IFMBE Proceedings

Volume 60

The International Federation for Medical and Biological Engineering, IFMBE, is a federation of national and transnational organizations representing internationally the interests of medical and biological engineering and sciences. The IFMBE is a non-profit organization fostering the creation, dissemination and application of medical and biological engineering knowledge and the management of technology for improved health and quality of life. Its activities include participation in the formulation of public policy and the dissemination of information through publications and forums. Within the field of medical, clinical, and biological engineering, IFMBE's aims are to encourage research and the application of knowledge, and to disseminate information and promote collaboration. The objectives of the IFMBE are scientific, technological, literary, and educational.

The IFMBE is a WHO accredited NGO covering the full range of biomedical and clinical engineering, healthcare, healthcare technology and management. It is representing through its 60 member societies some 120.000 professionals involved in the various issues of improved health and health care delivery.

More information about this series at http://www.springer.com/series/7403

Isnardo Torres · John Bustamante
Daniel A. Sierra (Eds.)

VII Latin American Congress on Biomedical Engineering CLAIB 2016, Bucaramanga, Santander, Colombia, October 26th – 28th, 2016

Springer

Editors
Isnardo Torres
Presidente
Asociación Colombiana de Bioingeniería
y Electrónica Médica - ABIOIN
Bucaramanga
Colombia

John Bustamante
Centro de Bioingeniería
Universidad Pontificia Bolivariana - UPB
Medellín
Colombia

Daniel A. Sierra
Escuela de Ing. Eléctrica, Electrónica
Universidad Industrial de Santander
Bucaramanga
Colombia

ISSN 1680-0737 ISSN 1433-9277 (electronic)
IFMBE Proceedings
ISBN 978-981-10-4085-6 ISBN 978-981-10-4086-3 (eBook)
DOI 10.1007/978-981-10-4086-3

Library of Congress Control Number: 2017934456

Printed on acid-free paper

This Springer imprint is published by Springer Nature
The registered company is Springer Nature Singapore Pte Ltd.
The registered company address is: 152 Beach Road, #21-01/04 Gateway East, Singapore 189721, Singapore

Preface

The CLAIB 2016 Organizing Committee, representing the Regional Council of Biomedical Engineering for Latin America (CORAL) with the Latin American and international scientific community has carried out the VII Latin American Congress on Biomedical Engineering (CLAIB2016) www.abioin.com/infoclaib2016.html. This event, which has taken place since 1998, had its seventh edition in Bucaramanga, Santander, Colombia, at Universidad Autónoma de Bucaramanga (UNAB).

The organization of the Latin American Congress on Biomedical Engineering was carried out by the "*Asociación Colombiana de Bioingeniería y Electrónica Médica- ABIOIN*". This congress was also supported by academic teams from Universidad Industrial de Santander, Universidad Pontificia Bolivariana, Universidad EIA, Instituto Tecnológico Metropolitano de Medellín, Universidad Javeriana and the Universidad Autónoma de Bucaramanga as the host University. Also, from the medical field, the FOSCAL, International Hospital of Colombia, FCV, among other organizations and international organisms have promoted this congregation of scientists, scholars and biomedical engineers from Latin America and other continents in an environment conducive to academic exchange and professional growth.

The Latin American congresses have provided a forum to present research findings, share experiences and coordinate activities between institutions and universities in the region to develop Bioengineering, Biomedical Engineering and related sciences.

The CLAIB 2016 in Bucaramanga (Santander – Colombia) followed previous meetings that were held in Mazatlan (Mexico) in 1998, Havana (Cuba) in 2001; Joao Pessoa (Brazil) in 2004, Margarita Island (Venezuela) in 2007, Havana (Cuba) in 2011, and Paraná (Entre Ríos – Argentina) in 2014. The congress program was designed so as to cover topics of regional and international interest and to meet scientific expectations.

Around 350 scientific articles were peer reviewed, each one by three independent reviewers and members of the CLAIB scientific committee, to select those meeting CLAIB's quality requirements. Moreover, 7 keynote talks from Europe, Asia and United States, and 11 keynote talks from Latin American countries were delivered by distinguished international scientists and academic and professional discussion panels with a high impact on bioengineering/biomedical engineering. Companies and firms also participated in CLAIB 2016 with an exposition of their innovations in the fields of medical products and equipment. Inside the activities of this congress, multiple meetings such as CORAL, IFMB, South-American working group, and social, cultural and recreation activities took place.

We thank you all for your participation in CLAIB2016 and you are always welcome to come back to Bucaramanga at any time.

Isnardo Torres
Chair CLAIB2016
John Bustamante
Co-Chair CLAIB2016
Daniel A. Sierra
Director of Publishing and Documentation CLAIB2016

Preface

Organizations

Organizing Institutions

Regional Council of Biomedical Engineering for Latin America/Consejo Regional de Ingeniería Biomédica para América Latina (CORAL)

Colombian Association of Bioengineering and Electronics Medicine/Asociación Colombiana de Bioingeniería y Electrónica Médica (ABIOIN)

Sponsors

International Federation for Medical and Biological Engineering (IFMBE)

The IEEE Engineering in Medicine and Biology Society (EMBS)

Supporting Organizations

"La Ingeniería Biomédica y la Bioingeniería en la visión integral de la salud humana y ambiental"
VII Latin American Congress On Biomedical Engineering Claib 2016
26 to 28 October 2016
PROCESAMIENTO DE SEÑALES BIOMÉDICAS
INGENIERÍA NEURAL Y REHABILITACIÓN
SISTEMA DE INGENIERÍA CARDIOVASCULAR RESPIRATORIO Y ENDOCRINO
BIOMECÁNICA
APARATOS MÉDICOS
BIOINFORMÁTICA Y BIOLOGÍA COMPUTACIONAL
BIOSENSORES Y BIOINSTRUMENTACIÓN
TELEMEDICINA
INGENIERÍA CLÍNICA Y EVALUACIÓN DE TECNOLOGÍA EN SALUD
INFORMÁTICA DE LA SALUD Y TELEMEDICINA
EDUCACIÓN EN BIOINGENIERÍA
BIOROBÓTICA
PROCESAMIENTO DE IMÁGENES BIOMÉDICAS
BIOMATERIALES
INGENIERÍA MOLECULAR, CELULAR Y DE TEJIDOS
TECNOLOGÍAS BIO-MICRONANO
CORAL
Organized By: ABIOIN ASOCIACION COLOMBIANA DE BIOINGENIERIA Y ELECTRONICA MEDICA
Registrations: www.abioin.com
Place: unab Universidad Autónoma de Bucaramanga (Avenida 42 No. 48 - 11)
Supported By IFMBE Springer IEEE EMB
www.springer.com
Short courses and workshops
UNIVERSIDAD EIA
UNIVERSIDAD CES
ITM
Universidad Pontificia Bolivariana
FOSCAL
fcv
JAVERIANA
Biomedical Engineering and Bioengineering: An integral vision of the human and environmental health

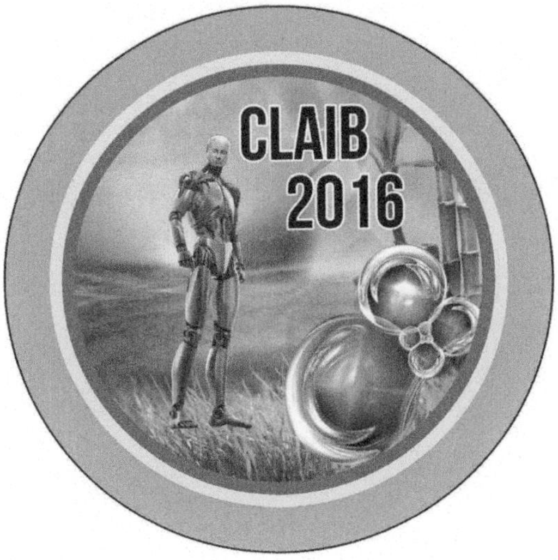
CLAIB
2016

VII LATIN AMERICAN CONFERENCE ON BIOMEDICAL ENGINEERING
CONGRESO LATINOAMERICANO DE INGENIERIA BIOMEDICA
CONGRESSO LATINO-AMERICANO DE ENGENHARIA BIOMÉDICA
BIENVENIDOS
Bucaramanga, Octubre 26-28 /2016
INGENIERÍA CLÍNICA Y EVALUACIÓN DE TECNOLOGÍA EN SALUD
BIOSENSORES BIOINSTRUMENTACIÓN
EDUCACIÓN EN BIOINGENIERÍA
APARATOS MÉDICOS
ABIOIN
CORAL
unab
IFMBE IEEE Springer EMB ACCE

Conference General Chair

Isnardo Torres Rivera
National President
Professor Medical Physics, Biophysics and Biomimetic Colombian Association of Bioengineering and Medical Electronics
Calle 48 # 39-59 Bucaramanga, Colombia Tel. 57 315 7267665
Isnardo26@yahoo.com
Web: www.abioin.com

Conference General Co-chair

John Bustamante Osorno
Director of the Center of Bioengineering, Head of the Cardiovascular Dynamic Research Group, and Coordinator of the Doctoral Programme in Medical Sciences
Universidad Pontificia Bolivariana
Circular 1ª No 73–76 Laureles, Bloque Externo 22C, Planta 2ª
Tel. (57/4) 448 83 88 Ext. 12400 Medellín, Colombia
john.bustamante@upb.edu.co
Web: www.upb.edu.co

Publications and Documentation Manager, Claib 2016

Daniel Alfonso Sierra Bueno
Planning Director, UIS
Electrical, Electronics and Telecommunications Engineering School Professor
Universidad Industrial de Santander, Bucaramanga, Colombia
dasierra@uis.edu.co
Web: www.uis.edu.co

Scientific Committee

Biosensors and Bioinstrumentation

Dr. Alejandro Weinstein, Chile
Dr. Alher Mauricio Hernández, Colombia
Mg. Andrés Valdez Díaz, Argentina
Dr. Angel Regueiro Gómez, Cuba
Prof. Dr. Anselmo Frizera Neto, Brasil
Dr. Arturo Vega González, México
Dr. Bertemes Filho Pedro, Brasil
Dr. César Antonio González Díaz, México
Dr. Daniel Alfonso Sierra Bueno, Colombia
Dr. Diego Sebastian Grafcaride, Argentina
Dr. Edith Pulido Herrera, Colombia
Mg. Eduardo Filomena, Argentina
Dr. Elisa Perez, Argentina
Dr. Fabiola Martínez, México
Dr. Franco Simini, Uruguay
M.sc. Guillermo Avendaño Cervantes, Chile
Dr. Henry Hermel Andrade Caicedo, Colombia
M.sc. Ing. Jaime Guillermo Barrero Pérez, Colombia
Mg. Javier Enrique González Barajas, Colombia
Dr. Jesus Enrique Chong Quero, México
Ph.D. Jhon Faber Archila Díaz, Colombia
Dr. Jorge Emilio Monzón, Argentina
Dr. José María Triano, Argentina
Ing. José Pirrone Puma, Venezuela
Dr. Juan Manuel Olivera, Argentina
Mg. Julio Alberto Aldonate, Argentina
Dr. Luis Enrique Llamosa Rincón, Colombia
Ing. Luis Roberto Barriere, El Salvador
Dr. Miguel Cadena Méndez, México
Dr. Myriam Cristina Herrera, Argentina
Dr. Natalia López, Argentina
Dr. Olimpia Arias De Fuentes, México
Dr. Omar Escalona, Reino Unido
Dr. Oscar Eduardo Gualdrón Guerrero, Colombia
Dr. Ricardo Silva, Venezuela/Ecuador
Ing. Ricardo Taborda, Argentina
Dr. Róbinson Alberto Torres Villa, Colombia
Dr. Rossana Madrid, Argentina
Mg. Rubén Darío Flórez Hurtado, Colombia
Mg. Sergio Conde Ocasionez, Brasil
Dr. Sonia Helena Contreras Ortiz, Colombia
Dr. Viviana Inés Rotger, Argentina
Dr. Whisner Fraga Mamede, Brasil
Dr. Willian Callejas, México

Biomedical Imaging and Image Processing

M.Sc. Adrián Carlos Ramón Salvatelli, Argentina
Dr. Agustina Bouchet, Argentina
M.Sc. Alejandro Javier Hadad, Argentina
Dr. Alejandro José Uriz, Argentina
Ph.D. Alexandra La Cruz, Ecuador
Dr. Alfredo Corniali, Brasil
Dr. Ana Lía Albarracín, Argentina
Dr. Ana Beatriz Ramírez Silva, Colombia
Dr. André Victor Alvarenga, Brasil
Mg. Andrés Valdez Dìaz, Argentina
Dr. Andrés Felipe Ruiz, Colombia
Dr. Ariel Braidot, Argentina
Post Dr. Catalina Tobón Zuluaga, Colombia
Dr. Cesar a. D. Teixeira, Portugal
M.Sc. Diana Estefy Gutiérrez Galvis, Colombia
Dr. Diego Sebastián Comas, Argentina
Dr. Diego Sebastián Graf Caride, Argentina
Dr. Domingo Rodríguez, Puerto Rico
Dr. Edgar Eduardo Romero Castro, Colombia
Dr. Eduardo Martínez Montes, Cuba
Dr. Elisa Pérez, Argentina
Dr. Fernando César Coelli, Brasil
Dr. Gustavo Meschino, Argentina
Bioing. Gustavo Horacio Bizai Tonello, Argentina
Dr. Javier Fernando Adur, Argentina
Mg. Javier Enrique González Barajas, Colombia
Dr. Jesús Enrique Chong Quero, México
Post Dr. João Carlos Mendes Carvalho, Brasil
Mg, José Maria Triano, Argentina
Dr. José Manuel Ferrer Villena, Perú
Dr. José Mauricio Santos Torres da Motta, Brasil
Prof. Juan Enrique Suen Díaz, Cuba
Dr. Juan Enrique Palomares Ruiz, México
Dr. Juan Ignacio Pastore, Argentina
Mg. Juan Manuel Olivera, Argentina
Dr. Juan Valentín Lorenzo Ginori, Cuba
Dr. Lola Xiomara Bautista Rozo, Colombia
M.Sc. Marco Antonio Coca Pérez, Chile
Prof. Marlen Pérez Díaz, Cuba
Dr. Miguel Cadena Méndez, México
Dr. Rossana Madrid, Argentina
Ph.D. Said David Pertuz, Colombia
Dr. Sonia Helena Contreras Ortiz, Colombia
Dr. Virginia Laura Ballarín, Argentina
Dr. Andrés Orozco Duque, Colombia
Dr. Eduardo Peón Aves, Cuba
M.Sc. Edwin Santiago Alférez Baquero, España
M.Sc. Javier Fernando Adur, Argentina
M.Sc. José Manuel Ferrer Villena, Perú
Ing. Juan Pablo Ángel López, Colombia

Cardiovascular Engineering, Lung and Endocrine

Dr. Alfredo Rosado, España
Dr. Ana María Rodríguez, Colombia
Dr. Ana Milena Herrera Torres, Colombia
Dr. David Alejandro Miranda Mercado, Colombia
Dr. Hyxia Villegas, Venezuela
Dr. Jesús Enrique Chong Quero, México
Dr. Jorge Emilio Monzon, Argentina
Dr. John Bustamante, Colombia
Dr. Lina Marcela Hoyos Palacio, Colombia
Dr. Mónica Karel Huerta, Venezuela
Dr. Pedro Arini, Argentina
Dr. Ricardo Luis Armentano Feijoo, Argentina
Mg. Sandra Milena Luna Osorio, Colombia
Dr. Vera Zalulich Pérez Ariza, Colombia

Molecular Engineering, Cell and Tissue

M.Sc. Ing. Alejandro Arboleda, Colombia
Dr. Ana María Rodríguez, Colombia
Dr. Andrés Mauricio Pinzón, Colombia
Ph.D. Carolina Londoño Peláez, Colombia
Dr. Diego Alejandro Velásquez Puerta, Colombia
Dr. Fabián Alejandro Buffa, Argentina
Dr. Ing. Florencia Montini Ballarín, Argentina
Mg. Guillermo Carlos Campiglio, Argentina
Dr. Gustavo A. Abraham, Argentina
Dr. Juan Manuel Olivera, Argentina
Dr. Juliana Velásquez Gómez, Colombia
Dr. Lizette Morejón Alonso, Cuba
Dr. Maria Fernanda Izaguirre, Argentina
Dr. Nydia Paola Rondón Villareal, Colombia
Ph.D. Pablo Cortez Tornello, Argentina
Mg. Sandra Milena Luna Osorio, Colombia
Dr. Ing. Sergio Lew, Argentina
Dr. Teresita Cuadrado, Argentina

Neural Engineering and Rehabilitation

Dr. Bioing. Alfredo Corniali, Brasil
Dr. Lizett Morejón Alonso, Cuba
Dr. Álvaro David Orjuela, Colombia
Mg. Ing. Ana María Echenique, Argentina
Dr. Andrés Felipe Ruiz, Colombia
Mg. Andrés Torres Velásquez, Colombia
Prof. Dr. Anselmo Frizera Neto, Brasil
Ph.D. Belkys Tibisay Amador Cáceres, Venezuela
Dr. Carlos Andrés Cifuentes García, Colombia
Mg. Carolina Tabernig, Argentina
Dr. Clara Isabel López Gualdrón, Colombia

Biomedical Signal Processing

Dr. Eduardo Pedro Serrano, Argentina
Dr. Edwin Alexander Cerquera, Colombia
M.Sc. Edwin Santiago Alférez Baquero, España
Dr. Eric Laciar Leber, Argentina
M. en IB. Fabiola Martínez, México
Dr. Franklin A. Sepúlveda Sepúlveda, Colombia
Dr. Gustavo Meschino, Argentina
Dr. Jairo Lenín Ramón Valencia, Colombia
Mg. Javier Enrique Gonzalez Barajas, Colombia
Dr. Jesús Enrique Chong Quero, México
Dr. Juan Carlos Perfetto, Argentina
Mg. Juan Pablo Tello, Colombia
Mg. Julio César Politti, Argentina
Ph.D. Julián Luciano Cárdenas Barrera, Canadá
Dr. Bioing. Lorena Correa Prado, Argentina
Mg. Ing. Luis Romero, Argentina
Mg. Luis Enrique Llamosa Rincón, Colombia
Dr. Luis E. Bergues Cabrales, Cuba
Dr. Bioing. María Paula Bonomini, Argentina
Dr. Miguel Cadena Méndez, México
Dr. Ing. Natalia López Celani, Argentina
Ph.D. Omar J. Escalona, Reino Unido
Ing. Óscar Rodolfo Vallena, Argentina
Dr. Óscar Eduardo Gualdrón, Colombia
Dr. Pedro Bertemes Filho, Brasil
Dr. Pedro D. Arini, Argentina
Dr. Pedro Federico Diez, Argentina
Dr. Rafael Torres Amarís, Colombia
Prof. Raul Correa Prado, Argentina
Dr. René Iván González Fernández, Cuba
Dr. Robinson Alberto Torres Villa, Colombia
Dr. Ricardo Silva, Ecuador/Venezuela
Dr. Rubén Carlos Acevedo, Argentina
Dr. Rubiel Vargas Cañas, Colombia
PhD. Said David Pertuz Arroyo, Colombia
Dr. Sergio Conde Ocazionez, Colombia
Dr. Ph.D. Sonia Helena Contreras Ortiz, Colombia
Dr. Verónica Medina Bañuelos, Mexico
Mg. Bioing. Victor Eduardo Martínez Abaunza, Brasil
Dr. Virginia Ballarín, Argentina
Dr. Whisner Fraga Mamede, Brasil
Mg. Bioing, Yanina Atum, Argentina
Dr. Yusely Ruíz González, Cuba

Bio-Micro-Nano Technologies

M.Sc. Adrian Salvatelli, Argentina
Dr. Alejandro José Uriz, Argentina
Dr. Alher Mauricio Hernández, Colombia
Dr. Carlos Julio Arismendi Pereira, Colombia
Dr. Diego Alejandro Velázquez Puerta, Colombia
Mg. Eduardo Filomena, Argentina

Dr. Eduardo A Albanesi, Argentina
Dr. Henry Hermel Andrade Caicedo, Colombia
Ph.D. John Bustamante, Colombia
Mg. Julio Alberto Aldonate, Argentina
Dr. Jorge Emilio Monzón, Argentina
Dr. Lina Marcela Hoyos Palacios, Colombia
Mg. Luis Romero, Argentina
Dr. Marlen Pérez Díaz, Cuba
Dr. Martha Elena Londoño López, Colombia
Dr. Myriam Cristina Herrera, Argentina
Ph.D. Olimpia Arias de Fuentes, México
Dr. Rafael Torres Amarís, Colombia
Dr. Rossana Elena Madrid, Argentina
Dr. Virginia Walz, Argentina
Mg. Bioing. Viviana Inés Rotger, Argentina

Medical Physics and Nuclear Medicine

Mg. Antonio Vélez Montoya, Colombia
Dr. César Antonio González Díaz, México
Dr. Clara Isabel López Gualdrón, Colombia
Mg. Consuelo Valera Corona, Cuba
Dr. David Alejandro Miranda Mercado, Colombia
Dr. Ernesto Suaste Gómez, México
Dr. Fernando Cesar Coelli, Brasil
Dr. Franco Simini, Uruguay
Dr. Gabriel Ruiz, Argentina
Dr. Graciela Vélez, Argentina
Prof. Mg. Guillermo Avendaño, Chile
Mg. Guillermo Carlos Campiglio, Argentina
Dr. Javier Fernando Adur, Argentina
Mg. José Luis Favant, Argentina
Dr. Ing. Josefina Ballare, Argentina
Dr. Juan Enrique Suen Díaz, Cuba
Dr. Leonardo Makinistián, Argentina
Dr. Luis E. Bergues Cabrales, Cuba
Mg. Luis Enrique Llamosa Rincón, Colombia
Dr. Marco Antonio Coca Pérez, Chile
Dr. María Fernanda Izaguirre, Argentina
Dr. Marlen Pérez Díaz, Cuba
Dr. Martha Elena Londoño López, Colombia
Dr. Miriam Marañón Cardonne, Cuba
Dr. Mónica T Millares Aranda, Argentina
Dr. Pedro Bertemes Filho, Brasil
Dr. Ricardo Luis Armentano Feijoo, Brasil
Dr. Rossana Madrid, Argentina
Dr. Virginia Walz, Argentina

Biomechanics

Dr. Andrés Felipe Ruiz, Colombia
Dr. Andrés Torres Velásquez, Colombia
Prof. Dr. Anselmo Frizera Neto, Brasil
Dr. Ariel Braidot, Argentina
Dr. Arturo Vega González, México
Dr. Belkys Tibisay Amador Cáceres, Venezuela
Dr. Carlos Andrés Cifuentes García, Colombia
Dr. Clara Isabel López Gualdrón, Colombia
Mg. Claudia Bonell, Argentina
Mg. Diana Estefy Gutiérrez Galvis, Colombia
Dr. Diego Martín Campana, Argentina
Dr. Edith Pulido Herrera, Colombia
Mg. Guillermo Carlos Campiglio, Argentina
Dr. Jesús Enrique Chong Quero, México
Ph.D. Jhon Faber Archila Díaz, Brasil
Dr. João Carlos Mendes Carvalho, Brasil
Dr. Jorge Ernesto Letechipia Moreno, Mexico
Dr. Jorge Mazzeo, Argentina
Dr. Jose di Paolo, Brasil
Dr. José Maurício Santos Torres da Motta, Brasil
Dr. Juan Enrique Palomares Ruiz, Mexico
Dr. Juan Pablo Ángel López, Colombia
Julio Cesar Politti, Argentina
Mg. Luis Eduardo Bautista Rojas, Colombia
Dr. Luis Eduardo Rodríguez Cheu, Colombia
Mg. Luis Romero, Argentina
Ph.D. Luis Vilcahuaman, Peru
Dr. Marco Henrique Terra, Brasil
Dr. Martha Zequera, Colombia
Dr. Mónica T Millares Aranda, Argentina
Dr. Ricardo Luis Armentano Feijoo, Argentina
Dr. Sonia Helena Contreras Ortiz, Colombia
Dr. Yolanda Torres Pérez, Colombia

Bio Robotics

Dr. Carlos Andres Cifuentes García, Brasil
Dr. João Carlos Mendes Carvalho, Brasil
Dr. John Archila Díaz, Colombia
Dr. Juan Enrique Palomares Ruiz, Mexico
Mg. Julio Cesar Politti, Argentina
Dr. Mario Luis Tronco, Brasil
Dr. Manuel Betancur, Colombia
Dr. Marco Henrique Terra, Brasil
Dr. Paulo Eigi Miyagi, Brasil

Bioinformatics – Computational Biology

M.Sc. Bio Ing. Alejandro Javier Hadad, Argentina
Dr. Andrés Mauricio Pinzón Velasco, Colombia
Dr. Ariel Fernando Amadio, Argentina
Dr. Bio Ing. Eduardo Martínez Montes, Cuba
Ph.D. Inti Anabela Pagnuco, Argentina
Dr. Julián Antonio Villamarín Muñoz, Colombia
Dr. Luciano Garelli, Argentina
Dr. Luis E. Bergues Cabrales, Cuba
Dr. María Fernanda Izaguirre, Argentina
Prof. María Margarita Goire Castilla, Cuba
Dr. Mauricio Corredor Rodríguez, Colombia
Dr. Nydia Paola Rondón Villareal, Colombia
Ph.D. Ricardo Silva, Venezuela/Ecuador
Dr. Rubiel Vargas Cañas, Colombia
Dr. Ing. Sergio Lew, Argentina
Dr. Victor Eduardo Martínez Abaunza, Brasil
Dr. Victor H. Castro, Argentina

Biomaterials

Dr. Alejandra Vera Álvarez, Argentina
Dr. Alher Mauricio Hernández, Colombia
Dr. Andrés Ozols, Argentina
Dr. Andrés Mauricio Pinzón Velasco, Colombia
Dr. Arturo Vega González, México
Dr. Carlos Oldani, Argentina
Dr. Carolina Londoño Peláez, Colombia
Dr. Diego Alejandro Velasquez Puerta, Colombia
Dr. Eduardo Peón Aves, Cuba
Dr. Ernesto Suaste Gómez, México
Dr. Fabián Alejandro Buffa, Argentina
Dr. Ing. Florencia Montini Ballarín, Argentina
Mg. Guillermo Carlos Campiglio, Argentina
Dr. Gustavo A. Abraham, Argentina
Dr. Jimena González, Argentina
Dr. Ing. Josefina Ballare, Argentina
Dr. Juan Manuel Olivera, Argentina
Dr. Lina Marcela Hoyos Palacios, Colombia
Dr. Lizett Morejón Alonso, Cuba
Dr. Luis Bergues Cabrales, Cuba
Dr. María Fernanda Izaguirre, Argentina
Dr. Martha Elena Londoño López, Colombia
Ph.D. Olimpia Arias de Fuentes, México
Dr. Pablo Cortés Tornello, Argentina
Dr. Silvia Ceré, Argentina
Dr. Teresita Cuadrado, Argentina
Mg. Walter Pardave Livia, Colombia
Dr. Whisner Fraga Mamede, Brasil

Control and Nonlinear Systems

Dr. Álvaro David Orjuela, Colombia
Dr. Andrade Kléber de Oliveira, Brasil
Prof. Dr. Anselmo Frizera Neto, Brasil
Dr. Carlos Andrés Cifuentes García, Colombia
Dr. Carlos Julio Arizmendi Pereira, Colombia
Mg. Carmen Telleria Prieto, Cuba
Mg. Claudia Bonell, Argentina
Dr. Franklin A. Sepúlveda Sepúlveda, Colombia
Ph.D. Jhon Faber Archila Díaz, Colombia
Dr. João Carlos Mendes Carvalho, Brasil
Dr. José Maurício Santos Torres Da Motta, Brasil
Dr. Juan Enrique Palomares Ruiz, Mexico
Dr. Juan Manuel Gómez González, Mexico
Ing. Juan Pablo Angel López, Colombia
Mg. Julio Cesar Politti, Argentina
Mg. Bioing. Luciano Schiaffino, Argentina
Dr. Luis E. Bergues Cabrales, Cuba
Dr. Luis Eduardo Rodriguez Cheu, Colombia
Dr. Manuel J. Betancur, Colombia
Dr. Marcio José Dacunha, Brasil
Dr. Marco Henrique Terra, Brasil
Dr. Mario Luiz Tronco, Brasil
Dr. Paulo Eigi Miyagi, Brasil
Dr. Vera Zalulich Perez, Colombia

Innovation- Education Standards

Dr. Alfredo Corniali, Brasil
Ph.D. Andrés Orozco Duque, Colombia
Dr. Angel Regueiro Gomez, Cuba
Dr. Antonio Miguel Cruz, Colombia
Mg. Carolina Tabernig, Argentina
Ing. Esp. Carolina Salazar, Colombia
Ing. Esp. Diana García Peñaranda, Colombia
Dr. Franco Simini, Uruguay
Dr. Gabriel Ruiz, Argentina
M.Sc. Prof. Guillermo Avendaño, Chile
Dr. Jairo Lenín Ramón Valencia, Colombia
Dr. Jesús María Soto, Colombia
Dr. Jose di Paolo, Brasil
Mg. José Luis Favant, Argentina
Dr. Juan Manuel Olivera, Argentina
Dr. Juan Pablo Graffigna, Argentina
Mg. Julio Alberto Aldonate, Argentina
Dr. Lizett Morejón Alonso, Cuba
Mg. Luis Alfredo Rocha, Argentina
Dr. Miguel Cadena Méndez, México
Dr. Miriam Marañón Cardonne, Cuba
Dr. Mónica Karel Huerta, Venezuela

Dr. Mónica T. Millares Aranda, Argentina
Ing. Óscar Vanella, Argentina
Dr. Ricardo Silva, Ecuador
M.Sc. Ph.D. Rossana Rivas, Perú
Dr. Virginia Walz, Argentina
Mg. Bioing. Viviana Inés Rotger, Argentina
Dr. Willian Ricardo Rodríguez Dueñas, Colombia
Dr. Mg. Yanina Atum, Argentina
Dr. Alfredo Corniali, Brasil
Ph.D. Andrés Orozco Duque, Colombia
Dr. Ángel Regueiro Gómez, Cuba
Dr. Antonio Miguel Cruz, Colombia
Mg. Carolina Tabernig, Argentina
Ing. Esp. Carolina Salazar, Colombia
Ing. Esp. Diana García Peñaranda, Colombia
Dr. Franco Simini, Uruguay
Dr. Gabriel Ruiz, Argentina
M.Sc. Prof. Guillermo Avendaño, Chile
Dr. Jairo Lenín Ramón Valencia, Colombia
Dr. Jesús María Soto, Colombia
Dr. Jose di Paolo, Brasil
Mg. José Luis Favant, Argentina
Dr. Juan Manuel Olivera, Argentina
Dr. Juan Pablo Graffigna, Argentina
Mg. Julio Alberto Aldonate, Argentina
Dr. Lizett Morejón Alonso, Cuba
Mg. Luis Alfredo Rocha, Argentina
Dr. Miguel Cadena Méndez, México
Dr. Miriam Marañón Cardonne, Cuba
Dr. Mónica Karel Huerta, Venezuela
Dr. Mónica T. Millares Aranda, Argentina
Ing. Oscar Vanella, Argentina
Dr. Ricardo Silva, Ecuador
Dr. Virginia Walz, Argentina
Mg. Bioing. Viviana Inés Rotger, Argentina
Dr. Willian Ricardo Rodríguez Dueñas, Colombia
Dr. Mg. Yanina Atum, Argentina

Bio Prosthetic Implants (Bionics)

Dr. Lizette Morejón Alonso, Cuba
Dr. Andrés Ozols, Argentina
Prof. Dr. Anselmo Frizera Neto, Brasil
Dr. Carlos Oldani, Argentina
Dr. Eduardo Peón Avés, Cuba
Dr. Fabián Alejandro Buffa, Argentina
Dr. Luis E. Bergues Cabrales, Cuba
Dr. Teresita Cuadrado, Argentina

Medical Informatics/Science of Computing in Health - TIC

M.Sc. Bioing. Adrián Salvatelli, Argentina
Dr. Alejandro Weinstein, Chile
M.Sc. Alejandro Javier Hadad, Argentina
Dr. Bioing. Alfredo Corniali, Brasil
Dr. Álvaro David Orjuela, Colombia
Dr. André Victor Alvarenga, Brasil
Mg. Ing. Andrés Valdez Dìaz, Argentina
Dr. Ariel Fernando Amadío, Argentina
Dr. Carlos Julio Arizmendi Pereira, Colombia
Prof. César Augusto Aceros Moreno, Colombia
Dr. Daniel Ulises Campos Delgado, México
Dr. Diego Sebastián Grafcaride, Argentina
Prof. Eduardo Martínez Montes, Cuba
Dr. Franco Simini, Uruguay
Dr. Gustavo Meschino, Argentina
Lic. Inti Anabela Pagnuco, Argentina
Dr. Juan Pablo Graffigna, Argentina
Dr. Julián Antonio Villamarín Muñoz, Colombia
Mg. Lola Xiomara Bautista Rozo, Colombia
Dr. Luciano Garelli, Argentina
Dr. Luis E. Bergues Cabrales, Cuba
Dr. Luis Enrique Sánchez Crespo, España
Ing. Luis Roberto Barriere, El Salvador
Mg. María Del Carmen Telleria Prieto, Cuba
Prof. María Margarita Goire Castilla, Cuba
Dr. María Paula Bonomini, Argentina
Prof. Maria Luiz Tronco, Brasil
Prof. Mónica Karel Huerta, Venezuela
Prof. Paulo Eigi Miyagi, Brasil
Dr. Pedro Federico Diez, Argentina
Dr. Ing. Raúl Correa Prado, Argentina
Prof. René Iván González Fernández, Cuba
Ph.D. Ricardo Silva, Ecuador
Prof. Roger Clotet Martinez, Venezuela
Dr. Rubiel Vargas Cañas, Colombia
Dr. Said David Pertuz Arroyo, Colombia
Dr. Willian Ricardo Rodríguez Dueñas, Colombia

Clinical Engineering, Technology Assessment Health, Medical Devices

Dr. Alexandra La Cruz, Venezuela
Dr. Bioing. Alfredo Corniali, Brasil
Dr. André Victor Alvarenga, Brasil
Mg. Andrés Valdez Dìaz, Argentina
Prof. Dr. Anselmo Frizera Neto, Brasil
Dr. Antonio Hernández, USA
Dr. Antonio Miguel Cruz, Colombia
Ms. Bioing. Atum Yanina, Argentina
Mg. Beatriz Janeth Galeano Upegi, Colombia

Ing. Esp. Carlos Mesa Duarte, Colombia
Dr. Carlos Román Vásquez Seisdedos, Cuba
Mg. Carmen Tellería Prieto, Cuba
Ing. Esp. Carolina Salazar, Colombia
Dr. Daniel Ulises Campos Delgado, México
Ing. Esp. Diana García Peñaranda, Colombia
Dr. Diego Sebastián Grafcaride, Argentina
Dr. Fernando Cesar Coelli, Brasil
Mg. Guillermo Avendaño, Chile
Mg. Guillermo Carlos Campiglio, Argentina
Ing. Isabel Angélica Morales Ledezma, Bolivia
Dr. Jairo Lenín Ramón Valencia, Colombia
Mg. Javier Enrique Camacho Cogollo, Colombia
Dr. Jesús María Soto, Colombia
Dr. Ing. Jorge Mazzeo, Argentina
Dr. Jose Pirrone Puma, Venezuela
M.Sc. Ing. José María Triano, Argentina
Dr. José Manuel Ferrer Villena, Perú
Ing. José Maria Flores, Argentina
Dr. Juan Enrique Suen Díaz, Cuba
Dr. Juan Manuel Olivera, Argentina
Dr. Julián Antonio Villamarín Muñoz, Colombia
Mg. Julio Alberto Aldonate, Argentina
Dr. Lizette Morejón Alonso, Cuba
Mg. Luis Romero, Argentina
Ph.D. Luis Vilcahuaman, Perú
Mg. Luis Alfredo Rocha, Argentina
Ing. Luis Roberto Barriere, El Salvador
Dr. Marcio Jose Dacunha, Brasil
Dr. Miguel Cadena Méndez, México
Dra. Miriam Marañón Cardonne, Cuba
Dr. Mónica T. Miralles, Argentina
Dr. Myriam Cristina Herrera, Argentina
Ing. Óscar Vanella, Argentina
Ing. Raúl Romo Doña, Argentina
Dr. Renato García, Brasil
Ing. René Iván González Fernández, Cuba
Dr. Ricardo Silva, Ecuador
Dr. Ricardo Taborda, Argentina
Dr. Roger Clotet Martínez, Venezuela
Ph.D. Rossana Rivas, Perú
Dr. Ing. Sergio Lew, Argentina
Mg. Viviana Inés Rotger, Argentina

Mems and Microelectronics

Dr. Luis E. Rodríguez Cheu, Colombia
Dr. Luis E. Bergues Cabrales, Cuba
Dr. Miguel Cadena Méndez, México
Dr. Víctor H. Castro, Argentina
Dr. Yolanda Torres, Colombia

Modelling/Simulation

Dr. Alher Mauricio Hernández, Colombia
Dr. Andrés Orozco Duque, Colombia
Dr. Andrés Felipe Hernández Marulanda, Colombia
Dr. Anselmo Frizera Neto, Brasil
Dr. Belkys Tibisay Amador Cáceres, Venezuela
Dr. Catalina Tobón, Colombia
Ph.D. Daniel Lorias Espinosa, Mexico
Dr. Daniel Alfonso Sierra Bueno, Colombia
Dr. David Alejandro Miranda Mercado, Colombia
Dr. Diego Martín Campana, Argentina
Dr. Domingo Rodríguez, Mayagüez, Puerto Rico
Dr. Fernando Cesar Coelli, Brasil
Dr. Ing. Florencia Montini Ballarín, Argentina
Dr. Franklin A. Sepúlveda, Colombia
Dr. João Carlos Mendes Carvalho, Brasil
Dr. Jose di Paolo, Brasil
Ing. Jose Pirrone Puma, Venezuela
Dr. Juan Enrique Palomares Ruiz, Mexico
Dr. Leonardo Makinistian, Argentina
Dr. Luciano Garelli, Argentina
Dr. Luis E. Bergues Cabrales, Cuba
Dr. Manuel J. Betancur, Colombia
Dr. Mauricio Corredor Rodríguez, Colombia
Dr. Nydia Paola Rondón Villareal, Colombia
Dr. Omar J. Escalona, Reino Unido
Dr. Paulo Eigi Miyagi, Brasil
Dr. Ricardo Silva, Venezuela/Ecuador
Dr. Ing. Sergio Lew, Argentina
Dr. Virginia Walz, Argentina

Telemedicine - Health Care

Dr. Alfredo Corniali, Brasil
Ing. Esp. Carolina Salazar, Colombia
Dr. Jose Pirrone Puma, Venezuela
Mg. Ing. Juan Pablo Tello Portillo, Colombia
Dr. Leandro di Presia, Argentina
Dr. Lola Xiomara Baustista, Colombia
Mg. Ing. Luis Alfredo Rocha, Argentina
Dr. Luis Enrique Sánchez Crespo, España
Dr. María Margarita Goire Castilla, Cuba
Dr. Mónica Karel Huerta, Venezuela
Dr. Paulo Eigi Miyagi, Brasil
Ph.D. René Iván González Fernández, Cuba
Dr. Ricardo Silva, Ecuador
Dr. Roger Clotet Martínez, Venezuela
Ph.D. Rossana Rivas, Perú
Mg. Rubén Darío Florez Hurtado, Colombia
M.Sc. Viviana Inés Rotger, Argentina
Dr. William Ricardo Rodríguez Dueñas, Colombia

Other Topics

M.Sc. Bioing. Adrian Salvatelli, Argentina (Imágenes)
Dr. Agustina Bouchet, Argentina (Logica Difusa, Matematicas)
Dr. Ing. Agustina Garcés Correa, Argentina (Dosimetria, Blindajes, Radioterapia)
Bioing. Alejandro Hadad, Argentina (Imag y Sistemas Expertos)
Dr. Ana Lía Albarracin, Argentina (Ultrasonido, Metrología; Computer-Aided Diagnostic Systems)
Dr. Ph.D. Ana Milena Herrera, Colombia (Anatomia, Fisiopatologia)
Dr. Antonio Hernández, USA (Aparatos Medicos, Evaluación de Tecnologias)
Dr. Bioing. Ariel Andrés Braidot, Argentina (Biomecánica E. Imagen)
Mg. Beatriz Janeth Galeano Upegi, Colombia (Gestión de Mantenimiento y Activos Hospitalarios)
Ing. Esp. Carlos Mesa Duarte, Colombia (Creación de Empresas de Salud)
Dr. Carlos Oldani, Argentina (Biomateriales)
Mg. Bioing. Carolina B. Tabernig, Argentina (Estimulacion Electric Funcional)
Bioing. Claudia Bonell, Argentina (Electromiografía)
Dr. Bioing. Diego Martín Campana, Argentina (Mecánica Fluidos)
Mg. Bioing. Eduardo Filomena, Argentina (Electrónica Digital)
Bioing. Eduardo Aldo Albanesi, Argentina (Imágenes)
Dr. Bioing. Gustavo Bizai, Argentina (Microscopia y Radiones No Inozante, Biología Molecular)
Mg. Bioq. Javier Fernando Adur, Argentina (Radiaciones Ionizantes)
Ing. José Luis Favant, Argentina (Ingeniería Hospitalaria)
Dr. Bioing. José María Flores, Argentina (Física Teoría de Modelo)
Mg. Bioing. Leonardo Makinistian, Argentina (Control de Proceso)
Dr. Luciano Schiaffino, Argentina (Microscopia, Biología Molecular)
Dr. Víctor Hugo Castro, Argentina (Microscopia, Biología Molecular)

Organizing Committe

Directive Board

Isnardo Torres
chair
John Bustamante
vice-chair

Honor Committe

Jorge Reynolds
Ignacio Escobar

Junta Ejecutiva

Economic and Financial Affairs: **Carlos Mesa**
Academic Oversight and Scientific Activities: **Jesus Maria Soto**
Area scientific communications, marketing and advertising: **Alejandro Arboleda**
Executive Secretary on Scientific, Legal and Health aspects Marketing: **Diana García**

Organizers Board

Infrastructure and Logistics: **Carlos Arizmendi**
Workshop and Commercial Staff: **Carolina Salazar**
Government and Business Management: **Organization Commitee**
Publications and Documentation Area: **Daniel A Sierra Bueno**
Protocol and Host Activities: **Francelly Bustamante**
Coordination of the Green Day and extra Congress Activities-Tour operator: **Delfín Albarracín**
Academic Network and Scientific Societies: **Martha Zequera**
Scientific Comunicator: **Nydia Paola Rondón**
Logistics and Administrative Affairs: **Pedro Albarracín**

Students Support Committee

Universidad Autónoma de Bucaramanga

Biomedical Engineering Program
Karen S Vega B
Daniel Figueroa
Carolina Carreño
Daniela Monsalve
Diana Garcia Rivero
Marly Galeano Blanco

Universidad Insdustrial de Santander

Andrés Suarez
Juan Carlos Sierra
Elizabeth Martinez
Maria Camila Pinilla
Mauricio Ballesteros
Angie Melissa Galindo

Keynote Conferences

Regeneration Technology for Functional Musculoskeletal Tissue

October 26th, 2016 – 10:00–11:00 a.m.
Auditorio Mayor Carlos Gómez Albarracín,
Universidad Autónoma de Bucaramanga, Colombia

Professor James GOH: President, International Federation of Medical and Biological Engineering (IFMBE)
PhD (1982), University of Strathclyde, Glasgow, UK; B.Sc (Hons) (1982), University of Strathclyde, Glasgow, UK; CEng (1993), The Engineering Council, UK
Appointment Status: Professor and Head, Department of Biomedical Engineering, NUS; Research Professor, Orthopedic Surgery, NUS.
Professional Societies: President, Biomedical Engineering Society (Singapore); President, International Federation of Medical and Biological Engineering. Council Member, World Council of Biomechanics. Secretary General, Asia-Pacific Association for Biomechanics. Treasurer, Executive Council, World Association for Chinese Biomedical Engineers. Secretary, TERMIS Asian Pacific chapter.

Abstract
Biomaterials including polymers, hydrogels, ceramics, metals and self-assembled materials have been widely used for bone, cartilage, ligament, tendon and dental tissue regeneration. Most importantly, micro/nano- biomaterials provide some important and interesting properties to the traditional biomaterials, including high ratio of surface area to volume, enhanced mechanical properties, high purity and homogeneity, and outstanding magnetic, optical and electrical properties. Taking advantages of these novel properties, micro/nano-biomaterials can be manipulated to stimulate the chemical and structural similarities to natural musculoskeletal tissues. For instance, cartilage is hard to be regenerated due to lack of an efficient vascular system, limited progenitor cells and chondrocyte mobility in the dense cartilage extracellular matrix. However, micro/nano-biomaterials can provide a biomimetic interface (or environment) to improve functions of chondrocyte, inhabiting and differentiation of progenitor cells. Moreover, the micro/nano-biomaterials will enhance the lifetime and performance of tissue engineered scaffolds for cartilage regeneration (such as improved mechanical properties and hierarchical structures). On the other hand, the development of micro/nano-biomaterials on bone regeneration is now focusing on two different directions: scaffolds for treatment of segmental defects (with structural functions) and cavity defects (such as fillers and injectable scaffolds). The utilization of micro/nano-biomaterials in orthopedic applications has been proved to be an effective and innovative technique to enhance osseointegration and bone regeneration. The micro/nano-biomaterials also have significant impacts on ligament/tendon regeneration. For example, nanostructured scaffolds are extremely helpful that can serve as structural and mechanical support for cellular function and tissue regeneration. In conclusion, countless opportunities and approaches have been brought by the development of micro/nano-biomaterials to solve the current difficulties of implant materials, providing lots of success for micro/nano-biomaterials in a variety of musculoskeletal tissue regeneration applications.

A Wellness-Centric Healthcare System Interoperable with Public Health: the Multidimensional Global Threats, Interdependences of the Critical Infrastructures and Geo Medicine

Keynote: October 26th, 2016 11:00 a.m. –12:00 a.m.
Auditorio Mayor Carlos Gómez Albarracín
Universidad Autónoma de Bucaramanga, Colombia

Luis Kun, Ph.D. Editor in Chief - Journal of Health & Technology - Springer - IUPESM - WHO, IFMBE Chairman Global Citizen Safety and Security WG.

Ex- Professor of National Security Affairs, CHDS/National Defense University, Fellow AIMBE, Lifetime Fellow, IEEE Editor in Chief - Journal of Health & Technology - Springer - IUPESM – WHO. IFMBE Chairman Global Citizen Safety and Security WG, Member Developing Countries WG, IEEE Society on Social Implications of Technology (SSIT) - Board of Governors, IEEE Life Science Technical Committee Member, IEEE - SSIT - Distinguished Lecturer (DL) IEEE - Computer Society - Distinguished Visitor Program (DVP), IEEE - Computer Society Ad hoc Public Policy Committee, Member - Healthcare Transformation Ad hoc Committee, Chair, IEEE EMBS, Editorial Board.

Abstract

In the Information Age, reliance on cyberspace pervades almost every aspect of life including healthcare. Private citizens have utilized digital systems – including the internet – to communicate with others around the globe, build social networks and share information. Businesses have taken advantage of online and ecommerce to expand their client base, manage supply chains and communicate with customers. The United States Government relies on digital systems to collect taxes, manage critical infrastructure, ensure national security and interact with constituents. As one of the most connected countries on earth, the United States has enormously benefited from the transformations brought about by the Information Age. However, the diverse benefits of the Information Age also come with enormous risks. The Internet and other digital infrastructure are lightly regulated and unsecure. As a "network of networks" where all types of secure government, open public and restricted private networks are located, security vulnerabilities in one internet network can be used to exploit vulnerabilities in another. A growing group of state and nonstate actors, including disgruntled employees, criminals, terrorists and foreign intelligence agencies, have exploited these vulnerabilities to maliciously target U.S. citizens, commerce and government.

For over 40 years researchers have written about applying computer technologies to improve daily medical care, and in the past decade, the goal of using genomic data to truly personalize care has been woven into those concepts. Since 2004, the US has focused on creating a complete personal health record for each citizen and great strides have been made in developing the technical standards to allow near-realtime health data acquisition either directly from medical devices, health practitioners, caregivers, and/or patients themselves. In this period government and private agencies have also been separately collecting and organizing vast quantities of scientific data that relate to our health. It has become clear also that many illnesses and injuries are directly caused or are significantly influenced by the food and water we consume, the air we breathe and the environment we live in as well as our education. Further, as human lifetimes are extended around the globe; environmental factors simply have more years' damage to human health and wellness. In 2009 Dr. Kun organized and conducted at the National Academy of Science and Engineering the Transformation of the American Health Care System (https://www.hawaii.edu/csati/summit/spvg/spvg.html) from mainly a diagnostic, prognostic and therapeutic approach, to one that fosters wellness, and incorporates information and practices which seeks to advance disease prevention, and the arrest of disease manifestation in populations. 23 different Federal Agencies or Departments were invited as well as 30 professional organizations to represent the highly inter-disciplinary and multi-disciplinary scientific infrastructure, methods and means, which at the heart will protect, and purvey timely information to enable the implementation of the most effective and efficient medical practices for the patient's benefit. This vision acknowledges and asserts that the scientific community has a very large role, and a responsibility toward creating the fertile environment, the conditions, the means, and the practices, which can bring about much needed Healthcare transformations.

The globally-interconnected digital information and communications infrastructure underpins almost every facet of modern society and provides critical support for the economy, civil infrastructure, public safety, and national security. This technology has transformed the global economy and connected people in ways never imagined. Yet cybersecurity risks pose some of the most serious economic and national security challenges of the 21st Century. A growing array of state and non-state actors are compromising, stealing, changing, or destroying information and could cause critical disruptions to U.S. systems and those of our allies. In the face of this expanding threat, traditional telecommunications and Internet networks continue to converge, and other infrastructure sectors are adopting the Internet as a primary means of interconnectivity. Within the domain of cyberspace, nations must not only maintain an environment that promotes efficiency, innovation, economic prosperity and free trade but also assure safety, security, civil liberties, and privacy rights. It is the fundamental responsibility of our governments to address strategic vulnerabilities in cyberspace and ensure that the world realize the full potential of the information technology revolution. Malicious cyber activity is occurring on an unprecedented scale with extraordinary sophistication. We are confronted by nation-states, terrorist networks, organized criminal groups, individuals, and other cyber actors with varying combinations of access, technical sophistication and intent. Daily, elements of the U.S. information infrastructure are being targeted for intelligence collection, intellectual property theft, or disruption. Criminal elements are operating a pervasive, mature on-line service economy providing illicit cyber capabilities and services to anyone willing to pay. Against this backdrop the U.S. Government (USG) is focusing its efforts on helping the United States achieve a more reliable, resilient, and trustworthy digital infrastructure for the future. It is working with all key players in U.S. cybersecurity, including state and local governments and the private sector, in a manner consistent with ensuring privacy rights and civil liberties.

E-Health: Notable Developments, - Who: the Temple of Public Health

Keynote: October 26th, 2016 2:00 – 3:00 p.m
Auditorio Mayor Carlos Gómez Albarracín
Universidad Autónoma de Bucaramanga, Colombia

Marc Nyssen Treasurer IFMBE International Federation for Medical and Biological Engineering Professor Medical Informatics - Medische Informatica Faculteit Geneeskunde en Farmacie
Vrije Universiteit Brussel, BELGIUM

Doctor in Applied Sciences in June 1983, after defending a thesis: "New architectures for opto-electronic signal processing" (Maxima Cum Laude).
Member bureau VLIR-UOS since August 2004, Chairman VLIR-UOS 2005–2009. Member of the Commission "Computer management" and of the Board of Directors of the "Data-center VUB-ULB". Since January 1st 2006 to March 2008 Chairman of the Board of Directors of the "Data-center VUB-ULB". Affiliated with the International Union for Physical and Engineering Sciences in Medicine. Member of the "Vlaamse Toezichtcommissie" 2010–2014 (as engineer) Since September 1st 2014 chairman of the "Vakgroep Geneeskunde Wetenschappen" (Public Health) at VUB, Faculty of Medicine and Pharmacy.

Abstract
The aim of this presentation is to give an overview of the operational features of e-health developments, linking the technical building blocks, available to today's developers together in ways that enable to satisfy the requirements of the operational public health systems.
Constraints from the professional and legal context of modern medicine and privacy are highlighted, as well as the information and communication tools that make it possible to realize usable systems. To have a chance for success, win-win situations must be created, as well for the health care professionals as for patients. Real examples illustrate successful and less successful implementations.
Finally, the future directions towards more interaction and control by patients themselves and the impact of mobile devices are put in perspective.

A Glimpse at Emerging Innovative Cardio-Tech Solutions

Keynote: October 27th, 2016 8:00 – 9:00 a.m
Auditorio Mayor Carlos Gómez Albarracín
Universidad Autónoma de Bucaramanga, Colombia

Shankar Krishnan, Ph.D. Elected President International Federation for Medical and Biological Engineering - IFMBE. Professor & Chair. Department of Biomedical Engineering. Wentworth Institute of Technology, Boston, USA.

Dr. Shankar Krishnan is the founding chair of the Biomedical Engineering program and an endowed chair Professor at Wentworth Institute in Boston since 2008. At NTU in Singapore, he was the founding director of the BME Research Center and the founding head of the Bioengineering division. He has served the administrative counsel of IFBME for the past ten years, and he is presently the Secretary General of IFBME. He was elected as a Fellow of American Institute of Medical and Biological Engineering and he was a member of a team which received the CIMIT Kennedy Innovation Award in Boston.

Abstract
Recent technological advancements tend to introduce innovative ways to provide better product performance in many fields, including health care applications. Heart disease and strokes cause nearly 750,000 deaths in the annually. Heart disease is considered to be the leading cause of death in both men and women, according to the CDC, with more 600,000 deaths annually in the U.S. The emerging new technological applications can provide early diagnosis and proper treatment which will be beneficial to patients, doctors, hospitals, payers and manufacturers. These new "cardiotech" solutions are of particular interest as they are providing state-of-the-art heart disease detection and therapy significantly impacting patients, healthcare providers and the society on a whole. The objective of the presentation is to offer a glimpse at recent and upcoming innovative cardiotech solutions.
To facilitate cardiovascular disease diagnosis and therapy, numerous devices and techniques have been developed: including ECG, ICU monitors, ultrasound, CT, MR, pacemaker, ICD, and LVAD. Despite these advances, cardiovascular diseases remain very lethal. However, recently approved devices and developing technologies aim to provide better diagnosis and therapy to reduce the devastating effects of the cardiovascular diseases.
In the following section, a few state-of-the-art emerging technologies based solutions for cardiovascular disease are highlighted. "The Katheter" utilizes ultrasound to construct a more accurate 3-D image of the heart and provides a map of electrical activation of the heart. It can locate the sources of irregular heartbeats. In addition, procedures that utilize the information from this technology have seen improved success rates and even reduces the length of time to perform A-Fib catheter ablation.
Better monitoring systems are becoming smaller, mobile, and more convenient for clinical and in home use than ever before. Medronic's Seeq Mobile Cardiac Telemetry system is allowing for longer monitoring, while St. Jude has developed the CardioMEMS HF system to allow for wireless monitoring of the heart. These systems are providing vital information, such as ECG and heart rate that could not have been achieved before.
The transformation to smart phone use allows for mobile phone to monito vital signs without needing to spend precious time waiting at the doctor's office. AliceCor's Mobile ECG Clips attachment can go directly onto the phone and record, store, analyze, and transfer single-channel ECGs directly through the phone in just 30 seconds. This information can then be easily shared with a doctor.
Other newly emerging diagnostic practices include noninvasive assessment of coronary artery blockages, lipoprotein associated CHD evaluation and the Trimethylamine-N-oxide test. These are then paired with novel monitoring technologies such as the Zio XT Patch, leadless pacemakers, and wearable ECG patches.

The emergent technologies are making diagnosis and proper treatment less expensive, less time consuming, and more accurate, reducing risks and costs for all parties involved. They are paving the way for even more innovative technologies. Radical and impossible ideas of the past may reach fruition in the future. Stem cells may be used for reconstructing myocardium and wireless charging could power the total artificial heart. All of these technologies are expected to emerge in the future and will save lives. The multi-segmental costs associated with innovative cardiotech solutions may impose constraints and challenges based on the health care coverage models in different countries across the globe. Efforts must be made to address and overcome the challenges with cohesive multi-organizational and international levels and participations.

In conclusion, recent cardiotech solutions prove to be promising techniques to diagnose, treat and monitor patients with cardiovascular diseases. It is anticipated that carefully coordinated interdisciplinary efforts around the globe may lead to greater efficacies of procedures, reduced costs, reduced lengths of stay in hospitals, and improved overall outcomes and biomedical engineers make their contributions toward inventions and implementations to enhance the quality of life.

Emerging Trends in the Signal and Image Processing and Computer-Aided Diagnosis

Keynote: October 27th, 2016 1:30 p.m. – 2:30 p.m.
Auditorio Mayor Carlos Gómez Albarracín
Universidad Autónoma de Bucaramanga, Colombia

Andrew Laine President IEEE/EMBS Professor of Biomedical Engineering, Chair Department of Biomedical Engineering, Columbia University, New York USA. Professor of Radiology (Physics), Department of Radiology; Director, Heffner Biomedical Imaging Laboratory Columbia University, New York, NY USA.

Andrew F. Laine is currently Chair of the Department of Biomedical Engineering and Director of the Heffner Biomedical Imaging at Columbia University and the Percy K. and Vida L. W. Hudson Professor of Biomedical Engineering and Professor of Radiology. He was the Program Chair for the IEEE EMBS annual conference in 2006 held in New York City and served as Program Co-Chair for IEEE ISBI in 2008 (Paris, France). He served as the IEEE EMBS Vice President of Publications 2008–2012, and currently the President of IEEE EMBS.

Abstract
Quantitative Imaging Informatics in Cost effective PET Imaging and Classification of lung disease. This talk presents a novel method for emphysema quantification, based on parametric modeling of intensity distributions in the lung and a hidden Markov measure field model to segment emphysematous regions. The framework adapts to the characteristics of an image to ensure a robust quantification of emphysema under varying CT imaging protocols and differences in parenchymal intensity distributions due to factors such as inspiration level. Compared to standard approaches, the present model involves a larger number of parameters, most of which can be estimated from data, to handle the variability encountered in lung CT scans. The method was used to quantify emphysema on a cohort of 87 subjects, with repeated CT scans acquired over a time period of 8 years using different imaging protocols. The scans were acquired approximately annually, and the data set included a total of 365 scans. The results show that the emphysema estimates produced by the proposed method have very high intra-subject correlation values. By reducing sensitivity to changes in imaging protocol, the method provides a more robust estimate than standard approaches. In addition, the generated emphysema delineations promise great advantages for regional analysis of emphysema extent and progression, possibly advancing disease subtyping, including COPD.
An important tool for studying brain disorders is positron emission tomography (PET), a nuclear imaging technology that allows for the in vivo functional characterization and quantification of blood flow, metabolism, protein distribution, and drug occupancy using radioactively tagged probes (tracers). Full quantification of PET images requires invasive arterial input function (AIF) measurement through online arterial blood sampling for the duration of the scan (1-2 hours). The AIF is used to correct images by accounting for the tracer bioavailability, which depends on an individual's physiological capacity for clearance, distribution and metabolism of the tracer. However, AIF measurement is invasive, risky, time consuming, uncomfortable for patients, and costly. Perhaps most importantly, it is impractical at the point-of-care and therefore limits clinical utility of PET. We believe an integrative multi-modal approach is possible via the amount of personalized information about the physiological and biochemical makeup of individuals available in their electronic health record (EHR). This talk will outline a novel approach to combine EHR and dynamic PET imaging data in an optimization framework based on simulated annealing to non-invasively estimate the AIF. Techniques that will be outlined are applicable across imaging modalities, organs and diseases, such as functional imaging of prostate cancer images where increasingly more complex tracers are utilized for assessment and require AIF measurement.

Improving Prevention and Treatment of Diabetes

Keynote: October 28th, 2016 8:00 – 9:00 a.m.
Auditorio Mayor Carlos Gómez Albarracín
Universidad Autónoma de Bucaramanga, Colombia

Ratko Magjarević. Past President IFMBE Faculty of Electrical Engineering and Computing, University of Zagreb, Croatia

Ratko Magjarević received his Ph.D. in Electrical Engineering in 1994 from the University of Zagreb, Faculty of Electrical Engineering. He is full professor teaching several courses in Electronic Instrumentation and Biomedical Engineering at undergraduate, graduate and at postgraduate studies. As visiting professor he was teaching in Stuttgart, Trieste, Ljubljana, Madrid and Bogota.

R. Magjarevic is elected the President of International Federation for Medical and Biological Engineering (IFMBE) for the term of office 2012-15. He is also the Editor in Chief of the IFMBE Proceedings series.

Abstract

Patient care and management in non-communicable chronic diseases can be significantly improved by introducing of ICT tools into management and self-management of the disease which is recognized in the new health policies of the European Union as empowerment of the patient. WHO defines patient empowerment as "a process through which people gain greater control over decisions and actions affecting their health". Empowerment process should be seen both, as an individual and as a community measure, bringing primarily benefits to individuals but also to the community through reduced costs.

Patients with diabetes are supposed to accurately collect data such as blood glucose level, insulin doses applied, weight, daily exercise and activity and/or blood pressure in order to be able to self-manage their disease. This paper describes a complete support system for monitoring of patients and for remote home care based on the Personalized Intelligent Mobile Health System (PIMHS) for diabetic patients. The system is based on a custom made wireless network and includes several types of medical, wellness or from the shelf devices that can provide information on the patient. Mobile unit is the main device for monitoring basic physiological functions, data storage and management. Peripheral devices from different manufacturers with connectivity (e.g. glucometer, blood pressure meters, scales) or custom made sensors can be connected to the network via interconnection devices.

Heart, Heart Rate, & Heart Rate Variability with Respiratory Sinus Arrhythmia

Keynote: October 28th, 2016 5:00 – 6:00 p.m.
Auditorio Mayor Carlos Gómez Albarracín
Universidad Autónoma de Bucaramanga, Colombia

Kang-Ping Lin, Ph.D. Secretary General IFMBE Distinguished Professor, Chung-Yuan Christian University

He is Distinguished Professor of Electrical Engineering at Chung-Yuan Christian University, Taiwan. He served as Director of Medical Device Technology Division of the Biomedical Engineering Center in Industrial Technology Research Institute in Taiwan (2000–2004). He was the president of Taiwanese Society of Biomedical Engineering (2007–2010) and the Editor-in-Chief of the Journal of Medical & Biological Engineering (1999–2007). He is now the Director of
He has several roles in IFMBE including the Chair of Publication Committee and Publicity Committee, the Co-Chair of Asia Pacific Working Group Committee, and the Editor of IFMBE Newsletter from 2009 to now.

Abstract
Aging and many diseased states have been shown to be associated with depressed vagal activity. The extent of vagal depression correlates with the severity and poor prognosis of the disease. It will be desirable if there is physiological vagal enhancement method based on a reliable self-evaluation technology, and a tool easy to use. The tool needs to be simple, remote monitoring capability, self-manageable and affordable. In this presentation, a handheld ECG device will be introduced for heart rate variability (HRV) analysis and respiratory sinus arrhythmia (RSA) application. Based on the device, results will show that heart rate varies during the respiratory cycle and slowing during inspiration, which is the feature of a healthy heart and autonomic nervous system and is known as RSA, and the vagal effect of slow deep breathing increases HRV and high frequency (HF) spectral power components in particular. Some studies concluded that the correlation between RSA and HR reflects the cardio-pulmonary coupling under parasympathetic control. In additions, because it is more convenient to measure photoplethysmography (PPG) than ECG, PPG is supposed as a surrogate of ECG for HRV analysis. In this presentation, a few comparisons measured the spectrum of PRV from PPG with HRV from ECG with different breathing conditions. The results showed that PRV might be used as an alternative to HRV depending on the applications and conditions.

Semi – Keynote Conferences

EL RETO DE LA INTEROPERABILIDAD EN TECNOLOGÍAS PARA LA SALUD

Hora: 6:00 p.m. – 6:30 p.m.
Fecha: Octubre 27
Lugar: Auditorio Mayor Carlos Gómez Albarracín

Dr. Antonio Hernandez, EE PE. USA

Antonio Hernández es un fundador y miembro del Consejo de Administración del Global Health International Advisors (GHIA).Trabajó por 29 años en la Organización mundial de la Salud (OMS) como mienbro sénior de Servicios de Salud, Infraestructura Física y Tecnología. Cuenta con más de 40 años de experiencia en el campo de la tecnología de la salud y ha sido miembro de At-Large para la Junta del American College of Clinical Engineering – ACCE.

Abstract

Healthcare has experienced a fast evolution due to the convergence of information technologies and communications with the health technologies. Medical technology in healthcare facilities is moving from standalone equipment to networks of medical devices (analog and digital). Technology is evolving and impacting public health and increasing used and servicing communities. Another current trend is the use of mobile communication platforms for personal monitoring and auto-management of health (people responsible of their own health). In parallel to this is the evolution of epidemiological information systems, the clinical information systems, and the information generated by the mobile technologies. All come together in the electronic Health records (EHR). We will highlight some aspects of the integration and convergence of these technologies and the standard that allows the communication among information systems and medical devices under development by the "Integrating the Healthcare Enterprise" (IHE) initiative. This evolution and its implementation is impacting the operation of healthcare delivery systems and dramatically changing the workflows in the healthcare institutions. Based on the change in the demographic indicators and epidemiological profile in Latin America and the Caribbean population, there are challenges and opportunities to better deploy, use, and maintain these evolving technologies.

DESAFÍOS DE LA INGENIERÍA CLÍNICA: NUEVOS ESCENARIOS DE GESTIÓN DE TECNOLOGÍA MÉDICA

Hora: 6:30 p.m. – 7:00 p.m.
Fecha: Octubre 27
Lugar: Auditorio Mayor Carlos Gómez Albarracín

Dr. Renato García Ojeda, Brasil.

Es miembro Sénior EMBS-IEEE, del American College of Clinical Engineering - ACCE y de la Sociedad Brasileña de Ingeniería Biomédica. De 2011 a 2013 fue Presidente del Consejo Regional de Ingeniería Biomédica para América Latina – CORAL. En 2015 recibe reconocimiento de la Inter-nacional Federation for Medical & Biological Enginnering por contribuciones como CORAL-Chair y por el IEB-UFSC el Clinical Engineering Manuscript Award de la División de Ingeniería Clínica de IFMBE; en el primer congreso Mundial de Ingeniería Clínica y Gestión de Tecnología en Salud. En la actualidad es Profesor Asociado, Investigador y Coordinador del Instituto de Ingeniería Biomédica (IEB-UFSC) de la Universidad Federal de Santa Catarina.

Abstract
Desafíos de la Ingeniería Clínica : Nuevos Escenarios de Gestión de Tecnología Medica. Las tendencias actuales de cambios en los sistemas organizacionales de Salud en el mundo y la modificación de perfiles demográficos y tecnológicos generan nuevos escenarios y desafíos.
Tendencias como modelos de salud personalizada, centradas en la seguridad del paciente, métodos de gestión de enfermedades crónicas, entre otras, incorporan necesidades de nuevas soluciones tecnológicas y de gestión, que modifican significativamente las actividades de Ingeniería Clínica.
Conceptos como Big Data, Interoperabilidad de sistemas, tecnologías pervasivas y modelos de sistemas ubicuos, determinan la necesidad de nuevas herramientas y modelos para la Ingeniería Clínica en sus programas de Gestión de Tecnología Medica. De esta manera, como ejemplo, herramientas como ingeniería de factor humano, modelos de toma de decisiones multiparametros como MCDA, gestión de procesos distribuidos, evaluación de tecnologías en salud se tornan fundamentales para la gestión de tecnologías medicas. Esta presentación analiza estos nuevos escenarios discutiendo los desafíos que la Ingeniería Clínica deberá enfrentar.

REGISTRO DE IMÁGENES MÉDICAS: APLICACIONES EN EPILEPSIA

Hora: 6:00 p.m. – 6:30 p.m.
Fecha: Octubre 27
Lugar: Auditorio de Ingenierías Jesús Alberto Rey Mariño

Dr. Juan Pablo Graffina, Argentina.

Ingeniero Electrónico, Doctor en Ingeniería, Magister en Ingeniería Biomédica y Especialista en Docencia Universitaria. En investigación ha dirigido y participado en proyectos de la Universidad Nacional de San Juan y de la Agencia de Promoción Científica y Tecnológica de Argentina, con publicaciones en su área disciplinar y dirección de tesis y tesinas. En gestión ha sido Director del Gabinete de Tecnología Médica (2003-2012) y autoridad del Departamento de Electrónica y Automática de la Facultad de Ingeniería de la Universidad Nacional de San Juan (2012 a la fecha). Actualmente es Presidente de la Sociedad Argentina de Bioingeniería (2015 a la fecha).

Abstract
The neurophysiology requires different equipment and studies for increasingly complex diagnosis. The pathologies are approached from anatomical, functional and metabolic information from different modalities. Epilepsy is a chronic neurological disorder that requires a variety of diagnostic techniques for the location of the epileptic area. In some cases, Magnetic Resonance Imaging (MRI) seems to be the technique of choice because it allows extraordinary precision to locate the anatomical area responsible of irritative focus. However, the seizure focus does not always coincide with a structural alteration, and in such cases functional imaging techniques such as Single Photon Emission Computed Tomography (SPECT) and Positron Emission Tomography (PET) are recommended. In addition, there are other complex techniques as Tractography, Functional Connectivity, Electrical Maps, etc.
Medical Image Modalities have different reference frameworks and theirs fusion requires adjust position and applied different transformations. Registration is defined as a process to transform mobile volume(s) in same framework of fixed volume. Maintz [1998-2016] proposed different categories for analysis of Registration Process.
Conference describes a basis of Registration and its application to different examples with epileptic patients. First part exposes theoretical description to calculate optimal transformation and classic evaluation functions. Calculation depends of different factors: Modalities, Anatomic parts, etc.
In second part, some cases are presented with epileptic patients. Different problem associated to registration and its solution is shown. These include: MRI-fMRI, MRI-Tractography, MRI-CT, MRI-Atlas, Medical Volumes-Patient applied to neuronavegation, MRI-SPECT, MRI-PET-Tractography, MR-Tractography-Electric Map, and so on.
As conclusion, references and main ideas are presented as summary of presentation.

LA FORMA DE LOS ÁTOMOS EXPLICARÍA EL FUNCIONAMIENTO DE ESTRUCTURAS ORGÁNICAS COMPLEJAS

Hora: 6:00 p.m. – 6:30 p.m.
Fecha: Octubre 27
Lugar: Auditorio Menor Alfonso Gómez Gómez

DE LA FORMA DE LOS ÁTOMOS AL CÓDIGO OCULTO DEL ADN

Hora: 6:30 p.m. – 7:00 p.m.
Fecha: Octubre 27
Lugar: Auditorio Menor Alfonso Gómez Gómez

Jaime Delgado Avendaño, Bucaramanga, Colombia.
Físico y matemático, Universidad de Pamplona. 40 años de experiencia como profesor de física Publicaciones: UNIFICACIÓN DE LA FÍSICA, (Primera Aproximación). 1976; MAS CERCA DEL CIELO, (Ecuación de Órbita). 2003 y FISICANOVA, (Una Aproximación a la Realidad). 2008

Abstract
FISICANOVA es un conjunto de teorías acerca de las fuerzas fundamentales, que intentan una aproximación a la realidad, basadas en la concepción de un Universo formado por minúsculas burbujas tangentes, cuyo interior e intersticios son vacío absoluto, que se colapsan para formar partículas materiales que deforman el espacio circundante originando campos de energía. Esta sencilla concepción del Universo permite explicar los fenómenos naturales que van desde lo más pequeño hasta lo más grande: Desde la más pequeña de las partículas subatómicas hasta el más grande de los agujeros negros, de una manera elegante y coherente sin recurrir a definiciones abstrusas e imposibles. Del estudio de estas teorías se desprende que los átomos tienen formas particulares que les confieren sus propiedades físicas y químicas, de tal manera que sus compuestos y cristales están organizados en arreglo a dichas formas. En su aplicación a la Bioquímica, FISICANOVA muestra cómo la forma de los átomos es decisiva al momento de explicar el funcionamiento de estructuras orgánicas complejas como la Hemoglobina, la Cobalamina, ADN y ARN entre otras. FISICANOVA abre una puerta para aplicaciones médicas y tecnológicas que permitirán el mejoramiento de la calidad de vida física, emocional y social del ser humano.

MEDICIONES PSICOFISIOLÓGICAS INSTRUMENTALES, UN CAMPO EMERGENTE DE LA INGENIERÍA BIOMÉDICA

Hora: 1:30 p.m. – 2:00 p.m.
Fecha: Octubre 28
Lugar: Auditorio Mayor Carlos Gómez Albarracín

Dr. Guillermo Avendaño, Chile.

Presidente de CORAL. Académico e investigador de la Escuela de Ingeniería Civil Biomédica de la UV. Ingeniero Ejecución Electrónico, U. Norte, Chile; Magister en Educación Ambiental, UPLA, Chile. Áreas de Desarrollo: Bioinstrumentación. Líneas de Investigación: Psicofisiológica, Radiometría, Simulación de procesos biofísicos y equipos médicos.

Abstract

Entre los equipos que en la actualidad se utilizan para realizar mediciones de procesos encefálicos, están los conocidos con herramientas psicofisiológicas, las cuales presentan como principal característica, ser de fácil utilización, baja o nula invasividad y un costo mucho menor que cualquier otro dispositivo bio-instrumental usados en la medición encefálicas como los electroencefalógrafos, potenciales evocados o los complejos equipos de imagenología, al mismo tiempo son más eficaces y eficientes que los test o escalas psicológicas convencionales.

Sus principales aplicaciones son, estudiar procesos encefálicos relacionados con el aprendizaje humano, todos los fenómenos encefálicos originados por las cargas psíquicas como el Stress, Burn Out y los trastornos del comportamiento afectivo social, trastornos del procesamiento cognitivo, detección de alteraciones de las llamadas funciones ejecutivas, F.E.

En psicofisiología se utilizan exitosamente, medidores de Tiempo de Reacción simples y complejos, medidores de Umbral de Discriminación Táctil (Estesiómetros), medidores de Frecuencia Crítica de Fusión (Flicker), sistemas de Biorealimentación (Bio Feedback), Medidores de tremor muscular mecánicos (Peeple), medidores de Coordinación motora (MAINS) y otros de naturaleza similar.

Se han diseñado, construido y aplicado a niños escolares, un conjunto de varios equipos que nos han permitido obtener útiles resultados verificando o refutando diagnóstica en Trastorno de Déficit Atencional con Hiperactividad y desorganización visoespacial

TENDENCIAS EN LA BIOMECÁNICA CLÍNICA Y DEPORTIVA

Hora: 1:30 p.m. – 2:00 p.m.
Fecha: Octubre 28
Lugar: Auditorio de Ingenierías Jesús Alberto Rey Mariño

Ph.D. Ariel Braidot, Argentina.

El PhD Bioingeniero Ariel Braidot se ha especializado en aplicaciones clínicas y deportivas de la biomecánica en los últimos años. Se desempeñó como Presidente, Vicepresidente y Comisión directiva de la Sociedad Argentina de Bioingeniería-SABI-.2003-2013. Director de la Investigador Asistente, Becarios de Posdoctorado y Doctorado del CONICET y Maestría. Evaluador de Proyectos Académicos Y Tecnológico. Director de Proyectos de Investigación Y Cooperación nacionales e internacionales. Responsable de Convenios de Colaboración con Instituciones nacionales e internacionales. Director Académico, Coordinador Académico, de la Maestría en Ingeniería Biomédica, FI-UNER. Presidente de la comisión de Comisión de Extensión e Investigación. 2006-2016. Director y Subdirector del Departamento Bioingeniería de la Carrera Bioingeniería y Bioinformática de la FI-UNER. Integrante del Banco de Evaluadores de CONEAU desde 2012, Par evaluador en las Carreras de Bioingeniería/Ingeniería Biomédica en 2015. Director del Laboratorio de Biomecánica de la FI-UNER. Posee publicaciones internacionales en revistas, libros, capítulos y eventos científicos. Participó como responsable del dictado de cursos de grado y posgrado para carreras de Grado, Maestría y Doctorado de Bioingeniería/Ingeniería Biomédica.

Abstract
Desde sus inicios el análisis biomecánico ha tenido su mayor desarrollo en el área de mejoras en patologías del movimiento. Específicamente, la marcha es el gesto más común que se ha estudiado desde hace varias décadas. También tiene un menor desarrollo el análisis del movimiento aplicado a patologías del miembro superior, en este aspecto uno de los gestos más comunes es alcanzar y agarrar un objeto con la mano, actividad ampliamente empleada en actividades de la vida diaria.
El análisis del movimiento de gestos deportivos es más reciente. En este aspecto se ha fortalecido el estudio de atletas de elite tanto para la mejora en la técnica específica como en la valoración de posibles lesiones futuras, empleando así el análisis biomecánico como una herramienta de predicción.
Para lograr un adecuado análisis biomecánico se requiere el desarrollo de herramientas de modelado y estimaciones antropométricas con la mejor precisión posible. Como ejemplo, se presentan las herramientas de desarrollo de modelos biomecánicos aplicados a gestos deportivos de interés. Se analizan estos nuevos ejemplos de aplicación considerando que una mejora en la calidad de los gestos tanto en deportistas profesionales de elite como jugadores amateurs permitirá fomentar un mejor desarrollo del deporte en forma continua. Además contribuirá a disminuir lesiones, lo que sin duda significa una mejora en la calidad de vida y social de la población.

IMPLANTES CARDIOVASCULARES A LA LUZ DE NUEVAS TÉCNICAS Y MATERIALES

Hora: 2:00 p.m. – 2:30 p.m.
Fecha: Octubre 28
Lugar: Auditorio de Ingenierías Jesús Alberto Rey Mariño

Dr. John Bustamante, Colombia

Médico y Cirujano de la Universidad Pontificia Bolivariana/Ph.D. en Cardiología de la Universidad Autónoma de Barcelona/Post-Ph.D. en Prótesis y Bioimplantes Cardíacos del Instituto Nacional de Cardiología de México. Director del Grupo de Investigación de Dinámica Cardiovascular de la Universidad Pontificia Bolivariana/Coordinador del Programa de Doctorado en Ciencias Médicas y Director del Centro de Bioingeniería de la misma Universidad. Investigador Sénior, vinculado a múltiples proyectos de investigación en el área cardiovascular, en los temas de: prótesis e implantes, biomateriales, biomecánica, instrumentación, modelización y simulación; de los cuales se han publicado varios textos y artículos científicos. Desarrollador de dos patentes de invención de dispositivos cardiovasculares. Miembro de la Sociedad Colombiana de Cardiología y Cirugía Cardiovascular: Comité Científico y Comité de Arbitraje de la Revista Colombiana de Cardiología. Miembro Honorario Federación Argentina de Cardiología: Moderador Mesa Temática "Foro Falla Cardiaca (Heart Failure Forum)". Fellowship de la Interamerican Society of Cardiology. Miembro de la Junta Directiva de la Asociación Colombiana de Bioingeniería ABIOIN, y ExPresidente del Capítulo ACOBIAN; Miembro del Comité Editorial, Científico y de Arbitraje, en varias revistas indexadas de carácter científico internacionales y nacionales.

Abstract
Múltiples alteraciones estructurales del corazón y el sistema vascular traen como consecuencia disfunciones en la acción circulatoria; dependiendo de la severidad de tal lesión llega a ser necesario el reemplazo o la intervención sobre algunas estructuras anatómicas por medio de diferentes implantes que pueden ser de tipo biológico o sintético. Las posibilidades abiertas con la incursión de diferentes biomateriales, con gran compatibilidad biológica y excelentes comportamientos biomecánicos, permiten disponer en la actualidad de una variedad de elementos para el manejo quirúrgico de pacientes con diversas patologías cardiovasculares, incrementando su calidad de vida.
Los biomateriales de uso cardiovascular surgen de los desarrollos de nuevos materiales, mediante procesos expeditos que brindan nuevas texturas, diversos comportamientos mecánicos y tolerancias al esfuerzo y a la agresión histoquímica, lo que garantiza un adecuado comportamiento al ser implantados en el organismo humano.
En la presente conferencia se presenta la dinámica de estas mejoras que propenden en alcanzar una mejor operatividad, así como una mayor adaptabilidad del tejido biológico, lo que redunda en la funcionalidad y duración de la prótesis, reduciendo la aparición de falla primaria. Se ilustran los medios de evaluación de dispositivos cardiovasculares, mediante ensayos de biocompatibilidad en modelos in vitro de células cardiacas valorando su capacidad proliferativa y generación de sincitios que interactúan con el implante, así como diferentes bancos de ensayos cardiovasculares, facilitando la valoración objetiva y permitiendo determinar si son aceptables para su uso clínico.
El impacto de los proyectos que se enmarcan en la disposición de dispositivos que permiten la introducción de nuevas prótesis cardiovasculares en el mundo de esta especialidad médica, han de repercutir también en el ámbito social hacia una alta disponibilidad, buena calidad y menores costos.

NEFRO-RED: UN NUEVO CONCEPTO DE GESTIÓN PARA HEMODIÁLISIS

Hora: 1:30 p.m. – 2:00 p.m.
Fecha: Octubre 28
Lugar: Auditorio Menor Alfonso Gómez Gómez

Dr. Miguel Cadena Méndez, México.

El Dr. Cadena trabaja en la Universidad Autónoma Metropolitana-Unidad Iztapalapa (UAM-I), desde 1974, adscrito desde el 2003 al Centro de Investigación en Instrumentación e Imagenología Médica, (www.ci3m.mx) en la Cd. de México. Egresó como Ingeniero (IPN-ESIME, 1965-1968, México), y realizó estudios a nivel maestría (WU, 1978-1980, EUA) y a nivel doctorado (UNAM, 2002-2005, México). Es profesor fundador de la licenciatura y posgrado en Ingeniería Biomédica de la UAM-I (1974 y 1982). Vicepresidente y Presidente SOMIB (1994–1997) y CORAL (2005–2010). Medalla al Mérito Universitario por la UAM-I (2012) y socio honorario por la IFMBE (Canadá, 2015).

A New Model for Hemodialysis Management

Miguel Cadena-Méndez, Emilio Sacristán-Rock, Fabiola Martínez-Licona and Joaquín Azpiroz-Leehan
Research Center for Medical Imaging and Instrumentation (www.ci3m.mx)
EE Department, Universidad Autónoma Metropolitana-Iztapalapa, México City.
mcm@xanum.uam.mx

RESUME

Purpose:

The aims in this semi-plenary are to explain: (a) how lethal rates (20-30%) in hemodialysis clinics is incremented when their management models are oriented to reduce operational costs in the treatment processes. The high operational costs and the low payment per treatment push clinics to reduce expenses in quality control (QC) such that they produce high rates of intradialytic hypotension (30-40%), and (b) how clinics using a different model based in business processes management (BPM), which is oriented to preserve patients' homeostasis, can reduce operational costs 35% and improve QC in order to reduce hypotension events (10%) and lethal rates (5%).

Background:

The problem is the unacceptable lethal rates (20-30%) in 65% in outpatient's clinics in México. The aims are to reduce intradialytic hypotension events and operational costs to finally reduce lethal rates. Today, the management models in clinics & hospitals for hemodialysis are oriented according to the QC suggested by the KDOQI guidelines (Kt/V> 1.2). However, this standard is not enough to assure high performance in clinics' processes. The health care delivery system in México together with the supplier companies have generated wrong business-rules. These rules are based in the "pay per treatment or pay per event" model. Public bidding contracts squeeze amounts of payment per treatment up to 65±10 USD. The indirect consequence is that public and private clinics unnecessarily work in unethical manner for profit: (1) reducing QC costs until $5 since supplies and overhead costs are approximately 50±5 USD/treatment, and (2) reducing treatment's time to 2.5 hours in average instead using 3.5 hours since they look for the treatment of the "third patient" in 8 hours shifts to generate positive return margin, and (3) reducing the hemodialysis services offer to only 13% of the total ESRD prevalence (200,000 patients/year) to unbalance demand vs offer index. The deplorable consequences are: (4) high rate of intradialytic hypotension events (30-40%), (5) high lethal rates of 200-300 deaths in 145,000 treatments per year, and (6) poor benefits for all stakeholders involved since an assessment study in 79 outpatient clinics (Malaquias et al, UNAM-2009) showed that 65% of the clinics run under improper QC performance.

Methodology Approach:

The targets in the BPM model were defined as: (a) to reduce up to 10% the hypotension rate which is equivalent to reduce up to 5% the lethal rate, (b) to reduce treatment operational costs 35%, and (c) to increase 300% the costs devoted to QC equivalent to 20 USD/treatment. Then, the new aim was defined as: "the more patient's homeostasis is preserved at long time, the less operational costs are generated". Then, the outputs and outcomes from the BPM model were defined as: less medical consults, less laboratory tests, less use of chronic pharmaceutical prescriptions (EPO, kelants, vitamins, anti-hypertensives, etc), better hemoglobin levels (10-12 gr/dl), better albumin levels (4-5 gr/dl) and the most important, more patient's quality of life.

The BPM modeling and operational processes were designed in three phases: first, the "what's" were defined in an assessment process "as is" in standard clinics. Second, the benefits were defining mapping decisions and actions in the "who's" and "how's" silos were defined in such a manner that QC process were based in a sequence of homeorhesis state variables measurement along each treatment: (a) energy expenditure with metabolic substrates, (b) cardiac simpatho-vagal index and (c) hydric composition. All these physiological variables were measured as a response to changes of multiple factors: (1) UFR by hemodiafiltration (HDF), (2) PO4 clearance, (3) dialysate temperature control, (4) intradialytic exercise, (5) reuse of dialyzer filters, and (5) nutritional support. The impact of this new BPM model was partially proved by multiple pilot studies at the National Cardiology Institute Ignacio Chavez, México City, in patients under preparation to receive kidney transplants.

Findings:

The costs for QC process were incremented up to $20 USD per treatment reducing up to 12% the hypotension events.

IMPACTO DE LAS NUEVAS TECNOLOGÍAS BIOMÉDICAS EN EL FUTURO DE LA MEDICINA

Hora: 2:00 p.m. – 2:30 p.m.
Fecha: Octubre 28
Lugar: Auditorio Menor Alfonso Gómez Gómez

Dr. Jorge Reynolds, Colombia.

El trabajo de Dr. Reynolds ha sido valorado mundialmente. Es miembro de 42 sociedades científicas en Colombia y el exterior; en algunas de ellas como miembro honorario. Es miembro de la Academia de Ciencias de Nueva York desde 1989, miembro de la Academia Colombiana de Ciencias Exactas, Físicas y Naturales también desde 1989 y miembro Asociado de la Academia Nacional de Medicina desde el año 2004. También es miembro fundador de varias sociedades científicas. En el 10 de noviembre de 2015 recibió el doctorado Honoris Causa por parte de la Universidad de La Sabana, en reconocimiento de sus 55 años investigativos en el funcionamiento eléctrico del corazón.

Precongress Courses

October 25th, 2016 Duration: 4 Hours

Curso	Profesor	Lugar/Hora
Aplicaciones Clínicas de la Bioimpedanciometría	**Prof. Dr. Ing. Franco Simini** 	Auditorio Facultad de Medicina UNAB 8:00 a.m. – 12:00 m
La Nanotecnología en la Bioingeniería	**Lina Marcela Hoyos Palacios** 	Universidad Pontifica Bolivariana Bloque K 8:00 a.m. – 12:00 m 2:00 p.m. – 6:00 p.m.

(*Continued*)

(*Continued*)

Curso	Profesor	Lugar/Hora
Prospectiva de la Ingeniería de rehabilitación y tecnología biomédica para la discapacidad	**Dr. Rafael Kohanoff**  **Prof. Mario Aguilar** 	Laboratorio de Electrónica UNAB 8:00 a.m. – 12:00 m
Análisis de Micro potenciales Cardíacos en Registros ECG de Alta Resolución	**Prof. Eric Laciar Leber** 	Universidad Autónoma de Bucaramanga UNAB 2:00 p.m. – 6:00 p.m.

(*Continued*)

(*Continued*)

Curso	Profesor	Lugar/Hora
Aplicaciones en la Biomecánica Clínica y Deportiva	**Dr. Ariel Braidot** 	Universidad Autónoma de Bucaramanga UNAB 2:00 p.m. – 6:00 p.m.
Diseño Curricular en Ingeniería Biomédica	**Dr. Joaquín Azpiroz Leehan** 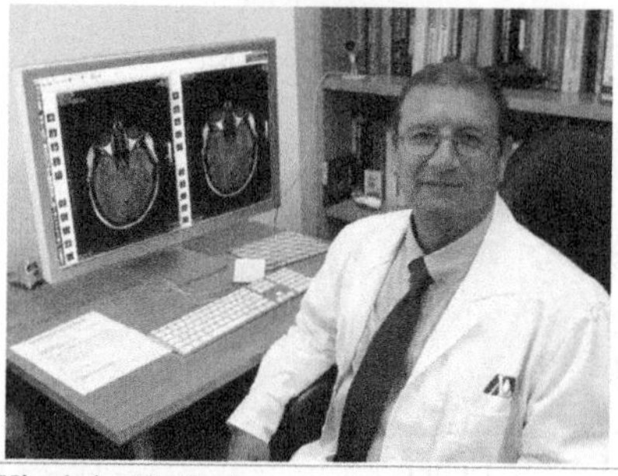	Universidad Autónoma de Bucaramanga UNAB 2:00 p.m. – 6:00 p.m.
Procesamiento de Imágenes de Microscopía	**Virginia Ballarin**  **Juan Ignacio Pastore** 	Universidad Industrial de Santander U.I.S 8:00 a.m. – 12:00 m

Academical Workshops

OCTOBER 26th, 2016 Duration: 4 Hours

Curso	Profesor	Lugar/Hora
Taller de Educación en Ingeniería Biomédica y Bioingeniería	**Jesús María Soto Director Del Taller** 	Universidad Autónoma de Bucaramanga UNAB 3:00 p.m. – 6:30 p.m.
Taller de Innovación y Desarrollo Empresarial	**Carolina Salazar Andres Orozco** 	Universidad Autónoma de Bucaramanga UNAB 3:00 p.m. – 6:30 p.m.

Clinical Workshops

October 26th, 2016

Engineering Workshop Hospital and/or Clinic

International Foscal
With the participation of Professor Marc Nyssen Treasurer IFMBE Universiteit Brussel, Belgium

Workshop of Social Projection of the Disability'S Technology

With the participation of Professor James GOH: President, IFMBE. Conference Jean Marc Van Hissenhoven, chief scientist Mobility Group "High technology and rehabilitation."

Clinical Engineering and/or Hospital Engineering

FCV INTERNATIONAL HOSPITAL.
Organized by Jaime Poveda and Carlos Mesa

Workshop on Education in Bioengineering and Biomedical Engineering

Organized by Jesús María Soto - Universidad EIA (Medellín, Antioquia)

Workshop on Innovation and Business Development

Organized by the Medellin Metropolitan Institute (ITM), led by medical equipment companies in UNAB

IFMBE – Industry Working Group Symposium at the CLAIB 2016

From 9:00 am to 12:00 m. Venezuela Room, UNAB Hostal

"Successful cases of research, innovation and transfer technology to promote a link between the scientific community, industry and health service providers"

Speakers:

- Prof. Piotr Ladyzynski, Poland.
- Prof. Martha Lucía Zequera Díaz, Colombia.
- Prof. Paulo Carvalho, Portugal.
- Prof. Ratko Magjarevic, Croatia.
- Ing. Bladimir Guzmán Páez, Colombia.
- Prof. Gina Paola Flórez, Colombia.
- Dr. Héctor Castro, Ministerio de Protección Social, Colombia.

Participants in Innovation and Bussiness Development Fair

1. Materialise Colombia S.A.S.
http://www.materialise.com/

2. Advanced Instruments S.A.S.
www.advanced.com.co

3. Medinistros S.A.S.
www.medinistros.com

4. Set & Gad S.A.S.
http://setgad.com/

Contents

A System to study the Evolution of Cardiac Arrhythmias at Home

R.I. Gonzalez-Fernandez[1], M.L. Mulet-Cartaya[1], J.D. Lopez-Cardona[1], A. Lopez-Reyes[1], O. Canto-Hernandez[1], E. Ledesma Valdes[1], V. Rodríguez-Peraza, P. Gonzalez-Acosta[1] and M. Gomez-Florido[1]

[1] ICID Digital Medical Technology/Division of Medical Devices, Havana, Cuba

Abstract— **The aim of this paper is to discuss the main features of a system to follow-up cardiac arrhythmias without people being hospitalized. Patients are registered in the system during their first visit to the physician. A battery-powered device was developed to digitize a one-lead ECG up to three minutes; physician should indicate the frequency to make the tests. The digital ECG is transmitted, using a Bluetooth connection, to an Android telephone running a mobile application developed by the authors. The incoming ECG is processed to detect and classify QRS complexes as an essential step for measuring RR intervals. Identification of bigeminism, trigeminy, asystole and compatible signals with atrial fibrillation is based on the measurements made. This data is uploaded to a web site using the HTTP protocol and a GSM/GPRS network as support; a web application stores this data in a SQL database. Physicians can log in this application to study the evolution of their patients using trend charts and others graphical tools. This operation mode allows medical treatment can be adjusted as soon as it is necessary. An alternative telephonic device is being developed by the authors to provide a low-cost solution for poor countries. Algorithms for ECG processing were tested with annotated database and voluntaries. Sensitivity in QRS complex detection was 98.77% using the MIT-BIH database as reference. The Bluetooth communication between the ECG device and several mobile telephone models was reliable, without data loss or data corruption. The same result was obtained testing the upload process with more than 600 ECG files. The ECG device was evaluated according to the IEC 60601-2-47 standard; results for all safety and parametric tests never exceed the permissible limits.**

Keywords— **Telemedicine, arrhythmia, ECG processing.**

I. Introduction

Public health systems of many countries face a serious challenge today: increased life expectancy of people is accompanied by an increasing consumption of resources or diagnosis and treatments of chronic diseases. Among these diseases, heart disorders classified as one of the three leading worldwide causes of mortality and morbidity.

Cardiac arrhythmias are increasingly affecting more people and many scientists around the world dedicate their efforts to find solutions for the prevention and treatment of these diseases [1].

Improvement and continuous expansion of public data transmission networks and the development of medical technology have created the necessary conditions for the establishment of Telemedicine as one of the specialties of health sciences with greater impact on last years. Health services have been established and traditional services have been refined to contribute to increase health indicators in countries that apply these technologies [2].

People suffering arrhythmia should be under medical surveillance to check the disease evolution and to correct medical treatment if it is necessary. This follow-up procedure causes agglomerations in hospitals and discomfort for patients and medical staff, conspiring against the medical service quality. If these people could receive the necessary medical care in their homes, without having to visit their doctor, the described problem would disappear. Multiple systems, based on mobile networks, have been developed to capture and measure different signals and parameters (ECG, blood pressure, blood glucose, weight, etc.). However, a unique and definitive solution for this kind of systems is not available on the market [2].

The aim of the proposed system is to discuss the main features of a system developed as solution to the situation described above.

II. Materials and Methods

The proposed system is composed by three layers defined clearly: an acquisition layer where ECG is digitized and transmitted via Bluetooth together with complementary information (pacemaker detection and electrode status). A second layer dedicated to the ECG processing based on the high computing power of the current mobile phones. Finally, a web application, as the third layer, stores all data associated to each ECG. This information can be analyzed by specialists using trend charts, reports, tables and other tools provided by the proposed system. This approach is better than the traditional procedure to study the arrhythmia evolution because patient's data, digital ECGs and electrocardiographic measurements can be analyzed exploiting database tools and the computer's graphical capabilities. Also, the medical treatment can be modified making a telephone call

I. Torres et al. (eds.), *VII Latin American Congress on Biomedical Engineering CLAIB 2016, Bucaramanga, Santander, Colombia, October 26th -28th, 2016*,
IFMBE Proceedings 60,
DOI: 10.1007/978-981-10-4086-3_1

to the patient, avoiding unnecessary visits to the hospital and reducing the stress of the medical staff.

The system operation is very simple for patients and physicians; this process can be summarized as follow:

- The physician registers the patient in the system during his first visit and gives him an ECG device with its accessories and the Android application for ECG processing is installed on the patient´s phone. Physician explains the patient how to acquire and transmit an ECG. Also, the patient receives the User Guide.
- The patient acquires the ECGs as physician indicates. He has to places the electrodes on his skin and turn on the ECG device. A bipolar ECG lead is amplified and digitized; this digital signal is transmitted to the telephonic terminal.
- The mobile application receives and processes the digital ECG to identify any arrhythmic event. Parameters like heart rate and premature beat rate are computed and each QRS complexes are classified as a previous step to the cardiac rhythm classification. All these data is stored in a database and uploaded to a HTTP server.
- The web application inserts all the upload information in a database.
- The physicians log in the web application to study a specific patient. Arrhythmic evolution is analysed and the treatment is corrected if necessary, a simple telephonic conversation with the patient is enough.

A. Layer 1: ECG acquisition and Bluetooth transmission

A small battery powered device was designed to amplify and to digitize an ECG lead. The electronic design was oriented to be a low-cost solution with a high reliability.

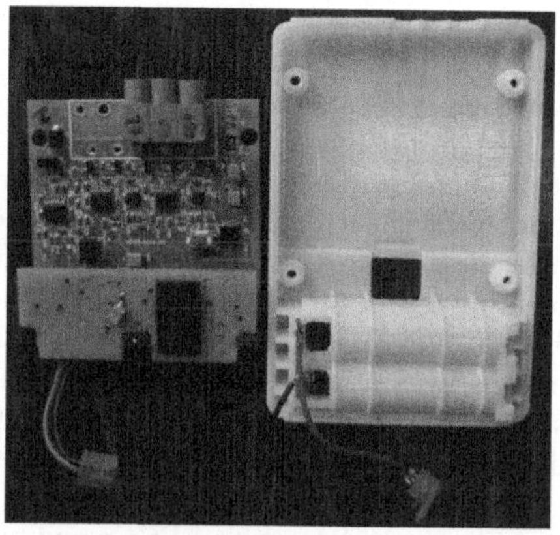

Fig. 1 ECG device: Electronic board and plastic case prototype.

The ECG device can be divided in the following blocks:

- The ECG amplifier: It is based on the OPA4236 operational amplifier and the mcropower single-supply INA826 instrumentation amplifier. The bandwidth is limited between 0.5 Hz and 40 Hz to improve the signal quality and to keep the required ECG frequency components for rhythm studies [3]. A right-leg circuitry was implemented to improve the signal to noise ratio of the ECG lead. Also, electrode status is checked continuously and a trace recovery circuitry was implemented to minimize the recovering time when an electrode is attached after a connection fail.
- The MSP430F5529 microcontroller: It is an ultralow-power microcontroller with a 16-bit RISC architecture from Texas Instruments. The microcontroller includes, in a single chip, all the resources necessaries for the application: serial ports for data interchange, programming port timers, program memory, data memory and a 12-bit A/D converter.
- The LMX9838 module: It is a fully integrated Bluetooth 2.0 base band controller. All hardware and firmware is included to provide a complete solution for this kind of wireless communication, helping to get a short time to market.
- The battery: The isolated power is supplied by two AAA NiMH batteries.

The user interface is easy. The ECG device has a single button for its operation. If the user presses the button for three seconds or more, the ECG device starts the pairing process with the Android phone running the ECG application. However, if the button is pressed for less than three seconds, the ECG device turns on automatically to acquire and transmit ECG samples until the Android application ends the communication and the ECG device turns off by itself. Sampling frequency always is set to 250 Hz. Detection of pacemaker spike and electrode status is packed with the ECG samples using a protocol defined by the authors.

B. Layer 2: The ECG processing

The ECG processing algorithms were implemented for Android platform and for a telephonic terminal developed by the authors to be used in poor countries. Java language was used to develop the Android application and C language for MSP microcontroller was the basic tool to develop the firmware of the specific telephonic terminal.

A FIR band pass filter proposed by Ligtenberg and Murat [5] was applied to minimize noise influence and to avoid baseline wandering; its implementation has been explained in previous researches [4, 6]. This filter implementation is easy because it is based on integer arithmetic only. Another significant advantage is that the cut-off frequencies can be easily modified by changes in the values of the K and L constants; the filter expression is the following:

$$y(k) = \frac{1}{K^2} \sum_{m=k-K+1}^{k} \sum_{n=m-K+1}^{m} x(n) - \frac{1}{L^2} \sum_{m=k-L+1}^{k} \sum_{n=m-K+1}^{m} x(n) \quad (1)$$

where:
x(n): input signal
y(k): output filtered signal
K, L: filter constant

QRS complex detection is based on an energy collector and two thresholds. Energy is computed as the sum of the Teager operator's values corresponding to a time's window of 150 milliseconds previous to the studied sample. The window width was defined taking into account the typical duration of the ectopic QRS complexes and previous experiences [3].

$$T(i) = x(i)^2 - x(i-1) * (xi+1) \quad (2)$$

where:
T(i): Teager operator
x(i): ECG samples.

A threshold is used to detect high-energy peaks because this kind of peaks is strongly associated to the QRS complexes. This threshold is computed as the 85% of the maximum energy value. For each detected QRS complex, the offset is defined as the sample corresponding to the 95% of the maximum energy value and from this sample a search backward process is made looking for the QRS complex onset. This process stops when the energy value is less or equal than the 0.5% of the maximum energy value.

RR intervals are computed after the identification of all R wave's peaks. Each cardiac beat is classified as premature or not premature according to their previous RR interval duration and width of QRS complex. The classification rule is as simple as the following: "the duration of the RR interval previous to an ectopic QRS complex will be less than 80% of the mean value of this interval. Also, the QRS width will be greater than the 120% of the mean value of this measurement".

When all QRS complexes have been classified, they are analysed to detect the following arrhythmic events:

- Arrhythmic ECG Strip: When ectopic beats are detected, but a high rate is not reached.
- Frequency Premature Contractions: Premature beat rate exceeds the programmed limit.
- Dangerous tachycardia: Heart rate exceeds the programmed limit.
- Extreme bradycardia: Heart rate drops below the programmed limit.
- Asystole: A QRS complex is not detected during more than five seconds. Except for the cases caused by bad signal quality.
- Atrial fibrillation: RR interval is variable without ectopic beat and a flat 100 millisecond segment is not detected for each RR interval.

At the end of ECG processing, all data is uploaded to the system website where is running the web application. HTTP protocol is used for this purpose and the information is uploaded as a file identified by a code generated when the patient was registered in the system.

C. Layer 3: The Web application

The web application was developed using Codeigniter framework; this open source tool is very useful to build dynamic web sites with PHP. This application is composed by a SQL database, a graphic interface to display ECGs and trend graphics and other complementary tools. ECG files are read and insert into a SQL database. The proposed application provides the following features:

- User login/logout: User name and password are used to organize the data access. Each physician can only check the data corresponding to his patients, so a physician is associated with N patients, but a patient is only associated with a physician.
- Patient registration: A patient is registered in the system when he visits a physician and arrhythmic disturbances are diagnosed; this process set a relationship in the database between the patient and the physician. All general patient data are introduced by the physician using an electronic form implemented into the proposed system. Also, the matching between patient and ECG device is registered.
- Database management (querying, sorting, updating, etc.): A SQL database engine provides all the mechanism to management measurements and digital signals associated to each patient. A friendly user interface was implemented to access the data.
- Several formats to display measurements: Several graphic tools are provided to facilitate measurement analysis. The values are shown as tables or several types of graphics. Trend graphs are available to study of the progression of parameters such as heart rate and arrhythmic events detected.
- Printed reports: This feature is implemented to allow data interchange between specialists and to create traditional patient paper-files.
- Export results: It allows the integration with other systems. The well-known PDF format is used.

This kind of system allows studying the evolution of patients and correcting their drug treatments in a simple way.

For example, a cardiologist instructs a patient to make himself three electrocardiograms daily, corresponding to morning, afternoon and evening. After ten days, the cardiologist has thirty ECGs of the specific patient and he can analyse in detail how it is influencing medical treatment in the disease studied. When cardiologist considers medical treatment should be changed, he only has to make a telephone call to the patient in order to tell him the changes. The authors think this is an innovative point of view for this problem.

III. Results and discussion

Two ECG device´s prototypes have been tested according to the IEC 60601-2-47 standard and all the tests were passed successfully. The proposed device is not classified as an ambulatory medical device because it is not its intended use, but it shares some technical characteristics with ambulatory ECG devices, so the cited standard can be applied.

The use of the proposed device can be considered safe for patients because all electrical safety parameters comply with the existing international standards.

The Bluetooth communication was tested with six hundred three-minute ECG strips; these signals were generated by an electronic ECG simulator and voluntaries. These strips were transmitted from the ECG device to a mobile phone running the proposed Android application. This test passed without errors, the simulated signals received by the Android application were identical to the original ECGs. Also, the communication process was never aborted by errors, so it demonstrates that Bluetooth standards and HTTP protocol are good choices to link the system layers.

The digital filter performance was not tested alone because this filtering scheme has been used and tested in previous researches with good results [3, 6].

The QRS detection process was based on previous experiences of the authors [3, 6]. This detection scheme was tested previously with twelve ECG strips from MIT-BIH database and results were published. QRS complex detection algorithm was tested with MIT-BIH database and the sensitivity was 98.77%; this result is enough for this kind of application.

The proposed Web application provides several graphical tools to analyse the electrocardiographic information captured. A desktop application with the same features was development for those scenarios when Internet connectivity is not available. In this situation, the ECGs would be stored in the local database created at the mobile phone. When the patient visits the cardiologist, ECG could be downloaded using an USB connection between the mobile phone and the desktop application. The same analysis could be made by cardiologist in a different scenario.

IV. Conclusions

A first version of the proposed system was completed and tested preliminary to guarantee patient safety, data security and a stable and reliable operation of all parts of the proposed system. The goal of the research project described in this paper has been achieved, but more tests should be done to be sure of the system performance.

The proposed system seems a useful tool to study arrhythmia disturbances with more details about the disease progression and less discomfort for patients than the way traditionally used in health services.

Acknowledgment

The authors of this paper would like to acknowledge the medical advice provided by cardiologists at the Institute of Cardiology and the "Hermanos Ameijeiras" Hospital, both in Havana.

Conflict of Interest

The authors declare that they have no conflict of interest.

References

1. World Health Organization. The World Health Statistics 2015 at http://www.who.int/gho/publications/world_health_statistics/2015/en/
2. Choi Y B, Krause J J, Seo H, Capitan K E and Chung K (2006) Telemedicine in the USA. Standardization through information management and technical applications. IEEE Communications Magazine
3. Campillo D, Torres H, Gonzalez R, Valdes K, Lopez R (2014) A portable Device for a Modular System of Patient ECG Monitoring. Computing in Cardiology 2014. Vol 41 pp 1077-1079
4. Gonzalez R I, Guardarrama R, Leon F, Rodriguez A, Jimenez D (2009) Two Low-Cost Solutions for Cardiac Mobile Monitoring. IFMBE Proc. Vol 25, World Congress on Me. Phys. & Biomed. Eng., Munich, Germany, 2009, pp. 188-191
5. Ligtenberg A and Murat K (1983) A Robust Digital QRS-Detection Algorithm for Arrhythmia Monitoring. Computing and Biomed. Res., 1983: 16 pp. 273-286
6. Gonzalez R (2003) A Spatial Study of the ST Segment. Proceedings of the 25th Annual International Conference of the IEEE Medicine and Biology Society, 2003:1 pp 2610-2614

Address for corespondence:

Rene Gonzalez-Fernandez
ICID Digital Medical Technology
Calle 202 #1704 entre 17 y 19, Playa
Havana, Cuba
rigf58@yahoo.com

Implementation of risk management activities within a quality management system. An osseous adhesive as case study

R.M. Guerra Bretaña[1], M.C. Pérez Álvarez[2], M.S. de Almeida[3], L.A. de Sena[4]

[1]Biomaterials Center, University of Havana, Havana, Cuba
[2]Dental Teaching Clinic/Surgical Department, Bauta, Artemisa, Cuba,
[3]Faculty of Dentistry, Federal Fluminense University, Niterói, RJ, Brazil
[4]Faculty SENAI, Rio de Janeiro, Brazil.

Abstract— **Medical device regulatory agencies have adopted quality and risk management approaches to demonstrate compliance with the essential principles of safety and performance. For this reason, specific standards are established in the medical device sector for the quality management system by regulatory purposes and a systematic process, in order to minimize risks throughout the life-cycle of the product. The objectives of this paper are to show how the risk management process can be inserted into the product design/development strategy within a quality management system and to apply such strategy to an osseous adhesive as a case study. Integration of a risk management system within a QMS can be an advantage for manufacturers of medical devices in order to comply with regulatory requirements. At the design and development stages the risk management principles should be applied and used to identify and address safety issues and to plan risk control activities. The risk assessment conducted shows that the Wollastonite / n-butyl-2-cyanoacrylate adhesive composite is safe by design. Adequate labeling and operating instructions have to be incorporated to the final product to minimize the risks or improper usage and storage outside prescribed environmental conditions. In addition, validation and control measures of the sterilization process and packaging have to be taken to maintain the acceptable levels for all identified risks.**

Keywords— **risk management, quality management, osseous adhesive, Wollastonite, butylcyanoacrylate.**

I. Introduction

The development of medical devices involves several activities for maximizing the benefit and minimizing risks in its intended use. Designers, manufacturers and testing services have to provide the required evidence of the safety and efficacy of the developed products. Quality and risk management approaches through the whole life cycle of the medical products have to be integrated to demonstrate compliance with the *Essential Principles of Safety and Performance of Medical Devices* [1]. In this context, a safe product is the one that has acceptable risks, in comparison with the benefit expected and the alternatives available [2]. For these reasons in the medical device industry there are specific standards, harmonized guides and regulations that establish a systematic process to minimize risks throughout the life-cycle of the product. Medical device regulations required that manufacturer implements a quality management system (QMS) and also a system for managing device related risks. The integration of the QMS and the risk management system may be advantageous because of reducing costs, the elimination of redundancies, and more effective management [3].

These international trends are incorporated in the medical devices regulations in Cuba [4] and Brazil [5], and despite the fact that these countries have different quality system regulations for medical products [6, 7] both explicitly include the requirement for risk management by the manufacturers. The objectives of this paper are to show how the risk management process can be inserted into the product design/development strategy within a QMS strategy and to analyze a case study of an osseous adhesive product.

II. Materials and Methods

A. Medical device risk management

Risk is *the combination of the probability of occurrence of harm and the severity of that harm, meanwhile risk management is a systematic application of management policies, procedures, and practices to the tasks of analyzing, evaluating, controlling, and monitoring risk* [8]. Risk management should be applied throughout the life cycle of medical devices and the international standard ISO 14971 *specifies a procedure by which a manufacturer can identify the hazards associated with medical devices and their accessories, including in vitro diagnostic medical devices, estimate and evaluate the risks, control these risks, and monitor the effectiveness of the control.*

The risk management process includes the following steps: risk analysis, risk assessment, risk control, production and postproduction information (Fig. 1). The requirements of the ISO 14971 standard can be certainly incorporated into an ISO 13485 QMS as shown in Table 1.

I. Torres et al. (eds.), *VII Latin American Congress on Biomedical Engineering CLAIB 2016, Bucaramanga, Santander, Colombia, October 26th -28th, 2016,*
IFMBE Proceedings 60,
DOI: 10.1007/978-981-10-4086-3_2

Table 1 Compatibility between high level requirement clauses of ISO 13485 and ISO 14071

ISO 13485:2003	ISO 14971:2007
4 Quality management system	3 General requirements for risk management 3.1 National or regional regulatory requirements 3.2 Risk management process 3.6 Risk management file 8 Risk management report
5 Management	3.3 Management responsibilities
6 Resource management	3.4 Qualification of personnel
7 Product realization	3.5 Risk management plan 4 Risk analysis 5 Risk evaluation 6 Risk control 7 Overall residual risk evaluation
8. Measurement, analysis and improvement	9 Post-production Information

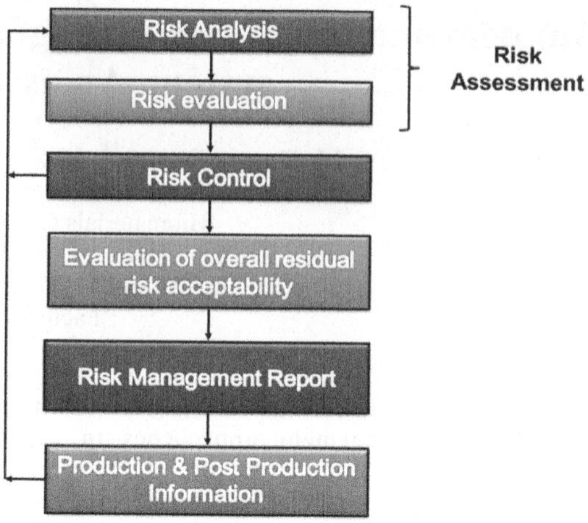

Fig. 1 Risk management process

The scope of the risk management plan developed in this study comprises the design and development stage of an experimental osseous adhesive. An overview of the relationships between risk management activities and design and development activities are shown in Figure 2. The risk management team was constituted by two professors in dentistry with expertise in maxillofacial surgery and dental materials, and two researchers in the field of biomaterials with knowledge and experience of the product, medical device regulations and risk management techniques.

Moreover the acceptability of risks is based on the estimation of their probability and consequence levels, codified to obtain a "scoring system", giving a risk priority "number" (RPN = Probability x Severity) as shown in Table 2. The estimation is performed under normal and fault conditions. After the team work to identify the hazards and the potential harm that they provoke, each expert done estimation and the final level was the mode of the overall values. For all fault conditions the experts were asked to value the probability at least in the level 2.

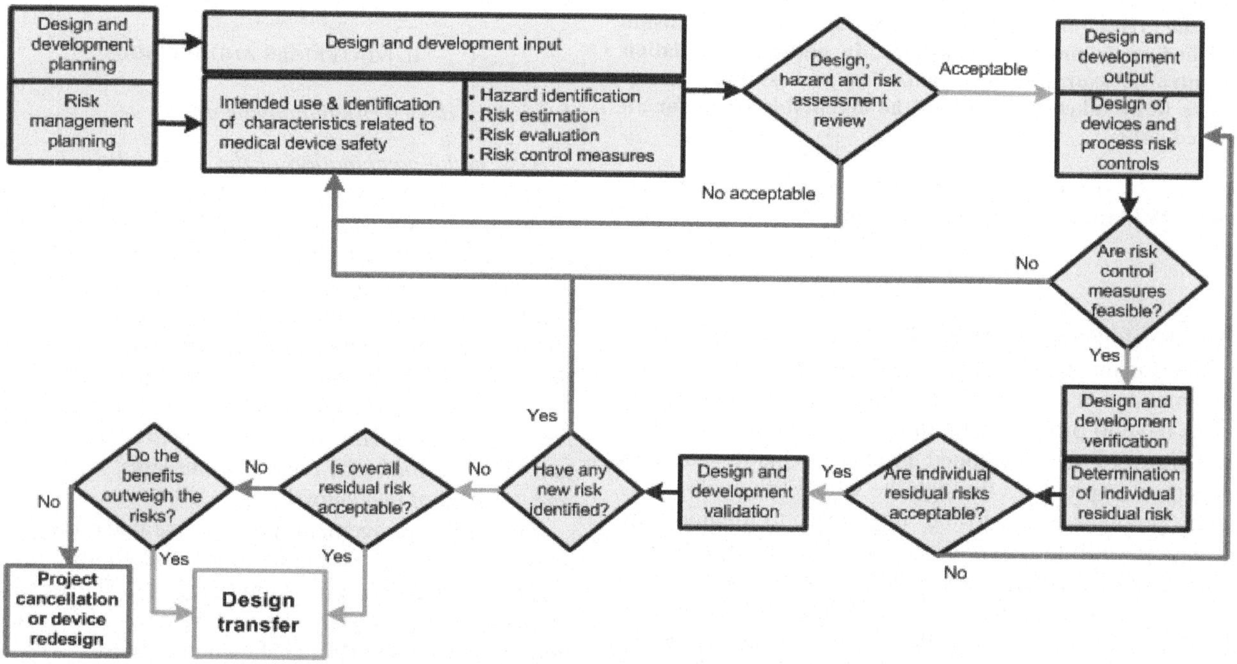

Fig. 2 Risk management activities in design and development. Modified from GHTF/SG1/N41, 2005 [1]

B. The studied osseous adhesive

A composite adhesive made of natural Wollastonite and n-butyl-2-cyanoacrylate (W-BCA) was chosen as case study. The adhesive is intended to be purchased as two-component system. Component I is natural Wollastonite powder with average particle size of 13.2 μm and 1.06 SiO2/CaO molar ratio. Component II is n-butyl-2-cyanoacrylate monomer, purity ≥ 97%. The adhesive is prepared in situ by mixing the two components in 60/40 solid/liquid (Wt/Wt) proportions. The polymerization occurs in less than 30 min.

Table 2 Risk estimation based on the risk priority number

Levels	Codification
Probability levels	1: Improbable
	2: Probable
	3: Frequent
Severity levels	1: Trivial
	2: Moderate
	3: Significant
Risk estimation	1-2: Acceptable
	3-4: Undesirable
	6-9: Inacceptable

III. Results

Design and development inputs include considerations about the intended use of the medical device, safety and regulatory requirements. The present study is focused on the design and development of a biocompatible adhesive to glue small bone fragments in craniofacial surgery. The bond strength value required to ensure bony healing in such application is still unknown. The fact remains that an adhesive used for bone bonding must possess several properties. It must hold the bone fragments together with adequate bond strength for the time required for the natural tissue healing and degrade at a rate which is consistent with its replacement. It must be accepted by the body, without toxic reactions and the degradation products must be able to be removed by natural processes. Also, the adhesive must be easy to handle and use by the surgeon [9].

Risk estimation for the W-BCA studied adhesive composite is based on data reported in the literature for similar products, expert opinions and data generated in chemical, physical and biological evaluation [10,11]. The adhesive composite is non-cytotoxic and have tensile bond strength to bone comparable to the medical n-butyl-2-cyanoacrylate adhesive. Also in vivo study shows that the composite is biocompatible when in contact with osseous and subcutaneous tissue. Table 3 gives the overall risk estimation done for the studied osseous adhesive and the residual risks after application of the control measures to reduce the probability of the occurrence of the hazard.

IV. Discussion

The risk assessment shows that W-BCA adhesive composite is inherently safe by design. Only one risk remains undesirable (RPN=3), that is the inflammatory reaction. However, the evaluated composite do not promote an exacerbated inflammatory reaction when in contact with bone [11]. It is not completely degraded after nine weeks of implantation in the femur of Wistar rats, delaying the formation of the new bone. It is important to notice that the biomaterial for osseous fixation has to be slowly biodegradable, to guarantee stability during the healing process. These findings should be confirmed in animal trials under loading patterns. For other identified hazards, adequate labeling and use instructions have to be incorporate to the final product to minimize the risks of improper usage and storage outside prescribed environmental conditions. Validation and control measures of the sterilization process and packaging have also to be taken to maintain the acceptable levels for all identified risks. This control measures are incorporated into the design and development outputs.

Besides ethical and regulatory compliance, medical device manufactures can obtain several benefits from an effective risk management process. This can also reduce operating costs and increase profits by identifying and preventing problems before products are marketed and providing a systematic framework for understanding the causes of problems, which allows more rapid and cost-effective solutions.

V. Conclusions

Integration of a risk management within a quality management system can be an advantage for manufacturers of medical devices in order to comply with regulatory requirements. At the design and development stage the risk management principles and activities allow to identify and address safety issues and to plan risk control activities. The risk assessment conducted shows that the Wollastonite / n-butyl-2-cyanoacrylate adhesive composite is safe by design. Adequate product and process control measures have to be taken to maintain the acceptable levels for all identified risks.

Conflict of Interest

The authors declare that they have no conflict of interest.

Table 3 Osseous adhesive risk management summary

Hazard	Harm	Probability	Severity	RPN	Risk control measures	Residual risk
Biological hazards						
Biological Contamination (Fault condition)	Infection	2	2	4	– Validation and control measures of the sterilization process and packaging. – Adequate labeling.	2
Bio-incompatibility /toxicity (Normal condition)	Inflammatory reaction	3	1	3	– Inherent safety by design (biological evaluation - ISO 10993)	3
Incorrect chemical formulation (Fault condition)	No osseous adhesion	2	2	4	– Process control, quality control of the final product.	2
Cross contamination (Normal condition)	Infection	2	2	4	– Adequate labeling (one use product).	2
Degradation (sub- products) (Normal condition)	Inflammatory reaction	3	1	3	– Inherent safety by design (biological evaluation - ISO 10993).	3
Environmental hazards						
Storage outside prescribed environmental conditions (Normal condition)	No osseous adhesion	3	2	6	– Adequate labeling and use instructions.	2
Accidental mechanical damage (Normal condition)	No osseous adhesion	2	2	4	– Adequate packaging.	2
Hazards related to the use of the product						
Reasonably foreseeable misuse (Normal condition)	No adhesion. Infection	2	2	4	– Adequate labeling and use instructions.	2
Inadequate labeling and operating instructions (Fault condition)	No adhesion. Infection	2	2	4	– Control of the labeling and packaging process.	2
Use by untrained personnel. (Normal condition)	No osseous adhesion.	2	2	4	– Adequate labeling and use instructions	2
Insufficient warning of hazards likely the re-use. (Normal cond.)	Infection	2	2	4	– Adequate labeling and use instructions	2
Incompatibility with consumables or other medical devices (Normal cond.)	Adhesion to instruments	2	1	2	– Adequate labeling and use instructions	1
Hazards arising from failure and aging						
Inadequate determination of the validity period (Fault condition)	No osseous adhesion.	2	2	4	Process control, quality control of the final product.	2
Inadequate packaging (contamination and/or deterioration) (Fault condition)	No adhesion. Infection	2	2	4	– Validation and control measures of the sterilization process and packaging. – Control of the labeling and packaging	2

References

1. GHTF/SG1/N41 Essential principles of safety and performance of medical devices, 2005.
2. ISO/TR 24971:2013 Medical devices — Guidance on the application of ISO 14971.
3. Guerra-Bretaña RM, Meizoso-Valdés MC. La gestión integrada de la calidad y los riesgos, bases para la efectividad y seguridad de las tecnologías médicas. Memorias Convención Internacional de Salud Pública. Cuba Salud 2012. La Habana 3-7 de diciembre de 2012. ISBN 978-959-212-811-8.
4. CCEEM. Reglamento para la Evaluación y el Control Estatal de Equipos Médicos. MINSAP Res No. 184/2008. Gaceta Oficial 1424-36, Cuba.
5. ANVISA RDC nº 185/2001 Registro de produtos médicos. Brasil.
6. ISO 13485:2016 Medical devices - Quality management systems - Requirements for regulatory purposes.
7. ANVISA RDC nº 16/2013 Boas Práticas de Fabricação. Brasil.
8. ISO 14971:2007 Medical devices—Application of risk management to medical devices.
9. Kandalam U, Bouvier AJ, Casas SB, et al. Novel bone adhesives: a comparison of bond strengths in vitro. Int J Oral Maxillof Surg, 2013, 42: 1054–9.
10. de Sena LA, de Almeida MS, de Oliveira Fernandes GV, et al. Biocompatibility of wollastonite-poly(N-butyl-2-cyanoacrylate) composites. J Biomed Mat Res Part B: Appl Biomat, 2014, 102(6): 1121–9.
11. de Sena LA, Guerra-Bretaña RM, de Almeida MS, et al. Functionality of wollastonite/n-butyl-2-cyanoacrylate bone adhesives: bond strength in vitro. XIV Braz Mat Res Soc, RJ, 27 Sept - 1 Oct, 2015.

Author: Rosa Mayelin Guerra Bretaña
Institute: Biomaterials Center, University of Havana
Street: Ave Universidad
City: Havana
Country: Cuba
Email: mayelin@biomat.uh.cu

Estimation of the optimal maintenance frequency of medical devices: A Monte Carlo simulation approach

Antonio Miguel Cruz[1]. William Ricardo Rodriguez Dueñas[1]

Programa Ingeniería Biomédica, Escuela de Medicina y Ciencias de la Salud. Universidad del Rosario
Bogotá, Colombia
e-mail: antonio.miguel@urosario.edu.co

Abstract—**This paper aims to implement and validate a Monte Carlo-based algorithm to determine the optimal interval of preventive maintenance of medical devices. The optimization criterion used was that which maximizes the equipment's achieved availability. The Monte Carlo algorithm was implemented and tested using maintenance data from infusion pumps, electrocardiographs, and ECG monitors from both primary and secondary data source of information. The performance of the algorithm behaved well; it had a 65-sec response time for 10,000 simulations. The accuracy of the calculations did not exceed 1%. In addition to that, the implementation of the Monte Carlo algorithm was able to determine the better availability curve for the interval of preventive maintenance "*tuned*" with the mean time to failure for each medical devices population type.**

Keywords-component: Monte Carlo simulation, maintenance optmization, preventive maintenance frequency, biomedical engineering.

I. INTRODUCTION

Maintenance can be defined as those actions (scheduled and unscheduled) for retaining a piece of equipment in, or restoring it to, a given safe and reliable condition [1]. The annual cost of maintenance (corrective and preventive), as a fraction of the total operating budget can be as high as 40%-50% for the mining industry [2], 20%-30% for the chemical industry [3], and 1% for medical devices in hospital settings [4]. Unavailability of devices causes economical loses in production lines. For example, in the oil industry, in a refinery a typical 10 days of downtime per year due to equipment failures causes an estimated economic loss of $20 000-$30 000 per hour [3]. For the case of industries such as the healthcare system, nuclear, and aviation, the question is not only how much preventive maintenance is economically advisable, but also how much safety as "*an added value*" these maintenance activities provide. As a result, many industries worldwide have begun to realize the importance of having an *effective maintenance policy*, not only because of the maintenance costs, but also an unexpected failure can cause unavailability in devices and safety issues [5]. A maintenance policy (e.g. standard periodic preventive maintenance (PM)) is understood as an implementation of a certain maintenance model to improve the reliability of a system [1] (see section II.A for more details).

In regard to the establishment of periodic PM intervals, in the maintenance of medical devices, there is a current debate that states on one hand some federal standards such as the Interpretative Guidelines Center of Medicare and Medicaid Services (CMS), Section 482.41 or The Laboratory Improvement Amendment (CLIA) which are claiming that equipment must be maintained "*periodically in accordance with the manufacturers' recommendations*" [6] cited by [7]. On the other hand, the clinical engineering and maintenance engineering community in general are questioning this claim because it seems to be arbitrary in nature. Some devices that appear to be similar both in function and complexity have manufacturer maintenance-recommended intervals with variations of one factor and two. As a result, the validity of these manufacturer maintenance-recommended intervals are in doubt.

The determination of maintenance-recommended intervals can be obtained from two sources, including the manufacturer and the in-house maintenance data of medical devices. However, these data most of the time are nonexistent. On one hand, sometimes manufacturers are not willing to share how maintenance-recommended intervals were determined (e.g. test data); as a result, there is no credible rationale to accept it. On the other hand, in-house maintenance or clinical engineering departments do not have a strong record maintenance history to determine "in the field" their own maintenance-intervals. Fortunately, when the data are nonexistent there are mathematical tools (e.g. Optimization methods based on mathematical statistics or stochastic) that help in determining maintenance-intervals. While these methods have been widely used in the electronic, chemistry, aviation, and nuclear industries, a recent literature review conducted by Jamshidi, Ait-kadi, & Bartolome [8] , revealed that "*not much research has been presented in the literature during 25 years to address proper strategies and the methods for implementing the [maintenance optimization models]*". In order to fill this gap in the existing knowledge in the determination of optimal maintenance intervals this study aims to:

Implement and validate a Monte Carlo algorithm to determine the optimal maintenance intervals of medical devices.

II. THEORETICAL FRAMEWORK AND MATHEMATICAL MODELS

A. Maintenance policies

As mentioned previously, a maintenance policy is understood as an implementation of a certain maintenance model to improve the reliability of the system [1]. Amongst the most common maintenance policies are age replacement policy, periodic preventive maintenance (PM) policy, and sequential

I. Torres et al. (eds.), *VII Latin American Congress on Biomedical Engineering CLAIB 2016, Bucaramanga, Santander, Colombia, October 26th -28th, 2016*,
IFMBE Proceedings 60,
DOI: 10.1007/978-981-10-4086-3_3

preventive maintenance policy. According to our experience, the maintenance policy more widely used in the maintenance of medical devices is the periodic PM policy. In the periodic PM policy, a unit is preventively maintained at fixed time intervals k*T (k =1, 2,...*n*) independent of the failure history of the unit, and repaired at intervening failures where T is a constant. Obviously, the decision variable in this policy is the T value. Ideally; a "*well tuned*" T value should be as close as possible to the equipment mean time between failures (MTBFs) or Mean Time to Failure (MMTF). As a result, the device will never fail because the PM is always performed first, then the availability of the device is maximized.

B. Mathematical model and optimizations methods.

The problem to be solved then, is what is the best T value that "optimizes" the availability of a certain device, knowing that the maintenance task has a cost. The mathematical model of achieved availability to be optimized is shown in Eq.1 [5].

$$A_a = \frac{1}{1+\lambda*(\mathrm{MCMT}+\mathrm{MCVT})+f_{pm}*(MPMT+MPVT)} \quad (1)$$

Where:

A_a: achieved availability

λ: Failure rate (MTBF= $1/\lambda$)

MCMT: Mean time to perform a corrective maintenance (in hours)

MCVT: Mean time to perform a verification or quality inspection of corrective maintenance (in hours)

f_{pm}: PM frequency (times/year, 1/T)

MPMT: Mean time to perform a PM (in hours)

MPVT: Mean time to perform a verification or quality inspection of PM (in hours)

It can be seen from Eq.1 that if the two components of maintenance (preventive and corrective) are present, the availability is lower. As mentioned previously, this scenario occurs when f_{pm} ($1/T_{pm}$) $\neq \lambda$ (1/MTTF), especially when $\lambda > f_{pm}$. One might think, then, that by increasing the PM frequency it would be a suitable solution, i.e. $f_{pm} >> \lambda$ because the equipment never fails. In other words, the more PM the better. However, the availability might decrease because the device is "always" in PM. Therefore, from Eq.1 the ideal λ value is when it is equal to the f_{pm} value ($1/T_{pm}$) value. Under this condition, ideally, the equipment never fails as a result, the factor λ*(MCMT+MCVT) = 0; thus achieving the highest availability possible.

The optimization method used in this study was based on the availability criterion only. In other words, the objective is to obtain for a specific $1/\lambda$, the f_{pm} ($1/T_p$) that maximizes the availability of the given system. To solve this problem we implemented and validated an algorithm using a Monte Carlo simulation. The concept of the Monte Carlo method comes from the gaming tables at the casinos of Monte Carlo. It is a useful tool for modelling phenomena with significant uncertainty in inputs (which is our case, because PM intervals are not available) [5]. To implement and run the Monte Carlo algorithm it is necessary to generate random variables that follow an arbitrary statistical distribution. In this research the Weibull distribution was used to generate the random numbers of MCMT, MCVT, MPMT, and MPVT (see Eqs. 1 and 2).

$$f(t) = \frac{\beta*(t)^{(\beta-1)}}{\eta^{\beta}} * e^{-(\frac{t}{\eta})^{\beta}} \; ; \; F(t) = 1 - e^{-(\frac{t}{\eta})^{\beta}} \quad (2)$$

Where:

β and η: are the scale and shape parameters of the distribution.

t: is the time (in hours).

Note: β=1, *f(t)* become in the exponential distribution

III. MATERIALS AND METHODS

A. Materials

To implement the Monte Carlo simulation algorithm the *Scilab* v.5.5.2 software package licensed by GPL [9] was used. The programming and the validation of the simulation was conducted on an Inter Core *i7* personal computer, 3,40Ghz, 6Gb RAM, with the Windows 7, 64 bits operative system.

B. Data sources

To validate the implemented Monte Carlo simulation algorithm the data were extracted from two sources. The first was a primary source of information of standardized maintenance data from the *American Hospital Association (AHA)* [10]. In this database preventive maintenance frequencies (f_{pm}) and the time consumption of PM tasks per equipment type are provided (e.g. electrocardiographs f_{pm} MPMT, and MPVT). The second source of data was a secondary source of maintenance records from the clinical engineering departments. From this secondary information source the corrective maintenance data, including λ and corrective time consumption in hours, were obtained (e.g. MCMT, MCVT).

C. Methods

In this study the next methodology was followed. First, the research team selected three types of medical devices to determine the optimum maintenance frequency, including (i.e. to validate the Monte Carlo algorithm) electrocardiographs, infusion pumps, and ECG monitors. Second, standard PM data from the *American Hospital Association (AHA)* database were extracted for device types selected. Next, corrective maintenance data were extracted from the secondary information data for the same device types. Fourth, with the maintenance data extracted, the Weibull probability distributions for MCMT, MCVT, MPMT, and MPVT variables were fitted. Next, the Monte Carlo algorithm was implemented by using *Scilab*. In this research the implemented Monte Carlo simulation algorithm was validated as follows:

1. Three experiments were run, one per type of device type (i.e. electrocardiographs, infusion pumps, and ECG monitors).
2. It was verified that for each availability curve obtained from different scenarios of f_{pm}, a parabolic curve was obtained (side-opening parabola, see Eq. 1). That is: the availability increases as the λ value (or MTTF increases) decreases.
3. It was verified that for a specific λ value, if $f_{pm} = \lambda$ the availability curve at this point was maximum (i.e. the f_{pm} is optimum or "*tuned*"). The Monte Carlo simulation model was compared with the deterministic model.

4. It was verified that the number of simulations guarantee an statistical error of no greater than 1%.

D. Monte Carlo algorithm implementation

As was previously mentioned, to run the Monte Carlo simulation, the implemented algorithm has to generate random variables that follow a statistical distribution. The inputs are randomly generated from probability distributions to simulate the process of sampling from an actual population; therefore, one has to choose a distribution for each input that best represents our current state of knowledge. Thus, a general Monte Carlo algorithm for the case of this study had the following steps:

Step1 1. Define the problem and the overall objectives of the study: *What is the f_{pm} ($1/T_p$) that maximizes the availability of the system*?

Step 2. Define the system and create a parametric model. Under the framework of this study, the mathematical model is represented by Eq. 1.

Step 3. Design the simulation.

3.1. Determine and fit the probability distributions for each of the inputs (i.e. Calibration of Weibull distributions parameters scale and shape of MCMT, MCVT, MPMT, and MPVT according to Eq. 2). The calibration of Weibull distribution parameters for all maintenance variables were determined by setting up a 25% of standard deviation (i.e. σ=0.25) in each maintenance variable data. Thus, the couple of (η,β) parameters for each device type and maintenance variable are shown in Table I. A standard value of shape parameter η=4.5 was selected because Weibull probability density function showed an exponential shape, a reasonable assumption for non-obsolete equipment.

3.2. Estimation of the number of simulation runs and the accuracy of results. In doing that Eq. 3 was used.

$$m = \sqrt{\frac{Z_{\alpha/2\sigma}}{\varepsilon}}, \quad (3)$$

Were

ε: standard error of the mean

σ: standard deviation of the output.

Accuracy: 1% 95% confidence level at α=0.05

Step 4. Generate a set of random inputs. In this study random inputs for MCMT, MCVT, MPMT, and MPVT according to a Weibull distribution were generated according to Eq. 4:

$$t = -(\eta * [\ln(rnd()]^{1/\beta}) \quad (4)$$

Where:

β and η: are the scale and shape parameters of the distribution.

t: is the time (in hours).

rnd: is a function to generate a random number.

Step 5. Run the deterministic model with the set of random inputs and assess the results of the model (according to Eq. 1).

Step 6. Repeat steps 4 and 5 for i = 1 to m, where m is the number of simulations calculated by Eq. 3.

Step 7. Analyze the results (i.e. statistics, confidence intervals, histograms, best fit distribution, or any other statistical measure.

IV. RESULTS

Tables II and III show the maintenance data per equipment type and the achieved availability versus mean time to failure in months for the Monte Carlo simulation for each medical technology type, respectively.

Table I. Calibrated Weibull distributions parameters of MCMT, MCVT, MPMT, and MPVT maintenance variables per medical device type population

Equipment type	Maintenance parameters			
	MCMT (η,μ)	MPMT (η,μ)	MCVT (η,μ)	MPVT (η,μ)
ECG	(4.5,10.9)	(4.5,3.3)	(4.5,3.3.)	(4.5,3.3.)
IP	(4.5, 87,7)	(4.5, 8.8)	(4.5, 5.5)	(4.5, 5.5)
MMP	(4.5, 186.3)	(4.5, 10.9)	(4.5, 6.6)	(4.5, 5.5)

Note: ECG, IP, and MMP stand for electrocardiographs, infusion pumps, and ECG monitors.

Table II. Maintenance parameters per medical device type population

Equipment type	Maintenance parameters, in hours			
	MCMT	MPMT	MCVT	MPVT
ECG	10	3	3	3
IP	80	8	5	5
MMP	170	10	6	5

Note: ECG, IP, and MMP stand for electrocardiographs, infusion pumps, and ECG monitors.

Table III. Achieved availability versus Mean Time to Failure (in months) for each medical device type

	f_{mp}	M 2	M 4	M 6	M 8	M 10	M 12
ECG	12/yr	**0.974**	0.976	0.976	0.976	0.976	0.976
MCMT = 10 hr	6/yr	0.972	**0.987**	0.988	0.988	0.988	0.988
MPMT = 3 hr MCVT = 3 hr	3/yr	0.966	0.985	**0.992**	0.993	0.994	0.994
MPVT = 3 hr	2/yr	0.968	0.982	**0.990**	**0.994**	**0.995**	**0.996**
	1/yr	0.970	0.984	0.988	0.991	0.993	0.995
IP	12/yr	**0.941**	0.948	0.949	0.949	0.949	0.949
	6/yr	0.886	**0.969**	0.973	0.973	0.974	0.974
MCMT = 80 hr MPMT = 8 hr	3/yr	0.833	0.939	**0.976**	**0.984**	0.986	0.986
MCVT = 5 hr MPVT = 5 hr	2/yr	0.838	0.907	**0.958**	0.979	**0.987**	**0.989**
	1/yr	0.838	0.910	0.936	0.951	0.964	0.977
MPM	12/yr	**0.930**	0.940	0.941	0.941	0.941	0.941
	6/yr	0.826	**0.963**	**0.969**	0.969	0.970	0.970
MCMT = 170 hr MPMT = 10 hr	3/yr	0.717	0.892	**0.967**	**0.980**	**0.983**	0.984
MCVT = 6 hr MPVT = 5 hr	2/yr	0.721	0.832	0.924	0.968	0.982	**0.986**
	1/yr	0.722	0.834	0.881	0.908	0.934	0.958

Note: M2, f_{mp}, ECG, IP, and MMP stand for month 2, 4,..,12, f_{mp} maintenance frequency (in years), electrocardiographs, infusion pumps and ECG monitors.

Fig. 1 and Fig. 2 show the trends of the mean achieved availability for different maintenance frequencies versus mean time to failure (MTTF, in months) for all equipment types (Fig. 1 deterministic model, Fig. 2 stochastic model or Monte Carlo simulation). Fig. 3 shows the error calculation and accuracy of the results for the Monte Carlo simulations of the developed algorithm.

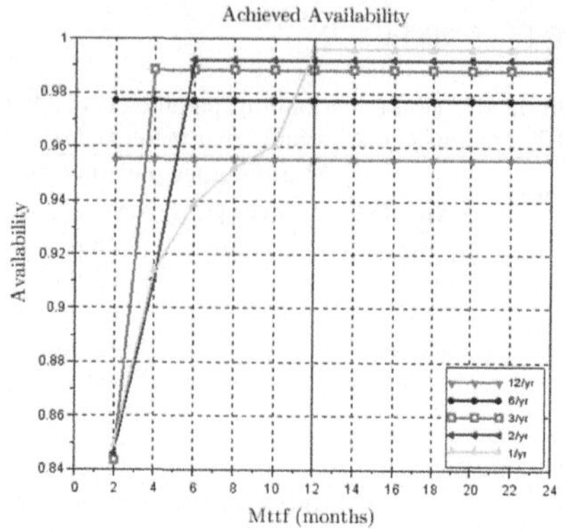

Figure 1. Deterministic model for all technologies. Achieved availability (mean) versus Mean Time to Failure (MTTF, in months). Simulation Window: 2 year. m=10 000 Simulations

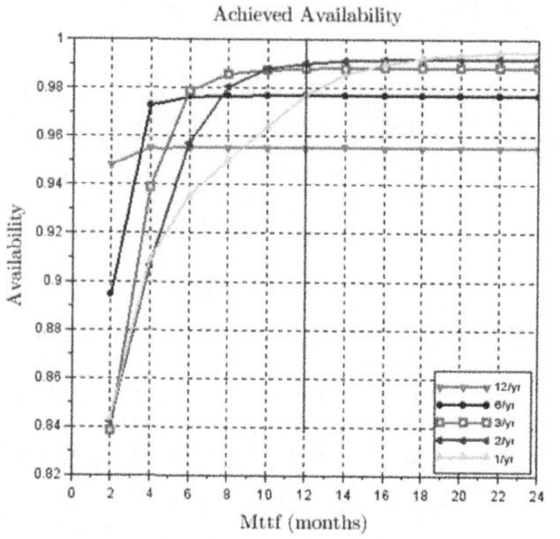

Figure 2. Stochastic model (Monte Carlo Simulation). Achieved availability versus Mean time to Failure (MTTF, in months). Simulation Window: 2 year. m=10 000 Simulations. Run time: 65 sec.

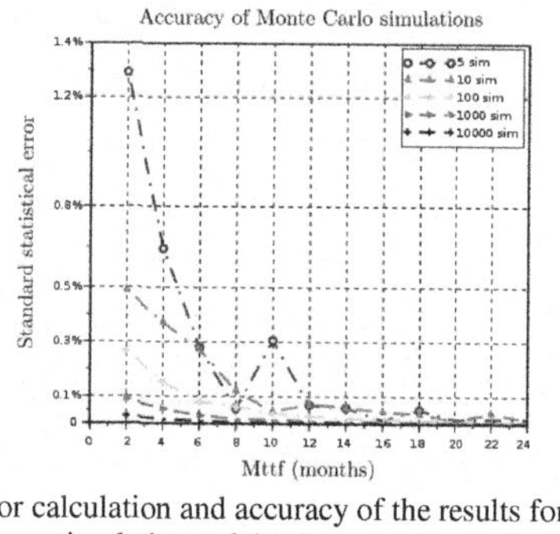

Figure 3. Error calculation and accuracy of the results for the Monte Carlo simulations of the developed algorithm.

V. DISCUSSION

Figs. 1 and 2 show that both package of curves are parabolic in nature (side-opening parabola). Fig. 1 also shows that when $f_{pm} < \lambda$, the achieved availability decreased, as the theoretical model predicted (i.e. preventive and corrective maintenance are present). Fig 3. shows that the calculated statistical error for 10 000 simulations becomes less than 0.1% and its variation over the MTTF range is minimized, which indicates that the results yield an acceptable level of accuracy, as detailed by [11]. More importantly, the predictive power of the results for the Monte Carlo simulations of the developed algorithm implemented was good. From Table II and Fig 3 it can be seen that the achieved availability is always maximum when f_{pm} is $\approx \lambda$. For example, for the case of electrocardiograph and infusion pump device types, when M=6 (i.e. a failure every 6 months, λ=0.17) the optimum maintenance frequency is 3-2 times per year (i.e. T=4-6 months), because the achieved availability has its maximum value. Future research should be conducted in real settings (i.e. healthcare organizations) order to determine whether the optimal maintenance frequencies calculated using Monte Carlo simulations differs or not from those suggested by manufacturers or third party maintenance organizations.

VI. CONCLUSIONS

According to the results shown after the implementation of the Monte Carlo algorithm one can reach the conclusion that it has acceptable levels of efficacy because it was able to determine the optimum values of maintenance frequency that minimize the effect of an unexpected failure.

ACKNOWLEDGEMENTS

We want to thank Oliver Jarvis for his proofreading of this paper. The authors assume overall responsibility for the manuscript.

REFERENCES

[1] H. Pham y H. Wang, Relaibility and optimal maintenance, Springer, 2007.

[2] D. Murthy, A. Atrens and J. Eccleston, "Strategic maintenance management," *2002,* vol. 8, no. 4, pp. 287-305, J. Qual. Maint. Eng..

[3] J. S. Tan and M. A. A. Kramer, "General Framework for Preventive Maintenance Optimization in Chemical Process Operations.," *Comput Chem.Eng.,* vol. 21, no. 12, pp. 1451-1469, 1997.

[4] B. Wang, Medical Equipment Maintenance: Management and Oversight, Princeton NJ.: Morgan & Claypool Publ, 2012.

[5] D. O. '. C. Patric and C. Cleiner, Practical Reliability engineering., 5th Ed. Wiley and Sons., 2012.

[6] CLIA, April 2016. [Online]. Available: https://www.cms.gov/.

[7] M. Ridgway, "Manufacturer-recommended PM intervals: It is time dor a change?," *Biomedical Instrumentation and Technology,* pp. 498-500, 2009.

[8] A. Jamshidi, A. RS, D. Ait-kadi and A. Bartolome, "Medical devices Inspection and Maintenance; A Literature Review," in *Proceedings of the 2014 Industrial and Systems Engineering Research Conference,* Montréal, Canada, 2014.

[9] S. Enterprises, "Scilab: Free and Open Source software for numerical computation (OS, Version 5.XX) [Software]. Available from: http://www.scilab.org". 2012.

[10] AHA, "Maintenance frequencies.," in *Maintenance manager for Medical equipment,* AMA, 1996.

[11] N. Negra, E. Fredericia, O. Holmstrom, B. Bak-Jensen and P. Sorensen, "Aspects of Relevance in Offshore Wind Farm Reliability Assessment," *IEEE transactions on energy conversion,* vol. 22, no. 1, pp. 159 - 166, 2007.

A visual EEG epilepsy detection method based on a wavelet statistical representation and the Kullback-Leibler divergence

A. Quintero-Rincón[1], M. Pereyra[2], Carlos D'Giano[3], H. Batatia[4] and M. Risk[1,5]

1 Instituto Tecnológico de Buenos Aires/Department of Bioengineering, Av. Eduardo Madero 399, C1106ACD, Buenos Aires, Argentine
2 University of Bristol, University Walk/Department of Mathematics, Clifton, Bristol, BS8 1TW, Bristol, U.K.
3 Fundaci´on contra las Enfermedades Neurológicas Infantiles (FLENI), Montañeses 2325. Buenos Aires, Argentina
4 University of Toulouse/IRIT - INP-ENSEEIHT, Toulouse, 2, rue Charles Camichel, B.P. 7122, 31071, Cedex 7, Toulouse, France
5 Consejo Nacional de Investigaciones Científicas y Técnicas (CONICET), Argentina

Abstract— **This paper presents a statistical signal processing method for the characterization of EEG of patients suffering from epilepsy. A statistical model is proposed for the signals and the Kullback-Leibler divergence is used to study the differences between *Seizure/Non-Seizure* in patients suffering from epilepsy. Precisely, EEG signals are transformed into multivariate coefficients through multilevel 1D wavelet decomposition of different brain frequencies. The generalized Gaussian distribution (GGD) is shown to model precisely these coefficients. Patients are compared based on the analytical development of Kullback-Leibler divergence (KLD) of their corresponding GGD distributions. The method has been applied to a dataset of 18 epileptic signals of 9 patients. Results show a clear discrepancy between *Seizure/Non-Seizure* in epileptic signals, which helps in determining the onset of the seizure.**

Keywords— **Kullback-Leibler divergence, Epilepsy, Seizure/Non-Seizure, Multivariate wavelet decomposition, Generalized Gaussian distribution.**

I INTRODUCTION

The International League Against Epilepsy (ILAE) [1] defines an *epileptic seizure as a transient occurrence of signs and/or symptoms due to abnormal excessive or synchronous neuronal activity in the brain*. The different types of seizures depend on the location in the brain where it originated and on how far and fast it spreads. The correct identification of this onset location is key to a proper treatment. Electroencephalography (EEG) is a non-invasive and widely available biomedical modality that is used to diagnose epilepsy and plan treatment; neurologists trained in EEG are able to identify visually the onset and subsequent seizure through analysis of characteristic waveforms associated with seizures. This problem has been addressed in various research works such as [2–8], but remains an open issue. In this work, we adopt a statistical approach to distinguish *Seizure/Non-Seizure* in epileptic signals. The data is represented using the generalized Gaussian distribution in the wavelet domain. The analytical development of Kullback-Leibler Divergence (KLD) or relative entropy, is used to measure the discrepancy between probability density functions (PDF), specifically among the PDFs of the generalized Gaussian distributions for *Seizure/Non-Seizure* signals. See [9–11] for some works on this topic in epilepsy and [12–14] for some applications in EEG signals. The remainder of this paper is structured as follows. Section 2 presents the proposed methodology and details the generalized Gaussian model and the analytical development of Kullback-Leibler divergence. Section 3 describes the experimentation on real EEG signals and the presents results obtained. Section 4 discusses the findings and provides perspectives for future work.

II METHODOLOGY

The methodology used to analyze the EEG signals has three stages. The first stage is to represent the signals using a time-frequency Dauchebies wavelet decomposition [15, 16] with 6 scales, this gives the bands delta (0.5-4Hz), theta (4-8Hz), alpha (8-13Hz), beta (13-30Hz) and gamma (>30Hz). The aim of this stage is to assess the distribution of the energy throughout the frequency spectrum. The second stage consists in summarizing the information contained in each group (band and scale) of wavelet coefficients. The approach adopted consists in fitting the generalized Gaussian distribution statistical model to each group. The parameters α and β are estimated for each PDF using (3) giving a parameter vector that represents each group [17]. The third stage consists in measuring the difference between *Seizure/Non-Seizure* in epileptic signals by calculating the Kullback-Leibler Divergence (7) between generalized Gaussian distribution PDFs obtained for each patient.
We now introduce the generalized Gaussian distribution and Kullback-Leibler Divergence.

A Generalized Gaussian distribution

The univariate generalized Gaussian distribution (GGD) is a flexible statistical model for one-dimensional signals that

I. Torres et al. (eds.), *VII Latin American Congress on Biomedical Engineering CLAIB 2016, Bucaramanga, Santander, Colombia, October 26th -28th, 2016*,
IFMBE Proceedings 60,
DOI: 10.1007/978-981-10-4086-3_4

has found numerous applications in science and engineering. Its application to epilepsy signal has been studied in [18–20]. Since the wavelet detail coefficients arise from high-pass filtering a zero-mean EEG signal matrix, it can be safely assumed that they also have mean value of zero [21]. Consequently, the wavelet coefficients can be modeled through the parameters of the GGD [22] whose probability density function (PDF) is given by

$$f(x;\mu,\sigma,\beta) = \frac{\beta}{2\alpha(\sigma)\Gamma(\frac{1}{\beta})} \exp\left(-\frac{|(x-\mu)|^{\beta}}{2\sigma^2}\right) \quad (1)$$

$$\alpha(\sigma) = \sigma\sqrt{\frac{\Gamma(\frac{1}{\beta})}{\Gamma(\frac{3}{\beta})}}, \quad \Gamma(z) = \int_0^{+\infty} t^{z-1}e^{-t}\mathrm{d}t, z>0 \quad (2)$$

where $\mu \in \mathbb{R}$ is a location parameter, $\alpha \in \mathbb{R}^+$ is a scale parameter and $\beta \in \mathbb{R}^+$ is a shape parameter that controls the shape of the density tail. In the case of a zero-mean GGD, (1) can be written as

$$f_{\mathrm{GGD}}(x;\alpha,\beta) = \frac{\beta}{2\alpha\Gamma(\beta^{-1})} \exp\left(-\left|\frac{x}{\alpha}\right|^{\beta}\right) \quad (3)$$

where α replaces the scale parameter σ.

It should be noted that the GGD parametric distribution family includes many popular distributions that are commonly used in biomedical signal processing. For example, setting $\beta = 1$ leads to a Laplacian or double-exponential distribution, $\beta = 2$ leads to Gaussian or normal distribution, and $\beta \to \infty$ leads to a uniform distribution.

The GGD was fitted using a window shift of two seconds with overlapping of one second in 18 epileptic signals, for each signal obtained 205 fits on average for each epoch and calculated the parameters related to the scale (α) and shape (β) for each rhythm band. We refer the reader to [23] for a comprehensive treatment of the mathematical properties of the GGD model and [17,22] for a detailed explanation on the estimation of the GGD parameters.

B Kullback-Leibler Divergence

Let p and q two PDFs, then a Kullback-Leibler Divergence (KLD) [24] is given by

$$D_{KL}(p||q) = \int_{-\infty}^{\infty} log\left(\frac{p_x(x)}{q_x(x)}\right)p_x(x)dx \quad (4)$$

$$D_{KL}(p||q) = -\int_{-\infty}^{\infty} log(q_x(x))p_x(x)dx + \int_{-\infty}^{\infty} log(p_x(x))p_x(x)dx = H_c(p,q) - H(x) \quad (5)$$

Notice that in general $D_{KL}(p||q) \neq D_{KL}(q||p)$, and that $D_{KL}(p,q) = 0$ if and only if $p = q$ [25].

Rewriting the equation (3), the probability density function of GGD is given by

$$p(x,\alpha,\beta) = \frac{e^{-|\frac{x}{\alpha}|^{\beta}}}{2\alpha\Gamma[1+\beta^{-1}]} \quad (6)$$

Here we consider the divergence between two generalized Gaussian models with parameters (α_1,β_1,μ_1) and (α_2,β_2,μ_2) subject to the constraint $\mu_1 = \mu_2 = 0$ because our signals have zero mean. This divergence is given by

$$KLD_{pdf}(p||q) = \int_{-\infty}^{\infty} p(x,\alpha_1,\beta_1) log\left(\frac{p(x,\alpha_1,\beta_1)}{p(x,\alpha_2,\beta_2)}\right)dx \quad (7)$$

$$= \int_{-\infty}^{\infty} p(x,\alpha_1,\beta_1)\left[\left|\frac{x}{\alpha_2}\right|^{\beta_2} - \left|\frac{x}{\alpha_1}\right|^{\beta_1} + log\left(\frac{\alpha_2\Gamma(1+\frac{1}{\beta_2})}{\alpha_1\Gamma(1+\frac{1}{\beta_1})}\right)\right]dx$$

$$= \frac{-\Gamma(\frac{1}{\beta_1}) + (\frac{\alpha_1}{\alpha_2})^{\beta_2}\beta_1\Gamma(\frac{1+\beta_2}{\beta_1}) + \beta_1\Gamma(\frac{1}{\beta_1})log\left(\frac{\alpha_2\Gamma(1+\frac{1}{\beta_2})}{\alpha_1\Gamma(1+\frac{1}{\beta_1})}\right)}{\beta_1\Gamma(\frac{1}{\beta_1})}$$

$$KLD_{pdf}(p||q) = -\frac{1}{\beta_1} + \frac{(\frac{\alpha_1}{\alpha_2})^{\beta_2}\Gamma(\frac{1+\beta_2}{\beta_1})}{\Gamma(\frac{1}{\beta_1})} + log\left(\frac{\alpha_2\Gamma(1+\frac{1}{\beta_2})}{\alpha_1\Gamma(1+\frac{1}{\beta_1})}\right)$$

We compare the PDFs obtained 18 epileptic signals, using the scales and the shapes of the GGD using (6), in two stages in steps of one second without overlapping for each rhythms brain bands

1. Between sliding window and the seizure onset

$$KLD_{pdf}(p^{(i)}||q_{onset}) = \mathbb{W}^{(i)}KLD_{pdf}(p||q)$$

2. Between continuous PDFs coupled with a 7-order one-dimensional median filter [26]

$$KLD_{pdf}(p^{(i)}||q^{(i+1)}) = \mathbb{W}^{(i)}\mathbb{F}^{(i)}KLD_{pdf}(p||q)$$

with

$$\mathbb{W}^{(i)} = [0^{L\times iL}, I^{L\times L}, 0^{L\times N-iL-L}]$$
$$\mathbb{F}^{(i)} = medianFilter(KLD_{pdf}(p^{(i)}||q^{(i)}))$$

where $0^{N\times M} \in \mathbb{R}^{N\times M}$ is the null matrix, $I^{N\times N} \in \mathbb{R}^{N\times N}$ is the identity matrix and L is the number of measurement obtained in one second. We refer the reader to [25, 27–29] for a comprehensive treatment of the mathematical properties of the KLD statistical theory.

III Results and Discussion

The performance of the proposed statistical method was evaluated using the Children's Hospital Boston database [30, 31], which consists of 36 EEG recordings from pediatric subjects with intractable seizures. In this work we used 18 seizures from 9 subjects. Datasets including two to five bipolar EEG recordings sampled at 256Hz were available for each subject. Each recording contained a seizure event with a labeled onset that was detected by an experienced neurologist, who worked backward from the observed clinical onset to find the epilepsy seizure onset. The signals used have one epoch focused on *Seizure/Non-Seizure* where the onset of the *Seizure* begins at two minutes.

We obtained a good performance of the KLD method via visual inspection in all 18 epileptic signals by an experienced neurologist. For illustration, Figures (1) and (2) depicts the different brain rhythms: delta, theta, alpha, beta and gamma, where the *Seizure* is 40 seconds of duration, we can see increase in activity between 2 minutes and 2.4 minutes in all brain rhythms. In Figure (1) we can notice how the signal have the *Seizure* onset begins at minute two; we can see clearly the discrepancy between *Seizure/Non-Seizure* in epileptic signals; while in Figure (2) the *Seizure* onset is detected clearly given by the highest peak which emerges from background of EEG showing a discrepancy between *Seizure/Non-Seizure*. Once the *Seizure* finished, there are several medical pathological factors that causes the signal takes a time in stabilize it again, this is the reason why the *Seizure* does not have an instantaneous change after the 2 minutes 40 seconds, however, is clear the discrepancy after the *Seizure*.

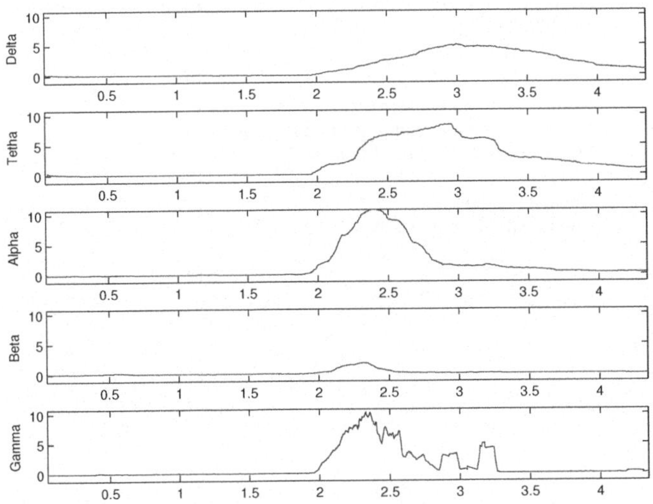

Fig. 1: KLD illustration between sliding window and the *Seizure* onset of the epileptic signal, showing a clear discrepancy between *Seizure/Non-Seizure*. In this example the *Seizure* onset begin at minute 2, and its duration is 40 seconds.

IV Conclusions

The preliminary results reported in this work in 18 epileptic signals suggest that the proposed method, is potentially useful for differentiating *Seizure/Non-Seizure* signals in epileptic signals and for onset detection.

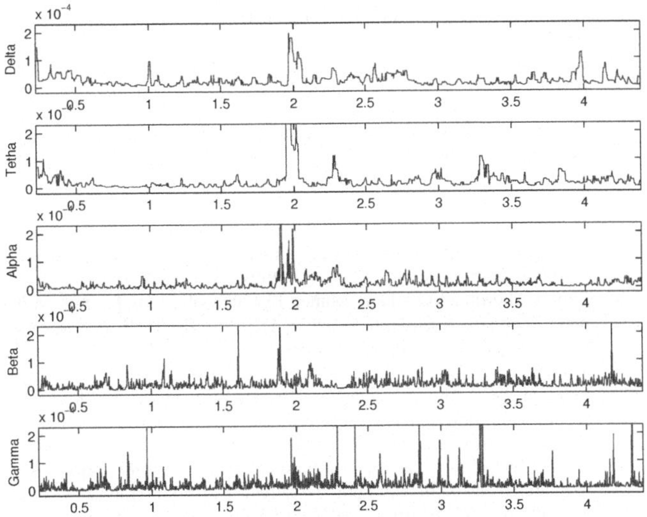

Fig. 2: KLD illustration between continuous PDFs coupled with a 7-order one-dimensional median filter, showing clearly the discrepancy given by the highest peak which emerges from background of EEG. In this example the *Seizure* onset begin at minute 2, and its duration is 40 seconds.

V Future Work

Perspective for future work include an extensive evaluation of the proposed methodology, as well as performing comparisons with other methods from the state of the art, and the development of fusion techniques to combine detections from several algorithms to increase robustness to noise and to artifacts.

It is also interesting to investigate how implement this method in real time for *Seizure/Non-Seizure* EEG classification using shift windows with different filters, optimizing display scales and calculating the onset delay by comparing the average amplitude with the background, similar to [19, 32].

Conflict of Interest

The authors declare that they have no conflict of interest.

Acknowledgements

Part of this work was funded by: a) the *SuSTaIN* program, EPSRC grant EP/D063485/1 at the Department of Mathematics, University of Bristol, b) *ITBACyT* grant DP/No.557/2015, Scientific and Technical Activities at Research Department, Instituto Tecnológico de Buenos Aires and c) *07/15Protocol* at FLENI .
Marcelo Pereyra holds a Marie Curie Intra-European Fellowship for Career Development.

REFERENCES

1. Fisher Robert S., Acevedo Carlos, Arzimanoglou Alexis, et al. ILAE official report: A practical clinical definition of epilepsy *Epilepsia.* 2014;55:475-482.
2. Qu Hao, Gotman Jean. A Patient-Specific Algorithm for the Detection of Seizure Onset in Long-Term EEG Monitoring Possible Use as a Warning Device *IEEE Transactions on Biomedical Engineering.* 1997;44:115-122.
3. Jerger Kristin K., Netoff Theoden I., Francis Joseph T., et al. Early Seizure Detection. In: Journal of Clinical Neurophysiology *Journal of Clinical Neurophysiology.* 2001;18:259-268.
4. Subasi Abdulhamit, Alkana Ahmet, Koklukayab Etem, Kiymik M. Kemal. Analysis Of Epileptic Seizure. Detection Methods Based on Parameter Estimation, Power Spectrum Density And Morlet Wavelet Transform *Neural Networks.* 2005;18:985?997.
5. Schelter Bjorn, Winterhalder Matthias, Drentrup Hinnerk Feldwisch, et al. Seizure prediction The impact of long prediction horizons *Epilepsy Research.* 2007;73:213-217.
6. Osorio Ivan, Frei Mark G. Realtime Detection, Quantification, Warning, and Control of Epileptic Seizures: The Foundations of a New Epileptology *US Neurology.* 2008;4:67-70.
7. Kharbouch Alaa, Shoeb Ali, Guttag John, Cash Sydney S.. An algorithm for seizure onset detection using intracranial EEG *Epilepsy Behavior.* 2011;22:29-35.
8. Alotaiby Turkey N, Alshebeili Saleh A, Alshawi Tariq, Ahmad Ishtiaq, El-Samie5 Fathi E Abd. EEG seizure detection and prediction algorithms a survey *URASIP Journal on Advances in Signal Processing.* 2014;183.
9. Quiroga R. Quian, Arnhold J., Grassberger P.. Kulback-Leibler and renormalized entropies: Applications to electroencephalograms of epilepsy patients *Physical Review E.* 2000;62:8380-8386.
10. Sabesan Shivkumar, Good Levi B., Tsakalis Konstantinos S., Spanias Andreas, Treiman David M., Iasemidis Leon D.. Information flow in coupled nonlinear systems: Application to the epileptic human brain *Data mining in biomedicine, Springer optimization and its applications series, P. Pardalos, V. Boginski, and A. Vazacopoulos, eds., . New York: Springer..* 2007:483-504.
11. Sabesan Shivkumar, Good Levi B., Tsakalis Konstantinos S., Spanias Andreas, Treiman David M., Iasemidis Leon D.. Information flow and application to epileptogenic focus localization from intracranial EEG *IEEE transactions on neural systems and rehabilitation engineering.* 2009;17:244-253.
12. Phothisonothai Montri, Nakagawa Masahiro. EEG signal classification method based on fractal features and neural network *30th Annual International IEEE EMBS Conference.* 2008:3880-3883.
13. Gupta Anjum, Parameswaran Shibin, Lee Cheng-Han. Classification of electroencephalography (EEG) signals for different mental activities using Kullback Leibler (KL) divergence *IEEE International Conference on, Acoustics, Speech, and Signal Processing.* 2009:1697-1700.
14. Wang Haixian. Harmonic Mean of Kullback-Leibler Divergences for Optimizing Multi-Class EEG Spatio-Temporal Filters *Neural Processing Letters.* 2012;36:161-171.
15. Daubechies Ingrid. *Ten Lectures On Wavelets Daubechies.* SIAM 1992.
16. Strang Gilbert, Nguyen Truong. *Wavelets and Filter Banks.* Wellesley-Cambridge Press 1996.
17. Pascal Frédéric, Bombrun Lionel, Tourneret Jean-Yves, Berthoumieu Yannick. Parameter Estimation For Multivariate Generalized Gaussian Distributions *IEEE Transactions on Signal Processing.* 2013;61:5960-5971.
18. Quintero-Rincón Antonio, Batatia Hadj, Pereyra Marcelo, Risk Marcelo. Detection of Onset in Epilepsy Signals using Generalized Gaussian Distribution *Fifth International Conference on Advances in New Technologies, Interactive Interfaces and Communicability.* 2014;2014. ISBN:978.88.96.471.37.1, DOI: 10.978.8896471/371.
19. Quintero-Rincón Antonio, Pereyra Marcelo, D'Giano Carlos, Batatia Hadj, Risk Marcelo. A new algorithm for epilepsy seizure onset detection and spread estimation from EEG signals *Journal of Physics: Conference Series.* 2016;705:012032.
20. Quintero-Rincón Antonio, Prendes Jorge, Pereyra Marcelo, Batatia Hadj, Risk Marcelo. Multivariate Bayesian Classification of Epilepsy EEG Signals *2016 IEEE 12th Image, Video, and Multidimensional Signal Processing Workshop (IVMSP).* 2016:1-5.
21. Schomer Donald L., Lopes.da-Silva Fernando. *Electroencephalography Basic Principles, Clinical Applications, and Related Fields.* Lippincott Williams and Wilkins 2010.
22. Do Minh N., Vetterli Martin. Wavelet-Based Texture Retrieval Using Generalized Gaussian Density and Kullback-Leibler Distance *IEEE Transactions on Image Processing.* 2002;11:146-158.
23. Nadarajah Saralees. A Generalized Normal Distribution *Journal of Applied Statistics.* 2005;32:685-694.
24. Kullback S., Leibler R.A.. On information and sufficency *The Annals of Mathematical Statistics.* 1951;22:79-86.
25. Cover Thomas M., Thomas Joy A.. *Elements of Information Theory.* John Wiley and Sons, Inc. 2006.
26. Pratt William K.. *Introduction to Digital Image Processing.* CRC Press 2013.
27. Kullback Solomon. *Information theory and statistics.* Dover Publications 1978.
28. Gray Robert M.. *Entropy and Information Theory.* Springer 2011.
29. Nafornita Corina, Berthoumieu Yannick, Nafornita Ioan, Isar Alexandru. Kullback-Leibler Distance Between Complex Generalized Gaussian Distributions *20th European Signal Processing Conference (EUSIPCO 2012).* 2012.
30. Goldberger Ary L., Amaral Luis A.N., Glass Leon, et al. PhysioBank, PhysioToolkit, and PhysioNet: Components of a New Research Resource for Complex Physiologic Signals *Circulation 101(23):.* 2000;101:e215-e220.
31. Shoeb Ali, Edwards Herman, Connolly Jack, Bourgeois Blaise, Treves S. Ted, Guttagf John. Patient-specific seizure onset detection *Epilepsy and Behavior.* 2004;5:483-498.
32. Gotman Jean. Automatic seizure detection: improvements and evaluation *Electroencephalography and Clinical Neurophysiology.* 1990;76:317-324.

Author: Antonio Quintero-Rincón
Institute: Instituto Tecnólogico de Buenos Aires
Street: Avenida Eduardo Madero 399
City: Buenos Aires
Country: Argenitna
Email: aquinter@itba.edu.ar

First Steps For A Web-Based Platform Oriented To The Study Of The Infant Neuro-Development And Disability

Sergio D. Cano Ortiz[1], Zoila Gonzales Videaux, Amilcar Borrás Gonzalez [2] Reinhardt Langmann, Hartmut Haehnel[3], Ludmila Regueiferos Diaz[4]

[1] CENPIS, Universidad de Oriente, Santiago de Cuba, Cuba, scano@uo.edu.cu
[2] Departamento de Informatica, Universidad de Oriente, Santiago de Cuba, Cuba, zoe@fie.uo.edu.cu, amilcarb@fie.uo.edu.cu
[3] Duesseldorf University of Applied Sciences, Germany, reinhard.langmann@fh-duesseldorf.de, hartmut.haehnel@fh-duesseldorf.de
[4] Servicio de Neonatología , Hospital Materno Sur Stgo de Cuba, Cuba

Abstract— **The first steps for a telemedicine network which connect the units and places within the Medical Consultation for Infant Neurodevelopment and Disability (CPNDI) in Santiago de Cuba are discussed in the paper. For that target a web-based system for communications within the CPNDI structure and also to access rural areas with limited communication infrastructures is finally proposed. The system consists of a patient management system and a medical device system that are available to the doctor and the patient-side assistant during treatment via an Internet browser. The combination of local and remote web applications enables a reliable integration of the medical devices required at the patient's location into the doctor's remote station based on browser technology.**

Keywords— **web-based telemedicine, infant neuro-development outcome, cry analysis.**

I. Introduction

A study of infant cry using advances from the field of cybernetics and signal processing was initiated by the Group of Speech Processing (GPV) from Universidad de Oriente in 1990. Just in 1991 the multidisciplinary group formed with specialists within the Medical Consultation for Infant Neurodevelopment and Disability (CPNDI of Southern Infant Hospital of Santiago de Cuba) was created, funding the basis of the research project Infant Cry Analysis Oriented for Diagnosis. In 2008 GPV researchers designed a cry-based methodology for newborn diagnose well commented through a PhD thesis successfully defended by a member of GPV in February 2010. As alternative results to that methodology several software tools were obtained for parametric characterization of the cry signal, also a fundamental frequency-based cry pre-classifier , as well as some theoretical contributions concerned with new algorithms for hybrid classification of cry signal. During the whole multidisciplinary research period the members of GPV has been established as a priority the study of infant neurodevelopment and disability holding the CPNDI as its main stone.

This entity assumes as a main goal to find out neurological and sensorial deviations precociously, and also to describe the related risk factors in newborns. In that way newborns who displayed some kind of sequel as neurological as motor or functional disability are properly studied and followed up.

The effective routine performance of the CPNDI is often limited by complexities of Cuban daily life as the urban mobility, problems with the communications, limitations in the access and use of the new communication technologies, existence of remote and mountain locations with difficult access where many newborns live, etc. In spite of the big efforts done by the Government and administration staff in order to extend people access to those technologies, until now is even insufficient and the CPNDI is affected too.

With the objective of contributing to the solution of this problem, the authors propose a low cost web-based system that should assume the communication via web of the nodes that conform the CPNDI and remote rural places located in the municipalities of the province of Santiago de Cuba.

The system allows not only the remote communication among the entities of the CPNDI but also the communication between a medical doctor (or multidisciplinary group located at the CPNDI) and an medical assistant with a patient (located in a remote place or a place with difficult access, e.g. a municipality town surrounding Santiago de Cuba).

In highly developed industrial countries with an extensive medical infrastructure, telemedical systems are mainly used to improve the quality of life for older people [1, 2] as well as emergency healthcare systems [2,4] and systems used in particular situations (on ships [5], in the space flight industry [6, 7] and in military applications [8, 9]).

The available telemedical systems are not suited for a regular and conventional treatment process with only untrained or semi-skilled medical staff being available at the patient's location. In addition, the costs of conventional systems available on the market exceed the cost limit for practicable systems in development

I. Torres et al. (eds.), *VII Latin American Congress on Biomedical Engineering CLAIB 2016, Bucaramanga, Santander, Colombia, October 26th -28th, 2016*,
IFMBE Proceedings 60,
DOI: 10.1007/978-981-10-4086-3_5

countries by far. The technical solutions are based on proprietary and supplier-specific components, while standard ITK technologies are hardly used.

Some web-based telemedical projects planned to be deployed in underdeveloped regions and countries already exist [10, 11, 12]. The respective concepts however only address the required telemedicine infrastructure in most cases, while not considering the actually required telemedical stations, for which conventional systems available on the market shall be used. In rural and isolated areas where there is a geographical limitation, the use for telemedical stations that take into account satellite connectivity, power supply such as solar panels, and ready to use teleconsultation equipments are very highly needed and not easily available or affordable.

II. MATERIALS AND METHODS

A. System Requeriments

For the design and implementation of the telemedical system named as TeleMedSys several requirements were established:

- A doctor station (DS), which will be able to cooperate with N patient stations (PS) respectively, will be set up.
- The PS deployment site has no power supply and communication infrastructure. The TeleMedSys will be transported through harsh terrain to the deployment site.
- Only medical assistants will be available at the PS site, who will have to be guided by the doctor during the treatment process.
- The TeleMedSys shall be web-based in order to integrate simple eLearning elements parallel (work-integrated learning).
- 95% of the treatment process is performed asynchronously between the patient/assistant and the doctor while the remaining 5% are performed synchronously via video conference and/or real-time chat.
- The weekly period of use for a PS amounts to 3-4 hours. The maximum distance between the PS and the respective DS is 2,000 km.
- The medical examination equipment is connected to the PS; it will be unlocked by the DS and operated by the assistants according to the doctor's instructions. The respective measuring data are automatically provided to the DS.

A particular challenge is the asynchronous treatment process because in the traditional medical practice such a procedure is not required and unbeknown..

B. Equipment structure of the TeleMedSys.

The device-related part of the DS can be realized with a standard computer, as the physician disposes of the required infrastructure (fixed power supply, Internet access) and no medical devices need to be connected. Developing the PS will be the biggest project challenge. Based on the above-mentioned requirements, the PS will therefore be equipped with the following components:

- A robust notebook for use under critical environmental conditions (ruggedized).
- A solar power station featuring a buffered battery operation for the power supply (solar output: 28W; battery:15Ah, 12,8V; module output: 15V, 1800mA).
- Satellite modem (Inmarsat satellites) for realizing an Internet access with 384 Kbit/s download and 240 Kbit/s upload.
- Medical examination devices with Bluetooth interface. The prototype will feature a blood pressure measuring device and a 6-channel ECG device only for the time being.

Figure 1 shows a photo of the PS device components.

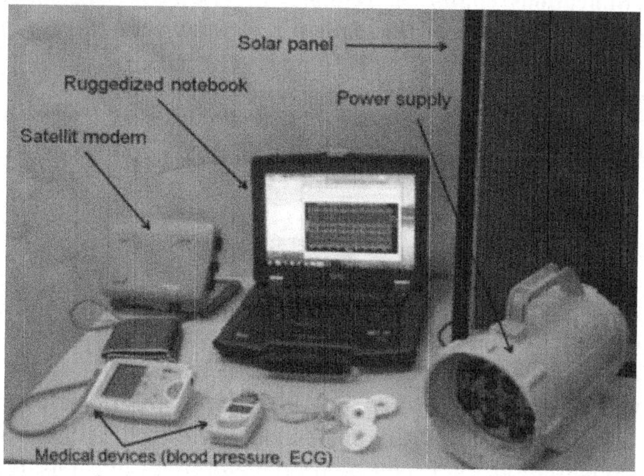

Figure 1. Device components of the Patient Station for the TeleMedSys

According to the primary health needs and common diseases of the area the appropriate technology devices will be implemented. For further development of the system should be connected at least the following additional equipment for medical examinations:

- ultrasonic diagnostic device;
- digital camera;
- body temperature measuring instrument

For all equipment components of the PS is planned also the developing of a special shipping box that is suitable for transport over rough terrain. After a field test of the prototype will also be re-examined the appropriateness of the used medical devices with Bluetooth connectivity. Although the wireless connection of the medical device with the PS is a flexible and elegant solution, but all units require an extra power supply with batteries, which may in some cases not be available. Moreover, for the communication interface of medical devices (Bluetooth, USB or serial), the respective communication protocols and APIs are required. This is usually not the case, since the devices usually come as a complete application solution. Exceptions are the devices of Corscience Company (Germany) which are used in the prototype of the PS.

The serial protocol is published by Corscience and can be used very well for development purposes. The protocol is designed for simple and memory-saving implementation in a microcontroller. The overhead was kept as low as possible. The principle of this protocol is basically modeled on the PPP (point-to-point protocol), which is often used to establish modem connections. Also, escape sequences are used to filter out reserved bytes (start flag, end flag, escape flag) from the data stream.

III. Results and Discussion

A. Functional structure.

The TeleMedSys software consists of the following three key components:

- Patient Management System (PMS)
- Medical Device System (MDS)
- Online Collaboration System (OCS)

The prototype requires the PMS and the MDS (only for asynchronous operation). Later, the prototype will be extended with the OCS.

Figure 2 depicts the functional structure of the TeleMedSys:

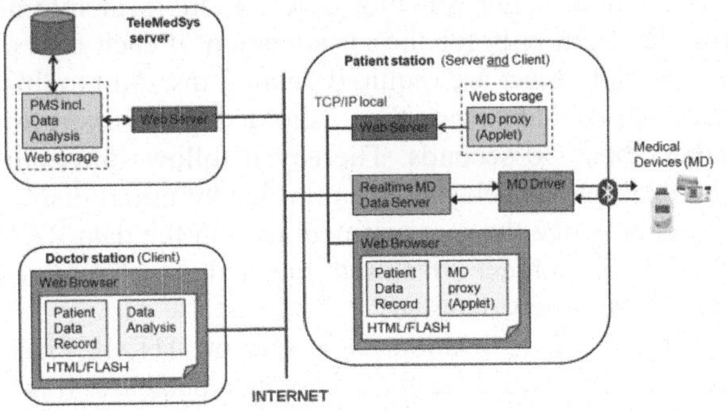

Figure 2. Functional structure of the TeleMedSys

All patient information is stored by the Patient Management System (PMS) on a TeleMedSys server. Both the DS and the PS have an access per web browser to the PMS database on the TeleMedSys server and can manipulate the information in a Patient Data Record (PDR). The DS is working as a client and uses only a web browser.

The asynchronous patient treatment is the main issue of the PMS (see Fig. 3). It requires separating the overall treatment session for a patient into partial steps for the assistant at the PS and the doctor at the DS. These partial steps are performed. alternately by the assistant and the doctor. The system automatically displays the status of all patient sessions and provides information about the next required step for each patient through an event screen.

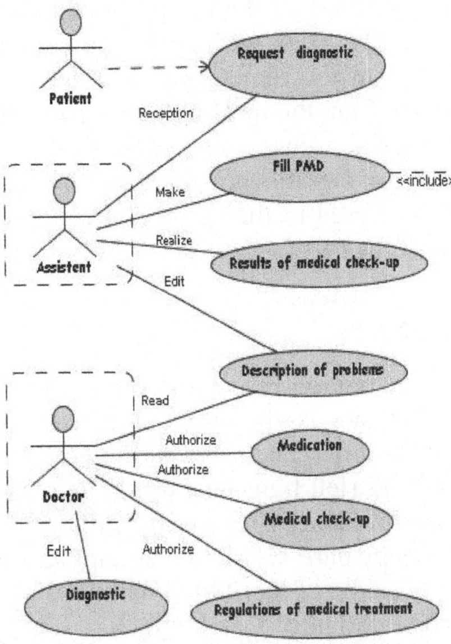

Figure 3. Graphic description of the asynchronous mode

In principle, the treatment is done in the asynchronous mode of operation similar to the remote diagnostic and service procedures in a technical system. The entire process is well structured and each sub-step ends with a conclusion. Finished partial steps are not repeatable in the current session.

A relatively stringent regulation of the treatment steps is required because only semi-skilled or non-professional medical personnel will be working at the patient's side. During a field test of the prototype under real conditions it will be shown whether this step-by-step process is suitable for the medical treatment and will also be accepted by the participating medical staff.

The doctor can unlock the required medical devices via the PMS at the PS site and provide the assistant with additional context-related support instructions. The assistant at the PS will be also here instructed by the doctor, in order to make no mistakes in carrying out the medical measurement.

The MDS consists mainly of the following components:

- MD driver: Implementation of the Corscience protocol interface to medical devices.

- MD data server: Recording and preprocessing of the measured medical data and implementation of a TCP/IP command interface.
- MD proxy: implementation of the TCP/IP command exchange between the MDS web page and MD data server.
- MDS web pages: The operation of each medical device is made via a web page with a MD device operator panel.

All MDS components are located on the PS and will be activated via the MDS web pages in the web browser of the PS. For this purpose on the PS is running a local web server by whom the MDS web pages will be started. The PMS (loaded from the remote server TeleMedSys) and the MDS (loaded from the local web server in the PS) are connected in a frame website according the MashUp principle. Only by this way it is possible that through this web page can be accessed to the peripheral components (medical devices) of the PS

B. Running the system.

Figure 4 shows the frame website of the PS. The visual design of the PS website is not yet optimized. At the moment it is a design difference between the MDS (right gray part) and the PMS (left light-blue part).The reason is the implementation of this two parts by different web technologies – MDS by pure HTML/JavaScript and PMS by Flash. The next version of the PS/DS will have also an unified user interface design.

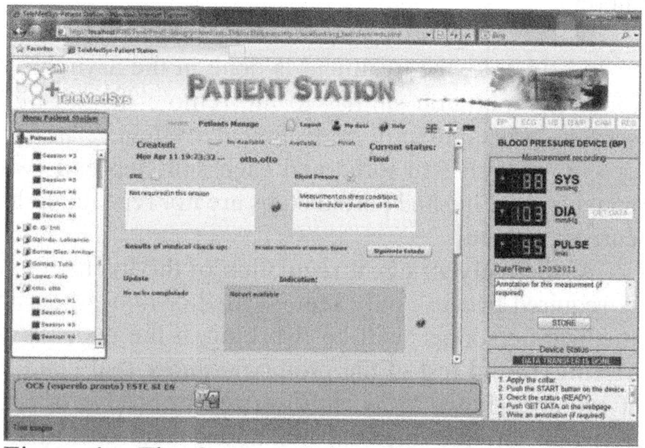

Figure 4. The frame website of the Patient Station in the prototype

In case of a limited data value volume, the MDS transmits the measuring data (e.g. blood pressure) in real-time to the PMS and stores the data in the database. In case of larger data volumes (e.g. ECG) the data are saved locally in a measuring file first and subsequently transferred remotely in a corresponding file of the TeleMedSys server. In this case, only the name and location of the data file (not the data itself) are stored in the database.

The measuring data can be analyzed at the DS. Figure 5 shows an analysis example for the 6-channel ECG in the DS web browser.

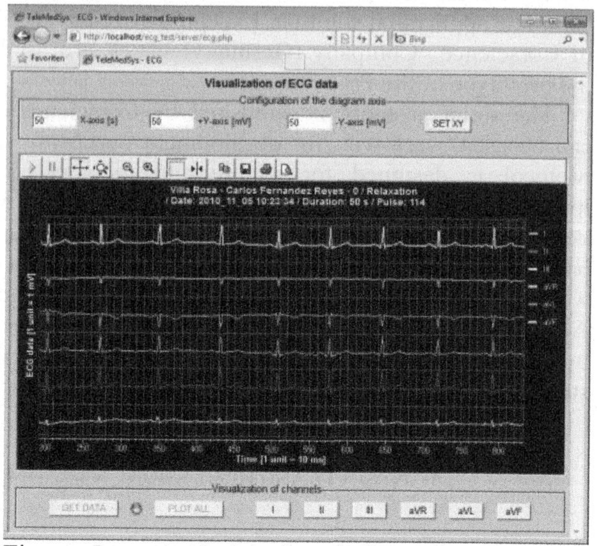

Figure 5. Analysis of a 6-channel ECG with a measuring time of 50 s in the Doctor Station's web browser.

The XY plot object (ActiveX object) from the Iocomp visualisation package is used for the graphical ECG data presentation [13]. Here, however, a not yet optimal solved problem arises for the visualization of long-term ECG measurement:

The recording of ECG data occurs at a clock rate of 100 Hz or 500 Hz. For a quite normal stress ECG with e.g. 20 min duration and a resolution of 10 ms (100 Hz) this results in 120,000 data values. The representation of these data in the XY plot object is in principle not a problem, but only for the visualization of each data point in the plot object are required about 2 ms. Add to this the retrieval of data from the TeleMedSys server, which takes about 30 seconds. Thereby it follows in result for the example a delay of 4.5 min for the initial display of the data. Once the measurement data in the data RAM of the XY plot object are stored, the analysis of data can be quickly and very detailed.

The PMS realisation is based on HTML/Flash in combination with Java for accessing a database. It is used mySQL as database server and the design of the relational database was made according with the specification of business applications. The MDS is developed with HTML, JavaScript and Java applets. The MD server is a Java application that is operated via a Bluetooth interface and controlled through TCP/IP commands by a Java applet (MD-Proxy, see Fig. 2).

For the ensuring of the persistence of the PMS the Hibernate framework is used. The main task of Hibernate is the Object-Relational Mapping (ORM). This makes it possible to store ordinary Java objects with attributes and methods in relational databases, and to create again objects from these records. Relations between objects are mapped to corresponding database relations. By this way it is certain that in case of connection abort or bad Internet connection no data are lost during treatment process, and the system can be put again on the last treatment date.

On the PS is running an Apache web server as a local web server. The PMS uses a Tomcat web server with a Spring Framework as a Java application support.

The DS will function as a TeleMedSys server simultaneously during a first field test of the prototype. Later, the TeleMedSys server will however operate on a dedicated computer, to provide the required safety and data security.

IV. Conclusions

The paper describes a low-cost telemedicine system supported by web technology which is used for connection of two types of nets: an urban one for remote connection with the Medical Consultation for Neurodevelopment and Dissability (CPNDI) in province Santiago de Cuba, as well as a a second one for access to rural locations. The combination of remote and local web applications facilitate a direct remote access to ambulatory medical equipments through browser without technological gaps.

The doctor station and N distributed patient stations are provided with a structured access to all patient data through a common web browser, enabling them to perform asynchronous treatment procedures.

A first field test for the prototype was carried out successfully in Santiago de Cuba, supplemented with a second test in Dusseldorf, where all the functionalities and technical features of the system were evidenced. Both tests were carried out with the participation of real patients.

The acquisition costs of the TeleMedSys should not exceed EUR 20,000. The operating costs should be less than EUR 1,000 per month.

V. Acknowledgments

This research has been partially supported by the *Webbasierte Be-Plattform Zur Signalanalyse (WebSA)* project in collaboration with the University of Applied Sciences of Dusseldorf, Germany and the Instituto Nacional de Astrofisica, Optica y Electronica, Puebla, Mexico

References

1. Zydacron: Information about CareStation 140. – www.zydacron.com, 2009
2. Bellazzi, R., et al.: Web-based telemedicine systems for home-care: technical issues and experiences. - Computer Methods and Programs in Biomedicine, 64 (2001), pp. 175–187
3. TMA Medical: Erfolg für Telemedizin und E-health durch das Mobile Care Unit. – www.tma-medical.com, 2010
4. MEYTEC: Ein neuesTelemedizinsystem für Echtzeitanwendungen in Rettungsfahrzeugen. – www.meytec.com, 2010
5. GS Elektromedizinische Geräte G. Stemple: Information about corpuls systems. - www.corpuls.de, 2009
6. Williams, D. R., et al: A Strategic Vision for Telemedicine and Medical Informatics in Space Flight. - Telemedicine Journal and e-Health, Vol 6, Nov 4, 2001, pp 441-448
7. Centre Spatial Guyanais: Telemedicine in French Guiana. – www.cnes-csg.fr, 2010
8. Cali, C.: The Italian Military Medical Services and Telemedicine. – Proc. of the Telemedicin Workshop and 3rd Meeting of the SIMIHO new Experts technical Group, Rome, 7 - 8 July 2003
9. Melcer, T., et al.: A Prospective Evaluation of ENT Telemedicine in Remote Military Populations Seeking Specialty Care. - Journal and e-Health, September 2002, 8(3), pp 301-311.
10. Logan, J. S.: Telemedicine 101: Innovations in Rural Health Care Delivery Systems. – Proc. of Cooperative Solutions to Rural Health Care Problems: Emergency Medical Services, University of California, Davis School of Medicine, June, 1993
11. Blue print: The Ohanyido Matrix: Development of a Telemedicine System for Nigeria. - http://ohanyido.tripod.com/id13.html, 2005
12. Pandian, P. S., et al: Store and Forward Applications in Telemedicine for Wireless Ip Based Networks. - Journal of Networks, Vol. 2, No. 6, Dec. 2007, pp 58 - 65
13. IOcomp: Description of the iPlot object. – www.iocomp.com, 2010

How do our students learn clinical engineering? A pilot study

Antonio Miguel Cruz[1], Daniel Alejandro Quiroga Torres[1], Ana Maria Presiga[1], Nestor Flórez Luna[1]

School of Medicine and Health Sciences, Universidad del Rosario/Biomedical Engineering Department. Calle 63D # 24-31, 7 de Agosto, Bogotá D.C, Colombia. Email's corresponding author: antonio.miguel@urosario.edu.co, miguelcr@ualberta.ca

***Abstract*—This paper aims to measure what are the students' perceived learning outcome achievements after finishing their clinical engineering major courses. This is a pre- post-test with no control group study design. Forty students were involved in this pilot study. A paper-based survey composed of a demographic section and a 5-point Likert ("1" is strongly disagree and "5" is strongly agree) section measured the students' perceived learning outcome achievements after exposing them to clinical engineering major courses. A Wilcoxon signed-rank and Mann-Whitney U test statistics were conducted to test the two hypotheses of this study. Our analysis showed statistically significant results between the pre-survey mean and SD: 21.10 SD 3.54; and between the post-survey mean and SD: 22.75 SD 3.68 (Z=-2.12, p<0.033), indicating that overall, students' perceived learning outcome achievements after exposing them to clinical engineering major courses had significantly improved by the end of the major. Also, statistically significant results were found between the post-survey mean and SD: 3.94 SD 0.61 learning outcome perceptions and between the students' actual marks mean and SD: 4.53 SD 0.22 (-5.00, p<0.000), indicating the students had low confidence in their learning outcomes after completing their clinical engineering major.**

Keywords-component: *clinical engineering, biomedical engineering, engineering education*

I. INTRODUCCIÓN

A. *Clinical engineering roles and training methods: a brief state of the art*

According to the American College of Clinical Engineers "*a Clinical Engineer is a professional who supports and advances patient care by* applying engineering and managerial skills to healthcare technology" *[1]*; this definition is also adopted by the International Federation for Medical and Biological Engineering *[2]* and by the U.S. Board of Examiners for Clinical Engineering Certification *[3]*. Simply put, a clinical engineer is a biomedical engineer settled in the clinical environment that serves as a "*facilitator*" between medical technology and its users by providing operational and technical support.

Although there is an indisputable need for the presence of clinical engineers in healthcare organizations gained across the years, legitimate concern on the part of the consumers (i.e. patients/clients) has arisen. Since most of the time these consumers are interacting with medical technologies, they want to be confident that the professionals providing the servicing (e.g. maintenance) for their medical devices are properly qualified. Therefore, universities and non-governmental bodies of countries (e.g. the U.S. led by the ACCE) and international federations (e.g. the International Federation for Medical and Biological Engineering (IFMBS), and the European Alliance for Medical and Biological Sciences (EAMBES)) are putting special emphasis on the development of guidelines containing protocols for professional education and training clinical engineers. For example the EAMBES released its own protocol for training clinical engineers in Europe. According to the EAMBES´s protocol, 3 years of education and training are needed to be a certified clinical engineer. Students' core training areas include management (32 %), technology assessment (15%), regulatory/QA issues (11 %), repair/systems thinking (6 %), risk management/safety issues (9%), education (8%), (product development (8 %), and miscellaneous topics (11 %). These core training areas are in conjunction with the ACCE [1]. The training and education protocol should be conducted in either specialized "*Training Centres*" or in "*Institutions*" that must be accredited with the national or the European Clinical Engineering Professional Development Panel CEPDP [2]. In the U.S. similar efforts have been made. One example is the cooperative training course offered by Trinity College, the Hartford Graduate Center, Hartford Hospital, the University of Connecticut Medical Center in Farmington, the Baystate Medical Center in Springfield, Massachusetts, and NOVAMED. This is a hospital-based two-year program in which students have to take a clinical engineering internship program followed by a normal Master's degree program in biomedical engineering.

Despite the advantages of the training protocols of the European Union and the U.S., for low-income countries such comprehensive, structured, and integrated programs rarely exist. As a result, the main responsibilities in the education and training of future clinical engineers are under the umbrella of the biomedical engineering programs of higher education centers/universities. For example, this is the case of the biomedical engineering program (BEP) offered in a partnership between the *School of Medicine and Health Sciences* at the Universidad del Rosario and the *School of Engineering Julio Garavito* in Bogotá. In brief, this program has 154 credits[1] and 9 semesters (4.5 years). To obtain a degree, the students have to do a major in any area of biomedical engineering covered by the program, and to have a certificate at B2 proficiency level in a foreign language (English) [4]. One of the majors is clinical engineering.

I. Torres et al. (eds.), *VII Latin American Congress on Biomedical Engineering CLAIB 2016, Bucaramanga, Santander, Colombia, October 26th -28th, 2016*,
IFMBE Proceedings 60,
DOI: 10.1007/978-981-10-4086-3_6

The authors of this pilot study carried out a rapid review of academic studies in both undergraduate and graduate programs aimed at measuring the impact of clinical engineering programs' implementation in terms of either the achievement of the students' intended learning outcomes, or employers' perceptions of program graduated skills. Unfortunately few studies exist [5, 6], indicating a gap in the knowledge in this area.

B. Clinical engineering training method: our proposal

The teaching method of the clinical engineering major on the BEP follows a well-structured and balanced number of courses and credits, the teaching methods and learning strategies with the course content. This is the so-called "*constructive alignment*" teaching philosophy [7]. Simply put, in the tasks of designing and implementing a particular course, a relationship must exist between the *intended learning outcomes (ILOs)*, the *teaching and learning activities (TLAs)*, and the *assessment of the actual students' learning (AL)*. In our experience, when this alignment exists, the students' anxiety decreases, while their perceived competence, motivation, and satisfaction with learning increases, because they know in advance what they have to learn and how they will be assessed. In addition, the above-mentioned alignment facilitates the process of accountability of the teacher to his/her audience: the students. Therefore, under the conditions of a constructive alignment teaching philosophy, this study proposes two hypotheses:

Hypothesis 1 (H_1): There are statistically significant differences in students' perceived learning outcome achievements after putting them through clinical engineering major courses.

Hypothesis 2 (H_2): There are no statistically significant differences between students' perceived learning outcome achievements and the actual students' marks after putting them through clinical engineering major courses.

Tables I, II, and III show the course's structure, the intended learning outcomes, and the constructive alignment of the clinical engineering major of the BEP.

Table I. Clinical engineering training major courses

Course name	Course id	Cr	Total hours	Type
Maintenance Management	MTM001	3	48	M
Clinical Engineering	CE001	3	48	M
Medical Devices I	MD001	3	48	E
Medical Devices II	MD002	3	48	E
Medical Devices III	MD003	3	48	E
Hospital Engineering	HE001	3	48	E
Thesis or Internship	GRADOPT	4	640	M

E: Elective course, M: Mandatory course. MD001: Cardiovascular system devices; MD002: Respiratory system devices; MD003: Imaging devices

The obtain the major in clinical engineering the students have to take at least MTM001 and CE001 mandatory courses, one elective course (see Table II) and Thesis or Internship course either in a healthcare institution or in a enterprice in the area of clinical engineering (e.g. maintenance, biomedical metrology, etc.)

Table II. Learning outcomes.

ILO code	Learning outcome (Name/code)
C13	Student is able to schedule the life cycle medical technologies, proposing a more feasible option for their introduction into the healthcare system.
C14	Student is able to conduct an economic analysis of health.
C15	Student is able to conduct an economic analysis of medical technologies during acquisition processes (cost benefit analysis).
C16	Student is able to perform scheduled and unscheduled maintenance tasks.
C17	Student is able to schedule and implement scheduled and unscheduled maintenance programs.
C18	In general student is able to design and implement the full process of technology management.

Table III. Constructuve aligment on the clinical engineering track courses biomedical engineering program.

Course code	Learning outcome (Name/code)	Teaching and learning activities	Assessment method
MTM001	C16-C17	Lectures (20%) Laboratories (80%)	Theory test Practical test in labs Rubrics
MD001, 002, 003	C16, C18	Lectures (20%) Laboratories (80%)	Theory test Practical test in labs Rubrics
CE001 HE001	C13-C15, and C18	Lectures (20%) Collaborative learning techniques (Colts) (80%)	Theory test Collaborative learning test (Rubrics)

II. MATERIALS AND METHODS

A. Study design

Pre- post-test with design with no control group.

B. Participants.

All students that enrolled on *MTM001* and *CE001* mandatory courses and the students that chose at least two elective courses in their clinical engineering major.

C. The instruments

In this study, a survey was designed to measure the students' perceived learning outcome achievements. The survey was composed of a demographic section (section A, 10 items) and a 5-point Likert (section B, 6 items) *"1" is strongly disagree and "5" is strongly agree*) section measuring the students' perceptions of their learning outcome achievements when passing the courses of the clinical engineering major. Also, the students' average marks on the clinical engineering major courses were recorded.

D. The measures

In this study two outcome variables were included. The first one was students' perceptions of their learning outcome achievements when passing the courses of the clinical engineering major, measured using the 6-item survey (C13-C18, see Table II). The second one was the average student's marks on the clinical engineering major courses.

Included among the demographics data recorded were: gender (male/female), the socio-economical strata (Low/Medium /High[2]), the type of high school the student graduated from in terms of whether the high school was private or public; and the type of the high school's degree profile (Academic/Technical).

E. Procedures and data collection.

During the first week of the first semester of the major (sixth semester of the program), the research team, with prior approval by the course instructors and the university Ethics Committee, administered the pre-survey to the students. When the students finished their major the post-survey was administered again (ninth semester). The post-survey was administered at a time when the students were not experiencing any anxiety, physiological pressure, or fatigue due to any exams. A research assistant who was not involved in the teaching sessions was assigned to assist the research team in administering the pre- and post-surveys.

F. Statistical analysis

Descriptive statistics were used to summarize the demographic data of the students. To test H_1, one paired Wilcoxon signed-rank test for two dependent samples was performed on the data gathered from the pre- and post-surveys (students' learning outcome perception achievements). To test H_2, a Mann-Whitney U test for two independent samples was performed between the students' perception achievements and the students' mean course marks. In addition to that, the Pearson correlation between the students' perception achievements and the students' mean course marks was calculated.

The alpha level of significance for all the tests was set at $p \leq 0.05$ (*two-sided*). The SPSS® V 22.0 statistics package was used to generate descriptive, univariate, and bivariate statistics, respectively.

III Results

A. Participants' description

Forty participants were involved in this study (female 45%: male 55%). Most of them (i.e. 77.5%) were within the age range of 19 to 21 years old, whereas the remaining 22.5% were within the age range of 22 to 24 years old. Most of them lived in families of medium socio-economical strata (i.e. 82.5%). Regarding the students' procedence baccalaureate school types, 82.5% had graduated from private schools and 17.5% from public schools. Finally, 87.5% had graduated from a school with an academic profile, 7.5% had graduated from schools that had a technical profile, and the remaining 5% had another type of profile.

B. What students' believe they learned and what they really learned: Testing hypotheses H_1 and H_2

Tables IV and V show the Wilcoxon signed-rank test statistic and Mann-Whitney U test results of the testing hypotheses H_1 and H_2. Table IV shows the pre- and post-Wilcoxon signed-rank test of the students' perceived learning outcome achievements (H_1). Fig. 1 shows the pre- and post-means of the students' perceived learning outcomes achievement (C16-C18). Table V shows the Mann-Whitney U test between the students' perceived learning outcome achievements (at the post-test) and the actual actual students' marks after putting them through clinical engineering major courses (H_2).

Table IV. Wilcoxon signed-rank test statistics: Students' perceived learning outcome achievements after putting them through clinical engineering major courses. (N=28)

Hypothesis 1

Learning outcomes	Pre-test		Post-test		Wilcoxon signed-rank test	
	Mean	S.D	Mean	S.D	(z, p)	Power Test
C13	3.39	1.06	3.50	1.13	-0.33, 0.73	6.6%
C14	3.42	0.99	3.50	0.96	-0.27, 0.78	6.1%
C15	3.17	1.05	3.67	0.90	**-2.0, 0.045**	48.1%
C16	3.92	0.85	4.10	0.73	-1.05, 0.29	13.6%
C17	3.89	0.91	4.07	0.66	-0.84, 0.39	13.5%
C18	3.28	0.93	3.89	0.83	**-3.11, 0.002**	73.5%
All*	21.10	3.54	22.75	3.68	**-2.12, 0.033**	40.1%

Notes: Sum of all the scores of the learning outcomes (C13-C18)

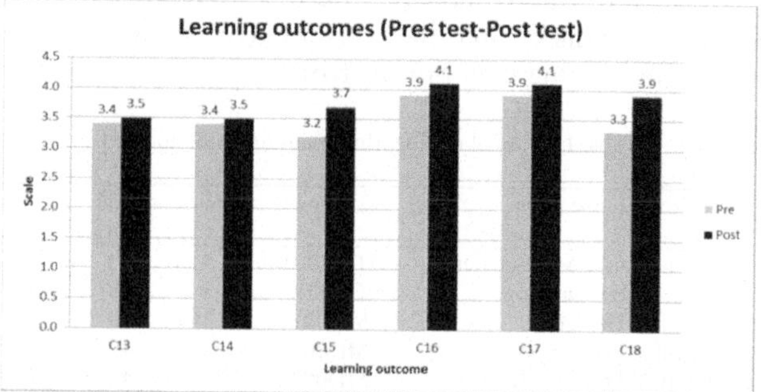

Fig 1. Students' learning outcome perception achievement: Pre-test versus Post-test

Table IV. Mann-Whitney U Test: Students' perceived learning outcome achievements versus actual students' marks after putting them through clinical engineering major courses. (N=40)

Hypothesis 2

Learning outcomes	Post-test	Students' marks	Mann-Whitney U Test

[2] Low< 800US Medium: 800-2500US; High >2500USD

	Mean	S.D	Mean	S.D	(z, p)	Power Test
All (C13-18)	3.94	0.61	4.53	0.22	**-5.00,0.00**	100%

A negative correlation between the students' perceived achievements and the students' course mark average was found (r_{xy}=-0.11, p=0.24), although it was not statistically significant.

III. Discussion

This paper aims to measure what the students' perceived learning outcome achievements were after finishing their clinical engineering major courses. From Table IV one can see that overall, our results support *Hypothesis 1* (H_1). The one paired Wilcoxon signed-rank test for two dependent samples study showed that there were statistically significant differences in the students' perceived learning outcome achievements after putting them through clinical engineering major courses for all the learning outcomes (i.e. all (C13-C18) sum scale means at the pre-test 21.11 SD 3.54 compared with the post-test 22.75 SD 3.68 (Z=-2.12, p=0.033)). Also, Table IV and Fig 1. show that all the mean values for all the learning outcomes at the post-test are higher compared to the pre-test. However, only two of the six learning outcomes (33%) were statistically significant (C15 and C18), indicating that after finishing the clinical engineering major courses, the students still perceived that they have had not achieved the learning outcomes C13-C14 and C16-C17. This seems to be contradictory because, as previously mentioned, they perceived they had achieved the learning outcome C18, i.e. *In general student is able to design and implement the full process of technology management*, which includes the C13, C14, C15, C16, and C17 learning outcomes. One possible explanation for this result is that the effect sizes of all the tests were low, indicating that the sample size in this study is still too low to achieve a power of 80% or higher; in fact for all the tests the power was <80%.

From Table V one can see that in spite of the actual students' marks, after putting them through clinical engineering major courses they were higher than the perceived learning outcome achievements, indicating they had actually achieved their learning outcomes; this result does not support *Hypothesis 2* (H_2) because statistically significant differences were found after conducting the Mann-Whitney U test (i.e. Z=-5.00, p<0.000). This result is consistent with previous research, which demonstrated that students had low confidence in their skills to successfully complete an engineering degree, especially women [8, 9]. In other words, students are not fully aware of what they have learned, indicating that instructors and staff professors should make a better effort to explain to their students what they are able to do and what they have actually learned.

IV. Conclusion

Under the conditions of a constructive alignment teaching philosophy, the results of this study indicate that the students' overall perceptions improved after their experience of taking clinical engineering major courses. However, the students still have low confidence in their skills, as measured by the students' perceived learning outcome achievements, compared to what they actually learned, as measured by the students' actual marks. An unresolved matter is still pending; what are the employers' perceptions about our students' learning outcome achievements? And to what extent are the employers' perceptions about our students' learning outcome achievements similar (or not) to what the students' actually learned or perceived they had learned? This will be covered by future research.

Conflicts of Interest

The authors report no conflicts of interest. The results reported in this manuscript were not sponsored by any grant.

Acknowledgements

We want to thank Oliver Jarvis for his proofreading of this paper. The authors assume overall responsibility for the manuscript.

References

[1] ACCE, "American College of Clinical Engineering," 1992. [Online]. Available: http://accenet.org/about/Documents/What's_a_Clinical_Engineer.pdf. [Accessed 09 March 2016].

[2] J. Nagel, "Protocol for the training of clinical engineers in Europe," BIOMEDEA PROJECT, Stuttgart, 2005.

[3] HTF, "2013 Candidate Handbook for Certification in Clinical Engineering by the Healthcare Technology Certification Commission," Healthcare Technology Certification Commission, Plymouth, 2013.

[4] A. Miguel-Cruz, A. Rios-Rincon, W. Rodríguez-Dueñas, N. Florez-Luna and D. Quiroga-Torres, "What is the effect of an introductory biomedical engineering course on the students' perceptions of the engineering profession?," *International Journal of Engineering Education,* vol. 32, p. 136–149, 1(A) 2016.

[5] B. Zeng, M. Huang, F. Chen and L. Chen, "Exploration on undergraduate curriculum reform of hospital-oriented biomedical engineering major," in *Proceedings of the 2009 2nd International Conference on Biomedical Engineering and Informatics, BMEI 2009*, Tianjin, 2009.

[6] N. Cram, "Educating a New Generation of Biomedical and Clinical Engineers," *Journal of Clinical Engineering,* vol. 39, no. 2, pp. 76–78, 2014.

[7] J. Biggs, Teaching for Quality Learning At University, Berkshire: England: The Society for Research into Higher Education and Open University Press, 2003.

[8] H.-S. Lee, L. N. R. Flores and M. Kanagui-Muñoz, "A longitudinal test of social cognitive career theory's academic persistence model among Latino/a and White men and women engineering students," *Journal of Vocational Behavior,* vol. 88, p. 95–103, 2015.

[9] A. Cruz, A. Rincon, W. Dueñas, N. Luna and D. Torres, "The impact of an introductory biomedical engineering course on students' perceptions of the engineering profession," *International Journal of Engineering Education,* vol. 32, no. 1, pp. 136–149, 2016.

An Embedded Hybrid BCI Speller

P.A. García[1], E.M Spinelli[1,2] and G.M. Toccaceli[1,2]

[1] Universidad Nacional de La Plata/Facultad de Ingeniería/Electrotecnia/LEICI, La Plata, Argentina
[2]Consejo Nacional de Investigaciones Científicas y Técnicas (CONICET), Argentina

Abstract— **The present work proposes an embedded real time hardware-software platform for brain computer interfaces (BCI) based on steady state visual evoked potentials (SSVEP) and alpha rhythm. The complete implementation of the embedded system is described, including the electroencephalogram (EEG) amplifier, signal acquisition and processing stages, and practical implementation. The device is a 25 characters hybrid-BCI speller that uses 5 visual stimulus and visual alpha waves for control instructions. The system focuses on simplicity and portability, using only two electrodes, a simple EEG amplifier and an embedded computer. The speller does not require any training from the user and provides real time biofeedback to increase attention on stimuli. The platform is based on an embedded system with a real time operative system, Windows CE. Experimental results are presented in order to show the feasibility of the proposed system.**

Working as a speller, the current average speed is around 7 characters per minute.

Keywords— **Alpha rhythm, brain computer interface, embedded system, speller, steady state visual evoked potential.**

I. Introduction

A brain computer interface (BCI) is a communication channel intended to translate voluntary brain activity into commands to control devices without the use of muscles [1]. The main application of BCIs are to provide a communication channel for severely disabled people [2], although in recent years its usage in entertainment applications has grown [3]. Non-invasive BCIs based on electroencephalography (EEG) are the most common implementations [4]. Several types of EEG signals have been used in these BCIs, such as alpha rhythms [5], slow cortical potentials [6], P300 event related potentials [7], steady state visual evoked potential [8], and motor imagery [9] among others.

In particular, visual evoked potential based BCIs are described as more accessible because they need little user training and achieve high information transfer rates (ITR) [10]. These devices use the brain response to a visual stimulus. During operation, the user is exposed to visual stimuli blinking at different frequencies, which are associated to different commands. When the user focuses on a command, the brain visual cortex generates an EEG evoked potential with the associated stimulus frequency (and its harmonics).

In recent years, an increasing number of researchers have begun to explore a new variant: hybrid BCIs. In this case multiple modalities are combined on a single BCI improving performance [11].

This paper presents a hybrid BCI based on SSVEP and alpha rhythm. BCIs based only on SSVEP cannot be turned on and off voluntarily because of their exclusive dependence on evoked potentials. In our hybrid implementation, by including spontaneous potentials, the user can turn on and off the stimuli avoiding visual fatigue and the generation of false commands. The proposed BCI is designed to be used independently and without supervision, unlike other implementations intended for research. This BCI aims to be a practical and autonomous implementation for an end user.

Thus, the system focuses on simplicity, robustness and on-demand operability, and should function outside the laboratory using as few EEG channels as possible. It uses only two electrodes on the occipital zone to capture SSVEP and two on the forehead to implement a Driven Right Leg circuit [12]. It works on-demand using the alpha rhythm to implement a brain switch for providing control commands without the need of visual stimulus. For instance, BCI users can turn on and off the SSVEP stimuli by means of time duration coded alpha wave activations.

As mentioned in [4] and [13], BCIs have proved to be useful and it is time to migrate the deployment from a laboratory prototype to a portable device reduced in size. In order to bring BCI technology closer to the end users, the computer should be replaced by more compact and reliable equipment. In this work an embedded system with a friendly graphical user interface (GUI) familiar to the user and fast start up is proposed. Any BCI or assistive device must be able to become operational as fast as possible without the need of a calibration process to be useful. A slow start up process can incline the user to choose another communication channel.

This work presents the complete implementation of the Hybrid BCI Speller system.

II. System overview

The proposed BCI is an embedded system on a single device. It implements a speller allowing to select among 25 characters using 5 visual stimuli plus an alpha-rhythm-

I. Torres et al. (eds.), *VII Latin American Congress on Biomedical Engineering CLAIB 2016, Bucaramanga, Santander, Colombia, October 26th -28th, 2016*,
IFMBE Proceedings 60,
DOI: 10.1007/978-981-10-4086-3_7

based switch for control instructions. It can be operated autonomously by the user, without external assistance or supervision.

A. Main board

The embedded system is implemented with a Micro2440SDK board from FriendlyArm, and a custom front-end for EEG amplification and digitalization. The main board is based on a 32 bit processor (Samsung S3C2440A) with ARM920TDI architecture and a 400 MHz processing speed. This board accepts different operative systems: Windows CE, Linux and Android. It has commonly used peripherals such as four USB host and one USB device ports, Ethernet, LCD screen with touch-screen input, 3 RS-232 serial ports, audio input and output, a socket for SD memory cards, and a real time clock, among others.

In Fig. 1, a diagram of the full system is presented.

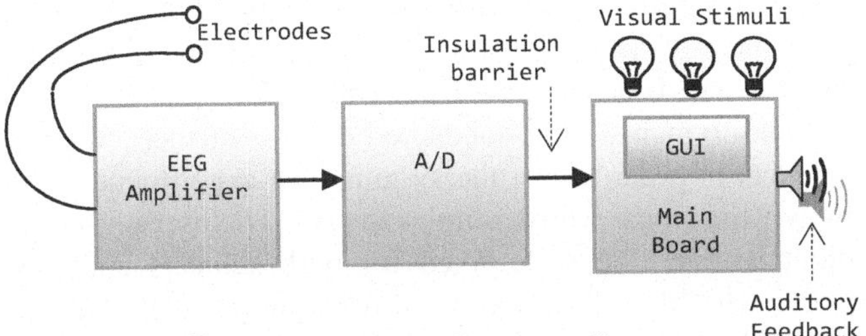

Fig. 1 Hybrid BCI Speller system diagram

B. Front-End: EEG Amplifier and digitalization

To acquire the EEG signals, a previously designed biopotential amplifier was used [12]. It is a low noise AC coupled amplifier with an overall gain of 5800, powered from a single 5V supply. It is complemented with a driven right leg (DRL) circuit to set the patient DC potential to V_{REF} (1.25 V). Two independent 3M auto-adhesive electrodes were used to place the DRL on the forehead avoiding problems to affix the circuit. An ADuc812 microcontroller from Analog Devices was used to digitalize the signal and impose the sampling rate. The microcontroller transfers the EEG signals digitally through an isolated SPI channel to the main board. Amplifier and microcontroller power and signals are isolated by an ADuM6401 DC/DC converter from the same company.

The A/D converter imposes the system time base at 1024 Hz sampling frequency. The A/D converter notifies the "sample ready" event to the main board by means of an external interrupt also routed through the ADuM isolator.

C. Low level software

Among the various operating systems (OS) options on the main board, Windows CE 6.0 was selected, because it is the only one with native hard real time features and presents a user interface based on windows, friendly and known to most users. In order to read the EEG signal from the front-end, a low level driver was developed to use the SPI port. This driver complies with real time constraints and is implemented into the operating system's image [14]. The information generated by the low level driver is stored on a circular buffer and available for any high level software implementation by means of a memory map file resource. This way, multiple applications can use the EEG data. The driver was programmed using C language.

D. High level software

Because of the implemented layered software scheme, any application can access the EEG information. The high level software runs in user mode and can be developed using the object oriented programming paradigm. The C# language was selected to program the speller.

E. Speller: Graphical interface

The graphical user interface (GUI) is shown in Fig. 2. The speller main window presents 25 characters grouped in 5 columns with 5 letters each.

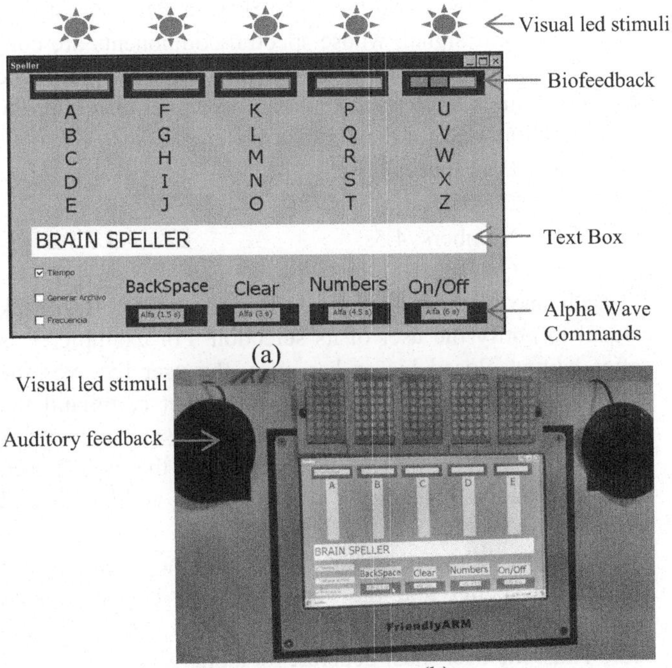

Fig. 2. Speller GUI. (a) The speller main window with the visual stimuli at the top of each letter column. (b) A photograph of the speller implementation showing the second window after the first column selection.

The five visual stimuli are implemented with 7x5 led arrays (Fig 2.b) and were located respectively at the top of each character column. In order to select a letter, the user must focus on the visual stimulus at the top of the corresponding column. The BCI detects the evoked potential and redraws the five letters of the selected column, but now, each letter is drawn in its own column (it can be seen in Fig 2.b that after the first column selection, the five letters are each in one column). Finally the user must focus on the visual stimulus corresponding to the target letter to select it.

The BCI processes the EEG signal in order to obtain its power spectral density (PSD) using a 0.5 s window. The algorithm estimates the stimulus frequency corresponding to the SSVEP potential with higher amplitude. Once a frequency stimulus is detected, the BCI activates one of three segments of a biofeedback progress bar. The command is selected after the three segments, i.e., three consecutive windows (1.5s), with activity of a particular frequency stimulus are detected. The frequencies used on the stimuli are 14, 16, 18, 20 and 22 Hz. They are 2 Hz apart because this is the frequency resolution on the fast Fourier transform (FFT) using a window size of 0.5 s.

At the bottom of the window, the "backspace", "clear", "numbers" and "On/Off" commands are displayed. All of them are activated using the alpha rhythm. The easiest way to generate this signal, whose spectral components are concentrated around 10 Hz, is by closing the eyes [15]. The four commands implemented using this signal differs in the time duration used for their activation:

- Backspace: 1.5s.
- Clear: 3s.
- Numbers: 4.5s.
- On/Off: 6s.

Each one of these commands has an auditory feedback signal to notify the user of its selection. For example, 1.5s after having closed his or her eyes, the user receives the Backspace notification. If this is the target command the user opens the eyes finishing the command selection. If this is not the target command, the user keeps the eyes closed until the auditory feedback corresponding to the desired command is received.

The most relevant command allows the user to turn on/off the speller and the visual stimuli. This command must be robust, and unintended activations should not occur. Thus, the user must generate an alpha wave signal during at least 6 s.

The "numbers" command is implemented to offer the user an alternative keyboard with numbers and unusual letters and symbols. The scheme is like the main keyboard with 25 alternative characters.

F. Algorithm: Power Spectral Density

In order to calculate the signal power spectrum, a class named "FFT" was implemented that returns a 512-point fast Fourier transform (FFT). A graphical interface was also implemented in the form of a spectrum analyzer as a debugging tool during the tests. The implemented algorithm calculates the 512-point FFT every 0.5 s since the sampling frequency is 1024 Hz. To detect the user's intention, the algorithm looks for the stimulus frequency band with highest energy and compares it with the spectrum mean value. The algorithm's decision may be for one of the five target stimulus or none of them in case the user is not focusing in any target. As mentioned before, three consecutive decisions for the same stimuli are needed to issue the command.

III. EXPERIMENTS

In order to test and characterize the speller, various experiments were conducted using only two standard EEG electrodes located on O_1 and O_2 (international 10-20 system). Two independent 3M^{TM} disposable auto-adhesive electrodes, placed on the forehead were used to implement the DRL circuit. The EEG data was acquired at a sampling frequency of 1024 Hz after being band-pass filtered between 1 and 100 Hz by the front-end.

Measurements were conducted on two of the authors as subjects in order to test the system. The experiment started by writing single words, similar to [10]. The user must write eight words: five were fixed for both subjects and three were variable. The fixed words were: BCI, BRAIN, SPELLER, INTERFACE and COMPUTER. The variable words were the user's first name, last name and football team. Each word must be written correctly without errors. The users could make corrections using the alpha command backspace. The user could have an optional rest between consecutive words turning off the SSVEP stimuli if desired.

IV. RESULTS

The parameter commonly used to measure and contextualize a BCI is the information transfer rate (ITR) [1][16] defined as:

$$ITR = \frac{Number_{of\,commands} * B}{T} \quad (1)$$

Where B is the channel capacity given by:

$$B = \log_2 N + P\log_2 P + (1-P)\log_2\left(\frac{1-P}{N-1}\right) \quad (2)$$

P is the probability to correctly detect a command, N is the number of commands and T is the time required to produce a certain number of commands. In our implementation with N=*5*, the best B is *2.32* bits/command. With any other P<*1* the channel capacity decreases.

Ideally, in the best situation, the user can execute a command in *1.5s* and the corresponding *ITR* with P=*1* would be *92.87* bpm. In this scenario, the speller speed would be *20* letters per minute (*2* commands/letter).

Tables 1 and 2 present the spelling results for authors 1 and 2 respectively.

TABLE I
ACCURACY AND BIT RATES FOR SUBJECT 1

Word	#Commands	Time[s]	Accuracy[%]	ITR[bpm]	[LETTERS/MIN]
BCI	6	20	100	41.76	9
BRAIN	10	34	100	40.94	8.82
SPELLER	14	59	100	33.03	7.12
INTERFACE	18	90	100	27.84	6
COMPUTER	16	85	100	26.20	5.64
PABLO	10	45	100	30.93	6.67
GARCIA	12	53	100	31.52	6.79
BOCA	8	40	100	27.84	6

TABLE II
ACCURACY AND BIT RATES FOR SUBJECT 2

Word	#Commands	Time[s]	Accuracy[%]	ITR[bpm]	[LETTERS/MIN]
BCI	6	28	100	29.83	6.43
BRAIN	10	36	100	38.67	8.33
SPELLER	15	134	93.3	12.22	3.35
INTERFACE	18	125	100	20.04	4.32
COMPUTER	16	185	100	12.04	2.59
ENRIQUE	14	120	100	16.24	3.50

V. CONCLUSIONS

A novel embedded hybrid BCI speller is introduced to enable people with disabilities to communicate. The system is robust, simple to use, and portable. It dispenses the use of a personal computer and presents a fast start up time (20 s). LED arrays are used to generate the visual stimuli, avoiding problems associated with screen size and refresh, but limiting the number of stimuli. The system implements the brain switch concept using alpha waves, which allows the user to turn on/off the visual stimulators if desired. The system implements a visual feedback using a progress bar which results very useful to enhance the SSVEP recovery, and an auditory feedback for the alpha wave commands.

Experimental results validate the complete system operation and its feasibility to be transferred to end users. Although the few conducted measurements show a good performance, the experimental results suggest that user training can significantly enhance the speller speed. With very little training, the average speeds obtained are 7 and 4.3 letters per minute for subject 1 and 2, respectively.

This work presents the complete implementation of the hybrid BCI speller. The system is fully functional and represents a good precedent for the effective implementation required by an end user. The hardware can be easily modified to implement other devices like wheelchairs controllers, or prosthesis controllers, among others.

ACKNOWLEDGMENT

This work was supported in part by the Universidad Nacional de La Plata (UNLP) under Grant 11/I167 and PPID-I003 and ANPCyT by project PICT-2012/0037

REFERENCES

1. J. R. Wolpaw, N. Birbaumer, D. J. McFarland, G. Pfurtscheller, and T. M. Vaughan, "Brain-computer interfaces for communication and control," *Clin. Neurophysiol.*, vol. 113, no. 6, pp. 767–791, 2002.
2. N. Birbaumer, "Breaking the silence: Brain-Computer-interfaces (BCI) for communication and motor control", *Psychophysiology*, vol. 43, pp. 517-532, 2006.
3. D. Marshall, D. Coyle, S. Wilson, and M. Callaghan, "Games, Gameplay, and BCI: The State of the Art", *IEEE Trans. on computational intelligence and ai in games*, vol. 5, no. 2, pp. 82-99, 2013.
4. J. E. Huggins, C. Guger, B. Allison, C. W. Anderson, A. Batista, A-M. Brouwer, C. Brunner, R. Chavarriaga, M. Fried-Oken, A. Gunduz, D. Gupta, A. Kübler, R. Leeb, F. Lotte, L. E. Miller, G. Müller-Putz, T. Rutkowski, M. Tangermann & D. E. Thompson, "(2014) Workshops of the Fifth International Brain-Computer Interface Meeting: Defining the Future, Brain-Computer Interfaces", 1:1, 27-49, DOI: 10.1080/2326263X.2013.876724.
5. P. A. García, M. Haberman and E. Spinelli, "A versatile hardware platform for brain computer interfaces", Engineering in Medicine and Biology Society (EMBC), 2010 Annual International Conference of the IEEE, pp. 4193 – 4196.
6. N. Neumann and A. Kubler, "Training locked-in patients: a challenge for the use of brain-computer interfaces", *IEEE Trans. on Neural Systems and Rehabilitation Engineering,* vol. 11, issue 2, pp. 169-172, 2003.
7. C. Postelnicu and D. Talaba, "P300-Based Brain-Neuronal Computer Interaction for Spelling Applications", *IEEE Trans. on Biomedical Engineering,* vol. 60, issue 2, pp. 534 – 543, 2012.
8. R. Ortner, B. Z. Allison, G. Korisek, H. Gaggl and G. Pfurtscheller, "An SSVEP BCI to Control a Hand Orthosis for Persons with Tetraplegia", *IEEE Trans. on Neural Systems and Rehabilitation Engineering,* vol. 19, issue 1, pp. 1-5, 2011.
9. Y. Tianyou, X. Jun, W. Fangyi, R. Zhang, G. Zhenghui, A. Cichocki and L. Yuanqing, "Enhanced Motor Imagery Training Using a Hybrid BCI with Feedback", *IEEE Trans. on Biomedical Engineering,* vol. 62, issue 7, pp. 1706 –1717, 2015.
10. H. Cecotti, "A Self-Paced and Calibration-Less SSVEP-Based Brain-Computer Interface Speller", *IEEE Trans. on Neural Systems and Rehabilitation Engineering,* vol. 18, no. 2, pp. 127-133, 2010.
11. G. Pfurtscheller, B. Allison, C. Brunner, G. Bauernfeind, T. Solis-Escalante, R. Scherer, T. Zander, G. Müller-Putz, C. Neuper, and N. Birbaumer, "The hybrid BCI," *Frontiers in Neuroscience*, vol. 4, art. 42, pp. 1-11.
12. E. Spinelli, N. Martinez y M. Mayosky, "A Single Supply Biopotential Amplifier". *Medical Engineering and Physics*, ISSN 1350-4533, Vol. 23/3, pp. 235-238, 2001.
13. J. Millán, R. Rupp, G. Müller-Putz, R. Murray-Smith, C. Giugliemma, M. Tangermann, C. Vidaurre, F. Cincotti,A. Kübler,R. Leeb, C. Neuper,K-R. Müller and D. Mattia, "Combining brain-computer interfaces and assistive technologies: state-of-the-art and challenges", *Frontiers in Neuroscience,* DOI: 10.3389, 2010.
14. P. A. García, E. M. Spinelli, and G. M. Toccaceli, "An embedded system for evoked biopotential acquisition and processing", *Int. J. Embedded Systems*, Vol. 6, No. 1, pp.86–93, 2014.
15. Y. Ishikawa, M. Takata and K. Joe, "Constitution and phase analysis of alpha waves", Biomedical Engineering International Conference (BMEiCON), 2012, DOI: 10.1109/BMEiCon.2012.6465482.
16. Gao X, Xu D, Cheng M and Gao S, "A BCI-Based Environmental Controller for the Motion-Disabled". *IEEE Transactions on neural systems and rehabilitation engineering*, vol. 11, Nº. 2, June 2003.

Teaching maintenance of medical devices in simulation centers: a pilot study

Daniel Alejandro Quiroga Torres[1],Antonio Miguel Cruz[1], Ana Maria Presiga[1], Nestor Flórez Luna[1]

School of Medicine and Health Sciences, Universidad del Rosario/Biomedical Engineering Department. Bogotá D.C, Colombia.

***Abstract*—This paper aims to measure what the students' perceived learning outcome achievements are after finishing medical devices courses I and II. This is a pre- post-test with no control group study design. *Twenty-four* students doing medical devices courses I and II with a test in the simulation center of a Biomedical Engineering Program participated in this pilot study. A paper-based survey composed of a demographic and 5-point Likert ("1" is strongly disagree and "5" is strongly agree) measured the students' perceived learning outcome achievements after exposing them to medical devices courses. A Wilcoxon signed-rank test and Mann-Whitney U Test statistics were conducted to test the two hypotheses of this study. Our analysis showed statistically significant results between the pre-survey sum mean and SD: 7.50, 1.31; and between the post-survey mean and SD: 8.56, 1.15, p=0.015, indicating the students' perceived learning outcome achievements after putting them through medical devices courses I and II had significantly improved by the end of the courses. Also, no statistically significant results were found between the post-survey mean and SD: 4.37, 0.57 learning outcome perceptions or between the students' actual marks mean and SD: 4.58, 0.18, p=0.28), indicating no differences between what the students perceived they had learned compared with what they had really learned.**

Keywords-component:* *biomedical engineering, engineering education, medical devices, simulation center.

I. Introducción

A. Clinical engineering and simulation centers: a brief state of the art

According to the American College of Clinical Engineers "a Clinical Engineer is a professional who supports and advances patient care by applying engineering and managerial skills to healthcare technology" [1]; this definition is also adopted by the international organizations [2] [3]. In other words, a clinical engineer is a biomedical engineer working in a healthcare organization whose function is to serve as a *"mediator"* or facilitator between the medical technology and its users by providing operational and technical support.

There is legitimate concern on the part of consumers (i.e. patients/clients) about the quality of medical devices or medical technology in the healthcare environment. Since most of the time these consumers are interacting with medical technology they want to be confident that the professionals providing the servicing (e.g. maintenance) for these medical devices are properly qualified. Therefore, the methods of the pedagogic used are important as to how these professional should be trained is a timely and important issue. One of these methods is the use of simulation centers as a practice scenario before introducing the students to real clinical settings. Teaching through simulation and virtual environments brings advantages such as the transfer of learning, practice, feedback, and the teaching process in a controllable environment and through trial and error [4]. In addition, some studies have reported that the use of simulations also helps in students' learning gains during education and training, technology assessment, product development, and the research and evaluation workflows process [5] [6]. Some tools used in the simulation for teaching are mannequins, role play, virtual labs, and computer modeling. An example of the use of simulation in clinical engineering teaching was performed at the Huntington hospital, in which the instructors used clinical simulations in the evaluation of the medical equipment. This study demonstrated from the assessment of the participants that the simulation allows a comprehensive assessment of the user interface of the device, the clinical application, and the overall usability without putting the patients at risk [7].

In countries like Israel [8], Scotland [9] and Germany [10] simulation centers are very common. In Israel, a medical simulation center was built. The simulation center contained emergency room, intensive care unit, operating rooms, among others. To Israel, the simulation center was a cost effective to provide a clinical simulation-based education solution. It also allowed the transfer of knowledge and learning in a health profession to another [8]. In the case of Germany and Scotland, the first simulation centers focused on training anesthesia residents. With these centers learning materials, forms of training and skills are improved as residents as they made use of simulators were designed [9] [10].

In spite of the aforementioned advantages of the use of simulation as a teaching method, the authors of this pilot study carried out a rapid review of academic studies in both undergraduate and graduate programs, aimed at determining the impact of simulation centers in the clinical engineering field in terms of either the achievements of the students' intended learning outcomes or the employers' perceptions of the program graduates' skills. Unfortunately only a few studies exist, indicating a gap in the knowledge in this area. Therefore, the motivation of this study was to explore how the use of a simulation center impacts the learning process of a population of students after their experience of taking a medical device course.

B. Medical devices training method: our proposal

The main responsibilities in the education and training of future clinical engineers are under the umbrella of the biomedical engineering programs of higher education centers/universities. For example, this is the case of the biomedical engineering program (BEP) offered in partnership with the School of Medicine and Health Sciences of Universidad del Rosario and the School of Engineering Julio Garavito of Bogotá. In brief, this program has 154 credits and 9 semesters (4.5 years). To obtain the degree, the students have to do a major in any area of biomedical engineering covered by the program and to have a certificate at B2 proficiency level in a foreign language (English) [11]. One of the majors is clinical engineering.

I. Torres et al. (eds.), *VII Latin American Congress on Biomedical Engineering CLAIB 2016, Bucaramanga, Santander, Colombia, October 26th -28th, 2016,*
IFMBE Proceedings 60,
DOI: 10.1007/978-981-10-4086-3_8

Our teaching method on the medical devices course at the BEP is based on two principles. The first one follows a well-structured and balanced course, and there is a link between the teaching methods and learning strategies and the course's content. This is the so-called "constructive alignment" teaching philosophy [12]. In other words, in the tasks of designing and implementing a particular course, a relationship must exist between the intended learning outcomes (ILOs), the teaching and learning activities (TLAs), and the assessment of the actual students' learning (AL). In our experience, when this alignment exists, the students' anxiety decreases, while their perceived competence, motivation, and satisfaction with learning increases because they know in advance what they have to learn and how they will be assessed. In addition, the above-mentioned alignment facilitates the process of accountability of the teacher to his/her audience: the students. The second one is based on the use of practical environments such as the use of simulation centers to provide students with hands-on experience in the resolution of the most typical failures that medical devices have.

Therefore, under the conditions of a constructive alignment teaching philosophy, this study proposes two hypotheses:

Hypothesis 1 (H_1): There are statistically significant differences in the students' perceived learning outcome achievements after putting them through the medical devices courses in simulation centers.

Hypothesis 2 (H_2): There are no statistically significant differences between the students' perceived learning outcome achievements and the actual students' marks after putting them through the medical devices courses in simulation centers.

Medical Devices I and II are elective courses on the clinical engineering major of the BEP. The courses have 3 credits, each one representing 144 working hours in the semester or term (i.e. 48 hours of work in a classroom and 96 hours of independent work) (See Table I). The content covered by Medical devices I is related to medical devices' functioning and operation, the basic principles of troubleshooting (i.e. failure diagnostic and resolution), the identification of electronic components in the electric schematics of the equipment used in the cardiovascular system (electrocardiographs, multi-parameter monitors, defibrillators, electrosurgical units, mechanical ventilators). In Medical Devices II course, the structure of the content is similar, but the equipment is for the respiratory system including mechanical ventilators, anesthesia units, surgical lighting, surgical tables, dental equipment, and sterilizing equipment.

Table I. Medical devices training courses

Course name	Course id	Cr	Total hours	Type
Medical Devices I	MD001	3	48	E
Medical Devices II	MD002	3	48	E

E: Elective course.
MD001: Cardiovascular system; MD002: Respiratory system devices.

The students are evaluated three times during each course; each evaluation has the weight distributed as 30 %/30 %/40 % of the overall mark. In the first evaluation, the students do a theoretical exam aimed at assessing the functioning, the operation, and the basic principles of troubleshooting medical devices. The second evaluation is a practical exam in which the students, under controlled laboratory conditions, have to demonstrate hands-on the diagnostic and resolution of standardized failures of certain medical devices provided in advance by the instructors. Ideally, under these conditions the student's anxiety or stress could be considered as minimal because he/she knows about the failure in advance and ideally has enough time to solve it. The third evaluation is also a practical exam, which is conducted in a simulation center. Under these conditions, the students also have to demonstrate hands-on the diagnostics and resolution of standardized failures of certain medical devices, but the exercise is as real as possible, as the students are in simulated real clinical settings. That is, the call is made by a healthcare professional, the failure to be diagnosed and resolved in the medical device is in a real ("simulated") clinical area; e.g., an intensive care unit and connected to real patients (i.e., volunteers are used that are real and not mannequins). In this exam, the student is under pressure because she/he has to solve the failure in finite time because the patient is at risk of death. A quantitative rubric was used to assess the students and to calculate their marks. The student was evaluated from the time he/she answered the call until he/she left the simulation center.

Tables II and III show the intended learning outcomes and the constructive alignment of the medical devices courses of the BEP.

Table II. Learning outcomes.

ILO code	Learning outcome (Name/code)
C16	Student is able to perform scheduled and unscheduled maintenance tasks.
C18	In general student is able to design and implement the full process of technology management.

Table III. Constructive alignment on medical devices courses on a biomedical engineering program.

Course code	Learning outcome	Teaching and learning activities	Assessment method
MD001, 002.	C16, C18	Lectures (20%) Laboratories (80%)	Theory test Practical test in labs Rubrics Practical test in simulation centers

II. MATERIALS AND METHODS

A. *Study design and Participants.*

Pre- post-test with design with no control group. The participants are all students enrolled on MD001 and MD002 elective courses.

B. *The instruments*

In this study, a survey was designed to measure the students' perceived learning outcome achievements. The survey was composed of a demographic section (section A, 10 items) and a 5-point Likert (section B, 6 items), "1" is strongly disagree and "5" is strongly agree, section measured the students' perceptions of the learning outcome achievements during their time on the Medical Devices I course. Also recorded were the students' average marks from the Medical Devices courses (I and II). Finally, the practical test was graded using a rubric in the simulation center. A quantitative rubric was used to assess the students' learning outcomes and to perform a 360-degree

trough four (4) dimensions including: failure call management, biosafety, failure diagnostics and resolution, and failure report documentation in a work order. Finally, a 5-point Likert (section B, 15 items), "1" is strongly disagree and "5" is strongly agree, survey was applied that aimed to measure the students' satisfaction with their learning when they used the simulation center.

C. The measures

In this study four outcome variables were included. The first one was the students' perceptions of the learning outcome achievements during their time on the Medical Devices courses I and II, measured using the 2-item survey (C16 and C18, see Table II). The second one was the average student's marks on Medical Devices courses I and II. The third one was the average student's marks obtained in the 360-degree assessment scored by the rubric. The last one was the students' satisfaction with their learning during their time in the simulation center.

Among the demographics data recorded included were: gender (male/female), the socio-economical strata (Low/Medium /High), the type of high school the student graduated from in terms of whether the high school was private or public; and the type of the high school's degree profile (Academic/Technical).

D. Procedures and data collection

During the first week of the first semester of the clinical engineering major (the sixth semester of the program), with previous approval from the course instructor and the university ethics committee, the pre-survey was administered to the students by the research team. When the students had finished their courses the post-survey was administered (eighth semester). The post-survey was administered at a time when the students did not feel any anxiety, physiological pressure, or fatigue due to any exams. A research assistant, who was not involved in the teaching session, was assigned to assist the research team in administering the pre- or-post surveys.

E. Statistical analyses

Descriptive statistics were used to summarize the demographic data of the students. To test H_1, one paired Wilcoxon signed-rank test for two dependent samples were performed on the data gathered from the pre- and post-surveys (students' learning outcome perception achievements). To test H_2, Mann-Whitney U Tests for two independent samples were performed between the students' perceived achievements and their course mark average. The alpha level of significance for all the tests was set at $p \leq 0.05$ (*two-sided*). An SPSS® V 22.0 statistics package was used to generate descriptive, univariate, and bivariate statistics, and PLS path modeling, respectively.

III. Results

A. Participants' description

Twenty-four participants were involved in this study (female 54.2%: male 45.8%). Most of them (i.e., 83.3%) were within the age range of 19 to 21 years old, whereas the remaining 16.7% were within the age range of 22 to 24 years old. Most of them live in families from medium socio-economical strata (i.e. 87.5%). Regarding the students' procedence baccalaureate school types, 83.3% had graduated from private schools and 16.7% from public schools. Finally, 83.3% had graduated from a school with an academic profile, whereas 16.6% were from schools that had either a technical profile or another type of profile (8.3% for each type of shool).

B. What students' believe they learned and what they really learned: Testing hypotheses H_1 and H_2

Tables IV and V show the Wilcoxon signed-rank test statistics and Mann-Whitney U test statistics, respectively, the results of the testing hypotheses H_1 and H_2. Table IV shows the pre- and post-Wilcoxon signed-rank tests of the students' perceived learning outcome achievements (H_1), whereas Table V shows the Mann-Whitney U test between the students' perceived learning outcome achievements (at post-test) and their actual marks after putting them through Medical Devices Courses I and II using simulation centers (H_2).

Table IV. Wilcoxon signed-rank test statistics: Students' perceived learning outcome achievements after putting them through the medical devices courses in simulation centers. (N=16)

Hypothesis 1

Learning outcomes	Pre-test		Post-test		Wilcoxon signed-rank test	
	Mean	S.D	Mean	S.D	(z, p)	Power Test
C16	4.06	0.85	4.31	0.60	-1.19, 0.23	16.1%
C18	3.43	0.96	4.25	0.57	-2.6, 0.008	83.6%
All*	7.50	1.31	8.56	1.15	-2.4, 0.015	68.2%

Notes: Sum of all scores of learning outcomes (C16 and C18)

Table V. Mann-Whitney U Test: Students' perceived learning outcomes achievement versus actual students' marks after putting them through medical devices courses in simulation centers. (N=24)

Hypothesis 2

Learning outcomes	Post-test		Students' marks		Mann-Whitney U Test	
	Mean	S.D	Mean	S.D	(z, p)	Power Test
All (C16&18)	4.37	0.57	4.58	0.18	-1.06 ,0.28	40.6%

Figure 1 shows the 360-degree assessment of the different categories during the last practical test in the simulation center.

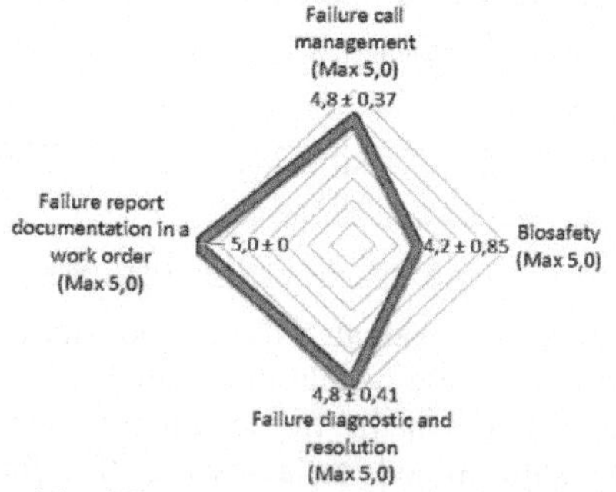

Fig. 1. Behavior of the different categories that are evaluated with the rubric.

The mean and SD of satisfaction with learning of the students' perceptions after exposing them to the simulation center was 46.4 SD 3.57. This indicated a high level of satisfaction with learning as the mean was located in the last third of the scale (the scale was divided into three parts including: First section: low satisfaction with learning= 0-16; second section: medium satisfaction with learning = 16-32; and high satisfaction with learning= 32-50).

IV. Discussion

This paper aimed to explore what the students' perceived learning outcome achievements were after finishing their Medical Devices Courses I and II.

From Table IV one can see that overall our results support Hypothesis 1 (H_1). One paired Wilcoxon signed-rank test showed that there were statistically significant differences in the students' perceived learning outcome achievements after putting them through Medical Devices Courses I and II for all the learning outcomes (i.e., all (C16 and C18), i.e., the sum scale mean at pre-test was 7.50 SD 1.31 compared with the post-test 8.50 SD 1.15 (Z=-2.4, p=0.015)). The results support expressed in [13] where it says that the simulation-based mastery learning improves students' skills and retention of those skills.

From Table V one can see that our results supported Hypothesis 2 (H_2), i.e., there were no statistical significant differences between the students' actual marks after putting them through Medical Devices Courses I and II and their perceived learning outcome achievements, (i.e, Z=-1.06, p=0.28). This indicates that there is a correspondence between what the students believed they learned compared with what they actually learned.

Despite our results supporting both hypotheses, they should be treated with caution; 75% of the power statistics of the tests were lower than 80 %, indicating not acceptable values to be protected against Type II error [14]. Therefore, in future studies a bigger sample size should be needed.

Finally, the 360° assessment showed that students had a good performance in the four dimensions of the assessment. However, biosafety was the dimension with lower results, a mean of 4.02 SD 0.85. This indicates that the BEP should include either courses covering these topics or these topics should be covered across the entire program in other subjects that might be taught.

V. Conclusion

Under the conditions of a constructive alignment teaching philosophy there were no differences between what the students perceived they had learned, as measured by the students' perceived learning outcome achievements, compared to what they had actually learned, as measured by their actual marks.

The use of simulation centers as a teaching strategy was demonstrated to be an effective method because the students were satisfied with their learning and there are differences in the students' perceived learning outcome achievements after putting them through the medical devices courses in simulation centers.

Conflicts of Interest

The authors report no conflicts of interest.

The results reported in this manuscript were not sponsored by any grant.

Acknowledgements.

We want to thank Oliver Jarvis for his proofreading of this paper. The authors assume overall responsibility for the manuscript.

References

[1] ACCE, "American College of Clinical Engineering," 1992. [Online]. Available: http://accenet.org/about/Documents/What's_a_Clinical_Engineer.pdf. [Accessed 09 March 2016].

[2] J. Nagel, "Protocol for the training of clinical engineers in Europe," BIOMEDEA PROJECT, Stuttgart, 2005.

[3] HTF, "2013 Candidate Handbook for Certification in Clinical Engineering by the Healthcare Technology Certification Commission," Healthcare Technology Certification Commission, Plymouth, 2013.

[4] G. H. Buck, "Development of simulators in medical education," *Gesnerus : Swiss Journal of the history of medicine and sciences,* vol. 48, no. 1, pp. 7-28, 1991.

[5] M. KORETSKY, C. KELLY and G. E, "Student Perceptions of Learning in the Laboratory: Comparison of Industrially Situated Virtual Laboratories to Capstone Physical Laboratories," *Journal of Engineering Education,* pp. 540-573, 2011.

[6] G. Alinier, B. Hunt, R. Gordon and C. Harwood, "Effectiveness of intermediate-fidelity simulation training technology in undergraduate nursing education," *J Adv Nurs,* pp. 359-369, 2006.

[7] I. Gieras, P. Sherman and D. Minsent, "Patient Safety Trilogy: Perspectives from Clinical Engineering," *Biomedical Instrumentation & Technology,* vol. 47, no. 2, pp. 137-142, 2013.

[8] D. E. a. H. B. A. Ziv, "38-Clinical Simulation on a National Level: Israel," in *Clinical Simulation*, Oxford, Academic Press, 2008, pp. 371-375.

[9] R. J. Glavin, "37 - A National Simulation Center Influences Teaching at a National Level: Scotland," in *Clinical Simulation*, Oxford, Academic Press, 2008, pp. 365-370.

[10] S. Mönk, J. Vollmer and W. Heinrichs, "34 - A National Simulation Program: Germany," in *Clinical Simulation*, Oxford, Academic Press, 2008, pp. 345-353.

[11] A. Miguel-Cruz, A. Rios-Rincon, W. Rodríguez-Dueñas, N. Florez-Luna and D. Quiroga-Torres, "What is the effect of an introductory biomedical engineering course on the students' perceptions of the engineering profession?," *International Journal of Engineering Education,* vol. 32, p. 136–149, 1(A) 2016.

[12] J. Biggs, Teaching for Quality Learning At University, Berkshire: England: The Society for Research into Higher Education and Open University Press, 2003, p. 54.

[13] I. Motola, L. A. Devine, H. Soo Chung, J. E. Sullivan and B. Issenberg, "Simulation in healthcare education: A best evidence practical guide. AMEE Guide No. 82," *Medical Teacher,* vol. 35, no. 10, pp. e1511-e1530, 2013.

[14] L. Portney and M. Watkins, Foundations of Clinical Research:Applications to Practice, New Jersey: Pearson/Prentice Hall, 2008, pp. 421-422.

Author: Antonio Miguel Cruz
Institute: Universidad del Rosario
Street: Carrera 24 No 63C – 69
City: Bogotá D.C.
Country: Colombia
Email: antonio.miguel@urosario.edu.co, miguelcr@ualberta.ca

Biofunctionalization Process of *a*-SiC:H Surfaces Applied to an Interdigitated Microelectrode Array to Detect Enterotoxigenic *Escherichia coli*

José Herrera-Celis[1], Claudia Reyes-Betanzo[1], Abdu Orduña-Díaz[2], Ana Pérez-Coyotl[2], Armando Hernández-Flores[1], Janet Morales-Chávez[2] and Arely Culebro-Gomez[1]

[1]Instituto Nacional de Astrofísica, Óptica y Electrónica/Departamento de Electrónica, Luis Enrique Erro #1, Santa María Tonantzintla, San Andrés Cholula 72840, Puebla, México

[2] Centro de Investigación en Biotecnología Aplicada del Instituto Politécnico Nacional, Ex-Hacienda San Juan Molino Carretera Estatal Tecuexcomac-Tepetitla Km. 1.5, Tepetitla 90700, Tlaxcala, México

Abstract— **A biofunctionalization process of *a*-SiC:H surfaces has been applied to an interdigitated microelectrode array (IMA) whose microelectrodes are covered by this material. The biofunctionalization process has been monitored stage by stage using Fourier transform infrared spectroscopy (FTIR), while its effects on the electrical behavior of the IMA were recorded in electrical impedance spectra. The process involves hydroxylation, silanization, generation of aldehyde groups, binding via protein A, immobilization of anti-*Escherichia coli* polyclonal antibodies, entrapping and detection of enterotoxigenic *Escherichia coli* (ETEC) in Luria-Bertani medium. The FTIR spectra confirm the success of the process. Regarding the performance of the IMA, although the detection of ETEC is successful and its percentage change in impedance reaches a value of 133.37% to 10^7 CFU/mL, some considerations may be taken into account to improve the sensitivity of the IMA by mean of the optimization of both the IMA design and the biofunctionalization process.**

Keywords— **Hydrogenated amorphous silicon carbide, biofunctionalization process, enterotoxigenic *Escherichia coli*, Fourier transform infrared spectroscopy, electrical impedance spectroscopy.**

I. Introduction

The biosensor development requires the simultaneous development of biocompatible materials and biofunctionalization processes [1]. The degree of exigency in terms of properties of materials and processes goes hand in hand with applications. For example, requirements for materials used in implantable biosensors are greater than those requested in biosensors working in disease diagnosis and in the food industry because the biosensor will be in direct and long term contact with the living system [2]. This work is focused to the food industry, where biosensors with high sensitivity, short response-time, high precision and label-free are necessary.

Among others biocompatible materials, hydrogenated amorphous silicon carbide (*a*-SiC:H) is chemically inert and feasible for *in situ* observation [3, 4]; it can be deposited at low temperatures compatible with manufacturing processes of silicon technology and its surface has already been biofunctionalized using self-assembled layers of organosilanes [5, 6]. Furthermore, the development of a biofunctionalization process for the surface of an amorphous network which in turn increases the sensitivity of a biosensor is not an easy task. The surface of an amorphous network is inhomogeneous, making difficult to obtain homogeneous self-assembled layers and oriented away from the surface [7, 8].

The main purposes of this work are to apply a biofunctionalization process of *a*-SiC:H surfaces to capture any strain of enterotoxigenic *Escherichia coli* (*E. coli*, ETEC), to functionalize the surface of an interdigitated microelectrode array (IMA), and to assess the effects of every step of the process on the electrical behavior of the IMA. To achieve them, this work uses Fourier transform infrared spectroscopy (FTIR) and electrical impedance spectroscopy (EIS).

II. Materials and Methods

A. Deposition of a-SiC:H thin film

Plasma enhanced chemical vapor deposition (PECVD) was used as technique to obtain *a*-SiC:H thin films on silicon (100) wafer (*Addison Engineering, #99CE-498/11*). The deposition was carried out in a capacitively coupled PECVD reactor (*AMP 3300 PLASMA II, Applied Materials*) operating at a power of 800 W and a plasma excitation frequency of 110 kHz. The precursor gases were silane (SiH_4) and methane (CH_4). Hydrogen (H_2) was used as a carrier and diluent gas. The gas flow ratio and the hydrogen dilution were 0.7 and 9.0, respectively. The electrode temperature was 350 °C and the chamber pressure was 1.5 Torr for 1 h of deposition.

B. Interdigitated microelectrode array

The IMA used in this work includes *a*-SiC:H on and between interdigitated microelectrodes (IM) as surface to be biofunctionalized [9]. This characteristic allows to capture bacteria of *E. coli* on whole surface of the array. The structure of the IMA was made on a square Si (100) substrate (*Addison Engineering, #99CE-498/11*) with side length of 4.4 mm and 300 μm of thickness, on which a silicon dioxide (SiO_2) of 700

I. Torres et al. (eds.), *VII Latin American Congress on Biomedical Engineering CLAIB 2016, Bucaramanga, Santander, Colombia, October 26th -28th, 2016*,
IFMBE Proceedings 60,
DOI: 10.1007/978-981-10-4086-3_9

nm of thickness was grown and three layers of undoped *a*-SiC:H, titanium and doped *a*-SiC:H with thicknesses of 450 nm, 50 nm and 50 nm were consecutively deposited. The last two layers were patterned for obtain the array. The finger spacing (S) and the finger width (W) are of 10 µm. The sensing area is a square of 10.24 mm^2, resulting in 160 IM of length 3.16 mm. The mean length and diameter of *E. coli* are 2.79 µm and 0.68 µm, therefore that sensing area can accommodate around 5x10^6 colony-forming unit (CFU).

C. Biofunctionalization process

a) Hydroxylation

The IMA and the *a*-SiC:H thin films on silicon substrates (samples) were immersed in a solution of potassium hydroxide and methanol (45% KOH/MeOH, 1:1) for 15 min, making hydrophilic the surfaces [8]. They were then rinsed in deionized (DI) water and dried under a stream of nitrogen gas. The second step of the hydroxylation of the surfaces was carried out in a Micro RIE system (*800 Series-RIE, Technics*). An oxygen plasma treatment (100 mTorr and 200 W of RF-power) was done for 10 min in order to improve the layer of –OH groups [7]. Finally, they were dried under nitrogen gas flow to remove any moisture present on the surfaces.

b) Silanization

The samples were immersed in 3-aminopropyltrimethoxysilane (APTMS) (*Sigma-Aldrich, #281778*) diluted at 1% (v/v) in 99.8% (v/v) anhydrous toluene (*Sigma-Aldrich, #244511*) for 1 h at room temperature. Then the samples were rinsed with anhydrous toluene and methanol, consecutively. Finally, the samples were dried in an oven at 110 °C for 1 h, removing water and forming siloxane bonds [10].

c) Generation of aldehyde groups

The silanized samples were dipped in a 2.5% solution of glutaraldehyde (GA) solution (*Sigma-Aldrich, #G7651*) in phosphate buffer (PB) solution (*Sigma-Aldrich, #63238*) for 1 h at room temperature. After this step, the samples were dipped in ~1.2 mg/mL of sodyum borohydride ($NaBH_4$) (*Sigma-Aldrich, #452882*) in PB solution for 1 h to reduce the imine bonds and to give rise to amine bonds [11]. This was followed by rinsing the samples with phosphate buffered saline (PBS, pH 7.4), solution prepared from one package of phosphate buffered saline powder (*Sigma-Aldrich, #P3813*) and 1 L of sterile water for injection (*PiSA, #SC-PI-AGIR1L/SG*). Finally, the samples were dried with nitrogen.

d) Binding via protein A

Protein A (PrA) from *Staphylococcus aureus* (*Sigma-Aldrich, #P6031*) was added to the GA/APTMS surface of the samples immersing them for 30 min in 1 mg/mL of PrA (at storage temperature) in phosphate buffered saline (PBS, pH 7.0) solution, which was prepared from PBS (pH 7.4) solution adding hydrochloric acid (HCl, 1.0 M) (*Hycel de México, #1348*). After that time, the samples were pulled out from solution and rinsed with PBS (pH 7.0). Finally, the samples were stored at about 4 °C in Axygen® 1.5mL MaxyClear microtubes (*Corning, #MCT-150-C*) containing PBS (pH 7.0) until the next stage.

e) Immobilization of anti-E. coli polyclonal antibodies

Rabbit serum with purified polyclonal antibodies (Ab) specific to Enterotoxigenic *Escherichia Coli (ETEC)* was diluted in PBS (pH 7.4) (1:4). The samples with PrA/GA/APTMS surfaces were changed from one tube to another, from PBS solution to this diluted serum, and allowed to react them for 20 h at 4 °C. Then, the samples were immersed in 2 mg/mL of bovine serum albumin (BSA) in PBS (pH 7.4) solution for 30 min at about 4 °C In this step, the BSA/PBS solution acts as a blocking agent for the terminal groups that did not bind to any antibody [8]. All samples were rinsed with PBS (pH 7.0) solution inside tubes to remove weak interactions. Upon completion of this step, the samples were stored in tubes at about 4 °C with PBS (pH 7.4) solution covering them fully.

f) Entrapping of ETEC in Luria-Bertani medium

The bacteria of *E. coli* were cultivated in PPLO agar at 37 °C for 24 h. Subsequently, a colony of bacteria was inoculated in 5 mL of Luria-Bertani medium (LB) and incubated at 37 °C under stirring for 4 h. The bacterial concentration measured using a spectrophotometer (*Evolution 300 UV-Vis, Thermo Scientific*) was 1x10^7 CFU/mL. The samples with Ab/PrA/GA/APTMS surfaces were taken out of storage, dried in nitrogen gas flow and arranged to come into contact with the bacterial medium.

D. FTIR and EIS measurements

The FTIR spectra corresponding to each stage of the process were obtained using a spectrometer (*Vertex 70, Bruker*) working in absorbance mode in middle infrared (500–4000 cm^{-1}). To achieve it, one sample of *a*-SiC:H on silicon substrate was assigned to each stage of the process. Whereas the changes in electrical impedance of the IMA, stage by stage, were registered with an impedance analyzer (*IM3570, Hioki*), applying a sinusoidal potential of 1 V in a frequency range from 1 kHz to 5 MHz across the IMA pads.

All samples were dried with nitrogen gas flow before measurements. The EIS measurement procedure was done as quickly as possible to achieve slightest affectation because of the environment (only data acquisition of each spectrum took about 5 min). The medium on the sensing area of the IMA was always air in the stages before entrapping of *E. coli*. In the last stage (entrapping and detection), a drop of bacteria in

LB of 30 μL was put on the sensing area of the IMA and two EIS measurements were performed. The first one was done straight away, and the other one 25 min after.

III. Results and discussion

A. Monitoring of the biofunctionalization process by FTIR

The peaks due to the APTMS layer can be seen in the S1 spectrum in the Fig. 1 by comparing with the S0 spectrum corresponding to the *a*-SiC:H thin films. These peaks are according with a previous work [10]. The low occurrence of the peak at 1200 cm^{-1} indicates that the majority of O–C_2H_6 terminal groups were hydrolyzed during the process.

The GA layer is used as a crosslinker between APTMS and PrA. The peaks in the S2 spectrum related to the stretching of C–N, C–O, C–C groups in the range 1200–1500 cm^{-1} are not visible [12], but there is a mean peak and shoulders at the amide II band (1510–1580 cm^{-1}), which can be attributed to C–N stretching and N–H bending [8]. In the amide I band (1600–1700 cm^{-1}) appears the characteristic peak at 1640 cm^{-1} corresponding to C=O groups on the surface [13].

The use of PrA to immobilize antibodies is based on a previous study [8], which shows that PrA improves orientation and does faster detection. A peak at 1600 cm^{-1} corresponding to C=N and C=C stretching can be seen in the S3 spectrum in Fig. 1 [13]. Moreover, there is a peak around 1650 cm^{-1}, which indicates that the time of reaction was not enough and some aldehyde groups were not linked. After Ab immobilization, one peaks appears at 1460 cm^{-1} corresponding to C–CH_3 and N–CH_3 deformations [13], other one at 1535 cm^{-1} corresponding mainly to N–H in-plane bending and the rest to the C–N and C–C stretching vibrations [14] (see S4 spectrum in Fig. 1). The absorption in the amide I band is due to stretching vibrations of C=O groups [13]. Besides, the absorption by amine groups of the Ab in 3200–3500 cm^{-1} proves the success of the Ab immobilization stage [14].

The capture of *E. coli* by Ab is backed up by the increasing of the absorption in the amide I and II bands (protein peaks of the bacteria) in the S5 spectrum in Fig. 1 [12]. Besides, the peaks at 1405 and 1460 cm^{-1} related to C–O–H in-plane bending and $C(CH_3)_2$ symmetric stretching, belonging to carbohydrates, glycoproteins and lipids, are also visible in the S5 spectrum [12].

B. Effects of the biofunctionalization process on IMA

The Fig. 2 shows the EIS spectra before and after hydroxylation. The thin layer of–OH groups decreases the electrical impedance magnitude (|Z|) in two orders. Accordingly, the formation of the layer of –OH groups leaves a negatively charged surface.

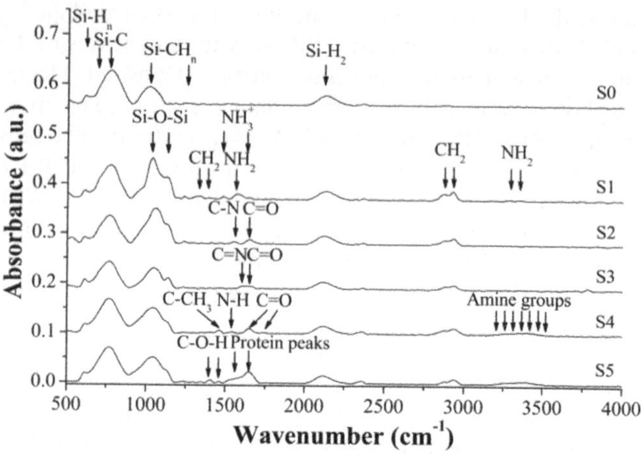

Fig. 1 FTIR spectra of biofunctionalization stages. S0: *a*-SiC:H thin film, S1: APTMS/*a*-SiC:H, S2: GA/APTMS/*a*-SiC:H, S3: PrA/APTMS/*a*-SiC:H, S4: Ab/APTMS/*a*-SiC:H, S5: ETEC/Ab/APTMS/*a*-SiC:H

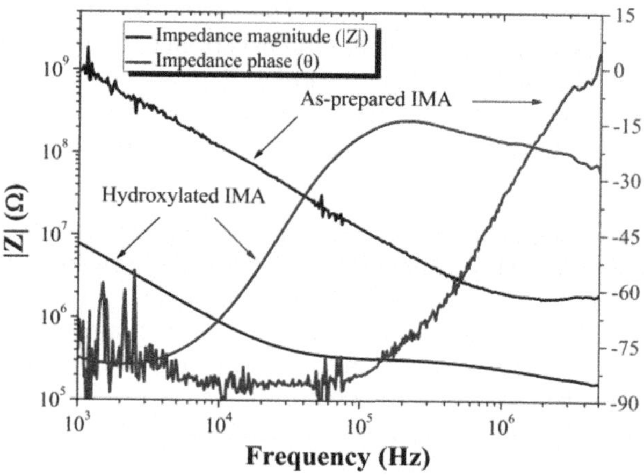

Fig. 2 Impedance spectra of as-prepared and hydroxylated IMA

The other EIS spectra of the process appear in the Fig. 3. The APTMS layer causes an increasing in |Z| because it form a covalent bond with the surface and its thickness decreases the lateral area of IM. Regarding the electrical behavior after generation of aldehyde groups, at frequency above 100 kHz, |Z| increases because the lateral area of IM decreases. Instead, in the region of 1–100 kHz, the GA/APTMS layer has lower electrical resistance than the APTMS layer. After PrA binding and Ab immobilization, the protein layers increase the permittivity of the medium between IM and disappear the resistive effect of aldehyde groups on the surface.

The Fig. 4 shows the change in the |Z| after 25 min in contact with the bacterial medium. Doing calculations, the IMA has a meaningful detection in the frequency range of 10–100 kHz and its percentage change in |Z| reaches its maximum of 133.37% (impedance response of 1.51 MΩ) at 28.5 kHz.

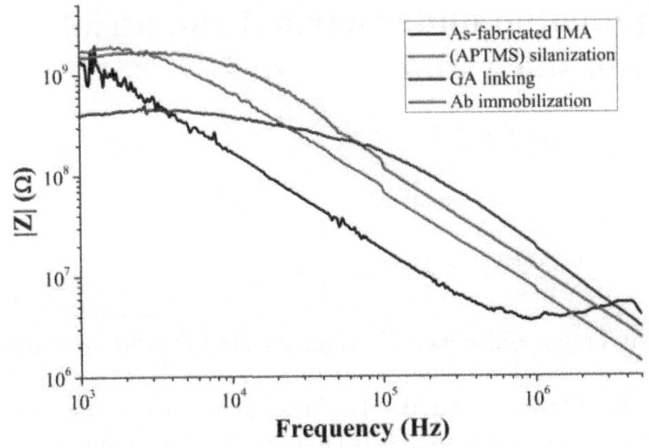

Fig. 3 Impedance magnitude spectra of biofunctionalization stages on IMA

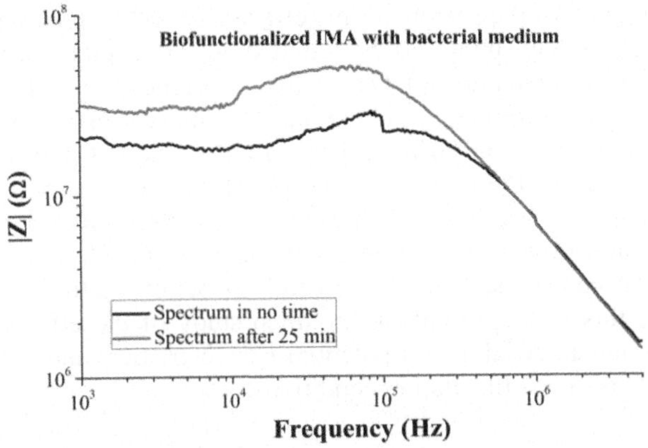

Fig. 4 Impedance magnitude spectra of IMA over time

Although the impedance response is above those presented in a similar study [15], the design of the IMA and the process can still be improved. The sensitivity may be better if the conductivity of both the a-SiC:H layer on IM and the metal are increased, and if thickness of Ti layer is such that the thickness of biofunctionalization layer does not affect.

IV. Conclusions

A biofunctionalization process was carried out on *a*-SiC:H surfaces of one thin film and one IMA. The FTIR spectra corresponding to each stage showed successful results, which proves that the layers were properly linked to each other. The EIS spectra allowed to know the effects of each layer on the electrical behavior of the IMA. Besides, this work proves that improve the biosensor sensitivity based on such arrays using the whole area on the IMA for sensing, along with a properly designed process, is possible. To achieve it, the next step will be to reduce the high IMA impedance increasing the electrical conductivity of the *a*-SiC:H layer on MI and adjusting the thickness of the Ti layer or changing the metal by other biocompatible metal with greater electrical conductivity.

Acknowledgment

The authors are thankful to *Hospital Infantil de México Federico Gómez* for the antibodies and to the National Council of Science and Technology under the project No. 242440.

Conflict of Interest

The authors declare that they have no conflict of interest.

References

1. Saddow S (2016) Silicon carbide biotechnology. Elsevier, Waltham
2. Nezafati M (2014) Biomaterial testing methodology for lon-term in vivo applications: Silicon carbide corrosion resistance, biocompatibility and hemocompatibility, Ph.D. thesis, USF, Tampa
3. Iliescu C, Chen B et al. (2008) PECVD amorphous silicon carbide membranes for cell culturing. Sens Actuators B 129:404–411
4. Yakimova R, Petoral R Jr et al. (2007) Surface functionalization and biomedical applications based on SiC. J Phys D: Appl Phys 40:6435–6442
5. Hijikata Y (2013) Physics and technology of silicon carbide devices. InTech, Rijeka
6. Galopin E, Touahir L et al. (2010) Amorphous silicon-carbon alloy for efficient localized surface plasmon resonance sensing. Biosens Bioelectron 25:1199–1203
7. Kumar N, Singh M et al. (2014) Optimization and characterization of biomolecule immobilization on silicon substrates using (3-aminopropyl)triethoxysilane (APTES) and glutaraldehyde linker. Appl Surf Sci 305:522–530
8. Huy T, Hong N et al. (2011) Characterization of immobilization methods of antiviral antibodies in serum for electrochemical biosensor. Appl Surf Sci 257:7090–7095
9. Herrera-Celis J, Reyes-Betanzo C, Orduña-Diaz A (2015) Design of an interdigitated microelectrode biosensor using *a*-SiC:H surface to capture *E. coli*, IEEE Conference Publications, SBMicro, Salvador, Brazil, 2015, pp. 1–4
10. Herrera-Celis J, Reyes-Betanzo C et al. (2015) *a*-Si_xC_{1-x}:H thin films with subnanometer surface roughness for biological applications. J Vac Sci Technol A 33(5):05E108-1–05E108-7
11. Olde Damink L, Dijkstra P et al. (1995) Glutaraldehyde as a crosslinking agent for collagen-based biomaterials. J. Mater. Sci.-Mater. M. 6(8):460–472
12. Mura S, Greppi G et al. (2012) FTIR nanobiosensors for *Escherichia coli* detection. Beilstein J Nanotechnol 3:485–492
13. Tolstoy V, Chernyshova I, Skryshevsky V (2003) Handbook of infrared spectroscopy of ultrathin films. Wiley, Boboken
14. Surewicz W, Mantsch H, Chapman D (1993) Determination of protein secondary structure by Fourier transfor infrared spectroscopy: A critical assessment. Biochemistry 32(2):389–394
15. Ghosh S, Barizuddin S et al. (2013) Impedance biosensor based on interdigitated electrode arrays for detection of low levels of E O157:H7, IEEE Conference Publications, MEMS, Taipei, Taiwan, 2013, pp. 955–958

Dofetilide effect on human atrial action potential under normal and atrial fibrillation conditions. In silico study

C. Tobón[1], S. Pérez[2], J.P. Ugarte[2] and J. Saiz[3]

[1] MATBIOM, Universidad de Medellín, Medellín, Colombia
[2] Grupo de Dinámica Cardiovascular, Universidad Pontificia Bolivariana, Medellín, Colombia
[3] CI²B, Universitat Politècnica de València, Valencia, Spain

Abstract— **Atrial fibrillation is the most common sustained cardiac arrhythmia. Dofetilide is an antiarrhythmic drug for the treatment of chronic AF that specifically blocks the rapid component of the delayed rectifier potassium current I_{Kr}. Dofetilide prolongs the action potential duration and QT interval in a concentration-dependent fashion, therefore, the risk of QT prolongation is dose related. It is important to study the electrophysiological effects of dofetilide at different concentrations in human atrial cells. For this, we simulated the effects of dofetilide on human atrial cell and studied its effect on atrial action potential under normal conditions and during cAF. We developed a model of dofetilide effects on I_{Kr} and I_{KACh}. Our results show that dofetilide blocks both currents in a fraction greater as the concentration increases, which results in an action potential duration lengthening. To our knowledge, this is the first work that has developed mathematical models of dofetilide effects on I_{Kr} and I_{KACh} currents to study its effect on human atrial action potential.**

Keywords— **Dofetilide, antiarrhythmic drug, atrial fibrillation, in silico model.**

I. Introduction

Atrial Fibrillation (AF) affects between 1 and 2 % of general population, with a peak prevalence of 10 % in those older than 80 years. It is estimated that by 2050 nearly 16 million US patients will have AF [1]. Paroxysmal AF is initiated by focal triggers localized usually in one or more pulmonary veins and can be cured by a catheter-based ablation procedure. However, in chronic AF, the prevailing theory is that multiple, random wavelets of activation coexist to create a chaotic cardiac rhythm, and therapy is more challenging [2]. Despite of this, currently available treatments for AF, radiofrequency ablation and antiarrhythmic drugs, are low effective and hampered by significant cardiac and extracardiac side effects [3].

Dofetilide is an antiarrhythmic drug that specifically blocks the rapid component of the delayed rectifier potassium current I_{Kr} [4] this block is voltage dependent. In 1999 dofetilide (Tikosyn) was approved by Food and Drug Administration (FDA) for the treatment of persistent (non-paroxysmal) AF and flutter [5]. Dofetilide induces the prolongation of action potential duration (APD) in reentrant arrhythmias and hence, causing the prolongation of the QT interval and the increment of refractory period [6]. However, the FDA cautioned: "Because Tikosyn can cause life-threatening ventricular arrhythmias, it should be reserved for patients in whom AF/flutter is highly symptomatic" [5]. While the efficacy of oral dofetilide as antiarrhythmic drug is related to the increment in refractoriness, this pharmacologic therapy should not be initiated out of hospital because of the prolongation of QT, that may trigger the TPD [7]. Dofetilide prolongs the APD and QT interval in a concentration-dependent fashion, therefore, the risk of QT prolongation and TDP is dose related [8, 9]. For this reason, it is important to study the electrophysiological effects of dofetilide at different concentrations in human atrial cells.

To contribute with the research of better treatments for AF, this work presents an in silico study of the effects on the human atrial action potential characteristics, under normal and atrial fibrillation conditions.

II. Methods

A. Electrophysiological model

We used the Courtemanche-Ramirez-Nattel-Kneller [10, 11] membrane formalism. A 0.005 uM of acetylcholine (ACh) concentration was simulated. The basic equation to calculate trasnmembrane voltage (V_m) is:

$$C_m \frac{dV_m}{dt} + I_{ion} + I_{stim} = 0 \quad (1)$$

where C_m is the specific membrane capacitance (100 pF), I_{ion} is the total ionic current that crosses the membrane cells, I_{stim} is the stimulus current. The model under physiological normal conditions corresponds to the control model.

B. Model of chronic atrial fibrillation

To reproduce the atrial electrical remodeling generated by cAF, changes in conductance of different ionic channels of human atrial cells observed in experimental studies of cAF [12, 13] have been incorporated in the electrophysiological model.

I. Torres et al. (eds.), *VII Latin American Congress on Biomedical Engineering CLAIB 2016, Bucaramanga, Santander, Colombia, October 26th -28th, 2016*,
IFMBE Proceedings 60,
DOI: 10.1007/978-981-10-4086-3_10

Several parameters were changed: the conductance for both I_{Kur}, I_{to} was decreased by 50 %, the conductance for I_{CaL} was decreased by 70 %, while the conductance for I_{K1} was increased by 100 %.

C. Model of dofetilide effects on I_{Kr} and I_{KACh}

To develop a basic model of the effect of the dofetilide on the I_{Kr} and I_{KACh}, we used the steady state fraction of block (b). In this model the kinetics of the channel would be considered unchanged in the presence of the drug.

$$b = \frac{1}{1+\left(\frac{IC_{50}}{D}\right)^{h}} \qquad (2)$$

where IC_{50} is the half maximal inhibitory concentration for the current block by dofetilide. D is the dofetilide concentration. For the IC_{50} to block I_{Kr} we used 0.32 uM, this value were found by [14] in HERG K+ channels. For the IC_{50} to block I_{KACh}, we used 6.6 uM, this value were found by [15] in isolated atrial myocytes. And Hill coefficient (h) of 1.

D. Simulation protocol and data analysis

We implemented the unicellular models to simulate the sinus rhythm under physiological and cAF conditions, using the Cellular Open Resource (COR) public CellML software. Forward Euler method with a time step 0.001 ms was implemented to solve the equations. A train of 10 stimuli was applied. The basic cycle length was 1000 ms. Dofetilide concentrations from 0 to 20 uM were implemented. The APD at 90 % of the repolarization (APD_{90}), ionic currents and the resting membrane potential (RMP) were measured. All measurements were made on the 10th beat using a program developed in MATLAB software.

III. Results

A. Model of dofetilide effects on I_{Kr} and I_{KACh}

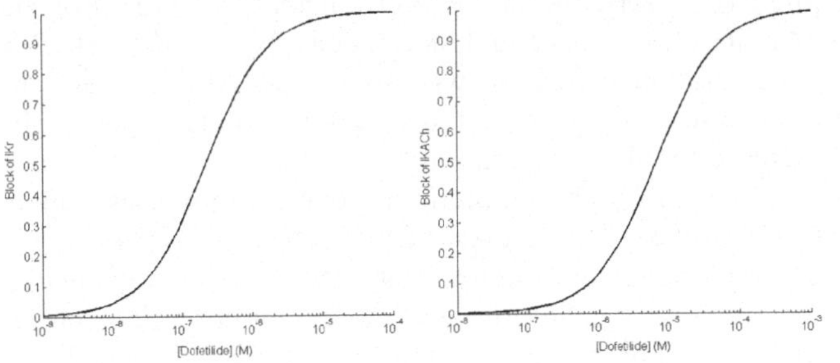

Fig. 1 Fraction of block (b) for I_{KI} and I_{KACh} by dofetilide.

Fig. 1 shows the fraction of block curves using (2) for the I_{Kr} and I_{KACh} currents, in function of the dofetilide concentration. We can observe that a higher concentration of dofetilide is necessary for blocking the I_{KACh} current than the concentration for blocking the I_{Kr} current.

B. Dofetilide effects on I_{Kr}, I_{KACh} and action potential

Fig. 2 and Fig. 3 show the effects of different dofetilide concentrations on I_{Kr} and I_{KACh} currents, and on atrial action potential, under control and cAF conditions.

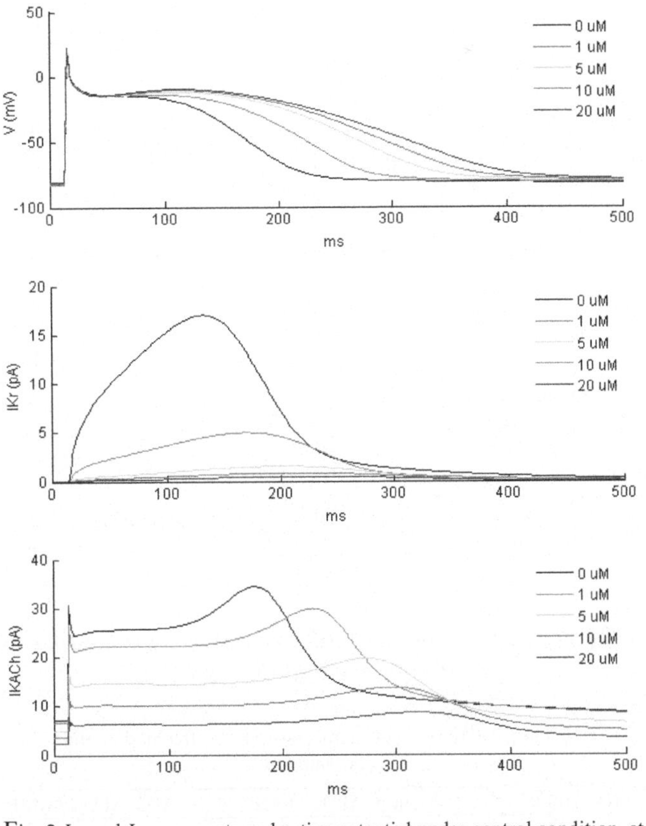

Fig. 2 I_{KI} and I_{KACh} currents and action potential under control condition, at different dofetilide concentration.

In control (Fig. 2) we can observe that I_{Kr} without dofetilide shows a peak of 17.0 pA at 117 ms far from depolarization and I_{KACh} shows a peak of 34.3 pA at 161 ms far from depolarization, these peaks contribute to the action potential repolarization, and they are reduced and shifted to the right over time as the dofetilide concentration increases.

Additionally, the steady-state values of these currents decreased. All of this caused an action potential lengthening and a slight RMP depolarization (see Table 1). When the highest dofetilide concentration was applied (20 uM), the

APD_{90} reached a value of 385 ms and a RMP depolarization of 1.6 mV was observed.

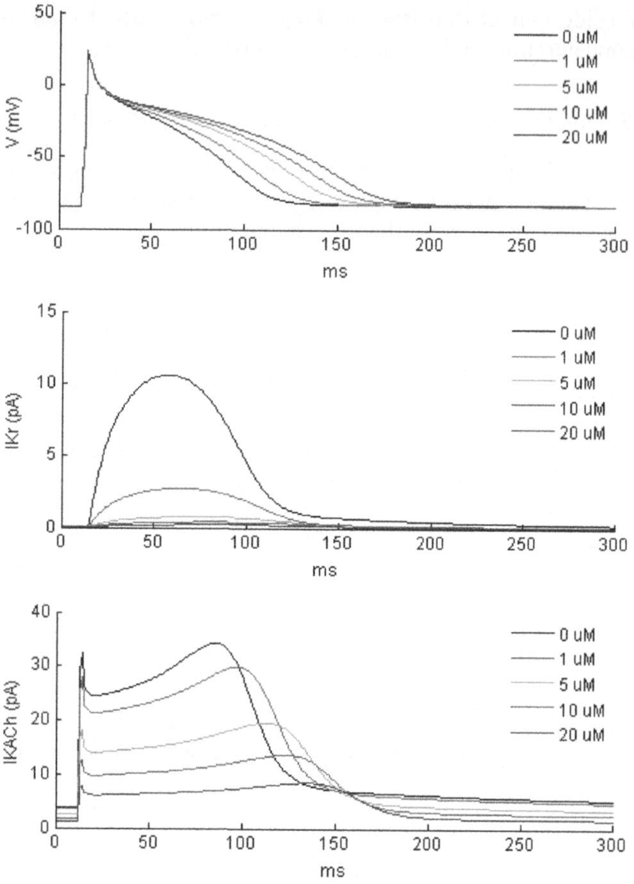

Fig. 3 I_{K1} and I_{KACh} currents and action potential under cAF condition, at different dofetilide concentration.

Table 1 APD_{90} during Control and cAF at different dofetilide concentration.

[Dofetilide]	Control APD_{90} / RMP	cAF APD_{90} / RMP
0 uM	211 ms / -82.8 mV	98 ms / -84.6 mV
0.5 uM	251 ms / -82.6 mV	107 ms / -84.6 mV
1 uM	269 ms / -82.5 mV	112 ms / -84.5 mV
5 uM	327 ms / -81.9 mV	130 ms / -84.4 mV
10 uM	357 ms / -81.5 mV	143 ms / -84.3 mV
20 uM	385 ms / -81.2 mV	155 ms / -84.3 mV

They were generated earlier in time (58 and 86 ms respectively), this caused an APD_{90} shortening of 54 % and a slight RMP hyperpolarization of 1.8 mV as shown in Table 1. The effect by increasing the dofetilide concentration on currents and action potential is in general the same, APD lengthening and slight RMP depolarization. When the highest dofetilide concentration was applied, the APD_{90} reached a value of 155 ms and a RMP depolarization of 0.3 mV. However, even if the action potential is lengthened, there is no recovery plateau phase.

Fig. 4 shows the APD_{90} increasing trend curves in function of the different dofetilide concentration for control and cAF. In both cases, a dofetilide concentration increase causes an APD lengthening.

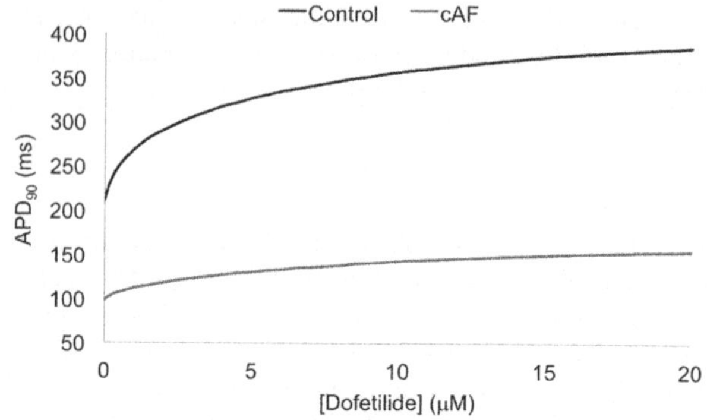

Fig. 4 APD_{90} increasing trend curves in function of dofetilide concentration for control and cAF.

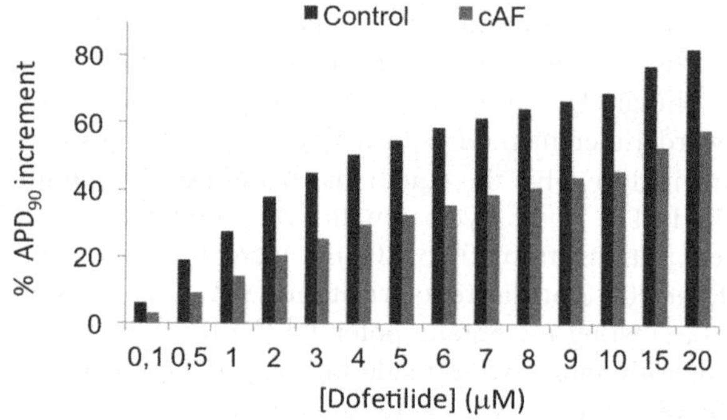

Fig. 5 Percentage increase in APD_{90} in function of the dofetilide concentration for control and cAF.

Fig. 5 shows the percentage increase in APD_{90} in function of the dofetilide concentration for control and cAF. We can observe when a low dofetilide concentration was applied (0.1 uM), the slight APD increases for control and cAF are close (6 % and 3 % respectively). As the dofetilide concentration increases, we observe an APD increase for both cases, being control that has the highest percentage increase for all concentrations.

When the highest dofetilide concentration was applied (20 uM) the APD_{90} increased 82 % in control and 58 % in cAF. These results suggest that dofetilide has a greater effect on control conditions and more high dofetilide concentrations will be needed to achieve similar effects in cAF.

IV. Discussion

The cAF model developed has an APD shortening as has been reported experimentally [12, 13]. It has been demonstrated that potassium currents as I_{Kr} and I_{KACh} contribute to action potential repolarization [15]. The lower the peak of these currents and further of the depolarization, APD lengthening and RMP depolarization are caused. Our models reproduced these physiological behaviors. I_{KACh} and I_{Kr} are currents of interest because they play a significant role during AF. It has been reported that inhibition of I_{KACh} effectively terminates or prevents AF [15]. Several groups have studied the ability of oral and intravenous dofetilide to convert sustained episodes of AF. Dofetilide blocks I_{Kr} in all myocardial tissues, an IC_{50} in the nanomolar range has been experimentally measured in HERG expressed in human cells, namely 0.32 uM [14]. Dofetilide is also a potent blocker of I_{KACh}, an IC_{50} of 6.6 uM was obtained from isolated atrial myocytes of patients with cAF [15]. Our results are similar to the observed experimentally [16, 17]. To our knowledge, there are no reported in silico study of the dofetilide effects on human atrial action potential.

V. Conclusions

Our results suggest that as the dofetilide concentration increases the APD increases. This in silico study of the effects on the human atrial action potential characteristics, under normal and atrial fibrillation conditions will contribute with the research of better treatments for AF. To our knowledge, this is the first work that has developed mathematical models of dofetilide effects on I_{Kr} and I_{KACh} currents to study its effect on human atrial action potential.

Acknowledgment

This work was partially supported by Departamento administrativo de ciencia, tecnología e innovación de la República de Colombia – COLCIENCIAS, projects # FP44842-075-2015 (2014-0401) and # 121065741044.

Conflict of Interest

The authors declare that they have no conflict of interest.

References

1. Goralnick E, Bontempo LJ. (2015) Atrial Fibrillation. Emerg. Med. Clin. North Am 33(3):597–612.
2. Aliot E, Haïssaguerre M, Jackman WM. (2008) Catheter Ablation of Atrial Fibrillation. Blackwell Publishing, Oxford.
3. Zipes DP, Jalife J. (2016) Cardiac Electrophysiology: From Cell to Bedside, Sixth. Elsevier, Philadelphia.
4. Jurkiewicz NK, Sanguinetti MC. (1993) Rate-dependent prolongation of cardiac action potentials by a methanesulfonanilide class III antiarrhythmic agent. Cir Res 72:75-83.
5. Lauer MR, Jose S. (2001). Dofetilide: Is the Treatment Worse Than the Disease? J Am Coll Cardiol 37(4):1106–1110.
6. Saiz J, Ferrero JM, Monserrat M, et al. (2003) Effects of the antiarrhythmic drug dofetilide on myocardial electrical activity: a computer modelling study. Comput Cardiol 291–294.
7. January CT, Wann LS, Alpert JS, et al. (2014) AHA/ACC/HRS Guideline for the Management of Patients With Atrial Fibrillation: Executive Summary: A Report of the American College of Cardiology/American Heart Association Task Force on Practice Guidelines and the Heart Rhythm Society. J Am Coll Cardiol 212.
8. Jaiswal A, Goldbarg S. (2014) Dofetilide induced torsade de pointes: Mechanism, risk factors and management strategies. Indian Heart J 66(6):640–648.
9. Zimetbaum P. (2012) Antiarrhythmic drug therapy for atrial fibrillation. Circulation 125(2):381–389.
10. Courtemanche M, Ramirez RJ, Nattel S. (1998) Ionic mechanisms underlying human atrial action potential properties: insights from a mathematical model. Am J Physiol 275(1):H301–H321.
11. Kneller J, Zou R, Vigmond EJ, et al. (2002) Cholinergic atrial fibrillation in a computer model of a two- dimensional sheet of canine atrial cells with realistic ionic properties. Circ Res 90(9):E73–E87.
12. Courtemanche M, Ramirez RJ, Nattel S. (1999) Ionic targets for drug therapy and atrial fibrillation-induced electrical remodeling: insights from a mathematical model. Cardiovasc Res 42(2):477–489.
13. Van Wagoner DR. (2003) Electrophysiological remodeling in human atrial fibrillation. Pacing Clin Electrophysiol 26(7):1572–1575.
14. Ficker E, Jarolimek W, Kiehn J, et al. (1998) Molecular Determinants of Dofetilide Block of HERG K+ Channels. Circ Res 82:386-395.
15. Voigt N, Rozmaritsa N, Trausch A, et al. (2010) Inhibition of IK,ACh current may contribute to clinical efficacy of class I and class III antiarrhythmic drugs in patients with atrial fibrillation. Naunyn-Schmied Arch Pharmacol 381: 251-259.
16. Tande PM, Bjornstad H, Yang T, et al. (1990) Rate-dependent class III antiarrhythmic action, negative chronotropy, and positive inotropy of the novel blocking drug, UK-68,798: Potent in guinea pig but no effect in rat myocardium. J Cardiovasc Phamacol 16: 401-410.
17. Ohler A, Amos GJ, Wettwer E, et al. (1994) Frequency-dependent effects of E-4031, almokalant, dofetilide and tedisamil on action potencial duration: no evidence for "reverse use dependent" block. Arch Pharmacol 349: 602-610.

Address of the corresponding author:

Author: Catalina Tobón
Institute: Universidad de Medellín
Street: Carrera 87 N° 30 - 65
City: Medellín
Country: Colombia
Email: ctobon@udem.edu.c

Architecture of an emotion recognition and video games system to identify personality traits

M. Callejas-Cuervo[1], L. A. Martínez-Tejada[2] and J. A. Botero-Fagua[1]

[1] Pedagogical and Technological University from Colombia/System engineering and computation School, Tunja, Colombia
[2] Saint Thomas University/Electronic engineering faculty, Tunja, Colombia

Abstract— **Emotional recognition is a groundbreaking technique currently used in research where it is necessary to evaluate the emotional reaction of an individual with respect to the content which is being exposed, thanks to advances in electronic devices and computing this technique is used in different fields as an interdisciplinary study. Physiological signals as emotive response and video games as a tool for stimulating these emotional responses, allows in an effective manner the recognition of emotional state carrying the individual to express these emotions through physiological changes emulating situations through virtual environments with which can interact.**

This article proposes an architecture for building a system of emotional recognition from physiological signals (ECG, EMG) and games, to obtain a technological tool that can supplement the results of psychometric personality test, and identify personality traits from the interaction of an individual with a virtual environment.

Keywords— **Architecture, Emotion recognition, Personality traits, Psychometric test, Video games.**

I. Introduction

The analysis and study of human emotions has been a complex field of research, which can be covered from different fields of knowledge starting from psychology, medicine and social sciences to evaluate the reaction and behavior of individuals with their environment, reaching currently fields such as engineering and computer science to build systems that can identify emotional reactions and respond in natural and empathetic way to the individual who manipulates the system. Emotional recognition systems have been growing over the past decade, since the consolidation of the term affective computing [1] [2] to the simulation of emotions for virtual agents [3][4], these systems have found a field development and action that can bring benefits applicable to different fields of development in social life.

From the emotions is possible to determine characteristics of reaction and behavior that are evoked in certain situations, emotional reactions are part of a person and can define the personality of an individual [5]. Psychology defines personality as "a particular pattern of behavior and thinking that prevails over time and context and differentiates one person from another" [6]. To understand patterns of behavior, psychology aims to identify individual characteristics of personality called personality traits, which are assumed as a durable and relatively constant characteristic, psychometric tests are most often used tool for identifying these characteristics. By linking emotions with the personality of an individual, it is possible to consider different techniques of emotional recognition to measure physiological responses and identify personality traits from emotional reactions, and can get to complement the results obtained through the psychometric test.

From the potential of these technologies, it is possible to link the tools and techniques of recognition in areas such as psychology and the application of psychometric personality test to identify personality traits. In this work an architecture for the development of a system using emotional recognition, physiological signals and video games to identify personality traits that may be related to the results obtained from psychometric tests, process and considerations are discuss for system construction.

II. Related works

Emotion is a mental and physiological state associated with a variety of feelings, thoughts and behaviors. The mental state of an individual is accompanied by physiological changes that may lead to the modification of the expressions of a person which are observable and measurable manifestations and can be perceived and evaluated by others as evidence of an emotional state [8]. Among the techniques used for emotional recognition we can find facial recognition [9], speech recognition [10] and measurement of physiological signals [11] [12], which are being used in different studies because the measured signals are controlled by the autonomic nervous system (ANS) of the human body, and therefore, are not affected by choice, leading to more accurate and consistent measured data that represent the emotional reaction of the individual.

The detection process emotional information begins with passive sensors that capture data of the state and behavior of the individual without interpreting the entrance where they

I. Torres et al. (eds.), *VII Latin American Congress on Biomedical Engineering CLAIB 2016, Bucaramanga, Santander, Colombia, October 26th -28th, 2016,*
IFMBE Proceedings 60,
DOI: 10.1007/978-981-10-4086-3_11

come from. The data collected are analogous to signals that humans use to perceive emotions in others [13]. Recognition of emotional information requires extracting meaningful patterns of data collected, which is done through machine learning techniques that process information from different sensors that together produce any emotional label. Physiological changes have been associated with the emotions experienced by a person, it is recognized that, in most cases, measurable physiological changes occur when an emotion is expressed or felt, for example, changes in heart rate [14] [15], galvanic skin response [16], muscle tension and facial expressions [17], respiration rate and skin temperature [18] [19] and electrical activity in the brain [20] [21]. Sensing these changes is expected to build computer systems that can automatically recognize emotions identifying patterns in these signals sensed that capture physiological responses. The methods used to capture these physiological signals depend on the degree of invasion required and have differences associated in the accuracy of the signals and the extracted features [6].

III. Methodological proposal

For the design of the system we part from two models which allow the identification and recognition of personality traits and emotions:

A. Classification model of personality traits:

One of the most commonly used test in the field of measuring personality traits is the PEN model of Eysenck psychometric test shown in Figure 1. Hans Eysenck [7], proposed that only two factors were necessary to explain individual differences in the personality, argued that these traits were associated with innate biological differences. Eysenck identified three independent dimensions of personality: neuroticism (N), extraversion (E) and psychoticism (P). The three dimensions are sufficient to describe the personality properly, because from these predictions, can be made prediction at physiological, psychological and social level

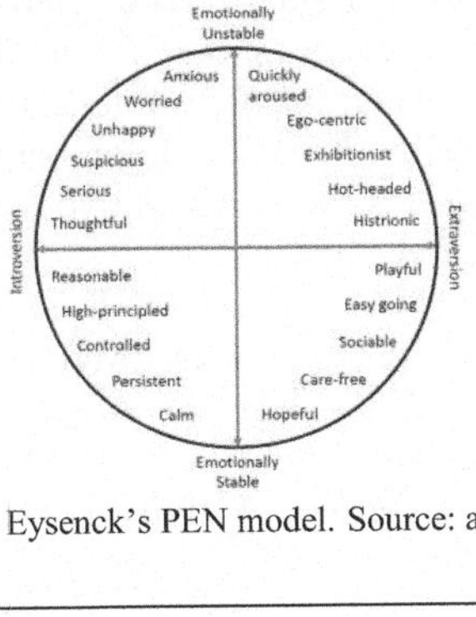

Fig. 1. Eysenck's PEN model. Source: authors.

B. Emotional classification model:

From the physiological signals acquired and the information measured by the acquisition system the behavior is classified by affective models, among which is found the affective model James Russell [22]. This model suggests that emotions are distributed in a space of two dimensions: excitation (vertical axis) and valence (horizontal axis), in the center of the model is shown the neutral valence and the medium arousal [23]. In this model, emotional states can be represented with any level of arousal and valence. The circular model has been used for the assessment of affective states in computer systems [24].

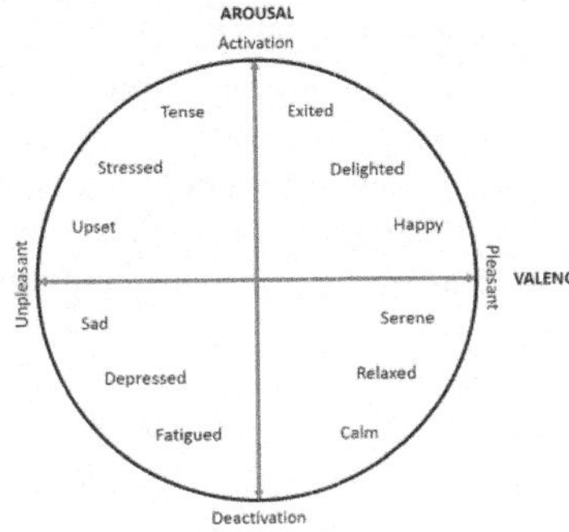

Fig. 2. Russell's affective model. Source: authors.

From the two models consulted from the emotional approach and the approach of personality, the aim of this work is the construction of a system for emotional recognition from the measurement of physiological signals used as stimulation technique video games, whose metrics can be associated with psychometric test values personality, in Figure 3 is shown the methodology proposal that consists of three major phases:

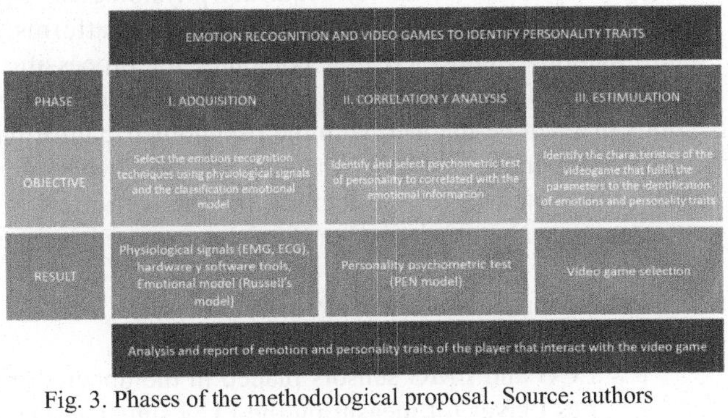

Fig. 3. Phases of the methodological proposal. Source: authors

Phase 1 - Acquisition of information to extract metrics for emotional recognition:

1. Selection of physiological signals measured in the player.

2. Selecting and placing sensors on the player for measuring physiological signals.
3. Selecting hardware system for acquisition and conditioning of physiological signals.
4. Selection the software platform for processing physiological signals acquired.
5. Obtaining metrics from physiological signals for later analysis.

Phase 2 - Mapping and analysis of psychometric information for determining personality traits:

1. Selection and design of the algorithm for analyzing metrics obtained.
2. Getting identifying emotions from emotional computational model proposed.
3. Correlation with model information for identifying personality traits chosen.
4. Report of the information obtained.

Phase 3 - Stimulation of emotion through video games:

1. Selection or video game design that allows stimulation of basic emotions in the player.
2. Selection of hardware tools for player interaction with the virtual environment.
3. Acquisition of information Player performance in the virtual environment for information regarding the physiological signals measured.

IV. Architecture proposal

The architectural proposal, shown in Figure 4, raises the integration of hardware for acquiring physiological signals (ECG and EMG) along with video game platforms, this information is collected in a computer for processing and extraction of metrics that allow algorithms to identify emotions that can be mapped with the psychometric information to detect personality traits in the individual who interacts with the system. We also apply a psychometric test before the test to evaluated the response of the proposed system.

Components and system elements:

1. ECG and EMG sensors placed in the upper extremities Player for measuring the ECG signal and muscle activation by stress.
2. Controlled micro platform for acquiring physiological signals.
3. Platform for interaction of the individual with the video game.
4. PEN psychometric personality test to obtain information about the personality traits of the individual.
5. Software application that allows for the processing and adaptation of physiological signals.
6. Software application that allows the extraction of metrics for analysis and identification of emotions through specialized algorithms.
7. Application software for capturing video player performance and annotation in it.
8. Application software for mapping metrics psychometric information and identification of personality traits.

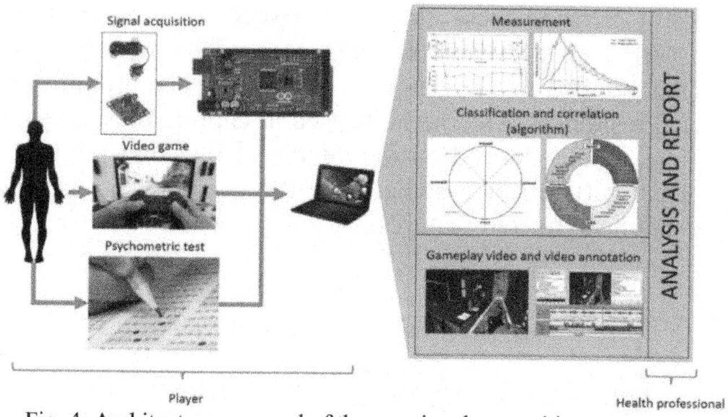

Fig. 4. Architecture proposal of the emotional recognition system. Source: authors

V. Conclusions

Systems involving emotional recognition techniques and video games are focused on solving problems in areas such as education, medicine, economics, politics and research in general. Psychometric personality tests are a very wide scope for these technologies, the results can improve processes in areas such as psychological applied to medicine, work psychology, providing relevant information, by providing environments immersive and interactive evaluation for evaluation of conscious and unconscious behavior of the individual.

Emotional recognition as an interdisciplinary line engineering and computer science is a relatively new field, due to technological advances that allow the acquisition and processing of signals that give relevant information on the behavior of an individual. For Colombia the development of these systems means the investigative foray into a field that is still growing and evolving.

At this stage of the research we have built the acquisition system with one ECG and one EMG sensor, and collected the data in LabVIEW software to adequacy and monitory these two variables. We expect to add a third sensor to measure Electro Dermal Activity (EDA), and implement different algorithms to identify the emotions and perform the correlation with personal traits in the players who interact with the system.

References

1. Picard R, M.I.T Media Laboratory Perceptual Computing Section Technical Report No. 321 at http://affect.media.mit.edu/pdfs/95.picard.pdf
2. Ward R D, Marsden P H, (2004) Affective computing: problems, reactions and intentions, Interacting with Computers, Volume 16, Issue 4, August 2004, pp 707-713, DOI 10.1016/j.intcom.2004.06.002
3. Madni A M, Boehm B, Spraragen M, Madni A M, (2014) Modeling of Emotional Effects on Decision-making by Game Agents, Conference on Systems Engineering Research, Procedia Computer Science, Volume 28, 2014, pp 736-743, DOI 10.1016/j.procs.2014.03.088
4. Leite I, Pereira A, Mascarenhas S, Martinho C, Prada R, Paiva A, (2013) The influence of empathy in human–robot relations, International Journal of Human-Computer Studies, Volume 71, Issue 3, March 2013, pp 250-260, DOI 10.1016/j.ijhcs.2012.09.005
5. Plutchik, R. E. and Hope R. C (1997) Circumplex models of personality and emotions. American Psychological Association.
6. American Psychological Association at http://www.apa.org/topics/personality/
7. H. J. Eysenck and S. B. G. Eysenck (1976) Psychoticism as a Dimension of Personality, London
8. R. A. Calvo and S. D'Mello, "Affect Detection: An Interdisciplinary Review of Models, Methods, and Their Applications," in IEEE Transactions on Affective Computing, vol. 1, no. 1, pp. 18-37, Jan. 2010 DOI 10.1109/T-AFFC.2010.1
9. Soladié C, Salam H, Pelachaud C, Séguier R, (2012) A multimodal fuzzy inference system using a continuous facial expression representation for emotion detection, ICMI '12 Proceedings of the 14th ACM international conference on Multimodal interaction, ACM New York, USA, 2012, pp 493-500 DOI 10.1145/2388676.2388782
10. Schipor O A, Pentiuc S G, Schipor M D, (2011) Towards a multimodal emotion recognition framework to be integrated in a Computer Based Speech Therapy System, Speech Technology and Human-Computer Dialogue (SpeD), 6th Conference on, Brasov, 2011, pp. 1-6 DOI 10.1109/SPED.2011.5940727
11. Kortelainen J et al., (2012) Multimodal emotion recognition by combining physiological signals and facial expressions: A preliminary study, Annual International Conference of the IEEE Engineering in Medicine and Biology Society, San Diego, CA, 2012, pp. 5238-5241 DOI 10.1109/EMBC.2012.6347175
12. Takahashi K, Namikawa S Y, Hashimoto M, (2012) Computational emotion recognition using multimodal physiological signals: Elicited using Japanese kanji words, Telecommunications and Signal Processing (TSP), 35th International Conference on, Prague, 2012, pp. 615-620 DOI 10.1109/TSP.2012.6256370
13. Tao J, Tan T, (2005) Affective computing: A review affective computing and intelligent interaction, Affective Computing and Intelligent Interaction, ser. Lecture Notes in Computer Science, J. Tao, T. Tan, and R. W. Picard, Eds. Berlin, Heidelberg: Springer Berlin / Heidelberg, 2005, vol. 3784, ch. 125, pp. 981–995.
14. Raaijmakers S F, Steel F W, de Goede M. van Wouwe N C, van Erp J B F, Brouwer A M, (2013) Heart Rate Variability and Skin Conductance Biofeedback: A Triple-Blind Randomized Controlled Study, Affective Computing and Intelligent Interaction (ACII), Humaine Association Conference on, Geneva, 2013, pp. 289-293 DOI 10.1109/ACII.2013.54
15. Mitkidis P, McGraw J J, Roepstorff A, Wallot S, (2015) Building trust: Heart rate synchrony and arousal during joint action increased by public goods game, Physiol-ogy & Behavior, Volume 149, pp 101-106, DOI 10.1016/j.physbeh.2015.05.033
16. Holmgård C, Yannakakis G N, Martínez H P, Karstoft K I, (2015) To rank or to classify? Annotating stress for reliable PTSD profiling, Affective Computing and Intelligent Interaction (ACII), 2015 International Conference on, Xi'an, pp. 719-725 DOI 10.1109/ACII.2015.7344648
17. Chittaro L, Sioni R, (2013) Exploring Eye-Blink Startle Response as a Physiological Measure for Affective Computing, Affective Computing and Intelligent Interaction (ACII), 2013 Humaine Association Conference on, Geneva, pp. 227-232 DOI 10.1109/ACII.2013.44
18. Chanel G, Rebetez C, Bétrancourt M, Pun T, (2011) Emotion Assessment From Physiological Signals for Adaptation of Game Difficulty, IEEE Transactions on Systems, Man, and Cybernetics - Part A: Systems and Humans, vol. 41, no. 6, pp. 1052-1063 DOI 10.1109/TSMCA.2011.2116000
19. Landowska A, Wróbel M R, (2015) Affective reactions to playing digital games," 2015 8th International Conference on Human System Interaction (HSI), Warsaw, pp. 264-270 DOI 10.1109/HSI.2015.7170678
20. S. A. Hosseini and M. A. Khalilzadeh, (2010) Emotional Stress Recognition System Using EEG and Psychophysiological Signals: Using New Labelling Process of EEG Signals in Emotional Stress State, 2010 International Conference on Biomedical Engineering and Computer Science, Wuhan, pp. 1-6 DOI 10.1109/ICBECS.2010.5462520
21. Z. Khalili and M. H. Moradi, (2009) Emotion recognition system using brain and peripheral signals: Using correlation dimension to improve the results of EEG, 2009 International Joint Conference on Neural Networks, Atlanta, GA, pp. 1571-1575 DOI 10.1109/IJCNN.2009.5178854
22. Russell J, (1980) A circumplex model of affect, Journal of Personality and Social Psychology 39, pp. 1161–1178.
23. Rubin D C, Talerico J M, (2009) A comparison of dimensional models of emotion, Memory 17, pp. 802–808.
24. Remington N A, Fabrigar L R, Visser P S, (2000) Re-examining the circumplex model of affect, Journal of Personality and Social Psychology 79, pp. 286–300.

Corresponding author:

Author: Mauro Callejas-Cuervo
Institute: Pedagogical and Technological University of Colombia/System engineering and computation School, Tunja, Colombia
Street: North Central Avenue 39-115, CP: 150003
City: Tunja, Boyacá
Country: Colombia
Email: mauro.callejas@uptc.edu.co

Classification of Weight Status Using Anthropometric and Clinical Indicators

A.E. Martinez-Licona, F.M. Martinez-Licona, J.C. Romero-Macias and L.A. del Castillo-Alfaro

Universidad Autonoma Metropolitana Iztapalapa /Electrical Engineering Department, Mexico City, Mexico

Abstract— **Obesity is a chronic disease that became a crisis on adult and child health worldwide, especially in Mexico where about 70 % of Mexican adults are overweight. Although several efforts have been made in order to take control over the problem, obesity as a disease is complicated to monitor and the solutions have had a limited success. In the other hand, disorder eating behaviors have been reported as an increasing health problem. Both clinical and anthropometric measurements can help on identify weight condition and in this paper we present an analysis of some of these attributes on a database devoted to identify the weight status of the subject. The data was collected from 600 records of a clinical laboratory in Mexico City and the attributes included cholesterol, LDL, HDL, body mass index (BMI) and waist measurement. We applied two classifiers to find out the way the classes are distributed and the possible relationship between the two categories of attributes. The classes were grouped in three categories (thinness, normal and obesity) and three levels of risk according to the experts for a total number of 28 classes. The results showed issues on severe thinness, medium and morbid obesity classes as well a strong impact of BMI on the classification of the obesity classes. The classification can be seen as an aid tool to help managing weight problems by an integral subject-centered strategy.**

Keywords— **Obesity, thinness, perceptron, BMI.**

I. Introduction

Obesity is a chronic disease that is characterized by a complex relationship among their causals, which are related to the interaction of genetic, social, environmental and life-style factors, as well as to underlying social and economic issues [1]. At present Mexico faces a crisis concerning adult and child obesity; about 70 % of Mexican adults are overweight or obese and this public health problem may cost up to 12 500 billion dollars by 2017 [2,3]. In the other hand, low weight as a result of bad nutrition and eating disorders is affecting the personal life of many people, especially the young ones [4]. Efforts have been made in order to take control over these problems but both are complicated to monitor and those solutions have had a limited success.

Some physiological variables that can lead to identify weight issues are related to cardiac functions and anthropometric measurements. The World Health Organization (WHO) defined obesity in terms of fat storage and health risks; it is the condition in which excess adipose tissue adversely affects the health and wellbeing. The same WHO defines a person as underweighted when he/she has low body weight relative to the height. According to these definitions the best indicators to assess weight status should be those that quantify the amount of adipose tissue or the body fat percentage that can be estimated through the body mass index (BMI) [5,6].

It has been shown that body fat distribution is correlated not only to obesity, but also to other high-risk metabolic and cardiac diseases [7]. Waist measurement is a common index that can be indirectly used to assess the cardiometabolic risk. Generally speaking, HDL-C levels are associated with both the degree and distribution of the fat in the body. Moreover, there is research on the effects of low-density lipoprotein cholesterol (LDL-C) on obesity [8] while higher levels of LDH-C may also lead to a cardiovascular disease [9]. In contrast, low weight issues are usually related to eating disorders, which are considered as psychiatric disorders that have been assessed with other kind of tools [10].

This paper presents an analysis of anthropometric and clinical indicators to assess weight status. A database of different body composition profiles was created from cholesterol levels, BMI and waist measurements in subjects of a clinical laboratory. The data were classified under two classifiers in order to find out the way the classes are distributed and to identify the impact of the attributes on the assigned classes.

II. Methods

A. Dataset

A dataset was collected from 600 records between 2012 and 2014, 233 males and 367 females which ages range from 20 to 80 years old, of a clinical laboratory in Mexico City. The clinical attributes were cholesterol, HDL and LDL; blood extraction was performed based on Mexican Official Standard NOM-253-SSA1-2012, for the disposal of human blood and blood components for therapeutic purposes. The anthropometric attributes of BMI and waist measurement were obtained following the NOM-008-SSA3-2010, for the integral treatment of overweight and obesity.

I. Torres et al. (eds.), *VII Latin American Congress on Biomedical Engineering CLAIB 2016, Bucaramanga, Santander, Colombia, October 26th -28th, 2016*,
IFMBE Proceedings 60,
DOI: 10.1007/978-981-10-4086-3_12

WHO classified according to the BMI index into thinness, normal, pre-obese and obese with different grades of severity [11]. Also three risk ranges were included in the classes, depending on the variables that are out of the ranges considered as normal, which are shown below:

- Cholesterol: < 200mg/dL
- LDL: >100 mg/dL
- HDL: <40 mg/dL in men, <50 mg/dL in women
- Waist measurements: <90 cm in men, <80 cm in women

The following resolution was taken to allocate risk: if the subject has one variable out of range a low risk is assigned, if he/she has two variables out of range a medium risk is given and three or more variables out of range equals high risk. Consequently, the classes of the database are shown in Table 1.

Table 1 Classes assigned to the subjects of the database

Class	Abbreviation
Severe thinness with high risk	BTHR
Severe thinness with medium risk	BTMR
Severe thinness with low risk	BTLR
Severe thinness with no risk	BTNR
Moderate thinness with high risk	MTHR
Moderate thinness with medium risk	MTMR
Moderate thinness with low risk	MTLR
Moderate thinness with no risk	MTNR
Slight thinness with high risk	STHR
Slight thinness with medium risk	STMR
Slight thinness with low risk	STLR
Slight thinness with no risk	STNR
Normal with high risk	NHR
Normal with medium risk	NMR
Normal with low risk	NLR
Normal with no risk	NNR
Pre-obesity with high risk	PHR
Pre-obesity with medium risk	PMR
Pre-obesity with low risk	PLR
Slight obesity with high risk	SOHR
Slight obesity with medium risk	SOMR
Slight obesity with low risk	SOLR
Medium obesity with high risk	MOHR
Medium obesity with medium risk	MOMR
Medium obesity with low risk	MOLR
Morbid obesity with high risk	BOHR
Morbid obesity with medium risk	BOMR
Morbid obesity with low risk	BOLR

B. Classification methods

We applied two classifiers to the data, Naive Bayes and Perceptron. We used Weka to implement and run the algorithms [12, 13].

A Naive Bayes Classifier is an algorithm that predicts a class value given a set of attributes using a probability error minimization criterion called the Bayes rule. For each known class value it calculates the probabilities for each attribute, conditional on the class value. Then it uses the product rule to obtain a joint conditional probability for the attributes. Finally it applies the Bayes rule to derive conditional probabilities for the class variable. Once this has been done for all class values, the output will be the class with the highest probability.

The perceptron is a linear classifier based on a neural network that uses supervised learning. For a data set classified according to binary categories (labeled +1 and -1) the classifier seeks to divide the two classes by a linear separator. The separator is a *(n-1)*-dimensional hyperplane in a *n*-dimensional space. The learning algorithm to find the optimal weights of the network starts with a guess for the weight vector, takes one example at a time to process and compares the predicted label to the actual value (the correct one) by applying the perceptron rule. If the output does not correctly classify the example the algorithm updates the weights vector in the correct direction until all examples are correctly classified.

C. Class clustering

After classifying the data by the techniques described above, we focused on the thinness and obesity categories in order to observe the behavior of the data and to identify the impact of the attribute on the classification of each subcategory. We used the perceptron to run these experiments separating the database into two data sets:

- Thinness which included the following classes: BTHR, BTMR, BTLR, BTNR, MTHR, MTMR, MTLR, MTNR, STHR, STMR, STLR and STNR.
- Obesity which included the Pre-obesity and Obesity classes: PHR, PMR, PLR, SOHR, SOMR, SOLR, MOHR, MOMR, MOLR, BOHR, BOMR and BOLR.

III. RESULTS

First we considered all classes and all attributes, Figure 1 shows the resulting classification percentages of the thinness, normal, pre-obese and obese classes using the machine learning methods selected.

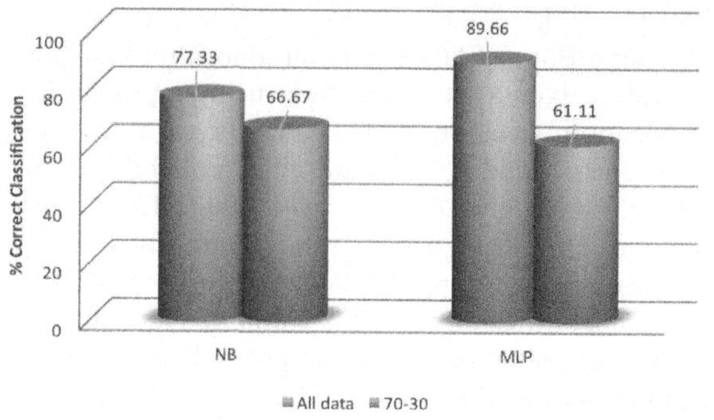

Fig. 1 General classification results using all data (blue) and 70% train 30% test data (red). NB: Naive Bayes, MLP: Multilayer Perceptron.

Figure 2 shows the classification results (percentages) of each of the 28 classes using the two classifiers. Looking at the results, we found that there are cases where class assignment places the person in a category where the associated care can result in a worse state than the original, for example that severe thinness high risk were classified as moderate thinness low risk. Then we proceeded to analyze the data by the thinness, pre-obese and obese categories.

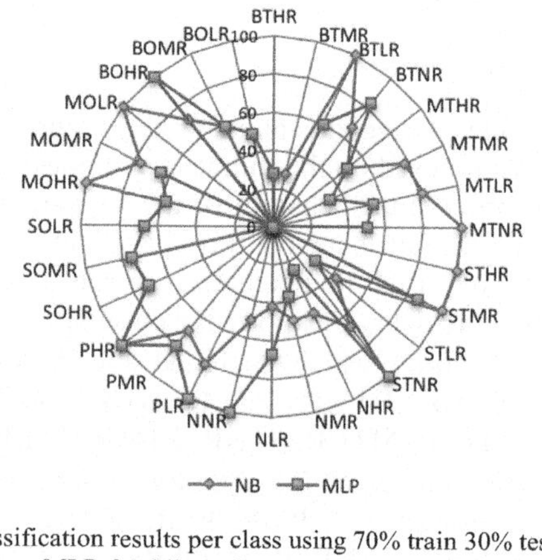

Fig. 2 Classification results per class using 70% train 30% test data. NB: Naive Bayes, MLP: Multilayer Perceptron. Classes are described in the methods section.

Figure 3 shows the confusion matrix of the 11 thinness classes using a 1-layer perceptron in a 70% train 30% test data set. We can observe that classes BTLR and STLR are spread into more classes and most of the data of classes MTNR, STHR and STLR are not correctly classified.

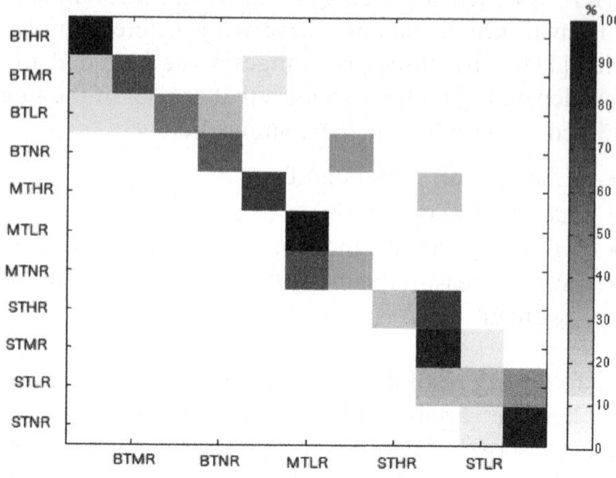

Fig. 3 Confusion matrix of thinness classes using a perceptron; classes are described in the methods section.

Figure 4 shows the confusion matrix of the 12 pre-obesity and obesity classes using the same neural network and data configuration as in the thinness classes. In this case we found that classes PMR, SOMR and BOHR are spread into more classes while class SOLR did not have any correctly classified data.

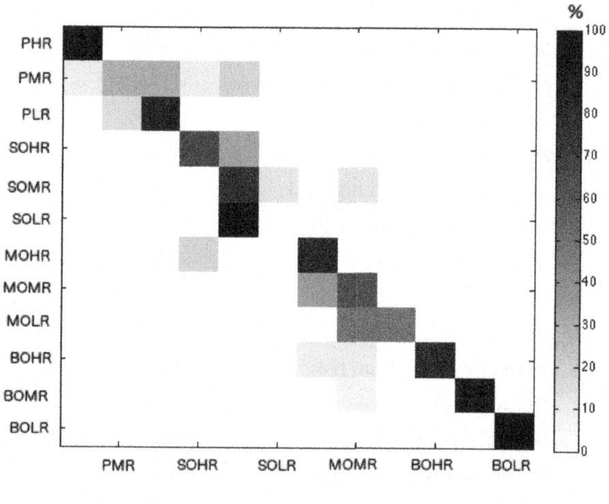

Fig. 4 Confusion matrix of thinness classes using a perceptron; classes are described in the methods section.

Taking a look at the results there is a case when a class spread leads to different treatments: the morbid obesity with high risk class (BOHR) spreads into its own class and medium obesity with medium risk (MOMR) and medium obesity with high risk (MOHR) classes. In order to observe the influence of the attributes on the classification of these particular classes we used a perceptron to run neural networks eliminating one attribute at a time, the results are shown in Table 2.

Table 2 Correct classification results (%) in the Obesity classes selected when one of the attributes is eliminated from the database

Attribute eliminated	MOMR	MOHR	BOHR
None	90	77.78	100
Gender	100	55.56	100
Waist Measurement	100	77.78	100
LDL	90	66.67	100
HDL	100	55.56	100
Cholesterol	100	66.67	100
Body Mass Index	80	22.22	40

IV. Discussion and Conclusions

Weight management is a health problem complex to manage, both thinness and obesity; although clinical and anthropometric are important variables to monitor, the interactions among them can't be established if there is not a considerable amount of data and, even so, it is not possible to guarantee a generalization of the results. Our database was conformed with such attributes and a high number of classes. This allowed us to identify from the groups of interest, in both thinness and obesity categories, what the level of risk was assigned if the subject were classified in an incorrect class.

According to the results, in the thinness category we found that the severe thinness low risk class (BTLR), data were distributed in the other severe thinness classes (BTMR and BTHR); this would imply a different strategy for treatment but, as we are talking about a severe condition, these are not going to have many changes. In the other hand, the obesity category presented the case of BOHR class data mentioned above. The possibility of a major procedure (as the bypass gastric balloon surgery) may signify an important economic expenditure to a misclassified patient that could have a less aggressive and less expensive nutrition treatment based on metabolic monitoring. BMI seems to be an important and quick way of measurement that tends to differentiate the risks levels when there is a doubt about the subject situation.

The classifiers showed different results in each of the 28 classes. Although Naive Bayes classifier got a slight better performance over the perceptron in general, more experimentation in a more balanced database must be done.

These findings showed that it's important to consider in more detail the interaction of both categories of attributes in the assessment of the weight status in order to focus on the treatment that best fit to each case. More important is the fact of having an integral approach subject-centered where these tools work as an aid to diagnosis and treatment; this will help to understand and optimally manage weight issues.

Acknowledgment

Authors thank Dr. Eduardo Mera Montufar and Chemical Amelia Briones Amador for providing clinical data.

References

1. Rivera J, Velasco A, Hernandez M, et al (2013) Obesity in Mexico: recommendations for a state policy. UNAM, Mexico
2. Barquera S, Campos I, Hernández L, Pedroza A, & Rivera J (2013). Prevalence of obesity in Mexican adults 2000-2012. *Salud Pública de México, 55*(Suppl. 2), S151-S160. http://www.scielosp.org/scielo.php?script=sci_arttext&pid=S0036-36342013000800012&lng=en&tlng=es
3. Forbes at http://www.forbes.com.mx/obesidad-un-problema-de-5500-mdd-para-mexico/
4. Vázquez C, et al. (2015) Prevalencia de conductas alimentarias de riesgo y su asociación con ansiedad y estado nutricio en adolescentes de escuelas secundarias técnicas del Distrito Federal, México. *Rev Esp Nutr Comunitaria* 21(1), 15-21.
5. Kaufer M, Toussaint G (2008) Anthropometric indexes to evaluate pediatric overweight and obesity. *Boletín médico del Hospital Infantil de México, 65*(6), 502-518. http://www.scielo.org.mx/scielo.php?script=sci_arttext&pid=S1665-11462008000600009&lng=es&tlng=es.
6. World Health Organization (1995) Physical status: the use and interpretation of anthropometry: Report of a WHOS Expert Committee. WHO Technical report Series, Geneva, pp. 312, 345.
7. Tchernof A, Després JP (2013) Pathophysiology of human visceral obesity: an update. Physiological reviews, *93*(1), 359-404.
8. Kesaniemi YA, Grundy SM (1983) Increased low-density lipoprotein production associated with obesity. Arteriosclerosis, Thrombosis, and Vascular Biology, 3:170-177 DOI: 10.1161/01.ATV.3.2.170
9. Grundy S, Pasternak R, Greenland P, Smith S, Jr, Fuster V (1999) Assessment of cardiovascular risk by use of multiple-risk-factor assessment equations: A statement for healthcare professionals from the American Heart Association and the American College of Cardiology1. J Am Coll Cardiol. 34(4):1348-1359 DOI:10.1016/S0735-1097(99)00387-3.
10. Morgan JF, Reid F, Lacey JH (1999) The SCOFF questionnaire: assessment of a new screening tool for eating disorders *BMJ* 319 :1467-1468.
11. WHO at http://apps.who.int/bmi/index.jsp?introPage=intro_3.html
12. Bishop C (2006) Pattern Recognition and Machine Learning, Springer
13. Hall M, Frank E, Holmes G, Pfahringer B, Reutemann P, Witten IH (2009); The WEKA Data Mining Software: An Update; SIGKDD Explorations, Volume 11, Issue 1.

Author: Fabiola Martinez Licona
Institute: Universidad Autonoma Metropolitana
Street: Av. San Rafael Atlixco 186, Col. Vicentina Iztapalapa
City: Mexico City
Country: Mexico
Email: fmml@xanum.uam.mx

How Can Biomedical Engineering and Health Science Students Learn Together?

W.R. Rodríguez[1], A.M. Ríos[2], A. Miguel Cruz[1]

[1] University of El Rosario, School of Medicine and Health Sciences, Biomedical Engineering Program. Bogota, Colombia.
[2] University of El Rosario, School of Medicine and Health Sciences, Occupational Therapy Program. Bogota, Colombia.

Abstract— **The aim of this study is to pilot test the active learning techniques we have developed to promote interdisciplinary work in students from Biomedical Engineering and Health Science programs. A combination of an only post-test one-group design using a survey with a descriptive analysis of the students' perceptions of the learning strategies was used during the course using a group interview. As a result, the students believed that the ASSI-TECH course contributed to the creation of communities of practice, they were satisfied with their learning processes, and they valued the active and collaborative learning activities.**

Keywords— **Engineering education, interdisciplinary work, professional skills, active learning.**

I. Introduction

The current practice of any professional involves interactions with other professionals from different disciplines. This situation requires the development of skills for multi and interdisciplinary interactions at university. In spite of this, few courses are designed to foster the development of a skill for integrating students from different disciplines, and it is not common to find courses that are designed with an interdisciplinary approach. In other words, students do not easily find courses where they carry out real interdisciplinary work with counter pairs from other careers; although in their professional lives they have to address problems in interdisciplinary teams, this situation made them feel unprepared for it [1].

Besides, it is common on elective courses to find students from different courses and semesters (Freshman, Sophomore, Junior, etc.). This is challenging for teachers as they have to adapt their lectures, labs and other learning activities to achieve a genuine integration of all the students in the proposed learning activities. If a teacher uses traditional 'passive' learning techniques, where students simply take notes or just listen to the teacher's lecture, the learning outcomes obtained at the end of the academic process are not exactly the best; moreover, the students' potential is not fully exploited. Despite sharing the same classroom, they usually fail to experience interdisciplinary work. On the other hand, if a teacher uses active learning techniques or collaborative learning techniques it is possible to create natural critical learning environments [2]–[6] to improve the students' performances. There is evidence that using active learning techniques is critical in two ways: the improvement of critical thinking and problem-solving skills, and increasing the likelihood that a student will finish his course (reducing attrition) [5]. These student-centered environments provide the opportunity to try, fail and try again before any evaluation, as well as providing the opportunity to close the gap between the current and desired student performance [7]. In addition, collaborative learning techniques foster the generation of Communities of Practice [8].

Another important context to highlight is that currently the professional market is increasingly competitive; however, on average worldwide 35% of business managers have difficulty filling vacancies (the average in Colombia is 30%) according to the results of the Manpower Talent Shortage Survey 2013 [9]. This study states that the main reasons for this difficulty are a lack of the candidates' technical skills (34%), shortages of candidates (32%) and a lack of experience (24%). According to the Labor Observatory for Education of the Ministry of Education in Colombia, when the Labor Observatory asked 2013 graduates the question what the main difficulty of getting the job they had been looking for was, 44.6% said they did not have the experience required for their target job [10].

The situations described above show an urgent need to improve the acquisition and strengthening of skills for problem solving in interdisciplinary groups, thus promoting the tangible application of the knowledge acquired by students. The present paper reports the advances in this direction by means of a strategy which promotes interdisciplinary work in Biomedical Engineering and Health Science students. The course selected for this purpose is called ASSI-TEC01 which main topic is assistive technology (AT). AT is any tool that an individual with a disability uses to accomplish something that would be difficult or impossible for him or her to do without using that tool [11]. Thus, AT increases the capabilities of people with disabilities. The importance of developing skills to work on interdisciplinary teams is overt in the field of assistive technology due to the fact that developing useful and usable assistive technologies often presents complex challenges that require input from multiple disciplines and sectors [11].

I. Torres et al. (eds.), *VII Latin American Congress on Biomedical Engineering CLAIB 2016, Bucaramanga, Santander, Colombia, October 26th -28th, 2016,*
IFMBE Proceedings 60,
DOI: 10.1007/978-981-10-4086-3_13

The aim of this study was to pilot test active learning techniques we developed to promote interdisciplinary work in students from Biomedical Engineering and Health Science programs.

II. Materials and Methods

A. *Study design and measures*

The constructs measured in this study were Satisfaction with Learning (SL) and Communities of Practice (CoP). For this study, the construct SL is defined as the extent to which student perceive how well a learning environment supports academic success in a given topic or field [12]. The construct CoP (or learning communities), is defined is the extent to which students perceive they are part of a community of peers who share interests and work together to learn and achieve professional growth [8].

For this pilot study the authors used a combination of an only post-test one-group design using a survey with a descriptive analysis of students' perceptions of the learning strategies used during the course through a group interview. A 7-point Likert scale with 36 items grouped into three sections was used. In the first section we asked for the participants' demographics and program. In the second section we inquired about whether or not the students' used the open access technologies we had used during the course in their clinical or hospital fieldwork with people with disabilities or on other courses. In the third section we created specific questions by adapting the scales and items already validated that measured the two constructs SL and CoP. SL included items related to students' satisfaction with the materials, equipment, activities and instructor's support, as well as motivation and sense of control on their learning process. CoP included items related to students' connectedness with peers and instructors. Each item had the same weight into each construct. This survey was used in previous research with students on a master program in Occupational Therapy [8]. Additionally, a semi-structured interview was designed for this study in order to assess the students' perceptions with the active and collaborative learning activities implemented in the ASSI-TECH course.

B. *Participants*

In this pilot study all the students enrolled on the ASSI-TEC01 course in one term (n=5) participated.

C. The ASSI-TEC01 course: our proposal

The ASSI-TECH is an elective course offered to students during the last two semesters of their courses. Although the University teaches in Spanish, this course is taught in English because to obtain a degree students have to achieve a certificate at B2 proficiency level in a foreign language (English). Thus, besides teaching the content of Assistive Technology, this course is intended to promote the use of English as a second

claib2016-wrodrig_et_all_FV3.doc

language with academic purposes. ASSI-TECH is an elective course with 3 academic credits and 48 hours of work per semester. It is offered by the Rosario University and it focuses on the implementation, adaptation and development of assistive technologies for people with a disability.

In this course we implemented strategies for the use of active learning techniques that promote interdisciplinary work. The activities were designed to improve skills such as: communication with professionals from different disciplines, hands-on skills, working under pressure, problem solving and communication in an academic environment. These active learning activities were accompanied by lectures and laboratory work with content on assistive technologies such as AT definitions and classifications, statistics about people with disabilities throughout the world and in Colombia, the International Classification of Functioning, Health and Disability and Human-Technology Interfaces.

The active learning activities used during the course were the following:

Introduction activity: *Noodle structure* is an activity used to integrate working groups and to discover skills like competence, leadership and others [13]. This activity was used to introduce students to the philosophy of interdisciplinary work. The course is divided into groups and each group has to build a stable structure with noodles and pieces of duct tape; the structure must be constructed in horizontal direction and attached to the edge of a table. The requirement is that the structure must be elevated without coming into contact with the floor.

Case Study: In the second session a case study was assigned to the students. This case study was about a person with a severe motor impairment who required assistive technologies. Students had to use this case study throughout the whole course. At the end of the course the students had to design an assistive technology device for the person in the case study.

Role-play: This was an activity in which students simulated different roles: patients with disabilities, health or engineering professionals, patients' families and social services professionals, all in a real-life situation. The students prepared their role-play based on the previously assigned case study. The play had to be as real as possible. The final aim is the students' awareness of 'real' professional experiences as well as the needs of the patients and their families.

Hands-on Technology: This is an activity where the students find out about, set up, test and compare different technologies with a specific aim. In this case, the students compared their performance by writing a text using different methods for accessing a computer such as head trackers, adapted keyboards and keyboards on the screen.

Hackathon: In this activity the students had to solve a real technical problem in limited time. Similar to the *Hands-on Technology* technique [14], with this activity the students could gain a more in-depth understanding of the concepts related to conceiving, designing, constructing and testing a low-cost assistive device (Fig. 1). This is an example: "*TASK: You will have the challenge of creating a device as an assistive technology. The device will allow you to emulate the left mouse click in a person with reduced mobility by means of a 'switch'. You have three hours to finish the activity. You can use the elements provided (e.g. springs, magnets, levers, aluminum bars, Arduino boards, and electronic elements and sensors). The design requirements are: the device cannot use batteries, the device should be as small as possible and the device should be as cheap as possible. You have three hours.*"

Fig. 1 Students working in a hackathon.

D. *S*tatistical analysis

In order to respond to our research question: What were students' perceptions of Satisfaction with Learning and Communities of Practice on the ASSI-TECH course?, descriptive statistics were used. The mean and standard deviation of the constructs SL and CoP were calculated. Mean values of these constructs higher than 5 points, indicate good perceptions with the active and collaborative learning activities on the ASSI-TECH course (i.e. seven point Likert Scale was used to measure Communities of Practice and Satisfaction with learning constructs students' perception). In addition to that, a Spearman correlation between CoP and SL constructs was calculated. We expected a positive correlation between these two constructs. The alpha level of significance for all the tests was set at $p \leq 0.05$ (two-sided). The SPSS V 22.0 statistics package was used to generate descriptive, univariate, and bivariate statistics, respectively. We used non parametric statistical analysis in this pilot study due to the ordinal nature of the variables and because our goal was to identify some hints about how the constructs SL and CoP behaved under the learning environment created in the course.

III. Results

A. Participants' description

All the students were female. Three of them were from Occupational Therapy, one from Psychology and one from Biomedical Engineering.

B. What were the students' perceptions of Satisfaction with Learning and Communities of Practice on the ASSI-TECH course? (from the survey).

The means and standard deviations of all the items or questions from both the Satisfaction with Learning and Communities of Practice constructs were calculated. Also, a summative scale with all the items or questions from both constructs were calculated, and then the mean and standard deviation of this new scale were calculated.

Regarding Satisfaction with Learning, the mean and SD of all the items were 6.49 (SD=0.74), while for the summative scale they were 110.4 (SD=5.98), which means that the Satisfaction with Learning construct measured by 17 items (maximum possible score 119) was located in the highest third of the summative scale. This indicates that students' satisfaction with learning on the ASSI-TECH course was high.

In relation to the Communities of Practice construct, the mean was 6.58 (SD=0.67). The mean and SD of the summative scale were 79 and 3.94, respectively, which means that this construct measured by 12 items (maximum possible score=84) was located in the highest third of the summative scale. This reveals that the students believed that the ASSI-TECH course contributed to the creation of communities of practice.

Finally, the correlation coefficient ($r_{xy}=+0.945$, $p<0.016$) indicates that as the perception of the Communities of Practice increases, the Satisfaction with Learning increases as well, and vice versa.

C. What were the students' perceptions with the active and collaborative learning activities implemented in the ASSI-TECH course? (from the semi-structured interview).

Introduction activity: The Health Sciences students perceived that their first natural reaction before this activity was to rely on the engineering students; however, as the activity progress they realized that their own ideas were valuable and useful for solving the problem.

Case Study: The students perceived that having a case study throughout the whole term increased the meaning of what they were learning because they had a tangible purpose associated with their learning process. The students valued the opportunity to contribute to the solution of this case study from their own careers.

Role-play: The students found this activity to be beneficial for their learning because they had the opportunity to reflect on the feelings, concerns and frustrations of a person with a disability and his or her family. The student in the role of the person with a severe motor impairment said that she felt she was able to understand better the needs of an individual with a disability after this class. The students commented on the importance of involving caregivers and relatives in the decisions about the design and development of an AT device. A student from the Biomedical Engineering program played the role of

Biomedical Engineer. She commented that she felt a big responsibility when the family of the person with the disability asked her about the best technical features of an AT device for their relative. The students also commented that they valued this learning strategy because they had the opportunity to put themselves in "*another's shoes*".

Hands-on Technology: The students valued the opportunity to use and experience different types of AT. They liked the method of comparing their own performance while doing the same activity using different types of input control interfaces.

Hackathon: The students commented that they felt pressed for time during the activity. The most difficult aspect of the activity was to decide quickly how to solve the problem. However, they valued working together to find a solution for the challenge.

IV. Discussion

The strong positive correlation between Satisfaction with Learning and Communities of Practice and the interview results indicate that the students were very satisfied with the learning process on the ASSI-TECH course. This is because they had opportunities such as creating a community of practice around the project they developed as they learned from and taught their peers, and playing the role of AT professionals (e.g. Occupational Therapists, Biomedical Engineers, Psychologists and Social workers), individuals with a disability, their caregivers and relatives. The students' comments during the course provided us with hints that they were working with an interdisciplinary approach, which is critical in the field of AT [11]; for instance, near the end of the course, one of the Occupational Therapy students said during a follow-up meeting "... *We just adjusted the sensor sensibility this morning and now our device is working*". This expression allowed us to see that Health Sciences students were committed to solving the technical problems at the same level as the Engineering students.

V. Conclusion

After conducting the pilot study the authors were able to reach the following conclusions: (1) The students believed that the ASSI-TECH course contributed to the creation of communities of practice; (2) the students were satisfied with the learning process during the ASSI-TECH course; and (3) the students were satisfied with the active and collaborative learning activities because they had the opportunity to create a community of practice around the project they were developing. These results encourage us to continue with this strategy in the future with larger courses.

Acknowledgment

The authors want to thank the students from the ASSI-TEC01 course for their participation in this project.

claib2016-wrodrig_et_all_FV3.doc

Conflict of Interest

The authors of this work declare that they have no conflict of interest.

References

[1] B. Williams, T. Brown, R. Scholes, J. French, and F. Archer, "Can interdisciplinary clinical DVD simulations transform clinical fieldwork education for paramedic, occupational therapy, physiotherapy, and nursing students?" *J. Allied Health*, vol. 39, no. 1, pp. 3–10, Jan. 2010.

[2] E. F. Barkley, K. P. Cross, and C. H. Major, Collaborative learning techniques: a handbook for college faculty. Jossey-Bass, 2005.

[3] UNC, "Classroom activities for active learning," Suggestions and reflections on teaching and learning, 2009. [Online]. Available: http://cfe.unc.edu/files/2014/08/FYC2.pdf.

[4] CEA, "Reflexiones pedagógicas No1 - Aprendizaje activo," Bogotá, 2015.

[5] T. A. Angelo and K. P. Cross, Classroom assessment techniques: a handbook for college teachers. Jossey-Bass Publishers, 1993.

[6] K. Bain, What the Best College Teachers Do. Harvard University Press, 2004.

[7] J. Charles, M.-D. Debra, M. Bob, N. David, R. David, and S. Brenda, "Enhancing student learning through effective formative feedback," 2004.

[8] L. Liu, A. M. Ríos Rincón, S. Esmail, E. Taylor, and A. Miguel Cruz, "Usability of mobile technologies in an occupational therapy" in CAOT, Banff, Canada, 2016.

[9] Manpower Group, "Talent shortage survey: research results," Research results, 2013. [Online]. Available: http://www.manpowergroup.com/wps/wcm/connect/587d2b45-c47a-4647-a7c1-e7a74f68fb85/2013_Talent_Shortage_Survey_Results_US_high+res.pdf?MOD=AJPERES [Accessed: 14-Mar-2014].

[10] MEN, "Observatorio Laboral para la Educación," Graduados Colombia, Ministerio de Educación Nacional., 20016. [Online]. Available: http://redes.colombiaaprende.edu.co/ntg/men/Observatorio/index.htm. [Accessed: 19-Apr-2016].

[11] J. Bogera, P. Jackson, M. Mulvenna, J. Sixsmith, A. Sixsmith, A. Mihailidis, and Kontos, "Principles for fostering the transdisciplinary development of assistive technologies," Disability And Rehabilitation: Assistive Technology, p. online 10 March, 2016.

[12] C.C. Lo. "How student satisfaction factors affect perceived learning." *Journal of the Scholarship of Teaching and Learning,* vol. 10, no. 1, pp. 47 – 54, January 2010.

[13] T. Willkomm, Assistive technology solutions in minutes. New Hampshire: ATECH Services. 2005.

[14] CRLT-Engin, "Teaching a Discussion," Center for Research on Learning and Teaching in Engineering. [Online]. Available: http://crlte.engin.umich.edu/gsitraining/webresources/. [Accessed: 26-Mar-2016].

Author: William Ricardo Rodríguez Dueñas
Institute: University of El Rosario
Street: AK 24 63C 74
City: Bogotá
Country: Colombia
Email: william.rodriguez@urosario.edu.co

Effect of Light Scattering due to Multilamellar Bodies in the Human Eye

Emilia M. Méndez-Aguilar[1], Ismael Kelly-Pérez[1] and L. R. Berriel-Valdos[1]

[1] Instituto Nacional de Astrofísica, Óptica y Electrónica, Departamento de Óptica, Santa María Tonatzintla, Puebla, México

Abstract— **A model of the human eye has been used to analyze the radial profiles of light pattern distributions formed in the retina due to scattering caused by cataracts with multilamellar bodies (MLBs) within crystalline lens. This computational model takes into account different parameters such as the particle diameter, the coat size, the particle number, the incident light wavelength and a specific refractive indices. The results show the effect on light scattering by MLBs according to the different parameters. It can be seen that at higher diameter, thinner coating and shorter wavelength, more light is scattered (this can be clearly seen in the results shown in Fig. 2 and Fig. 3).**

Keywords— **Forward scattering, Mie theory, Cataracts.**

I Introduction

Crystalline lens provides accommodation, this is the ability of the eye to change the focus between near and far objects (approximately 30% of the eye focusing power) [1, 2]. $\alpha-$, β- and γ-crystallins are crystalline proteins that contribute to get a refractive index which makes the lens transparent [1]. If the refractive index undergoes some changes then the crystalline lens can be visualized with opacity or clouding. Any opacity or clouding of crystalline lens caused by disruption of visual acuity is called cataract [2]. Cataracts are a problem of global health, approximately 20 million people suffering this affection are blind (51% of global blindness) while over 80 million people with cataracts have low vision [1]. Cataracts can be classified as subcapsular, nuclear, and cortical depending on the location of the opacity [3]. Ultrastructural analysis of lenses with nuclear cataracts (often age-related) shows that the appearance of the MLBs contributes to light scatter within the lens, for this reason this work is focused on analysing MLBs particles that appear in nuclear cataracts. MLBs have been observed in all nuclear cataracts examined from US and Indian people [4]. MLBs are lipid coated spheres (1-4 μm in diameter) found with major frequency in the nuclear region of age-related cataracts compared with human transparent lenses. Thus, opacity of this kind of cataracts is the consequence of the formation of lipid bilayers in the particles that are inside the lens. MLBs contain a core of cytoplasmic proteins surrounded by multiple lipid-rich bilayers forming a membrane, given rise to one of the major sources of scattering in the human nuclei [4–7]. Those coatings come to have 3 to 10 thin bilayers forming the coat of the MLBs and the thickness of each bilayer is about 5 nm, thus the thickness range of the coat is from 15 nm to 50 nm [7].

In a previous study of the model, we only focused on simulating the effect of light scattering on the image formation on the retina due to spherical particles without coating [8]. The main interest of the present study is to analyze the effect of the MLBs on the image formation. It is worth mentioning that MLBs are candidates for large scattering particles mainly by their form, size and number of particles. These attributes of MLBs affect the propagation path making them potential sources of forward light scattering.

As development tool, this work uses a modified computational model to obtain and to compare different radial profiles of light pattern distributions formed in the retina of human eyes under the presence of MLBs inside the crystalline lens. The simulations were carried out using different core and coat sizes for the MLBs, at wavelengths of 400 nm and 700 nm, using specific refractive indices for the medium, the coat and the core of the MLBs, all this with the purpose of knowing their effects on the retinal images by analyzing the radial profiles.

II Methods

The computational model of the optical system of the eye that was proposed by Kelly (for details see Ref. [8]) is used to analyze the effect of MLBs. In that model, there is a random distribution of particles within the crystalline lens, which are considered as sources of scattering of light inside the eye. These particles are spherical and have a refractive index which is different from the surrounding medium. When a beam intersects one particle, there is a change of direction of that beam which can be calculated using the scattering and azimuthal angles. Theory of exact ray tracing and Monte Carlo method were used to simulate the paths of the photons through the eye and to calculate those angles, respectively. These calculations are performed each time the photons intersect a MLBs, and so on, until they reach the retina. The scattering of light generated by these particles is simulated by applying the Mie theory [3, 4]. This work uses as probability density function and a phase function which gives the direction of a beam scattered. This function is obtained ap-

I. Torres et al. (eds.), *VII Latin American Congress on Biomedical Engineering CLAIB 2016, Bucaramanga, Santander, Colombia, October 26th -28th, 2016,*
IFMBE Proceedings 60,
DOI: 10.1007/978-981-10-4086-3_14

plying the Mie theory [8, 11, 12] and the normalized phase function is calculated by:

$$P(\theta) = \frac{4}{x_L^2 Q_{sca}} \frac{(S_1(\theta)^2 + S_2(\theta)^2)}{2} \quad (1)$$

Where x_L is the size parameter of the last layer, Q_{sca} is the scattering efficiency factor, and S_1 and S_2 are the scattering amplitudes. The size parameter is defined as $x_l = 2\pi n_m r_l/\lambda = kr_l$, where $l = 1,2,...,L$ is each layer, n_m is the refractive index of the medium, r_l is the outer radius of the lth layer, λ is the wavelength of the incident wave in vacuum, and k is the propagation constant. The scattering amplitudes are determined by [9–12]:

$$S_1(\theta) = \sum_{n=1}^{\infty} \frac{2n+1}{n(n+1)} [a_n \pi_n(\theta) + b_n \tau_n(\theta)] \quad (2)$$

$$S_2(\theta) = \sum_{n=1}^{\infty} \frac{2n+1}{n(n+1)} [a_n \tau_n(\theta) + b_n \pi_n(\theta)] \quad (3)$$

Where n denotes mode number; π_n and τ_n are the angular functions, which can be establish from recurrence relations [9, 13]; and a_n and b_n are the so-called Mie coefficients or scattering coefficients obtained by [9, 14]:

$$a_n = a_n^{L+1} = \frac{[H_n^a(m_L x_L)/m_L + n/x_L]\psi_n(x_L) - \psi_{n-1}(x_L)}{[H_n^a(m_L x_L)/m_L + n/x_L]\xi_n(x_L) - \xi_{n-1}(x_L)} \quad (4)$$

$$b_n = b_n^{L+1} = \frac{[m_L H_n^b(m_L x_L) + n/x_L]\psi_n(x_L) - \psi_{n-1}(x_L)}{[m_L H_n^b(m_L x_L) + n/x_L]\xi_n(x_L) - \xi_{n-1}(x_L)} \quad (5)$$

Where ψ_n and ξ_n are the Riccatti-Bessel functions; m_L is the relative refractive index of the last layer defined as $m_l = n_l/n_m$, where n_l is the refractive index of outside the particle with lth component; This work uses the scattering series of l- layered sphere in H_n^a and H_n^b, these are the determinants calculated according to reference [9].

Finally, the scattering efficiency factor Q_{sca} is readily obtained by the following equation:

$$Q_{sca} = \frac{2}{x_L^2} \sum_{n=1}^{\infty} (2n+1)[|a_n|^2 + |b_n|^2] \quad (6)$$

The following are some real parameters of MLBs used inside the previous equations of the model as constants. The diameter particles is in the range from 1 μm to 4 μm, whereas the refractive indices of the MLBs are 1.49, 1.50 and 1.42 for the interior core, the lipid bilayer coat and the surrounding cytoplasm, respectively (Fig. 1). Hence, resulting $m_1 = n_l/n_m = n_i/n_{lb} = 1.49/1.5$ and $m_2 = n_l/n_m = n_{lb}/n_c = 1.5/1.42$ (relative refractive indices of the first and the second layer, respectively) [4, 6].

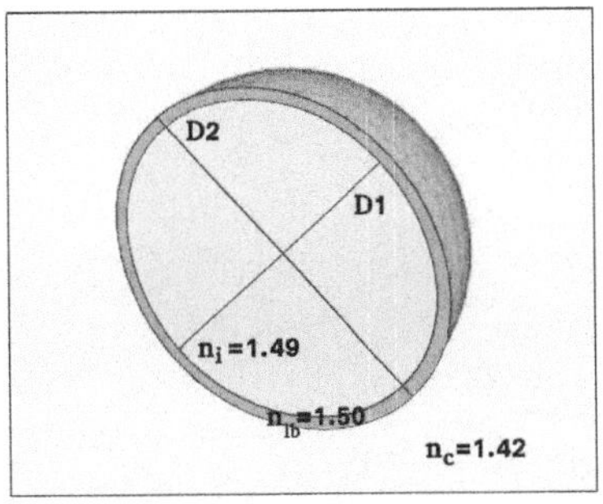

Fig. 1: Scheme MLBs, the refractive indices of the MLBs are 1.49, 1.50 and 1.42 for the interior core (n_i), the lipid bilayer coat (n_{lb}) and the surrounding cytoplasm (n_c), respectively. D_1 is diameter core and D_2 is diameter of the entire particle

The characteristics of scattering are essentially controlled by simple factors (r, λ, m_l). Thus, changes in these factors may involve significant variations in our results. Experimentally it is not possible to conduct scattering measurements such as size, wavelength or relative refractive index within a living eye; for that reason it is a good option to simulate in the computer the variations of those parameters and observe the effects on the scattering patterns formed on the retina. The computer simulations of the optical system of the human eye allow us to do different kinds of variations and to analyze the effects caused by the variation of a single factor or parameter in the formation of the image on the retina.

III RESULTS

By employing the model and the method presented in the previous section, simulations with different parameters were performed in order to know the effects of these parameters. One of them was the diameter of the particles, whose values were 1, 2, 3 and 4 μm; the other one was the light wavelength, which was 400 nm in some cases and 700 nm in others. In addition, 15, 30 and 50 nm were chosen as coat sizes of MLBs. The distributions of light corresponding to these simulations, considering 400,000 particles inside the crystalline lens and one million of photons, are shown in Fig. 2 and Fig. 3.

From the results obtained, it is possible to assert the following. Major light intensity distributions were obtained at the central region going from 0.1 mm to 1.9 mm, focusing in 2 μm diameter particle and 30 nm of coat size in the MBLs (Fig. 2b and Fig. 3b), when the simulations operate with wavelength of 400 nm than when working with the wavelength of 700 nm. By decreasing the wavelength, the scatter

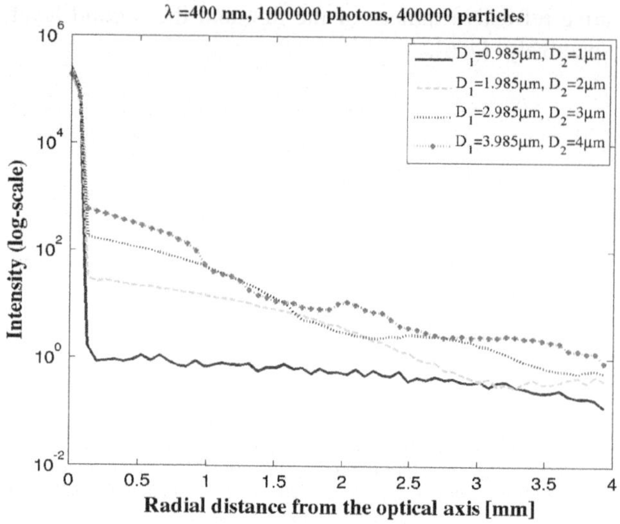

(a) BL=15nm λ=400 nm

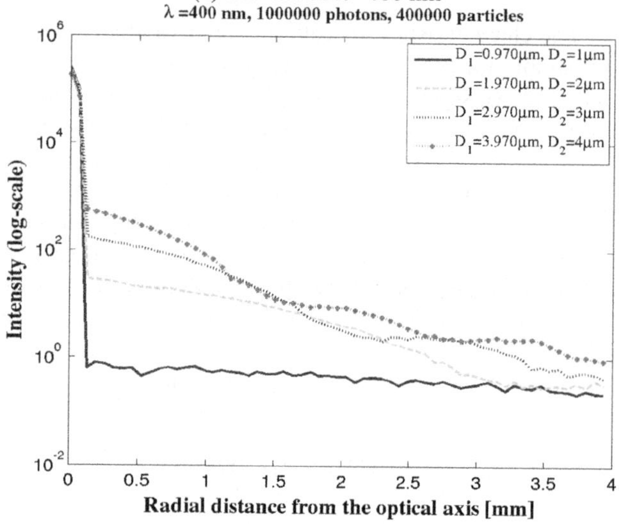

(b) BL=30nm λ=400 nm

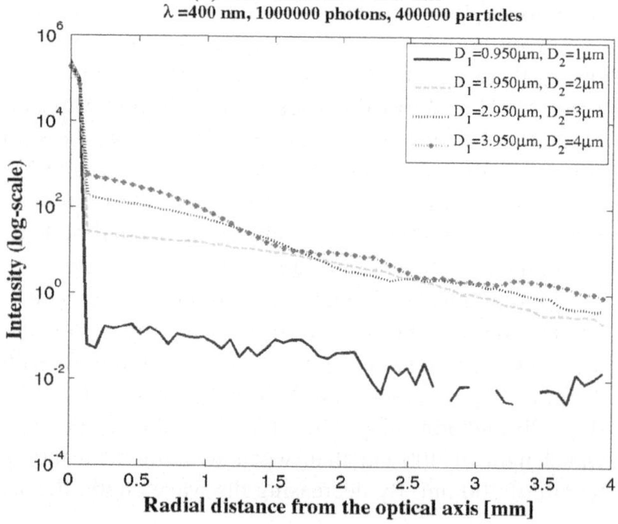

(c) BL=50nm λ=400 nm

Fig. 2: MLBs with coat size of 15, 30 and 50nm, λ=400nm

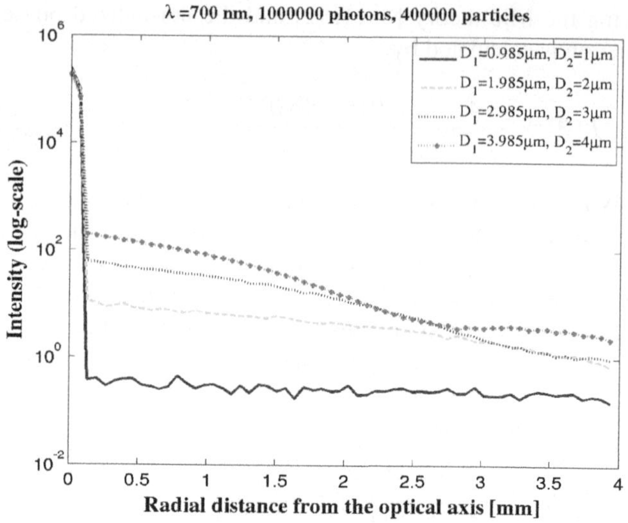

(a) BL=15nm λ=700 nm

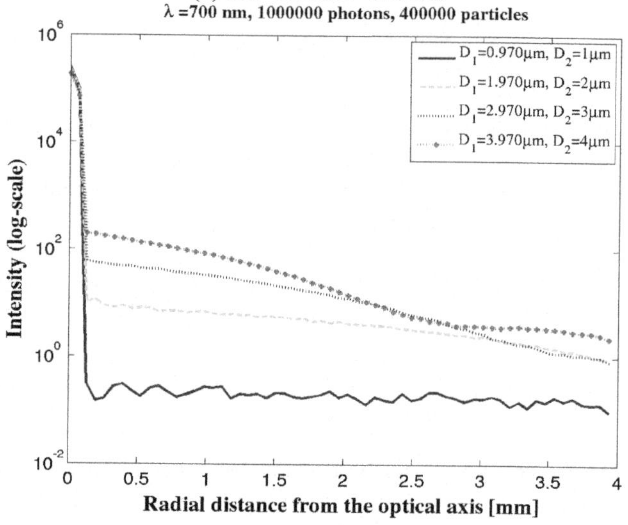

(b) BL=30nm λ=700 nm

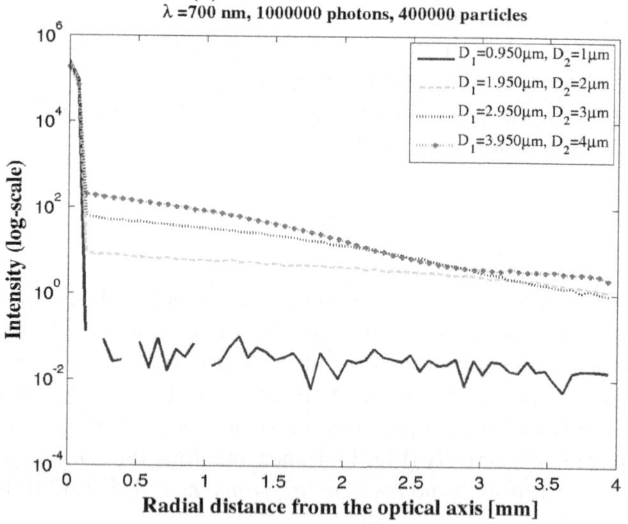

(c) BL=50nm λ=700 nm

Fig. 3: MLBs with coat size of 15, 30 and 50nm, λ=700nm

becomes mainly directed into smaller angles in forward direction, hence more scattered photons reach the central region on the plane of the retina. The intensity also changes when the particle diameters are varied. The light intensity distribution at the central region in all the cases is bigger as the particle size increases since larger particles scatter light at smaller angles in the forward direction while smaller particles scatter light at greater angles causing a decrease in the light scattered at the central region of the retina plane.

By focusing on the coat size, it can be seen that when the coating is increased from 0.13 mm to 4 mm, the intensity decreases at the central region (see Fig. 2 and Fig. 3). For example, the radial profiles with 1 μm diameter particle and 15 nm coat size in the MBLs (Fig. 2a and Fig. 3a) present an intensity in range of 10^{-1}-10^{0}, whereas with 50 nm, the intensity reaches values of the order of 10^{-2} (Fig. 2c and Fig. 3c). This effect which is due to the forward scattering is lower when the scattering angle is greater. Other factors to take into account are core size and particle size. Thus, when the particle s smaller it is more likely to have larger scattering angles and therefore less forward scattering. When photons pass through MLBs formed by layers with a refractive index different from the core of the MLBs, the scattering angle range increases even more and hence the forward scattering decreases. In research done by Costello [4, 6, 7], the MBLs are considered to be potential sources for excess light scattering in age-related nuclear cataracts and described likely candidates for large scattering particles. Thus, to be considered as large particles suggested that their major contributions would be low-angle scattering owing to their lipid-rich layer and fluctuations in the cytoplasm might be responsible for key fluctuations in the local refraction index essential to produce significant light scattering.

IV Conclusion

By means of this work, a computational model of the optical system of the eyes including MLBs with lipid bilayers was developed and used to analyze the effect of the MLBs in the formation of images on the retina. By varying parameters such as the core and the cover sizes it was found that MLBs particles produce more scattered light at the central regions when their diameter is greater, their lipid bilayer coat is thinner, and at shorter wavelengths. According to these results, the MLBs particles with lipid-rich outer layer surrounded by cytoplasm change the propagation path of the light. These kinds of particles generate small scattering angles which increases the forward light scattering. These results are consistent with this phenomenon of cataracts, and thus this work provides a model and possible interpretation of this phenomenon.

Conflict of Interest

The authors declare that they have no conflict of interest.

Acknowledgements

E.M acknowledges CONACYT for scholarship.

References

1. Pébay A. Regenerative Biology of the Eye edited by Alice Pbay *Q. Rev. Biol.* 2015:444–444.
2. Miller S K *Adult Nurse Practitioner Certification Review Guide*. New York:Jones and Bartlett Learning 2012.
3. Thomas van den Berg J T P and Henk S. Light scattering model for donor lenses as a function of depth *Vis. Res.* 1999:1437–1445.
4. Gilliland O K, Johnsen S, Metlapally S, Costello M J, Ramamurthy B, Krishna P V and Balasubramanian D. Mie light scattering calculations for an Indian age-related nuclear cataract with a high density of multilamellar bodies *Mol. Vis.* 2008:572–582.
5. Acton Q A *Advances in Intracellular Space Research and Application: 2011 Edition*. New York:ScholarlyEditions 2012.
6. Costello M J, Johnsen S, Metlapally S, Gilliland K O, Frame L and Balasubramanian D. Multilamellar spherical particles as potential sources of excessive light scattering in human age-related nuclear cataracts *Exp. Eye. Res.* 2010:881–889.
7. Costello M J, Johnsen S, Gilliland K O, Freel C D, Fowler W C. Predicted light scattering from particles observed in human age-related nuclear cataracts using mie scattering theory *IOVS.* 2007:303–312.
8. Kelly-Pérez I, Bruce NC, Berriel-Valdos L R, Werner A and Atencio J A . Computational model of the effect of light scattering from cataracts in the human eye *JOSA A.* 2013:2585–2594.
9. Pena O and Pal U .Scattering of electromagnetic radiation by a multilayered sphere *Comput. Phys. Commun.* 2009:2348–2354.
10. Hulst H C and Van De Hulst H C *Light scattering by small particles*. New York:Courier Corporation 1957.
11. Bohren C F and Huffman D R *Absorption and scattering of light by small particles*. New York:John Wiley & Sons 2008.
12. Boucher O. *Atmospheric Aerosols: Properties and Climate Impacts*. Netherlands:Springer Netherlands 2015.
13. Wiscombe W J. Improved Mie scattering algorithms *Appl. Opt.* 1980;1505–1509.
14. Yang W. Improved recursive algorithm for light scattering by a multilayered sphere *Appl. Opt.* 2003;1710–1720.

Formalization of Gene Ontology relationships with factor graph towards Biological Process prediction

F. Spetale*[1], P. Bulacio[1], F. Krsticevic[1], S. Ponce[2] and E. Tapia[1]

[1] CIFASIS-Conicet Institute, Bv. 27 de Febrero 210 Bis, Rosario, Argentina.
[2] Facultad Regional San Nicolás, Colón 332, Universidad Tecnolgica Nacional, Argentina.
* spetale@cifasis-conicet.gov.ar

Abstract— **Gene Ontology is a hierarchical controlled vocabulary for protein annotation. Its synergy with automatic classification methods, ensemble, has been widely used for the prediction of protein functions. Current classification methods use only the relation *is_a* and a few little *part_of* to generate prediction model. In this work we formalize the GO *part_of, regulates*; *negatively_regulates* and *positively_regulates* relationships through predicate logic. This formalization is incorporated within an ensemble method based on graph factor called *Factor Graph GO Annotation*. The proposed model is validated against four model organisms for GO Biological Process prediction.**

Keywords— **Gene Ontology, Factor Graph, Automatic function prediction**

I INTRODUCTION

The high-throughput of sequencing technologies provides huge amounts of data opening unlimited opportunities for better understanding of biological behavior of target organisms. The use of machine learning methods may achieve the initial approach for data analysis focalizing experiments, saving time and money. A central point of genomic research is to establish the biological functions of proteins, also called annotation. *Gene Ontology* (GO) provides a hierarchical architecture of biological functions [1] which may guide the automatic annotation of protein function. GO is composed of three sub-ontologies: Biological Process (BP), Molecular Function (MF) and Cellular Component (CC). Each of them is a Directed Acyclic Graph (DAG), where every node represents a GO-term (a biological function) and every edge represents a relationship between two GO-terms. The commonly used relationships in GO are: *is_a* (is a subtype of); *part_of*; *regulates*; *negatively regulates* and *positively regulates* [2]. Traditional ensemble methods for automatic function prediction based on GO consider the relationship *is_a* [3], [4], [5] and a few the relationship *part_of* [6].

In this paper, we propose the formalization of GO relationships beyond *is_a* for GO-BP prediction. Regarding inference process interpretability, a classification method based on factor graph [7] is considered. In particular, we use the *Factor Graph GO Annotation* (FGGA) [8] which models GO relationships with logical factor nodes. The formalization must consider TPG constraint, *"If the child GO-term describes the protein, then all its parent terms must also apply to that protein; and if a GO-term not describes a protein, then all its descendant GO-terms must not describe it"*, that governs the structure and inference within GO-DAG. The extension of logical factor nodes within FGGA model, hereafter FGGA$^+$, is able to infer functional predictions of proteins by using the adapted version of sum-product algorithm [8].

This paper is organized as follows. In Section II, GO relationships are formalized thought predicate logic to be included to FGGA$^+$. Section III discusses the results on *A. thaliana*, *D. melanogaster*, *D. rerio*, and *C. elegans* in BP-GO. In the last Section, conclusions are presented.

II METHOD

Given a GO subgraph, GO-terms *GO:i* are mapped to binary-valued latent variable nodes x_i of FGGA$^+$. Relationships between GO-terms are mapped to logical factor nodes f_k which describe valid *GO:i* configurations under the TPG constraint; and probabilistic factor nodes g_i which model statistical dependence between latent variable nodes (ideal) x_i and variable leaf nodes y_i modeling observable (real), i.e., uncertain in *GO:i* term predictions (see Fig. 1).

Practically, logical factor nodes f_k are implemented with truth tables of $2^{\#child+\#parents}$ entries. At each of these entries, the specific parent/child role and relationships of participating variable nodes are required to check the TPG constraint. As shown in Table 1, where 1/0 denotes positive/negative annotation, respectively. The logical factor f_4 in Fig. 1-b ensures that TPG constraint over variable nodes x_3, x_4 and x_5 is fulfilled whenever x_5 is a child node of x_3 (x_5 *regulates* x_4) and x_4 (x_5 *part_of* x_4), i.e., multiple inheritance over x_5.

Formally, logical factor nodes f_k over subsets of variable nodes x_i ensure the local satisfiability of TPG constraint. With this aim, two logical rules are repeatedly evaluated. Specifically, if a child GO-term is annotated positive, then its parent GO-term(s) must also be annotated positive. On the other hand, if a parent GO-term is annotated negative,

I. Torres et al. (eds.), *VII Latin American Congress on Biomedical Engineering CLAIB 2016, Bucaramanga, Santander, Colombia, October 26th -28th, 2016*,
IFMBE Proceedings 60,
DOI: 10.1007/978-981-10-4086-3_15

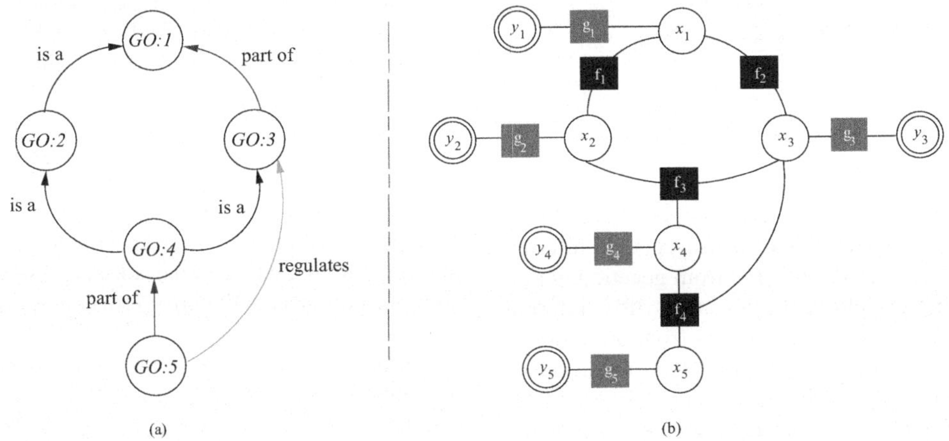

Fig. 1: (a) GO-DAG where *GO:i* nodes are GO-terms and edges are relationships (b) FGGA$^+$ model where x_i are latent variable nodes modeling actual positive/negative *GO:i* annotations and f_k are logical factor nodes modeling the TPG constraint over them, y_i are observable variable leaf nodes modeling real-valued *GO:i* predictions and g_i are probabilistic factor nodes modeling their statistical dependence on latent variable nodes x_i.

Table 1: The truth table of the logical factor node f_4. Positive/negative annotations of variable nodes x_3, x_3 and x_5 are depicted as 1/0. Parent variable nodes x_3 and x_4 are shown in blue.

x_3 [*regulates*]	x_4 [*part_of*]	x_5	$f_4(x_3,x_4,x_5)$
0	0	0	1
0	0	1	0
0	1	0	1
0	1	1	0
1	0	0	1
1	0	1	0
1	1	0	1
1	1	1	1

then its children GO-term must also be annotated negative. In addition, they must fulfill a requirement of transitive inference on their grandparents. Using predicate logic [9], let $part\_of(GO{:}j, GO{:}i)$ denotes *GO:j* (child) is part of *GO:i* (parent) and $is\_a(GO{:}i, GO{:}z)$ denotes *GO:z* is parent of *GO:i* (child). Similarly, let $annotated(\cdot)$ denotes the positive annotation of the target protein with a GO-term. As a result, at least one of the following rules (Eq.1 or Eq.2) must be active and fullfilled by any pair of GO-terms involved within a *part_of* relationship:

$$r_1 : \forall i,j,z \;\; part\_of(GO{:}j, GO{:}i) \wedge annotated(GO{:}j) \wedge [is\_a(GO{:}i, GO{:}z) \vee part\_of(GO{:}i, GO{:}z)] \rightarrow annotated(GO{:}i) \quad (1)$$

$$r_2 : \forall i,j,z \;\; part\_of(GO{:}j, GO{:}i) \wedge \neg annotated(GO{:}i) \wedge [is\_a(GO{:}i, GO{:}z) \vee part\_of(GO{:}i, GO{:}z)] \rightarrow \neg annotated(GO{:}j) \quad (2)$$

In the same way, we can extend the predicate logic to *regulates* relationships:

$$r_3 : \forall i,j,z \;\; reg\_GO(GO{:}j, GO{:}i) \wedge annotated(GO{:}j) \wedge [is\_a(GO{:}i, GO{:}z) \vee part\_of(GO{:}i, GO{:}z)] \rightarrow annotated(GO{:}i) \quad (3)$$

$$r_4 : \forall i,j,z \;\; reg\_GO(GO{:}j, GO{:}i) \wedge \neg annotated(GO{:}i) \wedge [is\_a(GO{:}i, GO{:}z) \vee part\_of(GO{:}i, GO{:}z)] \rightarrow \neg annotated(GO{:}j) \quad (4)$$

where $reg\_GO(GO{:}j, GO{:}i)$ can be just *regulates* or *positive/negative regulation.*

When multiple inheritance exists, multiple relationships must be considered in both, GO and FGGA$^+$ sides. For instance, Table 1 shows that "x_5 is the child of x_3" and "x_5 is also child of x_4", considering the *regulates* relation between *GO:3* and *GO:4*, and the *part_of* relation between *GO:4* and *GO:5*. For instance, row 1 shows the fulfillment of both relationships: *part_of* and *regulates*, by rule 2 and rule 4 activation, hence, f_4 is true. On the other hand, row 4 shows for these relationships, rule 1 and rule 3 are active but only rule 1 is fulfilled, hence, f_4 is false. Note that the modeling of GO relationships by predicate logic requires a detailed examination of cascade GO relationships to accomplish transitiveness.

III Results and Discussion

A Experimental Protocol

Four models organisms, *D. rerio* [10], *A. thaliana* [11], *C. elegans* [12] and *D. melanogaster* [13] are considered. For

Table 2: Datasets in the GO-BP

Organism	# GO-terms	# Samples
D. rerio	44	1002
A. thaliana	97	6032
C. elegants	112	3223
D. melanogaster	156	4189

Table 3: GO-BP prediction performance, Hierarchical Precision (HP), Hierarchical Recall (HR), Hierarchical F-score (HF)

Organism	HP	HR	HF
D. rerio	0.66	0.72	0.66
A. thaliana	0.52	0.68	0.57
C. elegants	0.56	0.76	0.63
D. melanogaster	0.59	0.75	0.64

each organism, GO-BP annotation datasets (see Table 2) are generated with experimental GO evidence codes[1]: inferred from mutant phenotype (IMP), inferred from genetic interaction (IGI), inferred from physical interaction (IPI), inferred from expression pattern (IEP) and inferred from direct assay (IDA), considering GO-terms with at least 300 positively annotated proteins. To balance the training dataset [14] of each GO-term, the number of positive and negative samples must be the same. The negative annotated samples are selected by the *inclusive* separation policy [15]. The protein characterization to a fixed number of input features is done by 457 physicochemical/secondary structure properties, Physicochemical^{+}, 453 of the physicochemical type [16] and 4 of the secondary structure [17]. Practically, protein characterization is implemented with the Bio.SeqsUtils [18] package. FGGA^{+} method is built from GO-term classifiers implemented with SVM default constant complexity C=1. The Gaussian assumption in FGGA^{+} is attained by real valued predictions of SVM soft-margin outputs (implemented with e-1071 R package [19]).

The FGGA^{+} is evaluated with 5-fold cross-validation test, computing per GO-term the AUC average scores [20]. Taking into account that GO annotation gets harder as deeper levels of the hierarchy [21], prediction performance was measured by the hierarchical precision (HP), the hierarchical recall (HR), and the hierarchical balanced F-score (HF) reflecting their trade-off.

B Prediction performance on model organisms

Whatever the organism, FGGA^{+} improves the SVM baseline classifiers. This is particularly evident in the annotation of *D. melanogaster* and *C. elegants*, see Fig. 2.

All relationships modeled in this paper are presented in the Fig. 3 and shows the GO-DAG the annotated sequence "ENSDARP00000061793" of the *NR1H4* gene in *D. rerio*. This gene is related to the hormone nuclear receptor family members and encodes a nuclear receptor for bile acids ENSDARP00000061793 protein which regulates the expression of genes involved in bile acid synthesis. The *is_a* consideration in the GO-BP activate two novel and specific terms, GO:0050794, regulation of cellular process, and GO:0044700, single organism signaling. By including relations *part_of* and *regulates* within GO-BP (see Fig. 4) allow the annotation of three new terms, the more specific term GO:0007165, signal transduction, which is part of terms GO:0007154, cellular communication, and GO:0051716, cellular response to stimulus.

Enrichment through this three new nodes indicates that probably the *NR1H4* gene is involved in the regulation of a cellular process, in this case a bile acid synthesis. It function is also related to the response to a stimulus, in this case the presence of bile, and to transduction signal within the cell, in this case expression of genes involved in bile production. The new prediction enriched of the GO term GO:0044700 results in the biological sense acquisition.

The performance of GO-BP prediction by FGGA^{+} is presented in Table 3. The results show a good F-score independent of the number of GO-terms and organism complexity.

IV CONCLUSIONS

The formalization of the GO relationships within FGGA^{+} allows a hierarchical and consistent prediction of GO-terms within any of the three sub-ontology GO (BP, MF or CC) achieving deeper, broader, and more jumping edges of predicted DAGs. This approach may be extended to another types of no transitive relationships which are in development, such as *capable of* and *occurs in*[2].

CONFLICT OF INTEREST

"The authors declare that they have no conflict of interest."

ACKNOWLEDGEMENTS

The authors were supported by the project PID INI 3600 - Facultad Regional de San Nicolás, Argentina.

REFERENCES

1. Consortium Gene Ontololy. Creating the gene ontology resource: design and implementation *Genome Res.*. 2001;11:1425-1433.
2. Consortium The Gene Ontology. The Gene Ontology in 2010: extensions and refinements *Nucleic Acids Research.* 2010;38:D331-D335.

[1] http://geneontology.org/page/guide-go-evidence-codes

[2] ftp://ftp.geneontology.org/pub/go/www/GO.draft-page.shtml

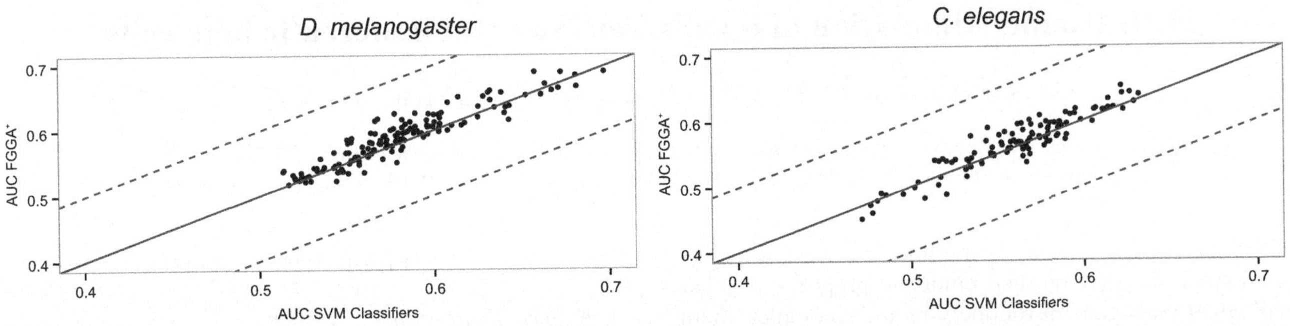

Fig. 2: Scatter-plot of the average AUC of base SVM vs. FGGA$^+$ GO-BP predictions on *D. melanogaster* (left) and *C. elegans* (right) with Physicochemical$^+$ characterization.

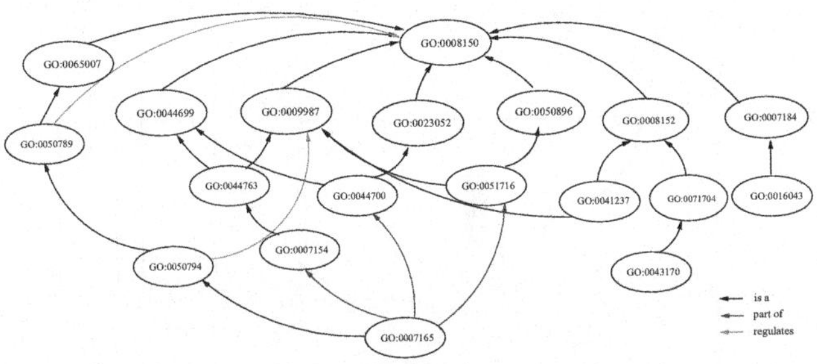

Fig. 3: GO-DAG of a *D. rerio* annotated sequence "ENSDARP00000061793"

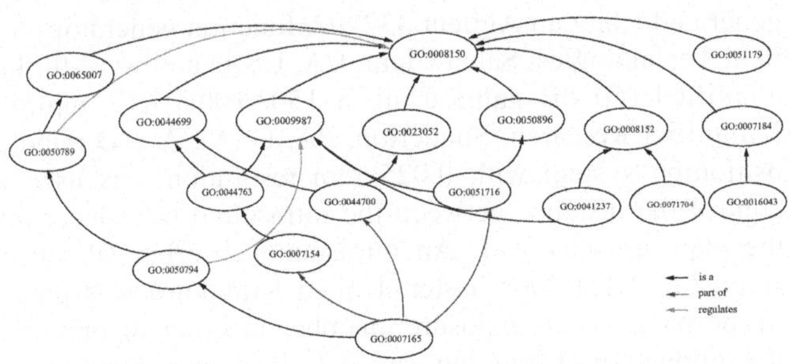

Fig. 4: GO-DAG of a *D. rerio* predicted sequence using FGGA$^+$

3. Barutcuoglu Zafer, Schapire Robert E., Troyanskaya Olga G.. Hierarchical multi-label prediction of gene function *Bioinformatics.* 2006;22:830-836.
4. Valentini Giorgio. True Path Rule Hierarchical Ensembles for Genome-Wide Gene Function Prediction *Computational Biology and Bioinformatics, IEEE/ACM Transactions on.* 2011;8:832-847.
5. Sykacek Peter. Bayesian assignment of gene ontology terms to gene expression experiments *Bioinformatics.* 2012;28:i603-i610.
6. Cheng Liangxi, Lin Hongfei, Hu Yuncui, Wang Jian, Yang Zhihao. Gene Function Prediction Based on the Gene Ontology Hierarchical Structure *PLoS ONE.* 2014;9:e107187.
7. Kschischang Frank R., Frey Brendan J., Loeliger Hans-Andrea. Factor Graphs and the Sum-product Algorithm *IEEE Trans. Inf. Theor..* 2001;47:498-519.
8. Spetale F.E., Tapia E., Krsticevic F., Roda F., Bulacio P.. A Factor Graph Approach to Automated GO Annotation *PLoS ONE.* 2016;11:1-16.
9. Burger Albert, Davidson Duncan, Baldock Richard A.. Formalization of mouse embryo anatomy *Bioinformatics.* 2004;20:259-267.
10. Carlson Marc. Genome wide annotation for Zebrafish 2016. Version: 3.0.0, Accessed: 2016-04-06.
11. Carlson Marc. Genome wide annotation for Arabidopsis 2016. Version: 3.0.0, Accessed: 2016-04-06.
12. Carlson Marc. Genome wide annotation for Worm 2016. Version: 3.0.0, Accessed: 2016-04-06.
13. Carlson Marc. Genome wide annotation for Fly 2016. Version: 3.0.0, Accessed: 2016-04-06.
14. Wei Qiong, Dunbrack Roland L.. The role of balanced training and testing data sets for binary classifiers in bioinformatics. *PloS one.* 2013;8.
15. Eisner Roman, Poulin Brett, Szafron Duane, Lu Paul, Greiner Russ. Improving protein function prediction using the hierarchical structure of the Gene Ontology in *Proc. IEEE CIBCB*:1-10 2005.
16. Lee Bum, Shin Moon, Oh Young, Oh Hae, Ryu Keun. Identification of protein functions using a machine-learning approach based on sequence-derived properties *Proteome Science.* 2009;7:27.
17. Chou Peter Y., Fasman Gerald D.. Prediction of protein conformation *Biochemistry.* 1974;13:222-245.
18. Sicheritz-Ponten Thomas, Alsmark Cecilia. Package SeqUtils 2002. Second Version, Accessed: 2015-09-02.
19. Meyer David, Dimitriadou Evgenia, Hornik Kurt, Weingessel Andreas, Leisch Friedrich. Misc Functions of the Department of Statistics (e1071), TU Wien 2014. Version: 1.6-4, Accessed: 2015-09-02.
20. Fawcett Tom. An Introduction to ROC Analysis *Pattern Recogn. Lett..* 2006;27:861-874.
21. Verspoor Karin, Cohn Judith, Mnizewski Susan, Joslyn Cliff. A categorization approach to automated ontological function annotation *Protein Science.* 2006;15:1544-1549.

Ultrasound stimulation of insulin release from pancreatic beta cells

I.M. Suarez Castellanos[1], B. Balteanu[1], Tania Singh[1], A. Jeremic[2] and V. Zderic[1]

[1] The George Washington University/Department of Biomedical Engineering, Washington DC, United States
[2] The George Washington University/Department of Biological Sciences, Washington DC, United States

Abstract— Type 2 diabetes mellitus is a complex metabolic disease that has reached epidemic proportions. Pharmacological management routinely requires complex therapy with multiple medications, and loses its effectiveness over time. Thus, new modes of therapy are needed that will target directly the underlying causes of abnormal glucose metabolism. The aims of this study were to explore a potential new treatment method that utilizes the non-invasive application of ultrasound energy to induce release of insulin from pancreatic β-cells. Our experiments consisted of assessing the safety, effectiveness and controllability of ultrasound-induced insulin secretion from pancreatic β-cells.

Keywords— **Ultrasound Therapy, Cell Stimulation, Insulin, Beta Cells, Type 2 Diabetes**

I. Introduction

Diabetes mellitus is a complex and heterogeneous metabolic disease that has reached epidemic proportions [1]. Controlling type 2 diabetes is often difficult as pharmacological management routinely requires complex therapy with multiple medications, and loses its effectiveness over time. Low-intensity therapeutic ultrasound has been utilized before in production of reversible changes in cell membrane permeability, modulation of neural tissues, and enhancement of release of the adiponectin from adipose cells, and epinephrine and norepinephrine from adrenal cells [2] [3] [4] [5]. However, ultrasound methods have not been explored before in treatment of secretory defects in type 2 diabetes. Furthermore, in vivo application of therapeutic ultrasound in modulation of cell secretion is a novel area of research, which can potentially lead to new treatments for various endocrine diseases. Clear advantage of using ultrasound to correct β-cell secretory deficiencies, as compared to surgical and pharmacological approaches, lies in its non-invasive and selective therapeutic targeting of the human pancreas. In this study, we assessed the effectiveness and safety of ultrasound stimulation in evoking secretory responses (insulin release) in pancreatic β-cells. We identified ultrasound parameters that showed to be safe and effective at enhancing insulin secretions from suspended β-cells, offering a potential novel method in correcting insulin deficiency in type 2 diabetes.

II. Methods and Materials

A. Experimental setup

The experimental setup was placed in a water bath maintained at 37°C (Thermo Haake DC10-P21, Fisher Scientific, Waltham, MA, USA) as shown in Figure 1.

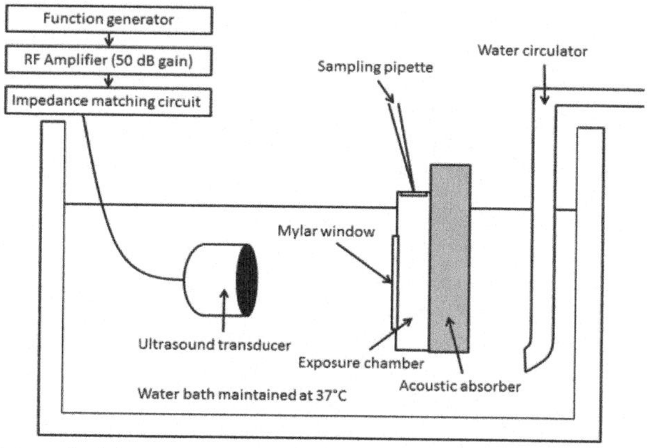

Figure 1. Schematic of experimental setup

Circular planar ultrasound transducers with an active diameter of 1.5 cm and center frequencies of 400 kHz, 600 kHz, 800 kHz and 1 MHz (Sonic Concepts, Inc. Bothell, WA, USA) were directed towards the exposure chamber at a distance of 1.5 cm, 2.25 cm, 3 cm and 3.75 cm, which corresponded to their respective near-field to far-field distances [6]. Ultrasound waveforms used as stimuli were generated using an Agilent 33220A function generator (Agilent Technologies, Santa Clara, CA, USA) and were further amplified (50 dB gain) using a 150A100B RF amplifier (Amplifier Research, Souderton, PA, USA). A 3-D micropositioning system with 0.025 mm resolution was used to control the distance between the ultrasound transducer and the exposure chamber. An ultrasound absorber (Precision Acoustics LTD, Dorchester, United Kingdom) was placed in the back of the exposure chamber in order to minimize the production of standing waves. Cell samples with density approximately of 2-4×10^6 cells/ml were suspended in 1 ml of glucose-free Krebs bicarbonate solution (KBS), placed inside the exposure chamber and treated for 5 min with ultrasound at intensity I_{SPTA} of 1 W/cm^2 using the previous-

I. Torres et al. (eds.), *VII Latin American Congress on Biomedical Engineering CLAIB 2016, Bucaramanga, Santander, Colombia, October 26th -28th, 2016*,
IFMBE Proceedings 60,
DOI: 10.1007/978-981-10-4086-3_16

ly mentioned center frequencies. Aliquots of 100 μL were acquired prior to the start of the treatment (t = 0 min), immediately after treatment (t = 5 min) and 30 minutes after treatment (t = 35 min) for analysis. The cells in the chamber were resuspended with a micropipette immediately before collecting each alliquot. To serve as positive controls, insulin release was measured from cells suspended in glucose-supplemented (12 mM) modified KBS (pH 7.4). In sham groups, the procedure was the same as for ultrasound treatment groups but ultrasound was not turned on.

B. Measurements of cell viability

Various studies have associated the stimulatory effects of ultrasound on different cell types to a direct interaction between ultrasound-induced bioeffects such as acoustic cavitation, acoustic streaming and acoustic radiation forces and the cell's plasma membrane [3] [7]. Therefore, cell membrane integrity was assessed as an indicator of cell viability using trypan blue dye exclusion test. Cell viability was measured for all tested frequencies, positive control groups and sham groups. Ultrasound-treated groups were compared to sham groups (n=7) using a two-tail Student t-test with unequal variances.

C. Measurements of ultrasound-induced secretory events

a) Insulin ELISA

Insulin concentration in cell suspensions was determined using enzyme-linked immunosorbent assay (ELISA) Insulin Kit (Millipore Corporation, Billerica, MA, USA) with a SpectraMax M5 Spectrometer (Molecular Devices, Sunnyvale, CA, USA). Measured insulin concentrations from samples acquired at t = 5 min and t = 35 min were normalized to their respective initial concentration measured at t = 0 min, and data was expressed as fold-change from their initial concentrations. Ultrasound-treated groups were compared to sham groups (n=6) using a two-tail Student t-test with unequal variances. Samples acting as positive controls were suspended in KBS supplemented with 12 mM glucose, a concentration shown to naturally induce insulin secretion in INS-1 cell lines [7].

In a separate set of experiments, intracellular insulin content in ultrasound- and sham-treated groups was determined. These experiments were performed to ensure that any released insulin detected in the extracellular space of the cells came from healthy cells and not from cells that might have been lysed as result of ultrasound exposure. Briefly, cell samples acquired from ultrasound-treated and sham groups were washed twice in modified KBS (pH 7.4), resuspended in RIPA 1X buffer for lysing and kept on ice for 90 min. Samples were then centrifuged for 10 min at 14000 rpm, and supernatants were collected and stored at -70°C for subsequent insulin quantification with ELISA. Ultrasound-treated groups were compared to sham groups (n=7) using a two-tail Student t-test with unequal variances.

b) Carbon fiber amperometry

Carbon fiber amperometry was used to detect ultrasound-induced secretory events from β-cells loaded with serotonin (5-HT) [8]. Neurotransmitters loaded into β-cells have been shown to be localized within the secretory vesicles of the cells [8]. Therefore, detection of 5-HT release was used as a surrogate of insulin release. This technique has advantage of detecting stimulated secretory events in real-time with very high temporal resolution. Cells were plated at a density of approximately 6×10^5 cells/ml and 4 to 6 hours prior to ultrasound treatment these cells were loaded with 5-HT and 5-hydroxy-L-tryptophan, a direct precursor to serotonin. The cells were then bathed in 2 ml of KBS and the carbon fiber electrode was placed in the solution as close as possible (> 1 mm) to the bottom of the well. The oxidation current was digitized and further processed in LabView 2015 (National Instruments, Austin, TX, USA).

III. RESULTS

Our studies showed that cell viability was not significantly affected during and for up to 30 min after treatment when β-cells were exposed to ultrasound at frequencies of 800 kHz and 1 MHz (Figure 2).

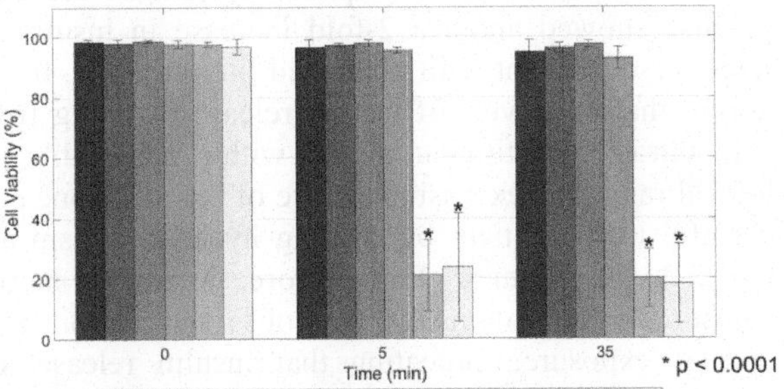

Figure 2. Cell viability measured with trypan blue dye exclusion test in ultrasound exposed cells (n=6), sham treatment (n=6), and positive control (12 Mm glucose, n=6).

Cell exposure to 800 kHz ultrasound resulted in approximately 4-fold increase in insulin release ($p < 0.005$), while exposure at 1 MHz also showed a potential increase in insulin release (around 50%) when compared to sham treatment (Figure 3B).

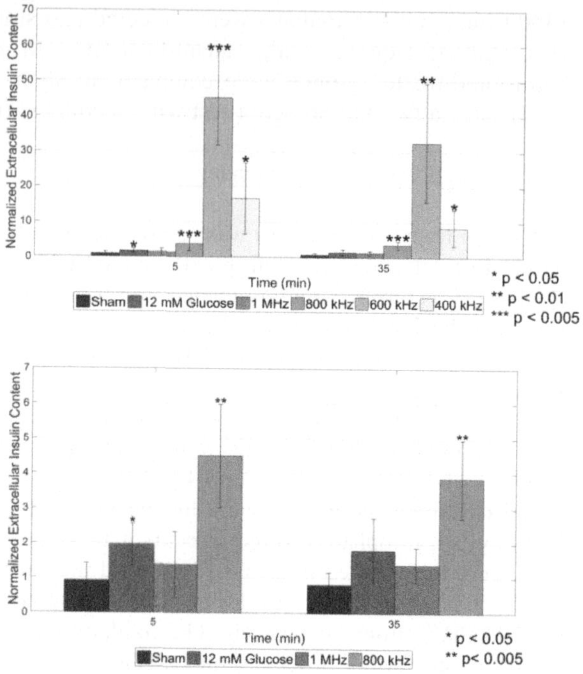

Figure 3. **(A - Top)** Insulin release from pancreatic β-cells exposed to ultrasound at different frequencies as compared to sham treatment and positive control (12 mM glucose). **(B - Bottom)** Zoom in on the results at ultrasound frequencies for which no loss in cell viability was observed. Ultrasound was able to release insulin within clinically-relevant ranges. No further release was observed at 30 min after the ultrasound treatment indicating that no irreparable damage was produced in the cell membranes.

In comparison, cells exposed to natural secretagogue glucose showed about a 2-fold increase in insulin release which is consistent with published literature [7]. It is worth noting that the levels of insulin release resulting from 800 kHz ultrasound stimulation fall within acceptable physiological ranges as excessive release of this hormone could be harmful to the patient by causing hyperinsulinism and potential hypoglycemia. Furthermore, no further increase in measured levels of insulin was observed 30 min after the end of exposure, indicating that insulin release stopped immediately after the end of treatment. Cell viability was highly reduced (by around 80-90%) when the cells were exposed to ultrasound at lower frequencies of 400 kHz and 600 kHz ($p < 0.001$) (Figure 3), which also resulted in significant amounts of insulin released from the cells ($p < 0.05$) (Figure 4). Furthermore, we did not detect a major insulin release and accumulation in the extracellular fluid up to 30 min after the end of ultrasound treatment (insulin levels remained constant at t = 5 min and t = 35 min), thus revealing the potent, inducible and reversible pattern of insulin release from β-cells using ultrasound. These results also suggested that ultrasound can be used in a safe and controlled manner.

We further measured intracellular insulin content in cells treated with 800 kHz ultrasound compared to sham group. Measurements of intracellular insulin content were consistent with results obtained from the extracellular fluid in our samples. Our results showed a reduction of approximately 20% at both t = 5 min ($p < 0.005$) and t = 35 min ($p < 0.005$) in the intracellular insulin content of the cells in ultrasound-treated samples as compared to the sham group (Figure 4), thus indicating increased insulin release from β-cells in response to ultrasound exposure.

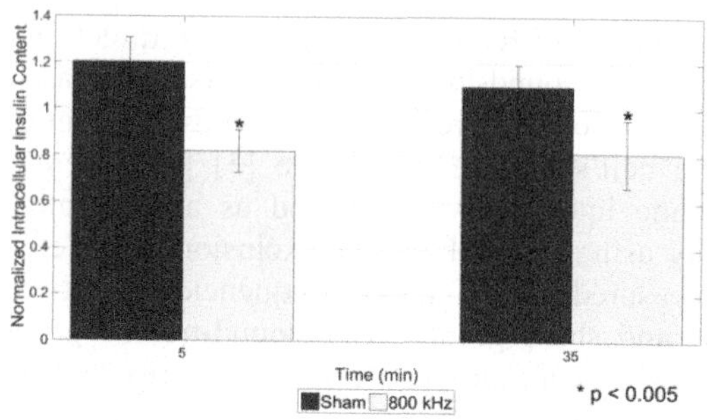

Figure 4. Intracellular insulin in pancreatic β-cells exposed to 800 kHz ultrasound (n=6) compared to sham treatment (n=6).

Amperometric recordings showed an immediate secretory response following 5-15 second long ultrasound stimulation at t = 2 min, t = 4 min and t = 6 min (black arrows in Figure 5). Moreover, the duration of the detected response matched the duration of the ultrasound stimulus. These results suggested that not only is ultrasound effective at stimulating insulin release from pancreatic β-cells, but that we could also potentially control such release by carefully selecting the appropriate ultrasound parameters to deliver a specific dose of insulin.

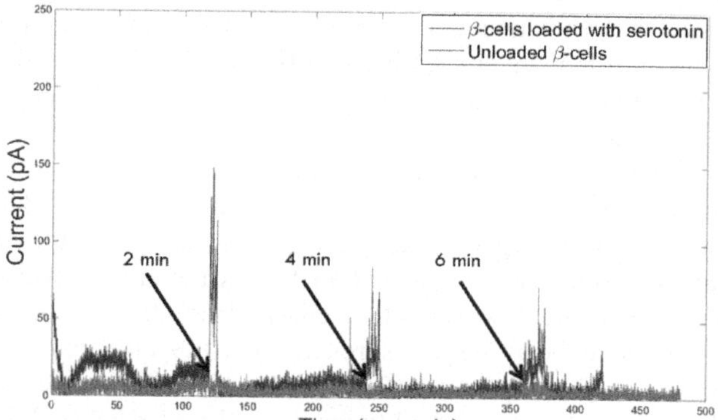

Figure 5. Amperometric detection of secretory events in ultrasound-stimulated pancreatic β-cells.

IV. DISCUSSION

The results of our experiments suggest that ultrasound exposure can stimulate insulin release from pancreatic β-cells in a safe and controlled manner. Our experiments showed that ultrasound applied at an intensity of 1 W/cm^2 and frequencies of 800 kHz and 1 MHz appears to have no significant effect on cell viability. However, 800 kHz ultrasound showed a significant (4-fold) increase in insulin release from the β-cells, whereas the cells exposed to 1 MHz ultrasound showed a lesser (50% on average) increase in insulin release, both of which could be useful in fighting hyperglycemia in type 2 diabetics. It is worth noting that no further increase in extracellular insulin was measured after the end of ultrasound exposure showing that insulin release eventually ceased after the end of treatment. Amperometric recordings showed immediate detection of released 5-HT (an indicator of insulin release) in response to ultrasound stimulation. Furthermore, the duration of the detected signal increased according to increased duration of the ultrasound stimulus, suggesting that ultrasound exposure caused reversible release of secretory products from pancreatic β-cells.

The exact mechanisms involved in ultrasound-stimulated insulin release from pancreatic β-cells are still unknown. However, it is possible that the process might be modulated by calcium dynamics [9]. Insulin is released from β-cells in response to natural secretagogue glucose in a calcium-dependent manner. Ultrasound-induced bioeffects have been shown to produce intracellular calcium transients in various cell types which triggered Ca^{2+}-dependent exocytosis of secretory vesicles [2]. These studies have attributed the creation of such transients to transient poration of the cell membrane caused by acoustic cavitation, referring to specific bioeffects resulting from oscillating gas bubbles in the extracellular medium [9] [10]. Another possible mechanism for ultrasound-enhanced Ca^{2+} influx and subsequent insulin release includes stimulation of mechanically sensitive proteins such as stretch-activated cation channels (SAC) that would allow Ca^{2+} to flow into the cell down its concentration gradient [11]. Our future studies will include characterization of the role of Ca^{2+} in ultrasound-enhanced insulin release from pancreatic β-cells, as well as application of optimized ultrasound parameters to more physiologically relevant models such as human pancreatic islets and animal models with type 2 diabetes.

V. CONCLUSION

If shown successful our proposed approach may open new strategies to combat secretory deficiencies in type 2 diabetes. Ultrasound has the advantage of being selectively targeted anywhere in the body thus potentially providing a localized therapeutic strategy for correcting secretory defects in the pancreas. Future work will include applying the parameters used in this work to more physiologically relevant models such as human islets and further experimentation to highlight the ultrasound-generated biological effects involved in ultrasound-enhanced insulin release from pancreatic β-cells. This approach may be used alone or in combination with existing pharmacological strategies, which makes ultrasound attractive as a potential novel method for treatment of type 2 diabetes.

VI. REFERENCES

[1] P. Zimmet, K. Alberti, J. Shaw, «Global and societal implications of the diabetes epidemic,» *Nature,* pp. 782-787, 2001.

[2] I. Robinson, R. Kinnick, J. Greenleaf, J. Fernandez, «Stimulation of secretion from bovine adrenal chromaffin cells by microsecond bursts of therapeutic levels of ultrasound.,» *The Journal of Physiology,* vol. 492, pp. 257-263, 1996.

[3] D. Miller, N. Smith, M. Bailey, G. Czarnota, K. Hynynen, I. Makin, «Overview of Therapeutic Ultrasound Applications and Safety Considerations,» *Journal of Ultrasound in Medicine,* vol. 31, n°14, pp. 623-634, April 2013.

[4] Y. Fujii, N. Taniguchi, M. Satoh, T. Irie, K. Itoh, «Effect of repeated US stimulation on adiponectin secretion by adipocytes of obese human subjects,» chez *AIP Proceedings 5th International Symposium on Therapeutic Ultrasound*, Boston, MA, 2006.

[5] A. Yudina, C. Moonen, «Ultrasound-induced cell permeabilisation and hyperthermia: Strategies for local delivery of compounds with intracellular mode of action,» *International Journal of Hyperthermia,* pp. 311-319, 2012.

[6] D. Christensen, Ultrasonic Bioinstrumentation, United States: John Wiley & Sons, Inc. , 1988.

[7] M. Hamid, J. McCluskey, N. McClenaghan, P. Flatt, «Comparison of the properties of four-insulin secreting cell lines,» *Endocrine Research,* pp. 35-47, 2002.

[8] J. Deeney, R. Branstrom, B. Corkeyb, O. Larsson, P. Berggren, «3H-serotonin as a marker of oscillatory insulin secretion,» *FEBS Letters,* vol. 581, p. 4080–4084, 2007.

[9] I. Suarez Castellanos, B. Balteanu, T. Singh, V. Zderic, «Modulation of calcium dynamics with ultrasound and other energy-based techniques,» *IEEE Reviews in Biomedical Engineering,* vol. 9, 2016.

[10] M. Hassan, P. Campbell, T. Kondo, «The role of Ca2+ in ultrasound-elicited bioeffects: progress, perspectives and prospects,» *Drug Discovery Today,* pp. 892-906, 2010.

[11] Y. Hu, J. Wan, A. Yu, «Cytomechanical perturbations during low-intensity ultrasound pulsing,» *Ultrasound in Medicine and Biology,* vol. 40, n°17, pp. 1587-1598, July 2014.

Lead (Pb^{++}) effect on human atrial action potential under normal and atrial fibrillation conditions. In silico study

C. Tobón[1], D. Pachajoa[2], J.P. Ugarte[3] and J. Saiz[4]

[1] MATBIOM, Universidad de Medellín, Medellín, Colombia
[2] GI^{2}B, Instituto Tecnológico Metropolitano, Medellín, Colombia
[3] Grupo de Dinámica Cardiovascular, Universidad Pontificia Bolivariana, Medellín, Colombia
[4] CI^{2}B, Universitat Politècnica de València, Valencia, Spain

Abstract— **Lead (Pb^{++}) is a toxic agent that can exert adverse effects on the cardiac human health. Pb^{++} blocks the L-type Ca^{++} channels. A decrease in L-type calcium current (I_{CaL}) is an important mechanism favoring atrial fibrillation. It is important to study the electrophysiological Pb^{++} effects on the atrial action potential in healthy people and those with AF. For this, we study the consequences of Pb^{++} on action potential, under normal and atrial fibrillation condition using in silico models. Our results suggest that Pb^{++} blocks I_{CaL} current in a fraction greater as the concentration increases, resulting in an action potential duration shortening, Pb^{++} has a greater action potential duration effect on control conditions. To our knowledge, this is the first work that has developed mathematical models of Pb^{++} effect on I_{CaL} current to study its effect on human atrial action potential.**

Keywords— **Lead (Pb^{++}), L-type Ca^{++} current, Atrial fibrillation, In silico models.**

I. Introduction

Air pollution causes 4.3 million premature deaths annually. The mortality rate increased 4 % around the world, 5 % in China and 12 % in India, between 2005 and 2010. The cost of the health impact of air pollution in Organization for Economic Co-operation and Development (OECD) countries (including deaths and illness) was about USD 1.7 trillion in 2010 [1]. The cost in 2009 of damage caused by emissions from European Pollutant Release and Transfer Register industrial facilities is estimated as being at least EUR 102–169 billion [2]. Environmental exposure to metal pollution like lead (Pb^{++}), contributes to cardiovascular disease [3]. Pb^{++} is a toxic agent that can exert adverse effects on the human health, according to United States Environmental Protection Agency (EPA) is one of the most dangerous pollutants in the air [4]. It has been amply used in the industrial sector in several countries [5]. Pb^{++} is a cumulative toxicant that affects multiple body systems and is particularly harmful to young children. In general, more than 143000 people died every year from Pb-related diseases, developing countries have the highest mortality rates [6]. Experimental and epidemiologic studies have reported Pb^{++} effects on cardiac system [7-14]. An experimental study shows that Pb^{++} blocks the L-type Ca^{++} channels [15]. A decrease in L-type calcium current (I_{CaL}) is an important mechanism favoring atrial fibrillation (AF) [16]. Studies by Van Wagoner et al [17] provide important information about the potential role of Ca^{++} channel abnormalities in clinical AF. AF is the most frequent cardiac arrhythmia, affects between 1 and 2 % of general population, with a peak prevalence of 10 % in those older than 80 years. It is estimated that by 2050 nearly 16 million US patients will have AF [18].

It is important to study the electrophysiological Pb^{++} effects on the atrial action potential (AP) in healthy people and those with AF. The main goal of our in silico study is to illustrate, using a current mathematical model of the AP in human atria, the consequences of Pb^{++} on AP, under normal and AF conditions. For this purpose, Pb-induced changes in I_{CaL} that are similar to those reported in the primary experimental data [15] were introduced into two different models of the atrial AP (control and AF).

II. Methods

A. Electrophysiological model

We used the Courtemanche-Ramirez-Nattel [19] membrane formalism. A 0.005 uM of acetylcholine (ACh) concentration was simulated. The basic equation to calculate transmembrane voltage (V_m) is:

$$C_m \frac{dV_m}{dt} + I_{ion} + I_{stim} = 0 \quad (1)$$

where C_m is the specific membrane capacitance (100 pF), I_{ion} is the total ionic current that crosses the membrane cells, I_{stim} is the stimulus current. The model under physiological normal conditions corresponds to the control model.

B. Model of chronic atrial fibrillation

AF is self-perpetuating, suggesting that it causes electrophysiological changes that contribute to the progressive

I. Torres et al. (eds.), *VII Latin American Congress on Biomedical Engineering CLAIB 2016, Bucaramanga, Santander, Colombia, October 26th -28th, 2016*,
IFMBE Proceedings 60,
DOI: 10.1007/978-981-10-4086-3_17

nature of the disease. This is mediated by shortening of AP termed electrical remodeling. To reproduce the atrial electrical remodeling generated by chronic atrial fibrillation (cAF), changes in conductance of different ionic channels of human atrial cells observed in experimental studies of cAF [20, 21] have been incorporated in the electrophysiological model. Several parameters were changed: the conductance for both the ultrarapid delayed rectifier potassium current I_{Kur} and the transient outward potassium current I_{to} were decreased by 50 %, the conductance for I_{CaL} was decreased by 70 %, while the conductance for the inward rectifier potassium current I_{K1} was increased by 100 % (see Table 1).

Table 1 Conductance (g) changes of currents for cAF

Conductance	cAF
gI_{K1}	Increased by 100 %
gI_{Kur}	Decreased by 50 %
gI_{to}	Decreased by 50 %
gI_{CaL}	Decreased by 70 %

C. Model of Pb^{++} effect on I_{CaL}

To develop a basic model of the Pb^{++} effect on the I_{CaL}, we used the steady state fraction of block (*b*). Hill equation has been used to fit the concentration-response relationships for ligand block. It describes the fraction of the macromolecule saturated by ligand as a function of the ligand concentration; it is used in determining the degree of cooperativeness of the ligand binding to the receptor. In this model the kinetics of the channel would be considered unchanged in the presence of the Pb^{++}.

$$b = \frac{1}{1+\left(\frac{IC_{50}}{D}\right)^h} \tag{2}$$

where IC_{50} is the half maximal inhibitory concentration for the current block by Pb^{++}. *D* is the Pb^{++} concentration. Hill coefficient (*h*) of 1 indicates completely independent binding. For the IC_{50} to block I_{CaL} we used 152 nM, this value was found by [15] in ventricular myocytes. There are no reported values of the IC_{50} for blocking channels by Pb^{++} on atrial myocytes.

D. Simulation protocol and data analysis

We implemented the unicellular models to simulate the sinus rhythm under physiological and cAF conditions, using the Cellular Open Resource public CellML software. Forward Euler method with a time step 0.001 ms was implemented to solve the equations. A train of 10 stimuli was applied. The basic cycle length was 1000 ms. Pb^{++} concentrations from 0 to 300 nM were implemented. The APD at 90 % of the repolarization (APD_{90}), ionic currents and the resting membrane potential (RMP) were measured. All measurements were made on the 10th beat using a program developed in MATLAB software.

III. Results

A. Model of Pb^{++} effect on I_{CaL}

Fig. 1 shows the fraction of block curve using (2) for the I_{CaL} current, in function of the Pb^{++} concentration (range 1 nM to 100 uM). The model of Pb^{++} effect on I_{CaL} fit well the experimental data [15].

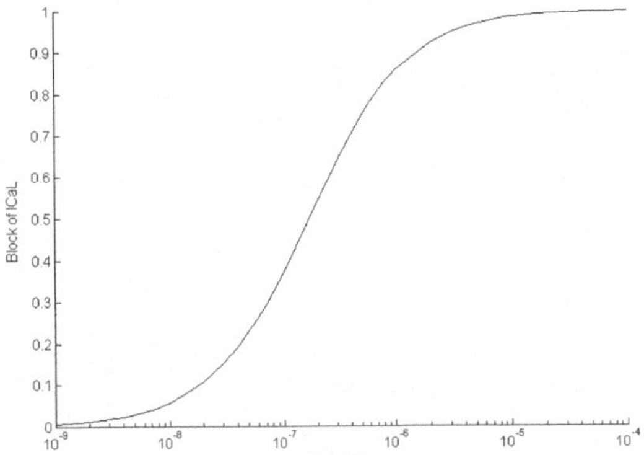

Fig. 1 Fraction of block (*b*) for I_{CaL} by Pb^{++}.

B. Pb^{++} effects on I_{CaL} and action potential

Fig. 2 and Fig. 3 show the effects of different Pb^{++} concentrations on I_{CaL} current and on atrial AP, under control and cAF conditions.

In control (Fig. 2) we can observe that I_{CaL} without Pb^{++} shows a peak of -454 pA, the current remains active during during the plateau phase of AP. As the Pb^{++} concentration increases, I_{CaL} downregulation is observed, which causes an AP shortening and loss of plateau phase (see Table 2). When the highest Pb^{++} concentration was applied (300 nM), the I_{CaL} peak decreased 58 % (-191 pA) and the APD_{90} reached a value of 115 ms. The RMP did not show significant changes (≈ -83 mV). cAF is characterized by a shortened AP without plateau phase (Fig. 3). In cAF without Pb^{++} we can observe that amplitude of I_{CaL} peak decreased 67 % (-152 pA), an APD_{90} shortening of 54 % and a slight RMP hyperpolarization of 1.8 mV, as shown in Table 2, all of these are caused by the atrial remodeling.

The effect by increasing the Pb^{++} concentration on the I_{CaL} current and AP is similar than in control case, I_{CaL} downregulation and APD shortening. When the highest Pb^{++}

concentration was applied, the APD_{90} reached a value of 77 ms.

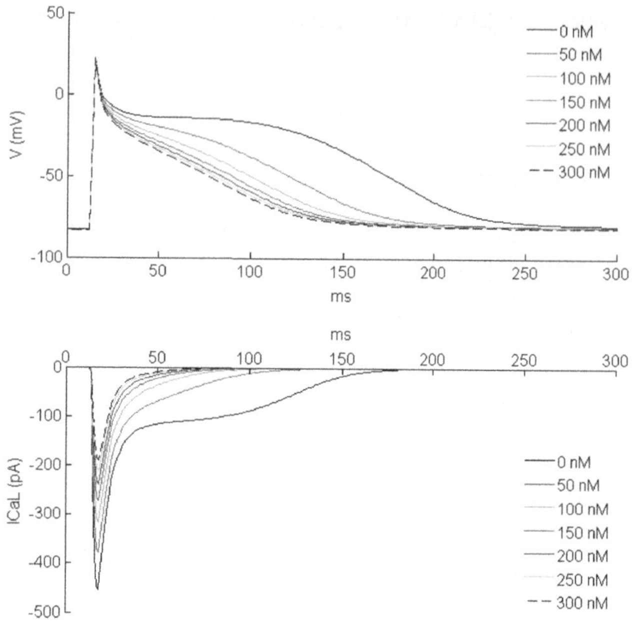

Fig. 2 Action potential and I_{CaL} current under control condition, at different Pb^{++} concentration.

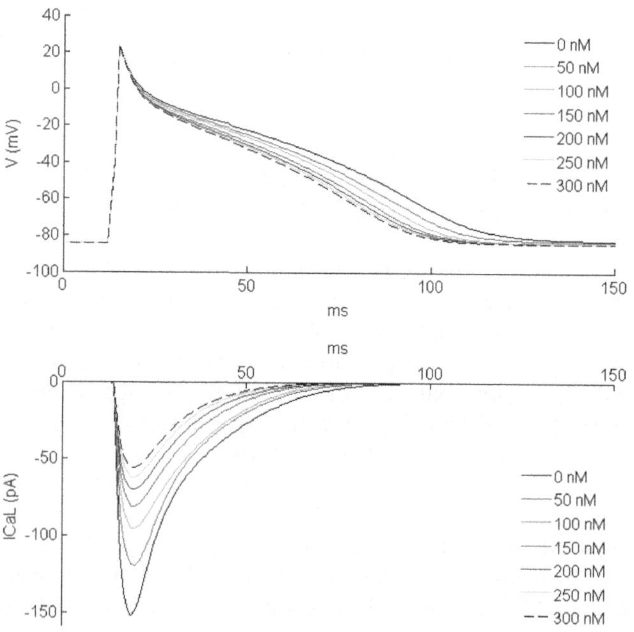

Fig. 3 Action potential and I_{CaL} current under cAF condition, at different Pb^{++} concentration.

Table 2 shows the APD90 decreasing values in function of the different Pb^{++} concentration for control and cAF. In both cases, a Pb^{++} concentration increase causes an APD shortening. In control we can observe a remarkable effect on the APD between Pb^{++} concentrations of 0 to 50 nM.

Table 2 APD_{90} during control and cAF at different Pb^{++} concentration.

$[Pb^{++}]$	Control APD_{90}	cAF APD_{90}
0 nM	211 ms	98 ms
50 nM	160 ms	91 ms
100 nM	139 ms	86 ms
150 nM	129 ms	82 ms
200 nM	122 ms	80 ms
250 nM	118 ms	78 ms
300 nM	115 ms	77 ms

Fig. 4 shows the APD_{90} percentage in function of the Pb^{++} concentration for control and cAF. As the Pb^{++} concentration increases, we observe an APD shortening for both cases, being control that has the most APD_{90} percentage decrease for all concentrations. When the highest Pb^{++} concentration was applied (300 nM) the APD_{90} shortened 45 % in control and 21 % in cAF. These results suggest that Pb^{++} has a greater APD effect on control conditions.

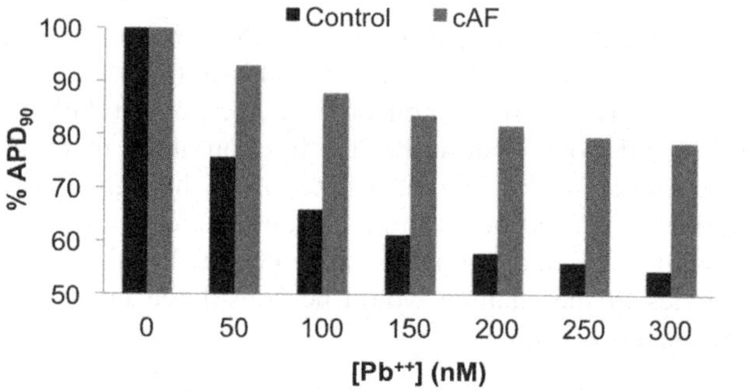

Fig. 4 APD_{90} percentage in function of the Pb^{++} concentration for control and cAF.

IV. Discussion

Experimental and epidemiologic studies have reported Pb^{++} effects on cardiac system. Acute and chronic Pb^{++} exposure can cause potentially lethal consequences [9, 10], because it acts in different places into the cardiovascular system, including endothelial dysfunction [11], direct effects on cardiac excitability and contractility [10], tissue damage by free radicals [12], cardiotoxicity and heart failure [13] and increasing blood pressure [14]. A study in 62 people exposed to Pb^{++} [7] showed a positive correlation between increasing inhibition of nitric oxide with increasing blood lead levels, which involves an association between Pb^{++} exposure and an increasing cardiovascular risk. Other study in isolated rat atria [8] showed that Pb^{++} has a bradycardia effect in the atria, which was antagonized by an increase of Ca^{++}.

An experimental study in ventricle myocytes of rats [15] showed that Pb^{++} blocks the L-type calcium channels, however, its mechanism is not known with clarity. The major mechanism of AP shortening during AF is the reduction of I_{CaL}, a decrease of 70 % in I_{CaL} has been consistently observed during AF in both human atrial myocytes and experimental models [17]. This current is the main depolarizing current that activates during the plateau phase of the AP. Our results are consistent with theses experimental studies, Pb^{++} blocks I_{CaL} current in a fraction greater as the concentration increases, prolonging its action in time, which results in an APD shortening as was demonstrated experimentally [15], Pb^{++} has a greater APD effect on control conditions. There are no reported in vitro and in silico studies of the Pb^{++} effect on human atrial action potential.

V. Conclusions

Our results suggest that Pb^{++} blocks I_{CaL} current in a fraction greater as the concentration increases, resulting in an APD shortening, Pb^{++} has a greater APD effect on control conditions. This in silico study contributes with the research on the electrophysiological Pb^{++} effects on the atrial AP in healthy people and those with AF. To our knowledge, this is the first work that has developed mathematical models of Pb^{++} effect on I_{CaL} current to study its effect on human atrial action potential.

Acknowledgment

This work was partially supported by Departamento administrativo de ciencia, tecnología e innovación de la República de Colombia – COLCIENCIAS.

Conflict of Interest

The authors declare that they have no conflict of interest.

References

1. OECD. (2014) The Cost of Air Pollution: Health Impacts of Road Transport, OECD Publishing, Paris. DOI: http://dx.doi.org/10.1787/9789264210448-en.
2. European Environment Agency. (2011) Revealing the costs of air pollution from industrial facilities in Europe. EEA Technical report, No 15. EEA, Copenhagen.
3. Bhatnagar A. (2004) Cardiovascular pathophysiology of environmental pollutants. Am J Physiol Heart Circ Physiol 286(2): H479–H485.
4. U.S. EPA. (2012) Integrated Science Assessment for Lead (Third External Review Draft). U.S. Environmental Protection Agency, Washington, DC, EPA/600/R-10/075C.
5. Rodrigue JP. (2013) Pollutants Emitted by Transport Systems (Air, Water and Noise), 3rd ed. Routledge, New York.
6. OMS. (2015) Intoxicación por plomo y salud, Nota descriptiva No. 379.
7. Barbosa F, Sertorio J, Gerlach R, et al. (2006) Clinical evidence for lead-induced inhibition of nitric oxide formation. Arch. Toxicol 80(12):811–816.
8. Acosta G, Rubio M. (1990) Efecto del nitrato de plomo en aurículas aisladas de rata. Acta Physiol Pharmacol Latinoam 40(2):137–148.
9. Kopp S, Baker J, D'Agrosa L, et al. (1978) Simultaneous recording of His bundle electrogram, electrocardiogram, and systolic tension from intact modified Langendorff rat heart preparations, I: Effects of perfusion time, cadmium, and lead. Env Heal Perspect 46:475–487.
10. Prentice R, Kopp S. (1985) Cardiotoxicity of lead at various perfusate calcium concentrations: Functional and metabolic responses of the perfused rat heart. Toxicol Appl Pharmacol 81:491–501.
11. Vaziri N, Liang K, Ding Y. (1999) Increased nitric oxide inactivation by reactive oxygen species in lead-induced hypertension. Kidney Int 56:1492–1498.
12. Patrick L. (2006) Lead toxicity part II: The role of free radical damage and the use of antioxidants in the pathology and treatment of lead toxicity. Altern Med Rev 11:114–127.
13. Ansari M, Maayah Z, Bakheet S, et al. (2013) The role of aryl hydrocarbon receptor signaling pathway in cardiotoxicity of acute lead intoxication in vivo and in vitro rat model. Toxicology 306:40–49.
14. Tsao D, Yu H, Cheng J, et al. (2000) The change of β-adrenergic system in lead-induced hypertension. Toxicol Appl Pharmacol 163:127–133.
15. Bernal J, Lee J-H, Cribbs LL, et al. (1997) Full Reversal of Pb^{++} Block of L-Type Ca^{++} Channels Requires Treatment with Heavy Metal Antidotes. J Pharm Exp Ther 282(1): 172–180.
16. Dinanian S, Boixel C, Juin C, et al. (2008) Downregulation of the calcium current in human right atrial myocytes from patients in sinus ehythm but with a high risk of atrial fibrillation. Eur Heart J 29: 1190-1197.
17. Van Wagoner DR, Pond AL, Lamorgese M, et al. (1999) Atrial L-type Ca^{2+} currents and human atrial fibrillation. Circ Res 85: 428–436.
18. Goralnick E, Bontempo LJ. (2015) Atrial Fibrillation. Emerg Med Clin North Am 33(3): 597–612.
19. Courtemanche M, Ramirez RJ, Nattel S. (1998) Ionic mechanisms underlying human atrial action potential properties: insights from a mathematical model. Am J Physiol 275(1): H301– H321.
20. Courtemanche M, Ramirez RJ, Nattel S. (1999) Ionic targets for drug therapy and atrial fibrillation-induced electrical remodeling: insights from a mathematical model. Cardiovasc Res 42(2): 477–489.
21. Van Wagoner DR. (2003) Electrophysiological remodeling in human atrial fibrillation. Pacing Clin Electrophysiol 26(7): 1572–1575.

Address of the corresponding author:
Author: Catalina Tobón
Institute: Universidad de Medellín
Street: Carrera 87 N° 30 - 65
City: Medellín
Country: Colombia
Email: ctobon@udem.edu.co

Acoustic spectrometer: resonant sensing platform for measuring volumetric properties of liquid samples

S. Villa-Arango[1, 2], R. Torres[1], P.A. Kyriacou[2] and R. Lucklum[3]

[1] Biomedical Engineering Research Group (GIBEC), EIA University, Envigado, Colombia
[2] Research Centre for Biomedical Engineering (RCBE), City University of London, London, UK
[3] Institute of Micro and Sensor Systems (IMOS), Otto-von-Guericke University, Magdeburg, Germany

Abstract— **A sensing platform for measuring volumetric properties of liquid samples using phononic crystals is presented in this paper. The proposed sensor concept is based on the transmission of elastic and acoustic waves through solids and liquids respectively to gather relevant information about the properties of the liquid under test. A major difference between this concept and the majority of current resonant sensors, like the well-known quartz crystal microbalance, is that the acoustic spectrometer proposed measures bulk properties and not interfacial properties of the liquid. The sensing platform uses a disposable analyte container to facilitate the measurement of hazardous substances and enable its use in biomedical applications. An electronic characterization system based on the acquisition of three mixed signals was developed to obtain the frequency response of the designed sensor. Finally, experimental and theoretical realizations were performed, using different analytes and showing characteristic transmission features that can be used as measures to determine the physical value speed of sound.**

Keywords— **Acoustic sensor, liquid sample, biomedical applications, phononic crystals, electronic characterization system.**

I. Introduction

During recent years, there has been a growing interest in sensing platforms based on resonant systems. The perhaps most known is the quartz crystal microbalance, QCM, with a broad application range reaching from film thickness monitor to advanced biosensors which work in liquid environments [1, 2].

Phononic crystals, PnCs, are a new platform that has been proposed to be used in several applications due to their characteristic capacity of generating frequency bands in which elastic waves cannot propagate, giving scientists a way of designing a selective transmission spectrum [3-5]. They have only recently been introduced as sensors and a few research groups have already started performing experimental realizations. [6, 7].

QCM is a resonant sensor with extraordinary (mass) sensitivity. In order to acquire high biochemical sensitivity, it is necessary to merge the QCM with a biological compound deposited onto the QCM surface that enhances a selective response of the system to the analyte of interest. QCM therefore only measures changes at the interface between the sensor and the analyte, making it impossible to acquire bulk properties of the liquid under test. Unlike the QCM, PnC sensors merge the properties of resonant sensors and ultrasonic sensors by measuring frequency changes of relevant transmission features of ultrasonic waves transmitted through the PnC. The liquid under investigation is confined in a cavity having a well-defined resonance which depends on volumetric properties of a liquid analyte. Similar to QCM, the PnC sensor determines frequency changes of the cavity resonance which appear as a shift in the maximum of transmission and a phase change. PnC sensors are still in a preliminary stage of development, however, a series of structures have been proposed that could be implemented in real sensor applications [6 - 9]. Besides fundamental requirements, real applications may also introduce additional constraints, specifically in biosensing [10, 12].

Commercial electronic characterization systems that are commonly used to measure the characteristic frequency features of resonant structures are vector network analyzers and high-frequency lock-in amplifiers. These conventional systems are very robust and expensive, making it very challenging to conduct tests in the field, thus, forcing the users to send the samples to specialized laboratories and, therefore, limiting the applicability of these sensing systems in various areas. However, a novel characterization system for measuring frequency changes of resonant structures like PnC has recently been introduced. This system is based on a double sideband modulation with suppressed carrier and a special demodulation process that involves a series of operations to obtain a signal that depends both on changes in the amplitude and phase induced by the resonant system under test. It enables, therefore, the use of phononic crystals in the field due to its portability [12].

The block diagram of the electronic system can be observed in Fig. 1.

I. Torres et al. (eds.), *VII Latin American Congress on Biomedical Engineering CLAIB 2016, Bucaramanga, Santander, Colombia, October 26th -28th, 2016*,
IFMBE Proceedings 60,
DOI: 10.1007/978-981-10-4086-3_18

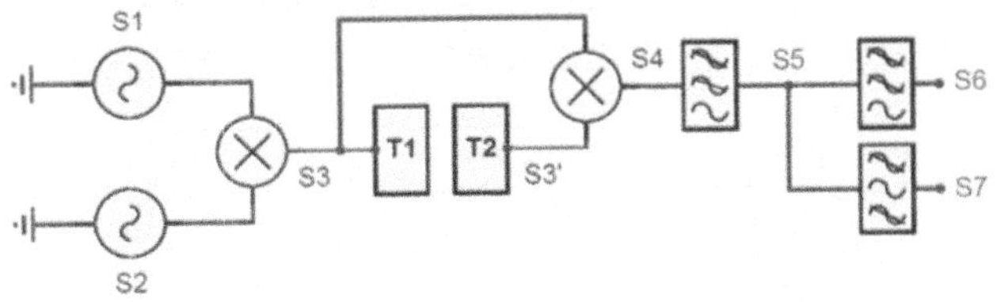

Fig. 1 Block diagram of the electronic characterization with a double sideband modulation. Source: [12].

A resonant sensing platform that measures volumetric properties of liquid analytes in a disposable container using an electronic characterization system based on the double sideband modulation system is presented on this paper. The approach used relies on the measurement of both gain and frequency of relevant transmission features present in frequencies in the order of MHz.

II. Materials and Methods

Biomedical applications often require the use of biological or hazardous samples and, therefore, it is important to take into account that all the elements of the system that are in contact with the analyte must be disinfected or discarded after the test is completed. The acoustic spectrometer designed has three main components: An electronic characterization system, a pair of piezoelectric ultrasonic transducers and a disposable analyte container made of glass that can be discarded or sterilized after each test. Fig. 2 shows a graphic representation of the container and the analyte, which can be understood as the very simple 1-D PnC. The ultrasonic waves travel from one transducer to the other generating a cavity resonance inside the container.

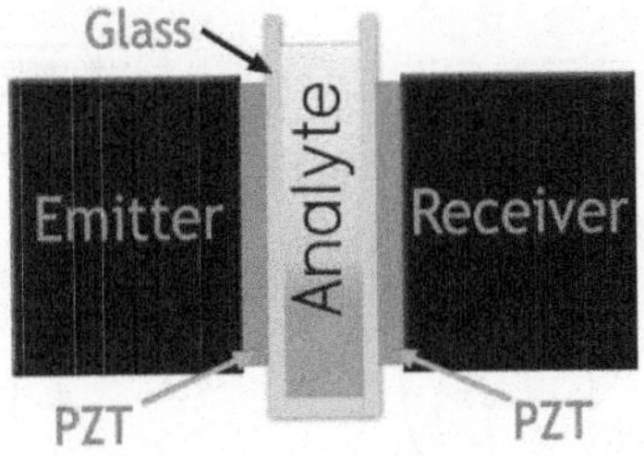

Fig. 2 Graphic representation of the acoustic spectrometer structure.

The electronic characterization system is based on obtaining three main signals, these signals are acquired by using the excitation signal that is fed to the transducer configured as the transmitter, S0, and the signal acquired by the transducer configured as the receiver, S1, same as in the system presented in Fig. 1, but without the double sideband modulation.

The three signals that the system obtains are: the square of the RMS value of S0, the square of the RMS value of S1, and the DC component of the signal obtained by multiplying S0 by S1. The signals are then passed through a series of mathematical operations to acquire the gain and phase of the system, thus generating a frequency spectrum with valuable information on the volumetric properties of the substance being analyzed. The block diagram of the proposed system is shown in Fig. 3.

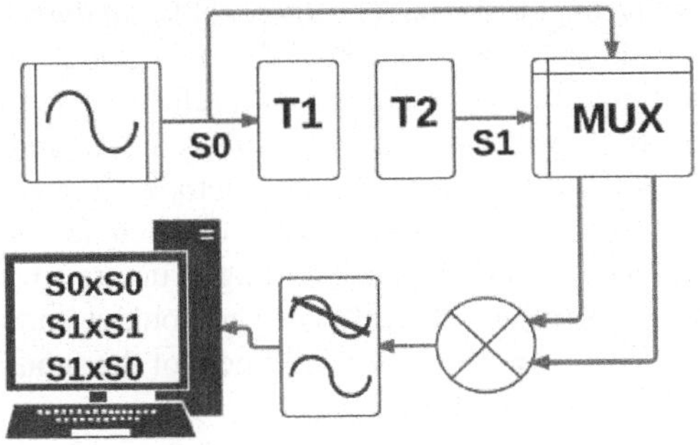

Fig. 3 Block diagram of the electronic characterization system designed.

As seen in the block diagram, Fig. 3, an analog multiplexer feeds a four quadrant multiplier with the two signals, S0 and S1, to perform three multiplications and to obtain 3 mixed signals, S2, S3 and S4. The signal acquired by the receiver, S1, experiences modifications when traveling through the resonant structure and the ultrasonic transducers. These variations are presented as changes in gain, G, and changes in phase φ2.

$$S_0(t) = A\sin(2\pi ft + \varphi)\,; S_1(t) = AG\sin(2\pi ft + \varphi + \varphi_2) \quad (1)$$
$$S_2(t) = A^2 sin^2(2\pi ft + \varphi) \quad (2)$$
$$S_3(t) = A^2 G^2 sin^2(2\pi ft + \varphi + \varphi_2) \quad (3)$$
$$S_4(t) = (A^2G^2/2)[\cos(\varphi_2 - \varphi) - \cos(4\pi ft + \varphi_2 + \varphi)] \quad (4)$$

The three mixed signals obtained by performing the multiplication between S0 and S1 are then passed through a low-pass frequency filter to obtain the DC component of each signal. The resulting signals are then passed through a digital to analog converter to be processed by a microcontroller.

$$S'_2(t) = A^2/2 \quad (5)$$
$$S'_3(t) = A^2G^2/2 \quad (6)$$
$$S'_4(t) = (A^2G/2)\cos(\varphi_2 - \varphi) \quad (7)$$

The three signals obtained contain relevant information to extract the value of the gain, G, and phase, φ, of the system. The gain is obtained by calculating the square root of the resulting value of dividing S'3 by S'2, while the phase is obtained by calculating the inverse cosine of the resulting value of dividing S'4 by G and S'2.

$$G(t) = \sqrt{(A^2G^2/2)/(A^2/2)} \quad (8)$$

$$\varphi(t) = \cos^{-1}\left(\frac{A^2G}{2}\cos(\varphi_2 - \varphi)/G\frac{A^2}{2}\right) \quad (9)$$

For the experimental realizations, wide bandwidth piezoelectric ultrasonic transducers with a central frequency of 1.5MHz and a half-peak band width of 1 MHz were used.

The electronic system was set to acquire a total of 400 points starting at 0.85MHz, with frequency steps of 3 kHz and ending at 2.05MHz. The disposable analyte container used was a 700uL glass cuvette and only 500uL of analyte were used per test. The cuvette was carefully rinsed and dried using acetone before introducing a new sample and the temperature was kept constant via room temperature control.

The ultrasonic transducers were coupled to the analyte container using glycerol and a holding structure specially designed to ensure an adequate surface contact and constant pressure. The experimental arrangement of the acoustic spectrometer can be observed in Fig. 4.

Fig. 4 Acoustic spectrometer experimental arrangement.

To evaluate the performance of the acoustic spectrometer, tests using different alcohol analytes were conducted. The properties of the analytes used can be found in Table 1. There are many methods to simulate phononic crystal structures, some of them are the Layer Multiple Scattering Theory (LMST), the Finite difference Time Domain (FDTD), the Finite Element Analysis (FEA) and the 1D Transmission Line Model (TLM). Simulations using the 1D Transmission Line Model to corroborate the experimental results were performed. This method was selected because it is widely used to simulate the performance of multilayered structures giving accurate results and it uses a reduction of the model to 1D that enhances the calculation speed and lowers the computation power required, that are commonly high when other methodologies are used. The TLM uses a chain matrix technique and an analogy between the electrical impedance and the acoustic impedance to perform the calculation of the transmission and reflection coefficients. The geometric and material properties of each layer were used to calculate the elements of the propagation matrix [10 - 14].

Table 1 Analytes used for the preliminary tests

Analyte	Density kg/m³	Speed of sound m/s
Distilled Water	998	1493
n-Propanol	786	1170
Methanol	792	1100

Source: Adapted from [15]

III. Results and discussion

The results obtained from the experimental realizations conducted are shown in Fig. 5. No further digital signal processing was made to enhance the quality of the signals given the signal to noise ratio exhibited by the electronic system.

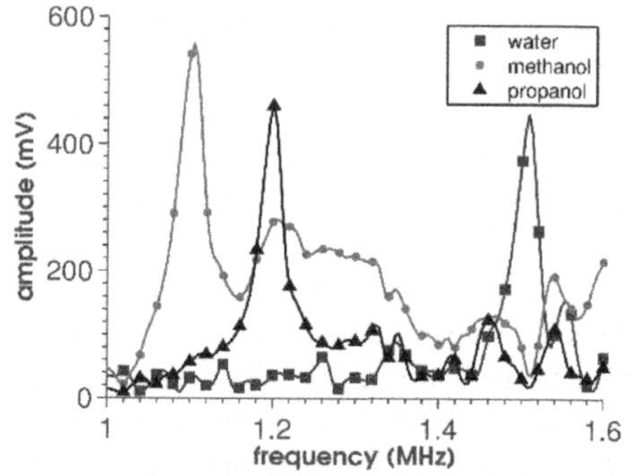

Fig. 5 Experimental results using different alcohols as analytes.

Each test showed a well differentiated maximum of transmission generated by the resonance in the cavity. This characteristic transmission feature has a good quality factor and was located at different frequencies on each experimental realization. The results from the simulations using the 1D transmission line model are shown in Fig. 6.

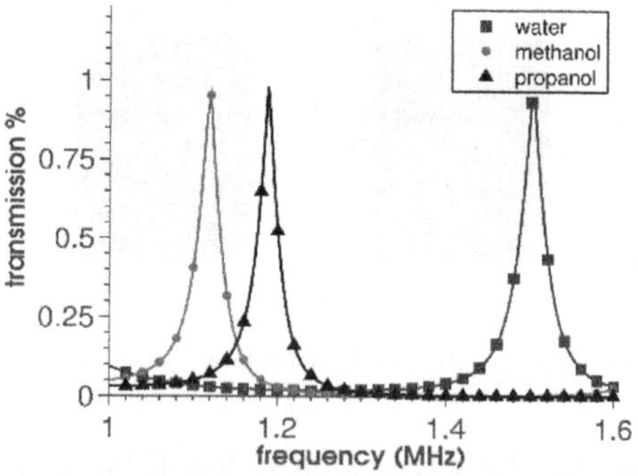

Fig. 6 Theoretical results using different alcohols as analytes.

The transmission curves obtained in both theoretical and experimental realizations show significant changes in the frequency and magnitude of the main resonant peaks. Variations in the acoustic properties of the liquid mixtures showed to be responsible of the frequency changes on the characteristic transmission features. The maxima of transmission of the experimental realizations agree well with the ones obtained in the simulations, the lateral miniaturization of the model in the TLM could explain the small differences present in the computational results.

The sensitivity obtained in these tests is very promising, variations of 1 m/s^{-1} on the longitudinal component of the speed of sound of the analyte produce variations of 1098 Hz in the frequency of the characteristic transmission peaks. This result is very important given the fact that the signal to noise ratio of the system is high and enables accurate measurements of the frequency of the transmission features without noise interference.

IV. Conclusions

An acoustic spectrometer with a disposable cavity that measures volumetric properties of liquid analytes was studied in this manuscript. The acoustic spectrometer relies on the determination of the frequency of relevant transmission features, in this case, maximums of transmission.

The 1D transmission line model showed to be useful to corroborate and even predict the results obtained with the system, being a reliable tool for designing new sensors.

The removable component of the sensing platform enables its use in applications where hazardous substances need to be analyzed, like point of care testing and other biomedical applications.

The electronic characterization system designed is portable and economic and makes it possible to use the acoustic spectrometer in the field. The designed platform could differentiate two different alcohols and distilled water with sufficient sensitivity and resolution, however, in order to use the designed system in label-free applications with extraordinary sensitivity and lower analyte availability, further optimization processes need to be done, like using much higher frequencies to enhance sensitivity and using a smaller analyte container to lower analyte consumption.

On-going investigations deal with making the acoustic spectrometer an absolute sensor, which can effectively differentiate target analytes without the need of a reference transmission spectrum by means of including a biochemical compound to the analyte container that enhances the sensitivity and selectivity of the system, like it is done with QCM biosensors. [2]

Conflict of Interest

The authors declare that they have no conflict of interest.

References

1. Arnau A (2008) Piezoelectric Transducers and Applications (2-ed.). Berlin: Springer, ISBN: 978-3-540-77507-2
2. Montagut Y, Narbon J G, Jimenez Y, March C, Montoya A, Arnau A, Serra P A (2011) QCM Technology in Biosensors. Biosensors - Emerging Materials and Applications, pp. 153-178. DOI: 10.5772/17991
3. Pennec Y, Vasseur J O, Djafari-Rouhani B, Dobrzyński L, Deymier P A (2010) Two-dimensional phononic crystals: Examples and applications. Surface Science Reports 65, pp. 229–291, DOI:10.1016/j.surfrep.2010.08.002
4. Sigalas M M, Economou E N (1992) Elastic and acoustic wave band structure, J. Sound Vibr. 158, pp. 377-382, DOI: 10.1016/0022-460X(92)90059-7
5. Kushwaha M S, Halevi P, Martinez G, Dobrzynski L, Djafari-Rouhani B (1993) Acoustic Band Structure of Periodic Elastic Composites, Phys. Rev. Lett. 71, 2022
6. Lucklum R, Li J (2009) Phononic crystals for liquid sensor applications, Meas. Sci. Technol. 20, 124014 (12pp). DOI:10.1088/0957-0233/20/12/124014
7. Jyun-Hong L, Dong-Po C, Cheng-Yi H, Fu-Li H, Chii-Chang C (2015) Liquid sensor composed of one-dimensional sonic Helmholtz resonator array. Appl. Phys. A. 120, pp. 509–517, DOI: 10.1007/s00339-015-9203-3
8. Lucklum R, Ke M, Zubtsov M (2012) Two-dimensional phononic crystal sensor based on a cavity mode. Sensors and Actuators B. 171– 172, pp. 271– 277, DOI: 10.1016/j.snb.2012.03.063
9. Ke M, Zubtsov M, and Lucklum R (2011) Sub-wavelength phononic crystal liquid sensor. J. Appl. Phys. 110, 026101. DOI: 10.1063/1.3610391
10. Lucklum R, Villa S, Zubtsov M (2014) Cavity resonance biomedical sensor, IMECE2014-38222: ASME 2014 International Mechanical Engineering Congress and Exposition, 2014, Montreal, Quebec, Canada. ISBN: 978-0-7918-4962-0
11. Lucklum R, Villa S, Zubtsov M, Grundmann R (2014) Phononic cystal sensor for medical applications. IEEE SENSORS 2014, Valencia, Spain, pp. 903 - 906. DOI: 10.1109
12. Villa S, Torres R, Lucklum R (2015) Electronic characterisation system for measuring frequency changes in phononic crystals. Electronics Letters, Volume 51, Issue 7, pp. 545-546. DOI: 10.1049/el.2014.4151.
13. Qiu C, Liu Z, Mei J, Ke M (2005) The layer multiple-scattering method for calculating transmission coefficients of 2D phononic crystals. Solid State Communications, 134, pp. 765-770.
14. Tanaka Y, Tomoyasu Y, Tamura I (2000) Band structures of acoustic waves in phononic lattices: Two-dimensional composites with large acoustic mismatch. Phys. Rev. B 62, pp. 7387-7392.
15. Kuhnkies R, Schaaffs W (1963) Untersuchungen an adiabatisch und isotherm aufgenommenen Schallkennlinien binärer Mischungen, Acustica 13, 407.

Author: Simón Villa Arango
Institute: EIA University
Street: Calle 25 Sur 42-73
City: Envigado
Country: Colombia
Email: simon.villa@eia.edu.co

Rehabilitation equipment for rotator cuff injuries shoulder

Hedrick ROBLES[1], Andrea SALAMANCA[1], Valentín MOLINA[1,2], Horderlin ROBLES[2,3],

[1]Ingeniería Biomédica, Escuela Colombiana de Carreras Industriales – ECCI.
[2] GINIC-HUS, Hospital Universitario de la Samaritana.
Bogotá, Cundinamarca, Colombia.
[3]DEMA, Universidad del Sinú – Elías Bechara Zainúm.
Montería, Córdoba, Colombia.

Abstract — **in this paper, the design of a biomedical device for rehabilitation of rotator cuff tendinitis and methodology for development is exposed. The first part is introduced into the rotator cuff anatomy and biomechanics, and then some epidemiological profiles found in the literature are presented. Subsequently the constituent parts of the proposed prototype and operation of electromechanical elements are described. Finally, the results obtained in designing the prototype, and the conclusions and future work are outlined.**

Key Works: **rotator cuff, rehabilitation, tendonitis, biomedical equipment.**

1. INTRODUCTION

The Rotator Cuff (MR) is an anatomical structure located on the shoulder and is make of four muscles and their tendons: supraspinatus, infraspinatus, teres minor and subscapularis [1]. His muscle complex enables movement of lifting, lowering and rotation arm also is prone to disease or inflammation due to sustained postures shoulder repetitive motion, force, vibration exposure and psychosocial factors such as stress, consumption cigarette and caffeine, life styles inadequate, leading to the involvement generating Rotator cuff tendinitis (TMR) [1].

In Colombia, the Musculoskeletal Disorders (DME) are the leading cause of occupational morbidity in the contributive regime of General System of Social Security in Health (SGSSS), representing 82% of all diagnoses made in 2004, in this regime there were 14,857,250 people affiliated to 11,885,800 that year and had some kind of DME [2]. The Ministry of Health and Social Protection between 2003 and 2005 showed that shoulder injuries had a 4% prevalence [2].

Treatment for TMR includes ergonomic modifications, education and information, pharmacological management orally, immobilization of the segment, physiotherapy management, infiltrations of corticosteroids [2], but in the rehabilitation stage, ie, after the acute stage (pain, inflammation, etc.), has not been incorporated a method using an automated biomedical device with manual control and easy guide to the professional in charge of the therapy, which allows the patient to perform all shoulder movements (lifting, lowering, bending external and internal) rotation without making any effort, and that, in turn, will provide regain normal biomechanical function of the limb.

2. BIOMEDICAL EQUIPMENT PROPOSED.

2.1. Biomechanics rotator cuff.

Biomechanically, the toracoescapulohumeral joint is the most complex joint in the body, with the broadest range of motion of all joints, but on the other hand, has little stability [3]. To improve the stability, are bony structures, the joint capsule and glenohumeral ligaments, labrum, rotator cuff muscles, long biceps tendon, and the two bags synovial [4]. It is consider that there are two types of stabilizers on the shoulder. On the one hand, we have the static stabilizers, which are the bony elements. This stability is consider very low because the humeral head is rounded and almost flat glenoid, with a much smaller area, therefore the true stabilizers are considered the dynamic elements. For many real dynamic stabilizer is the rotator cuff [4]. The function of the rotator cuff is center the humeral head in the glenoid (mainly the supraspinatus) and participate in abduction and external rotation movements. If the rotator cuff function is impaired, will be a displacement of the humeral head and negatively affect the intervening tissues, causing damage to the bursa and collagen fibers cuff tendons, leading to inflammatory changes and edema [3][5]. The shoulder is a joint that has three degrees of freedom and is composed functionally by-acromioclavicular [glenohumeral joint and [6].

2.2. Epidemiology rotator cuff.

The painful shoulder has a prevalence between 3 and 7% with an annual incidence of 6.6 to 25 cases per 1000 patients, reaching the highest odds of the fourth to sixth decade of life [5]. Among these causes of shoulder pain is rotator cuff pathology, especially the supraspinatus tendinitis, which represents 65% of the causes of shoulder pain [7]. This condition is directly associated with age and often appeared between the fourth and sixth decades of life. It has also been found that this condition tends to be more prevalent in females, with more frequent involvement of the dominant side of patients [8][9]. The increased prevalence of this disease, related to sports practices and poor ergonomic conditions in the workplace, has drawn attention to the significant costs to health systems and services. Situation is more complex for physiotherapy services, considering that these not only target the reduction of symptoms and recovery of joint mobility conditions, but to promote the scapular and shoulder muscle biomechanical balance, avoiding compensation posture, which is a higher investment involving both time and money in [8][9].

3. BACKGROUND.

The state of the art in relation to medical equipment to perform rehabilitation affectations shoulder and MR is very focus on

I. Torres et al. (eds.), *VII Latin American Congress on Biomedical Engineering CLAIB 2016, Bucaramanga, Santander, Colombia, October 26th -28th, 2016*,
IFMBE Proceedings 60,
DOI: 10.1007/978-981-10-4086-3_19

devices with large and complex physical structures. Mustafa et al, presented a robotic arm with an anthropocentric bio-inspired design with seven degrees of freedom for rehabilitation of the shoulder joint. Its advantages lie in high-skill and low weight [13].

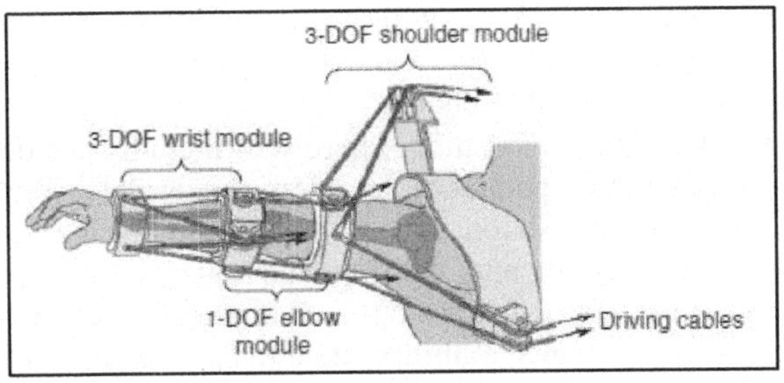

Figure 1. Conceptual design of an anthropocentric biologically inspired rehabilitator arm.

Moreira Nunes et al. (2011) designed a system based on manipulators cables shoulder joints and elbow equipment, using these mechanisms for good kinematic and dynamic characteristics in addition to its low acquisition cost and ease of transport (see Figure 2) [14].

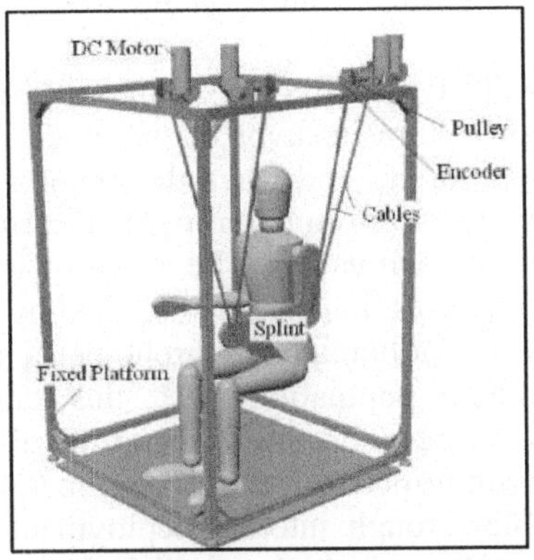

Figure 2. Manipulator based on parallel cables for rehabilitation of shoulder and elbow movements.

Galiana et al. (2012) made a similar to Mustafa robotic arm, this could be used at home for patient comfort by caregivers [15] In addition, Ball et al., In 2007, they had already submitted MEDARM, a robotic exoskeleton also for shoulder rehabilitation post-rupture [16]. Conventional therapy consists of physiotherapists passive movements, using 'ladder fingers' and the use of weights to strengthen and recovery of all range of motion [1,2].

4. MATERIALS AND METHODS.

This conceptual prototype was make using the software computer-aided design (CAD) Solidworks® Edition 2015 mechanical tool used for 3D modeling. Programming a microcontroller is performed via software MPLAB ™ Microchip®, program used to establish lines of programming code for this type of device, this was done in order to allow greater engine performance employees, lower consumption greater efficiency in electricity and automation.

4.1. Description of the proposed prototype. The equipment consists the following parts:

- Base Areas: These are in contact with the floor, the material is iron and stainless steel.
- Engine Cover: Iron is stainless steel and coated the engine from above and laterally.
- Horizontal inner tube: the tube is connected to the engine 1 to the rotational movements, externally has another tube for adjusting the length thereof.
- External standpipe: coats the inner tube for adjusting the lengths.
- Attach to rim and third tube: here the rim and standpipe for movements in that plane are assemble, this.
- Bearing: allows the mobilizations in all axes of the machine, this wheel has freedom of movement for all set-ups.
- Worm: This screw allows vertical movements, is connected with the engine number two on top is the basis for hand stand.
- Standpipe: covers the screw on all sides, has freedom of movement and which is connect to the motor base two.
- Rotational Base: it is connect from below with the worm, enables the coupling with this and so transmit motion.
- Pillow: this sponge plastic material will stand the patient's hand.

The proposed complete prototype is illustrated in Figure 3 team applied the principle of operation dictated by J. David Reinkensmeyer which states that recent research in animals and humans show that the optimal physical recovery is possible for upper and lower limbs by practicing repetitive movements [10] Pad: this sponge plastic material will stand the patient's hand (See Figure 3).

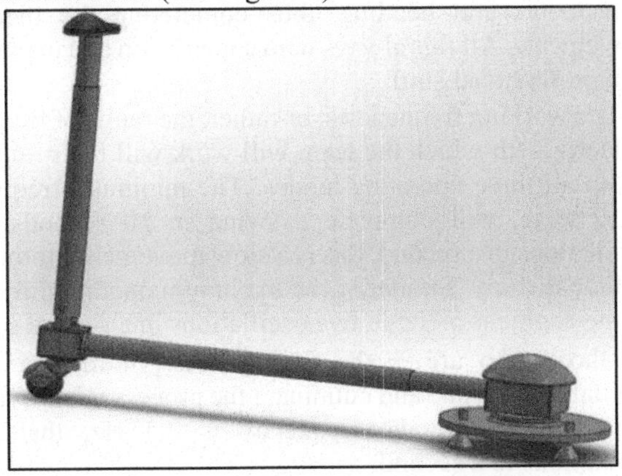

Figure 3. Proposed prototype.

The device will perform passive movements (movements where the patient does not move the joint at will [11]) of the affected arm in the three degrees of freedom (degree of freedom - DOF) and the five movements made by the shoulder joint where the muscles involved rotator cuff, allowing recovery - according to the intensity and duration of therapy will be discussed later - and recovering the arches of normal anatomical motion of the shoulder complex.

Two Step by step motors are used to regulate the speed of therapy, these, in their view, will be controlled by a microcontroller from Microchip. It was selected because stepper motors:

- They work better with analog signals.
- Manage good repeatability - the team will perform many repetitive movements in these directions simultaneously.
- Precise Positioning: 3-5%.
- Do not make periodic maintenance since they have brushes.
- Turn in different ways according to the input frequencies.
- They have simple control circuits.

The team has two motors located in such a way as to allow the above movements and studied in the preceding paragraphs:

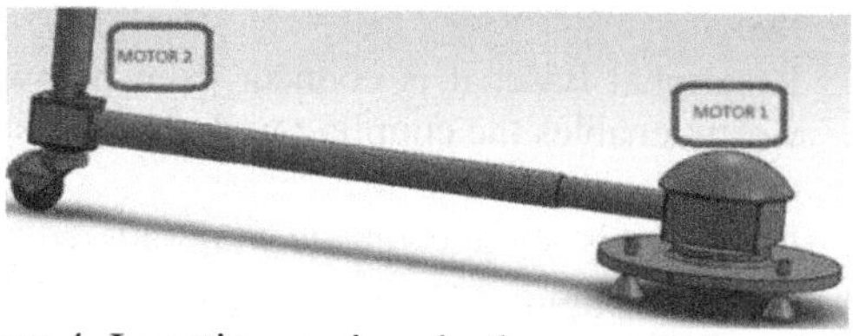

Figure 4. Location engines in the proposed prototype.

The motor 1 will allow the team to perform internal and external rotation, motion transmission is made by connecting the motor shaft with the metal housing. The motor 2 will allow the team and the patient to perform the movements in the vertical plane such as abduction and adduction. The simultaneous activation of both engines allow Circumduction movements and bending, thus completing the therapeutic movements. All this always with supervision and management of a professional staff.

The working frequencies, or rather, the range of frequencies in Hertz with which the team will work will be from 0.1 to 5 Hz in all three operating modes. The minimum frequency is 0.1 Hz, ie, will complete a swing in 10 seconds, or 0.1 oscillations in a second, this is a slow movement for those who initiate therapy. Similarly, the maximum operating frequency of the engine is 5 Hz, ie five oscillations in a second set speed for those who are in the final process of recovery; these oscillations are fast and culminate the process rehabilitation of patients. The average frequency is 2.5 Hz, that is, 2.5 oscillations in a second.

5. Results.

The designed device has three operating modes (MDF):

- Vertical mode: perform torque abduction-adduction movements in the same plane with the speed controlled.
- Rotational Mode: will the movements of internal and external rotation with controlled rate set by the professional in charge.
- Circumduction mode: here will include, besides the abduction-adduction, internal-external rotation, shoulder flexion, thus completing the three degrees of freedom and the five movements that activate muscles Rotator cuff successfully completing requirements mobility should have a team of this nature.

Mentioned throughout the project must have the ability to emulate or imitate equipment movements made by the shoulder complex, here, a more detailed description.

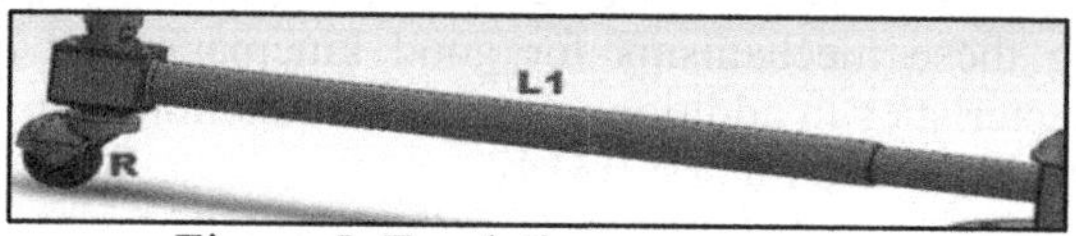

Figure 5. Emulation movement.

The L1 tube will have the freedom to move the rotational arm, when the motor 1 turn in favor of the clock hand, this will move from right to left, and when the motor 1 rotate against clockwise, this will go from left to right. The rim R allow this movement smooth and easily. The range of lengths L1 was determined with bases in a study conducted by researchers at the University of Antioquia on anthropometric measurements of the Colombian population [12], this range has been established as average men and women from different regions of Colombia with respect to the arm length: 0.45 m - 1.02 m. This is possible through internal deployment of two pipes located horizontally and which are adjustable according to the length of the arm of the patient being treated.

Figure 6. Emulation movement 2.

L2 receive the tube movement from the motor 2 and moves vertically. When the motor rotates in favor of clock hands, this will move from bottom to top and move up and down when the motor rotates in anti hourly basis. Similarly, considering the same study anthropometric measures, has determined the range of shoulder level [39], ie, the measure from floor to shoulder, this range is 0.45 m to 1.90 m.

Reaching this point is possible by deploying the auger when the motor rotates in favor of clock hands. Referencing the patient's hand (center of mass), emulation by Solidworks® showed that this moves along a path so nearly sinusoidal, thus activating the four muscles, thus generating a recovery of mobility in the 3 degrees of freedom.

6. CONCLUSIONS·

Repetitive movements in patients with acute illness physical damage which has been relegated generate positive effects, with good therapy, can achieve enable lost or attenuated joints and movements again. Physical rehabilitation requires a multidisciplinary approach, where biomedical engineering part of this by creating and technological innovation, offering alternatives to staff the area of health and patients.

It is important you tackle those diseases with high incidence and prevalence in our country through biomedical technologies, increasing the input fields of research and design of equipment for prevention, diagnosis, treatment and rehabilitation. The shoulder complex is one of the very prone to multiple pathologies body, so that prevention is always the best weapon to fight disease. The field of rehabilitation of upper limb still is a vast space for research, development, implementation and testing equipment from different mechanisms of action, so that here lies the importance of the development of this project, providing a different picture in the treatment of acute and chronic diseases such.

7. FUTURE WORK.

For now, this computer was designed completely in their electronic, mechanical and electromechanical components, such as future work interested in the following points:

- Perform physical implementation team, which is a financial component which although not as high, is significant.
- Perform tests or clinical trials once the equipment is manufactured in order to see real progress in patients anywhere in Colombia, remembering that this disease is highly prevalent in our country.
- Assess the feasibility and functionality of biomedical equipment in patients with pathologies of neuro-motor, neurological damage as hemiplegia, quadriplegia and paraplegia.
- To analyze the response of patients diagnosed with painful shoulder different from tendinitis, since initially is designed only to treat this, that although it is one of the diseases more present in diagnostics painful shoulder, not the one that occurs in this syndrome.

8. REFERENCES.

[1] Ministerio de protección social República de Colombia y Pontificia Universidad Javeriana, "Guía de Atención Integral de Salud Ocupacional Basada en la Evidencia para Hombro Doloroso (GATISO – HD) Relacionado con Factores de Riesgo en el Trabajo".

[2] Ministerio de Protección Social, República de Colombia, "Guías de atención integral de salud ocupacional basadas en evidencia (GATISO)", 2006.

[3] Ruiz Sánchez F, Ruiz Santiago F, Platero Rico D. Diagnóstico y tratamiento en la patología del manguito rotador. http://www.felipeisidro.com/recursos/diagnostico_tratamiento_patologia_mang

[4] Munuera, L. Introducción a la traumatología y cirugía ortopédica. McGraw-Hill Interamericana. 2002.

[5] Rene Cailliet MD. Síndromes dolorosos de hombro. 3ª ed. México: Manual Moderno; 1983.

[6] Carlos Rodríguez, Hugo Quintero y Helena Aschner, "Movimientos del brazo humano, de los tres planos a las tres dimensiones", Revista de ingeniería, Facultad de ingeniería – Universidad de los Andes, noviembre de 2005.

[7] Martínez Flores D. Hombro doloroso. Revista boliviana de Ortopedia y Traumatología. 2005; 13(1):43-44.

[8] Sánchez Sánchez F, Llinares Clausi B, Cruz Gisbert JM. Patología del manguito de los rotadores en el ambiente laboral. Master Universitario en medicina evaluadora. 2007.

[9] Pino Tapia G, Selmas Meñoz N. Comparación entre dos tipos de tratamiento kinésico en el síndrome de pinzamiento de manguito rotador. Instituto traumatológico de Santiago. Universidad de Chile. 2006.

[10] Reinkensmeyer DJ (2003) Rehabilitators, Standard Handbook of Biomedical Engineering & Design, McGraw-Hill, chapter 35, pp. 35.1 - 35.17.

[11] José Abad Querol, María del Carmen Ramos Serrano, Nuria Sánchez Labraca, "Técnicas de movilización". Disponible en: http://www.eccpn.aibarra.org/temario/seccion1/capitulo7/capitulo7.htm

[12] Jairo Estrada, Jesús Antonio Camacho, María Teresa Restrepo, "Parámetros antropométricos de la población laboral colombiana 1995 (acopla95)", publicado en la Rev. Fac. Nac. Salud Pública 1998; 15(2): 112-139.

An overview on biological effects of trace-element in substituted calcium phosphates

R.M. Guerra Bretaña[1], J.R. Guerra-López[2], L.A. de Sena[3]

[1]Biomaterials Center, University of Havana, Havana, Cuba
[2] Basic Sciences Department, National University of Lujan, Luján, Argentina,
[3]Faculty SENAI, Rio de Janeiro, Brazil.

Abstract—**Calcium phosphate materials have been widely used as bone substitute due to its good osteointegration. To improve their osteoinductive and osteoconductive properties several element-substitutions in the calcium phosphate matrices have been explored. In the present work, an overview of biological effects of some substitutions in calcium phosphates is done based on an extensive literature revision. Incorporation of elements such as magnesium, fluoride, manganese, zinc, silicon and strontium enhance the bone induction and conduction of calcium phosphate materials. However, this behavior cannot be attributed only to the elements release. The analyzed literature indicated that element substitutions in low concentrations increase the solubility of the calcium phosphate matrices, increasing their bioactivity. Doped materials result osteoinductive and osteoconductive.**

Keywords— **calcium phosphates, ionic substitutions, osteoinduction, osteoconduction.**

I. Introduction

Calcium hydroxyapatite (CaHap, $Ca_{10}(PO_4)_6(OH)_2$) and tricalcium phosphate (TCP) are widely recognized as potential biomaterial for clinical application due to their chemical similarity with inorganic component of bone and tooth mineral. In fact bone mineral phase is a carbonated apatite that contains trace elements such as magnesium (Mg), sodium (Na), fluoride (F), manganese (Mn), zinc (Zn), silicon (Si) and strontium (Sr) which are considered essential for bone metabolism [1, 2]. Carbonate, sodium, and magnesium ions are the major dopants of apatitic bone mineral.

Calcium phosphate (CaP) biomaterials exhibit osteointegrative properties, which mean that, after implantation, they allow deposition of bone cells on their surface followed by bone grow. However, when a bone defect exceeds a critical size, it is not enough to repair the defect and osteoinductive materials, which stimulate the osteogenesis of nearby undifferentiated mesenchymal stem cells, inducing bone formation, and osteoconductive properties, meaning the ingrowth of vascular tissue and mesenchymal stem cells into the scaffold structure of the graft material are needed [3].

To improve the biofunctionality of synthetic calcium phosphates several element substitutions have been explored. These substitutions can be easily obtained by adding the corresponding ions into the reactions media. The most common example is the preparation of carbonate-substituted [4] and fluoride-substituted CaHap [5]. Another approach is the incorporation of divalent cations (Mg(II), Sr(II), or Zn(II)) into its lattice [1, 5, 6]. TCP also has undergone element substitutions [7]. This paper presents an overview on biological effects of some substitutions in calcium phosphates.

II. Materialsand methods

Selected articles on the topic of interest were analyzed. The papers were searched on ScienceDirect and Scopus-Elseiver data bases, using the key words "biological effect" + "substituted calcium phosphates".

III. Results

A. Zn-doped calcium phosphate materials

Among mentioned substituents, Zn(II) deserves some attention because it is one of the essential trace elements in human bone and plasma [7] and due to their inhibitory activity to osteoclastic bone resorption and its capacity to induce apoptosis of mature osteoclasts [8]. Zinc is important in many biological functions, for instance, this element participates in the activity of enzymes like alkaline phosphates, which take part in bone metabolism.

It has been reported that in the concentration range of 10^{-8}–10^{-5} M (0.65 μg L^{-1} to 0.65 mg L^{-1}), Zn inhibits osteoclast like cell formation in mouse marrow cell cultures and, in concentrations as low as 10^{-14} M (0.65 pg L^{-1}) constraints rats osteoclastic bone resorption in vitro [9]. Moreover, zinc deficiency causes bone growth retardation and it is a risk factor of osteoporosis in humans [10]. Because these characterristics, it is expected that Zn-containing materials would release Zn(II), and influence the differentiation of osteogenic cells promoting the regeneration of adjoining host bone [11].

It was demonstrated that Ca-deficient apatite doped with Zn could be prepared up to Zn 9.1 wt%, because at higher concentrations, Zn(II) doping shows inhibitory effect on apatite formation [12]. Zn-doped CaHap, with Zn content below to 6.5 wt%, has a structure comparable to those ones of animal bones.

I. Torres et al. (eds.), *VII Latin American Congress on Biomedical Engineering CLAIB 2016, Bucaramanga, Santander, Colombia, October 26th -28th, 2016*,
IFMBE Proceedings 60,
DOI: 10.1007/978-981-10-4086-3_20

Supplementing zinc in the medium significantly increases the osteogenic differentiation of both rat and human bone marrow stromal cells (BMSc) [13]. It is also involved in the osteoclastic differentiation from macrophages [14].

Bhattacharjee et al. [15] studied the in vivo performance of Zn-doped porous CaHap scaffolds in critical sized tibial defects of rabbits. It was observed that Zn-substituted CaHap showed better osteointegration than undoped material. SEM, histopathological study, boxytetracycline labeling study and mechanical push out test further confirmed superior bone–implant attachment of Zn-doped CaHap.

Other in vivo experiment shows that Zn-TCP induced ectopic bone formation in canine muscle related to the concentration of zinc they contained; however, TCP without Zn did not formed bone in ectopic sites [16].

B. Mg-doped CaP materials

Among substituting cations, magnesium is widely studied, being the fourth most abundant cation in the human osseous tissues [17]. Enamel, dentin and bone contain, respectively, 0.44, 1.23, and 0.72 wt% of Mg [18]. It is well known that Mg is closely associated with the mineralization of calcified tissues, directly stimulating osteoblast proliferation. Mg depletion adversely affects all stages of skeletal metabolism, causing cessation of bone growth, decreased osteoblastic and osteoclastic activities, osteopenia and bone fragility. Consequently, the incorporation of Mg ions into the CaHap lattice is of great interest for the developing of artificial bone substitutes. It has been shown that the presence of Mg(II) within CaHap lattice sensibly affects apatite crystallization in solution and its thermal stability, promoting the formation of β-TCP and thus forming biphasic calcium phosphates (BCP). Besides, the incorporation of Mg(II) stabilizes the β-TCP phase, increasing its transition temperature to β-TCP above 1125 °C. These materials are considered promising substitutes for bone replacement due to their unique biological features.

Mg 0.55 wt% inhibits the crystal growth of synthetic apatites and it is associated with increased solubility of Mg-CaHap [8]. Higher cellular activity was observed in vitro with materilas containing high level of Mg(II) ions in CaHap lattice [18] in comparison with pure CaHap.

C. Si-substituted CaP materials

In the early 1970s, Carlisle [19] determined that silicon is an essential trace element for metabolic processes associated bone and connective tissues development. Schwarz and Milne [20], who recognized Si as the crosslinking agent in the connective tissues and its importance to vascular health, confirmed these findings. Since that date, one of the most frequently studied substitutions in CaPs is the anionic substitution of silicate for phosphate in the CaHap structure.

A silicon doped calcium phosphate materials have been reported as conducive to both osteoblast deposition and osteoclast resorption, allowing it to participate in the bone remodeling processes [21-25]. Khan et al. [24] presented a comprehensive overview about the role of silicon in enhancing the biological performance and bone forming capabilities of conventional calcium phosphate based bioceramics.

Porter et al. [23] observed that an increase in concentration of Si in hydroxyapatite (between 0 and 1.5 wt%) increase the dissolution and the bioactivity of CaHap implants after 6 and 12 weeks in vivo. Their observations confirmed that defects, in particular those involving grain boundaries, were the starting point of dissolution in vivo.

D. F-substituted CaHap

Fluoride ions added to cell culture media were shown to have effects on in vitro activities of osteoblasts and osteoclasts, influencing on cell proliferation, extracellular matrix production, and alkaline phosphate activity [26]. Furthermore, NaF not only acts as potent stimulator of ongoing osteogenesis from already differentiated osteoblasts, but also it influences in the initiation of osteogenesis from embryonic mesenchyme in vitro [27]. However, high F concentrations may exert a toxic effect on osteoblast function [28]. On this basis, it has been shown that F ion and F-substituted apatite promoted osteoblast proliferation and inhibited osteoclast cell activity. Inoue et al. [29] investigated the in vivo rat tibia activity on F-substituted unsintered calcium deficient apatite, with various F concentrations for 1 and 2 weeks. Results showed that low F concentration (0.48 wt%) induced better and faster new bone formation in vivo compared to calcium deficient apatite and high fluoride concentration material (2.23 wt%). Therefore, they concluded that F had a suitable effect on bone formation in vivo.

E. Sr-doped CaP materials

Strontium has known influence as promoting bone mass and bone strength agent, due to a dual mechanism, enhancing osteoblast differentiation and inhibiting osteoclast differentiation. Even when incorporated into hydroxyapatite Sr(II) stimulates osteoblast activity and exerts its inhibitory effect on osteoclast proliferation [30]. Marques et al. [31] studied the antibiotic release and biocompatibility of strontium and magnesium co-substituted biphasic hydroxyapatite/β-tricalcium phosphate powders with composition (Ca+Sr+Mg)/P = 1.62 impregnated with levofloxacin. The Sr-doped granules exhibited the best drug release profile and the highest proliferation yields and efficiency in osteoblastic maturation, including morphology differentiation and high levels of secreted alkaline phosphatase.

F. Doped calcium phosphate nanoparticles and coatings

Devanand et al. [32] developed a drug delivery system with zinc-doped CaHap nanoparticles loaded with Ciprofloxacin for use in bone infections. They found that the antimicrobial activity against *Pseudomonas aeruginosa* and *Staphylococcus aureus* increases with increase in the drug concentration and the amount of Zn and that the drug release from zinc-doped hydroxyapatite is higher than from pure hydroxyapatite. The implantation of these nanoparticles is expected to induce cell growth and antimicrobial activity.

Zinc substituted CaHap nanoparticles (1.6 wt% Zn) have been found to enhance bioactivity to human adipose-derived mesenchymal stem cells, along with the increase in the bone cell differentiation markers [33]. Zn-CaHap nanoparticles also exhibit antimicrobial capability against Gram-negative and Gram-positive bacterial strains [34, 35]

Due to good biocompatibility calcium phosphate coatings on metallic biomedical implants are also widely employed in orthopedic and dental applications, with the aim of improving osteointegration between bones and metallic implants [35]. Chung and Long [36] attempt to enhance the osseointegration of titanium implants by fabricating Sr-CaHap coatings. Obtained results indicate Sr-CaHap coating had higher osteoblast compatibility than raw titanium metal and the CaHap coating. Moreover, Sr(II) in low concentrations stimulates osteoblast adhesion and proliferation. However, high Sr concentrations (38.9 at.% Sr) significantly inhibits osteoclast differentiation.

Composite coatings, containing silicon substituted CaHap and poly-(ε-caprolactone) on titanium substrate, have improved bonding strength [37].After immersion in simulated body fluid for 8 days, the composite coatings have the ability to induce the bone-like apatite formation.

Mg-substituted and Sr-substituted octacalcium phosphate (OCP) was also studied as thin films on titanium substrates [38]. Human osteoblast like MG-63 cells were cultured on the studied biomaterials up to 14 days. MgOCP and SrOCP coatings promote osteoblast proliferation and differentiation with respect to OCP. The level of differentiation of the cells grown on the different coatings increased in the order SrOCP>MgOCP>OCP.

IV. Discussion

Studied ion-substitutions enhance the osteoinductive and osteoconductive properties of calcium phosphate materials. The in vitro cellular results and release behavior of the minor elements studied correlate with the in vivo responses. These effects are synergistically combined with the osteointegrative properties of the calcium phosphate materials, due to their ability for protein adsorption based on surface features, chemical composition and hydrophobic properties.

Despite that, the biological mechanisms of the interactions of bioceramics and proteins are still not completely understood [39]; the widely accepted theory for bone bonding to calcium phosphate based bioceramics is based on the dissolution–reprecipitation mechanism of apatite nucleation [40-42]. As it is explained in this theory, dissolution of calcium and phosphate ions from the bioceramic surface increases their concentration locally, leading to the precipitation of biological apatite heterogeneously on the surface of proteins present close to the bioceramic implant or coating or directly on its surface. Next, more proteins and osteoblasts are absorbed on the modified bioceramic surface promoting cell adhesion and bone formation.

Even though, Porter [43] determined that dissolution is not a prerequisite for osteointegration, he also confirmed that it accelerates nucleation of biological apatite. Thus, the generation of defect centers with elemental substitution in CaP implants or coatings enhances the in vitro dissolution of the materials with a corresponding increase in their osteoinductive capacity.

Although many studies of HA substitutions have already been made, and many papers have been published, there search results are in many cases so far from commercial implementation. Studies relating co-doped CaHap materials have also been conducted [44]. Much interdisciplinary work remains to be done to transfer interesting research findings into valuable materials for use in medical applications.

V. Conclusions

Doping of CaP materials, with elements important to bone metabolism such as Zn, Mg, Sr, Si, F, brings an increment in bone induction and conduction. This behavior cannot be attributed only to the elements release. The analyzed literature indicated that element substitutions in low concentrations increase the solubility of the calcium phosphate matrices, and play an important role in their bioactivity. This review contribute to deeper understanding that addition of mentioned elements is a useful strategy to convert calcium phosphates in really osteoinductive and osteconductive biomaterials.

Conflict of Interest

The authors declare that they have no conflict of interest.

References

1. Zofková I, Nemcikova P, Matucha P. Trace elements and bone health *Clin Chem Lab Med.* 2013; 51:1555-61.
2. Dorozhkin SV. Bioceramics of calcium orthophosphates *Biomaterials.* 2010;31:1465–1485.
3. Kao ST, Scott DD. A Review of Bone Substitutes *Oral Maxillofacial Surg Clin N Am.* 2007;19:513–521.
4. Ito A, Kawamurab H, Otsukac M et al. Zinc-releasing calcium phosphate for stimulating bone formation *Mat Sci Eng C.* 2002; 22:21–25
5. LeGeros RZ. Properties of osteoconductive biomaterials: calcium phosphates *Clin Orthop Relat Res.* 2002;395:81-98.
6. Lia ZY, Lama WM, Yang C et al. Chemical composition, crystal size and lattice structural changes after incorporation of strontium into biomimetic apatite *Biomaterials.* 2007;28:1452–1460.
7. Luo X, Barbieri D, Davison N et al. Zinc in calcium phosphate mediates bone induction: In vitro and in vivo model *Acta Biomater.* 2014;10:477–485.
8. Webster TJ, Ergun C, Doremus RH. Hydroxylapatite with substituted magnesium, zinc, cadmium, and yttrium. II. Mechanisms of osteoblast adhesion *J Biomed Mater Res.*2002;59: 312–317.
9. Li X, Sogo Y, Ito A et al. The optimum zinc content in set calcium phosphate cement for promoting bone formation in vivo *Mater Sci Eng C Mater Biol Appl.* 2009;29:969–75.
10. Ito A, Otsuka M, Kawamura H et al. Zinc-containing tricalcium phosphate and related materials for promoting bone formation *Current App Phys.* 2005;5:402-406.
11. Ishikawa K, Miyamoto Y, Yuasa T et al. Fabrication of Zn containing apatite cement and its initial evaluation using human osteoblastic cells *Biomaterials.* 2002;23:423–8.
12. Guerra-López J, Echeverría G, Güida J et al. Synthetic hydroxiapatites doped with Zn(II) studied by X-ray diffraction, infrared, Raman and thermal analysis *J Phys Chem Solids.* 2015;81:57-65.
13. Ikeuchi M, Ito A, Dohi Y et al. Osteogenic differentiation of cultured rat and human bone marrow cells on the surface of zinc-releasing calcium phosphate ceramics *J Biomed Mater Res Part A.* 2003;67:1115–1122.
14. Roy M, Fielding GA, Bandyopadhyay A et.al. Effects of zinc and strontium substitution in tricalcium phosphate on osteoclast differentiation and resorption *Biomater Sci.* 2013;1:74-96.
15. Bhattacharjee P, Begam H, Chanda A, Nandib SK. Animal trial on zinc doped hydroxyapatite: A case study *J Asian Ceramic Soc.* 2014;2:44–51.
16. Li X, Sogo Y, Ito A et al. The optimum zinc content in set calcium phosphate cement for promoting bone formation in vivo *Mater Sci Eng C Mater Biol Appl.* 2009;29:969–975.
17. Larson PR, Madden AS, Tas AC. Non-stirred synthesis of Na- and Mg-doped, carbonated apatitic calcium phosphate *Ceramics International.* 2013;39:1485–1493
18. Mehrjoo M, Javadpourl J, Shokrgozar MA et al. Effect of magnesium substitution on structural and biological properties of synthetic hydroxyapatite powder *Mater Express.* 2015;5:41-48.
19. Carlisle EM. Silicon: a possible factor in bone calcification *Science.* 1970;167:279–280.
20. Schwarz K, Milne DB. Growth-promoting effects of silicon in rats *Nature.* 1972;239:333–334.
21. Sayer M, Stratilatov AD, Reid J et al. Structure and composition of silicon-stabilized tricalcium phosphate *Biomaterials.* 2003;24:369–82.
22. Langstaff SD, Sayer M, Smith TJN et al. Resorbable bioceramics based on stabilized calcium phosphates, Part II: Evaluation of biological response *Biomaterials.* 2001;22:135–150.
23. Porter AE, Patel N, Skepper JN et al. Comparison of in vivo dissolution processes in hydroxyapatite and silicon-substituted hydroxyapatite bioceramics *Biomaterials.* 2003;24:4609-4620.
24. Khan AF, Saleem M, Afzal A et al. Bioactive behavior of silicon substituted calcium phosphate based bioceramics for bone regeneration *Mater SciEng C Mater Biol Appl.* 2014;35:245-252.
25. Botelho C, Brooks R, Kanitakahara M et al. Effect of Protein Adsorption onto the Dissolution of Silicon-Substituted Hydroxyapatite *J Encapsul Adsorp Sci.* 2011;1:72-79.
26. Farley JR, Wergedal JE, Baylink DJ. Fluoride directly stimulates proliferation and alkaline phosphatase activity of bone-forming cells *Science.* 1983;222:330-332.
27. Hall BK. Sodium fluoride as an initiator of osteogenesis from embryonic mesenchyme in vitro *Bone.*1987;8:111–116.
28. Pitt P, Berry H. Fluoride treatment in osteoporosis *Postgrad Med J.* 1991;67:323– 326.
29. Inoue M, Nagatsuka H, Tsujigiwa H et al. In vivo effect of fluoride-substituted apatite on rat bone *Dent Mater J.* 2005;24:398-402.
30. Capuccini C, Torricelli P, Boanini EJ et al. Interaction of Sr-doped hydroxyapatite nanocrystals with osteoclast and osteoblast-like cells *Biomed Mater Res.*2009;89A:594–600.
31. Marques CF, Lemos A, Vieira SI et al. Antibiotic-loaded Sr-doped porous calcium phosphate granules as multifunctional bone grafts *Ceram Int.* 2016; 42:2706–2716.
32. Devanand Venkatasubbu G, Ramasamy S, Ramakrishnan V et al. Nanocrystalline hydroxyapatite and zinc-doped hydroxyapatite as carrier material for controlled delivery of ciprofloxacin *Biotech.* 2011;1:173–186.
33. ThianE S, Konishi T, Kawanobe Y et al. Zinc-substituted hydroxyapatite: a biomaterial with enhanced bioactivity and antibacterial properties *J Mat Sci: Mat Med.* 2013;24:437-445.
34. Swetha M, Sahithi K, Moorthi A et al. Synthesis, Characterization, and Antimicrobial Activity of Nano-Hydroxyapatite-Zinc for Bone Tissue Engineering Applications *J Nanosci Nanotechnol.* 2012;12:167-172.
35. Sun G, Ma J, Zhang SH. Electrophoretic deposition of zinc-substituted hydroxyapatite coatings *Mat Sci Eng* C. 2014;39:67–72.
36. Chung CJ, Long HY. Systematic strontium substitution in hydroxyapatite coatings on titanium via micro-arc treatment and their osteoblast/osteoclast responses *Acta Biomater.* 2011;7:4081–4087.
37. Xiao X, Liu R, Tang X. Electrophoretic deposition of silicon-substituted hydroxyapatite/poly(ε-caprolactone) composite coatings *J Mat Sci: Mat Med.* 2009;20:691-698.
38. Boanini E, Torricelli P, Fini M et al. Magnesium and strontium doped octacalcium phosphate thin films by matrix assisted pulsed laser evaporation *J Inorg Biochem.* 2012;107:65–72.
39. Wang K, Zhou Ch, Hong Y et al. A review of protein adsorption on bioceramics *Interface Focus.* 2012;2:259–277
40. Ducheyne P, Radin S, King L. The effect of calcium phosphate ceramic composition and structure on in vitro behavior. I: dissolution *J Biomed Mater Res.* 1993;27:25–34.
41. Radin SR, Ducheyne P. The effect of calcium phosphate ceramic composition and structure on in vitro behavior. II: precipitation *J Biomed Mater Res.* 1993;27:35–45.
42. Ducheyne P. Bioactive ceramics: the effect of surface reactivity on bone formation and bone cell function *Biomaterials.* 1999;20:2287–2303.
43. Porter AE. Nanoscale characterization of the interface between bone and hydroxyapatite implants and the effect of silicon on bone apposition *Micron.* 2006;37:681–688.
44. Kaygili O, Keser S. Sol–gel synthesis and characterization of Sr/Mg, Mg/Zn and Sr/Zn co-doped hydroxyapatites *Mat Lett.* 2015;141:161–164.

Author: Rosa Mayelin Guerra Bretaña
Institute: Biomaterials Center, University of Havana
Street: Ave Universidad
City: Havana
Country: Cuba
Email: mayelin@biomat.uh.cu

Trading barriers in the medical devices industry. Are these barriers hindering the development of this sector in Cuba?

Y. Chaveco Salabarria[1], J.C. Rubio Romero[2] and R.M. Guerra Bretaña[3]

[1] Universidad de Oriente/Centro de Biofísica Médica, Santiago de Cuba, Cuba
[2] Universidad de Málaga/Dpto. de Economía y Administración de Empresas, Málaga, España
[3] Universidad de la Habana/Centro de Biomateriales, La Habana, Cuba

Abstract— **Cuba has developed innovative medical technologies for sale within the country and in other geographical areas. Why Cuban medical technologies have not reached the markets of developed countries? What factors might be slowing down their introduction? From this point will examine the barriers to commercialization in the foreign market of Cuban health technologies, particularly medical equipment and devices. The economic factor is presumed as an influential element but there are also other internal and external aspects influencing this business. To introduce more Cuban medical equipments in the foreign market, necessarily requires the implementation of legal and others standards established worldwide. Strengthen capacities within organizations, to develop more competitive products and better use of existing conditions in the country, will overcome the many obstacles of international trade in a sector so heavily regulated. The information shared in this article is part of a much extensive research currently in progress.**

Keywords— **medical devices and equipments, trade barriers, sanitary regulations**

I. Introduction

The Cuban health technologies, especially drugs, vaccines and monoclonal antibodies have achieved a significant development worldwide and provide millionaire revenue to the country. The institutions manufacturing these health technologies have been consolidated under a management system which incorporates scientific research and development of new products within the value chain and used them in the negotiations for commercial realization of value added [1]. The creation of this type of high tech companies is promoted in Cuba, encouraging science and innovation organizations to meet public health needs by means of self financing and performing the complete cycle of research-development-production-commercialization.

However, the health technologies industry in Cuba includes other products such as diagnostic equipment, biomaterials and others medical devices; and not all have been developed and commercialized with the same success.

The Cuban industry of medical devices has advanced moderately with high value added, created and developed mostly in research centers working in what is named full cycle scheme. Several of these products are intended to meet domestic demand in order to replace costly imports and have also been successfully introduced in foreign markets. The centre of attention in this study is the health technologies industry.

Cuban manufacturers who currently accumulate more experience exporting medical equipments are integrated to what is called BIOCUBAFARMA an integrated medical equipment company. Among these manufacturers stand out: Digital Medical Technology Company (ICID) - a spanish acronym-, leading the sector in the country, through its marketing COMBIOMED S.A has exported electrocardiography, cardiac rehabilitation, monitoring and electrostimulation equipment. Another particular Cuban entity that has completed exports of products for medical diagnosis is Immunoassay Center (CIE) with it sales agency TECNOSUMA S.A. The Center for Neurosciences of Cuba, this manufacturer using its mixed capital company in Spain and sale agency IC Neuronic S.L has introduced its products in the European market from within, with technology fundamentally linked to clinical neurophysiology. Some other manufacturers have been able to make punctual sales abroad. However, unlike the biotechnology and pharmaceutical industry, the field of medical devices still does not reach the production and export levels that could be achieved with the high scientific and technological innovation level of their organizations. Specifically exports to Europe and other developed countries are still unsatisfactory.

According with data published in 2013 by Center for State Control of Drugs and Medical Devices (CECMED), most of the requests for free sales certificate for exports of Cuban medical equipment are traditionally made towards Latin America [2]. Venezuela is the main destination with 62% of applications and there are also listed other countries like Peru with 22%, Ecuador and Mexico, both with 8%; and less significant amounts have requested certificates for

I. Torres et al. (eds.), *VII Latin American Congress on Biomedical Engineering CLAIB 2016, Bucaramanga, Santander, Colombia, October 26th -28th, 2016*,
IFMBE Proceedings 60,
DOI: 10.1007/978-981-10-4086-3_21

export of certain equipment to Colombia, Brazil and Argentina. Few have been processed for export outside the continent, mainly to Argelia and China. Marketing is virtually zero in most developed or also called first world countries. In this situation the question arises: What can be limiting the trades of Cuban medical equipment and devices in the international market?

It is known that many economic sectors have regulations for protecting companies already established in the market and prevent the entry of new competitors [3]. These market entry barriers arise from various sources [4, 5]. To mention some examples: the legal regulations, economies of scale and the learning curve of mature companies , highly differentiated products, capital requirements and costs of change to enter the sector, ability to access raw materials and channels distribution, favorable location, patents and knowledge of exclusive property, retaliation, government politics, etc.

Based on the above, the main purpose of this study is to identify the trading barriers for Cuban medical equipment and devices entering international market and verify manufacturer's perception of this matter. This study is an integrated part of a much large research currently in progress.

II. Materials and methods

For this investigation a comprehensive literature, concerning the state of the art in the subject, was examined. This is including databases Web of Science, original article and reference from identified magazine, also guidelines and other materials published through institutional repositories of recognized organizations such as WTO (World Trade Organization), WHO (World Health Organization), PAHO (Pan American Health Organization) and the Cuban regulatory agency CECMED; where information was obtained regarding trade barriers, regulations and directives for European Conformity (CE, by acronym in Spanish), sanitary regulations, among others.

This investigation is complemented with a study of exploratory cases following the Yin methodology (1984). The study is in progress with a number of companies. Our goal is to verify the presence of these barriers in the practice and identify possible corrective actions. The study result will be published in the near future.

III. Results

Within the health sector, the technology has an increasingly leading role in the diagnosis and medical therapy. Its influence in the quality of life of citizens has forced the health authorities to establish regulatory programs to ensure health equipment operate safely and efficiently. This sector is globally regulated as a result of potential and actual risks associated with the use of medical technologies. Therefore it requires a proper assessment, of the risks vs. the benefits of the new technology, before introducing a product in the market, which is only possible applying internationally harmonized industry standards and specific regulations such as ISO 13485 quality management systems for medical devices for regulatory purposes and ISO 14971 for risk management of medical equipment [6].

Failure to comply with regulatory requirements is the first barrier to health registration and marketing medical technologies, both domestically and internationally. In areas such as the European Union, United States and Japan, world leaders; and even in countries of the Latin American region, these requirements are already part of complex regulatory frameworks established to protect the public from unsafe and ineffective products.

In the European Union, for example, the procedure of conformity assessment and CE marking of health products is valid in all countries member and is based on compliance with harmonized European standards [7]. Complying with these procedures is the necessary process for the free movement of medical devices within the European market. As part of these procedures, the requirements that medical devices should meet are highly demanding in terms of demonstrating the performance and security, which translate to higher costs for conformity assessment. This aspect constitutes one of the possible technical barriers for the entry of Cubans products into European market.

Several authors have classified trade barriers into two groups: tariff and nontariff barriers. Tariff barriers refer to the value of the goods at customs which aims to raise the import duties. However, today tariff barriers tend to be reduced in a number of sectors, since the agreement concerning Technical Barriers to Trade of the WTO (World Trade Organization). The process of liberalization significantly eliminates tariffs that raise and hinder the exchange administratively, at the same time new measures such as approvals, certifications and standards are created on a regular basis [8]. Consequently, it has increased the relative importance of nontariff barriers, especially technical regulations which are including more challenging process and product requirements.

Nontariff barriers contain a wide range of elements, among the ones stand out quantitative aspects, administrative, technical, sanitary and fiscal requirements established in one country or another. Another element is the documentation, required properly authenticated and in a different language other than exporter such as: commercial invoices, certificates of origin, shipping company certificates, quality certificates issued by standards organizations, health certificates, content lists, and even company insurance certifications, all depending on the product in question [9].

Overall, many of the above barriers are applicable to any sector of the global economy. But the medical devices industry has its own peculiarities and strong restrictions, because logically, the characteristics, health implication and scope of use of the products. In addition to the general barriers to exports, there are a set of regulatory (and even ethical) factors that inhibit or mark the entrance of this technology to a particular country. In most countries not only the medical device should be registered in accordance with health regulations, but also should be registered the manufacturer to the relevant health authority, which must meet multiple, complex and frequently costly requirements.

Besides, in the sectors of the economy with high value added of the first world countries, companies use two main types of nontariff barriers: the intellectual property (patents) and technological. Both are preventing developing countries from producing goods and services of high technology. The protectionist bias is not only limited to entry barriers imposed by developed countries to protect their domestic markets; but also the claim of global harmonization of regulations allows them to be protected from all other countries of the world and preserve their domestic market for their own companies [1, 10]. Medical equipment, which most of them have high value added, do not escape from these trends as they are often subject to protectionist practices in industrialized countries.

To overcome the technical barriers associated to trade in the European Union, the costs are high and restrictive for Cuban medical equipment manufacturers. These costs are including both, the initial resulting costs, resulting from the regulation study and product adjustment for compliance and the ongoing costs corresponding to periodic tests to demonstrate a product complies with the standards. Among these costs there are the cost corresponding to product design adjustment, production system restructuring, technical testing, cost associated with the certification process required by exporting agencies , translation of regulations to foreign language, hire foreign technicians or relocate nationals to other country [11].

Other factors may be also limiting for these purposes, including: manufacturing costs, spare parts or parts, consumables and accessories, training and expertise of personnel, local infrastructure, regulations and social context and cultural [12].

IV. Discussion

According to WHO the medical devices industry is one of the most vital and dynamic sectors of the world economy. In the Americas region there is a highly competitive and growing market with manufacturers and distributors actively trading their products. With some exceptions, the countries of the Americas import most of their medical equipment and devices. But only a few reach have accomplished medical devices export. A high number of these countries do not have the institutional regulatory systems to ensure product conformity, safety and quality or the technical competence for implementing appropriate sanitary regulatory and assessment programs [13].

Cuba is one of the Latin American nations that have maintained a systematic policy regarding the evaluation activity and health records of medical equipment, continuously enriching its regulatory program. Since its foundation in 1992 the Cuban regulatory agency developed its regulations based on existing international harmonization efforts. Currently this regulatory agency is certified as Reference Regulatory Authority of the Americas Region, condition that gives and evaluates periodically the PAHO and WHO. This center has developed and significantly updated its regulatory program in collaboration with several foreign agencies such as Canadian and Spanish agencies; reviewing the proceedings for conformity assessment used by these agencies [14]. As follows, for the establishment of the Cuban regulations in the field of medical devices, the country has taken into account some of the most advanced experiences in this field. It is valid to ask then: if Cuba considers international regulations and standards in the equipment and medical devices industry, what is preventing the country to reach new and larger markets?

The economic factor is influential in the fact that few Cuban medical equipments have managed to achieve European conformity marking or other element leading market. The regulatory requirements concerning the safety and operation of medical devices are essentials but result in additional financial resources for the manufacturers. However, other elements must be considered in this situation. Many of the Cuban organizations in this field still suffer from significant internal capacities to cope with this reality. Complete assimilation and implementation of internationally approved standards is still missing. This is illustrated by the fact that only four Cuban manufacturers of medical devices have management systems certified by the generic standard ISO 9001. Only the most advanced manufacturers have implemented management systems that incorporate risk management as required in current regulations. Just recently the country implemented the certification system guided by ISO 13485 standards, only ICID achieved, in April this year, this standard certification. The company IC Neuronic SL based in Spain is also certified by this standard.

Overcome the trading barriers than limiting or preventing the export of medical equipment to other countries, specifically to the first world, is a pending task for this industry in Cuba. Achieve innovative products, in line with the world state of art and comply with regulations of the geographical areas where the equipment or device is intended to be introduced, is essential for achieving its certification within the most challenging markets. It is essential to take advantage of existing potential in terms of national infrastructure created, assimilated technology, expertise and qualification of human capital in the industry. Taking into consideration the identification of these barriers and the gradual incorporation of the aspect outlined in the daily life of the Cuban entities, will help to enhance the industry competency and will increase its exportable assets, contributing at the same time to the macroeconomics of the country. The implementation of coordinated strategies, concerning this affair, will help to transform these barriers into opportunities for improving this essential Cuban industry.

It will be necessary continue exporting under the scheme of intergovernmental agreements that open space to these highly innovative products, inside or outside of the American continent, as it has been in many of the Cuban cases. But it is also necessary strengthen the resources and internal capabilities of organizations throughout the product development cycle and adopt new marketing strategies to enter new markets. In any case, it is essential to be technically prepared. Domestic manufacturers have the maximum responsibility, without forgetting the role regulatory agency and all interested parties must meet.

V. Conclusions

The existing entry barriers (regulatory, technical, fiscal, or other) for accessing international markets, highly demanding for quality and safety of medical devices, are high, costly, complex and numerous. Regardless the presence of these barriers it is essential the creation of more competitive internal capabilities in the Cuban entities within the industry to modify, where suitable, existing organizational routines vetting its development.

Cuba should continue adopting regulatory, economic, productive and commercial actions in the medical equipment industry to satisfy primarily the needs of a social practice, especially the health system as well as working toward increasing exports of these goods which translates to increased revenue and development for the country.

Acknowledgment

The authors express gratitude to Agencia Española de Cooperación Internacional (AECID), Asociación Universitaria Iberoamericana de Posgrado (AUIP) and Universidad de Malaga, for all support provided for the completion of this project.

Conflict of Interest

The authors declare that they have no conflict of interest.

References

1. Lage, A. (2013) La economía del conocimiento y el socialismo. *Academia*.
2. CECMED. (2014). Boletín informativo InfoCECMED Año 18, No. 66. Edición Especial. ISSN 1684-1867.
3. Porter, M. E. (2009). Estrategia Competitiva: Técnicas para el análisis de las empresas y sus competidores. *Madrid: Piramide*.
4. Bain, J. S. (1956). Barriers to new competition, their character and consequences in manufacturing industries (No. HB771 B23).
5. Karakaya, F., & Stahl, M. J. (1989). Barriers to entry and market entry decisions in consumer and industrial goods markets. *The Journal of Marketing*, 80-91.
6. Guerra R.M., Meizoso M.C., Ramírez J.R. (2015) Influence of the Quality Infrastructure in the Development of Medical Technologies. VI Latin American Congress on Biomedical Engineering CLAIB 2014, Paraná, Argentina 29, 30 & 31 October 2014. IFMBE Proceedings; 49, 746-749.
7. Abad, M.C., Cantalapiedra, M. J. (2002). Productos Sanitarios: Aspectos Legales y Perspectivas de Futuro. Ministerio de Sanidad y Consumo de España.
8. World Trade Organization (WTO). Serie de acuerdos de la WTO. Obstáculos técnicos al comercio at http://www.wto.org
9. AFI. Análisis del mercado saudí para el exportador español at http://www.afi.es
10. Bergsland, J., Elle, O. J., & Fosse, E. (2014). Barriers to medical device innovation. Med Devices (Auckl), 7, 205-9.
11. Togan, S., & Dogan, S. (2009). La eliminación de obstáculos técnicos al comercio en el contexto de la unión aduanera entre Turquía y la Unión Europea. Información Comercial Española, ICE: Revista de economía, (846), 47-64.
12. World Health Organization. (2010). Barriers to innovation in the field of medical devices: background paper 6, August 2010.
13. World Health Organization. (2012). Dispositivos médicos: la gestión de la discordancia: un resultado del proyecto sobre dispositivos médicos prioritarios.
14. CECMED. Informe Anual 2013 at http://www.cecmed. sld.cu.

Author: Yariza Chaveco Salabarria
Institute: Medical Biophysics Center
Street: Ave Patricio Lumumba s/n. Universidad de Oriente
City: Santiago de Cuba
Country: Cuba
Email: chaveyary@yahoo.com

A risk-based integrated management for patient safety and quality in healthcare services

R. Roque González[1], R.M. Guerra Bretaña[2], Y. Chaveco Salabarria[3], K.F. Ortiz Jaya[4], O. Vera Cabezas[1]

[1]National Center for Minimal Access Surgery Havana, Cuba
[2] Biomaterials Center, University of Havana, Habana, Cuba
[3]Center of Medical Biophysics, Santiago de Cuba,
[4]Public Health Fac., High Polytechnic School of Chimborazo, Ecuador

Abstract—**Quality of health care services and patient safety has become a matter of interest to healthcare professionals, researchers and governments all over the world. The aim of this paper is to show the implementation of the risk-based integrated management for safety and quality at the National Center for Minimal Access Surgery in Havana, Cuba. Based on the structure, responsibilities and documents established for the health care service quality, thorough studies on patient safety and the risk-based thinking led to the development of an integrated system. This approach shifts the quality and safety management from reactively assessment of complications and incidents to proactive evaluation of the potential risks. Reporting of errors, injuries and complications can play an important role in the continuous health service improvement and as learning opportunities, may avoid future harm to patients.**

Keywords— **risk management, quality management, patient safety, healthcare services.**

I. Introduction

Errors and injuries are common and often very serious in the delivery of health care. For this reason hospitals have created several offices and committees for managing service quality and patient safety. However, in many healthcare institutions these structures work independently without the understanding that safety is just one of the dimensions of the quality of health care, together with access, timeliness, efficacy, efficiency, appropriate measures and acceptability [1]. In the context of patient safety, errors are defined as a failure of a planned action to be completed as intended (error of execution) or the use of a wrong plan to achieve an aim (error of planning) [2]. Errors may manifest by doing the wrong thing (commission) or by failing to do the right thing (omission), at either the planning or execution phase [3]. To have a valid, reliable, and meaningful error rate an accurate data has to be compiled. However, in the best situations, the rate of reporting or identifying medical errors in the review of the records may not reflect the true rate. In fact, engaging health care professionals and staff about reporting errors, reducing risk and improving the safety and quality is a crucial but difficult task [4].

To manage all dimensions of the health care quality, a risk-based approach may be adopted, focused on identifying the underlying hazards in the health care service that lead to risky situations and finally to errors and injuries [2]. Off course, the appreciation of risks has to be based on the available data about the occurrence and consequences of errors and injuries. Reports about injuries and errors are one of the sources of risk identification. Other potential sources of risk identification include the published literature [5], medical device recalls, consulting experts and patient safety research [6].

The World Alliance for Patient Safety [7] offers a definition of safety risk management for health care services, as activities or measures taken by an individual or a health care organization to prevent, remedy or mitigate the occurrence or reoccurrence of a real or potential patient safety event. Patient safety is the reduction of risk for needless harm associated to healthcare at an acceptable minimum. An acceptable minimum refers to the collective notions of given current knowledge, available resources and the context in which care was delivered weighed against the risk of non-treatment or other treatment.

Considering the drawbacks of having the quality management and patient safety functions in separated structures, the aim of this work is to show the implementation of the risk-based integrated management for patient safety and quality (IMS-SQ) at the National Center for Minimal Access Surgery in Havana, Cuba. The application of the IMS-SQ in biliary lithiasis surgery is demonstrated.

II. Materials and methods

The methodology for the implementation of the risk-based integrated management system for safety and quality includes five stages (Fig. 1) following the Plan-Do-Check-Act cycle. UNE 179003:2013 [8] and ISO 9001:2015 standards are employed as the current normative basis [9].

In the case studied, the organization at first implemented a quality management system and on this basis, an integral system for service quality and patient safety management is developed. In the first stage, "Leadership and commitment",

I. Torres et al. (eds.), *VII Latin American Congress on Biomedical Engineering CLAIB 2016, Bucaramanga, Santander, Colombia, October 26th -28th, 2016*,
IFMBE Proceedings 60,
DOI: 10.1007/978-981-10-4086-3_22

the higher echelons identify and assess the risk management to be relevant to a health care organization and makes it known the benefits of the IMS-SQ, decides its implementation and designates the project team.

Fig. 1 Integrated management system implementation project.

In the "Designing Stage", medical care activities are analyzed identifying and assessing risk through analysis and clinical interpretation. Then the next step is the reduction and elimination of the risk incorporating these activities in the already implemented QMS. For monitoring, review and improvement, the quality management tools are employed.

Data and qualitative information were collected from hospital records and documentation establishes in the QMS.

III. Results

The National Center for Minimal Access Surgery (CNCMA) - a spanish acronym- is the reference institution in Cuba for minimally invasive medical procedures. In the Center an ISO 9001 quality management systems (QMS) are implemented, consisting of four strategic processes, three key processes and two support processes, as shown in the Fig. 2. The healthcare process includes all the activities (Fig. 3) needed to satisfy the needs and expectations of the patients and their relatives.

In the Evaluation and Improvement process, the QMS incorporates several Committees which look after different aspects of hospital quality and patient safety, such as medical ethics, healthcare service, surgical activities, hospital mortality, prevention and control of infectious diseases, pharmaco-therapeutic practices and transfusion medicine.

Based on the structure, responsibilities and documents established for the health care service quality, thorough studies on patient safety and the risk-based thinking develops an integrated system. The IMS-SQ is based on prevention of risks on each subprocess involved in the medical care.

In general, surgical complications can be divided into preoperative, intraoperative and postoperative. The last two obviously occur while the patients are still in the hospital.

Preoperative risk factors in elective laparoscopic surgery are recognized prior to surgery in order to reduce complications and individualized treatment is implemented as soon as possible. However, some risk factors such as age, previous abdominal surgery and co-morbidities can obviously not be changed before surgery [5]. Regarding the patient's age, several factors, such as history of hypertension, pulmonary, neurologic and coronary artery diseases, can increase the odds of developing any postoperative adverse events in these patients. Other patient factors, such as obesity, anemia and general nutritional state can be dealt with before surgery.

Preoperative factors also include the adequate identification of the patient, experience of the surgeon and the number of surgical cases involved in the work load. These factors are prevented by means of organizational measures taken as part of the hospital management system.

In the protocols established for each medical procedure, quality indicators are measured including safety issues. Specifically, in the standard operational procedure (SOP) for biliary lithiasis, various process quality indicators are established to assure that the preoperative care adhered to the established protocols.

In the prevention of intraoperative complications, the laparoscopic technique and the medical devices chosen have a great influence and are detailed in the SOP. The anesthetic procedures and surgical preparation are also documented and controlled, including the prevention of acute venous thromboembolism, antibiotic prophylaxis in all surgical patients and sterile procedures while placing central intravenous catheters.

Examples of the safety related indicators in surgery for biliary lithiasis are the following:

- Surgical accidents.
- Injuries of the main biliary system.
- Index of conversion to open surgery.
- Morbidity associated to endoscopic retrograde cholangiopancreatography with endoscopic sphincterotomy (ERPC-ES).
- Mortality associated to ERPC-ES.

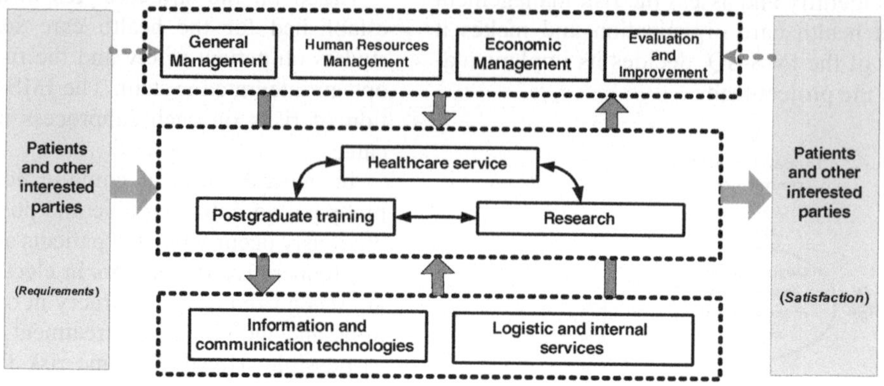

Fig. 2 Process map of the quality management system of the National Center for Minimal Access Surgery.

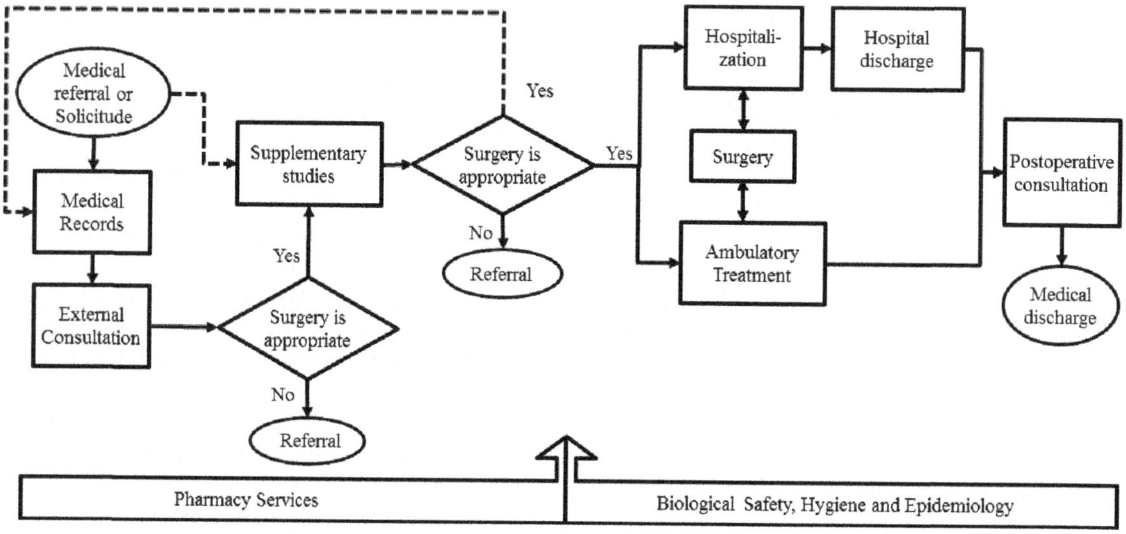

Fig. 3 Process diagram of the healthcare service in the National Center for Minimal Access Surgery.

A set of potential in-hospital postoperative complications that might represent patient safety events are identified, such as, adverse drug events, surgical site infection, central venous line infection, and ventilator-associated pneumonia. Other nosocomial infections may result from lack of proper hand hygiene.

Adverse events in the SOP are recorded. Several indicators in the post-surgical period at the Intensive Care Unit are established, such as:

- Complications in ventilated patients.
- Morbidity in ventilated patients.
- Mortality in ventilated patients.
- Morbidity associated to sepsis.
- Mortality associated to sepsis.

The safety outcomes are discussed in the corresponding Service and the Hospital Quality Committees, based on the detailed description of the events in the patient´s clinical record. Of course the first thing to do when a complication occurs is to respond adequately in order to minimize its effect on the patient, and then record what happened and/or what was done. The causes of the errors or injuries are analyzed and the corrective measures are taken. The purpose of these analysis are not only to know what happened and how, but also to utilize the incident to identify the faults of the medical care process. The discussion and measures taken are important feedback for the IMS-SQ improvement.

Registration, imaging, clinical laboratory and blood bank are very important subprocesses to assure the quality of the health care service and patient safety, in order to guarantee the proper inputs for the surgical act and the postsurgical monitoring. Also the patient flow to the operative theater and back to the hospitalization room is accompanied by nursery personnel to prevent accidental falls and to ensure the correct identification of the patient. The recount of instruments, gauzes, needles, etc. by the nurse is obligatory after each intervention before the patients leave the surgery room. All these activities are recorded and are strictly supervised.

In the CNCMA, not only the medical care process is involved in the risk management of patient safety, but also the research and postgraduate training process develops projects to predict morbidity and mortality of various surgeries, including for biliary lithiasis surgery, based on the predictive risk factors. Support processes related to maintenance and metrological assurance of medical devices, hospital cleaning and the diet are also involved in patient safety.

The survey of patient satisfaction includes safety issues and allows the perception of patients and their relatives about the medical service provided. Every patient complaint is informed to the corresponding service. Those related to satisfy the medical needs and comfort of patients and their relatives are analyzed by the board of directors and due action is taken.

Indicators reported in 2015 show a hospital mortality of 0.6%, a rate of infection of all surgical wounds of 0.9% and an urgent reoperations ratio of 0.5%. These quality outcomes are a result of the work done by all staff at the hospital in the framework of the risk-based integrated management for patient safety and quality.

IV. Discussion

Laparoscopic cholecystectomy remains the treatment of choice for symptomatic cholelithiasis, partially because of the shorter recovery period, decreased postoperative discomfort, improved cosmetic results and fewer complications compared to open surgery. However, approximately 5% of laparoscopic cases are converted to an open procedure secondary to difficulty in visualizing the anatomy or a complication. Even so, the risk management is necessary to improve the quality of the provided medical care, which includes the patient safety. Complications such as infection, bleeding, bile leakage, injury to the bile duct, injury to the intestine, bowel and blood vessels, deep vein thrombosis, risks of the general anaesthesia and the appearance of the post-cholecystectomy syndrome are recorded. International data suggest that the rate of trocar related complications is less than 3% [10].

A risk-based approach shifts the quality and safety management from response to the assessment of complications and incidents to proactive evaluation of the potential risks. This approach has the added value of involving frontline staff in this process [11], and also its positive connotation, due to the recording of measures taken to reduce or eliminate risks which results in enhanced safety. Health care staff is encouraged to actively seek out potential risks, even though they might not have led yet to an error or injury. The risks are then communicated and eliminated whenever possible. The accurate identification and recording of errors, injuries and complications can reflect an institution´s patient safety and quality record, and also they play an important role in continuous health service improvement because identified errors are learning opportunities that allow prevention of future harm to patients.

V. Conclusions

The integrated management system for safety and quality results in better performance of the National Center for Minimal Access Surgery and more efficient use of organizational resources, based on risk management approach.

Conflict of Interest

The authors declare that they have no conflict of interest.

References

1. Runciman WB, Williamson JAH, Deakin A, et al. An integrated framework for safety, quality and risk management: an information and incident management system based on a universal patient safety classification. QualSaf Health Care, 2006, 15:i82–i90.
2. Scanlon MC, Karsh B-T, Saran KA. Risk-Based Patient Safety Metrics. In: Henriksen K, et al. editors. Advances in Patient Safety: New Directions and Alternative Approaches Vol. 1 Rockville (MD): Agency for Healthcare Research and Quality; 2008 Aug.
3. Dückers et al. Safety and risk management in hospitals. The Health Foundation, London, 2009.
4. Leviton J, Valentine J. How Risk Management and Patient Safety Intersect: Strategies to Help Make It Happen March 24, 2015.
5. Kirchhoff P, Clavien P-A, Hahnloser D. Complications in colorectal surgery: risk factors and preventive strategies. Patient SafSurg 2010, 4:5.
6. Battles JB, Lilford RJ. Organizing patient safety research to identify risks and hazards. QualSaf Health Care, 2003; 12: 2-7.
7. WHO, World Alliance for Patient Safety 2009.
8. UNE 179003:2013 Servicios sanitarios. Gestión de riesgos para la seguridad del paciente. AENOR, Madrid; 2013.
9. ISO 9001:2015. Sistemas de gestión de la calidad. Requisitos. Organización internacional de Normalización, Ginebra; 2015.
10. Fuller J, Scott W, Ashar B, et al. Laparoscopic trocar injuries. A report fron U.S. Food and Drug Administration, Nov. 7, 2003
11. Morag I, Gopher D. A reporting system of difficulties and hazards in hospital wards as a guide for improving human factors and safety. Paper presented at the Human Factors and Ergonomics Annual Meeting. San Francisco, CA; 2006.

Author: Rosa Mayelín Guerra Bretaña
Institute: Biomaterials Center, University of Havana
Street: Ave Universidad
City: Havana
Country: Cuba
Email: mayelin@biomat.uh.cu

CAMACUA: Low Cost Real Time Risk Alert and Location System for Healthcare Environments

I. Decia[1], A. Farías[1], D. Szerman[2], L. Grundel[1], D. Briatore[2], M. Piñeyrúa[2], A.Villar[2] and F. Simini[1]

[1] Núcleo de Ingeniería Biomédica, Universidad de la República, Montevideo, URUGUAY
[2] Hospital Maciel, ASSE, Montevideo, URUGUAY

Abstract— **Medication error and other risky situations can lead to serious patient injuries or discomfort. We have designed a real time risk alert and location system to increase patient safety and to reduce medication administration errors. CAMACUA is based on barcode and passive RFID technology for real time location. First field test results show that 100% of patients, visitors and hospital assets are correctly located within hospital rooms or transitions > 1m. When closer than 1m from room entrance, 30% of objects need additional data to determine their location.**

Keywords— **Medication error, Patient safety, Radio frequency identification, Real time location system.**

I. Introduction

Medication error refers, as stated by the National Coordinating Council for Medication Error Reporting and Prevention, to "any preventable event that may cause or lead to inappropriate medication use or patient harm while the medication is in the control of the health care professional, patient, or consumer. Such events may be related to professional practice, health care products, procedures, and systems, including prescribing, order communication, product labeling, packaging, and nomenclature, compounding, dispensing, distribution, administration, education, monitoring, and use" [1].

When administering medication to a patient, the following aspects must be considered:

- The medication must belong to the patient.
- The medication, dose and route must be that of the prescription.
- The dose must be administered in due time.

A study conducted in 2008 in US hospitals concluded that 34% of the medication error occurred at the administration phase, and only 2% are intercepted (near misses). An additional 10% occurs while the prescription is transcribed and medication dispensed [2]. Some of the errors could be life-threatening [3]. These reports [2], [3] conclude that bar-code technology can help mitigate the risk of medication error. Despite the efforts to prevent medication errors, they are still reported in 2016 [4].

Other risk situations are related to patients and visitors wandering within hospital corridors or non-authorized zones. In some occasions wrong patients can even be operated on [5].

The goals of CAMACUA were defined as (1) to reduce medication error, (2) to notify when someone is in an unauthorized zone and (3) the system should be flexible enough to define new risk situations.

II. Real time location systems in healthcare environments

State-of-the-art real time location systems (RTLS) were tested by Clarke in a seminal publication [6] where the most relevant basic functions of such systems are identified, along with their error rate.

Wi-Fi: Based on a proprietary protocol over 802.11b/g [7], the system consists, at the physical level, of wearable devices called tags, and Wi-Fi access points (AP). Each tag periodically sends a "probe request" to the APs, which send in turn the received signal strength indication (RSSI) and tag identification to a central server (CS). Based on the gathered information, the CS computes each (x,y) tag coordinates. The most remarkable functionalities are: real time assets location in 2D coordinates, location history logging and alerts on location and state of tags [8].

Ultrasound: The hardware layer consists of ultrasound (US) mobile emitter tags and fixed receivers. Each tag sends a pulse containing its identification. The receiver demodulates each US pulse, extracting the tag identifier. Since US cannot go through walls [9], it is fair to assume that the identifier corresponds to a tag in the same zone as the receiver. The main functionalities are: real time assets location at zone level, alerts about tags locations, patient fall detection, hands hygiene compliance control [10].

Radio frequency and infrared: It relies on a proprietary protocol based on radio frequency (RF) and infrared (IR). Among the vendor declared functionalities, the most relevant are: real time assets location at zone level, alerts on tags locations, hands hygiene compliance control [11].

Ultra wide band: This system provides only real time location functionality. It consists of emitter tags and fixed receivers, communicating with each other in the ultra wide band (UWB) radio frequency spectrum (6 – 7 GHz). It can locate a tag with 15 cm accuracy [12].

Active radio frequency identification: Consists of self-powered radio frequency identification (RFID) tags emitting signals periodically to readers. They differ from passive RFID systems, where tags are remotely powered. The system main functionalities are: real time assets location at zone level; search assets by location and alerts if some asset exits an assigned zone [13]˙

I. Torres et al. (eds.), *VII Latin American Congress on Biomedical Engineering CLAIB 2016, Bucaramanga, Santander, Colombia, October 26th -28th, 2016*,
IFMBE Proceedings 60,
DOI: 10.1007/978-981-10-4086-3_23

Coordinate level active RFID location: This system is also based on active RFID technology, but the vendor claims it can locate an asset with (x,y) coordinate accuracy. The most remarkable functionalities are: real time assets location; medication near miss or error alerts; notify medical personnel if a patient enters an operating room (OR) and automatic bed management [14].

Table 1 shows some of the results by Clarke [6]. The "n/a" stands for "no test could be made", either due to system limitations or because it was not deployed in all the rooms. The experimental scenario consisted of two ORs separated by scrub, wet storage and autoclave areas. Each system was separately implemented, several tags placed in different locations and situations (i.e. covered by other object, over an instrumental tray, etc.) and the reported locations where compared with the real ones. Further description of the experiment can be found in [6]

Table 1 Error rates for commercial RTLSs

System	Error rates per zone (%)			
	OR1	Wet storage	Scrub	OR2
Wi-Fi	24.7	0.0	22.2	6.7
Ultrasound	2.9	62.9	78.5	70.1
RF + IR	1.3	100.0	100.0	2.5
UWB	n/a	n/a	n/a	14.9
Active RFID (zone)	5.7	95.0	85.1	41.4
Active RFID (x,y)	1.3	42.4	64.1	12.0

Table extracted from Clarke [6]

III. System specification

In order to detect risk situations, our design -which we called CAMACUA- must identify and track hospital entities (staff, patients, visitors, unit-doses, medicines and physical objects). For this, our system combines barcode and RFID technology.

RFID tags and barcodes are attached to entities to be tracked. Upon admission, patients are also provided with an ID barcode wristband. Fixed RFID readers are mounted in room entrances and corridors for receiving signals generated by passing tags. Each employee carries a mobile device, equipped with barcode reader, used to check entities. Visitors are given cards with passive RFID.

If a person or object is identified (by a sensor or employee) in a forbidden zone, the system sends and alert to the security staff, notifying the event. Zones can be rooms, OR, beds or hospital main entrances.

In order to prevent medication administration errors, some mobile function of CAMACUA should request the nurse to scan the patient barcode and unit-dose QR code. The application should check if the unit-dose matches the patient and if it is being administered at the right time. Should a mismatch occur, a warning message is displayed. Detailed information of medicines and patient should also be shown at the bedside, including patient photo, observations and dosages. Leftover medicines are notified to pharmacy in order to prevent waste and unauthorized use.

Patients and objects should also be located using a mobile application, with reports created from the "last seen" zone. CAMACUA should be freely configurable - using a flexible event specification language - to alert in case of entities and locations matches, which include drug mis-administration risks.

IV. System design

A. *Software architecture*

Main system components are (1) the back-end application, (2) mobile application, (3) web client and (4) RFID sensor middleware. The back-end application is the key component, which executes all the event processing logic and notifies when a risk situation has been detected. It is composed of the event engine and the alert system,

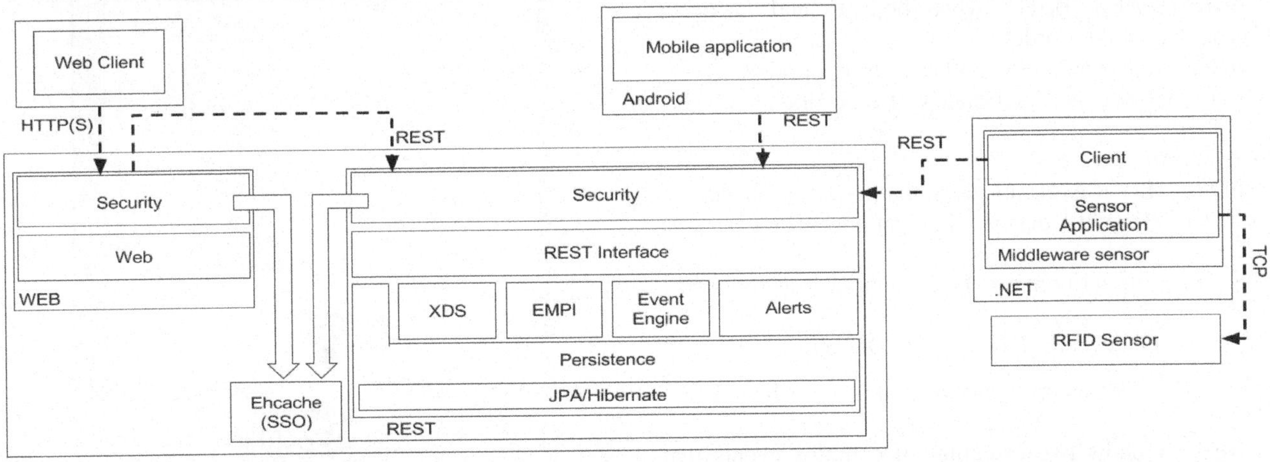

Fig 1: CAMACUA arquitecture. Main components are the back-end application, the mobile application, web client and RFID sensor middleware

and provides a lightweight interface, used by client side components. Web client for administrative staff manages such functions as entity registration and zone permission assignment. The mobile application runs on devices carried by medical and security staff. It is used for entity identification and to display notifications according to predefined priorities. Nursing staff use it as an assistant for unit-dose administration to prevent errors. Finally, the sensor middleware is responsible for receiving and filtering raw data from RFID sensors.

B. *Hardware components*

Hardware components include readers, antennas, tags and barcodes. We selected passive UHF RFID as location technology. As stated by Weis [15], it provides reading ranges up to 10m, doesn't require "line of sight" and it is supported by well known standards (EPC UHF Gen2). Furthermore, passive tags do not need power supply and are available in a variety of materials and shapes. Sensors are connected to circular polarized antennas which send and receive signals from/to tags. Finally, we use barcodes which provide an inexpensive way to identify entities.

C. *Event specification*

Events are classified as primitive or complex. Primitive events are generated by external actors and constitute the main input data. They can represent, for example, the following situations: a patient, unit-dose or object is identified; a unit-dose is packaged; there are unused medicines or a log-out. All triggered primitive events are persisted in an event log for further analysis and traceability. On the other hand, complex events are combinations of primitive events according to a set of operators applied to both primitive and complex events. In this paper we consider the following basic operators, based on the work of Wang et al. [16]:

a) Logical operators
- *AND* $(E_1 \wedge E_2)$: occurs when both E_1 and E_2 occur irrespective of their order.
- *OR* $(E_1 \vee E_2)$: occurs when either E_1 or E_2 occurs.
- *NOT* $(\neg E)$: occurs if no instance of E occurs.

b) Temporal operators
- *SEQ* $(E_1; E_2)$: if E_2 occurs given that E_1 has occurred.
- *WITHIN (E, t)*: if E occurs within an interval $\leq t$.

The syntax for defining events and actions is as follows:

ON event ; IF condition ; DO action

where *action* is executed iff *condition* is true when *event* occurs.

Next we show two examples of complex events that are freely defined by the administrator of CAMACUA. Let be *P*, *U* denote the following primitive events:

P: A patient is identified. *P.patientID* is the patient identifier, *P.zoneID* is the zone where the patient was detected and *P.userID* is the user who identified him/her.
U: A unit-dose is identified. *U.unit-doseID* is the unit-dose identifier, *U.patientID* is the patient whom the unit-dose belongs to & *U.userID* is the user who identified it.

(I) Patients in unauthorized zone:
ON *P*
IF is_in_unauthorized_zone(*P.patientID*, *P.zoneID*)
DO notify_security_staff();
(II) Wrong unit-dose-patient correspondence:
ON WITHIN(*P* ; *U* , 2 min)
IF (*P.userID*==*U.userID*) && (*U.patientID*!=*P.patientID*)
DO display_message("Wrong unit-dose for Patient!");

By launching these two safety rules the administrator is putting in place an error risk reduction functionality.

V. SYSTEM EVALUATION

Several test where conducted to evaluate the system capability to locate a tag within a zone.

Given a certain tag, it is said the system locates it correctly if only a sensor in the same zone as the tag detects it.

Figure 2 shows the RFID sensors placement schematics. One of them was installed in the room above the door (RS), and the other, in the corridor (CR). The tested scenarios are:

1. Person in corridor > 1m from door. Closed door.
2. Same as scenario 1, Open door.
3. Person in corridor < 1m from door. Closed door.
4. Same as scenario 3. Open door.
5. Person in room > 1m from door. Closed door.
6. Same as scenario 5. Open door.
7. Person in room < 1m from door. Closed door.
8. Same as scenario 7. Open door.

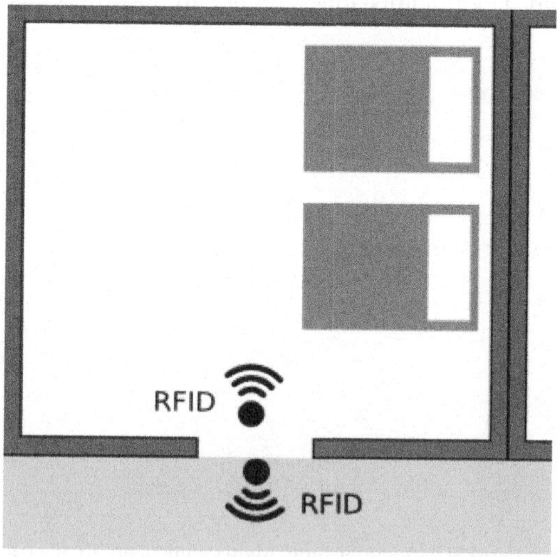

Fig 2: Experimental set up at the entrance door of a two bed Hospital room.

Detection statistics referring to 50 tries for each scenario are summarized in Table 2.

If a person is in the room and he stands more than 1 m from the door, the corridor sensor never detects him, and the room detector always does (scenarios 1, 2, 7 and 8). Closer than 1 m from the door, there is a 10% false negative by the corridor sensor (scenarios 3 and 4) and a 30% false positive by the corridor sensor. In other words the CS puts people in the corridor when they are in the room (30%) and does not see a person in the corridor when he actually is there (10%).

Table 2 CAMACUA Room RFID Tag Detection Error Rate

Scenario	Should detect tag?		Error rate (%)	
	CS	RS	CS	RS
1. corridor >1 m door. CD	Yes	No	0	0
2. corridor >1 m door. OD	Yes	No	0	0
3. corridor < 1 m door. CD	Yes	No	10	0
4. corridor < 1 m door. OD	Yes	No	10	0
5. room > 1 m door. CD	No	Yes	0	0
6. room > 1 m door. OD	No	Yes	0	0
7. room < 1 m door. CD	No	Yes	30	0
8. room < 1 m door. OD	No	Yes	30	0

CS corridor sensor; RS room sensor; CD closed door; OD open door.

On the other hand, regarding medication error mitigation functionality, not enough data were collected to give a quantitative evaluation on system performance yet, although - in testing scenarios - CAMACUA always prevented the user from administering a unit-dose to a wrong patient.

An approximated budget of USD 1500 was enough to buy the necessary equipment to build CAMACUA.

VI. CONCLUSIONS

A system with enough flexibility to define risk situations was built. RFID technology proved to suit the requirements successfully. Barcode as a complementary technology was a good choice, for the system to be built with a reduced budget. Future work includes refinement in the system user interface and evaluation of further field metrics on safety and error reduction impact. Nurses reacted positively to the use of CAMACUA as help to strengthen patient safety by way of assisting in medication administration.

CONFLICT OF INTEREST

The authors declare that they have no conflict of interest.

REFERENCES

1. "About Medication Errors | NCC MERP." [Online]. Available: http://www.nccmerp.org/about-medication-errors. [Accessed: 18-May-2016].
2. ISMP, "Pharmaceutical Bar Coding to Improve Patient Safety: Options for Technical Standards in the Canadian Environment" 2008.
3. E. G. Poon et al., "Effect of Bar-Code Technology on the Safety of Medication Administration," *N. Engl. J. Med.*, vol. 362, no. 18, pp. 1698–1707, 2010.
4. M. L. Durham et al., "Reducing Medication Administration Errors in Acute and Critical Care: Multifaceted Pilot Program Targeting RN Awareness and Behaviors.," *J. Nurs. Adm.*, vol. 46, no. 2, pp. 75–81, Feb. 2016.
5. S. SC and B. P, "Wrong-side/wrong-site, wrong-procedure, and wrong-patient adverse events: Are they preventable?," *Arch. Surg.*, vol. 141, no. 9, pp. 931–939, 2006.
6. D. Clarke, "Active-RFID System Accuracy and its Implications for Clinical Applications," 2006, pp. 21 – 26.
7. "IEEE 802.11 Specification." [Online]. Available: http://standards.ieee.org/getieee802/download/802.11-2012.pdf. [Accessed: 01-Jun-2016].
8. "Ekahau real-time location systems." [Online]. Available: http://www.ekahau.com/userData/ekahau/documents/solution-brochures/Ekahau-RFID-over-WiFi-RTLS-Solutions.pdf. [Accessed: 16-May-2016].
9. M. Mühlhäuser, *Handbook of Research on Ubiquitous Computing Technology for Real Time Enterprises*. Information Science Reference, 2008.
10. "Sonitor RTLS Technologies - Experience Real Time Intelligence." [Online]. Available: http://www.sonitor.com/. [Accessed: 15-May-2016].
11. "Healthcare: Real-Time Patient Status and Healthcare Solutions | Radianse." [Online]. Available: http://www.radianse.com/solution-overview/healthcare/. [Accessed: 17-May-2016].
12. "Ubisense :: Dimension4." [Online]. Available: http://ubisense.net/en/products/Dimension4. [Accessed: 17-May-2016].
13. "GE Healthcare Finds Opportunity, Acquires RTLS Firm - RFID Journal." [Online]. Available: http://www.rfidjournal.com/articles/view?7062. [Accessed: 18-May-2016].
14. "Exavera Technologies Incorporated - Insight Comes from Information™." [Online]. Available: http://www.exavera.com/healthcare/eshepherd.php#. [Accessed: 18-May-2016].
15. S. A. Weis, "RFID (Radio Frequency Identification): Principles and Applications," 2007. [Online]. Available: http://www.eecs.harvard.edu/cs199r/readings/rfid-article.pdf. [Accessed: 13-Jun-2016].
16. F. Wang et al., "Complex RFID event processing," *VLDB J.*, vol. 18, no. 4, pp. 913–931, 2009.

Author: Franco Simini
Institute: Núcleo de Ingeniería Biomédica
Street: Hospital de Clínicas, Av. Italia S/N, 11600
City: Montevideo
Country: URUGUAY
Email: simini@fing.edu.uy

Pupillary Latency in Chromatic High speed Video-oculography

Sánchez. S. Anabel, Suaste G. Ernesto

Centro de Investigación y de Estudios Avanzados del Instituto Politécnico Nacional, Departamento de Ingeniería Eléctrica, Sección Biolectrónica.

Abstract— **to know characteristics of pupillary reflex are useful in medical diagnosis. In this way High Speed Video-oculography (HSV) let to observe pupillary modifications in time more accurately than other video techniques. Differences till 7 ms are possible to measure with HSV in this application. The purpose of this study is to determine if there is a difference in Latency Time (LT) when pupillary reflex is stimulated with three isoluminant chromatic stimuli. The observations obtained from 15 people exhibit red color evoke a slower pupillary contraction than green and blue in a luminance level of 29 cd/m^2.**

Keywords— **HSV, Latency time, isoluminant chromatic stimuli.**

I. Introduction

Pupillary characteristics have been studied with flicker chromatic lights increasing luminance levels (blue, green and red) [1-4]. Kimura E and Young R. observed pupillary responses to six chromatic stimuli [5]. Rodriguez D and Suaste E. studied pupillary responses to chromatic stimuli, inside of visible spectrum from 400 nm to 650 nm during steady state [6].

On the one hand, video techniques have been used to observe pupillary responses some with a range of capture from 20 to 102.4 Hz. And others applications high speed video techniques given a higher sample time helping in many eye observations such as ocular movements, eye tracking systems and surgical monitoring [7-16].

On the other hand, many characteristics from pupillary responses can be studied for example Latency [17-22]. Biologists defined latency; as the spend time to obtain a response, considering it since a stimulus is applied. In this way, pupillary latency can be defined as the lag resulted from nervous transmission or neuromuscular excitation, when stimuli are applied (generally a light stimulus) until the initial contraction of iris muscles. It has been found that latency of pupil constriction did not change with an age between 14-45 years [23].

Many ways to know pupillary Latency Time (LT) have been proposed [22-25]. Bos J. E. described that there is discrepancy between detecting latency time employing the subjective way by hand, and determining it based on deflections from zero.

The **aim** of this study is to determine if there is a difference in LT, among three colors using High Speed Video-oculography (HSV), based on deflections from zero in the velocity signal. Because this meth is suggested for data sampled at a higher rate than 200 Hz [22]. Figure 1 describes this method.

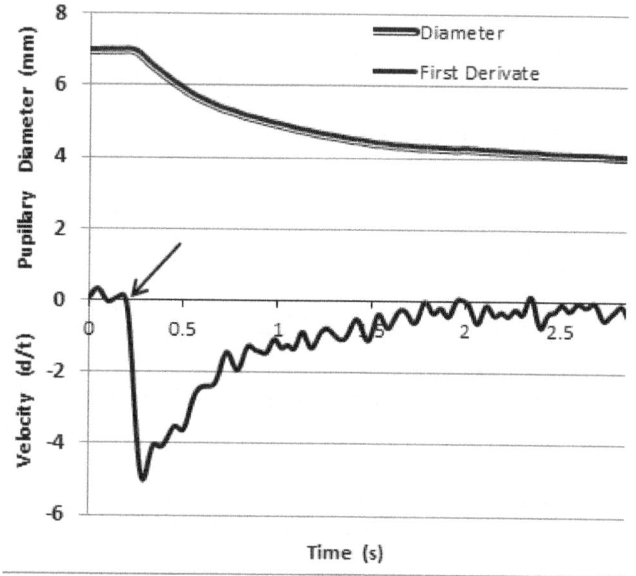

Figure 1 PUPILLARY REFLEX. UP. The model represents a pupillary reflex induced by blue color. DOWN. The first derivate, LT is indicated by the arrow considering the deflection from zero, in the velocity signal. Where: d= mm y t=ms, which corresponds to the image closer in size to the beginning of the stimulation.

II. Methodologhy

The mean age was 29 years in a range from 22 to 42 years. Tests of HS-VOG were realized in 15 subjects. In Vision Builder program 51, 805 images were processed to obtain Lt corresponding to each color (45 values).

A. High Speed Video-Oculography

This study consists in the application of High-Speed Video-oculography (HS-VOG). A High Speed DALSA

I. Torres et al. (eds.), *VII Latin American Congress on Biomedical Engineering CLAIB 2016, Bucaramanga, Santander, Colombia, October 26th -28th, 2016*,
IFMBE Proceedings 60,
DOI: 10.1007/978-981-10-4086-3_24

camera is used (model CA-D6-0256W-ECE), with a capability up 929 frames per second (fps) and a resolution of 260x260 pixels. A study to observe eye characteristics was done with a lower capacity [26]. An infrared light system is employed to capture pupillary images. Sample time was 150 fps (sample time 7 ms), for that pupillary response was not interpolated [13].

First a period of darkness adaptation (10 min), then it is showed the first stimulus (6 s). Images were captured during this time, but focusing in the aim of this paper we are depicting only the 3 first seconds. It means that around of 460 images per stimulus were processed to found their corresponding area. An example of it was illustrated in Figure 1. In Figure 2, they are showed only 4 pictures which were founded exactly in the interest points.

After the first stimulus, one more time darkness adaptation (5 min), and then, the next stimulus was projected. And, lastly the third was reflected.

The stimuli sequence was: red (615nm), blue (455 nm), green (530 nm). The darkness adaptation was to obtain the major pupillary dilatation and to avoid the effects of opponent chromatic perception.

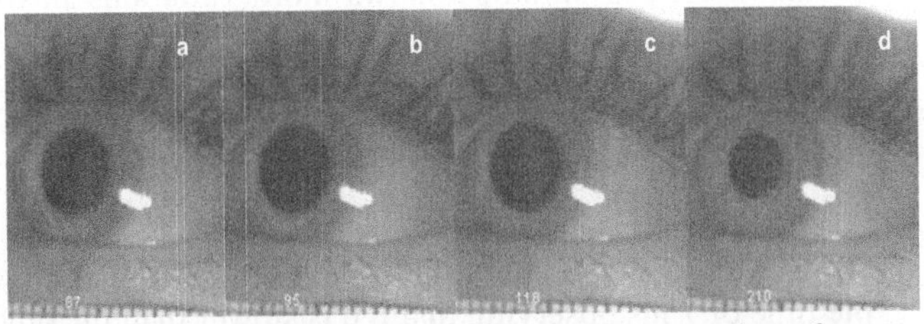

Figure 2 MAIN POINTS. These pictures are examples of one pupillary response in different times during the test. As it is possible to observe, paying attention, the pupillary areas among a, b and c are showing a scarcely difference. Because, a) Pupillary area in the darkness. It describes the pupillary dilatation before the stimulus. b) Pupillary response. It is the first image when the stimulus is activated. c) Latency area. It corresponds to the image linked to the deflection from zero before de maximum contraction velocity. Finally d) Pupillary response in stable state. It is only one image later than the third second. Dow it is showed an scale which let do calibration step in Image Processing.

B. Image Processing

Graphic programming is done with Vision Builder AI 3.5 software which let us to work with a folder of images from a specific address. The actions on the first image will be applied on the other 928. First, the spatial calibration is done by software, it establish the relation between a pixel and real units. In Figure 2 it is possible to observe a scale in millimeters to calibrate. Second, the Region Of Interest (ROI) is delimited with the purpose of to accelerate the process. It is the same for all the images in the processes and it is selected manually taking in account the biggest pupillary size and avoiding eyelashes. Third, smoothing filters are brushed over in order to avoid aberrations, the program contains four sorts: low pass, local, average, Gaussian, Median. It is selected depending of the type of images that we have, as it is programmed in the first picture, it will be established until the last. Contrast is modified to decrease the number of gray levels; it is umbralized, accentuating black and white to remark borders. As a result of it pupillary edge is highlighted, making easier to identify it (with locate feature). Furth, edges are identified in order to calculate pupillary areas as it is described in L. Villamar and E. Suaste 2011, with Hough transformed. Lastly, data is register in an excel file to be analyzed. Figure 3 describes image processing steps.

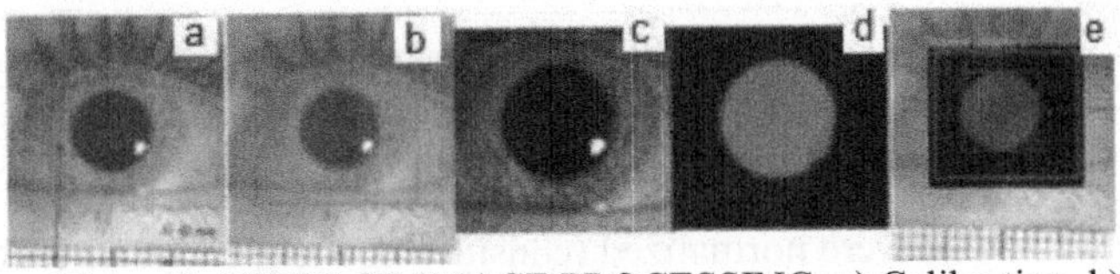

Figure 3. STEPS OF IMAGE PROCESSING. a) Calibration, b) ROI, c) Umbralized, d) Identify objects, and e) pupillary area.

From them, the first derivate is obtained. The time matched to the deflection from zero, closer to the beginning size in the test is chosen as Latency Time.

C. Isoluminant Stimuli

The isoluminant chromatic stimuli (red, green and blue) are adjusted to have a comfortable reading luminance **level** (29 cd/m^2).

The stimuli consist in three different color screens. They were selected, considering bending points showed in D. Rodriguez and E. Suaste [6]. They are the closest of maximum sensitivity of each photoreceptor. There are constituted as CIE 1978 stablished by RGB (Red, Green, Blue) colors. They are described in Table 1[6].

Table 1. Stimuli´s chromatic component.

	RED	GREEN	BLUE
AZUL (440 nm)	1	0	255
VERDE (530 nm)	70	252	0
ROJO (620 nm)	211	18	0

D. Sample Characteristics

Ethics Committee from Research Center and Advanced Studies (CINVESTAV) reviewed and authorized the procedure. **Subjects** were informed about the procedure and signed their **agreement** as Helsinki Declaration has established.

Inclusion criteria were

- Without systemic or ocular diseases.
- No presbyopia
- Normal Chromatic perception
    - It was evaluated with Farnsworth-Munsell test D15. It consists in 15 colored chips which must be ordered in decreased hue.
- Visual Capacity of 20/20.

  They were analyzed with a Snell chart

III. RESULTS

Later, ANOVA test and Means difference were done. To do it, data were normalized (considering the greatest of each group). One group consists of 15 values to each color. ANOVA was done among colors (α= 0.05%), value p=0.022. Homogeneity of variance test, Levene´s describes p=0.6; the population variations are not significantly different due it; it is possible to apply ANOVA.

Values from Means comparison is described in Table 2.

Table 2. Means comparison

	FISHER	TUKEY	BONFERRONI	SIDAK
Red VS **Blue**	0.008*	0.02*	0.02*	0.02*
Green VS **Blue**	0.51	0.78	1	0.88
Green VS **Red**	0.04*	0.1	0.12	0.12
*Statistically different				

LT among colors are statistically different. Mean and standard deviation values are depicted in Figure 3

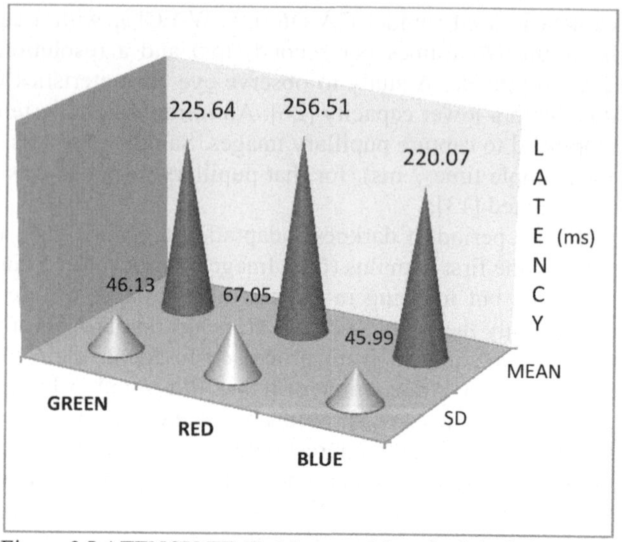

Figure 3 LATENCY TIME. Mean and Standard deviations

LT in response to red color is slower than blue and green. Green and blue responses do not show a statistical difference, they look with similar response times. The standard deviation for red stimulus shows more variations in dates than green and blue.

IV. DISCUSION

Studies describe chromatic changes in size [2-3] from red to blue. In contrast in this study, Lt is analyzed with a sample time of 7 ms observing that red is slower than the other.

Sobaszek A. et al. in their study of pupil response to light stimuli analyzed red, green and blue at three luminance levels. In the Figure 6 of that article, it is possible to observe longer latency times in red than in the other [2]. Also, another study done by Drew P. et. al. of pupillary response to chromatic flicker; in Figure 1b, it is possible to observe that red-black change was largest time to reach its maximum contraction[3]. It is in agreement with the results obtained in this article where red Lt is longer than green and blue, although not statistically. What is important to emphasize is: interpolation of pupil recording is not necessary due to sample time of HSV.

Boss J. E. in their observation suggest sample time higher than 200 Hz are able to determine LT directly. In the results obtained it can be confirmed [16].

Bergamin O. and Kardon Randy demonstrated "variability in latency as a function of intensity in normal subjects" , in this way it is no possible to discuss with it because luminance levels in this study where a control variable [19].

V. CONCLUSIONS

HSV is a very helpful tool to determine Lt, due to reduces errors in sample time and lets to know more precisely it mainly because of frames per second what are captured. It makes no necessary interpolation of pupil recordings.

LT depends of the color which the eye is looking. LT to red color is slower than blue.

ACKNOWLEDGMENT

Authors want to thank CONACYT-Mexico by support to project 141191.

CONFLICT OF INTEREST

The authors declare that they have no conflict of interest.

REFERENCES

1. Kohn M. and Clynes M. Color dynamics of the pupil. *Ann NY Acad Sci* 156(1969) 1.
2. Sobaszek A., Nowak W., Hachol A., Moron Z.Measurement of the eye pupil response to light stimuli with regulated waveform, wavelength and photopic level. *JPhys: Conference Series* 238(2010)1.
3. Drew P., Sayres R., Watanabe K., Shimojo S. Pupillary response to chromatic flicker. *Exp Brain Res* 136(2001)256.
4. Banks M., Munsinger H. (1974). Pupillometric measurement of difference spectra for three color receptors in an adult and a four-year-old. Vision Res. 4:813-817.
5. Kimura E., Young R. S. L. (1995). Nature of the pupillary responses evoked by chromatic flashes on a white background. Vision Res. 35:897-906.
6. Rodríguez D, Zuñiga A, Suaste E. Respuestas pupilares a estímulos cromáticos en el espectro de 400nm a 650nm, en el estado estable.*Rev Mex Fis* 51(2005)365.
7. Hammoud R. I. *Passive eye monitoring, Algorithms, applications and experiments,* (Springer, 2008)1.
8. Bond L. J., Cimino W.W. Physics of ultrasonic surgery using tissue fragmentation: Part II. *Ultrasound in Med. and Biol.* 22(1996)101.
9. Miyosi T., Yoshida H. J. Ultra-high-speed digital video images of vibrations of an ultrasonic tip and phacoemulsification *Cataract Refract Surg* 34(2008)1024.
10. Miyosi T., Yoshida H. J. Emulsification action of longitudinal and torsional ultrasound tips and the effect on treatment of the nucleus during phacoemulsification.*Cataract Refract Surg* 36(2010)1201.
11. Fernandez L. E. de Castro, Dimalanta L.C., Solomon K. D. J. *Cataract Refract Surg* 36 (2010) 1201.
12. Duchowski A. *Eye tracking methodology, Theory and practice,* (Second edition. Springer 2007), p.49-59.
13. Skatani T., Isa T. (2004). PC-based high speed video-oculography for measuring rapid eye movements in mice. Neurosci. Res. 49:123-131. A. Iijima, H. Minamitani, N. Ishikawa, *Med Biol Eng Comput* 39(2001)2.
14. Hachol A, Szczepanowska-Nowak W, Kasprzak H, Zawojska I, Dudzinski A, Kinasz R, Wygledowska-Promienska D. Measurement of pupil reactivity using fast pupillometry. Physiol Meas 2007;28:61-72.
15. Villamar L, Suaste E. High velocity videoculography to determination of the pupil dynamics. *Proceedings of The Tenth Mexican Symposium on Medical Physics,* Mexico 2008, edited by Am Institute of Phys, (Melville, New York, 2008), p.276.
16. Bos J. E. (1991). Detection of the pupil constriction latency. Med Biol Eng Comput. 529- 34.
17. Bitsios P., Prettyman R., Szabadi E. (1996). Changes in autonomic function with age: a study of pupillary kinetics in healthy young and old people. Age Ageing 25:432-438.
18. Barbur J.L., Wolf J., Lennie P. (1998) Visual processing levels revealed by response latencies to changes in different visual attributes.*Proc R. Soc. Lond. B.* 265(1998)2321.
19. Bergamin, O., Kardon R. Latency of the Pupil Light Reflex: Sample Rate, Stimulus Intensity, and Variation in Normal Subjects. *Invest Ophthalmol Vis Sci* 4(2003)1546.
20. Feinberg R, Podolak E (1965) Latency of pupillary reflex to light stimulation and its relationship to aging. In: Behavior, aging and the nervous system (Welford AT, Birren JE, eds), pp 1–14. Springfield, IL: Charles C. Thomas.
21. Bos J. E., Fock, J. M., Reulen J. P. H. Clinical quantitative pupillometry *Neuro-ophthalmology* 10(1990)187.
22. Kasthurirangan S., Glasser A. Age related changes in the characteristics of the near pupil response Vision Res 2006; 6:1393-1403
23. J. L. Barbur, J. Wolf, P. Lennie, *Proc R. Soc. Lond. B.* 265 (1998) 2321.
24. A. Migliaccio, H. MacDougall, L. Minor and C. Della Santana, J Neur Meth, 143 (2) pp. 141–150, 2005
25. Gunter Wyszecki y W.S. Stiles, Color Science: Concepts and Methods, quantitative data formulae, Second edition (John Wiley and sons INC, 2000)
26. Migliaccio A., MacDougall H., Minor L. and Della Santana C. Inexpensive system for real-time 3 dimensional video-oculography using a fluorescent marker array. J Neur Meth, 143 (2) pp. 141–150, 2005INC, 2000)

Pressure Monitor in Endotracheal Cuff for Effective Intubation of Patients

S. Villa-Arango[1, 2], E. Mejía-Mejía[1], J. M. Gómez[1], S. Uribe[1] and R. Torres[1]

[1] Biomedical Engineering Research Group EIA-CES (GIBEC), EIA University, Envigado, Colombia
[2] Biomedical Engineering Research Group (BERG), City University London, London, UK

Abstract—**Mechanical ventilation is usually used in patients found in the intensive care units in order to avoid the appearance of respiratory acidosis, rise the oxygenation levels and reduce the workload of breathing. Endotracheal tubes have a special cuff located in the tracheal area which is used to assure an efficient ventilation of the patient. These cuffs are generally inflated without measuring its pressure, which could cause several damages to tracheal tissue or under-oxygenation. An instrument for the measurement of the endotracheal cuff pressure is proposed in this study for its use in clinical environments. It consists on an analog and digital system which measures the pressure inside the cuff and displays it on an LCD. It is portable, easy to use, non-invasive, has suitable resolution, and in a future iteration will feature a wireless communication module for its use in intensive care units. Its mean appeal is the possibility of deploying it as a cheap, high-resolution and self-calibrating monitoring device for the measurement of the pressure on the endotracheal cuff, reducing the occurrence of complications that may arise from high pressure in the cuff.**

Keywords— **Mechanical ventilation, intubation, endotracheal cuff.**

I. Introduction

Human ventilation is the process in which the air travels from outside the body to the alveoli where the gas exchange process is performed, oxygenating blood and freeing it from carbon dioxide. Most of the times, human ventilation needs to be assisted during surgeries and in patients located in the intensive care units (ICUs), using mechanical ventilators. According to Rodríguez and Brochard, mechanical ventilation is generally used in patients in the ICU to reduce respiratory workload, increase oxygenation or amend respiratory acidosis [1].

Mechanical ventilators must be connected to an endotracheal tube through which air flows from the ventilator to the lungs. Generally, this tubes are positioned using a laryngoscope and are equipped with an endotracheal cuff. These endotracheal tube cuffs are used for 2 main reasons: they decrease the possibility of aspiration and they help to prevent air leakages during ventilation [2] .

The pressure inside the cuff must be monitored regularly and must be between 20 and 30 cmH_2O to assure an adequate perfusion pressure of the tracheal tissue [3]. It's been reported that hyperinflation of the cuff causes mucosal damage, destroying cilia and even ulcerating or perforating the trachea, whereas underinflating the cuff increases aspiration risk [4]. Moreover, prolonged pressures over 30 cmH_2O can cause serious complications as tracheoesophageal fistula [2].

In a study performed in a trauma unit, it was found that cuff pressures in patients admitted to the service ranged from 10 to 120 cmH_2O, and only 23% of the cases showed cuff pressures within the acceptable range [5]. Moreover, once inflated, the cuff pressure changes and therefore needs to be adjusted once in a while [6].

The measurement of this pressure is not always performed, although several devices to measure it are available [7]. In Colombia, these devices are costly and not available in every ICU and usually the inflation of the cuff is performed empirically by nurses using a syringe. Therefore, a system for the continuous, non-invasive measurement of cuff pressure is proposed in this paper, which features a digital visualization panel, an integrated syringe for the cuff inflation, a high-resolution sensor and low-cost materials, focused on usability in clinical environments.

II. Materials and Methods

A. Hardware

The block diagram of the circuit proposed in this paper is shown in Fig. 1. The sensor's reference is MP3V5010 (Freescale Semiconductor, USA) (Fig. 1 (a)). This sensor features a maximum error of 5%, is ideally suited for microprocessor or microcontroller-based systems, and has a pressure range from 0 to 101 cmH_2O [8]. The output of the sensor is passed through a first-order low-pass filter with cutoff frequency of 750 Hz (Fig. 1 (b)). An analog switch takes the pressure signal (Fig. 1 (c)). The switch selects whether to send the pressure measured by the sensor to the microcontroller, or an automatic calibration signal generated by the same microcontroller.

The automatic calibration function is applied every time the system is turned on. It consists on the modification of some parameters in the equation for acquiring the pressure value by the calculation of a known pressure value sent by the microcontroller. Once the system is calibrated, the switch allows the measured signal to pass and it is acquired by the

I. Torres et al. (eds.), *VII Latin American Congress on Biomedical Engineering CLAIB 2016, Bucaramanga, Santander, Colombia, October 26th -28th, 2016*,
IFMBE Proceedings 60,
DOI: 10.1007/978-981-10-4086-3_25

microcontroller using an analog-to-digital converter available in the microcontroller. For this system, an Arduino FIO (Arduino, USA) was used (Fig. 1 (d)). This system allows the implementation of an Xbee module (Digi, USA) which communicates wirelessly to another module connected to a computer. A 4-bits LCD display is used to indicate the pressure measured inside the cuff (Fig. 1 (e)).

The system is supplied by a 9-volts battery, making the system portable. The battery level is measured as well (Fig. 1 (f)) and displayed as a percentage in the LCD display.

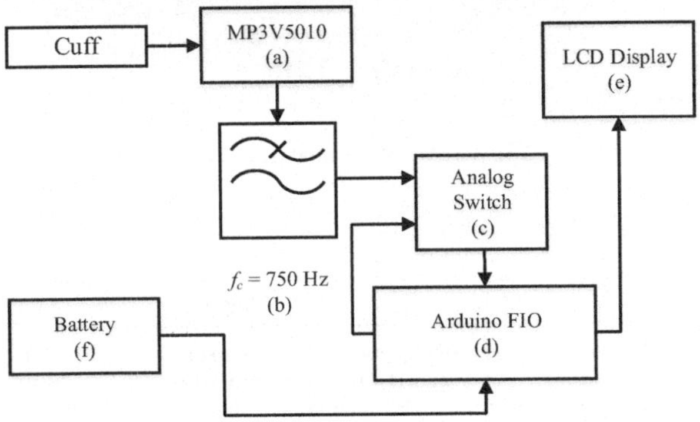

Fig. 1 Block diagram of the hardware of the system

B. Case

An enclosure in which the system can be allocated and transported as a whole was designed. This enclosure needs to be as small as possible to assure portability.

The selected method to inflate the cuff was a syringe, due to its portability and the high resolution it possess. Therefore, the designed enclosure must have an available space in which the syringe can be allocated. To connect the syringe to the cuff, a 3-way valve was used, connecting the pressure sensor both to the syringe and the cuff.

The case was designed using SolidEdge$_{TM}$ (Siemens, USA) and fabricated in ABS using a fused filament fabrication 3D printer. In order to allow the visualization of the LCD and a simple replacement of the battery, a couple of transparent acrylic lids were fabricated.

C. Firmware

The firmware was developed in the Arduino Development Environment (Arduino, USA). It consists mainly in the analog-to-digital conversion of the sensor output and the calculation of the pressure.

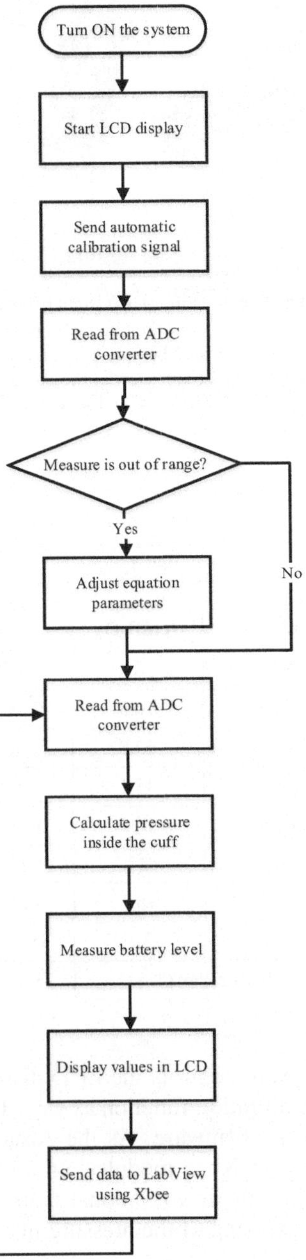

Fig. 2 Flow diagram of the firmware, implemented using Arduino Development Environment

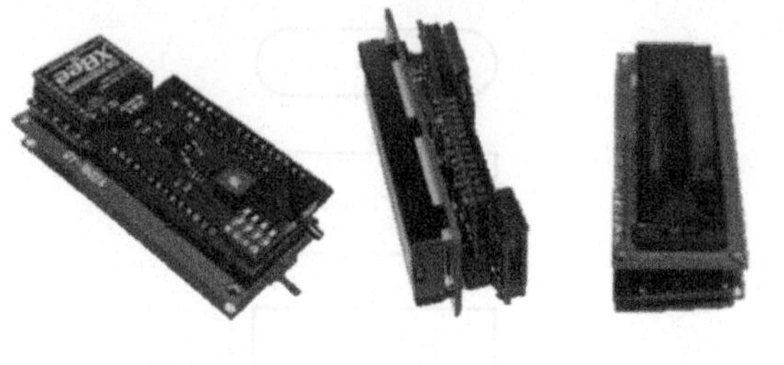

Fig. 3 Constructed circuit. (a) Isometric view. The first panel corresponds to the PCB of the circuit, on top of the Arduino FIO and the LCD. It has an Xbee module attached to it. (b) Lateral view of the circuit. (c) Top view of the circuit, corresponding to the LCD display

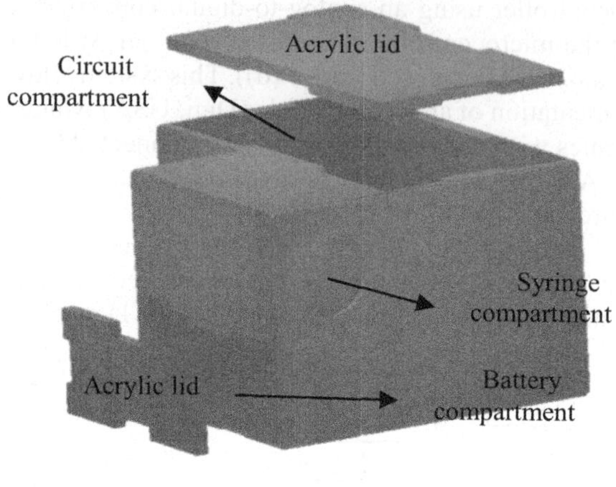

Fig. 4 CAD of the final design of the case

Table 1 Results of the comparison to a reference patron

	Set 1			Set 2			Set 3		
	Reference (cmH_2O)	**Sensor (cmH_2O)**	**Error (%)**	**Reference (cmH_2O)**	**Sensor (cmH_2O)**	**Error (%)**	**Reference (cmH_2O)**	**Sensor (cmH_2O)**	**Error (%)**
1	80.2	80.3	0.12%	86.5	87	0.58%	78.7	78.8	0.13%
2	70.3	70.5	0.28%	83.7	83.5	0.24%	66.8	67	0.30%
3	63.3	63.8	0.79%	75.2	75.6	0.53%	52	52.5	0.96%
4	45	45.1	0.22%	61.9	62.1	0.32%	41.5	41.5	0.00%
5	35.9	36.4	1.39%	48.6	48.6	0.00%	32.3	33	2.17%
6	26.7	27	1.12%	34.5	34.9	1.16%	21.1	21.9	3.79%
7	15.5	16.1	3.87%	18.3	19.1	4.37%	11.2	11.6	3.57%
8	2.8	3.8	35.71%	7.7	8.57	11.30%	2.1	2.69	28.10%
	Mean error (%)		5.44%	**Mean error (%)**		2.31%	**Mean error (%)**		4.88%

The firmware controls the LCD display and performs the automatic calibration function as well. Fig. 2 shows the flow diagram of the firmware. As the Arduino FIO features the connection with Xbee modules, an additional software was developed in LabView (National Instruments, USA) for the wireless monitoring of the pressure in a computer.

D. Performance

The pressure measured by the device is calculated using equation 1 as stated by the sensor manufacturer [8], which establishes a relationship between a static voltage source and a linear response of the sensor in a specific range of temperature for the output voltage and the pressure.

$$V_{out} = V_s(0.09P + 0.08) \pm (0.09V_sP_{error}) \quad (1)$$

V_s: Voltage supply

The performance of the system was evaluated comparing its output with the pressure measurement performed by a Fluke VT Mobile Gas Flow Analyzer (Fluke Biomedical, USA). This reference standard allows the measurement of the pressure of the air going through a conduct. It was connected to the cuff end of the 3-way valve, and the designed device was connected as it would be regularly.

Three sets of recordings were done by measuring 8 different, randomly chosen pressures, and the results were compared with the pressure indicated by the reference standard. The aim of this test was to estimate the performance of the designed device when pressures within the normal operating range are measured. Error was calculated as shown in equation 2.

$$\%Error = \frac{|P_{reference} - P_{sensor}|}{P_{reference}}(100\%) \quad (2)$$

III. Results

The constructed circuit can be seen in Fig. 3. Fig. 4 corresponds to the CAD of the final design of the case. Three sets of recordings were done using the reference standard and the designed system. Table 1 shows the results obtained. Fig. 5 shows the assembly of the hardware, the firmware and the case.

IV. Discussion

The system designed works as intended. The sensor has a resolution of 0.1 cmH_2O and has a mean error of 4.21%. However, in the region of interest between 10 and 40 cmH_2O, the mean error of the sensor is of 2.68%. This indicates that the system can perform well in the ICUs to help maintain an appropriate cuff pressure. Nevertheless, the error of the system can be decreased by improving the conditioning of the signal and by elaborating a more precise calibration curve.

It is important to improve and establish the wireless communication to a remote computer. If this is achieved, a central monitoring system could be used in the ICU where several patients are intubated. This would mean a real-time monitoring of the cuff pressure performed by a healthcare professional in most of the patients in the unit, which could improve their outcomes once they are extubated or generate alarms whenever the pressure is falling or increasing from a secure range.

The system is currently being tested by the Hospital San Vicente Fundación, so far with satisfying outcomes.

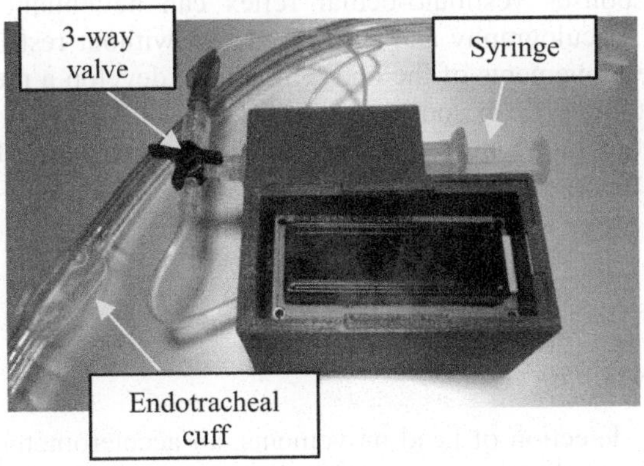

Fig. 5 Final assembly of the hardware, software and case

V. Conclusions

A completely functional device for measuring the pressure inside an endotracheal cuff was designed and fabricated. It features high resolution and portability, and is ideal for its use in ICUs. It is currently being tested and optimized by Hospital San Vicente Fundación, in Medellín, Colombia.

Further works involve the improvements of the case, in order to make it safe for its use in clinical environments; the settlement of a proper curve to measure the pressure inside the cuff, which is expected to reduce error; and the establishment of a wireless communication system for the acquisition of the signals from several different systems in a computer, allowing the monitoring of the pressure in the cuff of different patients at the same time.

Acknowledgment

The authors thank the Hospital San Vicente Fundación for their participation in the project.

Conflict of Interest

The authors declare that they have no conflict of interest.

References

[1] P. O. Rodriguez and L. Brochard, "Ventilación Mecánica Asistida: Hacia una mejor adaptación del respirador a las necesidades del paciente," *Med. Intensiva*, vol. 25, no. 1, pp. 28–38, 2008.

[2] J. Chapman, D. Pallin, L. Ferrara, S. Mortell, J. Pliakas, M. Shear, and S. Thomas, "Endotracheal tube cuff pressures in patients intubated before transport," *Am. J. Emerg. Med.*, vol. 27, no. 8, pp. 980–982, 2009.

[3] Y. Ramírez Tapia, "Rango de Seguridad de la Presión del Globo Endotraqueal con Técnica de Escape Mínimo o de Digitopresión," Instituto Politécnico Nacional, 2010.

[4] C. Ganner, "The accurate measurement of endotracheal tube cuff pressures.," *Br. J. Nurs.*, vol. 10, no. 17, pp. 1127–34, 2001.

[5] T. C. Hardcastle, M. Faurie, and D. J. J. Muckart, "Endotracheal tube cuff pressures and tube position in critically injured patients on arrival at a referral centre: Avoidable harm?," *African J. Emerg. Med.*, pp. 24–29, 2015.

[6] H. Kako, A. Goykhman, A. S. Ramesh, S. G. Krishna, and J. D. Tobias, "Changes in intracuff pressure of a cuffed endotracheal tube during prolonged surgical procedures," *Int. J. Pediatr. Otorhinolaryngol.*, vol. 79, no. 1, pp. 76–79, 2015.

[7] S. Gopalakrishnan, N. Barry, J. Rice, and J. D. Tobias, "Cuffed endotracheal tubes in infants and children: A technique to continuously measure the intracuff pressure," *Int. J. Pediatr. Otorhinolaryngol.*, vol. 77, no. 7, pp. 1135–1138, 2013.

[8] Freescale Semiconductor, "MP3V5010," *Integrated Silicon Pressure Sensor On-Chip Signal Conditioned , Temperature Compensated and Calibrated.* pp. 1–12, 2009.

Author: Elisa Mejía Mejía
Institute: Universidad EIA – Universidad CES
Street: Calle 25 Sur 42 - 73
City: Envigado
Country: Colombia
Email: elisa.mejia@eia.edu.co

Simultaneous detection of the position of eye movements and head for diagnostic purposes of the vestibular system

G. Palomino-Roldán[1], E. Suaste-Gómez[1], A. Castañeda-Galván[2]

[1] IE-Bioelectrónica, CINVESTAV, Ciudad de México, México
[2]UPIITA, IPN, Ciudad de México, México

Abstract— **A device for the simultaneous detection of the position of eye movements and head for diagnostic purposes of the vestibular system was developed. A sensor (3-axis accelerometer) attached to a helmet was used to detect the movements of the head, processing data from the accelerometer was held in LabVIEW. A video camera was used to detect eye movements and then image processing was performed. The tests were conducted on a mobile base in which a second sensor (3-axis accelerometer) was included to perform the tests while the subject was in movement. Experimental results show that it is possible to track eye movements and head without restriction in body movements, which is useful for detecting problems associated with loss of balance, especially under certain circumstances, for example when there is no reference point.**

Keywords— **Accelerometer, vestibular system, vestibulo-ocular reflex, head movements, eye movements.**

I. Introduction

There are three systems that allow us to properly orient in space: the organ of vision, the vestibular apparatus of the inner ear and proprioceptive system [1].

The eye is a highly reliable orientation mechanism when it has appropriate benchmarks. However, under certain conditions, for example for a pilot in flight, interpreting visual signals can be difficult.

Furthermore, the semicircular canals of the vestibular system are not responsible for maintaining balance, however, they are able to detect the movements of the head. Because each of the three semicircular canals is oriented in a different plane, they can detect rotation in three planes. Sensations of relative movement and relative position come from the sensors that are in the skin, joints and muscles. Otoliths provide information about the position [1, 2].

Disorientation in flight is considered as an important pathology in aviation medicine which is based on physiological mechanisms. A common way of inducing vestibular responses is through natural movements of the head and vestibulo-ocular reflex (VOR) [3, 4].

The movements of the eyeballs are coordinated with movements of the head, when attention is fixed on an object and the head is accelerated in vertical, longitudinal, lateral or angular direction, there is a compensation movement which rotates the eye in the opposite direction to the head and with the same magnitude This allows the eyes remain fixed on an object that deserves attention, despite the rapid movement of the body and head [5, 6].

Eye movements are controlled by the vestibular system, therefore, they can be influenced by stimulation of the vestibular system, but are also influenced by visual and neck proprioceptive inputs.

The vestibular nuclei are attached directly to the brainstem nuclei which interprets the signals as a movement [7].

To detect eye movements often it is used video - oculography, which is a non-invasive measurement technique based on video image processing [8, 9]. The study of head movements is vital when it comes to the relationship between them and the vestibulo - ocular reflex or the relationship between the head - neck. The head movements are important in the gaze, therefore, movements and eye muscles have been studied by many physiologists [10].

From the foregoing arises the interest in studying eye movements, head, and its relationship with the vestibular system to associate them to detecting problems such as disorientation in space, involuntary movements, posture, gait and loss of balance.

Responses vestibulo-ocular reflex can be obtained by video - oculography and accelerometers without restriction on the movements of the body, therefore, develop a mechanism to detect eye and head movements with a similar scenario to which a pilot in flight faces is necessary in order to study the vestibulo-ocular reflex.

II. Methodology

A. Positioning sensors.

For detection of head movements, an accelerometer was fixed in a helmet on subject's head, facing the subject a video camera was placed (about 28.5 cm) that recorded the eye movements. The video camera was mounted on a mobile base on which a second accelerometer was fixed, both accelerometers with sensitivity of 300 mV/g and range of ±3.6 g.

At a distance of approximately 1.50 m a screen was placed, where patterns were projected that the subject must

I. Torres et al. (eds.), *VII Latin American Congress on Biomedical Engineering CLAIB 2016, Bucaramanga, Santander, Colombia, October 26th -28th, 2016*,
IFMBE Proceedings 60,
DOI: 10.1007/978-981-10-4086-3_26

trace while the test was performed. Figure 1 shows the final assembly for performing tests.

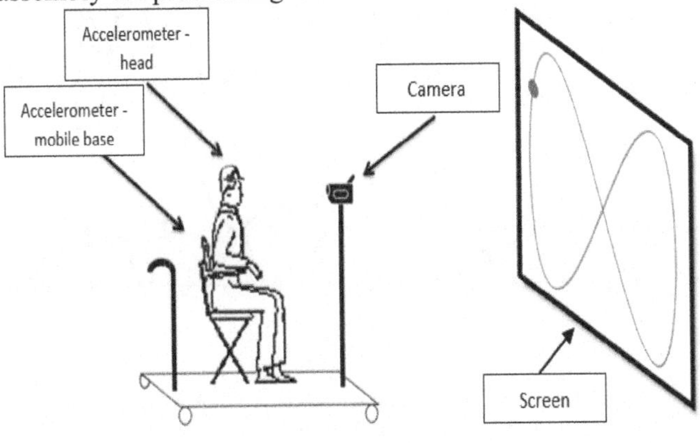

Fig. 1 Final assembly for performing tests.

B. Tests.

The tests were performed with the subject sitting on the mobile base. These consisted of tracking a red point which was projected onto the screen describing a familiar figure, while the mobile base was subjected to quasi-periodic movements. Twenty-four tests were performed. Variations in the tests were the projection of three different patterns and the mobile base was moved from left to right (between ± 28°) or from front to back. Figure 2 illustrates one of the patterns that were projected during the test.

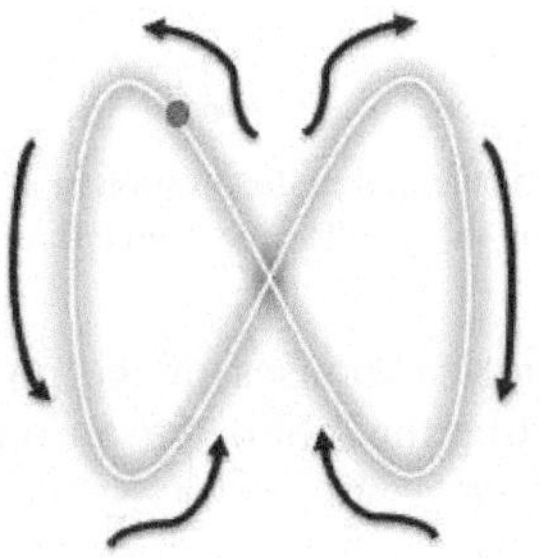

Fig. 2 Projected pattern during testing.

C. Data acquisition and processing.

The Arduino Uno was used as a DAQ through a graphical interface in LabVIEW in order to acquire data from both accelerometers, the data processing was carried out in Matlab to get the angle of displacement of the head and body (mobile base). Arduino and LabVIEW were used by presenting a graphical programming, interactive development and use of libraries that facilitate data acquisition. Figure 3 demonstrates graphically as it was done the acquisition and processing of data obtained from the accelerometers.

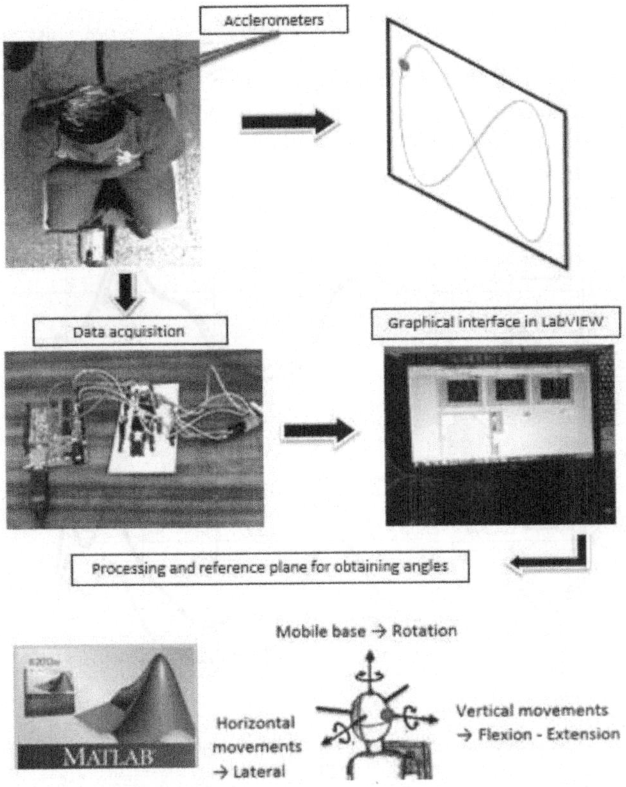

Fig. 3 Diagram of acquisition and data processing of accelerometers.

The image processing for detection of eye movements was performed through an interface in LabVIEW, which allows focus on the subject's eye and then process the data obtained in Matlab to determine the displacement angle of the eyes. Figure 4 depicts graphically as it was done processing for eye movements.

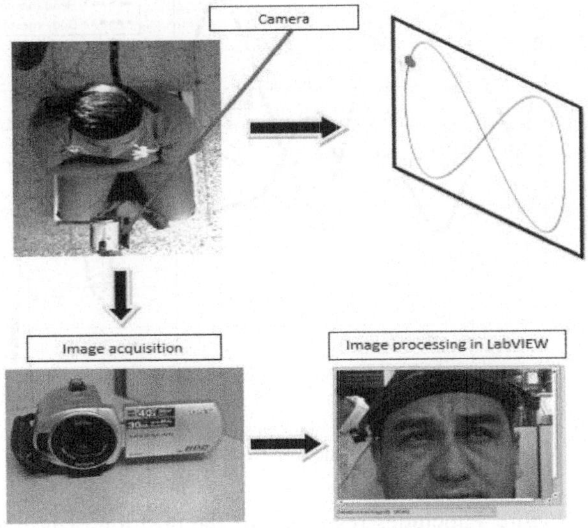

Fig. 4 Diagram of acquisition and processing of images.

III. RESULTS

To verify that the tracking eye movements were correct, the paths of the two eyes of a healthy subject were obtained, which should exhibit similar movements. Figure 5 exhibits the position of vertical and horizontal movements in both eyes.

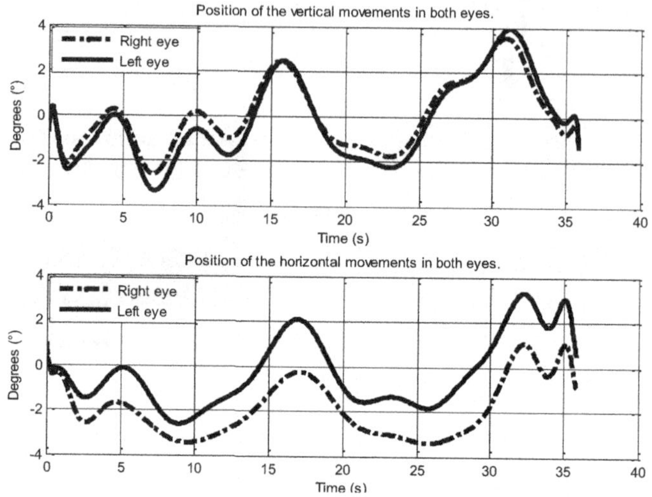

Fig. 5 Position of vertical (up) and horizontal (down) movements in both eyes.

Some tests were performed shifting the mobile base from left to right and others from front to back, then the responses of the eye movements, of the head and the body corresponding movement (mobile platform) were plotted. In Figure 6 can be observed responses of the position of vertical movements, shifting the mobile base from front to back.

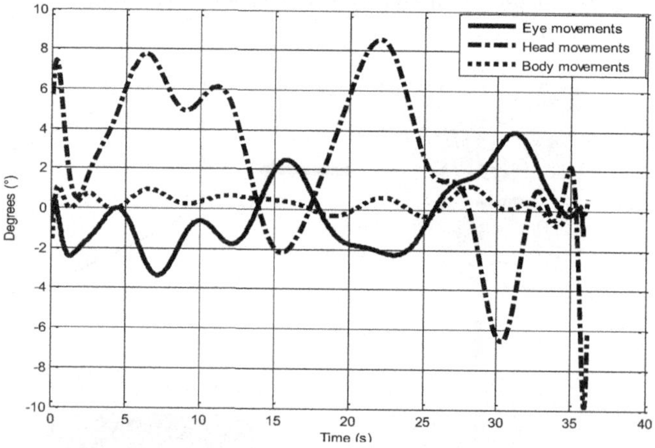

Fig. 6 Responses obtained of the position of vertical movements, shifting the mobile base from front to back.

Figure 7 includes responses of the position of horizontal movements, shifting the mobile base from front to back.

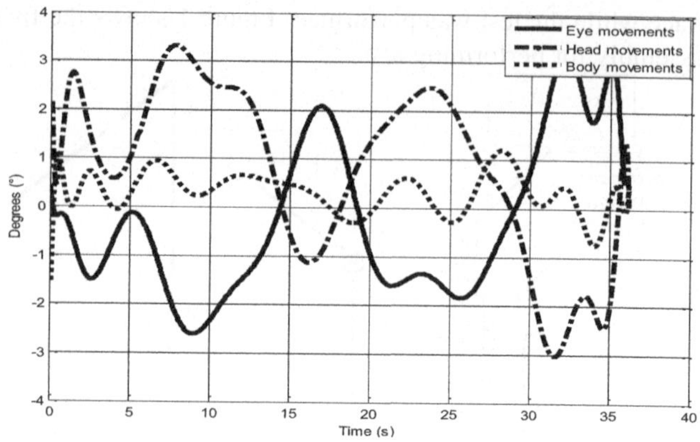

Fig. 7 Responses obtained of the position of horizontal movements, shifting the mobile base from front to back.

Figure 8 illustrates responses of the position of vertical movements, shifting the mobile base from left to right.

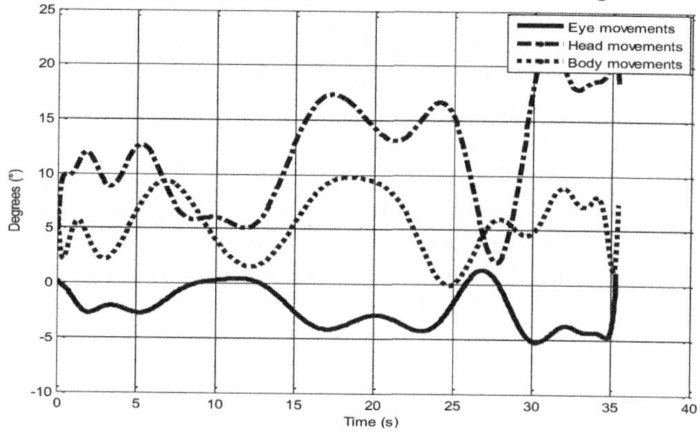

Fig. 8 Responses obtained of the position of vertical movements, shifting the mobile base from left to right.

Figure 9 depicts responses of the position of horizontal movements, shifting the mobile base from left to right.

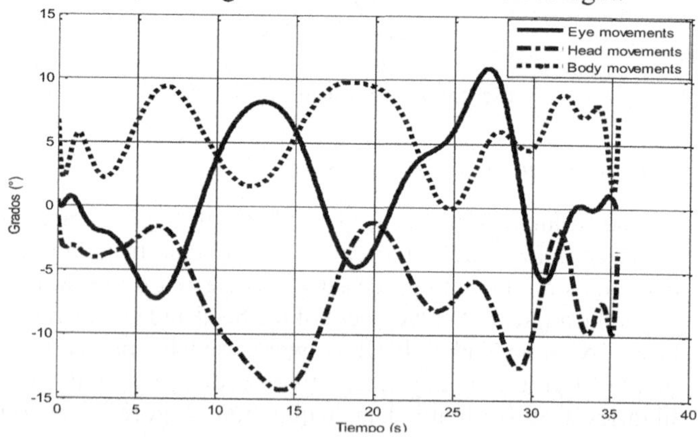

Fig. 9 Responses obtained of the position of horizontal movements, shifting the mobile base from left to right.

Finally, the velocities of the position of movements were obtained. Figure 10 corresponds to the velocities of the responses obtained in Figure 9.

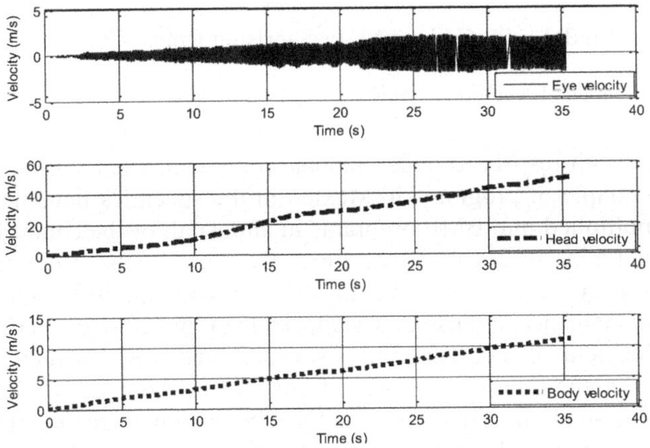

Fig. 10 Velocities of the responses obtained of the position of horizontal movements, shifting the mobile base from left to right.

IV. Discussion

In the figures shown can observe a very similar response to that obtained in the vestibulo-ocular reflex (VOR), however, the amplitude of head movements in most cases is usually higher unlike response vestibule-ocular reflex where the amplitude is equal in a healthy patient. The detected amplitude is greater in the head because all tests were performed with the subject in motion, so the head tends to compensate for body movements to maintain the gaze and track the projected pattern.

Figure 7 shows that the amplitude of movement of the head is very similar to the eye (VOR), this because the head receives less perturbation from front to back than from left to right and therefore the head movements to compensate for the movement of the body are minimal.

According to the velocities obtained, it demonstrates that velocity in the head is higher due to body movement and the required movement for tracking the pattern.

V. Conclusion

It is noted that the proposed method for position detection of movements in order to obtain the response of the vestibulo-ocular reflex is feasible, especially for cases where is required perform random movements or where there is no reference point, tests and records that can be of great help to areas such as aviation medicine, to observe the responses of the vestibulo-ocular reflex in the pilots. Also with these records of positions, it is possible to know velocities of the eye, head and body.

These tests can be used for early detection of the presence of different pathologies like disorientation in space, involuntary movements, posture, gait and loss of balance.

Acknowledgment

The authors would like to thank CONACYT for financial support through the project 141194.

Conflict of Interest

The authors declare that we have no conflict of interest.

References

[1] Arthur C. Guyton. (2005) Textbook of Medical Physiology, 11th edition. Elsevier Saunders. Philadelphia, Pennsylvania. ISBN: 0-7216-0240-1

[2] John A. Kiernan. (2009) The Human Nervous System an anatomical viewpoint, 9th edition. Wolters Kluwer. London, Canada. ISBN 978-0-7817-8256-2

[3] International Civil Aviation Organization – ICAO. (2012) ICAO Doc 8984 Manual of Civil Aviation Medicine – 3rd edition

[4] Wolfgang H Zangemeister et al (1987) Fixation suppression of the vestibular ocular reflex and head movement correlated EEG potentials. In book: Eye Movements from Physiology to Cognition, pp.247-256. DOI: 10.1016/B978-0-444-70113-8.50037-0

[5] Gabirel M. Gauthier et al (1987) Adaptive Optimization of eye-head coordination with degraded vision. In book: Eye movements: from Physiology to Cognition, Publisher: Elsevier Science Publishers BV, Editors: J.K. O'Regan and A. Lévy-Schoen, pp.201-210. DOI: 10.1016/B978-0-444-70113-8.50032-1

[6] M. Rodenburg (1987) Coordination of head and eye position during fixation. In book: Eye Movements from Physiology to Cognition, pp.211-218. DOI: 10.1016/B978-0-444-70113-8.50033-3

[7] Robert A. Moses. (2004) Adler's Physiology of the eye, clinical application, 10th edition. C.V. Mosby Company. USA. ISBN: 0-323-01136-5

[8] Fabiola M. Villalobos-Castaldi, Ernesto Suaste-Gómez (2013) A new spontaneous pupillary oscillation-based verification system, Expert Systems with Applications 40:5352-5362

[9] Ernesto Suaste Gómez and Anabel S. Sánchez Sánchez (2012). Biomedical Instrumentation to Analyze Pupillary Responses in White-Chromatic Stimulation and Its Influence on Diagnosis and Surgical Evaluation, Biomedical Engineering - Technical Applications in Medicine, Dr. Radovan Hudak (Ed.) DOI: 10.5772/48704

[10] L. Stark, W. H. Zangemeister (1981) Simulation of Head Movement Trajectories: Model and Fit to Main Sequence. Biological Cybernetics. 41, 19-32.

Biomedical Engineering in Latin America: A Survey of 90 Undergraduate Programs

J. Azpiroz-Leehan, F. Martínez-Licona, E.G. Urbina-Medal, M. Cadena M, E. Sacristán Rock

[1] Department of Electrical Engineering and CI3M, Universidad Autónoma Metropolitana-Iztapalapa

Abstract— **Undergraduate programs in BME started in Latin America over 40 years ago. From two programs in the seventies, the number has grown into at least 90 at present. This work presents a survey of all the programs and analyzes the curricular structure, aims and focus areas. It shows that the original aims that were to manage and maintain the technical structure in the health sectors still persist. The influence of early founders and the field of electrical engineering now hinder its transformation that would address more challenging problems in fields such as cellular, molecular and tissue engineering, for example. The region has an abundance of very similar programs at institutions that could be considered to be "Teaching Universities" Much work needs to be done to transform the aims and approaches into programs that stress Engineering Design, Innovation and Development.**

Keywords— **Curricular Design, Teaching Universities, Research Universities, Biomedical Engineering Programs.**

I. Introduction

The discipline of Biomedical Engineering (BME) in Latin America is not new. It was first established in the very early seventies by groups in Argentina, Brazil and Mexico by physiologists and electrical engineers who were able to start research groups dedicated to physiological research and biomedical instrumentation. Most of these pioneers had found their training in the U.S, while working at research institutions and universities. For example, among the first returnees from abroad in different countries we find Max Valentinuzzi in Argentina, who came from the Baylor school of medicine; Joaquín Remolina, from Mexico, who studied at the Grass Institute, while Drance de Matos Amorim was working at the Research Division of RCA Labs. In Brazil and Mexico, these groups started graduate programs at their institutions (Universidade Federal Do Rio de Janeiro, Instituto Politécnico Nacional) within a few years of their founding. In Argentina, graduate degrees were offered at Universidad Nacional de Tucumán on a personalized basis before the discipline was recognized officially. In Colombia, Jorge Reynolds, from Trinity College and Salomon Hakim were the first researchers who started the development of the discipline in their country, although BME programs were started several years after they were recognized internationally [1-6].

With respect to undergraduate programs, what started as a couple of programs in Mexico in the seventies have now multiplied into BME programs in almost all of the countries in Latin America. At the time, the aims of the programs were to "Train high-level professionals for the installation, maintenance and use of systems and equipment dedicated to the field of medicine and associated sciences". In this respect the aims were very different from those established almost at the same time in the US, which were oriented towards the "Design equipment and devices, such as artificial internal organs, replacements for body parts, and machines for diagnosing medical problems". In Mexico, initial funding for these programs and jobs for its graduates came from the public health sector, while in the US funding was provided by the National Institutes of Health, the National Science Foundation and private foundations such as the Whitaker Institute

At present, there are at least 85 BME undergraduate programs in the region. There are a wide variety of types in the region. For example, in Mexico, we find some programs that are only three years long, while in Chile there are programs that are six years in length. This difference leads us to carefully examine the types of programs in order to evaluate their relevance to the profession. The purpose of this paper is to analyze the aims, characteristics and outlook for the profession in Latin America, based on the analysis of these programs.
In view of these differences around the world, Krishnan [7,8] proposed a classification of university programs in BME, based on the qualifications, funding and needs of the different countries. He then proceeds to recommend some types of programs, based on his classification, for incorporation in different countries. The models he proposes are:

1. Research Universities with multiple focus areas.
2. Limited focus areas and co/internship in hospitals or medical companies to prepare students for work in those industries.
3. Universities that provide associate degrees or training similar to those for BM equipment technicians of BME technologists.

A secondary aim of this paper is to analyze the information we have gathered in light of this point of view.

I. Torres et al. (eds.), *VII Latin American Congress on Biomedical Engineering CLAIB 2016, Bucaramanga, Santander, Colombia, October 26th -28th, 2016*,
IFMBE Proceedings 60,
DOI: 10.1007/978-981-10-4086-3_27

II. METHODS

Information on all the BME curricula was obtained from several national BME societies' websites (in Mexico, SOMIB; in Brazil, SBEB; In Colombia, ABIOIN; in Argentina, SABI) [7-10], from publications such as the Colombian, Mexican and Brazilian Journal of Biomedical Engineering, and from consultations with colleagues from other institutions in Latin America. In all, the curricula from 90 institutions were analyzed. We did not attempt to find every program in the region but we are reasonably sure that the programs that were included represent over 95% of the programs on offer.

The countries and the number of institutions are noted in Table 1

Table 1 Countries and Number of Institutions offering BME programs

Country	Number of Programs
Mexico	48
Colombia	13
Brazil	12
Argentina	7
Chile	2
Honduras	1
El Salvador	1
Peru	1
Bolivia	1
Cuba	1
Venezuela	1
Paraguay	1
Ecuador	1

Following Krishnan's approach, the curricula were analyzed according to the type and number of courses they offered, the duration of the program and the focus areas in each program. In his analysis, he separates the curriculum into different areas. Table 2 shows these areas and the distribution for the three types of institutions. Additionally, the different focus areas were noted, in order to determine if these were in accordance to the stated aims of the programs.

Table 2 Grouping of subjects and classification into types of universities

Subject Area	% for Research	% for Teaching	% for Technical
Mathematics	13	12	10
Science	15	21	12
Engineering	18	17	27
Biomedical Engineering	37	29	29
Social Science/Humanities	18	21	22

In this work, we were interested in exploring additional details from the curricular analysis, so we divided the curriculum into:

1.Introduction/Capstone Projects
2. Mathematics
3.Science
4.Computing
5.Electrical Engineering
6.Physiology
7.Biomedical Engineering
8. Social Sciences and Humanities
9.Electives/Other Courses

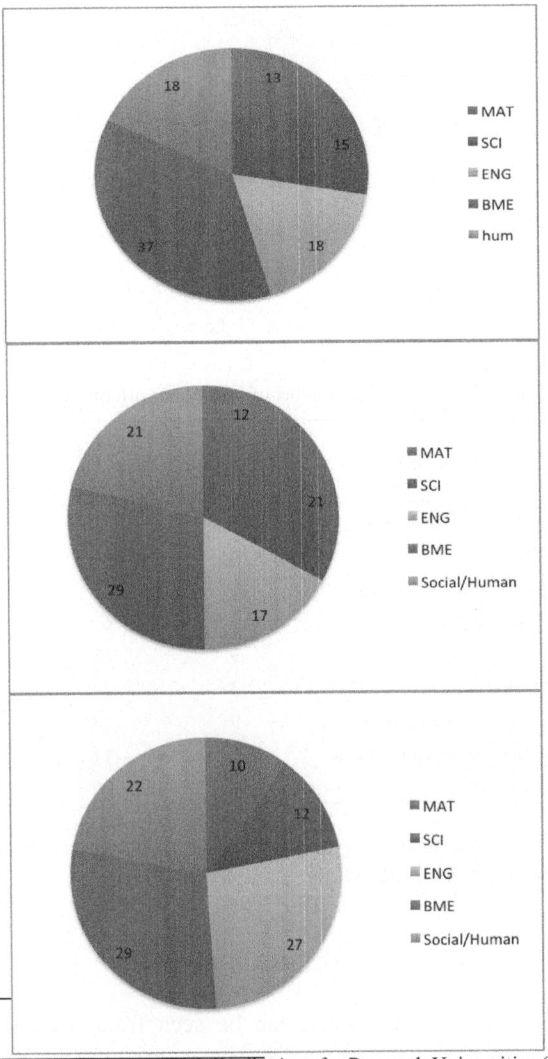

Figure 1: Krishnan's model distributions for Research Universities (top), Teaching Universities (middle) and Technical or technological colleges (bottom)

So the information that was gathered then was converted to Krishnan's models that are shown in figure 1.

III. Results

Figure 2 shows the average course distributions in Latin America. Table 3 shows the distributions of the different types of courses in number of subjects per type of course and also as a percentage of the curriculum. Other relevant information that was obtained was that the average number of courses was 51 and the average duration of the programs was 4.5 years.

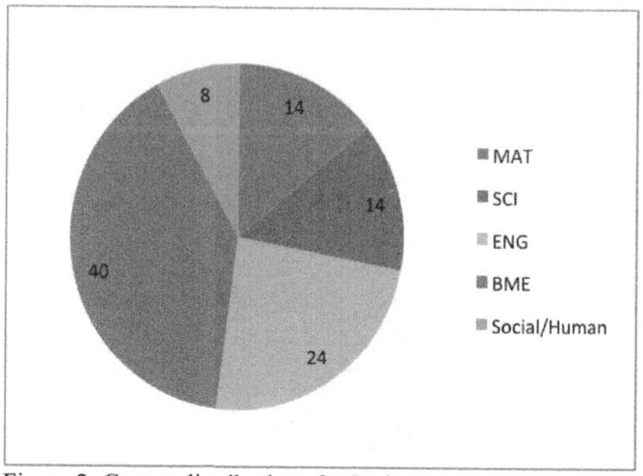

Figure 2: Course distributions for Latin American Universities

Table 3 Countries and Number of Institutions offering BME programs

Type of Course	Average Number of Courses	%Contribution to the program
Introduction/Capstone	4	7
Mathematics	6	12
Science	6	12
Computing	3	6
Electrical Engineering	9	18
Physiology	3	7
Biomedical Engineering	9	18
Social Sciences/Humanities	6	11
Electives/Other	4	9

The distribution by percentage of all the focus areas from the programs under study is shown in figure 3.

IV. Discussion

The first result which can be seen from the curricular analysis is that excepting the 3-year programs at Mexico's Polytechnic Universities and the 6-year programs at Chile's Universities, granting the degree of Biomedical civil engineer, the rest of the university programs are very similar, which is not surprising, given the cross-pollinating among faculties and the time and space distributions of the program starting dates. A second important result is the congruence between the aims of the programs and the distribution of focus areas: These programs are first and foremost designed to produce graduates that can maintain and manage the technological infrastructure in the respective countries' health sectors. It is not surprising to see many courses on clinical engineering, hospital equipment or health technology management, as well as classical medical instrumentation subjects.

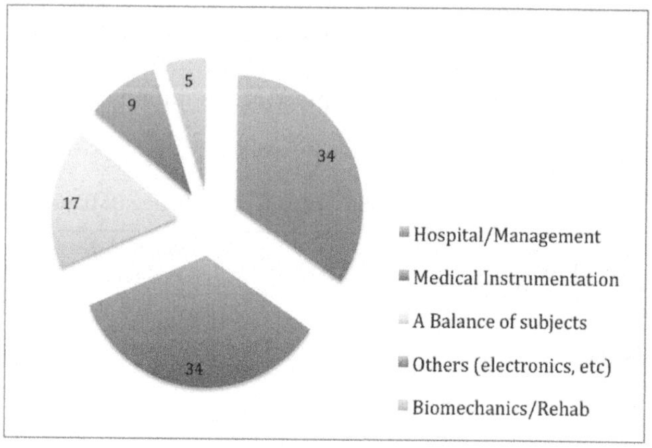

Figure 3.Distribution of focus areas in the BME programs in Latin America

It could be thought that these programs might resemble Krishnan's proposal for BME technicians, but the distribution is different. First, in all of the models, the percentage of SSH courses in the curricula is around 20% while in Latin America it is only 8%. This might underline the technical emphasis in all Latin American Universities to the detriment of a more balanced education. Second, discounting the contributions of SSH courses, the distribution of most programs in Latin America, resembles those of a Research University. This seems a bit contradictory, and indeed if the contribution of these 90 institutions to forums such as the last five IEEE BME conferences is taken into consideration, the universities' scientific production is negligible when compared to those from other parts of the world. Only about five percent of all Latin American institutions that offer the BME undergraduate degree can be considered to be Research Universities by its contributions to the field.

There is more relevant information to be found in the analysis: The number of courses (credits) that are required for graduation is high, compared with those required in countries such as the U.S., even when normalized with respect to the duration of studies. Focus areas are completely different from those outside the region, where cell, molecular and tissue engineering are as important as biomedical signal processing and imaging and both of these comprise almost 50% of focus areas at prominent universities, while what could be called health care (Hospital Engineering/Management in Latin America) is cited in 3% of the programs.

Another prominent difference is the excess influence of electrical engineering subjects in detriment to courses in quantitative physiology, systems biology, or cell and molecular biology. Whereas in some universities the EE requirements are significant, in the US it is common to find requirements for only one or two courses. This excess of EE subjects is evidence of resistance to change in Latin American institutions that prefer to address perceived current needs instead of looking to the future of the profession.

Regardless of the course distribution in the Latin American Universities' curriculum, the aims and the focus areas, it appears that most institutions could be considered to be Teaching Universities, where the emphasis is on engineering knowledge, but not engineering design, as has been documented in [13].

V. Conclusions

From the curricular analysis it can be seen that many universities provide the students with a good number of competencies, but they might be overqualified since the job market only offers positions at a lower technical level.

Technical colleges or technological universities which grant associate degrees are able to provide the "job market" with personnel that are able to solve satisfactorily many of the daily needs for maintenance of BM Equipment and the need to manage the medical infrastructure in these countries. In some cases, graduating engineers become service and applications engineers for sophisticated types of equipment, such as imaging systems, cardiac and surgical suites but have missed the opportunity to contribute to the field, and have not reached their full potential. In short, the state of the profession in the region is at a higher level of maturation than those that several authors considered to be necessary for a developing country [14,15], but the current structure in the programs, the emphasis on techniques, the need for health technology management and the overqualification of graduates will not help solve the real needs in the field of BME.

Only a Research University will allow for a personal and professional development to be fully reached. It will be necessary to pursue original research and development in the field of BME, even with modest means, and even if it means starting or incubating (almost) non existent BME industries in the region. Teaching Universities are leading us nowhere. The challenge in the region will be to transform the institutions into centers of Engineering Design and Innovation that work with industry via technology transfer to provide health solutions to populations. Without innovation it will be impossible to foresee a change from what are and will remain technical programs.

Conflict of Interest

The authors are with the Department of Electrical Engineering and CI3M, Biomedical Engineering and Instrumentation Center, at Universidad Autónoma Metropolitana Iztapalapa.

References

1. COPPE Universidade Federal do Rio de Janeiro Programa de Engenharia Biomedica: 40 anos de engenharia dedicados à saúde at: http://www.peb.ufrj.br/noticias/RevistaBiomedica40Anos.PDF
2. Muñoz J, Joaquín Remolina, in Ciencia y Tecnología en México en el siglo XXI, Biografías de personajes ilustres, Vol IV Academia Mexicana de las Ciencias 2005, pp 205- 234 fin: http://www.uam.mx/e_libros/biografias/REMOLINA.pdfSmith
3. Universidad de Valparaíso at http://icb.uv.cl/index.php/historia
4. Universidad Nacional de Tucumán at http://www.tucuman-conicet.gov.ar/Secciones.php?IdSeccion=17
5. Soto Castaño, J.M Ingeniería biomédica. Historia en construcción Revista Ingeniería Biomédica ISSN 1909-9762, Vol. 3, No. 5, January-June 2009, pp. 28-30
6. Soler López, F. A., La bioingeniería y la ingeniería biomédica en Colombia Revista Ingeniería Biomédica ISSN 1909-9762, Vol. 3, No. 5, January-June 2009, pp. 31-32
7. Krishnan, S. Challenges and Models for Baccalaureate BME Program Design, Proc IEEE EMBC 2015.
8. Krishnan, S. Academic Program Models for Undergraduate Biomedical Engineering Proc IEEE EMBC, pp. 5145-5148, 2014.
9. SBEB at http://www.sbeb.org.br/centros.php
10. SOMIB at http://www.somib.org.mx/universidades.html
11. SABI at http://www.sabi.org.ar/index.php?page=Regionales
12. ABIOIN at http://www.abioin.com
13. Martínez F, Azpiroz J, Urbina G, Sacristan E, Cadena M Critical Assessment and Outlook for the 50 Biomedical Engineering Undergraduate Programs in Mexico *Proc IEEE-EMBC* 2015, pp 3973- 3977.
14. Douglas, T.S. Biomedical Engineering education in developing countries: research synthesis Proc. 33 Ann. IEEE-EMBC pp.3628-3630.
15. Lustick, D.R.; Zaman, M.H. Biomedical Engineering Education and Practice Challenges and opportunities in improving health in developing countries Proc. IEEE Atlanta Conference on Science and Innovation, 2011 DOI:10.1109/ACSIP.2011.6064477

Author: Joaquín Azpiroz Leehan
Institute: CI3M, Dept. Electrical Engineering, UAM-I
Street: Av. San Rafel Atlixco 186 Col. Vicentina
City: Mexico
Country: Mexico
Email: jazpiroz@ci3m.mx

Continuous body temperature monitoring system based on a flexible PPy/PLA wristband.

G. Rodríguez-Roldán[1], E. Suaste-Gómez[1]

[1]CINVESTAV/Department of Electrical Engineering, Section of Bioelectronics, Mexico City, Mexico

Abstract—**A flexible wristband for continuously measuring body temperature was fabricated using a conductive organic polymer, polypyrrole (PPy) and a biodegradable and biocompatible polyester, polylactic acid (PLA). Changes in the PPy/PLA membrane resistivity were observed according to variations of temperature, thus, the response was characterized. Experimental results showed an almost linear and inversely proportional behaviour between the temperature and the resistance of the PPy/PLA wristband. In addition, the LabView based system for continuously monitoring and recording body temperature has been described. This proposed system is a promising tool for medical care due to its reliability for using over a long period of time so that medical staff will no longer have to manually take the patient's temperature, therefore, nursing workload is reduced as well as the quality of medical care is improved.**

Keywords— **Temperature, wristband, flexible, monitoring**

I. Introduction

Body temperature is one vital parameter which is important to measure accurately. A healthy body maintains its temperature within a narrow range using homeostatic thermoregulation mechanisms [1]. The average normal body temperature is between 36.1 ° C and 37.2 ° C [2]. Body temperature varies according to different factors such as age, activity and time of the day [3]. As certain diseases are accompanied by characteristic changes in body temperature, this parameter is measured at hospitals. Nurses generally take manually the patient's temperature from once every 15 minutes to once a day, however, this task, involves a lot of time and many staff members [4].

Nowadays, there are equipment such as thermal imaging cameras that successfully detect changes in body temperature without human intervention, nevertheless, these kind of devices are quite expensive [5].

Currently, new developments in materials field, have allowed the creation of wearable devices, modifying the way that the user can get information about himself such as temperature, blood pressure, pulse without human intervention [6].

Based upon the foregoing, a low-cost medical device to continuously measure the body temperature is required.

In this paper, the stages of fabrication and characterization of a temperature skin sensor based on a flexible PPy/PLA composite and the design of a LabView based continuous temperature monitoring system will be presented.

II. Experimental section

A. Fabrication

1. Flexible Polylactic acid (PLA) was solved in N, N-Dimethylformamide (from Sigma-Aldrich) at 60 °C. The concentration of the solution was about 33 wt%.
2. The solution was mixed by stirring and sonicating.
3. 0.15 grams of PPy were dispersed in the solution and stirred for 15 min at 60 °C
4. The mixture was put into an ultrasonic cleaner bath for 15 min until PPy powder was completely dispersed.
5. PPy/PLA composite was deposited on a glass substrate.
6. DMF was evaporated slowly at room temperature for 2 days.
7. Once the composite is dry, a rectangle was cut out. This rectangle is 18 cm long and 1 cm wide
8. Velcro tape was cut and placed at both ends of the rectangle.
9. Two electrodes made from conductive adhesive copper tape were placed 1 cm apart. Wires were welded over the copper tape.

B. Characterization

a) Temperature characterization:

The temperature sensor characterization was performed using an electric stove from 25 °C to 40 °C. Changes in electrical resistance of this composite were observed according to variations in temperature.

Temperature characterization was done in triplicate.

Finally, tests were validated on human skin by comparing with an infrared thermometer TM-969 from Lutron.

b) Scanning electron microscopy:

In order to obtain topographical, morphological and compositional information, the PPy/PLA membrane was evaluated by scanning electron microscopy (SEM).

I. Torres et al. (eds.), *VII Latin American Congress on Biomedical Engineering CLAIB 2016, Bucaramanga, Santander, Colombia, October 26th -28th, 2016*,
IFMBE Proceedings 60,
DOI: 10.1007/978-981-10-4086-3_28

C. LabView based temperature monitoring system.

The continuous body temperature system was developed in LabView. It displays and plots the body patient's temperature in virtual-real time. The information from a patient is also recorded in a database.

The analog signal was acquired and converted to a digital one through an Arduino board.

The circuit shown in Figure (1) was used to obtain a voltage value as a function of the wristband's resistance. R2 was selected for an intermediate value of the working range of the wristband, which is between 3 kΩ and 5kΩ.

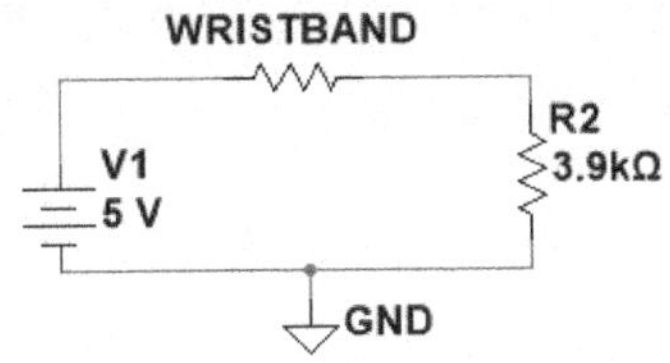

Fig. 1 Voltage divider circuit

Figure (2) shows the block diagram developed in Labview. The first part is the data acquisition using an arduino board. Then, a Mathscript node was used to calculate the resistance of the wristband as function of voltage, so, variations in temperature can be measured according to equation(1). Finally, temperature is displayed and plotted. The user can see the temperature data even in the indicator or the graph. The information is sent to a notepad. This database contains information about date, time and temperature, respectively. Furthermore, the interface contains an alarm, which is activated in case temperature reaches 38 °C.

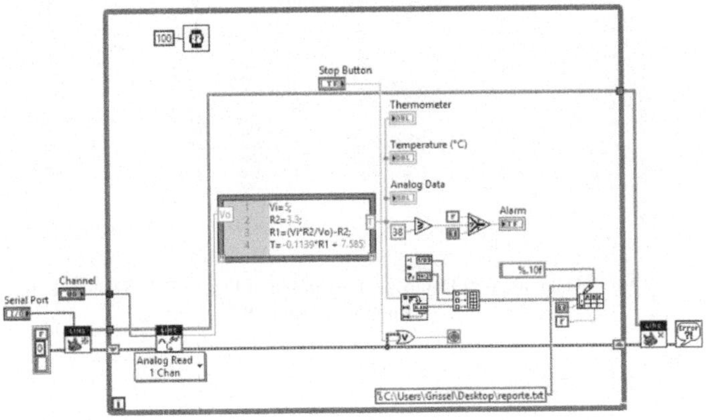

Fig. 2 Block diagram

III. Results

A. Fabrication

Figure (3) shows the wristband made of PPy and PLA.

As it is shown in Figure (4), due to the flexibility of the material, it is easily adaptable to any surface, in this case, skin.

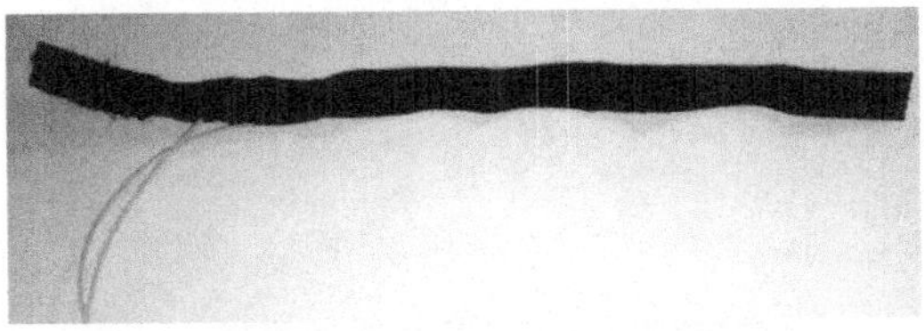

Fig. 3 PPy/PLA wristband

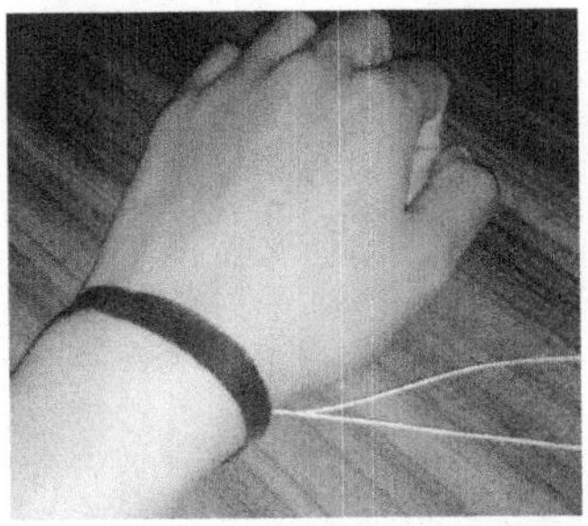

Fig. 4 Flexibility of the PPy/PLA wristband

B. Characterization

a) Temperature characterization

Temperature characterization was done in triplicate. Changes in electrical resistance were observed according to variations in temperature.

Figure (5) shows the response of the PLA/PPy/PVDF composite as a temperature sensor.

The equation that describes this phenomenon is given as,

$$T = -0.1139R + 7.5855 \tag{1}$$

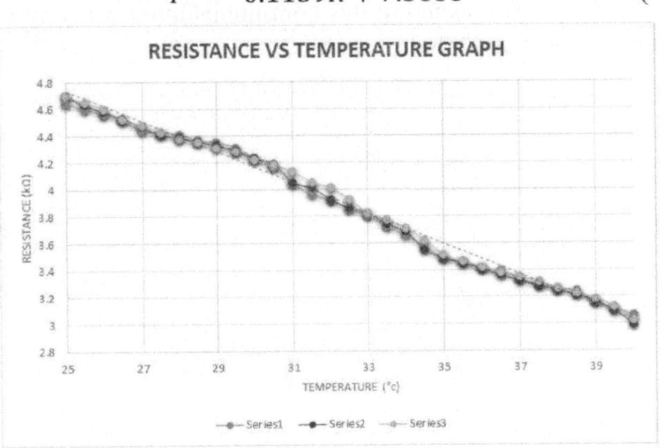

Fig. 5 Resistance versus temperature graph

b) SEM

The PPy/PLA membrane was evaluated by scanning electron microscopy (SEM). Images of the composite at high magnification (1000×, 5000×) are shown in figure (6) where homogeneity of the membrane are clearly seen.

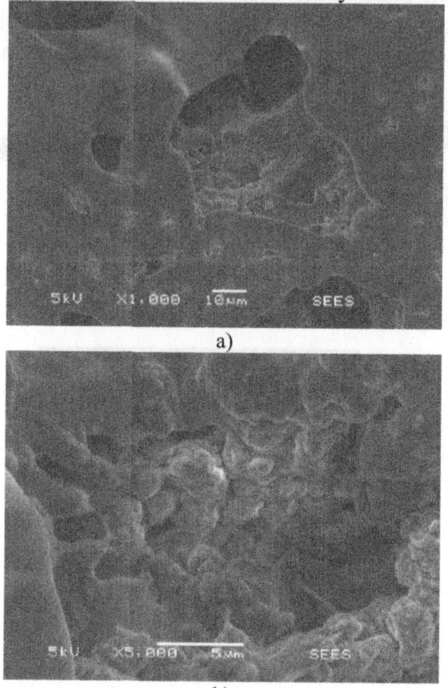

a)

b)

Fig. 6 SEM micrograph of the PPy/PLA composite. a) Magnification 1000x; b) Magnification 5000x

C. LabView based temperature monitoring system.

Figure (7) shows the interface created in Labview. A virtual thermometer displays the body temperature while variations of this one could be observed in the graph. The sampling rate was about twenty samples per hour. In this example, the patient was connected to this continuous monitoring system for one hour and a half.

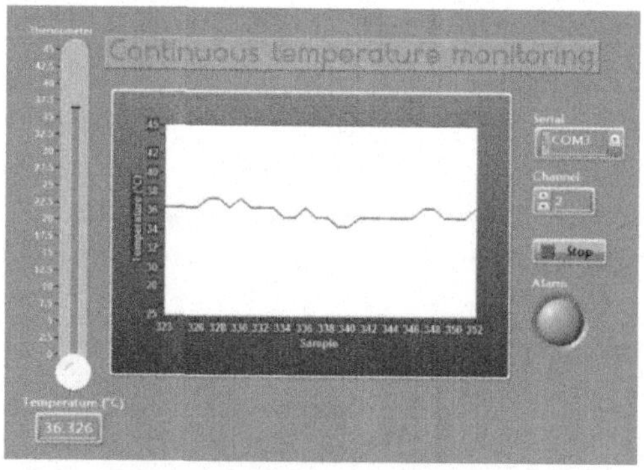

Fig. 7 Continuous temperature monitoring interface

Figure (8) shows a case when temperature reaches 38 °C, so, the alarm is activated.

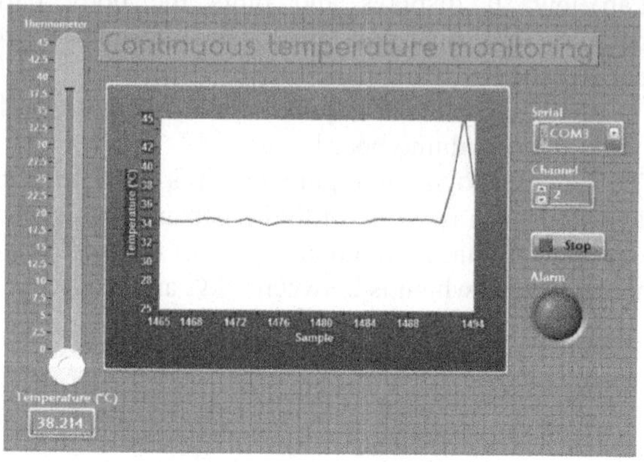

Fig. 8 Alarm system activated

Furthermore, a database containing information about time, date and temperature was created in a notepad. Figure (9) shows examples of databases corresponding to Figure (7) and Figure (8), respectively.

Finally, Figure (10) shows the complete continuous body temperature monitoring system.

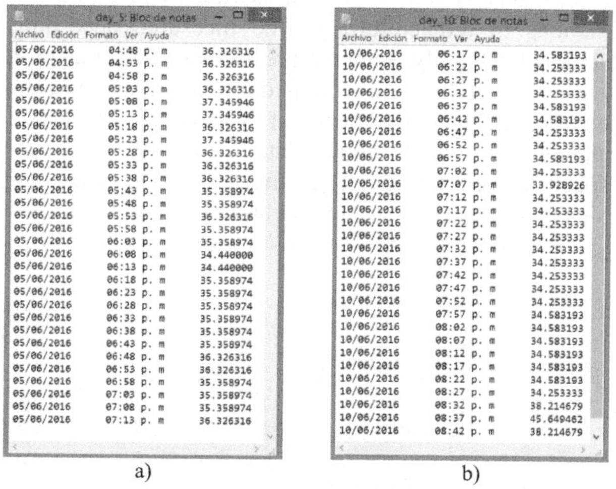

05/06/2016	04:48 p. m	36.326316
05/06/2016	04:53 p. m	36.326316
05/06/2016	04:58 p. m	36.326316
05/06/2016	05:03 p. m	36.326316
05/06/2016	05:08 p. m	37.345946
05/06/2016	05:13 p. m	37.345946
05/06/2016	05:18 p. m	36.326316
05/06/2016	05:23 p. m	37.345946
05/06/2016	05:28 p. m	36.326316
05/06/2016	05:33 p. m	36.326316
05/06/2016	05:38 p. m	36.326316
05/06/2016	05:43 p. m	35.358974
05/06/2016	05:48 p. m	35.358974
05/06/2016	05:53 p. m	36.326316
05/06/2016	05:58 p. m	35.358974
05/06/2016	06:03 p. m	35.358974
05/06/2016	06:08 p. m	34.440000
05/06/2016	06:13 p. m	34.440000
05/06/2016	06:18 p. m	35.358974
05/06/2016	06:23 p. m	35.358974
05/06/2016	06:28 p. m	35.358974
05/06/2016	06:33 p. m	35.358974
05/06/2016	06:38 p. m	35.358974
05/06/2016	06:43 p. m	35.358974
05/06/2016	06:48 p. m	36.326316
05/06/2016	06:53 p. m	36.326316
05/06/2016	06:58 p. m	35.358974
05/06/2016	07:03 p. m	35.358974
05/06/2016	07:08 p. m	35.358974
05/06/2016	07:13 p. m	36.326316

a)

10/06/2016	06:17 p. m	34.583193
10/06/2016	06:22 p. m	34.253333
10/06/2016	06:27 p. m	34.253333
10/06/2016	06:32 p. m	34.253333
10/06/2016	06:37 p. m	34.583193
10/06/2016	06:42 p. m	34.583193
10/06/2016	06:47 p. m	34.253333
10/06/2016	06:52 p. m	34.253333
10/06/2016	06:57 p. m	34.583193
10/06/2016	07:02 p. m	34.253333
10/06/2016	07:07 p. m	33.928926
10/06/2016	07:12 p. m	34.253333
10/06/2016	07:17 p. m	34.253333
10/06/2016	07:22 p. m	34.253333
10/06/2016	07:27 p. m	34.253333
10/06/2016	07:32 p. m	34.253333
10/06/2016	07:37 p. m	34.253333
10/06/2016	07:42 p. m	34.253333
10/06/2016	07:47 p. m	34.253333
10/06/2016	07:52 p. m	34.253333
10/06/2016	07:57 p. m	34.583193
10/06/2016	08:02 p. m	34.583193
10/06/2016	08:07 p. m	34.583193
10/06/2016	08:12 p. m	34.583193
10/06/2016	08:17 p. m	34.583193
10/06/2016	08:22 p. m	34.583193
10/06/2016	08:27 p. m	34.253333
10/06/2016	08:32 p. m	38.214679
10/06/2016	08:37 p. m	45.649462
10/06/2016	08:42 p. m	38.214679

b)

Fig. 9 Databases. a) Database of example of Figure 7; b) Database of example of Figure 8

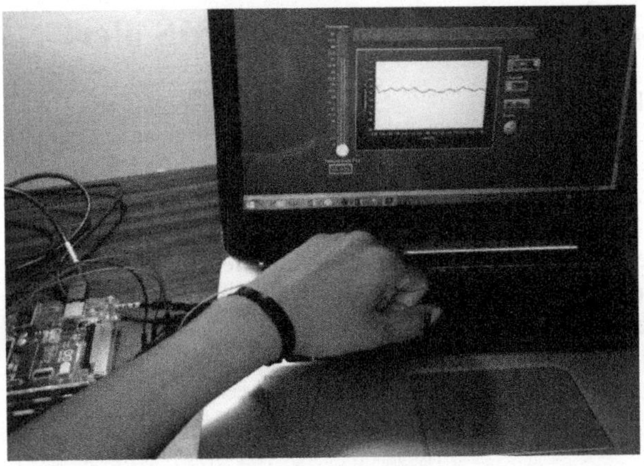

Fig. 10 Continuous body temperature monitoring system

IV. DISCUSSION

Body temperature is a vital sign that can be measured with different types of thermometers, such as temple, ear and oral thermometers or forehead thermometer strips[7]. The first ones are faster and more accurate than the last one, however, all of them require a person to take and register the patient's temperature. This sensor allows to measure temperature continuously. Also, as this wristband sensor is permanently in contact with the skin, it was made of biocompatible [9, 10, 11] and conductive [8,11] polymers. The repeatability of the results allows to evaluate the PPy/PLA composite as a reliable material. This is an important advantage in wearables manufacturing due to the fact that the whole wristband works as the temperature sensor, instead of having just one sensor placed on a wristband as many wearables do [12]. Besides, due to its composition, it can be easily washable and reusable, without losing accuracy.

V. CONCLUSIONS

In this work, a low-cost body temperature continuous monitoring system based on a flexible PPy/PLA wristband was presented. Results showed that this smart thermometer wristband made of PPy and PLA is flexible on the human skin and as well as accurate. A mathematical relationship between the temperature and resistance of the sensor was presented. The temperature-resistance curve demonstrated a linear trend within the range of 25°C to 40 °C.

The system was programmed satisfactorily in Labview for continuously monitoring body temperature data and it also demonstrated its ability of recording the information in a database for its medical analysis. Graphs showed the feasibility of fully using the temperature sensor over a long period of time so that nurses or medical staff will no longer have to manually take the patient's temperature, reducing the nursing workload and improving the quality of the medical care.

ACKNOWLEDGMENT

Authors express their gratitude to the Mexican National Council for Science and Technology (CONACYT) for financing this work through the project 151894.

CONFLICT OF INTEREST

The authors declare that they have no conflict of interest.

REFERENCES

1. Guyton, A. C. (1977). Basic human physiology: Normal function and mechanisms of disease. Saunders, Philadelphia
2. Benzinger, T. H (1969) Heat Regulation: homeostasis of central temperature in man., Physiol. Rev, Maryland
3. Tooley, M (2009) Design Engineering Manual. Butterworth-Heinemann, England
4. Smith, J; Johnson, J (2009). Nurses' Guide to Clinical Procedures. Lippincott, Philadelphia
5. Jiang, L.J; Ng, E.Y.K; Yeo, A.C.B; Wu, S; Pan, F; Yau, W.Y; Chen, J.H; Yang, Y. (2005) A perspective on medical infrared imaging. J. Med. Eng. Tech.
6. Stoppa, M.; Chiolerio, A. Wearable Electronics and Smart Textiles: A Critical Review (2014) Sensors 11957-11992 DOI 10.3390/s140711957
7. Ganio MS, Brown CM, Casa DJ, et al (2009). Validity and Reliability of Devices That Assess Body Temperature During Indoor Exercise in the Heat. Journal of Athletic Training 44:124-135.
8. R. Balint, N. J. Cassidy, and S. H. Cartmell (2014). Conductive polymers: Towards a smart biomaterial for tissue engineering. Acta Biomaterialia, 10 (6): 2341-2353.
9. A. Kotwal, and C. E. Schmidt (2001). Electrical stimulation alters protein adsorption and nerve cell interactions with electrically conducting biomaterials. Biomaterials, 22(10): 1055-1064.
10. J. Y. Lee, C. A. Bashur, A. S. Goldstein, and C. E. Schmidt (2009). Polypyrrole-coated electrospun PLGA nanofibers for neural tissue applications. Biomaterials, 30 (26): 4325-4335.
11. K. Dong-Hwan, S. M. Richardson-Burns, J. L. Hendricks, C. Sequera, and D. C. Martin (2007). Effect of immobilized nerve growth factor on conductive polymers: electrical properties and cellular response. Advanced Functional Materials, 17:79-86.
12. Zhu, Z.; Liu, T.; Li, G.; Li, T.; Inoue, Y (2015). Wearable Sensor Systems for Infants. Sensors, 15: 3721-3749.

Fall detection system for elderly by MEMS accelerometer and SMS alert

E. Oporto, L. A. Vilcahuaman

Pontifical Catholic University of Peru, Engineering Department, Biomedical Engineering Master Program, Lima, Peru

Abstract— **A major problem in the elderly are the negative effects of prolonged unattended stay after a fall. The proposed solution helps to reduce the negative consequences of falls through a portable noninvasive low-cost electronic system. The system, properly attached to the person, has the capacity to detect falls using MEMS accelerometers and generate an alert by sending a short text message through the public cellular network. The project includes the full design and the implementation of a biomedical portable prototype equipment using a reliable Atmel microcontroller, a medical-grade Bluetooth module and functional tests to evaluate the global effectiveness. The results were excellent, with sensitivity of 1 and specificity of 0.77 in several typical fall and not-fall sceneries.**

Keywords— **fall detection, MEMS, acceleration sensor, SMS alerts, the elderly.**

I. Introduction

In Latin America and the Caribbean the population over 60 years old is steadily increasing. Currently, there are in the region 63.1 million of older adults and it is estimated that for the next 38 years this figure will triple to 187 million, representing 25% of the regional population [1]. In old age, people face different ways of vulnerability by progressive biological deterioration associated with health problems, demonstrated as a deficit of functional abilities and autonomy. In this sense, among the most feared accidents in the life of an elderly are falls [2]. Currently, various electronic systems to detect falls in elderly individuals are available. These devices are commercialized worldwide. However, in Latin American countries, its use has not been sufficiently adopted because of its high cost [3]. The aim of this study was to design and develop a portable, low-cost, autonomous, wireless, lightweight, fully functional biomedical prototype for fall detection capable of sending SMS alerts by cellular network. Besides, the system's effectiveness was assessed by testing it in various typical scenarios of falls and daily activities. In Fig. 1 a diagram of the proposed system is shown. This consists of an electronic circuit attached to the thorax of the person, with a MEMS (Microelectromechanical Systems) acceleration sensor capable to detect falls and impacts. In addition, a processor to constantly read the acceleration data is available to determine fall events. In such cases, the processor sends AT commands via a Bluetooth link to a local mobile phone (near the elder). These commands instruct the local phone to generate SMS alerts by the public cellular network to previously selected recipients.

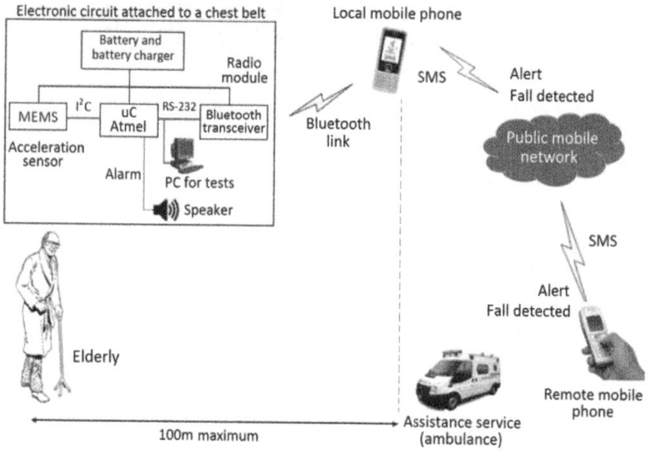

Fig. 1 Schematic of the fall detection system with sending alerts

II. Materials and methods

A. Data acquisition system

This comprises a three-axis MEMS accelerometer, Analog Devices ADXL345, and an Atmel ATMEGA88L microcontroller, both surface mounted and with low power consumption (Fig. 2). The MEMS measured accelerations in the range of +/- 16g ($g = 9.8 m/s^2$). The microcontroller used is mid-range, since the proposed system does not require complex computation or enormous storage capabilities. The connection between the sensor and microcontroller is through the I^2C digital serial bus. The sensor was programmed to acquire analog samples of acceleration, in each of its three axes, at a sampling frequency of 200Hz (every 5ms). All samples were converted to 13-bit digital values. The sensor includes a digital filter to reduce internal noise. Additionally, the circuit also contains noise decoupling capacitors. The programming and configuration of the microcontroller was performed in C language and included communication routines for the I^2C bus and to access the ADXL345 sensor. Other routines were also implemented for the RS-232 port communication required for sending acceleration data to an application developed in Matlab, during the system test stage.

I. Torres et al. (eds.), *VII Latin American Congress on Biomedical Engineering CLAIB 2016, Bucaramanga, Santander, Colombia, October 26th -28th, 2016*,
IFMBE Proceedings 60,
DOI: 10.1007/978-981-10-4086-3_29

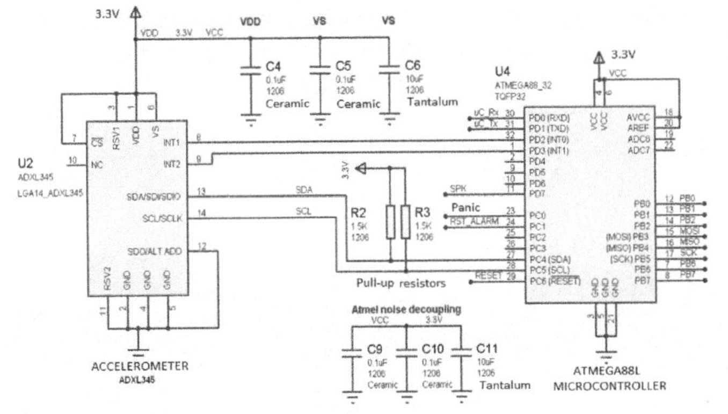

Fig. 2 Data acquisition circuit, consisting of ADXL345 MEMS acceleration sensor and the ATMEGA88L microcontroller

B. Fall detection logic

The fall detection algorithm was based on the use of an acceleration threshold in order to discriminate whether an event is or is not a fall. Continuous monitoring of acceleration was performed by the sensor taking into account the different positions of the person before, during and after a fall. The state of rest or ADL (activities of daily living) was considered as the normal state of the elderly (e.g., sitting in a chair or walking) with low magnitudes of acceleration (less than 4g). When the old man loses his balance, goes to a momentary state of free fall with an acceleration similar to Earth's gravity and then hit the ground. This impact is characterized by its duration (around 30ms) and the peak magnitude of acceleration (more than 4g). After the impact, it returns to a quiescent state with the person in a stunned or prolonged stationary position lying horizontally on the ground. The detection of this state also confirms the fall event.

C. Alert Notification

To ensure a system with a high reliability, a medical grade, Class 1, 100m range, highly immune to interferences, an Ezurio BTM402 Bluetooth module was included. This device was used for sending AT commands from the processor to the local phone during the fall events (Table 1). The local phone receives commands and automatically generates text messages to the programmed recipient. SMS is used because of its high reliability and fast delivery of messages, even with overloaded telephone networks.

D. Power supply system

Power consumption is critical in portable systems. That is why in this study Lithium ion 3.7 V batteries were used, suitable for electronic devices operating at 3.3V. The estimated current consumption was 3mA with an autonomy of approximately 14 days; the calculated battery capacity was 1000mAh. This system also included a recharging circuit based on the Maxim MAX1811 battery charger. The voltage of 3.3V was generated by the Microchip MCP1253 DC/DC converter.

Table 1 AT commands sent by the microcontroller ordering local phone to send an SMS to the programmed recipient's phone

AT+CMGF=1	Command to send a SMS
AT+CMGS="+985620246"	Destination telephone number (985620246) to send the SMS.
"ALARM: Fall detected"	Message to send. Can include basic data of patient.

E. Fall detection system tests

Two sets of tests were considered for the sensor placed at the chest of the person, clamped with a sash, in a similar form to that used by Jantaraprim et al [5]. The first test group determined peak accelerations in ADL and in various fall scenarios in order to define an acceleration threshold for both types of events. For all cases, the MEMS sensor digitalize at 13 bits and 200Hz sampling frequency. The data was received on a laptop, via a RS-232 link with the microcontroller and, by means of an application developed in Matlab, the acceleration curves were plotted and analyzed. ADL tests were conducted by an old man, 78 years old, in good physical and mental conditions. These tests included the following activities: walking, jogging, sitting, getting up from a chair, giving high jumps, lying on a bed and going downstairs. Tests were also conducted in fall scenery, made by a male adult, 44 years old, under controlled conditions using a pad or also using a dummy. In this case several types of falls were assessed: lateral from a chair, fall from initial standing position, drop from 2m high and falls downstairs. The second set of tests involved the whole system, subjecting it to five different scenarios between falls and ADL (side falls from an initial state chair and standing, sitting, high jumps and lying on a bed). All tests were carried out by repeating each event 10 times for each of the three different impact thresholds (4g, 5g and 6g). These tests enabled us to determine the sensitivity and effectiveness of the system as a whole using equations 1 and 2, respectively. For all cases, a fall event was considered as detected only if it was possible to receive SMS alert generated by the microcontroller.

$$\text{Sensitivity} = \frac{Detected\ falls}{Total\ real\ falls} = \frac{TP}{TP+FN} \quad (1)$$

$$\text{Specificity} = \frac{DLA\ detected}{Total\ ADL} = \frac{TN}{TN+FP} \quad (2)$$

III. Results

A prototype of the system (Fig. 3), completely wireless, 6cm x 4cm size, with a weight of 80g, and at an average cost of $ 70 in electronic devices (including manufacturing and assembly of the printed circuit board) was implemented. Performance tests of each stage: data acquisition, fall detection logic and sending SMS alerts were satisfactory. In Fig. 4 the acceleration curve of a side fall is shown from an initial standing position, which is representative of all tests performed with the device. This test lasted 3 seconds and its start was marked by a free fall of about 0.5s with a decrease in the acceleration of 0.5g. The initial contact with the ground occurs for a period of 0.7s with accelerations up to 2g. After that, the impact with a peak acceleration of 5.1g, 100ms duration occurs. Finally, due to rebounding and final movement of the person before lying motionless on the ground, acceleration variations less than 1g are recorded for a period of 1.2s. Acceleration peaks and the duration measured in various daily life activities and fall scenarios, for the second set of tests, are shown in Table 2. These results allowed us to determine that the 4g acceleration value corresponded to threshold separating the fall events of the ADL. Routine activities of the elderly recorded peak accelerations between 1.2g and 3g, except hops that showed values of 4.8g. In fall situations the peak accelerations exceeded 4g.

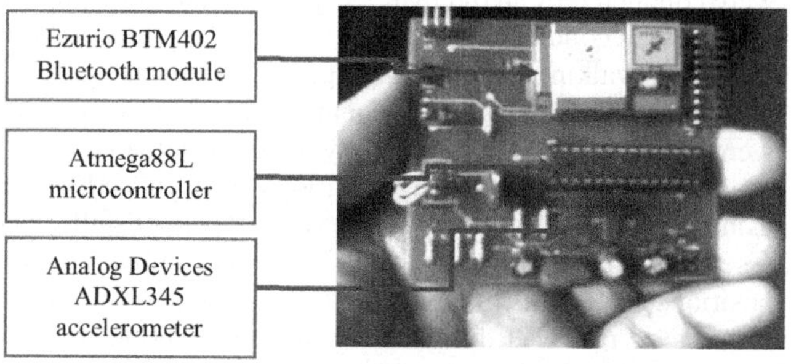

Fig. 3 Card prototype version implemented with Atmel microcontroller, Bluetooth module and the ADXL345 acceleration sensor

False positive, true positive (confirmed by SMS alert), false negatives and true negatives records, under the three acceleration thresholds during system testing are shown in Table 3. Regardless of the acceleration thresholds studied, in ADL events, such as sitting or lying down, no false positives were recorded. However, only on the 4g threshold, actions such as jumping, gave 7 false positives out of 10 trials executed. On the other hand, all fall events were detected as true positive with a 4g threshold. In Table 4, the evaluation results obtained by the system according to the criteria of sensitivity and specificity in the three thresholds impact studied are summarized. The highest sensitivity for detection of falls was obtained with a threshold of 4g and the highest specificity with thresholds of 5g and 6g.

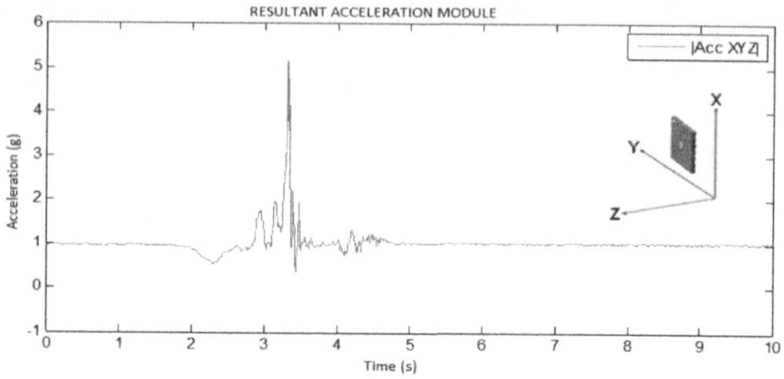

Fig. 4 The resultant acceleration module during lateral fall from a standing initial position

Table 2 Maximum and minimum acceleration in activities of daily life and fall scenarios

	Event type	Acceleration: Peak magnitude (g)	Acceleration: Duration (ms)
Daily life activities	Hike	1.5	150
	Trot	3	140
	Sit on a chair	1.2	230
	Up from a chair	1.3	300
	High jump, several	4.8	160
	Lying on a bed	1.5	100
	Down the stairs	3	120
Controlled tests	Lateral fall from a chair	4	140
	Falling passed out standing	5.1	100
	Fall from 2m height	27	35
	Fall staircase	21.6	30
	Fall back	4.1	35

Table 3 Results of system tests to measure their effectiveness

Event Type	Number of events detected by acceleration threshold (4g, 5g and 6g)											
	False positives			True positives			False negatives			True negatives		
	4g	5g	6g	4g	5g	6g	4g	5g	6g	4g	5g	6g
Lateral fall from chair				10	8	6	0	2	4			
Fall with lateral swoon				10	9	9	0	1	1			
Sit on a chair	0	0	0							10	10	10
Several high jumps	3	0	0							7	10	10
Lying on a bed	0	0	0							10	10	10

Table 4 Determining the sensitivity and specificity of the fall detection system according the impact threshold

Impact threshold	Sensitivity	Specificity
4g	1	0.77
5g	0.85	1
6g	0.75	1

IV. Discussion

The prototype hardware, design and implementation done in this work represents an excellent evidence of the feasibility of developing portable medical devices that meet the characteristics required for such systems. In this sense, experience was gained in this respect, but more comprehensive studies are needed to achieve compliance with international design standards such as AS4607 and EN60950 validation. The card designed in this study does not use generic modules or hardware, only the necessary components, low power and surface assembly, allowing the system to be small with low consumption. The prototype is totally software autonomous; the Matlab application was only used for testing and debugging stages. In addition, its high efficiency and an extended battery life were the outstanding features of the system that differentiate it from others [6].

In this study, it was observed that some ADL and fall events showed very close peak accelerations, as was the case in the high jump and lateral fall from a chair with 4.8g and 4g, respectively. It is likely that the accelerations obtained are systematically lower because test were performed using a mattress pad. This produced an attenuation of the impact; hence, we assume that the peak acceleration values can be greater than those recorded. However, the results show that fall events have peaks higher than 4g with typical durations of 30 to 35ms, similar to that reported by Jia [4] and Jantaraprim et al [5]. Moreover, it could be shown that the system's capacity to detect falls (true positives) decreases with increasing acceleration thresholds, especially for events with low peaks such as falls from a chair, but the specificity increases. Therefore, these results show that the 4g acceleration threshold, proved to be the value with the greater sensitivity for detection of falls in 100% of cases with a specificity of 77% in all ADL events. These results agree with Huynh, et al [7], who reported a sensitivity of 100% and specificity between 70 and 98%, using a system comparable to that proposed in this study.

V. Conclusions

The results in this study showed that the proposed system as a whole was effective. The fall detection algorithm, based on acceleration threshold, and the procedure for generating SMS alerts performed properly. However, validation of this prototype by using it in a long-term test on older people with different scenarios in order to perfect the design and expand its advantages is necessary. To end this, one of the possible future benefits of the device is to incorporate a GPS module that allows to know the location for immediate medical assistance, as well as to incorporate a system with air bags that protect the vulnerable anatomical areas from fractures, such as those of the hip. It is important to highlight that besides the elderly, users of this device include disabled people and patients with clinical conditions prone to increase the risk of falls, such as neurological patients. A robust and comfortable case design is also an important point to consider.

Acknowledgment

The authors thank Mr. Eleazar Oporto and Mrs. Marina Diaz, for their support as volunteers under informed consent IC for performing daily living activities tests.

Conflic of Interests

None declared.

References

1. Organización Mundial de la Salud (2015) Informe mundial sobre envejecimiento y la salud. Ed. Estados Unidos de América. pp 266
2. Izquierdo M et al (2008) Valoración de la capacidad funcional en el ámbito domiciliario y en la clínica. Nuevas posibilidades de aplicación de la acelerometría para la valoración de la marcha, equilibrio y potencia muscular en personas mayores, SCIELO, Navarra, 2008
3. Brickhouse Alert datasheets at http://www.brickhousealert.com
4. Jia N (2009) Detecting Human Falls with a 3-Axis Digital Accelerometer, Analog Dialogue vol. 43, 2009
5. Jantaraprim P et al (2009) Evaluation of Fall Detection for the Elderly on a Variety of Subject Groups, Proceedings of the 3rd International Convention on Rehabilitation Engineering & Assistive Technology, Singapore, 2009.
6. Huynh Q et al (2013) Fall Detection System Using Combination Accelerometer and Gyroscope, EDC 2013 Proceedings.
7. Huynh, Q et al (2015) Development of novel algorithm and real-time monitoring ambulatory system using Bluetooth module for fall detection in the elderly, IFMBE Proc. vol. 46, 2015, pp 184-189

Author: Edgard Oporto
Institute: Pontifical Catholic University of Peru, Engineering Department, Biomedical Engineering Master Program.
Street: 1801 Universitaria Avenue
City: Lima
Country: Peru
Email: edgard.oporto@pucp.edu.pe

Plastic Casing in Medical Equipment: Evaluation of 3D Design and Molding Simulation

M. Gómez[1], Y. H. Villalta[1], J. E. Suen[1], R. I. González[1]

[1]Digital Medical Technology Company/Medical Equipment Division, SOCBIO, Havana, Cuba

Abstract—**The aim of this paper is to present mechanical design solutions and verification techniques for the most complex plastic parts composing the digital electrocardiograph CARDIOCID T50/S100 and the vital sign monitor DOCTUS VIII. The most important manufacturing details were analyzed as well as the simulation of the mold injection process, in order to obtain the accurate vision of the problems that may arise. The details of the dynamic study of the different design solutions before the parts injection are also exposed. This practice optimizes time in the development of the mold, and consequently reduces the final product costs. By using Moldflow simulation package the 3D models were performed, testing the validity of technological parameters such as draft angles, possible areas of material shrinkage (shrinkage cavities) and others. The final results of the process are presented, from the injected parts, corroborating calculations made in the design stage. This methodology proves to be a highly effective and essential tool as a way to avoid mistakes and improve solutions. Modern designs are obtained, with quality, speed and safety.**

Keywords—**design, 3D models, simulation, medical equipments.**

I. Introduction

Nowadays, to make that new biomedical equipments be exclusives has become a key and basic premise. Each is distinguished from others similar products because of the greater added value and for being visually different and more attractive. Increasingly, the "design" of products is the main element that differentiates an innovator product of another that is not. To obtain this result is essential the development and effectively use of technologies for support the design and engineering functions [1].

During design and before the manufacturing of the plastic parts of casing of electrocardiographs and therapy's monitors, several iterations of simulation are required. It is undesirable to find in the final results failures that converge and create imperfections, such as a strains, sunken areas, etc. There is no process of rapid prototyping able to accurately predict the behavior of the proposed design when is being manufactured or removed from mold [2].

Using a tool for simulation of the injection process of the piece during design, it is possible to identify areas where these defects can occur and these errors can be corrected since the early stages. With this tools, is also possible to have the additional advantage of evaluate various design configurations for molds and select which manufacturing variant is better and cheaper [3].

This paper presents the main mechanical design solutions and details of the process that simulates the injection in molds of mechanical parts of the digital electrocardiograph CARDIOCID T50 / S100 and the physiological parameters monitor DOCTUS VIII, which are the latest models of this type of medical equipments that is being developed at Digital Medical Technology Company.

II. Materials and methods

A. Concurrent design of plastic pieces

The design of biomedical equipment begins with a concurrent and interactive process of exchange of ideas and requirements between electronic, mechanical and industrial designers with the final objective of to improve quality and product reliability, increase mark-up and reduction of total costs.

In the team, industrial designers define the main external aesthetic features taking into consideration the state of the art and the team agreements during conceptualization. The electronic team is responsible for specifying the characteristics of the interconnection wiring, keyboards, cards and other modules, in particular by establishing their needs for electromechanics component location. The mechanical team, gives the solution for where and how to place physically each element within the equipment, while sets the order for the assembly and disassembly thereof. This process is not sequential and many times each solution has to be changed or improved according to modifications in the requirements or because a new decision contradicts something previously established. For this reason there must be a constant exchange between members of different teams in project group. As a result of this exchange, the virtual assembly which simulates the physical location of all components is generated [4] [5].

At the design stage it is necessary to identify future defects that may occur during piece manufacturing. The first

I. Torres et al. (eds.), *VII Latin American Congress on Biomedical Engineering CLAIB 2016, Bucaramanga, Santander, Colombia, October 26th -28th, 2016*,
IFMBE Proceedings 60,
DOI: 10.1007/978-981-10-4086-3_30

parameter that must be checked is the variation of wall thickness. It is desirable that this variation be very small in the whole piece. It is important to define the nominal wall thickness value, which means wall thickness that will prevail in the piece. A thickness below the nominal value cause that filling process becomes very hard, because the liquid plastic must pass through very small ducts and the injection requires low-viscosity materials. Conversely, thicknesses greater than nominal value causes large solidification intervals in those areas. This will increase material contractions and will appear dimpling on surfaces that we name shrinkage cavities.

The second aspect to control during design is draft angle of the piece sides, regarding to the output shaft in the mold (hereinafter referred as draft angles). This parameter is closely linked to the surface finish.

If the piece has side holes or other details that limit its direct output from the mold, it will be necessary to include mechanisms that allow it.

In each design iteration of the pieces, the feasibility of making the proposed solutions should be checked. Program CAE (Computer Aided Engineering) Moldflow® helps carry out these verification, because it allows the simulation of each injection's parameters in a mold with the aim of obtaining a piece where the defects are minimal [6].

B. Design of plastic parts CARDIOCID T50/S100.

The upper cover of CARDIOCID T50/S100 has external dimensions of 327x242x47 mm. Here is where the thermal recorder, the membrane keyboard, screen and electronic card that allow connectivity during equipment operation are located. Because of the placement of all these electronic blocks, this piece has a very complicated configuration and requires great precision in its final dimensions. Figure 1 shows on the left the virtual assembly of all the above elements, and on the right, the piece without any element.

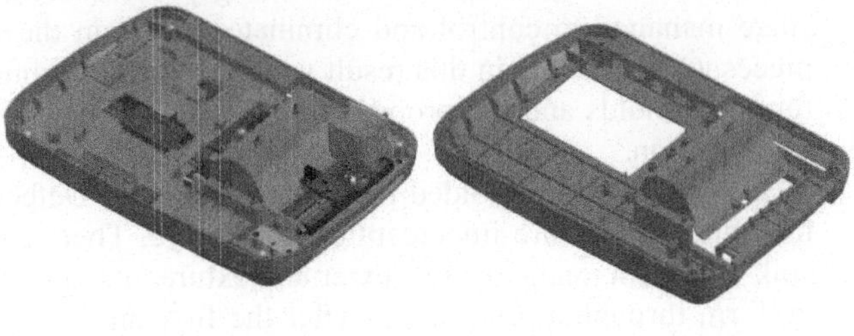

Fig. 1 Cover of CARDIOID T50 / S100 with its elements assembled (left) and the same piece empty (right)

The selected plastic material was ABS-PC (Acrylonitrile butadiene styrene with polycarbonate), as it has high resistant to chemicals, impact, and other attacks. Also it is an electrical insulator. ABS-PC offers better mechanical properties than the ABS alone and is much cheaper than the PC alone. Practice has shown that biomedical equipments are often exposed to shock and suffer breakages during use [7]. It is recommended for them ABS-PC whose thickness may vary in the range between 1 and 10 mm. In our case we adopt a nominal wall thickness value of 3 mm based on experience and the study of similar biomedical equipments [8].

Figure 2 shows on red color, the areas where there is a greater thickness than the nominal value. On the left, the design of the piece at an early stage is observed. In the areas circulated with ellipses in orange color, the greater thickness (4mm) causes the emergence of shrinkage cavities, and enlarges cooling time causing lower productivity of injection cycles.

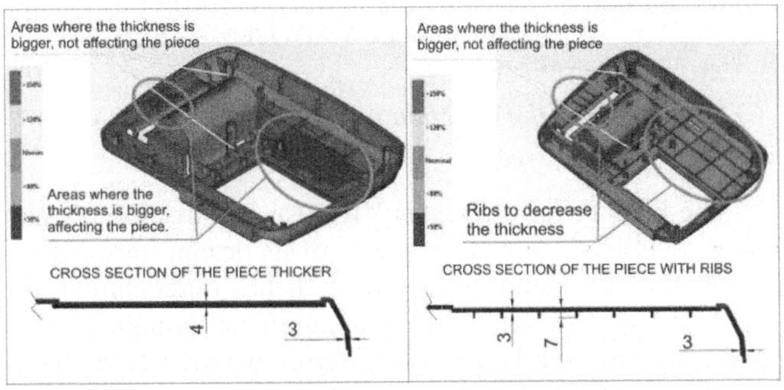

Fig. 2 Simulation with MOLDFLOW ADVISER showing thicknesses varying in the cover of CARDIOCID T50 / S100 (left: preliminary design, right: final design)

These areas of the equipment receive bigger external forces during operation and therefore require higher strength. The solution made for eliminating this problem was to reduce to nominal value the wall thickness of those areas and put ribs to increase their strength. It allows maintaining uniform the wall thickness in all piece. Additionally, future problems in the processes of injection and post-injection are reduced. If there are different wall thicknesses, changes from a dimension to another are performed abruptly and flow turbulences are formed during filling of the piece, provoking that not homogeneous mechanical properties be obtained.

On the right side of the same picture, the piece is observed with the new design (constant wall thickness of 3 mm). Is kept only the two areas marked in yellow color inside of the piece where the formation of shrinkage cavities does not affect the aesthetics or operation.

Regarding the relationship between draft angles and texture requirements of the piece, no difficulties arose. Figure 3 shows that the outer surfaces represented in light blue have a higher draft angle of 3°. The selected texture value on the external parts is YS1286 A [9] which has a requirement for a minimum draft angle of 1.5 °, (less than 3 ° set at design). Additionally, this value helps the proper removal of the piece, and prevents tearing in the surface due to friction between the metal cavity and the plastic itself.

The dark blue surfaces are parallel to the parting line of the mold and therefore it does have not an effect on the texture.

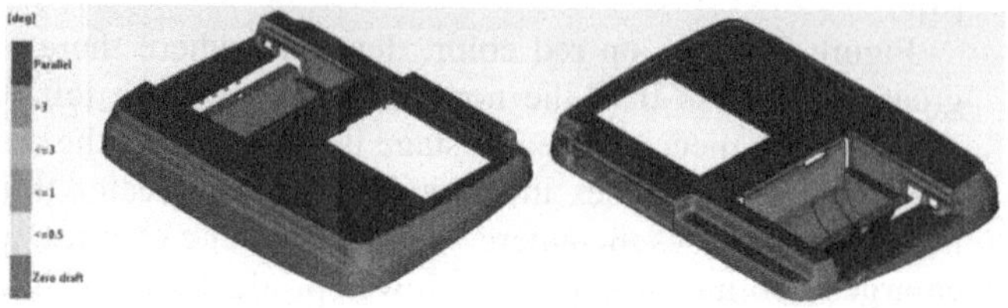

Fig. 3 Simulation of draft angles on the cover of CARDIOCID T50/S100 with MOLDFLOW ADVISER

C. Simulation of the injection process in the back cover of Doctus VIII

The back cover of Doctus VIII also requires the revision of all details since earliest stages of its design to avoid any difficulties at manufacturing time. It has outer dimensions of 307x266x137 mm. The most significant complexity lies precisely in their height (137 mm), so that liquid plastic must travel long distances during injection into the mold. This makes very important the correct definition of where to locate the points through which the liquid is injected in order to fill the mold easily.

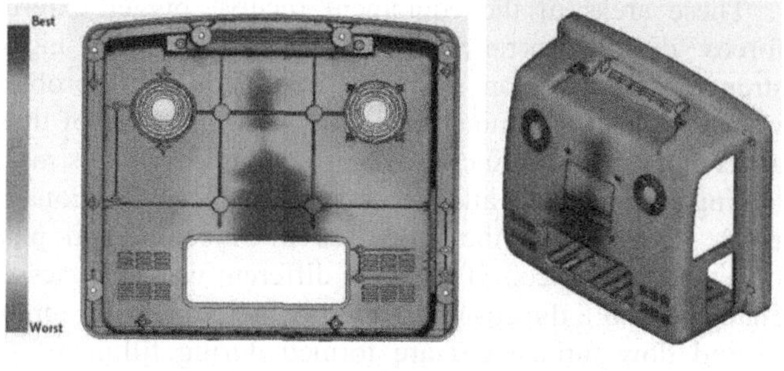

Fig. 4 Simulation of the location of the injection point in the cover of DOCTUS VIII with MOLDFLOW ADVISER

Figure 4 shows in dark blue the best location for the injection points, according to the simulation performed by Moldflow Adviser. In this geometrical area, elements (racks, posts, etc.) that may obstruct the correct entry and movement of fluid into the mold must be avoided. Three injection points were established to facilitate efficient filling of the mold and reduce the necessary pressure for injection.

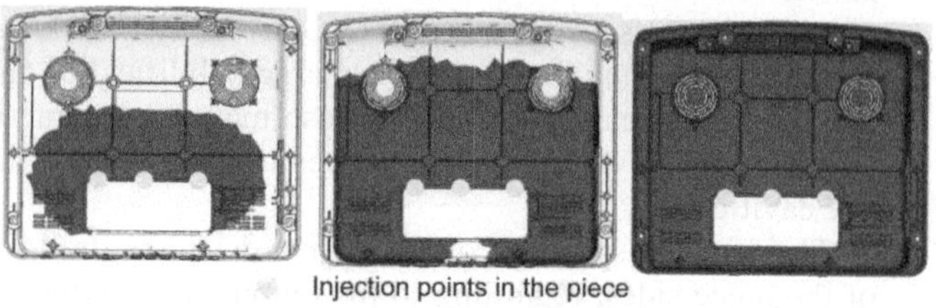

Fig. 5 Sequences of injection process simulation on DOCTUS VIII cover using MOLDFLOW ADVISER

Figure 5 shows how dark green liquid fills the mold cavity to complete it and finally get the cover. The three dots in yellow represent the injection points where the liquid penetrates the mold. They are located closest to the geometric center of the cover and at its thickest area. Thus is better for easy filling of cavity, while keeping open conduits through which the material moves as long as possible. Besides, good guidance and free flow movement is achieved, allowing good distribution of the polymer chains and improved mechanical properties of the cover.

In those areas where the walls are thicker, higher percentages of shrinkage will occur. That's why it is important keeping this areas near to the injection points in order to receive molten material as long as possible to compensate the contraction produced.

III. Results

The simulation of the manufacturing process at design stage managed to control and eliminates errors in the final pieces. Other factors in this result were the correct manufacturing of molds and the proper selection of parameters during injection.

In Figure 6 final molded pieces are observed. Defects in the outer surface are imperceptible to the eye. There are no sink mark, or tears on the external texture. Its quality is uniform throughout the piece. After the first injection, corrections were minimal and not major modifications to the initial design were made.

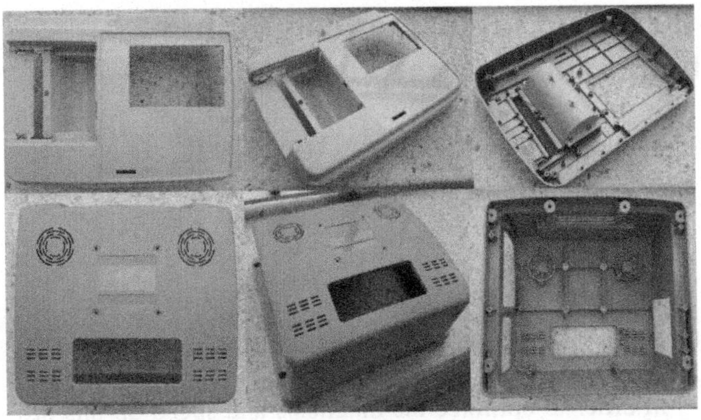

Fig. 6 Pictures of the injected covers of CARDIOCID T50/S100 (above) and DOCTUS VIII (below).

IV. Discussion

In the case of CARDIOCID's cover, the greatest difficulty presented was because of the large number of metal inserts for screws. This piece has 23 inserts which slows the injection process because having to place them on each plastic injection. Subsequently, it is applied inside the cover a layer of conductive paint to prevent electromagnetic interference.

Injection on DOCTUS VIII's cover, had problems with filling of the mold in the area with circular grids. This was resolved by increasing thickness of the ribs that separate the grids.

In both cases the obtained pieces have the precise dimensions, allowing its perfect assembly and the correct operation of biomedical equipment which are intended.

V. Conclusions

Modern technologies for the design, modeling and simulation are essential because allow to create virtual product models that simulate manufacturing and assembly. These tools should be properly coordinated and integrated into the product design and development cycle to achieve the ultimate goal: bring products to market with higher quality, lower cost and in less time. You must have an appropriate methodology of concurrent work that integrates all disciplines involved in product development.

Injection molding is a complicated process and the validation of any simulation model is also difficult. As discussed, the shrinkage and warpage depend on the pressure time history as well as the properties of the injection molded material. Consequently, any validation must first deal with prediction of pressure and temperature.

Simulating the process of pieces injection by suitable programs as Moldflow ®, areas where defects can occur are identified. This tool brings many options to correct problems in the design phase. Additionally it has the advantage of performing various design configurations to see what can be made better and more profitably. This simulation is a first step for understanding of the molecular orientation of the plastic into the piece. From this result, it is possible to study the piece when is subjected to loads by finite element simulation and calculate dynamic response of the piece to external stresses being able to avoid its breakage or fracture.

Conflict of Interest

The authors of this study declare that we have no conflicts of interest.

References

1. Maniscalco M. (2002) CAE for plastic getting up to speed. Injection Molding Magazine, Cleveland, USA.
2. Granchi M., Fantozzi C., Vettori E., (2007) Moldes para substancias plásticas: Optimización con técnicas CAE de un sistema de inyección, 8º Congreso Iberoamericano de Ingeniería Mecánica, Cusco, Perú.
3. Menges G, Michaeli W, Mohren P: (2001), How to Make Injection Molds, 3rd Edition, Carl Hanser Verlag Munchen Wien.
4. Albin S.L., Crefeld. P.J. (1994) Getting started: Concurrent engineering for a medium-sized manufacturer. Journal of Manufacturing Systems.
5. Chituc C-M, Azevedo A, Toscano C. (2009) A framework proposal for seamless interoperability in a collaborative networked environment. Computers in Industry.
6. Autodesk Simulation Moldflow®, Reference Manual, (2015).
7. Hoffman J.M., Ralston B., Chakravartula A., Reitman M. (2013) ESC Of Polycarbonate Exposed to Hospital Disinfectants, Proceedings, ANTEC 2013, Society of Plastic Engineers, Cincinnati, OH, April 2013.
8. Lamberti G., Peters G.W.M., and Titomanlio G. (2007) Crystallinity and linear rheological properties of polymers. Int. Polym. Processing.
9. Ying Sang Metal Plastic Mold Texturing Limited, No.22 Tsun Fat Street, G/F., to Kwa Wan, Kowloon.

Author: Miguel Gómez Florido
Institute: ICID-Empresa de Tecnología Médica Digital
Street: 202#1704, Siboney
City: La Habana
Country: Cuba
Email: mgomez@icid.cu

Automatic ABR detection at near-threshold intensities combining template-based approach and energy analysis

I.M. Cabana-Pérez[1], E. Velarde-Reyes[1], A. Torres-Fortuny[2], E. Eimil-Suarez[2], A. García-Giró[1]

[1]Department of Electronics, Cuban Neuroscience Center, Havana, Cuba
[2] Department of Audiology, Cuban Neuroscience Center, Havana, Cuba

Abstract— **This paper presents an algorithm for detecting auditory brainstem responses elicited by click stimuli at low intensities. Feature extraction is focused on peak V detection including analysis of the instantaneous energy and comparison with a template constructed from a reference dataset. All detection methods were validated and adjusted using ROC curve analysis. The algorithm was implemented in MatLab and evaluated in 135 ABR recordings. The combination of sensitivity, specificity and ROC area values exceeding 95 % with a 35 ms running time validates this approach for fast and accurate ABR detection.**

Keywords— **auditory brainstem response, infant hearing screening, instantaneous energy, ROC curves, template.**

I. Introduction

Hearing is the key for learning the spoken language. Without proper interventions, hearing loss is a barrier to education and social integration. Therefore, sooner a child is diagnosed with hearing loss, greater is the chance that, incorporating the appropriate treatment, he/she can master the spoken language and lower the likely adverse impact of hearing loss.

The Joint Committee on Infant Hearing recommends that all children with hearing loss should receive intervention by six months of age at the latest [1].

ABR (Auditory Brainstem Response) is an electrophysiological measure used to predict hearing sensitivity. Eight positives peaks compose the ABR signal (figure 1). They are generated by the auditory nerve and by structures in the auditory brainstem, in response to acoustic stimulation [2].

This response is very useful in infant's hearing screening because it is immune to subject state (so it can be reliably obtained in sleeping neonates) and it is recorded from surface electrodes. In addition, peak's latencies are stable within and across patients and the most robust component, the wave V, can be observed at levels very close to behavioral thresholds [3].

A conventional auditory testing strategy includes presenting click stimuli with a fixed intensity level, usually 30 to 40 dB nHL. Then, trained specialists visually determine whether a reliable response can be recorded. A critical issue arises if the number of infants who require screening far exceeds the highly skilled personnel available to carry out conventional ABR.

In order to optimize the diagnosis and speed up this procedure, several methods have been proposed for automatic ABR (aABR) detection [4-8]. The most widely used is to compare the recorded tracings with a template that represents expected results. Several approaches in this line have included filter techniques for emphasizing the morphology of ABR wave. Other methods are based on signal properties analysis for indirectly identifying the presence of ABR.

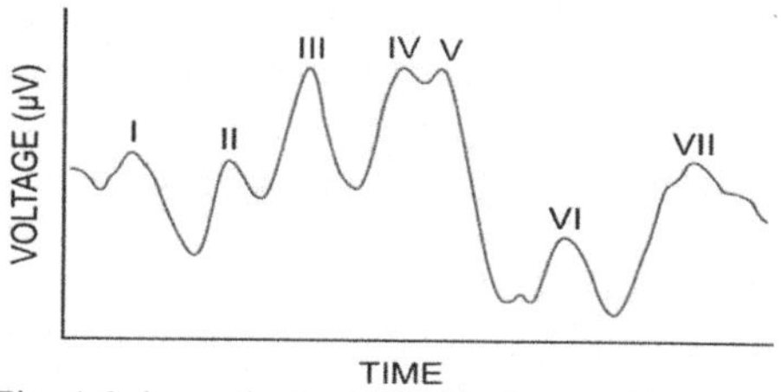

Fig. 1 Schematic Auditory Brainstem Response

When ABR is evoked at low, near-threshold, intensities, the signal is noisy and only wave V can be perceived. By the other hand, methods for aABR must be fast and low computational complexity for online applications.

Considering these requirements, an algorithm for automatic ABR detection at near-threshold hearing intensities based on template comparison and statistical analysis, is proposed.

II. Materials

The two mandatory steps for creating an ABR detection algorithm can be defined here as follows:

First, a reference dataset was created to explore the characteristics of ABR signals at near-threshold hearing intensities.

Second, an evaluation dataset was further created for quantifying the performance of designed algorithm.

In order to obtain the auditory evoked potentials in infants to construct these two datasets, a positive electrode was

I. Torres et al. (eds.), *VII Latin American Congress on Biomedical Engineering CLAIB 2016, Bucaramanga, Santander, Colombia, October 26th -28th, 2016,*
IFMBE Proceedings 60,
DOI: 10.1007/978-981-10-4086-3_31

placed on the Fpz derivation, a negative electrode was located on the ipsilateral mastoid (relative to stimulation) and a ground electrode on the contralateral mastoid. All stimulation and recording procedures were implemented in the Neuronic Audix system (Neuronic S.A.), keeping the electrode's contact impedance below 10 kΩ. Each record included 2000 sweeps, 15 ms long.

All operations were implemented on MatLab R2015b, with an Intel Core i5 3.30 GHz processor.

A. Reference Dataset

The reference dataset was compiled including 38 ABR screening records from 20 infants (6 female). Infant's age vary between 1 and 6 months. All ABR signals in this reference dataset were obtained via click stimulation at 30 dB nHL, applied to a subset of 22 recordings. During the remaining 16 recordings, the infants were not stimulated, thus used as an ABR signal-absent group.

These recordings were analyzed for selecting the most appropriate mathematical statistical measures to implement the feature extraction step.

B. Evaluation Dataset

The evaluation dataset contained 135 ABR infant screening records. Clicks with 21 Hz repetition rate and intensity levels in 30 dB and 40 dB (nHL) were used as stimuli. A signal subset with ABR presence encompassed 87 and 15 recordings, at 30 dB and 40 dB, respectively. The remaining 33 recordings were then labeled as the signal-absent subset.

The parameters sensitivity, specificity and ROC area were used in this evaluation dataset for quantitatively testing the detection program's performance. Sensitivity corresponds to the ability to detect the ABR absence in pathological subjects and specificity is the ability to sense ABR signal in normal subjects.

III. METHODS

Wave V tracking is the most prominent issue in infant hearing screening[3]. Figure 2 depicts how peak V in ABR signal is reduced in amplitude and increased in latency when click intensity level is reduced.

There are many ways to approach searching automatic ABR detection methods. In fact, at low intensities (30 dB nHL) only the peak V is distinguishable in most of ABR signals. Thus, here we are focused on detecting this waveform.

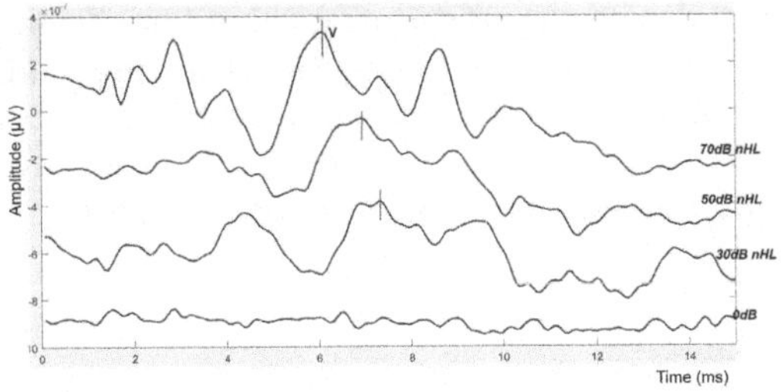

Fig. 2 Behavior of peak V for variations in intensity stimulation.

A. Feature Extraction

Wave V detection comprises comparing the signal with a pattern, such that it highlights the wave's morphology from the base activity. In order to do that, a pattern of the peak V was created using the average of a subset of recordings with the most representative signals from reference database. In figure 3 is presented graphically the effect of template used.

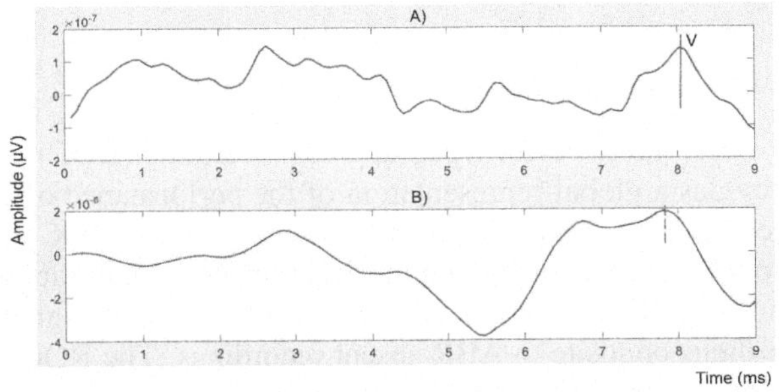

Fig. 3 Wave V detection method using template. A) First 9 ms from ABR signal at 30 dB with peak V marked. B) Effect of convolution with created template.

The time range used for template construction enclosed latencies between 4.95 ms and 10.05 ms. This interval was selected since peak V maximum amplitude is commonly described between 5.9 ms and 8.56 ms (for 30 dB nHL) and one of the essential criteria to visually identify it is the slope that follows the maximum amplitude value.

However, as observed in figure 2, peak V can be altered or distorted by base noise at 30 dB intensities. Therefore, statistical measures represent a complementary analysis method to characterize the signal.

Since wave V presence produce an increase in amplitude followed by a considerable decrease, we analyze the instantaneous energy and the variance in the interval of interest of the signals, looking for a statistical descriptor for ABR existence.

The instantaneous energy was determined simply according to equation 1.

$$IE(x) = (x)^2 \tag{1}$$

Figure 4 depicts the effects of equation 1 application in the ABR signal. As observed in graphic 4B, all amplitude variations are more conspicuous using this method. The analysis was limited to the specific time interval of peak V for 30 dB nHL using variance as another energy descriptor.

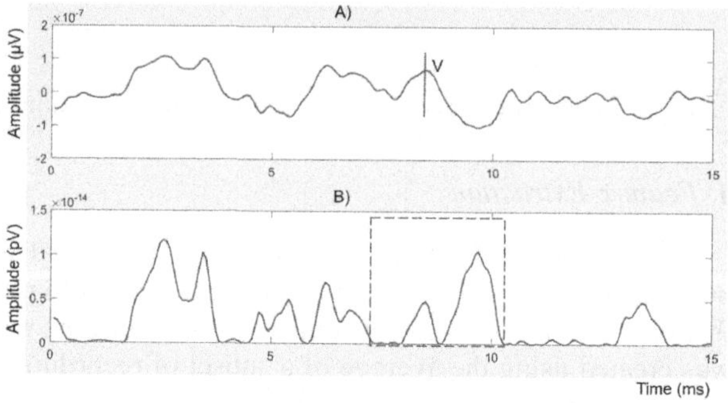

Fig. 4 ABR detection via energy analysis. A) ABR signal at 30 dB nHL. B) Instantaneous energy in period of interest for peak V appears.

B. ROC-based analysis

Here, the Receiver Operating Characteristic (ROC) curve provides a global representation of the performance of each method in reference dataset [9]. The area under ROC curve provides information about the accuracy of the measure used and quantifies the contrast of the results in ABR signal recordings opposite to ABR absent recordings. The ROC area values are qualifies as follows:

- from 0.90 to 1 is excellent
- from 0.80 to 0.90 is good
- from 0.70 to 0.80 is tolerable
- from 0.60 to 0.70 is poor
- from 0.50 to 0.60 is bad

Both methods, comparison with the template and energy analysis, were checked by using ROC curve analysis. The areas under the curve, as evaluated in reference dataset at the maximum sweep number for every recording, were 0.97443 and 0.97159, respectively. These values are classified as excellent descriptors for the contrast between distributions.

Taking advantage that ROC analysis allow cutoff or threshold's selection we determined thresholds for best classification in both methods, according to a predefined range (FVP = 90 %, FFP=10 %). In the particular case of template-based method, two threshold values were necessary. The first value localized the maximum amplitude after template convolution and the second defined the position of that maximum.

C. Designed Algorithm

The methods selected from the feature extraction procedure were combined in a general algorithm, presented in figure 5. The final value is logical: (1) in case of ABR-absent or possible hearing disorder and (0) in case of ABR identification or healthy child.

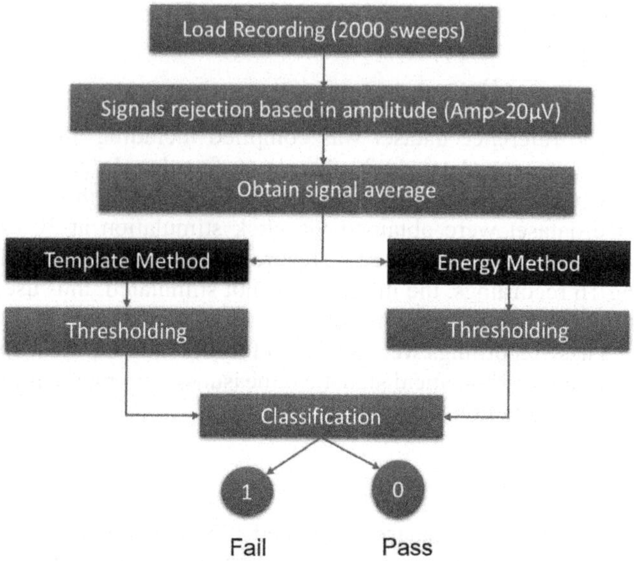

Fig. 5 General algorithm designed for ABR detection.

The classification process is described in table 1. When results from both methods are different, a third analysis is necessary. For this purpose, two simple stats are used: *Xaxis* and *IEmax*.

Xaxis relates to the latency of the maximum amplitude corresponding to peak V interval in the IE signal and is used when the energy method solely is positive to ABR presence. Otherwise, *IEmax* is the output value from energy analysis, but in this case was compared with a minimum amplitude noise-defining threshold when the template-based method solely is positive to ABR presence.

Table 1 Classification process in designed algorithm

Template-based method	Energy method	Further analysis	Final decision
1	1		1
0	0		0
1	0	Xaxis < 9ms	0
0	1	IEmax >low_threshold	0

IV. Results

The designed algorithm was tested using the evaluation dataset. The average execution time was 35 ms and the detection results involve 32 true positives and 97 true negative recordings. Performance appraisal was as follows:

$$Sensitivity = 96.9697\ \%$$

$$Specificity = 95.098\ \%$$

$$ROC\text{-}area = 0.96034$$

V. Discussion

We have presented a combination of elementary procedures for automatic ABR detection in infants at 30 and 40 dB nHL. The resulting low computational cost of the designed algorithm enables its implementation for hearing screening in newborns and infants, even using low-end equipment with limited storage capacity.

Several already published works on this topic have also introduced methods with low computational load. In [7], the authors used the estimation of first and second derivatives of the signals, but only for high intensity level (90 dB). Also, in [4] the instantaneous energy method was introduced for wave V detection at high stimulation intensities (80 dB). Our results showed that this statistical procedure can be used also at low stimulation intensities.

Our algorithm have achieved high accuracy values in terms of sensitivity, specificity and ROC area. Alternative methods reported (i.e. the continuous wavelets transform described in [5]) do not yield our sensitivity and specificity levels.

We must emphasize that all precision values exceeded 95 % as evaluated in a large set of ABR records from numerous infants. This peculiarity in the databases, for both the reference and the evaluation sets, guarantees a good level of reliability to the algorithm, as based on a very stable waveform. This feature becomes critical especially in infants.

A separate work [6] has reported the use of cross correlation with good results, but only in adult databases. Also, similar to us, another work [8] has showed a mathematical approach based on kernel methods for low stimulation level. The results are over 90 % accuracy, but in smaller recording datasets. The authors in [8] reduce the analysis to 200 sweeps, resulting in faster ABR testing.

We recommend to continue the work on the method's combination proposed here, extending its validity for lower number of sweeps.

VI. Conclusions

The combination of a template-based approach and the instantaneous energy analysis for detecting peak V, as a stable and replicable component of the ABR, at near-threshold intensities enable an accurate automatic ABR detection.

These detection methods require only a few simple mathematical operations, thus guarantying a low computational load.

Furthermore, the fast execution time enables this procedure to be included in online detection routines, as those mandatory for clinical and research recording software.

References

1. World Health Organization (2016) Childhood Hearing Loss: Strategies for prevention and care.
2. Schomer D.L., Silva (2011) Nierdermeyer´s Electroencephalography: Basic Principles, Clinical Applications and Related Fields. 6 ed.
3. Stach B.A. (2010) Clinical Audiology: An Introduction. Second Edition.
4. Rushaidin M.M. et al. (2009) Wave V Detection Using Instantaneous Energy of Auditory Brainstem Response Signal. American Journal of Applied Sciences. 6(9): 1669-1674.
5. Rushaidin M.M. et al. (2012) Wave V Detection Using Continuous Wavelet Transform of Auditory Brainstem Response Signal. Progress In Electromagnetics Research Symposium Proceedings. pp. 1889-1893
6. Berninger E., Olofsson Å, Leijon A. (2014) Analysis of Click-Evoked Auditory Brainstem Responses Using Time Domain Cross-Correlations Between Interleaved Responses. Ear & Hearing, 35:318-329.
7. Bradley A.P., W.J. Wilson (2004) Automated Analysis of the Auditory Brainstem Response. Proceedings of the 2004 Intelligent Sensors, Sensor Networks & Information Processing Conference: 541-545
8. Riedel H, Granzow M, Kollmeier B. (2001) Single-sweep-based methods to improve the quality of auditory brainstem responses. Part II: Averaging methods. Z Audiol; 40(2): 62-85.
9. Cabana I.M., Sánchez M., Galán L.(2015) Online Artifact Detection in EEG using Simple Digital Signal Processing's Procedures. IFMBE Proceedings

Comparative Study of Robust Methods for Motor Imagery Classification based on CSP and LDA

Ana Julia Villar[1]

[1]Instituto Tecnológico de Buenos Aires, Computer Engineering Department, Buenos Aires, Argentina

Abstract— **Common spatial patterns analysis and linear discriminant analysis are popular algorithms for spatial filtering and classifying in motor imagery. These algorithms are very sensitive to noise and artifacts which affect the classification accuracy. To deal with this issue, it is proposed to replace the usual estimators of covariance and scale used in the algorithms for robust versions. The performance of the methods are evaluated and compared on EGG data from BCI competition data sets; results show that robust methods outperformed classical techniques for subjects with poor classification accuracy.**

Keywords— **BCI, motor imagery, CSP, LDA, robustness**

I. Introduction

Motor imagery (MI) is a neuronal activity that occurs when a subject voluntarily imagine making a movement, for example, moving the right hand. To imagine a movement produces a neuronal activity that is spatio-temporally similar to the activity generated during the real movement. The Brain-Computer Interface (BCI) protocol usually employed involves asking the patient to imagine various types of movements while the EEG signals are recorded. The characteristics used to quantify the EEG activity are extracted and subsequently classifiers can be applied to discriminate between two or more imagined movements, allowing each imagined activity to be assigned to a particular control signal.

Common spatial patterns analysis (CSP) [1] is a supervised spatial filtering method that is used to find a transformation that maximizes the separability between the EEG data of two conditions. Usually, only a few of most discriminative filters obtained are used for classification. Linear discriminant analysis (LDA) [2] is a classifier that provides acceptable accuracy without high computation requirements. It is usually applied to classify patterns into two classes, although it is possible to extend the method to multiples classes.

EEG signals are widely affected by a variety of large signal contaminations or artifacts, such as eyes movements and blinkings, heart and muscle activities, head and body movements, as well as external interferences due to power sources. This causes the EEG signals contain outliers, which are observations that deviate from the general pattern of the data [3]. The outliers can adversely affect the results obtained from the conventional estimation methods such as CSP and LDA. These algorithms are strongly influenced by outliers because they involve the use of sample covariance that is highly non-robust.

In this work, methods based on the regularized common spatial patterns analysis (RCSP) algorithm proposed by Yong et al. [4] are evaluated for three datasets taken from the BCI competitions III and IV using several robust covariance estimators: Minimum covariance determinant (MCD) [5], Stahel-Donoho estimator (DS) [6,7], MM-estimator [8]. The proposed estimators for the covariance matrix provide robust estimates with respect to individual samples, i.e. they ignore trial structure and downweights outlier samples rather than whole trials. Not only the estimation of covariance matrices, but also the sample variance estimates used in extracting the features from the projected EEG signals are also easily affected by even a single outlier [3]. In order to deal with this problem, the scale estimate is replaced with the median absolute deviation and mean absolute deviation estimates.

II. Material and methods

In BCI design the goal is to process the EEG signal to translate into the mental state of the user. The two main steps in the processing system are feature extraction and classification.

In motor imagery based BCI can be distinguished imaginary movements of the right hand from the left hand. To identify these two mental states from EEG signals, band power features are usually extracted in the μ (about 8–12 Hz) and β (about 16–24 Hz) frequency bands, for electrode localized over the motor cortex areas of the brain. Such features are then typically classified using LDA.

A. Classical and robust CSP

One way to extract features from multiple EEG channels is to use spatial filtering, as CSP, which is a technique to analyze multi-channel data based on recordings from two classes.

I. Torres et al. (eds.), *VII Latin American Congress on Biomedical Engineering CLAIB 2016, Bucaramanga, Santander, Colombia, October 26th -28th, 2016*,
IFMBE Proceedings 60,
DOI: 10.1007/978-981-10-4086-3_32

Let $\mathcal{X} = \{X_1, \cdots, X_M\}$ be a sample of M training EEG trials corresponding to two different mental states (classes 1 and 2), where $X_j \in \mathbb{R}^{N\times C}$ is the data matrix which corresponds to a trial j $(1 \le j \le M)$ of imaginary movement, with N the number of observations in each trial and C the number of channels. Let $\Sigma_i \in \mathbb{R}^{C\times C}$ be the spatial covariance matrix of the band-pass filtered signals in $\mathcal{X}$ from class i $(i = 1, 2)$.

CSP yields a data-driven supervised decomposition of the signal parameterized by a matrix $W \in \mathbb{R}^{m\times C}$ that projects a signal $X \in \mathbb{R}^{N\times C}$ in the original sensor space $X_{\mathrm{CSP}} = XW^t$ where the m spatial filters (rows of W) $w_k \in \mathbb{R}^C$ $(1 \le k \le m)$ extremize the Rayleigh quotient $\mathrm{J}(w) = w^t\Sigma_1 w / w^t\Sigma_2 w$.

For a given trial matrix X_j (centered and scaled) the normalized sample covariance matrix is obtained as $S_j = X_j{}^t X_j / \mathrm{tr}(X_j{}^t X_j)$.

For each class i, an estimator of Σ_i is computed by averaging the covariances matrices of each trial as

$$\check{\Sigma}_i = \frac{1}{|\mathcal{C}_i|}\sum_{j\in\mathcal{C}_i} S_j, \qquad (1)$$

where $\mathcal{C}_i$ is the set of trials in $\mathcal{X}$ belonging to the class i and $|\mathcal{C}|$ denotes the cardinality of $\mathcal{C}$. Such computation of covariance matrices assumes that the signals have zero mean, which is true in practice for band-pass filtered signals.

The vector w in $\mathrm{J}(w)$ can be achieved by solving the generalized eigenvalue problem

$$\Sigma_1 w = \lambda\Sigma_2 w. \qquad (2)$$

Then the matrix W in X_{CSP} consists of the generalized eigenvectors w_k of Eq. 2 and $\lambda_{i_k} = w_k^t\Sigma_i w_k$ are the corresponding diagonal elements of Λ_i, while λ in Eq. 2 equals to $\lambda_{1_k}/\lambda_{2_k}$.

The sample covariance matrix is highly non-robust and has a breakdown point of 0, implying that is affected by even a single outlier. The simplest way to deal with this problem is to replace it with a robust estimate. In this study, the robust estimators used instead of S_j are MCD, DS-estimator and MM-estimator.

The MCD estimator is given by the subset of h $(> N/2)$ out of N data points with smallest covariance determinant. The location and scatter estimates are therefore the mean and a multiple of the covariance matrix computed on h such points. The DS-estimators of multivariate location and scatter are defined as a weighted mean and a weighted covariance matrix. The weights depend on a measure of outlyingness obtained by considering all univariate projections of the data. The idea of the multivariate MM-estimators is to estimate the scale by means of a very robust S-estimator, and then estimate the location and shape using a different loss function that yields better efficiency at the central model. The location and shape estimates inherit the breakdown point of the auxiliary scale.

The average estimates obtained in Eq. 1 can be substantially disturbed by outliers. It is possible to make them more robust by reducing the contributions of outlier covariances. An alternative iterative method is proposed by [9] using a weighted average where the weights controls the importance of each covariance depending on the distance from the center. Another weighted average, based on the number of outliers, can be obtained as follows. For a given X_j, let $\hat{\boldsymbol{\mu}}_j$ and $\hat{\Sigma}_j$ be robust estimates of location and scatter. Let $\{\boldsymbol{x}_1, \cdots, \boldsymbol{x}_N\}$ be the N rows of X_j, where $\boldsymbol{x}_i \in \mathbb{R}^C$ $(1 \le i \le N)$. A measure of outlyingness for a data point $\boldsymbol{x}_i$ is given by a robust version of the Mahalanobis distance $RD_i{}^2 = (\boldsymbol{x}_i - \hat{\boldsymbol{\mu}}_j)^t \hat{\Sigma}_j{}^{-1}(\boldsymbol{x}_i - \hat{\boldsymbol{\mu}}_j)$ with the usual cutoff value of $\chi^2_{C,0.975}$. For each row $\boldsymbol{x}_i$, RD_i is calculated to determine if $\boldsymbol{x}_i$ is an outlier; let n_j the number of rows of X_j which are not outliers. Thus, a robust estimator of Σ_i is given by $\tilde{\Sigma}_i = {}^1/_N \sum_{j\in\mathcal{C}_i} n_j\hat{\Sigma}_j$.

Then, replacing $\check{\Sigma}_i$ by $\tilde{\Sigma}_i$ in Eq. 1, the proposed robust version of CSP will generate a weight matrix $\widetilde{W}$ as output.

B. Feature extraction

The classical measure for the selection of CSP filters is based on the eigenvalues in Eq. 2. The m filters $(m = 2p)$, corresponding to the p largest and the p lowest eigenvalues are used.

Once these filters are obtained, a CSP feature f_k for a signal X is defined as $f_k = \ln(w_k{}^t X^t X w_k) = \ln(\mathrm{var}(w_k X))$, i.e., the m features used are simply the band power of the spatially filtered signals.

Each eigenvalue is the relative variance of the signal filtered with the corresponding spatial filter. This measure is not robust to outliers because it is based on simply pooling the sample covariance matrices in each class. A simple way to fix this issue proposed by Blankertz et al. [10]; they calculate the variance of the filtered signal within each trial and then calculate the corresponding ratio of medians: $\mathrm{score}(w) = \frac{med_1}{med_1 + med_2}$, with $med_i = \mathrm{median}(w^t X^t X w)$.

On the other hand, a robust scale estimate $\hat{s}$ of a sample $\mathcal{Y} = \{y_1, \cdots, y_N\}$ can be obtained using the median of the absolute deviations (MAD) of the sample from their median, given by $\hat{s} = \frac{1}{\Phi^{-1}(3/4)}\mathrm{median}(|y_i - \mathrm{median}(y_i)|)$.

Another scale estimator $\tilde{s}$ is the mean absolute deviation ($\overline{MAD}$), given by $\tilde{s} = \frac{\sqrt{\pi}}{N\sqrt{2}}\sum_{i=1}^N |y_i - \mathrm{mean}(y_i)|$.

Thus, using any of the proposed variance estimators, for a signal X, the vector $F = [f_1 \cdots f_m]$ of features used in classification is found.

C. Classification

Linear discriminant analysis is typically carried out using Fisher's method. LDA assumes that the two classes are linearly separable. According to this assumption, it defines a linear discrimination function which represents a hyperplane in the feature space in order to distinguish the classes. The class to which the feature vector belongs will depend on the side of the plane where the vector is found.

For each X_j in the sample $\mathcal{X}$, the features vector F_j is obtained. Let Φ_i the set of these vectors such signals are in $\mathcal{C}_i$, and let $\breve{\mu}_i$ and $\breve{\Sigma}_i$ be the sample mean and covariance obtained with the data in Φ_i. For a new signal $X \notin \mathcal{X}$, let π_i be the probability that the signal X to classify belongs to class i and F the features vector for X. Let $\breve{\Sigma}$ be the within groups covariance matrix given by the pooled version of the different scatter matrices $\breve{\Sigma} = \sum_{i=1}^{2} \pi_i \breve{\Sigma}_i$.

X is assigned to class l for which $D_l(F) = \min_{i=1,2} D_i(F)$, where $D_i^2(F) = (F - \breve{\mu}_i)^t \breve{\Sigma}^{-1} (F - \breve{\mu}_i) - 2 \ln \pi_i$.

D. Numerical experiments

a) Data sets

In order to compare the performance of the robust algorithms proposed with the classical CPS and LDA, three data sets of the BCI competitions, which contain motor imagery EEG signals, were used. The first two datasets were collected in a multiclass setting, with the subjects performing more than two different MI tasks; for them, the algorithms were evaluated on two-class problems by selecting only signals of left and right hand MI trials.

Datasets are: **IV-IIa** [11]: recorded using 22 electrodes from 9 subjects who performed left-hand, right-hand, foot and tongue MI. The training and testing sets containing 72 trials for each class; **III-IIIa** [12]: recorded using 60 electrodes from 3 subjects who performed left-hand, right-hand, foot and tongue MI. The training and a testing sets contain 45 trials per class for subject 1, and 30 trials per class for subjects 2 and 3; **III-IVa** [13]: recorded using 118 electrodes from 5 subjects who performed right hand and foot MI. A training set and a testing set were available for each subject, with different sizes (280 trials were available for each subject, among which 168, 224, 84, 56 and 28 composed the training set for subject A1, A2, A3, A4 and A5 respectively, the remaining trials composing the test set).

b) Data processing

For all data sets, EEG signals were band-pass filtered in 8-30 Hz, using a 5th order Butterworth filter. For each trial, the features are extracted from the time segment located from 0.5s to 2.5s after the cue instructing the subject to perform MI. The Matlab code of EEG signal processing/classification by Lotte, available at [14], was adapted to process the signals. Data were spatially filtered using the different variants of CSP presented in Section II.A. The robust statistical toolbox FSDA [15] was used to obtain the robust estimates of the covariance matrices by the functions "mcd" for MCD, "SDest" for DS-estimator and "MMmultcore" for MM-estimators. The filters were selected according to the p pairs of extreme eigenvalues and the p pairs of extreme scores. After spatially filtering the data, the log-variance was calculated in each trial using the different estimators of scale proposed in Section II.B, and LDA was trained on those features. The subjects' performance was calculated by filtering and classifying the testing data with the filters and LDA obtained in the manner mentioned before.

III. Results

Table 1 shows the best results of classifying the signals from the test sets using each of the techniques presented in Section II. The notation used is the following: CSP* for CSP, MCD* for MCD, DS* for Stahel-Donoho (using the weight functions: MDC (DSM*), Tukey's biweight (DST*) and Huber's (DSH*)) and MM* for MM-estimator; *1 for the scale estimator using MAD and *2 for $\overline{MAD}$; and *s when the m features are selected using scores instead of the extreme eigenvalues. Although in the BCI literature the usual number of features considered is between 2 and 6, in this work the number of pairs p (*pf*) was varied between 1 and 11 for the first dataset and between 1 and 30 for the last two ones; only the lowest p is reported in the case of ties for the best accuracies from the same method.

Results show that robust methods outperformed classical CSP for subjects with poor performances. In these cases, for most of them, the use of robust covariance matrices in CPS together the mean absolute deviation for scale and the features selected through the extreme eigenvalues produce the best classification accuracy. For the other subjects, this procedure provides an accuracy close to that of conventional method. This suggests that robust techniques are an alternative to improve the classification accuracy of CPS with LDA. However, determine the optimal method and the number of features suitable for classification is highly dependent on the individual.

IV. Conclusions

This study presents a comparison between the classical MI classification based on CSP and LDA and robust proposals. The methods are evaluated on MI data from 17 subjects. For all of them, better or similar performance in classification with classical CSP and RCSP (see [16]) are obtained. Results shows that for most subjects the best classifications are achieved using robust CSP.

The method with the best performance depends on the subject as it is seen from Table 1. Future work could be de-

Table 1 Classification accuracies obtained for each subject for the different methods. For each subject, the best result is displayed in bold characters

	BCI IV dataset IIa																	
	1		2		3		4		5		6		7		8		9	
	pf	%	pf	%	pf	%	pf	%	pf	%	pf	%	pf	%	pf	%	pf	%
CSP1	4	90.28	1	54.17	3	96.53	1	74.31	2	64.58	3	71.53	2	**84.03**	6	97.22	2	**93.75**
CSP1s	4	**92.36**	5	54.86	8	94.44	4	72.22	5	56.25	5	67.36	7	81.94	8	97.22	6	**93.75**
CSP2	3	88.19	1	54.17	5	96.53	1	75.00	2	65.97	3	69.44	5	81.94	5	96.53	1	92.36
CSP2s	4	90.28	2	57.64	3	95.14	4	76.39	4	56.94	5	68.75	3	76.39	9	96.53	11	93.06
MCD1	2	83.33	2	60.42	5	96.53	10	68.06	3	66.67	6	70.83	2	72.22	2	98.61	3	92.36
MCD1s	2	83.33	3	60.42	4	96.53	5	68.75	11	65.97	5	70.14	2	74.31	3	98.61	3	92.36
MCD2	3	88.19	3	**66.67**	7	**97.92**	1	76.39	6	67.36	6	67.36	2	79.17	8	97.92	3	93.06
MCD2s	3	88.19	3	**66.67**	4	**97.92**	2	74.31	3	67.36	6	68.06	2	79.17	7	98.61	2	93.06
DSM1	3	88.89	4	54.17	4	95.83	3	69.44	5	59.72	6	71.53	1	73.61	8	**99.31**	3	92.36
DSM1s	4	85.42	3	56.25	7	95.83	7	74.31	3	63.19	9	67.36	6	73.61	6	98.61	2	90.97
DSM2	3	89.58	2	55.56	5	96.53	10	70.83	3	61.81	4	**77.78**	2	78.47	8	98.61	3	93.06
DSM2s	5	88.89	7	58.33	3	96.53	6	70.14	3	61.11	8	73.61	4	73.61	8	98.61	2	92.36
DST1	1	88.19	3	57.64	6	96.53	8	72.22	2	64.58	9	70.83	3	75.69	6	**99.31**	3	91.67
DST1s	4	85.42	3	56.25	3	94.44	7	74.31	3	63.19	9	67.36	6	70.83	7	98.61	2	93.06
DST2	3	90.28	4	57.64	5	**97.92**	2	72.92	2	62.50	2	75.00	2	79.17	11	98.61	2	92.36
DST2s	5	88.89	7	58.33	3	96.53	6	70.14	7	61.81	8	73.61	4	76.39	8	**99.31**	4	**93.75**
DSH1	1	84.72	1	56.94	4	95.83	6	73.61	2	**73.61**	1	70.83	1	77.78	1	97.92	1	93.06
DSH1s	2	85.42	1	59.72	4	95.14	6	73.61	3	65.28	3	70.83	1	75.69	1	97.92	1	90.97
DSH2	3	88.19	1	56.25	3	95.83	6	75.00	2	65.28	1	75.69	11	75.69	5	97.22	1	92.36
DSH2s	2	90.28	1	63.89	4	96.53	6	76.39	1	63.19	2	69.44	4	75.69	6	97.22	1	92.36
MM1	2	85.42	3	56.25	5	95.83	1	72.92	2	67.36	2	66.67	1	75.69	5	98.61	2	91.67
MM1s	2	86.81	2	59.03	4	95.83	1	72.92	3	59.72	3	70.83	1	75.69	5	98.61	1	92.36
MM2	3	90.28	1	59.03	3	96.53	11	**77.78**	2	70.14	3	73.61	2	75.00	7	97.92	3	93.06
MM2s	2	91.67	2	57.64	4	97.22	11	**77.78**	10	62.50	4	62.50	3	77.08	6	**99.31**	1	93.06

	BCI III dataset IIIa						BCI III dataset IVa									
	1		2		3		1		2		3		4		5	
	pf	%	pf	%	pf	%	pf	%	pf	%	pf	%	pf	%	pf	%
CSP1	1	97.78	7	68.33	1	98.33	2	71.43	10	**100**	6	61.73	9	78.13	10	67.46
CSP1s	4	95.56	8	63.33	6	91.67	26	74.11	24	98.21	14	57.65	4	59.82	5	58.73
CSP2	2	97.78	6	73.33	2	**100**	2	74.11	10	**100**	5	57.14	3	78.13	10	76.19
CSP2s	5	95.56	16	61.67	7	91.67	12	74.11	8	98.21	6	55.61	6	61.61	5	58.73
MCD1	2	98.89	1	66.67	2	98.33	2	70.54	10	**100**	7	60.71	4	82.59	3	87.30
MCD1s	3	97.78	17	68.33	4	93.33	16	71.43	12	**100**	4	64.29	11	77.23	5	73.41
MCD2	2	98.89	14	73.33	2	**100**	2	74.11	8	**100**	8	64.29	6	80.80	7	82.94
MCD2s	3	98.89	16	68.33	4	91.67	7	67.86	22	**100**	4	64.29	10	78.13	5	74.21
DSM1	2	98.89	7	63.33	2	98.33	1	68.75	16	**100**	1	**73.98**	1	86.61	2	87.30
DSM1s	4	94.44	10	70.00	16	90.00	1	78.57	7	98.21	5	56.12	17	66.52	24	61.90
DSM2	3	**100**	10	66.67	2	98.33	1	69.64	27	**100**	3	71.94	3	85.71	2	82.54
DSM2s	7	94.44	7	66.67	11	86.67	1	**80.36**	16	**100**	7	57.14	20	71.88	9	56.35
DST1	2	98.89	1	61.67	2	98.33	1	73.21	8	**100**	2	69.90	1	**87.95**	1	**88.89**
DST1s	6	94.44	7	58.33	7	93.33	1	76.79	2	**100**	18	59.18	1	53.57	10	71.83
DST2	2	98.89	20	71.67	2	98.33	1	75.00	3	98.21	3	**73.98**	2	86.61	3	84.92
DST2s	4	93.33	8	61.67	10	88.33	1	77.68	6	98.21	17	58.67	8	54.91	11	67.46
DSH1	2	98.89	10	68.33	2	98.33	1	76.79	25	**100**	4	70.92	1	86.16	2	85.71
DSH1s	4	97.78	11	71.67	5	93.33	29	67.86	17	96.43	15	58.16	3	80.36	3	53.57
DSH2	1	97.78	13	**75.00**	2	98.33	1	72.32	26	**100**	2	66.33	1	80.36	2	82.94
DSH2s	3	96.67	12	73.33	9	93.33	2	67.86	6	98.21	12	61.22	3	81.70	21	57.54
MM1	2	96.67	17	68.33	2	**100**	9	74.11	1	96.43	6	71.43	2	86.61	2	79.37
MM1s	2	96.67	16	71.67	3	90.00	11	65.18	2	94.64	6	69.90	13	70.54	11	63.49
MM2	3	**100**	12	70.00	2	98.33	7	75.00	24	98.21	4	70.41	2	85.27	2	76.59
MM2s	2	96.67	13	73.33	3	95.00	7	67.86	4	96.43	17	67.86	12	73.66	15	62.70

veloping techniques to determine from the training data which is the most convenient method among the proposed in this paper for classifying MI signals.

Conflict of Interest

The author declare that she has no conflict of interest.

References

1. Hoffman, A., Farkaš, I. (2013) Using common spatial patterns for EEG feature selection. Technical report TR-2013-040. Comenius University in Bratislava.
2. Fisher, R. A. (1936) The use of Multiple Measurements in Taxonomic Problems. Annals of Eugenics **7** (2): 179–188.
3. Maronna, R.A., Martin D. et al (2006), Robust Statistics, Theory and Methods, Wiley, New York.
4. Yong, X., Ward, R. K., et al (2008) Robust Common Spatial Patterns for EEG Signal Preprocessing. 30th Annual International Conference of the IEEE Engineering in Medicine and Biology Society: 2087 - 2090
5. Rousseeuw, P. J. (1984) Least medians of squares regression. Journal of American Statistical Association, vol. 79: 851–857
6. Stahel, W.A. (1981) Breakdown of covariance estimators. Research Report 31, Fachgruppe für Statistik, E.T.H. Zürich, Switzerland.
7. Donoho, D.L. (1982) Breakdown Properties of Multivariate Location Estimators, Ph.D. dissertation, Harvard University
8. Lopuhaä, H. P. (1992) Highly efficient estimators of multivariate location with high breakdown point. The annals of Statistics 1992 Vol 20 No. 1: 398-413.
9. Kawanabe, M., Vidaurre, C. (2009) Improving BCI performance by modified common spatial patterns with robustly averaged covariance matrices. World Congress on Medical Physics and Biomedical Engineering, Munich, Germany Vol. 25/9 IFMBE Proceedings: 279-282.
10. Blankertz, B., Tomioka, R. et al (2008) Optimizing Spatial Filters for Robust EEG Single-Trial Analysis. IEEE signal processing magazine, vol. XX: 41-56.
11. Naeem, M., Brunner, C. et al (2006) Separability of four-class motor imagery data using independent components analysis, Journal of Neural Engineering, Vol. 3, no. 3: 208-216.
12. Schlögl, A., Lee, F. et al (2005) Characterization of four-class motor imagery EEG data for the BCI-competition 2005. Journal of Neural Engineering, Vol. 2, no. 4: 14-22.
13. Blankertz, B., Muller, K. et al (2006) The BCI competition III: Validating alternative approaches to actual BCI problems. IEEE Trans. Neur. Syst. Rehab, Vol. 14: 153-159.
14. RCSP toolbox at sites.google.com/site/fabienlotte/code-and-softwares
15. FSDA at ec.europa.eu/jrc/en/scientific-tool/fsda-matlab-code/downloads
16. Lotte, F. Guan, C. T. (2011) Regularizing Common Spatial Patterns to Improve BCI Designs: Unified Theory and New Algorithms. IEEE Transactions on Biomedical Engineering, vol. 58, no. 2: 355-362

CINARTRO: Clinical Tool to Assess Knee Kinematics by Videofluoroscopy

Williams Olivera[1], Marcio Rodriguez[1], Darío Santos[1,2] and Franco Simini[1]

[1] Núcleo de Ingeniería Biomédica, Universidad de la República, Montevideo, Uruguay.
[2] Clinical Biomechanics, Department of Rehabilitation, Hospital de Clínicas, Universidad de la República.

Abstract— **A new software application has been added to CINARTRO, a clinical tool to study knee kinematics from videofluoroscopic images during extension. The user selects, for each image, a set of points located on the Tibial Plateau and Condyle projections to determine the value of the Tibiofemoral Contact Point and a proxy of the Quadriceps Ligament Moment Arm. The software generates reports in PDF, Excel and CDA formats. CDA (Clinical Document Architecture) reports are uploaded into an XDS repository, as part of the patient clinical record. To solve the problem of "pincushion" effect in images, a gray tone recognition algorithm and distortion correction algorithm were implemented. The software was used to analyze data of injured and healthy knees of 3 patients. Precision errors introduced during the location of the points were 0.235mm. The Tibiofemoral Contact Point shows an average displacement of 59% in terms of tibial plateau length for healthy knees and 49% for injured knees. The average Moment Arm proxy was 50mm for healthy and 45mm for injured knees. The reduction in Quadriceps Muscle strength after ACL injury is compensated by a 11% increase in the Quadriceps Moment Arm.**

Keywords— ***Knee Kinematics, Anterior Cruciate Ligament (ACL), Tibio-Femoral Contact Point (TFCP), Quadriceps Moment Arm, Computer Document Architecture (CDA).***

I. Introduction

The knee of a person is subject to continuous movements and exercises in everyday life. After a serious lesion such as the rupture of the Anterior Cruciate Ligament (ACL) the knee joint kinematics are altered [1]. Postoperatively, functional instabilities may develop which require careful and evidence-based rehabilitation.

One of the most common ways to check the state of a reconstructed knee is with the Lachman test [2]. This evaluation depends greatly upon the perception and experience of the physician. We have developed a way to enhance and supplement this test with quantitative methods [3]. The aim of CINARTRO (such is the name of our instrument) is the collection of data and images to be processed into numerical results to help determine the condition of the patient after an ACL surgery [4]. The set of images obtained from a study by videofluoroscopy (VFC) is the source of input data to the software which interacts with the user to produce a relevant clinical report for the patient's medical record.

The procedure follows the theoretical methods by Baltzopoulos [5]. In particular, CINARTRO calculates the instantaneous position of the tibio-femoral contact point (TFCP) during extension movements [6]. Once the TFCP is known for every image, its posterior-anterior migration during extension is an indication of knee kinematics [4] and we add in the present paper the estimation of a proxy of the Moment Arm to further refine knee joint kinematics evaluation. This proxy is based on the TFCP rather than the instantaneous center of rotation, for simplicity.

II. Design and Architecture

CINARTRO is a software application which includes usability, integration with external components and generation of medical reports based on the specifications of Table 1.

The design and architecture for CINARTRO is shown in Figure 1 to fulfill the following functions:

- Acquisition of VFC images
- Phantom image processing for distortion correction
- Decimation of VFC to 15 images for extension
- Management of patient/VFC/healthy-or-injured knee
- Interaction to select anatomical repères in each image
- Geometric calculations of TFCP and moment arm
- Report generation

Table 1 CINARTRO software specification

Characteristic	Description
Calculation	TFCP and Moment Arm for every VFC image
Platform	Multiplatform: Windows, MacOS, Linux
Internationalization	English, Spanish; easy adoption of other languages
Target user	Physicians, specially physiotherapist, exercise physiologist
External agents	EMPI [7], XDS [8], AGESIC [9], Salud.uy [10]
Output information	CDA [11] and PDF clinical reports. Excel spreadsheets containing raw data

I. Torres et al. (eds.), *VII Latin American Congress on Biomedical Engineering CLAIB 2016, Bucaramanga, Santander, Colombia, October 26th -28th, 2016,*
IFMBE Proceedings 60,
DOI: 10.1007/978-981-10-4086-3_33

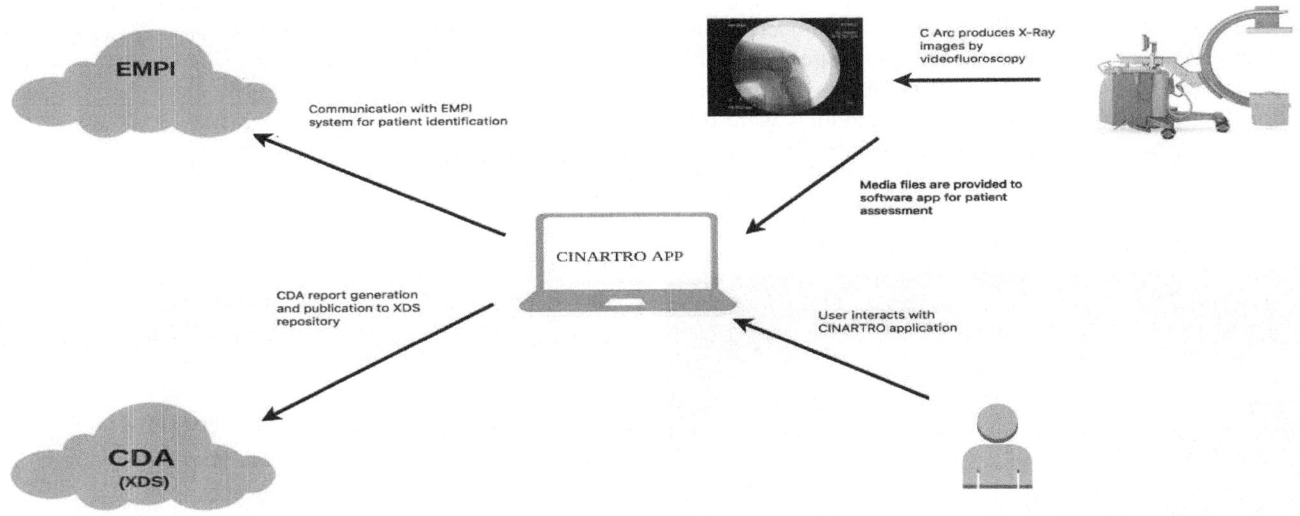

Fig. 1 General Architecture of CINARTRO. C arm sequence is analyzed to create a CDA report, with patient identified by EMPI web service.

Operation of CINARTRO begins when the user enters patient data, which is checked against a National Health Users EMPI (Enterprise Master Patient Index) index [6]. This ensures a reliable patient identification. Then the patient is subjected to X-ray with a C arm to obtain a VFC showing a knee extension movement. The media files of the VFC are presented to the user as individual frames and reduced to 15 images, at regular intervals during the leg extension movement.

The software guides the interactive identification of anatomical points following the method of Baltzopoulos [5], and our enhancements [4]. Once the process is completed, a report in Clinical Document Architecture (CDA) format [10] is created and published in an XDS repository [7] in the cloud for later use and access by Electronic Clinical Record (ECR) systems. Note that the Enterprise Master Patient Index (EMPI) is responsible for the identification of patients within either the Health institution or the Health Sector in case interoperability is available.

These two features (patients identified by an EMPI and CDA reports stored in XDS) give CINARTRO the ability to smoothly interact with any modern clinical record system. Additionally this feature is fully compatible with Uruguayan state standards under the SALUD.UY initiative [8][9].

III. WORKFLOW INSIDE THE APPLICATION

To obtain results from VFC images, the user must place a total of 11 points on each image as shown in Figure 2:

- Two points along the posterior Femur rim
- Two points along the posterior Tibia rim
- Three points along the internal Femur condyle
- Two extremities of the Tibial plateau
- Inferior Patella and anterior Tibial Tuberosity

CINARTRO calculates with these points the TFCP, a proxy of the Moment Arm and the bending angle, as also shown in Figure 2. Once processed, the user can see the fluctuation of these values in X-Y graphs where X is the frame number and Y the parameter such as TFCP position or Moment Arm proxy length. We call it "Proxy of Moment Arm" because the TFCP is not an instantaneous center of rotation.

Raw data may be output in Excel spreadsheets for further analysis. Since the CDA clinical reports (automatically uploaded to an XDS repository) are only readable by Electronic Clinical Record applications, CINARTRO also outputs regular PDF printable reports containing the same information. This information (both CDA and PDF) is of clinical value since it allows to compare different kinematics parameters (TFCP migration, MA proxy) of injured, contralateral and reconstructed knees.

Image distortion is corrected by using a phantom. To do this, the user must upload the rX image of the phantom, under the same conditions as the VFC, and select the visible parts of the grid points, as seen in Figure 3, so that the software identifies the matrix points to implement the corrections explained by Baltzopoulos [5]

IV. IMPLEMENTATION

Table 2 summarizes the programming elements used for a compact implementation of CINARTRO. We selected libraries to solve several programming issues. Java was the basic coding language following the paradigm of Object Orientation. JavaFX framework was chosen to operate the graphics application. The Singleton design pattern was used to maintain a single instance of classes. Other modules were dedicated to generate reports with external agents.

Pin cushion distortion error using the phantom was 0.235mm for 40% utilization of phantom points, which is close to 0.246mm, obtained by Baltzopoulos.

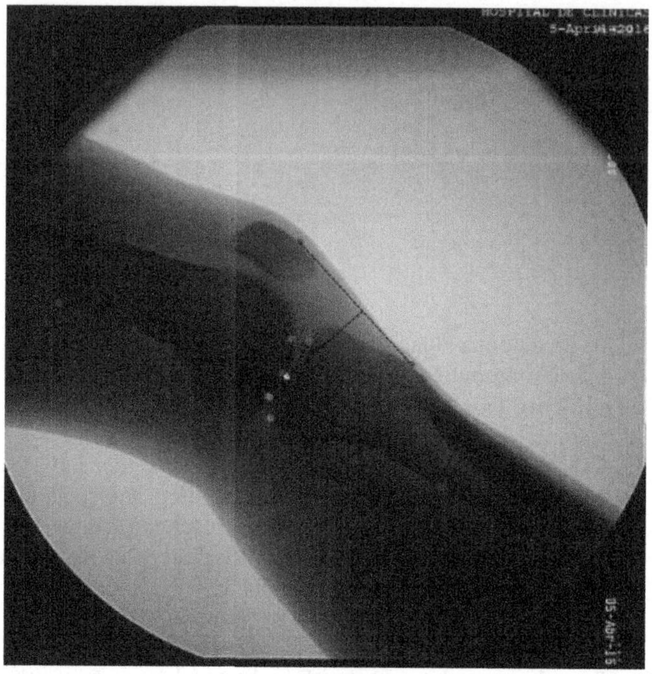

Fig. 2 Anatomical points marked by the user with CINARTRO. The line between patella an anterior tuberosity represent quadriceps ligament.

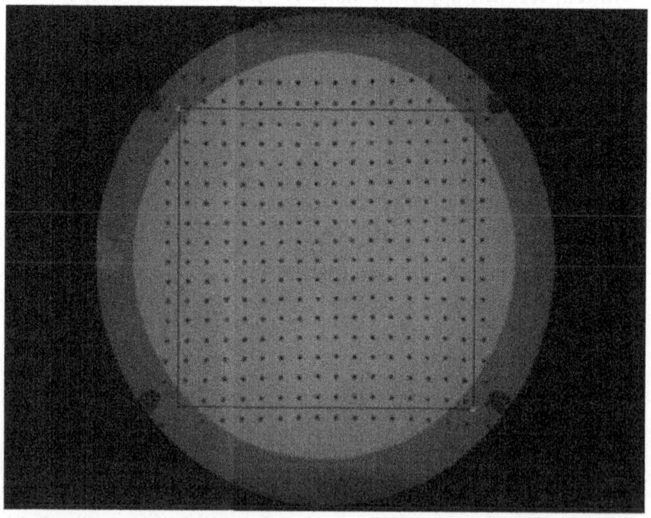

Fig.3 Equidistant lead sphere regular pattern seen by C-arm & CINARTRO. Note the pin cushion effect distorting shapes.

Table 2 CINARTRO Software Implementation

Feature	CINARTRO
Installation	Standalone (Desktop)
Programming Language	Java [12]
Graphic interface	JavaFX [13]
Graphic design tool	Scene Builder
Patient identification	EMPI (OpenEmpi) [7]
Studies output format	CDA (XML)
Math library	Apache Commons Math [14]
Data output / reports	Excel, PDF
PDF library	iText PDF [15]
Excel library	No, String CSV
XML management library	JAXB [16]
Video manipulation library	Xuggler [17]
HTTP messages exchange	HTTP Client (Java Core)
Architecture patterns	POO, Singleton

V. TESTS AND RESULTS

The software obtained fulfills the goal set in the specifications: we obtained a clinical tool to process VFC of injured, healthy and reconstructed knees during extension, with direct ECR documentation. We tested CINARTRO with VFC images obtained from 3 patients. The tests correspond to healthy knee and ACL injured knee for each patient.

Table 3 shows the results of TFCP with an average of 59% displacement for healthy knee and an average of 49% for injured knee. Additionally the Moment Arm proxy was 50 mm in average for healthy knees and 45 mm for injured knees.

Table 3 Moment Arm from TFCP to Quadriceps Ligament

Patient	Hanging 100° -110°		(135° - 145°)		Extended 170°-180°		Average Moment Arm (mm)	
	H	I	H	I	H	I	H	I
1	43	43	47	51	44	46	45 ± 2	48 ± 4
2	43	46	51	57	41	54	47 ± 5	52 ± 5
3	44	47	45	54	43	49	44 ± 1	51 ± 4
mean	**43**	45	**48**	**54**	**43**	**50**	**45 ± 3**	**50 ± 4**

Table 4 Tibio-Femoral Contact Point Migration

Patient	Hanging 100° -110°		(135° - 145°)		Extended 170°-180°		Average migration % TFCP	
	H	I	H	I	H	I	H	I
1	57	48	63	46	60	58	60 ± 3	50 ± 6
2	52	55	54	45	67	51	57 ± 8	50 ± 5
3	54	46	68	46	65	51	62 ± 7	47 ± 2
mean	**54**	**49**	**61**	**45**	**64**	**53**	**60 ± 3**	**49 ± 2**

VII. DISCUSSION AND CONCLUSION

For the first time we could easily process VFC images to obtain a quantitative evaluation of the kinematics of the knee. This follows the pioneering works of Baltzopoulos who defined parameters in the research setting, which we are now extending to potential routine clinical assessments with the new CINARTRO instrument [6]. Using the TFCP as a proxy for an equally migrating instantaneous center of rotation, we have defined a proxy of the Moment Arm seen by the quadriceps ligament during knee extension movement.

The measurement of a proxy of the Quadriceps Moment Arm (QMA) during extension is valuable to assess the load of the quadriceps during rehabilitation [3]. We found in this paper hat after ACL injury, the QMA increases with respect to the pre-injury knee, as measured in the contralateral joint. Additionally, clinical practice [18]observes that quadriceps muscle strength (QMS) decreases after ACL injury. We may therefore be in a position to describe a natural compensation mechanism by which the product of QMS x QMA (Moment) is kept as constant as possible after ACL injury, MS decreasing approximately 30% [18] compensated by 11% increase in QMA.

This new finding needs further research to be confirmed by careful protocols and patient follow-up with CINARTRO.

Conflict of Interest

The authors declare that they have no conflict of interest.

References

[1] S. Brandsson, J. Karlsson, B. I. Eriksson, and J. Kärrholm (2001) Kinematics after tear in the anterior cruciate ligament: dynamic bilateral radiostereometric studies in 11 patients. Acta Orthop. Scand., vol. 72, no. 4, pp. 372–8.
[2] M. Collette, J. Courville, M. Forton, and B. Gagnière (2012) Objective evaluation of anterior knee laxity; comparison of the KT-1000 and GNRB® arthrometers. Knee Surgery, Sport. Traumatol. Arthrosc., vol. 20, pp. 2233–2238.
[3] F. Simini and D. Santos (2014) Anterior Cruciate Ligament Reconstruction Follow-up Instrumentation based on Centre of Rotation Videofluoroscopy Determination: Development of an original equipment, CINARTRO, and first clinical use," IEEE Conf. proceedings- Int. Instrum. Meas. Technol. I2MTC., 2014.
[4] D. Santos, F. Massa, and F. Simini (2015) Evaluation of anterior cruciate ligament reconstructed patients should include both self-evaluation and anteroposterior joint movement estimation? Phys. Ther. Rehabil., vol. 2, no. 1, p. 3.
[5] V. Balzopoulos (1995) A videofboroscopy method for optical distortion correction and measurement knee-joint kinematics," Clin. Biomech., vol. 10, no. 2, pp. 85–92.
[6] F. Simini, D. Santos, and L. Francescoli (2016) Videofluoroscopy instrument to identify the tibiofemoral contact point migration for anterior cruciate ligament reconstruction follow-up: CINARTRO J. Phys. Conf. Ser., vol. 705 (2016).
[7] Sysnet International Open Enterprise Master Patient Index (OpenEMPI) (2016) [Online]. Available: http://www.openempi.org/
[8] Integrating the Healthcare Enterprise (IHE) (2016) Cross-Enterprise Document Sharing XDS [Online]. Available: http://wiki.ihe.net/index.php/Cross Enterprise_Document_Sharing
[9] Presidencia de la República. Agencia de Gobierno Electrónico y Sociedad de la Información de Uruguay (AGESIC) (2016). [Online]. Available: http://www.agesic.gub.uy/
[10] Agencia de Gobierno Electrónico y Sociedad de la Información de Uruguay (AGESIC), SALUD.UY (2016). [Online]. Available: http://salud.gub.uy/
[11] Health Level Seven International (HL7) (2005) Clinical Document Architecture (CDA) Release 2.0, HL7 Normative Edition.
[12] Oracle Corporation, Java Software (2016). v 8. [Online]. Available: https://www.java.com/es/about/
[13] Oracle Corporation, JavaFX (2016). v 8. [Online]. Available: http://docs.oracle.com/javafx/
[14] Apache Softfware, Apache Commons Math (2016). v 3.6.1.[Online].Available:https://commons.apache.org/proper/com
[15] NV iText Group, iText PDF (2016). v 5.5.7. [Online]. Available: http://www.itextpdf.com/
[16]Oracle Corporation, Java Architecture for XML Binding (JAXB) (2016). v 2.2. [Online]. Available: https://jaxb.java.net/.
[17] LLC ConnectSolutions, Xuggle/Xuggler (2016) v 5.2. [Online]. Available: http://www.xuggle.com/xuggler/
[18] W. Micheo, L. Hernández, and C. Seda (2010) Clinical Review : Current Concepts Evaluation , Management , Rehabilitation , and Prevention of Anterior Cruciate Ligament Injury : Current Concepts Phys. Med. Rehabil., vol. 2, no. 10, pp. 935–944.

Author: Dario Santos
Institute: Hospital de Clínicas - NIB
Street: Avda Italia s/n
City: Montevideo
Country: Uruguay
Email: dsantos@hc.edu.uy

Test Bench to Validate Audio CODEC Kit as EIT Complex Voltage Measurement Circuit

M. Arregui, E. Santos and F. Simini

Núcleo de Ingeniería Biomédica, Universidad de la República, Montevideo, URUGUAY

Abstract— **A simple bench test instrument was developed to measure the real part of the impedance of a patient simulated by a phantom (variable R 350-600 ohm in parallel with C=47nF). The instrument is part of an EIT design which includes a EVM-L137 evaluation kit to enhance previous designs. The DSP included in the kit generates the waveform to control current sources. We use this original signal as a time reference to measure voltage phase, thus minimizing unknown delays. Once data are corrected by a linear fit to remove systematic errors in measurement, the values obtained deviate on average 0.9% from the expected. This instrument will allow to further compare design options in EIT circuit development.**

Keywords— **Electrical Impedance Tomography, Audio Codec, Digital Signal Processor, Howland Current Source, Test Bench**

I. Introduction

Stemming from clinical practice, IMPETOM is an instrument to solve the problem of obtaining simple and reliable graphical information on liquid occupancy of the thorax of intensive care patients[1][2]. Following the seminal papers of the Sheffield School[3] the first prototypes of IMPETOM included a 16 electrodes chest strap, a multiplexed 30KHz current source and 16 voltage meters. The resulting matrix was processed based on "back projection" algorithms to show impedance moduli color coded areas. Different architecture were devised and put in operation [4] including three rows of electrodes to better control current dispersion[5]. In 2014 a new processing alternative, based on better finite elements mesh shapes, was associated with enhanced hardware solutions, taken from "off the shelf" circuit elements rather than in house development [6].

DSP technology and audio CODEC circuitry were adopted for IMPETOM when an EVM-L137 evaluation board was used [7]. The first results of such tomographic reconstruction were calibration measurements, published in 2015 [8], the layout of which is shown in Figure 1.

While at least one commercial option is available on the market [9], we are faced here with the problem of developing a bench test to compare strategies and design options. Our goal is a tool to measure complex voltage resulting from different body impedance simulators.

II. System Description

We build a simple phantom with a parallel RC circuit, where R is variable from 350 ohms to 600 ohms and fixed C= 47 nF. This circuit is fed by a modified Howland current source [10], controlled by one of the CODEC output signals. This signal is proportional to a DSP originated 16 KHz sinus wave. In parallel to feeding the current source, the 16 KHz wave is measured by the oscilloscope, allowing to have a phase reference for the current measurements.

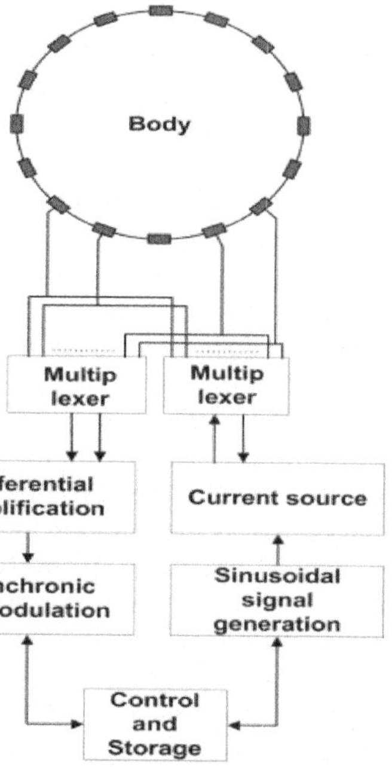

Fig. 1: EIT system block structure. Current source and resulting 16 voltage measurements are both referenced to the same DSP generated 16 KHz sinusoidal signal. Taken from Santos et al. [8].

The CODEC was configured to work with a 48 kHz sampling frequency in slave mode, the data transfer protocol was "I2S to DSP" and 16 bits word length. No offset is added to the data. The PGA gain of each channel was set to 0db. Communication between DSP and CODEC is through McASP serial port communication settings synchronized with 2 slots TDM format.

In parallel, an oscilloscope is connected to the output of the phantom circuit and to the input of the current source, where the voltage is in phase with the current

I. Torres et al. (eds.), *VII Latin American Congress on Biomedical Engineering CLAIB 2016, Bucaramanga, Santander, Colombia, October 26th -28th, 2016*,
IFMBE Proceedings 60,
DOI: 10.1007/978-981-10-4086-3_34

injected into the phantom. This allows to compare measurements made by the system under test implemented with a reliable instrument such as a laboratory oscilloscope.

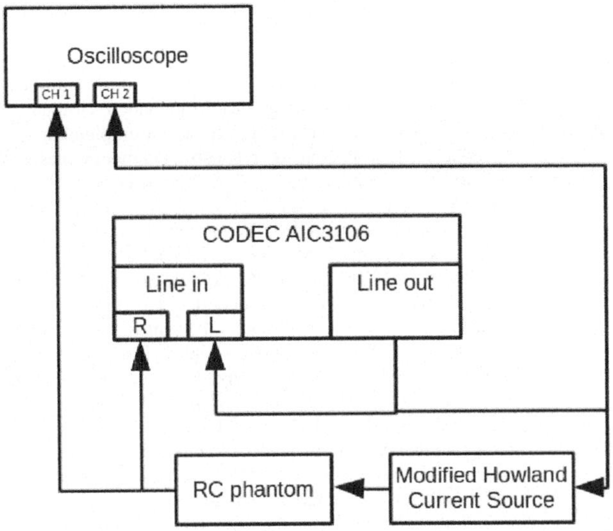

Fig. 2: Connection diagram of the test bench. The modified Howland current source feeds a phantom (parallel RC: C= 47nF and a variable resistor). The oscilloscope allows to compare the original CODEC voltage signal to the phantom-shaped voltage.

The software we developed uses the phase of the reference voltage signal to compare it to the phase of the resulting phantom voltage. The output voltage signal of the phantom circuit allows to calculate the real part of the phantom impedance. We do not use the original DSP generated signal which is fed to the CODEC to control the current source, because it would introduce a lag from the time the DSP sends the signal to the time the CODEC effectively outputs the signal at its output channel. To cancel this delay we chose to use a second CODEC input channel to measure the signal of its own output port. In this way we reduce the time overhead derived from causes other than the wanted phase change between the input and output signals caused by the RC phantom.

Once sampled signals and data are sent from the CODEC to the DSP, the software routine behaves according to the diagram in Figure 3.

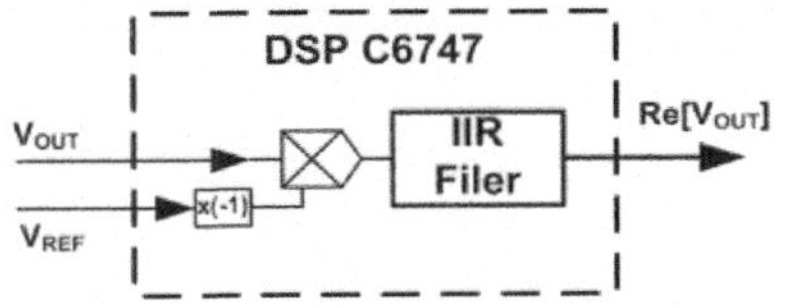

Fig. 3: DSP Implemented Synchronous Demodulator. Real part of V_{OUT} is output.

III. Results

We have tested our circuit with several values of R and a fixed C value, all reported in Table 1.

Table. 1: Real part of the phantom impedance.

R PHANTOM (Ω)	Impedance R//C real part @ 16KHz (Ω)	Oscil. measurement (Ω)	IMPETOM measurement (Ω)	IMPETOM adjusted R//C (Ω)	Relative Error
350	93.7	107.6	108.3	92.6	1.23%
400	87.5	111.3	104.5	88.2	0.82%
450	81.5	101.3	98.8	81.8	0.38%
500	76.0	94.7	94.6	77.1	1.43%
550	70.9	85.5	88.7	70.4	0.74%
600	66.4	78.9	84.7	66.0	0.66%
				Average Error	0.876%

Note: The first column shows the R value of the parallel RC circuit.

IV. Discussion

Our previous IMPETOM designs, along with other projects, include current control by a single analog signal multiplexed into 16 current sources. When a decision was made to use digital circuits and audio kits to cut costs and ease reliable duplication of IMPETOM instruments, real part impedance measurement errors were high. Debugging gave as a result an unexpected phase lag between signals. A theoretical internal digital signal as a vector in CODEC memory was initially used as a reference to measure real world hardware signal phases. This gave rise to uncontrolled errors. By using the same CODEC generated D/A signal as a reference for measurement as well as to create the signal to submit to the "patient simulator" or phantom, reliable values were expected. This was actually the case and the present paper describes therefor the error found with the suggested circuit. An error of approximately 1% was measured, which insures a reliable enough test bench.

We found that the implemented system manages to make consistent and comparable measurements. Once the data are corrected by a linear fit to remove systematic errors in measurement, the values obtained deviate 1% on average from the expected. This shows that our system inherits the precision of a laboratory instrument. Without our system, the impedance real part would only be estimated with an error due to lack of common time reference. This special measurement is necessary because modern designs include software signal generation, as opposed to traditional layouts with a central analog oscillator to command all current sources.

V. Conclusions

We obtained a test bench that allows us to evaluate the circuit configurations for EIT design as well as for future modifications and enhancements. The ability to test the behaviour of hardware and software components on a simple phantom is a great advantage for design purposes and decision making. Several improvements can be made to the system, such as reducing the test impedance vulnerability to noise derived from the lack of electrical shielding. Additionally the inclusion of measuring bridges to improve system sensitivity should be considered. We also plan to explore the use of 32-bit digital encoding to reduce quantization errors.

VI. Acknowledgment

The authors thank the work of all IMPETOM team workers (students, technicians and staff) since 2002, and specially the recent contribution of Nicolás Alfaro and Fernanda Martinucci.

VII. Conflict of Interest

The authors declare they have no conflict of interest.

VIII. References

[1] A. Rodríguez, A. Ferreira, and F. Simini, 2003 "IMPETOM-C Tomógrafo de Impedancia Eléctrica," Thesis, Universidad de la República, Montevideo, Uruguay.

[2] R. Hartmam, J. Lobo, M. Ruétalo, and F. Simini, 2002 "IMPETOM-I Tomógrafo de Impedancia Eléctrica," Thesis, Universidad de la República, Montevideo, Uruguay.

[3] D. C. Barber and B. H. Brown, 1984 "Applied potential tomography," *J. Phys. E.*, vol. 17, no. 9, p. 723.

[4] S. González, A. Liguori, and F. Simini, 2005 "IMPETOM: Tomógrafo de Impedancia Eléctrica," Thesis, Universidad de la República - Uruguay.

[5] W. Quinteros and F. Simini, 2007 "IMPETOM-48: Tomógrafo de Impedancia Eléctrica con tres hileras de electrodos," Thesis, Universidad de la República, Montevideo, Uruguay.

[6] E. Santos, 2014 "Alternativas de proyecto e implementación de circuitos y de programas de reconstrucción tendientes a un tomógrafo por impedancia eléctrica para la presentación compacta del estado edemático de cortes torácicos en tiempo real," Thesis, Universidad de la República - Uruguay.

[7] N. Alfaro, M. Arregui, F. Martinucci, E. Santos, and F. Simini, 2015 "IMPETOM-B: Tomógrafo de Impedancia Eléctrica," Universidad de la República, Montevideo, Uruguay.

[8] N. Alfaro, M. Arregui, F. Martinucci, E. Santos, and F. Simini, 2016 "Audio Codec and Digital Signal Processor for an Electrical Impedance Tomography System," in *II Latin American Conference on Bioimpedance*, F. Simini and P. Bertemes-Filho, Eds., Springer, pp. 16–19.

[9] E. Teschner and M. Imhoff, 2010 *Tomografia por Impedancia Electrica: A realização da monitorização da ventilação regional*, Drager tec. Lubeck, Alemania.

[10] G. Saulnier, 2005, "EIT Instrumentation," In Electrical Impedance Tomography: Methods, History And Application, D. Holder, Ed. London: Institute Of Physics, P. 67.

Author: Franco Simini
Institute: Núcleo de Ingeniería Biomédica
Street: Hospital de Clínicas, Av. Italia S/N, 11600
City: Montevideo
Country: URUGUAY
Email: simini@fing.edu.uy

Images Digitization and Characterization of Surface and Fundus obtained through a Slit Lamp Adapted

I. Cassi[1], A. Salvatelli[1], G. Bizai[1], A. Hadad[1], D. Ramirez Arduh[2], B. Drozdowicz[1]

[1] Facultad de Ingeniería, Universidad Nacional de Entre Ríos, Lab. Sistemas de Información, Oro Verde, Entre Ríos, Argentina
[2] Sociedad Entreriana de Oftalmología, Paraná, Entre Ríos, Argentina

Abstract— **Slit Lamp (SL) is the equipment commonly used in most eye clinics. This instrument allows to observe ocular structures of both the anterior segment and posterior segment with different magnification values. Currently is very common obtain images from SL with a Smartphone (for documentation purposes) by means of a simple adapter in the eyepiece. These images lose their diagnostic quality due to the information lose to correct the lack correction of optical aberrations and noise introduced by this system. This paper analyzes the hardware settings and image characterization to need diagnostic quality image in SL. For this purpose a universal adapter for SLR cameras were developed. To adquisition process image the camera software capture were use. They were captured 128 images: 73 of anterior segment, 25 fundus and the rest with a magnifying glass and panfunduscope. These images were characterized by type and noise power. For de full optical system the point spread function PSF and the consequent modulation transfer function MTF were determined. For the new image postprocessing the Lucy-Richardson deconvolution algorithm were applied. Finally in the SL the numerical aperture and F number (#f) were obtained. Those data, uncommon in the technical data of manufacturers, are of fundamental importance in future adaptations.**

Keywords: **Slit Lamp, Lucy Richardson, universal adapter, SLR camera, f number.**

I. Introduction

Based on the lateral or oblique and focal illumination, Gullstrand, has devised the slit lamp (biomicroscope). This lamp consists of an illumination arm system containing the lamp emitting light through a variable diaphragm (slit-shaped), and an optical system of positive lenses which allows to focus that light slit on segment site above to be examined. The slit lamp (SL) or biomicroscope is a fundamental tool in ophthalmological practice [1, 2]. With it can be considered the anterior segment of the "naked eye". In addition is possible examine the vitreous body, the retina (vascular structures, macula and fovea) and optic nerve. To observe it, is necessary to place in front of the eye a lens (typically 90 diopters) air or contact (panfundoscope), which make these structures visible.

In the 1950s, with the advent of electronic flashes and 35 [mm] cameras, the fundus photography became a common practice. The main aim of this photography mode was documentation, rather than post-processes or diagnostic consultation.

However, digital age has facilitated sending images for remote diagnosis. In this sense, there is a increased interest in digital fundus photography with slit lamps [3].

Moreover today is very common in the field of ophthalmology, obtain images from the slit lamp using a smartphone to placed with a simple adapter in the eyepiece. With images obtained, only documentation purpose, due to lose their diagnostic quality. This loss are related with optical aberrations and noise introduced with that system. Then thiss images can not be used for disease control and monitoring with involving, for example, distances and áreas measurements.

For this reason, this paper analyzes the necessary configuration and hardware settings to achieve a 2D image of diagnostic quality in SL, including characterization of the images obtained. The development of a universal adapter for cameras SLR by the capture system is also described. The noise model, point spread function PSF for the total system with all its gains were obtaining. In addition, Lucy Richardson deconvolution algorithm were applied only to posterior pole photograph. Finally, focal length, numerical aperture and F number (#f) were obtained. Those data, uncommon in the technical data provided by manufacturers, are of fundamental importance in future adaptations.

The topics addressed here is part of one of the research lines within the Research and Development Project "Plenoptic Information System as a diagnostic tool applied to Slit Lamp". The purpose project is to add diagnostic value to current slit lamps, by means three-dimensional reconstruction of plenoptics images (obtaining depth map with a single photographic shot). These plenoptics images need to have reference data from planar images (2D).

II. Materials y Methods

A. Slit Lamp (SL)

The conventional SL consists of three basic: elements, lighting system, stereo-biomicroscope, and mechanical system. [4]

I. Torres et al. (eds.), *VII Latin American Congress on Biomedical Engineering CLAIB 2016, Bucaramanga, Santander, Colombia, October 26th -28th, 2016*,
IFMBE Proceedings 60,
DOI: 10.1007/978-981-10-4086-3_35

The "HGM Laser systems" SL used in this project is attached at the top of laser photocoagulation system and its corresponding eye filter.

This SL has the next technical especification:

- Magnifications of 6X, 10X, 16X, 25X, 40X, in the Galileo system,
- A focal length of 125 [mm].
- Adjustable eyepieces 12.5X between -6 to 6 diopters.
- Variable halogen lighting in three levels.
- Filters green and blue light.
- Slit adjustable in shape and dimensions.

B. Materials

The digital SLR camera is the Canon EOS Rebel T3i/600D. This model were selected because of its good signal noise ratio at high ISO. The more relevant technical characteristics are:

- CMOS sensor with 18 effective megapixel size 22.3 x 14 mm,
- Maximum resolution of 5,184 x 3,456 px,
- Pixel pitch 4.3μm.
- Canon EOS Utility software with USB connection.

The eyepieces used for the construction of universal adapter for SLR cameras were the Ramsden type with a 10X magnification and a focal length of 20[mm]. This adapter replaces one of the original eyepieces.

To capture the fundus images, a magnifying glass model VOLK Superfield (Field of View: 95 ° / 116 °, magnification : 0,76X , working distance : 7 [mm]) were used . Another magnifying panfunduscopio volk quadraspheric (contact view: 120 ° / 144 °, magnification : 0,51X) were used too.

C. Methodology

The images of anterior and posterior pole by a specialist ophthalmologist according to techniques that can be reviewed in [1] and [2] without changing the daily clinic were taken. In this work 128 images: 73 previous segment, of which 3 were obtained with Goldmann lens, 25 fundus, of which 8 were taken with Goldmann lens, 9 with panfunduscope and 8 magnifying glass, were captured.

Determining the focal distance (DF) of the eyepieces and all lenses used it was based on the principle of geometrical optics which states that all beams incident on the surface of a lens, parallel to its main axis, is diffracted so such that intersect at the focal point thereof [5].

The noise power (NPS) of the images by a variation proposed by J.H. et al Siewerdsen algorithm in [6] implemented in MatLab program R2013a (MathWorks, Inc. USA) were determined. In this case, regions of interest (ROI) were selected according to Gonzalez and Woods [7].

The method proposed by U. Neitzel in [8] has been used for PSF system determination. Finally the MTF calculus uses the formula (1) where F expresses the Fourier transform.

$$MTF(u,v) = F|PSF(x,y)| \qquad (1)$$

The selected algorithm to perform deconvolution (aberration correction) for ease of application was proposed by Lucy - Richardson (LR). For more details about this algorithm, reading [9], [10] and [11] is recommended .

The numerical aperture (NA) of an optical system is a dimensionless number that characterizes the range of angles for which the system accepts light rays. Especially in microscopy, the NA lens is defined by equation (2) :

$$NA = n.\sin\theta \qquad (2)$$

Where n is the refractive index of the medium in which the lens is placed (1 for air, 1.33 for pure water), and θ is half of maximum angle of acceptance that can enter or leave the lens. The NA is generally measured with respect to an object or a point of an image and varies with the position of the point. In microscopy, the NA is important because of indicates the resolving power of a lens. It is also known that the NA directly affects the functions, PSF and MTF [12].

Through the assembly described in [5], focal distance and the diameters of opening of the images projected onto the lens (aperture) were obtained. We can then estimate the *number #f*, for photography by the expressions (3) and (4)

$$\#f\ number = \#\boldsymbol{f} = 1/(2 * \mathrm{NA}) \qquad (3)$$

$$\#\boldsymbol{f} = DF/\emptyset FOV \qquad (4)$$

Were ØFOV is the diameter of the field of view

III. Results

In order to obtain images of anterior and posterior eye pole, was made a universal adapter. This development fulfilled the premise of not compromising the structure of the slit lamp, conected at eyepiece level.

The distance between the CCD camera and relay lens of adapter was calibrated by the ophthalmologist. Then, the image seen in the free eyepiece is the same image that the captured by the camera (both fields of vision are matched).

The adapter was designed based on a 10x eyepiece and a adjustable relay lens, coupled to the camera via a ring mount T2 EF, Figure 1.

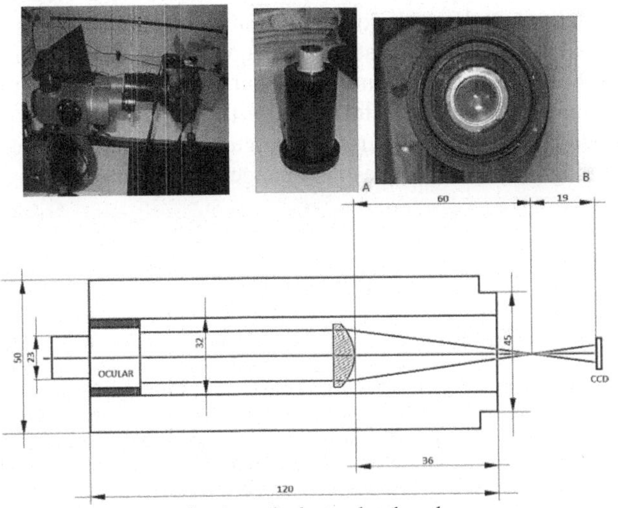

Fig. 1: Measures and universal adapter developed.

From the statistical analysis of the histogram of the noise present in the photographs of posterior pole, for each channel (red, green and blue) the standard deviation values were obtained. This values fluctuate within the range [4; 8], symmetry values located in the range [-0.1; 0.23] close to zero and kurtosis whose values fluctuate in the range [3; 4] close to three. Based on the morphological analysis of the histogram for each ROI and statistics presented above, Gaussian noise was determined.

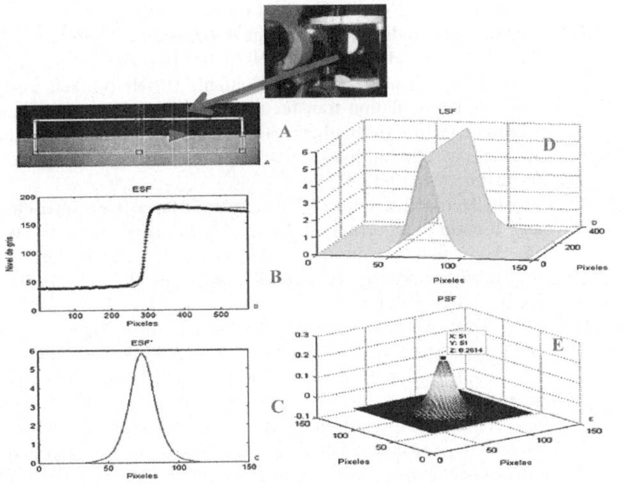

Fig. 2. Example of Calculation of the PSF. (A) step-like phantom. (B) ESF average of ROI column. (C) derived ESF. (D) LSF of ROI. (E) PSF

There are few differences between the statistical characteristics of noise in images of the anterior pole and posterior pole. It is possible to mention that the histograms obtained on fundus images are slightly sharper than those obtained on the images of anterior pole. The average noise power of the red and green channels reaches a maximum of 0.06 dB/mm2, while the blue channel reaches a maximum of 0.075 dB/mm2. Numerous filters were tested with the purpose of attenuate this noise, media filtering with a 5x5 window were selected.

For each of the magnifications including a magnifier, the PSF were obtained, Figure 2. From those PSF the MTF of this SL were obtained too, as shown in Figure 3.

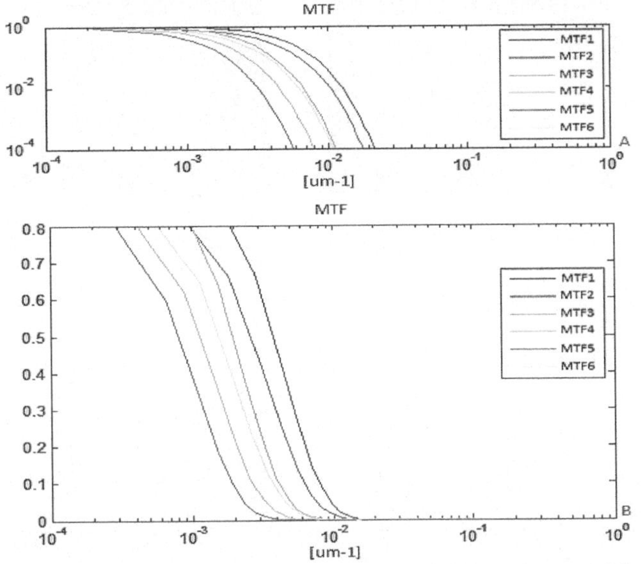

Fig . 3 (A) detail of the top of the MTF (B) detail of the bottom of the MTF for different magnifications system. MTF1: 6X. MTF2: 10X. MTF3: 16X. MTF4: 25X. MTF5: 40X. MTF6: 6X + LUPA VOLK SUPERFIELD NC.

We preset a complete set of NA and #f values of the SL, obtained by the expressions (3) and (4) in Table 1.

Finally it is shown in Figure 4 a posterior pole image, which was noise filtered and then deconvolved with LR method with five iterations. This number of iterations was selected by an ophthalmologist after seeing the pictures. The criteria chosen were contrast and sharpness enhancement. Nevertheless more iterations increase the total time of obtening image.

IV. Discussion And Conclusions

By designing the universal adapter for SLR cameras and applying techniques of geometrical optics we obtain high-quality diagnostic images. Thus, we reach the goal of providing diagnostic value to SL with very low cost and with a universal adapter. In free aberration images, fig. 4C for example, is possible to make measurements to monitor and control changes in response to treatments. This measure opens the possibility for medical interconsultations more accurately.

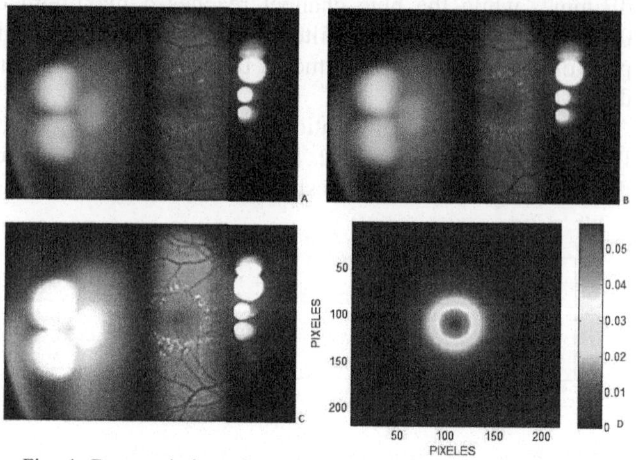

Fig. 4: Deconvolution of an image fundus by the LR algorithm. (A) original image documentation. (B) filtered with a median filter with a window size of 5x5 image. (C) deconvolved and aberration-free image. (D) image of the PSF.

Tabla 1 : NA and f-number calculated for each magnification system

Magnificación	Ø FOV[mm]	DF [mm]	NA	#f
6X	9	215	0,0209	23,89
10X	11	150	0,0366	13,63
16X	15	145	0,0517	9,67
25X	16	135	0,0592	8,44
40X	18	125	0,0719	6,95

In particular, from the results observed in Figure 3A and 3B, and data that expresses in Table 1, it is possible to infer that the gradient of the various ESF (one for each optical magnification) decreases as the greater the magnification system. Therefore the maximum peak of the different PSF is calculated. This result corresponds to Table 1 as with increasing the numerical aperture of the system decreases the cutoff frequency and light that reaches the CCD camera is diminishing (high f number) and the reflective field area vision is becoming smaller and the diameter of the diaphragm (Ø FOV) on the lens increases. That is, further details will be obtained at low magnification images. This corresponds with the daily clinical since increases are mostly used, 6X and 16X.

Lucy Richardson (LR) deconvolution algorhythm working properly to restore ophthalmological images captured with a slit lamp. The experimental results suggest do not use higher of five iterations.

The diagnostic quality of the reconstructed images with LR was assessed by expert ophthalmologists; achieving very good results refer to enhancement of hidden details and correction of aberrations caused by the optical system. However, a negative aspect is that much enhances the reflections present in the images which often can degrade the quality of them.

To calibrate the images acquisition process for measurement we focusing on a pattern, in this case a Neubauer chamber (used in cell count) 0.1 mm resolution. Finally it is possible to mention that this capture system, which can be calibrated to see the same image on the free eyepiece in the computer screen. This form to use the SL adapted provides freedom to other tasks routine doctor and gives confidence to diagnose because does not change clinical practice.

Referencias

1. Susana Figueiredo e Silva, Development of an optical confocal module for a slit lamp microscope. Master tesis, Faculdade de Ciências e Tecnologia, Universidade de Coimbra, 2011.
2. Gabriela González, Andrés Faisano, y Rodrigo Torres, The Slit-Lamp Survey, VI Latin American Congress on Biomedical Engineering CLAIB 2014, Paraná, Argentina 29, 30 & 31 October 2014, Volume 49 of the series IFMBE Proceedings pp 117-119.
3. UmitYolcu, Omer Faruk Sahin y Fatih C. Gundogan, Imaging in Ophthalmology., IntechOpen, 2014, Available on: http://cdn.intechopen.com/pdfs-wm/46615.pdf
4. B. Drozdowicz, A. Salvatelli, G. Bizai, A Hadad, D. Evin, R.Torres, Análisis y propuesta de trabajo para la reconstrucción tridimensional de imágenes maculares obtenidas mediante lámparas de hendidura, JAIIO-CAIS 2015, 6° Congreso Argentino de Informática y Salud, Set. 2015, Proceedings online http://44jaiio.sadio.org.ar/?q=node/226.
5. Sears, Zemansky, Young y Freedman, Física Universitaria Vol. 2; 2da Edición. Ed. Addison-Wesley, pag. 896, 2009. ISBN: 978-607-442-307-5.
6. J. H. Siewerdsen, I. A. Cunningham, D. A. Jaffray, A framework for noise-power spectrum analysis of multidimensional images, Med Phys. 2002 Nov;. DOI: 10.1118/1.1513158.
7. Gonzalez &Woods, Digital Image Processing., 3° Edición, Ed. Pearson Prentice Hall, 2008, ISBN-10: 013168728X.
8. Egbert Buhr, Susanne Günther-Kohfahl, Ulrich Neitzel, Simple method for modulation transfer function determination of digital imaging detectors from edge images. Proc. SPIE vol. 5030, Medical Imaging 2003: Physics of Medical Imaging, 877 (June 9, 2003); doi:10.1117/12.479990.
9. Eymard Hernandez López, Un acercamiento a la deconvolución ciega usando el algoritmo de Lucy – Richardson, tesis de Maestría, Universidad Autónoma Metropolitana 2010. On line: http://www.remeri.org.mx/portal/REMERI.jsp?id=oai:tesiuami.izt.uam.mx:UAMI14709.
10. Maja Temerinac-Ott, Tile-based Lucy-Richardson Deconvolution Modeling a Spatially-Varying PSF for Fast Multiview Fusion of Microscopical Images, Technical Report 260, University of Freiburg(260), 2010, Available from: http://cia.informatik.unifreiburg.de//Publications/2010/Tem10a
11. Arijit Dutta, Aurindam Dhar, Kaustav Nandy, Image Deconvolution by Richardson Lucy Algorithm,Indian Statistical Institute, Nov. 2010. http://www.isid.ac.in/~deepayan/SC2010/projectsub/RLA/report.pdf
12. Kenneth Spring, Michael Davidson, Microscopy U, Modulation Transfer Function, Available on: https://www.microscopyu.com/articles/optics/mtfintro.html.

Non-homogeneous multichannel electroencephalographic dynamic forward modeling of epilepsy

P.A. Munõz-Gutiérrez[1] and E. Giraldo[2]

[1] Universidad del Quindío/Electronic Instrumentation Technology, Assistant professor, Armenia, Colombia
and Universidad Tecnológica de Pereira/Doctorade in Engineering, Student, Pereira, Colombia
[2] Universidad Tecnológica de Pereira/Department of Electrical Engineering, Associated professor, Pereira, Colombia

Abstract— **This paper shows signals of brain activity per zones, which were obtained through a dynamic forward model that considers non- homogeneous temporal activity in a group of regions. This activity could be observed in a multichannel electroencephalography. The forward model combines the reduction of source space and a model of spatio-temporal propagation of neural activity, in which the independent non-homogeneous brain activity is introduced for each zone. This model allowed us to simulate brain activity similar to epilepsy in a specific area, which could be observed in the multichannel electroencephalography, and is therefore a first step to carry on a focused analysis of the brain activity and the inter-connectivity among zones.**

Keywords— **non-homogeneous, dynamic modeling, multichannel, epilepsy.**

I INTRODUCTION

The electroencephalographic (EEG) and magnetoencephalographic (MEG) signals are frequently used in the analysis or diagnosis of different neural disorders (epilepsy, tumors, brain damage) with the purpose of rebuilding the brain activity; and thus, to localize the specific current sources that could be generating these disorders. The EEG based neural activity estimation through dynamic inverse problems assumes that the region in which the brain activity is estimated, is homogeneous. Therefore, the parameters obtained from the temporal restriction are the same for this region. This may prevent the analysis of inter-connectivity and at the same time, may prevent to observe the temporal evolution of the brain activity. In [1] it is pointed out that the inverse solution can improve if the temporal structure observed in the neural recordings is not ignored in the entire levels.

In general, most methods incorporate temporal structures in their models, which specify spatio-temporal interactions that are highly restricted. For instance: space-time separability or spatial independence. Yet this does not necessarily reflect the accuracy in the dynamic relations among the different regions of the brain. Therefore, when considering a non-homogeneous dynamical model for neural activity, the focus spatially currents with temporal restrictions could be reconstructed [2].

In this article a novel non-homogeneous multichannel electroencephalographic dynamic forward modeling of epilepsy is presented. Proposed model considers non-homogeneous dynamics that allows simulation of several types of pathologies as well as analysis connectivity of brain zones. The simulated EEG signals are compared with homogeneous EEG signals during normal and pathological activity. This article is structured as follows: in section II a physiological model for simulation of multichannel EEG with a homogeneous dynamic model is presented, in section III a multichannel EEG with a non-homogeneous dynamic model is proposed, and in section IV simulation results are presented.

II PHYSIOLOGY-BASED DYNAMIC MODELS

A key component for realistic EEG modeling is the use of an adequate set of features which represent the system dynamics [3]. In general, the EEG can be described in terms of the neural activity as follows:

$$\boldsymbol{y}_k = \boldsymbol{M}\boldsymbol{x}_k + \boldsymbol{\varepsilon}_k \tag{1}$$

being $\boldsymbol{y}_k \in \mathbb{R}^{d\times 1}$ the EEG measured signals, $\boldsymbol{M} \in \mathbb{R}^{n\times n}$ the leadfield matrix, $\boldsymbol{x}_k \in \mathbb{R}^{n\times 1}$ the neural activity and $\boldsymbol{\varepsilon}_k \in \mathbb{R}^{n\times 1}$ the zero mean uncorrelated additive Gaussian noise.

Several models have been used to describe the neural activity in the whole brain. For example, a homogeneous multichannel EEG model is proposed in [3], grounded in a telegrapher's equation as follows:

$$\boldsymbol{x}_k = \boldsymbol{A}_1\boldsymbol{x}_{k-1} + \boldsymbol{A}_2\boldsymbol{x}_{k-2} + \boldsymbol{\eta}_k \tag{2}$$

being $\boldsymbol{A}_1 = a_1\boldsymbol{I}_n$ and $\boldsymbol{A}_2 = a_2\boldsymbol{I}_n$ with $\boldsymbol{I}_n \in \mathbb{R}^{n\times n}$ an identity matrix, and $\boldsymbol{\eta}_k$ a zero mean uncorrelated additive Gaussian noise. This model successfully describes the spatio-temporal

I. Torres et al. (eds.), *VII Latin American Congress on Biomedical Engineering CLAIB 2016, Bucaramanga, Santander, Colombia, October 26th -28th, 2016*,
IFMBE Proceedings 60,
DOI: 10.1007/978-981-10-4086-3_36

propagation of neural activity x_k under the assumption that the whole brain is represented by the same dynamic. In [4, 5, 6] the following homogeneous model grounded in [7] is used:

$$\begin{aligned} x_k &= A_1 x_{k-1} + A_2 x_{k-2} \\ &+ A_3 x_{k-\tau} + A_4 x_{k-1}^2 + A_5 x_{k-1}^3 + \eta_k \end{aligned} \tag{3}$$

being $A_i = a_i I_n$ with $i = 1,\ldots,5$. This models includes nonlinear and delayed terms that improve the underlying dynamics of neural activity according to the physiological behavior. However, a drawback of the model is the assumption that the whole brain is described by the same dynamics (homogeneous modeling). On the other hand, in [2] a non-homogenous first order model is proposed as follows:

$$x_k = A_1 x_{k-1} + \eta_k \tag{4}$$

being $A_1 = \operatorname{diag} a$, with $a \in \mathbb{R}^{n\times 1}$. This work proposes a model in which the regions are characterized for being non-homogeneous and for having their own temporal evolution, but the model is restricted to a first order model that can not adequately describe the underlying dynamics of the neural activity.

III PROPOSED NON-HOMOGENEOUS MULTICHANNEL EEG MODEL

Grounded in the models proposed by [2, 6] the following model is proposed where the activity is assumed to be represented for the same dynamic for a reduced number of zones [6]. Consequently, a spatial reduction can be included in Eq. (1) as below:

$$x_k = \Phi_s c_k, \tag{5}$$

where the vector $c_k \in \mathbb{R}^{s\times 1}$ weights the matrix of spatial coefficients, noted as $\Phi_s \in \mathbb{R}^{n\times s}$.

Considering the spatial basis (see (5)) the following model is obtained:

$$y_k = M\Phi_s c_k + \varepsilon_k, \tag{6}$$

$$c_k = f(c_{k-1}, c_{k-2}, c_{k-\tau}, w_k) + \eta_k, \tag{7}$$

A discrete state space representation based on the structure of Eq. (3) and Eq. (6) is proposed for describing the non-homogeneous activity into the brain. The proposed model assumes that each zone evolves independently from others, and its dynamical behavior can be defined as time-varying. These features allow that the proposed model describes adequately normal and pathological non-homogeneous activity, even for localized epilepsy events. This model can be defined as follows:

$$\begin{aligned} f(c_{k-1}, c_{k-2}, c_{k-\tau}, w_k) &= A_1 c_{k-1} + A_2 c_{k-2} \\ &+ A_3 c_{k-\tau} + A_4 c_{k-1}^2 + A_5 c_{k-1}^3, \end{aligned} \tag{8}$$

being $A_1 = \operatorname{diag}(a_1), A_2 = \operatorname{diag}(a_2), A_3 = \operatorname{diag}(a_3), A_4 = \operatorname{diag}(a_4)$ and $A_5 = \operatorname{diag}(a_5)$, where $A_i \in \mathbb{R}^{s\times s}$ and $a_i \in \mathbb{R}^{s\times 1}$ are the parameters matrices which described the dynamics of the model. The vector function f is time varying since $w_k \in \mathbb{R}^{p\times 1}$ can change at each sample k. The set of parameters associated with the dynamics of (8) is w_k with $p = 5s$ defined as

$$w_k^{\mathsf{T}} = \begin{bmatrix} a_1^{\mathsf{T}} & a_2^{\mathsf{T}} & a_3^{\mathsf{T}} & a_4^{\mathsf{T}} & a_5^{\mathsf{T}} \end{bmatrix}. \tag{9}$$

This model involves high complexity and flexibility which allows to describe any behavior of the brain.

IV RESULTS AND DISCUSSION

In this section, an analysis of the proposed model is performed considering homogeneous and non-homogeneous activity. In addition, estimation of models parameters is performed for homogeneous and non-homogenous models parameters.

A *Homogeneous epileptic activity*

To emulate either homogenous normal or epileptic activity the model of Eq. (3) and Eq. (8) are used where the parameter values are fixed to $\tau = 20$, $a_1 = 1.0628$, $a_2 = -0.42857$, $a_3 = 0.008$, $a_4 = 0.000143$, $a_5 = -0.000286$, and $\|\eta_k\| \leq 0.05$. These values were used in [5, 6]; there, this model was compared with real EEG. The epileptic seizure is simulated at sample $k = 125$ ($t = 0.5\,s$) by modifying the values of a_1 from 1.0628 to 1.3, while a_2 from -0.428 to -1 over the entire diagonal. The simulated EEG y_k is obtained from x_k using $y_k = Mx_k + \varepsilon_k$, where ε_k is set to achieve the Signal-to-Noise Ratio (SNR) equals $7\,dB$. The resulting EEG signal can be seen in Fig. 1.

B *Non-homogeneous epileptic activity*

The proposed model of Eq. (6) and Eq. (8) can be also used to simulate the EEG with non-homogeneous activity assuming that the parameters are fixed in $\tau = 20$, $a_1 = 1.0628$, $a_2 = -0.42857$, $a_3 = 0.008$, $a_4 = 0.000143$, $a_5 = -0.000286$, and $\|\eta_k\| \leq 0.05$. The simulated EEG y_k is obtained from x_k using $y_k = M\Phi_s c_k + \varepsilon_k$, where ε_k is set to achieve the Signal-to-Noise Ratio (SNR) equals $7\,dB$. The

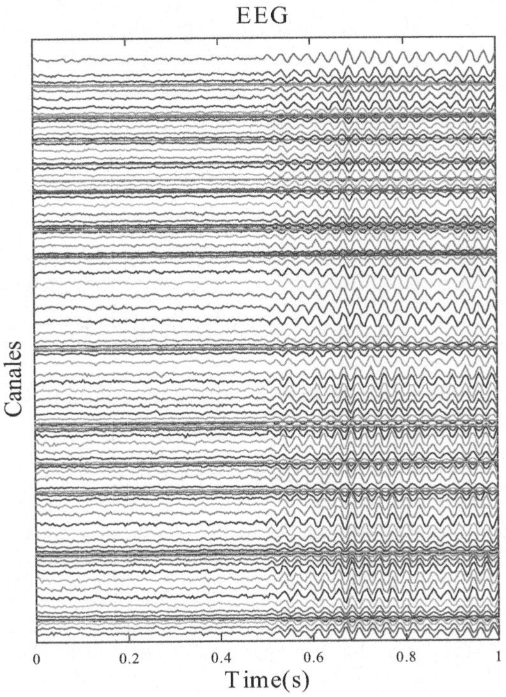

Fig. 1: Homogeneous EEG simulation

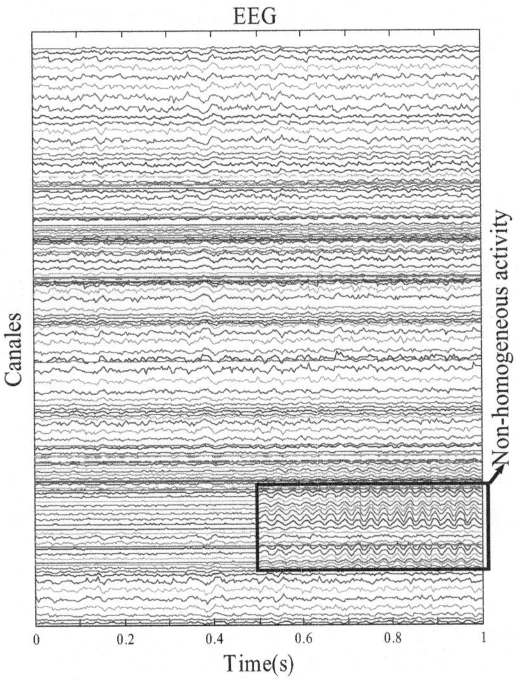

Fig. 2: Non-homogeneous EEG simulation

epileptic seizure is simulated at sample $k = 125$ ($t = 0.5\ s$) by modifying the values of a_1 from 1.0628 to 1.3, while a_2 from -0.428 to -1, but this values were fixed over 5 regions c_k to activate a group the zones in the brain. The resulting EEG with SNR of 7 dB is shown in Fig. 2.

C Estimation of the model parameters

With the purpose of observing the evolution in time of the parameters of both homogeneous and non-homogeneous activity models, a least square multivariable method was implemented according to [8]. This let us estimate the values of the parameters a_1 and a_2 through the temporal values obtained for the different regions c_k. Fig. 3 shows the parameters a_1 and a_2 or the case of homogeneous activity and we can observe how the parameters reveal a trend until 0.5s; however at the time that the instability the simulates epilepsy is introduced, these values tend to adopt the values of the instability.

For the analysis of the non-homogeneous activity, it was necessary to divide the estimation of the parameters into two parts. The first involved only the 5 regions in which the instability was applied. Fig. 4 shows the estimation of a_1 and a_2 for these regions, and we can observe how the trend of these values adopts the new values established to simulate epilepsy in half of the time.

On the other hand, Fig 5 reveals the estimated values for a_1 and a_2 in the rest of the regions, and we can observe how the values of a_1 and a_2 remain within some constant values.

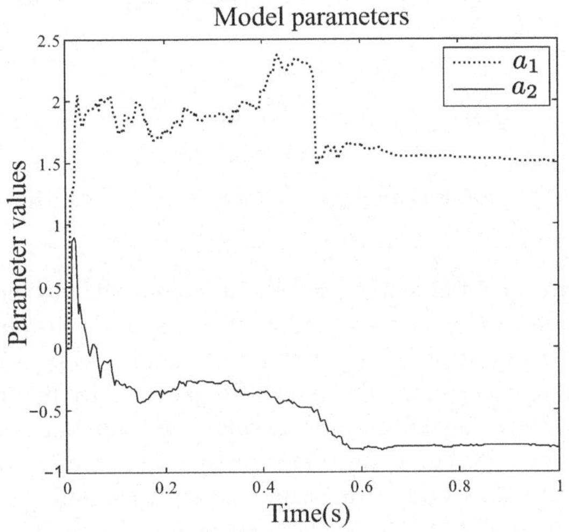

Fig. 3: Model Parameters estimated for Homogeneous Activity

V Conclusions

In this work we could establish that it is possible to observe the brain activity generated in a specific zone by using the

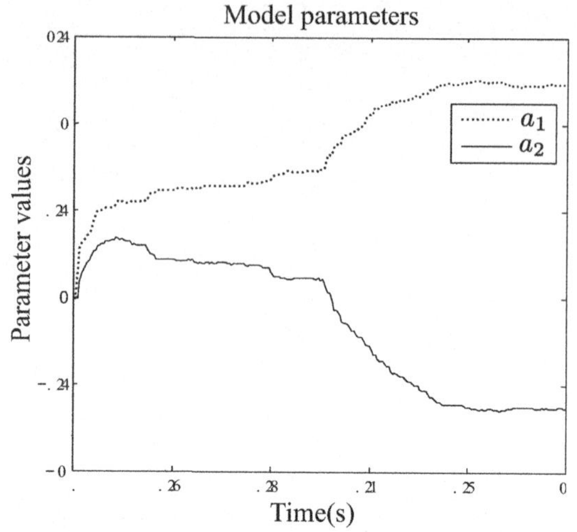

Fig. 4: Model Parameters estimated at a Non-Homogeneous activity zone

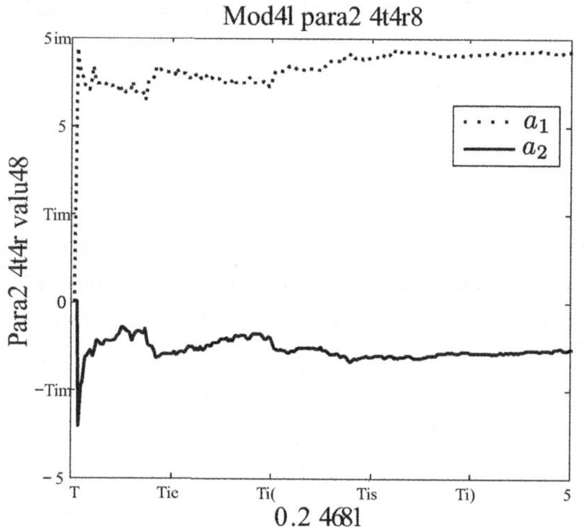

Fig. 5: Model Parameters estimated at a Homogeneous activity zone

signals of EEG obtained for our model. It was possible through a dynamic forward model and by considering a group of cortical regions with a non-homogeneous temporal evolution. Therefore, this is a first step to analyze the focused behavior of the current sources and their temporal evolution in a specific region since the trendy assumption is that all current sources follow the same temporal evolution over the entire brain or over a specific cortical area [2]. That is to say, Brain activity is usually considered as homogeneous in the whole brain or in cortical regions.

This dynamic forward model can be useful to generate focused simulations in different regions of the brain. Additionally, this will be of great use to evaluate approaches that solve the neuromagnetic inverse problem applied to the location of current sources or neural activity reconstruction.

We can also consider the possibility of analyzing the inter-connectivity among zones by using models of interrelation of brain zones that allow us to observe the influence that some regions have over others, and their possible temporal evolution.

In future work involving the estimation of the model parameters, EEG signals will be used to evaluate the accuracy of the model.

CONFLICT OF INTEREST

The authors declare that they have no conflict of interest.

REFERENCES

1. Lamus Camilo, Hmlinen Matti S., Temereanca Simona, Brown Emery N., Purdon Patrick L.. A spatiotemporal dynamic distributed solution to the {MEG} inverse problem *NeuroImage.* 2012;63:894 - 909.
2. Fukushima M., Yamashita O., Kanemura A., Ishii S., Kawato M., Sato M.. A State-Space Modeling Approach for Localization of Focal Current Sources From MEG *IEEE Transactions on Biomedical Engineering.* 2012;59:1561-1571.
3. Barton M. J., Robinson P. A., Kumar S., et al. Evaluating the Performance of Kalman-Filter-Based EEG Source Localization *IEEE Transactions on Biomedical Engineering.* 2009;56:122-136.
4. Giraldo E., Castaño-Candamil J. S., Castellanos-Dominguez G.. A Weighted Dynamic Inverse Problem for electroencephalographic current density reconstruction in *Neural Engineering (NER), 2013 6th International IEEE/EMBS Conference on*:521-524 2013.
5. Giraldo-Suárez E., Castellanos-Dominguez G.. *Artificial Computation in Biology and Medicine: International Work-Conference on the Interplay Between Natural and Artificial Computation, IWINAC 2015, Elche, Spain, June 1-5, 2015, Proceedings, Part I*ch. Neural Activity Estimation from EEG UsinganIterative Dynamic Inverse Problem Solution, :388–397. Cham: Springer International Publishing 2015.
6. Giraldo E., Martinez-Vargas J. D., Castellanos-Dominguez G.. Reconstruction of neural activity from EEG data using dynamic spatio-temporal constraints *International Journal of Neural Systems.* 2016;26:1-15.
7. Robinson P.;, Kim J.;. Compact Dynamical Model of Brain Activity *Physical Review E.* 2007;75:701 -710.
8. Goodwin Graham C., Sin Kwai Sang. *Adaptive Filtering Prediction and Control.* Mineola, NY, USA: Dover Publications, Inc2nd ed. 2009.

Author: P.A. Munõz-Gutiérrez
Institute: Universidad del Quindío
Street: Carrera 15 Calle 12 Norte
City: Armenia
Country: Colombia
Email: pamunoz@uniquindio.edu.co

Author: E. Giraldo
Institute: Universidad Tecnolǵica de Pereira
Street: Carrera 27 N. 10-02 Barrio Alamos
City: Pereira
Country: Colombia
Email: egiraldos@utp.edu.co

Discovery of novel dihydroorotate dehydrogenase inhibitors in trypanosomatids through a molecular docking and molecular dynamics approach

Rodrigo Ochoa[1], Carlos Muskus[1] and Maria Luisa Serrano[2]

[1] Programa de Estudio y Control de Enfermedades Tropicales –PECET, Universidad de Antioquia, Medellín, Colombia
[2] Unidad de Química Medicinal, Facultad de Farmacia, Universidad Central de Venezuela, Caracas 1041A, Venezuela

Abstract— **The classical treatments for the diseases caused by trypanosomatids are associated to severe side effects, including death. Furthermore, drug resistant parasites are a major health problem in different endemic countries. For those reasons, there is an urgent need for new, safe and inexpensive anti-trypanosomatid drugs. One strategy is through the structure-based design of inhibitors against essential molecular targets. Our study focuses in one enzyme, the dihydroorotate dehydrogenase from *Leishmania major* and *Trypanosoma cruzi*. Both parasite's protein crystals were used to find novel active compounds through a virtual screening approach, combining a molecular docking analysis followed by a refinement stage of the binding score based on molecular dynamics simulations. An initial set of ~600.000 compounds were tested, obtaining at the end 16 promising molecules ranked by the predicted binding energies and the interactions created after visual inspection. The refinement protocol prioritized the molecules for further validation using *in vitro* experiments.**

Keywords— **Drug Discovery, *Leishmania*, Molecular docking, Molecular dynamics, Dihydroorotate dehydrogenase.**

I. Introduction

Trypanosomatids are kinetoplastid protozoans associated to different diseases commonly categorized as neglected (e.g. trypanosomiasis, leishmaniasis), due to the population affected that is located mainly in developing countries, and the lack of resources for improving the control and treatment alternatives [1]. One of the main problems is the limited number of drug options and the adverse effects caused by their use, which sometimes could generate worst consequences to the patients [2, 3]. In addition, there are reports of treatment failures due to increased parasite resistance [4, 5]. Because of that, during the last three decades some solutions have been designed and implemented to accelerate drug discovery protocols through computational models and simulations [6–8].

Protein 3D models allow the simulation of potential interactions against other kind of molecules (e.g. peptides, small molecules, DNA) based on physical theories and empirical knowledge [9, 10]. One important limitation of these techniques in trypanosomatids is the small set of protein structures available. Despite that, some essential proteins have been structurally characterized due to their potential roles as drug targets, as the dihydroorotate dehydrogenase (DHODH) [11].

DHODH has been previously discussed as a molecular target for leishmaniasis and trypanosomiasis therapeutics due to its critical role in oxidizing dihydroorotate to orotate, and reducing fumarate to succinate during pyrimidine biosynthesis [12]. The structures of dihydroorotate dehydrogenase from prokaryotic and eukaryotic organisms are significantly different; in trypanosomatids the enzyme is homodimeric whereas in humans the enzyme is monomeric [11]. In addition, binding site differences are being studied to design more selective inhibitors through techniques such as molecular docking and molecular dynamics (MD) [13].

Here, we present a virtual screening approach for the discovery of novel inhibitors against the DHODH enzyme in *Leishmania major* and *Trypanosoma cruzi* using computational structure-based methodologies.

II. Material and methods

A. Molecular docking screening with LmDHODH and TcDHODH

Both the DHODH structures from *L. major* (PDB:4EF8) and *T. cruzi* (PDB:3W23) were parametrized based on the functionalities available in the AutoDock Tools package [14]. The binding site was delimited according to previous studies reporting key residues conformations from these kind of trymanosomatids DHDOH.

With regard to the ligands, a subset of commercial drug-like molecules was obtained from the ZINC database [15].This subset contained 579.890 compounds with tridimensional coordinates. All the ligands were parametrized using AutoDock Tools, with the additional assignment of flexible torsions. The molecular docking was run with AutoDock Vina [16]. The results were prioritized according to the predicted free energy of binding in kcal mol^{-1}.

B. Preparation of LmDHODH and TcDHODH complexes for MD simulations

The MD simulations involved the enzymes LmDHODH and TcDHODH of the best ranked protein-ligand complexes

I. Torres et al. (eds.), *VII Latin American Congress on Biomedical Engineering CLAIB 2016, Bucaramanga, Santander, Colombia, October 26th -28th, 2016*,
IFMBE Proceedings 60,
DOI: 10.1007/978-981-10-4086-3_37

obtained previously by docking analysis. The complexes were embedded in a 100 Å water sphere using VEGA software [17] and were preliminarily minimized to optimize the relative position of the water molecules with the protein backbone atoms fixed to preserve the folding of the enzymes.

C. Molecular dynamics simulations and energy analysis

The obtained systems underwent a one nanosecond (ns) MD simulation with the following characteristics: (a) Periodic Boundary Conditions (PBC) were introduced to stabilize the simulation space; (b) the long-range electrostatic potential was treated by the Particle Mesh Ewald summation method (PME) [18]; (c) Newton's equation was integrated using the r-RESPA method [19] (every 4 fs for long- range electrostatic forces, 2 femtoseconds (fs) for short-range non-bonded forces, and 1 fs for bonded forces); (d) the temperature was maintained 300±5 K by means of Langevin's algorithm [20]; (e) Lennard-Jones (L-J) interactions were calculated with a cut-off of 14 Å, the switching distance was 10 Å, and the non-bonded pair list distance was 14 Å; (f) a frame was stored every 5picoseconds (ps), yielding 2000 frames; (g) protein backbone atoms fixed.

The simulations were carried out in three faces: an initial period of heating from zero to 300 K over nine ps; an equilibration MD simulation of 0.1 ns, and the monitored MD phase of simulation of one ns. Equilibration MD runs had also the aim to optimize the interactions between ligands and the enzyme. Only the frames memorized during the monitored MD phase were considered. The lowest energy frame of the MD simulation was further minimized and represented the final structure of the LmDHODH and TcDHODH complexes. All MD simulations described in the study were performed with NAMD 2.8 [21] using the CHARMM force field [22] and the Gasteiger charges. All reported minimizations were performed using the conjugated gradient algorithm until a Root-Median Square (RMS) gradient was smaller than 0.01 kcal mol^{-1} $Å^{-1}$.

III. Results and discussion

The LmDHODH and TcDHODH are Class 1A dihydroorotate dehydrogenase in comparison with the human homolog (HsDHODH) which belongs to Class 2A. Topographical analysis on the active site reported a set of regions (S1 to S5) that potentially could be used for the design of new specific inhibitors against the parasite's enzyme [11]. The complete hydrophobic active site is illustrated on Fig. 1.

The main region (S1) is the one involved on the interactions with substrates such as orotate, which is conserved among the DHODH homology family. Near to this site, there is a larger cavity (S2) composed of amino acids connecting the active loop α4-βA, essential for the catalytic activity through the pocket opening to S1 from the surface. S2 is a key site for enzyme activity, and have been demonstrated that there is not an equivalent region on HsDHODH [23].

Fig. 1 Structural characterization of the LmDHODH hydrophobic active site in complex with a binder obtained from the molecular docking results. According to the hydrophobicity scale, most of the residues that conforms the binding site reports a hydrophobic nature. The visualization was created with Discovery Studio.

The other subsites report small differences that could benefit specificity for trypanosomatids, despite their locations on deeper regions inside the protein. In addition, the presence of the flavin mononucleotide (FMN) is essential to preserve the active folding of the enzyme.

The virtual screening approach allowed us to rank a diverse set of compounds based on the AutoDock Vina scores. For the case of TcDHODH, the best complex predicted a binding energy of -11.8 kcal mol^{-1}. For LmDHODH the best was -10.5 kcal mol^{-1}. Based on the scores, we selected the best eight protein-ligand complex for each trypanosomatid structure, which were consequently re-evaluated and visually inspected using the MD protocols. This procedure looked for theoretically analyze if their mechanisms of action are suitable for finding disease-specific inhibitors, and also to include new parameters that are usually skipped by docking protocols such as the role of the solvent and the protein intrinsic flexibility [24].

MD simulations were carried out on LmDHODH and TcDHODH complexed with the selected ligands and the prostetic FMN group, with the aim to investigate the binding properties of the ligands and the influence of the water molecules on the binding mode at the S1 active site. The analysis of the resulting complexes clearly reveals that the ligands remained close to the prostetic FMN group and some waters were involved in hydrogen bonds with the ligands in the active site. The energy results from the molecular docking and the MD analysis obtained for the best eight compounds against LmDHODH are described in Table 1.

Table 1 Energy scores ranked by Vina for the best eight compounds against LmDHODH

Complex Structure	Vina Score	MD minimum energy	Re-minimization energy
LmDHODH-C1	-10.5	-37609.16	-138482.41
LmDHODH-C2	-10.4	-38107.76	-140740.6
LmDHODH-C3	-10.3	-38998.84	-140252.14
LmDHODH-C4	-10.1	-38275.25	-140976.9
LmDHODH-C5	-9.7	-38475.19	-141172.56
LmDHODH-C6	-9.2	-37996.15	-140878.68
LmDHODH-C7	-9.1	-39171.46	-141709.7
LmDHODH-C8	-9	-38227.25	-141065.7

According to the predicted energies, the compound LmDHODH-C7 was initially ranked in the seventh position (based on docking score), but later repositioned as the energetically best complex based on the MD re-scoring function. This analysis is important in order to prioritize more accurately compounds with potential activity for further *in vitro* and *in vivo* validations.

In detail, in the LmDHODH-C7 (Fig. 2), the ligand molecule clearly occupied de S2 binding site and interact with residues belonging to this site, Leu72, Leu102, but also interacts with some residues from S1 and S3 site, Leu72, Pro132, Asn195, Sr196, Leu72, Leu102 and Pro132, Lys215, respectively.

Molecules with scaffolds formed by several rings, as the one depicted on Fig. 2, tend to create multiple π-stacks interactions, and consequently are more stable during the interaction. In fact, according to the MD trajectories the ligands conserved most of their bonds with LmDHODH during the simulated time.

Similarly, the energy results for the best eight compounds against TcDHODH are described in Table 2.

Table 2 Energy scores ranked by Vina for the best eight compounds against TcDHODH

Complex Structure	Vina Score	MD minimum energy	Re-minimization energy
TcDHODH-C1	-11.8	-39285.96	-140422.25
TcDHODH-C2	-11.8	-38998.84	-140252.14
TcDHODH-C3	-11.6	-38886.94	-140652.7
TcDHODH-C4	-11.6	-38959.17	-140141.7
TcDHODH-C5	-11.5	-38568.65	-139845.75
TcDHODH-C6	-11.3	-38890.11	-139957.38
TcDHODH-C7	-11.3	-38794.72	-139898.25
TcDHODH-C8	-10.9	-39124.64	-140627.48

In the TcDHODH-C8 (Fig. 3), the ligand molecule also occupied de S2 binding site interacting with Gln138, but also interacts with residues from S1, Asn67, Met69 and Pro134, and with Asn198 from the S3 site. This molecule in particular interacts with several water molecules that could confer stability within the complex. Interesting it has been reported that molecules that concomitantly exploit the fully conserved S1-S3 sites can be considered attractive lead compounds for further development.

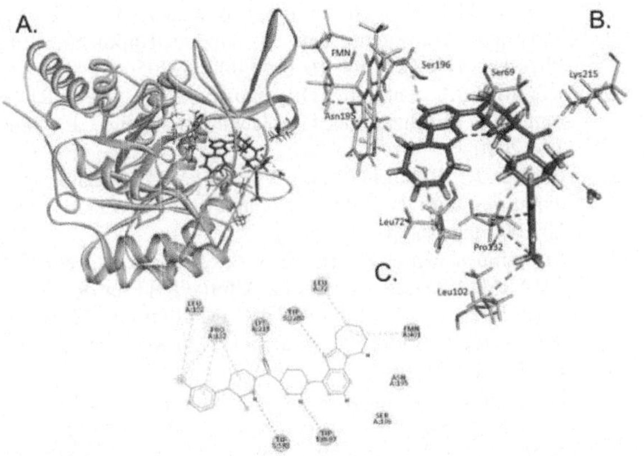

Fig. 2 Diagrams of the protein-ligand interaction between LmDHODH and C7. (A) Complete structure of the complex with the compound bound in the active site. (B) 3D and (C) 2D representations of the complex interactions categorized by hydrogen bonds (green dashed lines), π-stacks (pink dashed lines) and solvent (water) bonds (blue dashed lines).

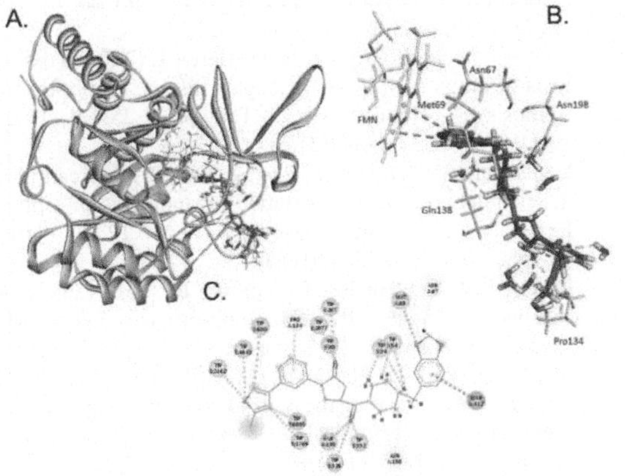

Fig. 3 Diagrams of the protein-ligand interaction between LmDHODH and C7. (A) Complete structure of the complex with the compound bound in the active site. (B) 3D and (C) 2D representations of the complex interactions categorized by hydrogen bonds (green dashed lines), π-stacks (pink dashed lines) and solvent (water) bonds (blue dashed lines).

IV. CONCLUSION

Regarding the structure-based design of active molecules, the combination of techniques in different stages of the prediction is useful to complement the limitation associated to each technique. In this case, the MD approach provides an explicit way to theoretically analyze the protein flexibility, and consequently obtain energy predictions more aligned with biological behaviors. Moreover, the computational time spent during the virtual screening was reduced (1 week) thanks to the use of the Texas Advanced Computer Center (TACC) infrastructure. Finally, the prioritized list of molecules were built based on drug-like properties, and according to the results are potential candidates for the optimization of novel chemical treatments against complex living organisms such as the trypanosomatids.

Acknowledgment

This work was supported by the IBM's World Community Grid initiative.

Conflict of Interest

The authors declare no conflict of interest.

References

1. H. Lopes A (2010) Trypanosomatids: Odd Organisms, Devastating Diseases. Open Parasitol J 4:30–59. doi: 10.2174/1874421401004010030
2. Oliveira LF, Schubach AO, Martins MM, et al (2011) Systematic review of the adverse effects of cutaneous leishmaniasis treatment in the New World. Acta Trop 118:87–96. doi: 10.1016/j.actatropica.2011.02.007
3. Castro J, Montalto de Mecca M, Bartel L (2006) Toxic side effects of drugs used to treat Chagas' disease (American trypanosomiasis). Hum Exp Toxicol 25:471–479. doi: 10.1191/0960327106het653oa
4. Vanaerschot M, Dumetz F, Roy S, et al (2014) Treatment failure in leishmaniasis: drug-resistance or another (epi-) phenotype? Expert Rev Anti Infect Ther 12:937–946. doi: 10.1586/14787210.2014.916614
5. Barrett MP, Vincent IM, Burchmore RJ, et al (2011) Drug resistance in human African trypanosomiasis. Future Microbiol 6:1037–1047. doi: 10.2217/fmb.11.88
6. Florez AF, Park D, Bhak J, et al (2010) Protein network prediction and topological analysis in Leishmania major as a tool for drug target selection. BMC Bioinformatics 11:484. doi: 10.1186/1471-2105-11-484
7. Katsuno K, Burrows JN, Duncan K, et al (2015) Hit and lead criteria in drug discovery for infectious diseases of the developing world. Nat Rev Drug Discov. doi: 10.1038/nrd4683
8. Velásquez V, Ochoa R, Muskus C (2015) Detección de blancos moleculares en la ruta de señalización del fosfatidilinositol en Leishmania spp. a través de herramientas bioinformáticas y de modelado matemático. Biomédica 35:235–246. doi: 10.7705/biomedica.v35i2.2298
9. Irwin JJ, Shoichet BK (2016) Docking Screens for Novel Ligands Conferring New Biology. J Med Chem 59:4103–4120. doi: 10.1021/acs.jmedchem.5b02008
10. Meng X-Y, Zhang H-X, Mezei M, Cui M (2011) Molecular docking: a powerful approach for structure-based drug discovery. Curr Comput Aided Drug Des 7:146–157. doi: 10.2174/157340911795677602
11. Cheleski J, Rocha JR, Pinheiro MP, et al (2010) Novel insights for dihydroorotate dehydrogenase class 1A inhibitors discovery. Eur J Med Chem 45:5899–5909. doi: 10.1016/j.ejmech.2010.09.055
12. Reis RAG, Ferreira P, Medina M, Nonato MC (2016) The mechanistic study of Leishmania major dihydro-orotate dehydrogenase based on steady- and pre-steady-state kinetic analysis. Biochem J 473:651–660. doi: 10.1042/BJ20150921
13. Pinto Pinheiro M, da Silva Emery F, Cristina Nonato M (2013) Target sites for the design of anti-trypanosomatid drugs based on the structure of dihydroorotate dehydrogenase. Curr Pharm Des 19:2615–2627.
14. Morris GM, Huey R, Lindstrom W, et al (2009) AutoDock4 and AutoDockTools4: Automated docking with selective receptor flexibility. J Comput Chem 30:2785–2791. doi: 10.1002/jcc.21256
15. Irwin JJ, Shoichet BK (2005) ZINC - A free database of commercially available compounds for virtual screening. J Chem Inf Model 45:177–182. doi: 10.1021/ci049714+
16. Trott O, Olson AJ (2009) AutoDock Vina: Improving the speed and accuracy of docking with a new scoring function, efficient optimization, and multithreading. J Comput Chem. doi: 10.1002/jcc.21334
17. Pedretti A, Villa L, Vistoli G (2004) VEGA - An open platform to develop chemo-bio-informatics applications, using plug-in architecture and script programming. J Comput Aided Mol Des 18:167–173. doi: 10.1023/B:JCAM.0000035186.90683.f2
18. Toukmaji A, Sagui C, Board J, Darden T (2000) Efficient particle-mesh Ewald based approach to fixed and induced dipolar interactions. J Chem Phys 113:10913–10927.
19. Masella M (2006) The multiple time step r-RESPA procedure and polarizable potentials based on induced dipole moments. Mol Phys 104:415–428. doi: 10.1080/00268970500404414
20. Izaguirre JA, Catarello DP, Wozniak JM, Skeel RD (2001) Langevin stabilization of molecular dynamics. J Chem Phys 114:2090.
21. Phillips JC, Braun R, Wang W, et al (2005) Scalable molecular dynamics with NAMD. J Comput Chem 26:1781–1802. doi: 10.1002/jcc.20289
22. Vanommeslaeghe K, Hatcher E, Acharya C, et al (2010) CHARMM general force field: A force field for drug-like molecules compatible with the CHARMM all-atom additive biological force fields. J Comput Chem 31:671–90. doi: 10.1002/jcc.21367
23. Feliciano PR, Cordeiro AT, Costa-Filho AJ, Nonato MC (2006) Cloning, expression, purification, and characterization of Leishmania major dihydroorotate dehydrogenase. Protein Expr Purif 48:98–103. doi: 10.1016/j.pep.2006.02.010
24. Bauer MR, Ibrahim TM, Vogel SM, Boeckler FM (2013) Evaluation and Optimization of Virtual Screening Workflows with DEKOIS 2.0 – A Public Library of Challenging Docking Benchmark Sets. J Chem Inf Model 53:1447–1462. doi: 10.1021/ci400115b

Space Adventures: a serious game for childhood obesity prevention

L. M. Parra Navarro[1], D. R. Paez Ardila[1], M. M. S. Pires[2] and J. L. B. Marques[1]

[1] Federal University of Santa Catarina, Institute of Biomedical Engineering, Florianópolis - SC, Brazil
[2] Federal University of Santa Catarina, Department of Medicine, Clinical Research Laboratory, Florianópolis - SC, Brazil

Abstract— **Prevention of childhood obesity is a global public health priority. Obesity is associated with dyslipidemia, type 2 diabetes and long-term vascular complications. Therefore, obesity prevalence remains high making it imperative to continue surveillance and alternatives or plans to pre-vent it. We developed an innovative population strategy to support initiatives that work towards reducing the obesity epidemic in Brazil. The educational game developed - Space Adventures - is aimed to promote the consumption of healthier food by children. The theme resembles a galaxy and different idealized planets corresponding to the daily meals. To evaluate the educational game developed, test were conducted with children from the ages of 5 to 10; the food choices of each volunteer were analyzed during the execution of the game. Two institutions tested the educational game developed and it was observed that children of a primary school (Group 1) showed significantly better learning-acquisition scores than children of a children's hospital (Group 2). Group 1 performed better in the educational game than Group 2, based in healthy eating choices. The results of this study provide evidence to support the importance of innovative strategies in health.**

Keywords— **Educational Game, Digital Game-Based Learning, Serious Games, Childhood Obesity, Nutrition.**

I. Introduction

In the United Kingdom, the epidemic of pediatric obesity began in the late 1980s. Prevalence has continued to increase rapidly, and obesity is now the most common disorder of childhood and adolescence. The health survey for England 2004 showed that 14% of 2-11 year olds and 25% of 11-15 year olds were obese [1].

Childhood obesity is a worldwide issue, especially in developed countries like the United States. More than one-third of adults and 17% of children in the United States are obese, although the prevalence continued stable between 2003-2004 and 2009-2010 [2].

Latin America has experienced epidemiologic and nutritional transitions that have contributed to the high rates of overweight and obesity. By some estimates, more than 50% of males and 60% of females will be obese by 2030. Likewise, in children between the ages of 5 and 11 the prevalence of obesity ranged from 18.9% to 36.9% and from 16.6% to 35.8% in adolescents [3].

Obesity is the driving force for the epidemic of diabetes mellitus which has, similarity, extended to Latin America, and the increasing frequency of these two metabolic problems runs in parallel together with the metabolic syndrome [4].

Thereby, obesity in children is currently an international salient public health crisis that can no longer continue to be overlooked. To substantiate, the prevalence of childhood obesity has significantly increased over the last few years throughout the world, engendered by an imbalance between the calorie intake and the actual calories utilized. One or more factors (e.g., genetic, behavioral, and environmental) have been known to cause obesity in children, ultimately leading to physical and psychological health problems.

Nonetheless, much can be done in the struggle to counteract and prevent the adverse effects of obesity. Acquiring good nutrition habits is the first step in preventing these conditions; so it is very important for people to know how to implement a daily diet through an authentic understanding of the nutritional content in every food intake. [5].

Consequently, intervention strategies are crucial to prevent and control obesity in children. With this in mind, a serious and measurable method is employed and, as a result, an immersive experience is engendered. In search of obtaining sustained changes in nutrition behavior and habits while preserving the essential entertainment aspects for the target group, video games come into play. In order to substantiate the alternative method, a serious video game is designed seeking to improve children's food choices. [6].

These intervention strategies involve video games focused on childhood nutrition. Namely, educational video games, a proven effective learning tool considered advantageous and suitable for children at any age. Videogames have induced positive health behavioral changes among children [7,8], such as increased fruit and vegetable intake [9,10]. In order to motivate obese people to have a daily diet and do physical exercise regularly, one possible approach suggested in recent years is to utilize serious gaming concepts, which teach health issues [11,12].

Nowadays, the use of the technologies, such as video games, supposes an important tool to get an active learning under a concept known as Serious Games.

These videogames and the potential application area of Serious Games come from Zyda [13]. He defines serious games as a "mental contest, played with a computer in ac-

I. Torres et al. (eds.), *VII Latin American Congress on Biomedical Engineering CLAIB 2016, Bucaramanga, Santander, Colombia, October 26th -28th, 2016,*
IFMBE Proceedings 60,
DOI: 10.1007/978-981-10-4086-3_38

cordance with specific rules that uses entertainment to further government or corporate training, education, health, public policy, and strategic communication objectives". According to this definition, the main intent of a serious game is to impart knowledge or skills through direct experience of carrying out a task. The playful environment promotes learning because children face challenges, test their limits and solve problems. This is a way to acquire knowledge in a pleasant form, spontaneous and at the same time with rules.

This paper reports the development and results of the nutrition education game "Space Adventures". The overall aim of this work is to encourage children between 5 and 10 years old to consume nutritious foods through a game based on Digital Game-Based Learning.

II. MATERIALS AND METHODS

The methodology follows the essential features for learning and witty games development; these include imaginary situation, control by rules, progressive difficulty, sound effects, dynamic graphics, user control and feedback.

The current concern is that by 2050, obesity may become the primary health issue of Santa Catarina (Brazil). Reason why we developed this game seeking to eating practices to children in Florianópolis, Brazil.

A. Brazilian Food Guide Pyramid for preschool and school-age children

Nutritional monitoring by the pediatrician during obesity treatment and prevention is fundamental. The food pyramid is a tool often used in nutrition education and it is important to spread the concepts of variety, moderation and proportionality. The food groups are organized in the pyramid in the following respective seven food groups: grains, greens and vegetables, fruits, dairy (milk, cheeses and yogurt), meat, beans, sugar and oils [14].

The daily requirement for energy is defined as the Estimated Energy Requirement (EER). The EER is based on calculations that account for an individual's energy intake, energy expenditure, age, sex, weight, height, and physical activity level (PAL). The EER is the estimated number of daily calories an individual requires in order to maintain the current weight [15] According to the calculations mentioned above the EER for boys and girls is presented in Table 1, both for ages from 5 to 10 years.

Table 1 Energy requirements, considering the moderate level of PAL.

Age (yrs)	Boys (Kcal per day)	Girls (Kcal per day)
5	1.475	1.325
6	1.575	1.425
7	1.700	1.550
8	1.825	1.700
9	1.975	1.825
10	1.475	1.325

The values presented in Table 1 were used to do the activities in the educational game. Each level in "Space Adventure" corresponds to the daily meals and the goal for each level is to eat the corresponding quantity of food to the day and to the percent corresponding to each meal (Fig 1).

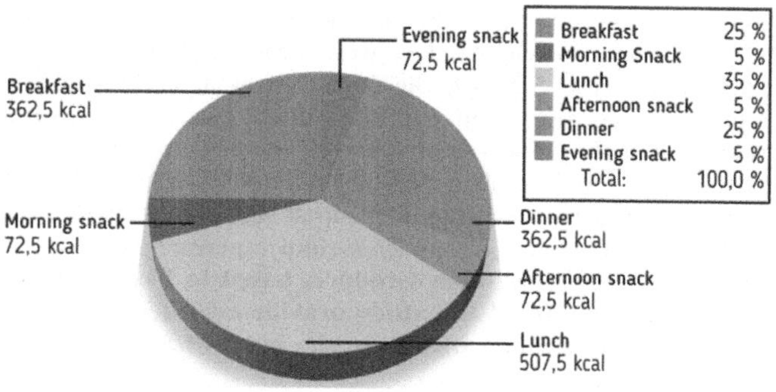

Fig. 1 Distribution of daily used for 1300 Kcal per day.

B. Game Design

The development of serious games is a topic rising in the scientific community, particularly in the area of education, due to its complying function that targets educational purposes through innovative entertainment and constant motivation. One of the strengths of using games for learning is that they are tools to connect students with knowledge, key concepts and processes [16].

As cited by Prensky "Digital Game-Based Learning is precisely about fun and engagement, and the coming together of and serious learning and interactive entertainment into a newly emerging and highly exciting medium" [16]. Taking a cognitive approach to game development will give developers the tools to create theories and methods for modelling and simulating computer characters, story and emotion; innovating new game genres and play styles; and integrating pedagogy with story in the interactive game scenario [13], [16].

In the scope of game design literature, we highlight the work by Jesse Schell [17] which defines the basic tetrad of a game composed by four main elements that make up a set.

Mechanics: These are the procedures and rules of the game. Mechanics describe the goal of the game, how players can and cannot try to achieve it, and what happens when they try an action or movement.

Story: This is the sequence of events that unfolds throughout the game. The game has a narrative behind it, presented by scene cuts (animations with the central character of the narrative), making the user's experience a significant one through a set of questions. One of the animations also has an instructional character, in charge of explaining the mission of the game and accompanies the players to do their job.

Aesthetics: This is how your game looks, sounds, smells, tastes, and feels. Aesthetics are an incredibly important aspect of game design since they have the most direct relationship with a player's experience. When you have a certain look that you want players to experience and become immersed in, you will need to select the top notch technology that will not only allow the aesthetics to come through, but also amplify and reinforce it.

Technology: It refers to any material and interaction that makes your game possible. The technology you select for your game enables it to do certain things and prohibits it from doing other things.

According to Schell [28] "The technology is essentially the medium in which the aesthetics take place, in which the mechanics will occur, and through which the story will be told". According to him, all these elements are essential for a game, and all have the same importance.

We developed the serious game using the Unity game engine (Unity Technologies, USA), the inputs and interactions of the player were captured by the keyboard and mouse, the game was built as a multiplatform software to supports different operating systems (Windows, OS X, Linux) and its final output is a database that contains information about player's performance over the game.

III. Results

The pilot tests were conducted by inviting children between the ages of 5 and 10 years old, from a Hospital and a Primary School in Florianopolis, southern Brazil, between September and November 2015. The evaluation involved a sample of 75 participants (31M, 44F).

"Space Adventures" is divided in different parts (Fig. 2). At first, the nutrition questions to make a profile about the user. In the second place, five levels which goal is to promote healthy feeding, they are distributed as three main meals (breakfast, lunch and dinner) and the final level is reserved for drinks. Finally, there is a nutrition question created for reassessment.

Fig. 2 Screenshots and theme of the Space Adventures game.

Results of children's food choices during the execution of the game can be found in Fig. 3. This result was obtained during the active participation of the children with the game designed, passing the serious game levels and, it ultimately, showed a tendency of the school children's food choices inclining (M= 25.83, SD = 9.27) to be healthier, because their choices registered significantly less sugar and sweets over the hospital children (M= 25.61, SD= 7.76). The t-value is 2.70838 and p-value is 0.026814 ($p < 0.05$).

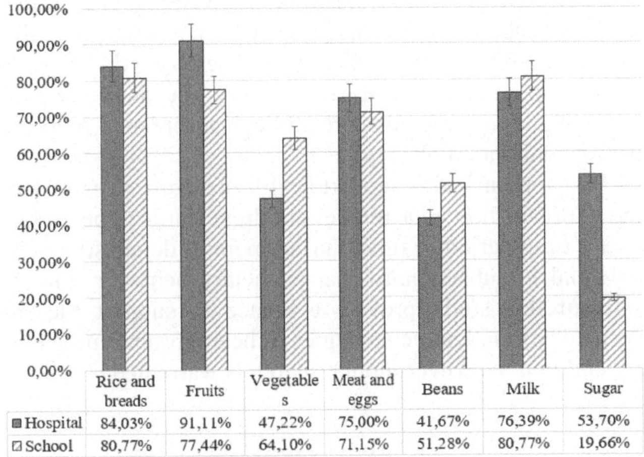

	Rice and breads	Fruits	Vegetables	Meat and eggs	Beans	Milk	Sugar
Hospital	84,03%	91,11%	47,22%	75,00%	41,67%	76,39%	53,70%
School	80,77%	77,44%	64,10%	71,15%	51,28%	80,77%	19,66%

Fig. 3 Results obtained of the choices in the educational game between the two groups of children. (Hospital *vs* School).

Regarding the results of the evaluation section (Final Nutrition Question) the choice of fruits and vegetables were highlighted in the two institutions. Candies, ice cream and soda were rejected in the choices of children in both institutions. Results revealed significant differences in healthy food choices frequency between hospital children (M= 1.49, SD = 0.6) and school (M= 0.64, SD = 0.49) at $p<0.05$.

IV. Discussion

A similar work to this project developed is found in *DigesTower* [18] the main focus is on comprehension and differentiation of where and how each food is digested in the body and its composition. However, the educational game "Space Adventures" is a platform which purpose is to easily explain, through active learning, how to maintain a healthy diet.

In the case of the child population, we ought to pay special attention in choosing the game method, as this age group generally requires a respondent (parent or guardian), especially when dealing with preschoolers, in relation to questions for food record.

In summary, using the technology and the methodology described in this work, and taking into account its results, it may be helpful in supporting children to learn nutritional concepts and adequate healthy-eating habits, offering a new approach that could be interesting to slowdown the increasing rates of children obesity and overweight in our society.

V. Conclusions

This game associated with educational aspects is a complementary tool to carry out prevention and help in the efforts against childhood obesity and associated health issues. The new generation of children is involved and accustomed to using technology, so they usually prefer new approaches than a traditional education system. Therefore, the proposal to develop an educational game is very relevant be-cause it is presented as an innovative strategy in order to help and prevent childhood obesity.

This approach will support healthcare professionals dealing with nutrition education of children, if it is to be applied over a long period of time. Furthermore, it demonstrates the potential to substantially change dietary behavior. The results of this study provide evidence to support the importance of innovative strategies in health promoting games for children. However more research is warranted.

Acknowledgment

We thank the Brazilian National Council for Scientific and Technological Development (CNPq) for the financial support provided.

Conflict of Interest

The authors declare that they have not no conflict of interest.

References

[1] Reilly JJ, Wilson D. ABC of obesity: Childhood obesity. *BMJ* 2006;333:1207–10. doi:10.1136/bmj.39048.503750.BE.
[2] Ogden CL, Carroll MD, Kit BK, Flegal KM. Prevalence of Childhood and Adult Obesity in the United States, 2011-2012. *J Am Med Assoc* 2014;311:806. doi:10.1001/jama.2014.732.
[3] Obesity Society at www.obesity.org/obesity-latin-america.
[4] Aschner P. Obesity in Latin America. *Metab. Syndr., Cham*: Springer International Publishing; 2016, p. 33–9. doi:10.1007/978-3-319-11251-0_4.
[5] World Health Organization. Obesity and overweight at. http://www.who.int/mediacentre/factsheets/fs311/en/
[6] Thompson D. What serious video games can offer child obesity prevention. *JMIR Serious Games* 2014;2:e8. doi:10.2196/games.3480.
[7] Baranowski T, Buday R, Thompson DI, Baranowski J. Playing for real: video games and stories for health-related behavior change. *Am J Prev Med* 2008;34:74–82. doi:10.1016/j.amepre.2007.09.027.
[8] Lu AS, Thompson D, Baranowski J, Buday R, Baranowski T. Story Immersion in a Health Videogame for Childhood Obesity Prevention. *Games Health J* 2012;1:37–44. doi:10.1089/g4h.2011.0011.
[9] Bahia A, Berndt A, Bordignon G, Takase E. Nutrition at play technology promoting alimentary behavior modification. IEEE, *Int. Conf. Serious Games Appl. Heal.*;3(Rio de Janeiro, Brazil):1–8 2014. doi:10.1109/SeGAH.2014.7067085.
[10] Baranowski T, Baranowski J, Cullen KW, Marsh T, Islam N, Zakeri I, et al. Squire's Quest! Dietary outcome evaluation of a multimedia game. *Am J Prev Med* 2003;24:52–61.
[11] Alamri A, Hassan MM, Hossain MA, Al-Qurishi M, Aldukhayyil Y, Hossain MS. Evaluating the impact of a cloud-based serious game on obese people. *Comput Human Behav* 2014;30:468–75.
[12] Scarle S, Dunwell I, Bashford-Rogers T, Selmanovic E et al. Complete Motion Control of a Serious Game against Obesity in Children in *Third Int. Conf. Games Virtual Worlds Serious Appl.*, IEEE: 178–9 2011. doi:10.1109/VS-GAMES.2011.48.
[13] Zyda M. From visual simulation to virtual reality to games. IEEE Comput Soc 2005;38:25–32. doi:10.1109/MC.2005.297.
[14] Brazilian Society of Pediatrics. *Infant Feeding Guidelines (in Portuguese).* Rio de Janeiro: Department of Nutrition Science 2012.
[15] National Academy of Sciences. *Dietary Reference Intakes.* Washington, D.C.: The National Academies Press; 2005.
[16] Prensky M. *Digital Game-Based Learning.* USA: Paragon House 2007.
[17] Schell J. *The Art of Game Design.* USA: Elsevier 2008.
[18] Dias J, Silva G, Beder DM, Mekaro MS, Santiago DL, Bordini RA. Digestower: Educational game to help prevent childhood obesity in *Int. Conf. Comput. Educ.*;10(Fortaleza, Brasil): 309-317 2014.

Cardiac dynamic assessment through entropy proportions and probability

Javier Rodríguez[1], Leonardo Ramírez[2], Signed Prieto[1] and Catalina Correa[1]

[1]Insight Group, Universidad Militar Nueva Granada. Research center, Clínica del country, Bogotá, Colombia
[2] Universidad Militar Nueva Granada. Eng, MSc and PhD in Biomedical Enginnering.
TD&I Director of Development and Innovation Office, Bogotá, Colombia

Abstract—**A new predictive methodology of diagnosis aid of cardiac dynamic was developed from dynamical systems theory, probability and entropy, allowing to differentiate normal states from pathological ones, and also allowing the quantification of evolution between normality and disease. Mathematical characteristics of this methodology and its simplicity may indicate an easy automation in the future. For this study, 520 holters were selected, 50 normal and 470 with different pathologies; through maximum and minimum values of heart rate (HR) and total beats per hour, a simulation of the whole dynamic was developed, then with these values a numerical attractor was generated. Probability, entropy, S/k proportion and proportions of entropy were evaluated. Sensitivity, specificity and Kappa coefficient were obtained.**

The values obtained for entropy, S/k proportion and entropy proportions are within ranges of normality and disease established previously. The sensitivity and specificity values were 100%, and Kappa coefficient was 1, confirming the diagnostic and predictive capability of methodology for normal and pathological dynamics, showing new predictive algorithms for Holters and demonstrating its value as a monitoring tool for daily clinical practice that may contribute in decision-making on specific therapeutic interventions.

***Keywords*—Heart Rate, Biomedical Engineering, nonlinear dynamics, ambulatory electrocardiography, mathematics.**

I. INTRODUCTION

The objective of dynamic systems theory is the study of the state and evolution of systems, building for it abstract spaces, where obtained attractors are of three types; periodic, punctual and chaotic [1]. The two first types of attractors represent predictable dynamics, while the last of them is associated to unpredictability. The irregularity of attractors has been measured through fractal geometry [2], which had allowed the characterization of various physiological systems, and between them of the cardiac dynamics [3]. In this way a new conception of health and disease arises [3], and establishes that excessive periodicity and increased randomness are associated to disease, while an intermediate behavior corresponds to health, on the contrary of the conventional homeostatic regard.

Not only dynamic systems theory has been useful in physiology or cardiac dynamics. In general, mathematical and physical perspective has led to models that may study physiological mechanisms [4], through the analysis of several variables with mathematical and physiological implications [5]. Symbolic dynamics for the cardiac dynamic variability have been studied [6, 7]. In the other hand, multifractal analysis has suggested an apparent multiscale loose of complexity in life-threatening conditions, possibly useful in a clinical context [8]. Huikuri et al. [9] applied fractal analysis to the study of cardiac dynamics, establishing more precise predictive parameters of mortality than the actual ones in post-Acute Myocardial Infarction patients with an ejection fraction lower than 35%. There have been analyzed short-time ECG series through Chaos Theory and nonlinear dynamics in a study with patients with coronary disease [10], also nonlinear properties of heart dynamic have been studied in patients with advanced age [11] as well as the clinical impact of the assessment of cardiovascular control through different methods in heart rate dynamic [12]. Fourier analysis, chaos theory, entropy as well as other concepts and methods have been used to analyze ventricular fibrillation [13]. It is remarkable how in a study, patients that were diagnosed with ventricular fibrillation and after that were successfully defibrillated, exhibited ECG recordings with an increased irregularity, showing a chaotic behavior of heart dynamics, [14] which is consistent with previous studies that show that cardiac chaotic behavior is related to health [3]. Fractal and complexity measures of heart rate variability have also been assessed, nevertheless the clinical applicability of these methodologies has not been established yet [15].

A new methodology was developed in the context of dynamical systems theory, probability and entropy, which differentiates acute disease from normality and chronic disease, as well as evolution between states [16]. Its clinical applicability was confirmed in previous studies [17-20]. The purpose of this paper is to apply the methodology to 520 holters to confirm its clinical applicability.

I. Torres et al. (eds.), *VII Latin American Congress on Biomedical Engineering CLAIB 2016, Bucaramanga, Santander, Colombia, October 26th -28th, 2016*,
IFMBE Proceedings 60,
DOI: 10.1007/978-981-10-4086-3_39

II. MATERIALS AND METHODS

A. Definitions

Map delay: plot, which is generated from each heart rate graphed as a function of prior heart rate

Ordered pairs of Heart Rate: It is composed by the consecutive values of HR represented as (x, y).

Probability of ordered consecutive pairs range: Ratio between the amount of ordered pairs in a specific range and the total of ordered pairs of the record. The ranges are of 5 beats/min.

Entropy of attractor

$$S = k\sum_{x=1}^{n}\sum_{y=1}^{n} P(X,Y) \times Ln\, P(X,Y) \qquad \text{Eq.2}$$

Where $P(X,Y)$ is the probability for range (X, Y), k is Boltzmann constant, 1.38x10-23 (Joules/Kelvin) and S is entropy.

Attractor S/k ratio: $\frac{S}{k} = \sum_{x=1}^{n}\sum_{y=1}^{n} P(X,Y) \times Ln\, P(X,Y)$

This expression is decomposed in terms corresponding to different orders of magnitude of occupation frequency: units (U), tens (Tn), hundreds (H) and thousands (Th), this terms are sums which subsequently are analized regarding to totality (T) with the proportions:

U / T; Te/ T; H/ T; Th / T; H / Th; Te /H

Physical-mathematical diagnosis evaluation:

According to the developed methodology [16-20], abnormality can be differentiated from normality quantifying the number of proportions outside the normality limits established previously; abnormality is characterized for more than two values outside those limits.

In order to quantify the degree of the abnormality respect to the normality, the following subtraction was made: the values less than minimum value of normality were subtract from this limit, and the superior normality limit was subtract from the proportions values that were greater than normality limits. Those values are organized and sum in four groups: 1) the subtractions of the proportions associated to occurrence frequencies of thousands, 2) the subtractions correspondent to hundreds, 3) the subtractions associated to tens, and 4) the subtractions correspondent to units. By this method, it is obtained a unique value associated to the four orders of magnitude, which allows quantifying the distance of the dynamic respect to normality, so that major values will imply a high grade of abnormality of the dynamics, and lower values will imply evolution to normality [16-20].

B. Population

A total of 520 holter were taken of older patients to 21, for a minimum of 21 hours will be analyzed; of which 50 are from normal cases and 470 from patients diagnosed with different pathologies. The records will be taken from Insight group database. The conditions of each patient are provided by an expert cardiologist and the findings of each study, from the conventional parameters, were verified by an expert cardiologist.

C. Procedure

Based on the values of minimum and maximum frequencies each hour and the number of beats per hour, a computational simulation of the whole dynamic was made in order to build an attractor in the phase space according to the method previously developed [15]. Each pair of cardiac frequencies is assumed as an event inside probability space. After, for each region and each holter the occupancy probability regarding the totality was evaluated and the probability for each of the ranges of 5 beats/min was quantified in the phase space. Later, the entropy of each attractor was evaluated.

Finally, from the entropy formula the Boltzmann, the S/ k ratio (see definitions) was obtained in order to group the addends corresponding to probabilities associated to occupation frequencies of units (U), tens (Te), hundreds (H) and thousands (Th). Then sums were made (equation 4) for evaluating the proportions (Eq.5). Then the mathematical parameters of diagnosis were applied (see definitions).

D. Ethic aspects

This study follows the requirements of Helsinki declaration and according to Colombian Health Ministry laws, it is a research without risk, because the measures are performed of medically prescribed tests from an anonymous database retrospectively evaluated, making it unnecessary informed consent. The privacy of research subjects and the confidentiality of their personal information were fulfilled.

E. Statistical Analysis

Subsequently the clinical diagnosis was taken as a Gold-Standard, and this was compared with the mathematical methodology, calculating the specificity and sensitivity. Such measures were carried out through a binary classification where true positives (VP) are the number of holter clinically diagnosed as abnormal and with mathematical values of abnormality. False positive (FP) is the number of holter with mathematical values of abnormality whose clinical diagnosis is normal. False negative (FN) is the number of holter mathematically diagnosed as normal but with pathological clinical values. Finally, true negatives (VN) defined as the number of records clinically diagnosed as normal and whose mathematical values also correspond to normal.

In order to evaluate the correlation between physical-mathematical values and conventional clinical diagnosis, the Kappa coefficient was calculated.

III. RESULTS

For cardiac dynamics studied entropy values (Table 1) showed values between $7{,}058\text{x}10^{-23}$ and $4{,}328\text{x}\ 10^{-23}$. The values of the proportions of entropy found for dynamic studied varied between 0 and 0.00731 for the U/T, between 0 and 0.1345 to Te/T, between 0 and 0.5528 for H/T, between 0 and 0.901 to Th/T, between 0 and 3.3513 to H/Th and 0 to 4.0239 for Te/H. The maximum values of thousand were in the dynamics with acute disease.

Table 1. Clinical diagnosis and mathematical measures of some examples of holter studied. U, Te, H and Th have same meaning of Eq.4.

Clinical Diagnosis	**S/k Proportion**	**Entropy**	**U**	**Te**	**H**	**Th**
Atrial fibrillation disorder intraventricular conduction. Atrial extrasystoles frequent	-4,4967448	$6{,}21\text{x}\ 10^{-23}$	0,01	0,0076	0,724	0
Sinus tachycardia inappropriate	-4,3235898	$5{,}97\text{x}\ 10^{-23}$	0	0,0182	0,0404	0
Acute coronary disease	-3,1511188	$4{,}35\text{x}\ 10^{-23}$	0,001	0,0343	0,4509	1,8554
Arrhythmia	-4,7144408	$6{,}51\text{x}\ 10^{-23}$	0	0,0037	0,22	0
Normal	-5,1146311	$7{,}06\text{x}\ 10^{-23}$	0	0	0	0
Atrial fibrillation	-4,3446415	$6\ \text{x}\ 10^{-23}$	4E-04	0,0188	1,5473	0,0319
Normal	-4,700169	$6{,}48\text{x}\ 10^{-23}$	0	0	0	0
Normal	-5,0794418	$7{,}01\text{x}\ 10^{-23}$	0	0	0	0
Acute myocardial infarction	-3,2138363	$4{,}44\text{x}\ 10^{-23}$	0,002	0,0164	3,0931	1,7877
Acute coronary disease	-3,3210591	$4{,}58\text{x}\ 10^{-23}$	6E-05	0,0378	3,0581	1,6591

The results of the clinical conclusion of cardiac dynamics evaluated were unmasked for determining values for specificity and sensitivity, which were 100%, and comparing the mathematical diagnosis with the standard Gold, the Kappa coefficient was equal to 1. Thus the clinical applicability and reproducibility of the method is confirmed.

IV. DISCUSSION

This is the first work in which the diagnostic ability of the methodology developed based on dynamic systems, probability and entropy in a sample of 520 cases, 50 normal and 470 with different pathologies is confirmed. This study demonstrates that the methodology allows to differentiate normal and abnormal cardiac dynamics, as well as evolution between these two states by assessing the proportional entropy of each attractor. In this methodology, the concept of entropy is reinterpreted, giving it a new geometric context that could show new perspectives in medicine and physics.

With this methodology, it is not only possible to assess the state of cardiac dynamics, it can also assess the evolution of dynamics from normal to chronic illness, and from chronic disease to acute disease, allowing to prevent the evolution to chronic and acute dynamics, which would help in making clinical decisions about the time required to monitor patients, since those patients that show values tending to acute disease should be monitored more continuously.

As this is a physical and mathematical methodology, its automation is easier than in the case of qualitative parameters of conventional evaluation. The predictions developed do not depend on population consensus, statistical or epidemiological parameters such as sex, pathology, risk factors, etc., but it is applicable to each particular case, if the age is older than 21 years, allowing a correct clinical following for each individual and making possible the assessment of the individual impact of different surgical or pharmacological interventions.

V. CONCLUSION

Usefulness and clinical applicability of the methodology based on probability, dynamical systems theory, and entropy were confirmed in the assessment of the state and evolution of cardiac dynamics. This methodology is new, objective and reproducible; it is applicable to each specific dynamic regardless statistical factor like age (if is older than 21 years), sex, risk factors or pathology.

ACKNOWLEDGMENT

We thank to Universidad Militar Nueva Granada for funding through the research fund the project INV-ING-2096 with which it could be developed this work. Also to university research fund, to research vicerrectorate and to engineering faculty, for their support to our work. We also thank to Research Center of Clínica del Country, especially to dr. Tito Tulio Roa, director of medical education; to Dr. Jorge Ospina, medical director, to Dr. Natalia Malaver, to Dr. Alfonso Correa, director of research center; to Dr. Lizbeth Ortiz, epidemiologist and to Silvia Ortiz, chief nurse of the research center, for their support to our work. Also to nurse Sandra Rodríguez.

REFERENCES

1. Devaney R. (1992) A first course in chaotic dynamical systems theory and experiments. Addison- Wesley, Massachusetts, pp 1-48
2. Peitgen H. (1992) Length area and dimension. Measuring complexity and scalling properties In: Chaos and Fractals: New Frontiers of Science. NY, Springer-Verlag, pp 183-228
3. Goldberger A, Amaral L, Hausdorff J. (2002) Fractal dynamics in physiology: alterations with disease and aging. Proc Natl Acad Sci USA 99(suppl1):2466–72
4. Prudat Y, Madhvani RV, Angelini M. (2016) Stochastic pacing reveals the propensity to cardiac action potential alternans and uncovers its underlying dynamics. J Physiol 594(9):2537-53
5. Eastman J, Sass J, Gomes J. (2016) Using delay differential equations to induce alternans in a model of cardiac electrophysiology. J Theor Biol 404:262-272
6. Guzzetti S, Borroni E, Garbelli PE, et al. (2005) Symbolic dynamics of heart rate variability: a probe to investigate cardiac autonomic modulation. Circulation 112:465–470
7. Maestri R, Pinna GD, Balocchi R. (2006) Clinical correlates of non-linear indices of heart rate variability in chronic heart failure patients. Biomed Tech 51:220–223
8. Norris PR, Anderson SM, Jenkins JM. (2008) Heart rate multiscale entropy at three hours predicts hospital mortality in 3,154 trauma patients. Shock 30:17–22
9. Huikuri HV, Makikallio T, Peng CK et al. (2000) Fractal correlation properties of R – R interval dynamics and mortality in patients with depressed left ventricular function after and acute myocardial infarction. Circulation 101:47–53
10. Krstacic G, Krstacic A, Smalcelj A. (2007) The "Chaos Theory" and nonlinear dynamics in heart rate variability analysis: does it work in short-time series in patients with coronary heart disease? Ann Noninvasive Electrocardiol 12(2):130-6
11. Schmitt DT, Ivanov PCh. (2007) Fractal scale-invariant and nonlinear properties of cardiac dynamics remain stable with advanced age: a new mechanistic picture of cardiac control in healthy elderly. Am J Physiol Regul Integr Comp Physiol 293(5):R1923-37
12. Huikuri HV, Perkiömäki JS, Maestri R. (2009) Clinical impact of evaluation of cardiovascular control by novel methods of heart rate dynamics. Philos Trans A Math Phys Eng Sci 367(1892):1223-38
13. Mezentseva LV. (2000) Ventricular fibrillation: the current methods for analysing the degree of irregularity of the process. Usp Fiziol Nauk 31(3):70-8
14. Jagric T, Marhl M, Stajer D. (2007) Irregularity test for very short electrocardiogram (ECG) signals as a method for predicting a successful defibrillation in patients with ventricular fibrillation. Transl Res 149(3):145-51
15. Perkiomaki J, Mäkikallio T, Huikuri H. (2005) Fractal and complexity measures of heart rate variability. Clin Exp Hypertens 27(2-3):149-158
16. Rodríguez J. (2010) Entropía Proporcional De Los Sistemas Dinámicos Cardiacos: Predicciones físicas y matemáticas de la dinámica cardiaca de aplicación clínica. Rev Colomb Cardiol 17:115-129
17. Rodríguez J, Prieto S, Domínguez D. (2013) Mathematical-physical prediction of cardiac dynamics using the proportional entropy of dynamic systems. J Med Med Sci 4(8):370-381
18. Rodríguez J, Prieto S, Bernal P. (2015) Entropía proporcional aplicada a la evolución de la dinámica cardiaca. Predicciones de aplicación clínica. La emergencia de los enfoques de la complejidad en América Latina. Tomo I. Comunidad del pensamiento complejo, Argentina, pp 247-264
19. Rodríguez J, Prieto S, Correa C. (2015) Predicciones de la entropía proporcional en la dinámica cardiaca. Rev Fac Med 23(2):60-70
20. Rodríguez J, Prieto S, Correa C. (2016) Proporciones de la entropía aplicadas a la unidad de cuidados intensivos: predicciones en infarto agudo de miocardio. Rev Acad Colomb Cienc Ex Fis Nat 40(155):192-199

Automatic Detection of the Retroareolar Region in X-ray Mammography Images

Germán F. Torres[1] and S. Pertuz[1]

[1] Universidad Industrial de Santander/Electric, Electronic and Telecomunications Engineering School, Bucaramanga, Colombia

Abstract— **Mammographic image analysis is an important tool for the detection and assessment of breast cancer. Previous studies have shown that the performance of image analysis algorithms can be improved by applying them in the retroareolar (RA) region of the breast. However, previous works have relied on subjective, manual segmentation of the RA region. This paper presents a method for the fully-automated detection of the RA region in x-ray mammography images. The method is based on a curvilinear coordinate system that automatically adapts to the breast shape and size. Experiments using logistic regression analysis on images from a publicly available dataset show that the proposed method outperforms the traditional approach in the task of cancer detection.**

Keywords— **mammography, retroareolar region, breast coordinate system, image analysis.**

I INTRODUCTION

Breast cancer is the most common cancer in women worldwide. Diagnostic in early stages is fundamental for increasing the probability of full recovery and for reducing the associated mortality rate. At present, screening mammography is one of the most widely used, effective and low cost techniques for early detection of breast cancer [1].

Computerized mammographic image analysis has been studied for aiding breast cancer detection and assessment. For instance, several researchers have applied image-based biomarkers on specific regions of interest (ROI) in order to examine parenchymal patterns [2]. In this scope, previous studies have demonstrated that the association of image-based biomarkers is higher in the zone immediately behind to the nipple, namely the *retroareolar* (RA) region[3]. Subsequently, several researchers have also developed methods for the analysis of parenchymal tissue in the RA region [4].

Despite its promising results, the main limitation of the aforementioned works is the need of human interaction for manually segmenting the RA region [3, 4]. On the one hand, subjectively human-annotated ROIs imply certain limitations in terms of reproducibility and scalability (e.g. the application of these methods to large datasets). On the other hand, fixed squared ROIs, as used in the literature, cannot adapt to a wide variety of breast shapes and sizes.

In this work we propose a methodology for automatic segmentation of the RA region in mammographic images. The proposed method takes into account the breast geometry and automatically adjusts to diverse shapes and sizes. For this purpose, we have built upon recent developments in the creation of anatomical coordinate systems in mammographic images [5, 6, 7]. Experimental results in publicly available datasets show that the proposed method outperforms manually selected ROIs for the task of cancer detection using logistic regression analysis.

The remainder of this paper is organized as follow. Firstly, the proposed method for the automatic segmentation of the RA region is presented in section II. Then, experiments and results are described in section III. Finally, conclusions are summarized in section IV.

II PROPOSED METHOD

The method proposed here can be divided in three steps: breast detection, ST-based anatomical mapping and automatic RA region detection.

A *Breast segmentation*

Breast segmentation is arguably the first preprocessing step in mammographic image analysis algorithms. In this work breast segmentation is performed in two steps: *Removal of scanning artifacts* and *Breast contour detection*.

Tape artifacts are markings left by tapes, or other shadows that appear as horizontal or vertical running strips in the image. Since these artifacts are commonly defined by straight lines, the first step for breast segmentation uses the Hough line transform for the their detection and removal [8]. For the second step, breast contour detection, a statistical technique was used in this work. Bearing in mind the rich texture in the breast and the homogeneity of pixels in the background, this approach suggests that pixels in the breast region can be treated as samples from a normal distribution. Therefore, the Anderson-Darling test is applied in order to test the hypothesis that the observations are samples from a normal distribution [9]. Pixels with p-values below a predefined threshold are marked as background. The obtained binary mask is post-processed by means of morphological operators in order to remove spurious artifacts and the breast contour is smoothed.

I. Torres et al. (eds.), *VII Latin American Congress on Biomedical Engineering CLAIB 2016, Bucaramanga, Santander, Colombia, October 26th -28th, 2016*,
IFMBE Proceedings 60,
DOI: 10.1007/978-981-10-4086-3_40

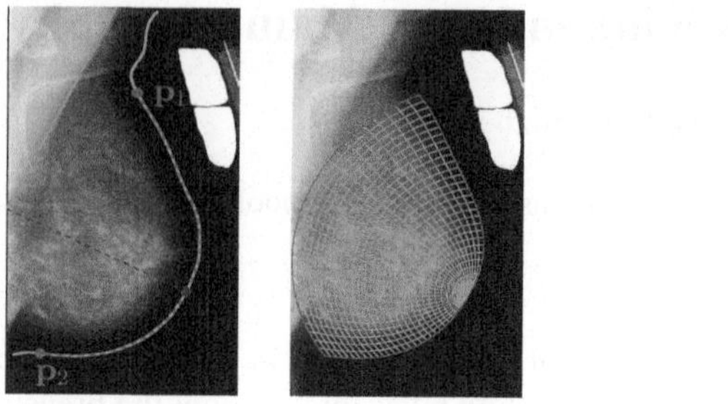

Fig. 1: ST coordinate system. Left: Location of the reference points Right: Coordinate system generated from de breast contour and key-points.

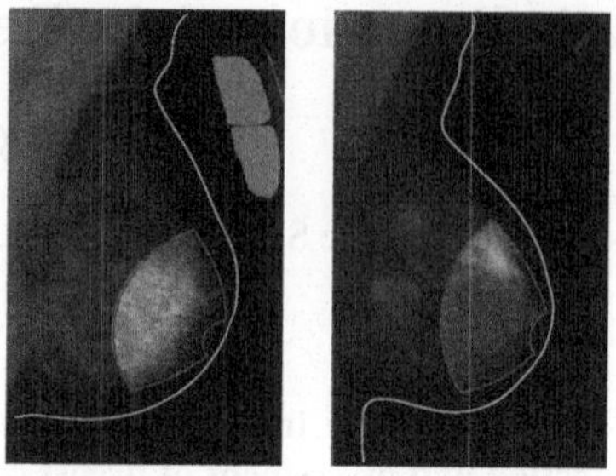

Fig. 3: Automatic RA region detection for two different cases. This figure shows how the detection adapts to the breast shape.

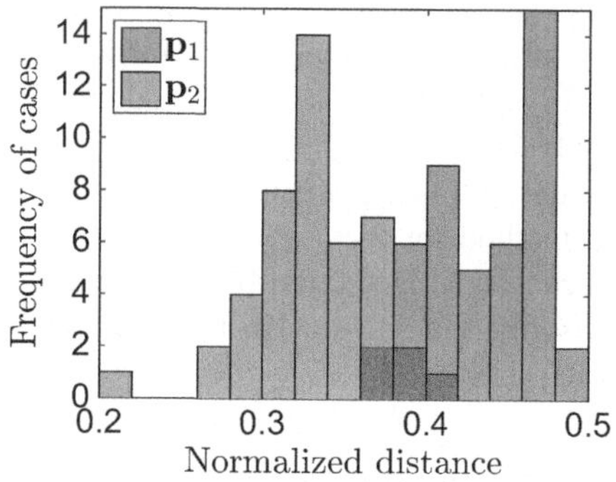

Fig. 2: Histograms of normalized distance of the extreme points. Based on this results, $\mathbf{p}_1$ and $\mathbf{p}_2$ are set to 0.44 and 0.33, respectively.

B ST-based anatomical mapping

The proposed methodology for automatic RA region detection is based on a coordinate system that adapts to breast shape. Anatomy-based breast coordinate systems have been studied previously for image registration [5, 6].

In this work, we have adapted the system proposed by Pertuz et al., namely the *ST coordinate system* [6], for automatic RA region detection. This coordinate system was chosen due to its flexibility and fast computation. The ST coordinate system can be interpreted as a mapping rule that assigns a coordinate pair (s,t), to each pixel in Cartesian coordinates (x,y). This process can be interpreted as a fully invertible, closed-form transformation represented as:

$$(x,y) \xrightarrow{\mathscr{M}} (s,t) \xrightarrow{\mathscr{M}^{-1}} (x,y) \quad (1)$$

where $\mathscr{M}$ and $\mathscr{M}^{-1}$ are the *direct* and *inverse* mapping operators, respectively.

The mapping in (1) requires the specification of three reference points: the origin, $\mathbf{p}_0$, and the extreme points $\mathbf{p}_1$ and $\mathbf{p}_2$ (Fig.1). In the original work [6], these points were manually selected. In order to make the process fully-automatic, we define these reference points in two steps.

Firstly, following the work by Nasser et al.[7], $\mathbf{p}_0$ is set at the nipple of the breast. As shown in Fig. 1, the nipple can be identified as the farthest point from the pectoral line [10] which, in turn, can be detected by using the Hough line transform. The location of the extreme points $\mathbf{p}_1$ and $\mathbf{p}_2$ is an important parameter of the ST coordinate system. These extreme points indicate what portion of the mammogram is considered as breast tissue. For instance, if $\mathbf{p}_1$ and $\mathbf{p}_2$ are localized too distant from the nipple, the resulting ST coordinate system includes information of the pectoral muscle and outside the image limits. In contrast, extreme points too close to the nipple will yield the exclusion of relevant information of the image. In order to determine the optimum location of $\mathbf{p}_1$ and $\mathbf{p}_2$, we have conduced an experiment with the aim of statistically describing their location. For nearly 10% of our correctly segmented dataset (45 images), the extreme points were manually established onto the contour. By measuring the distance from the reference point $\mathbf{p}_0$ (normalized over the entire length of the breast contour), $\mathbf{p}_1$ and $\mathbf{p}_2$ are located at normalized distances of $0.44(\pm 0.03)$ and $0.33(\pm 0.03)$, respectively. For illustration, Fig. 2 shows the histogram of the location on the extreme points in the test set.

C Automatic RA region detection

Since the ST coordinate system automatically adapts to the breast shape, the RA region can be segmented in Cartesian space by placing limits on the s and t coordinates. Thus, the RA region $\mathscr{R}$ is defined as the set of pixels given by:

$$\mathscr{R} = \{(x,y) \in \mathscr{M}^{-1}\{s,t\} | s \in [0.1, 0.5], t \in [0.1, 0.9]\} \quad (2)$$

For illustration, two examples of the extracted RA-ROIs with this methodology are shown in Fig. 3.

III Experiments and Results

Three experiments are conducted in order to assess the performance of breast segmentation, the robustness of the de-

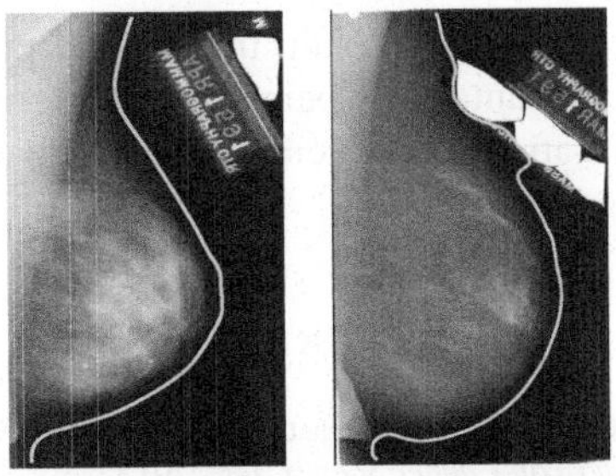

Fig. 4: Examples of correct and wrong breast segmentation. Left: correct breast segmentation. Right: deficient breast segmentation

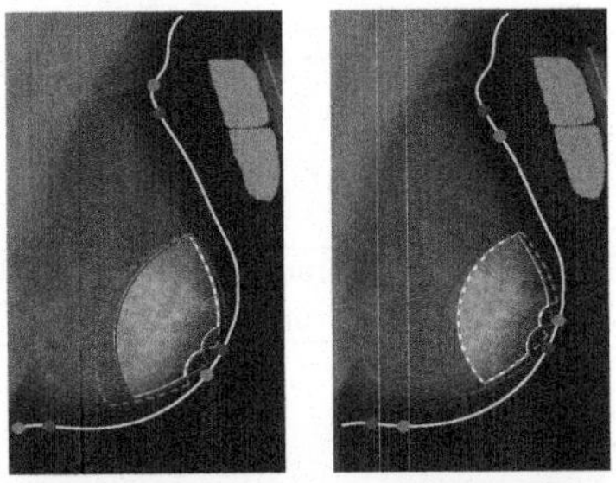

Fig. 5: Overlap of the detected RA region as a function of changes in the reference points. Left: $\varepsilon = +5\%$. Right: $\varepsilon = -5\%$.

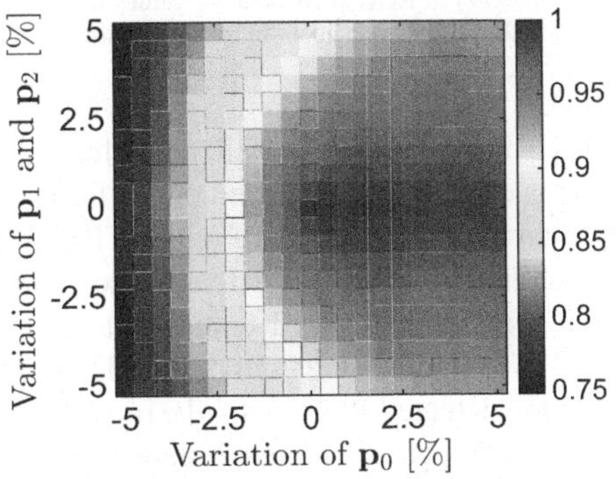

Fig. 6: Average map of ROIs overlapping

tection of the RA region and the accuracy in breast cancer detection. These experiments are detailed below.

The image dataset used in this study was taken from the Digital Database for Screening Mammography (DDSM) [11]. The DDSM database contains approximately 2500 cases of mammograms from several medical institutions in the USA. We have considered images with MLO view, which were scaled by a factor of 8 and converted in a 8-bits unsigned integer format to speed up the processing.

A Performance on breast segmentation

Notice that, although the proposed method is independent of the breast segmentation approach used, the segmented breast is the starting point of the algorithm. Therefore, this initial experiment is aimed at detecting the cases where the segmentation is not accurate in order to discard them and avoid biasing the results of subsequent experiments.

The breast segmentation algorithm described in section II-A was applied to a large subset of the DDSM dataset. The results on both normal and malignant mammograms are summarized in table 1. Most of the issues found in the segmentation were related to the to inaccurate detection of the breast contour due background noise and artifacts in the scanned mammograms. For illustration, Fig. 4 shows different cases with successful and deficient breast segmentation. For remaining experiments in this section, only the images with correct segmentation were considered.

Table 1: Performance of breast segmentation method in a large dataset

Type	Analyzed	Correct	%
Normal	982	354	36.04
Cancer	579	150	25.90
Total	1561	504	32.28

B Robustness of the detection of the RA region

An erroneous estimation of the reference points ($\mathbf{p}_0$, $\mathbf{p}_1$ and $\mathbf{p}_2$) can yield an inaccurate detection of the RA region. In order to assess the robustness of the method to changes in these parameters, we measure the area overlap between a reference ROI and a test ROI. The reference ROI is computed using predefined values for these points. The test ROI is then computed by changing the location of the reference points to $\mathbf{p}_i = (1+\varepsilon)\mathbf{p}_i$, for $i = 0, 1, 2$. This process is illustrated for a given image with $\varepsilon = \pm 5\%$ in Fig. 5. In this figure, the reference retroareolar ROI is depicted in blue, the test ROI is depicted in red, and the overlapping region is enclosed by the yellow contour.

Experiments were conducted for values of $\varepsilon \in [-5, +5]\%$ and repeated for 150 images randomly selected from our dataset. Fig. 6 summarizes the average overlap by means of a colormap. Results shown that the average overlap is 90% with a minimum value (worst case scenario) of 75%.

C Comparative tests

In order to compare the proposed methodology for automatic detection of the RA region with the traditional approach (manually selected squared ROI), we have imple-

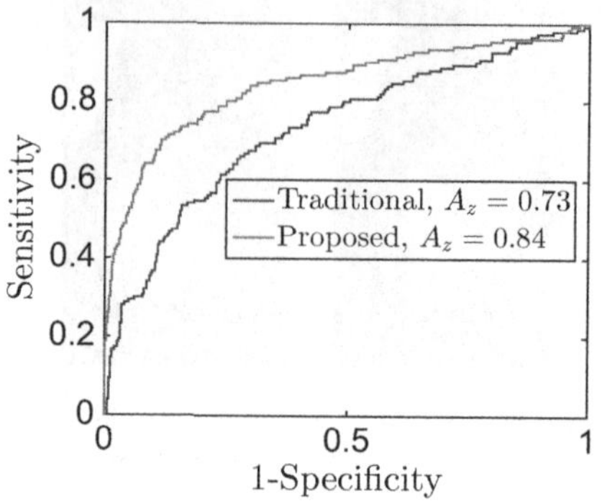

Fig. 7: Comparison of ROC curves and A_z values for image cancer detection experiment in the RA region

mented a logistic regression classifier for the task of cancer detection. This experiment is carried out as follows.

Based on an extensive literature review, 101 image-based features were implemented. The feature selection process, based in the sequential forward feature selection algorithm [12], yielded 12 features in the final model. These features are measured on the detected ROIs using the proposed method on our dataset (504 images). The performance of the classifier is then measured by using ROC analysis and leave-one-out cross validation. This process is repeated using the traditional method for the segmentation of the RA region. Specifically, square ROIs were manually localized for each image in the RA region. The size of the ROI was changed depending of the breast size according to the work by Wei et al. [4]. For our dataset, the size of square ROIs were 128x128, 112x112, 96x96, and 64x64 pixels for 101, 201, 183 and 19 cases, respectively.

ROC curves and areas under the ROC curve (A_z values) are shown in Fig. 7. Based on these results, it is clear that the proposed method for automated ROI detection outperforms the traditional method that uses manually selected squared ROIs. This result can be explained since, in contrast to manual ROIs, the proposed approach can adapt to the breast anatomy.

IV CONCLUSIONS

A method for the automatic detection of the RA region in breast images has been proposed. To the best of our knowledge this is the first fully-automated method for this task. The relevance of this contribution stems from the fact that previous studies have shown that the performance of mammographic image analysis be can improved by extracting features on the RA region. Interestingly enough, previous work relied on the manual selection of squared regions of interest by human readers. As a result, the method proposed in this work has the advantage of being automatic and more flexible, thus facilitating reproducibility and allowing to conduct studies in large datasets.

REFERENCES

1. American Cancer Society. Breast Cancer Facts and Figures 2015-2016 *Atlanta:American Cancer Society, Inc.* 2015.
2. Tahoces P. G., Correa J., Soutos M., Gomez L., Vidal J. J.. Computer-assisted diagnosis: the classification of mammographic breast parenchymal patterns *Physics in Medicine and Biology.* 1995;40:103.
3. Li Hui, Giger Maryellen L., Huo Zhimin, et al. Computerized analysis of mammographic parenchymal patterns for assessing breast cancer risk: Effect of ROI size and location *Medical Physics.* 2004;31:549–555.
4. Wei Jun, Chan Heang-Ping, Wu Yi-Ta, et al. Association of computerized mammographic parenchymal pattern measure with breast cancer risk: a pilot case-control study *Radiology.* 2011;260:42–49.
5. Brandt S. S., Karemore G., Karssemeijer N., Nielsen M.. An Anatomically Oriented Breast Coordinate System for Mammogram Analysis *IEEE Transactions on Medical Imaging.* 2011;30:1841–1851.
6. Pertuz S., Julia C., Puig D.. A Novel Mammography Image Representation Framework with Application to Image Registration in *2014 22nd International Conference on Pattern Recognition (ICPR)*:3292–3297 2014.
7. Abdel-Nasser Mohamed, Moreno Antonio, Puig Domenec. Temporal mammogram image registration using optimized curvilinear coordinates *Computer Methods and Programs in Biomedicine.* 2016;127:1–14.
8. Ballard Dana H. Generalizing the Hough transform to detect arbitrary shapes *Pattern recognition.* 1981;13:111–122.
9. Liu Li, Wang Jian, Wang Tianhui. Breast and Pectoral Muscle Contours Detection Based on Goodness of Fit Measure in *(iCBBE) 2011 5th International Conference on Bioinformatics and Biomedical Engineering*:1–4 2011.
10. Karssemeijer Nico. Automated classification of parenchymal patterns in mammograms *Physics in medicine and biology.* 1998;43:365.
11. Heath Michael, Bowyer Kevin, Kopans Daniel, Moore Richard, Kegelmeyer W Philip. The digital database for screening mammography in *Proceedings of the 5th international workshop on digital mammography*:212–218Citeseer 2000.
12. Devijver Pierre A, Kittler Josef. *Pattern recognition: A statistical approach*;761. Prentice-Hall London 1982.

Author: Germán F. Torres
Institute: Universidad Industrial de Santander
Street: Cra. 27 No. 9
City: Bucaramanga
Country: Colombia
Email: german.torres@correo.uis.edu.co

Author: S. Pertuz
Institute: Universidad Industrial de Santander
Street: Cra. 27 No. 9
City: Bucaramanga
Country: Colombia
Email: sdpertuz@uis.edu.co

Improved Particle Swarm Optimization algorithm applied to rigid registration in medical images

Ramiro Isa Jara[1], Francisco J. Buchelly[1], Gustavo J. Meschino[1] and Virginia L. Ballarin[1]

[1] ICYTE, CONICET – UNMDP Universidad Nacional de Mar del Plata, Argentina

Abstract— **Image registration is a space-temporary correlation process that allows comparison and/or image matching. This process has value into medical area when it comes to compare images acquired by different modalities or in different times. In this work, we present a method based on computational intelligence techniques from Particle Swarm Optimization (PSO) algorithm, to seek the best answer (particle) exploring a solutions set (swarm). The algorithm we developed includes an original idea for starting and for avoiding local minimum values, in order to achieve good rigid registration results (scale, rotation, translation). The mono and multi modal registration are performed on 2D magnetic resonance imaging (MRI) in T1-T2 sequences and single-photon emission computed tomography (SPECT) images. Experimental results show better optimal solution and decrease in convergence time compared to PSO original algorithm.**

Keywords— **Image registration, Particle Swarm Optimization, Swarm Intelligence.**

I. Introduction

Medical image registration allows specialists obtaining full anatomy information through image comparison at the same or different modality and acquisition times. This process is used in medical treatment verification, illness monitoring (vigilance) and for constructing anatomic atlas from image information databases, among other applications [1, 2].

In this work, the registration process follows these characteristics: Dimensionality: 2D, Nature of transformation: rigid, Modalities: mono-modal, multi-modal and Optimization procedure: Swarm Intelligence [2, 3, 4].

We present a new perspective to improve image registration results through guided search and different similarity measures as fitness function. There are many works in PSO applied to registration [6, 7, 8, 11, 12].

This work is structured as follows: Section 2 describes the theoretical basis used. Section 3 sets out the comparison between PSO-original and PSO-improved algorithms. Section 4 presents and discusses the results obtained. Finally, Section 5 presents the main contributions and future work.

II. Image registration

A. Rigid Registration

Registration is the process for finding the best geometric alignment between two or more images at the same scene, the same or different modality, different time and/or different viewpoints. The involved images are called: *source* (target) and *floating* (reference) [2].

Registration aim is to find and apply geometric transformations on the reference image in order to modify pixels' coordinates, until the difference between the floating and target images is minimized through the definition of some measure of comparison. It can be written as:

$$I_R = T * I_F, \qquad (1)$$
$$M = \min(||I_O - I_F||)$$

where T is the geometric transformation, I_R is the modified image, I_F is the reference image, I_O is the target image and M is the measure of comparison.

In rigid registration, deformation occurs considering three parameters: scale, rotation and translation. To compute scale and translation, equation 2 shows the transformation matrix in 2D homogenous coordinates, where e_x and e_y is the scale and t_x and t_y defines the translation factor:

$$\begin{bmatrix} x' \\ y' \\ 1 \end{bmatrix} = \begin{bmatrix} e_x & 0 & t_x \\ 0 & e_y & t_y \\ 0 & 0 & 1 \end{bmatrix} * \begin{bmatrix} x \\ y \\ 1 \end{bmatrix} \qquad (2)$$

In the same way, equation 3 shows the rotation transformation matrix, where α is the rotation angle:

$$\begin{bmatrix} x' \\ y' \\ 1 \end{bmatrix} = \begin{bmatrix} Cos\,\alpha & Sin\,\alpha & 0 \\ -Sin\,\alpha & Cos\,\alpha & 0 \\ 0 & 0 & 1 \end{bmatrix} * \begin{bmatrix} x \\ y \\ 1 \end{bmatrix} \qquad (3)$$

B. Particle Swarm Optimization (PSO)

PSO is an algorithm that is part of the computational intelligence paradigm, which was initially inspired by the

I. Torres et al. (eds.), *VII Latin American Congress on Biomedical Engineering CLAIB 2016, Bucaramanga, Santander, Colombia, October 26th -28th, 2016*,
IFMBE Proceedings 60,
DOI: 10.1007/978-981-10-4086-3_41

behavior of birds. PSO was developed by Kennedy and Eberhart in 1995, where important terms are defined such as a swarm and particles "flying" into a solution space [4].

Each particle is a parametric potential solution. PSO is an iterative algorithm where particles change their position in the parameter space through velocity. It is based on the best swarm behavior position and the best particle. To update the particle values, PSO seeks to minimize or maximize a target function called Fitness. Equation 4 defines PSO iterative operation:

$$V_i^{t+1} = WV_i^t + C_1R_1(P_{best} - X_i^t) + C_2R_2(g_{best} - X_i^t) \quad (4)$$
$$X_i^{t+1} = X_i^t + V_i^{t+1}$$

C. Similarity Measures

In registration, similarity measures allow quantizing difference between images. These measures should be based on probabilistic paradigms and they avoid analyzing pixel's information (intensity) directly, so it is an advantage because previously segmentation or feature extractions are not necessary [9].

Cross-Correlation (CC): is the ratio of the covariance to the product of standard deviations of two images. In this work, as usually done, we use the CC in mono-modal registration. The ideal alignment will produce CC = 1.

Mutual Information (MI): quantifies the statistical dependence between two variables based on individual entropies with the join entropy. MI is applied in multi-modal registration.

III. Proposed Algorithm

Let us consider two images A and B, mathematical defined as $A, B : Z^2 => [0,255]$, where A is the target image and B is the reference image.

The proposed algorithm uses cross-correlation and mutual information for *Fitness* function. The procedure is summarized in the next stages:

1. Estimate initial Fitness function (FP) between inputs images A, B.
2. Initialize the swarm (X) in the initial solutions space composed by scale (E), translation (T) and rotation (R).

   $E = [E_{MIN}, E_{MAX}]$, $T = [T_{MIN}, T_{MAX}]$, $R = [R_{MIN}, R_{MAX}]$

   $X_{i \to 1:N} = (X_i = rand[e,t,r] => e \in E \wedge t \in T \wedge r \in R)$
   $B^T{}_{i \to 1:N} = X_{i \to 1:N}\, B$
   $F_{i \to 1:N} = F\,(A, B^T{}_{i \to 1:N})$
   $If\ F_i > F_P => g_{Best} = X_i$

   where B^T is the image after applying the geometric transformation, F is the Fitness function and g_{BEST} is the best particle among all swarm found by PSO.
3. According to particle $g_{Best} = [e_g, t_g, r_g]$, modify initial solutions in order to optimize the search. The next process is referred to scale but is similar in the case of translation and rotation:

   $E_M = prom(E_{Min}, E_{Max})$
   $Si\ e_g > E_M => E_Y = [E_M, E_{Max}]$
   $else,\ E_Y = [E_{Min}, E_M]$
   $X_{i \to 1:N} = (X_i = rand[e,t,r] => e \in E_Y \wedge t \in T_Y \wedge r \in R_Y)$
4. Considering the swarm X, the PSO algorithm avoids local minimum values, during the execution at specific iterations among randomize particles positions in solutions space.
5. Stop the algorithm after a given number of iteration.

Normally, PSO algorithm initializes with random particles of the swarm at the whole solutions space, which influences directly in the final solution, because searching is performed from the best parameters found according to algorithm heuristic. Oppositely, the proposed algorithm carries out guided search following the particle to improve initial *fitness*.

Introducing the previous modification, the heuristic of PSO algorithm is performed in a new reduced solution space, which makes possible obtaining a better result with low computational cost than in the original algorithm. Another change introduced is the randomization the particles after a specific number of iterations, which allow avoiding multiple local minimum values.

These proposed changes improve significantly the results of medical image registration as it will be presented in the next section.

IV. Results and Discussion

In order to test the proposed algorithm, MR images in T1-T2 sequences are used for monomodal and multimodal registration with SPECT modality (Fig. 1). Reference images are generated with known values of scale, translation and rotation, so the performance of the algorithms can be estimated.

In the tests, we analyzed fitness values, processing times and errors achieved between target and registered image. In Fig. 2 we show mono modal registration results compared the original PSO and the proposed algorithm. Fitness values (CC) are shown, achieved with different particles and number of iterations. In the case of PSO algorithm, fitness reached around 0.4 and 0.7, as long as with proposed algorithm the fitness is constant in 0.7, being independent of particles values.

Fig. 3 shows computation times. In test 1 and 2, time for PSO algorithm is directly proportional to the number of iterations and particles' values. It is greater than 10 minutes, but with the proposed algorithm, time is significantly reduced. In test 1, PSO algorithm with 20 particles, and the proposed algorithm with 40 particles was considered, both using 50 iterations. In test 2, PSO algorithm with 50 particles, and the proposed algorithm with 60 particles was considered, always taking 50 iterations.

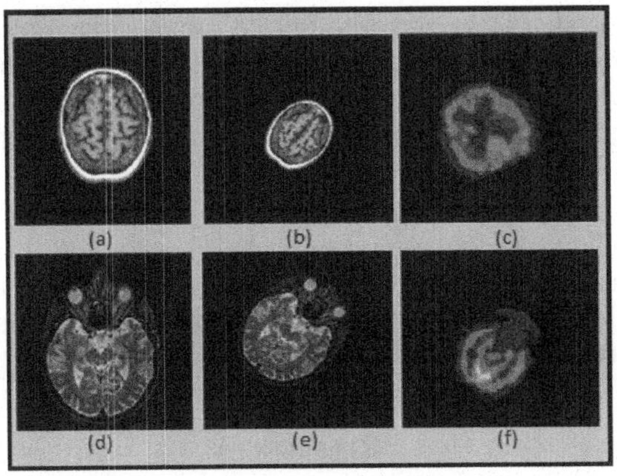

Fig. 1 Head images in MR modality **a)** Target image in T1 sequence; **b)** Reference image in T1 sequence **c)** Reference image in SPECT; **d)** Target image in T2 sequence **e)** Reference image in T2 sequence; **e)** Reference image in SPECT.

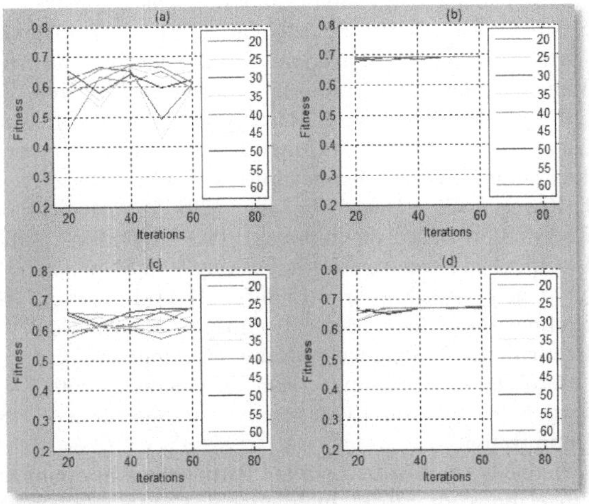

Fig. 2: Fitness vs. Iterations. a) PSO original, T1 image; b) Proposed algorithm; c) PSO original, T2 image; d) Proposed algorithm.

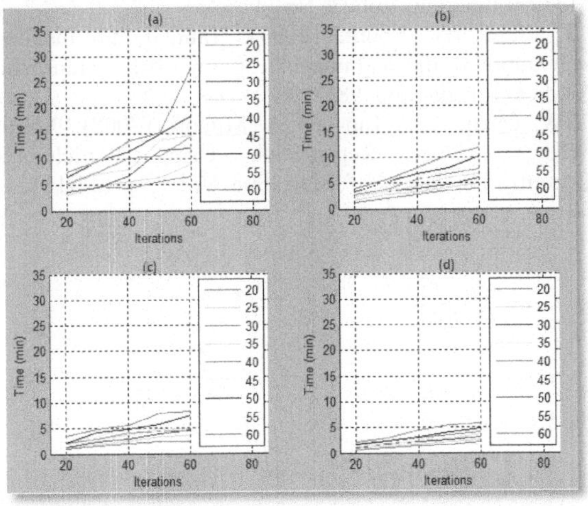

Fig 3: Time vs. iterations. a) PSO original, T1 image; b) Proposed algorithm; c) PSO original, T2 image; d) Proposed algorithm.

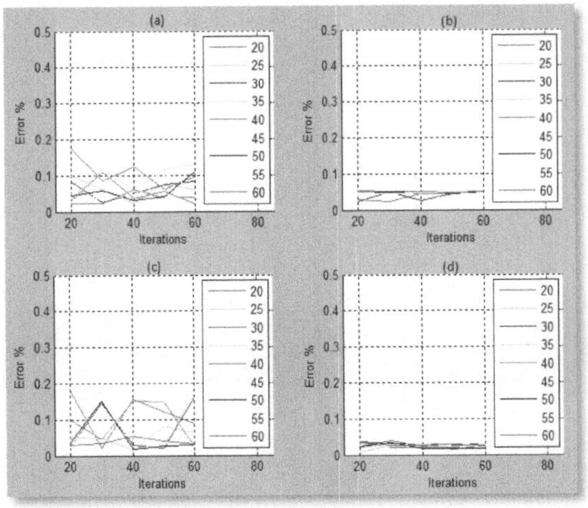

Fig. 4: Error vs. Iterations. a) PSO original, T1 image; b) Proposed algorithm; c) PSO original, T2 image; d) Proposed algorithm.

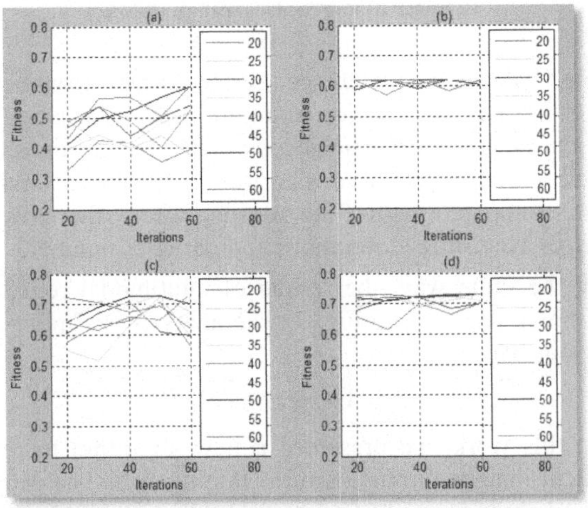

Fig 5: Fitness vs Iterations. a) PSO algorithm, T1-Spect; b) Proposed algorithm, T1-Spect; c) PSO algorithm, T2-Spect; d) Proposed algorithm, T2-Spect.

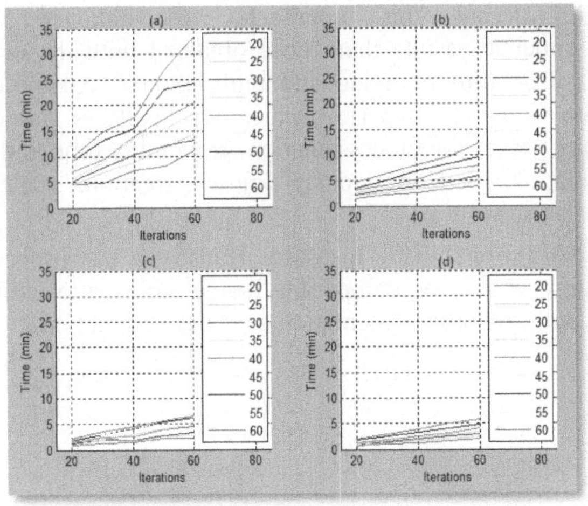

Fig 6: Time vs. iterations. a) PSO algorithm, T1-Spect; b) Proposed algorithm, T1-Spect; c) PSO algorithm, T2-Spect; d) Proposed algorithm, T2-Spect.

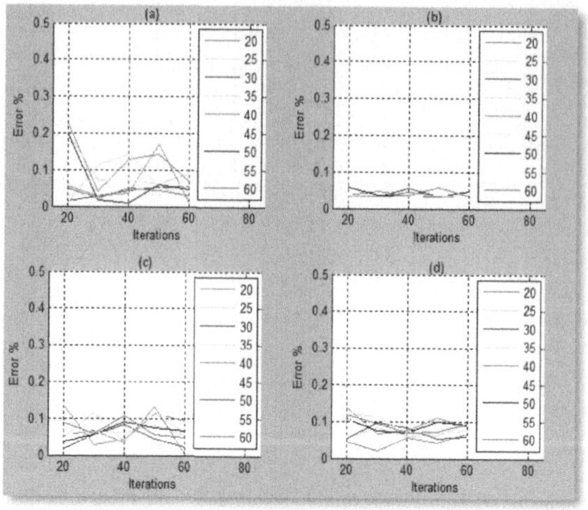

Fig 7: Error vs Iterations. a) PSO algorithm, T1-Spect; b) Proposed algorithm, T1-Spect; c) PSO algorithm, T2-Spect; d) Proposed algorithm, T2-Spect.

Fig. 4 shows the difference (hereby called Error) between target values of scale, rotation and translation and references, considering achieved values. The results show in the case of the proposed algorithm a difference less than 0.1%, compared with PSO results, where the difference changes according to iterations and particles number.

In the same way, the results of multimodal registration are shown in Fig. 5 to Fig. 7. The analysis is similar as the previous part.

V. Conclusion

In this work, we presented a new algorithm to register medical images through Swarm Intelligence. The proposed method based on PSO improves the rigid registration results; therefore we have shown that computational algorithms with collective behavior are useful to optimization procedures in images registration applications.

Considering this approach, we obtained better experimental results than those obtained with the original PSO algorithm in mono and multi modal registration. It improves the PSO performance, which allows stable and proper values for fitness function, computing time and error estimation, with previously known rotation, scale and translation values.

Using probabilistic measures leads to a fast processing, because it allows comparing image information directly without previous segmentation or feature extraction.

It is always necessary to reduce computational resources and processing time. Swarm Intelligence paradigm has shown to be useful in optimization problems and in medical image registration, as it provided significant improved results.

Future work is focused in 2D and 3D registration including also affine and projective registration, which is required in muscular and abdominal imaging. We aim to develop faster and "smarter" algorithms to be used in different applications.

References

1. A. Maintz, A. Viergever. (1998) "A survey of medical image registration", Medical Image Analysis Vol. 2, N 1, pp. 1–36.
2. L. Brown. (1992) "A survey of image registration technique", ACM Comput. Surv., Vol. 24, pp. 325-376.
3. B. Zitova, Jan Flusser. (2003) "Image registration methods: a survey". Image and Vision Computing Vol. 21 pp. 997 – 1000.
4. S. D. Chavan, Nisha P. Adgokar. (2005) "An Overview on Particle Swarm Optimization: Basic Concepts and Modified Variants", International Journal of Science - Research, Vol. 4.
5. J.Pramanik, S. Dalai, D. Rana. (2015) "Image Registration using PSO and APSO: A Comparative Analysis", International Journal of Computer Applications, Vol. 116 – No. 21.
6. L. Zheng, G. Qian, L. Lin. (2011) "Medical Image Registration Based on Improved PSO Algorithm", Advanced Electrical and Electronics Engineering Vol. 87 pp. 487-494.
7. Y. Chen, A. Mimori. (2009) "Hybrid Particle Swarm Optimization for Medical Image Registration". Fifth International Conference on Natural Computation.
8. C. Grosan, A. Abraham, S. Han, A. Gelbukh. (2003) "Hybrid Particle Swarm–Evolutionary Algorithm Search Optimization".
9. J. Kennedy, R. Eberhart.(1995)"Particle swarm optimization". Proceedings of IEEE International Conference on Neural Networks IEEE, Piscataway, NJ, Vol. 4, pp.1942-1948.
10. J. Kim, J. Fressler. (2004) "Intensity-based Image Registration using robust correlation coefficients", IEEE Transactions on Medical Imaging, Vol.23, No 11.
11. J. Pluim, A. Maint. (2003) "Mutual information-based registration of medical image: a survey", IEEE Transactions on Medical Imaging, pp. 986-1004.
12. Y.W. Chen, C.-L. Lin and A. Mimori. (2008) "Multimodal Medical Image Registration Using Particle Swarm Optimization", Proc. of ISDA2008, Vol.3, pp.127-131.

Assessment of Surface Electromyography During Orofacial Praxis in Healthy Subjects

S. Cadavid-Arboleda[1], L.M. Ramirez-Arbelaez[1], E. Perez-Giraldo[1], S. Restrepo-Agudelo[2], S. Roldan-Vasco[2], J.C. Suarez-Escudero[3], G. Cantillo-Mackenzie[4], C.L. Bedoya-Londono[1,5] , L. Martinez-Moreno[5] and A. Orozco-Duque[1]

[1] Department of Biomedical Engineering, Instituto Tecnologico Metropolitano, Medellin, Colombia
[2] Faculty of Engineering, Instituto Tecnologico Metropolitano, Medellin, Colombia
[3] Grupo de Salud Publica, Universidad Pontificia Bolivariana, Medellin, Colombia
[4] Student, Faculty of Pharmacy and Biochemistry, Universidad de Buenos Aires, Buenos Aires, Argentina
[5] IPS Organizacion Fonoaudiologa E.U. (OFA), Medellin, Colombia

Abstract— **Traumatic Brain Injury or injuries involving the posterior region of the brain can cause orofacial apraxia. Rehabilitation and diagnosis of pathologies associated with apraxias require better tools to evaluate and quantify the muscles activation during praxis performing. Developing a protocol to acquire surface electromyography (sEMG) from small muscles associated with orofacial praxis is necessary for the quantitative valuation of apraxias. In this paper we aim to test the use of Ag/AgCl electrodes with 20 mm diameter respect to Au electrodes with 7 mm diameter to acquire sEMG from orofacial muscles. Our findings show that there is not statistically significant difference between medians of Ag/AgCl and Au measures. We recommend the use of Ag/AgCl in the masseter region and Au electrodes in the orbicularis oris and other small muscles.**

Keywords— **apraxia, orofacial praxis, surface electromyography, signal processing.**

I INTRODUCTION

Praxis are voluntary and planned movements or tasks, executed in defined sequences, with a determined purpose [1]. The orofacial apraxia refers to an inability to execute praxis related to facial, lingual, pharyngeal and masticatory actions [2] [3]. Apraxias are higher cognitive function disorders which affect patients after a stroke or lesions in posterior region of the brain [3], i.e. head trauma. Deglution disorders could be involved in case of orofacial apraxia [4]. The treatment for apraxia consists mainly in rehabilitation based on movements repetition where the patient evolution is determined by clinical considerations [3]. Evaluation using surface electromyography (sEMG) from masseter and temporalis muscles has been considered in praxis related to chewing [5] [6]. sEMG is able to measure muscular activation in a non-invasive way. In case of deglutition disorders, some authors have suggested the use of sEMG to assess the patients, nevertheless, there is not a gold standard for the diagnosis [7]. Problems with sEMG acquisition from orofacial muscles are related to both the small size of the muscles, and the overlying of the fibers [8]. Although activation of masseters muscles can be easily measured by sEMG using commercial Ag/AgCl electrodes, other muscles, as orbicularis oris, show more limitations for the electrodes placement [8]. In this paper, the preliminary results from muscular activation measurements of healthy subjects, in two muscular groups during orofacial praxis, are presented. The aim of this project is to develop a sEMG acquisition protocol for quantitative diagnosis of apraxias. For this purpose, it is necessary to identify the muscular activation pattern using commercial electrodes and a correct placement. The use of Ag/AgCl are compared with Au electrodes to determine the appropriate electrode for acquisition of sEMG signals in small orofacial muscles.

II MATERIALS AND METHODS

A Subjects

We recruited eight healthy subjects (four male and four female) aged between 21 and 30 years old (23.5 ± 2.88). The subjects were selected using the following exclusion criteria: dental pathologies, congenital oral malformations, active inflammatory processes (mouth, head or neck), strange elements in mouth (like piercing), diagnosed cognitive disorders (motor or sensorial), chronic obstructive pulmonary disease, head or neck cancer antecedents or facial aesthetic surgery. Ethical approval was obtained and informed consent was used.

B Signal acquisition

The differential bioamplifier ML138, connected to polygraph PowerLab 16/35 (AD Instruments Inc.) was used for

I. Torres et al. (eds.), *VII Latin American Congress on Biomedical Engineering CLAIB 2016, Bucaramanga, Santander, Colombia, October 26th -28th, 2016*,
IFMBE Proceedings 60,
DOI: 10.1007/978-981-10-4086-3_42

the sEMG acquisition. Two type of superficial electrodes were tested: disposable Ag/AgCl electrodes (Ref. 2228, 3M - 30 mm x 35 mm, 15 mm diameter in gel area); and gold electrodes (Au) for electroencephalography (Ref. MLAWBT9 EEG, AD Instruments Inc.) with 7 mm diameter, using conductive gel (Nuprep, Weaver and Company) and fixed to the skin with adhesive tape.

Two muscular groups were studied: masseter (left and right) and orbicularis oris (superior and inferior). The electrode positions in the muscular groups were split in two stages: first, a pair of Ag/AgCl electrodes were placed in the right masseter (RM) and another pair in inferior orbicularis oris (IO); Au electrodes were placed in the left masseter (LM) and superior orbicularis oris (SO) (Figure 1).

A second experiment was performed switching the type of electrodes in each subject (second electrode configuration). An Ag/AgCl electrode placed in the forehead was used as reference.

Subjects were asked to clenching teeth to detect the anatomical position of masseter muscles contraction. Electrodes were placed parallel to the muscles fibers as shown in Figure 1. The electrodes for superior orbicular oris were placed in the upper lip at both sides of the philtrum. For the inferior orbicularis oris the electrodes were placed in the lower lip in mentolabial sulcus (see Figure 1). For Au electrodes the inter-electrode distance was 20 mm. However, for Ag/AgCl case, the inter-electrode distances depended of the electrodes size.

Two orofacial praxis were tested: clench the teeth (CT) and blow a kiss (BK); each subject was asked through a verbal command to execute the both praxis for 10 times (five with the first and five with the second electrode configuration).

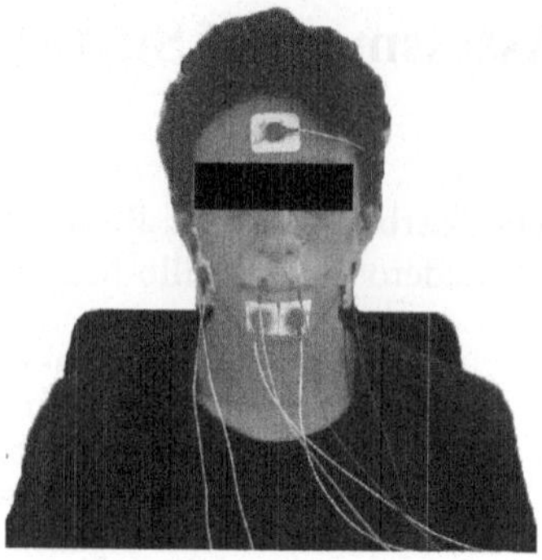

Fig. 1: Initial electrode configuration

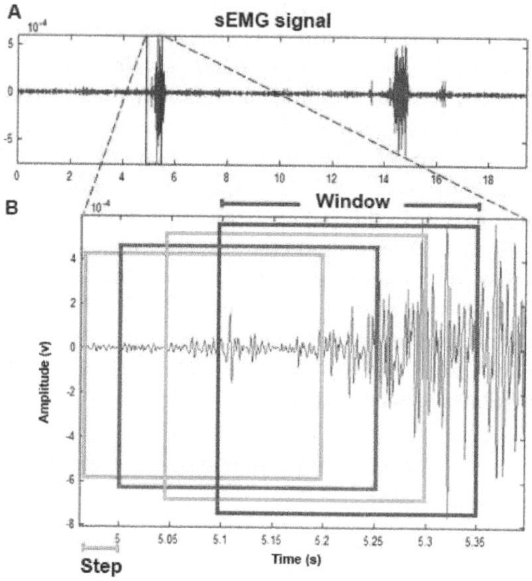

Fig. 2: A. Raw sEMG signals for two sequential praxis. B. Zoom of a section from raw signal and the computation of RMS by windowing.

C Signal processing

The sEMG signals were acquired with sampling frequency of 2 kHz. The pre-processing, storage and visualization of the raw sEMG data was carried out with LabChart Pro (AD Instruments Inc.). The offline analysis and the computation of the root mean square value (RMS) was done in MatLab environment (MathWorks Inc.). A 10th order Butterworth band-pass filter was designed with high and low cutoff frequencies at 25 Hz and 500 Hz, respectively. Moreover, a 20th order notch filter in cascade was employed to eliminate the 60 Hz line noise and its harmonics until 360 Hz [9]. The RMS value was estimate for each muscle at each praxis (Figure 2).The proposed procedure estimates the RMS value for time windows of 500 samples (250ms) with step time of 100 samples (50ms) (see Figure 2).

For each subject, the peak RMS value (RMS^p) of each muscle at each praxis was computed. A total of 80 values (RMS^p) per subject was extracted per session (five tests with Au electrode and five with Ag/AgCl). The normalization of the RMS values ($nRMS^p$) by subject and by praxis was performed, using the following equation:

$$nRMS^p = \frac{RMS^p - RMS^p_{min}}{RMS^p_{max} - RMS^p_{min}} \tag{1}$$

where RMS^p_{min} corresponds with the lowest RMS^p value along 10 repetitions (five Au and five Ag/AgCl), and RMS^p_{max} is the highest RMS^p value. The $nRMS^p$ value eliminates the inter-muscular variability in sEMG amplitude and allows a comparison of activation levels between the muscles evaluated at each subject using different types of electrodes.

Two statistical sample groups were compared: Au vs.

Ag/AgCl electrodes measurements according to praxis and muscular group. Table 1 shows the mean $nRMS^p$ values of each muscle group during each praxis. The U-Mann Whitney test (U-test) was applied to evaluate the null hypothesis of equal medians ($p = 0.05$).

III RESULTS

The Figure 3A shows an example of sEMG signal (blue) and the corresponding RMS value (red line). The Figure 3B illustrates the RMS value of the sEMG signal in masseter muscular group during the praxis CT, comparatively acquired with Ag/AgCl electrodes (red and blue lines) and Au electrodes (yellow and green lines). The red and green lines represent the activity in right hemisphere whilst the blue and yellow lines show the left hemisphere activity.

The U-test accepted the null hypothesis for equality of medians for both electrodes in every muscle and praxis. Although a difference of amplitudes was observed between the raw signals measured with Au and Ag/AgCl electrodes (Figure 3B), the $nRMS^p$ value did not have such significant statistical differences in the involved muscles.

The $nRMS^p$ values yield the patterns of activation of praxis. In the CT praxis, a high intensity of contraction for masseter muscles was observed simultaneously with low intensity in orbicularis oris (Table 1A). On the other side, the pattern activation during the praxis BK was inverted (Table 1B).

Figures 3A and 3B show that both Ag/AgCl and Au electrodes are able to acquire EMG signals. Both electrode configurations display the same activation pattern by the same praxis performed in different patients.

Table 1: Mean values $nRMS^p$. Null hypothesis accepted (A) mean that Au and Ag/AgCl electrodes belong to the same population. Bold numbers correspond to muscles activation.

Clenching teeth								
Subjects	RM		LM		SO		IO	
	Au	Ag	Au	Ag	Au	Ag	Au	Ag
1	**0.76**	**0.50**	**0.68**	**0.84**	0.07	0.06	0.04	0.07
2	**0.54**	**0.88**	**0.59**	**0.65**	0.02	0.01	0.05	0.03
3	**0.54**	**0.50**	**0.37**	**0.41**	0.09	0.05	0.39	0.10
4	**0.74**	**0.79**	**0.90**	**0.86**	0.02	0.01	0.03	0.05
5	**0.28**	**0.54**	**0.30**	**0.88**	0.00	0.05	0.03	0.04
6	**0.41**	**0.71**	**0.77**	**0.93**	0.09	0.01	0.09	0.07
7	**0.43**	**0.66**	**0.54**	**0.56**	0.06	0.09	0.21	0.09
8	**0.65**	**0.86**	**0.52**	**0.66**	0.02	0.03	0.06	0.11
U-test	A	A	A	A	A	A	A	A

Blow a kiss								
Subjects	RM		LM		SO		IO	
	Au	Ag	Au	Ag	Au	Ag	Au	Ag
1	0.11	0.08	0.09	0.02	**0.50**	**0.53**	**0.69**	**0.27**
2	0.07	0.03	0.01	0.07	**0.66**	**0.56**	**0.85**	**0.85**
3	0.01	0.12	0.16	0.01	**0.36**	**0.54**	**0.94**	**0.82**
4	0.02	0.02	0.01	0.01	**0.30**	**0.31**	**0.79**	**0.57**
5	0.02	0.04	0.02	0.03	**0.38**	**0.71**	**0.51**	**0.68**
6	0.01	0.01	0.01	0.01	**0.32**	**0.34**	**0.50**	**0.83**
7	0.03	0.07	0.03	0.01	**0.41**	**0.37**	**0.87**	**0.57**
8	0.04	0.04	0.03	0.04	**0.30**	**0.45**	**0.77**	**0.45**
U-test	A	A	A	A	A	A	A	A

IV DISCUSSION

In this research, we evaluated commercial Au electrodes with only 7 mm diameter respect to Ag/AgCl with 20 mm diameter. This approach is consistent with other authors who have carried out comparisons between Au and Ag/AgCl electrodes in other applications [10]. Authors in [8] have reported the use of non-commercial small Ag/AgCl electrodes to record sEMG in facial muscles. Small electrodes allow the correct configuration in the muscle and simplify the placement considering the inter-electrode distance. Our findings show also that there is not statistically significant difference between medians of Ag/AgCl measures and Au measures (see Table 1).

Normalized RMS value is recommended to avoid the variability of amplitude in sEMG. Author in [11] reported that different electrode positions, or different materials, can affect the sEMG amplitude in the raw signal. We used the normalized RMS value in order to compare different muscles contraction during a specific praxis according to [11] and [12]. Using this approach is possible to compare several muscle independently of the used electrodes.

Our experimental results (summarized in Table 1) suggest that exist an activation pattern which allow the discrimination between different praxis using sEMG signals. Other authors have reported that sEMG from the masseter muscle could be used to detect clenching in bruxism studies [5]. Activation patterns in orbicularis oris muscle have been studied during movements like depressing and protruding the lower lip to analyze oralized deaf individuals [13]. Adhesive properties of commercial Ag/AgCl electrodes make them suitable for their usage in facial extensive areas, such as masseter region.

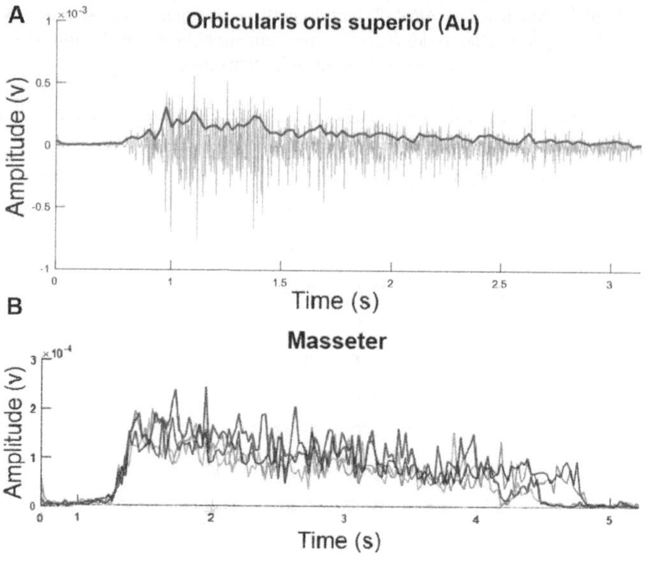

Fig. 3: A. Typical EMG signal (blue) and the correspond RMS (red) in "BK". B. RMS in masseter muscles during "CT"; using Ag/AgCl electrodes (red and blue lines) and Au electrodes (yellow and green lines).

However, their use in small areas such as orbicularis oris region are not appropriate because it is difficult to get a correct placement. Our findings show that commercial Au electrodes with 7 mm diameter are suitable for acquisition of sEMG in small regions. Authors in [12] recommend the use of electrodes with maximum 10 mm diameter and 20 mm of interelectrode distance placed along the fiber direction. Au electrodes facilitate this configuration in the orbicularis region.

V CONCLUSION

In this paper, a protocol for quantitative assessment of orofacial praxis based on sEMG signals and RMS values is presented. The normalization of the RMS values ($nRMS^p$) allows a correct comparison among sEMG signals and the detection of muscular activation patterns. The comparative results obtained with Au and Ag/AgCl electrodes show that there are not significant differences in the medians of $nRMS^p$ value among the two evaluated statistical samples. We propose the use of Au electrodes in orbicularis oris region due to its size, which makes easy the positioning; and the use of Ag/AgCl electrodes in masseter muscles, because the area is bigger and the adhesive of the electrodes fits easily. The assessed muscular groups (bilateral masseter and orbicularis oris) are of interest in swallowing studies, so the proposed protocol constitutes the first step in the development of an assessment system of apraxias and swallowing disorders through sEMG signals.

CONFLICT OF INTEREST

The authors declare that they have no conflict of interest.

ACKNOWLEDGEMENTS

This work has been supported by COLCIENCIAS-Republica de Colombia, project No. 115071149746.

REFERENCES

1. Schrager O.L., O'Donnell C.M.. Actos motores oro-faringo-faciales y praxias fonoarticulatorias. *Fonoaudiológica.* 2001;47:22–32.
2. Canzano L., Scandola M., Pernigo S., Aglioti S.M., Moro V.. Anosognosia for apraxia: Experimental evidence for defective awareness of one's own bucco-facial gestures *Cortex.* 2014;61:148–157.
3. Selnes O.. Apraxia, Handedness and Language Laterality *Encyclopedia of Language & Linguistics.* 2006:420–423.
4. Drozdz D., Mancopes R., Silva A.M.T., Reppold C.. Analysis of the level of Dysphagia, anxiety, and nutritional status before and after speech therapy in patients with stroke. *International archives of otorhinolaryngology.* 2014;18:172–7.
5. Moreno I., Sanchez T., Ardizone I., Aneiros F., Celemin A.. Electromyographic comparisons between clenching, swallowing and chewing in jaw muscles with varying occlusal parameters *Medicina Oral, Patologia Oral y Cirugia Bucal.* 2008;13:207–213.
6. Schumann N.P., Bongers K., Guntinas-Lichius O., Scholle H.C.. Facial muscle activation patterns in healthy male humans: A multi-channel surface EMG study *Journal of Neuroscience Methods.* 2010;187:120–128.
7. Vaiman M.. Standardization of surface electromyography utilized to evaluate patients with dysphagia. *Head & face medicine.* 2007;3:26.
8. Lapatki B.G., Stegeman D.F., Jonas I.E.. A surface EMG electrode for the simultaneous observation of multiple facial muscles *Journal of Neuroscience Methods.* 2003;123:117–128.
9. Mello R. G.T., Oliveira L.F., Nadal J.. Digital Butterworth filter for subtracting noise from low magnitude surface electromyogram *Computer Methods and Programs in Biomedicine.* 2007;87:28–35.
10. Tallgren P., Vanhatalo S., Kaila K., Voipio J.. Evaluation of commercially available electrodes and gels for recording of slow EEG potentials *Clinical Neurophysiology.* 2005;116:799–806.
11. Huckabee M., Low I.S., McAuliffe M. J.. Variability in Clinical Surface Electromyography Recording of Submental Muscle Activity in Swallowing of Healthy Participants *Asia Pacific Journal of Speech, Language and Hearing.* 2012;15:175–186.
12. Stepp C.E.. Surface electromyography for speech and swallowing systems: Measurement, analysis, and interpretation *Journal of Speech, Language, and Hearing Research.* 2012;55:1232–1247.
13. Regalo S.C.H., Vitti M., Moraes M.T.B., et al. Electromyographic analysis of the orbicularis oris muscle in oralized deaf individuals *Brazilian Dental Journal.* 2005;16:237–242.

Author for correspondence: Lina Marcela Ramirez Arbelaez
Institute: Instituto Tecnologico Metropolitano
Street: Calle 73 No 76A - 354
City: Medellin
Country: Colombia
Email: linaramirez@itm.edu.co

Analyzing Multiple Accelerometer Configurations to Detect Falls and Motion

J.D. López[1], C. Ocampo[1], A. Sucerquia[1] and J.F. Vargas-Bonilla[1]

[1] SISTEMIC, Facultad de Ingeniería, Universidad de Antiquia UDEA, Calle 70 No. 52-21, Tel.: +57-4-2198561, Medellín, Colombia

Abstract— **The use of wearable devices with accelerometers developed to detect falls and motion has been continuously growing because of their small size, low weight, energy efficiency, and low price. However, the number of sensors, their position in the body and estimation methodologies are still open issues when tested in uncontrolled conditions. In this paper, we perform a discriminant analysis to determine which combinations of feature extraction characteristics and accelerometer positions are best fitted to estimate falls and particular movements. A dataset of 33 activities recorded with eight accelerometers distributed along the body of six participants is released as part of this work. Our analysis concludes that a waist-arm combination with a statistical feature is best fitted for most cases. But this work, rather than a single conclusion, is intended to provide a benchmark to other authors. The specific tests explain for example that the foot is not a good location for the accelerometer, or that static features are less relevant than dynamic ones.**

Keywords— **Triaxial accelerometer, wearable devices, mobile health-care, elderly fall detection, motion capture.**

I INTRODUCTION

There is a current impulse in healthcare services for constant monitoring of independent elderly people with minimum intrusion. Real-time fall detection and automatic assistance on physical therapy exercises are good examples of the current needs in the field [1]. Additionally, the demand of wearable devices developed for improving sports performance has increased with the introduction of smart-watches and similar gadgets. However, fall detection and motion capture are complex processes and still an open issue when implemented under uncontrolled conditions.

Falls in elderly people have been traditionally detected with environment based (cameras, pressure sensors, etc.) and wearable devices (mainly accelerometers, see [2–4] for reviews). But the indoor constrain of environment based systems makes them infeasible for most activities of daily living (ADL). The most popular sensors used for wearable devices are capacitive accelerometers because of their small size, low weight, energy efficiency, and low price [5]. These characteristics have popularized accelerometers in several electronic devices for a wide range of uses (stabilizing cameras, video-game pads, smartphone positioning, etc.), and recently in smart-watches or directly attached to clothes.

Motion capture with accelerometers has not been broadly studied. Some works have characterized walk and jog by detecting the periodicity of the signal (see [6] and the references therein). Cola et al. [7] detected gait deviation as a fall risk feature. An overview of the field [1] highlights classification of posture and transition, gait, balance and sway, unsupervised monitoring, and general movements (sit, sleep, etc.) as important human motion activities for monitoring clinical conditions, quality of life, sports performance, and aging.

Some authors have proposed to use various accelerometers distributed along the body. Although using a single accelerometer provides enough information for detecting falls, the algorithms used for classification are getting more complex in order to improve the accuracy [2]. But due to this complexity, their implementation increases the energy consumption, and in most cases they must be permanently active. The use of multiple accelerometers may allow simpler algorithms with reduced energy consumption [8]. Moreover, Gao et al. [9] analyzed multiple vs. single sensor systems. They demonstrated that elaborated classifiers and feature sets are not required to obtain high accuracies on a multi-sensor system.

There are not standardized methodologies to determine the best fitted location of the sensors, and few works address the problem. In [5] the authors analyzed different locations for placing a triaxial accelerometer. They concluded that the most common positions in the body (waist, chest, etc.) provide similar information, suggesting the waist as the best one. But as an example of the widespread of works in this field, Yuan et al. [10] proposed the wrist as the best placement for detecting falls, despite the higher false positive rates they acknowledge this position implies. For a wider overview, the review presented in [3] resumes the most common locations of the sensors, reviewing 12 works using the chest, 8 the head, 29 the waist, 13 the thigh, 6 the wrist, 3 the back, and 5 the ankle or foot.

The objective of this paper is to analyze possible locations and combinations of accelerometers and features to discriminate between and within falls and ADL. To accomplish this objective, a fall and ADL dataset acquired with in eight accelerometers was generated with a self developed device

I. Torres et al. (eds.), *VII Latin American Congress on Biomedical Engineering CLAIB 2016, Bucaramanga, Santander, Colombia, October 26th -28th, 2016*,
IFMBE Proceedings 60,
DOI: 10.1007/978-981-10-4086-3_43

(Section II). We used a Fisher score to discriminate among sensors and features (Section III). Results and conclusions of this analysis are shown in Sections IV and V.

II System and Experimental Set-up

In order to determine the best number, location and combination of accelerometers for detecting different ADL and falls, four devices, each with two triaxial accelerometers (an ADXL345 from Analog Devices, and a MMA8451Q from Freescale) were used for acquiring and recording acceleration data. Fig. 1 (center) shows one of the four devices. Each one consisted on a Kinets MKL25Z128VLK4 microcontroller with 128 kB of flash memory and 16 kB of SRAM, the two accelerometers, a SD card for recording, and a 1000 mA/h battery. The size of each device was $53 \times 81 \times 25$ mm, and its weight 92 g.

The eight accelerometers were placed as shown in Fig. 1 (left): head (A1), chest (A2), hip (A3), waist (A4), foot (A5), thigh (A6), wrist (A7), and arm (A8). These locations were selected for being the parts with higher mobility in the body (limbs: A5–A8), or for being in the center of mass (trunk: A1–A4). The ADXL345 (sensors A1, A3, A5, A7) was configured for ± 16 g, 13 bits of ADC; and the MMA8451Q (sensors A2, A4, A6, A8) was configured for ± 8 g, 14 bits of ADC. All recordings were performed with a sampling frequency of 300 Hz.

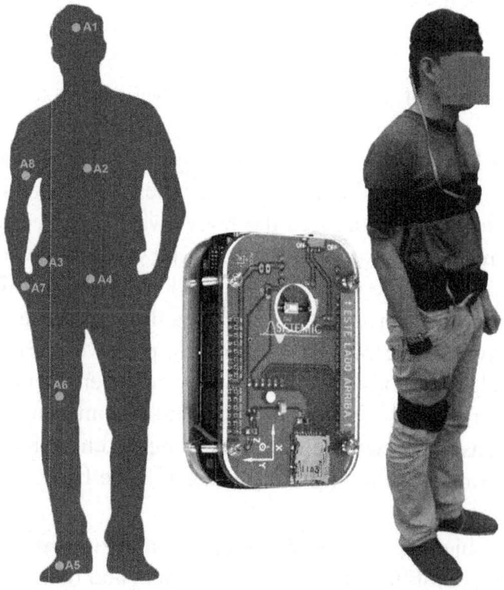

Fig. 1: Location of accelerometers (left), example of the device used for acquisition (center), and placement of the devices on the participants (right).

Six participants: 3 male (age 21–23 years, height 1.69–1.77 m, weight 55–61 kg), and 3 female (age 24–26 years, height 1.58–1.65 m, weight 48–53 kg) performed 5 repetitions of 18 types of ADL and 15 types of falls wearing 8 accelerometers in different parts of the body (see Fig. 1). 1002 total trials (544 ADL and 458 falls) were recorded. The dataset is publicly available as part of this paper (`http://sistemic.udea.edu.co/investigacion/proyectos/multiple-accelerometers/?lang=en`). The list of activities performed by the participants is included in the Readme file (clic here).

III Methods

We performed a discriminant analysis to determine the capability of each selected location to discriminate among activities and/or falls compared with the other locations, and the best features for each case. No pre-processing was performed in order to avoid bias caused by specific selections (as every author has particular filters).

The most popular features used by authors to classify between ADL and falls are those based on acceleration peaks [3], but they are not expected to be feasible for differentiating among ADL (or even among falls). A simple and not computationally intensive dynamic estimation consists of obtaining statistical moments from a sliding-horizon window. Other option not too explored by authors is to analyze the frequency domain with a fast Fourier transform (FFT). Given that this work is expected to help authors taking decisions for their own works, in this paper (Table 1) we analyzed two widely used peak-based features, two windowed statistical-based, and the FFT (see [3, Table 4] for a complete list).

Table 1: Selected feature extraction characteristics.

No.	Feature	Equation
P1	Sum vector magnitude	$\lvert a\rvert_{\text{RMS}} = \frac{1}{3}\sqrt{a_x^2 + a_y^2 + a_z^2}$
P2	Signal magnitude area	$SMA = \frac{1}{t}\left(\sum_{i=1}^{3}\int_0^t \lvert a_i(t)\rvert dt\right)$
P3	RMS of mean over each axis, 1 s window	$\lvert\mu\rvert = \frac{1}{3}\sqrt{\mu_x^2 + \mu_y^2 + \mu_z^2}$
P4	Standard deviation magnitude, 1 s window	$\lvert\sigma\rvert = \sqrt{\sigma_x^2 + \sigma_y^2 + \sigma_z^2}$
P5	FFT of P4	$FFT(\lvert\sigma\rvert)$

The sliding-horizon window for the dynamic features (P2–P5) was selected of 1 s. Different time windows between 0.2 s and 2 s were tested, being 2 s better for some activities (it can contain a full period of walk) but not enough for accepting the higher computation burden.

Fisher developed a simple to implement discriminant score for separating data between two classes (see [11] for a survey on the field). This analysis can be used to discriminate among pairs or sets of activities (fall vs. ADL, with- vs. without displacement ADL, etc.), with the following algorithm:

1. Compute the features of Table 1 for each trial and accelerometer. It will result in a set of N_f feature matrices $\mathscr{F} = \{F_1, \ldots, F_{N_f}\}$, where each matrix $F_i \in \Re^{N_s \times N_r}$ contains the scalar values of the computed features, in N_r recorded files over the N_s sensors.
2. Compute the Fisher discriminant score $J \in \Re^{N_f \times N_s}$ for each feature ($i = 1, \ldots, N_f$) and sensor ($j = 1, \ldots, N_s$):

$$J_{ij} = \frac{(\bar{f}_1 - \bar{f}_2)^2}{\sigma_1^2 + \sigma_2^2} \quad (1)$$

where $\bar{f}_k$ is the mean over N_r recorded files for the i-th feature, j-th sensor, and characteristic $k = 1, 2$ (fall vs. ADL for example). Sub-indices i and j were intentionally omitted in f_k for simplicity in the equation writing.

IV RESULTS

A Fall vs. ADL analysis

Fig. 2 shows the Fisher discriminant score between falls and ADL for the eight accelerometers and five features selected. Note how the head (A1), waist (A4), and arm (A8) have higher capabilities of discriminating among falls and ADL with most features. However, given that the head and waist are in the same axis, and that it is easier to attach accelerometers to the waist and arms than to the head, the best combination for this discrimination is sensors A4 and A8 with feature P4, or a combination of A4–P3 and A8–P5. Note that the overall best features (P3, P4 and P5) are dynamic (over a 1 s window) suggesting that the well-known sum vector magnitude (P1) might not be a good feature.

This analysis gives other interesting results. The FFT feature presented high discriminant differences between head, arm, and the other sensors. This result is quite unexpected given that most features are at least consistent between all sensors based in the trunk. Other interesting result is that sensors A5 (foot) and A6 (tight) presented almost none discriminant capacity. This could be expected because walk or jog may have high accelerations making it difficult to separate them from falls. The other sensors (A2, A3, and A7) presented mixed results.

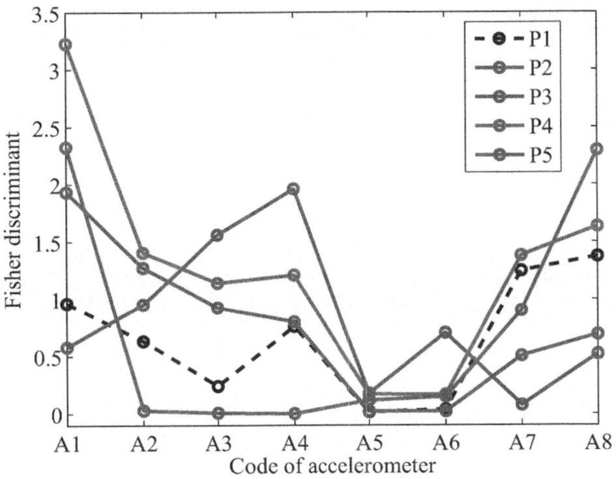

Fig. 2: Capability of features and sensors to discriminate between falls and ADL. The head (A1), waist (A4) and arm (A8) presented higher values with most features, being P3 and P4 (statistical ones) the best fitted ones.

B Discriminating among activities

Fig. 3 presents the comparison of two Fisher scores: ADL with displacement (D01–D06, periodic signals, see the Readme for details –clic here–) vs. ADL without displacement (D7–D19, sit, lie in a bed, etc.) in dotted red; and quickly sitting (D08, D10) vs. slowly sitting (D07, D09) in gray. All five features were added as a single one on each score. It is clear that the high discriminant capability of the foot sensor (A5) between displacement-based and quiet ADL goes down to the worst fitted sensor in activities where the foot is usually quiet, such as sitting down. This result may look obvious, but it provides useful conclusions: (i) Discriminating among groups of activities may be tricky, and (ii) The variety of ADL human activities is large enough for limiting to a single sensor or feature for all cases.

V CONCLUSIONS

This paper presents a discriminant analysis for analyzing positions and combination of sensors to differentiate between falls and ADL, and within ADL. This analysis is based on five features, four commonly used in the literature and a FFT analysis. A dataset with six participants performing 18 ADL and 15 falls was generated for this work. This dataset consisted of 1002 trials of acceleration data from eight accelerometers in different parts of the body acquired with a self developed device.

A combination of sensors in the arm and waist seems ideal for separating falls and ADL, and both sensors seem to be

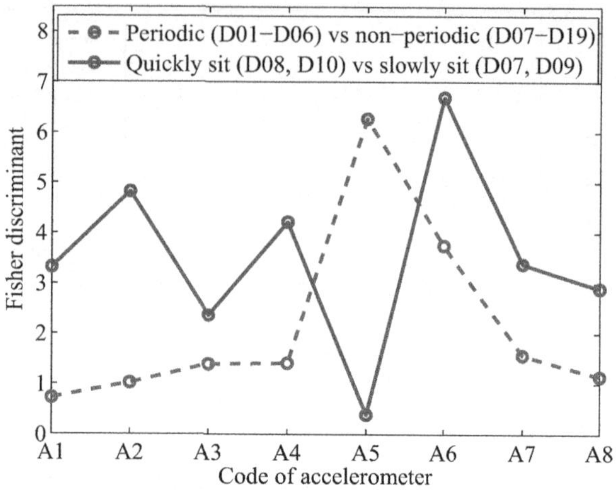

Fig. 3: Comparison between the sum of all features in the Fisher score for: (dotted red) with vs. without displacement ADL, and (gray) sitting quickly vs. slowly. Note how almost all sensors change their discriminant capability demonstrating that there is not a gold standard for motion recognition.

feasible for discriminating among ADL; this because they are consistent among features in all tests. In the other hand, the leg (foot and tight) did not show consistent results, it was only relevant in specific situations. The head and the wrist were good but not outstanding, and the head has several drawbacks for being useful in commercial wearable devices. Curiously, the wrist did not show good discriminating capabilities between falls and ADL, or among ADL.

Among features, the sum vector magnitude was consistently the worst feature. This result suggests that it is time to increase efforts in developing dynamic algorithms; and they do not need to be complex, as demonstrated in this work where the standard deviation based feature was relevant in all tests. However, it was relevant in different sensors for different analysis, suggesting that it would be necessary a multi-sensor array.

Finally, this work only intended to show a methodology for selecting the best configuration of sensors and features depending on the objective each author has. Knowing the lack of public datasets in the literature (to our knowledge, none with task specific devices), we published this dataset with the intention of encouraging authors to test new ideas with it. No pre-processing was added as a way to reduce biases in the analysis; but filtering may highly improve the performance of novel features. Additionally, several sub-group tests as the shown in Fig. 3 may be performed in order to answer a wide number of research questions.

Ethical approval

All activities performed by the participants were approved by the Bio-ethics Committee of the Medicine Faculty, Universidad de Antioquia UDEA (Medellín, Colombia). Additionally, all participants were evaluated by a sports specialized physician.

The authors declare that they have no conflict of interest.

Acknowledgements

The authors thank the participants for their insightful help and patience while recording the dataset.

References

1. Shany Tal, Redmond Stephen J., Narayanan Michael R., Lovell Nigel H.. Sensors-Based Wearable Systems for Monitoring of Human Movement and Falls *IEEE Sensors Journal.* 2012;12:658 - 670.
2. Igual Raul, Medrano Carlos, Plaza Immaculada. Challenges, issues and trends in fall detection systems *BioMedical Engineering OnLine.* 2013;12:1–24.
3. Pannurat Natthapon, Thiemjarus Surapa, Nantajeewarawat Ekawit. Automatic fall monitoring: A review *Sensors.* 2014;14:12900-12936.
4. Casilari Eduardo, Luque Rafael, Morón María-José. Analysis of Android Device-Based Solutions for Fall Detection *Sensors.* 2015;15:17827–17894.
5. Cleland Ian, Kikhia Basel, Nugent Chris, et al. Optimal Placement of Accelerometers for the Detection of Everyday Activities *Sensors.* 2013;13:9183-9200.
6. López José David, Sucerquia Angela, Duque-Muñoz Leonardo, Vargas-Bonilla Francisco. Walk and jog characterization using a tri-axial accelerometer in *IEEE International Conference on Ubiquitous Computing and Communications (IUCC)*:1406-1410 2015.
7. Cola Guglielmo, Avvenuti Marco, Vecchio Alessio, Yang Guang-Zhong, Lo Benny. An On-Node Processing Approach for Anomaly Detection in Gait *IEEE Sensors Journal.* 2015;15:6640 - 6649.
8. Gao Lei, Bourke Alan K, Nelson John. A comparison of classifiers for activity recognition using multiple accelerometer-based sensors in *11th International Conference on Cybernetic Intelligent Systems (CIS), IEEE*:149-153 2012.
9. Gao Lei, Bourke A.K., Nelson John. Evaluation of accelerometer based multi-sensor versus single-sensor activity recognition systems *Medical Engineering & Physics.* 2014;36:779–785.
10. Yuan Jian, Tan Kok Kiong, Lee Tong Heng, Koh Gerald Choon Huat. Power-Efficient Interrupt-Driven Algorithms for Fall Detection and Classification of Activities of Daily Living *IEEE Sensors Journal.* 2015;15:1377 - 1387.
11. Li Tao, Zhu Shenghuo, Ogihara Mitsunori. Using discriminant analysis for multi-class classification: an experimental investigation *Knowledge and Information Systems.* 2006;10:453-472.

Corresponding author: José David López
Institute: Universidad de Antioquia
Street: Calle 70 No. 52-21
City: Medellín
Country: Colombia
Email: josedavid@udea.edu.co

MC3C3-E1cell response to zirconium (Zr) implants with different surface characteristic by digital image processing analysis

M.R. Katunar[1], A. Bouchet[2], J. Ballarre[1] and J.I. Pastore[2]

[1] Instituto en Ciencia y Tecnología de los Materiales (INTEMA) CONICET, Universidad Nacional de Mar del Plata, División Electroquímica Aplicada, Mar del Plata, Argentina.
[2] Instituto en Científicas y Tecnológicas en Electrónica (ICYTE), CONICET, Universidad Nacional de Mar del Plata, Argentina.

Abstract— **In this work we propose investigate attachment and proliferation behavior of the osteoblastic cell line MC3T3-E1 on Zr implants with different anodization treatments as surface modifications. To achieve the objective it is important count the number of cell nucleus three and five days after the cells were seeded over control and anodized samples. We presented an automatically method that combines different techniques for image processing. The algorithms used for analysis were implemented in a toolbox for Matlab, to facilitate reuse and distribution. The results showed an interesting relationship between cell differentiation and material surface.**

Keywords— **Cell proliferation, zirconium, anodization, image analysis.**

I. INTRODUCTION

In normal bone remodeling or bone turnover, osteoblastic bone formation and osteoblastic bone resorption is coupled in a precise and orchestrated manner. Osteoblastic differentiation is an essential part of bone formation especially after a metallic orthopaedic implant is placed [1,2]. Surface topography has profound impact on osteoblast cell attachment, proliferation, and differentiation as the first events after the material is placed in biological medium. Zirconium (Zr) is an ideal metal for intra-osseous implants due to its properties such as corrosion resistance, its inert nature, its ability to adsorb proteins readily onto its surface, and low cytotoxic. Zr is considered as a potential good biocompatible material for orthopaedic implantation [3,4].

The use of image processing analysis for cell proliferation and viability result in an excellent tool to automatize and standardize the nuclei number in a determined substrate. Then the comparison between different substrates and times of incubation results more efficient and rapid. The ability to detect objects with certain characteristics within an image is a fundamental appearance for the automation of various kinds of applications. Such procedures, known as image segmentation, allow the decomposition of an image into regions of interest, according to each specific application [5]. This stage is of paramount importance in any automated vision system because it is the first level of the task of understanding the image and severely affects the subsequent process of image interpretation, providing useful structures such as regions and borders.

In this work, we investigated attachment and proliferation behavior of the osteoblastic cell line MC3T3-E1 on Zr implants with different anodization treatments as surface modifications. To achieve the objective we use different techniques for image processing.

II. MATERIALS AND METHODS

A. Base material and surface treatment

Commercially pure zirconium cylinders (99.5% Roberto Cordes S.A., Argentina) of 40-50 mm length and 1 mm diameter were used for the in vivo test and plane samples of 1 cm2 for the in vitro assays . Three surface conditions were compared: as received pure zirconium (Zr0, control), zirconium anodised at a constant potential of 30 V (Zr30) and at 60V (Zr60) during 60 minutes in 1 mol L-1 H3PO4. All samples were mechanically polished with 600 grit emery paper, degreased with ethanol and rinsed with deionized water. The sample conditioning and oxide growth details have been previously reported [6].

B. Cell culturing

Zirconium plates of 1 cm^2 were autoclaved and washed with completed medium previous to cell seeding. Three procedures of seeding were performed.

Seeding with MC3T3-E1 cell line (ATCC® CRL-2593™). This is a mouse preosteoblastic cell line. Cells were maintained in completed α-MEM (A10490, Gibco, UK) without ascorbic acid supplemented with 10% fetal bovine serum (FBS, Hyclone®, Thermo Scientific) plus antibiotics (100 U/mL penicillin and 100 μg/mL streptomycin sulphate, Sigma-Aldrich, St. Louis, MO) incubated at 37°C with 5% CO_2. For differentiation, cells were plated at 1x104 cells per cm^2 in completed maintenance medium with 50 ug/mL of ascorbic acid, 10 nM dexamethasone and 10 mM β-glycerolphosphate, complete differentiation medium was changed every 3 or 4 days.

I. Torres et al. (eds.), *VII Latin American Congress on Biomedical Engineering CLAIB 2016, Bucaramanga, Santander, Colombia, October 26th -28th, 2016,*
IFMBE Proceedings 60,
DOI: 10.1007/978-981-10-4086-3_44

C. Cell Cytotoxicity Assays

Cells were fixed with 4% paraformaldehyde (PFA) solution for 15 min. After PFA was removed, cells were rinsed with PBS twice and permeabilized with 0.1% (v/v) Triton X-100, washed with PBS. Hoechst staining (Invitrogen, Molecular Probes ®) was used for nuclei visualization. Finally fluorescent-labelled cells were observed using an inverted fluorescence microscope (Olympus IX51) with a DAPI filter for Hoechst ($\lambda ex/\lambda em$ = 380/455nm) using CellD analysis software (Olympus). The sizes of the images were 2070×1548 pixels. The cell cytoxicity assays-test for proliferation assay was repeated twice, always with three parallel samples for each surface type.

D. Automatic segmentation of the proliferation cell images

The segmentation is typically used to locate objects and boundaries within an image. This stage is of paramount importance in any automated vision system because it is the first level of the task of understanding the image and severely affects the subsequent process of image interpretation.

We propose the following algorithm to achieve the objective proposed in this paper:

 # Step 1: Segmentation of cell proliferation images:
In this step the segmentation of the cell proliferation images is extremely important because it affects the next stage of analysis. The success of the subsequent stage depends of an appropriate segmentation. The following steps are performed:

a. *Binarization:* The original image, a color image, is converted into a binary image, where the objects (cell nuclei) are separated from the background.

The binarization was performed using a local Otsu method, to adapt the spatial hue variations in the blue component of the original image. The image is divided into a 10x10 nonoverlapping windows. For each window the optimal threshold was calculated using the Otsu algorithm [7, 8] on another 20x20 enlarged window centered on the original window. Determine the threshold value on a larger window allows a smooth transition between windows [9].

b. *Filtering and separation:* A morphological filtering on the resulting image of the previous step is proposed because there are objects detected in the binarization that not correspond to cell nucleus.

This step is applied for filter the objects that do not correspond cell nucleus and for the separation of objects superimposed. In the filtering process, the objects whose sizes are below or above to the rank of typical values corresponding to cell nuclei are removed. A proposed to achieve this algorithm is described:

i. The closing area operator is applied to fill any internal region of a cell that not fully detected [10,11].
ii. The internal gaps are eliminated using opening area operator [10,11].
iii. Finally, the cell nuclei superimposed are divided using the Watershed Transform [10-12].

Figure 1 show the results of this first stage. Figure 1-(b) shows an enlarged section of a binary image and the result of the filtering, which have been removed unwanted objects.

 # Step 2: Determination of spacial position

This analysis is designed to determine the spatial position of proliferated cell nuclei. The center of area of each cell nucleus corresponding to the center of mass of the objects detected in the binarization is determined. The information is stored as vectorized and the image of Figure 1-(d) is obtained. Then, the numbers of nuclei were counted.

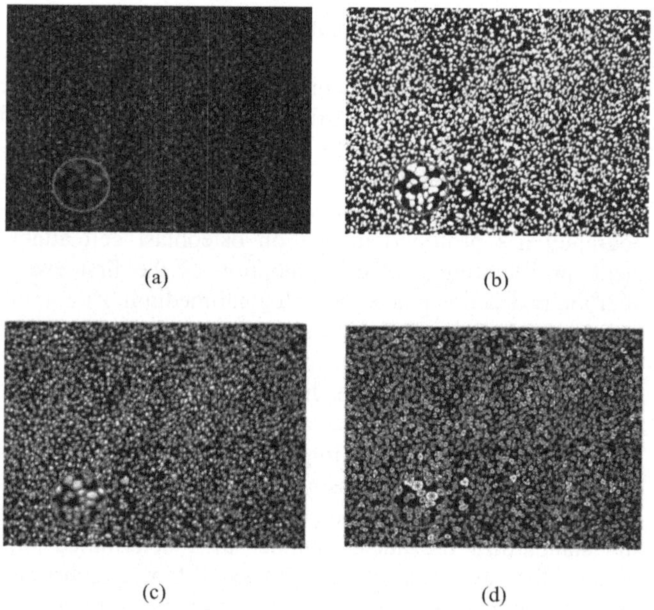

Fig. 1 Segmentation of the proliferation cell images. (a) Original image. (b) Binary image. (c) Overlapping of the segmentation. (d) Binary image with the center of nucleus.

III. RESULTS AND DISCUSION

Figure 2 shows MC3T3-E1 attachment and proliferation behavior three and five days after the cells were seeded over the materials (control and anodization ones) as a measure of cytotoxicity. There is a homogenous nuclei distribution along the implants Zr0 (without treatment), Zr30 (with anodized 30V) and Zr60 (with anodized 60V). It is possible to note a significant increased in the number of fluorescent nuclei five days after the cells were seeded over the materials. The quantitative analysis of the number of cell nuclei can be seen in Figure 3 and 4. Three days after cells were seeded over the materials, it is possible to note that anodization treatment affect the first contact events with the cells, showing a reduction in the number of cell nuclei attached to the surface both in Zr30 and Zr60v materials. Nevertheless, five days after cell were seeded, the number of nuclei increased significantly in control and anodized materials reaching comparable among them. At this time, there is a notorious dispersion of the results obtained over Zr 0V samples which could be awared to the material surface conditions and this dispersion is increased as the culture growth showing that cells behavior is strongly dependent of the surface characteristics.

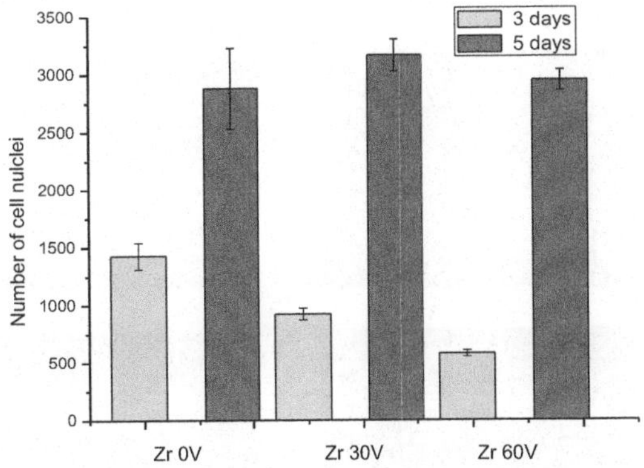

Fig. 2 Column plot of the anodized Zr surfaces with the MC3C3-E1 cell proliferation after 3 and 5 days

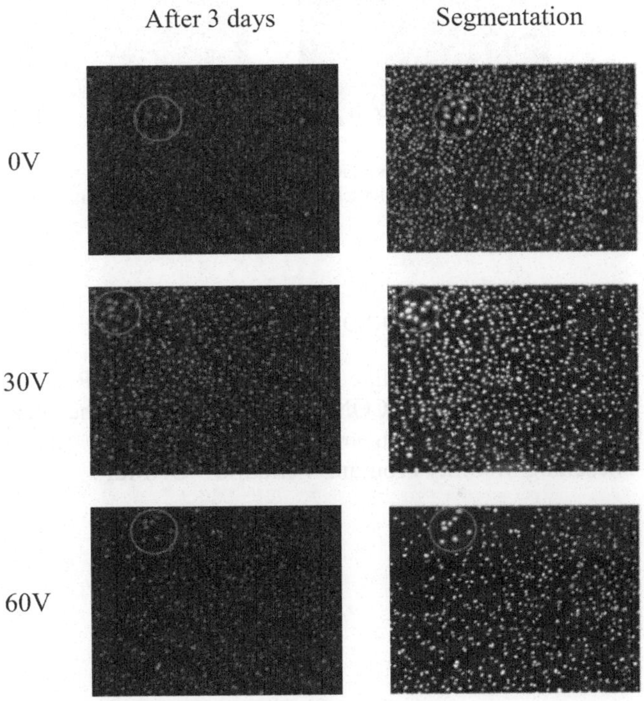

Fig. 3 Images of the metallic surface of anodized zirconium with the proliferation of MC3C3-E1 cell after 3 days and the nuclei segmentation.

IV. CONCLUSIONS

Our in vitro results highlight that anodization treatment has a direct effect over cell attachment and proliferation at early stage. Nevertheless five days after MC3T3-E1 cells were seeded over the materials, proliferation increased significantly on control and anodization materials. The results showed an interesting relationship between cell differentiation and material surface.

The results obtained by the techniques of Processing Digital Images indicate that it is possible to obtain the number of nucleus automatically and accurately. The algorithms used for analysis were implemented in a toolbox for Matlab, to facilitate reuse and distribution. The continuity of this work involves the application of separation steps or very close overlapping objects, and further analysis of the spacial distribution of the nucleus. It also raises generate a graphical interface to facilitate the work of the experts.

As a future work we propose compare our method in front of other segmentation strategy available in the literature.

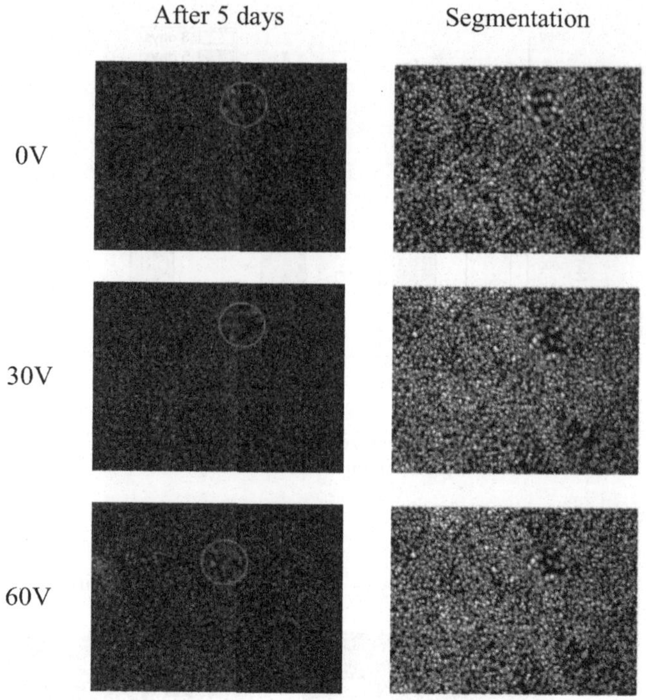

Fig. 4 Images of the metallic surface of anodized zirconium with the proliferation of MC3C3-E1 cell after 5 days and the nuclei segmentation.

Acknowledgment

The authors would like to thank the National Research Council of Argentina (CONICET) PIP 0434, for the financial support. Dr. Katunar would like to thank the 6th Theaching Mobility Program - Madrid – UNMdP.

References

1. Keller JC, Schneider GB, Standford CM, Kellog Bradley. (2003) Effects of implant microtopography on osteoblast cell attachment. Implant Dentistry, 12:175-181.
2. Galli C, Piemontese M, Ravanetti F, Lumetti S, Passeri G, Gandolfini M, Macaluso GM. (2012) Effect of surface treatment on cell responses to grades 4 and 5 titanium for orthodontic mini-implants. Am J Orthod Dentofacial Orthop, 141(6):705-14. doi: 10.1016/j.ajodo.2011.12.022.
3. Hoerth RM, Katunar MR, Gomez Sanchez A, Orellano JC, Ceré SM, Wagermaier W, Ballarre J. (2014) A comparative study of zirconium and titanium implants in rat: osseointegration and bone material quality. J Mater Sci Mater Med, 25(2):411-22. doi: 10.1007/s10856-013-5074-3.
4. Katunar MR, Gomez Sanchez A, Ballarre J, Baca M, Vottola C, Orellano JC, Schell H, Duffo G, Cere S. (2014) Can anodised zirconium implants stimulate bone formation? Preliminary study in rat model. Progress in Biomaterials, 3(24):1-10.. doi: 10.1007/s40204-014-0024-9.
5. Pham DL, Xu C, Prince J L. (2000) Current methods in medical image segmentation. Annual Review of Biomedical Engineering, 2:315-37. doi: 10.1146/annurev.bioeng.2.1.315.
6. Katunar MR, Gomez-Sanchez A, Pellejero J, Ballarre J, Santos-Coquillat A, Martinez-Campos E, Civantos A, Baca M, Vottola C, Haddad K, Ramos V and Cere S. (2016) The effect of anodised treatment on zirconium implant at early stages: A preliminary in vivo and in vitro characterization. 10th World Biomaterials Congress (WBC), which will be held in Montreal May 17-22.
7. Otsu N. (1979) A threshold selection method from gray-level histogram. EEE Trans. Syst.Man Cybern, SMC-9(1): 62-6.
8. Mehmet S and Bulent S. (2004) Survey over image thresholding techniques and quantitative performance evaluation. Journal of Electronic Imaging, 13:146-65.
9. Brun M, Pastore JI, Gonzalez MA, Cacace JP, Rapacioli M, Flores V, Ballarin VL. (2013) Caracterización de la dinámica de la proliferación celular en el sistema nervioso central en desarrollo utilizando procesamiento digital de imágenes. XIX Congreso Argentino de Bioingeniería. San Miguel de Tucumán, Tucumán, Argentina. ISBN: 978-987-23950-7-0.
10. Serra, J. (1982) Image analysis and mathematical morphology, vol 1, Academic Press, London.
11. Serra J. (1988) Image analysis and mathematical morphology, vol 2, Academic Press, London.

Author: Juan Ignacio Pastore
Institute: National University of Mar del Plata
Street: Juan B. Justo 4302
City: Mar del Plata
Country: Argentina
Email: jpastore@fi.mdp.edu.ar

Cost Estimate Methodology in procurement processes of ME

Fagundes, V. O.[1]; Zaniboni, R.[1]; Garcia, R.[1]

[1] Instituto de Engenharia Biomédica, Universidade Federal de Santa Catarina, Florianópolis, Santa Catarina, Brasil

Abstract: **For procurement process in Health Technology Management (HTM), after the step of planning, design and specification that will meet the clinical demand, we seek to set the amount to be spent on its procurement. Following, in public institutions, after the application approval, bidding process starts for the procurement, which must fully comply with the laws governing the Public Bidding. For the technology incorporation in the health unit applicant, usually the process is quite lengthy due to its complexity, however, if well planned, budgeted and defined, its chances of achieving success within a reasonable time is much higher. In this paper, we will consider in the budget definition of Medical Equipment (ME) to be acquired, which in most cases becomes the villain of delays justifications. In this context, this paper aims to present a new methodology for defining the estimated value / cost that will be used in a future bidding process, aiming at greater effectiveness of these processes and the reduction of the implementation period, which cause a lot of damage to the bidder body, the health unit and the citizen.**

Keywords — **Clinical engineering, Health Technology Management, Medical Equipment, cost estimation, Health Care Facility.**

I. Introduction

In present days due to the dynamism and volume that the activities are performed, a control over management action is required within a group that seeks an organization dedicated to increase the profit on a determined product. Thus, there are cost methods that aims to minimize costs and maximize profits, according to Porter [5] this means "deliberately choose a different set of activities to deliver a unique mix of values."

Usually when developing a technical specification (TS) of Medical Equipment (ME) to meet clinical demand of a Health Care Facility (HCF), are sought-in the market technologies required in TS. Then is requested to the suppliers to submit a budget of their equipment and by considering these budgeted amounts it is presented the estimated cost for use in the public notice. Another way often used in the Health Technology Management (HTM) is the calculation performed by the arithmetic mean of the values quoted in a previous bidding process; however, the result is very inaccurate because of the wide range of values and usually carries old process addictions. In both methodologies the estimated value information does not take into consideration the other requirements that incorporate the bidding process and especially the necessary time from the estimated value definition up to the procurement conclusion with the payment to the homologated provider in the bidding process.

Given this finding and the experience in HTM in the HCF [9,10] along with the Santa Catarina State of Health Bureau (SES/SC), it was identified that undoubtedly the definition of the estimated value of ME to be acquired is one of the main factors that have influenced the huge delay in the processes and in many cases making it desert or finalized without the completion of the good procurement.

So with the objective of developing a methodology for the definition of ME cost that would bring benefits in the agility of the bidding process, to be employed in the Dimensioning and Incorporation of Technology (DIT) for the field of Clinical Engineering (CE) in the Institute of Biomedical Engineering (IEB-UFSC), it was conducted the present study.

In this paper, initially it presents traditional methods and defined by the study to meet the definition of the value of ME. In sequence, it compares the value estimated by the methods used before and after the study to low technology and high added value, comparing percentage according to the methodology used, ending with the benefits of this new methodology and the need for further comparative studies seeking further refine the estimate of the future value.

II. Methodology

Furthermore, is a brief definition of the methods traditionally established, such as: Absorption Costing, Variable Costing, Standard Costing, Activity-based Costing (ABC) to analyze which of the methods would best apply or adapt to the Clinical Engineering activities at IEB-UFSC.

Absorption Costing: "Characterized by the appropriation of all the internal operational cycle costs to the final carriers' costs" [2], i.e. all production efforts are distributed to all products.

Variable Costing: Attributes to each cost a specific classification in the form of fixed or variable costs. Lopes de Sá says that the variable costing is "the cost counting process that excludes fixed costs" [3].

Standard Costing: "The standard cost is a pre-assigned cost, taken as basis for the production registration before determination of the effective cost" [8]. So it would be the cost that should be pursued in full efficiency and maximum yield.

I. Torres et al. (eds.), *VII Latin American Congress on Biomedical Engineering CLAIB 2016, Bucaramanga, Santander, Colombia, October 26th -28th, 2016*,
IFMBE Proceedings 60,
DOI: 10.1007/978-981-10-4086-3_45

Activity-based Costing (ABC): This costing system aims to supply the need for more accurate information on the cost of each activity involved in the production processes and the generation of services. "The ultimate goal of ABC is to create information for a decision to improve the competitiveness of enterprises" [4].

A. Applied costing at DIT on IEB-UFSC

Applied costing at DIT of IEB-UFSC: The fact that all methods have advantages and disadvantages between them, and none of them would be directly applied or adapted to this reality; key ideas of all methods were extracted. For example, the idea of involving total costs (Absorption Costing), focus on unit cost (Variable Costing), use the maximum efficiency (Standard Costing) and a constant database updates (ABC Costing).

The methodology employed aims to separate in the database the ME initially approved or rejected, as long as acquired under the same technical specification. For example, assuming that a public notice for procurement of General Surgical Table there are three approved proposals, assuming that all of them will work at its maximum efficiency, only one will be the winner of the bidding process/public notice, the lowest cost compared to the others. While on the other hand, there is a General Surgical Table that was disapproved for some reason, such as: excessive cost; incompatibility of technical specifications; lack of sample (in case of brand and/or equipment unknown by the medical staff); or still if it failed the clinical judgment. Thus, this method to separate the approved and disapproved, serves as the first filter to enable the estimated value for the public notice, because in this universe of approved and disapproved, only the approved ones will be considered.

To be able to estimate a more accurate value, it will be examined the bidding processes history in which the studied equipment had the declared winning bid and it will be made a filter which will consist in stopping the entry of devices approved with cost higher than 30% of the winning product. Therefore, it will be used for calculation purposes of this methodology, the value of equipment which are in the database with costs up to 30% higher than the value of the winning bid.

After the insertion of these two filters, it is necessary to weigh in this information to be able to apply to all equipment. Therefore, mathematical methods will be used, such as geometric mean, because then no data will dominate the sample, normalizing the scope of the estimated value. This mean is applied only to approved and that do not exceed the amount of 30% of the winning equipment value.

To increment and reach an estimated value closer to the reality of the technology supplier market, that notwithstanding seek to increase their profit margins, sums up to the average of the dollar variation and the inflation of the period following the closing of the bidding process (from which the cost data were taken – Table 1) until the estimated cost of ME setting date, both applied on the geometric mean of the product values under the conditions stipulated in this methodology.

The necessity to increase the value with inflation is due to the fact that this "consists of the persistent and widespread increases in costs of goods and available services to society" [1]. And to try to correct or tend to neutralize the cost discrepancy is performed monetary correction, which results in increasing the cost of medical equipment.

Finally, as the dollar exchange rate varies from day to day, to streamline the quotation update, is computed an average of the period so that this variation is considered in the estimated final cost.

III. Results and Considerations

To verify the accuracy of the methodology with respect to the first quoted model, a comparison will be made of these methodologies.

The equation of the methodology of mean cost:

$$P_f = \frac{P_1+P_2+P_3+P_{(n-1)}}{n-1} \tag{1}$$

Being:

- $\mathbf{P_f}$: Final cost.
- $\mathbf{P_1, P_2, P_3, P_{(n-1)}}$: Equipment Cost;
- **n-1:** Quantity of equipment.

The equation developed for the new methodology applied:

$$P_f = M_g + (INF_v \times M_g) + (DOL_v \times M_g) \tag{2}$$

$$P_f = M_g x(1 + INF_v + DOL_v)$$

Being:

- $\mathbf{P_f}$: Final cost;
- $\mathbf{M_g}$: Geometric mean with the costs within the stipulated conditions;
- $\mathbf{INF_v}$: Inflation variation in the period;
- $\mathbf{DOL_v}$: Dollar exchange rate in the period.

Using two cases for comparison, the first in which costs have a great variation, but with a low earned value, and the second when there is little cost difference, with a high earned value equipment.

Case I: Great cost variation, low earned value.

Analyzing the following data (Table 1) taken from a procurement made in December 2014, and applying the filter using the value of the winning bid plus 30%, we have a limiting value of R$ 260.00 (R$ = Brazilian Real), therefore it will only be considered for the purpose of calculation values up to R$ 260.00.

Table 1 Table of costs for products with low earned value and great variation. (R$ = Brazilian Real)

Brand	Cost (R$)	Situation
A	200,00	Approved/Winner
B	230,00	Approved
C	270,00	Approved
D	1.000,00	Approved
E	1.055,00	Approved
F	1.100,00	Approved

In addition to this application condition, we still have two data to be used in the formula for calculating the estimated value of the ME. Thus, using the dollar exchange rate extracted data in the period between the quotation/bidding process and the definition date of the estimated value (January 2015) the variation is "6.73%" [6] and also the inflation index in the same period - IPCA in the average period of 1.24% [7], so:

$$P_f = (\sqrt{200 \times 230})\, x(1 + 0{,}0124 + 0{,}0673) \quad (3)$$

$$\mathbf{P_f = 214{,}47\ x\ 1{,}0797}$$

$$\mathbf{P_f = R\$\ 231{,}57}$$

While using the overall average, so:

$$M_a = \frac{(200+230+270+1000+1055+1100)}{6} \quad (4)$$

$$\mathbf{M_a = R\$\ 642{,}50}$$

When comparing the methodologies, it is observed an extremely large reduction in the estimated value between the methodologies: R$ 231.57 to R$ 642.50, i.e. almost twice the highest value (1.8x).

As the equipment has a low earned value, normally several units are bought, and with this difference in the definition of the item's estimated value to be purchased would have a fairly high amount at the end of the process as it would allow companies to quote their costs above the actual market value. Thus, applying this new methodology this fact would not happen generating more competition between companies and savings for the health system.

$$E_{\%} = \frac{(231{,}57-642{,}50)}{642{,}50} \times 100 = -63{,}96\% \quad (5)$$

Case II: Little cost difference, with a high earned value.

Analyzing the following data (Table 2) and by applying the limiting filter using the winning bidder value increased by 30%, thereby the border value will be R$ 1,443,000.00.

Table 2 Table of costs for products with high earned value and little variation

Brand	Cost (R$)	Situation
X	1.110.000,00	Approved /Winner
Y	1.170.000,00	Approved
Z	1.200.000,00	Approved

Using the extracted data from 2015 variation of dollar exchange rate "6.73%" and the inflation index - IPCA of 1.24%.

$$P_f = M_g + (INF_v \times M_g) + (DOL_v \times M_g) \quad (6)$$

$$P_f = M_g x\ (1 + INF_v + DOL_v)$$

$$M_g = \sqrt[3]{1.110.000 \times 1.170.000 \times 1.200.000}$$

$$\mathbf{M_g = 1.159.391{,}28}$$

$$\mathbf{P_f} = 1.159.391{,}28\ x(1 + 0{,}0124 + 0{,}0673)$$

$$\mathbf{Pf = R\$\ 1.251.794{,}76}$$

Comparing with the methodology of overall average:

$$M_a = \frac{1.110.000+1.170.000+1.200.000}{3} \quad (7)$$

$$\mathbf{M_a = R\$\ 1.160.000{,}00}$$

Differently from the previous case, the arithmetic mean showed a lower value, however, the estimated cost does not contain information referring to the economy dynamism, fact that a device with cost of this magnitude (more than R$ 1 million), would result in relevant cost changes, for higher or lower depending on the variation of the American currency and inflation. It can be noticed a percentage difference of 7.91%, although it is a difference of average magnitude, due to a high equipment value which would generate a cost difference of R$ 91,794.76 being a fairly high cost variation, from the estimated costs.

$$E_{\%} = \frac{(1.251.794{,}76-1.160.000{,}00)}{1.160.000{,}00} \times 100 \quad (8)$$

$$E_{\%} = 7{,}91\%$$

IV. Final considerations

By analyzing the first case with low earned value equipment, but with a high variation it was observed a percentage reduction of -63.96% in estimated value, having an estimated final cost variation of R$ 410.93. Therefore, to estimate the value of the equipment by the simple average in relation to the costs quoted in the last bidding process on the same product, allows companies to participate with high costs proposals for such equipment what may cause huge damage to health facilities.

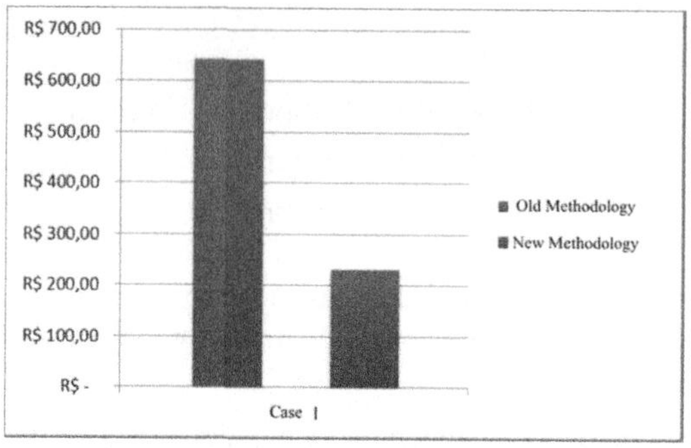

Graph 1 - Methodology comparison, on the first case

While in the second case resulted in a much smaller percentage change than in the first case, but still considerable of 7.91%, and the difference in the estimated cost between methodologies R$ 91,794.76, i.e. a high magnitude value, therefore if not taken into consideration in a public notice, this value with the weighted average may result that the process do not have participants or their costs be above the estimated, making it deserted and the process ineffective, and also the hospital unit that needs the equipment functionality will have to wait much longer to acquire it.

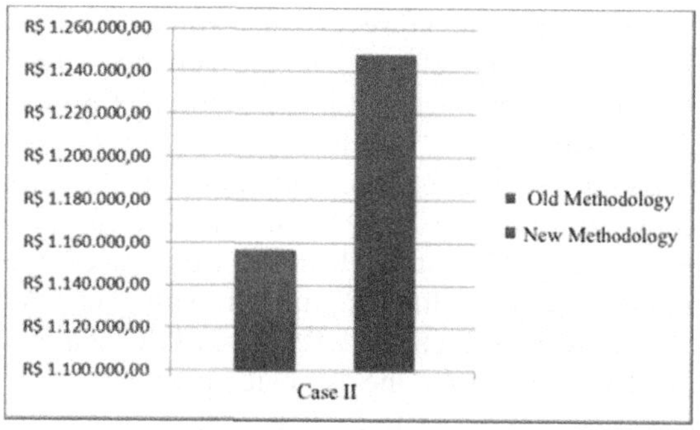

Graph 2 - Methodology comparison, on the second case

V. Conclusions

Considering the presented results, it can be observed that the new methodology will be able to streamline and qualify advisory activities in DIT, specifically the appropriate setting of the estimated value of ME. So with this new method based on the database of quoted costs in past bidding processes, will be available to the institute as a tool to minimize a major problem in the effectiveness of the bidding process which is related in the estimated value of ME definition, thereby reducing the time taken currently in the process of procurement through public notices and assisting in reducing the time to provision of the required technology to health facilities.

The estimated value of ME depends on several factors involving the market, such as: economic, financial, political (internal and external) and social, among others. Therefore, is important to seek for continuous improvement in the presented method, based on new comparative case studies and research from other models of financial evaluation, as well as new formulas combining and adding other parameters to the definition of the estimated cost, since the purpose is to seek more and more the approach of the product estimated cost (forecast) with the quoted market cost in the bidding process (actual).

References

1. BERTOLO, L. INFLAÇÃO. Available at <www.bertolo.pro.br/AdminFin/HTML/INFLACAO.htm>. (Accessed on March 3, 2016).
2. KOLIVER, O. Os Custos dos Portadores Finais e os Sistemas de Custeio. [S. I.: s.n.], 2000.
3. LOPES DE SÁ, A. Dicionário de Contabilidade. 8. ed. São Paulo: Atlas, 1990.
4. MAUSS, C, V, COSTI, R, M. O Método de Custeio ABC como instrumento de gestão. Available at: <aedb.br/seget/arquivos/artigos06/784_Artigo%20Abc_Seget1.pdf>. (Accessed on February 26, 2016).
5. PORTER, M. E. Competição = On competition: estratégias competitivas essenciais. Rio de Janeiro: Campus, 1999.
6. UOL. Câmbio. Available at: <economia.uol.com.br/cotacoes/cambio/dolar-comercial-estados-unidos/?historico>. (Accessed on February 27, 2016).
7. UOL. Inflação de 1,24% em janeiro é a maior em 12 anos e estoura limite da meta. Disponível em: <economia.uol.com.br/noticias/redacao/2015/02/06/inflacao-de-1,24-em-janeiro-e-a-maior-em-12-anos-e-estoura-limite-da-meta.htm>. (Accessed on February 27, 2016).
8. ZANLUCA, J, C. Custos Padrão. Available at: <www.portaldecontabilidade.com.br/tematicas/custospadrao.htm>. (Accessed on February 27, 2016).
9. SANTOS, F.; GARCIA, R. Supporting Decision Making Tool within the process of Replacement and incorporation of New Technology to the Health Area. IFMBE Proceedings, 2009, v.25, p.453-456.
10. HAYASHIDE, M.; ZANIBONI, R.; GARCIA, R. Metodologia de avaliação após a incorporação de Novas Tecnologias em EAS. In: XXII Congresso Brasileiro de Engenharia Biomédica, 2010, Tiradentes/MG. ISSN 2179-3220, 2010. p. 214-217.

Author: Renato Zaniboni
Institute: Instituto de Engenharia Biomédica - UFSC
Street: Campus Universitario
City: Florianópolis/SC
Country: Brasil
Email: zaniboni@ieb.ufsc.br

Evaluation of methods based on conventional videography for detection of gait events

A. Arcila Cano[1], D. Ewins[2], A. Shaheen[2] and P. Catalfamo Formento[1,2,3]

[1]LIRINS. Facultad de Ingeniería, Universidad Nacional de Entre Ríos, Ruta Prov. 11, Km 10, CP 3101, Oro Verde, Entre Ríos.
[2]Centre for Biomedical Engineering, Faculty of Engineering and Physical Sciences, University of Surrey, Guildford, GU2 7XH U.K
[3]CITER Ruta Prov. 11, Km 10, CP 3101, Oro Verde, Entre Ríos

Abstract— **The detection of gait events plays an important role in gait analysis. The initial foot contact (IC) and the end of foot contact or foot off (FO) represent the start of stance and swing, respectively. They are commonly used to compute parameters that offer information for the characterization of gait.**

To determine these events, kinematic methods that implement optical systems, including conventional video cameras, have been suggested.

The key objective of this study was to evaluate the performance of two event detection methods, one visual (VM) and one semiautomatic (SAM), which use conventional video recordings, for the visual detection of IC and FO, during level ground and stair walking. The methods were tested in 10 healthy subjects, wearing confortable shoes and walking at a self-selected speed. The accuracy of the proposed methods was compared with force plate data (gold standard). Furthermore, the VM and the SAM were compared with an alternative kinematic method: The Velocity-Algorithm (VA), previously described in the literature.

The results of the level ground test showed that the VM and the SAM detected IC and FO with an absolute mean difference (AMD) lower than 30 ms.

For the stair climbing test, it was found that the VM and the SAM detected IC and FO with an AMD lower than 55 ms.

In general terms, the VM and the SAM produced differences within the ranges reported in the literature for both terrains.

These results provide preliminary evidence supporting the use of both methods for gait event detection on level walking as well as stair climbing.

Keywords— **Gait analysis, gait event, detection methods, level ground walking, stair climbing.**

I. Introduction

The detection of gait events plays an important role in gait analysis. The initial foot contact (IC) and the end of foot contact or foot off (FO) are commonly used to compute spatio-temporal parameters that offer information for the characterization of gait [1], [2].

Force platforms represent the gold standard method for determining gait events. However, despite their recognized accuracy, the method is usually restricted to gait laboratory environments and the number of force platforms available limits the number of steps per trial that can be recorded [2].

The investigation of alternative methods that could provide portable and reliable measurements is an ongoing area of research. A number of methods, including kinematic methods using optical systems [3-6], foot switches [7], pressure matrices [8], pressure insoles [2], accelerometers [9] and gyroscopes [1], have been suggested and evaluated.

Conventional video cameras are commonly used to provide a qualitative recording of walking during gait analysis. They are easy to use, portable, can be installed in small physical areas, and their cost is relatively low, therefore they could be considered for gait event detection in the clinical setting. A previous pilot study showed promising results for the use of conventional video cameras in the detection of gait events [10] on level ground. That study, investigated the accuracy and reliability for different raters of two event detection methods, the visual (VM) and the semiautomatic (SAM), using conventional video cameras in subjects walking on level ground.

While level ground walking usually represents the basic scenario for evaluation, rehabilitation programs for lower limb amputees normally include stair climbing training as the final stage of the protocol [11]. So, for a detection algorithm to be used in clinical settings, its performance on stair walking should also be considered.

Despite its importance for the clinical practice, few studies have used stairs for gait event detection [11], [12] and the evaluation of methods, using conventional video cameras, for detection of gait events on stairs is still pending. The key objectives of the present study were to extend the evaluation of both event detection algorithms presented in [10], for the visual detection of IC and FO to a larger number of subjects, while walking on level ground, and to evaluate the performance of the algorithms for subjects walking on stairs.

I. Torres et al. (eds.), *VII Latin American Congress on Biomedical Engineering CLAIB 2016, Bucaramanga, Santander, Colombia, October 26th -28th, 2016*,
IFMBE Proceedings 60,
DOI: 10.1007/978-981-10-4086-3_46

II. Materials and Methods

i) Subjects

Ten healthy subjects, 6 men and 4 women (age 32.2 ± 10.5 years, weight 71.7 ± 17.2 kg, height 1.69 ± 0.09 m) without motor impairments, participated in the study. All participants were informed of the objectives and methods of the study and gave their informed consent for participation. The study protocol was approved by the University of Surrey Ethics Committee.

ii) Data Acquisition Protocol

All subjects wore comfortable shoes and walked at a self-selected speed, both on level ground and stairs

Kinematic data was acquired using the ProReflex system, (ProReflex, Qualisys Medical AB, Sweden) operating at frequencies of 200 Hz for level ground tests and 100 Hz for stair climbing tests. Markers were placed on the heel, between the second and third metatarsal head and on the sacrum.

On level ground tests, kinematic data was acquired simultaneously with data measured by two force plates (AMTI 400600HF-2000, Advanced Mechanical Technology Inc., U.S.).

A video camera (Panasonic model NV DS 15), with a sampling frequency of 25 Hz, positioned on a tripod one metre high and three metres away from the walking path recorded the level ground trials. The video camera was positioned at a height of 1.5 metres and three metres away from the sagittal plane of the stairs.

Time synchronization of the equipment was performed visually using data from the ProReflex system (synchronized with the force platform) and from the conventional video. For synchronization, the IC previous to the first step taken on the force platform was used for level ground trials and the IC previous to the first step on the stairs was used for stair trials.

During level ground trials, participants were instructed to walk on a path that included two force platforms (AMTI, U.S). The participants performed at least six successful walks on the path, with clean force plate contacts. During stair climbing tests, the subjects went up and down a staircase, which included 6 steps, performing at least twelve walks.

iii) Methods used for detection

Event detection was accomplished using two reference algorithms and the proposed methods. The reference methods used data from the force platform (gold standard), and kinematic data (Velocity-based Algorithm (VA)), previously described in the literature [6]. The proposed methods used conventional videos from video cameras.

All data was analyzed using MATLAB® (Mathworld Inc. Student Version 7.12.0).

1. Gold Standard

For detection of the events, the vertical force recorded from the force platforms was used, with a threshold set at 10 (N) [4], [5]. When the vertical force signal exceeded this threshold, the routine detected IC and when it fell below this threshold, FO was detected.

2. Velocity-based Algorithm (VA)

The VA uses the change in the direction of the velocity of the foot markers, to determine the IC and FO [6]. The algorithm uses the displacement in the direction of progression of the markers of the heel, toe and sacrum. It determines the coordinates for the markers of the heel and the toe relative to the sacrum, such that:

$$Heel_{relative} = X_{heel} - X_{sacrum}$$
$$Toe_{relative} = X_{toe} - X_{sacrum} \quad (1)$$

Linear velocities for $Heel_{relative}$ and $Toe_{relative}$ are then calculated.

To determine the IC, all zero crossings of the velocity vector of $Heel_{relative}$, are detected. IC is calculated as the first sample after the signal exceeds the zero value.

To determine the FO, all zero crossings of the velocity vector of $Toe_{relative}$, are detected. FO is calculated as the first sample after the signal falls below the zero value.

For stair walking evaluation, caution should be taken especially during IC detection. This is important because unlike walking on level surfaces, on stairs IC tends to occur with the forefoot.

3. Visual Event Detection Method (VM)

Visual detection was performed by direct observation of the video, without using any tools provided by the software (VirtualDub version 1.9.11) [10]. The frame in which the foot makes contact with the floor after the swing phase was recorded as the IC and the frame after which the foot leaves the ground after the stance phase was recorded as FO. All events were evaluated by the same researcher, although good intra and inter rater repeatability has been established for this method [10].

4. Semi-Automatic Event Detection Method (SAM)

Using tools provided by the software (Kinovea version 0.8.5), the positions of two markers were tracked [10]. The markers were located on the heel and on the toe of the shoes used by the subject. The software provides the possibility of reconstructing automatically or manually the trajectory of the markers. The trajectories were then exported to a data sheet as position coordinates (x, y) for each frame. The time of the first minimum value of "y" (height) of the heel marker (indicating the lowest vertical position of the trajectory of the marker) was used to define IC. The time immediately after the minimum value of "y" from the toe marker (indicating the upward movement of the foot after stance) was used to define FO.

During stair walking, IC tends to occur with the forefoot, so only the information of the toe marker is used.

iv) Data Analysis

The same number of events was used for each subject, to avoid any bias due to the difference in the number of events detected.

Once the events were determined for each method, during level ground trials, the comparison between the kinematic methods (m: VA, VM or SAM) and the gold standard (platform), was carried out by calculating the difference in time (ms) between detections for each step analysed and each subject (s: subject) (equation 2).

$$\Delta IC_{m,s} = IC_{m,s} - IC_s\,plat \qquad (2)$$
$$\Delta FO_{m,s} = FO_{m,s} - FO_s\,plat$$

During stair climbing tests, the differences between the proposed methods (m: VM or SAM) and the reference method (VA) were calculated for each event and for each subject (s: subject) (equation 3).

$$\Delta IC_{m,s} = IC_s\,ref - IC_{m,s} \qquad (3)$$
$$\Delta FO_{m,s} = FO_s\,ref - FO_{m,s}$$

The mean differences (MD) and absolute mean differences (AMD) were calculated for each of the algorithms.

III. Results and discussion

A. Accuracy evaluation for level ground tests

During level ground trials, 6 IC and 6 FO from each subject were analysed. Data from Subject 2 could not be used due to technical problems in data collection.

Table 1 shows the MD and AMD for both events and methods, for subjects walking on level ground. The results showed that all the methods detected IC and FO with an AMD lower than 30 ms.

For IC, results show that the reference method (VA) showed larger differences with respect to the platform than the proposed methods (VM and SAM). Also, there is a negative tendency in detection of IC for all methods, indicating that they detected the event earlier than the platform.

For FO, there is a positive tendency in detection, indicating that all methods detected the event later than the platform. MD and AMD were similar for VM and SAM for both events.

The results obtained for the VA, were comparable to those obtained in the original publication [6]. The original article [6] reported that the VA detected IC 23 ms after the force platform and FO 8 ms before. In the present study, although the magnitudes were very close, the IC is detected before the reference method and the FO later (25 ms before the force platform for the IC and 12 ms later for FO). Those differences may be due to methodological factors that varied in the original study and in the present investigation.

Furthermore, for all cases, the differences were within the ranges reported in the literature for other kinematic methods [3-6].

Table 1. Mean and standard deviation (SD) of MD and AMD in the determination of gait event times (IC and FO) for the three algorithms (VA, VM and, SAM) applied to the steps of healthy subjects (n= 9, steps= 54), during level ground tests.

		IC			FO		
		VA	VM	SAM	VA	VM	SAM
MD [ms]	Mean	-25	-5	-7	12	12	12
	SD	12	16	16	7	16	17
AMD [ms]	Mean	26	14	15	13	16	16
	SD	12	10	10	6	12	13

B. Accuracy evaluation for stair climbing tests

During stair climbing tests, 12 IC and 12 FO were analysed per subject, so a total of 120 IC and FO events were detected.

Table 2 shows the MD and AMD for both events and methods evaluated. The results showed that the VM and the SAM detected IC with an AMD lower than 55 ms and FO with an AMD lower than 20 ms, related to VA.

Also, there is a negative tendency in detection of IC, indicating that the VA detected the events earlier than the VM and SAM, which is consistent with the results obtained on level ground (Table 1).

The differences for FO were smaller than for IC, with a negative tendency in detection.

The MD and AMD, in particular during IC detection, were higher for stair climbing tests, than level ground. This may be due to the reference methods used (the VA during stair walking and the Gold Standard, using data from the force plate, during level ground walking); also the VA was not originally proposed for the evaluation of gait events during stair walking.

Table 2 shows that VM and SAM show similar differences with respect to the reference method, for both events.

The proposed methods presented MD and AMD within the ones reported in the literature [11], [12]. They showed similar behavior for both terrains, so the choice of using one or the other will depend on the application. VM is a simple method that will provide basic information for the calculation of temporal parameters. SAM is more time consuming but can provide the temporal trajectories of the markers, which could be used for calculation of other gait parameters.

Table 2. Mean and standard deviation (SD) of MD and AMD in the determination of gait event times (IC and FO) for the VM and SAM, applied to the steps of healthy subjects (n= 10, steps= 120), during stair climbing tests.

		IC		FO	
		VA-VM	VA-SAM	VA-VM	VA-SAM
MD [ms]	Mean	-50	-46	-6	-5
	SD	31	29	17	17
AMD [ms]	Mean	50	46	14	14
	SD	31	29	11	11

IV. Conclusions

The present work is part of a research line that aims at implementing Human Movement Analysis tools in the clinical practice.

The final purpose of the present study is then focused on the implementation of the proposed systems to calculate parameters that facilitate the characterization of the gait and as a tool to follow the process of rehabilitation of patients with lower limb amputations in public and private institutions. In this regard, the evaluation of gait events on different surfaces, including stairs, implies a starting point for further research.

Conventional video cameras are relatively inexpensive, easy to use and portable, making them suitable for clinical environments.

The proposed methods for event detection using conventional video cameras have generated results (MD and AMD) that are within the ranges reported in the literature for kinematic methods using optical systems, for both terrains, despite the low sampling rate of the video camera.

Both methods showed similar results for both terrains, indicating that, although further work is required, they could be used interchangeably for the detection of gait events and the method of choice will depend on the application.

The VM provides with basic information regarding only time occurrence of events and the process of detection is faster than for the SAM. If the final objective of the procedure is the calculation of temporal parameters, then this method would represent the best choice. The SAM method provides also with the paths of the markers during the trial. Although it is more time consuming, it could be used if the final objective also includes other biomechanical parameters such as joint angles.

References

1. Catalfamo P, Ghoussayni S et al. Gait event detection on level ground and incline walking using a rate gyroscope. *Sensors (Basel)*. 2010; 10:5683-5702.
2. Catalfamo P, Moser D et al. Detection of gait events using an F-Scan in-shoe pressure measurement system. *Gait Posture*. 2008; 28:420-426.
3. Desailly E, Daniel Y et al. Foot contact event detection using kinematic data in cerebral palsy children and normal adults gait. *Gait Posture*. 2009; 29:76-80.
4. Hreljac A and Marshall R N. Algorithms to determine event timing during normal walking using kinematic data. *J Biomech*. 2000; 33:783-786.
5. Mickelborough J, van der Linden M L et al. Validity and reliability of a kinematic protocol for determining foot contact events. *Gait Posture*. 2000; 11:32-37.
6. Zeni J A, Richards J G et al. Two simple methods for determining gait events during treadmill and overground walking using kinematic data. *Gait Posture*. 2008; 27:710-714.
7. Hanlon M and Anderson R. Real-time gait event detection using wearable sensors. *Gait Posture*. 2009; 30:523-527.
8. Webster K E, Feller J A et al. Validity of the GAITRite walkway system for the measurement of averaged and individual step parameters of gait. *Gait Posture*. 2005; 22:317-321.
9. Sant'anna A and Wickstrom N. A symbol-based approach to gait analysis from acceleration signals: identification and detection of gait events and a new measure of gait symmetry. *IEEE Trans Inf Technol Biomed*. 2010; 14:1180-1187.
10. Peterson M V, Ewins D et al. Evaluation of methods based on conventional videography for detection of gait events. *IFMBE Proceedings*. 2015; 49: 234-237.
11. Catalfamo P, Acevedo R et al. Gait event detection during stair walking using a rate gyroscope. *Sensors*. 2014; 14:5470-5485.
12. Bergmann J H M, Mayagoitia R E et al. A novel method for determining ground-referenced contacts during stair ascent: comparing relative hip position to quiet standing hip height. *Gait Posture*. 2010; 31:164-168.

Author: Andrea Arcila Cano
Institute: FIUNER
Street: Ruta Prov. 11, Km 10
City: Oro verde
Country: Argentina
Email: andrea.arcila.c@hotmail.com

Brain Functional Connectivity in Parkinson's disease – EEG resting analysis

J. Carmona[1], J. Suarez[2] and J. Ochoa[2]

[1] Grupo de Neurociencias de Antioquia - GNA, Universidad de Antioquia UdeA, Sede de Investigación Universitaria - SIU; Calle 62 No. 52-59, Medellín, Colombia

[2] Bioinstrumentation and Clinical Engineering Research Group - GIBIC, Bioengineering Department, Engineering Faculty, Universidad de Antioquia UdeA; Calle 70 No. 52-21, Medellín, Colombia

Abstract— **Brain functional connectivity evaluates the statistical dependencies between spatially distributed brain regions. The patterns obtained from this analysis have been related to different cognitive processes and are altered by neurodegenerative diseases. Parkinson's disease (PD) is the second most frequent neurodegenerative disease and presents a combination of motor and cognitive disturbances. In this study the early changes in connectivity patterns in PD are evaluated. EEG was recorded at resting state with eyes closed in twenty-three patients with PD without cognitive decline (PD-CogNL) and twenty three healthy controls. Spectral coherence was estimated between electrode pairs in fronto-parietal and inter-hemispheric regions. Rhythms of interest were: delta (1–4 Hz), theta (4–8 Hz), alpha1 (8–10 Hz), alpha2 (10–13 Hz), beta1 (13–20 Hz), and beta2 (20–30 Hz). Compared to controls, PD-CogNL had an increased coherence in frontal inter-hemispheric electrodes in delta and theta bands. In the fastest bands were found correlations between connectivity and executive function measured by INECO test. The results of this paper show early changes in frontal inter-hemispheric coupling in PD.**

Keywords—**Parkinson's disease, EEG, Functional Connectivity, Resting State, Coherence.**

I. Introduction

Normal cognitive functions are based on the integrated activity between distinct brain regions, and this is attained in essence by synchronization of distributed neuronal populations [1]. These events can be examined by studying interdependencies among signals from different brain regions [2], designated as functional connectivity [3]. Changes in brain connectivity patterns may be identified by different activation arrays within structural and functional networks linked to different cognitive processes [4]. Alterations in cortico-cortical synchronization, obtained by means of neurophysiological measures of functional connectivity, has been found in central nervous system diseases as Alzheimer's disease [5] multiple sclerosis [6], and schizophrenia [7].

Parkinson's disease (PD) is the second most frequent neurodegenerative disease and affects significantly life quality with a heterogeneous combination of motor and cognitive disturbances [8]–[10]. The physiopathological mechanisms in PD lead to a disruption in segregated cortico-thalamic circuits, which would be directly related to the various clinical features of the disease [11]. Changes in the activity of the cortical components of these circuits should alter the normal connectivity patterns with other cortical areas and might be expressed in modulation of cortico-cortical coupling indices. Although scarce, EEG studies in PD subjects show that resting state cortical synchronization measures can differentiate not demented PD individuals from healthy controls [12], and suggest that connectivity indices could be sensitive to different therapeutic interventions [13], [14]. Nevertheless, there is a lack of evidence about the early EEG connectivity changes in PD populations without objective cognition impairments, and it is unclear the association of the electrophysiological signature and the early cognitive dysfunction related to PD [15]. In this paper we analyze functional connectivity at rest in patients with PD classified as cognitively preserved and healthy controls. We sought to determine whether there are early changes in cortico-cortical coupling, and its relationship with executive function, which has been seen to be affected from the beginnings of the disease.

II. Methodology

A. Subjects

Twenty-three patients with PD without cognitive decline (PD-CogNL) and twenty three healthy controls matched for age, gender, and educational level were recruited. PD was diagnosed by an expert neurologist according to the criteria of the United Kingdom PD Society Brain Bank [16]. PD progression stage was determined with the Hoehn & Yahr scale (H&Y) [17], and motor disability was assessed with section III of the Unified Parkinson's Disease Rating Scale (UPDRS-III) [18]. Moreover, all participants underwent an extense neuropsychological evaluation to appraise global cognitive efficiency and to exclude the coexistence of mild cognitive impairment (MCI) or dementia. The patient's cognitive screening was performed using the Montreal Cognitive Assessment (MoCA) [19], a reliable psychometric instrument which has shown to be capable to identify MCI in PD [20], [21], and that has been validated in the Colombian population [22]. The executive function was evaluated with the INECO frontal screening test [23]. Exclusion criteria were the presence of Parkinson-plus symptomatology, other neurological disorders, major psychiatric conditions or current use of psy-

I. Torres et al. (eds.), *VII Latin American Congress on Biomedical Engineering CLAIB 2016, Bucaramanga, Santander, Colombia, October 26th -28th, 2016*,
IFMBE Proceedings 60,
DOI: 10.1007/978-981-10-4086-3_47

choactive drugs that could alter the EEG brain activity. Inclusion criteria for the PD-CogNL group were spared functional independence and a MoCA score of 23 or above. All PD patients were taking antiparkinsonian medication and were evaluated during the "on" phase of medication. Subject characteristics are listed in Table 1. Informed consent for participation was obtained from all subjects according to the protocol approved by the Human Subjects Committee of the Universidad de Antioquia.

Table 1 Subject characteristics (Mean ± Standard Deviation)

Item	PD-CogNL	Control	P-value
N	23	23	
Age (years)	60.74 ± 9.39	61.17 ± 7.93	0.4409
Gender (F/M)	8/15	8/15	
Education (years)	11.78 ± 5.05	12.35 ± 4.58	0.3490
Disease duration (years)	4.6 ± 2.84	N.A	
H&Y	2.1 ± 0.38	N.A	
UPDRS-III	28.3 ± 12.32	N.A	
MoCA	26.57 ± 1.56	27.26 ± 1.51	0.0686
INECO	20.89 ± 2.46	23.13 ± 2.78	0.0030

H&Y: Hoehn and Yahr rating scale. UPDRS-III: motor part of the Unified Parkinson's Disease Rating Scale. MoCA: Montreal Cognitive Assessment. INECO: Instituto de Neurología Cognitiva test. NA: not applicable.

B. EEG acquisition

EEG signals were recorded at awake resting (eyes closed, 5 minutes). The EEG was recorded with 58 tin electrodes (positioned according to the international 10-10 system) using the software and amplifiers Neuroscan (Scan 4.5, SynAmps2). The signal was digitized at a sampling rate of 1000 Hz and filtered online (bandpass filter: 0.05 to 200 Hz, and band reject filter 60 Hz to eliminate noise from the power supply). The reference acquisition consisted of an electrode located on the right earlobe, and an electrode located between Cz and Fz was used as ground.

C. EEG data pre-processing

It was implemented a semi-automated pre-processing pipeline using two MATLAB toolbox: EEGLAB [24] and the standardized early-stage EEG processing pipeline (PREP) [25], which includes a robust reference to average, where bad channels are excluded, and detection and interpolation of bad channels relative to this reference. The data were segmented into 2s epochs; and independent component analysis enhanced by wavelet was used to correct muscular and eye blinks artifacts [26]. Remaining bad epochs were rejected by a procedure based on linear trend, joint probability and kurtosis approach [27].

D. Functional connectivity analysis

Functional connectivity was estimated from spectral coherence function using the multi-taper spectral estimation method [28], available in Chronux MATLAB toolbox [29]. Despite coherence is affected by volume-conduction of remote EEG sources, and only captures the linear component of the functional coupling [30], we decided to use this analysis because EEG coherence is the most common methodological approach used in the study of several clinical applications [31]. The EEG coherence was calculated at all electrode pairs in six frequency bands: delta (1–4 Hz), theta (4–8Hz), alpha1 (8–10Hz), alpha2 (10–13Hz), beta1 (13–20Hz), and beta2 (20–30Hz), for 50 free artifacts epochs, which were randomly selected from each recording. Following previous studies [12], [15], we analyzed only the fronto-parietal functional coupling (intra-hemispheric) and callosal functional coupling (inter-hemispheric). The fronto-parietal electrode pairs of interest were F3–P3, Fz–Pz, and F4–P4, while the inter-hemispheric electrode pairs of interest were F3–F4, C3–C4, and P3–P4 (Fig. 1).

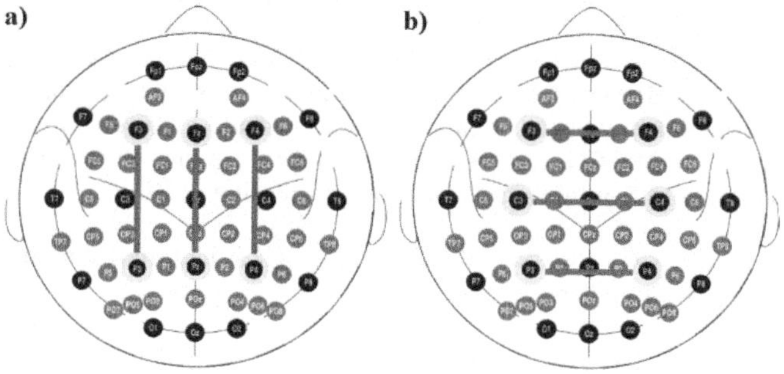

Fig. 1 a) Fronto-parietal (F3–P3, Fz–Pz, F4–P4), and b) Inter-hemispheric (F3–F4, C3–C4, P3–P4) pairs of electrodes. Electrodes positioned according to the International 10–10 System.

E. Statistical analysis

Differences between PD-CogNL and Control groups in connectivity measures for each frequency band were analyzed by means a non-parametric two-sample T-test, that uses permutations of group labels to estimate the null distribution [32]. Statistical significance used was $\alpha = 0.05$. P values were corrected for multiple comparisons using the false discovery rates method [33]. In order to complement P-values, we also estimated measures of effect size to calculate the magnitude of the difference between the groups [34], using the Hedges' g (standardized mean difference) [35]. In addition, we determined if there was any correlation between the functional connectivity and executive function using Spearman's rank correlation coefficient. All statistical analysis were implemented in MATLAB.

III. Results

There were no differences in demographic features (age, gender and education) or MoCA test, which indicates the good general cognitive state of the patients (Table 1). Difference in executive function between groups was found. PD-CogNL patients have a lower INECO scale compared to healthy controls ($p<0.01$). Regarding the connectivity analysis, no differences were found in fronto-parietal connectivity. In inter-hemispheric coupling there were significant differences in delta, theta and alpha2 bands (Table 2). Relative to the Control group, PD-CogNL had an increased coherence in frontal region in delta and theta bands, and a decreased coherence in parietal region in alpha2 band. Table 3 reports the Spearman's rank correlation coefficient between coherence and INECO score in all subjects considered as a whole group. There is a positive relationship with the fronto-parietal coupling for alpha1, alpha2 and beta1 bands. Conversely there is a negative relation between the inter-hemispheric coupling and INECO in parietal region for beta2 band.

Table 2 Control vs PD-CogNL - Inter-hemispheric connectivity

Band	Pair	P-value	Hedges	CI Bootstraped
Delta	F3-F4	0.0086	-0.8492	-1.5233 -0.2942
Theta	F3-F4	0.0135	-0.7844	-1.3805 -0.2501
Alpha2	P3-P4	0.0197*	0.6275	0.0529 1.3594

CI: confidence interval. *: P-value not corrected.

Table 3 Spearman's rank correlation coefficients

Connectivity	Band	Pair	P-value	r
	Alpha1	Fz-Pz	0.049	0.29
		F3-P3	0.024	0.33
Fronto-parietal	Alpha2	Fz-Pz	0.014	0.36
		F4-P4	0.011	0.37
	Beta1	F4-P4	0.048	0.29
Inter-hemispheric	Beta2	P3-P4	0.023	-0.33

r: correlation coefficient.

IV. Discussion

In this work we evaluated functional connectivity at rest in PD-CogNL patients and healthy controls. We found that PD-CogNL group showed mainly an increased inter-hemispheric coherence in frontal region in delta and theta rhythms. Also, a decrease is observed in parietal inter-hemispheric coherence in alpha2 band. In addition to these differences, we found relationship with executive function for the fronto-parietal and parietal inter-hemispheric connections. Previous studies have reported an increased coherence in non-demented PD patients (PD-ND) compared to controls mainly in theta and beta bands. Moazami et al [12], found enhanced coherence in PD-ND in frontal region in theta, high beta and gamma bands, and lower coherence in the parietal region around 10 Hz. Fonseca et al [15], only found an increase in inter and intra-hemispheric coherence in beta rhythm in demented PD patients (PDD) compared to PD-ND and controls. It has to be noted that the classification of PD-ND patients in previous studies included patients without cognitive impairment and patients with MCI. Our results confirm the increased coherence in theta band in frontal regions, which could be one of the first cortical locations disturbed in early stage of disease in patients, without the presence of cognitive impairment. Our work excludes patients with MCI and shows that neurophysiological changes occurs before cognitive impairment. It is possible that changes in higher rhythms are more related to cognitive decline with involvement of posterior components of cognitive large scale networks, as shown by decreased connectivity in inter-parietal alfa2 band and the correlations performed with executive function.

V. Conclusion

The results of this paper show early changes in frontal inter-hemispheric coupling in PD using EEG in resting state, a noninvasive methodology, easy to be carried out in clinical environment. This is a preliminary result of a broader analysis which involving the qEEG analysis in patients with PD with MCI, in order to search for markers to aid in the diagnosis, prognosis, and therapeutic control of PD patients.

Acknowledgment

This work was supported by Vicerrectoría de Investigación of Universidad de Antioquia (CODI), Project "Neurofisiología y Neuropsicología en Enfermedad Ganglio Basal", code PRG2014-768.

Conflict of Interest

The authors declare that there is no conflict of interest regarding the publication of this paper.

References

[1] F. Varela, J.-P. Lachaux, E. Rodriguez, and J. Martinerie, "The brainweb: Phase synchronization and large-scale integration," *Nat. Rev. Neurosci.*, vol. 2, no. 4, pp. 229–239.

[2] E. Pereda, R. Q. Quiroga, and J. Bhattacharya, "Nonlinear multivariate analysis of neurophysiological signals," *Prog. Neurobiol.*, vol. 77, pp. 1–37, 2005.

[3] K. J. Friston, "Functional and effective connectivity: a review," *Brain Connect.*, vol. 1, no. 1, pp. 13–36, 2011.

[4] S. L. Bressler and V. Menon, "Large-scale brain networks in cognition: emerging methods and principles," *Trends Cogn. Sci.*, vol. 14, no. 6, pp. 277–290, 2010.

[5] C. Babiloni, R. Lizio, N. Marzano, P. Capotosto, A. Soricelli, A. I. Triggiani, S. Cordone, L. Gesualdo, and C. Del Percio, "Brain neural synchronization and functional coupling in Alzheimer's disease as revealed by resting state EEG rhythms," *Int. J. Psychophysiol.*, 2015.

[6] K. S. Cover, H. Vrenken, J. J. G. Geurts, B. W. van Oosten, B. Jelles, C. H. Polman, C. J. Stam, and B. W. van Dijk, "Multiple sclerosis patients show a highly significant decrease in alpha band interhemispheric synchronization measured using MEG.," *Neuroimage*, vol. 29, no. 3, pp. 783–788, Feb. 2006.

[7] E. Carlino, M. Sigaudo, R. Rosato, S. Vighetti, and P. Rocca, "Electroencephalographic connectivity analysis in schizophrenia.," *Neurosci. Lett.*, vol. 604, pp. 145–150, Sep. 2015.

[8] T. Pringsheim, N. Jette, A. Frolkis, and T. D. L. Steeves, "The prevalence of Parkinson's disease: A systematic review and meta-analysis," *Mov. Disord.*, vol. 29, no. 13, pp. 1583–1590, 2014.

[9] Y. Bocanegra, N. Trujillo-Orrego, and D. Pineda, "Demencia y deterioro cognitivo leve en la enfermedad de Parkinson: Una revisi??n," *Rev. Neurol.*, vol. 59, no. 12, pp. 555–569, 2014.

[10] L. V Kalia and A. E. Lang, "Parkinson's disease.," *Lancet (London, England)*, vol. 386, no. 9996, pp. 896–912, Aug. 2015.

[11] R. M. Bonelli and J. L. Cummings, "Frontal-subcortical circuitry and behavior," *Dialogues Clin. Neurosci.*, vol. 9, pp. 141–151, 2007.

[12] M. Moazami-Goudarzi, J. Sarnthein, L. Michels, R. Moukhtieva, and D. Jeanmonod, "Enhanced frontal low and high frequency power and synchronization in the resting EEG of parkinsonian patients," *Neuroimage*, vol. 41, pp. 985–997, 2008.

[13] P. Silberstein, A. Pogosyan, A. a. K??hn, G. Hotton, S. Tisch, A. Kupsch, P. Dowsey-Limousin, M. I. Hariz, and P. Brown, "Cortico-cortical coupling in Parkinson's disease and its modulation by therapy," *Brain*, vol. 128, pp. 1277–1291, 2005.

[14] J. S. George, J. Strunk, R. Mak-Mccully, M. Houser, H. Poizner, and A. R. Aron, "Dopaminergic therapy in Parkinson's disease decreases cortical beta band coherence in the resting state and increases cortical beta band power during executive control," *NeuroImage Clin.*, vol. 3, pp. 261–270, 2013.

[15] L. C. Fonseca, G. M. a S. Tedrus, P. N. Carvas, and E. C. F. a Machado, "Comparison of quantitative EEG between patients with Alzheimer's disease and those with Parkinson's disease dementia," *Clin. Neurophysiol.*, vol. 124, no. 10, pp. 1970–1974, 2013.

[16] A. J. Hughes, S. E. Daniel, L. Kilford, and A. J. Lees, "Accuracy of clinical diagnosis of idiopathic Parkinson's disease: a clinico-pathological study of 100 cases.," *J. Neurol. Neurosurg. Psychiatry*, vol. 55, no. 3, pp. 181–184, Mar. 1992.

[17] M. M. Hoehn and M. D. Yahr, "Parkinsonism: onset, progression and mortality.," *Neurology*, vol. 17, no. 5, pp. 427–442, May 1967.

[18] S. Fahn and R. Elton, "UPDRS Development Committee. The Unified Parkinson's Disease Rating Scale," in *Recent Developments in Parkinson's Disease*, 2nd ed., S. Fahn, C. Marsden, D. Calne, and M. Goldstein, Eds. Florham Park: Macmillan Healthcare Information, 1987, pp. 153–163.

[19] Z. S. Nasreddine, N. A. Phillips, V. Bedirian, S. Charbonneau, V. Whitehead, I. Collin, J. L. Cummings, and H. Chertkow, "The Montreal Cognitive Assessment, MoCA: a brief screening tool for mild cognitive impairment.," *J. Am. Geriatr. Soc.*, vol. 53, no. 4, pp. 695–699, Apr. 2005.

[20] J. C. Dalrymple-Alford, M. R. MacAskill, C. T. Nakas, L. Livingston, C. Graham, G. P. Crucian, T. R. Melzer, J. Kirwan, R. Keenan, S. Wells, R. J. Porter, R. Watts, and T. J. Anderson, "The MoCA: well-suited screen for cognitive impairment in Parkinson disease.," *Neurology*, vol. 75, no. 19, pp. 1717–1725, Nov. 2010.

[21] N. Kandiah, A. Zhang, A. R. Cenina, W. L. Au, N. Nadkarni, and L. C. Tan, "Montreal Cognitive Assessment for the screening and prediction of cognitive decline in early Parkinson's disease.," *Parkinsonism Relat. Disord.*, vol. 20, no. 11, pp. 1145–1148, Nov. 2014.

[22] L. Gil, C. Ruiz de Sánchez, F. Gil, S. J. Romero, and F. Pretelt Burgos, "Validation of the Montreal Cognitive Assessment (MoCA) in Spanish as a screening tool for mild cognitive impairment and mild dementia in patients over 65 years old in Bogotá, Colombia," *Int. J. Geriatr. Psychiatry*, vol. 30, no. 6, pp. 655–662, 2015.

[23] T. Torralva, M. Roca, E. Gleichgerrcht, P. López, and F. Manes, "INECO Frontal Screening (IFS): a brief, sensitive, and specific tool to assess executive functions in dementia.," *J. Int. Neuropsychol. Soc.*, vol. 15, no. July, pp. 777–786, 2009.

[24] A. Delorme and S. Makeig, "EEGLAB: An open source toolbox for analysis of single-trial EEG dynamics including independent component analysis," *J. Neurosci. Methods*, vol. 134, pp. 9–21, 2004.

[25] N. Bigdely-Shamlo, T. Mullen, C. Kothe, K.-M. Su, and K. a. Robbins, "The PREP pipeline: standardized preprocessing for large-scale EEG analysis," *Front. Neuroinform.*, vol. 9, no. June, pp. 1–20, 2015.

[26] N. P. Castellanos and V. a. Makarov, "Recovering EEG brain signals: Artifact suppression with wavelet enhanced independent component analysis," *J. Neurosci. Methods*, vol. 158, pp. 300–312, 2006.

[27] A. Delorme, T. J. Sejnowski, and S. Makeig, "Enhanced detection of artifacts in EEG data using higher-order statistics and independent component analysis," *Neuroimaging*, vol. 34, no. 4, pp. 1443–1449, 2007.

[28] P. Mitra and H. Bokil, *Observed Brain Dynamics*. New York: Oxford University Press, 2008, p. 408.

[29] P. Mitra, H. Bokil, H. Maniar, C. Loader, S. Mehta, D. Hill, S. Mitra, P. Andrews, R. Baptista, S. Gopinath, H. Nalatore, and S. Kaur, "Chronux." [Online]. Available: http://chronux.org/. [Accessed: 15-Mar-2016].

[30] V. Sakkalis, "Review of advanced techniques for the estimation of brain connectivity measured with EEG/MEG.," *Comput. Biol. Med.*, vol. 41, no. 12, pp. 1110–7, Dec. 2011.

[31] S. M. Bowyer, "Coherence a measure of the brain networks: past and present," *Neuropsychiatr. Electrophysiol.*, vol. 2, p. 1, 2016.

[32] E. Glerean, R. K. Pan, J. Salmi, R. Kujala, J. M. Lahnakoski, U. Roine, L. Nummenmaa, S. Leppämäki, T. Nieminen-von Wendt, P. Tani, J. Saramäki, M. Sams, and I. P. Jääskeläinen, "Reorganization of functionally connected brain subnetworks in high-functioning autism," *Hum. Brain Mapp.*, vol. 37, no. November 2015, pp. 1066–1079, 2016.

[33] J. D. Storey, "A direct approach to false discovery rates," *J. R. Stat. Soc. Ser. B (Statistical Methodol.*, vol. 64, no. 3, pp. 479–498, 2002.

[34] H. Hentschke and M. C. St??ttgen, "Computation of measures of effect size for neuroscience data sets," *Eur. J. Neurosci.*, vol. 34, no. July, pp. 1887–1894, 2011.

[35] L. V Hedges, "Distribution theory for Glass's estimator of effect size and related estimators," *J. Educ. Stat.*, vol. 6, no. 2, pp. 107–128, 1981.

Corresponding author
Author: John Fredy Ochoa Gomez
Institute: Universidad de Antioquia
Street: Calle 67 No. 53-108
City: Medellín
Country: Colombia
Email: john.ochoa@udea.edu.co

Modeling Transient Otoacoustic Emissions in children with hearing impairment

Y. Torné-Cabrera[1], L. M. Alvero[1], E. Martínez-Montes[2]

[1] Electronics Department, Cuban Neuroscience Center (CNEURO), Havana, CUBA
[2] Neuroinformatics Department, Cuban Neuroscience Center (CNEURO), Havana, CUBA

Abstract— **Due to the importance of hearing in the development of speech, language, communication and learning, the early detection of disorders associated to the auditory pathway is essential to reduce their negative impact in the long term. Among the techniques for auditory evaluation are Otoacoustic Emissions (OAE) and Auditory Brainstem Response. Both of them allow the fast, objective and non-invasive study of the integrity of the ear canal, which is essential in neonatology testing. The OAE is a simpler and easier tool to use, but it is very sensitive to noise and statistical processing is necessary for a reliable diagnosis. In this paper we propose a model for simulating OAE signals from both healthy and hearing impaired children, with the objective of creating a reliable database for a thorough assessment of OAE detection methods. We present a preliminary evaluation of the model with classical methods and with other methodologies based on time-frequency analysis.**

Keywords— **otoacoustic emissions, time-frequency distributions, short-term Fourier transform, wavelet transform, detection of biological signals**

I. Introduction

Hearing is fundamental for the normal development of a child, being in the familiar as in the social environment. Children with hearing loss or problems in the processing of acoustic information have usually a late development of expression skills, which leads to learning disabilities and corresponding problems in school. Together with other negative consequences in the long term, shows the importance of the early detection of hearing problems, especially in the first 3 years of life. The diagnosis and study of hearing impairments, as well as the rehabilitation, should be carried out in the first 6 months of age [1].

Recording the Oto-Acoustic Emissions (OAE) is one of the techniques for auditory assessment and diagnosis. They are a quick, cheap, non-invasive and relatively simple method to obtain information about the correct processing of sound. The OAE are very low-intensity sounds produced by the movement of external ciliated cells in the cochlea. Several researches have shown that these transient and complex signals are present in people with normal hearing [2].

In general, the OAE are classified in "evoked" and "spontaneous" OAE, according to whether they are generated using stimulation or not. The spontaneous OAE are narrowband signals, mostly in a frequency range of 1 to 2 kHz, with a sinusoid shape similar to that of a pure tone. The evoked OAE are the physiological response to a stimulus and does not appear if the hearing loss is over 20-40 dB [3]. According to the type of stimulation, there are three types of OAE. The most used in the clinical practice is the transient OAE (TOAE), which are generated by short stimuli such as the click or brief tones. The presence of TOAE, by itself, does not indicate the type and degree of the hearing loss. However, clinical and research studies based on their morphology and reproducibility, usually consider that the lack of a response about 3 to 6 dB over the noise level suggests that ciliated cells for the corresponding frequency range are affected [3].

A number of methods have been used to analyze the TOAE, in order to diagnose whether there is an impairment or not of the auditory canal. Since the evoked OAE are signals with a very low amplitude (usually between -20 y 20 dB SPL), it is quite difficult to detect them within the noisy measurements. That´s why one strategy commonly followed is to reduce noise in the temporal and/or frequency domain. Among the useful measures to detect these responses are the mean reproducibility, the signal-to-noise ratio (SNR) and the amplitude. Detection is based on empirical threshold values for these parameters [4].

However, these biological signals are non-stationary and therefore, some authors have started using time-frequency (TF) analysis for detecting them, such as the Short-Term Fourier Transform (STFT), the Continuous Wavelet Transform (CWT), and the Wigner–Ville and Choi–Williams distributions [6]. Noteworthily, there is no perfect method able to detect these signals in the great variety of real OAE scenarios. Moreover, to our knowledge, there are no validation studies showing a fair comparison of the performance of all methods in the detection of OAE in the same set of data. For carrying out such study, it is crucial to have a database of simulated and real OAE recordings, together with some reliable expert criteria as a gold standard.

Precisely, the objective of our study is the exploration and assessment of the use of TF methods for the analysis of OAE, in comparison with the classical methodology. In this paper, we present as a first step, the modeling of TOAE for healthy and hearing-impaired children, in order to create a simulated database that allows a subsequent thorough comparison among the different methods. As a preliminary validation of these models, we explore here the information provided by

I. Torres et al. (eds.), *VII Latin American Congress on Biomedical Engineering CLAIB 2016, Bucaramanga, Santander, Colombia, October 26th -28th, 2016*,
IFMBE Proceedings 60,
DOI: 10.1007/978-981-10-4086-3_48

the classical methodology and TF methods, such as the STFT and the CWT. This is an indispensable step for the development of new and more reliable methodologies for detecting OAE, allowing for large scale screening and auditory assessment of children, especially neonates.

II. Methods

A. Model for the TOAE and simulated database

For simulating the TOAE signals $s(t)$ (in mPa), we use the model that has been proposed for the response of the external ciliated cells. This consists in the sum of 5 gamma tones $g_i(t)$ [5]:

$$s(t) = \sum_{1=1}^{5} g_i(t);\;\; g_i(t) = a_i \cdot t^3 \cdot e^{-2\pi\beta f_i t} \cos 2\pi f_i t \quad (1)$$

Where the amplitude factor depends non-linearly on the central frequency f_i, $a_i = (2\pi f_i)^{3.5}$, and the relaxation factor β is a constant that determines the amplitude-falling rate of each gamma tone. This model allows simulation of normal TOAE signals (Fig. 1A and C) but not those signals from subjects with hearing impairments, which do not only have smaller amplitudes but also different morphologies (Fig. 1B). We propose here to introduce a different exponential amplitude factor, $a_i = e^{\frac{(f_i - \min(f_i))}{\alpha}}$, where small values for α lead to signals similar to those from hearing-impaired subjects (using a relaxation factor β in the range [0.008, 0.120]. As a first option, we fixed α=0.001. Gaussian noise is added for a predetermined SNR value ("true SNR"). We can also control the final amplitude in dB SPL of the noisy simulated signal (i.e. to set a maximum amplitude SPL_{max}), by normalizing $s(t)$ to have rms=1 and then multiplying it by a gain factor $A = 0.02 * 10^{\frac{SPL_{max}}{20}}$. This model allows generating realistic signals from subjects with hearing impairment (Fig. 1D).

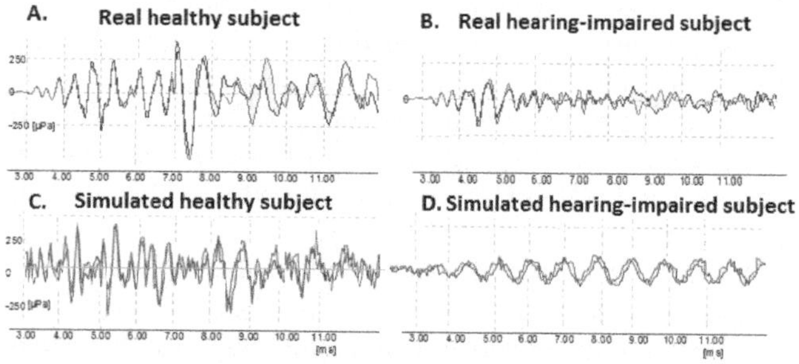

Fig. 1. Real TOAE signals from healthy (A) and (B) subjects. Simulated TOAE signals for healthy (C) and hearing-impaired (D) subjects.

For evaluating the models and subsequently the detection methods, we prepared a database with 3 different data sets (sampling frequency of 20 kHz). The first data set would allow to explore which central frequencies f_i are more appropriate (we test 2 sets from the literature: 1.1, 1.53, 2.10, 2.90, 4 kHz and 1.0, 1.5, 2.2, 3.3, 5 kHz [5]. The second data set is to study the influence on results of the selected value for β (using 3 values that permit simulating realistic signals for both healthy and impaired subjects: 0.05, 0.045 and 0.0325). Finally, we simulated TOAE signals with varying the true SNR (1, 5, 10 and 20 dB). The first 2 data sets contain 10 signals (repetitions with different sampled noise) for each parameter value and the last one contains 100 repetitions for each true SNR. In all cases, we simulated signals using random maximum amplitudes in 3 different ranges corresponding to i) healthy subjects: 15-25 dB SPL; ii) hearing impaired subjects: 2-6 dB SPL; iii) intermediate amplitudes: 5-20 dB SPL. In total, the database contains 1350 simulated signals.

B. Methods of analysis of the OAE

a) *Classical methods*

The traditional analysis for detecting TOAE is based on computing parameters such as the level of reproducibility, the signal-to-noise ratio and the estimated amplitude of the response [6]. In practice many repetitions of the OAE signals are obtained using trains of stimuli. The odd and even repetitions are averaged separately for creating two buffer responses A and B (of n time points). The correlation between these (centered) buffers is known as *reproducibility*:

$$\gamma = \frac{\sum_n A(n)B(n)}{\sqrt{\sum_n A^2(n)}\sqrt{\sum_n B^2(n)}} \quad (1)$$

Another feature is based on estimating the level of response as the power cross-spectrum between A and B and the level of noise as the power auto-spectrum of the mean difference $\frac{A-B}{2}$. The difference between these values (in dB) is an estimate of the SNR of the OAE signal, for any particular frequency band. From these two features, the detection/classification of a biological OAE signal is done using a heuristic classifier: reproducibility higher than a pre-determined threshold (usually between 50 and 70%) and SNR≥3 dB in two or more frequency bands (usually centered in 1, 2, 3, 4, 5 kHz, with a bandwidth of 1 kHz each) [4].

b) *Short Term Fourier Transform (STFT) and Continuous Wavelet Transform (CWT)*

The STFT is a powerful general tool for Fourier analysis of non-stationary signals, such as OAE (x(t)), in smaller time segments in which the signal can be considered as stationary. These segments are defined by the non-null time support of a window function ω(t-τ):

$$\mathrm{STFT}(\tau, f) = \int_{-\infty}^{\infty} x(t)\omega(t-\tau)\, e^{-j2\pi ft} dt \qquad (2)$$

On the other hand, the CWT consists in a similar decomposition of the signal but using wavelet functions ψ instead of windowed Fourier basis:

$$\mathrm{CWT}(\tau, f) = \int_t x(t) \cdot \sqrt{\frac{f}{f_0}} \cdot \psi^*\left(\frac{f}{f_0} \cdot (t-\tau)\right) dt \qquad (3)$$

where ψ^* is the conjugate of the mother wavelet (in our case we used Morlet), f/f_0 is the wavelet dilation factor, f_0 is the central frequency and τ is the time shift. In both cases, the amplitude/energy distribution of the signal are estimated for each frequency f and time point τ (time-frequency maps).

The STFT and CWT coefficients obtained from the mean sum and the mean difference of A and B offer new estimates of response and noise levels. These can now be computed for all frequencies (narrow band) and the whole temporal window, which offers a more complete map of the SNR variability in time and frequency. From this more general features we can define new and more flexible criteria for the heuristic detection/classification of OAE responses. In this work, we will consider a successful detection if the SNR time-frequency map shows amplitudes over 3dB for any 3 bands of frequency in a time interval larger than 2 ms inside the windows of analysis.

III. Results and discussion

For a preliminary validation of the model introduced in section II.A, we simulated 130 TOAE signals (buffer responses A y B) of children with normal hearing ("healthy subject", amplitudes between 15 and 20 db SPL) and 130 TOAE of hearing impaired children ("impaired subject", amplitudes between 2 and 6 db SPL). We used true SNRs of 5 dB (60 signals) and 1 dB (200 signals), the three values of factor β and the first set of central frequencies. We explored the results from the 3 methods using two slightly different windows of analysis: from 4 to 12 ms and from 4 to 10 ms. This parameter is highly variable in the literature (between 2.5 and 12.5 ms) and there are no studies on its influence on the OAE detection. In practice, defining this window is crucial for eliminating the part of the signal not related with the cochlear physiological response but with artifacts from different sources. However, this selection is done visually and is highly dependent on the expertise of the specialist.

For each signal we computed the reproducibility and the SNR for frequency bands centered in 1, 2, 3, 4, 5 and 6 kHz (bandwidth of 1 kHz). Fig. 2 shows the results for the classical analysis of two typical simulated signals of a healthy and an impaired subject (subjects 1 and 2, respectively). All processing was done with Matlab R2014a.

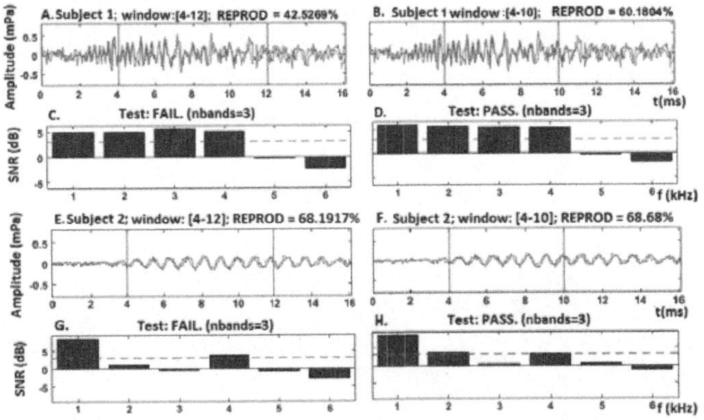

Fig. 2. Reproducibility (REPROD) and SNR for the two windows (4-12 ms, left and 4-10 ms, right) for simulated signals of a healthy subject (A-D) and an impaired subject (E-H). Simulations used true SNR=5dB and β=0.0325.

A clinical neurophysiologist, expert in hearing problems, considered a successful detection of the biological response if reproducibility was above 60% and the SNR above 3 dB in at least 3 out of the 6 frequency bands. Fig 2 shows that according to that criteria, a healthy subject failed the test (i.e. was considered as without response) for one of the windows and the impaired subject passed the test for the other window. Both cases illustrate the dependence of results on the selected windows of analysis.

The TF distributions of the SNR, obtained with the STFT and CWT for the same two virtual subjects are shown in Fig. 3 and Fig. 4 (left panels). From them, we extracted the temporal evolution for the 6 frequency bands, which are compared with the threshold of 3 dB (Fig. 3 and Fig. 4, right panels). With these figures one can better assess whether there are time windows and frequency bands with large enough amplitudes to consider a successful detection of OAE. In this case, we observed that the healthy subject has a lot more activity above the threshold, with more than two bands consistently over the threshold in the whole segment, while the impaired subject has 2 or 3 bands over-threshold simultaneously only for small time periods of less than 2 ms. In general, the TF methods offer more information and allow us to explore in which time window and frequencies the level of the signal is above the noise level. In this way, there is no need of an arbitrary (expert-dependent) selection of time-frequency windows for the analysis.

For the evaluation of the reliability of the proposed model with this simulated data subset, we computed the sensitivity (Sens) and specificity (Spec) of each method by using our simulating criteria as the gold standard (Table 1). Results showed that the STFT and CWT offered higher accuracy (Acc), although the three methods had a very high specificity. This suggests that our model is simulating signals for healthy and impaired subjects that can be well discriminated with the TF methods. The results were similar for the two windows of analysis.

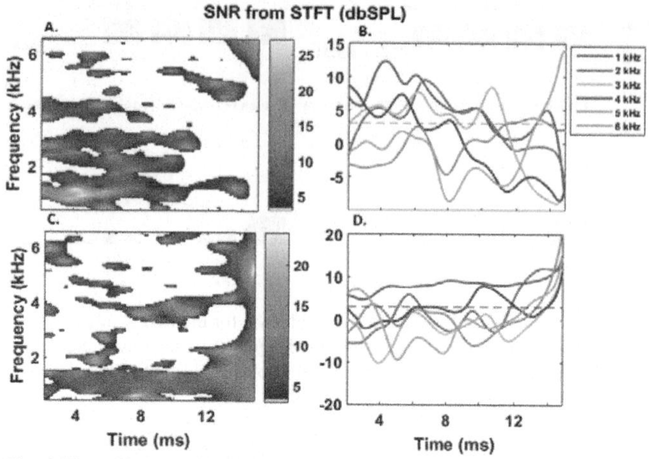

Fig. 3 Time-frequency SNR maps (left panels) and time courses for the selected 6 bands (right panels) obtained with the STFT for a healthy (A and B) and an impaired (C and D) subjects (true SNR=5dB and β=0.0325).

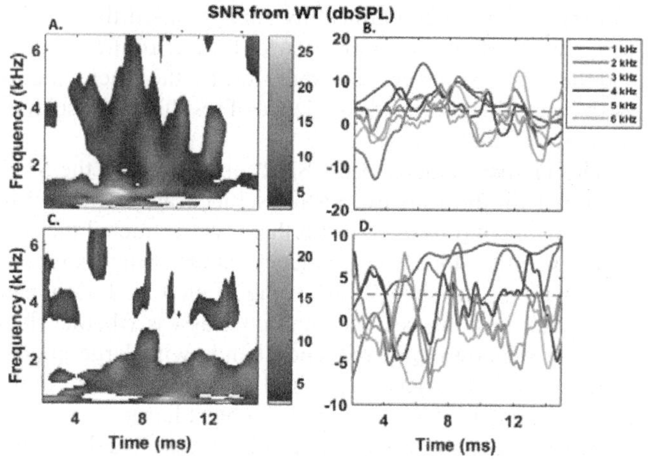

Fig. 4 Time-frequency SNR map (left panels) and time courses for the selected 6 bands (right panels) obtained with the CWT for a healthy (A and B) and an impaired (C and D) subjects (true SNR=5dB and β=0.0325).

Table 1 Classification results for the three methods in a subset of the database: 60 and 200 signals with true SNR=5db and 1db, respectively.

	window: 4-12 ms			window: 4-10 ms		
Method	Sens(%)	Spec(%)	Acc(%)	Sens(%)	Spec(%)	Acc(%)
Classical	21.5	**100**	60.8	23.3	**100**	66.5
STFT	74.6	**100**	87.3	70.8	**100**	85.4
CWT	**81.5**	99.2	**90.4**	**81.5**	99.2	**90.4**

III. Conclusions

In this article we introduced a new model for simulating TOAE in both healthy and hearing impaired children. We defined a first attempt of a simulated database for a thorough study of methods for detecting OAEs. We explored the information obtained with the use of TF methods of analysis, in comparison with the traditional analysis. We found that the arbitrary selection of (even slightly different) time windows can lead to totally different results with the classical method, and the TF methods offer high sensitivity and specificity in a very inhomogeneous database. This suggest that the TF methods are useful for a reliable detection of TOAE since they do not need an *ad hoc*, before-hand selection of the most appropriate segment of the signal (in time and frequency) for the detection. This work provides the grounds for a more thorough assessment of the methods for detecting OAE and for the development of more automatic methodologies that reduce the need for an experienced neurophysiologist imposing arbitrary time windows and other analysis parameters.

Conflict of Interests

Authors declare no conflicts of interests.

References

1. Vashistha I, Aseri Y. (2015) Prevalence of Hearing Impairment in High Risk Infants. Indian Journal of Otolaryngology and Head & Neck Surgery: 1-4 10.1007/s12070-015-0869-9
2. Engdahl B, Tambs K et al. (2013) Otoacoustic emissions, pure-tone audiometry, and self-reported hearing. International Journal Audiology 52(2): 74-82 10.3109/14992027.2012.733423
3. Akinpelu O, Peleva Ea et al. (2014) Otoacoustic emissions in newborn hearing screening: a systematic review of the effects of different protocols on test outcomes. International journal of pediatric otorhinolaryngology 78(5): 711-717.
4. Zhang Vicky Wei., Zhang Zhi-Guo. et al. (2011) Detection improvement for neonatal click evoked otoacoustic emissions by time-frequency filtering. Computers in Biology and Medicine 41(8): 675-86 10.1016/j.compbiomed.2011.06.003
5. Tognola Gabriella, Grandori Ferdinando et al. (1998) Time-frequency distribution methods for the analysis of click-evoked otoacoustic emissions. Technology and Health Care 6(2-3): 159-75
6. Jedrzejczak W Wiktor, Kochanek Krzysztof et al. (2012) Otoacoustic emissions for the evaluating the low-frequency hearing of patients considered for partial deafness treatment. Journal of Hearing Science 2(2): 10.17430/882768

IoT Protocol Model on Healthcare Monitoring

Edward Guillén[1], Jeisson Sánchez[2] and Leonardo Ramírez López[3]

[1]IEEE member, leader of GISSC Group of Nueva Granada Military University, Bogotá, Colombia
[2]IEEE member, researcher of GISSIC Group of Nueva Granada Military University, Bogotá, Colombia
[3]IEEE-EMBS member, leader of TIGUM Group of Nueva Granada Military University, Bogotá, Colombia

Abstract— **different technologies such as cloud computing, Internet of Things –IoT, big data or machine learning have been used to improve the quality of human life. The quality of human life often depends on healthcare. In telemedicine, one approach of these technologies is the remote control of patients in real - time. Although multiple science fields are working on telehealth technologies, there are different medical parameters in technical level to evaluate. In this paper we analyze the IoT protocol behavior and its architecture applied to medical data transmission. Security and availability of patient's data, among others parameters are organized through this model by improving the general proposed models in telehealth for information acquisition.**

Keywords— **Healthcare, Internet of Things, Cloud Computing, Protocol.**

I. INTRODUCTION

Nowadays emergent technologies are implemented in Telemedicine field. The main feature in the digital era to monitor different health diagnostics is the Electronic Health Record - HER. According to American Telemedicine Association- ATA the Telemedicine impact on Healthcare is growing [1]. Over half of U.S hospitals use some telemedicine technology. Additional approximately one million Americans are using remote cardiac monitors. Further, the telemedicine solutions in Latin America are exposed by Enlace Hispano Americano de Salud – EHAS with multiple projects [2]. Telemedicine services through communication networks were deployed in different isolated areas. The main scope was a Tele-stethoscope, Tele-microscopy and Health pregnancy project to improve the healthcare of Latin America isolated areas.

Nevertheless, in telemedicine the communication protocols, interfaces, devices and medical equipment have evolved with Internet of Things –IoT technology. The main parameters are real-time monitoring, security protocols, data transmission, simple architecture to remote control, among others. In 2015, Gerdes M. et al proposed a reference design for telehealth and telecare services on upcoming IoT technologies [3]. The system architecture contains a secure health cloud HER, based on two points of information. The first one is the care as home context information to monitor the patient and the last one is the health and care sources including hospital specialist. Then W. Ren et al. presented a scheme with cipher-policy attributes to solve the security problems based on transmission medical information to cloud computing systems. The scheme has a general framework to control the measurements at Body Area Network –BAN architecture. They compared the computational cost of IoT cloud architecture with their scheme [4]. The analysis improves the data security and robustness between Hospital Server and Home controller. However nowadays there are multiple security protocols and cloud computing resources to optimize the data transmission.

A social inclusion project is presented by Duretti S.et. al. in 2015. Mainly, ALL4ALL system improves the elderly and disabled population healthcare. They analyzed the economic crisis and the consequent over the health situation resources [5]. ALL4ALL system enhances the quality of life through care services platform. Multiple actors are working on this platform such as administrators, social community, service center personnel, sanitary professionals, among others. Different modules as gaming system, social video network, user profiling and clinical management interacts with the user to solve medical attention. Then, Lamprinakos et al. proposed an integrated remote monitoring platform towards Telehealth and Telecare services interoperability [6]. Model inCASA platform enables the deployment of services to follow-up the patient's health status. InCASA integrates Telehealth services while also encompassing prevention, treatment, cure and health promotion aspects of care. The benefit is access for relevant new diseases information, take decision at the right time, among others.

Recently, Park et al. present an elderly health monitoring system approached into brain signals [7]. The main problem is the increasing of the incidence of stroke in Korea. The study based on the problem improves the prevention on a disease with the highest index. The most important in engineering science is enabling the number of pilot systems in the real-world scenarios. On thc other hand, Prakash et al. present a real-time remote monitoring of human vital signs using IoT and GSM Connectivity [8]. The study has an essential component, technical device connectivity analysis. The measurement of pulse rate and body temperature is controlled by the calibration of the sensor and the configuration of media channel. The system is proposed to be used in hospitals with high accuracy.

Different models are proposed to monitor remotely patients with telehealth approaches as rural scenarios, areas without specialized medical attention, patients with chronic pathology, elderly people, among others. The main idea with IoT technology and cloud computing inclusion in telemedicine field is generate solutions to prevent diseases. Monitoring solutions improve the response-time of the medical physicians, ambulances, nurses and other medical

I. Torres et al. (eds.), *VII Latin American Congress on Biomedical Engineering CLAIB 2016, Bucaramanga, Santander, Colombia, October 26th -28th, 2016*,
IFMBE Proceedings 60,
DOI: 10.1007/978-981-10-4086-3_49

nodes. Related researches in different scenarios as presented are important to help the medical work with pervasive computing. The proposed model in this paper is intended to improve these systems by optimizing results. New models deployed with IoT architecture have not been analyzed through a protocol of data transmission. The quality of service in telemedicine is diverse and important to analyze. The introduction of new technologies as IoT has changed the wired test. IoT architecture allows the patient to do normal activities during the testing stage. In this paper we present an analysis based on IoT advances and the secure architecture to control medical parameters from home patients. A model of data transmission between cloud server and different IoT controllers is analyzed to determine the main parameters in a Telemedicine scheme.

II. INTERNET OF THINGS ARCHITECTURE

A. IoT taxonomy

In general, different models have data acquisition modules through an IoT network. The main IoT components are the sensors. The sensors are professional medical hardware developed or software applications validated by a healthcare professional. In a telemedicine model based on IoT architecture there are essential components as mentioned by Intel blueprint solution [9], such as systems to acquired data, gateways, computing infrastructure and healthcare providers. In order to typical IoT architecture in Telemedicine as shown in figure 1, the data transmission is between server/ broker and user devices. The user device sends constantly messages based on clinical parameters. The most common clinical parameters are heart rate, temperature, percentage of oxygen saturation, among others.

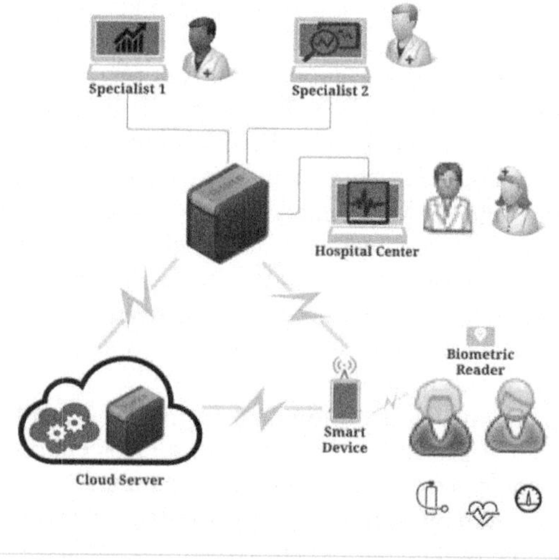

Fig. 1 IoT Network in Telemedicine architecture

The accelerated growth of devices as IoT technology is changing the common health assistance remotely. According to Intel for 2020 will be 198 billion devices rather than in 2006 [10]. IoT architecture in telemedicine presents an additional node, professional physician center. In the next section we proposed an IoT protocol model to evaluate different medical variables in optimum conditions.

B. IoT Proposed Model

The objective in medical monitoring is the control of clinical variables. There are many factors to analyze in IoT communication. The first feature is the device selection based on the study objective. Multiple devices can take the variables, but not all are certified and precise. For example, the heart rate has different devices and measures to check the information about the patient. The analysis on the device selection is proposed according to the specific datasheet of device manufacturer. The main features are output values, conversion analog schemas of electronic hardware, ports and communication protocols integration. The environment conditions are considered during the device operation. One factor in heart rate is based on the sensor location in the body to monitor the medical feature. A medical validation explain us the difference to monitor the patient finger or chest.

The model measurement IoT for telemedicine proposed is presented in following steps:

1. Study the medical providers datasheet and compatibility of software architecture with other medical systems.
2. A technical analysis of media channel as wired or wireless transmission, noise, delay, among others.
3. Determine the environmental conditions during testing stage and real function. The measurement could be affected the patient status and the diagnostic reason of possible diseases.
4. Analog-Digital Conversion parameters in electronic devices to analyze variations of measures.
5. Data acquisition by mobile apps or intelligent cards.
6. Sampling rates optimized depending on the clinical parameter tested. A pulse oximetry test has different samples per day in a controlled patient that a critical. One heart rate testing during few time compared with a physical activity that has samples per minute in a long time.
7. Generate alarms before the IoT device start.

The features selection based on the proposed IoT protocol take advantage with other models to consider the conditions during the testing stage. The tolerance error considers the real measure of clinical variables with different devices. Further an Analog-Digital conversion in IoT devices changes based on environmental conditions, media channel, and data storage on memory device or server, among others.

III. Comparison performance of standard IoT model and proposed

The study is based on computational resources, test time, patient benefit, among others features. The cloud computing server stores data and deploy techniques to process information with analytics algorithms. The conventional IoT model and proposed IoT protocol are compared in figure 2. The proposed model reduces computational resources as Virtual Machine –VM hours, gateway use, devices energy, memory and storage disk space. Although these limitations are considered in cloud computing solutions, the main idea is assist the diagnostic performed by healthcare professionals. The

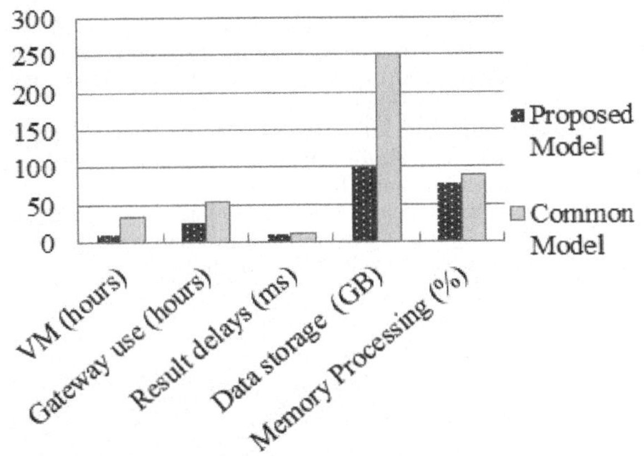

Fig. 2 Proposed model vs. conventional model of IoT measurements

The IoT proposed model was tested in a pilot scenario with a telemedicine server in cloud computing. IoT broker is the central node of data interchange with 3 topics of medical information such as temperature, measure oxygen saturation and heart rate. All the devices are certified equipment controlled by the same protocol of data transmission. The tests were developed in laboratory environment with different programmable cards to control the sample rates, time acquisition and incorporate the analytics algorithms. The devices were connected to the programmable cards and interconnected through a secure media channel with the cloud server.

The difference with Virtual Machines –VM that stores the databases is based on the usage. The connection processing time with VPN connection and gateway decreases with the proposed model. The main difference is data storage due to the sample rate in each variable. The optimization of the variables is according to the relevance of quantity of samples per patient. For example, a patient heart rate is monitored in regular times based on daily activities. But the measure of oxygen saturation is acquired only in long time periods. The same reason in memory of processing is necessary on multi-tasks when the variables are adjusted to the model. Multiple cases requiring this model as elderly patient at home, child care, critical health patients, among others. In order to healthcare cases the patient information is secure and just can be analyzed with previously authorization.

The synchronization to publish a message by the devices and the reception in the server is almost in real-time. The media channel was interconnected with a direct protocol of communication. Also the server was exclusive dedicated to process the data of IoT protocol model. Although there are asynchronous scripts programs to decrease the processor performance. Additional data reliability the software validation based on result delays were promising. The IoT callbacks of messages are synchronized between clients and data storage systems. Multiple medical devices were compared through the measures acquisition, features range and capacity maximum of measurement. The transmission on wired conditions has been configured with the same parameters as bitrate, coverage, transmission medium with similar electrical conductance, among others. IoT transmission was tested on TCP and SSL protocol of communication. The difference was the reliability of data, important in telehealth. Proposed model are tested with quality of service of 2 due to the bidirectional subscribe and publish methods.

IV. Conclusions

IoT proposed model has reached the main objective in telemedicine infrastructure to monitor remotely patients. The difference with other models is the optimization of resources as computational processing, data storage and data acquisition. The standard model has been tested in a pilot project. Medical measurements were tested based on sampling rate, then an analytics algorithm generate rules according to patient diagnostic.

The analysis of technical devices during the test was a significant part in the study. Engineering field could determine the tolerance error of equipment calibration, type of connectors, relevance of software updates in programmable cards and devices, among others. Although the technology transmission channel in IoT proposed model was wireless, some medical equipment often works properly in wired conditions with the patient. Another approach in the model was the satisfaction of the patient. The devices selected were appropriate in the pilot scenario due to the function and comfort. The patient comfort to do the common activities without limitations as space monitoring or mobility with wires is other advantage in this model.

Acknowledgment

This work was possible in part with funds of project code INV-ING-2108 at Military University Nueva Granada with GISSIC and TIGUM research groups

Conflict of Interest

"The authors declare that they have no conflict of interest".

References

1. The American Telemedicine Association. How typical is telemedicine? at http://www.americantelemed.org/about-telemedicine/faqs#.V18cabsrLIU. 2012
2. Prieto I, Simó J, Liñán L, García V, Martínez A. (2015) Telemedicine networks of EHAS Foundation in Latin America. Front Public Health. 2: 188. DOI 10.3389/fpubh.2014.00188
3. Gerdes M, Reichert F, Nytun J.P, Fensli R. (2015) Reference Design for Smart Collaborative Telehealth and Telecare Services based on IoT Technologies, International Conference on Computational Science and Computational Intelligence (CSCI), IEEE, Las Vegas, United States, 2015, pp 817-820.
4. Ren W, Ren Y, Wu ME, Lee CJ. (2015) A Robust and Flexible Access Control Scheme for Cloud-IoT Paradigm with Application to Remote Mobile Medical Monitoring, Third International Conference on Robot, Vision and Signal Processing (RVSP), IEEE, Kaohsiung, Taiwan, 2015, pp 130-133.
5. Duretti S, et al (2015) ALL4ALL: IoT and telecare project for social inclusion, 1st International Forum on Research and Technologies for Society and Industry (RTSI), IEEE, Torino, Italy, 2015, pp 17-22.
6. Lamprinakos GC, et al (2015) An integrated remote monitoring platform towards telehealth and telecare services interoperability. Information Sciences, 308, 23-37.
7. Park SJ, Subramaniyam M, Kim SE, Hong S, Lee JH, Jo CM, Seo Y. (2017) Development of the Elderly Healthcare Monitoring System with IoT. In Advances in Human Factors and Ergonomics in Healthcare. Springer International Publishing, 2017, pp 309-315.
8. Prakash R, Girish SV, Ganesh AB. (2016) Real-Time Remote Monitoring of Human Vital Signs Using Internet of Things (IoT) and GSM Connectivity. In Proceedings of the International Conference on Soft Computing Systems, Springer India, 2016, pp 47-56.
9. Dell and Intel IoT solutions Alliance, Felix M (2016) Transforming Healthcare with Telemedicine Solutions based on the Internet of Things (IoT). Intel-Dell 1:10
10. Intel Corporation. A Guide to the Internet of Things infographic at http://www.intel.com/content/www/us/en/internet-of-things/infographics/guide-to-iot.html

Force Plate Calibration and Setup for Assessments of Human Balance

I. Ghersi[1,2], C. F. Castro Arenas[1,3], P. D. Borsoi[1], M. T. Miralles[1,2]

[1] Pontificia Universidad Católica Argentina. Facultad de Ciencias Fisicomatemáticas e Ingeniería. Laboratorio de Biomecánica e Ingeniería para la Salud. Ciudad Autónoma de Buenos Aires, Argentina.
[2] Universidad de Buenos Aires. Facultad de Arquitectura, Diseño y Urbanismo. Centro de Investigación en Diseño Industrial de Productos Complejos. Ciudad Autónoma de Buenos Aires, Argentina.
[3] Universidad de Buenos Aires. Facultad de Arquitectura, Diseño y Urbanismo. Instituto de la espacialidad Humana, Centro Laboratorio de Morfología. Ciudad Autónoma de Buenos Aires, Argentina.

Abstract—**Repeatable, quantitative evaluations of human balance rely on thorough characterizations of the devices and sensors used in measurements. Given the growing concern on this field, the research group has developed dedicated devices for the assessment of gait and balance, including a force plate. In this work, a calibration procedure for this force plate is devised and implemented, based on a calibration device which combines versatility, ease of use, and low-cost, and on reference weights. The procedure is detailed, and its results are shown, indicating a positive response for the device, and proving a superior fit compared to the theoretical estimation of its behavior. Based on this improved knowledge on the designed force plate, a dedicated signal-analysis tool is shown, incorporating calibration results and curves. Finally, examples of use of this tool are detailed, with standing balance registers from a control subject, indicating the resulting capabilities of the whole system as a promising tool for the assessment of human balance.**

Keywords—**Balance, force plate, calibration, curve-fitting.**

I. Introduction

Gait and balance disorders are a growing concern for healthcare institutions. While many devices and tests have been elaborated over the past decades, the quantification of human balance remains decentralized, with no unified standard or method for the acquisition or analysis of results. As populations grow older [1] the reliable evaluation of risk of falls becomes a necessity. In this context, force plates have been used for research and clinical purposes over the past years [2], while differences between signal processing and mathematical models [3, 4] applied to their signals results in a difficult-to-trace, large number of non-uniform results. As part of a research effort into accurately quantifying gait and balance, the research group has worked in the development of specialized devices for the quantification of motions and ataxias. Among these, a wearable acceleration sensing device was developed [5], as well as a force plate. Given the need to validate results coming from the latter, a low-cost, versatile calibration device and procedure was developed, which is shown in this work. The calibration procedure is detailed, from its requisites to its implementation, and compared to other research efforts in this field. As a result, the response of the force plate is characterized. Finally, records from a test subject are shown, as obtained through a dedicated application, incorporating the calibration results and specific filters, illustrating the performance capabilities of the system.

II. Force plate calibration

A. Background

Force plates rely on multiple load sensors (usually four) to dynamically estimate the weight of a subject (W_O), and the relative position of the center of pressure (CoP) in two orthogonal axes (i.e. Anterior Posterior-AP and Medial Lateral-ML). These sensors support a stable, flat surface, capable of holding large weights and sustaining their motions without experiencing deformations during tests. The actual weight of the support surface (W_S) is therefore added as an offset value in such a way that, when considered ideally distributed, it is equivalent to placing W_S at the center of the force plate [6]. When adding the weight of another object or subject, the actual, measured CoP (CoP^M), will be affected by this force, so that the ratio between the coordinates of the CoP of the object (CoP^R) and CoP^M will be, for the AP and ML directions:

$$r_C = CoP^R / CoP^M = (W_O + W_S)/W_O \qquad (1)$$

Therefore, for a fixed W_S the ratio between real and measured coordinates (r_C) becomes a function of body weight. While this background allows estimating CoP coordinates, a validation for the real, precise behavior of force plates is needed, especially for dedicated designs, and for the assessment of the stability of the behavior of any device over time.

B. Requisites

The basic premise for calibrating a force plate requires the fixed positioning of a number of known weights at known positions on the device, resulting in calibration functions or matrices. For this purpose, specific devices have been developed as part of research solutions [6, 7]. While some of these

I. Torres et al. (eds.), *VII Latin American Congress on Biomedical Engineering CLAIB 2016, Bucaramanga, Santander, Colombia, October 26th -28th, 2016*,
IFMBE Proceedings 60,
DOI: 10.1007/978-981-10-4086-3_50

require versatile, elaborate mechanics and active closed-loop control, others rely mainly on knowledge about the behavior of the force plate to determine an overall response characteristic. A dedicated device, combining characteristics of these, was developed as a low-cost, simple to use reference for the creation of calibration curves and analysis of force plates.

III. Materials and methods

A. Force Plate

A dedicated force plate (figure 1), developed by the group, was designed in order to assess static balance and specific gestures over a broad range of subjects (older adults, young athletes, etc.). This device includes four uniaxial load cells (Soehnle SEB4A [8]), with a capacity of 490.3 N in the vertical direction (combined error ±0.017%), resulting in a maximum capacity of 1961.2 N for the plate over a flat surface of 0.50 m by 0.50 m. Outputs from the load cells are amplified and acquired at a sampling rate of 60 Hz, through a 12-bit Analog-to-Digital converter, and later accessed by a dedicated application that updates the CoP plot in real time, while saving records on a text file. CoP coordinates, CoP offset values (which are independent of W_O), raw signals and event indicators are calculated and saved into this file, which is, as a result, available for further analysis. This dedicated force plate was developed by the group as a research effort, in order to promote a higher flexibility and upgraded capabilities for the system, in combination with other devices that have been developed for the study of human balance: its architecture allows incorporating combined trigger functions for these devices, and programmable sample rate, with dedicated filters, analysis tools and user interfaces, and the further possibility to upgrade and embed signal processing algorithms (such as the calibration function) into the device in the future.

B. Calibration device and accessories

The proposed calibration device (figure 2) consists on a spinning surface, weighing 44.7 ± 0.7 N, which smoothly rotates over a strong axis supported by three equidistant bases, guaranteeing the constant distribution of a rotating weight over its full rotation. As a consequence of this setup, and considering that the calibration device adds a specific weight, equivalent weight and radius values have to be calculated, through a similar transformation to the one defined in equation (1), in order to estimate the real CoP coordinates to be fitted in defining the calibration function.

This layout allows tracing curves (i.e. circles, under ideal conditions), following the path of the CoP^M in both axes, while weights are rotated. Through the tracing of these shapes, a further understanding concerning the fine, joint response of the load cells, and the CoP coordinates, can be achieved. This layout also allows positioning weights on top of the rotation axis, as well as weights over the rotating surface. In this way, this device can be used under a significant number of combinations, achieving both large distances from the center of the device and sustaining significant weights. An adaptable motor can also be added to the device, in order to provide slow, constant rotations of the equivalent weight.

Grading marks, specific to this device, have been developed with the aid of CAD tools, and fitted to the device and platform, in order to guarantee the fixed placement of the calibrating device, and of each weight, at different distances from the axis of rotation (resolution ±1.0 mm). Known weights were combined, resulting in a large number of curves for different equivalent weight values and distances from the center of the device, as shown in figure 2. The force platform was leveled during testing.

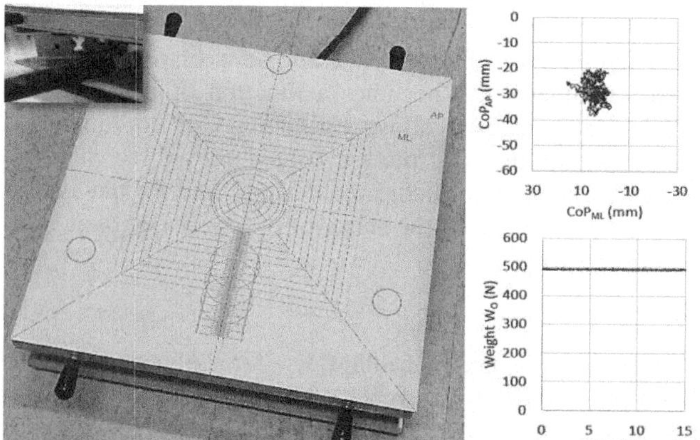

Fig. 1 Force plate layout, SEB4A load cell (top-left), and derived signals (right: Center of Pressure Coordinates CoP_{AP}, CoP_{ML}, and Weight W_O)

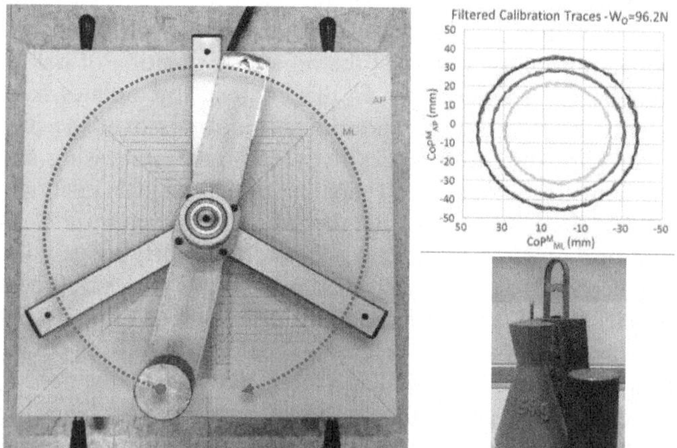

Fig. 2 The calibration device is positioned over the plotted surface of the platform. Rotation of combined weights allows tracing calibration patterns.

C. Calibration curve estimation, analysis and system setup

Table 1 specifies the traces and static measurements involved in the calibration procedure. The relationship between measured and reference W_O values, as well as the value of W_S (offset), were estimated by placing static weights on the plate. Measurements from a single-sensor force plate (Vernier FP-BTA, resolution ± 0.3N) were used as reference in determining this response. In order to define the calibration functions, force plate measurements were performed with the calibration device for up to five distances from its axis and different weights. Slow rotations were repeated three times in each case, and signals were filtered with a Hamming window digital filter (F_{cutoff}=1.0 Hz), for analysis. Based on these traces, ellipses were fitted with a modified least-squares method [9] solving equation 2:

$$\left(\frac{CoP^M{}_{AP}-Z_{AP}}{S_{AP}}\right)^2+\left(\frac{CoP^M{}_{ML}-Z_{ML}}{S_{ML}}\right)^2=1 \qquad (2)$$

where parameters Z_{AP} and Z_{ML} are the central coordinates of the ellipse (center of the calibration device), and S_{AP} and S_{ML} are the respective axis values of the fitted shapes. This model assumes no significant rotation for the ellipses. Results from this method allowed inferring ratios between CoP^R and CoP^M as functions of W_O, and therefore calibration curves of r_C in both axes. These were compared to the ideal curve (equation 1) resulting from the calculated W_S value. Error curves, eccentricity and linearity were characterized, and the derived functions were implemented on a dedicated analysis tool.

Table 1: Traces and measurements used for calibration, both with the calibration device (rotation) and with static weights on the force plate

Weight rotation over axis						*Static weight on force plate surface*		
Equivalent Weight (N)	Calibration Traces: Equivalent Radius (mm) *					Weight (W_O ± 0.6) N		
73.0 ± 0.7	36.4	45.6	54.7	63.8	72.9	26.6	47.8	56.5
101.3 ± 0.7	55.8	66.9	78.1	89.2	100.4	96.2	145.2	196.3
141.5 ± 0.7	68.4	75.3	82.1	88.9	95.8	291.5	388.3	486.8
184.5 ± 0.9	52.1	57.4	62.6	67.8	73.0	*Propagated uncertainty specific to each value		
246.8 ± 0.5	14.4	18.3	22.2	26.1	30.0			
294.6 ± 1.1	15.3		18.6		21.9			

IV. Results

A. Calibration results

The relationship between averaged static measures of W_O by the device (measurement time: t_m>30 s), and the acquired reference values, proved to be linear (offset W_S = 262.1 N) with a coefficient of determination R^2=1.0 for a linear fit.

Rotation of the equivalent weights from Table 1 resulted in adequate traces by the force plate (example in Figure 2). Relative error between traces and fitted ellipses was plotted as a function of rotation angle, showing no significant trends compared to noise levels, reinforcing the assumption of elliptic traces with dominant AP and ML axes. Given the multiple traces acquired for each weight, a measure of linearity (optimal value: 1.000) was tested for changes in S_{AP} and S_{ML}, in relation to the corresponding relative increases in equivalent radius. This resulted in values of (0.999 ± 0.009) for the AP axis and (0.994 ± 0.009) for the ML axis.

Considering the theoretical background from section II.A, r_C values for CoP coordinates in the AP and ML directions (estimated as the ratio between each equivalent radius and corresponding S_{AP} or S_{ML} values), were fitted as potential functions in W_O. Results were weighted by a factor of $1/(M\ \Delta r_C)^2$, where M is the number of traces plotted for each weight (3 or 5, see Table 1) and Δr_C is the propagated uncertainty of each r_C value (prioritizing results with a higher degree of confidence). As a result, the calibration curves and equations shown in Figure 3 were obtained. Combined with the previously calculated CoP offset values, these functions characterize the spatial response of the force plate.

Similarity between AP and ML curves proved to be significant, with a low eccentricity (e = 0.095 ± 0.044) for the fitted ellipses. Results show, however, a difference between real and theoretical calibration curves, and this finding is most probably due to device-specific factors, such as differential traction between load cells when fixed to the support surface, as well as tolerances in dimensions and amplification stages between cells. This response has proved to be consistent and stable between calibration procedures.

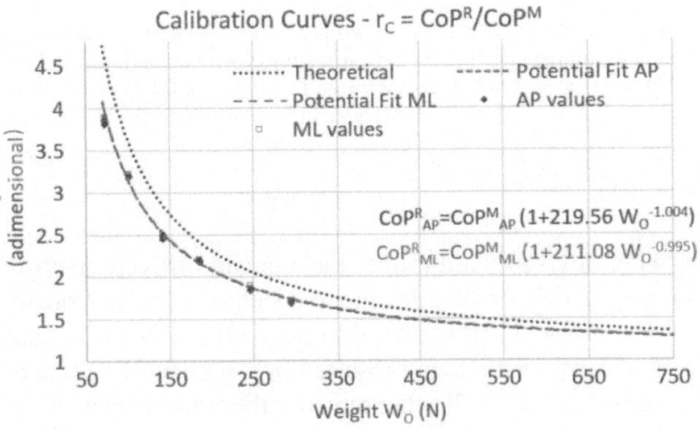

Fig. 3 Acquired and theoretical calibration curves of the force plate.

B. Force Plate Analysis Tool

Acquired calibration functions were incorporated into a dedicated analysis tool developed in Matlab, resulting, as a whole, in the completion of a comprehensive system for the assessment of human balance. Figure 4 shows results from tests on a young adult, a professional ballet dancer (19 years old), while performing standing balance assessments without feet separation, both with open and closed eyes, as shown by the dedicated interface. This subject gave specific informed consent before testing, and performed these tests while standing on the platform with no shoes on.

The application allows selecting portions of the acquired signals, and choosing specific filters for standing balance analysis. Furthermore, it automatically calculates biomechanical values associated to gestures (sway-path length, deviation and range in both axes, areas covered by the gesture, etc.), and saves test results. CoP plots are prominently shown by the interface, allowing for comparisons between segments of different tests, as shown in Figure 4. For this subject, results showed a 12.6 % larger sway path for a test with closed eyes (t_m = 60 s), as opposed to the same test when performed with the subject's eyes opened.

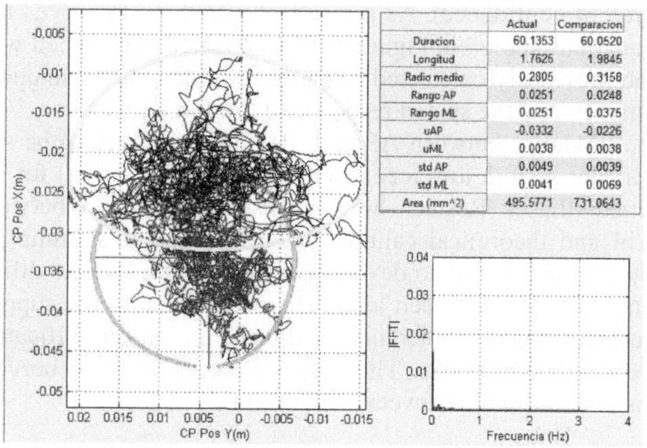

Fig. 4 Section of the developed interface for balance assessments. Standing balance CoP registers (t_m=60 s) from a subject with open (blue) and closed (black) eyes are filtered, displayed, and compared on the table (left: eyes open, right: eyes closed), showing significant differences between cases.

V. CONCLUSIONS

The devised calibration mechanism proved useful for gaining a deeper insight into the behavior of the designed force plate. Calibration results opened the way for validating its derived registers, and resulted in a superior fit for the device, compared to the theoretical calibration curve. As a result, a comprehensive, programmable system was set up, able to generate biomechanical evaluations associated to static human balance, as well as for dynamic gestures within the range and capabilities of this device. Further work includes the periodic reiteration of the calibration procedure, and the integration of the force plate with other devices developed by the group (wearable accelerometers and gyroscopes), so as to obtain synchronized data during gait and balance assessments.

ACKNOWLEDGMENT

This work was supported in part by Universidad de Buenos Aires, Programación Científica 2013-2016, Proyecto Interdisciplinario PIUBAMAS 20621201000011BA: "Evaluación del riesgo de caídas en adultos mayores y en pacientes geriátricos".

This work was supported in part by Pontificia Universidad Católica Argentina, Facultad de Ciencias Fisicomatemáticas e Ingeniería: "Modelos biomecánicos y desarrollos de dispositivos para la cuantificación del movimiento humano en individuos normales y/o disfuncionales".

Authors would like to thank Gonzalo Fernández for his insight and assistance during the setup of the device.

CONFLICT OF INTEREST

The authors declare that they have no conflict of interest.

REFERENCES

1. United Nations DoEaSA (2013) World Population Ageing. New York. ST/ESA/SER.A/348
2. Duarte M, Freitas SMS (2010) Revision of Posturography based on force plate for balance evaluation, *Rev Bras Fisioter*, 2010 May-Jun; Vol. 14(3), pp. 183-92, ISSN 1413-3555
3. Paillard T and Noé F (2015) Techniques and Methods for Testing the Postural Function in Healthy and Pathological Subjects *Biomed Res Int.* 2015; 2015:891390. doi: 10.1155/2015/891390.
4. Han J, Moussavi Z, Sztrum T, Goodman V (2005) Application of nonlinear dynamics to human postural control system, *Conf Proc IEEE Eng Med Biol Soc.* 2005; 7, pp. 6885-8.
5. Miralles M, Vecchio R et al (2012) Estudios de equilibrio en pacientes con riesgo de caída a partir de datos de acelerometría en tres ejes, *Proc. XIV JIIC*, Paraná, Argentina. http://bioingenieria.edu.ar/grupos/geic/biblioteca/j2012/Documentos/Trabajos/T12TCAr14.pdf
6. Baratto M, Cervera C, Jacono M (2004) Analysis of adequacy of a force platform for stabilometric clinical investigations, *IMEKO, IEEE, SICE 2nd International Symposium on Measurement, Analysis and Modelling of Human Functions, June 14-16, 2004*, Italy. http://ada-posturologie.fr/home-a.htm
7. Blanchard Sanhueza G (2010) Test system for clinical force platforms, *Conf Proc IEEE Eng Med Biol Soc* 2010; 2010:5772-5. doi: 10.1109/IEMBS.2010.5627840.
8. Soehnle SEB4A. See http://www.soehnle-professional.com/
9. Ghersi I, Miralles MT, Mariño M (2014) Augmented-Reality and Alternative-Input System based on Simultaneous Head Position and Pose Estimation, *IFMBE Proceedings*, Vol 49, pp 222:225.

Corresponding author:	Ignacio Ghersi
Institute:	LaBIS-FI-UCA
Street:	1500 Moreau de Justo Ave.
City:	Buenos Aires
Country:	Argentina
e-mail:	igghersi@gmail.com

Medical Device by Health Care Facility Interoperability in Alarm Management

J. Loureiro[1] and R. Garcia[1]

[1]Biomedical Engineering Institute, Federal University of Santa Catarina (IEB-UFSC), Florianópolis, Brazil

Abstract— **This paper presents a proposal to incorporate the concept of medical device interoperability in the healthcare technology management as an evaluation tool of clinical engineering, contributing to the risk management of the technological process in health. The methodology consists of researching the interrelation between each level of interoperability with each domain of the technological process in health, identifying the common attributes that may be evaluated in a structural model. The proposed model was used to classify the medical device interoperability stage of health care facility and also alarms management evaluation of interoperability solutions in ICU simulation environment. It was verified that interoperability may further enhance alarm management, contributing to propitiate increasing patient safety, through appropriate connectivity, clearly defined processes as well as trained human resources.**

Keywords — **Medical device by health care facility interoperability, alarm management, healthcare technology management, clinical engineering, process technology in health.**

I. Introduction

The incorporation and use of Medical Device by Health care Facilities (MDHcF) [1] from different manufacturers are one of the main challenges faced by Health care Facilities (HcF) [1] in the proper use of their resources directed to patient care.

The Emergency Care Research Institute (ECRI) highlights the importance of the alarms of the ten hazards to health technology [2] and the World Health Organization (WHO) has studies indicating that one of the main causes of MDHcF inefficiency is caused by inadequate management [3].

An appropriate policy of alarm management could reduce adverse events related to monitoring, increasing confidence in the urgency and relevance of the alarms [4]. Implementation of such a policy could be made with the use of interoperability between MDHcFs of different manufacturers, by open standards, providing a reduction of false alarms, reducing cognitive load of the medical staff and hence greater safety for the patient [5].

Interoperability of MDHcF may be understood as the ability of MDHcFs and clinical systems communicate with each other safely and with a specific purpose [6].

However, the simple adoption of standards may generate no guarantee of safety and reliability, because these patterns tend to be multi-purpose general rules not designed to address specific clinical needs. As a consequence, organizations that act as certification entities, such as the Integrating the Healthcare Enterprise (IHE), specify integration profiles with precise functions of how the standards should be implemented [7].

These problems arising from an MDHcF multivendor environment could be avoided with an appropriate policy of Healthcare Technology Management (HTM) having interoperability of MDHcF as one of its assessment tools. The Clinical Engineering (CE) has made efforts based on the HTM, aimed to generate safety, reliability and effectiveness for the Health Technological Process (HTP). Thus, the CE must make use of assessments tools for the operation of its activities, which are based on three domain of HTP: infrastructure, human resources and technology [8].

In this context, the paper presents a methodology for incorporating the MDHcF interoperability concept in HTM, as an assessment tool to be used by the CE to improve the HTP.

II. Materials and Methods

The structure of the methodology was systematized in four stages - definition of a structural model of interoperability MDHcF, application and validation of the model in HcFs, analysis of integration solutions of MDHcF in the HTP laboratory of the IEB-UFSC and recommendation actions.

The stage 1 is founded on a stratified structural layered model, based on the context of different visions for interoperability. The structural model is divided into a first layer supporting the others and related to aspects of infrastructure; one following layer is concerned with establishing communication and make it safe; a forthcoming layer addresses the aspects related to the nomenclature of clinical terms and finally a layer organizing the data exchanged in processes and well-defined actors.

The methodology in this stage 1 consists in researching the interrelation between each level of the proposed model of MDHcF interoperability with each domain HTP, identifying attributes that are common to both domain of HTP as the interoperability levels adopted in the model.

I. Torres et al. (eds.), *VII Latin American Congress on Biomedical Engineering CLAIB 2016, Bucaramanga, Santander, Colombia, October 26th -28th, 2016*,
IFMBE Proceedings 60,
DOI: 10.1007/978-981-10-4086-3_51

Has been found that the levels of communication and model semantics are related to technological domain of the HTP, the model data level is related to infrastructure domain of the HTP and model organizational level is related to human resources domain of the HTP.

The application of the structural model requires specific attributes to each of its levels, as shown in Table 1. It is emphasized that the model uses an adaptation of the general requirements of IHE PCD to the Brazilian reality, but this fact is in accordance with the fourth volume of the document Technical Framework (TF 4) IHE which provides for the use of national extensions [9].

Table 1 Attributes of the structural model of MDHcF interoperability

Interoperability level	HTP Domain	Recommended attributes
Data	Infrastructure	NBR14565, IEEE802
Communication	Technology	HL7 version 2.3 or 2.4
Semantics	Technology	SNOMED-CT
Organizational	Human Resources	Eng. Human Factor [10]

Stage 2 deals with the application of the structural model in HcFs having MDHcF integration solution of different manufacturers in order to identify the MDHcF interoperability scenario and suggest an action plan which is expected to improve the interoperability of MDHcF.

The MDHcF interoperability assessment method used in this research model is an adaptation of the method used in the Levels of Information Systems Interoperability (LISI) model [11], according to the protocol below:

1. Stage Independent: Inability to interoperate. HcF does not have minimal infrastructure.
2. Initial Stage: Has infrastructure to interoperate. The HcF meets only at the level of the model data.
3. Connected Stage: Has connectivity. HcF meets the levels of data and communication of the model, using standard protocol of interoperability in healthcare (HL7).
4. Integrated Stage: Has integration. HcF meets the levels of data, communication and model semantics, using standard of clinical terminology.
5. Interoperable Stage: Has interoperability. HcF serves all levels of the interoperability model, also worrying about organizational processes and human aspects.

Once known some of MDHcF integration solutions within the clinical context in HcF, we verified the importance that these solutions have in alarm management for patient safety. Stage 3, then consists of testing solutions in the laboratory of IEB-UFSC under alarm management approach.

MDHcF integration solutions were evaluated by clinical alarms simulations and technicians on the multiparameter monitors, pulmonary ventilators and infusion pumps, checking the record of these alarms in a central of monitoring and at one Hospital Information System (HIS) as shown in Fig. 1. The results from the case study were compared with the attributes of levels communication and semantics of the structural model of MDHcF interoperability.

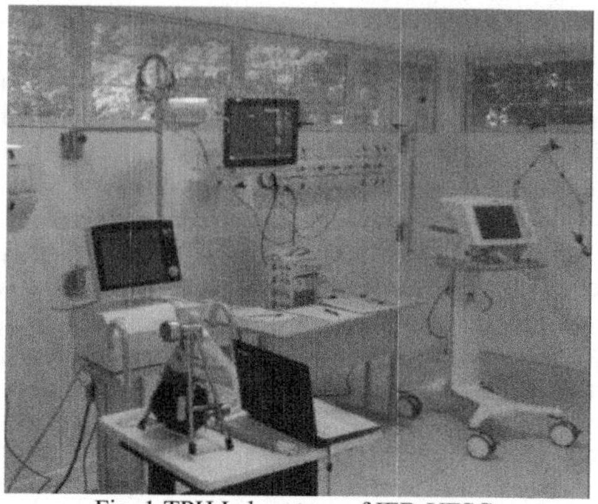

Fig. 1 TPH Laboratory of IEB-UFSC.

The actions provided in the Test Procedures and MDHcF Testing Protocols developed by the EC team of IEB-UFSC, seek to replicate the critical alarm situations to a patient in adult Intensive Care Unit (ICU) of a HcF, such as:

1. No alarm indication of a lost connection due to disconnected wire;
2. No indicator, audible and visual on the monitor screen and/or central, due to an alarm problem caused on the fan and/or pump;
3. Conditions of high priority alarms the fan and/or pump displayed on the monitor and/or central as medium or low priority;
4. Inability to display waveforms of physiological signals of the fan and / or pump on the monitor and central of monitoring.

The physiological parameters under test on the monitor are related to no invasive pressure, oxygen concentration, heart rate, respiratory rate, temperature, apnea and capnography. Whereas on the fans were tested pressure limit, tidal volume limit, PEEP limit, apnea and tachypnea. Technical alarms are related to failures in sensors of MDHcF, leaks, AC supply absence, and dead battery on the monitors and the fans.

Considering the result of the evaluation of MDHcF interoperability stage of HcF, it goes to stage 4 of the methodology in which are made recommendations for improvement in MDHcF interoperability of HcF.

III. RESULTS

This section presents the results obtained by the methodology applied in both Health care Facilities (HcF) analyzed and TPH laboratory of IEB-UFSC.

A. Case study in HcF

The studies were conducted in the adult ICU from two HcF of Santa Catarina, where an HcF is in the installation phase of MDHcF integration solution and the other in the use phase of the integration solution.

The physical infrastructure of both HCF agrees to the attributes established in the structural model data-level interoperability MDHcF, therefore attending the preconized in NBR 14565 and IEEE 802 standards.

However, the solution of one of the HcF requires the gateway installation in its data center to communicate in HL7 with the HIS. This generates more consumption of Information Communication Technology (ICT) resources, and more points of failure that the solution installed in other HCF in which communication by HL7 is made between the central of monitoring and HIS in the ICU.

The communication level of the model is attended by both HcF because they use HL7 standard message in version 2.4 in sending vital signs to the HIS server.

The naming clinical terms adopted by each hospital is of a proprietary nature, not considering the semantic level of the structural model of interoperability MDHcF, which indicates the SNOMED-CT.

At the organizational level of the model we sought to understand the change brought about in routine clinical and CE and ICT professionals, as well as the benefits for the patient using the integration solution in the ICU, of hospital 1 (H1). Whereas in the hospital 2 (H2) we aimed to realize the expectations of these professionals in relation to automatic collection of the patient's vital signs, since there is only monitoring alarms.

It was found that there is high expectation in H2 professionals with future automation of data collection, whereas H1 did not show the same optimism; although they agree that automation makes the job easier.

The alarm management had significant importance for improving patient safety in both hospitals, however H2 had more significant result. The solution in H2 has an alarm panel and vital signs in place visible to everyone anywhere in the ICU, whereas in H1, these data are presented only in a central of monitoring.

Regarding training, the professionals of both hospitals agree that it is suitable for the operation of its activities.

It was also found that the incorporation of the MDHcF integration solution improved the management of the location of MDHcF and scheduling of preventive and corrective maintenance. However, there is a lack in the use of quality indicators in MDHcF maintenance policy.

B. Case study in the HTP laboratory of IEB-UFSC

This test aimed to validate the levels of communication and semantics of the structural model, which correlates with the HTP technology domain. The other levels were validated in the case studies in the HcF.

The choice of MDHcF manufacturers for testing was based on the importance of the brand's market share, and could test so far two of the four solutions that dominate the Brazilian market for MDHcF integration.

It was observed that, relative to the level of communication of the structural model, the solutions using the standard HL7 versions 2.3 and 2.4 for the messages exchanged between central monitoring and HIS. Communication between the ventilator and the monitor and between the monitor and the central of monitoring is through the monitor manufacturer's proprietary protocol.

As for the semantic level of the structural model, the research found that the HIS database allows the HcF fill it with their own data in different nomenclatures.

Another observed factor was that the integration solution of each manufacturer can monitor a distinct relationship of physiological ventilation parameters using a particular set of acronyms to represent the physiological parameters and alarm messages.

Regarding on the registration of alarms, one of the solutions could not provide technical alarms in the central of monitoring. HIS managed to store clinical alarms, however did not have the possibility of registration of technical alarms.

The MDHcF integration solutions tested, fail to monitor alarms and physiological parameters of infusion.

IV. DISCUSSION

The Brazilian reality is dominated by integration solutions of own MDHcFs manufacturers, with the use of an external module acting as a "converter" to compatibilize the communication between MDHcFs from different manufacturers. Interoperability aspects can be found only on the messages between the central of monitoring and HIS made through the HL7 standard.

The focus on MDHcF integration solutions is in monitoring alarms and physiological patient parameters. The focus in the technical data of MDHcF, important for proper management of the technological park in the Health care Facilities by the CE, still lacks development.

It was found that the HcF, analyzed according to criteria adopted in the structural model of MDHcF interoperability, are on the "connected" stage of the MDHcF interoperability, because exclusively meet the attributes of data levels and communication model.

In the semantic level of the model it was found the possibility of improvements in interoperability MDHcF of the HcF with the adoption of standard medical terminology SNOMED-CT provided in the ordinance 2073 from the Ministry of Health (MOH).

At the organizational level of the model, it was noticed that the improvement in the quality of record maintenance service orders, with well-defined terminology for the failure modes, could allow the CE to make use of reliability and performance indicators to improve risk management of Health Technological Process.

Tests conducted in the laboratory of IEB-UFSC were able to highlight potential improvements in alarm messages leaving the meaning of more evident alarm, and the possibility of HIS store technical alarms in order to improve risk management by the CE.

The ECRI has been developing tests with similar purpose in order to collaborate with the MDHcFs manufacturers and make the solution more reliable and safe for the health of the patient [6]. These tests allow the CE and ICT professionals know the MDHcF integration solutions and act in support of the decision of the HcF managers in the acquisition of MDHcFs and ICT systems that are more compatible with each other, bringing greater safety, effectiveness and reliability to Health Technological Process.

V. CONCLUSION

The experimental data obtained from the application of the structural model developed in the methodology, indicated that a suitable alarm management in a heterogeneous environment of MDHcF may increase confidence in the relevance of the alarms, opening the possibility of expanding the management without increasing fatigue for clinical staff.

Thus, the preliminary results of incorporating of the methodology in the Healthcare Technology Management showed promising to increase patient safety, through a risk management concerned to reduce the number of adverse events caused by alarm management limitation.

Finally, the methodology will be applied in other MDHcF integration solutions in other Health care Facilities, in order to expand the evaluation of the MDHcF interoperability scenario and propose improvements to the Health Technological Process.

ACKNOWLEDGMENT

The authors would like to acknowledge financial support from CAPES.

CONFLICT OF INTEREST

The authors declare that they have no conflict of interest.

REFERENCES

1. WHO. World Health Organization. Available at www.who.int.
2. ECRI. (2014) Top 10 Health Technology Hazards for 2015. Health Dev.
3. Nunziata E. et al. (2004) Application of the EHTP Methodology to plan Primary Health Care Services in Mozambique at http://www.who.int/medical_devices/publications/en/EHTP_Mozambique.pdf.
4. Joint Commission. (2011) Sound the Alarm: Managing Physiology Monitoring Systems. The Joint Commission Perspectives on Patient Safety, v.11, issue 12.
5. Arney D. *et al.* (2011) Biomedical Devices And Systems Security. In: Engineering In Medicine And Biology Society, EMBC, 2011 Annual International Conference Of The IEEE. IEEE, pp. 2376-2379.
6. Association for the Advancement of Medical Instrumentation; U.S. Food and Drug Administration. (2012) Medical Device Interoperability: a Safer Path Forward - Priority Issues from the 2012. AAMI-FDA Interoperability Summit.
7. Rhoads J et al. (2010) Medical Device Interoperability and Integrating the Healthcare Enterprise (IHE) Initiative. Biomedical Instrumentation & Technology Journal, pp. 21-27.
8. Moraes L, Garcia R. (2004) Proposta de um Modelo de Gestão da Tecnologia Médico-Hospitalar. In: Anais do III Congresso Latino-Americano de Engenharia Biomédica CLAIB'2004, 2004, v.4, pp.309-312.
9. IHE PCD TF-1. (2015) IHE Patient Care Device (PCD) Technical Framework v.1 at http://www.ihe.net/uploadedFiles/Documents/PCD/IHE_PCD_TF_Vol1.pdf.
10. Carayon P. (2011) Handbook of Human Factors and Ergonomics in Health Care and Patient Safety. [S.l.]: CRC Press.
11. Rezaei R et al. (2014) Interoperability evaluation models: A systematic review. Computers in Industry, v.65, n.1, pp. 1-23.

Author: Jorge Luiz da Costa Loureiro
Institute: Instituto de Engenharia Biomédica - UFSC
Street: Campus Universidade Federal de Santa Catarina - UFSC
City: Florianópolis
Country: Brazil
Email: jorgeloureiro@outlook.com

Characterization Framework for Ex-combatants Based on EEG and Behavioral Features

Andrés Quintero-Zea[1], Lina M. Sepúlveda-Cano[2], Mónica Rodríguez Calvache[1], Sandra Trujillo Orrego[3], Natalia Trujillo Orrego[3,4] and José D. López[1]

[1] SISTEMIC, Engineering Faculty, Universidad de Antioquia UDEA, Calle 70 No. 52 - 21, Medellín, Colombia
[2] ARKADIUS, Engineering Faculty, Universidad de Medellín UDEM, Cra. 87 No. 30 - 65, Medellín, Colombia
[3] Mental Health Group, School of Public Health, Universidad de Antioquia UDEA, Cra. 53 No. 61-30, Medellín, Colombia
[4] Neuroscience Group, Faculty of Medicine, Universidad de Antioquia UDEA, Medellín, Colombia

Abstract— **This paper presents a framework to characterize the emotional processing of Colombian ex-combatants from illegal groups. The classification process is performed using EEG-ERP data and behavioral features from psychological tests. The results show that ex-combatant and civilian populations can be automatically separated using supervised techniques. With this, we can provide a decision support system for psychologists to improve current interventions aimed to help ex-combatants to make a successful reintegration to civilian life.**

Keywords— **Emotional Processing, Emotional Recognition Task, ERP, Ex-combatants, Supervised Learning.**

I Introduction

Ex-combatants from illegal groups in Colombia manifest an increased expression of aggression. A previous work demonstrated that they present alterations on their emotional processing [1]. However, this study did not analyze early stages of physiological processing. Veterans mainly manifest differences on the processing of unpleasant or violent emotions [2]. On ex-combatants, we hypothesize an atypical functioning on similar mechanisms. In order to improve the current interventions, we need biological markers to characterize atypical functioning of emotional processing.

In this line, Electroencephalography (EEG) is widely used as an index of neurophysiological activity associated to electrical activations in the brain, measured with electrodes placed on the scalp [3]. Given that event-related potentials (ERP) capture neural activity related to both sensory and cognitive processes, it is the most used approach to characterize EEG changes [4].

Several studies have reported relations between ERP components and the behavior or personality of healthy subjects. Results in [4] show an overview of the different ERP waveforms and the major findings in various psychiatric conditions. Relations between impulsivity and P3 amplitude/latency were studied in [5] with subjects suffering of high anxiety. While [2] showed that combat veterans with PSTD exhibit greater ERPs.

This paper presents a framework based on ERP components and psychological tests for automatic characterization of ex-combatants, in order to design strategies that are useful for their reintegration into civilian life. For feature selection and extraction, Partial Least Squares (PLS) method is used as a supervised projection, aiming to preserve the components that are maximally related with the labels. Results show that we indeed can characterize ex-combatants from civilians. The proposed methodology can be used as a decision support system to develop efficient intervention protocols.

II Materials

A Participants

The participants were 30 Colombian ex-combatants (two female) from *Agencia Colombiana para la Reintegración* (ACR) reintegration program, and 20 Colombian individuals (paired by sex, age and school level). All subjects participated voluntarily and signed an informed consent in agreement with the Helsinki declaration. Demographic information is provided in Table 1.

Table 1: Demographic information.

	Ex-combatants	Civilians	
	$n = 30$	$n = 20$	p
Gender	2:28	2:18	0.678
Age (years)	M = 37.50 SD = 8.22	M = 36.15 SD = 9.17	0.589
Educational level (years)	M = 10.33 SD = 3.10	M = 11.05 SD = 2.14	0.373

I. Torres et al. (eds.), *VII Latin American Congress on Biomedical Engineering CLAIB 2016, Bucaramanga, Santander, Colombia, October 26th -28th, 2016*,
IFMBE Proceedings 60,
DOI: 10.1007/978-981-10-4086-3_52

B Emotional Recognition Task Procedure

For this study, we implement a modification of the Dual Valence Association Task (DVAT) [6]. A two-alternative, forced-choice task, in which participants are asked to classify words or faces displayed on a computer screen according to their valence, into one of two categories (positive or negative) as quickly as possible [6]. Our modifications consisted in the inclusion of *Neutral* as a third valence level, and the removal of the simultaneous stimuli block described in [6].

Fig. 1 shows the pipeline of a single trial. Block trials are presented one-by-one with strict alternation between words and faces, and no more than two consecutive stimuli with the same valence.

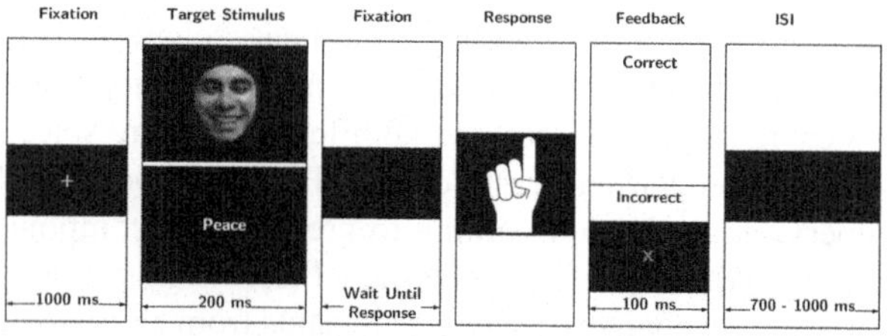

Fig. 1: **Experimental design.** Trials start with a fixation cross, followed by a target stimulus: single stimulus face or word. Feedback is provided only in error trials. Time between trial completion and onset of the subsequent trial (ISI) varies between 700 and 1000 ms.

C ERP Recordings

EEG signals were sampled at 1000 Hz from a 64-channel Neuroscan SynAmps2 amplifier. They were band-pass filtered between 0.1 and 30 Hz. We selected 60 electrodes for this study (HEO, VEO, CB1 and CB2 were excluded as they do not record neural activity). Acquired EEG signals were re-referenced off-line to average electrodes, and downsampled to 500 Hz. Continuous EEG data were segmented from 200 ms prior to the stimulus to 800 ms after. All segments contaminated with eye movement were removed from further analysis using Independent Component Analysis (ICA) and visual inspection. Artifact-free segments were averaged to obtain ERPs. The preprocessing stage was performed using EEGLAB Toolbox [7]. Unless DVT presents faces and words, only word stimuli were considered for this work.

D Psychological Tests

Both ex-combatants and civilians completed a neuropsychological evaluation. It included two psychological tests: the Social Ability Scale of Gismero (EHS) [8], and the Reactive-Proactive Aggression Questionnaire (RPQ) [9].

The EHS measures the assertion of individual's everyday social interaction. It consists of 33 items, 28 redacted in negative sense (absence or lack of abilities), and 5 in positive sense. It explores 6 dimensions of the social skills. This scale allows obtaining a global score and punctuations for each of the six dimensions [8].

The theoretical foundation of the RPQ suggests that the exploration of reactive and proactive aggression has to include the motives associated with behavior, context and type of reaction; such as physical and verbal aggression [9]. It has 23 items divided in two dimensions rated as *never*, *sometimes* and *often* for frequency of occurrence [9].

III Methods

A Features Extraction

The main ERP components and their properties are summarized in [10]. In accordance with the nature of the DVAT, two peaks are identified as dominants: (*i*) N170, a a member of the N2 family with latency between 156 and 189 ms, which reflects expert object recognition; and (*ii*) P300 (or P3), which occurs in response to an unexpected stimulus type approximately 300 ms after the stimulus onset.

In this work, we use the following procedure to identify the lag and amplitude of the N170 and P300 ERP components: *i*) a multilevel 1-D non-decimated Haar wavelet decomposition is calculated for the ERP signal, *ii*) the minimum (N170) and maximum (P300) values of the first approximation coefficient are found between 150 and 350 ms, *iii*) the occurrence time of both peaks is brought to the original ERP signal, and *iv)* the peak amplitudes are calculated. Fig. 2 shows a random ERP signal, its first approximation coefficient, and the detected peaks.

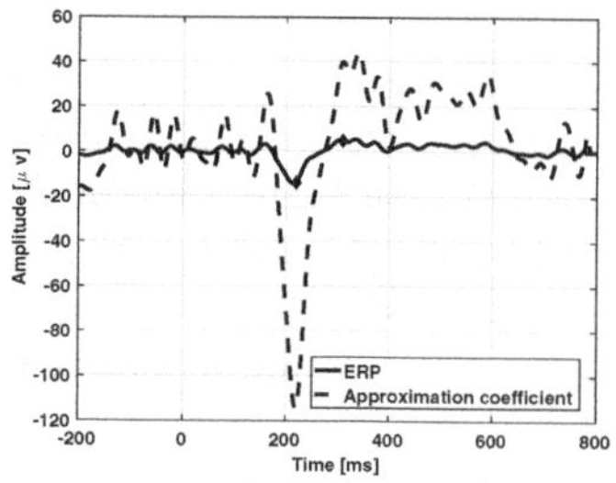

Fig. 2: **ERP component detection.** The peaks N170 and P300 are detected from the first approximation coefficient of the non-decimated Haar wavelet decomposition of the ERP. The peaks amplitudes are then obtained from the original ERP signal.

Finally, we complete the feature vector with the psychological scores. Specifically, for each participant we take the global score from the EHS, the punctuations in each dimension (reactive and proactive) of the RPQ, and the total RPQ punctuation.

B Partial Least Square Regression

The main idea of PLS regression is to find a linear model to describe some predicted variables $\mathbf{X}$ in terms of other observable variables $\mathbf{Y}$. For a L classification problem as in [11], with $(\mathbf{x}_i, y_i) \in \mathbf{X} \times \{C_1, \ldots, C_L\}$, $\mathbf{x} \in \mathbb{R}^p$, with Q the number of features and n the number of observations. The sample vectors $\mathbf{X}$ and response $\mathbf{Y}$ matrices are given by:

$$\mathbf{X} = \begin{bmatrix} x_{11} & \ldots & x_{1Q} \\ \vdots & \ddots & \vdots \\ x_{n1} & \ldots & x_{nQ} \end{bmatrix} \qquad \mathbf{Y} = \begin{bmatrix} 1 & 0 & \ldots & 0 \\ \vdots & \vdots & \ddots & \vdots \\ 0 & 0 & \ldots & 1 \end{bmatrix} \tag{1}$$

where each row of $\mathbf{Y}$ contains ones in positions denoting class labels. PLS regression searches for a set of components (called latent vectors) that perform a simultaneous decomposition of $\mathbf{X}$ and $\mathbf{Y}$ as in Eq. (2). These components explain as much as possible of the covariance between $\mathbf{T}$ and $\mathbf{U}$ [12].

$$\begin{cases} \mathbf{X} & = \mathbf{T}\mathbf{P}^\top + \mathbf{E} \\ \mathbf{Y} & = \mathbf{U}\mathbf{Q}^\top + \mathbf{F} \end{cases} \tag{2}$$

Here, matrices $\mathbf{E}$ and $\mathbf{F}$ are the error terms, assumed to be i.i.d. normal, $\mathbf{P}$ and $\mathbf{Q}$ are orthogonal loading matrices, and $\mathbf{T}$ and $\mathbf{U}$ are the projections of $\mathbf{X}$ and $\mathbf{Y}$. The algorithm used in this work for PLS implementation is the well–know Non-Linear Iterative Partial Least Squares Algorithm (NIPALS).

C k-Nearest Neighbors Classifier

The k-Nearest Neighbors (kNN) algorithm starts from a training set $\mathbf{X}_e = \mathbf{x}_1, \ldots, \mathbf{x}_m \subset \mathbf{X}$, $m < n$, labeled with a class label $y_j \in \mathbf{Y}$. Its objective is to classify an unknown sample χ. According to [13], for each $\mathbf{x}_i \in \mathbf{X}_e$ the distance between χ and $\mathbf{x}_i$ could be calculated as follows:

$$d\{\chi, \mathbf{x}_i\} = \sum_{q \in Q} w_q \delta(\chi_q, x_{iq}), \tag{3}$$

where δ is any distance metric, i.e., Mahalanobis, Euclidean, Minkowski, Hamming, among others; and w_q is the distance weighting function. The k nearest neighbors are selected based on this distance metric. The class of χ could be assigned to the majority class among the nearest neighbors; nevertheless, there are several ways to select the χ label, according to the weighting factor.

D Support Vector Machine Classifier

Support vector machines (SVM) are linear classifiers developed on statistical learning theory (SLT) by Vapnik [14]. SVM are supervised learning models that aim to find an optimal hyperplane to separate the points of two classes. C-SVM is a widely used type of SVM (see [15] for implementation details).

For linearly separable data, the maximum margin hyperplane is determined by the constrained optimization problem

$$\begin{aligned} \underset{\alpha}{\text{minimize}} \quad & \frac{1}{2} \sum_{i=1}^{m} \sum_{j=1}^{m} y_i y_j \alpha_i \alpha_j \kappa(\mathbf{x}_i, \mathbf{x}_j) - \sum_{j}^{m} \alpha_j \\ \text{subject to} \quad & \sum_{i=1}^{m} \alpha_i y_i = 0, \\ & 0 \leq \alpha_i \leq C, i = 1, \ldots, m, \end{aligned} \tag{4}$$

where $\kappa(\mathbf{x}_i, \mathbf{x}_j)$ represents a kernel function, and C is the penalty factor that controls the complexity of the SVM. The decision function is

$$f(\mathbf{x}) = \operatorname{sgn}\left(\sum_{i}^{m} \alpha_i y_i \kappa(\mathbf{x}_i, \mathbf{x}) + b \right) \tag{5}$$

For the C-SVM model specification, the Gaussian (or radial basis) kernel is defined as

$$\kappa(\mathbf{x}, \mathbf{y}) = \exp\left(\frac{-\|\mathbf{x} - \mathbf{y}\|^2}{2\sigma^2} \right) \tag{6}$$

The penalty factor C and the parameter σ of the kernel must be tuned by minimizing the estimated generalization error.

IV Results and Discussion

We used the classification accuracy to tune the parameters of both the number of PLS components, and the number of neighbors for the kNN classifier. The highest results were obtained with two components and three neighbors. The C-SVM model selection was made by grid search. The best performance was reached with $C = 1.6$ and $\sigma = 1.0$.

Results show that although both classifiers achieved over 75% of accuracy (see Table 2), the SVM-based classifier presented better performance, achieving 80.00% of accuracy with a confidence interval of (63.87 - 90.88)%. Despite that evidence points out the existence of alterations in the emotional processing of ex-combatants, the confidence intervals are wide due to the limited size of the sample, and because some of the features found also reflect non-emotional cognitive processes that may be common to both populations.

The sample size of our study was modest because neuropsychological assessment combined with ERPs in illegal ex-combatants is difficult and uncommon.

Table 2: Accuracy, sensitivity and specificity reached with the proposed methodology.

Method	Accuracy (%) (Conf. Interval)	Sensitivity (%) (Conf. Interval)	Specificity (%) (Conf. Interval)
KNN	77.50 (61.14 - 89.03)	85.00 (69.48 - 94.39)	70.00 (53.29 - 83.21)
SVM	80.00 (63.87 - 90.88)	85.00 (69.48 - 94.39)	75.00 (58.48 - 87.14)

Regarding to the population grouping, Fig. 3 shows that ex-combatants conform a well defined group, while civilians tend to share some features with the other group, *i.e.* some civilian samples mix with the ex-combatants group of samples. This complicates their identification as a group.

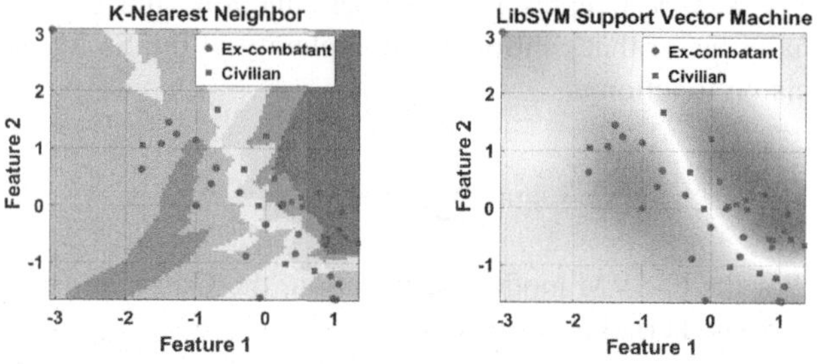

(a) (b)

Fig. 3: **Classification results for (a) *k*NN and (b) SVM.** Note that civilians are spread, while ex-combatants are more compact. This drive to lower specificity values.

V CONCLUSION

This paper introduced a framework to characterize ex-combatants from civilian people, as a first step to design interventions to treat their rage issues. This study showed that supervised techniques with EEG and behavioral features may differentiate emotional processing from ex-combatants with high accuracy and sensitivity.

ACKNOWLEDGEMENTS

The authors appreciate the assistance of *Agencia Colombiana para la Reintegración*. This work was partially supported by Colciencias Grants [122266140116 and 111556933399], CODI-UDEA INV518-16, doctoral fellowship call 647 (year 2014), and research project 762 (*Universidad de Medellín* and *Neurocentro de Pereira*).

CONFLICT OF INTEREST

The authors declare that they have no conflict of interest.

REFERENCES

1. Tobón Carlos, Ibañez Agustín, Velilla Lina, et al. Emotional processing in Colombian ex-combatants and its relationship with empathy and executive functions *Social Neuroscience.* 2015;10:153-165. PMID: 25302548.
2. Shu I-Wei, Onton Julie A., Prabhakar Nitin, O'Connell Ryan M., Simmons Alan N., Matthews Scott C.. Combat veterans with PTSD after mild TBI exhibit greater ERPs from posterior-medial cortical areas while appraising facial features *Journal of Affective Disorders.* 2014;155:234 - 240.
3. Light Gregory A, Williams Lisa E, Minow Falk, et al. Electroencephalography (EEG) and Event-Related Potentials (ERPs) with Human Participants *Current protocols in neuroscience.* 2010;CHAPTER:Unit–6.2524.
4. Sur Shravani, Sinha VK, others . Event-related potential: An overview *Industrial psychiatry journal.* 2009;18:70.
5. Pascalis Vilfredo De, Strippoli Enrica, Riccardi Patrizia, Vergari Fabiola. Personality, event-related potential (ERP) and heart rate (HR) in emotional word processing *Personality and Individual Differences.* 2004;36:873 - 891.
6. Ibáñez Agustín, Urquina Hugo, Petroni Agustín, et al. Neural Processing of Emotional Facial and Semantic Expressions in Euthymic Bipolar Disorder (BD) and Its Association with Theory of Mind (ToM) *PLoS ONE.* 2012;7:1-12.
7. Delorme Arnaud, Makeig Scott. EEGLAB: an open source toolbox for analysis of single-trial EEG dynamics including independent component analysis *Journal of Neuroscience Methods.* 2004;134:9 - 21.
8. Gismero E. EHS Escala de habilidades sociales *Madrid: TEA Publicaciones de Psicología Aplicada.* 2000.
9. Raine Adrian, Dodge Kenneth, Loeber Rolf, et al. The Reactive–Proactive Aggression Questionnaire: Differential Correlates of Reactive and Proactive Aggression in Adolescent Boys *Aggressive Behavior.* 2006;32:159–171.
10. Key Alexandra P Fonaryova, Dove Guy O., Maguire Mandy J.. Linking brainwaves to the brain: An ERP primer *Developmental Neuropsychology.* 2005;27:183–215.
11. Blaszczyk Pawel. Semi-supervised Feature Extraction Method Using Partial Least Squares and Gaussian Mixture Model *Lecture Notes in Engineering and Computer Science..* 2015.
12. Abdi Herv. Partial least squares regression and projection on latent structure regression (PLS Regression) *Wiley Interdisciplinary Reviews: Computational Statistics.* 2010;2:97–106.
13. Cunningham Padraig, Delany Sarah Jane. k-Nearest neighbour classifiers *Multiple Classifier Systems.* 2007:1–17.
14. Vapnik Vladimir N. *The Nature of Statistical Learning Theory*ch. Methods of Pattern Recognition, :138–155. New York, NY: Springer–Verlag2nd ed. 1995.
15. Shi Guo-You, Liu Shuang. Model selection of RBF kernel for C-SVM based on genetic algorithm and multithreading in *2012 International Conference on Machine Learning and Cybernetics*;1:382-386 2012.

Address of the corresponding author:
Author: Andrés Quintero Zea
Institute: Universidad de Antioquia
Street: Calle 70 No. 52 - 21
City: Medellín
Country: Colombia
Email: andres.quintero@udea.edu.co

Auditory Brainstem Responses with AEP_AUDIX System Using an Optimized Broadband Chirp Stimulus

L.M. Alvero[1], J.A. Gaya[2], C. Miret[1], E. Velarde[1], A. Torres[3], E. Eimil[3], Y. Torne[1], and E. Martinez[4]

[1] Cuban Neuroscience Center/Department of Electronics, Havana, Cuba
Cuban Neuroscience Center/Department of Neurophysiology, Havana, Cuba
[3] Cuban Neuroscience Center/Department of Audiology, Havana, Cuba
[4] Cuban Neuroscience Center/Department of Neuroinformatics, Havana, Cuba

Abstract— **A recent alternative in the generation of Auditory Evoked Response (ABR) has been the use of chirp stimuli developed to compensate for the traveling wave delay in the cochlea. This study examines the usefulness of a rising broadband chirp stimulus (modified version based on chirp stimulus development by Dau et al, 2000) implemented in the AEP-AUDIX system (Havana, Neuronic SA). Two experiments in normal-hearing subjects were performed: in the first, ABR elicited by the click and the broadband chirp were obtained at single moderate stimulus level (60 dB nHL) and repetition rate of 21/s. Wave-V (peak-to-peak) amplitude with the chirp-evoked ABR was larger (on the order of 0.20 µV, chirp-to-click amplitude ratio: 1.62, gain in amplitude: 54%) and significant than the obtained with click-evoked ABR (p<0.001, Wilcoxon matched-pairs test). In the second, the effect of different repetition rates (17, 21, 33, 55, 71, 77, 89/s) on wave-V amplitude for the chirp-evoked ABR was investigated. It was found that the wave-V amplitude remained constant at stimulus rate up to 89/s (higher significant: *t*-value=24.926, p=0.000), suggesting that the synchronicity was still large enough to generate stable components (the systems behaves essentially linearly at stimulus rate up to about 100/s). The overall results indicate that a rising broadband chirp is a more efficient stimulus than a click for the recording of early ABRs in normal hearing young adults. The use of chirp-evoked ABR may be particularly interesting in neonatal hearing screening devices (larger amplitudes might facilitate the automatic detection and reducing the testing time) or many other clinical testing situations.**

Keywords— **broadband chirp stimulus, Auditory Brainstem Response, cochlear delay model, wave-V, click**

I. Introduction

The auditory brainstem response (ABR) is considered a synchronized potential evoked by the simultaneous discharge of a large number of nerve cells in the auditory brainstem [1]. The ABR has been traditionally evoked with a click stimulus because their rapid onset and broad-frequency characteristics support higher neural synchrony and larger waveform amplitude.

However the response to a click is not entirely synchronous as a result of the traveling wave delay (change of stiffness along the basilar membrane): as there is an approximately exponential rise in frequency with that delay, so the response is dominated by neural activity from cochlear high-frequencies.

A recent advance in the generation of AEP has been the use of chirp stimuli which compensate for frequency-delay characteristics of the basilar membrane. The first comprehensive description and mathematical formulation of the chirp was presented by Dau et al., 2000 [2] and since then several experiments have been carried out about its use within the auditory field in the international literature. In several publications it has been shown that an upward chirp is more efficient than a click when recording the auditory compound action potential, ACAP [3]; the ABR [4-10], and the auditory steady-state responses, ASSR (narrow-band chirps) [11-14]. Still, the chirp is a relatively novel stimulus and there is a need for more systematic studies (commercial accessibility in a medical device has less than 10 years). There are a few reports about the use of chirp stimuli in children [15-18] or individuals with hearing losses [19-20].

The aims of this study were: *i*) to designed and generated a modified broadband chirp (on the basis of equations described in Dau et al., 2000) and implemented in the AEP_AUDIX system (Havana, Neuronic SA) recommended in national hearing screening program; *ii*) preliminary investigated about their feasibility for evoking ABR in normal hearing subjects.

II. Methods

A. Design of a broadband chirp

The general idea is to generate a stimulus in which the high-frequency components are delayed relative to the low-frequency components and produced synchronous displacement maxima and neural discharges resulting from all frequency components. The chirp used in this study was derived from the ′optimizing′ stimulus described by Dau et al., 2000 [2]. Because hearing thresholds information generally is not available in newborn hearing screening programs or other clinical testing situations, the present study selected a modified flat spectrum (or ′M′) chirp [2, 10].

I. Torres et al. (eds.), *VII Latin American Congress on Biomedical Engineering CLAIB 2016, Bucaramanga, Santander, Colombia, October 26th -28th, 2016*,
IFMBE Proceedings 60,
DOI: 10.1007/978-981-10-4086-3_53

The resulting chirp (Figure 1) is directly given by the general form:

$$s(t) = A(t)\sin(\varphi(t) - \varphi(t_0)) \quad (1)$$

When:

t (as a function of frequency) is the needed time to arrive at the place of resonance, $\varphi(t)$ is the instantaneous phase and is given by integrating over time the frequency instantaneous (inverse function of time), and $A(t)$ is formulated to obtain a flat amplitude spectrum (modified for this work):

$$A(t) = \sqrt{\frac{1}{\alpha} * \left(\frac{-\gamma * e^{-L*\frac{\alpha}{2}}}{\beta * \left(\left(e^{-L*\frac{\alpha}{2}} \right) * \left(1 + \frac{t_0 - t}{\beta} \right) \right)^{\gamma - 1}} - 1 \right)} \quad (2)$$

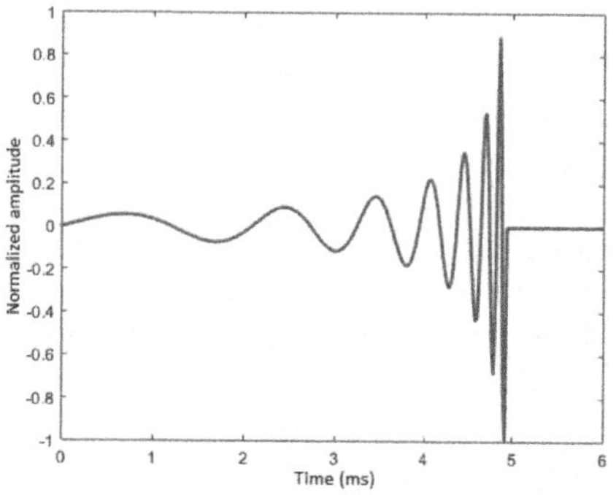

Figure 1 The chirp stimulus is design electrically to have a *flat amplitude spectrum* within five octave-bands ranging from 350-11300 Hz. The stimulus was generated in Matlab and inserted in AEP_AUDIX system with 48 KHz sampling rate, resulting in durations of 4.958 ms.

B. Data Acquisition

Experiments: Two experimental series were performed: *in the first*, ABR elicited by click and broadband chirp were obtained (stimulus level: ~60 dB nHL /repetition rate: 21/s) and compared wave-V amplitude for the two stimuli; *in the second*, chirp-evoked ABR were obtained at different repetition rates (17, 21, 33, 55, 71, 77 and 89/s) with respect to investigated their possible influence in the temporal processing.

Subjects: Nine normal-hearing (NH) volunteered subjects (5 females) ranging in age from 24-32 years (mean=25.8 years, SD=2.9) with pure-tone audiometric thresholds equal to or better than 20 dB HL for octave frequencies from 250-8000 Hz (in each ear) participated in the experiments. Participants were placed on a comfortable couch in a sound-isolated and electrically shielded booth and were instructed to relax and, if possible to sleep during the recording session. Participants review and signed informed consent approved by the Research Committee of the Cuban Neuroscience Center (in accordance with the Helsinki Declaration).

Equipment: The recordings are obtained with a commercial AEP software module running on AUDIX-5 system (Havana, Neuronic SA). Surface (gold) electrodes were placed on the head at Cz, M_1/M_2 (ipsilateral/contralateral mastoid), and high forehead as ground. Inter-electrode impedance was maintained below 5 kΩ. Clicks and chirps were presented monaurally via ER-3A insert earphones.

C. Data Analysis

Procedure: The individual ABR waveform was first determined by visual identification by a trained bioengineering and supervised and verified by a medical neurophysiologist. The most prominent response peak within the first 10-12 ms was identified and referred to as wave-V.

Wave-V (peak-to-peak) amplitude was measured from the (maximum) peak to the largest negativity direct following it and evaluated separately.

Statistics: The Wilcoxon test compared wave-V latency and amplitude results between right and left ears. The differences between ears were not significant (chirp–latency: p=0.090 and amplitude: p=0.441; click–latency: p=0.635 and amplitude: p=0.411). For this reason, the following data analysis considered the values of both ears (totalizing 18 ears). In the first experiment, a Wilcoxon matched-pairs signed-rank test (α=0.01) was performed to testing whether the response (wave-V amplitude) differed significantly for the two comparison stimuli. In the second, a student *t*-test (Test of means against reference constant, Ho: mean value=´0´) was applied to identify the larger amplitude when all repetition rates were compared.

III. Results

A. Comparison of ABR with click and broadband chirp.

Figure 2 provides an example of ABR waveforms obtained (females subjects, left ears) with a click (left panel) and broadband chirp (right panel) respectively and shows how wave-V latency and (peak-to-peak) amplitude were marked for both stimuli (small vertical bars). In the case of the chirp stimulus, it can be seen that the wave-V latencies, relative to stimulus onset, are shifted by their total duration (≈10.5ms).

The mean (maximum-minimum), standard deviations, confidence interval and amplitude ratio (chirp/click mean amplitude ratio) were calculated for the wave-V amplitude (Table 1). Wave-V amplitude with the chirp-evoked ABR was larger (on the order of ~0.2 μV) and significant (p=0.000196) than the amplitude obtained with click-evoked ABR.

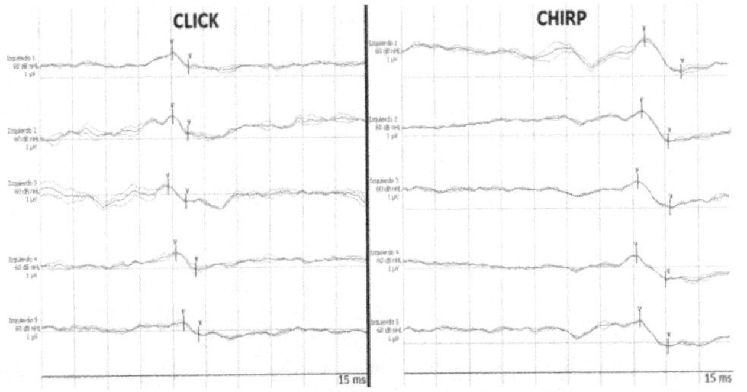

Figure 2 ABRs recording with click and chirp stimuli

The chirp-to-click ABR (wave-V) amplitude ratio showed that there is higher mean ABR amplitude to chirp than click (a value greater than 1 indicates that ABR to chirps are larger than ABR to clicks).

Table 1 Comparison of wave-V amplitude ABR with click and chirp stimuli (level: 60 dB nHL, repetition rate: 21/s)

Descriptive Statistic	Wave-V amplitudes (μV)		Amplitude Ratio
	Click	Chirp	
No. ears	18	18	18
Mean	**0.382**	**0.591**	**1.619**
Minimum	0.246	0.458	1.065
Maximum	0.475	0.724	2.365
Std. Dev.	0.086	0.085	0.412
CI	0.429	0.638	1.847

B. Wave-V amplitude with chirp-evoked ABR at different repetition rates

Figure 3 shows the individual chirp-evoked ABR waveform (one female subject, left ear) at different stimulus repetition rates used in the experiment.

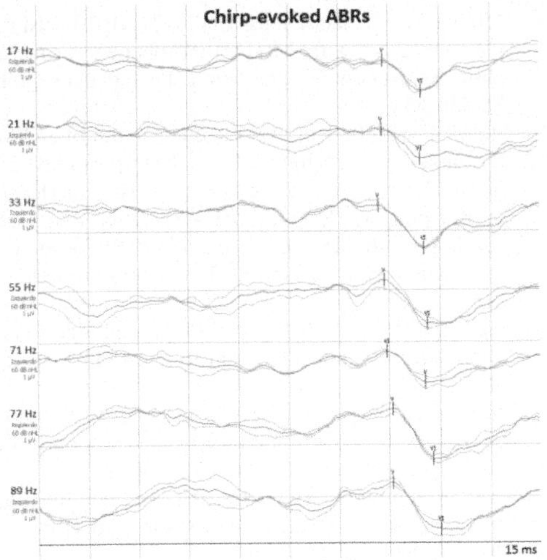

Figure 3 Chirp-evoked ABRs at different repetition rates. From top to bottom, 17, 21, 33, 55, 71, 77, and 89/s stimulation rates are represented.

The mean, standard deviations and confidence interval for the wave-V amplitude were calculated (Table 2).

Table 2 Comparison of wave-V amplitude with chirp-evoked ABR at different repetition rates

	17/s	21/s	33/s	55/s	71/s	77/s	89/s
Ears	18	18	18	18	18	18	18
Mean	**0.7542**	**0.7502**	**0.7654**	**0.7704**	**0.7693**	**0.7698**	**0.7713**
S.D	0.1673	0.1300	0.1268	0.1519	0.1618	0.1265	0.1148
CI	0.8074	0.7855	0.8485	0.9119	0.8709	0.9527	0.9484
***t*-value**	**23.356**	**23.514**	**23.273**	**24.715**	**23.348**	**23.838**	**24.926***

* Higher significant (*t*-test, Ho: mean=´0´)

With respect to wave-V amplitude there is no systematic and significant amplitude changes at stimulus rate up to 89/s. Thus, the mean wave-V amplitude remained constant suggested that the synchronicity was still large enough to generate stable amplitude values. The largest wave-V amplitude was founded for repetition rate of 89/s.

IV. Discussion

In the present study, the differences obtained in the comparison between wave-V amplitude on the ABR recording with click and broadband chirp stimuli were analyzed in NH young adults. The averaged wave-V amplitude is 0.382 μV for click-evoked ABR and 0.591 μV for chirp-evoked ABR, indicating an increase in amplitude of 54% (chirp/click mean amplitude ratio of 1.62 at 60 dB nHL). The increase in amplitude associated with the chirp stimulus are significant ($p<0.001$; $n=18$). These results are expected and similar to report in previous studies that compared chirp and click stimuli in the ABR recording [2, 4-10].

Most electrophysiological studies that employed amplitude and latencies measures of ABR used the traditional single-stimulus paradigm. In those studies, ABR latencies generally increased and ABR amplitude decreased as click rate increased [21-23]. In the second experiment, with respect to temporal processing, the influence of repetition rate on wave-V amplitude for the chirp-evoked ABR was investigated. It was found that, at stimulus rate up to 89/s, the wave-V amplitude remained constant, suggesting that the synchronicity was still large enough to generate stable amplitude values (overall neural activity remained unchanged).

This is consistent with the results obtained in earlier studies [2, 10, 14]. For very high rates (>100/s), synchronicity might be diminished further, as reflected the reduction of wave-V amplitude because the temporal smearing of the neural activity might ´exceed´ the duration of the effective integration window at the stage of processing where wave-V is generated [9]. Wave-V amplitude might therefore less sensitive to changes in synchronization, so in that terms, the system behaves essentially linearly at stimulus rate up to about 100/s. However, for a more quantitative understanding of the relation between the neural processing in the brainstem and the evoked far-field AEP, modeling work is needed.

A. Additional Considerations

Stimulus Level: In the present study, one signal level was used for two reasons: *i*) overall level and loudness of the stimuli remain within a comfortable range (previous studies reports better chirp/click amplitude ratios at moderate-intensity levels) [8, 10-12, 19]; *ii*) to represent a level high enough to yield a stable waveform with good identifiable components.

Clinical Applications: The quick evolution of different chirp model is evident in the literature but the availability of little tested stimuli still in development in commercialized equipment is a worrying factor. The use of chirp-evoked ABR (and ASSR) may be particularly interesting in neonatal hearing screening devices since larger amplitudes might facilitate the automatic detection of responses reducing the testing time. Satisfactory results for this end have already been reported [19-20].

V. Summary and Conclusions

AEP to broadband chirp stimuli (implemented in the AUDIX system, Neuronic SA) can be recorded and detected reliably. Chirp-evoked ABR have larger wave-V amplitude than click-evoked ABR, due to presumably better synchronization of the firing of a larger population of neural components. This yield improves the detectability of the response.

The wave-V amplitude remain constant at stimulus repetition rate up to 89/s, suggesting that the synchronicity was still large enough to generate stable amplitude values. In that term, the system behaves essentially linearly at stimulus rate up to about 100/s.

Conflict of Interest

The authors declare that they have no conflict of interest.

References

1. Jewett DL, Williston JS (1971) Auditory-evoked far fields averaged from the scalp of humans. Brain 94:681-696.
2. Dau T, Wegner O, Mellert V et al. (2000) Auditory brainstem responses with optimized chirp signals compensating basilar-membrane dispersion. J Acoust Soc Am 107:1530-1540
3. Shore SE, Nuttall AL (1985) High-synchrony cochlear compound action potentials evoked by rising frequency-swept tone burst. J Acoust Soc Am 78:1286-1295
4. Elberling C, Kristensen SG, Don M (2012) Auditory brainstem responses to chirps delivered by different insert earphones. J Acoust Soc Am 131:2091-2100
5. Kristensen SG, Elberling C (2012) Auditory brainstem responses to level-specific chirps in normal-hearing adults. J Am Acad Audiol 23:712-721
6. Rodrigues GRI, Lewis DR (2012) Comparison of click and CE-chirp® stimuli on brainstem auditory evoked potential recording. Rev Soc Bras Fonoaudiol 17:412-416
7. Elberling C, Callo J, Don M (2010) Evaluating auditory brainstem responses to different chirp stimuli at three levels of stimulation. J Acoust Soc Am 128:215-223
8. Elberling C, Don M (2010) A direct approach for the design of chirp stimuli used for the recording of auditory brainstem responses. J Acoust Soc Am 128:2955-2964
9. Junius D, Dau T (2005) Influence of cochlear traveling wave and neural adaptation on auditory brainstem responses. Hear Res 205:53-67
10. Fobel O, Dau T (2004) Searching for the optimal stimulus eliciting auditory brainstem responses in humans. J Acoust Soc Am 116:2213-2222
11. Elberling C, Don M (2008) Auditory brainstem responses to a chirp stimulus designed from derived-band latencies in normal-hearing subjects. J Acoust Soc Am 124:3022-3037
12. Elberling C, Don M, Cebulla M et al. (2007) Auditory steady-state responses to chirp stimuli based on cochlear traveling wave delay. J Acoust Soc Am 122:2772-2785
13. Bell SL, Allen R, Lutman ME (2002) An investigation of the use of band-limited chirp stimuli to obtain the auditory brainstem response. Int J Audiol 41:271-278
14. Wegner O, Dau T (2002) Frequency specificity of chirp-evoked auditory brainstem responses. J Acoust Soc Am 111:1318-1329
15. Cebulla M, Lurz H, Shehata-Dieler W (2014) Evaluation of waveform, latency and amplitude values of chirp ABR in newborns. Int J Pediatr Otorhinolaryngol 78:631-636
16. Rodrigues GRI, Lewis DR (2014) Establishing auditory steady-state response thresholds to narrow band CE-chirps® in full-term neonates. Int J Pediatr Otorhinolaryngol 78:238-243
17. Zirn S, Louza J, Reiman V et al. (2014) Comparison between ABR with click and narrow band chirp stimuli in children. Int J Pediatr Otorhinolaryngol 78:1352-1355
18. Cebulla M, Shehata-Dieler W (2012) ABR-based newborn hearing screening with MB11BERAphone® using an optimized chirp for acoustical stimulation. Int J Pediatr Otorhinolaryngol 76:536-543
19. Maloff ES, Hood LJ (2014) A comparison of auditory brainstem responses elicited by clicks and chirp stimuli in adults with normal hearing and sensory hearing loss. Ear Hear35:271-282
20. Xu Z, Cheng W, Yao Z (2014) Prediction of frequency-specific hearing threshold using chirp auditory brainstem response in infants with hearing losses. Int J Pediatr Otorhinolaryngol 78:812-816
21. Jiang ZD, Wu YY, Zhang L (1991) Amplitude change with click rate in human brainstem auditory-evoked responses. Audiology 30:173-182
22. Burkard R, Hecox K (1983) The effect of broadband noise on the human brainstem auditory evoked response. II. Frequency-specificity. J Acoust Soc Am 74:1214-1223
23. Pratt H, Sohmer H (1976) Intensity and rate functions of cochlear and brainstem evoked responses to click stimuli in man. Arch Otorhinolaryngol 212:85-92

Protein Network Related to Unfavorable Prognosis in Acute Myeloid Leukemia

L.F. Restrepo[1] and S. Röthlisberger[1]

[1] Instituto Tecnológico Metropolitano, Facultad de Ciencias Exactas y Aplicadas,
Grupo de Investigación e Innovación Biomédica - GI^{2}B, Medellín, Colombia

Abstract— **Acute myeloid leukemia (AML) is a cancer of the myeloid line of blood cells, characterized by the rapid growth of abnormal white blood cells that accumulate in the bone marrow and interfere with the production of normal blood cells. Identifying proteins involved in AML and understanding their molecular mechanisms is important for proposing biomarkers that can aid in the clinical management of the disease. The aim of this study was construct a protein-protein interaction (PPI) network based on proteins associated with unfavorable prognosis in AML and analyze biological pathways underlying the molecular complexes in the network. The proteins associated with unfavorable prognosis were identify from studies performed in bone marrow samples of AML patients. The PPI network was construct with Cytoscape using association networks from IntAct, String, and Biogrid databases. We carried out a Gene Ontology (GO) enrichment analysis with DAVID and Gene Ontology Consortium online tools. Finally, we construct a biological map of proteins associated with unfavorable prognosis using PathVisio, showing their roles, interactions and the possible combined influence of these on AML prognosis. The PPI contain 1922 nodes and 5905 edges, the pathways related with AML pathogenesis were VEGF and VEGFR signaling network, signaling evens mediated by focal adhesion kinase, IGF1 pathway, IL3-mediated signaling, Class I PI3K signaling events mediated by Akt, EGF receptor (ErbB1) signaling pathway, among others. Our results represent a bioinformatic approach to understand of the molecular mechanisms of AML development and may even provide additional targets for clinical prognosis and treatment.**

Keywords— **Acute myeloid leukemia, protein interaction network, prognosis and gene ontology.**

I. Introduction

Acute myeloid leukemia (AML) is a malignant clonal disorder characterized by chromosomal, gene alterations, and accumulation of poorly differentiated hematopoietic cells. Genetic alterations inhibit differentiation of cells and induce blast proliferation. The accumulation of these immature cells begins in the bone marrow, but in most cases, they quickly build up in the blood, and sometimes spread to other parts of the body such as the lymph nodes, spleen, liver, testes and the central nervous system.

Proteins are directly responsible for cell function; analysis of differentially expressed proteins in AML is valuable for gaining a better understanding of the disease and how a particular protein signature can be tied to a specific clinical feature, such as unfavourable prognosis.

Two systems are commonly used for AML classification: The French-American-British (FAB) system is based on the type of cell from which the leukemia develops and how mature the cells are (M0 – M7), while the World Health Organization (WHO) classification includes genetic, immunophenotypic, biological, and clinical features.

II. Material and Methods

A. Literature search of proteins associated with AML

We identified candidate proteins associated with unfavorable prognosis in AML using a text mining system, Polysearch, and Pubmed publications. We searched for proteins associated with AML, query type 'Disease-Gene/Protein' and query keyword 'Acute myeloid leukemia'. Due to the wide range of publications available, we manually confirmed whether these proteins are associated with unfavorable prognosis in AML, and only included the studies that met the following criteria: 1) proteomic studies examining bone marrow samples of AML patients, 2) reports of differentially expressed proteins related to unfavorable prognosis in AML (in terms of poor response to treatment, poor overall survival, short complete remission, and relapse), and 3) studies that included a verification/validation methodology.

B. Construction of protein-protein interaction network

The proteins related to unfavorable prognosis identified in the literature searches were used to create a protein-protein interaction (PPI) network. For this, we used three public databases: IntAct, String, and Biogrid, including only PPI with experimental evidence, and complemented with curated relationships parsed from Literature was carried out with the Agilent Literature Search plug-in in Cytoscape 3.4.0. The duplicate edges, self-loops and false-positive interaction information was removed from the results.

C. Network topology analysis

The network analysis was carried out using the Network Analysis plug-in. The parameters evaluated included clus-

I. Torres et al. (eds.), *VII Latin American Congress on Biomedical Engineering CLAIB 2016, Bucaramanga, Santander, Colombia, October 26th -28th, 2016,*
IFMBE Proceedings 60,
DOI: 10.1007/978-981-10-4086-3_54

tering coefficient, topological coefficient and degree distribution. MCOMD (Clusterviz plug-in) was used for the correlation analysis of biological networks; CytoCluster was used to identify protein clusters in the subnetworks, and the biggest maximal cliques were detected with MClique. Finally, we conducted an enrichment analysis with DAVID Bioinformatics Resources 6.7 and Gene Ontology Consortium. In order to explore the overrepresented biological terms, we set the threshold of false discovery rate to P 0.05.

D. Construction of a map showing the roles and interactions of the altered proteins

To illustrate the possible combined influence of the differentially expressed proteins in AML on unfavorable prognosis, we constructed a functional map with PathVisio 3.2.2, showing their roles and interactions, using the Pathway common analysis (WebGestalt) to complement the GO results. For the map construction we only used the proteins reported in subtypes M4 and M5, corresponding to acute myelomonocytic leukemia and acute monocytic leukemia, as these were the most reported subtypes in our literature search (Table 1). For each protein of interest, we searched for analyses carried out in bone marrow samples of AML patients; if the protein has not been reported in patient studies, then we searched for studies in leukemic cell lines.

III. Results

A. General survey of literature search

Data mining and analysis yielded 16 differentially expressed proteins, related to unfavorable prognosis, reported in bone marrow samples from AML patients (Table 1). These proteins were described in proteomic studies based on two-dimensional gel electrophoresis (2DGE), MALDI-TOF, SELDI-TOF, western blot, flow cytometry or Inmunostaining methodologies. For 9 of these proteins there is a reported correlation with a FAB subtype (1/M1, 2/M2, 5/M4, 8/M5) and for 8 proteins there are reports of correlation to unfavorable karyotype. Fifteen of the proteins were found to be upregulated in AML, while only one protein (RhoG12) is reported as downregulated.

B. Network analysis

With the 16 proteins identified, we constructed a network with 1922 nodes and 5905 edges. In this network, 8 molecular complexes were detected. The topological coefficients shows a tendency of the nodes in the network to have shared neighbors (R^2 value = 0.951 and a correlation= 0.936). The average clustering coefficient of the network was 0.394 of the nodes, which decreased as the number of interactions per protein increased; the R^2 value was 0.847 and a correlation of 0.917. The highly connected subnetwork included, among others, a subnetwork that includes Hsp90, AKT1, TP53 and EGFR, with 8 nodes and 27 edges, the GO analysis show a central role of the VEGF pathway in AML pathogenesis (Figure 1).

Table 1. Proteins found in literature search related to unfavorable prognosis in AML.

Gene symbol	Uniprot ID	Methodology	Up / Down	FAB	Ref.
ACTG	P63261	2DGE, MALDI-TOF ESI-MS, Western Blot	Up	M1-M2	[1], [2]
BCL-2	P10415	Western blot	Up	M4-M5	[3]
BIRC5	O15392	Western blot/ Inmunostaining	Up	M4-M5/ M2	[4], [5]
HNRNPH1	P31943	2D-DIGE MALDI-TOF , Western blot	Up	NS	[6]
HO-1	P30519	Western Blot	Up	NR	[7]
HSP110	Q92598	Inmunostaining, Flow cytometry, Western blot	Up	M5	[8]
HSP27	P04792	Western Blot, Inmunostaining, Flow cytometry	Up	M4-M5/ NR	[8], [9]
HSP60	P10809	Inmunostaining, Flow cytometry, Western blot	Up	NR	[8]
HSP90	P08238	Inmunostaining, Flow cytometry, Western blot	Up	M5	[8]
MCL-1	Q07820	Western blot	Up	NS	[10]
NM23H-1	P15531	2DGE, ESI-MS/MS	Up	M5	[11]
NRP-1	O14786	Inmunostaining	Up	NR	[12]
RhoG12	P52566	2DGE, MALDI-TOF	Down	All sub-types WHO	[13]
S10A8	P05109	SAX/SCX SELDI-TOF, MALDI-TOF/ TOF, LC-MS/MS, Western blot	Up	NS	[14]
SHPS1	P78324	Western Blot	Up	M4-M5	[15]
XIAP	P98170	Western blot	Up	M4-M5	[4], [16]

UP: Upregulation. Down: Downregulation. NS: Not specified, data not reported in the article. NR: Not related, the authors did not find a relationship between subtype and prognosis.

C. Map of proteins related to unfavorable prognosis in AML

A functional map of the proteins reported in subtypes M4 and M5 was constructed (Figure 2) based on reports found in the literature together with pathway common analysis, and is summarized in Table 2. BIRC5, HSP27, HSP90 and

Table 2. Function in cancer of proteins related to unfavorable prognosis in AML subtypes M4 and M5.

Protein	Leukemic cells	Function in cancer	Ref.
BCL-2	BM-MNCs	Bcl-2 and bax represent both sensitive indicators of clinical outcome and potential targets of novel proapoptotic molecules in order to circumvent chemoresistance.	[17]
BIRC5	HL-60, OCI-AML3 and BM-MNCs of AML patients.	MEK and PI3K pathways upregulate survivin (BIRC5) expression. Survivin is a critical down-stream effector of the PI3K/Akt pathway. Has antiapoptotic and mitogenic effects.	[18] [19]
	NB4	Survivin-2B accumulated IKK alpha in the nucleus, which further stabilized P73, a transcription factor of UVRAG. P73 in the nucleus upregulated UVRAG, an initiator of autophagy.	[20]
HSP27	U937	HSP27 binds to cytochrome c released from the mitochondria and prevents cytochrome c mediated interaction of Apaf-1 with procaspase-9.	[21]
HSP90	BM-MNCs of AML patients.	Constitutive activation of PI3K/AKT and ERK pathways.	[22]
	HL-60, THP-1, and HEL.	VEGF increased the expression of Hsp90 by interacting with KDR and activating the MEK cascade. Hsp90 in turn promotes survival of leukemic cells by binding APAF-1 and Bcl-2.	[23]
	NB4, HL-60 and jukart cells	Sodium selenite-decreased Hsp90 expression, and attenuated Hsp90/IKK interaction, which suppressed the activation and nuclear translocation of NFκB. The inactivated NFκB then suppressed becn1 transcription and anti-apoptotic genes, leading to a cell signaling switch from autophagy to apoptosis and, finally, irreversible cell death.	[24]
NM23H1	Primary cultured human acute myelogenous leukemia cells	Promotes growth and survival of primary AML cells mediated by MAPK activation, signal transducers and activators of transcription (STAT) activation and cytokine release.	[25]
SHPS1	KG1a, Kasumi-1 and HL-60	SIRPa inhibits growth and causes caspase-independent cell death.	[15]
XIAP	OCI M2, HL60, K562, and U937	Is an anti-apoptotic protein and a potent inhibitor of active caspases 9 and 3.	[26]
	OCI-AML3, Acute myeloid leukemia cells from bone marrow AML patients.	XIAP expression was regulated through the PI3K and, to a lesser degree, the MAPK/ ERK pathways.	[27]

BM-MNCs: Bone marrow-derived mononuclear cell; MEK: Mitogen-activated protein kinase kinase; PI3K: Phosphatidylinositol-3 kinase

XIAP are predominantly localized in the cytoplasm and are mainly involved in the inhibition of the intrinsic apoptosis pathway in mitochondria [18], [20], [21], [27]. Accordngly, VEGF increases the expression of Hsp90 and promotes APAF-1 binding, inhibiting apoptosis [23]. Additionally, stimulation of the PI3k/Akt pathways in the mitochondria is dependent on HSP90 chaperoning activity in the cytoplasm [22].

In the Endoplasmic Reticulum (ER), phoshosphorylation of HSP27 at Ser 15, 78 and 82 sites causes the inhibition of Cytochrome C activity, mediated by BCL-2 [21]. NM23H-1 is localized both in the extracellular and cytoskeleton components. Extracellular NM23H-1 stimulates MAPK and STATs to promote growth and survival [25]. On the other hand, findings indicate that a growth-suppressive and caspase-independent mode of programmed cell death is induced via SHPS1 in AML [15]. Interestingly, studies explaining the molecular mechanism of HSP110 in terms of AML poor prognosis were not found. According to the pathway common analysis, the pathways related to poor prognosis in AML were the VEGF and VEGFR signaling networks, as well as PI3K signaling events mediated by Akt.

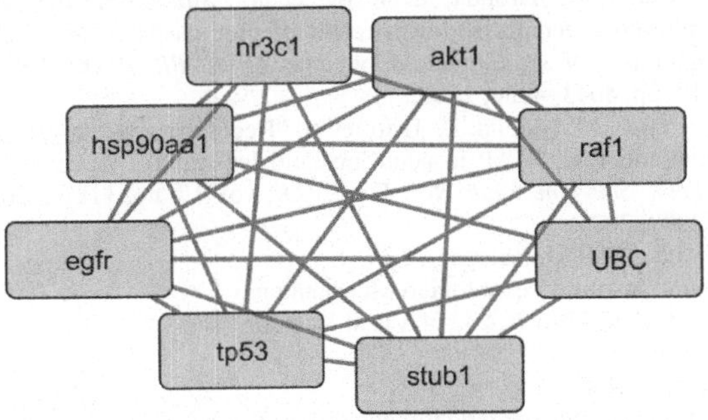

Fig.1 Protein LMA subnetwork of the biggest maximal clique.

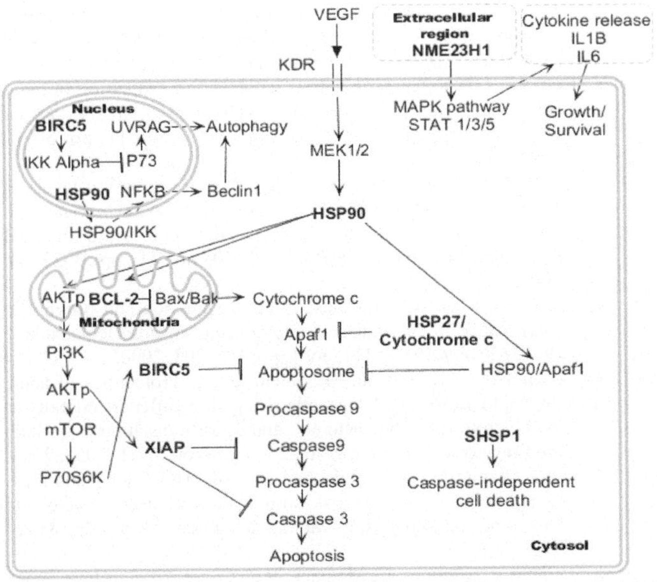

Fig.2 Map of proteins related to unfavorable prognosis, reported in AML subtypes M4 and M5. "→" indicates activation or enhancement, and "⊣" indicates inhibition.

IV. Conclusions

A literature search of proteomic studies of bone marrow samples of AML patients yielded 16 proteins reported to be associated with unfavorable prognosis. PPI and GO analysis revealed pathways related to stress response and apoptosis, which is coherent with their role in cancer. The map of proteins related to unfavorable prognosis reported in AML subtypes M4 and M5 represents an approach to understanding the molecular mechanisms involved in AML and gives an idea of targets that could be considered for treatment.

Acknowledgment

We acknowledge support received from Colciencias (Grant call 657-2014; Project Code: 115065743926), as well as from the Instituto Tecnológico Metropolitano.

Conflict of Interest

The authors report no conflicts of interest in this work.

References

[1] M. Luczak, M. Kaźmierczak, L. Handschuh, et al, "Comparative proteome analysis of acute myeloid leukemia with and without maturation," *J. Proteomics*, vol. 75, no. 18, pp. 5734–5748, 2012.

[2] M. Kazmierczak, M. Luczak, K. Lewandowski,et al, "Esterase D and gamma 1 actin level might predict results of induction therapy in patients with acute myeloid leukemia without and with maturation," *Med. Oncol.*, vol. 30, no. 4, pp. 1–9, 2013.

[3] L. Campos, J. P. Rouault, O. Sabido, et al, "High expression of bcl-2 protein in acute myeloid leukemia cells is associated with poor response to chemotherapy.," *Blood*, vol. 81, no. 11, pp. 3091–3096, 1993.

[4] I. Tamm, S. Richter, D. Oltersdorf, et al, "High expression levels of x-linked inhibitor of apoptosis protein and survivin correlate with poor overall survival in childhood de novo acute myeloid leukemia.," *Clin. Cancer Res.*, vol. 10, no. 11, pp. 3737–44, 2004.

[5] C. Adida, C. Recher, E. Raffoux, et al, "Expression and prognostic significance of survivin in de novo acute myeloid leukaemia.," *Br. J. Haematol.*, vol. 111, no. 1, pp. 196–203, 2000.

[6] M. Y. Balkhi, a K. Trivedi, M. Geletu, et al, "Proteomics of acute myeloid leukaemia: Cytogenetic risk groups differ specifically in their proteome, interactome and post-translational protein modifications.," *Oncogene*, vol. 25, no. 53, pp. 7041–7058, 2006.

[7] M. Yu, J. Wang, D. Ma, S. Chen, X. Lin, et al, "HO-1, RET and PML as possible markers for risk stratification of acute myelocytic leukemia and prognostic evaluation," *Oncol. Lett.*, pp. 3137–3144, 2015.

[8] X. Thomas, L. Campos, C. Mounier, et al, "Expression of heat-shock proteins is associated with major adverse prognostic factors in acute myeloid leukemia.," *Leuk. Res.*, vol. 29, no. 9, pp. 1049–58, 2005.

[9] L. Yang, L. Cao, M. Yang, et al, "Hsp27: A novel therapeutic target for pediatric M4/M5 acute myeloid leukemia," *Oncol. Rep.*, vol. 29, no. 4, pp. 1459–1466, 2013.

[10] S. H. Kaufmann, J. E. Karp, P. a Svingen, S. Krajewski, P. J. Burke, S. D. Gore, and J. C. Reed, "Elevated expression of the apoptotic regulator Mcl-1 at the time of leukemic relapse.," *Blood*, vol. 91, no. 3, pp. 991–1000, 1998.

[11] J.-W. Cui, J. Wang, K. He, et al, "Two-dimensional electrophoresis protein profiling as an analytical tool for human acute leukemia classification.," *Electrophoresis*, vol. 26, no. 1, pp. 268–279, 2005.

[12] M. Kreuter, K. Woelke, R. Bieker, et al, "Correlation of neuropilin-1 overexpression to survival in acute myeloid leukemia.," *Leukemia*, vol. 20, no. 11, pp. 1950–4, 2006.

[13] C. López-Pedrera, J. M. Villalba, E. Siendones,et al, "Proteomic analysis of acute myeloid leukemia: Identification of potential early biomarkers and therapeutic targets.," *Proteomics*, vol. 6 Suppl 1, pp. S293–S299, 2006.

[14] E. Nicolas, C. Ramus, S. Berthier, et al, "Expression of S100A8 in leukemic cells predicts poor survival in de novo AML patients.," *Leuk. Off. J. Leuk. Soc. Am. Leuk. Res. Fund, U.K*, vol. 25, no. 1, pp. 57–65, 2011.

[15] M. Irandoust, J. Alvarez Zarate, I. Hubeek, et al, "Engagement of SIRP alpha Inhibits Growth and Induces Programmed Cell Death in Acute Myeloid Leukemia Cells," *PLoS One*, vol. 8, no. 1, pp. 1–13, 2013.

[16] I. Tamm, S. Richter, F. Scholz, et al, "XIAP expression correlates with monocytic differentiation in adult de novo AML: impact on prognosis.," *Hematol. J.*, vol. 5, no. 6, pp. 489–495, 2004.

[17] G. Del Poeta, A. Venditti, M. Ilaria, et al, "Amount of spontaneous apoptosis detected by Bax/Bcl-2 ratio predicts outcome in acute myeloid leukemia (AML)," *Ratio*, vol. 101, no. 6, pp. 2125–2131, 2003.

[18] B. Z. Carter, M. Milella, D. C. Altieri, et al, "Cytokine-regulated expression of survivin in myeloid leukemia," *Blood*, vol. 97, no. 9, pp. 2784–2790, 2001.

[19] J. Serrano-López, J. Serrano, V. Figueroa, A. Torres-Gomez, et al, "Cytoplasmic localization of wild-type survivin is associated with constitutive activation of the PI3K/Akt signaling pathway and represents a favorable prognostic factor in patients with acute myeloid leukemia," *Haematologica*, vol. 98, no. 12, pp. 1877–1885, 2013.

[20] K. Shi, J. An, L. Shan, Q. Jiang, et al, "Survivin-2B promotes autophagy by accumulating IKK alpha in the nucleus of selenite-treated NB4 cells," *Cell Death Dis.*, vol. 5, no. e1071, pp. 1–10, 2014.

[21] J. M. Bruey, C. Ducasse, P. Bonniaud, et al, "Hsp27 negatively regulates cell death by interacting with cytochrome c," *Nat. Cell Biol.*, vol. 2, no. 9, pp. 645–652, 2000.

[22] P. Flandrin, D. Guyotat, A. Duval, et al, "Significance of heat-shock protein (HSP) 90 expression in acute myeloid leukemia cells.," *Cell Stress Chaperones*, vol. 13, no. 3, pp. 357–64, 2008.

[23] S. Dias, S. V Shmelkov, G. Lam, et al, "VEGF 165 promotes survival of leukemic cells by Hsp90-mediated induction of Bcl-2 expression and apoptosis inhibition," *Blood*, vol. 99, no. 7, pp. 2532–2540, 2002.

[24] Q. Jiang, Y. Wang, T. Li, et al, "Hsp90-mediated inactivation of NF κ B switches autophagy to apoptosis through becn1 transcriptional inhibition in selenite-induced NB4 cells," *Mol. Biol. Cell*, 2011.

[25] J. Okabe-Kado, T. Kasukabe, Y. Honma, et al, "Extracellular NM23 protein promotes the growth and survival of primary cultured human acute myelogenous leukemia cells," *Cancer Sci.*, vol. 100, no. 10, pp. 1885–1894, 2009.

[26] B. Z. Carter, M. Gronda, Z. Wang, et al, "Small-molecule XIAP inhibitors derepress downstream effector caspases and induce apoptosis of acute myeloid leukemia cells," *Blood*, vol. 105, no. 10, pp. 4043–4050, 2005.

[27] B. Z. Carter, M. Milella, T. Tsao, et al, "Regulation and targeting of antiapoptotic XIAP in acute myeloid leukemia.," *Leuk. Off. J. Leuk. Soc. Am. Leuk. Res. Fund, U.K*, vol. 17, no. 11, pp. 2081–2089, 2003.

Author:Sarah Röthlisberger
Institute: Instituto Tecnológico Metropolitano
Street: Calle 73 No 76A -354 Vía al Volador
City: Medellin
Country: Colombia
Email: sarahrothlisberger@itm.edu.co

Green synthesis of silver nanoparticles using green coffee bean extract

Lukas Cardeño Calle[1] and Martha Elena Londoño López[1-2]

[1]EIA University – [2]CES University, Biomedical Engineering, Envigado, Colombia

Abstract- **This investigation report a simple green synthesis for stabilizing silver nanoparticles (AgNPs) using green coffee beam extract. Silver nitrate was used as the metal precursor and green coffee extract as a reducing agent. The AgNPs UV-Vis spectrum was recorded, the morphology of AgNPs was determinate by transmission electron microscopy (TEM) and atomic force microscopy (AFM). Antibacterial activity of the synthetized nanoparticles was analyzed by means well diffusion method against three different microorganisms (*Escherichia coli* ATCC 25922, *Pseudomonas aureginosa* ATCC 27853 and *Staphylococcus aureus* ATCC 25923). The silver nanoparticles showed high stability and antimicrobial activity against both gran positive and gram negative bacteria.**

Key Words - **Green synthesis, silver nanoparticles, coffee extract, antibacterial activity, nanotechnology.**

I. Introduction

According to the National Institute of Health of America, nanotechnology is defined as "the creation of functional materials, devices or systems through control of matter at a scale of 1-100 nm and the exploitation of novel properties and phenomena at this scale" [1]. Applications in this field have spread exponentially due to their relevance on diverse areas of knowledge such as physics, electronics, chemistry, biology, medicine and optics [2].

Amongst nanostructures studied today, metallic nanoparticles are considered promising because of their antibacterial properties attributed to their surface to area ratio. Nowadays, a great number of harmful organisms have developed resistance towards several antibiotics used [3], it is because of this that the need to develop new materials with better antibacterial properties has become one of the principal focuses of research. Because of its high electrical conductivity, optical properties and antibacterial effect, one of the most currently used materials to create nanoparticles is silver.

One of the drawbacks of the synthesis of nanoparticles through chemical routes is the possibility of contamination of the environment or the colloidal solutions with various chemical byproducts of the reactions that take place in order to reduce silver. For these reasons, in order to develop environmentally friendly protocols to synthesize silver nanoparticles, natural reducing agents have been thoroughly studied. Reducing agents including plants such as garlic [4], [5], Hibiscus [6], oranges [7], bananas [8], bacterial and many others [9]–[14] have been studied.

Another common inconvenience when synthesizing nanoparticles is the necessity of stabilize the particles in order to avoid agglomerates and the loss of chemical and physical properties of the colloidal solutions. This is achieved by coating the particles with a capping agent that in most cases requires the addition of secondary reagents after the reduction has been carried [15], [16]. The autors Nadagouda and Varma presented a green approach to generate nanocrystals of noble metals such as palladium and silver using coffee extracts obtained from commercial instant coffee. This method yielded sphere-like particles in the size range of 20-60 nm [17]. The single-pot method described used no surfactant, capping agent, and/or template to synthesize the nanocrystals.

In order to improve the silver nanoparticle synthesis and to avoid chemical in instant coffee from altering the reduction process, we propose a single-step reaction by using green coffee bean extract instead of commercial coffee as both reducing and capping agent. This method yields spherical nanoparticles with controlled size distribution and presents a "green" approach to reduce nanoparticles.

II. Materials And Methods

A. Chemical

Silver nitrate ($AgNO_3$) was purchased from Chemí (Colombia) and used as received, green coffee bean was obtained by a Colombian plantation. The antibiotics were purchased from the company Oxoid (United Kingdom), Mueller-Hinton agar and BHI broths were purchased from the CES University and Instituto Colombiano de Medicina Tropical production center.

B. Synthesis of silver nanoparticles

100 g of Colombian green coffee beans were washed thoroughly with distilled water and then ground in 400 mL of milliQ water using a kitchen blender. Afterwards the solution was heated to 60°C for 20 minutes and left to cool down to be stored at 4°C overnight. Finally, to obtain the green coffee beans extract –GCBE- the solution was filtered. The solutions containing 50 mL of milliQ water and 85, 24 and 20 mg of silver nitrate (solution 1, 2 and 3 respectively) were heated up to 55°C and 5 mL of GCBE was added drop wise while maintaining the temperature constant. The reaction was

I. Torres et al. (eds.), *VII Latin American Congress on Biomedical Engineering CLAIB 2016, Bucaramanga, Santander, Colombia, October 26th -28th, 2016*,
IFMBE Proceedings 60,
DOI: 10.1007/978-981-10-4086-3_55

carried out for 2 h under white light and constant stirring (300 rpm). Within seconds of exposure to the light, the light orange colored solution started changing to dark brown indicating the formation of silver colloid. The color intensity kept increasing progresssively until it reached a plateau at 2 h. The synthesized nanoparticles were purified by washing four times with milliQ water by centrifuging (5.000 rpm for several cycles).

C. Characterization

Uv-Vis spectra of silver nanoparticles solution was recorded as a function of wavelength using a Perkin-Elmer (USA) spectrophotometer from 200-800 nm operated at a resolution of 1nm. Atomic force microscopy images were taken using a Nanosurf Easy Scan 2 Atomic Force Microscope in tapping mode. The size and morphology of the synthesized nanoparticles were recorded by using a transmission electron microscope (TEM model: JEOL 1200EX, (Japan)) at an accelerating voltage of 120 kV. Antibacterial activity of the synthetized nanoparticles was analyzed by well diffusion method against three different microorganisms (Escherichia coli ATCC 25922, Pseudomonas aureginosa ATCC 27853 and Staphylococcus aureus ATCC 25923) using the antibiotics erythromycin, penicillin, tetracyclin and ampicillin as controls. 5 mm wells where punctured in the agar with the help of a stainless steel cork borer in order to deposit 40 μL of the nanoparticle solutions. Then the plates containing both the nanoparticles and the controls were incubated at 37°C and then examined in order to determine the inhibition zones.

III. Result and discussion

A. UV-visible spectroscopy

The Uv-Visible spectra of solutions containing 85, 24 and 20 mg of $AgNO_3$ is shown in the figure 1. It is widely known that solutions containing silver nanoparticles exhibit a yellowish-brown color when the particles present in the solution have spherical shape [18]. The presence of a peak centered around 422 nm (shown in Fig. *1*) and the final yellowish-brown colors of the coffee/$AgNO_3$ solutions give the first indication of silver nanoparticle formation. The solution containing 85mg of silver nitrate shows a well-defined peak with high intensity; this according to Slistan et al. corresponds to an increase of the radii of particles up to a maximum of 20 nm, threshold after which the peak maxima decreases as the size of the particles increases surpassing the 20 nm range [19]. To evaluate the stability of the solutions the SPR spectra was measured over different periods of time as shown in Fig. *2*. It demonstrate how silver nanoparticles obtained at 85 mg of $AgNO_3$ has better characteristics than others.

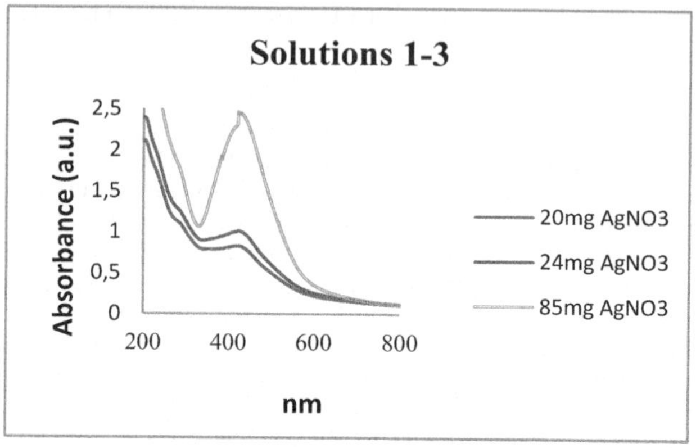

Fig. 1 Uv-Visible spectra of silver nanoparticles solutions at different concentrations of AgNO3.

B. Atomic Force Microscopy

The AFM results display the surface morphology of silver nanoparticles obtained through reduction of silver nitrate by GCBE as shown in Fig. *3*. The morphologies of the nanoparticles obtained using this method are mostly spherical and close in size, presenting both agglomerates and fine-dispersion areas.

C. Transmission Electron Microscopy

Fig. *4* shows TEM analysis of the silver nanoparticles obtained at 85 mg of $AgNO_3$; image revealed the presence of spherical nanoparticles with sizes ranging mostly between 5-10 nm (for solutions 1 and 2) and in the case of solution 3, polydispersed and fewer particles with sizes between 8 and 44 nm, no shwon. As minus quantity of extract is used more polydispersed particles were obtained.

A. Antibacterial activity

The inorganic nanoparticles have a distinct advantage over conventional chemical antimicrobial agents. The most important problems caused by the chemical antimicrobial agents is multidrug resistance. Generally, the antimicrobial mechanism of these agents depends on the specific binding with the surface and the metabolism of agents into the microorganism. Various microorganisms have evolved to be resistant to drug compounds over many generations [10]. The antimicrobial activity of silver nanoparticles against various pathogenic organisms was investigated and the sizes of inhibition were recorded for all test pathogens. The solutions of silver nanoparticles evaluated showed antibacterial activity against all pathogens, as shown in figures 5, the antibacterial efficiency of the solutions evaluated was higher against the microorganism *Pseudomona aureginosa* with values of 14 to 16 mm.

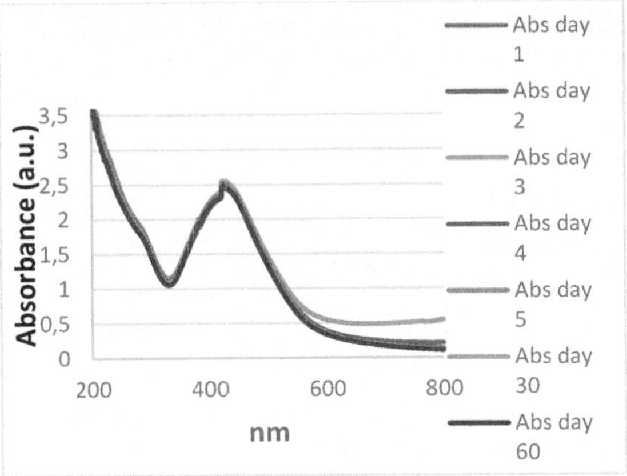

Fig. 2 Uv-Visible spectra of silver nanoparticles solutions at different periods of time - 85 mg AgNO3.

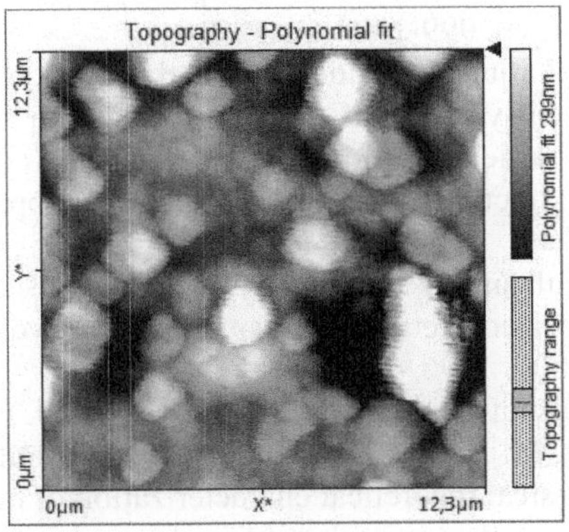

Fig. 3 AFM images of nanoparticle solutions at of AgNO3 20mg.

IV. Conclusion

An eco-friendly method to synthesize silver nanoparticles using coffee bean extract has been development by metal reduction, without using of any special capping or surfactant agents. The silver nanoparticles showed high stability and antimicrobial activity against both gram positive and gram negative bacteria. Silver nanoparticles obtained at 85 mg of $AgNO_3$ shown better characteristics than others. This green method may find its use as a drug delivery vehicle, sensors, and other medical and technological applications.

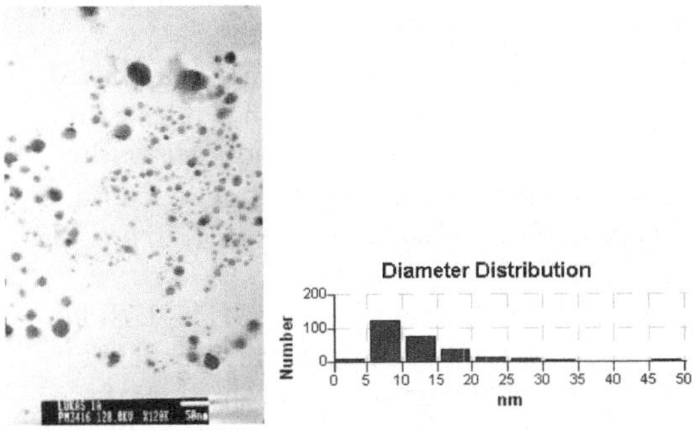

Fig. 4 TEM image of synthesized silver nanoparticles at 85mg of silver nitrate and their respective histogram and diameter distribution.

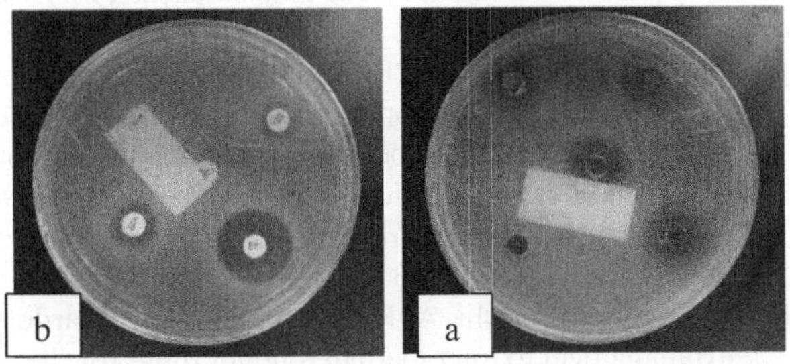

Fig. 5 Antibacterial activity assay against Pseudomona aureginosa a) antibiotic controls, b) Silver nanoparticles solutions.

Acknowledgements

This research has been sponsored by Colciencias under the category "joven investigador" and EIA University and CES University.

Conflic Interest

The authors declare that they have no conflict interest.

References

[1] J. Han, «Biomedical Nanoengineering For Nanomedicine», *Tissue Eng. Artif. Organs*, pp. 289-290, 2006.

[2] N. Misra, V. Kumar, L. Borde, and L. Varshney, «Localized surface plasmon resonance-optical sensors

based on radiolytically synthesized silver nanoparticles for estimation of uric acid», *Sens. Actuators B Chem.*, vol. 178, pp. 371-378, 2013.
[3] M. Rai, A. Yadav, and A. Gade, «Silver nanoparticles as a new generation of antimicrobials», *Biotechnol. Adv.*, vol. 27, pp. 76-83, 2009.
[4] L. Rastogi and J. Arunachalam, «Sunlight based irradiation strategy for rapid green synthesis of highly stable silver nanoparticles using aqueous garlic (Allium sativum) extract and their antibacterial potential», *Mater. Chem. Phys.*, vol. 129, pp. 558-563, 2011.
[5] M. Ahamed, M. A. MajeedKhan, M. K. J. Siddiqui, M. S. AlSalhi, and S. A. Alrokayan, «Green synthesis, characterization and evaluation of biocompatibility of silver nanoparticles», *Phys. E*, vol. 43, pp. 1266-1271, 2011.
[6] D. Philip, «Green synthesis of gold and silver nanoparticles using Hibiscus rosa sinensis», *Phys. E*, vol. 42, pp. 1417–1424, 2010.
[7] R. Konwarh, B. Gogoi, R. Philip, and M. A. Laskar, «Biomimetic preparation of polymer-supported free radical scavenging, cytocompatible and antimicrobial "green" silver nanoparticles using aqueous extract of Citrus sinensis peel», *Colloids Surf. B Biointerfaces*, vol. 84, pp. 338-345, 2011.
[8] A. Bankar, B. Joshi, A. R. Kumar, and S. Zinjarde, «Banana peel extract mediated novel route for the synthesis of silver nanoparticles», *Colloids Surf. Physicochem. Eng. Asp.*, vol. 368, pp. 58-63, 2010.
[9] S. Vivekanandhan, D. Tang, M. Misra, and A. K. Mohanty, «Novel Glycine Max (Soybean) Leaf Extract Based Biological Process for the Functionalization of Carbon Nanotubes with Silver Nanoparticles», *Nanosci. Nanotechnol. Lett.*, vol. 2, pp. 240-243, 2010.
[10] M. Vijayakumar, K. Priya, F. T. Nancy, A. Noorlidah, and A. B. A. Ahmed, «Biosynthesis, characterisation and anti-bacterial effect of plant-mediated silver nanoparticles using Artemisia nilagirica», *Ind. Crops Prod.*, vol. 41, pp. 235-240, 2013.
[11] M. Thakur, S. Pandey, A. Mewada, R. Shah, G. Oza, and M. Sharon, «Understanding the stability of silver nanoparticles bio-fabricated using Acacia arabica (Babool gum) and its hostile effect on microorganisms», *Spectrochim. Acta. A. Mol. Biomol. Spectrosc.*, vol. 109, pp. 344-347, 2013.
[12] G. M. Sulaiman, W. H. Mohammed, T. R. Marzoog, A. A. A. Al-Amiery, A. A. H. Kadhum, and A. B. Mohamad, «Green synthesis, antimicrobial and cytotoxic effects of silver nanoparticles using Eucalyptus chapmaniana leaves extract», *Asian Pac. J. Trop. Biomed.*, vol. 3, pp. 58-63, 2013.
[13] R. S. Patil, M. R. Kokate, and S. S. Kolekar, «Bioinspired synthesis of highly stabilized silver nanoparticles using Ocimum tenuiflorum leaf extract and their antibacterial activity», *Spectrochim. Acta. A. Mol. Biomol. Spectrosc.*, vol. 91, pp. 234-238, 2012.
[14] A. J. Kora, R. B. Sashidhar, and J. Arunachalam, «Gum kondagogu (Cochlospermum gossypium): A template for the green synthesis and stabilization of silver nanoparticles with antibacterial application», *Carbohydr. Polym.*, vol. 82, pp. 670-679, 2010.
[15] D. Guo, L. Zhu, Z. Huang, H. Zhou, Y. Ge, W. Ma, J. Wu, X. Zhang, X. Zhou, Y. Zhang, Y. Zhao, and N. Gu, «Anti-leukemia activity of PVP-coated silver nanoparticles via generation of reactive oxygen species and release of silver ions», *Biomaterials*, vol. 34, n.º 32, pp. 7884-7894, oct. 2013.
[16] S. M. Ghaseminezhad and S. A. Shojaosadati, «Data on the role of starch and ammonia in green synthesis of silver and iron oxide nanoparticles», *Data Brief*, vol. 7, pp. 999-1003, jun. 2016.
[17] Mallikarjuna N. Nadagouda and Rajender S. Varma, «Green synthesis of silver and palladium nanoparticles at room temperature using coffee and tea extract», *Green Chemestry*, vol. 10, pp. 859-862, 2008.
[18] P. Mulvaney, «Surface Plasmon Spectroscopy of Nanosized Metal Particles», *Langmuir*, vol. 12, pp. 788-800, 1996.
[19] A. Slistan-Grijalva, R. Herrera-Urbina, J. F. Rivas-Silva, M. Ávalos-Borja, and F. F. Castillón-Barraza, «Classical theoretical characterization of the surface plasmon absorption band for silver spherical nanoparticles suspended in water and ethylene glycol», *Phys. E*, vol. 27, pp. 104–112, 2005.

Surface Electromyographic Characterization of Five Orofacial Ideomotor Praxis in 20 Healthy Individuals

G. Cantillo-Mackenzie[1;5], L. Martinez-Moreno[2], C.L. Bedoya-Londoño[2;3], E. Perez-Giraldo[3], L.M. Ramirez-Arbelaez[3], S. Cadavid-Arboleda[3], S. Restrepo-Agudelo[4], S.Roldan-Vasco[4], A. Orozco-Duque[3] and J.C. Suarez-Escudero[5]

1 Student International Master in Biomedical Science, Universidad de Buenos Aires, Buenos Aires, Argentina
2 IPS Organizacion Fonoaudiologa E.U. (OFA), Medellin, Colombia
3 Department of Biomedical Engineering, Instituto Tecnologico Metropolitano, Medellin, Colombia
4 Faculty of Engineering, Instituto Tecnologico Metropolitano, Medellin, Colombia
5 Grupo de salud p´ublica, escuela de ciencias de la salud, Universidad Pontificia Bolivariana, Medellin, Colombia

Abstract— **Praxis is a voluntary nonverbal sequence of movements that allows the completion of programmed and organized movements in defined sequences. A type of praxis are the orofacial ideomotor, that allows to perform movements such as open and close the mouth, smiling, make tongue gestures, among others. In the execution of any praxis, a cortical cerebral component and a neuromuscular peripheral structure are involved. There is a limited literature related with the orofacial praxis recording using surface electromyography (sEMG) signals. The purpose of this paper is to describe the electromyographic characteristics of ideomotor orofacial praxis in 20 healthy individuals. Eight electrodes were placed during, the execution of five praxical orofacial movements and we evaluated the root mean square (RMS) of the sEMG signals for the orbicularis oris and masseter muscles. In the five executed praxis we found significant differences in the median values of RMS data, except for the smile praxis where median RMS values were similar. The evaluation of praxis is mainly used in the clinical neuropsychology field, done through tests highly dependent of the clinical expertise as a key element in determining the alteration or normal praxic funtions. This EMG characterization forecasts a progress in terms of the objective study of the praxical disorders with future implementation in diagnosis of apraxias secondary to brain lesions.**

Keywords— **apraxia, ideomotor apraxia, motion, central nervous system, electromyography.**

I Introduction

Nonverbal praxis sequences, are defined as programmed, organized and voluntary movements in defined sequences, with a particular purpose, which are executed intentionally and coordinated [1]. They are part of the higher cognitive functions, concept described from the observation of patients with neurological injuries since 1871 [2]. The alteration of these movements is called apraxia.

In voluntary motor behavior, where praxis are included, two components are distinguished: a cognitive (cortical cerebral component) and a motor (neuromuscular peripheral structure). The cognitive component is responsible for processing information that facilitates the execution of motor plans and contains internal and unobservable aspects of the action, such as planning, initialization or completion of the motor activity. Moreover, the motor component has parameters associated with the execution of the motor action in effector muscles through precision, strength, performance, movement, among other variables that can be quantified [3]. So the motor component is common to several actions such as praxis, verbal movements, breathing, chewing, swallowing and phonation. In other words, the muscle is the common denominator, since it is the final pathway effector. Therefore, the changes come from the cortical program that stores and manages the type of movement to execute. Evaluation of praxis are used to assess higher cognitive functions, which at present are evaluated by clinical methods based on the expertise of a professional and include a series of neuropsychological tests.

The electromyography (EMG) is an objective diagnostic test, based on the extracellularly recording of superimposed action potentials generated by muscle fibers, subjected to mechanical work [4]. Surface electromyography (sEMG) is a non-invasive electrophysiological measure of the muscles activity. The sEMG provides quantitative information about the neuromuscular activity associated with muscle contraction with higher activation mediated by the central nervous system. Therefore, sEMG becomes a tool for evaluating the motor component, which is used by praxis, by way of a non-verbal voluntary movement sequence.

One type of praxis is the orofacial ideomotor that allows the execution of movements such as the opening and closing of the mouth, smile, make tongue gestures, among oth-

I. Torres et al. (eds.), *VII Latin American Congress on Biomedical Engineering CLAIB 2016, Bucaramanga, Santander, Colombia, October 26th -28th, 2016*,
IFMBE Proceedings 60,
DOI: 10.1007/978-981-10-4086-3_56

ers; those praxis have social and survival implications of the species. Orofacial praxis are part of a type called ideomotor praxis, since at present praxical disorders (apraxia) of orofacial type, which usually coexists with apraxia of a limb [5]. Limb apraxia has been studied in depth since it is of easy characterization, as in clinical and physiological studies of sEMG. However, the scientific literature regarding the evaluation of praxis recording through surface sEMG is sparse. The purpose of this paper is to describe the sEMG characteristics of orofacial ideomotor praxis in 20 healthy individuals.

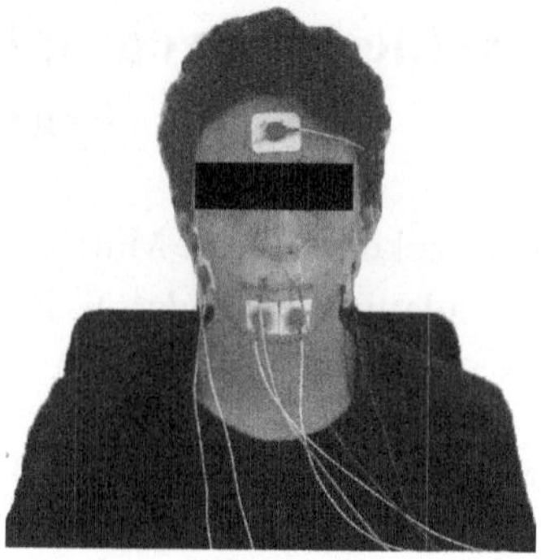

Fig. 1: Distribution of electrodes. The forehead electrode acts as reference.

II Materials and Methods

Twenty healthy individuals were voluntarily recruited in the city of Medellin,Colombia, where surface EMG signals were measured in perioral, zygomatic and mandibular areas. A differential bioamplifier ML 138 (AD Instruments Inc.) was used with sampling frequency of 2 kHz; the recording and storage was performed with the preprocessing software LabChart Pro (AD Instruments Inc.). The signals were stored in a datasheet format and later processed in Matlab (MathWorks Inc.). We applied a bandpass filter with cutoff frequencies of 25 and 500 Hz, and a notch filter with central frequency of 60 Hz. Four channels of acquisition were used. The target muscles were the right and left masseters and the upper and lower orbicularis oris. The upper orbicularis oris was measured with gold electrodes, whilst the masseter region and lower orbicularis oris were assessed with conventional Ag/AgCl electrodes. The Figure 1 shows the distribution of electrodes. Each individual was asked to execute five praxical orofacial movements: open mouth, close mouth, clench teeth, smile and give a kiss (protrusion of lips). For sEMG signal analysis we used the Root Mean Square or RMS (Figure 2), i.e. the square root of the arithmetic mean of the squares of the values. The RMS value was estimated for each muscle at each praxis and the procedure estimates this value for time windows of 500 samples (250ms) with step time of 100 samples (50ms). All individuals signed an informed consent approved by the Ethical Committee of the Universidad Pontificia Bolivariana (UPB) and the Instituto Tecnolgico Metropolitano (ITM).The research pilot phase was approved by Colciencias and carried out by the Biomedical Research and Innovation Group and the Research Group in Advanced Materials and Energy of the ITM, the Public Health Research Group of the UPB (disability research line) and the clinical institution Organizaci6n Fonoaudiol6gica E.U. (Medellín, Colombia).

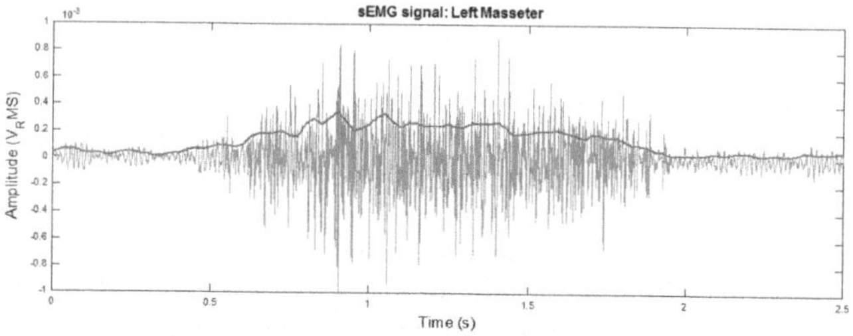

Fig. 2: Typical EMG signal (blue) and the correspond RMS (red).

III Results

Ten women and ten men with an average age of 24.5 5.12 years old (minimum: 19, maximum: 37) were studied. Figure 2 shows the normalized RMS (nRMS) of both muscles for each evaluated praxis with its respective p values.

In all praxis we found significant differences in the median values of the normalized RMS feature, except for smile praxis (0.35 in masseter and 0.25 in orbicularis oris). Statistical analysis was performed by t-student test. In clenching teeth and smile praxis, the interquartile range of the masseter muscle (0.655, 0.445 and 0.465, 0.23 respectively) has higher values than the interquartile range of the orbicularis oris muscle (0.05 and 0.24 respectively). In the other three praxis, the values of the orbicularis oris were higher than those of the masseter muscle. In the clenching teeth praxis, the highest RMS value was for the masseter muscle as expected (0.55). The lowest nRMS value of the masseter was measured during the open mouth (0.06) and blow a kiss praxis (0.03). The highest nRMS value of the orbicularis oris was in blow a kiss (0.56), followed by closing mouth, opening mouth and smile. The lowest nRMS value of the orbicularis oris was in the clenching teeth praxis (0.06 RMS).

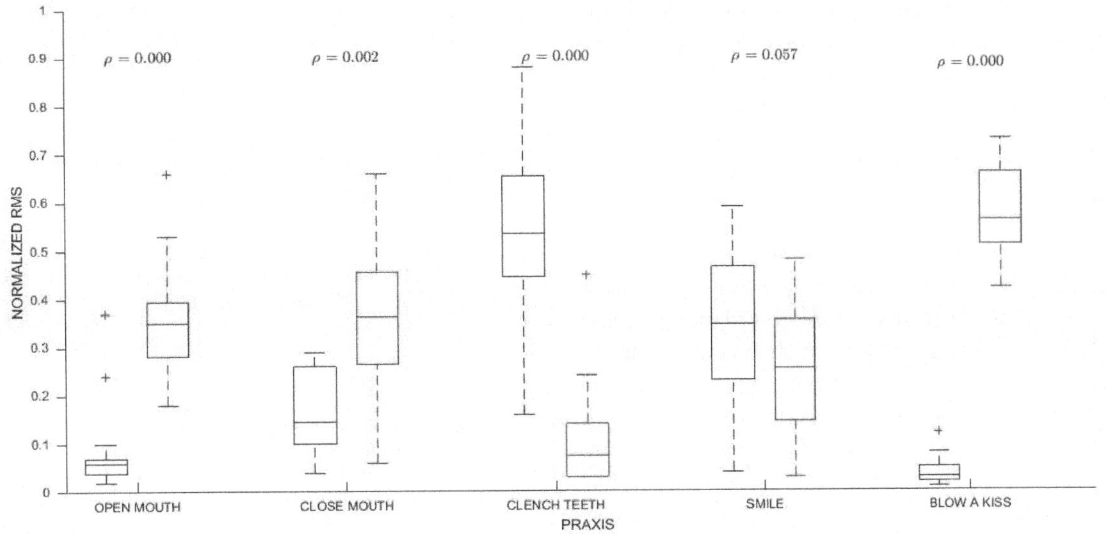

Fig. 3: Box and whiskers diagram of the median of RMS values in five praxis. The blue boxes show the masseter muscle and the red boxes show the Orbicularis oris activity.

IV DISCUSSION

In the open- and close mouth praxis (essential movements in deglutition, breathing and speaking), the dominant muscle according to the nRMS value is the orbicularis oris. Such result comes from the sphincter function and its great cortical representation in the motor primary area [6]. In the clench teeth praxis we noticed that the muscle dominance was for the masseter muscle, structure that behaves as principal agonist in chewing movements. In these three praxis the computed nRMS differences between orbicularis oris versus masseter muscle were statistically significant, indicating the agonist or lead pattern of a muscle over the other depending on the executed praxis. In the blow a kiss praxis, the dominant muscle is the orbicularis oris, with a significant difference with the masseter. This is related to the sphincteric activity of the lips [7]. On the smile praxis, we did not find a predominant muscle ($p=0.057$). This praxis has an importance in social behavior, and reflects the degree of peripheral complexity, by the action of several muscles including zygomatic, risorius (Santorinis muscle), elevators of oral commissure, as well as the highest degree of cortical representations in various neural structures (cortex primary motor, hypothalamus, limbic system) [8] [9] [10].

The surface electromyography (sEMG) is a reliable method for the focused assessment of muscular function [11] not only in healthy individuals but also in patients. Although the sEMG activity of swallowing muscles has been studied in clinical context and research, there are few studies of electromyographic behavior of the muscular structures involved in the realization of orofacial praxis, especially in healthy population. The study of praxis and apraxias has been carried out since the late nineteenth century [12], however, it remains as a subjective topic, with few clarification and no consensus about definitions and objective methods for evaluation. Our findings suggest that using sEMG is possible to develop a method for objective assessment of ideomotor orofacial praxis that will allow an objective diagnosis of this higher cortical function in normal subjects and in pathological conditions. The praxia assessment is based mainly on the clinical trials expertise, as mentioned above, using mostly neuropsychological test, which does not determine a universal character of the score, laying on the clinical expertise the diagnosis of praxical alteration. It is then that neuropsychological assessment tests of motor performance, usually of the upper extremities, not emphasizing the orofacial praxis, evidenced in numerous neurological disorders. There is a number of useful measures for the identification of motor impairment and thus make inferences about the functional integrity of the cerebral hemispheres and particularly of voluntary motor activity. These measures include different aspects of laterality, such as manual preference, strength, speed and skill [13] [14] [15] [16], in the absence of an objective technique applied to praxic processes. The current work, which is preliminary for the establishment of a diagnostic tool for orofacial disorders, raises an advance in the development of quantitative methods with utility in the clinical setting, since the sEMG quantifies the muscle activity involved in the orofacial praxical processes, through two main muscles, orbicularis oris and masseter, allowing the pattern detection in normal subjects.

V Conclusion

This sEMG characterization forecasts a progress in the objective study of the praxis with future diagnostic application for secondary apraxis cause by neurological lesions. This work is a preliminary study that contributes to work in a bigger challenge which is the establishment of the sequential pattern of cortical activation and neuromuscular that organizes, stores, edits and executes voluntary movements, in the connection between the peripheral motor systems (through EMG) and central motor systems (through electroencephalography or functional neuro-images) of praxis.

Conflict of Interest

The authors declare that they have no conict of interest.

Acknowledgements

This work has been supported by COLCIENCIAS Republica de Colombia, project No. 115071149746.

References

1. O Schrager. Actos motores oro-faringo-faciales y praxias fonoarticulatorias 2001;47:22–32.
2. Osiurak F Le Gall D. Apraxia: Clinical Types, Theoretical Models, and Evaluation *N Engl J Med.* 2012.
3. S Fernndez-Guinea. *Apraxias y agnosias. In: Rehabilitacin neuropsicolgica Intervencin y prctica clnica.* Espaa: Elsevier Masson 2011.
4. Caballero K Duque LM Ramrez JC Pelaez Vargas A. Conceptos bsicos del anlisis electromiogrfico *Ces Odontologa.* 2002;15:41–50.
5. Gross RG Grossman M. Update on apraxia *Curr Neurol Neurosci Rep.* 2008.
6. DE Haines. *Fundamental neuroscience for basic and clinical applications.* Elsevier, 2013.
7. Yazar M Kozanolu E Karsidag S. Functional Sharing of the Upper Orbicularis Oris Muscle for the Reconstruction of the Lower Lip *J Craniofac Surg.* 2015.
8. Ramachandran V. S.. The neurology and evolution of humor, laughter, and smiling: The false alarm theory *Medical Hypotheses.* 1998;51:351–354.
9. Vrticka PascalBlack Jessica M Reiss Allan L. The neural basis of humour processing *Nature Reviews Neuroscience.* 2013;14:860–868.
10. Wild Barbara, Rodden Frank A., Grodd Wolfgang, Ruch Willibald. Neural correlates of laughter and humour *Brain.* 2003;126:2121–2138.
11. Kant P., Bhowate R. R., Sharda N.. Assessment of cross-sectional thickness and activity of masseter, anterior temporalis and orbicularis oris muscles in oral submucous fibrosis patients and healthy controls: An ultrasonography and electromyography study *Dentomaxillofacial Radiology.* 2014;43.
12. Lehr RP Coulson R. Electromyographic evidence of a multiple motor system: implications for apraxia *Eur Neurol.* 1992.
13. Annett M. A classification of hand preference by association analysis. *British journal of psychology (London, England : 1953).* 1970;61:303–321.
14. Anstey KJ Smith GA. Interrelationships among biological markers of aging, health, activity, acculturation, and cognitive performance in late adulthood *Psychol Aging.* 1999.
15. Benton AL Polder GJ. Some aspects of handedness *Psychiatr Neurol.* 1962.
16. Haaland K Y, Temkin N, Randahl G, Dikmen S. Recovery of simple motor skills after head injury. *Journal of clinical and experimental neuropsychology.* 1994;16:448–56.

Correspondence: Juan Camilo Suarez Escudero
Institute: Universidad Pontificia Bolivariana
Street: Calle 78b No 72a−109
City: Medellín
Country: Colombia
Email: juanca.suarez@upb.edu.co

New technique for determining age of coastal skates from Argentinian sea by digital image processing analysis: A preliminary study

P.A. Cristini[1], J.I. Pastore[2], S.A. Barbini[3], J. Ballarre[1] D. Sabadín[3] and A. Bouchet[2]

[1] Instituto en Ciencia y Tecnología de los Materiales (INTEMA) CONICET, UNMdP, División Electroquímica Aplicada, Argentina.
[2] Instituto de Investigaciones Científicas y Tecnológicas en Electrónica (ICyTE), CONICET, UNMdP, Argentina.
[3] Instituto de Investigaciones Marinas y Costeras (IIMyC) CONICET, UNMdP, Lab. de Ictiología, Mar del Plata, Argentina.

Abstract— **To understand dynamic populations of cartilaginous fishes age and growth studies have essential importance in the demographic analysis application, especially to estimate vulnerability to mortality by fishing. However, knowledge age of cartilaginous fishes have a serious inconvenient since in many opportunities the calcified structures, as vertebrae, must be stained with different techniques to identify rings of growth, and the effectiveness of each technique varies according to the study species. This problem would be solved using methods that allow identifying rings of growth with an objective technique such as digital image processing analysis. In the present work we performed a preliminary study using as model the Smallnose skate *Sympterygia bonapartii* (Rajidae). Corpus calcareum from small specimen does not have growth bands while the medium one has four bands and the large, six. The number of bands is related directly with the age of the specimens and is in good agreement with the visual counting made by two readers. The use of the algorithm to determine the number of bands in skates allows to be independent of the operator and to have a systematic method of measure.**

Keywords— **Cartilaginous fishes, calcium phosphate, age determination, image analysis.**

I. Introduction

Cartilaginous fishes (sharks, skates and chimeras: Chondrichthyes) are characterized by having life histories which make them very vulnerable to intensive fishing activities and destruction of habitats in short time [1-3]. Such vulnerability is because of their life histories have low growth rates, late sexual maturation and low fecundity compared to bone fishes [4]. There is a strong evidence of recent decline populations as a result of over fishing [5], being Argentina the fifth country in the world which more exploit these species with an annual average of 35089 tonnes between 2000 and 2008 [6]. The increase of captures and international market, add to the low population productivity of cartilaginous fishes, have produced a concern about their conservation.

To understand dynamic populations of cartilaginous fishes age and growth studies have essential importance in the demographic analysis application, especially to estimate vulnerability to mortality by fishing [7]. When knowing growth population rates of a given species results difficult (a very common situation around the world because of the lack of fisheries statistics in cartilaginous fishes), estimate the age of sexual maturity can have a strong implication in the design of management strategies for conservation. This is because of species with late sexual maturation ages will result in populations with higher risk of extinction than those with early sexual maturation ages, and, therefore are critical for conservation [4]. For this, determining age and growth parameters are of vital importance. However, knowledge age of cartilaginous fishes have a serious inconvenient since in many opportunities vertebrae must be stained with different techniques to identify rings of growth, and the effectiveness of each technique varies according to the study species. This make that many times result impossible to compare curves of growth among species (or populations of the same species) whose rings where identified with different techniques. This problem would be solved using methods that allow identifying rings of growth objectively. The aim of this work is to determine age of skates by using and objective technique such as digital image processing analysis. For this work, we used as model study the Smallnose skate *Sympterygia bonapartii* (Rajidae).

II. Materials and methods

A. Field sampling and storage

Of three individuals of different body size (small, medium and large) of *Symterigia bonapartii* a section numbering between 10 and 15 of the largest vertebrae from the thoracic region were removed from the individuals. The vertebral section were bagged, labeled and frozen at -20°C until ready for preparation.

B. Cleaning, cutting and mounting

Vertebral samples were thawed and thoroughly cleaned with scalpel to remove excess of tissue and separated into individual centra. Then samples were storage in alcohol 70% at least during 24hs. Vertebral centra were included in

I. Torres et al. (eds.), *VII Latin American Congress on Biomedical Engineering CLAIB 2016, Bucaramanga, Santander, Colombia, October 26th -28th, 2016*,
IFMBE Proceedings 60,
DOI: 10.1007/978-981-10-4086-3_57

p-methyl methacrylate (PMMA) and storage in heater at 32°C during four days. After that each vertebral centra was sagitally sectioning with a low speed diamond-bladed saw (Buelher Co) and gridded until obtain thickness minor than 80 microns. Each section was mounting to slides and optical images from corpus calcareous were taken with an Sony Camara with 5.1MPixels and zoom variable between 1.9 and 2.4x depending on the vertebral size (Fig. 1) The microscope used was a stereoscopic microscope (10X) magnification.

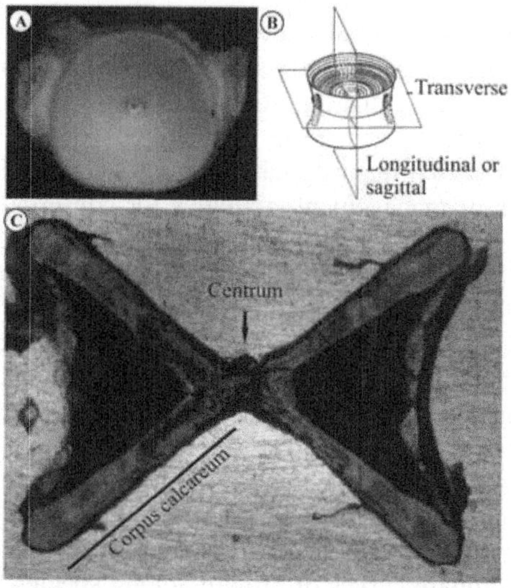

Fig. 1 A: Vertebral centra. B: Two sectioning planes that can be used on vertebral centra. C: Sagitally cut vertebral section showing centrum and corpus calcareum containing bands.

C. Image analysis

Image segmentation is the process of dividing an image into multiple parts. Its objective is to simplify or change the representation of an image in a more meaningful one to facilitate the analysis. Image segmentation is typically used to locate objects and boundaries.

The Mathematical Morphology (MM) has been studied and successfully applied in biomedical image segmentation. MM is based on concepts of geometry, algebra and set theory and was created to characterize physical and structural properties of materials, studying the geometrical structures of the components present in the images using nonlinear operations. This theory allows to analyze the shape, size, orientation and overlapping objects in a digital image. The key aspect of this approach lies in the structuring element (SE), a set with know geometry, which is compared with the image from translations. The shape and size of the SE allow test and quantify how that element is or is not contained in the image.

The basic operations of the MM are erosion and dilation. By the composition of these basic operations it is possible to build new operators such as, for example, opening and closing. By combining the basic operations, we can define the basic morphological filters opening and closing [8-10].

The opening is useful to remove small luminous details in relation to the SE, leaving the rest of the image relatively unchanged, while the closing is useful to eliminate small dark details in relation to the SE, leaving the rest of the image relatively unchanged. The main disadvantage of applying basic filters is the distortion produced by the SE on the original image structures. Basic filters can be combined with reconstruction operators thus defining the opening by reconstruction and closing by reconstruction [11].

The opening by reconstruction $\gamma^{rec}_{bero,bc}$ of the image f by the marker image g is defined as:

$$\gamma^{rec}_{bero,bc}(f) = \gamma_{bc,g}(f) \quad (1)$$

where $g = \varepsilon_{bero}(f)$ is the erosion of the image f by the SE $bero$ and bc is the SE using for the reconstruction. The image f is reconstructed for the marker image g by recursive iterations to the idempotency of the dilation of g conditionate to f.

The closing by reconstruction $\phi^{rec}_{bdil,bc}$ of the image f by the marker image g is defined as:

$$\phi^{rec}_{bdil,bc}(f) = \phi_{bc,g}(f) \quad (2)$$

where $g = \delta_{bdil}(f)$ is the dilation of the image f by the SE $bdil$ and bc is the SE using for the conditional erosion. The image f is reconstructed for the marker image g by recursive iterations to the idempotency of the erosion of g conditionate to f.

By sequential combination of the operators defined in equations (1) and (2) with SEs of increasing size, can be defined Alternating Sequential Filters (ASFs) for reconstruction. Formally they can be defined as follows:

The sequential filter "OC" (opening-closing) of n iterations:

$$nb, nc - \gamma^{rec}\phi^{rec}(f) = \gamma^{rec}_{nb,bc}(\phi^{rec}_{nb,bc}(\ldots\gamma^{rec}_{b,bc}(\phi^{rec}_{b,bc}(f)))) \quad (3)$$

The sequential filter "CO" (closing-opening) of n iterations:

$$nb, nc - \phi^{rec}\gamma^{rec}(f) = \phi^{rec}_{nb,bc}(\gamma^{rec}_{nb,bc}(\ldots\phi^{rec}_{b,bc}(\gamma^{rec}_{b,bc}(f)))) \quad (4)$$

where $n \in \mathbb{N}$ and nb is:

$$nb = \underbrace{\delta_b(\delta_b(\ldots(\delta_b(b))))}_{(n-1)\text{ veces}} \quad (5)$$

The ASFs for reconstruction allows homogenize the regions of the image, eliminating structures that are not of interest and producing less distortion at the edges of the image structures than traditional ASFs.

D. Segmentation Method

In this section, the algorithm proposed to segment the rings of growth in the vertebrae of the skates is described. To show the effects of the algorithm, as an example it was applied to some images. Figure 2 shows the different objects that compose the image and they are important to detect with the algorithm. Rv is the vertebral radius while R1, R2, R3 and R4 are the radius to the different bands (i.e. years). Each measure of radius obtained by this algorithm is very important to validate periodicity of ring formation and to estimate back-calculation of length-at-age data.

Then the proposed algorithm is explained in detail.

Fig. 2 Vertebrae composition.

 # Step 1: Background extraction

When segmenting the vertebrae of the skates images, a wide range of useless information arises, which has to be discarded as a prior step for the detection of the rings areas. To obtain effective results during the segmentation process, it is necessary to work with images containing the vertebrae only and eliminate the background and the noise. The segmentation of the vertebrae is the first step in the quantification of the distances of the rings. A correct segmentation will end, finally, in the success or failure of the analysis curves of growth among species.

To obtain this segmentation, we use an automatic algorithm in which a preprocessing is applied. In a first step, this preprocessing consists in apply Alternating Sequential Filters (ASFs) by reconstruction. In this way, we obtain a new image in which the corpus calcareum is homogenize and then regions labeling criterion, using geodesic distance, were applied, which can evolve adapting to the topology of the objects in the image [11].

Once the work area has been limited, a bandpass filter was applied that passes frequencies within a certain range and rejects (attenuates) frequencies outside that range. The parameters that defined this filter were: 151 (high size), 20 (low size) and 8 (strength). Finally, an Otsu binarization was applied [12, 13].

Figure 3 show the result of the classification after the extraction of the corpus calcareum by the algorithm proposed. Figure 3(a) show the original image, (b) show the result after the application the geodesic distance for obtain the corpus calcareum and (c) show the image result after Otsu binarization to the bandpass filter.

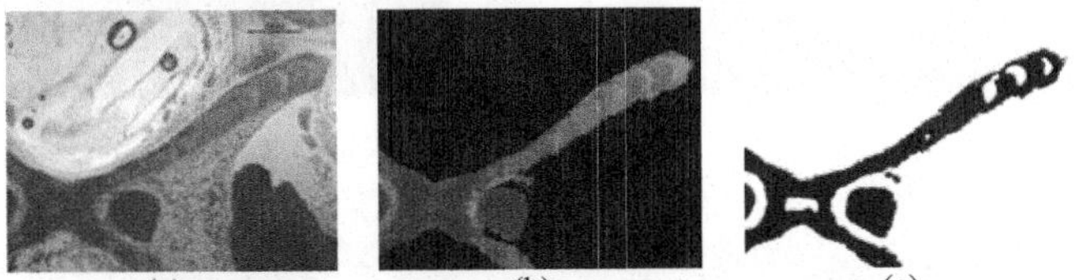

(a) (b) (c)

Fig. 3 (a) Original image. (b) Corpus calcareum. (b) Image result after Otsu binarization to the bandpass filter.

 # Step 2: Semi-automatic segmentation of rings in the vertebrae

In this second step, the method for segmentation of the rings in the vertebrae is described. First the connected components of the filter image are labeled. Then, the valid bands are manually selected.

The center of the vertebra is obtained as the center of mass of the region corpus calcareum and the end of the vertebra is calculated by the external edge of the corpus.

In the next section, the results obtained through the proposed method are described.

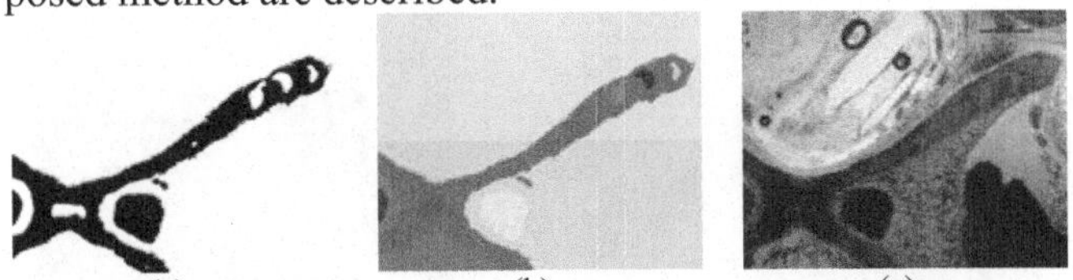

(a) (b) (c)

Fig. 4 (a) Image obtained in the previous step. (b) Tagged image region. (c) Overlap with the original image.

III. Results and Discusion

The images obtained for the optical microscope of the different specimen were analyzed. An algorithm in Matlab was developed and the number of bands and the distance be-tween the centrum and the end of the corpus calcareum were measured. Also the distance between the centrum and each band was measured. Table 1 show the different distance for three specimen: large, medium and small.

Figures 5, 6 and 7 show the segmentation of the rings areas for each specimen analyzed, the reference point in the ring area and the vertebral radius, which they were obtained according to the steps described in the previous section.

From the image processing analysis of the corpus calcareum of the analyzed samples can be extracted that the small specimen does not have growth bands, the medium one has four bands and the large, six. The number of bands is related directly with the age of the specimens and is in good agreement with the visual counting made by two readers.

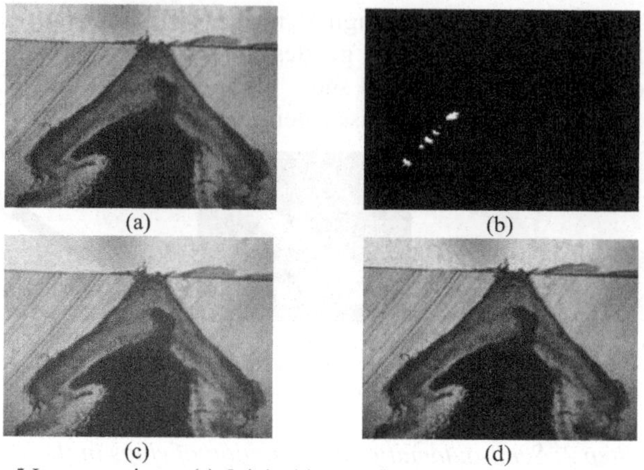

Fig. 5 Large specimen. (a) Original image. (b) Segmentation of the growth bands. (c) Reference points in the ring area. (d) Rv.

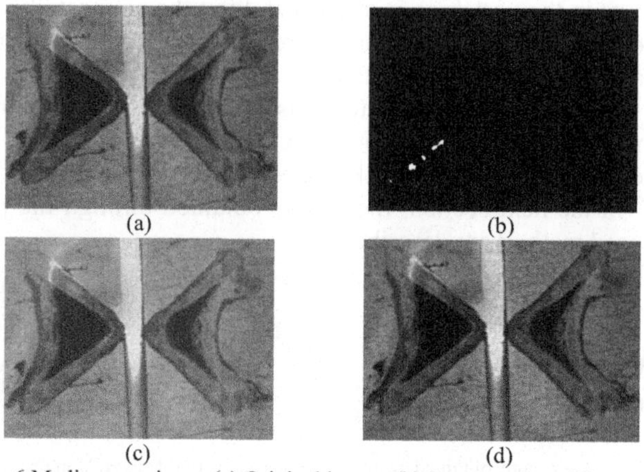

Fig. 6 Medium specimen. (a) Original image. (b) Segmentation of the growth bands. (c) Reference points in the ring area. (d) Rv.

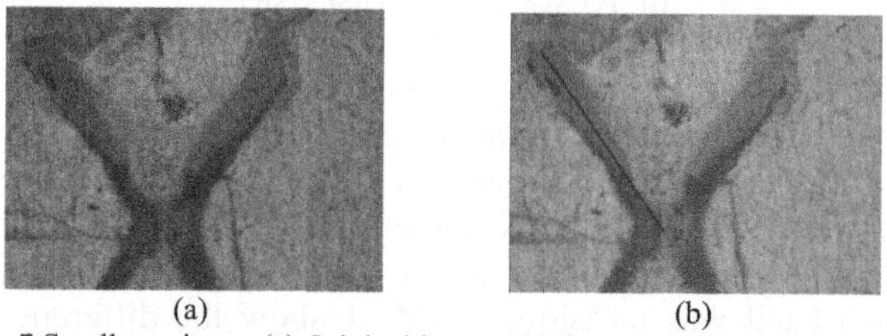

Fig. 7 Small specimen. (a) Original image. (b) Rv.

Table 1 Vertebral radius are radius in pixels to the different bands in the three specimen analyzed.

	Rv	R1	R2	R3	R4	R5	R6
Large	3657	3467	2960	2765	2541	2356	2124
Medium	2807	2696	2218	1848	1561		
Small	1855						

IV. Conclusions

The estimation of the number of bands in vertebrae of cartilaginous fishes is a way to determine age and growth of them. The impact of the observator counting affects the measurement since it is a subjective method. In this work, we developed an algorithm to detect the different growth bands of each specimen and measure the vertebral radius are the radius to the different bands. The use of the algorithm to determine the number of bands in skates allows to be independent of the operator and to have a systematic method of measure.

Acknowledgment

The authors would like to thank Consejo Nacional de Investigaciones Científicas y Técnicas (CONICET) and Fondo para la Investigación Científica y Tecnológica (FONCyT) PICT 14 N° 819, for the financial support.

References

1. Stevens JD, Bonfil R, Dulvy NK and Walker PA. (2000) The effects of fishing on shaks, rays, and chimaeras (chondrichthyans), and the implications for marine ecosystems. ICES Journal of Marine Science, 57:476-494.
2. Myers RA and Worm B. (2005) Extinction, survival or recovery of large predatory fishes. Philosophical. Transactions of the Royal Society B, 360:13-20.
3. Dulvy N. K. et al. (2014) Extinction risk and conservation of the world's sharks and rays. eLife 3: e00590.
4. Hutchings JA, Myers RA, García VB, Lucifora LO and Kuparinen A. (2012) Life-history correlates of extinction risk and recovery potential. Ecological Applications, 22:1061-1067.
5. Ferretti F, Worm B, Britten GL, Heithaus MR and Lotze HK. (2010) Patterns and ecosystem consequences of shark declines in the ocean. Ecology Letters, 13:1055-1071.
6. Lack M and Sant G. (2011) The Future of Sharks: A Review of Action and Inaction. TRAFFIC International and the Pew Environment Group.
7. Cortés E. (2004) Life history patterns, demography, and population dynamics. En: Biology of sharks and their relatives (Carrier J.C., J.A. Musick y M.R. Heithaus eds.), 449-469. CRC Press, Boca Raton, United State.
8. Serra, J. (1982) Image analysis and mathematical morphology, vol 1, Academic Press, London.
9. Serra J. (1988) Image analysis and mathematical morphology, vol 2, Academic Press, London.
10. R. González and R. Woods. (1996) Digital Image Processing. Addison-Wesely Publishing Company.
11. Pastore J, Moler E and Ballarin V. (2005) Segmentation of brain magnetic resonance images through morphological operators and geodesic distance, Digital Signal Processing, 15:153-60.
12. Otsu N. (1979) A threshold selection method from gray-level histogram. EEE Trans. Syst.Man Cybern, SMC-9(1): 62-6.
13. Mehmet S and Bulent S. (2004) Survey over image thresholding techniques and quantitative performance evaluation. Journal of Electronic Imaging, 13:146-65.

Author: Agustina Bouchet
Institute: National University of Mar del Plata
Street: Juan B. Justo 4302
City: Mar del Plata
Country: Argentina
Email: abouchet@fi.mdp.edu.ar

Electrodes based on PPy polymer for electrocardiography and impedance plethysmography

O. Teran-Jiménez[1], D. Hernández-Rivera[1] and E. Suaste-Gómez[1]

[1] Center for Research and Advanced Studies of the National Polytechnic Institute, Engineering Electrical Department, Bioelectronics Section, Av. IPN No. 2508 Mexico City, 07360, Mexico

Abstract— **Polymer electrodes based on Polypyrrole (PPy) were fabricated for their use in electrocardiography and impedance plethysmography applications. The fabrication process was also described, which had the advantage of produce disposable or even reusable electrodes. The electrodes were fabricated as films or bulks elements according to the requirements of their use and they were compared with nickel clamp electrodes in order to evaluate their use in electrocardiographic records. Measurements of impedance of the PPy electrodes were done and compared with the impedance of clamp electrodes to evaluate their behavior. Also, impedance pletismographic measures were done to probe the potential uses of this electrodes in films form.**

Keywords— **Electrodes, ECG, polymer, Polypyrrole.**

I. Introduction

The material research has created new opportunity areas for engineers and scientist in the search of new solutions to typical demands in health care areas. All the novel technologies associated to this area have increased in the same magnitude. The miniaturization and the improvement in the capability and velocity of many medical devices is evidence of that. In the same way, a basic element, like the bioelectrodes, has acquired a great complexity in the design due to the specific requirements of their application. More comfortable and reliable designs have been created to reduce skin-electrode impedance or to improve their behavior for long term recordings and so ensure an acceptable signal noise ratio [1, 2].

Recent studies of polymer composites based on conductive polymers have demonstrated a great field of application on medicine as a result of improving properties like a larger sensing area, weight reduction, enhanced strength and flexibility, visible or infrared light sensibility and many other properties that open the application areas [3-5]. These polymers possess an excellent control of the electrical stimulus of cells, very good electrical and optical properties, and high conductivity/weight ratio. Also, they can be fabricated to be biocompatible, biodegradable, and porous [6-8]. Another advantage of conductive polymers is that their chemical, electrical, and physical properties can be adapted to the specific requirements of the application [6, 9]. In this context, with the polymer electrodes, the contact area has been increased and improved as a consequence of their flexibility and adaptability to a large number of flat or curved surfaces. Typically the non-metallic bioelectrodes presented now a days have probed their efficacy in electrocardiography (ECG) and electroencephalography (EEG) records [10-13]. However, bioelectrodes are required in other kind of useful non-invasive medical test. One of them is the impedance plethysmography (IPG), which has been an useful technique to reflect blood volume changes and can indirectly indicate the presence or absence of nowadays thrombosis [14] or where the bioelectrical impedance changes need to be monitored [15]. The technique uses an arrangement of electrodes that need to fit in an extremity of the body to receive and stimulate it with an enough small current at high frequency signal to correlate the impedance of the extremity with its blood volume.

The electrodes presented here are based on polypyrrole which is a conductive polymer. It has a good conductivity and can be processed easily. Therefore, this material can be used in solution or in powder form to be mixed with other polymeric substrates. The versatility of this material optimizes the fabrication process of the electrodes, the time involved of stages of its development and reducing significantly the production cost. The two types of elaborated electrodes employ a polymeric agglutinant that performs a double function. It contributes to the mechanical resistance of the electrode and, it keeps joined the PPy as a single block or as a uniform layer. In this way, the body of the electrode is completely elaborated of polymers. The experimental results show the functionality through some ECG and IPG records acquired with them.

II. Methodology

A. ECG Electrodes

The fabricated ECG electrodes were designed with a rounded shape. Mixed compound of PPy (Aldrich No.

I. Torres et al. (eds.), *VII Latin American Congress on Biomedical Engineering CLAIB 2016, Bucaramanga, Santander, Colombia, October 26th -28th, 2016*,
IFMBE Proceedings 60,
DOI: 10.1007/978-981-10-4086-3_58

577030, USA) and agglutinant (multipurpose adhesive spray, 3M, USA) with a ratio 9:1 respectively. The mixture was pressed into a die with a 980.7 MPa to obtain discs (2 cm in diameter with and 1.5 mm thick) during 30 minutes.

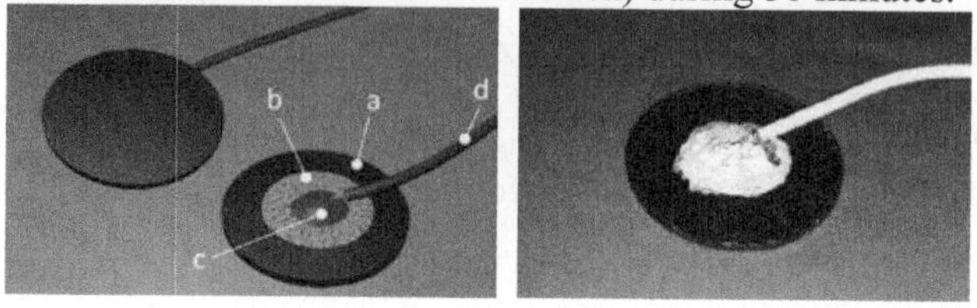

Fig. 1: Schematic electrode of ♷ PPy: a) Pressed PPy disc, b) Silver paint, c) Solder, d) Cu wire. (Right). Fabricated ECG Electrode (Left).

One side was painted with a layer using silver paint (SPI Supplies, Figure 1). The conductive layer makes possible the interface between the plastic electrode and a Cu wire. Only a section of the electrode was painted to reduce the parasite capacity effect due to the silver paint layer. The wires were joined to paint with a droplet of lead free solder. The other side of the wire where connected to the ECG equipment (Contec ECG 1200G, China). For these probes were selected 10 subjects with a range of age from 26 to 28 without heart disease, pacemaker or under medical treatment.

B. IPG Electrodes

The IPG electrodes were also based on PPy polymer. A shape of rectangular bars deposited over an acetate sheet and then stick to Velcro strip were chosen to ease their placement over the finger of each subject0.

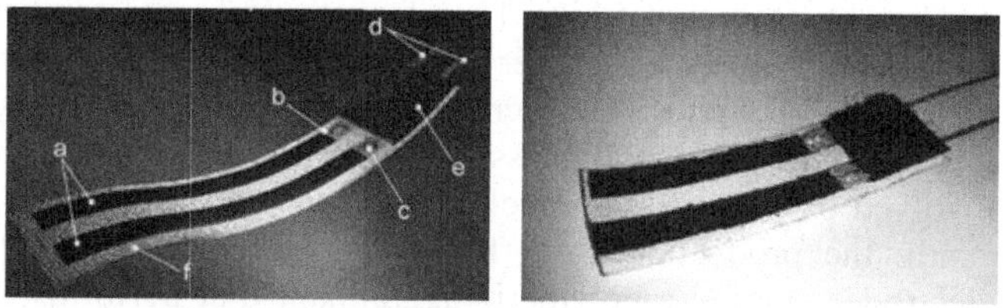

Fig. 2: Schematic adjustable sheet electrode. a) ♷ PPy flexible sheets, b) Silver paint layer, c) Solder droplet, d) Cu wires, e) velcro loops strip, f) velcro hook strip (Right). Fabricated adjustable bar electrodes (Left).

The same mixture of ECG electrodes were made by coating rectangular areas of acetate sheet, and then stick to Velcro strip were chosen to ease their placement on subject finger of each.

They were reduced in thickness to 1 mm and pressed under 0.19 MPa, in order to increase the electrodes flexibility. Pressed mixture was cut into rectangle sheets of 0.1 x 0.5 x 4 cm^3.

These sheets were adhered in a Velcro strip disposed and placed in parallel (separated by 1 cm). The electrodes were covered in one of their borders with silver paint and soldered to the wires, for connection to the designed current amplifier (Figure 2).

A controlled current source was also required to generate a potential difference between the sensing electrodes when they were placed in the finger of the subject. This source was coupled to a voltage sinusoidal generator to obtain a voltage to current (V-I) converter which produce a sinusoidal current with a frequency of 5 kHz and constant peak amplitude of 90 μA. The amplitude of current signal was small enough to be safe and imperceptible by the subject.

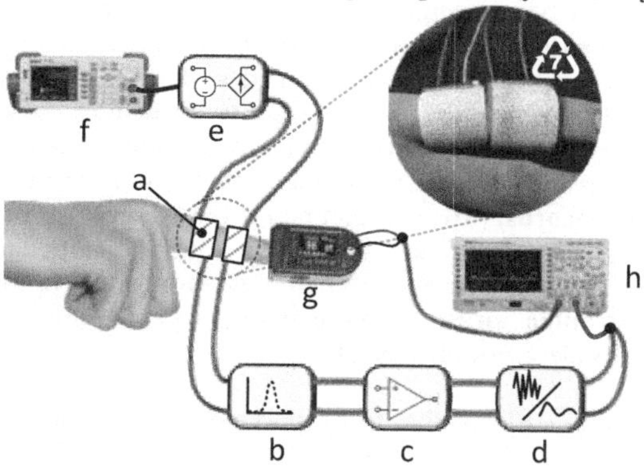

Fig. 3: Connection of plethysmography electrodes and fingertip oximeter. a) Flexible ♷ PPy Electrodes, b) Tuned filter, c) Amplifier, d) Demodulator, e) Voltage to current converter, f) Waveform generator, g) Fingertip oximeter h) Oscilloscope.

To avoid electric shock risk, all the electronic equipment was connected to an isolated power supply based on the integrated circuit MAX256 (Maxim Integrated, USA). The arrangement of electrodes and circuit blocks used to recover and demodulate the voltage signal present in the finger due to the signal of V-I converter is shown in Figure 3. The modulated signal fluctuates according the impedance in the finger. The first block connected to electrodes was a tuned filter at 5 kHz which reduce the level of noise and another kind of parasite frequencies far of the tuned applied signal. A voltage amplifier with gain of 100 to increase the amplitude of small filtered signal was placed at the output of the filter. Next, a demodulator was also required to get the amplitude variations modulated by the instant impedance changes as a result of the blood flow. Simultaneously, a fingertip oximeter (SB220, Rossmax, Taiwan) was used in the sensed finger to extract the real time signal from the infrared sensor photoplethysmographic (PPG) signal which is related with the blood flow) and compare it with the IPG signal captured with the plastic PPy electrodes. These signals were observed and saved with a digital oscilloscope (Agilent DSO3062A, USA). The same 10 subjects were used in this part of the test. The records of IPG and PPG were captured simultaneously to ensure that IPG signal was directly correlated with the blood flow.

III. Results

A. ECG Electrodes

Previous to get the register of the electrodes, the electric impedance of the ECG fabricated electrodes were compared with the nickel clamp electrodes used in many ECG equipment. It was measured with a typical arrangement of three electrodes [16].

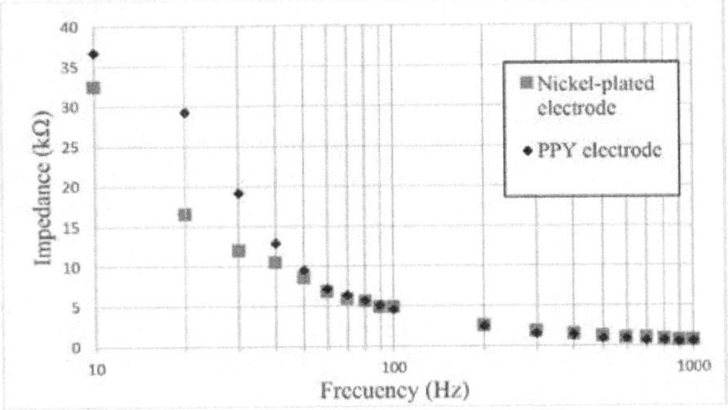

Fig. 4: Electrode impedance of PPy and nickel clamp electrodes.

Ten measures were developed for each electrode (PPy and nickel clamp electrodes) at each frequency and the impedance measures were averaged to get the curves on Figure 4. The results show that polymer electrodes have a similar impedance behavior respect to the clamp electrodes.

The polymer and clamp electrodes were placed in the same manner to record the ECG signals. The register shows low noise ECG (I derivation) signal from the same subject using PPy electrodes and the Nickel clamp electrodes (Figure 5). As can be observed, the registers show a good morphology that do not exhibit evidence of distortion or the presence of noise due to the electrode impedance.

B. IPG Electrodes

The IPG electrodes were connected according to the typical connection (Figure 3). Two electrodes supply AC current and two electrodes acquire the voltage generated by it.

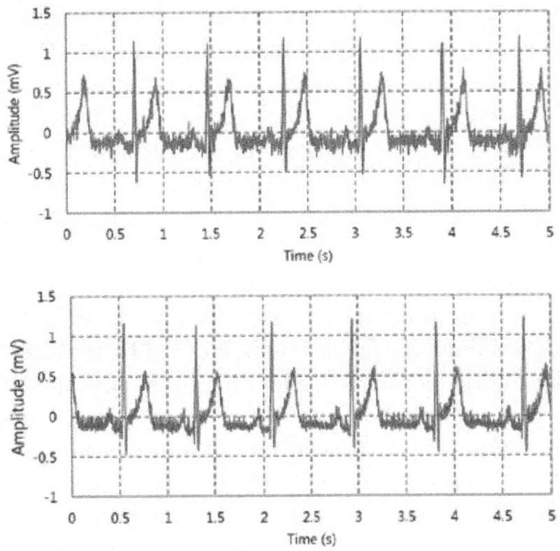

Fig. 5: ECG derivation I recorded with PPy (Top) and Nickel (Bottom) electrodes.

The signals are signal recorded with the PPy-IPG electrodes and other recorded by the commercial oximeter with similar frequencies (1.33Hz) are synchronized (Figure 6). The PPG signal of fingertip oximeter is only used to get a reference for the changes in volume due to the blood flow. An increment of blood flow corresponds to an increment in the impedance of the measured section as a result of the volume increment. This is established by the Nyboer's formula [17]: $\Delta V=\rho \cdot L^2 \cdot Z_0^{-2} \cdot \Delta Z$ where ΔZ and ΔV are the impedance and volume changes respectively, L is the length between the voltage sensing electrodes, Z_0 is the initial value of impedance limited by the voltage sensing electrodes and ρ are the blood resistivity.

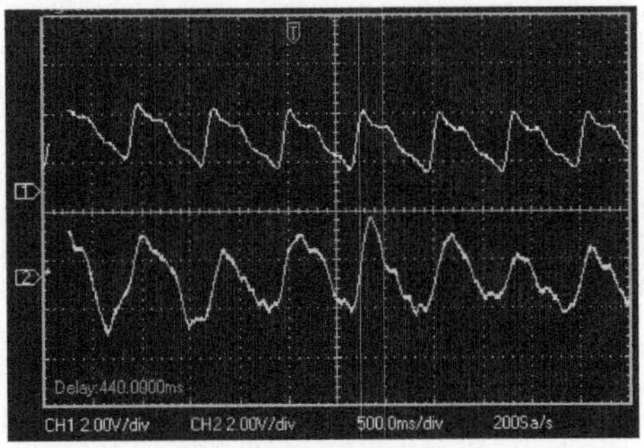

Fig. 6: PPG signal (Top trace) and IPG signal (Bottom trace) recorded with bar electrodes.

IV. Discussion

Currently, the ambulatory monitor of physiological signals is increasing because of the processing power of the mobile processors and the miniaturization of many integrated circuits, that have been made this fact possible. An essential part of this technology is the electrodes used to sense the involved signals. The bio-electrodes still keeping their utility in a variety of studies where a bio-signal needs to be registered. In this context, the polymer electrodes proposed in this work contribute, not only to the actual alternatives of polymer electrodes development but to an innovative and ease way to elaborated reusable and flexible electrodes. They have the potentiality to be in a bulk or film form or even deposited over fabrics, which could be recycled.

The microelectrodes are another area were the PPy electrodes can assume the same roll of other used substance like Indium oxide (ITO), typically used in micro arrays as occur with other conductive polymers [13]. Moreover, brain-machine interfaces or recent active bioprosthesis depend on the electrode tissue contacts. Some of them require electrodes which are in contact with the neural tissue. To keep the low impedance is a decisive factor to ensure in this applications the record of signals and in some cases, produce microstimulation by the same electrodes [18].

References

[1] G. I. Petrova, "Influence of electrode impedance changes on the common-mode rejection ratio in bioimpedance measurements," Physiological Measurement, vol. 20 (4), pp. N11-N19, 1999.

[2] D. J. Hewson, J. Y. Hogrel, Y. Langeron, and J. Duchene, "Evolution in impedance at the electrode-skin interface of two types of surface EMG electrodes during long-term recordings," Journal of Electromyography and Kinesiology, vol. 13 (3), pp. 273-279, 2003.

[3] R. Balint, N. J. Cassidy, and S. H. Cartmell, "Conductive polymers: Towards a smart biomaterial for tissue engineering," Acta Biomaterialia, vol. 10 (6), pp. 2341-2353, 2014.

[4] Z. Khatri, S. Ali, I. Khatri, G. Mayakrishnan, S. H. Kim, and I.-S. Kim, "UV-responsive polyvinyl alcohol nanofibers prepared by electrospinning," Applied Surface Science, vol. 342 pp. 64-68, 2015.

[5] G. Buchberger, P. Bartu, R. Schwoediauer, B. Jakoby, W. Hilber, and S. Bauer, "A Flexible Polymer Sensor for Light Point Localization," 26th European Conference on Solid-State Transducers, Eurosensor 2012, vol. 47 pp. 795-800, 2012.

[6] B. Lakard, L. Ploux, K. Anselme, F. Lallemand, S. Lakard, M. Nardin, et al., "Effect of ultrasounds on the electrochemical synthesis of polypyrrole, application to the adhesion and growth of biological cells," Bioelectrochemistry, vol. 75 (2), pp. 148-157, 2009.

[7] A. Kotwal and C. E. Schmidt, "Electrical stimulation alters protein adsorption and nerve cell interactions with electrically conducting biomaterials," Biomaterials, vol. 22 (10), pp. 1055-1064, 2001.

[8] J. Y. Lee, C. A. Bashur, A. S. Goldstein, and C. E. Schmidt, "Polypyrrole-coated electrospun PLGA nanofibers for neural tissue applications," Biomaterials, vol. 30 (26), pp. 4325-4335, 2009.

[9] K. Dong-Hwan, S. M. Richardson-Burns, J. L. Hendricks, C. Sequera, and D. C. Martin, "Effect of immobilized nerve growth factor on conductive polymers: electrical properties and cellular response," Advanced Functional Materials, vol. 17 (1), pp. 79-86, 2007.

[10] C. Yuanfang, P. Weihua, C. Sanyuan, W. Xian, Z. Shanshan, W. Huan, et al., "Poly(3,4-ethylenedioxythiophene) (PEDOT) as interface material for improving electrochemical performance of microneedles array-based dry electrode," Sensors and Actuators: B Chemical, vol. 188 pp. 747-56, 2013.

[11] P. Leleux, J.-M. Badier, J. Rivnay, C. Benar, T. Herve, P. Chauvel, et al., "Conducting Polymer Electrodes for Electroencephalography," Advanced Healthcare Materials, vol. 3 (4), pp. 490-493, 2014.

[12] C. Yun-Hsuan, M. Op de Beeck, L. Vanderheyden, E. Carrette, V. Mihajlovic, K. Vanstreels, et al., "Soft, Comfortable Polymer Dry Electrodes for High Quality ECG and EEG Recording," Sensors, vol. 14 (12), pp. 23758-80, 2014.

[13] E. Castagnola, L. Maiolo, E. Maggiolini, A. Minotti, M. Marrani, F. Maita, et al., "PEDOT-CNT-coated low-impedance, ultra-flexible, and brain-conformable micro-ECoG arrays," IEEE Transactions on Neural Systems and Rehabilitation Engineering, vol. 23 (3), pp. 342-50, 2015.

[14] E. Irzmanska, G. Padula, and R. Irzmanski, "Impedance plethysmography as a tool for assessing exertion-related blood flow changes in the lower limbs in healthy subjects," Measurement, vol. 47 pp. 110-115, 2014.

[15] A. Rebeca Gutierrez-Cruz, B. Soto-Rivera, B. Alicia Leon-Chavez, E. Suaste-Gomez, D. Martinez-Fong, and J. Antonio Gonzalez-Barrios, "Active core rewarming avoids bioelectrical impedance changes in postanesthetic patients," BMC Anesthesiology, vol. 11 (2), pp. 1-6, 2011.

[16] S. Grimnes, "Impedance measurement of individual skin surface electrodes," Medical and Biological Engineering and Computing, vol. 21 (6), pp. 750-755, 1983.

[17] J. Nyboer, "Regional pulse volume and perfusion flow measurement: Electrical impedance plethysmography," Archives of Internal Medicine, vol. 105 (2), pp. 264-276, 1960.

[18] B. Ghane-Motlagh and M. Sawan, "Design and implementation challenges of microelectrode arrays: a review," Materials Sciences and Applications, vol. 4 (8), pp. 483-495, 2013.

Institute: Center for Research and Advanced Studies of the National Polytechnic Institute, Engineering Electrical Department, Bioelectronics Section
Street: Av. IPN No. 2508
City: Mexico City
Country: Mexico
Email: omartjmz@hotmail.com, esuaste@cinvestav.com

Gesture Recognition and Machine Learning Applied to Sign Language Translation

Luis A. Estrada Jiménez[1], Marco. E. Benalcázar[2] and Nelson Sotomayor[1]

[1]Escuela Politécnica Nacional/Departamento de Automatización y Control Industrial, Quito, Ecuador
[2] Escuela Politécnica Nacional/Departamento de Informática y Ciencias de la Computación, Quito, Ecuador

Abstract— **In this paper we propose an intelligent system for translating sign language into text. This approach consists of hardware and software. The hardware is formed by flex, contact, and inertial sensors mounted on a polyester-nylon glove. The software consists of a classification algorithm based on the k-nearest neighbors, decision trees, and the dynamic time warping algorithms. The proposed system is able to recognize static and dynamic gestures. This system can learn to classify the specific gesture patterns of any person. The proposed system was tested at translating 61 letters, numbers, and words from the Ecuadorian sign language. Experimental results demonstrate that our system has a classification accuracy of 91.55%. This result is a significant improvement compared with the results obtained in previous related works.**

Keywords— **Sign language translation, gesture recognition, machine learning, pattern classification**

I. Introduction

Worldwide, at least 360 million people are likely to be deaf and dumb (or simply deaf) [1]. Most of these people do not have access to public services. In the case of hearing impairment, mutism has created the need of using sign language. A sign language consists of a set of manual gestures that use fingers, palms, arms, and even body movements to represent letters, numbers, and words. This type of language allows people who cannot speak or hear to communicate with others. However, most of the non-disabled people do not know this type of language. Therefore, for deaf people performing daily activities turns out to be hard, especially in public areas. Additionally, it is difficult and expensive to make non-disabled people learn sign language. For these reasons, automatic systems that translate sign language into text are required.

Some systems that recognize gestures and translate sign language into text have been proposed. Most of these systems use image processing techniques to perform the translation [2]. Other approaches use different kinds of sensors to acquire the orientation of the hand and fingers [3] [4]. The main drawback of all these systems is that they cannot distinguish similar gestures neither patterns that consist of movements (i.e., dynamic patterns).

In [3], the authors propose a sensorized glove combined with Euclidean classifiers capable of translating up to 10 static patterns (i.e., patterns that have no motion). A similar system is proposed in [4]. This system is able to recognize the Malaysian sign language. This approach is based on flex sensors and an accelerometer, and is able to recognize 9 static patterns. Even though these approaches show good classification accuracy, they were not developed to provide a wide range of recognition of words and expressions.

In [5], [6], and [7], prototypes of sign language translators based on the Kinect sensor are proposed. In [7], a system based on depth-estimates from the Kinect sensor is presented. To compute the angles of the fingers, this system estimates the position of both hands and some features of their contours. The dynamic time warping (DTW) algorithm is used for a pattern recognition task. This system is able to identify 52 static and dynamic gestures. Since these approaches are based on artificial vision systems, their performance is highly dependent on the light conditions where the system is used. In [8], the sign language translation is based on a sensor called leap motion controller. This approach uses decision trees (DT) and genetic algorithms to recognize 24 static characters with 82.7% of classification accuracy. Besides this relatively good performance, this approach is not capable of recognizing dynamic gestures.

Because of the problems described above, the field of gesture recognition and sign language translation is still open for research. In this paper, we propose an intelligent system to translate sign language into text. This approach can learn to recognize different gestures (i.e., static and dynamic) using machine learning and pattern recognition techniques.

The hardware of the proposed system was mounted on a polyester-nylon glove which has good properties in terms of durability, elasticity, and comfort. The hardware is composed of 3 types of sensors: flex, contact, and inertial. Combining the information of these sensors, we estimated the relative orientation and movement of the hand and its fingers. These estimates are used as the inputs of a gesture recognition system. This system is based on 3 classification algorithms: k-nearest neighbors (k-NN), decision trees, and the DTW algorithm [9]. The first two algorithms were used to recognize static gestures, whereas the DTW algorithm was used to identify dynamic gestures. We tested the proposed system at translating 61 gestures from the Ecuadorian sign language [10]: 30 letters, 10 numbers, and 21 expressions (Fig. 1).

The proposed system for gesture classification is based on a learning process. This process makes possible that our system can be trained to work with other sign languages different from the Ecuadorian one. Moreover, the proposed system could be trained to recognize signs and gestures from other domains different from the sign language

I. Torres et al. (eds.), *VII Latin American Congress on Biomedical Engineering CLAIB 2016, Bucaramanga, Santander, Colombia, October 26th -28th, 2016*,
IFMBE Proceedings 60,
DOI: 10.1007/978-981-10-4086-3_59

translation. These domains can include gaming, robotics, and assistive technology.

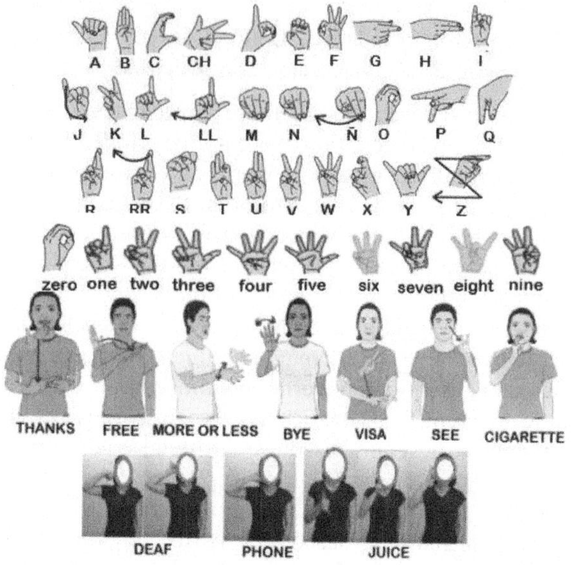

Fig. 1 Patterns and gestures that represent letters, numbers, words, and expressions of the Ecuadorian Sign Language [10]

This paper is organized as follows. In the first section, we have presented an introduction about the problem of sign and gesture recognition applied to sign language translation. The second section provides information about the materials and methods we used to develop the proposed approach. In the third section, we present and discuss the results obtained. Finally, in the fourth section, we present some concluding remarks of this work.

II. Materials and Methods

A. *Types of hand movements*

This work is based on the theory developed by Rouviere and Delmas [11]. These authors studied the following 4 types of movements of the hand: flexion and extension of the fingers, flexion and extension of the wrist, supination and pronation of the forearm, and radial and ulnar deviation of the hand. Fig. 2 shows some of these movements.

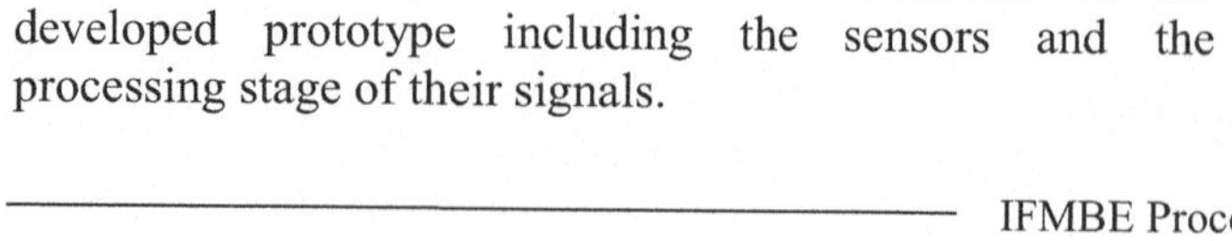

Fig. 2 Movements of the hand according to Rouviere and Delmas [11]

B. *Description of the Developed Hardware*

In this section, we describe the structure of the developed prototype including the sensors and the processing stage of their signals.

All the sensors used in this work were mounted on a polyester-nylon glove and their distribution is according to the graph that is shown in Fig. 3. The flexion and extension signals of the fingers were acquired using flex sensors. Two flex sensors were attached to each finger, except in the case of the thumb where we used only one sensor. We used two flex sensors because the two phalanges are needed for representing the movement of a finger [11]. The hand movements were measured by an inertial measurement unit (IMU) placed on the back of the hand. The IMU used for this work is the MPU-6050. This sensor measures the pitch, roll, and yaw angles (i.e., Euler angles). These angles were used to estimate the orientation of the hand in a 3D space.

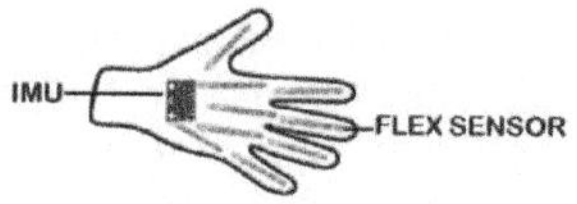

Fig. 3 Distribution of the sensors on the glove

The signals from the flex sensors were preprocessed by using a Wheatstone bridge followed by a differential amplifier. Then, all these signals were digitalized, with 10 bits of resolution, using an Arduino mega board. The data from the MPU-6050 IMU sensor was sent to the Arduino board using I^2C communication. Then, this information was processed in the Arduino to obtain the corresponding Euler angles. In this work, we only used the pitch and roll angles. We did not use the yaw angle since it is required a magnetometer to provide an accurate estimate of this angle. An aluminum contact sensor was placed between the middle and index fingers. This was done because the flex sensors and the MPU-6050 did not allow differentiating accurately the letters U and V. For each gesture, we obtained 12 digital signals: 9 from the fingers, 1 from the contact sensor, and the 2 Euler angles. Finally, all these signals were sent through a radio-frequency module to a personal computer for the classification task.

The proposed system is able to classify the movements and gestures of the hand into 61 different classes. Some gestures can only be differentiated from others based on the context where they are used. An example of this situation is number 2 and letter U. Because of this reason, the set of gestures to be classified was divided into 3 groups: numbers, letters, and words. Before testing the proposed system, the user needs to select the group inside which the classification will be performed.

C. *Training Process*

The proposed system needs to learn to recognize the movements and gestures for each user. For this process, we used a 12-dimensional feature vector containing the information from all the sensors on the glove. For training the system, the user must perform 5 repetitions for each gesture to be recognized. This value was selected as a tradeoff between the minimum number of samples for training accurately the classifiers and the time it takes to acquire each training sample. As a result of this process, for each user we obtain a training dataset composed of a total of 5*61 = 305 12-dimensional feature vectors.

D. Classification Process

The first step of the gesture recognition process is deciding whether a given gesture is static or dynamic. This was based on the classification performed by the k-NN and the decision tree algorithms. If the result is static, then the recognition process ends returning the result from these algorithms. Otherwise, if the result is dynamic, the algorithm identifies the movement using the DTW algorithm. In Fig. 4, we summarize the flowchart of the proposed classification algorithm.

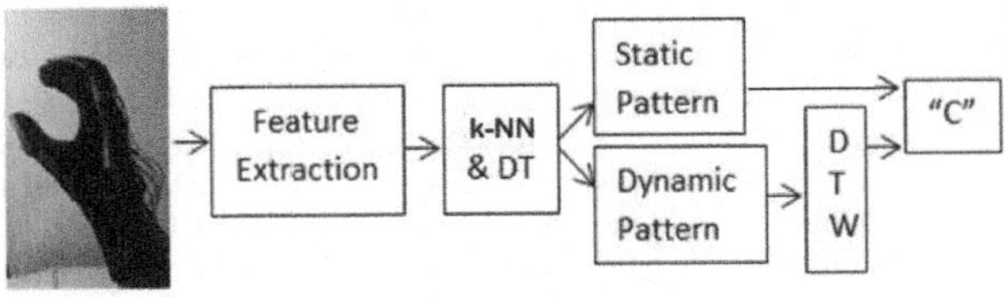

Fig. 4 Flowchart of the gesture classification process

E. k-Nearest Neighbors (k-NN) Algorithm

k-NN is a supervised classification algorithm based on finding the k closest training samples to a test sample **x**. For this task, we computed the Euclidean distance $d_E(\mathbf{x},\mathbf{x}_j)$ between **x** and each training sample $\mathbf{x}_j$, with $j = 1,2,\ldots,N$. We labeled the test sample **x** with the most frequent label among all the labels of the k closest training samples to **x**.

The k-NN algorithm is universally consistent if $k \rightarrow \infty$ and $k/N \rightarrow 0$ when $N \rightarrow \infty$ [12]. This means that the rate of grow of k must be much slower than the rate of growth of N. Based on this fact, in this work, we selected the number of neighbors, k, to be the closest odd integer that is greater than or equal to $log_2(N)$. Based on this rule, we obtained $k = 7$ for the 3 groups of gestures we used in this work (i.e., alphabet, numbers, and words).

F. Decision Trees

Few patterns of the Ecuadorian sign language could not be recognized accurately by the k-NN algorithm. This occurs because some gestures have very similar features (e.g., letters F and T). To solve this issue, we used decision trees. Decision trees classify a given pattern based on the value of each of its features in a given order [13]. In the case of the letters F and T, after the response of the k-NN algorithm, we refined the decision by observing the value of the index lower sensor. In Table 1, we show the signs for which we applied the decision tree algorithm to improve the classification. As an example, Fig. 5 shows the decision process when translating the signs of the letters F and T.

Table 1 Classification using decision trees

Signs	Feauture sonsidered
F, T	Index lower sensor
G, L, LL	Orient. and movement of the hand
H, U, V	Hand orientation and contact sensor
I, J	Hand movement
K, CH	Hand orientation
N, Ñ	Hand movement
R, RR	Hand movement
"Thumbs up" and "Thumbs down"	Hand orientation
"Coin" and "Hospital"	Hand orientation
"One moment" and "license"	Hand orientation

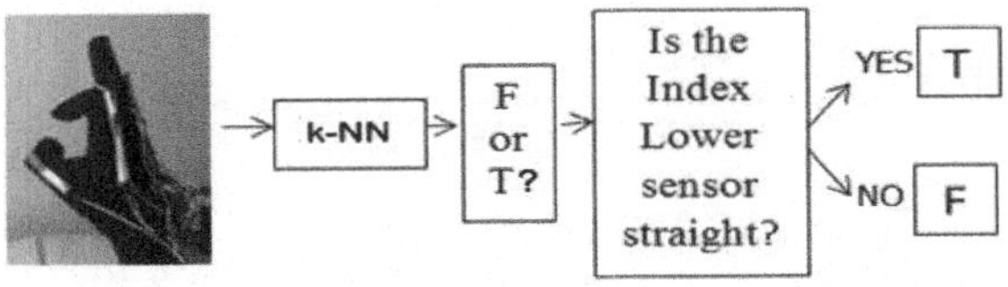

Fig. 5 Decision process for classifying the letters F and T

G. Dynamic Time Warping (DTW) Algorithm

After processing a dynamic pattern using the k-NN and decision tree classifiers, we analyzed its subsequent movement using the DTW algorithm. This algorithm was applied to the variation of the pitch and roll angles of the hand. The decision was based on the minimum of the sum of the DTW distances of these angles with respect to their reference signals. The DTW algorithm consists of finding an optimal alignment between two time series [9]. In our case, the inputs of the DTW algorithm are the variations of the pitch and roll angles of the hand measured during 2 seconds. These measurements result in 2 time series (i.e., test and reference series) composed of 20 values each.

The structure of the DTW algorithm is as follows. Let $\mathbf{A} = a_1, a_2, \ldots, a_m$ and $\mathbf{B} = b_1, b_2, \ldots, b_n$ be two sequences of length m and n, respectively. First, a $m \times n$ distance matrix **C** is computed, where the $c_{i,j}$ element is given by $c_{i,j} = |a_i - b_j|$, with $i = 1, 2,\ldots,m$ and $j = 1, 2,\ldots,n$. Then, a matrix of minimum distances **D** is obtained from **C**. The $d_{i,j}$ element of **D** is given by the recursive equation $d_{i,j} = c_{i,j} + min(d_{i-1,j-1}; d_{i-1,j}; d_{i,j-1})$, where the base case is $d_{1,1} = c_{1,1}$. In this equation, if the index of any element inside the minimum operator is out of **C**, then we set that element to be infinity. Finally, the optimal warping path is obtained from **D**. This path defines an optimal alignment between the sequences **A** and **B**. Formally, the warping path is a sequence $\mathbf{W} = \mathbf{w}_1, \mathbf{w}_2, \ldots, \mathbf{w}_l, \ldots, \mathbf{w}_L$ of L elements, where $\mathbf{w}_l = (i_l, j_l) \in [1,\ldots,m] \times [1,\ldots,n]$, that satisfies the following rules:

Boundary Condition: $\mathbf{w}_1 = (1,1)$ and $\mathbf{w}_L = (m,n)$.

Monotonicity: $i_1 \le i_2 \le \ldots \le i_L$ and $j_1 \le j_2 \le \ldots \le j_L$.

Continuity: $\mathbf{w}_{l+1} - \mathbf{w}_l \in \{(0,1),(1,0),(1,1)\}$ for $l = 1,2,\ldots,L-1$.

Finally, the DTW distance between **A** and **B**, DTW(**A**,**B**), is given by $\sum_{i=1}^{L} c_{\mathbf{w}_i} = d_{m,n}$, where $c_{\mathbf{w}_i} \in \mathbf{C}$ and $d_{m,n} \in \mathbf{D}$.

An example of the classification of a dynamic gesture between the classes "free", "bye", and "so so" is shown in Fig. 6. The signals in blue, black, and pink show the variation of the pitch angle of the reference "bye", "free", and "so so" gestures, respectively. The signal in red shows the variation of the pitch angle of the gesture to be classified (i.e., test gesture). The sum of the DTW distances between the reference and test pitch and roll signals are the following: $d(Bye,Test)$ = 13.57 + 9.66 = 23.23, $d(SoSo,Test)$ = 32.08 + 30.12 = 62.12, $d(Free,Test)$ = 18.6 + 19.38 = 37.98. Therefore, the test gesture is classified as "Bye" since it has the minimum sum of DTW distances.

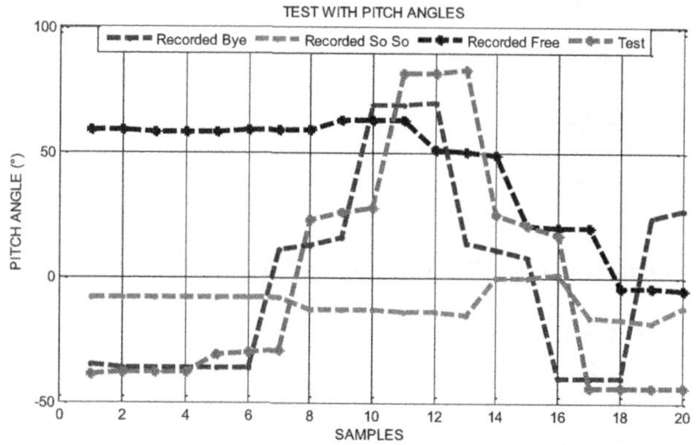

Fig. 6 Pitch angles for the reference and test patterns

III. Experimental Results

To evaluate the performance of the proposed system, we tested it with 6 non-disabled people. For each gesture to be classified, each person performed 50 repetitions. Table 2 summarizes the classification accuracies we obtained.

Table 2. Classification accuracies for the 3 groups of gestures

Group	Number of Gestures	Average Accuracy
Alphabet	30	89.01%
Numbers	10	95.90%
Words	21	89.75%

In the case of the alphabet, the letter with the lowest classification accuracy was the M with 82%. This letter was sometimes confused with the O or the N. On the other hand, the W had the best classification accuracy with 98.87%. This was because the W had a gesture pattern that is very different from the other letters. A similar situation occurs for the case of the numbers. This is the reason why their classification accuracy exceeds 95%. Finally, the group of words had relatively high classification accuracy with 89.75%. Static patterns like "thumbs up" and "thumbs down" resulted in the best performance with an accuracy of 97.33% and 98.33%, respectively. On the other hand, dynamic patterns like "free" (80.33%) and "thank you" (81%) were the most difficult to be classified.

IV. Conclusion

In this paper, we have proposed an intelligent system to translate sign language into text. This system is composed of hardware and software. The hardware is composed of flex, contact, and inertial sensors. The software consists of a classification algorithm based on the k-NN, decision tree, and DTW algorithms. This system has been able to learn to recognize 61 static and dynamic gestures of the Ecuadorian sign language. Experimental results show that the overall classification accuracy of the proposed system is 91.55%. Compared to similar approaches described in this work, this system has a better accuracy of classification. Future work includes testing the proposed system on deaf people and other sign languages different from the Ecuadorian one.

References

[1] Organización Mundial de la Salud (2011) Informe mundial sobre la discapacidad, Malta

[2] Kumar V, Goudar R, and Desai V (2015) Sign Language Unification: The Need for Next Generation Deaf Education. International Conference on Intelligent Computing, Communication and Convergence, Odisha, India, 2015, pp 673-678

[3] Kanwal K, Abdullah S, Ahmed Y, and Jafri A (2014) Assistive Glove for Pakistani Sign Language Translation. IEEE 17th International Multi-Topic Conference (INMIC), Pakistan, 2014, pp 173-176

[4] Shukor A, Miskon M, and Jamaluddin M (2015) A New Data Glove Approach for Malaysian Sign Language Detection. Procedia Computer Science 76:60-67

[5] Microsoft Research (2013) Kinect sign language translator expands communication possibilities for the deaf

[6] Li K, Lothrop K, Gill E, and Lau S (2011) A Web-Based Sign Language Translator Using 3D Video Processing. The 14th International Conference on Network-Based Information Systems, Tirana, Albania, 2011, pp 356-361

[7] Plouffe G and Cretu A (2016) Static and Dynamic Hand Gesture Recognition in Depth Data Using Dynamic Time Warping. IEEE transactions on instrumentation and measurement 65:305-316

[8] Funasaka M, Yu I, Takata M, and Joe K (2015) Sign Language Recognition using Leap Motion Controller. The 21st International Conference on Parallel and Distributed Processing Technques and Applications, Las Vegas, Nevada, USA, 2015, pp 263-269

[9] Cong G and Tuan D (2016) Similarity search for numerous patterns over multiple time series streams under dynamic time warping which supports data normalization. Vietnam Journal of Computer Science 3:181-196

[10] Consejo Nacional de Igualdad de Discapacidades de Ecuador (2016) Diccionario de Lengua de Señas Ecuatoriano "Gabriel Román"

[11] Rouviere H and Delmas A (2005) Anatomía humana, descriptiva, topográfica y funcional: miembros, Elsevier, España

[12] Dong Y (2014) Universal Consistency of k-NN classifier, University of Ottawa, Canada

[13] Duda R, Hart P, and Stork D (2012) Pattern Classification, Wiley, New York, USA

Authors: Luis A. Estrada, Marco E. Benalcázar, and Nelson Sotomayor
Institute: Escuela Politécnica Nacional
Street: Ladrón de Guevara E11-253
City: Quito
Country: Ecuador
Emails: {luis.estrada, marco.benalcazar, nelson.sotomayor}@epn.edu.ec

An IoT Approach to an ECG Online Monitor System in an Android Application

F.M. Machado[1], E.G. Bertogna[1] and M.A. Sovierzoski[1,2]

[1] Federal University of Technology – Paraná – DAELN - UTFPR, Curitiba, Brazil
[2] Graduate Programm in Biomedical Engineering – PPGEB-UTFPR, Curitiba, Brazil

Abstract— **We currently live in a world with many paradigms and being subject to changes such as the ones related to mHealth. With the advances in Mobile Cloud Computing (MCC), new concepts have been introduced, such as the Internet of Things (IoT) and the use of bluetooth for low power data transmission. On the other hand, sensors like the ones used in mHealth can monitor vital signals and send them over the internet for many different purposes. This paper shows an approach of MCC and Internet of Things (IoT) applied to mHealth, which suggests a new architecture developed to an application for mobile devices that can receive electrocardiographic signals from a 3-lead ECG sensor, using cloud data storage and a dedicated frame architecture between the ECG sensor and the mobile device that allows error correction. The proposed approach focuses on an mHealth scenario, where a secure connection algorithm between the bluetooth module and the local android device must be established, so that the system as a whole knows if there is too much error within the local communication, or if the signal is up to be sent to the cloud.**

Keywords— **IoT, Cloud, Computing, Bluetooth, ECG monitoring, Error Control.**

I. Introduction

Nowadays smartphones are playing an important role in our day-to-day lives, breaking paradigms not only concerning mHealth, but also in many other applications as the ones for monitoring and controlling residences, booking hotels, calling the most nearby taxis and so on. As mentioned above, almost all of the new applications for mobile devices have to solve problems in different areas. Smartphones have become part of an increasing revolution in everyone´s lives, such as remotely monitoring elderly people´s health conditions like the idea proposed in this paper. This work proposes a new way to provide means for using the available hardware present at most mobile devices in order to have a secure link connection to an ECG node. This paper also makes use of new services recently made available on the internet that allow remote users to access any previous stored data from a remote server. In order to uniquely identify the signal sent to the cloud, information such as the date when the examination was done, the sampling rate, the name of the patient, the signal data itself, and a unique identification of each package are sent to the cloud.

II. Related Works

There are many applications for the use of Internet of Things. Some of them for instance are related to controlling a whole city, while others are being developed for the industrial domain. This paper concerns in the use of it for mHealth among all the others. According to [1] medical and healthcare sector will be affected by IoT by monitoring medical parameters and vital functions. For different scenarios, many recent studies have proposed solutions involving IoT and Cloud Computing like those for the so called Smart City [2,3,4]. Gachet et al. developed in [5] an application for mobile health monitoring called the Virtual Cloud Carer Project, which presents an overview of an architecture developed by using intelligent mobile devices to integrate with many biosensors. The cloud takes care of the data for further processing. Pereira et al. in [6] developed a system based on a Mobile Cloud Computing in which there is an interaction among sensors and a mobile device by bluetooth. Mohammed et al. in [7] developed an ECG monitoring system using the cloud as a FTP server.

The difference this work intends to present does not only concern in the bluetooth communication improvements between the sensor node and the local application. The difference is also present in the way the ECG signal is sent remotely to the cloud server, having its own identity, keeping which is one of the features of IoT systems. This identity is provided by the cloud functionality to every single received package. A second ID is present among the data within the package sent to the cloud. One part of the data refers to the ECG signal itself while the other part refers to general examination information, as date and name of patient.

III. Cloud Computing and Internet of Things

As we know, Internet of Things has not been around for long. Everyday new services are influencing the way new businesses models are currently run. Since approximately 1999, when the first idea of IoT was introduced by MIT [1], the main concept has changed. In 2005, a first paper published by the ITU referred to this theme [8]. In 2009 another

I. Torres et al. (eds.), *VII Latin American Congress on Biomedical Engineering CLAIB 2016, Bucaramanga, Santander, Colombia, October 26th -28th, 2016*,
IFMBE Proceedings 60,
DOI: 10.1007/978-981-10-4086-3_60

definition was made by the Cluster of European Research projects on the internet of things (CERP-IoT) that said: „a dynamic global network infrastructure with self capabilities based on standard and interoperable communication protocols where physical and virtual things have identities, physical attributes, virtual personalities and use intelligent interfaces and are seamlessly integrated into the information network“ [9].

The existence of internal hardware such as for NFC, bluetooth, WiFi and 3G allow today´s smartphones and tablets to perform complex tasks, from smart home control [1] to patient monitoring [10] by using cloud services.

Another concept that has emerged together with IoT refers to Cyber Physical Systems (CPS). They represent a new generation of embebbed systems and have been developed combining the mobile device computation power and networking capabilities, as well as all physical processes so as to enhance the way the process of receiving, processing and sending data to the cloud can be reliable, efficient and secure [1]. Over time many applications related to medical and healthcare systems began to use this concept.

IV. SCENARIOS FOR IMPLEMENTATION

This paper firstly concerns in the ECG signal and proposes a solution to improve the reliability for the communication between an ECG sensor and a local mobile device over bluetooth, taking into account the local device must be aware whenever a transmission error occurs in order to preserve the ECG signal in time domain. For this purpose a more secure connection control over the data flowing between the two parts had to be developed. A sort of oriented connection is established between the local android device and the microcontroller. Two routines are responsible for that. One of them controls the data being sent by Bluetooth and the other one remains in the android code. After the bluetooth reception, the data are sent to a cloud. It stays there until someone erases them. The remote mobile device is the one responsible for getting the data back for online or on demand displays. Figure 1 shows the main idea better.

Among the devices on the market equipped with android operation systems, most of them have an embedded bluetooth hardware. Although bluetooth is capable of flow control, the existing bluetooth software cannot implement a flow control over the data being sent and received in an asynchronous transmission mode. The first task to be implemented by this work is an algorithm that does this job.

Two data flow control algorithms run simultaneously. One runs on the sensor side and the other at the android data reception routine.

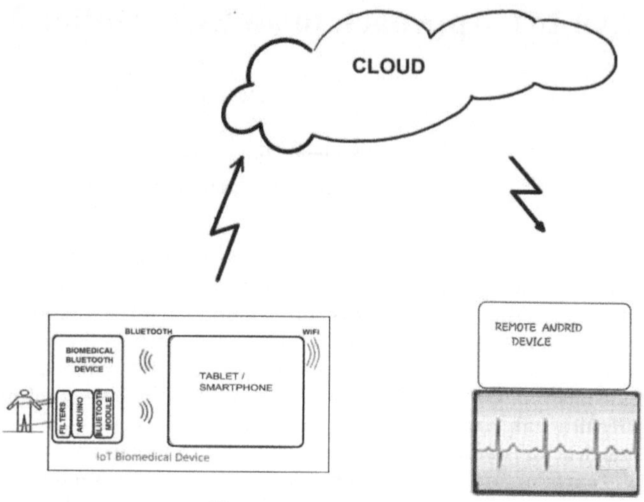

Fig. 1 Cloud use approach.

Those two routines communicate with one another whenever every single ECG data frame is or is not received. If it is not received it is asked for that specific lost frame to be retransmitted. With such algorithms it is possible for the remote android application to be aware of an eventual package transmission lost along the timeline of the ECG signal. It also allows the microcontroller to resend the lost frame within its proper time interval, among samples one to four as shown below. At last, it makes for a specific frame to be resent more than once, if the baud rate set allows it. In this work the baud rate was set to 115000 bits per second.

A. The ECG Sensor

The ECG waveform is amplified, filtered and conditionned to a 1 Vpp signal before applied to an arduino internal analog-to-digital converter (ADC) sample circuitry. The sampling rate is fixed at 2400 samples per second. The microcontroller gathers every five samples, adds an index frame number, a sampling rate and its checksum before sending the ECG signal over bluetooth to a local tablet or smartphone running its application named app 1. The frame format is shown in figure 2. Frame transmission errors are supposed to be reduced, although errors may always occur in spite of the possible frame retransmissions.

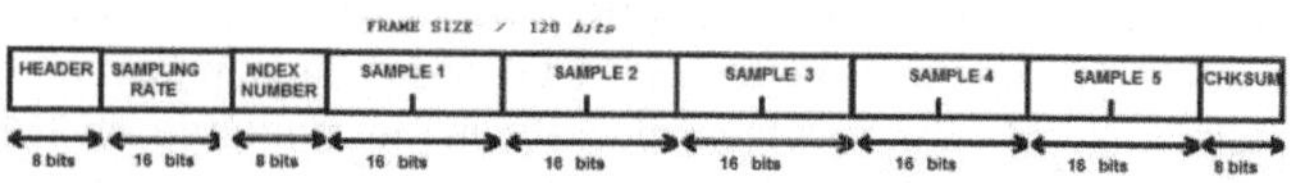

Fig. 2 Format of frame.

In extreme situations the android application may stop receiving frames from the ECG node. For these cases, a time-out interruption signal is provided in the android application, which restarts the bluetooth connection all over again. Those are the cases when the sensor is either far from the mobile device or with some kind of problem in the transmission itself. A small number of errors can be easily handled by the algorithm. Nevertheless, if many consecutive errors occur, they must be signalized to the packages being sent to the cloud.

The sample rate was established at 240 samples per second.. As samples are collected, they are internally stored, and their checksum is calculated. The algorithm in the form of a state diagram is presented in figure 3. There are five states at the edges and one state represented in the middle. In each of the five states the microcontroller is sampling data and mounting the frame. From samples one to four, they are only stored for further transmission and the algorithm verifies if any frames that still have not been appropriately received by the android device are still to be transmitted. In the fifth sample one frame is produced for transmission. This frame contains a preamble, the five sampled data, an index that corresponds to the frame number, the sampling rate and a checksum.

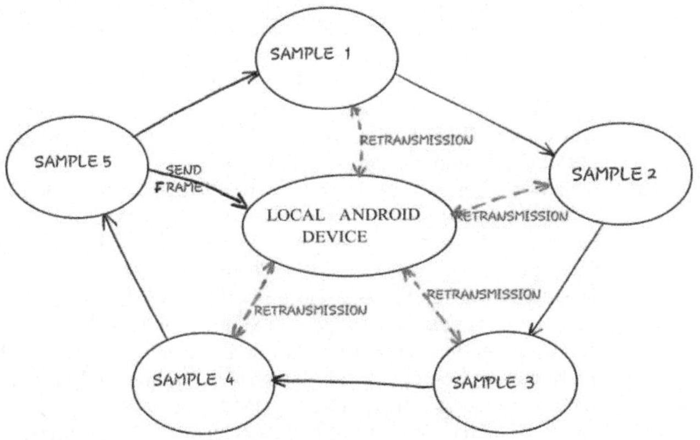

Fig. 3 Scheme of transmission.

B. The Android Application Side

Figure 4 presents the whole cycle, from sending the frame until it is received in the android side. On the top left part of figure 4 there is the sampling routine which keeps on mounting the frame. At every fifth sample, the frame is sent by the Arduino board. In the android side, the received frame is decoded. Its checksum is verified and if it is incorrect, an error message is sent back to the Arduino board. This error code corresponds to the frame index number itself. It is sent back from the android application to the arduino board that puts it into a buffer asking for further retransmission. From the first to the fourth sample, if the buffer is not empty, which means there is a frame to be resent, the main routine resends the proper frame.

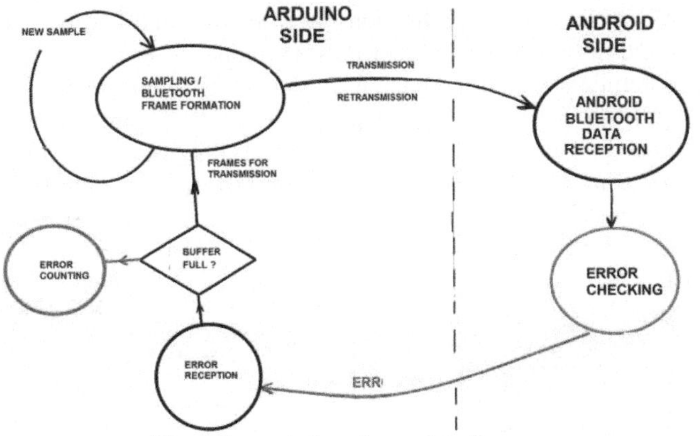

Fig. 4 Transmission scheme state diagram.

The idea was to implement a sort of oriented connection communication between both parts.

A proper thread to receive data by bluetooth hardware was developed. After decoding 48 frames it sends the samples to the main activity which displays them as an ECG signal representing a one-second period signal.

C. The Cloud

The choice for parse.com as the data storage cloud was due to its simplicity and ease of use. Data is stored there as parse objects containing many variables, such as information from the patient, date of examination and raw data from the ECG signal itself. A parse object named „ECG41" was created. It has the following fields: name of patient, sampling rate, year, month, day, hour, minute and second of examination, number of sequence of the package, ECG data, an unique identification of the package and time of creation of the object in the cloud.

It is possible to abstract the object „ECG41" as a huge table in the cloud. Every 240 samples, one package is sent to the cloud. This package is converted into a new line to be inserted in the table named „ECG41". Hence, every second a package is sent to the cloud creating a new line in the table named „ECG41". Each new line has its own read and write permissions. In order to have access to the ECG41 parse object, the application must also have two other key permissions that are implemented in the application. Both local and remote applications have those keys.

D. The Remote Android Application

The remote application was implemented so as to read new data coming from the cloud, as queries. Whenever there are new data to be displayed, they are retrieved by the application. If it does not, it means that no new data related to the name of the patient has have been sent to the cloud.

When the remote application starts, it waits until some data come in otherwise an error message appears on the display. When data are pulled back from the parse server, the remote application starts displaying one second of samples at a time. In other words, the application pulls one parse object at a time, corresponding to 240 samples of ECG data to be displayed. A flow diagram of the online application running at the remote mobile device is shown in figure 5.

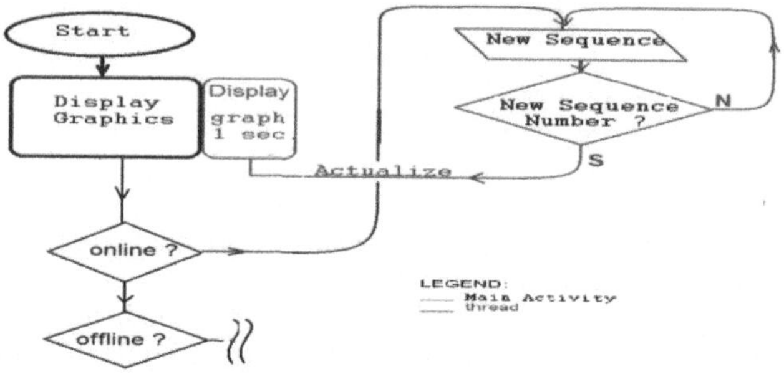

Fig. 5 Online application flow diagram.

V. Results

Figure 6 presents the two devices side by side. A tablet from Samsung model P7500 was used as the local device and a four-inch model smartphone AN400 as the remote device. The figure size of the AN400 is distorted because of the size of the ECG signal at the display. Once the signal was sent to the cloud by the tablet, it could be received by the AN400. The AN400 could be started anytime by pressing the "Conectar" button.

A natural delay was seen due to the many delays related to the internet network.propagation. It was also observed a small amount of errors because of the electromagnetic interference coming from all around the experiment. The experiment took place at room Q109 at Technological University of Paraná.

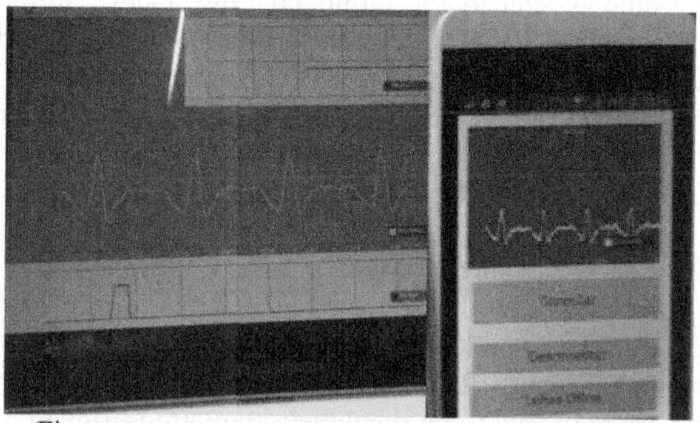

Fig. 6 The ECG transmitted signal and the received signal.

VI. Conclusions

A new scheme of data transmission to a controlled communication link between an ECG sensor node and an android application device was developed. Flow and error control in an asynchronous transmission was implemented. Local data coming from the sensor node were transformed into packages sent to the cloud as objects. In our approach, it was established a secure connection between the local and remote applications with the cloud, so that only allowed applications may access the ECG data. The convergence of mobile cloud computing, the IoT and mobile health systems was were highlighted. The screen of the local device displayed the ECG signal, and also left the ECG data available at a cloud server from Parse for online readings by a remote doctor´s application.

References

1. Borgia E (2014) The Internet of Things Vision: Key features, applications and open issues, Comp. Comm., v. 54, pp. 1–31.
2. Wang H, Yu Y, Zhu P, Yuan Q (2011) Cloud computing based on internet of things, In: 2nd International Conference on Mechanic Automation and Control Engineering, pp. 1106-1108.
3. Fox GC, Kamburugamuve S, Hartman RD (2012) Architecture and Measured Characteristics of a Cloud Based Internet of things, In: 2012 International Conference on Collaboration Technologies and Systems (CTS), pp. 6-12.
4. Soldatos J, Serrano M, Hauswirth M (2012) Convergence of Utility Computing with the Internet-of-Things. In: 2012 Sixth International Conference on Innovative Mobile and Internet Services in Ubiquitous Computing (IMIS), pp. 874-879.
5. Gachet D, Buenaga M, Aparicio F, Padrón V (2012) Integrating Internet of Things and Cloud Computing for Health Services Provisioning: The Virtual Cloud Carer Project. In: 2012 Sixth International Conference on Innovative Mobile and Internet Services in Ubiquitous Computing (IMIS), pp. 918-921.
6. Pereira PP, Eliasson J, Kyusakov R, Delsing J, Raayatinezhad A, Johansson M (2013) Enabling Cloud Connectivity for Mobile Internet of Things Applications. IEEE 7th International Symposium on Service Oriented System Engineering (SOSE), pp. 518-526.
7. Mohammed J, Lung CH, Ocneanu A, Thakral A, Jones C, Adler A (2014) Internet of Things: Remote Patient Monitoring Using Web Services and Cloud Computing. IEEE International Conference on Internet of Things (iThings), and Green Computing and Communications (GreenCom), IEEE and Cyber, Physical and Social Computing (CPSCom), pp. 256-263.
8. International Telecommunication Union: ITU Internet Report (2005) The Internet of Things, International Telecommunication Union, Geneva.
9. Jain, AK, Hong, L, Pankanti, S (2014) Internet of Things - Strategic Research Roadmap, Tech. Report, Cluster of European Research projects on the Internet of Things, pp. 9-52.
10. Lau SL, König I, David K, Parandian B, Carius-Düssel C, Schultz M (2010) Supporting patient monitoring using activity recognition with a smartphone. In: 7th International Symposium on Wireless Communication Systems (ISWCS), pp. 810-814.

Noninvasive approach to estimate ventilatory mechanics in spontaneous breathing with different PEEP and pressure support values: validation with mechanical simulation

I.C. Muñoz[1] and A.M. Hernández[1]

[1]University of Antioquia/Bioengineering Department, GIBIC, Calle 70 No. 52-21, Medellín, Colombia

Abstract—**Measuring or estimating the state of the ventilatory mechanics, expressed in airway resistance and compliance of ventilatory system, and muscles effort are useful in mechanically ventilated patients to identify the level of obstruction or restriction in lung diseases and if the presence of respiratory muscle weakness is evident, the decision making concerning right time for extubation is a clinical issue of interest. Traditionally, the ventilatory mechanics has been estimated in sedated patients under controlled ventilation modes. However the available approaches are not appropriate in the case of spontaneous ventilation during the weaning test. This article presents an approach to evaluate ventilatory mechanics and respiratory muscle effort under spontaneous ventilation in different levels of positive end-expiratory pressure and pressure support, that has been tested using an advanced breathing mechanical simulator. This procedure allowed proper estimation of ventilatory mechanics and respiratory muscle effort in line with simulator settings.**

Keywords— **Mechanical ventilation, ventilatory mechanics, static compliance, airway resistance, respiratory muscular effort, work of breathing.**

I. Introduction

The mechanical properties of the respiratory system include the airway resistance (R_{aw}) and elastic recoil forces of the lung tissues associated with both lung (C_l) and chest wall (C_{cw}) compliances (static compliance), which must be successfully overcome by the respiratory muscles in each breath [1]. Whenever these properties are modified, the respiratory process is affected. An abnormally low compliance is reflected in lung stiffness and additional effort is required to insufflate the lungs with an adequate volume of air [2]. When the airway resistance is abnormally high, a reduction of airflow during ventilation occurs, particularly during expiration, causing pulmonary hyperinflation because it avoids ejecting all inhaled air, as a consequence a reduction in lung compliance is presented, producing an extra load to be overcome by respiratory muscles [3].

In advanced stages of diseases, the muscle weakness due to the altered ventilatory mechanics leads to a reduction in alveolar lung volume, leading to the use of mechanical ventilation (MV) [4], [5].

The MV treatment could be improved if the respiratory pathophysiology and changes in the ventilatory mechanics of the patient were known and used to optimize the mechanical ventilator settings. In addition to a proper configuration of mechanical ventilator and the evaluation of the effectiveness of ventilatory support are required to increase the patient comfort and reduce the time of intubation. Estimating the state of the ventilatory mechanics is an strategy that provides complementary information to decide the right time for extubation [1], [6]. Furthermore, positive end-expiratory pressure (PEEP) and pressure support (PS) are common parameters of mechanical ventilation and due to they modify the pressure column in the airways, the estimation of airway resistance and lung compliance is particularly interesting when different levels of PEEP and PS are applied [7].

Ventilatory mechanics of a patient under mechanical ventilation has been traditionally estimated by parameter fitting of a mathematical model, using fluidic variables such as airflow, pressure and volume, which are generally controlled by the mechanical ventilator, assuming that the patient is totally controlled and relaxed. In spontaneous breathing the last assumptions are not correct, which leads to devise a new approach including the respiratory muscle activity in the model [8].

This article presents a preliminary non-invasive strategy for evaluating ventilatory mechanics and respiratory muscle effort under spontaneous ventilation, tested at different levels of PEEP and PS using a mechanical simulator connected to a mechanical ventilator.

II. Material and methods

A. Experimental Design

ASL5000 simulator of Ingmar Medical is a mechanical breathing simulator that provides control over parameters of ventilatory mechanics and breathing pattern [9].

In order to test the estimation approach, the ASL5000 simulator was configured to simulate a healthy adult using reported reference of static compliance (C) and airway resistance (R) values of 100 mL/cmH_2O and 3 cmH_2O/L/s

I. Torres et al. (eds.), *VII Latin American Congress on Biomedical Engineering CLAIB 2016, Bucaramanga, Santander, Colombia, October 26th -28th, 2016*,
IFMBE Proceedings 60,
DOI: 10.1007/978-981-10-4086-3_61

respectively [10]. To guarantee a tidal volume between 500 and 1000 mL, which depends of the body weight [11], the muscular pressure was configured in 12 cmH_2O. C was changed from its reference value to 50, 150 and 200 mL/cmH_2O with healthy reference value of R. Finally, R was modified in the simulator to 5, 10 and 15 cmH_2O/L/s with a healthy reference value of C.

For each of the above configurations, the simulator was connected to the Hamilton C1 mechanical ventilator in spontaneous mode. Under conditions of a healthy subject, changes in the level of PEEP from 0 to 10 cmH_2O were performed with steps of 2 cmH_2O. After this test, the PEEP value was restored to 0 cmH_2O and the PS was changed from 0 to10 cmH_2O with steps of 2 cmH_2O. Each pressure level was maintained for 3 minutes. Throughout the test, ventilatory signals and parameters were recorded with Hamilton datalogger.

B. *Ventilatory Mechanics Estimation*

The equation of motion of respiratory system (1) relates the respiratory system pressure, airway flow and volume [8].

$$P_{aw} + P_{mus} = \frac{1}{C} * V + R * Q + PEEP_{total} + I * Q'' \quad (1)$$

Where P_{aw} is the pressure of the respiratory system measured in mouth, P_{mus} the muscular pressure, Q is airflow, V is volume, C is respiratory system compliance, R is the resistance of the system and $PEEP_{total}$ is total positive end-expiratory pressure, which is defined as the sum of the applied positive end-expiratory pressure (PEEP) and the intrinsic positive end-expiratory pressure (PEEPi). The inertance (I) represents the inertia of gas, which is considered negligible [12].

In controlled ventilation the patient is making no respiratory efforts and therefore the pressure exerted by the respiratory muscles (P_{mus}) is zero[8]. While in spontaneous ventilation Pmus, C and R, are unknown and the equation has multiple solutions [13].

To calculate the mechanical properties of the respiratory system, occlusion maneuvers are generally necessary, which consist in pauses during expiration, where is possible to calculate both static compliance and the airway resistance [8], [14].

$$C = \frac{V_t}{P_{Plat} - PEEP}. \quad (2) \qquad R = \frac{PIP - P_{Plat}}{Q}. \quad (3)$$

Where V_t is tidal volume, P_{Plat} is the plateau pressure during the pause, PIP is peak inspiratory pressure and Q is airflow (see Fig. 1).

The occlusion maneuvers are associated only with controlled modes. In order to create the breathing pattern shown in Fig. 1 a manual occlusion was necessary, establishing conditions for calculating the compliance by (2). Three manual occlusion maneuvers were performed for all tests.

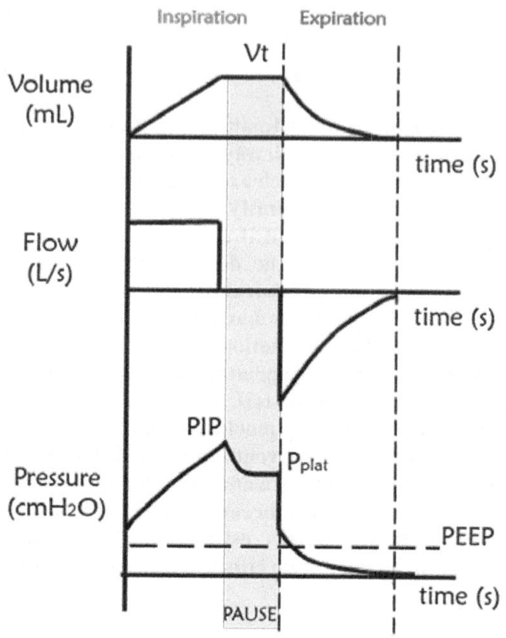

Fig. 1 Volume, flow and pressure curves. A pause at the end of inspiration is shown with a shadow.

Due to the difficulty to get a peak pressure (PIP) with a maneuver to calculate R with (3), an optimization algorithm based on the equation of motion (1) was used. The cost function to be minimized was:

$$J = \sum_{n=1}^{n=N} \frac{1}{N} \left(P_{aw}(n) - P_{eq}(n)\right)^2. \quad (4)$$

Where Paw is the pressure measured at mouth and Peq is the sum of pressure measured at mouth (P_{aw}) and muscular pressure (P_{mus}) obtained from (1) using Q and V signals acquired from Hamilton C1, PEEP was configured and C was obtained by (2). In some cases the PEEPi did not reach zero, so if the minimum volume was greater than zero, the value of PEEPi was optimized too. The optimization was based in a sequential quadratic programming algorithm with Matlab® and it was applied only during expiratory cycle, where the ventilation is passive and the assumption of muscular pressure equal to zero is proper [7].

Some authors suggest that differences between expiratory resistance and total resistance are not significant in healthy subjects [15]. This is why, once R and C were estimated, the muscular effort was deduced from (1).

Finally, the WOB for each respiratory cycle (t) was estimated as follows [13]:

$$WOB = \frac{1}{Vt}\int_0^t P_{mus}(t) \times Q\, dt. \quad (5)$$

III. Statistic analysis

A descriptive analysis of the data through the mean and standard deviation was implemented. For compliance, the measurements were calculated between the values obtained in each occlusion maneuver, and for resistance it was between the values obtained in each expiratory cycle. Finally, a comparison between the configured and calculated values was performed through the nonparametric statistical test of Wilcoxon Mann Whitney in order to know if statistically significant differences were evidenced between set and estimated parameters.

IV. Results and discussion

The Table 1 shows the C, R and P_{mus} values for each test. The compliance values obtained by occlusion maneuver are very close to those set in the simulator for all tests, with low standard deviation and porcentual error lower than 10%. For resistance, porcentual errors greater than 10% are presented, mainly in incremental PEEP test, with an R value set to 3 $cmH_2O/L/s$ where the highest difference reached 0.9 $cmH_2O/L/s$. Although the error, no significant statistical differences occurred.

C and R values were replaced in the equation of motion (1) to calculated P_{mus}. Significant statistical differences were found in incremental PEEP and PS tests, where estimated values were higher than set values, reaching a difference of 3.4 cmH_2O. This difference could be attributed to the absence of an assumption of a specific shape for the muscular pressure signal that distributes the energy between expiration and inspiration in a real way.

Concerning WOB, estimated from (5), values between 0.4375 J/L and 0.8832 J/L were found, values that are in agreement of some reports for healthy subjects, where WOB is in the range of 0.35 J/L to 1.2J/L [16], [17].

Some authors have found the WOB in patients with invasive mechanical ventilation using an esophageal balloon and relaxed patient [13], however this technique due to its invasive nature and a required catheter is difficult to apply [18].

Table 1 Comparison between set and estimated values of ventilatory mechanics. Values of estimated Pmus (muscular effort) for each test is shown.

Test	PEEP (cmH_2O)	PS (cmH_2O)	Configured Values			Calculated Values			Error (%)		
			C (mL/cmH_2O)	R ($cmH_2O/L/s$)	Pmus (cmH_2O)	C (mL/cmH_2O)	R ($cmH_2O/L/s$)	Pmus (cmH_2O)	C	R	Pmus
1	0	0	100	3	12	108.8±2.8	2.7±0.0	11.56	8.77	10.11	3.65
2	0	0	50	3	12	52.4±0.9	3.5±0.11	12.46	4.85	15.56	3.80
3	0	0	150	3	12	148.0±7.1	2.7±0.1	11.68	1.32	9.17	2.66
4	0	0	200	3	12	200.3±3.1	2.6±0.0	11.43	0.15	14.60	4.76
5	0	0	100	5	12	104.7±1.2	4.3±0.1	11.43	4.70	13.85	4.73
6	0	0	100	10	12	100.4±1.4	8.4±0.1	10.82	0.39	16.20	9.82
7	0	0	100	15	12	94.0±1.1	14.2±0.2	11.58	6.03	5.62	3.53
8	2	0	100	3	12	104.4±1.1	3.0±0.0	12.33*	4.44	0.58	2.73
9	4	0	100	3	12	105.3±2.4	3.6±0.0	13.56*	5.27	20.60	13.02
10	6	0	100	3	12	99.0±8.0	3.3±0.4	13.46*	1.04	10.65	12.13
11	8	0	100	3	12	100.0±4.3	3.4±0.4	13.29*	0.22	11.52	10.78
12	10	0	100	3	12	94.9±6.7	3.9±0.1	14.25*	5.15	30.84	18.79
13	0	2	100	3	12	96.0±3.7*	3.2±0.1	13.65*	4.04	6.34	13.75
14	0	4	100	3	12	96.7±1.4*	3.1±0.1	14.63*	3.33	4.21	21.92
15	0	6	100	3	12	95.2±2.0*	3.2±0.1	15.42*	4.81	7.48	28.46
16	0	8	100	3	12	96.2±0.6*	3.0±0.1	14.56*	3.80	1.14	21.30
17	0	10	100	3	12	93.6±1.9*	3.0±0.1	15.09*	6.43	1.30	25.77

* The group presents significant statistical differences between set and estimated values with $P<0.05$

Other studies have calculated WOB with non-invasive techniques using optimization algorithms without maneuvers, however due to the equation of motion has multiple solutions it is necessary to apply different constraints that only work with specific breathing patterns [13]. The strategy presented in this article allowed to estimate both ventilatory mechanics and WOB in different simulated conditions in spontaneous ventilation without constraints in the algorithm. Although, the proposed technique has the limitation that an occlusion maneuver must be done.

V. Conclusions

This article presented an approach to estimate airway resistance and static compliance in a noninvasive manner, which in turn, allowed to estimate the muscular effort in terms of muscular pressure exerted by respiratory muscles during spontaneous breathing. From simulation results with ASL5000 simulator, it can be inferred that the proposed technique could allow the estimation of the ventilatory mechanics in both restrictive and obstructive conditions.

These findings lead to future studies focused in computational estimation of mechanical ventilation parameters in mechanically ventilated patients. This kind of techniques will allow the optimization of the diagnosis and treatment of mechanically ventilated patients.

Future extension of this work involves recording a database of healthy subjects performing controlled modifications of airway resistance and ventilatory system compliance, in order to represent some of the conditions commonly presented on respiratory diseases. A complementary database of mechanically ventilated patients will be needed in the future to test the proposed approach.

Acknowledgment

Research supported by "PRG13-2-08 Protocolo para destete de la ventilación mecánica en pacientes intoxicados con organofosforados". Grant by University of Antioquia (Medellín, Antioquia, Colombia), 2013-2016

Conflict of Interest

The authors declare that there is no conflict of interest regarding the publication of this paper.

References

1. Bates JH (2009) Lung Mechanics. An Inverse Modeling Approach. Cambridge University Press. New York.
2. Shoemaker (2002), Tratado de medicina crítica y terapia intensiva. Editorial Médica Panamericana.
3. Hernández AM (2014), Sistema de control respiratorio ante estímulos y patologías: Análisis, Modelado y Simulación (Spanish Edition). Publicia.
4. Troosters T, Gosselink R, and Decramer M (2005) Respiratory muscle assessment. Eur. Respir. Monogr 31:57.
5. Carlucci A, Pisani L, Ceriana P et al (2013) Patient-ventilator asynchronies: may the respiratory mechanics play a role? Crit. Care 17(2): R54.
6. Shaikh H, Morales D, and Laghi F (2014) "Weaning from mechanical ventilation. Seminars in respiratory and critical care medicine 35(4): 451–468.
7. van Drunen EJ, Chiew YS, Chase JG et al (2013) Model-based respiratory mechanics to titrate PEEP and monitor disease state for experimental ARDS subjects. In Conf. Proc. Annu. Int. Conf. IEEE Eng. Med. Biol. Soc. IEEE Eng. Med. Biol. 2013 (5), pp. 5224–7.
8. García-Prieto E, Amado-Rodríguez L, Albaiceta G et al. (2014), Monitorization of respiratory mechanics in the ventilated patient, Med. Intensiva (English Ed.) 38(1):49–55.
9. IngMar Medical. (2015), User´s Manual ASL 5000, Active Servo Lung. IngMar Medical, 1:1-152.
10. Tobin M.J (2012). Principles And Practice of Mechanical Ventilation, Third Edition. McGraw-Hill Education, Chicago, Illinois.
11. Brower RG, Matthay MA, Morris A et al. (2000) Ventilation With Lower Tidal Volumes As Compared With Traditional Tidal Volumes for Acute Lung Injury and the Acute Respiratory Distress Syndrome N. Engl. J. Med 342(18):1301–1308.
12. Hess D.R (2014) Respiratory mechanics in mechanically ventilated patients. Respir. Care 59(11): 1773–94.
13. Vicario F, Albanese A, Wang D et al (2015) Constrained optimization for noninvasive estimation of work of breathing, in Engineering in Medicine and Biology Society (EMBC), 2015 37th Annual International Conference of the IEEE, Milan, Italy, 2015, pp. 5327–5330.
14. Polese G, Serra A and Rossi A (2005) Respiratory mechanics in the intensive care unit. Eur. Respir. Monogr., 31:195.
15. Paredi P, Goldman M, Alamen A et al (2010) Comparison of inspiratory and expiratory resistance and reactance in patients with asthma and chronic obstructive pulmonary disease. Thorax 65:263–267.
16. Cabello B, Mancebo J (2006) Work of breathing. Intensive Care Med. 32: 1311–1314.
17. Dellweg D, Haidl P, Siemon K et al (2008), Impact of breathing pattern on work of breathing in healthy subjects and patients with COPD. Respir. Physiol. Neurobiol. 161(2): 197–200.
18. Iglesias NR (2012) Protocolo para el destete de pacientes acoplados a ventilación mecánica. Universidad de Ciencias Médicas de Camagüey. 1:1-158

Author: A.M. Hernández.
Institute: University of Antioquia
Street: calle 70 No. 52 - 21
City: Medellín
Country: Colombia
Email: alher.hernandez@udea.edu.co

Characterization of silver nanoparticles for potential use as antimicrobial agent

J. Zapata-Giraldo[1], P. Mena[1], B. Galeano[1], N. Escobar[1], M. Mejía[1], I.C. Ortiz[2], D. Cuesta[2], L.E. Botero[2] and L.M. Hoyos-Palacio[1]

[1]Centro de Bioingeniería. Universidad Pontificia Bolivariana, Medellín, Colombia
[2] Facultad de Medicina. Universidad Pontificia Bolivariana, Medellín, Colombia

Abstract—**The aim of this study was produced silver nanoparticles using chemical reduction method and controlling size through pH variation. The nanoparticles had been characterized by studying their structural and morphological properties through UV-vis spectroscopy, atomic absorption spectroscopy, transmission electron microscopy and dynamic light scattering. Our results indicate that with the variation of pH levels can be obtained smaller silver nanoparticles. These nanoparticles could be evaluated as antimicrobial agents for potential use as a barrier control of hospital acquired infections.**

Keywords—**Chemical synthesis; nanoparticles; cross infections; infection control.**

I. Introduction

Health-care associated infections (HAIs) represent a major health and financial problem in the world due to the frequency which they occur, morbidity and mortality they cause and the burden they impose on patients, health workers and health systems [1].

The problem of HAIs is notoriously higher in developing countries compared with develop countries. In develop countries, the prevalence of hospitalized patients who acquired at least one HAIs is ranging between 3.5 % and 12 %, while in developing countries varies between 5.7 % and 19.1 % [2].

With the emergence and increase of microbial microorganism resistant to multiple antibiotics and the widespread of those in hospital environment, many researchers have tried to develop new, effective antimicrobial reagents free of resistant and use these agents as a reinforce of polymer fibers which acts as a barrier control for HAIs [3][4].

Nanotechnology has emerged as a rapidly growing multidisciplinary field with wide applications in science and technology and with the aim of manufacturing new materials at the nanoscale. Silver nanoparticles or nanosilver (AgNP), are cluster of silver atoms, which have an attracting interest as antibacterial and antimicrobial reinforcements for polymers involved in care of hospital environment [5][6].

Production of silver nanoparticles with different morphologies and sizes using different routes has been reported [7][8][9][10]. The aim of this research was synthesized AgNPs by chemical reduction and with pH adjustment. These nanoparticles could be used as antimicrobial agents for infection control and as a reinforcement of polymer fibers of hospital use.

II. Materials and Methods

A. Materials

To prepare Ag nanoparticles (AgNPs), silver nitrate ($AgNO_3$), sodium citrate ($C_6H_5O_7Na_3$), sodium borohydride ($NaBH_4$) and sodium hydroxide (NaOH) were used. All the chemicals were of analytical reagent grade and used without further purification.

B. Synthesis method

The AgNPs were synthesized by chemical reduction method using water Milli Q® as a solvent. Monodispersed citrate-stabilized silver nanoparticles were performed by the reduction of $AgNO_3$ using sodium borohydride [7]. The mixture contained water Milli Q®, 0.05 M sodium citrate, and 0.05 M of $AgNO_3$, which was stirred for 3 minutes at 10°C. Then, 0.05 M of $NaBH_4$ was added drop wise to the reaction mixture. Immediately after addition of $NaBH_4$ stirring was stopped and was made an adjustment of the pH adding drops of 1.25 M NaOH [8]. AgNPs were stored in the dark at 4°C.

C. Characterizations

In order to evaluate physical and chemical characteristics of silver nanoparticles, structural characterizations were made by an UV-visible spectrophotometer (Shimadzu) with a 1-cm quartz cell; It was analyzed the concentration of nanoparticles obtained by atomic absorption spectroscopy (AAS); Mean particle size was examined through dynamic light scattering analysis (DLS - Zetasizer Nano Series Malvern Instruments), this technique exhibit a Z average, based on consecutive measures of particles sub-groups; For studying nanoparticle morphology, transmission electron microscopy (TEM) analysis was performed at an accelerating voltage of 200 kV (Tecnai F20 Super Twin TMP). The

I. Torres et al. (eds.), *VII Latin American Congress on Biomedical Engineering CLAIB 2016, Bucaramanga, Santander, Colombia, October 26th -28th, 2016*,
IFMBE Proceedings 60,
DOI: 10.1007/978-981-10-4086-3_62

samples were prepared by placing a drop of homogeneous suspension on a copper grid with a lacey carbon film and allowing it to dry in air. All the characterizations analyses were made using the obtained aqueous dispersion of silver nanoparticles.

III. Results

UV-visible spectroscopy is one of the most widely used techniques for structural characterization of silver nanoparticles. The absorption spectrum of the dark yellow silver colloids prepared by NaBH4 reduction and with the adjustment of pH from 9 to 11 it is showed in Fig. 1. The surface plasmon absorption band for silver nanoparticles with a pH of 9 has a maximum of 395.5 nm (Fig 1a.) and with a pH of 11 has a maximum pf 399.5 nm (Fig 1b.). This indicate the presence of lone spherical or roughly spherical silver nanoparticles.

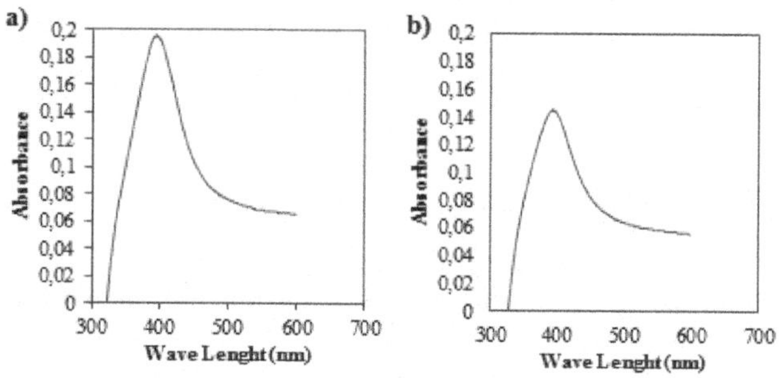

Fig 1. Absorption spectrum of silver nanoparticles solution. a) Without the adjustment of pH; b) With the adjustment of pH.

TEM imaging confirmed the shape of silver nanoparticles produced (Fig. 2).

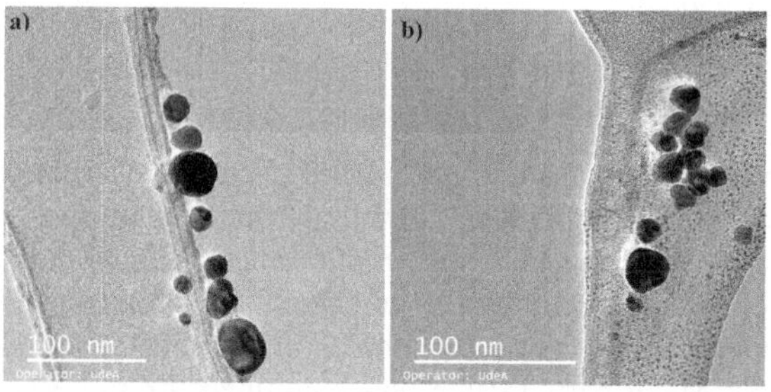

Fig 2. TEM images of silver nanoparticles. a) pH 9; b) pH 11.

The mean particle size for silver nanoparticles with a pH of 9 was 36 nm and with a pH of 11 was 18 nm. The concentration of the silver nanoparticle with a pH of 9 produced 140 ppm, while AgNPs with a pH of 11 produced 130 ppm.

IV. Discussion

This research modified the methodology of Brown *et al* of synthesis of silver nanoparticles in order to produce AgNPs varying their size through pH adjustment [7]. Ajitha *et al* [8] found that pH has a direct relation with the mean size of silver nanoparticles, and with an increase in pH, nanoparticle size reduced.

The AgNPs of our study were monodisperse citrate-stabilized with a variation of their pH from 9 to 11. All the particles seem to be spherical, homogenous and are without any substantial agglomeration.

Sodium citrate play an important role by preventing them to form aggregation. Agglomeration is due to the high surface energy and thermodynamic instability exhibited by the surface of the particles [9]. When the reaction took place, it was observed that AgNPs colloid had an initial pH of 9. By adding drops of NaOH, we adjusted this pH to 11. Then, it was observed that those nanoparticles synthesized under a pH of 9 had grater sizes than those with a pH of 11, indicating by DLS 36 nm and 18 nm sizes respectively.

These behaviors could be explained because in the alkaline pH range, the stability of cluster distribution and colloid formation is increased with a declined tendency for aggregation of the particles due to complete charging of the clusters which maximize the repulsive electrostatic/electrosteric interactions. During the elevated pH, the reaction rate will be increased with subsequent crystallization into smaller particles, which involved the nucleation and growth processes of smaller particles from silver ions (Ag^0) nuclei.

Direct comparisons between our results and those from other studies are difficult because the physiochemical characteristics of the nanoparticles used differs, and the protocols used for their synthesized are not the same [9][11][12].

Currently, we are evaluating these synthesized nanoparticles as antimicrobial agent against gram positive and gram negative bacteria, susceptible and resistant to antibiotics.

V. Compliance With Ethical Requirements

The research is experimental with physical and chemical basis, which did not involve humans or animals as an object of study.

The object of study involved the development of nanostructures, which chemical composition was not used in humans or animals. The risk of manipulation of this nanostructures by the laboratory workers was eliminated by using appropriate protective barriers, considering its possible ability to cause some irritation.

VI. Conclusions

We showed here that AgNPs prepared by chemical reduction method exhibited smaller sizes with an increase of their solution pH. In the future, we will determine their antimicrobial properties and use them as a reinforcement of polymer matrix solutions with a potential use as a barrier control for HAIs.

Acknowledgment

This research work was supported by Universidad Pontificia Bolivariana, Medellín-Colombia under grant number 438B-08/15-65 and by COLCIENCIAS, Colombia under grant number 121071149742.

Conflict of Interest

The authors declare that they have no conflict of interest.

References

1. Centers for Disease Control and Prevention. Healthcare associated infections [Internet]. Atlanta: Centers for Disease Control and Prevention; 2016 Mar [cited 2016 Jun 14]. Available form: http://www.cdc.gov/hai/

2. World Health Organization. The burden of health care-associated infection worldwide [Internet]. Geneva: World Health Organization; cited [2016 Jun 14]. Available from: http://www.who.int/gpsc/country_work/burden_hcai/en/.

3. Pan American Health Organization, World Health Organization. Antimicrobial resistance [Internet]. Geneva: World Health Organization; cited [2016 Jun 14]. Available from: http://www.paho.org/hq/index.php?option=com_topics&view=article&id=7&Itemid=40740&lang=en.

4. World Economic Forum. The Global Risks Report 2016 [Internet]. cited [2016 Jun 14]. Available from: http://www.weforum.org/reports/the-global-risks-report-2016.

5. Huh AJ, Kwon YJ (2011) "Nanoantibiotics": a new paradigm for treating infectious diseases using nanomaterials in the antibiotics resistant era. J Control Release 2:128–145.

6. Pelgrift RY, Friedman AJ (2013) Nanotechnology as a therapeutic tool to combat microbial resistance. Adv Drug Deliv Rev 13–14:1803–1815.

7. Brown AN, Smith K, Samuels TA et al (2012) Nanoparticles functionalized with ampicillin destroy multiple-antibiotic-resistant isolates of *Pseudomonas aeruginosa* and *Enterobacter aerogenes* and methicillin-resistant *Staphylococcus aureus*. Appl Environ Microbiol 8:2768–2774.

8. Ajitha B, Ashok Kumar Reddy Y, Sreedhara Reddy P (2015) Enhanced antimicrobial activity of silver nanoparticles with controlled particle size by pH variation. Powder Technol 269:110–117.

9. Guzman M, Dille J, Godet S (2012) Synthesis and antibacterial activity of silver nanoparticles against gram-positive and gram-negative bacteria. Nanomedicine Nanotechnol Biol Med 1:37–45.

10. Pal S, Tak YK, Song JM (2007) Does the Antibacterial Activity of Silver Nanoparticles Depend on the Shape of the Nanoparticle? A Study of the Gram-Negative Bacterium Escherichia coli. Appl Environ Microbiol 6:1712–1720.

11. Martínez-Castañón GA, Niño-Martínez N, Martínez-Gutierrez F et al (2008) Synthesis and antibacterial activity of silver nanoparticles with different sizes. J Nanoparticle Res 8:1343–1348.

12. Kim JS, Kuk E, Yu KN et al (2007) Antimicrobial effects of silver nanoparticles. Nanomedicine Nanotechnol Biol Med 1:95–101.

Author: Jenniffer Zapata Giraldo
Institute: Universidad Pontificia Bolivariana
Street: Circular 1 No. 70 – 01
City: Medellín
Country: Colombia
Email: ferzapatagiraldo@gmail.com

Learning Tool for Mechanical Ventilation during Spontaneous Breathing Test on Patients Intoxicated with Pesticides

M. B. Salazar Sánchez[1], A. M. Hernández Valdivieso [1], A. F. Botero Ospina [1], C. C. Cortés Daza[2]

[1] Bioinstrumentation and Clinical Engineering Research Group - GIBIC, Bioengineering Department, Engineering Faculty, Universidad de Antioquia UdeA; Calle 70 No. 52-21, Medellín, Colombia.
[2] Hospital San Vicente Fundación, Calle 64 # 51D-154, Medellín, Colombia

Abstract— **Organophosphorus compounds (OC) are widely used as pesticides and produce neuromuscular compromise after acute exposure causing paralysis in rib cage muscles. Given the acute respiratory failure in these patients, it is necessary to provide assistance for breathing with mechanical ventilation. According to the World Health Organization, about 20% of poisonings in the world involves organophosphate pesticides [1]. The management of this kind of patient is difficult due to the lack of information concerning the evolution of poisoning and the long time under mechanical ventilatory support. The flaw in determining the ideal moment for patient disconnection of ventilator, not only increases the time attendance but also the probability of death [2]. Therefore, the development of tools to facilitate the dissemination of knowledge about the effect on ventilatory mechanics of poisoning with these compounds contributes significantly to patient treatment. This article describes a learning tool on effect of intoxication over the mechanical ventilation, through of the interaction with different clinical cases of poisoned patients under mechanical ventilation. This tool allows knowing different variables (heart rate, respiratory rate, arterial blood pressure, rapid shallow breathing index, Richmond agitation sedation scale index, PaO_2, $PaCO_2$, QT interval, corrected QT interval) that are important in decisions-making during the spontaneous breathing tests.**

Keywords— **Organophosphates, Mechanical Ventilation, Spontaneous Breathing Test, Clinical Cases.**

I. Introduction

Organophosphorus compounds are used as pesticides and the main action is inhibiting irreversibly the function of the cholinesterase. The management of this intoxication is difficult because after acute exposure to these compounds the patient requires ventilatory support in Intensive Care Unit (ICU). In developing countries are considered as one of toxic more involved in voluntary, occupational and accidental poisoning by its high toxicity, wide availability and for the bad practice in use and storage [3].

According to the World Health Organization, the exposure to toxic (per year) are divided into one million accidental cases and two million suicide attempts, of which two hundred thousand individuals die, most in developing countries [3]. The mortality for poisoning with OC in the world is about 20% but can be as high as 46%, due to the lack of research to support treatment protocols, despite the high number of poisoning per year [1]. According to the epidemiological bulletin reported by the National Institute of Health of Colombia, in 2015 occurred 29,839 cases of poisoning, of which 8,738 (29.28%) were secondary to pesticide, being the leading cause of poisoning in our country [4].

The morbidity and mortality increases with the cases where the process of withdraw ventilatory support (weaning test) is not successful because there is an increment times of ventilatory assistance [2], [5]. Therefore the process of ventilation in these patients is high complexity due to lack of information about of evolution of the poisoning and the recovery of the autonomous breathing capacity, also to the long and costly periods of mechanical ventilation.

Biomedical engineering is a discipline which principles and techniques are applied to the medical field and clinical practice. This allows joining knowledge from engineering and needs in the medical field. It has proven that an efficient tool allows the access to knowledge simple and interactive way [6], [7]. An interactive tool provides the possibility that the user can make decisions to understand the importance of variables or parameters in the system behavior, without the need of wait until the clinical practice to acquire knowledge about the topic of interest. Several studies have demonstrated the utility of types tools in the diagnosis and clinical treatment [8], [9].

This article describes a learning tool implemented in IOS and aimed at the diffusion of knowledge about the effect on ventilatory mechanics of poisoning with organophosphorus compounds. The medical students (*user*) can interact through decision-making in the management of patients under mechanical ventilation during spontaneous breathing test (SBT).

II. Materials and Methods

The structure of the application has two modules, as shown in Figure 1. The first module allows to medical students to use the *training* module to learn of the clinical management of different real clinical cases. The second module provides the possibility of *interaction* through decision-making during a weaning attempts. The graphical interface as

I. Torres et al. (eds.), *VII Latin American Congress on Biomedical Engineering CLAIB 2016, Bucaramanga, Santander, Colombia, October 26th -28th, 2016*,
IFMBE Proceedings 60,
DOI: 10.1007/978-981-10-4086-3_63

IOS app displays all variables and parameters common during weaning test and it is totally interactive, making possible its use by clinical personal.

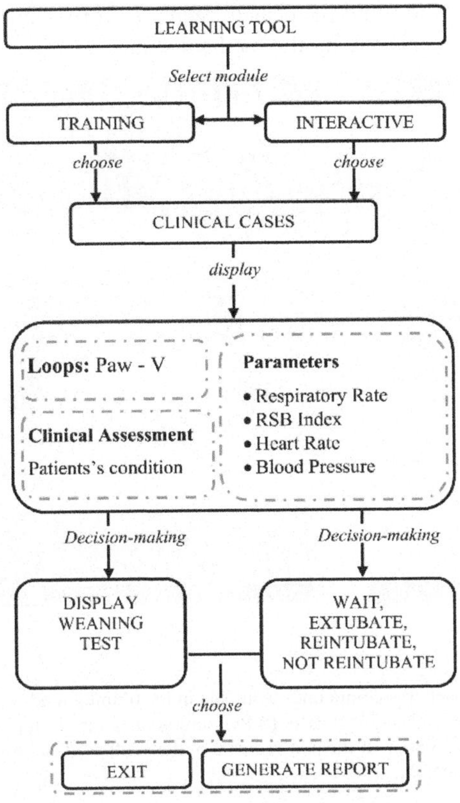

Fig. 1 Block diagram of the learning tool on ventilatory mechanics during spontaneous breathing test in intoxicated patients with pesticides.

A. Database

We obtained the clinical cases of this application through the monitored of weaning attempts by ventilatory signals in patients admitted and managed in the Hospital San Vicente Foundation (Medellin, Colombia). The patients were included according the following inclusion criteria: i) intoxication with OC, ii) requirement of mechanical ventilation, iii) older than 16. This study was conducted under the approval of the Ethical Review Committee of the Hospital and written informed consent was obtained from the patient's family for publication of this study.

The variables available in the two modules of the tool as a source of daily information in ICUs: heart rate (HR), respiratory rate (RR), systolic/diastolic blood pressure (SBP/DBP), Rapid Shallow Breathing index (RSB), partial pressure of oxygen (PaO_2) and carbon dioxide ($PaCO_2$), oxygen saturation (SpO_2), QT interval (QT), corrected QT interval (QTc) and Richmond Agitation Sedation Scale index (RASS). The tool has documentation related with the toxic involved in the clinical cases and guidelines for the use of learning tool.

B. Training Module

The training module presents a group of real clinical cases of patients poisoned with OC and under mechanical ventilation during SBT. Each clinical case shows the clinical condition of the patient during the spontaneous ventilation, so that the medical students can analyze the situation from the symptoms, through evolution, until the decision of the medical staff at the end of test. As shown in Figure 1, once the user has selected a clinical case, in the panel will have access to:

- *Airway Pressure – Volume (Paw-V) curve.* The inspiratory and expiratory phase are showed, from which one can infer information about the variability of compliance and respiratory system resistance [10]. Moreover, it shows setting of the ventilator during spontaneous breathing test: pressure support (Psupp), positive end-expiratory pressure (PEEP) and fraction of inspired oxygen (FiO_2).
- *RSB index,* which quantifies whether the breaths are rapids and shallows [11].
- *Parameters*: HR, RR, SaO_2, FiO_2, PaO_2, $PaCO_2$, SBP, DBP. The trend in these variables is available on the *panel trend*, in order to have an overview of the clinical condition of patient in ICU.
- *Clinical Assessment*, in which relates the clinical manifestations of the patient during the test and the possible causes of weaning failure. Understood as weaning failure, when the physician determines that the patient is not ready to withdraw ventilatory support.

All weaning test for each patient are available when it select the clinical case of interest.

C. Interactive Module

This module has two clinical cases. Around the selection of the clinical case it is possible to analyze the clinical condition of the patient during the first weaning test. The medical student through decision-making (wait, extubate, reintubate or not reintubate the patient) builds the patient's progress. Each decision must be justified on the tool to later be evaluated by the teacher:

- *Wait*, patient is not ready to withdraw ventilatory support, the medical student must decide how many hours will wait until the next weaning test.
- *Extubate,* patient is ready to withdraw ventilatory support.

- *Reintubate*, patient will be reintubate, the medical student must decide how many hours will wait until the next weaning test.
- *Not Reintubate*, patient is discharged from the ICU by the user.

When patient has been extubated, the *Patient's Progress* panel is enabled. The medical student can see the patient outcome on the following hours before to decide either required or not reintubate. Unlike the display panel in the training module, the trend of measurements is built according to the decisions-making by the medical student.

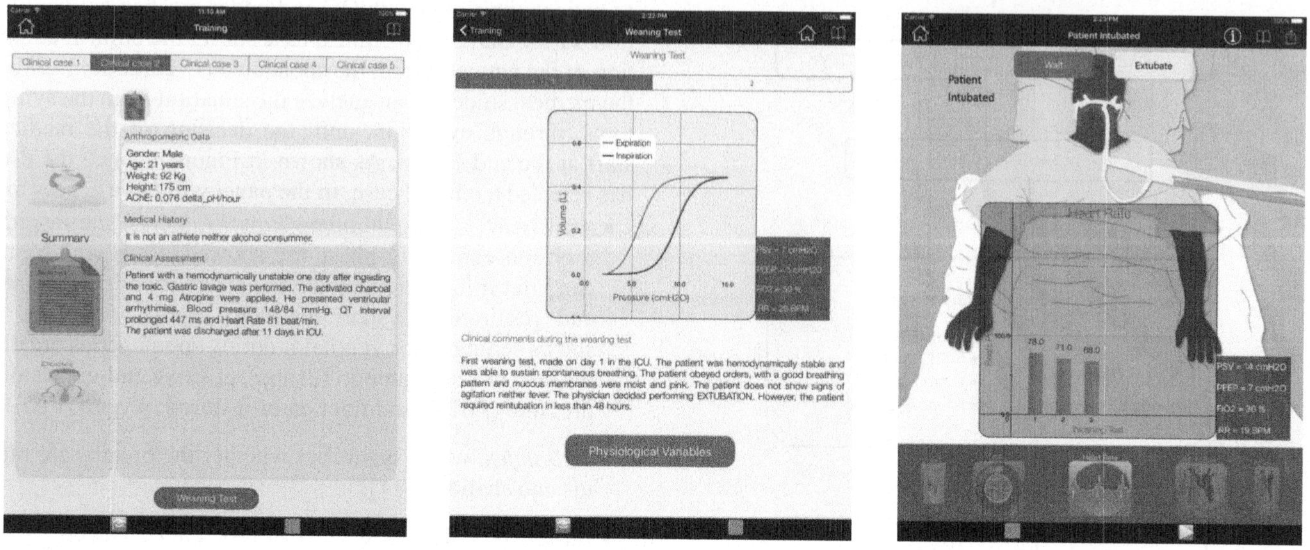

(A) (B) (C)

Fig. 2 Graphic interface: **A)** Set of real clinical cases of poisoned patients with organophosphorus compounds available in the training module. The anthropometric information, medical history and the clinical condition of patient are available for each clinical case. **B)** Pressure-volume curve on the first weaning test of the clinical case selected by the user. **C)** Interactive module, the extubation, reintubation or the non-removal of the ventilatory support is dependent of the decision-making by the user.

In each clinical case is presented the clinical condition of the patient on admission to ICU by the *nurse* and *doctor* (left slider, see Fig 2A). Also in the *summary* icon the general information of the clinical case is presented. Besides, the application has *panel notes*, to write all that the medical student considers relevant for the analysis of the case. The notes can be saved in Word2007 or pdf format.

III. Results

The following results were obtained of random testing of application functionality:

A. Training Module

In Figure 2A shown the training module at the beginning when the *clinical case 2* and *summary* icon has been selected. The anthropometric information, medical history and the clinical condition of the patient upon arrival at the emergency room are available. The *clinical case 2* is about a male patient who was admitted to the emergency department 6 hours after of drink an unknown amount of Neguvón®, he presented respiratory distress. The patient was discharged from the ICU after 11 days and 2 weaning tests were performed. The Figure 2B shows the results and ventilator settings on the first weaning test (WT1), moreover the clinical condition that led to withdraw ventilatory support. The patient required to be reintubated in less than 48 hours. The variability in the pressure-volume curve allows inferring information about the ventilatory capacity of the patient during the test.

B. Interactive Module

In Figure 2C is presented the interactive module when has been selected the *clinical case 1* and has decided to wait until the third weaning test (WT3). The patient was not extubated in any of these tests (WT1-WT3). The *clinical case 1* is about a male patient who was admitted to the emergency department an hour after of the self-poisoning with parathion. He drank an unknown amount of pesticide and presented respiratory failure. The tendency of variables, shown in Figure 2C (bottom slider), presents the values of the variables during the three weaning test and reflect the improvement of the critical condition of the patient.

Once it has been decided to end the interaction with tool, pressing *generate report*, the document will be created. This report has all justifications that had been written by user when the tool asked him the reason of his decision, therefore it may be used by the teacher as evaluative test. Moreover, the report presents suggestions for the user, if the decision has not been in line with expectations according to the clinical condition of the patient.

IV. Discussion

This learning tool allows the access to knowledge about mechanical ventilation of simple and interactive way. Several studies have demonstrated the utility of types tools in the diagnosis and clinical treatment[9]. Serna et al. worked with mechanically ventilated patients but not with patients poisoned with organophosphorus compounds [8].

This interactive tool provides the possibility that medical students can acquire knowledge about the weaning test, mechanical ventilation and clinical management of this kind of patient without the need of wait until the clinical practice. The student can repeatedly analyze each clinical case in detail, to know the variability of all the clinical variables used and to learn of each medical decision made during the patient's progress.

Although there are tools that allow to set the patient and to know the response of the respiratory and cardiovascular system [12], these are robust tools that run on platforms whose license carries high costs and are difficult to implement in mobile tools easily accessible to both students and teachers. Unlike of learning tool proposed in this work, which can be downloaded for free in the lite version.

V. Conclusions

Interactively, this application allows the user to know the response of the ventilatory mechanics of an intoxicated patient during spontaneous breathing tests. Through graphical interface is transmitted the outcome of each therapeutic decision, as well as the differences and similarities with clinical progress of the real clinical case. Knowledge, that otherwise, could only be acquired in clinical practice.

Acknowledgment

This work was partially supported by "PRG13-2-08 Protocolo para destete de la ventilación mecánica en pacientes intoxicados con organofosforados". Grant by Universidad de Antioquia (Medellín, Antioquia, Colombia), 2013-2016.

Conflict of Interest

The author(s) declare(s) that there is no conflict of interest regarding the publication of this paper.

References

[1] M. J. Abedin, A. A. Sayeed, A. Basher, R. J. Maude, G. Hoque, and M. a Faiz, "Open-label randomized clinical trial of atropine bolus injection versus incremental boluses plus infusion for organophosphate poisoning in Bangladesh," *J. Med. Toxicol.*, vol. 8, no. 2, pp. 108–117, Jun. 2012.

[2] J.-M. J. Boles, J. Bion, A. Connors, M. Herridge, B. Marsh, C. Melot, R. Pearl, H. Silverman, M. Stanchina, A. Vieillard-Baron, and others, "Weaning from mechanical ventilation," *Eur. Respir. J.*, vol. 29, no. 5, pp. 1033–1056, 2007.

[3] J. Jeyaratnam, "Acute pesticide poisoning: a major global health problem," *World Heal. Stat Q*, vol. 43, no. 3, pp. 139–144, 1990.

[4] S. N. de V. en S. P. SIVIGILA., "Informe epidemiológico nacional 2013," 2013. [Online]. Available: http://www.ins.gov.co/lineas-de-accion/Subdireccion-Vigilancia/sivigila/Paginas/documentos.aspx.

[5] M. Meade, G. Guyatt, D. Cook, L. Griffith, T. Sinuff, C. Kergl, J. Mancebo, A. Esteban, and S. Epstein, "Predicting Success in Weaning From Mechanical Ventilation," *CHEST J.*, vol. 120, pp. 400–424, 2001.

[6] S. Dormido, S. Dormido-Canto, R. Dormido, J. Sánchez, and N. Duro, "The role of interactivity in control learning," *Int. J. Eng. Educ.*, vol. 21, pp. 1122–1133, 2005.

[7] J. Sánchez, S. Dormido, and F. Esquembre, "The learning of control concepts using interactive tools," *Comput. Appl. Eng. Educ.*, vol. 13, no. 1, pp. 84–98, 2005.

[8] L. Y. Serna, A. M. Hernandez, and M. A. Mañanas, "Computational tool for modeling and simulation of mechanically ventilated patients.," *Conf. Proc. IEEE Eng. Med. Biol. Soc.*, pp. 569–572, Jan. 2010.

[9] A. M. Hernández Valdivieso, M. B. Salazar Sánchez, D. A. Urrego Higuita, R. Costa-Castelló, and M. Á. Mañanas Villanueva, "Virtual laboratory for simulation and learning of cardiovascular system function in BME studies," *Rev. Fac. Ing. Univ. Antioquia*, vol. 60, pp. 194–201, 2011.

[10] B. Cabello and J. Mancebo, "Work of breathing," *Intensive Care Med.*, vol. 32, no. 9, pp. 1311–1314, Sep. 2006.

[11] E.-J. Kang, S.-J. Seok, K.-H. Lee, H.-W. Gil, J.-O. Yang, E.-Y. Lee, and S.-Y. Hong, "Factors for determining survival in acute organophosphate poisoning.," *Korean J. Intern. Med.*, vol. 24, no. 4, pp. 362–367, Dec. 2009.

[12] L. Cheng and M. C. K. Khoo, "Modeling the autonomic and metabolic effects of obstructive sleep apnea: a simulation study.," *Front. Physiol.*, vol. 2, no. January, pp. 1–20, Jan. 2012.

Author: Alher Mauricio Hernández
Institute: University of Antioquia
Street: Calle 70 No. 52 – 21, AA 1226
City: Medellín
Country: Colombia
Email: alher.hernandez@udea.edu.co

A Comprehensive System for Healthcare Technology Management HTM

L. Vilcahuaman[1], M. Cordova[1], J. Kalafatovich[1], R. Rivas[1]

[1] Pontifical Catholic University of Peru / Engineering Department, Bioengineering Laboratory, Lima, Peru

Abstract— **Current models to manage and organize technology in hospitals have evolved over the established 40or 60 years ago according to totally different circumstances of now, therefore, have a failure of origin. In the context of new technologies, it is not enough improve the outdated model, but a new model needs to be raised by re-engineering to achieve appropriate levels of clinical effectiveness, efficiency, safety, cost control and quality that users expect of the technology used in hospitals. Many of the current premises on good practices in healthcare technology management HTM provided by specialized institutions on health are impractical, because in the current processes do not have neither the human resources, nor the liability, nor the appropriate procedures for its implementation. The aim of this work is to provide a process map and their components to implement an effective HTM system in hospitals. Our initial results obtained with the support of a computer platform makes evident that the parameters of clinical effectiveness, efficiency, safety, cost control and quality can be improved, as well as, to reflect on the need to rethink the hospital organization for decision-making on technology.**

Keywords— **Clinical Engineering, Healthcare Technology Management, HTM Process map.**

I. Introduction

Peruvian health sector as other ones from developing countries, has not understood neither manages adequately Health Technology HT and its relevant influence in safety, effectiveness, quality and sustainability of health organizations [1]. This evidence determines low-quality of health services, lack of human resources experts, absence of HT policies and regulation, plans and strategies with no accuracy/level of quality required, etc. [2]. Not developed countries traditionally focus mainly on equipment maintenance; according to surveys: the budget aimed to equipment maintenance is less than 2% of the annual operative budget, while the operability of medical equipment is under 57% in a hospital classified as "high-level", in other side the regulation related to medical equipment ignores the metrological verification; all of these factors are against the implementation of a functional clinical service.

This article presents the design of a healthcare technology management-HTM system, in a comprehensible model. The initial information obtained primarily [3], supported the evaluation of HTM in Peruvian hospitals, based on the current processes, the authors adjust and propose new ones.

II. New perspective in CE, HTM, HTA and others

A. Healthcare Technology HT

HT is not only medical devices, that in the new concept of medical devices they include implants, implements, software, applications, etc. In this sense, HT include as well: clinical procedures, information systems, materials and medical supplies, energy systems, hospital equipment and infrastructure, further organizational models and medicines. New prospects should also be considered in health services such as telemedicine and technologies for prevention, promotion and protection.

B. Clinical Engineering CE and Healthcare Technology Management HTM

XXI century requires a multidisciplinary approach to health sector: physicians, engineers, physicists, architects, managers, economists, etc. because of the importance and complexity of technology [1]. This fact is even more urgent in developing countries. Given the limited number of engineers in the hospital, health sector's network requires clinical engineers or biomedical engineers capable of addressing the functionality of technology in health services; the success of the interaction depends on a strong interaction between the engineers, the medical staff and the administration.

Partial Clinical Engineering criteria [4] have been implemented in health organizations. In Peru the implementation of CE criteria with emphasis on HTM, is highly needed. The article emphasizes on the interaction between the stakeholders as one of the key factors to facilitate the implementation and emphasis stated above.

C. Healthcare Technology Assessment HTA

HTA reports in health institutions, supports the decision-making to be more objective and transparent not only related to acquisition but also related to HTM´s issues [5][6]. Peru have started efforts around HTA but they´re

I. Torres et al. (eds.), *VII Latin American Congress on Biomedical Engineering CLAIB 2016, Bucaramanga, Santander, Colombia, October 26th -28th, 2016,*
IFMBE Proceedings 60,
DOI: 10.1007/978-981-10-4086-3_64

basics and not enough, there is a need for policies, guidelines, training to respond adequately to the country´s goals. Encouraging the development of HTA demands the sustained participation of experts and also permanent training in HTA methods and procedures. The inclusion of HTA in research activities at the hospitals is one of the key-aspects to be considered.

D. Risk Management RM in health

Risk management focuses on identifying and reducing risks, follows the health regulations regarding the use of technology in humans. The danger is inherent in technology also it is not acceptable to admit any risk, consequently it is required to reduce and if possible to eliminate damage to people. In this regard the following processes are considered: techno-surveillance, analysis of adverse events, application of technical standards, identification of the benefit by diagnostic of quality, productivity, cost of acquisition, operating cost, compatibility between technologies, identifying of hazards and risks, etc. [7].

III. MAIN PROCESS FOR COMPREHENSIVE SYSTEM OF HTM

After an extensive analysis of the reality of the health system, and considering the main objectives which are improve clinical effectiveness, improve efficiency in the use of technology resources, increase security, better cost control and higher quality of health services with the compliance of the premises of the world's major health organizations; and considering the rules, regulations and guidance given to health systems, we propose six main processes for HTM based in ISO9000 and EFQM [4]:

A. Planning Process

The planning process elaborates and follows up on corporate strategic plan, annual operating plans and generates HT policies and is the reason why targets and indicators are posed. A major sub-process is the assessment of technology needs. It requires the development of HTA reports of types: medical, economic and development of clinical services in the medium and long term device. With the information gathered is possible to analyze the situational status and plan new activities. Another product is the technology procurement annual plan. Another major sub-process is the economic assessment oriented to have the reference price to buy, annual budgets and others to medium and long term. These comprise directors and multi annual plans [7][9].

B. Acquisition Process-AP

Acquisition process goal is to acquire equipment that matches technical specification at a lower cost. The most important input of acquisition process is the list of future acquisition equipment that as mention before is one of the outputs of planning process.

The acquisition process aims to programming technology purchases and contracts based on the points made in the planning process. This process includes the preparation of purchase and call for technology providers. After that follows the evaluation of tenders, awarding and receipt of purchased technology. In the acquisition process the technical specifications are developed by market and according to HTA reports. Also, committees are formed for evaluation and monitoring of the whole process. The selected suppliers sign contracts and acquired technology is inspected upon arrival at the hospital being issued a certificate of conformity [10].

C. Management and maintenance of technology assets

The process of management and maintenance of technology assets aims to manage technology goods as medical equipment to ensure its operation as part of the functionality of clinical services. It has a sub-process of document management and information system, where the inventory of medical and hospital devices [11] is continuously updated, as well as the historical record of work orders. Another sub-process is the asset management and inspection which apply procedures for installation of new equipment [7], the contract supervision, scheduling and execution of inspections and preventive and corrective maintenance [12]. Besides the decommissioning of medical devices are evaluated. Additionally it has a sub-process for analyzing costs and waste of economic resources (results assurance) aimed at improving the efficiency of technological resources.

D. Risk Management

The process for managing technology risks complements and strengthens risk management oriented to occupational health-OSHA, at the hospital. A major sub-process is the Techno-Surveillance Management, that includes developing the matrix of hazard identification, risk assessment and risk control HIRARC for clinical services and their technologies [13], also manage of adverse events from registration until analysis and verdict. In the end the annual risk reduction plan is developed, it includes disaster mitigation. Another sub-process is the metrological verification and programming [7], which includes the development of metrological testing procedures and supervision of these.

E. Acquisition Process AP

The process for managing research aims to establish in the hospital the ability of continuous quality improvement and analysis of results in technology [2]. A major sub-process is to improve the technological design and functionality of clinical services [4] by re-design processes and procedures HTM, support for the development of clinical practice guidelines and reports HTA, as well as the conceptual and comprehensive re-design of clinical services considering architectural requirements, power systems, hospital equipment, medical devices, information systems and organization. Another major sub-process is the project management in the perspective of PMBOOK [14]. Projects can be research itself, or investment projects under internal or external funds, also we include the health technology transfer to others hospitals or entities.

F. Human Resources for HTM

The process of human resources development aims to a sustainable management of training programs for medical assistance, engineering, managers and technical personnel. It includes the identification of training needs, the design of academic proposals [2], the management of teachers and teaching assistants, the inter-institutional management for dissemination and financing of programs, budget management , logistics management and control of activities. In addition, internship management and pre-professional practices are included.

IV. RESULTS ANS DISSCUSSION

A. Process map for HTM

The comprehensive system of HTM is presented in Figure 1. It shows the main processes described above and the interaction between them. The sub-processes and detailed procedures were developed in accordance with the criteria established by the standard ISO9000 and EFQM model.

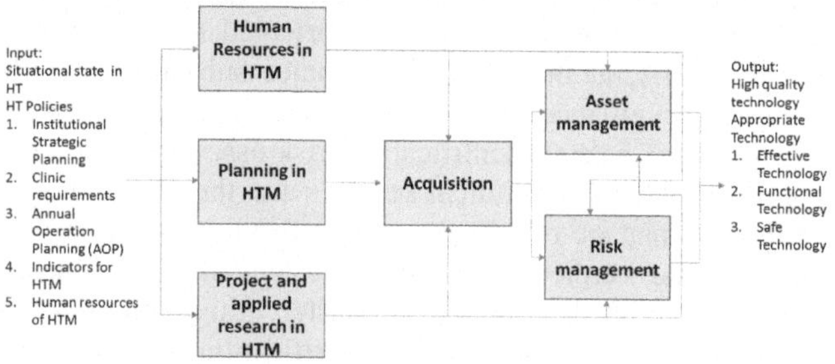

Fig. 1 Processes Map of HTM comprehensive system

B. Computer platform for HTM comprehensive system

For the application of the HTM process map computing platform was developed with scalable features. This is on test period at the Maternal Prenatal National Institute INMP. The software was developed in the programming language PHP Hypertext Pre, very versatile, widely used in web servers, it allows to create complex applications and has great capacity to handle dynamic content. Thus, all HTM processes mentioned above are very easy to use.

The development of this platform is an instrument to put in practice the proposed process map. However, other programing methods could be used for its development and implementation.

C. Indicators for HTM

The application of the new map of processes in the hospital INMP (Maternal Prenatal National Institute in Lima-Peru) is ongoing for 6 weeks. We present the initial results for the processes of management and maintenance of technology assets, particularly with respect to medical devices. Response time for maintenance request went from 11.3 hours to 8.5 hours on average. Figure 2 shows the response times in various clinical services.

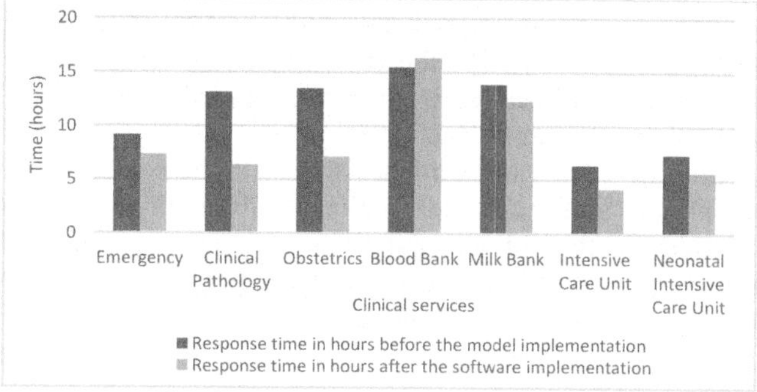

Fig. 2 Response time for maintenance request

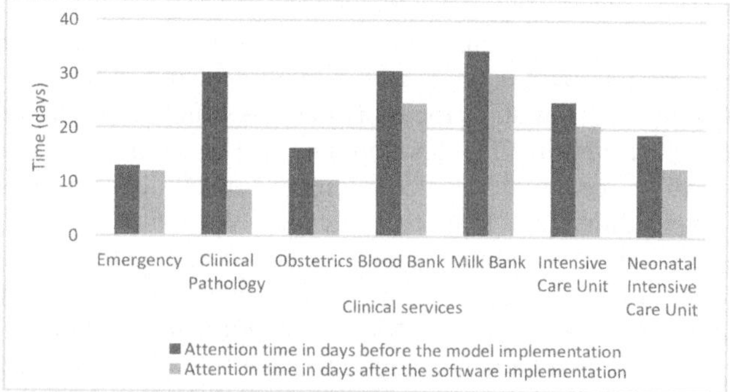

Fig. 3 Maintenance attention time

Another indicator is the repair maintenance attention time, the average time went of 24.5 days to 17 days. That includes the time for repairing and the time to get replacement for broken pieces. Figure 3 shows a comparison between times; all times have been reduced at least one day.

The frequency of corrective maintenance is a critical indicator. The average value was reduced from 4.4 times to 3.2 times per month. Figure 4 shows these results in each clinical service.

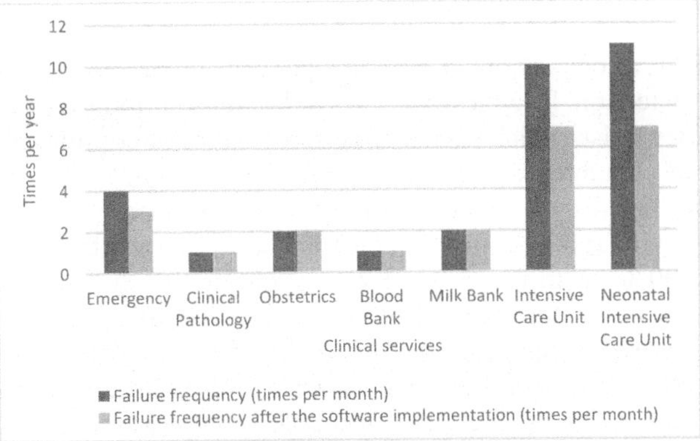

Fig. 4 Corrective maintenance frequency

The corrective maintenance cost depends on the frequency of work done. Applying the new process map the average value decreased from 3273 to 2576 PEN per equipment.

V. Conclusions

Process map developed in the Comprehensive System of healthcare technology management has proven to be suitable to incorporate gradually and without major disturbs to the current processes of the hospital, both aspects determines a feasible application, whether partial or complete. In addition the Process Map is a guide to effectively update HT it also improves the functionality of clinical services. One of the findings shows that the hospital tends to re-think their organization according to the proposed processes; this fact supports the proposal of an organizational structure of higher quality than before. The indicators feedback on-time the state of technology in the hospital, they also facilitate the effective and on-time correspondent action.

The computer platform developed in HTM Comprehensive System proves to be flexible and scalable, allowing high feasibility of application due to the easy adaptability to particular environments in hospitals. Like any software, it requires continuous monitoring to ensure operability or improve procedures.

This paper presents the preliminary implementation of HTM Comprehensive System software, a report with complete results of validation will be elaborated later.

Acknowledgment

This work was supported by project 209 FINCyT – IA – 2013. FINCyT is a part of INNOVATE PERU founded by Peruvian government for science and technology research.

Conflict of Interest

The authors declare that they have no conflict of interest.

References

[1] Vilcahuaman L, Rivas R, (2006). Ingeniería Clínica y Gestión de Tecnologías en Salud: Avances y Propuestas. Lima: CENGETS-PUCP.

[2] Vilcahuaman, L, Rivas, R., Portella, J., Paz, D., Luis, J., & Mezarina, E. (2013). Unidad de Gestión de Tecnología en Salud e Ingeniería Clínica en Hospitales Peruanos: Excelencia y Calidad de Tecnología en el INMP. In V Latin American Congress on Biomedical Engineering CLAIB 2011 May 16-21, 2011, Habana, Cuba (pp. 579-582). Springer Berlin Heidelberg.

[3] Vilcahuaman L, Talla M, Cordova M, (2014, Sept) , "Method for evaluating healthcare technology management in Peruvian hospitals", Presented: VI Latin American Congress on Biomedical Engineering (CLAIB)[Online] Available: http://link.springer.com/chapter/10.1007%2F978-3-319-13117-7_187

[4] Dyro J, (2004) The clinical engineering handbook, Setauket, New York: Academic Press.

[5] Finn, Helga SIGMUND, Copenhagen S. Denmark: Danish Centre for health technology assessment. National Board of Health. 2008.

[6] Health technology assessment of medical devices. World Health Organization (WHO), Geneva, Suisse: WHO Press. 2012.

[7] Lenel A, Temple-Bird C, Kawohl W, Kaur M (2005) ´How to manage´ series for healthcare technology, Guide5: How to organize the maintenance of your healthcare technology, East Sussex, UK: Ziken International, 2005

[8] IAT. Guía para una gestión basada en procesos. Instituto Andaluz de Tecnología. ISBN 84-9234464-7-7.

[9] Development of medical device policies. World Health Organization (WHO), Geneva, Suisse: WHO Press. 2012.

[10] Procurement process resource guide. World Health Organization (WHO), Geneva, Suisse: WHO Press. 2012.

[11] Introduction to medical equipment inventory management. World Health Organization (WHO), Geneva, Suisse: WHO Press. 2012.

[12] Medical equipment maintenance program overview. World Health Organization (WHO) , Geneva, Suisse: WHO Press. 2012.

[13] Conceptual Framework for the International Classification for patient safety. World Health Organization (WHO), Geneva, Suisse: WHO Press. 2009.

[14] A Guide to the Project Management Body of Knowledge (PMBOK Guide), Project Management Institute, Inc, 5th ed., Pennsylvania.. 2013.

Author: L. Vilcahuaman
Institute: Engineering Department -PUCP
Street: 1801 Universitaria Ave.
City: Lima
Country: Peru
Email: lvilcah@pucp.edu.pe

Development and Evaluation of a Method for Fall Detection Based on a Wrist-Located Device

T. de Quadros, A.E. Lazzaretti and F.K. Schneider

Federal University of Technology – Paraná/CPGEI, Curitiba, Brazil

Abstract— **The consequence for a fall in elderly people ranges from a soft psychological insecurity and fear of falling to a more physical and severe complication, leading in many cases the victim to death. However, as quick as the subject receives help and medical care, lower is the risk for serious consequences and sequelae. Therefore, several fall detectors have been developed, aiming at detecting fall through wearable devices located on neck, waist, chest, etc. However, for devices located on user wrist, the results are commonly worse, since its detection becomes harder on a body member that moves unimpeded in space. Also, threshold-based methods, whose requires low processing resource and, consequently, becomes easier to be implemented in a wearable embedded device, are not efficient enough to detect falls with high accuracy. This paper presents a threshold based method for fall detection using motion signal variables, evaluating different configurations and their contributions to improve the final fall accuracy, compared to state-of-the-art wrist-based devices.**

Keywords— **Elderly, falls, detection.**

I. Introduction

According to World Health Organization (WHO), approximately 28%-35% of people aged of 65 and over fall each year [1]. This fraction increases to 32%-42% when the age is 70 and over. Further, the average cost for every fall injury related hospitalization range from US$6646 on Ireland to US$17483 on USA. These elevated costs justify why elderly falls must be considered a public health problem. About falls recurrence, for every three elderly victims of a fall, two of them will suffer a new fall in six months [2]. The possible consequences of falls for elderly people are diverse, both physiological and psychological. Depression and fear of fall are commonly seen in elderly fall victims, reducing their quality of life [3].

In order to reduce fall consequences, immediate relief is crucial. Studies performed with reports about elderly falls identified a great influence of time-to-help on fall results: the premature obit increased from 8% to 40% when the help happened after one hour [4]. For this reason, many fall detectors have been developed, based on different technologies. The goal of fall detection is to alert family and medical services quickly, reducing the risk of fall sequelae. As consequence of the wearable devices popularity, a lot of solutions have been adopted motion signals as input variable for fall detection [5]. However, considering solutions located on user wrist, which seems to be more discreet, comfortable and acceptable (e.g. smartwatches, smartbands), the efficiency seems to be far from the ideal purpose: 100% of sensitivity (no falls undetected) and a high specificity (at least 90%), with few false alarms, avoiding user to consider the device as unreliable, leading to rejection [6]. Since human wrist translates and rotates in different axis constantly, detect a fall motion becomes harder on this approach. Also, for embedded devices like wearable, complex processing algorithms leads to high battery consumption, which means that a great solution for this hardware model must be simple and efficient, defining a challenge and an open problem in the literature.

This work aims to develop a solution based on different motion signals acquired with a device located on user wrist. The selected method for fall detection used a threshold-based approach, allowing it to be easily embedded on a commercial wearable device. Further, exhaustive tests were performed, in order to identify the best configuration and threshold for every signal and their combinations. The results were compared to those presented on literature.

II. Materials and methods

The development and evaluation of the presented work were done in five steps, as follows.

A. Raw information acquisition

For the motion signal acquisition, an IMU (Inertial Measurement Unit) was used, comprised of a 3-axis accelerometer ADXL345 (Analog Devices, USA) and a 3-axis gyroscope L3G4200D (ST, Switzerland) [7][8]. The raw data from IMU was acquired by a development board Arduino UNO (Arduino, Italy), and processed in a MatLab (MathWorks, USA) environment, where the samples from the accelerometer and the gyroscope can be further analyzed. The acquired signals were divided into four fall activities: forward, backward, sideward to the side of the wrist with the device, sideward to the opposite side of the wrist with the device; and four non-fall activities: clapping hands,

I. Torres et al. (eds.), *VII Latin American Congress on Biomedical Engineering CLAIB 2016, Bucaramanga, Santander, Colombia, October 26th -28th, 2016*,
IFMBE Proceedings 60,
DOI: 10.1007/978-981-10-4086-3_65

sitting in a chair and standing up again, walking, opening and closing a door. Every volunteer performed five times each one of the activities, resulting in 20 fall and 20 non-fall acquisitions. The learning set (training set) and test set was comprised of two and four volunteer's information, respectively (80 and 160 acquisitions). Every signal was acquired at 100Hz sampling rate, in a range of 4G for accelerometer and 500 degrees/sec for gyroscope.

B. Definition of configuration variables and their range

Acquiring acceleration from the IMU device, three signals were selected to be measured: the magnitude of the resultant acceleration after gravity removal (AN, acceleration-normal), the velocity after AN window-integration (VN) and the displacement after VN window-integration (DN). The window-integration was performed because a simple integration would accumulate error over time. Integrating data only in a fixed-size window avoids this problem.

When the motion intensity is low, the device inclination may be estimated by accelerometer-only information. Using the angular velocity measured by the gyroscope, this estimation can also be performed when the device is translating and rotating [9]. Making use of this information, the vertical component of AN can be calculated, resulting in an estimated vertical acceleration (AV). Applying a window-integration to AV, a vertical velocity is found (VV), and applying a second window-integration, the vertical displacement is calculated (DV).

These six signals were selected to be evaluated over time. This work presupposes that isolated signals are not reliable enough to detect falls, but their combinations can increase sensitivity, without compromising specificity. Studies performed using only acceleration or velocity as input data did not reach satisfactory results to fall detection, when the device was located at user wrist [10-12]. Adding displacement to the evaluation, and calculating the vertical component for motions signals, it is expected to achieve results of high accuracy. Considering the mentioned signals, four configuration variables were selected in order to have different results, and a range for each one was defined as follows:

- four options for the window time and five for the limit value to consider acceleration magnitude low enough to reset the device angular position measured using accelerometer-only data;
- five options for window-time to perform acceleration integration and
- five options for window-time to perform velocity integration.

Therefore, 500 combinations for these configuration variables were possible. Running the algorithm to each one of these signals, 500 different thresholds were selected for every acquired signal, starting on the threshold that results on 100% of sensitivity and ending on the one that results on 100% of specificity. These tests were performed using only the learning set.

C. Single signals evaluation

Combining the 500 different configuration variables with the 500 different thresholds, an exhaustive test was performed for all the 500x500 possibilities with each one of the six signals. For this test, the acquired data from learning set was read and all the moments when the signal was above the threshold were registered, recording the time information. Thus, a large signal evaluation report was generated, which can be later analyzed, and different combinations can be tested. Using the results from these tests, the sensitivity and specificity were calculated for each one of the 500x500 combinations. Then, the confusion matrix was calculated [13] and the results were sorted by their sensitivity, specificity and classification accuracy.

Whereas the results would be combined, some points must be considered. Firstly, if two different signal configurations provide 100% and 0% of specificity, respectively, the result for their combination will not necessarily be near to 0% of specificity. Rather, it could be as high as 100%. So, signal combinations are able to increase specificity when evaluated together. Secondly, if two different signal configurations provide 100% and 0% of sensitivity, then the highest sensitivity allowed for a combination of these signals is 0%. Thus, it is not worthwhile to combine signals with low sensitivity, because their results will surely be unsatisfactory. Finally, the accuracy information for every signal configuration allows a fast identification whether the signal has a great sensitivity without saving in specificity, whose combination results could be more stable and efficient. For every signal combination, the best accuracy and specificity were selected from each configuration variable set, which means two combinations. Also, eight different time window sizes were selected to identify when the detected fall happened for the same event along the signal. Therefore, 500x2x8 tests were done for every combination, in order to estimate the best solution. This cross validation was performed in order to select the configurations combination with the most promising results [14].

D. Combined signals evaluation

Considering the six signals acquired and previously evaluated, 15 combinations were possible for a 2-by-2 set, 20 for a 3-by-3 set, 15 for a 4-by-4 set, 6 for a 5-by-5 set and, finally, a single combination with all the six signals. Performing the tests for them, the best accuracy case for each one was identified, considering the configurations for both accuracy and sensitivity best cases of the previously tests.

E. Evaluation with test set

Comparing the results for each signals combination, the best configuration (considering configuration variables, thresholds and time window size) were selected for each one. Then, a final test was performed using the data from test set. This process was relevant to identify how every signal contributes to increase sensitivity and specificity. Also, it allows an estimation of the best signals combination and variable configuration for a final solution.

III. Results

The tests were performed for the learning set with the best thresholds and configuration variables, including both sensitivity and accuracy selection criteria. When the signals combination was composed by two or three signals, the accuracy standard represented the most part of the better cases. In this situation, 25 from 35 combinations produced better results with the accuracy selection criteria. In contrast, when four or five signals were combined, sensitivity standard appears to be more efficient: 15 from 21 combinations. For a single signal or for all of them, this evaluation is not worthwhile. The best results for the test set are presented in Table 1. According to them, displacement is not present, reducing its relevance on the proposed fall detection method. On the other hand, velocity is present in all them, reinforcing the results presented by other works on literature, which suggest velocity as a relevant information for fall detection [10,15]. When no signals combination is performed, the obtained results do not confirm the relevance of the vertical components of the motion signals. Considering only the single signal tests results, the vertical components (AV, VV and DV) presented worse results than the others, as shown in Table 2.

Table 3 presents the average accuracy for combinations with two and three signals, discerning the results by every single signal contribution. The normal and vertical acceleration and the normal velocity presented better results, reinforcing the results presented on Table 2. For combinations with four or more signals, there is a lot of interference on results from each other, constraining a proper evaluation. Comparing the results between the tests performed on learning set and test set, combinations using a higher number of variables presented a lower variation, despite being not related to the best results from Table 1. On Table 4, the average accuracy for every combination quantity with the learning set and test set is presented (except 6-by-6, since it contains just one possibility). The 5-by-5 combination presented the worst average accuracy for the learning set, but the best for the test set.

Table 1 Best accuracy results for test set

Signal Combination	Sensitivity	Specificity	Accuracy
AN – VN	98.8%	82.5%	90.6%
AV – VN	97.5%	82.5%	90.0%
AN – AV –VN	96.3%	82.5%	89.4%
VN	100.0%	77.5%	88.8%

Table 2 Test set results for single signals

Signal	Sensitivity	Specificity	Accuracy
AN	98.8%	67.5%	83.1%
AV	93.8%	70.0%	81.9%
VN	100%	77.5%	88.8%
VV	92.5%	53.8%	73.1%
DN	92.5%	66.3%	79.4%
DV	86.3%	50.0%	68.1%

Table 3 Average accuracy for every single signals contribution

Signal included	Average accuracy	
	2-by-2 combination	3-by-3 combination
AN	85.5%	85.1%
AV	84.3%	84.4%
VN	83.9%	82.8%
VV	79.0%	81.4%
DN	79.9%	81.4%
DV	80.3%	82.0%

Table 4 Average accuracy for learning and test set results

Combination	Average accuracy	
	Learning set	Test set
1-by-1	92.3%	79.1%
2-by-2	95.5%	82.1%
3-by-3	95.4%	82.3%
4-by-4	95.1%	83.7%
5-by-5	95.0%	84.4%

IV. Discussion

Evaluating the obtained results for all configuration combinations, velocity is revealed as a key variable for fall detection through a threshold based method. Its relevance is increased when it is combined with normal and/or vertical acceleration. Intuitively, a higher velocity in a fall may be related to a most injurious consequence, allowing the development of a fall severity identifier. Displacement, against the hypothetical prediction, presented great specificity for many cases (100% in many of the learning set). However, its sensitivity is unstable, becoming unreliable for a commercial application.

The results for vertical components of the acquired signals were not relevant. This does not mean that a true decomposition is not relevant, but the method developed is not able to decompose the signals properly enough for a device located on user wrist. If an ideal decomposition be performed, a vertical displacement becomes a relevant information, since every fall must present a body height variation. Increasing the number of variables in the signals combinations (i.e., 4-by-4 and 5-by-5 configurations) makes the system be more predictable. For a 1-by-1 combination, the standard deviation of the accuracy for the learning set and test set was 6.7% and 7.4%, respectively – against 0.8% and 1.0% of the 5-by-5 configuration.

The best result (98.8% of sensitivity, 82.5% of specificity), even presenting accuracy higher than 90%, is far from the ideal model proposed in this paper. However, it is considerably higher than those presented on literature. Degen et al. achieve a sensitivity of 65% (specificity was not calculated) with a device similar to a watch [10]. Kang et al., with a system able to measure several biomedical signals, presented a 91.3% of fall detection rate [11]. Bagnasco et al. although achieving great results for device located at chest and head, presented only 71% of sensitivity and 89% of specificity when the fall detector was located on wrist [16]. Similarly, Kangas et al. achieve 100% of accuracy for a threshold based method with a device located on head, but 75% of sensitivity when it was placed on wrist [17].

V. Conclusions

Fall detection by motion signal acquisition using a threshold-based method is a challenge. In this work, three different signals were evaluated, as well as their vertical compositions. The obtained results, although being not ideal, promote a continuation for wrist located fall detector devices research, aiming at embedded solutions for this purpose. For future works, displacement and vertical velocity may be discarded, and acceleration and velocity may be better evaluated. A mathematical combination between them is also suggestive, like a cross product of their vectors or a time window convolution. In order to obtain better results for the vertical component of the signals, a magnetometer may be used to increase reliability of the device angular position, since the gyroscope data is not enough for this purpose.

Conflict of Interest

The authors declare that they have no conflict of interest.

References

1. World Health Organization (2007) WHO Global Report on Falls Prevention in Older Age. WHO Press, Switzerland
2. Tinetti ME et al. (1988) Risks factors for falls among elderly persons living in the community. N Engl J Med 319(26):1701-1707 DOI 10.1056/NEJM198812293192604
3. Aschkenasy MT1, Rothenhaus TC (2006) Trauma and falls in the elderly. Emerg Med Clin North Am 24(2):413-32 DOI 10.1016/j.emc.2006.01.005
4. Gurley RJ, Lum N, Sande M, Lo B, Katz MH (1996) Persons found in their homes helpless or dead. N Engl J Med 334:1710-1716 DOI 10.1056/NEJM199606273342606
5. Igual R, Medrano C, Plaza I (2013) Challenges, issues and trends in fall detection systems. Biomed Eng Online 12:66 DOI 10.1186/1475-925X-12-66
6. Noury N, Rumeau P, Bourke AK, Ólaighin G, Lundy JE (2008) A proposal for the classification and evaluation of fall detectors. IRBM 29(6):340-349 DOI 10.1016/j.irbm.2008.08.002
7. Analog Devices (2015) ADXL345 3-axis Digital Accelerometer Datasheet. Rev. E. © 2009–2015 Analog Devices, Norwood.
8. ST Microelectronics (2010) L3G4200D Ultra-Stable Three-Axis Digital Output Gyroscope Datasheet. Doc ID 17116 Rev 3
9. Pedley M (2013) Tilt Sensing Using a Three-Axis Accelerometer. Freescale App. Note. Document number: AN3461. Rev. 6
10. Degen T, Jacckel H, Rufer M, Wyss S (2003) Speedy: a fall detector in a wrist watch. In Proc. Seventh. IEEE International Symposium on Wearable Computers, Washington, USA, 2003, pp. 184-187 DOI 10.1109/ISWC.2003.1241410
11. Kang JM, Yoo T, Kim HC (2006) A wrist-worn integrated health monitoring instrument with a tele-reporting device for telemedicine and telecare. IEEE Trans Instrum Meas 55(5):1655-1661 DOI 10.1109/TIM.2006.881035
12. Pannurat N, Thiemjarus S, Nantajeewaraw E (2014) Automatic fall monitoring: a review. Sensors 14(7):12900-12936 DOI 10.3390/s140712900
13. Fawcett T (2006) An introduction to ROC analysis. Pattern Recognit Lett 27(8):861-874 DOI 10.1016/j.patrec.2005.10.010
14. Bishop CM (2006) Pattern recognition and machine learning. Springer-Verlag New York, Secaucus
15. Wu G (2000) Distinguishing fall activities from normal activities by velocity characteristics. J Biomech 33:(11)1497-1500 DOI 10.1016/S0021-9290(00)00117-2
16. Bagnasco, A., Scapolla, A.M., Spasova, V. (2011) Design, implementation and experimental evaluation of a wireless fall detector. In Proc. 4th International Symposium on Applied Sciences in Biomed. and Comm. Technologies, Barcelona, Spain, 2011, pp 26–29 DOI 10.1145/2093698.2093763
17. Kangas, M., Konttila, A., Winblad, I., Jamsa, T. (2007) Determination of simple thresholds for accelerometry-based parameters for fall detection. In Proc. 29th Annual International Conf IEEE Eng Med Biol Soc, Lyon, France, 2007, pp 1367–1370 DOI 10.1109/IEMBS.2007.4352552

Extending the horizon of biomedical engineering to help other species

A.M. Gonzalez-Vargas[1]

[1] Departamento de Automática y Electrónica, Universidad Autónoma de Occidente, Cali, Colombia

Abstract— **It is important for engineers to become aware of how we affect the ecological environment around us. Therefore, we must understand how engineering can be not only at the service of humans but of other animal species as well. Prosthetics, orthotics and similar devices are intended to replace or improve the functions of lost limbs and organs. Traditionally, they have been used only in humans, but in recent years there has been a growth in the development of these devices for animals. They are particularly important when they help to restore the locomotion abilities to their users, in which case they are called ambulation devices. For companion animals, this is important because it improves the welfare of the animal, avoiding the use of euthanasia, and improving also the psychological well-being of its owner. In this paper we discuss the generalities of veterinary ambulation devices, and present the idea that the *maker* movement can be a great aid to introduce this concept to new generations of biomedical engineers, helping to create a new research line on veterinary engineering. We illustrate the concept with an example of an activity of this type carried out at the Universidad Autónoma de Occidente.**

Keywords—**Ambulation devices, Veterinary, Maker Movement, Biomedical Engineering.**

I. Introduction

Biomedical Engineering applies the knowledge of diverse engineering areas, such as mechanics, electronics and computer science to solve problems of the medical field. Therefore, the biomedical engineer acquires not only a thorough knowledge of the technological principles, but also solid knowledge in the field of biology. As many of the principles of biology areas are equally applicable to a large number of species, this leads us to think that biomedical engineering does not have to be limited to human applications.

Two of the research lines of biomedical engineering in universities are bioinstrumentation and rehabilitation engineering. Thanks to the development of these two lines, in recent decades there have been great advances in the development of prostheses for upper and lower limbs in humans. The knowledge gained in the development of such devices can be extrapolated (with special considerations) to domestic animals; particularly quadrupeds such as dogs, cats or horses, and has allowed the creation of different devices to improve the mobility of these animals. However, this practice has had a noticeably smaller diffusion than its human counterpart, and it is often performed heuristically, without many biomechanical considerations or high investment in materials and components, as there is not a wide offer of these products for animal purposes [1].

The advances of the last decade in the understanding of animal biomechanics, as well as the development of inexpensive mass manufacturing technologies, and the growing interest in animal welfare from a scientific and ethical view, suggest that it is time to extend the horizons of biomedical engineering beyond our species, and develop a new field of research to improve the quality of life of other animals. For it to happen, however, we must raise awareness in the new generation of engineers and, from an early stage of their education, familiarize them with the knowledge and technologies that will be useful to this purpose.

Fab-labs are small-scale workshops that offer (personal) digital fabrication capabilities [2]; these spaces are being integrated into many universities, due to the popularization of the *maker* movement. These facilities give us an excellent opportunity to carry out activities that involve building small ambulation devices for veterinary purposes.

This article is organized as follows: Section II is a brief introduction to veterinary ambulation devices (VAD). In section III we describe the history and generalities of the *maker* movement. In section IV we present a first experience with biomedical engineering students on building VADs, which was carried out at the Universidad Autónoma de Occidente (UAO). Finally, we present some conclusions and propose future activities.

II. Veterinary ambulation devices.

Animal limbs are subject to the same diseases and accidents that affect humans, with the same consequences of reduced functionality and locomotion. For wild animals, this type of dysfunction makes them more likely to die victims of predators and, for centuries, domestic animals suffered a similar fate, since in most cases euthanasia was the only remedy offered to their situation. However, in recent decades, the interest in animal welfare and animal rights has been constantly increasing, and this has led the owners of pets that have suffered a crippling injury to find solutions that improve their quality of life. The types of devices developed for this purpose are [3]:

Boots: Used to protect the animal's pads and feet from abrasion, and to provide support.

I. Torres et al. (eds.), *VII Latin American Congress on Biomedical Engineering CLAIB 2016, Bucaramanga, Santander, Colombia, October 26th -28th, 2016*,
IFMBE Proceedings 60,
DOI: 10.1007/978-981-10-4086-3_66

Prosthesis: Replace lost limbs fully or partially, helping to regain some level of functionality and mobility

Orthotics: Used to provide support and help in locomotion where the individual has not lost his limb, but its functionality has been altered.

Straps, bands and slings: Used to help locomotion, by limiting the movement of the affected limbs.

Carts: Also called veterinary wheelchairs. They allow the animal to move around and provide stability to the body and support to the affected limbs.

Some of these devices can be seen in Fig. 1.

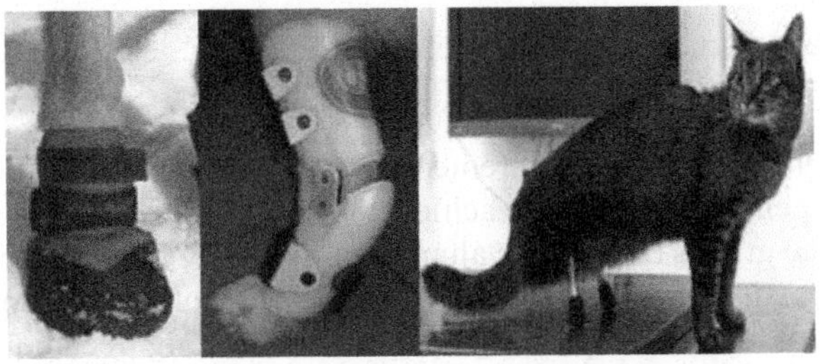

Fig. 1. Examples of veterinary ambulation devices. Boots that allow dogs to walk on snow (left) [4], ankle orthosis for a dog (center) [5] and hindlimb prosthesis on a cat (right) [6]

III. The maker movement

The maker movement describes a community of individuals who promote the "do it yourself" (DIY) and tinkering philosophy, encouraging active learning and free experimentation when building objects for purposes of entertainment, science, or art. This concept was popularized around 2005 and 2006 with the publication of first issue of the "Make" magazine [7] and the organization of the first "Maker Faire" festival [8], which opened a space for sharing and collaboration between large groups of individuals enthusiastic to develop themselves the products they needed or wanted. Some of the principles that characterize the maker movement involve sharing your work and knowledge, learn constantly and having fun in the process of making [9].

In recent years, a lot of manufacturing technologies, such as 3D printing or laser cutting have reduced their costs, becoming affordable to a large number of people. This and the spread of social networks to share experiences and knowledge have led to an accelerated growth of the interest of people to become makers themselves rather than just being users. Creativity in product development has gone from being a property of large corporations to be a common good for all people.

Engineering education has also been revolutionized by this way of thinking, and it has seen increased efforts to develop innovative and problem-solving skills in students [10] by designing and manufacturing devices. Many programs are integrating design experiences at an early stage in their program, and everyday more universities build their own Fablabs where students can turn their ideas into tangible innovations.

IV. A first experience for biomedical engineering students

A. Complementary Activity

In the Faculty of Engineering at UAO there are ten undergraduate degree programs, of which nine share a similar course structure during the basic stage (1st-6th semester). Within this stage, between fourth and fifth semester, the Complementary Activity course has been included. The purpose of this course is to provide engineers in training the opportunity to complement their education with activities that enhance skills such as teamwork, independent learning, critical thinking, management skills, leadership and creativity. These activities are freely chosen by students from a wide variety presented by teachers of different programs and carried out independently. The activities are supervised by a team of tutors, who monitor and evaluate the goals, outcomes and products of learning, whereas a team of mentors carry out the assessment and thematic guidance [11].

In the first semester of 2016, the complementary activity "Design of veterinary prosthesis" was offered. This activity sought to develop a sense of belonging and understanding of other animal species, using engineering as a tool to improve the quality of life of these species. Inspired by the maker movement, the idea was that students would help animals with disabilities using open technologies and inexpensive materials.

The two objectives of the activity were:

1. Find cases of animals with deformities or amputations in which it was feasible to develop a low cost ambulation device.

2. Design and build the device using CAD techniques, 3D printing technologies or other low cost manufacturing techniques and materials.

The activity was carried out by seven groups of two or three students each. Each group worked on one case. In addition to promoting awareness of the needs of other species, the project sought that students improved their skills to work in a group, learn to use different computer design software and become familiar with basic principles of biomechanics.

Table 1 Cases treated

Case	Animal	Type of problem	Place	Solution	Materials
A	Dog	Paralysis	Right Forelimb	Orthosis	Fiberglass
B	Dog	Paralysis	Both Hindlimbs	Wheelchair	Polypropylene, fabric
C	Dog	Paralysis	Both Hindlimbs	Wheelchair	PVC, fabric
D	Dog	Amputation	Left Hindlimb	Prosthesis	ABS, Aluminium
E	Dog	Amputation	Left Hindlimb	Prosthesis	Polypropylene
F	Cat	Amputation	Right Hindlimb	Prosthesis	MDF, polypropylene, fiberglass
G	Dog	Amputation	Right Hindlimb	Prosthesis	Aluminium, fabric

B. Results

Seven cases were selected: six dogs and one cat. Of these, three had paralyses and the rest were amputees. In six cases, the problem was presented in the hind limbs. The list of cases is presented in the Table 1.

The design process was based on a simplified version of the methodology proposed in [12], consisting of the following steps:

1. Identification of user needs
2. Establishing target specifications
3. Concept generation
4. Concept selection
5. Design
6. Prototyping
7. Testing and final adjustments

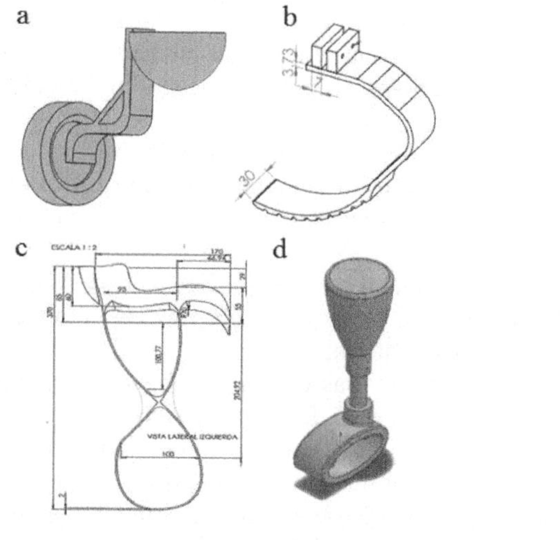

Fig. 2. Preliminary designs of the devices built a) 3D model of a prosthesis for case E. b) Trimetric view of prosthesis for case F. c) Lateral view of orthosis for case A. d) 3D model of prosthesis for case D.

One of the key requirements was that the materials were inexpensive and easy to achieve. Among the materials used by the groups there were: aluminum, ABS, PVC, polypropylene, MDF, fiberglass, and different types of fabric. Some groups used 3D printing and thermoforming techniques, while others did all the manufacturing by hand. Most of the students used the equipment available at the UAO Fab-lab. The devices built were two wheelchairs, four prostheses and one orthosis. Some examples of preliminary designs can be seen in Fig. 2.

The success was variable; the most satisfactory results were obtained with wheelchairs, which had greater stability and ease of adaptation. The orthoses and three of the prostheses had good adaptation by the animal, although some improvements should be made in the stability and flexibility. The most difficult case of adaptation was the cat, probably due to the complexity of its movement patterns, when compared to dogs. Fig. 3 shows some of the animals using the devices.

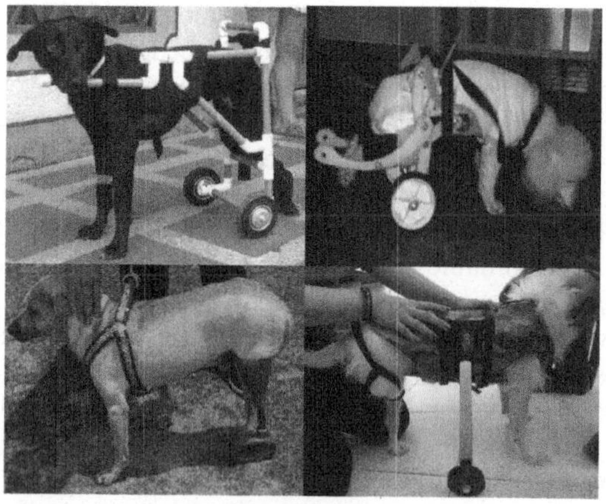

Fig. 3 Final results in users of cases C (upper left), B (upper, right), D (bottom left) and G (bottom right)

The final results were presented to all other participating students from different complementary activities. The activity was well received by students and teachers alike, and several participants were interested in participating of this project. The dog owners were satisfied with the results, although in some cases it is necessary to implement further improvements to enhance the functionality of the device.

V. CONCLUSIONS

Everyday more people become aware of the importance of treating well other species and take care of the world that we inhabit. Biomedical engineers play a huge role because their knowledge can make use of the constantly evolving technology to improve the well-being of humans and other animal species. This kind of activities promotes a new way of seeing the world, in which students learn the importance of creating things to help others. The good results achieved during this project motivate us to continue. Some improvements we plan for the future include developing course material of animal biomechanics and animal orthotics, creating a database of the cases treated to follow up their progress, improving the designs to make them more robust and promoting a network of cooperation with animal shelters and other interested institutions.

ACKNOWLEDGMENT

Thanks to the owners of the animals, the tutors and professors who provided their guidance during the activity and especially to all the students who participated in this experience: D. Restrepo, M. Sanchez, A. Hurtado, L. Ospina, J. Escobar, A. Cachimbo, L. Gomez, J. Vergara, C. Thorp E. Cabrera, O. Jimenez, G. Garcia, M. Ramirez, R. Peña, D. Marin, M. Rodriguez.

REFERENCES

1. Patrice M, The Emerging Role of Veterinary Orthotics and Prosthetics (V-OP) in Small Animal Rehabilitation and Pain Management, Topics in Companion Animal Medicine, Volume 29, Issue 1, March 2014, Pages 10-19, ISSN 1938-9736, http://dx.doi.org/10.1053/j.tcam.2014.04.002.
2. Menichinelli M, "Business Models for Fab Labs" at http://www.openp2pdesign.org/2011/fabbing/business-models-for-fab-labs/.
3. Denis J. Marcellin-Little and David Levine, 17 - Devices for Ambulation Assistance in Companion Animals, In Canine Rehabilitation and Physical Therapy (Second Edition), W.B. Saunders, St. Louis, 2014, Pages 305-311, ISBN 9781437703092, http://dx.doi.org/10.1016/B978-1-4377-0309-2.00017-X.
4. Ultra paws durable dog boots. At http://www.ultrapaws.com/ ultra-paws-durable-dog-boots/
5. Hock (ankle) orthosis. At http://orthodesign.ca/photos.html
6. Scott C, Vincent the Cat Gets New Legs Thanks to Iowa State University and BioMedtrix at https://3dprint.com/ 109075/ vincent-the-cat-prosthetic/
7. MAKE. vol. 1, February 2005. O'Reilly Media Inc.,
8. Maker Faire, a bit of history. At http://makerfaire.com/ makerfairehistory/
9. Hatch, M. (2014). The maker movement manifesto: Rules for innovation in the new world of crafters, hackers, and tinkerers.
10. M. M. Hynes and W. J. Hynes, "Appearances matter: The perceived complexity, coherence, mystery, and legibility of makerspace labs," 2014 IEEE Frontiers in Education Conference (FIE) Proceedings, Madrid, 2014, pp. 1-7. doi: 10.1109/FIE.2014.7044473
11. Facultad de Ingeniería, UAO. Actividad Complementaria, instructivo AC-02: Orientaciones para mentores(as). 2016
12. Ulrich, K. T., & Eppinger, S. D. (2012). Product design and development. New York: McGraw-Hill/Irwin.

Corresponding author:

Author: Andrés Mauricio González Vargas
Institute: Universidad Autónoma de Occidente
Street: Cll 25#115-85 Km 2 Vía Cali-Jamundi, Postal Code: 760030,
City: Cali
Country: Colombia
Email: amgonzalezv@uao.edu.co

Platonic Tensegrities: dynamic aspects and characterization

C. Castro Arenas[1,2], I. Ghersi[1,3], P. D. Borsoi[1], M. Miralles[1,3]

[1] Pontificia Universidad Católica Argentina/Facultad de Ciencias Fisicomatemáticas e Ingeniería. Laboratorio de Biomecánica e Ingeniería para la Salud. Ciudad Autónoma de Buenos Aires, Argentina.
[2] Universidad de Buenos Aires/Facultad de Arquitectura, Diseño y Urbanismo. Instituto de la espacialidad Humana, Centro Laboratorio de Morfología. Ciudad Autónoma de Buenos Aires, Argentina.
[3] Universidad de Buenos Aires/Facultad de Arquitectura, Diseño y Urbanismo. Centro de Investigación en Diseño Industrial de Productos Complejos. Ciudad Autónoma de Buenos Aires, Argentina.

Abstract— **Biotensegrity is the natural extension of the tensegrity principles, developed by Fuller in the early 60's, into the biologic field. With the singular properties of these structures, these principles have gained renewed attention in the scientific and technologic fields, both as models for theorizing the behavior of biological structures at different scales, and as useful tools in a wide variety of applications, ranging from the matrix biomaterials to extreme bionics. Therefore, understanding the dynamic behavior of these structures becomes a fertile field, enabler of future applications in the field of Bioengineering. Platonic Solids represent recurrent morphologies in nature, with well-studied symmetries. This work is part of a broad research dedicated to the study of Biotensegrity. Among the Platonic Solids, the Tetrahedron, Octahedron and Icosahedron were translated into tensegrities (Platonic Tensegrities), using the Rot-Umbela Manipulation and Symmetry methods. In this work, we present the results obtained after instrumentation and processing of the aqcired signals from homologues and non-homologues nodes of these Platonic Tensegrities, when subjected to vibrations in the vertical axis, between 1.0 Hz and 60.0 Hz. These results indicate a close correspondence between Tetrahedral and Octahedral Tensegrities, which present resonant frequencies in the ranges of (8.0 ± 0.1) Hz; (15.0± 0.1) Hz to (21.0 ± 0.1) Hz and (40.0 ± 0.1) Hz to (45.0 ± 0.1) Hz. The Icosahedral Tensegrity presents more resonant frequencies compared to the previous two Tensegrities: (9.0 ± 0.1) Hz; (13.0 ± 0.1) Hz to (14.0 ± 0.1) Hz; (26.0 ± 0.1) Hz to (27.0 ± 0.1) Hz; (30.0 ± 0.1) Hz, to (35.0 ± 0.1) Hz and (41.0 ± 0.1) Hz to (42.0 ± 0.1) Hz. These differences are discussed from the perspective of symmetry groups associated to each of the Platonic Solids.**

Keywords— **Tensegrity, Biotensegrity, Platonic Solids, Vibration.**

I. Introduction

Tensegrity is a principle of structural relationship, as described by Fuller in 1961 [1]. This principle applies both to static and dynamic systems of different types, achieving the balance of whole structures through the interaction of continuous tension and discontinuous compression forces, respectively materialized in tensors and compressors of different materials and forms [2, 3, 4, 5]. As a result of these interactions, tensegrity structures have specific properties, among which self-stability, lightness and adaptability, are some of the most relevant to the field of bioengineering [6]. Since the 80's, this principle has been used in modeling of biological structures at different scales. Molecules, viruses, tissues and even some aspects of the complex human biomechanics (neuro-muscle-skeletal system) are some examples of these models. These findings have brought forward the field of Biotensegrity [7] which, together with the development of new materials and a biomimetic approach, have revolutionized design strategies, leading to innovative products [8].

Platonic Solids are convex polyhedra whose faces are equal, regular polygons, and for which each vertex anchors the same number of faces, with many derived properties, which are further discussed in [9]. The recurrence of platonic morphologies in both organic and inorganic natural systems [10,18], evidences a strategy of material and energy optimization featured in the nature. The possible conformation of "basic pieces" (all configurable from triangles), with simple patterns and various and numerous spatial symmetries, favors self-assembly and easy integration to other structures [4, 19, 20].

In this paper, we present the frequency response of three out of the five Platonic Solids, modeled as tensegrities (Platonic Tensegrities), namely the Tetrahedron, the Octahedron and the Icosahedron. The selection of these Platonic Solids, is based on the fact that all of them have equilateral triangular faces (the Tetrahedron has 4, the Octahedron has 8 and the Icosahedron has 20), resulting in

I. Torres et al. (eds.), *VII Latin American Congress on Biomedical Engineering CLAIB 2016, Bucaramanga, Santander, Colombia, October 26th -28th, 2016*,
IFMBE Proceedings 60,
DOI: 10.1007/978-981-10-4086-3_67

structures displaying maximum stability with minimal components. Triangular patterns are synonymous of equilibrium of forces and maximum strength. From an atomic viewpoint, a triangle can be determined from the centers of three packed spheres (atoms). Their successive stacking give place to crystalline systems (Face-Centered Cubic or Hexagonal Close-Packed), according to the order of successive atomic stacking. In the field of biology, the generation of icosahedral capsids of some viruses, whose triangular faces are conformed by capsomers [17], are paradigmatic examples of Platonic Solids.

Within Tensegrity systems, Platonic Solids can be classified as "cells", because they are among the first instance of a tridimensional configuration in this system, whose grouping can generate structures in 1, 2 or 3 dimensions, according to the spatial relationship that dominates the guidelines or groupings [21, 22]. Research on these cells is of interest as a novel basis to the study of the structural behavior of Tensegrity and Biotensegrity tissues of different spatial dimensions.

II. MATERIALS AND METHODS

A. Platonic Tensegrity form-finding

A form-finding method is a methodology applied to find stable forms corresponding to some specific known structural system. In the case of Tensegrity structures, several methods with different approaches have been developed for this purpose, including mathematical, geometric, kinetic and algorithmic methods, among others [23].

In order to transform the Tetrahedron and the Octahedron to Platonic Tensegrities, the Rot-Umbela Manipulations method, developed by Gomez-Jauregui et al [24], was applied. For the case of the Icosahedron, we have opted towards a geometric method that favors symmetry. In this case, compression members (i.e. aluminum bars) are placed in symmetrical pairs inside the polyhedron, parallel to the cartesian axes. This configuration ensures structural and morphological stability and optimization of materials, which is the reason for which it is of interest in the development of robotic systems used in space exploration [25, 26]. This configuration has also been used to simulate the behavior of biological systems at micro [19, 27, 28] and macro scales [29, 30].

The presented methods allow reinterpreting any polyhedral configuration, and even the creation of Double-Layer Tensegrity Grids (DLTGs). The latter have been of interest, mainly for engineering, architecture and other fields related to these. A similar behavior and configuration is also observed in the external wall of algae colonies called Volvox, and this correlation is being used in the field of robotics [31].

B. Construction of Platonic Tensegrities

In order to manufacture the three Platonic tensegrities shown in Figure 1 (below), hollow aluminum bars were used as compressors, and springs were used as tensors. Aluminum bars with a thickness of (1.0 ± 0.1) mm, a length of (200.0 ± 0.1) mm and an external diameter of (6.0 ± 0.1) mm were used. The tensor springs have a diameter of (4.0 ± 0.1) mm, and were used in three different lengths in order to accommodate to the geometry of each solid: (35.0 ± 0.1) mm, (60.0 ± 0.1) mm and (85.0 ± 0.1) mm. As a result, their combination resulted in variable weights for each model: Tetrahedron (101.7 ± 0.1) g; Octaedron (206.3 ± 0.1) g; Icosahedron (113.3 ± 0.1) g. Figure 1 shows the three selected Platonic Solids and their respective tensegrities.

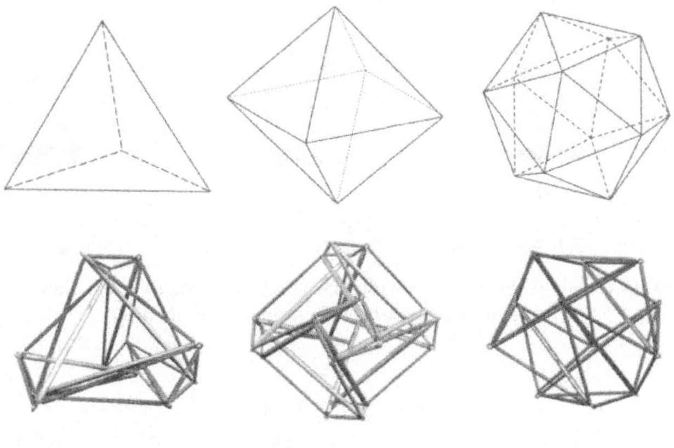

a) b) c)

Fig. 1 Platonic Solids (top) and Platonic Tensegrities (bottom) a) Tetrahedron, b) Octahedron, c) Icosahedron respectively

C. Study method of vibrational modes for Tensegrity structures

In order to assess the dynamic properties of these particular structures, they were instrumented with LED markers (in this case, positioned at the ends of the compressors, representing "nodes" of interest), and subjected to controlled mechanical oscillations in the vertical axis. These oscillations are adjustable in amplitude and frequency, and the acquisition of the response of these nodes across a frequency sweep (fixed amplitude), allowed determining the frequency-response for each node, characterizing the structure.

The details of the applied method, developed for the study of the dynamic behavior of tensegrities, is described in [32]. For this study, a new function was included in the acquisition algorithm, which allows calculating the amplitude of the motion of markers, independently of its oscillation direction, more directly responding to the needs of this work. Signal analysis was performed through a dedicated algorithm designed for the study of oscillations of the markers, as obtained from video files, captured with high-speed cameras at 120 fps (LED marker segmentation and centroid-estimation). Figure 2 shows the three models, instrumented with LED markers in homologous and non-homologous nodes. The vibration frequencies to which the structures were subjected in the vertical axis, were in the range of 1.0 Hz to 60.0 Hz. One additional node in the base of the oscillator was used as reference, detecting the motion of the base.

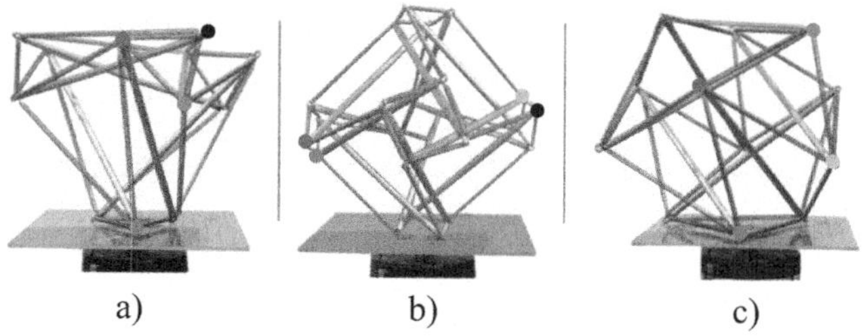

Fig. 2 Platonic tensegrities with instrumented nodes over the plattform: a) Tetrahedron, b) Octahedron, c) Icosahedron.

III. RESULTS

Vibration-mode results for the detailed structures, obtained through the video-processing algorithm, are presented in Table 1. Standout peak frequency values, for each node, are presented in succession in the third column.

Results from Table 1 indicated a close correspondence between Tetrahedral and Octahedral Tensegrities. Both structures present the main resonant frequencies for all studied homologous nodes (N1, N3, N4), for the values and ranges of (8.0 ± 0.1) Hz; (15.0 ± 0.1) Hz to (21.0± 0.1) Hz and (40.0 ± 0.1) Hz to (45.0 ± 0.1) Hz. Both structures exhibit greater stiffness than the Icosahedral tensegrity, due to the greater amount of nodes (compressors) corresponding to each vertex. Similarly, this comparative response stands out since it corresponds a two structures with a greatly different number of components and weights.

The Icosahedral Tensegrity, however, is more complex, presenting a larger number of standout frequencies (≥ 4 peak values), with intermediate peak frequencies in addition to the ones covered by the aforementioned ranges: (9.0 ± 0.1) Hz; (13.0 ± 0.1) Hz to (14.0 ± 0.1) Hz; (26.0 ± 0.1) Hz to (27.0 ± 0.1) Hz; (30.0 ± 0.1) Hz to (35.0 ± 0.1) Hz; (41.0 ± 0.1) Hz to (42.0 ± 0.1) Hz.

This observation can be explained by the different symmetry groups that apply to each Platonic Solid: Tetrahedron are congruent to A4 (set of permutations, combined with the identity) and S4 (group of rotations and reflections); Octahedron is S4, but the rotational symmetries of Icosahedron are isomorphic to A5, as defined by [33, 34].

Table1: Vibrational modes of Platonic Tensegrities.

Polyhedron	Node	Frecuency (± 0.1) Hz
Tetrahedron Amplitude 50%	N1	8.0, 15.0, 40.0
	N3	8.0, 15.0, 40.0
	N4	8.0, 15.0, 40.0
Octahedron Amplitude 50%	N1	8.0, 21.0, 45.0
	N2	8.0, 22.0, 43.0
	N4	8.0, 22.0, 42.0
	N5	8.0, 22.0, 44.0
Icosahedron Amplitude 50%	N1	9.0, 27.0, 30.0, 41.0
	N3	9.0, 13.0, 26.0, 35.0, 42.0
	N4	9.0, 13.0, 26.0, 34.0, 41.0

IV. CONCLUSION

In this paper, an approach to the dynamic aspects of three Platonic Tensegrities was presented, obtaining:

a) Equivalency between Tetrahedron and Octahedron tensegrities for the specified parameters of construction: vibration frequencies are clearly measurable and similarly fall between the specified ranges.
b) The Icosahedral Tensegrity modes shows a higher number of standout vibrational frequencies, with an overall differentiated response. The geometry of this construction has different symmetry features to the previous ones.

Based on these results, a model that predicts the response of certain tensegrity structures based on their geometric features can be conceived. Just like accepted by the modern theoretical physics: *symmetry defines structure*.

Future work includes the study of the two remaining Platonic Solids, not covered by this paper (Cube and Dodecahedron), and the further study of transformations of Platonic Tensegrities and their effects on their symmetry groups, in light of the developed method of analysis.

Acknowledgment

This work was supported in part by Universidad de Buenos Aires, Programación Científica 2013-2016, Proyecto Interdisciplinario PIUBAMAS 20621201000011BA: "Evaluación del riesgo de caídas en adultos mayores y en pacientes geriátricos".

This work was supported in part by Pontificia Universidad Católica Argentina, Facultad de Ciencias Fisicomatemáticas e Ingeniería: "Modelos biomecánicos y desarrollos de dispositivos para la cuantificación del movimiento humano en individuos normales y/o disfuncionales".

Authors would like to thank Gonzalo Fernández for his insight and assistance during the setup of the device.

Conflict of Interest

The authors declare that they have no conflict of interest.

References

1. Fuller B, Applewhite E, (1975, 1979) Synergetics: Explorations in the Geometry of Thinking. Macmillan Publishing Co. Inc.
2. Burkhardt R, (2008) A Practical Guide to Tensegrity Design. 2nd edit. Cambridge, MA
3. Motro R. (2003). Tensegrity: Structural Systems for the Future. London and Sterling, VA: Krogan Page Science.
4. Fuller B, (1961) Tensegrity, Portfolio and Art News Annual,No. 4 pp. 112-127, 144, 148.
5. Gómez-Jáuregui V, (2010) Tensegrity structures and their application to architecture, Universidad de Cantabria. Servicio de Publicaciones, Santander.
6. Cretu S-M, (2009) Tensegrity as a Structural Framework in Life Sciences and Bioengineering, In: Modeling, Simulation and Control of Nonlinear Engineering Dynamical Systems. State-of-the-Art, Perspectives and Applications, Ed. J. Awrejcewicz, published by Springer-Verlag, Heidelberg, ISBN978-1-4020-8777-6: 301-312.
7. Randel L, Swanson I, (2013) Biotensegrity: A Unifying Theory of Biological Architecture With Applications to Osteopathic Practice,Education, and Research—A Review and Analysis. J Am Osteopath Assoc. 113(1):34-52
8. Ananthanarayanan A, Azadi M, and S Kim (2012)Towards a bio-inspired leg design for high-speed running, Bioins.&Biom., v 7 (4) DOI 10.1088/1748-3182/7/4/046005
9. Sutton D (2002) Platonic and Archimedean Solids. Walker Books, London
10. Levin S (1981)The Icosahedron as a Biologic Support System. Proc.vol. 23, 34th An. Conf. on Engin.in Med. and Bio., Huston. 1981, p 404.
11. Levin S. (1990) Primordial Structure. Proc. vol 2, 34th An.Meet. Int. Soc. Syst. Sci., Portland.
12. Morgan G (2010) Laws of biological design: A reply to John Beatty. Biol. Philos, 25:379–389DOI 10.1007/s10539-009-9181-y
13. Wells, A(1969) The geometrical basis of crystal chemistry. X. Further study of three-dimensional polyhedral. ActaCryst. B25: 1711-1719. DOI:10.1107/S0567740869004626
14. Twarock R (2006) Mathematical virology: a novel approach to the structure and assembly of viruses. Phil. Trans. R. Soc. A. 364, 3357–3373 DOI:10.1098/rsta.2006.1900.
15. Morgan G (2006)Why there was a useful plausible analogy between geodesic domes and spherical viruses. Hist. Phil. Life Sci., 28 (2): 215-236.
16. Johnson J, Speir J (1997) Quasi-equivalent Viruses: A Paradigm for Protein Assemblies. J. Mol. Biol., 269: 665-675.
17. Caspar D, Klug A (1962) Physical principles in the construction of regular viruses. Cold Sp. Symp. Quant. Biol., 27: 1 – 24. DOI 10.1101/sqb.1962.027.001.005.
18. Caspar D, Klug A (1963) Structure and assembly of regular virus particles. Virus, nucleic acids and cancer. Williams and Wilkins, Baltimore, Maryland, USA, 2-39
19. Ingber D, (1993) Cellular Tensegrity: Defining New Rules of Biological Design That Govern the Cytoskeleton. Journal of Cell Science, Vol. 104, No. 3: 613–627
20. Pearce P (1978) Structure in Nature Is a Strategy for Design. MIT Press, Cambridge.
21. Reissig P (2012) Tecnomorfología como estrategia de diseño: entre la teoría y la práctica.PhD Thesis FADU – Universidad de Buenos Aires.
22. Motro R (2009) Structural morphology of tensegrity systems, As J. Civ. Eng. (build. and hous.) 10 (1): 1-19
23. Tibert A, Pellegrino S (2003) Review of form-finding methods for tensegrity structures. Roy.Ins.Tech., Dep. Str.Eng. Cambridge, UK
24. Gómez-Jáuregui V (2012) Mallas tensegríticas de doble capa y manipulaciones de Rot-Umbela. Inf. Const. 64 (527) : 331-344. DOI:10.3989/ic.11.053
25. Andrew P. et al (2015) System Design and Locomotion of SUPERball, an Untethered Tensegrity Robot. Proc. Int. Conf. Rob.Aut. ICRA2015. Seattle, Washington
26. Agogino A, SunSpiral V, Atkinson D (2013)Super Ball Bot - Structures for Planetary Landing and Exploration - Final Report. NASA Innovative Advanced Concepts (NIAC) Program. NASA Ames Research Center. Intelligent Systems Division.
27. Ingber D (1998) The Architecture of the Life. Sci. Am. Mag. 278: 48-57.
28. Ingber D (2003) Tensegrity I, Tensegrity II, J. Cell Sci. 116: 1157-1173, 1397-1408.
29. Ingber D (1995) The importance of soft tissues for structural support of the body at http://goo.gl/bxO4ex
30. Ingber D (1997) Putting the shoulder to the wheel: a new biomechanical model for the shoulder at http://goo.gl/avpO21
31. Yu C-H, Haller K, Ingber D, Nagpal R (2008) Morpho: A Self-deformable Modular Robot Inspired by Cellular Structure. IEEE/RSJ Proc. Int. Conf. Intell. Robots Syst, 2008, pp. 3571-3578
32. Castro Arenas C, Ghersi I, Miralles M (2016) Biomechanics and Biotensegrity: Study Method and Frequency Response of the Simplex and 3-bar-SVD Tensegrity Configurations.IOP Proc. J. Phy: Conf. S., vol. 705. SABI 2015. San Nicolás, Argentina.
33. Wo W (2004) The platonic solids. Berkeley, at https://goo.gl/PwmqOR
34. Lim Y, (2008) Symmetry Groups of Platonic Solids at http://goo.gl/BmOXlt

Corresponding author:	Cristhian Castro Arenas
Institute:	LaBIS-FI-UCA
Street:	1500 Moreau de Justo Ave.
City:	Buenos Aires
Country:	Argentina
e-mail:	ccastroarenas@gmail.com

Wiimote-based Infrared System for Instrumental Tracking in Minimally Invasive Spine Procedures for Training

M. Domínguez[1], D. Lorias[2] and R. Martínez[3]

[1,2] CINVESTAV/Departamento de Ingeniería Eléctrica, Sección Bioelectrónica, Ciudad de México, México
[3] UAEM/Facultad de Ingeniería, Toluca, Estado de México, México

Abstract— **In this paper we present a Wiimote-based infrared tracking system to estimate the position of a single infrared marker placed in surgical instruments used for spine fixation procedures in an image-guided simulation system for training. The purpose of our research is the objective assessment of surgical skills and abilities during the simulation-based approach. A stereovision setup is used for calculating the position of the instrumental. In order to implement the triangulation algorithms, the horizontal and vertical fields of view need to be known, lens distortion is not taken into consideration due that the camera represents a linear homogenous system in this project. Moreover the tracking system has been integrated into a graphic user interface that allows the visualization of the spine anatomy in Postero-anterior, Lateral and Axial planes for visual feedback in pedicle screw insertion accuracy task.**

Keywords— **Infrared Tracking, Training System, Wii remote, Stereovision, Minimally Invasive Spine Procedures.**

I. Introduction

The most common fractures of the spine occur in the thoracic and lumbar spine or at the connection of the two. A severe fracture may require a pedicle screw insertion surgery to return the bones to their normal position, but it may be associated with complications such as screw malpositioning [1]. Due to the impact of misaligning one or more pedicle screws that can directly influence patient safety, learning and development of skills and abilities for spine fixation procedures are extremely important to properly perform a satisfactory surgery, for this reason, it is important to effectively learn these complex techniques before clinical exposure. Simulation-based education plays a great role for these purposes. The Simulation-training field in medicine has evolved rapidly over the past three decades and training systems have been developed to help medical students to acquire experience through them avoiding practicing with patients. There's also emerging evidence that skills learned in the simulation environment carry over to improved clinical performance thus, learning a practical skill becomes easier with time, with an initial period of difficulty followed by a rate of improvement and stabilization in performance [2].

In order to evaluate the learning process and development of surgeon's abilities, a system for tracking the kinematics of the instrumental it's needed. Different solutions have been employed in tracking systems. A first solution consists in using visible markers along with cameras. Such a solution may still be complex and computer demanding for real time tracking. A solution based on infrared (IR) tracking allows capturing only the markers without the remaining environment and simplifies target localization. Passive or active marker technologies can either be used for such a purpose. In the active solution, IR sources are directly placed on the body to be tracked, while in the passive solution the body is equipped with reflective markers to be illuminated by IR LEDs [3]. For this reason, we propose an active Infrared (IR) tracking system using Nintendo Wii remotes. By using multiple Wiimotes we can apply stereovision techniques to acquire 3D tracking data from individual IR sources [4]. Each Wii remote has an IR camera sensor, which provides tracking of up to four simultaneous IR light sources. This camera has a built-in image processor that can analyze the raw camera image, identify bright spot and compute their positions (x, y) in the camera's image plane and transmit the data to the host computer through Bluetooth [5].

II. Materials and Methods

A. General system setup

There are many possible setups for calculating the position in space of a marker. An orthogonal arrangement with two Wiimotes can be used. This setup has a small coverage area, and the degrees of freedom that handle the instrument for the correct placement of the screws inside the pedicle, are within the field of view. Nonetheless, a stereovision setup has a greater coverage area and a wider range of angles to be seen by the Wiimotes. For this reason, a setup presented in Fig. 1 has been used. In this setup two IR cameras located at a fixed distance between them and at a fixed distance between the Wiimotes and the synthetic model of the spine, moreover, one IR source is needed to calculate the position of the marker [6,7]. Since the distance between the Wiimotes is measured by hand, and any small error in the positioning of the Wiimotes will introduce an

I. Torres et al. (eds.), *VII Latin American Congress on Biomedical Engineering CLAIB 2016, Bucaramanga, Santander, Colombia, October 26th -28th, 2016*,
IFMBE Proceedings 60,
DOI: 10.1007/978-981-10-4086-3_68

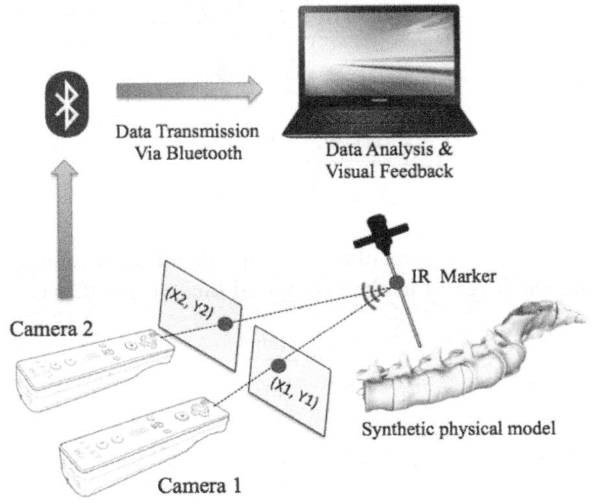

Fig. 1. General setup of the system.

error, it is important to fix the Wiimotes into a device to always keep the same distance between them. Moreover, the Wiimotes must be at a fixed distance from the synthetic physical model of the spine to ensure an accurate calculation of the position of the marker. In this way, the system knows where is the spine in relation with the Wiimotes, and at the same time knows the position of the marker in space. The goal of the system is to improve the skills needed for spine fixation procedures within a free-risk environment using a physical synthetic training model

B. Extrinsic and intrinsic properties of the cameras

These intrinsic parameters are typically stored as a matrix, and include data such as the focal length, image center and size of the image sensor. Typically a distortion skew matrix is also used to account for lens distortion that exists in all lens cameras. Nevertheless, for this project the Wiimote camera is assumed to be a linear homogeneous system, thus some intrinsic parameters are needed. These values are the horizontal and vertical fields of view of the camera. Lourenco [8] and Wronski [9] experimentally determined these properties. They found that viewing angles were 41° horizontally and 31° vertically and the reported x and y coordinates ranges from 0 to 1023 and 0 to 767 respectively.

Different values of these two angles can be found in the literature, in addition they may vary from one Wiimote to another one, for this reason is necessary to estimate the value of the angles corresponding to each device. Experiments have been conducted to estimate the angles of the two Wiimotes used for our system by measuring the changes in the horizontal and vertical viewing ranges and by simply finding the limits of the field of view with IR sources located at different distances from the camera. From these experiments the mean values 0 to 1016 and 0 to 760 for the horizontal and vertical fields of view respectively have been found. Thus, the viewing angles are, 40.6° horizontally and 30.7° vertically. For the aims of this investigation this parameters are used as constants in the programing code as described in the next section.

In order for both cameras to be referenced using the same coordinate system, the extrinsic parameters of the camera system need to be resolved [10]. These include the translation and rotation vectors of one camera in relation to the other. In our model, the cameras are pointed parallel outward and are located a fixed distance apart from each other, resulting in zero effective rotation and a simple translation in one axis.

C. Stereoscopic triangulation: Intersection of two 3D rays.

In this section, estimation of position is described considering that two IR cameras are used and only a single IR source is tracked. The main tool used for the reconstruction of the 3D position is an estimation technique for the near intersection of two rays [11]. If a Wiimote is selected as the reference frame $o_1 = [0,0,0]^T$ then the second Wiimote lies along the x-axis at a known distance d from the reference at $o_2 = [d,0,0]^T$ as manually measured when the system was constructed. In the parallel configuration of Fig. 3 the coordinates of the point to be tracked is $x = [x,y,z]^T$. A vector from the camera position to the detected point can be obtained using the fields of view. These vectors can be viewed as rays from the IR source to the camera; in this way any detected point on a ray will report the same two-dimensional coordinates. Let the two rays r_1 and r_2 be defined as:

$$r_1 = o_1 + t_1 d_1 \tag{1}$$

$$r_2 = o_2 + t_2 d_2 \tag{2}$$

where d_1 and d_2 are direction vectors from each sensor to the source, while t_1 and t_2 are the unknown distance between the source and each Wiimote in these directions. To compute the direction vectors it's necessary to calculate the normalized pixel values for each camera that measures two coordinates in its two-dimensional focal plane denoted (x_1, y_1) and (x_2, y_2).

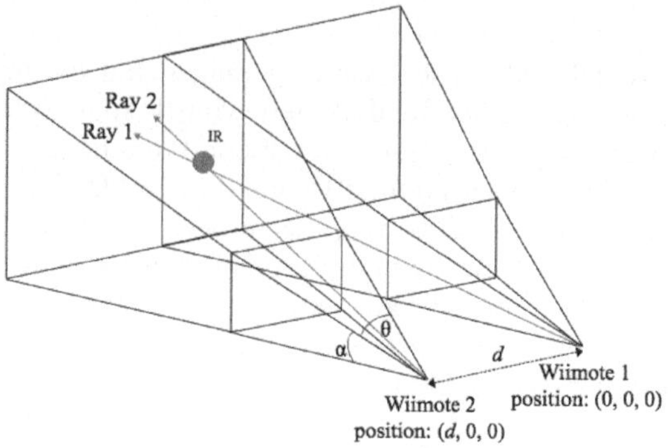

Fig 3. Stereoscopic triangulation. With two parallel cameras two separated rays are constructed from one IR marker.

The resolution of the focal plane is 1016 x 760 for the Wiimotes used in this project, and mapping these coordinates to [-1; 1]:

$$\hat{x}_{1,2} = 2\frac{x_{1,2}}{1016} - 1 \tag{3}$$

$$\hat{y}_{1,2} = 2\frac{y_{1,2}}{760} - 1 \tag{4}$$

This now allows to calculate the direction vectors d_1 and d_2. The Wiimote is assumed to point in the direction $[0,0,-1]^T$, straight down the z-axis, in its own coordinate system. The coordinates of each ray direction are given by

$$d_1 = \begin{bmatrix} \hat{x}_1 \cdot \tan\left(\frac{\alpha}{2}\right) \\ \hat{y}_1 \cdot \tan\left(\frac{\theta}{2}\right) \\ 1 \end{bmatrix}, d_2 = \begin{bmatrix} \hat{x}_2 \cdot \tan\left(\frac{\alpha}{2}\right) \\ \hat{y}_2 \cdot \tan\left(\frac{\theta}{2}\right) \\ 1 \end{bmatrix} \tag{5}$$

where α and θ are the horizontally and vertically fields of view respectively. The remaining two unknown constants are estimations of the closest points on each ray to the other ray. Since this is a physical system, there are inaccuracies in the equipment and measurements, as a consequence, the two rays will never intersect perfectly and the linear system would have no real solution, for this reason we need to compute t_1 and t_2 by equating r_1 and r_2:

$$o_1 + t_1 d_1 = o_2 + t_2 d_2 \tag{6}$$

Computing the cross product of this expression for d_1 and d_2 yields

$$t_1 = \frac{\det(o_2 - o_1, d_2, d_1 \times d_2)}{\left\| d_1 \times d_2 \right\|^2} \tag{7}$$

$$t_2 = \frac{\det(o_2 - o_1, d_1, d_1 \times d_2)}{\left\| d_1 \times d_2 \right\|^2} \tag{8}$$

A final coordinate can be calculated using these lengths along each vector as the average of the two points as:

$$\hat{x} = \frac{r_1 + r_2}{2} \tag{9}$$

D. System coverage

The setup was designed considering the maximization of the area where the IR markers would be detected. It's evident from Fig. 4 that in order to maximize the area, Wiimotes need to be as close together as possible, to allow the user to have a greater space to interact with the system [8]. Fig. 4 shows that for the horizontal field of view the overlapping area where the user can interact with the system at the distance of H from the origin is represented by the length L and width W for the vertical field of view. The length and the width can be calculated using Pythagoras theorem as:

$$L = 2H \tan\frac{\alpha}{2} - d \tag{10}$$

$$W = 2H \tan\frac{\theta}{2} \tag{11}$$

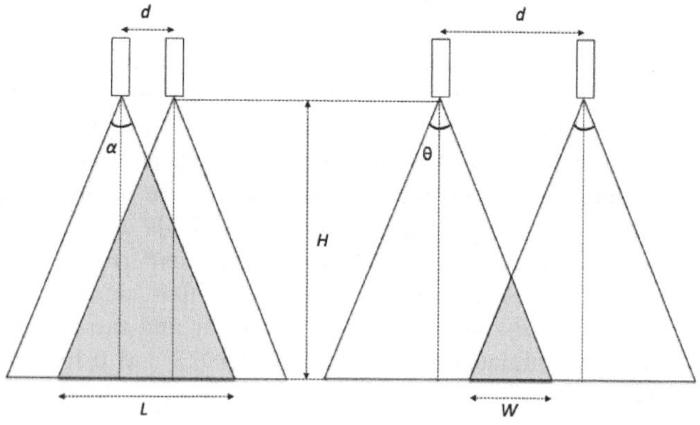

Fig 4. Wiimotes arrangement, blue area is where the vision of both remotes intersects. Left: shows a greater area to interact with the system when the Wiimotes are as close as possible. Right: Shows that with a greater distance between the Wiimotes the area to interact with the system is smaller.

However, the nearest the Wiimotes, any small error in measuring the distances will give as a result an inaccurate position of the marker.

III. Preliminary results

Acquiring information for the means of this system has been achieved. Two of the most important parameters during spine fixation procedures are: the needle depth and angulation to guide the safe placement of the screw. For this reason an interface able to plot the position of the marker placed on the instrumental was developed in C# in Visual Studio 2015. A trajectory line referenced to the long axis of the probe is superimposed on the Postero-anterior, Lateral and Axial views as shown in Fig. 5. As the probe is moved through the pedicle, the position of the trajectory line will change accordingly. The user interface displays the coordinates of the marker as well as the needle depth inside the pedicle. In addition, lateral and posterior angle measurements needed to guide the placement of the screws inside the pedicle are shown. These angles may vary between vertebras of the same region. For this reason, the system has prior information of some parameters such as, the angles and needle depth in order to evaluate the correct placement of the screw by comparing these parameters with the parameters measurements during the task. In any case, the degrees of freedom that handle the instrument for a lumbar pedicle fixation are within the field of view of the system.

IV. Discussion

One of the major challenges to effectively teach and learn how to perform complex spine fixation procedures, based on simulation, is to minimize the learning curve of the students, nevertheless is necessary to evaluate the progress throughout registration and comparison of parameters such as the trajectory of the instrumental. In order to prevent errors in the calculation of the position of the instrumental, it is extremely important to maintain the static parameters of the system such as: the distance between Wiimotes and the distance between the Wiimotes and the synthetic physical trainer. For this reason, it was necessary to fix the Wiimotes in a physical device, and to fix the device in a fixed place in the training system. Moreover, to maximize the field of view of the system, the Wiimote should be as close together as possible, however the nearest the Wiimotes, any error in measuring the constants will be maximized. Furthermore, the use of IR markers as well as IR cameras reduced considerably processing time and ambient noise produced by light.

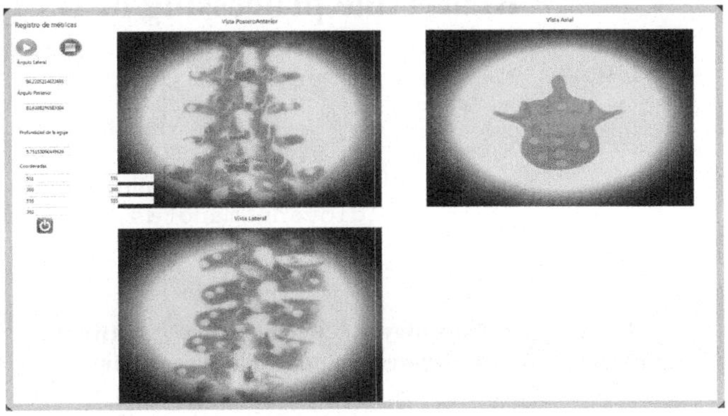

Fig 5. Graphic User Interface.

Conflict of Interest

The authors declare that they have no conflict of interest.

References

1. Manbachi A, Cobbold, Ginsberg. (2014) Guided pedicle screw insertion: techniques and training. Spine J. 14(1): 165-79
2. James D. Michelson, Lance Manning. (2008) Competency assesment in simulation-based procedural education," The American Journal of Surgery, pp. 609–615.
3. Sylvain Bertrabd et al. (2011) A low-cost system for indoor motion tracking of unmanned aerial vehicles. ISSNIP.
4. Johnny Chung Lee. (2008) "Hacking the Nintendo Wii Remote," Pervasive Computing IEEE, vol. 7, pp. 39-54, doi: 10.1109/MPRV.2008.53
5. Wei Wang. (2010) A Case Study of Interactive Teaching and Learning Practices on Information Technology Course in Primary School by Using Low-cost Wiimote Whiteboard.
6. Zhen Duo and Wenhui Zhang, (2012) Constructing 3D Position Tracking System with Twin-Wiimote Structure," International Conference on Information Management, Innovation Management and Industrial Engineering, vol. 2, pp. 466–469, doi: 10.1109/ICIII.2012.6339879
7. T. Petric, A. Gams, A. Ude and L. Zlajpah. (2010) "Real-time 3D marker tracking with a WIIMOTE stereo vision system: aplication to robotic throwing," Robotics in Alpe-Adria-Danube Region (RAAD), 2010 IEEE 19th International Workshop, IEEE Press, pp. 357-362, doi: 10.1109/RAAD.2010.5524560.
8. Joa Lourenco. (2010) "Wii3D: Extending the Nintendo Wii Remote into 3D," Bachelor's Degree, Rhodes University,
9. Michael Wronski. (2008) "Design and Implementation of Hand Tracking Interface using the Nintendo Wii Remote, Master's thesis, University of Cape Town.
10. Z. Zhang. (1999) Flexible camera calibration by viewing a plane from unknown orientations. Technical report, Microsoft Corporation,
11. Eric Haines Tomas Moller and Naty Hoffman. (2008) Realtime Rendering, Third Edition. A.K. Peters, Ltd., Natick, MA, USA, third edition.

Real-Time processing of DVI signals on a FPGA as a Telemedicine modular solution

G.F. Manotas[1], L.A. Rodríguez[1] and J.M. Velandia[1]

[1] Fundación Cardiovascular de Colombia FCV, Bucaramanga, Colombia
{giovannimanotas,leonardorodriguez,josevelandia}@fcv.org

Abstract— **Nowadays, telemedicine networks are implemented with limited compatibility regarding data and signal transmission. This hinders their implementation as each medical-device manufacturer has its own set of transmission systems and methods, frequently unable to talk among themselves. On the other hand, medium tier and higher class medical devices have digital or/and analog video outputs (DVI,HDMI,VGA), which are commonly used for improving visibility of the device on medical institutions by using larger screens. Said devices usually lack of any other form of data output, and those who have it follow their own protocol of medical data transmission (only comprehensible by other devices of the same family, or manufacturer), thus making the use of video signals for the establishment of a telemedicine network a feasible option. An earlier implementation of a video-processing architecture using the Nios II softcore processor, along with custom dedicated processing modules for scaling, color conversion, and compression (M-JPEG) for a 22kLE FPGA has been developed to allow video signals from VSMs with DVI output to be transmitted by WWAN, taking into account the bandwidth limitations of each transmission channel.**

Keywords— **FPGA, DVI, M-JPEG, WWAN, DSP, WWAN, Biomedical telemetry.**

I. Introduction

Nowadays reconfigurable computing is one of the most evolving digital technologies along with the custom-architecture approach, which is designed in order to accelerate certain time-intensive tasks in a video processing system. An embedded system, for which power and size optimization are basic requirements of design, can notably benefit from these approaches resulting in the reduction of system clock speeds, and for the case of Programmable Logic, the implementation of various independent signal-processing subsystems in the same device, being the latter the main focus of this paper.[1]

Performance and reliability on programmable logic devices have seen a significant change since their appearance as commercial (Field Programmable Gate Arrays) FPGAs, noting that previous versions of (Programmable Logic Devices) PLDs, (Programmable Array Logic) PAL and similar devices granted very low amounts of Logic Elements or Cells, (One-Time Programmability) OTP or very high costs and complexity; and thus were not considered to be used as main processing units.

Due to its parallel nature and implementation of almost any digital function, an FPGA device may provide DSP-Like capabilities, only limited by its number of logic elements, with a higher grade of customization in comparison of a DSP dedicated device. This includes but isn't limited to: the instantiation of a complete (System on Chip) SoC or the more appropriate term (System on Programmable Chip) SoPC, integrating multiple CPUs, memory controllers, peripheral interfaces, interconnect logic, and acceleration circuitry.

A SoPC implementation is presented on this paper, in which a NIOS II[2] Soft-Core architecture is programmed on an FPGA as the controller of custom video DSP modules[3], which process video signals from medical devices with DVI output, in order to allow its transmission through WWANs, thus giving them connectivity to telemedicine networks.

II. State of the art

A diagram of the current connectivity architecture in commercial VSMs is shown in Figure 1, where the device has to send video data to screens and others visual devices, as well as processing vital data internally to send tele-monitoring packets through a network device.

Basically each manufacturer uses several propietary protocols or network frames[4]. Even between same manufacturer devices, there are additional differences between frames. Earlier versions of these protocols usually carry more data or are more stable than older ones, however, the coexistance of said versions add complexity and incompatibility to operative networks. This fosters the

I. Torres et al. (eds.), *VII Latin American Congress on Biomedical Engineering CLAIB 2016, Bucaramanga, Santander, Colombia, October 26th -28th, 2016*,
IFMBE Proceedings 60,
DOI: 10.1007/978-981-10-4086-3_69

development of "not so" standard translators of network frames by software or hardware application.[5]

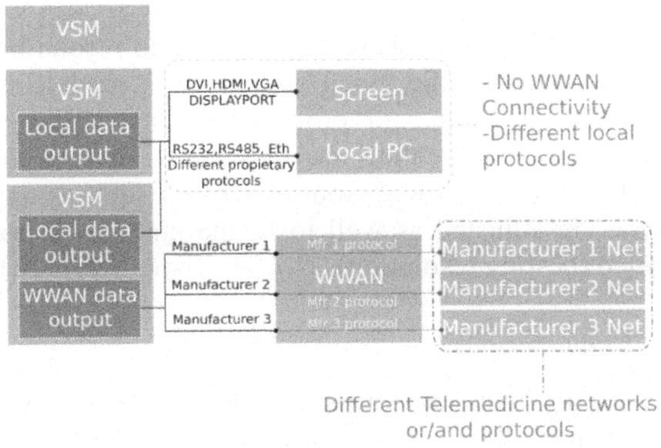

Fig. 1 Diagram of the current connectivity architecture between VSMs.

The implementation of such translations is known to bring unreliability to the device or network, since for these translators to work properly, they must be updated constantly, not even allowing for a proper validation of their stability. Differences of data volume, data synchronism, sample precission, and parameters measured, could also bloat the payload to non-practical dimensions for its use in WWAN networks [6].

The current direction in the VSM market shows a higher degree of integration among devices, proposing local networks interconnected by dedicated servers, and making alliances between manufacturers to extend their compatibility [7]. Legacy devices however aren't able to benefit from this new approaches, and as not all medical institutions have the capability of renewing them, they would stay without connectivity, limiting the expansion of the telemedicine network[8].

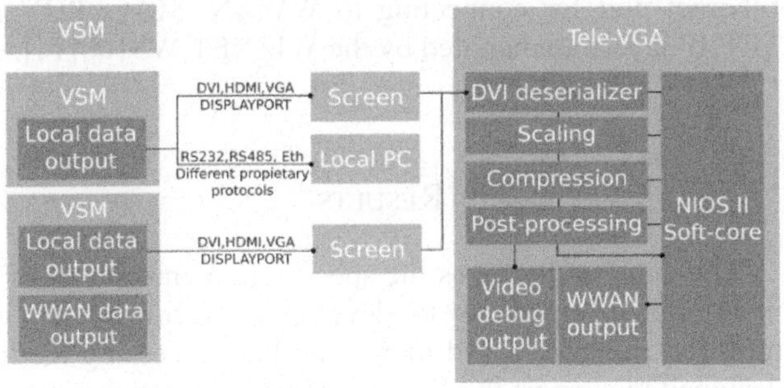

Fig. 2 Proposed architecture for telemedic modular solution

III. ATRCHITECTURE

The proposed architecture takes several common data signals and gives telemedicine capabilities to any device connected. Figure 2, is a teardown representation of the different scenarios (VSMs). This representations also shows the different stages of video adapting for making it suitable for transmision, and how they would connect devices to a telemedic network using the Tele-VGA module based on a 22kLE Cyclone IV FPGA and a 32MB SDRAM.[9]

A. Local data outputs

It is common for Vital Signs Monitor to have local data outputs, in the form of serial data or video data. Serial data is principally used for interfacing with client propietary applications running in PCs, for downloading patient data, parameter registers, and mantainance. Video data is used when connecting VSMs to surgery rooms screens, in order to improve parameter visibility. The use of video signals as telemedic data source allow us to guarantee all relevant data will be available on the RX side of the communication.

B. WWAN data outputs

Representation of all internet connectivity that may be available on the device. It is less common than local connectivity, and uses propietary protocols from end to end[10]. Usually it brings Ethernet connectivity, Wifi, and mobile networks in high-end devices (although they aren't common because of high power consumption, and possible noise emission, which may interfere with other devices[11][12]).

C. Tele-vga

This device takes video signals from any VSM with DVI output, which then is processed and compressed in order to meet WWAN transmission criteria (bandwidth limits) and reasonable Round trip times for telemedic devices. (Previous devices developed at FCV Colombia have shown 8 seconds as a feasible value). All modules are implemented in Verilog, on a 22kLE Altera Cyclone IV FPGA, with a soft-core Nios II Processor running at 120 MHz for control of hardware modules and accelerators for speciphic video calculations.

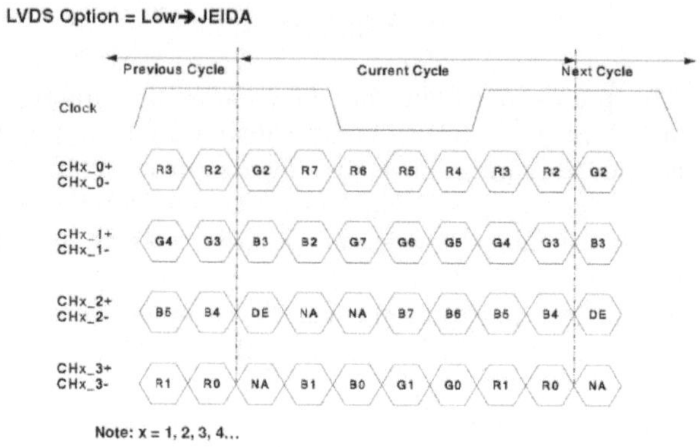

Fig. 3 Serialization of 4 channel video data on DVI and LVDS busses

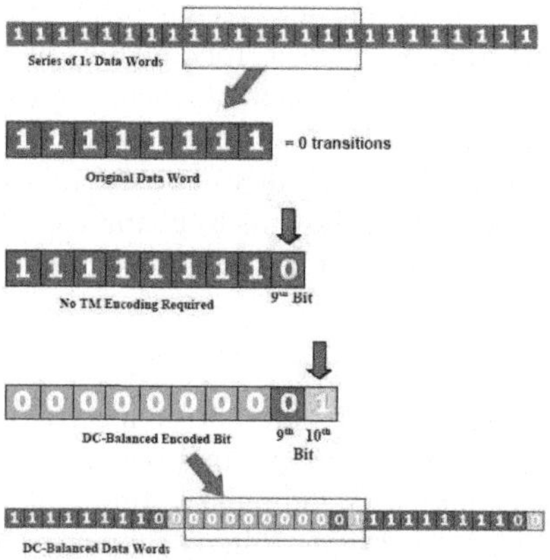

Fig. 4 8b10b encodinf and DC balancing

a) DVI Deserializer:

DVI signals are transmitted using TMDS (Transition minimized differential signaling), which basically is LVDS (Low voltage differential signaling) with an 8b10b coding and bit balancing[13]. This codification allows, by using two aditional bits for each transmitted byte to minimize bit transitions using codewords, thus improving data transmission rates. Figure 3 shows serialized differential data from a parallel video input without coding, Figure 4 shows how stream is 8b10b encoded, and Table 1 the code words for DVI. The deserializer decodes DVI TMDS signals first as 4 serial signals, which are then finally converted to parallel RGB for further manipulation.

Table 1 TMDS Code words

C0	C1	Output codeword
0	0	0010101011
0	1	0010101010
1	0	1101010100
1	1	1101010101

b) Scaler:

Since all VSMs don't share the same video resolution, it is necessary to scale the video stream. This module is in charge of accelerating said process, by interpolating in-between pixels when scaling up, and averaging continuous pixels when scaling down. Control signals such as vsync and hsync are adjusted as well to fit the number of pixels transmitted (destiny resolution).

c) Compressor:

Signals are then compressed using MPEG-1 Video format, which does color conversion to YCbCr and subsamples Cb and Cr components (the human eye is more sensitive to Y component, so changes in other components are less perceived). I-Frames, P-Frames, B-Frames and D-Frames are computed by hardware accelerators. Frame sequence is then rebuilt in SDRAM.

d) Nios II Soft-Core processor:

A Nios II Processor is in charge of controlling all hardware accelerators, as well as creating the packets to be sent to the Ethernet PHY.

e) Video debug output:

In order to provide a debug output, processed data is converted to VGA by using the FPGA parallel outputs and a Video DAC, to show it in a standard resolution monitor.

f) WWAN Output:

Data is sent into packets by the NiosII processor using the websocket protocol for transmission.The device uses a Ethernet port for connecting to WWAN. MAC, PHY and TCP/IP layers are handled by the WIZNET W5100[14].

IV. RESULTS

This section describes the specific implementation of the mentioned components to develop a Telemedicine video transmitter, and focused mainly on the processing modules designed to support the Video real-time operations in a way that performance and reliability are not affected.

Figure 5. Shows a commercial VSM with DVI output, connected to a ECG Simulator and the TeleVga Prototype device. Sampled video data from the VSM is compressed and transmitted via Ethernet to a local service within the FCV, which makes it available to the end user through a web-platform previously developed by FCV DTICS Department. Figure 6 shows how the video streaming is fully integrated within the medical platform of SAHIWeb.

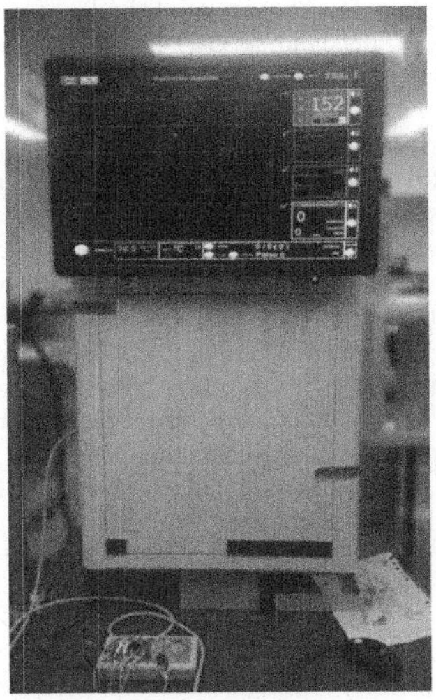

Fig. 5 MSV transmiting

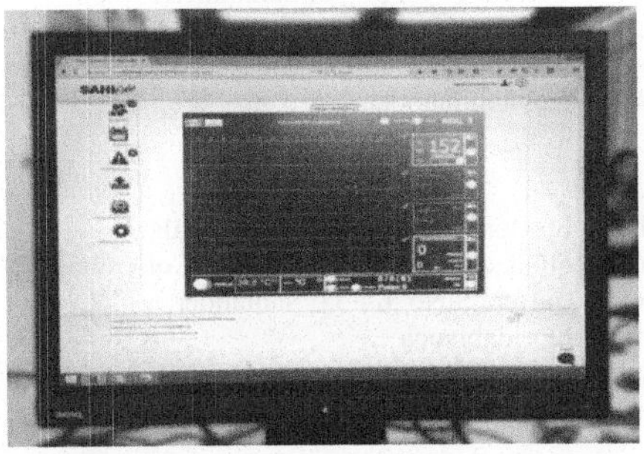

Fig. 6 Reception of MSV video data on test web platform

For verification purposes, each waveform available on the device was enabled sepparately, in order to take 5 different samples of the delay in seconds between the waveform in the VSM and the waveform in the PC. Table 2 summarizes sample times and averages them at approximately 8s, at 29 fps.

There are however, details that must be improved. Some small texts have proven to get unreadable due to pixelization while resizing the image to fit the available space within the web plataform. This development is currently on execution.

Table 2 Delay times from VSM to PC

		Delay (s)					Average
		1	2	3	4	5	
Enabled Waveform	1	8.7	7.1	7.1	8.5	8.5	8
	2	9.1	7.5	7	8.8	8.4	8.2
	3	9	7.5	7.1	8.6	8.1	8
	4	8.4	7.4	7.3	9.5	7.9	8.1
	5	8.5	7.8	7.7	9	7.9	8.1

A. Characteristics and Performance

A summary of the resource usage for the current implementation on the Altera FPGA EP4CE22F17C6 device is presented on Table 2. The device has proven to be almost enough for fitting designment specifications. It is possible that ongoing developments may rise resource consumption, even beyond this device available elements. A Cyclone V SoC 5CSEMA5F31C6 Device will be used in said case.

V. Conclusions

A functional Telemedicine transmiter device prototype was constructed using the described architecture implementation in a FPGA design with custom hardware video processing modules, capable of interacting with the existing server platform and web application. This paper presents the implementation of the required modules to handle acquisition, compression and optimization of input data, using a proposed Custom video processing architecture on an FPGA, along with custom hardware description.

This approach has allowed the integration of systems that would otherwise require the implementation of multiple Video DSPs, and the logic interconnect among them, therefore decreasing the size, complexity and power consumption of the device [15][16], which reflects significatively on lower (Non-Recurring Engineering) NRE and production costs.

Acknowledgment

First, thank COLCIENCIAS and program of technological development and innovation, which in its call 669 of 2014 favored the participation of Bioengineering FCV in the project whose results are the subject of this publication.

Conflict of Interest

The authors declare that they have no conflict of interest.

References

1. Quinnel R, Mannion P, 2014 Embedded Market Study | Then, Now: What's Next, 2014. Available: http://bd.eduweb.hhs.nl/es/2014-embedded-market-study-then-now-whats-next.pdf, Accessed on June, 2016.
2. Nios II (Gen 2) Processor Reference Handbook Altera Corporation, NII5V1-13.1, EE.UU, 2015. Avalible: https://www.altera.com/en_US/pdfs/literature/hb/nios2/n2cpu-nii5v1gen2.pdf, Accessed on June, 2016.
3. Video and Image Processing Suite User Guide Altera Corporation, 2016. Avalible: https://www.altera.com/content/dam/altera-www/global/en_US/pdfs/literature/ug/ug_vip.pdf, Accessed on June, 2016.
4. Galarraga M, Serrano L, Martinez I, de Toledo P, Reynolds M, Telemonitoring Systems Interoperability Challenge: An Updated Review of the Applicability of ISO/IEEE 11073 Standards for Interoperability in Telemonitoring. 2007 Conf Proc IEEE Eng Med Biol Soc, 2007, pp.1-5.
5. Mansoor M GholamHosseini H, Connolly M, Integrated vital signs monitoring system using ubiquitous devices: Multiple physical signs detection and decision support for hospitalized older adults, 2015, DOI 10.1109/EMBC.2015.7318586
6. Dialogic Corporation, Considerations for creating streamed video content over 3G-324M mobile networks, 2009, Avaliable: https://www.dialogic.com/~/media/products/docs/whitepapers/11489-video-streaming-wp.pdf, Accessed on June, 2016.
7. Lantronix Inc, Medical Device Networking for Smarter Healthcare: Part 3 Next-Generation WLAN Deployments in Hospitals, Avaliable: https://www.lantronix.com/wp-content/uploads/pdf/Part3-Lantronix-Medical-Device-Networking-Next-Gen-Wireless-Deployments_WP.pdf, Accessed on June, 2016.
8. Capsule Tech Inc, Using enterprise connectivity to make medical devices IHE compliant, 2013, Available: http://www.interoperabilityshowcase.org/himss13/documents/Capsule.pdf, Accessed on June, 2016.
9. Altera Corporation, "Cyclone IV FPGA device family overview. 2014, Available: https://www.altera.com/en_US/pdfs/literature/hb/cyclone-iv/cyiv-51001.pdf, Accessed on June, 2016.
10. Ekkebus S, "Telemonitoring services for home care ventilators, using web services", pp. 17-18. Master Thesis, Universtity of Twente, The Netherlands, 2006.
11. Medical design, "Eliminating RF interference in healthcare wireless networks", Available: http://medicaldesign.com/electronics/eliminating-rf-interference-healthcare-wireless-networks, Accessed on June, 2016.
12. Aruba networks, "Secure Wi-Fi For Healthcare Applications", Available:http://www.arubanetworks.com/pdf/technology/whitepapers/wp_Healthcare_WLAN.pdf , Accessed on July, 2016.
13. BICSI, Understanding DVI-D, HDMI And DisplayPort Signals, Available: https://www.bicsi.org/pdf/Regions/charlotte_se_4_11/Understanding%20DVI-D,%20HDMI%20and%20Display%20Port%20Signals.pdf Accessed on June, 2016.
14. WIZnet, W5100 Datasheet, Available: https://www.sparkfun.com/datasheets/DevTools/Arduino/W5100_Datasheet_v1_1_6.pdf , Accessed on June, 2016.
15. Altera Corporation, Adding Hardware Accelerators to Reduce Power in Embedded Systems, 2009, Available: http://www.altera.com/literature/wp/wp-01112-hw-reduce-power.pdf, Accessed on June, 2016.
16. Altera Corporation, AN 531: Reducing Power with Hardware Accelerators, 2008, Available: http://www.altera.com/literature/an/an531.pdf, Accessed on June, 2016.

Author: Giovanni Francisco Manotas Rodriguez
Institute: Fundación Cardiovascular de Colombia FCV
Street: Carrera 5 No. 6 - 33, Santander.
City: Floridablanca
Country: Colombia
Email: giovannimanotas@fcv.org

Author: Leonardo Andres Rodriguez Salazar
Institute: Fundación Cardiovascular de Colombia FCV
Street: Carrera 5 No. 6 - 33, Santander.
City: Floridablanca
Country: Colombia
Email: leonardorodriguez@fcv.org

Author: Jose Manuel Velandia Rojas
Institute: Fundación Cardiovascular de Colombia FCV
Street: Carrera 5 No. 6 - 33, Santander.
City: Floridablanca
Country: Colombia
Email: josevelandia@fcv.org

Construction of arterial networks considering the Fahraeus-Lindqvist effect

P. F. Brito[1], L. D. M. Meneses[1], B. M. Rocha[1], R. W. Santos[1] and R. A. B. Queiroz[1]

[1] Graduate Program in Computational Modelling, Federal University of Juiz de Fora, Brazil

Abstract— **Arterial networks can be generated from lumped parameter, anatomical, fractal and optimized models. Inside the optimized class, the computational method of Constrained Constructive Optimization (CCO) is able to generate models that mimic important properties of real coronary arterial trees. This work contributes with an algorithm based on CCO method to create arterial networks taking into account the Fahraeus-Lindqvist effect during the growth process. This effect is associated with the viscosity of the blood changes with the diameter of the vessel nonlinearly. Similarly to the CCO, in this algorithm, a arterial tree is represented as dichotomously branching system of straight cylindrical tubes, with resistance and flow conditions calculated by Poiseuille's law. At bifurcations, the radii of parent and daughter segments obey a power law and the process of growing the tree is governed by minimizing total intravascular volume. The models obtained by the algorithm presented here have shown to adequately reproduce the distribution of the vessel radii of real coronary arterial trees when compared with experimental data.**

Keywords— **Arterial network, computational model, optimization.**

I. INTRODUCTION

Arterial trees serve the purpose of conveying blood to all sites of a tissue. Hemodynamic simulation studies uses computational models of arterial trees as their geometrical substrate. These studies have been used to gain a better understanding of all aspects related to blood flow, from wave propagation and analysis of pressure pulse can be used for diagnosis and surgical planning applications. To date these simulations can employ one of following class of models of arterial trees: lumped parameter models [1], anatomical models [2], fractal models [3, 4] and optimized [5, 6, 7, 8, 9, 10, 11, 12].

In particular, this work is interested in optimized models generated by Constrained Constructive Optimization (CCO) [5]. Arterial tree models generated by CCO are able to mimic important properties of real arterial trees, such as segment radii, branching angle statistics and pressure profiles. However, the CCO method is not able to take into account the Fahraeus-Lindqvist effect during the generation of the models, which is an effect where the viscosity of the blood changes with the diameter of the vessel it travels through. For example, there is a decrease of viscosity as the vessel's diameter decreases when the vessel diameter is between 10 and 300 micrometers.

In this context, the purpose of this work is provide an algorithm based on CCO method that generates arterial tree models considering the Fahraeus-Lindqvist effect [13]. The models obtained by the algorithm are compared with morphometric data of real coronary arterial trees.

The remainder of this paper is organized as follows. In Section II, the algorithm proposed based on CCO method is described. In Section III, results obtained using the algorithm developed are presented. Section IV contains our conclusion and discusses the future direction of this work.

II. THE ALGORITHM BASED ON CCO METHOD

The algorithm here proposed is based on the assumptions and constraints listed below [5, 6, 11]:

- The concept associated with the construction is to minimize the total intravascular volume

$$V = \pi \sum_{i=1}^{K_{tot}} l_i r_i^2, \qquad (1)$$

  where l_i and r_i are the length and radius of the segment i, K_{tot} is the number of segments of the tree in growth stage;
- the tree model is generated on a fixed 2D or 3D domain non necessarily convex that represents a tissue and organ, respectively;
- the proximal localization $\mathbf{x}_{prox}$ of the root segment (feeding artery) is known and fixed in the boundary of the domain at the beginning of the simulation;
- the arterial tree is modeled as a dichotomously branching (binary) system of straight cylindrical tubes (vessel segments);
- the model tree starts at the root segment (main feeding artery) and it is truncated in the form of terminal segments on prearteriolar level;
- the model tree should take up the space of the perfusion domain as homogeneously as possible without intersection of segments;

I. Torres et al. (eds.), *VII Latin American Congress on Biomedical Engineering CLAIB 2016, Bucaramanga, Santander, Colombia, October 26th -28th, 2016*,
IFMBE Proceedings 60,
DOI: 10.1007/978-981-10-4086-3_70

- the blood is modeled as an incompressible, homogeneous Newtonian fluid at steady state and laminar flow conditions;
- at bifurcations the radii of parent (r_0) and daughter segments (r_1, r_2) are forced to exactly fulfill a bifurcation law derived from real coronary trees [14]

$$r_0^\gamma = r_1^\gamma + r_2^\gamma, \quad (2)$$

where γ is a constant exponent with ranging between 2.55 and 3, governing the shrinkeage of radii across bifurcations;
- flow resistance R of each segment of the tree is assumed to follow Poiseuille's law

$$R = \left(\frac{8\eta}{\pi}\right)\frac{l}{r^4}, \quad (3)$$

where η is the blood viscosity. The method CCO adopts $\eta = 3.6cP$ (constant). In this work the blood viscosity $\eta(d_i)$ used is described as a non-linear function given by [13]:

$$\eta_p\left[1+(\eta_{0,45}-1)\left(\frac{d_i}{d_i-1,1}\right)^2\right]\left(\frac{d_i}{d_i-1,1}\right)^2, (4)$$

where d_i is the diameter of the segment i, $\eta_p = 1,1245$ cP is the plasma viscosity, and $\eta_{0,45}$ is the apparent viscosity of the plasma for a discharge hematocrit of 0,45 given by

$$6\exp(-0,085d_i)+3,2-2,44\exp(-0,06d_i^{0,645}); (5)$$

- the pressure drop Δp_s along each segment is given by $\Delta p_s = RQ$, where Q is the flow through segment;
- each terminal segment supplies an identical and equal amount of blood flow Q_{term} into the microcirculatory network, which is not modeled in detail;
- the resistance of the resulting model tree induces a prespecified perfusion flow Q_{perf} across the overall pressure drop $\Delta p = p_{perf} - p_{term}$, where p_{perf} is the perfusion pressure at the inlet of the root segment, p_{term} is the pressure at the outlet of all the terminal segments and Δp is constant for the CCO method;
- the radius of the root segment r_{iroot} is scaled during the growth of the tree model by CCO method as follows:

$$r_{iroot} = \left[R^*_{sub,iroot}\frac{Q_{perf}}{\Delta p}\right], \quad (6)$$

where $R^*_{sub,iroot}$ denotes the reduced hydrodynamic resistance of the whole tree (see details in [6]) and $Q_{perf} = k_{term}Q_{term}$ which k_{term} denotes the number of terminal segments to be supplied.

A. Algorithm of tree generation

The algorithm of tree generation based on CCO has previously been described in detail [5, 6, 11]. The optimized tree model generation starts by planting the root segment with its proximal end $\mathbf{x}_{prox}$ fixed at the perfusion domain Ω_{perf} and the distal $\mathbf{x}_{inew}$ selected randomly within this domain. If this position is not too close to $\mathbf{x}_{prox}$, $\mathbf{x}_{inew}$ is connected to $\mathbf{x}_{prox}$, resulting in a root segment length (l_{iroot}). The radius r_{iroot} of the root segment is such that the hydrodynamic resistance $R^*_{sub,iroot}$ yields the flow $Q_{perf} = Q_{term}$ through one terminal segment ($k_{term} = 1$) into the microcirculatory network. At this moment, $R^*_{sub,iroot} = \frac{8\eta_i}{\pi} l_{iroot}$ and η_i assumes 3.6 cP to calculate radius. In following, η_i is updated using Equation (4) with $d_i = 2r_{iroot}$ and again the radius r_{iroot} is computed. This iterative procedure is repeated until convergence is achieved with precision $\varepsilon = |r^{n+1}_{iroot} - r^n_{iroot}| < 10^{-5}$ for example.

Given a tree with k_{term} terminal segments, the stepwise growing of the tree is as follows. First, the location $\mathbf{x}_{term}$ for a new terminal is selected from a pseudorandom number sequence, uniformly distributed inside the perfusion domain. The prospective location $\mathbf{x}_{term}$ is accepted as a candidate for a new terminal site only if $\mathbf{x}_{term}$ satisfy a distance criterion (see [5]).

Since $\mathbf{x}_{term}$ has been accepted as a distal end of a new terminal segment, it is temporarily connected to each of the neighboring segments, one after the other. Connecting the new terminal segment to a preexisting segment, consequently cause violation in the boundary condition regarding the terminal flows. In order to return the proper terminal flows, the flow resistance of the tree must be adjusted for each temporary connection. This can only be performed by rescaling of the segments' radii [6] equipped with an iterative procedure that estimates and corrects the non-linear viscosity in each segment of the tree.

The bifurcation site resulting in each temporary connection is optimized in order to minimize the total intravascular volume and dissolved again. After assessing all possible connections in the neighborhood of $\mathbf{x}_{term}$, the connection that provided the lowest optimization target is adopted as permanent for the new terminal site $\mathbf{x}_{term}$. Thus, the tree is grown to $k_{term}+1$ terminal segments. The process of growing the tree summarized above is repeated until $k_{term} = N_{term}$, i.e., the preset number of terminals N_{term} is achieved. The steps described previously are systematized in the Algorithm 1.

Algorithm 1: Tree generation inspired in the CCO.

Data: Ω_{perf}, $\mathbf{x}_{prox}$, Q_{perf}, N_{term}, γ.

Fix the proximal position $\mathbf{x}_{prox}$ of the root segment into domain Ω_{perf};
repeat
  Generate the distal position $\mathbf{x}_{inew}$ for the root segment into domain Ω_{perf};
  Distance criterion: check if this position is not too close to $\mathbf{x}_{prox}$;
until *(distance criterion is met)*;
Connect $\mathbf{x}_{inew}$ to $\mathbf{x}_{prox}$ *(planting the root segment)*;
Call iterative procedure (non-linear viscosity);
$k_{term} \leftarrow 1$;
while $(k_{term} < N_{term})$ **do**
  repeat
    Generate the distal position $\mathbf{x}_{inew}$ for a new terminal segment;
    Distance criterion: check if this position is not too close to any of the existing segments;
  until *(distance criterion is met)*;
  Obtain the neighboring segments N_{con} of $\mathbf{x}_{inew}$ for temporary connection;
  for $j \leftarrow 1$ *to* N_{con} **do**
    Connect $\mathbf{x}_{inew}$ to the midpoint $\mathbf{x}_{ibif}$ of segment j;
    Call iterative procedure (non-linear viscosity);
    Optimize the bifurcation position $\mathbf{x}_{ibif}$;
    Perform restriction checks (e.g. segment intersection, segments traversing forbidden domains);
    Save value of target function, position $\mathbf{x}_{ibif}$ and results of check in line j of the *Connection Evaluation Table (CET)*;
    Remove the bifurcation created;
  Restrict *CET* to allowed connections: CET^v;
  if *(CET^v is not an empty set)* **then**
    Find optimal connection j_{opt} from CET^v *(structural optimization)*;
    Make connection from $\mathbf{x}_{inew}$ to j_{opt} permanent;
    Call iterative procedure (non-linear viscosity);
    $k_{term} \leftarrow k_{term} + 1$;
  else
    Refuse the position $\mathbf{x}_{inew}$;
Obtain computed quantities (length, radius, resistance, pressure drop);

III. RESULTS: COMPARISON WITH REAL CORONARY ARTERIAL TREES

For morphometric comparison with real coronary arterial trees, the Algorithm 1 was applied to generate arterial trees with 250 terminal segments (499 in total) in order to represent the tree of the left anterior descending (LAD) coronary artery. In Figures 1 and 2 observe that the LAD orifice has level 1 in a tree starting with the main left coronary artery (level 0). In this figure, the bifurcation level indicates the number of proximal bifurcations of a segment.

The arterial tree models were generated under the following conditions [5]: total perfusion flow $Q_{perf} = 500$ mL/min, terminal flows $Q_{term} = 2$ mL/min, bifurcation exponent $\gamma = 3$, circular area $\Omega_{perf} = 78.54$ cm^2 representing tissue to be perfused (LAD region). Two simulations were performed using constant viscosity blood $\eta = 3.6$ cP or viscosity given by Equation (4).

For each simulation, ten replicates of the tree with 250 terminal segments were generated on the same predefined parameters and the same optimization target function (total intravascular volume). Each tree was generated using a different sequence of pseudorandom numbers for casting the distal ends of its terminal segments.

Figures 1 and 2 display the mean diameter and standard deviation of this mean diameter (SDM) of all vessel segments at a certain bifurcation level that indicates the number of proximal bifurcations of a segment. Close agreement with the experimental data acquired from corrosion casts from two human hearts by Zamir and Chee [15] is found when the models are generated with viscosity given by Equation (4) for the levels 3-8 (see Figure 2).

Table 1 shows the absolute mean error calculated by

$$E_{abs} = \frac{1}{N_{bif}} \sum_{i=1}^{N_{bif}} \mid d_i - \tilde{d}_i \mid, \qquad (7)$$

where N_{bif} is the number of bifurcation levels of the tree model, d_i is the mean diameter in the bifurcation level i [15] and $\tilde{d}_i$ is the mean diameter in the bifurcation level i of the tree model. For hearts A and B adopt $N_{bif} = 34$ and $N_{bif} = 30$, respectively. One can conclude of this quantitative result that consider variable viscosity in the simulations provide tree models more close to real coronary arterial trees.

IV. CONCLUSION AND FUTURE WORK

In this work is presented an algorithm developed based on CCO method that is capable of generating tree models taking into account the Fahraeus-Lindqvist effect. These arterial tree

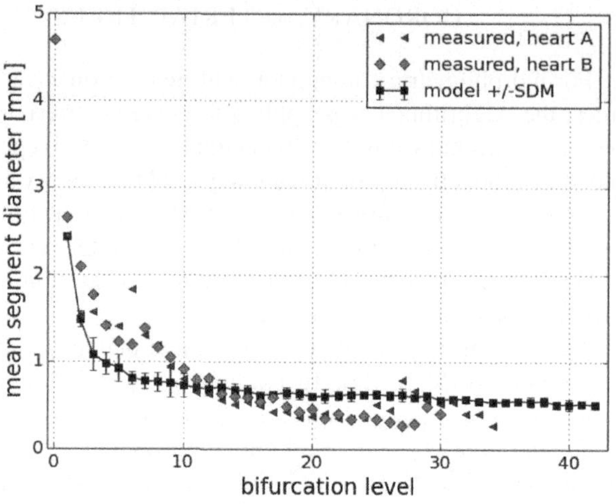

Fig. 1: Morphometric comparison between tree models and real left coronary arterial trees of two humans – constant viscosity 3.6 cP.

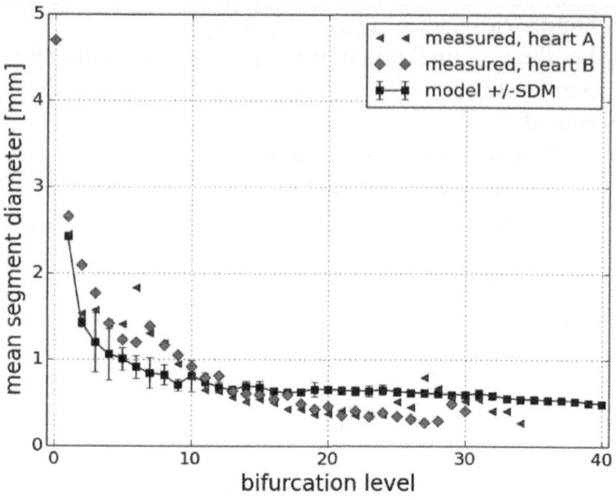

Fig. 2: Morphometric comparison between tree models and real left coronary arterial trees of two humans – variable viscosity given by Equation (4).

Table 1: Absolute mean error (E_{abs})

viscosity	heart A	heart B
constant (3.6 cP)	0.220697	0.256466
variable (Equation (1))	0.214034	0.245019

models are in agreement with real vascular trees regarding morphometric parameter. It should be emphasized that the described algorithm can be improved. In future work, we plan to incorporate elastic representation and collateralization of the vessels. Furthermore, hemodynamic simulations will be performed considering as geometric substrate the tree models here constructed.

Conflict of Interest

The authors declare that they have no conflict of interest.

Acknowledgement

This work was partially supported by the Brazilian agency FAPEMIG which is gratefully acknowledged (Proc. Num. 00795-14).

References

1. Mates R.E., Klocke F.J., Jr. J.M. Canty. Coronary capacitance *Prog Cardiovasc Dis.* 1988;31:1–15.
2. Onuki T., Nitta S.. Computer simulation of geometry and hemodynamics of canine pulmonary arteries *Ann Biomed Eng.* 1993;21:107–115.
3. Beek J.H. Van, Roger S.A., Bassingthwaighte J.B.. Regional myocardial flow heterogeneity explained with fractal networks *Am J Physiol.* 1989;257:H1670–H1680.
4. Yang J., Wang Y.. Design of vascular networks: A mathematical model approach *Int. J. Numer. Meth. Biomed. Engng.* 2012.
5. Schreiner W., Buxbaum P.F.. Computer-optimization of vascular trees *IEEE Trans Biomed Eng.* 1993;40:482–491.
6. Karch R., Neumann F., Neumann M., Schreiner W.. A tree-dimensional model for arterial tree representation, generated by constrained constructive optimization *Comput Biol Med.* 1999;29:19–38.
7. Schreiner W., Karch R., Neumann M., Neumann F., Szawlowski P., Roedler S.. Optimized arterial trees supplying hollow organs *Med Eng & Phys.* 2006;28:416–429.
8. Blanco P.J., Watanabe S.M., Queiroz R.A.B., Trenhago P.R., Fernandes L.G., Feijóo R.A.. *Trends in the computational modeling and numerical simulation of the cardiovascular system.* Petropólis–RJ: Scientific Computing Applied to Medicine and Healthcare 2012.
9. Blanco P.J, Queiroz R.A.B., Feijóo R.A.. A computational approach to generate concurrent arterial networks in vascular territories *Int J Numer Method Biomed Eng.* 2013;29:601–614.
10. Queiroz R.A.B., Blanco P.J., Feijóo R.A.. Multiple vascularization of spatial domains using CCO-based algorithms in *WCCM Proc, 10th World Congress on Comput. Mech*(São Paulo–SP, Brazil) 2012.
11. Queiroz R.A.B., Blanco P.J., Feijóo R.A., Ulysses J.N.. Constructive Algorithm of Optimized Arterial Networks in *IFMBE Proceedings. 1ed.: Springer International Publishing*;49:884-887 2015.
12. Schwen L. O., Wei W., Gremse F., et al. Algorithmically generated rodent hepatic vascular trees in arbitrary detail *Journal of Theoretical Biology.* 2015;365:289-300.
13. Pries A.R., Secomb T. W., Gaehtgens P., Gross J. F.. Blood flow in microvascular networks. Experiments and simulation *Circulation Research..* 1990;67(4):826-834.
14. Zamir M.. Distributing and delivering vessels of the human heart *J G Physiol.* 1988;91:725-735.
15. Zamir M., Chee H.. Segment analysis of human coronary arteries *Blood Vessels.* 1987;24:76–84.

Optical microscopy and autofluorescence recognition of *Toxoplasma gondii* and *Cryptosporidium parvum* oocysts

Alejandra Alba[1], Laura Baker[1], Graciela Juez[1] and Andrés Ramírez[1]

[1] Universidad El Bosque/Bioingenería, Bogotá, Colombia

Abstract— **Oocysts *Toxoplasma gondii* and *Cryptosporidium parvum* were visualized using conventional optical microscopy (Olympus microscope CX31), allowing recognize their main morphological characteristics. In addition, its ability to autofluorescence when were exposed to certain range of ultraviolet light was confirmed. *Cryptosporidium parvum* presented autofluorescence under the FITC filter (excitation 450-490 nm and emission 515 nm) and *Toxoplasma gondii* under (excitation 330-380 nm and emission 397 nm) UV filter. The above was done to develop an autofluorescence experimental protocol for recognition of *C. parvum* and *T. gondii* oocysts in samples of raw cow's milk.**

Keywords— ***Toxoplasma gondii*, *Cryptosporidium parvum* parasite, autofluorescence, microscopy**

I Introduction

Protists parasites such as *Cryptosporidium parvum* and *Toxoplasma gondii* pose a high-risk to the health of the human population, due to their biological mechanisms of resistance to environmental conditions in their state of oocyst and its means of transmission which are determined by direct consumption oocysts parasites in food and water sources [1]. For example, in foods like unpasteurized milk and goat cheese, some studies suggest that *T. gondii* may be a potential contaminant [2]. Other studies have reported the presence of *C. parvum* oocysts in unpasteurized milk [3].

According to the "quality monitoring of water for human consumption" in 2013, the National Institute of Health in Colombia, reports *C. parvum* and *T. gondii* as parasites of high importance for health and long persistence in water supply systems, indicating that these microorganisms are considered high epidemiological relevance to the country [4].

Cryptosporidium is an intracellular parasite that affects human health causing diarrheal diseases especially in people with immune deficiency, malnutrition, and patients vulnerable children. Routes of transmission are determined by the consumption of contaminated water and food with oocysts of the parasite.

T. gondii is a protozoan that causes toxoplasmosis in humans, characterized by affecting immunocompromised individuals who can establish chronic infection and congenital toxoplasmosis responsible for contributing to the presence of disturbances such as macrocephaly, hydrocephalus, correoretinitis, among others. Routes of transmission are determined by the consumption of fruits, vegetables and water in the presence of oocysts of the parasite [5].

For detection of *C. parvum* oocysts and *T. gondii*, different techniques are currently used as optical microscopy using staining as thionine, Wright-Giemsa and MIF (merthiolate, iodine and formaldehyde) [6] and fluorescence microscopy using antibodies specific fluorochrome conjugated [7]. In order to develop these techniques it is important to implement purification protocols that allow a high concentration of oocysts of the parasite and removing pollutants that affect the specificity of the technique agents.

It has now been shown that some microorganism have the property of autofluorescence, which is an intrinsic property of certain molecules or organisms that are characterized by transient emission of light when stimulated by an exogenous source of a length shorter wavelength [8]. Therefore, autofluorescence can be used as an alternative to recognition methods *C. parvum* oocysts and *T. gondii*, as evidenced in [9] where successfully detected with *T. gondii* oocysts in samples of fresh water and estuarine water in the Galapagos islands, Ecuador. Therefore, this study aims to develop an experimental protocol for the recognition of *C. parvum* oocysts autofluorescence and *T. gondii* in samples of raw cow's milk.

II Materials and Methods

A Optical microscopy

Visualization of *C. parvum* and *T. gondii* oocysts was performed by optical microscopy on samples purifieddas in 1X PBS.

A sample of 10 μl of oocysts purified in PBS 1X *C. parvum* was taken, placed on a slide by adding 10 μl lugol 7% to stain dimly, then the sample is passed through capillarity Neubauer chamber and placed in a optical microscope Olympus CX 31 model, 100X.

To view *T. gondii* oocysts, 10 μl of purified in PBS 1X it was taken and placed on a slide adding 10 μl lugol 7%,

I. Torres et al. (eds.), *VII Latin American Congress on Biomedical Engineering CLAIB 2016, Bucaramanga, Santander, Colombia, October 26th -28th, 2016*,
IFMBE Proceedings 60,
DOI: 10.1007/978-981-10-4086-3_71

to stain faintly, then the sample was passed by capillarity Neubauer chamber and it was visualized in the light microscope Olympus CX31, 100X.

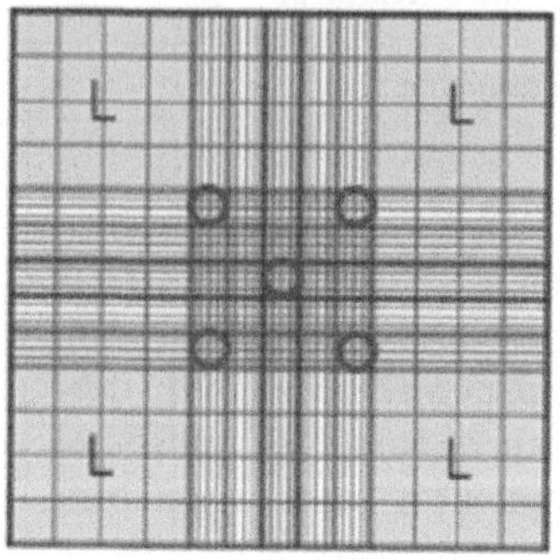

Fig. 1: Neubauer chamber's counting grid [10]

B Quantification of the sample

It is important to quantify the microorganisms present in the sample to estimate the number of parasites that are to be analyzed, for this process the quantification was performed in 1 *ml* sample of oocysts purified in PBS 1X, using a Neubauer chamber (Boeco) and optical microscope Olympus CX31, 100X.

Initially 10 μl of sample and 10 μl of lugol 7% were mixed, subsequently display and counting was performed using the upper and lower Neubauer chamber's panel.

T. gondii, the count was conducted in the four quadrants marked with the letter L (see Figure 1) and counting was repeated at the bottom of the chamber's grid. Finally, the concentration of cells in the sample (1 *ml*) was calculated, taking into account equation 1 [11]. The procedure was repeated ten times using different sample for every of the counts and the data obtained were averaged to obtain a final concentration.

$$C\left(\frac{cells}{ml} = \frac{N^{\circ}CC}{N^{\circ}squares} f_d \times p_c\right) \quad (1)$$

where C=concentration, $N^{\circ}CC$=number of counted cells, $N^{\circ}squares$=number of counted frames, f_d=dilution factor and p_c=chamber depth.

To perform the quantification of *C. parvum* in a 0.5 *ml* sample, the count was carried out in the five sections of central quadrant (marked with a red circle, see Figure 1) and counting was repeated in the same pictures of the lower chamber's grid.

The concentration of cells in the sample (0.5 *ml*) was calculated based on Equation 1. The last procedure was repeated ten times using different samples for every of the counts and the data obtained were averaged to obtain a final concentration.

C Autofluorescence microscopy

For visualization of autofluorescence *C. parvum* and *T. gondii* oocysts purified samples and samples of raw cow's milk contaminated with oocysts of the parasites were used.

The Axio Imager microscope A2, Zeiss and ultraviolet filters was used to *T. gondii* (330-380 *nm* excitation, 397 *nm* emission) suggested by Lindquist, *et al*, [12] and FITC for *C. parvum* (Arousal 450-490 *nm*, emission 515 *nm*) recommended by Varea, *et al* [13].

Initially, four sheets were mounted: the first with 20μl of *T. gondii* oocysts in PBS 1X, the second with 20μl of a 1 : 15 dilution of the sample *T. gondii* oocysts in PBS, with 20μl third: Initially, four sheets were mounted in sample of *C. parvum* oocysts in PBS and 20 μl fourth with a 1 : 15 dilution of the sample of *C. parvum* oocysts in PBS and allowed to a drie at room temperature, then was added to each sheet ten drops to set absolute methanol and allowed to dry at room temperature amment [13].

Subsequently, each sheet covered with one or two coverslips (depending on the size of the sample), which were attached to the blade with clear nail polish and the sheets were stored in a plastic container in a place protected from light and covered with foil.

Prior to visualization the sheets were cleaned with 50% ethanol and placed placed on the fluorescence microscope using a 40X objective with the respective filters, the corresponding images captured by camera Axiocam is HRm.

Table 1: Quantification of *Toxoplasma gondii* oocysts

N° Count	Count upper chamber	Counting bottom chamber	Quantification average (cells/ml)	Standar deviation
1	146	126	680.000	110.131,9
2	136	111	617.500	65.937,7
3	135	110	612.500	62.402,2
4	129	111	600.000	53.563,3
5	82	84	415.000	77.251,4
6	83	88	427.500	68.412,6
7	98	102	500.000	17.147,3
8	94	106	500.000	17.147,3
9	84	88	430.000	66.644,8
10	84	100	460.000	45.431,6
Quantification average (cells/ml)				524.250

III Results and Discussion

Table 1 show ten counts was performed to calculate the amount of *T. gondii* oocysts present in the sample of 1 *ml* and the final result was 5.2×10^5 in 1 *ml*.

Table 2: Quantification of *Toxoplasma gondii* oocysts

N° Count	Count upper chamber	Counting bottom chamber	Quantification average (cells/ml)	Standar deviation
1	12	15	1.350.000	77781,7
2	10	11	1.050.000	134350,3
3	11	12	1.150.000	63639,6
4	12	12	1.200.000	28284,3
5	13	14	1.350.000	77781,7
6	12	13	1.250.000	7071,1
7	10	15	1.250.000	7071,1
8	14	12	1.300.000	42426,4
9	11	14	1.250.000	7071,1
10	12	13	1.250.000	7071,1
Quantification average (cells/ml)				1.240.000
Quantification end (cells/ml) in 0,5 ml				620.000

Table 2 show ten counts was performed to calculate the amount of *C. parvum* oocysts present in 1 *ml* and the final result was 6.2×10^5 oocysts in 0.5 *ml*.

In Figure 2 it is seen in a 100X objective, *C. parvum* oocysts with its double cell wall and inside some peripheral sporozoites using conventional optical microscopy. In the right image, the autofluorescence of *C. parvum* oocysts viewed with a 40X objective under a microscope immunofluorescence is observed.

Figure 3 *T. gondii* oocysts are observed. In the image on the left side are observed in an objective of 100X using a conventional microscope to show their morphological characteristics as the thick cell wall and two sporocysts each. In the image on the right side autofluorescence of *T. gondii* oocysts seen in a 40X objective displayed on a fluorescent microscope.

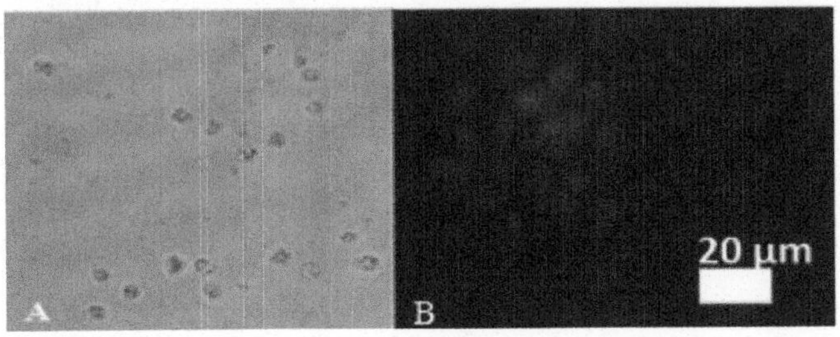

Fig. 2: *Cryptosporidium parvum*, left: visible light, right: ultraviolet light

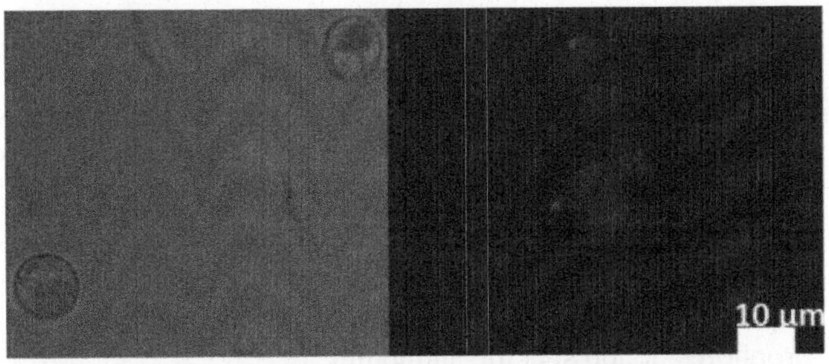

Fig. 3: *Toxoplasma gondii*, left: visible light, right: ultraviolet light

Figure 4 *T. gondii* and *C. parvum* oocysts are observed in samples of raw cow's milk. In the image on the left side of *T. gondii* autofluorescence it was observed using a fluorescence microscope to analyze their morphological characteristics as its cellular wall and two sporocysts each. In the right image autofluorescence of *C. parvum* oocysts observed in sample of raw cow's milk.

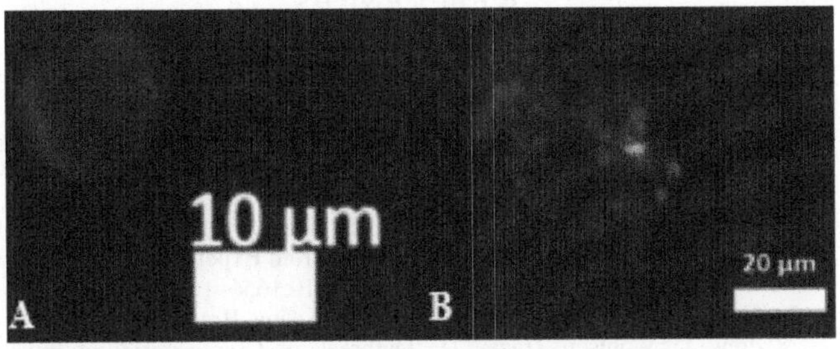

Fig. 4: Samples oocysts in raw cow's milk. (A) Autoflores-cencia of *T. gondii* (B) autofluorescence of *C. parvum*

IV Conclusions

Recognition of the morphology of both parasites in their oocyst forms using optical microscopy and fluorescence microscopy was achieved because, both meet the characteristics studied in the literature, and its double cell wall, the diameter of oocyst *C. parvum* (4-6 μm) and *Toxoplasma gondii* (10-12 μm), the two sporocysts inside the oocyst *T. gondii* and peripherals within sporozoites *C. parvum* oocysts.

It was possible counts for quantifying ooquísticas forms of each parasite in samples, although the results throw a concentration value on the scale 10^5, which is rather low for counts normally done in chamber Neubauer.

It was confirmed that both *T. gondii* and *C. parvum* have the property of autoflourescencia (low light emission wavelength which excites the cell, without using any staining) under UV light, to be displayed successfully in the fluorescence microscope under ultraviolet filters and FITC respectively. Contrary to what is stated in the book Atlas of Human Parasitology, where it is mentioned that Cryptosporidium oocysts have a lack of autofluorescence under UV light. [14].

Autofluorescence property opens the way to an alternative method of effective and simple recognition of these parasites oocysts in food samples, without the need for dyes, fluorophores or add specific antibodies.

ACKNOWLEDGEMENTS

Special thanks to GEPAMOL Research Group at the University of Quindio (Armenia, Quindio, Colombia) by providing us the respective samples of *T. gondii* oocysts and *C. parvum*. As the Research Laboratory of the University El Bosque for letting us use their facilities to take pictures of autofluorescence under the microscope Axio Imager A2.

REFERENCES

1. Bigot-Clivot Aurélie, Ladeiro Mélissa Palos, Lepoutre Alexandra, et al. Bioaccumulation of Toxoplasma and Cryptosporidium by the freshwater crustacean Gammarus fossarum: Involvement in biomonitoring surveys and trophic transfer *Ecotoxicology and Environmental Safety.* 2016;133:188–194.
2. Dubey J. P., Verma S. K., Ferreira L. R., et al. Detection and Survival of Toxoplasma gondii in Milk and Cheese from Experimentally Infected Goats *Journal of Food Protection.* 2014;10:1656–1833.
3. Harper Katherine M., Cowell Noel A., Adams Brad C., Langley Andrew J., Wohlsen Tracey D.. Outbreak of Cryptosporidium linked to drinking unpasteurised milk *Communicable Diseases Intelligence.* 2002;26:–.
4. Barragan Blanca Lisseth Guzman, Tovar Gerardo Nava, Bustacara Angela Liliana Mejia. Enfermedades vehiculizadas por agua-EVA e Índice de riesgo de la calidad en Colombia IRCA *Instituto nacional de salud, Bogotá, D.C., Colombia.* 2014;1:1-3.
5. Authority New South Wales. Toxoplasma and Cryptosporidium in the food supply *Food authority.* 2014;–:1–34.
6. Importancia de la referencia médica en el diagnóstico de parasitosis intestinales por métodos coproparasitologicos at http://www.usfx.bo/nueva/vicerrectorado/citas/salud_10/Medicina/69.pdf
7. Bushkin G. G., Motari E., Carpentieri A., et al. Evidence for a structural role for acid-fast lipids in oocyst walls of Cryptosporidium, Toxoplasma, and Eimeria *mBio.* 2013;4:1–8.
8. Utilidad de la autofluorescencia en las formas atróficas de la degeneración macular asociada a la edad at http://www.oftalmo.com/studium/studium2009/stud09-4/09d-04.html
9. Verant M. L., D'ozouville N., Parker P., Shapiro K. G, Vanwormer E.. Attempted Detection of Toxoplasma gondii Oocysts in Environmental Waters Using a Simple Approach to Evaluate the Potential for Waterborne Transmission in the Galápagos Islands, Ecuador *EcoHealth.* 2014;11:207–214.
10. Preparación de la muestra at http://www.celeromics.com/es/resources/Technical%20Notes/Como-contar-celulas-con-camara-de-Neubauer-Paso1.php
11. Camara de recuento at http://www.brand.de/fileadmin/user/pdf/GK900/Zaehlkammern/GK900_05_Clinical_Lab_Zaehlkammern_s.pdf
12. Lindquist H D Alan, Bennett Jason W, Hester Jeff D, Everson William V. Autofluorescence of Toxoplasma gondii and Related Coccidian Oocysts *Journal of Parasitology.* 2003;89:865–867.
13. Varea M, Clavel A, Doiz O, Castillo FJ, Rubio MC, Gómez Luis R.. Fuchsin fluorescence and autofluorescence in Cryptosporidium, Isospora and Cyclospora oocysts *International journal for parasitology.* 1998;28:1881–1883.
14. Lawrence Ash R., Orihe Thomas C.. *Cryptosporidium hominis, parvum y otras especies: Diagnóstico.* Los Angeles, Estados Unidos: Atlas de Parasitología Humana 2007.

Author: Francy Alejandra Alba Patiño
Institute: Universidad El Bosque
Street: Cra. 9 # 131a-2, Bogotá, Colombia
City: Bogotá
Country: Colombia
Email: falba@unbosque.edu.co

Author: Laura Viviana Baker Leon
Institute: Universidad El Bosque
Street: Cra. 9 # 131a-2, Bogotá, Colombia
City: Bogotá
Country: Colombia
Email: lbaker@unbosque.edu.co

A human gait temporal parameters calculation algorithm

P.E. Caicedo-Rodríguez[1], C.F. Rengifo-Rodas[2] and L.E. Rodríguez-Cheu[3]

1 Corporación Universitaria Autónoma del Cauca, Grupo investigación en tecnologías y ambiente, Popayán, Colombia
2 Universidad del Cauca, Grupo investigación en automática industrial, Popayán, Colombia
3 Escuela Colombiana de Ingeniería, Grupo investigación en ingeniería biomédica, Bogotá, Colombia

Abstract— **This paper presents a new algorithm to calculate three temporal parameters of human gait: cadence, swing time and stride time. These values are computed from: heel-strike time, toe-off time and mid-swing point time which can be in turn obtained from the first time derivate of the medio-lateral axis angular velocity of both shanks. The results generated by the proposed algorithm were validated with those of the commercial software *Tech MCS Studio* of *Technaid*. The comparison gives a mild difference for cadence and stride time. Aditionally, the new algorithm is a good alternative to the thresholding-based algorithms previously reported in the literature which require a great amount of work dedicated to the manual tuning of their multiple parameters.**

Keywords— **Inertial sensor, Gait events, Heel-strike, Toe-off, Stride time, Stance time, Swing time.**

I Introduction

The analysis of the human gait allows to evaluate the motion quality of a person [1]. Some of the parameters obtained from these gait studies are cadence, stride time and swing stance time [2]. In most cases, these analyses require a well-instrumented laboratory with optical, electromagnetic or electronic-walkway-based movement capture systems [3]. However, due to the importance of of patients' diagnosis in their natural environment, some authors propose portable motion capture systems which are mostly based on inertial devices such as accelerometer and gyroscopes. Jasiewicz et al [4] have identified step events by measuring linear accelerations in the horizontal and vertical axis of motion. Aung et al [5] report the detection of heel-strike and toe-off events through a multiscale and frequency analysis of acceleration signals. Some other authors describe gyroscopes-based analysis systems. In most cases, gyroscopes are located in the shanks of the patient [6, 7]. Greene et al [6] use an adaptive threshold based algorithm to process the medio-lateral angular velocity of both shanks. Fraccaro et al [7] change the threshold obtention process for their algorithm; besides, Fraccaro et al [7] compare their results with those of Greene. In particular, Greene's method requires six adaptive thresholds, to match the local maxima and minima of the medio-lateral angular velocity (MLAV) with heel-strike, toe-off and midswing point events. These thresholds are highly signal-dependent and sensor-dependent making more susceptible to erroneous signal sources. The proposed algorithm was designed to use the MLAV signals of both shanks for event detection; however the theoretical basis of the algorithm are noise-cancelling techniques for the MLAV digital signal. This approach requires less qualitative information than the threshold-based approach; because it uses the signal noise characteristics, which are only sensor-dependent; making the algorithm error immunity higher. The validation of the algorithm was done with the *Technaid* gait analysis software

II Temporal gait parameter algorithm

A Experimental platform

For the algorithm development, the *Technaid* Motion Capture System (MCS) was used. The MCS has four different type of sensors: accelerometers, gyroscopes, magnetometers and a thermometers, which should be located at the eight positions recommended by *Technaid*: on the two legs, on the two shanks, on the two ankles, one on the trunk and one on the lumbar zone. For the validation of the proposed algorithm only the data obtained from the gyroscopes located at shanks were used.

The data acquisition sampling frequency was 40Hz for every MCS sensor. Figure 1 shows the location of the sensors in both legs and their coordinate system. The x, y and z axis of the sensors correspond to the longitudinal, anteroposterior and mediolateral axis of the human body.

B Algorithm description

The proposed algorithm processes the MLAV digital signals of both shanks to detect the following gait events: (i) heel-strike time, (ii) toe-off time and (iii) mid-swing point time. The computation of these values involves six main steps where the signal obtained in the step i is used as the input to the step $i+1$

I. Torres et al. (eds.), *VII Latin American Congress on Biomedical Engineering CLAIB 2016, Bucaramanga, Santander, Colombia, October 26th -28th, 2016*,
IFMBE Proceedings 60,
DOI: 10.1007/978-981-10-4086-3_72

1. *Noise-cancelling filter:* This is done by using a 10th order causal moving average filter passed in both forward and reverse directions to avoid phase delays. In *Matlab* this operation can be effectuated with the function `filtfilt`.
2. *Signal derivate calculation:* In this step the lead-lag network described by the transfer function (1) is applied. The pole limits the derivative gain for frequency components greater than 22 Hz. The system tuning was done through a heuristic process based on a frequency response analysis

$$G_d(s) = \frac{s}{0.0007s+1} \tag{1}$$

3. *Smoothing signal derivative:* The algorithm applies a 5th order causal moving average filter to the signal obtained from $G_d(s)$. This filter is similar to the one applied in step one.
4. *Sign calculation:* To detect the time intervals where the signal given by the gyroscopes is increasing or decreasing with respect to time, the signum function is applied to the signal obtained in the previous step. This function returns $+1$ when the signal is increasing and -1 when the signal is decreasing.
5. *First order difference:* To detect when the signal changes from increasing to decreasing or viceversa, the first order difference described by the following input-output relationship is applied: $y_k = u_k - u_{k-1}$. At the sampling time when the signal u_k is transitioning from increasing to decreasing y_k is -2. For the transition from decreasing to increasing $y_k = +2$. If the signal is not transitioning y_k remains equal to 0.
6. *Local maxima and minima detection:* A local maximum value of the angular velocity occurs at the sampling time when the signal y_k obtained from the previous step is equal to -2. Converseley, there exists a local minimum in the angular velocity when $y_k = +2$. To avoid spurius local inflection points, the values of y_k beloging to the set $\{-2, 2\}$ and ocurring when the absolute value of the angular velocity is smaller than 10 rad/seg, are excluded.
7. *Gait event detection:* The events detected by the algortihm are presented in the Figure 2a, 2b, 2c
   (a) *Toe-off points* are the moments when the restless leg leaves the floor. In the algorithm, a toe-off point is a local minimum before a midswing point.
   (b) *Midswing points* are the maximum angular velocities of the restless leg during the swing phase. In the algorithm, a mid-swing point is a local maximum with positive angular velocity.
   (c) *Heel-strike points* are the events when the restless leg touches the floor. In the algorithm, a heel-strike point is a local minimum after a midswing point.
8. *Gait parameter calculation:*
   (a) *Stride time* is defined as the elapsed time between two consecutive gait events. In the proposed algorithm the stride time is calculated as the average of three values. The first value is the elapsed time interval between two consecutive midswing points. The second and last one are similar to the first one but for the heel-strike points and the toe-off points.
   (b) *Swing time* is computed as the elapsed time between a toe-off event and the next heel-strike event.
   (c) *Cadence* is defined as the rythm of walking. It is calculated as the quotient of the number of steps and the gait time. In the algorithm, the number of steps is the number of the complete gait cycles (successive heel-strike, midswing and toe-off events); hence, the number of steps is equal to the number of midswing events. The gait time is the elapsed time (in seconds) between the first toe-off and the last heel-strike.

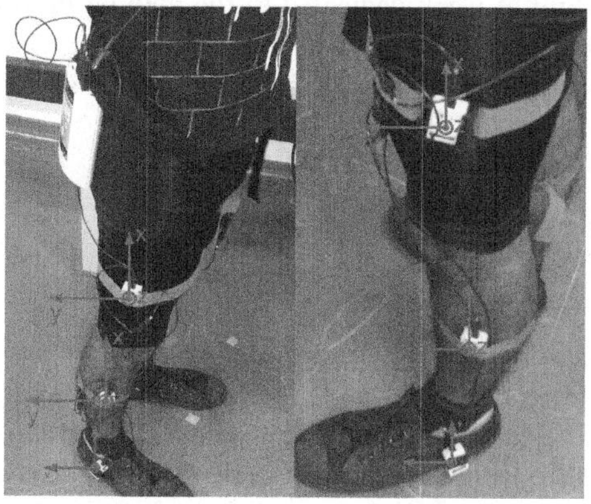

Fig. 1: Sensor locations for both legs

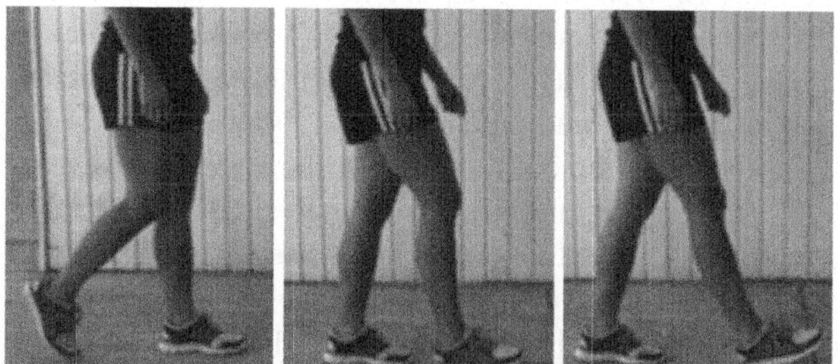

(a) Toe-off point (b) Mid-Swing point (c) Heel-strike point

Fig. 2: Gait cycle

III Results

The human gait temporal parameters calculation algorithm was used to calculate the gait parameters of a 21 year old subject with an apparently normal gait. He performed six different experiments, each one consisting in a straight line walking composed of 21 steps. Figure 3 presents the events midswing, heel-strike and toe-off points for the first experiment. The *Alg.* column of the Table 1 contains the cadence values given by the proposed algorithm. The *Tech.* column shows the same variable but obtained with the *Technaid* system. The third column is the percentage difference between these two values. Tables 2 and 3 present the comparisons for the swing time and the stride time, respectively.

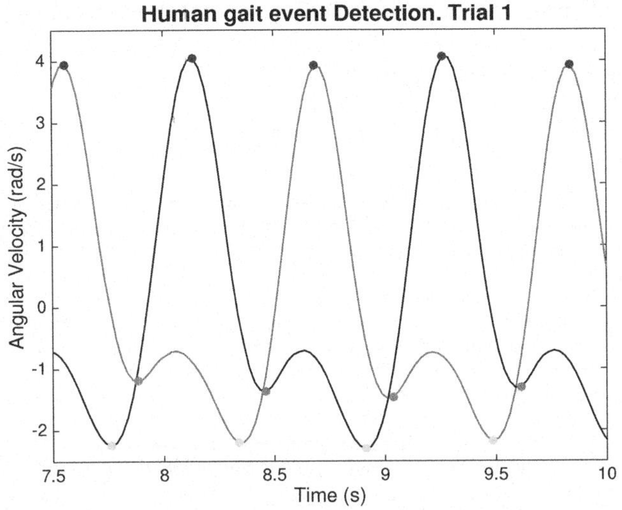

Fig. 3: *Human gait event detection for trial 1.* In red, the figure presents the MLAV signal for the left shank; while, in black, it shows the MLAV signal for the right shank. Blue dots represents the mid-swing event detection. Cyan marks are the heel-strike event detection, and magenta dots are the toe-off detection.

Table 1: Cadence calculation statistics (steps/min)

Tr.	Alg.	Tech.	%Diff.
1	105.3624	104.3	1.0186
2	107.8353	105.5	2.2136
3	105.7927	103.2	2.5123
4	107.9473	107.9	0.043824
5	104.7981	104.3	0.47754
6	100.9127	100	0.91268

Table 2: Swing stance time statistics (seconds)

	Right leg			Left leg		
Tr.	Alg.	Tech.	% Diff.	Alg.	Tech.	% Diff.
1	0.75322	0.64	17.6907	0.69588	0.65	8.7305
2	0.72829	0.6	21.3824	0.70599	0.63	17.665
3	0.72589	0.64	13.4208	0.70587	0.65	10.2919
4	0.74195	0.6	23.659	0.7085	0.59	18.083
5	0.74642	0.63	18.4788	0.7009	0.67	11.2545
6	0.76455	0.65	17.6227	0.71836	0.54	10.5163

Table 3: Stride stance time statistics (seconds)

	Right leg			Left leg		
Tr.	Alg.	Tech.	% Diff.	Alg.	Tech.	% Diff.
1	1.1807	1.15	2.6654	1.1598	1.15	0.85149
2	1.1533	1.13	2.0605	1.1451	1.15	0.42395
3	1.1756	1.18	0.37181	1.1911	1.15	3.5735
4	1.15	1.13	1.7658	1.1466	1.1	4.2377
5	1.1832	1.15	2.8862	1.1842	1.15	2.9749
6	1.2306	1.2	2.5528	1.2153	1.2	1.2781

IV Discussion

The gyroscope-based gait event detection algorithms, widely reported on the literature [6–8], are used in the design of human gait temporal parameter calculation systems.

The event detection results given by the proposed algorithm are similar to those presented by Greene et al [6] and Fraccaro et al [7], the difference with those works lies the nature of the algorithms. The reported ones are adaptive threshold based algorithms while this is a derivative-based algorithm.

In order to show how well the algorithm behaves on real data, Figure 4 aims to display the coincidence between the human gait events and the derivate sign changes. In this figure are presented the local maxima and minima calculated through the signum function.

The proposed algorithm requires only two sensors instead of the eight sensors required by the *Technaid* software for the same task. As shown in Tables 2 and 3, the proprietary software was not able to process the data obtained from the third experiment. On the other side, the new algorithm computes the required values for all the six trials. The results of the algorithm present a low mean error in cadence and stride time parameters; but, in the swing time, this mean-error increases. Standard deviation shows a low dispersion over the data, which is an encouraging fact. Table 4 is a statistics sum-

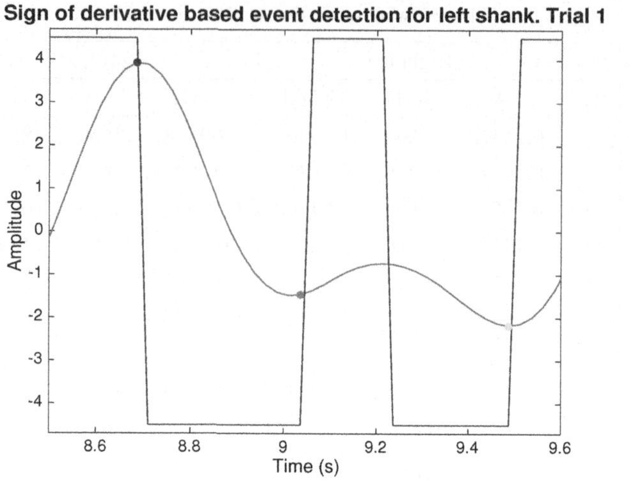

Fig. 4: *Event detection process for trial 1.* The red line represents the MLAV of the left shank. The blue line shows the signum of the MLAV signal derivative. The blue, cyan and magenta dots respectively represents the midswing, heel-strike, and toe-off points.

mary for measurement errors. The nature of these differences can not be determined because there is no information about the algorithms supporting the *Technaid* software. Nevertheless, in the aim to compare the *Technaind* algorithm and the proposed one, an optical motion capture system can be used due to its great accuracy.

Table 4: Statistics summary of human gait temporal parameters

Parameter	Mean	Standard Deviation
Cadence	1.1964	0.9720
Right stride time	2..0504	0.9196
Left stride time	2.2233	1.5785
Right swing stance time	18.709	3.5181
Left swing stance time	12.7569	4.0502

V Conclusion

The results generated by the proposed algorithm were validated with those of the commercial software *Tech MCS Studio* of *Technaid.* The comparison gives a mild difference for cadence and stride time. The new algorithm is a good alternative to the thresholding-based algorithms previously reported in the literature which require a great amount of work dedicated to the manual tuning of their multiple parameters.

Conflict of Interest

The authors certify that they have NO affiliations with or involvement in any organization or entity with any financial interest or non-financial interest in the subject matter or materials discussed in this manuscript.

Acknowledgement

The authors wish to express their gratitude to: Corporación Autónoma de Occidente, Universidad del Cauca, Escuela Colombiana de Ingeniería Julio Garavito, Proyecto Innovación Cauca for the academic goods necessary to develop the research described in this paper.

References

1. Herran Alvaro, García-Zapirain Begoña, Méndez-Zorrilla Amaia. Gait analysis methods: An overview of wearable and non-wearable systems, highlighting clinical applications *Sensors (Switzerland).* 2014;14:3362–3394.
2. Bugané F., Benedetti M. G., Casadio G., et al. Estimation of spatial-temporal gait parameters in level walking based on a single accelerometer: Validation on normal subjects by standard gait analysis *Computer Methods and Programs in Biomedicine.* 2012;108:129–137.
3. Greene Barry R, Foran Timothy G, McGrath Denise, Doheny Emer P, Burns Adrian, Caulfield Brian. A Comparison of Algorithms for Body-Worn Sensor-Based Spatiotemporal Gait Parameters to the GAITRite Electronic Walkway *Journal of Applied Biomechanics.* 2012;28:349–355.
4. Jasiewicz Jan M., Allum John H J, Middleton James W., et al. Gait event detection using linear accelerometers or angular velocity transducers in able-bodied and spinal-cord injured individuals *Gait and Posture.* 2006;24:502–509.
5. Aung Min S H, Thies Sibylle B., Kenney Laurence P J, et al. Automated detection of instantaneous gait events using time frequency analysis and manifold embedding *IEEE Transactions on Neural Systems and Rehabilitation Engineering.* 2013;21:908–916.
6. Greene Barry R., McGrath Denise, O'Neill Ross, O'Donovan Karol J., Burns Adrian, Caulfield Brian. An adaptive gyroscope-based algorithm for temporal gait analysis *Medical and Biological Engineering and Computing.* 2010;48:1251–1260.
7. Fraccaro Paolo, Walsh Lorcan, Doyle Julie, O'Sullivan Dympna. Real-world Gyroscope-based Gait Event Detection and Gait Feature Extraction *eTELEMED 2014, The Sixth International Conference on eHealth, Telemedicine, and Social Medicine.* 2014:247–252.
8. Sabatini Angelo M., Martelloni Chiara, Scapellato Sergio, Cavallo Filippo. Assessment of walking features from foot inertial sensing *IEEE Transactions on Biomedical Engineering.* 2005;52:486–494.

Autor: M.Sc. Pablo Eduardo Caicedo R.
Universidad: Corporación Universitaria Autónoma del Cauca
Dirección: Calle 5 # 3-85, Popayán, Cauca, Colombia
Email: pablo.caicedo.r@uniautonoma.edu.co

Autor: Ph.D. Carlos Felipe Rengifo R.
Universidad: Universidad del Cauca
Dirección: Carrera 2 # 4N-140, Popayán, Cauca, Colombia
Email: caferen@unicauca.edu.co

Autor: Ph.D. Luís Eduardo Rodríguez C.
Universidad: Escuela Colombiana de Ingeniería
Dirección: AK 45 (Autonorte) # 205-59, Bogotá, Cundinamarca, Colombia
Email: luis.rodriguez@escuelaing.edu.co

Health Technology Management for Digital Medical Scales in Primary Healthcare

F.S. Rosa[1] and R. Garcia[2]

[1,2] Instituto de Engenharia Biomédica (IEB), Departamento de Engenharia Elétrica (DEEL), Universidade Federal de Santa Catarina (UFSC), Florianópolis, Brasil

Abstract— **This paper presents a study on the management of Digital Medical Scales in primary health care facilities. Poor assessment of calibration status by the medical staff and the costs involved in this type of activity hinder the release of financial resources in public healthcare systems, limiting the management actions of health technology. The methodology used calculations of the total cost of the park to assess the maximum percentage cost for budget prediction, costs prevailing in the market and a logistics solution to enable the realization of this activity. From the obtained results it can be observed that in many situations the reading errors presented by the equipment went beyond the maximum permitted tolerances, evidencing the low perception of the clinical staff as for the equipment calibration. They are presented some of the origins of non calibration of equipment identified during the execution in the field. The implementation of a program of preventive maintenance by the Clinical Engineering structure is an interesting solution to improve the quality of services provided in basic healthcare.**

Keywords— **Health Technology Management, Primary Healthcare, Clinical Engineering.**

I. Introduction

The Biomedical Engineering Institute at UFSC (IEB-UFSC) has a methodology for application of Clinical Engineering (CE) consolidated through its Center of Management for Medical and Hospital Technology Development (CEGED-TMH). This methodology of CE application, in which the focus is the technological process composed by three domains: Technology, Infrastructure and Human resources. The concern to keep these three basic domains in a proportional manner, therefore contribute with quality in technological processes, reaching greater reliability, safety and effectiveness.

Indicators and data processing tools are important to evaluate the different aspects around these three pillars, and where studies are always necessary to explore distinct indicators. Programs of Preventive Maintenance (PM) in equipment can generate several indicators, reducing the need for corrective interventions.

The development of methodologies HTM is consolidated through the application in extension projects by CEGED-TMH, being one of IEB-UFSC projects focused on Primary Healthcare (PH) [1]. This project is a partnership between the IEB-UFSC and the Florianopolis' City Health Bureau (SMS-PMF), a project is being developed since 2007 [2].

The medical scale is used to measure the body weight of patients in general, it is indispensable for prenatal care, nutritional monitoring, evaluation of physical development of neonates and children patients, monitoring in patients with a body weight fluctuation history, and proper dosage of medicine to patients, are some of the significant cases of its application and is directly related to its construction quality and equipment calibration.

The CE has a key role in the management of these technologies and their calibrations because the medical staff has no knowledge and/or appropriate instruments to assess the situation, and as it will be noted in this paper, are induced to believe that the displayed values by the equipment are appropriate. A PM program allows proper evaluation of these conditions.

The objective of this study was to evaluate the impact on the management and financial benefits of the application of a preventive maintenance program in Body Weight scales in Primary Health Care.

II. Materials e Methods

The PM program implantation theoretically provides an improvement in the availability of equipment for use, reduction of activities and Corrective Maintenance (CM) costs, as well as providing constant monitoring to the use of the equipment and its technology life cycle.

In legislation terms, currently the ordinance of Inmetro 236/1994 defines calibration scales throughout the country [3]. This ordinance, together with the Guidance Document DOQ-CGCRE-036 of Dec/2012 (Guidelines for intermediate verification on scales) from Inmetro, define the procedure and the calibration periodicity in up to 12 months (ca-

I. Torres et al. (eds.), *VII Latin American Congress on Biomedical Engineering CLAIB 2016, Bucaramanga, Santander, Colombia, October 26th -28th, 2016*,
IFMBE Proceedings 60,
DOI: 10.1007/978-981-10-4086-3_73

libration of scales), and also the calibration of the weights standards (every 60 months). [4]

The PM program implementation was necessary in the scope of Primary Health care (PHC), and a study analyzing a Digital Medical Scales inventory (DS) composed of approximately 550 equipment, with about 55% Digital Physical Scales (Adult) and 45% Digital Pediatric Scales (Neonates).

The removal of each DS for evaluation in third party companies, as well as costly, is a excessively time-consuming process, making this kind of activity impossible technically and economically. The most appropriate method would be to carry out the activities on site where the equipment is in use.

The costs definition for execution of these activities was based on the scales park total cost and individual calibration costs on third party enterprises, with an acceptable spending limit of up to 5% of the park total cost a year to implement the PM program (methodology applied by IEB-UFSC in HTM forecast an annual investment of up to 10% in accumulated external maintenance in 12 months). Thus the following data were obtained.

Table 1 Contracted park Estimated cost (R$ = Brazilian Real)

Item	Qty.	Unit Cost	Total Cost
Digital Pediatric Scales	259	R$ 500,00	R$129.500,00
Digital Physical Scales	328	R$ 900,00	R$ 295.200,00
Total Park	587	---	R$ 424.700,00
5% of the Park			R$ 21.235,00

The equipment estimated cost is related to the beginning of the Primary Healthcare Management, since 2007.

Table 2 Individual calibration cost of equipment in third party enterprises

Item	Qty.	Unit Cost	Total Cost
Digital Pediatric Scales	259	R$ 65,00	R$ 16.835,00
Digital Physical Scales	328	R$ 68,00	R$ 22.304,00
Total Park Cost	587	---	R$ 39.139,00
Monthly Cost			R$ 3.261,58

The cost was extremely high and therefore was obtained a new cost to the activities execution directly in the locations where the equipment was allocated, and obtained the following budget forecast.

Table 3 Individual equipment calibration cost by third party enterprises on site

Item	Qty.	Unit Cost	Total Cost
Digital Pediatric Scales	259	R$ 23,50	R$ 6.086,50
Digital Physical Scales	328	R$ 32,50	R$ 10.660,00
Total Park Cost	587	---	R$ 16.746,50
Monthly Cost			R$ 1.395,54

Therefore, costs became acceptable to perform the activities and also with a safety margin to perform corrective maintenance activities with parts replacement when needed. The PM activities were planned in partnership with a company specialized in weight adjustments and DS calibration of various brands, possessing accreditation in Inmetro and standard operating procedures by issuing fully traceable reports with all data for each scale.

The standard operating procedure (SOP) applied to Digital Pediatric Scales recommended the growing inclusion of standards loads of 1000 g, 5000 g, 10000 g and 15000 g, followed by decreasing removal of equipment with capacity up to 15 kg, or extra loads of 20000 g and 25000 g in equipment with capacity up to 25 kg. The maximum tolerated errors in equipment were typically ± 15 g, where the equipment weight and calibration adjustment that exceeded this tolerance.

The standard operating procedure applied in Digital Physical Scales recommended similar procedure to the used in Pediatric Scales, being used standards loads of 5000 g, 50000 g, 70000 g, 90000 g, 115000 g and 150000 g, in equipment with capacity up to 200 kg. The maximum tolerated errors were typically ± 300 g, the weight and calibration adjustments being carried when exceeding this tolerance.

The Clinical Engineering Structure has a total of 56 Health Facilities (HF). The geographical disposition and distance between these Facilities (Figure 1) is one of the most important factors for defining the assistances planning and execution of activities by the CE structure, which performs weekly at least one visit to each health facility. The assistance logistics depends on some other factors such as: unit degree of complexity, the call degree of risk and stop services.

Fig. 1 HF geographical disposition of Primary Healthcare in Florianópolis.

Therefore, for the PM activity in DS planning logistics and execution of activities were developed, predicting a 12-month cycle for full coverage of the facilities, collection and refining the data produced. The planning forecasted an average service of 2-3 facilities biweekly, reaching an average of approximately 50 equipment monthly.

III. RESULTS

During the execution of this study, some facilities were closed for renovation and therefore their DS, about 100 equipment, had their calibrations postponed for the facilities reopening. Some devices were taken to CM before the visit to the sectors, so there were not collected data for this study.

Health facilities were concatenated into five regions due to the geographical disposition region of the study, thus providing the following data collection:

Table 4 Percentage of DS verified out of calibration

Region	Qty DS evaluated	Qty DS not calibrated	Percentage of Calibration
A	71	40	56,3%
B	37	12	32,4%
C	148	95	64,2%
D	96	47	48,9%
E	105	42	40%
Total	457	236	51,6%

The economic advantage of the PM implementation in all equipment (including equipment that showed no calibration failure) in relation to performing calibration only in equipment that had called for CM is what is called economic equilibrium point.

The economic cutoff point of DS points toward a percentage of approximately 37.8% of the equipment requiring calibration for economic compensation compared to the previous method.

It can be observed below that this point has been surpassed in almost all regions.

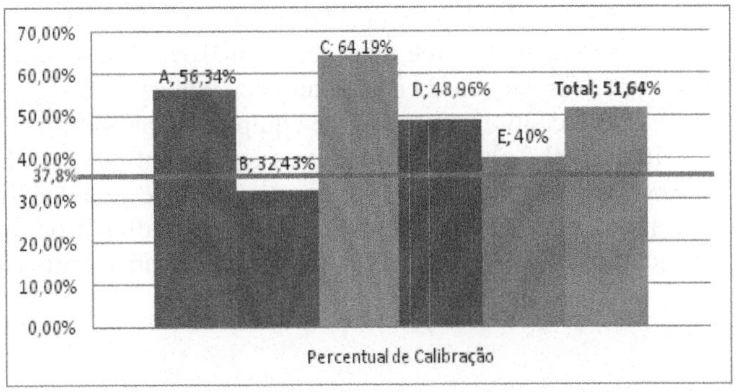

Fig. 2 Calibration economic equilibrium point.

The equipment calibration problems origins are the most diverse. During this study, the operational error most frequently identified was the placement of objects on the equipment' weighing platforms. This action causes changes in load cells, including permanently damaging it. Other observed errors were: inadequate transportation between offices (carrying the Digital Pediatric Scales of the tray), children playing and jumping on the platform (Digital Physical Scales), power outage causing short circuit of the power supply and interruption of power supply while scale' startup.

The facilities infrastructure had some flaws, which were corrected. The main occurrences were the floor unevenness in the facilities causing measurement errors, even with calibrated equipment (solved by regulating the shallow foundation), no standard electrical outlet, equipment exposed to the sun and little space for allocating the equipment.

IV. CONCLUSIONS

Operating costs for performing the activity were significantly reduced with the methodology adoption of a calibration implemented on site. The logistics employed during the study also provided other economic advantages to the method, since there was a significant reduction in calls to CM and the need to forward equipment to third party companies.

The maximum tolerable errors for Digital Pediatric Scales are 15 g while in Column is 300 g. This sensitivity degree ends up producing a lack of awareness on users in the calibration status of the equipment. The amount of equipment out of calibration identified during the study shows that this perception is practically null.

However, a more thorough analysis of the data points out that the amount of equipment with error way above the measurement tolerance (considering the triple tolerance) was low compared to the total, less than 10%. The absence of comparison standards is also another important factor in the users perception regarding the equipment calibration.

Although the alarming amount of equipment out of calibration, the risks to patients were not as significant since the tolerances were not overly extrapolated, but after program application of PM, these risks were considerably reduced. The medical staff confidence and security for the use of equipment has been improved.

CONFLICT OF INTEREST

The authors declare that they have no conflict of interest.

REFERENCES

1. Garcia SJA, Santos RAL, Avelar PS, Zaniboni R, Garcia RO, Health care technology management applied to public primary health care. PAHCE 2011; 2011 Mar 28-Apr 1; Rio de Janeiro, Brasil.
2. Santos RAL, Junior CDMSM, Garcia RO. Health technology management model applied in primary healthcare in Brazil. 2º Who Global Forum on medical devices; 2013 Nov 22-24; Genebra, Suiça.
3. INMETRO. (1994) Portaria INMETRO nº 236. Brasília, Brasil. http://www.inmetro.gov.br/rtac/pdf/rtac000180.pdf
4. INMETRO. (2012) DOQ-CGCRE-036 Orientações sobre verificação intermediária das balanças. Brasília, Brasil. http://www.inmetro.gov.br/Sidoq/pesquisa_link.asp?seq_tipo_documento=5&cod_uo_numeracao=00581&num_documento=036

Corresponding author:

Author: Felipe dos Santos Rosa
Institute: IEB-UFSC
Street: Campus Universitário – UFSC, S/N.
City: Florianópolis
Country: Brasil
Email: felipedsr@gmail.com

Selective cytotoxicity of a novel compound based on Ruthenium II in a Gallbladder carcinoma cell line

Villota, Hernán[1,2]., Pizarro, Sebastian[3]., Gajardo, Francisco[3]., Delgadillo, Álvaro[3]., Cortés-Mancera, Fabián[2]., Bernal, Giuliano [2].

[1]Ciencias Biomédicas, Facultad de Medicina, Universidad Católica Del Norte. Coquimbo, Chile
[2] Facultad de Ciencias Exactas y Aplicadas- GI^{2}B, Instituto Tecnológico Metropolitano. Medellín, Colombia
[3]Departamento de Química, Facultad de Ciencias, Universidad de La Serena. Serena, Chile

Abstract— **Gallbladder cancer (GBC) is highly prevalent in Chile and some southern regions of Colombia. This cancer presents low survival rate and poor prognosis in advanced stages. At this respect, the search for new drugs with better selective activity is necessary, among them ruthenium complexes represent an alternative for the development of more efficient drugs for the treatment of GBC. The objective of this study was to determine the *in vitro* cytotoxic activity of a novel complex based on Ruthenium (II) on the GBC cell line G-415 vs the non-tumor cell line HEK239 using MTT assay. The cell viability of the experiments showed a significant activity of the treatments in tumor cells.**

Keywords— **chemotherapy, ruthenium complexes, selectivity index, gallbladder cancer, specificity.**

I. Introduction

The gallbladder cancer (GBC) is highly prevalent in Chile surpassing the levels of incidence worldwide [1], while Colombia has reported significant levels in the incidence of this disease in the southern part of the country [2]. The survival rate of GBC is low due to the lack of early detection and poor prognosis in advanced stages. Indeed, current drugs have limited effects for this type of cancer, including Cisplatin medication approved by the American agency Food and Drug Administration (FDA) in 1978. This drug and ruthenium are members of the platinum group, although Cisplatin has had excellent results in the treatment of some types of cancer, its efficacy is limited due to high toxicity and treatment resistance development [3].

In this context is necessary the development of new and more effective anti-cancer agents, such as complexes based on ruthenium, which represent a more effective and less toxic alternative for treatment [4]. Different studies have proven anticancer activity *in vivo* and *in vitro* of different complexes based on ruthenium [5], conducting clinical trials in humans, however the mechanism of action of the different synthesized complexes still unknown. Some theories suggest that ruthenium complexes are more specific and less toxic due to the advantages that represent their various oxidation states, as well as their resemblance to elements such as iron (Fe), and the union to biomolecules of transport as transferrin [6]. The selectivity of drugs are very important in cancer treatment because if this is low the consequences of treatment are reflected in serious side effects because they affect both tumor cells and non-tumor cells. In this context, it is important research a new drug that provides a high level of specificity against the tumor cells and thus enhance chemotherapeutic treatment [7], ruthenium complexes appear to have a more specific mechanism of action than cisplatin, and other drugs currently used in the clinic.

II. Materials and Methods

A. Ruthenium complex

For the synthesis of the ruthenium complex [(pbbzim) Ru (tpy-phCOOCH3)](Cl)2 were mixed in amounts stoichiometries; ruthenium trichloride (III) (RuCl3xH2O) with the respective ligand pbbzim and heated for 4 to 8 hours, The reaction mixture was cooled and filtered, later concentrated and precipitated with ethyl ether. The solid obtained was purified by column chromatography. The characterization was carried out using the technique of Proton Nuclear Magnetic Resonance (1H-NMR), the complex present a concordance between the area of integration of the signals of protons with the number of protons that presents the complex. The resonance spectrum indicated the correct synthesis of the new complex developed [8].

The synthesis and characterization were performed in the laboratory of inorganic chemistry at the Universidad de la Serena, Serena, Chile. (Data not shown)

B. Cell culture

Epithelial cell line of a Human gallbladder carcinoma G-415 was used. As non-tumoral cell line was cultured the human

I. Torres et al. (eds.), *VII Latin American Congress on Biomedical Engineering CLAIB 2016, Bucaramanga, Santander, Colombia, October 26th -28th, 2016,*
IFMBE Proceedings 60,
DOI: 10.1007/978-981-10-4086-3_74

kidney embryonic cells HEK-293 (©RIKEN BRC Cell Bank). These cell lines were grown under controlled conditions of 5% CO2, 70% humidity and a temperature of 37 ° C. The G-415 was cultured in medium RPMI 1640 supplemented with 10% FBS, LC5 pen / strep and 125µl Fungizone, while HEK 293 cells were cultured in medium (Dulbecco's Modified Eagles medium) DMEM supplemented with 5% FBS, LC5 pen / strep and 5ML Fungizone.

C. Cell viability assay

For evaluation of cellular viability, G-415 and HEK-293 cells were seeded in 96-well plates at 1x10^4 cells/well. After 24 h, cells were treated with different concentrations of the ruthenium complex up to 100µM. The complex was dissolved in dimethyl sulfoxide (less than 1 % final concentration) and incubated for 24–48-72 h (37 °C, 5 % CO2). Cisplatin was used as control in the experiments. After incubation with MTT reagent, formazan crystal were dissolved with DMSO. Reading of the absorbance was conducted at 570 nm (NOVOSTAR 700-0130); the percentage of living cells was given by dividing the optical density (OD) of the cells treated with the (OD) from the control cells * 100.

$$Cell\ viability\ (\%) = \frac{\text{OD sample}}{\text{OD control}} X\ 100 \quad (1)$$

D. Statistical analysis

Statistical calculations to determine the maximum concentration that cause 50% of cytotoxicity in cells (CC_{50}) were performed in the software program GraphPad Prism. To determine the selectivity index of the new ruthenium complex was used the CC_{50} values of HEK-293 and CC_{50} of G-415 using the following equation: [9].

$$SI = \frac{\text{CC50 nontumoral cells}}{\text{CC50 tumoral cells}} \quad (2)$$

To assess differences between data Kruskal-Wallis method was applied, using SPSS windows software. Significance differences were consider when p value was ≤ 0.05.

III. Results

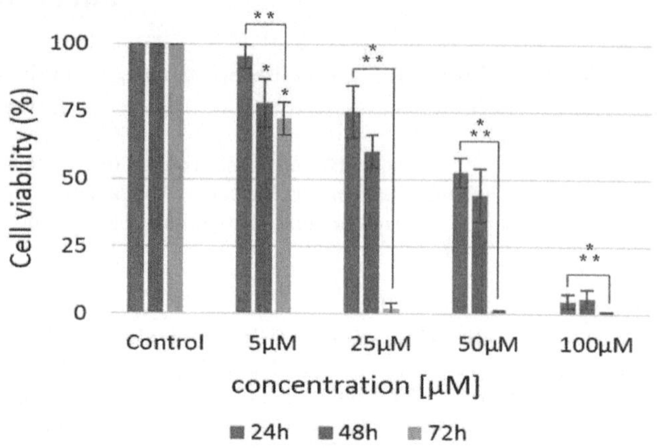

Figure 1. Antiproliferative activity of the Ruthenium complex (Ru) in the cancer cell line of gallbladder G-415.
* p=<0.05 compared to the control, ** p=<0.05 in the same concentration but different time.

The results of the cell viability showed an evident inhibition of proliferation dose/time dependent with the novel compound (Fig 1).

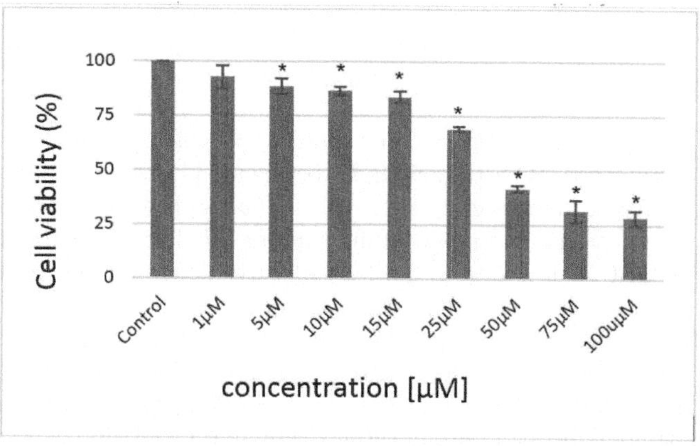

Figure 2. Antiproliferative activity of the Cisplatin in the cancer cell line of gallbladder G-415 after 48h. * p=<0.05 compared to the control,

The results of MTT showed an inhibition of proliferation of dose dependent with Cisplatin (Fig 2). The CC_{50} value in G-415 cells corresponded to 41,11µM at 48 h of treatment. In contrast ruthenium complex (Ru) preparation had a more potent activity against G-415 cells with a CC_{50} value of 28.07 µM to 48h treatment (Fig 1).

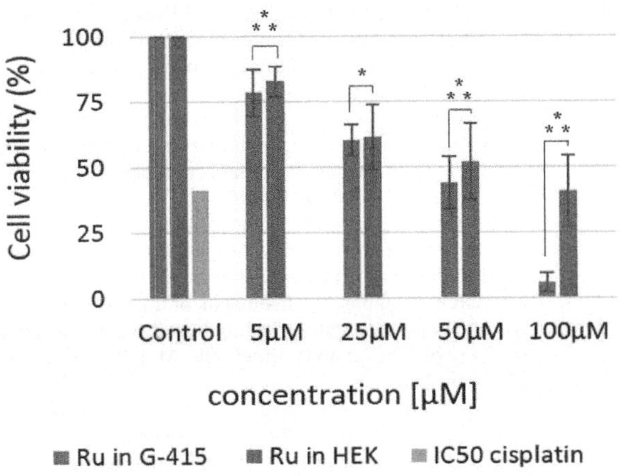

Figure 3. Comparison of antiproliferative activity of Ru in G-415 and HEK cells during 48h of treatment. * p=<0.05 compared to the control, ** p=<0.05 in the same concentration but different time.

As you can observe in figure 3, the results of cell viability indicated a greater inhibition of proliferation of Ru in G-415 cells compared to the activity in HEK-293 cells which were less sensitive to the treatment (CC_{50} value of 118,01 µM). Statistical analysis revealed a significant difference in the inhibition of viability of HEK-293 cells against G-415, starting 5 µM. Using calculated CC_{50} values, the selectivity index value was ≥ 4.

IV. Discussion

The properties that have some metals have drawn attention to research in chemotherapeutic agent for cancer treatment. The main complex used today correspond to Cisplatin, which has been effective in treating some cancers such as testicular cancer in the early diagnosis of this. However for other cancers with late diagnosis, such as gallbladder cancer (GBC), most patients (90%) are candidate for chemotherapy and radiotherapy [10]. In the case of GBC, chemotherapy protocol has not been fully defined, although some clinical trials testing the effectiveness of treatment indicates that combination of Cisplatin and Gemcitabine appear more effective than monotherapy [11-12]. However, the use of agents such as cisplatin is limited by problems of toxicity and resistance given mainly by the unspecificity in the mechanism of action of these complexes. In this line of research, the ruthenium complexes (Ru) represent a possible alternative for a most effective and safe in the chemotherapy treatment. The properties of Ru, such as its wide range of oxidation states and the binding affinity with cell molecules like transferrin, would be related to the benefits in the selectivity of these complexes towards cancer cells. In this work it was evaluated the anti-proliferative activity of a new complex based on Ru II in the gallbladder cancer cell line G-415 and the non-tumoral cells HEK-293. The results of cytotoxicity assays with MTT indicated a significant difference in the activity of the complex dependent of the cell line, being carcinoma cells G-415 more sensitive to the treatment compared to non-tumoral cells HEK-293, with values of CC_{50} to 48h of 41,11uM for cisplatin in G-415, 28, 07 for the ruthenium complex in G-415 cells and of 118, 01 for the Ru complex in HEK-293 cells, indicating a good selectivity index.

Although no previous data exist about ruthenium complexes selectivity in the cell model used in the present study, some reports have found heterogeneous results in other tumoral lines, such as DU-145 (prostate cancer) and HepG2 cells (hepatocellular carcinoma) [13-14]. In overall, our results with a novel Ru complex indicates that this compound would correspond to an important alternative in cancer treatment.

V. Conclusions

The results showed that the ruthenium II complex presented evident of higher antiproliferative activity in tumor cells, with an important index of selectivity when compare with control cells. Our results indicate that the synthesized ruthenium complex would be a potential agent for the treatment of GBC.

Acknowledgment

Authors thank to Group of molecular and cellular biology of cancer. Universidad Católica Del Norte (UCN), Coquimbo, Chile.

Conflict of Interest

The authors declare that they have no conflict of interest.

References

1. MINSAL. (2004). Registro Poblacional del Cáncer. Provincia de Valdivia, Chile.
2. Anderson Rocha-Buelvas, Elizabeth Trujillo-Montalvo, Carlos Hidalgo-Patiño, Angela Hidalgo-Eraso, (2010). Carga de Cáncer del Departamento de Nariño y Subregiones. Revista de la Escuela Nacional de Salud Pública.

3. Bruijnincx, Pieter C. A. and Sadler, P. J.. (2008) *New trends for metal complexes with anticancer activity*. Current Opinion in Chemical Biology, Vol.12 (No.2). pp. 197-206. ISSN 1367-5931
4. William M. Motswainyana and Peter A. Ajibade.(2015) "Anti-cancer Activities of Mononuclear Ruthenium(II) Coordination Complexes," Advances in Chemistry, vol. 2015, Article ID 859730, doi:10.1155/2015/859730
5. Irena Kostova. (2006). Ruthenium Complexes as Anticancer Agents, Current Medicinal Chemistry, vol 13 Page: [1085 - 1107], doi: 10.2174/092986706776360941
6. Claire S. Allardyce and Paul J. Dyson. (2001). ruthenium in medicine current clinical uses and future prospects. Platinum Metals Review
7. Ramesh B. Badisa, Selina F. Darling-Reed, Patrick Joseph, John S. Cooperwood, Lekan M. Latinwo And Carl B. Goodman. (2010) Selective Cytotoxic Activities of Two Novel Synthetic Drugs on Human Breast Carcinoma MCF-7 Cells. anticancer research 2993–2996. PMCID: PMC2885965
8. Chanchal Bhaumik, S. D. (2010). Synthesis, Characterization, Photophysical, and Anion-Binding Studies of Luminescent Heteroleptic Bis-Tridentate Ruthenium(II) Complexes Based on 2,6-Bis(Benzimidazole-2-yl)Pyridine and 40. Inorganic Chemistry, 5049 – 5062, doi: 10.1021/ic100138s.
9. Badisa RB, Lambert AT, Ikediobi CO, Walker EH. (2006) Selective anticancer activity of pure licamichauxiioic-B acid in culture cell lines. Pharmaceut Biol., 141–145.
10. Paula María Jaramillo y Sergio Iván Hoyos. (2009). Carcinoma de vesícula biliar en el Hospital Pablo Tobón Uribe de Medellín. Revista Colombiana de Cirugia, 250-257
11. Juan Valle, M.D., Harpreet Wasan, M.D et al. (2010) Cisplatin plus Gemcitabine versus Gemcitabine for Biliary Tract Cancer, New England Journal of Medicine. doi: 10.1056/NEJMoa0908721
12. Thongprasert S, Napapan S, Charoentum C, Moonprakan S. (2005) Phase II study of gemcitabine and cisplatin as first-line chemotherapy in inoperable biliary tract carcinoma. Annals of Oncology; 279-81. PMID: 15668284
13. Ariadna Grau-Campistany, Anna Massaguer, Dolors Carrion-Salip, Barragan Flavia, Gerard Artigas, Paula Lopez-Senin, Virtudes Moreno, and Vicente Marchán. (2013) Conjugation of a Ru(II) arene complex to neomycin or to guanidinoneomycin leads to compounds with differential cytotoxicities and accumulation between cancer and normal cells, Mol. Pharmaceutics, doi: 10.1021/mp300723b
14. Xin Li, Kirsten Heimann, Fangfei Li, Jeffrey M. Warner F. Richard Keene and J. Grant Collins. (2015) Dinuclear ruthenium(II) complexes containing one inert metal centre and one coordinatively-labile metal centre: syntheses and biological activities, DaltonTransactions; doi: 10.1039/c5dt04885k

Author: Hernán Darío Villota Bolaños
Institute: Instituto Tecnológico Metropolitano
Street: 73 No 76A – 354 vía al volador
City: Medellin.
Country: Colombia.
Email: Hernanvillotah@hotmail.com

Apoptosis induced by a novel Ruthenium II complex in a Gallbladder carcinoma cell line

Villota, Hernán[1,2]., Pizarro, Sebastian[3]., Gajardo, Francisco[3]., Delgadillo, Álvaro[3]., Cortés-Mancera, Fabián[2]., Bernal, Giuliano [2].

[1]Ciencias Biomédicas, Facultad de Medicina, Universidad Católica Del Norte. Coquimbo, Chile
[2] Facultad de Ciencias Exactas y Aplicadas- GI[2]B, Instituto Tecnológico Metropolitano. Medellín, Colombia
[3]Departamento de Química, Facultad de Ciencias, Universidad de La Serena. Serena, Chile

Abstract— **Some metals with striking properties have been used for the development of different complexes with activity against malignant cells. Due to difficulties in early diagnosis of gallbladder cancer (GBC) the main treatment option is chemotherapy. In this line of research, ruthenium complexes represent a less toxic alternative that others metal complexes such as cisplatin. The aim of this study was to evaluate the effect of a novel ruthenium complex in activation of apoptosis in a cell lineG-415. Our results show an activity of inducing cell death and overexpression of pro-apoptotic transcripts.**

Keywords— **chemotherapy, ruthenium complexes, apoptosis, gallbladder cancer, qPCR.**

I. Introduction

The gallbladder carcinoma (GBC) is very difficult to detect due to its anatomical location and nonspecific symptoms [1]. Surgical resection is the only potentially curative treatment but only 10% of cases are detected in the early state to be considered candidates for this intervention. In this context, the chemotherapy is used as auxiliary treatment after surgical resection and as the main procedure for patients with locally advanced disease [2]. However, the chemotherapy for GBC has not been defined totally, and most of the current knowledge is derived from some few phase II studies that have indicated better responses and survival rate with drugs such Gemcitabine, Cisplatin, 5-Fluorouracil, Capecitabine and Oxaliplatin [3].

Current metal complexes for the treatment of cancer, such as Cisplatin, offer good results in some types of cancer, however its use is very limited by its high toxicity. In this line, Ruthenium (Ru) complexes seem to have a mechanism that improves the specificity against tumor cells; the interaction of the ruthenium complexes with some molecules, such as proteins and mainly with DNA, causes irreversible damage and activation of apoptosis [4]. Specifically, some studies have evidenced interactions of the ruthenium complexes with cytochrome c protein which is located in the intermembrane space of mitochondria, affecting electron transportation which is important in cell respiratory chain [5]. Also, this protein is key the route of apoptosis called mitochondrial pathway, where the family of Bcl-2 regulates the process of apoptosis on the mitochondria. It has also been shown that Ru complexes has an anticancer through interactions with proteins related with DNA replication or directly to Nuclei acids [6-7].

According to the evidence discussed above, the purpose of this study was to evaluate the effect of treatment with a novel complex of ruthenium (II) on the activation of mechanisms of programmed cell death in the cell line G-415.

II. Materials and Methods

A. Ruthenium complex synthesis and characterization

For the synthesis of the ruthenium complex *[(pbbzim) Ru (tpy-phCOOCH$_3$)](Cl)$_2$* were mixed in amounts stoichiometries; ruthenium trichloride *(III) (RuCl$_3$xH$_2$O)* with the respective ligand pbbzim and heated for 4 to 8 hours, the reaction mixture was cooled and filtered, later concentrated and precipitated with ethyl ether. The solid obtained was purified by column chromatography. The characterization was carried it out using the technique of Proton Nuclear Magnetic Resonance (1H-NMR) presenting a concordance between the area of integration of the signals of protons with the number of protons that presented the complex, suggesting the correct synthesis of the new complex developed [8]. The synthesis and characterization were performed in the laboratory of inorganic chemistry at the Universidad de la Serena, Serena, Chile. (Data not shown)

B. Cell culture

Epithelial cell line of human gallbladder carcinoma G-415 was used. The cells were cultured under controlled conditions of 5% CO2, 70% humidity and a temperature of 37 ° C.

I. Torres et al. (eds.), *VII Latin American Congress on Biomedical Engineering CLAIB 2016, Bucaramanga, Santander, Colombia, October 26th -28th, 2016*,
IFMBE Proceedings 60,
DOI: 10.1007/978-981-10-4086-3_75

The G-415 was cultured in medium RPMI 1640 supplemented with 10% FBS, LC5 pen / strep and 125μL Fungizone.

C. *Real Time Quantitative PCR*

In previous experiments performed by our group the cytotoxic activity of the complex in G-415 cells was evaluated, the results showed a strong antiproliferative activity. Based on these results, we selected two concentrations for evaluation of gene expression. G-415 cells were treated with 25 and 75 μM of Ru complex for 24h and 48h. Total RNA of $4x10^5$ cells/well was extracted with the RNA SV Total RNA Insolation System of Promega following the manufacturer's protocol. Total RNA was used to produce complementary DNA (cDNA) with the AffinityScript QPCR cDNA Synthesis Kit of agilent technologies® with oligo dt primers in a total of 20 μL reaction mixture according to the manufacturer's protocol. Real-time PCR reactions were then carried out in a total of 10 μL reaction mixture: 2.2 μL of cDNA, 5 μL of SYBR Green PCR Master Mix, 2.6 μL H2O and 0,1 μL of each forward and reverse primer, the genes evaluated in this study were p53, Bax and Caspase 3. The PCR amplification conditions comprised 35 cycles of 95°C for 10 sec, and 57° or 62°, as appropriate, for 30 sec. To calculate the relative quantification of gene expression, the method $2^{(-\Delta\Delta Ct)}$ was used [9] with B2M as housekeeping gene with a relatively stable expression in normal and cancerous gastrointestinal tissue [10].

D. *Statistical analysis*

For data analysis SPSS software was used. To assess differences Kruskal-Wallis, ANOVA and Tukey methods were implemented. Statistical significance was considered when *p* value was ≤ 0.05.

III. RESULTS

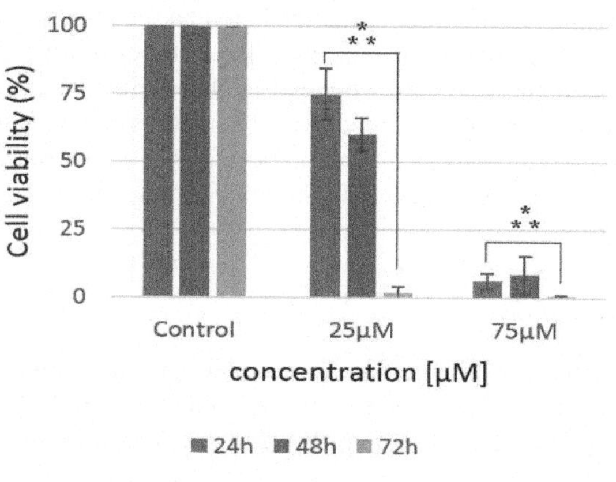

Fig 1. Antiproliferative activity of the Ruthenium complex in the cancer cell line of gallbladder G-415. * $p \leq 0.05$ compared to the control, ** $p \leq 0.05$ in the same concentration but different time.

The results of the cell viability show an inhibition of proliferation of dose/time dependent, the concentrations of 25μM and 75μM had an effect that could be comparative to perform gene expression analysis (Fig 1).

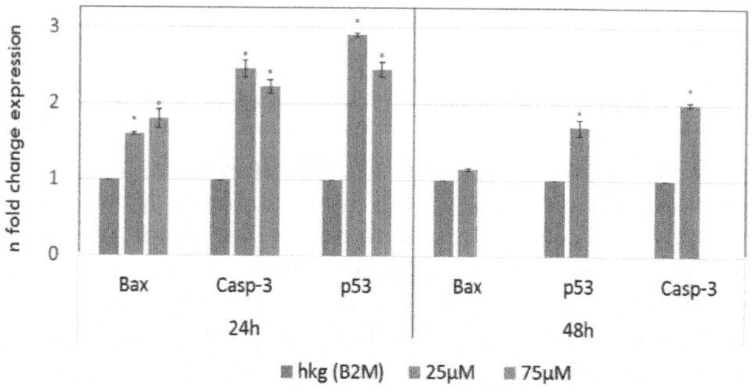

Fig 2. Effect of the Ru complex in the expression of Bax, p53 and caspase -3 in cancer cells gallbladder G-415 at 24h and 48h of treatment. * $p \leq 0.05$ compared to the control.

Gene expression was measured after 24 and 48 hours of treatment with 25 and 75 μM of Ru complex. B2M was used as housekeeping gene (fig.2). The results showed upregulation of pro-apoptotic genes up to three times in relation to the control. Activity in gene expression to different treatment concentration was similar for p53 and caspase-3, wherein the concentration of 25μM had a slightly greater effect than the concentration of 75μM. For Bax the behavior at this two different concentrations was the opposite, the concentration of 75μM had a slightly greater effect on the expression.

IV. DISCUSSION

The activity of the ruthenium complexes is dependent on its oxidation state as well as the structure of its ligands. Some hypotheses indicate that ruthenium complexes are more specific and less toxic due to the advantages that represent their various states of oxidation in addition to its similarity with iron (Fe), favoring the binding to biomolecules for transportation such as apotransferrin. It would facilitate the introduction of the complex to the tumor cells, changing the balance among anti and pro-apoptotic proteins. When the pro-apoptotic proteins of the Bcl-2 family or Bax are activated it is produced an opening in the mitochondrial membrane resulting in the release of proteins such as cytochrome c, inducing activation of a protein complex called the apoptosome, which activates caspase protease 9 and finally induces activation of effector caspases such as caspase 3 leading final stages of apoptosis [11].

The results of real time PCR in this study showed a clear overexpression of p53 to 24h treatment. p53 is the gene of response to DNA damage, inducing the expression of proapoptotic genes and preventing cell replication [12]. Genes such as PUMA or NOXA are expressed for activating proapoptotic proteins like BAX gen; this marker was increased in the present study. The activation of Bax induces releasing cytochrome c of the mitochondria, and subsequently trigger caspase cascade whose expression was also increased in our results [13].
In overall, our preliminary results confirmed that the evaluated complex is able to induce apoptosis in G-415 cells.

V. Conclusions

The results indicate that the ruthenium (II) complex has evident antiproliferative activity and induces an increasing expression of pro-apoptotic genes in the cell line G-415. Our results indicates that the novel ruthenium complex synthesized can be a potential agent for the treatment of GBC.

Acknowledgment

Authors thank to Group of molecular and cellular biology of cancer. Universidad Católica Del Norte (UCN), Coquimbo, Chile.

Conflict of Interest

The authors declare that they have no conflict of interest.

References

1. Rajveer Hundal and Eldon A Shaffer. (2014) Gallbladder cancer: epidemiology and outcome. Clinical Epidemiology 6: 99–109. doi:10.2147/CLEP.S37357
2. Andrew X. Zhu,Theodore S. Hong, Aram F. Hezel, y David A. Kooby. (2010). Current Management of Gallbladder Carcinoma. The Oncologist, 168–181 doi:10.1634/theoncologist.2009-0302
3. Dr. Mario Uribe M, Dr. Claudio Heine T, Dr. Freddy Brito M, Dra. Diana Bravo L. (2013). Gallbladder Cancer: An Update Revista Médica Clínica Las Condes. 638-643
4. Claire S. Allardyce and Paul J. Dyson. (2001). ruthenium in medicine current clinical uses and future prospects. Platinum Metals Review.
5. L Trynda-Lemiesz. (2004) Interaction of an anticancer ruthenium complex HInd[RuInd2Cl4] with cytochrome c. 199-205. Acta Biochimica Polonica.
6. Brabec V, and Nováková O. (2006) DNA binding mode of ruthenium complexes and relationship to tumor cell toxicity. *Drug Resistance Updates*. PMID: 16790363
7. Yadav A, Janaratne T, MacDonnell FM et al. (2013) Regression of lung cancer by hypoxia-sensitizing ruthenium polypyridyl complexes. Molecular Cancer Therapeutics 643-53. doi: 10.1158/1535-7163.MCT-12-1130.
8. Chanchal Bhaumik, S. D. (2010). Synthesis, Characterization, Photophysical, and Anion-Binding Studies of Luminescent Heteroleptic Bis-Tridentate Ruthenium(II) Complexes Based on 2,6-Bis(Benzimidazole-2-yl)Pyridine and 40. Inorganic Chemistry, 5049 – 5062, doi: 10.1021/ic100138s.
9. Kenneth J. Livak and Thomas D. Schmittgen. (2001) Analysis of Relative Gene Expression Data Using RealTime Quantitative PCR and the $2^{(-\Delta\Delta Ct)}$ Method. Methods. 402–408 doi:10.1006/meth.2001.1262
10. Rubie C, Kempf K, Hans J, Su T, Tilton B, Georg T, Brittner B, Ludwig B, Schilling M. (2005) Housekeeping gene variability in normal and cancerous colorectal, pancreatic, esophageal, gastric and hepatic tissues. Molecular and Cellular Probes 101-9 PMID: 15680211
11. Susan Elmore. (2007) Apoptosis: A Review of Programmed Cell Death. Toxicologic Pathology, 495-516 doi: 10.1080/01926230701320337
12. Linlin Li, Wenqiang Cao, Wenjie Zheng, Cundong Fana and Tianfeng Chen. (2012). Ruthenium complexes containing 2,6-bis(benzimidazolyl)pyridine derivatives induce cancer cell apoptosis by triggering DNA damage-mediated p53 phosphorylation. Dalton Transactions, 41, 12766-12772, doi: 10.1039/C2DT30665D
13. Lima AP , Pereira FC , Almeida MA , Mello FM , Pires WC , Pinto TM , Delella FK , Felisbino SL , Moreno V , Batista AA, de Paula Silveira-E Lacerda. (2014). Cytoxicity and apoptotic mechanism of ruthenium(II) amino acid complexes in sarcoma-180 tumor cells. PLoS One, doi: 10.1371/journal.pone.0105865.

Author: Hernán Darío Villota Bolaños
Institute: Instituto Tecnológico Metropolitano
Street: 73 No 76A – 354 vía al volador
City: Medellin.
Country: Colombia.
Email: Hernanvillotah@hotmail.com

Optimization of spectral analysis of electrophysiological recordings of the subthalamic nucleus in Parkinson's disease: A retrospective study

S.E. Valderrama-Hincapié[1], A.M. Hernández[1], F. Sánchez[1], S. Roldán-Vasco[2], A.L. López-Ríos[3], W. D. Hutchison[4].

[1] Bioinstrumentation and clinical engineering Research Group – GIBIC, Bioengineering Department, Engineering Faculty, Universidad de Antioquia UdeA; Calle 70 No. 52-21, Medellín, Colombia
[2] Instituto Tecnológico Metropolitano, Research Group in Advanced Materials and Energy, Medellín, Colombia
[3] Fundación Hospital San Vicente de Paúl, Rionegro, Colombia
[4] University of Toronto, Department of Surgery and Physiology, and Krembil Reserch Institute, Toronto, Canada

Abstract— **Parkinson's disease (PD) is a neurodegenerative disorder that is diagnosed in over 60.000 people per year in the USA alone. Deep brain stimulation of the STN has been implemented to ameliorate the motor symptoms, being highly effective. PD has been associated with increased power in the beta frequency band (13-35 Hz) in the STN's stereo electroencephalography signals (sEEG). Several studies have estimated the spectrum of the sEEG signals in order to identify spectral behavior according to anatomical structures and pathologies; however, the estimation methods do not give enough sensitivity. In the present study, we hypothesize that parametric methods can have a better performance to correctly estimate the spectrum of a sEEG signal in the beta band using the right model order. AR models were estimated, with four different information criteria to choose the proper order, where the Akaike's information criterion corrected (AICc) gives the best estimation with an order of 17.**

Keywords— **Autoregressive models, Parkinson's disease, Spectral analysis, Stereo-electroencephalography, Subthalamic nucleus.**

I. INTRODUCTION

Parkinson´s disease (PD) is a neurodegenerative disorder that affects basal ganglia [1] and is associated with symptoms of bradykinesia, dyskinesia and tremor [2]. One of the existing treatment consists on a high-frequency deep brain stimulation in the subthalamic nucleus (STN) [3,4], which is highly effective in reducing the motor symptoms. Nevertheless, the efficacy of this treatment is sensitive to the correct targeting of the STN [4] with microrecordings, due to its small size and the possible displacement of the brain after the incision in the cranium is made.

Several studies had found that PD is associated with an increase in the oscillatory activity [2,5]. Local field potential's recordings (LFP) within the STN show a prevalent band in patients with PD, which is called the beta band (13-35 Hz) [4]. It has been found that exist a phenomenon of phase-amplitude coupling between the beta band and the LFP during akinesia and rigidity episodes in PD [6]. It has been hypothesized that evaluating the energy in the beta band allows the correct STN's targeting [4,7,8]. The spectral estimation of the studies is often made with the fast Fourier transform (FFT) or with the Welch periodogram, which are non-parametric estimation methods. It is well known that the FFT introduces discontinuities, generating perturbations in the spectrum [9]. On the other side, the Welch approach is highly sensitive to the overlapping and windowing where the frequency resolution is reduced when the length of the signal is shorted [10]. A high resolution estimation method is recommended in this kind of signal processing because the relevant information is often associated with the occurrence of specific rhythms [9].

The parametric modeling allows the generation of a mathematical model that can represent with good approximation the behavior of the signal with the help of the indexed parameters of the model [11,12]. The auto-regressive structure (AR) is defined by equation (1):

$$x_t = \sum_{i=1}^{p} a_i\, x_{t-i} + \varepsilon_t \qquad (1)$$

Where ε_t is white noise, x_t is the signal to be modeled, a_i are the parameters of the model and p is the order of the model [12].

The information criteria are widely used to correctly estimate the order of the models and some of them are Akaike's information criterion (AIC), more often used although it has a tendency to overestimate the order [13], Bayesian information criterion (SBC) is an extension of the minimum AIC procedure and is less likely to overestimate the order of the models [13], Hannan–Quinn information criterion (HQIC) which can be considered as a variation of the SBC criterion with a small penalty of the sample size [14], and the corrected Akaike's information criterion (AICc), which has been recently explored, because it has shown a high precision in determining the order of the model with a less quantity of parameters as well as the generation of stable models [15].

In order to guarantee stable models, is necessary that the

I. Torres et al. (eds.), *VII Latin American Congress on Biomedical Engineering CLAIB 2016, Bucaramanga, Santander, Colombia, October 26th -28th, 2016*,
IFMBE Proceedings 60,
DOI: 10.1007/978-981-10-4086-3_76

modeled signal is stationary. Due to the non-stationary nature of biosignals [9], finding a specific windows size is necessary to provide the stationarity [16]. Is also necessary to verify that the roots of the characteristic polynomic of the model's transfer function are also within the unit circle [17], and check the quality of the model by its goodness of fit.

Estimating the spectrum of a signal with a precise order model is highly suitable, because it reduces the problems that the non-parametric estimation methods have [18].

The aim of this study consisted on determining a precise order of AR models of the STN's sEEG in subjects with PD, in order to obtain a high resolution spectrum. To achieve such a goal, a cluster analysis of the cutoff frequencies of the filter in the acquisition system was made, with the purpose of only processing the signals that have information in the beta band. Then, all signals were filtered in this band and down-sampled, also the stationarity was checked. The AR models were generated, the stability was tested and the higher order model was chosen.

II. METHODS

A. Data

Signals from 18 patients were obtained, where 15 were bilaterally implanted and 3 unilaterally. The micro targeting system was the Guideline 4000 LP+ with up to five recording channels. The STN was localized using MRI, a stereotactic frame and intraoperative microrecordings.

The microrecordings were sampled at 48000 Hz. To all of the sEEG signals a hardware based filtering was applied, where the cutoff frequencies in most of the patients was different and dependent of the necessities of the surgical procedure, which lead to make a retrospective study.

B. Cluster analysis

A cluster analysis was done according to the cutoff frequencies of filtering. High and low cutoff frequencies of all the recordings were acquired, differentiating the hemispheres of the brain. The hierarchical cluster analysis was calculated with Euclidean distance, shortest distance grouping technique and it was defined with a distance of 0.25 in order to ensure a difference no superior to 50 Hz in the low cutoff frequency.

C. Signal preprocessing

A digital FIR low band pass filter with order of 1500 and cutoff frequency of 45 Hz was applied in Matlab. It was implemented also a down-sampling at 192 Hz. This process guaranteed the content of information in the band of interest.

D. Modelling

In order to find an appropriate window's length with weak stationarity and check the normality of the signals, epochs of 1, 2 and 5 seconds were tested. The median, standard deviation and normal distribution were calculated in each epoch. To verify normality Q-q graphics and histograms were obtained. In Fig. 1 a summary of the followed process is shown.

Fig. 1 Summary of the implemented methodology to obtain the AR models.

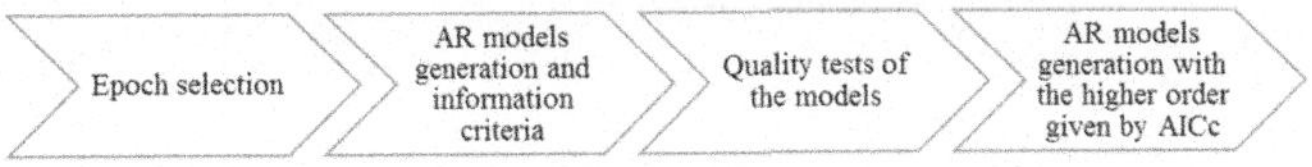

The AR models of the selected epoch of each signal were generated and were accepted exclusively if their fit was superior to 80%. The order model was determined by four information criteria: AIC, SBC, HQIC and AICc. The fit, stability and order model of every information criterion were compared. The AICc showed stability and high fit with a minor order model. Finally, the higher order model given by the AICc was chosen among all the models and all the recordings were modelled again with the order found, where were checked again the stability and fit.

E. Spectral and mean energy analysis

The energy of each epoch in the beta band over the complete surgery's trajectory in the patients was calculated, in order to check if a difference of energy exists when the electrodes are approaching the STN.

Three different spectral estimation methods were compared: Welch's periodogram, FFT and Yule-Walker. The first mentioned was implemented with a Hamming window.

III. RESULTS

The cluster analysis provided a classification of the signals according to the cutoff frequencies. As was previously described, the two hemispheres of the brain were analyzed separately. Fig. 2 shows the groups obtained by the cluster analysis in both hemispheres. According to the results obtained from the cluster analysis, the groups that had relevant information in the beta band in the left hemisphere were 4, 5, 8, 9 and 11. In the right hemisphere the groups were 5 and 6.

After the cluster analysis was made, the number of patients was reduced to 11. When the final patients were selected and all the signals were filtered and down-sampled, normality and stationary tests were carried out (described in section D), where a window's length of 2 seconds was the one that presents a reasonable amount of data and a weak stationarity.

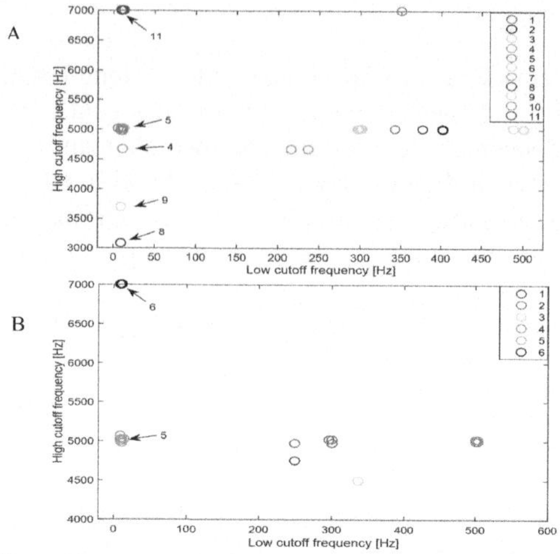

Fig. 2 Cluster analysis in left (A) and right (B) hemispheres. The groups with beta band information are indicated.

Fig. 3 shows the energy of every recording at different depths of the trajectory in one of the patients. It was found that the major part of the trajectories showed an increased energy in the proximities of the STN, as is shown in Fig. 3 where the target marked in the surgery corresponds to the electrode depth 0 mm.

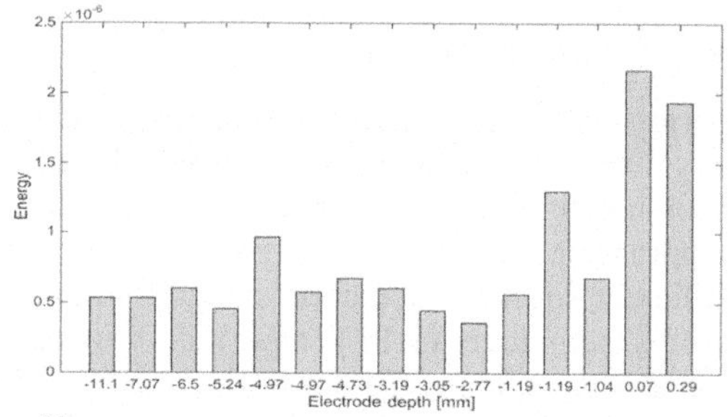

Fig. 3 Energy calculated of a complete trajectory over a patient.

It was evaluated the spectrum in diverse sEEG recordings processed in the beta band, in Fig. 4 an example of one recording is shown.

In Table 1 are shown the average values and standard deviation from all the models generated for one patient with the order models given by the information criteria.

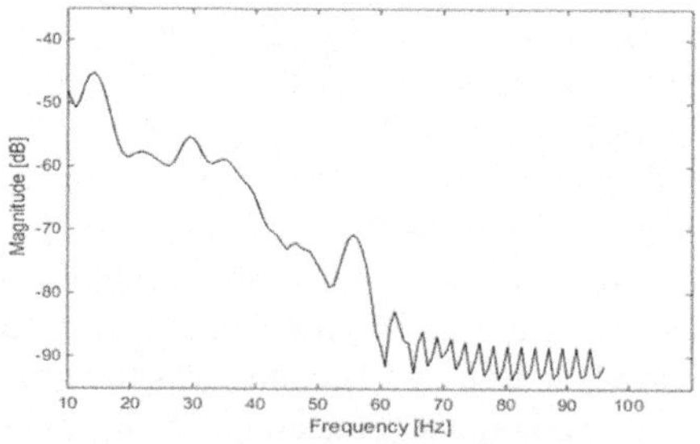

Fig. 4 Spectrum of a signal calculated with the Welch's periodogram.

Table 1 Comparison of 4 information criteria in one patient. An average fit and order model of all the recordings of a patient was obtained.

Information criterion	Order model	Fit (%)
AIC	55 ± 11	95 ± 1.8
HQ	38 ± 12	94 ± 2.1
SBC	27 ± 6	94 ± 2.2
AICc	9 ± 2	91 ± 2.6

The AR models were generated with a 17 order model. In Table 2 a summary of the AR models is shown.

Table 2. Mean and standard deviation of the fit per patient.

Patient	Right hemisphere	Left hemisphere
2	92,317 ± 2,390	-
5	-	95,177 ± 2,312
7	-	92,122 ± 2,735
9	95,089 ± 3,540	94,824 ± 2,361
10	91,380 ± 3,882	-
12	95,301 ± 1,124	-
13	92,769 ± 3,689	91,017 ± 5,100
14	87,321 ± 3,716	92,591 ± 3,058
15	93,946 ± 2,137	-
17	-	89,214 ± 3,462
18	90,522 ± 3,393	93,897 ± 2,439

Fig. 5 shows a comparison between the FFT, Welch's periodogram and Yule-Walker.

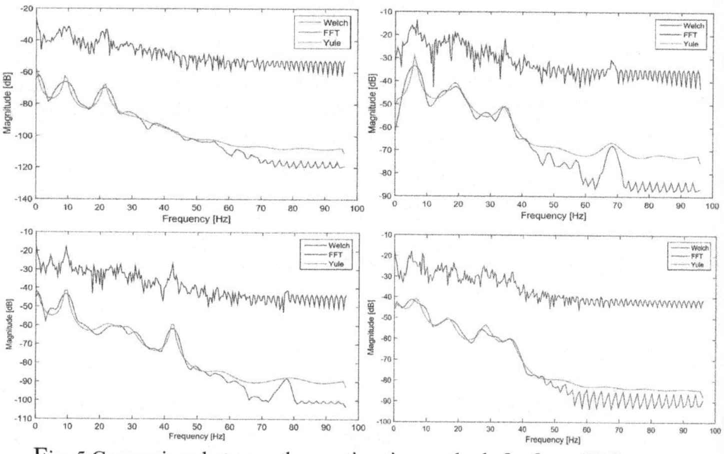

Fig. 5 Comparison between three estimation methods for four sEEGs.

IV. Discussion

From Fig. 4 is possible to observe that the frequency components of interest in this study remain unchanged, meanwhile the other bands were attenuated. The down-sampling process was also necessary due to the high original sampling frequency (48 kHz). Studies support frequencies near to 200 Hz [9,10]. The spectrum shown in Fig. 4 is not a generalized behavior for all the sEEG signals. It was found that in some cases the high beta band was the predominant but in other patients the opposite situation was given. The results in the present study are congruent with those showed in different studies, where the difference in the spectrum of STN's signals over diverse patients is remarked [7].

It was found that the AICc can accurately define the order of the model for sEEG signals. The difference between the fit of the evaluated information criteria is minimum. However, a substantial difference in the orders of the models is clear, where the smaller order model was given by the AICc, as is seen in Table 1. All the models generated with an order model of 17 showed an accuracy close to 90% (Table 2), i.e. the models represent highly the beta band coupling in the sEEG signals. Also, it was found that all the models were stable.

As is seen in Fig. 5, the parametric estimation method presents a higher resolution than the other two non-parametric methods because a better definition of the peaks is seen, which is concordant with other studies [11], as well as this estimation method does not present windowing problems, as is shown in Fig. 4 and Fig. 5.

With the AR models of sEEG signals it is hypothesized that is possible to find the entry to the STN through the calculated parameters, however a further investigation is needed.

Conclusions

In the present study was found that the AICc, generate stable models with a high accuracy and estimates a lower order in comparison with the other information criteria evaluated. It was stablished also that a 17 order model, is sufficient to simulate and estimate the spectrum of a sEEG signal in the beta band. In the same way, it was determined that the parametric estimation method, Yule-Walker, is more sensitive than the FFT and Welch's periodogram.

The correct placement of the stimulation electrode is essential, because a misplacement can generate adverse effects. Likewise, a new intervention in the patient is risky due to the high possibilities of hemorrhage and infection. Due to the above mentioned, every attempt that help to correctly identify the zone to stimulate is relevant.

Acknowledgment

This work was supported partially by Sistema General de Regalías, Republic of Colombia through the project "Fortalecimiento de la plataforma tecnológica para la formación especializada en el área de la salud y el desarrollo de tecnología biomédica", RutaN-139C and Instituto Tecnológico Metropolitano through the Project "Desarrollo de un simulador computacional de señales eléctricas de ganglios basales en coordenadas oblicuas afines de entrenamiento en neurocirugía estereotáctica", code p14222.

Conflict of Interest

The authors declare that they have no conflict of interest.

References

1. Albin RL, Young AB, Penney JB. (1995) The functional anatomy of disorders of the basal ganglia. Trends Neurosci.18:63–4.
2. Alavi M, Dostrovsky JO, Hodaie M, Lozano AM, Hutchison WD. (2013) Spatial extent of beta oscillatory activity in and between the subthalamic nucleus and substantia nigra pars reticulata of Parkinson's disease patients. Exp. Neurol. 245:60–71.
3. Obeso J A, Rodríguez-Oroz MC, Rodríguez M, Lanciego JL, Artieda J, Gonzalo N, et al. (2000) Pathophysiology of the basal ganglia in Parkinson's disease. Trends Neurosci. 23:S8–19.
4. Chen CC, Pogosyan A, Zrinzo LU, Tisch S, Limousin P, Ashkan K, et al. (2006) Intra-operative recordings of local field potentials can help localize the subthalamic nucleus in Parkinson's disease surgery. Exp. Neurol. 198:214–21.
5. Hutchison WD, Allan RJ, Opitz H, Levy SR, Dostrovsky J, Lang a E, et al. (1998) Neurophysiological Identification of the Subthalamic Nucleus in Surgery for Parkinson's Disease. Ann. Neurol. 44:622–8.
6. Beudel M, Brown P. (2016) Adaptive deep brain stimulation in Parkinson's disease. Parkinsonism Relat. Disord. 22:S123–6.
7. Kühn AA, Trottenberg T, Kivi A, Kupsch A, Schneider GH, Brown P. (2005) The relationship between local field potential and neuronal discharge in the subthalamic nucleus of patients with Parkinson's disease. Exp. Neurol. 194:212–20.
8. Guridi J, Rodriguez-Oroz MC, Lozano AM, Moro E, Albanese A, Nuttin B, et al. (2000) Targeting the basal ganglia for deep brain stimulation in Parkinson's disease. Neurology. 55:S21–8.
9. Senhadji L, Shamsollahi MB, Bouquin-jeannks R Le. (1998) Representation of sEEG signals using time-frequency signatures. IEEE. 20:1454–7.
10. Foffani G, Bianchi A., Baselli G, Priori A. (2005) Movement-related frequency modulation of beta oscillatory activity in the human subthalamic nucleus. J. Physiol. 568:699–711.
11. Shiman F, Safavi SH, Vaneghi FM, Oladazimi M, Safari MJ, Ibrahim F. (2012) EEG Feature Extraction using Parametric and Non-Parametric Models. IEEE. 25:66–70.
12. Pardey J, Roberts S, Tarassenko L. (1996) A review of parametric modelling techniques for EEG analysis. Med. Eng. Phys. 18:2–11.
13. Wei W. (2006) Time series analysis: Univariate and multivariate methods. 2nd ed. Pearson education.
14. Hannan EJ, Quinn BG. (1979) The Determination of the Order of an Autoregression. J. R. Stat. Soc. Ser. B. 41:190–5.
15. Song PX-K, Freeland RK, Biswas A, Zhang S. (2013) Statistical analysis of discrete-valued time series using categorical ARMA models. Comput. Stat. Data Anal. 57:112–24.
16. Blinowska K, Zygierewicz J. (2012) Practical biomedical signal analysis using Matlab. Boca Raton: CRC Press.
17. Kuo BC. (1996) Automatic control systems. 7 ed. Prentice Halll.
18. Alkan A, Yilmaz AS. (2007) Frequency domain analysis of power system transients using Welch and Yule-Walker AR methods. Energy Convers. Manag. 48:2129–35.

Author: A.M. Hernández
Institute: University of Antioquia
Street: Calle 67 No. 53 - 108
City: Medellín
Country: Colombia
Email: alher.hernandez@udea.edu.co

Influence of the Immune System on the Biological Dynamics of the Interstitial Fluid Pressure

R.F. Reis[1], R.W. dos Santos[1] and M. Lobosco[1]

[1] Federal University of Juiz de Fora, Graduate Program in Computational Modeling, Juiz de Fora, Brazil

Abstract— **The increase of the interstitial fluid pressure (IFP) can be influenced by an immunological defense action followed by an edema formation depending on the inflammation level. Some causes that may lead to an abnormal IFP rising include the increasing of the capillary permeability and the extrapolation of the maximum lymphatic flow. These situations can lead to an accumulation of interstitial fluid because the incoming flux becomes greater than the outgoing one. This paper presents the first modeling steps that in the future will allow us to capture this dynamic completely. The simplified model presented in this work considers an inflammation due to unspecified bacteria, that are removed by neutrophils. All immunological process is modeled using an multi-phase model in a porous media and its interaction with the IFP.**

Keywords— **Computational Immunology, Computational Modeling, Edema Formation, Finite Volume Method**

I INTRODUCTION

The presence of excess fluid in body tissues may induce an edema formation. The edema can be caused by extracellular or intracellular fluid accumulation. One of the causes that may lead to an extracellular edema is a bacterial infection. When the body starts an inflammatory reaction, it also increases the capillary filtration (i.e., the fluid that leaves the capillary)[1]. As an attempt to counterpoise the excess of fluid, the lymphatic system increases the lymph flow until a given threshold value [2].

This study aims to model a bacterial infection along with the immune system reaction and its consequences for the interstitial fluid pressure (IFP) dynamics. First of all, an unspecified bacteria was considered as the pathogen that enters in the interstitium. The bacterial dynamics include its diffusion, growing and death due to the immune system response. The neutrophils were the only immune system cell considered in the model due to its major presence among the circulating white blood cells [3]. Its dynamics include diffusion, chemotaxis, growing due to extravasation, death due to interaction with bacteria, and induced apoptosis (a process of programmed cell death). Finally, all these interactions between bacteria and neutrophil change the plasma flow dynamics between capillaries and interstitium which also alter the IFP. In other words, neutrophil-bacteria iterations trigger a chain of cytokine reactions which in turn change the vessel wall permeability, increasing also the capillary filtrating rate, which impacts the IFP. Moreover, the lymph flow is also considered.

All these interactions are mathematically modeled using a partial differential equation (PDE) system based on a porous media approach. The resulting PDE system is solved using a numerical strategy known as finite volume method (FVM) [4] combined with a technique called first order upwind (FOU) scheme [5].

While some previous studies model the IFP dynamics due to cancer growing [6] or the bacteria-neutrophil iterations [7], to the best of our knowledge, the first study which models the coupling of bacterial infection dynamics with the dynamics of IFP was presented at [8]. This works extends the previous one[8], considering the lymph flow and the reduction of osmotic reflection coefficient.

This paper is divided into 5 sections, starting with this introductory section, followed by a brief overview of immune system, the proposed mathematical model, the results of our simulations, and finally, the conclusions and future works.

II IMMUNOLOGY BACKGROUND

While pathogens try to attack the organisms, specialized cells are used to seek and destroy them. These cells are called leukocytes and are classified as mononuclear (T lymphocytes, B lymphocytes, natural killer, monocytes, macrophages and dendritic cells) or polymorphonuclear (neutrophils, eosinophils, basophils and mast cells). Usually neutrophils are the first defense cells to achieve the inflammation site and represent 70% of leukocytes[9]. Cells and molecules of the immune system encounter pathogens in the interstitium. The immune cells phagocyte (engulfe and absorbe) the pathogens and start producing specific proteins called cytokines that triggers a chain of reactions which finally attract other immune cells for help them.

The vascular endothelium acts as an important mediator which regulates the movement of bloodstream molecules and

I. Torres et al. (eds.), *VII Latin American Congress on Biomedical Engineering CLAIB 2016, Bucaramanga, Santander, Colombia, October 26th -28th, 2016*,
IFMBE Proceedings 60,
DOI: 10.1007/978-981-10-4086-3_77

leukocytes into the tissue. A process known as extravasation consists of the movement of blood cells from the capillaries to the inflamed site. During an inflammatory response, cytokines produced by immune cells as well as other inflammatory mediators act upon the local blood vessels, inducing the increase of endothelial cell-adhesion molecules (CAMs) expression [9]. In this situation, the vascular endothelium is called activated, or inflamed. Thus, it allows immune cells, like neutrophils, to leave the circulatory system, enter into the tissue and migrate to the infection site [3]. The extravasation allows not only immune cells, but also fluids, to enter into the tissue[2].

About one-sixth of the total body volume consists of space among cells which is called interstice. Interstitial fluid is the liquid phase that fills this space among cells [2]. When the interstice is excessively filled occurs an extracellular edema formation. Usually, there are two possible reasons to the edema formation: abnormal plasma liquid leaking from capillaries to the interstitial space; and lymphatic system fail[2].

III MATHEMATICAL MODEL

Eq. (1) is the well-posed problem, with proper boundary conditions, which describes the bacteria dynamics in the interstitial tissue [8]. Model parameters and their values utilized in this study can be found at Tab. 1.

$$\begin{cases} \dfrac{\partial \phi \rho_b s_b}{\partial t} = \nabla \cdot (D_b \nabla(\rho_b s_b)) - r_b + q_b & \text{in } \Omega \times I \\ \alpha s_b + \beta \nabla s_b \cdot \mathbf{n} = f_b & \text{on } \partial\Omega \times I \\ s_b(\cdot, 0) = s_{b0} & \text{in } \Omega, \end{cases} \quad (1)$$

where the source and the sink are given by $q_b = c_b \rho_b s_b$ and $r_n = \lambda_{nb} \rho_n s_n \rho_b s_b$, respectively.

Eq. (2) describes the neutrophils dynamics in the interstitial tissue [8], as well as the boundary conditions.

$$\begin{cases} \dfrac{\partial \phi \rho_n s_n}{\partial t} = \nabla \cdot (D_n \nabla(\rho_n s_n) - \chi_{nb} \rho_n s_n \nabla(\rho_b s_b)) & \\ -r_n + q_n & \text{in } \Omega \times I, \\ \alpha s_n + \beta \nabla s_n \cdot \mathbf{n} = f_n & \text{on } \partial\Omega \times I \\ s_n(\cdot, 0) = s_{n0} & \text{in } \Omega, \end{cases} \quad (2)$$

where the source and the sink are given by $q_n = \gamma_n \rho_b s_b \rho_n (s_{n,max} - s_n)$ and $r_n = \lambda_{bn} \rho_n s_n \rho_b s_b + \mu_n \rho_n s_n$, respectively. Again, model parameters and values used in this study can be found at Tab. 1.

Finally, Eq. (3) couples the bacteria-neutrophils dynamics with IFP. This equation is a modified version of the pressure equation presented in our previous work [8]. In this work, it has been included the influence of lymph flow and the reduction of osmotic coefficient.

$$\begin{cases} \nabla \cdot \dfrac{\mathbf{K}}{\mu} \nabla P = q_c + q_l & \text{in } \Omega \\ \alpha P + \beta \nabla P \cdot \mathbf{n} = f_p & \text{on } \partial\Omega, \end{cases} \quad (3)$$

where $q_c = -k_f(P_c - P - \sigma(\pi_c - \pi_i))$ is the influence of capillary fluid exchange according to the Starling's hypothesis and $q_l = \nabla(\frac{-\delta P}{\gamma + P})$ is the influence of lymph flow. To couple Eq. (1) with Eq. (3), $k_f = L_p(S/V)$, where $L_p = L_{p0}(1 + c_{bp}\rho_b s_b)$. Furthermore, $\sigma = \frac{\sigma_0}{(1 + c_{bp}\rho_b s_b)}$. All model parameters name and their values can be also found at Tab. 1.

The above partial differential equation system was solved using the finite volume method with first order upwind scheme to ensure a stable numerical solution.

IV MODEL RESULTS

This section presents the numerical model solution as well as the parameters used in the PDE system to reproduce the modeled phenomenon. All simulations are presented in an one-dimensional domain of 1 cm length, simulating the equivalent of $30s$ of infection.

The simulated scenario is a bacterial infection seated at $x \in [0.48, 0.52]$ with $s_{b0} = 0.001$. For all other points outside this interval, we consider that $s_{b0} = 0.0$. In the case of the neutrophils population, we consider that initially there is no cells in the domain, so $s_{n0} = 0.0\ \forall x \in \Omega$. This simulated scenario represents an initial bacterial infection.

On the boundary of the domain, Neumann boundary condition was imposed to the bacteria and neutrophil, so $\nabla s_b \cdot \mathbf{n} = 0$ and $\nabla s_n \cdot \mathbf{n} = 0$, respectively. On the other hand, the boundary of Eq. (3) was modeled as lymph vessels, thus, it was applied a combination of Dirichilet and Neumann boundary conditions $-\delta P + (\gamma + P)\frac{\mathbf{K}}{\mu}\nabla P \cdot \mathbf{n} = 0$, which results $\frac{\mathbf{K}}{\mu}\nabla P \cdot \mathbf{n} = \frac{\delta P}{\gamma + P}$ on $\partial\Omega$.

The diameter of the blood capillaries are between 5 e 10 μm. The diameter of lymphatic capillaries are about $0.2mm$, so the caliber of lymphatics capillaries are much greater than blood ones. According to [10], the lymphatic vessels represents about 2.9% of an tissue. Thus, this work models lymphatics capillaries as points randomly seated with 2.9% of probability to be found over the domain. The blood capillaries were distributed uniformly, except at points where lymphatics capillaries are found.

All parameters names, symbols, units and their respective value are shown at Tab. 1, which gather data from distinct works [6, 11, 8]. μ and $\mathbf{K}$ parameters were marked with "*" because it was not found their individual values, just their relationship: $\mathbf{K}/\mu = 2.5\text{x}10^{-7} cm^2/s/mmHg$ and $L_{p0}(S/V) = 626.4(1/s/mmHg)$.

Table 1: Model Parameters Values

Name	Symbol	Unit	Value
Fluid velocity	v_f	$\frac{cm}{s}$	-
Interstitial Pressure	P	$mmHg$	-
Capillary Pressure	P_c	$mmHg$	20.0
Viscosity	μ	$\frac{g}{cms}$	*
Permeability	$\mathbf{K}$	cm^2	*
Saturation	s	–	-
Porosity	ϕ	–	0.2
Density	ρ	$\frac{g}{cm^3}$	1.0
Filtering coefficient	k_f	$\frac{1}{smmHg}$	-
Neutrophil diffusion coefficient	D_n	$\frac{cm^2}{s}$	0.0001
Bacteria diffusion coefficient	D_b	$\frac{cm^2}{s}$	0.0001
Chemotaxis rate	χ_{nb}	$\frac{cm^5}{sg}$	0.001
Bacteria reproduction rate	c_b	$\frac{1}{s}$	0.15
Induced apoptosis rate	λ_{bn}	$\frac{cm^3}{sg}$	0.1
Phagocytosis rate	λ_{nb}	$\frac{cm^3}{sg}$	1.8
Capillary permeability to neutrophil	γ_n	$\frac{cm^3}{sg}$	0.1
Neutrophil source	$s_{n,max}$	–	0.55
Apoptosis rate	μ_n	$\frac{1}{s}$	0.2
Hydraulic permeability	L_{p0}	$\frac{cm}{smmHg}$	$3.6\text{x}10^{-8}$
Osmotic reflection coefficient	σ_0	-	0.91
Capillary oncotic pressure	π_c	$mmHg$	20.0
Interstitial oncotic pressure	π_i	$mmHg$	10.0
Bacterial influence in hydraulic permeability	c_{bp}	$\frac{cm^3}{g}$	1.0
Weight of lymph vessel	δ	$\frac{cm^2 smmHg}{g}$	0.000001
Pressure Shift	γ	$mmHg$	0.1

Although the values used in the simulations must be adjusted to reflect a real infection scenario, the qualitative values shows that the model can reproduce some key aspects related to an initial edema formation as Fig. 1 shows. Figs. 1(a) and 1(b) show the bacteria-neutrophil values obtained from Eqs. (1) and (2), respectively. As one can observe, the bacteria starts to proliferate in the middle of the domain, consequently neutrophils are attracted initially from blood vessels to the interstice at this point. Fig. 1(c) shows the influence of a bacterial infection in the IFP dynamics as a result of Eq. (3). It is important to highlight that the lymph vessels are seated randomly at $x = 0.06cm$ and $x = 0.74cm$ (represented by 2 black points at the x-axis). A higher concentration of bacteria induces the increase of hydraulic permeability along with the reduction of osmotic reflection coefficient, which results in the rise of the filtering coefficient. Moreover, the lymph flow keeps the solution smaller than in the neighborhood, inducing a pressure gradient. As a result, the IFP still increases at the infection site, despite of the presence of lymphatic capillaries.

The initial reaction of the body to the increase of the capillary permeability is to counterpoise its effect: the excess of plasma entering the interstitial space. This is done by rising the lymph flow. An accumulation of fluids takes place if the incoming flux become grater than the outgoing one, which results in an edema formation. So, the presence of bacteria and the respective immune system response are responsible for changing the plasma fluxes, resulting in an increase of pressure at the infection site. The increase of IFP shown at Fig.1(c) is the initial scenario that may give rise to an extracellular edema formation. The accumulation of fluids, which may result in an edema formation, is not yet included in our model and will be subject of future works.

V CONCLUSIONS AND FUTURE WORKS

This paper presented a mathematical model that couples bacteria-neutrophil dynamics with a pressure equation. This papers included the influence of lymph flow and modified the osmotic reflection coefficient, which now reduces its value as the inflammatory response evolutes. Despite of some simplifications and limitations of this model, the results were able to reproduce some key aspects related with an initial edema formation.

As future works, it is planed to analyze the influence of the increase of IFP in the immune system cells due to the pressure gradient created in the interstice. Moreover, we would like to evaluate the impact of lymphangiogenesis in the edema formation[10].

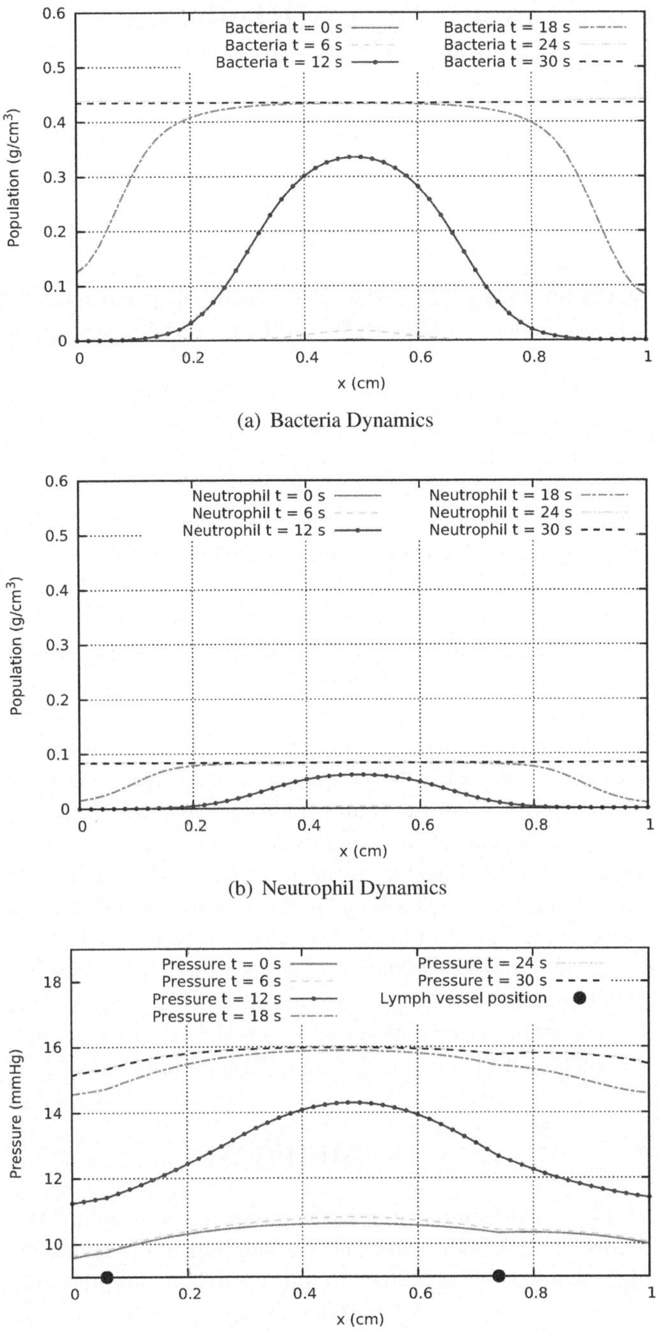

(a) Bacteria Dynamics

(b) Neutrophil Dynamics

(c) Interstitial Fluid Pressure Dynamics

Figure 1: Simulation Results

Conflict of Interest

The authors declare that they have no conflict of interest.

Acknowledgments

The financial support by CNPq, CAPES, UFJF and FAPEMIG is greatly acknowledged.

References

1. Scallan Joshua, Huxley Virginia H, Korthuis Ronald J. *Capillary fluid exchange: regulation, functions, and pathology*;2 . 2010.
2. Guyton A.C., Hall J.E.. *Textbook of Medical Physiology*. Guyton Physiology SeriesElsevier Saunders 2006.
3. Sompayrac Lauren. *How the Immune System Works*. Wiley-Blackwell 2012.
4. Versteeg H.K., Malalasekra W.. *An Introduction to Computational Fluid Dynamics: The Finite Volume Method*. Prentice Hall2 ed. 2007.
5. Alves J.R., Queiroz R.A.B., Santos R.W.. Simulation of cardiac perfusion by contrast in the myocardium using a formulation of flow in porous media *Journal of Computational and Applied Mathematics*. 2016;295:13 - 24. VIII Pan-American Workshop in Applied and Computational Mathematics.
6. Phipps Colin, Kohandel Mohammad. Mathematical model of the effect of interstitial fluid pressure on angiogenic behavior in solid tumors *Computational and mathematical methods in medicine*. 2011;2011.
7. Pigozzo Alexandre Bittencour, Macedo Gilson Costa, Santos Rodrigo Weber, Lobosco Marcelo. On the computational modeling of the innate immune system *BMC bioinformatics*. 2013;14:S7.
8. Reis Ruy Freitas, Santos Rodrigo Weber, Lobosco Marcelo. *Bioinformatics and Biomedical Engineering: 4th International Conference, IWBBIO 2016, Granada, Spain, April 20-22, 2016, Proceedings*ch. A Plasma Flow Model in the Interstitial Tissue Due to Bacterial Infection, :335–345. Cham: Springer International Publishing 2016.
9. Abbas A.K., Lichtman A.H.. *Basic Immunology Updated Edition: Functions and Disorders of the Immune System*. Elsevier Health Sciences 2012.
10. Rahier J-F, De Beauce S, Dubuquoy L, et al. Increased lymphatic vessel density and lymphangiogenesis in inflammatory bowel disease *Alimentary pharmacology & therapeutics*. 2011;34:533–543.
11. Basser Peter J. Interstitial pressure, volume, and flow during infusion into brain tissue *Microvascular research*. 1992;44:143–165.

Author: Ruy Freitas Reis
Institute: Universidade Federal de Juiz de Fora
Street: José Lourenço Kelmer
City: Juiz de Fora
Country: Brasil
Email: ruyfreis@gmail.com

Comparing Myocardium Perfusion Data Acquired by a MRI-Phantom and a Mathematical Model

J. R. Alves[1], R. A. B. de Queiroz[1] and R. W. dos Santos[1]

[1] Federal University of Juiz de Fora, Graduate Program in Computational Modeling, Juiz de Fora, Brazil

Abstract— ***Myocardial Perfusion* refers to how cardiac tissue is being supplied by nutrients and oxygen. There are several exams used to verify if myocardial perfusion is taking place properly. These exams, for instance, can verify if there are regions in the heart with ischemia (lack of oxygen and nutrients). One exam that is widely used for this purpose is contrast-enhanced *Magnetic Resonance Imaging* (MRI). Unfortunately, the images provided by this exam usually offers a more qualitative assessment of myocardial perfusion than a quantitative one. Different techniques that aim to better understand and quantify the information acquired by contrast-enhanced MRI have been developed. Among this techniques, a MRI-phantom device was developed in the framework of an European project. This complex hardware device enables the simulation of contrast-enhanced MRI exams under controlled conditions. In this paper, we present another technique that also enables a similar quantitative analysis of this exam, a mathematical model based on Partial Differential Equations that treats cardiac tissue as a porous media. The results obtained by our mathematical model were compared to different results obtained by the MRI-Phantom. Different parameters such as Myocardial Perfusion Rates (MPR) and Cardiac Output (CO) were tested. For all the simulations we observed a good agreement between the results obtained by the mathematical model and the data obtained by the MRI-Phantom, in terms of the temporal dynamics of the contrast agent (CA). Therefore, we conclude that the presented mathematical models can be taken as promising tools for the quantification of cardiac perfusion.**

Keywords— **Myocardial Perfusion; Magnetic Resonance Imaging; Porous Media; Computational Modeling; Biomedical Engineering**

I INTRODUCTION

Cardiac diseases are the main cause of deaths around the world. Clinical imaging has been one important tool for the diagnosis of several cardiovascular diseases [1]. Among these imaging tools, contrast-enhanced MRI (Magnetic Resonance Imaging), or first-pass MRI, is highly used for the early detection of myocardial ischemia, infarct or any related cardiac perfusion anomaly. Cardiac or myocardium perfusion is the phenomenon associated to the supply of oxygen and nutrients to cardiac cells.

To improve and further explore the technique of contrast-enhanced MRI, several clinical, multi-center projects are continuously under way. In addition, in order to better understand this complex phenomena, in particular, how MRI images are related to ischemia, models have been also developed. With this respect, both hardware models, Phantoms [2], as well as mathematical models [3, 4] were also developed and tested.

In this work, we compare our previous computer simulations of contrast in myocardium to data acquired from a sophisticated Phantom for contrast-enhanced MRI. It is important to note that quantitative data regarding contrast concentration (typically Gadolinium is used as contrast in these exams) is not available for the physicians. Quantitative assessments of contrast Signal Intensity (SI), which is measured as the sum of intra and extravascular contrast concentration, can provide significant information of cardiac perfusion. Therefore, any tool, whether hardware-like (Phantoms), or software-like (Computer models), that can contribute to a more quantitative evaluation of myocardium perfusion is of great interest.

Is important to mention that our model considers the relation between SI and contrast concentration to be linear.

II METHODS

During a Phantom simulation pumps are used to quantitatively control both water and Gadolinium influx to the system. An entry tube follows to four cavities that represent the heart atria and ventricles. After the cavities, the fluid flows through another tube that represents the aorta. From the simulated aorta a portion of the fluid flows to the right and left coronary arteries and another portion goes out of the system (simulating the blood that flows to the other portions of the body). From the coronaries, the fluid goes to two cylindrical compartments that represent the right and left myocardial tissue that are perfused. These two cylindrical compartments have many small pipes within them, imposing the flow to be near laminar.

I. Torres et al. (eds.), *VII Latin American Congress on Biomedical Engineering CLAIB 2016, Bucaramanga, Santander, Colombia, October 26th -28th, 2016*,
IFMBE Proceedings 60,
DOI: 10.1007/978-981-10-4086-3_78

The *Arterial Input Function* (AIF) can be used for the quantification of myocardial perfusion. In the Phantom, it is measured in the tube that represents the aorta. For more details, see the original paper that describes the Phantom [2].

In our previous work [3, 5], we proposed a mathematical model for cardiac perfusion based on the framework of continuous porous media. The model is general enough to represent the complex 3D geometry of the heart as well as the heterogeneous and anisotropic nature of cardiac tissue. In this work, we will verify if the same model, using some simplifications driven by the features of the hardware model, can reproduce data acquired by Phantom simulations. The simplifications include the hypothesis of homogeneous 1-D flow, as described in the next section.

A Modeling the contrast dynamics as a single-phase flow

Different works have already modeled the myocardium structure as a porous media [3, 5, 6]. For this case, classical Darcy's formulation gives us:

$$\nabla \cdot \mathbf{K}\nabla p = 0, \; in \; \Omega \tag{1}$$

where $\mathbf{K}$ is the permeability coefficient; p is the pressure; and Ω is the domain. The boundary conditions for Eq. 1 are given by Dirichlet boundary conditions. $\vec{v} = -\mathbf{K}\nabla p$ is the Darcy velocity.

Given the contrast concentration (C), the total porosity (ϕ), advective ($\vec{v}C$) and diffusive flux ($-D\nabla C$), the CA dynamics can be modeled by the following advective-diffusive equation:

$$\frac{\partial(\phi C)}{\partial t} + \nabla \cdot \vec{v}C - \nabla \cdot (\mathbf{D}\nabla C) = 0, \tag{2}$$

where $\mathbf{D}$ is the diffusion coefficient. In this work, for the influx boundary condition (Γ_i, see Fig. 1), we use the concentration in the aorta obtained in [2].

We use the Neumann boundary condition ($\mathbf{D}\nabla C \cdot \vec{\mathbf{n}} = \gamma$) in Γ_o. Here, $\vec{\mathbf{n}}$ is the normal vector. As initial condition, we have $C(x,0) = C_0(x)$ in Ω

B Numerical Methods

Fig. 1 presents the cylinder compartment of the Phantom that simulates the myocardial compartment together with the associated 1D simulation domain of our mathematical model.

The analytic solution of Eq. 1 in 1D is given by $p(x) = ax + b$. Considering the boundary conditions $p = p_i$ in Γ_i, $p = p_o$ in Γ_o, and $v = -\mathbf{K}\frac{\partial p}{\partial x}$, we have $v = -\mathbf{K}\frac{p_o - p_i}{L}$. Once knowing the pressure distribution, this last equation is used to obtain the velocity field, which is used in the second term of the left side in Eq. 2.

As highlighted in [7], an analytic solution for the Eq. 2 is not simple. It can be as computational expensive as a numerical solution. Thus, we apply in our work numerical methods for solving this part of the model.

For the discretization of the advection-diffusion equation, the FVM formulation was used, which consists in the evaluation of influx and outflux of a control volume around each node of the mesh. The control volumes were taken as lines of length equal to h.

We used the Explicit Euler method in the temporal part of the advection-diffusion equation. For the diffusive term, the central differences scheme works very well, without any major problems. But for the advective term it is necessary to make the decision of which scheme must be used to ensure accuracy and convergence of the numerical method. For the numerical approximation of this term, we implemented the third-order polynomial upwind scheme (TOPUS). For more details, see [3].

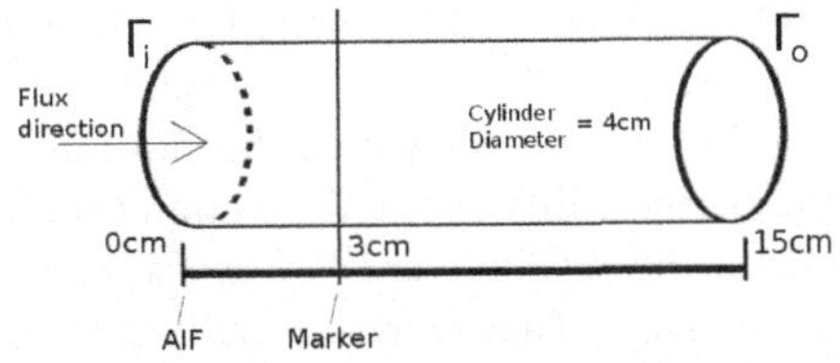

Fig. 1: Basic scheme for both Phantom and mathematical model simulations: in the Phantom, after the fluid pass through the coronary circulation pipes, it flows to the myocardial compartments. A marker presents were the MRI captures the images. In our 1D model, the AIF is taken as a boundary condition (left point).

C Fermi Deconvolution

The results obtained by our model were evaluated in the context of the Fermi Deconvolution, a simple mathematical model widely used for the quantification of cardiac perfusion [2, 4]. This method assumes that concentration can be approximated by the convolution $C_{in} \otimes h(t)$ where $h(t) = \frac{F}{exp(k(t-\tau_0-\tau_{onset})+1}H(t-\tau_{onset})$ is the Fermi function. F, k and τ_0 are parameters to be determined. $H(t-\tau_{onset})$ is the Heaviside function. This parameters are obtained by minimizing the difference between a given myocardial concentration C_{out} and $C_{in} \otimes h(t)$. For more details, see [2, 4]. We solved the minimization problem using as entries the first initial condition (AIF 1, 3 L/min) and the corresponding concentration obtained via our mathematical model.

D Numerical Experiments

In [2], different types of experiments were performed to evaluate the sensitivity of SI curves capture by the MRI-Phantom under different conditions of flux velocity and contrast concentration. In particular, one of the tests evaluated the sensitivity of SI to different values for myocardial perfusion rate (MPR) whereas another test evaluated the sensitivity of SI to different values cardiac output (CO). MPR can be seen as the rate of contrast flow per unit of water in the Phantom (mL/mL/min). CO is the volume of the fluid been pumped through the hardware (L/min).

To simulate different MPR values using our model we changed the velocity values v, which are obtained for different pressure boundary conditions. For this purpose, we keep the initial condition of the contrast, the AIF distribution, and varied the fluid velocity as follows: fix $p_i = 2.0kPa$ and use $p_o = 5.0$, 4.2, 3.5, 2.75 and 2.3 kPa, obtaining $v = 0.15$, 0.1125, 0.075, 0.0375 and 0.015 mm/s, respectively.

The different CO values in the Phantom results in different AIF measured distributions. To simulate this test we used two different acquired AIF as boundary conditions in our model. As shown in Fig. 1, the simulated SI curves were captured 3 cm away from where the fluid arrives. In all scenarios, the permeability K and diffusion D coefficients that best calibrated the results were 5.0 $mm^2kPa^{-1}s^{-1}$ and 0.5 mm^2s^{-1}, respectively. They were manually adjusted. The porosity $\phi = 0.14$ was taken from the literature [3].

The code implementation has been done in C, and the simulations were run on a Intel(R) Core(TM) i7 3.07GHz, with 8 Gb RAM.

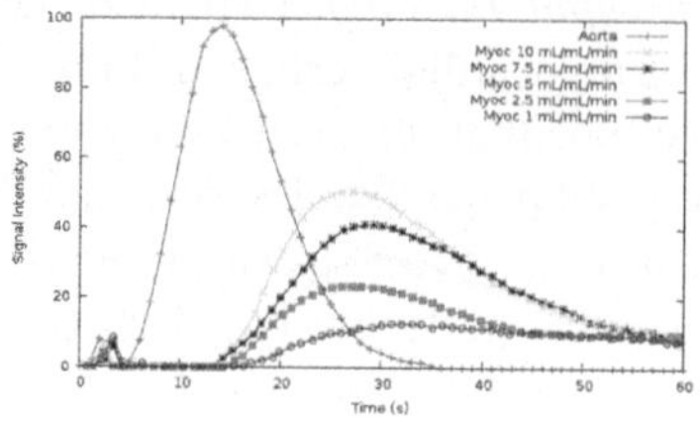

(a)

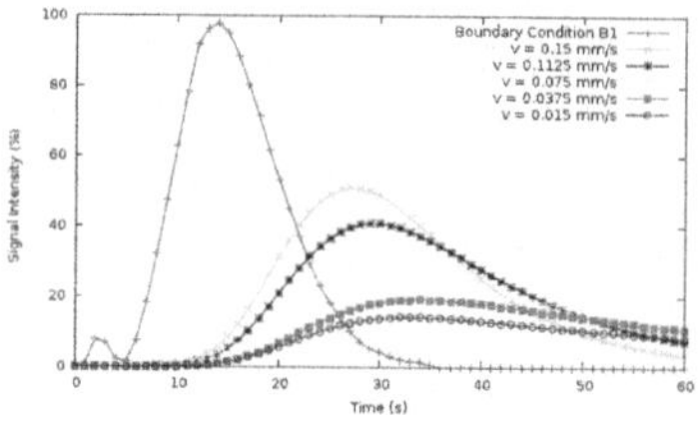

(b)

Fig. 2: SI values obtained for different perfusion rates (a) by the Phantom in the myocardium (1, 2.5, 5, 7.5 and 10 mL/mL/min) and (b) by our model.

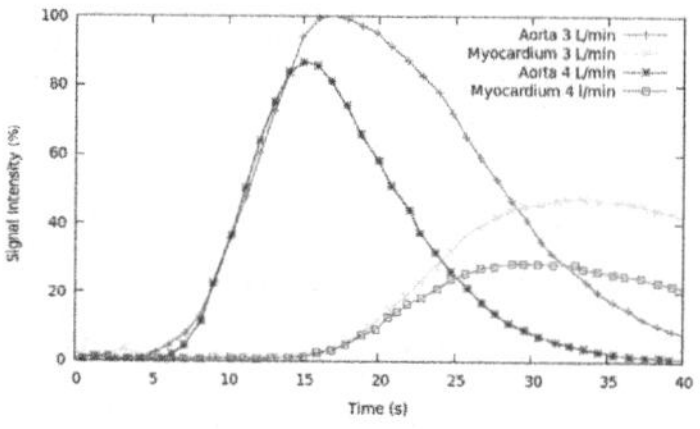

(a)

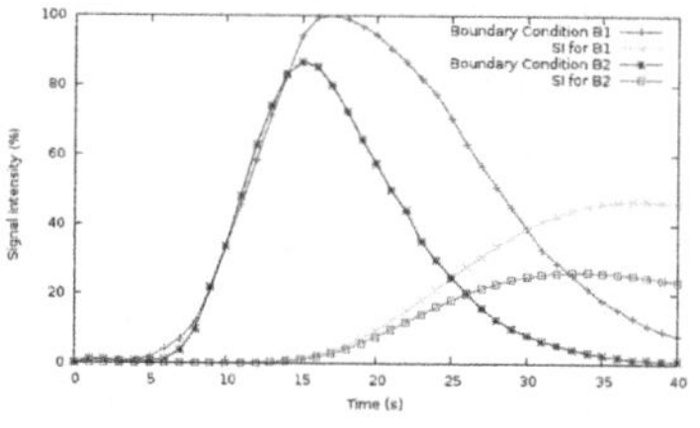

(b)

Fig. 3: SI values obtained for different cardiac output values (3 and 4 L/min) (a) by the Phantom in the myocardium and (b) by our model.

III Results

In this section we present the results obtained in the simulations. Fig. 2 presents the normalized SI (normalized with respect to AIF) for different MPR values obtained by the MRI-Phantom (2(a)) and by our mathematical model (different velocities, 2(b)). In Fig. 3 we have the results of normalized SI curves for different AIF distributions as obtained by ther Phantom (3(a)) and by our mathematical model (using different boundary conditions, 3(b)). From the figures we can observe that the dynamics of SI obtained by the Phantom and by our mathematical model are very similar for the different setups and tests performed in this study.

The obtained values for the Fermi distribution were $F = 1.222$, $k = -0.219$ and $\tau = 27.418$. Fig. 4 presents the results of the concentrations obtained by the Fermi deconvolution and the concentrations computed via our mathematical model. By only using a single pair C_{in}, C_{out}, based on AIF 1, the Fermi deconvolution underestimates the concentration for a different input data, AIF 2, whereas our model reproduces well both myocardial concentrations obtained by the Phantom (from AIF 1 and AIF 2) without the need of any change of parameters.

IV Discussions and Future Works

The development of both hardware and software tools that can help physicians in the development of new protocols for

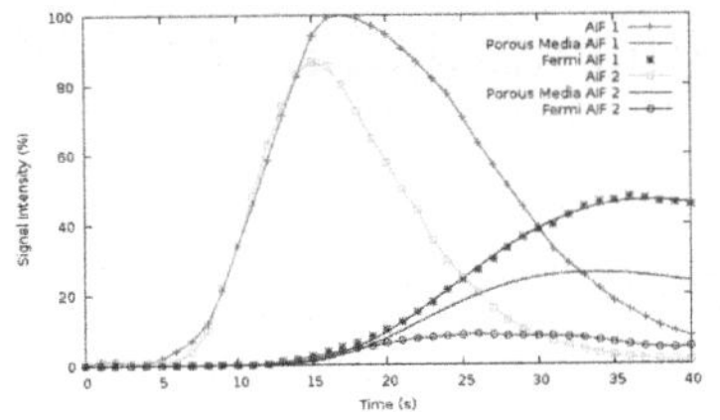

Fig. 4: The parameters F, k and τ_0 are adjusted to minimize the difference between AIF 1 and the corresponding simulated myocardial concentration. With this same Fermi model we present the prediction of the concentration for another input, AIF 2. We observe that the Fermi model does not reproduce the result obtained by the PDE-based mathematical model.

acquiring quantitative data that reflects cardiac perfusion is a great challenge. In this work we show that both hardware solutions, as Phantoms, and software ones, as the presented mathematical model can be used to quantitatively evaluate one of the most used protocols for measuring myocardium perfusion, contrast-enhanced MRI.

In terms of the Fermi deconvolution tests, we observed that after adjusting the Fermi distribution with one input, AIF 1, and its corresponding output concentration (Fig. 4) the Fermi model underestimated the concentration for a second input, AIF 2. This shows how sensitive are the parameters of the Fermi model. On the other side, we have adjusted only once the parameters of the PDE-based mathematical model, and with the same set of parameters it was enough to reproduce many different scenarios observed by the Phantom. Therefore, the proposed model seems to be more robust than the Fermi deconvolution model. Nevertheless, this particular comparison between the two mathematical models deserves further attention and tests. In addition, as the proposed PDE-based model admits an analytic solution for the mentioned 1D homogeneous case, in the near future we plan to compare the Kernel of this analytic solution and the Fermi distribution in the perspective of improving its general use.

It is know that not always the relation between SI and contrast concentration is linear [1]. In future works, we intend to take this issue into account.

Another important limitation that was not approached by neither the Phantom nor the computational model is the leak that the contrast substance suffers from the intravascular media to the extravascular media. Other computational models have recently taken this into account [4]. In order to obtain more accurate and precise results we also intend to included this leak in our model. The mentioned leak happens in the micro-circulatory (capillaries) level. Thus, it is important to work in a domain that captures the circulatory tree structure. For this purpose, we intend to develop a hybrid model. For the extravascular media, we will use the continuum model proposed in this work. For the intravascular media, we will use discrete models proposed in [8] for the construction of the myocardium arterial tree.

V CONCLUSION

Given all the simplifications and the simulated scenarios, we can conclude that the proposed model reproduced very well the data obtained by the MRI-Phantom tests. Furthermore, having in mind the substantial information that quantitative methods can provide for physicians, we remark that our mathematical model and its future planned extensions to 2D and 3D can be taken as promising tools for the further understanding of the complex phenomena and relations between the acquired clinical MRI data and cardiac perfusion.

ACKNOWLEDGEMENTS

The authors would like to thank CAPES, CNPQ, UFJF and FINEP; and the fruitful discussions with profs. Markus Baer and Tobias Schaeffter, from PTB - Berlin.

CONFLICT OF INTEREST

The authors declare that they have no conflict of interest.

REFERENCES

1. Daly C, Kwonj R Y: Cardiac MRI for Myocardial Ischemia. Methodist Debakey Cardiovasc. 2013;9:123-131
2. Chiribiri A, Schuster A, Ishida M et al. Perfusion Phantom: An Efficient and Reproducible Method to Simulate Myocardial First-Pass Perfusion Measurements with Cardiovascular Magnetic Resonance. Magnetic Resonance in Medicine. 2013;69:698-707
3. Alves J R, Queiroz R A B, dos Santos R W: Simulation of cardiac perfusion by contrast in the myocardium using a formulation of flow in porous media. Journal of Computational and Applied Mathematics. 2015;295:13-24 (2015)
4. Cookson A N, Lee J, Michler C et al: A spatially-distributed computational model to quantify behaviour of contrast agents in MR perfusion imaging. Medical Image Analysis. 2014;18:1200-1216
5. Alves J R, dos Santos R W, Novaes G M, de Queiroz R A B: Simulating the Perfusion of Contrast in Cardiac Transmural Wall. IFMBE Proceedings. 2015;49:837-840
6. Michler C, Cookson A N, Chabiniok R et al: A computationally efficient framework for the simulation of cardiac perfusion using a multi-compartment Darcy porous-media flow model. Int J Numer Method Biomed Eng. 2013;29:217-232
7. Mazaheri M, Samani J M V, Samani H M V: Analytical Solution to One-dimensional Advection-diffusion Equation with Several Point Sources through Arbitrary Time-dependent Emission Rate Patterns. Journal of Agricultural Science and Technology. 2013;15:1231-1245
8. R.A.B. de Queiroz, Construção automática de modelos de árvores circulatórias e suas aplicações em hemodinâmica computacional. PhD Thesis. 2013;LNCC/MCTI

Color morphological reconstruction as a segmentation tool for microscope cell images

J.I. Pastore[12], M. Brun[12], A. Bouchet[12] and V.L. Ballarin[23]

[1] ICYTE, CONICET - UNMDP, Universidad Nacional de Mar del Plata, Argentina.
[2] Department of Mathematics, School of Engineering, UNMDP, Argentina
[3] Department of Electronics, School of Engineering, UNMDP, Argentina

Abstract— **There are numerous segmentation methods for gray level microscope cell images, however in some situations the texture features and roughness are not as relevant as the color for the segmentation task. For example, for detection of histological analysis, the relative sizes of nucleus and cytoplasm, as well as their shapes, are the relevant features, while other characteristics such as texture and roughness have no value in the diagnosis.**

In this context, geodesic reconstruction is one of the image operators, of mathematical morphology, that facilitates image segmentation of individual objects. This operator uses markers to highlight, in the image, objects of interest, in order to separate them from the rest of the scene.

This paper presents a new segmentation method, for microscope cell images, based on mathematical morphology color reconstruction, where the markers can be obtained automatically or semi-automatically. Automatically, when you want to detect those cells that were not removed at the generation of the markers. Semiautomatically when the expert manually selects one pixel for each cell of interest.

Keywords—**Mathematical morphology, color, segmentation, reconstructions, cells.**

I. INTRODUCTION

Modern technological development has enabled a huge advance in the histological knowledge. Currently, the use of segmentation algorithms aids the detection and treatment of different lesions. Thus, lots of researchers devote their efforts to develop more powerful algorithms that support diagnostics issued by medical specialists [1]. However, they are still insufficient achievements in this regard. This paper proposes a new segmentation method, for microscope cell images, based on mathematical morphology color reconstruction, where the markers can be obtained automatically or semi-automatically.

The automatic segmentation of cells microscopic images can be considered to be one of the major hurdles for a robust analysis. Automatic or semiautomatic segmentation techniques attempt to improve sensitivity and specificity by detecting, segmenting and classifying the cells and later obtain the cytoplasm or nuclei shape. To solve this problem, in this work, color mathematical morphology reconstruction technique was developed based on a new local defined ordering [2]. The marker image can be obtained automatically or semi-automatically. Automatically, when you want to detect those cells that were not removed at the generation of the markers. Semiautomatically when the expert manually selects one pixel for each cell of interest. In the first case, the marker image is obtained performing successive dilations with a structuring element which size is related to the size of the nucleus of the cell that wants to be removed. In that way we obtained the image mask automatically. Also the expert can also select by clicking a particular cell to segment it.

Experimental results show that the proposed color morphological reconstruction can be efficiently used in the segmentation of cells.

II. MATHEMATICAL MORPHOLOGY IN COLOR SPACE

Mathematical Morphology (MM) has a strong theoretical basis supported on concepts of geometry, algebra, topology and set theory. The principal idea of this theory is to compare the objects of interest with a set of predefined and known geometry, called "structuring element". The use of different shapes and sizes for the structuring element (SE) allows testing and quantifying how the structuring element "is, or is not contained" in the image [3-4].

The MM has been studied and successfully applied in biomedical image segmentation. The process of segmenting an image consists in generating a partitioning of the image set in groups of pixels. Its objective is to simplify or change the representation of an image in a more meaningful one to facilitate the analysis. The segmentation is used either to locate objects of interest or to determine the boundaries within an image [5].

The MM had been introduced as a processing technique for binary images, which were regarded as sets; therefore, its elementary operations are based on the set theory [3-4]. However, the extension to sets of grayscale images, using the umbra concept [3], introduced a generalization of the basic morphological operations. The grayscale morphology is based on the lattice theory, which implies a partial ordering of the data within the grayscale images. Therefore

I. Torres et al. (eds.), *VII Latin American Congress on Biomedical Engineering CLAIB 2016, Bucaramanga, Santander, Colombia, October 26th -28th, 2016*,
IFMBE Proceedings 60,
DOI: 10.1007/978-981-10-4086-3_79

erosion and dilation, the fundamental operations of MM, are defined within a complete lattice as the operations which distribute over the infimum and the supremum [6].

The erosion and the dilation of an image f, using a flat SE g, are defined as follows [6]:

$$\varepsilon_B^{\leq_3}(f) = \inf_{s\in B}^{\leq_3}\{f \circ \tau_s\} \tag{1}$$

$$\delta_B^{\leq_3}(f) = \sup_{s\in B}^{\leq_3}\{f \circ \tau_{-s}\} \tag{2}$$

It can be noticed that the basic morphological operations involve finding an infimum and a supremum for the points within a local region, given by the SE positioning. For a general account on mathematical morphology the interested reader should refer to the two pioneer books by Serra [3-4]. Fundamental references to works which have studied the theory of vector morphology theory are [7-11].

While the extension from binary to grayscale images is a natural one, the extension to color or multivariate images is not straightforward, because of the vectorial nature of the data and the difficulty in finding a suitable ordering for it.

A color can be represented by different algebraic structures. The color spaces provide a way to specify order and manipulate colors. These representations correspond to a subset of a three-dimensional coordinate system in which each color is represented by a point univocally. A color image can be modeled as a function $f:\Omega\subset\mathbb{R}^2\rightarrow\Im\subset\mathbb{R}^3$ where $\Im$ represents a color space. As mentioned above, the definition of morphological operators needs a complete lattice structure to spatial structures [6], i.e., the possibility of defining an ordering relationship among the points to be processed.

Being $\Im\subset\mathbb{R}^3$ a color space and $\leq_\Im$ a proposed order that gives to $\Im$ a structure of complete lattice. The space of functions $f:\Omega\subset\mathbb{R}^2\rightarrow\Im\subset\mathbb{R}^3$ with the order $\leq$ defined as: $f,g:\Omega\subset\mathbb{R}^2\rightarrow\Im\subset\mathbb{R}^3$, $f\leq g$ if and only if $f(x)\leq_\Im g(x)\ \forall x\in D$, has a lattice structure (Serra, 1982; Talbot et al., 1998). Indeed, $\forall f,g:\Omega\subset\mathbb{R}^2\rightarrow\Im\subset\mathbb{R}^3$, $\exists \inf(f,g),\sup(f,g):\Omega\subset\mathbb{R}^2\rightarrow\Im\subset\mathbb{R}^3$ defined by:

$$\left.\begin{aligned}\inf(f,g)(x) &= \inf_{(\Im,\leq_\Im)}\left(f(x),g(x)\right)\\ \sup(f,g)(x) &= \sup_{(\Im,\leq_\Im)}\left(f(x),g(x)\right)\end{aligned}\right\}\forall x\in\Omega$$

This allows defining the basic operations of erosion and dilation for a color image $f:D_f\subset\mathbb{R}^2\rightarrow\Im\subset\mathbb{R}^3$ by a SE B following Eq.1 and Eq.2.

A central aspect of the MM is the construction of any operator as a combination of basic operators: erosion, dilation and the operations supreme and infimum. The formal definition of this decomposition is given by a formal language called Morphological Language [7-8]. Particularly, the reconstruction operator, which is a very useful tool provided by mathematical morphology, is employed not only in binary images but also in grey levels. Besides, it can be applied to different stages of image processing such as filtering, segmentation or features extraction. In general, it is introduced as part of a set of operators called geodesic operators [12-13]; and it basically consists in extracting the connected components of an image from a marker image.

A. Geodesic morphological operators

The geodesic reconstruction is one of the operators of mathematical morphology that facilitates image segmentation. This operator allows, by using markers, highlighting an image objects of interest, in order to remove them from the rest of the scene. That means geodesic operators are useful when you want to process a subset of the space analyzed. The geodesic reconstruction operator employs successive dilations (erosions) to a marker image whose result is delimited by a mask image [12-13].

The definitions of geodesic erosion and dilation are closely related to the geodesic distance [14].

The *geodesic erosion* of an image f (called mark), by an structuring element b, conditioned to g (called mask), $g\leq f$, is defined as:

$$\varepsilon_{b,g}^{(1)}(f) = \varepsilon_b(f)\vee g \tag{3}$$

being $\vee$ the supremum.

First, the mark image f is eroded. Then the supremum between the eroded image and the mask is calculated. The visual effect of this type of erosion is that the mask retains the marker so that it does not disappear (the contraction of the marker is limited). In this case the geodesic erosion is greater than or equal to the mask, it is also an increasing and anti-extensive operation.

The geodesic erosion of size n, with $n\geq 1$, of an image f, by an structuring element b, conditioned to g, is defined as the iteration of geodesic erosions of increasing size, i.e.:

$$\varepsilon_{b,g}^{(n)}(f) = \underbrace{\varepsilon_{b,g}^{(1)}(\varepsilon_{b,g}^{(1)}(\ldots..\varepsilon_{b,g}^{(1)}(f)))}_{n-times} \tag{4}$$

The *geodesic dilation* of an image f (mark), by a structuring element b, conditioned to g (mask), $f\leq g$, is defined as:

$$\delta_{b,g}^{(1)}(f) = \delta_b(f)\wedge g \tag{5}$$

being $\wedge$ the infimum.

The mask acts as the limit of the marker image dilated therefore $\delta_{b,g}^{(1)}(f) \leq f$. The geodesic dilation, like classical dilating, is a growing and extensive operator.

The *geodesic dilation* of size n, $n \geq 1$, of an image f, by an structuring element *b*, conditioned to g, like the geodesic erosion, is defined as the iteration of geodesic dilations of increasing size, i.e.:

$$\delta_{b,g}^{(n)} = \underbrace{\delta_{b,g}^{(1)}(\delta_{b,g}^{(1)}(.........\delta_{b,g}^{(1)}(f)))}_{n-times} \quad (6)$$

The geodesic erosion and dilation have the particularity that they allow, when iterated until stability, the definition of powerful algorithms of morphological reconstruction. Both the geodesic erosion and dilation converge in a finite number of iterations [15].

Reconstruction by dilation of a mark image f from a mask image g, both with the same domain $f \leq g$, is define as the geodesic dilation of f conditioned to g until stability and it denotes $\rho_g(f)$.

The *gray level reconstruction* $\rho_g(f)$ of an image g with respect a marker image f, $f < g$, is obtained iterating successively geodesic dilations in gray levels, by a structuring element b, until new changes do not occur. That means:

$$\rho_g(f) = \bigvee_{n \geq 1} \delta_{b,g}^{(n)}(f) \quad (7)$$

As g acts like constraint, $\rho_g(f) \leq g$, so at the end of the geodesic dilation produces no change in the image. Reconstruction by dilation is an anti-extensive operation.

Each geodesic dilation of the reconstruction is performed from the geodesic dilation of the previous iteration. Thus the marker can progressively reducing its intensity in the spread under the mask

Similarly the dual reconstruction $\rho_g^*(f)$ with respect to an image g is defined $g \leq f$, as:

$$\rho_{b,g}^*(f) = \bigwedge_{n \geq 1} \varepsilon_{b,g}^{(n)}(f) \quad (8)$$

The idea is to iterate several times to spread the minimum (maximum) value of each component till the entire component is homogenized. That minimum (maximum) represents the gray level of the reconstruction of the object, so that the reconstruction of an image in gray levels will not be perfectly reconstructed, as it happens in the binary case.

B. Proposed Method

The proposed method to segment the cells can be summarized in the following four steps:

Step 1: First we define a total order in the RGB color space. In order to define the morphological operators involved in the segmentation, we use the previously defined order in [16]. This order have three stages: The order between two colors is defined first observing the average intensities, if it matches we take the distance to the color of the central element of the structural element and finally if there is a new match we used the order predefined in the decision window.

Step 2: we define the geodetic operators in the complete lattice generated by the former order.

Step 3: the marker image is determined.

Step 4: the dual reconstruction is applied using the marker image of the previous step.

Figure 1 show the proposed color morphological reconstruction applied to a synthetic image. The marker image was obtained performing a dilation with a structuring element. The size of the structuring element is related to the size of the object that wants to be removed by reconstruction. In that way we obtained the marker image automatically.

III. RESULTS

As an example of application, we applied the proposed color segmentation method to cell images from different sources.

Figure 2 shows an example of an image of avian erythrocyte [17], or red blood cells, which also show their nuclei.

We also tried our segmentation approach on these images, first with manual selection of cells, and then with automatic marker detection, with an structuring element of size 11. Figures 2 and 3 show the respective results.

(a) (b) (c)

Figure. 1: Result of the propose method. (a) Original image. (b) Marker image. (c) Result of the dual reconstruction.

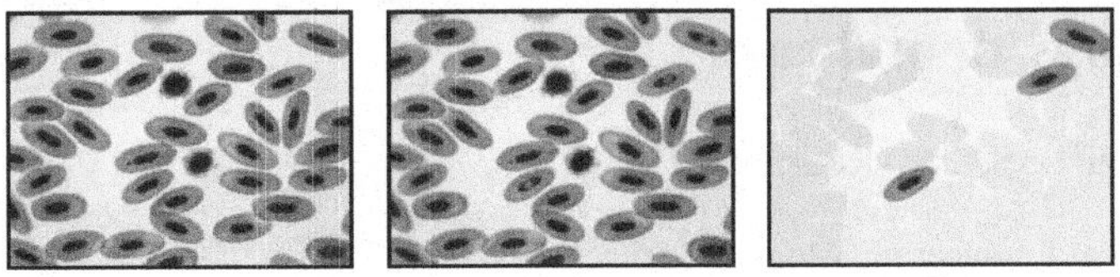

(a) (b) (c)

Figure 2. Avian erythrocyte sample. b) Manually selected markers in red. c) Segmented image, where we can see the three selected cells.

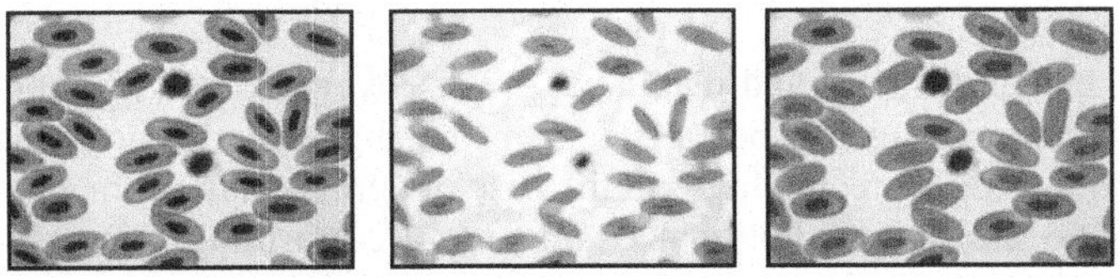

(a) (b) (c)

Figure 3. Avian erythrocyte sample b) Automatic selection of markers. c) Reconstructed image.

We can see that in all cases we obtain a good recovery of the marked cells. In the last case, because of the overlap between cells, the reconstructed image contains additional cell regions. This issue is not present on the previous examples, where the cells were shaped in a more consistent shape.

We also applied the proposed segmentation method to frog red blood cells, called also erythrocytes and also to nasal mucosa in allergic rhinitis obtained from a nassal cytology. Both cases were successfully segmented.

IV. CONCLUSIONS

We proposed here a new method for color image segmentation based on color reconstruction, for application on segmentation of cells images. The algorithm works directly on color images, avoiding the need to convert them to gray scale images, with the loss of information that it produces, and the final results are the full selected cells, reconstructed from original markers. Another advantage of this approach is that, because of the use of the new color image operators, no false colors are created in the process, and the resulting image contains exactly the same color present in the original image. As we can see in the result, the algorithm can perform properly in different situations, where cell nucleous are well defined, with option to select the cells manually, or using an automatic marker algorithm.

REFERENCES

1. Glasbey C. and Horgan G. (1995). Image Analysis for the Biological Sciences. Chichester, England.
2. Pastore J., Bouchet A., Brun M. and Ballarin V. (2015). New Windows based Color Morphological Operators for Biomedical Image Processing.. XIX Congreso Argentino de Bioingeniería. San Miguel de Tucumán, Tucumán, Argentina. ISBN: 978-987-23950-7-0.
3. Serra J.: Image analysis and mathematical morphology, Vol. I. Academic Press, London, 1982.
4. Serra J.: Image analysis and mathematical morphology, Vol. II. Academic Press, London, 1988.
5. González R. and Woods R.: Digital Image Processing. Addison-Wesely Publishing Company, 1996.
6. Angulo J. Morphological colour operators in totally ordered lattices based on distances: Application to image filtering enhancement and analysis. Computer vision and image understanding, 2007, 107(1-2):56-73.
7. Banon, G. J. F. and Barrera, J.: 1993, "Decomposition of mappings between complete lattices by Mathematical Morphology". Signal Processing Vol. 30, pp. 299–327.
8. Banon, G. J. F. and Barrera, J.: 1993, "A decomposition theorem in Mathematical Morphology". Proceedings of the International Worshop on Mathematical Morphology and its Application to Signal Processing, Barcelona, Spain, pp. 234–238.
9. Goutsias J., H.J.A.M. Heijmans H.J.A.M., Sivakumar K., Morphological operators for image sequences, Computer Vision and Image Understanding 62 (3) (1995) 326–346.
10. Serra J., Anamorphoses and Function Lattices (Multivalued Morphology), in: Dougherty (Ed.) Mathematical Morphology in Image Processing, MarcelDekker, 1992, pp. 483–523.
11. Talbot H., Evans C., Jones R.. Complete ordering and multivariate mathematical morphology: algorithms and applications, in: Proceedings of the International Symposium on Mathematical Morphology (ISMM'98), Kluwer, 1998, pp. 27– 34.
12. Lantuéjoul, C. and Beucher, S. (1981). "On the use of the geodesic metric in image analysis", J. Microsc., 121, 39-49.
13. Lantuéjoul, C. and F. Maisonneuve (1984). "Geodesic methods in Quantitative image Analysis", Pattern Recognition, 2, 177-187.
14. Pastore J, Moler E & Ballarin V. "Segmentation of brain magnetic resonance images through morphological operators and geodesic distance", Digital Signal Processing, Vol. 15, pp 153-60, 2005.
15. Vincent L.. "Morphological Grayscale Reconstruction in Image Analysis: Applications and efficient Algorithms", IEEE Transactions On Image Processing, vol 2, 1993. 176-201.
16. Pastore J., Bouchet A., Brun M. and Ballarin V. New Windows based Color Morphological Operators for Biomedical Image Processing. Journal of Physics: Conference Series. ISSN 1742-6596 (On-Line). ISSN 1742-6588 (Print). doi:10.1088/1742-6596/705/1/012023. 04/2016.
17. http://veterinaryonline.blogspot.com.ar/2013/01/erythrocyte-in-animals-erythrocyte.html.

Virtual environment as complementary tool to learning of the Imaginology dental care

P.B. Neto [1], F.D.L. Abreu[1] e S.C.M. Rodrigues[1]

[1] University of Mogi das Cruzes/Biomedical Engineering, Mogi das Cruzes, Brazil.

Abstract— **The study of dental radiographic image requires, in addition to knowledge of anatomy, injuries of knowledge and their interpretations. This work aims to develop a Virtual Learning Environment (VLE) for knowledge tests with focus on the image through the assistance of information technology as a facilitator to access a virtual learning environment. The main idea is that environment presents a greater number of radiographic images, thereby allowing greater learning in the interpretation of images possible. The software used for the development of AVA was Notepad ++, which is free and HTML, JavaScript, jQuery, CSS. 18 screens were prepared with different levels of difficulty and based on the ISO 9126 interface. AVA was analyzed by a group of graduate professors from the University of Mogi das Cruzes resulting in an excellent evaluation.**

Keywords— **Computer-assisted learning; software in oral radiology, technology assistance.**

I. Introduction

In recent years, studies using computer resources as a tool for improving teaching and learning have shown positive results [1]. Medical students have received, with approval, the use of computer resources as a tool for learning [2]. Normally, the exploitation of computer resources is accomplished through multimedia in presentations or the World Wide Web.

Subjects presenting a computational tool as a complement to education tends to attract more students to search for information and knowledge, thus improving the ability of learning and not just of the conventionally presented statically as books and handouts.

The virtual learning environments used as tools to support face-to-face teaching assist users/students to achieve the goals of your informational behavior as mentioned by [3][4][5].

The discipline of dental radiology in Brazil is part of the curriculum of pedagogic project undergraduate program in dentistry since 1940, being of utmost importance for the future professional field of different radiographic techniques, processing steps and interpretation of images. The discipline is usually provided to users/students of 2nd year with total hour load of 160 hours/year in that amount of time the user/student is introduced to all theoretical and practical content through lectures, seminars, practices of various radiographic techniques and interpretation of images.

In dentistry and medicine, the discovery of new diagnostic imaging has become bigger and bigger and I need the use of this technology. In the case of dentistry, the film aims to assist, as additional examination, confirming, discover, define and classify an injury [6], being, therefore, an essential element in dental clinic [7].

Among the many advantages we can mention the interactivity and involvement of the student front of the technology. The progress in the area of health are strictly linked to technological developments.

In recent years, use of computer technology in dentistry has gradually increased in all specialties. Even with few published works on computing and its relationship / insertion in dentistry, the vast majority of these show that this method of aid to learning is equally or more effective than the classic and formal education. The global trend has been to seek alternative ways of teaching. The development of information and knowledge technologies are promoting the possibility of transforming the teaching-learning process.

II. Methods

After defining the educational content to be addressed, we define a detailed roadmap on environmental structure, with the consequent establishment of a planning of the various stages of the work and development of visual programming and design appropriate to the target audience.

A. *Scanning of images*

The images used were obtained from Oral *Radiology book: foundations and interpretation* [8]. For scanning of dental radiographic images using the scanner of HP G4050 Scanjet model, in 300 dpi resolution, jpeg format and without any type of treatment or manipulation of images.

I. Torres et al. (eds.), *VII Latin American Congress on Biomedical Engineering CLAIB 2016, Bucaramanga, Santander, Colombia, October 26th -28th, 2016*,
IFMBE Proceedings 60,
DOI: 10.1007/978-981-10-4086-3_80

B. Selected Images

Forty five selected images containing cystic lesions. We opted for this kind of injury, because they are the most frequent cases in patients in clinical care being images of difficult interpretation, are they:

A) Apical Cyst;
B) Residual Cyst
C) Dentigerous Cyst
D) Keratocyst
E) Periodontal Cyst Side
F) Cyst Duct Nasopalatine
G) Nasolabial Cyst
H) Bleeding Cyst
I) Mucus Retention maxillary Sinus Cyst

C. Development environment

The software used for the development of AVA was Notepad ++ (free), the languages were the HTML, JavaScript, JQuery and CSS.

In all 18 screens were made with standard-based interface ISO/IEC 9126 [9].

Settled 3 blocks of questions with different difficulty levels, easy, medium and hard, figure 1. In each level there are 3 questions each of them with image randomization and alternatives.

The intention is that each image the user can identify the type of cystic lesion that is being presented. Each hit will be a score of 0.2 total 3.0 points at the end of each group of 5 images the level of difficulty. The final score of the user / student, after finishing a certain level of difficulty can be a maximum of 3.0 points. At the end of the three proposed levels, the user can add up to maximum 9.0 points. In this way, we will seek review in a future work, which is the most difficult of the user who answered the questions.

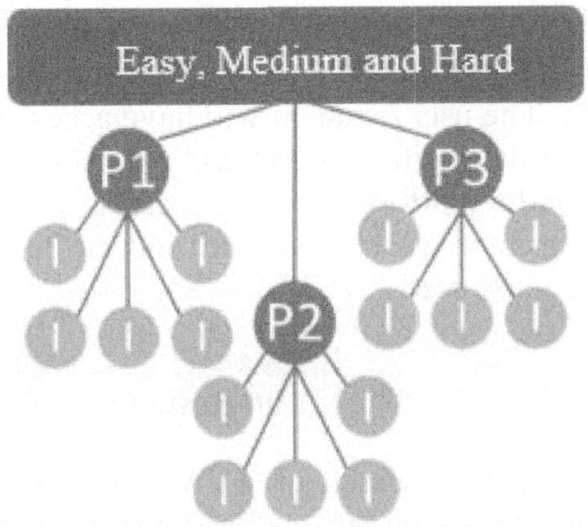

Figure 1*- Difficulty levels and random images.
*where: P1, P2 and P3 questions, and the alternatives to the images.

D. Development of interface

- First screen: Composed by the questionnaire name and the buttons to read and / listen to instructions as well, the start button.
- Second screen: Also known as ID screen is formed by the name and RGM (Matricula registration), which are required for initiation of the tests with the user.
- Third screen: Represented by the questionnaire, where the first level of difficulty is easy. There are 3 questions with random images, and for this level of difficulty has 15 images, so it is possible that the tutor add more images at any time. Similarly occur with other levels of difficulty (Medium and Hard).
- Fourth screen: will demonstrate the result in points obtained by the user / student through questionnaires at each level of difficulty. It is important to note that this screen are two buttons to trigger: Save (in. ".pdf" Through the memory drive on your computer) and Print (will be sent a report to the printer with the result obtained by the feedback).

E. Evaluation of the Virtual environment

- An environment satisfaction questionnaire was prepared, which was answered by four teachers (teachers and doctors) who teach disciplines related to imaging exams and diagnostics in dentistry.

The virtual environment can be installed on a MOODLE platform where the user / student will be able to follow the course of the activities on the Internet. The user / student will have access to the platform with the login as a user and a personal password. The Moodle platform can be accessed from any computer with internet. The platform indeed be free today it is the most used.

III. Results

The first screen of AVA, shows the survey name and the buttons to read and / or listen to the instructions and the start the application, Figure 2.

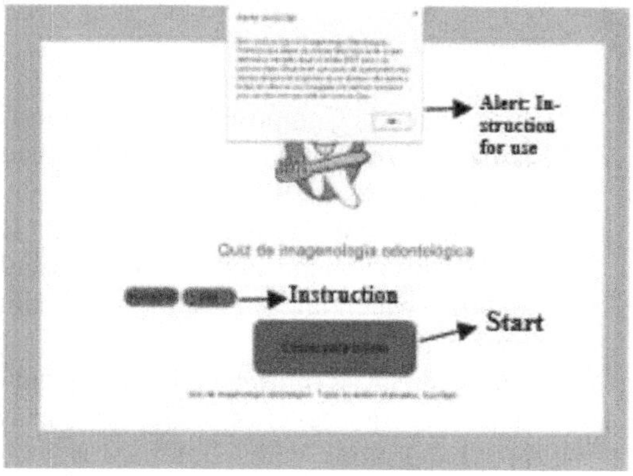

Figure 2 - Home screen with instructions.

When pressing the button, the user / student will be directed to screen identification/registration, Figure 3.

Figure 3 - ID screen.

On the third screen you can see the level of difficulty (easy), the progress bar, the question and the image. The responses following a standard randomization as well as images, Figure 4.

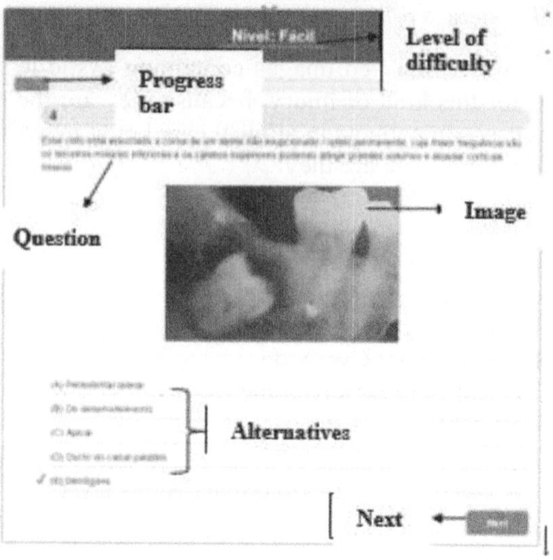

Figure 4 - Screen with the question, progress bar, difficulty level, image and alternatives.

This type of pattern was used on all 15 screens, and changed the questions, alternatives, images and level of difficulty.

The last screen is the finish in which the user / student will be able to see their learning as well as name, number (RGM) Points at each level of difficulty and also the questionnaire with questions, pictures and the answers given by him. On this screen it has the option to save or print, Figures 5 and 6. In the latter screen is also observed the whole progress bar filled in orange.

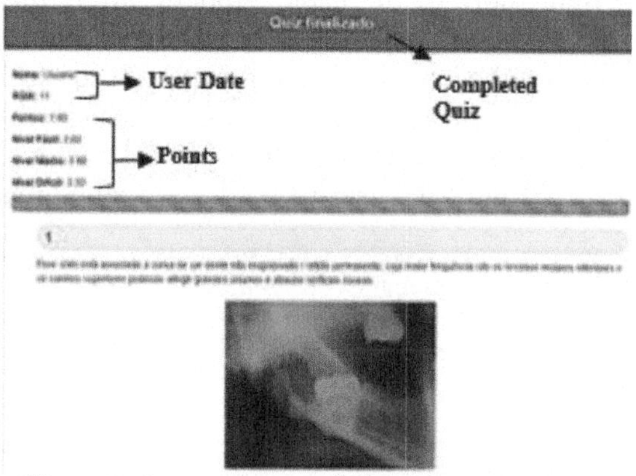

Figure 5 - Last test screen with performance results.

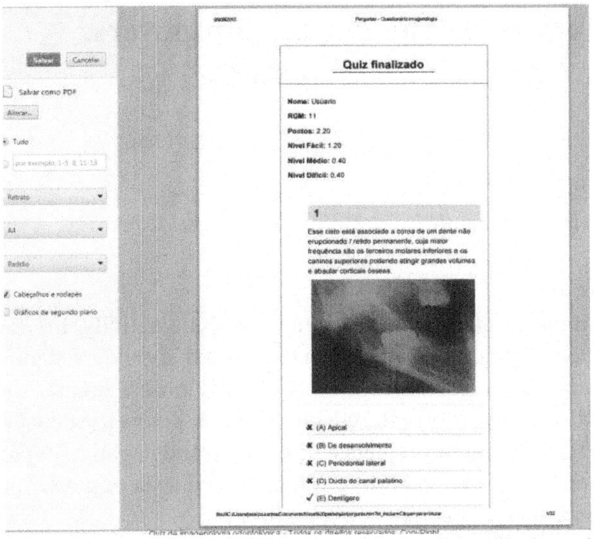

Figure 6 - Screen printing and creation of the file in .pdf format.

IV. Conclusions

The Virtual Learning Environment as an auxiliary tool in learning Dental Imaging was very highly rated by teachers (respondents) who tested and analyzed. two were lauded important points: ease of use and the quality of the questions and images as they were presented.

With the virtual environment, it could assist in learning Dental Imaging of a more concrete and based on new technologies which are of great interest to the young. We believe that the student will have a much greater responsibility in the teaching-learning process, because the income will depend largely on their willingness and interest in wanting to learn. However, it is self-learning does not rule out in any way, the teacher's role (tutor).

Another important factor of application development is that when it is allocated on a server it can be accessed by a link through the Moodle platform for people with disabilities or in a state of disability (patients) without going to university.

References

1. Parrish K.L. (1996) Using interactive multimedia to teach the cognitive e psychomotor skills of cardiopulmonary resuscitation (CPR) and advanced cardiac life support (ACLS). Resuscitation, 31:3:2-2(1).
2. Devitt S, Jones, J.,M Jr, Houghton L et al (1999) Future of health insurance. N Engl J Med 965:325–329 DOI 10.10007/s002149800025
3. Poline T., Bartalo L. (2011) O Comportamento Informacional e a Aprendizagem no Ensino Superior. In: III SBA. Salvador, BA.
4. Fernandes J. C. L., Rodrigues S. C. M., Netto O. L. D. C. A (2013) Virtual environment to help deaf students with written language. International Journal of Biomedical Engineering and Technology, 12:1: pp 26-37.
5. De Castro M. V. et al. (2014) Effect of a virtual environment on the development of mathematical skills in children with dyscalculia. PloS one, 9(7): e103354. doi:10.1371/journal.pone.0103354.
6. Yacovenco A. A. (2001) Análise dos problemas mais frequentes da radiografia na prática odontológica. Revista ABRO, 2: 1: 29-39.
7. Brettle D. S., Worman A., Ellwood, R., Launders J. H., Horner K., Davies R.m.(1996) The imaging performance of a storage phosphor system for dental radiography. The British Journal of Radiology, Vl. 69, pp. 256-261.
8. White & Pharoah, (2015) Radiologia Oral: fundamentos e interpretação, 7ª Ed., ISBN-10: 85-352-7841-9.
9. ABNT at http://www.google.com.br/url?sa=t&rct=j&q=&escr=s&source=web&cd=1&ved=0CFgQFjAA&url=http%3A%2F%2Fprof.swai.com.br%2Fdownload.php%3Ffile%3D1012%26idProf%3D4&ei=wQe1T4zzKYyO8wTew_TqDw&usg=AFQjCNGgEWfb2IGBVhlcOG7tMomPpOGS8Q

Ergonomic and Biomechanical Evaluation of the use of Computers, Tablets and Smart Phones by Children. A pilot study

Holman Ospina-Mateus[1], Benilda Niño-Prada[1], Keyla Tilbe-Ayola[1] and Sonia Contreras-Ortiz[2]

[1] Department of Industrial Engineering, Universidad Tecnológica de Bolívar, Cartagena, Colombia
[2] Department of Electrical and Electronic Engineering, Universidad Tecnológica de Bolívar, Cartagena, Colombia

Abstract— **The use of technological devices has widespread around the world. It is known that inappropriate or prolonged use of electronic devices can cause musculoskeletal discomfort or even illness. However, there are not many studies that evaluate the associated risks on the children's health. This paper describes a pilot ergonomic study for the analysis of the postures of a five-year old girl while using a laptop computer, a tablet, and a smart phone. The analysis was done using the software Delmia to generate 3D models of the child body, that were created based on photographs. We considered five positions: desk sitting, sitting cross legged, lying supine, lying prone and standing. The results suggest that the safest positions are sitting on a desk and standing, and the greatest risk is in the area of neck-trunk. However, further investigation is needed to validate the musculoskeletal risks associated to the use of electronic devices by children.**

Keywords— **Computer, Tablet, Smart phone, Posture, Musculoskeletal discomfort, Children, Ergonomics, Biomechanics**

I INTRODUCTION

Thanks to the advances in the technologies of information and communication (ICT), the users of technological aids such as computers, tablets and smart phones have increased in the last years. Since the invention of computers, their performance has improved significantly while their price has reduced. As a result, its use for educative and entertaining purposes has increased. The growing use of computers among children elevated the concern about long-term consequences that they may have on the children's musculoskeletal (MSK) health, as they are still physically developing. Several studies have been conducted to evaluate the physical impacts of the use of computers on children [1, 2, 3, 4]. Harris [3] aplied a questionnaire to 1351 children to identify MSK outcomes associated to computer use. He found that the most common sympthoms were neck and lower back discomfort and its severity depended on frequency and duration of computer use. Straker et. al. [1] used photogrammetry and superficial electromiography (sEMG) to assess posture during tablet computer, desktop computer and paper use by young children. The results show that tablet computers and laptops are associated with greater neck muscle activity, more trunk flexion, and more flexed and elevated shoulders than desktop computers. Therefore, they likely pose a greater risk for MSK discomfort. Based on previous studies, Staker et. al. [5] reviewed the evidence of positive and negative impacts of computer use in children and proposed a set of guidelines for wise use of computers by children.

On the other hand, smart phones and tablets began to widespread in the first years of the 21st century. Pereira et. al. [6] evaluated the effects of holding a tablet with one hand in relation to usability, fatigue and biomechanics among users with small hands. Young et. al. [7] studied head and neck postures during four common tablet user configurations. Ning et. al. [8] evaluated the neck posture and kinematics of the cervical spine during the operation of a touch screen tablet and a smart phone. They use a wireless sEMG system and inertial sensors to record muscle activity and posture data. The results showed that users maintain deep neck flexion when using touchscreen mobile devices.

It is thought that high prevalence of MSK discomfort increases the risk of musculoskeletal disorders (MSD). MSDs are injuries in soft tissues like tendons, ligaments, cartilage, muscles and nerves [9]. However, there is a lack of research on quantitative analysis of the effects in the MSK system of children when using technological devices. The purpose of this study is to evaluate the posture of a child during the use of a laptop computer, a tablet and a smart phone; to assess the associated MSK risks. In order to do that, we performed qualitative and quantitative analysis from the viewpoints of ergonomics and biomechanics. Ergonomics can help us to study the interactions among children and the devices, and with biomechanics the effects of energy and forces on the human body can be evaluated.

II METHODS

A Questionnaires

A questionnaire was designed to evaluate habits, postures, timing and discomfort in children when using mobile devices. A total of 60 questionnaires were applied to parents of chil-

I. Torres et al. (eds.), *VII Latin American Congress on Biomedical Engineering CLAIB 2016, Bucaramanga, Santander, Colombia, October 26th -28th, 2016*,
IFMBE Proceedings 60,
DOI: 10.1007/978-981-10-4086-3_81

dren with ages between 2 and 12 years old. Some of the questions are given next.

- What kind of technological devices do children use?
- How much time a day they use the devices?
- What are the usage positions?

B Data acquisition

A five-year old girl was recruited for this pilot study. Her parents signed an informed consent to authorize her participation. Twenty six anthropometric measurements were taken according to the norm NTC 5649: Basic measurements of the human body for technological design[1]. The girl was asked to play games using three devices: a laptop computer (Microsoft Surface Pro 4, China), a tablet (IPad Mini 2, Apple, Cupertino, CA, USA), and a smart phone (Galaxy A5, Samsung Electronics, China). The positions used were: desk sitting, sitting cross legged, lying prone, lying supine, and standing. The postures of the girl in each case were captured using a digital camera (NEX-F3, Sony) and described in terms of positions and angles using photogrammetry.

C Digitalization

There are software tools to create digital human models for ergonomics and biomechanical analysis. One of these programs is Delmia V5 (Dassault Systemes, Waltham, MA, USA). We used Delmia to create a basic 3D model of the girl. This model was adjusted to the different postures using the data obtained from the pictures.

D Analysis

D.1 Ergonomics analysis

Among the methods for ergonomics evaluation are the Job Strain Index (JSI), Occupational Repetitive Action (OCRA), Rapid Upper Limb Assessment (RULA), and Rapid Entire Body Assessment (REBA). These methods can estimate the risks associated to postures and tasks. We used the RULA method [10], as other studies on the physical impacts of the use of computers by children [4]. RULA evaluates the risk of upper limb disorders based on posture, muscle use, weight of loads, task duration, and frequency. It assigns a score that determines the risk of an upper limb MSD as follows:

- Score 1 - 2: negligible risk, acceptable posture if not maintained or repeated for long periods.
- Score 3 - 4: low risk, changes may be needed
- Score 5 - 6: medium risk, further investigation, changes required soon
- Score 7: very high risk, changes required immediately

In RULA, the human body is divided in two groups: group A includes upper arm, lower arm, and wrist; and group B includes neck and trunk. The data required by the RULA analysis are joint angles and twisting of the arms, wrist, neck, trunk and legs; ease in load handling, whether it is static or repeated, and number of (task) repetitions per minute.

D.2 Biomechanical analysis

Delmia was used to perform the biomechanical analysis of each of the postures to estimate forces (tension and compression) and moments. The National Institute for Occupational Safety and Health (NIOSH) equation determines the load limits that can be considered acceptable (AL) as 3400N, and hazardous (MPL) as 6400N [11].

III Results

The results assess the biomechanics and ergonomics conditions when technological devices are used in different environments: sitting, standing and lying. The positions considered for the evaluation and digitalization of posture habits were selected according to the results of the questionnaires. The questionnaire allowed us to assess the perception about 60 children and their relationship with electronic devices, which are used mostly for gaming. The average time that children use devices is one to three hours per day. Within the conditions presented by children, there are pain in the neck with a percentage of 22%, pain in the lower back with 18%, shoulder 15%, chest and wrist 8% and 6% respectively. The percentages for frequent visual fatigue discomfort is 26%, blurred vision is 21% and dryness 15%.

A representative child (five-year old girl) was digitalized in order to simulate the positions of use of electronic devices as seen in Fig. 1. The human model, also called manikin, is an essential part of human modeling tools. The 3D models of different body parts were generated using a parametric CAD system (Delmia). The appearance of the human models in the human modeling tools varies from simple stick figures to realistically scanned human images based on the anthropometric measurements.

A Ergonomic analysis with RULA

A model was generated for five positions: sitting cross legged, lying prone, lying supine, and standing to evaluate

[1]NTC 5649. ICONTEC

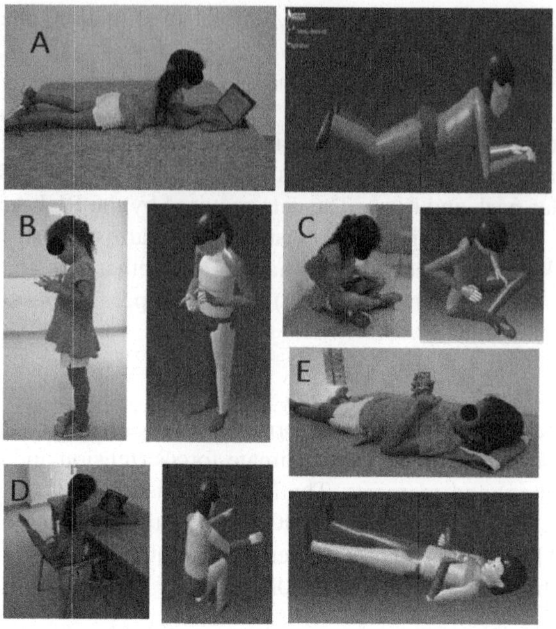

Fig. 1: Photographs and human models in different postures: A- Lying prone, B- Standing, C- Sitting cross legged, D- Desk Sitting, and E Lying supine.

Table 1: RULA scores

Parameter	Lying prone	Standing	Desk sitting	Sitting cross legged	Lying supine
Global	7	4	4	7	5
Upper arm	3	1	3	3	2
Forearm	2	3	3	2	3
Wrist score	2	2	2	3	3
Wrist twist	2	1	2	1	1
Posture A	4	3	4	4	4
Muscle	1	1	1	1	1
Wrist and arm	5	4	5	5	5
Neck score	4	3	2	2	3
Trunk score	1	1	2	4	1
Leg score	1	1	1	1	1
Posture B	5	3	2	5	3
Neck trunk leg	6	4	3	6	4

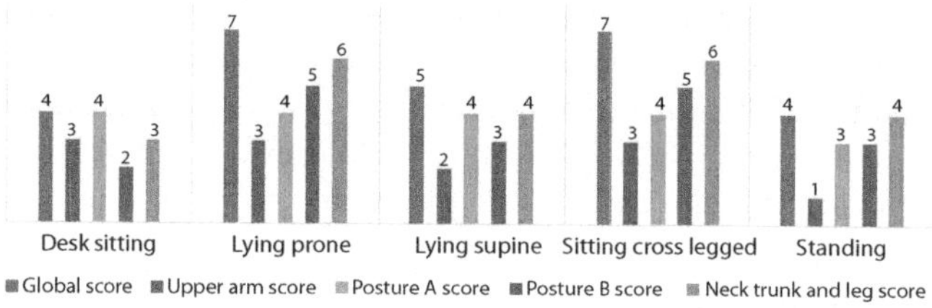

Fig. 2: Scores in different areas for the positions analyzed.

the risk of upper limb injuries. The RULA tool available in Delmia was used for this purpose. This analysis evaluates a posture and rates it on a scale of one to seven, being one the most comfortable. The results show a global score of 7 for lying prone and sitting cross legged, 5 for lying supine, 4 for desk sitting, and 5 for standing position. The high scores for lying prone and sitting cross legged positions suggest immediate revisions. The analysis showed that the primary reasons for the high scores were uncomfortable conditions in the area of neck-trunk.

Among the areas most affected by postural conditions are the arm and wrist, with scores between 4 and 5. The most awkward postures assessed by RULA are in the group A: composed arm, wrist and forearm. The most critical positions are lying prone and sitting cross legged for the group B (neck, trunk and legs). A score of 5 is given for discomfort in legs. In case of sitting cross legged, it violates the degrees of freedom in natural movements. Table 1 and Fig. 2 show the results.

B Biomechanical analysis

The biomechanical analysis was done with the aim of assessing compression forces and moments in the different postures. The L4-L5 moment is seen in Fig. 3. Extensor moments are expressed as positive and flexor moments as negative. The lying supine position presents a bending moment, all other moments show extension. This resistive moment is referred to as the support moment.

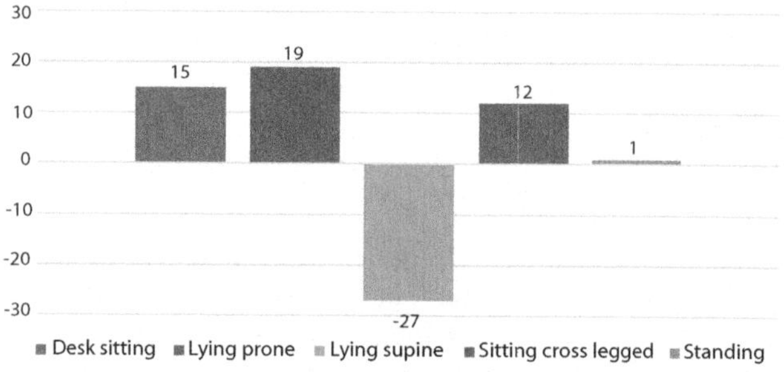

Fig. 3: L4-L5 moment in all positions.

The L4-L5 compression value represents the compressive force acting upon the L4-L5 intervertebral joint. It is seen in Fig. 4 with respect to the lumbar spine, the L4-L5 vertebrae are compressed together by the forces due to the mass of the body, the forces acting upon the hands and the trunk muscles that are used to generate the support moment. The L4-L5 compression value in the lying prone position is the highest: 597 N, and the lowest score is in the condition of desk sitting with 119 N.

The results in compression body load are shown in Fig. 5.

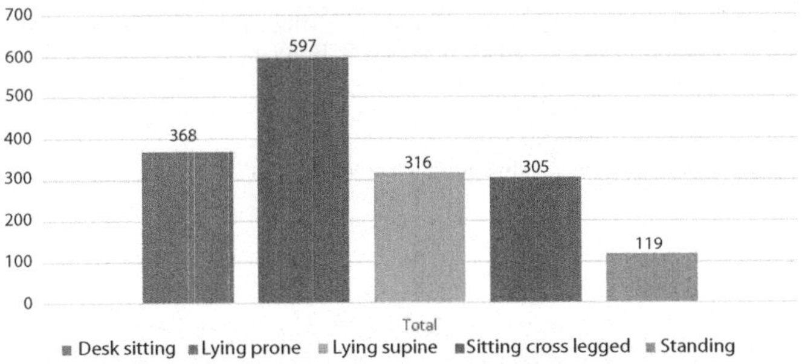

Fig. 4: L4-L5 compression in all positions.

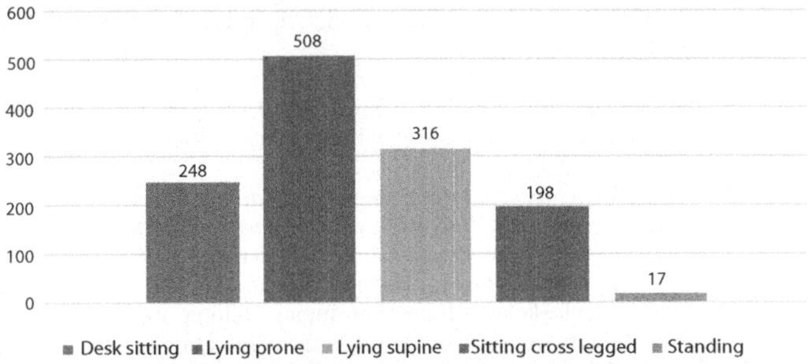

Fig. 6: Flexion-extension compression in all positions.

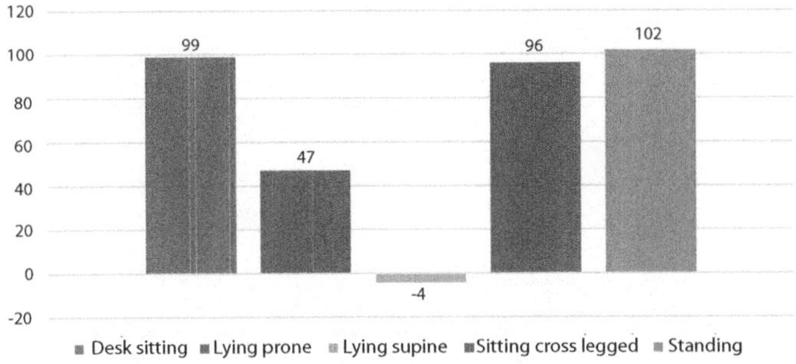

Fig. 5: Body load compression in all positions.

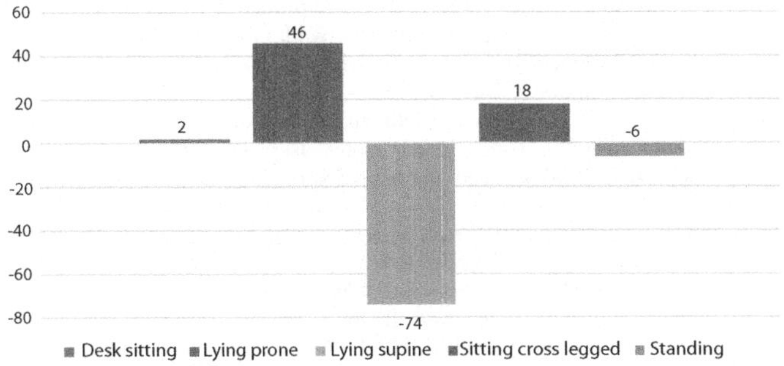

Fig. 7: Joint shear in all positions.

On standing positions, sitting cross legged and lying prone compression values over 90 N values. The lying supine has the lowest value, as the child is resting on a flat surface.

Flexion and extension in the lying supine position are related to the efforts between the trunk and neck. This position has a critical value in extension due to abdomen pressure to hold the weight of the trunk. The healthiest position is standing because of the sagittal and natural body posture, as seen in Fig. 6.

The L4-L5 joint shear represents the resultant shear force due to the sum of the reaction shear and the muscle/ligament shear. This value includes the effect of muscle/ligament forces, which represents the actual shear experienced at the L4-L5 joint. The critical values of pressure occurs in positions lying prone and supine with 46N and -74N respectively. They can be considered inadequate and may cause discomfort in the L4-L5 joint.

IV Conclusion

This pilot study evaluated the ergonomic and biomechanical conditions of the use of electronic devices by a young child. The postures analyzed were desk sitting, sitting cross legged, lying supine, lying prone and standing. The results showed that the most comfortable positions are desk sitting and standing. However, it is important to consider that maintaining the standing position for long periods of time may produce MSK discomfort because of the moments and compression forces required to hold stability and equilibrium. The best position is desk sitting, where the hips, low back and trunk rest comfortably. The results of the biomechanical analysis with respect to the joint shear suggest that the lying positions present higher risks. Children technology usage habits should be evaluated considering their anthropometric and biomechanical conditions. They usually choose as preferred positions the ones that can be adopted in a spontaneous manner, like lying supine, lying prone and sitting cross legged. However, they are not aware of consequences such as MSK discomfort and MSD. The results agree with the most common conditions reported in the questionnaires, such as shoulder and low back pain. Future work includes increasing the number of subjects of different ages and the types of technological devices considered. We also plan to include factors related to the visual field such as scope, visual angle and focusing.

References

1. Straker L. M., Coleman J., Skoss R., Maslen B. A., Burgess-Limerick R., Pollock C. M.. A comparison of posture and muscle activity during

tablet computer, desktop computer and paper use by young children *Ergonomics.* 2008;51:540–555.
2. Maslen Barbara, Straker Leon. A comparison of posture and muscle activity means and variation amongst young children, older children and young adults whilst working with computers *Work.* 2009;32:311–320.
3. Harris C. *Musculoskeletal outcomes in children using computers: a model representing the relationships between user correlates, computer exposure and musculoskeletal outcomes.* PhD thesis 2010.
4. Binboga Elif, Korhan Orhan. Posture, musculoskeletal activities, and possible musculoskeletal discomfort among children using laptops or tablet computers for educational purposes: A literature review *Journal of Science Education and Technology.* 2014;23:605–616.
5. Straker L, Pollock C, Maslen B. Principles for the wise use of computers by children. *Ergonomics.* 2009;52:1386–1401.
6. Pereira Anna, Miller Tevis, Huang Yi-Min, Odell Dan, Rempel David. Holding a tablet computer with one hand: effect of tablet design features on biomechanics and subjective usability among users with small hands. *Ergonomics.* 2013;56:1363–75.
7. Young Justin G., Trudeau Matthieu, Odell Dan, Marinelli Kim, Dennerlein Jack T.. Touch-screen tablet user configurations and case-supported tilt affect head and neck flexion angles *Work.* 2012;41:81–91.
8. Neck kinematics and muscle activity during mobile device operations *International Journal of Industrial Ergonomics.* 2015;48:10–15.
9. Freivalds A. *Biomechanics of the Upper Limbs: Mechanics, Modeling and Musculoskeletal Injuries.* Taylor & Francissecond ed. 2011.
10. McAtamney Lynn, Corlett E. Nigel. RULA: a survey method for the investigation of work-related upper limb disorders *Applied Ergonomics.* 1993;24:91 - 99.
11. Waters Thomas R, Putz-Anderson Vern, Garg Arun, Fine Lawrence J. Revised NIOSH equation for the design and evaluation of manual lifting tasks *Ergonomics.* 1993;36:749–776.

Authors: Ospina-Mateus, Niño-Prada, Tilbe-Ayola, Contreras-Ortiz
Institute: Universidad Tecnológica de Bolívar
Street: Km 1 Via Turbaco
City: Cartagena de Indias
Country: Colombia
Email: hospina@unitecnologica.edu.co

Cardiovagal Reflex Blood Pressure Set Point Determination by Time Domain Analysis

Juan Carlos Perfetto[1], G.A. Ruiz[2]

[1] Biomedical Engineering Institute, Buenos Aires University, Av Paseo Colon 850, Buenos Aires, Argentina
[2] Fernández Hospital, Cerviño 3356, Buenos Aires, Argentina

Abstract— **The baroreflex is not only an important marker of cardiac risk, but also can serve as a monitor of improvements in baroreflex sensitivity and consequences of drug treatments, physical training, etc. Sensitivity is not the only parameter of baroreflex function, but another very important is the set point. This allows to know the value of blood pressure that baroreflex tries to stabilize. A method to obtain the Blood Pressure Set Point corresponding to vagal baroreflex curve, based on the analysis of blood pressure and heart pulse interval variability is presented. The Eurobavar, which is a heterogeneous dataset, extracted from subjects with different reflex response is used to test the method. Systolic Blood Pressure (SBP) and interval time between beats (RR) series are constructed using records from that Dataset. From both series baroreflex sequences are extracted using traditional Sequence Method. Instead of computing the average of the slopes obtained from all arterial pressure points, we segment the pressure space in non-overlapping windows of 10 mm of Hg wide averaging the gain values corresponding to pressure values within each window (data binning). The Blood Pressure corresponding to the maximum of histograms curves, constructed from SBP of the baroreflex sequences, is used to contrast the result obtained with other method. Results were 119 ± 17,7mmHg for method 1 and 118 ± 21,9 mmHg for method 2. In conclusion, with only rearranging the data, an important parameter, such as Set Point of the baroreflex, can be achieved.**

Keywords— **Time domain, Spontaneous Baroreflex, RR Interval and Blood Pressure Variability, Blood Pressure Set Point, Eurovabar.**

I. Introduction

Vagal Baroreflex function (VBF) can be analyzed by RR intervals changes induced from variations of systolic blood pressure (SBP). The baroreflex is a well-known system to stabilize arterial blood pressure by means of negative feedback. The first baroreflex studies were done in early 1930 [1], and as many other biologic functions, its shape is sigmoid (Figure 1). We can therefore distinguish tree important points, the threshold, the saturation and the Set Point (Blood Pressure to be stabilized).

The baroreflex has a minimal value of pressure to work (threshold), and beyond a determined value (saturation) the response is not increased. The slope at the centering point is the maximum slope or gain of the curve. For this reason it is accepted to use that value as a unique characteristic of the baroreflex function. This value is called baroreflex sensitivity (BRS) and is measured in ms /mmHg.

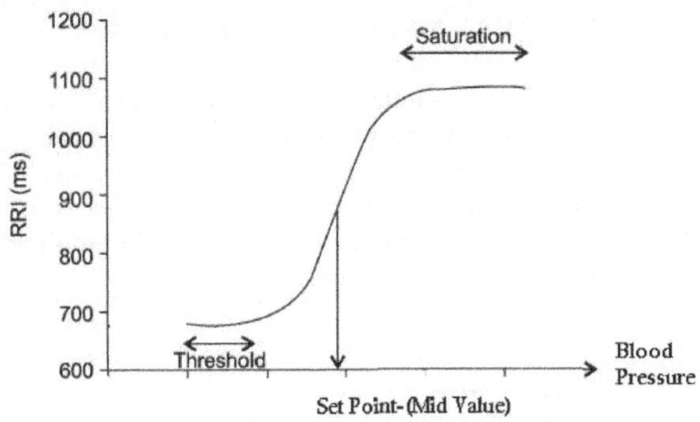

Fig. 1-Baroreflex Idealized Curve function

The methods available for baroreflex sensitivity estimation can be divided in two main groups. The first one includes methods that cause artificial increments and decrements of the blood pressure, while the second one involves spontaneous variations of arterial blood pressure.

Due to its non-invasive nature, the last group became more popular.

BRS estimated by spectral methods is biased by the simultaneous presence of regulatory mechanisms other than baroreflex, and cannot discriminate between positive and negative feedback, as time domain methods can. Thus the latter is frequently chosen.

In time domain, the most popular non-invasive method to study the VBF is the sequence method [2, 3]. The main criticism made to Sequence method is the small number of baroreflex sequences obtained by this technique. Arterial Stiffness due to aging and cardiac diseases results in baroreflex impairment. In these situations sequential methods usually fails. Others methods, like Cross Correlation (xBRS) [4] and Baroreflex Events (BE) [5] seem to have improved this issue, but differs only in the technique used to find local gains, and finally all of them average the gains.

In Sequence method, baroreflex function can be evaluated from spontaneous and concordant variations in RR intervals and systolic arterial blood pressure time series. For each sequence baroreflex, gain is estimated computing the ratio between slopes from both series. Instead of computing the average of the slopes obtained from all arterial pressure points, we first segment the pressure space in non-overlapping

I. Torres et al. (eds.), *VII Latin American Congress on Biomedical Engineering CLAIB 2016, Bucaramanga, Santander, Colombia, October 26th -28th, 2016*,
IFMBE Proceedings 60,
DOI: 10.1007/978-981-10-4086-3_82

windows of 10 mm of Hg wide. We then average the gain values corresponding to pressure values within each window (data binning).
The Set Point corresponds to the greater value taking into account all the bins. As a check, this peak is compared to that obtained from the histograms of pressure series formed with sequences detected.

II. Methods and Materials

A. Experimental Data

All of the records used in this study were from the EUROBAVAR [6] data set, acquired from 21 patients (four men and seventeen women) in supine and standing positions. The patients were noninvasively monitored. Continuous arterial blood pressure (CBPC) and Electrocardiogram (ECG) are the signals acquired. The sample rate used is 500 samples per second. Both signals are available in text format. The studies elapsed 10 minutes aproximately.

B. Methods

Preprocessings:

Two series are needed: beat to beat interval and beat to beat systolic BP. They are obtained respectively from ECG and CBPC of the Eurobavar dataset. This implies detection of R wave on ECG to compute intervals between them, and determination of maximum of blood pressure for every cardiac cycle on CBPC. Pam Tompkins [7] is used for the detection of R wave on ECG. The intervals between them were used to form the RR event series. Another method [8] is used to detect systolic arterial blood pressure from continuous arterial blood pressure. This method was developed to be used in these studies were stationarity cannot be assured. From these points, the SBP series can be constructed.

Baroreflex Sequences Detection:

We adopt the criteria of the sequential method [9]. Using the time series created three or more consecutive cycles of blood pressure increments coupled with RR interval concordant changes, are identified. Only values of correlation coefficient greater than 0.85 between both point series are included, to assure linear relationship between point serial pairs. The quotient between segments is the gain. The minimum increment to be considered is 1 mmHg for the arterial pressure. For each ramp not only the gain is registered but also the corresponding arterial pressure. A lag of one beat between ramps is allowed.

Data binning:

As the number of sequences that can be detected are limited, and have great variance, the solution that sequential method brings out, is to average all the gains points obtained from the selected ramps.
One alternate procedure is to cluster points on the basis of similar arterial pressure, and average only those points which belong to the same class. The pressure range from 60 to 170 mmHg is then segmented into non-overlapping windows of 10 mmHg wide. To choose the range of pressures, possible values from a physiological point of view have been taken into account. Twelve gain values are then computed within each window. To register the blood pressure associated with each sequence detected two series are created, one with partial gain values, and the other with corresponding systolic blood pressure. Both of them are indexed with the sequence number. The index of the maximum gain from all the bins is used to find the systolic arterial blood pressure of the corresponding bin. This pressure value corresponds to the systolic arterial pressure set point. Depending on the subject only a few of the classes have elements, so the others are allowed to be empty. For comparing purpose with the sequence method all the slopes were also averaged.

Blood Pressure Set Point:

The maximum from gains of different classes becomes the Blood Pressure Set Point for each patient, and for each situation. The justification comes from theory, as the slope for the baroreflex curve at the set point is maxima.

Systolic Arterial Pressure set point using histogram:

The goal of baroreflex is to stabilize blood pressure at a given value, therefore, if it works as it should, blood pressure should have values close to this value, more often than to any other value. For this reason, histograms are obtained, using systolic arterial blood pressure of baroreflex sequences, and the pressure which corresponds to the maximum of each histogram is taken a second estimate of the systolic arterial blood pressure set point, and used to validate the other estimate. If both estimates do not match, conclusion would be that baroreflex is impaired.

III. Results

Using twenty one records from Eurovabar dataset, corresponding to supine position, all the processes explained in the previous section were done. The results are show in the Table 1. Each row corresponds to a different subject, and each column shows different results. The first column, the systolic blood pressure Set Point obtained from Data Binning the baroreflex sequences detected. The second column, the peak of the histogram of the systolic pressure series of the baroreflex sequences. In one case no baroreflex sequence was detected, and this case is removed.

Table 1- Systolic Pressure Set Points from the two methods.

Set Point1	Set Point2
100	101
90	104
110	116
140	130
130	132
80	86,7
100	102
120	115
90	97,7
110	115
140	116
110	108
140	132
140	148
160	144
140	144
120	121
90	102
130	142
120	117

Another case is probably corresponding to a patient with impaired baroreflex, because gain in this case is extremely low, and only one bin has sequences. In all the others cases, three or more bins have been involved in the calculus. Four records corresponding to two patients were duplicated to test reproductibility.

As expected, results from each pair of duplicaded records were identical. For this reason records named B014 and B015 were excluded from statistical results. B005 cannot be evaluated and was also excluded.

The correlation coefficient between estimates of systolic arterial pressure set point was 0.91.

Results were 119mmHg ± 17,7mmHg for method 1 and 118 ± 21,9 mmHg for method 2.

The Bera–Jarque test on method 2 rejected 2 registers (A006LC and B007LC) as normally distributed ($p<0.05$).

For method 1 the number of elements of each bin was not enough to run statistical tests.

The Eurobavar Dataset is non-homogenous: one subject is diabetic with autonomic neuropathy and another is heart transplanted. The one transplanted has no baroreflex sequences, for that reason this case can not evaluated. The remaining 19 subjects are 12 normotensive, 3 hypertensive, and 4 healthy volunteers. From the table 1 we can identify those subjects suffering hipertension.

VI. Discussion

Spontaneous methods have inherent limitations: they cannot do full range analysis, but it does not mean they only can estimate gain at mid point. They are limited to the range of excursion that blood pressure naturally has. Blood pressure variations, from Eurobavar Dataset, in most cases, allow computing gains at different operating points of the baroreflex curve, and therefore, determining the set point, looking for the maximum between them.

Our technique provides information about the blood pressure set point. This is really important, if we want to compare different treatments, for example the response of antihipertensive drugs. On the other hand, the use of histograms of the pressure series involved in baroreflex sequences, allow to check normality of the reflex. If it is healthy the histogram will exhibit one peak that will correspond to the peak computed with the other technique. This could be useful when assessing patients with impaired baroreflex functions. This theoretical conclusion must be corroborated in studies including impaired baroreflex and healthy subjects.

V. Conclusions

A novel method is proposed to improve time domain assessment of baroreflex function. It is very simple, and was able to process records for all the patients, except for the one transplanted.

Conflict of Interest

"The authors declare that they have no conflict of interest".

References

1. Koch, E., *Die reflektorische Selbststeuerung des Kreislaufes*. Vol. 1. 1931: Theodor Steinkopff.
2. Bertinieri, G., et al., *Evaluation of baroreceptor reflex by blood pressure monitoring in unanesthetized cats.* Am J Physiol, 1988. 254(2 Pt 2): p. H377-83.
3. Bertinieri, G., et al., *A new approach to analysis of the arterial baroreflex.* J Hypertens Suppl, 1985. 3(3): p. S79-81.

4. Westerhof, B.E., et al., *Time-domain cross-correlation baroreflex sensitivity: performance on the EUROBAVAR data set.* J Hypertens, 2004. 22(7): p. 1371-80.
5. Gouveia, S., et al., *Time domain baroreflex sensitivity assessment by joint analysis of spontaneous SBP and RR series.* Biomedical Signal Processing and Control, 2009. 4(3): p. 254-261.
6. Laude, D., et al., *Comparison of various techniques used to estimate spontaneous baroreflex sensitivity (the EuroBaVar study).* American Journal of Physiology-Regulatory, Integrative and Comparative Physiology, 2004. 286(1): p. R226-R231.
7. Pan, J. and W.J. Tompkins, *A real-time QRS detection algorithm.* Biomedical Engineering, IEEE Transactions on, 1985(3): p. 230-236.
8. Perfetto, J.C., et al., *Pressure-detection algorithms.* IEEE Eng Med Biol Mag, 2009. 28(5): p. 35-40.
9. Parati, G., M. Di Rienzo, and G. Mancia, *How to measure baroreflex sensitivity: from the cardiovascular laboratory to daily life.* J Hypertens, 2000. 18(1): p. 7-19.

Experimental Study of Apatite Layer Formation on Chitosan/Bioactive Glass Scaffolds for Bone Tissue Regeneration

L.A. Quintero[1*], D. O. Grajales[1], D. M. Escobar[1]

[1] Biomaterials Research Group, Engineering Faculty, University of Antioquia, St 70 Nº 52-21, Medellin, Colombia.

* lindsey.alejandra@gmail.com

Abstract--**In this study, sol-gel derived bioactive glass (BG), in the ternary SiO_2-CaO-P_2O_5 system, was synthesized and added as a reinforcement material in Chitosan (CH)-matrix scaffolds with ratios of CH/BG 70/30, 50/50 and 30/70. The scaffolds were prepared by freeze-drying process for 24 hours. The addition of BG in the CH-matrix improved the composite scaffolds bioactivity, as seen by the precipitation of bone like apatite layer after immersion in simulated body fluid (SBF) for 7 days. The apatite layer formation on the surface of each scaffold was confirmed by Scanning Electron Microscopy (SEM), X-ray Powder Diffraction (XRD) and Fourier Transform Infrared Spectroscopy (FTIR). XRD result for BG indicated net formation using alternative calcium and phosphorus sources. SEM, XRD and FTIR results demonstrated that CH/BG scaffolds had higher bioactivity in comparison with CH scaffolds. These results advised that synthesized BG had a potentiality as reinforcement material in CH scaffolds and better bioactive behavior.**

Keywords--**composite scaffold, chitosan, bioactive glass, bioactivity test, apatite**

I. INTRODUCTION

Three-dimensional porous structures, named scaffolds, are developed to contribute to formation of functional tissue displaying properties that mimics natural, healthy tissues [1].

In this sense, scaffolds for bone tissue engineering (BTE) are required to be bioactive, biodegradable and provide an appropriate environment for cell attachment, proliferation and differentiation to finally form a new bone tissue [2]. Given that bone tissue is a mineralized composite, it is well know that one biomaterial type is not sufficient for bone tissue regeneration [3]. Therefore, composite scaffolds based on flexible biodegradable biopolymers and inorganic ceramics have been developed.

CH is a linear polysaccharide composed by glucosamine units extracted from crustaceans. Also, this is a biodegradable polymer, which allows osseointegration and presents an intrinsic antibacterial activity [4]. However, CH itself is not the ideal material for BTE due to its low mechanical strength and poor bioactive capabilities [5]. For this reason, the addition of bioactive glasses (BG), as reinforcement in CH scaffolds, could improve mechanical and bioactive properties without decreasing CH qualities. BG capability increase apatite layer formation giving a suitable and safe chemical bonding between composite scaffolds and living bone [6]. The bioactivity has been studied through apatite layer formation on the biomaterial surface after being soaked in SBF solution [7]. In this paper, composite scaffolds made of CH and sol-gel derived BG with three different ratios were analyzed in order to identify bioactive performance by in-vitro tests for 7 days using SBF solution.

II. MATERIALS AND METHODS

A. *Materials*

Chitosan (medium molecular weight) and glutaraldehyde 25% were purchased from Sigma-Aldrich, Germany. The N-deacetylation degree of chitosan was found to be 80.22%. Tetraethylorthosilicate (TEOS: $C_8H_{20}O_4Si$) and calcium acetate ($Ca(CH_3COO)_2$) were purchased from Merck Inc. P_2O_5 was purchased from Alfa Aesar. Ethanol, Acetic acid and Lactic acid were purchased from Panreac. All chemicals from SBF preparation were obtained from Sigma-Aldrich, Germany.

B. *Bioactive glass synthesis*

Sol-gel prepared BG consisting of $65SiO_2$-$5P_2O_5$-30CaO (based on mol.%) was synthesized and characterized, as it is following described: TEOS was added into an ethanol/distilled water solution, with a molar ratio 1:4 both TEOS/distilled water and TEOS/ethanol according to Vaid, et al. [8]. The mixture reacted for 1 h assisted by agitation. P_2O_5 and an ethanol/distilled water solution with a molar ratio 1:4 both P_2O_5/distilled water and P_2O_5/ethanol were added to react completely in stirring for 45 min. Later on, calcium acetate was added to react for 11 min. Finally, an acetic acid/distilled water solution (6:1) was added into mixture (distilled water/BG 1:4). The final solution was kept in agitation until the gel was formed (gel point).

I. Torres et al. (eds.), *VII Latin American Congress on Biomedical Engineering CLAIB 2016, Bucaramanga, Santander, Colombia, October 26th -28th, 2016*,
IFMBE Proceedings 60,
DOI: 10.1007/978-981-10-4086-3_83

BG gel was kept in a container for 3 days at room temperature. The gel was heated at 120°C for 2 days to remove all water content. The final product was crushed in order to obtain a dry powder.

C. Preparation of the composite scaffolds

With the aim of preparing highly porous and bioactive composite scaffolds, CH was dissolved in 1% (v/v) aqueous lactic acid in order to obtain a CH solution at 2% (w/v). Composite scaffolds with different ratios of CH/BG (70/30, 50/50, 30/70) were elaborated. Stirring for 30 min and sonication for 1h were needed to assure the BG homogenization into CH solution [2]. In addition, Glutaraldehyde (1% w/w, respect to CH amount) was added to the mixture with constant stirring. The mixtures were frozen with dry ice for 24 h. Lastly, the scaffolds were taken to a freeze dryer at -80°C and 0.02 mbar for 24 h in order to produce 3D porous structure.

D. Sample characterization

The synthesized BG were analyzed by XRD with XPert PANalytical Empyrean Series II diffractometer. This instrument worked with voltage and current settings of 45 kV and 40 mA respectively and used Cu-Kα radiation (1.5405980 Å). Regarding SEM analysis the morphology and microstructure of the composite scaffolds were evaluated using a scanning electron microscope (SEM-JOEL-JSM 6490 LV) that operated at the acceleration voltage of 20 kV. The functional groups present on the composite scaffolds were examined by Fourier Transform Infra-Red (FTIR) analysis with Perkin Elmer Spectrum One model with DTGS detector spectrometer.

E. In vitro bioactivity study in SBF solution

The SBF solution was prepared according to Kokubo et al. [7]. The composite scaffolds were immersed in the SBF solution and incubated at 37 °C in close tubes for 7 days. Afterwards, the scaffolds were removed from the SBF solution and washed with distilled water to removed adsorbed minerals, and finally they were analyzed by FTIR and XRD.

III. Results and Discussion

A. XRD analysis of the synthesized BG

Fig. 1 presents narrow and differentiable peaks, which are characteristics of augmented crystallinity due to thermal treatment (1050 °C) carried on the XRD analysis. Primary and secondary peaks, even the hump centered close to 2θ =21°, present important consistencies regarding angle and intensity, as it was reported by other authors for comparable systems [9,10].

B. SEM observations

The low and high magnifications from the top view of scaffolds are shown in Fig. 2. According to the observations, the CH scaffolds, 70/30 and 50/50 (CH/BG) scaffolds showed high porosity with lamella structure (Fig. 2.a, c and e). The micrographs of these sort of scaffolds show a network of interconnected pores, having different shapes, in a size range from 100 to 200 μm. This is desirable to achieve osteogenic cell ingrowth (Fig. 2.b, d and f). [11,12].

These results allow to identify the possible bioactive behavior according to its porosity of each scaffold on in vivo conditions. Nevertheless, 30/70 (CH/BG) scaffolds show less porosity and less interconnection due to the high amount of BG incorporated into the composite (Fig. 2.g and h).

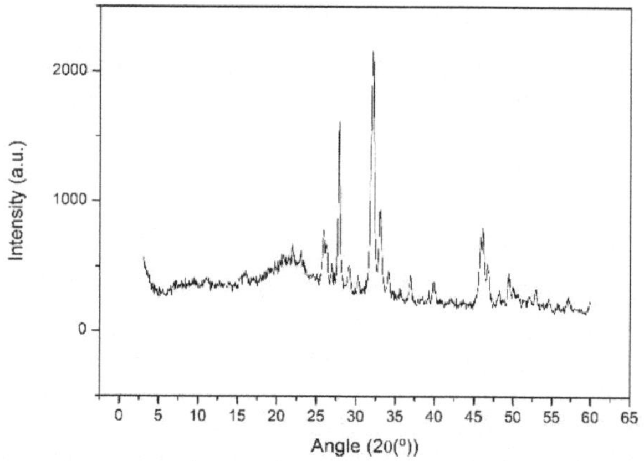

Figure 1. XRD patterns of synthesized BG after 1050°C treatment.

C. Sample characterization after in vitro assays

The two samples with appropriate pore size were immersed in SBF solution. FTIR and XRD were used for apatite layer study. XRD spectra showed in Fig. 3 and Fig. 4 compare 70/30 and 50/50 (CH/BG) scaffolds before and after soaking in SBF. The main peaks on both samples after SBF immersion match with the main peaks of Hydroxyapatite pattern spectra on HighScore Plus software (Reference code 96-901-3628) as Pourhaghgouy et al [2] reported. Furthermore, Fig. 5 and 6 depict FTIR spectra which were performed on 70/30 and 50/50 (CH/BG) scaffolds before and after immersion in SBF. As it can be seen on Fig. 6 for 50/50 (CH/BG) scaffold, the IR spectrum after soaking in SBF revealed two phosphate bonds at 576 and 669 cm^{-1} related to P-O bonding vibration on PO_4^{-3} group in calcium phosphate crystalline phases and the decrease of the 794 and 1085 cm^{-1} peaks related to Si-O-Si bonding vibration indicate the Hydroxyapatite layer formation on the surface of the scaffold [13]. However, this same result on 70/30 (CH/BG) scaffold proves the low capacity of forming a mineralized layer due to the poor amount of BG on the composite.

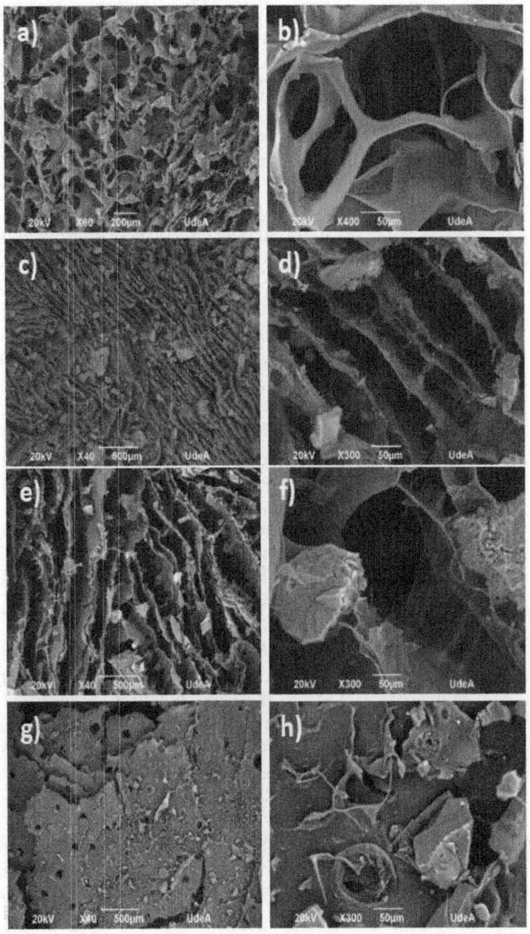

Figure 2. SEM micrographs of CH scaffold (a-b), 70/30 CH/BG scaffold (c-d), 50/50 CH/BG scaffold (e-f) and 30/70 CH/BG scaffold (g-h).

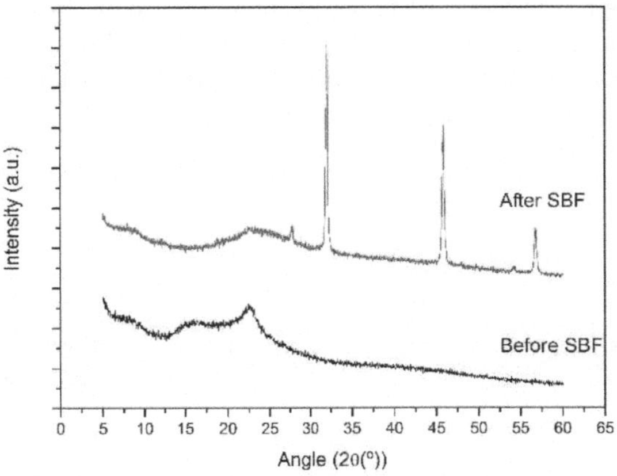

Figure 3. XRD patterns of 70/30 CH/BG scaffolds before and after immersion in SBF.

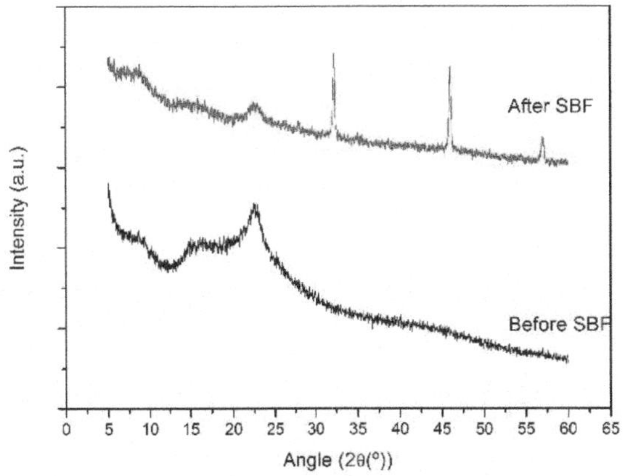

Figure 4. XRD patterns of 50/50 CH/BG scaffolds before and after immersion in SBF.

IV. Conclusions

The synthesized BG demonstrated appropriate network formation for bioactive glasses from SiO_2-CaO-P_2O_5 ternary system and an excellent bioactivity. The BG synthetized with calcium acetate and P_2O_5 (Phosphorus (V) oxide) reagents showed a similar response in comparison of bioactive glasses with tetrahydrate calcium nitrate ($Ca(NO_3)_2.4H_2O$) and TEP (triethyl phosphate). CH/BG scaffolds presented desirable porous morphology and bioactivity. Using BG as reinforcement material CH scaffold improved bioactivity without reducing its ability of forming 3D porous structures.

The pore size of scaffolds decreased due to the large amounts of BG; however, lower quantities decreased scaffold bioactivity. 50/50 scaffolds were proved to be appropriate for BTE applications due to its pore size and interconnectivity as it shown in Fig. 2.e and 2.f and bioactivity in Fig. 4. This study has shown that desirable pore structure and interconnected porosity of the composite scaffolds can be achieved through procedure and CH/BG ratio control.

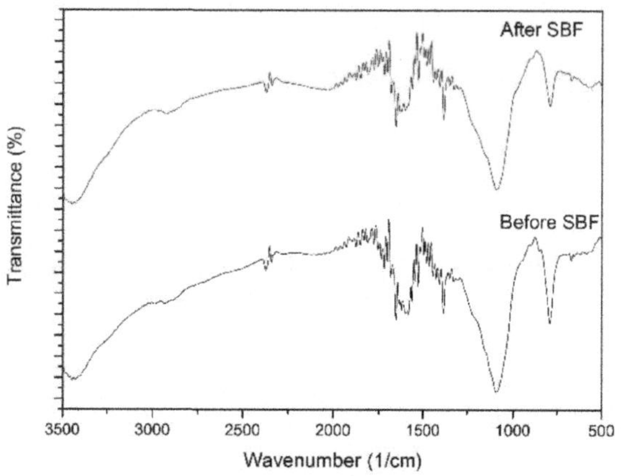

Figure 5. FTIR patterns of 70/30 CH/BG scaffolds before and after immersion in SBF.

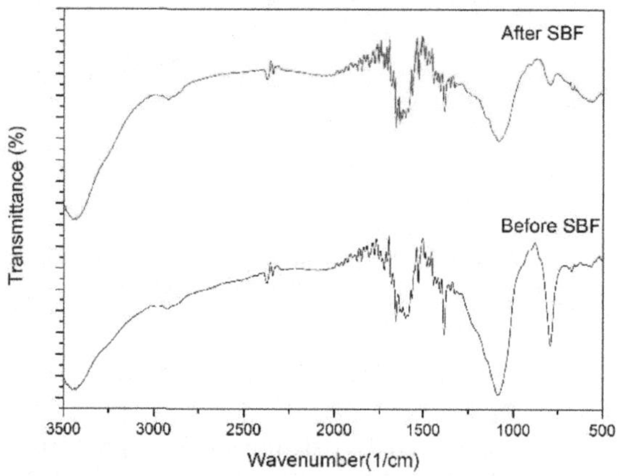

Figure 6. FTIR patterns of 50/50 CH/BG scaffolds before and after immersion in SBF.

Acknowledgment

The authors are thankful with Biomaterials Research Group for providing the necessary reagents and studies during the development of this project.

Conflict of Interest

The authors declare that they have no conflict of interest.

References

1. Marelli, B.; Ghezzi, C. E.; Mohn, D.; Stark, W. J.; Barralet, J. E.; Boccaccini, A. R.; Nazhat, S. N. Accelerated mineralization of dense collagen-nano bioactive glass hybrid gels increases scaffold stiffness and regulates osteoblastic function. *Biomaterials* **2011**, *32* (34), 8915–8926 DOI: 10.1016/j.biomaterials.2011.08.016.
2. Pourhaghgouy, M.; Zamanian, A.; Shahrezaee, M.; Masouleh, M. P. Physicochemical properties and bioactivity of freeze-cast chitosan nanocomposite scaffolds reinforced with bioactive glass. *Mater. Sci. Eng. C* **2016**, *58*, 180–186 DOI: 10.1016/j.msec.2015.07.065.
3. Luz, G. M.; Mano, J. F. Mineralized structures in nature: Examples and inspirations for the design of new composite materials and biomaterials. *Compos. Sci. Technol.* **2010**, *70* (13), 1777–1788 DOI: 10.1016/j.compscitech.2010.05.013.
4. Baldino, L.; Cardea, S.; De Marco, I.; Reverchon, E. Chitosan scaffolds formation by a supercritical freeze extraction process. *J. Supercrit. Fluids* **2014**, *90*, 27–34 DOI: 10.1016/j.supflu.2014.03.002.
5. Anitha, a.; Sowmya, S.; Kumar, P. T. S.; Deepthi, S.; Chennazhi, K. P.; Ehrlich, H.; Tsurkan, M.; Jayakumar, R. Chitin and chitosan in selected biomedical applications. *Prog. Polym. Sci.* **2014**, *39* (9), 1644–1667 DOI: 10.1016/j.progpolymsci.2014.02.008.
6. Peter, M.; Binulal, N. S.; Nair, S. V.; Selvamurugan, N.; Tamura, H.; Jayakumar, R. Novel biodegradable chitosan–gelatin/nano-bioactive glass ceramic composite scaffolds for alveolar bone tissue engineering. *Chem. Eng. J.* **2010**, *158* (2), 353–361 DOI: 10.1016/j.cej.2010.02.003.
7. Kokubo, T.; Takadama, H. How useful is SBF in predicting in vivo bone bioactivity? *Biomaterials* **2006**, *27* (15), 2907–2915 DOI: 10.1016/j.biomaterials.2006.01.017.
8. Vaid, C.; Murugavel, S. Alkali oxide containing mesoporous bioactive glasses: Synthesis, characterization and in vitro bioactivity. *Mater. Sci. Eng. C* **2013**, *33* (2), 959–968 DOI: 10.1016/j.msec.2012.11.028.
9. Dziadek, M.; Zagrajczuk, B.; Menaszek, E.; Wegrzynowicz, A.; Pawlik, J.; Cholewa-Kowalska, K. Gel-derived SiO2–CaO–P2O5 bioactive glasses and glass-ceramics modified by SrO addition. *Ceram. Int.* **2016**, *42* (5), 5842–5857 DOI: 10.1016/j.ceramint.2015.12.128.
10. Desogus, L.; Cuccu, A.; Montinaro, S.; Orrù, R.; Cao, G.; Bellucci, D.; Sola, A.; Cannillo, V. Classical Bioglass® and innovative CaO-rich bioglass powders processed by Spark Plasma Sintering: A comparative study. *J. Eur. Ceram. Soc.* **2015**, *35* (15), 4277–4285 DOI: 10.1016/j.jeurceramsoc.2015.07.023.
11. Francois, E.; Dorcemus, D.; Nukavarapu, S. *Biomaterials and scaffolds for musculoskeletal tissue engineering*; Elsevier Ltd, 2015.
12. Liu, Y.; Lim, J.; Teoh, S. H. Review: Development of clinically relevant scaffolds for vascularised bone tissue engineering. *Biotechnol. Adv.* **2013**, *31* (5), 688–705 DOI: 10.1016/j.biotechadv.2012.10.003.
13. Luz, G. M.; Boesel, L.; Campo, A. Del; Mano, J. F. Micropatterning of bioactive glass nanoparticles on chitosan membranes for spatial controlled biomineralization. *Langmuir* **2012**, *28* (17), 6970–6977 DOI: 10.1021/la300667g.

A Simple Physical Model of Human Gait Using Principles of Kinematics and BTS GAITLAB

C. A. Collazos, H. E. Castellanos, J. A. Cardona, J. C. Lozano, A. Gutiérrez, M. A. Riveros

Manuela Beltrán University, Bogotá, Colombia
cacollazos@gmail.com

Abstract— **The aim of this work is to present the one-dimensional gait kinematic principle, in order to identify the kinematic parameters for each type of gait associated with uniform linear motion and uniformly accelerated motion. This paper presents the one-dimensional gait kinematic principle, in order to identify the kinematic parameters for Normal and Pathological (transtibial amputation) Gait of two subjects with similar anthropometry. The developed physical model complements the information of the data acquisition system and is used for teaching Mechanics Physics in Biomedical Engineering and Physiotherapy.**

Keywords— **Gait, kinematic, modeling, physics.**

I. Introduction

Gait analysis is the measurement and assessment of human locomotion which includes both walking and running [1]. These movements, known as stereotyped reflexes, are characterized for being repetitive in time when the velocity and the acceleration are constant [2]. Therefore, it is possible to obtain reference curves at each movement phase that could help to determine abnormalities or pathologies which are related to the musculoskeletal system and modify normal behavior [3].

The different tissues involved during walking namely: muscles, tendons, cartilage, ligaments, connective tissue (fascia) and the bone component, perform different functions such as motion generation, power transmission, buffer loading, joint stabilization of segments, among others. These functions are the basis of motion and therefore are constantly analyzed and evaluated to determine alterations that modify their performance [4], [5].

Subjects with lower limb amputations, have compensatory adjustments in gait where the soft tissues and the mechanical stress of the body must adjust to the structural and functional changes in this one. This suggests an increased muscular demand, energy expenditure, the alignment of gravity center and mass center, the static and dynamic postural alignment among others. These parameters are relevant for measuring the risk factors presented by these subjects, such as falling due to alterations in the stability and balance control of the body [6] [7],[8].

The aim of this work is to compare the gait analysis to healthy and pathological subjects. In this research, pathological gait refers to prosthetic gait transtibial amputation. The motion analysis techniques used to measure accurately kinematic curves are obtained through skin markers, which record the position, velocity and acceleration of a body segment. These measurements provide quantitative information about the movement [9].

A complete review of human walking modeling and simulation is presented in [10], [11], [12], [13]. This research review focuses on physics-based human walking simulations in biomechanics literature and robotics. The gait synthesis methods are broadly divided into five types: inverted pendulum model; passive dynamics walking; zero moment point methods; optimization-based methods; and control-based methods [11]. The above- referenced models are highly complex by the methods and principles involved.

This paper describes a simple Physical Model of Human Gait of the Normal and pathological gait at a constant velocity and acceleration. We identify the kinematic parameters for each type of gait and compare the kinematics curves presented in each of the cases.

The article is structured as follows: Section II shows the fundamentals in kinematics and Physics Principles. Section III shows the instrumentation used and the associated markers for gait evaluation. Section IV presents the identification of the kinematic parameters related to the gait and the mathematical tools. Finally Section V is dedicated to the discussions and conclusions.

II. Fundamentals of the One-Dimensional Kinematics

For a position linear model in uniform linear motion, the first-degree polynomial is given by: ($\hat{y} = a + bx$). For N pairs of data, the regression parameters a and b are refered to (1) and (2) respectively as:

I. Torres et al. (eds.), *VII Latin American Congress on Biomedical Engineering CLAIB 2016, Bucaramanga, Santander, Colombia, October 26th -28th, 2016,*
IFMBE Proceedings 60,
DOI: 10.1007/978-981-10-4086-3_84

$$a = \frac{\sum_{i=1}^{N} x_i^2 \sum_{i=1}^{N} y_i - \sum_{i=1}^{N} x_i \sum_{i=1}^{N} (x_i y_i)}{N \sum_{i=1}^{N} x_i^2 - \left(\sum_{i=1}^{N} x_i \right)^2} \quad (1)$$

$$b = \frac{N \sum_{i=1}^{N} (x_i y_i) - \sum_{i=1}^{N} x_i \sum_{i=1}^{N} y_i}{N \sum_{i=1}^{N} x_i^2 - \left(\sum_{i=1}^{N} x_i \right)^2} \quad (2)$$

For a position no linear model in uniformly accelerated motion , in a parabolic model, the second-degree polynomial is given: ($\hat{y} = a + bx + cx^2$). For N pairs of data, the regression parameters a , b and c are refer to (3), (4), (5), (6), (7), (8), (9) and (10) as follow:

$$a = \frac{\left[S(x^2 y) S(xx) \right] - \left[S(xy) S(xx^2) \right]}{\left[S(xx) S(x^2 x^2) \right] - \left[S(xx^2) \right]^2} \quad (3)$$

$$b = \frac{\left[S(xy) S(x^2 x^2) \right] - \left[S(x^2 y) S(xx^2) \right]}{\left[S(xx) S(x^2 x^2) \right] - \left[S(xx^2) \right]^2} \quad (4)$$

$$c = \left[\frac{\sum_{i=1}^{N} y_i - b \sum_{i=1}^{N} x_i - a \sum_{i=1}^{N} x_i^2}{N} \right] \quad (5)$$

where:

$$S(xy) = \sum_{i=1}^{N} x_i y_i - \frac{\sum_{i=1}^{N} x_i \sum_{i=1}^{N} y_i}{N} \quad (6)$$

$$S(xx^2) = \sum_{i=1}^{N} x_i^3 - \frac{\sum_{i=1}^{N} x_i \sum_{i=1}^{N} x_i^2}{N} \quad (7)$$

$$S(x^2 y) = \sum_{i=1}^{N} x_i^2 y_i - \frac{\sum_{i=1}^{N} x_i^2 \sum_{i=1}^{N} y_i}{N} \quad (9)$$

$$S(x^2 x^2) = \sum_{i=1}^{N} x_i^4 - \frac{\left(\sum_{i=1}^{N} x_i^2 \right)^2}{N} \quad (9)$$

We use the fundamentals of the one-dimensional kinematic to identify the kinematic parameters for each type of gait associated with uniform linear motion and uniformly accelerated motion [9], [15].

III. INSTRUMENTATION

Manuela Beltrán University Biomechanics Laboratory was used for data logging. We use BTS GAITLAB [16]. This acquisition system of high precision for motion analysis has six optoelectronic cameras that measures the displacement ($\pm 10^{-7} m$) of body segments in time ($\pm 10^{-2} s$).

The device has 3 markers placed strategically as Fig. 1 indicates. The markers involved in the gait were the sacrum (marker 6), the right greater trochanter (marker 7) and the left greater trochanter (marker 8). The study of movement in this work is restricted to the X axis only.

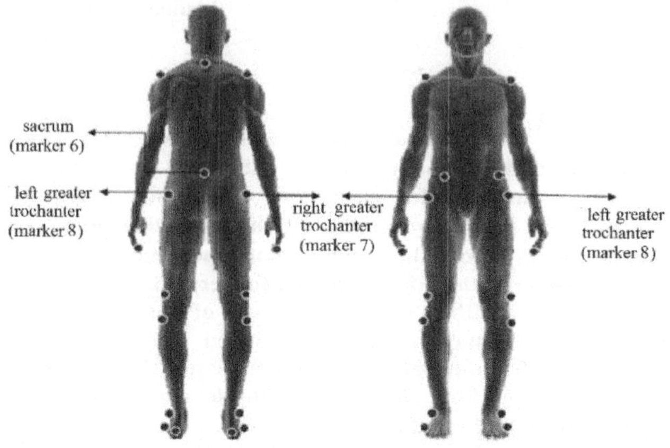

Fig. 1. Disposition of the cutaneous markers in the human body [16]

IV. METHOD, RESULTS AND ANALYSIS

For the data analysis the least squares and correlation coefficient methods were used. The test was applied for two subjects of (34 $\pm$ 1) years old male, height (1.64 $\pm$ 0.01) m and mass of (64 $\pm$ 0.1) kg. The first subject has unilateral transtibial amputation of right lower extremity. The second subject has a normal gait and an anthropometry like that of the first subject. The study was approved by the Ethics Committees of the Manuela Beltrán University.

The figures 2 and 3 show experimental data in dotted line and models in continuous black line. We use squares for Normal Gait and points for Pathological Gait. The units for all variables and parameters are represented in the International Units System.

A. Normal and Pathological Gait with constant velocity

Fig. 2 illustrates the trajectories for Normal and Pathological Gait. This figure indicates the position registered towards the sacrum (marker 6) and the linear interpolation of the two trajectories. Here, it is observed that there is a high correlation between the model and the experimental data for Normal Gait. This correlation is significantly statistical ($r = 0.997$). The identification of the model allows determining the initial position (x_0 =-2.13m) and the average velocity on the walk (v_0 =1.00m/s). The model of the position in function of time for Normal Gait is therefore $x(t) = -2.13 + 1.00t$. We observed that there is a high correlation between the model and the experimental data for Pathological Gait. This correlation is statistically significant ($r = 0.997$). The identification of the model allows determining the initial position (x_0=-2.12m) and the average velocity on the walk (v_0=0.60m/s). The model of the position in function of time for Pathological Gait is therefore $x(t) = -2.12 + 0.60t$. We use a percentage difference defined as: $\%E = \left[\frac{control - experimental}{control}\right].100\%$, to compare Normal Gait (control value) and Pathological Gait (experimental value) average velocity. In this case we obtain a percentage difference of $\%E$=39.60%.

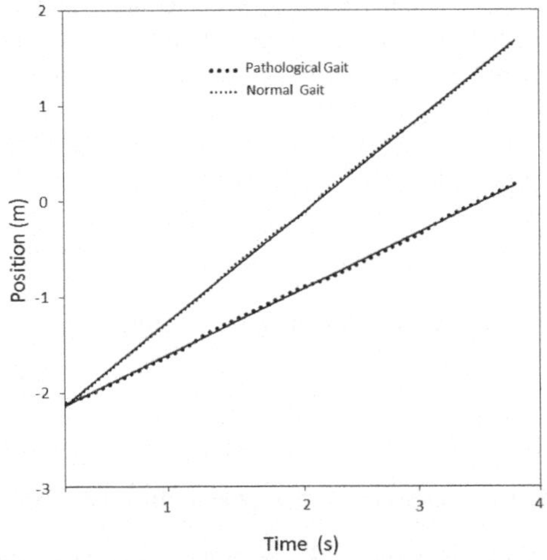

Fig. 2 The position–time graph on the X axis for the sacrum with constant velocity.

The position registered towards the right greater trochanter (marker 7) and the linear interpolation for Normal and Pathological Gait show that the initial position (x_0=-213m) and the average velocity in the normal gait (v_0= 1.02 m/s). Here can be seen that there is a high correlation between the position and the experimental measurements with $r = 0.996$. The model of the position in function of time for Normal Gait is therefore $x(t) = -2.13 + 1.02t$. The identification of the model for Pathological Gait determines the initial position (x_0=-2.12m) and the average velocity in the pathological gait (v_0= 0.61 m/s). It can be seen that there is a high correlation between the position and the experimental measurements with $r = 0.996$. The model of the position in function of time for Pathological Gait is therefore $x(t) = -2.12 + 0.61t$. In this case we obtain a percentage difference of $\%E$=39.60% for average velocity.

The position register to the left greater trochanter (marker 8) and the linear interpolation for Normal and Pathological Gait show that identification of the model for Normal Gait in initial position (x_0=-2.13m) and the average velocity on the gait (v_0=0.99 m/s). In this case the correlation coefficient is $r = 0.996$.The model of the position in function of time for Normal gait is consequently $x(t) = -2.13 + 0.99t$. The identification of the model for Pathological Gait determines the initial position (x_0=-2.12m) and the average velocity on the gait (v_0=0.59 m/s). In this case the correlation coefficient is $r = 0.996$. The model of the position in function of time for Pathological Gait is consequently $x(t) = -2.12 + 0.59t$. In this case we obtain a percentage difference of $\%E$=40.00% for average velocity.

B. Normal and Pathological Gait with constant acceleration

Fig. 3 illustrates the trajectories for Normal and Pathological Gait. The Figure 5 indicates the position registered towards the sacrum (marker 6) and quadratic interpolation of the two trajectories. The identification of the model allows determining the initial position (x_0=-2.13 m), initial velocity (v_0=-1.09 m/s) and half of the average acceleration (0.90 m/s^2) for Normal Gait. There is a high correlation between the model and the experimental data ($r = 0.997$). The model of the position in function of time for Normal Gait consequently is $x(t) = -2.13 - 1.09t + 0.90t^2$. The identification of the model for Pathological Gait determines the initial position (x_0=-2.27 m), initial velocity (v_0=-0.31m/s) and half of the average acceleration (0.29 m/s^2). There is a high correlation between the model and the experimental data ($r = 0.997$). The model of the position in function of time for Pathological Gait consequently is $x(t) = -2.27 - 0.31t + 0.29t^2$. In this case we obtain a percentage difference for average acceleration of $\%E$=71% and of $\%E$=67% in initial velocity.

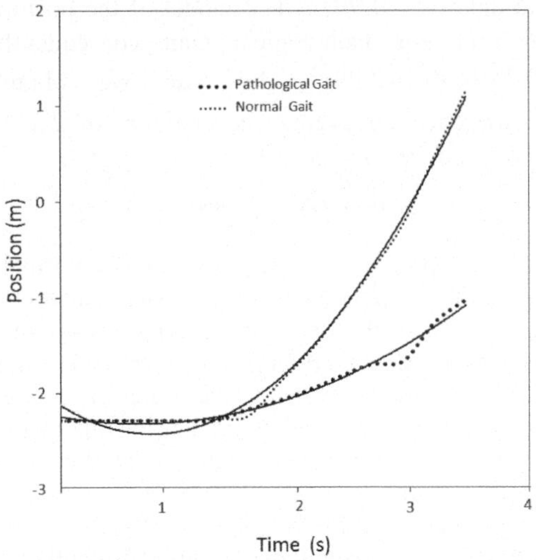

Fig. 5 The position-time graph on the X axis to the sacrum with constant acceleration

The position registered towards the right greater trochanter (marker 7) and the quadratic interpolation for Normal and

Pathological Gait show that the identification of the model for Normal Gait allows determining the initial position (x_0 =-2.10m), initial velocity (v_0 =1.18 m/s) and half of the average acceleration (0.91 m/s^2). Like we can see there is a high correlation between the position and the experimental measurements ($r = 0.996$). Hence, the model of the position in function of time for Normal Gait is $x(t) = -2.10 - 1.18t + 0.91t^2$.The identification of the model for Pathological Gait determines the initial position (x_0 =-2.27 m), initial velocity (v_0 =-0.33 m/s) and half of the average acceleration (0.31 m/s^2). There is a high correlation between the model and the experimental data ($r = 0.997$). The model of the position in function of time for Pathological Gait consequently is $x(t) = -2.27 - 0.33t + 0.31t^2$. We obtain a percentage difference for average acceleration of *%E* =72% and of *%E*=65% for initial velocity.

The position register to the left greater trochanter (marker 8) and the quadratic interpolation for Normal and Pathological Gait show that the identification of the model for Normal Gait allows determining the initial position (x_0 =-2.15m), initial velocity (v_0 =1.06 m/s) and half of the average acceleration (0.89 m/s^2) for Normal Gait. Here can be seen that there is a high correlation between the model and the experimental data ($r = 0.997$). The model of the position in function of time for Normal Gait is $x(t) = -2.15 - 1.06t + 0.89t^2$.

The identification of the model for Pathological Gait determines the initial position (x_0 =-2.27 m), initial velocity (v_0 =-0.30 m/s) and half of the average acceleration (0.29 m/s^2). There is a high correlation between the model and the experimental data ($r = 0.997$). The model of the position in function of time for Pathological Gait consequently is $x(t) = -2.27 - 0.30t + 0.29t^2$. In case we obtain a percentage difference for average acceleration of *%E*=70% and of *%E*=67% for initial velocity.

IV. Conclusion

This work has presented a case of study to Normal and Pathological Gait. Experimental results were theoretically validated for the physical models and parameters found. The technique used involves three reference markers (sacrum, right and left greater trochanter,) related to the center of mass of the human body. The identified models predict in time quantities such as position, velocity and acceleration at the different types of motion with constant velocity and acceleration. Orders of magnitude found for the physical models of position are within the range of magnitudes reported by authors like Winter in [9].

This work compare physical models between Normal and Pathological Gait. It is known that the normal gait may be affected in subjects with unilateral transtibial lower limb aputacion where the musculoskeletal system and soft tissues that help maintain the dynamic mechanism of the body are compromised. It is for this reason that the compesatory in pelvic floor and lower limbs are notorious in gait of people with amputation as you can see in the graphics of trajectory. In this sense our purpose in future is to establish a three-dimensional kinematic modeling involvement other markers such as hip, knee and ankle depending on the level of amputation for determining characteristic patterns in each study subject.

It is important to note that the normal gait pattern modeling can be affected by many causes, such as size, age, footwear, terrain, load, activity of the subject, which are not necessarily pathological but are related to the alteration or adaptation of musculoskeletal structures for movement. In this case we can generate in future works to do comparisons of the gait in different pathologies.

CONFLICT OF INTEREST

The authors declare that they have no conflict of interest.

References

1. K. Luttgens, K. Wells, *Bases científicas del movimiento humano*. Madrid, 1982, pp. 21–34.
2. K. Hainaut, *Introducción a la Biomecánica*, Barcelona, 1982, pp. 31–34.
3. B. Le Veau. *Biomecánica del movimiento humano*. México, 1991. pp.51–54.
4. F. Plas, E. Viel, Y. Blanc. *La marcha humana. Cinesiología dinámica, biomecánica y patomecánica*. Barcelona, 1984, pp.12–24
5. P. Rasch, R. Burke, *Kinesiología y Anatomía aplicada*, Buenos Aires, 1986, pp.25–27.
6. Natalie Vanicek , Siobhan Strike , Lars McNaughton, Remco Polman. Gait patterns in transtibial amputee fallers vs. non-fallers: Biomechanical differences during level walking , UK, 2009 pp.415–420.
7. Ava D. Segal, Michael S. Orendurff, Joseph M. Czerniecki ,Jason Schoen, Glenn K. Klute. Comparison of transtibial amputee and non-amputee biomechanics during common turning task. USA, 2011,pp. 41–47.
8. Yoshimasa Sagawa Jr et al. Biomechanics and physiological parameters during gait in lower-limb amputees: A systematic review. France, 2011 pp. 511–526
9. D. Winter, *Biomechanics of Human Movement*, USA, 1979, pp.9–22.
10. Y. Xiang , JS Arora ,K. Abdel-Malek, *Physics-based modeling and simulation of human walking: a review of optimization-based and other approaches*, Struct Multidisc Optim 42, 2010,pp 1-23.
11. C. Detrembleur , F. Dierick F, A. van den Hecke A, *Motion of the body centre of gravity as a summary indicator of the mechanics of human pathological gait*, Gait Posture 12(3),2010,,pp. 243-250.
12. Y. Kalyan Chakravarthy, et al, *Haptics based Robotic Prosthetic Leg*, Advances in Robotics , Mechatronics and Circuits, Proceedings of the 18th International Conference on Circuits (part of CSCC '14), Santorini Island, Greece, July 18-20, 2014,pp 296-302.
13. Nurfarahin Onn, et al, *Considerations of Biped Locomotion for Lower Limb Exoskeleton Designs*, Latest Trends in Circuits, Control and Signal Processing, Proceedings of the 13th International Conference on Robotics,Control and Manufacturing Technology (ROCOM '13), Kuala Lumpur, Malaysia ,April 2-4, 2013, pp. 59-64.
14. M. Spiegel, *Estadística*, España, 1991, pp.102-104.
15. R. Serway, J. Jewetter, Física para ciencias e Ingenierías, México, 2005, pp.25-30.
16. BTS Gait-Lab, www.btsbioengineering.com/

Glu4Pred: A computational tool for design and testing of insulin therapies for patients with type 1 diabetes based on interval simulation

M. García-Jaramillo[1], J.S. Delgado[1] and F. León-Vargas[2]

[1] Laboratory for Advanced Computational Science and Engineering Research (LACSER), Universidad Antonio Nariño, Bogotá, Colombia
[2] Research Group in Energy and Materials (REM), Universidad Antonio Nariño, Bogotá, Colombia

Abstract— **Diabetes management is a key factor in preventing clinical complications in type 1 diabetes (T1D) patients, where the use of manual or automatic insulin therapies are designed and implemented to counteract the effects of diabetes. However, a good glucose control may perform poorly if any therapy condition changes suddenly or over time, without the patient's awareness. This phenomenon can seriously compromise the patient's health and is usually attributed to uncertainty in the T1D models used to this end.**

Some simulation tools have been successfully designed to develop strategies for automated control in T1D patients; one of them even has the Food and Drug Administration's (FDA) approval to be used for pre-clinical testing replacing animal experiments. However, none have explicitly included uncertainty as part of their simulation results.

This work proposes a simulation tool named Glu4Pred that includes different sources of uncertainty as well as intra-patient variability to predict a more comprehensive dynamics of the glucose-insulin system.

The Glu4Pred tool allows to define a complex scenario of several days or meals, where the patient's variability and several sources of uncertainty are simulated obtaining as a result an envelope of glucose traces. This is achieved using different mathematical interval models of the glucose-insulin system. Furthermore, the tool includes a risk index outcome that quantifies the risk of experiencing different grades of hypo- or hyperglycemia in the postprandial state from interval simulation. As a result, a new computational tool integrating several sources of uncertainty through interval simulation, that can support the design of decision-aid systems for insulin therapy and glucose control in T1D, was developed.

Keywords— **Computational tool, Glucose control, Interval simulation, Type 1 diabetes, Uncertainty.**

I INTRODUCTION

Diabetes is a metabolic disease characterized by elevated plasma glucose levels, corresponding to acute or chronic hyperglycemia, which can lead to long-term micro- or macrovascular complications. This is so due to the lack of insulin secretion by the beta-cells in the islets of Langerhans in the pancreas (type 1 diabetes) or a combination of resistance to insulin action and an inadequate compensatory insulin secretory response (type 2 diabetes). Diabetes is one of the most serious diseases that must be regulated artificially. According to the latest data from the *International Diabetes Federation* (IDF), it is estimated that diabetes and its complications are major causes of early death in most countries, approximately 5.0 million people died from diabetes in 2015 of the world's adult population aged between 20 and 79 years [1].

Diabetes management is a key factor in preventing clinical complications in type 1 diabetes (T1D) patients, where the use of manual or automatic insulin therapies, usually based on subcutaneous insulin injections or insulin pump, are designed and implemented to counteract the effects of diabetes. However, an insulin therapy well established for glucose control may perform poorly if any therapy condition changes suddenly or over time, without the patient's awareness. This may occur when uncertainty in the carbohydrate (CHO) estimation of food intake to calculate the corresponding insulin dose is not considered [2], or when unaccounted physical or mental states that affect the glucose-insulin system happen (e.g. illness, stress, physical activity, etc.) [3, 4]. In addition to this, there is a large intra-individual patient variability, as diurnal changes in the patient's insulin sensitivity that affects the insulin needs along the day [5]. These sources of uncertainty can seriously compromise the glucose control performance and consequently the patient's health if they are not considered in the design of the control algorithm.

Some simulation tools have been successfully developed to design and testing insulin therapies for T1D patients, most of them related to automated control strategies which have been further extensively evaluated in clinical trials [6]. The Food and Drug Administration (FDA) has approved one simulation tool to be used for pre-clinical testing replacing animal experiments [7] which are very time consuming and expensive. However, according to state of the art of related simulators [8], none have explicitly included uncertainty features as part of their simulation results.

I. Torres et al. (eds.), *VII Latin American Congress on Biomedical Engineering CLAIB 2016, Bucaramanga, Santander, Colombia, October 26th -28th, 2016*,
IFMBE Proceedings 60,
DOI: 10.1007/978-981-10-4086-3_85

This paper presents a computational tool for design and testing of insulin therapies for T1D patients based on different mathematical models of the glucose-insulin system, where the parameters and initial states related to several sources of uncertainty are evaluated using Modal Interval Analysis (MIA) to obtain an optimal rational computation. The tool performs an interval simulation for a given testing scenario, such as one or more meals per day or several days. The resulting glucose response is presented as an envelope enclosing all glucose values obtained at each simulation step considering a given uncertainty in model parameters of the glucose-insulin system. Furthermore, a risk index outcome based on the resultant envelope of glucose that quantifies the risk of suffering different grades of hypo- and hyperglycemia episodes in the postprandial state [9] was also included.

II Materials and Methods

The design concept of the computational tool, referred here as Glu4Pred, is based on three key modules: inputs of CHO, bolus insulin for food intake and the percentage of uncertainty in model parameters; processing of interval models of the glucose-insulin system to simulate the resultant envelope of glucose; and an outcome of estimated risk of hypo- or hyperglycemia episodes in the postprandial state. Figure 1 shows the interaction of the key modules of the Glu4Pred tool. They are detailed in following subsections.

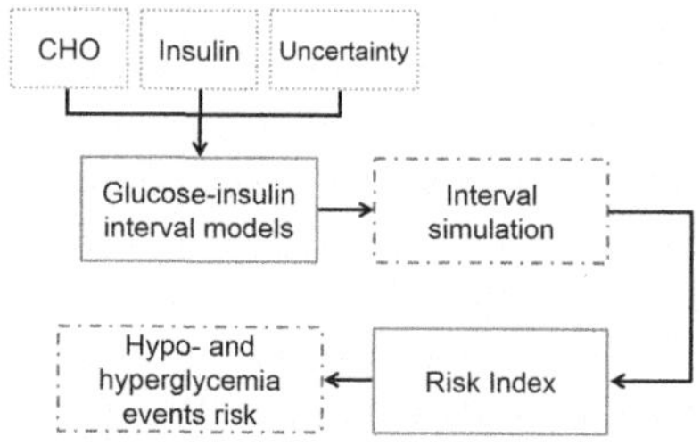

Fig. 1: Interaction of key modules of the Glu4Pred tool. Dotted line indicates the simulation inputs, solid line indicate the processing modules for interval models and the risk index outcome respectively, and dashed line indicate outputs for interval simulation and outcome values respectively.

A Glucose-insulin interval models

To perform a model-based *in silico* simulation and for clinical diabetes decision support, a pharmacodynamic model of the interaction between glucose and insulin is required. The engineering effort in modelling the glucose-insulin system over the past 50 years, starting with the minimal model and including subsequent models, has been revisited by Cobelli et al. [10]. Most of them have been used for glucose prediction in decision-aid systems for insulin therapy optimization and glucose control strategies for T1D patients. However, the intra-individual variability in patient behavior and the different sources of uncertainty are not accounted for in these models.

Recently, several interval models of the glucose-insulin system including different sources of uncertainty and intra-patient variability were presented and compared by García-Jaramillo et al. [11]. They applied MIA to allow the computation of the tight enclosure of the envelope of glucose that includes all possible behaviors of the system, and concluded that under variability, simple glucose-insulin models may be sufficient to describe patient dynamics in most situations. Accordingly, the interval model of the Bergman et al. model [12] was selected to be implemented in the Glu4Pred tool to represent the insulin action and glucose kinetics. Similarly, the interval model of the Hovorka et al. model [13] and Dalla Man et al. model [14] were selected to represent the subcutaneous insulin absorption; and the interval model of the Hovorka et al. model [13] and the Dalla Man et al. model [15] were selected to represent the gastric emptying, digestion, and absorption.

B Interval simulation

Interval simulation is performed using Matlab software to obtain the envelope of glucose (i.e. the range of the trajectories) for the above interval models using a class named *vector interval arithmetic* [16], which incorporates the so-called extended interval arithmetic, or Kaucher arithmetic, and allows using the MIA theory.

To do this, interval values for parameters, inputs, and/or initial states associated to uncertainty are required in order to obtain the envelope of glucose. The corresponding parameters and inputs that can be considered as uncertain in each of the interval models included in the Glu4Pred tool are presented in table 1.

C Risk index outcome

Several metrics to analyze the risk of suffering hypoglycemia or hyperglycemia events from glucose traces have been proposed. Some of the most used are the low blood glucose index (LBGI), the high blood glucose index (HBGI), the average daily risk range (ADRR) [17] or the lability index (LI) [18], among others. However, any attempt to calculate the risk of postprandial hypo- or hyperglycemia should consider the different sources of uncertainty and

Table 1: Models parameters uncertainty

Model	Parameter uncertainty
Carbohydrate digestion and absorption	
Hovorka et al. (2014)	$D, A_G, t_{max,G}$
Dalla Man et al. (2006)	D
Subcutaneous insulin absorption	
Hovorka et al. (2004)	u
Dalla Man et al. (2007)	u
Insulin action and glucose kinetics	
Bergman et al. (1981)	p_1, p_2, p_3, p_4, n

patient variability, and none of the aforementioned methods deals with this problem.

The Glu4Pred tool incorporates an outcome that quantifies the risk of experiencing different grades of hypo- or hyperglycemia in the postprandial state induced by insulin therapy [9]. This outcome is computed from a quantification of the excursions, provided by the glucoregulatory model with uncertainty, based on mild and severe hypo- and hyperglycemia ranges and their relative importance. Figure 2 shows an example of risk assessment from interval simulation where a severe hyperglycemia is identified between 245 and 340 min, while two mild hyperglycemia are presented between 121 and 244 min and between 341 to 420 min. A detailed explanation of the index calculation can be found in [9].

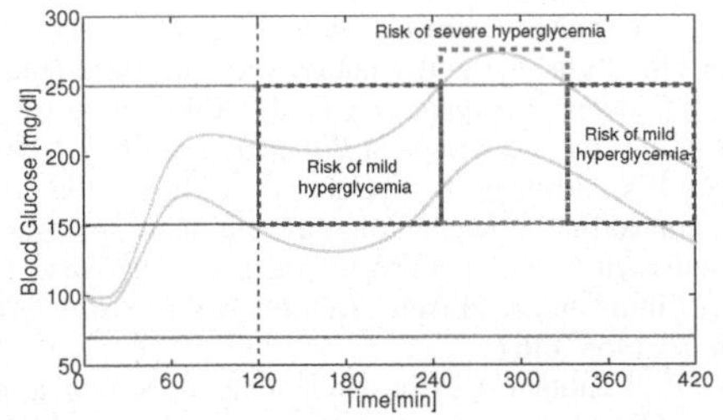

Fig. 2: Example of risk index basis. Severe and mild hyperglycemia events are highlighted by red and purple rectangles, respectively. The vertical dashed line indicates the relevance of hyperglycemia 2 h after a meal.

III Results

The interface of the Glu4Pred tool was divided into five steps to be friendly and easy to use. Figure 3 shows the main interface, where red circles highlight the five steps needed to set a simulation protocol.In the first step the user selects the model for each subsystem using a listbox and, accordingly, inputs related to uncertainty are enabled. Likewise, specific virtual patients are enabled according to the model selected (step 2). The model parameters used in the Glu4Pred tool were taken from literature [11, 7]. The tool allows to setting different scenarios such as one or more meals per day or for several days. Accordingly, the third step enables the user to specify the timing (min) and size (gr) for each meal and the corresponding timing (min) and bolus insulin units (IU). Finally, the Glu4Pred tool provides options for selecting the risk index outcome and displaying the results in graphical format, which can be selected by clicking on the check box button (step 4).

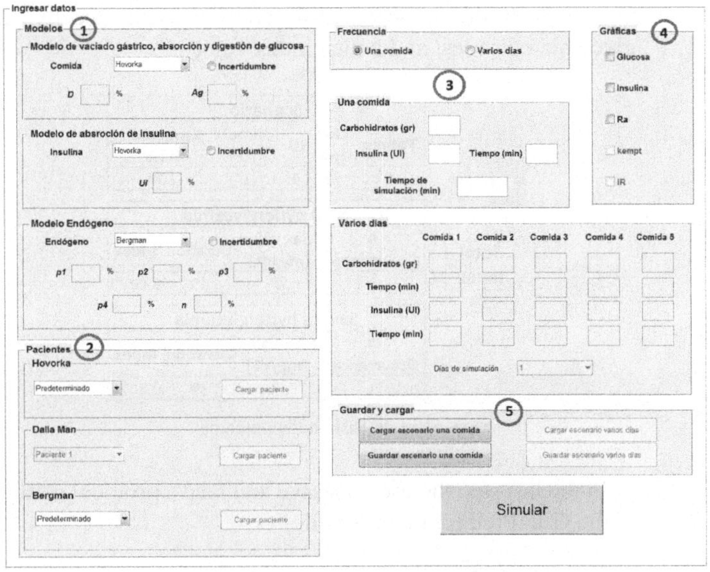

Fig. 3: Glu4Pred: Home screen

The resulting available graphs of the Glu4Pred simulation are related to blood glucose, carbohydrate digestion and absorption, and subcutaneous insulin absorption. Once all of the above have been set up, the simulation starts by clicking on "Simular".

In case the uncertainty and variability are considered in the simulation protocol, an envelope of glucose is obtained representing the set of all possible glucose responses. An interval simulation example of an envelope of glucose excursion for a 6 h period after a meal using 10% variation in meal intake and 2% variation in insulin sensitivity is presented in Figure 4.

In addition to above options, a risk index of suffering hypo- and hyperglycemia episodes can be performed when the simulation results correspond to an envelope of glucose. This outcome results in an Excel file with data related to hypo- and hyperglycemia episodes classified as severe or

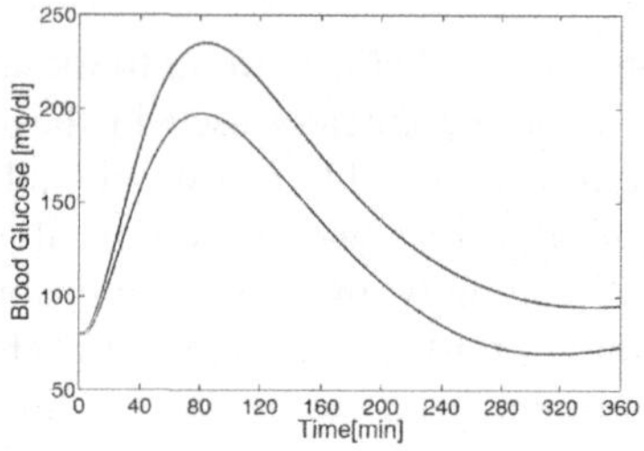

Fig. 4: Example of interval simulation of glucose in the presence of intra-individual variability and uncertainty in food intake.

mild (see Figure 5). The file shows a summary of the testing scenario, the number and duration of hypo- and hyperglycemia episodes, and a total index value.

Total index
3,945

Scenario				
Patient	CHO (gr)	IU	Time IU	Simulation time
1	70	4	1	300

Mild hyperglycemia			
# Occurrences	Max value (mg/dL)	Duration	Index
2	0	19	3,0779

Severe hyperglycemia			
# Occurrences	Max value (mg/dL)	Duration	Index
1	275,90674	39	0,8678

Fig. 5: Risk index outcome

In the last step the user can save and load a protocol setting of the Glu4Pred tool in a .xls file.

IV Conclusion

A computational tool which implements different mathematical models of glucose-insulin system for T1D that includes different sources of uncertainty and intra-patient variability was developed. This tool can be used as a key decision-aid system for insulin therapy optimization in T1D and to support the automated glucose control design.

However, future improvements in this tool are required in order to cover additional realistic scenarios for automated glucose control design. Some of these improvements are focused on modifying the basal insulin dose and the inclusion of explicit profiles of circadian variability of insulin sensitivity in addition to corresponding uncertainty.

Conflict of Interest

The authors declare that they have no conflict of interest.

Acknowledgments

This work was supported by the Universidad Antonio Nariño through project No. 20141062.

References

1. Federation International Diabetes. *IDF Diabetes Atlas 7th edition* . 2015.
2. Graff M. R., Gross T. M., Juth S. E., Charlson J.. How Well Are Individuals on Intensive Insulin Therapy Counting Carbohydrates? *Diabetes Research and Clinical Practice.* 2000;50:S238.
3. Barnett A. H.. A review of basal insulins *Diabetic Medicine.* 2003;20:873-885.
4. Guerci B., Sauvanet J. P.. Subcutaneous insulin: pharmacokinetic variability and glycemic variability *Diabetes and Metabolism.* 2005;31(4):S7-S24.
5. Scheiner Gary, Boyer Bret A.. Characteristics of basal insulin requirements by age and gender in Type-1 diabetes patients using insulin pump therapy *Diabetes Research and Clinical Practice.* 2005;69:14-21.
6. Forlenza G, Buckingham B., Maahs D.M.. Progress in Diabetes Technology: Developments in Insulin Pumps, Continuous Glucose Monitors, and Progress towards the Artificial Pancreas *The Journal of Pediatrics.* 2015;169:13-20.
7. Dalla Man C., Micheletto F., Lv D., Breton M., Kovatchev B., Cobelli C.. The UVA/PADOVA Type 1 Diabetes Simulator New Features *Journal of Diabetes Science and Technology.* 2014;8:26-34.
8. Colmegna Patricio H.. *Simulation & Control in Type 1 Diabetes.* PhD thesisInstituto Tecnologico de Buenos Aires 2015.
9. García-Jaramillo M., Calm R., Bondia J., Tarin C., Vehi J.. Computing the Risk of Postprandial Hypo- and Hyperglycemia in Type 1 Diabetes Mellitus Considering Intrapatient Variability and Other Sources of Uncertainty *Journal of Diabetes Science and Technology.* 2009;3(4):895-902.
10. Cobelli C., Dalla Man C., Sparacino G., Magni L., De Nicolao G., Kovatchev B.. Diabetes: models, signals and control *IEEE Reviews in Biomedical Engineering.* 2009;2:54-96.
11. García-Jaramillo M., Calm R., Bondia J., Vehi J.. Prediction of postprandial blood glucose under uncertainty and intra-patient variability in type 1 diabetes: a comparative study of three interval models *Computer Methods and Programs in Biomedicine.* 2012;108(1):224-233.
12. Bergman RN., Phillips LS., Cobelli C.. Physiologic evaluation of factors controlling glucose tolerance in man: measurement of insulin sensitivity and beta-cell glucose sensitivity from the response to intravenous glucose *The Journal of Clinical Investigation.* 1981;68(6):1456-1467.
13. Hovorka R., Canonico V., Chassin L. J., et al. Nonlinear model predictive control of glucose concentration in subjects with type 1 diabetes *Physiological Measurement.* 2004;25:905-920.
14. Man C. Dalla, Raimondo D., Rizza R., Cobelli C.. GIM, Simulation Software of Meal Glucose-Insulin Model *Journal of Diabetes Science and Technology.* 2007;1(3):323-330.
15. Man C. Dalla, Camilleri M., Cobelli C.. A System Model of Oral Glucose Absorption: Validation on Gold Standard Data *IEEE Transactions on Biomedical Engineering.* 2006;53:2472-2478.
16. Herrero P, Georgiou P, Toumazou C, Delaunay B, Jaulin L. An Efficient Implementation of the SIVIA Algorithm in a High-Level Numerical Programming Language. *Reliable Computing.* 2012;16:239–251.
17. Kovatchev B., Cobelli C.. Glucose Variability: Timing, Risk Analysis, and Relationship to Hypoglycemia in Diabetes *Diabetes Care.* 2016;39:502510.
18. Ryan E. A., Shandro T., Green K., et al. Assessment of the Severity of Hypoglycemia and Glycemic Lability in Type 1 Diabetic Subjects Undergoing Islet Transplantation *Diabetes.* 2004;53:955-962.

Tribological properties study of lubricants for possible nanoclays reinforced biomedical applications

J. Pardo Bernal[1,*], L. Peña Parás[2], R. Tamayo Ramirez[3]

1. El Bosque University. Av. Cra 9 No. 131 A – 02. Bogotá, D.C, Colombia.
2. Department of Engineering, University of Monterrey. Av. Morones Prieto 4500 Pte. San Pedro Garza García, N.L. 66238, México.
3. Department of Basic Sciences, University of Monterrey. Av. Morones Prieto 4500 Pte. San Pedro Garza García, N.L. 66238, México.

Abstract— **This research focuses on the use of nanoparticles halloysite nanotube (HNTs) and montmorillonite (MONT) to improve lubricating properties of polyethylene glycol (PEG) for total knee arthroplasty (TKA). Tribological tests nanofluids under Tribotester T-02U were performed using the ASTM D4172 standard changing the test temperature at 36 ° C-37 ° C, the analysis was made from the coefficient of friction (COF) and diameter worn (WSD) provided by the inspection worn mark on the optical microscope Alicona. There was a clear reduction of COF by 27% at 0.05% HNTs and WSD by 10% at 0.1% HNTs. However in MONT implementation COF is reduced by 6% at 0.05% concentration and WSD 2% at 0.01% concentration. In conclusion we obtained that nanoparticles HNTs improve the properties of PEG in greater proportion than nanoparticles MONT, although emphasis on further studies in tribological applications in bioengineering industry with both materials, due to their biological and physicochemical properties, is made. Cytotoxicity tests that ensured the biocompatibility of the nanoparticles with C6 rat glioma cells were performed.**

Keywords— **Tribology, C6 cells, Halloysite, Montmorillonite, Total Knee Replacement, Polyethylene Glycol**

I. Introduction

The knee is the largest, most important joint for walking and running in the human body because they provide stability to the locomotor system [1]. There are knee conditions that can be expressed in any population, without any precedent in age [2]. Processes that can trigger some kind of damage to the knee joint movements or strokes are carried out without caution; largely these injuries can be treated from processes physiotherapy and drugs prescribed by the doctor [2]. There is a condition known as osteoarthritis (OA) which is defined by the National Center of Technological Excellence in Health in Mexico as a degenerative joint condition characterized by progressive loss of articular cartilage, marginal bone hypertrophy (osteophytes) and changes in the synovium [3]. OA is a disease that generates major impact on global health because of its prevalence and costs associated with rehabilitation [4].

According to the study organized by the National Center of Technological Excellence in Health in Mexico, patients with severe OA had no positive response to drugs and exercise, resulting in body mass loss. A clinical diagnosis from biomechanical tests recommends the need for surgical replacement process, referred to as total knee arthroplasty (TKA) [5]. The ATR is the artificial replacement of the knee joint in its three behaviors by introducing metallic components and a surface of ultra-high density polyethylene [5].

The materials do not differ much from the implemented between the last 15 or 20 years, which is the use of CoCr base alloy (ASTM F75) and Ti6Al4V alloy (ASTM F136) for the manufacture of both the femoral component on the tibial. The insert generally present in these prostheses is composed of ultra-high-molecular-weight polyethylene (UHMWPE). Moreover, the CoCr has improved and polished smoothness that allows improved and longer slipping lasting on the UHMWPE [6].

According to the study by Madl et al. [9], there are certain toxicological concerns caused by orthopedic implants. Infers that good surgical implant procedure provides system stability leaving residues volumetric wear under wear composed of Cr nanoparticles (<100nm) with no content of Co. For implants with poor surgical implant position, distribution wear parti-cles are very large (> 1000 nm) and contains a higher con-centration of Co [8], which generate migration of wear debris which induce an inflammatory response, causing toxicity, osteolysis and subsequent loosening of the implant. [9]

Trying to provide a solution to the problem this study is focused on the prevalence of new systems of friction and wear in the ATR and new technologies applied to nano-technological processes. In a study by Alnaimat et al. [10], polyvinyl alcohol (PVA) in different concentrations ex-posed to tribological tests with different combinations of CoCr, UHMWPE and PEEK generates COF reduction [9], but in a study by Kobayashia et al. [11], using the PEG lubricant has high viscosity characteristics of high density; We can prolong the longevity of the ATR, wear processes, making the lubricant tribological tests to check its efficiency in material exposure Co-Cr [10]. In this project the PEG lubricant is reinforced with HNTs nanoparticles because of their potential to

I. Torres et al. (eds.), *VII Latin American Congress on Biomedical Engineering CLAIB 2016, Bucaramanga, Santander, Colombia, October 26th -28th, 2016*,
IFMBE Proceedings 60,
DOI: 10.1007/978-981-10-4086-3_86

work in biomedical applications as anti-corrosive, heat resistant agents, nanocomposites and bio-materials [11]. On the other hand, in pharmaceutical preparations, MONT and catalysts [12] are implemented.

II. Materials and Méthods

Selected nanoparticles HNTs and MONT were supplied by Sigma-Aldrich and described in Table 1. The nanoparticles were incorporated into the PEG 400 (polyethylene glycol) at various concentrations of filling: 0.01%, 0.05% and 0.10% wt %. The nanoparticles were treated with ultrasound for 5 min using a Cole Parmer 500W ultrasonic probe with a fixed frequency of 20 kHz. C6 glioblastoma cells of rat were cultured in Dulbecco's Modified Eagle's medium (DMEM) medium, fetal bovine serum (FBS), to perform a cytotoxicity assay with the kit Celltiter Promega (Promega) with the concentrations mentioned to check its compatibility with the environment.

Table 1. Properties of Material

Materials	Properties
Base fluids	
Polietilenglicol 400 (PEG)	Viscous liquid
Nanoparticles	
MONT	< 25 μm, sheets
HNTs	30-70 nm × 1-3 μm nanotubes
Pellets test	**Chemical composition**
AISI 52100	0.98-1.1% C, 0.15-0.30% Si, 0.25-0.45% Mn, 1.30-1.60% Cr, <0.025% S Diameter : 12.7 mm, 60 HRC

I. Experimental

A. *Tribological experiment*

The tribological properties were measured by a Four-Ball T-02U Tribotester. As shown in Figure 1. The Tribotester has three stationary lower balls that are fixed in the pot ball and presses against the upper ball or cone with a constant load P, the ball at the top is fixed to the mandrel ball and rotates at a constant speed n. The material is an AISI 52100 Ball [14] steel. The tribological system is immersed in the test lubricant, which generates friction and wear at the points of contact with the pellet that is moving. [15]. The method used to obtain the wear scar diameter (WSD) and COF was a modified ASTM D4172 that follows these conditions: load of 392 N, the rotation speed of 1200 rpm, and time of 60 min at 36-37 °C (modified from 75°C) [16]. The WSD were calculated from the average diameter of the three lower balls, using an optical microscope Alicona.

Fig 1. Design Tribotester T-02U in Fusion 360.

B. *Preparation of samples and evaluation of crop*

Thawing and culture of C6 rat glioblastoma cells in monolayer

C6 glioblastoma cells were obtained from rat American Type Culture Collection (cat. No. CCL-107, Manassas, VA, USA). C6 cells rat glioma were thawed and cultured in cryovials T boxes 25, by seeding button. After 1 h of adhesion in the incubator, 5 mL of medium are added and kept in incubation at 37 ° C, humidified atmosphere and 5% CO2, until 70% cell confluency.

Culture observation was performed in the microscope Axiovert 40C to verify the proliferation of C6 rat glioblastoma cells. Then, we worked on the removal of conditioned medium and proceeded to implement 5 ml trypsin for enzymatic digestion of the crop, and then incubate at 37 ° C for 2 min. The sample was centrifuged at 1290 rpm for 10 min, the supernatant where can appreciate the pellet C6 glioma cells rat was removed. 5 ml of Advanced D-MEM (Dulbecco's Modified Eagle Medium) (1X) were added, and the pellet disaggregated, and then centrifuged at 1290 rpm for 10 min. Finally, the pellet was disaggregated with 5 ml of medium in a T26 box, adding 5 ml of the medium to incubate at 37 ° C for 24 hours.

C. Cell viability

The feasibility assessment was made every 24 h for 2 days. Viability was assessed by the method of trypan blue exclusion, performing enzymatic digestion of the culture with 0.025% trypsin and cell counting in a Neubauer chamber.

D. Test cytotoxicity kit Promega Celltiter

The amount of ATP is directly proportional to the number of cells present in culture. The CellTiter-Glo assay Luminescent is a homogeneous method of determining the number of viable cells in culture based on quantification of ATP present, which signals the presence of metabolically active cells. It is implemented according to the protocol provided by the commercial, which works as follows:

CellTiter-Glo buffer was added to the CellTiter-Glo substrate generating a homogeneous solution. The reagent was carried to the culture dish of 96 wells which work in a low light environment. Agitation is performed for 5 minutes and the luminometer read afterwards. [7]

II. Results/Discussion

Figure 1 and Figure 2 represent the WSD of steel shot and Figure 3 depicts the results COF base lubricant nanoparticles, according to the ASTM D4172 method, modifying the standard 75 ° C to 36 ° C-37 ° C simulating body temperature of the human body. The WSD is obtained from an average between the diameters of the footprint of the bottom three pellets.

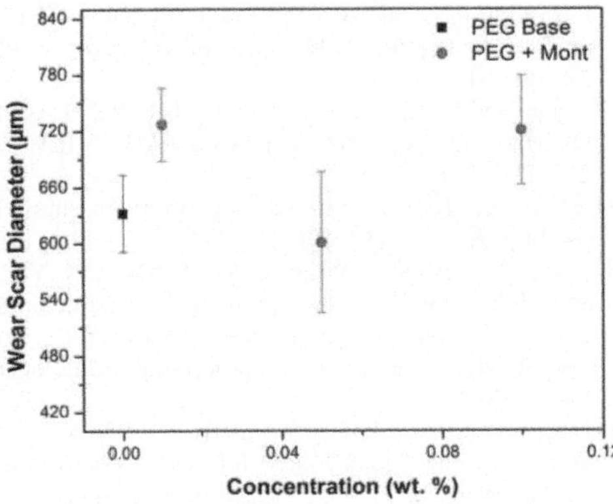

Fig 2. Results of the lubricating base WSD PEG and PEG + MONT in concentrations of 0.01%, 0.05% and 0.1%

Figure 2 and Figure 3 show the diameter of the wear scar PEG base lubricant is 0.6322 mm, which provides the comparative factor with the addition of nanoparticles. Larger signs of wear belongs to PEG + MONT at concentrations of 0.01 and 0.1%. This representation ensures that the nanoparticles are not sufficient in very small amounts do not increase in lubricant properties and higher amounts are agglomerated abrasion generating system. Moreover, at 0.05% concentration tends to optimize the lubricant. The PEG + HNTs at higher concentrations increase the base lubricant properties as shown in Figure 2, thus ensuring a higher yield in the ATR.

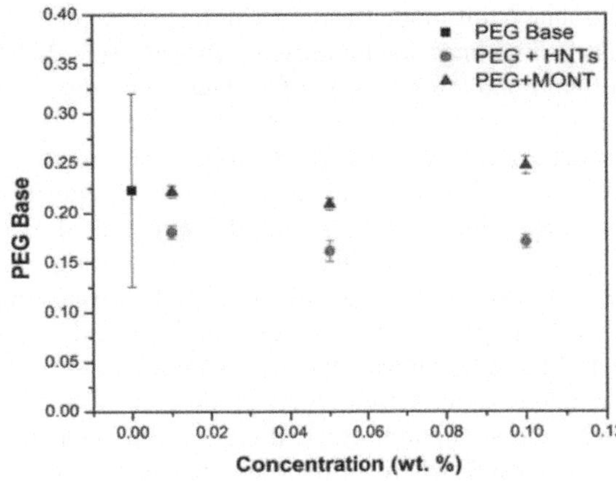

Fig 3. Results of the lubricating base WSD PEG and PEG + HNTs at concentrations of 0.01%, 0.05% and 0.1%.

Figure 4 shows the behavior of the coefficient of friction of the various tests of the lubricant with nanoparticles, where it can be appreciated that the participation of HNTs nanoparticles did provide an automation nano the par participation of nanoparticles HNTs if ensured automation of lubricating base PEG, as the results show in Figure 2. However, although HNTs nanoparticles bore the coupling with lubricant system under the ASTM D4172 standard, MONT nanoparticles could be also be used in precise concentrations for the improvement in mechanical properties in surgical replacements properties.

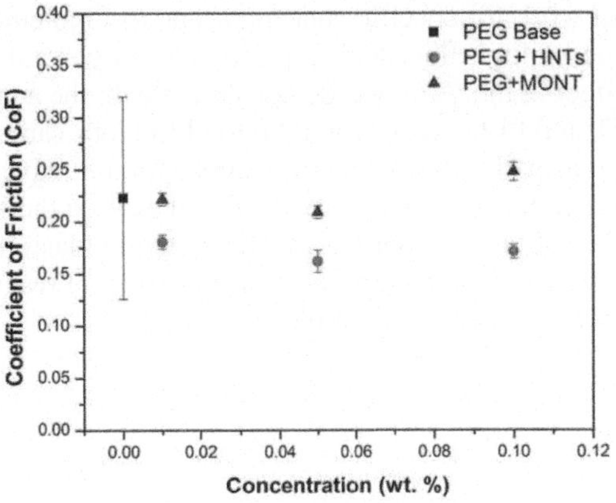

Fig 3. Results of the lubricating base COF PEG, and PEG +MONT PEG+ HNTs at concentrations of 0.01%, 0.05% and 0.1%. 0.05% HNTS (e) Adding 0.01% mont (f) Adding 0.1% mont (g) Adding 0.05% mont, based on ASTM D4172 standard.

The population doubling (PDT) C6 rat glioma cells was determined by the CellTiter-Glo assay. 3x103 C6 (3 replicates control) were seeded in a 96 wells box. The measurement of fluorescence intensity was performed in the GloMax®-Multi + Microplate Multimode Reader With Instinct® reader. The population doubling time for the C6 cells was every 24 hours, with a final population of 6x103 C6. The same procedure for treatments with PEG 400 + 0.05% HNTs and PEG 400 + 0.05% montmorillonite in different volumetric amounts of 5 ul, 10 ul, 20μL (3 replications) in 96 wells box is performed. From testing, a guaranteed cell growth was obtained with the exposed media being biocompatible for the grafting process in TKA, however behavior population doubling was compromised by the experimental technique implemented as shown in Figure 2, due to the test objective being focused to represent their cytotoxicity from ATP production in a 24 hour time, with the simulation of a critical breaking sequence of PEG into the ATR, where the nanolubricant is exposed on the tissue.

III. Conclusions

Throughout this research, nanofluids composed of PEG +HNTs nanoparticles and PEG + MONT nanoparticles. Their tribological properties of WSD and COF were tested in accordance with ASTM D4172 method with variation in temperature to 36 ° C-37 ° C to simulate body temperature of the human body, using the T-02U four ball tribotester. An improvement in COF for PEG + HNTs was found in higher concentrations, with increased lubricant performance by 27% at 0.05% concentration. In the case of PEG + MONT, it does not provide an automation system in all its concentration because they can cause abrasion on the ATR by agglomerates of nanoparticles between surfaces. However, it generates a field of work for specific concentrations for implementation in some biological systems replacement. In general, a positive effect in the COF was demonstrated with the addition of MONT and PEG nanoparticles to the PEG lubricant.

Similarly, these nanoparticles can be used in bio compatible lubricants, as cytotoxicity results guarantee the biocompatibility of the nanoparticles with C6 rat glioblastoma cells. HNTs and MONT nanoparticles are innovative resources to optimize processes lubrication.

IV. Acknowledgements

The authors acknowledge the support of Ing. Patricio García Pineda, Luis Alberto Carreon and Carlos Rafael Martínez Chacín.

V. References

[1] D. Ledia, H. G. García, D. Cruz, M. R. García, D. Isabel, and G. Fuentes, "Laboratorios de Anticuerpos y Biomodelos Experimentales ARTICULACIÓN DE LA RODILLA Y SU MECÁNICA ARTICULAR," vol. 7, no. 2, pp. 100–109, 2003.

[2] I. Nacionales, "¿ Cuáles son los problemas de las rodillas ?," *NIH*, pp. 1–7, 2014.

[3] Mexico: Secretaria de Salud, *Guía de Práctica Clínica Diagnóstico y Tratamiento de la Osteoartrosis de rodilla*. 2009.

[4] F. Poblete, "Osteoartritis (artrosis) de rodilla," *Rev. Chil. Ortop. y Traumatol.*, vol. 56, no. 3, pp. 45–51, 2015.

[5] Mexico. Secretaria de salud. 2011, "Tratamiento con Artroplastia de Rodilla en pacientes mayores de 6O años," pp. 1–55, 2011.

[6] R. S. Cerveró, "Biomecánica de las prótesis de rodilla," *Patol. del Apar. ...*, vol. 46400, no. 4, pp. 242–259, 2005.

[7] M. Ortega Andreu, R. Barco Laakso, and E. C. Rodríguez Merchán, "Artroplastia total de rodilla," *Rev Ortop Traumatol*, vol. 5, pp. 476–484, 2002.

[8] A. K. Madl, M. Liong, M. Kovochich, B. L. Finley, D. J. Paustenbach, and G. Oberdörster, "Toxicology of wear particles of cobalt-chromium alloy metal-on-metal hip implants Part I: physicochemical properties in patient and simulator studies.," *Nanomedicine*, vol. 11, no. 5, pp. 1201–15, 2015.

[9] F. A. Alnaimat, D. E. T. Shepherd, and K. D. Dearn, "The effect of synthetic polymer lubricants on the friction between common arthroplasty bearing biomaterials for encapsulated spinal implants," *Tribol. Int.*, vol. 98, pp. 20–25, 2016.

[10] M. Kobayashi, T. Koide, and S. H. Hyon, "Tribological characteristics of polyethylene glycol (PEG) as a lubricant for wear resistance of ultra-high-molecular-weight polyethylene (UHMWPE) in artificial knee join," *J. Mech. Behav. Biomed. Mater.*, vol. 38, pp. 33–38, 2014.

[11] D. Rawtani and Y. K. Agrawal, "MULTIFARIOUS APPLICATIONS OF HALLOYSITE NANOTUBES : A REVIEW," vol. 30, pp. 282–295, 2012.

[12] J. M. Hartwell, "T h e d i v e r s e u s e s of m o n t m o r i l l o n i t e," no. July, pp. 111–118, 1965.

[13] P. Krajnik, A. Rashid, F. Pušavec, M. Remškar, A. Yui, N. Nikkam, and M. S. Toprak, "Transitioning to sustainable production - part III: Developments and possibilities for integration of nanotechnology into material processing technologies," *J. Clean. Prod.*, 2015.

[14] Matweb, "No Title," 2016. [Online]. Available: http://www.matweb.com/search/DataSheet.aspx?MatGUID=d0b0a51bff894778a97f5b72e7317d85&ckck=1. .

[15] R. Michalczewski and W. Piekoszewski, "The New Methods for Scuffing and Pitting Investigation of Coated Materials for Heavy Loaded , Lubricated Elements," 2010.

[16] M. T. Unit, "Tribology and Mechanical Testing Unit Tribology and Mechanical Testing Unit," vol. 91, pp. 1–2.

Robotic kinematics applied to human biomechanics

J.F. Archila-Diaz[1], I.L. Argote-Pedraza[2], R. Bortholin[3], R. Rubin[3], J.N. Archila-Diaz[4] and M.L. Tronco[2]

[1] Industrial University of Santander/Industrial Design, Titular Professor, Bucaramanga, Colombia
[2] University of São Paulo/ Department of Mechanical Engineering, São Carlos, Brazil
[3] Federal Institute of Education, Science and Technology of Sao Paulo/ Professor, Araraquara, Brazil
[4] Maternal And Child Health Center San Luis, Physician, Bucaramanga, Colombia

Abstract— **This work presents two applications of the kinematic robotics in human biomechanics. The first one of these applications is about analysis of upper limb joints in lifting weights; the second one is in human gate. These applications were developed using homogenous matrix and were simulated using the software MatLab®. With the development work was possible to get two mathematical models applied in the area of occupational health and movement analysis. The purpose of these models is to be applied as a tool that will improve the health at work and life quality.**

Keywords— **Kinematics, Biomechanics, Technical helps, Health at work, Rehabilitation.**

I. Introduction

The human body can be considered the most perfect machine; can be adapted to many environments without changes in its configuration, however, like any machine can fail, if this happens, you need to replace parts and take special care.

Many factors affect the human health in this article we work with lifting and correct movement, giving a new tool from kinematics robotics, to health fields. The paper starts with; robotic kinematics techniques, subsequently it presents the human limbs descriptions and mathematical model, finishing with simulation and conclusions about lifting weight and correct movement.

A. Robotics Kinematics Techniques

There are many mathematical techniques to model the 3D kinematics, that is the case of human body, it is a kinematic chain open and closed in different times, but the more accurate mathematical models are the robotic techniques, between more commons are: Vectorial analyses, rotation and translation matrix, quaternions, Euler angles and homogenous matrix, below presents a description of homogeneous matrix to be used in this work.

Homogenous Matrix: Are arrays of 4x4 dimensions; they represent a linear transformation of a vector from a coordinated system to other. This matrix represents the position and orientation in one array; it also represents scales and perspectives. Equation (1).

$$A = \begin{bmatrix} [R(3\times 3)] & [T(3\times 1)] \\ [P(1\times 3)] & [E(1\times 1)] \end{bmatrix} \quad (1)$$

Where R (3x3) it's a rotation matrix of three row by three columns ;T (3x1) it's a vector of three row by one column and represents a translation; P (1x3) it´s a vector of one row by three columns and represents the perspective; E (1x1) it´s a scalar and represent the scale of transformation. In this case P= [0, 0, 0] and E=1.

Principal Homogeneous Matrix: A spatial movement is represented by rotations and translations. It can be represented as a multiplication of homogenous matrix and the principal arrays can be appreciated in Figure 1 and Figure 2.

$$\begin{bmatrix} Cos(\theta) & -Sin(\theta) & 0 & 0 \\ -Sin(\theta) & Cos(\theta) & 0 & 0 \\ 0 & 0 & 1 & 0 \\ 0 & 0 & 0 & 1 \end{bmatrix}$$

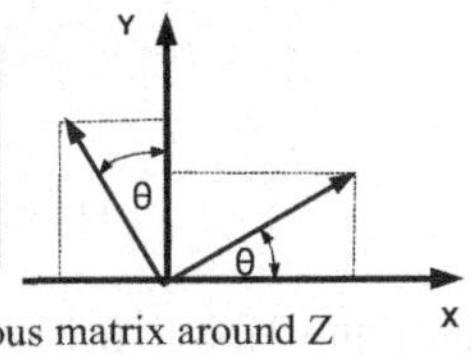

Fig. 1 Rotation Homogeneous matrix around Z

$$\begin{bmatrix} 1 & 0 & 0 & P_x \\ 0 & 1 & 0 & P_y \\ 0 & 0 & 1 & P_z \\ 0 & 0 & 0 & 1 \end{bmatrix}$$

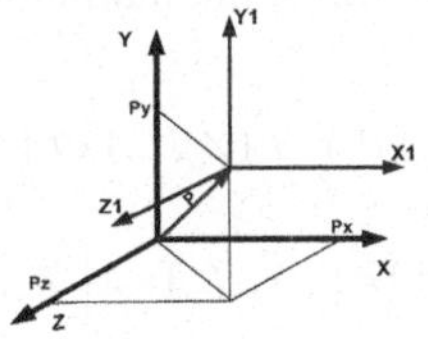

Fig. 2 Translation Homogeneous matrix

Direct Kinematics: The direct kinematics is a mathematical model that describes the robot movements, in this model are known the articular coordinated of degrees of freedom (DOF), and is found the final position of robot. To find the direct kinematical model is used the homogenous matrix. To find the homogeneous matrix, that represents the successive movements between the articular coordinated systems (ϴ1, ϴ2,…, ϴn) is determinates with each homogenous matrix and the final result is the product of this homogeneous matrixes. As shown in Figure 3.

I. Torres et al. (eds.), *VII Latin American Congress on Biomedical Engineering CLAIB 2016, Bucaramanga, Santander, Colombia, October 26th -28th, 2016*,
IFMBE Proceedings 60,
DOI: 10.1007/978-981-10-4086-3_87

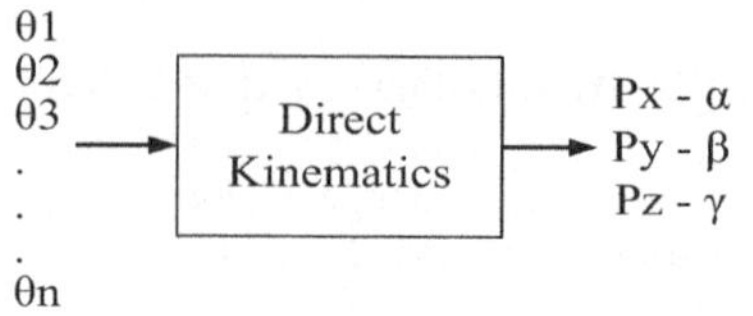

Fig. 3 Direct kinematics

Mathematical Model to Direct Kinematics: The movements between coordinated systems are represented by transformations matrix, following the robot geometry. The kinematical model corresponds to multiply the transformation matrix an example is the KUKA ROBOT KR6, presented in Figure 4. [1]

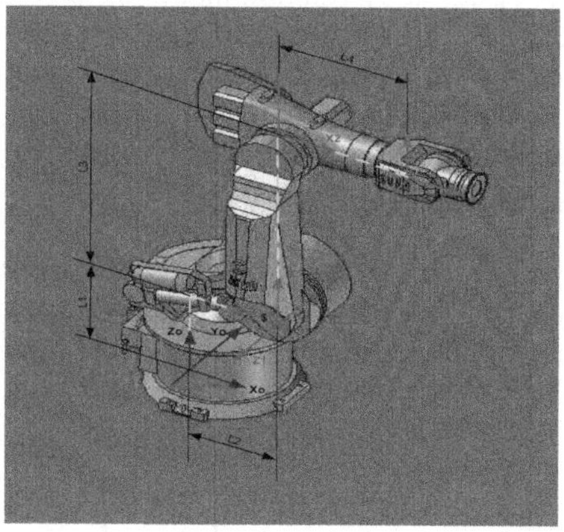

Fig. 4 Kuka Robot [1]

The figure above presents KUKA KR6 CAD model with associate links, the mathematical model is presented at equation (2).

$$T = R(Z_o,\theta_0)\times T(Z_o,L_1)\times T(X'_o,L_2)\times R\left(X'_o,\frac{\pi}{2}\right)\times R\left(Z'_o,\frac{\pi}{2}\right) \quad (2)$$
$$\times R(Z_1,\theta_1)\times T(X_1,L_3)\times R\left(Z'_1,-\frac{\pi}{2}\right)\times R(Z_2,\theta_2)\times T(X_2,L_4)$$

II. Kinematical Model of Human Body

A. Kinematical Model of Human Arm

The human arm can be represented by a kinematics open chain, is conformed by four segments they are: Segment 1: shoulder (Clavicle and scapula), segment 2: arm (humerus), Segment 3: forearm (Ulna and Radio), Segment 4: hand (Carpal, metacarpus, phalanges). The present kinematical model is composed by eight degree of freedom (DOF's). Other models as the one presented by pons et al [2] in 2008 consider nine DOF. The length of links are presented at Table 1.

Table 1 Link length

LINK LENGTH (H=Height)	
LINK	**LENGTH**
Shoulder (L1)	0.094 H
Arm (L2)	0.186 H
forearm(L3)	0.146 H
Hand (L4)	0.108 H

The coordinate system, for each human articulation are represented in the Fig 5, each articulation has the DOF associated [3].

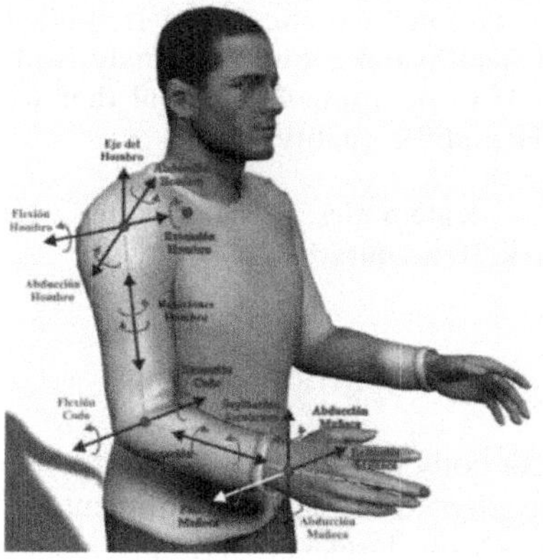

Fig. 5 Upper limb movements [4]

Each articulation connects human body parts and has an associated movement. The shoulder articulation connects the scapula with clavicle and humerus, the associated movements are: Flexion – extension between 130° - 80° to flexion and 30° - 80° to extension, the other movement is Abduction – adduction 180° - 50° and circumduction it's a movement around humerus, between 60° to 90°. The elbow connect the humerus with ulna and radio, the associated movements are flexion – extension and pronation – supination, there are other rotation between ulna -radio and wrist but this movement is not consider in the elbow articulation. The wrist articulation connects the ulna and radio with the wrist. The associated movements are: Flexion – extension reaching 90°, and abduction – adduction 30° - 40° and 15° respectively.

Taking the recommendation of Veslin [5] the kinematic model are described by the Equation (3)

$$T_1^0(A_1) = A_1^0(X_0, A_1)$$
$$T_2^1(\alpha_1, L_1) = R_x(X_1,\alpha_1)\cdot T_x(X_1,L_1)$$
$$T_3^2(\alpha_2,\beta_1,L_2) = R_x(X_2,\alpha_2)\cdot R_y(Y_2,\beta_1)\cdot T_y(Y_2,L_2)$$
$$T_4^3(\alpha_3,L_3) = R_x(X_3,\alpha_3)\cdot T_y(Y_3,L_3)$$

$$T_5^4(\beta_2) = R_y(Y_4, \beta_2)$$
$$T_5^0(q) = T_1^0 \cdot T_2^1 \cdot T_3^2 \cdot T_4^3 \cdot T_5^4 \quad (3)$$

B. Kinematical Model to Human Gait

The biomechanics of human displacement describes the behavior of different elements that conform the legs during the normal progress.

Progress Cycle: The normal locomotion is an alternative series of movements, the rhythms of legs and trunk determine the displacement to front of gravity center [6] the cycle of displacement, start when a foot take contact with floor and finish with the next contact of the same foot.

The cycle is divided by two principal components: The contact phase and the balance phase. One leg is in contact and the other is in balance. This two phases alterning between legs during the progress. In a complete step the contact is refered to a period, where only one leg is in contact with floor. The period of double contact is when two foot are in contact with floor simultaneously. The difference between walk and run is the absent of the double period of contact.

The contact phase is divided by five sub phases, Figure 8: Initial contact phase (CI): is the initial contact of foot with floor. Initial support phase (AI); is between the initial contact and the start to leave the foot of the floor. The mean support phase (AM); Final support phase (AF): in this phase the body is ahead of support foot, fall to front; oscillation previous phase (OP): is the preparation, to oscillation phase

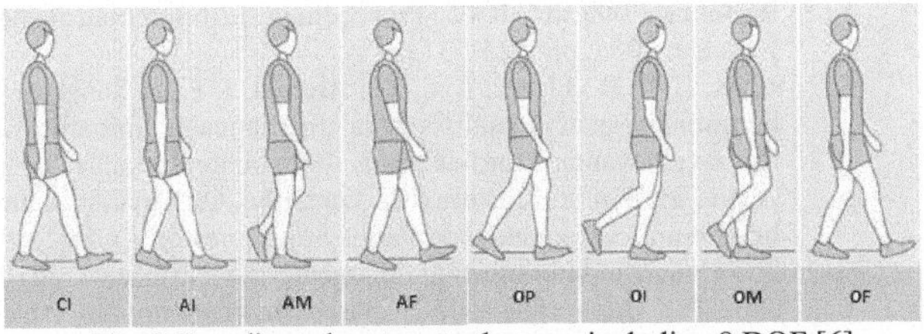

Fig. 8 Coordinated system to the arm, including 8 DOF [6]

The oscillation phase is divided in tree sub phases: Initial phase of oscillation (OI) starts with leave the foot to the floor and achieve the other leg. Medium oscillation phase (OM) starts when both legs are cross, and finish when the tibia achieve the vertical position. Final oscillation Phase (OF) It is limited by next contact with the floor and give the start to a new step.

Mathematical Model: The mathematical model presented in this article represents the legs, each leg was modelled with five DOF and was linked to pelvis link the DOF total are 11, the length of legs are obtained with anthropometric data. The scope of model is describe the phases of human progress. Equation 4 to 16.

$$link0 = T(X_0, Y_0, Z_0) \quad (4)$$
$$link1 = link0[R(Z_1, \emptyset_1) \times T(X_1, d2)] \quad (5)$$
$$link2 = link1[R(X_2, \emptyset_2)] \quad (6)$$
$$link3 = link2[R(Z_3, \emptyset_3) \times T(X_3, d3)] \quad (7)$$
$$link4 = link3[R(Z_4, \emptyset_4) \times R(Y_4, \beta_4) \times T(X_4, d4)] \quad (8)$$
$$link5 = link4[R(Z_5, \pi/2) \times T(X_5, d5)] \quad (9)$$
$$link6 = link5[R(Z_6, \emptyset_6) \times T(X_6, d6)] \quad (10)$$
$$link7 = link6[R(Y_7, \beta_7) \times R(X_7, \alpha_7) \times T(Z_7, 36)] \quad (11)$$
$$link8 = link7[R(Z_8, \emptyset_8) \times T(X_8, d7)] \quad (12)$$
$$link9 = link8[R(Z_8, \emptyset_8) \times T(X_8, d8)] \quad (13)$$
$$link10 = link9[R(Z_9, \emptyset_9) \times R(Y_9, \beta_9) \times T(X_9, d9)] \quad (14)$$
$$link11 = link10[R(Z_{10}, \pi/2) \times T(X_{10}, d10)] \quad (15)$$
$$link12 = link11[R(Z_{11}, \emptyset_{11}) \times T(X_{11}, d11)] \quad (16)$$

III. Results

A. Arm Models

To the arm model was studied the biomechanics to lifting weight to upper the head, this movement has the next phases: phase 0, is the lifting weight to height of waist. Is the start of the analyses of movement, phase 1, lifting the weight to height of the chest. Phase 2, leave the weight from chest to final position. As presented in Figure 9.

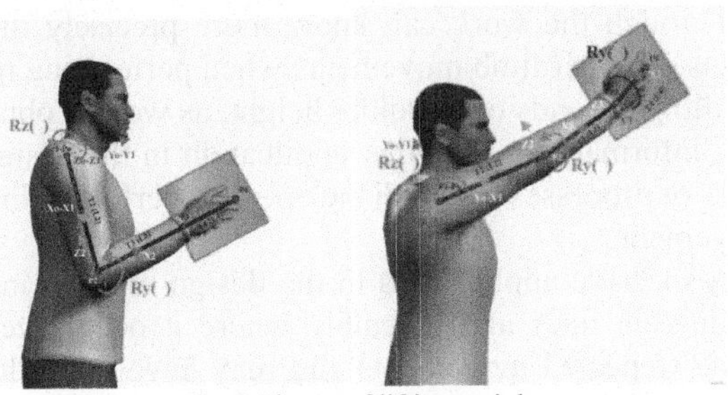
Fig. 9 Phases of lifting weight

During the development of movement was studied the influence of weight in each articulation, especially in shoulder articulation. The weight influence was analysed by Jacobian propagation method over the kinematical model. The results are showed at Figure 10.

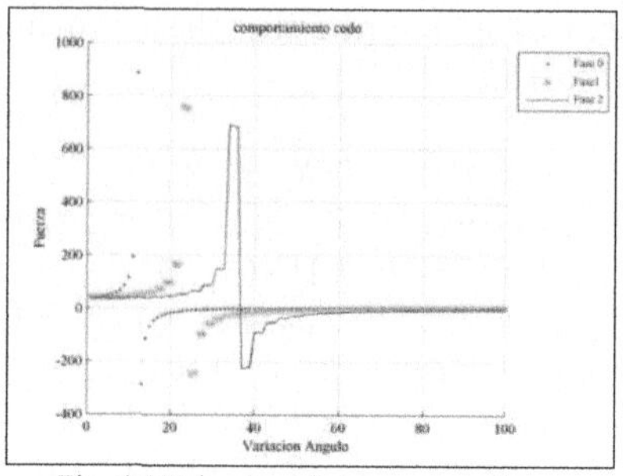

Fig. 10 Behavior of stress over elbow

B. *Kinematical Analyses of Human Progress*

In a robot [7] the kinematics studies the movement robot respect to an inertial coordinated system. In this way the kinematic is interested by the description of movement in function of space and time in an inertial coordinated system. In the human displacement the kinematics model is interested in the movement in respect a coordinated system in the body.

In the Figure 11 are presented the simulation of mathematic model describe above in the equation 4 to 16, the simulation was made in sagittal, front and transversal planes. [8]

Sagittal Plane	Frontal Plane	Transversal Plane
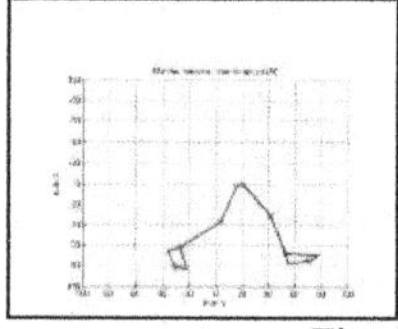	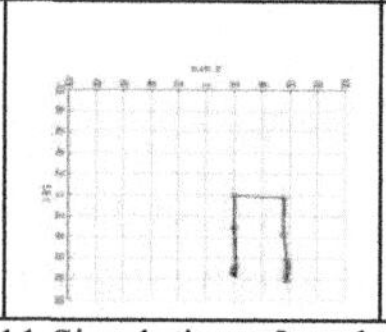	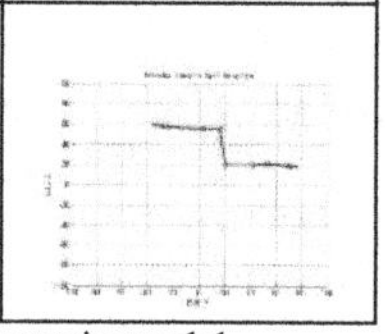

Fig. 11 Simulation of mathematic model

IV. Conclusions

Through the work can know more precisely the movements of upper limb movements when performing repetitive handling of loads on shoulder height, as well as obtain valuable information for further application in the design of devices or processes in which the operator performs this type of movement.

Also, have applications in the design of jobs, industries, production lines and assembly where it occurs very commonly repeated movements and may have musculoskeletal injuries that can be converted illnesses.

This tool is the first step in designing a more efficient workplace that can be predicted and controlled manner possible health consequences of avoiding and reducing operator fatigue and possible muscle disorders.

The separation distance between the cargo handled and the body has a direct influence on the mechanical stress of the upper extremities and more intensely on the shoulder. We found that the shoulder joint is significantly affected by the stresses generated in the movement in phase 2. Although not expected, also highlighted the overload on the wrist joint present at all stages of motion.

The kinematic modeling of a lower extremity, which is implicit in the model of the human walk, provides information about the direct relationship between the rotations of each joint, taking into account their movements; in some phases of human walking have different senses. Moreover, achieving emulates the movements described in the phases of human gait, described by specialists.

Acknowledgment

The authors express its acknowledgment to UIS Industrial Univiersity of Santander, IFSP Federal Institute of Education, Science and Technology of Sao Paulo – Araraquara, EESC-USP Engineering School of São Carlos of São Paulo University and CNPq.

References

1. Lazinica A, Hiroyouki Kawai (2010) Robot manipulators New Achievements: InTech. 601 p.
2. Pons, J. L. (Ed.). (2008). Wearable robots: biomechatronic exoskeletons.
3. Carignan, C., Liszka, M., "Design of an Arm Exoesqueleton whith Scapula Motion for Shoulder Rehabilitation", Advanced Robotics, 2005, ICAR'05, Proceedings, 12th International Conferencie, 2005, pp. 524-531.
4. Rojas, L. E. B., Florez, L. C. G., Archila, J. F., & Sanabria, J. J. herramienta para el analisis de la cinematica del miembro superior en el levantamiento de pesos sobre el hombro. 2010
5. Veslin. E., Slama. J., Dutra. M., Lengerke, O. "Análisis Cinemático de un Exoesqueleto de Partes Superiores de 7 GDL", 2007. 2do Congreso Internacional de Ingenieria mecatrónica, 2010
6. Vera. P., "Biomecánica de la marcha humana normal y patológica", Instituto de Biomecánica de Valencia. Capítulo 3, 1999.
7. Barrientos. A., "Fundamentos de robótica". Ed. McGraw Hill, pp. 15-38, Barcelona, 1997.Smith J, Jones M Jr, Houghton L et al. (1999) Future of health insurance. N Engl J Med 965:325–329
8. Hernández, A. E. G., & Díaz, M. J. F. kinematic model for direct analysis of human progress modelo cinemático directo para el análisis de la marcha humana. Revista Colombiana de Tecnologías de Avanzada V2. N 18, 2011 pg 63 - 69

Author: John Faber Archila-Diaz
Institute: Industrial University of Santander
Street: 27th Av. 9th St.
City: Bucaramanga
Country: Colombia
Email: jfarchid@uis.edu.co

SAFER: A Context-Aware Ubiquitous Assistance Platform for Elderly Care

J.F. Bravo-Torres[1], H.S. Redrován-Parra[1], A.F. Soto-Sarango[1], J.A. Andrade-Padilla[1], E.F. Ordoñez-Morales[1], M. López-Nores[2] and Y. Blanco-Fernández[2]

[1] Grupo de Investigación en Telecomunicaciones y Telemática, Universidad Politécnica Salesiana, Cuenca, Ecuador
[2] AtlantTIC Research Center, University of Vigo, Vigo, Spain

Abstract— **We present the SAFER platform, a ubiquitous and smart assistance platform for elderly care. The idea is to provide, in an integrated way, a set of smart and context-aware services that improve the quality of life and social integration of elderly. This paper shows the design of the platform from the lowest levels for sensing health parameters of elderly and their location, up to the highest level of smart assistance services. Furthermore, we describe a smart health services for detection of falls and assistance to elderly, as an example of application of our platform that we are already experimenting with.**

Keywords— **Smart Health, Ubiquitous Health, Healthcare, Pervasive Elderly Care, Assistance Services.**

I Introduction

The rapid growth of the elderly population in the world is generating more demanding challenges to social and health systems to address the needs and requirements of this population: social participations, health care, leisure, culture, among others [1]. Dealing with these challenges demands cities that provide services able to dynamically adapt to the needs of citizens and making urban environments more friendly spaces with elderly.

In this context, the development of electronic devices, wireless communications and communications networks is changing the way we interact with our environment; and specifically in the cities, it is taking us to the concept of smart cities. The citizens of these cities, travelling by museums, parks and shopping centers, can enjoy interactive services, enriching their experiences and/or facilitating their tasks. Similarly, the growing popularity of smartphones and new patterns of interaction-driven by Web allow to enable a new era of information services adapted to the physical and social context of people [2].

These services, related to the ubiquitous social computing [2], generates many opportunities for the elderly population to improve their quality of life, increasing their opportunities for social interaction and entertainment, and improving health care [3]. So, several approaches have been proposed in the literature about how to generate digital platforms that leverage different technologies to assist to elderlies in their daily activities [4,5], leisure [6], or encourage their social interaction with relatives or acquaintances [1]. However, these platforms address each of these issues separately and without considering the information generated in the city (museums, health centers, theaters, shopping centers, ...) for planning entertainment activities and their social integration.

This paper is about applying technologies to enable an integral care of the physical and social health of elderly. Specifically, we are building a platform called SAFER (*"Smart Assistance platform For EldeRly care"*) that aims at facilitating automatic and proactive deployment of different smart services-oriented to the health care (monitoring vital signs, detection of falls, medication management, ...), orchestration of entertainment activities (recommendation cultural agenda, cultural content (video or text), sports activities, ...), and social integration of elderly (integration in social networks, management of communication with family and friends, forming sporadic groups with people around them with common interests , ...). Each of these services are developed independently, but are triggered automatically in accordance with the context and user needs.

The paper is organized as follows. In Section II, we will provide an overview of the architecture of our platform. Next, Section III and Section IV describe a smart health services for detection of falls and assistance to elderly —as an example of application of our platform that we are already experimenting with—, its implementation and evaluation. Finally, Section V concludes the paper and points out some lines of further work.

II The SAFER Platform

The SAFER platform is being developed as an extension of the technology that is already available to people, aiming to improve the quality of life and autonomy of elderly. It relies on an fully interactive multi-layer architecture, organized in several layers, applications and services. Conceptually, this architecture has four levels (see Fig. 1) that will be described below.

I. Torres et al. (eds.), *VII Latin American Congress on Biomedical Engineering CLAIB 2016, Bucaramanga, Santander, Colombia, October 26th -28th, 2016*,
IFMBE Proceedings 60,
DOI: 10.1007/978-981-10-4086-3_88

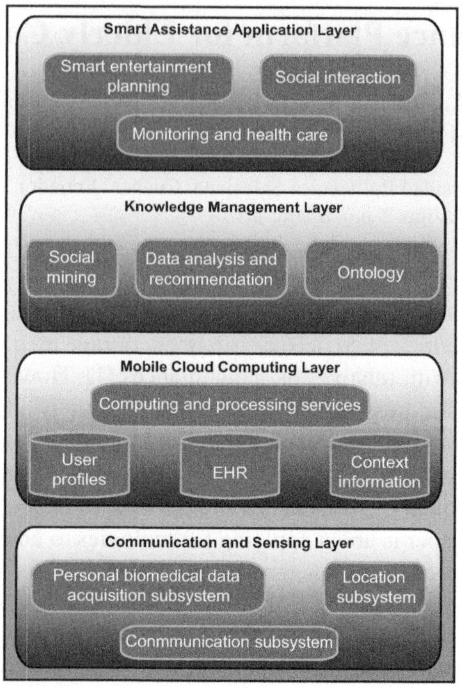

Fig. 1: Four layers of the SAFER platform

A Communication and Sensing Layer

This layer is the responsible for collecting key information about the health state, location, and the current activities that are being developed by elderly. For this, our system relies on three subsystems: (i) a personal biomedical data acquisition subsystem, consisting in a set of biomedical sensorial devices (such as blood-glucose meter, accelerometers for detection of falls, oximeter,...) to capture the biomedical data from old people; (ii) a location subsystem to track the location of elderly and triggering different services accordingly; and, finally, (iii) the communication subsystem, which enables wireless connections (links to hotspots or 3G/4G connections) that are necessary to send and/or receive information or services. All protocols and mechanisms for establishing links and maintaining the necessary QoS levels are housed in this subsystem, too.

B Mobile Cloud Computing Layer

The second layer aims at providing and sharing the resources for the storage, analysis and processing of the information of elderly (health information, profiles, interests, preference, activities, habits,...) that are necessary to support the different kind of smart services provided from our platform. The core ideas here are borrowed from *Mobile Cloud Computing* paradigm, which has become a choice to support the great data demand coming from an explosion of handheld devices thanks to its rapid scalability, ubiquitous network access, and on-demand services [7]. The tandem between mobile devices and cloud computing works perfectly because handheld terminals are constrained by their processing power, battery life and storage capabilities, whereas cloud computing provides the illusion of "infinite" computing resources. In our platform, the extra resources required by the handheld devices are provided by centralized services in the cloud. So, our cloud computing layer provides the following services:

- Storing information of elderlies in spaces in the cloud.
- Synchronising multiple flows of information coming from the connected devices.
- Pooling data from various sensors on multiple devices to achieve greater precision in geolocation.
- Delegating complex tasks on remote machines, to overcome the limitations of the mobile devices in terms of battery, memory and/or computing power.

C Knowledge Management Layer

Our platform uses information of elderly derived from very diverse and varied sources (personal biomedical sensorial devices, location sensors, electronic health records, personal profiles, social networks data,...) to offer different services focused on health care, entertainment, and information for elderly. In order to tackle this overload of information, the knowledge management layer of SAFER brings together solutions from the areas of data mining, machine learning, pattern recognition, recommender systems, Semantic Web to support the monitoring and automatic diagnosis of health emergencies, automatic orchestration of entertainment among elderlies with common interests and located close to each other, selecting pieces of information (music, videos, health news, ...) or cultural activities offered in the city of interest of elderly, among other. Making this vision reality requires to rely on: (i) data analysis strategies applied to health-related data to decision support and prognosis, (ii) techniques for modeling elderly preferences via personal profiles and, (iii) recommendation strategies that select the most appropriate activities and content.

D Smart Assistance Application Layer

The focus of the topmost layer is on the application interface, so this layer refers to a variety of applications that can be triggered manually or automatically depending on the context and needs of users. The aim is to harness the information and tools provided in the lower layers to deploy a range of services to improve the autonomy and quality of life of elderly, both in health care, social integration, and entertainment. To do this, different services and applications are supported for the monitoring of the patient's health, automation of healthcare in emergencies, the recommendation of cultural and sports activities and contents (both video and text), the formation of social and face-to-face interaction groups, among others. All this, according to the interests and needs of elderly.

III A SMART APPLICATION FOR MONITORING, DETECTION, REPORTING AND CARE OF ELDERLY AGAINST FALLS

In this section, we describe a first application of our platform, aimed at monitoring elderly in nursing homes for the fall detection and automated support to assistance from medical staff. Thus, after the detection of the fall, the system sends a notification message to the mobile device of the nurse located at the nearest position to the place of the event. This message contains a summary of the health information of the injured elderly, as well as his/her location and the shortest path to him/her. Figure 2 shows the functional diagram of our application.

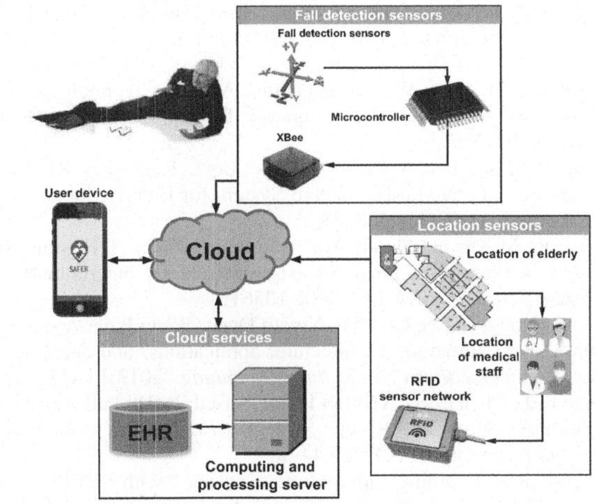

Fig. 2: Functional diagram of the smart application for monitoring, detection, reporting and care of elderly against falls

For the detection of falls, we designed a portable device, which is located in the old man's chest. Its operation is based on the use of a triaxial accelerometer to detect sudden movements; a microcontroller, for the acquisition of data; and XBee wireless module, for sending data to the cloud for storage and analysis.

The triaxial accelerometer provides a vector signal magnitude (VMS), S[n], which is constituted by the acceleration values of the movements of elderly in the axes x, y and z, for the sample n (see equation 1). For this, the device takes periodic samples at a frequency of 20Hz.

$$S[n] = \sqrt{a_x^2[n] + a_y^2[n] + a_z^2[n]} \quad (1)$$

Depending on the activity of elderly (lying down or getting out of bed, go up or down stairs, walking, etc.), values of gravity S[n] will be within $0.5G \leq S[n] \leq 2.2G$. These values are used by the detection algorithm (on the server in the cloud) to determine whether or not there has been a fall [8].

For the monitoring and detecting the location of elderly within the nursing home, we use the technology of radio frecuency identification (RFID) to deploy a network of sensors [9]. Different RFID readers and tags are used to personal identification (both elderly and medical staff) and to detect their current location. For this, RFID readers are deployed in each nursing home environments (dining room, games room, bathrooms, ...), and each of the elderly residents and medical staff are identified through their respective RFID tags. Whenever a user accesses a space managed by a different RFID reader, it sends the location information to the location database in the cloud.

The data from the RFID readers and the accelerometer devices are sent to the cloud through of wireless routers. These information is used in the processes of care of elderly that are triggered during an emergency. Thus, the location data of elderly and medical staff, stored in the location database, allow us to determine which is the support staff that is closer to each elder in every moment. In this way, when an event of fall of an elderly is detected by our system, an alarm message is sent to the nurse who is closer to the location of the injured old man. This message contains a summary of the main elderly health information (illnesses, medications, treating doctor, ...) and a map with the shortest route to the elderly (see Figure 3). In addition, an SMS message is sent to a close relative, reporting the event, the place and time.

An external server, located in the cloud, allow us implementing the different algorithms which are necessaries to determine if a fall of an old man happened, what the support staff is whose location is closest to each elder in every moment, what the shortest path is to the crash site, and what information should be given to each user during emergency management.

IV EVALUATION AND RESULTS

A prototype of our system to monitoring, detection, reporting and care of elderly against falls was implemented and tested in the nursing home "Casa de Residencia Geriátrica

Fig. 3: A snapshot of the user interface, the notification message of the fall of an old map and the shortest route to the crash site

San Andrés" of the city of Cuenca-Ecuador. The user interface was implemented on mobile devices with IOS operating system. A set of RFID sensor was placed in bedrooms, games room and dining room. Each member of staff health and elderly was assigned an RFID tag.

With the aim to set the fall detection parameters used in our algorithm, several tests were developed with elderlies. For example, variations of acceleration were analyzed when performing activities as: stand or sit on the chair, walking up or down stairs, and getting in and out of bed. Furthermore, we developed several simulations scenarios of falls: lateral fall from a chair, fainting from a vertical position, fall back, stumbling and falling, and fall from bed. The developed simulations allowed us to verify the correct operation of processes for detecting falls, the location of the accident and notifications to the support staff and relatives of the injured old man.

V Conclusions

In this paper, we have proposed a new four-layer architecture to elderly care, with the aim of providing a set of services for integral care for physical and social health of elderly. This approach extends the classical view of care of elderlies as a separate aspect of their social activity and entertainment without integration services provided by the city. Nowadays, we are working on the development of each layer of our platform; specifically, we are proposing new strategies to integrate information available in cities (From museums, theaters, shopping centers, ...) to orchestrate different social activities and entertainment for elderlies based on their interests and needs.

Conflict of Interest

The authors declare that they have no conflict of interest.

Acknowledgements

The authors from the University of Vigo were supported by the European Regional Development Fund (ERDF) and the Galician Regional Government under agreement for funding the Atlantic Research Center for Information and Communication Technologies (AtlantTIC), and by the Ministerio de Educación y Ciencia (Gobierno de España) research project TIN2013-42774-R (partly financed with FEDER funds).

References

1. Skouby Knud Erik, Kivimäki Anri, Haukiputo Lotta, Lynggaard Per, Windekilde Iwona Maria. Smart Cities and the Ageing Population 2014.
2. Zhou Jiehan, Sun Junzhao, Athukorala Kumaripaba, Wijekoon Dinesh, Ylianttila Mika. Pervasive Social Computing: augmenting five facets of human intelligence *Journal of Ambient Intelligence and Humanized Computing.* 2012;3:153–166.
3. Mann W. C., Helal S.. Pervasive computing research on aging, disability and independence in *Applications and the Internet Workshops, 2004. SAINT 2004 Workshops. 2004 International Symposium on*:244-246 2004.
4. Kötteritzsch Anna, Weyers Benjamin. Assistive Technologies for Older Adults in Urban Areas: A Literature Review *Cognitive Computation.* 2016;8:299–317.
5. Pires Pedro, Mendes Luís, Mendes Jorge, Rodrigues Rúben, Pereira António. Integrated e-Healthcare System for Elderly Support *Cognitive Computation.* 2016;8:368–384.
6. Hossain M Anwar, Alamri Atif, Almogren Ahmad S, Hossain SK, Parra Jorge. A framework for a context-aware elderly entertainment support system *Sensors.* 2014;14:10538–10561.
7. Dinh Hoang T., Lee Chonho, Niyato Dusit, Wang Ping. A survey of mobile cloud computing: architecture, applications, and approaches *Wireless Communications and Mobile Computing.* 2013;13:1587–1611.
8. Kau L. J., Chen C. S.. A Smart Phone-Based Pocket Fall Accident Detection, Positioning, and Rescue System *IEEE Journal of Biomedical and Health Informatics.* 2015;19:44-56.
9. Want R.. Enabling ubiquitous sensing with RFID *Computer.* 2004;37:84-86.

Author: Jack Fernando Bravo-Torres
Institute: Universidad Politécnica Salesiana
Street: Calle Vieja 12-30 and Elia Liut
City: Cuenca
Country: Ecuador
Email: jbravo@ups.edu.ec

Time-course reconstruction of neural activity for multiples simultaneous sources

P.A. Munõz-Gutiérrez[1] and E. Giraldo[2]

[1] Universidad del Quindío/Electronic Instrumentation Technology, Assistant professor, Armenia, Colombia and Universidad Tecnológica de Pereira/Doctorade in Engineering, Student, Pereira, Colombia
[2] Universidad Tecnológica de Pereira/Department of Electrical Engineering, Associated professor, Pereira, Colombia

Abstract— **In this paper a new methodology for time-course reconstruction of neural activity is proposed. Proposed approach allows finding the locations and waveforms (time-course activity reconstruction) of inner brain sources, that could be used for source-based noninvasive brain computer interface technologies. Proposed framework for source-reconstruction considers a dynamic state space model which allows successfully the reconstruction of time-course activity at each source. A comparative analysis is performed against dynamic Thikonov methods, where the proposed approach obtain the minimum reconstruction error.**

Keywords— **time-course reconstruction, neural activity, inverse problem.**

I Introduction

The use of estimated brain activity or source signals instead of using their surface projections, namely, electroencephalographic (EEG) signals, for Brain Computer Interfaces (BCI) is a new area of development [1]. In this context, finding the locations and waveforms (time-course activity reconstruction) of inner brain sources represents a crucial task for advancing source-based noninvasive BCI technologies. Several methods have been proposed to estimate the EEG brain source spatial locations and their corresponding waveforms [2, 3], however, this methods can successfully reconstruct only a limited number of sources and their corresponding waveforms.

In BCI applications a detailed knowledge of the activity into the brain is highly required. Instead of estimated a limited number of sources, which is useful for analysis epileptic seizures, BCI applications require the estimation of multiple simultaneous activity in several sources into the brain, and it also requires its temporal behavior. Temporal behavior of brain activity considering key features of EEG dynamics as described in [4, 5]. Neural activity estimation under the aforementioned conditions also provide tools for analyzing a potential dipole correlation and connectivity, which is a central issue in neuroscience [1].

In this article a methodology for multiple simultaneous time-course source reconstruction is proposed. The results, based on simulated EEG data show that the proposed framework reconstruct correctly the brain activity at each source and its corresponding waveform.

This paper is organized as follows: In section II a forward modeling for EEG is described. In Section III the proposed time-course reconstruction for multiple simultaneous sources is outlined including a framework for regularization parameter tuning. In Section IV the proposed approach is applied to simulated EEG data and compared with alternative solutions, based on the Tiknonov dynamic methods where the proposed approach successfully reconstruct time-course activity over multiple sources.

II Forward EEG modeling

The following forward modeling is proposed to describe the EEG dynamics as follows:

$$\boldsymbol{y}_k = \boldsymbol{M}\boldsymbol{x}_k + \boldsymbol{\varepsilon}_k \tag{1}$$

$$\boldsymbol{x}_k = \boldsymbol{\Phi}_s \boldsymbol{c}_k \tag{2}$$

$$\boldsymbol{c}_k = \boldsymbol{f}(\boldsymbol{c}_{k-1}, \boldsymbol{c}_{k-2}, \boldsymbol{c}_{k-\tau}) + \boldsymbol{\eta}_k, \tag{3}$$

where $\boldsymbol{x}_k \in \mathbb{R}^{n\times 1}$ is the neural activity at sample k, $\boldsymbol{M} \in \mathbb{R}^{d\times n}$ is the leadfield matrix, $\boldsymbol{y}_k \in \mathbb{R}^{d\times 1}$, $\boldsymbol{c}_k \in \mathbb{R}^{s\times 1}$ weights the matrix of spatial coefficients, noted as $\boldsymbol{\Phi}_s \in \mathbb{R}^{n\times s}$, $\boldsymbol{\varepsilon}_k \in \mathbb{R}^{d\times 1}$ and $\boldsymbol{\eta}_k \in \mathbb{R}^{s\times 1}$ the zero mean uncorrelated additive Gaussian noises, and being the neural activity model defined, according to [5], as

$$\begin{aligned}\boldsymbol{f}(\boldsymbol{c}_{k-1}, \boldsymbol{c}_{k-2}, \boldsymbol{c}_{k-\tau}) = {} & \boldsymbol{A}_1\boldsymbol{c}_{k-1} + \boldsymbol{A}_2\boldsymbol{c}_{k-2} \\ & + \boldsymbol{A}_3\boldsymbol{c}_{k-\tau} + \boldsymbol{A}_4\boldsymbol{c}_{k-1}^{\circ 2} + \boldsymbol{A}_5\boldsymbol{c}_{k-1}^{\circ 3},\end{aligned} \tag{4}$$

being $\boldsymbol{A}_1 = a_1\boldsymbol{I}_s$, $\boldsymbol{A}_2 = a_2\boldsymbol{I}_s$, $\boldsymbol{A}_3 = a_3\boldsymbol{I}_s$, $\boldsymbol{A}_4 = a_4\boldsymbol{I}_s$ and $\boldsymbol{A}_5 = a_5\boldsymbol{I}_s$, where $\boldsymbol{I}_s \in \mathbb{R}^{s\times s}$ is an identity matrix and $a_i \in \mathbb{R}$ are the model parameters which describe the dynamics of the brain activity, where $\boldsymbol{c}_{k-1}^{\circ 2}$ denotes the Hadamard Power.

I. Torres et al. (eds.), *VII Latin American Congress on Biomedical Engineering CLAIB 2016, Bucaramanga, Santander, Colombia, October 26th -28th, 2016*,
IFMBE Proceedings 60,
DOI: 10.1007/978-981-10-4086-3_89

III Time-course Reconstruction of Neural Activity

A time-course reconstruction of neural activity can be obtained by including temporal constraints that describe adequately the temporal dynamics of the neural activity [5]. This can be represented by the following functional:

$$J(\mathbf{c}_k,\rho,\lambda) = \|\mathbf{y}_k - \mathbf{M}\mathbf{\Phi}_s\mathbf{c}_k\|_2^2 + \rho^2\|\mathbf{c}_k\|_2^2 + \lambda^2\|\mathbf{c}_k - \mathbf{c}_k^-\|_2^2 \quad (5)$$

where $\rho \in \mathbb{R}^+$ and $\lambda \in \mathbb{R}^+$ are the regularization parameters and $\mathbf{c}_k^- = \mathbf{f}(\widehat{\mathbf{c}}_{k-1},\widehat{\mathbf{c}}_{k-2},\widehat{\mathbf{c}}_{k-\tau})$ from Eq. (4). This functional is similar to the one proposed in [6] but including a complex dynamic temporal constraint based in a realistic physiological brain model [7].

Therefore, $\widehat{\mathbf{c}}_k$ can be obtained from Eq. (5) as follows:

$$\widehat{\mathbf{c}}_k = \arg\min_{\mathbf{c}_k}\{J(\mathbf{c}_k,\rho,\lambda)\} \quad (6)$$

An analytical solution for Eq. (6) can be obtained as follows:

$$\begin{aligned}\frac{\partial J}{\partial \mathbf{c}_k} &= -\mathbf{\Phi}_s^\top\mathbf{M}^\top\mathbf{y}_k + \mathbf{\Phi}_s^T\mathbf{M}^\top\mathbf{M}\mathbf{\Phi}_s\mathbf{c}_k \\ &\quad + \rho^2\mathbf{c}_k + \lambda^2\mathbf{c}_k - \lambda^2\mathbf{c}_k^- \\ &= \mathbf{0}\end{aligned} \quad (7)$$

Consequently, the solution for $\widehat{\mathbf{c}}_k$ that can describe the time-course evolution of multiple sources simultaneously is obtained as follows:

$$\widehat{\mathbf{c}}_k = \left(\mathbf{\Phi}_s^\top\mathbf{M}^\top\mathbf{M}\mathbf{\Phi}_s + \rho^2\mathbf{I}_s + \lambda^2\mathbf{I}_s\right)^{-1}\left(\mathbf{\Phi}_s^\top\mathbf{M}^\top\mathbf{y}_k + \lambda^2\mathbf{c}_k^-\right) \quad (8)$$

Regularization parameters are computed by generalized cross-validation method described in [5] as follows:

$$v(\rho,\lambda) = \frac{\|\mathbf{M}_s\mathbf{c}_k(\rho,\lambda) - \mathbf{y}_k\|_2^2}{\mathrm{tr}(\mathbf{I}_d - \mathbf{M}_s\mathbf{M}_s^\dagger(\rho,\lambda))^2} \quad (9)$$

where notation $\mathrm{tr}(\cdot)$ denotes the trace of argument matrix, $\mathbf{M}_s = \mathbf{M}\mathbf{\Phi}_s$ and $\mathbf{M}_s^\dagger(\rho,\lambda)$ is the regularized inverse of the matrix $\mathbf{M}_s$, and the matrix $\mathbf{M}_s^\dagger$ is defined as follows:

$$\mathbf{M}_s^\dagger = \left(\mathbf{M}_s^\top\mathbf{M}_s + \rho^2\mathbf{I}_s + \lambda^2\mathbf{I}_s\right)^{-1}\mathbf{M}_s^\top \quad (10)$$

IV Results and Discussion

In order to evaluate the performance of the proposed method, a comparison between the estimated source and the simulated source neural activity is performed (for the following figures will be x145 or source 145). The neural activity is simulated by using the model described in Eq. (1), Eq. (2) and Eq. (3) where the model parameter are set to $\tau = 20$, $a_1 = 1.0628$, $a_2 = -0.42857$, $a_3 = 0.008$, $a_4 = 0.000143$, $a_5 = -0.000286$, and $\|\boldsymbol{\eta}_k\| \leq 0.05$. The model was changed the stable to unstable at sample $k = 125$ ($t = 0.5\,s$) by modifying the values of a_1 from 1.0628 to 1.3, while a_2 from -0.428 to -1 over the entire diagonal. The simulated EEG $\mathbf{y}_k$ is obtained from $\mathbf{x}_k$ using $\mathbf{y}_k = \mathbf{M}\mathbf{x}_k + \boldsymbol{\varepsilon}_k$, where $\boldsymbol{\varepsilon}_k$ is set to achieve the Signal-to-Noise Ratio (SNR) equals $7\,dB$. The resulting simulated EEG signal can be seen in Fig. 1.

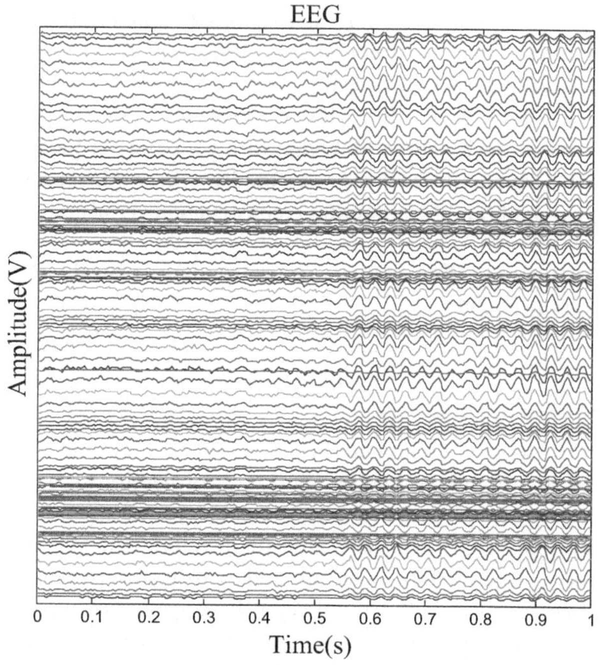

Fig. 1: Simulated EEG signals with a generalized epilepsy seizure. It can be seen that the epilepsy seizure starts at time $t = 0.5$ s.

A comparison analysis of the proposed method described in Eq. (8) is evaluated, in terms of the time-course reconstruction, against the following cases:

1. Tikhonov method [8] without the inclusion of a dynamical model:

$$\widehat{\mathbf{c}}_k = \left(\mathbf{\Phi}_s^T\mathbf{M}^T\mathbf{M}\mathbf{\Phi}_s + \lambda^2\mathbf{I}_s\right)^{-1}\mathbf{\Phi}_s^T\mathbf{M}^T\mathbf{y}_k \quad (11)$$

2. Dynamic Tikhonov method considering a linear dynamical model:

$$\widehat{\mathbf{c}}_k = \left(\mathbf{\Phi}_s^T\mathbf{M}^T\mathbf{M}\mathbf{\Phi}_s + \lambda^2\mathbf{I}_s\right)^{-1}\left(\mathbf{\Phi}_s^T\mathbf{M}^T\mathbf{y}_k + \lambda^2\mathbf{c}_k^-\right) \quad (12)$$

being $\mathbf{c}_k^- = \mathbf{A}_1\mathbf{c}_{k-1} + \mathbf{A}_2\mathbf{c}_{k-2}$.

3. Dynamic Tikhonov method considering a nonlinear dynamical model, being $c_k^- = A_1 c_{k-1} + A_2 c_{k-2} + A_3 c_{k-\tau} + A_4 c_{k-1}^{\circ 2} + A_5 c_{k-1}^{\circ 3}$.

It is worth noting that the methods are qualitative evaluated during the epileptic stage ($0.5 \leq t \leq 1$).

In Fig. 2 a comparison between the ground-truth simulated activity for a particular state and the estimated time-course activity is performed by using a Tikhonov method without the inclusion of a dynamical model considering additive noise of 7dB.

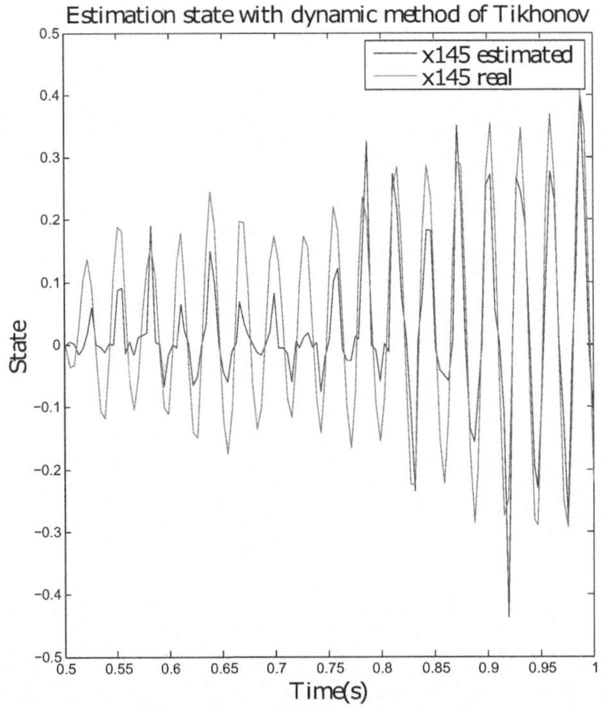

Fig. 2: Time-course comparison using Tikhonov method without dynamical model.

In Fig. 3 a comparison between the ground-truth and the estimated time-course activity is performed by using the Tikhonov method with the inclusion of a dynamical linear model considering additive noise of 7dB.

In Fig. 4 a comparison between the ground-truth an the estimated time-course activity is performed by using the Tikhonov method with the inclusion of a dynamical nonlinear model considering additive noise of 7dB.

In Fig. 5 a comparison between the ground-truth an the estimated time-course activity is performed by using the proposed method that includes a nonlinear dynamical constraint.

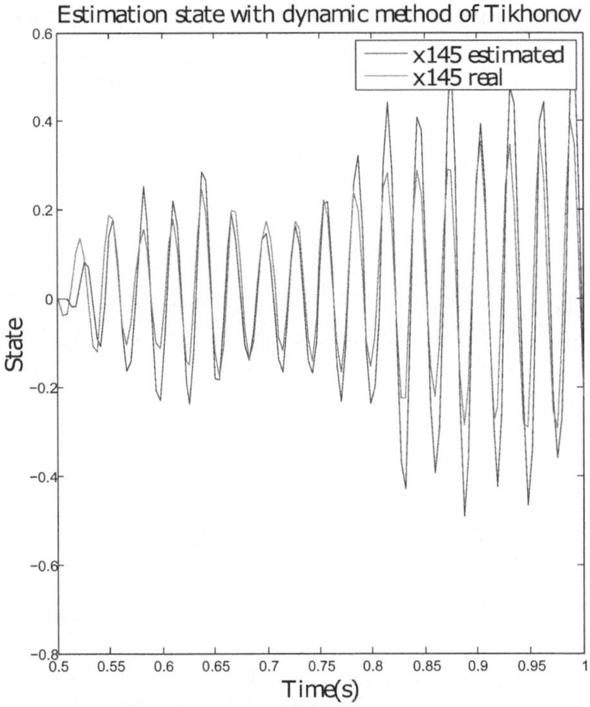

Fig. 3: Time-course comparison using the Thikhonov method with a dynamical linear model.

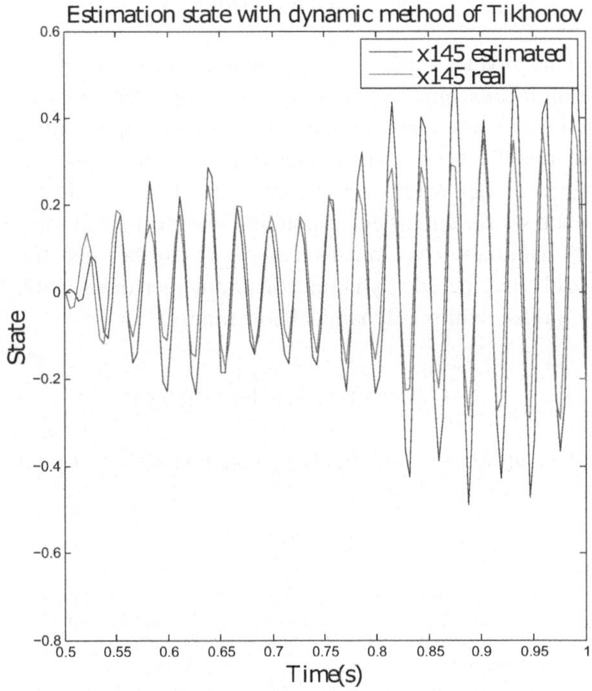

Fig. 4: Time-course comparison using the Thikhonov method with a dynamical nonlinear model.

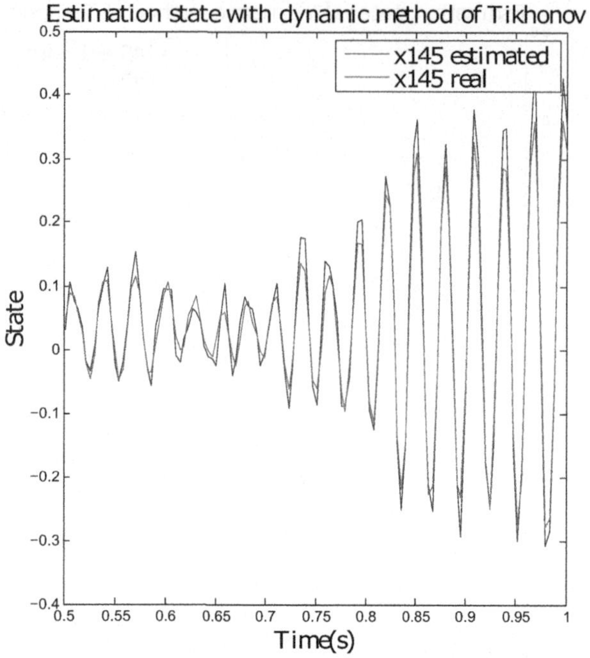

Fig. 5: Time-course comparison with dynamical model using the proposed method that includes a nonlinear dynamical constraint.

V Conclusions

This paper address a methodology for multiple simultaneous time-course source reconstruction that can be used for a source-based noninvasive brain computer interface. The results, based on simulated EEG data show that the proposed framework reconstruct correctly the brain activity at each source and its corresponding waveform in comparison with Tikhonov dynamical methods. It can be seen that the models that include dynamic constraints performs better that the models without dynamical constraints.

Conflict of Interest

The authors declare that they have no conflict of interest.

References

1. Georgieva P., Bouaynaya N., Silva F., Mihaylova L., Jain L. C.. A Beamformer-Particle Filter Framework for Localization of Correlated EEG Sources *IEEE Journal of Biomedical and Health Informatics.* 2016;20:880-892.
2. Brookes Matthew J., Stevenson Claire M., Barnes Gareth R., et al. Beamformer reconstruction of correlated sources using a modified source model *NeuroImage.* 2007;34:1454 - 1465.
3. Jonmohamadi Yaqub, Poudel Govinda, Innes Carrie, Jones Richard. Source-space {ICA} for {EEG} source separation, localization, and time-course reconstruction *NeuroImage.* 2014;101:720 - 737.
4. Barton M. J., Robinson P. A., Kumar S., et al. Evaluating the Performance of Kalman-Filter-Based EEG Source Localization *IEEE Transactions on Biomedical Engineering.* 2009;56:122-136.
5. Giraldo E., Martinez-Vargas J. D., Castellanos-Dominguez G.. Reconstruction of neural activity from EEG data using dynamic spatio-temporal constraints *International Journal of Neural Systems.* 2016;26:1-15.
6. Schmitt U, Louis A K. Efficient algorithms for the regularization of dynamic inverse problems: I. Theory *Inverse Problems.* 2002;18:645.
7. Robinson P.;, Kim J.;. Compact Dynamical Model of Brain Activity *Physical Review E.* 2007;75:701 -710.
8. Fukushima M., Yamashita O., Kanemura A., Ishii S., Kawato M., Sato M.. A State-Space Modeling Approach for Localization of Focal Current Sources From MEG *IEEE Transactions on Biomedical Engineering.* 2012;59:1561-1571.

Author: P.A. Munõz
Institute: Universidad del Quindío
Street: Carrera 15 Calle 12 Norte
City: Armenia
Country: Colombia
Email: pamunoz@uniquindio.edu.co

Author: E. Giraldo
Institute: Universidad Tecnolǵica de Pereira
Street: Carrera 27 N. 10-02 Barrio Alamos
City: Pereira
Country: Colombia
Email: egiraldos@utp.edu.co

Biomedical IoT Device for Self-Monitoring Applications

Tatiana Huertas[1] and Diego Mendez[2]

[1] Pontificia Universidad Javeriana - Department of Electronics Engineering. Bogota, Colombia
[2] Pontificia Universidad Javeriana, Department of Electronics Engineering - Center of Excellence in IoT (CEA-IoT). Bogota, Colombia

Abstract— **Currently in Colombia, the coverage of medical services is not enough to fulfil the growing demand of patients in medical institutions. Furthermore, the access to good medical facilities is limited to certain communities in the country. For these reasons, the aim of this project is to design and implement a system, which could be used by the patient to monitor her own health without having to go to a hospital. This system integrates different sensors of a physiological monitor and visualizes the data acquired by these sensors in a mobile device. This device could allow the possibility to send the data through the Internet (enabling a remote monitoring) for future projects. The project is made up of a software (Android application and microcontroller firmware), as well as a hardware platform (eHealth systems and Bluno board). In order to ease the integration of the acquisition system with the mobile phone, a BLE connection is required. Preliminary testing of the developed system has shown that all the acquired data from the different sensors can be displayed simultaneously in the mobile phone (in a text and graphical manner), achieving a real-time local visualization platform.**

Keywords— **physiological monitor, remote monitoring, telemedicine, biomedical, IoT.**

I Introduction

Nowadays, there are many Colombians who do not have the opportunity to attend to a hospital or medical center to get a constant monitoring of their vital signals. This situation arises for economical reasons or for not having any nearby medical facilities. The elderly is the most affected population at the moment. As presented in [1], three out of five seniors are emotionally affected by diseases, including heart disease, pneumonia, hypertension, arthritis, epilepsy and cancer, among others.

Additionally, there are many hospitals that do not have enough space to cover the high demand of patients. Some of these patients require a hospitalization, while others do not need to stay permanently at the health center. In January 2013 the number of beds available in 22 different hospitals in Bogota, Colombia, was 414, but the high quantity of patients in hospitals of levels II and III, exceeded this number. According to the District Government, hospital services of gynaecology are occupied 145%; internal medicine are 154%; surgical, 109%; of Paediatrics, 106%; basic neonatal care, 102%; symmetry, 124% and emergency, 137% in third level hospitals [2]. This high demand of beds is caused by the patients who are not necessarily in danger but have to be under observation with physiological monitors.

The goal of this project is to help relieve congestion in hospitals and, at the same time, to help people who (for different reasons) are not able to attend to a medical center. The aim of this project is to design and build a data acquisition system, which integrates the sensors of a physiologic monitor into a first level integrating digital board, and additionally, to transmit the acquired information to a mobile phone in order to centralize and visualize the data.

The rest of the paper is organized as follows: Section II presents the overall system architecture to provide the reader with the big picture. Section III details about the development process of the system, including the hardware and the software components. Section IV presents the results and corresponding analysis, while Section V concludes the document.

II Proposed Architecture

For the development of this project the following functional specifications and restrictions were proposed.

- The creation of an acquisition system in and Arduino-like Platform that integrates different sensors that measure physiological signals.
- The data collection process can be done once a day, over a period of 30 minutes in total (using the acquisition system).
- The Bluetooth (LE) communication module between the acquisition system and a local mobile device (Smartphone) must have a sampling rate of at least 100 Hz.
- The creation of an Android application for local visualization of the data received from acquisition system, in order to let the patient study the current status of her vital signals.
- The Android application must be compatible with as many mobile devices as possible so that more people have access to it.

I. Torres et al. (eds.), *VII Latin American Congress on Biomedical Engineering CLAIB 2016, Bucaramanga, Santander, Colombia, October 26th -28th, 2016*,
IFMBE Proceedings 60,
DOI: 10.1007/978-981-10-4086-3_90

- The acquisition system should always be connected to a power supply; this means that the device is not completely portable and the patients cannot be doing exercise while using it. In the next version of the system, a rechargeable battery will be included.
- In order to reduce power consumption, the mobile device (Smartphone) should support Bluetooth 4.0 connection (BLE). This restriction implies that the mobile device operating system should be either Android 4.3 (API 18) or higher, or IOS 7.0 or higher.

A General Description

The project is focused on the implementation of a local monitoring system for patients, in order to provide a better medical service by keeping track of their vital signals while they stay at home. This was done with a data communication network, which, starts from a digital acquisition system of physiological signals.

To meet the mentioned specifications, the overall architecture shown in Figure 1 was proposed. As shown in the diagram, the biomedical signals from the patients are acquired through a detection device (first level integrator). Once this digital systems has collected the data coming from the different sensors, the information is transmitted to the mobile phone via a Bluetooth LE connection. The mobile phone is in charge of centralizing the data and visualize the information in a convenient way.

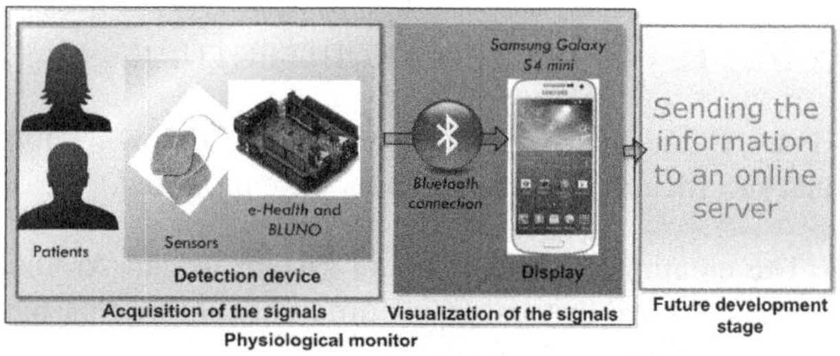

Fig. 1: General System Architecture.

B Signal acquisition

The acquisition system for this project will be the e-Health Sensor Shield 2.0[1] (e-HSS) complete kit and the BLUNO board. The e-Health Sensor Shield is the revamped version of the first biomedical board for Arduino and Raspberry Pi [3]. Thanks to the versatility of e-HSS it is possible to study different body signals (see Figure 2).

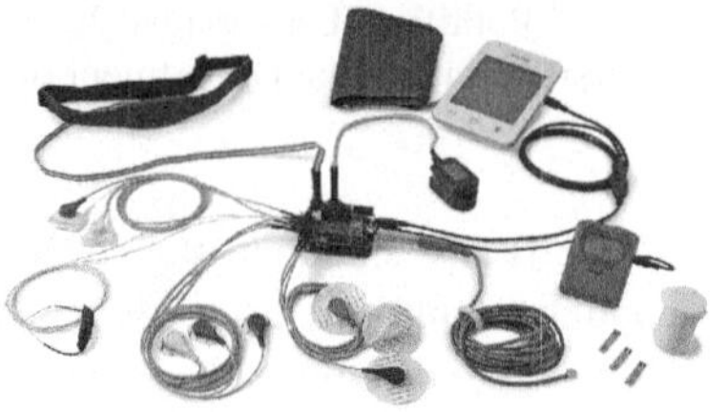

Fig. 2: The e-HSS connected to the 9 compatible sensors. Taken from [3]

The sensors shown in Figure 2 are (in clockwise direction starting from the top):

1. Blood pressure sensor (sphygmomanometer).
2. Pulse and percentage of oxygen in blood (SPO_2).
3. Glucometer.
4. Body temperature sensor.
5. Electrocardiogram (ECG).
6. Electromyography (EMG).
7. Airflow.
8. Galvanic skin response (GSR).
9. Patient position sensor (Accelerometer).

These sensors, except the Glucometer, are non-invasive.

Along with the e-HSS, the development board BLUNO was also used for the acquisition of the signals. The BLUNO board was the first board in the market that integrated a Bluetooth 4.0 modem on an Arduino UNO-like board. Its dimensions and pins distribution are exactly the same to an Arduino UNO, which allow us to develop Arduino projects with real-time communications and low energy consumption [4]. When used together, the e-HSS and the BLUNO board allow us to acquire all the data coming from the biomedical sensors, and subsequently transmit it to the visualization device (Smartphone).

C Bluetooth connection

A Bluetooth connection is useful for transmitting information between two or more devices (1-to-N paradigm) that are within walking distance. This technology is best suited for low bandwidth applications, like transferring sound from a mobile device to a headset or transferring data (bytes) from a wireless keyboard to a computer [5]. Also, as explained in [6], the Bluetooth technology provides sufficient channel bandwidth to transmit the data measured by physiological sensors. After preliminary tests, we could verigy that both Bluetooth

[1] As explained by [3] the e-HSS does not have medical certifications, meaning that the device can not be used to monitor critical patients who must be accurately measured for professional diagnosis. The interval of current used in this device (20mA-50mA) could cause muscle contraction/pain if the pins are touched directly with a wire but that is not the case for this project.

and WiFi communication had no loss of information. These results would let us conclude that the Bluetooth connection is compatible for this project since the two devices will be near to each other, and the ease of use of the Bluetooth communication is fairly simple compare to WiFi (not to mention power consumption).

D Signal Visualization

In order to conveniently present the acquired data to the user/patient, the easiest way to locally display this information is through the screen of a mobile phone (Samsung Galaxy S4-mini), considering also the wide spread use of Smartphones (almost everyone has at least one). This Signal Visualization block is characterized by the development of an Android application, enabling not only the BLE connection with the physiological acquisition system, but also the local visualization (graphical and text mode) of the data collected by the nine biomedical sensors.

III System Development

First of all, before integrating the acquisition system and the visualization system, the nine sensors were tested to verify that all of them were measuring and providing data from the patient. These tests were made using the example codes provided by [3]. Two of the sensors, the Accelerometer and the Blood Pressure sensor, were not working correctly; the former was not providing the right position of the patient, and the latter was not giving any information at all. These problems were solved by correcting and modifying the original libraries provided by the e-HSS developer.

Depending on the physiological signal, different visualization modes must be provided:

1. Text type: Static numbers that show the current measured data (Temperature, SPO_2, Blood pressure, GSR, Glucometer and Airflow).
2. Real-time (RT) graphs: Data that change rapidly in time and have to be shown in graphical form (ECG, EMG and Airflow).
3. Accelerometer graphs: The actual position of the patient and the acceleration values of each axis (x, y, z).

The selected sampling frequency for the first mode was the same sampling rate of the SPO_2 sensor [2], which is 2 Hz. For the RT-graphs, the sampling frequency was 300 Hz because that is the minimum sampling rate for the EMG sensor. Lastly, the sampling frequency for the Accelerometer graphs was 100 Hz.

In order to establish the Bluetooth connection between the BLUNO board and the Samsung Galaxy S4-mini (which supports Bluetooth 4.0), a code based on an application provided by [4] (BlunoBasicDemo[3]) was designed. This code not only enables the reception and transmission of the data, but also verifies if the devices were compatible for the communication. Tests were made to prove that the data were being sent without loss.

After proving that the data measured by the sensors were correctly sent via Bluetooth, the Android application was designed according to the distinctive style of an Android Operating System. As shown in Figure 3, three different sections were created in order to display the data in text (left) and graphical mode (center), and to show the position of the patient with the acceleration values of the 3 axis (right).

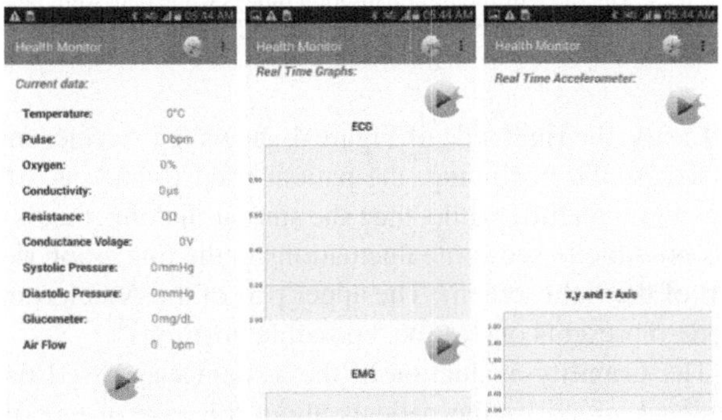

Fig. 3: Sections of the Android App

IV Results

The experiments were performed with healthy university students from 20 to 25 years old. Each student was given knowledge about the tests and the project, they were explained why they were fit for this procedure and they were asked to sign the informed consent before they were tested, and a specialised companion guided the whole process of data acquisition. Each patient (student) showed different data for each sensor.

The test was passive, each student was tested separately and remained seated almost the entire process. Also, the students were able to deny the usage of any sensor that made them feel uncomfortable. Additionally, the students interacted with the Android app when the sensors were placed on their bodies, and were asked to move only while using the Accelerometer.

[2]Commonly pulsations (bpm) can be sampled from 0.5 Hz to 5 Hz

[3]This app can be downloaded from [7]

Overall, the results were consistent to those of a healthy person. The temperature was a little higher than normal because the sensor was placed under the clothing. This patient did not allow the use of the Glucometer (it displayes 0 mg/dL on the left screenshot of Figure 4). The graphs for the ECG and EMG signals are shown in the center of Figure 4.

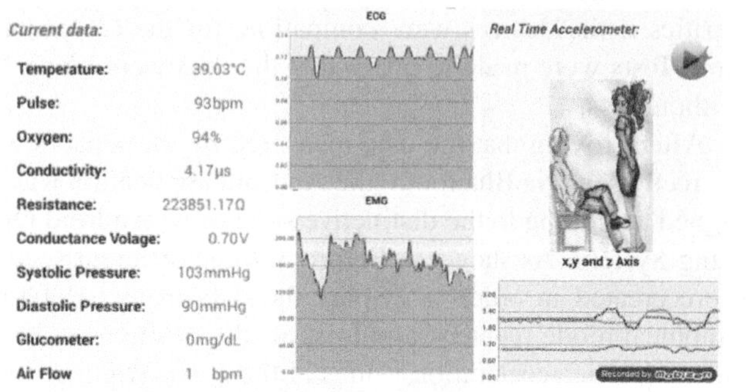

Fig. 4: The three different visualisation modes while data were being collected from the patients. Left (text), Center (graph), Right (accelerometer).

Lastly, the right side of Figure 4 shows the Accelerometer mode. At the beginning, the patient tried to stay as still as possible for a little while, then she started shaking, that is why it is possible to see some fluctuations in the line graph (lower part of the right screen). The upper part of the Accelerometer mode represents one of the 5 possible positions[4].

The usability evaluation of the system was based on the feedback given by the patients about the ease of use of the app, how they felt using the physiological monitor and if they understood the visualised data.

V Conclusion

Current monitoring systems have a wide number of functionalities, but commonly they have quite complex systems for a non-specialist public. Therefore, new smart monitors have been created, trying to reduce this complexity. Similarly, the number of low price physiological sensors available to the public have increased thanks to the development of new scientific disciplines such as Telemedicine.

The system developed in this project meets the general goal to locally acquire and visualize biomedical data allowing patients to have a better understanding of their vital signals without having to attend to a hospital. Additionally, this system was designed so that patients could use it in their own home without the need of specialized help. Preliminary results show that the system is robust, easy to use, and effectively acquire and visualize the biomedical data received from the e-HSS platform. The patients were able to study and understand their current health status and see the behavior of their physiological signals with no complication.

Our current and future work focuses on the extension of this system to a completely mobile device including rechargeable batteries as its main supply source, and including the transmission and centralization of the data in a cloud-based server.

[4]Prone, Supine, Seating or Standing, Left lateral recumbent and Right lateral recumbent

Acknowledgements

The authors would like to acknowledge the cooperation of all partners within the *Centro de Excelencia y Apropiación en Internet de las Cosas (CEA-IoT)* project. The authors would also like to thank all the institutions that supported this work: the Colombian Ministry for the Information and Communications Technology (*Ministerio de Tecnologías de la Información y las Comunicaciones - MinTIC*) and the Colombian Administrative Department of Science, Technology and Innovation (*Departamento Administrativo de Ciencia, Tecnología e Innovación - Colciencias*) through the *Fondo Nacional de Financiamiento para la Ciencia, la Tecnología y la Innovación Francisco José de Caldas* (Project ID: FP44842-502-2015).

References

1. PROFAMILIA. "Situación de los Adultos mayores en Colombia". Encuesta Nacional de Demografía y Salud. Available from: `www.profamilia.org.co/encuestas/Profamilia/Profamilia/documentos/boletines/prensa/SituaciondelosAdultosMayoresenColombia.pdf` Accessed July 2015.
2. Malaver, C. La falta de camas hace colapsar salas de urgencia en red hospitalaria. "Periódico El Tiempo". Available from: `http://www.eltiempo.com/archivo/documento/CMS-12822365`. Accessed May 2016.
3. Cooking-Hacks, "e-Health Sensor Platform V1.0 for Arduino and Raspberry Pi [Biometric / Medical Applications]". Available from: `https://www.cooking-hacks.com`. Accessed August 2015.
4. DFRobot, "Bluno - An Arduino with Bluetooth 4.0". Available from: `http://www.dfrobot.com/index.php?route=product/product&product_id=1044#.V2IcyGIrK00`. Accessed May 2016.
5. Kate T., Poonam S. , Pooja Sehgal , Nikhilesh Jasuja , Bert Ishita and Kyle Gauthammanoj. "Bluetooth vs Wi-Fi." Available from: `http://www.diffen.com/difference/Bluetooth_vs_Wifi`. Accessed May 2016.
6. Xiaoming Zhao , Ding-Yu Fei , Charles R. Doarn , Brett Harnett and Ronald Merrell, "A Telemedicine System for Wireless Home Healthcare Based on Bluetooth and the Internet" Telemedicine Journal and e-Health. November 2004.
7. DFRobot/BlunoBasicDemo. Available from: `https://github.com/DFRobot/BlunoBasicDemo`. Accessed May 2016.

Physical and Chemical Emulation of a Cornea

R. Jaramillo Diaz[1], H. Dávila Torres[1, 2], A.V. Molina Mojica[1] and L.J. Martínez Guerrero[1]

[1] Universidad ECCI /Coordinación de Biomédica, research professor, Bogotá, Colombia
[2] Universidad Distrital Francisco José de Caldas /Coordinación de Biomédica, professor, Bogotá, Colombia

Abstract— *Abstract*— **the corneas lack for transplant and acquiring surgical experience system are areas that need constant advances both medical and scientist to supply maximal approach to the physical and chemical patient's reality, for this reason a possible solution are the organs emulators modules. Based on the physical and chemical corneal characterization and an inverse engineer process which model and generate its artificial counterparty by means the intelligent biopolymers constitution. The polyacrylamide, cyanoacrylate (minimal quantity) and chitosan three layered biopolymer is generated with a half layer 200% swelling index, 80 % environment absorption, semi-permeable aspheric geometry and temperature work window among 15°C to 37°C in vitro. This method contributes to an artificial organ generation that allows the possible prosthetics implantation previously the medical-ethic commitment proving and accepting to start an experimental surgical module due to the physics and biochemical similarities to the ophthalmic tissues.**

Keywords— **cornea, emulation, biochemistry, prostheses, transplant.**

I. Introduction

The ophthalmic surgery residency students work by means of observation joined by the specialist during a real surgery or an experimental surgery in which animal specimens like pigs or cow's necrotic heated tissues are used. However, a complication with this procedure is that the eye has lost some of its natural characteristics. Other alternatives are didactic elements that produce damage in the phacoemulsifier or the virtual practice which does not let the user acquire the expertise with the necessary care and force in the real structure. The other possibility is the technique used in countries like India in which the procedure is offered at a discounted price in order to be performed by an ophthalmology resident on what usually amounts to low income populations.

In Latin-American countries, the number of patients on transplantation lists are very long, for example: Mexico (6952) [1], Argentina (3043) [2] and Colombia (1300) [3]. The amount of people in need of organs cannot be fulfilled with the available organs and organs importation laws requiere for very rigorous adjustments that require biosecurity for evaluation and transplants tissue transports.

On the other hand in order for a patient to get approval for transplantation, the diagnostics needed are HLA antigens, blindness, generals grafts or lamelars and age associated factors in order to generate prioritization for transplant execution and it could still not be successful. Complementing the donated tissues rate vs. processed and transplanted tissues has a very low reception rate in Europe contrary South American statistics [4].

Mimetic tissues is a promissory field in the artificial organs conformation needed for surgical medical practice [5] to supply the transplant tissue needed which are made by complex system models and precision medicine to synthetize prostheses that can replace the organ according to biochemical, physiological, anatomical and its surroundings.

Based on embryology and histogenesis working through reverse engineering to the molecular level that constitute by means of permutation and combination the more basic anatomy and physiology. The corneal structure forms at the fifth and sixth week of pregnancy when the epiblast appears from the ectoderm and is then molded beginning from the placode to the ocular globe which is divided by the vitreous humor and crystalline capsule. The mesenchyme distribution which provides the strong protoestroma-like long structure and the stability point to form the barrier and the last part of the corneal epithelium. A corneal structure depth analyses shows the carrying of a proteoglycans chains and basics unit glicosaminoglycans. The chondroitin sulfate leads to understand the connective tissues bases.

The crystalline emulation is the initial step when going about developing the emulate systems ophthalmological constitution [6] which corresponds to a corneal diffraction of about twenty percent and a regeneration inability that allows for time adaptations, process and closure mechanism. The fucose and manose are key molecules to the constitutive adequate process as a result from the keratan substrate which generates the diffraction, the dissemination, change substance, leucocytes interaction adequate environment by aqueous medium [7], the hyaluronic acid is one of the most important glycosaminoglycan scaffold agents.

The next stage is based on the knowledge that collagen interaction, depending on the substrate, either collagen two, three or five, disappears in the conformation process leaving

I. Torres et al. (eds.), *VII Latin American Congress on Biomedical Engineering CLAIB 2016, Bucaramanga, Santander, Colombia, October 26th -28th, 2016*,
IFMBE Proceedings 60,
DOI: 10.1007/978-981-10-4086-3_91

collagen four as constant raw matter, for experimentation purposes, it was obtained by means of spider webs and avian placenta [8]. The collagen fibrilles works in the light diffraction process that is evident in the wavelength dispersion [9] in three different difraction levels helping the retina performance to the absorbed chromatic gamma.

The corneal layering begins from the corneal epithelium, the external tissue is composed by a central zone and the limbo, a joined and parallel structure with differentt collagen layers. There are 7 biochemical layers and the basement membrane with a similar internal structure near to the aqueous humor and join together with the blood vessel due to its week abrassion [10]. The corneal stroma is the biggest size structure having ninety percent of the corneal volume [11] this tissue transfers substances from the corneal epithelium to the corneal endothelium with a high macrofage population density. The stroma possesses a differential absorbed pressure of 20 mmHg joined to 60 mmHg cavities of external pressure which are supported by a different conformation and lamela size that includes a width range from 0.5 to 30 μm and thickness range from 0.2 to 1.2 μm, different dimensions in a rear part with a width range from 100 to 200 μm and a thickness range from 1 to 2.5μm. The lamellas alignment are according to the temporal axis and the nasal sinus with distribution angles from 90° to 45° [12]. This layer has keratin sulfate enrichment, this sulfate is originated by the D-galactosa residues that are subdivided in ks-1 from the n-asparagine and O-serine links which are found in the manosa, o-serine and tronine obtained by hydrolilation that let an area components high cohesion [13].

II. Methodology

The emulator is generated according to the natural biochemical and physical characteristics, in this way it exhibits their properties in a different nature ambience. The mechanical characteristics are set according to the material analyses versus material tissue.

According to the standard aspheric geometry measures: the vertical ratio is 5.75mm, the horizontal ratio is 6 mm, the distal lumen thickness is 1 mm, and the medium thickness is 0.58mm, this thicknesses are subdivided: the corneal endothelium 18-20μm, 38μm fibrils corneal stromal composition 500μm and a keratinocytes medial layer, corneal epithelium 58-64μm. The Solidworks designed model took the measurements and the mechanical proprieties to analyses the eyes exposed forces and the pressure (table 1).

Table 1 Material Properties

Properties	Values	Units
Elastic Modulus in X	0.25	N/mm^2
Poisson's Ratio in XY	0.00049	
Shear Modulus in XY	0	
Mass Density	0.001	Kg/m^3
Tensile Strength in X	30	N/mm^2
Thermal Conductivity in X	0.2256	W/(K-m)
Specific Heat	4.186	J/(Kg-K)

The polyacrylamide and chitosan solgel fibrils had cuts with 0.5 μm minimal determinate forming a 7.6 μm height structure. In distilled water there was four, five and seven collagen constituted and hydrated controlling the size so as not to exceed 65.8 initial size to conserve the corneal stromal thickness. Water at 50°C was added to obtain the aspheric shape. All previous procedures were searched to emulate the original corneal embryology [14]. The upper and bottom layers were created by means of an airflow in the drying process. The first layer was cut to 0.25 μm made by double cuts (by a histopathology expert) according to a metrology pattern measurement. The last cut was performed smoothly. The wrapping layer was created to generate a multilayer system [15] the size presented a discrepancy of 10 times in the basement membrane but this was compensated with a diffraction reduced coefficient by reduced doping.

III. Results

The force and pressure Solidworks data entry propitiated the observation of the material interaction and reaction. The iris and sclerotic lack support generated a corneal base depression (fig. 1) and the material possible appearance (fig. 2).

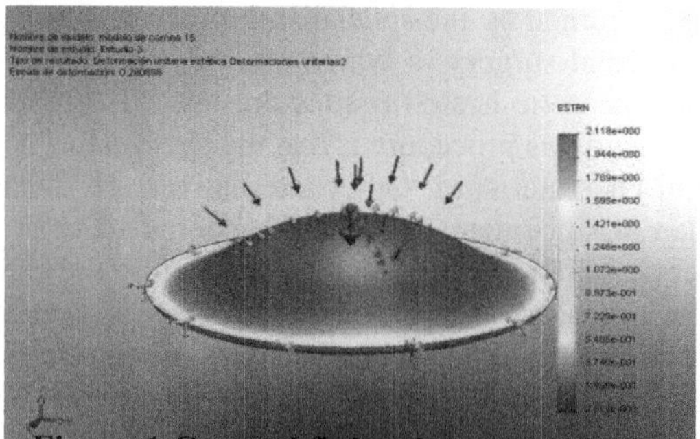

Figure 1 Corneal finite element analysis.
Source: Author

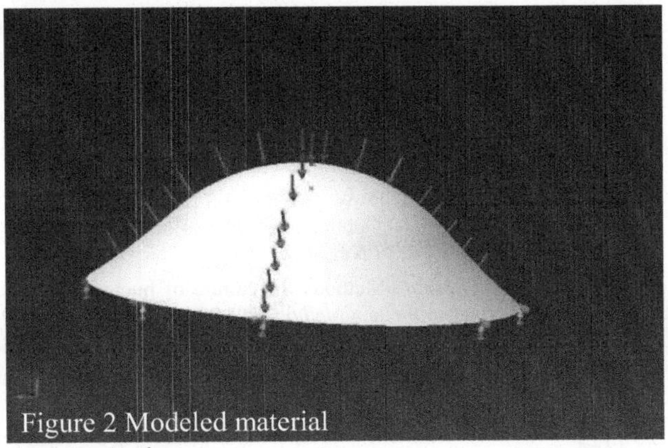

Figure 2 Modeled material

Source: Author.

A mold is necessary according to the layer resulting deformation with the corresponding adjustments and the posterior 3D performance and the specified size methodology microtome cuts. The enveloping created polymer had a porous upper and bottom layer. The exceeded pressure and impermeability incorrect fractured enveloping tissue reducing its crystallinity (Fig. 3)

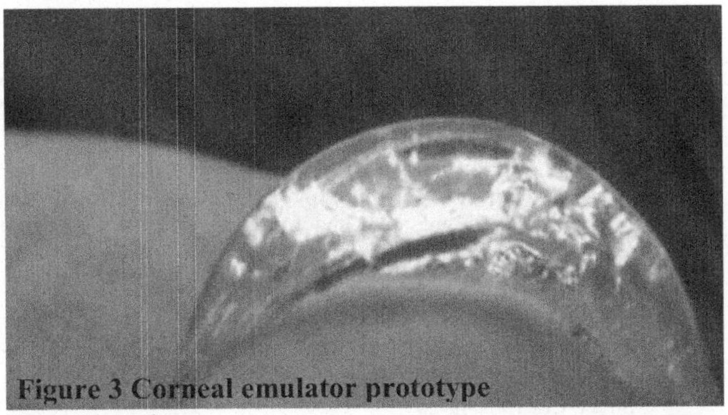

Figure 3 Corneal emulator prototype

Source: Author

The tissue conservation endured 15 days without apparent changes, oxidation or fungi. The prototype was isolated in a lenses container constantly hydrated with collagen and water, afterwards the prototype was exposed to the environment having a size reduction due to thermic changes and the fractures increased when the example was sent to laboratory for diffraction tests. The diffraction coefficient was 1.367.

IV. Discussion

The near corneal tissues function works to thermal regulation and the accompanying tissue reconstruction. The aqueous humor feeds the internal tissue through the aqueous porosity producing an electro-osmosis, which needs solutes and solvents transport. The aqueous humor introduction process begins in the ciliars bodies with a ciclic volume variation of 164 μL in a 90 minutes lapse [16].

For the studied investigation the hydration continuity of the eye according to the lacrimal composition, its process and the fifth cranial nerve interaction [17] achieving a constants cyclic modeled in Matlab (Fig. 4 & Fig. 5).

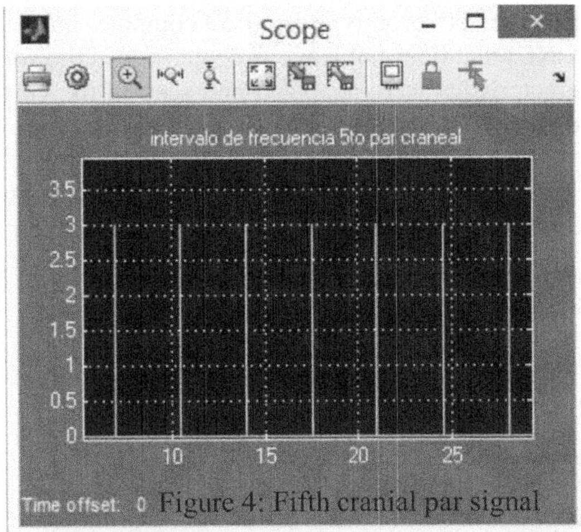

Figure 4: Fifth cranial par signal

Source: Author

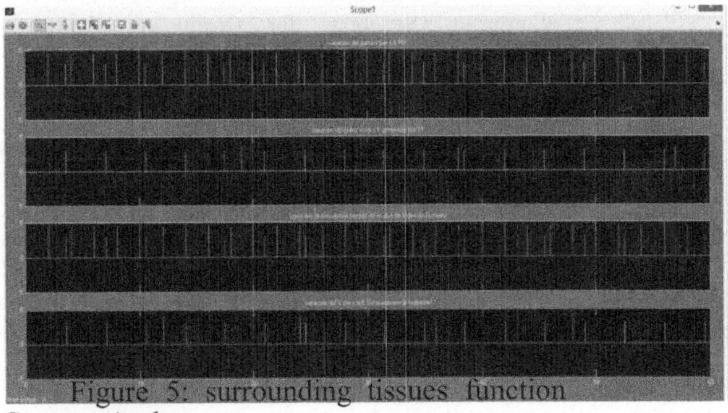

Figure 5: surrounding tissues function

Source: Author.

V. Conclusions

- The hydrogel absorption coefficient is about 80% which allows for the achievement of biocompatibility in the necessary limit for transplants.

- The dry failure and conservation times are evidence of the possible functionality capability of the prototype in order for medical personnel to carry out their procedures.
- Based on the Solidworks model, the need to carry out other material structure analyses is observed.

VI. Recommendation

The modeled and cut prototype require a computer numerical control (CNC) that does not leave irregularities in the solgel, these characteristics were analyzed and seen in the corneal embryology needing shaping and a cut that provides the exact size, form and structure combination.
Humidification is a necessary process creating near structures like ciliary bodies.

VII. Referencias

[1] C. Contreras, «Corneas encabezan listas de transplantes,» *Excelsiour,* p. 5, 22 Abril 2014.

[2] M. Garcia, «Por día, tres personas logran recueprar la vista en el pais,» *El Clarin,* 01 Marzo 2010.

[3] S. Perilla Santamaria, «A Colombia le va a tocar traer córneas desde otros paises,» *El Tiempo,* 24 Octubre 2013.

[4] «European Directorate for Quality of Medicine & Healthcare,«International Figure on Donation and Transplantation-2013,» *Newsletter transplant,,* vol. 9, nº 1, pp. 34-58, 2014.

[5] D. Barbieri, C. A. Bruijin y H. Y. van Blitterswijk, «Instructive nanomaterials hold the potential to mimic tissue complexity,» *IEEE PULSE,* vol. 5, nº 2, pp. 44-49, 2014.

[6] R. Jaramillo Diaz y C. H. Pinilla, «Ocular Surgery Emulator,» de *2015 PAN AMERICAN HEALTH CARE EXCHANGES (PAHCE)*, Chile, IEEE, 2015, pp. 159-165.

[7] M. P. Shafiquzzaman, «Fucosylated chondroitin sulfate diversity in sea cucumbers: A review,» *Carbohydrates Polymers,* vol. 112, pp. 173-178, 4 November 2014.

[8] A. J. Balley y N. C. Avery, «quantitative determination of collagen cross-links,» *Methods in molecular biology,* vol. 522, nº 1, pp. 103-121, 2009.

[9] K. M. Meek, ,. D. W. Leonard y S. Khan, « Corneal cross-linking – a review» Ophthalmic & Physiological Optics, vol. 30, nº 1, pp. 927-936, 17 November 2013.

[10] I. K. Gipson, «Methods for culture of human corneal and conjunctival epithelia,» *Methods in Molecular Biology,* vol. 945, pp. 31-43, 2013.

[11] A. Wilson y J. Marshal, «The role of light in measuring ocular biomechanics,» *Eye,* vol. 30, nº 2, pp. 234-240, 2016..

[12] J. W. Ruberti, «Corneal Structure and Function,» *CornealBiomechanics and Biomaterials,* vol. 13, pp. 269-295, 2011.

[13] Y. Zang, A. H. Conrad y E. S. Trasheva, «Detection and Quantification of Sulfated Dissacharides from Keratan Sulfates, Chondroitin/Dermatan Sulfates during Chick Corneal Development,» *Investigative Ophthalmology 6 Visual Science,* vol. 46, nº 5, pp. 1604-1613, 2005.

[14] S. J. Tuft y D. J. Coster, «The Corneal Endothelium,» *Eye,* nº 4, pp. 389-424, 1990.

[15] G. Decher y J. B. Schlenoff, « Layer-by-Layer Assembly: from Conventional to Unconventional Methods,» de *Multilayer Thin Films* , Strasbourg, Wiley Vch Verlag GmbH & Co. KGaA, 2016, pp. 43-64..

[16] J. Fishbarg, «Water Channels and Their Roles in the Some Ocular Tissues,» *Molecular Aspects of Medicine,* vol. 33, nº 5-6, pp. 638-641, October 2012.

[17] J. Huang y H. B. Hindman, «In vivo thickness dynamics measurement of tear film lipid and aqueous layers with optical coherence tomography and maximum-likelihood estimation,» *Optic letter,* vol. 41, nº 9, pp. 1981-1984, 2016.

Author: Ricardo Jaramillo Diaz
Institute: Universidad ECCI
Street: Cra 19 N 49-20
City: Bogotá D.C., Colombia
Email: rjaramillod@ecci.edu.co

Development of an Equipment to Prepare Nanofibers for Tissue Engineering Since a Standpoint of the Industrial Design

S.L. Rúa Jiménez[1], J.D. Villate Lagos[1], E.Y. Gómez-Pachón[1], Y. Torres Perez[2], E. Muñoz Prieto[3]

[1]Universidad Pedagógica y Tecnológica de Colombia. Escuela de Diseño Industrial. Facultad seccional Duitama, Colombia.
[2] Facultad de Ingeniería Mecánica, Universidad Santo Tomas. Seccional Tunja, Boyacá, Colombia.
[3]Universidad Pedagógica y Tecnológica de Colombia. Escuela de Ciencias Químicas. Facultad de Ciencias, Tunja, Colombia.

Abstract- **The use of electrospinning equipment built empirically for preparation of nanofibers and for research processes has confirmed the existence of design problems. Some of them are: the user interaction, equipment and environment; the location and calibration of injectors; the design, location and handling of collectors; the wiring and instrumentation equipment, etc. It has become evident to some users. The foregoing, has made palpable the need to perform an assessment from the Industrial Design point of view of this type of equipment and the generation of a design proposal for one more functional electrospinning equipment.**

As a case study, we evaluated and studied from different points of view (ease and safety of usability, user interaction, equipment and environment, quality of manufactured of fibers, etc). Initially an equipment of electrospinning was designed and manufactured by researchers from Universidad Pedagógica y Tecnológica de Colombia-UPTC. This equipment allows experimentation with fibers in the micro and nano scale. It is being used for the synthesis and preparation of biomaterials potentially useful for tissue regeneration by their nanometric size; for cell adhesion and the generation of pores for fluid ingress.

Subsequently, an analysis of usability and the main problems in handling were identified, so that the new design would have more security and a correct interaction between user and computer environment. The design of a new equipment electrospinning incorporating a positioning system injector-collector in the X, Y, Z and seeks greater precision, dimensional variety and order in the fibers was proposed. It also will facilitate and enable researchers to experiment with different ensembles team to achieve better mechanical, physical and chemical properties of the fibers. Also, in this new design involved people in industrial design, mechanical and electronics engineering and chemistry working daily with this equipment electrospinning.

Keywords: **Industrial design, equipment, Nanofibers, electrospinning, technological development.**

I. Introduction

Nanotechnology is the science that manipulates the matter to atomic dimension scales molecular and macromolecular in range nanometric. "These nanoparticles have important applications in the medical field, especially in the diagnostic and therapy areas. We can already see progress in the development of devices with nanotechnology to the early detection of infectious diseases and even cancer" [1].

One of the technics that allows the manipulation of nanometric scale is the electrospinning, which consist of remove from a cone injector polymeric particles in liquid state electro-statically loaded by a high voltage source, to a surface or collector that solidifies the substances generating walls of membranes of these polymeric fibers but in nanometric dimensions (Fig. 1) [2-4]. The process of electrospinning allows the preparation of nanofiber almost of 50 kind of different polymers for preparation of membranes for tissue engineering. The size of fibers in nanometers facility the cell adhesion and grow of the tissue.

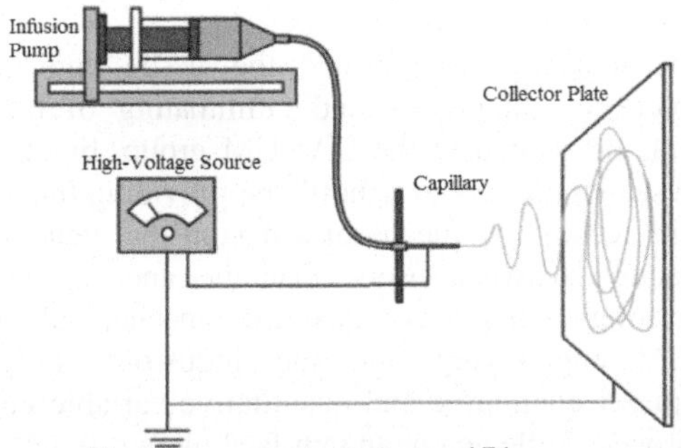

Fig.1 Electrospinning technique [2].

The new systems for positioning between injector-collector of electrospinning equipment seek to evolve the technical of manipulating the morphology of fibers pressure gauge by means of the experimentation and evolution of their components as the collector, injector, source, etc [3, 4].

The DANUM research group of Universidad Pedagógica y Tecnológica de Colombia-UPTC elaborated first an equipment of electrospinning base that allows preparation of fibers into micro scale and has facilitated learning and advances in researching field in the University [5]. In this previous equipment, the source of power produced fibers to scale nano but with its maximum power of 20 kV which reloading the process and generated more risks for the user.

The last experiences with the first equipment generated the initiative for design a electrospinning equipment since a

I. Torres et al. (eds.), *VII Latin American Congress on Biomedical Engineering CLAIB 2016, Bucaramanga, Santander, Colombia, October 26th -28th, 2016*,
IFMBE Proceedings 60,
DOI: 10.1007/978-981-10-4086-3_92

standpoint of the Industrial Design for solve the problems and research other kind of configurations.

This equipment has a new source of power of 15 to 30 kV of power was developed, that in addition to get to scale nano easily, decreases them damage of radiation electrostatic in the user which has been problematic of equipments of electrospinning similar, "makes missing remember that the process of spinning of ectrospinning is works with voltages very high, solvents that can be toxic [4]".

For electrospinning equipment design, the task force was looking for the way to incorporate a positioning system injector- collector in three exes X, Y, Z which can be found with more specific, dimensional variety and order and to elaborate the nanometric fibers, with the end of strengthen the research process of new materials and its possible applications in addition to generating further technological development in the area.

A fundamental facet in the creation of the equipment was the problems analysis that arises from the interactions between users and the previous equipment environment, which will be a starting point to improve and use the security safety. To achieve the development, all design Industrial process is taken as a pillar the Bernd Lobach [6]. Which identify and look for solutions to each problem according to the required conditions of users.

The industrial design and the tissue engineering will intervene as integrators and culminating of technological process developed in the DANUM group, by strengthening the research process of nanofibers, providing the team access and its technic by means of a design that generated proper interaction between users, machine and environment, by integrating applications fields of the mechanical engineering. Tissues engineering and the industrial design as an interpreter qualitative and quantitative variable connector of the process, reflected in an interface more efficient.

The DANUM research has developed researches of electrospinning and creation of nanometric fibers, among them the design of electrospinning machine to the processing of polymeric nanofibers project. This equipment seeks to strengthen the research process integrating and encouraging the experimental research as part of university.

Fig.2 First electrospinning equipment developed in UPTC.

The figure 2 shows the first equipment electrospinning designed by DANUM research group which elaborate fibers to micro scale (Fig. 2). This equipment was analyzed carefulness with the objective to propose a new technique for preparation of nanofibers for tissue engineering considering the relation of parameters of equipment and the morphology of nanofibers to obtain.

This electrospinning equipment offers 11 components such as: 1) source of power of 30 kV; 2) probe bk pr-28ª to track the varying electrical voltages; 3) voltmeter 4) material injector ne-1002x y ne-4002x that synchronize the application of polymeric material; 5) a syringe of 5mm to inject in this case the polymeric material; 6) cabin made in (MDF) laminated of 1cm with an acrylic of 5mm; 7) earth pole to dial the electric power disruption of power; 8) power stabilizers to the voltage shall be stabilized; 9) electric cables to make the connections among the machine components; 10) a collector to the material produced and 11) a placement to allocate the collector (Fig. 3).

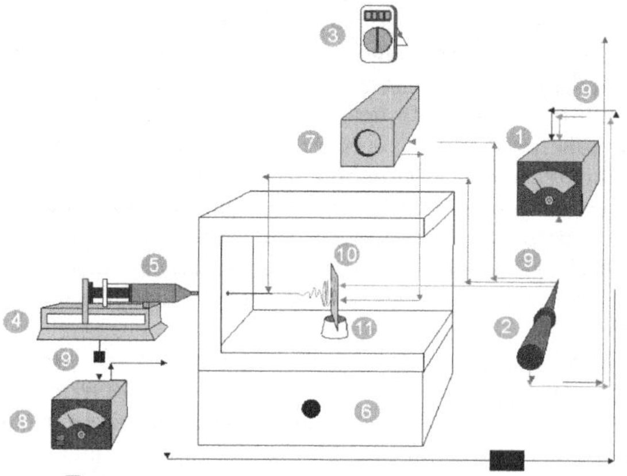

Fig.3 Electrospinning equipment components

This experimental prototype allocate the electrospinning technique into position horizontal and vertical. To each position, it must be change of positions the injector and the collector what instills complications and increase the time when it`s developed each process.

The first equipment as conducted with academic purposes in order that chemistry students of the university achieve advanced experiment learning about the electrospinning. Where get excellent outcome of several polymeric membranes and the composite materials, which were studied and analyzed achieving extensive documentation and assumptions of its possible uses. Take into account that the process was assertive that it initiated a stage of design and collection of electrospinning equipment which is the necessary requirement to develop the technic to a professional level.

II. Methodological Process

Initially, it was identified the available problems in the first electrospinning equipment that generally are exposed by similar equipment for research and preparation of the nanofibers in the world. The users and students that handled the equipment indicated necessity such as: interaction, security precision of data, easy learning, and some others; later is done a data collection of the problems caused by user and environment, it is done its proper analysis and by means of a hierarchy resolver each one of them by means the designing requirements, alternative formulation, experimentation on sketch and functional models. Finally, develop the proposal of equipment that provide the best solution to the problematic of the previous equipment correcting the minimum details that arise of the tests to define the final design.

A. *Problems Analyses:*

- *Users:* Researchers showed a long time to learn to handle the equipment.
- *Environment:* The environment where is located the electrospinning usually generate visual and physical congestion, electric risk, chemical, locatives etc.
- *Disconnection of equipment:* The communication of the majority of electrospinning equipment doesn´t send properly signals that allows to user a coherent operation. The components usually send information separately, which increases the steps in the process.

B. *Design Requirements*

Afterwards the analyses of problems of majority of electrospinning equipments of laboratory and its hierarchy that generate requirements of design that help to guide the final design according to each side that has a product such us the requirement uses, functionally, technic productive, structures, forms, economics from the marker (Table 1).

Table 1. Requirements Design

#	Requirements of design
1	Allowing the visualization of the injection process.
2	The equipment must have a support or fixed base that allows the holding of the actual injector and/or a future one.
3	The contention cabin of the principal elements process (injector, collector, water) must be hermetically sealed
4	The internal walls must have with a material formed of aluminum that simulates the principle of Faraday box to avoid damage by radiation.
5	The elaboration of internal walls from must integrate acrylic sheet the equipment of 6mm and aluminum sheet of 2mm.
6	The equipment must have a casing that cover the internal structure, mechanisms and components, creating the general morphology and external design.

C. *Creative Process*

The requirements of designs starts a process of proposals using technics such as brainstorm, sketches, and models. Actually the work team is in the process of establishing alternatives of the functional components and internal of the equipment, such us the positioning system, injector, collector in axes x, y, z and source of power. The first render is showing an external design (Fig. 4); it begins in the spilling machine, it can have adjustment according to the internal components.

Fig 4. Initial render of electrospinning equipment

III. RESULTS

The design and appropriate adjustment of the electrospinning equipment was defined, it worked with experimental methods of research, where was done with an observation of the phenomena participant in the process and with the abstract and creative thinking, where designed the necessary hypothesis to experience to study and thus generate validity of hypothesis. To the development of the equipment, a mechanical system was designed and set up based on three axes that generate the movement in three axes of the injection needle and the collector. The mechanical system is integrated by three independent motors, where all transmissions of movement are power screws with roll bars. It was taken as concerning existing equipment and technological printing technology 3D. A first approach of the positioning system is showing in the Fig. 5 y 6. The figure 5 shows the disposition of the external parts such as voltage, the control panel, insulating box and the external formal aspects .The figure 6 shows the mechanics of the positioning control.

The system is controlled and set up by an electronic system of software and hardware looking for more specific in each position.

Fig 5 External positioning system

This first approach of the system was able to obtain a positive outcome with respect to the movements that must generate in the three exes of the positioning system, by ensuring greater efficient in the electrospinning process.

To the formal development was taken into account concepts about ergonomics cognitive [7] and physical to give equipment that fits to the context of laboratory and nanometric science and give to the user's colors signals that provide the understanding and interaction with the equipment.

The first results in the test show a fulfillment of the requirements such as precision in the position, easy of the manipulate the system and high level of the security.

Fig 6 Mechanical and electronic system

The electrostatic radiation and magnetic fields generated in the process of electrospinning specially in the voltage source, can affect the electronic system and the users, it was applied the principle of the cage of Faraday [8], reflected in plates elaborated with material compound to a polymethylmethacrylate (acrylic and aluminum sandwich type, as structural base of the case and internal walls of the equipment.

IV CONCLUSIONS

The technological development achieved in this project has allowed developing different fields in a mutual benefit where each one plays a role essential and interconnected in the process. The industrial design proposes a great for design of this kind of equipment that allows research the optimal preparation of polymer nanofiber for tissue engineering. This system allows a major result in the research in engineering tissue.

Acknowledgment

The present study was carried out under the cooperation agreement UPTC-IIM-UNAM and supported by the projects with codes SGI 1769 and 1960 in the UPTC-Colombia.

Conflict of Interest

The authors declare that they have no conflict of interest.

References

1. Ortiz M, Albert N (2009) Nanotecnología y biomedicina Publicación Científica en ciencias biomédicas. Nova, New York.
2. Gómez-Pachón E, Vera-Graziano R, Montiel R. (2014) Structure of poly(lactic-acid) PLA nanofibers scaffolds prepared by electrospinning. IOP Conf. Series: Materials Science and Engineering 59:1-10.
3. Duque S, Rodriguez L, Lopez M. (2013) Electrospinning: La era de las nanofibras. Revista Iberoamericana de polímeros 14: 10-27.
4. Huang Z-M et al (2003) A review on polymer nanofibers by electrospinning and their applications in nanocomposites Composites Science and Technology. 63: 2223-2253
5. Cárdenas W, Gómez-Pachón E, etal. (2016) Preparation of potato starch microfibers obtained by electro wet spinning. IOP Conference Series: Materials Science and Engineering, Volume 138:1-10.
6. Lobach B (1981). Diseño Industrial. Bases para la configuración de los productos industriales. Editorial Gustavo Gili, Barcelona.
7. Romero A. (2006) Ergonomía cognitiva y usabilidad at http://www.um.es/docencia/agustinr/Tema6-0607a.pdf
8. Faraday M., (1847) The Bakerian lecture: experimental relations of gold (and other metals) to light', Philosophical Transactions of the Royal Society of London, Vol. 147: 145-181

Corresponding author:
Author: Edwin Yesid Gómez-Pachón
Institute: Universidad Pedagógica y Tecnológica de Colombia
Street: Kra 2 B No 46-30
City: Tunja
Country: Colombia
Email: edwin.gomez02@uptc.edu.co

Acquisition of Lower Limb Joint Variables by an Inertial Card System

C. Ramos[1], C. A. Collazos[1] and A. Maldonado[1]

[1] Universidad Manuela Beltrán, Bogotá, Colombia

Abstract— **The high cost and low portability of joint variables acquisition systems that are commercialy used nowadays, make this technology usually unavailable for the rehabilitation of orthopedic problems in the lower limb. This work searches implementing a more economic, portable and efficient alternative for this kind of instrumentation, by using commercial inertial cards (IMUs: Inertial Measurement Unit) and the Lab-VIEW software for registering the joint variables related to knee and ankle during human gait.**

Keywords— **IMU, human gait, gait lab, joint variables, lower limb.**

I INTRODUCTION

This work presents the implementation of an acquisition system of lower limb joint variables using IMUs.

The fundamental goal of this work is adapt and evaluate a portable system that is able of capturing kinetic and kinematic variables, by using commercial inertial cards, in order to detect gait patterns, by analysing the biomechanical behavior of a lower limb kinematic chain [1]. In [2] and [3] a comparative summary is made of different acquisition techniques for joint variables, showing their main advantages and disadvantages.

These different movement acquisition systems can be found for different purposes. In [4] a system that registers movement in three dimensions through the reconstruction of video sequences is presented. This system is applied particularly for injuries in anterior cruciate ligament. [5] and [6] present an accelerometer and inertial sensor for analysis of sports gesture in elite athlets. [7] makes a systematic review of wireless inertial sensors for biomechanical studies of lower limb. [8] also makes a review, this time of vision based systemas for recognizing actions and gestures in sports. Finally, in [9] a classification is presented for different mechanic, optic and electromagnetic systems, with the goal of refining the golf gesture.

This paper has been structured in the following manner: section II presents the hardware design. Section III shows the software design. Section IV displays the tests and results of the implemented acquisition system and Section V finalizes with the conclusions.

II HARDWARE DESIGN

The card IMU GY-85, shown in figure 1 [10], takes care of extracting the data of each one of the IMUs, sending it to the computer for processing and visualization. In this application, neither wireless transmission of data nor card integrated microcontroller usage is necessary because I2C protocol allows for the extraction of the data in a simple, fast and safe manner, by using 4 lines: two for polarization (the card works at 5V or 3.3V) and two for control (SCL and SDA, which are part of I2C protocol).

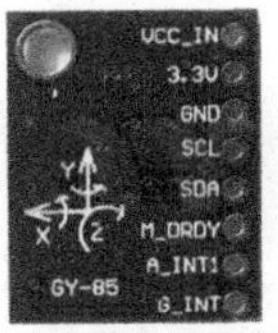

Fig. 1: IMU GY-85 Card

By having four IMUs GY-85 with I2C transmission, each working independently, it is necessary to use a signal hub, a central device that receives all the information of the peripherals. In this case, an ARDUINO UNO card was chosen, due to its compatibility with ATMEL microcontrolers and the compatibility of ARDUINO's IDE programmation platform with I2C communication protocol [10], [11], [12], [13], [14], [15]. It was necessary to include an analog multiplexor, specifically the CD4052 chip [15], to handle the data of the four IMUs on the I2C bus. The final implemented solution is shown in figure 2.

The inertial sensor has three sensors: an accelerometer (ADXL345), a gyroscope (PS-ITG-3205A) and a magnetometer (HMC5883L) which deliver data independently, in the three space axes (x,y,z). These sensors use an I2C communication protocol, which must be adequately configured in each sensors' own registers, in order to obtain the desired resolutions and sensibilities. The configuration of these registers is shown in [16], [17] and [18].

III SOFTWARE DESIGN

Two programs are being used: ARDUINO's IDE and LabVIEW, which handle specific tasks. In figure 3, each software

I. Torres et al. (eds.), *VII Latin American Congress on Biomedical Engineering CLAIB 2016, Bucaramanga, Santander, Colombia, October 26th -28th, 2016*,
IFMBE Proceedings 60,
DOI: 10.1007/978-981-10-4086-3_93

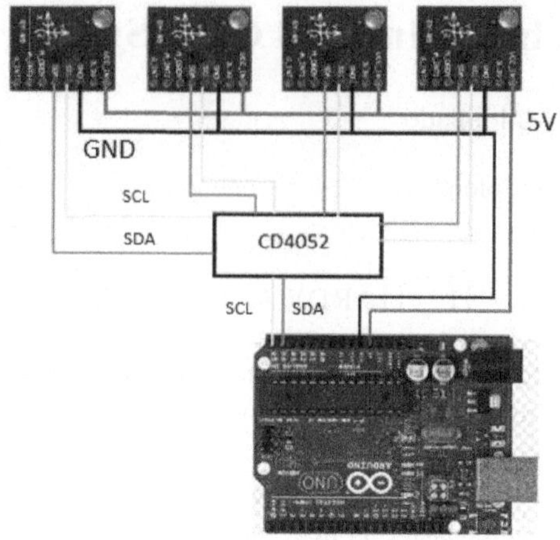

Fig. 2: Data acquisition hardware final conection diagram with 4 GY-85 IMUs.

and its main task is shown.

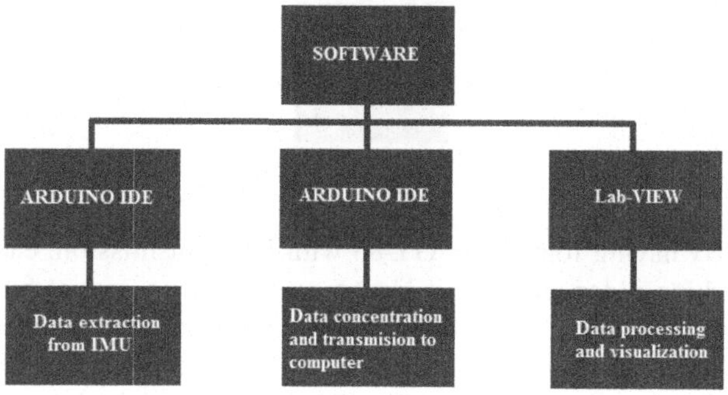

Fig. 3: Software Task Division

The sensors and the data hub use the same programming platform, which is ARDUINO's IDE. This software has the logic that differentiates and organizes the data that comes from the IMUs. It also transmits the data to the computer through serial communication and in the computer, data processing and visualization take place [19], [20],[21].

In order to extract the data from the four IMUs in a controlled manner, digital pins 11 and 12 in the ARDUINO UNO card are enabled. These pins control the selectors in the CD4052 multiplexor. The selectors choose one of the four IMU cards, adjust each of their own parameters and register the 9 values of the sensors. The program finishes by sending the information to the computer. This process is inside a loop, making a continuous register of the data given by all four cards, allowing real time visualization [22],[23].

In the Lab-VIEW tool, the tasks shown in figure 4 are performed, beginning by the serial data reception from the ARDUINO at 115200 bps. Here, the information trace is received from the sensors and each of the sensors' 9 degrees of freedom are distributed in a matrix, for a total of 36 data. The storage matrix has a size of 4 rows x 9 columns, making it easy to read and visualize each sensor individually, for posterior numeric analysis [21], [22], [23].

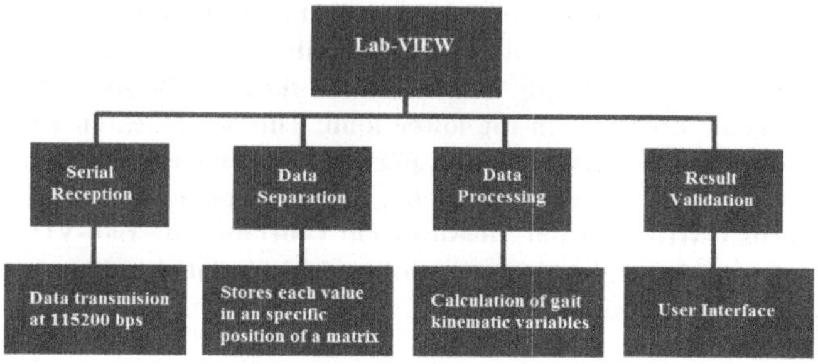

Fig. 4: Lab-VIEW tasks.

In the final user interface, the following results are given for each axis and IMU separately:

- Acceleration values normalized to $9,8m/s^2$.
- Spin speed values in [°/s].
- Normalized magnetic field in Gauss.
- Graphs of the previously listed values for better visualization.

In the kinematic data analysis, the following values are shown:

- Joint angle measurement for each card, with respect to the vertical.
- Joint angle measurement for hip, knee and ankle, with respect to the vertical (which is the reference).
- Joint angle measurement for hip in a graph, with positive and negative values with respect to the vertical.
- Joint angle measurement for knee, with positive values with respect to the vertical.
- Joint angle measurement for ankle, with positive and negative values with respecto to the horizontal [24].

IV Implemented System Validation

In order to confirm the quality of the data obtained by the implemented system, validation tests were run through the Biomechanics lab BTS-GaitLab of the Universidad Universidad Manuela Beltrán [25], [26].

The evaluation was done on the data delivered by both

joint angle measurement equipments, in knee and angle. BTS System markers (green) and inertial sensors (red) were used simultaneously, as shown in the figure 5.

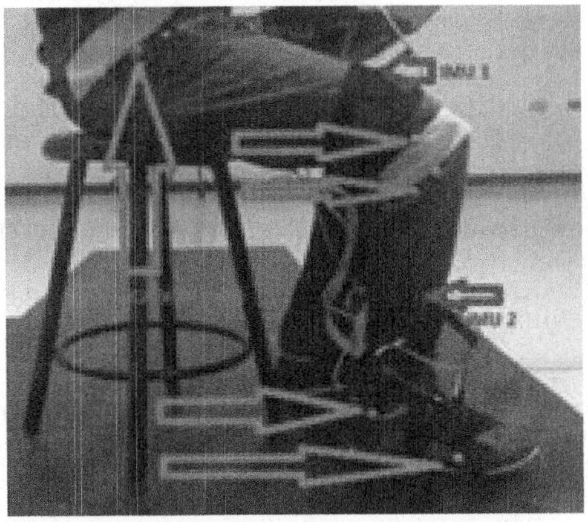

Fig. 5: Location of markers for both systems (Inertial and BTS).

Figure 6 shows the geometry that was used for the joint angle measurement, which are measured simultaneously by both systems [24].

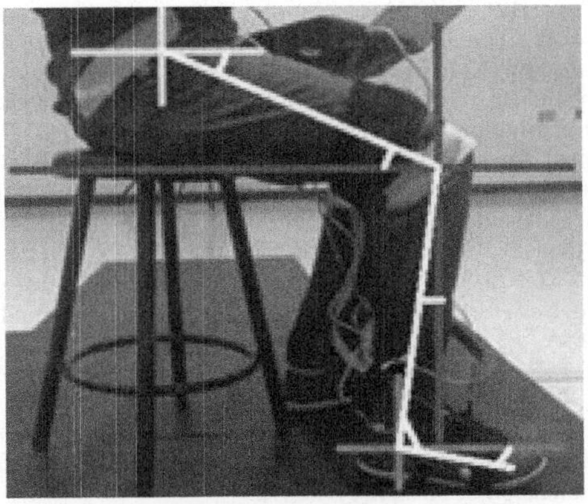

Fig. 6: Taken joint angles.

Figures 7 and 8 show the variation between knee and ankle joint angles during the measurement. Red cirles show the points where the comparison between both technologies was made.

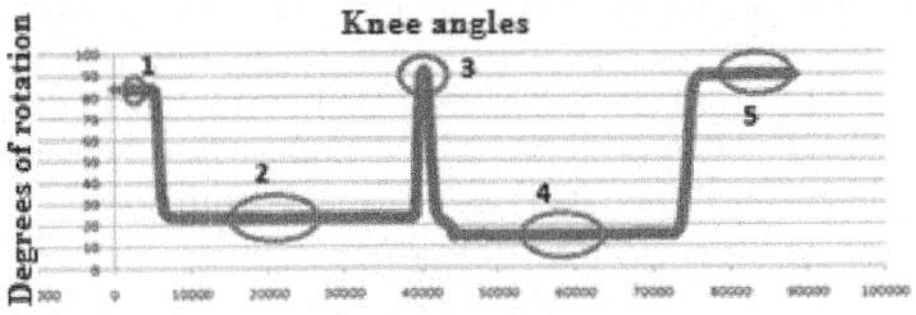

Fig. 7: Variation in knee angles graph. Taken with the BTS system.

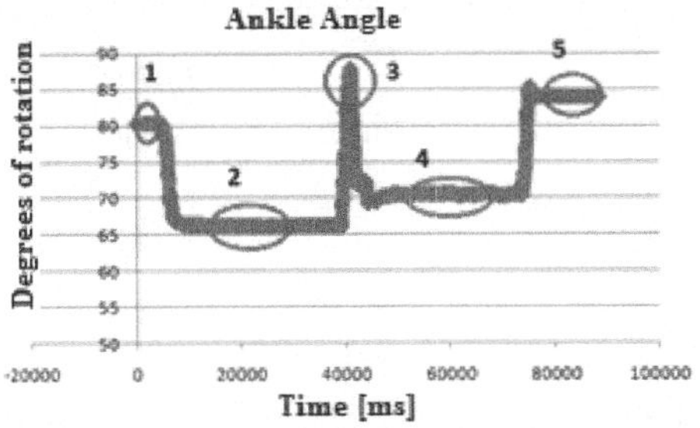

Fig. 8: Variation in ankle angles graph. Taken with the BTS system.

For analysis, the following time lapses and their respective subject movements are taken into account:

- From 0 to 5.19 s. Flexed knee. Beginning of the measurement and 1st. evaluation point.
- From 5.19 to 6.68 s. Change from flexion to extension.
- From 6.68 to 3.00 s. Knee Extension. 2nd. evaluation point.
- From 38.00 to 39.00 s. Change from knee extension to flexion.
- From 39.00 to 42.00 s. Change from knee flexion to extension. 3rd. evaluation point.
- From 42.00 to 73.00 s. Knee extension. 4th. evaluation point.
- From 73.00 to 76.00 s. Change from knee extension to flexion.
- From 76.00 to 88.00 s. Knee flexion. 5th. evaluation point.

In table 1 the values measured with both technologies (BTS and inertial) are shown. This measurements were taken in the five points shown in figures 7 y 8. The extracted data are compared from statistic parameters, like absolute error, average error and standard deviation of the measurements [27]. In the taken data of table 1, it is observed that the average error is not greater than 10%, which gives a validation of the data obtained with the implemented system.

V CONCLUSIONS

In order to make the reconstruction of the kinematic chain, when using the BTS technology, the location of the markers is of vital importance. On the other hand, when using inertial technology, its location must be respected but it is not necessary for it to be so specific. This is due to the fact that each marker delivers its own variables and any linearity issue can be eliminated in the algorithm, thanks to the rotation matrices. These matrices allow for the calculated angles for each

Table 1: BTS vs. Inertial Systems Measurement Validation

Joint	BTS Angle	Inertial Angle	Absolute Error
Knee flexion (sitting). Position 1	83.8	89.5	6.80%
Knee extension (sitting). Position 2	24.0	22.7	5.33%
Switch from flexion to extension. Position 3	80.0	87.0	7.81%
Knee extension (standing). Position 4	15.3	13.9	8.63%
Knee flexion (sitting). Position 5	89.5	98.1	9.65%
Ankle. Position 1	80.3	87.3	8.67%
Ankle. Position 2	69.6	75.6	8.56%
Ankle. Position 3	84.5	87.8	3.91%
Ankle. Position 4	70.4	76.9	9.23%
Ankle. Position 5	84.0	91.0	8.38%
Average error			7.70%
Error standard deviation			1.83

IMU to be located in any referential system.

When comparing the data that is delivered by a commercial gait laboratory (with BTS technology) with the data obtained with inertial equipment (with IMU technology), the observed variations do not surpass 10% in average error. It is important to note that the delivered data with the BTS technology are not absolute, despite the fact that their fidelity is very high, specially when working with all the markers, since the system itself generates Euler angles in the center of the joints. However, the inertial system presents great advantages, not only in price, size and easiness of test application, but also, allows the combination of stationary and mobile data.

CONFLICT OF INTERESTS

The authors declare that they do not have any conflict of interest.

REFERENCES

1. Martínez F, Gómez F, Romero E. Desarrollo de un laboratorio de marcha con integración sincrónica mediante una arquitectura en módulos. *Acta Biológica Colombiana.* 2010;15:235-250.
2. Sistemas adaptativos y bioinspirados en inteligencia artificial <http://sabia.tic.udc.es/gc/Contenidos%20adicionales/trabajos/Peliculas/Mocap/tecnol.htm> Consulted October 1st 2013, from Sabia.
3. Vera P. *Biomecnica de la marcha normal y patolgica.* Espaa: IBV 2000.
4. Krosshaug T, Slauterbeck J R, Engebretsen L, Bahr R. Biomechanical analysis of anterior cruciate ligament injury mechanisms: three-dimensional motion reconstruction from video sequences. *Scandinavian Journal of Medicine & Science in Sports.* 2007;17:508-19.
5. James D A, Davey N, Rice T. An accelerometer based sensor platform for insitu elite athlete performance analysis. in *Sensors*:1373-76 2004.
6. James D A. *The Engineering of Sport* 6ch. The Application of Inertial Sensors in Elite Sports Monitoring., :289-94. Springer 2006.
7. Fong D, Chan Y. The Use of Wearable Inertial Motion Sensors in Human Lower Limb Biomechanics Studies: A Systematic Review. *Sensors.* 2010;10:11556-65.
8. Poppe R. Vision-based human motion analysis: An overview. *Computer Vision and Image Understanding.* 2007;108:4-18.
9. Noiumkas S, Tirakoat S. Use of Optical Motion Capture in Sports Science: A Case Study of Golf Swing. in *Informatics and Creative Multimedia (ICICM), 2013 International Conference*:310-3 2013.
10. Spark Fun Electronics <https://www.sparkfun.com/search/results?term=imu> Consulted October 1st 2013, from Spark Fun Electronics.
11. Vesga J, Sinsel M. *Microcontroladores Motorola-Freescale: programación, familias y sus distintas aplicaciones en la industria.* México: Alfaomega 2008.
12. PHILIPS . The I2C Bus tecnhnologies specifications. 2000.
13. Angulo J, Romero S. *Microcontroladores PIC: diseño práctico de aplicaciones. Segunda parte: PIC16F87X, PIC18FXXXX.* España: McGraw-Hill 2006.
14. García E. *Compilador C CCS y Simulador PROTEUS para Microcontroladores PIC.* España: Marcombo 2008.
15. Semiconductor National. CD4051BM/CD4051BC Single 8-Channel Analog Multiplexer/Demultiplexer CD4052BM/CD4052BC Dual 4-Channel Analog Multiplexer/Demultiplexer CD4053BM/CD4053BC Triple 2-Channel Analog Multiplexer/Demultiplexer datasheet.
16. Devices Analog. Digital Accelerometer ADXL345L datasheet.
17. InvenSense . PS-ITG-3205A datasheet.
18. Honeywell . 3-Axis Digital Compass IC HMC5883L datasheet.
19. Sabrie S. *Sensors Handbook.* USA: McGraw-Hillsecond edition ed. 2000.
20. Wolf P R. *Elements of Photogrammetry with Applications in GIS.* USA: McGraw-Hillfourth edition ed. 2014.
21. Ruiz J. *Utilización de Labview para la Visualización y Control de la Plataforma Open Hardware Arduino.* Herramientas gráficas para la programación de Arduino. NI 2012.
22. Monk S. *30 arduino projects for the evil genius.* USA: McGraw-Hill 2010.
23. Barckhahn F, Sebastián D. *Interconectividad local de sistema de posicionamiento autónomo para una comunidad de dispositivos compactos de arquitectura modular.* Chile: Ed. Santiago de Chile 2011.
24. Díaz C A, Torres A, Ramírez J I. Descripción de un dispositivo destinado al análisis de la marcha en dos dimensiones, CineMED. *Revista EIA.* 2006;5:85-92.
25. BTS Gait-Lab <www.btsbioengineering.com>
26. Laboratorio de Biomecnica Digital, Universidad Manuela Beltrán <http://umb.edu.co/biomed/tecno.html.> Consulted October 1st 2013, from Laboratorio de Biomecnica Digital UMB.
27. Kawamura C M, Morais M C, al . Comparison between visual and three-dimensional gait analysis in patients with spastic diplegic cerebral palsy. *Gait & Posture.* 2007;25:18-24.

Analysis of the Improvement on Textural Information in Human Iris Recognition

Eduardo Garea Llano[1], Mireya S. García-Vázquez[2], Luis M. Zamudio-Fuentes[2], Juan M. Colores Vargas[3] and Alejandro A. Ramírez-Acosta[4]

[1] Advanced Technologies Application Center -CENATAV, Cuba
[2] Instituto Politécnico Nacional- CITEDI, México
[3]Universidad Autónoma de Baja California- ECITEC, México
[4]MIRAL R&D&I, USA
[1]egarea@cenatav.co.cu, [2]{msarai, lzamudiof}@ipn.mx, [3]dr.jcolores@gmail.com, [4]ramacos10@hotmail.com

Abstract— **Recent works in the area of biometrics have shown that fusion at segmentation level (FSL) has contributed to the robustness in iris recognition compared with the recognition obtained from a single segmentation. Different segmentation algorithms can produce different iris textural information from the same image. Considering FSL and combining a method for quality evaluation of images, in this paper we present the analysis of the improvement on textural information in human iris images. The images set for the experiments were four international databases, MBGC-V2 (iris video database), CASIA-V3-Interval, CASIA V4 Thousands and UBIRIS v1 (iris image databases). The Equal Error Rate is used as metric to show the improvement of the recognition rates and hence the textural information improvement.**

Keywords— **Video iris, segmentation fusion, iris texture enhancement, verification and quality.**

I. Introduction

At present time, people recognition by iris constitutes one of the main applications of the biometrics. The biometrics applications focus on increase the iris recognition rates. For this reason localize the textural information in human iris images has received attention in many researches. The errors occurred at segmentation stage are propagated on the rest of stages of the recognition process. Therefore, this stage turns out to be more complex, especially when the system acquire eye images under less controlled environments, for example: at a distance, on the move, under visible light, among others. Some international evaluations have shown the importance of the robustness of the iris segmentation in the last generation of biometric systems [1].

In 2013, is introduced the concept fusion at segmentation level FSL to combine the results of several segmentations. The aim of FSL is to improve the accuracy of the iris recognition systems [2]. The authors evaluated two fusion algorithms using the CasiaV4-Interval database; the segmentation of the iris region was manually obtained [2]. On the other hand, a previous research evaluated seven fusion methods to enhance the texture information in normalized iris templates from each video sequence of the MBGC iris video database [3]. Another approach, presented the results from the fusion of three automatic iris segmentation. They used the sum-Rule fusion algorithm in FSL after the normalization step to increase the recognition rate [4]. Based on these works, in our previous research we evaluated four fusion methods in FSL. During this evaluation, we took into account three iris images from different sensors and we used two iris segmentation algorithms. From diverse iris image source, the recognition rate increase using FSL [5]. Finally, some authors revealed auto corrective properties of augmented model fusion on masks before FSL in most of the tested cases. They proposed a scanning iris masks before the merging process. Although, the recognition rates were also increased, the authors did not take into account the quality of the mask fusion when boundaries are frequently overestimated. Furthermore, non-convexity of the mask can lead to sample points that are attributed to the wrong boundary [1].

We believe that combining a method for quality evaluation of images, automatic segmentation methods and a robust fusion method; it is possible to obtain iris images with higher content of textural information. To validate this combination, in this paper we present the analysis of the improvement on textural information in human iris recognition during an experiment of biometric verification.

The remainder of this paper is organized as follows. Section 2 presents the improvement of the textural information in human iris images; section 3 describes the experimental design, section 4 presents the results and discussion of the experimental evaluation. Finally, section 5 concludes this work.

II. Improvement of the Textural Information in Human Iris Images

All iris recognition begins with the video/image acquisition. At this point, the user typically is lead to a certain spot and position. In contrast, other users could not be oriented during this step. Either, video or images both need preprocess

I. Torres et al. (eds.), *VII Latin American Congress on Biomedical Engineering CLAIB 2016, Bucaramanga, Santander, Colombia, October 26th -28th, 2016*,
IFMBE Proceedings 60,
DOI: 10.1007/978-981-10-4086-3_94

stage to locate the iris region and extract the iris texture to recognize a person. In some cases, the captured iris texture is not appropriate for recognition. This situation affects directly to the recognition rates rate [1, 2]. Figure 1 shows the general scheme of the experiment.

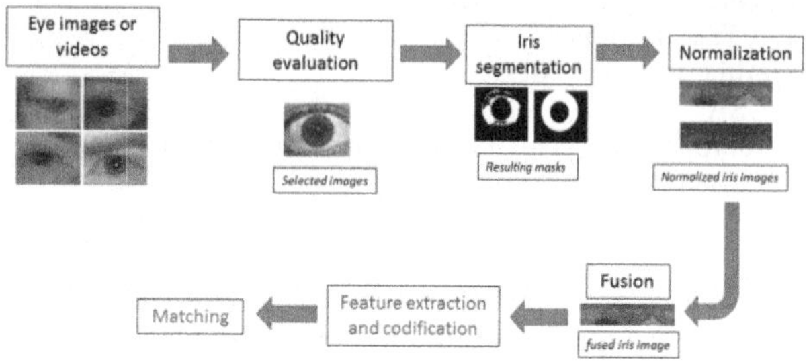

Fig. 1 Experimental iris recognition scheme

From the two images of the quality evaluation stage, the following steps extract the iris region of each image with a different segmentation method and normalize a template. These four templates separately are the inputs of the FSL.

A. Quality Evaluation Stage

The stage of quality assessment, allows debugging those images with low quality that adversely affect the recognition performance. This stage uses the convolution kernel for measuring defocus level in images. The segmentation is not required because the operator is applied to the entire image.

Due, the result from the analysis and comparison and the determined superiority in our proposal the quality evaluation stage uses the Kang & Park method, which is capable to evaluate the whole image and detect the high frequencies of the iris texture [6, 7]. As result, this step easily rejects the images with less quality and obtain the best two images

B. Segmentation Fusion level Stage

The fusion at segmentation level is able to produce images with greater information content than those obtained by a single method of segmentation.

The Laplacian pyramid fusion method used in this stage obtains a fused image by convolution a band past kernel, then each resulting image is a band pass convolved copy of its predecessor [8]. The difference between images low-pass (Gaussian filter) in successive levels of a Gaussian pyramid produce the Band-pass copies.

We propose the use only of the first level of the Gaussian Pyramid for a given image $I(x,y)$such is the normalized template. Secondly, we obtain the first level of the Laplacian Pyramid. The stage fuse the low pass coefficients by the mean value of the previously obtained Gaussian pyramids concatenated into a matrix. Thirdly, calculate the high-pass coefficients using the biggest absolute value of the obtained by the Laplacian pyramids. Finally, the fused image is the sum of the resulting images.

III. EXPERIMENTAL DESIGN

In this section, we present the results, discussion and the analysis of the improvement on textural information in human iris recognition combining image evaluation method and robust fusion method. During an experiment of biometric verification. The experiments were conducted over four international iris datasets and focused on to assess how impact into an iris recognition system. The iris collections used in our experiments were:

MBGC V2 [9] provided 986 near infrared eye videos. The acquisition of all videos were the LG2200 EOU iris capture system. The camera uses NIR illumination. Database presents noise factors: reflections, contrast, luminosity, eyelid and eyelash iris obstruction and focus characteristics. These facts make it the most appropriate for the objectives of real iris systems for uncontrolled environments. We processed the database by the experimental iris recognition scheme (fig1) obtaining our own database of 2,000 iris frames from the MBGC iris video database V2 that represents 100 subjects.

CASIA-V3-Interval [10]. All iris images are 8-bit gray-level JPEG files, collected under near infrared illumination. The CASIA-V3-Interval 1 database is composed of high quality NIR illuminated indoor images with 320×280 pixel resolution (2,639 images, 395 classes). For the experiments, we used the whole database.

CASIA V4-Thousands [10] is a collection that contains 20 000 irises images from 1000 subjects. The main sources of intra-class variations in this dataset are the eyeglasses and specular reflections.

UBIRIS.v1 [11] is a dataset comprised of 1877 images collected from 241 persons in two distinct sessions. This database contains images with several noise factors (contrast, reflections, luminosity, focusing, occlusion by eyelids and eyelashes), simulating less constrained image acquisition environments.

In the experiments, for each of the 26516 images in the iris collection we used the following automatic segmentation algorithms to obtain the two initial segmentations: Contrast-Adjusted Hough Transform (CHT) [12], a traditional sequential (limbic-afterpupillary) method based on circular Hough Transform (HT) and Viterbi-based Segmentation Algorithm (VIT) [13] that use a circular HT-based method with boundary refinement. The motivation for selecting these algorithms were their public availability as open source software for reproducibility and besides, they provide good results in the iris segmentation.

For the fusion stage, we analyze the results of three algorithms to fuse iris images and iris video frames: EM-Exponential Mean and PCA-Principal Component Analysis [14] and LP-Laplacian Pyramid.

After fusion, the well-known Daugman's rubber sheet model normalize the iris images. Then, for feature extraction and coding n the experiments, we used two state of the art methods [15]: Daugman (1993), and. Masek (1997).

Finally, we computed the Hamming distances at the verification task to obtain the results in terms of equal error rate (EER) and compared these values to analyze of the improvement on textural information.

For each results on the experimented databases, figure 2 to 5 report the results of the ERR. This shows the improvement of the textural information in the iris verification. In contrast, the EER decrease compared with the results when the single segmentation algorithms are used.

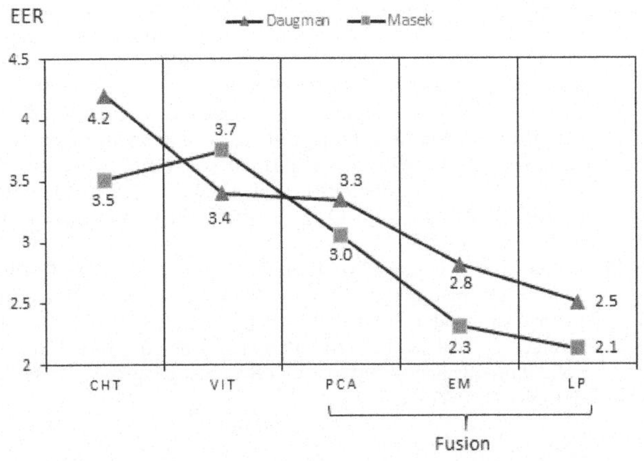

Fig. 2 Comparison of the EER values using fusion method on MBGC V2 database

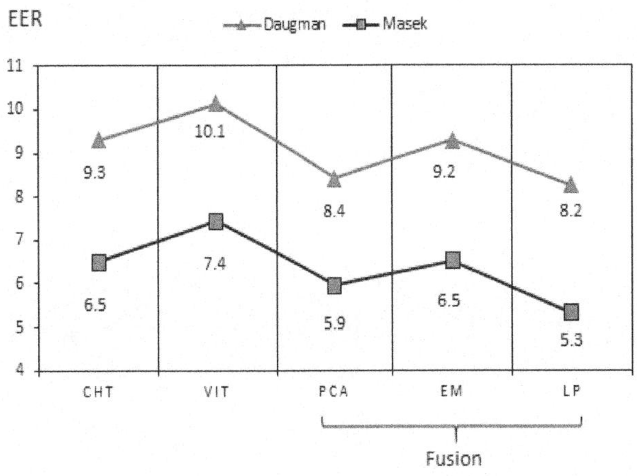

Fig. 3 Comparison of the EER values using fusion method on CASIA V3 Interval database

By analyzing the experiments, we demonstrate the improvement in the textural quality by decreasing the EER. These results compared with those obtained by individual segmentation methods confirm the best textural quality.

From the evaluated fusion methods, it is possible to see that the LP method obtains a less Equal Error Rate for the four databases.

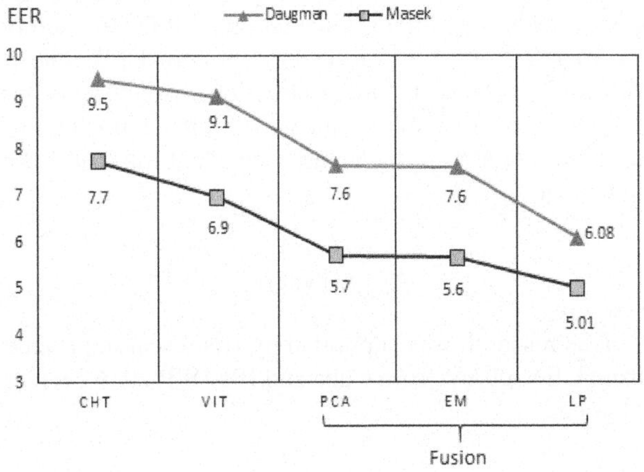

Fig. 4 Comparison of the EER values using fusion method on CASIA V4 thousands database

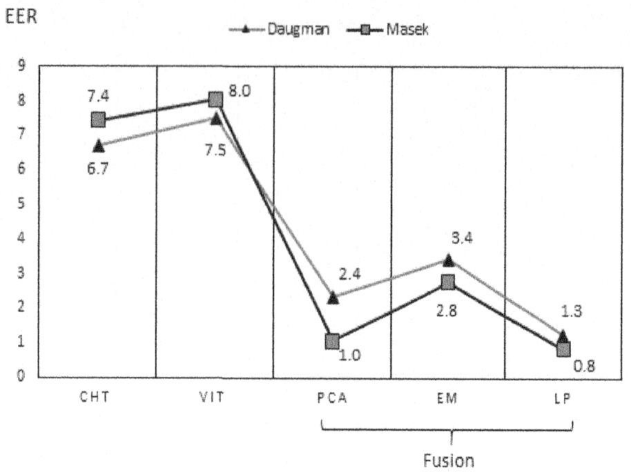

Fig. 5 Comparison of the EER values using fusion method on UBIRIS v1 database

IV. CONCLUSIONS

The analysis on the experimental results shows enriched iris texture in processed images by combining image evaluation method and robust fusion method. Consider that UBIRIS database images contains several noise factors the results shows the most significant iris texture improvement on this database. This is due to the combination of the information contained in the initial segmentations. LP fusion method was the most robust showing lesser degree of EER. The further work will focus on the new extraction characteristic methods between the different elements of human iris texture and evaluate combination of clustering algorithms at segmentation stage.

ACKNOWLEDGMENT

This research was supported by Iris research project grant from CENATAV from Cuba and IPN-SIP 2016 project from Mexico.

CONFLICT OF INTEREST

The authors declare that they have no conflict of interest.

REFERENCES

1. Wild P., H. Hofbauer, J. Ferryman, and A. Uhl. (2015) Segmentation-level fusion for iris recognition, in BIOSIG.2015 International Conference of the. IEEE, 1–6.
2. Uhl A., Wild P. (2013) Fusion of Iris Segmentation Results. CIARP 2013. Part II, LNCS 8259, 310–317.
3. Colores J. M., Garcia V. M., Ramirez A. A., Perez M. H. (2011) Iris image evaluation for non-cooperative biometric iris recognition system. MICAI 2011, Part II, LNAI 7095:499–509.
4. Sanchez-Gonzalez Y., Chacon-Cabrera Y., Garea-Llano E. (2014) A Comparison of Fused Segmentation Algorithms for Iris Verification. CIARP 2014, LNCS 8827:112–119.
5. Garea. E., Colores J.M., García M S. Zamudio L.M., Ramírez. (2015) Cross-sensor iris verification applying robust fused segmentation algorithms, IEEE. Proceedings of. ICB 2015.
6. Colores-Vargas J. Miguel, Mireya S. García-Vázquez, and Alejandro A. Ramírez-Acosta. (2010) Measurement of Defocus Level in Iris Images Using Different Convolution Kernel Methods. In JMCPR 2010, LNCS 6256:125–133, 2010.
7. Kang, B.J., Park, K.R. (2005) A study on iris image restoration. In: International Conference on Audio- and Video-Based Biometric Person Authentication, 31–40.
8. Burt, P., Kolczynski, R.. (1993) Enhanced image capture through fusion. In: Proc. Fourth Int. Conf. on Computer Vision, 173–182.
9. Multiple Biometric Grand Challenge at http://face.nist.gov/mbgc/
10. CASIA Iris databases. The Center of Biometrics and Security Research, CASIA Iris Image Database at http://biometrics.idealtest.org/
11. Proenca H. and L. A. Alexandre. Ubiris, (2005) A noisy iris image database. In Image Analysis and Processing–ICIAP 2005, 970–977.
12. Masek, L. (2003) Recognition of human iris patterns for biometric identification. Master's thesis, University of Western Australia.
13. Sutra G., S. Garcia-Salicetti, and B. Dorizzi, (2012) The viterbi algorithm at different resolutions for enhanced iris segmentation, in Biometrics (ICB), 2012 5th IAPR International Conference on. IEEE, 310–316.
14. Colores J.M., Garcia V. M., Ramirez A. A., Perez M. H., Nakano M. M. (2013) Video images fusion to improve iris recognition accuracy in unconstrained environments. MCPR2013. LNCS 7914: 114-125.
15. Rathgeb C., Uhl A., and Wild. P. State-of-the-Art in Iris Biometrics, vol. 59 ofAdvances in Information Security, chapter 3: 21–36. Springer New York.

Author: Luis Miguel Zamudio Fuentes
Institute: Instituto Politécnico Nacional, Centro de Investigación y Desarrollo de Tecnología Digital.
Street: Av. Instituto Politécnico Nacional #1310 Col. Nueva Tijuana, Baja California, México. C.P. 22435.
City: Tijuana
Country: México
Email: lzamudiof@ipn.mx

Evaluation of computer vision based objective measures for complementary balance function description and assessment in Multiple Sclerosis

Germán D. Sosa[1], Juanita Sánchez[2], Xiomary Bermúdez[3], Angélica Ramírez[1] and Hugo Franco[1]

[1] Universidad Central, 110311 Bogotá, D.C. Colombia
[2] Universidad del Rosario, Facultad de Ciencias de la Salud, 111221 Bogotá, D.C., Colombia
[3] Fundación para la Esclerosis Mútltiple (FUNDEM), 111221 Bogotá, D.C., Colombia

Abstract— **Multiple Sclerosis is a neurodegenerative–autoimmune disorder caused by a demyelination process of the axonal tracts within the Central Nervous System. This condition will increasingly affect cognitive, perceptual, motor and even vital life functions at different rates. Sensorimotor impairments have an increasing impact in the patient functionality, altering basic abilities e.g. static and dynamic equilibrium, whose preservation is an important therapeutic goal. The assessment of the state and progression of the associated disabilities is a relevant issue in the election and adjustment of a rehabilitation pathway.**

This work presents an exploratory study on the use of angular kinematic variables as objective descriptors for Multiple Sclerosis diagnostic support, comparing its behavior against the score values for a subset of five equilibrium tests within the Berg Balance Scale. These values were estimated using a computer vision-based framework integrating data from a Kinect sensor and the NiTE skeleton model. This version of the framework provides angular measures for mediolateral and anteroposterior balance. To evaluate this quantitative approach, a sample of six patients with diagnosis of Multiple Sclerosis, and able to maintain the standing position, performed all five balance tests while both mediolateral and anteroposterior angles were registered along each of them and compare its behaviour against the corresponding evaluation using the standard Berg Balance Scale scores assigned by a physiotherapist.

I INTRODUCTION

Multiple Sclerosis (MS) is a neurodegenerative autoimmune disorder of the Central Nervous System (CNS) related to disability [1] caused by the emergence of focal areas of inflammatory demyelination causing symptoms like limb weakness, gait ataxia, pharestesia, neck flexion, etc [1]. Despite MS does not significantly reduces life expectancy, it demands for life physical rehabilitation in order to mitigate increasing impairments which usually reduce the ability of the patient to carry out common-day tasks involving motor faculties. Indeed, MS is the major cause of non-traumatic neurological disability [2], with an average prevalence of 30 cases per 100.000 habitants with ages between 5 and 80 years. In Colombia, this prevalence is around 5 cases per 100.000 habitants [3]. MS is then considered an orphan disease with relevant challenges s.a. the difficulties to obtain a prompt yet precise diagnosis, lack of information for patients and families, limited or expensive therapeutic options, etc.

It has been shown that regular therapy sessions on MS patients has a positive impact on disability progression along time [4], normally requiring between 1 to 4 interventions per week depending on the current patient condition. However, it is usually accepted that as MS progresses, sessions become longer and more frequent, turning the rehabilitation process into an exhausting procedure affecting the patient adherence, which compromises its own impact. So, medical community is adopting technological tools to support both diagnosis and therapeutic intervention, integrating Natural User Interfaces (NUI) –e.g. Kinect– into the rehabilitation pathway.

The design and adjustment of therapeutic pathways require the adoption of a reliable set of anthropometric and biomechanical descriptors allowing the clinical staff to provide a correct diagnosis and to determine the current state of a specific patient along the disease progression. Up to date, physiatrists and physical therapists use several indexes and scales to describe the patient performance specifically, for equilibrium and balance. Thus, It is usual to find implementations built upon the Tinneti Scale, the Timed "Up-and-Go" or the Berg Balance Scale (BBS [5]). Such descriptors have been validated and adopted as standard clinical procedures,but their design involves subjective ordinal scores make them prone to a high inter-specialist variability and lack of precision and detail to support medical decisions.

This work presents a low-cost computer-vision based framework using the Kinect©sensor designed to acquire –in real-time– angular measures for mediolateral balance and anteroposterior balance during the execution of a subset of balance tests from the BBS. This system is intended to provide an objective description of the patient performance for each test and, then, evaluate its reliability according to its correspondence to the evaluation carried out by an expert (physiotherapist) using the BBS standard protocol.

I. Torres et al. (eds.), *VII Latin American Congress on Biomedical Engineering CLAIB 2016, Bucaramanga, Santander, Colombia, October 26th -28th, 2016*,
IFMBE Proceedings 60,
DOI: 10.1007/978-981-10-4086-3_95

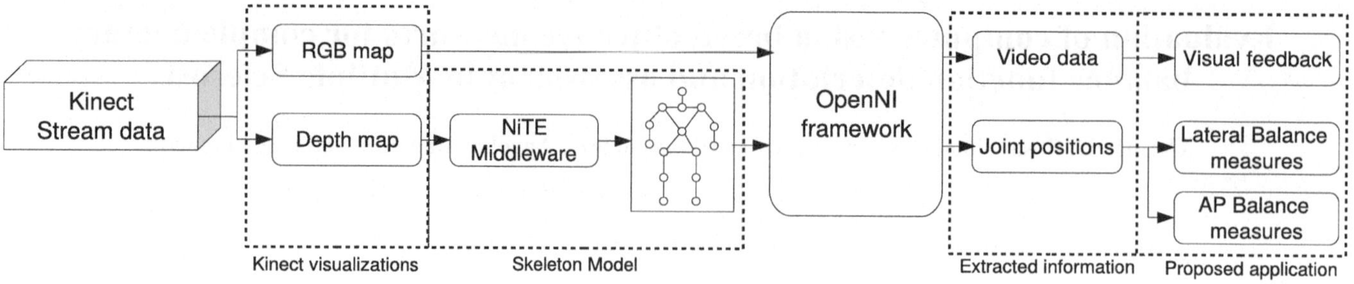

Fig. 1: Overview of the proposed method: Application developed oevr OpenNI framework takes both RGB and depth information. RGB video data is used to provide visual feedback whereas depth video is used to built the NiTE skeleton model and extract balance objective measures. Source: own elaboration

II Materials and Methods

A. Kinect for acquisition of motion data:

Kinect™ is a motion sensing device originally developed for Microsoft© as a NUI for Xbox360™ gaming console in 2010 [6], allowing players to interact directly with video games by using only their own body (and some spoken commands). Kinect design integrates a variety of sensors including a depth sensor, a RGB camera and an array of four microphones, making it capable of sensing its environment with a three-dimensional notion of present objects[7].

Kinect SDK skeleton model have proved to be robust, yielding a reasonable accuracy for gesture recognition in a wide variety of applications[8]. Given the proposed application is intended be cross-platform, the proposed framework was built using software libraries and tools released under the GNU General Public License (GPL [9]), specifically the NiTE middleware –developed by PrimeSense™– as well as the OpenNI Application Programming Interface (API). Figure 1 summarizes the architecture of the proposed framework, integrating OpenNI and NiTE to get the skeleton model of user during each test execution.

B. The Berg scale for balance assessment:

The Berg Balance Scale (BBS) [10] is a model and protocol test initially designed to assess balance ability among older people with impairments in balance function by assigning an ordinal score that describes the patient performance in 14 different movement tasks, involving several seat and stand postures, transfers picking up objects, head rotations, 360°turns, etc.

Every balance test has an ordinal score of five levels to assess the patient performance, from 0 (incapable to perform the test with no support) to 4 (complete execution of test with no exhibition of body compensations nor difficulties). The score assigned to each balance test is determined by the maximum duration achieved by the patient during the task execution whereas, for other balance tests (e.g. transferring and picking up objects), this grade is related to the level of support he/she needs to complete the assigned task.

The time required to perform all BBS tests should not exceed 15 minutes, and the total score is calculated by summing all the individual scores for each task demanding no more than 5 minutes. The outcome yielded by BBS test score may vary depending on the interpretation of physician. However, standard practices assign a low fall risk level for patients with score between 41-56 points, medium fall risk for scores between 21-40 points and high fall risk for scores between 0-20 points. Several studies [11] have shown that an improvement of 8 points in BBS reveal a genuine change in motor functions and may lead to rehabilitation.

C. Balance measurement with Kinect:

Following the proposed approach, as it is shown in Fig. 1, this framework uses the visual information provided by Kinect (i.e. RGB color and depth information) discarding sound and tilt signals. To track and register the patient movement, it is employed the skeleton model of its corresponding NiTE module. This model brings a set of 15 body joints[12] and implements real-time user tracking as well as gesture recognition –up to 30 frames per second (fps) in optimal configurations. Relevant joint positions (for angle estimation) are then recorded in raw text files, one per each test instance and patient. Video data is presented in a digital TV monitor to improve overall patient performance by including the *mirror effect* [13] during the activity execution.

Given that the estimated three dimensional position of shoulders, neck, torso and hip joints behaves like an unified and rigid structure in the NiTE model, the rotation of its projection on the mediolateral and anteroposterior planes (in frontal view) yields an approximated estimation of balance of the full body trunk. Angular values for MLB and APB are then obtained by using the estimated position of torso and neck joints in the application three dimensional space, and recorded along the duration of each balance test. Thus, balance angular measurement results are acquired as a set of time-series of angle vs time for different types of balance for every test executed by patient.

From the original set of 14 tests in the BBS, those that can be executed standing in frontal view respect to the Kinect sensor with no trunk rotations were selected for the system evaluation and variable analysis: i) unsupported standing up to 2 minutes; ii) unsupported standing with closed eyes up to 10 seconds; iii) unsupported standing with feet together up to one minute; iv) unsupported standing with one foot in front of the other (tandem position) up to 30 seconds; and v) unsupported standing on one leg for 10 seconds or more.

III EVALUATION: A CASE STUDY WITH MULTIPLE SCLEROSIS PATIENTS

To evaluate the reliability and compatibility of the obtained measures, a group of six patients diagnosed with MS from the *Fundación para la Esclerosis Múltiple* (FUNDEM, Bogotá, Colombia) were selected by applying the inclusion criteria: ambulatory people able to keep standing position without support during 10 seconds or more and able to walk by itself for several minutes, even with use of mobility aids like crutches or walkers.

Every patient included in the study had previously signed up the informed consent in the version approved by the institutional ethics committee of the Universidad Central (Bogotá, Colombia). After applying the inclusion and exclusion criteria, the resulting group comprises 4 female and 2 male subjects with ages ranging between 35 and 62 years with an average of 50.8 yo, weight range between 47 and 70 kg with an average of 59.83 kg and height range between 146 and 168 cm with an average of 158.67 cm. All those patients satisfied the inclusion criteria before test execution.

A. Balance reported measures:

Balance variation while performing different task may be observed in two directions: mediolateral balance (MLB) and anteroposterior balance (APB). Every joint location yielded by skeleton model has three components x, y (coplanar to the projection plane) and z (orthogonal into the projection plane). Given the location of torso and neck joints, MLB is measured by taking x and y components of both joints, tracing a line between them, and calculating its deviation respect to y coordinate normal vector; similarly, APB takes y and z components and calculate the deviation of the line between them against same y normal vector. Both, MLB and APB are calculated in every frame and recorded into a file during balance test realization. As the frame rate of Kinect application was consistent at 30 fps, reported results consist of five graphs containing two time series of angle vs time for MLB and APB respectively.

B. Testing protocol:

To evaluate the validity of objective measures of APB and MLB in comparison with BBS the following protocol was implemented: i) every patient had to perform all 5 five balance tests in frontal position with respect to Kinect; ii) the computer vision application records APB and MLB variations with time into a raw file; and iii) at the same time, the physiotherapist assigns the BBS-based score achieved by each patient for each test, and annotates the presence of remarkable APB or MLB events, as it is shown in Table 1, in order to check the validity of the reported measures.

C. Comparison with Berg Scale test:

Due to the different character of reported measures for BBS and Kinect angles assessing postural balance of patient for the selected balance tests, a direct, quantitative comparison requires an specific model (section V). Instead, this work presents a qualitative evaluation on the behavior and potential use of objective measures of changes of angle in APB and MLB (vs. time) as complementary assessing descriptors for diagnosis of current state of patient.

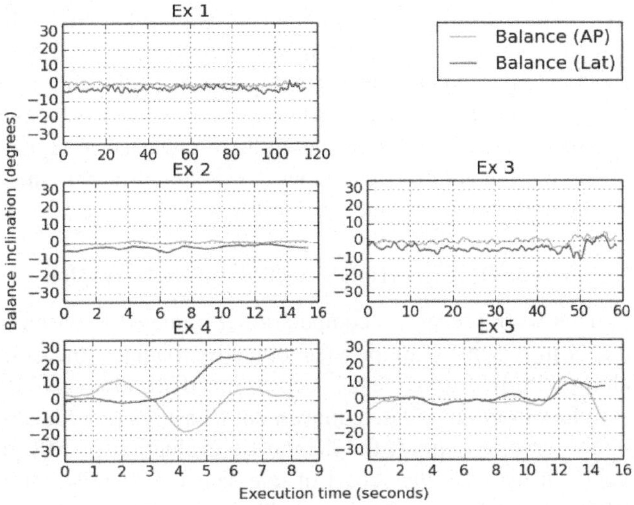

Fig. 2: Balance objective measures of APB and MLB for patient P1

IV RESULTS AND DISCUSSION

A. Berg Balance Scale scores:

Table 1 summarizes the diagnostic achieved by BBS test. During balance tests, physiotherapist used a chronometer to determine the maximum duration achieved by patients, nonetheless BBS scores does not contemplate such time value but is used to determine the performance level from 0 to 4, so

Table 1: Experimental results according to the Berg Balance Scale for each patient in the study: standing (Ex 1); standing, eyes closed (Ex 2); standing, feet together (Ex 3); standing, tandem position (Ex 4); and one-foot standing (Ex 5). Balance variation (BV) was registered when observed by physiotherapist, blank spaces are considered as holding a straight position during a balance test

Patient	Ex 1		Ex 2		Ex 3		Ex 4		Ex 5	
	Score	BV	Score	BV	Score	BV	Score	BV	Score	BV
P1	4	–	4	–	3	MLB	1	APB, MLB	3	MLB
P2	4	MLB	4	MLB	3	MLB	3	MLB	4	MLB
P3	4	–	4	–	4	MLB	3	MLB	3	MLB
P4	4	APB	4	APB	4	APB, MLB	0	–	0	–
P5	4	APB	4	APB	4	APB, MLB	1	APB, MLB	0	–
P6	4	MLB	4	AP	4	APB, MLB	1	MLB	3	MLB

cannot be considered as an full objective measure. Regarding to balance variation,taking into account that was assessed only by observation, we only consider such value as the presence or not of APB or MLB.

B. Balance objective measures:

For every patient, balance variations with time were recorded and used to plot the patient balance performance (angle vs time graphs). A visual inspection was carried out on these graphs, in contrast with the BBS-based scores for every patient and exercise in the study. Such inspection showed that graphs are useful to quantify the grade and direction of unbalance exhibited by a patient during BBS tests. Fig.2 present the angle vs time graphs exhibited for Patient P1 (See Table 1); each plot includes the variation of APB as well as MLB for an unique balance test, corresponding to the five plots (one for each test in the selected subset). Note that balance variations in Figure 2 coincide with the presence of balance in Table 1, but additionally allows to the clinical staff to quantify spatiotemporal balance variability.

V Conclusions and Future Work

This work presents a computational framework, based on the Kinect sensor, to provide objective balance measures along with traditional tests, such as the BBS. Objective descriptors from quantitative movement measures provide a reliable input to support the diagnostic evaluation of MS patients, at least for the subset of five selected exercises from BBS, alongside the ordinal score. To support this affirmation, an exploratory study was carried out with six patients diagnosed with MS, and evaluated according to the results obtained by execution of all five selected balance tests for every patient in the study.

A qualitative inspection of the BBS scores against the objective measures achieved by proposed framework shows that Kinect measures are fully compatible with scores assigned by the physiotherapist who assisted this work, regarding to variations of balance. However, a systematic offset on some measures was observed, probably induced by uncontrolled tilted positions of the sensor. Incoming versions of this framework should consider a more exhaustive preparation and calibration protocol, an implementation for the complete set of BBS exercises and a quantitative model to fusion further objective variables with ordinal scores.

VI Conflict of Interest

The authors declare that they have no conflict of interest.

References

1. Wingerchuk Dean M, Lucchinetti Claudia F, Noseworthy John H. Multiple sclerosis: current pathophysiological concepts *Laboratory investigation.* 2001;81:263–281.
2. Trapp Bruce D, Nave Klaus-Armin. Multiple sclerosis: an immune or neurodegenerative disorder? *Annu. Rev. Neurosci..* 2008;31:247–269.
3. Sáncheza J, Aguirreb C, Arcos-Burgosc O, et al. Prevalencia de la esclerosis múltiple en Colombia *Rev Neurol.* 2000;31:1101–1103.
4. Solari A, Filippini G, Gasco P, et al. Physical rehabilitation has a positive effect on disability in multiple sclerosis patients *Neurology.* 1999;52:57–57.
5. Bohannon Richard W. Objective measures *Phys Ther.* 1989;69:590–593.
6. Jana Abhijit. *Kinect for Windows SDK Programming Guide.* Packt Publishing Ltd 2012.
7. Zhang Zhengyou. Microsoft kinect sensor and its effect *MultiMedia, IEEE.* 2012;19:4–10.
8. Shotton Jamie, Sharp Toby, Kipman Alex, et al. Real-time human pose recognition in parts from single depth images *Communications of the ACM.* 2013;56:116–124.
9. Stallman Richard, others . Gnu general public license *Free Software Foundation, Inc., Tech. Rep.* 1991.
10. Berg Katherine O, Wood-Dauphinee Sharon L, Williams J Ivan, Maki Brian. Measuring balance in the elderly: validation of an instrument. *Canadian journal of public health= Revue canadienne de sante publique.* 1991;83:S7–11.
11. Steffen Teresa M, Hacker Timothy A, Mollinger Louise. Age-and gender-related test performance in community-dwelling elderly people: Six-Minute Walk Test, Berg Balance Scale, Timed Up & Go Test, and gait speeds *Physical therapy.* 2002;82:128–137.
12. Gonzalez-Pacheco Victor, Malfaz Maria, Fernandez Fernando, Salichs Miguel A. Teaching human poses interactively to a social robot *Sensors.* 2013;13:12406–12430.
13. Ramachandran Vilayanur S, Altschuler Eric L. The use of visual feedback, in particular mirror visual feedback, in restoring brain function *Brain.* 2009;132:1693–1710.

Biomedical computational reproducible research using Madagascar

Jorge E. Monsegny[1]

[1] Universidad Industrial de Santander, Bucaramanga, Colombia

Abstract— **Madagascar is an open-source collaborative software project for multidimensional data analysis and reproducible computational experiments. The purpose of this work is to show how Madagascar can be used to provide a platform to process biomedical data, like diagnostic medical images, among others, in a reproducible research fashion. Although Madagascar has been used mainly to deal with geoscientific data it can also be used as a powerful and complete framework to create biomedical scientific literature with attached data and computer recipes that simplify the verification and repetition of computational experiments by peer reviewers, fellow colleagues and even by the original authors and their succesors.**

Keywords— **Computational reproducibility, image processing, magnetic resonance image**

I Introduction

Computational experiments have become one of the main pillars of the scientific method and current scientific publications base much of their discoveries in computer simulation, data analysis and processing. Although experimentation must be reproducible in order to make the scientific community confident about the validity of scientific research results, computational experiments are hard or even impossible to reproduce by third parties, or even by the original authors, due to the lack of precise and practical documentation about the setup of these experiments, and the un-availability of the original datasets that were used to perform the experimentation.

The more practical documentation about a computational experiment are the programs used to produce their results, the complete specification of the parameters in a recipe-like form and the datasets used. However, when there are issues like floating point calculations the objective of complete reproducibility is difficult because of round off errors. As a result, a more plausible aim is to make easier for the reader to verify and extend published research [1].

However there has been some hesitancy to publish code and data accompanying research documents because of a variety of reasons including that the code is not designed for others to use, the code is a valuable intellectual property or it uses some forbidden proprietary software, among others [2]. However, there has been found that researchers that share their codes and datasets have the tendency to be more cited, at least in image processing research [3]. It also seems that reproducible research is a kind of extra-work for the authors and thus a burden, but the main beneficiaries of this effort are the authors themselves because improving the way to communicate research to other persons inevitable improves the way the researchers communicate with their future selves [1].

The Madagascar software was designed for analysis of large-scale multidimensional data and provides a framework for computational research [4, 5]. It contains a collection of computational modules, data-processing scripts and research papers [4]. Although is mainly used for exploration geophysics research it has the potential to be applied in research areas where multidimensional data processing is used like in biomedical signal processing research. It provides their users with a versatile data format and allows them to add programs and develop computational recipes that other users can easily reproduce. It also allows the creation of high quality documents where the code that make the images is embedded.

II Architecture

Madagascar Architecture is composed of four layers as Figure 1 shows. Each layer builds upon the layer immediately below. At the base the Regularly Sampled Format (RSF) is a very general and flexible data format used to store and interchange all the data that Madagascar can process. Above this level the visualization is based on a device independent data format called Vplot that can be interpreted and converted to a variety of visual representations allowing portability between platforms. At the same level, a set of Application Programming Interfaces (APIs) with very popular programming languages let the users to contribute with their own processing programs. Just above this section is the computational recipes stage where the experiments with data are defined. The very versatile tool Scons, which use the Python language, is used to organize the data dependencies between experiments. At the apex of the architecture is the publishing module based on LaTeX that generates documents with reproducible computational experiments and data attached to them.

The Regularly Sampled Format (RSF) stores n-dimensional arrays of information on disk. The number of dimensions ranges from 1 to 9 and each one of them

I. Torres et al. (eds.), *VII Latin American Congress on Biomedical Engineering CLAIB 2016, Bucaramanga, Santander, Colombia, October 26th -28th, 2016*,
IFMBE Proceedings 60,
DOI: 10.1007/978-981-10-4086-3_96

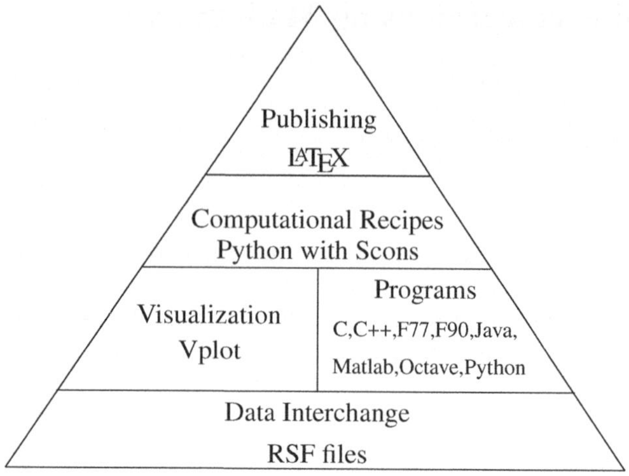

Fig. 1: Madagascar architecture.

represents a discrete or continuous domain that is sampled regularly.

The Vplot file format is capable of describing in a platform independent way the information stored in a RSF file in a variety of forms depending on the nature of this information. For example, if the information is an image, the program `sfgrey` can transform it in a raster display form.There are also programs that convert RSF files to contours, polylines, and dot diagrams, to name a few. Once there is a Vplot file display of a RSF file information, a set of platform dependent programs called pens are used to produce the actual display that the user can view. Some pens display an interactive window that the user can interact to while others write to graphics format files like pdf, tiff and svg, for example.

A large base of academic and industrial users have taken advantage of the programming APIs that Madagascar offers to contribute with processing programs. When a processing technique is lacking in Madagascar because it is new, for example, a user can easily add it to the official Madagascar release so that other users can start using it. There is a philosophy of making Madagascar programs that perform only one action but that perform it well and to build the complexity in the computer recipe. The programing APIs allow the users to read and write RSF files from the standard input and output so the programs can be connected between themselves with Unix pipes.

Computer recipes can have very complex dependency relationships with experiments that need a very precise sequence of other experiments to be executed before they can produce their results. These relationships are described by triplets composed of a set of targets, a possible empty set of sources and a command that builds the targets based on the sources. The relationships are expressed as a Python file called Sconstruct that the software construction tool Scons can interpret. Scons avoids the execution of commands when their targets have been build before and there is no change in their sources or the command itself. This way, expensive calculations are not repeated when there is no necessity to do it.

The commands themselves are composed of computer programs joined together using Unix pipes so that the output of one of them is connected directly to the input of the following one. Not only Madagascar programs are allowed to form commands, but external programs and other utilities are also allowed. As an example, in the next section, some external programs are used to compose commands that read Dicom files.

During the publishing part a LaTeX document that includes the required figures generated by the computational recipes is written by the user. Using a small Sconstruct file, Scons performs the compilation of this document and transforms it to a pdf file or a html file.

III Computer recipes

Computer recipes in Madagascar are stored in files called Sconstruct that are executed with the Scons tool. The recipe downloads an online Dicom dataset, performs the conversion from this format to RSF files, applies a very simple processing step and displays the resulting dataset.

A Getting the dataset

The next code fragment is the beginning of a Sconstruct file. The first line uses the Python directive `import` to get the proper Madagascar header definitions. They are located in the package `rsf.proj`. After a few constant definitions that will be clear later it is invoked the first Madagascar function, `Fetch`.

```
from rsf.proj import *

n1 = 256
n2 = 256
first=1
last=41
step=2

Fetch('ANONYMIZE.zip',server='http://www.osirix-
     viewer.com/',top='/',dir='datasets/DATA/')
```

The purpose of Fetch is to get from an on-line repository the data that is going to be used by the script. If this data is always available anyone with access to this Sconstruct will be able to execute the computer recipe. In this example a public dataset provided by Osirix Imaging Software is used

[6]. The complete path of the dataset is http://www.osirix-viewer.com/datasets/DATA/ANONYMIZE.zip

B Generating the RSF files

The downloaded dataset is a zip file that contains various subdirectories and Dicom files inside of them. In this exercise we are interested in files contained inside subdirectory `ANONYMIZE/BRAIN/Ax PD&T2 FSE - 4/`. The computer recipe first extracts a subset of these Dicom files, then converts them to raw data files and finally creates the RSF files. Currently, Madagascar does not have tools to read zip files or Dicom files, but the computer recipe can invoke other tools to perform these tasks. This is accomplished with the Madagascar function `Flow` that appears in the next code fragment.

```
for i in range(first,last,step):
    stri = "%02d"%i
    Flow('IM-0001-00' + stri + '.dcm', 'ANONYMIZE
        .zip',
         '''
         unzip -jo ${SOURCE}
         'ANONYMIZE/BRAIN/Ax PD&T2 FSE - 4/IM
             -0001-00''' + stri + '''.dcm'
         ''')
    Flow('IRM' + stri + '.dcm', 'IM-0001-00' +
        stri + '.dcm', 'gdcmconv -w ${SOURCE} ${
        TARGET}')
    Flow('IRM' + stri + '.png', 'IRM' + stri + '.
        dcm', 'dcm2pnm ${SOURCE} ${TARGET}')
    Flow('IRM'+ stri + '.gray', 'IRM' + stri + '.
        png', 'convert - ${TARGETS}')
    Flow('IRM'+ stri,'IRM'+ stri + '.gray',
         '''
         echo
         in=$SOURCE
         data_format=native_uchar
         esize=1
         n1=%d
         n2=%d
         d1=1
         d2=1
         o1=1
         o2=0 | sftransp | sfdd type=float
         ''' % (n2, n1))
```

The code fragment is a Python for loop that iterates between `first` and `last` using intervals of size `step`. These constants were defined in the first code fragment as 1, 41 and 2, respectively, because only files with names `IM-0001-00AB.dcm`, where `AB` is two digit odd number between 1 and 41 with an optional leading zero are going to be extracted.

Inside the loop the Madagascar function `Flow` is invoked several times. This function is in charge of performing the computational experiments and has three parameters separated by commas. The first is a set of target files that are going to be generated, the second is a set of source files and the third is the command that uses the sources to generate the targets. By way of illustration, the first `Flow` function invocation in the code fragment extracts the Dicom files from the zip file using the Unix application unzip. In each iteration the target is a file called `IM-0001-00AB.dcm`, the source file is the downloaded dataset ANONYMIZE.zip and the command that produces the target file from the source file is:

```
unzip -jo ANONYMIZE.zip \
'ANONYMIZE/BRAIN/Ax PD&T2 FSE \
- 4/IM-0001-00AB.dcm'.
```

Following the same idea, the next `Flow` function invocation decompresses each Dicom file using the tool gdcmconv [7] producing files `IRMAB.dcm`:

```
gdcmconv -w IM-0001-00AB.dcm IRMAB.dcm
```

where parameter `${TARGET}` is also replaced by its value.

The third `Flow` converts each decompressed Dicom file to a png file using the program dcm2pnm [8]:

```
dcm2pnm IRMAB.dcm IRMAB.png
```

The fourth `Flow` employs the software convert [9] to transform the png file to a raw data file:

```
convert - IRMAB.gray < IRMAB.png
```

Now each IRMAB.gray file contains the raw data of one of the original Dicom files. The last `Flow` invocation is a combination of three programs joined by Unix pipes. The first is the Unix echo command which creates a text file with a pointer to the raw data file. Due to the fact that this text file is a RSF file, it must also contain info about the dimensions of the Dicom file. These info is specified by the `n1`, `o1`, `d1`, `n2`, `o2` and `d2` keywords. Keywords `n1` and `n2` take their values from the respective constants. The second program joined by the pipe is the Madagascar program sftransp that is called to transpose the two axes of the RSF file. The third is the Madagascar sfdd program that transforms the data to float type for further processing. At the end of the loop there is a RSF file for each extracted Dicom file.

C Processing

With the Dicom files transformed to RSF files it is possible to perform on them some processing tasks. For example, a Sobel filtering can be applied to a single image using the Madagascar program `sfgrad2` as the following code fragment shows. The next subsection will show how to visualize the filtering result.

```
Flow('IRM21sobel', 'IRM21', 'sfgrad2')
```

D Visualization

Visualization can be accomplished by using two functions, `Plot` and `Result`. The difference is that the first one generates a temporary image while the second one creates a final image that can be inserted in a document. The following fragment shows how to visualize one of the extracted Dicom files and the Sobel filtering result using function `Result`.

```
Result('IRM21', 'sfgrey pclip=100 gainpanel=a
    allpos=y screenratio=1 scalebar=y title="
    IRM21"')

Result('IRM21sobel', 'sfgrey pclip=100 gainpanel=
    a allpos=y screenratio=1 scalebar=y title="
    Sobel IRM21"')
```

The `Result` function needs three parameters separated by commas. In a similar way to `Flow`, the first parameter is the target, the second is the source and the third is the command that generates the target using the source. When the source is omitted, as in the code above, it is supposed to have the same name as the target. The command to generate the image uses the Madagascar program `sfgrey` with some parameters to control the image that is shown in Figure 2.

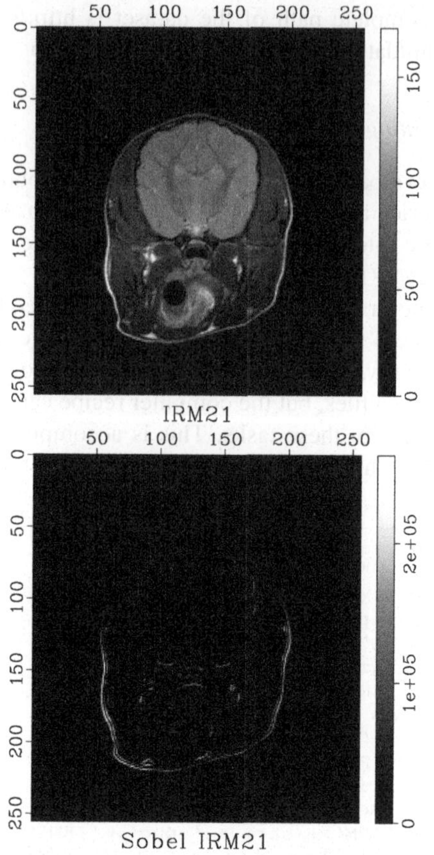

Fig. 2: Visualization of one of the extracted Dicom files and its Sobel filtered version.

IV CONCLUSIONS

This paper showed how to use Madagascar, a software for multidimensional data analysis mainly used in geoscientific research, as a reproducible computational research platform that can be used in biomedical processing research. The main reason behind using a software platform made for a seemingly unrelated research area is that it is well established and the problems that solves are not very far from the problems in biomedical image processing.

Although Madagascar does not have tools to process biomedical data format files like Dicom files, it was showed how to incorporate other tools into Madagascar computer recipes to access the data inside the Dicom files and transform them into RSF files that Madagascar can process. This also allowed the processing and displaying of the original Dicom files using Madagascar utilities.

The more important characteristic is that because the data used is a public online dataset and the computer recipes are available all computational experiments performed in this paper are easily reproducible by a reader who has Madagascar and the other open tools.

CONFLICT OF INTEREST

The author declare that they have no conflict of interest.

REFERENCES

1. Fomel S.. Reproducible Research as a Community Effort: Lessons from the Madagascar Project *Computing in Science Engineering.* 2015;17:20-26.
2. LeVeque R.. Top Ten Reasons To Not Share Your Code (and why you should anyway) *Siam News.* 2013.
3. Vandewalle P.. Code Sharing Is Associated with Research Impact in Image Processing *Computing in Science Engineering.* 2012;14:42-47.
4. Fomel S., Sava P., Vlad I., Liu Y., Bashkardin V.. Madagascar: open-source software project for multidimensional data analysis and reproducible computational experiments *Journal of Open Research Software.* 2013;1:e8.
5. Madagascar . Madagascar http://www.ahay.org/ 2016.
6. Osirix . Osirix Imaging Software http://www.osirix-viewer.com/ 2016.
7. Gdcm . Grassroots DICOM library http://gdcm.sourceforge.net/ 2016.
8. Dcmtk . DCMTK - DICOM Toolkit http://www.dcmtk.org/dcmtk.php.en 2016.
9. ImageMagick . Image Magick http://www.imagemagick.org/ 2016.

Muscle strain field estimation using object tracking in high definition video sequences

M.A. Zamora[1], H. Franco[2], Q. Fang[3], S. Shefelbine[3], A.Taylor [4,5] and A. Ramirez-Martinez[1]

[1] Departamento de Ingeniería Mecánica, Universidad Central, Bogotá, Colombia
[2] Departamento de Ingeniería de Sistemas, Universidad Central, Bogotá, Colombia
[3] Department of Biomedical Engineering, Northeastern University, Boston, USA
[4] Department of Physical Medicine and Rehabilitation, Harvard Medical School, Boston, Massachusetts, Boston, USA
[5] Cardiovascular Research Laboratory, Spaulding Rehabilitation Hospital, Boston, Massachusetts

Abstract— **Measuring muscle activation is a valuable tool for the determination of diagnostic-focused parameters and the assessment of therapeutic intervention effectiveness. This paper presents a low cost, portable system for estimating muscular strain by measuring local superficial deformation of muscle tissues. Using stereo vision, the 3D positions of a set of markers, placed on the belly of a particular muscle, are estimated to calculate the relative distance between neighbor markers in order to approximate the superficial muscle strain field. This technique was tested on a female subject to measure the deformation of her right quadriceps during a electrical stimulation applied as a 20 Hz train of pulses. The current intensity is from 5 mA to 45 mA, with increasing steps of 5 mA. To validate the system results, the obtained strain fields are evaluated according to their consistency to the force signals resulting from the contractions induced by applied stimuli.**

Keywords— **Stereo vision, video tracking, Superficial strain, GoPro Action Camera, Electrical stimulation, Muscle deformation.**

I INTRODUCTION

In vivo characterization and assessment of muscle function and its contractile properties is of critical importance in the diagnosis and therapeutic intervention of patients of a wide range of diseases and conditions involving motor impairment.

Objective measurement of muscular strain is a relevant value in order to comprehensively describe the muscle elasticity. Classical methods quantify muscular deformation using isolated samples of animal tissue, measuring the distance between terminal clamps [1, 2]. In vivo experimental studies with animals [3] use image based-techniques s.a. tracking of reflective marker arrays, which are directly attached to the surface of a peeled muscle. This approach provides values for spatial elastic displacements representing the muscular strain.

Nevertheless, for clinical applications those invasive techniques are not suitable and non–invasive measurement approaches have been proposed. In this context, several works [4, 5] used ultrasound imagery to measure 2D and 3D tendon and muscle deformation. However, in the case of 3D deformation assessment, the tight field of view of common ultrasound transducers could significantly increase the complexity of the underlying 3D reconstruction methods. On the other hand cine-phase contrast magnetic resonance imagery (cine PC-MRI) yields accurate spatial measurements for dynamic soft tissue tracking and deformation measurement [6], but it is constrained to the limitations of magnetic resonance acquisition (limited space, absence of metallic instruments, etc.).

Some optical approaches based on stroboscopic illumination, use structured light imaging to reconstruct the shape of 3D objects. Particularly, by means of structure light 3D surfaces of skin, muscle, and fat tissues are obtained accurately in [7], and a patient-specific foot model is obtained in [8]. Furthermore, using 3D reconstructions, deformations can be estimated along time by calculating the difference between local features of the 3D surfaces. This approach was applied to study the heart deformation [9]. In fact, those methods become robust when a 3D photogrametry correlation technique is added [10], allowing an accurate estimation of 3D deformation of complex objects (e.g. thighs or shins).

3D photogrametry is based on stereo vision, where at least two cameras capture the same stage from different viewpoints to represent the 3D scene. Therefore, this non-contact technique provides a full-field 3D measurement of displacement or deformation of the studied sample giving an accurate approach in case of non-homogeneous and anisotropic materials typical in soft tissues [2].

The aim of this work is to estimate the full strain field of the Quadriceps muscle in isometric contractions induced by electrical stimulation. For this purpose, a low cost, portable, simple and non invasive system was implemented based on object tracking of high definition video sequences.

II METHODOLOGY

The experimental setup consists in two main tasks: i) muscular deformation video acquisition and ii) 3D stereographic-based measurement of local muscular stretching.

I. Torres et al. (eds.), *VII Latin American Congress on Biomedical Engineering CLAIB 2016, Bucaramanga, Santander, Colombia, October 26th -28th, 2016*,
IFMBE Proceedings 60,
DOI: 10.1007/978-981-10-4086-3_97

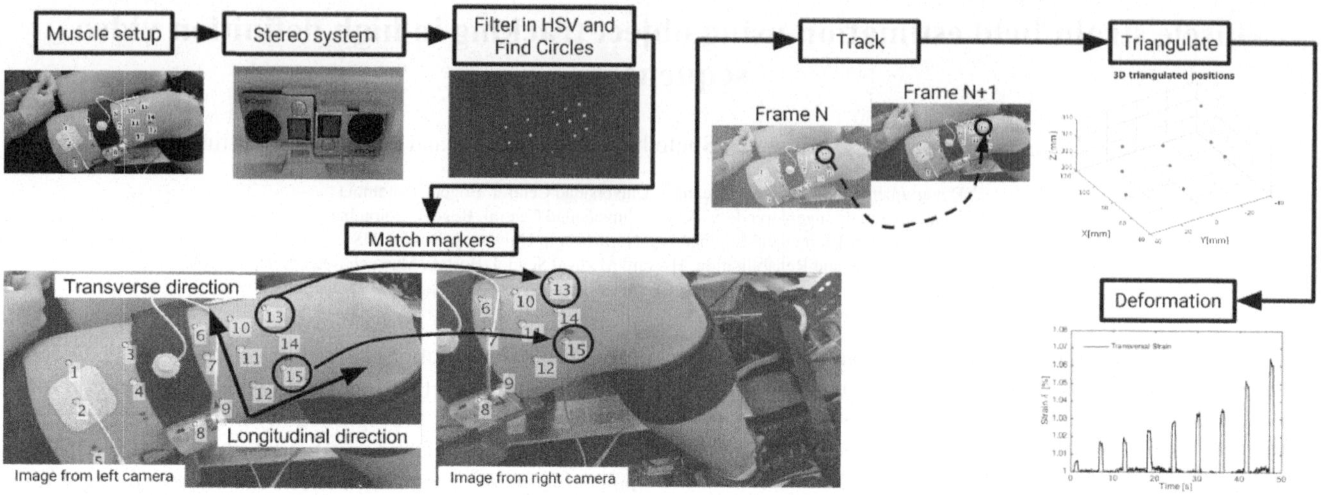

Figure 1: Stereo vision algorithm starts from the acquisition system, and passes through a preprocessing stage before using the matching and tracking strategies, to produce the 3D position of each marker and the strain curves of neighbor markers. Lower limb images have been horizontally clipped

A Muscular deformation video acquisition

To acquire video sequences of muscular deformation, a subject (female) is asked to seat on the adapted chair with a seat locked in a fixed position (knee angle: 120°, hip angle: 100°). In this position, the lower leg is strapped to the chair structure with a brace whose axis coincides with the subject's shin axis, as it is shown in Figure 1 (Muscle setup).

After the subject leg is strapped to the chair, a sequence of electrically induced isometric contractions is applied to the quadriceps by two active electrodes placed over its motor points. Such sequence is implemented as train of pulses of 20 Hz with increasing current intensities starting at 5 mA and stopping at 45 mA, with steps of 5 mA. Each train of pulses has a duration of 350 ms, while the resting time between two consecutive trains of pulses is 5 s long.

A set of circular–shaped green markers are placed in the belly of the muscle, aligned in a grid spanning along the longitudinal and transverse directions, with a minimum spacing of 2 cm between each row or column. Given the rectangular topology of the grid, each marker is assigned with 4 neighbors, two in its same row (left and right markers) and two in its same column (top and bottom markers). This setting allows the estimation of local strains describing the geometric distortion yield by the electrical stimuli applied to the muscle.

The superficial strain of the muscle is then measured through a custom-built stereo video system composed of a pair of action cameras, to record at 1080p 120 fps (HD 4K Action Camera - GoPro Hero 4 BLACK). Both cameras are mounted on a 3D printed support (Figure 1) to avoid relative motions of the cameras. This custom made stereo system is firmly supported by a tripod located at 40 cm in front of the thigh to provide a full coverage of the thigh surface and to maintain the cameras steady. In order to synchronize the two video sequences, the turn-on of a led is recorded at the beginning of the trial.

B 3D muscular deformation measurement by video tracking

Stereo calibration is a fundamental step to build a stereo system. For this application, such procedure was carried out with a checkerboard pattern using the technique proposed in [11]. The accuracy of the calibrated stereo system was validated by computing the mean re-projection error, which is shown to be lower than 2 pixels.

A second step towards building a stereo system is to match the images of the two cameras. Matching is the process of identifying the same object in two different images that are captured at the same time from different perspectives. Conventional matching approaches using a descriptor (feature vector) such as SURF [12] or DAISY [13] failed in this application due to the homogeneous texture of the thigh skin.

To overcome this issue, a set of green circular stickers placed on the surface of the muscle are selected to facilitate streamlined image processing. The centroids of these markers are estimated using a circle Hough transform. Exploiting the pattern of the markers, a heuristic matching process is performed. Computing a radius from an arbitrary origin to the center of each marker, the markers are paired according to the length of such a radius. Assuming that the tracking of each marker is done robustly enough, so that no marker will be lost, the described heuristic matching is executed only once at the beginning of the video sequence. In Figure 1 radii where computed relative to the position of marker 1 (leftmost

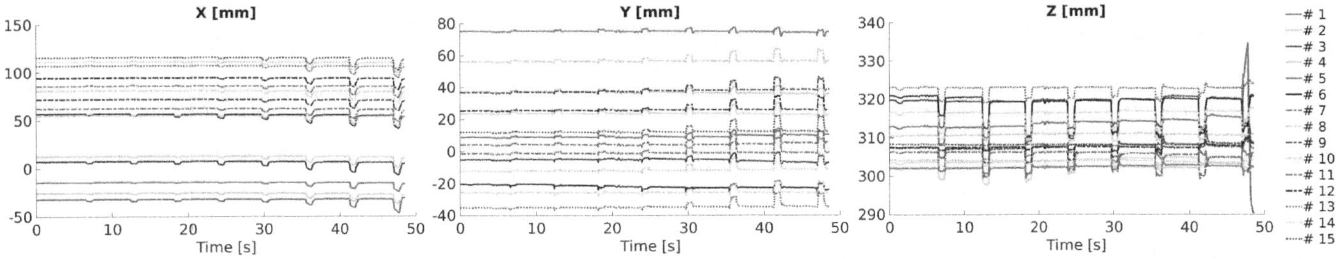

Figure 2: Triangulated x,y, and z coordinates of each marker, with respect to the frame of reference of the left camera, for 20 Hz trains of pulses and current intensities ranging from 5 mA to 45 mA with 5 mA steps.

marker).

The KLT algorithm was used to track the markers and together with the previously described matching process it is possible to use epipolar geometry to triangulate the 3D position of each marker [14]. By tracking stationary skin markers, a metric to quantify the position noise inherent to the trackers is computed, obtaining a standard deviation lower than 0.15 mm in the x, y and z directions.

The muscle force produced by isometric activation of the quadriceps is transmitted to the foot where a force plate is located. The signal from the force plate (sample rate of 100 Hz) is acquired by a USB Data Acquisition External Bus (Model USB-6211DAQ, National Instruments). This signal supports the validation of the muscular deformation estimation.

III RESULTS

By stereo triangulation, the 3D position of each marker was obtained respect to a frame of reference whose origin is located at the center of the left camera field of view and whose z–axis points away while its y–axis points downwards. For a stimulus frequency of 20 Hz and an increasing intensity (from 5 to 45 mA) the peak displacements in x, y and z directions are 14.4 mm, 10.5 mm, and 9.6 mm, respectively. Furthermore, Figure 2 presents the 3D tracking algorithm performance, which exhibits a good behavior for most of the experimental configurations.

Using the tracking algorithm described in section B, the 3D position of each marker was estimated for each video frame. The euclidean distance between neighbor markers was computed and normalized in terms of their initial distance, to obtain a strain curve for each pair of neighbor markers and for each stimulus. Figure 3a,b shows the deformation between representative marker pairs for both longitudinal (#10–#11) and transverse strain (#12–#15). In this Figure, the positive direction corresponds to a contraction displacement. The muscle force measured by the load cell shows an asymptotically exponential response starting at 5 N for 10 mA and raising to 190 N for 45 mA, as it is shown in Figure 3c.

IV DISCUSSION

The proposed method presents satisfactory results for all but one experimental configuration (Figure 2), proving that 3D photogrametry is a suitable approach to implement video tracking techniques supporting non–invasive muscular strain in clinical, therapeutic and training applications, among others. The muscular strains estimated with this method (Figure 3a,b) are consistent with the Poisson's effect and the expected behavior for an isometric muscle contraction.

The system implementation yields a temporal signal whose strain peaks accurately fit the force responses induced by the electrical stimuli on the muscle, as they are acquired into the system from the dynamometer (Figures 3a,b,c). On the other hand, and in consistency with previous works [2], the strain increases at an asymptotically exponential rate with the electrical stimulus intensity (Figure 3e), resembling the behavior of the force curves (Figure 3f). It is also worth noticing that for an isometric contraction, significantly greater transverse strains than longitudinal strains are observed for lower current stimuli (Figure 3f), but the longitudinal strain increases at a faster rate for stimuli above 35 mA, surpassing the transverse strain response.

V CONCLUSIONS AND FUTURE WORK

This work presents a low-cost, simple, and noninvasive stereo–imaging system to approximate the non-rigid muscular strain field due to electrical stimuli, based on video tracking of local muscle surface displacements.

In this study, the sliding motion between the muscle-fat tissue and the skin is neglected. This limitation should be addressed in future versions of the system. Furthermore, long local displacements (for high intensity stimuli) could induce tracking errors (specifically in the z coordinate for markers #1 and #13, Figure 2), pointing to further triangulation algorithm enhancements.

The experimental results for the strain field could support the validation of simulation models of the muscular activation [15].

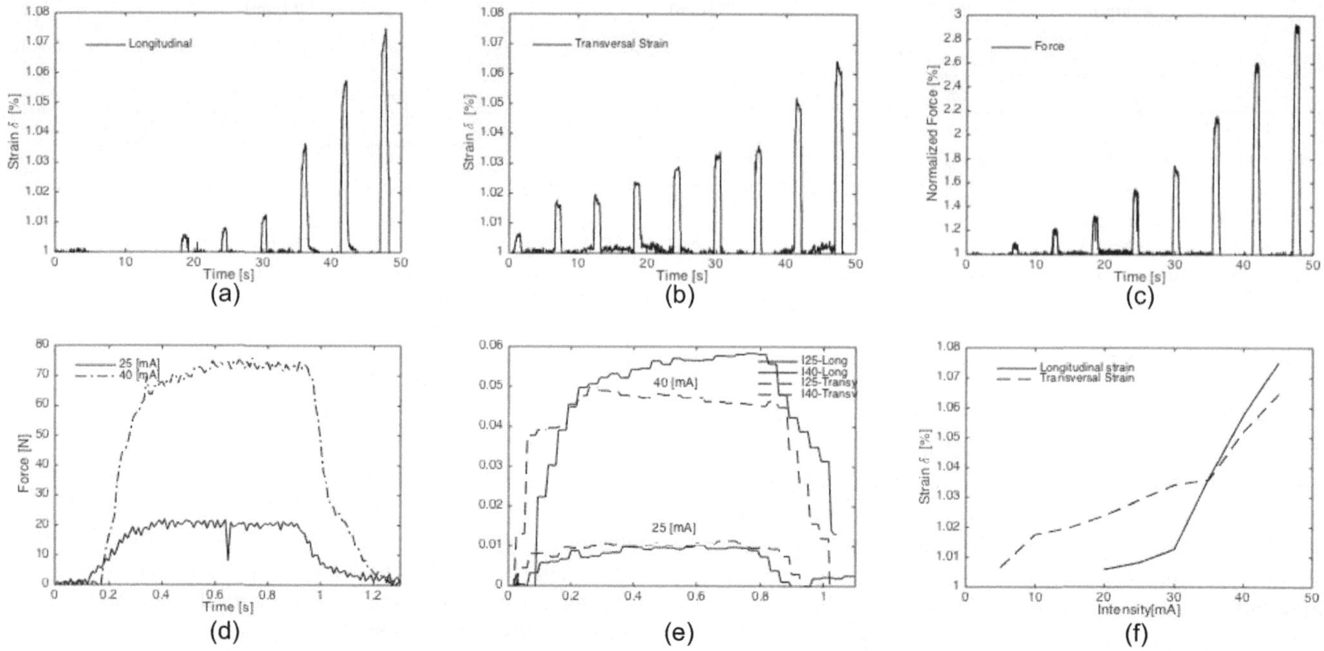

Figure 3: (a) Strain in the longitudinal direction. (b) Strain in the transverse direction. (c) Force registered by the load cell. (d) Force detail for the 25 mA and the 40 mA stimuli. (e) Strain in the longitudinal and transverse direction for the 25 mA and the 40 mA stimuli. (f) Longitudinal and transverse strain vs stimulus intensity.

Conflict of Interest

The authors declare that they have no conflict of interest.

Acknowledgments

This research is part of the project *computational model of the skeletal muscle adaptation during functional electrical stimulation* (20306009006), supported by Cluster of Convergent Technologies of the Central University, Colombia. In addition, We thank Adina Draghici and Glen Piccard for their assistance with the methods and patient recruitment.

References

1. Gordon a. M., Huxley a. F., Julian F. J.. The variation in isometric tension with sarcomere length in vertebrate muscle fibres. *The Journal of Physiology.* 1966;184:170–192.
2. Ramírez A., Grasa J., Alonso A., et al. Active response of skeletal muscle: In vivo experimental results and model formulation *Journal of Theoretical Biology.* 2010;267:546–553.
3. Donkelaar C C Van, Drost M R. Skeletal muscle transverse strain during isometric contraction at different lengths *Science.* 1999;32:755–762.
4. Lopata Richard G P, Dijk Johannes P, Pillen Sigrid, et al. Dynamic imaging of skeletal muscle contraction in three orthogonal directions. *Journal of applied physiology (Bethesda, Md. : 1985).* 2010;109:906–915.
5. Prager R W, Ijaz U Z, Gee a H, Treece G M. Three-dimensional ultrasound imaging. *Proceedings of the Institution of Mechanical Engineers. Part H, Journal of engineering in medicine.* 2010;224:193–223.
6. Sheehan F T, Zajac F E, Drace J E. In vivo tracking of the human patella using cine phase contrast magnetic resonance imaging. *Journal of biomechanical engineering.* 1999;121:650–6.
7. Wilm J, Jensen SHN, Aanæs H. Structured Light Scanning of Skin, Muscle and Fat 2015.
8. Lochner Samuel J., Huissoon Jan P., Bedi Sanjeev S.. Development of a patient-specific anatomical foot model from structured light scan data *Computer Methods in Biomechanics and Biomedical Engineering.* 2014;17:1198-1205. PMID: 23181631.
9. Laughner Jacob I, Zhang Song, Li Hao, Shao Connie C, Efimov Igor R. Mapping cardiac surface mechanics with structured light imaging. *American journal of physiology. Heart and circulatory physiology.* 2012;303:H712–20.
10. Schmidt T, Tyson J, Galanulis K. Full-field dynamic displacement and strain measurement using advanced 3D image correlation photogrammetry: part 1 *Experimental Techniques.* 2003;27:47–50.
11. Zhang Z.. A flexible new technique for camera calibration *IEEE Transactions on Pattern Analysis and Machine Intelligence.* 2000;22:1330-1334.
12. Bay Herbert, Tuytelaars Tinne, Van Gool Luc. *SURF: Speeded Up Robust Features*:404–417. Berlin, Heidelberg: Springer Berlin Heidelberg 2006.
13. Tola E., Lepetit V., Fua P.. DAISY: An Efficient Dense Descriptor Applied to Wide-Baseline Stereo *IEEE Transactions on Pattern Analysis and Machine Intelligence.* 2010;32:815-830.
14. Hartley R. I., Zisserman A.. *Multiple View Geometry in Computer Vision.* Cambridge University Press, ISBN: 0521540518second ed. 2004.
15. Robertini Nadia, Neumann Thomas, Varanasi Kiran, Theobalt Christian. Capture of arm-muscle deformations using a depth-camera *Proceedings of the 10th European Conference on Visual Media Production.* 2013:12.

Design and Construction of a Wearable Wireless Electrogoniometer for Joint Angle Measurements in Sports

Isabela M. Mercado-Aguirre[1] and Sonia H. Contreras-Ortiz[1]

[1] Department of Electrical and Electronic Engineering, Universidad Tecnológica de Bolívar, Cartagena, Colombia

Abstract— **Advances in the field of biomedical engineering have allowed the development of biometric devices to evaluate sport performance. This article describes the design and construction of a wearable electrogoniometer to monitor joint angle in real time for sports applications. The electrogoniometer uses four accelerometers connected to a microcontroller. The joint angle is estimated and transmitted via Bluetooth Low Energy. This prototype is a low-cost, low-power, and comfortable to wear solution for joint angle measurement.**

Keywords— **Electrogoniometer, accelerometer, wireless, bluetooth, joint angle.**

I Introduction

The use of technological devices for biosignal measurement in sports has increased in the last years. Nowadays there are biometric devices to perform quantitative measurements on the athletes during the execution of exercises. The acquired data can be used to determine optimal techniques, improve performance and prevent injuries [1, 2]. Measuring kinematic variables is fundamental to perform inverse dynamics analysis of the athlete's movements. One of these variables is joint angle, and its estimation is useful to evaluate the efficiency during the execution of exercises in sports such as weightlifting [3] and to classify movements during sports training sessions [4]. Joint angles can be measured using two approaches: optical methods and goniometers.

Optical methods are the most common methods to record human movement. They use cameras and specialized software to make kinematic measurements on the athletes from their pictures and video-recordings. This technique is called photogrammetry. To improve accuracy, markers are placed on anatomical landmarks to indicate their position in space and aid in the tracking process [5]. Optical methods have some disadvantages. They require expensive equipment and software, and because of the movements, the markers may disappear of the camera's field of view and affect the accuracy of the measurements.

There are also the goniometric systems, which allow direct measurement of the joint angles. Electrogoniometers convert angle to voltage. They can be based on resistive sensors like potentiometers and strain gages, or on microelectromechanical inertial sensors (MEMS). Electrogoniometers have been used for gait assessment and rehabilitation [6, 7, 8], and sports [9, 10, 11, 12]. Petushek et. al. compared optoelectronic systems and electrogoniometers and concluded that electrogoniometers are a cost-effective and time-efficient alternative to video analysis for the assessment of joint angles when the sensor is precisely attached. Among the major challenges of electrogoniometers are data logging and signal processing to ensure portability and robustness [13].

Optical methods can be very expensive and usually require laboratory environments. On the other hand, goniometric-based systems within body area networks are cost-effective and can be used outdoors with less space limitations. This paper describes the design and construction of an electrogoniometer based on accelerometers for joint angle measurement. The system is a new version of a previous work [14, 15] with added wireless communication, reduced size, and improved performance. The electrogoniometer can estimate joint angles and transmit data in real time to a mobile phone or computer. The system is intended for use in sports, so it was designed to be unobtrusive, lightweight and wireless, to avoid interfering with the movements of the user. Additionally, it provides a low-cost solution to perform kinematic analysis and evaluate sport performance.

II Materials and Methods

Four three-axis accelerometers (MMA7361L, Freescale Semiconductor, Austin, TX) connected to a Bluno microcontroller (Nano, DFRobot, Shanghai, China) were used for joint angle measurement. The accelerometers have two operating ranges: $1.5g$ and $6g$, where g the is standard gravity acceleration equivalent to $9.806\ m/s^2$ [16]. The $1.5g$ range was selected for this application.

A Calibration of the accelerometers

The output voltage of the accelerometers for each axis can be related to the acceleration in the corresponding direction

I. Torres et al. (eds.), *VII Latin American Congress on Biomedical Engineering CLAIB 2016, Bucaramanga, Santander, Colombia, October 26th -28th, 2016*,
IFMBE Proceedings 60,
DOI: 10.1007/978-981-10-4086-3_98

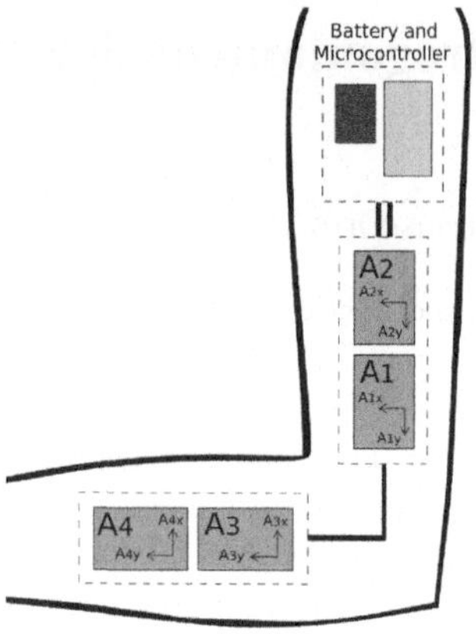

Fig. 1: Basic diagram of the system

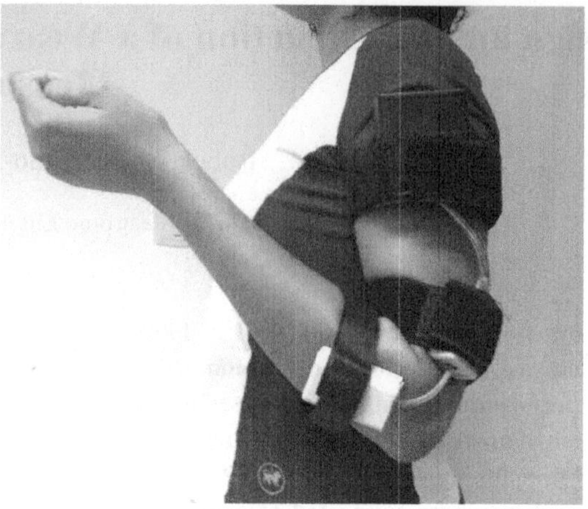

Fig. 2: Volunteer wearing the electrogoniometer

using equation 1.

$$V_{axis} = V_{off} + S \cdot A_{axis} \tag{1}$$

Where V_{off} is the voltage produced for $0g$ acceleration, S is the sensor sensitivity, and A_{axis} is the detected acceleration for the axis. The analog to digital converter (ADC) has a voltage reference V_{ref} of 3.3 V, and a resolution of 10-bits A/D [17]. Equation 2 relates the converted digital value DV with the original analogical value given by the accelerometer.

$$DV = \frac{V_{axis} \cdot 1023}{V_{ref}} \tag{2}$$

The calibration of the X axis for each accelerometer was done as follows. The accelerometer was placed on a flat surface with the X direction pointing up (perpendicular to the surface). One hundred readings were acquired and averaged to obtain the voltage that corresponds to acceleration $+1g$. Then, the accelerometer was rotated 90° so that the X direction was parallel to the surface, to obtain the voltage value that corresponds to acceleration $0g$. These two values were used to estimate the slope (m) and intercept (b) of equation 3

$$A_{axis} = m \cdot DV + b \tag{3}$$

Equation 3 calculates the acceleration in each axis based on the digital value read with the microcontroller.

B Angle calculation

Joint angle was estimated through the *Distributed Common Mode Rejection* method (DCMR). Two accelerometers were located in each segment of the joint as seen in Fig. 1. They were used to simulates two "virtual accelerometers" located at the center of the joint. The acceleration values of the virtual accelerometers were estimated from the sensors' readings [18]. DCMR is preferable to the *Common Mode Rejection* (CMR) method, as CMR needs two accelerometers physically placed at the center of the joint. This approach can be used for measuring angles in mechanical arms, but it is not advisable for measurements of joint angles of the human body as high precision is required in the accelerometer placement [19].

The acceleration values for the X and Y axis of the virtual accelerometer A'_1 can be calculated from equation 4

$$\begin{aligned} A'_{1x} &= \frac{(r_2 \cdot A_{1x}) - (r_1 \cdot A_{2x})}{r_2 - r_1} \\ A'_{1y} &= \frac{(r_2 \cdot A_{1y}) - (r_1 \cdot A_{2y})}{r_2 - r_1} \end{aligned} \tag{4}$$

Where r_1 and r_2 are the distances from the center of accelerometers A_1 and A_2 to the joint vertex. A similar procedure can be used to calculate the acceleration components of the virtual accelerometer and A'_2, using the readings from accelerometers A_3 and A_4. The joint angle φ is given by the difference between the angles of the acceleration vectors A'_1 and A'_2 [18].

$$\varphi = \arctan\left(\frac{A'_{1y}}{A'_{1x}}\right) - \arctan\left(\frac{A'_{2y}}{A'_{2x}}\right) \tag{5}$$

C Description of the system

The accelerometers were located inside two small boxes that can be attached to the arm and forearm with a velcro type stripe. The acceleration values were estimated with the Bluno microcontroller. A 7.4 V lithium-polymer rechargeable battery was used for power supply. The battery and the microcontroller were located in other unit that can be attached at the upper part of the arm. Fig. 2 shows a volunteer wearing the system. The volunteer expressed that the system is comfortable and does not limit her movements. The components of the system are commercially available and low cost. The dimensions of the electrogoniometer were chosen according to the elbow joint of an average adult, however, it can be adjusted for other body joints.

D Wireless communication

The Bluno microcontroller transmits data wirelessly through an embedded *Bluetooth Low Energy* (BLE) module. The BLE technology was created for applications requiring low power consumption, and it has a maximum transmission speed of 1 Mb/s. Data can be received by multiple devices, being smart phones the most popular. However, most of the Bluetooth receivers of computers are not compatible with the BLE protocol, so an adapter is required. We used the USB BLE-Link (DFRobot, Shanghai, China) for data acquisition with a computer. The baud rate was configured to 115200 bps, which is the value recommended by the manufacturer. The transmission range in open space can be up to 70 m [17].

E Data visualization

The microcontroller acquires data from the sensors and calculates the angle. Angle measurements are transmitted to a computer or smart phone, and an application for data acquisition and display was developed with Processing (Version 2.2.1, Processing Foundation). Fig. 3 shows an screenshot of the application.

III Results

The average computational time for angle calculation and data transmission was measured as $T_c = 36$ ms, what allows a sampling frequency of the angle signal equal to 27.78 Hz.

A Accuracy of the system

A manual goniometer was attached to the electrogoniometer to evaluate its accuracy. Twenty readings were taken and averaged for each angle. We considered angles from 0°,

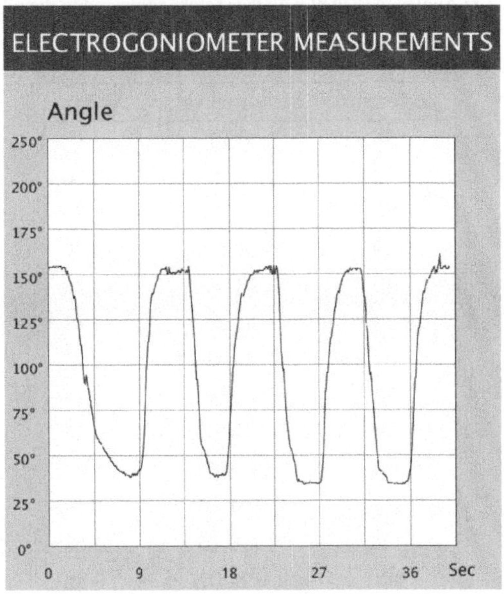

Fig. 3: Screenshot of the Processing application. The data was acquired while a volunteer was doing flexion/extension exercises of the elbow joint.

which correspond to total extension of the joint to 160°, which is the maximum passive flexion angle of the elbow joint. Table 1 shows the results and error percentages.

Note that error percentages were lower than 5%. Possible sources of error includes insecure attachment of the electrogoniometer, and electric noise in the circuitry.

B Power consumption

The LiPo battery provides a voltage of 7.4 V and has a capacity of 500 mAh. The current drawn by the system was measured as 39.5 mA during data transmission. The instantaneous power can be estimated as 0.29 W, and the expected runtime with a full charged battery is approximately 9 hours.

IV Conclusion

An electrogoniometer based on accelerometers was implemented using low-cost components. The system was evaluated with respect to accuracy, power consumption, processing time and comfort. The electrogoniometer proved to be effective for angle measurement and wireless transmission in real time. It has a battery autonomy of about 9 hours, and can transmit angle data at a frequency of 27.78 Hz using the BLE protocol. This system may be used for joint angle measurement in sport applications as it has a compact design, it is comfortable to wear and does not restrain the users movements. Further research should be done to evaluate the perfor-

Table 1: Angles measured with the system compared to the reference. Values expressed in degrees.

Reference	Measured value	% Error
0	0.14	–
10	9.69	3.20
20	19.86	0.71
30	29.86	0.46
40	41.79	4.28
50	48.86	2.33
60	58.41	2.72
70	70.52	0.73
80	79.31	0.87
90	89.07	1.05
100	101.10	1.09
110	110.76	0.68
120	121.07	0.88
130	129.97	0.03
140	139.34	0.47
150	150.59	0.39
160	159.62	0.24

mance of the system during the execution of sports exercises. Future work includes improving the accuracy of the system through more precise sensors' attachment and signal processing.

CONFLICT OF INTEREST

The authors declare that they have no conflict of interest.

REFERENCES

1. Mertz Leslie. Technology comes to the playing field: new world of sports promises fewer injuries, better performance. *IEEE pulse.* 2013;4:12–17.
2. Waltz E.. The quantified olympian *Spectrum, IEEE.* 2015;52:44-45.
3. Harbili Erbil. A gender-based kinematic and kinetic analysis of the snatch lift in elite weightlifters in 69-kg category *Journal of Sports Science and Medicine.* 2012;11:162–169.
4. Ahmadi Amin, Mitchell Edmond, Richter Chris, et al. Toward automatic activity classification and movement assessment during a sports training session *IEEE Internet of Things Journal.* 2015;2:23–32.
5. Kutz M.. *Standard Handbook of Biomedical Engineering and Design.* McGraw-Hill Handbooks SeriesMcGraw-Hill 2003.
6. Willemsen A Th M, Van Alste JA, Boom HBK. Real-time gait assessment utilizing a new way of accelerometry. *Journal of Biomechanics.* 1990;23:859–863.
7. Piriyaprasarth Pagamas, Morris Meg E, Winter Adele, Bialocerkowski Andrea E. The reliability of knee joint position testing using electrogoniometry. *BMC musculoskeletal disorders.* 2008;9:1.
8. Walker CRC, Myles C, Nutton R, Rowe P. Movement of the knee in osteoarthritis The Use Of Electrogoniometry to Assess Function. *Journal of Bone & Joint Surgery, British Volume.* 2001;83:195–198.
9. Lotfian Sara, Cherati Afsaneh Safar, Jamshidi Aliashraf, Sanjari Mohammad Ali. Assessment of Correlation between Electrogoniometer Measurements and Sports-Specific Movement in Karate Elites *Asian journal of sports medicine.* 2014;5:115.
10. Pietraszewski BOGDAN, Struzik Artur. Evaluation of selected biomechanical parameters in female team sports players *Acta of Bioengineering and Biomechanics.* 2013;15.
11. Fonda Borut, Sarabon Nejc, Li François-Xavier. Validity of Different Kinematical Methods for Assessing Knee Angle during Cycling *1. Specific Rhytmic Gymnastics Skills Acquisition Conditionallity In Preschool.* 2013:129.
12. Waltz Emily. A wearable turns baseball pitching into a science [News] *Spectrum, IEEE.* 2015;52:16–17.
13. Fong Daniel Tik-Pui, Chan Yue-Yan. The use of wearable inertial motion sensors in human lower limb biomechanics studies: A systematic review *Sensors.* 2010;10:11556–11565.
14. Haye Chamorro G.I., Mercado Aguirre I.M., Contreras-Ortiz S.H.. Design of an electrogoniometer based on accelerometers for the evaluation of sports gesture in weight lifting in *Engineering Mechatronics and Automation (CIIMA), 2014 III International Congress of*:1-3 2014.
15. Mercado-Aguirre Isabela M., Mercado-Medina Edgardo L., Chavarro-Hernandez Zulay D., Dominguez-Jimenez Juan A., Contreras-Ortiz Sonia H.. A wearable system for biosignal monitoring in weightlifting *Sports Engineering.* 2016:1–8.
16. Semiconductor Freescale. MMA7361L at https://www.nxp.com/files/sensors/doc/data_sheet/MMA7361L.pdf
17. DFRobot . Bluno Nano at http://www.dfrobot.com/wiki/index.php/Bluno_Nano_SKU:DFR0296
18. Dong Wei, Chen I, Lim KY, Goh YK, others . Measuring uniaxial joint angles with a minimal accelerometer configuration in *Proceedings of the 1st international convention on Rehabilitation engineering & assistive technology: in conjunction with 1st Tan Tock Seng Hospital Neurorehabilitation Meeting*:88–91ACM 2007.
19. Cheng Peng, Linnarsson Fredrik, Oelmann Bengt. Joint Angular Sensor Based on Distributed Biaxial MEMS Accelerometers in *Industrial Electronics Society, 2007. IECON 2007. 33rd Annual Conference of the IEEE*:2242–2247IEEE 2007.

Authors: Mercado-Aguirre, Contreras-Ortiz
Institute: Universidad Tecnológica de Bolívar
Street: Km 1 Via Turbaco
City: Cartagena de Indias
Country: Colombia
Email:

Improve Image Quality for Minimally Invasive Surgery Simulator Using Lenses and Image Processing.

Edgar M. Ramírez-Rodríguez [1], Daniel Lorias-Espinoza [2], José Antonio Gutierrez-Gnecchi [3]

[1, 2] Department of Electric Engineering, Section Bioelectronic, CINVESTAV, México City, México, 07360.
[3] Instituto Tecnológico de Morelia, Department of Electronic Engineering, Avenida Tecnológico 1500, Col. Lomas de Santiaguito, Morelia, Michoacán, México.

Abstract: ***A video camera for minimally invasive surgery is an important tool and should be smaller, lighter and provide high quality images but these cameras are very expensive, so use in simulators is to raise costs. The objective of this work is to implement image processing techniques and a system of magnifying lenses to improve the quality and realism obtained from commercial use webcams and replace the high-resolution camera on minimally invasive surgery simulators. An application of image processing is shown for a closed environment aimed at learning medical operations of minimally invasive surgery.***

Keywords: *Minimally invasive surgery, Image processing, Magnifying lenses, High resolution camera, Simulator.*

I. Introduction

The equipment needed to create a clinical simulation center is high technology, and the purchase is an expensive proposition for most hospitals and medical training centers. Hospitals, universities and training centers that do not have enough funds for normal administration may be impossible to invest large sums of money in a clinical simulation center [1].

Learning for minimally invasive surgery (MIS) is a fundamental aspect in its development and introduction in daily practice. For its realization several psychomotor skills quite different from open surgery are necessary. Besides the time spent on the learning curve it is still very long [2].

One of the essential tools is the high-end camera, this has a high cost depends on the number of sensors containing.

The project is to design and build a system that can replace the high-definition camera. This proposal has two parts: a physical part that it comprising a focus control and a lens arrangement to amplify the image, and a virtual part of processing the image to prevent image distortion, as well as get color reproduction close to the real.

II. Materials

Image capture is performed through a webcam s-vision of the company perfect choice model PC-320425. Targeted a textile that simulates stomach distension and the system is illuminated by light-emitting diodes are used.

As computational tool is used the Graphical User Interface (GUI) of the program MATLAB ® in its 2014 version.

Once the camera is selected it is need a coupling to align with the laparoscopic optics. This coupling is designed in *"SolidWorks 2014 x64 Edition"* with the camera and arthroscopy measures.

Although the image is sharp, very small, so it seeks to expand by physical means, for this purpose a magnifying glass was used, the best option was 10x lens.

III. Methods

A. Image Acquisition.

The quality of the images obtained by a digital camera is given by the CCD chip (charge-coupled device), which are state image sensors.

The amount of light is measured in lux, usually they need 6-10 lux to have the minimum necessary amount of light for a quality picture, but there are models they only need 3 lux SIS (suspended image system) [3].

The lighting system is independent and is provided by an array of LEDs with adjustable intensity. The target is placed at least 2 cm from the camera, this is considered by the distance that MIS are performed.

The part of the optics focuses on the use of lenses to concentrate light beams. Using different lenses with Macro function and 10x magnification which combine to increase the image.

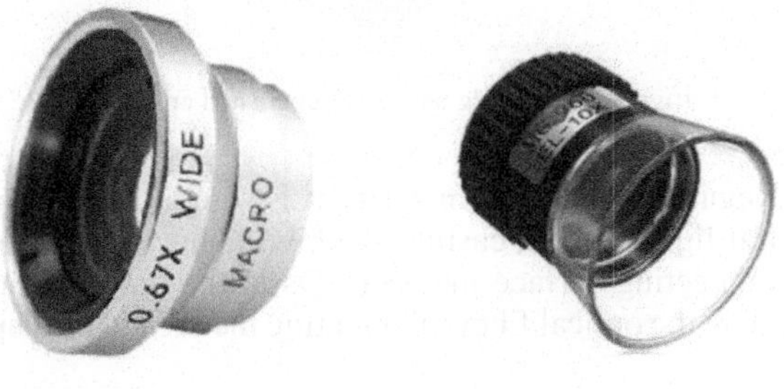

(a) (b)

Figure 1. Lenses used (a) Macro (b) Increase 10x.

The image provides an arthroscope is sharp, but very small. By bringing the camera is misses the point of focus and blurs the captured image. a lens macro is placed in the output of the camera, which provides a clear picture.

A lens is used with 10x magnification. A standard lens is a lens that provides a perspective image and a field of view (FOV) adjusted to the human visual system. They usually have a focal length equal to the diagonal of the format where

I. Torres et al. (eds.), *VII Latin American Congress on Biomedical Engineering CLAIB 2016, Bucaramanga, Santander, Colombia, October 26th -28th, 2016*,
IFMBE Proceedings 60,
DOI: 10.1007/978-981-10-4086-3_99

the image is recorded, although there is considerable variation.

The objectives are combinations of individual lenses, grouped into elements, normally used to form images of distant objects. Each element is modeled as a thin lens with optical characteristics (focal length, position of the planes and main points) of the combination.

A zoom lens is a lens that has varifocal according to their relative distance (t) maintaining a fixed image plane where the film or detector is positioned. The simplest case consists of two elements φ1 and φ2 optical powers whose power and BFD see in equation 1 and 2.

$$\varphi = \frac{1}{f} = \varphi_1 + \varphi_2 - \varphi_1\varphi_2 t \quad (1)$$

$$BFD = f - \left(\frac{\varphi_1}{\varphi}\right) \quad (2)$$

To align the camera with optics proposal used a connection between the camera and medical instruments. This coupling is designed in SolidWorks with the measures of the camera, lenses and arthroscope.

Figure 2. Coupling between camera and arthroscope.

Regarding the light intensity it is sought to be similar to natural light without casting shadows. This effect is obtaine by connecting surface mount LEDs in the ureteroscope and aided by the optical fiber illuminating the area to be displayed

B. Image processing

The frequency histograms are bar graphs used to summarize and illustrate the variation that occurs in a data set. They are used to investigate how you can solve a problem or improve a process [4].

The histogram equalization is used as a tool for adaptive enhancement. This option automatically determines the transformation function and seeks to produce an output image having a uniform histogram [5].

The histogram is a graphical representation of the distribution of the different tones of an image. Understand helps control exposure in our images, as well as to correct colors.

The histogram will help us get an idea of how correct the exposure of an image. Thus, in an underexposed image graphics you tend to move to the left and in an overexposed image tends to move to the right. An image containing all shades of gray will have a more or less uniform throughout the horizontal axis histogram. The histogram has a certain way does not have to mean that the image is incorrectly exposed. It depends on what you look at the time of capture.

It is defined as false color transformation process which seeks to make visible small changes in gray level palette changing output and transforming each gray value in the palette.

When you want to change an image with few pixels to a larger blur effect happens, this is precisely the problem we face when proposing the lens system. Besides the blur effect it is emphasized by the digital approach.

A set of digital filter, each to solve a specific problem is proposed. A Gaussian filter (3) to smooth noise, Sobel to detect edges.

The Gaussian filter is elected on the other that is separable, that is, instead of performing a convolution elected bidimensional we can perform two dimensional convolutions: a horizontally and one vertically.

$$g(x,y) = \sum_{s=-a}^{a} \sum_{t=-b}^{b} w(s,t) f(x+s, y+t) \quad (3)$$

Being a closed environment where the tests are performed, and control the light intensity, the filter parameters are fixed and are chosen for the characteristics of lighting system.

The Sobel operator calculates the gradient of the intensity of an image at each pixel (4). It can be implemented by simple definitions.

Only eight points around the point image are used to analyze to calculate the corresponding pixel of the resulting image. The separate calculations give good advantage because they involve less arithmetic operations for each pixel.

$$G_x = \begin{bmatrix} -1 & 0 & +1 \\ -2 & 0 & +2 \\ -1 & 0 & +1 \end{bmatrix} * A \; ; \; G_y = \begin{bmatrix} -1 & -2 & -1 \\ 0 & 0 & 0 \\ +1 & +2 & +1 \end{bmatrix} * A \quad (4)$$

Being a closed environment there is a dark outline and center images to study. Keep the contours of this center to add them to the image obtained from the Gaussian filter and highlight even more intensity changes.

The option to change the color image components, emphasizing the red component (R) and reducing the blue component (B). It changes to enhance image contrast.

C. Implementation of the Graphical User Interface

Once it has scheduled actions are brought together in a graphical user interface (GUI), this in order to facilitate its handling in a single window. The GUI is divided into three sections according to their function:

Video Options
Image Options and
General Operations.

The GUI consists of an area where the image is displayed, the area of the histogram, an indicator of size, an indicator of the option being underway and a menu of available options.

It also has options to zoom in or out of the image obtained.

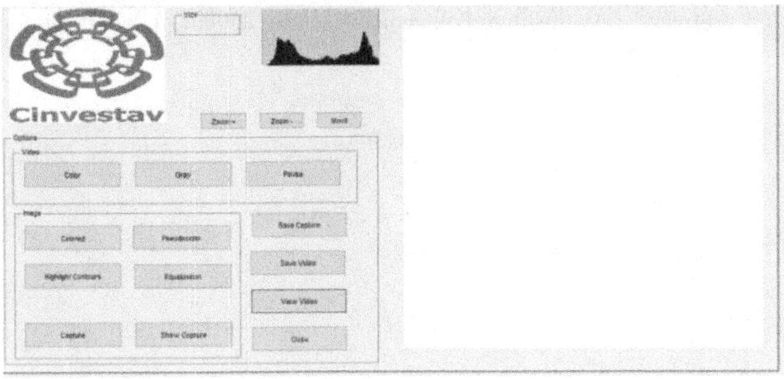

Figure 3. GUI window.

The area of options is divided into three sections: video preview, image processing and capture options. The first section is to display images without any processing, here is where we must use the manual zoom in the camera installed.

In the image processing section is where changes to the histogram and filters are applied, it is in this part where you choose happens to the image. I n the last section it is to save the images and video to be analyzed later.

IV. Results

When you start the GUI this shows how the image is viewed through the camera, this is where you can adjust the camera zoom. The GUI shows the original image in real time.

Histogram equalization in the video doing the contrasts between light and dark is shown. Recall that acts as an adaptive improvement.

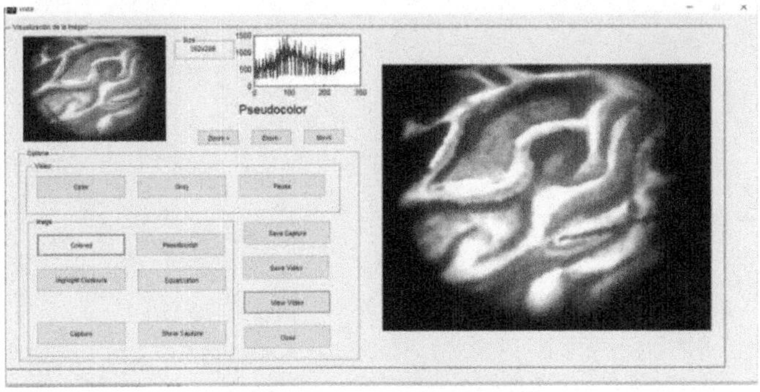

Figure 4. Colored Option.

Pseudocolor results in an image in which the colors were changed to a color palette in which the blue represents areas with lower intensity and red the bright areas.

V. Discuccion

The information gives us the image histogram indicates the degree of the necessary lighting. Being a closed environment and to control the intensity of illumination no conflict is generated by the acquisition. It helps to use manual zoom of the selected camera.

What must be taken into account is the alignment lenses and lens cleaning, if any of these factors is altered reflected in the final result. The use of lenses before the image processing is for the changes that have to be made in the image are lower and, therefore, consumption of time by the algorithm is less.

The main problem was detected when very high or very low values of few frequencies, which can make the adjustment of the histogram is very bad.

In histogram equalization we took the video as a sequence of frames, each of these are taking the time and frequency intensity. Arithmetic operations of multiplication and addition are applied to the histogram has equidistant values and a sharp image samples.

Coloring a picture is taken and the values of each layer are modified by multiplying different constants.

The Sobel operator calculates the gradient of the intensity of an image at each pixel. Thus the operator gives the greatest magnitude of change, its direction and the direction from dark to light.

Image processing is a task that handles a lot of data, so you spend a lot of resources of the computer on which it is used, you have to assess the cost-benefit of using this type of processing in buying a specialized camera MIS.

The work is focused for the training of surgeons, mainly inexperienced. The simulator is simple and inexpensive, they can be used anywhere, so that training can be performed outside the hospital environment, further increasing training time. Not only it limited to a single anatomical area and has the option to share progress with other surgeons through internet.

VI. Conclusion

This article is intended to show the advantages of working with crisp images similar to the environment minimally invasive surgery but without the cost of buying a specialty camera for such surgeries.

The concept of contrast resolution concerns the accuracy comparing the brightness of a pixel and corresponding to that of the original image in the same place. This is achieved by varying the masks of each template color, which does not give extra information on what happens to the patient.

They are still some problems with the filters mainly due to motion blur which generates, blur and different illuminated areas. It is intended to implement adaptive filters to overcome this problem.

Conflict of Interest

The authors declare that they have no conflict of interest.

References

[1] Poggi Machuca, L. (1999). Cirugía Laparoscópica. In: O. Salaverry García, ed., Cirugía : I cirugía general, 1st ed. [online] Lima: Fondo Editorial de la UNMSM, pp.58-76.

[2] Hanna M. I, Hanna J. I. Fundamentos de Cirugía Endoscópica Rev. Med. Electrón. vol.34 no.1 Matanzas ene.-feb. 2012.

[3] J.E. Greivenkamp: Field Guide to Geometrical Optics. Society of Photo-Optical. Instrumentation Engineers (SPIE), 2004.

[4] Anil K. Jain. Fundamentals of Digital Image Processing. Prentice-Hall, 1989.

[5] Esqueda Elizondo, José Jaime. Fundamentos de Procesamiento de Imágenes. Universidad Autonoma de Baja California. 2010.

Author: Edgar M. Ramírez-Rodríguez
Institute: Centro de Investigación y de Estudios Avanzados
Street: Av. Instituto Politécnico Nacional 2508, Col. San Pedro Zacatenco.
City: Ciudad de México
Country: México
Email: emramirez@cinvestav.mx

Fabrication of piezoelectric PVDF/Graphene membranes by electrospinning for respiratory rate and temperature sensing

D. Hernández-Rivera[1], E. Suaste-Gómez[1]

[1] Electrical Engineering Department, Bioelectronic Secction,
Centro de Investigación y de Estudios Avanzados del IPN, Cd. de México, México.

Abstract— **Graphene (G) based materials are highly attractive materials in biomedical area because of its biocompatibility and excellent physicochemical properties acquired by adding G. The purpose of this paper was to show the application potentiality of piezoelectric PVDF/G membranes. In this context, a methodology to fabricate PVDF/G membranes by electrospinning was developed. A preliminary comparison between PVDF/G and PVDF membranes was realized. Finally, it was possible to test the capacity of PVDF/G membranes as frequency respiratory and temperature sensor.**

Keywords— **PVDF, Graphene, Electrospinning, Sensors.**

I. Introduction

The medical field demands high technology to physiological measurements such as respiration, heartbeat, temperature, arterial pressure, among others. This factor has contributed to growth of technological areas such as material science, sensors and instrumentation.

Polymers have been making a profound impact in material science and biomedical applications due to their mechanical, electrical and thermal properties, biocompatibility and cost [1-3]. In recent years, there has been an increasing interest in Graphene (G) because of its excellent electrical conductivity, strong mechanical strength, thermal conductivity, ease of functionalization and biocompatibility. Therefore, G based materials are highly attractive materials [4-6]. PVDF is a biocompatible ferroelectric polymer, which possesses piezoelectric and pyroelectric effects when it is crystallized in β phase [7, 8]. The addition of carbon allotropes such as carbon nanotubes or G improves piezoelectric and pyroelectric properties of PVDF [9-11].

Taking into account these technological advances, it is necessary to develop alternative applications where the novel features can be exploited. This paper presents a real opportunity for application of PVDF/G composite within the field of biomedical engineering.

II. Materials y methods

A. Preparation of PVDF/G solution

This paper will focus on a mixture of PVDF with G. Recently; studies have shown that the piezoelectric and pyroelectric properties are improved by the influence of G addition in PVDF. [10-11].

PVDF is typically dissolved in polar solvents such as DMF [1-8]. Recent evidence shows that crystallization from DMF solution at <70°C results predominantly in formation of the phase β [8].

PVDF powder and N,N-dimethylformamide (DMF) were purchased from Sigma Aldrich. Ultra-high Concentration Dispersion of Graphene Nanoplatelets was purchased from Graphene Supermarket.

Two solutions were used; PVDF/DMF and PVDF/G/DMF. The first solution was prepared by dissolving 2 g of PVDF powder in 8g of DMF at 60°C in order to ensure a homogeneous mixture. The detailed procedures to prepare PVDF/G solution of 0.5 wt% are described as follows: firstly, 4g of PVDF powder was dissolved in 16 g of DMF to give a concentration of 4:1, then the solution was warmed at 60°C in order to ensure a homogeneous mixture, on the other hand, G was dispersed in DMF by using an ultrasound dispersion machine during three 30 min sessions to generate a homogenous solution. After, PVDF/DMF and G/DMF solutions were mixed, stirred and sonicated during two 30 min sessions. Finally, uniform solution of appropriate concentration to electrospinning was obtained.

B. Membrane fabrication

Polymeric fibers were fabricated by electrospinning. In the electrospinning process, high voltage induces a charged jet of a polymer solution; this jet is moved toward the collector plate where fibers are deposited [12-14]. A DC power supply 0-25kV, a syringe (volume of $3\,cm^3$, hypodermic needle) and an aluminum collector plate (15 x 10 cm) were used. The distance between tip and collector plate was 15 cm.

I. Torres et al. (eds.), *VII Latin American Congress on Biomedical Engineering CLAIB 2016, Bucaramanga, Santander, Colombia, October 26th -28th, 2016*,
IFMBE Proceedings 60,
DOI: 10.1007/978-981-10-4086-3_100

Figure 1 shows a scheme of technique used. PVDF/G/DMF solution was deposited in syringe, needle was connected to positive voltage (25 kV) and aluminum plate was grounded. Ultimately, high voltage induced a jet of PVDF/G toward the aluminum plate.

The two kinds of membranes shown in Figure 2 were produced with this process. PVDF membrane was used as reference sample and PVDF/G was the main sample.

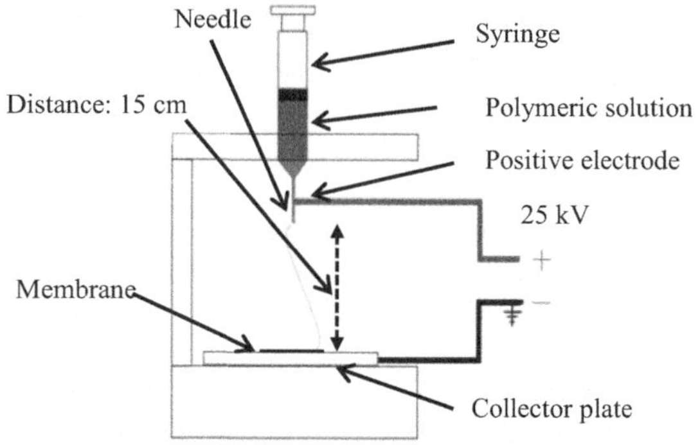

Fig. 1 Electrospinning technique

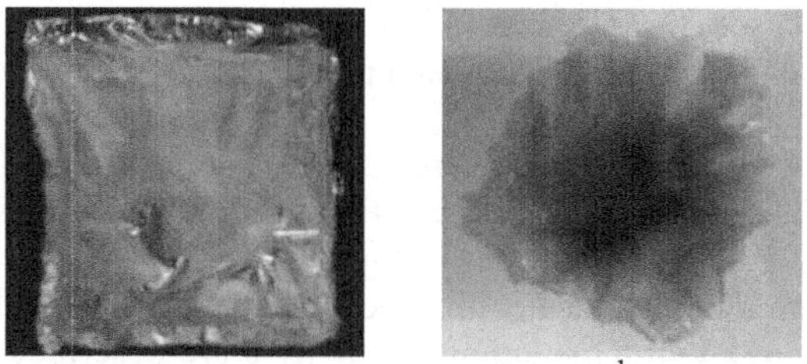

a b

Fig. 2 PVDF (a) and PVDF/G (b) membranes

Micrographs of membranes were obtained by a scanning electron microscope JEOL (JMS-6360LV). PVDF fibers with diameters ranging from 50 nm to 700 nm and PVDF/G fibers with diameters ranging from 130 nm to 600 nm were achieved. Figure 3 illustrates the polymeric fibers produced by electrospinning.

a

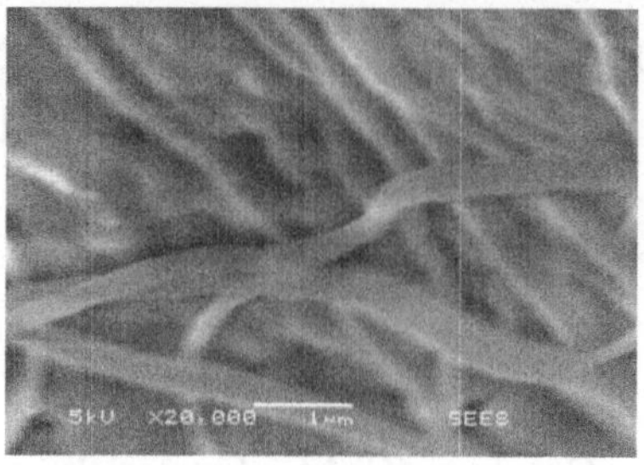

b

Fig. 3 SEM micrographs of membranes PVDF (a) and PVDF/G (b)

C. Experimental tests

In this stage, the fabricated membranes by electrospinning were analyzed. Two possibilities of application in biomedical field were considered.

Firstly, membranes were used as respiratory rate sensors. In this case, a membrane was fixed on a section of an elastic strip which should be attached to the chest wall where the maximum changes during inspiratory phase are noticed. Inspiration and exhalation of air generates volume changes of the ribcage, which induce pressure changes in the polymer membrane. These variations of pressure generate electric charges according to respiration cycle. The next stage consists in a signal conditioning process; first the signal is amplified by a factor of 500 and filtered with a low pass filter of 1 Hz. Figure 4 illustrates how the respiratory rate sensor is placed on a subject test.

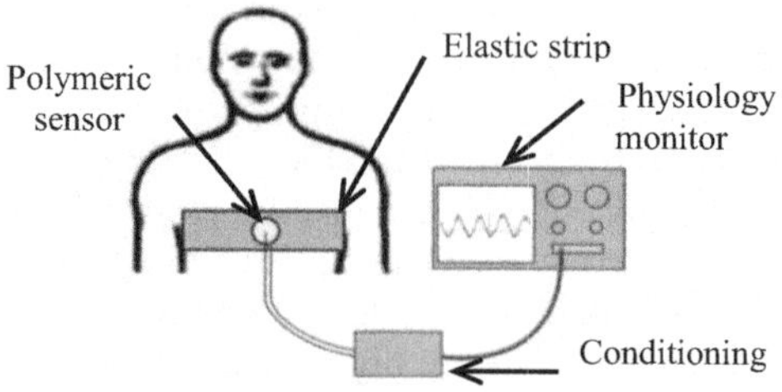

Fig. 4 Respiratory rate sensing

The polymer membranes were proved as temperature sensor in the second stage. In this experiment, the temperature dependence of the dielectric constant in piezoelectric material was exploited [2]. Following this, the samples were placed between two parallel copper plates in order to measure the temperature-dependent dielectric constant as shown in Figure 5.

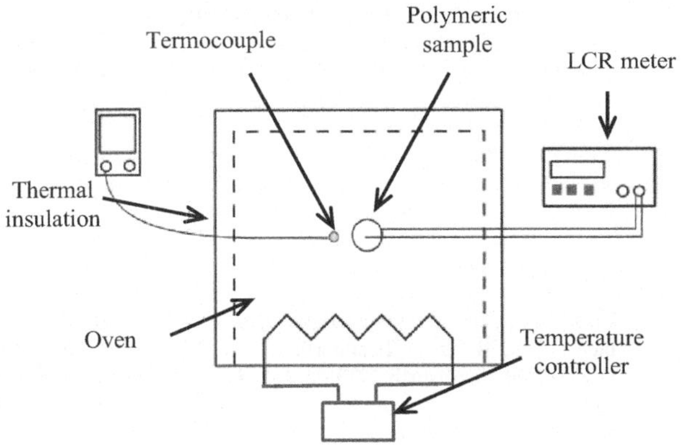

Fig. 5 Temperature sensing

III. Results

Circular samples of 1 cm diameter were obtained of PVDF and PVDF/G membranes. The average thickness of PVDF membranes was around 124,4 μm and PVDF/G membranes was around 127 μm. Copper electrodes were placed on both faces of membranes as illustrated in Figure 6.

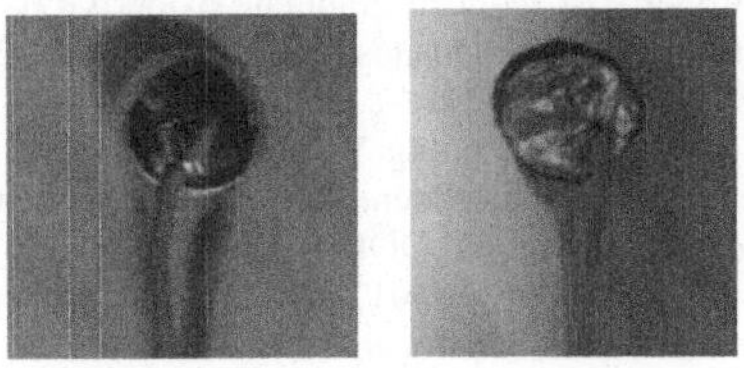

a b

Fig. 6 PVDF (a) and PVDF/G (b) samples with cupper electrodes

Firstly, dielectric constant (ε_r) was measured by using a LCR meter at a 1 kHz and considering equation 1, where C is capacitance, d is distance between plates, A is the contact area and $\varepsilon_0 \left(8.854 \times 10^{-12} \frac{F}{M}\right)$ is the vacuum permittivity. Dielectric constants were $\varepsilon_r = 6.32$ (PVDF) and $\varepsilon_r = 6.05$ (PVDF/G).

$$\varepsilon_r = \frac{C \times d}{\varepsilon_0 \times A} \quad (1)$$

Pre-testing was performed with the purpose of checking the response of the membranes and assessing the viability as a respiratory rate and temperature sensor. Tests of both membranes (PVDF and PVDF/G) were realized in order to compare their responses.

Once the conditioning process was applied, typical waveforms of respiratory cycle were acquired. Figure 7 shows signals of 2 test subjects where respiratory signals were obtained by using PVDF and PVDF/g membranes as respiratory rate sensors. The respiratory rates obtained were approximately from 20 to 40 breaths per minute, matching with typical range in adults from 4 to 50 breaths per minute under certain conditions [15].

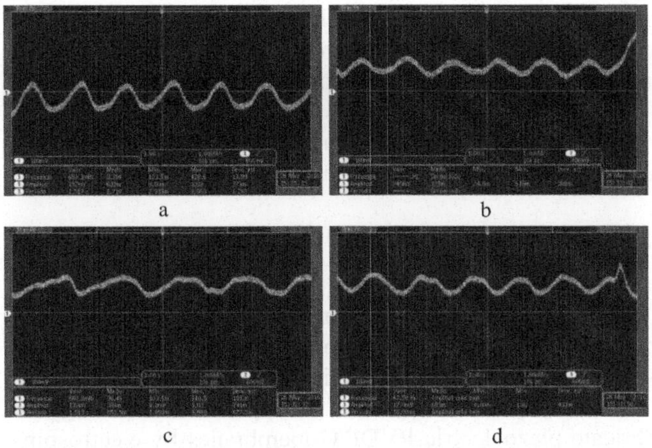

Fig. 7 Response of PVDF membrane (a) subject 1, (b) subject 2 and PVDF/G membrane (a) subject 1, (b) subject 2, as respiratory rate sensors.

Finally, membrane response to controlled temperature variations was analyzed. PVDF and PVDF/G samples were subjected to oven heating from 30°C to 90°C with a rate of 5°C/min, simultaneously capacitive impedance was measured at 1 kHz. The thermal behavior is clearly illustrated in Figure 8, where the graphs show the capacitance response to temperature variations.

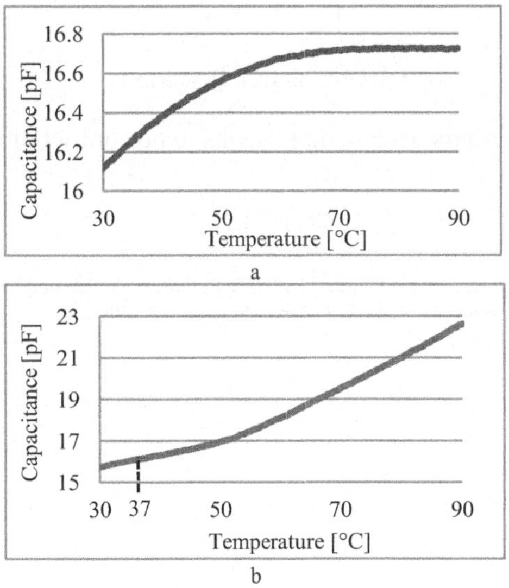

Fig. 8 Thermal response of PVDF (a) and PVDF/G (b) membranes

IV. Discussion

Analyzing waveforms and amplitude of signals in respiratory rate sensing, the results show similar responses of PVDF and PVDF/G membranes. In all these cases, the amplitude was around 100 mV_{PP}.

About of thermal response of membranes, there was a notable difference between behaviors of both membranes. In the case of body temperature (37°C) sensing is apparent that PVDF/G shows lineal behavior and an improved sensitivity.

Preliminarily, fabricated membranes exhibit a proper functioning as respiratory rate and temperature sensor.

V. Conclusions

In recent studies, it has been demonstrated that G based materials have excellent electrochemical and optical properties. Nevertheless, real applications are necessaries.

In this paper, was possible to apply a methodology to fabricate piezoelectric PVDF/G membranes by electrospinning, however, it is interesting to note that these membranes can function as respiratory rate and temperature sensor. Moreover, the PVDF/G membrane can be used for moisture, light and pressure sensing due to their ferroelectric properties. Other possible applications can be filtration, growth of living tissue and prosthesis.

Acknowledgment

This study was supported in part by the Mexican Agency CONACYT through grant 151894.

Conflict of Interest

The authors declare that they have no conflict of interest.

References

1. Kataoka K, Scholz C (2007) Polymers in Biomedical Applications. Progress in Polymer Science 32: 876-921.
2. Suaste-Gómez E, Rodríguez-Roldán G, Reyes-Cruz H et al. (2015) Developing an Ear Prosthesis Fabricated in Polyvinylidene Fluoride by a 3D Printer with Sensory Intrinsic Properties of Pressure and Temperature. Sensors 16:332 DOI:10.3390/s16030332
3. Kwang J K, Satoshi T (2013) Electroactive Polymers for Robotic Applications (Artificial Muscles and Sensors). Springer, USA
4. Avouris P, Dimitrakopoulos C (2012) Graphene: synthesis and applications. Materials Today 15: 86-97 DOI:10.1016/S1369-7021(12)70044-5
5. Pinto A M, Gonçalves I C, Magalhães F D (2013) Graphene-based materials biocompatibility: A review. Colloids and Surfaces B: Biointerfaces 111: 188-202 DOI:10.1016/j.colsurfb.2013.05.022
6. Yang Y, Asiri A M, Tang Z et al. Graphene based materials for biomedical applications. Materials Today 16:365-373 doi:10.1016/j.mattod.2013.09.004
7. Benz, M, Euler W B (2002) Determination of the Crystalline Phases of PVDF Under Different Preparation Conditions Using Differentail Scanning 99 Calorimetry and infrared Spectroscopy. Journal of Applied Polymer Science 89:1093 -1100 DOI: 10.1002/app.12267
8. Rinaldo G (2006) Determination of the α, β, γ and Crystalline Phases of Poly(vinylidene fluoride) Films Prepared at Different Conditions. Journal of Applied Polymer Science 100:3272–3279 DOI: 10.1002/app.23137
9. Ramaratnam A, Jalili N (2006) Reinforcement of Piezoelectric Polymers with Carbon Nanotubes: Pathway to Next-generation. Sensors Journal of Intelligent Material Systems and Structures 17: 199-208 DOI:10.1177/1045389X06055282
10. Alamusi, Xue J, Wu L K et al. (2012) Evaluation of piezoelectric property of reduced graphene oxide(rGO)- poly(vinylidene fluoride) nanocomposites. Nanoscale, 4:7250-7255 DOI: 10.1039/C2NR32185H
11. Hu Y C, Hsu W L, Wang Y T et al. (2014) Enhance the pyroelectricity of polyvinylidene fluoride by graphene-oxide doping. Sensors 14:6877-6890 DOI:10.3390/s140406877
12. Jayakumar R, Nair S (2012) Biomedical Applications of Polymeric Nanofibers. Springer, New York
13. Cozza E S, Monticelli O, Marsano E et al. (2013) On the electrospinning of PVDF: influence of the experimental conditions on the nanofiber properties. Polym. 62:41–48. DOI: 10.1002/pi.4314
14. Moradi R, Karimi-Sabet J, Shariaty-Niassar M, et al. (2015) Preparation and Characterization of Polyvinylidene Fluoride/Graphene Superhydrophobic Fibrous Films. Polymers 7: 1444-1463 doi:10.3390/polym7081444
15. Guyton C, Hall J (2010) Textbook of Medical Physiology. Elsevier, Philadelphia

Educative ECG Platform for Undergraduate Courses in BME

Armando S. Hernández-Delgado[1], Arturo Vega-González[2], Juan M. Gomez-González[3]

[1, 3] Facultad de Ingeniería, DCR, Universidad Nacional Autónoma de México, Ciudad de México
[2] División de Ciencias e Ingenierías, DIQEB, Universidad de Guanajuato, León Guanajuato
[1]armando.salomon_h@yahoo.com.mx, [2]a.vega@ugto.mx, [3]juan@dctrl.fi-b.unam.mx

Abstract— **In the area of Biomedical Instrumentation and Physiology, the teaching process is commonly supported by commercial instrumentation systems, frequently used with minimal understanding of the fundamental principles of registering and processing physiological signals. In BME-closing module at Faculty of Engineering, National Autonomous University of Mexico, the Fundamentals of Biomedical Instrumentation and Introduction to Physiology courses, are in need to put in practice theoretical subjects, using accessible equipment. Therefore, an instrument to register and display ECG signals was developed for educational purpose, training students in topics such as: identify main problems when acquiring biosignals, basic medical electronics, designing customized instrumentation systems and biosignal processing. Development was conducted using free licensed software such as IDE Arduino for hardware controller and, Phyton, Processing and Octave for signal display and processing data, making accessible the use of this platform in both courses. Theoretical and practical knowledge obtained will allow students to continue his/her career as designer/developer of instrumentation systems –hardware and software- for research, as well as in medical equipment maintenance.**

Keywords— **biomedical instrumentation, ECG, open source.**

I. Introduction

New Technologies are continuously employed in developing new teaching methods to improve the quality of Education. In the area of Biomedical Instrumentation and Physiology, teaching is frequently supported by commercial instrumentation systems -hardware and software-, tools frequently used with minimal understanding of the fundamental principles in registering physiological signals and placing main emphasis on interpretation, for example in ECG [1].

Worldwide, public and private healthcare systems need professionals specialized in the area of preventive and corrective maintenance of biomedical equipment. Thus, knowledge of specific functions of medical instrumentation systems is required in order to perform a correct failure detection and correction [2].

Highly used medical equipment in education and in hospitals is electrocardiograph or ECG machine, which frequently suffers from several issues as noise, interferences and artifacts, and even broken components, in addition to incorrect practices when registering signals. These factors may suggest a malfunction in equipment and if technical staff or service engineers are uninstructed or inexperienced, issues may be misidentified and diagnosed incorrectly.

Knowing how an instrumentation system works, the elements and devices that contains, the type of signal that is going to be measured and its characteristics, symptoms for common issues and the correct use of such instrument, helps isolate failures and then, to perform corrective actions and even modifications or to design a new improved system.

In BME-closing module at Faculty of Engineering, National Autonomous University of Mexico, the Fundamentals of Biomedical Instrumentation and Introduction to Physiology courses are in need to set up in practice theoretical subjects, only possible with high cost commercial equipment, such as a commercial ECG alone, or, included in a patient monitor, or, a Biopac® System, frequently used in teaching and research [3, 4]. Also is possible implement a very simple ECG circuit, then process and display the signal with licensed (not free) software such as Matlab® or LabView® [4, 5], which is a drawback if we are looking for an accessible and low cost system for teaching/learning and we avoid software piracy.

Considering these factors, a low cost instrument to register and display ECG signals for educational purpose was developed covering two needs:

a) Bring to students a comprehensive and practical understanding of electronic instrumentation -commonly found in any modern ECG machine- involved in registering bioelectrical signals; set in practice theoretical aspects of signal processing, programming techniques for a standalone digital controller for the system, as well as understanding the origin of electric biopotentials and its utility in medical diagnosis with further signal processing, preparing students in developing custom instrumentation systems and in research on biosignal processing.

b) Prepare students in technical aspects involving detection of malfunction in common medical equipment such as the ECG if he/she decides the area of medical equipment main-

I. Torres et al. (eds.), *VII Latin American Congress on Biomedical Engineering CLAIB 2016, Bucaramanga, Santander, Colombia, October 26th -28th, 2016*,
IFMBE Proceedings 60,
DOI: 10.1007/978-981-10-4086-3_101

tenance, or other career paths such as professional designer and tester of new devices in market.

The system contains the basics circuits to sense and coupling the bioelectrical signal, denoising and removing offset artifacts from it, including reduction of common mode voltage using Right Leg Driven (RLD) circuit and patient protection, performing isolation of Man-Instrument System from processing and displaying stages.

Part of processing stage is the analog to digital conversion, performed by an Analog to Digital Converter (ADC), contained on a microcontroller unit (MCU). A broad diversity in MCU manufacturers, architectures, capabilities, cost, and even programming languages of these devices may be overwhelming when choosing one. To keep the system simple, low cost, accessible and with ease to get support from developer community, hardware platform Arduino UNO is used, which contain an *ATmega328* MCU.

This reduces time developing digital controller for the system, allowing direct proof of concepts, using C language syntax, which is a programming language common on engineering courses. A previous introductory course in microcomputers/microcontrollers or embedded systems may be useful but not mandatory for students.

Hardware for displaying signal is any computer running the User Interface created in different scripting languages: Python, Processing, Octave and MatLab, leaving to the student or the professor wide options about write off-line and even on-line processing algorithms.

Advantages on using open source, free licensed software, are several and already known: possibility to modify existing code to own needs with proper credits to original developer; to develop systems at low cost; it can be shared publicly, improved and maintained by a growing community, and also, is an important issue in public education, allowing students learn and study code from other developers, creating new software without rely on commercial ones or commit piracy.

II. Description of the system.

The platform is divided on three main stages named *sensing*, *processing* and *displaying*, according to Webster [6]. Extended description of a generalized instrumentation system includes a *feedback* stage, which is also included in this work.

A. Sensing ECG signal.

Integrated on one board, is performed by the following elements and their functions are described briefly.

Ag/AgCl electrodes: sense the ionic current transducing it to electronic current, supplying a proper working signal.

Shielded cables: coaxial cable type is used to connect electrodes with circuitry through core conductor and, braid shield is used to reflect EMI or conduct it to ground system; also is used in drive shield technique.

Protection Diodes: are present on each lead. Provide protection to patient and system from high voltage surge, for example, from electrostatic discharge or in clinical environment, from defibrillator pulse, which is drained to ground system, avoiding dangerous currents returns.

Instrumentation amplifier (IA): high impedance inputs of AD620 IA are suitable to be coupled to the biopotential-low current signal, brings a moderated amplification (G=10), configurable via a potentiometer or a fixed resistor, performs reduction of common mode voltage (mainly 60 Hz hum), and its reference terminal is further used to adjust dynamically the zero volts level, reducing baseline wander (offset) of ECG.

Offset compensator: an active one pole low pass filter (LPF) or analog integrator, cut off frequency 0.5Hz, its output is applied to the IA reference terminal and its purpose is to counteract low frequency amplitude variations due to movement of electrodes or cables. This circuit can also be seen as a part of processing stage because it uses the signal already acquired.

Sensing stage is powered with four 1.5 Volts coin cell batteries in order to source enough voltage to sensing circuits but small amounts of current back to the patient via a third electrode in Right Leg, used as ground loop, featuring IA as double-ended differential.

RLD circuit: Other reason to use batteries is the implementation of a feedback loop, which drains a small amount of current back to the subject. Consist on a buffer and an inverting OpAmp, whose input is an average of common voltage and, its output is applied on subject's right leg, intending to reduce the amplitude of interference. Design of RLD is according to [7]. This circuit provides also a high impedance path to ground system, protecting patient from possible high currents.

B. Analog signal processing.

A separate board contains processing stage. Connection between sensing and processing boards is made with headers in a stackable fashion. Here, the output of the IA is coupled to an Isolation Amplifier, in order to separate the subject from possible high currents from this stage, which uses a regulated symmetric power supply of 15V, therefore, it provides isolation security to patient. Then, signal goes to analog processing. Elements and its function are described next.

Band pass filter: a LPF and a high pass filter (HPF) acting as anti-alias filter, limiting interest signal's band width (BW). For ECG, common BW is 0.01-250Hz [7]. Both filters are second order Sallen-Key with Butterworth characteristic, in order to provide an output with flat response in pass band (G=1).

Notch filter: rejecting 60Hz interference, using two active double tee notch filters in cascade, G=1, in which resistor values are variable via double potentiometers to adjust cut off frequency.

Amplifier: a non inverter amplifier to provide gain to processed signal (G=100-500), spanning its amplitude to dynamic range of ADC; maximum signal peak-to-peak amplitude must be adjusted and lower than Vref of ADC.

Non inverting summing amplifier: only adds without gain (G=1) ECG signal and offset voltage Vref/2 in order to provide only positive voltage to a single ended ADC.

A microcontroller: it samples analog signal periodically according to Nyquist theorem, and sends data to a computer running user interface. ADC resolution is 10 bits, meaning *4.8 mV/LSB.*

C. Displaying signal.

Displaying is performed by a host computer running a script to display received data in a plot. Scripting languages used are Phyton, Processing, Octave and Matlab.

Using first three languages make suitable for the student: using free licensed, well documented software, which include package/functions for signal processing and interfacing hardware through the Serial Port. Using Matlab is frequent in engineering courses, for this reason we also develop a script for it.

III. WORKING WITH THE ECG IN LAB.

In both courses, all students provide written informed consent prior to its participation using the system and recording data. The experimentation procedure was performed according the Helsinki Declaration of 1975 and approved by the ethical committee of the institution.

In order to get a good understanding of how an ECG works diverse activities are recommended to experience in Lab in two sections.

A. Lab first section: Analog.

Is Focused on working with hardware shown in Figure 1, an oscilloscope and a power supply. All analog design and modifications are referenced on current regulation and compliance with AHA.

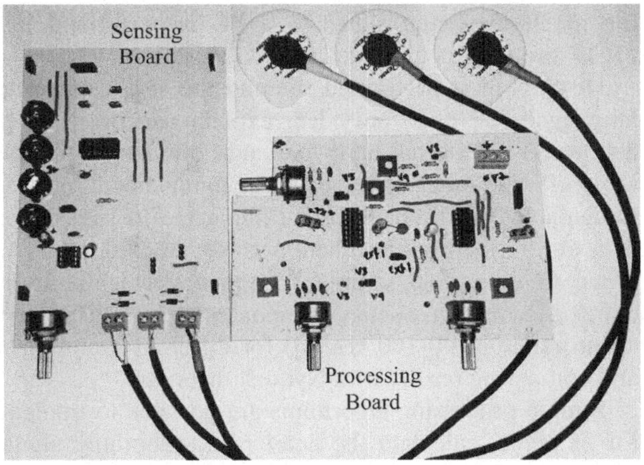

Fig. 1 ECG Educative Boards; left side is the Sensing and decoupling board with electrodes connected using short length cables (0.5m) , right side is the analog processing board.

In Sensing stage, relevant test points are accessible on board trough male headers terminals, such as the common voltage coming from each electrode via two averaging resistors, the output of the IA, the output of the offset compensator, as well as the output of RLD and drive shield circuits.

These signals can be first displayed on an oscilloscope and then modify some parameters of the system, identifying its characteristic behavior. For example, eliminate third electrode, grounding it, ground the reference terminal from IA, change values of resistor and capacitance in offset compensator, using different electrode cable lengths, implement RLD and/or shield drive circuits and observe its effect on signal's common voltage.

Real artifacts can be induced and its characteristic shape/waveform or visual features registered, getting a realistic experience that sometimes is assumed with virtual instrumentation. For example, to produce movement on cables, induce electrostatic discharge on subject, change electrode temperature, or exaggerate respiration movement.

Processing stage is useful to reduce persistent noise as 60Hz interference. On board, filters inputs and outputs are also accessible, being able to test its performance using solely an external signal. Moderate level of adjustment to cut off frequencies can be made via trimpots or potentiometers. Also is possible to verify how conditioning circuit works coupling ac signal to a single-ended ADC, adding gain and dc offset.

B. Lab second section: Digital.

Conducted using hardware and software involved in acquiring signal. Once the signal has been processed, digitizing and transferring signal to a PC is performed by the MCU, programmed using C syntax language.

Displaying is performed running the script in preferred language. The code provided, explain use of serial ports connected to external hardware and how data is plotted on screen. Some other activities are recommended once signal is displayed. A first one is writing code to save received data to a formatted file. Then, perform an exploratory technique to determine sampling frequency calculating FFT, extracting main frequency components for an ECG (which are in range of 40-100 Hz [8]) for different sampling rates and comparing results with Nyquist theorem.

Simple processing algorithms are possible to implement, for example, calculate the heart rate, smoothing signal or even to implement digital filters on line.

Analyze relation between resolution in bits of sampled data, sampling rate, number of sampled channels and bit rate during data transference are of big importance, because based on the technique used, polling or interrupt to sampling and transfer data, the system behavior could be adequate or erratic.

IV. RESULTS.

A basic ECG platform was developed as a prototype to be used in our courses. Although the subject is not new, is very common at undergraduate course on Biomedical Instrumentation, in order to provide a first insight registering physiological signals. The use of this ECG is suitable in real situations with actual subjects due to be provided with proper protection devices.

The whole system cost is less than 100 U.S. dollars.

Regarding basic ECG, once basic learning needs are covered, new topics need to be addressed in a new platform, so further work need to be done, for example: include resistor network to generate Wilson's Central Terminal and Goldberger Reference to measure ventricular and augmented leads; if only one IA and one ADC is used, then, design and implement a multiplexing stage is needed along with its control via the microcontroller used; once this stage is implemented, a calibration pulse can be included.

Recommendations from students are also received, for example, to shield the full system from electromagnetic noise and grounding the system to earth ground, to use commercial available lead cables, possibility to change the order in filters and the sequence they are connected, implementation of different types and high quality notch filters.

V. CONCLUSIONS

Experiences in Lab are important aspects for students, because they want to set in practice theoretical knowledge, using real equipment, sometimes restricted or unavailable,

Availability on hardware and software encourage students to develop their own ECG or new and diverse instruments, and also for teachers, developing their own instrumentation systems fitting their needs on research and teaching.

Biomedical research may benefit from using open source software for signal processing due its availability and a growing community doing research using these tools.

ACKNOWLEDGMENT

Authors would like to thank to UNAM-PROYECTO PAPIME PE104715 for support in this project.

CONFLICT OF INTEREST

The authors declare that they have no conflict of interest.

REFERENCES

1. Jacqui Crawford, Linda Doherty (2012). *Practical Aspects of ECG Recording*. M&K Publishing
2. Laurece Street (2008) *Introduction to Biomedical Engineering Technology*. Taylor & Francis Group, CRC Press.
3. A. J. Macy (2003) *Inquiry-based biomedical signal processing laboratory: from practice to simulation*. Engineering in Medicine and Biology Society. Proceedings of the 25th Annual International Conference of the IEEE, pp. 3513-3516 Vol.4. DOI: 10.1109/IEMBS.2003.1280909
4. M. A. B. S. Akhanda, S. M. F. Islam et al (2014) *Monitoring the performance of computer user by analyzing physiological nals.* Computer and Information Technology (ICCIT), 16th International Conference on, Khulna, pp. 120-125. DOI: 10.1109/ICCITechn.2014.6997300
5. Marcs, D. M. (2006). *Developmental laboratories for biomedical instrumentation and digital signal processing with virtual instrument technology and diverse software techniques*. Available from ProQuest Dissertations & Theses Global. http://search.proquest.com/docview/304979454?accountid=14598
6. J. G. Webster-editor (2010) *Medical Instrumentation Application and design*. Fourth edition. John Wiley and Sons Inc.
7. B. B. Winter and J. G. Webster (1983) *Driven-right-leg circuit design*. IEEE Transactions on Biomedical Engineering, BME-30, no. 1, pp. 62-66, doi: 10.1109/TBME.1983.325168
8. B H Brown, R H Smallwood et al (1999). *Medical Physics and Biomedical Engineering*. IOP Publishing Ltd.

Biomechanical analysis of a cranial Patient Specific Implant on the interface with the bone using the Finite Element Method

J.M. Díaz[1], O.A. González-Estrada[1] and C.I. López[2]

[1] Universidad Industrial de Santander, School of Mechanical Engineering, Bucaramanga, Colombia
[2] Universidad Industrial de Santander, School of Industrial Design, Bucaramanga, Colombia

Abstract— **New advance technologies based on reverse engineering, design and additive manufacturing, have expanded design capabilities for biomedical applications to include Patient Specific Implants (PSI). This change in design paradigms needs advanced tools to assess the mechanical performance of the product, and simulate the impact on the patient. In this work, we perform a structural analysis on the interface of a cranial PSI under static loading conditions. Based on those simulations, we have identified the regions with high stress and strain and checked the failure criteria both in the implant and the skull. We evaluate the quality of the design of the implant and determine their response given different materials, in order to ensure optimality of the final product to be manufactured.**

Keywords— **Biomechanics, Finite Element Method, Patient Specific Implant, Cranial Implant.**

I. Introduction

Biomechanics studies the mechanical behaviour of biological systems. It helps to provide solutions to the injuries the body is not capable of healing itself, i.e., cases in which natural recovery is not possible. This could be accomplished by locating orthopaedic implants where the tissue did not heal to help with the recovery of a patient [1].

The study of these biomedical implantable devices involves areas such as materials, medicine and engineering [2]. These disciplines cannot be mutually exclusive if we want to ensure that the final product is in agreement with the expectations of the design. Moreover, it is very important to evaluate the quality of the designs before manufacturing them, in order to find any possible existing flaws, optimise the implant prototype and guarantee the patient is getting a product that is going to enhance his quality of life [3] [1].

This testing process can be done through non-destructive tests by means of numerical simulation, using finite element analysis (FEA) [4] to evaluate the implant-bone interface behaviour [5]. The model for the skull-implant assembly is considered, subjected to given boundary conditions. The structural analysis is performed in order to evaluate the stress and strain fields. Failure criteria for the materials are also considered. It is expected that critical zones are located in the interface between the skull and the implant, and that will be the main subject of this work.

II. Methods

A. Model identification and geometry definition

The models of the study are 3D skull reconstructions obtained with a multislice spiral CT scanner, Toshiba Aquilon, in DICOM format. Using the software Invesalius 3, with semiautomatic segmentation, they are converted to STL format. Later, they receive CAD treatment with Rhino®. Based on the reconstruction, two different implants are designed and all models imported to ANSYS 15.0.

For optimisation purposes, we removed some of the bone tissue, leaving only the bone portion of the cranial vault near to the defect, thus simplifying the model.

B. Material models

Mechanical properties of the materials to use in the numerical models are defined. The software used in the CAD reconstruction did not allow displaying the entire Hounsfield scale [6]; therefore, the values used to represent the mechanical properties are taken from the literature. In the case of bone, Galicer [7] suggests that the value of the bone matrix density is $\rho_{mo} = 2.02\ \mathrm{g/cm^3}$. The bone elastic constants, Young modulus and Poisson ratio, are defined by the following equations [8]:

$$E{=}1763\rho 3{,}2 \text{ if } \rho{>}1{,}2\ \mathrm{g/cm^3} \tag{1}$$

$$\upsilon{=}0{,}32 \text{ if } \rho{>}1{,}2\ \mathrm{g/cm^3} \tag{2}$$

By replacing the bone matrix density in (1), we have $E =$ 16725 MPa.

Table 1 shows the Polyether ether ketone (PEEK) and titanium alloy Ti6Al4V properties, suggested by Safi [9]:

Table 1. Young modulus E and Poisson ratio υ for PEEK and Ti6Al4V.

Material	E [MPa]	υ
PEEK	4000	0,4
Ti6Al4V	110300	0,36

I. Torres et al. (eds.), *VII Latin American Congress on Biomedical Engineering CLAIB 2016, Bucaramanga, Santander, Colombia, October 26th -28th, 2016*,
IFMBE Proceedings 60,
DOI: 10.1007/978-981-10-4086-3_102

C. Contact type definition

Many numerical models are available to define the contact conditions. For this problem we have chosen a linear contact condition, defined using Multi-Point Constraint (MPC). Linear contact models are faster to solve and less computational demanding. Extension of this work to include nonlinear models is expected. The MPC condition defines a state of no separation between the implant and the bone tissue. MPC are recommended for non-conforming contact surfaces, as it is the case for the implant-bone assembly due to features of the design introduced in the CAD model.

D. Meshing

Meshing is a fundamental part in the FEA, since a coarse mesh can generate results with low accuracy, whereas a fine mesh can generate a high computational cost and poor numerical conditioning. Safi [9] recommends the use of quadratic tetrahedral elements (SOLID 187).

Two nonconforming meshes were used: one for the implant and one for the skull. Both were generated with the *Patch Independent* method to overcome issues related to the STL features imported from the CAD. This process is ideal for models with complex geometries, with very sharp angles, as the algorithm ensures that the mesh is refined where needed, but maintains larger elements in areas where possible. Table 2 shows the parameters used to generate the meshes for implant 1, with a total of 529113 elements. The resulting meshes are shown in Figure 1.

Table 2. Parameters used for the mesh of implant 1.

Parameter	Implant	Skull
Element's maximum size	2 mm	5 mm
Element's minimum size	0,75 mm	1,5 mm
Number of cells across a gap	8	3
Curvature angle	7,5°	
Smooth transition	Yes	Yes
Number of elements	180303	167450

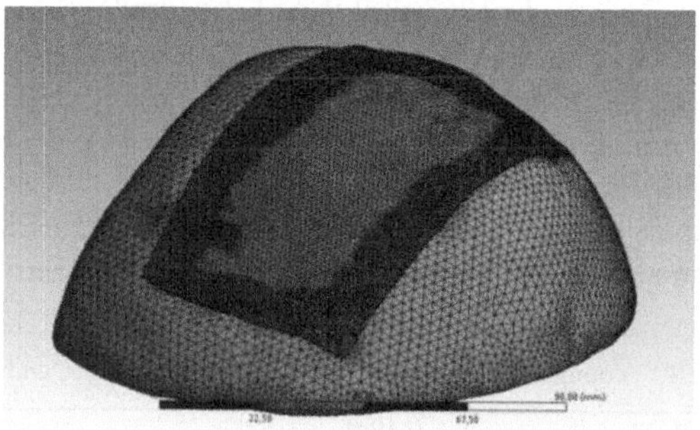

Figure 1: Final mesh for the model of the skull with implant 1.

Table 3 shows the values of the parameters used to generate the meshes for the implant 2, with a total of 681643 elements. Figure 2 shows the mesh for the implant and bone tissue.

Table 3. Parameters used for the mesh of implant 2.

Parameters	Implant	Skull
Element's maximum size	2 mm	3 mm
Element's minimum size	0,75 mm	1 mm
Number of cells across a gap	3	3
Smooth transition	Yes	No
Number of elements	199685	250001

Figure 2: Final mesh for the skull with implant 2.

E. Boundary conditions and applied loads

In the skull part, displacement in all directions is restricted along the boundary defined by the cutting planes. For the implants, displacement is restricted in the x and y directions of the surfaces representing the perforations aimed for the location of the screws, thus simulating the screw clamping.

The applied load is a pressure of 1 MPa which will be applied on an circular area with a radius of 15 mm in the central zone of the implant, this is equivalent to statically applying 70 kg, as used previously in [9]. Additionally, to simulate the pressure exerted by the screws, pressure is applied around the holes aimed to hold the screws. The value is defined in terms of the torque recommended by ASTM F- 543 [6] for HA 2.0 screws:

$$P = 4T/\pi d^3 \quad (3)$$

Where P is the applied pressure, T is the torque, d is diameter of the head of the screw, giving a value of $P =$ 6,96 MPa.

III. Results

The quality of the implant designs will be assessed taking into account the stresses obtained for the implant and the skull, which should be below the material limits.

A. Skull with implant 1 using PEEK

The obtained results are analysed under the von Mises criterion, with an applied load $P = 1$ MPa. In Figure 3, we observe the obtained stress distribution. The greatest stress is located in the implant, on the region in contact with the fracture, and has a value of 96.291 MPa. This may be because the PSI implant contour adjusts tightly to the shape of the fracture, and the interference is large. It is recommended to change the CAD implant so that when the two parts come into contact interference is smaller.

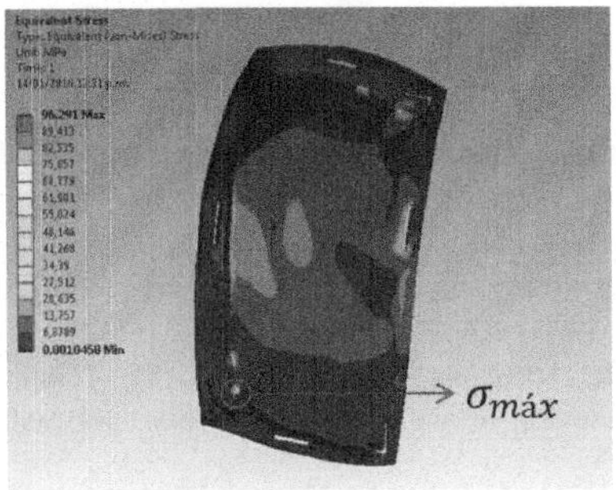

Figure 3: Stress distribution in the implant 1 using PEEK.

Figure 4 shows the strain and displacements, indicating that there are not large deformations. Structural stability is an important condition to ensure the health of the patient.

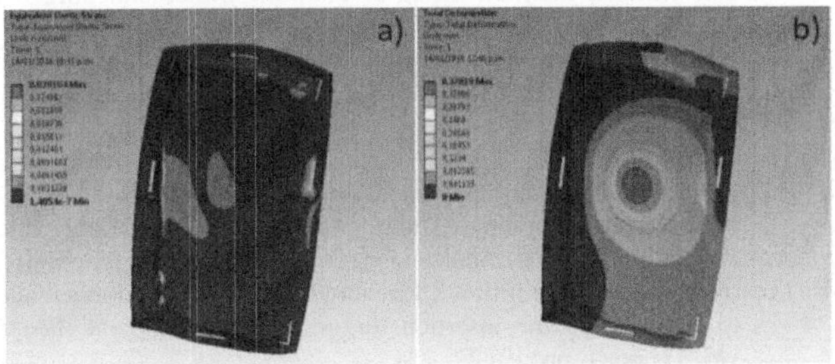

Figure 4: Strain (a) and displacements (b) in implant 1 using PEEK.

B. Skull with implant 1 using titanium alloy (Ti6Al4V).

The simulation is performed changing the material assigned to the implant part to titanium alloy Ti6Al4V. The maximum stress for this model is 125.83 MPa. Compared to the stress for the PEEK model, this one is slightly larger due to the higher elastic modulus. Using Ti6Al4V, it can withstand a pressure $P = 5$ MPa, with a maximum stress of 629.16 MPa, below the elastic limit of $S_y = 675$ MPa. The maximum displacement for the Ti6Al4V.is 0.01543 mm, which compared with the PEEK model is much lower. This behaviour is ideal as it is intended that the implant is deformed as little as possible once it has been placed on the patient. However, notice that the more rigid Ti6Al4V has a stronger stress shielding effect, bad for bone resorption. Table 4 contains the results of the analysis for both implants.

Table 4: Results obtained for the model of the skull with implants 1 and 2.

Material	Max. stress [MPa]	Max. strain	Max. displacement [mm]
Implant 1			
PEEK	96,291	0,0281	0,3701
Ti6Al4V Alloy	125,83	0,0013	0,0154
Implant 2			
PEEK	32,575	0,0092	0,1448
Ti6Al4V Alloy	47,034	0,0004	0,0050

C. Skull with implant 2 using PEEK.

Considering the previous results we generate an enhanced implant model. We use the same boundary conditions for the implant 2. Figure 5 shows the stress distribution obtained for this model using PEEK.

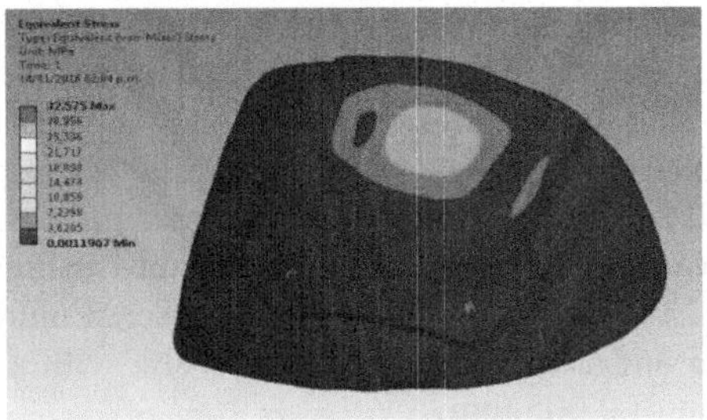

Figure 5: Stress distribution for the skull with implant 2 using PEEK.

The maximum stress is located on the face that comes into contact with the bone, with 32.575 MPa. This implant model has a smoother contour in the area, which contact the fracture. Comparing the results obtained for the implant 1 using PEEK, and given that the behaviour of the stress of this model is linear, it can resist three times more load. Figure 6 shows the strain and displacements obtained, which are much lower compared to the implant 1 using PEEK. The maximum displacement has a value of 0.1448 mm.

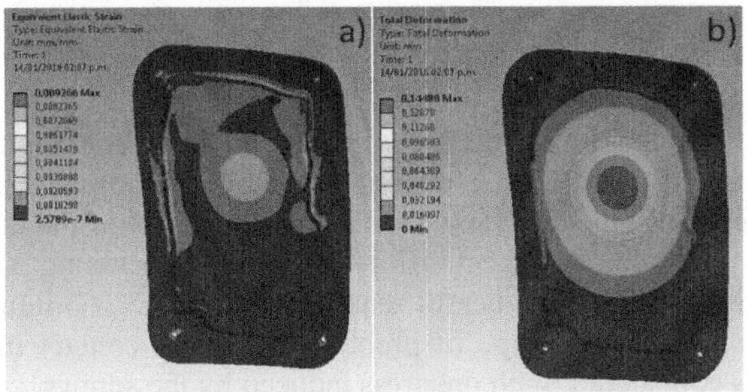

Figure 6: Strain (a) and (b) displacements in implant 2 using PEEK.

D. Skull with implant 2 using titanium alloy (Ti6Al4V).

Simulation is performed using the titanium alloy Ti6Al4V. Figure 7 shows the stress distribution for this case.

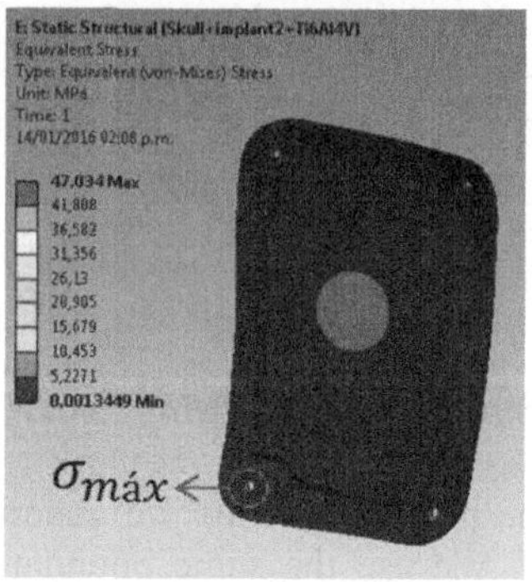

Figure 7: Stress distribution in the implant 2 using Ti6Al4V.

Maximum stress is located in the holes provided to secure the implant with screws and have a value of 47 MPa. Different behaviour from the previous model can be observed because the titanium alloy is more rigid than PEEK and the contour that gets in contact with the bone is practically not deformed so the holes act as stress concentrators. Comparing this value with the obtained in implant 1 using Ti6Al4V, implant 2 resists a load three times higher. Figure 8 shows the strain and displacements respectively, with a max. displacement of 0.005 mm.

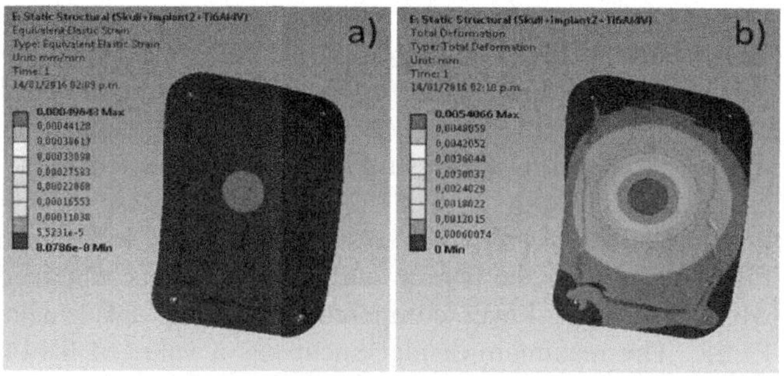

Figure 8: Strain (a) and (b) displacements in implant 2 using Ti6Al4V.

IV. Conclusions

For this work it was assumed that the bone behaves as an isotropic linear elastic material, which is an approximation to simplify the analysis. Meshing is a key part of the analysis and requires great effort in the preprocessing stage. Manual interaction is still required to produce good quality meshes for such type of problems. A linear contact model was considered, with prestress induced by the screws.

In the analysis performed, implant 2 is superior to implant 1 model, primarily because implant 2 is a redesign of the implant 1, in which the geometry is redefined to improve the contour. PEEK exhibits lower values of stress that the more rigid Ti6Al4V, which is in agreement with the known stress shielding effect of Titanium alloys. We show the feasibility of virtual prototyping for modelling PSI implants.

Extension of this work could include nonlinear models for contact conditions, and more complex material models, including varying bone density obtained from the scans.

Acknowledgment

This work was supported in part by VIE-UIS under Code No. 1361.

V. References

[1] Schmitt K, Niederer P, Walz F. (2013). Trauma biomechanics: introduction to accidental injury, Springer Science & Business Media.

[2] Juanola-Feliu E, Colomer-Farrarons J, Miribel-Català P, et. al. (2012). Market challenges facing academic research in commercializing nano-enabled implantable devices for in-vivo biomedical analysis. Technovation, vol. 32, no. 3–4, pp. 193–204.

[3] Sumida T, Otawa N, Kamata Y, et. al.(2015.) Custom-made titanium devices as membranes for bone augmentation in implant treatment: Clinical application and the comparison with conventional titanium mesh. J. Cranio-Maxillofacial Surg., vol. 43, no. 10, pp. 2183–2188.

[4] Diaz Lantada A, (2013) Handbook on Advanced Design and Manufacturing Technologies for Biomedical Devices. Springer New York.

[5] Harrysson O, Hosni Y, J. F. Nayfeh (2007) Custom-designed orthopedic implants evaluated using finite element analysis of patient-specific computed tomography data: femoral-component case study, BMC Musculoskelet. Disord., vol. 8, no. iv, pp. 91.

[6] López C, Moreno J, Pinillos J, Galeano C (2014) Experimental study to establishing the insertion torque on dental implant, Iteckne, vol. 11, no. 1, pp. 17–27.

[7] Galicer Sánchez M (2012) Simulación del Bruxismo y su Efecto en la Estructura Ósea de la Mandíbula.Universidad de Sevilla pp 184.

[8] Taylor W, Roland E, Ploeg H, et al. (2002) Determination of orthotropic bone elastic constants using FEA and modal analysis, vol. 35, pp. 767–773.

[9] Safi Y, Hohenberger S, Robbenmenke C, Banki F, et. al. (2010) Investigation of the failure behavior of a cranial implant-skull model under different load conditions using FEM, in *SIMULIA Customer Conference*, 2010, no. Spetzger, pp. 1–16.

Author: O. A. González-Estrada; C.I. López2
Institute: Universidad Industrial de Santander
Street: Cra 27 street 9
City: Bucaramanga
Country: Colombia
Email: agonzale@uis.edu.co; clalogu@uis.edu.co

Numerical simulation and fitting of tumor growth kinetics models using Python

E.E. Ramirez Torres[1], L.E. Bergues Cabrales[2], R.E. Rivero Labrada[1] and J. Lambert Cause[1]

[1] Departamento de Ingeniería Biomédica, Universidad de Oriente, Santiago de Cuba, Cuba
[2] Centro Nacional de Electromagnetismo Aplicado, Universidad de Oriente, Santiago de Cuba, Cuba

Abstract— **Tumor growth kinetics may be mathematically described by ordinary differential equations. Fitting experimental data to different models often leads to non-convex functions optimization with multiple local minima. The aim of this paper is to present numerical methods for simulation and fitting of ordinary-differential-equation models of malignant tumors implemented in Python. The suggested protocol combines the generation of 1000 random initial parameter values with the Nelder-Mead simplex direct search for data fitting in order to escape from local minima. Sums of squared residuals obtained are compared with those achieved by the Levenberg-Marquardt method.**

Keywords—**Tumor growth kinetics, ordinary differential equations, Nelder-Mead method, Levenberg-Marquardt method, Python.**

I. Introduction

For decades, mathematical models of tumor growth kinetics (TGK) have provided an important tool for testing hypotheses and treatment planning. In particular, models based on ordinary differential equations (ODE) have been useful to describe TGK [1].

Benzekry et al. [2] outlined, using Matlab®, a protocol based on the Nelder-Mead simplex method for data fitting although details of how to avoid local minima, a common problem in nonlinear and non convex functions optimization, are not given. For this purpose, it has been suggested the Levenberg-Marquardt method of nonlinear least-squares minimization for 1 000 random initial parameter values [3]. Nevertheless, we are not aware of any reference of the combination of multi-start optimization with Nelder-Mead method using free software built-in functions. This paper describes how to fit TGK to ODE models with Nelder-Mead method for 1 000 random initial parameter values using free software Python for an improvement in local-minima rejection. The results are compared with the Levenberg-Marquardt optimization.

II. Methods

A. Mathematical models

The first ODE model used is the conventional Gompertz growth for unperturbed tumors [4], named model I:

$$\frac{dc}{dt} = -\lambda c \log\left(\frac{c}{K}\right) \tag{1}$$

where c is the number of tumor cells, λ is the net population growth rate and K is the carrying capacity.

Hahnfeldt et al. [5] considered a modification with a variable carrying capacity:

$$\frac{dK}{dt} = \phi c - \varphi K c^{(2/3)} \tag{2}$$

where ϕ and φ are constants that represent stimulation and inhibition rates of angiogenesis respectively. The combination of equations (1) and (2) is one of the most studied models [6], hereinafter model II.

B. Curve-fitting techniques and numerical implementation

Equations (1) and (2) are implemented in Python (version 3.4.3), which is a free, interpreted and high-level language. Python has gradually become the unofficial standard of scientific computing and is being used successfully in applications for life sciences [7]. It offers a free alternative to Matlab and other scientific computing programs without practical limitations for fitting and simulation of ODE models for TGK.

In order to fit and simulate models I and II, a methodology is established:

1. Import the data from file to Python.
2. Generate 1 000 random initial parameter values with uniform distribution in order to avoid local minima.
3. Fit data to models I and II starting from all initial parameter states using **fmin**.

I. Torres et al. (eds.), *VII Latin American Congress on Biomedical Engineering CLAIB 2016, Bucaramanga, Santander, Colombia, October 26th -28th, 2016*,
IFMBE Proceedings 60,
DOI: 10.1007/978-981-10-4086-3_103

4. Choose the fit with lower sum of square residuals (SSR) and plot numerical simulation with these parameters.

For the numerical integration of models I and II, the **odeint** function is used, which is included in the integration routines of the high level pack of scientific computing **scipy**. The **fmin** function is used for data fitting and its basis is the Nelder-Mead method. It is included in the **scipy.optimize** module, which provides useful algorithms for function minimization.

Levenberg-Marquardt method is widely applied for SSR minimization. It is the basis of the Python function **curve_fit**. This method is also used to fit models I and II in order to compare its results with those obtained by the Nelder-Mead method, starting from 1 000 random initial parameter values. It must be noted that these initial points are not be confused with the prior of Bayesian parameter estimation.

SSR is chosen as cost function. Minimizing SSR coincides with finding maximum likelihood as scattered data follow a Gaussian distribution with constant standard deviation [8].

C. *Experimental data*

In order to fit models I and II, experimental data of Ehrlich and fibrosarcoma Sa-37 tumors [9] are used.

III. Results and discussion

Results for model I, with two freedom degrees, are very similar when both fitting algorithms are used (Table 1). However, the results differ as these two algorithms are applied to model II, with four freedom degrees (Table 2).

Table 1 Parameters obtained for model I

Parameters.	Nelder-Mead method	Levenberg-Marquardt method
	Ehrlich Tumor	Fibrosarcoma Sa-37 tumor
λ	0.2619	0.2460
K	3539.8001	3092.9095

Table 2 Parameters obtained for model II

Parameters.	Nelder-Mead method	Levenberg-Marquardt method
	Ehrlich Tumor	Fibrosarcoma Sa-37 tumor
λ	0.1539	0.0708
K0	15187.0137	136147.8016
ϕ	1.1684	1.5336
φ	0.0042	0.0051

The fit obtained with the Nelder-Mead method corresponds with the minimum SSR found that makes sense from biophysical point of view. Negative values of ϕ and φ are not considered.

Simulations carried out with Nelder-Mead method are shown in Figures 1 and 2 for Ehrlich and fibrosarcoma Sa-37 tumors, respectively.

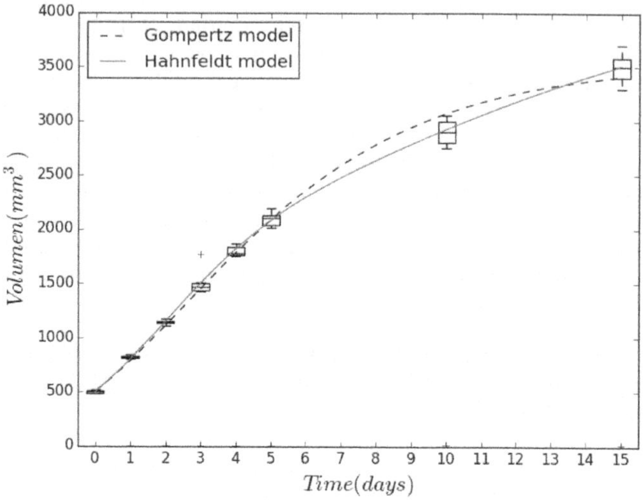

Fig. 1 Fit of Ehrlich tumor data with Nelder-Mead method.

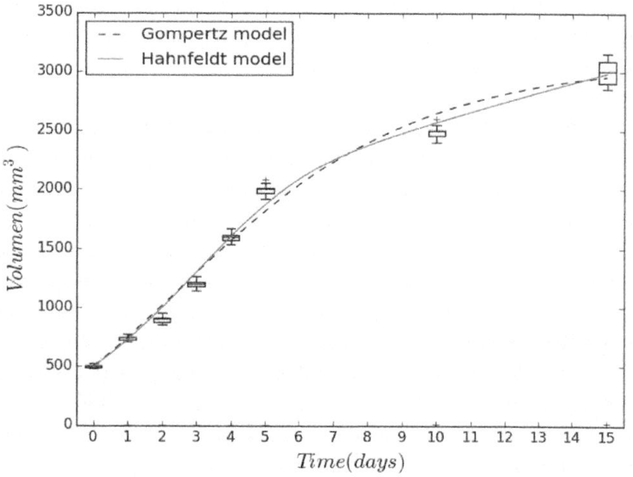

Fig. 2 Fit of fibrosarcoma Sa-37 tumor data with Nelder-Mead method.

Simulations carried out with Levenberg-Marquardt fit are shown in Figures 3 and 4 for Ehrlich and fibrosarcoma Sa-37 tumors, respectively.

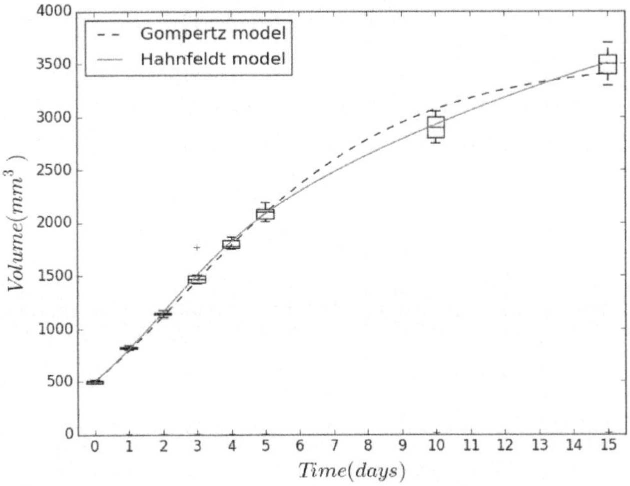

Fig. 3 Fit of Ehrlich tumor data with Levenberg-Marquardt method.

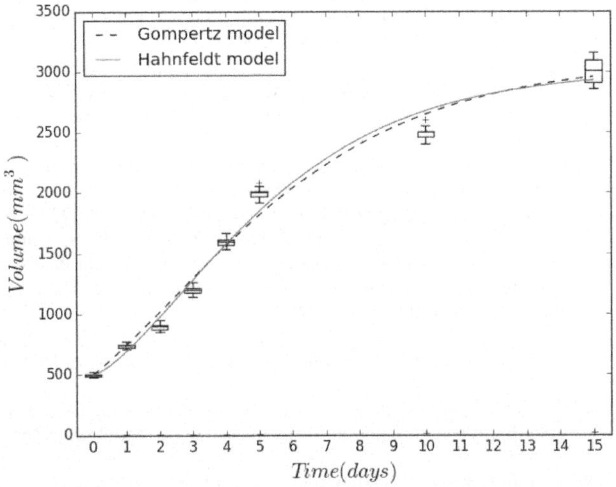

Fig. 4 Fit of fibrosarcoma Sa-37 data with Levenberg-Marquardt method.

Table 3 shows SSR values obtained from fitting data of Ehrlich and fibrosarcoma Sa-37 tumors to models I and II for 1 000 random initial parameters values, using Nelder-Mead and Levenberg-Marquardt methods. Both methods reveal the same minimum-SSR values, except for model II on fibrosarcoma Sa-37, in which the Levenberg-Marquardt method finds a local minimum. This is no guarantee that Nelder-Mead fit has found a global minimum.

Table 4 shows minimum SSR values obtained with Nelder-Mead method for both tumor types, models I and II, and different numbers of random initial parameters. For both tumor types, the minimum-SSR value obtained with model II depends on the number of start points, which is not the case for model I, reaching its lower value for the highest number of the initial random parameters. Despite results shown in Tables 3 and 4, Figures 1-4 do not show great differences on Ehrlich and fibrosarcoma Sa-37 TGK when models I and II are used. This may indicate that both models are adequate to fit experimental data. Nevertheless, the selection of the best model to describe TGK still remains in debate [6, 10].

Table 3 Minimum SSR found in fitting with 1000 initial random values of parameters

Tumor/Model	Minimum SSR	
	Nelder-Mead method	Levenberg-Marquardt method
Ehrlich/I	39039.0209	39039.0209
Ehrlich/II	1425.7286	1424.7143
Fibrosarcoma Sa-37/I	84999.6556	84999.6556
Fibrosarcoma Sa-37/II	43324.6941	79693.2052

Table 4 Minimum SSR obtained from fitting with Nelder-Mead method for different amounts of random initial parameters

Tumor/Model	Minimum SSR		
	5 initial random parameters	50 initial random parameters	500 initial random parameters
Ehrlich/I	39039.0209	39039.0209	39039.0209
Ehrlich/II	1971.6492	1426.0796	1426.0796
Fibrosarcoma Sa-37/I	84999.6556	84999.6556	84999.6556
Fibrosarcoma Sa-37/II	77924.9286	59881.3470	46309.2313

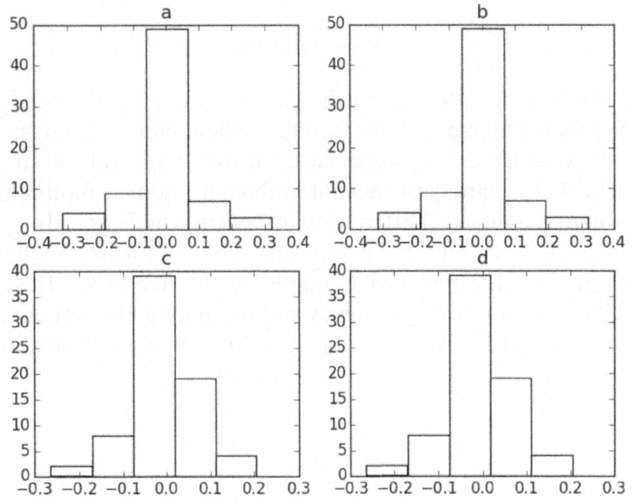

Fig. 5 Histograms of residuals for Ehrlich tumor data. (a) Model I/Nelder-Mead fit (b) Model I/Levenberg-Marquardt fit (c) Model II/Nelder-Mead fit (d)) Model II/Levenberg-Marquardt fit.

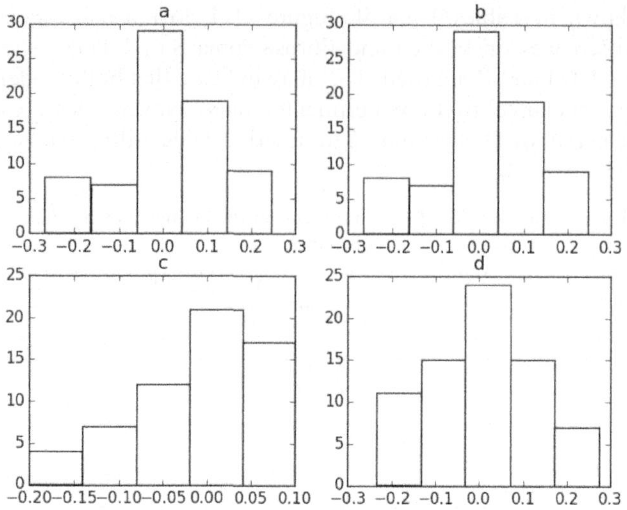

Fig. 6 Histograms of residuals for fibrosarcoma Sa-37 data. (a) Model I/Nelder-Mead fit (b) Model I/Levenberg-Marquardt fit (c) Model II/Nelder-Mead fit (d)) Model II/Levenberg-Marquardt fit.

Error histograms between the data and the fit for Ehrlich and fibrosarcoma Sa-37 tumors are shown in Figures 5 and 6, respectively along with their dependence with models I and II, and data-fitting methods. An asymmetry of these histograms suggests that Gaussian model of error with constant standard deviation is, at least, questionable [1]. Nelder-Mead method makes easier to choose a customized function for minimization, being an important advantage of this method.

This methodology can be extended to any ODE model.

IV. CONCLUSIONS

Fitting of tumor growth kinetics using Python with Nelder-Mead method for 1 000 random initial parameter values achieves in some cases lower SSR values than Levenberg-Marquardt method, although a global minimum is not guaranteed. Differences in both methods are always obtained with Hahnfeldt but not with Gompertz model, suggesting that the model complexity influences on fit results. The number of iteration starting points also can alter the minimum SSR found, being lower for more points, as expected.

ACKNOWLEDGMENT

Authors wish to thank Sébastien Benzekry and Manuel Rodríguez Bebelagua for their technical assistance. Also, we thank in a special way the reviewers for their invaluable recommendations and suggestions.

CONFLICT OF INTEREST

The authors declare that they have no conflict of interest.

REFERENCES

1. Benzekry S, Lamont C, Beheshti A, Tracz A, Ebos JML, et al. (2014) Classical Mathematical Models for Description and Prediction of Experimental Tumor Growth. PLoS Comput Biol 10(8): e1003800. doi:10.1371/journal.pcbi.1003800.
2. Benzekry S, Beheshti A, Hahnfeldt P et al. (2015) Capturing the Driving Role of Tumor-Host Crosstalk in a Dynamical Model of Tumor Growth. Bio-protocol, 2015, 5 (21), <http://bio-protocol.org/e1644>. <hal-01222068>.
3. Poleszczuk J, Hahnfeldt P, Enderling H (2015) Therapeutic Implications from Sensitivity Analysis of Tumor Angiogenesis Models. PLoS ONE 10(3): e0120007. doi:10.1371/journal.pone.0120007.
4. Akanuma A (1978) Parameter analysis of Gompertzian function growth model in clinical tumors. Eur J Cancer 14: 681–688. Available: http://www.ncbi.nlm. nih.gov/pubmed/658092.
5. Hahnfeldt P, Panigrahy D, Folkman J et al. (1999) Tumor development under angiogenic signaling: a dynamical theory of tumor growth, treatment response, and postvascular dormancy. Cancer Res 1999; 59(19): 4770-5.
6. Enderling, H, Chaplain M A J (2014) Mathematical Modeling of Tumor Growth and Treatment. Current Pharmaceutical Design, 2014, 20, 000-000.
7. Libeskind-Hadas R, Bush E (2014) Computing for Biologists: Python Programming and Principles, 1st ed., Cambridge Univ. Press.
8. Motulsky H, Christopoulos A (2004) Fitting models to biological data using linear and nonlinear regression. Oxford University Press.
9. Ciria H M C, Quevedo M C S, Cabrales L E B et al (2004) Antitumor effectiveness of different amounts of electrical charge in Ehrlich and fibrosarcoma Sa-37 tumors, BMC Cancer 4 87–91.
10. Murphy H, Jaafari H, Dobrovolny H M (2016) Differences in predictions of ODE models of tumor growth: a cautionary example, Murphy et al. BMC Cancer (2016) 16:163. DOI 10.1186/s12885-016-2164-x.

Author: Erick Eduardo Ramirez Torres
Institute: Universidad de Oriente
Street: Ave. Las Américas S/N
City: Santiago de Cuba
Country: Cuba
Email: eramirez@uo.edu.cu

A Low-Cost Methodology for Biomechanical Analysis of Martial Arts Using Videography and Accelerometers

J.G. Vejar-Robles[1], A. Vega-Gonzalez [2] and R.P. Duarte-Zamorano[1]

[1]Universidad de Sonora/Departamento de Física, Hermosillo, Sonora, México
[2] Universidad de Guanajuato/División de Ciencias e Ingenierías, León, Guanajuato, México

Abstract— **To achieve this, in this paper is made a study of motion using videographic techniques and using accelerometers to make a comparison between different individuals and identify differences and similarities, which is shown in the figures and graphs that are included. The advantage of this method is the low cost and potential for analysis presenting the data obtained in this way.**

Keywords— **Videography, accelerometers, martial arts, Tang Soo Do, motion.**

I. Introduction

Biomechanics is the study of continuum mechanics of biological systems (i.e., it studies load, movement, tension, etc.) and mechanical effects resulting in the movement, size, shape and body structure at the molecular level to the cellular level. One area of particular interest where biomechanics can contribute significantly is in sports. Here, the specific objectives of biomechanics include improved performance, prevent injury and promote safety of individuals. Motion analysis is key to professional sports training, in order to optimize and improve athletic performance. Therefore, this paper aims to develop a low-cost methodology to study the biomechanics of martial arts with the particularity of being able to study it in the participant´s environment.

Currently, there is not a standard for the biomechanical study of martial arts [8,10]. However, there is a consensus that quantitative data are required and it is necessary to study the kinematics and kinetics. Most systems currently used in martial arts, only contemplate the kinematics. This paper considers a methodology to study the kinematics and kinetics in martial arts.

II. Motivation

The interest in carrying out a systematic study of the movements in martial arts arises when, in a review for the next martial degree, for one of the authors, that includes, as part of the test, to make a break of wood (which is 12in x 12in side and 1in thick). It was noted that all grades and almost all ages could break them (although performers 3 to 8 years old had difficulty). From the above observation raises the question: does the martial degree, physiology or physics behind it, to achieve the successful movement to make the break?

III. Methodology and Results

In order to answer the previous question, a first analysis consisted of the following methodology: several practitioners were filmed (of different degrees or ribbons, aged between 3 and 48 years old), to be analyzed frame by frame and thus study the kinematics of the movement, and the relative positions of the extremities according to blow that was made for a successful break. The techniques used for breakups are presented with this methodology: *Choon Dan Kong Kyuck* (right fist in the middle) [4]. For the study of the movements the participant was recorded with two home video cameras (30 frames per second – fps-) with the idea that a camera captured the vertical movement and another captured the horizontal movement. The recorded videos were edited using VideoPoint Capture™ program, and then were analyzed using the VideoPoint Physics Fundamentals™ program for the positions of the tips on each stroke. With these position data, and by employing an applet that makes calculations using least squares adjustment, we obtained velocity and acceleration.

Some of the results obtained with this method are shown in Figure 1.

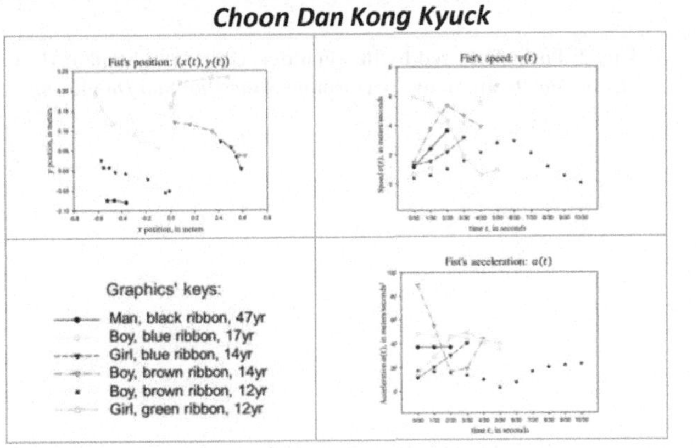

Fig. 1: Graphs obtained using videography with a home video camera recording at 30 frames per second (fps).

I. Torres et al. (eds.), *VII Latin American Congress on Biomedical Engineering CLAIB 2016, Bucaramanga, Santander, Colombia, October 26th -28th, 2016,*
IFMBE Proceedings 60,
DOI: 10.1007/978-981-10-4086-3_104

In a second part of this study the coup was reported in 3D. Practitioners were recorded with two home video cameras (30 fps), which were placed at the same height and axes perpendicular, so that the planes in each chamber have a coincident axis, the y axis.

The obtained videos were edited with VideoPoint Capture™ [6] program, and then were analyzed using the VideoPoint Physics Fundamentals™ [5] program for positions on every shot. With these position data, and making the necessary adjustments to "match" the $y(t)$ data, we obtained the 3D path followed by hands, elbows and shoulders. It is also important to mention that the tied axes coincide with the coordinate axes of the human body. In Figures 2, 3 and 4, are shown shoulder's, elbow's and hand's path of 6 *Tang Soo Do* [4] amateurs performing the blows called *Sang Dan Soo*, *Had Dan Soo Do* and *Yuk Soo* [4], respectively.

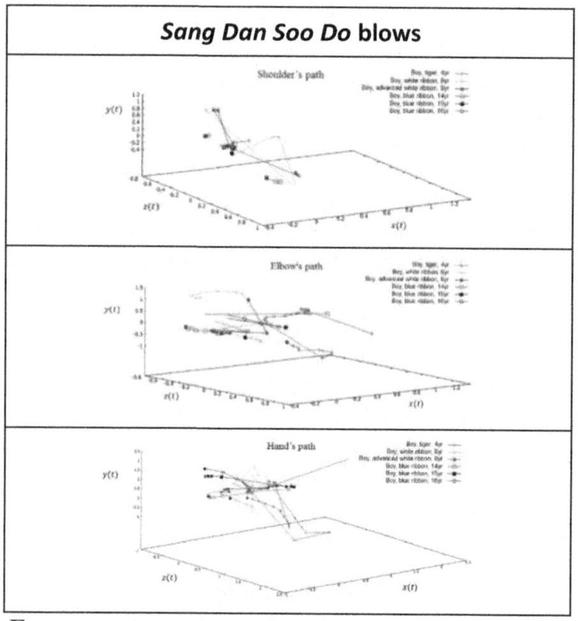

Fig. 2: Paths followed by the shoulder, elbow and hand of six *Tang Soo Do* amateurs performing *Sang Dan Soo Do* blows.

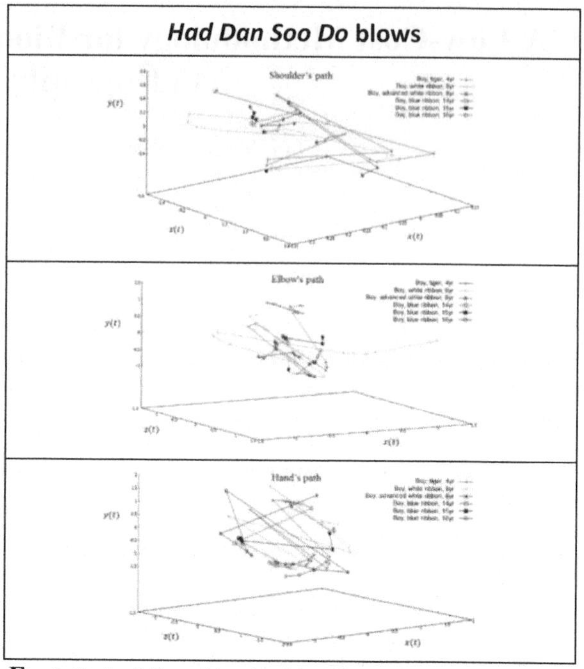

Fig. 3: Paths followed by the shoulder, elbow and hand of six *Tang Soo Do* amateurs performing *Had Dan Soo Do* blows.

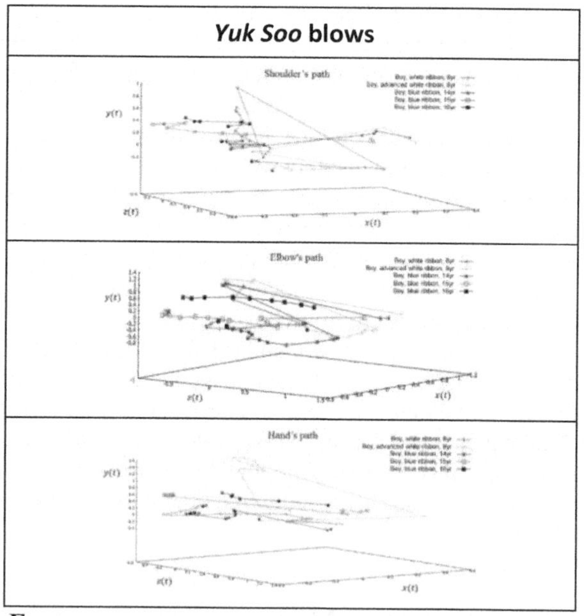

Fig. 4: Paths followed by the shoulder, elbow and hand of six *Tang Soo Do* amateurs performing *Sang Dan Soo Do* blows.

Finally, in a third phase the use of accelerometers was implemented and the set designed and the position of the markers for videography was improved [9].

A. Kinematics (Artificial Vision)

Artificial vision was used for kinematics, which consisted of three video cameras and open source software.

Video cameras were set to cover least two orthogonal planes. The recording was made at 60 fps (maximum speed of the Sony HandyCam commercial camera).

During the evaluation, different types and positions of the markers were considered. Figure 5 shows the proposed camera arrangement and example in a *Dojo* blow.

Fig. 5: proposed camera arrangement and example in a *Dojo* blow.

B. Kinetics (accelerometers)

A set of accelerometers was used with LabVIEW and Arduino for kinetics. They use the triaxial accelerometer ADXL345 (Analog Devices), which can support up to $16g$ accelerations [7].

The accelerometers were placed at following points: distal radioulnar (wrist), humeral epicondyle (elbow) and clavicular acromion (shoulder), as shown in Figure 6. The sampling frequency was 60Hz, and "ad-hoc" programs for calibration and use were developed.

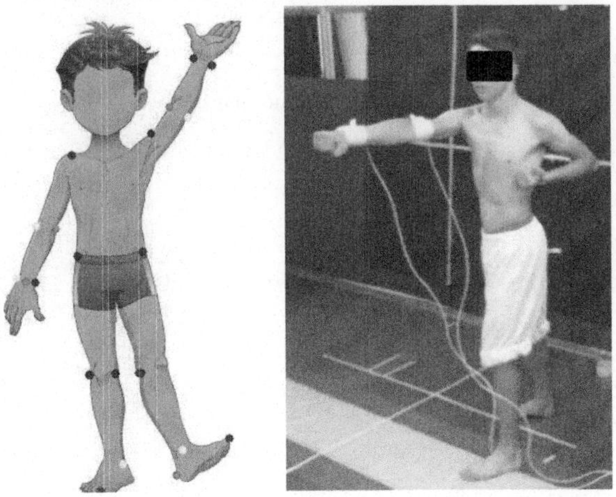

Fig. 6: Accelerometers and markers positions for evaluating the coup.

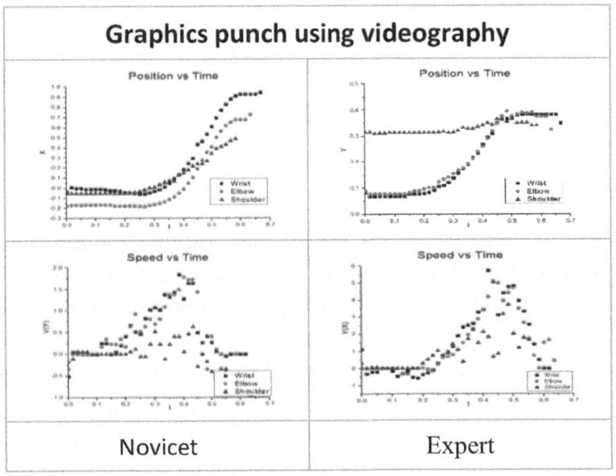

Fig. 7: Results obtained by videography for the performance of an expert and a novice.

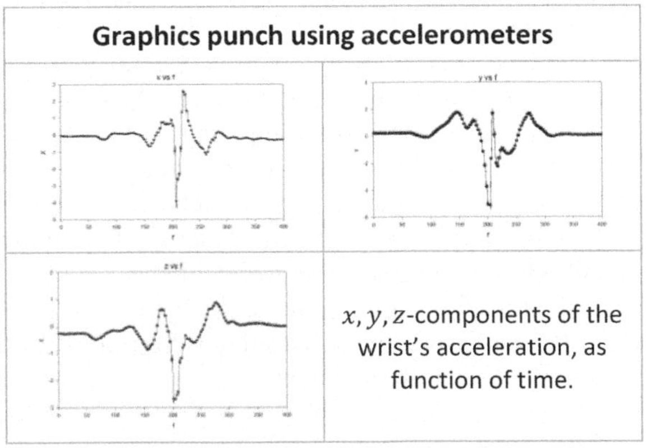

Fig. 8: Results obtained by using accelerometers considering wrist's movement.

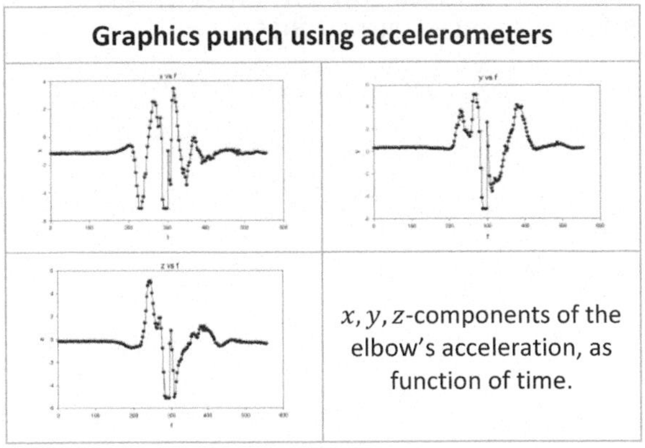

Fig. 9: Results obtained by using accelerometers considering elbow's movement.

IV. Conclusions

- The trajectory of the strokes of each practitioner is similar, however for participants with experience, it was noted that this is more uniform and stable.
- The system and methodology proposed for the study of martial arts provides qualitative and quantitative data, and it is flexible as it can be transported to different places, not to mention that it's a low-cost implementation (common commercial chambers). The cost of accelerometer is approximately US$20 for which it is planned to place accelerometers in strategic anatomical points, and one external, which will serve as a benchmark. This will apply techniques such as the direction cosines resulting tracking in the space of any point, as a result we could calculate rotations and other useful parameters in sports biomechanics.
- One of the disadvantages of the method of accelerometers is the connection because it limits the movement of the participant; this is due to the long cable, which is needed to maintain the connection. No overwhelms has considered the possible inclusion of a wireless system.
- Data from the sensors allow applying physical equations for speed and position is a reverse form of videography
- The proposed system should continue to be refined to achieve establish a low-cost methodology to provide relevant information to the coach and athlete.
- The methodology used allow us to distinguish a person who has never practiced martial arts from a martial arts practitioner regardless of the time it has practiced

Acknowledgment

DAIP-UGTO 1, 146/2016 PAPIME PE104715
PIFI 3.0/UNISON (2014 y 2015)

Conflict of Interest

The authors declare that they have no conflict of interest.

References

1. Le Veau, Barney (1991) Biomecánica del movimiento. Editorial Trillas, México
2. Kapanji, A.I. (2012) Fisiología articular (Tomo 1). Editorial Médica Panamericana, México.
3. Griffin, Michael J (2012) Enciclopedia de la salud y seguridad en el trabajo. Editorial Chantal Dufresne, BA, España.
4. Delcole, Larry (1991). Manual del Tang Soo Do. WTSDA.
5. VideoPoint Fundamentals at http://vpfundamentals.com
6. VideoPoint Capture at http://www.lsw.com/videopoint/capture
7. ADXL345 Accelerometer Quickstart Guide at https://www.sparkfun.com/tutorials/240
8. Kerstin Witte, Peter Emmermacher, and Marion Lessau. Biomechanical Measuring Stations To Solve Practical Problems In Karate Sport, ISBS Conference 2008, July 14-18, 2008, Seoul, Korea
9. A. Kolahi, M. Hoviattalab, T. Rezaeian, M. Alizadeh, M. Bostan, and H. Mokhtarzadeh, "Design of a marker-based human motion tracking system," Biomedical Signal Processing and Control, vol. 2, no. 1, pp. 59 – 67, 2007.
10. Fernandes, FM., Wichi, RB., Silva, VF., Ladeira, APX. and Ervilha, UF. Biomechanical methods applied in martial arts studies. J. Morphol. Sci., 2011, vol. 28, no. 3, p. 141-144

Corresponding author:

Author: Roberto Pedro Duarte Zamorano
Institute: Departamento de Física, Universidad de Sonora
Street: Luis Encinas y Rosales S/N
City: Hermosillo, Sonora
Country: México
Email: roberto.duarte@fisica.uson.mx

Identifying the needs in the integration of disciplines in the hospital infrastructure management in Colombia

M. Madroñal Ortiz[1], B. Galeano Upegui[2], N. Escobar Mora[3], L. Cruz Parra[4] and I. Rios Cuartas[5]

[1] Universidad Pontificia Bolivariana, PhD. Engineering Student, Medellín, Colombia
[2] Universidad Pontificia Bolivariana, Mechanical Engineering Faculty, Medellín, Colombia
[3] Universidad Pontificia Bolivariana, Agroindustrial Engineering Faculty, Medellín, Colombia
[4]Universidad Pontificia Bolivariana, Msc. Bioengineering Student, Medellín, Colombia
[5]Universidad Pontificia Bolivariana, Msc. Technologic Management Student, Medellín, Colombia

Abstract — **This paper presents an overview of the state of the art of Hospital Physical Asset Management for Health Service Institutions (IPS) in Colombia. These institutions have the need, within the requirements of the sector, to optimize the management of their physical asset as a fundamental part of the Strategic Plan of the Organization for the health services they provide, relying on disciplines that can be aligned to the Objectives, by Processes, Strategies and Policies of the providing health services institution. The international standards linked to such disciplines have generally evolved in organizations of production, manufacturing or energetic character and less on those organizations that provide services such as the hospital case. Therefore, the aim of this document is to identify and know the different management disciplines, as Building Maintenance Management (BMM) Facility Management (FM) and Asset Management (AssM), and the associated international standards (ISO 55000, UNE-EN 15221, UNE-EN 16646, ISO 31000) in order to meet the state regulations in the health field. That being the case, this review has established the main needs for the integration of management disciplines in the hospital infrastructure in the IPS in Colombia.**

Keywords — **Asset Management, Maintenance Management, Infrastructure, Clinical Engineering, Facility Management.**

Introduction

Hospital physical asset may represent the most important investment for the IPS, hence engineering and architectural disciplines are fundamental to reform and transform the organizational structure in order to come to the aid of the patients recovery in the organizational climate and the community in general. In the case of hospital infrastructure, this represents the mission, values and principles of the organization, therefore it must be built, maintained and operated intentionally to achieve their specific objectives, working on the maintenance, the investments and the internal building management [1].

The term (Physical) Asset Management is a concept in development, so there is not yet an established definition, even less in the hospital sector. The use of this term for maintenance managers in the USA has been used to gain greater credibility in their activities. As maintenance was understood as a job without importance, the term "asset management" seems to be more appealing and professional, even though maintenance is a fundamental component in the asset management policy of the company. The term that we encompass in this study shows the evolution of the overall concept of physical asset management in enterprises [2].

Taking into consideration that the IPS are increasingly forced worldwide to seek a competitive advantage to get a foothold in the market, they should work on the optimization of their services, status and business continuity. To achieve this goal, we have checked that some IPS in Colombia are unionized to defend their corporate interests in the form of cluster health, this has favored that Colombia has 50% of the 43 best clinics and hospitals in Latin America [3].

Linking this study to the doctoral thesis of the main author, it has been considered the evolution of disciplines such as Facility Management, which is defined as *"A management model of real estate asset of companies that aim to adapt them permanently to the organization and the companies personnel at the lowest possible cost, by integrating all the management responsibilities on those resources"*, according to the Spanish Society of Facility Management (SEFM) and International Facility Management Association (IFMA). This discipline has been focused on the physical asset management (infrastructure), which become more complex with the past of days and contain more technology, sparking a major competitiveness, cost, demand, health expectations, consumers′ safety and welfare, in addition to the environment. Considering that the institutions add new responsibilities and challenges in their work environments to the social purpose of the organization [4], the FM has a very wide scope of work, also aligned with the Strategic Plan of the Organization [5].

Moreover, due to the increase of the importance of the building maintenance, the BMM [6] has generated a growing awareness of the need to manage more effectively the building condition. Despite this, a great part of the maintenance is carried out in a context that does not create a fully integrated performance of the buildings management, and therefore buildings are not used as is due.

I. Torres et al. (eds.), *VII Latin American Congress on Biomedical Engineering CLAIB 2016, Bucaramanga, Santander, Colombia, October 26th -28th, 2016*,
IFMBE Proceedings 60,
DOI: 10.1007/978-981-10-4086-3_105

Instead, the AssM is defined as the *"coordinated activity of an organization to generate value of the asset"*, being the definition of an asset as *"an item, thing or entity that has a potential or actual value in an organization"*. In the case of physical asset, they assume an important role in the organizations, within the five types of assets (financial, human, intangible, information and physical) [7]. Physical asset must be managed throughout the lifecycle, covering processes such as design, construction, operation, maintenance and replacement of asset and infrastructure in order to maximize their value; leading the efforts to improve their performance, reduce their costs, extend their useful life and improve the return on investment of asset for the organization [8].

Several factors appear to distinguish hospitals from many other businesses. Firstly, they are facilities that are open 24 hours a day, 7 days a week; secondly, they offer particularly complex services; and, thirdly, an error in a hospital can cost a life. This hospital infrastructure must constantly update their assets in order to meet the highest technical and safety standards, even though this can bring about high levels of investment [9].

The ISO 55000 standard [10] , preceding of the PAS 55 British Standards Institution (BSI), provides an overview of asset management and the systems of asses management through common practices that can be applied to the widest range of asset and organizations across different cultures. The ISO 55000 enables an organization to achieve its objectives through the effective and efficient management of its asset. Applying an asset management system ensures that these goals can be achieved consistently and with sustainability over the time. This standard is associated with others as the UNE-EN 16646, offering the vision of maintenance in the physical asset management, the UNE-EN 15221-1: 2012 for Facility Management, and the ISO 31000 for Risk Management in all the operations within the organizations.

Considering this is a sector with a great deal of responsibility, it should focus on achieving Operational Excellence (OPEX), that according Amendola [11] it is the pursuit of conducting business in a way that improve continuously the quality of goods and services, trying to achieve competitive superiority from the core of the company "Processes – People – Technology - Networks". Though, within these factors there are three pillars of OPEX: • Planning and Production Control • Manufacturing process (or services) • Operational Effectiveness of People, Processes and Physical Asset Management as it can be seen in Figure 1.

The first hurdle that companies have not implemented a Business Continuity Planning (BCP) face is the lack of knowledge about this and the magnitude of the effort, and the second difficulty is overcome budget to run it. Within corporations must be sized to BCP as safe and not as an expense *"must be taken to not use it sometimes, but the day that required the company will succeed"* [12].

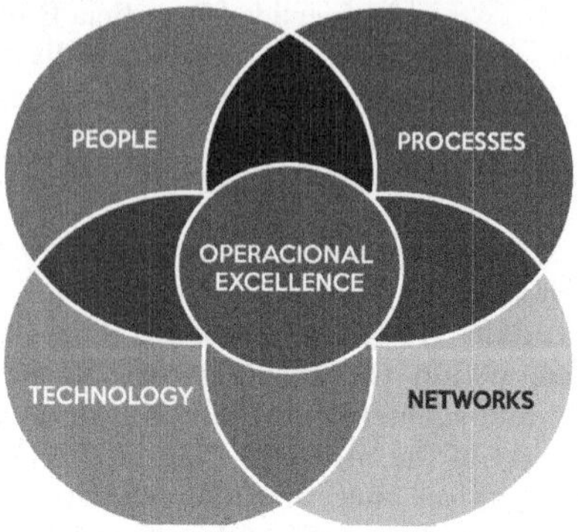

Fig.1 Core of the organization on OPEX. [11]

Objetives

Given the above mentioned, the proposed objectives of this study are to conduct a literature search on standards, disciplines, buildings or infrastructure and international health organizations that are involved in the proper management of the hospital infrastructure for its contextualization in Colombia. Then, evaluate and analyze the gathered data taking into account the model of the IPS in Colombia, identifying the needs that IPS present in order to integrate disciplines and international standards for the infrastructure management in Colombia.

Methodology

To meet the objectives described above, the procedure involves a literature search as the first step in the process of scientific research, through access to scientific literature in digital databases, selecting the right terms for the drafting of the study considering those latest and relevant references related to the topic addressed. This allows to capture information about science and technology, select and analyze it, in order to make decisions identifying which countries has been investigated, which authors have written about and in what areas they have worked the selected topics.

Scopus is today one of the best tools for bibliometric studies and scientific production assessments, inasmuch it encompasses a multidisciplinary vision of science and integrates all relevant sources for basic research, applied research and technological innovation through patents, sources of Internet of scientist content, free access scientific journals, congresses and conferences reports. This search tool is not only unique for its updated daily content with more than 3,000 journals, but also as the sole basis that brings the right tools, such as: profile author profile institution, cites tracker, analyzer h and index scientific journals [13].

The three initial search equations used in the database took into account the disciplines "Asset Management", "Facility Management" and "Building Maintenance Management" associated with building management, infrastructure and organizations physical asset; performing them in individual equations, finding a sum of 127,555 results adding the three initial equations.

However, these are too many results to make a realiable analysis, for that reason a more precise equation was undertaken and new particular searches were carried out in each discipline linking them to the rest of the terms, such as "building, hospitals, Healthcare, World Health Organization (WHO), Pan American Health Organization (PAHO), Joint Commission International (JCI) and the international standards", obtaining 20 different equations with positive solutions that added the amount of 1,648 results, and then a final search combining the disciplines "Asset Management AND Facility AND Building Maintenance", resulting in 63 documents as well as the defined criteria as shown in Table 1.

Table 1 Results of the search equations (Authors)

SEARCH EQUATION	RESULTS
Asset Management (AssM)	112.385

SEARCH EQUATION	RESULTS
Facility Management (FM)	13.462

SEARCH EQUATION	RESULTS
Building Maintenance Manag. (BMM)	1.708

SEARCH EQUATION	RESULTS
AssM AND ISO 55000	19
AssM AND ISO 31000	3
AssM AND Hospitals	114
AssM AND Healthcare	96
AssM AND WHO	19
AssM AND PAHO	1
AssM AND Joint Commission Intern.	2
AssM AND Hospitals AND FM	26

SEARCH EQUATION	RESULTS
FM AND ISO 55000	1
FM AND Hospitals	691
FM AND Healthcare	508
FM AND WHO	36
FM AND PAHO	1
FM AND Joint Commission Intern.	4
FM AND Hospitals AND AssM	26
FM AND Healthcare AND AssM	17
FM AND EN 15221	1

SEARCH EQUATION	RESULTS
BMM AND Hospitals	46
BMM AND Healthcare	35
BMM AND Joint Commission Intern.	2

SEARCH EQUATION	RESULTS
AssM AND BMM AND FM	63

Source	Scopus
Fields	Article title, abstract, keywords
Subject areas	Engineering, Business, Management and Accounting
Date range	2006 - 2016

The terms related with the disciplines "AssM, FM and BMM" are associated with the buildings management, infrastructure and physical asset organizations, as well as the international standards associated with the disciplines "ISO 55000, EN 15221, UNE -IN 16646, ISO 31000".

After gathering the data of these equations, the authors who have written about hospital physical asset management internationally are analyzed. The findings provide an overview of the number, location and volume of production of those authors who do research in this field, as well as international organizations that work on biomedical equipment management, industrial equipment management and infrastructure management, plus studies conducted to the health sector, since most of work done so far has focused on the production, manufacturing and energy sectors.

Likewise, the international health organizations that are responsible for recognizing, promote, regulate and investigate in order to reach levels of quality in health services, "WHO, PAHO and JCI"; and finally, keywords of the clinical maintenance sector: "Hospitals, Healthcare, Building and Infrastructure" that reduce the results to an adequate amount of documents that facilitate its study.

All this information represents a theoretical basis, which has gone through an evaluation and analysis process that shed light on the needs that the organizations present for the proper infrastructure management with the current international standards. Thus, we proceed to make the adaptation to the colombian hospital environment, being necessary to have a wide view of the IPS in Colombia; how the physical asset management is carried out in the institutions, what the typical organizational structure is like, as well as the human management responsible for the maintenance and heritage, and the integration of the bioengineering area with the rest of the entity.

During the completion of the development of this study and due to the preliminary results at the time, the Universidad Pontificia Bolivariana has financed this research project to carry out (through a web application to manage surveys) a Diagnosis of the State of Hospital Physical Asset Management in the IPS (as a pilot in the Aburrá Valley) based on the ISO 55000.

Results

In summary, once done the analysis and the evaluation of the information obtained from the literature search of such documentation, we get the results that lead to establish the main needs in the integration of the disciplines of hospital infrastructure management in Colombia. As a result of this study, the hospital infrastructure management is not aligned with the international standards nor with disciplines such as AssM, FM and BMM. Another point worth mentioning it is the lack of knowledge and training of the personnel working in hospital infrastructure in such disciplines.

Moreover, this study reveals the existence of incompatibilities between departments, leaders and organizational structure to achieve objectives; besides, the processes and methodologies used currently by the IPS are confused and do not allow to maintain properly the infrastructure considering it a world-class service.

Additionally, there is a visible lack of integrated systems with adequate tools that lead to make objective decisions in order to manage the hospital infrastructure in a proper way. This shows that the physical asset hospital management does not work in conjunction with such asset life cycle. It related studies [14], it is proposed that the Asset Management and Facility Management intertwine in the physical asset (infrastructure) with Operation and Maintenance Service of Building Maintenance Management.

Thus, these three disciplines (AssM, FM and BMM) are complementary with each other in quest of the accomplishment of the objectives of the organization, as shown in Figure 2.

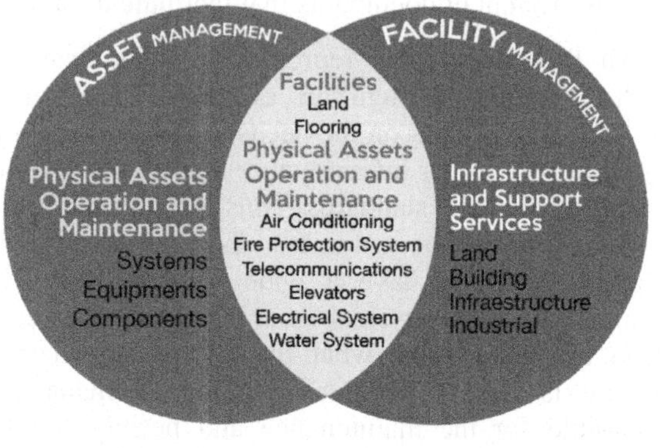

Fig.2 Modified, Relation between FM, AssM and BMM [14].

Conclusions

This work reveals certain shortcomings that IPSs in Colombia present in order to manage their infrastructure. Likewise, it has been identified that disciplines such as AssM, FM and BMM are highly related to infrastructure management, and that the international standards associated with these (ISO 55000, EN 15221, EN 16646, ISO 31000) are required to align infrastructure management to the objectives of the institution, through the processes, policies and strategies previously planned.

Physical asset management is the best way to reach the proper management of the asset lifecycle, in order to accomplish and back up the objectives, strategies and policies that lead to the Operational Excellence of the organization.

Being relevant to think about the development of a methodology for managing infrastructure based on physical asset management, facility management and maintenance management for hospital entities, aligned with international standards (ISO 55000, EN 15221, EN 16646, ISO 31000), seeking to reach a compromise from the administrative management to the workers in order to cover the found needs in the Colombian context, also supported by those ones coming from the IAM, IFMA.

Given these points, this study and further researches will aim at coming up with information that complement advanced training in universities, training of future professionals in hospital engineering, and also the development of information for the doctoral thesis of the main author, in search of the development of methodologies that lessen the IPS needs for a proper infrastructure management.

This work is aligned with what it has been done by the research group GIBIOING (Universidad Pontificia Bolivariana), based on the PAS 55, and developed from the approval of the legal requirements, national and international accreditation and the recommendations coming from the WHO, PAHO, the Joint Commission International and the Ministry of Social Protection, for its implementation in Colombia [15].

Conflict of Interest

The authors declare that they have no conflict of interest.

References

1. Negrotti, A. (2015). Conference in VII World Congress of Maintenance and Asset Management: "Curso Fundamentos para el Mantenimiento de Edificios Industriales y Corporativos. Facility Management (FM)", 2015.
2. Cabrera C. y Martinez I. (2008), Conference: "Business Continuity Plan: Es el momento de asegurar la continuidad del Negocio", Administración de Riesgos KPMG, 2008.
3. Revista América Economía (2015): http://bit.ly/2bBJNyE
4. Feingold V. et alts. (2012). Chapter Origen de la disciplina of Book "El libro del Facility Management" pp. 13-19. SLFM, Argentina, 2012.
5. Feingold V. et alts. (2012). Chapter Real Estate Planning of Book "El libro del Facility Management" pp. 92-123. SLFM, Argentina, 2012.
6. Chanter B. and Swallow P. (2007), Chapter 1 of Book "Building Maintenance Management". Blackwell Publishing, 2007.
7. The Institute of Asset Management – IAM. Accessed November 7, 2015. Defining Asset Management: http://bit.ly/2ccX82I , 2015.
8. Candelario, J.M. (2014), Chapter 1: Introducción of Doctoral Tesis. Diseño e implementación de un modelo de gestión integral de activos físicos para la gestión de cadena de suministro en la industria. España, UPV, 2014.
9. Lennerts, K (2009), Chapter 9 of Book, "Investing in hospitals of the future". World Health Organization, 2009.
10. ISO 55000:2014, Asset management -- Overview, principles and terminology, International Standard Organization. 2014.
11. Amendola, L. (2015), Article: Camino a la Excelencia Operacional, http://bit.ly/2bBJbsO, 2015.
12. Ferreira, L.A. (2012) Article: LCC Life Cycle Cost, su importancia para la política de Physical Asset Management, 2015.
13. ELSEVIER. Accessed July 17, 2015: http://bit.ly/2bF8f3s
14. PMM Business School. (2016) Conference: Estándares en Facility Management: http://bit.ly/2bTxoDM ,2016.
15. Galeano B. et als. (2015), Article: Modelo integrado de gestión de activos hospitalarios basado en la PAS 55, Revista Ingeniería Biomédica, 2015.

Author: Manuel Madroñal Ortiz et als.
Institute: Universidad Pontificia Bolivariana
Street: Bloque Externo 22C, Planta 2ª Circular 1ª No 73 – 76
City: Medellín
Country: Colombia
Email: manuel.madronal@upb.edu.co

Approximations of Pupillary Shape in High-Definition Video-Oculography Registers

R. Mora-Martínez, E. Suaste-Gómez,

Center for Research and Advanced Studies of the National Polytechnics Institute,
Department of Electrical Engineering, Bioelectronics Section, Mexico City, Mexico

Abstract—**Two approximations of the pupillary shape (in laboratory environment) were developed: a nonlinear elliptic fit with outliers rejection and the assessment of the realistic amorphous shape via radial distances measurement, for the more diaphanous registers. The present methodology achieved a resolution of 10 μm/pixel. The algorithm of nonlinear elliptic fit with outliers rejection resulted a robust option for infrared reflex artifacts, whereas, radial distance measurement could lead us towards a new way of spatiotemporal analysis of pupillary dynamic.**

Keywords— **Pupil, Video-Oculography, Elliptic Fit, Radial Distances, High-Definition.**

I. Introduction

Pupil is a structure located over the iris, which plays the role of a control diaphragm of the amount of light that arrives to the retina. Its contraction (miosis) is effectuated by the sphincter muscle (mediated by the parasympathetic system) and the responsible of its dilatation (mydriasis) is the dilator muscle (sympathetic activity) [1].

The study of pupillary dynamic has been a source of information for a variety of clinical and engineering applications. Some of the pupillary event of interest are the *pupil cycle time* [2], *pupillary latency* [3], *hippus* [4], also in an indirect form, *ocular movements* [5]. Some examples of clinical applications are studies of *Type I-II Diabetes* [6], *Nervous system diseases* [7] etc. whereas engineering application include but are not limited to *Biometric systems* [4] and *Eye-trackers* [8].

For the pupillary dynamic assessment, a variety of methodologies (pupilometers) have been developed as well as commercial and for research purposes. The proposed algorithms for this second group solve specific needs, for example, the real-time detection of the pupillary centroid for eye trackers [8] (which means very fast and robust detection), high sample rate for eye movements [9] (which means a high frame rate, in other words, high speed cameras) or high spatial resolution for analysis of differences in consensual and direct pupillary reflex (which means HD cameras) [10].

Typically, pupillary shape is fitted to a circular or elliptic approximation. In [11] more realistic analysis of the pupillary shape was introduced. The present work deals with the most general case of ellipse fitting (centered out of origin and rotated) besides, an alternative of realistic assessment is performed. The last implementation is based on the measurement of radial distances to the edge of the pupil.

II. Methodology

For the present experiment, two modules were implemented. The first one involves the development of the control interface of the luminous stimuli and video recording, whereas the second refers to the techniques of digital image processing employed in the measurement of the pupillary characteristics. The next sections describe, in detail, all particularities of the two modules.

All registers were processed off-line in the Matlab platform. As it was early mentioned, all registers were taken in a laboratory environment. The room for the tests was conditioned in such way that the external interferences were reduced, and the ambient illumination level, constant (6.1 cd/m^2).

Since the conditions of the experiment (in the sense of the subjects under study) are completely cooperative, and the length of the stimuli is short, typical artifacts such as blinks and eye movements are not present. Nevertheless, in some records the infrared reflexes occlude a portion of pupillary region. This fact lead us to a degradation of the signal. For this purpose, two algorithms were proposed.

When pupil is partially occluded, the measurement was implemented via a *nonlinear ellipse fit with outliers rejection* (NEFOR) algorithm; in the case where the register are free from occlusions, *radial distance measurement* (RDM) was performed.

A. Control Interface

All system was mounted on an ophthalmological table, where the subject rests his chin on a headrest. The illuminations array consists in two infrared lamps, which are disposed at the sides of the camera (in order to increase the image contrast with no pupil stimulation) and a white LED lamp for the smooth ambient light. For the recording operation, a modified reflex camera plus a macro lens was employed (the IR

I. Torres et al. (eds.), *VII Latin American Congress on Biomedical Engineering CLAIB 2016, Bucaramanga, Santander, Colombia, October 26th -28th, 2016*,
IFMBE Proceedings 60,
DOI: 10.1007/978-981-10-4086-3_106

filter was removed). The model of the camera is Canon EOS T3i and the recordings was taken at a resolution of 1280x720 pixels and frame rate of 60 fps. The stimulus was created by an ultra-brilliant white LED (8,500 cd/m^2 measured over its surface).

User can control the amount of ambient and infrared light and the length of the luminous stimuli through a graphic interface programed in the LabView platform. Current drivers provide the power to the sources of light, being governed by an Arduino Leonardo board. The whole system is shown in the Figure 1.

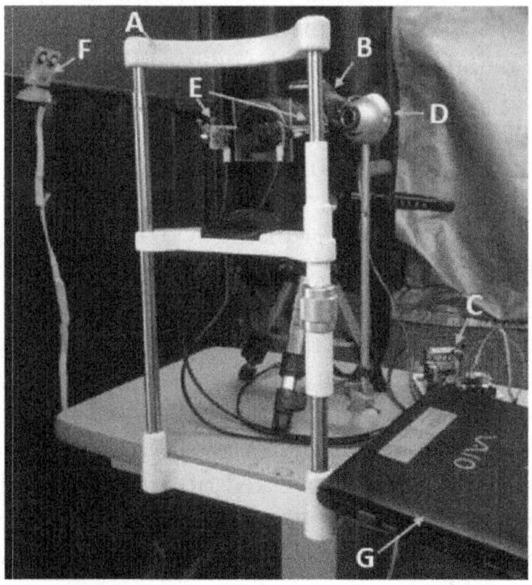

Fig. 1 Montage of the whole system. (A) Chin rest, (B) HD Camera (C) Control Circuit (D) Ultra brilliant white LED (E) IR Lamps (F) White Lamp (G) PC

B. Digital image processing

For the processing of partially occluded pupils, NEFOR algorithm is implemented. It consists in the next steps:

-Frame extraction: The video is segmented in frames, and only those which are useful remain for the next steps.

-Preprocessing: A camera correction is performed, then, a procedure to enhance the pupil. Typically, this last task consists in some kind of filtering [10, 12]. In this case, morphological operations was done (opening and close) [4].

-Pupil segmentation: In the same way like its predecessor, standard procedure is performed, a threshold segmentation [4]. Because of the homogeneity of laboratory condition, the threshold level was adjusted to a normalized value of 0.035. In addition, a hole-fill operation is executed, in order to get a better edge extraction and reduce the presence of outliers which may causes failures with the ellipse fitting (such as erroneous adjustment or no convergence [9,12-13]).

-Edge extraction: Canny edge-detector was employed to obtain the edge of the pupil. This technique has been widely employed in similar papers [9]. The resultant image is mapped to a table of x-y coordinates.

-Nonlinear ellipse fitting: In this step, the algorithm described in [13] is implemented. The task deals with the nonlinear problem of minimizing the perpendicular distances from the adjusted ellipse.

-Rejection of outliers: In this point, the focuses are located and the sum of the distances from here to every point of the data set (fourth step) are calculated. Those points whose distance are out from the range (mean ± std. desv.), are labeled as outliers. A new data set is created ignoring the outliers.

-A second nonlinear ellipse fitting: Is a new fitting with no presence of outliers.

After the execution of all the above steps, the result of the algorithm is a vector of five elements, which are: *the x-coordinate of the ellipse center, the y-coordinate of the ellipse center, length of the semiaxis-a, length of the semiaxis-b* and *the rotation angle*. The steps mentioned above are illustrated in the Figure 2. Considering all frames which integrates the register, a set of four 2D curves and a 3D curve are create.

It is important to highlight the next statement: all nonlinear method of adjustment needs an initial solution which must be close to the real solution. For this purpose, the starting parameters, associated with the first frame, are initiated to an ellipse with origin in the center or the image and values of 480 pixels for semiaxis-a, 490 for semiaxis-b and a rotation angle of 0.1 rad. For the next frames, the initial solution will be the parameters of the previous one.

In the case of the diaphanous registers, the algorithm of RDM is executed. This algorithm is compounded by the next steps:

Frame extraction, Preprocessing, Pupil segmentation and *Edge extraction* are the same as NEFOR algorithm.

Calculus of radial distances: In this scenario, the centroid of the pupil is located and then, at certain angle θ, a line departs from here until it locates the edge of the figure. The ray goes forward one pixel by step (in the 'x' axis) until, at least, five elements of a five-by-five neighborhood has value of one (Figure 3). When this requirement is satisfied, the center of this mask is labeled as the end of the line. The radial distance is calculated as the Euclidian distance between the centroid and the end of the ray. The complete figure is fully analyzed when the sweep of the angle θ is finished in the range 0-2π. The increase of θ, for this case, is ten grades. This is a variation of the first stage in StarBurst algorithm [12] where the edge of the pupil is founded with aim of gradient value.

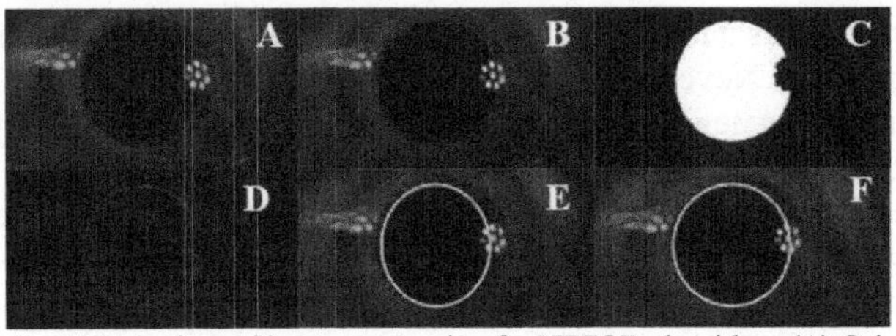

Fig. 2 Procedure of image processing for NEFOR algorithm. (A) Original image, (B) Preprocessed image, (C) Segmented pupil, (D) Pupillary edge detection, (E) First ellipse fitting, (F) Second fit without outliers

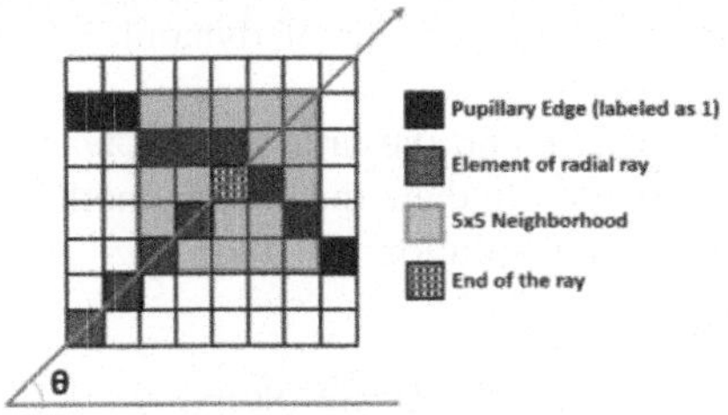

Fig. 3 Schematic of the use of 5x5 mask to identify the end of the radial line in the RDM algorithm

After that, the result of the algorithm is a curve which describes the radial distance (from center to the edge of the pupil) versus angle. An example of this procedure is illustrated in the Figure 4. Considering all frames which integrates the register, a surface with axis angle, time and radial distance is created.

III. Results

Six voluntary and healthy subjects (age 27.3 ± 1.5) were tested. All procedures were done in accordance with the Declaration of Helsinki and letters of consent were signed by every voluntary.

In order to get the maximum dilatation at the ambient illumination, the voluntary under test was disposed in a seat for a period of five minutes, then, was asked to assume the position for the procedure on the headrest. Once the camera has been focused, the recording starts. The frames of interest are those which are located after a period of absence of infrared illumination. The stimuli had a length of 200 ms and starts when has passed 1000 ms.

Results from NEFOR are summarized in four graphs versus time (Figure 5), which represent the length of semiaxis-a, semiaxis-.b, rotation angle, and the location of the ellipse center.

In addition, the differences between the two semiaxis (Figure 6) were calculated in order to visualize their tendency and evaluate, in a qualitative way, how elliptical or circular is the pupil. A circumference is a particular case of an ellipse, where both semiaxis are equal.

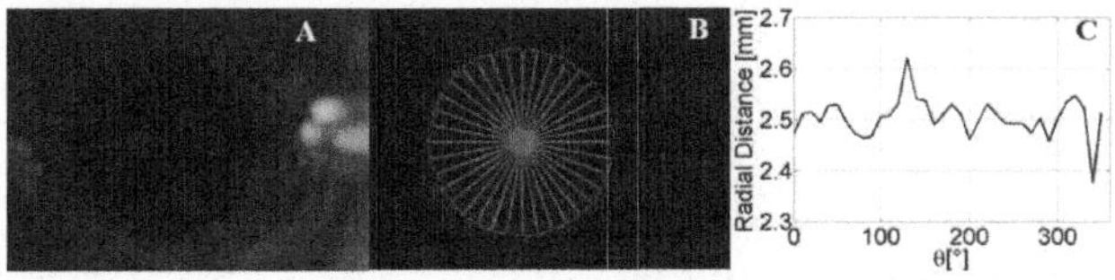

Fig. 4 Example of the procedure in the RDM algorithm. (A) Original image, (B) Contour extracted and radial lines drawn (C) Plot of radial distance versus angle

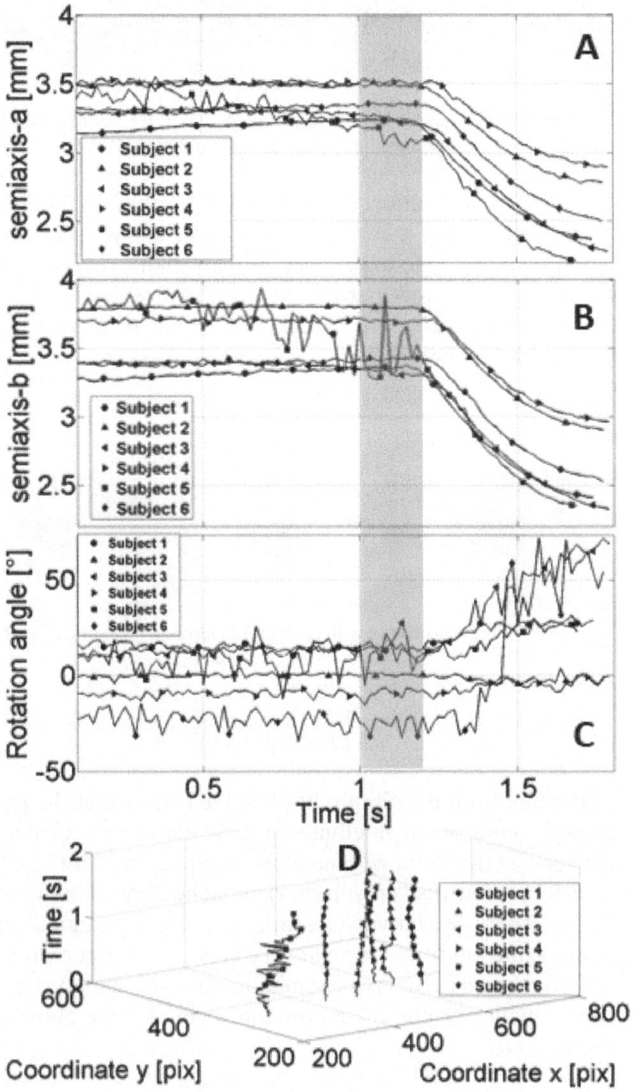

Fig. 5 Graphs from NEFOR algorithm. (A) Semiaxis a, (B) Semiaxis b, (C) Rotation angle, (D) Center of the ellipse. Shaded region represents the period when the stimulus is active.

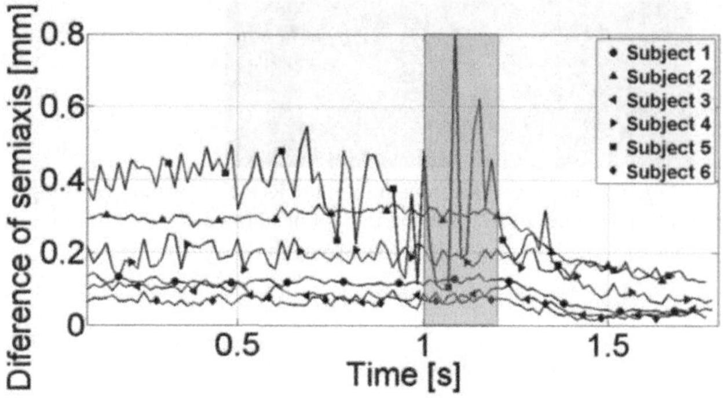

Fig. 6 Differences in both semiaxis length for all the subjects. Shaded region represents the period when the stimulus is active.

As it was mentioned, the RDM algorithm applied over all frames lead to a surface with axis of angle θ, time and radial distance. Due to the need of diaphanous registers, this implementation was performed only in 3 subjects, where large infrared occlusions did not appear. The results are summarized in the Figure 7.

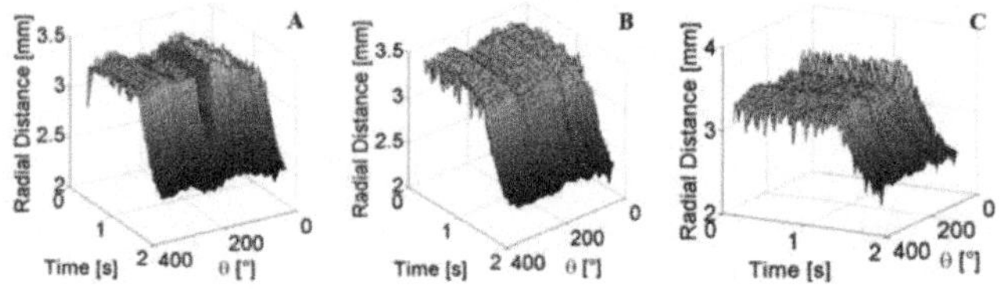

Fig. 7 Surfaces created with the RDM algorithm. (A) Subject 1, (B) Subject 3, (C) Subject 6

IV. Discussion

Results from the elliptic fit (NEFOR) show that the pupil could be considered an ellipse with the major axis located in the vertical direction with no particular pattern in rotation.

Effects of artifacts (eyelashes) could be viewed in the subject 5, where the curves of semiaxis presents irregularities, besides, the pupil was located very close to the left extreme.

Differences between the semiaxis-a and the semiaxis-b become smaller when time go forward (and also the pupil size decreases).

Surface from RDM algorithm for the subject 1 exhibits a dent, which is the product of a slightly pupillary occlusion

V. Conclusions

Both algorithms included in this work are a special implementation for video oculography registers in HD cameras, with the lowest amount of artifacts. The goal of this paper never was a highly-robust or fast algorithm for real time processing. Nevertheless, the NEFOR algorithm was able of avoid the negative effect of the infrared occlusions in the pupil.

The shape of the pupil could be considered, in a general way, an ellipse with its major axis in the vertical position, but, the smaller the pupil becomes, the more it looks like a circumference.

The RDM algorithm is, in principle, quite sensitive to noise but, techniques of smoothness or surface fit could be implemented to increase its robustness. This could be an option towards the assessment of dynamic in regions of the muscles in iris.

For this experiment, the ratio μm/pix was in the range of 9.97388-12.9863 μm, then, the present methodology is capable of achieves a resolution of 10 μm.

References

1. Loewenfeld IE, Lowenstein O. (1999) The Pupil: Anatomy, Physiology, and Clinical Applications.Butterworth-Heinemann. Boston
2. Howarth PA, Heron G, Whottaker L (2000) The Measurement of Pupil Cycling Time. Graefe's Arch Clin Exp Ophthalmol. 238:826–832
3. Bergamin O, Kardon RH (2003) Latency of the Pupil Light Reflex: Sample Rate, Stimulus Intensity, and Variation in Normal Subjects. IOVS Vol. 44, No. 4
4. Villalobos-Castaldi FM, Suaste-Gomez E (2013) A new spontaneous pupillary oscillation-based verification system. Expert Systems with Applications. Vol 40, Issue 13, pp 5352–5362 ISSN: 0957-4174; DOI: 10.1016/j.eswa.2013.03.042.
5. Van der Geest JN, Frens MA (2002) Recording eye movements with video-oculography and scleral search coils: a direct comparison of two methods. Journal of Neuroscience Methods. Vol 144 pp 185-195
6. Taranpreet K, Suchitira-Devi A, Satyabhama-Devi M (2014) A comparative study of Edge light pupil cycle time in type-II diabetes mellitus patients and normal subjects. Journal of Dental and Medical Sciences. Vol 13 Issue 1 pp 19-23
7. Stergiou V et al (2009), Pupillometric findings in patients with Parkinson. International Journal of Psychophysiology. Vol 72 pp 97-101
8. Swirski L, Bulling A, Dodgson N (2012) Robust real-time pupil tracking in highly off-axis images. Proceedings of the Symposium on Eye Tracking Research and Applications. March 28-30, 2012, Santa Barbara, California [DOI: 10.1145/2168556.2168585]
9. Jansen SMH, Kingma H, Peeters RLM (2009) A Confidence Measure for Real-time Eye Movement Detection in Video-oculography. ICBME 2008 Proceedings 23, pp. 335–339
10. Espinosa J, Roig AB, Pérez J, Mas D (2015) A high-resolution binocular video-oculography system: assessment of pupillary light reflex and detection of an early incomplete blink and an upward eye movement. BioMedical Engineering Online. 14:22 DOI: 10.1186/s12938-015-0016-6
11. Wyatt HJ (1995) The Form of the Human Pupil. Vision Research. Vol. 35, No. 14, pp. 2021-2036
12. Li D, Winfield D, Parkhurst DJ (2005) Starburst: A hybrid algorithm for video-based eye tracking combining feature-based and model-based approaches. Proceedings of the 2005 IEEE CVPR pp 79-86
13. Ahn SJ, Rauh W, Warnecke HJ (2000) Least-squares orthogonal distances fitting of circle, sphere, ellipse, hyperbola, and parabola. Pattern Recognition. Vol. 34 pp 2283-2303

Motion Artifacts Recognition in Electrocardiographic Signals through Artificial Neural Networks and Support Vector Machines for Personalized Health Monitoring

F. A. Castaño[1] and A. M. Hernández[1]

[1] Bioinstrumentation and Clinical Engineering Research Group - GIBIC, Bioengineering Department, Engineering Faculty, Universidad de Antioquia UdeA; Calle 70 No. 52-21, Medellín, Colombia

Abstract— **Nowadays a personalized approach is being giving to health care concerning the prevention of diseases, improving diagnosis and treatment of patients, for this, equipment to measure ambulatory vital signs are used, allowing to get large volumes of information. Nevertheless, the obtained information from ambulatory electrocardiography has no largely clinical validity because it is contaminated with motion artifacts, for this reason, it is necessary to determine what information is useful and what information can be ruled out. This paper presents a comparison between two different classification methods of electrocardiography signals: Artificial Neural Networks and Support Vector Machines. Database includes electrocardiography signals of volunteers and some important features of these signals are extracted to train both classification methods. Also, performance of methods is assessed verifying the generalization capabilities. The best performance was presented by the Radial Basis Function kernel.**

Keywords— **Health Information Management, Artificial Neural Networks, Support Vector Machines, Electrocardiography**

I INTRODUCTION

Currently, the main medical paradigm that rules the determination of diagnoses and clinical treatments is the Evidence Based Medicine (EBM), that consists in identifying symptoms based on the obtained results on researches performed in a determinate population [1]. Although it is well accepted and discussed because it does not consider specific aspects of the patient, which can lead to obtain inappropriate diagnoses and treatments [2].

The increase in the elderly population, the wide variety of chronic diseases and economic costs have led to a paradigm shift in health care toward a predictive, preventive and personalized medicine, able to identify diseases efficiently and overcome consequences of the generic treatments, which do not work for all people [3].

The development of researches and solutions has been prompted to make use of physiological information acquired through vital signs monitors, mainly wearable devices [4]. The wearable devices allow the measurement during continuous and over long time intervals in everyday environments such as home or work, providing an alternative for better monitoring and identification of the patient's health status [5].

However, data acquired by these devices simply indicate a possible trend of the vital signs, because the information recorded do not have clinical significance due to including artifacts generally produced by the movements of people while using these devices [6].

In the case of ambulatory electrocardiography (ECG), its recording produces large amounts of information that could be used for patient diagnosis [7]. However, ECG presents artifacts caused by the movement of the person, which can be misinterpreted by medical specialists and lead to inappropriate diagnosis of his condition [7].

The automatic analysis of ECG signals is a problem that has focused efforts worldwide [8], however, the non-linearity of the signal and the presence of contaminate noise make necessary the use of pattern recognition methods, among which some of the most used are the Artificial Neural Networks (ANN) [9], and Support Vector Machines (SVM) with Radial Basis Function kernel (RBF) [10].

This paper presents a method for classification of ECG signals that allows classify between ECG signals useful for diagnosis and motion artifacts-contaminated ECG signals, so that useful information can be presented to medical specialist to track his patient, besides being useful to study the tendency of cardiovascular diseases using big data methods. The presented method uses ANN and SVM for classification of signals in addition to compare the performance presented by both techniques.

II METHODOLOGY

Database used in this study was recorded from a physiological study of 10 healthy volunteers with ages of 24 ± 6 and Body Mass Index (BMI) of 22 ± 6. Each subject was asked to stay at rest and to make controlled movements during two minutes while their ECG signals were recorded and sampled at 100 Hz. The exercise consisted in raising and lowering the arms repeatedly in order to obtain normal ECG signals and motion artifacts-contaminated ECG signals

I. Torres et al. (eds.), *VII Latin American Congress on Biomedical Engineering CLAIB 2016, Bucaramanga, Santander, Colombia, October 26th -28th, 2016*,
IFMBE Proceedings 60,
DOI: 10.1007/978-981-10-4086-3_107

respectively. The ethics committee for human studies of the Universidad de Antioquia approved the recording protocol.

2000 signal epochs of 1.2 s corresponding to a heartbeat with a resolution of 11 bits were taken from database. In Figure 1, an example of the two signals of different quality are shown.

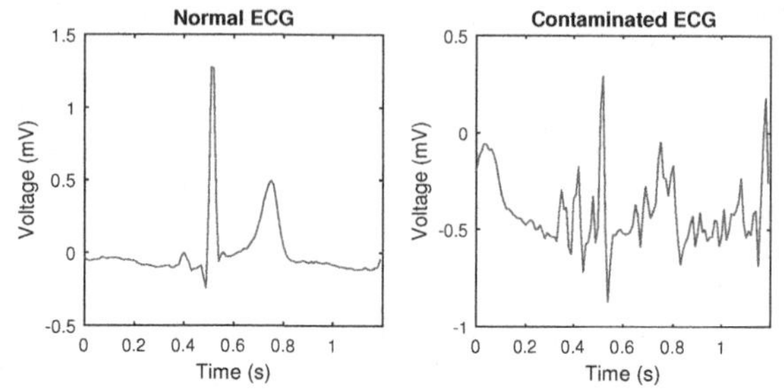

Fig. 1: ECG signals in database: 1) normal ECG signal of a patient at rest, 2) contaminated ECG signal by motion artifacts

Database had 500 normal *ECG* signals and 1500 contaminated *ECG* signals. To train both methods, 1000 samples were used, and these were chosen at random from the database, the remaining 1000 samples were used to measure the performance of the method against unknown information to evaluate the level of generalization. The selection of samples were performed watching the balance of the database. In both cases, 25% of clean samples and 75% of contaminated samples were used, to avoid overtraining and biased information.

Some authors propose six different characteristics that can be extracted from a heartbeat [11]; the R peak value, the mean power spectral density, the area under the QRS complex, the energy of the signal, the distance between $Q-S$ waves and the autocorrelation value.

A Stationarity analysis

Stationarity of normal *ECG* signals and contaminated *ECG* signals was analyzed using the Bootstrap method [12]. This method allows to know the probability of non-stationarity inside a time series.

B ANN design

An Feed-Forward Backpropagation *ANN* topology was implemented. The descent gradient method was used as the training technique and two layers were used in the network to have a topology with one hidden layer and one output layer. The hidden layer had 10 neurons with tan-sigmoid activation function. Network inputs were the 6 features extracted from the ECG signal epochs and 2 possible states were used as outputs.

C ANN training

The learning process of *ANN* involves training, validation and test stages. In the training stage, the 50% of the data was used. The *ANN* adjusts the connection weights between neurons, once it finds the minimum error. In validation stage, the connection weights are adjusted with unknown data, in this stage, the 25% of the database was used. Finally, test stage is applied to evaluate the performance of the net with the other 25%.

D Performance test of ANN

1000 samples were used for a more rigorous performance test of the *ANN*. These samples were evaluated with the *ANN* and the error obtained between the predicted outputs and expected results was determined.

E SVM design

SVM were designed to perform a binary classification of information [11], the method requires a group of training dataset $S=\{(x_1,y_1),(x_2,y_2),...,(x_m,y_m)\}$. Where x_i are the vectors of attributes and $y_i \in \{-1,+1\}$, is the result set. In this work, a $C-SVM$ was used to perform the analysis. Also, the performance of the linear kernel, polynomial kernel and *RBF* kernel functions was tested.

F Sensitivity and specificity analysis

Cross-validation was performed with 10 folds for each of kernel functions. The sensitivity (equation 1), specificity (equation 2), precision (equation 3) and efficiency (equation 4) of the method were evaluated.

$$Sensitivity = \frac{T_P}{T_P + F_N} \tag{1}$$

$$Specificity = \frac{T_N}{T_N + F_P} \tag{2}$$

$$Precision = \frac{T_P}{T_P + F_P} \tag{3}$$

$$Efficiency = \frac{T_P + T_N}{T_P + T_N + F_P + F_N} \tag{4}$$

Where T_P represents the number of samples that are normal and were classified as normal (Positive True), F_P represents the number of samples that are contaminated and were classified as normal (Positive False), F_N represents the number of samples that are normal and were classified as contaminated (Negative False) and T_N represents the number of contaminated samples that were classified as contaminated (Negative True).

III RESULTS

A Stationarity analysis

From the test with the Bootstrap method, it was found that normal *ECG* signals and contaminated *ECG* signals with artifacts presented non-stationarity with a probability greater than 95% for a signal time of 1.2 s. That means, classifying from stationarity condition is a not suitable approach.

B ANN training

The histogram of the error obtained in the learning stage is shown in Figure 2. The 99% of data had an error below ±0.1 indicating that the *ANN* can identify the difference between normal *ECG* signals and contaminated *ECG* signals. Besides, the error presents a normal distribution.

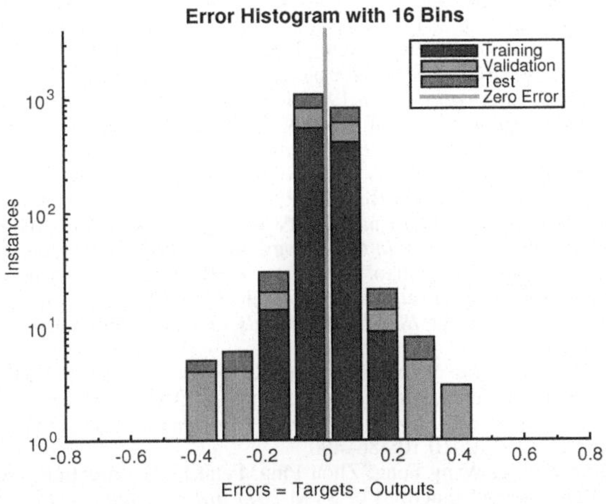

Fig. 2: Histogram of the error obtained in the training stage of ANN, the y axis is presented in logarithmic scale

C Performance test of ANN

The histogram of error obtained in the testing stage is shown in Figure 3. The test was conducted with 1000 samples to measure the performance of *ANN*, and it was observed that most of the data had a maximum error of ±0.1.

D Sensitivity and specificity analysis

Table 1 shows the results of the sensitivity and specificity test for *ANN* and *SVM* with each kernel evaluated. For each estimation, the mean and the confidence interval are presented; these were calculated with a confidence level of 95%.

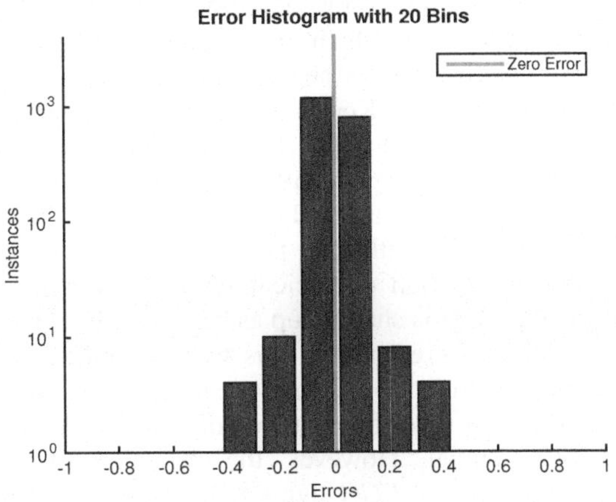

Fig. 3: Histogram of the error obtained in the test stage of ANN, the axis y is presented in logarithmic scale

Table 1: Results of the sensitivity and specificity test for ANN and SVM, (*MSE* - Mean Squared Error, R^2 - Squared Correlation Coefficient, *CI* - Confidence Interval)

RESULTS				
Method	ANN	SVM		
Kernel	–	Linear	Pol 2	RBF
Sens (%)± CI	99.60±0.31	96.49±0.79	98.80±0.37	99.06±0.37
Spec (%)± CI	99.86±0.11	95.19±0.47	96.62±0.33	99.74±0.09
Prec (%)± CI	99.60±0.36	86.99±1.12	90.70±0.83	99.24±0.28
Effi (%)± CI	99.80±0.13	95.51±0.44	97.16±0.27	99.57±0.11
MSE ± CI	0.025±0.003	0.179±0.017	0.113±0.011	0.016±0.004
R^2± CI	0.956±0.004	0.786±0.019	0.862±0.012	0.977±0.006

IV DISCUSSION AND CONCLUSIONS

Stationarity analysis does not allow to classify between normal *ECG* signals and motion artifacts-contaminated *ECG* signals [13], this is due to both signals are not stationary for the selected epoch, then the use of other techniques is necessary for classification of these signals.

The trained *ANN* as classifiers are a very useful tool for pattern recognition, these facilitate the task of identifying *ECG* signals by extraction of relevant signal characteristics, reaching fit percentages greater than 80% to classify arrhythmias in ischemic heart disease [8].

The power of the *SVM* method was subjected to tests with data that were not used in the training stage obtaining a fit of 99.7% in the classification. It has been found that the *SVM* is used to identify strange forms in some segments of the *ECG* signals with fit of 95%, higher than the presented by *ANN* [14].

The sensitivity and specificity analysis presented estimation values between 90% and 100%, indicating that the evaluated methods allow to classify the *ECG* signals within the groups to which it belongs, minimizing the presence of positive false or negative false [15].

SVM with different Kernel functions as linear kernel, polynomial kernel and *RBF* kernel were used and its performance was compared with respect to the *ANN*. The best performance was presented by the *RBF* kernel function, this is comparable with the performance of *ANN*, but the squared correlation coefficient (R^2) is lower in *ANN*. Therefore, the *SVM* is shaping up as best classifier [16].

The techniques presented in this work have been widely used in the classification of *ECG* signals, especially for the determination of different types of arrhythmias present in the *ECG* signal. However, the detection of motion artifacts-contaminated *ECG* signal remains an open topic to research. This paper shows the utility of the *ANN* and *SVM* to determine motion artifacts in *ECG* signals.

Future works will explore the multiclass methods of *SVM* to identify not only whether it is a normal *ECG* signal or motion artifacts-contaminated *ECG* signal, but also determine the type of movement present in the signal. In addition, it is proposed to use *ANN* for the classification of different types of artifacts that are affecting the *ECG* signal like a method to determine the type of filtering that is necessary to reduce the artifact.

ACKNOWLEDGMENT

This work has been supported in part by "Convocatoria 727 – Programa de Becas de Doctorado Nacionales 2015 de Colciencias" and the project "Fortalecimiento de la plataforma tecnológica para la formación especializada en el área de la salud y el desarrollo de tecnología biomédica", code Ruta N 139C of Fondo Nacional de Regalías, Republic of Colombia.

CONFLICT OF INTEREST

The authors declare that there is no conflict of interest regarding the publication of this paper.

REFERENCES

1. Sarraf Elie. Quality of Life *IEEE PULSE.* 2010;1:45–50.
2. Darlenski Razvigor Borislavov, Neykov Neyko Valentinov, Vlahov Vitan Dakov, Tsankov Nikolaï Konstantinov. Evidence-based medicine: Facts and controversies *Clinics in Dermatology.* 2010;28:553–557.
3. Golubnitschaja Olga, Kinkorova Judita, Costigliola Vincenzo. Predictive, Preventive and Personalised Medicine as the hardcore of 'Horizon 2020': EPMA position paper. *The EPMA journal.* 2014;5:1–29.
4. Pantelopoulos Alexandros, Bourbakis Nikolaos. A health prognosis wearable system with learning capabilities using NNs *Proceedings - International Conference on Tools with Artificial Intelligence, ICTAI.* 2009:243–247.
5. Jung Sang-Joong, Kwon Tae-Ha, Chung Wan-Young. A new approach to design ambient sensor network for real time healthcare monitoring system *2009 IEEE Sensors.* 2009:576–580.
6. Pandian P. S., Mohanavelu K., Safeer K. P., et al. Smart Vest: Wearable multi-parameter remote physiological monitoring system *Medical Engineering and Physics.* 2008;30:466–477.
7. Vullings E, Krijgsman A J, Verbruggen H B. Validating ECG Signals using a single-layer ANN *18th Annual International Conference of the IEEE Engineering in Medicine and Biology Society, Amsterdam 1996.* 1996;18:919–920.
8. Maglaveras Nicos, Stamkopoulos Telemachos, Diamantaras Konstantinos, Pappas Costas, Strintzis Michael. ECG pattern recognition and classification using non-linear transformations and neural networks: A review *International Journal of Medical Informatics.* 1998;52:191–208.
9. Vargas Fabian, Lettnin Djones, De Castro Maria Cristina Felippetto, Macarthy Marcello. Electrocardiogram pattern recognition by means of MLP network and PCA: A case study on equal amount of input signal types *Proceedings - Brazilian Symposium on Neural Networks, SBRN.* 2002;2002-Janua:200–205.
10. Han Liang, Pu Xiujuan, Zhou Na, Zhang Bowei, Jiang Wenhao. Extraction of Fetal Electrocardiogram Using Online Least Squares Support Vector Machines *Journal of Information & Computational Science.* 2011;8:2045–2057.
11. Liu Bo, Hao Zhifeng, Tsang Eric C C. Nesting one-against-one algorithm based on SVMs for pattern classification *IEEE Transactions on Neural Networks.* 2008;19:2044–2052.
12. Barnett Adrian G, Wolff Rodney C. A time-domain test for some types of nonlinearity *IEEE Transactions on Signal Processing.* 2005;53:26–33.
13. Kauffmann F., Maison-Blanche P., Cauchemez B., et al. A study of non stationary phenomena of HRV during 24-hour ECG ambulatory monitoring *Computers in Cardiology, 1988..* 1988;2:303–306.
14. Zheng Qian, Chen Chao, Li Zhinan, et al. A novel multi-resolution SVM (MR-SVM) algorithm to detect ECG signal anomaly in WE-CARE project *ISSNIP Biosignals and Biorobotics Conference, BRC.* 2013.
15. Kohli Narendra, Verma Nishchal K., Roy Abhishek. SVM based methods for arrhythmia classification in ECG *2010 International Conference on Computer and Communication Technology, ICCCT-2010.* 2010;10:486–490.
16. Ding Zining, Wang Feng, Zhou Ping. Fetal ECG extraction based on different kernel functions of SVM *2011 3rd International Conference on Computer Research and Development.* 2011;3:205–208.

Author: F. A. Castaño
Institute: Universidad de Antioquia
Street: Calle 67 No. 53 - 108
City: Medellín
Country: Colombia
Email: fabian.castano@udea.edu.co

Author: A. M. Hernández
Institute: Universidad de Antioquia
Street: Calle 67 No. 53 - 108
City: Medellín
Country: Colombia
Email: alher.hernandez@udea.edu.co

Thermographic comparative study between an acupuncture point and sham acupuncture for cervical pain

D.V.Q. Moreira[1] and P. Nohama[1]

[1] Pontifical Catholic University of Paraná - PUCPR/Graduate Program on Health Technology, Curitiba, Brazil

Abstract — **Thermography may facilitate the analysis of therapeutic effects of acupuncture (ACP) in cervical pain. The goal of this research was to assess thermal variations in TA15 acupuncture point in subjects with cervical pain when submitted to different stimulation forms. The sample was composed of twenty volunteers of both genders and mean age of 31 years old, divided in two groups: group 1, who received stimuli in the acupuncture point; and group 2, who received sham stimuli. Thermal images obtained before and after acupuncture application in TA15 and in sham showed reduction in skin temperature: 0.1 and 0.4 °C, 0.11 and 0.48 °C, respectively, for test side and in contralateral one. Therefore, both the true as the sham acupuncture produce a reduction in the skin temperature, even though greatly reduced.**

Keywords — **Acupuncture, Thermography, Cervical pain.**

I. Introduction

Acupuncture (ACP) is an ancient science based on stimulation of certain body points with filiform needles in order to restore or maintain health promoting mobilization, circulation and the strengthening of human energies [1]. According to Svedberg, acupuncture is a mode of sensory stimulation that evokes strong vasodilator effect. The skin temperature rises as sympathetic activity is reduced, presumably due to the mediation of opioids [2].

Science comes every day trying to understand acupuncture mechanisms of action, its effectiveness and the use of noninvasive resources, which has motivated the growth of research on this therapy [3].

Several studies have been shown ACP efficacy in different fields of health such as: acute and chronic pain [4] in knee [5], sleep disturbs [6], women health [7], obesity [8], sports [9], stomach disorders [10, 11].

As a technology for healthcare, infrared thermography may indicate thermal skin reactions to needle´s application, and maybe explain how the acupuncture points (meridians) work. It measures body electromagnetic radiation emitted with precision, without invasive procedure, quickly, painless, safe and with non-ionizing radiation [12, 13].

In clinical practice, a minimum difference in temperature of the skin surface around 0.5 °C may be considered as an indication of a clinically significant alteration or disorder.

Omura [14, 15], Thomas et al., Kozlowski [16], and Ipólito [3] used infrared thermography to assess local and systemic effects of ACP, showing contradictory results about thermal changes caused by ACP application in the acupuncture points. Kozlowski et al. has studied the muscle needling resulting in decrease in temperature, but the needle insertion at point in IG4 has produced an increase in temperature [16]. Others have measured possible changes in temperature on skin surface when applying laser on Hegu acupuncture point and has verified its elevation [17].

In this way, the goal of this paper was to evaluate the thermographic profile in the acupuncture point TA 15 with a non-acupuncture point (sham acupuncture), in subjects with cervical pain.

II. Methodology

A. Sample

In this research, twenty volunteers participated of the clinical study. They had mean age of 31 years old; cervical pain for more than three weeks, and they had not taken pills for pain in the survey day nor other pain relief therapies and had no kind of body infections.

B. Materials

It was used disposable needle, dimensions 0.25 x 30 mm, stainless, alcohol for cleaning, cotton, measuring tape, a thermographic camera (Flir Systems, USA, model A325, 320 x 240 pixels, 60 Hz, temperature range from -20ºC to + 350º C).

C. Protocol

The experimental protocol was approved by the Research Ethics Committee of the Pontifical Catholic University of Parana (PUCPR), under registration number 761176. All volunteers were attended in the Medical Thermography Laboratory at Technological Federal University of Paraná, Brazil. The environment was held stable in the range of 21 to 23 ºC, with 60% humidity. Before tests, the volunteers remained 15 minutes for temperature stabilization. Afterward, the volunteers seated with camera behind, 1 m distance, for 10 minutes unclothed upper body for thermal image acquisition. Meanwhile,

I. Torres et al. (eds.), *VII Latin American Congress on Biomedical Engineering CLAIB 2016, Bucaramanga, Santander, Colombia, October 26th -28th, 2016*,
IFMBE Proceedings 60,
DOI: 10.1007/978-981-10-4086-3_108

marks were made on skin for anatomical orientation and imaging processing, as illustrated in figure 1. Both sides of all volunteers served to the study: one side for study group and the other for the control one, as shown in figure 1.

D. Thermography and Acupuncture

Both sides of the upper back were used to stimulate and to acquire thermal images: during the first 10 minutes of ACP application, and along 5 minutes after needle had been taken off. The ACP application stopping time was called "De CHI", indicating the end of needle insertion.

In the *sham* group (fake acupuncture point), application was done at 3 cm away from AT 15 point [18].

The camera remained activated all session. Its temperature range was adjusted through the software named Thermacam Researcher 2.9 (Flir Systems, USA), in the range of 26 to 38 ºC. After outline the interest area by means of a circle (one of the analysis functions of the camera software), as illustrated in figure 1, the software provided the mean value of temperature as well as its maximum, minimum, and the standard deviation.

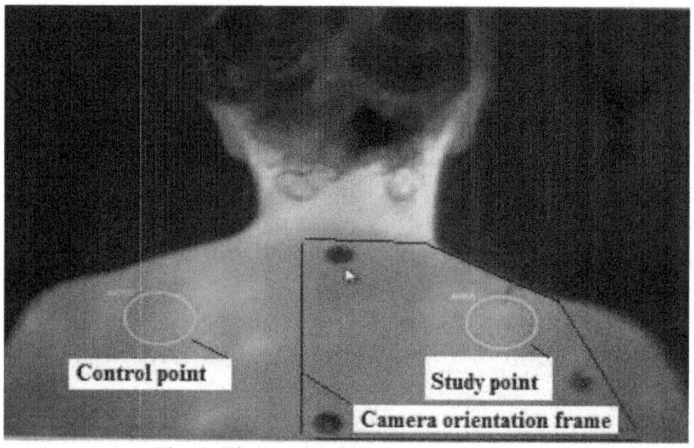

Fig. 1. Thermographic areas for analysis.

E. Statistical analysis

The analysis of data was made using Excel Microsoft and Statistical Package for the Social Science v. 20.0. After analysis of normality and homogeneity, it was applied the Student´s t-test for correlation of the pairs for the assessed moments. Values of $p< 0.05$ pointed out statistical significance.

III. RESULTS

Table 1 shows demographic data from the volunteers and thermographic values such as mean, maximum, minimum and standard deviation.

Table 2 presents temperature mean values and standard deviation for the moments before needle insertion and 10 minutes after it, between study group and *sham* group.

TABLE I
VOLUNTEERS' DEMOGRAPHIC AND THERMOGRAPHIC DATA. MIN: MINIMUM, MAX: MAXIMUM, BMI: BODY MASS INDEX, TEMP: TEMPERATURE, SD: STANDARD DEVIATION

	N	Mean	Min	Max	*SD*
Age (years)	20	31.15	20	50	±9.78
BMI (kg/m^2)	20	22.12	18.16	24.85	±2.10
Temp. (°C)	20	36.05	35	36.8	±0.51

TABLE II
TEMPERATURE MEAN VALUES AND SD IN TA 15 ACUPOINT AND IN *SHAM* FOR THE MOMENT BEFORE AND 10 MINUTES AFTER STIMULI

Group	**N**	Before (°C)	SD (°C)	After (°C)	SD (°C)
ACP_1	10	34.10	±0.78	34.00	±0.74
$SHAM_1$	10	34.00	±0.47	33.60	±0.69
ACP_2	10	34.15	±0.72	34,04	±0.59
$SHAM_2$	10	34.17	±0.52	33.69	±0.63

$_1$ test side; $_2$ control side.

In table 3, it is listed the mean temperature variation for moment 10 min after needle insertion with relation to the moment before insertion, for TA 15 acupoint and sham, and its corresponding p-value. Differences statistically significant were obtained only for ACP_2.

TABLE III
TEMPERATURE VARIATION DATA 10 MINUTES AFTER NEEDLE INSERTION WITH RELATION TO THE MOMENT BEFORE INSERTION

Group	Variation (°C)	P-value
ACP_1	-0.06±0.21	0.13
SHM_1	-0.45±0.38	0.39
ACP_2	-0.11±0,33	0.015
SHM_2	-0.48±0,40	0.39

According to data described in table 4, it is possible to check that sham acupuncture point showed a variation p =

0.005, on both studied sides. It represents significant results for the comparison 10 minutes after versus before the experiment.

TABLE IV
P-VALUES FOR TA 15 ACUPOINT AND *SHAM*, FOR THE TEST SIDE (TS) COMPARED TO THE CONTROL SIDE (CS), FOR 10 MINUTES AFTER NEEDLE INSERTION VERSUS THE MOMENT BEFORE INSERTION

Group	P-value	Group	P-value
ACP_TS	0.40	SHAM_TS	0.005*
ACP_CS	0.32	SHAM_CS	0.005*

Figure 2 illustrates comparative temperature (in °C) values for volunteers before and after the essay for acupuncture and sham tests.

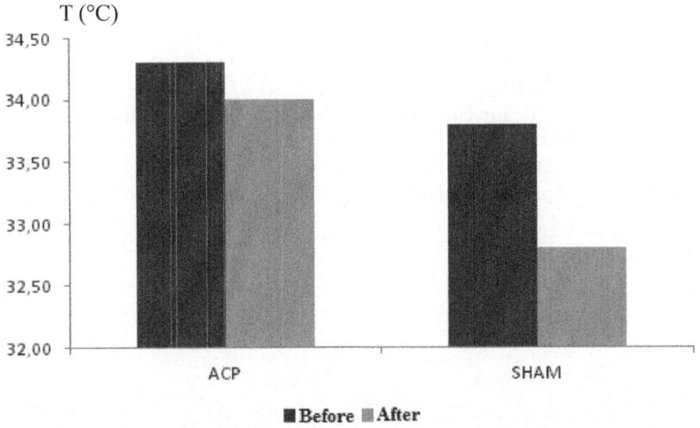

Fig. 2. Comparative temperature values between ACP and sham sessions before and after the experimental protocol.

IV. DISCUSSION

Thermography may help to study and understand acupuncture. It may be applied for accomplishment of protocols based on response time of needle´s placing, on mechanoreceptors activation and skin physiologic responses. Depending of deepness of acupuncture needle's placing, Ruffini's warming receptors located deeply inside of skin are activated and Krause receptors for cold, as well.

In the present paper, we assessed the viability of the thermography to measure temperature variation on skin after needles insertion in the skin in two situations: the group who was stimulated with needle in the acupoint TA 15 and the one whose needle was placed outside the acupoint (*sham*). The results showed that in both groups, there is a skin temperature variation after some time.

In table 2, the acupuncture group showed a difference in the mean values of 0.10 °C in the acupoint TA 15, while *sham* group had a difference of 0.40 °C after needle's placing.

In table 3, the mean temperature variation is negative in both groups ACP and sham. Also, the results listed in the table 2, for both groups temperature, decreased if considered the beginning and the end of experimental session. However, the *sham* group showed temperature variation of -0.45 °C while in the acupuncture group occurred a much lower variation, whose mean did not surpass -0.06 °C. This can be explained by reaction after material inserted superficially or deeply in the skin.

In table 4, the results for ACP_TS showed p-value of 0.4 comparing to 0.32 ACP_CS; and sham_TS p=0.005 versus sham_CS p=0.005, which indicates that stimuli outside to acupoint (sham) have denoted statistical significance.

Figure 2 shows temperatures in the moments before and after needle insertion for ACP and sham acupuncture, indicating decrease of temperature after insertion. The findings of Kozlowiski *et al.* [16], with relation to sham tests, corroborate with our research results when comparing false acupuncture with the true one.

Narongpunt *et al.* [19] assessed the results of an experimental protocol using infrared radiation to detect warming points on acupuncture in the bladder acupoint by pressure, while Raith et al. [17] used laser as stimulus on aupuncture point IG4 (Hegu) yielding a greater amount of warm emission comparing to the point that had not received laser stimulation.

However, there was no significant statistical difference in the acupuncture groups for the measured temperatures, between the test side and the control one. A small difference was perceived in both sides for the *sham* group, taking a p-value <0.05. Some studies use *sham* acupuncture to test skin sense for discovering real and false sensations [20].

V. CONCLUSION

The present results showed reduction in mean values of temperature, when comparing the moments before and after of the ending of session between effective acupuncture and false (*sham*) acupuncture.

By means of high thermal resolution thermography, it was possible to observe that there is a difference in temperature using needle's application in the skin but false acupuncture showed a psychological influence after touch perception even without needle.

However, there are many challenges using thermography during acupuncture application, mainly assessing small temperature variations on the skin. In the future, these quite soft alterations will help to understand the greater reactions that may occur in the tissues under the skin. Then, it will be used in clinical applications.

ACKNOWLEDGMENT

The authors are grateful to CAPES and CNPq, Brazil, for scholarships.

CONFLICT OF INTEREST

The authors declare that they have no conflict of interest.

REFERENCES

1. Yamamura Y. Acupuntura Tradicional: A Arte de Inserir. 2°ed.rev. e ampl. São Paulo: Roca, 2001.
2. Svedberg LE, Nordahl UEG, Lundeberg TCM. Effects of acupuncture on skin temperature in children with neurological disorders and cold feet: an exploratory study. Complementary Therapies in Medicine, 2001.
3. Ipólito AJ. Efeitos Térmicos da Acupuntura no Ponto Taixi (Rim 3), Avaliados Mediante Teletermografia Infra Vermelha. [Dissertação do Programa de Pós Graduação Interunidades em Bioengenharia]. São Carlos, 2010.
4. Menezes CRO, Moreira ACP, Brandão, WB. Base neurofisiológica para compreensão da dor crônica através da Acupuntura. Rev Dor 2010; 11(2):161-168.
5. Berkovitz, S., Cummings, M., Perrin,C., Ito, R. High volume acupuncture clinic (HVAC) for chronic knee pain – audit of a possible model for delivery of acupuncture in the National Health Service. ACUPUNCTURE IN MEDICINE 2008; 26(1):46-50.
6. Adamowicz T. Estudo dos efeitos da acupuntura sobre os padrões de sono e a secreção de cortisol em mulheres com dor crônica. [Dissertação em Biologia Celular e Molecular]. Universidade Federal do Paraná. Curitiba, 2010.
7. Luca ACB. Medicina Tradicional Chinesa – Acupuntura e Tratamento da Síndrome Climatérica. [Tese em Obstetrícia e Ginecologia]. Universidade de São Paulo 2008.
8. Haddad ML, Marcon SS. Acupuntura e apetite de trabalhadores obesos de um hospital universitário. Acta Paul Enferm 2011; 24(5):676-82.
9. Nascimento RC, Hossri CAC, Berwanger O, Carvalho, VO. Acupuncture and exercise capacity: A case report. CLINICS 2012; 67(2):193-194
10. Enblom A, Johnsson A, Hammar M, Onelo E, Steineck, G, Borjeson S. Acupuncture compared with placebo acupuncture in radiotherapy-induced nausea—a randomized controlled study. Annals of Oncology Advance Access published September 23, 2011.
11. Fernandes MH, Graziani SR. Acupuntura na prevenção da náusea e do vômito decorrentes do tratamento da quimioterapia antineoplásica. Revista PIBIC, Osasco, v. 3, n. 2, 2006, pag. 49-58
12. Carvalho AR, Medeiros DL, Souza FT, et al. Variação de Temperatura do Musculo Quadríceps Femoral Exposto a Duas Modalidades de Crioterapia por Meio de Termografia. Rev. Bras. Med Esporte v. 18, n° 2 – Mar/Abr, 2012.
13. Sanches IJ. Sobreposição de Imagens de Termografia e Ressonância Magnética: Uma Nova Modalidade de Imagem Médica Tridimensional. Tese [doutorado] Universidade Tecnológica Federal do Paraná. Programa de Pós-Graduação em Engenharia Elétrica e Informática Industrial, Curitiba/PR, 2009.
14. Omura Y. Acupuncture, infra-red thermography and Kirlian photography. Acup Electrother Red Int J (Pergamon Press) 1976, 1 e 2 pag 43-86.
15. Thomas D, Collins S, Strauss S. Somatic Sympathetic Vasomotor Changes Documented by Medical Thermographic Imaging During Acupuncture Analgesia. Clinical rheumatology, 1992, 11, n°1 pag.55-59
16. Kozlowski AK, Lange AC, Beck H. Contact-free Infrared Thermography for Assessing Effects during Acupuncture: A Randomized, Single-blinded, Placebo-controlled Crossover Clinical Trial. Anesthesiology v. 111, 632–9, 2009.
17. Raith W, Litscher G, Sapetschnig, I, Bauchinger, S.,Ziehenberger E, Muller W, Urlesberger B. Thermographical Measuring of the Skin Temperature Using Laser Needle Acupuncture in Preterm Neonates. Evidence-Based Complementary and Alternative Medicine, v. 2012, Article ID 614210, 5 pages.
18. Carneiro MN. Acupuntura Médica Contemporânea. Os resultados das pesquisas recentes: Gonartrose (osteoartrite do joelho), 2008. Instituto de Pesquisa e Ensino de Terapêutica Contemporânea – IPETC. Disponível em: www.acupunturacomtemporanea.blogspot.com/ Acesso em: 12/10/2014
19. Narongpunt V, Cornillot P, Attali J-R, Molinier F, et al. Infrared Thermographic Visualization Of The Traditional Chinese Acupuncture Meridian Points. Medical Acupuncture. A Journal for Physicians by Physicians, v.16, n 2 s/ano.
20. Moffet HH. Sham Acupuncture May Be as Efficacious as True Acupuncture: A Systematic Review of Clinical Trials. The Journal of Alternative And Complementary Medicine Volume 15, Number 3, pp. 213–216, 2009.

Author: Percy Nohama
Institute: PUCPR
Street: Rua Imaculada Conceição 1155, CEP 8025-901
City: Curitiba
Country: Brazil
Email: percy.nohama@pucpr.br

Iterative joint dynamic brain mapping and neural activity modeling from electroencephalographic signals

E. Giraldo[1], P.A. Munõz-Gutiérrez[2] and G. Castellanos-Dominguez[3]

[1] Universidad Tecnológica de Pereira/Department of Electrical Engineering, Associated professor, Pereira, Colombia
[2] Universidad Tecnológica de Pereira/Doctorade in Engineering, Student, Pereira, Colombia and Universidad del Quind´ıo/Electronic Instrumentation Technology, Assistant professor, Armenia, Colombia
[3] Universidad Nacional de Colombia, sede Manizales/Department of Engineering and Architecture, Full Professor, Manizales, Colombia

Abstract— **A novel joint iterative dynamic inverse problem solution for brain mapping based on electroencephalographic (EEG) signals is presented. Proposed approach considers linear and nonlinear time-varying state space models of the brain as dynamic constraints in the solution of the dynamic inverse problem where the brain mapping and the neural activity model are estimated simultaneously from EEG signals. The method performance is evaluated in terms of standard error, projection error, and residual error for several SNRs by using simulated EEG signals. As a result, a considerable improvement over Low Resolution Tomography (LORETA) and Dynamic LORETA approaches is found.**

Keywords— **dynamic inverse problem, joint estimation, time-varying parameters.**

I Introduction

Among brain mapping techniques, neuroimaging based on EEG signals is widely used due to its high temporal resolution and low implementation cost [1, 2, 3]. However, EEG-based neuroimaging spatial resolution is very low, since it reflects activity estimation triggered by a few thousands of neural generators, but having just a very reduced set of scalp measures. This estimation problem is termed the EEG inverse problem [4].

In order to improve inverse solutions, the dynamics of the brain activity can be included in the form of a dynamical estimation model constraining the solution to some predefined geometric or physiological restrictions [5, 6]. Particularly in [7], the solution is constrained by considering maximum temporal smoothness. On the other hand, a physiologically motivated model allows an adequate system dynamics description and also increases the information provided by the inverse solution. To this end, a better statement of global EEG source dynamics is accomplished grounded on the first and second order models, as proposed in [5, 2, 1].

This article presents an alternative iterative joint dynamic inverse problem solution for brain mapping and neural activity modeling from EEG signals. The proposed method, considers linear and nonlinear state space models with two, three an five time-varying parameters in order to estimate adequately normal and pathological activity into the brain. The method performance is evaluated in terms of standard error, projection error, and residual error for several SNRs by using simulated EEG signals. As a result, a considerable improvement over LORETA and dynamic LORETA approaches is found.

II Methods

Consider that the EEG can be described by the following state space model:

$$z_k = \boldsymbol{F}_k z_{k-1} + \boldsymbol{\upsilon}_k \tag{1}$$

$$\boldsymbol{y}_k = \boldsymbol{C} z_k + \boldsymbol{\zeta}_k \tag{2}$$

being

$$z_k = \begin{bmatrix} \boldsymbol{c}_k \\ \boldsymbol{w}_k \end{bmatrix}, \boldsymbol{F}_k = \begin{bmatrix} 0 & \boldsymbol{G}_k \\ 0 & \boldsymbol{I}_p \end{bmatrix}, \boldsymbol{\upsilon}_k = \begin{bmatrix} \boldsymbol{\eta}_k \\ \boldsymbol{\varepsilon}_k \end{bmatrix} \tag{3}$$

and

$$\boldsymbol{C} = \begin{bmatrix} \boldsymbol{M\Phi}_s & \boldsymbol{0} \end{bmatrix} \tag{4}$$

where the following definitions are considered: $\boldsymbol{y}_k \in \mathbb{R}^{d\times 1}$ is the EEG measurement of d channels at sample k, $\boldsymbol{c}_k \in \mathbb{R}^{s\times 1}$ is the neural activity at s zones, $\boldsymbol{w}_k \in \mathbb{R}^{p\times 1}$ is the vector of model parameters, $\boldsymbol{z}_k \in \mathbb{R}^{(s+p)\times 1}$ is the augmented state space vector, $\boldsymbol{F}_k \in \mathbb{R}^{(s+p)\times(s+p)}$ is the augmented transition matrix, $\boldsymbol{I}_p \in \mathbb{R}^{p\times p}$ is an identity matrix, $\boldsymbol{M} \in \mathbb{R}^{d\times n}$ is the lead field matrix being n the number of current sources, $\boldsymbol{\Phi}_s \in \mathbb{R}^{n\times s}$ the spatial basis that maps the neural activity at each source into zones [7], $\boldsymbol{C} \in \mathbb{R}^{d\times(s+p)}$ is the augmented output matrix. $\boldsymbol{\zeta}_k \in \mathbb{R}^{d\times}1$, $\boldsymbol{\eta}_k \in \mathbb{R}^{s\times 1}$ and $\boldsymbol{\varepsilon} \in \mathbb{R}^{p\times 1}$ are the uncorrelated zero mean Gaussian noises with covariances matrices $\boldsymbol{\Sigma} \in \mathbb{R}^{d\times d}$, $\boldsymbol{\Lambda} \in \mathbb{R}^{s\times s}$ and $\boldsymbol{\Gamma} \in \mathbb{R}^{p\times p}$ respectively,

I. Torres et al. (eds.), *VII Latin American Congress on Biomedical Engineering CLAIB 2016, Bucaramanga, Santander, Colombia, October 26th -28th, 2016*,
IFMBE Proceedings 60,
DOI: 10.1007/978-981-10-4086-3_109

$\boldsymbol{G}_k \in \mathbb{R}^{s\times p}$ is the dynamical model matrix, $\boldsymbol{\upsilon}_k \in \mathbb{R}^{(s+p)\times 1}$ is the augmented noise vector and $\boldsymbol{\Upsilon} \in \mathbb{R}^{(s+p)\times(s+p)}$ the augmented noise covariance defined as

$$\boldsymbol{\Upsilon} = \begin{bmatrix} \boldsymbol{\Lambda} & \boldsymbol{0} \\ \boldsymbol{0} & \boldsymbol{\Gamma} \end{bmatrix} \tag{5}$$

The dynamical model of neural activity is defined as $\boldsymbol{c}_k = \boldsymbol{G}_k\boldsymbol{w}_k + \boldsymbol{\eta}_k$ and the parameter evolution $\boldsymbol{w}_k = \boldsymbol{w}_{k-1} + \boldsymbol{\varepsilon}_k$, and where the following structures for $\boldsymbol{G}_k$ are considered:

$$\boldsymbol{G}_k = \begin{cases} [\ \widehat{\boldsymbol{c}}_{k-1} \quad \boldsymbol{L}\widehat{\boldsymbol{c}}_{k-1}\ ], p=2 \\ [\ \widehat{\boldsymbol{c}}_{k-1} \quad \boldsymbol{L}\widehat{\boldsymbol{c}}_{k-1} \quad \widehat{\boldsymbol{c}}_{k-2}\ ], p=3 \\ [\ \widehat{\boldsymbol{c}}_{k-1} \quad \boldsymbol{L}\widehat{\boldsymbol{c}}_{k-1} \quad \widehat{\boldsymbol{c}}_{k-2} \quad \widehat{\boldsymbol{c}}_{k-1}^{\circ 2} \quad \widehat{\boldsymbol{c}}_{k-1}^{\circ 3}\ ], p=5 \end{cases} \tag{6}$$

where $\boldsymbol{c}_{k-1}^{\circ 2}$ denotes the Hadamard Power, being $\boldsymbol{L} \in \mathbb{R}^{s\times s}$ the spatial coupling among sources, and with the following set of time-varying parameters [6]:

$$\boldsymbol{w}_k^\top = \begin{cases} [\ a_{1,k} \quad b_{1,k}\ ], p=2 \\ [\ a_{1,k} \quad b_{1,k} \quad a_{2,k}\ ], p=3 \\ [\ a_{1,k} \quad b_{1,k} \quad a_{2,k} \quad a_{3,k} \quad a_{4,k}\ ], p=5 \end{cases} \tag{7}$$

The joint estimation of states and parameters is performed by estimating $\boldsymbol{c}_k$ and $\boldsymbol{w}_k$ simultaneously through $\boldsymbol{z}_k$. This can be done by defining the following functional

$$\Phi(\boldsymbol{z}_k, \lambda_k) = \|\boldsymbol{y}_k - \boldsymbol{C}\boldsymbol{z}_k\|_{\boldsymbol{\Sigma}}^2 + \lambda_k^2 \|\boldsymbol{z}_k - \boldsymbol{F}_k\boldsymbol{z}_{k-1}\|_{\boldsymbol{\Upsilon}}^2 \tag{8}$$

Accordingly, an estimate of the $\widehat{\boldsymbol{z}}_k$ can be obtained by minimizing (8) with respect to $\boldsymbol{z}_k$ as follows:

$$\widehat{\boldsymbol{z}}_k = \arg\min_{\boldsymbol{z}_k} \{\Phi(\boldsymbol{z}_k, \lambda_k)\} \tag{9}$$

where the minimization of (8) can be obtained by applying the Jacobian with respect to $\boldsymbol{z}_k$ and equaling to $\boldsymbol{0}$, where $\widehat{\boldsymbol{z}}_k$ can be obtained analytically for each k as follows:

$$\widehat{\boldsymbol{z}}_k = \left(\boldsymbol{C}^\top\boldsymbol{\Sigma}^{-1}\boldsymbol{C} + \lambda_k^2\boldsymbol{\Upsilon}^{-1}\right)^{-1}\left(\boldsymbol{C}^\top\boldsymbol{\Sigma}^{-1}\boldsymbol{y}_k + \lambda_k^2\boldsymbol{\Upsilon}^{-1}\boldsymbol{F}_k\widehat{\boldsymbol{z}}_{k-1}\right) \tag{10}$$

Consequently, the states $\boldsymbol{c}_k$ and parameters $\boldsymbol{w}_k$ can be estimated for each k by applying Eq. (10).

III RESULTS AND DISCUSSION

The effectiveness and performance of the proposed method is evaluated by estimating the neural activity and its dynamics from simulated EEG signals. In order to obtain a realistic EEG data set, a model which displays complex spatio-temporal behavior for the brain dynamics as proposed in [5] is used, where the temporal dynamics of $\boldsymbol{x}_k$ are modeled using a linear combination of sine functions whose components are evenly spaced in the alpha band (8 to 12 Hz). The epileptic seizure is simulated for each of the 30 trials for one randomly located active source at time $t = 0.5$ seconds. The simulated EEG $\boldsymbol{y}_k$ is obtained from $\boldsymbol{x}_k$ using $\boldsymbol{y}_k = \boldsymbol{M}\boldsymbol{x}_k + \boldsymbol{\zeta}_k$, where $\boldsymbol{\zeta}_k$ is set to achieve the Signal-to-Noise Ratio (SNR) equals $5\,dB$, $10\,dB$, $15\,dB$ and $20\,dB$ with a sample rate of 250Hz.

The performance of the proposed method is compared with LORETA [8] and with a Dynamic LORETA presented in [9], where we also include the Laplacian operator as weighting matrix. The comparison was carried out in terms of the following measures:

$$\text{RE (Residual Error)} = \|\boldsymbol{M}\widehat{\boldsymbol{X}} - \boldsymbol{Y}\|_{FRO}^2$$

$$\text{SE (Standard Error)} = \|\widehat{\boldsymbol{X}} - \boldsymbol{X}\|_{FRO}^2$$

$$\text{PE (Projected Error)} = \left\|\widehat{\boldsymbol{X}} - \sum_{i}^{\text{rank}(\boldsymbol{M})} \langle \boldsymbol{X}, \boldsymbol{v}_i\rangle \boldsymbol{v}_i\right\|_{FRO}^2$$

where $\|\cdot\|_{FRO}$ is the Frobenius norm, $\boldsymbol{v}_i$ are the right singular vectors of $\boldsymbol{M}$ and $\boldsymbol{X}$ and $\boldsymbol{Y}$ are $n\times K$ and $d\times K$ matrices respectively that contain the activity on the dipoles and the sensor measurements for all the time instants. Furthermore, the projected error describes the error relative to the recoverable solution i.e. the part of the solution that is in the space rank of $\boldsymbol{M}$, which is the best reconstruction possible.

Figure 1 shows spatial visualization of the simulated and reconstructed activity with the considered methods, using measurements with a SNR of 20dB. It can be seen that there is spurious activity in the reconstruction by using the LORETA approach. This spurious activity is not observed in estimated brain activity by using the dynamic LORETA or the proposed models with $p = 2$, $p = 3$ and $p = 5$ parameters.

Figures 2(a), 2(b) and 2(c) show that the lowest error is obtained using the most complex dynamic model. It can be seen that the difference of the performance can be seen clearly only in the projected error. For high SNRs, the performance of the proposed dynamic models is very similar, however, when the SNR decreases, the performance of the simpler models decays faster than the performance of the model that contains five parameters.

In general, the proposed method proved to be robust to the SNR of the signal. More specifically, the model with five parameters is the most robust to noise.

This method can estimate changes the model parameters, when brain activity changes the normal to pathological, figures 3(a), 3(b) and 3(c) show the changes in the dynamics of the brain (normal to pathological states) and is successfully modeled. Therefore, this method can be used to detect pathological signals.

It is worth noting that by using a dynamic approach for the regularization parameters (update them at every time instant), adds noise to the solution and greatly increase the computational cost of the algorithm. However, the inclusion of additional parameters in the model does not increase considerably the computational cost of the proposed method.

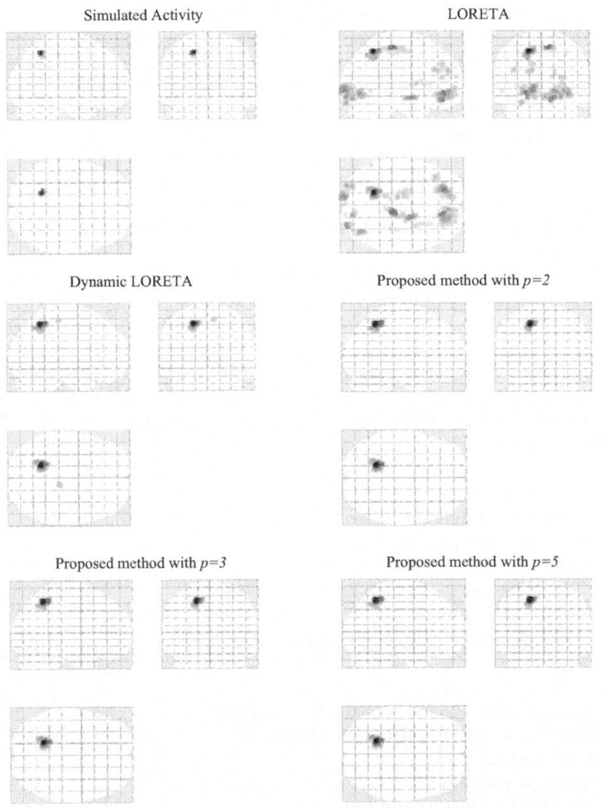

Fig. 1: Spatial visualization of the simulated and reconstructed activity with the considered methods, using measurements with a SNR of 20dB.

IV CONCLUSIONS

We address a joint iterative dynamic inverse problem solution for brain mapping and neural activity modeling from electroencephalographic signals. Proposed approach considers linear and nonlinear time-varying state space models of the brain as dynamic constraints in the solution of the dynamic inverse problem where the brain mapping and the neural activity model are estimated simultaneously from EEG signals. The obtained results demonstrate that the dynamic inverse problem with time-varying parameters perform better than LORETA and dynamic LORETA. Also, it can be seen that the changes in the dynamics of the brain (normal to pathological states) is successfully described by the model parameters. As future work, the authors plan to expand the discussed methodology to detect different mental states using the parameters of the dynamical model.

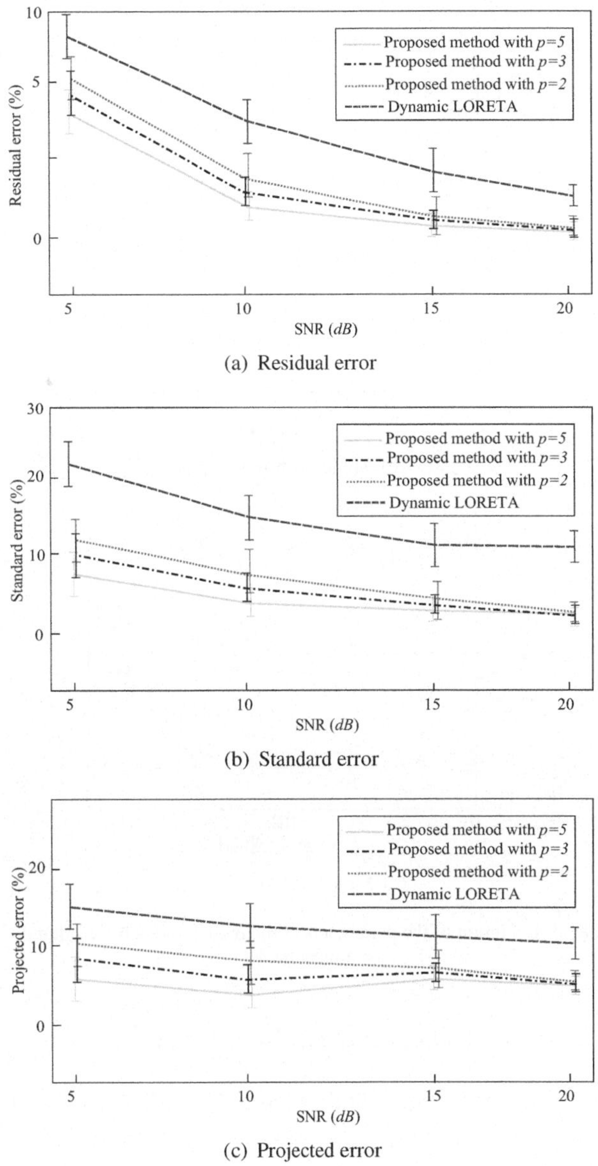

Fig. 2: Estimation errors under several SNRs for each considered method.

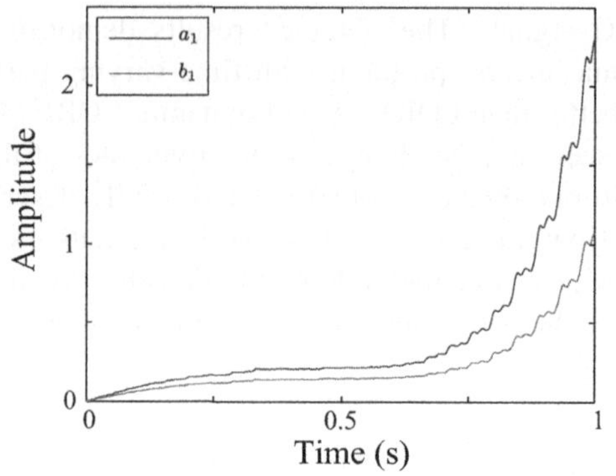

(a) Model with $p = 2$.

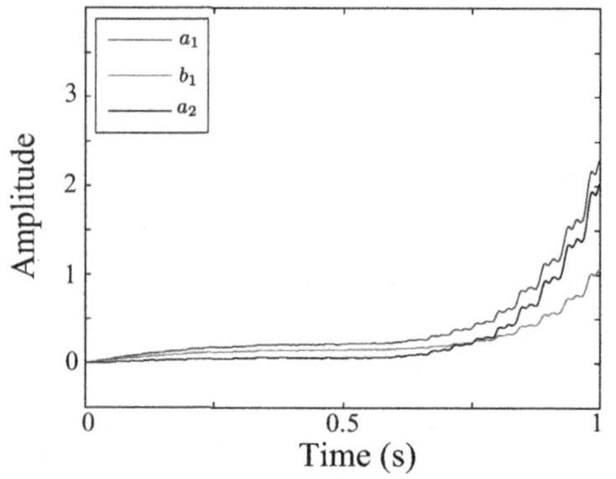

(b) Model with $p = 3$.

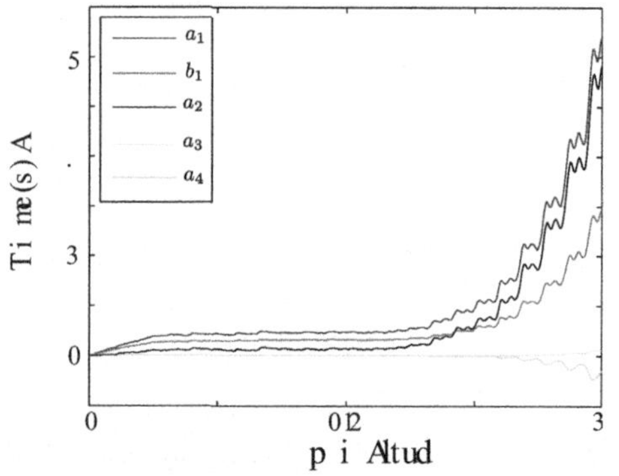

(c) Model with $p = 5$.

Fig. 3: Evolution of the parameters of the model for each considered dynamical model.

Conflict of Interest

The authors declare that they have no conflict of interest.

References

1. Fukushima Makoto, Yamashita Okito, Kanemura Atsunori, Ishii Shin, Kawato Mitsuo, Sato Masa. A State-Space Modeling Approach for Localization of Focal Current Sources From MEG *IEEE Trans. Biomed. Engineering.* 2012;59:1561-1571.
2. Giraldo E.;, Castellanos-Dominguez G.;. Current Density Reconstruction from EEG Based on a Time Varying Nonlinear Physiological Model *Proceedings of the 5th International IEEE EMBS Conference on Neural Engineering.* 2011:2914 -2917.
3. Giraldo E.;, Padilla-Buriticá J.I.;, Castellanos-Dominguez G.;. EEG Source Localization Based on Multivariate Autoregressive Models using Kalman Filtering *Proceedings of the 33rd Annual International Conference of the IEEE Engineering in Medicine and Biology Society, EMBC 2011.* 2011:2914 -2917.
4. Plummer C.;, Simon-Harvey A., Cook M.;. EEG source localization in focal epilepsy: Where are we now? *Epilepsy.* 2008;49:201 -218.
5. Barton M. J., Robinson P. A., Kumar S., et al. Evaluating the Performance of Kalman-Filter-Based EEG Source Localization *IEEE Transactions on Biomedical Engineering.* 2009;56:122-136.
6. Giraldo-Suárez E., Castellanos-Dominguez G.. *Artificial Computation in Biology and Medicine: International Work-Conference on the Interplay Between Natural and Artificial Computation, IWINAC 2015, Elche, Spain, June 1-5, 2015, Proceedings, Part I*ch. Neural Activity Estimation from EEG UsinganIterative Dynamic Inverse Problem Solution, :388–397. Cham: Springer International Publishing 2015.
7. Giraldo E., Martinez-Vargas J. D., Castellanos-Dominguez G.. Reconstruction of neural activity from EEG data using dynamic spatio-temporal constraints *International Journal of Neural Systems.* 2016;26:1-15.
8. Pascual-Marqui R D, Michel C M, Lehmann D. Low resolution electromagnetic tomography: a new method for localizing electrical activity in the brain. *International Journal of Psychophysiology.* 1994;18:49–65.
9. Yamashita O.;, Galka A.;, Ozaki T.;, Bizcay R.;, Valdez-Sosa P.;. Recursive Penalized Least Squares Solution for Dynamical Inverse Problems of EEG Generation *Human Brain Mapping.* 2004;21:221 -235.

Author: P.A. Munõz-Gutiérrez
Institute: Universidad del Quindío
Street: Carrera 15 Calle 12 Norte
City: Armenia
Country: Colombia
Email: pamunoz@uniquindio.edu.co

Author: E. Giraldo
Institute: Universidad Tecnolgica de Pereira
Street: Carrera 27 N. 10-02 Barrio Alamos
City: Pereira
Country: Colombia
Email: egiraldos@utp.edu.co

Author: G. Castellanos-Dominguez
Institute: Universidad Nacional de Colombia, sede Manizales
Street: Cra 27 N 64-60
City: Manizales
Country: Colombia
Email: cgcastellanosd@unal.edu.co

Autoregressive Models of Electrocardiographic Signal Contaminated with Motion Artifacts: Benchmark for Biomedical Signal Processing Studies

F. A. Castaño[1] and A. M. Hernández[1]

[1] Bioinstrumentation and Clinical Engineering Research Group - GIBIC, Bioengineering Department, Engineering Faculty, Universidad de Antioquia UdeA; Calle 70 No. 52-21, Medellín, Colombia

Abstract— **Motion artifacts are one of the biggest challenges in analysis of electrocardiographic signals in outpatients, different signals processing techniques have been developed to significantly reduce these artifacts, however, there is not a standard signal that consider artifacts and allows to compare the proposed algorithms to reduce these, as is done in other areas such as acoustic or seismic. In this paper, a mathematical model for the representation of the electrocardiographic signal with motion artifacts is proposed; Fourier analysis and adjustment parameters are shown to mathematically represent an electrocardiography lead. Identification of motion artifacts in ambulatory electrocardiography was performed by autoregressive models. Additionally, the obtained signal from the mathematical model and its performance are shown.**

Keywords— **Electrocardiography, Mathematical Model, Computer Simulation, Electrophysiology, Artifacts Reduction**

I INTRODUCTION

Motion artifacts are one of the biggest challenges in ambulatory electrophysiological information recording, especially in those signals acquired through portable monitoring devices. The greatest difficulty in the treatment of motion artifacts is its dynamic nature and often its amplitude is greater than the amplitude of the interest signal [1].

Ansari, Ward and Najarian by reviewing the state of the art in terms of strategies to reduce motion artifacts, showed that it is possible to classify them into three groups of techniques [1]:

a). The first group is focused on instrumentation, the main purpose of these studies is to look for alternative hardware designs to acquire the signals, improved electrodes, sensors or cables as well as different locations for these that are less susceptible to movement [2].

b). Computational analysis techniques as independent component analysis (*ICA*) and principal component analysis (*PCA*) to estimate motion artifacts as an independent source of signal variation [3]. The objective of ICA is to estimate the linear coefficients group through which the interest signal and motion artifacts have been mixed. This requires acquisition of multi-channel signal, which is performed on the electrocardiogram (*ECG*).

c). Motion artifact reduction using adaptive filters that modeled the artifact to be adaptively removed. Different adaptive filters types have been applied to this problem including Least Mean Square (*LMS*) [4], Recursive Least Square (*RLS*) [5], and others. Despite the popularity of adaptive filters, there are some disadvantages associated with their use. In many cases these not only adapt to motion artifacts, but also they adapt to the signal of interest, it is due to the adaptive filters do not have a mechanism to distinguish between body motion artifacts and interest components [1]. Also, it can happen that a higher order artifact could not be identified by the modeling stage of the filter. However, this technique has been reported as one of the most efficient in the motion artifacts reduction [6].

Simulated signals with different degrees of realism or real signals are used in the proposals of filtering techniques to reduce motion artifacts in *ECG* signals, but often it is difficult to compare the performance of these filtering strategies because the synthesis of the used signals differs between authors.

In this paper, a mathematical model to simulate an *ECG* signal contaminated with movement artifacts is presented. The model presented may be used in the analysis of methodologies to reduce motion artifacts.

II METHODOLOGY

The $x(n)$ signal represents the artifact-contaminated *ECG* signal. Signal consists in a linear combination of a normal synthetic *ECG* signal, $ecg(n)$ (see Appendix 1 for a reference of *ECG* model [7]) and a synthetic artifacts $g(n)$, the expression is shown in equation1.

$$x(n) = ecg(n) + g(n) \tag{1}$$

In most devices that acquire the *ECG* signal, leads I, II and III are measured, however, lead II is commonly used in clinical practice as a standard to do a preliminary analysis of the electrical heart condition and heart rhythm. For this reason, the signal presented in this paper as example simulates the lead II.

A Acquisition of ECG signal

Lead II of the *ECG* signal of a healthy volunteer at rest was recorded with a sampling frequency of 100 Hz. From

I. Torres et al. (eds.), *VII Latin American Congress on Biomedical Engineering CLAIB 2016, Bucaramanga, Santander, Colombia, October 26th -28th, 2016*,
IFMBE Proceedings 60,
DOI: 10.1007/978-981-10-4086-3_110

this signal, amplitude a_j, duration d_j, time of occurrence t_j of each event $j \in J$ and heart rate (N_{hb}) were extracted to perform the *ECG* signal synthesis according to Javed and Ahmad [7].

Additionally, the volunteer was asked to perform a controlled movement of its arms while the *ECG* signal was recorded in order to obtain motion artifact-contaminated signal. This signal was processed and identified to complete the mathematical model.

B Motion artifacts model identification

The function $g(n)$ presented in equation 1 represents the artifacts that can significantly affect the *ECG* signal, it can simulate different types of artifacts, either motion artifacts, breathing artifacts or electric noise.

From motion artifacts-contaminated *ECG* signal, the *ECG* component was removed using a lowpass *FIR* digital filter with 256 tags and 2 Hz cutoff frequency.

Once the motion artifact signal was obtained, an autoregressive model (*ARX*) was identified [8]. This model has the structure shown in equation 2.

$$A(q)y(n) = B(q)u(n) + e(n) \quad (2)$$

Where $A(q)$ and $B(q)$ represent the polynomials to adjust the autoregressive model, $y(n)$ is the system output, $u(n)$ is the system input and $e(n)$ is random noise that affects the system. The system output is given by equation 3.

$$y(n) = -a_1y(n-1) - ... - a_{na}y(n-na) + b_0u(n-nk) + b_1u(n-nk-1) + ... + b_{nb}u(n-nk-nb) + e(n) \quad (3)$$

Where a_n are the coefficients of the polynomial $A(q)$, b_n are the coefficients of the polynomial $B(q)$, na is the number of coefficients taken for the function $A(q)$, nb is the number of coefficients taken for the function $B(q)$ and nk is the delay relative to input function $u(n)$. The model order was selected according to the Akaike Information Criterion (*AIC*) [9].

The motion artifact signal extracted from *ECG* signal was used as output signal for the *ARX* model identification. As input signal, a vector time from 0 s to 10 s was used, this vector has a sample frequency of 100 Hz.

III Results

A Simulation of ECG signal

The signal characterization is shown in Table 1, the value of the parameters a_j, d_j, t_j, and each measured event $j \in J$ are presented. Heart rate N_{hb} was 66 bpm. In Figure 1, the similarity between the acquired signal of a patient at rest and the synthesized *ECG* signal is shown; the synthesized signal was sampled at 100 Hz.

Table 1: Parameters for the synthesis of lead II *ECG* signal

CHARACTERIZATION						
Parameters	P	Q	R	S	T	U
Amplitude	0.10	−0.10	1.20	0.10	0.34	0.01
Duration	0.0900	0.0660	0.1100	0.0660	0.1820	0.0476
Time	0.160	0.250	0.300	0.370	0.525	0.733

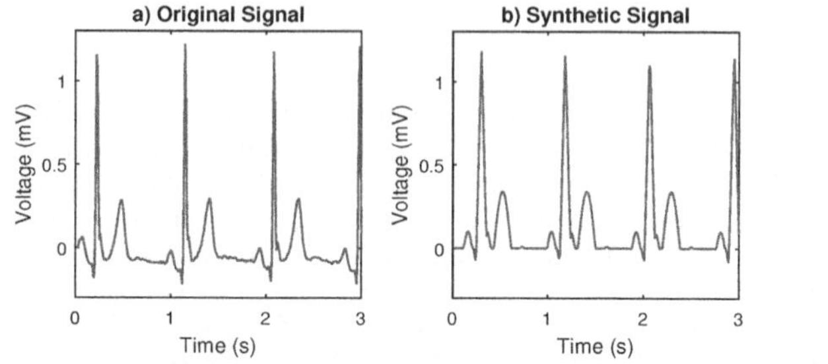

Fig. 1: a). Acquired *ECG* signal from a volunteer at rest and b). synthesized signal from the mathematical model of Appendix 1

In Figure 2, Power Spectral Density (*PSD*) of the acquired *ECG* signal from a volunteer at rest and the synthesized *ECG* signal by the mathematical model are shown.

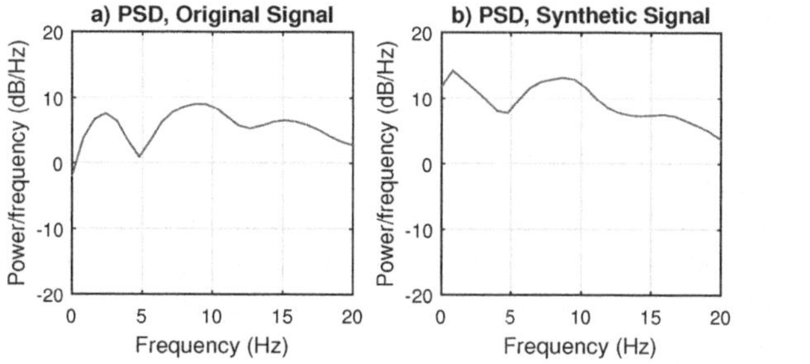

Fig. 2: Power Spectral Density (*PSD*) of: a). *ECG* signal of a healthy volunteer at rest and b). *ECG* signal synthesized from mathematical model of Appendix 1

B Identification of motion artifacts component

In Figure 3, the acquired *ECG* signal from healthy volunteer in movement and the motion artifact signal resulting from signal filtering are shown.

The obtained order of the model of the artifact signal was 10, $na = 8$ and $nb = 2$, additionally, the estimated delay was $nk = 3$. In equations 4 and 5 the polynomial $A(q)$ and the polynomial $B(q)$ are shown respectively.

$$A(q) = 1 - 4.993q^{-1} + 9.31q^{-2} - 6.977q^{-3} + 2.325q^{-5} - 0.9978q^{-7} + 0.333q^{-8} \quad (4)$$

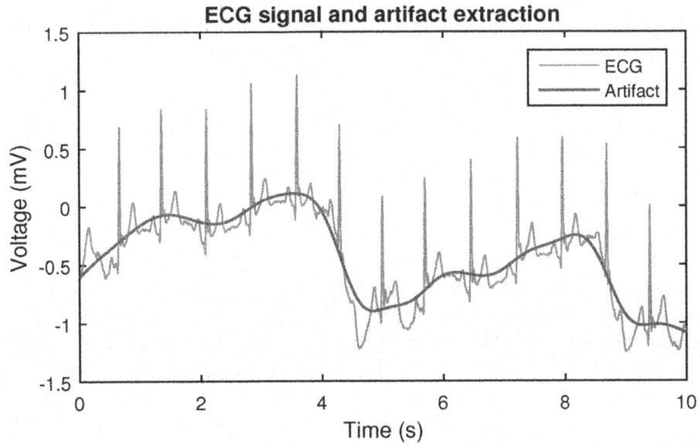

Fig. 3: *ECG* signal with motion artifacts (gray) and signal corresponding to motion artifact partially extracted (blue)

$$B(q) = -4.527x10^{-7}q^{-3} + 4.517x10^{-7}q^{-4} \quad (5)$$

Therefore, the following function $g(n)$ represents motion artifacts in *ECG* (equation 6):

$$\begin{aligned} g(n) &= 4.993g(n-1) - 9.31g(n-2) + 6.977g(n-3) \\ &- 2.325g(n-5) + 0.9978g(n-7) - 0.333g(n-8) \\ &- 4.527x10^{-7}u(n-3) + 4.517x10^{-7}u(n-4) + e(n) \quad (6) \end{aligned}$$

Where $g(n-1)$, $g(n-2)$, $g(n-3)$, $g(n-5)$, $g(n-7)$ and $g(n-8)$ represent the past values of motion artifacts signal. $u(n-3)$ and $u(n-4)$ represent the past values of input signal.

The fitting of the *ARX* model to experimental data was 98% with a Mean Square Error (*MSE*) of $6.26x10^{-17}$. A prediction horizon of 100 samples was used. In Figure 4, a comparison between the *ECG* signal contaminated by motion artifact and synthesized *ECG* signal from equation 1 is shown.

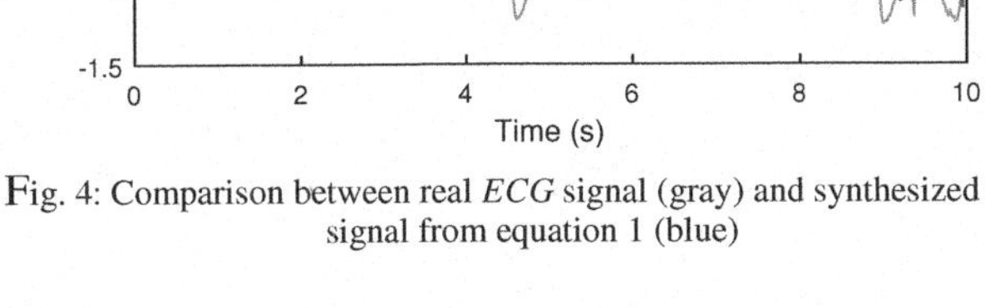

Fig. 4: Comparison between real *ECG* signal (gray) and synthesized *ECG* signal from equation 1 (blue)

IV Discussion and Conclusions

In this paper an approach to use autoregressive models to generate synthesized versions of artifact-contaminated *ECG* signals was presented. From this signal, different conditions of artifacts to study biomedical signal processing techniques and reducing motion artifacts may be generated and compared using this signal or similar as a benchmark [10].

From synthesized *ECG* signal, it is possible to determine heart rate, R-peak signal voltage and when added noise, it could be explored linear and non-linear techniques for artifacts reduction, which will remain a future work.

Slight differences between both frequency spectra are observed, a lower curl is shown in the real signal due to some characteristics of smoothing that has the original signal and need a complex model to be simulated. However, the spectrum generally has a great similarity between the two signals along the frequency axis, which allows to conclude that the synthesized signal reliably represents the real *ECG* signal [11].

Changes in the value of the parameters for each of the events, can produce other derivations of *ECG* or synthesize signals with different morphologies, a situation that is useful when doing analysis by *ICA* or *PCA*, because it is possible to simulate signals from different sources [12].

As future works, identification of different types of artifacts produced by a volunteer during walking or running is proposed. Also, the use of input signals different like accelerometer signals captured by inertial sensors will be analyzed to determine possible sources of the motion artifacts. Also, the autoregressive model implemented in this work was tested with a single volunteer ECG signal, to validate the model, it will plan a test with signals of different people that can prove the robustness of propose model.

Appendix 1

Javed and Ahmad presented a mathematical model of the *ECG* signal based on simple functions and Fourier series $ecg(n)$ [7]. The waveform of the *ECG* signal is composed of a group of peaks and undulations named as events $j \in J = \{P, Q, R, S, T, U\}$. These events are associated with atrial and ventricular depolarization and polarization.

The mathematical model was designed through the use of periodic waveforms based on three key parameters: amplitude, duration and time of occurrence of the event. Heart rate or number of beats per minute was denoted by N_{hb}. Equation 7 expresses the waveform of each one of the events $j \in J$ at any instant of time t.

$$E_j(t) = A_j + \sum_{n=1}^{N_{hb}} U_j(n) \cos(\omega_{hb}(t - t_j)n) \quad (7)$$

Where $\omega_{hb} = \pi N_{hb}/30$ is the fundamental frequency of the ECG signal, A_j and $U_j(n)$ are Fourier coefficients, while the parameter t_j is the time of occurrence of the event $j \in J$. Under these conditions, ECG signal is represented as the sum of all events belonging to the set J, as shown in equation 8.

$$ECG(t) = \sum_{j \in J} E_j(t) \tag{8}$$

The P, T and U events (group 1) have a semi-elliptical shape, while the Q, R and S events (group 2) have a conic shape. A function for the A_j and $U_j(n)$ coefficients from the groups 1 and 2 were obtained respectively. Equations 9 and 10 show the coefficients for $j \in \{P,T,U\}$ and the equations 11 and 12 show the coefficients for $j \in \{Q,R,S\}$ [7].

$$A_{j \in \{P,T,U\}} = \frac{a_j d_j N_{hb}}{30\pi} \tag{9}$$

$$U_{j \in \{P,T,U\}}(n) = \frac{\frac{240a_j}{d_j N_{hb} \pi}}{\left(\frac{60}{d_j N_{hb}}\right)^2 - 4\pi^2} \cos\left(\frac{n\pi d_j N_{hb}}{60}\right) \tag{10}$$

$$A_{j \in \{Q,R,S\}} = \frac{a_j d_j N_{hb}}{120} \tag{11}$$

$$U_{j \in \{Q,R,S\}}(n) = \frac{120a_j}{d_j N_{hb}(n\pi)^2} \left(1 - \cos\left(\frac{n\pi d_j N_{hb}}{60}\right)\right) \tag{12}$$

Where a_j is the amplitude of event $j \in J$ and d_j is the duration of event $j \in J$.

ACKNOWLEDGMENT

This work has been supported in part by "Convocatoria 727 – Programa de Becas de Doctorado Nacionales 2015 de Colciencias" and the project "Fortalecimiento de la plataforma tecnológica para la formación especializada en el área de la salud y el desarrollo de tecnología biomédica", code Ruta N 139C of Fondo Nacional de Regalías, Republic of Colombia.

CONFLICT OF INTEREST

The authors declare that there is no conflict of interest regarding the publication of this paper.

REFERENCES

1. Ansari Sardar, Ward Kevin, Najarian Kayvan. Epsilon-tube filtering: Reduction of high-amplitude motion artifacts from impedance plethysmography signal *IEEE Journal of Biomedical and Health Informatics.* 2015;19:406–417.
2. Liu Yan, Pecht Michael. Reduction of Skin Stretch Induced Motion Artifacts in Electrocardiogram Monitoring Using Adaptive Filtering *Annual International Conference of the IEEE Engineering in Medicine and Biology - Proceedings.* 2007;28:6045–6048.
3. Milanesi M., Martini N., Vanello N., et al. Multichannel techniques for motion artifacts removal from electrocardiographic signals *Annual International Conference of the IEEE Engineering in Medicine and Biology - Proceedings.* 2006;28:3391–3394.
4. Jeong Do Un, Kim Se Jin. Development of a Technique for Cancelling Motion Artifact in Ambulatory ECG Monitoring System *Third 2008 International Conference on Convergence and Hybrid Information Technology Development.* 2008;1:954–961.
5. Raya Mary Anne D., Sison Luis G.. Adaptive noise cancelling of motion artifact in stress ECG signals using accelerometer *Proceedings of the Second Joint 24th Annual Conference and the Annual Fall Meeting of the [Biomedical Engineering Society] [Engineering in Medicine and Biology Society].* 2002;2:1756–1757.
6. Eilebrecht B., Wartzek T., Willkomm J., Schommartz A., Walter M., Leonhardt S.. Motion artifact removal from capacitive ECG measurements by means of adaptive filtering *5th European IFMBE Conference, IFMBE Proceedings.* 2011;37:902–905.
7. Javed Shazia, Ahmad Noor Atinah. An Adaptive Noise Cancelation Model for Removal of Noise from Modeled ECG Signals *Region 10 Symposium, 2014 IEEE, Kuala Lumpur.* 2014:471–.
8. Mendez Martin O., Ruini Davide D., Villantieri Omar P., et al. Detection of sleep apnea from surface ECG based on features extracted by an autoregressive model *Annual International Conference of the IEEE Engineering in Medicine and Biology Society- Proceedings.* 2007;29:6105–6108.
9. Snipes Michael, Taylor D. Christopher. Model selection and Akaike Information Criteria: An example from wine ratings and prices *Wine Economics and Policy.* 2014;3:3–9.
10. Devuyst S, Dutoit T, Stenuit P, Kerkhofs M, Stanus E. Removal of ECG artifacts from EEG using a modified independent component analysis approach. *Conference proceedings : Annual International Conference of the IEEE Engineering in Medicine and Biology Society..* 2008;30:5204–5207.
11. Hamprecht Fred A., Achleitner Ulrich, Krismer Anette C., et al. Fibrillation power, an alternative method of ECG spectral analysis for prediction of countershock success in a porcine model of ventricular fibrillation *Resuscitation.* 2001;50:287–296.
12. Chawla M. P. S.. PCA and ICA processing methods for removal of artifacts and noise in electrocardiograms: A survey and comparison *Applied Soft Computing Journal.* 2011;11:2216–2226.

Author: F. A. Castaño
Institute: Universidad de Antioquia
Street: Calle 67 No. 53 - 108
City: Medellín
Country: Colombia
Email: fabian.castano@udea.edu.co

Author: A. M. Hernández
Institute: Universidad de Antioquia
Street: Calle 67 No. 53 - 108
City: Medellín
Country: Colombia
Email: alher.hernandez@udea.edu.co

Feature selection for KNN classifier to improve accurate detection of subthalamic nucleus during deep brain stimulation surgery in Parkinson's patients

L Schiaffino[1], A Rosado Muñoz[2] , J Francés Villora[2], M Bataller[2], A Gutiérrez[3], I Martínez Torres[3], V Teruel-Martí[4] and J Guerrero Martínez[2]

[1] Engineering Faculty, LIRINS, National University of Entre Ríos, Oro Verde, Argentina.
[2] Digital Signal Processing Group, ETSE, Department of Electronic Engineering, University of Valencia, Valencia, Spain.
[3] Functional Neurosurgery Unit. La Fe Hospital, Valencia, Spain.
[4] Department of Anatomy and Embryology, Faculty of Medicine, University of Valencia, Valencia, Spain

Abstract— **The tremor and dystonia associated with Parkinson's disease can be treated with deep brain stimulation (DBS) implanted into the subthalamic nucleus (STN). The accurate STN detection is a complex neurosurgeon task during a DBS surgery since a proper fixing of stimulating electrodes will impact on the patient's future life. The brain electrical signals obtained with Micro Electrodes Register (MER) are acquired at different depths of the brain during DBS surgery to detect STN. In our previous work, we found good accuracy performance to improve the localization of STN using K-Nearest Neighbours (KNN) supervised learning algorithm. However, for real-time classification, it is essential to reduce the feature dimension without loss of classification accuracy so as to reduce computation time. In this study, we compared the KNN classification trained with 16 features with other KNN with a reduced number of features resulting from applying four feature selection techniques. We obtained similar classification results compared to the classifier using all features, with 6 feature resulting from Branch & Bound (KNN+B) algorithm, giving the best performancewith 86.13% accuracy and 92.72% AUC. From the study, we infer that the KNN+B classifier has the potential to be used in detecting the STN in real time**

Keywords— **feature selection, K-Nearest Neighbors (KNN) classifier, deep brain stimulation (DBS), Parkinson**

I. Introduction

Deep brain stimulation (DBS) is a well- established treatment for Parkinson's disease (PD) movement disorders, such as essential tremor and dystonia. The best clinical effect depends on precise placement of the electrode in the target nucleus by stereotactic neurosurgery [1] . The most common DBS target is the subthalamic nucleus (STN), a little structure with an approximated size of 5.9×3.7×5 mm [2].

During DBS surgery, a set of parallel MER microelectrode recording (MER) are inserted into patient's brain. The detection of the target area to fix the stimulation electrodes in the STN is a complex task moving the electrodes in the brain which is performed by the neurosurgery team during a surgery to determine whether the recorded signal comes from the STN and then, fix the electrodes in there. Previous magnetic resonance images (MRI) help the neurosurgeon to estimate and plan the trajectory of MER from the brain surface to the target [1]. The typical trajectory of MER in the DBS surgery is following the midbrain structures: the thalamus, zona incerta, STN, and substantia nigra. These structures have different neurophysiological spiking and spike background patterns [2].

In this study, we explored the benefits of using a microelectrode recording (MER) data base to improve the localization of STN using supervised learning algorithms. To train this models it is necessary to calculate temporal [2, 3] and spectral [3, 4] features from MER. An important aspect to consider in the process of real-time classification is the possibility of reducing the feature dimension.

Feature selection (FS) for classification can be formalized as a combinatorial optimization problem, finding the feature set maximizing the quality of the hypothesis learned from these features. For the sake of the learning performance, it is highly desirable to discard irrelevant features prior to learning. Because of its computational intractability, the FS problem has been tackled by means of heuristic algorithms based on various machine learning techniques [5]. By selecting only a subset of features, the prediction accuracy can improve and more insight in the nature of the prediction problem can be gained.

Chaovalitwongse W. et al [2] proposed three classifiers: Naive Bayes, K-nearest neighbours (KNN) and decision tree using 14 features obtained from MER. From the three proposed classifiers, those obtaining better ratios were with decision tree, achieving a sensitivity and specificity around 85% for the STN brain area. They did not study a reduction of feature dimension. In [2] and [6], the classification results were provided without statistical comparative analyses.

Rajpurohit V. et al [7] worked with four classification methods: Logistic Regression, Gaussian Naive Bayes, K-Nearest Neighbors and SVM. They extracted 13 features from 65 MER tracks for classifier training and then used sequential forward and backwards feature selection. They

I. Torres et al. (eds.), *VII Latin American Congress on Biomedical Engineering CLAIB 2016, Bucaramanga, Santander, Colombia, October 26th -28th, 2016*,
IFMBE Proceedings 60,
DOI: 10.1007/978-981-10-4086-3_111

suggested that this task should be routinely used to eliminate redundant features for optimal automated classification of MER data. However, other authors use heuristic strategies only, which have a limitation because they may result in local optima. There is other interesting feature selection techniques that could avoid this limitation [5].

In our previous work [8], we studied that K-nearest neighbors algorithm (KNN) provides good classification accuracy for STN. We trained the KNN with 16 features obtained from MER signals from 15 patients. As the goal of the classifier is working in real time in the operating room, it is necessary to study the reduction of features dimension without losing accuracy.

In this study we experimentally compared four state-of-the-art feature selection techniques [5]: Branch and Bound, ReliefF, LASSO and ELASTIC NET. The results of these techniques were used to train and validate a KNN classification model. Finally, we present the statistical comparative study of the performance for the obtained models.

II. Materials

A. Data recordings

The MER acquisitions are from 15 patients of La Fe Hospital in Valencia (Spain) who signed informed consent and selected from an IRB-approved.

In this study, we included MER registers from unmedicated patients aging 57± 6 (9 male/ 6 female) from bilateral DBS surgery to implant DBS stimulator in the STN nucleus.

During a surgery the neural activity recording was made in both brain hemispheres with 2 parallel MER electrodes per hemisphere (2 mm separation) using a Medtronic recorder model 3389. Recordings were obtained at 12 kHz sampling frequency using a 12-bit A/D converter at a total gain of 10000x with a 50 Hz mains filter. All recordings were acquired with the Micro Guide ProTM system from Alpha Omega Company.

After acquisition, the entire register for each patient was divided in non-overlapping windows of 1 s. In total, 1200 registers of 1 s were obtained from each patient. MER signals were labeled by specialists in neurosurgery and neurophysiology as 'STN' or 'non STN'. This classification was also validated by the acquired coregister images done at the end of surgery. This strategy minimized the human error in the classification procedure and allowed a reliable database for training the classifiers. Those registers showing outliers were discarded. In total, there were 16.999 useful neural recordings divided in two classes: 11.058 (65.08 %) like 'non STN' and 5.933 from 'STN' (34.92 %).

B. Features from MER and Data Base

A total of 16 features were obtained from 1 s registers to be use as input parameters to the classifiers. The obtained parameters were based on spikes and background neural activity, according to [2, 8]:

1. *Background neural activity features:*

•1-basal: Basal amplitude value assessing the envolvent using the Hilbert transform.

•2-kurt: Signal kurtosis.

•3-CL: Sum of absolute value for the signal derivative.

•4-TH: Threshold in the signal amplitude.

•5-PK: Peaks number.

•6-RA: RMS value of amplitude.

•7-NE Non-linear mean energy.

•8-ZC: Zero crossings.

2. *Spike neural activity features:*

•9-SBI: Ratio between the number of interspike time intervals lower than 10 ms and intervals higher than 10 ms.

•10-SPI: Ratio between the number of interspike time intervals higher than 50 ms and intervals lower than 50 ms.

•11-SPR: Ratio between the accumulated interspike time in a time higher and lower than 50 ms.

•12-SC: Spikes per second.

•13-SMA: Mean spike differential amplitude as the mean value for the amplitude in two consecutive spikes.

•14-SSD: Standard deviation of interspike intervals.

•15-SCR: Spike count ratio.

•16-SF: Mean spike trigger frequency.

III. Methodology

A. Classifier Algorithm

In our previous work [8] the K nearest neighbors algorithm had an acceptable accuracy to detect STN when the input to the model were the 16 features defined in the previous section. We found the best performance in the KNN classifier with K=9, city block distance metric and square inverse distance weight. We used the KNN classification with cost of misclassification matrix to make the algorithm less sensitive to class imbalance.

B. Feature selection techniques

Supervised feature selection methods can further be broadly categorized into filter models, wrapper models and embedded models [5].

In this work we consider a filter technique called ReliefF, a wrapper technique, called Branch and Bound, and two embedded approaches called LASSO and ELASTIC NET.

Table 1 shows the results obtained when applying the feature selection algorithms detailed above.

Table 1 Feature selection for each algorithm

B&B	ReliفF	LASSO	ELASTIC NET
5-PK	16-SF	5-PK	5-PK
14-SSD	15-SCR		8-ZC
16-SF	12-SC		
8-ZC	14-SSD		
3-CL	10-SPI		
2-kur	2-kur		

C. Performance of classifiers and statistical analysis

The same sets of training data and testing resulting from 10-fold cross validation were used for KNN classification with total features (KNN) and with the features resulting from applying ReliefF (KNN+R), B&B (KNN+B), LASSO (KNN+L) and ELASTIC NET (KNN+E). Thus, the results obtained of each classifier are comparable. We analyzed KNN, KNN+R, KNN+B, KNN+L and KNN+E as five different classifiers.

For each classifier, the sensitivity, specificity, accuracy and STN class area under curve ROC (AUC) was obtained. We carried out non parametric Friedman test followed by Tukey-Kramer post-hoc test for pairwise comparisons if the results of the Friedman test indicated overall significance.

IV. Result and discussion

Table 2 shows the mean 10-fold cross validation mean performance measure of KNN classifiers using all features in the first column and other columns show the classification results obtained by those using reduced features according to Table 1. We observe that KNN+B provides the best performance with 86.13% accuracy and 92.72% area under the ROC (AUC).

Friedman test showed significant differences for all groups. Fig. 1 shows the graphical information for KNN+B compared against the other classifiers. In Fig. 1 we observed the following: KNN and KNN+B performed the best in sensitivity, KNN, KNN+B and KNN+E performed best in specificity, KNN and KNN+B performed best in accuracy and AUC. Therefore, it is explicit under the conditions of this study that KNN and KNN+B models are giving best results. The substantial difference between both is that KNN+B used 6 features instead of 16 features used by KNN. A limitation of the study is the total number of patients included (15) but the present work is a promising first step towards that long-term goal. To generalize our results, we performed a non-parametric test. In comparison with other studies who used support vector machines [7], decision trees [2] and K-nearest neighbors [2, 6, 7] we achieved similar performance with only 6 features.

Table 2 Performance measure for each model

Performance Measures	KNN	KNN+R	KNN+B	KNN+L	KNN+E
Sensitivity (%)	93.39±1.02	83.94±1.14	94.4±1.07	90.91±1.96	81.19±2.13
Specificity (%)	81.73±1.45	72.73±1.66	81.69±1.59	66.49±1.89	74.72±1.46
Accuracy (%)	85.8±0.95	76.65±1.10	86.13±1.20	75.02±1.26	76.98±0.72
AUC (%)	92.46±0.57	81.98±1.11	92.72±0.72	80.19±1.29	82.8±0.94

V. Conclusions

In DBS surgery, the correct implantation of stimulating electrodes in the STN area is a neurosurgeon's crucial task. In the present work we obtained 16 features from MER signal used to train and validate KNN classifier with different number of features. The contribution of this study is the following: we have developed a model (KNN+B) using 6 features only to detect the STN area. In particular, and under the conditions of this work, KNN+B had a superior significant performance relative to the rest of classifiers. Unlike previous works, we also validated statistically the models.

From the study, we infer that the KNN+B classifier has the potential to be used in real time to detect the STN area using 6 input features from MER signal. The KNN+B proposed model can then become a tool to assist the neurosurgeons for correct implantation of stimulation electrode in DBS surgery.

Acknowledgment

The authors wish to thank the Carolina Foundation, the University of Valencia, the Faculty of Engineering of the National University of Entre Ríos and the Autonomous University of Entre Rios for the aid received for the doctoral scholarship of Ms. Bioing. Luciano Schiaffino. This study was made possible by funding received by the University of Valencia through the "VLC-BIOMED 2015 Subprogram B, project "02-Parkinson-2015-B".

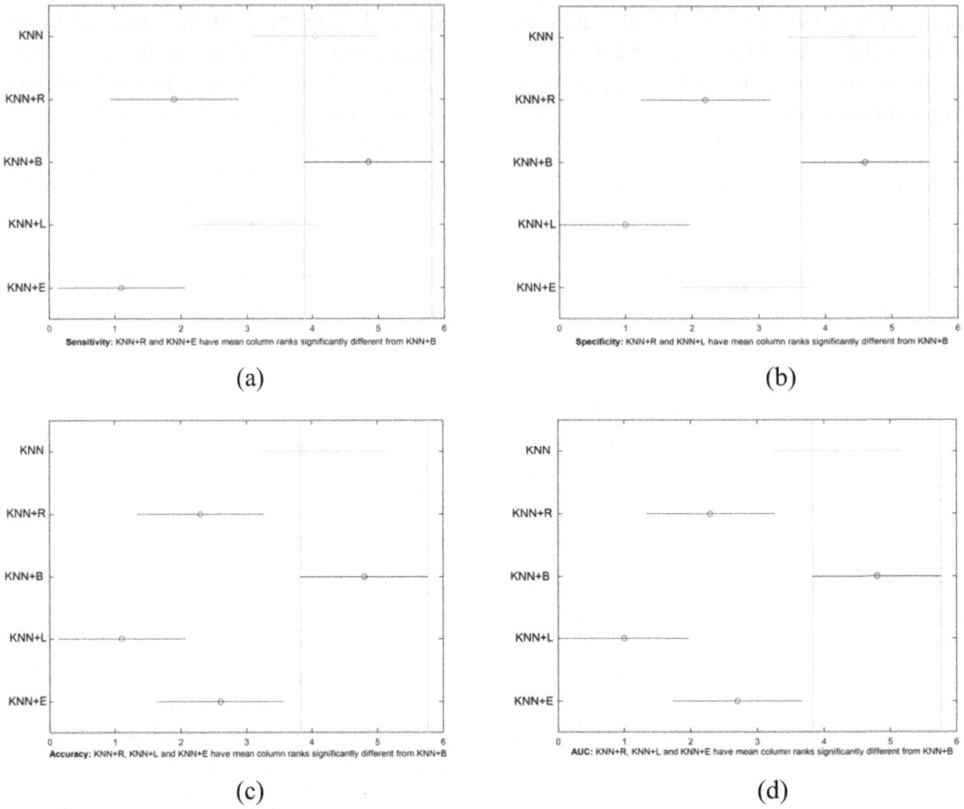

Fig. 1. Tukey-Kramer graphics for each performance measures and all classifiers: (a)Sensitivity (b)Specificity (c)Accuracy (d)AUC

Conflict of Interest

The authors declare that they have no conflict of interest.

References

1. Itakura T (2014) Deep Brain Stimulation for Neurological Disorders: Theoretical Background and Clinical Application, 2015 edition. Springer, New York

2. Chaovalitwongse W, Jeong Y, Jeong MK, et al (2011) Pattern Recognition Approaches for Identifying Subcortical Targets during Deep Brain Stimulation Surgery. IEEE Intell Syst 26:54–63. doi: 10.1109/MIS.2011.56

3. Ciecierski K, Raś ZW, Przybyszewski AW (2012) Foundations of Recommender System for STN Localization During DBS Surgery in Parkinson's Patients. In: Proc. 20th Int. Conf. Found. Intell. Syst. Springer-Verlag, Berlin, Heidelberg, pp 234–243

4. Pinzon-Morales R-D, Orozco-Gutierrez A-A, Carmona-Villada H, Castellanos C-G (2010) Towards high accuracy classification of MER signals for target localization in Parkinson's disease. Conf Proc Annu Int Conf IEEE Eng Med Biol Soc IEEE Eng Med Biol Soc Annu Conf 2010:4040–4043. doi: 10.1109/IEMBS.2010.5628014

5. Jiliang Tang, Salem Alelyani, Huan Liu (2014) Feature Selection for Classification: A Review. In: Data Classif. Chapman and Hall/CRC, pp 37–64

6. Guillén P, Martínez-de-Pisón F, Sánchez R, et al (2011) Characterization of subcortical structures during deep brain stimulation utilizing support vector machines. In: 2011 Annu. Int. Conf. IEEE Eng. Med. Biol. Soc. pp 7949–7952

7. Rajpurohit V, Danish SF, Hargreaves EL, Wong S (2015) Optimizing computational feature sets for subthalamic nucleus localization in DBS surgery with feature selection. Clin Neurophysiol 126:975–982. doi: 10.1016/j.clinph.2014.05.039

8. Schiaffino L, Muñoz AR, Martínez JG, et al (2016) STN area detection using K-NN classifiers for MER recordings in Parkinson patients during neurostimulator implant surgery. J Phys Conf Ser 705:12050. doi: 10.1088/1742-6596/705/1/012050

Ensemble Kalman filter for state estimation of brain activity by considering a large scale nonlinear dynamical model

P.A. Munõz-Gutiérrez[1] and E. Giraldo[2]

[1] Universidad Tecnológica de Pereira/Doctorade in Engineering, Student, Pereira, Colombia and Universidad del Quindío/Electronic Instrumentation Technology, Assistant professor, Armenia, Colombia
[2] Universidad Tecnológica de Pereira/Department of Electrical Engineering, Associated professor, Pereira, Colombia

Abstract— **Data assimilation techniques, like the Ensemble Kalman filter (EnKF), have been successfully used for weather forecasting or in general for state space estimation tasks that involve large scale nonlinear complex dynamical models. In this paper a novel application of the EnKF is presented for estimation of neural activity into the brain considering a large scale time-varying nonlinear model to describe its dynamics. A comparative analysis is performed in terms of brain activity reconstruction and time-course reconstruction. As a result, the EnKF outperforms methods like the Multiple Sparse Priors (MSP) and Low Resolution Tomography (LORETA).**

Keywords— **ensemble Kalman, dynamic modeling, brain activity.**

I INTRODUCTION

Data assimilation techniques allows the estimation of complex dynamical models for forecasting [1]. These methods have been specially used for weather modeling and forecasting [2, 3]. Kalman filtering (KF) is a data assimilation technique for linear systems; Extended KF (EKF) and EnKF are for nonlinear systems. In the KF framework, the propagation of model error covariance matrix is computed in an analytical closure form which is challenging for highly nonlinear models; EnKF can capture the covariance matrix by an ensemble of forecasts using Monte Carlo methods [2]. The advantages of the EnKF in comparison with the EKF are the low computational cost, the quality of the estimation of error covariance by Monte Carlo methods (instead of linear approximations) and the memory storage [4, 5].

The neural activity into the brain can be described as a large scale nonlinear model [6, 7], and therefore the estimation of brain activity from electroencephalographic (EEG) signals can be achieved by the application of data assimilation techniques for nonlinear systems. Moreover, considering the advantages of the EnKf in comparison with the EKF [5], the EnKF should be preferred for the aforementioned task. In this paper an novel application of EnKF is presented to estimate the neural activity into the brain. In this case, a large scale nonlinear time-varying model is used to describe the neural activity, and the EnKF is modified according to that nonlinear model for neural activity estimation. The paper is structured as follows: in section II the nonlinear state space model and its measurement equation are presented, in section III the EnKF is presented in the context of brain activity estimation, and in section IV a comparison analysis with LORETA and MSP methods is presented for several Signal-to-Noise-Ratio (SNR) and active sources by using simulated EEG signals.

II LARGE SCALE NONLINEAR MODEL FOR BRAIN ACTIVITY

Consider the following state space nonlinear model of brain activity:

$$\boldsymbol{x}_k = \boldsymbol{f}(\boldsymbol{x}_{k-1}, \boldsymbol{x}_{k-2}, \boldsymbol{x}_{k-\tau}) + \boldsymbol{\eta}_k \tag{1}$$

with

$$\begin{aligned}\boldsymbol{f}(\boldsymbol{x}_{k-1}, \boldsymbol{x}_{k-2}, \boldsymbol{x}_{k-\tau}) = \boldsymbol{A}_1\boldsymbol{x}_{k-1} + \boldsymbol{A}_2\boldsymbol{x}_{k-2} + \boldsymbol{A}_3\boldsymbol{x}_{k-\tau} \\ + \boldsymbol{A}_4\boldsymbol{x}_{k-1}^{\circ 2} + \boldsymbol{A}_5\boldsymbol{x}_{k-1}^{\circ 3}\end{aligned} \tag{2}$$

being the transition matrices $\boldsymbol{A}_i = \operatorname{diag} \boldsymbol{a}_i$, $i = 1, \ldots, 5$ and being $\boldsymbol{a}_i \in \mathbb{R}^{n\times 1}$, with $\boldsymbol{x}_k \in \mathbb{R}^{n\times 1}$ the neural activity at sample k, and where the term $\boldsymbol{x}_k^{\circ 2}$ denotes the Hadamard power, and being $\boldsymbol{\eta}_k \in \mathbb{R}^{n\times 1}$ a zero mean uncorrelated Gaussian noise. It is worth noting that the number of states n is large. In addition, the measurement equation is given by

$$\boldsymbol{y}_k = \boldsymbol{M}\boldsymbol{x}_k + \boldsymbol{\varepsilon}_k \tag{3}$$

being $\boldsymbol{y}_k \in \mathbb{R}^{d\times 1}$ the d EEG measurements and $\boldsymbol{M} \in \mathbb{R}^{d\times n}$ the leadfield matrix [7], where $n >> d$, and being $\boldsymbol{\varepsilon}_k \in \mathbb{R}^{d\times 1}$ a zero mean uncorrelated Gaussian noise.

III ENKF FOR BRAIN ACTIVITY ESTIMATION

The state estimation of the brain activity $\boldsymbol{x}_k$ can be obtained by applying the EnKF in two stages [4]. First,

I. Torres et al. (eds.), *VII Latin American Congress on Biomedical Engineering CLAIB 2016, Bucaramanga, Santander, Colombia, October 26th -28th, 2016*,
IFMBE Proceedings 60,
DOI: 10.1007/978-981-10-4086-3_112

the forecast stage, where an ensemble of q forecasted state estimates is computed

$$\boldsymbol{x}_k^{f_i} = \boldsymbol{f}\left(\boldsymbol{x}_{k-1}^{a_i}, \bar{\boldsymbol{x}}_{k-2}^{a}, \bar{\boldsymbol{x}}_{k-\tau}^{a}\right) + \boldsymbol{\omega}_k^i \quad (4)$$

being $i = 1, \ldots, q$, $\boldsymbol{\omega}_k^i \in \mathbb{R}^{n\times 1}$ is a zero mean random variable with normal distribution an covariance $\boldsymbol{\Omega}_k \in \mathbb{R}^{n\times n}$. The sample error covariance matrix computed from $\boldsymbol{\omega}_k^i$ converges to $\boldsymbol{\Omega}_k$ as $q \to \infty$. The forecasted measurement is given by

$$\boldsymbol{y}_k^{f_i} = \boldsymbol{M}\boldsymbol{x}_k^{f_i} \quad (5)$$

where f_i denotes the $i-$th forecast ensemble member. Then, the ensemble mean for states $\bar{\boldsymbol{x}}_k^f \in \mathbb{R}^{n\times 1}$ is defined by

$$\bar{\boldsymbol{x}}_k^f = \frac{1}{q}\sum_{i=1}^{q} \boldsymbol{x}_k^{f_i} \quad (6)$$

and the ensemble mean for measurements $\bar{\boldsymbol{y}}_k^f$ is defined by

$$\bar{\boldsymbol{y}}_k^f = \frac{1}{q}\sum_{i=1}^{q} \boldsymbol{y}_k^{f_i} \quad (7)$$

The ensemble error matrix around the ensemble mean is defined as

$$\boldsymbol{E}_k^f = \left[\begin{array}{ccc} \boldsymbol{x}_k^{f_1} - \bar{\boldsymbol{x}}_k^f & \cdots & \boldsymbol{x}_k^{f_q} - \bar{\boldsymbol{x}}_k^f \end{array}\right] \quad (8)$$

being $\boldsymbol{E}_k^f \in \mathbb{R}^{n\times q}$, and the ensemble output error

$$\boldsymbol{E}_{y_k}^f = \left[\begin{array}{ccc} \boldsymbol{y}_k^{f_1} - \bar{\boldsymbol{y}}_k^f & \cdots & \boldsymbol{y}_k^{f_q} - \bar{\boldsymbol{y}}_k^f \end{array}\right] \quad (9)$$

being $\boldsymbol{E}_{y_k}^f \in \mathbb{R}^{d\times q}$, and where the forecast covariances are approximated as

$$\begin{aligned} \boldsymbol{P}_k^f &= \frac{1}{q-1}\boldsymbol{E}_k^f\left(\boldsymbol{E}_k^f\right)^\top \\ \boldsymbol{P}_{xy_k}^f &= \frac{1}{q-1}\boldsymbol{E}_k^f\left(\boldsymbol{E}_{y_k}^f\right)^\top \\ \boldsymbol{P}_{yy_k}^f &= \frac{1}{q-1}\boldsymbol{E}_{y_k}^f\left(\boldsymbol{E}_{y_k}^f\right)^\top \end{aligned}$$

being $\boldsymbol{P}_k^f \in \mathbb{R}^{n\times n}$, $\boldsymbol{P}_{xy_k}^f \in \mathbb{R}^{n\times d}$ and $\boldsymbol{P}_{yy_k}^f \in \mathbb{R}^{d\times d}$.

The second stage is the analysis stage, where an ensemble of perturbed observations $\boldsymbol{y}_k^i$ is obtained as follows

$$\boldsymbol{y}_k^i = \boldsymbol{y}_k + \boldsymbol{\upsilon}_k^i \quad (10)$$

where $\boldsymbol{\upsilon}_k^i \in \mathbb{R}^{d\times 1}$ is a zero mean random variable with normal distribution an covariance $\boldsymbol{\Upsilon}_k \in \mathbb{R}^{d\times d}$. The sample error covariance matrix computed from $\boldsymbol{\upsilon}_k^i$ converges to $\boldsymbol{\Upsilon}_k$ as $q \to \infty$. Also, the EnKF gain matrix is approximated as

$$\boldsymbol{K}_k = \boldsymbol{P}_{xy_k}^f\left(\boldsymbol{P}_{yy_k}^f\right)^{-1} \quad (11)$$

where it is worth noting that $\left(\boldsymbol{P}_{yy_k}^f\right)^{-1}$ is an inverse of $d\times d$, being $d << n$.

Therefore, an ensemble of data assimilation cycles is obtained as follows:

$$\boldsymbol{x}_k^{a_i} = \boldsymbol{x}_k^{f_i} + \boldsymbol{K}_k\left(\boldsymbol{y}_k^i - \boldsymbol{M}\boldsymbol{x}_k^{f_i}\right) \quad (12)$$

where a_i denotes the $i-$th analysis ensemble member. And finally, the ensemble mean of the analysis stage is computed as

$$\bar{\boldsymbol{x}}_k^a = \frac{1}{q}\sum_{i=1}^{q} \boldsymbol{x}_k^{a_i} \quad (13)$$

where $\bar{\boldsymbol{x}}_k^a$ is the estimated brain activity at sample k.

IV RESULTS AND DISCUSSION

A non-homogeneous EEG simulation is used in order to evaluate the performance of the EnKF. The neural activity is simulated by using a nonlinear model as follows:

$$\boldsymbol{y}_k = \boldsymbol{M}\boldsymbol{x}_k + \boldsymbol{\varepsilon}_k \quad (14)$$

$$\begin{aligned} \boldsymbol{x}_{k+1} =& \boldsymbol{A}_1\boldsymbol{x}_k + \boldsymbol{A}_2\boldsymbol{x}_{k-1} + \boldsymbol{A}_3\boldsymbol{x}_{k-\tau} \\ &+ \boldsymbol{A}_4\boldsymbol{x}_k^{\circ 2} + \boldsymbol{A}_5\boldsymbol{x}_{k-1}^{\circ 3} + \boldsymbol{\eta}_k \end{aligned} \quad (15)$$

being $\boldsymbol{A}_i = \operatorname{diag}\boldsymbol{a}_i$, $i = 1,\ldots,5$ and being $\boldsymbol{a}_i \in \mathbb{R}^{n\times 1}$. The parameters values are fixed to $\tau = 20$, $a_1 = 1.0628$, $a_2 = -0.42857$, $a_3 = 0.008$, $a_4 = 0.000143$, $a_5 = -0.000286$, and $\|\boldsymbol{\eta}_k\| \leq 0.05$, this parameters were used in [6, 7]. The epileptic seizure is simulated at sample $k = 125$ ($t = 0.5\,s$) by modifying only the parameters of some sources (active sources) of a_1 from 1.0628 to 1.3, while a_2 from -0.428 to -1. The active sources are modified considering the spatial coupling with the neighbor sources (which means that the model parameters are also changed in the neighbor sources). In this paper, a comparison is performed for one and five active sources. A visual comparison is performed using the MSP and LORETA [6, 7] methods in order to evaluate the spatial and temporal resolutions of the EnKF. The analysis is achieved for SNRs of: 1dB and 30dB. A model with $n = 8196$ and $d = 128$ is used. The number of ensemble members used is $q = 20$.

Figure 1 shows the simulated EEG for one active source with an SNR of 1dB, where the epileptic activity begins at $t \geq 0.5$s.

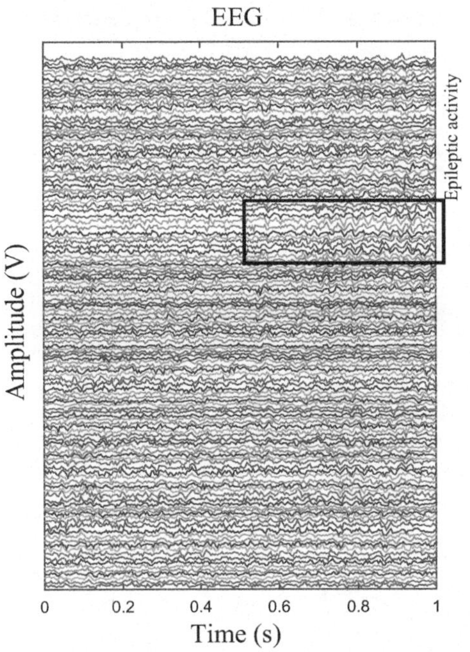

Fig. 1: Simulated EEG for one active source with an SNR of 1dB.

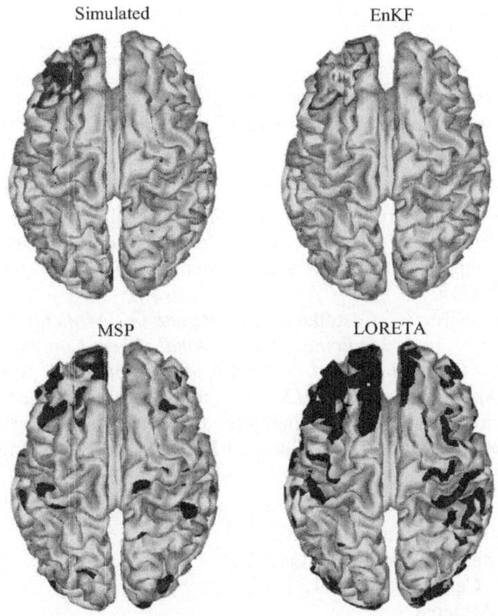

Fig. 2: Simulated and estimated brain mapping with one active source for each considered method with an SNR of 1dB.

Figure 2 shows the simulated and estimated brain mapping with one active source for each considered method with an SNR of 1dB. It can be seen that the MSP method exhibit some spurious activity while LORETA method exhibits a large amount of activity all over the brain. In contrast the EnKF method shows the reconstructed activity in the same

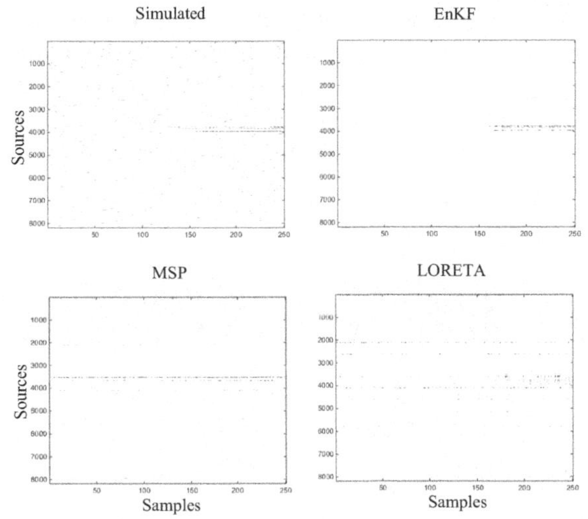

Fig. 3: Time-course evolution of each source for each considered method with an SNR of 1dB.

area than the simulated one.

Figure 3 shows the time-course evolution of each source for each considered method with an SNR of 1dB. It can be seen that the EnKF method reconstruct the activity in to the source from $t = 0.5$ s, while the MSP method reconstructs the activity for $0 \leq t \leq 1$ since the MSP computes a covariance considering the complete data set.

Figure 4 shows the simulated and estimated brain mapping with one active source for each considered method with an SNR of 30dB. It can be seen that the MSP and the EnKF method show the reconstructed activity in the same area than the simulated one, while the LORETA method shows a large amount of spurious activity.

Figure 5 shows the brain mapping and time-course evolution of the reconstructed activity for five sources with an SNR of 1dB for the MSP and EnKF methods. It can be seen that even when the MSP and the EnKF methods reconstruct the activity adequately, the time evolution of the activity is not estimated adequately in the case of the MSP. However, the estimated activity by the EnKF method shows an adequate behavior both in time and space.

V CONCLUSIONS

In this paper a novel application of EnKF is presented to estimate the neural activity into the brain. As shown in section IV the EnKF method outperforms the MSP and LORETA methods for spatial and temporal resolution under several noise conditions (SNRs: 1dB and 30dB), where the EnKF turns into a more robust estimation method for the considered experimental setup. It is worth noting that the EnKF shows better time resolution than the MSP since it takes into consideration the nonlinear model that describes the neural activity behavior. In future works, dual methods for the EnKF that estimate states and parameters simultaneously will be considered.

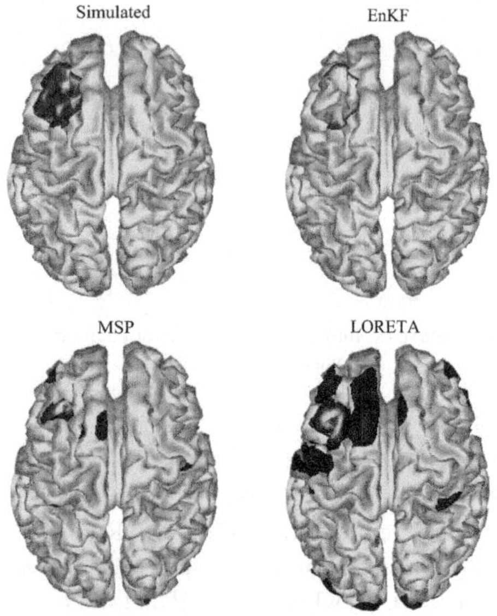

Fig. 4: Simulated and estimated brain mapping with one active source for each considered method with an SNR of 30dB.

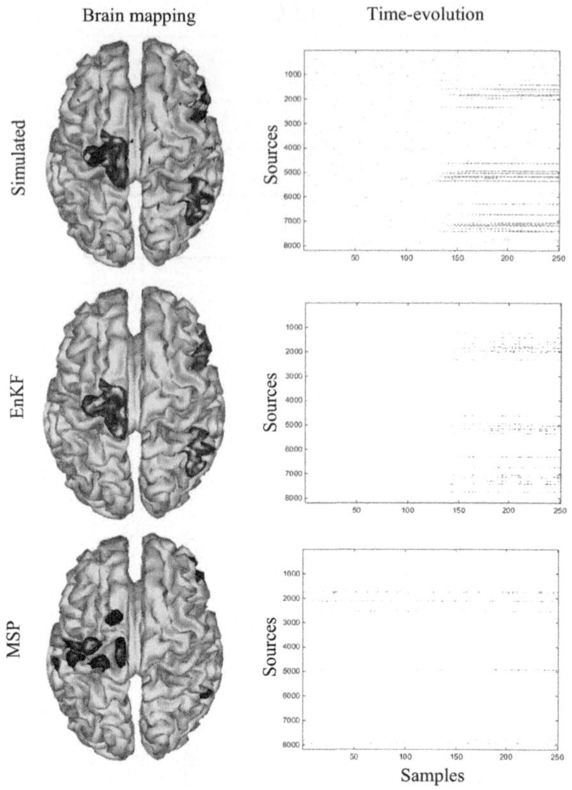

Fig. 5: Brain mapping an time-course evolution of the reconstructed activity for five sources with an SNR of 1dB for the MSP and EnKF methods

Conflict of Interest

The authors declare that they have no conflict of interest.

References

1. Dumedah Gift, Coulibaly Paulin. Evaluating forecasting performance for data assimilation methods: The ensemble Kalman filter, the particle filter, and the evolutionary-based assimilation *Advances in Water Resources.* 2013;60:47 - 63.
2. Tamura Hitoshi, Bacopoulos Peter, Wang Dingbao, Hagen Scott C., Kubatko Ethan J.. State estimation of tidal hydrodynamics using ensemble Kalman filter *Advances in Water Resources.* 2014;63:45 - 56.
3. Zuluaga Carlos D., Álvarez Mauricio A., Giraldo Eduardo. Short-term wind speed prediction based on robust Kalman filtering: An experimental comparison *Applied Energy.* 2015;156:321 - 330.
4. Gillijns S., Mendoza O. B., Chandrasekar J., Moor B. L. R. De, Bernstein D. S., Ridley A.. What is the ensemble Kalman filter and how well does it work? in *2006 American Control Conference*:6 pp.- 2006.
5. Hut Rolf, Amisigo Barnabas A., Steele-Dunne Susan, Giesen Nick. Reduction of Used Memory Ensemble Kalman Filtering (RumEnKF): A data assimilation scheme for memory intensive, high performance computing *Advances in Water Resources.* 2015;86, Part B:273 - 283. Data assimilation for improved predictions of integrated terrestrial systems.
6. Giraldo E., Martinez-Vargas J. D., Castellanos-Dominguez G.. Reconstruction of neural activity from EEG data using dynamic spatio-temporal constraints *International Journal of Neural Systems.* 2016;26:1-15.
7. Giraldo-Suárez E., Castellanos-Dominguez G.. *Artificial Computation in Biology and Medicine: International Work-Conference on the Interplay Between Natural and Artificial Computation, IWINAC 2015, Elche, Spain, June 1-5, 2015, Proceedings, Part I*ch. Neural Activity Estimation from EEG UsinganIterative Dynamic Inverse Problem Solution, :388–397. Cham: Springer International Publishing 2015.

Author: P.A. Munõz-Gutiérrez
Institute: Universidad del Quindío
Street: Carrera 15 Calle 12 Norte
City: Armenia
Country: Colombia
Email: pamunoz@uniquindio.edu.co

Author: E. Giraldo
Institute: Universidad Tecnolgíca de Pereira
Street: Carrera 27 N. 10-02 Barrio Alamos
City: Pereira
Country: Colombia
Email: egiraldos@utp.edu.co

A CT-based and mechanobiologic model for the simulation of rotation of tibia deformities during patient´s immobilization treatment.

R.A. González-Carbonell[1], A. Ortiz-Prado[2], V.H. Jacobo-Armendáriz[2], Y.A. Cisneros-Hidalgo[1] and L. Morales-Acosta[2]

[1] Universidad de Camagüey/Mechanical Engineering, Professor, Camagüey, Cuba
[2] Universidad Nacional Autónoma de México/UDIATEM, Professor, Coyoacán, DF, México

Abstract — **Tibial torsion deformity produces a change in the alignment of the planes of motion of the articulations of the knee and ankle. Lower limb immobilization by cast is one of the corrective treatments. The objective of this paper was to study the bone remodeling of rotation of tibia deformities during patient´s immobilization treatment. The Finite Element (FE) method was used in the simulation of bone remodeling of the correction process of tibial torsion. A mechanobiologic CT-based FE model of the tibia was defined. An user material subroutine was used to define the bone constitutive material model. Bone density increased its value when the torque was bigger than 12 Nm and it decreased when the torque was lower than 8 Nm. Stress during the iteration process for all load conditions increased. Except for a torque of 4 Nm, the strain decreased its value during the iteration process. At the final iteration step, strain was equal to 0.025 approximately for all load conditions. As it was expected, the variation of bone density during simulation indicates that the bone adapts its structure to load conditions. The bone showed the same density distribution at the end of the simulation for all torque values applied. The highest density was located at the mid-shaft of the diaphysis and at the epiphysis density decreased, adopting a circular ring arrangement.**

Keywords — **bone remodeling, tibial torsion, bone density, torque.**

I. Introduction

Bones serve as a body support and due to the changes in severity and complexity of loads, they adapt their internal and external shape (Wolff´s Law). According to this theory, it is considered that mechanical stimulus is the cause of bone remodeling [1, 2]. Biomechanics studies the mechanical behavior of bone´s stress, strain and displacement under a load. It also studies the bone remodeling under different load conditions [3].

The 3D reconstruction of bone geometry is a complex process, it can be simplified with the help of a software for the manipulation of Computed Tomography (CT) data retrieved in DICOM format (Digital Imaging and Communication in Medicine) [4].

The CT-based finite element (FE) model is used in orthopedic medicine as non-destructively preparatory surgery [5], locating bone defects [6] and in the prediction of remodeling of bone-implant integration and bone healing fractures [7, 8].

A tibial torsion deformity is defined as any twisting of the tibia on its longitudinal axis which produces a change in alignment of the planes of motion of the articulations of the knee and ankle [6].

Children can correct this deformity spontaneously, but it is not possible to predict when it will happen. In some tibial torsion treatments, the lower limb of the patient is immobilized and the tibia is twisted. The predominant load during this type of treatments is torque.

When the patient´s leg is accommodated into the corrective appliance, the physician should collocate the foot in a position that might favor the tibia rotation in an opposite direction to the deformity. The magnitude of the corrective torque and the stress condition are related to the position of the foot angle into the appliance. When orthopedic treatment is indicated, the magnitude of the corrective torque is unknown and an empirical value of torque is applied [9].

Patient-specific FE model of the tibia under torque has been used to determine the stress and strain of the bone as mechanical stimulus that induced the bone remodeling [10]. Besides, the change of bone density and strain magnitude on the whole bone due to the action of a torque load were simulated by the use of a mechanobiologic model [11]. However, there are differences between the tibial torsion treatments and the bone remodeling processes related to it that have not been studied yet. The objective of this paper is to study the bone remodeling of rotation of tibia deformities during patient´s immobilization treatment.

II. Materials and methods

The FE method was used to simulate the bone remodeling of the twisting of the tibia. In the pre-processing step the bone geometry and the initial density computed from CT was defined to the model. Then, the bone readapts its structure during the simulation. The methodology used in this research is shown in the Fig. 1.

I. Torres et al. (eds.), *VII Latin American Congress on Biomedical Engineering CLAIB 2016, Bucaramanga, Santander, Colombia, October 26th -28th, 2016*,
IFMBE Proceedings 60,
DOI: 10.1007/978-981-10-4086-3_113

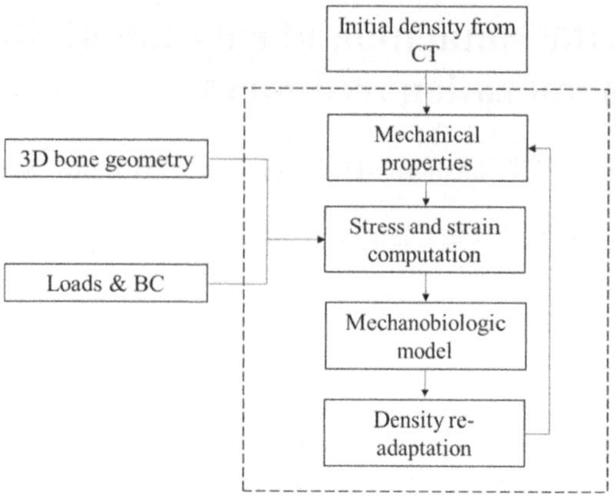

Fig. 1. CT-based and mechanobiologic model for the simulation of bone remodeling.

A. 3D bone geometry.

The 3D bone geometry reconstruction employed images in DICOM format from the Computed Tomography Scanner GE LightSpeed VCT (120kV/89.40mAs, pixel 0.773 mm, 512x512, slice 5 mm) installed at the National Institute of Rehabilitation, of México City. 3D-Slicer software (http://www.slicer.org) was used for processing the medical images and 3D geometry reconstruction. The superficial mesh with triangular element was generated, and a quality control of triangular elements (shape factor base/ height, aspect ratio, and element size) was used [12, 13]. The geometry was exported in STL (STereo Lithography) file extension.

The STL file generated in 3D-Slicer was imported into Abaqus (SIMULIA, Dassault Systems, RI, USA), and a mesh with tetrahedral elements was obtained. The size of tetrahedral elements (2.5 mm) was chosen based on the mesh refinement study in which differences in stresses with higher-density meshes were less than 3% [13]. The number of elements of each mesh used in the mesh convergence study were 392126 (1 mm), 155588 (1.5 mm), 79746 (2 mm), 50890 (2.5 mm), 34705 (3 mm) and 30171 (3.5 mm).

B. Loads & Boundary conditions.

The correction treatment of tibial torsion where the patient was immobilized during the derotation of the tibia deformity was selected. The twisting of the tibia was defined as load condition. A set of nodes was created on the surface of the distal epiphysis, and later it was coupled to a reference point. The reference point was used to apply the moment (in the axial direction) that represents the bone twist. All nodes with couple restriction rotated together to the reference point. A fastening restriction was applied to the proximal epiphysis surface and the movement in all direction was removed [11].

C. Mechanical properties

The mechanical properties were defined as a function of density [14]. The bone density was calculated through a linear regression that converted the Hounsfield Units into apparent density. The mechanical properties of the bone were recalculated after each iteration during the simulation. Ten levels of bone density were selected [11, 15].

$$E = \begin{cases} 2042 \cdot \rho^{2.5} & \rho \le 1.2 \, g/cm^3 \\ 1798 \cdot \rho^{3.2} & \rho > 1.2 \, g/cm^3 \end{cases} \quad (1)$$

D. Mechanobiologic model.

The bone remodeling theory was used [16]. The mechanical stimulus needed to initialize the bone remodeling was calculated by:

$$\Psi = \left[\sum_{i=1}^{N} n_i \sigma_i^m \right]^{\frac{1}{m}} \quad (2)$$

where

$n_i = 1$ denotes the number of load cycles per day, (one cycle per day because the torque was considered as static)

σ_i the tissue level stress for each load case i, and

$m = 4$ the stress exponent that weights the relative importance of the stress magnitude and the number of load cycles in the mechanical stimulus.

The stress at continuum-level was defined as [17]:

$$\sigma_c = \sqrt{2EU} \quad (3)$$

where E and U are the modulus of elasticity and the strain energy density at the continuum level, respectively. The continuum-level stress and the tissue level stress were related as [17]:

$$\sigma = \left(\frac{\rho_{cb}}{\rho} \right)^2 \sigma_c \quad (4)$$

where:

ρ_{cb} was the maximum density of the fully mineralized with null porosity, and

ρ was the apparent density.

The non-linear remodeling law was represented as the surface remodeling rate as [17]:

$$\dot{r} = \begin{cases} c_1 \; [\Psi - \Psi_{AS}(1+w)] \; when \; \Psi > \Psi_{AS}(1+w) \\ 0 \; when \; \Psi_{AS}(1+w) \le \Psi \le \Psi_{AS}(1-w) \\ c_2 \; [\Psi - \Psi_{AS}(1-w)] \; when \; \Psi < \Psi_{AS}(1-w) \end{cases} \quad (5)$$

where:

$c_1 = 0.02\mu m / MPa$ was the coefficient of bone formation.

and $c_2 = 3c_1$ was the constant of bone resorption, and

$w = 0.25$ was the mid-half of the dead zone, where the remodeling difference was not great enough to trigger either bone resorption or formation.

The change of bone density was expressed as:

$$\Delta\rho = k \cdot S_v \cdot \rho_t \cdot \dot{r} \cdot \Delta t \qquad (6)$$

where k was the fraction of local area which was actively remodeling, ρ_t was the density of the bone tissue that was being remodeled and Sv was the bone-specific surface described by Martin [18] to quantify the available remodeling surface of a bone volume as:

$$Sv = 0.03226p - 0.09394p^2 + 0.13396p^3 - 0.10104p^4 + 0.02876p^5 \qquad (7)$$

The bone remodeling algorithm was written by using FORTRAN user material subroutine (UMAT subroutine), on the FE Software Abaqus. Each iteration step of the analysis represents a time period equivalent to one day and 100 iterations were defined in this simulation.

III. RESULTS AND DISCUSSION

At initial iteration step of all simulations, the highest value of bone density was 1.77 g/cm^3. For all torque magnitudes, the bone reduced its density at the epiphysis and redistributed it at the diaphysis (Fig. 2). The bone density increased its value when the torque was higher than 12 Nm and it decreased when the torque was lower than 8 Nm (Fig. 3). When the torque was between 8 Nm and 12 Nm the maximum value of density was approximately equal to the maximum of the density at the beginning of the simulation.

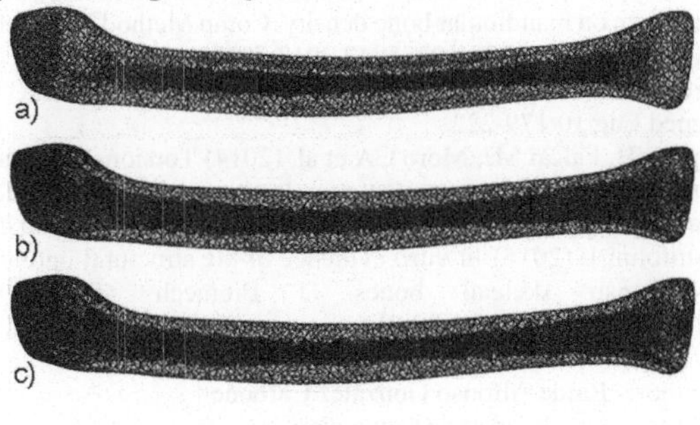

Fig. 2. Bone density distribution: a) 1st iteration step, it is the same as bone density from CT, b) 100th iteration step when 4 Nm was applied, and c) 100th iteration step when 15 Nm was applied.

Stress was increased during the iteration process for all torque magnitudes. This result confirms that bone adapted its structure to the new load condition. On this way, the bone can support new load conditions avoiding stress concentration and fracture. The surface area with maximum density was located at the shaft diaphysis nearest the distal epiphysis. This result is coincident with the reported by González-Carbonell et al. [11]. They conducted a comparison of bone remodeling simulation by using two different initial density (homogeneous and density computed form CT).

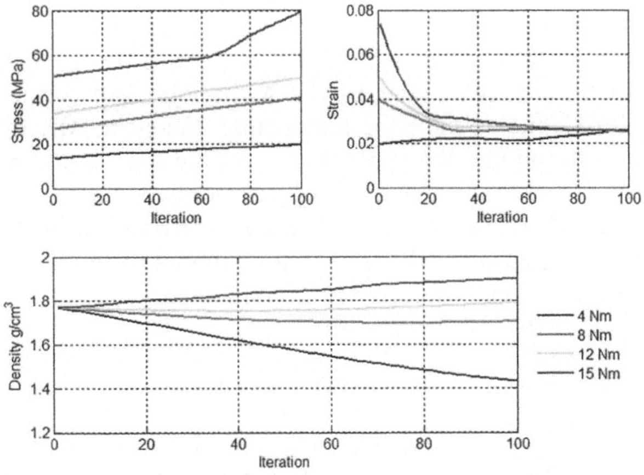

Fig. 3. Stress, strain and bone density behavior during simulation for 4, 8, 12 and 15 Nm. The stress increased for all torque loads, the strain decreased except for 4 Nm, and the bone density increased.

Except for a torque of 4 Nm, strain decreased its value during the iteration process. At 100th iteration step, it was equal to 0.025 approximately for all load conditions. Maximum principal strain occurred at the diaphysis nearest to the distal epiphysis, and it is supposed that the deformity rotation occurs in this region of the bone. This result showed an agreement with the experimental measurements reported by Batista-Volpon et al. [19]. These authors fixed the tibia with Schanz screws at the distal and proximal epiphysis. This fastened configuration do not allow to study the stress and strain behavior at epiphysis. However, they obtained that a controlled twisting applied to the immature tibia caused a progressive rotation of the distal epiphysis.

In everyday activities, the tibia is subjected to complex load condition. It is a long bone with no major muscle insertion in a large portion of the diaphysis, the main proximal insertion is the patellar tendon. Compression and flexion are the predominant loads during gait [20]. Compression loads are transmitted to the bone through the two main articulations (knee and ankle), while a moment is produced in the sagittal plane by the patellar tendon. The treatments of tibial torsion represent a modification to this load condition. The bone tissue formation or resorption was associated with the stress behavior. At the epiphysis, stress was lower, density decreased, and density showed a circular ring arrangement. The highest modification of the internal structure of the bone was observed in this region of the bone. A CT Scan at the final of the treatment of tibial torsion is required to compare the bone density and the results of the simulation.

IV. CONCLUSIONS

In this paper, the tibia under torque was simulated. The model was sufficiently refined to accurately represent the geometry. The bone density was computed from CT data to study the change in bone density. An in-house subroutine was used to simulate the bone remodeling. Independently of the foot position into the appliance (magnitude of the corrective torque), the bone density was transformed mainly at the epiphysis, and the bone should readapt to the torque load conditions. It is known that after a long immobilization period, the limb must be subjected to rehabilitation.

ACKNOWLEDGMENT

The authors would like to thank Dr. Armando Alpizar Aguirre (Instituto Nacional de Rehabilitación de México) for providing image data. The authors acknowledge the assistance of UDIATEM. This research was supported by the Universidad Nacional Autónoma de México through the projects PAPIIT (IN115015, IN115415).

CONFLICT OF INTEREST

The authors declare no competing interests.

REFERENCES

1 Fang J, Gong H, Kong L et al. (2013) Simulation on the internal structure of three-dimensional proximal tibia under different mechanical environments. Biomed Eng OnLine 12:1-17 DOI 10.1186/1475-925X-12-130
2 Wolff J (1986) Concept of the Law of Bone Remodelling. Springer Berlin Heidelberg DOI 10.1007/978-3-642-71031-5_1
3 Cisneros Hidalgo YÁ, González Carbonell RA, Camue Corona E et al. (2014) Aplicación de los modelos mecanobiológicos en los procesos de regeneración ósea. Rev Cubana Ortop Traumatol 28:214-222
4 Cisneros Hidalgo YA, González Carbonell RA, Puente Alvarez A et al. (2014) Generación de imágenes tridimensionales: integración de tomografía computarizada y método de los elementos finitos. Rev Cubana Inv Biomèd 33:313-321
5 Chernikov A, Foteinos P, Liu Y et al. (2013) Tetrahedral Image-to-Mesh Conversion Approaches for Surgery Simulation and Navigation. In Y Zhang (Ed.), Image-Based Geometric Modeling and Mesh Generation vol. 3, pp 69-84: Springer Netherlands DOI 10.1007/978-94-007-4255-0_5
6 Agnihotri G, Rath G, Kullar J et al. (2014) Human tibial torsion - Morphometric assessment and clinical relevance. Biomed J 37:10-13 DOI 10.4103/2319-4170.117890
7 Hambli R, Lespessailles E, Benhamou C-L (2013) Integrated remodeling-to-fracture finite element model of human proximal femur behavior. J Mech Behav Biomed Mater 17:89-106 DOI 10.1016/j.jmbbm.2012.08.011
8 Son D-S, Mehboob H, Jung H-J et al. (2014) The finite element analysis for endochondral ossification process of a fractured tibia applied with a composite IM-rod based on a mechano-regulation theory using a deviatoric strain. Compos Part B-Eng 56:189-196 DOI 10.1016/j.compositesb.2013.08.004
9 González R, García E, Moya J (2013) Tacón de Torque para uso Ortopédico: Propuesta de un Nuevo Diseño. In J Folgueras Méndez, TY Aznielle Rodríguez, CF Calderón Marín et al. (Eds.), V Latin American Congress on Biomedical Engineering CLAIB 2011 May 16-21, 2011, Habana, Cuba vol. 33, pp 912-915: Springer Berlin Heidelberg DOI 10.1007/978-3-642-21198-0_232
10 González-Carbonell RA, Ortiz-Prado A, Jacobo-Armendáriz VH et al. (2015) 3D patient-specific model of the tibia from CT for orthopedic use. J Orthop 12:11-16 DOI 10.1016/j.jor.2015.01.009
11 González-Carbonell RA, Ortiz-Prado A, Cisneros-Hidalgo YA et al. (2015) Bone Remodeling Simulation of Subject-Specific Model of Tibia under Torque. In A Braidot, A Hadad (Eds.), VI Latin American Congress on Biomedical Engineering CLAIB 2014, Paraná, Argentina 29, 30 & 31 October 2014 vol. 49, pp 305-308: Springer International Publishing DOI 10.1007/978-3-319-13117-7_79
12 Bright JA, Rayfield EJ (2011) The Response of Cranial Biomechanical Finite Element Models to Variations in Mesh Density. Anat Rec 294:610-620 DOI 10.1002/ar.21358
13 González Carbonell RA, Ortiz Prado A, Jacobo Armendáriz VH et al. (2015) Consideraciones en la definición del modelo específico al paciente de la tibia. Rev Cubana Inv Bioméd 34:157-167
14 Beaupré GS, Orr TE, Carter DR (1990) An approach for time-dependent bone modeling and remodeling—theoretical development. J Orthop Res 8:651-661 DOI 10.1002/jor.1100080506
15 Cisneros Hidalgo YA, González Carbonell RA, Ortiz Prado A et al. (2015) Modelo mecanobiológico de una tibia humana para determinar su respuesta ante estímulos mecánicos externos. Rev Cubana Inv Bioméd 34:54-63
16 Chen G, Schuetz M, Pearcy M (2010) Mechanobiology of Bone Development and Computational Simulations. In F Bronner, MC Farach-Carson, HI Roach (Eds.), Bone and Development vol. 6, pp 279-295: Springer London DOI 10.1007/978-1-84882-822-3_17
17 Chou H-Y, Satpute D, Müftü A et al. (2013) Influence of mastication and edentulism on mandibular bone density. Comp Method Biomech Biomed Eng:1-13 DOI 10.1080/10255842.2013.792916
18 Martin RB (1984) Porosity and specific surface of bone. Crit Rev Biomed Eng 10:179-222
19 Volpon JB, Falcai MJ, Moro CA et al. (2014) Torsional force applied to the tibia of living lambs in an attempt to change the bone rotational axis. Acta Cir Bras 29:193-200 DOI 10.1590/S0102-86502014000300008
20 Cristofolini L (2015) In vitro evidence of the structural optimization of the human skeletal bones. J Biomech 48:787-796 DOI 10.1016/j.jbiomech.2014.12.010

Author: Raide Alfonso González Carbonell
Institute: Universidad de Camagüey
Street: Camagüey
City: Camagüey
Country: Cuba
Email: raide1977@gmail.com

A Comparison between Solar Radiation and Skin Cancer in South Brazil

P. Bertemes-Filho[1] and F. Imai[2]

[1]Dept. Electrical Eng., State University of Santa Catarina, Joinville, Brazil
[2] Federal Institute of Santa Catarina, Jaragua do Sul, Brazil

***Abstract* — Skin cancer has been highly detected in Brazil over the last decade. It causes genetic mutations in cells, which are mostly related to family health history and exposure to ultraviolet (UV) ray. It has not found in hospitals data a correction between UV radiations to cancer incident. The objective of this work is to investigate the correlation level between skin cancer cases and the solar UV radiation levels from 2006 to 2014. It was considered two South Brazilian cities: Curitiba at 800 meters above the sea level and the island of Florianopolis. Data were obtained from INCA (National Institute of Cancer) and INPE (National Institute of Space Research). Pearson's correlation coefficient was used for the numerical analysis, according to the city, cancer type (non-melanoma and melanoma) and sexual gender. According to Student "t" curve and using the 5% meaningfulness level for a bilateral test, results showed a minimum absolute value of 0.88 for the correlation. The cases of non-melanoma cancer in women could not be corrected, accordingly. It can be concluded that the geographical factor also shows to be relevant in the skin cancer incidence, as well as the solar UV radiation.**

Keywords— **UV Radiation, Skin Cancer, Pearson Correlation.**

I. INTRODUCTION

In terms of cases, the skin cancer type is the most one register in Brazil over the last decades [1]. It is important to take into account that cancer is a multifactorial problem. It causes genetic mutations in cells, which are mostly related to family health history and exposure to ultraviolet (UV) ray [2,3]. UV radiation is the main risk factor for most type of skin cancer, such as melanoma type. 5% of skin cancer cases in Brazil are melanoma, which causes 75% of deaths [3]. On the other hand, the non-melanoma types have much lower mortality rate but can cause irreversible injury to epithelial tissues, cartilage and bone [3].

The UV radiation is an electromagnetic wave ranging from 100 to 400 nm, which is divided into UV-A (315 to 400 nm), UV-B (280 to 315 nm) and UV-C (100 to 280 nm). Both UV-A and UV-B may cause mutation in human skin cells and then increasing the number of cancer diagnosis [4].

Caucasians are more likely to develop skin cancer [5] due to the skin pigment composition. Furthermore, albinos and people with white skin color are more likely to develop skin cancer under prolonged exposure to UV rays.

84% of the population in Florianopolis and Curitiba city are considered to be white skin color, and then cancer incidence tends to be high [6].

The amount of UV exposure a person can receive on their skin or eyes during an 8-hour period varies with the wavelength of the UV radiation, according to the Canadian Centre for Occupational Health and Safety. Wischermann *et al* (2008) reported that different UV-A exposure regime (4 x 5 J/cm2 per week or 1 x 20 J/cm2 per week) caused tumorigenic conversion (tumors in nude mice) [7].

According to the National Institute of Cancer (INCA), it were expected 20 new cases of melanoma among the male population in Florianopolis in 2014 whereas 70 in Curitiba [7]. However, according to INCA's prediction applied to 100,000 women, 11 ones would develop melanoma in 2014 in Florianopolis whereas 7.8 in Curitiba [8]. Besides the ethnic classification, socio-cultural factors and locality are other factors which may affect the solar radiation exposure. Robinson *et al* (1997) found that people with higher financial income expose themselves more frequently in the weekends whereas the poor ones during the week [9].

By investigating the difference in the solar radiation intensity and relating it to the cancer incidence statistics may help to understand how relevant these factors are in the melanoma occurrence [10], for example. Duzen and Aydin (2012) have described a method to estimate the global solar radiation in a place of 41.800 km², by using pyrometers in weather forecast stations. However, this study was only applied to solar power generation [11], but not for medical purpose.

The objective of this article is to investigate the correlation level between skin cancer cases estimated for Curitiba and Florianopolis and the solar UV radiation level. It was considered the collected data in 2006, 2008, 2010, 2012 and 2014.

II. METHODOLOGY

INCA develops ever 2 years the national estimation for 19 types of cancer in the Brazil. The estimation is done by all of States from Brazil, and then data are separated by

I. Torres et al. (eds.), *VII Latin American Congress on Biomedical Engineering CLAIB 2016, Bucaramanga, Santander, Colombia, October 26th -28th, 2016*,
IFMBE Proceedings 60,
DOI: 10.1007/978-981-10-4086-3_114

gender. Data are indicated in absolute value and also per 100,000 inhabitant rate. Equation (1) shows how the cancer incidence rate (T_i) is calculated, where T_m is the cancer mortality rate estimated by the local historic series, I_r is the cancer incidence rate (data from 2002 to 2011) and M_o is the cancer mortality rate. The database cases are presented in the period from 2002 to 2011.

$$T_i = T_m \times \frac{I_r}{M_o} \quad (1)$$

The cancer databases are provided by 282 hospitals all over Brazil. While the mortality rates are obtained from the Brazilian Mortality Information System (IML). Both INCA and IML data are updated and trustable for scientific analysis [8].

On the other hand, the incidences of UV radiation were obtained from the GL 1.2 model, which consist of the solar irradiance level reflected by the atmosphere and the earth ground [12]. These data are acquired from the GOES 12 geostationary satellite's VIS sensor. Data are recorded by the National Institute for Space Research (INPE), together with the National Environmental Prediction Center (NCEP), and then discloses graphics containing the daily solar UV radiation curves [13] in terms of mW/cm² unit.

Data were extracted from the database registered at 11:00, 13:00, 15:00 and 17:00. This is due to the fact that they comprehend the highest values for the solar incidence at the earth surface [14]. Annual averaging and then overall averaging were calculated.

Due to the fact that rain and fog occurrence, besides cloud presence, affect the UV radiation levels directly [14], it was considered in the model a factor called "Cloud Modification Factor" (CMF). Therefore, data shows two curves: one with cloudless sky (maximum irradiance) and other under a described weather conditions (real irradiance) [15]. The data considered in this article is only related the real irradiance values. CMF is obtained by computer simulations using experimental measures [15], taking into consideration clouds positioning (high, medium or low ones). This may vary from 0.2 to 1.0. The UV radiation corrected by CMF [mW/cm²] is equal to the product between cloudless sky UV radiation and the cloudiness broker factor.

Due to the fact that radiation levels are presented graphically, a visual analysis was made in order to determine the correlated data for the analyzed periods. It was considered 5 decimals when collecting data from graphics. UV radiation data are only available from 2011 onwards for both city of Curitiba and Florianopolis, resulting in a time lag. Therefore, it was made a linear regression [16] in order to estimate the irradiance values from 2006 to 2010.

The mathematical method used to analyze the data was Pearson's correlation coefficient [16] for 2006, 2008, 2010, 2012 and 2014.

III. Results

New skin cancer cases are divided in melanoma and non-melanoma. Figure 1 shows the cancer incidence from 2006 to 2014 per 100,000 inhabitants in Curitiba and Florianopolis, according to sexual gender [8,18-21]. Non-melanoma cases are show in the left side of the figure whereas the melanoma ones are in the right, where City 1 is Florianopolis and 2 is Curitiba.

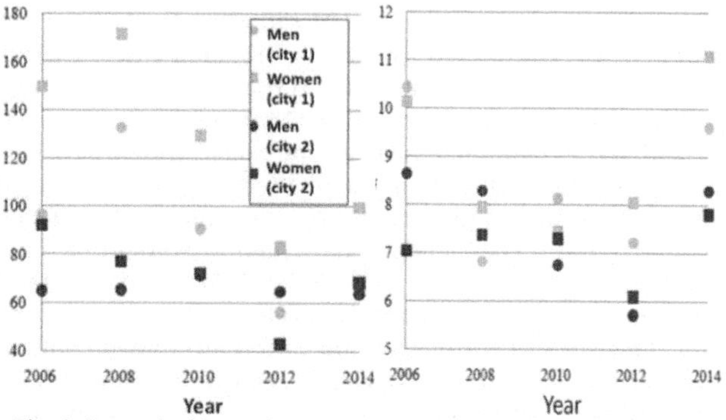

Fig. 1 Cancer incidence from 2006 to 2014 per 100,000 inhabitants in Curitiba and Florianopolis, according to sexual gender [8,18-21].

The average annual UV radiation incidence in Curitiba and Florianopolis from 2011 to 2015 are 4.06 and 3.96 [13], respectively, according to the visual calculation using the graphics.

The linear regression used in this work is presented in equation (2), where *IUV(Y)* is the average estimated UV radiation as a function of time [mW/cm²], a is the equation's linear coefficient, b is the equation's slope, Y is the year and N is the number of data sample.

$$IUV(Y) = a + (b \times Y) \quad (2)$$

where

$$b = \frac{N \times \sum(Y \times IUV) - (\sum Y) \times (\sum IUV)}{N \times \sum Y^2 - (\sum Y)^2} \text{ and } a = \frac{\sum IUV - b \times \sum Y}{N}$$

The expected radiation mean values from 2006 to 2010 are 3.9 for Curitiba whereas 3.5 for Florianopolis.

Pearson's correlation coefficient p was applied according to equation (3) [16], where T_i is the cancer incidence average whereas *IUV* is the *UV* irradiance average. Calculations were performed over 5 years of data analyzed here.

$$\rho = \frac{\sum\left[T_i - \overline{T_i}\right]\times\left(IUV - \overline{IUV}\right)}{\sqrt{\sum\left(T_i - \overline{T_i}\right)^2}\times\sqrt{\sum\left(IUV - \overline{IUV}\right)^2}} \quad (3)$$

It was performed 8 correlation analyses in order to verify the interaction between the values analyzed. The results are presented in Table 1.

Table 1 Pearson's correlation coefficient (-*p*) according to the city, cancer type and sexual gender, where MCN and MC stand for non-melanoma and melanoma, respectively.

Curitiba				Florianopolis			
Men		*Women*		*Men*		*Women*	
NMC	MC	NMC	MC	NMC	MC	NMC	MC
0.26	**0.57**	**0.88**	**0.29**	**0.81**	**0.35**	**0.92**	**0.07**

The Student "*t*" curve distribution and using the 5% meaningfulness level for a bilateral test, it can be observed in Table 1 that it is necessary an minimum absolute value of 0.88 for the correlation between data [16]. It was not found a minimum for the case of non-melanoma cancer in women.

IV. DISCUSSIONS

Both cancer incidence and UV radiation were collected from graphics with a time lag of few years. However, time recreation was performed in order to use statistic models. A time lag of one year might be considered a reasonable time for skin cancer incidence appear due to UV exposures.

In order to verify the real correlation, the lower number of UV radiation levels used here was a key factor in the analysis, since the lower the observation number the bigger the minimum absolute value [16].

Difficulties were faced in the extraction of the UV radiation levels by using the information system from INPE. It was necessary to convert graphic information into a number in order to perform the calculations. Besides that, it was needed to establish an approximation pattern when the radiation level did not match an integer value, since auxiliary decimal gridlines are not presented.

Based on the analysis performed in this work, it was only possible to establish a correlation between the level of UV radiation and the skin cancer incidence for non-melanoma cancer cases for women. This fact shows that UV radiation must not be interpreted as the single cause in skin cancer development.

Curitiba registered the highest irradiance values over the period analyzed, but the cancer incidence was lower (=18) than Florianopolis city (=20). Higher non-melanoma cancer cases in the male population was found in Florianopolis, which is in contrast to the UV radiation levels observed in the period analyzed in this work.

The overall results showed that the higher the exposure to UV radiation the higher the skin cancer incidence, especially to those city close to the beaches. Habitants from coastal cities, such as Florianopolis, tend to expose themselves more frequently to solar rays, due to the specific characteristics of leisure activities, professional and domestic perceived in these regions [17].

V. CONCLUSION

It can be concluded that the geographical factor also shows to be relevant in the skin cancer incidence, as well as the solar UV radiation. Care should be taken when interpreting these results, as data over all cities in Brazil need to be collected in order to show some evidences from this study within the framework of population habits, Government Programmes, medical recommendations and public health results.

ACKNOWLEDGMENT

The authors thank the institutional support of the State University of Santa Catarina (UDESC) and the financial support of the Technological Institute Foundation of Joinville (FITEJ).

CONFLICT OF INTEREST

The authors declare that they have no conflict of interest.

REFERENCES

1. Castilho I, Souza M, Sousa M (2010) Fotoexposição e fatores de risco para câncer da pele: uma avaliação de hábitos e conhecimentos entre estudantes universitários. Anais Brasileiros de Dermatologia 85(2):173-178
2. Carvalho A, Cunha M, Giugliani R et al. (2004) Melanoma hereditário: prevalência de fatores de risco em um grupo de pacientes no Sul do Brasil. Anais Brasileiros de Dermatologia 79(1):53-60
3. Souza R, Rezende M, Duarte E et al. (2009) Estimativa do custo do tratamento de câncer de pele tipo melanoma no Estado de São Paulo – Brasil. Anais Brasileiros de Dermatologia 84(3):237-43
4. Okuno E, Nakajima T, Yoshimura M et al. (1996) Radiação ultravioleta solar em São Paulo, Chiba, Calafate e Ilha de Páscoa. Caderno de Engenharia Biomédica 12(3):143-53

5. Battisti R., Weber A, Sgrott I et al. (2009) Avaliação do perfil epidemiológico e da mortalidade dos pacientes com diagnóstico de melanoma cutâneo primário no município de Florianópolis - SC, Brasil. Anais Brasileiros de Dermatologia 84(4):335-42
6. Instituto Brasileiro de Geografia e Estatística. Estatísticas de Gênero: Censo de 2010 at http://www.ibge.gov.br/apps/snig/v1/?loc=0&cat=
7. Wischermann K, Popp S et al. (2008) UVA radiation causes DNA strand breaks, chromosomal aberrations and tumorigenic transformation in HaCaT skin keratinocytes. Oncogene 27: 4269–4280.
8. Instituto Nacional do Câncer José Alencar Gomes da Silva. Incidência de Câncer no Brasil: Estimativa at http://www.inca.gov.br/estimativa/2014/estimativa-24042014.pdf
9. Robinson K, Rigel D, Amonette R (1997) Trends in sun exposure knowledge, attitudes, and behaviors: 1986 to 1996. Journal of the American Academy of Dermatology 37(2):179-86
10. Osterlind A et al. (1988) The Danish case-control study of cutaneous malignant melanoma. II - Importance of UV light exposure. Int.J.Cancer 42: 319-24
11. Duzen H, Aydin H (2012) Sunshine-based estimation of global solar radiation on horizontal surface at Lake Van region (Turkey). Energy Conversionand Management 58:35-46
12. Instituto Nacional de Pesquisas Espaciais. Estimativa de Radiação Solar Mediante Satélite at http://satelite.cptec.inpe.br/uv/docs/GLweb1_modelo1.pdf
13. Instituto Nacional de Pesquisas Espaciais. Acervo de Imagens at http://satelite.cptec.inpe.br/acervo/acervo.formulario.logic?idProd=4000
14. Oliveira L, Glauss N, Palma A (2011) Hábitos relacionados à exposição solar dos professores de educação física que trabalham com atividades aquáticas. Anais Brasileiros de Dermatologia 86(3):445-50
15. Instituto Nacional de Pesquisas Espaciais. As nuvens e a radiação UV at http://pyata.cptec.inpe.br/radiacao/docs/OIUV.pdf
16. Barbetta P, Reis M, Bornia A (2004) Estatística para Cursos de Engenharia e Informática. Atlas, São Paulo
17. Szklo A, Almeida L, Figueiredo V et al. (2007) Comportamento relativo à exposição e proteção solar na população de 15 anos ou mais de 15 capitais brasileiras e Distrito Federal, 2002-2003. Caderno de Saúde Pública 23(4):1-7
18. Instituto Nacional do Câncer José Alencar Gomes da Silva. Incidência de Câncer no Brasil: Estimativa 2006 at http://bvsms.saude.gov.br/bvs/publicacoes/versaofinal_estimativa2006.pdf
19. Instituto Nacional do Câncer José Alencar Gomes da Silva. Incidência de Câncer no Brasil: Estimativa 2008 at http://bvsms.saude.gov.br/bvs/publicacoes/estimativa_incidencia_cancer_2008.pdf
20. Instituto Nacional do Câncer José Alencar Gomes da Silva. Incidência de Câncer no Brasil: Estimativa 2010 at http://bvsms.saude.gov.br/bvs/publicacoes/estimativa_2010_incidencia_cancer.pdf
21. Instituto Nacional do Câncer José Alencar Gomes da Silva. Incidência de Câncer no Brasil: Estimativa 2012 at http://portal.saude.sp.gov.br/resources/ses/perfil/gestor/homepage/estimativas-de-incidencia-de-cancer-2012/estimativas_incidencia_cancer_2012.pdf

Author: Pedro Bertemes-Filho
Institute: The State University of Santa Catarina
Street: Paulo Malschitzki 200 - Zona Industrial Norte
City: Joinville
Country: Brazil
Email: pedro.bertemes@udesc.br

Inhibition of Erythrocyte Hemolysis induced by H_2O_2 with *Mangifera indica* L. extract

E. Pareja [1], Y. Salazar [1], S.S. Arango[1], M.E. Maldonado[2], J.G. Zuluaga[1], S. Restrepo[1]

[1] Instituto Tecnológico Metropolitano/Facultad de Ciencias Exactas y Aplicadas- GI²B. Medellín, Colombia
[2] Universidad de Antioquia/Escuela de Nutrición y Dietética– ICAS, Medellín, Colombia

Abstract- **oxidative stress is originated by a high production of ROS (pro-oxidant process), along with a impaired antioxidant mechanisms in the organism, which leads to the damage of different macromolecules. One of the most important targets of ROS are the polyunsaturated fatty acids (PUFA) of cell membranes. These damages produce changes in the structure and function of the cells, favoring process like premature aging and the appearance of different degenerative and autoimmune diseases. In this study it was evaluated the effect of *Mangifera indica* L. extracts treatments over the Human Erythrocyte Hemolysis caused by the pro-oxidant agent H_2O_2. On the other hand, in order to analyze cell membrane changes induced by the treatments, erythrocytes sediments were observed and photographed with an inverted microscope, using the phase contrast tool. Erythrocytes were obtained from healthy donors, and the test samples were prepared with 250 µl of erythrocyte suspension, 250 µl of H_2O_2. (1000mM or 100mM) and 1000 µl of mango extracts (30, 50, 100, 150, 200µg/ml); curcuma extracts at 50 µg/ml were used as control. Percentages of hemolysis inhibition over 49% were observed with mango treatments of 30 and 50 µg/ml. Likewise, the mango treatments shown a protective activity, which was evidenced by the conservation of the erythrocyte membranes treated with H_2O_2. Our results suggest that the damages suffered by the erythrocyte membrane by the oxidant action of H_2O_2 can be reduced by the action of compounds that are present in the Mango extracts.**

Keywords— **Hemolysis, Erythrocytes, membrane, lipid peroxidation, ROS.**

I. INTRODUCTION

Reactive oxygen species (ROS) are molecules produced by the cells in different processes like mitochondrial electron transport, physical activity, environmental stimulus, cigarette smoking, and alcohol consumption abuse, among others. The ROS are characterized for having an unpaired electron, which have the ability to interact with other molecules to transform them in other reactive species as: (O2-) superoxide anion ($O_2\cdot-$) singlet oxygen, and hydroxyl radical ($HO\cdot$) [1] [2]. H_2O_2 do not have unpaired electrons and it is not considered as a standard free radical; however, it is a molecule that can react with O_2 and cupric or ferric ions to produce OH, one of the most injurious oxidant species for its short half-life and high reactivity [3]. The oxidative stress is originated by a high production of these ROS and by deficient antioxidant mechanisms, which lead to macromolecules damage. One of the most important targets in the oxidation process the polyunsaturated fatty acids (PUFA) of cell membranes. It starts with the loss of an H atom of the fatty acid, generating a peroxyl radical (ROO-) at the same time, these reactions are known as lipid peroxidation [4]. These damages produce changes in the structure and function of the cells, favoring process like premature aging and the appearance of different degenerative and autoimmune diseases [5].

Blood is an important biological fluid, it fulfills different functions and acts like substrate in various intracellular biochemical reactions resulting in a high production of H_2O_2 and superoxide. Erythrocytes carry oxygen in the organism and circulate repeatedly through the lungs and capillaries, where are continuously exposed to intracellular ROS derivatives of autoxidation of oxyhemoglobin. These ROS have a high interaction with PUFA and proteins in their membranes, causing changes in the fluidity and the morphology of the membranes, enzymes inactivation, hemolysis, fragmentation and degradation of proteins. The erythrocytes hemolysis assays have already been used for a long time in measuring free radical damages [6] [7].

A strategy to remove or protect cells from ROS is the consumption of dietary sources of natural antioxidants. Experimental *in vitro* and *in vivo* studies, have demonstrated that the damage may be stopped, reduced or reverted, with an antioxidant-rich diet [8]. Mango (*Mangifera indica)* belong to the Anarcardiaceae family, it was established as one of the most important fruits around the world in terms of production and consume [9]. Furthermore, it is an important resource of micronutrients, vitamins and others phytochemicals, becoming as a potential source of anticancer, anti-diabetics and anti-inflammatory compounds and also it have shown antimicrobial and cardiovascular protector activities [10][11]. Mango it is recognized as a nutraceutical food, for its antioxidant capacity, its (poly) phenols content and its total carotenoids. [12], [13]. For this reason, the goal of our

I. Torres et al. (eds.), *VII Latin American Congress on Biomedical Engineering CLAIB 2016, Bucaramanga, Santander, Colombia, October 26th -28th, 2016*,
IFMBE Proceedings 60,
DOI: 10.1007/978-981-10-4086-3_115

research was to assess the inhibition of human Erythrocyte Hemolysis due to oxidative damage induced by H_2O_2 with *Mangifera indica* L. extract.

II. MATERIALS AND METHODOS

A. Mango extract preparation

Mango was grown by a Colombian company and ripe fruits were used in agreement with NTC-5139 norm. Fruits were selected, cut into pieces and stored at -20°C. The extract was prepared with the dilution of the pulp in water (1:4), with the addition of unflavored gelatin (stabilizer), sucralose, sucrose until to reach between 12 and 13° Brix. Finally, it was pasteurized between 80 and 85°C, packed and stored at 4°C. The nectar is lyophilized in a vacuum chamber to pressure 0.427 ± 0.5 mm Hg, temperature at 50°C, stored at 25°C ± 2°C and protected from light in a low density polyethylene packaging.

B. Erythrocyte suspension Preparation

Blood samples were obtained from two healthy volunteers, who previously read and signed the Informed Consent. According to the methodology explained by Duran et al 2013[14], 6ml of heparinized peripheral blood were obtained by venipuncture, centrifuged at 2500 rpm for 10 minutes and the supernatant discarded. 3 ml of PBS pH 7.4 were added to each tube for washing the cell pellet and centrifuged again at 2000 rpm for 10 minutes. The washing step was repeated two times to ensure the removal of other cells present in the blood. Finally, erythrocytes sediment was re-mixed whit (PBS) pH 7.4 until reach a 6 ml volume.

C. Erythrocyte hemolysis inhibition assay

These assay were performed according to the procedure described by Aguillon et al 2013 [15]. Briefly, samples were prepared with 250 µl of erythrocyte suspension, 250 µl of H_2O_2 1000mM or 100 mM (final concentration is 134 mM and 13,4mM respectively and 1000 µl of mango treatments (30, 50, 100, 150, 200µg/ml) in PBS; curcuma extract at 50 µg/ml were used as control. The preparations were put in an incubator at 37°C for 3 hours, immediately after adding the H_2O_2. After the incubation step, the preparations were centrifuged at 5.000 rpm for 10 minutes and supernatants collected and read in the Glomax TM Promega equipment to 540 nm. An erythrocyte sample treated with H_2O_2 500mM were used as hemolysis control. The results were reported as % hemolysis Inhibition = [(Ac - A) / Ac] x 100. Where Ac is the absorbance of the hemolysis control and A is the absorbance of the sample.

D. Observation of the effect on erythrocytes structure/ morphology

30 µl of erythrocytes-treated sediment were resuspended in 50 µl of PBS to analyze the cell membranes changes induced by the treatments, using a Nikon microscope and the phase contrast tool. Images were acquired using a Nikon DsFi1c digital camera with magnification of 60x (numerical aperture: 0.25) and image resolution of 1280 x 960 pixels; every image was captured at 40ms of exposure and gain of 1.4x, with linear contrast. Additionally, in all registers the microphotography was balanced selecting the auto white function of the camera interface.

E. Image processing

MATLAB Software was used to filter and detect cell membrane damage. It is employed a mean filter to remove image noise and it is made point by point constituting the median of a pixel with a matrix 10x10 around of this. It is designed as an algorithm to detect damage in membrane edges.

F. Statistical analysis

Data were presented as mean ± standard deviation (SD) of least four independent experiments. Linear Regression Analysis was used to calculate effective concentration or IC_{50} and inhibition percentage correlation (doses-response). Statistical differences between groups were evaluated using ANOVA with repeated measures and p values ($P < 0.05$) using Statgraphics Centurion XVI.

III. RESULTS

A. Erythrocyte hemolysis inhibition assay

The Hemoglobin released into medium was measured by optical density (OD) and it was used to estimate the percentages of hemolysis inhibition. The figures 1 and 2 show the percentages found for every H_2O_2 concentration.

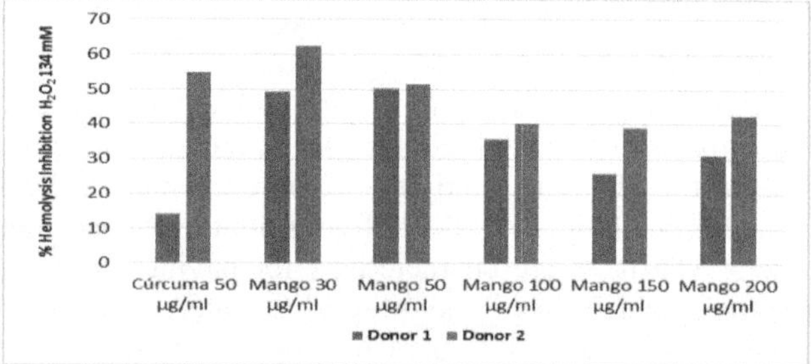

Fig. 1: Comparison of % Inhibition of erythrocyte hemolysis treats with H_2O_2 134mM, for every treatment. Significant differences between treatments and the negative control* ($P < 0.05$). A multiple range analysis (Fisher) was performed.

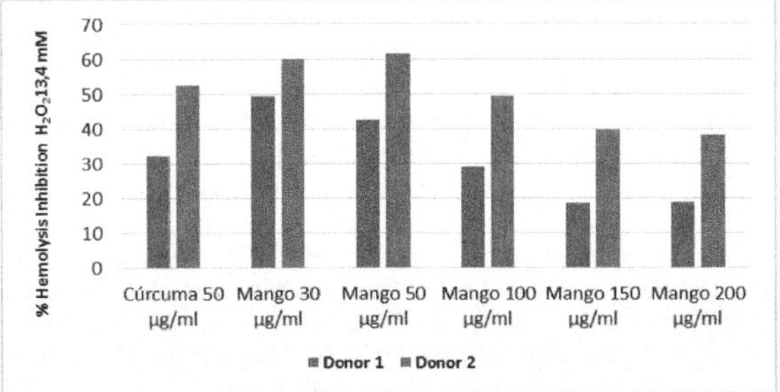

Fig. 2: Comparison of % Inhibition of erythrocyte hemolysis treats with H_2O_2 13,4mM, for every treatment. Significant differences between treatments and the negative control* ($P < 0.05$). A multiple range analysis (Fisher) was performed.

The percentages of hemolysis inhibition over 49% were observed with mango treatments of 30 and 50 µg/ml for both donors and independent of the concentration of H_2O_2 employed (134mM and 13.4mM). in comparison they were less that for treatments to the curcuma extract used as positive control in consequence the effect of antioxidant activity that has been reported for this plant. Furthermore, it was observed that increasing the mango extract concentrations about 100 ug / ml, the protection activity of erythrocyte hemolysis decreased.

B. Determination of hemolysis inhibitory concentration (IC_{50})

The IC_{50} were determined with the percentages of hemolysis inhibition for each treatment using a Linear Regression Analysis (see Table 1).

Table 1. Linear Regression Analysis

	Regression equation	Coefficient Determination (R2)	Value-P	CI_{50} (µg/ml)
% Inhibition 134 Mm H_2O_2 Donador 1	Y = 53,1154 – (0,139164)X	78,1594 %	0,0465	IC_{50} = 22.4 µg/ml
% Inhibition 134 Mm H_2O_2 Donador 2	Y = 58,6308 – (0,109101)X	60,5703 %	0,1211	IC_{50} = 62 µg/ml
% Inhibition 13.4 Mm H_2O_2 Donador 1	Y= 51,7025 – (0,189514)X	90,753 %	0,0123	IC_{50} = 9 µg/ml
% Inhibition 13.4 Mm H_2O_2 Donador 2	Y= 65,8596 – (0,150978)X	93,0846 %	0,0079	IC_{50} = 105 µg/ml

A statistically significant relationship among extract concentration and the percentages of hemolysis inhibition was found. The IC_{50} were lower of 100 µg/ml, which is the reference concentration established by the National Cancer Institute of the United States to consider an extract with promissory bioactivity [16].

C. Effect on erythrocyte membrane structure/ morphology

The H_2O_2 effect on erythrocytes cell membranes is visible microscopically at 40X magnification. The figure 3 and 4 show treated erythrocytes. The difference in cell membranes and morphology in general of the treated erythrocytes with H_2O_2, what evidence the oxidative damage generated in membranes.

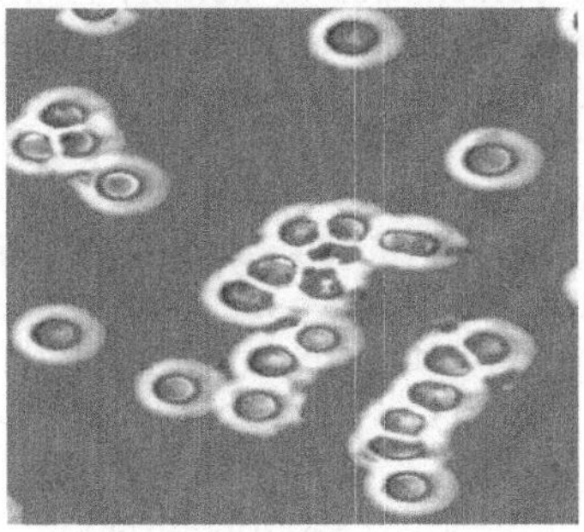

Fig 3: Changes in cell membrane obtained from 50 µg/ml of mango extract as treatment.

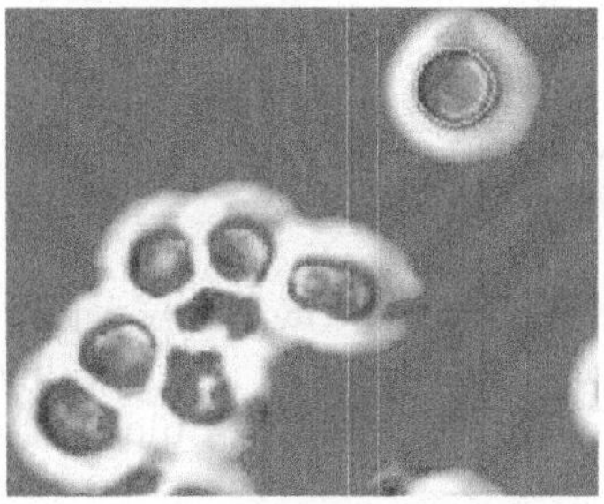

Fig 4: Segmentation of mango sample (50 µg/ml), where changes on erythrocytes cell membranes.

IV. Discussion

The stability of the erythrocyte membrane, is a cellular model widely used as an indicator of oxidative damage in the biomembranes [6]. The oxidation process affects the components responsible for membrane stability, such as phospholipids, cholesterol and transmembrane proteins, in combination with a network of cytoskeletal proteins. The alteration of the basic form of the erythrocytes and the loss of membrane integrity impairs its functions and can originate several disorders [17]. When the membranes lost their integrity are more susceptible to suffer hemolysis, for this reason after centrifugation, the hemoglobin is released into medium and can be an indicator of erythrocyte damage [18]. The mechanism of erythrocyte hemolysis by H_2O_2 oxidation occurs when the hemoglobin suffers degradation of the heme group with the released of catalytically active

Fe ions that trigger free radical production, membranes lipid peroxidation and OH- group production [19].

In this research, it was found a protective activity with the mango extracts that was evidenced by the conservation of the membranes of erythrocyte treated with H_2O_2. Several studies have demonstrated the mango antioxidant activity associated with its polyphenol content, anthocyanins, carotenoids and vitamin C, among others [19], [20]. The effect on the percentages of hemolysis inhibition can be related to the presence of antioxidant compounds that have been reported. In general, IC_{50} lower 100 µg/ml were found, these is desirable for compounds with promising biological activity; it is important to note that with concentrations of mango extract above of 100 µg/ml, this activity decreased. Previous reports have shown that mango peel extract, could inhibit the oxidative hemolysis of rat erythrocytes induced by H_2O_2 [3]

V. Conclusions

Our academic work shows that the damages suffered by the erythrocyte membrane by the oxidant action of H_2O_2 can be reduced through the effect of the Mango extract compounds (*Mangifera indica)*. It is suggested to continue with further studies to determine the mechanisms of the antioxidant activity of these extracts.

Acknowledgment

This work was supported by Instituto Tecnológico Metropolitano (ITM), Medellin, Colombia.

Conflict of interest

The authors declare that they have no conflict of interest.

References

1 Sies H (2015) Oxidative stress: A concept in redox biology and medicine, *Redox Biol.*, 4:180–183,.
2 Elejalde J (2001) Estrés oxidativo, enfermedades y tratamientos antioxidantes, *an. Med. Interna*, 18, no. 6:326–335,.
3 Cárdenas-Rodríguez N, Pedraza-Chaverri J (2005) Especies reactivas de oxígeno y sistemas antioxidantes : aspectos básicos, *Educ. Química*, 12, no. 2:164–173,.
4 Ajila C, Prasada U (2008) Protection against hydrogen peroxide induced oxidative damage in rat erythrocytes by Mangifera indica L. peel extract," *Food Chem. Toxicol*, 46, no.1: 303–309,.
5 Martínez C, Vargas C, Arancibia S (2003) Estrés oxidativo y neurodegeneracion, *Rev. la Fac. Med.*, 46, no. 6:229–235
6 Ibrahim I, Sallam S, Omar H, Rizk M (2006) Oxidative hemolysis of erythrocytes induced by various vitamins, *Int. J. Biomed. Sci.*, 2, no. 3:295–8,.
7 Włodek L, KusiorD (2006) Oxidative hemolysis of erythrocytes. *Biochem. Mol. Biol. Educ.*, 34, no. 6:438–443,.
8 Rajendran P, Nandakumar N, Rengarajan T, Palaniswami R, Gnanadhas R, Lakshminarasaiah U, Gopas J, Nishigaki I (2014) Antioxidants and human diseases. *Clin. Chim. Acta*, 436:332–347,.
9 Parvez G, (2016) Pharmacological Activities of Mango (Mangifera Indica): A Review. 5, no. 3:1–7,.
10 Corrales-Bernal A, Maldonado M, Urango L, Franco M, Rojano B (2014) Mango de azúcar (Mangifera indica), variedad de Colombia : características antioxidantes , nutricionales y sensoriales *Rev. Chil Nutr*, 41, no. 3:312–318,
11 Maldonado M, (2014) Propiedades quimiopreventivas del mango y la manzana en el cáncer de colon Chemopreventive properties of mangoes and apples in colon cancer 20:614–618.
12 Kumar S, (2014) The importance of antioxidant and their role in pharmaceutica science - a review, *Asian J. Res. Chem. Pharm. Sci.*, 1, no. 1, pp. 27–44,.
13 Jahurul M, Zaidul J, Ghafoor K, Al-Juhaimi F, Nyam K, Norulaini N, Sahena F, Mohd A (2015) Mango (Mangifera indica L.) by-products and their valuable components: A review, *Food Chem.*, 183:173–180,.
14 Durango M, Montero P, Marrugo Y (2013). Extractos metanólicos de corteza de guayaba (Psidium guajava l.), Revista U.D.C.A. Actualidad & Divulgacion Científica vol 16 2:327–334.
15 Aguillón J, Maldonado M, Loango N, Arango S, Landázuri P (2013) Antioxidant and antiproliferative activity of ethanolic and aqueous extracts from leaves and fruits juice Passiflora edulis. *Perspect. en Nutr. Humana*, 15, no. 1:13–25,.
16 WHO, (2000) General Guidelines for Methodologies on Research and Evaluation of Traditional Medicine World Health Organization, Geneva.

Author: Sandra Sulay Arango Varela
Institute: Instituto Tecnológico Metropolitano -ITM
City: Medellín-Colombia
Email: sandrarango@itm.edu.co

DelphiCare 4.0: Web Biotelemetry System

R.J. Díaz[1], R. Villalpando[1], J.C. Hinojo[2], G.E. Ramírez[2] and L.A. Retana[2]

[1] ITESM / Deparment of Engineering, Professor, Torreón, México
[2] ITESM / Deparment of Engineering, Student, Torreón, México

Abstract— **Nowadays, health monitoring systems, that comprise electronic devices; patients, doctors and appointments registry, supported by Internet, tend to be very costly. For this reason, its massive presence in hospitals is reserved for private health institutions, that have high economic resources. In this paper we propose an instrument that allows measurement, processing and transmission of patient vital signs, via Internet to a mobile device at a low cost. This instrument, is based, on three modular applications working in four functional blocks, to transmit the information. The first software module consists of an application programed in Node.js, that receives information from a mini computer (PCDuino) where the vital signs sensors are connected. Information from sensors is transmitted through UDP protocol to a second application, that uses Websocket connection. Later this information is shown in an administrative interface.This process allows doctors and attention personnel to know and update patient vital signs in real time. This tool represents a main contribution for the medical sector.**

Keywords— **Biotelemetry , vital signs , mobile applications**

I INTRODUCTION

Today, many patients around the world do not have economic resources to pay periodical medical treatment in specialized institutions. This forces them to take medical treatment by home. These people constantly require monitoring of their vital signs, where a small variation could represent a severe health issue. Patients usually request immediate remote assistance by cellphone or other electronic device, when symptoms arise.

New existing technologies can be applied to solve both problems (but are rarely used) which are detailed below.

II APPLIED TECHNOLOGIES

A New technologies

At the early 2005, there were many services that facilitated communication in everyday life. The Internet and new technologies in web, enabled many companies on Silicon Valley, to create applications that were impossible to achieve, prior to applications like Google Fit, which allows the use of accelerometers and motion sensors that detect heart rate and footsteps done by the user, in order to make a report of the physical activity done by people during the day.

E-mail services like Gmail and Hotmail had an incredible interactivity with the user, giving real-time notification when a new mail is received, in later years they allowed the develompment of social networks like Facebook and Twitter, allowing the user share content and watch it at any time and anywhere.

B Changing the Paradigm

In MID-2005, a Danish programmer, David Heinemier Hannson, designed the source code of an entire set of tools and libraries , for application development in the cloud,this changed the paradigm of web systems and applications. These tools and libraries are named Ruby on Rails, as they were programed with the Ruby programming language [1].

At the end of the past decade, in the European Conference JSConf.eu, the programmer Ryan Dahl released the source code of Node.js, a platform that allows developers to implement real-time web applications. In a very short time Node.js reached a wide acceptance as a tool for software development. It allowed creation of applications such as Netflix and Uber, which send large amounts of data in real time [2].

These technologies allowed the development of the proposed real-time health monitoring technique Delphicare, as will be explained further.

III DELPHICARE

A Software Architecture

The proposed system consists of four main blocks which are illustrated in Figure 1. The first block is a program developed in Node.js that allows to sample the vital sign delivered by the electronic sensors , attached to aPCDuino [3] through different I/O ports. The data obtained from the different sensors is packaged and sent through a connection called User Datagram Protocol (UDP). This protocol is commonly used to send a large amount of data to a remote server in real-time.

The second block, named Node.js, receives and process the data from the UDP port, and then replaces it through

I. Torres et al. (eds.), *VII Latin American Congress on Biomedical Engineering CLAIB 2016, Bucaramanga, Santander, Colombia, October 26th -28th, 2016*,
IFMBE Proceedings 60,
DOI: 10.1007/978-981-10-4086-3_116

socket connection.io to the next block.

The third block uses the information received in real time to create an HTML application within the HTTP protocol. The HTML page is created within an administrative system developed in Ruby on Rails. Through this block we keep a record of patients, doctors, general data, in addition it can be access by software administrators. Approved users can see real-time data of patients/family. This may be useful for patient monitoring and therefore is relevant in the DelphiCare tool.

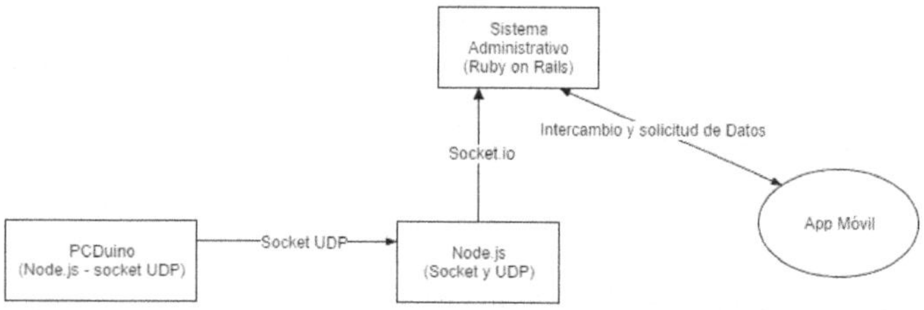

Fig. 1: DelphiCare block diagram.

B Mobile platform

A mobile app was developed for the Android platform, programed in the Java environment, which uses libraries for Android 4.1 JellyBean [4]. This application communicates with the central management system of Delphicare (referring to the third block) using a HTTP protocol. This system encrypts the data in JSON format, and delivers it to the mobile application so that it can be processed and displayed on the mobile device. The real-time data display is able to adapt to any screen; therefore, a user can watch the system via web from any mobile device. This application has communication with a service developed by Google, called Google Cloud Messaging. This service sends notification messages to any place in the world, this lets medical personnel to receive emergency notifications at any time.

C Hardware

As shown in subsection III.A, the first block the DelphiCare system is a group of sensors responsible for measuring three of the most important vital signs indicated by specialists. These three variables are: oxygen concentration in blood, temperature and electrocardiographic signal. Each variable was measured through the corresponding sensor, that were build according to the standard IEC 60364-7-710 [5], on equipment for hospitals and clinics.

D Electrocardiographic signal

There is a large amount of circuits for measuring electrocardiographic signal. One of the goals of the DelphiCare was to obtain a system that, is low cost and has a large portability with regard to similar systems. After a search conducted, it was found the AD8232 [6] of Analog Devices, which is a cardiograph integrated circuit. The standard used is the 60364-7-710 of the International Electrotechnical Commission, that refers to the electrical installations and electronic devices used in hospitals or clinics. The standard requires that the voltage in a medical complex is 25V, establishing a limit of electrocution of 30mA. The aforementioned circuit is powered from 2 to 3.8V, operating with a current of 170 micro-amp; their electrical qualities and size make it ideal for the DelphiCare. According to the data sheet of the AD8232 circuit, there are two ways to connect the electrodes to the patient (Figure 2).

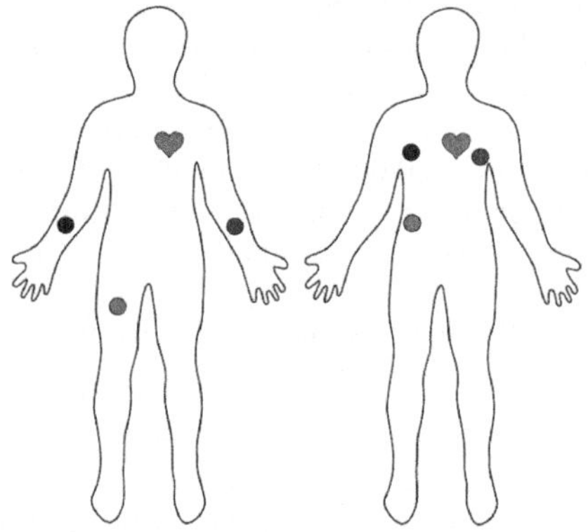

Fig. 2: Electrode position on the body.

The ECG signal can be read directly by one of the analog ports of the pcDuino, where it gets digitalized and displayed.

Testing of the ECG sensor was done with an Arduino UNO [7], due to its architecture, is very similar to that of the pcDuino. By testing, it was possible to find an appropriate rate of sampling, approximately 60 Hz. This was necessary because the sampling frequency suggested by the manufacturer of the device does not gave satisfactory results by plotting the signal, since these results do not coincide with the medical standards.Testing is still in progress. Figure 3 is a graph of the ECG sensor values already processed of a test subject.

E Temperature Measurement

The temperature module of the system consists of the MLX90614 sensor[8], which is an infrared thermometer suitable for measuring temperatures in the range of -70 C to +382 C, with digital output of 17 bits. The sensor provides two

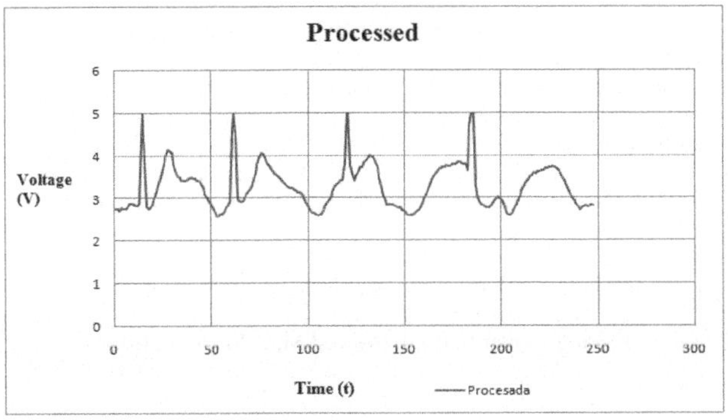

Fig. 3: Signal processed by software.

types of outputs: PWM for a resolution of 10 bit (0.14C) and I2C for the maximum resolution (0.02C), which was chosen. The sensor has a default setting range of -40 C to +85 C. The temperature value given is the average temperature of the objects included in the field of vision of the sensor. The accuracy of the sensor is 0.5 C at room temperature of 25 C. In Figure 4 shows the sensor operating range. Because the

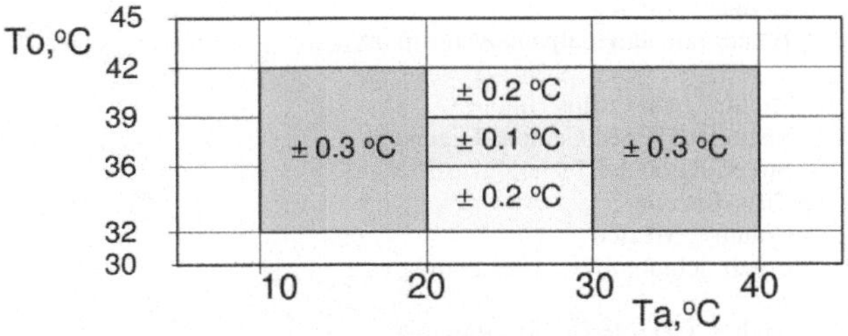

Fig. 4: Operating temperature graph.

sensor requires a special serial communication, you need a software library for it to work, not available in pcDuino, so it was necessary to use the Arduino Micro, to serve as the interface between the sensor and the pcDuino. Table I shows values obtained with the DelphiCare compared with those of a digital thermometer.

Table 1: Practical results obtained (in Centigrade).

DelphiCare	Thermometer
37.48	37.1
37.24	37
36.94	36.8
36.7	36.6
36.52	36.5
36.3	36.2
36.14	36.5

F Measurement of Oxygen Concentration in Blood

The oxygen sensor was designed from scratch, on the basis of the digital electronics and the pulse oxymetry method. It consists of four main parts: Signal conditioning circuit (constituted by a I/V converter),pulse generator (designed from two digital outputs of pcDuino), switching LED circuit (constant current source and LED switching driver) and thimble (formed by an emitter and a photoreceptor). The SpO2 sensor (oxygen concentration in blood) [9] is based on the principle of pulse oxymetry, it generates two beams of light, one in the red light spectrum (wavelength: 600nm - 750nm) and another in the infrared spectrum (wavelength: 850nm - 1000nm) and measures the amount of light that is transmitted through the index finger (Figure 5).

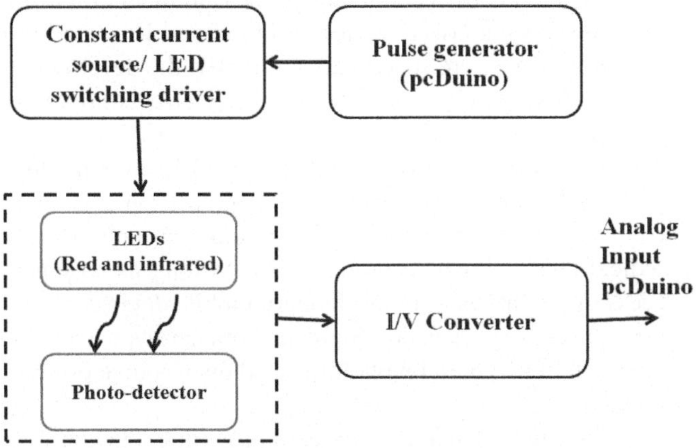

Fig. 5: Block diagram of Sp02 saturation sensor.

The constant current source is implemented using an H-bridge L293D [10], which together with the pulse generator is the LED switching driver. Red and infrared LEDs are used [11] to create the beams of light, that pass through the finger and arrives at the photo-detector, in this case a photodiode connected to a current/voltage converter made of operational amplifiers. In Table II shows the readings obtained by the oxygen sensor from the DelphiCare, compared with the readings obtained by a commercial sensor.

IV CONCLUSIONS

This paper presents the design of DelphiCare system in its fourth version , reaching the following conclusions:

1- The DelphiCare answers a specific and important issue for the health sector.

2- It is supported by different medical experts in the region, for device testing and standards approval, so it can be considered reliable and could be used in the field of medicine.

Table 2: Oxygen sensor results.

Oxygen saturation %SpO2			
DelphiCare	Comercial Sensor	DelphiCare	Comercial Sensor
96	97	98	98
96	97	99	98
97	97	98	98
97	97	98	99
97	97	99	99
97	97	99	99
97	98	98	99
98	98	99	99
97	98	99	99
98	98	98	99
98	98	99	99

3- The tests corresponding to the measurement of temperature and oxygen concentration in blood and its integration with the system were successful. Currently, the development team is working in the integration of the electrocardiographic sensor.

4- The system software was developed in four modular blocks, which are responsible of the acquisition and signal processing,user management and their quick reception by the medical staff; anywhere in the world, as long as they have access to the Internet in most modern mobile devices.

5- The entire project development, both hardware and software system was built on platforms and open source program languages.

6- Now, this system is in process of getting a patent from the OTT from ITESM.

7- There are development kits [12] that have a larger number of devices and sensors for monitoring vital signs that can be adapted to the system. Which they were not known at the beginning of development.

8- The cost of the components was around 200 US dollars. Regardless aesthetic design and manufacturing process. The cost of software development is estimated at 24,000 dollars.

From a social and human stand point, DelphiCare will bring services, that can give follow up to a group of patients with less financial resources, closer; services that mainly are offered in private clinics.

REFERENCES

1. Ruby on Rails (n.d.). Retrieved May 26, 2016, from `http://rubyonrails.org/`
2. Node.JS (n.d.). Retrieved May 26, 2016, from `https://nodejs.org/en/`
3. PCDuino User Guide (n.d.). Retrieved June 15, 2016 `http://bit.ly/21ndsvF`
4. Android Developers (n.d.). Retrieved May 26, 2016 `https://developer.android.com/`
5. IEC 60364-7-710 normative (n.d.). Retrieved May 26, 2016 `http://bit.ly/291a18L`
6. Heart Rate Sensor (n.d.). Retrieved May 26, 2016, from `http://bit.ly/24RurXG`
7. Arduino UNO (n.d.). Retrieved June 15, 2016, from `https://www.arduino.cc/en/Main/ArduinoBoardUno`
8. Sensor Temperatura Infrarrojo (MLX90614) (n.d.). CRetrieved May 26, 2016, from `http://bit.ly/1BDBLwE`
9. Oxmetro de Pulso (n.d.). Retrieved June 15, 2016, from `http://bit.ly/24RuO4t`
10. L293D Motor Driver (n.d.). Retrieved May 26, 2016, from `http://www.ti.com/lit/ds/symlink/l293.pdf`
11. LED Datasheet (n.d.). Retrieved June 15, 2016, from `http://bit.ly/1trG4ZM`
12. e-Health Sensor Platform (n.d.). Retrieved June 15, 2016, from `http://bit.ly/2cg31vU`

Author: René Joaquín Díaz Martínez
Institute: ITESM Campus Laguna
Street: Paseo del Tecnológico 751
City: Torreón
Country: México
Email: renejdm@itesm.mx

Author: Rafaela Villalpando Hernández
Institute: ITESM Campus Laguna
Street: Paseo del Tecnológico 751
City: Torreón
Country: México
Email: rafaela.villalpando@itesm.mx

Author: Juan Carlos Hinojo
Institute: ITESM Campus Laguna
Street: Paseo del Tecnológico 751
City: Torreón
Country: México
Email: jchinojob7@gmail.com

Author: Gerardo Ernesto Ramírez
Institute: ITESM Campus Laguna
Street: Paseo del Tecnológico 751
City: Torreón
Country: México
Email: gerc_89@hotmail.com

Author: Luis Alonso Retana
Institute: ITESM Campus Laguna
Street: Paseo del Tecnológico 751
City: Torreón
Country: México
Email: luis.alonso.retana@hotmail.com

Cardiorespiratory Interaction Using Nonlinear Data Processing Techniques In Patients Undergoing Test Tube T

J.I. Trapero[1], C.J Arizmendi[1], C.A. Forero[1], S.K. Lopez[2] and B.F. Giraldo[3]

[1] Autonomous University of Bucaramanga, Dept, of Physicomechanical, Bucaramanga, Colombia
[2]Clinica Materno Infantil, Bucaramanga, Colombia
[3] Polytechnic University of Catalunya, Dept. of ESAII, Barcelona, España

Abstract—**The estimate of the optimal time to remove the ventilator from a patient in intensive care remains critical in clinical practice. This study analyzes the breathing pattern from cardiorespiratory signals extubation patients undergoing performing resampling the signal, then the Symbolic Dynamics technique for data processing is implemented, together with the techniques of Support Vector Machines (SVM) and Linear Discriminant Analysis (LDA) for classifying 154 patients conglomerates in the Group Success and Failure Group classification , obtaining the best result obtained from 93.87 ± 0.01 % prediction, for SVM .**

Keywords - Extubation Process , Symbolic Dynamics , Support Vector Machines .

I. INTRODUCTION

It is essential to try to maintain the life of severe acute patient without the aid of Mechanical Ventilation (MV), more so when the disease in its extreme cases the patient has Acute Respiratory Failure (ARF), choosing the optimal time to remove the MV depends on the stage of evolution of the patient , so there is a possibility that has to be reintubated for not choosing the optimal time of disconnection. This is an aspect that can not miss when it comes cessation [1]. There are multiple etiologies of Weaning Failure and a bad decision regarding when to start it,should not be identified as a cause of failed Weaning all this imposes a burden on the inspiration of patients who are being released from the MV that the patient can not beat and muscle fatigue appears, this being one of the causes of the failed weaning due to load imbalance being ventilatory and respiratory pump [1,2] .

In the present study it is implemented using nonlinear analysis of dynamic systems [3] for the stage classification is proposed to use the LDA and SVM obtaining new indices based on the respiratory signal that allow us to determine the optimum time for removing MV to a patient.

II. MATERIALS AND METHODS

A. Database

The study was conducted with the WEANDB database, respiratory signals formed by patients connected to the MV through endotracheal tube. From the result of the test tube T, patients were classified into 2 groups : Success Group (SG) with 94 patients who successfully passed the test and regained spontaneous breathing ; Failure Group (FG) with 38 patients who failed the test and were reconnected to the ventilator.
Patients were recorded in the Departments of Intensive Care Hospital de la Santa Creu i Sant Pau , Barcelona, and the Hospital of Getafe , Spain , according to approved protocols ethical committees .
The respiratory flow signal was obtained using a pneumotachograph (Datex-Ohmeda monitor variable reluctance transducer) connected to an endotracheal tube. They were recorded 30 minutes at a sampling frequency of 250 Hz with a National Instruments data acquisition card dAQCard6024E [1]. Once processed cardiac and respiratory signals, the corresponding time series were obtained: duration of the cardiac cycle (RR) , inspiratory time (TI), expiratory time (TE) , breath duration (TTot) , tidal volume (VT) , inspiratory fraction (IT / TTot) , flow medium inspired (VT / TI) and the relationship between frequency and tidal volume (f / VT) .

III. PROCESSING TECHNIQUES AND DATA

A. Symbolic Dynamics

Symbolic Dynamics (SD) has shown to be a method of analysis it provides important information to characterize the complexity of the Heart Rhythm Variation information. It is based on the transformation of a signal into a sequence of symbols that takes the value of an alphabet, eliminating the detailed information and keeping in the new series clinically useful information present in the cardiac signal, but not available or observable directly [2]. The main idea of SD is divided into outlines the space in which a signal can fluctuate [3]. Each contour will be represented by a symbol the set of all symbols representing the contours in which a signal is divided is called the alphabet. When making this conversion, it can be build a sequence of symbols from a time series or any other opening sequence; this process is called "symbolization ".
The above it can be formally expressed as follows [4]. Assume

I. Torres et al. (eds.), *VII Latin American Congress on Biomedical Engineering CLAIB 2016, Bucaramanga, Santander, Colombia, October 26th -28th, 2016*,
IFMBE Proceedings 60,
DOI: 10.1007/978-981-10-4086-3_117

the time series $\{x_n\}_{n=0}^{\infty}$ in $\mathcal{U}\epsilon\mathbb{R}$, , and $\mathcal{A} = \{A_0, A_1, \ldots, A_{k-1}\}$ is a partition that satisfies $\mathcal{U} = \bigcup_{i=0}^{k-1} A_i$, and $A_i \cap A_j = \emptyset, \forall i \neq j$. If each region A_i is denoted by a symbol $\theta(A_i)$, then the original time series can be rewritten as a symbolic sequence simply determining the regionA_i including each individual value x_n. In Figure 1 a representation of the SD process is shown.

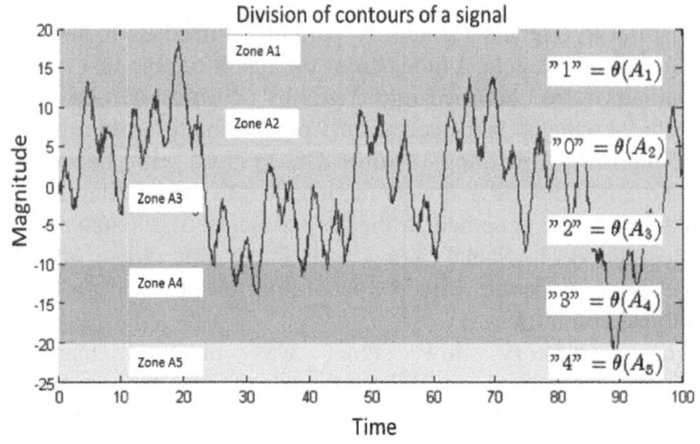

Fig. 1. Division sequences of symbols according to variablity of the signal [2]

In Figure 1 a signal is converted into a sequence of symbols, an alphabet of four symbols A = {0,1,2,3}. Implementing equation 1. This transformation of a signal x considers three levels given, where it refers to the average of a signal and is used as an special parameter normally assigned a value of 0.1 [4] . It is noteworthy that small changes in the values of μ and α do not influence considerably the partition being a constant value alpha between 0 and 1.

$$\begin{cases} 1\ si\ (1+\alpha)\mu < xi < \infty \\ 0\ si\ \ \mu < xi \leq (1+\alpha)\mu \\ 2\ si\ (1-\alpha)\mu < xi \leq \mu \\ 3\ si\ 0 < xi \leq (1-\alpha)\mu \end{cases} \quad n = 1,2,\ldots\ldots N \qquad (1)$$

Once obtained the sequence of symbols $s(x)$, it must to be fragmented the sequence in " words over the alphabet A " with long ℓ . A word on an alphabet A is defined as a finite sequence of symbols (possibly empty) [4]. For example, the study of an electrocardiogram (ECG) of 30 minutes which corresponds to an approximate value of 1800 RR intervals in the tachogram [5] . Was chosen for this case using words of length 3, and constant tau (τ) consisting of symbols of an alphabet $A = \{0,1,2,3\}$, whereby 64 types of different words (bins) [5] are obtained . In Figure 2 the conversion process in a time series illustrates a symbolic sequence .

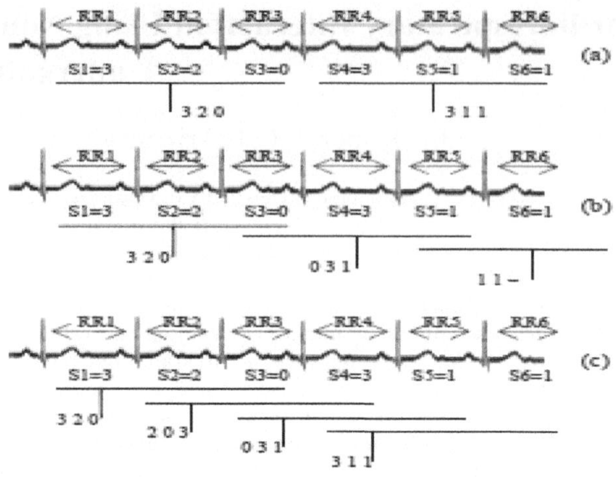

Fig. 2. How to build words ℓ = 3 and overlaps . (a) τ=0; (b) τ=1; (c) τ=2 [2]

B. Forward Selection

The technique starts evaluating the classifier using a single variable which changes with each of the existing variables; the result of this classification is ranked being selected the variable with the highest percentage of classification. Then a new set of classifications is carried out taking as input the selected variable with each of the remaining variables; then , the pair of variables that gets the best result is chosen. This process is repeated until finishing all the variables, each time selecting the set of variables with which the best ranking was obtained.

C. Moving Window with Variance Analysis (MWVA)

The MWVA is a feature selection method type filter, consisting in the creation of a variable moving window , which comparing two or more signals by the computation of the intragroup, the inter group variance and the energy criterion, find the most relevant variables and its relevance in the system [6].

D. Linear Discriminant Analysis

The LDA is a technique that identifies the most relevant characteristics to differentiate two or more groups and to create a function able to distinguish as accurately as possible the members of two or more groups. The belonging to groups, previously known, is used as the dependent variable. The final goal of LDA is to find the linear combination of independent variables, that allows differentiated the groups. Once is found the discriminant function, it can be used to classify new cases.

E. Support Vector Machines

A SVM learns the decision surface of two different clases, independent of entry points . As a classifier of a single class , the description given by the vector data support is capable of forming a decision boundary around the domain of the training data with little or no knowledge of the data outside this border. Data are mapped by a Gaussian kernel or other kernel to a higher dimensional feature space, looking for the the maximum separation between classes. This boundary fucntion, when is brought back to the input space, can separate the data on all different classes [6].

The SVM first maps the entry points to a feature space of greater dimension (i.e. if the entry points are in $\Re 2$ then are mapped by the SVM to $\Re 3$) and finds a hyperplane that separates and maximize the margin m between classes in the space.

IV. DATA PROCESSING AND CLASSIFICATION

In studies based on Weandb data base is illustrated that the resampling to 2 Hz guarantees the nonexistence of loss of signal energy [8], by computing the power spectrum of the resampled signal. Because the cardiac signal has a different sampling compared with the respiratory signals, is made a resampling of the signal of 2Hz, and linear interpolation is performed to equalize the number of samples for cardiac and respiratory signals [8].

In order to obtain indexes to improve the optimal time of weaning the SD is used, where the transformation of the series in an alphabet of four consecutive words with words of length $\ell = 3$ and a value of tau = 0 is used, obtaining a total of 64 words per series. Next the alpha value that allows us to have all the words with a frequency of occurrence greater than 0 is chosen. Due to the large cardinality of the system, we proceed to perform a dimensionality reduction by implementing the FS and MWVA techniques, for data classification and the LDA and SVM is implemented, the results were divided into 4 stages. In all the stages were implemented 150 runs, together with 4 Fold- Crossvalidation, being the mean and standard deviations of the test group averaged to all the classification results, in the SVM was used a radial kernel with different values of alpha.

In the first stage are implemented the overall 560 variables of Success Group and Failure Group, processed with SD toguether with the LDA and SVM classifiers, the results are illustrated in Table 1.

Table 1. Classification percentages of SG Vs FG without dimensionality reduction

Classifiers	Classification Percentage	Variables	Time (Hours)
LDA	60.12% ± 2.42%	560	1 H
SVM	69.12%±1.71	560	2 H

In the second stage, is reduced the cardinality of the system implementing FS technique in conjunction with LDA and SVM, the computational time cost was 144 and 720 hours respectively, Table 2 illustrates the mean and standard deviation of the best classifications results implementing the FS and SVM.

Table 2. Classification percentages of SG Vs FG implementing FS technique

Classifiers	Classification Percentage	Variables	Time (Hours)
LDA	85.51% ± 1.17%	44	600 H
SVM	90.33±1.52	81	620 H

In the third stage in order to reduce the cardinality of the system, MWVA technique is implemented [6], the optimal window width was equal to one, the stop criterion chose the number of variables containing 99% of the total energy of the system and by adding more variables it does not represent a significant increase in the total energy of the system. Table 3 illustrates the results obtained.

Table 3. Classification percentages of SG Vs FG implementing MWVA technique

Classifiers	Classification Percentage	Variables	Time (Hours)
LDA	71.26% ± 1.28%	170	0.6H
SVM	74.68% ± 0.0119%	170	0.65H

In the fourth stage the MWVA is implemented as variable preselection technique, obtaining 170 variables then FS and SVM techniques are implemented, Table 4 illustrates the results obtained.

Table 4. Classification percentages of SG Vs FG implementing MWVA-FS techniques

Classifiers	Classification Percentage	Variables	Time (Hours)
LDA	88.87 ± 1.45	51	360H
SVM	93.87±0.01	42	360H

V. CONCLUSIONS

The SVM's are algorithms with good performance against the data without any processing, indicating the great capacity in classifying.

Comparing the two classification techniques used in this study, is illustrated that the performance of classifiers increases considerably when processing techniques and dimensionality reduction are used.

By implementing the 560 system variables, the best percentage of 69.12% classification was obtained, using the dimensionality reduction by technical Forward Selection, the best percentage of 90.23% classification with 81 variables was obtained, however this method has a very high computational cost.

Using the MWVA technique to preselect the variables before implementing FS, the best overall classification percentage of 93.87% was obtained with 42 variables with the lower the computational cost in all the experiments.

Acknowledgment

The authors would like to thank to Doctors of Cardiac Unit and his collaborators of the Santa Creu I Sant Pau Hospital, Barcelona, Spain, by their collaboration in the signal database acquisition. This work was supported in part by the Secretariat of Universities and Research of the Department of Economy and Knowledge of the Government of Catalonia (Consolidated research group GRC 2014 SGR 1569) and by the Ministry of Economy and Competitiveness of the Spanish Government (DPI2015-68820-R).

Conflict of Interest

The authors reports no conflicts of interests.

VI. References

[1] Preciado J. A. C, and Giraldo Giraldo B (2011). Análisis y clasificación del patrón respiratorio de pacientes en proceso de retirada del ventilador mecánico, *Ingeniería Biomédica,* vol. 5, nº 1909-9762, p. 7,.

[2] Brunel, H., Vallverdú, M., Perera, A., Caminal, P., Voss, A., Cygankiewicz, I., ... & de Luna, A. B (2011). Estudio de la complejidad de la variabilidad del ritmo cardiaco en pacientes con cardiomiopatia dilatada.

[3] Hadamard J. S (1898). *Les surfaces à courbures opposées et leur lignes geodésiques,* vol. 5e série, Paris: Journal de mathématiques pures et appliquées,.

[4] Voss A, (1996). T*he application of methods of non-linear dynamics for the improved and predictive recognition of patients threatened by sudden cardiac death,* vol. 31, Berlin: Cardiovascular research Elsevier.

[5] Yazlle J, and Jadur C (2006). Dinámica Simbólica: Apuntes de Cátedra, Salta: Departamento de Matemática de la Universidad Nacional de Salta.

[6] Arizmendi C, Romero E, Sierra D and Vellido A (2014). Automated Classification of Brain Tumours from Short Echo Time in Vivo MRS Data Using Gaussian Decomposition and Bayesian Neural Networks. *Expert Systems with Applications,* vol. 11, pp. 5296-5307.

[7] Betancourt, G. A. (2005). Las máquinas de soporte vectorial (SVMs). *Scientia et Technica, 1*(27).

[8] Pandit, D. (2010). Respiratory weaning. In *Core Topics in Critical Care Medicine*(pp. 241-245). Cambridge University Press New York, NY.

[9] González-Acevedo, H., Arizmendi-Pereira, C. J., & Giraldo-Giraldo, B. (2015). Classifier design for patients on weaning process. *Iteckne, 12*(2), 131-137..

[10] Arizmendi, C., Hernández-Tamames, J., Romero, E., Vellido, A., & del Pozo, F (2010). Diagnosis of brain tumours from magnetic resonance spectroscopy using wavelets and Neural Networks. In *2010 Annual International Conference of the IEEE Engineering in Medicine and Biology* (pp. 6074-6077). IEEE..

[11] Arizmendi C (2012). *Signal processing techniques for brain tumour diagnosis from magnetic resonance spectroscopy data* (Doctoral dissertation, Universitat Politècnica de Catalunya).

Author: Carlos Julio Arizmendi
Institute: Universidad Autónoma de Bucaramanga
Street: Avenida 42 -48-11
City: Bucaramanga
Country: Colombia
Email: carizmendi@unab.edu.co

Incidence and risks associated with nasal injury in newborns undergoing non-invasive ventilation through the binasal prong

D.F. Camillo[1], F.S. Barros[1], T.D. Costa[1] and P. Nohama[2]

[1]Universidade Tecnológica Federal do Paraná/ PPGEB, Curitiba, Brasil
[2] Pontifícia Universidade Católica do Paraná/PPGTS, Curitiba, Paraná, Brasil

Abstract— **Nasal injuries due to the use of non-invasive ventilation (NIV) prongs have become increasingly common adverse events in Neonatal Intensive Care Units (NICU). The goal of this study was to evaluate the incidence and risk factors associated with this type of nasal injury in newborns (NB). It was collected data in the patient records, and in the Assistance Management System of Unified Health System (UHS) of NB who have required NIV from 01/01/2013 to 12/31/2014, in a public hospital from the south of Brazil. The parameters analyzed were: gender, birth body mass, Apgar index and length of stay with NIV. In total, 158 NB have used a NIV along that period. The incidence of injury reached 41.14%, and the main responsible variables were: gestational age, body mass at birth and length of stay in the NIV were statistically significant (p <0.05) for the development of the nasal injury. It was concluded that these factors constitute the main risk factors in developing nasal injuries.**

Keywords— **Risk factors, newborn, non-invasive ventilation**

I. Introduction

Prematurity is considered a relevant public health problem due to the high rates of infant mortality [1]. Premature NB are those NB with gestational age less than 37 weeks [2]. Usually, they present risks of developing respiratory complications resulting of the respiratory system immaturity, being necessary in most cases ventilatory support [3].

Non-invasive ventilation is defined as a technique that provides ventilatory support without creating an artificial airway, avoiding possible complications inherent to the presence of endotracheal tube and the use of invasive mechanical ventilation [4].

In general, NIV devices involve a gas source, a pressure generator and the patient interface [5]. In neonatology, the most commonly used devices as interface are the nasal prongs [6]. They are classified as short binasal, simple or nasopharyngeals [4].

The short binasal prongs are more effective to prevent reintubation, to improve oxygenation, for successful weaning and for improving respiratory rate when compared to the exclusive use of nasal and nasopharyngeal prongs. [7].

While nasal prongs have advantages and are the most widely used devices, they can cause nasal injury, an adverse event increasingly present in the NICU [8, 9]. Such injury varies from simple hyperemia of the nasal mucosa, bleeding, crusting, necrosis until total destruction of the nasal septum [9, 10].

The nasal injury can provoke aesthetic and functional consequences and represent a source of discomfort for NB [11]. Because of the painful condition due to nasal injury, it has been observed in increase of blood pressure and intracranial pressure, an increase in the risk of interventricular bleeding and, consequently, alterations in motor development of NB [12].

Thus, considering the presented scenario, the goal of the research described in this paper was to evaluate the incidence and the risk factors associated with nasal injury in NB submitted to NIV through short binasal prongs.

II. Methodology

This is a retrospective, descriptive and quantitative study, accomplished in the NICU of the Waldemar Monastier Public Hospital, localized in southern of Brazil.

The study population consisted of NB admitted to the NICU during the period from 01/01/2013 up to 31/12/2014, requiring non-invasive ventilatory support through short binasal prong.

Data collection was conducted by means of the researching in the patient records at the Assistance Management System of Unified Health System, Brazil, in the period from October 2015 to January 2016.

The following data were analyzed: gender, birth body mass, gestational age, apgar at first and fifth minute of life, length of stay in the NICU, total time of NIV, reason for the NIV institution, necessity to reuse the ventilatory support and presence of nasal injury during the permanence in the NIV.

In addition to these data, we observed in the notes of the patient records, which health professional reported the injury and if there was suspension of NIV or alternation of this support with the oxygen therapy through the oxygenation helmet due to nasal injury.

I. Torres et al. (eds.), *VII Latin American Congress on Biomedical Engineering CLAIB 2016, Bucaramanga, Santander, Colombia, October 26th -28th, 2016*,
IFMBE Proceedings 60,
DOI: 10.1007/978-981-10-4086-3_118

The study was approved by the UTPFR's Research Ethics Committee, CAAE number 42718915.4.0000.5547.

Data were tabulated and submitted to statistical analysis, which was performed using univariate analysis to check the minimum value, maximum value, mean, mode, and standard deviation. To check the sample normality, we applied the Shapiro-Wilk test, and for the variables correlation, the Wilcoxon test, chi-square and Fisher, according to the sample's requirements, considering the confidence limit of 95% with 0.05 significance level.

III. Results

During the research period, 369 NB have been admitted in the NICU studied; of these, 158 have needed NIV support.

Table 1 shows the demographic data of the NB who have required NIV during hospitalization in the NICU.

Table 1 Demographic data of the newborns who have required NIV in the NICU studied (n=158)

Variables	n	%
Gender		
Female	67	42,40
Male	91	57,60
Gestational Age		
Preterm	97	61,39
The term	61	38,61
Body mass at birth		
<1000g	22	13,92
1000 a 1499g	27	17,09
1500 a 2499g	45	28,48
≥ 2500g	64	40,51
Apgar first minute		
0 a 3	34	21,52
4 a 6	38	24,05
7 a 10	81	51,27
Uninformed	5	3,16
Apgar fifth minute		
0 a 3	6	3,80
4 a 6	23	14,56
7 a 10	124	78,48
Uninformed	5	3,16
NIV total time		
Up to 24 hours	24	15,19
From 25 up to 48 hours	34	21,52
From 49 up to 72 hours	23	14,56
More than 72 hours	77	48,73

Among the NIV users, 67 (42.40%) were female and 91 (57.60%) were male. The mean gestational age of the NB studied was 33.91±4.75 weeks and the average body mass at birth was 2142.61±955.47g. About the apgar, the average first minute was 5.90±2.71 and fifth minute was 7.84±2.02. The mean time of NB hospitalization who were treated with NIV was 39.04±24.00 days and average length of stay in this ventilatory support was 139h.

The reasons for the use of the non-invasive ventilator support were: elective post extubation 92 in NB (58.23%), respiratory effort in 23 NB (14.56%), apnea 10 NB (6.33%), fall saturation oxygen in 7 NB (4.43%) and others, in 17 NB (10.76%).

The nasal lesion was observed in 65 NB (41.14%) who required NIV. Table 2 shows the correlation between nasal injury and the characteristics of NB who required NIV: 60 NB (37.97%) needed reuse of the non-invasive ventilatory support at some moment during their hospitalization in the NICU.

Table 2 Relationship between nasal injury and characteristics of newborns who required NIV

Variables	With injury (n = 65)		Without injury (n= 93)		p value
	n	%	n	%	
Gender					
Female	28	41,79	39	58,21	0,43*
Male	37	40,66	54	59,34	
Gestational Age					
Preterm	49	50,52	48	49,48	0,001*
The term	16	26,23	45	73,77	
Body mass at birth					
<1000g	12	54,55	10	45,45	0,003**
1000 a 1499g	20	74,07	7	25,93	
1500 a 2499g	14	31,11	31	68,89	
≥ 2500g	19	29,69	45	70,31	
Apgar first minute					
0 a 3	15	44,12	19	55,88	0,43**
4 a 6	14	36,84	24	63,16	
7 a 10	32	39,51	49	60,49	
Apgar fifth minute					
0 a 3	1	16,67	5	83,33	0,059**
4 a 6	9	39,13	14	60,87	
7 a 10	51	41,13	73	58,87	
NIV total time					
Up to 24 hours	1	4,17	23	95,83	0,04**
From 25 up to 48 hours	6	17,65	28	82,35	
From 49 up to 72 hours	4	17,39	19	82,61	
More than 72 hours	54	70,13	23	29,87	

* Fisher test
** Chi-square test

The appearance of the lesions occurred in 34 NB (52.31%) during the first use of NIV, in 18 NB (27.69%) in the reuse of this support and in 13 NB (20.00%) when the appearance of the lesion occurred in both first use and during reuse.

About the occurrence of nasal injury, 27 (41.54%) were described only by physiotherapists, 21 (32.31%) by nurses and physiotherapists, 2 (3.08%) by physicians and physiotherapists and 15 (23.08%) concomitantly by physicians, nurses and physiotherapists.

Due to the severity of the injury, ventilatory support was suspended in 12 NB (18.46%) who had nasal injury. Of these, 7 NB (10.77%) underwent oxygen therapy through the oxygenation helmet, 1 NB (1.54%) was reintubated and 4 NB (6.15%) alternated periods of NIV and oxygen therapy through the oxygenation helmet, in order to reduce such injury.

IV. Discussion

The nasal injury was found in 41.14% NB who required NIV. Fisher *et al.* [11] evaluated the incidence and severity of nasal lesions in 989 NB with a mean gestational age of 34 weeks and mean body mass of 2142 g at birth, hospitalized in the NICU with NIV for longer than 24 hours. They observed occurrence of nasal injury in a very similar percentage, because resulted in 42.5% of their patients. Comparing the demographic data of NB evaluated in both studies, it was observed that the mean gestational age was similar and both had the same mean body mass at birth.

Bonfim *et al.* [13] evaluated the incidence and risk factors associated with nasal injury in 70 premature newborns who required NIV for longer than 24 hours and found that the incidence of injury reached 62.9% and the length of stay with NIV is a determining factor for the development of nasal injury. The evaluation of only premature NB submitted to the NIV in the study accomplished by Bonfim *et al.* [13] could be the cause of the higher incidence of nasal injury when compared to the present study, which included both premature newborns as newborns with gestational age longer than 37 weeks.

About the risk factors Bonfim *et al.* [13] evaluated the type of prong, gender, birth body mass, gestational age and length of stay in the NIV. And they found that gestational age and length of stay in NIV were determinants in the development of nasal injury. The findings of this study corroborate with Bonfim *et al.* [13], because both gestational age and the length of stay with NIV showed statistically significant values ($p=0.001$ and $p=0.04$ respectively) for developing nasal lesions.

In the present study it was observed that association of gestational age, the length of stay in the NIV and body mass at birth produces significant risk factors ($p < 0.05$) for the development of nasal injury.

The findings of this study corroborate with the results of Primo *et al.* [14] who found that gestational age and body mass at birth are risk factors associated with nasal injury when evaluated 122 NB who used continuous positive airway pressure (CPAP) at an university hospital in Vitória, Brazil.

Fischer *et al.* [11] verified that the shorter the gestational age and body mass at birth, and the higher the residence time in CPAP and in the NICU, greater the severity of nasal lesions resulting from non-invasive ventilatory support.

In the present study, it was not evaluated the severity of the lesion, since in the injuries' reports of the NICU studied, there is no description of the stage in which they were. According to Fischer *et al.* [11], nasal injuries resulting of the use of NIV, characterized initially by persistent hyperemia (grade I), are aggravated to superficial ulceration (grade II) and later to necrosis and loss of the nasal tissue (grade III).

Nasal injury is caused by the pressure applied by the prong on the nostrils, columella, and nasal septum, which induces to a reduction of local blood circulation, with consequent tissue perfusion damage, ischemia development and tissue damage [15].

Due to the immaturity of the integumentary system, premature NB and with low body mass are more susceptible to developing nasal injury [16]. Premature NB have epidermal barrier and immune systems less developed, which facilitates the occurrence of skin lesions when certain areas are compressed [17].

In this study with 65 NB who had nasal injury, 56.92% were preterm with low body mass at birth, showing that injuries using prongs affect more often preterm and low body mass.

Another important factor that showed a statistically significant difference in the development of nasal injury was the length of stay with NIV. It was found that the use of this support for longer than 72 hours is a determining factor for the occurrence of nasal injury.

Souza *et al.* [9] evaluated the prevalence and risk factors in 47 preterm NB undergoing NIV in a NICU of a University Hospital in Recife, Brazil, and identified that premature who used non-invasive ventilatory support for more than 72 hours were more susceptible to develop nasal injury. Nascimento *et al.* [10] verified that the stay in NIV for longer than 48 hours is already a risk factor for developing injuries.

In this study, the main reason to prescribe NIV was its use as support after elective extubation (58.23%). NIV has recommendation level A as ventilatory support for weaning and extubation after use, minimizing extubation failures which may reach levels of 25 to 40% [4].

In this study, the variables gender and Apgar at first and fifth minutes did not constitute significant risk factors for the development of nasal injury, which indicates that there is no association between these variables and the occurrence of injury by the use of NIV through binasal prongs. Most NB had good conditions of vitality both in first minute (51.27%) and in fifth minute (78.48%) of life. Primo *et al.*

[14] and Ota *et al.* [17] also did not find correlation between gender, Apgar at first and fifth minutes and the development of nasal injury.

The nasal injury can limit the use of NIV in NB who need this kind of ventilatory support [14], and also may be the cause of septicemia [18]. In this study, the non-invasive ventilatory support was suspended in 18.46% of the NB who had nasal injury.

The identification of the injury by the multidisciplinary group is of great importance for the development of procedures to prevent and/or minimize nasal injuries. And in the cases whose lesion has occurred, the identification of this problem implies on therapeutic procedures to avoid the deterioration thereof.

It was observed that the registration of the majority of nasal lesion occurrences were done by physiotherapists, probably related to the fact that in the NICU studied, the physiotherapy service provides 24-hour assistance and is the physiotherapist the professional responsible by the NIV's prescription.

According to Jerre *et al.* [19], the physiotherapist establishes and monitors the NIV in the intensive care environment. But this does not reduce the responsibility of the entire multidisciplinary staff to assess early the problems related to the use of NIV, and developing preventive procedures to reduce the incidence and severity of nasal lesions.

V. Conclusion

The gestational age, body mass at birth and length of stay in the NIV are the main risk factors for developing nasal injury in NB who required non-invasive ventilatory support through short binasal prong.

Gender and the Apgar at first and fifth minutes do not seem to be correlated with the development of nasal injury.

From the NICU's multidisciplinary staff assessed, the physiotherapist was the main professional to record the occurrence of nasal injuries.

The results of this study will contribute to the awareness of the multidisciplinary staff about the incidence and risk factors associated with the development of nasal injury, as well as will assist decision-making to carry out procedures to prevent or minimize such injuries and their consequences.

We suggest conducting further studies to evaluate beyond the neonatal risk factors, such as technological and assistance risk factors associated with the development of nasal injury.

Conflict of Interest

The authors declare that they have no conflict of interest.

References

1. Oliveira C H Y, Moran C A (2009) Estudo descritivo: ventilação mecânica não invasiva em recém-nascidos pré-termo com síndrome do desconforto respiratório. ConScientiae Saúde 8: 485-489
2. World Health Organization at http://www.who.int/mediacentre/factsheets/fs363/en/
3. Flenady V J, Gray P H (2002) Chest physical therapy for preventing morbidity in babies being extubated from mechanical ventilation. Cochrane Database Syst Rev 2: DOI: 10.1002/14651858.CD000283
4. Nicolau C M, Andrade L B (2012) Programa de atualização em Fisioterapia Pediátrica e Neonatal: Cardiorrespiratória e Terapia Intensiva. Artmed/Panamericana, Porto Alegre
5. Mahmound R A, Roehr C C, Schmalisch G (2011) Current methods of non-invasive ventilatory support for neonates. Paediatr Respir Rev 12:196-205
6. Sarmento G J V (2011) Princípios e práticas de ventilação mecânica em pediatria e neonatologia. Manole, Barueri
7. Chowdhury O, Wedderburn C J, Duffy D et al. (2012) CPAP review. Eur. J. Pediatr 171:1441-1448
8. McCoskey L (2008) Nursing care guidelines for prevention of nasal breakdown in neonates receiving nasal CPAP. Adv. neonatal care 8:116-124
9. Sousa N F C, Bonfim S F S F, Vasconcelos M G L et al. (2013) Prevalência de lesão do septo nasal em prematuros no uso de prongas nasais. Rev Esc Enferm USP 47:1285-1290
10. Nascimento R M, Ferreira A L C, Coutinho A P F P et al. (2009) The frequency of nasal injury in newborns due to the use of continuous positive airway pressure with prongs. Rev. Latino-Am. Enfermagem 17:489-494
11. Fischer C, Bertelle V, Hohlfeld J et al. (2010) Nasal trauma due to continuous positive airway pressure in neonates. Arch. Dis. Child. Fetal Neonatal Ed 95:447-451
12. Brasil (2013) Atenção à saúde do recém-nascido: guia para os profissionais de saúde. Ministério da Saúde, Brasília
13. Bonfim S F S F, Vasconcelos M G L, Sousa N F C et al. (2014) Lesão de septo nasal em neonatos pré-termo no uso de prongas nasais. Rev. Latino-Am. Enfermagem, 22:826-833
14. Primo C C, Baratela M S, Valladares M L et al. (2014) Fatores de risco associados à lesão nasal por dispositivo de pressão positiva em recém-nascidos. Rev enferm UERJ 22:16-21
15. Cartlidge P (2000) The epidermal barrier. Semin Neonatol 5:273-280
16. Robertson N J, McCarthy L S, Hamilton P A et al. (1996) Nasal deformities resulting from flow driver continuous positive airway pressure. Arch Dis Child 75:209-212
17. Ota N T, Davidson J, Guinsburg R (2013) Lesão Nasal precoce pelo uso da pronga em recém-nascidos prematuros de muito baixo peso: estudo piloto. Rev. Bra. Ter. intensiva. 25:245-250
18. Graham L, Begg M D, Larson E et al. (2006) Risk factors for late onset gram-negative sepsis in low birth weight infants hospitalized in the neonatal intensive care unit. Pediatr Infect Dis J 25:113-117
19. Jerre G, Silva T J, Beraldo M A et al. (2007) Fisioterapia no paciente sob ventilação mecânica. J Bras Pneumol 33:142-15

Study of Medical Device Purchasing Cycles through Temporal Series Analysis

J.C. Guerrero[1], J.H. García[1], A.M. Hernández[1]

[1] Bioinstrumentation and Clinical Engineering Research Group - GIBIC, Bioengineering Department, Engineering Faculty, Universidad de Antioquia UdeA; Calle 70 No. 52-21, Medellín, Colombia

Abstract— **Production of medical devices is subject to economic factors which vary according to the stability of regional markets. Manufacturers, distributors and other market participants do not always reach to commercialize medical products easily because the timing according to the market behavior is not always clear. The study of the buying cycles of the largest medical device markets worldwide using frequency analysis and techniques to identify markets coupling could serve as a guide to plan the interaction with the market or at least to synchronize the production schedule. This article presents an approach to study market behavior through time series that allows to decompose all the information facilitating analysis, using purchasing processes data of medical devices in the United States and the European Community obtained from the FDA and the WHO databases, the data comprises a period from 1996-2015 with a sampling frequency of 1 month. Singular Spectrum Analysis was made to extract the main components of each data set to identify which of the components coupled better both markets. Once the components were identified, a crossed correlation was applied over the reconstructed data to find the possible coupling between markets, separated by trend and oscillatory components. The main finding of this work is that there is no coupling between European and American markets, but both markets have in common purchasing cycles and can be affected by each other by non-immediate changes.**

Keywords— **medical device life cycle, temporal series analysis, singular spectrum analysis**

I. Introduction

There are about 27,000 medical device manufacturers in the world dedicated to develop about 10,000 different categories of products for clinical use, main of these belonging to the US and Europe markets [1]. The demand of devices does not meet the supply costs of these in developing countries mainly because timing and cost issues [2].

Global production of medical devices increased due to scientific and technological sustained industry progress in addition to the demand for quality health services, amounting to 635 billions of dollars each year. From this quantity the participation in production is leaded by north America with 38.7% and the European Union with 20.3% while Latin America has a stake of 1.4% [3], hence the country with the largest medical device market is the United States followed by the European Community. However the device production is subject to economic factors which vary depending on the stability of regional markets.

Manufacturers, distributors and other market participants are not always able to commercialize medical products easily because they do not always make the sale of its technology in times where the health institutions are able to acquire. Technology acquisition depends on the life cycle of the devices, macroeconomic factors, innovation processes and population demand, among others. Because of this it is necessary to study what are the times where most purchases of medical technology are made and its relationship with those factors affecting each region.

The duration of the marketing cycles for medical devices may allow to indicate the optimal time to enter into a market, which is why the focus of this article is to identify the relationship between the two major markets of medical devices, set cycles of purchasing devices to verify if there is any link or synchronization and to identify potential factors that affect them using frequency analysis of purchase data of United States and the European Community and if they can be affected by each other by both a linear and a non-linear way treating data as signals.

II. Methodology

A. Database

From a database generated by the FDA [4] recordings of average purchases of medical technology made by health facilities in the United States in the period from January 1996 to March 2015 were performed. The data were recorded on a monthly basis. Using data from the World Health Organization (WHO) [5] a monthly equal number of samples for the European market that similarly covers a period from 1996 to 2015 was established in order to facilitate comparison.

To find linear coupling each temporal serie was separated into trend, oscillatory and noise components applying singular spectrum analysis (SSA). Then, the correlation coefficient was calculated as linear coupling indexes between both trend and oscillatory components of the behavior of medical technology markets in Europe and the US. The following describes each of the signal processing techniques applied.

I. Torres et al. (eds.), *VII Latin American Congress on Biomedical Engineering CLAIB 2016, Bucaramanga, Santander, Colombia, October 26th -28th, 2016*,
IFMBE Proceedings 60,
DOI: 10.1007/978-981-10-4086-3_119

B. Singular Spectrum Analysis (SSA)

SSA is used to make a decomposition of the original series in the sum of a small number of independent and interpretable components such as a slightly varying trend, oscillatory components.

1. Decomposition:

Is a mapping that transfers series of one-dimensional time to multidimensional through a matrix of trajectories described as a matrix of Henkel in which the rows of size L are written, for the size of the divisions of the time series such that $2 \leq L \leq T/2$.

Once the trajectory matrix is obtained it can be written as $X = X_1 + \cdots + X_d$, and a decomposition is made in 3 different matrixes $X_i = \sqrt{\lambda_i}\, U_i V_i^T$. In the matrix $\sqrt{\lambda_i} V_i^T$ the principal components of the time series are saved [6].

2. Reconstruction:

Reconstruction is based on two main steps, first grouping where the components of interest are selected and diagonal averaging is made by a dehankelization process.

Once obtained all components for reconstruction, the logarithm of the eigenvalues is used to determine components. In the literature it has been reported, different criteria to infer which components correspond to trend, oscillatory and noise [6]. The components are selected until it ceases to be a significant change between the logarithms of the corresponding eigenvalues to each of the components [6], that is, until the slope is flat. According to the theory, the first component refers to the trend and the others correspond to oscillations. Then the reconstruction was applied using the number of components that explained similar percentages of variance.

C. Cross Correlation

The Pearson correlation coefficient was calculated between original time series describing the American and European purchasing. Furthermore, this was calculated between different components obtained from SSA. This coefficient was estimated by (1) [6]:

$$R_{xy}^2 = \frac{C_{xy}^2}{C_{xx}C_{yy}} \qquad (1)$$

Where C represents the corresponding covariance matrix.

III. RESULTS

The behavior of the U.S. and European markets are presented in Fig. 1, where can be observed differences between regions presenting a low Pearson correlation coefficient of 0.35. The American market presents a more dynamic behavior than European market, however it is possible that the dynamics of one market affects the other one. To find this underlying relationship to the original series were decomposed using the SSA technique.

Fig 1. Purchases for American and European markets

Fig.2 shows the logarithm of the explained variance by each of the components found by SSA. In the case of United States, an inflexion point where the slope changes is shown around the 20 component, i.e. the explained variance by following components is low; in the European case the inflexion point is found around the 7 component. Fig. 3 shows the trend series, equivalent to the first component and Fig. 4 shows oscillatory components reconstruction.

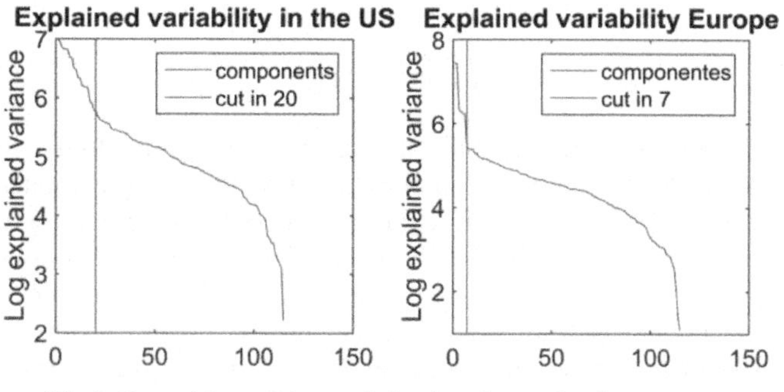

Fig 2. Logarithm of the explained variance for the components found

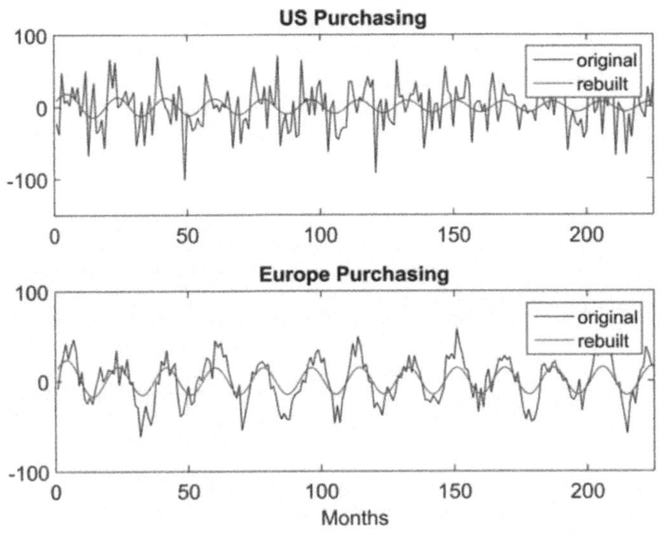

Fig 3. Reconstructed Signals from trend component

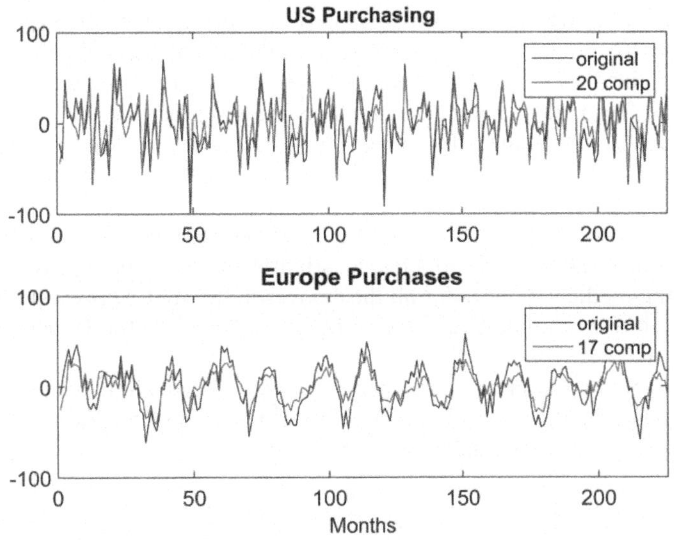

Fig 4. Reconstructed Signals from oscillatory components

The components for the reconstruction of the signal were chosen according to the explained variance. For the American market with 20 components an explained variance of 50% was achieved while for the European market using the inflexion point criteria, an explained variance of 37.2% was achieved with 7 components. To compare both markets 17 components are used on the European market, amount needed to achieve an explained variance of 50%.

As it can be seen in figures 3 and 4, the trend components of each market have high synchrony obtaining a high Pearson correlation coefficient of 0.97, while the oscillatory components present high variability with a correlation coefficient of 0.17.

IV. DISCUSSION

To find the relationship between markets it is necessary to use advanced techniques that allow to decompose the market behaviors in trend and variability and to analyze each of them separately.

Markets behavior shown in Fig 1. relates the differences from each market, while American market has a high variability, the European market looks more predictable, indicating to the naked eye completely different purchasing cycles showing the importance of decomposing the series to obtain the appropriated information. Furthermore, the explained variance of the components of each serie in Fig. 2 shows the level of redundant information that can disturb the analysis of the markets. The components used for this analysis were the first 20 for the American market and the first 17 for the European, indicating the dynamism of purchases in the U.S. while for both markets the oscillatory components did not have a linear relationship.

The trend component shown in Fig. 3 suggests a natural frequency of purchases of the two markets that look similar near to 18 months, being the last peak around month 224 of the series (July of 2014). However this is one component of all that the data sets may have, unable to explain the whole market behavior, while oscillatory components are more likely to rebuild the original series as shown in Fig 4. Thus, the trend component can be described as the behavior markets have at general level, while the oscillatory components can describe the internal behavior of the markets every month.

Data from the United States have a greater variability month to month, which could suggest that the life cycle of the devices is shorter in this market, a possible research topic for the future.

The trend component of both markets is quite similar, however, oscillatory components show a large difference. Thus indicating that there is a relationship between the two markets, even with a purchasing cycle or frequency close to each other's, also the main difference between the two markets are mainly due to regional influences.

With a Pearson coefficient of 0.35 for the original series it is easy to discard any linear relationship between both purchasing cycles, but taking into account that for the trend component data sets present a very high, close to 1 correlation coefficient, while oscillatory components have a low correlation coefficient it results more likely to understand both markets have similar trend. This allows to infer that most of linear coupling for the two sets of data is explained by the trend component, i.e. the behavior of the two markets are explained by the natural frequency of each and not by the behavior market of each month. This meaning that there is no coupling between the variability of the markets, but both markets present similar trend and can be affected by each other, not in immediate times but in larger epochs.

The limitant of this study refers to the fact that the database does not allow to relate health institutions or the type of the medical devices for all the devices in the study, however future work will focus on specific types of devices for deeper analysis.

Similar studies can be obtained with other techniques of signal analysis which might help for this kind of studies relating the delay time between both markets

V. CONCLUSIONS

European and the American markets are affected in a linear way, having the same trend near to 18. Monthly behavior is ruled by oscillatory components which in this case have no linear coupling, meaning that both markets are not governed by the same internal processes, thus the amount of purchases of medical devices in various markets does not depend solely on the market size, but also by demographic need for more specialized services, and for the market system itself which can give a lesser life cycle of medical devices.

The cycles of the two major markets of medical devices have common purchasing cycles, important information for any manufacturer, supplier and importer of technology, which is why it is necessary to find the relationship between markets that establish the best epochs to participate actively in a specific economic process.

There is a relationship between the two markets studied, first they share purchasing frequency, and the relationship of the two markets is linear which allows to recognize the relationship with simplicity, suggesting signal analysis as an important tool that might help to study market behavior for medical devices.

ACKNOWLEDGMENT

This work was supported by Sistema General de Regalías, Republic of Colombia through the project "Fortalecimiento de la plataforma tecnológica para la formación especializada en el área de la salud y el desarrollo de tecnología biomédica", code RutaN-139C.

CONFLICT OF INTEREST

The authors declare that there is no conflict of interest regarding the publication of this paper.

REFERENCES

1. MDDI, "Top 40 Medical Device Companies" at http://www.mddionline.com/article/top-40-medical-device-companies
2. C. L. Ventola (2008) Challenges in evaluating and standardizing medical devices in health care facilities. *P T*, vol. 33, no. 6, pp. 348–359.
3. Obregón, A. La Industria de los dispositivos Médicos. http://mim.promexico.gob.mx/work/sites/mim/resources/Local-Content/68/3/130815_DS_Dispositivos_Medicos_ES.pdf
4. FDA, 510 (k) Clearances at: http://www.fda.gov/MedicalDevices/ProductsandMedicalProcedures/DeviceApprovalsandClearances/510kClearances/ucm089428.htm
5. World Health Organization, et al. (2012). Dispositivos médicos: la gestión de la discordancia: un resultado del proyecto sobre dispositivos médicos prioritarios.
6. Hassani, H. (2007). Singular spectrum analysis: methodology and comparison. Journal of Data Science, 5(2), 239-257.

Author: A.M. Hernández

Institute: University of Antioquia

Street: Calle 67 No. 53 - 108

City: Medellín

Country: Colombia

Email: alher.hernandez@udea.edu.co

Creation of the Biomedical Systems Engineering undergraduate program, School of Engineering, UNAM

J.M. Dorador-Gonzalez [1], L. Báez-Rivas[2], L. Gonzalez-Gonzalez[3], M. Guillen-Mandujano[4]

[1,2,3,4] School of Engineering, National Autonomous University of Mexico (UNAM)
[1] Corresponding author: dorador@unam.mx

Abstract— **The proposal for the creation of the Biomedical Systems Engineering undergraduate program is presented by the School of Engineering and the School of Medicine of the National Autonomous University of Mexico (UNAM).**

This Engineering degree covers three main areas: Biomechanics, Biomedical Instrumentation and Hospital Logistics. It is originated by the experiences obtained in research and technology development conducted by the School of Engineering during the past twenty years.

The proposed curriculum is planned to be fulfilled within ten semesters, with a total of 430 credits or units, and a 3936 hours pensum. Out of the 54 mandatory courses, 8 would be imparted by the School of Medicine and, during the ninth semester, there is only one course denominated *Internship,* which would consist on the students' practical instruction in a full-time stay either at a secondary or tertiary Healthcare Center or a National Healthcare Institute.

Key words—**Biomedical Systems, academic plan.**

I. INTRODUCTION

Taking into consideration that a great deal of the research and technological innovation around the world is developed within the medical scope. And that the population's health, the amount of medical infrastructure, its level of innovation and development, along with the access to health service provided to the entire population, are some of the main indicators of a country's development, the School of Engineering at UNAM identified the need of creating an undergraduate program in Biomedical Systems Engineering.

The concept of Biomedical Systems Engineering differs from the concept of Biomedical Engineering primarily because the first does not include the aspects of cellular and tissue biology and engineering, that represent one of the main study areas of Biomedical Engineering in the world.

Biomedical Systems Engineering will play an important role within the National Health Sector (public and private), and will promote the technological development of medical devices, alongside with the incorporation of hospital planning, logistics and administration. According to professional profile, the engineer in Biomedical Systems will be able to design, implant, maintain and develop technology within the medical-hospital, industrial and service environments.

Likewise, it will have repercussions on the Industrial Sector (biomedical technology industry), where the generation of companies that would aim to satisfy those necessities currently fulfilled through imports [1] will be fostered.

II. BACKGROUND

In the early 20th century, engineers designed and built equipment for the hospitals, but they were not in charge of operating or maintaining them. The personnel who was in charge of such equipment were technicians, electricians or medical personnel who proved to have ability with the machines. Over time, the hospital equipment became more specialized and complex and the personnel in charge required deeper training in order to operate it. By the middle of the 20th Century, the medical equipment field was so specialized that the ones who operated were awarded with the degree of *Biomedical Equipment Technician, BMET.* Since then, the existence of a professional who is in charge of not only operating and giving maintenance to medical equipment, but also designing it has been proposed [3]. Nowadays, bachelor and technical degrees and study plans regarding the link between Engineering and medical equipment are commonly found among diverse institutions worldwide.

In 1965, the Mexican Institute of Social Security (IMSS) founded the first department dedicated to the maintenance and service of medical equipment in Mexico, denominated Office of Conservation of Medical Equipment. Under the direction of Miguel E. Schulz, a work team that consisted mainly of electronic engineers [2] was formed.

In 1973, the Iberoamerican University (Universidad Iberoamericana) provided, at first, an option of specialization in Biomedical Engineering as terminal branch of the last year of the degree in Electronic Engineering, and the degree on biomedical engineer afterwards [2].

In 1976, Enrique Hernández Matos, who returned from Germany with a diploma in Biomedical Engineering, founded, along with several collaborating doctors, the first Biomedical Engineering Department at the November 20th National Medical Center (Centro Médico Nacional 20 de Noviembre), in Mexico City. Being this the first time that a hospital in Mexico counted with biomedical engineers within its personnel. By the late sixties, all public hospitals had followed its example, as later did the private hospitals [2].

I. Torres et al. (eds.), *VII Latin American Congress on Biomedical Engineering CLAIB 2016, Bucaramanga, Santander, Colombia, October 26th -28th, 2016,*
IFMBE Proceedings 60,
DOI: 10.1007/978-981-10-4086-3_120

III. PROPOSAL OF THE CREATION OF THE DEGREE ON BIOMEDICAL SYSTEMS ENGINEERING

Background in Biomedical Systems Engineering in UNAM

The UNAM has several institutes and schools that promote the generation of projects and research areas for research and technological development. The Institutes of Physics, Biomedical Research and Cellular Physiology, to name some, count with lines of investigation such as biomedical instrumentation, molecular engineering of biomaterials (biopolymers), image processing, micromechanics for medical technologies, synthesis and characterization of biomaterials, among others. The Schools of Engineering, Sciences, Laws, Administration and Medicine, together, sum up to over 40 projects, whose main subjects are biomechanics, biomaterials, biomechatronics, biomedical instrumentation, medical informatics, imaging, hospital logistics, design of medical equipment, bioethics, regulation and normativity of health sector, among others.

In the School of Engineering, there has been work on projects related to the Biomedical Systems Engineering for over twenty years. Some examples are: the design of a gastric drainage pump, a microcomputer-controlled sterilizer for hospital use, a system for manufacturing surgical gloves, an infrapatellar prosthesis, an artificial respiratory system for intensive care, a baby incubator for intensive care, designed by the Center of Mechanical Design and Technological Innovation (CDMIT) and the Mechatronics Engineering Department. The School of Engineering in UNAM has also with the lines of investigation and technological development in engineering for welfare and design of smart mechatronic prosthetics. The Unit of Research and Technical Assistance in Materials (UDIATEM) has worked on the design of hip, knee and spine implants, and counts with research lines on the characterization of biomaterials and gait biomechanics. The Electrical Engineering Division has worked on projects within the medical imaging, biologic signals processing and sensor characterization areas. Additionally, since 2001, the careers of Electric-Electronic and Computer Engineering count with a terminal modulus on Biomedical Engineering, and, since November of 2005, there is a student chapter of Biomedical Engineering, associated to the Engineering in Medicine and Biology Society (EMBS) and the Institute of Electrical and Electronics Engineers (IEEE) [5].

Current Status and trends of the disciplines involved in the curriculum [1]

In México, there is an increase on the demand of engineers capable of designing and improving the diverse devices and processes employed by hospitals for the diagnosis, treatment and rehabilitation of the diverse ailments that attack the population. The areas that demand professionals with the profile of an engineer on Biomedical Systems are: services, research and development, shown in Figure 1.

Fig 1: Areas that demand engineers in Biomedical Systems

Career objectives [1]

The objective of the undergraduate degree on Biomedical Systems Engineering is to prepare engineers with solid knowledge and abilities on the principles of Engineering in the areas of biomechanics, hospital logistics and biomedical instrumentation, in order to integrate and apply them within the biomedical area, according to the professional profile.

Knowledge areas

The Biomedical Systems Engineering program has three knowledge areas. The student will take the courses of basic sciences and mathematics, engineering sciences and applied engineering according to the chosen area. The specialization areas are Hospital Logistics, Biomechanics and Biomedical Instrumentation.

The graduate from the Hospital Logistics area will be able to manage and administrate in an efficient and effective way medical units that provide health and hospital care, ensuring better quality in satisfying the specific needs of both the medical units and the users, with a competitive, productive and human approach.

The graduate from the Biomedical Instrumentation area will be the professional specialized in the design, integration, operation, maintenance, improvement and generation of equipment for biomedical applications; as well as the acquisition, conditioning, processing and monitoring of the human body biological signals.

The graduate from the Biomechanics area will be able to design, improve and create technology for medical assistance, such as prosthetic devices, medical instruments, hospital equipment and rehabilitation appliances, among others, and will have the capacity to improve the means of work, based on the study of anatomic, physiological and technological data, applied to the resolution of adaptation problems between humans and their environment, as well as promoting the optimal development of human activity. The graduate from the Biomechanics area will also do research within the areas of biomaterials, bio-thermofluids, and anthropometric analysis, to be applied on the development of biomedical systems.

Structure of the curriculum [1]

The curriculum structure of the degree offered by the School of Engineering, includes the formation of the students on five main areas, including courses of: Basic Sciences and Mathematics, Social Sciences and Humanities, Engineering Sciences, Applied Engineering and specific subjects of Biomedical Systems Engineering Sciences, as well as additional subjects known as Other Courses. The proportional credit distribution for each area is shown in Table 1.

Table 1: Credits by area

Classification	Percentage of credits from the curriculum total	Compulsory credits	Elective credits
Basic Sciences	30.2%	122	8
Social Sciences and Humanities	6.5%	22	6
Engineering Sciences	27.9%	64	56
Applied Engineering	14%	28	32
Biomedical Systems Eng. Sciences	17.2%	66	8
Other courses	4.2%	18	0

Students will have to complete at least 42 credits from subjects of Biomedical Systems Engineering Sciences, according to the chosen area (Biomechanics, Biomedical Instrumentation or Hospital Logistics), and choose subjects of Biomedical Systems Engineering Sciences from the other two areas, in order to complete at least 60 credits.

The students will have to course elective subjects of Applied Engineering in Biomedical Systems, until they complete at least 32 credits. The structure of the curriculum includes subjects focused on design, bioethics, regulatory legal framework and logistics, as well as a research internship.

A subject named Stay (internship), is included on the 9th semester. This course will be taught in a Hospital or National Health Institute. In order to take this course, the student must:

- Have completed at least 80% of the total credits.
- Not course another subject during the same semester of the stay.
- Stick to a schedule of 8 hours a day in a second or third level hospital or national health institute, as well as private hospitals.
- Have a project to work on.

IV. DISCUSSION

At the national level, Biomedical Engineering has been developing predominantly within the instrumentation area; some universities have entered the logistics area and development of research competence, and a scarce intervention within the areas of biomechanics and biomaterials can also be observed. It should be noted that there are no national programs that address the area of medical informatics, even though it is one of the major national and international needs.

In the case of the proposal of bachelor's degree in Biomedical Systems Engineering, there is a balance among the areas of instrumentation, biomechanics and logistics, the core subjects of the curriculum cover the profile of biomedical engineer currently offered by other national universities. It is important to note that although the instrumentation area is competitive, the areas of biomechanics and logistics are the ones that add a particular stamp to this program and distinguish it from the national supply.

V. CONCLUSION

Biomedical Systems Engineering does not intend to replace other programs on Biomedical Engineering, but contribute on the solution of national and international needs regarding the biomedical-hospital environment, within the Biomechanics, Biomedical Instrumentation and Hospital Logistics areas, capitalizing the extensive experience that the UNAM already has in the area, in order to provide a holistic and comprehensive formation of professionals who will be able to perform in multi and transdisciplinary environments.

This undergraduate program started in 2015 and nowadays we have students coursing the first, third and fifth semesters, due to the fact that the students started it by changing the program in which they were enrolled. The students were enrolled in the Mechanical, Industrial, Computing and Electric Engineering programs.

This paper was supported with the DGAPA PAPIIT IT101915 project.

REFERENCES

1. *Proyecto de creación del plan de estudios de la licenciatura en Ingeniería en Sistemas Biomédicos*, School of Engineering, National Autonomous University of Mexico (UNAM), Mexico, 2014
2. Cadena, Mendez M., Azpiroz Leehan J., "Overview of the Biomedical Engineering History in Mexico, a Personal Point of View", Autonomous Metropolitan University-Iztapalapa (UAM-I), *Proceedings of the 25' Annual International Conference of the IEEE EMBS Cancun, Mexico*, September 17-21, 2003.
3. Street, Laurence J. *Introduction to biomedical engineering technology*. CRC Press. USA, 2008
4. (*Plan de Desarrollo 2012* [Digital version], National Autonomous University of Mexico (UNAM). Consulted on April 20th, 2012 at: http://www.planeacion.unam.mx/consulta/Plan_desarrollo.pdf
5. Gomez J., Vega A. *BME Education Program and Research Activities at the National Autonomous University of Mexico (UNAM)*, National Autonomous University of Mexico, 32nd Annual International Conference of the EEE EMBS Buenos Aires, Argentina, August 31 - September 4, 2010.

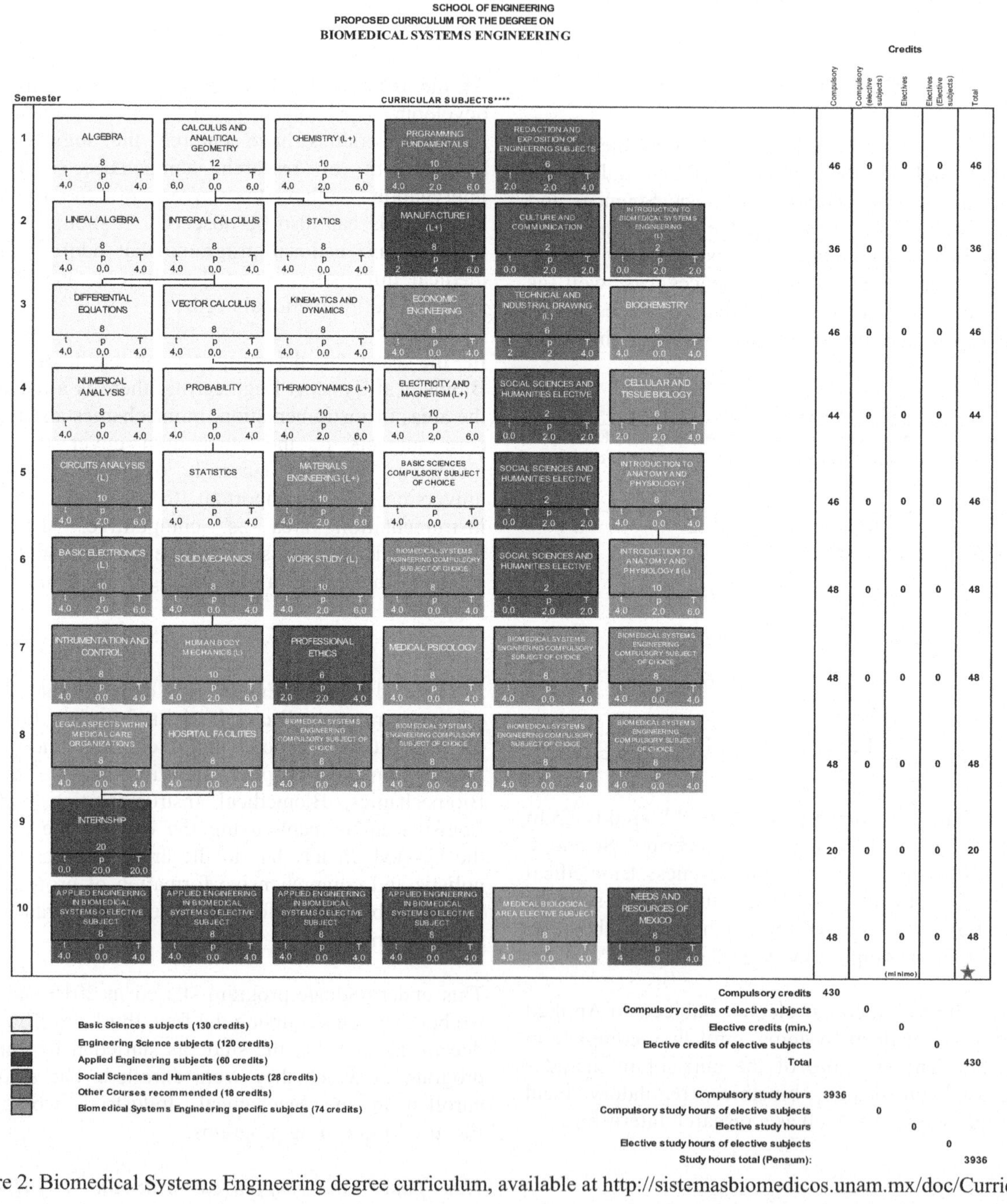

Figure 2: Biomedical Systems Engineering degree curriculum, available at http://sistemasbiomedicos.unam.mx/doc/Curricula.pdf

Analysis of the respiratory flow signal for the diagnosis of patients with chronic heart failure using artificial intelligence techniques

J.C. Rodríguez[1], C.J. Arizmendi[1], C.A. Forero[1], S.K. Lopez[2], and B.F. Giraldo[3,4,5]

[1] Universidad Autónoma de Bucaramanga (UNAB)/Dept. Ingenierías físico-mecánicas, Ingeniería Mecatrónica, Bucaramanga, Colombia
[2]Clinica Materno Infantil, Bucaramanga, Colombia
[3]Universitat Politècnica de Catalunya, Dept. ESAII, Barcelona, Spain
[4]Institute for Bioengineering of Catalonia, Barcelona, Spain
[5] CIBER de Bioingeniería, Biomateriales y Nanomedicina (CIBER-BBN), Madrid, Spain

Abstract— **Patients with Chronic Heart Failure (CHF) often develop oscillatory breathing patterns. This work proposes the characterization of respiratory pattern by Wavelet Transform (WT) technique to identify Periodic Breathing pattern (PB) and Non-Periodic Breathing pattern (nPB) through the respiratory flow signal. A total of 62 subjects were analyzed: 27 CHF patients and 35 healthy subjects. Respiratory time series were extracted, and statistical methods were applied to obtain the most relevant information to classify patients. Support Vector Machine (SVM) were applied using forward selection technique to discriminate patients, considering four kernel functions. Differences between these parameters are assessed by investigating the following four classification issues: healthy subjects versus CHF patients, PB versus nPB patients, PB patients versus healthy subjects, and nPB patients versus healthy subjects. The results are presented in terms of average accuracy for each kernel function, and comparison groups.**

Keywords— **Support vector machine, forward selection, chronic heart failure, periodic breathing, non-periodic breathing.**

I. Introduction

CHF is a serious public health problem for its increasing prevalence and the high number of hospital admissions and mortality [1]. This disease is associated with major abnormalities of autonomic cardiovascular control, and is characterized by enhanced sympathetic nerve activity and cardiorespiratory disarrangement. It is also a growing problem and its progressive increase is attributed to an aging population [2]. CHF patients often develop breathing anomalies such as various forms of oscillatory breathing patterns characterized by rises and falls in ventilation, called PB. Cheyne-Stokes respiration (CSR) is a more severe form of a PB pattern in which apneas and hypopneas alternate with repetitive gradual increases and subsequent decreases in ventilation [3]. PB has a prevalence as high as 70% in CHF patients [4], and is associated with increased mortality [5], especially in CSR patients [6]. It is therefore crucial to establish accurate risk stratification of CHF patients so as to optimize the allocation of limited resources. The number of available treatment options have increased, but this increase has rendered clinical decision making far more complex.

The respiratory pattern can be characterize by the following time series: inspiratory time (T_I), expiratory time (T_E), breathing cycle duration (T_{Tot}), tidal volume (V_T), inspiratory fraction (T_I/T_{Tot}), half inspired flow (V_T/T_I), and rapid shallow index (Fr/V_T), where f is respiratory rate. The analysis of respiratory pattern variability has been studied through the approximation and detail coefficients extracted from WT technique. In this study, we analyze the groups of healthy subjects and CHF, PB and non-periodic breathing (nPB) patients. CSR patients are considering into PB group. SVM classification method is used to extract the best characteristics when comparing each two groups. In this work, we present the best results using the functions kernel: linear, quadratic, third degree polynomial, and Gaussian Radial Basis function (RBF).

II. Material and Methods

A. Databases

Respiratory flow signal were recorded from 27 CHF patients (7 female, 65 ± 9 years, respiratory frequency 19.6 ± 3.4 breaths/min) and 35 healthy volunteers subjects (23 female, 27 ± 7 years, 15.5 ± 3.7 breaths/min) at Santa Creu i Sant Pau Hospital, Barcelona, Spain. All subjects were studied according to a protocol approved by the local Ethics Committee. The respiratory flow signals were acquired using a pneumotachograph, consisting of a Datex-Ohmeda monitor with a Validyne Model MP45-1-871 Variable-Reluctance Transducer. The pneumotachograph was connected to a mask. The signals were recorded at 250-Hz sampling rate and 12-bit resolution, during 30 min. The classification of the CHF patients, PB, nPB, and CSR was developed by the medical doctors at the Santa Creu i Sant Pau Hospital, Barcelona, Spain.

I. Torres et al. (eds.), *VII Latin American Congress on Biomedical Engineering CLAIB 2016, Bucaramanga, Santander, Colombia, October 26th -28th, 2016*,
IFMBE Proceedings 60,
DOI: 10.1007/978-981-10-4086-3_121

The CHF patients were classified into two groups: 8 patients with PB patterns (1 female, 71 ± 7 years, 18.4 ± 2.2 breaths/min), and 19 patients with nPB pattern (5 female, 62 ± 9 years, 22.5 ± 4.3 breaths/min). Within the PB group, three patients were classified as CSR (1 female, 68 ± 6 years, 21.7 ± 4.2 breaths/min) and five patients as PB without apnea (no females, 73 ± 8 years, 23.0 ± 4.7 breaths/min). Fig. 1 illustrates the different flow patterns observed in CSR, PB, nPB patients, and a healthy subject. For 62 subjects, 7 respiratory time series were extracted (Table 1).

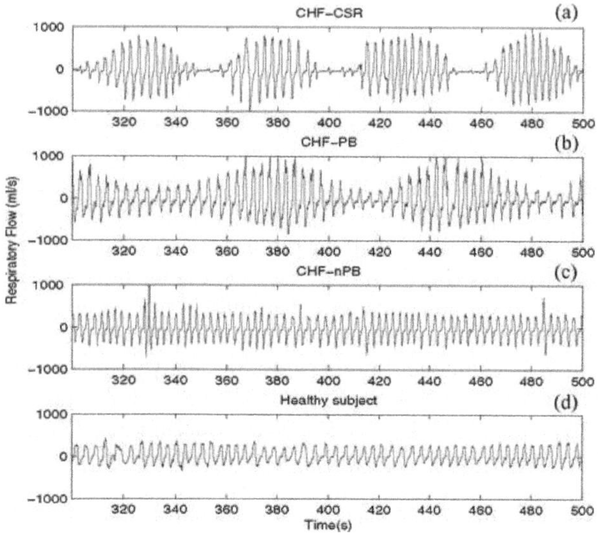

Fig. 1. Respiratory flow signal exemplified for (a) CSR patient, (b) a PB patient, (c) a nPB patient, and (d) a healthy subject.

Table 1. Respiratory time series

Serie	Description
T_{Tot}	Breathing cycle duration
T_I	Inspiratory time
T_E	Expiratory time
V_T	Tidal volume
T_I / T_{Tot}	Inspiratory fraction
V_T / T_I	Half inspired flow
Fr	Respiratory rate
Fr / V_T	Rapid shallow index

B. Wavelet Transform

Continuous Wavelet Transform (WT) of a signal x(t) with mother wavelet $\psi(\cdot)$ is defined as (1):

$$W(\tau, s) = \frac{1}{\sqrt{|s|}} \int_{-\infty}^{\infty} x(t)\psi\left(\frac{t-\tau}{s}\right) dt \tag{1}$$

being the transformed signal $W(\tau, s)$ a function of the translation parameter τ and the scale s. The signal energy is normalized at every scale by dividing the wavelet coefficients by $1/\sqrt{|s|}$.

The original signal can be reconstructed with the inverse continuous WT, defined by (2):

$$x(t) = \frac{1}{C_\psi^2} \int_{-\infty}^{\infty} \int_{-\infty}^{\infty} W(\tau, s) \frac{1}{s^2} \psi\left(\frac{t-\tau}{s}\right) d\tau ds\,, \tag{2}$$

with

$$C_\psi = \int_0^\infty \frac{|\psi(f)|^2}{f} df < \infty.$$

An important development for the application of Wavelet theory in discrete signal processing was presented in [7], using Multiresolution Analysis (MA). The Discrete Wavelet Transform (DWT) is implemented via an octave filter bank, as a cascade of low-pass L(z) and high-pass H(z) filters, followed by sub-sampling. Every pair of filters represents a decomposition level. The reconstruction of the original signal is possible using the synthesis filter bank where the signals are up-sampled and passed through the filters L'(z) and H'(z) [7].

The DWT is a process widely used in problems with temporal series and biomedical signals [8]. In [9] the DWT is applied to the original signal data, making the decomposition to the fifth level. In this study, DWT it was implemented with different mother wavelets, ranging from different orders of Biorthogonal (1.1, 1.3, 1.5, 2.2, 2.4, 2.6, 2.8, 3.1, 3.3, 3.5, 3.7, 3.9, 4.4, 5.5, 6, 8), Coiflet (1 to 5), Daubechies (1 to 43), and Symlet (1 to 25). For every mother wavelet, the absolute values of each decomposition coefficient were sorted in descending order, and the signal of each spectrum was reconstructed by adding consecutive coefficients. The average Mean Square Error (MSE) and Signal-to-Noise Ratio (SNR) were calculated over the whole set of patients for each Wavelet order r, together with the Number of Decomposition Coefficients (NDC). Finally, Q1 the index was computed as follows (3) [10, 11]:

$$Q1 = \frac{SNR}{MSE+NDC} \tag{3}$$

C. Support Vector Machine Classifier

SVM are supervised learning methods effective in high dimensional spaces that provide a good out-of-sample generalization even when the training sample has some bias [8]. SVM is a non-probabilistic binary linear classifier that divides the feature space in a hyperplane that can be defined by a Kernel function [9]. By introducing the kernel, SVMs gain flexibility in the choice of the form of the hyperplane separating two different sets of data, as it contains a non-linear transformation no assumptions are made. In this study four different kernel functions are used: linear, quadratic, third degree polynomial, and Gaussian Radial Basis function (RBF). These functions were applied using MATLAB toolbox ®. Forward selection technique was used to evaluate the parameters extracted from each classification.

D. Parameters used from classification

The parameters used are obtained from the time series of the respiratory flow signal, characterized by statistical values as mean (M), standard deviation (Std), kurtosis (K), interquartile range (IQR), and variance (V). Differences between groups are tested by Mann-Whitney *U* test, and a p value < 0.05 was considered statistically significant.

For each comparison groups, 150 run and 4-fold cross-validation were performed. For each kernel function and parameters extracted, the average classification correctness was obtained. The accuracy of predicting the different groups of subjects is evaluated for four classifications: healthy subjects versus CHF patients, PB versus nPB patients, PB patients versus healthy subjects, and nPB patients versus healthy subjects.

III. Results

First, a statistical analysis of the time series is performed. For 62 subjects, total of 35 parameters were extracted (M, Std, K, IQR and V from T_I, T_E, T_{Tot}, V_T, T_I/T_{Tot}, V_T/T_I, Fr/V_T). Tables 2 to 5 illustrates the average accuracy obtained for each classification when SVM classification technique is applied.

In the second part, when applied Wavelet processing, Daubechies 1 was the optimal Wavelet with the highest Q1 index. After, DWT method was computed to all parameters, and Approximation Coefficients (AC) and Detail Coefficients (DC) for five levels of decomposition were obtained (a total of 35 parameters were obtained for each coefficient). The best results were obtained with the AC parameters ($p < 0.05$). Tables 6 to 9 illustrates the average accuracy obtained with the best parameters in each case, considering different kernel functions applying SVM classification method and Forward Selection for dimensionality reduction.

Table 2. Average accuracy when comparing healthy subjects versus CHF patients using different kernels functions to SVM classification

Kernel	Accuracy (%)
Linear	96.3 ± 2.2
Quadratic	85.6 ± 3.3
Third degree polynomial	92.7 ± 2.0
RBF	68.8 ± 2.0

Table 3. Average accuracy when comparing PB versus nPB patients using different kernels functions to SVM classification

Kernel	Accuracy (%)
Linear	94.9 ± 3.7
Quadratic	71.3 ± 4.9
Third degree polynomial	79.8 ± 2.8
RBF	70.2 ± 0.0

Table 4. Average accuracy when comparing PB patients versus healthy subjects using different kernels functions to SVM classification

Kernel	Accuracy (%)
Linear	95.4 ± 2.6
Quadratic	92.8 ± 3.2
Third degree polynomial	96.8 ± 1.5
RBF	81.4 ± 0.0

Table 5. Average accuracy when comparing nPB patients versus healthy subjects using different kernels functions to SVM classification

Kernel	Accuracy (%)
Linear	95.4 ± 2.3
Quadratic	88.8 ± 3.2
Third degree polynomial	92.8 ± 1.9
RBF	72.0 ± 2.1

Table 6. Average accuracy of the best parameters obtained with AC coefficients, when comparing healthy subjects versus CHF patients using different kernels functions to SVM classification

Kernel	Parameters	Accuracy (%)
Linear	Std AC5-V_T/T_I M AC1-T_I	100.0
Quadratic	M AC2 -V_T/T_I M AC1 - T_I / T_{Tot}	100.0
Third degree polynomial	M AC2 - VT/TI Std AC4- TI	100.0
RBF	Std AC5-VT/TI M AC2 - T_I / T_{Tot}	100.0

Table 7. Average accuracy of the best parameters obtained with AC coefficients, when comparing PB versus nPB patients using different kernels functions to SVM classification

Kernel	Parameters	Accuracy (%)
Linear	IQR AC1 - Fr/V_T IQR AC1 - T_I IQR AC1 - V_T	100.0
Quadratic	IQR AC1 - Fr/V_T IQR AC1 - T_I IQR AC2 - V_T	98.8 ± 1.2
Third degree polynomial	IQR AC1 - Fr/V_T K AC2 - V_T V AC2 - T_I	99.4 ± 1.4
RBF	IQR AC1 - Fr/V_T IQR AC1 - T_I IQR AC2 - Fr/V_T	98.6 ± 1.4

Using the MWVA technique to preselect the variables before implementing FS, the best overall classification percentage of 93.87% was obtained with 42 variables with the lower the computational cost in all the experiments.

ACKNOWLEDGMENT

The authors would like to thank to Doctors of Cardiac Unit and his collaborators of the Santa Creu I Sant Pau Hospital, Barcelona, Spain, by their collaboration in the signal database acquisition. This work was supported in part by the Secretariat of Universities and Research of the Department of Economy and Knowledge of the Government of Catalonia (Consolidated research group GRC 2014 SGR 1569) and by the Ministry of Economy and Competitiveness of the Spanish Government (DPI2015-68820-R).

CONFLICT OF INTEREST

The authors reports no conflicts of interests.

VI. REFERENCES

[1] Preciado J. A. C, and Giraldo Giraldo B (2011). Análisis y clasificación del patrón respiratorio de pacientes en proceso de retirada del ventilador mecánico, *Ingeniería Biomédica,* vol. 5, nº 1909-9762, p. 7,.

[2] Brunel, H., Vallverdú, M., Perera, A., Caminal, P., Voss, A., Cygankiewicz, I., ... & de Luna, A. B (2011). Estudio de la complejidad de la variabilidad del ritmo cardiaco en pacientes con cardiomiopatia dilatada.

[3] Hadamard J. S (1898). *Les surfaces à courbures opposées et leur lignes geodésiques,* vol. 5e série, Paris: Journal de mathématiques pures et appliquées,.

[4] Voss A, (1996). T*he application of methods of non-linear dynamics for the improved and predictive recognition of patients threatened by sudden cardiac death,* vol. 31, Berlin: Cardiovascular research Elsevier.

[5] Yazlle J, and Jadur C (2006). Dinámica Simbólica: Apuntes de Cátedra, Salta: Departamento de Matemática de la Universidad Nacional de Salta.

[6] Arizmendi C, Romero E, Sierra D and Vellido A (2014). Automated Classification of Brain Tumours from Short Echo Time in Vivo MRS Data Using Gaussian Decomposition and Bayesian Neural Networks. *Expert Systems with Applications,* vol. 11, pp. 5296-5307.

[7] Betancourt, G. A. (2005). Las máquinas de soporte vectorial (SVMs). *Scientia et Technica, 1*(27).

[8] Pandit, D. (2010). Respiratory weaning. In *Core Topics in Critical Care Medicine*(pp. 241-245). Cambridge University Press New York, NY.

[9] González-Acevedo, H., Arizmendi-Pereira, C. J., & Giraldo-Giraldo, B. (2015). Classifier design for patients on weaning process. *Iteckne, 12*(2), 131-137..

[10] Arizmendi, C., Hernández-Tamames, J., Romero, E., Vellido, A., & del Pozo, F (2010). Diagnosis of brain tumours from magnetic resonance spectroscopy using wavelets and Neural Networks. In *2010 Annual International Conference of the IEEE Engineering in Medicine and Biology* (pp. 6074-6077). IEEE..

[11] Arizmendi C (2012). *Signal processing techniques for brain tumour diagnosis from magnetic resonance spectroscopy data* (Doctoral dissertation, Universitat Politècnica de Catalunya).

Author: Carlos Julio Arizmendi
Institute: Universidad Autónoma de Bucaramanga
Street: Avenida 42 -48-11
City: Bucaramanga
Country: Colombia
Email: carizmendi@unab.edu.co

Sudden Cardiac Death Prediction Based On a Nonlinear Estimation

R. Urda Benitez[1] and A. Orozco Duque[1]

[1] Grupo de Investigación e Innovación Biomédica, Instituto Tecnológico Metropolitano, Medellin, Colombia.

Abstract— **Sudden Cardiac Death (SCD) associated with ventricular tachyarrhythmia, is one of the main causes of death worldwide. The Left Ventricle Ejection Fraction (LVEF) is the predictive parameter used in clinical practice for the stratification of risk of SCD. However, this has a poor predictive value. Several authors have proposed methods seeking to estimate characteristics that can be used as predictors of SCD from an analysis of the nonlinear dynamics of the signal of heart rate variability (HRV). These authors have shown the great potential of fractal analysis for this purpose. In this article, we worked under the hypothesis that if there is an underlying non-linear dynamic to the HRV signal, this dynamic should be best described by the multifractal analysis, which by fractal indices. Therefore, a comparison of fractal and multifractal features for SCD prediction was made and implemented a classifier that will combine this type of characteristics. The results show that h-fluctuacion multifractal index shows a 94.44% sensitivity and 87.50% of specificity compared to the exponents of scale with Detrend Fluctuation Analysis (DFA) an 88.89% and 87.50% respectively. Combining these two features as input for a Support Vector Machine (SVM) is achieved an accuracy of the 97.06%, 100% sensitivity and specificity 94.44%, surpassing the results that are reported in the literature so far.**

Keywords— **Detrend Fluctuation Analysis, Hearth Rate Variability, Multifractal, Support Vector Machine.**

I INTRODUCTION

The Sudden Cardiac Death (SCD), is one of the leading causes of death worldwide, is related to ventricular tachyarrhythmias such as ventricular fibrillation (VF) and ventricular tachycardia (TV) [1, 2]. The relationship between Left Ventricle Ejection Fraction (LVEF) and arrhythmic events that are tied to the SCD has been studied and evaluated, obtaining 56-71% sensitivity, 74-83% specificity, 11-22% positive predictive value and negative predictive value 96-98% [3, 4]. Therefore, given the low positive predictive value of LVEF, this isolated test is insufficient to detect arrhythmic high-risk patients leading to SCD, so it is necessary to combine it with other diagnostic tests, such as electrocardiographic features removed by non-linear methods that correlated if they can become an independent predictor of SCD [1, 5].

Some recent studies show advances in the prediction of SCD, four minutes prior to the event of VF. In 2014, Elias [6] with temporal characteristics, frequency and non-linear as the exponents of Hurst and Poincare plot, got an accuracy of 83.93% and a sensitivity of 83.75% prediction, using neural networks and K-Nearest Neighbors (KNN). In the year 2015 Acharya et al. used data bases whit SCD and Normal Sinus Rhythm (NSR), estimating non-linear features, such as the fractal dimension, the exponents of Hurst, correlation dimension, approximate entropy and sample entropy and energy of the coefficient of the Discrete Wavelet Transform (DWT) and using Support Vector Machine (SVM) obtained 92.11% accuracy and 92.50% in sensitivity [7]. The same authors in the year 2016 employ the 4 best characteristics estimated from entropies, Hjorth's parameters and energies of DWT, achieving 94% of accuracy to 95% sensitivity with SVM [8].

Most of the described studies used fractal features from heart rate variability (HRV) signal. In this artıcle, we worked under the hypothesis that if there is an underlying non-linear dynamics to HRV signal, this dynamic should be best described by the multifractal analysis for fractal indices [9]. From this arises a comparison of indexes both fractal and multifractals. It evaluates the relevance of these indexes and features space is generated to evaluate the performance of a classifier based on SVM.

II MATERIALS AND METHODS

A DataBase

Two available databases of PhysioNet were used. The MIT-BIH Normal Sinus Rhythm database (NSR-DB), contain 18 long-term ECG records, taken at a frequency of sampling 128Hz, in this registration apparently there are no significant arrhythmias. The Sudden Cardiac Death Holter Database (SCDH-DB) is composed of 23 complete Holter recordings with a sampling frequency of 250Hz. All the patients had a ventricular tachyarrhythmia sustained and most presented a real heart attack, were discarded 3 records, which correspond to records of patients who did not have cardiac arrest (40, 42 and 49 records) [10].

I. Torres et al. (eds.), *VII Latin American Congress on Biomedical Engineering CLAIB 2016, Bucaramanga, Santander, Colombia, October 26th -28th, 2016*,
IFMBE Proceedings 60,
DOI: 10.1007/978-981-10-4086-3_122

B Selection of Windows and Segmentation

For the analysis of the signals, windows of 4, 8, 12, 20, 32 and 42 minutes are generated, which contain information from 4 minutes before the event, as illustrated in Figure 1.

Afterwards a detection process is performed of the R peaks from the algorithm proposed by Pan and Tompkins [11]. This detection shows failures of the algorithm in records 35, 41, 43 and 52 due to artifacts. They were removed from the analysis. From the 16 remaining records of SCD and 18 of NSR, it generates a signal with the HRV.

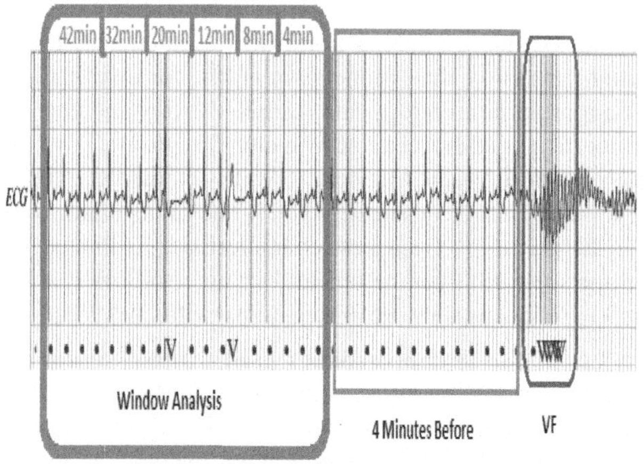

Figure 1: Selection Windows Analysis

C Fractals

The time series that have fractal behavior, present a statistical similarity at different scales. Since the Euclidean lengths present difficulty for the estimation of fractal measures, it is necessary to introduce the concept of fractal dimension.

To understand fractal dimension, suppose that we cover a set with balls of diameter ε; the fractal dimension D characterizes the dependence on ε of the number of covering elements $N(\varepsilon)$. $N(\varepsilon)$ is denoted as the number of balls required to coat the whole unit and satisfies the law of power as $\varepsilon \to 0$:

$$N(\varepsilon) \approx \varepsilon^{-D} \tag{1}$$

Where D is a constant called fractal dimension or the Hausdorffs dimension [12].

Correlation dimension (CD) estimates fractal dimension of the attractor that forms in the space. It is a type of dimension probabilistic, which depends on a dimension of embedding (m) and a time delay τ, to form the attractor [13]. In this work the parameter $\tau = 9$ is obtained from the average mutual information (average mutual information) and $m = 5$ as shown in [14, 15].

The exponent of scale is linked to the fractal dimension and quantifies the dynamics of time series, allowing weighing up to that point data can be persistent to long term [16]. The estimation of this characteristic was conducted with the method Detrended Fluctuation Analysis (DFA), with power scales of 2, from 16 to 512 [17].The behavior of the series of time at different scales is commonly called, exponent of Hurst [16].

The Lyapunov exponents (LE) can quantify the unpredictability of a series of time, under certain initial conditions, estimating the convergence or divergence of the trajectories of the attractor. The presence of positive exponents indicates signal chaos, and periodic signals are characterized by negative exponents [13].The greatest exponents of Lyapunov was estimated based in the method proposed by [18].

D Multifractals

The multifractal term is used for systems that are characterized by a range of different fractal dimensions, to which is attached a function called multifractal spectrum or singularities, in this sense, the equation 1 takes the form:

$$N_\alpha \sim \varepsilon^{-f(\alpha)} \tag{2}$$

where ε is the diameter of the balls that covered the set, α singularity index, and $f(\alpha)$ corresponds to the generalised Hausdorff dimension, that goes from being a constant to a function, which represents the local dimensions of subsets of a multifractal set. The spectrum of singularities measures the global distribution of singularities with various regularities [12, 19], and is related to the scaling exponent $\tau(q)$ via a Legendre transform, equation 3. where, the index q can take any real value except zero [20].

$$\alpha = \tau'(q) \qquad\qquad f(\alpha) = q\alpha - \tau(q) \tag{3}$$

The scaling exponents also called q-order mass exponents, leave the partition function in which they are based several methods, can reveal aspects of cardiac dynamics [21]. Multifractal detrend fluctuation analysis (MF-DFA) method proposed by [20], was used to calculated $f(\alpha)$. From the spectrum, determines the width (SW), which represents the degree of multifractality of the time series, the asymmetry index of the spectrum (AI) and the multifractal h-fluctuation index (*hFI*) is implemented and used in this work according with the parameters defined in [19]. The hFI extracts information for the shape of the singularity spectrum $f(\alpha)$ which is

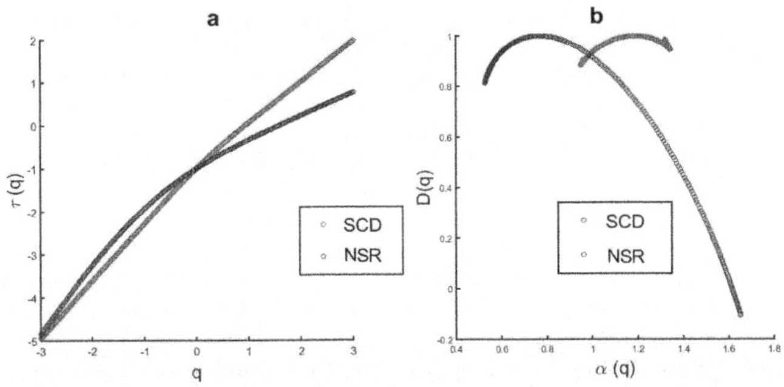

Figure 2: a) The q-order Mass Exponent. b) Multifractal spectrum

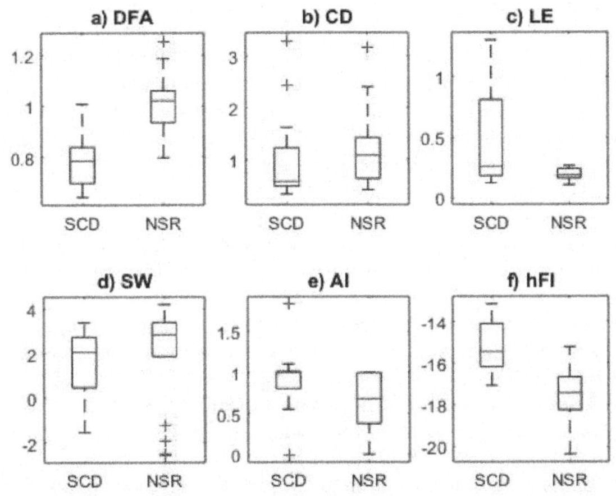

Figure 3: Boxplot of fractal features a, b, c and multifractal d, e, f

a generalization of fractal dimension when statistical scaling is characterised by different Hurst exponents.

E Features and Classification space

Using the Receiver Operating Characteristic (ROC) curves the sensitivity and specificity of each characteristic to discriminate between healthy subjects and subjects that are going to suffer SCD were evaluated. ReliefF was use as feature selection method and is based on the nearest neighbor paradigm [22]. Two features with the better performance were selected to build a 2-dimensional feature space and a SVM with a sigmoid kernel was used, a gamma = 1 and c = 8, these parameters were selected heuristically for the construction of the classifier with leave-one-out cross validation.

Table 1: Results of ROC curves with a window of 32 minutes of analysis. Area under the curve (AUC), sensitivity (sens), specificity (spec) and Standard Error of the Area (SE).

	DFA	CD	EL	SW	AI	hFI
(AUC)	93,4	63,54	73,26	65,63	74,83	93,4
sens	88,89	75	88,89	72,22	72,22	94,44
spec	87,5	61,11	62,5	62,5	75	87,5
SE	4,46	9,56	8,62	9,4	8,42	4,46

III RESULTS AND DISCUSSION

As shown in Figure 2a the q-order Mass Exponent (qME) demonstrate that the SCD signal tend to have a linear behavior, which indicates a loss of multifractality in the spectrum shown in Figure 2b, where the width of this is reduced, in comparison with the signal of NSR. These results are consistent with studies that have given evidence of HRV signals from healthy people present multifractals features with a wide spectrum and a non-linear qME, but these characteristics are lost in patient pathological [21, 23].

Figure 3 shows the boxplot for SCD and NSR samples, comparing the fractal indices (Figure 2a-2 c) concerning the multifractals (figure 2d-2f). Table 1, shows the quantitative results of sensitivity and specificity in discrimination between SCD and NSR of each index. These results belong to the window of analysis of 32 minutes, since it has the number of samples required for the good estimation of non-linear features (approximately 2284 ± 540 R peaks).In general you can see that multifractals indices have a better performance than the fractal indices. Particularly the best index is the hFI. Mul-

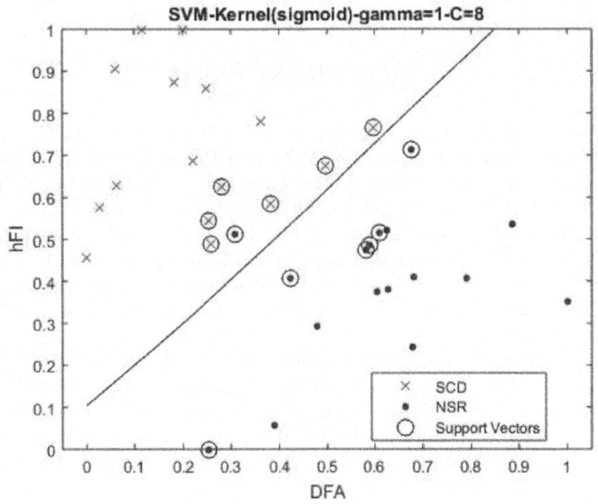

Figure 4: Classification with SVM, Kernel sigmoid, gamma = 1 and C = 8.

tifractals features show representative change in signals, SW shows that the NSR contain a high degree of multifractality and the SCD tend to lose the multifractality and present monofractal behavior. In the boxplot of the AI and hFI, you can see that NSR signals presented greater symmetry and few fluctuations, while the SCD has less symmetry and more fluctuations.

According to ReliefF and boxplot analysis, the most relevant characteristic after the h-fluctuations is DFA, which is monofractal. So these two features can be combined into an SVM classifier for the prediction of VF with an accuracy of 97.06%, 100% sensitivity and specificity 94.44%. This results exceeds those reported in the literature so far, where they only contemplate the monofraltal dynamic [6–8]. In Figure 4 you can see the diagram of dispersion of data, where there is evidence of a border's decision clearly established by the support vectors, and the need to increase the penalty C factor, to have a smooth margin.

IV CONCLUSIONS

The results achieved in this work are able to accurately classify two types of dynamics of the HRV, surpassing the works that are reported in the literature so far. Our findings present evidence to confirm the hypothesis that the multifractal analysis manages to best describe the nonlinear dynamics of the HRV, however combining these with fractals features, can generate an important descriptor which helps prediction the risk of VF associated to SCD.

CONFLICT OF INTEREST

The authors declare that they have no conflict of interest.

REFERENCES

1. Murukesan Loganathan, Murugappan M, Iqbal Mohammad. Sudden cardiac death prediction using ECG signal derivative (heart rate variability): a review in *Signal Processing and its Applications (CSPA), 2013 IEEE 9th International Colloquium on*:269–274IEEE 2013.
2. Sachdev Molly, Fetics Barry J, Lai Shenghan, Dalal Darshan, Insel Jerald, Berger Ronald D. Failure in short-term prediction of ventricular tachycardia and ventricular fibrillation from continuous electrocardiogram in intensive care unit patients *Journal of electrocardiology.* 2010;43:400–407.
3. De Chillou Christian, Sadoul Nicolas, Bizeau Olivier, et al. Prognostic value of thrombolysis, coronary artery patency, signal-averaged electrocardiography, left ventricular ejection fraction, and Holter electrocardiographic monitoring for life-threatening ventricular arrhythmias after a first acute myocardial infarction *The American journal of cardiology.* 1997;80:852–858.
4. Villacastín Julián, Bover Ramón, Castellano Nicasio P, Moreno Javier, Morales Ricardo, García-Espinosa Arturo. Estratificación del riesgo y prevención de la muerte súbita en pacientes con insuficiencia cardíaca *Revista española de cardiología.* 2004;57:768–782.
5. Fam Jiang Ming, Ching Chi Keong. Review on Non-invasive Risk Stratification of Sudden Cardiac Death *Proceedings of Singapore Healthcare.* 2011;20:263–278.
6. Ebrahimzadeh Elias, Pooyan Mohammad, Bijar Ahmad. A novel approach to predict sudden cardiac death (SCD) using nonlinear and time-frequency analyses from HRV signals *PloS one.* 2014;9:e81896.
7. Acharya U Rajendra, Fujita Hamido, Sudarshan Vidya K, et al. An integrated index for detection of Sudden Cardiac Death using Discrete Wavelet Transform and nonlinear features *Knowledge-Based Systems.* 2015;83:149–158.
8. Fujita Hamido, Acharya U Rajendra, Sudarshan Vidya K, et al. Sudden cardiac death (SCD) prediction based on nonlinear heart rate variability features and SCD index *Applied Soft Computing.* 2016;43:510–519.
9. Sassi Roberto, Signorini Maria Gabriella, Cerutti Sergio. Multifractality and heart rate variability *Chaos: An Interdisciplinary Journal of Nonlinear Science.* 2009;19:028507.
10. Goldberger Ary L, Amaral Luis AN, Glass Leon, et al. Physiobank, physiotoolkit, and physionet components of a new research resource for complex physiologic signals *Circulation.* 2000;101:e215–e220.
11. Pan Jiapu, Tompkins Willis J. A real-time QRS detection algorithm *Biomedical Engineering, IEEE Transactions on.* 1985:230–236.
12. Calvet Laurent E, Fisher Adlai. *Multifractal volatility: theory, forecasting, and pricing.* Academic Press 2008.
13. Parker Thomas S, Chua Leon O. Chaos: A tutorial for engineers *Proceedings of the IEEE.* 1987;75:982–1008.
14. Cao Liangyue. Practical method for determining the minimum embedding dimension of a scalar time series *Physica D: Nonlinear Phenomena.* 1997;110:43–50.
15. Fraser Andrew M, Swinney Harry L. Independent coordinates for strange attractors from mutual information *Physical review A.* 1986;33:1134.
16. Bassingthwaighte James B, Raymond Gary M. Evaluation of the dispersional analysis method for fractal time series *Annals of biomedical engineering.* 1995;23:491–505.
17. Ihlen Espen AF. Introduction to multifractal detrended fluctuation analysis in Matlab *Fractal Analyses: Statistical And Methodological Innovations And Best Practices.* 2012:97.
18. Wolf Alan, Swift Jack B, Swinney Harry L, Vastano John A. Determining Lyapunov exponents from a time series *Physica D: Nonlinear Phenomena.* 1985;16:285–317.
19. Orozco-Duque A, Novak D, Kremen V, Bustamante J. Multifractal analysis for grading complex fractionated electrograms in atrial fibrillation *Physiological measurement.* 2015;36:2269.
20. Kantelhardt Jan W, Zschiegner Stephan A, Koscielny-Bunde Eva, Havlin Shlomo, Bunde Armin, Stanley H Eugene. Multifractal detrended fluctuation analysis of nonstationary time series *Physica A: Statistical Mechanics and its Applications.* 2002;316:87–114.
21. Ivanov Plamen Ch, Amaral Luis A Nunes, Goldberger Ary L, et al. Multifractality in human heartbeat dynamics *Nature.* 1999;399:461–465.
22. Robnik-Šikonja Marko, Kononenko Igor. Theoretical and empirical analysis of ReliefF and RReliefF *Machine learning.* 2003;53:23–69.
23. Mercy Cleetus Helen Mary, Singh Dilbag. Multifractal application on electrocardiogram *Journal of medical engineering & technology.* 2014;38:55–61.

Author: Andrés Felipe Orozco Duque
Institute: Instituto Tecnológico Metropolitano
Street: Cl. 54a #1 - 30
City: Medellín
Country: Colombia
Email: andresorozco@itm.edu.co

Mechanical Characterization of a Breast Phantom

D.A. Triana[1,2], K.L. Cristiano[1], J.C. Gutiérrez[1] and D.A. Miranda[1]

[1] Universidad Industrial de Santander/Physics Department, Professor, Bucaramanga, Colombia
[2] Universidad Autónoma de Bucaramanga/Mathematics Department, Professor, Bucaramanga, Colombia

Abstract— **A Breast Phantom is an appropriate way to study early detection of breast cancer because the Phantom let to model and perform experiments of mechanical and optical properties of the breast that are ethically impossible to do in a woman. In this paper, we report the mechanical characterization of a breast Phantom. The Phantom preparation was a water based solution and glycerine mixed with phenol and pectin. At the elaboration process, the mix proportions were modified seeking varied textures, colors, and form which simulated mechanics and optics properties a mammary tissue. Stress and deformation measurement were done to perform mechanic characterization. The elasticity module of refined and unrefined pectin Phantom was compared with breast fibrous and glandular tissues achieving values in the same range.**

Keywords— **Breast phantom, tension compression proofs, breast tissue.**

I. Introduction

Early detection of cancer is a subject of study because this disease is associated with high incidence, prevalence, morbidity, and mortality. Some experiments to study early detection of cancer are ethically prohibitory e.g. the mechanical characterization in the nonelastic limits of a woman breast. The mechanics and optical characterization were aimed to model normal and abnormal breast tissue since real breast compression is invasive, painful and would need request approbation from an ethic committee.

Optical characterization of breast phantom at NIR range shown similar curves of absorbance like human breast [1, 2]. The elastic properties of breast tissue depend on the microscopic and macroscopic structural organization of their molecules [3]. The phantom is deformed when external forces are applied, and phantom interior forces reduce the accelerated motion. The restore forces as several point mass attached by springs were modeled [4]; it has constant forces then deformations are invariant along its axis and can be modeled by Hooke law (see equation 1).

$$\sigma = E\varepsilon \tag{1}$$

where σ is the stress in N/m^2 and ε, dimensionless relative deformation. These quantities have a linear relation with the Youngs module E as proportional constant.

In this document, we report the mechanical characterization of a phantom developed to study early detection of breast cancer.

II. Materials and Methods

In this study, a pectin named poly-galacturonic acid was selected; this solid substance has dispersed phase; the dispersion medium was water and glycerin (polyhydroxy alcohol) solution. The phase dispersed and dispersing medium interactions produce associations through hydrogen bonds improving texture and structure to the phantom.

The procedure used to fabricate the phantom is summarized as follow: first, standard water (99.46 g) and glycerin (177.88 g) were mixed to do the dispersion medium, then the pectin (111.56 g) was added to obtain the dispersed phase. Second, the mixture in an oil bath was heated for 14 hours to 83° Celsius into petri dish until the liquid state was achieved, then liquid phantom in syringe pump was poured. Finally, it was cooled down about 12 hours. Following this steps the dehydration of the phantom was delayed during heating, i.e., the water didn't evaporate as quickly the Phantom retains its structure for longer.

The phantom was cut in a cross section so that length is twice the diameter of the cylinder [5] (see figure 1). Finally, the obtained samples were placed into benchtop tester: Tinius Olsen H25KS, for tension compression to 25 kN. Several tests were made until the phantom achieved rupture limit (see figure 2).

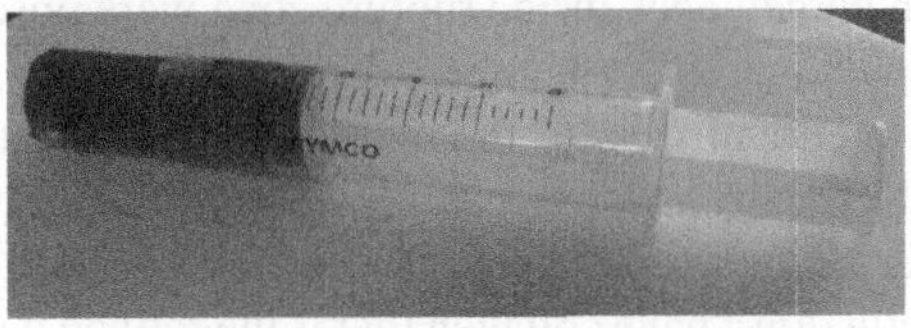

Fig. 1: Sample mold for stress measurements.

I. Torres et al. (eds.), *VII Latin American Congress on Biomedical Engineering CLAIB 2016, Bucaramanga, Santander, Colombia, October 26th -28th, 2016*,
IFMBE Proceedings 60,
DOI: 10.1007/978-981-10-4086-3_123

Fig. 2: Phantom rupture limit.

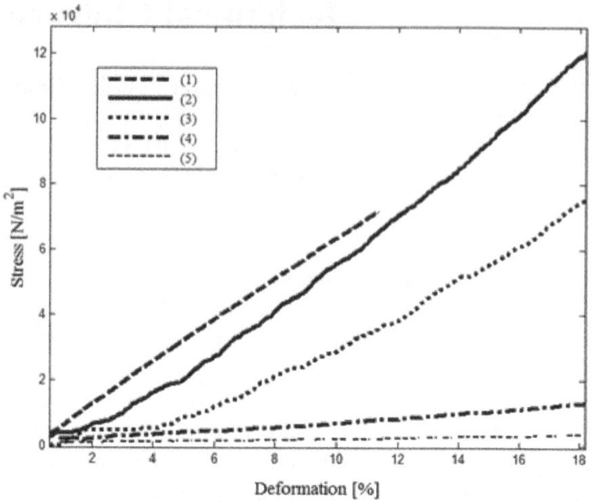

Fig. 3: Stress of fabricated phantom (2), (3) compared with heterogeneous breast tissue (1), (4), (5) reference data from [4].

III. RESULTS AND DISCUSSION

Mechanical characterization by tension-compression proofs was performed to the fabricated phantoms. The maximum velocity of compression was 0.5 mm/min for 40 minutes by each measure. The mean temperature of the laboratory was 17° Celsius to avoid an abrupt change in the sample. It is important to note that room temperature can affect the fabrication process of the phantom.

Compression measurements of the phantom as stress as a function of deformation percent (see figure 3) shown a similar behavior than fibrous and glandular tissues: the curves (2) and (3) are the efforts measured on the phantom of refined and unrefined pectin, respectively. The curve (1) is stress on fibrous tissue until 11 percent deformation. The curve (4) shown the stress on a non-fat glandular tissue by 30 percentage deformation. The curve (5) has stress information of breast fat tissue, this curve has less inclination than the other, then the Youngs modulus (breast fat tissue) will be lower than others (see table 1).

A linear regression analysis was applied to quantify the Youngs modulus for unrefined and refined pectin phantom to compare them with the Youngs modulus of each one soft breast tissue [6]. To calculate the Youngs modulus the linear curve zone until 23 deformation percent was analyzed (see Table 1), other relevant conditions like phantom cylindrical geometry (initial longer 52 mm and diameter 26 mm) were used [5]. If the deformation increase, the material entered the plastic zone between 24-50 deformation percent. Finally, deformation percentages inside rupture zone were observed: 51 to refine pectin and 58 to unrefined pectin.

In the chemical view of the material, the fabricated phantom exhibits a strength associated with hydrogen bonds: water, glycerin, and pectin make the phantom more resistant when pectin don't have collagen that is the refined pectin. Unrefined pectin has high collagen percent shaped by the amino acid with nitrogen as the main component, whereas the chemical composition of the refined pectin included carbon, hydrogen, and oxygen as a woman breast tissue.

Table 1: Calculated Youngs modulus for refined and unrefined pectin compared with data from [4].

Material	Young's modulus (KN/m^2)
Refined pectin	757
Unrefined pectin	493
Breast fibrous tissue	232
Breast glandular tissue	57
Breast fat tissue	20

IV. CONCLUSIONS

Two phantoms that model human breast tissue were fabricated and characterized by mechanical proofs (tension-compression). The mechanical characterization has shown similarities into the behavior of phantom and a woman breast that could be exploited in studies of mechanical properties of woman breast to develop new techniques to detect, treat or prevent cancer.

Relevant aspects of fabricates Phantom are their not toxicity and the fact that have biodegradable and recycled components.

Acknowledgments

We are grateful to CIMBIOS and GUQUIP groups at the Universidad Industrial de Santander for his collaboration during experiments development and measure time.

References

1. Miranda D A, Cristiano K L, Gutiérrez J C. Breast phantom for mammary tissue characterization by near infrared spectroscopy *Journal of Physics: Conference Series, IOP Publishing.* 2013;466:012018.
2. Miranda D A, Cristiano K L, Triana D A. Design assembly and calibration of an optoelectronic device that measures three wavelengths in the near infrared for breast cancer early detection *IEEE Xplore Circuits and Systems IEEE 5th Colombian Workshop.* 2014:1-4.
3. Fung Y C. *Biomechanics.* New York: Springer 1990.
4. Strelkóv S. *Mecánica.* Moscú: Mir Moscú 1978.
5. ASTM International. 1012-Standard Practice for Verification of Test Frame and Specimen Alignment Under Tensile and Compressive Axial Force Application
6. Krouskop T A, Wheeler T M, Kallel F, Garra B S, Hall T. Elastic moduli of breast and prostate tissues under compression *Ultrasonic imaging; SAGE Publications.* 1998;20:260–274.

Author: Daniel Andrés Triana Camacho
Institute: Universidad Autónoma de Bucaramanga UNAB
Street: Avenida 42 No. 48 11
City: Bucaramanga
Country: Colombia
Email: dtriana331@unab.edu.co

Theoretical study of the interaction between carbon nanotubes and the linoleic acid, an atherogenic polyunsaturated fatty acid

Ana M. Torres[1], John Bustamante[1], Andrés M. Garay-Tapia[2] and Tapas Kar[3]

[1] Universidad Pontificia Bolivariana/Centro de Bioingeniería, Grupo de Dinámica Cardiovascular, Medellín, Colombia
[2] Centro de Investigación en Materiales Avanzados/Grupo de Termodinámica Computacional, Monterrey, México
[3] Department of Chemistry and Biochemistry/Utah State University, Logan, UT 84321, USA

Abstract— **This investigation explores the interaction between single walled carbon nanotubes and the linoleic acid, a polyunsaturated fatty acid, present in low density lipoprotein, which is highly susceptible to oxidation. Molecular models were built and theoretical studies were performed using the PM6 semiempirical method and the density functional theory. The strength of the interaction was estimated from the interaction energy calculation. Negative values of E_{int} indicated that all evaluated systems are stable. Although, the interaction occurs through weak forces, CH$\cdots\pi$, OH$\cdots\pi$ and CH$\cdots$N, collectively such interactions are strong enough and may be considered as chemisorption.**

Keywords— **Carbon nanotubes, linoleic acid, lipid peroxidation, density functional theory.**

I Introduction

Experimental evidence has attributed to lipid peroxidation an important role in the pathogenesis of atherosclerosis. The structural changes experienced by low-density lipoprotein (LDL) after oxidation can significantly increase their atherogenic characteristics [1, 2]. About half of the content of fatty acids in LDL consists of polyunsaturated fatty acids (PUFAs), mainly linoleic acid (LA) [3, 4]. As consequence, oxygenated derivatives of LA have been found in human atherosclerotic lesions, as main peroxidation products [5, 6].

The hypothesis that antioxidants may inhibit LDL oxidation and reduce the incidence of coronary events has encouraged the development of various investigations. Some studies have been conducted using antioxidant vitamins (beta-carotene, vitamin C and vitamin E) and a synthetic antioxidant (Probucol), obtaining results in favor and against of its use [7, 8]. On the other hand, statins are drugs widely used in the treatment of hypercholesterolemia [9]. However, there is some concern because a relationship between statins use and the risk of developing other pathologies has been found [10].

Therefore, it is important the search of alternative treatments to prevent or inhibit lipid peroxidation. On this direction, the use of certain nanomaterials such as carbon nanotubes (CNTs) could be considered. Doping and functionalization enhance biocompatibility and solubility of these nanomaterials, and can alter its cellular interaction pathways, resulting in a reduction in the cytotoxic effects [11], which has opened the possibility of its application in medicine field. In addition, antioxidant activity has also been reported in CNTs [12, 13], which could be useful for the treatment of different diseases, including the atherosclerosis.

Modeling and simulation have been well established as instrumental in the study of CNTs [14, 15, 16]. Theoretical research is based on the development of a molecular model that allows the description, research and prediction of properties of interest. Electronic structure methods are the fundamental level of theory used for the description of the systems at nanoscale. These methods use the laws of quantum mechanics and are characterized by different levels of approach to the solution of the Schrödinger equation. Semiempirical methods, *ab initio* and the density functional theory (DFT) are classified as electronic structure methods [17].

As a first approach to the study of the inhibition of lipid peroxidation applying CNTs, we evaluated theoretically the interaction of the LA and perfect single walled carbon nanotubes (SWCNTs) and doped with nitrogen (SWCNT4N), using the PM6 semiempirical method [18] and the DFT [19, 20]. We took advantage of the highly cost effective semiempirical method to perform geometry optimization of SWCNT-LA and SWCNT4N-LA. The optimized coordinates were used to perform single point calculations at DFT level. Interaction energies were estimated and considered as a criterion for evaluating the stability of the systems. End-functionalization of LA with SWCNT has been reported using molecular dynamics [21], but electronic calculations are required for an adequate estimation of the interaction.

II Methodology

To study the interaction of SWCNTs with LA, we selected a (10,0) SWCNT, which has been commonly studied [22, 23]. Model of perfect SWCNT was built using 280 carbon atoms.

I. Torres et al. (eds.), *VII Latin American Congress on Biomedical Engineering CLAIB 2016, Bucaramanga, Santander, Colombia, October 26th -28th, 2016*,
IFMBE Proceedings 60,
DOI: 10.1007/978-981-10-4086-3_124

Doped nanotubes include four nitrogen atom (SWCNT4N) located at the edge of a divacancy, which was generated by removing two adjacent carbon atoms. SWCNT4N were built using 274 and 114 carbon atoms. The ends of the nanotubes were passivated with hydrogen atoms. For perfect and doped nanotubes, two orientations of LA were evaluated (Fig. 1).

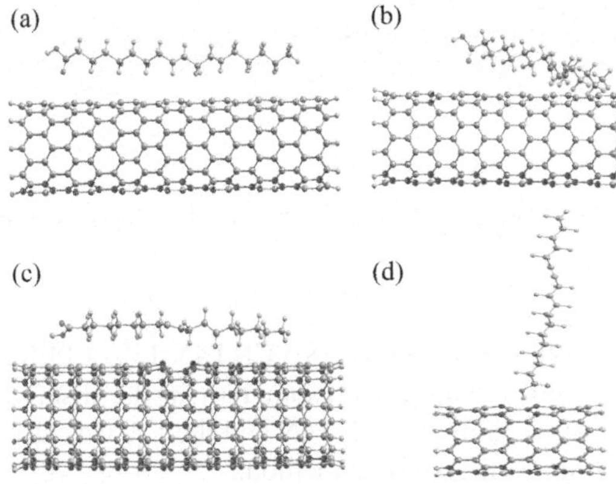

Fig. 1: Structure of orientations of LA. (a) SWCNT-LA-O1, (b) SWCNT-LA-O2, (c) SWCNT4N-LA-O1, (d) SWCNT4N-LA-O2. Color scheme: gray-carbon, light blue-hydrogen, red-oxygen, pink-nitrogen. O1 and O2: orientation 1 and 2, respectively.

LA ($C_{18}H_{32}O_2$) is a carboxylic acid with an 18-carbon chain and two C=C bonds at positions 9 and 12. Geometry optimization of the four conformers of LA was performed using PM6 and the DFT with the functionals B3LYP [24, 25] or wB97XD [26] and the 6-31G* basis set. The wB97XD functional includes an empirical dispersion correction. With most stable form of LA, the SWCNT-LA and SWCNT4N-LA were relaxed at PM6 level. Relaxed coordinates of the complex were used for single point calculations at DFT level. All calculations were performed using Gaussian 09 program [27]. The stability of the systems was estimated from the relative energy (RE) and the interaction energy (E_{int}) (Eq. 1).

$$E_{int} = E_{(SWCNT-LA)} - E_{(SWCNT)} - E_{(LA)} \quad (1)$$

where, $E_{(SWCNT-LA)}$ is the total energy of SWCNT-LA or SWCNT4N-LA, $E_{(SWCNT)}$ is the total energy of SWCNT or SWCNT4N, $E_{(LA)}$ is the total energy of LA.

III Results and Discussion

A Linoleic acid stability

According to relative energy (RE, relative to the most stable structure) values (Table 1), the PM6 semiempirical method predict a similar trend in the results to those obtained with a higher level of theory, such as DFT. The most stable conformer corresponds to *t*9-*t*12 and the less stable to *c*9-*c*12.

Table 1: Relative energy (RE) of LA

Conformer	RE (kcal/mol)		
	PM6	B3LYP/ 6-31G*	wB97XD 6-31G*
*t*9-*t*12	0	0	0
*t*9-*c*12	0.54	1.52	1.03
*c*9-*t*12	0.56	1.62	1.02
*c*9-*c*12	1.22	3.15	1.94

The *t*9-*t*12 conformer was selected to evaluate the interaction with a (10,0) SWCNT perfect and doped with nitrogen.

B Geometry optimization of SWCNT-LA and SWCNT4N-LA using PM6

The values estimated for the interaction energy are shown in Table 2

Table 2: Interaction energy (E_{int})

Orientation	E_{int} (kcal/mol)	
	SWCNT-LA	SWCNT4N-LA
O1	-5.88	-7.70
O2	-5.76	-9.56

Negative E_{int} equates to a stable complex, otherwise unstable complex. For SWCNT-LA, the E_{int} values differ by only 0.12 kcal/mol, indicating that two orientations got almost the same final state. In the case of SWCNT4N-LA, the interaction between the doped nanotube and LA seems to occur preferably through the carboxyl group, E_{int} was slightly lower for Orientation 2.

In optimized structures, shown in Fig. 2, several C-H bonds of LA are directed to the surface of the tube, indicating CH$\cdots\pi$ interactions between adsorbate and adsorbent. It may be worth mentioning that some of the C-H bonds are directed toward the center of hexagons of the tube surface. Stretching of those C-H bonds compared to others away from the tube surface, confirms such interaction [28]. Such intermolecular CH$\cdots\pi$ distances of about 2.7–3.4 Å, predicted by PM6 method, are slightly longer than the standard CH$\cdots\pi$ distance of 2.3–2.9 Å [29]. Such a discrepancy is quite reasonable as PM6 omits explicit correlation as well as dispersion correction required for proper description of weak CH$\cdots\pi$ interaction. To correct such effects, we employed the wB97XD functional. O-H$\cdots\pi$ interactions were established

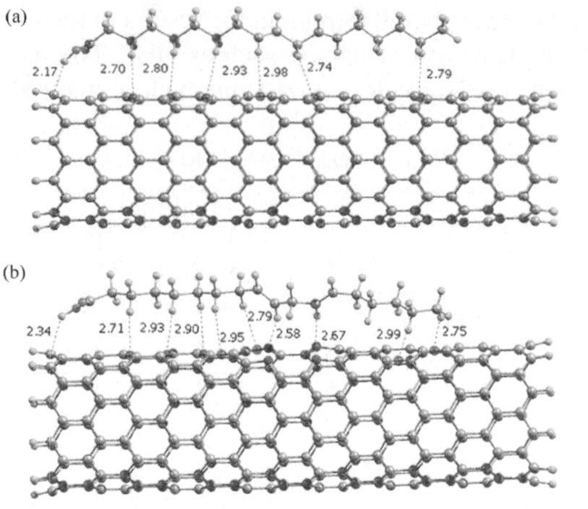

Fig. 2: Structures and key geometric parameters. (a) SWCNT-LA-O1, (b) SWCNT4N-LA-O1. Closest C-H, O-H and N-H distances are shown (in Angstrom Å). Color scheme: gray-carbon, light blue-hydrogen, red-oxygen, pink-nitrogen, violet-dummy atom at the center of C=C bond closest to C-H bond, or at the center of the ring.

through carboxylic group. For SWCNT4N-LA, also CH···N interactions were observed.

C Single point calculations using DFT

We performed a potential energy surface (PES) scan by bringing LA in Orientation-1 from 3.0 to 2.4 Å toward the (10,0) nanotube by freezing the PM6 optimized geometry of both LA and the SWCNT at each intermolecular distance R. It is worth noting that such a weak interactions has little or no effect on the internal geometrical parameters, which supports the freezing of LA and SWCNT geometries during the PES scan. The variation of interaction energy is reported in Table 3.

Table 3: Variation of interaction energy (E_{int}) with distance R between SWCNT and LA

R (Å)	E_{int} (kcal/mol)	
	PM6	wB97XD/3-21G
3.0	-5.70	—
2.9	-5.76	-29.68
2.8	-5.69	-30.45
2.7	-5.47	-31.01
2.6	-5.04	-31.28
2.5	-4.34	-30.28
2.4	-3.30	-28.67

It can be seen that PES (Fig. 3) exhibits a minimum around 2.6 Å and 2.9 Å at the wB97XD/3-21G level and PM6, respectively, which indicates that PM6 tends to overestimate the distance of interaction between the SWCNT and the LA.

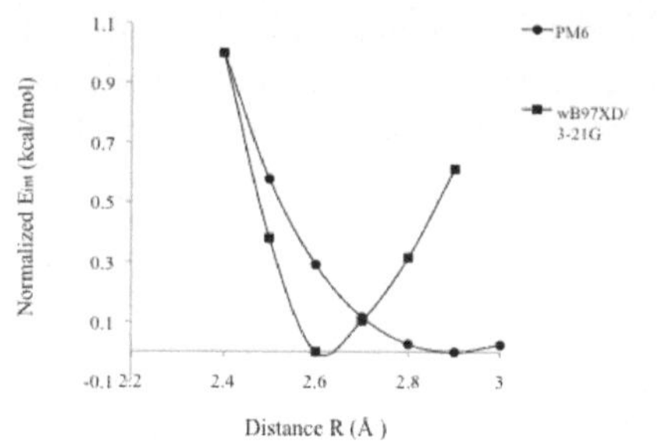

Fig. 3: Normalized interaction energies as a function of distance R for SWCNT-LA-O1

Also for both orientations of SWCNT4N-LA, a PES scan were performed. A slight overestimation in the distance of interaction between the SWCNT4N and the LA was also noticed with PM6 semiempirical method.

The minimum E_{int} reached with wB97XD/3-21G for each system evaluated are shown in Table 4. As mentioned earlier, a negative value for E_{int} indicate that the interaction between LA and SWCNT or SWCNT4N is thermodynamically feasible. E_{int} values are within the range reported for a chemisorption process (-23 to -230 kcal/mol) [30]. Then, although the interactions present are weak type, in conjunction they help to the stability of the systems, resulting in a comparable interaction with a chemical bond.

Table 4: Interaction energy (E_{int} predicted at DFT level, using wB97XD/3-21G

Complex	E_{int} (kcal/mol)
SWCNT-LA-O1	-31.28
SWCNT4N-LA-O1	-37.37
SWCNT4N-LA-O2	-29.92

According to the results, we can say that the interaction between LA and (10,0) SWCNT perfect or doped with nitrogen is of attractive nature. This could serve as a mechanism of inhibition of the oxidation of acid. However, other theoretical studies, such as oxidation mechanisms, are necessary.

IV CONCLUSION

Using the PM6 semiempirical method, we obtained optimized coordinates, which were used for single point calculations at higher theory level. Interactions energies indicate that the SWCNT-LA and SWCNT4N-LA complex were stable. We verified that interaction between LA and nanotubes is of weak nature. CH$\cdots\pi$, OH$\cdots\pi$ and CH$\cdots$N interactions were established. In conjunction, all weak interactions give stability to the complex and generate an attractive interaction equivalent to a chemical bond. This interaction could be useful as mechanism to avoid the lipid oxidation, but additional studies are required.

CONFLICT OF INTEREST

'The authors declare that they have no conflict of interest.

ACKNOWLEDGEMENTS

Ana M. Torres and John Bustamante thank to Colciencias (Colombia) for financial support, under project 1210569 33678. Computational facilities at the Computational Thermodynamics Laboratory at CIMAV (Monterrey, Mexico) and DoD HPC at AFRL (Dayton, Ohio) are gratefully acknowledged.

REFERENCES

1. Singh R., Mengi S., Xu Y., Arneja A., Dhalla N. Pathogenesis of atherosclerosis: A multifactorial process. *Experimental and Clinical Cardiology.* 2002;7:40–53.
2. Maiolino G., Rossitto G., Caielli P., Bisogni V., Rossi G., Calò L. The role of oxidized low-density lipoproteins in atherosclerosis: The myths and the facts. *Mediators of Inflammation.* 2013:1-14.
3. Wang T., Yu W., Powell W. Formation of monohydroxy derivates of archidonic acid, linoleic acid, and oleic acid during oxidation of low density lipoprotein by copper ions and endothelial cells. *J. Lipid. Res.*. 1992;33:525–537.
4. Spiteller G. Linoleic acid peroxidation—the dominant lipid peroxidation process in low density lipoprotein—and its relationship to chronic diseases. *Chemistry and Physics of Lipids.* 1998;95:105—162.
5. Waddington E, Sienuarine K, Puddey I, Croft K. Identification and quantitation of unique fatty acid oxidation products in human atherosclerotic plaque using high-performance liquid chromatography. *Analytical biochemistry.* 2001;292:234–244.
6. Waddington E., Croft K., Sienuarine K., Latham B., Puddey I. Fatty acid oxidation products in human atherosclerotic plaque: An analysis of clinical and histopathological correlates. *Atherosclerosis.* 2003;167:111–120.
7. Gotto A. Antioxidants, statins, and atherosclerosis. *Journal of the American College of Cardiology.* 2003;41:1205–1210.
8. Katsiki N., Manes C. Is there a role for supplemented antioxidants in the prevention of atherosclerosis?. *Clinical Nutrition.* 2009;28:3–9.
9. Rosenson R. Statins in atherosclerosis: Lipid-lowering agents with antioxidant capabilities. *Atherosclerosis.* 2004;173:1–12.
10. Navarese E., Szczesniak A., Kolodziejczak M., Gorny B., Kubica J., Suryapranata H. Statins and risk of new-onset diabetes mellitus: Is there a rationale for individualized statin therapy?. *American Journal of Cardiovascular Drugs.* 2014;14:79–87.
11. Vardharajula S., Ali S., Tiwari P., et al. Functionalized carbon nanotubes: Biomedical applications. *International Journal of Nanomedicine.* 2012;7:5631–5374.
12. Lucente R., Moore V., Leonard A., et al. Antioxidant Single-Walled Carbon Nanotubes. *Journal of the American Chemical Society.* 2009;131:3934–3941.
13. Amiri A., Memarpoor-Yazdi M., Shanbedi M., Eshghi H. Influence of different amino acid groups on the free radical scavenging capability of multi walled carbon nanotubes. *Journal of Biomedical Materials Research Part A.* 2013;101A:2219–2228.
14. Zhao J., Park H., Han J., Lu J. Electronic Properties of Carbon Nanotubes with Covalent Sidewall Functionalization. *The Journal of Physical Chemistry B.* 2004;108:4227-4230.
15. Orellana W., Fuentealba P. Structural, electronic and magnetic properties of vacancies in single-walled carbon nanotubes. *Surface Science.* 2006;600:4305–4309.
16. Gracia E, López F, Terrones H, Terrones M. Doping (10,0)-Semiconductor Nanotubes with Nitrogen and Vacancy Defects. *Materials Express.* 2011;1:127–135.
17. Yates, J. and Johnson, J. . *Molecular Physical Chemistry for Engineers.* University Science Books 2007.
18. Stewart J. Optimization of parameters for semiempirical methods V: Modification of NNDO approximations and application to 70 elements. *Journal of Molecular Modeling.* 2007;13:1173–1213.
19. Hohenberg P., Kohn W. Inhomogeneous Electron Gas. *Phys. Rev.*. 1964;136:B864–B871.
20. Kohn W., Sham L. Self-Consistent Equations Including Exchange and Correlation Effects *Phys. Rev.*. 1965;140:A1133–A1138.
21. Azevedo D., Nero J. Molecular Dynamics of Fatty Acid Interacting with Carbon Nanotube as Selective Device. *Internationl Scholarly and Scientific Reserch & Innovation.* 2010;4:474–477.
22. Yu S., Wen Q., Zheng W., Jiang Q. Effects of doping nitrogen atoms on the structure and electronic properties of zigzag single-walled carbon nanotubes through first-principles calculations *Nanotechnology.* 2007;18:1–7.
23. Min Y., Bae E., Kim U., et al. Unusual transport characteristics of nitrogen-doped single-walled carbon nanotubes. *Applied Physics Letters.* 2008;93:1–3.
24. Becke A. Density-functional exchange-energy approximation with correct asymptotic behavior. *Phys. Rev. A.* 1988;38:3098–3100.
25. Lee C., Yang W., Parr R. Development of the Colle-Salvetti correlation-energy formula into a functional of the electron density. *Phys. Rev. B.* 1988;37:785–789.
26. Chai J., Head M. Long-range corrected hybrid density functionals with damped atom-atom dispersion corrections. *Physical Chemistry Chemical Physics.* 2008;10:6615–6620.
27. Frisch, M. and Trucks, G. and Schlegel, H. and Scuseria, G. and Robb, M. and Cheeseman, J. and Scalmani, G. and Barone, V. and Mennucci, B.; Petersson, G. and Nakatsuji, H. and Caricato, M. and Li, X. and Hratchian, H. and Izmaylov, A. and Bloino, J. and Zheng, G. and Sonnenber, D . Gaussian 09. *Gaussian, Inc. Wallingford CT.* 2009:2–3.
28. Scheiner S., Grabowski S., Kar T. Influence of hybridization and substitution on the properties of the CH center dot center dot center dot O hydrogen bond. *J. Phys. Chem. A.* 2001;105:10607–10612.
29. Suezawa H., Ishihara S., Umezawa Y., Tsuboyama S., Nishio M. The Aromatic CH/π Hydrogen Bond as an Important Factor in Determining the Relative Stability of Diastereomeric Salts Relevant to Enantiomeric Resolution A Crystallographic Database Study. *European Journal of Organic Chemistry.* 2004;2004:4816–4822.
30. Yoder B. *Steric Effects in the Chemisorption of Vibrationally Excited Methane on Nickel.* Springer 2012.

Author: Ana María Torres López
Institute: Universidad Pontificia Bolivariana
Street: Circular 1ª No 73 – 76 Laureles
City: Medellín
Country: Colombia
Email: anatorreslopez@gmail.com

VCG and ECG indexes for classification of patients with Myocardial Infarction

R. Correa[1, 2, 3], P.D. Arini[3, 4], L.S. Correa[1, 3] and E. Laciar[1, 3]

[1]Gabinete de Tecnología Médica, Facultad de Ingeniería, Universidad Nacional de San Juan (UNSJ), San Juan, Argentina; rcorrea@gateme.unsj.edu.ar; lcorrea@gateme.unsj.edu.ar; laciar@gateme.unsj.edu.ar.
[2]Departamento de Física, Facultad de Ingeniería, Universidad Nacional de San Juan (UNSJ), San Juan, Argentina;
[3]Consejo Nacional de Investigaciones Científicas y Técnicas (CONICET), Buenos Aires, Argentina
[4]Instituto de Ingeniería Biomédica (IIBM), Facultad de Ingeniería (FI), Universidad de Buenos Aires (UBA), Buenos Aires, Argentina
pedro.arini@conicet.gov.ar

Abstract—**We proposed a classification technique for patients with Myocardial Infarction (MI), based on an Electrocardiographic (ECG) and Vectorcardiographic (VCG) signals analysis. We suggest and statistically analyze two VCGs and nine orthogonal ECG indexes, i.e., a) QRS loop Perimeter, b) Angle between QRS and T loops, c-e) the area under the QRS, T-wave and ST segment in X, Y and Z leads.**

For classification, the population was divided into two groups according to the infarcted area, that is, anterior or inferior sectors (MI-ant and MI-inf, respectively). The results indicate that combining eight indexes, we could separate out the MI patients in MI-ant vs MI-inf with a sensitivity = 89.8%, 84.8%, respectively, and an accuracy = 89.8%.

We conclude that the proposed technique could be suitable to estimatethe infarcted area localization.

Keywords— **ECG, VCG, MI.**

I. Introduction

The Myocardial Infarction (MI) is a portion of cardiac tissue necrosis that usually happens when the coronary vessels have been completely or partially blocked. The myocardial necrosis (infarction) produces an electromechanical dissociation, which could generate arrhythmias, ventricular fibrillation and sudden death. Thus, the MI is the main cause of death worldwide [1].

The heart electrical conduction is usually modified by MI. These changes can be recorded in the Electrocardiogram (ECG) and the Vectorcardiogram (VCG).

Several techniques have been developed that use the VCG constructed from ECG for monitoring patients after a MI episode, based on the new technological advances in digital ECG analysis. In the same way, several investigators have also used ECG and VCG techniques for MI detection [2], [3] and for the necrotic myocardial area localization.

In this way, Bakul et al. [3], have developed a set of features called Relative Frequency Band Coefficient for automatic identification of MI risk. Moreover, Keshtkar et al. [4] have proposed the evaluation of the wavelet coefficients set computed on the average ECG signal through neural networks as indices to detect MI. Furthermore, Muhammad et al. [2] presented automatic detection and localization of MI using k-Nearest Neighbor (KNN) classifier and Time domain features of each beat in the ECG signal, such as T-wave amplitude. In a recent work, Safdarian et al. [5] proposed two new features, i.e., T-wave integral and total integral as extracted features from one cycle of the ECG to detect and localize the infarcted area.

In this way, in previous works [6], [7] we have proposed two ECG and VCG signal processing techniques, for early identification of patients with Acute Myocardial Infarction and to differentiate them from healthy subjects. As result, it was demonstrated it is possible to classify the AMI and healthy subjects with an approximately sensitivity = 96%, a specificity = 94%, applying a linear discriminant classifier method.

In this paper, we present a technique for processing ECG and VCG signals for automatic classification of patients with MI according to the infarcted area. The hypothesis of this study is to determine approximately MI area of the heart by ECG and VCG analysis. The great advantages of the proposed technique are: very low cost; non-invasive and it can be carried out repeatedly without causing any harm to the patient.

II. Materials

We used ECG records from 95 patients with MI (69 men, age 58 ± 10 yrs, and 26 women, age 64 ± 12 yrs) obtained during the week after the MI episode. The areas of myocardial necrosis shown by the ECGs of the MI group were: anterior (n=14), antero-lateral (n=10), antero-septal (n=24), antero-septo-lateral (n=1), inferior (n=23), infero-lateral (n=17), infero-postero-lateral (n=6). All data were obtained from the *Physikalisch-Technische Bundesanstalt* (PTB) database of the National Metrology Institute of Germany, available at *Physionet*. Each record includes 15 simultaneously measured signals: the conventional 12 leads (I, II, III,

I. Torres et al. (eds.), *VII Latin American Congress on Biomedical Engineering CLAIB 2016, Bucaramanga, Santander, Colombia, October 26th -28th, 2016*,
IFMBE Proceedings 60,
DOI: 10.1007/978-981-10-4086-3_125

aVR, aVL, aVF, V1-V6) together with the 3-Frank lead-ECGs (X, Y, Z). Each signal was digitized at 1000 Hz, with 16 bits of amplitude resolution [8]. It is underlined that only the orthogonal ECG leads (X, Y, Z) were used to derive the VCG. Since the number of patients for each infarcted area is small, we grouped them into two sets, a) patients that have the anterior area affected (n=49), denoted as MI-Anterior, and b) those who have infarcted the inferior zone (n=46), denoted as MI-Inferior.

III. Methods

All ECG records (X, Y, Z) were pre-processed as follows: a) A band-pass filter (Butterworth, 4th order, 0.2-100 Hz, bidirectional) to reduce low and high frequency noise and a notch filter (Butterworth, 2th order, 50/60 Hz, bidirectional) to minimize the powerline interference were applied. b) A cubic spline interpolation filter was used to attenuate ECG baseline drifts and respiratory artifacts. c) When all records were filtered, the QRS-complex, T-waves and their end points were detected for each ECG using a wavelet-based technique [6]. d) Excessive noisy beats (with a RMS noise level >40μV, measured within a 40 ms window located at 2/3 of the RR interval) were excluded. e) In addition, ectopic beats were also eliminated by comparing incoming signals against a previously established template with the use of a cross-correlation technique. Besides, a visually low noise sinus beat extracted from the ECG record was selected as template (or reference).

After such ECG signal adaptation (or pre-processing), we carried out the feature extraction step. Figure 1 shows a general block diagram of the procedure.

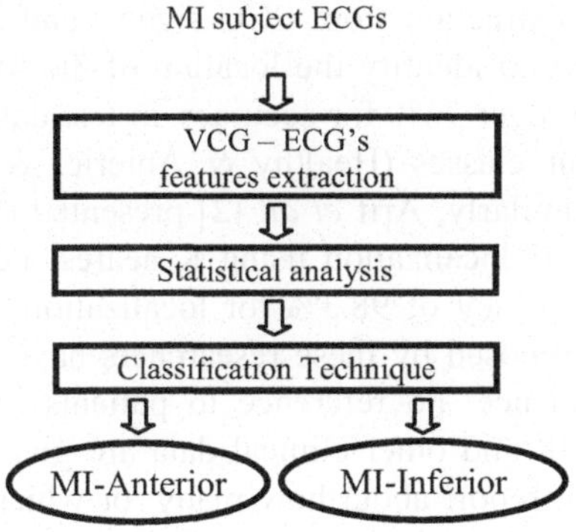

Figure 1-General block diagram of the proposed technique

A. Parameter Calculation

The VCG was obtained by drawing simultaneously in a 3-D plot the instantaneous amplitudes of XYZ orthogonal leads for each time sample in the temporal interval corresponding to the QRS complex. Herein, we analyzed 2 VCG parameters, 1 assessed on the QRS-loop and 1 between the QRS and T-loops. Also, 9 from the orthogonal ECG leads (X, Y, Z) were studied: 3 evaluated on the QRS complex waves and 6 on the ST-T segment. In the same way as in previous works, we have analyzed beat-to-beat morphological changes of the VCG and ECG [9], [10]. We defined,

- *QRS-loop Perimeter* ($\boldsymbol{QRS}_{P}$) [mV]: Computed around the QRS-loop projected over the Optimum Plane. It measures the loop total length and can detect loop contour changes [9].
- *Angle between the depolarization maximum vector (or QRS-complex waves) and the Repolarization (T-wave) in frontal, sagittal and transversal planes* ($\boldsymbol{aQRS\text{-}T}_{F}$,$\boldsymbol{aQRS\text{-}T}_{S}$,$\boldsymbol{aQRS\text{-}T}_{T}$): It reflects the angular difference between the maximum QRS and T vectors and estimates the Ventricular Gradient. This parameter is a strong and independent predictor of cardiac mortality [11], [12].
- *Area under the QRS complex in X, Y and Z leads* ($\boldsymbol{aQRS}_{X}$, $\boldsymbol{aQRS}_{Y}$, $\boldsymbol{aQRS}_{Z}$) [mVs]: It is defined as the difference area between the ECG signal at the current QRS complex and the abscissa. It estimates depolarization energy.
- *Area under the T-wave in X, Y and Z leads* ($\boldsymbol{aT}_{X}$, $\boldsymbol{aT}_{Y}$, $\boldsymbol{aT}_{Z}$) [mVs]: It is defined as the difference area between the ECG signal at the current T-wave and the abscissa. It estimates repolarization energy.
- *Area under ST segment in X, Y and Z leads* ($\boldsymbol{aST}_{X}$, $\boldsymbol{aST}_{Y}$, $\boldsymbol{aST}_{Z}$) [mVs]: It is defined as the difference area between the ECG signal at the current ST-segment and the abscissa. It estimates repolarization energy.

B. Statistical Analysis

The described indexes were computed for each detected sinus beat in each ECG record. Comparisons between mean values were computed for both populations using the non-parametric Mann-Whitney test.

The feature selection was performed with the Wilks Lambda (WL) stepwise method for identifying the variables that improve classification and reduce the number of variables in the Linear Discriminant Analysis (LDA) [13], [14].

Briefly, we selected eight ECG-VCG features that produced the lowest WL for the classification between MI-Anterior and MI-inferior and denoted them as Best Combination or BC-Ant-Inf.

C. Classification technique

After the descriptive analysis, linear discriminant analysis (LDA) was applied to the parameters obtained for each patient. A *leave one out cross validation* (*loo-cv*) method was applied to validate the model. The selected ECG-VCG features subsets are used for training, while one is saved and left out, using it to validate the discriminant function. Thereafter, the procedure is repeated until all of the ECG-VCG features subsets are used up in the validation.

IV. Results

From two VCG parameters, and 3 (XYZ)-ECG features, the study planned to quantify and assess the morphological differences between VCG-ECG records of patients with MI-anterior and MI-Inferior. Descriptive analysis results are shown in Figure 2 where mean values (M) and standard deviations (SD) are plotted. They were computed for each ECG-VCG record MI groups. The statistical significance p-values comparing both populations are also included. In patients with more than one ECG record during the first week after the MI episode, we averaged out the parameter values, so that each one corresponded to a single value for each feature.

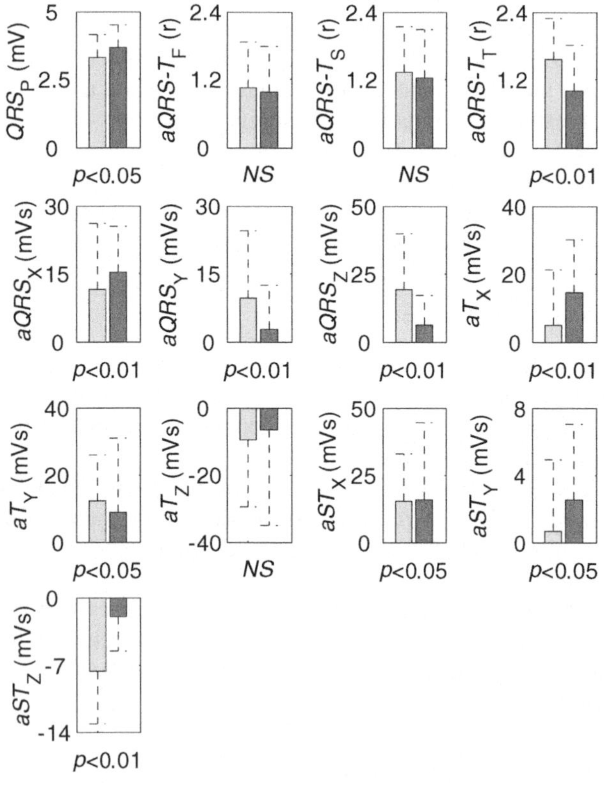

Figure 2- Mean values, standard deviations and statistical significance (p-value) of each parameter for MI-Ant (cyan) vs MI-Inf (magenta) population

To evaluate the classification performance, we determined 3 statistics that compare the results predicted by the proposed algorithm against the classification given by the database. These indices were sensitivity (Sen), specificity (Spe) and accuracy (Acc). Table 1 shows the results of the LDA classification using each parameter and the best combination (BC) of them. The BC of parameters was determined using the Lambda Wilks minimization technique, which determines the smallest set of discriminant variables that contributes most to the group differentiation [13], [15]. Hence, it was found that the BC was $\boldsymbol{QRS_P}$, $\boldsymbol{aQRS_X}$, $\boldsymbol{aQRS_Y}$, $\boldsymbol{aQRS_Z}$, $\boldsymbol{aQRS\text{-}T_S}$, $\boldsymbol{aT_X}$, $\boldsymbol{aST_Y}$ and $\boldsymbol{aST_Z}$.

Table 1 Classification outcomes for each computed index and for the BC

	QRS_P	$aQRS\text{-}T_F$	$aQRS\text{-}T_S$	$aQRS\text{-}T_T$	$aQRS_X$	$aQRS_Y$
Se (%)	58,70	58,70	52,17	76,09	79,45	65,62
Sp (%)	65,31	42,86	51,02	67,35	84,00	72,50
Ac (%)	62,11	50,53	51,58	71,58	81,07	68,07

	$aQRS_Z$	aT_X	aT_Y	aT_Z	aST_X	aST_Y	aST_Z	BC
Se (%)	58,70	54,35	60,87	43,48	6,52	58,70	78,26	**95,65**
Sp (%)	65,31	61,22	59,18	55,10	69,39	59,18	69,39	**91,84**
Ac (%)	62,11	57,89	60,00	49,47	38,95	58,95	73,68	**93,68**

V. Discussion and conclusion

In recent years, several researchers have proposed different ways to address the identification of the infarcted heart area, using the same PTB database. So, Safdarian *et al.* [5] used feature extraction from ECG signals and applied several classifiers to identify the location of the MI; these authors reached over 76% for accuracy in test data for localization in four classes (Healthy & Anterior & Inferior & Posterior). Similarly, Arif *et al.* [2] presented an automatic method for MI localization using K-nearest neighbor, obtaining an accuracy of 98.3% for localization. Though the techniques proposed by these researchers have shown very good performance, no reference to patients' age, gender, days since MI, and other clinical data are given. As far as we know, not report about how many, or which records of each patient were used.

This work proposes a new technique for estimating the localization of the infarcted area in two groups (MI-ant and MI-inf). With this aim, we only used the ECG's obtained during the week after MI for each patient and proposed two VCG and 3 (XYZ)-ECG indexes, such as: (a) perimeter of the QRS loop assessed on 3-D VCG representation ($\boldsymbol{QRS_P}$);

(b-d) Area under QRS complex, T wave and ST segment in X, Y, Z, leads (***aQRS***, ***aT*** and ***aST***, respectively) and (e) The spatial angle between QRS complex and T wave (***aQRS-T***) was also analyzed (see Figure 2). It can be seen that M's have a large dispersion because their SD's are often close to the M-values. However, the statistical comparison produced significant differences (p-value <0.05) in 10 of the 13 analyzed indexes (Figure 2), suggesting that they may be used to correctly differentiate between MI inferior and anterior.

In Table 1, it can be seen that the parameters with greater individual discriminating power between patients with MI inferior and Anterior are $aQRS_Z$ and aST_Z, also when combined with QRS_P, $aQRS_X$, $aQRS_Y$, $aQRS\text{-}T_S$, aT_X and aST_Y, with a Sen = 95,65%, Spe = 91,84% and Acc=93,68%.

We conclude that the new multivariable MI patient localization technique based on ECG and VCG indexes could be used to estimate the localization of the infarcted area (in two groups MI-ant and MI-inf). It could be used as an alternative diagnostic technique in the emergency room.

Acknowledgment

This study was supported by grants from *Consejo Nacional de Investigaciones Científicas y Técnicas* (CONICET – PIP 2014-2016 11220130100552CO and 2016-2018 11220150100059CO) and *CICITCA-UNSJ* (Res. 3635-CS)

References

[1] K. Thygesen, J. S. Alpert, A. S. Jaffe and others "Third Universal Definition of Myocardial Infarction," *Journal of the American College of Cardiology*, vol. 60, no. 16, pp. 1581–1598, 2012.

[2] M. Arif, I. A. Malagore, and F. A. Afsar, "Detection and localization of myocardial infarction using K-nearest neighbor classifier," *Journal of medical systems*, vol. 36, no. 1, pp. 279–289, 2012.

[3] G. Bakul and U. Tiwary, "Automated risk identification of myocardial infarction using relative frequency band coefficient (RFBC) features from ECG," *The open biomedical engineering journal*, vol. 4, p. 217, 2010.

[4] A. Keshtkar, H. Seyedarabi, P. Sheikhzadeh, and others, "discriminant analysis between myocardial infarction patients and healthy subjects using Wavelet Transformed signal averaged electrocardiogram and probabilistic neural network," *Journal of Medical Signals and Sensors*, vol. 3, no. 4, 2013.

[5] N. Safdarian, N. J. Dabanloo, and G. Attarodi, "A New Pattern Recognition Method for Detection and Localization of Myocardial Infarction Using T-Wave Integral and Total Integral as Extracted Features from One Cycle of ECG Signal," *Journal of Biomedical Science and Engineering*, vol. 2014, 2014.

[6] R. Correa, P. Arini, L. Correa, M. Valentinuzzi, and E. Laciar, "Identification of Patients with Myocardial Infarction: Vectorcardiographic and Electrocardiographic Analysis," *Methods of Information in Medicine*, vol. 55, pp. 242–249, 2016.

[7] R. Correa, P. Arini, L. Correa, M. Valentinuzzi, and E. Laciar, "Clasificación automática de pacientes con Infarto de Miocardio," *XX Congreso Argentino Bioingeniería y IX Jornadas Ingeniería Clínica (SABI2015), 28-30 October 2015, San Nicolás-Buenos Aires-Argentina*, 2015.

[8] R. Bousseljot, D. Kreiseler, and A. Schnabel, "Nutzung der EKG-Signal daten bank CARDIODAT der PTB über das Internet," *Biomedizinische Technik, Band 40, Ergänzungsband*, vol. 1, p. 317, 1995.

[9] R. Correa, P. D. Arini, L. Correa, M. E. Valentinuzzi, and E. Laciar, "Acute myocardial ischemia monitoring before and during angioplasty by a novel vectorcardiographic parameter set," *Journal of electrocardiology*, vol. 46, no. 6, pp. 635–643, 2013.

[10] R. Correa, P. D. Arini, L. S. Correa, M. Valentinuzzi, and E. Laciar, "Novel technique for ST-T interval characterization in patients with acute myocardial ischemia," *Computers in biology and medicine*, vol. 50, pp. 49–55, 2014.

[11] I. Kardys, J. A. Kors, I. M. van der Meer, A. Hofman, D. A. . van der Kuip, and J. C. M. Witteman, "Spatial QRS-T angle predicts cardiac death in a general population," *European Heart Journal*, vol. 24, no. 14, pp. 1357–1364, 2003.

[12] A. Rubulis, L. Bergfeldt, L. Rydén, and J. Jensen, "Prediction of cardiovascular death and myocardial infarction by the QRS-T angle and T vector loop morphology after angioplasty in stable angina pectoris: an 8-year follow-up," *Journal of electrocardiology*, vol. 43, no. 4, pp. 310–317, 2010.

[13] D. Altman, *Practical statistics for medical research*. Chapman and Hall/CRC; 1st ed edition.

[14] J. G. Flores, E. G. Jiménez, and G. R. Gómez, *Análisis discriminante*. La Muralla, 2001.

[15] J. Flores, E. Jiménez, and G. G, *Análisis Discriminante*. Editorial La Muralla, S.A.; Edición: 1, 2001.

Corresponding author:
Author: Raúl Correa
Institute: [1]Gabinete de Tecnología Médica, Facultad de Ingeniería, Universidad Nacional de San Juan (UNSJ), San Juan, Argentina
Street: Av Libertador Gral San Martín 1109 (O)
City: San Juan
Country: Argentina
Email: rcorrea@gateme.unsj.edu.ar

Incorporation of Current Density Map Method in the Epilepsy Surgery Protocol

A.R. García[1], J.P. Graffigna[1] and R. Otoya[2]

[1]Gabinete de Tecnología Médica, Facultad de Ingeniería, Universidad Nacional de San Juan
[2]Neuromed, Mendoza, Argentina

Abstract— **A non-invasive method is developed to identify brain regions compatible with epileptogenic foci. It combines different programming techniques such as free software, electroencephalographic signal processing, MRI and CT image processing.**

This method has been included in a protocol of epilepsy, in order to evaluate its performance compared with most commonly used techniques such as ECoG and SEEG. This method has been validated with seven patients against the diagnosis of a medical specialist in the subject.

We found that this method is consistent with the results obtained through ECoG and SEEG.

Keywords— **Epilepsy, electroencephalogram, seizure focus, surgery, inverse method.**

I. Introduction

This work aims to improve the accuracy of location techniques seizure focus on pre-surgical studies, including it in the epilepsy surgery protocol.

Epilepsy is a chronic neurological disorder that affects people of all ages. It is estimated that there are approximately 50 million people with epilepsy worldwide, representing about 1% of the world population [1].

Recent studies have revealed that up to 70% of patients diagnosed with epilepsy can be successfully treated, i.e. have their seizures completely controlled, with anticonvulsant drugs. For patients that respond poorly to drug treatment - the focus of study in this work and approximately 30% of epileptic patients- surgical treatment may be useful [1].

From the neuropsychological studies to the latest imaging techniques, there is a variety of diagnostic techniques for the location of the epileptic area. In some cases Magnetic Resonance Imaging (MRI) seems to be the technique of choice because it allows extraordinary precision to locate the anatomical area responsible of irritative focus. However, the seizure focus does not always coincide with a structural alteration, and in such cases functional imaging techniques such as Single Photon Emission Computed Tomography (SPECT) and Positron Emission Tomography (PET) are recommended.

The limitation of these techniques is the temporal resolution. On the other side, neuroimaging techniques that have a higher temporal resolution, e.g. electroencephalography (EEG), is deficient in spatial resolution. Which is why this work presents a noninvasive technique [2] based on software called *Brainstorm* [3], which combines the excellent temporal resolution of EEG with the very good spatial resolution of MRI in order to identify the current density area compatible with a possible seizure focus. In turn, this study seeks to demonstrate the integration of this technique in the epilepsy protocol, defining its aid in the diagnosis for different types of cases, and contrasting the results with currently used invasive searches such as deep (SEEG) and cortical (ECoG) electrodes.

II. Materials and methods

Fig. 1 shows the method developed [2]. For the execution, volumes of CT and MRI used, and EEG records of 128 electrodes.

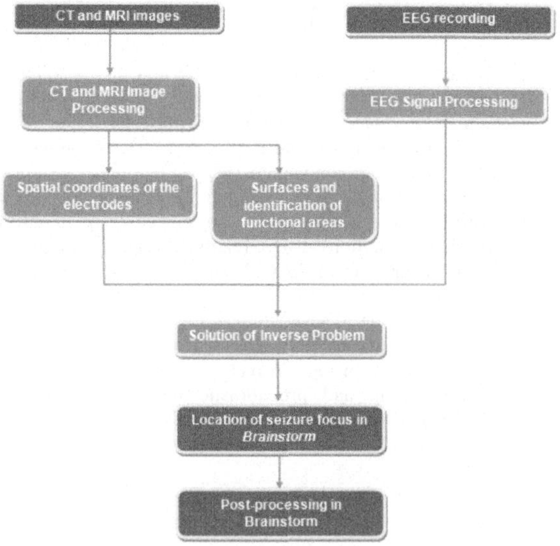

Fig.1. Steps followed in the identification of the epileptogenic focus.

I. Torres et al. (eds.), *VII Latin American Congress on Biomedical Engineering CLAIB 2016, Bucaramanga, Santander, Colombia, October 26th -28th, 2016*,
IFMBE Proceedings 60,
DOI: 10.1007/978-981-10-4086-3_126

The first essential step is to obtain images and signals with the following studies:

• A head CT, with EEG cap to determine the spatial location of the electrodes.

• A head MRI with sequences that the physician deems appropriate.

• A Video-EEG of 128 electrodes, to document a set of common seizures, defined by a relative or witness.

Then, a procedure is performed to extract features specific characteristics. Among these processes it is to obtain the spatial coordinates of the electrodes, which is a technique that is characterized by locating the electrodes over tomography volumes. Using image processing algorithms, electrodes descriptors are obtained, such as its centroid, determining its spatial position, among others. Then these points are transformed into the coordinate space of the MRI through a process called registration [4].

Then surfaces and identification of functional areas, with software called *BrainSUITE* [5] in which MRI images are inserted and extraction occurs four main surfaces are obtained. These are the cortex, inner and outer skull and scalp of the patient.

For the solution of the inverse problem we worked with distributed models based on mathematical constraints. A comparison was made between different models [6] concluding that *sLORETA* (standardized low resolution brain electromagnetic tomography) [7] was the best choice.

III. RESULTS

Studies have been compiled from seven patients under informed consent, including four males and three females. Each of these patients was subjected to epilepsy protocol designed, which was included in the noninvasive method proposed.

For each of the patients the method was adapted in a right way, finding with great precision the epileptogenic focus area compatible with higher current density in the cerebral cortex of the patient. Fig. 2, 3, 4 show epileptogenic foci of different patients.

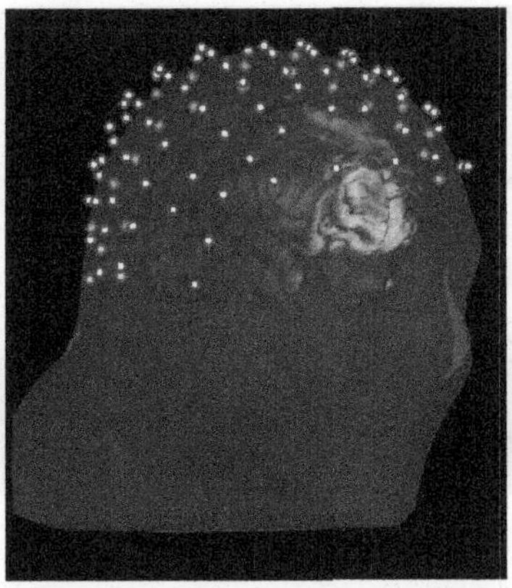

Fig. 2a. Higher current density area, in patients with seizure focus in the front area. EEG electrodes placed on the patient's head.

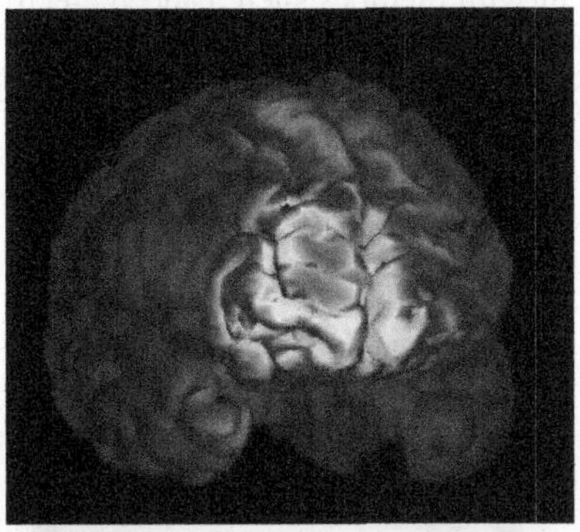

Fig. 2b. Higher current density area in the cerebral cortex of patients with seizure focus in the front area.

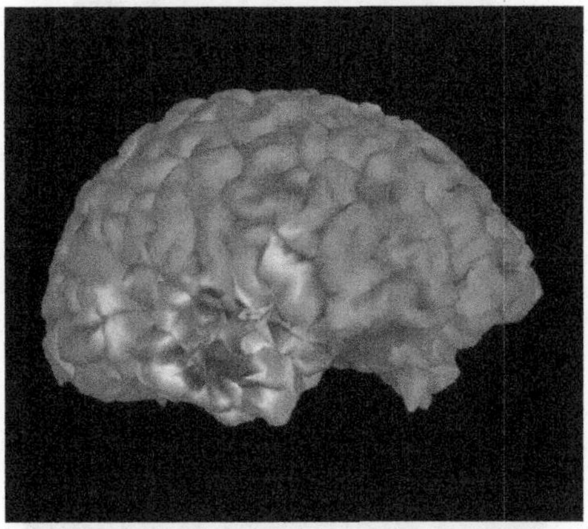

Fig.3. Higher current density area in the cerebral cortex of patients with seizure focus in the temporal area.

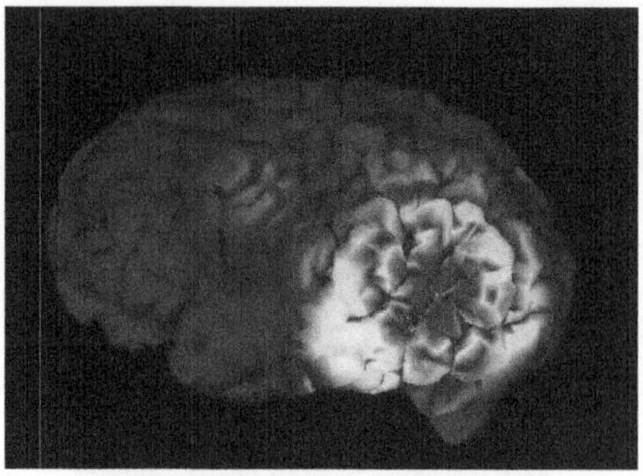

Fig.4. Higher current density area in the cerebral cortex of patients with seizure focus in the left temporal area.

The proposed method was applied to patients who have had treatment with methods such as deep electrodes (SEEG) and electrocorticography (ECoG). Fig. 5 shows an example in which SEEG was used and deep electrodes it can be identified.

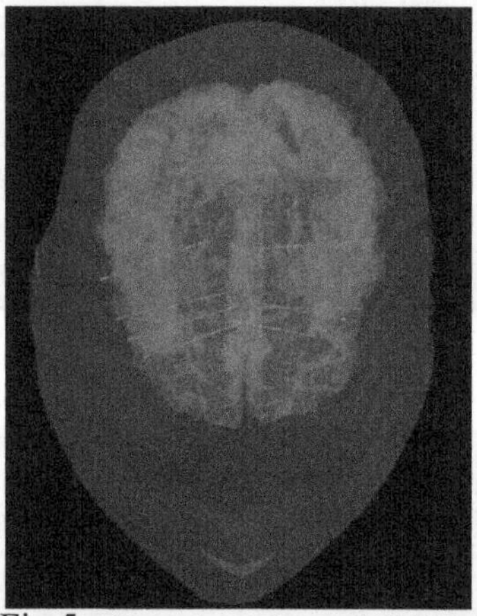

Fig.5a. Deep electrodes used (SEEG).

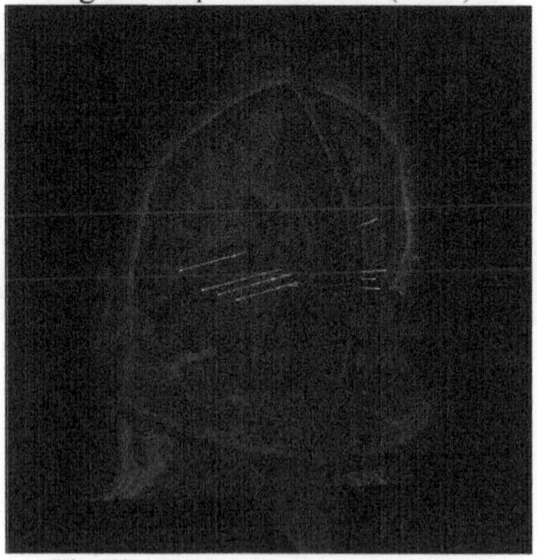

Fig.5b. Deep electrodes used (SEEG).

Following the analysis of the relevant area to the SEEG system, a seizure focus in the left temporal area is manifested, which was consistent with the information provided by the non-invasive method. In Fig. 6 it can be seen the higher current density area in the cerebral cortex of the patient, which matches the position expressed by the deep electrodes.

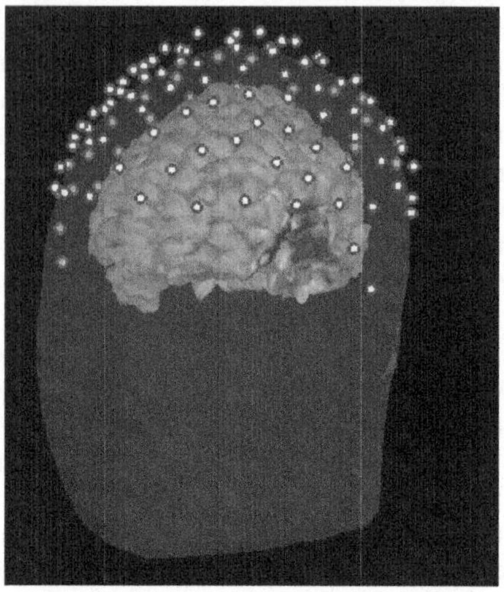

Fig.6. Higher current density area in patients with seizure focus in the left temporal area, which coincides with information provided by SEEG.

Finally the proposed non-invasive method was compared with ECoG. The location of the electrodes in the cortex of the patient is shown in Fig. 7.

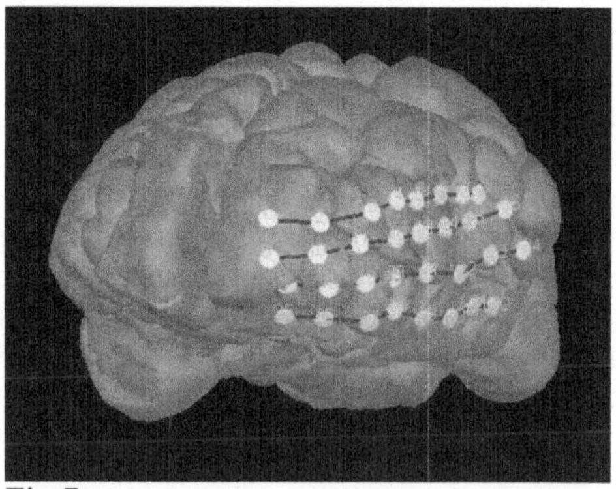

Fig.7. ECoG used for a detailed analysis of the case

The results provided by the non-invasive method is consistent with those reported by the ECoG. Fig. 8 exhibits the greatest current density area compatible with seizure focus located in the left temporal area.

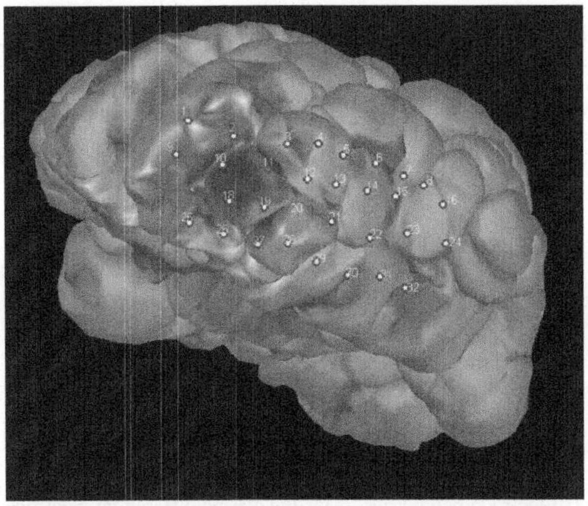

Fig. 8a. Higher current density area in patients with seizure focus in the temporal area. ECoG electrodes seen in the cerebral cortex.

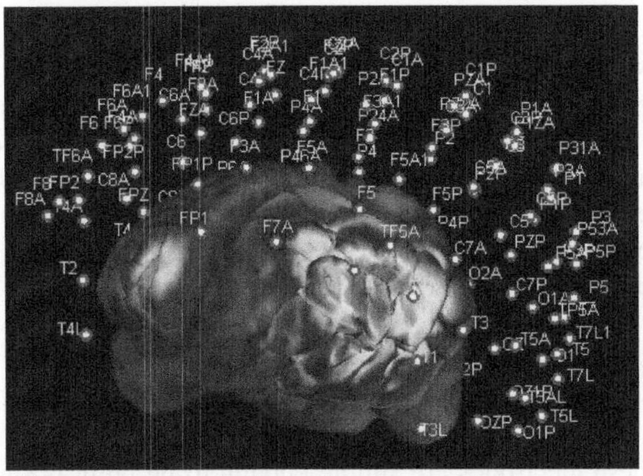

Fig. 8b. Higher current density area in patients with seizure focus in the temporal area. EEG electrodes displayed.

IV. Conclusion

This work has consisted in the integration of a non-invasive method for localization epileptogenic foci in epilepsy protocol. Within this protocol, there are different types of studies that aim to achieve a complete reading of the situation of each patient and try to establish a diagnosis as accurate as possible.

Furthermore, this method was adapted for each patient taking into account the anatomical variations and particular circumstances.

It can be noted that this non-invasive method has been able to exhibit areas of higher current density in the cerebral cortex of different patients, coinciding with the remaining areas of diagnostic methods manifested as potential epileptogenic foci.

In one case it coincided exactly with SEEG, establishing a potential epileptogenic focus in the same area. In another patient it matched perfectly with the information provided by the ECoG. The nature of use of these diagnostic methods is their high efficiency, however they are surgeries that carry all risks associated to any surgical intervention [8]. Hence, there is benefit of the information provided by the hereby proposed non-invasive method, which exhibits equally accurate performance when compared to invasive methods.

V. References

1. World Health Organization (2016). Epilepsy. Note No. 999
2. García A.R., Laciar Leber E., Graffigna J.P., Otoya R. (2014). Aplicación clínica de un método para análisis de mapas de densidad de corriente en pacientes epilépticos. CLAIB 2014.
3. Tadel F., Baillet S., Mosher J.C., et al (2011). Brainstorm: A User-Friendly Application for MEG/EEG Analysis. Computa-tional Intelligence and Neuroscience.
4. Bustos, J.J (2011). Validación de Técnica de Detección de Electrodos basada en Imágenes de MR y CT. Instituto Balseiro. Comisión Nacional de Energía Atómica-Universidad Nacional de Cuyo.
5. Official Website of BrainSuite http://brainsuite.loni.ucla.edu/
6. Pascual-Marqui R.D. Sitio oficial de LORETA. http://www.uzh.ch/keyinst/NewLORETA/ComparingLORETA/Comparing LORETA.htm
7. Pascual-Marqui R.D (2002). Standardized low resolution brain electro-magnetic tomography (sLORETA): technical details. Methods & Findings in Experimental & Clinical Pharmacology
8. Kochen S., Thomson A.E., García M.C (2006). Guía de cirugía de la epilepsia. Revista Neurológica Argentina, Vol. 31, pp. 117-122.

Signal modes for design-oriented analysis of active sEMG spatial filter electrodes

F.N. Guerrero[1], P.A. García[1] and E.M. Spinelli[1]

[1] LEICI, CONICET-UNLP, La Plata, Buenos Aires, Argentina

Abstract— **In this work, the signal modes necessary for the analysis of amplifiers for double differential and normal double differential electrodes are obtained for the first time. Superficial electromyographic (sEMG) electrodes may have multiple dry contacts with particular geometric arrangements in order to better pick up signals originated in motor units just below the electrode while rejecting other EMG signals. The amplifiers used for these electrodes are thus not differential and in order to evaluate their performance, the commonly used decomposition into common mode and differential mode is not applicable. Hence, a set of orthogonal signal modes were derived considering linear combinations of the input signals of a selection of spatial filter amplifiers. The first signal mode was associated with the desired output and the second to common mode electromagnetic interference, but there are other modes present in the amplifier that account for EMG crosstalk interference. The rejection coefficients associated with these modes were derived and the obtained laws were applied to a comparison of two double-differential electrode circuit designs.**

Keywords— **Superficial electromyography, double differential electrodes, dry electrodes, signal modes, biopotential amplifier**

I INTRODUCTION

Superficial electromyography (sEMG) signals are useful in medicine for clinical diagnosis, in rehabilitation applications as a quantitative evaluation of the muscle and nervous system progress[1], and in neuroprostheses as input reference signals for human-machine interfaces [2].

Usually, a differential pair of electrodes with an inter-electrode distance between 0.5 cm to 4 cm is placed on the skin in the closest possible locations above the muscles which activity it is desired to know. However, in general, superficial electrodes measure a superposition of signals from various EMG sources. Hence, EMG signals from muscles or motor units that are not of interest can overlap the desired components originated in nearby sources and are known as *crosstalk* [3].

Unwanted EMG signals can be rejected by use of multi-electrode configurations that create spatial filters[4][5], i.e., the gain for signals originated within a specific volume directly below the electrode is higher relative to those outside this volume, compared with the differential pair electrode.

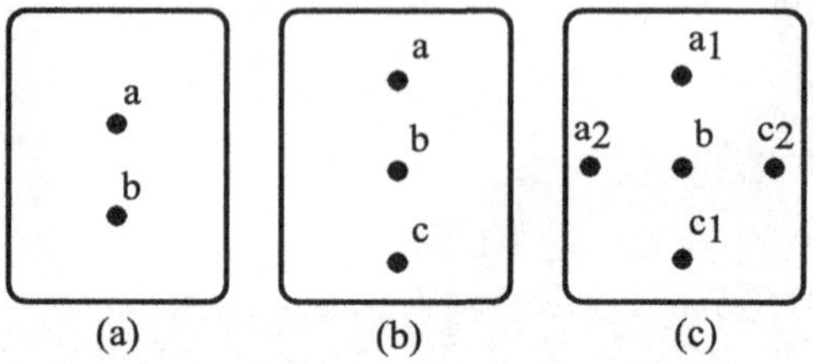

Figure 1: Electrode geometries. (a) is the classic differential electrode pair, (b) is the double differential or DD electrode, and (c) the Normal double differential or NDD electrode.

Spatial filters of varying degrees of complexity can be constructed [6]. The simplest configurations are the double differential (DD) and normal double differential (NDD) electrodes shown in Fig. 1. They are characterized by the number of contacts, their geometric disposition, and the weights assigned to each electrode, so their output is:

$$V_{o,DD} = V_a - 2V_b + V_c \quad (1)$$

$$V_{o,NDD} = V_{a1} + V_{a2} - 4V_b + V_{c1} + V_{c2} \quad (2)$$

These electrodes are simple enough to be built as single standalone electrodes, and it is possible to do so because contacts are close to each other. There are commercially available devices such as the DD Ottobock 13E200 and Delsys DE-3.1, and the NDD Delsys dEMG. Other multi-electrode spatial filter configurations are more selective but because of their complexity, they are usually implemented on EMG arrays.

The best crosstalk rejection is achieved with inter-electrode distances in the order of 5 mm to 10 mm [7], so contacts must have a small area and should be of the dry type to avoid short-circuiting electrodes with paste. Moreover, dry electrodes are preferred in many applications because they enable comfortable and faster installation, according to the "wearable" paradigm [8].

Small area, dry electrodes however, have a higher impedance and variability compared with wet electrodes [9] and it is standard practice to include an amplifier on the electrode itself, conforming *active electrodes*, to help reject electromagnetic interference (EMI) [10].

When designing the amplifiers for active electrodes, their performance must be evaluated and compared with alterna-

I. Torres et al. (eds.), *VII Latin American Congress on Biomedical Engineering CLAIB 2016, Bucaramanga, Santander, Colombia, October 26th -28th, 2016*,
IFMBE Proceedings 60,
DOI: 10.1007/978-981-10-4086-3_127

tives. In a typical differential amplifier with inputs V_a and V_b, useful parameters can be obtained through an orthogonal decomposition of signals into *signal modes* suitable for analysis: the differential mode (DM) and the common mode (CM) [11], that are defined through the following transformation:

$$\begin{bmatrix} V_{\mathrm{DM}} \\ V_{\mathrm{CM}} \end{bmatrix} = \begin{bmatrix} 1 & -1 \\ 1/2 & 1/2 \end{bmatrix} \begin{bmatrix} V_a \\ V_b \end{bmatrix} \quad (3)$$

In biopotential measurements, this decomposition is useful because the desired signal is usually the potential difference between two electrodes. Meanwhile, well-established EMI models (e.g. [12]) consider impedances inside the body to be sufficiently small so as to consider the body an equipotential volume; therefore EMI potentials are equally present on all electrodes, i.e., they are mostly CM signals.

Ideally, the output of an amplifier with single-ended output would be $V_o = G_D V_{DM}$. However, a practical amplifier will cause *mode transformaions* and as a result undesired signals will reach the output. This effect can be quantified considering gains from all modes to the output:

$$V_o = V_{\mathrm{DM}} G_{\mathrm{DD}} + V_{\mathrm{CM}} G_{\mathrm{CD}} = G_{\mathrm{DD}} \left(V_{\mathrm{DM}} + \frac{G_{\mathrm{CD}}}{G_{\mathrm{DD}}} V_{\mathrm{CM}} \right) \quad (4)$$

Where the quotient $G_{\mathrm{DD}}/G_{\mathrm{CD}}$ is called *common mode rejection ratio* (CMRR), and is profusely used as a figure of merit for amplifiers.

Active sEMG DD and NDD electrodes have multiple inputs, and hence equation 3 cannot be applied to their analysis. Moreover, as the signal space has a higher order, new modes must be defined and their mode transformation to the output analyzed.

II Method

Signal modes were obtained considering a signal space stemming from the N input electrode potentials of the amplifier $V_{e1} \ldots V_{eN}$, and a linear transformation to obtain the N modes $V_{m1} \ldots V_{mN}$.

The rationale to define the modes was first to obtain a mode corresponding to the output signal, and then orthogonal modes that could identify signals originated in independent sources. This is analogous to the DM-CM decomposition in a differential-input amplifier.

III Results and Discussion

A DD electrode signal modes

The DD electrode has 3 contact with voltages V_a, V_b , and V_c as shown in Figure 1.

The first mode was obtained directly as the output equation and hence was called double differential mode or DDM. The second mode should account for EMI. EMI is present equally on the 3 input contacts, analogous to the CM of a differential amplifier, and so it can also be called common mode.

The purpose of the DD electrode compared with a differential electrode is to increase spatial selectivity. Then, differential EMG signals contains crosstalk components that should be rejected by a DD electrode. The third mode of the DD electrode can then account for this component. Effectively, a differential mode (DM) defined as the difference of 2 input nodes is orthogonal to the DDM and CM previously defined. Hence, the modes and transformation matrix for the DD amplifier result:

$$\begin{bmatrix} V_{\mathrm{DDM}} \\ V_{\mathrm{CM}} \\ V_{\mathrm{DM}} \end{bmatrix} \begin{bmatrix} 1 & -2 & 1 \\ 1/3 & 1/3 & 1/3 \\ 1 & 0 & -1 \end{bmatrix} \begin{bmatrix} V_a \\ V_b \\ V_c \end{bmatrix} \quad (5)$$

The output of a practical amplifier can now be written in terms of these modes:

$$V_o = G_{\mathrm{DD}} V_{\mathrm{DDM}} + G_{\mathrm{C}} V_{\mathrm{CM}} + G_{\mathrm{D}} V_{\mathrm{DM}} \quad (6)$$

$$= G_{\mathrm{DD}} \left(V_{\mathrm{DDM}} + \frac{G_{\mathrm{C}}}{G_{\mathrm{DD}}} V_{\mathrm{CM}} + \frac{G_{\mathrm{D}}}{G_{\mathrm{DD}}} V_{\mathrm{DM}} \right) \quad (7)$$

Hence, two figures of merit can be defined for the DD amplifier: a common mode rejection ratio CMRR $:= \mathrm{G_{DD}}/\mathrm{G_C}$ and a differential mode rejection ration $DMRR := G_{DD}/G_D$. The CMRR as well as in the differential amplifier case should be in the range of 90 dB to 110 dB for good quality EMG measurements [7]. The DMRR however, is an indication of rejection of crosstalk EMG components. Because this signal has the same order of magnitude than the signal of interest, a DMRR in the order of 20 dB to 40 dB is sufficient.

B NDD electrode signal modes

The NDD electrode has five inputs as shown in Figure 1.c:

$$\vec{V_{in}} = \begin{bmatrix} V_{a1} & V_{a2} & V_b & V_{c1} & V_{c2} \end{bmatrix}^T \quad (8)$$

The first mode is the Normal Double Differential mode (NDDM), and a common mode can also be defined:

$$\begin{bmatrix} V_{\mathrm{NDDM}} \\ V_{\mathrm{CM}} \end{bmatrix} = \begin{bmatrix} 1 & 1 & -4 & 1 & 1 \\ 1/5 & 1/5 & 1/5 & 1/5 & 1/5 \end{bmatrix} \vec{V_{in}} \quad (9)$$

As in the DD electrode, differential signals picked up by this electrode would have crosstalk components to be rejected, so they can be identified as interference modes. This electrode has two such modes, determined by pairs of electrodes in directions normal to each other:

$$\begin{bmatrix} V_{\mathrm{DM1}} \\ V_{\mathrm{DM2}} \end{bmatrix} = \begin{bmatrix} 1 & -1 & 0 & 0 & 0 \\ 0 & 0 & 0 & 1 & -1 \end{bmatrix} \vec{v_{in}} \tag{10}$$

Finally, following the same rationale, the NDD electrode is an improvement over the DD electrode in that it can better reject crostalk components not only when it is placed along muscle fiber direction but also across. Hence, components that would be picked up by a DD electrode must be rejected. The weights assigned to the fifth mode effectively create a DD electrode with the NDD geometry:

$$V_{\mathrm{DDM}} = \begin{bmatrix} 1 & -1 & 0 & -1 & 1 \end{bmatrix} = \vec{v_{in}} \tag{11}$$

The output equation of a practical NDD amplifier is:

$$\begin{aligned} V_o &= G_{\mathrm{NDD}} V_{\mathrm{NDDM}} + G_{\mathrm{C}} V_{\mathrm{CM}} \\ &+ G_{\mathrm{D1}} V_{\mathrm{DM1}} + G_{\mathrm{D2}} V_{\mathrm{DM2}} + G_{\mathrm{DD}} V_{\mathrm{DDM}} \end{aligned} \tag{12}$$

And thus the figures of merit are the CMRR,DMRR_1 and DMRR_2, and the DDMRR. With the exeption of the CM, the interfering signals that the amplifier must reject are EMG signals and a value between 20 dB to 40 dB is appropriate.

C Comparison of two DD amplifier circuits

Figure 2 shows two circuits that implement the amplifier for a DD active electrode. In Figure 2a the amplifier is implemented with 3 instrumentation amplifiers (IAs) as found in [13], and in Figure 2b with a simpler 4 operational amplifiers (OAs) circuit found in [14]. Figures of merit for these amplifiers can be found according to the calculated modes and coefficients fom equations 5 and 7.

As a first approximation, a CM signal can be applied to the 3 IA circuit disregarding possible unbalances, with gains from the three amplifiers $G_1 = G_2 = G_3 = G$. This yields:

$$V_{o,\mathrm{CM}} = G\left(\frac{1}{\mathrm{CMRR}_{AI_1}} - \frac{1}{\mathrm{CMRR}_{AI_2}}\right) V_{\mathrm{CM}} \tag{13}$$

where CMRR_{AI_n} stands for the standard CMRR of instrumentation amplifier labeled IA_n.

When gain unbalances amongst the IAs are considered the output is:

$$V_o = G_3 \left(G_1 \left(V_a - V_b\right) - G_2 \left(V_b - V_c\right)\right) \tag{14}$$

Renaming gains to reflect an unbalance between G_1 and G_2 such that:

$$G_1 = G + \delta G \; ; \; G_2 = G - \delta G$$

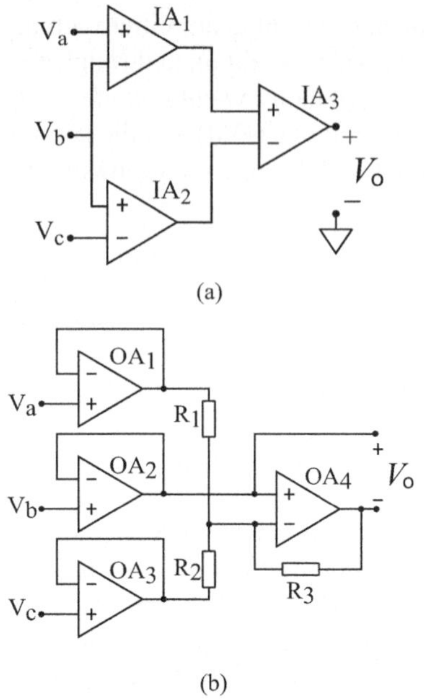

Figure 2: Amplifier circuits that implement the double differential function.

the expression results:

$$\begin{aligned} V_o &= G_3 \left(G\left(V_a - 2V_b + V_c\right) - \delta G \left(V_a - V_c\right)\right) \\ &= G_3 G \left(V_{\mathrm{DDM}} + \frac{\delta G}{G} V_{\mathrm{DM}}\right) \end{aligned} \tag{15}$$

Because of the modes definition, it is easily seen in Equation 15 that an unbalance in amplifier gain would yield a mode transformation producing crosstalk components to interfere with the DDM.

In the circuit from Figure 2b a perfect resistor balance would produce a gain $G_{DDM} = R_3/R_1$. A CM signal applied to the input would produce an output depending on the CMRR and open loop gain (A_{OL}) of operational amplifier OA_4, such that, disregarding second-order effects:

$$V_{o,\mathrm{CM}} = \left(1 + \frac{R_3}{R_1}\right)\left(\frac{1}{\mathrm{CMRR}_{\mathrm{OA}_4}} + \frac{1}{A_{\mathrm{OA}_4}}\right) \tag{16}$$

Hence, if $R_3/R_1 >> 1$:

$$\mathrm{CMRR} \approx \left(\mathrm{CMRR}_{\mathrm{OA4}}^{-1} + \mathrm{A}_{\mathrm{OL,OA4}}^{-1}\right)^{-1} \tag{17}$$

If unbalances in resistors are considered, the output of the amplifier is:

$$V_o = V_a \frac{R_3}{R_1} + V_b \frac{R_3}{R_2} - V_c \frac{R_3}{R_1 || R_2} \tag{18}$$

This equation does not allow to gain insight into the circuit function, so it is convenient to elaborate a so-called "low entropy expression" [15].

Table 1: Figures of merit of DD amplifiers

Amplifier	CMRR	DMRR
3 IA	$\left(\mathrm{CMRR}_{\mathrm{IA1}}^{-1}-\mathrm{CMRR}_{\mathrm{IA2}}^{-1}\right)^{-1}$	$G/\delta G$
4 OA	$\left(\mathrm{CMRR}_{\mathrm{OA4}}^{-1}+\mathrm{A}_{\mathrm{OL,OA4}}^{-1}\right)^{-1}$	$\bar{R}/\delta R$

If the unbalance is expressed such that $\bar{R}=(R_1+R_2)/2$ and $\delta R=R_1-R_2$ the following expression can be written:

$$V_o=\frac{R_3}{R_1R_2}\left(\bar{R}(V_a-2V_b+V_c)+\delta R(V_c-V_a)\right) \tag{19}$$

where the obtained modes can be recognized by simple inspection arriving to the more useful equation:

$$V_o=\frac{R_3}{2(R_1||R_2)}\left(V_{\mathrm{DDM}}+\frac{\delta R}{\bar{R}}V_{\mathrm{DM}}\right) \tag{20}$$

The DMRR for the 4 OA circuit is then $\delta R/\bar{R}$.

The obtained figures of merit for the circuits are summarized in table 1. This comparison would not be possible if the modes proposed in this work had not been defined.

IV Conclusion

Usual differential input amplifier circuits can be analyzed using the usual decomposition of the input into a common mode and a differential mode signal. This decomposition is useful because each mode can be associated with signals from independent sources: the EMG signal of interest for the DM and and unwanted EMI for the CM. Double differential and Normal double differential sEMG standalone active electrodes however require amplifiers with 3 and 5 input nodes respectively and hence these modes are not suitable for its analysis.

Signal modes for the DD and NDD electrodes were deduced using an analogy with the DM-CM decomposition. Orthogonal transformations of the inputs to the signal modes were found yielding the expected modes associated with the desired electrode output (DDM and NDDM) and with EMI (CM for both electrodes). But most importantly, a differential mode is shown (2 for the NDD electrode) that must be rejected because it contains crosstalk components. Additionally, the NDD electrode has a double differential mode that detracts from its capacity to reject unidirectional crosstalk.

Hence, besides providing a proper definition for the CMRR of DD and NDD electrodes, further figures of merit were defined, necessary for the evaluation of amplifiers for DD electrodes (the DMRR) and NDD electrodes (DMRR and DDMRR). A sample application of the obtained coefficients was demonstrated analyzing two DD amplifier circuits.

Conflict of Interest

The authors declare that they have no conflict of interest.

Acknowledgements

This work was funded by UNLP project I-167 and ANPCyT project PICT-2012/0037.

References

1. Zwarts Machiel J, Stegeman Dick F. Multichannel surface EMG: basic aspects and clinical utility. *Muscle Nerve.* 2003;28:1–17.
2. Zecca M, Micera S, Carrozza M C, Dario P. Control of multifunctional prosthetic hands by processing the electromyographic signal *Crit. Rev. Biomed. Eng.*. 2002;30:459–85.
3. De Luca Carlo J., Merletti Roberto. Surface myoelectric signal crosstalk among muscles of the leg *Electroencephalogr. Clin. Neurophysiol.*. 1988;69:568–575.
4. Lynn P a, Bettles N D, Hughes a D, Johnson S W. Influences of electrode geometry on bipolar recordings of the surface electromyogram. *Med. Biol. Eng. Comput.*. 1978;16:651–60.
5. Dimitrov George V., Disselhorst-Klug Catherine, Dimitrova Nonna a., Schulte Elke, Rau Günter. Simulation analysis of the ability of different types of multi-electrodes to increase selectivity of detection and to reduce cross-talk *J. Electromyogr. Kinesiol.*. 2003;13:125–138.
6. Reucher H, Silny J, Rau G. Spatial filtering of noninvasive multielectrode EMG: Part II–Filter performance in theory and modeling. *IEEE Trans. Biomed. Eng.*. 1987;34:106–113.
7. Merletti Roberto, Parker Philip A.. *Electromyography: Physiology, Engineering, and Noninvasive Applications*;53. Hoboken: John Wiley & Sons 2004.
8. Ha Sohmyung, Kim Chul, Chi Yu M., et al. Integrated circuits and electrode interfaces for noninvasive physiological monitoring *IEEE Trans. Biomed. Eng.*. 2014;61:1522–1537.
9. Searle a, Kirkup L. A direct comparison of wet, dry and insulating bioelectric recording electrodes. *Physiol. Meas.*. 2000;21:271–283.
10. Fernandez M., Pallas-Areny R.. A simple active electrode for power line interference reduction in high resolution biopotential measurements *Proc. 18th Annu. Int. Conf. IEEE Eng. Med. Biol. Soc.*. 1996;1:97–98.
11. Pallás-Areny Ramón, Webster John G. *Analog signal processing*. John Wiley & Sons 1999.
12. Huhta J C, Webster J G. 60-HZ interference in electrocardiography. *IEEE Trans. Biomed. Eng.*. 1973;20:91–101.
13. Rieger Robert. Variable-gain, low-noise amplification for sampling front ends *IEEE Trans. Biomed. Circuits Syst.*. 2011;5:253–261.
14. Guerrero Federico Nicolás, Member Student, Spinelli Enrique Mario, Haberman Marcelo Alejandro. Analysis and Simple Circuit Design of Double Differential EMG Active Electrode *IEEE Trans. Biomed. Circuits Syst.*. 2015;10:1–9.
15. Middlebrook R.D.. Low-entropy expressions: the key to design-oriented analysis *Proc. Front. Educ. Twenty-First Annu. Conf. Eng. Educ. a New World Order.* 1991:399–403.

Author: Federico Nicolás Guerrero.
Institute: Instituto de investigaciones en electrónica, control y procesamiento de señales LEICI (CONICET-UNLP).
Street: Calle 116 y 48.
City: La Plata.
Country: Argentina.
Email: federico.guerrero@ing.unlp.edu.ar

An Age-based Multiscale Mathematical Model of the Hepatitis C Virus Life-cycle During Infection and Therapy: Including Translation and Replication

B.M. Quintela[1], J.M. Conway[2], J.M. Hyman[3], R. F. Reis[1], R.W. dos Santos[1], M. Lobosco[1] and A.S. Perelson[4]

[1] Universidade Federal de Juiz de Fora, Pós-Graduação em Modelagem Computacional, Juiz de Fora, Brazil
[2] The Pennsylvania State University, Department of Mathematics, University Park, USA
[3] Tulane University, Mathematics Department, New Orleans, USA
[4] Los Alamos National Laboratory, Theoretical Biology and Biophysics, Los Alamos, USA

Abstract— **The dynamics of hepatitis C virus (HCV) RNA during translation and replication were added to an age-based multiscale mathematical model of HCV infection and treatment. The model allows the study of the production of HCV inside infected cells, which are later released as virus in the plasma, available to infect other cells. This is the first model to our knowledge to consider both positive and negative strands of HCV RNA with an age-based multiscale approach. Simulation of the effectiveness of direct acting antiviral (DAA) drugs in blocking HCV RNA intracellular production and release of new virions, and enhancing their depletion were performed. Changes on the set of parameters allowed the validation of the model comparing it to distinct experiments: in vitro and in vivo under therapy.**

Keywords— **Mathematical Model, Differential Equations, Computational Biology, HCV, RNA.**

I INTRODUCTION

Chronic hepatitis C virus (HCV) infection affects about 130-150 million people worldwide and is the primary cause of liver cirrhosis and liver cancer [1]. HCV has a linear positive strand RNA molecule as its genome with ~ 9600 nucleotides and was classified as the genus *Hepacivirus* in the *Flaviridae* family [2]. For many years HCV infection was not completely understood due to the inability to culture virus *in vitro*. However, the recent development of a HCV cell culture (HCVcc) system allows better investigation of the processes that govern the HCV life cycle [3, 4]. Moreover, new forms of distinguishing and quantifying the RNA strands have been developed and improved [5].

There is no vaccine approved for HCV infection and until recently treatment used a combination of pegylated interferon (PEG-IFN), which acts as replication inhibitor, with ribavirin (RBV), which is been shown to enhance the action of IFN. This treatment was inadequate for a significant number of patients. The approval of antiviral agents that target small molecules, DAA agents, represents an advance in the treatment due to improved efficacy and tolerability, shorter treatement durations and the option of PEG-IFN and RBV-free threapy [6, 7, 8].

The remainder of this paper presents a short introduction to mathematical models of HCV, followed by the new age-based multiscale model developed based on previous works [9, 10] and a summary of the results.

II MATERIAL AND METHODS

A Related Works

Mathematical models have been used to help understand the dynamics of HCV [11, 12, 13, 14] and valuable insights can be obtained from modeling [15]. The effect of treatment has also been represented in the context of HCV infection with models for therapy with IFN-α [16] and in combination with ribavirin [17] and also therapy with DAAs [18, 19, 9]. A multiscale model including the dynamics of intracellular viral RNA coupled to a biphasic model of viral kinetics during treatment was also presented [9, 19].

Age-based models are widely used to study the epidemiology of infectious diseases, such as HIV [20], hepatitis C [21] and tuberculosis [22]. An age-structured model to represent the dynamics of within host HIV was presented in [23]. One advantage of using such an approach is the possibility of considering that individuals with distinct ages could behave differently [24]. Using that approach in modeling the dynamics of virus in the host allows a realistic representation of infection biology in which the rate of production and release of new virus is not constant but rather depends on the length of time a cell has been infected. Moreover, the model could also account for an infected cell death rate that depends on the time the cell has been infected.

An age-structured multiscale model using partial differential equations (PDE) was able to fit patient viral load data for 2 days. However, for longer times with therapy the model was

I. Torres et al. (eds.), *VII Latin American Congress on Biomedical Engineering CLAIB 2016, Bucaramanga, Santander, Colombia, October 26th -28th, 2016*,
IFMBE Proceedings 60,
DOI: 10.1007/978-981-10-4086-3_128

not able to reproduce the data due to drug resistance [9]. Including more viral strains in the model was pointed as one method to study those features not reproduced by the model with one positive-strand of viral RNA. Another work studied the intracellular dynamics of both positive- and negative-strand viral RNA [10]. They represented the number of intracellular viral genomic units or positive-strands of viral RNA, available for transcription and translation, and the number of negative-strands of viral RNA or the replication units.

B The Intracellular model of HCV RNA in Translation and Replication

We initially developed a mathematical model using ordinary differential equations to represent intracellular HCV infection based on the life-cycle of HCV [25]. The model allows the study of aspects such as: translation of positive-strand HCV RNA after cell entry, transfer of the positive strand to an environment such as the membranous web where it is used for replication, production of negative- and positive-strand HCV RNA in the replication complex and secretion of positive-strand RNA as virions. We then extended it to a PDEs set to include the age-based approach and the parameters to study the effects of therapy.

Age-based Intracellular model with therapy:

$$\begin{aligned}
&\frac{\partial}{\partial t}R_t(a,t)+\frac{\partial}{\partial a}R_t(a,t)=\theta R_c-\\
&\qquad(\sigma+(1-\varepsilon_s)\rho(a)+\kappa_t\mu_t)R_t,\\
&R_t(0,t)=R_{t_0},R_t(a,0)=\overline{R}_t(a),\\
&\frac{\partial}{\partial t}R_c(a,t)+\frac{\partial}{\partial a}R_c(a,t)=\\
&(1-\varepsilon_\alpha)\alpha R_m+\sigma R_t-(\theta+(1-\varepsilon_s)\rho(a)+\kappa_c\mu_c)R_c,\\
&R_c(0,t)=R_{c_0},R_c(a,0)=\overline{R}_c(a),\\
&\frac{\partial}{\partial t}R_m(a,t)+\frac{\partial}{\partial a}R_m(a,t)=\\
&\qquad(1-\varepsilon_r)r(1-\frac{R_m}{R_{max}})R_c-\kappa_c\mu_cR_m,\\
&R_m(0,t)=R_{m_0},R_m(a,0)=\overline{R}_m(a).
\end{aligned} \tag{1}$$

in which, R_t represents the positive-strand HCV RNA in translation, R_c the positive-strand HCV RNA in replication complex and R_m the negative-strand HCV RNA in replication complex. After cell entry, the cell machinery is used by the virus to translate the polyprotein from the HCV RNA genome released in the cytoplasm [26]. We assume that the positive-strand HCV RNA used for translation can form a replication complex at rate σ day^{-1}, be assembled and exported as a virus at rate ρ day^{-1}, and has a natural decay rate of μ_t day^{-1}. Positive-strand HCV RNA in the replication complex (R_c) can move out of the replication complex and become accessible for translation at a rate θ day^{-1} or be assembled and exported at rate ρ day^{-1}.

A negative-strand of HCV RNA (R_m) is formed in the replication complex by copying the positive strand R_c at rate r day^{-1}. It is assumed that there is a limitation on the number of negative-strand RNAs that can be replicated (R_{max}) with a logistic term. R_c replicates at a rate α day^{-1} and depends on the existence of R_m. We consider that both R_c and R_m are in the replication complex and decay at the same rate μ_c day^{-1}.

Experimentally, a delay in the secretion of new virus of approximately 12 hours after infection has been observed [27]. To consider this biological delay, ρ, the viral secretion rate, was implemented as a function of time. We assume that given a time delay τ, $\rho=0$ for $time<\tau$ and $\rho=(1-e^{-k(a-\tau)})*\rho$, otherwise, meaning that ρ increases exponentially until it reaches the predefined value to avoid discontinuity.

In order to study the effects of therapy, we consider that: ε_α represents the effectiveness of therapy in decreasing or blocking positive-strand RNA replication, ε_r represents the effectiveness of therapy in decreasing or blocking negative-strand RNA replication, κ_t is a factor to increase degradation of positive-strand RNA in translation, κ_c is a factor to increase degradation of both strand RNA in replication complex and ε_s represents the effectiveness of therapy in decreasing or blocking secretion of positive-strand RNA. When we are simulating therapy, $\overline{R}_t(a)$, $\overline{R}_c(a)$ and $\overline{R}_m(a)$ are the steady state distributions after the infection.

We then coupled the intracellular model to a established HCV infection mathematical model (Equation 2) [28, 14], modified to perform the coupling.

Modified Infection Model:

$$\begin{aligned}
&\frac{d}{dt}T(t)=s-\beta V(t)T(t)-dT(t),\\
&\frac{\partial}{\partial t}I(a,t)+\frac{\partial}{\partial a}I(a,t)=-\delta(a)I(a,t),\\
&I(0,t)=\beta V(t)T(t),I(a,0)=\bar{I}(a),\\
&\frac{d}{dt}V(t)=(1-\varepsilon_s)\int_0^\infty\rho(a)R_s(a,t)I(a,t)da-cV(t),
\end{aligned} \tag{2}$$

in which, T are the target cells, I, the infected cells and V the HCV in the plasma. Target cells become infected at rate β, have a constant source rate s and a natural decay rate of d. The parameter δ represents the probability of a infected cell survive the age "a". R_s represents the exported positive-strand HCV RNA (R_t + R_c) and the effects of therapy on the virus export are given by ε_s. The virus in the plasma could be cleared by the immune response at a rate c. $\bar{I}(a)$ is the steady-state distribution of infected cells $\bar{I}(a) = \beta V(t)T(t)e^{-\delta x}$. The coupling between the model is clear in the integral on the equation representing the virus in the plasma (V). The number of virus depends on the amount of infected cells and number of virions being packaged and exported per infected cells. That coupling was realized before by [9] considering only one equation for intracellular HCV RNA.

III RESULTS

The three-equation intracellular model was compared to experiments of transfection in vitro from the reference [27]. They quantified the number of positive and negative HCV RNA in Huh7.5.1 cells with real-time RT-PCR analysis of the HCV genome. Our model was able to reproduce the behavior of HCV RNA replication over 5 days since infection. Figure 1 shows the result of parameters fitting. We set $\rho = 0.1$ day^{-1} with a 12h delay and no therapy. Other parameters are: $t_0 = 0$, $R_{t_0} = 12.8$, $\alpha = 30$ day^{-1}, $\mu_t = 24$ day^{-1}, $r = 3.18$ day^{-1}, $\mu_c = 1.05$ day^{-1}, $R_{max} = 100$ molecules, $\sigma = 0.1$ day^{-1}, $\theta = 1.2$ day^{-1},

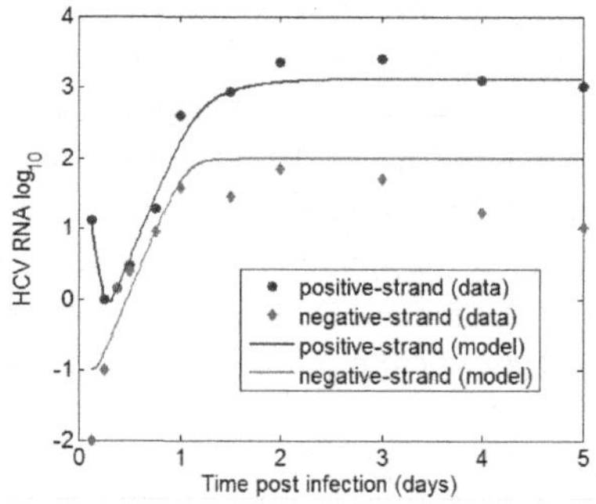

Fig. 1: Comparison of model results to in vitro infection data. Data points were extracted from Keum et al. (2012) and the lines were obtained with the translation and replication intracellular model.

The coupled multi-scale model was validated by fitting the Equations 1 and 2 data obtained from patients treated with one dose of 10 or 100 mg of Daclatasvir (DCV) [19]. DCV inhibits the action of NS5A, which has been shown to play an important role in HCV RNA replication and secretion [29, 30]. We used a distinct set of parameters to represent *in vivo* infection dynamics assuming there is no super-infection, with only one virus infecting each cell. The baseline *in vivo* set of parameters was based on the literature: $\alpha = 30\ d^{-1}$, $\rho = 8.18\ d^{-1}$, $\delta = 0.14\ d^{-1}$, c $= 22.3\ d^{-1}$, from [9] and $\kappa_t = \kappa_c = 1\ d^{-1}$, $\sigma = 1.3\ d^{-1}$, $\theta = 1.2\ d^{-1}$ $\varepsilon_\alpha = 0.99$, $\varepsilon_s = 0.998$, from [19].

Figures 2 - 3 depict the results obtained with the multi-scale model for one patient. We fixed the replication of positive HCV RNA $\alpha = 30\ d^{-1}$ and considered no enhancement on decay with therapy $\kappa_t = \kappa_c = 1\ d^{-1}$. We observed from the expected number per cell (Figure 3) that initiation of therapy affects the replication of both strands and there is a small increase in the number of positive strand HCV RNA in translation. Therapy blocks the appearance of new replication complexes which only decrease with the presence of the drug.

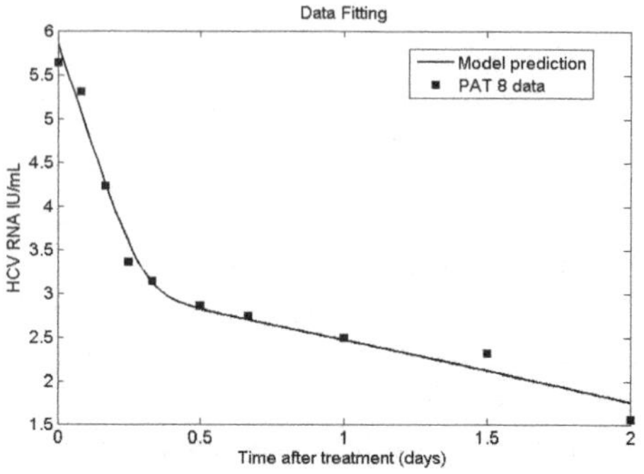

(a) PAT 8

Fig. 2: Parameters fitting of coupled multiscale model to the viral load of patient data from [19]. The model prediction is shown with lines and data points in squares. Patients were treated with one dose of 10 or 100 mg of daclatasvir.

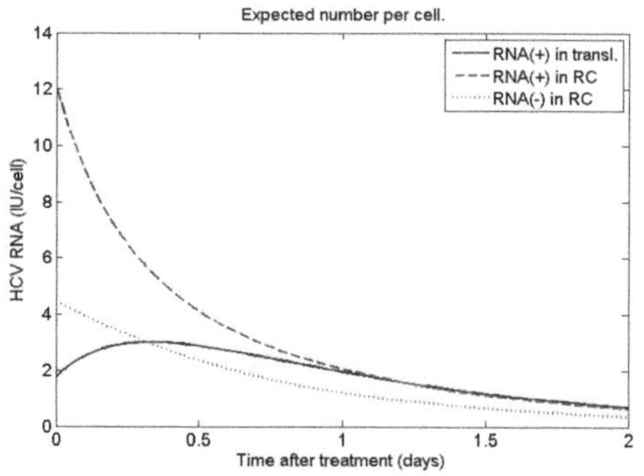

(a) PAT 8

Fig. 3: Intracellular HCV RNA obtained with parameters fitting of patient data from [19].

IV CONCLUSION

The dynamics of the negative-strand of HCV during replication and a new state for the positive-strand during translation were added to a multi-scale age-based mathematical model of HCV infection. The results showed the ability to represent the intracellular dynamics of HCV RNA translation, replication and export compared to transfection experiments and also the viral load compared to in vivo experiments.

Further work is still required to study other aspects such as the superinfection assumption [31], the influence of host factors and also the high mutation rate of the HCV [32] that were not included yet on this study.

Conflict of Interest

The authors declare that they have no conflict of interest.

Acknowledgements

This work was developed with the support of CAPES Proc. number 99999 .002789 / 2014-00 and the Center for Nonlinear Studies, Los Alamos National Laboratory.

References

1. WHO . Guidelines for the screening, care and treatment of persons with chronic hepatitis C infection 2016. Date accessed: July, 2016.
2. Gastaminza P., Cheng G., Wieland S., Zhong J., Liao W., Chisari F. V.. Cellular Determinants of Hepatitis C Virus Assembly, Maturation, Degradation, and Secretion *Journal of Virology.* 2008;82:2120–2129.
3. Appel N., Schaller T., Penin F., Bartenschlager R.. From Structure to Function: New Insights into Hepatitis C Virus RNA Replication *Journal of Biological Chemistry.* 2006;281:9833–9836.
4. Afzal M. S., Alsaleh K., Farhat R., et al. Regulation of Core Expression During the Hepatitis C Virus Life Cycle *Journal of General Virology.* 2015;96:311–21.
5. Craggs J K, Ball J K, Thomson B J, Irving W L, Grabowska AM. Development of a strand-specific RT-PCR based assay to detect the replicative form of hepatitis C virus RNA *J Virol Methods.* 2001;94:111–20.
6. Goutagny N., Inchauspe G.. *Emerging Therapies for Hepatitis C Infection*ch. 12. Human Virus GuidesInternational Medical Press 2009.
7. Liang T. Jake, Ghany Marc G.. Current and Future Therapies for Hepatitis C Virus Infection *The New England Journal of Medicine.* 2013;368:1907–1917.
8. Myers R. P., Shah H., Burak K. W., Cooper C., Feld J. J.. An update on the management of the chronic Hepatitis C: 2015 consensus guidelines from the Canadian Association for the Study of the Liver *Can J Gastroenterol Hepatol.* 2015;(In press).
9. Rong L., Guedj J., Dahari H., et al. Analysis of Hepatitis C Virus Decline during Treatment with the Protease Inhibitor Danoprevir Using a Multiscale Model *PLOS Computational Biology.* 2013;9:e1002959.
10. Guedj J., Neumann A.U.. Understanding hepatitis C viral dynamics with direct-acting antiviral agents due to the interplay between intracellular replication and cellular infection dynamics *Journal of Theoretical Biology.* 2010;267:330–340.
11. Dixit N. M.. Advances in the mathematical modelling of Hepatitis C virus dynamics *Journal of the Indian Institute of Science.* 2008;88:37–43.
12. Dahari H., Layden–Almer J. E., Kallwitz E., et al. A Mathematical Model of Hepatitis C Virus Dynamics in Patients With High Baseline Viral Loads or Advanced Liver Disease *Gastroenterology.* 2009;136:1402–1409.
13. Ribeiro R. M., Li H., Wang S., et al. Quantifying the Diversification of Hepatitis C Virus (HCV) during Primary Infection: Estimates of the In Vivo Mutation Rate *PLOS Pathogens.* 2012;8:e1002881.
14. Canini L., Perelson A. S.. Viral kinetic modeling: state of the art *J Pharmacokinet Pharmacodyn.* 2014;41:431–443.
15. Chaterjee A., Guedj J., Perelson A. S.. Mathematical Modelling of HCV infection: what can it teach us in the era of direct-acting antiviral agents? *Antiviral Therapy.* 2012;17:1171–1182.
16. Dahari H., Sainz B., Perelson A. S., Uprichard S. L.. Modeling Subgenomic Hepatitis C Virus RNA Kinetics during Treatment with Alpha Interferon *Journal of Virology.* 2009;83:6383–6390.
17. Banerjee S., Keval R., Gakkhar S.. Modeling the dynamics of Hepatitis C virus with combined antiviral drug therapy: Interferon and Ribavirin *Mathematical Biosciences.* 2013;245:235 – 248.
18. Guedj J., Perelson A. S.. Second-phase Hepatitis C Virus RNA Decline During Telaprevir-Based Therapy Increases With Drug Effectiveness: Implications for Treatment Duration *Viral Hepatitis.* 2011;53:1801–8.
19. Guedj J., Dahari H., Rong L., et al. Modeling Shows that the NS5A inhibitor dacatasvir has two modes of action and yields a shorter estimate of the hepatitis C virus half-life *PNAS.* 2013;110:3991-3996.
20. Thieme H., Castillo-Chavez C.. How may infection-age-dependent infectivity affect the dynamics of HIV/AIDS? *SIAM J. Appl. Math..* 1993;53:1337–79.
21. Martcheva M., Castillo-Chavez C.. Diseases with chronic stage in population with varying size? *Math. Biosci..* 2003;182:1–25.
22. Thieme H., Castillo-Chavez C.. A two-strain tuberculosis model with age infection *SIAM J. Appl. Math..* 2002;62:1634–56.
23. Nelson P. W., Gilchrist M. A., Coombs D., Hyman J. M., Perelson A. S.. An age-structured model of HIV Infection that Allows for Variations in the Death Rate of Productively Infected Cells *Mathematical Biosciences.* 2004;1.
24. Li Jia, Brauer Fred. Continuous-Time Age-Structured Models in Population Dynamics and Epidemiology in *Mathematical Epidemiology* (Brauer Fred, Driessche Pauline, Wu Jianhong. , eds.);1945 of *Lecture Notes in Mathematics*:205–227Springer Berlin Heidelberg 2008.
25. Moradpour D., Penin F., Rice C. M.. Replication of hepatitis C virus *Nature Reviews Microbiology.* 2007;5:453–463.
26. Shi S. T., Lai M. M. C.. *HCV 5' and 3'UTR: When Translation Meets Replication*ch. 2. Norfolk (UK): Horizon Bioscience: in: Hepatitis C Viruses: Genomes and Molecular Biology. Tan SL, editor 2006. Available from: http://www.ncbi.nlm.nih.gov/books/NBK1624/.
27. Keum S.J., Park S.M., Park J.H., Jung J.H., Shin E.J., Jang S.K.. The specific infectivity of hepatitis C virus changes through its life cycle *Virology.* 2012;433:462–470.
28. Neumann A. U., Lam N.P., Dahari H., Gretch D.R., Wiley T. E., al. . Hepatitis C viral dynamics in vivo and the antiviral efficacy of interferon-alpha therapy *Science.* 1998;282:103–107.
29. Lee C.. Daclatasvir: potential role in hepatitis C *Drug Des Devel Ther..* 2013;7:1223–1233.
30. Scheel T. K. H., Rice C. M.. Understanding the hepatitis C virus life cycle paves the way for highly effective therapies *Nat. Med..* 2013;19:837–849.
31. Webster B., Ott M., Greene W. C.. Evasion of Superinfection Exclusion and Elimination of Primary Viral RNA by an Adapted Strain of Hepatitis C Virus *Journal of Virology.* 2013;87:13354–13369.
32. Sanjuán R., Nebot M. R., Chirico N., Mansky L. M., Belshaw R.. Viral Mutation Rates *Journal of Virology.* 2010;84:9733–9748.

Vibrotactile System for the Replication of Textures

M. Santís[1 4], D. Jaramillo[2 4] and V. Z. Pérez[3 4]

[1] Electronic engineer, Masters Student in Engineering.
[2] General Practitioner. Specialization student in Biomedical Engineering.
[3] Electronic Engineer, PhD. in Engineering. Associate Professor. Facultad de Ingeniería Electrónica.
[4] Centro de Bioingeniería. Universidad Pontificia Bolivariana. Medellin, Colombia.

Abstract— **This article explores a tactile interface prototype made to recreate simulated textures using tactile sensations. The device works with vibration patterns found in a small area of the hand. It was tested on 30 people who evaluated the difference between 3 artificial surfaces presented to them and found similarities with known textures. Using a grading system of 1 to 10, considering 5 as the mean value, the difference between texture 1 and 2 and between 2 and 3 was valued at an average of 7, with a p-value < 0.01. An average of 9 was given to the difference between texture 1 and 3 with a p-value < 0.01. The above results indicates that the three textures were distinguishable from each other. Additionally, the majority of the participants interpreted the first texture as a corrugated or gravel surface (70%), the second texture as a wavy texture similar to uneven paving (64%), and the third texture was interpreted as a smooth surface like that of fabric or a very fine granite (60%). A development of this type has multiple applications of different areas of Bioengineering since it increases the effect of immersion in a virtual environment of simulation and it can be linked to kinesthetic interfaces to compliment the force feedback with the effect of superficial texture in the interaction with a virtual object.**

Keywords— **Tactile Interface, Kinesthetic Interface, LRA (linear resonator actuator), simulated textures.**

I. Introduction

In daily tasks, haptic information, which involves tact and proprioception, is fundamental to interact with the surrounding environment. The most common commercial haptic interface is aimed to recreate the interaction of virtual bodies from a kinesthetic perspective [1]. However, tactile information, such as texture, is not available.

To simulate texture, small scale electromechanical devices are used [3], such as inertia type of resonator-vibrators, linear, direct current motors with eccentric rotating mass [2][3], and impact actuators [4]. Also, medium devices, such as voice coils, are used [5]. Finally, there are big scale devices which offer better resolution such as graphic tactile screens, based on needles activated by solenoids [6], servomotors [7], or large piezoelectric dimorphic bars in an interface like Braille [8].

Additionally, electrocutaneous interfaces have been developed using the principle of electro stimulation of nerve endings on the surface of the skin. These interfaces tend to be small, durable, efficient and free of mechanical resonance [9][10]. This method of electro tactile stimulation can produce a wide range of sensations in the skin, from a light ant to painful thumps, by varying frequency and amplitude of the applied pulses [11]. Electrostatic interfaces have been implemented, which recreate the sensation of friction through the generation of normal forces between the skin of the finger and the screen, similar to capacitor plates [12].

High resolution tactile interfaces have been achieved [6] increasing the level of realism that the user can experience in a simulated environment, but their construction is complex due to the dense distribution of nerve endings on the skin, which requires stimulation into specific zones. Tactile, portable interfaces, which integrates the kinesthetic systems to perceive and detect texture, are scarse. It could increase the realism in the interaction with a virtual environment or in a Minimally Invasive Surgery procedure.

This work presents the development of a vibro-tactile system prototype to replicate textures. The prototype description and its components are described. A methodology to evaluate the difference among three surfaces is proposed. A test with 33 subjects was performed. Finally, the results are discussed.

II. Prototype Description

Figure 1 shows the developed device and a block diagram of its constitutive elements, which are described as follows.

--A vibro-tactile device, which is a Linear Resonator Actuator, LRA, generates the vibro-tactile stimulus to recreate skin texture of the hand and is based on the vibration frequency at 175 Hz. It works like a voice coil speaker oscillating at frequencies close to this value.

--The micro controller system, based on an Arduino R3, is responsible for generating stimulus patterns and connections of the actuator.

--The driver, DRV2605L, which receives the commands from micro controller via I2C, and enables the activation of

I. Torres et al. (eds.), *VII Latin American Congress on Biomedical Engineering CLAIB 2016, Bucaramanga, Santander, Colombia, October 26th -28th, 2016*,
IFMBE Proceedings 60,
DOI: 10.1007/978-981-10-4086-3_129

default patterns, stored in its nonvolatile memory, to deliver voltage and current levels to the vibro-tactile actuator. Through software programming, three different textures were created by concatening the available patterns, with the objective of generating low, medium and high levels of roughness.

--Infrared barriers that generate the activation signal when thimble passes over one of three possible regions (see Figure 1a)

When the user puts the thimble on the dominant hand and passes over regions 1, 2 or 3, it is sensed by the respective infrared barrier, the microcontroller sends the activation signal to the driver, which generates the vibrations patterns according to that texture, to produce the desired texture effect. The driver sends the stimuli to the vibro-tactile actuator. These stimuli are electric square signals which are amplitude and pulse-width modulated. Stimulation is paused when the user removes the finger from the region.

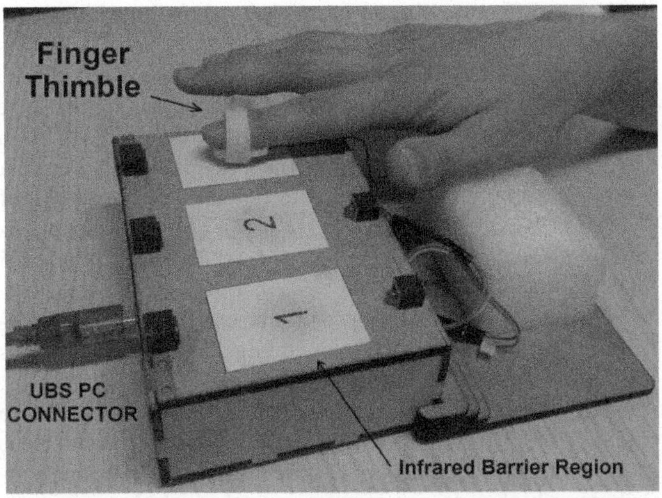

(a)

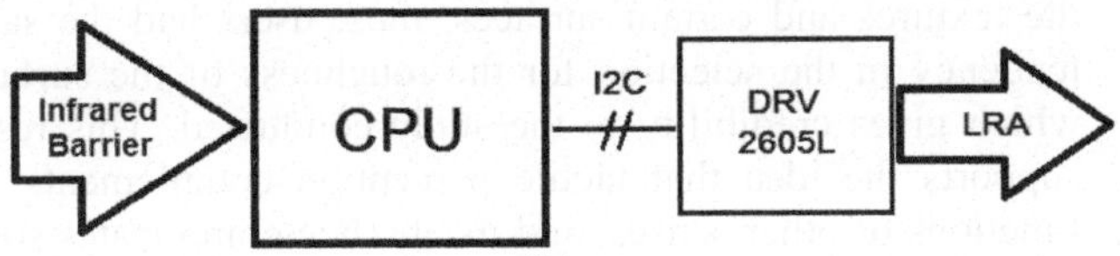

(b)

Figure 1. (a) Haptic System and (b) block diagram

III. METHODOLOGY

According to the device design described in section II, three textures were simulated whose differences were assessed on a group of subjects aged between 18 and 60 years old. A sample number of 60 was obtained using the program GPower 3.1, with a basis of Effect=0.4, Power=80, Confidence=95%, Measurement=3. Evaluation of a 33 subjects from the sample was carried out as a preliminary analysis. Tests were supported by a physician, who evaluated the skin condition and neurological response of the upper limbs at the beginning and end of the test.

Exclusion criteria includes hand skin conditions, scars, burns, dermatological conditions, traumas, neurological, psychiatric or immunological diseases, mellitus 1-2, dyslipidemia, or surgery that affects the sensitivity component or neurological condition, prolonged medication or stimulants which affects the sensitivity of the skin; exposition to toxic substances of diverse nature. Order of stimuli application was randomized and test subjects are not professionally related with biomedical field in order to avoid bias in the sample selection and detections of results [13][14].

The project was approved by Ethics Committee associated to the institution where test was carried out. Subjects were instructed about the appropriate position of the hand and test execution, allowing a correct evaluation of the functioning of the interface and avoiding interferences between the cables and the sensors. A manual was provided in order to guarantee that each subject receive the same information and perform the same tests. Additionally, an informed consent document were signed by each subject before starting the test.

Three tests were completed. In the first test, texture 1 and 2 were compared; in the second, test texture 2 and 3 were compared, and in the third test, texture 1 and 3 were compared. Each subject was asked to make exploratory, circular or zigzag movements over the two surfaces to compare and grade the differences between them in each test using a score between 1 to 10, with 1 indicating no difference between textures and 10 indicating entirely different textures. Thus, a matrix for each person was obtained, that allows comparison among textures.

Additionally, each participant was asked to assign to each surface a physical characteristic or texture. Perception of usage comfort was inquired. Differences were evaluated by means of t-student analysis. We considered 5 as the reference value corresponding to the case when no similarities or differences (equivalent to middle of the 1-10 score). $\bar{X}$ is the average of the 30 observations; S, the standard deviation and n, the number of participants, with the aim of obtaining $t > 2.39$ and a p-value < 0.05.

IV. *RESULTS*

From the 33 evaluated subjects, 3 were excluded due to irregular clinical exam results. The remaining 30 individuals (16 women and 14 men) were evaluated. At the end of the test, none of the participants suffered any type of pain or discomfort with the device. Three subjects experienced a tingling sensation or vibration lasting a few seconds after the test finished. One person reported a feeling of coldness for a few seconds on the finger used during the test, and another referred an improvement in sharp movements.

Figure 2 (top) shows percentages discrimination of people that scored the difference between textures in each test. Figure 3 shows the percentage of individuals who gave a score above 5 in each texture.

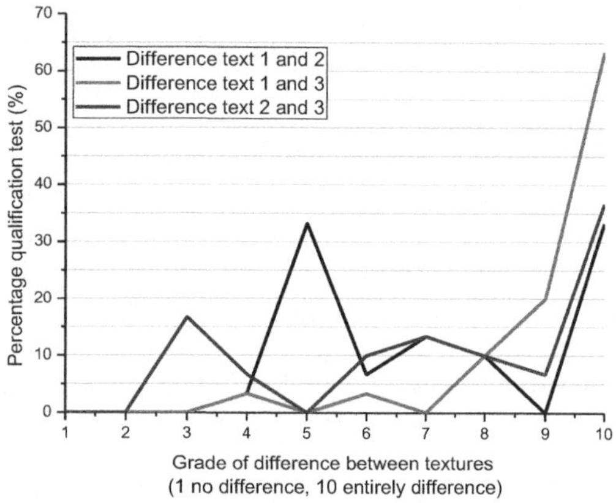

Figure 2. Percentage of people with their assigned grading to differentiate the textures

According to the scores matrix, for the differences between texture 1 and 2, t =5.613 (with a $\bar{X}$ =7.266 and S=2.211) and a $p < 0.01$; between texture 2 and 3, t=4.851 (with a $\bar{X}$ = 7.366 and S=2.671) $p < 0.01$; and between texture 1 and 3, t=9.266 (with a $\bar{X}$ = 9.266 and S=1.362) $p < 0.01$. Thus, all textures were distinguishable among each another.

Texture 1, was related with a corrugated or gravel surface (70% of subjects). Texture 2 was related with a wavy, uneven surface similar to a bumpy pavement (64% of subjects). Texture 3 was related with a smooth surface such as fabric or very fine granite (60%). This qualification evinces the distinction between the recreated textures. More than 13 subjects (43.3%) gave a grading greater than 5 to the 3 tests, from these subjects, 4 were male (13.3%) and 9 were female (30%). 15 subjects (50%) gave a grading greater than 5 to 2 of the tests, 8 were male (26.6%) and 7 were female (23.4%), and only 2 (6.7%) participants, both male, graded one test greater than 5.

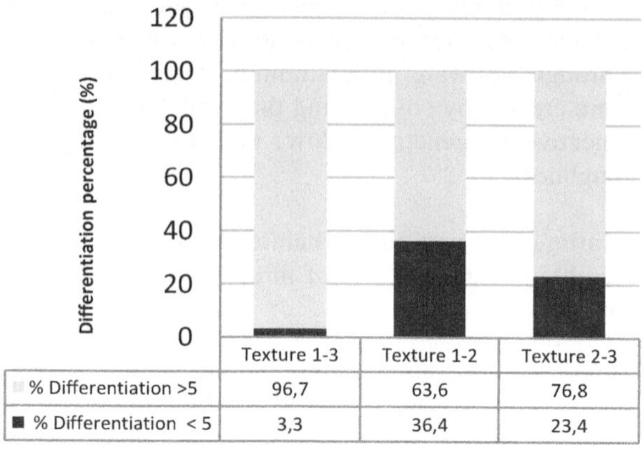

Figure 3. Percentage of individuals with a scoring of above 5

V. DISCUSSION

The developed system recreates vibro-tactile patterns that emulate different textures. This system uses a driver which includes different vibratory waves by default, which were configured to generate different patterns. The system was tested on 30 individuals who could differentiate among the three presented textures with statistical significance ($p<0.01$). The skin surface of the hand, can achieve high distinction of textures through the Pacinian and Meissner corpuscles, which are designed to perceive the sensation of vibration between 8 and 400 Hz [16]. The above explains the good perception of the textures which was around 175 Hz.

Tactile systems based on pressure enables the identification of borders. Other systems attempt to recreate textures through different vibrating systems such as speakers [15]. Our device, besides, simulates different surfaces, allows to identify them through tactile perception. A tactile interface complements the effects of a kinesthetic interface and contributes a higher degree of immersion through the user.

When looking for patterns of similarity between each of the textures and certain surfaces, most users had the same tendency in the selection for the roughness of the surface, which gives credibility to the study conducted. This result supports the idea that tactile perception complements the functions of other senses, and its absence complicates some labors and processes.

Our system enables the configuration of new textures that can be recreated at any time as additional function. The recreation of textures has multiple applications such as games based on the identification of textures for children, adult games with or without sight, as with teaching strategies which enable better immersion for the user through means of texture identification. Moreover, the system is portable, and of low risk for the user. It could be linked to current commercial kinesthetic systems, for virtual reality simulation or in other areas such as rehabilitation, tele-operation and entertainment.

VI. CONCLUSIONS

A system was developed which recreates vibro-tactile patterns which simulates three of different and distinguishable textures.

For future studies, the comparison test will be done with a real texture pattern and the unit of tactile interface will be added to a commercial kinesthetic interface to evaluate the performance. A larger subjects sample will be considered to improve the statistical significance.

ACKNOWLEDGMENT

The authors want to acknowledge Juan Pablo Ugarte whose help in the writing process was really important, and Lucelly López López, who guided the statistical analysis with assertiveness and willingness.

CONFLICT OF INTEREST

The authors declare that they have no conflict of interest.

REFERENCES

1. Worldhaptics at http://www.worldhaptics.org
2. Yang, Y., 2013. Design and Control of an Integrated Haptic Interface for Touch Screen Applications (Doctorat). Université Lille 1 [Accessed November 9, 2015]. Available at: http://goo.gl/9J8s4R.
3. Wang, F., 2014. Haptic Energy Consumption. Application. Report., Dallas, Texas. Available at: http://goo.gl/78bICu.
4. Pyo, D., Yang, T.-H., Ryu, S. & Kwon, D.-S., 2015. Novel linear impact-resonant actuator for mobile applications. Sensors and Actuators A: Physical, 233, pp.460–471.
5. Richter, H., Blaha, B. & Wiethoff, A., 2011. Tactile feedback without a big fuss: simple actuators for high-resolution phantom sensations. Proceedings of the 13th International Conference on Ubiquitous Computing - UbiComp '11, pp.85–88. Available at: http://goo.gl/tzaYpm.
6. Simeonov, S. & Simeonova, N., 2014. Graphical Interface for Visually Impaired People Based on Bi-stable Solenoids. arXiv preprint arXiv:1401.5289. Available at: http://goo.gl/Jlu3GT.
7. Wagner, C.., Lederman, S.J. & Howe, R.D., 2004. Design and performance of a tactile shape display. Haptics-e, p.6.
8. Lee, J.S. & Lucyszyn, S., 2005. A micromachined refreshable Braille cell. Journal of Microelectromechanical Systems, 14(4), pp.673–682.
9. Kajimoto, H., Kawakami, N., Maeda, T. & Tachi, S., 2004. Elec-tro-tactile display with tactile primary color approach. Proceed-ings of International Conference on Intelligent Robots and Sys-tems.
10. Sato, K. & Tachi, S., 2010. Design of electrotactile stimulation to represent distribution of force vectors. 2010 IEEE Haptics Sym-posium, HAPTICS 2010, pp.121–128.
11. Pamungkas, D. & Ward, K., 2015. Tactile sensing system using electro-tactile feedback. ICARA 2015 - Proceedings of the 2015 6th International Conference on Automation, Robotics and Applications, p.8.
12. Xu, C., Israr, A., Poupyrev, I., Bau, O. & Harrison, C., 2011. Tac-tile display for the visually impaired using TeslaTouch. Proc. CHI EA '11, pp.317–322. Available at: http://goo.gl/cDLu5c
13. Glantz, S. A. Bioestadística. 6ta Edición. Ed. McGraw Hill. ISBN 970-10-5701-5. 2006
14. Alvarez, F. Alvarez. A. Auditoría Médica y Epidemiología. Ecoe Ediciones. ISBN 978-958-648-578-4. 2009.
15. Smith, S., Smith, G. C., & Lee, J. (2015) The effects of realistic tactile haptic feedback on user surface texture perception. Journal of Vibroengineering, 17(2), 1004-1016.
16. Griffin, Michael. Vibraciones. Riesgos generales. Enciclopedia de salud y seguridad en el trabajo. Tomo 2, 50.1. Sumario. http://goo.gl/H0P7it Consultado en junio 2016.

Author: Mauricio Santís Chaves
Institute: Universidad Pontificia Bolivariana
Street: Circular 1# 73-76
City: Medellín
Country: Colombia
Email: mauricio.santis@upb.edu.co

Nonlinear estimators of human movement in biomechanical signals: Comparison between Extended Kalman Filter and Unscented Kalman Filter

M. Callejas Cuervo[1], M.A. Vélez Guerrero[1] and A.C. Alarcón-Aldana[1]

[1]Pedagogical and Technological University of Colombia, Engineering Faculty, Tunja, Colombia

Abstract— **In the capture of biomechanical signals, the application of Kalman filters allows a correction of the error caused by disturbances or noise in the system. This article presents a comparison of the performance of an Unscented Kalman Filter and an Extended Kalman Filter with the objective of identifying which of the two has a better performance for use in the area of the analysis of human movement. The results of the analysis of the measurement of flexion and extension of the arm, taken with inertial and magnetic sensors, show a Root Mean-Square Error (RMSE) of 2.43° in the Unscented Filter versus a 3.88° in the Extended one. The analysis was done with angular velocities between 1 and 15 rad/s, ideal for biomechanical applications.**

Keywords— **Kalman Filter, Movement capture, Biomechanical signals, inertial and magnetic sensors, Digital Processing of signals.**

I. Introduction

Within the electronic instrumentation that makes the capture of biomechanical data possible, there are inertial measurement and movement processing units based on Microelectromechanical Systems (MEMS), which are used for the diagnosis and rehabilitation of diseases which involve the locomotive system. These systems are relevant in the medical field in where they have advantages, such as the portability of the elements due to their reduced size and weight and, in addition, the electric and electronic characteristics, where the low power consumption and the interoperability with diverse platforms is an earth-shattering topic in this area [1].

The importance of the treatment of the data to obtain information lies in that, before making any kind of analysis of the signals, it is necessary to subject them to certain transformations: filtering, sundering, mixing, selecting, conditioning, detecting, classifying, or improving the information coming from the capture system that is used. In this particular case, the signal conditioning process will be focused on the filtering, which is done digitally through an algorithm that discriminates defined ranges of data based on specific characteristics of the signal, and generates variations in the amplitude or phase of the incoming signal. Among the signal processing filters used which involve movement and biomechanical signals, the Unscented Kalman Filter (UKF) and the Extended Kalman Filter (EKF) are found. Both can be used for the estimation of future states in a nonlinear system where exists white noise and other defects of the signal.

This article is structured as follows: first, some works related with the use and comparison of the UKF and the EKF are presented; second, the testing platform and the protocols used to run tests and evaluate the filters are described; third, the results are presented and an analysis of the results is carried out; and finally, the conclusions are presented.

II. Related works

Julier and Uhlmann [2] implemented the first UKF, which tried to solve the problem of the generation of estimative errors of the EKF by making a distribution of the states of estimation through an approximation of the constant of the Gaussian variation and selecting specific samples of the estimated values, which are known as sigma points. Khoder and Jida [3] used a variation of this filter to be employed in the experimental correction of the error produced in fused navigation systems, such as GPS/INS, obtaining satisfactory results in its application, but without comparing the results with other filters. Roetenberg, et al [4], developed an application which is intended to eliminate the cross-track error by fusing the data of sensors such as accelerometers, magnetometers, and gyroscopes. Moreover, it applies a Complementary Kalman filter achieving good results that can be proved through their comparison with an optical reference system which provides more exact measurements. As regards the comparison of the UKF and the EKF, the work proposed by LaViola [5] reports a biomechanical application where an analysis of the movement of the head and hands was made. The results highlighted the fact that, even when both filters have an equivalent and optimum level of performance, the use of the extended Kalman Filter is put forward as the best, given its lower computational cost. For objects in motion, the comparison between UKF and EKF is broad. D'Alfonso, et al [6], concluded that the Extended Kalman Filter is as efficient as the Unscented

I. Torres et al. (eds.), *VII Latin American Congress on Biomedical Engineering CLAIB 2016, Bucaramanga, Santander, Colombia, October 26th -28th, 2016*,
IFMBE Proceedings 60,
DOI: 10.1007/978-981-10-4086-3_130

version for the fusion of data from the ultrasonic sensors of a robot. The author is inclined towards using the UKF because the properties of approximation to the real values are much better than in the EKF. Toloey and Niazi [7] confirm that, under analysis of real data, the UKF works better due to a numerical approximation and stability in time, ideal characteristics for use in dedicated motion tracking systems. In aerial and navigation systems, Giannitrapani, et al [8], highlight that, even when the results of the two filters are satisfactory, the UKF shows better results in terms of the average localization of the error and the consistency of the estimated states. Allota, et al [9], use the comparison of both filters in an unmanned aquatic vehicle and state that the UKF deserves great merit for its superiority and easy implementation as it does not require complicated calculations, something that can occur with the EKF.

From the previous studies, it is concluded that although there is a considerable amount of information about the use of UKF and EKF to achieve the elimination of the errors present in the data captured in diverse applications, it is necessary to carry out more analysis for the capture and filter of biomechanical information for the purpose of diagnoses and rehabilitation of human beings.

III. MATERIALS AND METHOD

A previously built motion capture platform is employed that is made up of two IMMU (inertial and magnetic measurement unit), a control unit which applies capture algorithms, processing, and transmission of the biomechanical signals acquired from the human arm to a device [10] (Figure 1); furthermore, a testing platform is built. The platform has a head, a mobile element that on its surface holds a movement processing unit: MPU 9150 and allows for movements to be made simulating the flexion or extension of an upper limb of the human body (extremity used to carry out the biomechanical capture in this study), representing the segment of the forearm. The body of the platform which gives support to the moving parts and also serves as an axis of support for the second MPU 9150 sensor represents the segment of the arm. And, finally, the base supports the entire system on a solid surface (Figure 2).

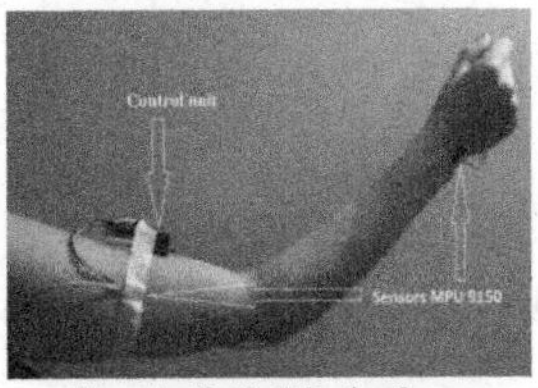

Fig. 1. IMMU platform

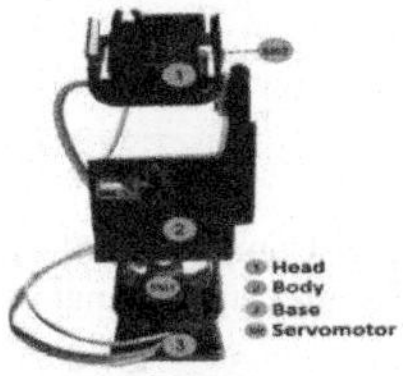

Fig. 2. Test platform

A series of test protocols were established which simulate sinusoidal movements, given that they represent the movement of the flexion and extension of the arm. The angular velocities of the movement analysed are between 1 and 15 rad/s, taking into account that the movements of the human body are produced within that range. An amplitude of movement of 20º is finally adopted, a value that can be changed at any time, given that the articular amplitude between the forearm and the arm is between 0º and 165º approximately. The purpose of the different tests presented in the results section is to show the value of the RMSE for each filter when exposed to different periods of use, diverse velocities, and with different covariant values in the process and the measurement.

The root mean-square error (RMSE) was used to quantify the difference of the value of the angles between the reference information (test platform) and the estimate using the inertial and magnetic measurement unit (IMMU). See equations (1) and (2), where k represents the number of samples taken with both systems and n represents a sample in a time slot.

$$e(n) = \theta_{test\ platform}(n) - \theta_{IMMU}(n) \quad (1)$$

$$\text{RMSE} = \sqrt{\frac{1}{k}\sum_{n=1}^{k} e^2(n)} \quad (2)$$

Below, three scenarios are presented, which shows the research progress until the achievement of the comparison of the filters with biomechanical signals of the human arm.

A. Scenario 1. Initial calibration of the filters

With the objective of carrying out the calibration of the filters, the capture of data was done in five captures, each one of them at specific angular velocities, being 1, 5, 10, and 15 rad/s, using diverse values of standard deviation in the measurement (r) and in the process (q) to determine the most adequate behaviour for the UKF and the EKF in each velocity. The range of values for the parameters r and q was established through the analysis of the behaviour of the systems and the sensors (Table 1).

B. Scenario 2. Optimal Parameters of the filters

An average standard deviation value for the measurement ($\bar{x}$r) and an average standard deviation for the process ($\bar{x}$q) are found using an arithmetic average of the parameters of optimal standard deviation for each specific angular velocity. The objective is to be able to use one value of r and q that is valid for any velocity between the range of 1 and 15 rad/s. Obtaining the optimal parameters in each filter for a range of velocities is justified within biomechanical research, as the movements made by the upper limbs of the human body are not always constant velocities, but they are within a certain range. After the arithmetic operations, an average is obtained for r and q= 0.0013250. With these values, a simulated movement of flexion and extension is established: of amplitude A=20°, offset=0°, sampling frequency Fs=35 Hz, shape of sinusoidal wave and capture time Tc=30 s (Table 2).

C. Scenario 3. Biomechanical Capture of the arm

Five captures of data at different velocities for a movement of flexion and extension are made at a sampling frequency Fs=35 Hz with variable amplitude and capture time. The purpose is to analyse the values of maximum flexion/extension, given that in the record of biomechanical data in applications such as physical rehabilitation of human beings, a record of this values is kept. It is determined which filter obtained the data closest to the real estimation (Table 3).

IV. Results

The results with the selection of the parameters r and q which gave less RMSE in the signals are shown in Table 1.

Table 1 Selection of the parameters r and q which gave less RMSE in the signals, together with the RMSE value obtained.

rad/s	UKF			EKF			RAW
	r	q	RMSE	r	q	RMSE	RMSE
1	0.0115	0.0015	0.9767	0.0155	0.00110	2.2598	1.5969
5	0.0125	0.0014	1.6714	0.0115	0.00150	2.1716	2.2015
10	0.0135	0.0013	1.9749	0.0115	0.00150	2.9150	3.9121
15	0.0155	0.0011	2.1974	0.0145	0.00120	2.6614	5.1637

From the *RAW* signal (without any filtering), it is observed that the quantity of errors grows as the angular velocity of the movement increases, finding a direct relation between the quantity of errors and the angular velocity. In the same way, a relation between the parameters of standard deviation and the efficiency of the filter for each measured velocity is established. The UKF as well as the EKF respond better to certain values of q and r than to others, and this process of comparison allows for the tuning that establishes the value of the optimum parameters. In general, the filter with better observed behaviour regarding the reduction of the RMSE is the UKF. The values that show the result of greater efficiency for each filter is always less for the UKF versus the EKF in all the chosen angular velocities.

The results of Scenario 2 are summarized in Table 2.

Table 2 RMSE obtained using the averaged standard deviation for the measurement (r) and the process (q).

rad/s	RMSE - UKF	RMSE - EKF	RMSE - RAW
1	1.6219	2.5189	1.8019
5	1.8536	2.7221	2.3946
10	2.8211	4.9549	4.1503
15	3.4374	5.3494	5.1651
$\bar{x}$	2.4335	3.8863	3.3780

In Figure 3, it is observed that, independently of the parameters used for the tuning of the filter, the UKF still shows less RMSE in any velocity of biomechanical movement, versus the EKF.

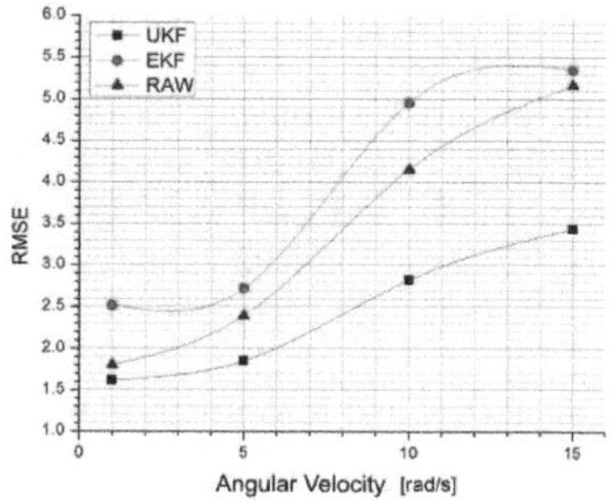

Fig. 3 RMSE value obtained with the parameters of averaged standard deviation.

The biomechanical values captured can be observed in Table 3.

As is evident, the UKF still has a better performance than the EKF, even with biomechanical signals. In addition, in Figure 4, it is possible to observe that in the values of maximum flexion the filters respond in a similar way when getting closer to the value of the estimated signal whereas in the values of maximum extension (negative values) the UKF achieves a better approximation, concluding that it has a better performance.

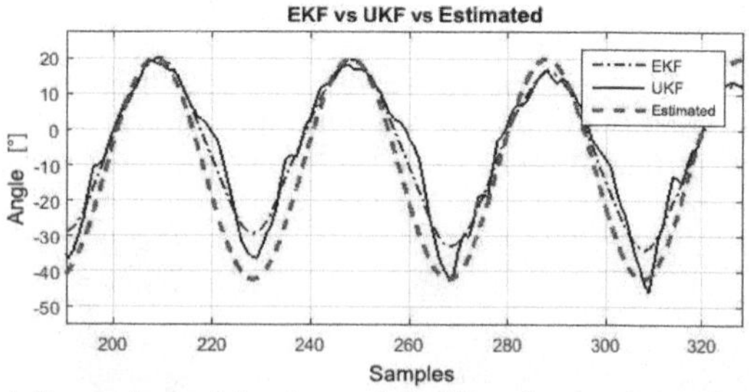

Fig. 4 Comparison of the shape of the filtered wave versus the estimated signal, approximate angular velocity of 6 rad/s and 10 s capture time.

Table 3 Values of maximum flexion and extension of the biomechanical data captured.

rad/s	Estimated Amplitude [°]	Estimated Offset [°]	Estimated Values		UKF		EKF		RAW	
			Ext.	Flex.	Ext.	Flex.	Ext.	Flex.	Ext.	Flex.
~0.5	~23.15	~-5.09	-28.23	18.06	-16.72	26.89	-9.21	38.43	-18.67	29.87
~2	~24.23	~-0.54	-24.77	23.69	-22.97	24.05	-24.03	25.14	-26.86	26.53
~4	~35.19	~-19.76	-54.95	15.43	-56.48	16.96	-51.39	17.45	-64.80	21.27
~6	~31.11	~-11.07	-42.18	20.05	-45.60	23.47	-33.88	24.92	-51.32	26.82
~8	~28.66	~-10.42	-39.08	18.24	-46.66	25.82	-18.89	30.85	-78.65	34.02

V. Conclusions

The use of recent technologies for the capture of biomechanical movement, such as the MPU used in the project, allows for a less subjective evaluation when it comes to the valuation of movements related with the human body.

It was possible to see a better performance of the Unscented Kalman Filter in comparison with the Extended Kalman Filter for the treatment of biomechanical data, specifically regarding flexion and extension, taking as a parameter the reduction of the RMSE in the signal given with respect to the reference platform.

There is a better approximation to the real values of the biomechanical signal in the Unscented Kalman Filter in comparison to the Extended Kalman Filter, taking into consideration the values of maximum flexion and extension estimated in a biomechanical movement of the upper limbs.

Acknowledgments

To the Universidad Pedagógica y Tecnológica de Colombia (Pedagogical and Technological University of Colombia - UPTC) and the Dirección de Investigaciones – DIN and a special thanks to the Grupo de Investigación en Software – GIS (Software Research Group).

Conflict of Interest

The authors declare that they have no conflict of interest.

References

1. Šprdlík O (2012) Detection and Estimation of Human Movement Using Inertial Sensors: Applications in Neurology. Czech Technical University in Prague, Prague
2. Julier S, Uhlmann J (1997) New Extension of the Kalman Filter to Nonlinear Systems, SPIE Proc. vol. 6, Sig. Process., Sens. Fusion & Target Recognit. VI, Orlando, USA, 1997
3. Khoder W, Jida B (2014) A Quaternion Scaled Unscented Kalman Estimator for Inertial Navigation States Determination Using INS/GPS/Magnetometer Fusion, J. Sens. Technol. vol. 4, Tripoli, Lebanon, 2014, pp 101-117
4. Roetenberg D, et al. (2005) Compensation of Magnetic Disturbances Improves Inertial and Magnetic Sensing of Human Body Segment Orientation. IEEE Trans Neural Syst Rehabil Eng 3:395-405
5. Laviola J (2015) A Comparison of Unscented and Extended Kalman Filtering for Estimating Quaternion Motion, Am. Control Conf. Proc. vol. 3, Adv. Sci. Comput. & Visualization, Brown Univ. Technol. Center, Providence, USA, 2003, pp 2435-2440
6. D'alfonso L, et al. (2015) Mobile robot localization via EKF and UKF: A comparison based on real data, Rob. Auton. Syst. vol. 74, Università della Calabria, Rende, Italy, 2015, pp 122-127
7. Toloei A, Niazi S (2015) Estimation of LOS Rates for Target Tracking Problems using EKF and UKF Algorithms- a Comparative Study, Int. J. Eng. Trans. B. vol. 28, Department of Aerospace, ShahidBeheshtiUniversity, Tehran, Iran, 2015, pp 172-178
8. Giannitrapani A,et al. (2011) Comparison of EKF and UKF for Spacecraft Localization via Angle Measurements, IEEE Trans. Aerosp. Electron. Syst. vol. 47, Universita di Siena, Siena, Italy, 2011, pp 75-84
9. Allota B, et al. (2015) A comparison between EKF-based and UKF-based navigation algorithms for AUVs localization, OCEANS-Genova, ISME - Interuniversity Center of Integrated Systems for the Marine Env, Florence, Italy, 2015
10. Callejas-Cuervo M, Alvarez J.C, Alvarez D (2016) Capture and analysis of biomechanical signals with inertial and magnetic sensors as support in physical rehabilitation processes, Proc. IEEE EMBS 13th Annu. Int. Body Sensor Networks Conf., San Francisco, USA, 2016, pp 119–123

Corresponding author:

Author: Mauro Callejas-Cuervo
Institute: Pedagogical and Technological University of Colombia/System Engineering and Computer School, Tunja, Colombia
Street: Avenida Central Norte 39-115, Post Code: 150003
City: Tunja, Boyacá
Country: Colombia
Email: mauro.callejas@uptc.edu.co

Voice Controlled Prosthetic Hand with Predefined Grasps and Movements

Juan Pablo Ángel-López[1,2] and Nelson Arzola de la Peña[2]

[1] Universidad Autónoma de Manizales (UAM®), Grupo de Investigación en Automática, Colombia
[2] Universidad Nacional de Colombia Sede Bogotá, Grupo de Investigación en Diseño Óptimo Multidisciplinario - OptimUN, Colombia

Abstract— **This paper shows the development of a robotic prosthetic hand controlled by voice commands, according to several grasps and movements. Drawing upon the kinematic definition and analysis of different grasps and movements from a previous work, the prototype prosthesis is able to achieve actions like fully hold or release objects, bi-digital gripping pincer, or postures like thumb up and pointing with the index finger. The robotic hand is controlled with a Graphical User's Interface (GUI) made upon the Matlab® environment, which recognizes a set of five voice commands by means of an Artificial Neural Network (ANN) and according to each command, the prosthesis develops a certain movement.**

Keywords— **Artificial limbs, electronics, voice, biomechanics, programming languages, motion capture.**

I. Introduction

Replicating the human hand has become one of the most desired achievements in robotics and prosthesis, but also one of the most difficult goals ever set in engineering. In the pursuit of this goal, there have been several researches aiming to reproduce the mechanical behaviour of the hand and to get control signals from the remaining muscles or nerves near the amputation, in order to create an ultimate hand prosthesis that can retrieve part or full functionality to an amputee.

As a Master's Degree Thesis in 2006, Alonso [1] explains in detail the attainment of the kinematic and dynamic models of an anthropomorphic robotic hand of four fingers and four Degrees of Freedom (DOF). All the fingers are actuated by means of bands connected to DC motors. This work aims to demonstrate that the control law, based on passivity has a better behaviour in respect to the error, compared to the calculated error. In the end, this reserach achieved an optimum control of four finger of four DOF each, resulting in a total of 16 DOF. Another Thesis in 2009 [2], was oriented to develop a robotic hand, controlled with a sensorized glove. The goal of this device was to follow the movements of the hand inside the glove with a certain level of accuracy. The research was able to determine a total positioning error of 1.55% when controlled with the glove, and 0.46% of error when controlled with direct angular parameters with a computer's GUI.

In 2004, Yang et. al. [3] showed a hand with multiple actuated fingers, four of them with two joints each, and the thumb with three joints. Each joint was designed with a flexible mechanism based on the load of a compression string on the axial and transversal directions, using a guidewire system. As a result, a five finger actuated hand was obtained, each finger with three DOF. In 2008, Pérez and Mendoza [4] worked on a research that looked for the develop of a robotic hand the could imitate the same movements of a human hand when playing some musical tracks with an electronic piano. In the end, the researchers obtained a device composed by three fingered hand moving along a rail, in order to be placed in front the keys to be pressed.

From a control strategy point of view, in 2009, Sarmiento et al. [5] conducted a research in order to evaluate and compare the efficiency and effectiveness of the interaction of a robotic prosthetic hand and its user, by controlling it with myoelectric signals or voice commands. The myoelectric signals were measured as electric pulses, and by counting these pulses the system recognized an action to be executed. For the voice commands, five different words were defined, each one related to a particular task of the prosthesis. As a result, they found that the voice command routine presented more efficiency and effectiveness. Also in 2011, Asyali et. al. [6] show the advantages of using a voice command control routine over a myoelectric control approach. The researchers designed and build a three DOF prosthetic hand with functions of opening and closing for grasping several objects. In the process, they developed an electronic control system based on voice commands as "pick-up" and "release". The authors claim that voice commands allow the prototype to have a larger range of actions and comfort than a myoelectric control system, besides the advantage of a short and easy training process.

According to the previously referenced works, this research aims to develop a voice-command based control system for a five DOF hand by some predefined and bio-inspired grasps and movements. The target is to create a routine that allows the user to manipulate objects or interact with others by a set of five actions, each related to a certain voice command.

I. Torres et al. (eds.), *VII Latin American Congress on Biomedical Engineering CLAIB 2016, Bucaramanga, Santander, Colombia, October 26th -28th, 2016*,
IFMBE Proceedings 60,
DOI: 10.1007/978-981-10-4086-3_131

II. Equipment and Methods

From an electronic point of view, the main goal of this research is to control the angular evolution of the different joints of the fingers in a seven DOF robotic hand (Fig.1), designed by [7] and improved by [9], according to the kinematic studies shown in [8], also expanded in [9]. This prosthetic hand prototype, designed in a previous research, was manufactured in a rapid prototyping process and its main material is ABS. The movement of the fingers two, three, four and five (index, middle, ring and little in this order) is given mainly on a sagittal plane by a guidewire attached to a pulley in a DC motor, considering the bi-dimensional model presented by [10]. The first finger (the thumb) has movements on a sagital and a transversal plane, as a simplification of the three plane movement noted by [11]. The prototype's DOF, considering the finger enumeration shown in the Fig.2, are divided as follows:

1. Second finger closing.
2. Third finger closing.
3. Simultaneous closing of the fourth and fifth fingers.
4. First finger closing.
5. First finger opposition.
6. Forearm rotation (180°).
7. Flexion and extension of the wrist (range of 180°).

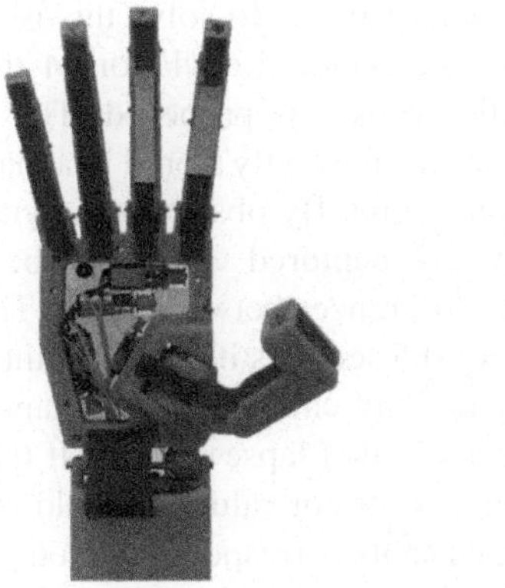

Fig. 1: Robotic prosthetic hand prototype improved by this research

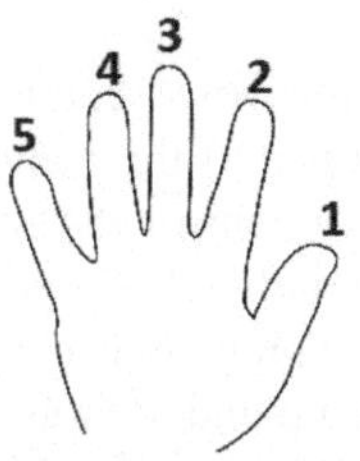

Fig. 2: Open hand (palmar view).

The fingers of the prosthetic prototype are actuated by DC motors, and are equipped with bending sensors (flex sensors onwards) placed on each DOF related to the fingers' movement (five out of the total seven DOF of the robotic hand). The use of the flex sensors responds to the need to control the angular evolution of the robotic device and the lack of space to place any other kind of sensor with less noise and more accuracy. Because of the their noisy response, the measure of the flex sensors had to be calibrated and filtered. First, a softening filter [12] was used in order to adjust the flex sensors' signals to an stable behaviour, and after conditioning the signals, it was necessary to create equations to convert the voltage given by the sensors into angular information. This conversion was achieved by using a Motion Capture System (MoCap) as shown in [8] so the received information from the sensors corresponds to the previously acquired kinematic data for different grasps and postures. The selected movements to be achieved by the prosthetic hand are:

1. Fully closing.
2. Fully opening.
3. Bi-digital clamp grasp.
4. "Ok" posture (thumb up).
5. Index finger pointing.

As the device shown in this work corresponds to a prototype, all the control routines were implemented under the Matlab® environment. All the algorithms written for this research were created with the purpose to later be transcribed to an embedded system like Arduino® or Raspberry Pi®, so all this routines have a very low level of complexity. The voice command control was achieved by adapting a simple ANN created by [13], used to control the movement of a mobile robot by making it go forward, reverse, left and right and to stop, also for the Matlab® environment, using a standard microphone connected to a standard PC audio card. This ANN works by a Back Propagation Algorithm that takes as entries a set of LPC[1], 12 for each saved command, and the output is a certain action. The idea for further improvements on the prosthetic hand is to use voice detection modules for Arduino® like EasyVR (VeeaR©) or a voice detection system for Raspberry Pi®.

The control interface between the device and the PC was achieved by means of an Arduino® Mega board, connected via USB to the computer and communicated with the Matlab® environment using the Hardware Package provided by Mathworks® for this purpose.

[1]Linear Predictor Coefficients

III. RESULTS

The prosthetic hand prototype mentioned above is actuated by five DC micromotors with reduction, under a nominal voltage supply of $5V$. In order to control the movement of these motors, five flex sensors were placed on the back of the hand, as shown in Fig.3. Given that the first finger possesses two DOF (OP[2] and flexion-extension), two flex sensors were placed on it, the fingers two to four have one flex sensor each (the fifth finger moves simultaneously with four finger) according to the one DOF each one has.

Fig. 3: Flex sensors placed on the Prosthetic Hand

Each sensor's signal is sent to an analog input of the Arduino® board, it means, to the ADC[3] inputs. Each ADC has a resolution of 10 bits (1024 different values) transforming each signal from $0V$ to $5V$ to an integer between 0 and 1023 bits, where each bit is approximately equal to $4.9mV$. This value in bits had to be converted to a number in degrees, between 0° and 180°. In order to get this equivalence, a MoCap system was used. Each DOF measured by a flex sensor was analysed, and one equation for each sensor (Eq.1 to Eq.5) was obtained. It has to be noted that even when the sensors are the same manufacturer and reference, each one has a different behaviour so that's why it was necessary to obtain a single equation for every flex sensor.

$$\theta_1 = 0.0005 f_1^2 - 0.3631 f_1 + 172.50 \tag{1}$$

$$\theta_2 = 0.0014 f_2^2 - 1.8395 f_2 + 730.34 \tag{2}$$

$$\theta_3 = 0.0015 f_3^2 - 1.7354 f_3 + 620.05 \tag{3}$$

$$\theta_4 = 0.0007 f_4^2 - 0.7797 f_4 + 336.75 \tag{4}$$

$$\theta_5 = 0.0015 f_5^2 - 1.9322 f_5 + 719.88 \tag{5}$$

Where f_1, f_2, f_3, f_4 and f_5 are the values in bits delivered by their respective flex sensor for each joint, and θ_1, θ_2, θ_3, θ_4 and θ_5 are the angles of each prosthesis' DOF. The enumeration of the θ and f variables corresponds to the robotic hand's DOF enumeration showed before in this paper.

[2]Opposition.

[3]Analog-Digital Converter

For achieving a closed loop control for the θ variables, a GUI under the Matlab® environment was created (Fig.4). Through the PC audio card and a standard microphone, the adapted ANN is trained, saving different voice commands for a posterior reading and detection of these commands. This voice control strategy is user- dependant, so only the prosthesis' user can control it. Each predefined action has maximum and minimum (or initial and final, if preferred) angle values for the joints when the hand is not grasping any object, and under a proportional control with hysteresis logic, it is wanted to match the measured angle with its desired value.

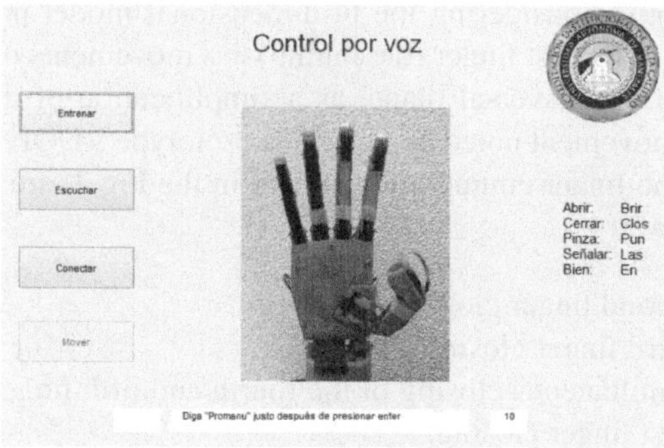

Fig. 4: GUI created for controlling the prototype.

Taking into count that the main idea is to manipulate objects, the joints of the hand not always will be able to reach their minimum values. To solve this issue, a stop criteria by calculating the standard deviation of the last sensed values for each flex sensor is proposed. The measured angles by each sensor are constantly stored in a matrix which can be accessed at any time. By observing a dynamic window of this matrix, the last captured values can be analysed and detect significant differences between them. The idea is to detect in a window of 0.5 seconds if a given joint has stopped because of an obstacle, by calculating the standard deviation of the captured data in that lapse of time. If this standard deviation is lower than a certain value, it would be considered that the joint stopped so its corresponding motor is turned off in order to avoid damages inside the robotic hand or to the manipulated object. This stop criteria showed an average of 83.97% of success in a set of 60 tests, with and without grasping objects.

After assembling the voice command detection and the flex sensors calibration into the same GUI, it was possible to achieve a system to control the movements of the prosthetic hand by five predefined actions such as: full opening, full closing, pitch grasping, index pointing and thumb up. For each action, a command was associated, so every time the user pronounces "Promanu" + one of five commands, the robotic device develops one of the said actions. If there is an obstacle during the closing or the opening of the fingers, the standard-deviation-based stopping criteria takes action and the movement of the given finger is finished. The commands and their related movements developed by the prosthetic hand are:

- "Promanu" + "brir" = open hand (all fingers in extension).
- "Promanu" + "clos" = close hand (all fingers in flexion).
- "Promanu" + "pin" = bidigital clamp (with fingers 1 and 2).
- "Promanu" + "en" = "OK" sign (thumb up).
- "Promanu" + "nas" = pointing with index finger (all the other fingers in flexion).

The training of the ANN required a recording of 10 repetitions for each word (Promanu, brir, clos, pin, en, nas) and 10000 iterations for the relative error calculation. Once the network is trained, the different commands are easily detected. It is necessary to clarify that any word with a high level of similitude to those listed above could be wrongly detected as an action and that's the reason of using the security command "Promanu".

IV. Conclusions

Voice commands detection obeys only to one user (the one who trained the ANN) due to the synaptic weights of the neurons change according to the tone, frequency and voice pitch of the user. Despite of this, the need to implement a security voice command was noted, in order to achieve a higher level of accuracy. So, before pronouncing any action command like "Open" or "Pitch", it was necessary to say "Promanu", so the robotic hand could be able to understand that the following word was going to be an action to do.

By increasing the training parameters of the ANN like the amount of times a command is trained, the number of LPC for each recorded command and the error calculation iterations, the voice command detections become more accurate.

Noting that there are several more robust and complex control strategies, like for example a PID control, the proportional control with hysteresis approach proposed in this research is showed as low-cost option (in terms of computer use), that could be easily transcribed to an embedded board like Arduino® or Raspberry Pi®, consequently decreasing the cost of the possible final prosthetic product.

In the pursuit of a better mechanical behaviour of the robotic hand, it is proposed to create an electronic control system based on servomotors, and mechanisms actuated with different approaches to the guidelines used in this prototype. The guideline system showed low efficiency and low reliability because of its susceptibility to fail.

It was also evident the need to design three DOF thumbs (like the human thumb), and two DOF fingers, dividing the action of the MCP[4] joint from the IF[5] joint, so movements like holding a book could be reached.

V. Acknowledgement

The authors thank the support given by Universidad Autónoma de Manizales, providing the facilities of the Biomechanics Laboratory, and Universidad Nacional de Colombia Sede Bogotá, by which it was possible to develop this research.

References

1. L. Alonso Juan, "Análisis y control de un efector reproduciendo un movimiento circular de una mano," Centro Nacional de Investigación y Desarrollo Tecnológico CENIDET, Cuernavaca, México, 2006.
2. D. Pilaquinga, "Diseño y Construcción de una Mano Robótica Controlada Mediante un Guante Sensorizado," Escuela Politécnica Nacional, Quito, Ecuador, 2009.
3. J. Yang, E. Pena Pitarch, K. Abdel-Malek, A. Patrick y L. Lindkvist, "A multi-fingered hand prosthesis," Mechanism and Machine Theory, n°39, pp. 555–581, 2004.
4. I. Pérez and H. Mendoza, "DESARROLLO DE UNA MANO ROBÓTICA CON CAPACIDAD DE MANIPULACIÓN," Télématique: Revista Electrónica de Estudios Telemáticos, vol. 7, n°1, pp. 180–207, 2008.
5. L. C. Sarmiento Vela, J. J. Páez y J. F. Sarmiento, "Prótesis mecatrónica para personas amputadas entre codo y muñeca," Tecné, Episteme y Didaxis, n°25, pp. 22–40, 2009.
6. M. H. Asyali, M. Yilmaz, M. Tokmakçi, K. Sedef, B. H. Aksebzec y R. Mittal, 2Design and implementation of a voice-controlled prosthetic hand," Turkish Journal of Electrical Engineering and Computer, vol. 19, n° 1, pp. 33-46, 2011.
7. J. L. Loaiza Bernal, "Diseño y simulación de un prototipo de prótesis de mano bioinspirada con cinco grados de libertad," Universidad Nacional de Colombia, Bogotá, 2012.
8. J. Angel and N. Arzola, "Kinematic Hand Analysis Using Motion Capture Technology," IFMBE Proceedings, vol. 49, pp. 257–260, Springer International Publishing Switzerland, 2015.
9. J. Angel, "Control de una prótesis de mano para varios tipos de prensión empleando comandos de voz," Universidad Nacional de Colombia, Bogotá, 2016.
10. Freivalds, Andris. "Biomechanics of the Upper Limbs: Mechanics, Modeling and Musculoskeletal Injuries," vol. 16, n°4. pags. 215–221. 2004.
11. Kapandji, Adalbert I. Fisiología Articular, Tomo 1, 6ª edición, pp. 250–325. 2006.
12. Matlab Documentation Center. Savitzky-Golay filtering "sgolayfilt". `http://www.mathworks.com/help/signal/ref/sgolayfilt.html`. 2013.
13. V. Samant, "Voice Controlled Robot," 11/01/2014. [Online]. Available: `https://developer.mbed.org/users/vsamant3/notebook/mini-project-1--voice-controlled-robot/`.

[4]Metacarpal Phalangeal
[5]Interphalangeal

ICA and SVM Clustering Applied to Remove Ocular Artifacts from Electroencephalography

J. Pena-Rodriguez[1], D. A. Sierra[1] and C. A. Conde-Cotes[2]

[1] Universidad Industrial de Santander/School of Electrical, Electronic and Telecommunication Engineering, Bucaramanga, Colombia
[2] Universidad Industrial de Santander/School of Medicine, Bucaramanga, Colombia

Abstract— **Brain behavior study has taken attention by the scientific world in last years. Comprehension of cerebral processes is made by means of data obtained through several techniques such as: electrocorticography, magnetoencephalography, functional and structural imaging, and electroencephalography (EEG). This latter is the most used due to its low cost and minimal risk. However, recording of cerebral information through EEG is affected by different artifactual sources, which influence their posterior data processing. Inside non-cerebral sources, the potentials caused by ocular movements during tracking and fixing tasks have the greatest impact. For this reason, a procedure to identify and subtract the ocular artifacts from EEG signals is needed. In this work, an automatic method to remove artifacts is presented. The algorithm is compounded of four main stages: a) the separation of the EEG and artifacts sources by means of Independent Component Analysis (ICA); b) the characterization of the EEG and artifact components using complexity features; c) the components classification through a Hybrid Support Vector Machine (SVM)-External clustering; and d) the reconstruction of the EEG free of artifacts. The method has an overall accuracy of 85.9%, the elimination of ocular artifacts is 85.68% and the preservation of EEG information is 85.9%. The entire algorithm was written in Python.**

Keywords— **Ocular artifacts, Electroencephalography, ICA, SVM clustering, feature extraction.**

I INTRODUCTION

Currently, one of the biggest goals in science is to understand the brain. To do it, researchers need to get information about how the brain reacts to stimuli, situations and pathologies. Consequently, several tools to register and to study neural dynamics have been developed. EEG is the most common technique owing to its simplicity and non-invasive nature. EEG signals have small amplitudes and they are very sensible to artifacts produced by non-cerebral sources such as: breathing, heart rate, muscular and ocular movements.

Artifacts caused by ocular movement are the main source of contamination in EEG recording. Thus, several methods to remove ocular artifacts in EEG signals have been developed in the last decade; a complete study of those methods is presented in [1]. Due to the similarity in the spectrum between EEG and artifacts and the unpredictable temporal behaviour of the artifacts, temporal, spectral or multiresolution analysis are not sufficient to remove them. For that reason, most of the current techniques use Blind Source Separation methods which search possible components of the mixture between artifacts and EEG. Besides, to avoid the need of having components being manually classified, several machine learning techniques have been applied. As a result, two methodologies have been taken into account: supervised and unsupervised learning.

In supervised learning, a classification model is trained with a labeled samples, then the model is validated with another portion of the data; in this manner, accuracy of the classifier is assessed. In particular, usually SVM predictors are used instead of another methods as neuronal networks (ANN) due to their better performance in this kind of problems as shown in [2].

By contrast, unsupervised learning searches clusters in the data according to a similarity measure. The model does not need a priori information about how much samples there are in each class, neither which label would belong to a sample. Some algorithms to remove artifacts based on hierarchical, k-means and fuzzy clustering are implemented in [3], [4] and [5] respectively.

In this work, an automatic method to separate ocular artifacts from EEG based on SVM External clustering is proposed. The system has a high performance in the removal task and preserves the most of EEG information.

II METHODS

The stages of the method are described in the following way:

Independent Component Analysis (ICA) decomposes the n recorded EEG channels $X = [x_1, x_2, \ldots, x_n]^T$, called *observations*, in a lineal mixture of m independents components (ICs),

$$X = A \cdot S \tag{1}$$

I. Torres et al. (eds.), *VII Latin American Congress on Biomedical Engineering CLAIB 2016, Bucaramanga, Santander, Colombia, October 26th -28th, 2016*,
IFMBE Proceedings 60,
DOI: 10.1007/978-981-10-4086-3_132

where $S = [s_1, s_2, ..., s_m]^T$ are the ICs, with $m <= n$, and A is the mixture matrix.

Then, several features are extracted from each independent component, i.e. $F(S) = [f(s_1), f(s_2), ..., f(s_m)]^T$ is the feature vector for IC, with $f(s_i) \in R^k$, where k is the number of features extracted.

Afterwards, the clustering algorithm sets up a decision boundary that allows to assign the samples to each group. Here g_1 is the group of artifact ICs and g_2 is the group of EEG ICs.

Finally, the EEG signals reconstruction is made in two steps. First, ICs corresponding to the artifacts are removed, in the following way:

$$\tilde{s}_i = s_i \frac{-g_i + 1}{2}, \ \forall i \tag{2}$$

where

$$g_i = \begin{cases} 1 & if \quad g_i = g_1 \\ -1 & if \quad g_i = g_2 \end{cases} \tag{3}$$

Second, observations without ocular artifacts $\tilde{X}$ are computed as,

$$\tilde{X} = A \cdot \tilde{S} \tag{4}$$

where, $\tilde{X} = [\tilde{x}_1, \tilde{x}_2, ..., \tilde{x}_n]^T$ and $\tilde{S} = [\tilde{s}_1, \tilde{s}_2, ..., \tilde{s}_m]^T$.

A EEG dataset

EEG dataset was taken from the *Institute for Knowledge Discovery* in the Graz University of Technology for the BCI Competition IV. The data was recorded from 9 subjects using 22 EEG and 3 EOG channels, with a sampling rate of 250 Hz. Sessions are divided in four blocks: (1) two minutes with closed eyes, (2) two minutes with opened eyes, (3) one minute with ocular movement and (4) six registers of four imaginary tasks [6]. Here, only the third block was used.

B ICA

Decomposition of EEG is made by means of the *FastICA* algorithm [7]. This method consists in finding the separation matrix coefficients that maximize the non-gaussianity of observations X.

C Feature extraction

Particular behaviour of the mixture between EEG and ocular artifacts limits some common techniques of feature extraction. In this case, frequency features are not efficient because EEG signals and ocular artifacts share the same spectral range. On other hand, a windowing method to extract temporal features is not an option because the algorithm should detect the exact moment when artifacts occur by means of a reference signal, e.g. an extra EOG channel.

For these reasons, a set of complexity features were randomly chosen, due to their capability of discrimination between ICs from EEG and artifacts. Then, an algorithm to chose the best mixture of features was implemented and the resultant subset was composed by: Kurtosis, Fisher Information, Detrended Fluctuation Analysis, Hjorth Parameters, Higuchi Fractal Dimension, Hurst Exponent, and Petrosian Fractal Dimension. Features were computed by means of the *PyEEG* library [8].

For the fitting and testing of the clustering model, a feature vector was extracted from the EEG dataset. The record sessions from subjects are chopped in l parts of 8 seconds, then an ICA algorithm runs on each part and features are extracted from the obtained ICs. The final feature set contains 1540 samples.

D Clustering

To identify the ICs belonging to ocular artifacts, the SVM External-Hybrid clustering proposed in [9] was used. The SVM External-Hybrid clustering works as an optimized decision tree. Firstly, the algorithm labels the dataset samples through a K-means clustering, later a SVM binary classifier is trained with the labeled dataset.

After that stage, the algorithm improves the result by flipping the labels of mislabeled data, which have a confidence factor greater than a set threshold, then the SVM model is trained again with the re-labeled data. This process continues until the difference between the expected labels and the estimated labels is under one stop condition.

The *Mean Square Error* (MSE) was used as stop criterion here. This is defined as [10]:

$$MSE = \frac{1}{N} \sqrt{(\hat{t} - t)^2} \tag{5}$$

where t is the target vector, $\hat{t}$ is the estimated vector and N is the length of each one.

The efficiency of the clustering method changes according to the SVM parameters, e.g. if a Gaussian kernel is used, the parameter C is the penalty for misclassified samples and γ is the kernel width, in this case the model was assessed for a range of parameters and finally the best of them were chosen.

III RESULTS

In this section, the results obtained from the implementation of a system to detect and remove ocular artifacts from EEG are presented. The system is split in two parts, the first one is the clustering model training stage, Algorithm 1, and the second one is the test stage, Algorithm 2.

Algorithm 1 Training stage of the clustering model

Require:
- Feature vector: $\mathbf{f}$
- Penalty parameter: C
- Kernel width γ
- Stop criterium: MSE

$\theta \leftarrow Clustering(\mathbf{f}, C, \gamma, MSE)$

Algorithm 2 Test stage of ocular artifact removal system

Require:
- EEG data: X
- Trained model: θ

$X \leftarrow$ Standardization of X
$S, A \leftarrow FastICA(X)$
for each $s_i \in S$ **do**
  $f_i \leftarrow$ Feature extraction of s_i
  $g_i \leftarrow Prediction(\theta, f_i)$
end for
$X \leftarrow Reconstruction(S, A, \mathbf{g})$

The algorithm has as input only the feature vector $\mathbf{f}$ and the model parameters C, γ and MSE. The first stage provides a trained model represented by a decision boundary θ used to assign each independent component to one cluster.

The validation of clustering method was made though 10000 runs. In each run the algorithm was trained with 1000 samples and tested with 540 samples; the average results are shown in the Table 1.

Table 1: Confusion matrix of the clustering validation with $C = 2.5$, $\gamma = 0.009$ and $MSE = 1e-6$

	Predicted class	
True class	g_1 (Artifact ICs)	g_2 (EEG ICs)
g_1	**85.68** %	14.3 %
g_2	14.01 %	**85.9** %

In the second stage of the algorithm, the EEG signals for processing are uploaded, standardized and decomposed in ICs by means of *FastICA*. From each IC, a feature vector is extracted, then the prediction is made through the trained model. Finally, the removal of ocular artifacts and the reconstruction of the EEG signals are implemented following (2) and (4).

Now, one example is shown. The algorithm was applied on the EEG recording from the subject A07E in the temporal range (412-420s). In Fig. 1, the EEG and HEOG channels forming the register are exposed. Apart from this, the influence of ocular artifacts in the EEG signals are indicated with a gray fringe.

The ICA decomposition of the registered EEG is displayed on Fig. 2. Here, the artifact components are indicated with a gray fringe too. In this case, the component IC1 describe clearly the ocular artifact behavior shown in the EEG record. Furthermore, the scalp mapping of the coefficients from the identified ocular artifact IC confirms the origin of the source of noise, Fig. 3. Once the algorithm identifies the artifact ICs using the features extracted and the trained clustering model, it eliminates them. That is to say, attenuate its influence over the recorded EEG, Fig. 4.

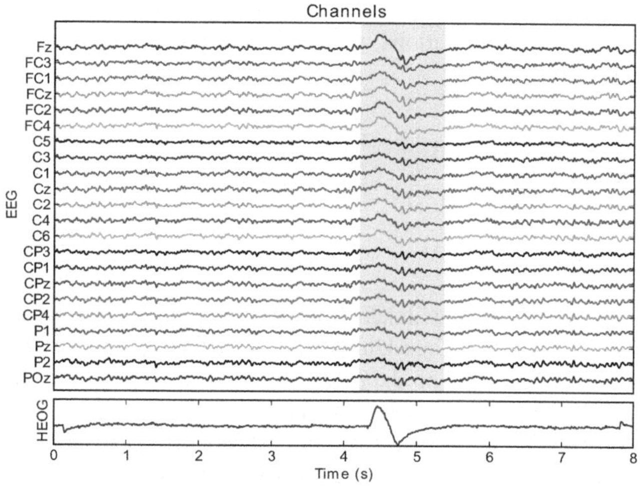

Fig. 1: EEG channels from the subject A07E from 412s to 420s. Top: the 22 EEG channels; Bottom: The horizontal EOG

IV CONCLUSION

In this work, an automatic removal system for ocular artifacts in EEG using ICA and SVM clustering was presented. This tool proposes a novel method to remove ocular artifacts avoiding the manual selection and elimination of artifacts from EEG signals. Furthermore, it allows a reduction in time and costs of the process. The method has an overall accuracy of 85.9%, the elimination of ocular artifacts is 85.68% and the preservation of EEG information is 85.9%. A new set

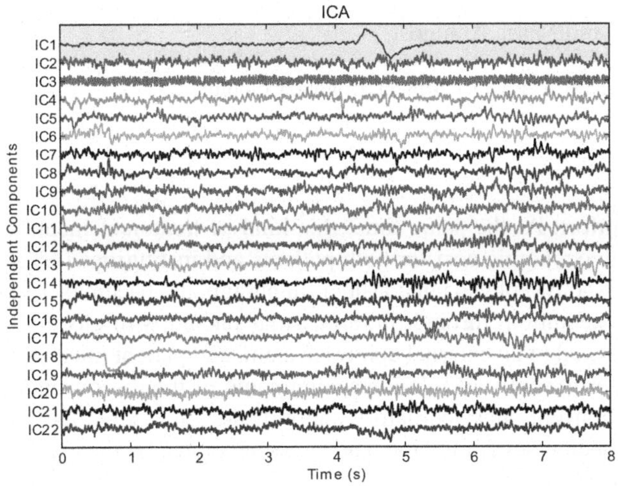

Fig. 2: ICA decomposition of the EEG channels

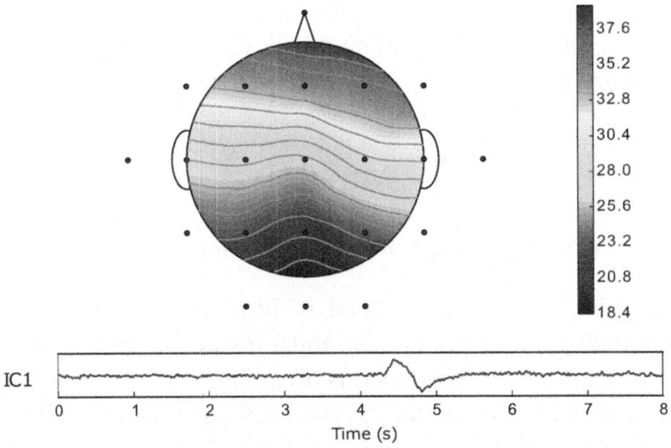

Fig. 3: Scalp mapping for the IC1 coefficients

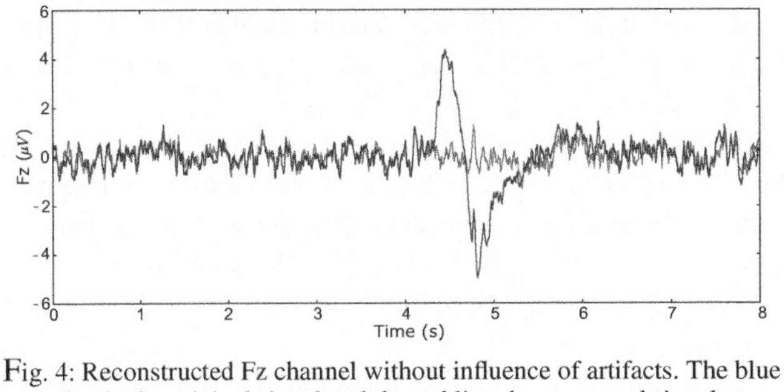

Fig. 4: Reconstructed Fz channel without influence of artifacts. The blue line is the original signal and the red line the processed signal.

of features will be proposed to improve the accuracy of the method in a future work.

CONFLICT OF INTEREST

The authors declare that they have no conflict of interest.

ACKNOWLEDGEMENTS

This work was developed at CEMOS Research Group in collaboration with NYC Research Group at Universidad Industrial de Santander.

REFERENCES

1. Urigüen J A, Garcia-Zapirain B. EEG artifact removal—state-of-the-art and guidelines *J. Neural Eng.*. 2015;12:031001.
2. Singla R, Chambayil B, Khosla A, Santosh J. Comparison of SVM and ANN for classification of eye events in EEG *Journal of Biomedical Science and Engineering.* 2011;04:62–69.
3. Zou Y, Hart J, Jafari R. Automatic EEG artifact removal based on ICA and Hierarchical Clustering in *2012 IEEE International Conference on Acoustics, Speech and Signal Processing (ICASSP)*;5(Kyoto, Japan):649–652 2012.
4. Otero-Millan J, Alba-Castro J L, Macknik S L, Martinez-Conde S. Unsupervised clustering method to detect microsaccades *Journal of Vision.* 2014;14:18–18.
5. Patidar U, Zouridakis G. A hybrid algorithm for artifact rejection in EEG recordings based on iterative ICA and fuzzy clustering. *Conference proceedings : ... Annual International Conference of the IEEE Engineering in Medicine and Biology Society. IEEE Engineering in Medicine and Biology Society. Annual Conference.* 2008;2008:50–3.
6. Brunner C, Leeb R, Müller-Putz G R, Schlögl A, Pfurtscheller G. BCI Competition 2008 - Graz data set A 2008.
7. Hyvärinen A, Karhunen J, Oja E. *Independent Component Analysis.* Wiley 2004.
8. Bao F S, Liu X, Zhang C. PyEEG: An open source python module for EEG/MEG feature extraction *Computational Intelligence and Neuroscience.* 2011;2011.
9. Winters-Hilt S, Yelundur A, McChesney C, Landry M. Support vector machine implementations for classification & clustering. *BMC bioinformatics.* 2006;7 Suppl 2:S4.
10. Bishop C M. *Pattern Recognition and Machine Learning (Information Science and Statistics).* Secaucus, NJ, USA: Springer-Verlag New York, Inc. 2006.

Author: Jesús Peña Rodríguez
Institute: CEMOS, Universidad Industrial de Santander
Street: 9 Carrera 27
City: Bucaramanga
Country: Colombia
Email: jesus.pena@correo.uis.edu.co

Diaphragmatic pacemaker prototype with wireless communication

Jorge Reynolds[1], Jose Forero[2] and Maria Arango[2]

[1] Shaio Clinic Foundation/ Electronic and Nanoelectronic Research Department, Bogota, Colombia
[2] Manuela Beltran University/ Biomedical Engineering, Bogota, Colombia

Abstract— **Purpose: To describe the design and development of a diaphragmatic pacemaker prototype with wireless control, which allow the user to modify the stimulation parameters and some physiology parameters related with respiration process.**

Method: Begins with a bibliographic review of the main concepts of respiration and the diaphragm muscle, and a background research focused on diaphragm stimulation parameters. Followed of the design of user interface and hardware in charged of generate the pulse train with the needed specifications. Finally, the construction and coupling of the analog system with the interface, and conducting pertinent tests that allow evaluation and analysis of the device.

Results: It has been made a prototype with a LabView interface and an analog system in charged of the digital to analog conversion of the information, to become the signal into a biphasic one, and a power stage, that allows to have a linear control of the current depending on the input voltage. The device had good results, beginning with and excellent transmission of the information from the user interface to generator system without data loss, in the other hand, the reception of the data and the execution of modifications is effective, with a little difference between the value entered by user and the response of the device.

Keywords— **Diaphragm pacing, diaphragmatic paralysis, electric parameters.**

I Introduction

The diaphragm is considered the main muscle in human respiration, due to it is responsible for 60-70% in inspiration process. It is innervated by phrenic nerves for contraction function, the main nerve branches originate from fourth cervical nerve and the accessory branches from third and fifth nerve [1,2,3].

Diaphragmatic paralysis caused by muscle or nerve pathologies, which forces patients to resort to several kinds of ventilatory support in order to supply respiratory function. However, even when it is essential for patient support, it has some shortcomings hampering with quality of life, and procedures like endotracheal intubation end up injuring vocal cords, disabling patient in basic actions such as speaking or swallowing, and representing a high sepsis risk. This is an inherent disadvantage in other procedures like tracheotomy, considered an open wound, subjecting the patient to discomfort. At the same time, this kind of treatment incurs in high economic cost for both patients and institutions [4,3,5]. Nowadays, physician consider diaphragmatic pacemaker implementation, through intramuscular pacing with electrodes attached to an external control system, and patient must carry it permanently. While improving some of the disadvantages of conventional ventilatory support, does not solve the risk of open wound infection, and stress generated by the external device [6].

For those reasons, we raised the development of a diaphragmatic pacemaker prototype using wireless technology, allowing the user change pacing parameters and some physiologic parameters related to respiration, through an user interface on a computer; taking the first step to achieve, in future researches, a fully implantable device pacing supporting respiratory function.

II Methods

A Electrical parameters for diaphragmatic pacing

*I*n order to achieve artificial respiration using diaphragmatic pacing, some parameters must be accomplished to ensure an imitation of spontaneous breathing. Those parameters are implemented in modern respiratory pacemakers with intramuscular electrodes showing a great effectiveness, and its allowable ranges are shown in Table 1 [7,8,9,10,11]. This information is applied to this investigation, and every single parameter's values are limited for avoiding unwanted effects such as tissue damage for high current application, or high frequencies in pacing signal producing tetanization of the excited area or even damage on the electrode because of electrolysis.

The waveform of the designed signal must be on ramp, where the gradual increase in amplitude of every pulse represents a more similar stimulation to the natural diaphragm innervation; moreover, pulses must be biphasic using the negative current to reverse unwanted chemical reactions over tissue or the electrode [12].

Moreover, the prototype's development presented in this article, and contrary to current electroestimulation electrodes, call for the incorporation of sigh, feature that should be ap-

I. Torres et al. (eds.), *VII Latin American Congress on Biomedical Engineering CLAIB 2016, Bucaramanga, Santander, Colombia, October 26th -28th, 2016*,
IFMBE Proceedings 60,
DOI: 10.1007/978-981-10-4086-3_133

plied to this kind of pacemaker, to evaluate in the future its effects against the alveolar collapse or atelectasis, taken into account in mechanical ventilation [13]. Furthermore, the designed signal must be generated four times simultaneously, two for each hemidiaphragm, through four independent electrodes, due to it can be considered that a single electrode cannot generate pacing enough to induce diaphragm movement [6].

Table 1: Intramuscular diaphragmatic pacing parameters

Parameter	Range	Unit
Amplitude	5 - 25	mA
Frequency	4 - 50	Hz
Pulse Width	10 - 200	u sec
Respiratory frequency	8 - 18	breaths/ minute
Inspiration Time	0.8 - 1.5	sec
sigh rate	2 -60	sigh/minute

B Wireless control software

The built prototype requires several orders and settings from a control equipment, which is represented by a software installed on computer; this software was designed with the development tool LabVIEW. The interface allows the user to turn on and off the device, and setup parameters such as amplitude, frequency, pulse width, respiratory frequency, inspiration time and sigh rate (Figure 1) taking into account the possible range of values.

Before starting configuration, patient data must be entered and then, the software can export a file with this information, settings made to parameters, and modification date and hour. The communication will be done under Bluetooth protocol, indicating the module address installed on hardware.

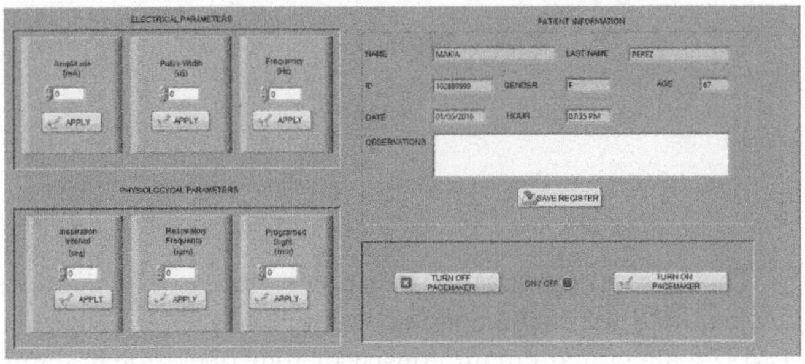

Fig. 1: Pacemaker Control Panel

C Hardware

Communication and signal generation generating the signal required a microcontroller PIC16F877A, which was programmed whit a C++ language code, made in CCS Compiler. This algorithm allows generate two digital 8 bits signals trough different ports, both signals are the same, but with a lag of 180 in order to generate an only biphasic signal based in those two. At this point, amplitude is voltage controlled, from 0 to 5 volts.

At this point, amplitude is voltage controlled, form 0 to 5 volts. The algorithm allows create a pulse train forming a ramp, increasing the twice the value of the previous pulse, up to the maximum amplitude configured from the interface. The train duration is directly associated with the inspiratory time, because while pulses are generated it will be produced tissue stimulation, and exhalation occurs in the absence thereof. Meanwhile, the sigh is generated by modifying the base signal, increasing the number of repetitions of the pulse of greater amplitude after generating the ramp, over the same duration of the ramp; that is to say, the PROGRAMMED SIGH lasts twice the time of inspiration and it's presented with the programmed frequency (4 Hz-50Hz) from the control software, through Timers. Respiratory rate is related to the number of pulse trains that the microcontroller generates in a minute, and the signal frequency is programmed through the period of the pulse, which is equal to the time high (pulse width) plus the time low.

All parameters are received through serial communication pins, integrated withing the microcontroller; the pins were connected to Bluetooth modules, testing tow references: HC-05 and HM-10

Signal processing After the Digital- analog conversion, we inverted the advanced phase signal using an inverting amplifier and we add it with the original wave, obtaining a biphasic pulse train; this was made using an LM324 operational amplifier.

Voltage to Current converter For obtaining a current controlled signal (form 0 to 25 mA), we implemented a voltage to current converter whit float charge, using operational amplifiers. This configuration allows control the output current proportionally to voltage in the input, keeping the programmed stimulation waveform.

This stage is a summing amplifier, and through nodal analysis is easy to establish the current and voltage limits determining the range of impedance of the charge where is possible to use this configuration, and calculate saturation voltage and reference voltage. To do this, we made an analogy with a linear equation, in order to obtain a linear response current versus voltage.

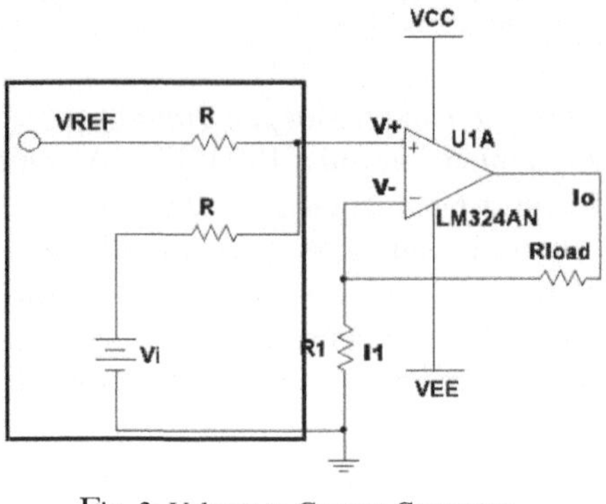

Fig. 2: Voltage to Current Converter

In figure 2 we can see the voltage to current converter circuit, where it is indicated by a black box the passive averager thereof, which, as its name indicates, is in charge of averaging the input voltages of amplifier noninverting input. Voltage of noninverting input correspond to add input voltages divided the number of inputs, at the same time, it is equal to noninverting input voltage, on the other hand, output current *Io* is equal to *I1* and this correspond to noninverting input voltage *V-* divided *R1* resistor. Thus, output current in terms linear equation is:

$$Io = \frac{1}{R1} * Vi + \frac{VREF}{2R1} \tag{1}$$

Due to voltage *Vi* isn't constant, takes x value in linear equation.

Knowing that output current *Io* is directly proportional to input voltage *Vi*, and that we know both variation ranges, we can obtain linear equation graph, see figure 3.

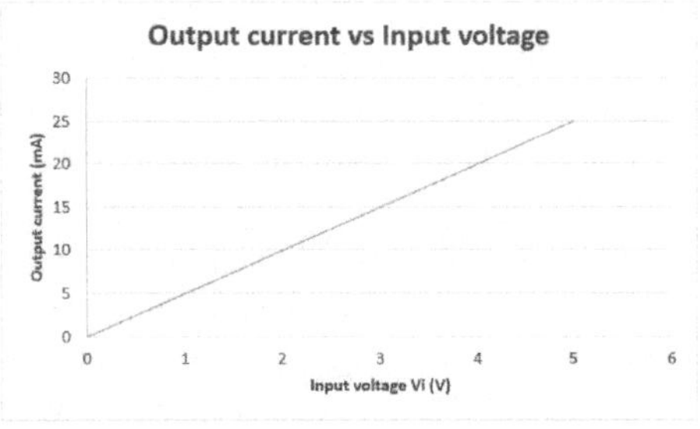

Fig. 3: Output current *(Io)* vs Input voltage *(Vi)*

Based on the above, calculate the slope of the line m and cut-off point in y axis (b) obtaining:

$$m = \frac{25-0}{5-0} = 5 = \frac{1}{2R1} \tag{2}$$

$$b = \frac{VREF}{2R1} = 0mA \tag{3}$$

In orther to find all the values of figure 2 scheme, isolate *R1* variable of four equation and taking into account that cut-off point in axes y is 0, VREF will take this value.

$$R1 = \frac{1}{2*5} = \frac{1}{10}m = 100\Omega \tag{4}$$

As R can takes every value as long as it has a high impedance bearing in mind *VREF* and *Vi* voltages, it was selected to 1Kohm.

Finally, we have to calculate saturation voltage *Vsat* of circuit, knowing that output voltage *Vo* is minor or equal to input voltage *Vi* and load resistor *Rload* is approximately 300ohms.

$$(Rload + R1) * Io \leq Vsat \tag{5}$$

We replace the variables and taking the maximum value of *Io* current (25mA), so:

$$Vsat = (300 + 100) * 25mA = 10V \tag{6}$$

That means, VCC should be higher than 10V. Due to input signal of power stage is biphasic, amplifier supply must be bipolar, and negative supply value equal to Vcc but with opposite sign. Thus, complete scheme can be seen down bellow:

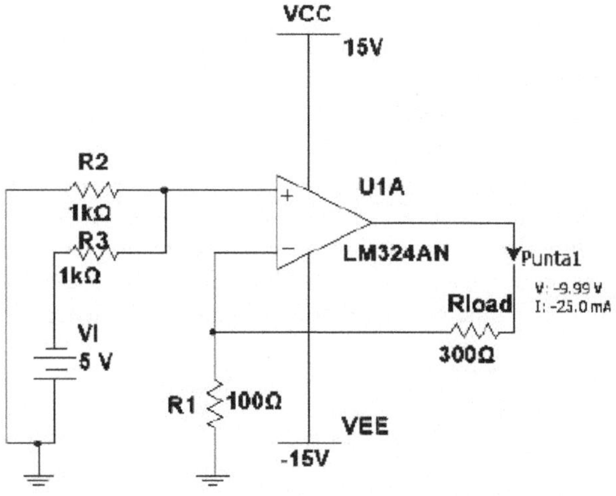

Fig. 4: Voltage to Current converter with *VREF* of 0v and *Vi* of 5V

This process is repeated four times, because 2 electrodes are required for each hemidiaphragm, to do so, it is used an LM324 integrated circuit consist of four operational amplifiers, through that it is possible to cover all the OPAMPs needed.

III RESULTS

After construction of the design system it was obtained the shown circuit in figure 5, in which is possible to observe Bluetooth module (1), microcontroller (2), In circuit Serial Programming connection (3), conversion stage to negative voltage (4), DAC (5), DAC of out of phase signal (6), biphasic signal generator (7), voltage to current converter stage (8) and the four output ports for electrodes.

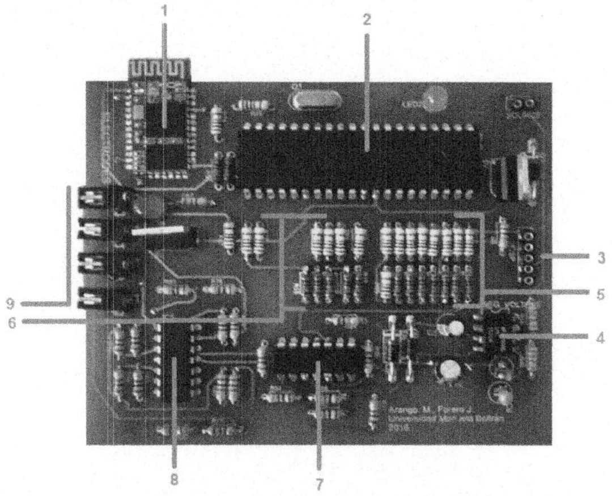

Fig. 5: Diaphragmatic Pacemaker Prototype

After the output signal obtaining with the desired features, they were performed comparative tests between theoretical and measured values by Fluke 199C oscilloscope, finding minimum or even null differences for all parameters save pulse width, which in configurations smaller than 150us, presents an error of 10 units more in comparison with the expected value. However, the error could be attributed to measured instrument implemented, which, due to it is a reduced time, it isn't possible to obtain a entirely correct measure. The remaining cases presented very precise and accurate results, after correlation with the programmed values on the user interface. Figure 3 show the output signal obtained.

Furthermore, wireless communication was test with HM-10 module, which has incorporated a Bluetooth 4.0 or low power BT, representing a considerable reduction in the circuit current consumption around 55.8, however, due it is an advance protocol version, communication with the computer software wasn't possible because of the incompatibility between the devices, therefore, tests had to be development with a conventional module (HC-05). Results show an easy connection without loss of information or data transmission while wireless programming of prototype. Even so, the following prototypes of the investigation will be commanded by smart phone applications, which, mostly could get communicated through a 4.0 Bluetooth profile, achieving a low current consumption in the next device.

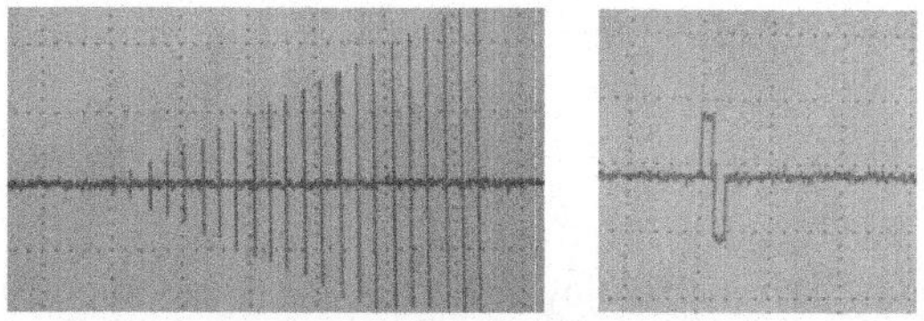

Fig. 6: Biphasic Stimulation Signal

Under this premise, conventional battery life of a cardiac pacemaker, commonly 2.5Ah, connected to supply the designed system, could supply it around just 17- 15 hours, in way that, is necessary to seek consumption reduction of all analog system components.

IV CONCLUSIONS

Based on obtained results, it was determined, that such electronic components for configuration and signal generation, as technology used for wireless communication, should be replaced by elements of low consumption, because of, current registered for prototype functioning is to high to ensure a long device autonomy with an small size battery; in addition; should be implemented SMD (Surface Mount Device) technology, in order to reduce prototype size in way that ensures it is fit for its implantation. Thus, through optimization of the presented circuit on this investigation, it is possible to create a wireless device with a battery capable of supplying the consumption thereof, for full implantation and that is controlled by an external system. It is expected that future created prototypes will be controlled with smart phone technology for a better portability and access (nowadays there isn't any other system with this features), also, it is necessary to establish a security protocol to annul any unexpected interference and avoid unauthorized person to accede patient information. It is recommended the use of more precisely electronic components, such as elements for military or medical purposes, due to conventional components have an error percentage between 5 and 15%.

INTEREST CONFLICT

Authors declare that there is no interest conflict.

ACKNOWLEDGMENTS

Our sincere thanks are with Electronic and Nanoelectronic research department of Shaio Clinic Foundation and it's working group, to Dr. Albert Borau of Guttman institute in Barcelona-Spain, as well as educational of Manuela Beltran University.

REFERENCES

1. Parsons, P.E., Heffner, J.E.: Pulmonary/ respiratory therapy secrets. Hanley & Belfus, inc, Filadelfia (2001)
2. Iglesias, B.G., Gallardo, M.L., Martn, A. I., Prieto, J.: Bases de la Fisiologa. Tebar, Madrid, (2007)
3. Cristancho, W.: Fisiologa Respiratoria: lo esencial en la prctica clnica. Manual moderno, Mxico (2007)
4. Kaufman, M.R., Elkwood, A.I., Rose, M.I., Patel, T., Ashinoff, R., Saad, A., Caccavale, R., Bocage, J., Cole, J., Soriano, A., and Fein, E.: Reinnervation of the Paralyzed Diaphragm: Application of Nerve Surgery Techniques Following Unilateral Phrenic Nerve Injury. Chest. 140, 191-197 (2011)
5. Dimarco, A.F.: Neural prostheses in the respiratory system. J Rehabil Res Dev. 38, 601607 (2001)
6. NeuRx Diaphragm Pacing System. Patient/Caregiver Information and Instruction Manual, www.synapsebiomedical.com/support/PDFs/006_ 77-0055_A Patient Caregiver Manual-ALS.pdf
7. Taira, T., Hori, T.: Diaphragm pacing with a spinal cord stimulator: current state and future directions. Acta Neurochir Suppl. 97, 289292 (2007)
8. Hogan, J. F.: Evaluation of Electrical Parameters for Diaphragm Pacing: An Experimental Study, J Surg Res. 153, 142153 (1981)
9. Universidad de Costa. Facultad de Medicina, http://163.178.103.176/Fisiologia/respiratorio/pracb_7/respi_pracb_14.html
10. Dobelle, W.H., D'Angelo, M.S., Goetz, B.F., Kiefer, D.G., Lallier, T.J., Lamb, J.I., Yazwinsky, J.S.: 200 Cases with a new breathing pacemaker dispel myths about diaphragm pacing. ASAIO (1994)
11. Ducko, C.T.: Clinical Advances in Diaphragm Pacing. Innovations. 6, 289 297 (2011)
12. Uranga, A.: Desarrollo de un estimulador elctrico integrado implantable para el control de la miccin en lesionados medulares. PhD. Thesis, Dept. Ing. Electron., UAB, Barcelona, Espaa.
13. Holstege, H.G., Beers, C.M.: The Central Nervous System Control of Respiration. Elsevier, Amsterdam (2014)

Three Dimensional Reconstruction and Airflow Simulation in a Realistic Model of the Human Respiratory Airways

A. E. Ruiz[1] and J. K. Aristizábal[1]

[1] Bioinstrumentation and Clinical Engineering Research Group - GIBIC, Bioengineering Department, Engineering Faculty, Universidad de Antioquia UdeA; Calle 70 No. 52-21, Medellín, Colombia

Abstract— **Respiratory tract diseases are a common cause of medical consultation. Indeed, most of this patients need an adequate treatment to improve their airway conditions. In order to understand how the airflow works through the lower airway and how disease condition may affect it, this paper presents an airflow simulation based on a three dimensional model from a real human airway constructed from a Computed Tomography scan.**

Methodology used to perform the simulation, include the implementation of Materialise's medical image software MIMICS® (v. 18.0), in order to segment the lower airway using region based segmentation procedures. Moreover, a computational fluid dynamic simulation was carried out by using the finite element software Comsol® (v. 5.2); to this end, this paper shows a comprehensive workflow integrating both software to obtain a realistic airflow simulation through the lower airway.

Velocity and pressure profiles are presented to assess the effectiveness of the simulation, showing a parabolic non-compressive flow and high pressure in complex anatomical geometric areas, validating the expected physiologic behavior.

Keywords— **Human lower airway, realistic model, computational fluid dynamics**

I. Introduction

Human respiratory system is a complex structure consisting of two main anatomical divisions: upper airways and lower airways, whose primary function is oxygen supply to the body cells and removing carbon dioxide, through a process called gas exchange [1]. Lower airways geometry is not homogeneous due to peculiarities in its own structure, as anatomical constriction zones, asymmetric branching patterns, distinct radius of curvatures and constant change of branching angles and orientation; which generate rapid changes in the flow direction [2]. Therefore, it is necessary to carry out different analysis to allow understanding flow behaviour within the airway under different conditions including normal physiologic function [3–5], particle deposition [4], [6], some pathologies as obstructive sleep apnea [7] and congenital glottic web [8] and situations under mechanical ventilation [9–11]; in the latter case, this is a procedure that can involve multiple complications such as airway edema, bleeding, obstruction, collapse, respiratory failure and arrest [12]. For this purpose, Computational Fluid Dynamics (CFD) simulation has been widely applied [2–9, 13–15], since fluid flow characteristics can be predicted in detail, under controlled conditions and noninvasive way. Nevertheless, some of these works have been performed in idealized and simplified models of human airways defined by only diameters and lengths of cylindrical structures. In these cases, particularities derived from asymmetry are ignored and the complexity of real airway is not considered. In this paper an airflow simulation in a realistic model of the human respiratory airway is presented. For this end, a methodology consisting in four stages was carried out: 1) Computed Tomography scans aqcuisition; 2) Airways segmentation and 3D modelling were performed through the commercially available software MIMICS® (v. 18.0), specially developed by Materialise for medical image processing and accurate creation of anatomy 3D models; 3) Volumetric model was imported into Comsol® (v. 5.2), wherein a convergence analysis was developed and 4) Velocity and pressure profiles were obtained from a CFD simulation, in which was possible to analyse fluid behaviour trough tracheobronchial tree.

II. Methodology

A. CT data acquisition

Computed tomography (CT) scan of a healthy volunteer without any problems in the respiratory system was obtained. Simple axial slices were taken from the oropharynx to the upper abdomen including the tracheobronchial tree, while volunteer was in a supine position. Images were obtained using a Philips Brilliance 64 CT scanner with a resolution of 512x512, 0.9 mm slice thickness, 120 kV peak and 333 mA.

I. Torres et al. (eds.), *VII Latin American Congress on Biomedical Engineering CLAIB 2016, Bucaramanga, Santander, Colombia, October 26th -28th, 2016*,
IFMBE Proceedings 60,
DOI: 10.1007/978-981-10-4086-3_134

B. Airway model construction

A segmentation of the region of interest (ROI) which includes the lower airways from larynx to fourth generation of bronchioles was implemented in order to create the 3D model; to do so, image segmentation techniques as dynamic region growing, morphologic operations and manual volumetric segmentation were carried out with MIMICS®. Different layers were created inside of the 3D model, allowing the volumetric surfaces been imported into 3-Matic®(v.10.0) module, to clean, adjust and obtain a preliminary mesh from the volumetric model.

C. Mesh generation

The preliminary mesh created with 3-Matic® was imported into Comsol®, where a convergence analysis was carried out in order to determinate the adequate number of tetrahedral elements in which the velocity magnitude and pressure do not depend of sizes of elements. Convergence analysis was performed evaluating several element size between (19.3 *mm* - 1.21 *mm*) and assessing that velocities and pressures values in critical points such as middle of trachea duct and tracheal distal bifurcation do not present variations above 5%.

D. Computational Fluid Dynamics Simulations

Due to equation for developing flow simulation cannot be solved analytically, a numerical solution using software Comsol® was considered. Airflow was assumed to be incompressible and steady state, the boundaries were located in bottom of larinx as inlet, and the bronchioles as the outlets, walls of airway were defined to be no-slip condition; fluid velocity was selected to be 2.66 m/s [5] and pressure was set to be 0 Pa at outlet in order to simulate normal breathing during inhalation [16]. Specifically, fluid flow modeling was carried out through Navier-Stokes equations [17], in which a steady, incompressible, isothermal and laminar flow were assumed in order to obtain a continuity equation as shown in (1), and a momentum equation as shown in (2).

$$\nabla \cdot \mathbf{v} = 0 \tag{1}$$

$$\rho(\frac{\partial \mathbf{v}}{\partial t} + \mathbf{v} \cdot \nabla \mathbf{v}) = -\nabla p + \mu \nabla^2 \mathbf{v} \tag{2}$$

Where, $\mathbf{v}$ is the flow velocity vector field, t is time, p is hydrodynamic pressure, ρ is fluid density and μ is fluid viscosity. Air properties are $\rho = 1.2 kg/m^3$ and $\mu = 1.78 \times 10^{-5}\ Pa \cdot s$.

III. Results and Discussion

Figure 1 shows the airway segmentation carried out with MIMICS® from volunteer CT scan. Respiratory tract was successfully delimited from lower hipopharynx to fourth bronchiole generation (with trachea as 1st generation).

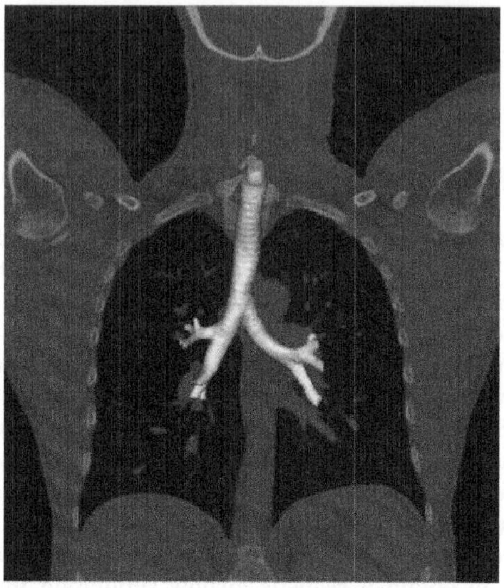

Fig. 1: Airway segmentation in MIMICS® software

Figure 2 shows 3D model that was performed in 3-Matic software, in which a preliminary mesh was carried out in order to export the volumetric geometry.

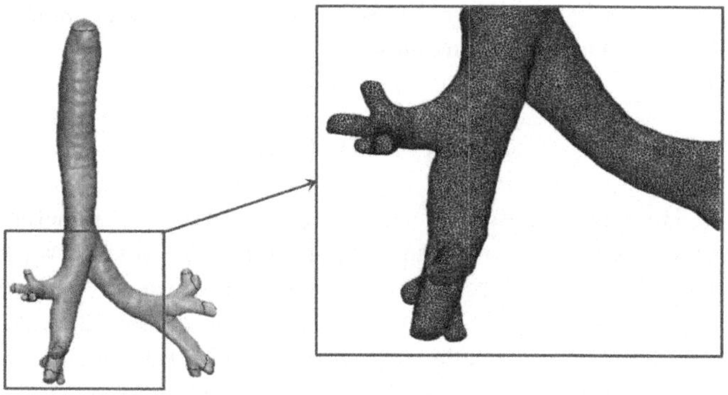

Fig. 2: 3D airway model and the preliminaary mesh obtained from 3-Matic

Volumetric mesh was imported into Comsol ® (v. 5.2). Several element sizes were evaluated and it was defined a volumetric mesh where parameters shown in Table 1 proved to be those in which velocities and pressure magnitudes do not depended of element size. 3D geometry and its corresponding mesh created in Comsol are shown in Figure 3.

Table 1: Mesh domain elements

Item	Value
Number of Elements	481478
Minimum quality element	7.77×10^{-4}
Average quality element	0.7692
Mesh Volume	61830 mm^3
Maximum growth rate	3.745
Average growth rate	1.614

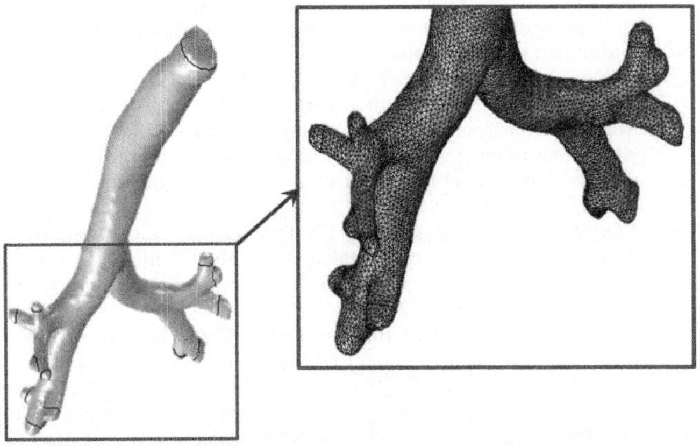

Fig. 3: 3D airway model after refined mesh in Comsol Multiphysics

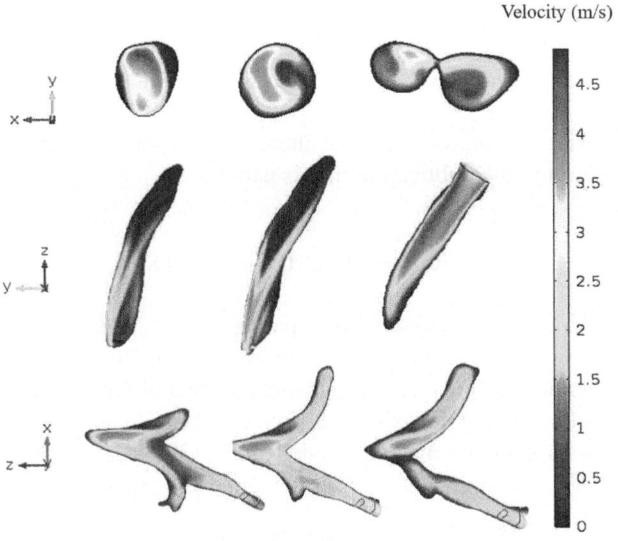

Fig. 4: Top to bottom: Velocity profiles in axial, sagittal and coronal planes

In Figure 4 are shown velocity magnitude contours (m/s) on three different views for axial, sagital and coronal planes, and finally in Figure 5 can be observed level curves for pressure trough airways in different views.

According to results, the process implemented to obtain realistic 3D models of human airways is appropriated to evaluate airflow profiles due to the coherence with a real airflow behavior under the same boundary conditions considered in the simulation [5], as the velocity profile shown a parabolic profile increasing from the airway walls to the center. In addition, the maximal pressure value is located in the carina being the critical point of bronchi bifurcation.

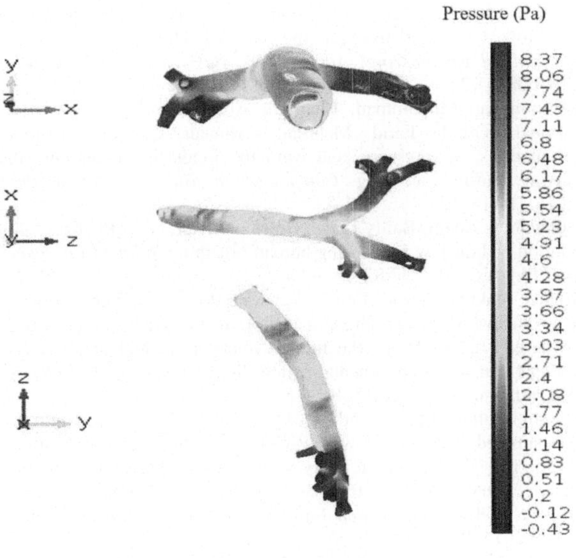

Fig. 5: Top to bottom: Pressure curve levels through airways in axial, coronal and sagittal planes

IV. CONCLUSIONS

This paper presents a series of steps implemented in order to obtain a realistic 3D model of human airways in which can be simulated and predicted some characteristics of airflow profiles inside tracheobronquial tree under different conditions. The results obtained provide a starting point of reference for further simulations; specifically, this results will be used to perform a complete analysis of airflow state in Intensive Care Unity patients which required endotracheal intubation.

Conflict of Interest

The authors declare that there is no conflict of interest regarding the publication of this paper.

Acknowledgements

This work was supported partially by Sistema General de Regalías, Republic of Colombia through the project "Fortalecimiento de la plataforma tecnológica para la formación especializada en el área de la salud y el desarrollo de tecnología biomédica", code RutaN-139C.

References

1. Lewis David J., McKevitt Tom P.. Respiratory System in *Toxicologic Pathology: Nonclinical Safety Assessment* (Sahota Pritam S., Popp James A., Hardisty Jerry F., Gopinath Chirukandath. , eds.)ch. 11Boca Raton, FL: CRC Press Tylor & Francis Group 2013.
2. Johari Nasrul Hadi, Osman Kahar, Helmi Nor Harris N, Abdul Kadir Mohammed a Rafiq. Comparative analysis of realistic CT-scan and simplified human airway models in airflow simulation. *Computer methods in biomechanics and biomedical engineering.* 2013;5842:37–41.
3. Yu Chi, Chen Shu-hua, Wang Gang, Wang Ying. Numerical Analysis of Flow Characteristics for the Normal Human Upper Airway *Biomedical Engineering: Applications, Basis and Communications.* 2016;28:1–12.
4. Rahiminejad Mohammad, Haghighi Abdalrahman, Dastan Alireza, Abouali Omid, Farid Mehrdad, Ahmadi Goodarz. Computer simulations of pressure and velocity fields in a human upper airway during sneezing *Computers in Biology and Medicine.* 2016;71:115–127.
5. Scheinherr Adam, Bailly Lucie, Boiron Olivier, et al. Realistic glottal motion and airflow rate during human breathing *Medical Engineering and Physics.* 2015;37:829–839.
6. Elcner Jakub, Lizal Frantisek, Jedelsky Jan, Jicha Miroslav, Chovancova Michaela. Numerical investigation of inspiratory airflow in a realistic model of the human tracheobronchial airways and a comparison with experimental results *Biomechanics and Modeling in Mechanobiology.* 2016;15:447–469.
7. Cisonni Julien, Lucey Anthony D., King Andrew J C, Islam Syed Mohammed Shamsul, Lewis Richard, Goonewardene Mithran S.. Numerical simulation of pharyngeal airflow applied to obstructive sleep apnea: effect of the nasal cavity in anatomically accurate airway models *Medical and Biological Engineering and Computing.* 2015;53:1129–1139.
8. Gokcan M Kursat, Gunaydinoglu Erkan, Kurtulus D Funda. Effect of glottic geometry on breathing: three-dimensional unsteady numerical simulation of respiration in a case with congenital glottic web. *European archives of oto-rhino-laryngology : official journal of the European Federation of Oto-Rhino-Laryngological Societies (EUFOS) : affiliated with the German Society for Oto-Rhino-Laryngology - Head and Neck Surgery.* 2016.
9. Trabelsi O, Malve M, Mena Tobar A, Doblare M. Simulation of swallowing dysfunction and mechanical ventilation after a Montgomery T-tube insertion *Comput Methods Biomech Biomed Engin.* 2015;18:1596–1605.
10. Ruiz A. E., Arboleda D. C., Aristizabal J. K.. Non-rigid registration of airway region in computed tomography of ICU patient using B-splines in 3D-slicer *Pan American Health Care Exchanges, PAHCE.* 2015;2015-July:6968.
11. Longest P Worth, Azimi Mandana, Hindle Michael. Optimal Delivery of Aerosols to Infants During Mechanical Ventilation. *Journal of aerosol medicine and pulmonary drug delivery.* 2013;27:1–15.
12. Youn Ann Misun, Yoon Seok-hwa, Park Soo-yong. Failed intubation of an unanticipated postintubation tracheal stenosis *Korean Journal of Anesthesiology.* 2016:2–5.
13. Rahimi-Gorji M., Pourmehran O., Gorji-Bandpy M., Gorji T.B.. CFD simulation of airflow behavior and particle transport and deposition in different breathing conditions through the realistic model of human airways *Journal of Molecular Liquids.* 2015;209:121–133.
14. Calmet Hadrien, Gambaruto Alberto M., Bates Alister J., Vázquez Mariano, Houzeaux Guillaume, Doorly Denis J.. Large-scale CFD simulations of the transitional and turbulent regime for the large human airways during rapid inhalation *Computers in Biology and Medicine.* 2016;69:166–180.
15. Kongnuan S., Na-Thakuatung U., Pholuang J.. a Comparative of Analytical and Numerical Simulations for the Oscillating Airflow in a Human Oral Airway *International Journal of Pure and Applied Mathematics.* 2014;90:321–333.
16. Sarasaen Chompunuch, Bhongmakapat Thongchai, Hemtiwakorn Khaisang, Laothamatas Jiraporn, Pintavirooj Chuchart. Computational Fluid Dynamic (CFD) in Thai patient with Obstructive Sleep Apnea Syndrome (OSAS) - A case report: Comparative study between healthy and OSA subject *5th 2012 Biomedical Engineering International Conference, BMEiCON 2012.* 2012:0–5.
17. Chase Michael. Simulation of Flow and Heat Transfer Through Packed Beds of Spheres Tech. Rep. AugustWorcester Polytechnic InstituteWorcester 2011.

Author: J.K. Aristizábal
Institute: Universidad de Antioquia
Street: Calle 67 No. 53-108
City: Medellín
Country: Colombia
Email: jenny.aristizabal@udea.edu.co

Heart Models in 3D Print

Jorge Reynolds[1] and Nasdly Diaz[1]

[1] Electronic and Nanoelectronic Research Department of Shaio Clinic , Bogota, Colombia

Abstract— **3D printing is one of the innovations of the century, because it allows us to see objects created virtually on a tangible medium, without further production processes. One technique used to create models, was fused deposition modeling (FDM), which is a technique layer by layer addition, a thermoplastic material such as PLA, ABS, HIPS or TPU, allowing well the reproduction of parts of the human body, where they can study the different morphologies, as is the study of the heart, through the import of images of CT (computed tomography) or MRI (magnetic resonance imaging), which are processed later to be printed, thereby obtaining a model of the organ in actual size. This makes especially science and medicine, progress in understanding the workings of the human body and its organs.**

Keywords— **3Dprinting,FDM,heart,models,materials,magnetic resonance.**

I INTRODUCTION

Time ago was hard to hear the term '3D printing', which is the reproduction of functional volumetric objects, from a prototype created virtually. Its origins come from the creation in 1976 of the ink jet printing. The first technique reported 3D printing was in an article published in 1981 by Hideo Kodama of Nagoya Industrial Research Institute Municipal in Japan[1], speaking on a rapid prototyping system for photopolymer[2]; later, on March 19, 1983, making some adjustments to the printer inkjet, Chuck Hull an engineer working photopolymer to cover furniture, [3][4] he prints a cup using sensitive resin ultraviolet light. These adaptations made by Hull opened the way to printing materials, thus inventing the technique of stereolithography (SLA), which is patented in March 1986 and that same year he adopted the term 3D printing[5] [5]; based on this, two years later it created: 3D Sys- tems, a pioneer in the field of 3D printing.

Later in 1988, 3D Systems creates and markets the first three-dimensional printer that uses SLA, which is given to know the process of 3D printing. At the end of the same year they created two new techniques: selective laser sin- tering the (SLS) developed by Carl Deckard of the Univer- sity of Texas at Austin [6]; and fused deposition modeling (FDM by Stratasys Inc. trademarked terms), the latter cre- ated by S. Scott Crump who develops and markets in 1990. In the same year Rep Rap (Replicating Rapid-Prototyper - Fast Prototipador Replicable-) European company created by Adrian Bowyer[7]records the term Fused Filament Fabrication (FFF) to have a terminology that could legally be used without limitations, since the patent of this technique has Stratasys. It is clear that the FDM and the FFF are the same printing technique.

A 3D PRINTING IN MEDICINE

The beginnings of 3D printing in medicine are largely reflected in the creation of prostheses since 2008. One of the pioneers was the South African Richard Van As with Robohand which are semi-functional hand prosthesis designed for children, spite of this, it is not until 2012, with the impression of a prosthetic chin titanium [8] that the race of 3D printing in medicine begins, extending to other fields such as manufacturing drugs.

In 2014Jason Kirk [9] Drexel University student [10] prints a CT images based on heart and says very textually in his thesis *"replicas of cardiac anatomy can be used to facilitate medical communication with the patient and complement contemporary visualization techniques providing accurate three-dimensional data that provides additional haptics and specific feedback to the anatomy and pathology of the patient "*.

As organ 3D printing have this science called Biorreprografia[11], which is defined as the method of creating printed alive and functional organs for a 3D printer; one of the first bodies and implanted is the kidney of Dr. Anthony Atala[12]. Today, with the advances that have occurred in science, we observe that arises using the 3D printing as a pre helps the doctor to perform surgery or as a guide for stem cells in case of implants and they make use of their pluripotent characteristics. Examples of this are:

- March 2012, a team from the University of Michigan, were able to reproduce a piece of trachea [13] biocompatible with a baby in order to save his life.[14]

- In February 2013, researchers at Cornell University [15]announced the creation of a complete auditory chamber filled collagen to be colonized by cartilage cells[16]

- The company Organovo [17] In April 2013, announced

I. Torres et al. (eds.), *VII Latin American Congress on Biomedical Engineering CLAIB 2016, Bucaramanga, Santander, Colombia, October 26th -28th, 2016*,
IFMBE Proceedings 60,
DOI: 10.1007/978-981-10-4086-3_135

the creation of liver tissue to achieve reproduce small livers including hepatocytes and capillaries, thus being fully functional. [18]

- In mid 2014, China's first transplant of a vertebra created by 3D printer[19]] is achieved. This has allowed a child of 12 years to recover faster from bone cancer who suffered through the creation of a vertebra plastic with accurate measurements, replacing the usual standard titanium prosthesis.
- In October 2015 researchers at the University of Carnegie Mellon took magnetic resonance imaging (MRI) of the coronary arteries and 3D images of embryonic hearts and made a bioimpresin 3D materials such as collagen and alginates, for reconstruction heart tissue, this printing technique called FRESH (reversible embedding supported by hydrogels) [20], and expected soon work with heart cells to make implants for stimulating contractile muscle.[21]
- In November 2015, doctors at the hospital of Mount Sinai in association with New York Eye and Ear Infirmary of Mount Sinai [22] created a nose 3D fully functional impression for a child of 15, who at nine years he lost his nose for a burn with high voltage cables. [23]

B SOFTWARES

Some of the free software used to perform segmentation DICOM images are *3D Slicer* and Brazilian software *In Vesalius*, in which images are edited, allowing to obtain only the body part you want to see, but one of the versatilities of these programs is the option to view 3D rotation bodies and then to make an export in STL (*STereo Lithography*) format so that the organ or area chosen to be printed. Other software that can also be used and has enhanced tools are MIMICS, 3D DOCTOR, ANALYZE PRO, OSIRIS, among others, but these are not free software.

II MATERIALS AND METHODS

taking into account that now gives necessary the creation of new materials to open more the landscape of 3D printing by molten filament Deposition technique (FDM) working with the following materials are:

A PLA(Polylactic Acid)

PLA (polylactic acid) [24] for strand material print fused deposition (FDM); which is extracted from starches, (extracts of wheat, maize, sugar beet etc.), this material does not produce toxic gases when printing is not recyclable, but if it is biodegradable, as to the physical properties, presents hardness, and resistance shock and fractures, but when it is done for some friction polishing material, tends to fray. As for the chemical treatment to improve the appearance, work well with acetone, the pieces still suffer from effect "warping" (the recording layer shrinks) in printing at medium speed.

B HIPS (High Impact Polystyrene)

HIPS (high impact polystyrene)[25] production base is petroleum, has characteristics similar to ABS, therefore has high hardness, stiffness and strength, in terms of physical treatments to improve appearance, is very good material can be sanded and no degradation with respect to chemical treatments with methylene chloride degrades very fast and if it penetrates the pores may have a total damage to the workpiece. With the D-limonene can be improved, but it is very slow, even after this piece is printed at medium or low speed, the details are very good and does not need further treatment.

C ABS (Acrylonitrile Butadiene Styrene)

The ABS (Acrylonitrile Butadiene Styrene) is a thermoplastic polymer derived from petroleum that melts at about 230°C [26], but has medium toxicity when heated, is highly resistant to physical impact , print leftovers can be recycled and can be re new fibers, which does not hap- pen with the PLA, this is one of the most used materials for prosthetic arms 3D printing because of its hardness. Regard- ing physical treatments can be sanded and buffed by hand, as this wear does not damage the piece, but if left exposed to the sun can have loss of color, you can also be chemically treated with acetone.

D TPU (Thermoplastic Polyurethane)

thermoplastic polyurethane, is a material for flexible and transparent, robust and average elasticity printing, after stretching its fibers back to the original form, it has a high resistance to fatigue use when printing if the model has many layers loses elasticity and transparency [27].

Printing technique: Fused Deposition Modeling (FDM)

It is an additive layer technique [28], that part of a thermoplastic material in the form of a filament, which is inserted into a nozzle, which is heated above the melting temperature of the material, this is deposited on a surface, where many individual layers overlap and are fused together, forming a model [29].

III Results

For printing hearts as learning models, FDM technique was used, a calibration first impressions was made an print on the scale of 1:10 in HIPS, whereby the printing speed was adjusted and performed chemical treatments for a good finish (figure 1), the next print was made larger, at a scale 1:5 in PLA as print media, with which it was denoted that the details were better in HIPS, with these 2 printed hearts rota- tion model was adjusted for printing and thus reduce material support, these models are digitally printed hearts and created a segmented MRI (magnetic resonance imaging).

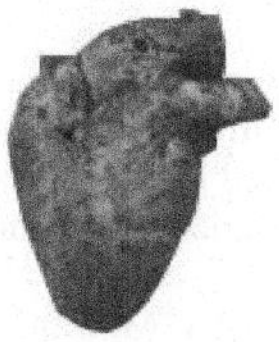

Fig. 1: Heart model from MR printed in HIPS, rear view Heart, after treatment with Methylene Chloride

In Figure 1 the result after chemical treatment with methylene chloride is, with which much weakened structures inside the model by the high penetration of the chemical, there was also a degradation of black color.

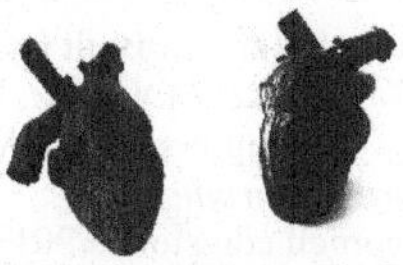

Fig. 2: Digital heart, printed on HIPS, scale 5: 1 with respect to Figure 1, virtual model side and front view

In Figure 2 is the impression of a slightly bigger heart, size 1:5 was treated with D-limonene which gives a shine for a better finished and there is no degradation of the material, but the material is hard to try to make a cut and see inside.

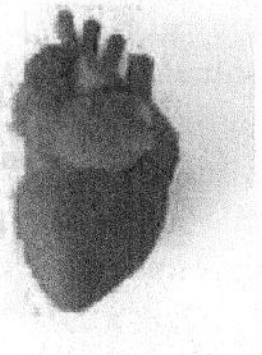

Fig. 3: Digital model of heart printed on PLA, scale 1:10, with respect to the big heart of figura4

Figures 3 and 4 impressions PLA, which have problems with details at high speeds, also have hardness, which pre-

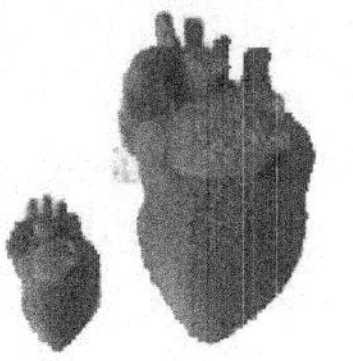

Fig. 4: Comparison of size and finishing details virtual heart model, printed in PLA, rear view

vents making a cut with a scalpel on the piece are; great details of the piece, such as arteries, are denoted very well, but as using support materials you need to make a physical and chemical to get a good finished treatment, this can be evidenced in the small heart, because on the surface not meshing see the typical material support, which is observed in the bigger heart.

Fig. 5: Heart printed on TPU, images from MRI, front view, scale 1: 4 with respect to the original size heart pattern

Fig. 6: Digital model of heart printed on TPU, internal view, with appreciation of chambers and valves

The small heart printed on TPU (Figure 5) comes from MRIs that were segmented, converted into a 3D model and then passed to STL format for later printing, for this reason the aortic arch is cut ; the second heart (Figure 6) and in actual size, is a digital model of the heart with its valves, this model has greater flexibility in their atria, but as the heart has a considerable number of layers in their ventricular walls , there is a loss of flexibility, however, it does allow this material cut with a scalpel on the surface.

The heart printed on ABS, also obtained resonance imaging is lifesize (Figure 7), printing was conducted in 2 parts to denote the internal cavities where chordae are and connection veins and arteries to the heart chambers, repairs that are evident in the model to the contraction suffered by the ABS must when in contact with a colder than the print area, better known as effect environment *warping*.

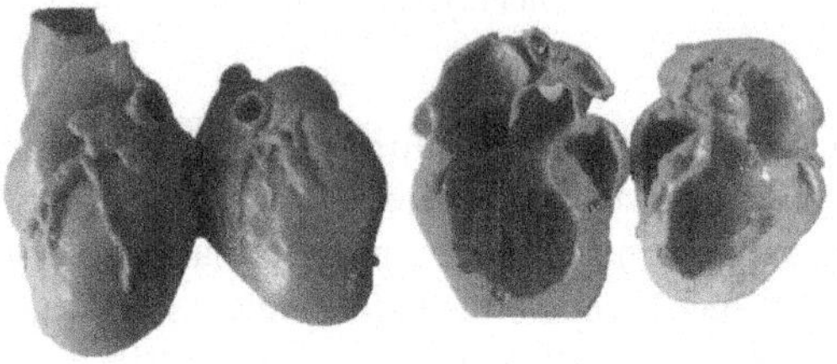

Fig. 7: Heart model from MR images printed on ABS, view of the rear and front faces (left) and internal view with appreciation of the heart chambers (right)

IV Discussion

The quality of prints from MRI (magnetic resonance imaging) depends significantly on the quality and distance with which the cuts or taking imaging tests because the greater the distance between cuts, there will be a loss considerable anatomy of the organ. Another factor to keep in mind is the technique of 3D printing, as used in this project an accessible art for equipment and materials, but the ideal or recommended would use a technique photopolymerization as Stereolithography (SLA) and Laser Sintering selective (SLS), as solidifying liquid resins they are light and have better detail, but do not offer the use of flexible materials to make simulation organs.

V Conclusion

To achieve optimal printing models, must have images in systole and diastole of the heart, and thus have a better view of the chambers and internal structures, and thus make educational and helps surgeries models; one of the best materials for education is the HIPS, for being a material high impact resistant to drops and cuts, and also allows internal structures l heart as are chordal, cavities and valves can see well and feel, regarding support for surgeries the best material of serious evaluated the TPU, allowing to make cuts with a scalpel and maintain the shape of the original models, does not lose rigidity to be cut, the only drawback that has this material is that the speed printing should be 50 % lower than with other materials, so that the model stay with good definition, which generates the printing time is increased twice, still remains a good material; It is expected to continue testing with other flexible materials to achieve similar to the heart muscle flexibility and thus have a better method of learning and medical support using advanced technologies

Conflict of Interest

The authors declare that they have no conflict of interest

References

[1] D. Goldberg, "HISTORY OF 3D PRINTING: IT'S OLDER THAN YOU ARE (THAT IS, IF YOU'RE UNDER 30)," at http://lineshapespace.com/history-of-3d-printing/.

[2] AVPLASTICS, "3D PRINTING HISTORY." at http://www.avplastics.co.uk/3d-printing-history.

[3] Impresoras3d, "Breve historia de la impresión 3D,".at http://impresoras3d.com/breve-historia-de-la-impresion-3d/.

[4] Impresoras-3d., "historia de la impresión 3D.": at http://www.impresoras-3d.info/historia-de-las-impresoras-3d/.

[5] S. Soman, "3D printing," . at Http://www.slideshare.net/Sun1249x/3d-printing-32088902.

[6] T. U. of T. at Austin., "Selective Laser Sintering, Birth of an Industry," at http://www.me.utexas.edu/news/news/selective-laser-sintering-birth-of-an-industry.

[7] RepRapPro., "RepRap."At Https://reprappro.com/about/.

[8] Andrei Vazhnov., "Impresión 3D: Cómo va a cambiar el mundo," in *Impresión 3D: Cómo va a cambiar el mundo*, Firts., B. Editorial, Ed. 2014.

[9] M. MOODY, "How Patient Specific 3D Printed Organ Replicas Help Patients Reach Informed Decisions.," *3d print.com*, Aug-2014.

[10] J. T. Kirk., "3D Printed Medical Imaging Data: A Tool to Supplement Doctor/Patient Communication," Drexel University, 2014.

[11] D. A. Zopf, S. J. Hollister, M. E. Nelson, R. G. Ohye, and G. E. Green, "Bioresorbable Airway Splint Created with a Three-Dimensional Printer," *N. Engl. J. Med.*, vol. 368, no. 21, pp. 2043–2045, May 2013.

[12] TED, "Anthony Atala: Printing a human kidney," 2011.

[13] M. . David A. Zopf, M.D., Richard G. Ohye, M.D. Of Akron Children's Hospital: Marc E. Nelson, "Baby's life saved with groundbreaking 3D printed device from University of Michigan that restored his breathing.," Health System University of Michigan, Michigan, May-2013.

[14] M. Molitch-hou, "trachea 3d printed with ordinary makerbot pla." 2015.

[15] J. Stromberg, "An Artificial Ear Built By a 3D Printer and Living Cartilage Cells," *SMITHSONIAN.COM*, Feb-2013.

[16] Cornell University, "Bioengineers, physicians 3-D print ears that look, act real," *Cornell Chronicle*. at http://www.news.cornell.edu/stories/2013/02/bioengineers-physicians-3-d-print-ears-look-act-real.

[17] Mava., "Organovo anuncia hígado impreso en 3D funcional," *Impresoras3D*. at Http://impresoras3d.com/organovo-anuncia-higado-impreso-en-3d/.

[18] B. KRASSENSTEIN, "ExVive3D 3D Printed Human Liver Tissue Now Commercially Available by." 2014.

[19] J. PRATS., "Cartuchos de células, impresoras de órganos," *El pais*, Valencia, 12-Nov-2013.

[20] U. CARNEGIE MELLON, "CARNEGIE MELLON RESEARCHERS HACK OFF-THE-SHELF 3-D PRINTER TOWARD REBUILDING THE HEART," Oct-2015.

[21] Infosalud, "bioimpresion en 3D para reconstruir el corazón," . at Http://www.infosalus.com/asistencia/noticia-bioimpresion-reconstruir-corazon-20151024102733.html.

[22] C. Scott, "3D Printing, Health 3D Printing," http://3dprint.com/106556/dalan-jennet-surgery/.

[23] J. WAGNER, "3d printing helps doctors give badly burned boy a new NOSE," *CBS NEWS.*, 2015.

[24] L. Serna C., A. Rodríguez de S., and F. Albán A., "Ácido Poliláctico (PLA): Propiedades y Aplicaciones," *Rev. Ing. y Compet.*, vol. 5, no. 1, pp. 16–26, 2011.

[25] D. Burón, "Qué materiales usar y dónde comprarlos,". at http://www.siliconweek.es/workspace/impresion-3d-que-materiales-usar-y-donde-comprarlos-50135.

[26] Mava., "El material de impresión ABS y sus Características.," .at http://impresoras3d.com/el-material-de-impresion-abs-y-sus-caracteristicas/.

[27] Americanchemistrycouncil, "Thermoplastic polyurethane."

[28] S. Sánchez., "Modelado por deposición de fundido, les explicamos todo," at http://www.3dnatives.com/es/modelado-por-deposicion-fundida29072015/.

[29] Stratasys, "Modelado por deposición fundida."at Http://www.dedutel.com/pdf/STRATASYS.pdf,.

Nonlinear measures characterize atrial fibrillatory dynamics generated using fractional diffusion

J.P. Ugarte[1], S.I. Duque[1], A. Orozco-Duque[2], C. Tobón[3], J. Bustamante[1] and H. Andrade-Caicedo[1]

[1] Grupo de Dinámica Cardiovascular, Universidad Pontificia Bolivariana, Medellín, Colombia
[2] Grupo de Investigación e Innovación Biomédica, Instituto Tecnológico Metropolitano, Medellín, Colombia
[3] Grupo de Investigación en Materiales Nanoestructurados y Biomodelación, Universidad de Medellín, Medellín, Colombia

Abstract— **Computational simulations are used as tool to study atrial fibrillation and its maintaining mechanisms. Phase analysis has been used to elucidate the mechanisms by which a reentry is generated. However, clinical application of phase mapping requires a signal preprocessing stage that could affect the activation sequences. In this work we use the fractional diffusion equation to generate fibrillatory dynamics, including stable and meandering rotors, and multiple wavelets, by varying the order of the spatial fractional derivatives obtaining different complexity levels of propagation in a 2D domain. We applied nonlinear measures to characterize the propagation patterns from electrograms. Our results show that electroanatomical maps constructed using approximate entropy and multifractal analysis, are able to detect the tip of stable and meandering rotors, and to mark the occurrence of collisions and wave breaks. Application of these signal processing techniques to clinical practice is feasible and could improve atrial fibrillation ablation procedures.**

Keywords— **atrial fibrillation, nonlinear measures, rotors, phase analysis, fractional diffusion**

I Introduction

Atrial fibrillation (AF) is the most common tachyarrhythmia observed in clinical practice [1]. There is experimental and theoretical evidence that complex propagating patterns during an AF episode, are derived from spiral waves of self-organized and high-frequency drivers known as rotors [2]. Phase maps have been widely used to elucidate the mechanisms by which a rotor is generated. However phase analysis requires high spatial resolution and an additional signal conditioning stage that could lead to changes in activation sequence [3].

Computational models are used as tool to study rotors and other arrhythmogenic mechanisms [4], including ectopic foci and multiple wavelets. In order to simulate complex propagation patterns, electrophysiological heterogeneity is implemented by varying the parameters of the cellular model, and coupling the system using standard difussion equation, which implies an elaborated design of the cardiac model. Fractional calculus generalizes the standard diffusion equation by letting the order of the spatial derivative be a real number. It has been reported that through this approach, different propagation dynamics can be obtained [5]. However, studies of systems involving cardiac membrane formalisms are scarce.

In this work, we simulated episodes of fibrillatory propagation using fractional reaction-difussion equation and Fenton-Karma cardiac celular model. Irregularity maps were built applying nonlinear measures, such as approximate entropy (ApEn) [6] and multifractal analysis (MF) [7], calculated from virtual unipolar electrograms (EGM). By applying a dynamical approach, our results evinced that nonlinear measures were able to characterize complex propagating patterns using phase analysis as gold standard method.

II Materials and Methods

We designed a cardiac excitable tissue through a 2D regular domain of cardiomyocytes. Action potential (AP) propagation was implemented using fractional diffussion operator which involves spatial fractional derivatives. Complexity of fibrillatory dynamics was controlled varying the order of the spatial derivative. We performed phase singularity (PS) tracking to quantify complexity of propagation. EGM from each simulated episode were calculated and their degree of irregularity was measure using ApEn and MF.

A Electrophysiological computational model

Propagation of AP was modeled using 2D isotropic fractional diffusion equation:

$$\frac{\partial U}{\partial t} = D\left(\frac{\partial^\alpha U}{\partial x^\alpha} + \frac{\partial^\alpha U}{\partial y^\alpha}\right) - (I_{ion} + I_{stim}) \qquad (1)$$

where U is the transmembrane potential, I_{stim} is the stimulus current, and x and y the spatial variables. The ionic current i_{ion} was defined by Fenton-Karma membrane formalism [8]. Fractional derivative order α was varied within the interval

I. Torres et al. (eds.), *VII Latin American Congress on Biomedical Engineering CLAIB 2016, Bucaramanga, Santander, Colombia, October 26th -28th, 2016,*
IFMBE Proceedings 60,
DOI: 10.1007/978-981-10-4086-3_136

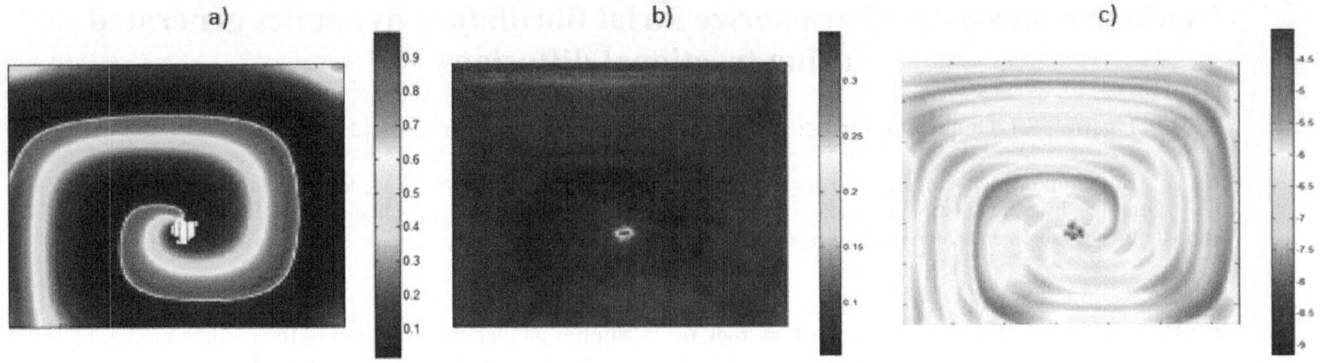

Fig. 1: Atrial Fibrillatory episode (AF1P3) is sustained by a stable clockwise rotor. a)Simulated episode, the white path represents the phase singularity. b) Irregularity map of approximate Entropy. c) Irregularity map of Multifractal index.

[1,2]. Diffusion coefficient D was fixed to obtain a conduction velocity of 60 cm/s for $\alpha = 2$.

B Tissue domain and stimulation protocol

Atrial tissue was represented by a $2x2\ cm^2$ domain, discretized into a $130x130$ mesh. Fibrillatory episodes were generated through $S1-S2$ protocol. $S1$ was a train of five pulses at the left boundary of the tissue. $S2$ was a single pulse applied after $S1$, at the inferior half of the middle vertical line of the domain.

C Phase singularity detection

Phase analysis provides information about the propagation dynamics by representing the state of the system as a phase angle, revealing mechanistic aspects of fibrillatory events [9]. If a functional reentry is present, the spatial distribution of phase will contain one or more singularities, which are points surrounded by tissue with phases ranging from $-\pi$ to π.The phase singularity (PS) marks a region where the phase is undefined. We obtained the phase maps from membrane potentials and singularity tracking was implemented using the topologic charge formulation [10].

D Electrograms, approximate entropy and multifractal analysis

Unipolar EGM were calculated as the extracellular potential Φ_e registered at the point r':

$$\Phi_e(r') = -\frac{1}{4\pi}\frac{\sigma_i}{\sigma_e}\iiint \nabla U \cdot \nabla\left(\frac{1}{r'-r}\right) dxdydz \qquad (2)$$

where r corresponds to a source point, and σ_i y σ_e are the intra and extracellular conductivities, respectively. Calculated EGM have temporal resolution of 1 ms. 16900 registry points (130x130) were located 0.1 mm above the tissue.

Fractionation levels in EGM were assessed using approximate entropy (ApEn) and multifractal analysis (MF). ApEn is a nonlinear statistic proposed by Pincus [6]. Calculation of $ApEn(N,m,r)$ depends on three parameters: number of data points N, embedding dimension m and threshold r. $ApEn(500,3,0.30)$ was used as reported in a previous work [11].

Multifractal h-fluctuation index (hFI) was implemented accordingly with the parameters defined in [12]. The hFI extracts information for the shape of the singularity spectrum $f(\alpha)$ which is a generalization of fractal dimension when statistical scaling is characterised by different Hurst exponents.

EGM irregularity maps were generated considering a 500 points window applied to each EGM signals in order to calculate ApEn and MF-a. Maps of both measures were built applying their respective ranges to a color scale.

E Numerical methods

Fractional diffusion equation was numerically solved using a semi-spectral approach, in which spatial variables in equation 1 were transformed into the frequency domain, time derivative was discretized using Euler method and inverse-transforming to the spatial domain [13]. Time step was fixed to 0.01 ms.

III Results

A Simulated fibrillatory episodes

Three fibrillatory episodes were generated using values of α equal to 1.3, 1.4 and 1.5, Referred to hereinafter as AF1p3,

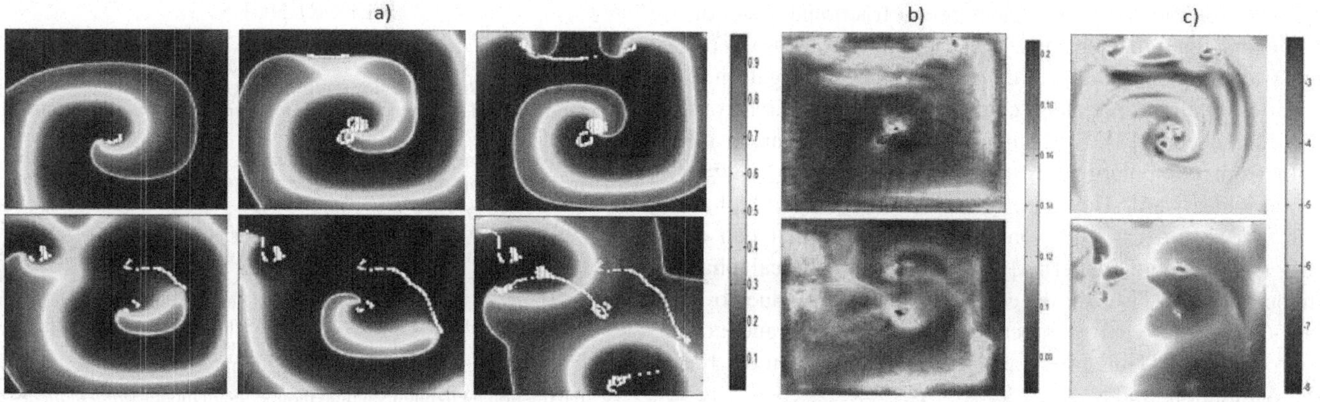

Fig. 2: Atrial fibrillatory episode (AF1P4) initiate with a clockwise rotor which breaks in several waves leading to new microreentrants. Top and bottom rows correspond to two consecutive 500 ms of simulation. a)Simulated episode, the white path represents the phase singularity. b) Irregularity map of approximate Entropy. c) Irregularity map of Multifractal index.

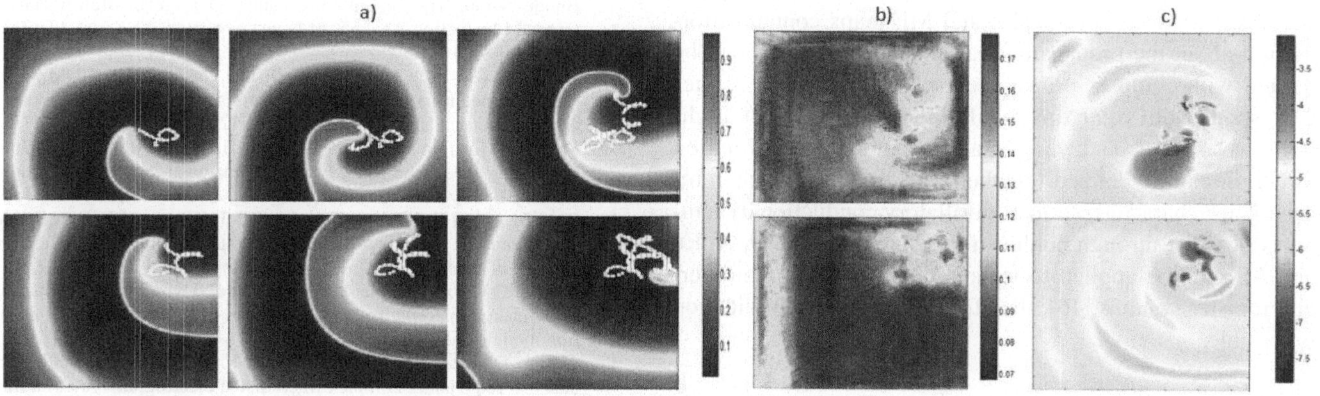

Fig. 3: Atrial Fibrillatory episode (AF1P5) is sustained by a meandering rotor. Top and bottom rows correspond to two consecutive 500 ms of simulation. a)Simulated episode, the white path represents the phase singularity. b) Irregularity map of approximate Entropy. c) Irregularity map of Multifractal index.

AF1p4 and AF1p5, respectively. AF1p3 is sustained by a stable clockwise rotor (Figure 1a). AF1p4 initiate with a clockwise rotor which breaks in several waves leading to new microreentrants (Figure 2a). AF1p5 is sustained by a meandering rotor. (Figure 3a). Phase maps were generated for each episode. Figures 1a, 2a and 3a mark the location of the PS motion as a white path. AF1p4 contains a scattered distribution of PS paths, compared to AF1p3 and AF1p5; marking fibrillatory dynamics of greater complexity. In Figures 2 and 3, top and bottom rows correspond to two consecutive 500 ms of simulation.

B Nonlinear meausures maps

Irregularity maps were generated for 1000 ms of each simulation. For episode AF1p3: ApEn detects the core of stable rotor through highest ApEn values (Figure 1b); while high hFI values mark the core and the profile of spiral propagation (Figure 1c). For episode AF1p4: within the first 500 ms interval, highest ApEn and hFI values (top Figure 2b-c) detect the cores of microreentrants waves (top-right Figure 2a); within the second 500 ms interval, the core of a microreentrant wave is also highlighted by high ApEn values, while middle ApEn values (bottom Figure 2b) correspond with the path of the singularity (bottom-right Figure 2a). For episode AF1p5: highest ApEn and hFI values (Figure 3b-c) highlight the meandering path of the rotor core (Figure 3a). Moreover, hFI depicts the spiral propagation of the AP.

IV DISCUSSION

In this study, we performed computational simulations to generate three fibrillatory episodes with different AP propagation complexity, by implementing fractional reaction-diffusion equation. Currently, phase mapping come up as primary technique for characterizing fibrillatory propagation. In order to obtain proper phase calculation, monophasic recordings are needed [3]. Since membrane potentials are available from computational simulations, phase map analysis is straightforward. However, for clinical procedures, phase analysis requires a EGM preprocessing stage that can affect the activation patterns [3, 9]. A recent clinical phase mapping study [14] shows controversial results, due to its non-disclosed signal preprocessing methods [15] and non-reproducibility. Thus phase analysis feasibility requires further validation.

In this work, we applied phase analysis for relating nonlinear measures outcomes with arrhythmogenic mechanisms. We show that quantification of EGM complexity using nonlinear measures can reveal mechanistic aspects of fibrillatory events. We showed that ApEn and MF maps contain information about the propagating patterns: ApEn values can be related with complex propagation where the highest values correspond with rotor tips; while intermediate values highlight waves shock and wave breaks. From MF maps, rotation direction can be defined together with the location of rotor tips. Thus, characterization of fibrillatory conduction dynamics is possible through nonlinear measures. Moreover, ApEn and MF can be applied to signals obtained using commercial register systems [16], therefore their clinical application is feasible.

V Conclusions

Our findings evince that fractional calculus can be used to generate arrhythmogenic mechanisms and these propagation patterns can be characterized trough nonlinear measures such as ApEn and MF. Future studies are needed to translate these results to clinical procedures.

Conflict of Interest

The authors declare that they have no conflict of interest.

Acknowledgements

This work was supported by the government of Colombia through COLCIENCIAS, research projects #121065741044 and #121056933647.

References

1. January Craig T, Wann L Samuel, Joseph S Alpert, et al. 2014 AHA/ACC/HRS Guideline for the Management of Patients With Atrial Fibrillation: Executive Summary *Journal of the American College of Cardiology.* 2014.
2. Berenfeld Omer, Jalife José. Mechanisms of Atrial Fibrillation: Rotors, Ionic Determinants, and Excitation Frequency *Heart Failure Clinics.* 2016;12:167–178.
3. Clayton Richard H., Nash Martyn P.. Analysis of cardiac fibrillation using phase mapping *Cardiac Electrophysiology Clinics.* 2015;7:49–58.
4. Clayton R. H., Bernus O., Cherry E. M., et al. Models of cardiac tissue electrophysiology: Progress, challenges and open questions *Progress in Biophysics and Molecular Biology.* 2011;104:22–48.
5. Liu F, Turner I, Anh V, Yang Q, Burrage K. A numerical method for the fractional FitzhughNagumo monodomain model *Mathematical Soc.* 2012;54:608–629.
6. Pincus S M. Approximate entropy as a measure of system complexity. *Proceedings of the National Academy of Sciences of the United States of America.* 1991;88:2297–301.
7. Kantelhardt Jan W, Zschiegner Stephan A, Koscielny-bunde Eva, Bunde Armin, Havlin Shlomo, Stanley H Eugene. Multifractal Detrended Fluctuation Analysis of Nonstationary Time Series *Physica A.* 2002;316:87–101.
8. Fenton Flavio H., Cherry Elizabeth M., Hastings Harold M., Evans Steven J.. Multiple mechanisms of spiral wave breakup in a model of cardiac electrical activity *Chaos.* 2002;12:852–892.
9. Shenasa Mohammad, Hindicks Gerhard, Borggrefe Martin, Gunter Breithardt, Josephson Mark E.. , eds.*Cardiac Mapping.* Blackwell4th ed ed. 2013.
10. Bray M a, Lin S F, Aliev R R, Roth B J, Wikswo J P. Experimental and theoretical analysis of phase singularity dynamics in cardiac tissue. *Journal of cardiovascular electrophysiology.* 2001;12:716–722.
11. Ugarte JP, Orozco-Duque A, Tobón C, et al. Dynamic approximate entropy electroanatomic maps detect rotors in a simulated atrial fibrillation model *PloS one.* 2014;9:e114577.
12. Orozco-Duque A., Novak D., Kremenr V., Bustamante J.. Multifractal analysis for grading complex fractionated electrograms in atrial fibrillation *Physiological measurement.* 2015;36:2269-2284.
13. Bueno-orovio Alfonso, Kay David, Grau Vicente, Rodriguez Blanca, Burrage Kevin, Interface J R Soc. Fractional diffusion models of cardiac electrical propagation : role of structural heterogeneity in dispersion of repolarization *Journal of the Royal Society Interface.* 2014;11.
14. Narayan SM, Krummen DE, Shivkumar K, Clopton P, Rappel WJ, Miller JM. Treatment of atrial fibrillation by the ablation of localized sources: CONFIRM (Conventional Ablation for Atrial Fibrillation With or Without Focal Impulse and Rotor Modulation) trial *Journal of the American College of Cardiology.* 2012;14:628-36.
15. Baumert Mathias, Sanders Prashanthan, Anand Ganesan. Quantitative-Electrogram-Based methods for guiding catheter ablation in atrial fibrillation *Proceeding of the IEEE.* 2016;104:416-431.
16. Orozco-Duque Andres, Bustamante John, Castellanos-Dominguez German. Semi-supervised clustering of fractionated electrograms for electroanatomical atrial mapping *BioMedical Engineering OnLine.* 2016;15:1–19.

Author: Juan Pablo Ugarte Macías
Institute: Universidad Pontificia Bolivariana
Street: Circular 1 No 73-76, Bloque22C
City: Medellín
Country: Colombia
Email: juan.ugartem@upb.edu.co

Construction of arterial networks considering a power law with exponent dependent on bifurcation level

L.D.M. Meneses[1], P.F. Brito[1], B.M. Rocha[1], R.W. Santos[1] and R.A.B. Queiroz[1]

[1] Graduate Program in Computational Modelling, Federal University of Juiz de Fora, Brazil

Abstract— **The structure of an arterial tree model is generated on the computer using an algorithm based on method of constrained constructive optimization(CCO). During the growth process of the model, the CCO method generates arterial tree model taking into account a power law at bifurcation with a constant exponent in all bifurcations throughout the tree. In this work we propose a different approach where the algorithm is able to adopt an exponent dependent on the number of proximal bifurcations along the path from the respective segment to the root segment, i.e. bifurcation level. The resulting model tree is analyzed regarding the relations between segment radii and bifurcation levels. Our results show satisfactory agreement with morphometric measurements taken from corrosion casts of human coronary arteries reported in the literature.**

Keywords— **Arterial network, computational model, optimization.**

I. Introduction

Frequently, computational models of arterial trees have been employed as geometrical substrates for hemodynamic simulation studies. Different class of arterial trees models have been reported in the literature such as: lumped parameter models [1], anatomical models [2], fractal models [3, 4] and optimized models [5, 6, 7, 8, 9, 10, 11, 12].

In particular, this work is focused in optimized models generated by the Constrained Constructive Optimization (CCO) method [5]. Arterial tree models generated by CCO are able to mimic important properties of real arterial trees, such as segment radii, branching angle statistics and pressure profiles.

Also, our group contributes with an algorithm based on CCO method that generates arterial tree models considering the Fahraeus-Lindqvist effect [13], which is an effect where the viscosity of the blood changes with the diameter of the vessel it travels through.

Here we describe an algorithm based on the CCO which is able to generate an arterial tree model with bifurcation exponent taking into account properties of the tree during the growth process. In particular, the proposed algorithm considers the bifurcation level, which is the number of proximal bifurcations along the path from the respective segment to the root segment (main feeding artery).

The remainder of this paper is organized as follows. In Section II, the proposed algorithm based on the CCO method is described. In Section III, results obtained using the algorithm developed here are presented. Finally, section IV contains our conclusions and discusses the future directions of this work.

II. The algorithm based on CCO method

The algorithm proposed here is based on the assumptions and constraints listed below [5, 6, 11]:

- The concept associated with the construction is to minimize the total intravascular volume

$$V = \pi \sum_{s=1}^{K_{tot}} l_s r_s^2, \tag{1}$$

  where l_s and r_s are the length and radius of the segment s, K_{tot} is the number of segments of the tree in growth stage;
- the arterial tree is modeled as a binary branching system of straight cylindrical tubes (vessel segments);
- the model tree starts at the root segment and it is truncated in the form of terminal segments on prearteriolar level;
- the blood is modeled as an incompressible, homogeneous Newtonian fluid at steady state and laminar flow conditions are considered;
- flow resistance R_s of each segment of the tree is assumed to follow Poiseuille's law

$$R_s = \left(\frac{8\eta}{\pi}\right)\frac{l_s}{r_s^4}, \tag{2}$$

  where $\eta = 3.6cP$ is the blood viscosity.
- the pressure drop Δp_s along each segment is given by $\Delta p_s = R_s Q_s$, where Q_s is the flow through the segment;
- the resistance of the resulting model tree induces a prespecified perfusion flow Q_{perf} across the overall pressure

I. Torres et al. (eds.), *VII Latin American Congress on Biomedical Engineering CLAIB 2016, Bucaramanga, Santander, Colombia, October 26th -28th, 2016*,
IFMBE Proceedings 60,
DOI: 10.1007/978-981-10-4086-3_137

drop $\Delta p = p_{perf} - p_{term}$;

- the perfusion flow is Q_{perf} which is given in the proximal localization $\mathbf{x}_{prox}$ of the root segment;
- each terminal segment supplies an identical and equal amount of blood flow $Q_{i,term}$ into the microcirculatory network, which is not modeled in detail. Thus, $Q_{perf} = \sum_{i=1}^{N_{term}} Q_{i,term}$, where N_{term} is the number of terminal segments of the tree;
- the overall pressure drop in the tree is given by $\Delta p = p_{perf} - p_{term}$,, where p_{perf} is the perfusion pressure at the inlet of the root segment, p_{term} is the pressure at the outlet of all the terminal segments.

A. Bifurcation law

Next, we describe our contribution to the proposed CCO algorithm that takes into account the bifurcation level for the power law exponent. At each bifurcation the radii of parent (r_{parent}) and daughter segments (r_{left}, r_{right}) are forced to exactly fulfill a bifurcation law

$$r_{parent}^{\gamma} = r_{left}^{\gamma} + r_{right}^{\gamma}, \tag{3}$$

where γ is a exponent that govern the shrinkage of radii across bifurcations. This power law was derived from morphometric measurements of coronary artery corrosion casts [14, 15] and also from theoretical considerations [16].

However, several values for the parameter γ have been reported in the literature. Measurements on corrosion casts of human coronary arteries suggested that a value of $\gamma = 3.0$ should be used, which would allow for uniform shear stress over all the tree [16, 17], since shear stress is proportional to flow/r^3. Conversely, it has also been argued that minimum reflection of pulse waves [18, 5] would be achieved with $\gamma = 2.55$.

Regardless of the particular value of γ given, the CCO method generates an arterial tree model maintaining this exponent constant in all bifurcations throughout the tree.

Here, we propose an algorithm based on CCO which is able to generate an arterial tree model with bifurcation exponent dependent on the bifurcation level. This tree property is defined as the number of proximal bifurcations along the path from the respective segment to the root segment.

Different mathematical relations between exponent γ and bifurcation level can be created. As it is reasonable to accept that this exponent is between 2.55 and 3 for the coronary arterial tree, here we consider equation (3) with γ satisfying

$$\gamma = \gamma_k(n) = \begin{cases} 2.55, & \text{para} \quad n \leq k, \\ 3.00, & \text{para} \quad n > k \end{cases}, \tag{4}$$

where k is the bifurcation level that the user can choose in order to simulate the algorithm and n denotes the level of the bifurcation in the tree model.

During the generation of model, if adding the terminal segment affects the level of bifurcation of some segments of the tree, then the segments' radii must be rescaled as follows: the tree has to be traversed up to the root segment, starting at all terminal segments, to recalculate the corresponding bifurcation ratios (see Equations (9) and (10) in [6]).

B. Generation of the arterial tree model

The algorithm of tree generation based on the CCO has previously been described in detail [5, 6, 11]. Here we only summarized the main ideas.

The optimized tree model generation starts by planting the root segment with its proximal end $\mathbf{x}_{prox}$ fixed at the perfusion domain Ω_{perf} and the distal $\mathbf{x}_{inew}$ selected randomly within this domain.

Given a tree with k_{term} terminal segments, the stepwise growing of the tree is as follows. Three fundamental steps are performed:

- Adding the new terminal segment
  Firstly, the location $\mathbf{x}_{term}$ for a new terminal is selected from a pseudorandom number sequence, uniformly distributed inside the perfusion domain. The prospective location $\mathbf{x}_{term}$ is accepted as a candidate for a new terminal site only if $\mathbf{x}_{term}$ satisfy a distance criterion (see [5]). Since $\mathbf{x}_{term}$ has been accepted as a distal end of a new terminal segment, it is temporarily connected to each of the neighboring segments, one after the other. Connecting the new terminal segment to a preexisting segment, consequently causes violation in the boundary condition regarding the terminal flows and it can affect the bifurcation level of the segments of the tree. In order to return the proper terminal flows, the flow resistance of the tree must be adjusted for each temporary connection. This can only be performed by rescaling of the segments' radii [6] equipped with a procedure that adjusts the exponent γ in accordance with the bifurcation level, see equation (4).
- Geometric optimization
  The bifurcation site resulting in each temporary connection is optimized in order to minimize the total intravascular volume and dissolved again. When this connection is dissolved, the exponent γ associated the bifurcation level can be adjusted.
- Structural optimization
  After the investigation of all possible connections in the neighborhood of $\mathbf{x}_{term}$, the connection that provided the

lowest optimization target is adopted as permanent for the new terminal site $\mathbf{x}_{term}$. Thus, the tree is grown to $k_{term}+1$ terminal segments.

The process of growing the tree summarized above is repeated until $k_{term} = N_{term}$, i.e., the preset number of terminals N_{term} is achieved. The steps described previously are systematized in the Algorithm 1.

Algorithm 1: Tree generation inspired in the CCO.

Data: Q_{perf}, Δp, N_{term}

Planting the root segment;
$k_{term} \leftarrow 1$;
while $(k_{term} < N_{term})$ **do**
    Adding the new terminal segment;
    Geometric optimization;
    Structural optimization;
    $k_{term} \leftarrow k_{term}+1$;
Obtain computed quantities (length, radius, resistance);

III. Results & Discussion

For morphometric comparison with real coronary arterial trees, the Algorithm 1 was applied to generate arterial trees with 250 terminal segments (499 in total) in order to represent the tree of the left anterior descending (LAD) coronary artery.

The arterial tree models were generated under the following conditions [5]: total perfusion flow $Q_{perf} = 500$ mL/min, terminal flows $Q_{term} = 2$ mL/min, bifurcation exponent given by Equation (4), circular area $\Omega_{perf} = 78.54$ cm^2 representing tissue to be perfused (LAD region).

For each simulation with k fixed in Equation (4), ten replicates of the tree with 250 terminal segments were generated on the same predefined parameters and the same optimization target function (total intravascular volume). Each tree was generated using a different sequence of pseudorandom numbers for casting the distal ends of its terminal segments.

Figure 1 shows the mean diameter and standard deviation of this mean diameter (SDM) of all vessel segments at a certain bifurcation level that indicates the number of proximal bifurcations of a segment. Close agreement with the experimental data acquired from corrosion casts from two human hearts by Zamir and Chee [14] is found when the models are generated with the paramter $k = 10$ in Equation (4) for the levels 5-15.

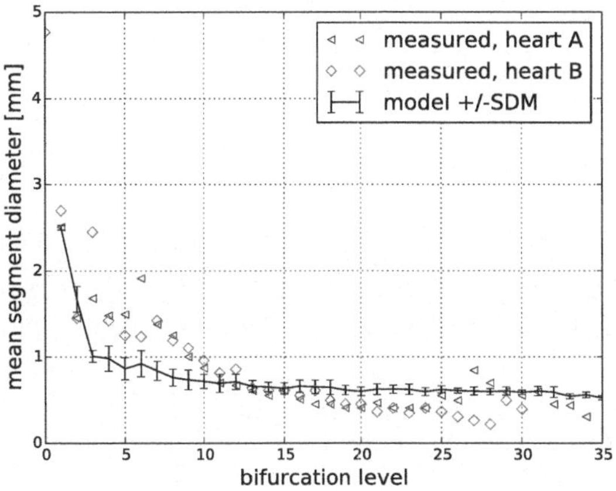

(a) $k = 2$

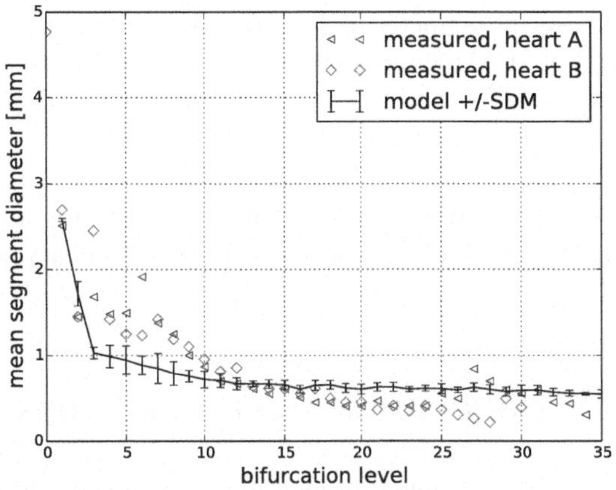

(b) $k = 6$

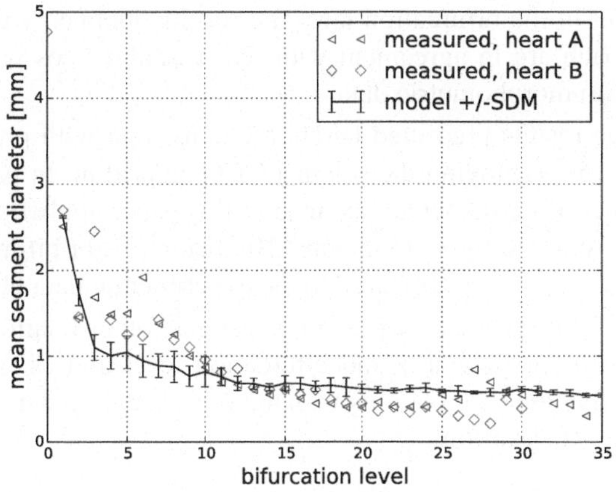

(c) $k = 10$

Fig. 1: Morphometric comparison between tree models and real left coronary arterial trees of two humans – the bifurcation exponent $\gamma = \gamma_k(n)$ is given by Equation (4).

Table 1: Relative mean error (E_{abs}) between tree models and real left coronary arterial trees of two humans.

	Parameter k in Equation (4)				
Heart	2	4	6	8	**10**
A	0.2137	0.2072	0.2137	0.2074	**0.2057**
B	0.3197	0.3289	0.3099	0.3113	**0.2954**

Table 1 depicts the relative mean error calculated by

$$E_{rel} = \frac{1}{N_{bif}} \sum_{i=1}^{N_{bif}} \frac{\min\{| d_i - \tilde{d}_i^{sup} |, | d_i - \tilde{d}_i^{inf} |\}}{d_i}, \quad (5)$$

where N_{bif} is the number of bifurcation levels of the tree model, d_i is the mean diameter in the bifurcation level i [14], $\tilde{d}_i$ is the mean diameter in the bifurcation level i of the tree model. Let SDM_i denotates the standard deviation of $\tilde{d}_i$. Thus $\tilde{d}_i^{sup} = \tilde{d}_i + \text{SDM}_i$ and $\tilde{d}_i^{inf} = \tilde{d}_i - \text{SDM}_i$. For hearts A and B we have $N_{bif} = 34$ and $N_{bif} = 30$, respectively.

One can conclude of this quantitative result that the consider the bifurcation exponent with $k = 10$ in Equation (4) produced tree models more close to real coronary arterial trees.

IV. Conclusion and future works

This work presented an algorithm developed which was based on CCO method that is capable of generating arterial tree models taking into account different values for the exponent in the bifurcation law. The results obtained with this approach are in agreement with real vascular trees with respect to morphometric data.

The results presented herein are consistent with previous work [5] employing the original CCO algorithm. In particular, the mean diameters obtained in this paper are better than those reported by Schreiner and Buxbaum [5] for bifurcation levels 5-11 when compared to the experimental data [14].

In future works, we plan to investigate other relations between the exponent γ and bifurcation level and as well as hemodynamic simulations with the tree models generated by proposed algorithm.

Conflict of Interest

The authors declare that they have no conflict of interest.

Acknowledgement

This work was partially supported by the Brazilian agency FAPEMIG which is gratefully acknowledged (Proc. Num. 00795-14).

References

1. Mates R.E., Klocke F.J., Jr. J.M. Canty. Coronary capacitance *Prog Cardiovasc Dis.* 1988;31:1–15.
2. Onuki T., Nitta S.. Computer simulation of geometry and hemodynamics of canine pulmonary arteries *Ann Biomed Eng.* 1993;21:107–115.
3. Beek J.H. Van, Roger S.A., Bassingthwaighte J.B.. Regional myocardial flow heterogeneity explained with fractal networks *Am J Physiol.* 1989;257:H1670–H1680.
4. Yang J., Wang Y.. Design of vascular networks: A mathematical model approach *Int. J. Numer. Meth. Biomed. Engng.* 2012.
5. Schreiner W., Buxbaum P.F.. Computer-optimization of vascular trees *IEEE Trans Biomed Eng.* 1993;40:482–491.
6. Karch R., Neumann F., Neumann M., Schreiner W.. A tree-dimensional model for arterial tree representation, generated by constrained constructive optimization *Comput Biol Med.* 1999;29:19–38.
7. Schreiner W., Karch R., Neumann M., Neumann F., Szawlowski P., Roedler S.. Optimized arterial trees supplying hollow organs *Med Eng & Phys.* 2006;28:416–429.
8. Blanco P.J., Watanabe S.M., Queiroz R.A.B., Trenhago P.R., Fernandes L.G., Feijóo R.A.. *Trends in the computational modeling and numerical simulation of the cardiovascular system.* Petrópolis–RJ: Scientific Computing Applied to Medicine and Healthcare 2012.
9. Blanco P.J, Queiroz R.A.B., Feijóo R.A.. A computational approach to generate concurrent arterial networks in vascular territories *Int J Numer Method Biomed Eng.* 2013;29:601–614.
10. Queiroz R.A.B., Blanco P.J., Feijóo R.A.. Multiple vascularization of spatial domains using CCO-based algorithms in *WCCM Proc, 10th World Congress on Comput. Mech*(São Paulo–SP, Brazil) 2012.
11. Queiroz R.A.B., Blanco P.J., Feijóo R.A., Ulysses J.N.. Constructive Algorithm of Optimized Arterial Networks in *IFMBE Proceedings. 1ed.: Springer International Publishing*;49:884-887 2015.
12. Schwen L. O., Wei W., Gremse F., et al. Algorithmically generated rodent hepatic vascular trees in arbitrary detail *Journal of Theoretical Biology.* 2015;365:289-300.
13. Pries A.R., Secomb T. W., Gaehtgens P., Gross J. F.. Blood flow in microvascular networks. Experiments and simulation *Circulation Research..* 1990;67(4):826-834.
14. Zamir M., Chee H.. Segment analysis of human coronary arteries *Blood Vessels.* 1987;24:76–84.
15. Zamir M.. Distributing and delivering vessels of the human heart *J G Physiol.* 1988;91:725-735.
16. Sherman T.F.. On connecting large vessels to small: the meaning of Murray's law *J G Physiol.* 1981;78:431–453.
17. Schreiner W., Neumann F., Neumann M., End A., Muller M. R.. Structural Quantification and Bifurcation Symmetry in Arterial Tree Models Generated by Constrained Constructive Optimization *J Theor Biol..* 1996;180:161–174.
18. Arts T., Kruger R. T. I., Gerven W., Lambregts J. A. C., Reneman. R. S.. Propagation velocity and reflection of pressure waves in the canine coronary artery *American Journal of Physiology.* 1979;237:H469–H474.

Leveraging Wireless Communications and Biomedical Devices to Support Prehospital Trauma Care in Cuenca, Ecuador

R.D. Contreras-Chacón[1], J.F. Bravo-Torres[1] and M.K. Huerta[1,2,3]

[1] Universidad Politécnica Salesiana, Grupo de Investigación en Telecomunicaciones y Telemática, Cuenca, Ecuador
[2] Prometeo Project Researcher (SENESCYT), Ecuador
[3] Network and Apply Telematic Group, Simón Bolívar University, Caracas, Venezuela

Abstract—

Advancement of information and communication technologies, along with the development of medical sensing devices and wireless communications systems are changing the way in which citizens can access to health services. Moreover, in treatment of trauma, the use of technology is playing an important role in the care of the wounded during transport to hospital. In this paper we advance in the development of our SINATRA platform for comprehensive care of trauma. Specifically, we present the processes needed for the acquisition of vital signs of wounded and its display through a Web service. Our system has been tested in a controlled laboratory environment and in real tests from an ambulance moving around the city of Cuenca-Ecuador. The results show the feasibility of our proposal.

Keywords— **Telemedicine, Clinical Decision Support Systems, Emergency Medicine, Injuries, Mobile Health Applications.**

I Introduction

Trauma is one of main requirements in calls to emergency services [1]; and, specifically, vehicle accidents are a major cause of death in the world [2–4]. In this context, trauma care systems and, mainly, pre-hospital emergency medical assistance systems play a crucial role in order to reduce death and disability resulting from traumatic injury —more than half of the deaths due to trauma occur outside the hospital, while the emergency medical service (EMS) arrives to the incident location or during treatment and transport between that location and the receiving health center [1].

This type of systems should be designed to act in coordination with local health care systems (e.g., system 911) and local governments to integrate the community. So, besides quick access of specialized health teams to the scene, it is necessary to maintain a trained local people to provide first aid to the injured person and contacting relief agencies. Similarly, during the transfer of the patient to the hospital, constant communication between ambulance and emergency department is required. Valuable data about his/her health status can be shared to assess in advance the seriousness of the wounded, to prepare the necessary surgical equipment, to instruct the medical staff in the ambulance to supply medicines or to develop additional procedures to stabilize the patient and, if it is necessary , refer the patient to another medical center with more equipment or more trained personnel.

To do the above, technologies can play an important role in the efficient deployment of this type of services-oriented trauma care. So, the advance of biomedical sensors devices [5] (such as blood-glucose meters, oximeters, ECG monitors, respiratory rate meters, temperature sensors...), wireless technologies and communication networks are paving the way towards the development of monitoring and health care systems, and support systems for medical emergencies [6]. With respect to trauma care, several approaches have been proposed in the literature about how to generate communication platforms that leverage different technologies to assist to emergency medical personnel in their work, receiving or transmitting information to/from the emergency department [2,4,7]. Other researchers have focused on training of medical staff and the community (especially in rural areas) to support first aid trauma events, taking advantage of wireless communications, virtualization and the proliferation of smart phones [8]. However, these platforms address each of these issues separately, without considering the problem in a holistic manner: (i) first aid training medical staff and community, and (ii) technological assistance to support specialized health teams.

The proposal of this paper is about applying technology to support the deployment of a digital platform that addresses in a holistic manner issues related to assistance to traumatic injuries. In [9], is presented the SINATRA (Intelligent System to Support Trauma, for its acronym in Spanish) platform, which integrates knowledge management of health information about trauma, acquisition and transmission processes of health information, and training of emergency personnel and the community to provide first aid. This platform is being developed by the Telemedicine Research Group of CEDIA (Red Nacional de Investigación y Educación del Ecuador) [1], which

[1] https://www.cedia.org.ec/

I. Torres et al. (eds.), *VII Latin American Congress on Biomedical Engineering CLAIB 2016, Bucaramanga, Santander, Colombia, October 26th -28th, 2016*,
IFMBE Proceedings 60,
DOI: 10.1007/978-981-10-4086-3_138

we are part. In this work, we focus on the design of communication and data acquisition processes.

This paper is organized as follows. In Section II, we will provide an overview of the architecture of our platform, focusing on the necessary mechanisms for the transmission of health information from the ambulance to the hospital. Next, Section III, we will discuss the settings and results of our preliminary implementation we have carried out at the Cuenca-Ecuador. Finally, Section IV concludes the paper.

II SYSTEM DESCRIPTION

This section briefly describes the hardware and software architecture of the system and the functionality developed to visualize the trauma information of a patient in real-time.

A Background on the SINATRA Platform

The SINATRA platform is a fully interactive multi-layer architecture, aimed to provide a complete support for trauma care. It relies on MLST (mechanism of injury, suspected lesions, vital signs and effected treatment) system, which is able to manage the critical information from the prehospital care [9]. Conceptually, this architecture is organized in five levels (see Fig. 1) that are described below.

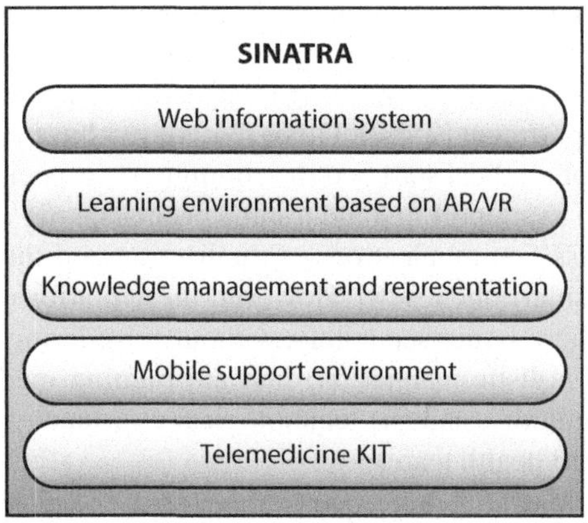

Fig. 1: Five layers of the SINATRA platform

On the lower level of our platform, the *telemedicine KIT* includes the set of biomedical sensors and electronic devices that collect information from patient's vital signs (respiratory rate, blood pressure, pulse and oxygen saturation) and the preliminary assessment of the paramedics on the state of the injured's health. In addition, the *mobile support environment* layer is the responsible for enabling the wireless connections (bluetooth, RFDI, WiMAX, 3G/4G...) that are necessary to send/receive information or health services from/to the ambulance. All protocols and mechanisms for establishing links

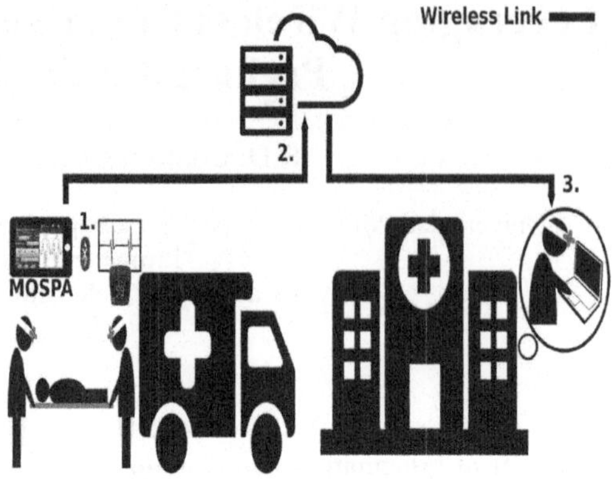

Fig. 2: Process for streaming the patient's vital signs

and maintaining the necessary QoS levels are housed in this layer, too.

The next level is the *knowledge management and representation*, this layer brings together solutions from areas of data mining, machine learning, recommender systems, semantic Web, to support monitoring and automatic diagnosis and decision about the health of the injured. Moreover, as SINATRA is an ecosystem that integrates trauma care training of medical staff and the community, this platform incorporates an *information system for academic tutoring* within the *knowledge management* layer and a *learning environment* based on augmented and virtual reality. Finally, the *web information system* layer manages the information through the services provided by the previous layer.

B Visualization of trauma situations in real-time

Functionality of visualization of trauma situations in real-time, enables us to monitor remotely a patient's information and his/her medical condition while he/she is been transported to the nearest hospital facility. To do this, this functionality leverages the services provided by the knowledge management and representation layer, which stores the current patient information that is sent from the mobile support environment layer and the telemedicine KIT layer (see Fig. 2).

So, different sensors and handheld devices allow to obtain vital signs to be transmitted. This information is sent to the cloud for storage and processing. For its part, the visualization of trauma situations in real-time is provided by the web information service, which is composed of two main process: the first one, allows to receive the vital signs delivered by the sensor device (through an application called MOSPA, which runs on handheld devices) and continuously upload them to the system's database; the second procedure, manages the clients' requests to a web service for the visualization of a

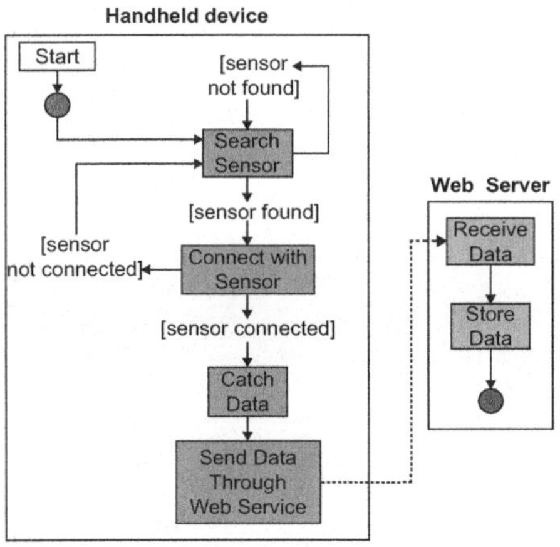

Fig. 3: Activity diagram for the data upload from the sensor device to the web server

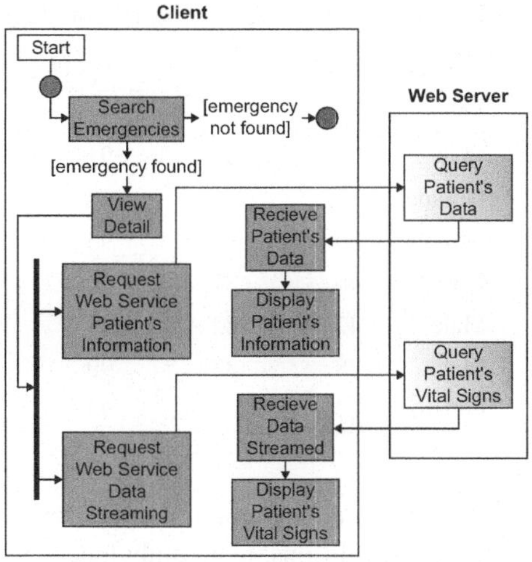

Fig. 4: Activity diagram for data streaming from the web server to the client

patient's information and his/her medical condition. In the remainder of this section we discuss each of these processes.

B.1 *Data upload from the sensors devices to the web server*

This procedure starts when the handheld application does the inquiry of sensor devices. All found devices are listed on the screen and then the user selects one to do pairing and connection. Once the connection is established, the handheld application continuosly catches the sent data and requests the upload data through a web service hosted in the web server. On the web server, the received data are stored in the system's database. Figure 3 illustrates the activity diagram of this procedure.

B.2 *Visualization request of the patient's information and medical condition*

This procedure starts when a client requests the current emergencies. All found emergencies are listed on the screen so the user selects one to view its detail. For this, two concurrent requests are sent to the web server to get the patient's information and to start live streaming of the patient's vitals signs. On the web server side, services do the corresponding queries and send the results as a response. On the client side, data is received and the patient's information is displayed once, while the patient's vital signs are displayed every time that new data is received. Figure 4 illustrates the activity diagram of the procedure.

III RESULTS

Tests were planned to be performed in the laboratory and in an ambulance. For this, it was necessary to analyze the communication technologies and ambulance route before the tests execution. So, firstly, we studied the available technologies in the city of Cuenca-Ecuador: LTE, WiMAX, and CDMA / EvDO; considering parameters such as transmission velocity (Mbps), coverage, compatibility with other communication systems, and current state of deployment worldwide. LTE technology was chosen because its characteristics were superior to other technologies discussed (see Table 1).

Table 1: Analysis of communication technologies

Technology	Tx. velocity (down / up)	Coverage (area)	Compatibility	Worldwide deployment
LTE	300 / 75	Urban	3G/2G	High
WiMAX	70 / 70	Urban	None	Medium
CDMA/EvDO	3.1 / 1.8	Rural	None	Low

Next, for the establishing of the test route, we analyzed the map of the urban area of Cuenca, supplied by the emergency services of Ecuador (ECU911), and the fact that usually the data communication network of LTE is overcrowded between 14 to 16 hours. Devices used for the tests were: Pulse Oximeter, Samsung Galaxy Tab 3 and Mobile WiFi Huawei E5377. The tests were conducted in two scenarios: the first test was executed under controlled conditions, in the laboratory, and

Table 2: Test results in the laboratory under controlled conditions

Patient	Gender	Age	HR	SpO2%
1	Male	57	82	91
2	Male	29	63	91
3	Female	57	79	98
4	Female	29	82	93
5	Female	46	87	90
6	Male	42	82	93
7	Male	28	82	90
8	Female	23	85	93

with the web server in debug mode; the second test was performed under real conditions, in an ambulance, and with the web server in production mode.

In the first scenario, we monitored the HR (Heart Rate) and SpO2 rate of eight patients, specifically four women and four men. The mean of the received values are shown in Table 2. The results indicate a reception rate of vital signs of 100%. In the second scenario, tests were done in an ambulance under real conditions. HR and SpO2 samples of a 25-year-old man patient were taken. In addition, each sample was identified with the timestamp in which it was taken. This value was used to determine the average latency of the data synchronization between the pulse oximeter and the web application (3 seconds). Figure 5 shows a slice of the vital signs of the 25-year-old man patient tested.

IV Conclusions

In this paper, we were designed and implemented a system for monitoring vital signs of patients, which is part of the development of our SINATRA platform. The system minimizes care time of trauma patient in emergency centers of the city of Cuenca-Ecuador due the data handling in real-time. The test, performed under real conditions, indicates that the system remains connected over the urban area of Cuenca, and the data acquired by the pulse oximeter are reflected in the web application with a latency of 3 seconds.

Conflict of Interest

The authors declare that they have no conflict of interest.

Acknowledgements

This work has been sponsored by the Consorcio Ecuatoriano para el Desarrollo del Internet Avanzado (CEDIA) and the Universidad Politécnica Salesiana.

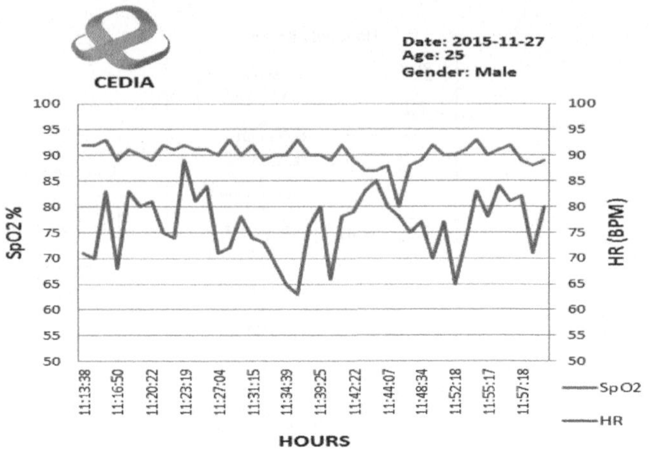

Fig. 5: Vital Signs of the 25-year-old man patient

References

1. BeeBe Richard, Jeffrey Myers Do. *Trauma Care & EMS Operations*;volume 3. DELMAR, Cengage Learning 2012.
2. Blomberg Hans, Svennblad Bodil, Michaelsson Karl, Byberg Liisa, Johansson Jakob, Gedeborg Rolf. Prehospital Trauma Life Support Training of Ambulance Caregivers and the Outcomes of Traffic-Injury Victims in Sweden *Journal of the American College of Surgeons.* 2013;217:1010 - 1019.e2.
3. Varghese Mathew, Sasser Scott, Kellermann Arthur, Lormand Jean-Dominique, others . *Prehospital trauma care systems.* Geneva: World Health Organization 2005.
4. Lavariega Juan C., Avila Alfonso, Gómez-Martínez Lorena G.. *Software Architecture for Emergency Remote Pre-hospital Assistance Systems*:453–471. Cham: Springer International Publishing 2015.
5. Nikita Konstantina S.. *Sensing Technologies for Biomedical Telemetry*:736-. Wiley-IEEE Press 2014.
6. Touati Farid, Tabish Rohan. U-Healthcare System: State-of-the-Art Review and Challenges *Journal of Medical Systems.* 2013;37:1–20.
7. Anantharaman V, Han Lim Swee. Hospital and emergency ambulance link: using {IT} to enhance emergency pre-hospital care *International Journal of Medical Informatics.* 2001;61:147 - 161.
8. Youngblood Patricia, Harter Phillip M, Srivastava Sakti, Moffett Shannon, Heinrichs Wm LeRoy, Dev Parvati. Design, development, and evaluation of an online virtual emergency department for training trauma teams *Simulation in Healthcare.* 2008;3:146–153.
9. Timbi-Sisalima Cristian, Rodas Edgar, Salamea Juan, Sacoto Hernán, Monje-Ortega Diana, Robles-Bykbaev Vladimir. An Intelligent Ecosystem for Providing Support in Prehospital Trauma Care in Cuenca, Ecuador *Studies in health technology and informatics.* 2015;216:329–332.

Author:Jack Fernando Bravo-Torres
Institute: Universidad Politécnica Salesiana
Street: Calle Vieja 12-30 and Elia Liut
City: Cuenca
Country: Ecuador
Email:jbravo@ups.edu.ec

Electronic Stethoscope With Wireless Communication to a Smart-phone, Including a Signal Filtering and Segmentation Algorithm of Digital Phonocardiography Signals

Reynolds Jorge[1], Forero José[1], Botero Juan[2], Leguizamón Vivian[2], Ramírez Luis[3] and Lozano Carlos[3]

[1] Shaio Clinic Foundation/Electronic and Nanoelectronic Research Department, Bogotá DC., Colombia
[2] Manuela Beltrán University, Bogotá DC., Colombia
[3] Asesoftware SAS, Bogota DC., Colombia

Abstract— **This paper describes the construction of an electronic stethoscope prototype, able to amplify and filter several sounds in order to obtain heart sounds captured by a microphone attached to a conventional stethoscope chest piece, using the diaphragm; sending wirelessly, over Bluetooth, this data to an android smartphone app, capable to play and plot the sound, acting as a Phonocardiogram(PCG); and parallel, the development of a software designed to denoise sound signals through Wavelet Filters, in order to migrate this algorithm to the phone app. We obtained an acceptable sound acquisition to identify cardiac activity, however, without an optimum audio fidelity and contaminated signals due to noise addition in the wireless transmission to the phone. On the other hand the development of the signal processing is divided on 3 main steps: the fist one is the normalization of the signal, the second is the PCG denoising and the last one is the identification of different sounds with the envelope. We obtain the peaks of the normal heart sounds, base for the sound segmentation. This project must continue under research in order to improve transmission quality, and adaptation of filtering software into the app, moreover, integrate to the developed software a comparison algorithm for the detection of pathologies and achieve an application of suggested diagnosis, as final target of this investigation on long term.**

Keywords— **electronic stethoscope, wavelet denoising, phonocardiogram, smartphone.**

I Introduction

In 2016, the bicentenary of the invention of the stethoscope by Dr. Rene Laennec in Paris,France is celebrated [1]. This device is the flagship of medicine and crucial for the daily routine of a doctor, commonly used to mechanically amplify the body sounds generated by: the Cardiovascular System, gastrointestinal system and lungs; being very useful for diagnostic of the health staff. The stethoscope has undergone major changes through history, to improve auscultation and facilitate the processes of diagnosis. Due to technological advances, this kind of devices can optimize his performance through an electronic treatment of the sound, to amplify what was meant to hear and attenuate unwanted ambient noise, through filtering processes. Thus, in the present article we expose the development of an electronic stethoscope, with the target of a cardiac auscultation, moreover of processing the signal, achieves wireless transmission of the sound to a mobile application developed on Android, for playback and phono-cardiogram display; which can work from the patient's home or inaccessible areas (rural zone) due to the unlikely implementation of a complex equipment as eco-Cardiographs. This coupled of a digital treatment of this type of signals, through a software developed in Matlab, which will be migrated to the application on the mobile device, so it can do it itself.

The digital processing software initially works with a database of classical cardiac pathologies recordings, to identify important parameters in signals classification [2,3,4].

The main objective of this research is to bring the profits of the mobile application to a suggested diagnosis of heart diseases , from an auscultation performed with an electronic stethoscope, by a software comparison between the parameters of the sounds captured and the characteristics of pathological or non-pathological signals; the current document represents the first stage of this process.

During cardiac auscultation, should normally encounter two main sounds, known as S1 (or R1) and S2 (or R2). They are usually related with the opening or closing of the two groups of heart valves: Atrioventricular (AV) valves (mitral and tricuspid) and semilunar valves (aortic and pulmonary). S1 refers to the closure of the AV valves and the beginning of systole and S2 the closure of the semilunar valves and end of diastole. Table 1 describes the characteristics in time and frequency of these noises. There is also a third noise, most of the time very difficult to hear, is weak and rumbling, and other sound called atrial cardiac tone or forth sound, which must be registered by phone - cardiogram due to its low frequency, less than 20 Hz. The third tone can only be register one third of the half of the entire population, and the fourth noise on a quarter [5].

I. Torres et al. (eds.), *VII Latin American Congress on Biomedical Engineering CLAIB 2016, Bucaramanga, Santander, Colombia, October 26th -28th, 2016*,
IFMBE Proceedings 60,
DOI: 10.1007/978-981-10-4086-3_139

Table 1: First and second cardiac sound Characteristics. [3]

Sound	length (s)	Frequency (Hz)
S1	0,14	30-45
S2	0,11	50-70

The bandwidth that contains sounds and murmurs is from 25 Hz to 2000 Hz.

II METHODS

A Hardware

This section will explain the most important phases of the physical design which obtains cardiac sounds to be processed, in Figure 1 a block diagram of the circuit and the sequence of steps are exposed.

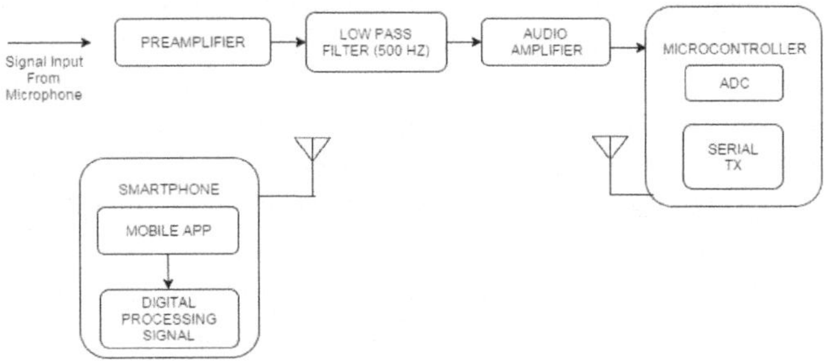

Fig. 1: Block diagram of the designed system

Pre-amplification: To capture sound, we used an Electret condenser microphone, built with a FET transistor. When is polarized in common emitter, it behaves as preamplifier; the gain is given in relation to a series of resistance between the commutator and the power supply. This microphone is omnidirectional, however when it is attached to a conventional stethoscope bell, it will be mechanically isolated from sound coming of all areas, except forward.

Analog filter: Given the frequency characteristics of a healthy and pathological heart sounds, a series of low-pass filters are develop, with a Butterworth topology and Sallen key type, which operates in cascaded twice, improving the quality factor, being a 4th order filter. The desired cutoff frequency was of 500 Hz, using values R1 = R2 = 2.2k Ohm C1 = 100 nF, C2 = 200nF calculated with (1).

$$Fc = 1\frac{1}{2\pi\sqrt{R1C1R2C2}} \quad (1)$$

Amplification: In the amplification step, an LM386 integrated is used, that is specially designed for audio processing, and configured for a gain of 200. At the output of this stage is possible to connect a headphone jack to listen to the signal directly.

Transmission: In this stage, a PIC16F873A microcontroller (of the Microchip enterprise) was used, programmed in C++ language; the CCS Compiler is used to program the microcontroller. The algorithm performs an analog - digital conversion (ADC) from the input audio with a sampling frequency of 50 kHz, and a digitalization of 10 bits, which sends the resulting values through a transmission under the RS - 232 protocol, using the Transmission Pin (TX). This pin, is connected to the Bluetooth module HC -05, responsible for conducting the transmission data wirelessly under the Bluetooth protocol to the mobile device.

B mobile app

The main application's features are capture, view, store, transmit, and process signals related to heart sounds. Those functions will be executed on a portable device such as smartphone or tablets.

In order to achieve these requirements, we used native functions that support access hardware device and allow the application of mathematical algorithms for classification of the acquired signals.

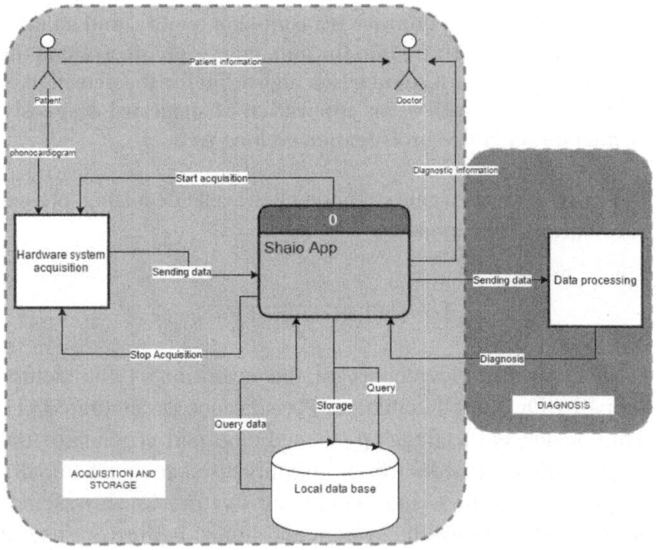

Fig. 2: Mobile application context Diagram

Figure2 shows the application context, and all components that will interact between different systems.

C Software

Database: The database used for sounds processing, starts whit acquisition of recordings of cardiac auscultation of the following discs of 33 revolutions: "Interpreting heart sounds" by Roche Laboratories, "Auscultation of the heart" by J.B. Barlow and W.A. Pocock, "Grabación cardiofónica para audición estetoscópica" by Dr. Hector P. Redondo Rampírez and "Manual de auscultación y percusión" by Dr.K.Holldock.

Database contains phonocardiography records taken from normal patients, and subjects who presents cardiac murmurs caused by heart valve disease (stenosis and insufficiency of the mitral valve, aortic, pulmonary and tricuspid).This database includes normal records and different pathological records.

Signal Standardization this step is responsible for balancing the number of samples and define a constant amount of gain for each recording of cardiac auscultation as part of the audio processing.

Denoising Using Wavelet transform, orthogonal deubechies 5 (db5) with a descomposition level 2 [6] and a low pass IIR (Infinite Impulse Response) filter whit cutoff frequency of 1.5 KHz, deleting the noise from 33 revolutions discs and other physiological sounds added in phonocardiography signal.

Signal Segmentation At this point, we calculated the signal envelope, which works from the recognition of the positions of S1 and S2 over time, defining a threshold as reference to find the limits of heart sounds, based in an algorithm of peak detection[7]. Later, the length of S1 and S2 are identified, along with the period of S1-S1 and S1-S2[8].

III Results

After building the prototype (Figure 3) Bluetooth transmission tests were performed, using a Hc-05 module communicated whit the mobile application, under the SSP (Synchronous Serial Port) protocol; however when connecting step of scanning and sending the analog circuit, audio output is contaminated with electrical noise produced by the microcontroller and the Bluetooth module. Under these conditions, bypass capacitors have been implemented, that while attenuate audio distortion, do not disappear it. We consider the possibility of using other bluetooth, such as 4.0 version -reference HM -10-, which in addition to reducing consumption of the circuit due to its BLE (Bluetooth Low Energy) chip, also reduced the noise considerably, however this version is not compatible even with the type of communication programmed in the mobile application, which could be replaced.

Fig. 3: Electronic stethoscope prototype

The display on the phone is successful, achieving appreciation of S1 and S2 Sounds (Figure ref figure4) due to dedicated space for the graphical representation of the acquired information . The application provides a simplified and comprehensive navigation, which allows the physician to move between options quickly and easily without the need for connections to other devices, always managing and processing the information locally.

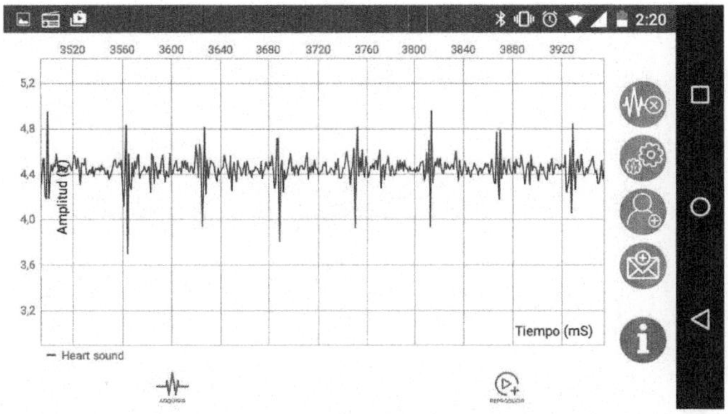

Fig. 4: view and acquisition system

In addition, the application has a help system to advice the user whenever requires information related to the app options, this characteristic improves the experience for the user, focusing efforts on the analysis of the patient's heart.

On the other hand, results obtained in the filtration step in the Matlab software are shown in Figure ref figure6, where the reduction of noise frequency components unwanted in the digital recording of cardiac auscultation is appreciated. The algorithm is qualitatively validated, through hearing appreciation of specialists in cardiology.

Figure 7, section a, shows the PCG when is filtered and normalized, where coefficients are calculated using the continuous wavelet transform, which allows to observe the variations of the signal shown in section b, where is no signal variation in the red zone, while green-yellow color indicates the

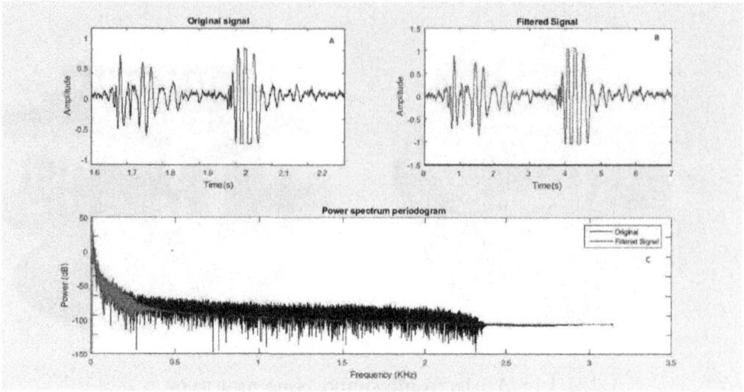

Fig. 5: comparison between healthy phonocardiography signal versus signal after digital processing. (a) Original healthy PCG (b) Filtered healthy PCG (c) Power spectrum Periodogram of signals above.

existence of peaks, this method allows to note that these positions contains S1 and S2 sounds. Another procedure useful for finding the peaks is bound by the absolute value envolve of the signal. The average of the signal is used as a threshold, and peaks exceeding this value will denominated as S1 and S2 sounds, as shown in Section C.

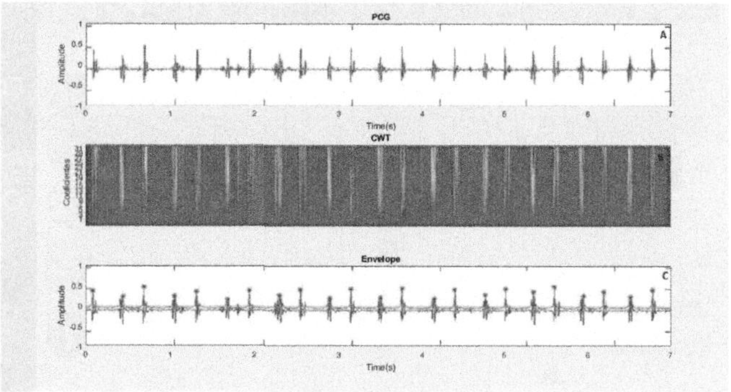

Fig. 6: Phonocardiography Signal segmentation. (a) Healthy PCG (b) rbior Continuous wavelet transform coefficients (c) PCG signal envolve and S1,S2 peak detection.

IV Conclusions

After the results we obtained, we conclude the initial part of the investigation, in which the first hardware version can be submitted to changes and improvements in order to correct the quality of the audio. Similarly, we develop the first part of the processing of heart sounds, the algorithm filtering and segmentation, this process was developed in Matlab, that allows the acquisition of the different characteristics of the processed signal and from this, in the next phase of the investigation, it must process the audio sent in real time from the hardware, and classify between pathological and non-pathological. The app is prepared to adapt processing developed in Matlab and migrate to the mobile device.

Conflict of Interest

The authors declare no conflict of interest.

Acknowledgment

We thank the support for the development of this research to: the company Asesoftware SAS , Manuela Beltran University and Sound Engineering program from San Buenaventura University.

Referencias

1. Reynolds, J.Cardona, A.: , BICENTENARIO DE LA INVENCIÓN DEL FONENDOSCOPIO. Editorial Carrera Séptima, Bogotá (2015).
2. Barlow, J.Pocock, W.: Auscultation of the heart. FFRR, Londres (1958).
3. Holldack, K.: Manual de Auscultacion y percusion. Teide (1961).
4. Geckeler, G.: Grabacion Cardiofonica para Audicion Estetoscopica. Columbia Records (1968).
5. Hall, J.Guyton, A.: Válvulas y tonos caríacos; cardiopatías valvulares y congénitas. Tratado de fosiología médica. pp. 265-272. ElServier, Barcelona (2011).
6. Mishra, G., Biswal, K., Mishra, A.: Denoising of heart signal using Wavelet Transform. International Journal of research in engineering and technology. 2, 719-723 (2013).
7. Li, T., Qui, T., Tang, H.: A heart sound segmentation method based on cyclostationary envelope. Proceeding of the 2nd international conference on computer science and electronics engineering(ICCSE),. 1, 2536-2539 (2013).
8. Singh, M.Cheema, A.: Heart Sounds Classification using Fature Extraction of Phonocardiography Signal. International Journal of Computer Applications. 77, 13-17 (2013).

Permanent Magnets to Enable Highly-Targeted Drug Delivery Applications: A Computational and Experimental Study

M. Mercado-M[1], A.M. Hernandez[1] and J.C. Cruz[1]

[1] Bioinstrumentation and Clinical Engineering Research Group - GIBIC, Bioengineering Department, Engineering Faculty, Universidad de Antioquia UdeA; Calle 70 No. 52-21, Medellín, Colombia.

Abstract— **Nanoparticle-based therapies have gained considerable attention over the past few years as potential candidates for the development of robust treatments for degenerative conditions such as Parkinson's or Alzheimer's. A major challenge to increase the specificity of such therapies is the limited control over their fate and transport upon administration. An avenue to overcome this issue is to immobilize the therapeutic molecules on magnetically responsive nanostructures. This allows a direct control via magnetic fields generated through either permanent or electromagnets. Permanent magnets offer technical and economic advantages over electromagnets but their spatial disposition needs to be dynamically adjusted to assure proper magnetic gradient directionality. The engineering of such a dynamic system requires a detailed understanding of the impact of changing the number and relative positions of magnets in close proximity to the particles, which can be tedious to achieve experimentally. Alternatively, electromagnetic modeling and simulation provides a rather expedite route to explore multiple configurations. Here we evaluate, against experimental data, the robustness of electromagnetic models incorporated in the software COMSOL® to predict field gradients and intensities. We tested two of the incorporated meshing approaches in a single magnet and an array of two magnets. Our findings indicate that the two approaches fitted the experimental data as evidenced by the statistical significance of various correlation indexes.**

Keywords— **Drug Delivery, FEM Analysis, Magnetic Fields, Permanent Magnets, Model Validation.**

I. INTRODUCTION

The delivery of drugs is always challenged by the fact that the delivered therapeutics come in contact with both healthy and diseased tissues. As a result, most treatments may induce a number of undesirable side effects upon administration. For this reason, scientists have invested much effort toward the development of new alternatives to target specific regions within affected tissues upon the delivery of the therapeutic molecules [1]. One attractive avenue to enhance the specificity of drugs relies on the use of nanoparticles as carriers. The main advantage of this vehicles is the availability of relatively simple chemical pathways to conjugate biocompatible polymers, which in turn can be further functionalized with therapeutic agents [2]. Additionally, nanoparticles offer large surface-to-volume ratios, and exhibit sizes similar to those of various biological molecules including antibodies, receptors and nucleic acids [3]. Consequently, nanoparticles are well suited to pass through highly-selective biological tissues such as the blood-brain barrier (BBB), which consists of tight junctions that confer selectivity towards water, some gases and lipid-soluble molecules [4] [5]. Therapies targeting brain cells need to be able to pass the BBB. Accordingly, nanoparticle-based therapies have gained considerable attention as potential candidates to develop robust treatments for conditions such as Parkinson's or Alzheimer's [6].

An interesting strategy to enhance directionality to therapeutic nanocarriers is to incorporate magnetically actionable nanostructures as supports [7]. Recent studies have demonstrated that magnetic fields produced by both electromagnets [4] [8] and permanent magnets [9] [10] are well suited to manipulate nanoparticles. Particularly, permanent magnets made of rare earths alloys have demonstrated higher magnetic field strength and gradient than systems based on electromagnets of the same size [11]. Perhaps the permanent magnets most widely used are those made of Neodymium-Iron-Boron (NdFeB) and Samarium-Cobalt (SmCo) alloys. Compared with traditional magnets, the NdFeB and SmCo magnets exhibit improved magnetic properties, and are relatively inexpensive [12]. Additionally, as opposed to electromagnets, permanent magnets do not produce Foucault parasite currents and can be set up with no need for voltage sources or refrigeration coils.

Despite the availability of these magnetic manipulation systems, complete control for nanoparticle transport continues to be viewed as a major experimental challenge. This has been attributed to absence of instrumentation to manipulate the magnetic gradient, which is directly responsible for the developed magnetophoretic force acting on magnetic nanoparticles [13] [14]. An avenue to determine the optimal spatial distribution of a system of magnets for nanoparticle manipulation is the *in silico* experimentation. This approach allows a rapid testing of different conditions at a reduced cost. Computational modeling is a powerful tool to simultaneously couple several physical and chemical phenomena and thereby simulate the interplay of different variables.

I. Torres et al. (eds.), *VII Latin American Congress on Biomedical Engineering CLAIB 2016, Bucaramanga, Santander, Colombia, October 26th -28th, 2016*,
IFMBE Proceedings 60,
DOI: 10.1007/978-981-10-4086-3_140

With this approach is possible to estimate *in silico* the impact of changing the relative levels of various parameters on properties of interest [15]. In this case, we are interested in studying the impact of changing the spatial location of permanent magnets on the intensity and 3D distribution of the associated magnetic fields. The results are critical to define the needs for an automatically-actuated system to assure that magnetic nanocarriers can be transported to regions of interest within biological systems.

Perhaps the major challenge of *in silico* experiments is to assure that the results truly represent what is observable experimentally. Thus far, numerous *in silico* studies are available with complete descriptions on the magnetization process of isotropic permanent magnets [16] and the nanoparticle magnetic manipulation using either permanent magnets [10] or a combination of permanent and electromagnets [1]. These studies failed however to show the statistical robustness and experimental validity of their models. This information is critical to move from laboratory-scale experiments to preclinical biological models. This work is aimed at modeling and simulating the magnetic field generated from commercially available Neodymium N35 permanent magnets in different spatial arrangements. Additionally, the predicted flux densities are validated experimentally using physical replicas of the simulated magnets.

A. Magnetic Field from Permanent Magnets

The magnetic form of Gauss' law (1) models a system of non-moving permanent magnets as unique sources of magnetic fields. This is described by the divergence of magnetic flux density equals cero [17].

In linear materials, the constitutive relationship between magnetic flux density **B** and magnetic field intensity **H** is given by (2). From (1) and (2), **H** can be expressed as the gradient of a magnetic scalar potential (3). In some non-linear materials such as permanent magnets, the constitutive relationship between **B** and **H** is given by (4), where $\mathbf{B}_r$ is the remanent magnetic flux density. This represents the magnetic flux density in the absence of magnetic fields [17] [18].

$$\nabla\cdot\mathbf{B}=0 \quad (1)$$

$$\mathbf{B}=\mu_0\mu_r\mathbf{H} \quad (2)$$

$$\mathbf{H}=-\nabla V_m \quad (3)$$

$$\mathbf{B}=\mu_0\mu_r\mathbf{H}+\mathbf{B}_r \quad (4)$$

II. Materials and Methods

Multiphysics modeling and simulation of the magnetic field generated from Neodymium N35 permanent magnets was conducted via COMSOL®. This software package uses the Finite Element approach to discretize the calculation domains. The experimental validation was completed with the aid of magnetic field meters.

A. Computational Domain and Material Properties

The computational domain (Fig. 1) was comprised by a cylindrical magnet of 53mm diameter and 33mm thickness and the surrounding air medium, which was confined by a 1m-diameter sphere. The isolation of the system was achieved by a 100mm layer adjacent to the limit of the confining air sphere as defined by an infinite element domain in COMSOL®.

Material properties are respectively 1.0 and 1.05 for air and Neodymium relative permeability. $\mathbf{B}_r$ magnitude is 1.2T and its direction is given by the magnet thickness direction.

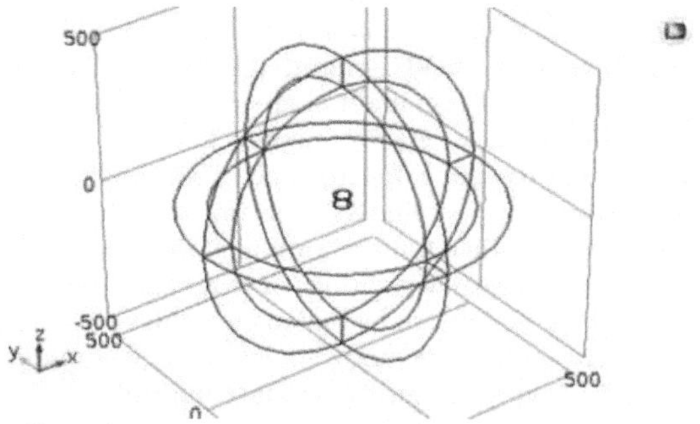

Fig. 1 Computational Domain: cylindrical magnet of 53mm diameter, 33mm thickness surrounded by a 1m diameter air sphere with a 100mm infinite element domain layer.

B. Governing Equations, Boundary Conditions and Solver

The magnet was modeled with the aid of *Magnetic Fields, No Currents* COMSOL® interface. *Magnetic flux conservation* was applied to all domains. Air constitutive relationship was set as *Magnetic Flux Density* while for the Neodymium magnet the *Remanent Flux Density* was considered. For all boundaries the *Magnetic Isolation* was implemented, and the initial value for magnetic scalar potential was set to zero.

No time dependence was considered and consequently, the stationary version of governing equations was implemented. The involved system of equations was resolved by the conjugated gradients iterative method.

C. Mesh and Convergence Analysis

Two different meshing approaches were considered, i.e., normal and adaptive mesh refinement. The size of the tetrahedral finite elements was controlled with the *Parametric Sweep* tool, using as parameter the maximum element size for permanent magnets domain (*max*) (Table 1). The minimum element size for magnets was set to *max*/6. For the air domain, minimum element size was set to *max/2* and maximum element size to *max**4. The convergence criterion was that upon duplicating the Degrees of Freedom (DOF), the magnetic flux density for two different locations remained within 1% of the previous value. Experiments taken into consideration were only those where convergence was reached.

Table 1 Degrees of Freedom and Their Increments

max (mm)	Normal Mesh		Adaptive Mesh Refinement	
	DOF	*DOF Increment Factor*	*DOF*	*DOF Increment Factor*
8	358237	3.21	1367603	3.38
6	870097	2.43	3453783	2.53
5	1505240	1.73	5842518	1.69
4	2947364	1.96	-	-

D. Two Permanent Magnets System.

A second study considered an array of two permanent magnets with sizes identical to the single magnet studied firstly. The two magnets were facing each other at a separation of 100mm in the thickness direction.

In this case, $\mathbf{B}_r$ for the magnets in the array oppose to each other. Computations for the locations previously defined were run using the same material properties, mesh parameters and solver. The model was validated in light of the same criteria used for single magnet.

E. Experimental Measurements and Model Validation.

The magnitude of the magnetic flux density was measured with a Lakeshore® 460 3-Channel Gaussmeter in 12 different locations around the Neodymium N35 magnet. At each location, six different measurements were conducted in a random manner. The collected values were averaged for direct comparison with those estimated via simulation.

The Goodness of Fit analysis was conducted in light of the coefficient of determination (R^2), Nash-Sutcliffe Efficiency (NSE), Index of Agreement (d), Kling-Gupta Efficiency (KGE) and Volumetric Efficiency (VE). All indexes were estimated using the 'hydroGOF' package, which is incorporated in the statistics software R. A complete validation of the model requires 95% in at least two of the indexes above.

III. RESULTS

Mesh convergence analysis results for normal mesh and adaptive mesh refinement are shown in Table 2. The results are presented as a function of parameter *max*. The percentage difference is calculated with respect to the previous mesh.

Table 2 Convergence Analysis Results for Single Magnet

max (mm)	B - Normal Mesh Difference (%)		B - Adaptive Mesh Refinement Difference (%)	
	Location 1	*Location 2*	*Location 1*	*Location 2*
8	9.13	0.18	2.25	1.03
6	4.96	5.51	**0.44**	**0.72**
5	3.52	4.89	0.22	0.47
4	**0.57**	**0.79**	N.A	N.A

Fig. 2 and Fig. 3 compare simulated and measured magnetic flux densities for single magnet and two magnets array, respectively. For the single magnet, the mean relative errors between the measured and simulated magnetic flux densities are 3.21% for the normal mesh and 2.92% for the adaptive mesh. For the two magnets array, the mean relative errors are 3.79% for the normal mesh and 4.28% for the adaptive mesh.

Table 3 contains the efficiency indexes for testing the robustness of the studied models.

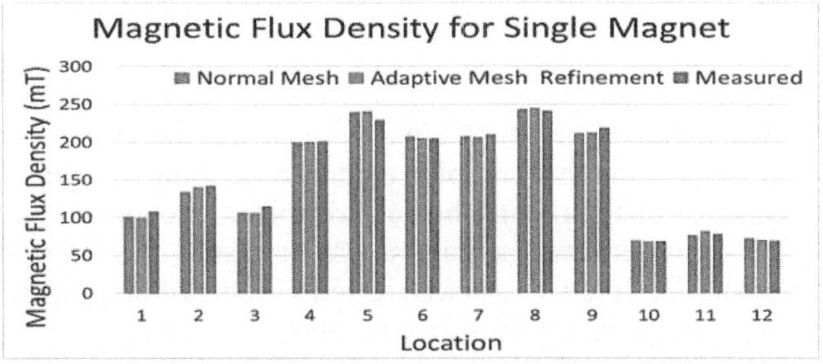

Fig. 2 Magnetic flux density for single magnet at 12 locations.

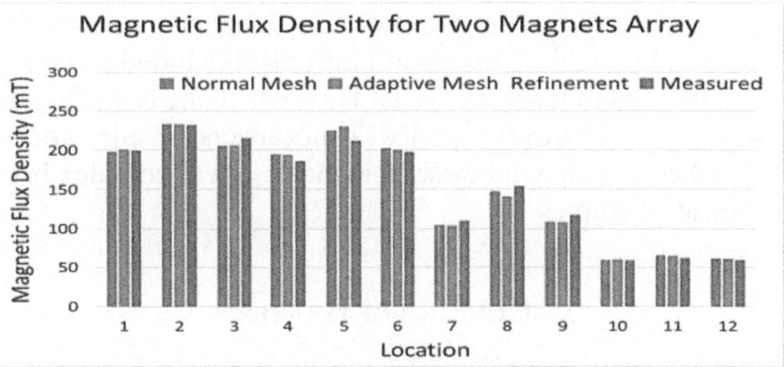

Fig. 3 Magnetic flux density for two magnets array at 12 locations.

Table 3 Model Efficiency Indexes

Model Efficiency Indicator	Normal Mesh		Adaptive Mesh Refinement	
	Single Magnet	Two Magnets Array	Single Magnet	Two Magnets Array
R^2	99.35%	98.97%	99.36%	98.46%
NSE	0.99	0.99	0.99	0.98
d	1.00	1.00	1.00	1.00
KGE	0.97	0.98	0.98	0.97
VE	0.97	0.96	0.97	0.96

IV. Discussion

As shown in Table 2, convergence is reached when max=5 for normal mesh and max=8 for adaptive mesh refinement (see rows in bold). For the two magnets study, convergence was reached using the same values for max (data not shown).

Fig. 2 and Fig. 3 graphically confirm that experiments and simulations agree well. The mean relative errors for both normal mesh (3.21% and 3.79%) and adaptive mesh refinement (2.92% and 4.28) are small, suggesting a high agreement between the models and the experiments. This is further confirmed by the model efficiency indexes shown in Table 3, whose values are at least at a 96% level.

Slightly higher R^2 and KGE indexes for the adaptive mesh refinement approach appear to suggest a better goodness of fit for the single magnet. Nevertheless, for the two magnets array, the normal mesh approach seems to provide a better agreement with the experimental data.

V. Conclusions

The rigorous statistical tests applied here prove to be useful to determine that complex models describing the electromagnetics of interacting bodies are robust enough to accurately represent real situations. As a result, the *in silico* approach explored here is well suited to help prototyping new devices based on interacting magnets.

A suitable model to be used as a base of simulation of magnetic sources interacting with materials and particularly with nanocarriers has been validated. Future work includes the design of a model incorporating magnetic nanocarriers and combining the physics of biofluids like blood as well as the topology of human vessels. This will be the basis for simulating magnetically guided nanocarriers inside arterial branches, which will be useful in the design of complex brain diseases therapies.

Conflict of Interest

The authors declare that there is no conflict of interest regarding the publication of this paper.

Acknowledgment

This work was supported by Colciencias through the project: col13-1-06 - "Lateral Association of BACE1 and BACE2 Transmembrane Domains in Model Lipid Bilayers to Enable Highly Efficient SIRNA-Based Therapies at the Cellular Level". Cto.458-2012.

References

1. L. Agiotis, I. Theodorakos, S. Samothrakitis, S. Papazoglou, I. Zergioti, and Y. S. Raptis, "Magnetic manipulation of superparamagnetic nanoparticles in a micro fluidic system for drug delivery applications," *J. Magn. Magn. Mater.*, vol. 401, pp. 956–964, 2016.
2. Wahajuddin and S. Arora, "Superparamagnetic iron oxide nanoparticles: Magnetic nanoplatforms as drug carriers," *Int. J. Nanomedicine*, vol. 7, pp. 3445–3471, 2012.
3. R. Tietze, J. Zaloga, H. Unterweger, S. Lyer, R. P. Friedrich, C. Janko, M. Pöttler, S. Dürr, and C. Alexiou, "Magnetic nanoparticle-based drug delivery for cancer therapy," *Biochem. Biophys. Res. Commun.*, vol. 468, no. 3, pp. 463–470, 2015.
4. T. D. Do, Y. Noh, M. O. Kim, and J. Yoon, "An Electromagnetic Steering System for Magnetic Nanoparticle Drug Delivery.," *12th Int. Conf. Ubiquitous Robot. Ambient Intell.*, no. Urai, pp. 528–531, 2015.
5. I. Khalin, R. Alyautdin, N. M. Ismail, M. H. Haron, and D. Kuznetsov, "Nanoscale drug delivery systems and the blood–brain barrier," *Int. J. Nanomedicine*, vol. 9, no. 1, p. 795, Feb. 2014.
6. T. Tagami, M. Taki, and T. Ozeki, *Nanomaterials in Pharmacology*, vol. 39. New York, NY: Springer New York, 2016.
7. S. Wang and Y. Zhou, "Magnetic nanoparticles: Synthesis, physicochemical properties and role in biomedicine," in *Advances in Nanotechnology*, vol. 13, Nova Science Publishers, Inc., 2015.
8. E. J. Kastner, R. Reeves, W. Bennett, A. Misra, J. D. Petryk, A. A. Petryk, and P. J. Hoopes, "Alternating magnetic field optimization for IONP hyperthermia cancer treatment," in *Progress in Biomedical Optics and Imaging - Proceedings of SPIE*, 2015, vol. 9326, p. 93260M.
9. H.-C. Roth, A. Prams, M. Lutz, J. Ritscher, M. Raab, and S. Berensmeier, "A High-Gradient Magnetic Separator for Highly Viscous Process Liquors in Industrial Biotechnology," *Chem. Eng. Technol.*, vol. 39, no. 3, pp. 469–476, 2016.
10. A. Garraud, C. Velez, Y. Shah, N. Garraud, B. Kozissnik, E. G. Yarmola, K. D. Allen, J. Dobson, and D. P. Arnold, "Investigation of the capture of magnetic particles from high-viscosity fluids using permanent magnets," *Biomedical Engineering, IEEE Transactions on*, vol. PP, no. 99. p. 1, 2015.
11. S. Erni, S. Schürle, A. Fakhraee, B. E. Kratochvil, and B. J. Nelson, "Comparison, optimization, and limitations of magnetic manipulation systems," *J. Micro-Bio Robot.*, vol. 8, no. 3–4, pp. 107–120, 2013.
12. W. N. L. W. Mahadi, S. R. Adi, and K. M. Nor, "Application of the rare earth permanent magnet in linear generator driven by an internal combustion engine," *Natl. Power Eng. Conf. Pecon 2003, Proc.*, pp. 256–261, 2003.
13. S. Erni, "Design and Optimization of Magnetic Manipulation Systems for Biomedical Applications presented by," University of Basel, 2015.
14. T. B. Jones, *Electromechanics of Particles*. Cambridge University Press, 1995.
15. H. Lin and J. Wooley, "Computational Modeling and Simulation as Enablers for Biological Discovery," in *Catalyzing inquiry at the interface of computing and biology*, 2005, p. 468.
16. F. Bruckner, C. Abert, C. Vogler, F. Heinrichs, A. Satz, U. Ausserlechner, G. Binder, H. Koeck, and D. Suess, "Macroscopic simulation of isotropic permanent magnets," *J. Magn. Magn. Mater.*, vol. 401, pp. 875–879, 2016.
17. J. D. Kraus and Carver Keith R., *Electromagnetics*, 2d ed. McGraw-Hill New York, 1973.
18. COMSOL Multiphysics, "AD/DC Module User's Guide," 2013.

Author: A. M. Hernández.
Institute: University of Antioquia
Street: calle 67 No. 53 - 108
City: Medellín
Country: Colombia
Email: alher.hernandez@udea.edu.co

Analysis of the stability control of motors used in biomechanical prostheses

C. Alvarez Picaza[1], M.I. Pisarello[1] and J.E. Monzón[1]

[1] Universidad Nacional del Nordeste (UNNE). Corrientes, Argentina

Abstract— **We present an alternative analysis for the energy optimization and system stability in servomotors that drive biomechanical prostheses using a LQR (Linear Quadratic Regulator) controller. The model is designed in the state space, and we use the Lyapunov Function as a parameter of the system stability. From the observation of the dynamical response and transition of the system variables, angular position, angular velocity and armature current of the model, we found that the use of LQR with its respective controlling action generates in the system a convergence to an equilibrium point x_e in less than 3 seconds, ensuring stability with a shorter trajectory.**

Keywords—**Prosthesis, Servomotor, Stability, LQR.**

I. Introduction

The design of artificial limbs requires full knowledge not only of the mechanics of the mechanisms, but also a clear understanding of electromechanical devices, among which driving motors play a key role in the area of prostheses. The maximal speed, force and stability of the anatomical limb are still unparalleled by the artificial prosthesis. These limitations are due to physical restraints of current technology to achieve the properties which the natural limb exhibits. Matching muscle speed and force with technological actuator is not an easy task, mainly when choosing a driving motor with the appropriate speed-torque relationship [1]. Instability of driving motors adds to the complexity of proper prosthesis design.

It is known that many theories in physical sciences are based or expressed in terms of optimality. In the field of motor control, optimality also plays a key role. The optimization processes give rise to a specific motor system under investigation (adaptation, development, evolution, recovery). These processes cause the system to perform better and better. In the area of theoretical investigations, it is natural to search for limits of optimal performance of motor control [2].

Several methods are used to control speed of DC mo-tors [3]. Neenu [4] reports that Proportional-Integral-Derivative (PID) controllers have been widely used for speed position control. Selecting PID parameters using genetic algorithms has lead to a more efficient controller [5]. Other authors like Boumediene [6], used a Particle Swarm Optimization (PSO) instead of GA.

They presented a PID controller based on PSO. They found that PID-PSO controller gives good performance and minimal rise time. Sharaf [7] presented a novel PID dual loop controller for a solar photovoltaic (PV) powered industrial permanent mag-net DC motor drive. However, in spite of the robustness and apparently simple structure of PID control strategy, optimiz-ing the gains of PID controller is still a difficult task [8].

An alternative method is the Linear Quadratic Regulator (LQR). Its performance index is found by using a mathematical algorithm that minimizes a cost function. This function is often defined as a sum of the deviations of key measurements from their desired values. In fact, this algorithm finds those controller settings that minimize the undesired deviations. Although LQR gives a near optimal response, the major drawback of this quadratic method lies in its lengthy mathematical calculations [8]. In addition, the advantages of using LQR are related to an easy design and increased accuracy of the state variables by estimating the state. The attractive feature of the LQR control as compared to pole placement is the intuitive specification of a set of performance weighting instead of having to specify where *n* eigenvalues should be placed [9].

Molavi and Khaburi [10] introduced the optimal strategies for speed control of permanent magnet synchronous motor (PMSM) through the linear quadratic regulator (LQR) and linear quadratic Gaussian (LQG) methodologies.

In previous works [11] [12] we found the desired energy optimization for a motor used to drive the movement of a robotic arm with the highest possible efficiency.

The objective of this paper is to analyze, based on LQR control, the stability of motors used in biomechanical prostheses.

II. Methodology

A. Dynamical model of the servomotor system.

In a previous work [11] we presented the model of the DC servomotor used in a robotic manipulator. The motor drives a load through a rigid axis. The estate space model is presented in equations (1), (2) and (3). Figure 1 shows the schematic diagram. In this analysis we assume that the input reference to the system is a step function.

I. Torres et al. (eds.), *VII Latin American Congress on Biomedical Engineering CLAIB 2016, Bucaramanga, Santander, Colombia, October 26th -28th, 2016*,
IFMBE Proceedings 60,
DOI: 10.1007/978-981-10-4086-3_141

The state variables of the system are:

$$x_1 = \theta(t)$$
$$x_2 = \dot{\theta}(t) = \omega(t) \quad (1)$$
$$x_3 = i_a(t)$$

for x_1: angular position; x_2: angular velocity and x_3: armature current.

The corresponding representation in state variables is as follows,

$$\begin{bmatrix} \dot{\theta} \\ \dot{\omega} \\ \dot{i}_a \end{bmatrix} = \begin{bmatrix} 0 & 1 & 0 \\ 0 & \frac{-b}{J} & \frac{K_t}{J} \\ 0 & \frac{-K_b}{L_a} & \frac{-R_a}{L_a} \end{bmatrix} \begin{bmatrix} \theta \\ \omega \\ i_a \end{bmatrix} + \begin{bmatrix} 0 \\ 0 \\ \frac{1}{L_a} \end{bmatrix} u(t)$$

$$y = \begin{bmatrix} 1 & 0 & 0 \end{bmatrix} \begin{bmatrix} \theta \\ \omega \\ i_a \end{bmatrix} \quad (2)$$

Given a control system,

$$\dot{\mathbf{x}} = \mathbf{A}\mathbf{x} + \mathbf{B}u$$
$$y = \mathbf{C}\mathbf{x} + Du \quad (3)$$

The servomotor system is shown in Figure 1.

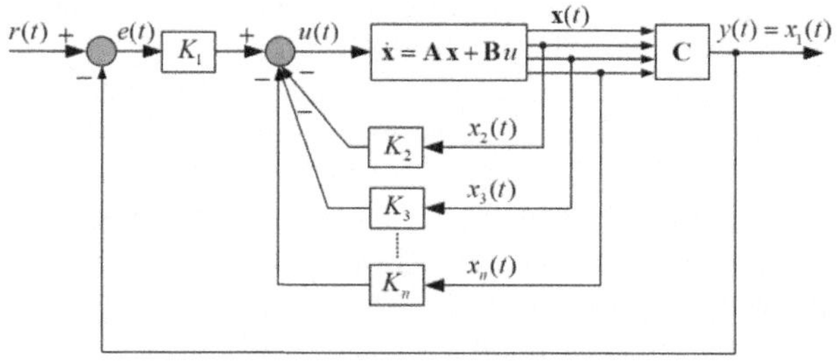

Figure 1. Block diagram of the system.

Mathematically,

$$u(t) = -\ \mathbf{K}\mathbf{x}(t) - \ K_1 r(t) \quad (4)$$

Therefore, in a permanent state, we have

$$\dot{x}(\infty) = (\mathbf{A} - \mathbf{BK})x(\infty) + \mathbf{BK_1}r(\infty) \quad (5)$$

We can get the error equation of this system

$$\dot{x}(t) - \dot{x}(\infty) = (\mathbf{A} - \mathbf{BK})[x(t) - x(\infty)] \quad (6)$$

where,

$$e(t) = x(t) - x(\infty) \quad (7)$$

Hence,

$$\dot{e}(t) = (\mathbf{A} - \mathbf{BK})e(t) \quad (8)$$

And its solution is,

$$\mathbf{e}(t) = e^{(\mathbf{A}-\mathbf{BK})t}\,\mathbf{e}(0) \quad (9)$$

The previous equation describes the dynamics of the system error of Figure 1. From the equation (9), it follows that this error **e***(t)* tends to zero for a given initial condition **e**(0) if the eigenvalues of the matrix (**A-BK**) are placed at the left semi plane of the *s* plane. Thus the closed loop system results asymptotically stable.

In permanent state, we have that,

$$\begin{aligned} x(t) - x(\infty) &= x_e(t) \\ e(t) - e(\infty) &= e_e(t) \\ u(t) - u(\infty) &= u_e(t) \end{aligned} \quad (10)$$

Equation (10) can be re-written as,

$$\begin{bmatrix} \dot{x}_e(t) \\ \dot{e}_e(t) \end{bmatrix} = \begin{bmatrix} \mathbf{A} & 0 \\ -\mathbf{C} & 0 \end{bmatrix} \begin{bmatrix} x_e(t) \\ e_e(t) \end{bmatrix} + \begin{bmatrix} \mathbf{B} \\ 0 \end{bmatrix} u_e(t) \quad (11)$$

Finally,

$$u_e(t) = -\ \mathbf{K}\mathbf{x}_e(t) + K_i e_e(t) \quad (12)$$

B. The Linear Quadratic Regulator (LQR).

One of the systematic and more used methods to determine the values of **K** is throughout the application of optimal control techniques. It is done by minimizing the indexes of quadratic performance, which can be achieved by using the following scalar equation:

$$\mathbf{M} = \int_0^\infty \left[\mathbf{x}^T\mathbf{Q}\mathbf{x} + \mathbf{u}^T\mathbf{R}\mathbf{u} \right] dt \quad (13)$$

This is the *cost function*. It is directly related with the energy involved in the control process. In equation (13), matrices **Q** and **R** are hermit or real positive symmetric defined, **x** is the state vector and **u** is the input vector. If the state equation $\dot{\mathbf{x}} = \mathbf{Ax} + \mathbf{B}u$ is controllable, then the state feedback gain matrix is given by the next equation.

$$\mathbf{K} = \mathbf{R}^{-1}\mathbf{B}^{T}\mathbf{S} \tag{14}$$

Where **S** is a positive defined and symmetric matrix satisfying the following expression:

$$\mathbf{SA} + \mathbf{A}^{T}\mathbf{S} - \mathbf{SBR}^{-1}\mathbf{B}^{T}\mathbf{S} + \mathbf{Q} = 0 \tag{15}$$

This is the well-known *Riccati algebraic equation.* Matrices **Q** and **R** are chosen commonly as diagonal matrices, where the elements q_{ii} of **Q** are associated to the energy of the states. Elements r_{ii} of **R** are associated to the energy de the control action.

C. Stability criteria

The equation

$$\mathbf{A}^{T}\mathbf{P} + \mathbf{PA} = -\mathbf{Q} \tag{16}$$

is the *Lyapunov equation*. The stability criteria is based on find **P** in (16), choosing previously **Q**. In general, **Q=I**
$x_e = 0$ is the equilibrium state of the linear invariant system described by the equation

$$\dot{\mathbf{x}}(t) = \mathbf{Ax}(t) \tag{17}$$

The state x_e is asymptotic and globally stable if and only if, all the eigenvalues of **A** have a negative real part.

The quadratic shape $\mathbf{V(x)} = \mathbf{x}^{T}\mathbf{Px}$, called *Lyapunov Function,* must accomplish with the Sylvester criteria and must be positive defined.

III. Results and Discussion

Data for the servomotor simulation model, developed in a previous section , were obtained from an actual engine (RE model Maxon® 40-40 mm) [11] generally used to drive anthropomorphic myoelectric prostheses [13].
The following values were taken from the manufacturer catalog:
R_a = Armature Resistance = 1,16Ω.
L_a = Armature Inductance = 0,329 mH.
K_t= Torque Constant = 60,3 mNm/A.
K_b = Constant Rate = 158 rpm/V.
b = Friction coefficient of the bearings = 3,04 rpm/mNm.
J = Inertia moment of the motor and load= 138 gcm^2.

The design specifications are:
Time constant T = 0,25 sec. We consider the criterion of 2% (Dumping tolerance).

This implies that if $T = 0,25;\ \sigma = \frac{1}{T} = 4$

System poles are:
-1,77 + j 14,4 ; -1,77 - j14,4 ; 0 ; 0

Poles (lqr):
-5,1723 + j13,6456 ; -5,1723 - j13,6456 ;
-1,7921 + j 1,0947 ; -1,7921 - j 1,0947
The gain from the quadratic optimal regulator is
K (lqr) = [609,0428 33,3480 3,4153 -707,1068].

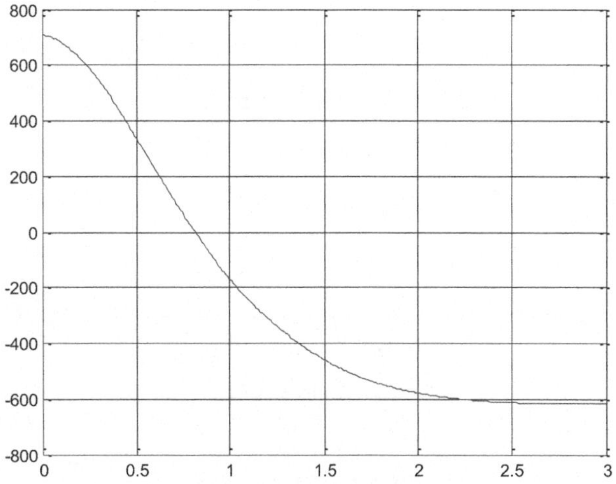

Figure 2. Control Action (LQR)

Figure 3 shows the step response of the state variables. It can be observed that the servomotor system achieves an equilibrium level, reaching stability after 3 seconds.

Figure 3 shows the step response of the state variables, angular position, angular velocity and armature current as defined in equation (1).

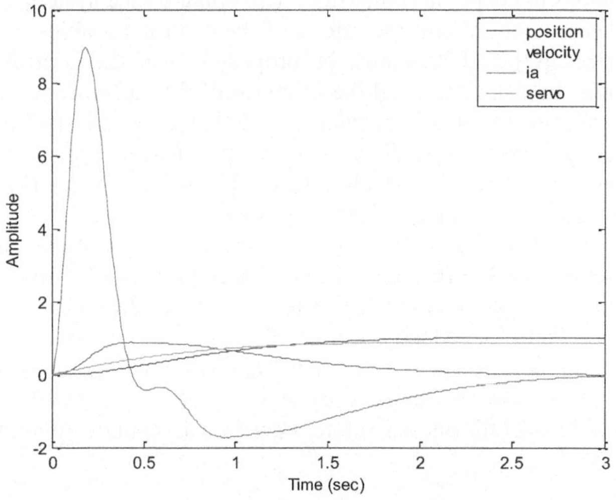

Figure 3. Step response of the state variables

The Lyapunov equation of the original system is

$$V(x)=0{,}0024021x_1^2+80{,}7601x_1x_2+10x_1x_3+155{,}6708x_2^2+0{,}1560x_2x_3+0{,}1515x_3^2$$

and the optimized system,

$$Vs(x)=5.0474*10^3x_1^2+926.5027x_1x_2+14.5705x_1x_3-1.6973*10^4x_1x_4+89.6971x_2^2+1.5879x_2x_3-1.3395*10^3x_2x_4+0.0609x_3^2-23.2638x_3^2-23.2638*x_3x_4+3.7191*10^4$$

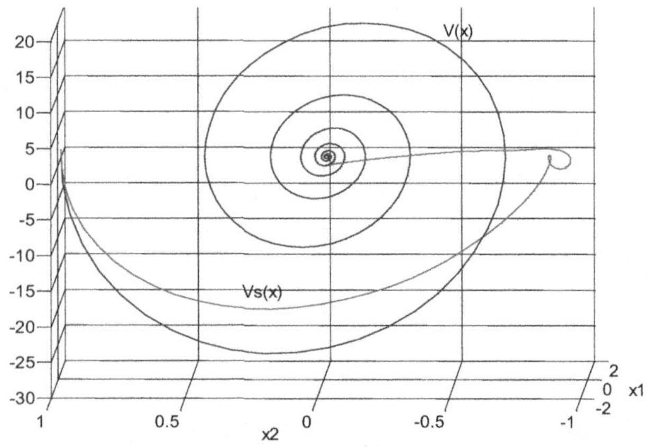

Figure 4. Evolution of x_e

IV. Conclusions

The use of the LQR with its respective control action generates in the system a convergence to an equilibrium point *xe* ensuring stability with a lower trajectory. To obtain a driver LQR it is necessary to define the cost function matrices **[Q]** and **[R]**. Unfortunately there is not a simple analytical methods that allows the designer to define the values to shape these matrices depending on the system, the desired control and the efforts of the control variables. Matrices **[Q]** and **[R]** must be proposed, then the controller must be calculated and the behavior of the closed-loop system checked. It is important to note that the matrix **[K]** it is not unique to a particular system, but depends on the desired positions of the closed-loop poles (which determine the speed and damping of the response).

It should be noted that there exists a compromise between speed of response and sensitivity to perturbations and noise measurement in the selection of the desired closed-loop poles, or property desired equation.

As there is a scalar function ***V(x)*** with continuous partial first derivatives, then steady state $x_e = 0$ is asymptotically stable overall, and therefore ***V(x)*** is a Lyapunov function. The use of LQR method allows us to optimize this function observing a shorter path to the equilibrium point x_e.

V. Acknowledgment

This work was performed within the framework of the Research Project PICTO UNNE 2011 - 249 funded by the National Agency for Scientific and Technological Promotion (AGENCIA) and the General Secretary of Science and Technology, National University of the Northeast

Conflicts of Interest

We haven't had any conflict of interest.

References

1. Weir R. F. ff, (2003) "*Design of Artificial Arms and Hands for Prosthetic Applications*" in "*Standard Handbook of Biomedical Engineering and Design*". Myer K. Editor, Mcgraw-Hill Publisher. New York..
2. Todorov E. and Li W., (2003) "Optimal Control Methods Suitable for Biomechanical Systems". *EMBS- IEEE Proceedings of the 25th Annual International Conference*. Vol 2. Pp 1758 – 1761.
3. Ruba M.K. and Hummadi A-M. (2012) "Simulation of Optimal Speed Control for a DC Motor Using Linear Quadratic Regulator (LQR)". Journal of Engineering. Num 3, vol 18, pp 340-346.
4. Neenu T. and Poongodi P., (2009) "Position Control of DC Motor Using Genetic Algorithm Based PID Controller", *Proceedings of the World Congress on Engineering.* Vol. II, WCE2009.
5. Delavari H., Noiey A.R., Minagar S., (2008) "Artificial Intelligent Controller for a DC Motor". *Advances in Computer Science and Engineering*, pp. 842-846.
6. Allaoua B., Gasbaoui B. and Mebarki B, (2009) "Setting Up PID DC Motor Speed Control Alteration Parameters Using Particle Swarm Optimization Strategy", *Leonardo Electronic Journal of Practices and Technologies*, Issue 14, pp. 19-32.
7. Sharaf, A.M., Elbakush, E., Altas, I.H., (2007) "Novel Control Strategies for Photovoltaic Powered PMDC Motor Drives', IEEE-Electrical Power Conference. EPC 2007.
8. Pal P., Dey R., Biswas K.R., Bhakta S. (2015), "Optimal Pid Controller Design For Speed Control Of A Separately Excited Dc Motor: A Firefly Based Optimization Approach", *International Journal of Soft Computing, Mathematics and Control IJSCMC.* Vol 4, No 4.
9. Isa N.I.A.B.M., (2008) "DC Motor Controller Using Linear Quadratic Regulator LQR) Algorithm Implementation On Pic." Thesis. Faculty of Electrical & Electronics Engineering. University Malaysia Pahang.
10. Molavi R, and Khaburi D.A (2008), "Optimal Control Strategies for Speed Control of Permanent-Magnet Synchronous Motor Drives", World Academy of Science, Engineering and Technology 44, pp.428-432.
11. Álvarez Picaza C, Pisarello MI and Monzón JE (2014) "Model of optimal energy regulation for the articulation of a robotic manipulator". *VI Latin American Congress on Biomedical Engineering CLAIB* (Berlin: Springer Berlin Heidelberg) Pp 908-911.
12. Álvarez Picaza C., Pisarello M.I., Monzón J.E. (2015) "Análisis teórico de la regulación de energía de servomotores utilizados en manipuladores biomecánicos". *Annals of the XX Congreso Argentino de Bioingeniería. IX Jornada de Ingeniería Clínica.*
13. RE 40 ∅ 40mm, Graphite Brushes, 150 Watt. Data Sheet. [On line. 6/2016].

Atrial fibrillation detection through heart rate variability using a machine learning approach and Poincare plot features

J.P. Sepulveda-Suescun[1], J. Murillo-Escobar[1], R.D. Urda-Benitez[1], D.A. Orrego-Metaute[1] and A. Orozco-Duque[1]

[1] Instituto Tecnológico Metropolitano, Grupo de Investigación e Innovación Biomédica Gi2B, Medellin, Colombia

Abstract— **Atrial Fibrillation (AF) is a common cardiac arrhythmia, and it has a high rate of morbidity and mortality. In this paper, an algorithm for automatic AF episodes detection based on novel low computational cost features is proposed. The features are based on Poincare plots calculated from heart rate variability signal. A supervised classification technique, Support Vector Machines, optimized with Particle Swarm Optimization, was implemented. The data was obtained from MIT-BIH Atrial Fibrillation and Normal Sinus Rhythm Databases. This method shows an accuracy of 92.9% to detect AF spontaneous episodes in signals from AF patients, and 97.8% to classify between AF episodes from AF patients and episodes from subjects with normal sinus rhythm. The proposed method can be employed in real time applications due to its performance as well for its low computation time around 8.8 ms per episode.**

Keywords— **Atrial fibrillation, SVM, Poincare plot, PSO, heart rate variability.**

I. Introduction

Atrial fibrillation (AF) is one of the most common cardiac arrhythmia worldwide, having a significant mortality rate, because AF may increases the blood coagulation, which cause 15% of strokes [1, 2]. The AF in the electrocardiogram (ECG) is characterized by irregularity present in the RR intervals and the replacement of the P wave by rapid oscillations and fibrillatory waves that vary in time, amplitude and shape [3], therefore, the diagnosis of paroxysmal AF presents a major challenge for clinicians and researchers because of its asymptomatic nature and spontaneous, especially in its early stages[4]. Do to this, there is an urgent need to develop methods for accurate and rapid detection of paroxysmal AF.

Different methods have been proposed to detect AF from the ECG signal using wavelet transform, Shannon entropy sequence analysis of the ECG signal and machine learning methods as genetic programming optimized simulated annealing and artificial neural networks[5, 6, 7, 8, 9], however, this strategy represents a high computational cost since it is sometimes necessary to segment the P wave or analyzing raw signal where there is information that is not relevant. Therefore, it has been proposed AF detection by RR intervals, because this offers a more efficient strategy from the computational point of view, in this context have been developed methodologies based on statistical characteristics of the Heart Rate Variability (HRV), which is a more robust method because the time intervals R-R are less affected by noise [10, 11].

In this paper, we present a method based on the estimation of novel features from RR interval and HRV to detect the presence of AF and Normal sinus rhythm using signals from MIT-BIH Atrial fibrillation database and MIT-BIH Normal Sinus Rhythm Database. We proposed a method based on Poincare plot with low computational cost in a support Vector machine optimize by particle swarm optimization.

II. Materials and Methods

The proposed methodology to the AF detection is show in Fig 1. As it can be seen, the detection system comprises three steps: preprocessing, characterization and classification, each step are detailed in the following subsections. With this methodology we intend to differentiate short events of AF to Normal Sinus Rhythm (NSR), in this sense we conduce two experiments, the first one using only signals that had at least one episode of AF or NSR, given that in literature is the most common approach, the second experiment we employ only signals from AF patients during NSR with spontaneous episodes of AF, having a greater proximity to real application due to the spontaneous nature of this affection.

A. *Data Base Description*

MIT-BIH Atrial Fibrillation Database (AFDB) is the most popular and often used publicly available database for AF detection, this data base has 23 ECG recording of subjects such had suffered significant arrhythmia each arrhythmia. Each record has a duration of about 10 hours. A total of 291 AF episodes, 14 of atrial flutter, 12 of junctional rhythm and 288 of NSR were found. [7]. MIT-BIH Normal Sinus Rhythm Database (NSRDB) includes 18 ECG recordings of subjects

I. Torres et al. (eds.), *VII Latin American Congress on Biomedical Engineering CLAIB 2016, Bucaramanga, Santander, Colombia, October 26th -28th, 2016*,
IFMBE Proceedings 60,
DOI: 10.1007/978-981-10-4086-3_142

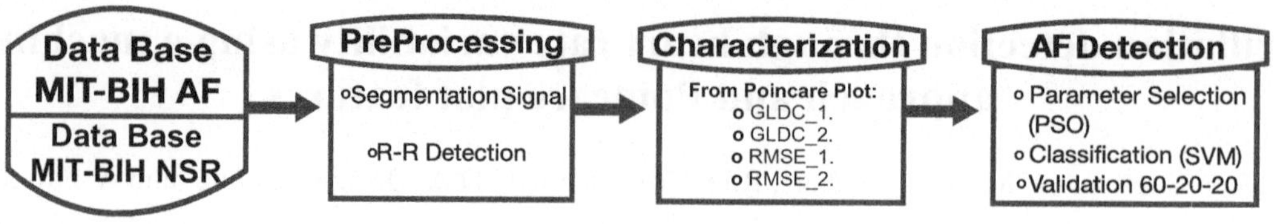

Fig. 1. Proposed Methodology

such had not suffered significant arrhythmia [12].

The analysis of developed algorithm for AF detection was made in two experiments. The first one using signals with at least one event of AF or NSR, in this regard we employ AFDB to extract signals with AF and the NSR signals were get from NSRDB. For the second experiment we use only signals that has both rhythms, which were found only in AFDB.

B. Preprocessing

The preprocessing begin, with the extraction the RR intervals from the ECG signals within the database. For this task, the entries in the database referenced with (.ATR) were used, then was assessed the duration of the events of atrial fibrillation and normal sinus rhythm, for the AF and NSR events greater to 30 seconds were selected from the signals. A windows with length of 30 seconds were construct for each event, centered around the middle point. For calculate the Heart Rate Variability (HRV) we start detecting R peaks employing the Pan-Tompkins method, due to is capability of been implemented in real time [13].

C. Characterization

The *Poincare Plot* is a scatter-plot of the pairs ($HRV[n]$, $HRV[n+m]$), showing the relationship among m consecutive beats [14]. The Poincare plot in NSR episode looks like a cluster of points grouped linearly unlike, the Poincare plot in AF episode evidenced a complete dispersal of the points (Fig 2), this occurs because of the higher erratic variability of the HRV in AF episode with respect to the NSR episode.

From the Poincare plot, linear dependence between $HRV[n]$ and $HRV[n+1]$ is analyzed, signals with AF present greater degree of dispersal, while NSR segments shows greater linear dependence, for this was used the generalized linear dependence coefficient (**GLDC_1**) proposed in [15] which is calculated as:

$$D = 1 - |R|^{1/(p-1)} \tag{1}$$

Where p is the number of involved variables and $|R|$ represents the determinant of correlation matrix, R is obtained from the covariance matrix S between $HRV[n]$ and $HRV[n+1]$ as $R = D^{-1/2}SD^{-1/2}$, D is a square matrix that retains only the main diagonal of S.

Moreover we extend this analysis to measure the generalized linear dependence coefficient between five consecutive heart beats (**GLDC_2**) using (1), in this case the correlation matrix S is calculated between $HRV[n]$, $HRV[n+1]$ $\cdots$ $HRV[n+5]$. Now, assuming that in NSR is possible find

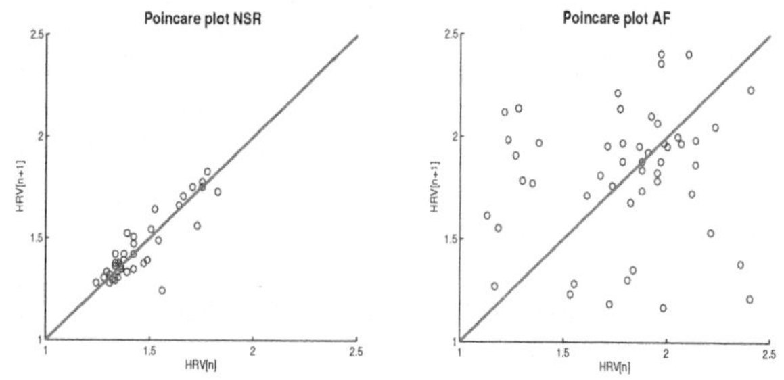

Fig. 2. Poincare plots of NSR and AF

$HRV[n+1]$ as linear function of $HRV[n]$ using regression, we propose measures such capacity through the difference between the actual $HRV[n+1]$ and the predicted one, following the next procedure. Let L be a vector that contain $HRV[n]$ for $n = 1, \cdots l-1$, and l is the number of heart beats of the signal, P is a vector with $HRV[n+1]$, if we estimate P as $\widehat{P} = \omega L$ is possible measure the Root Mean Square Error (RMSE) between $\widehat{P}$ and P (**RMSE_1**). The weight coefficient ω can be found using the normal equation as:

$$\omega = \left(L^T L\right)^{-1} L^T P \tag{2}$$

Is possible extend this analysis for five consecutive heart beats, for this case P will be a vector with $HRV[n+5]$, for $n = 1, \cdots l-5$ and L be a matrix in which every column represents $HRV[n]$, $HRV[n+1]$ $\cdots$ $HRV[n+4]$, repeating the above formulation is possible find the RMSE between the actual P and the predicted (**RMSE_2**).

Aditionally to Poincare plot featues, standard deviation of HRV (**STD_HRV**) was implemented due to it has been widely study in different papers [9, 16, 11].

D. AF Detection

The goal of this paper is to classify each signal segment into AF or NSR, using the proposed extracted features. To achieve this, we use Support Vector Machines (SVM) as our classification algorithm. SVM is supervised classifier proposed for Vapnik in 1995, the idea behind this methods is find an hyperplane that separate the data in their corresponding classes. Given a training data $\{x_i, y_i\}$, where $x_i \in \mathbf{R}^d$, $i = 1, 2 \cdots l$ and $y_i \in \{-1, 1\}$, this way the hyperplane can be found solving the dual problem formulation (3), this formulation allow us for non-linear solutions introducing a kernel function $\phi(x_i, x_j)$, note that in its dual form the optimization problem appears only in terms of the Lagrange multipliers α_i [17]. In this paper we employ the Radial Basis Kernel (RBK) due to it is the dot product in the infinite dimensional space which minimizing both the estimation and approximation errors of classifier in this regard the RBK function is $\phi(x_i, x_j) = e^{-\gamma \|x_i - x_j\|^2}$.

$$\max \sum_{i=1}^{l} \alpha_i - \frac{1}{2} \sum_{i,j=1}^{l} \alpha_i \alpha_j y_i y_j \phi(x_i, x_j)$$
$$\text{Subject to:} \begin{cases} 0 \le \alpha_i \le C \\ \sum_{i=1}^{l} \alpha_i y_i = 0 \end{cases} \tag{3}$$

For the well performance of the SVM is necessary a good selection of the parameter, we selected the C and γ of RBK through Particle Swarm Optimization (PSO) which algorithm is described in [18].

The data obtained in the characterization stage were subjected to a linear normalization between 0 and 1, to avoid redundancy in the data and increase the performance. Subsequently the data were randomly split into three subsets, one with 60% of the data to train the SVM, and two sets more, each with 20% of the data, we use the first subset (validation) for PSO to get the parameter combination that maximize the accuracy, the reminder data were employ to assessed the overall performance of the proposed system in terms of the accuracy, sensitivity and specificity.

III. Results and Discussion

A. Preprocessing

For the first experiment in which is necessary had 30 seconds of signal that has at least one event of AF or NSR, we obtain from AFDB 226 segments of signal with AF and 234 of NSR from NSRDB. In the second setup we extract from AFDB signals that has both of the study rhythms getting 226 segments of signal with AF and 264 of NSR.

B. Characterization

We extracted five features of the HRV, one of these commonly used in literature and the other four calculated from the Poincare plot were proposed in this paper for detecting AF. Our studies show that these features can be used as good markers for detection of AF as can be seen in Fig 3, from the boxplot is clear that features based in the analysis of five consecutive heart beats represent a most robust approach due can be less sensible to anomaly beats like premature ventricular contractions in contrast to statistical methods as the used in [11, 7].

C. AF Detection

The performance of the methodology is listed in Table 1, the proposed methodology on the experiment two presents performance values about 4 points below the results reported by other approaches [7], but these have the disadvantage of working with complete ECG signal and use characteristics with a very high computational cost, allowing these methodologies presented better performances but away from the possibility to be implemented in real time, contrasting with this experiment of our methodology, which were estimated characteristics of the same database and with a low computational cost, meanwhile in the experiment one the obtained performance is equivalent to another approaches [6, 9]. The PSO gave for both experiments a parameter $C = 131$ and $\gamma = 9.5$.

Table 1. Classification Performance

	Sen	Spe	Acc	F1
Exp 1	97.9%	97.8%	97.8%	97.9%
Exp 2	91.3%	94.3%	92.9%	92.3%

Sen: Sensitivity, Spe: Specificity, Acc: Accuracy, F1: F1 score, Exp 1: Experiment 1, Exp 2: Experiment 2

The mean time for the characterization of each of the signal segments for both experiments of our methodology, with Matlab 2014a on an Intel Core CPU to 2.0GHz, 8 RAM and Windows of 64 bits, was of 8.8 *mS*.

IV. Conclusion

In this paper, we presented novels features from Poincare plot, based on the HRV signal extracted from ECG recordings, focused to detect spontaneous atrial fibrillation events. These features show good difference leading to better classification using supervised techniques. The developed experiments show the feasible implementation of a low computational cost methodology for discrimination of paroxysmal AF episodes from NSR in the same HRV signals. The proposed

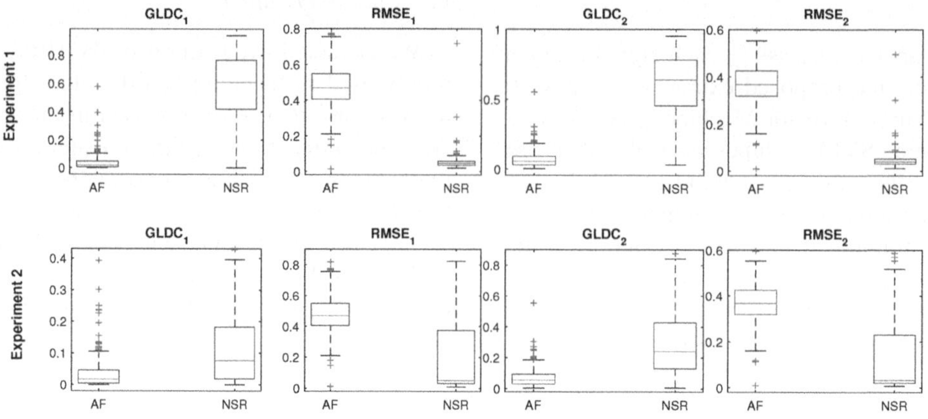

Fig. 3. Boxplots proposed features

method will be implemented in embedded systems in futures developments due its factible implementation in real time.

Conflict of Interest

The authors declare that they have no conflict of interest.

References

1. Chen J, Yang Z G, Xu H Y, Shi K, Long Q H, Guo Y K. Assessments of pulmonary vein and left atrial anatomical variants in atrial fibrillation patients for catheter ablation with cardiac CT *Eur Radiol.* 2016.
2. Mohebbi Maryam, Ghassemian Hassan. Prediction of paroxysmal atrial fibrillation based on non-linear analysis and spectrum and bispectrum features of the heart rate variability signal *Computer Methods and Programs in Biomedicine.* 2012;105:40–49.
3. Fuster Valentin, Rydén Lars E, Cannom David S, et al. ACC/AHA/ESC 2006 Guidelines for the Management of Patients with Atrial Fibrillation: a report of the American College of Cardiology/American Heart Association Task Force on Practice Guidelines and the European Society of Cardiology Committee for Practice *Circulation.* 2006;114:257–354.
4. Yoon Kwon Ha, Thap Tharoeun, Jeong Chang Won, et al. Analysis of Statistical Methods for Automatic Detection of Congestive Heart Failure and Atrial Fibrillation with Short RR Interval Time Series in *2015 9th International Conference on Innovative Mobile and Internet Services in Ubiquitous Computing*:452–457 2015.
5. Ghodrati Alireza, Murray Bill, Marinello Stephen. RR Interval Analysis for Detection of Atrial Fibrillation in ECG Monitors in *Annual International Conference of the IEEE Engineering in Medicine and Biology Society*;2008:601–604 2008.
6. Lian Jie, Wang Lian, Muessig Dirk. A simple method to detect atrial fibrillation using RR intervals. *American journal of cardiology.* 2011;107:1494–1497.
7. Asgari Shadnaz, Mehrnia Alireza, Moussavi Maryam. Automatic detection of atrial fibrillation using stationary wavelet transform and support vector machine *Computers in Biology and Medicine.* 2015;60:132–142.
8. Petrenas Andrius, Marozas Vaidotas, Lukosevicius Arūnas, Sörnmo Leif. A Noise-Adaptive Method for Detection of Brief Episodes of Paroxysmal Atrial Fibrillation in *Computers in Cardiology*:739–742 2013.
9. Yaghouby Farid, Ayatollahi Ahmad, Bahramali Reihaneh, Yaghouby Maryam, Alavi Amir Hossein. Towards automatic detection of atrial fibrillation: A hybrid computational approach *Computers in Biology and Medicine.* 2010;40:919–930.
10. Tateno K., Glass L.. A method for detection of atrial fibrillation using RR intervals in *Computers in Cardiology 2000. Vol.27 (Cat. 00CH37163)*:391–394 2000.
11. Ruan Xiuhua, Liu Changchun, Liu Chengyu, Wang Xinpei, Li Peng. Automatic detection of atrial fibrillation using R-R interval signal in *The 4th International Conference on Biomedical Engineering and Informatics (BMEI 2011)*:644–7 2011.
12. Goldberger A. L., Amaral L.A.N., Glass L., et al. PhysioBank, PhysioToolkit, and PhysioNet: Components of a New Research Resource for Complex Physiologic Signals *Circulation.* 2000;101:215–220.
13. Pan J, Tompkins W J. A real-time QRS detection algorithm. *IEEE transactions on bio-medical engineering.* 1985;32:230–236.
14. Brennan M., Palaniswami M., Kamen P.. Do existing measures of Poincar?? plot geometry reflect nonlinear features of heart rate variability? *IEEE Transactions on Biomedical Engineering.* 2001;48:1342–1347.
15. Peña Daniel, Rodríguez Julio. Descriptive measures of multivariate scatter and linear dependence *Journal of Multivariate Analysis.* 2003;85:361–374.
16. Maji U., Mitra M., Pal S.. Differentiating normal sinus rhythm and atrial fibrillation in ECG signal: A phase rectified signal averaging based approach in *International Conference on Control, Instrumentation, Energy and Communication, CIEC 2014*:176–180 2014.
17. Burges Christopher J.C.. A Tutorial on Support Vector Machines for Pattern Recognition *Data Mining and Knowledge Discovery.* 1998;2:121–167.
18. Kennedy J, Eberhart R. Particle swarm optimization *Neural Networks, 1995. Proceedings., IEEE International Conference on.* 1995;4:1942–1948.

Author: Juan Pablo Sepúlveda Suescún
Institute: Instituto Tecnológico Metropolitano
Street: Calle 73 No 76A - 35
City: Medellín
Country: Colombia
Email: jsepulveda1331@hotmail.com

Mathematical Modeling of Human Eye Affected by Increased Intraocular Pressure as a tool for the Prevention of Glaucoma

Eduardo Pinos-Vélez[1], Rocio Alvarez-Cardenas[2], Sebastian Torres-Ríos[2], Dr. Carlos Luis Chacón[3], PhD William Ipanaqué-Alama[4], Luis Serpa-Andrade[5]

[1]GIIATA, Research Group on Artificial Intelligence and Assistive Technologies at Universidad Politécnica Salesiana, Cuenca - Ecuador, DICOP- University of Piura - Peru. e-mail: epinos@ups.edu.ec
[2] Electronic Engineering, from Universidad Politécnica Salesiana, Cuenca – Ecuador. e-mail: nalvarez@est.ups.edu.ec; ctorresri@est.ups.edu.ec
[3]Ophthalmologist Surgeon, Glaucoma and Cataract Specialist, Board Member of the Ecuadorian Society of Glaucoma, Clinical Santa Lucía, Quito – Ecuador. e-mail: clchacon@clinicasantalucia.com.ec
[4]PhD of Engineering Computer Science and Control, Universidad de Piura – Perú. e-mail: william.ipanaque@udep.pe
[5]GIIATA, Research Group on Artificial Intelligence and Assistive Technologies at Universidad Politécnica Salesiana, Cuenca - Ecuador, e-mail: lserpa@ups.edu.ec

Abstract— **This article is intended to provide information about the development of a project, in order to assist the diagnosis of glaucoma through mathematical modeling of the human eye affected by increased Intraocular Pressure IOP. For this purpose a database was obtained from "Santa Lucía" Medical Clinic in Quito, Ecuador. This database is not only formed by the values of intraocular pressure, but also by the tests that detect damage to the optic nerve in the vision field, and as extra data, the value of the thickness of the cornea. This complements the IOP measurement, since, the greater the thickness, the smaller the pressure reading will be, and when it is thinner, the IOP tends to rise. the difference in corneal thickness according to the type of glaucoma where indicates the central corneal thickness is reduced in patients with normal-pressure glaucoma, suggesting that the IOP is underestimated in the presence of thin corneas, however, the specific finding of values paquimetry consistently lower in patients with normal IOP and glaucomatous damage involves the hypothesis that there may be a relationship between structural alterations (cribriform plate) patients with thin corneas and increased susceptibility to glaucoma progression. The mathematical modeling is divided into three equations which relate the increase in IOP with damage to the nerve fiber layer and with the decrease of the visual field.**

Keywords— Glaucoma, intraocular pressure, modeling, nerve fibers, visual field, aqueous humor, optic nerve.

I. INTRODUCTION

Currently, at Universidad Politécnica Salesiana, UPS, several researches related to biomedicine are being carried out in order to prevent any kind of silent disease in the human body. Therefore, this paper aims to help people who might suffer from Glaucoma or those at risk of having it.

Glaucoma is a disease that causes irreversible damage to the optic nerve by increased intraocular pressure, IOP, thus decreasing the thickness of the nerve fiber layer and subsequently affecting the vision field causing blindness in patients [10][11]. In Fig. 1 the internal structure of the eye is shown.

The increase of IOP occurs by the accumulation of aqueous humor in the anterior chamber [2]. This can happen when the production of this fluid is greater than the amount that is drained through the trabecular meshwork or it can also happen when the drainage pathways clog.

Normal pressure ranges from 10 mmHg and 20mmHg. Although there is no strict boundary, 20mmHg is considered to be the upper limit of normal, and higher Fig. 1 are considered suspicious. [3]

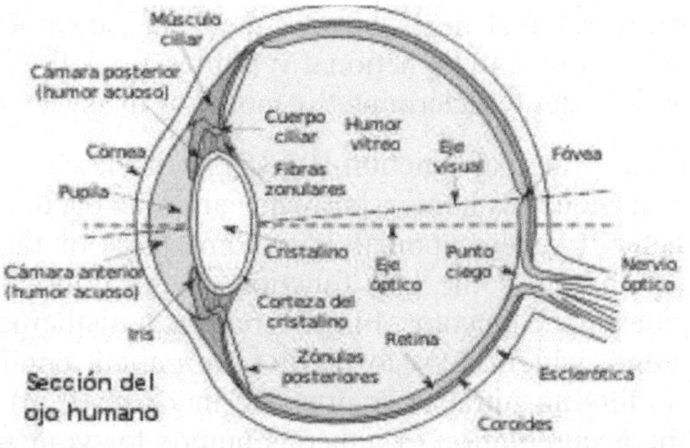

Fig. 1. Internal structure of the human eye [1]

The vision of those who have glaucoma [14] decreases when the pressure increases and the thickness of the nerve fiber layer is reduced; this happens immediately. Normal thickness value is 70 to 110μm lower values; this means that the person has already lost nerve fibers. These analyzes can be done by optical coherence tomography, OCT.

After five years of increased IOP, damage can be shown in the vision field, Fig. 2. Its state can be detected by the campimeter whose results show the sensitivity of the visual field. As the disease progresses, the damage begins in the periphery, and over time this increases until it finishes in a tubular vision and finally in a completely vision loss.

II. MATHEMATICS MODELING.

The Methodology that was used in order to find the mathematics model equations was based in the process follows Santa Clara Medical Clinic´s Ophthalmologist to detect the

I. Torres et al. (eds.), *VII Latin American Congress on Biomedical Engineering CLAIB 2016, Bucaramanga, Santander, Colombia, October 26th -28th, 2016*,
IFMBE Proceedings 60,
DOI: 10.1007/978-981-10-4086-3_143

Glaucoma. That is why; the obtained database belongs to the examination of IOP value measurement, the value of thickness of nerve fibers layers were obtained by the OCT, and finally, the examination that allows to determine the sensibility of the visual range which analysis has two different parameters: the one of average deviation, and from the standard deviation of the model.

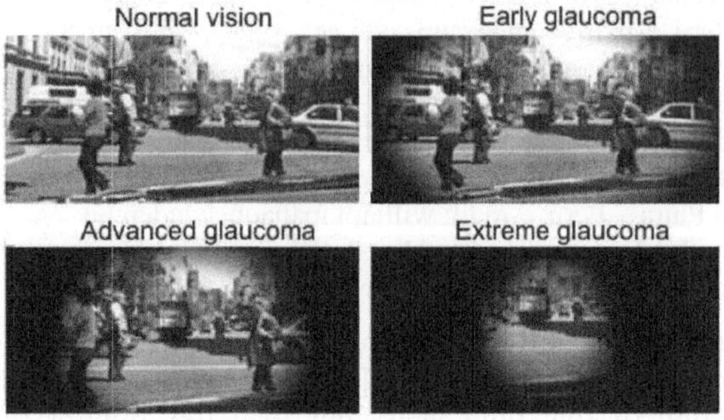

Fig. 2.The stages of Vision Loss for a person that suffers Glaucoma [18].

A. Human eye modeling affected by increase of Intraocular Pressure.

The most convenient mathematical model for this case would be the deterministic, since is possible to predict what is going to happen in the future when the value of the independent variable changes, besides the database values are real and no aleatories, the functional variables are defined in a perfect way in order to determine the model of the system .

Part of the principal function [9][8][17] of the eye is the drainage of its aqueous humor, but when an issue occurs with this parameter it causes anomalies in some parts of the eye anatomy, in this case the mal function of this fluid is the principal cause for Glaucoma. For this reason, a mathematical model is done, which helps to predict the patient condition when he suffers an intraocular pressure increase (IOP) as a result of the accumulation of aqueous humor for your paper size.

B. Mathematical model for Thickness of the Retinal Nerve fibers layer in function of the Intraocular Pressure.

For the first Mathematical model is necessary to relate the IOP with the thickness of the retinal nerve fibers layers [15] [16], because when a pressure increase exists, the damage of the nerve fibers is going to be immediate, although there is no evidence of direct relation between these two parameters, the purpose is to determine the damage in the decrease of the thickness in this layer when an increase of the pressure occurs. This equation was found trough a sixth level polynomial regret ion by the use of Matlab [19], as a result the polynomial coefficient were obtained, as shown in Equation (1) and (2) of the mathematical model.

$$G=a_0+a_1*p+a_2*p^2+a_3*p^3+a_4*p^4+a_5*p^5+a_6*p^6 \quad (1)$$

Where:

G: Is the thickness of the nerves fiber layer.

$a_0 \ldots a_6$: Are the polynomial coefficients.

p: Is the intraocular pressure.

Replacing the coefficient values the equation (2) is obtained which is the mathematical model that represents this relation.

$$G=-17187+7270.9*p-1258.1*p^2+114.21*p^3-5.7315*p^4+0.15063*p^5-0.0016191*p^6 \quad (2)$$

A second mathematical model refers to the existent relation between the thicknesses of the retinal nerve fibers layers with the visual field. As was mentioned before, the damage caused by the increase of the intraocular pressure (IOP) can be detected after 5 years approximately from the beginning of the illness.

C. Mathematical Model for the Average Deviation in function of the thickness of the retinal nerve fibers layers.

This model aims to determine the loss of the visual field which is designated by the average deviation index after having a value of the thickness of the retinal nerve fibers layers [12] [13]; in this case a polynomial regression of seventh order is done. The reason of why this order was chosen is because this is the closest curve to the real values from the database. Likewise, the prior equation, the parameters and equations were obtained by Matlab.

The mathematical representation is shown in equations (3) and (4), corresponding to the mathematical model of average deviation according to the thickness of the retinal nerve fibers layer.

$$D=a_0+a_1*g+a_2*g^2+a_3*g^3+a_4*g^4+a_5*g^5+a_6*g^6+a_7*g^7 \quad (3)$$

Where:

D: Is the Average deviation.

g: Represents the thickness of the retinal nerve fibers layer.

$a_0 \ldots a_7$: Are the polynomial coefficient.

By replacing the coefficient values on the equation (4), the mathematical model for this relation is obtained.

$$D=-57193+6534*g-314.6*g^2+8.2691*g^3-0.12815*g^4+0.00117*g^5-5.848x10^{-6}*g^6+1.2313x10^{-8}*g^7 \quad (4)$$

The third mathematical model is also related to the thickness of the retinal nerve fibers layers with the visual field.

D. Mathematical model of the standard deviation of the model as a function of layer thickness of retinal nerve fibers.

This second parameter determines the stage of the visual field during the advanced stage of the disease, with this result is possible to analyze the standard deviation of the model that depends on the thickness of the retinal nerve fibers [4][5][6]. Like the previous model, the polynomial order is seven, by the use of Matlab was possible to obtain the parameters and the equation.

Like the last model the polynomial order is seven, Matlab was also used in order to obtain the equation (5) parameters.

The mathematical representation is shown on the equation five corresponding to the mathematical model of the standard deviation of the model in function of the thickness of the retinal nerve fibers layers.

$$S=a0+a_1*g+a_2*g^2+a_3*g^3+a_4*g^4+a_5*g^5+a_6*g^6+a_7*g^7 \quad (5)$$

Where:

S: Is the standard model deviation

g: Represents the thickness of the retinal nerve fibers layer.

$a_0 \ldots a_7$: Are the polynomial coefficient.

Replacing the values of the coefficients equation (6) is obtained which is the mathematical model for this relation.

$$S=-3879+496.56*g-26.213*g^2+0.74577*g^3-0.012394*g^4+0.00012*g^5-6.38x10^{-7}*g^6+1.4179x10^{-9}*g^7$$
(6)

The three equations have been obtained and they represent different mathematical models in the relation between these three core parameters that are used for the detection of Glaucoma, somehow they help in the prevention of this illness. Despite, there is not an exact relation; there is an approximation that allows predicting a value.

III. Discussion and Further Work.

The equations corresponding to three mathematical models allow to predict values of the thickness of the retinal nerve fibers layers, average deviation and standard deviation of the model that corresponds to visual field once IOP has been measured [7]. This is due to the fact that having the values of pressure is possible to calculate the thickness of the nervous fibers layer by applying the first mathematical model, and with this data the visual field is of this two parameters are obtained. Normally, these values are detected within five years of the illness has been presented in the patient. However, this project development wills to reduce the time of Glaucoma diagnostic, by the early detection of eye damage.

The following step of this project is the readiness check of the system by an ophthalmology specialist in the Medical Clinic to a group of patients. The specialist will gather the test results and will determine that the found equations reaffirm the diagnostic of the illness. In this way, the mathematical models will be used as a test or diagnostic method for Glaucoma.

IV. Conclusions.

When was carried out to obtain the database, it was necessary not only to perform the measurement of IOP, although it is a fact that gives evidence that the person has the disease, but rather complete the diagnosis made other important tests, because not always a person who has elevated IOP mean you have glaucoma therefore taken as fact the thickness of the cornea, because if this value is too old or too less than the established reading IOP is wrong.

The values of RNFL thickness is an important thing, even though the subject is not mentioned; patients that this test is performed must have advance preparation as this analysis requires a lot of concentration by the patient and to expedite this process applies a salve which makes dilates faster pupil and outcome results are top.

Mainly three exams that patients are made are those that allow detection of Glaucoma, once he had all the results it was found that this disease affects as many older people to 40 years, both men and in women.

Acknowledgment

The execution of this project would not be possible without the cooperation of the Research Group on Artificial Intelligence and Assistive Technologies at Universidad Politécnica Salesiana, Cuenca - Ecuador, Dr. Carlos Luis Chacón from Santa Lucía Clinic in Quito, The Universidad de Piura – Perú, PhD William Ipanaqué and its Engineering Doctorial Program, DICOP.

References

1. Martins Joao C., Sousa Leonel A. "Bioelectronic Vision, Retinal Models, Evaluation Metrics and System Design". Series on Bioengineering & Biomedical Engineering. Vol. 3. World Scientific, ISBN-13 978-981-279-430-7. 2009.
2. P. B. Martín, "Estudio de la Dinámica del humor acuoso mediante fluorofotometría en el glaucoma de ángulo abierto", Departamento de Oftalmología, Facultad de Medicina, Universidad Complutense de Madrid, Madrid 2991
3. E. Aguilar R., S. Reyes G., "Glaucoma: La Enfermedad Silenciosa", Centro optometría internacional, Junior 2006.
4. Tombran-Tink J., Barnstable C. J., Shields M. B. "Mechanisms of the Glaucoma, Disease Process and Therapeutic Modalities". Ophthalmology Research. Human Press. ISBN: 978-1-58829-956-7. 2008
5. U.S. Department of Health and Human Services, "El Glaucoma", National Institutes of Health, National Eye Institute, NIH Publication No: 03-6515
6. Dr. D. Pérez, "Glaucoma", *Centro Universitario Oftalmológico Universidad Nacional de Tucumán*, 2005.
7. Dr. Cesar Larrain Alzamora, capitulo 9 ''Glaucoma''. Cirugia IV Oftalmologia. Obtenido el 10 de septiembre del 2014. Available: www.rinconmedico.com
8. Dr. Eugenio Maul de la Puente. "Glaucoma". *Escuela de Medicina Departamento de Oftalmologia Curso de Oftalmologia* 2011. 5º año Medicina.
9. Prof. Dr. Juan R. Sampaolesi, ''Efecto Hipotensor de las Prostaglandinas versus la Asociación Fija Timolol-Dorzolamida en el perfil diario de presión intraocular en pacientes con glaucoma''. Glaucoma Forum junio 2007.
10. Dr. Enrique Aramedía. "Glaucoma también conocido como neuropatía óptica glaucomatosa". Policlinica Gipuzkoa. Servicio de oftalmología
11. SBG-Sociedad Boliviana de Glaucoma. "Tipos de Glaucoma". Available: glaucomabolivia.com/tipos-de-glaucoma
12. Dr. Javier Fernando Casiraghi, dr. Juan Roberto Sampaolasi. Capítulo 9 ''Glaucoma''. Carlos Agento-oftalmologia General-Introduccion para el especialista.
13. ACG-Asociacion Colombiana de Glaucoma. "Que es la curva de la presión ocular y como se realiza". Available: www.glaucoma.org.co/curva%20de%20presion.html
14. La pérdida de la visión o también conocida como la agudeza visual. Reportaje. Blog de farmacia. Available: ec.globedia.com/perdida-vision-tambien-conocida-agudeza-visual.
15. David Carbonel Murillo. ''Aplicaciones clínicas de la paquimetría". Grado en opyica y optometría universidad de Zaragoza. 2013.
16. SEAGIG. South East Asia Glaucoma Interest Group. Supported by ALLERGAN."Gonioscopy", Available: www.apglaucomasociety.org/toc/Gonioscopy(NM).pdf
17. Ferreras Antonio, "Glaucoma Imaging" Springer ISBN: 978-3-319-18958-1. DOI: 10.1007/978-3-319-18959-8, 2016.

18. National Eye Institute. *Caption: Series of four photos demonstrating typical progression of vision loss due to glaucoma.* Available: http://www.visionaware.org/info/your-eye-condition/glaucoma/patients-guide-to-living-with-glaucoma/125
19. Álvarez Rocío, Torres Sebastián, "Simulación y modelamiento del ojo humano como herramienta para la prevención del Glaucoma a través de la medición de la Presión Intraocular". Universidad Politécnica Salesiana, 2015

Model Fitting and Simulation of the Respiratory Control System under Incremental Exercise and Altitude in Healthy Subjects

C.A. Sarmiento[1], A.M. Hernández[1] and L.Y. Serna[2,3]

[1] Bioinstrumentation and Clinical Engineering Research Group - GIBIC, Bioengineering Department, Engineering Faculty, Universidad de Antioquia UdeA; Calle 70 No. 52-21, Medellín, Colombia
[2] Department of Automatic Control, ESAII, Biomedical Engineering Research Center, CREB, Technical University of Catalonia, Barcelona, Spain
[3]Biomedical Research Networking Center in Bioengineering, Biomaterials and Nanomedicine, CIBER-BBN, Barcelona, Spain

Abstract— **The modeling and simulation of the respiratory control system allows a better understanding of the physiological system to different stimuli and pathologies. Although different models have been validated under different situations, it is unusual to perform a fitting of their parameters based on environmental characteristics such as altitude, which can significantly affect the accuracy of the simulation results. The purpose of this work is to fit the respiratory control model proposed by Fincham and Tehrani with experimental data obtained under exercise. The experimental test include carbon dioxide production between 0.4 and 2.5 l / min at an altitude of 1538 m above sea level. The variables compared between simulation without fitting and with fitting regarding experimental data were total minute ventilation, arterial partial pressure of oxygen, arterial partial pressure of carbon dioxide and Frequency of breathing. The results evidence the altitude as a stimulus that should be considered in simulations of respiratory models in order to obtain more accurate results.**

Keywords— **Computer simulation, Biological system modeling, Parameter estimation.**

I. Introduction

The respiratory control system is recognized as a highly nonlinear and dynamic system that is constantly disturbed by physiological and pathological changes [1]. It has been proposed different respiratory control models that have allowed a better understanding of the roles and physiological relations based on the simulation of their response to different stimuli and pathologies [2][3], in addition to promote the development of training and medical education tools [2][4], clinical diagnosis and treatment [5] and the medical equipment design [6][7].

Although several respiratory control models have been validated and studied under different situations, the parameter values for their computational simulations usually correspond to the specific experimental conditions used to develop the model [2][8], limiting its application to populations with the same characteristics and environmental conditions.

In this context, the altitude variation can be considered as a significant stimulus for human physiological system, mainly because of its direct relationship with the partial pressure of inspired oxygen and therefore its gas concentration in blood and organs such as the brain [9]. Variations of this stimulus are typically simulated by reducing the fraction of inspired oxygen at sea level [2], but this results are only valid to simulate the response of subjects adapted to sea level pressures, not for those acclimated to different altitude conditions [9].

In this paper a sensitivity analysis of the respiratory control model to variation on its plant parameters was initially performed in order to select which of them are the most representative regarding the system outputs. Then, the selected parameters were fitted with experimental data obtained in an exercise test where carbon dioxide production rate was between 0.4 and 2.5 l/min in all sample population at an altitude of 1538m above the sea level. Finally, simulation results for the same model with fitted and nominal parameters regarding the experimental data were compared in terms of variables of interest: total minute ventilation ($\dot{V}_E$), partial pressure of arterial oxygen (PaO_2), partial pressure of arterial carbon dioxide ($PaCO_2$) and breathing rate (f).

The implemented model of respiratory control system corresponds to the proposed by Fincham and Tehrani [10], which is one of the most studied and validated, being also suitable for simulating stimulus such as exercise and altitude in population belonging to sea level behavior [2].

II. Methodology

A. Respiratory control system model

The respiratory control system model implemented corresponds to the proposed by Fincham and Tehrani [10]. This model is based on a self-adaptive control that allows to calculate the respiratory pattern variables every respiratory cycle, and understand the results of peripheral processes as

I. Torres et al. (eds.), *VII Latin American Congress on Biomedical Engineering CLAIB 2016, Bucaramanga, Santander, Colombia, October 26th -28th, 2016*,
IFMBE Proceedings 60,
DOI: 10.1007/978-981-10-4086-3_144

gas exchange in the lungs, in the tissues and in the brain, delays in the transportation due the blood circulation and the blood-gas dissociation relationships [11][12]. This model is composed by two main compartments corresponding to the plant and the controller, whose elements are shown in the diagram of the Figure 1.

The chemical behavior of the system is comprised by the model plant, that integrates the activity in the lungs, brain and physiological tissues to changes in the concentrations of oxygen and carbon dioxide, feeding back their results to the controller via peripheral and central chemoreceptors [12].

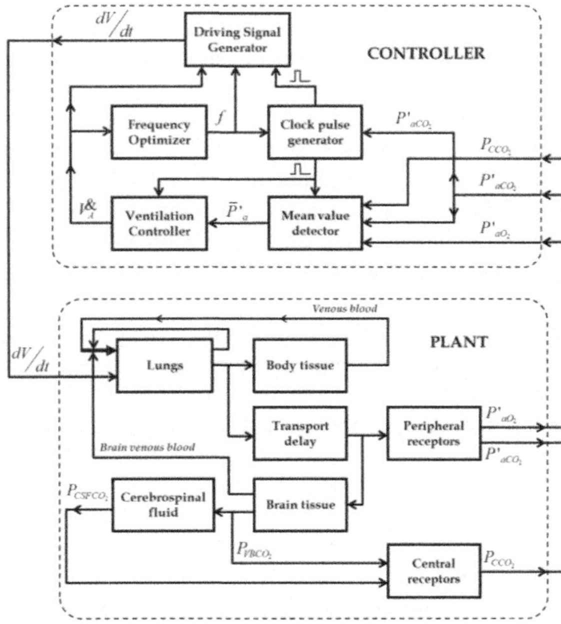

Fig. 1 Block diagram of the respiratory control system model proposed by Fincham and Tehrani [2].

In this sense, $d, K_1, K_2, K_3, K_4, K_5, S_T$ and S_B are the plant parameters in [10], related with the blood-gas dissociation relationships and carbon dioxide transduction in the central receptors, and are associated with the feedbacks of peripheral and central chemoreceptors. Parameter d represents the depth of central receptor below surface of the medulla, S_T and S_B correspond to the equivalent gas storage in the tissue and brain, K_1, K_2, K_3, K_4 and K_5 are constants that represent oxygen blood-gas dissociation, carbon dioxide blood-gas dissociation, central receptor and carbon dioxide diffusion time.

The respiratory controller implemented corresponds to a self-adaptive control that acts discreetly even though the plant does continuously, self-adjusting the breathing pattern, breathing frequency and dead space volume only at the end of each respiratory cycle, from the average value of the signals received from the chemoreceptors and based on the optimization of work breathing [12].

The breathing frequency is calculated based on the minimum respiratory work according to Otis et al [13], although there are variations of this equation that could be more appropriate for the case of moderate exercise, as the equation proposed by Mead [14]. The latter is based on the optimum integral time-inspiratory pressure as a measure of the cost of inspiration energy performed by the respiratory muscles. Equation 1 shows the Mead's equation [14], where E_{rs} is the pulmonary elastance, R_{rs} is the airway resistance, $\dot{V}_A$ is the alveolar ventilation and V_D is the dead space volume.

$$f = \sqrt[3]{\frac{E_{rs}\dot{V}_A}{4\pi^2 R_{rs} V_D}} \quad (1)$$

B. Experimental trial

The experiment consisted of a test of moderate exercise at an altitude of 1538 m above sea level. The test was performed by 13 healthy male volunteers with a mean age of 27 ± 4.0 years, weight of 78 ± 6.2 kg and height of 174 ± 8.6 cm. The measurements were taken at rest and under different levels of exercise by changes in the programmed load of a cycle ergometer, for a total experimental time of 30 minutes divided in 10 stages of three minutes each one, as described in Table 1. The information measured corresponds to the production rate of carbon dioxide ($\dot{V}CO_2$), consumption rate of oxygen ($\dot{V}O_2$), respiratory frequency (f), tidal volume (V_T), total minute ventilation ($\dot{V}_E$), end tidal oxygen tension ($PETO_2$), end-tidal end tidal carbon dioxide tension ($PETCO_2$) and barometric pressure (PB), registered with the Oxycon Mobile metabolic monitoring equipment [15].

Table 1. Description of the test stages

Stage	Description	Test Duration (min)
1	Rest	3
2	Warming	3
3 - 9	Increase of exercise load	21
10	Recovery	3

The measurements of $PETO_2$ and $PETCO_2$ in units of mmHg can be related with PaO_2 and $PaCO_2$ according to the equations 2 [2] and 3 [16].

$$PaCO_2 = PETCO_2 - 3.6 \quad (2)$$

$$PaO_2 = (PETO_2 * 1.5) - 6.53 \quad (3)$$

C. Sensitivity Analysis

In order to identify which parameters of the model plant are the most representative to be fitted, a sensitivity analysis of the system outputs regarding parameter variations was performed. Thereby, independent simulations of parameter variations between ± 50% of their nominal values were implemented, and the combined sensitivity was measured using the equation 4:

$$J = \left(PRD(f) + PRD(\dot{V}_E) + PRD(PaCO_2) + PRD(PaO_2)\right)^2 \quad (4)$$

The Equation 4 relates the variables of interest and additionally corresponds to a part of the cost function that should be minimized in the fitting process. PRD is the Percentile Root-mean-square Difference value that is shown in Equation 5, where $x_1(n)$ correspond to the reference experimental variable, $x_2(n)$ to the evaluated variable when the parameter changes and N is the number of compared elements.

$$\mathrm{PRD} = \sqrt{\frac{\sum_{n=1}^{N}(x_1(n)-x_2(n))^2}{\sum_{n=1}^{N}x_1{}^2(n)}100} \quad (5)$$

D. Model Fitting

The fitting of the selected parameters was implemented using the optimization technique of Sequencial Quadratic Programming (SQP). This is a deterministic algorithm based on the gradient technique and provides a method of nonlinear optimization applied to problems where the objective function and constraints are continuously differentiable [2].

The Equation 6 shows the total cost function that should be minimized in the fitting process, were J_i corresponds to the cost function shown in the Equation 4 applied to the N sets of the experimental data.

$$J_T = \left(\sum_{i=1}^{N} J_i\right)^2 \quad (6)$$

Then the simulation model with the fitted and nominal parameters was performed, in order to compare with experimental data the results for $\dot{V}_E, f, PaCO_2$ and PaO_2 in function of $\dot{V}CO_2$.

The Equation 1 was implemented as the system control equation and the values of barometric pressure, $\dot{V}_A$ at rest and basal metabolic rate were determined from the experimental data. The values of the other implemented parameters correspond to nominal values [10].

III. Results

A. Sensitivity Analysis

Figure 2 shows the results for the parameter sensitivity analysis.

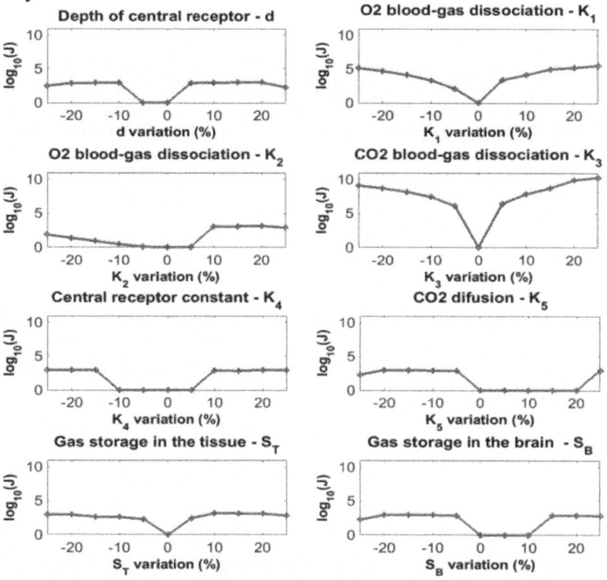

Fig. 2 Sensitivity analysis results for $d, K_1, K_2, K_3, K_4, K_5, S_T$ and S_B

B. Model Fitting

Table 2 shows the results of the fitting process, whose sensitivity analysis was the most representative (K_1 and K_3).

Table 2. Fitting of the parameters

Parameter	Nominal value [10]	Fitted value
K_1	0.200	0.0258
K_3	0.016	0.0120

Table 3 shows the mean value of PRD regarding to the experimental data for the fitted and unfitted model.

Table 3. Mean value of PRD for the fitted and unfitted model

Parameter	Fitted model	Unfitted model
f	32.8 ± 13.2	29.2 ± 10.2
$\dot{V}_E$	38.7 ± 4.7	43.2 ± 4.8
PaO_2	18.7 ± 2.8	32.6 ± 3.2
$PaCO_2$	28.0 ± 14.7	26.8 ± 14.1

Figure 3 shows the interest variables ($\dot{V}_E, f, PaCO_2, PaO_2$) in function of different levels of exercise ($\dot{V}CO_2$) for the fitted model, the unfitted model and the average of the experimental data.

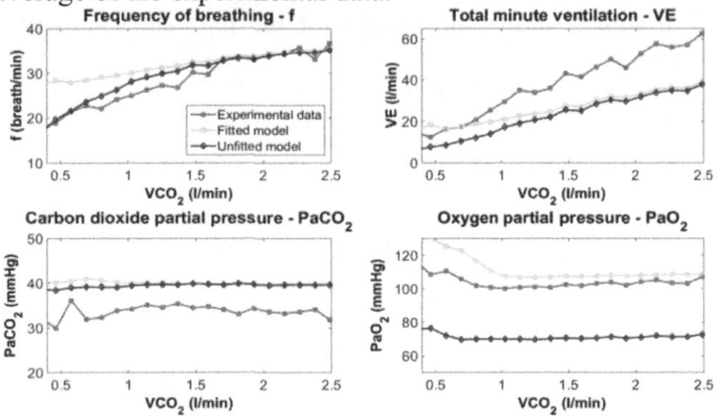

Fig. 3 Values of $\dot{V}_E, f, PaCO_2, PaO_2$ in function of $\dot{V}CO_2$. Green correspond to the fitted model results, blue correspond to the unfitted model results and red correspond to the average of the experimental data

IV. Discusion and Conclusions

Concerning the comparison of results, stands out mainly the improved of fitted model referring to the prediction capability of PaO_2, getting a reduction of approximately 42% of the average error compared to the unfitted model. This result evidence the altitude as a stimulus that should be considered in simulations of respiratory models in order to obtain more accurate results.

As for the other variables, there are not significant improvements, result that could be related to the selection of K_1 and K_3 as fitting parameters. Even though these parameters have the most significant effect on the system regarding the analyzed outputs, their fitting is related mainly to the blood-gas regulation.

Additionally, it was also observed that the results of the fitted model are more similar to the unfitted model with increasing of $\dot{V}CO_2$, which may indicate relationship of K_1 and K_3 with exercise changes.

The error reduction, particularly in the case of PaO_2, and the differences between the simulations results and the experimental data according to the exercise levels, show an open research field concerning the structure of chemical plant at altitude and the respiratory control system under changes of exercise stimulus.

Acknowledgment

This work has been supported in part by "Convocatoria 647 – Programa de Becas de Doctorado Nacionales 2015 de Colciencias" and the project "Fortalecimiento de plataforma tecnológica para la formación especializada en el área de la salud y el desarrollo de tecnología biomédica", code Ruta N 139C of the Fondo Nacional de Regalías, Republic of Colombia.

Conflict of Interest

The authors declare that there is no conflict of interest regarding the publication of this paper.

References

1. Poon C (2000) Respiratory Models and Control. In Bronzino J (ed) The Biomedical Engineering Handbook, 2nd Edn. Boca Ratón, United states
2. Hernández M (2014) Sistema de control respiratorio ante estímulos y patologías: Análisis, Modelado y Simulación. Alemania
3. Tehrani F (1998) Optimal control of respiration in exercise. Eng Med Biol 6–9
4. Serna L, Hernández A, Mañanas M (2010) Computational tool for modeling and simulation of mechanically ventilated patients, Proc. vol. 2010, Annual International Conference of the IEEE Engineering in Medicine and Biology 2010, pp. 569–572
5. Ellwein L, Pope S, Xie A et al. (2013) Patient-specific modeling of cardiovascular and respiratory dynamics during hypercapnia. Mathematical biosciences 56–74
6. Tehrani F (1999) Automatic control of mechanical ventilation and the inspired fraction of oxygen in the premature infant: A simulation study, Proc. vol. 1, Proceedings of the First Joint BMES/EMBS Conference., pp. 339
7. Tehrani F (2001) A control system for oxygen therapy of premature infants, Proc. vol. 2, 23rd Annu. Int. Conf. IEEE Eng. Med. Biol. Soc., pp. 2059–2062
8. Mananas M, Hernandez A, Romero S et al. (2003) Analysis of respiratory models at different levels of exercise, hypercapnia and hypoxia, Proc. vol. 3, 25th Annu. Int. Conf. IEEE Eng. Med. Biol. Soc., pp. 2754–2757
9. Lumb B (2010) Nunn's Applied Respiratory Physiology
10. Fincham W, Tehrani F (1983) A mathematical model of the human respiratory system. J. Biomed. Eng. 125–133
11. Fincham W, Tehrani F (1983) On the regulation of cardiac output and cerebral blood flow. J. Biomed. Eng. 73–75
12. Tehrani F (1993) Mathematical analysis and computer simulation of the respiratory system in the newborn infant. IEEE Trans. Biomed. Eng. 475–481
13. Otis B, Fenn W, Rahn H (1950) Mechanics of breathing in man. J. Appl. Physiol. 592–607
14. Mead J (1960) Control of respiratory frequency. J. Appl. Physiol. 325-336
15. Oxycon Mobile at http://www.carefusion.com/our-products/respiratory-care/metabolic-carts/oxycon-mobile-device
16. Bengtsson J, Bake B, Johansson A et al (2001) End-tidal to arterial oxygen tension difference as an oxygenation index. Acta Anaesthesiol. Scand 357–363

Author: A.M. Hernández.
Institute: University of Antioquia
Street: Calle 67 No. 53 - 108
City: Medellín
Country: Colombia
Email: alher.hernandez@udea.edu.co

Supporting Diabetic Patients with a Remote Patient Monitoring Systems

S. Zulj[1], G. Seketa[1], D. Dzaja[1], F. Sklebar[1], S. Drobnjak[2], L. Celic[1] and R. Magjarevic[1]

[1] University of Zagreb, Faculty of Electrical Engineering and Computing, Zagreb, Croatia
[2] S. D. Informatika, Velika Gorica, Croatia

Abstract—**In order to achieve better self-control of their diabetes and to decrease long-term risks of complications, diabetic patients monitor several different parameters on a day-to-day basis. These parameters, such as blood glucose level, insulin intake, weight, diet, exercise and physical activity, blood pressure, can be easily acquired using IT technologies without oppressing the patient with handwritten diaries. At the same time, healthcare professionals are provided with the data on time and can intervene without delay, if necessary. The system collects data on blood glucose level using special device for communication between glucose meter and either a smartphone or a PC. A person's weight is acquired using a modified body scale which, while weighing, at the same time scans patient's feet and stores images for further processing and comparison with the images from pressure measurement platform. Patients enter their own diet and therapy using a smartphone app. Information on exercise and physical activity is provided using WBAN (Wireless Body Area Network) system with complementing software, both developed at our University. The system supports patients while performing previously determined exercise plan without the immediate presence of trainer, and also acquires data on regular daily activity (walking, running, sitting,…). The parts of the system providing some of the parameters are either developed in our previous studies or commercial, and they are integrated into our e-system. After collecting, the data is stored in a database for further use by diabetic patients themselves or by their healthcare providers.**

Keywords—**diabetes management, remote monitoring, monitoring system.**

I. Introduction

WHO reports that the number of diabetic patients has almost quadrupled since 1980 to 422 million adults. Almost 50% of adults with diabetes are undiagnosed. Complications caused by diabetes can lead to heart attack, stroke, kidney failure, leg amputation, vision loss and nerve damage. In addition, people with diabetes also have a higher risk of developing infections [1]. Diabetes and its complications bring substantial economic loss to health systems and national economies through direct medical costs and loss of work and wages [2].

Regardless of the type of diabetes, WHO recommends cost-effective interventions which can improve patients' outcomes through a combination of diet, physical activity and medication. Also, it is recommended for patients to control blood pressure and lipids to reduce cardiovascular risks and complications, as well as screening for damage to the eyes, kidneys and feet. Early diagnosis and good diabetes management are a key factor in reducing the risk of complications.

Patients should monitor their blood glucose levels and achieve a specific level of glycemic control, to prevent hypoglycemia and to help healthcare professionals to adjust their treatment and advise correctly. Handwritten diabetic diaries are proven to often have incorrect and/or lacking data, and are therefore useless for clinical use [3].

In our earlier feasibility study [4], we have shown the benefits of integration of different measurement devices into one system which will increase the accurate number of data for clinicians and at the same time assist patients in self-managing their disease on day-to-day basis.

II. Monitoring and Acquisition e-System

The monitoring and acquisition e-system recently developed in our group has several modules described in the next chapters.

A. Day-to-day blood glucose levels acquisition

Module for acquisition of self-monitoring of blood glucose measurements consists of two specialized devices, both developed at our University, which connect most available commercial glucose meters and either a PC or a smartphone, and corresponding software. Both iOS and Android applications have been developed.

The two devices for connecting glucose meters have interfaces to allow connection with different glucose meters which can vary in appearance, memory capabilities, hardware interface and communication protocol. Therefore, communication protocols are implemented for each supported glucose meter. Currently, more than 25 different types of glucose meters are supported for use with the device for PC, and around 10 types for smartphone use. Although over 25 types are supported in PC app, we have shown in previous work that around 70% of the data were obtained from only three types of glucose meters [5].

The device which connects a mobile phone with a glucose meter primarily supports patients in personal use and it is called *mobile version*, whereas the device which connects to

I. Torres et al. (eds.), *VII Latin American Congress on Biomedical Engineering CLAIB 2016, Bucaramanga, Santander, Colombia, October 26th -28th, 2016,*
IFMBE Proceedings 60,
DOI: 10.1007/978-981-10-4086-3_145

PC and corresponding software are used by physicians and it is called *clinical version*. Physicians and patients are using the devices to acquire data from the glucose meter to PC/smartphone. The data are sent and permanently stored in the database after the acquisition. This allows physicians to have access to the patients' measurement without immediate appointment.

Figure 1 is showing smartphone application module which guides patients through the process of blood glucose data acquisition.

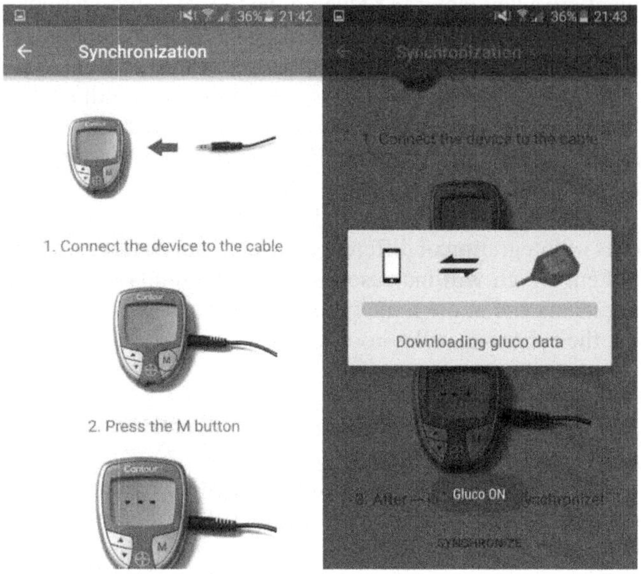

Fig. 1 Smartphone app module for blood glucose data acquisition

All stored data can be accessed and viewed by authorized personnel and patients themselves in visualization tools in each version of the app. Statistics such as total number of blood glucose measurements, minimal, maximal and average values, etc. and important glycemic variability and the number of hypoglycemic episodes as recommended by ADA [6] are offered for reviewing. Two most important graphic tools show *standard day graph* (Fig. 2), which shows measurements overlapped over 24-hour period, and *trend graph* (Fig. 3), which shows measurement on a timeline. Both graphs are widely used by clinicians as assistance when adjusting therapy and/or assessing patients' glycemic control, and are available in *the clinical version* of the app. Patients use smaller smartphone screens and only *trend graph* is available in the app.

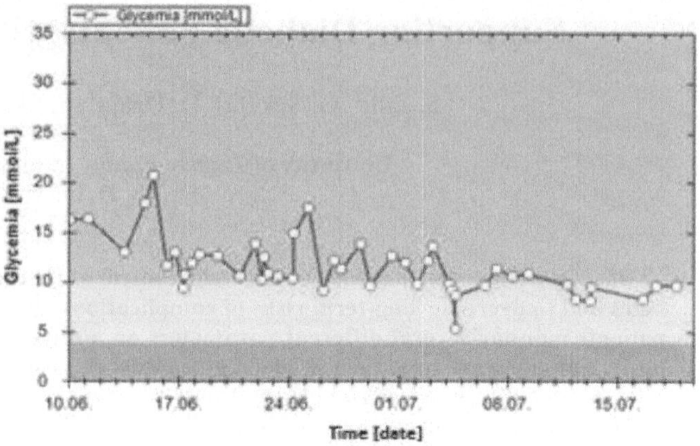

Fig. 2 *Trend graph* in *clinical version*

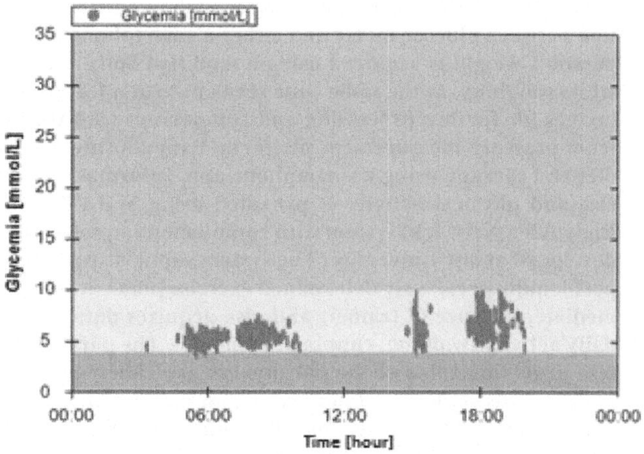

Fig. 3 *Standard day graph* in *clinical version*

B. Exercise module

The module for exercising developed at our University comprises WBAN and corresponding software [7]. Each node in WBAN has integrated sensor chip InvenSense MPU-9150, which contains an accelerometer, gyroscope and magnetometer in a single package, microprocessor Nordic Semiconductor nRF51822 for data gathering with Bluetooth LE V4.1 module for communication and Li-Ion battery. Nodes are low power and rechargeable.

The system is based on body segment orientation estimation using a rotation estimation algorithm on the microprocessor. Raw data from the sensors are processed and a quaternion description of orientation is given as an output. Each node in WBAN is positioned on the body corresponding segments in a Hanavan's model of the human body which consists of 15 rigid segments. It is possible to use different number of sensor nodes depending on the exercise, e.g. arm strength exercise can be performed by using only two sensor nodes, whereas at least 9 sensor nodes are necessary for the whole body motion capture.

Exercises and exercise plans are predefined and prerecorded by clinicians or trainers. Patients can perform exercise using the exercise module in the mobile app. The virtual scene is composed using the Unity game engine and it contains two virtual characters set side by side or one above the other (Fig. 4); *virtual trainer,* which guides patient during exercise and therefore compensate immediate presence of actual trainer, and patient's *avatar* – a personal representation in a virtual scene, which allows patients to compare their position and movement in real time [8].

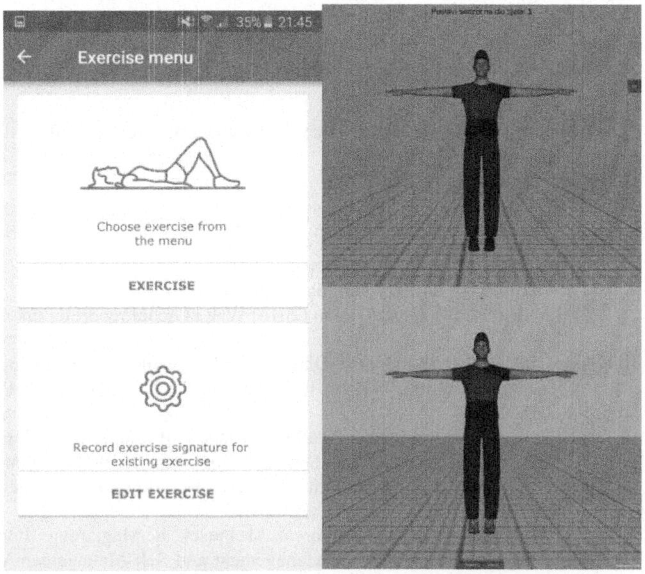

Fig. 4 Smartphone app module for assisted exercise

At the beginning of exercise series, the patient is guided through the process of positioning sensor nodes on the required position on the body one sensor node at the time. Sensor node which needs to be place signals with a vibration and visual guidance shows the patients where it needs to be placed on the body. After the positioning of last sensor node, the patient needs to stand in an initial position for a second for segment orientation estimation, and the assisted exercise is then ready to start.

Virtual trainer has prerecorded exercises and assists patients during the exercise. Every movement is qualitatively assessed comparing the movements of the two characters. After every movement patient is shown evaluation mark in the form of color; *green* for correct movement, *red* for incorrect and *yellow* for satisfactory. At the end of each cycle (e.g. 12 arm lifts in a set make one cycle), the patient is given a quantitative assessment as a number of correct movements in the entire cycle. That assessment is also stored in a database for further evaluation and to show progress over time.

Qualitative assessment takes into consideration the level of the patients in terms of ability to perform certain movements. The threshold is preset for beginners, intermediate and expert users in such manner that it allows beginners to deviate more from the prerecorded exercises performed by *the virtual trainer* than expert users. Threshold changes as the patient progresses in performing defined exercises.

C. Physical activity module

During the periods that the patient is not performing an exercise, one sensor node described in the previous chapter can be worn on the belt or in a pocket to capture the level of physical activities. The data is stored as the highest level of activity in the set intervals (e.g. 5 minutes).

D. Diet and therapy module

Diet and therapy can be entered using mobile app. The app contains lists of meals as well as the medications for users to choose from, or they can enter new meal/therapy themselves.

E. Foot scanner and body weight module

The modified body weight scale includes conventional body weight scale, body composition measurement and scanner. Body composition measures body fat, muscle, amount of water and bones percentage, and calculates BMI. Body weight scale has glass surface and is therefore suitable for placing image scanner underneath. A person's weight is acquired using a modified body scale which, while weighing, at the same time scans patient's feet and stores images for further processing and comparison with the images from pressure measurement platform.

F. Blood pressure

Both PC and mobile app offers patients to enter measured blood pressure. There are a number of blood pressure meters which enable Bluetooth connectivity and the option to connect to the system they will be included in the further development of the system. Also, there is a limited number of devices which integrate blood glucose and blood pressure measurements.

G. ECG monitor

The developed ECG monitor has similar features as aforementioned sensor nodes modified to accommodate module with ECG recorder. It is based on the 16-bit dual-channel sigma-delta analog-to-digital (A/D) converter Texas Instruments ADS1192. The A/D converter has a specific ECG functions such as detection of separation of the electrode from the patient and active right leg drive, which increases the common mode rejection ratio signal at the input. The A/D converter itself has a declared rejection ratio of 105 dB of the interference caused by the power network.

A segment of recorded signals during the walk up the stairs with the ECG node is shown in Figure 5. The digitized signals are stored on a microSD card for further processing.

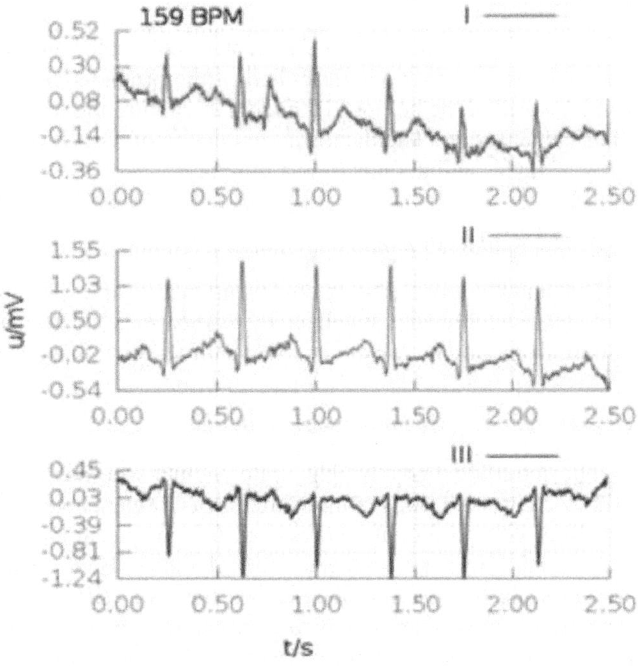

Fig. 5 Recorded ECG signal (I, II and III lead)

III. Conclusions and Future Work

We have integrated different measurement devices, either commercial or developed at our University, into the e-System for monitoring diabetic patients. This system has a great potential benefit for both healthcare professionals and patients. It facilitates entering of the data for patients, which may lead to more accurate and more frequently entered data. Also, physicians are provided with better quality data which has benefits for patients' therapy. Moreover, patients can enter the data, which immediately becomes available for physicians, between the appointments and thus shorten the time of needed reaction in case of unusual behavior.

In our future work, we are going to consider adding additional devices which could bring more useful data into the consideration. Smart algorithms for prediction of changes in data in order to alarm patients and physicians on time are also considered.

Acknowledgment

Development of a module for assisted exercise and device for the transfer of blood glucose levels to smartphone and complementing software has been partially supported by the European Regional Development Fund under the project "Technology platform for new ICT strategies in diabetes therapy and control" agreement n° RC.2.2.08 0010.

Conflict of Interest

The authors declare that they have no conflict of interest.

References

1. Complications of Diabetes. Retrieved July 4, 2016, at http://www.idf.org/complications-diabetes
2. World Health Organization (2016) WHO Global report on diabetes.
3. M. Prasek, "Self Control Diary – Challenges of New Technological Possibilities", IFMBE Proceedings 37, Budapest, 2011, p.972-973
4. L. Celic; D. Trogrlic; I. Paladin; M. Prasek; R. Magjarevic. "Integration of Measurement Devices Supporting Diabetic Patients into a Remote Care System", IFMBE Proceedings 37, Budapest, 2011, p.39-42.
5. S. Zulj, L. Celic, M. Grgurevic, M. Prasek, R. Magjarevic: Pilot Project: ICT System for Management and Self-Management of Diabetes, International Conference on Biomedical and Health Informatics, Haikou, China, October 2015, in press
6. Glycemic Targets, Diabetes Care 2015; 38 (Suppl. 1):S33–S40, American Diabetes Association, 2015.
7. D. Dzaja, M. Varga, G. Seketa, S. Zulj, L. Celic, I. Lackovic, R. Magjarevic: System for Assisted Exercising and Qualitative Exercise Assessment, 6th European Conference of the International Federation for Medical and Biological Engineering, IFMBE Proceedings vol 45, p 682-686, 2014.
8. S. Zulj, G. Seketa, D. Dzaja, L. Celic, R. Magjarevic: Virtual Reality System for Assisted Exercising Using WBAN, 6th European Conference of the International Federation for Medical and Biological Engineering, IFMBE Proceedings vol 45, p 719-722, .2014

Author: Sara Zulj
Institute: University of Zagreb
Street: Unska 3
City: Zagreb
Country: Croatia
Email: sara.zulj@fer.hr

Climbing/Descending Stairs Detection Using Inertial Sensors and Implementing PCA and a SVM Classifier

R. Alvarez[1], E. Pulido[1] and D.A. Sierra[1]

[1] Universidad Industrial de Santander, School of Electrical,
Electronics and Telecommunications Engineering, 680002, Bucaramanga, Colombia
roberth.alvarez@correo.uis.edu.co, epulidoh@saber.uis.edu.co, dasierra@uis.edu.co

Abstract— **Activity classification have been used in different fields such as energy expenditure measurement or health monitoring. Many combinations of different sensors and machine learning techniques have been proposed in order to do this kind of classification. The aim of this paper is to introduce an activity classification approach for Climbing/Descending stairs detection divided in two phases. In the first phase the signals from accelerometer and gyroscope are filtered, then implementing step detection allows us to extract the relevant features from these signals. The second phase consists of a principal component analysis (PCA) for reducing dimensionality, and a support vector machines (SVM) classifier to identify the motion. Using this methodology, an accuracy of 98.76% is achieved. The data used for classification were taken from an inertial measurement unit carried by three users in their ankles, which was provided by a database from the UCI machine learning repository.**

Keywords— **Support vector machines, Principal component analysis, Activity recognition, Inertial measurement unit.**

I. Introduction

In the last twenty years, inertial measurement units (IMU) have gained popularity due to its low invasiveness, size and cost, giving an useful alternative for obtaining information about human motion. These devices are compounded of accelerometers and gyroscopes, and in some cases magnetometers and barometers are included for an additional analysis. These sensors have been used in areas such as human activity recognition, pedestrian localization, physical activity monitoring, rehabilitation and health monitoring, among others [1].

In the case of the Human Activity Recognition (HAR), several activities have been studied such as walking, jogging, running, climbing/descending stairs, playing sports, cooking and irregular events such as falling. For this purpose, machine learning and pattern recognition techniques have been used to obtain high accuracy in classification [2].

We propose the support vector machines (SVM) as classification method, due to its capacity of low computational cost when high dimensions of data or non linear separation are required. This is possible because SVMs work with dot products and not with other complex operations, creating linear or non linear planes for separating the data. It is necessary to modify the data extracted before training the SVM, because it works with discrete samples and IMU data are a sequential signal, then we do a step detection for extracting the features every time the user takes a step.

The IMU data used in this work were taken from the UCI machine learning repository [3]. The IMU was placed on the user's ankle and the sensor was oriented as it can be seen in Figure 1. Data from three different people were taken while they were walking forward, climbing and descending stairs.

The paper begins with a short review of the related work, then the methodology proposed is presented, showing the application of principal component analysis (PCA) and the designing process of the SVM classifier. By last, the results of the classification are shown and discussed.

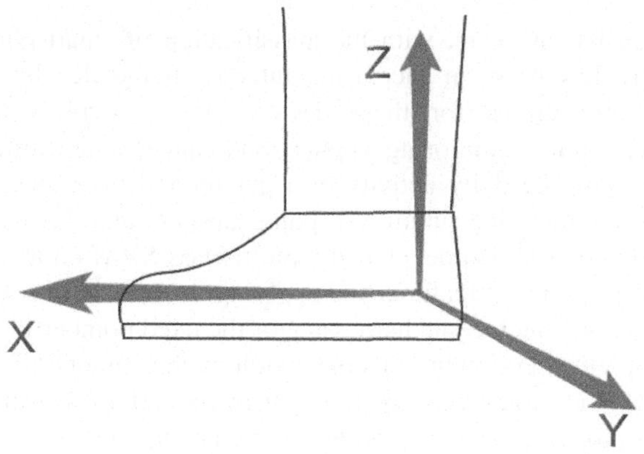

Fig. 1 IMU placement on user's foot

I. Torres et al. (eds.), *VII Latin American Congress on Biomedical Engineering CLAIB 2016, Bucaramanga, Santander, Colombia, October 26th -28th, 2016*,
IFMBE Proceedings 60,
DOI: 10.1007/978-981-10-4086-3_146

II. Related work

Several HAR methods and features sets have been explored for detecting when a person is climbing stairs. One of the biggest challenges in HAR is selecting the features that best describe the desired activities. In order to achieve a reduced set of features, techniques such as principal component analysis (PCA), linear discriminant analysis (LDA) and independent component analysis (ICA) are implemented [4].

Activity classification can be conducted using the features extracted and then comparing them with thresholds, in a process that gives information about the person's motion. Lee et al. propose an algorithm based in thresholds for first responder localization combined with an HAR system [5], this method can identify walking, running, climbing stairs or crawling. For the climbing stairs detection they use six features (energy, mean value, variance, product of signals, among others), obtaining a 90% of accuracy. Machine learning techniques as Hidden Markov models (HMM) are also popular for classifying activities due to its capacity for analyzing sequential signals [6]. In the work presented in [7], HMM is used for estimating the likelihood of the acceleration data obtained from an IMU, then SVMs are used to classify ten different activities (walking, running, sitting to a chair, among others). By using this fusion, a 99% of accuracy is achieved and the data were all labeled "by hand". Furthermore, in the work presented by [8] they compare a Bayes classification method (BCM) with SVM for detecting stairs climbing and descending. They make a 2D projection of 16 features (statistical parameters, and magnitude values) using Sammong's mapping, obtaining a better result using BCM but requiring a priori information that is not available in all cases.

In recent years, with the massification of smart-phones, there has been an increasing interest in the development of applications for these devices. For example, energy expenditure monitoring applications can be interesting for mapping the daily activity of a person and his capacity to do activities like running or going upstairs and downstairs. Anguita et al. [9] developed a multi-class SVM on a smartphone using 17 features and applying the hardware friendly concept, which is the harnessing of the fixed point arithmetic properties. Following this methodology the authors obtained a 89% accuracy classifying six different activities (walking, upstairs, downstairs, standing, sitting and laying).

A classification system as SVM requires the processing of a big group of features, so it will be important to reduced these groups providing an advantage in time processing. We propose a PCA and SVM classifier implementation to detect walking in stairs using the features suggested by Lee et al [5]. This is done by looking for a dimensionality reduction through PCA, and the implementation of a robust classification method as SVM to classify the data accurately.

III. Methodology

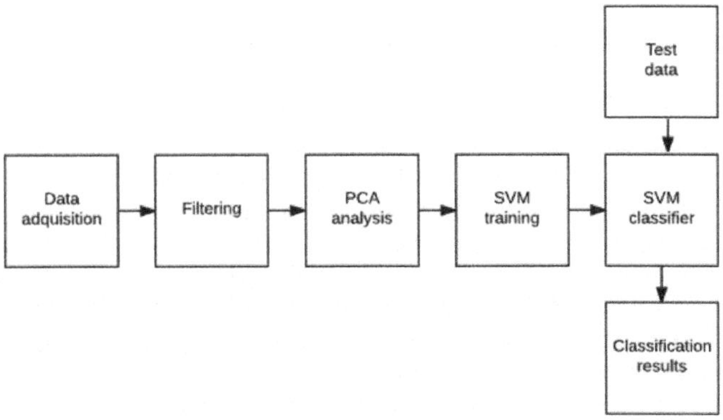

Fig. 2 Methodology followed in this work

In this section, we describe a detailed methodology (as seen in Figure 2) of stairs climbing recognition starting with the preprocessing of the signal from accelerometer and gyroscope. We also describe the PCA and the SVM implementation.

A. *Signal preprocessing*

Once the signal from accelerometer and gyroscope are recorded, it is necessary to filter the signal eliminating undesired high frequency noise. Ladetto [10] establishes that all the frequencies associated with human walking are in the 20 Hz range. Then, a 20 Hz filter proposed by Ying et al. [11], is applied:

$$Hz = \frac{1(1-z^{-4})^2}{16(1-z^{-1})^2} \tag{1}$$

B. *Feature selection and extraction*

Lee et al. [5], suggest six features for the detection of climbing or descending stairs, these features are listed in Table 1. In this work, these features are normalized for correcting the scale of the data values. The signals obtained by the IMU are sequential and cannot be used directly with Support Vectorial Machines (SVM), consequently for extracting the features it is necessary to have discrete samples. For this purpose, a step detection module is implemented to calculate the features required every time the user takes a step. There are different forms to detect steps such as flat zone detection, zero cross detection and peak detection [12]. We use a peak detection method setting a restriction on how near a step could be from another, avoiding the detection of false peaks.

Table 1 Features defined for classifying

Item	Feature	Axis
X1	Max. angular velocity	Y
X2	Acceleration variance	Y
X3	Mean Energy value	X, Z
X4	Energy variance	X, Z
X5	Acceleration variance	Y
X6	Mean acceleration value	Z

C. Principal component analysis

A good description of the activities could be provided using the features in Table 1, but it implies computational cost and cannot be visualized due to the high dimensionality of the data. Our proposal consist of using the PCA for reducing the dimensionality of the data from six to two features. This reduction allows to observe the relation between the samples of different classes in 2D, wich is relevant to select the classifier that has to be implemented.

For implementing the PCA, the first step consists of obtaining the dataset covariance matrix. It is necessary to select the two eigenvectors associated with the biggest eigenvalues of this matrix, these ones are related with the variability of the data, and the principal components will be oriented by this eigenvectors. The 2D representation of the data corresponds to its projection over the principal components and can be seen in Figure 3.

D. SVM classifier

An SVM classifier is implemented for classifying the data obtained after the PCA. As a result, it is possible to see in Figure 4 that the data can be separated linearly, therefore no transformation or linear kernel implementation is required.

IV. RESULTS

For validating the results, cross validation (CV) with four groups was conducted. The accuracy of the classifier

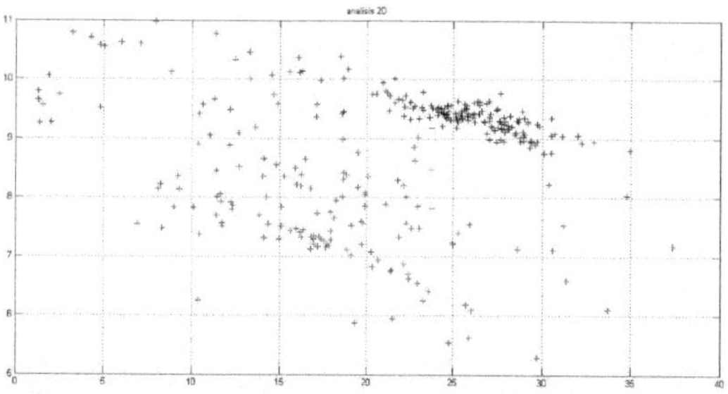

Fig. 3 2D visualization of walking forward (black) and stairs (red) using PCA

is calculated as amount of successes divided on the total number of test data. The accuracy obtained in every iteration of the CV is obtained for having a measurement of the classifier stability.

$$Acc = \frac{TP + TN}{Tdata} \tag{2}$$

Where TP are the true positives, TN the true negatives and Tdata the total data used for the testing.

When the CV is conducted, an accuracy of 98.76 % is obtained. The values of accuracy in each iteration can be observed in Table 2. These results show a maximum variation of 4% in the accuracy when the training and test data changes. The results of classification can be observed in Figure 4

Table 2 Accuracy in CV iterations

Iteration	Accuracy
1	100%
2	98.2%
3	96.3%
4	100%

V. CONCLUSIONS

In this paper, we propose a novel approach for detecting when a person is climbing stairs, using PCA and SVM showing a high accuracy.

The PCA allows us to obtain a 2D graphical visualization of the relation between the data, which is not possible having the six features. Moreover, reducing the dimensionality can be convenient to reduce the computational cost aiming to

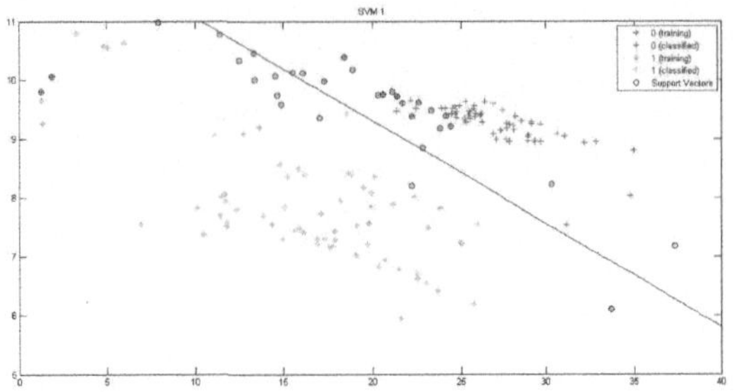

Fig. 4 SVM classification results for walking forward (red and pink) and walking in stairs (cyan and green)

a multi class implementation. Normalizing the data is an important step for avoiding scale disproportions between the features; if this is ignored, the eigenvalues of the covariance matrix of the data will not give the required information to calculate the principal components.

The SVM classifier does not required a kernel transformation, because the results indicate that the data can be separated linearly. From cross validation an accuracy of 98.76% was obtained, and the variation between the four iterations made is around 4%. These results indicate that the classifier is very stable as changes on training and test data do not show a significant effect on its performance.

For future work, an online multi-class robust algorithm for classifying others additional activities will be implemented.

CONFLICT OF INTEREST

The authors reports no conflicts of interests.

ACKNOWLEDGEMENTS

The authors would like to thank Juan Camilo Bravo and Clara Cañon for their support and witting assistance.

This research was party founded by the Colombian Administrative Department of Science, Technology, and Innovation (COLCIENCIAS) and the Universidad Industrial de Santander (UIS) (Grant FP44842-507-2014; Project 1817)

REFERENCES

1. Lowe S, ÓLaighin G. Monitoring human health behaviour in one's living environment: A technological review *Med. Eng. Phys.*. 2014;36:147–168.
2. Mannini A, Sabatini A. Machine learning methods for classifying human physical activity from on-body accelerometers *Sensors.* 2010;10:1154–1175.
3. Reiss A, Stricker D. Introducing a new benchmarked dataset for activity monitoring in *Proc. - Int. Symp. Wearable Comput. ISWC*no. June 2012(Newcastle, UK):108–109 2012.
4. Avci A, Bosch S, Marin-Perianu M, Marin-Perianu R, Havinga P. Activity Recognition Using Inertial Sensing for Healthcare, Wellbeing and Sports Applications: A Survey *Archit. Comput. Syst. (ARCS), 2010 23rd Int. Conf.*. 2010:1–10.
5. Lee M, Park C, Shim C. A Movement-Classification Algorithm for Pedestrian using Foot-Mounted IMU in *Proc. 2012 Int. Tech. Meet. Inst. Navig.*(Newport Beach, USA):922–927 2012.
6. Dietterich T. Machine learning for sequential data: A review *Struct. syntactic, Stat. pattern Recognit.*. 2002;2396:15–30.
7. Wan W, Liu H, Wang L, Shi G, Li W. A Hybrid HMM/SVM Classifier for Motion Recognition Using uIMU Data in *Proc. 2007 IEEE Int. Conf. Robot. Biomimetics*:115–120 2008.
8. Schmid M, Riganti-fulginei F, Bernabucci I, et al. SVM versus MAP on Accelerometer Data to Distinguish among Locomotor Activities Executed at Different Speeds *Comput. Math. Methods Med.*. 2013;2013:24–30.
9. Anguita D, Ghio A, Oneto L, Parra X, Reyes-Ortiz J. Human activity recognition on smartphones using a multiclass hardware-friendly support vector machine *Int. Work. Ambient Assist. Living.* 2012;7657:216–223.
10. Ladetto Q. On foot navigation : continuous step calibration using both complementary recursive prediction and adaptive Kalman filtering in *Proc. 13th Int. Tech. Meet. Satell. Div. Inst. Navig.*;2000(Salt Lake City, USA):1735–1740 2000.
11. Ying H, Silex C, Schnitzer A, Leonhardt S, Schiek M. Automatic Step Detection in the Accelerometer Signal in *4th Int. Work. Wearable Implant. Body Sens. Networks*;13(Aachen, Germany):80–85 2007.
12. Park J, Kim Y, Lee J. Waist mounted Pedestrian Dead-Reckoning system in *2012 9th Int. Conf. Ubiquitous Robot. Ambient Intell.*(Daejeon,Korea):335–336 2012.

Assessment protocol of wrist flexion and extension to support processes in occupational health using Myo Armband

V. Montoya-Leal[1], A. Orozco-Duque[2], J.P. Ugarte[1], M.A. Portela[1], J.C. Franco[1] and V. Z. Perez[1,3]

[1] Centro de Bioingenería, Universidad Pontificia Bolivariana, Medellín, Colombia
[2] Grupo de Investigación e Innovación Biomédica, Instituto Tecnológico Metropolitano, Medellín, Colombia
[3] Grupo de Investigaciones en Bioingeniería, Facultad de Ingeniería Electrónica, Universidad Pontificia Bolivariana, Medellín, Colombia

Abstract— **Carpal tunnel syndrome, epicondylitis and Quervain's disease are the main upper limb muscleskeletal disorders related with occupational health events. During the clinical practice, assessment of these pathologies includes subjectives methods and parameters. There are quantitative approaches for improve the analysis, however they do not consider dynamic conditions. This paper presents a protocol to perform flexo-extension wrist motion assessment during dynamic contractions. The protocol includes force estimation using *Thera-Bands*, and electromyography (EMG) signals acquired using a commercial, low-cost device (MYO-Armband). A test of feasibility with five healthy persons was carried out. Activation of muscular groups during wrist flexo-extension motions were identified and analyzed respect to the anatomic and physiologic theory. A quantitative assessment of activation patterns was performed using Root Mean Square calculation from EMG. Our results evinced that this protocol could be projected to be applied for occupational therapy.**

Keywords— **Flexo/extension Motion, Muscular Function, Upper limb, surface electromyography, Myo Armband.**

I Introduction

Between 2013 and 2015, 723,836 work accidents and 9,583 work diseases have been reported [1] in Colombia; being the upper limb events the most common. The continued execution of manual activities are related with musculoskeletal disorders of labor origin, which influences the functionality of people. The most common disorders are carpal tunnel syndrome *(CTS)*, epicondylitis and Quervain's disease. Principal symptoms include strength decrement when executing motion with the hands, such as flexion and extension. In order to recover strength, an initial assessment is necessary to support subsequent rehabilitation.

Traditional physical therapy performs a qualitatively assessment of the strength, by observing and palpating the muscles, and by applying a test known as Daniels grading system [2] . The latter classifies the strength as grades from 0 to 5, however it tends to be subjective because it depends on the perception of professional and patient cooperation [3].

There are studies that looked for relating the force exerted by hands with muscle behavior [4] by applying techniques such as dynamometry and surface electromyography *(sEMG)* [5, 6, 7]. However, muscle strength assessment is not possible given that they performed the evaluation of static grip [8].

This paper presents a protocol of dynamical assessment of wrist strength during flexo-extension. Muscular groups associated with wrist extension and flexion were identified, sEMG signals were acquired using the comercial device Myo Armband, and, a quantitative assessment of activation patterns was performed using Root Mean Square (RMS) calculation from sEMG. A proof-of-concept test was carried out aiming to a potential application in occupational therapy.

II Methods

A *Dynamical strength assessment protocol*

To assess the upper limb muscular function, we considered wrist flexion/extension motions taking into account a biped position, with shoulder adduction, 90° elbow flexion, neutral-supportless forearm and palmar grip (Figure 1A.), as suggested by the American Society of Hand Therapist. We recorded and processed *sEGM* under this setting in two stages described bellow.

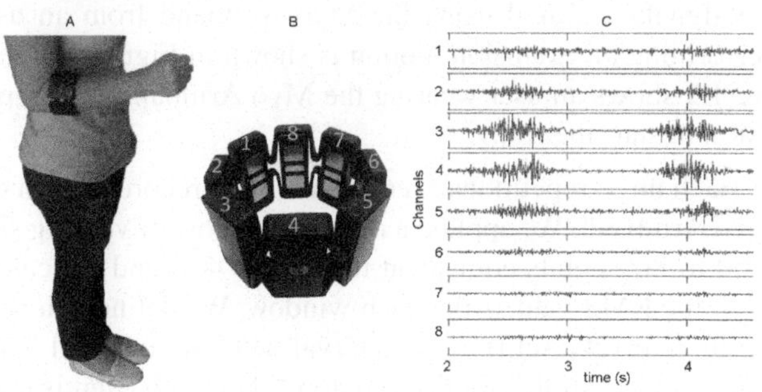

Figure 1: Myo Armband channels and recorded sEMG signals

Stage 1. Motion without resistance: 3 repetitions of wrist

I. Torres et al. (eds.), *VII Latin American Congress on Biomedical Engineering CLAIB 2016, Bucaramanga, Santander, Colombia, October 26th -28th, 2016*,
IFMBE Proceedings 60,
DOI: 10.1007/978-981-10-4086-3_147

flexo-extension are executed, with a rest period of one minute to avoid muscular fatigue.

Stage 2. Resistive motion: Resistance was implemented to the exercise through *TheraBands* of 40 cm in length. Black, silver and gold *TheraBands* were used. Each color represents different levels of resistance per percentage of elongation, being the black the lower level and the gold the highest. The *TheraBands* were fixed at a distance of 85.75 cm. The routine of stage 1 is repeated. The objective is to find the three-repetition maximum (3MR) of wrist extension/flexion, according to the subject capacity. If the person is able to perform more than 3 repetitions with the black *TheraBand*, resistance is increased by using the silver *TheraBand* and then the golden *TheraBand.* Finally, if the person executes 3 repetitions with the golden *TheraBand*, the exercise is repeated at a distance of 112.82cm (Gold Plus).

B EMG acquisition and processing

We used the comercial device Myo Armband from Thalmic Labs. It is composed by 8 bipolar medical grade stainless steel EMG electrodes. EMG channel distribution and their numeration is shown in Figure 1B. Channel 4, marked with Thalmic Labs logo, was used as position reference and it was placed over the lateral epicondyle of each arm. Anatomical points were identified, according with the muscular groups, and they were related with each channel. Previous skin conditioning is not required.

The Myo Armband acquires the *sEMG* signal sat a sampling frequency of 200 Hz and then transfers the data via Bluetooth to a desk computer or laptop. A Matlab® library called MyoMex, developed by Mark Tomaszewski was used to receive the 8 channels signal. We developed an graphical user interface for the sEMG recording process. A sample of the signals acquired using the Myo Armband from an user performing an extension motion is shown in Figure 1C. Figure 1A shows an user wearing the Myo Armband and in position to start the test.

Each three-repetitions exercise was fully recorded with the Myo Armband. We applied a moving window of varying size (100 to 800 points equivalent to 0.5 s to 4 s) and we calculated the RMS value from each window. We defined an activation interval (AcIn) as the interval with the maximal RMS value among all the applied windows. For each exercise, we recovered three AcIn and their respective RMS values, and the mean value of the three was the representative feature of each exercise.

Table 1: Maximum strength for each subject in the stage 2. S: Subject-URM: Upper Right Member-ULM: Upper Left Member Black: 4.4kg-Silver: 6kg-Gold: 9.8kg-Gold plus: 13.8kg

S	Extension		Flexion	
	URM	**ULM**	**URM**	**ULM**
1	Gold	Gold	Gold Plus	Gold Plus
2	Gold	Gold	Gold Plus	Gold Plus
3	Silver	Black	Gold	Gold
4	Gold Plus	Gold	Gold Plus	Gold Plus
5	Gold	Gold Plus	Gold	Gold

C Protocol test

In this work, 5 healthy people participated (3 men and 2 women, between 20 and 50 years old). One subjects is left-handed, and no one have some injury or disease in the upper member. Every subject were informed about the process and the possible risks that could arise. All of this, was wrote in the informed consent.

We assessed both upper extremities, starting with the right upper member, and using the channel 4 as a reference to localize the channels around the forearm, as described above. During execution of the protocol, a physiotherapist, gave indications to the subjects, and avoided postures compensations during the exercises.

III Results and Discussion

A Protocol implementation

The protocol was applied according t o the defined instructions. At the end of *Stage 1*, the subjects do not express any kind of annoyance in their wrists. At the end of *Stage 2* the subjects exhibited difficulty for sustain the posture, mostly by executing the wrist extension motion and a light feel of muscular fatigue.

Table 1 presents the 3MR results for stage 2 in terms of the TheraBand color. Resistance in *kg* is also indicated. For extension and flexion, 80% and 100% of the subjects, respectively, determine their 3MR in both arms while using at least the Gold TheraBand. 60% of the subjects reach the Gold Plus TheraBand with both arms.

B Muscle groups associated with Myo channels

We identified the main muscular groups that each Myo channel register, by relating anatomical points and muscular action with palpation. Table 2 shows the correspondence of the channels and muscles for the right and left arm. Channel correspondence varies for both arms with the exception of channel 4, which is the reference, and channel 8, which is diametrically opposed to channel 4. Most channels are influenced by more than one muscular group, excepting channels 1, 3 and 8 for URM, and 5, 7 and 8 for ULM. Note that the distribution could be affected by the perimeter of the forearm, as it may vary depending on the body composition of each person or may be differences according to the anthropometric between men and women.

The obtained resistance strength, according to 3RM test, is suitable for men and women. During wrist flexion, it is observed that both extremities were able to overcome resistance, while during wrist extension prevails the dominant limb. Additional studies are needed in order to determine differences while executing wrist flexion/extension exercises, in men and women aiming to identify the dominant hand influence.

Table 2: Muscle identification according to 8 channels of Myo Armband C: Channel-URM: Upper Right Member- ULM: Upper Left Member-F: Muscles Flexor-E: Muscles extensor

C	URM	ULM
1	F. carpi ulnaris	Brachioradialis F. digitorum superficialis
2	Anconeus F. carpi ulnaris	E. carpi radialis longus E. carpi radialis brevis Brachioradialis
3	E. digiti minimi	E. digitorum E. pollicis longus
4	E. digitorum E. carpi ulnaris	E. digitorum E. carpi ulnaris
5	E. digitorum E. pollicis longus	E. digiti minimi
6	E. carpi radialis longus E. carpi radialis brevis Brachioradialis	Anconeus F. carpi ulnaris
7	Brachioradialis F. digitorum superficialis	F. carpi ulnaris
8	F. carpi radialis	F. carpi radialis

C Quantitative assessment of muscular strength

Normalized values of the mean RMS from AcIn during wrist flexo-extension exercise were found for Stages 1 and 2 of the protocol. Boxplots were constructed, by grouping the normalized RMS of each channel from all subjects. Figure 2 shows the eight boxplots for all exercise variations assessed. A similar activation pattern, among the 8 channels, for both flexion and extension can be observed when comparing the exercises with and without resistance. The pattern of activation is different when comparing wrist extension and flexion.

Through this activation patterns, and the information of muscular group/sEMG channel relation (table 2), we infered that RMS from channels 3, 4 and 5 are higher for extension exercises, given that they are related with extensor carpi ulnaris muscles, whose main function is the extension of the wrist extensor and digiti minimi muscles, extensor digitorum longus and extensor pollicis which contribute to the extension of the wrist [2]. Similarly, we observed that channels 1, 7 and 8 correspond to muscles flexor carpi ulnaris and flexor carpi radialis which have, as main function, the wrist flexion, flexor digitorium superficialis muscles and the brachioradialis which contribute with the wrist flexion [2]. Moreover, activation in channels 2 and 6 for both exercices, is due to the relation of various muscular groups with these channels, being anconeus, extensor carpi radialis and extensor carpi brevis muscles contributors to the wrist extension and flexor carpi ulnaris and brachioradialis contributors to the wrist flexion [2].

IV CONCLUSIONS AND FUTURE WORKS

We presented a protocol for quantitative assessment of flexo-extension wrist motion during dynamic contractions using sEMG. The Myo Armband is a commercial low-cost device that has been used mainly for gesture recognition to control computer applications, including research and development in medical rehabilitation areas [9, 10, 11]. However, to the best of our knowledge, there are not studies related with assessment of muscle strength using this device. Therefore this study proposes a protocol, using the Myo Armband, that allows a quantitative approach in assessing muscle function during wrist flexion and extension in occupational health processes. In order to demonstrate the feasibility of its application, we performed a test with 5 healthy subjects considering full range of motion and higher restive loads. We highlight the versatility that the Myo Armband contributes to the execution of the protocol. This mades the Myo Armband an attractive option for occupational health applications.

We reported a quantitative analysis, which, given the small sample used, are not statistical significant. However, they evinced that useful information can be extracted by applying the proposed methodology. As a future work, the protocol will be tested using a larger subjects sample, in order to obtain more conclusive results and to establish parameters that allow objectify assessment processes. Furthermore, the protocol will be enhanced aiming to assess other motions such as pronation, supination, radial deviation and deviation ulnar, which are also affected by carpal tunnel syndrome, epicondylitis and Quervain's disease. Application of the protocol on real patients will need to address special issues: Range

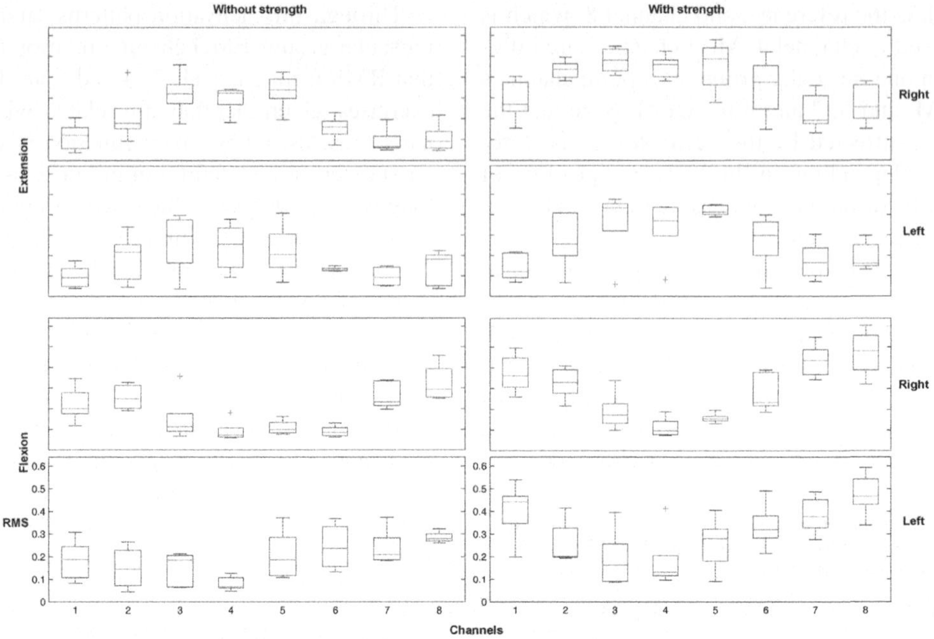

Figure 2: Activation of the different channels of Myo Armband for all subjects in wrist flexion and extension

of motion must be took as variables and lower resistive loads might be needed, given that patients could have limited motion and muscular weakness, due the the respective physiopathology.

CONFLICT OF INTERESTS

The authors declare that there is not potential conflict type of interest related to the paper or research.

ACKNOWLEDGEMENTS

The authors thank the physiotherapist and specialist in hand Alexis Polo Medina for their support in the development of the paper.

REFERENCES

1. Ministerio de trabajo . *Sistema de riesgos laborales estadísticas de afiliados y eventos ATEL años 2013 - 2014 - 2015* . 2015.
2. Daniels; Worthingham . *Pruebas funcionales musculares. Técnicas de exploración manual.* Madrid - España: Marbán6 ed. 2002.
3. Ottenbacher Kenneth J., Branch Laurence G., Ray Laura, Gonzales Vera A., Peek M. Kristen, Hinman Martha R.. The reliability of upper- and lower-extremity strength testing in a community survey of older adults *Archives of Physical Medicine and Rehabilitation.* 2002;83:1423–1427.
4. Ministerio de trabajo . *Guías de atención integral seguridad y salud en el trabajo - Desorden músculo esquelético de miembro superior.* Instituto de Evaluación Tecnológica en Salud 2015.
5. Judith Boscheinen-Morrin; W Bruce Conolly . The Hand: Fundamentals of Therapy in *Rehabilitation* (Heidi Allen . , ed.):1–12Reed Educational and Professional Publishingthird edit ed. 2001.
6. C J. De Luca . The use of surface electromyography in biomecanics *Journal of Applied Biomechanics.* 1997;13(2):135–163.
7. Disselhorst-Klug Catherine, Schmitz-Rode Thomas, Rau Günter. Surface electromyography and muscle force: limits in sEMG-force relationship and new approaches for applications. *Clinical biomechanics (Bristol, Avon).* 2009;24:225–35.
8. Ngo Binh P.T., Wells Richard P.. Evaluating Protocols For Normalizing Forearm Electromyograms During Power Grip *Journal of Electromyography and Kinesiology.* 2016:1–7.
9. Abduo Mahmoud, Galster Matthias. Myo Gesture Control Armband for Medical Applications 2015.
10. Lipovsky Rastislav, Ferreira Hugo Alexandre. Hand therapist: a rehabilitation approach based on wearable technology and video gaming *Bioengineering (ENBENG), 2015 IEEE 4th Portuguese Meeting on.* 2015:1–2.
11. Qamar Ahmad M., Khan Ahmed Riaz, Husain Syed Osama, Rahman Mohamed Abdur, Baslamah Saleh. A Multi-Sensory Gesture-Based Occupational Therapy Environment for Controlling Home Appliances *Proceedings of the 5th ACM on International Conference on Multimedia Retrieval.* 2015:671–674.

Author: Julieth Vanessa Montoya Leal
Institute: Universidad Pontificia Bolivariana
Street: Circ. 1 N 73-76, Bloque 22C
City: Medellín
Country: Colombia
Email: julieth.montoyal@upb.edu.co

In vitro study of proliferation and cellularisation on electrospun membranes for vascular prosthesis

Y. Montoya[1], R.A. Valencia[1,2], I.C. Ortiz[3], L.M. Hoyos[1] & J. Bustamante[1,*]

[1] Grupo de Dinámica Cardiovascular, Centro de Bioingeniería, Universidad Pontificia Bolivariana, Medellín, Colombia.
[2] Grupo de Automática y Diseño A+D, Universidad Pontificia Bolivariana, Medellín, Colombia.
[3] Grupo de Biología de Sistemas, Universidad Pontificia Bolivariana, Medellín, Colombia.

Abstract— **Tissue engineering applied to new therapies with synthetic vascular grafts, represents a breakthrough to improve the results of biocompatibility and functionality of implants for use in patients with cardiovascular disease. The manufacturing of vascular implants with electrospinning technique has gained interest for small-diameter blood vessels, thanks to its ability to generate micro-porous structures with large surface area. The objective of this work was to study the proliferation and generation of cellular environment, and how it influences the growth rate on the permeability, and porosity of electrospun membrane for use as vascular prostheses.**

Polyurethane membranes with shape memory (Irogran) were manufactured considering two thicknesses by electrospinning: Sample 0, between 0,2mm and 0,9mm, and sample +1, between 0,9mm and 1,0mm. In an *in vitro* model, cardiac fibroblasts were cultivated for a period of up to 10 days of incubation. The cell proliferation was evaluated by means of optical and scanning electron microscopy (SEM), and the porosity and permeability were evaluated by mean of hydrostatic pressure and gravimetric technique, according to ISO 7198 international standard.

It was found that samples +1 have an average permeability of 55,5% less than samples 0, and a reduction of porosity of 10,24%, associated to higher cellular growth evidenced by cell syncytium. This paper concludes that the variation of microporous structures with large surface area, affects the cell growth and subsequently the permeability and porosity, opening a great opportunity for its potential use in vascular applications.

Keywords— **Vascular implants; Cardiac fibroblast; Cellular proliferation; Tissue engineering; Permeability; Porosity.**

I. Introduction

Vascular diseases include multiple treatments, ranging from the use of medications up to the execution of surgical procedures that use the implantation of different vascular grafts [1,2]. When it becomes necessary to replace a segment of blood vessel; autologous, homologous or heterologous implants; are taken into consideration [3]; however, in most cases this is not possible [4], so one could resort to the use of synthetic implants [5]. These seek to be functionally similar to the blood vessel [6,7]. The most used materials in vascular prostheses are the polytetrafluoroethylene PTFE (Teflon), the polyethylene terephthalate or PET (Dacron®), and the synthetic polymers [3].

Meanwhile, tissue engineering has focused its attention on methods for manufacturing synthetic vascular implants with polymeric matrices in order to interact with blood cells, with the purpose to proliferate and create an extracellular matrix, that leading to the formation of new endothelial tissue, so that the graft can be accepted and integrated in the body without side effects [8,9]. One of the most attractive alternatives for the manufacture of polymer matrix is the electrospinning technique, given its ability to generate micro-porous structures, similar to an extracellular tissue, and to obtain large superficial areas where the cells can adhere and proliferate [5, 10,11].

Likewise, the electrospinning technique is highlighted as one of the options for the manufacture of small-diameter (<5mm) synthetic vascular implants [12,13]. However, it is necessary to highlight that in this type of vascular prosthesis remains unknown how the thickness of the membrane and other properties related with the mass transport in the wall, affect the generation of cell syncytia for the implant endothelialization.

The objective of this research was to study the proliferation and cellularisation of electrospinning implants of shape memory polyurethane, and how the growth rate of cardiac fibroblasts influences the permeability and porosity to different thicknesses as possible use in vascular prostheses.

II. Methodology

A. Manufacture of implants

In the Laboratory of Nanomaterials Synthesis, Universidad Pontificia Bolivariana, were manufactured electrospun membranes with different thicknesses (E), taking into account the thickness of real vessels and commercial vascular implants. The membrane thickness were selected as follows: Sample 0: 0,2mm$\leq$E <0.9mm, and Sample +1: 0,9mm$\leq$E

I. Torres et al. (eds.), *VII Latin American Congress on Biomedical Engineering CLAIB 2016, Bucaramanga, Santander, Colombia, October 26th -28th, 2016*,
IFMBE Proceedings 60,
DOI: 10.1007/978-981-10-4086-3_148

<1.0mm, these values represent thicknesses wall equivalent to vascular prostheses of small diameter vessels. For the manufacture of the samples, as base solution, it was used shape memory polyurethane (Irogran) 21% w/w in a solution in tetrahydrofuran (THF) / dimethylformamide (DMF) to 50% w/w. To determine the average thickness of the samples, a digital micrometer Mitutoyo was used. To each sample was taken 10 measurements of thickness, and the average was determined.

B. Samples preparation

The membranes were cut in circular pieces of 1.2 cm and 2.2 cm diameter, based so much in the requirements of the assemblage for the tests of porosity and permeability, according to ISO 7198 standard [14], as in the diameters of wells for the cell cultures. The total numbers of samples was 12 for each thickness referenced.

C. In vitro model: NHCF-A cardiac fibroblasts

The cell cultures were carried out in the Laboratory for Processing of the Cultures, Universidad Pontificia Bolivariana. The *in vitro* model used during the experiments were cultures of human cardiac fibroblasts of atria NHCF-A (Supplier Lonza Group Ltd.). This cell line was isolated from healthy adult atrial tissue, expressing about 90% of type I collagen.

Cell cultures of fibroblasts were maintained in FGM medium, 10% fetal bovine serum (FBS), supplemented with penicillin (1000U/ml) and streptomycin (1000g/ml), at a temperature of 37 ° C, 1 atm, 5% CO_2, 95% O_2 and 95% Hr. The medium was changed after those cells reached confluence approximate of the 88 to 90% and then incubated at 37°C (Fig. 1)

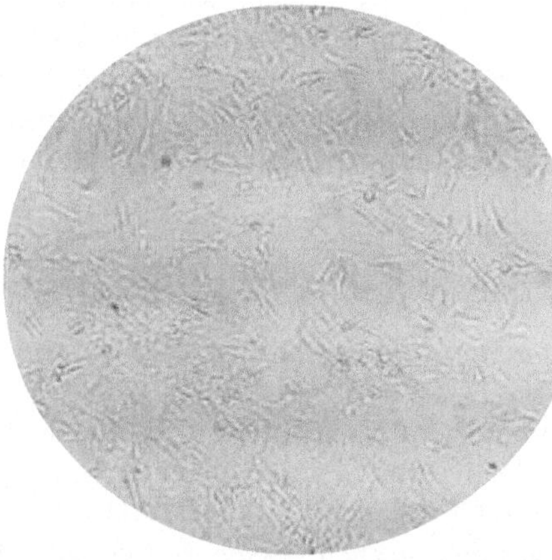

Figure 1. Cultures of cardiac fibroblasts. Cell confluence approximate of 85% (Objective: 100X. Source: Proteomic Laboratory, Universidad Pontificia Bolivariana).

D. Seeding cells into membranes

In culture's dishes of 6 and 24 wells, in which FGM culture medium supplemented with fetal bovine serum and antibiotics was added, were deposited electrospun membranes of the samples 0 and +1, and left suspended for 2 days. On the third day, the medium was discarded and a solution of 3500 cells/cm3 was added in the culture dishes. Finally, the samples were carried under controlled conditions at a temperature of 37°C, 5% CO_2, and 95% Hr. They were left in incubation for a period up to 10 days.

E. Analysis of cell growth

The membranes were deposited in a tissue processor for a period of 48 hours, immersed in alcohol at 50%, 70% and 80%, formalin at 10%, xylenes and liquid paraffin at 60°C. Then, they were passed to the central of inclusion to form paraffin blocks that were cut on a microtome (Leyca series RM2235). Finally, the plates were fixed and stained with hematoxylin-eosin and observed by optical microscopy.

F. Analysis by optical and scanning electron microscopy (SEM)

Membranes samples 0 and +1 were subjected to incubation cell periods and fixed in a solution of 10% formaldehyde in phosphate buffer solution for 24 hours. The samples were stained with hematoxylin-eosin, and analyzed in a microscope (Motic series ECO-Bino). Finally, it was carried out by SEM analysis, which was performed with a scanning electron microscope (JEOL JSM-6490 LV).

G. Measurement of permeability and porosity

For measuring the static permeability it was used the procedure described in the ISO 7198 standard [14]. The setup consisted in a column of hydrostatic pressure, with a height of liquid equivalent to a pressure from 0 to 150 mmHg. Then, the porosity was measured by gravimetric technique through the procedure indicated in ISO 7198 standard. In these tests the samples 0 and +1 were used at 48, 96, 192 and 240 hours cell incubation.

H. Statistical analysis

The data analysis of permeability and porosity in the samples was performed using ANOVA statistical analysis, to determine the influence of each variable, and their relation with the interest properties of the study.

III. RESULTS

A. Cell proliferation

Figure 2, shows micrographs by scanning electron microscopy of membranes with human cardiac fibroblasts proliferation, for both samples 0 and +1. In these images adherent cells can be seen forming syncytia in the surface of the electrospun membrane.

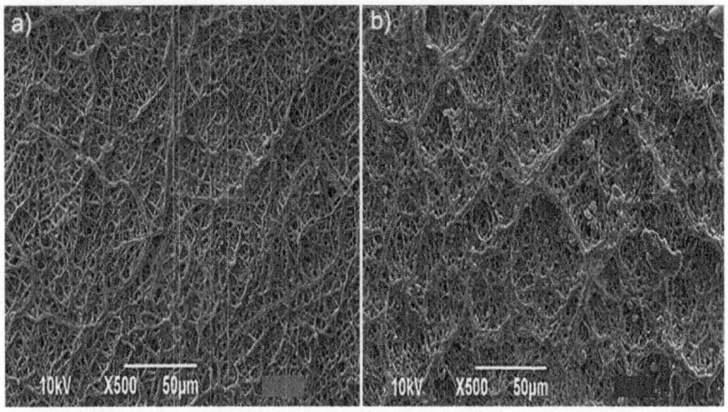

Figure 2. SEM images of electrospun membranes. a) Sample 0 with 240 hours cell incubation. b) Sample +1 with 240 hours cell incubation. Highlighted in red, cell growth indicated in the samples.

Figure 3a, shows electrospun membranes with 240 hours of incubation, stained with hematoxylin-eosin, which allows to observe the presence of intact cell nucleus on the sample, an element indicator of membrane viability and cell proliferation. Additionally, in Figure 3b are presented cell colonies inside of the implant wall, showing penetration and cellular interaction in the empty spaces of the membrane.

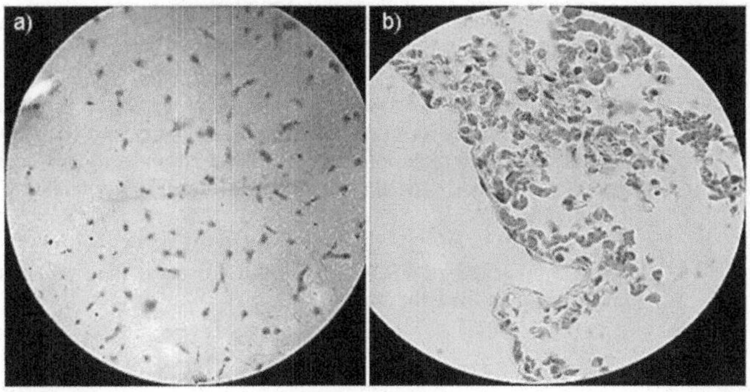

Figure 3. a) Electrospun sample with 240 hours of cell incubation, stained with hematoxylin-eosin in the implant surface and b) Cellularisation in the implant wall.

B. Permeability and porosity

The results of measurements of permeability and porosity for samples 0 and +1, according to the hours of incubation, are presented in Figure 4.

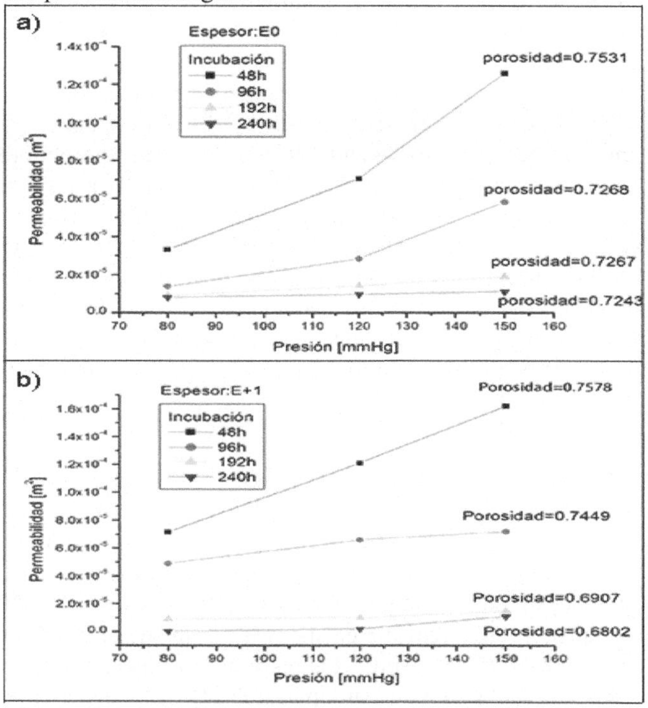

Figure 4. Permeability and porosity data at different pressures and hours of cell incubation for a) Samples 0. b) Samples +1.

Comparing these figures, it is noted that for the samples +1, incubated for 48 and 96 hours, the permeability average was higher in 35.2% and 46.3% respectively, compared to the samples 0 with the same incubation period. However, when the results for the samples incubated for 192, and 240 hours was observed otherwise, the samples +1 had average permeability lower to 20.3% and 55.5% respectively, compared to samples 0.

Then, porosity of the samples +1, showed that the reason for decreasing the incubation with days was higher, reaching up to 10.24%, when was compared in an incubation period between 48 and 240 hours. While for samples 0, the decrease presented was of 3.86% in the same incubation period.

IV. DISCUSSION

Vascular prosthesis can promote the generation of cells syncytia by manufacturing electrospun membranes with specific pore diameters according to the intended functions. The results of this study showed that in the samples +1, the growth of the tissue cell had higher proportion in comparison with the samples 0, leading to a drop in permeability values. One possible cause of the high cellular proliferation in thicker membranes would be the high values of porosity of the samples +1 before being covered by cells, which promoted a significant cell growth in the incubation period analyzed, generating an upholstered cellular.

In addition, was observed a marked difference between the values of membrane permeability for both samples +1 and 0 to pressures of 80 and 150 mmHg, when the periods were contrasted before and after that the cellularisation in membranes was presented with cardiac fibroblasts. However, at 240 hours, the effect of the thickness and the pressure on the permeability of the membrane does not change with regard to the previous; this behavior is probably due to the free space of the membrane that had been occupied by fibroblasts in the early hours and there is not marked growth later, on the other hand.

V. CONCLUSIONS

Electrospun membranes of the samples +1 showed a major proliferation and generation of cell syncytia. At 10 days of incubation cell culture occurs an upholstered of cell culture making apparent the creation of a cellularized layer on the inside of the implant. This response is translated into a decrease in the porosity and permeability with the incubation hours.

The observation of cell colonies showed that this type of membranes predisposes to the pseudo-endothelialization of vascular implant, a theme that opens therapeutic expectations in cardiovascular diseases.

ACKNOWLEDGMENT

Thanks to Simon Sanchez, for his participation in the operational support of experimental tests related to mass transport properties. Also, thanks to Departamento Administrativo de Ciencia, Tecnología e Innovación Colciencias, Universidad Pontificia Bolivariana and Universidad EAFIT (Colombia), for the academic and financial support in the formation of human resources involved in this researching work.

REFERENCES

1. Benrashid E, McCoy C, Youngwirth L, Kim J, Manson R et al. Tissue engineered vascular grafts: Origins, development and current strategies for clinical application. Regenerative Medicine. 2016:99:13-19.
2. Machy D, Carteron P, Jozefonvicz J. A New Vascular Polyester Prosthesis Impregnated with Cross-Linked Dextran. J. Biomater. Sci. Polym. 2002:13(8):963-975.
3. Tura, A. (Ed.) (2003). Vascular grafts. Experiment a modelling. Southhampton, United Kingdom; WIT Press, 422p.
4. Bergemeister H, Strobl M, Grasl C, Liska R, Schima H. Tissue engineering of vascular grafts. European Surgery. 2013:45(4):187-193.
5. Hirt M, Hansen A, Eschenhagen T. Cardio Vascular Tissue Engineering: Estate of Art. Circulation Research. 2014: 114(2):354-67.
6. FDA. Guidance for Industry and FDA Staff. Obtenido de Guidance document for Vascular Prostheses 510(k). 2012. Submissions: http://www.fda.gov/Medic.
7. Greenhalgh, E. and Dunn, M. Modelling blood flow through vascular grafts C98-P01. Philadelphia College of Textile and Science. 1998.
8. Szentivanyi A, Chakradeo T, Zemetsch H, Glasmacher B. Electrospun cellular microenviromental: Understending controlled release and Scaffold structure. Advanced Drug Delivery Review. 2011: 63(4-5):209-220.
9. Wade RJ, Burdick, JA. Advances in nanofibrous scaffolds for biomedical applications: From electrospinning to self-assembly. Nanotoday. 2014:9(6):722-742.
10. Chung S, Ingle NP, Montero GA, Kim SH, King MW. Bioresorbable elastomeric vascular tissue engineering scaffolds via melt spinning and electrospinning. Acta Biomaterialia. 2010:6(6):1958-1967.
11. Nabovati A, Llewellin EW, Sousa AC. Through-thickness permeability prediction of three-dimensional multifilament woven fabrics. Composites Part A: Applied Science and Manufacturing 41 (2010), 453 – 463.
12. Hasan A, Memic A, Annabi N, Hossain M, Paul A, Dokmeci MR, et. al. Electrospun scaffolds for tissue engineering of vascular grafts. Acta Biomaterialia. 2014:10(1): 11–25.
13. Kucinska-Lipka J, Gubanska I, Janik H, Sienkiewicz M. Fabrication of polyurethane and polyurethane based composite fibers by the electrospinning technique for soft tissue engineering of cardiovascular system. Material Science and Engineering:C. 2015:46:166-176.
14. International standard ISO/DIS 7198, Cardiovascular Implants; Tubular vascular prostheses. 1998.

Use macro [*] to enter the address of the corresponding author:

Author: Y. Montoya
Corresponding author: *J. Bustamante
Institute: Universidad Pontificia Bolivariana/ Grupo Dinámica Cardiovascular.
City: Medellín
Country: Colombia
Email: bustamante.john@gmail.com

Height Difference Effects Between the Standard and the Equipment Under Test in Calibration Process for Sphygmomanometers in Colombia

M.A. Castro-Leal[1], M.A. Castro-Cortés[2]

[1] Biosancta/Laboratorio de Calibración, Director, Bogotá, Colombia
[2]Universidad Nacional de Colombia/Ingeniería de Sistemas, Estudiante, Bogotá, Colombia

Abstract— **Height difference effects between the standard and the equipment under test, in calibration process for sphygmomanometers in Colombia is contemptible unto one-meter height differences within them. When is applied for standard lecture corrections and uncertainty measures effects with a 95,45% probability with T Student distribution, calculated for Colombian's cities with most and less gravitational acceleration, relative humidity (RH) among 20% and 80% and temperatures within 15°C and 25°C according to the R 16-1 Annex A.1 of International Organization of Legal Metrology OIML. The demonstration was realized with simplified calculus of air density given by the International Committee for Weights and Measures 2007, which are based on RH, temperature and atmospheric pressure, and the gravitational accelerations given by the national gravimetric network of first and second order SIGNAR in the thesis Determination of the vertical reference surface for Colombia made by Laura Sánchez Rodríguez Technische Universität Dresden Germany 2003.**

Keywords— **Calibration, Sphygmomanometers, Height Differences, Effects.**

I. Introduction

Sphygmomanometer's calibration is essential in the quality assurance for diagnostics and medical treatments, process in which the reliability of the equipment measures is achieved. This of vital importance in clinic medical process from the daily medical consult to the hospital intra and extra care.

The process of calibration of sphygmomanometers, being part of the legal metrology, has it recommendation from the International Organization of Legal Metrology OIML expressed in the R 16-1 Annex A.1, in which is clearly specified the calibration process, and other general aspects in the measure uncertainty. However, when a deep study of the elements that contribute to the uncertainty measure is realized, fundamental aspects are found that, also, are an impediment to realize the calibrations at site. One of many elements to consider, is the correction and uncertainty that gives the height difference between the standard and the sphygmomanometer under calibration. When it is take into account Colombian's geography and different conditions of the environment, the correction and uncertainty measures concludes in dispendious calculus, long and highly different for each one of the places where the calibration is needed. Reason for which is necessary (and much more efficient) in the reliability of the calibration process, *to demonstrate that height difference effects between different Colombian cities is not relevant for the sphygmomanometer's calibration process and results.* Additionally, it is important to recall two features. First, the standard and the sphygmomanometer under test have a one-meter height differences within them. And second, for this Colombian's cities the measure of the gravitational acceleration is known, which is comprehended between 0 and 2900 meters above the sea level, which involves more than 98% of the habitable Colombian territory and a 100% of the Colombian territory where there are more than 20.000 people.

II. METHODS

A. Height difference between the standard and the sphygmomanometer in calibration process.

The fluid that exerts pressure: Generally, when calibrating manometers, height difference between the standard and the manometer under calibration is important, because height differences generates errors which must be corrected. This errors are calculated in function of the density of the fluid that is used to exert pressure in the system, the gravitational acceleration of the place where the calibration is realized and the mentioned height difference. Normally, for high tensions, mineral oil is utilized, which has a high density compare to air density produced by the atmospheric pressure. For low pressures dry nitrogen gas is used, even if density values do not differ greatly from atmospheric air, it results relevant at errors calculations generated by pressure in low pressures and manometers of classes below 0,5. To calculate the error generated by height difference it should be taken into account the next variables in the following equation:

$$\mathbf{E = \delta P_{\Delta h} = (\rho_f - \rho_a) g \Delta h} \qquad \textbf{equation 1}$$

Where

I. Torres et al. (eds.), *VII Latin American Congress on Biomedical Engineering CLAIB 2016, Bucaramanga, Santander, Colombia, October 26th -28th, 2016,*
IFMBE Proceedings 60,
DOI: 10.1007/978-981-10-4086-3_149

$\delta P_{\Delta h}$ is the correction due to height difference between the standard and the manometer to be calibrated.

ρ_f is fluid's density that is being used as means to generate pressure.

ρ_a is air density at atmospheric pressure.

g is gravitational acceleration from the place from which the calibration is realized.

Δh is height difference between the standard and the manometer to be calibrated.

The calibration method for sphygmomanometers determined by the guide R 16-1 Annex A.1 of International Organization of Legal Metrology OIML stablished that the fluid to exert pressure is the ambient air, so it is required to calculate atmospheric air's density in each one of the calibration points stablished in the mentioned method.

To calculate air's density, the simplified formula of the International Committee for Weights and Measures CIPM of the year 2007 was utilized [2]

$$\rho_a = \frac{0.34848\,p - 0.009\,h_r\,e^{0.061\,t}}{273.15+t} \qquad \textbf{equation 2}$$

Where:

ρ_a = Air density in Kg/m^3

p = Atmospheric Pressure in hPa

h_r = RH in %

t = Temperature in °C

As density is in function of the atmospheric pressure, in sphygmomanometers cases, that uses ambient air as element to generate pressure, it must be taken into account that to atmospheric pressure it should be added the pressure at which the system is in each one of the measure points p_{mi} which are 0 mmHg, 50 mmHg (66.6612hPa), 100 mmHg (133.3224 hPa), 150 mmHg (199.9836 hPa), 200 mmHg (266.6448 hPa), 250 mmHg (333.306 hPa) y 300 mmHg (399.9672 hPa), so the fluid density that is used for generating pressure is:

$$\rho_f = \frac{0.34848(p+p_{mi}) - 0.009\,h_r\,e^{0.061\,t}}{273.15+t} \qquad \textbf{equation 3}$$

B. Conditions at which it is going to be calculated the error generated by height difference in the Colombian Territory:

As calibration method R 16-1 Annex A.1 of OIML express that conditions at which the calibration must be made it ought to be a temperature among 15°C and 25°C and a RH between 20% and 80%, measures conditions of higher and lower gravitational acceleration in the Colombian territory were given by the first and second order web SIGNAR [3] which in order they are *San Andrés Islas* at 0 meters above the sea level with a gravitational acceleration of g_1= 9.78385701 m/s^2 and an atmospheric pressure of 1013 hPa [4], and city of *Ipiales* at 2900 meters above the sea level with an gravitational acceleration of g_2=9.77238153 m/s^2 and an atmospheric pressure of 718 hPa [5]. For calculations it was defined a height difference between the standard and the sphygmomanometer under calibration of 1 meter, which is large enough to cover the comprehended cases in the field.

Uncertainty contributed to the process by height difference in the Colombian Territory: In this uncertainty it must be calculated sensitivity coefficients in respect of each one of the variables [6], therefore it remains:

According to equation 1 where it is expressed the error due to height difference, if we call (ρ_f-ρ_a) = Δρ then the correction due to height difference remains

$$\delta P_{\Delta h} = \Delta\rho g \Delta h \qquad \textbf{equation 4}$$

Calculating sensitivity coefficients for equation 4 we have:

$$C1 = \frac{\partial \delta P\Delta h}{\partial \Delta\rho} = g*\Delta h*1*\Delta\rho^{(1-1)} = g*\Delta h \quad \textbf{equation 5}$$

$$C2 = \frac{\partial \delta P\Delta h}{\partial g} = \Delta\rho*\Delta h*1*g^{(1-1)} = \Delta\rho*\Delta h \quad \textbf{equation 6}$$

$$C3 = \frac{\partial \delta P\Delta h}{\partial \Delta h} = g*\Delta\rho*1*\Delta h^{(1-1)} = g*\Delta\rho \quad \textbf{equation 7}$$

Therefore uncertainty [6] contributions are:

$$u_{1(\Delta\rho)} = C1 *u\,\Delta\rho = g*\Delta h*u\,\Delta\rho \quad \textbf{ecuación 8}$$

$$u_{2(g)} = C2 *u\,g = \Delta\rho*\Delta h*u\,g \quad \textbf{ecuación 9}$$

$$u_{3(\Delta h)} = C3 *u\,\Delta h = g*\Delta\rho*u\,\Delta h \quad \textbf{ecuación 10}$$

Consequently, to calculate uncertainty [6] due to height difference $u_{(\delta P\Delta h)}$ we have:

$$u_{(\delta P\Delta h)} = \sqrt{\sum (c_i * u_i)^2} = \sqrt{(g\Delta h u_{\Delta\rho})^2 + (\Delta\rho\Delta h u_g)^2 + (g\Delta\rho u_{\Delta h})^2}$$

equation 11

III. DISCUSSION

Each one of the errors were calculated for each one of air's density and flow's density in each condition of RH and temperature in both cities of higher and lower gravitational acceleration and the following values were obtained for each one of the defined cases: Case 1 San Andrés Islas at

15°C and 20% RH; Case 2 San Andrés Islas at 15°C and 80% RH; Case 3 San Andrés Islas at 25°C and 20% RH; Case 4 San Andrés Islas at 25°C and 80% RH; Case 5 Ipiales at 15°C and 20% RH; Case 6 Ipiales at 15°C and 80% RH; Case 7 Ipiales at 25°C and 20% RH; Case 8 Ipiales at 25°C and 80% RH.

Realizing the calculus for each one of the cases, the obtained errors are summarized in the following table

Table 1 Errors in mmHg due to height difference for the 8 cases.

Pressure	Case 1	Case 2	Case 3	Case 4	Case 5	Case 6	Case 7	Case 8
50 mmHg	0.0059	0.0059	0.0057	0.0057	0.0059	0.0059	0.0092	0.0056
100 mmHg	0.0118	0.0118	0.0114	0.0114	0.0118	0.0118	0.0114	0.0114
150 mmHg	0.0177	0.0177	0.0171	0.0171	0.0177	0.0177	0.0171	0.0171
200 mmHg	0.0236	0.0236	0.0228	0.0228	0.0236	0.0236	0.0228	0.0228
250 mmHg	0.0295	0.0295	0.0285	0.0285	0.0295	0.0295	0.0285	0.0285
300 mmHg	0.0354	0.0354	0.0321	0.0343	0.0354	0.0351	0.0342	0.0342

The maximum error calculated was 0.0354mmHg.

For the calculus of uncertainty contributed by height difference that is described in equation 11, it takes the following values:

g= maximum gravity equivalent to San Andrés Islas 9.78385701 m/s^2

Δh = 1 meter

$\Delta\rho$ = 0.483708381 kg/m^3 being the máximum value between ρ_f-ρ_a calculated for the 8 cases described.

$u_{\Delta g}$ = 0.098 m/s^2 [3]

$u_{\Delta h}$ = 0.005 m

$u_{\Delta\rho}$= 0.09050966799 kg/m^3 according to the next calculus

$$\mathbf{u_{\Delta\rho}} = x = \sqrt{u_{\rho f}{}^2 + u_{\rho a}^2}\ \textbf{equation 12}$$

as f and a are from the same air at different pressure, it assumes uncertainty equals and it is take as 0.064kg/m^3 [7], therefore the uncertainty contributed by height difference u($\delta P_{\Delta h}$) remains:

$$u(\delta P_{\Delta h}) = \sqrt{\begin{matrix}0.8855336496^2 + 0.04740342133^2 \\ +0.02366266817^2\end{matrix}}$$

u(δP$_{\Delta h}$)= 0.88711716 Pa = 0.0066 mmHg equation 13

for a probability of 65%, therefore when transiting it to a probability of 95,45% with T Student distribution and effective degrees of freedom near ∞, the uncertainty contributed by height difference is **u(δP$_{\Delta h}$)= 0.0132 mmHg**.

IV. RESULTS

Gravitational acceleration difference between San Andrés Islas and Ipiales, does not present appreciable differences of generated pressures. Being a 1-meter height difference between the standard and the sphygmomanometer under calibration, utilizing as system pressure fluid for the calibration process ambient air which was about 0.0354 mmHg.

The correction value plus the uncertainty contributed by height difference between the standard and the sphygmomanometer under calibration is 0.0486 mmHg for the studied case and it represents 2.43% of the resolution value of the sphygmomanometer which has a resolution of 2 mmHg.

The typical appreciation of a sphygmomanometer is of 4 mmHg, therefore values readings are made in 0.5 mmHg multiples and the value 0.0486 mmHg only represents 1.62% and 1.215% respectively of the accepted tolerances [1] which are of 3mmHg for new sphygmomanometers and 4 mmHg for sphygmomanometers in use.

V. CONCLUSIONS

Due to the circumstance that sphygmomanometers have a resolution of 2mmHg and their class is 1, an error due to height difference between the standard and the sphygmomanometer in a calibration process of 0.0354mmHg is contemptible.

In the same manner, the uncertainty contributed to the system by height difference with a probability of 95.45%, with 0.0132mmHg it is also not significant.

The error plus the uncertainty contributed by height difference is about 0.0486mmHg, which is contemptible compare to the resolution and possible appreciation in the lecture of the sphygmomanometer.

From the medical point of view, the noninvasive blood pressure values taken by the medical and paramedical staff, it is approximate to multiples of 1mmHg, for which the error plus the uncertainty contributed by height difference is also contemptible.

According to the previously exposed, for Colombia, between 0 meter above the sea level and 2900 meter above the sea level, with a maximum height difference of 1 meter between the standard and the sphygmomanometers in sphygmomanometer's calibration processes under the OIML R 16-1 Annex A1 the error and uncertainty contributed by this height difference is contemptible.

VI. REFERENCES

1. International Recommendation OIML R 16-1 Edition 2012 Non invasive mechanical sphygmomanometers.
2. A Pickard, R S Davis, M Gläser y K Fujii Revised formula for the density of moist air (CIPM-2007) Metrologia 45 (2008) Pg 149 – 155.
3. Sánchez R, Laura Determinación de la superficie vertical de referencia para Colombia Technische Universität Dresden, junio 2003, apéndice A páginas 79 a 82.
4. http://www.tutiempo.net/tiempo/San_Andres_Isla_Sesquicentenario/SKSP.htm?datos=detallados
5. http://rp5.ru/Tiempo_en_Ipiales,_San_Luis_(aeropuerto)
6. JCGM 100:2008 Evaluation of measurement data — Guide to the expression of uncertainty in measurement.
7. SIM MWG7/cg-01/v0.0 pág 63

Author$_1$: Miguel Ángel Castro Leal
Institute: Biosancta SAS
Street: Kra 16A No. 162-37
City: Bogotá
Country: Colombia
Email: macastrole@hotmail.com / direccionlaboratorio@biosancta.com

Parametric Modeling of Kinetic-Kinematic Polycentric Mechanical Knee

A.M. Cárdenas[1], J. Uribe[1] and A.M. Hernández[1]

[1] Bioinstrumentation and Clinical Engineering Research Group - GIBIC, Bioengineering Department, Engineering Faculty, Universidad de Antioquia UdeA; Calle 70 No. 52-21, Medellín, Colombia
(andresm.cardenas, alher.hernandez, juliana.uribep)@udea.edu.co

Abstract— **The transfemoral amputation involves the loss of the knee joint, which is recognized as a common and complex case. The knee is replaced by a polycentric mechanism, which is exposed to high levels of structural stress. Therefore, mathematical models of the mechanics knees are commonly used to kinetic analysis and simulation and determine possible failures. This paper describes the procedure for determining a kinematic model of a four-bars polycentric knee using a geometric analysis and the Grashof Law for a double rocker. The kinetic model was found using parametric, linear and nonlinear identification techniques, for this were used knee force and angle data supplied by the free database Orthoload. The model couples the kinetics and kinematics ARX structure, these can represent bending angles 110 ° and the total force exerted on the instantaneous center of rotation of the knee.**

Keywords— **Biomechanics, kinematics, nonlinear dynamical systems, prosthetics, prosthetic limbs, system identification.**

I. Introduction

The amputation is a progressive problem associated with different causes like work or traffic accidents [1], diseases [2] and war [3]. Regardless of the cause, unilateral lower limb amputation is a common case [4], assorted according to amputation height [5].

The transfemoral prosthesis is a mechanical assistance device for above knee (AK) amputees. This prosthetic system consists of five parts: prosthetic foot, shank, knee unit, socket and suspension [6]. Each of these elements is subjected to different mechanical stresses decreasing the life cycle of elements, affecting mobility and patient confidence [7]. The mechanical knee facilitates patient mobility, improves gait performance and stability of the amputee during their activities of daily living [8]. The prosthetic knees are characterized by monocentric and polycentric mechanisms, providing greater or lesser flexion and extension of the joint [9].

The polycentric knees have different rotation axes, which converge in the instant center of rotation (ICR). These knees are used by it's biomechanics versatility, presenting good performance in the stance and swing phase gait, allowing greater confidence, stability, bending and factly.

A typical four bar knee is shown in Fig 1. This is characterized by four center cranks A, B, OB y OA linked by OA-A, A-B, B-OB, OA-OB, conform 4 links a, b, c, d respectively. Mechanical linkage configuration is based on Law of Grashof mechanisms, where the length of each link is determined to have a full joint revolution [10].

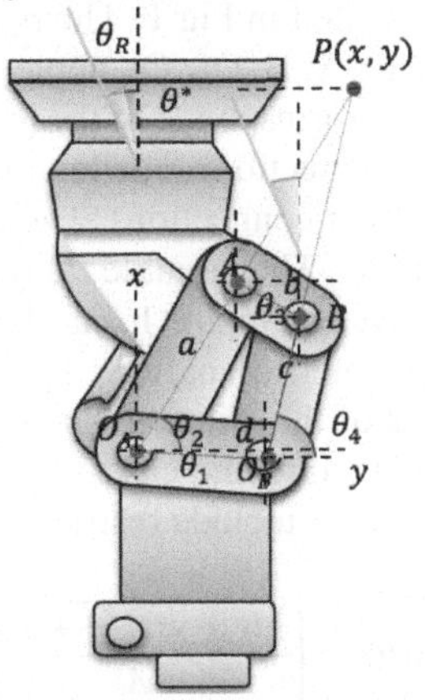

Fig 1. Extended knee geometry: links, articulation angles and instant center of rotation.

The knee should support the weight of the body and additional loads during gait cycles. The resulting forces stabilize the body and achieve the displacement during the phases of gait. If known the knee force magnitude and direction, is possible to choose a correct construction material, simulate the gait performance and estimate possible structural problems.

Two forms of analysis are discussed in this paper, kinematic and kinetic. To find kinetic model, the study is conducted with data from healthy patients, as suggested in [11]. Data is collected from 8 test subjects with an average weight of 100kg and different system identification techniques are evaluated. To find kinematic model, a geometric analysis of four- bar mechanism is done. The resulting model allows provides information about the ICR, marked with a *P(x,y)*.

Mohsen and colleagues [12] designed a method to parameterize a polycentric knee four-bars using the ICR and the floor reaction forces of (FRF) during gait (the initial contact, the support phase and voluntary bending before takeoff foot). To determine an appropriate measures and angles, a genetic algorithm uses FRF data of an amputee patient to parameterize the model of the knee.

The four-bars knee model and the optimal geometry of the joints are presented in [13]. They use the Grashof Law, the

I. Torres et al. (eds.), *VII Latin American Congress on Biomedical Engineering CLAIB 2016, Bucaramanga, Santander, Colombia, October 26th -28th, 2016*,
IFMBE Proceedings 60,
DOI: 10.1007/978-981-10-4086-3_150

transmission angle criteria and the angle sequence. Has been used an optimization method known as musical harmony, which is based on better state of harmony.

A general model for the simulation of polycentric knee has not been studied considerably. Therefore, the paper also suggests applicate a system identification.

II. Methodology

A. Kinematic Analysis

The analysis of four-bar mechanism is performed following the geometry presented in Fig 1. The *a*, *b*, *c*, *d*, variables are lengths of the links, A, B, O_B, O_A depict joints.

The lower link d, corresponds of the mechanical knee distal view part (shank connection). The top link b connected to the proximal part (socket connection). The alignment angles are defined by θ_1 and θ^*. The variable θ_R represents the angle of the knee or the system input. It can be calculated as $\theta_3 = \theta_R + \theta^*$. By a geometric analysis, θ_2 and θ_4 angles can be describes in terms of the joints and some intermediate variables as (1) and (2) . In some cases, these equations may be indeterminate, so must include restrictions in evaluating simulation.

$$\theta_2 = 2tan^{-1}\left[\frac{-B \pm \sqrt{B^2 - 4AC}}{2A}\right] \tag{1}$$

$$\theta_4 = 2tan^{-1}\left[\frac{-B \pm \sqrt{E^2 - 4DF}}{2D}\right] \tag{2}$$

Where,

$$A = \frac{K_2}{2ab} - \frac{dK_1}{a} - cos(\theta_3) + \frac{dcos(\theta_1)}{b} \tag{3}$$

$$B = 2\left[sin(\theta_3) - \frac{dsin(\theta_1)}{b}\right] \tag{4}$$

$$C = \frac{K_2}{2ab} - \frac{dK_1}{a} - \frac{dcos(\theta_1)}{b} + cos(\theta_3) \tag{5}$$

$$E = 2\left[\frac{dsin(\theta_1)}{b} - sin(\theta_3)\right] \tag{6}$$

$$D = \frac{K_3}{2bc} - \frac{dK_1}{c} - \frac{dcos(\theta_1)}{b} + cos(\theta_3) \tag{7}$$

$$D = \frac{K_3}{2bc} - \frac{dK_1}{c} + \frac{dcos(\theta_1)}{b} - cos(\theta_3) \tag{8}$$

$$K_1 = cos(\theta_1)\,cos(\theta_3) + sin(\theta_1)\,sin(\theta_3) \tag{9}$$

$$K_2 = a^2 + b^2 - c^2 + d^2 \tag{10}$$

$$K_3 = b^2 - a^2 + c^2 + d^2 \tag{11}$$

Finally, joint points A, B, O_A and O_B can be calculated as,

$$O_A(X_{O_A}, Y_{O_A}) = [0,0] \tag{12}$$

$$O_B(X_{O_B}, Y_{O_B}) = [dcos(\theta_1) dsin(\theta_1)] \tag{13}$$

$$A(X_A, Y_A) = [acos(\theta_2) - asin(\theta_2)] \tag{14}$$

$$B(X_B, Y_B) = [X_{O_B} + ccos(\theta_4), Y_{O_B} + csin(\theta_4)] \tag{15}$$

The $P(x, y)$ point is located where the projections of the *a* and *c*, converge.

$$P_x = X_B + \frac{K_4}{K_5} - dcos(\theta_1) - ccos(\theta_4) \tag{16}$$

$$P_y = Y_B + \frac{K_4}{K_5}\tan(-\theta_2) - dsin(\theta_1) - csin(\theta_4) \tag{17}$$

A double swing is simulated on Matlab, making the shortest link "*b*" and the base "*d*" are fixed, fulfilling the condition of the second law of Grashof ($b+a \leq c+d$). For simulation, $a=6cm$, $b=1cm$, $c=5.5cm$ and $d=3cm$ were chosen. An initial upper alignment angle (top of the knee) was considered by $\theta^* = -5°$ and lower alignment angle by $\theta_2 = -5°$. The model was tested with θ_1 = [-5°,-10°,-20°], simulating alignment changes on the socket and shank, then *P(x,y)* movements were evaluated.

B. Systems Identification

For model identification, the Orthoload data have been used without commercial use [14]. The data include information of 8 patients six men and two women, aged about 70 years, height 172cm, weight 91kg. Each patient flexed knee, the angle (°) and forces (N) were measured in each coordinate axis, then the joint total force was estimated. To eliminate weight patient variability, a proportional value of the force was found. The procedure of data processing used by Orthoload, includes a standardizes medial forces F_X, anterior F_Y, distal F_Z and resultant force F_{Res} about 100kg following the "dynamic time warping (DTW)" [15]. This procedure is a normalized time signal, a distorted of time signal, so that the summed squared errors between all of them become a minimum. Then, resulting signals are statistically treated obtaining different output signals: arithmetic mean, median, minimum, maximum, 25 percentiles, 75 percentiles. In this paper averages of one individual subject is used. Details of procedure are described in [16]. Data of resulting force and bending angle are seen in Fig 2.

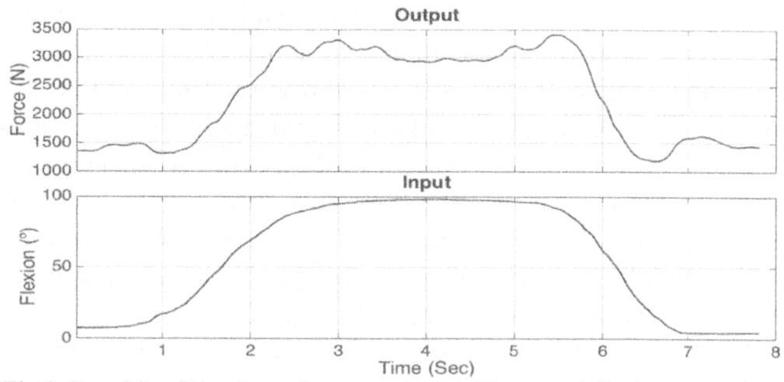

Fig 2. Resulting force is used as experimental input and flexion-extension used as experimental output.

Treatment of the data is performed before making the identification. First, an analysis is performed to determine frequency noise effects. In order to eliminate offset, the average values of the output signal and input were removed.

The linear polynomial structures tested were: ARX, ARMAX, OE and BJ. Then, a nonlinear Hammerstein-Wiener and nonlinear ARX. Nonlinear ARX model consists of a regressors block, a linear block and non-linear block. The Hammerstein-Wiener model consists of a series of linear block between two non-linear blocks. The best structure and order is performed from minimization of final prediction error criteria (18), the method of normalized Akaike criteria (19) and the adjustment percentage.

$$FPE = det\left(\frac{1}{N}\sum_{1}^{N} EE^{T}\right)\left(\frac{1+d/N}{1-d/N}\right) \quad (18)$$

$$AICc = log\left(det(E^{T}E/N)\right) + 2\,d/N \quad (19)$$

Where E is the prediction error matrix, d is the number of estimated parameters, N is the number of estimation data samples.

III. Results

A. Kinematic Model.

A knee flexion between $\theta_R = 10°$ and $\theta_R = 140°$ is applied. Six transitions of kinematic model simulation are showed in Fig 3. The P(x,y) describe an rotation angle between 10° and 140° is. This allowed to estimate the movement of the knee.

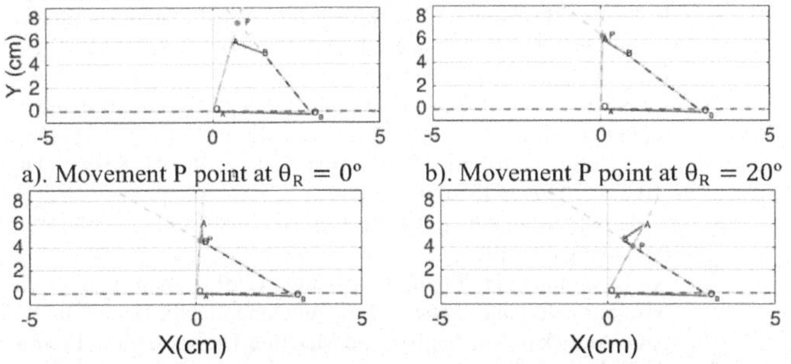

a). Movement P point at $\theta_R = 0°$ b). Movement P point at $\theta_R = 20°$

c). Movement P point at $\theta_R = 60°$ d). Movement P point at $\theta_R = 140°$

Fig. 3. Flexion knee on θ_3 joint. Displacement of instantaneous center of rotation.

B. System Identification Techniques

a) Linear parametric identification

Concerning linear structures, 1600 different ARX models varying parameters for $n_a = (1{:}10)$, $n_b = (1{:}10)$ and $n_k = (0{:}15)$ were tested. The models that minimized the two criteria (equations 18 and 19), FPE=0.69 and AIC=-0.29.

In the case of ARMAX structure, 2000 combinations were evaluated: $n_a = (1{:}5)$, $n_b = (1{:}5)$, $n_c = (1{:}5)$ and $n_k = (0{:}15)$. The minimizing of FPE = 0.74 and AIC = -0.30.

In the Output Error (OE) model, 1600 combinations of parameters $n_b = (1{:}10)$, $n_f = (1{:}10)$ and $n_k = (0{:}15)$. The FPE and AIC values were 1.49 y 9.61 respectively.

Concerning Box Jenquins (BJ) structure, 10.000 combinations were evaluated: $n_b = (1{:}5)$, $n_c = (1{:}5)$, $n_k = (1{:}5)$, $n_f = (1{:}5)$ and $n_k = (0{:}15)$, the minimization was achieved for FPE=0.74 and AIC=-0.31.

The minimizing criteria was achieved for each model (ARX, ARMAX, OE and BJ) with the same order for each one. Simulation of each model are presented in the figure 4.

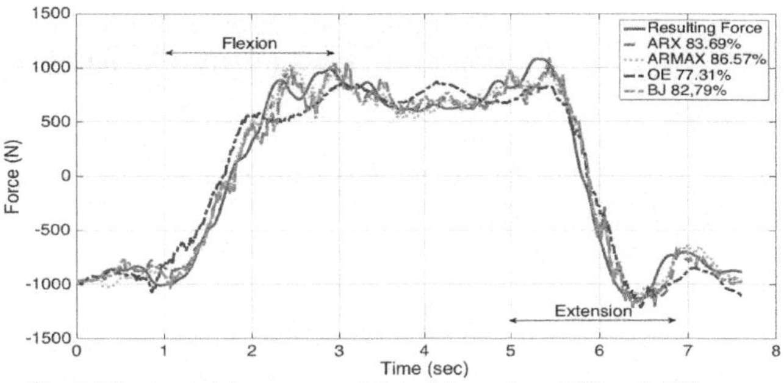

Fig 4. Linear models response. Minimizing values EFP and AIC.

b) Nonlinear identification

Model parameter of nonlinear ARX were varied in 3100 combinations for $n_a = (1{:}10)$, $n_b = (1{:}10)$ and $n_k = (0{:}30)$. The FPE=0.71 and AICc = -0.34. For the the Hammerstein-Wiener model was evaluated for 170 combinations of n_a=n_b=(1,3,5,7,9) and n_k=(0,5,10,15,20,25,30). The FPE=5.18 and AIC=8.55. The simulation models are present in the fig 5.

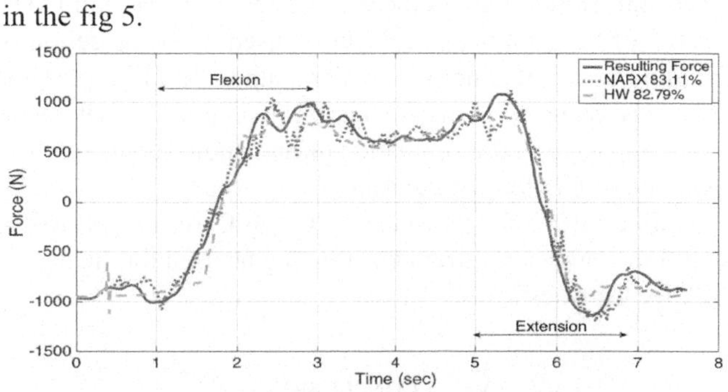

Fig 5. Non linear models response. Minimizing values EFP and AIC.

c) Model selection

The models were compared as shown in Table 1.

Table 1 Evaluation of models identified

MODELS	FPE	AIC	Mean error	Parameters*	FIT (%)
ARX	0,69	-0,29	1,72	(4,2,-,-,-,2)	83,69%
ARMAX	0,74	-0,30	-3,87	(5,3,1,-,-,2)	86,57%
OE	1,49	9,61	-2,05	(-,9,-,-,4,0)	77,31%
BJ	0,74	-0,31	-0,51	(-,1,1,5,3,2)	84,52%
NARX	0,71	-0,34	-2,65	(10,7,-,-,-,13)	83,11%
Ham.-Wiener	5,18	8,55	24,97	(-,5,-,-,5,30)	82,79%

The parameters row present the values of *na*, *nb*, *nc*, *nd*, *nf* and *nk*. The model that best adjusted all selection criteria in relation to minimizing information criterion, the average error of the order and maximizing the fit to the data was the ARX.

The continuous time model that describing the force exerted on the knee joint is presented in equation 20.

$$\frac{F_{Res}(s)}{\theta_R(s)} = e^{-0.013s}\frac{10.5s^3 + 2767s^2 + 232000s + 589600}{s^4 + 60.6s^3 + 1739s^2 + 9710s + 32100} \quad (20)$$

To found the ICR force (F_P), the P point angle is calculated by $\theta_P = cos(X_P, Y_P)$. This is the input to dynamic model, then ICR force is estimated.

IV. Discussion

The resulting model simulates a prosthetic knee joint permitting a range of performance of 110º, with variations between 10º and 120°. The angle of knee $\theta_R(t)$ is the input of the model, generating the instantaneous center of rotation (ICR), which, added to the estimate of the value of the force exerted on the joint (found by identification), provides information of a *P(x,y)* angle and force.

The resulting overall model can simulate knee flexion of 110° and estimating parameters of strength in the joint. The results of this simple model can be used in the selection of building materials knees (in patients up to 120 kg), possible structural damage, in the simulation of amputee's gait deviations, and even in the simulation of humanoid robots, with an estimate of 83.69%. of the data.

Unlike other models found by Ortho Care, this allows to determine angle and strength, allowing be used for the design of knee prosthesis.

V. Conclusions

The resulting model allows simulations of possible movement made by the amputee and determine the forces that will undergo construction materials of mechanical knee.

The right balance between the different criteria for selecting a parametric model for system identification, to find a tight model representing multiple system states.

Should properly choosing the links lengths for proper knee simulation.

The use of two sub-models in series allows adjustments to each of these, facilitating the simulation work.

Acknowledgment

This work has been supported in part by "Convocatoria 727 – Programa de Becas de Doctorado Nacionales 2015 de Colciencias" and the project "Fortalecimiento de plataforma tecnológica para la formación especializada en el área de la salud y el desarrollo de tecnología biomédica", Ruta N code 139C of the Fondo Nacional de Regalías, Republic of Colombia.

Conflict of Interest

The authors declare that there is no conflict of interest regarding the publication of this paper.

References

1. Ministerio del http://www.mintrabajo.gov.co/abril-2015/4398-disminuyen-muertes-por-accidentalidad-laboral.html.
2. Gil-Chang (2007) Fundamentos de Medicina de Rehabilitación, San José: Universidad de Costa Rica, 157-158.
3. J. P. Daniels (2016) Helping Colombia's landmine survivors. The Lancet 387(10033):2079–2080 DOI:10.1016/S0140-6736(16)30597-9.
4. J. E. Restrepo, E. A. Yara, J. Cano Betancur, L. N. Tavera (2014) Perfil emocional de un grupo de militares colombianos víctimas de minas antipersona o artefactos explosivos improvisados. Rev. Col. de Psiquiatría 87–95 DOI 10.1016/j.rcp.2014.02.004.
5. J. E. Edelstein, A. Moroz (2011) Lower-Limb Prosthetics and Orthotics: Clinical Concepts. SLACK Incorporate, New York:
6. R. Meier y D. Melton (2013) Ideal Functional Outcomes for Amputation Levels. Amputee Rehabilitation 25(1):199-212 DOI 10.1016/j.pmr.2013.09.011.
7. J. Andrysek (2010) Lower-limb prosthetic technologies in the developing world: A review of literature from 1994–2010. Prosthetics and Orthotics International 34(4):378-98 DOI 10.3109/03093646.2010.520060.
8. Calumet Orthopedic & Prosthetics Co (2007) Selecting the Right Mechanical Knee.
9. N. Palastanga, D. Field, R. Soames (2000) Anatomía y Movimiento Humano: Estructura y Funcionamiento, Barcelona.
10. A.G. Ambekar, Mechanism and Machine Theory (2007). Prentice Hall, New Delhi.
11. Belkys, C. Müller, E. Casanova, R. Torrealba (2013) Structural analysis during the design of polycentric prosthetic knee. 24th CISB. RJ, Brazil, 2013, DOI:10.13140/RG.2.1.1079.3124
12. R. Amador, R. Torrealba, M. Rojas, J. Cappalletto, C. M. Müller Karger (2012) Metodología para dimensionamiento de mecanismo policéntrico de rodilla utilizando análisis de marcha y algoritmos genéticos. Rev Ing. Biomédica 6(11):30-45.
13. S. Mohsen Alhajj (2011) Path Synthesis of Four Mechanism Using Harmony Search Optimization. National Institute of Technology, Rourkela.
14. Orthoload at http://www.orthoload.com/test-loads/standard-loads-knee-joint/.
15. G. Bergmann, A. Bender, F. Graichen, et al (2014) Standardized Loads Acting in Knee Implants. PLoS ONE.
16. Bender, G. Bergmann (2012) Determination of typical patterns from strongly varying signals. Biomed Eng., 15(7):761–769 DOI 10.1080/10255842.2011.560841

Usability Evaluation for a Vital Signs Monitor Prototype

L.E. Arenas[1], P.J. Bedoya[1], L. Correa[1], J.G. Barreneche[1], A.M. Hernández[1]

[1]Bioinstrumentation and Clinical Engineering Research Group - GIBIC, Bioengineering Department, Engineering Faculty, Universidad de Antioquia UdeA; Calle 70 No. 52-21, Medellín, Colombia

Abstract— **The usability test performed as part of the processes of biomedical equipment development has great importance for safety use of health care technology, between 50% and 70% of adverse events with medical devices are related with assembly mistakes, wrong connections and incorrect operation. Usability test requires detailed processes with rigorous methodologies that guarantee test to the user, who measures several aspects of the use, interface and process in the equipment. This article compares the usability of a prototype of vital signs monitor against a commercial product in terms of some aspects of the human factor engineering that were applied in order to test the usability of a Prototype, the commercial vital signs monitor is used as a reference. This paper integrates a methodology composed by 3 phases that allows an analysis of users` behavior interacting with medical devices. Time and steps were analyzed through statistics and heuristic analysis.**

Key words— **Usability, adverse event, vital signs monitor, human factor engineering, patient safety, prototype.**

I. Introduction

According to the World Health Organization (WHO) adverse events in patients related with medical devices in the United States are more than one million every year. [1] In Colombia, in 2010, 29 adverse events related with medical devices were identified in clinics or hospitals, and 21 of them were clearly predictable. [1]

In accordance with the Food and Drug Administration (FDA), between 45% and 50% retirements of medical devices in hospitals were a consequence of a bad design of the product. [2]

Data given by the WHO and the Global Harmonization Task Force (GHTF) show that between 50% and 70% of adverse events or incidents with medical devices are related with assembly mistakes, wrong connections and incorrect operation. [3]

Comfort, user friendliness, simplified steps, learnability, effectiveness, safety, visual appeal and ergonomics are crucial aspects to consider when design process is on, in order to ensure a successful marketing process.

Biomedical devices designers are focused on create products to satisfy the medical industry. At the same time, for hospitals is important that people who operate the devices feel comfortable with them, so a correct and optimal use can be made; otherwise, new technology will be underused and old technology will be overused, until it get damaged, generating losses.

In order to reduce that possibility, a usability test which can measure the quality of users' experience with the product is needed. Usability tests main purpose is to give useful information to designers, showing product performance and how reliable and safe it is.

In this paper, a methodology based on human factor engineering to design usability tests for medical devices is developed. Nine expertise nurses tested the main indicators of two vital signs monitors: heart rate, non-invasive pressure and oxygen saturation; showing how friendly each device was.

II. Methodology

The Human factor engineering, according to the *Guidance for Industry and FDA Premarket and Design Control Reviewers* [4], has three phases: task analysis, heuristic analysis and usability tests.

Human factor engineering was applied to nine head nurses. They tested two vital signs monitors (VSM): The first one is commercially used. It has a traditional design: conventional screen with selection mode through knobs and hard keys. The other one is a VSM in design process, or a prototype. Its interface might be a tablet, a cellphone or a computer. Another fact is that it only has two bottoms: one for turn on and another for NIBP measurement. The selected interface was a tablet.

The Purpose of this research was to make a comparison between the prototype and the commercial vital signs monitors in some aspects of usability such user friendliness, design, accessories connections, VSM startup among others.

The phases of the research are:

I. Torres et al. (eds.), *VII Latin American Congress on Biomedical Engineering CLAIB 2016, Bucaramanga, Santander, Colombia, October 26th -28th, 2016*,
IFMBE Proceedings 60,
DOI: 10.1007/978-981-10-4086-3_151

A. Phase 1: task analysis

The first step was to consider, hypothetically, which errors may occur, and which of them might trigger a negative impact, by pinpointing potential causes and drawing conclusions about consequences and appropriate design solutions.

B. Phase 2: usability tests

In this step, a flux diagram for each task was built (turn on, ECG, NIBP and SPO_2 measurements), in order to record the steps and the time taken during the whole test.

C. Phase 3: heuristic analysis

At this point, possible use-related hazards were identified in the interaction of the user with the vital signs monitor. 10 principles were included in Jakob Nielsen's Heuristic analysis: [5]

1. Visibility of system status.
2. Match between system and the real world.
3. User control and freedom.
4. Consistency and standards.
5. Error prevention.
6. Recognition rather than recall.
7. Flexibility and efficiency of use.
8. Aesthetic and minimalist design.
9. Help users recognize, diagnose, and recover from errors.
10. Help and documentation.

According to these heuristic analysis principles, and after testing both vital signs monitors, experts were allow to answer the question below, related to usability:

1. Is there any kind of feedback for each action or operation?
2. Response time is appropriate for each task?
3. Terminology, icons and colors. Do they answer to expected values according to the user's code?
4. Function keys are clearly tagged and are easily distinguishable.
5. Is it easy for users to change from one window to another in systems with overlapped windows?
6. Is it asked to the users to confirm actions which may have negative, drastic or destructive consequences?
7. Can users undo their actions in a simple way?
8. Sound: low tones for regular feedback devices and high tones for critical situations?
9. Does the system alert users if they are about to make a potentially serious error?
10. Browsing: is the information easy to find?
11. Is the device easy to use?
12. Are the instructions easy to remember?

Then, the answers in the previous step were ranked based on heuristic principles as follows:

1 point: The device doesn't obtain a proper score in the heuristic analysis.

2 points: The device obtains a medium score and it is necessary to improve it.

3 points: the device obtains a high score in the heuristic analysis.

III. Results

Personnel involved in the test was a group of nine head nurses, three of them were men between 24 and 27 years of age and six were women between 25 and 51 years of age.

Initially, a 15 minutes training was given to the head nurses, teaching them how to identify ECG, SPO_2 and NIBP modules and how to measure these indicators; while they were using both VSMs, time and steps to complete the tasks were recorded, then, the twelve heuristic questions were answered.

In Tables 1 and 2 results of the test are presented, pointing the number of people who rated each one of the heuristics with 1, 2 or 3.

Table 1 Data summary for VSM prototype

PROTOTYPE

HEURISTICS		**1**	**2**	**3**	**4**	**5**	**6**	**7**	**8**	**9**	**10**	**11**	**12**
SCORE	**1**	0	0	1	0	0	3	0	1	3	0	0	0
	2	2	2	1	1	2	2	2	1	3	1	1	2
	3	7	7	7	8	7	4	7	7	3	8	8	6

Table 2 Data summary for commercial VSM

COMMERCIAL

HEURISTICS		**1**	**2**	**3**	**4**	**5**	**6**	**7**	**8**	**9**	**10**	**11**	**12**
SCORE	**1**	0	0	1	2	2	3	2	2	1	0	0	1
	2	1	1	2	1	2	1	3	2	3	2	2	4
	3	8	8	6	6	5	5	4	5	5	7	7	4

The time and steps taken for each person and the expert is shown in Table 3.

Table 3 Time taken with prototype and commercial VSM

SAMPLE	PROTOTYPE		COMMERCIAL	
	Time (min)	Steps	Time (min)	Steps
1	2,6	19	5,9	18
2	2,7	23	6,6	19
3	2,5	21	7	22
4	3,1	23	7,2	21
5	2,2	20	5,5	20
6	2,6	23	5,8	21
7	3,1	26	7	21
8	2,8	22	6	20
9	2,4	19	5,7	21
Expert	2,1	19	5	18

Also, in Tables 4 and 5, averages and standard deviation are shown respectively.

Table 4 Steps and time average

AVERAGE			
		Step	Time (min)
Sample	Prototype	22	3
	Commercial	20	7
Expert	Prototype	19	2,1
	Commercial	17	5

Table 5 Steps and time standard deviation

	Standard deviation	Desviation coeficient	%
Prototype	2,149	0,099	9,87
Commercial	1,15	0,06	5,68

With Tables 1 and 2 data, the results of heuristic questions were sorted out in levels of satisfaction by evaluating the total number of people who rated the device with 3 points in each heuristic, as follows:

Low satisfaction rate: 1, 2 or 3 people.
Medium satisfaction rate: 4, 5 or 6 people.
High satisfaction rate: 7, 8 or 9 people.

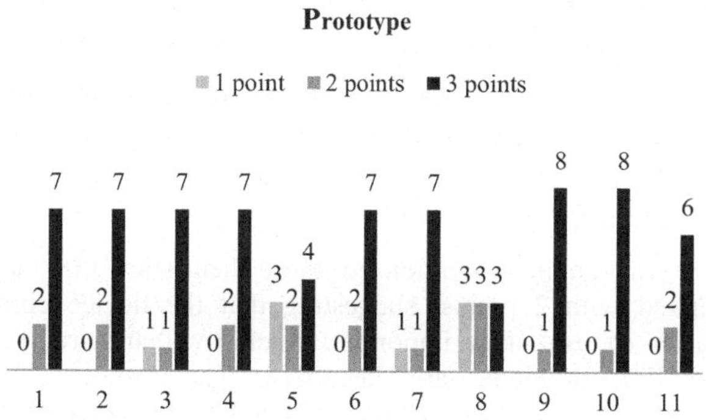

Graphic 3 Heuristic achievements in the VSM prototype

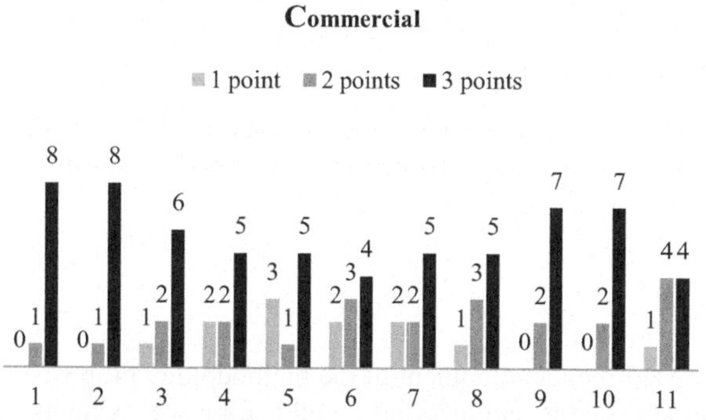

Graphic 4 Heuristic achievements in the commercial VSM

IV. Analysis

According to table 4, it is clear that the nurses took less time to complete the task with the VSM prototype than the time they took with the commercial one.
At the same time, the nine people involved in the test took more time to complete the task than the experts with both VSMs: assistance personnel took 2.66 minutes in average with the prototype, 33 seconds more than the experts. With the commercial VSM, they took 6.3 minutes, 78 seconds more than the experts.

Also, in table 5 head nurses and experts did two more steps with the prototype than with the commercial VSM. Analyzing the standard deviation, both VSMs were in the allowed range. It can be deduced that the prototype got a lower deviation percentage because the behavior of the nine users were similar, in contrast to the situation with the commercial VSM, which had a higher percentage because its proper use depends directly on how acquainted is the user with a tablet.

Also, it is clear that in despite of MSV prototype operation needs two more steps than the commercial one, it takes less time to complete the tasks, because browsing with a tablet is faster than with the commercial one, which has knobs and bottoms.

In a deeper analysis, and identifying the heuristics with higher and lower score, it can be regarded in graphic 3 that the results are favorable to the prototype, because nine of the twelve heuristics are highly ranked.

Now, paying attention to these heuristics the majority ranked with 2 points, suggesting that the device could be easier to use, it is important to clarify that here the real usability problems are identified. The prototype shows failures in heuristics 6 and 9 related to the device's alerts which prevent user's wrong decisions.

On the other hand, graphic 4 shows that 8 of the 12 heuristics for commercial VSM, were ranked in the medium satisfaction rate.

It is clear that for both VSMs, none of the heuristics were ranked in the low satisfaction rate for the majority of head nurses.

V. Conclusion

A correct development of the methodology proposed by the human factor engineering, could improve patients safety indicators, because it can show potential errors. It also may offer valuable information not available in other tests.

Usability should become an important part of the technology validation process, because of what it may generate in equipment's design; for this, the acceptance in the market might be higher. Every prototype should be validated directly by the users.

In the technology acquisition process, an evaluation by usability tests should be demanded, given that this guarantees that technology will be easy to use for the medical staff. The consequence will be a decrease in the number of adverse events caused by the incorrect use of the devices. Frequently, this is due to a low usability index of technology.

Acknowledgment

This work was supported partially by Sistema General de Regalías, Republic of Colombia through the project "Fortalecimiento de la plataforma tecnológica para la formación especializada en el área de la salud y el desarrollo de tecnología biomédica", code RutaN-139C.

Conflict of Interest

The authors declare that there is no conflict of interest regarding the publication of this paper.

References

1. Alfonso L P, Salazar C et al. (2010) Incidencia de eventos adversos asociados a dispositivos médicos en una institución de salud en Colombia. *Rev. Ing. Biomédica*, vol. 4, N° 8, 2010, págs. 71-84.

2. Gregory W. O'Connell, "Human Factors in the GMP Inspection. Food and Drug Administration. Regulatory Authority for Design Controls." at http://www.fda.gov/medicaldevices/deviceregulationandguidance/humanfactors/ucm128186.htm

3. Hernández A. (2008) Tecnovigilancia, perspectiva de la Organización Panamericana de la Salud, consultado el 20 de febrero de 2016, Bogotá, Colombia.

4. D. C. (2000) Reviewers, "Applying Human Factors and Usability Engineering to Optimize Medical Device Design", FDA Guid. Ind., no. 301, pp. 1–37, 2011.

5. Nielsen Norman Group (1995) Evidence-Based User Experience Research, Training, and Consulting. 10 Usability Heuristics for User Interface Design at https://www.nngroup.com/articles/ten-usability-heuristics/.

Author: A.M. Hernández.
Institute: University of Antioquia
Street: Calle 67 No. 53 - 108
City: Medellin
Country: Colombia
Email: *alher.hernandez@udea.edu.co*

New insights into the scoring of respiratory events based on alternative sensors: A comparative effectiveness study

C R Dell'Aquila, L S Correa, R Correa, G E Cañadas, E Laciar

Universidad Nacional de San Juan/Facultad de Ingeniería, Gabinete de Tecnología Médica, San Juan, Argentina.
Consejo Nacional de Investigaciones Científicas y Técnicas/Ministerio de Ciencia,
Tecnología e Innovación Productiva de la Nación, Argentina

Abstract— **In this study we compare the traditionally scoring of apnea hypopnea events (based on the respiratory flow with oxygen desaturation) with the obtained by applying an alternative methodology (based on the Respiratory Inductance Plethysmography flow estimation and oxygen saturation variability features: MAD and IQR), in 23 polisomnographic recordings. Two new models were proposed and compared with the standard one. The Apnea-Hypopnea Index was measured and root mean square was computed with the expert's scores. Patients were classified with the American Academy of Sleep Medicine recommended rules. Results indicate an improvement in AHI estimation using the model based on the RIPflow and the SpO_2MAD.**

Keywords— **Apnea, AHI, RIPflow, SpO_2 variability.**

I. INTRODUCTION

The sleep apnea-hypopnea syndrome (SAHS) is mainly characterized by the recurrence of both total breathing cessation (apnea events) and significant airflow reduction (hypopnea events) during sleep time [1]. It has been reported that SAHS affects severely to 1 in 15 adults, can induce a great amount of damage to all body systems and is associated with increased secondary cardiovascular morbidity and mortality [2].

The recommended rules for scoring respiratory events are detailed into the task force of the American Academy of Sleep Medicine [3]. Apnea in adults is scored when there is a drop in the peak respiratory signal excursion by $\geq 90\%$ of pre-event baseline for ≥ 10 seconds. Hypopnea in adults is scored when the peak signal excursions drop by $\geq 30\%$ of pre-event baseline for ≥ 10 seconds in association with either $\geq 3\%$ arterial oxygen desaturation or an arousal.

Currently, oronasal airflow sensors (OAS), and blood oxygen saturation sensor are considered as the standard sensors for apnea and hypopnea diagnosis.

The respiratory inductance plethysmography (RIP) signals to be used as alternative sensors for apnea and hypopnea detection specified in the task force [3] included the RIPsum which is the sum of the thorax and abdomen belt signals and the RIPflow (the time derivative of the RIPsum), an estimation of the airflow signal. During apnea these two signals show absent or minimal excursions, and during hypopnea, the excursions are diminished compared with baseline breathing.

As regards the detection of blood oxygen, the desaturation associated with a respiratory event is defined as a drop from a baseline SpO_2 preceding the event. However, no "baseline SpO_2" has been considered as standard since it is a no linear signal: the same drop percentage in the SpO_2 implies different variations in the arterial oxygen pressure depending on the baseline considered [3].

Studies have evaluated the accuracy of RIPsum or RIPflow to detect apneas and hypopneas. The work of Thurnheer et al showed good agreement in the measured with these signals with the pneumotachograph [4]. Another study proposed a method based on ensemble learning to estimate the respiratory flow, the thoracic respiratory effort and the abdominal respiratory effort from acceleration of suprasternal notch, the thorax and the abdomen, respectively [5].

In a recent research, Kogan *et al* showed that an improvement in sensitivity and specificity could be obtained when scoring hypopneas by RIPsum channel when compared with both the recommended and acceptable criteria of the American Academy of Sleep Medicine [6].

In this work, we propose an alternative way to measure the alterations of the oxygen saturations during respiratory events in different models using the respiratory flow and the RIPflow, and compare them by the AHI and the patient classification, validating the results by comparing with the physicians scores available at the St. Vincent's University Hospital / University College Dublin Sleep Apnea Database [7].

II. MATERIALS AND METHODS

A. Database

The database used is freely available in Physionet's web site [8]. It contains 25 full overnight PSG records of patients (age: 50±18 years, 21M and 4F) with suspected sleep-

I. Torres et al. (eds.), *VII Latin American Congress on Biomedical Engineering CLAIB 2016, Bucaramanga, Santander, Colombia, October 26th -28th, 2016*,
IFMBE Proceedings 60,
DOI: 10.1007/978-981-10-4086-3_152

disordered-breathing. The recording lasts from 5.9 to 7.7 hours. It also includes sleep stages and respiratory annotations done by experts with the respiratory event start and last; and event classification (Central Apnea/Hypopnea, Obstructive Apnea/Hypopnea and Mixed events).

In this study, the Respiratory Flow, RIPSum and SpO_2 signals, sampled at 128Hz, were used. Respiratory Flow is a calibrated signal expressed in liters/seconds and SpO_2 is the percentage of blood oxygen saturation. The RIPSum signal is the sum of Thorax and Abdomen respiratory effort, which is a non-calibrated signal.

The records UCDDB006 and UCDDB011 were not used because they have problems in the recording of the signals of interest for our purpose.

B. Signal Pre-processing

A digital filter is applied to Flow and RIPSum signals, to obtain the spectral range that defines the respiratory process, which comes up to 0.5Hz [9]. The bidirectional 4th order Butterworth low-pass filter has been selected with cut-off frequency in 0.5Hz. It has the zero-phase distortion advantage.

In order to obtain the RIPFlow, a five points Differentiator-Integrator filter is applied to the RIPSum. It derives the input signal up to the desired 0.5 Hz frequency and attenuates the rest.

The filtered Flow, RIPSum and RIPFlow signals are shown in Fig. 1.

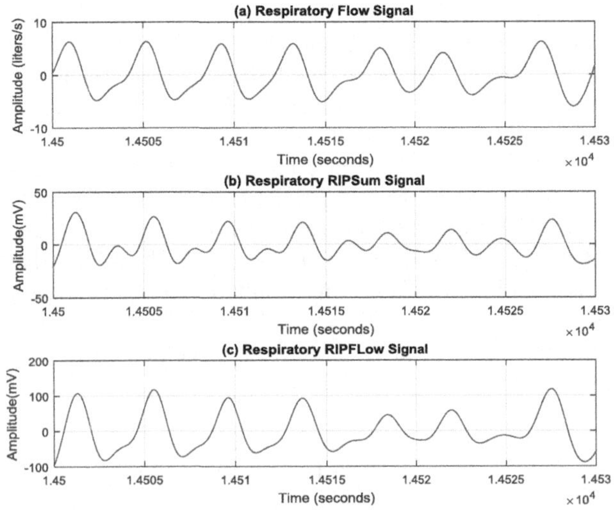

Fig. 1: (a) Respiratory Flow Signal; (b) Respiratory RIPSum Signal, and (c) Respiratory RIPFlow signal.

The SpO_2 signal is resampled to 32 Hz to decrease the computational cost, according to AASM – Manual for the Scoring of Sleep and Associated Events that recommends a minimum of 25 Hz [10].

C. Breathing and SpO_2 signal processing

The breathing baseline (BB) is defined as 3 largest breaths in the 2 minutes preceding onset of the event, in individuals without a stable breathing pattern, and it is valid for Flow and RIPFlow signals [3]. In the present work, BB is estimated to determine Thresholds (*Th*) that are used to convert a respiratory signal (RS) in square signals (SQS).

The RS is divided into 1 minute last segments and each one is associated with a *Th*. In order to compute it for a present segment (PS), the 2 minutes previous are analyzed. This previous segment (PRS) is divided into 3 sub-segments (SS) and for each one the mean of the 3 largest local maximum is calculated. Then the *Th* for PS is computed as the median of the 3 means of the 3 SS.

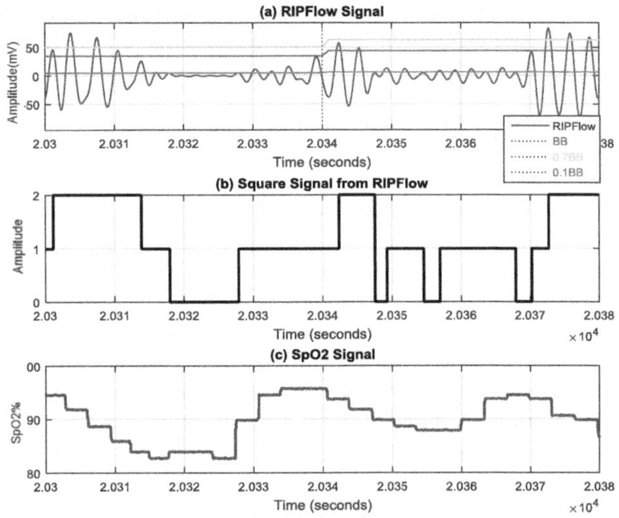

Fig. 2: (a) RIPFlow with Breathing baseline (BB), 0.7BB and 0.1BB; (b) Square signal from RIPFlow and SpO2 signal.

The SQS is generated for each RS, based on definition of apnea and hypopnea. If RS was lower than 70% of *Th*, then SQS was assigned with 1, and with be 0 if the RS is lower than 10% of *Th*. The SQS was assigned with 2 if the RS is larger than 70% of *Th*. This process and the estimated BB are shown in Fig. 2 (a) y (b).

The intervals where SQS is 1 or 0 were computed. If an interval is 0 for 10 or more seconds, it is scored as respiratory event (RE). On the other hand, if an interval is 0 or 1 for 10 or more seconds, it is scored as possible respiratory event (PRE) and the SpO_2 is then analized. In this sense, the following 3 approaches were used.

The first is based on the traditional method, which is the SpO_2 drop of 3% of pre-event SpO_2 baseline (SB). The SB is estimated to obtain a *Th* for each minute segment using the same method that in RS. Then, the detected event is scored as RE, if SpO_2 signal drop 3% of *Th* during a PRE ± 2 seconds.

The second is the Mean Absolute Deviation (MAD) of SpO_2 and the third is Interquartile Range (IQR), of SpO_2 signal during a PRE ± 2 seg. These two features estimate the variability of the SpO_2 signal. Then a fixed *Th*, previously set in the algorithm, is used to score the event. The Fig. 1 (c) shows the SpO_2 signal.

Table 1 describes the signals and features used in each analyzed model.

Table 1 Models for Respiratory Events

Model	Respiratory Signal	SpO_2 Analysis
Model 1 (M1)	Flow	Standard
Model 2 (M2)	Flow	MAD
Model 3 (M3)	Flow	IQR
Model 4 (M4)	RIPFlow	Standard
Model 5 (M5)	RIPFlow	MAD
Model 6 (M6)	RIPFlow	IQR

To compare the models performance, the apnea/hypopnea per hour index (AHI) is computed as in eq (1) [10].

$$AHI = \frac{\text{Respiratory Events (apneas/hypopneas)}}{\text{Total Sleep Time (in seconds)}} \cdot 60 \qquad (1)$$

III. Results

Respiratory Apnea/Hypopnea events were scored for each record using all 6 studied models and the AHI was computed. The results were contrasted with the physician's score available from database and the Root Mean Square Error (RMSE) was computed for each model with eq (2).

$$RMSE = \sqrt{\frac{1}{n}\sum_{i=1}^{n}\left(AHI_{DB(i)} - AHI_{E(i)}\right)^2} \qquad (2)$$

n: Total Records
AHI_{DB}: AHI from database
AHI_E: AHI Estimated with algorithm

The AHI is used to diagnose patients with breathing disordered. The index values < 5/h are considered as Normal, AHI ≥ 5/h but ≤ 30/h are considered Moderate and ≥30/h are Severe (3). Based on these criteria, patients were classified using AHI with the proposed models and physicians' database.

In order to compare models' performance patients were grouped as apneic or not apneic and the following parameters were calculated for each model: Specificity (Sp); Sensitivity (Se); and Accuracy (Acc).

Table 2 shows these results. The RIPFlow combined with SpO_2-MAD model (M5) present the lowest RMSE value (5.4), as well as highest Se, Sp and Acc combination.

Table 2 RMSE, Sensitivity, Specificity and Accuracy

	M1	M2	M3	M4	M5	M6
RMSE (RE/h)*	14.32	9,84	10,84	10,70	5,49	5,67
Se%	100	94,736	94,736	100	100	100
Sp%	25	25	75	25	75	75
Acc%	86,956	91,304	91,304	86,950	95,652	95,652

*RE/h: Respiratory Event per hour

IV. Discussion and Conclusions

The AASM recommended methodologies to detect respiratory events are based in the usage of the oxygen desaturation or arousals to determine a hypopnea event.

In this work, 6 different models (Table 1) to detect respiratory events have been computed and the results have been compared by the AHI in 23 subjects. The model M1 uses the recommended methodology based in the drop of respiratory flow signal and 3% oxygen desaturation, needing the baseline measures for both signals, with the inconvenient that there is no consensus on how to measure the SpO_2 baseline. We have introduced a new alternative method to detect respiratory events changing this measure by an estimation of the SpO_2 variability. In models M2 and M3 the respiratory flow signal is combined with the MAD and IQR of the SpO_2, respectively. We also studied what happens when the respiratory flow signal is changed by the RIPflow estimation. In this sense, the models M4 – M6 represent the RIPflow (also defined in the AASM) combined with the 3% oxygen desaturation, the SpO_2MAD and the SpO_2IQR, respectively.

As it can be seen in the first line of Table 1, the RMSE for models based on the flow signal are generally higher than the ones based on the RIPflow, except for M4, in which the arterial oxygen is measured in the traditionally 3% drop. Lower RMSE values are obtained when the arterial oxygen variations are estimated with MAD and IQR. As regards patient diagnostic, when severe and moderate SAHS are grouped, the Sensitivity, Specificity and Accuracy values for all models are acceptable and consistent with the indicated by physicians, but the higher values are obtained with M5 and M6, which used the RIPflow and SpO_2 variability features. On the other hand, only 4 subjects had AHI<5, so, the lower values in Specificity reflect the false negatives misclassifications. The best model may identify 3 of the 4 normal subjects. However, we consider for that knowledge more normal recordings are needed.

We supposed that the improvement observed between M1 and M2 is related to the apnea/hypopnea events are associated with desaturation and re-saturation changes in arterial oxygen [3].

Based on these results we may assure that using the RIPflow with the MAD of SpO_2 signal instead of the baseline 3% drop may improve the AHI and consequently, the patient diagnostic. The RIPflow improvement is consistent with the recently published Kogan's work [6], who showed better results by using RIPsum.

Another important advantage of the proposed methodology is the simplicity in the signal acquisition; it also is no invasive and does not affect patient comfort, as opposite to the oronasal flow sensor, which comfortless produce a lower quality in the polisomnography study.

Besides, the threshold analysis of the RIPflow and the variability SpO_2 measurements implies a low computational cost; it is easily computed and may be implemented in any processor. These qualities make it suitable for wireless applications. In this sense, the following expectations will focus in the event classification and the implementation in a portable device.

Acknowledgment

All authors are supported by Consejo Nacional de Investigaciones Científicas y Técnicas (CONICET). The research project is supported by grants of Secretaría de Ciencia, Tecnología e Innovación – Universidad Nacional de San Juan (SECITI-UNSJ – Res 3635-R-15), Argentina

Conflict of Interest

The authors declare they have no conflict of interest.

References

[1] G. C. Gutiérrez-Tobal, D. Alvarez, J. Gomez-Pilar, F. del Campo, and R. Hornero, "Assessment of Time and Frequency Domain Entropies to Detect Sleep Apnoea in Heart Rate Variability Recordings from Men and Women," *Entropy*, vol. 17, pp. 123–141, 2015.

[2] L. Ma, J. Zhang, and Y. Liu1, "Roles and Mechanisms of Obstructive Sleep Apnea-Hypopnea Syndrome and Chronic Intermittent Hypoxia in Atherosclerosis: Evidence and Prospective," *Hindawi*, vol. 2016, pp. 1–10, 2016.

[3] R. B. Berry, R. Budhiraja, D. J. Gottlieb, D. Gozal, C. Iber, V. K. Kapur, C. L. Marcus, R. Mehra, S. Parthasarathy, S. F. Quan, S. Redline, K. P. Strohl, S. L. D. Ward, and M. M. Tangredi, "Rules for Scoring Respiratory Events in Sleep: Update of the 2007 AASM Manual for the Scoring of Sleep and Associated Events," *Journal of Clinical Sleep Medicine*, vol. 8, pp. 597–619, 2012.

[4] R. Thurnheer, X. Vie, and K. E. Bloch, "Accuracy of Nasal Cannula Pressure Recordings for Assessment of Ventilation during Sleep," *American Journal of Respiratory and Critical Care Medicine*, vol. 164, pp. 1914–1919, 2001.

[5] P. Dehkordi, M. Marzencki, K. Tavakolian, M. Kaminska, and B. Kaminska, "Monitoring torso acceleration for estimating the respiratory flow and efforts for sleep apnea detection," in *2012 Annual International Conference of the IEEE Engineering in Medicine and Biology Society*, 2012, pp. 6345–6348.

[6] D. Kogan, A. Jain, S. Kimbro, G. Gutierrez, and V. Jain, "Respiratory Inductance Plethysmography Improved Diagnostic Sensitivity and Specificity of Obstructive Sleep Apnea," *RESPIRATORY CARE*, vol. 6, 2016.

[7] A. L. Goldberger, L. A. N. Amaral, L. Glass, J. M. Hausdorff, P. C. Ivanov, R. G. Mark, J. E. Mietus, G. B. Moody, C. K. Peng, and H. E. Stanley, "PhysioBank, PhysioToolkit, and PhysioNet: Components of a new research resource for complex physiologic signals," *Circulation*, vol. 101, pp. e215–e220, 2000.

[8] "St. Vincent's University Hospital/University College Dublin Sleep ApneaDatabase. (2008). [Online]. Available: http://www.physionet.org/pn3/ucddb/."

[9] S. J. Redmond and C. Heneghan, "Cardiorespiratory-Based Sleep Staging in Subjects With Obstructive Sleep Apnea," *IEEE TRANSACTIONS ON BIOMEDICAL ENGINEERING*, vol. 53, pp. 485–496, 2006.

[10] C. Iber, S. Ancoli-Israel, A. L. Chesson, and S. F. Quan, *The AASM Manual for the Scoring of Sleep and Associated Events*, 1st ed. Westchester, Illinois: American Academy of Sleep Medicine, 2007.

Author: Carlos R. Dell'Aquila
Institute: Universidad Nacional de San Juan - GATEME
Street: Av. Libertador 1109 (oeste) – CPA: J5400ARL
City: San Juan
Country: Argentina
Email: carlos.dellaquila@unsj.edu.ar

Prediction of Critical Air Quality Events Using Support Vector Machines and Particle Swarm Optimization

J.C. Zapata-Hernandez[1], Y.K Rojas-Idarraga[1], D.A Orrego[1] and J. Murillo-Escobar[1]

[1] Instituto Tecnologico Metropolitano, Grupo de Investigación e Innovación Biomédica Gi2B, Medellin,Colombia

Abstract— **In recent years several investigation prove the effects of air pollutants over human health, in this regard is necessary develop systems that allow people reduce the exposure to unhealthy air quality conditions. In This paper we propose a methodology to predict Critical Air Quality Events (CAQE) in Aburrá Valley based on Support Vector Machines (SVM) optimized with Particle Swarm Optimization (PSO) and a characterization scheme to asses the current and past tendencies of pollutants and weather behavior, analyzing the statistical behavior at different time intervals. We use a three stage methodology consisting in prepossessing, characterizations and CAQE prediction. The proposed method shows the better result for ozone CAQE prediction with an error of 30%. Due to low sensitivity among the pollutants is necessary use another machine learning technique that warranty a robust behavior working with unbalanced data.**

Keywords— **SVM, PSO, Air Quality, Feature Selection,**

I. INTRODUCTION

The origin of urban air pollution is mainly caused due to sources of emissions lead of the human activities, such as cars, industries and domestic fuel burning [1]. Several studies have provided evidence of adverse effects associated with air pollution over human health and environment [2]. The International Agency for Research on Cancer which found sufficient evidence to classify air pollution as a leading environmental cause of cancer deaths, also according to the US agency for environmental protection, exist vulnerable groups to high levels of contamination such people with respiratory diseases as asthma, chronic obstructive pulmonary disease, bronchitis [3, 4].

The air quality information gives cities a support tool in decision-making to protect citizens, allowing a reduction in exposure to unhealthy levels of pollution, this is the case of the Po Valley (Northern Italy), they measure thirteen air pollutants concentration hourly for 13 years, they investigated the different cause as, human habits, photochemical processes of each pollutant, they processed the data statistically hourly for detect the long-term trends in relation to the changes in emission scenarios, providing direct information on the levels and trends of key pollutants [5]. Air quality models are increasingly being used not only for research but also in a working context by national weather centers and environment institutes for air quality prediction for designing emission control policies and environmental impact assessment, the regional model ensemble established in the European Union within the Mainstreaming Adaptation to Climate Change project which provides working daily air quality forecasts for Europe [6].

This paper intends to predict 24 hours early Critical Air Quality Events (CAQE) in the Aburrá Valley, using a characterization strategy that allows asses the current and past tendencies of pollutants and weather behavior through the analysis of the statistical behavior at different time intervals. The predictions was perform using a hybrid machine learning strategy based on Support Vector machines and Particle Swarm Optimization.

II. MATERIALS AND METHODS

A. *Support Vector Machines*

Given a training vectors $x_i \in R^n, i = 1, ..., l$, in two classes and an indicator vector $y \in R^l$ such that $y_i \in \{1-1\}$, C-SVC [7] solves the following dual optimization problem which is the key to extend SVM for nonlinear functions:

$$\max \sum_{i=1}^{l} \alpha_i - \frac{1}{2}\sum_{i,j=1}^{l} \alpha_i\alpha_j y_i y_j \phi(x_i,x_j)$$
$$\text{Subject to:} \begin{cases} 0 \le \alpha_i \le C \\ \sum_{i=1}^{l} \alpha_i y_i = 0 \end{cases} \quad (1)$$

where $\phi(x_i)$ is a kernel function that maps x_i into a higher-dimensional space and $C > 0$ is the cost parameter. In this sense the optimization problem is expressed in terms of the lagrange multipliers α_i only, finally the decision function is $sgn(\sum_{i=1}^{l} y_i\alpha_i\phi(x_i,x_j)+b)$.

I. Torres et al. (eds.), *VII Latin American Congress on Biomedical Engineering CLAIB 2016, Bucaramanga, Santander, Colombia, October 26th -28th, 2016*,
IFMBE Proceedings 60,
DOI: 10.1007/978-981-10-4086-3_153

B. *Particle Swarm Optimization*

Swarm-based algorithms emerged as a powerful family of optimization techniques, inspired by the collective behavior of social animals. In particle swarm optimization (PSO) the set of candidate solutions to the optimization problem is defined as a swarm of particles flowing through the parameter space defining trajectories which are driven by their own and neighbors' best performances [8], the algorithm 1 resume the employed PSO.

Inputs: $f(x)$, stop criterion, α, β, θ & n
Output: g^*
Initialization: positions x_i and speed v_i of the n particles
$x_i^* = x_i$, ${}_{val}x_i^* = f(x_i)$
$g^* = \text{argmin} f(x_i)$, ${}_{val}g^* = f(g^*)$
while *stop criterion* **do**
    for *every n* **do**
        $v_i^{t+1} = \theta v_i^t + \alpha\varepsilon_1 \cdot [g^* - x_i^t] + \beta\varepsilon_2 \cdot [x_i^* - x_i^t]$
        $x_i^{t+1} = x_i^t + v_i^{t+1}$
        if $f(x_i^{t+1}) <_{val} x_i^*$ **then**
            ${}_{val}x_i^* = f(x_i^{t+1})$, $x_i^* = x_i^{t+1}$
        end
        if $f(x_i^{t+1}) <_{val} g^*$ **then**
            ${}_{val}g^* = f(x_i^{t+1})$, $g* = x_i^{t+1}$
        end
    end
end

α & β constants of group and individual behavior $f(x)$ objective function, θ inertia parameter, ε_1 & ε_2 Gaussian random variables, v_i particle speed, x_i particle position, x_i^* best particle position, g^* best position reached between all particles

Algorithm 1: PSO

C. *Proposed Methodology*

The fig 1 shows the proposed methodology to predict CAQE 24 hours early, initially a pre-processing stage was carried out for outliers detection, database selection and CAQE identification, moreover we conduct the data characterization and finally we use SVM optimized with PSO (SVM-PSO) to predict critical air quality events in urban and suburban areas.

D. *Data Base Description*

For this study we used the information provided by the Air Quality Laboratory (CALAIRE) of the National university of Colombia (UNAL). The data correspond to five stations in different sites of The Aburra Valley including Suburban and urban areas, in the urban areas we find 3 stations: Museum of Antioquia (MED-MANT), UNAL Medellín Campus (MED-NAL) and Itagui's House of Justice (ITA - HJ), the suburban areas include San Buenaventura university (BEL-USBV) and Liceo Council Itagui (ITA-CONC). They collect this data between January 2013 and December 31 of the same year, including measures hour after hour of six weather variables (air speed (WS),air direction (WD), air temperature (AT), humidity relative (AH), solar radiation (SR)) and six pollutants (nitrogen monoxide (NO), nitrogen dioxide (NO_2), Ozone (O_3), Carbon monoxide (CO), particulate matter 2.5 μm ($PM_{2.5}$), particulate matter 10 μm (PM_{10})). Although to clarify that monitoring of these six pollutants were not taken simultaneously in the five stations, but was taking in combinations of three or four of these. Allowing in this way to get a representation space between ten and twelve features and a total samples of 8760.

E. *Data Preprocessing*

The used data present missing values and outliers, occasioned for damage or maintenance in the sensors, taking into account that this values can affect how mathematical or statistics methods perform, we conduce a pre-processing stage. First we make the outliers detection using a box plot and then replacing as a missing values. Finally, the samples with missing data was removing. In order to make a comparative analysis between a urban station that is characterized by high levels of pollutants and a suburban station where its levels of pollutants are moderate, we just use the data base with the lower number of missing values. Then, we evaluated the pollutants to identify if their concentration levels corresponding to a CAQE, we get the mean (μ) and the standard deviation (σ) of every pollutant, if the concentration of the pollutant is greater than $\mu + 0.8\sigma$ was labeled as CAQE.

F. *Characterization*

With the aim to shape a feature space of effective representation for CAQE prediction 24 hours early, we characterize the pollutants concentration and the weather variables in the time domain, in this way we calculate features that allows us to asses the current and past tendencies of pollutants and weather, the first group of features consisting in the measure of the pollutants concentration and weather variables 24 hours before of the prediction, for evaluate the past tendencies, we split the analysis in pollutants and weather variables, given that the pollutants concentration varies mainly as function of the human activity, we get the pollutant concentration the same hour and day in the past seven weeks and calculate the mean, maximum

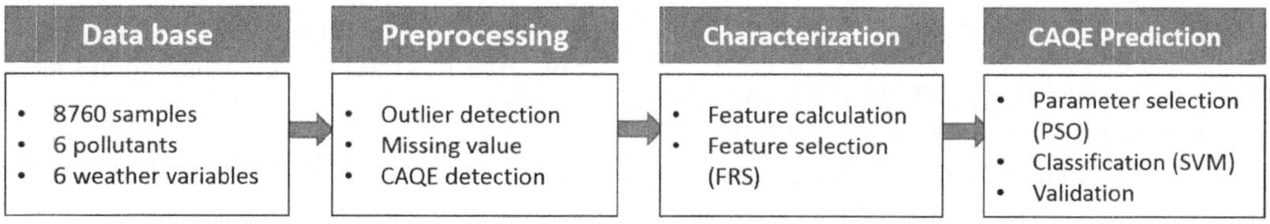

Figure 1. Proposed Methodology

and minimum concentration, in other hand the weather varies as function of region's climatic patterns, for this reason we get the mean, maximum and minimum of the weather variable in several time intervals before the CAQE prediction (12 Hours, 24 Hours and 7 days). Later we make feature selection using an algorithm based on Fuzzy-Rough set theory propose in [9], with the aim of reduce the problems associated with high dimensional space.

G. *Prediction*

For prediction, we performed an algorithm of classification based on SVM-PSO . In this step, we randomly split the data in three groups: training (60 %), validation (20%) and test (20%). The SVM learning process is conducted using the training data and the parameters C and γ is given for the PSO algorithm, is important note that PSO choose the setting that maximize the accuracy over the validation data, in this process we executed 200 iterations using 15 particles, $\theta = 0.9$ and $\alpha, \beta = 2$. Finally we assessed the obtained model through the test data, measuring the accuracy, sensitivity and specificity in the classification process.

III. Results and Discussion

A. *Data Preprocessing*

In this stage we realize the outlier detection, missing value removing, data base selection and CAQE identification. The urban station MED-MANT, MED-NAL and ITA-HJ has a proportion missing values of 12.7 %, 15.5 % and 15.8 % respectively, for suburban areas ITA-CONC and BEL-USB has 10.8 % and 12.7 % of missing values. The selected urban station was MED-MANT and the sub-urban station were ITA-CONC.

The monitor station MED-MANT measure the concentration of four pollutants: NO, NO_2, $PM_{2.5}$ and CO, now we identify the CAQE for every pollutant and find that for $PM_{2.5}$ was 3781 CAQE this correspond to the 43.2% of samples, the same way for NO was the 44.7%, NO_2 40.7% and for CO the 34.4%. The ITA-CONC station measure three pollutants $PM_{2.5}$, PM_{10} and O_3, for $PM_{2.5}$ we found a 39.4% of CAQE, PM_{10} has a 39.7% and O_3 with 44.6%. The mean CAQE portion for the 7 pollutants is 42%, this evidence that balance between classes is not ideal and could lead in a poor performance due to increase the probability that the classification algorithm label the test instance in the group with more samples.

B. *Characterization and Feature Selection*

Following the proposed characterization scheme, we find a total of 47 features split in three groups (table 1). The feature selection stage was conducted to find the optimal subset of features for CAQE prediction in each pollutant, in this regard for predict CAQE in the urban area of MED-NAL associated with NO we employ only 11 features, for NO_2 12 features, CO 8 features and $PM_{2.5}$ 11 features, in the suburban area ITA-CONC for $PM_{2.5}$, PM_{10} and O_3 the algorithm select a total of 11 different features for every pollutant.

Table 1. Proposed Features

First group	Second group	Third group
$WD^{\oplus\ominus\otimes}$	M-$PM_{2.5}$-24h$^{\otimes}$	M-AT-7w $^{\diamond}$
$AT^{\bigcirc\oplus}$	M-$PM_{2.5}$-7w	Max-AT-7w
$AH^{\diamond\ominus\otimes}$	Max-$PM_{2.5}$-7w$^{\oplus\ominus\otimes}$	Min-AT-7w
$SR^{\bigcirc\triangle\diamond}_{\oplus\ominus}$	Min-$PM_{2.5}$-7w $^{\diamond\oplus}$	M-AT-24h
$NO^{\triangle}$	M-NO-24h $^{\diamond}$	Max-AT-24h $^{\otimes}$
NO_2 $^{\triangle\diamond\bigcirc}$	M-NO-7w	Min-AT-24h *
$PM_{2.5}{}^{\bigcirc\triangle\diamond}_{\oplus\ominus\otimes}$	Max-NO-7w $^{\triangle\diamond\bigcirc}$	M-SR-12h $^{\triangle\diamond\bigcirc}$
CO $^{\bigcirc\triangle\diamond}$	Min-NO-7w	Max-SR-12h $^{\triangle\diamond\bigcirc}_{\oplus\ominus\otimes}$
$PM_{10}{}^{\oplus\ominus\otimes}$	M-NO_2-24h$^{\bigcirc}$	Min-SR-12h
$O_3{}^{\otimes}$	M-NO_2-7w	M-AH-7w
$WS^{\triangle\diamond\bigcirc}_{\oplus\ominus\otimes}$	Max-NO_2-7w	Max-AH-7w
	Min-NO_2-7w	Min-AH-7w$^{\oplus\ominus}$
	M-CO-24h$^{\triangle}$	
	M-CO-7w	
	Max-CO-7w$^{\bigcirc\triangle}$	
	Min-CO-7w	
	M-PM_{10}-7w	
	Max-PM_{10}-7w$^{\oplus}$	
	Min-PM_{10}-7w $^{\ominus\otimes}$	
	M-O_3-24h$^{\ominus}$	
	Min-O_3-7w	

MED-NAL: $\triangle$ selected for NO, $\diamond$ selected for NO_2, $\bigcirc$ selected for CO

ITA-CONC: selected $\oplus$ for O_3, selected $\otimes$ for $PM_{2.5}$, selected $\ominus$ for PM_{10}

C. *Critical Air Quality Events Prediction*

Using the proposed SVM-PSO algorithm we conduce the CAQE prediction in urban and suburban areas, the selected parameter for PSO in urban areas pollutants were $C = 88.3$ and $\gamma = 2.9$ for NO CAQE prediction, for NO_2 $C = 0.74$ and $\gamma = 10$, for $PM_{2.5}$ $C = 233.3$ and $\gamma = 3.6$, for CO $C = 0.8$ and $\gamma = 5.1$. In suburban areas PSO gave $C = 1000$ and $\gamma = 0.09$ for O_3, $C = 779.14$ and $\gamma = 0.07$ for PM_{10} and finally for $PM_{2.5}$ we obtain $C = 10.25$ and $\gamma = 0.78$.

Table 2 show the system performance for urban and suburban areas, the best result in urban areas was obtained for the NO and the poorest were reach for $PM_{2.5}$, while in suburban areas the best performance where reach for O_3, the poorest agree with urban area for $PM_{2.5}$. The sensitivity in the CAQE detection in the 7 contaminants is low, due to the unbalance between classes, moreover the feature selection stage in $PM_{2.5}$ for MED-NAL show that the proposed features are uncorrelated with CAQE given that only one feature was selected.

Table 2. CAQE prediction performance

Pollutant	Spe	Sen	Acc
MED-NAL			
NO	0.72	0.64	0.69
NO_2	0.86	0.38	0.66
$PM_{2.5}$	1	0	0.56
CO	0.85	0.44	0.68
ITA-CONC			
O_3	0.8	0.59	0.7
PM_{10}	0.89	0.2	0.63
$PM_{2.5}$	0.89	0.2	0.61

IV. Conclusion

This paper attempts to predict Critical Air Quality Events (CAQE) using Support Vector Machines (SVM) Optimized with Particle Swarm Optimization (PSO) through a characterization scheme to asses the current and past tendencies of pollutants and weather behavior through statistical features, obtaining a low performance in terms of sensibility and accuracy. The unbalance between classes lead in low sensibility since most of the label correspond to normal air quality conditions, this is confirmed through the high specificity, for these reason is necessary employ a machine learning techniques developed to work with unbalanced data. For some pollutants the feature selection algorithm chose a few features in the case of $PM_{2.5}$ in MED-NAL only one feature this can suggest that the proposed set of features are not so related to CAQE and is necessary extend the analysis to non linear features.

Conflict of Interest

The authors declare that they have no conflict of interest.

Acknowledgements

We would like to thank the Air Quality Laboratory of the National university of Colombia and Metropolitan Area of the Aburrá Valley for provide the data used in this study.

References

1. Bhaskar B. Vijay, Mehta Vikram M.. Atmospheric particulate pollutants and their relationship with meteorology in Ahmedabad *Aerosol and Air Quality Research.* 2010;10:301–315.
2. Bruno Francesca, Cameletti Michela, Franco-Villoria Maria, et al. A survey on ecological regression for health hazard associated with air pollution *Spatial Statistics.* 2016.
3. Buczyńska Anna J., Krata Agnieszka, Van Grieken Rene, et al. Composition of PM2.5 and PM1 on high and low pollution event days and its relation to indoor air quality in a home for the elderly *Science of the Total Environment.* 2014;490:134–143.
4. Kim Ki Hyun, Jahan Shamin Ara, Kabir Ehsanul. A review of diseases associated with household air pollution due to the use of biomass fuels *Journal of Hazardous Materials.* 2011;192:425–431.
5. Masiol Mauro, Agostinelli Claudio, Formenton Gianni, Tarabotti Enzo, Pavoni Bruno. Thirteen years of air pollution hourly monitoring in a large city: Potential sources, trends, cycles and effects of car-free days *Science of the Total Environment.* 2014;494-495:84–96.
6. Brunner Dominik, Savage Nicholas, Jorba Oriol, et al. Comparative analysis of meteorological performance of coupled chemistry-meteorology models in the context of AQMEII phase 2 *Atmospheric Environment.* 2015;115:470–498.
7. Boser Bernhard E., Guyon Isabelle M., Vapnik Vladimir N.. A Training Algorithm for Optimal Margin Classifiers *Proceedings of the Fifth Annual ACM Workshop on Computational Learning Theory.* 1992:144–152.
8. Marini Federico, Walczak Beata. Particle swarm optimization (PSO). A tutorial *Chemometrics and Intelligent Laboratory Systems.* 2015;149:153–165.
9. Orrego Diana a, Becerra Miguel a, Delgado-Trejos Edilson. Dimensionality reduction based on fuzzy rough sets oriented to ischemia detection *Annual International Conference of the IEEE Engineering in Medicine and Biology Society..* 2012;2012:5282–5285.

Author:J. Murillo-Escobar
Institute: Instituto Tecnológico Metropolitano
Street: Calle 73 No 76A-354
City: Medellín
Country: Colombia
Email: jpmurillo22@gmail.com

SIRUMED©, software for wheelchair selection. A preliminary report.

J. Letechipia[1] and A. Arredondo[1]

[1] Universidad Iberoamericana/Instituto de Investigación Aplicada y Tecnología, Mexico DF, Mexico

Abstract— **Rehabilitation organizations have recognized that personal mobility plays a significant role in the lives of many people with disabilities. It has been estimated that the number of people with disabilities in developing countries who require wheelchairs is approximately 1% of the population. In Latin America medical insurance, socialized or private, does not prescribe nor provides wheelchairs, therefore the only way to acquire one is through charity or through direct purchase. This represents a major obstacle particularly if considering that the economic resources of the population with disabilities throughout this region are very limited. Given that they are not prescribed, wheelchair users receive minimal advice from clinical personnel and end up buying a device based on salesman recommendation. With the purpose of facilitating the provision of properly fitted wheelchairs for Latin America, a software program, to be used by clinical personnel, was developed. Two versions were developed, a PC stand-alone and a Web based version. Fifty recommendations were made and compared to experienced wheelchair manufacturer representatives. Software recommendations coincided with the manufacturer's suggested sizing in 49/50 cases within +/- 1 cm. This paper presents the functions of the software and illustrates individual wheelchair recommendations. With this information, wheelchair users and their families can acquire a better fitted wheelchair for their patients.**

Keywords— **software, wheelchair selection, Latin America**

I. Introduction

Based on the 2010 population estimates and the 2004 disability prevalence estimates [1], including children, over a billion people (or about 15% of the world's population) were estimated to have a disability [2]. This is higher than the 1970s estimates that suggested a prevalence of 10% [3]. According to the WHO 80% of people with disabilities, particularly children, live in low income countries [4] where Assistive Technology (AT) is rarely available. AT device means any item, piece of equipment, or product system, whether acquired commercially off the shelf, modified, or customized, that is used to increase, maintain, or improve the functional capabilities of a person with a disability [5]. Wheelchairs are considered AT. Global data on the need for rehabilitation services (including mobility devices) and estimates of unmet need are very limited [6]. However, it has been estimated that the number of people with disabilities in developing countries who require a wheelchair is approximately 1% of the population [7].

The number of people with disabilities increases mainly for two reasons: the aging population and the increase in chronic degenerative diseases, especially common non-communicable diseases (NCD) such as diabetes, stroke and cancer. This suggests a correspondence increase in the need for mobility devices.

Exceptional rehabilitation organizations have recognized that personal mobility plays a significant role in the lives of many people with disabilities. They have focused their efforts exclusively in the provision of wheelchairs to low income countries. Organizations like Wheelchairs for the World [11], Joni and Friends [12], and pioneer organizations like the Wheelchair Foundation [13] have been providing wheelchairs since 1979. As meaningful, helpful, well structured, and funded as these organizations are, their lack of long term involvement with local organizations has rarely resulted in development and sustainability [14]. Their work is perceived mainly as charity.

When considering the reasons for not having commercially available quality wheelchairs specifically in Latin-America, literature comes short of answers. One potential explanation may be that as long as medical insurance, socialized or private, will not provide for these devices, the only technology that will be available are those received through charity or devices that people can acquire out of pocket. This condition represents a major obstacle for procuring quality wheelchairs, particularly if considering that the economic resources of the population with disabilities throughout this region are very limited. Given that wheelchairs are not recommended, therapists and physicians are not familiar with the recommendation process.

In consequence, not recommending, not providing and the lack of availability leave the wheelchair users to their own resources, and forces them to choose by price.

Therefore, if quality personal mobility devices are needed to improve the lives of people with disabilities in Latin America, it is necessary to develop alternate sustainable methods of designing, producing and distributing these technologies.

With the purpose of facilitating the provision of quality wheelchairs for Latin America, the Rehabilitation Engineering and Technology Center (CITeR® for its name in Spanish) at Universidad Iberoamericana in Mexico City, em-

I. Torres et al. (eds.), *VII Latin American Congress on Biomedical Engineering CLAIB 2016, Bucaramanga, Santander, Colombia, October 26th -28th, 2016*,
IFMBE Proceedings 60,
DOI: 10.1007/978-981-10-4086-3_154

barked in the development of a wheelchair recommendation software (Section 1: SIRUMED Ligera©, 03-2013-111413315900-01, and Section 2: SIRUMED Estándar©, 03-2013-111413282300-01) that would assist therapists and physicians in Latin America to properly recommend a customized wheelchair.

II. System requirements

A. General requirements

SIRUMED© would be integrated by two separate sections: a. rigid frame design, and b. folding frames design. Software would require patient upper and lower limb measurements (measured by therapist or physician) and will result in a wheelchair recommendation that will indicate: 1. seat length, 2. seat width, 3. seat height, 4. footrest height and 5. back height.

Software should be user friendly and easy to use by clinical personnel. They will learn to use it through an online tutorial.

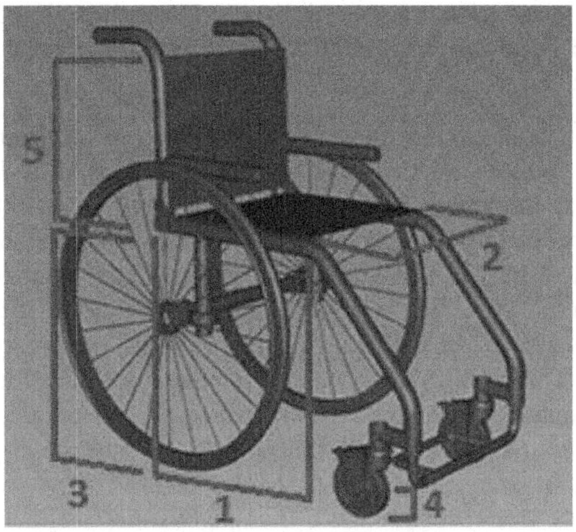

Fig. 1 Wheelchair dimensions

Software would recommend wheelchair parameters procuring to keep a propulsion angle (elbow) between 100 and 120 degrees. Wheel diameter would be limited to 24" given that other adult sizes wheels are rarely available. Hand breaks, push handles and arm rests are treated as accessories.

B. Special requirements

SIRUMED© will provide four possible configurations: three obtained automatically by the algorithm and a fourth that considers a manual mode where specific parameters can be entered by clinical personnel. These four options take into account therapist and physician clinical experience and individual considerations specific to each patient. Figure 2 illustrates the block diagram of the software.

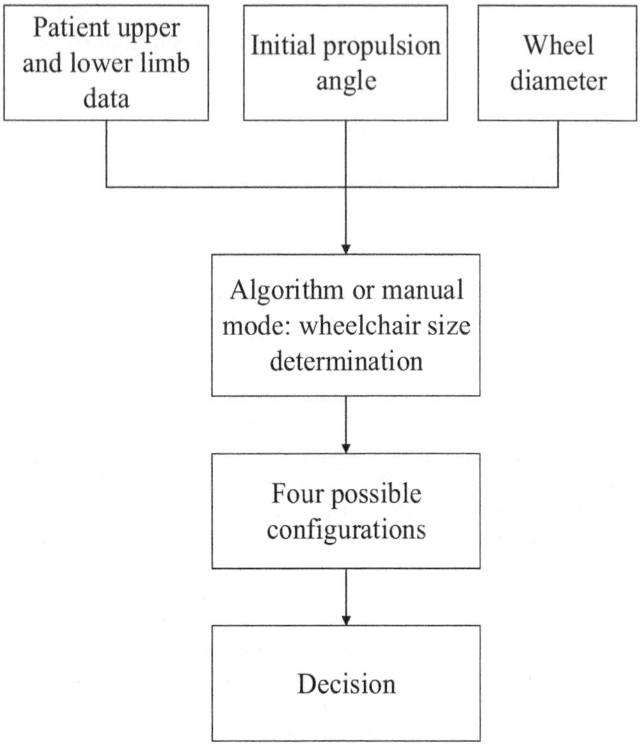

Fig. 2 Wheelchair recommendation process

III. Results

A software program was developed to facilitate recommending of manual wheelchairs, both, rigid and folding frames. Software is Windows® based. Two versions were developed, the stand alone version for PC was written in JAVA®, and for the Web based version, PHP® was used. Algorithm function is to calculate proper sizing of the wheelchair attending to the individual's upper trunk and leg measurements. If both calculations coincide a recommendation is made. If they do not coincide, software recalculates wheelchair dimensions by lowering (reducing the propulsion angle) or elevating (increasing the propulsion angle) the seat height. Software allows the clinician to select the final recommendation, taking into account the wheelchair user level of expertise.

Software is intuitive to use and was easily learned and applied by clinical personnel. Figure 3 illustrates the software results shown on a table where four options are presented. From this table, clinical personnel will select the most appropriate option for the client based on client information and condition.

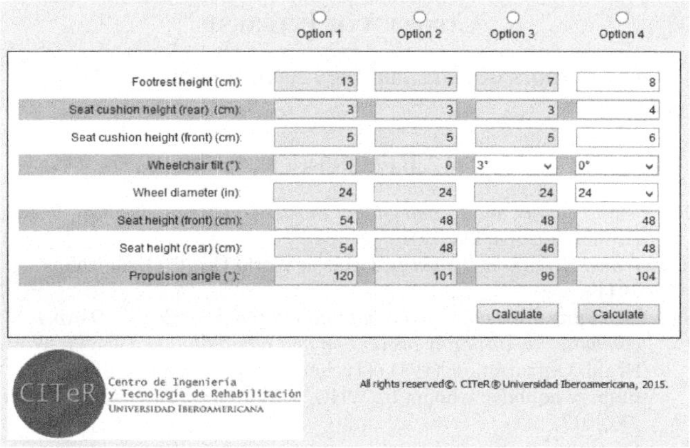

	Option 1	Option 2	Option 3	Option 4
Footrest height (cm):	13	7	7	8
Seat cushion height (rear) (cm):	3	3	3	4
Seat cushion height (front) (cm):	5	5	5	6
Wheelchair tilt (°):	0	0	3°	0°
Wheel diameter (in):	24	24	24	24
Seat height (front) (cm):	54	48	48	48
Seat height (rear) (cm):	54	48	46	48
Propulsion angle (°):	120	101	96	104
			Calculate	Calculate

Fig. 3 Wheelchair recommended options (labels are originally in Spanish for use in most Latin American countries)

Figure 4 illustrates the final recommendation report where all required dimensions to custom fabricate the wheelchair are provided.

Fifty wheelchair recommendations were made to volunteer "expert" male adult wheelchair users. Disabilities included spinal cord injury at different cervical and thoracic levels. Age range was from 21 to 45 years. While entering the users data to the software, two experienced wheelchair manufacturer representatives were present and were asked to produce their own recommendations.

The suggested wheelchair recommendations by the software program coincided with the manufacturer's suggested sizing in 49/50 cases within +/- 1 cm on each of the reported parameters. In one case, the recommendation made by the manufacturer representatives had a better fit for the client.

Twenty five wheelchairs were fabricated locally by manufacturer Fundación Bertha O de Osete A.C., sizing them according to the dimensions recommended by the software. Volunteers that received these customized wheelchairs recognized a better fit and ease of use than their existing wheelchair

IV. Conclusions

The results of this qualitative study suggest that a recommendation made by the software is quite similar to those issued by an expert provider. Therefore in most clinical setting in Latin America where no expert is available, the use of this program could enhance wheelchair recommendation practices.

More wheelchairs would have to be prescribed by the software in conjunction with the experts to assess its true clinical value.

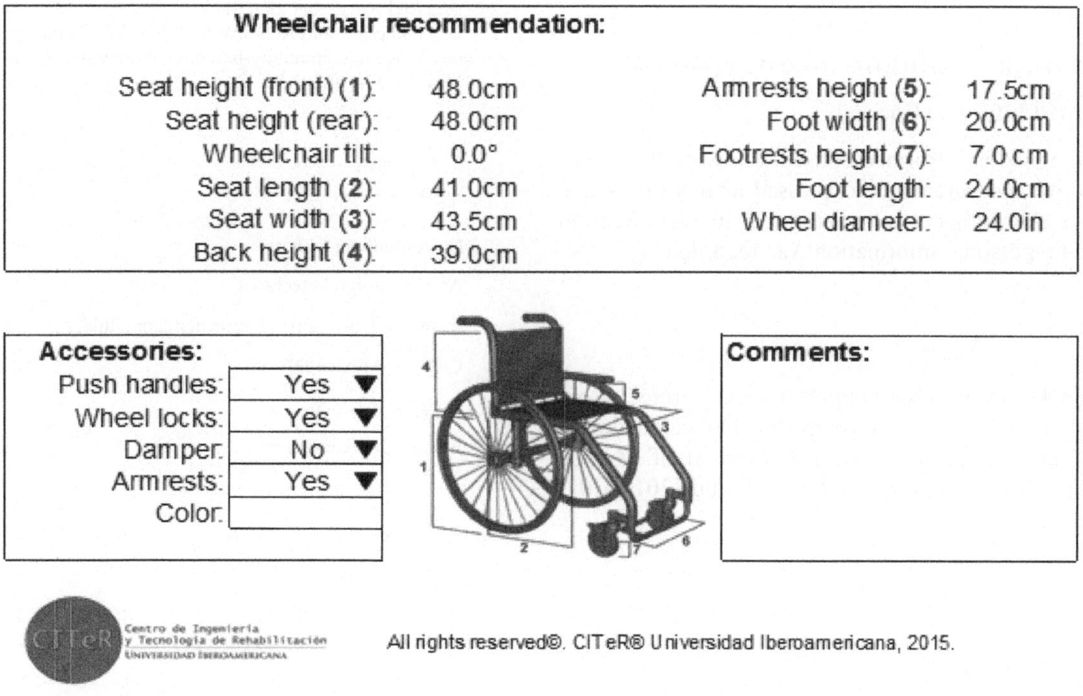

Fig. 4 Recommended Wheelchair

The recommendation generated by SIRUMED© could be used by the patient or his/her family to order a customized wheelchair, when available. Or at least, can be used as a guide to purchase one commercially available that is close to the proper size. This would facilitate individual propulsion for the active user and provide better comfort.

However, proper wheelchair size recommendation by itself is not a complete solution. For that reason, CITeR has also developed a seating fabrication system that in concert with the properly fit wheelchair would enhance mobility and comfort. For that purpose, a Web based application is being developed with the interest of providing a clinical tool to Latin American therapists and physicians.

As long as wheelchairs are not provided by the public or private health care system in Latin America, therapists and physicians will not see the need to prescribe them. Individuals with disabilities will rely entirely on the opinion of vendors to acquire them, a system that has proven not to be in the best interest of the wheelchair user, especially if he/she is paying out of pocket.

This software will provide interested clinical personnel with a tool to guide the patient in acquiring as close to their best fit wheelchair as possible, and in some instances, local manufacturers could fabricate the wheelchair to the specific needs of the user.

V. Compliance with Ethical Requirements

A. Statement of Informed Consent

Participants signed an informed consent allowing the use of their ergonomic measurements to be used as input data into the software. All subjects were assigned an identification number and no personal information was recorded.

Acknowledgment

Authors acknowledge the support for this project from Universidad Iberoamericana, through the Direccion de Investigacion and the National Science Council of Mexico (CONACYT) through grant number: Salud 2011-C01-161827.

Conflict of Interest

"The authors declare that they have no conflict of interest".

References

1. *World Health Survey and Global Burden of Disease*. Geneva, World Health Organization 2004.
2. *World Report on Disability*. Geneva, World Health Organization 2011.
3. *Disability prevention and rehabilitation report of the WHO expert committee on disability prevention and rehabilitation*. Geneva, World Health Organization, (1981) (Technical Report Series 668) (http://whqlibdoc.who.int/trs/WHO_TRS_668.pdf, accessed October *28, 2011)*.
4. *Disability, including prevention, management and rehabilitation*. The Fifty Eight World Health Assembly. WHA58.23 Ninth plenary meeting 25 May 2005.
5. *Public Law 10-407, The Technology-Related Assistance Act for Individuals with Disabilities*. Commonly known as the Tech Act, 1988 passage.
6. WHO, 2011, World Report on disability, Geneva World Health Organization
7. ISPO/USAID/WHO (2006). Report of consensus conference on wheelchairs for developing countries. Bengaluru, India, 6–11 November 2006 (http://www.who.int/disabilities/ technology/Wheelchair%20Consensus%20Conference%20 Report_Jan08.pdf accessed 4 July 2016), Bengaluru
8. Wheelchairs for the World (http://wheelchairsfortheworld.org.uk/, accessed November 29, 2011).
9. Joni and Friends. International Disability Center (http://www.joniandfriends.org/wheels-for-the-world/, accessed November 29, 2011).
10. Wheelchair Foundation (http://www.wheelchairfoundation.org/, accessed November 29, 2011).
11. G. Mukherjee and A. Samanta. "Wheelchair charity: a useless benevolence in community-based rehabilitation." *Disabil Rehabil* 2005, May 20, 27(10):591-6.

Corresponding author:

Author: Jorge Letechipia

Institute: Universidad Iberoamericana / InIAT

City: Mexico DF

Country: Mexico

Email: citer@ibero.mx

Upper-Limb Kinematics During Feeding and Drinking

Parra-Sánchez Sergio[1], Gómez-González Juan Manuel[2], Quintero-Ortega A. Iraís[1], Castellano Laura E.[1], Mendoza-Novelo Birzabith[1], Delgado-García José Jorge[1], Cuéllar-Cruz Mayra[3] and Vega-González Arturo*[1]

[1] División de Ciencias e Ingenierías DIQEB, Universidad de Guanajuato, León, Guanajuato, México
[2] Facultad de Ingeniería, DCR, Universidad Nacional Autónoma de México, México DF
[3] Divisi´on de Ciencias Naturales y Exactas, DB, Universidad de Guanajuato, Guanajuato Guanajuato, México

Abstract— **Feeding and drinking are Activities of Daily Living which can be used to assess the motor control and functional ability of the upper limb. This paper presents the upper-limb kinematics during the execution of feeding and drinking activities, such analysis consisted in the measurement of angles of flexion for trunk and arm. Eight healthy subjects performed these activities in a simulated-environment while they were video recorded. Markers on anatomical landmarks were used to analyze the kinematics of the upper limb in the sagittal plane. Additionally an electro-hydraulic sensor was attached to each upper limb to assess the vertical position of the wrist relative to the shoulder. Results showed a difference on the angles of the elbow and trunk. The electro-hydraulic sensor showed to be an efficient way to record the vertical position of wrist.**

Keywords— **upper limb, ADL, joint angles**

I INTRODUCTION

Kinematic analysis of task-oriented activities plays a fundamental role to asses functional movement. Functional evaluations refer to those evaluations that use real life activities, such as feeding and drinking. Task-oriented evaluations examine aspects of motor function through simulating a useful or familiar task. The kinematic analysis can help to depict the physical performance component of functional movement such as range of motion, strength, joint angles, etc. [1, 2]. Accurate quantification of joint motion of upper limbs gives important information about the differences between healthy subjects and people with joint injuries or limited movement. Such information can be used to improve the rehabilitation interventions [3]. There are a numerous quantity of analysis techniques for upper limb movements, which can be classified in two paradigms: one approach is based on markers and video-cameras [4, 5, 2, 6]. Other approaches are based on wearable sensors, such as magnetometers and inertial sensors [7, 8, 9, 10].

Assessing upper-limb function is a complex problem. There are reports studying cyclic movements, such as the reaching and grasp cycle [11], or a cyclic modification of the patterns during the execution of some ADL [5, 12]. There has been suggested a variety of ADL in the literature [13, 14]. Therefore, a standard set of ADL to assess upper-limb function has not been defined. Moreover, the upper limb kinematics during the performance of ADL might show differences even on non-pathological participants. Recently, it has been suggested the Reaching and Hand- to-Mouth Evaluation Method to assess upper-limb function in neurological patients [15, 16]. The aim of this study is to analyze the kinematics of hand-to-mouth movement through task-oriented activities (feeding and drinking) and by using a minimal technology. The latter will help to translate the methodology to a clinical environment. Additionally, a brief comparison with a portable sensor was done. This will help to establish a method based on wearable sensors.

II MATERIALS AND METHODS

A *Participants*

Eight healthy subjects (5 males and 3 females) were included in the study. Participants signed an informed consent prior to start the measurements. Measurement protocol is in line with the Helsinki declaration.

B *Measurements*

This is a pilot study based on the characterization of drinking and eating using fork and knife. Subjects were then evaluated through kinematic analysis using videography in the sagittal plane. Briefly, subject wore a set of markers on: wrist, elbow, epicondile, acromion, sternum and temple (see figure 1(a)) and a commercial video camera (Sony-HDR-CX250, 60 hz) was used for recording. Participants also wore a electro-hydraulic sensor on each upper limb[17]. The sensor measures the vertical displacement of the wrist in relation to the shoulder. Then they performed the activities of drinking and eating using fork and knife.

I. Torres et al. (eds.), *VII Latin American Congress on Biomedical Engineering CLAIB 2016, Bucaramanga, Santander, Colombia, October 26th -28th, 2016*,
IFMBE Proceedings 60,
DOI: 10.1007/978-981-10-4086-3_155

The tasks were performed in a sitting position. Before to start every activity, the participant maintain a position with the back in vertical and hands on the table (this was known as the initial position). When the instructor indicate, the participant began to perform an activity in a natural way. Measurements were made in two steps, in the first (control trial) the instructor indicates initial and final times for executing activities, during the second step, (free trial), participant initialize and finished the activity when he/she wanted. In both trials, activities were real and not simulated. The interval time during trials was of 8 minutes.

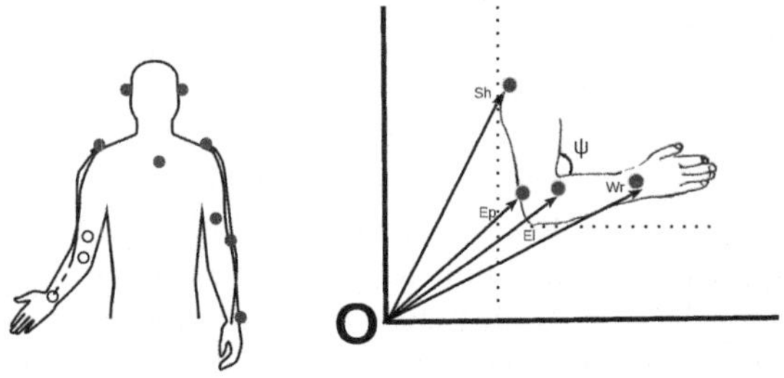

(a) (b)

Fig. 1: Marker positions to record trajectories followed by body segments during the execution of activities. a) Markers are represented by Red and white balls. They were placed at: wrist, elbow, epicondile, shoulder, sternum and temple. The Vertical displacement sensors are shown too, they were mounted in both shoulders and a tube connected to them, were placed along the arm until wrist. b) Elbow flexion angle calculated.

C Calculation of kinematics of movement

Trajectories for every one of the markers were obtained using the software Tracker ©[18]. Then, the trajectories were filtered with an 11-order Butterworth low-pass filter with a cutoff frequency of 10 Hz.

Arm flexion was calculated using vectorial dot product. Figure 1(b) show a diagram indicating the coordinates for markers on the arm. furthermore, inclination of trunk was measured respect to initial position of the shoulder considering it as a rigid body with only an effective degree of freedom (rotation):

$$\theta_{it} = \omega_t - \omega_i \tag{1}$$

where ω_i is the angle formed by the shoulder and sternum at initial position and ω_t is the angle formed by the sternum at time t (St_t) and the initial position of the shoulder (Sh_i):

$$\theta_{it} = Acos\left(\frac{\vec{St_t}\cdot\vec{Sh_i}}{|St_t||Sh_i|}\right) - Acos\left(\frac{\vec{St_i}\cdot\vec{Sh_i}}{|St_i||Sh_i|}\right) \tag{2}$$

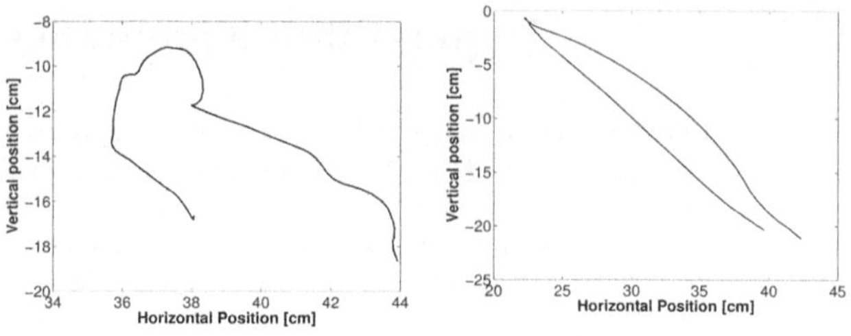

(a) (b)

Fig. 2: Arm trajectories for a) Picking food with a fork and b) Drinking from a cup. Trajectories were tracked with respect to the shoulder.

During the extraction of trajectories, time duration of every activity was normalized and an interpolation was made using cubic splines. Based on this set of trajectories and angles, mean and standard deviations were calculated.

III Results

Trajectories for activities are shown in fig. 2. Such trajectories were tracked with respect to the position of the shoulder. Furthermore, trunk and elbow flexion, and vertical displacements of the wrist are shown in figures 4, 3 and 5 respectively. Table 1 shows a summary of some kinematic variables as the maximum angular velocity and maximum angular acceleration.

Table 1: Maximum values of kinematic parameters.

Activity	Displacement	Angular vel.	Angular accel.
Eating	47.5±4.2cm	$69.0\pm18.7^0/s$	$429.4\pm142.4^0/s^2$
Drinking	26.4±5.5cm	$105.2\pm18.6^0/s$	$601.6\pm221.5^0/s^2$

IV Discussion

Trajectories show that both feeding activities have the same maximum displacement $(max(position) - min(position))$. Nonetheless drinking from a cup might be considered as a cyclic activity, whereas feeding do not present this feature because picking with a fork is only one step of a more complex activity.

Inclination of trunk (fig 3) shows an interesting difference between the two activities analyzed in this study: drinking is mainly an extension of the trunk and then return to the initial position, but for feeding a flexion is needed followed by a return to the initial position. For feeding a larger variance is observed because of the differences in body heights of par-

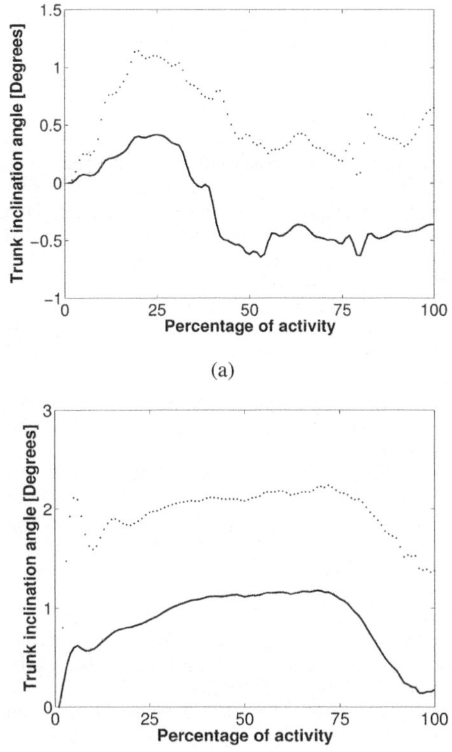

(a)

(b)

Fig. 3: Inclination of trunk during the performance of activities: a) picking food with a fork and b) drinking from a cup.

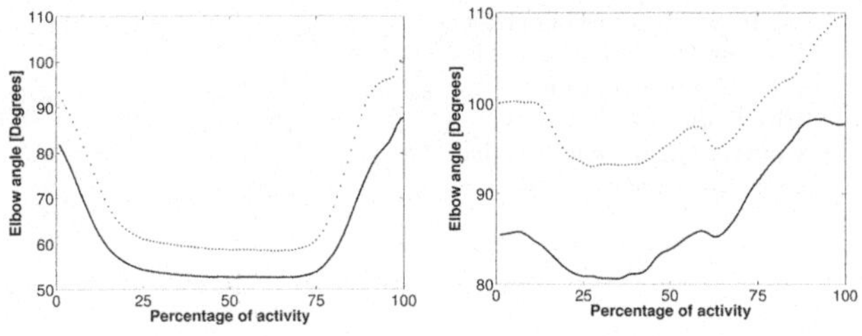

(a) Drinking (b) Eating

Fig. 4: Elbow flexion during the performance of a) drinking from a cup and b) picking food with a fork.

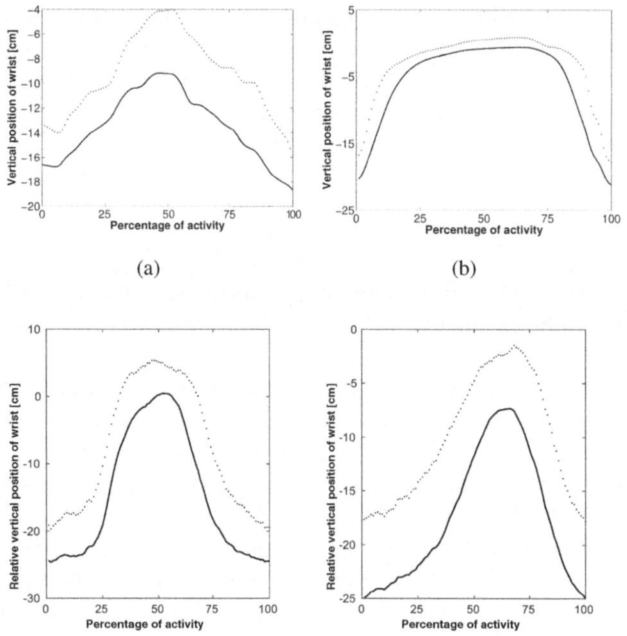

(a) (b)

(c) (d)

Fig. 5: Vertical displacement of wrist during the performance of the activities: picking food with a fork, and drinking from a cup; tracked with video (a) and b)) and recorded with the pressure sensor (c) and d)).

ticipants. In this sense, the environment slightly modifies our activities in different degrees depending of the activity.

Figure 5 shows relative vertical displacement of the wrist. It is possible to observe that the waveforms are considerably different. This could lead to the proposal of pattern recognition algorithms to differentiate among these activities [19]. In particular when assessment of the upper-limb function must be performed in the real-life situation, thus, outside the laboratory environment.

In the case of elbow flexion, results are in line with reports in other studies [12]. In this case, it is necessary to note that the angle reported in this study is the supplement to the one calculated in ref [12]. Elbow flexion showed a bigger variance for eating activity than for drinking. This is an expected result since the variance for the position of wrist is bigger for eating than for drinking, as shown in figure 5. This differences would be originated by all the possibles forms to eat using fork and knife, even when analysis presented here only covers the picking food, this can begin or end in different initial conditions according to the whole activity itself. This would be the mark to analyze real activities.

Table 1 shows that displacement for eating activity is bigger than for drinking. But drinking is an activity performed more quickly, as shown the greater values for angular velocity and acceleration. This is because eating requires more dexterity to manipulate a fork or a spoon. This table shows a great variances for angular velocity and acceleration because velocity and acceleration at returning to the table is bigger than the step to bring cup and fork to mouth.

V CONCLUSIONS

The contribution of this study was to detect the inclination of trunk for feeding activities, which usually is not considered in the studies of upper limbs because most rehabilitation interventions fix the trunk for rehabilitation, nonetheless, this study shown the role of trunk during the execution of some ADL related with feeding. Our results would suggest that trunk inclination must be included in the analysis of ADL performed with upper limbs.

Finally, we consider that information of vertical displacement of wrist (relative to shoulder) can be enough to differentiate feeding activities and it can be used to assess upper-limb function.

CONFLICT OF INTEREST

The authors declare that they have no conflict of interest.

ACKNOWLEDGEMENTS

DAIP-UGTO 1,146/2016, CONACYT CB2011/164440, PAPIIT IT101713 and PAPIME PE104715.

REFERENCES

1. Okkema Kathleen A., Culler Kathleen H.. Functional Evaluation of Upper Extremity Use following Stroke: A Literature Review *Topics in Stroke Rehabilitation.* 1998;4:54-75.
2. Rab George, Petuskey Kyria, Bagley Anita. A method for determination of upper extremity kinematics *Gait and Posture.* 2002;15:113 - 119.
3. Son Kwon, Park Junghong, Park Seonghun. Kinematic Analysis of the Neck and Upper Extremities During Walking in Healthy Young Adults *Journal of Bionic Engineering.* 2011;8:305 - 312.
4. Yang C., Kerr A., Stankovic V., Stankovic L., Rowe P.. Upper limb movement analysis via marker tracking with a single-camera system in *2014 IEEE International Conference on Image Processing (ICIP)*:2285-2289 2014.
5. Murgia Alessio, Kyberd Peter J., Chappell Paul H., Light Colin M.. Marker placement to describe the wrist movements during activities of daily living in cyclical tasks *Clinical Biomechanics.* 2004;19:248 - 254.
6. Pellegrino L., Stranieri G., Tiragallo E., et al. Analysis of upper limb movement in Multiple Sclerosis subjects during common daily actions in *2015 37th Annual International Conference of the IEEE Engineering in Medicine and Biology Society (EMBC)*:6967-6970 2015.
7. Ramanathan R., Eberhardt S. P., Rahman T., Sample W., Seliktar R., Alexander M.. Analysis of arm trajectories of everyday tasks for the development of an upper-limb orthosis *IEEE Transactions on Rehabilitation Engineering.* 2000;8:60-70.
8. El-Gohary M., Holmstrom L., Huisinga J., King E., McNames J., Horak F.. Upper limb joint angle tracking with inertial sensors in *2011 Annual International Conference of the IEEE Engineering in Medicine and Biology Society*:5629-5632 2011.
9. Ertzgaard Per, Öhberg Fredrik, Gerdle Björn, Grip Helena. A new way of assessing arm function in activity using kinematic Exposure Variation Analysis and portable inertial sensors–A validity study *Man Ther.* 2016;21:241-9.
10. Bouvier Brice, Duprey Sonia, Claudon Laurent, Dumas Raphaël, Savescu Adriana. Upper Limb Kinematics Using Inertial and Magnetic Sensors: Comparison of Sensor-to-Segment Calibrations *Sensors (Basel).* 2015;15:18813-33.
11. Butler Erin E., Ladd Amy L., Louie Stephanie A., LaMont Lauren E., Wong Wendy, Rose Jessica. Three-dimensional kinematics of the upper limb during a Reach and Grasp Cycle for children *Gait and Posture.* 2010;32:72 - 77.
12. Kim Kyung, Song Won-Kyung, Lee Jeongsu, et al. Kinematic analysis of upper extremity movement during drinking in hemiplegic subjects *Clinical Biomechanics.* 2014;29:248 - 256.
13. Buckley M. A., Yardley A., Johnson G. R., Carus D. A. Dynamics of the upper limb during performance of the tasks of everyday living a review of the current knowledge *Journal of Engineering in Medicine (Proc. Instn Mech. Engrs, Part H).* 1996;210:241-247.
14. Aizawa Junya, Masuda Tadashi, Koyama Takayuki, et al. Three-dimensional motion of the upper extremity joints during various activities of daily living *J Biomech.* 2010;43:2915-22.
15. Caimmi Marco, Guanziroli Eleonora, Malosio Matteo, et al. Normative Data for an Instrumental Assessment of the Upper-Limb Functionality *Biomed Res Int.* 2015;2015:484131.
16. Molteni F, Caimmi M, Maini ES, Sabatini AM, Rossini M, Franzini A. Instrumental kinematic approach for assesing upper limb motor control in Parkinson's disease (PD): A case report in *Movement Disorders*;17:S254–S254WILEY-LISS DIV JOHN WILEY & SONS INC, 605 THIRD AVE, NEW YORK, NY 10158-0012 USA 2002.
17. Vega-González Arturo, Granat Malcolm H. Continuous monitoring of upper-limb activity in a free-living environment *Arch Phys Med Rehabil.* 2005;86:541-8.
18. Douglas Brown. `http://physlets.org/tracker/` May, 2016.
19. Sergio Parra-Sánchez, Manuel Gómez-González Juan, Iraís Quintero-Ortega A., et al. *Automatic Pattern Recognition of Functional Upper-Limb Activities Using Markov Chains*:588–591. Cham: Springer International Publishing 2015.

Author:Arturo Vega González
Institute:División de Ciencias e Ingenierías
Street:Loma del bosque
City:León.
Country:Guanajuato, México
Email:a.vega@ugto.mx

The Output Circuit of a Biphasic Constant Current Electrical Stimulator

R.R. Nogueira[1], D.C. Souza[2], J.C. Palma[3], G.N. Nogueira-Neto[3] and P. Nohama[1,2,3]

[1] Pontifical Catholic University of Paraná/PPGTS, Curitiba, Brazil
[2] Federal University of Technology - Paraná/CPGEI, Curitiba, Brazil
[3] Pontifical Catholic University of Paraná/Computer Engineering Dept., Curitiba, Brazil

Abstract— **An Electrical Stimulator (ES) can promote physiological benefits during rehabilitation of patients with spinal cord injury. One of the main circuits in an ES system is its output stage. This paper outlines a transformer-based output circuit topology and shows the results obtained during the development of the ES biphasic output. The current in the circuit is controlled by a voltage-to-current converter whose input is driven by amplitude modulated pulses generated by a microcontroller. Darlington configuration transistor switching is responsible for regulating the current flow in the transformer's primary coil. The pulse width, frequency and amplitude can be set accordingly. Battery-supplied power enhances patient safety by avoiding use of high voltage sources. Tests with the transformer-based prototype indicated efficiency on pulse generation, transfer of data packets between stages, and amplification of signals. The circuit provided output currents of 200 mA and a 200 V on a 1 kΩ load. The pulse frequency varied from 500 to 5000 Hz, and pulse width from 30μs to 400μs. The burst frequency ranged from 10 to 100 Hz. The designed system represents a useful and versatile tool for functional and therapeutic applications since it has resources for patient safety and can be altered and expanded.**

Keywords— **Electrical Stimulator, Biphasic, Output Circuit, Constant Current.**

I. Introduction

The spinal cord can suffer trauma resulting from accident or disease, which can lead to various degrees of loss of voluntary limb control. Although physiologically intact, muscles are compromised by spinal cord injury (SCI), tending to atrophy due to inactivity, even when partial control remains [1,2]. Neuromuscular electrical stimulation (NMES) has been used to strengthen the muscles and reduce physical limitations and weakness that affect SCI patients and improve their condition [2–4]. The basic principle of the ES is based on the conversion of stimulatory low power pulses into high power signals that will be applied to neuromuscular tissues of patients with SCI. These stimulatory circuits aim to generate electrical impulses in order to artificially evoke physiological action potentials (APs) which are responsible for triggering muscular contractions [1].

Some have developed ESs make use of a transformer to amplify the generated low power signals. However, they operate on single-phase mode. Such a mode is not ideal for the application of electrical stimuli since it can cause ionic charge accumulation in the tissue causing damage to the patient's skin by means of electrochemical process [5,6].

Thus, a single-channel ES was developed with transistorized output in an H-bridge topology powered by high voltage and capable of operating in two-phase mode [7]. Another feature used in the development of ESs was to use microcontrollers in order to generate stimulatory patterns [8] [9]. The use of a microcontroller that supports I²C communication allows expanding the number of channels easily, being only limited to the number of available addresses on the bus. The expansion in the number of channels is required in order to generate functional movements of better quality. This is done with the intent of increasing the selectivity in muscle fibers recruitment during electrical stimulation. The stimulatory circuit proposed in this work makes use of an embedded system to generate configurable stimulatory patterns. In its output circuit a transformer is employed with three functions: elevate low power voltages, generate symmetric and biphasic signals, and electrically isolate the patient from the ES's circuit. High voltage is restricted to the secondary coil of the transformer, which is connected the electrode. The new design allows up to 127 biphasic stimulatory channels and is portable and battery powered.

The goals of this paper are to describe and discuss the design of the output circuit of a biphasic and constant current multichannel ES based on a voltage-to-current converter using a step-up transformer as the transferring element.

II. Materials and Methods

The ES was developed aiming to generate biphasic pulses, have at least eight stimulatory channels, and be a portable device for research and therapeutic applications using NMES. Its constructive topology is based on transformers with bipolar junction transistors (BJT) in Darlington configuration and battery powered, which provides galvanic isolation between the volunteer and the ES circuit. Such features prevent the use of high voltage sources and symmetric power supply, thereby enhancing patient safety.

I. Torres et al. (eds.), *VII Latin American Congress on Biomedical Engineering CLAIB 2016, Bucaramanga, Santander, Colombia, October 26th -28th, 2016*,
IFMBE Proceedings 60,
DOI: 10.1007/978-981-10-4086-3_156

Each stimulation channel has an 8-bit ATmega328 microcontroller which receives data packets containing pulse configuration parameters. The configuration is done by an embedded software in a Raspberry Pi 2 model B microcomputer and is transmitted to the microcontroller by the I²C bus. Due to the use of I²C communication, the increase in stimulation channels becomes quite simple, only requiring the replication of channel hardware and the change of devices' addresses. Addressing is done by attributing a 7-bit number to a device, which provides a total of 128 distinct addresses, i.e. theoretically the system could be expandable to up to 128 channels of stimulation. However, the total number is a somewhat lower.

The pulse generation and modulation stages run on low power. The output circuit requires high power due to the need for high voltage to be applied to the patient and promote nerve stimulation. An ATmega328 microcontroller generates pulses and sends them to two parallel pulse amplitude modulators (PAM). The digital-to-analog conversion data are sent to the Digital-Analog Converter (DAC) via SPI bus, resulting in modulated amplitude pulses. Fig. 1 presents an overview of the developed ES.

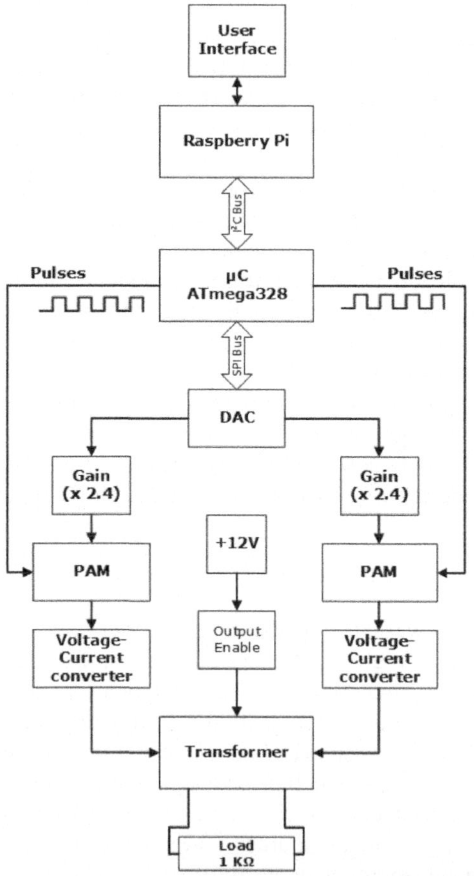

Fig. 1 Block diagram of the developed Electrical Stimulator

A. Pulses

The pulses are generated in different digital outputs of the microcontroller and delayed to at least 180°. They will switch the transistors in the voltage-to-current converter (Fig. 2).

B. Output Enable

This control signal is responsible for enabling the supply of energy in the transformer's center tap. For safety purposes, energizing the center tap is disabled while it is not in use.

C. DAC

Controlled by microcontroller via SPI bus, the DAC converts digital values to analog with a 12-bit resolution, i.e. pulses adjusted are adjusted into 4096 analog voltage levels.

D. Gain

This stage amplifies (2.4x) analog voltage pulses at the output of the DAC from 0 to 12 V in addition to increasing the input impedance of the PAM.

E. PAM

The PAM circuitry is composed of a BJT BC547 configured as an electronic enveloper, as shown in Fig. 2. In the collector terminal of the BC547, the modulating signal is applied through a 10 kΩ resistor. This signal represents the waveform profile and comes from the DAC. In the base terminal, the pulses generated by the microcontroller are applied.

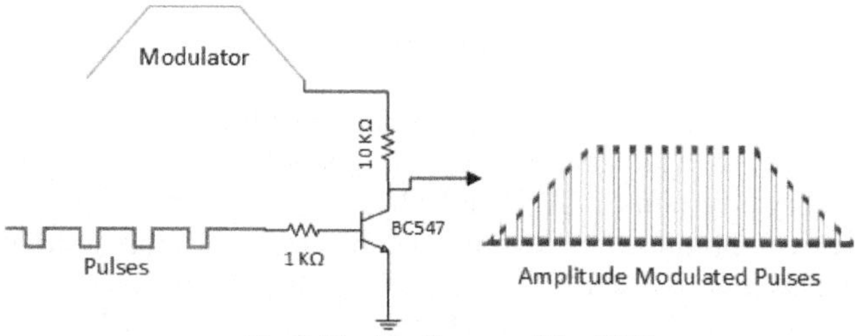

Fig. 2 Electric diagram of the PAM

F. Voltage-to-current Converter

The voltage-to-current converter circuit converts the amplitude of the voltage pulse into a current pulse with proportional intensity. This current will be drained by the primary coil of the transformer for transferring the energy to the secondary coil in order to magnify the pulses applied to its primary coil. This module consists of two op-amp and two Darlington BJT to provide the required current to each primary coil of the transformer. Resistors R_1 and R_2 (Fig. 3) are responsible for regulating the current I_e that flows through the primary coil. V_{in}' and V_{in}" are the amplitude of the pulse. As R_1 and R_2 are equal, they were called R (1).

$$I_e = \frac{V_{in}}{R} \quad (1)$$

G. Transformer

The transformer designed for the output stage elevates the output voltage at appropriate levels. It has two primary coils with 10 turns each, center tap derivation and 217 turns in the secondary coil. The transformer, together with the voltage-to-current converter, forms the output stage of the ES, as shown in Fig. 3. Table 1 shows the technical specifications of the designed transformer. First, the number of turns was calculated using (2-4),

$$N_1 = \frac{V_{in} \cdot 10^8}{4.44 \cdot B_m \cdot S_m \cdot f} \quad (2)$$

where N_1 is the number of turns in the primary coil; B_m is the magnetic induction in silicon iron, typically used 11,300; S_m is the magnetic core section and f is the frequency.

$$S_m = 7.5 \cdot \sqrt{\frac{1.25 \cdot W_2}{f}} \quad (3)$$

Equation (3) was used to calculate the magnetic section S_m, where W_2 is the desired power at the transformer's secondary. Finally, the number of turns in the primary coil was estimated, where V_1 is the voltage in the primary and V_2 is the voltage in secondary coil. N_2 is the number of turns in the secondary (4).

$$\frac{N_1}{V_1} = \frac{N_2}{V_2} \quad (4)$$

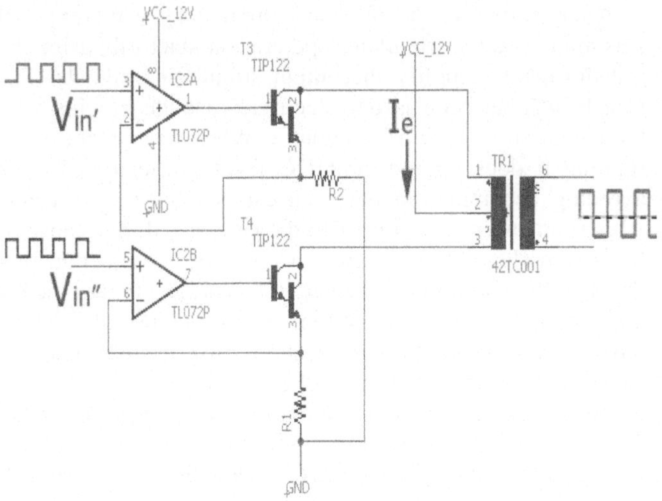

Fig. 3 Schematic of the developed ES output stage

Table 1 Technical specifications of transformer

Parameter Description	Parameter Value
Power in secondary coil	62.5 W (250 V – 250 mA on 1 KΩ load)
Magnetic section	2.09 cm^2
Frequency	1000 Hz
Number of turns in primary coil	10 + 10 turns, with center tap
Voltage in secondary coil	250 V
Number of turns in secondary coil	217 turns

III. RESULTS

The proposed ES generates biphasic pulses and high voltages, according to the transformer construction and microcontroller sequencing. The microcontroller generates two-monophasic pulse sequentially 180º out-of-phase one another (Fig. 4), aiming to activate the switching transistors in each of the primary coils. The collector current in each transistor feeds each half of the primary coil separately. The transistors are connected to the primary coil so that they allow producing a biphasic current in the secondary coil.

Each half of the primary coil operates like a monophasic stimulation channel: one winding generates anodic pulses whereas the other one generates cathodic pulses that, once combined, produce biphasic pulses (Fig. 5).

The transformer amplifies and inverts the current direction of pulses out of the digital outputs. The transformer's central derivation was supplied by a 12V, 5A power supply and the pulses switch a transistor making the current to flow through a primary winding each time, transferring energy to the secondary winding. Working at 1 kHz frequency and pulses of 40 μs pulses achieved 176 V in the positive phase and 146V in the negative phase, reaching a peak of 322 Vpp. Since the transformer power is 62.5W and the primary voltage is 12 V a maximum current of 5.21A flows through the winding. In order to provide the required energy, we have used a Darlington BJT (TIP 122) in the voltage-to-current conversion module.

A small pulse in the cathodic direction can be seen in Fig. 5 which is generated by the parasitic capacitance effect in the transistors. For patient's safety a relay couples the central derivation supply only when the stimulator is used. Moreover, the transformer does not transmit DC signals; it therefore has no need for connecting electrodes to reference because there is not going to be potential difference and no current will be flowing through the patient's tissue. The transformer will also provide galvanic isolation between patient and ES circuitry.

The DAC utilizes the SPI bus to transmit digital data to registers. Each transfer takes 128us each. Using a 12-bit resolution, a resolution of 1.22 mV was obtained. The output of the DAC is responsible for modulating pulse amplitude. To reach this goal, the DAC has in its output an amplifier with a 2.4x voltage gain, resulting in pulses that can vary amplitude from 0 to 12V.

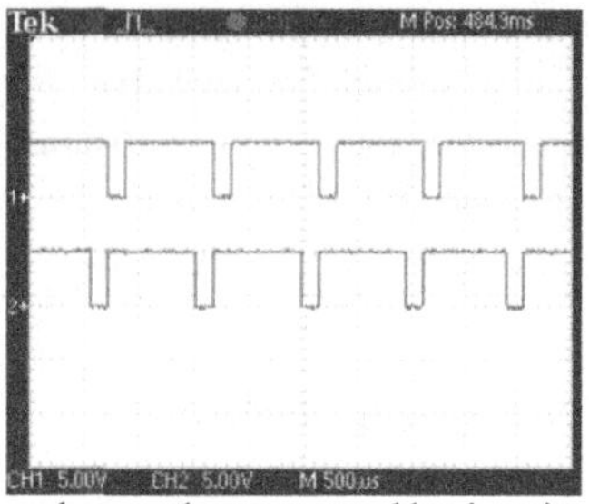

Fig. 4 Low voltage pulses generated by the microcontroller

Table 2 shows electrical characteristics of the developed ES.

Table 2 Electrical characteristics of ES

Parameter definition	Min. Value	Max. Value
Current	0 A	5 A
Supply voltage	12	18 V
Pulse frequency	50 Hz	5 KHz
Pulse width	30 µs	400 µs
Max. biphasic amplitude	0 mA	322 mA
Max. monophasic amplitude	0 mA	176 mA
DAC Resolution	-	12 bits

IV. DISCUSION

The flexibility in the generation of the stimulatory signal profiles was a goal of this project. We chose very small hardware like Raspberry Pi 2 and ATmega328 to make this possible. The consulted literature has used conventional PC, DSP [10] and timer [3]. We employed a transformer to increase voltage up to 200V and current up to 200mA on a resistive load of 1kΩ. Another study used a DC-DC converter to generate peak voltages of 300V [7]. The galvanic isolation of the transformer provides safety by leaving the patient isolated from the rest of the circuit. This safety may be lost if the feedback path is inadequate, however, Gaiotto et al. [7] ensures safety using optocouplers implemented in the feedback.

Another objective of this project was to generate biphasic signals in order to avoid ionic charge accumulation in the tissue of volunteers. Such accumulation does not occur when there is no charge equilibrium as with monophasic waveform stimuli [5,6]. Such goal was achieved with transformers, whereas Gaiotto et al. [7] created a biphasic output circuit using another type of topology: H bridge. It is also possible to achieve with high voltage operational amplifiers however, it requires high voltage power supply. Masdar [1] used an additional circuit controlled by DAC. One of the consulted projects used switch-capacitors to neutralize the waste electrical charges without the need of generating two-phase current [11].

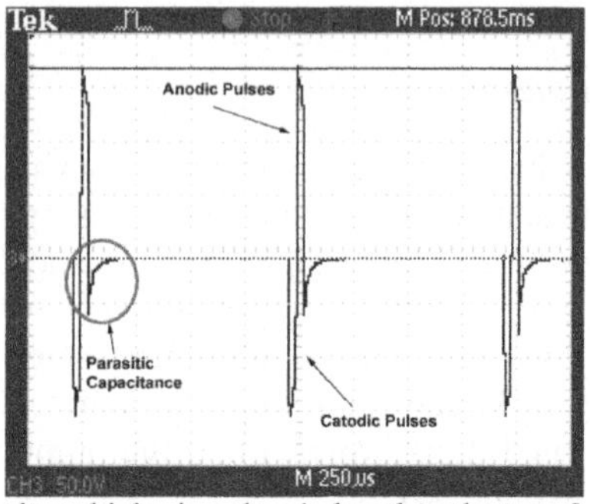

Fig. 5 High voltage biphasic pulses induced on the transformer secondary

This project used I²C bus for allowing the expansion in the number of channels using just a pair of wires (e.g., the SPI bus uses three wires for data transit and one wire for each device added). The number of devices is limited to the number of pins available on master devices [12]. RS-232 serial communication allows a unique point-to-point connection.

As alluded to earlier, another important point that promotes patient safety is the battery power. Other ES projects used the power grid to supply electrical circuits which, therefore, required that some safety procedures must be implemented to isolate the patient from the rest of ES hardware [5–7].

V. CONCLUSION

After analyzing the obtained stimulatory waveforms, it was found that the stimulator operates satisfactorily using the transformer to amplify the output stimulating signals. The transformer has been used to electrically isolate patients from the electrical circuits of stimulator. When the system is not running it does not transmit DC level to electrodes, thus avoiding accumulation of electrical charges in the tissue which can cause some injury to the patient's skin because of the electrochemical effect. It can therefore be asserted that the applied techniques described in this project turned the ES into a considerably versatile tool. Due to the use of I²C bus, the number of channels in the system can be easily expanded. The use of Raspberry Pi 2 model B, instead of a conventional PC, provides a substantial volume reduction, making it portable. Additionally, the incorporation of new features or experimental routines is easily performed with the Raspberry Pi 2 model B.

Acknowledgment

The authors thank Fundação Araucária and CNPq for financial support and scholarships and Dr. Ron Martinez for English revision.

References

1. A. Masdar, B. S. K. K. Ibrahim, and M. Mahadi Abdul Jamil, "Development of wireless-based low-cost current controlled stimulator for patients with spinal cord injuries," *Biomedical Engineering and Sciences (IECBES), 2012 IEEE EMBS Conference on*. pp. 493–498, 2012.
2. M. Bélanger, R. B. Stein, G. D. Wheeler, T. Gordon, and B. Leduc, "Electrical Stimulation: Can It Increase Muscle Strength and Reverse Osteopenia in Spinal Cord Injured Individuals?," *Arch Phys Med Rehabil*, vol. 81, 2000.
3. K. W. E. Cheng, Y. Lu, K. Y. Tong, a. B. Rad, D. H. K. Chow, and D. Sutanto, "Development of a Circuit for Functional Electrical Stimulation," *IEEE Trans. Neural Syst. Rehabil. Eng.*, vol. 12, no. 1, pp. 43–47, 2004.
4. M. Alfaro, L. Chairez, and L. N. Rivera, "Adaptive multi-channel portable stimulator based on PWM: A Tool for micro-stimulation using multi-array electrodes," in *Electrical Engineering Computing Science and Automatic Control (CCE) 8th International Conference on Merida City*, 2011.
5. G. N. Nogueira-neto, "Viabilidade do emprego de mecanomiografia no controle motor artificial em lesados medulares empregando estimulação elé-trica neuromuscular," UNICAMP, 2013.
6. A. L. Zagheni, "Sistema EENM multicanal controlado por computador para aplicações em locomoção artificial," Centro Federal de Educação Tecnológica do Paraná, 1998.
7. M. C. C. Gaiotto, G. N. Nogueira-Neto, E. Krueger, C. E. F. Magnani, and P. Nohama, "Uma contribuição ao projeto de saída de estimuladores elétricos neuromusculares bifásicos," in *XXIII Congresso Brasileiro em Engenharia Biomédica*, 2012.
8. D. T. O'Keeffe and G. M. Lyons, "A versatile drop foot stimulator for research applications," *Med. Eng. Phys.*, vol. 24, no. 3, pp. 237–242, 2002.
9. S. Koçer, M. R. Canal, and I. Guler, "Design of Low-Cost General Purpose Microcontroller Based Neuromuscular Stimulator," *J. Med. Syst.*, vol. 24, no. 2, pp. 91–101, 2000.
10. H. Wu, S. Young, T. Kuo, and S. Member, "A Versatile Multichannel Direct-Synthesized Electrical Stimulator for FES Applications," *IEEE Trans. Instrum. Meas.*, vol. 51, no. 1, pp. 2–9, 2002.
11. S. C. Huerta, M. Tarulli, A. Prodic, M. R. Popovic, and P. W. Lehn, "A universal functional electrical stimulator based on merged flyback-SC circuit," *Power Electronics and Motion Control Conference (EPE/PEMC), 2012 15th International*. pp. LS5a.3–1–LS5a.3–5, 2012.
12. F. Leens, "An introduction to I^2C and SPI protocols," *IEEE Instrumentation & Measurement Magazine*, vol. 12, no. 1. pp. 8–13, 2009.

Author: Raphael Ramos Nogueira
Institute: PUCPR
Street: Rua Imaculada Conceição 1155, CEP 80215-901
City: Curitiba
Country: Brazil
Email: rn.rapha@gmail.com

Facial Movements Detection Using Neural Networks and Mpeg-7 Descriptors Applied to Alternative and Augmentative Communication Systems

Alexandre Felippeto Henzen[1], Percy Nohama[2]

[1,2] Federal University of Technology/CPGEI, UFPR, Curitiba, Brazil

Abstract— **Assistive technologies are helping people without speech or writing functional, fully or partially, to communicate through methods and alternative and augmentative communication (CAA) systems, involving emulators keyboard and mouse. Most of these solutions use an input device as function activator. The drives are sensors connected to a part of the user's body with available effective or residual motor control. The aim of this paper is to present a new technique to detect facial movements applicable to a drive device in computer boards of alternative communication. The proposed algorithm uses neural networks in conjunction with MPEG-7 descriptors for detection of eye blinking, ranking the state of the eye between open and closed. The tests were performed using an application developed for simulation. The results reached 99% of accuracy using the descriptor Edge Histogram Descriptor (EHD). The developed method is efficient in environments with less than 60 lm. The article focused to eye blinking, but the technique can be easily adapted to any other movement, such as opening and closing the mouth.**

Keywords— **Alternative and Augmentative Communication, Neural Networks, MPEG-7, Sensors, Descriptors.**

I. Introduction

Communication requires the ability to transmit and receive information through codes and systems managed by general rules, so that a person may interact with others and be included in this society. By means of speech, combined with gestures, facial expressions and body, that the interaction with other person, communities and cultures occurs, forming social bonds which characterize the human condition [1].

The Alternative and Augmentative Communication aims at expanding communication skills to assist people without speech or functional writing or with communicative gap between the need and ability to speak and/or write [2]. As a result of such limitations, alternative methods and communication systems make it necessary [3].

Sensory, cognitive or motor impairments may compromise partly or fully human communicative skills, which can be complemented with support methods and systems for Alternative and Augmentative Communication (CAA). These alternative mechanisms are less efficient than natural, with significantly lower typing speeds [4]. However, in many cases, it may be the unique solution for persons with difficulties to access the computer and mobile devices in order to communicate with others.

Artificial intelligence has been incorporated to alternative communication techniques systems in areas such as word prediction, recognition of movements, speech, etc. [5]. Artificial neural networks (ANN) consist of a method for solving problems of artificial intelligence, developing systems containing circuits that simulate the human brain, including its behavior, for instance, learning, making mistakes and making discoveries. They allow acquiring and maintaining knowledge (based on the input information), and can be defined as a set of processing units, characterized by artificial neurons which are interconnected by a large number of connections (artificial synapses) [5].

The aim of this paper is to present an eye blink's detector based on image acquisition through video cameras to actuate as a drive device for CAA systems, such as virtual keyboards or communication boards. In this project, neural networks, in conjunction with MPEG-7 descriptors, are used to classify the image by differentiating the movement, when opening or closing the eyes even in environments where occurs brightness variation.

II. Materials and Methods

The development of the proposed solution follows the steps described below.

A. images Acquisition

The images were acquired by web camera or portable camera connected to the computer. There is no specificity in relation to the camera. One of the virtues of the developed solution is to use the more generic hardware possible, becoming irrelevant the camera quality.

B. Extraction of the image area for detecting movement

For extraction of the eye region, we used the Opencv library [6]. The Opencv was developed by Intel library, and is free to develop applications in computer vision area. The Opencv is not responsible for the detection of gestures. Its function is only to focus on the region where one intends to detect motion and extract the region of the image for

I. Torres et al. (eds.), *VII Latin American Congress on Biomedical Engineering CLAIB 2016, Bucaramanga, Santander, Colombia, October 26th -28th, 2016*,
IFMBE Proceedings 60,
DOI: 10.1007/978-981-10-4086-3_157

analysis. In the case of detection of eye blinks, it is possible to adapt a camera directly to glasses without the use of Opencv.

C. Extraction of MPEG-7 descriptors

The images extraction features for the samples uses the MPEG-7 descriptors, which is an emerging standard used in image classification systems. Its purpose is not to create a method of encoding and decoding video and audio, but to establish a pattern of how to describe the contents of a media [7].

Descriptors extract image features such as color and texture which are used to identify that image. The descriptors used in this project are: color and texture.

Color: is one of the most visual characteristics used in image and video retrieval. Color features are relatively robust for viewing angle, translation and rotation of the regions of interest. Color descriptors represent different aspects of color feature, including the distribution of colors, spatial arrangement of color and spatial structure color [8].

SCD (Scalable Color Descriptor): the scalable color descriptor is derived from a color histogram of color space defined in HSV (hue-saturation-value, i.e., hue, saturation, and value). HSV is also known as HSB (hue, saturation and brightness - hue, saturation and brightness, respectively). This color system sets the color space as described below, using the three parameters:

• Tint (hue): it checks the type of color, covering all colors of the spectrum from red to violet, more magenta. It reaches values 0 to 360, but for some applications, this value is normalized from 0 to 100 %;

• Saturation, also called "purity." The lowest this value, the greatest gray tone appears in the image. The highest the value, the most "pure" image. It reaches values from 0 to 100 %;

• Value (brightness): it sets the color brightness and reaches values between 0 and 100 %.

DCD (Dominant Color Descriptor): it allows the specification of a small number of dominant color values, as well as their statistical properties such as distribution and variance. It aims to provide an efficient, compact and intuitive representation of colors present in a region or image.

CLD (Color Layout Descriptor): it acquires the spatial arrangement of the representative colors on an array superimposed on a region or image. The representation is based on the coefficients of the discrete cosine transform. This is a very compact and very efficient descriptor in navigation applications and quick search. It can be applied to still images as well as video segments.

Texture: it corresponds to the visual patterns homogeneity properties, resulting from the presence of multiple colors or multiple intensities in the image. The texture descriptor contains important structural information of surfaces and the relationship with the environment. The description of the texture images by MPEG-7 provides a powerful mode to find similarity and to recover both the homogeneous and non-homogeneous textures.

EHD (Edge Histogram Descriptor): it is a texture descriptor for detecting image edges. The image is divided into 4×4 sub-images. Each sub-picture is divided into blocks of smaller images (typically 4×4 pixels). The pattern has four directional edges: vertical, horizontal, diagonal (45 and 135°) and non-directional. An histogram of 5 bars (one for each type of edge) is calculated on all image blocks in the sub-image. This procedure is repeated for all sub-images, resulting 80 coefficients histogram.

For extraction of image features, we used a library in C # developed by Chatzichristofis [9]. The technique is to generate a vector containing the image characteristics by concatenating the individual results of each descriptor. The result of the concatenation is used as input to the neural network.

D. Samples classification

For the image classification containing samples with open and closed eyes, we opted for the artificial neural network.

An ANN can be considered as a meta-function that accepts a fixed number of numeric inputs and produces a fixed number of numeric outputs. In most situations, a neural network has a layer of hidden neurons, where the neurons of this layer are connected between the input and output neurons. Associated to each individual hidden neuron and each individual output neuron there is a set of weight values and the so called "trend value". Weights and trends determine the output values for a given set of input values [10].

The need of a neural network-based classifier system is due to the nonlinearity of the image classification. Many images are similar, with characteristics that differ close to each other. So, they are not easily separable by a linear algorithm. The degree of variability is very diverse.

For ANN, we used an algorithm for the multilayer perceptron model (Multilayer Perceptron - MLP) with back propagation error. It is employed if only one hidden layer exists, ensuring that the effectiveness of the training will be dependent on the number of neurons in the hidden layer. The choice of a unique layer stems from the fact that many layers in a MLP induce to a retro error propagated very large [5].

In the network entry, it is applied the descriptors extracted from the acquired frames. Fifty open eyes samples and 50 closed eyes samples are captured. This number of samples does not follow any criterion and can be set to other amounts.

The network's objective is to classify the images and determine if the image corresponds to an open or closed eye. In this case, the network will have only one output. The output of the network uses a sigmoid function which generates a value between 0 and 1 determining the probability of the output represents one eye closed. The closer to 1, the more likely the image correspond to a closed eye. To minimize false positives, the algorithm has a trigger threshold setting (Threshold), whose goal is to eliminate areas where there is a classified image that does not correspond to a closed eye. For example, it is considered a totally closed eye only if the output value is equal to or greater than 0.7, and fully opened when the output is less than 0.3 for each acquired frame. The sample frequency was 20 Hz.

To avoid involuntary eye closure problem, it is considered as a voluntary movement of an output sequence of 10 ratings equal or approximately to 500 ms with the eye closed. To perform the neural network training, we used 25 frames for open eye image and the same for the closed eye.

The prototype implemented for tests execution was developed in C# language. For extraction of image features, we used a library in C# developed by Chatzichristofis [9]. The technique consists in generating a vector containing the image characteristics in order to concatenate the individual results of each descriptor. The result of the concatenation is used as input to the neural network.

Figure 1 shows some sample frames acquired and analyzed by the software with the highlighted probability with the number in red color at the top of each image.

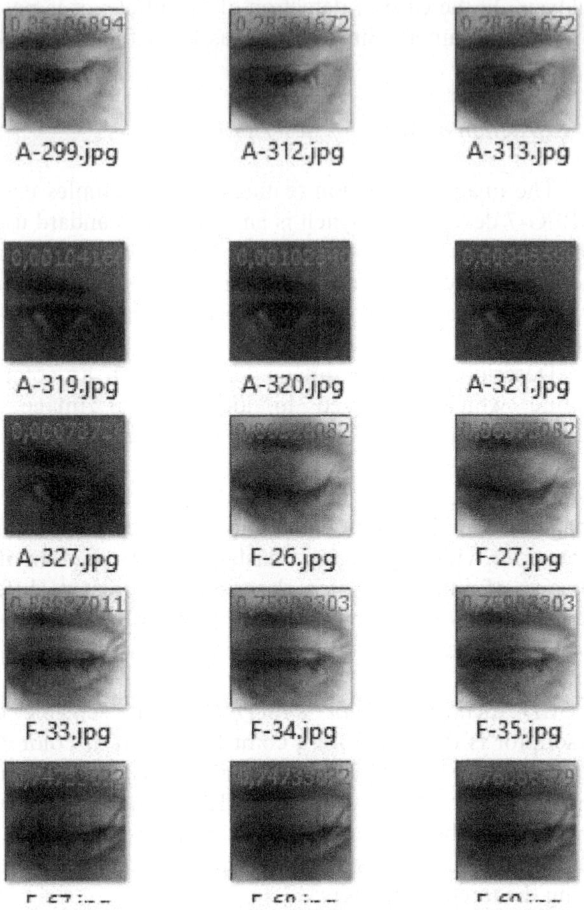

Figure 1 – Captured samples processed by neural network.

III. RESULTS

The developed solution had achieved a profitable solution as shown in Table 1. In environments with low-level light, less than 60 lm, between the camera and the eye at a distance of 50 cm, it obtained better results by increasing the number of neurons in the intermediate layer between 50 and 200.

To alleviate the problem of increasing the number of neurons in the hidden layer, we added in the simulator, brightness controls, contrast, gain and a control to convert the image to grayscale before extracting features.

Assays were performed with varying numbers of samples for training the network. It was observed that quantities between 40 and 100 samples had no significant change in result.

Table 1 shows the results of tests performed on combinations of descriptors using the prototype. It shows the number of 300 samples used for classification. Each combination of descriptors generates a number of input neurons, for example, the EHD has 80 input neurons. Fifty neurons were used in the hidden layer. This number has been changed to other values but they are not presented in this article. The number of epochs was resultant of the neural network processing. The accuracy was determined based on samples used in the tests.

Table 1 Results for descriptors combination.

Results	EHD	SCD	CLD	EHD SCD	EHD SCD CLD
Samples	300	300	300	300	300
Epochs	680	700	Error	2100	3749
Time Extraction Features	<1s	<1s	<1s	<1s	1s
Time Classification	<1s	<1s	-	<1s	<1s
Neurons (Input)	80	256	192	112	176
Neurons (Hidden)	50	50	50	50	50
Neurons (Output)	1	1	1	1	1
Accuracy	99.9%	80%	-	90%	50%

The combination of several descriptors does not guarantee, in the scenario evaluated, a better image classification. The best results were obtained using EHD alone and with the combination of EHD/SCD descriptors.

IV. DISCUSSION

The feature extraction techniques used in this work can be replaced by others as input to the neural networks. The descriptor MPEG-7 was used due to the small number of features, unlike the use of pure image (bitmap), which need a large number of inputs and neurons, creating instability and increases the processing time.

The prototype has being tested with volunteers with Amyotrophic Lateral Sclerosis (ALS) using the CAA software named Emulator for Keyboard and Mouse (ETM) [11] whose results will be published in future article.

V. CONCLUSION

In this article, we described a design algorithm using neural networks together with MPEG-7 descriptors to constitute an efficient method for feature extraction and classification of images. The use of MPEG-7 image descriptors with MLP proved effective when used descriptor EHD. The results reached 99 % of accuracy using the descriptor Edge Histogram Descriptor (EHD).

The method developed as a drive mechanism for alternative communication applications has proved effective in tests being conducted with volunteers and new tests should be conducted with other applications.

REFERENCES

1. Miranda, Leidy e Gomes, Ivone. Contribuições da comunicaçãoalternativa de baixa tecnologia em paralisia cerebral sem comunicação oral.
2. Sartoretto, Mara e Bersh, Rita. 2014. http://www.assistiva.com.br/ca.html. *Assistiva Tecnologia e Educação.*[Online] 2014. http://www.assistiva.com.br/ca.html.
3. Jordan, Monica. 2007. Predição de Palavras baseada em Modelos Ocultos de Markov. s.l. : Universidade Tecnologica Federal do Parana, 2007.
4. Garcia, Luis e Luis, Oliveira. *Sistema de CAA com Adaptação ao Contexto.*
5. Nunes Ivan, Hernane Danilo, Andrade Rogério. Redes Neurais Artificiais para Engenharia e ciências aplicadas, Artiliber, 2010.
6. Opencv, www.opencv.org
7. Wonil Kim, Sanggil Kang, Juyoung Lee Image Identification System Using MPEG-7 Descriptors
8. B. S. Manjunath, Philippe Salembier, Thomas Sikora. Introduction to MPEG-7: Multimedia Content Description Interface, Volume 1, 2002.
9. M. Lux and S. A. Chatzichristofis, "LIRE: LUCENE IMAGERETRIEVAL – AN EXTENSIBLE JAVA CBIR LIBRARY", «ACM International Conference on Multimedia 2008»,(ACM MM), Open Source Application Competition, pp.1085-1087, October 27 to 31, 2008, Vancouver, British Columbia.
10. McCaffrey James. Retropropagação de rede neural para programadores.https://msdn.microsoft.com/ptbr/magazine/ee532098(en-us).aspx?sdmr=JamesMcCaffrey&sdmi=authors.
11. Henzen Alexandre. ETM – Keyboard and Mouse Emulator. www.projetoetm.com.br

Main Effects of Energy Drinks on Mood, Reaction Time and Brain Regions

C. Martínez-Torres[1], M. Calvillo[2], C. Romero-Rebollar[1], D. Martínez-Cancino[1], M. Flores-Leal[1], and L. Jiménez-Angeles[3]

[1] Centro Nacional de Investigación en Imagenología e Instrumentación Médica, Universidad Autónoma Metropolitana-Iztapalapa, México.
[2] Graduate School of Arts and Sciences, New York University, New York, NY, USA.
[3] Departamento de Sistemas Biomédicos, Facultad de Ingeniería, Universidad Nacional Autónoma de México, México.

Abstract— **Effects of energy drinks (ED) are attributed to the combination of their three main ingredients: caffeine, taurine, and glucose. The primary objective of this study was to evaluate the main effects and interactions of these ingredients on mood, reaction time during a working memory task and to identify the specific brain regions involved using functional Magnetic Resonance Images (fMRI). In a double-blind, mixed factor and repeated measurement design, 12 healthy volunteers (6 male, 6 females) received one of four treatments (200 mg caffeine/0 mg taurine, 0 mg caffeine/2000 mg taurine, 200 mg caffeine/2000 mg taurine, 0 mg caffeine/0 mg taurine) on each of four separated days. Between-participants treatment was a glucose (11 g of glucose) or placebo (3 g of Stevia). All participants were evaluated with a mood questionnaire in three times and underwent a working memory task, where reaction time was registered, while fMRI were acquired. Functional images were analyzed considering treatment and task difficulty as within-participants in a repeated measurements general linear model. Our results shown that glucose had the most consistent effects on mood, decreasing feelings of tension, depression, anger and fatigue and increasing feeling of vigor. However, reaction time had significant differences only for task difficulty without interactions due to the treatment or glucose intake. Significant activations of brain regions were located at frontal, temporal and cingulate gyrus but main effects and interactions with treatment and glucose were explained only for the task difficulty.**

Keywords—. **fMRI, Energy drinks, Cognition, Working memory.**

I. Introduction

In Mexico, the consumption of energy drinks (ED) has raised 64% from 2003 to 2008 while per capita consumption is approximately 3.4 liters annually [1], and more health concerns are increasingly unfolded about their excessive consumption [2]. ED have become a popular beverage and the most common reason for consumption include counteracting sleepiness and increasing energy, maintaining alertness while study and driving and reducing symptoms of a hangover [3]. The cognitive effects of ED usually attribute changes in performance and mood to the combination of their three main ingredients: caffeine, taurine, and glucose.

Caffeine is the most commonly used behaviorally-active substance in the world and can reliably enhance vigilance and psychomotor performance. It has been suggested that the beneficial effects of caffeine on cognitive performance are actually due to the reversal of caffeine withdrawal [4] or the reversal of environmental-induced cognitive impairments, such as sleep deprivation, physical fatigue and psychological stress [5].

Taurine is highly concentrated in the heart and liver as well as the central nervous system including the brain stem and hippocampus [6], where it plays a role in osmoregulation, membrane stabilization, neuroprotection, neuromodulation, and regulation of cellular calcium level [7]. The cognitive effects of taurine studied in animals suggest that taurine may prevent or reverse neurotoxin-induced deficits in learning, memory, and long-term potentiation, but does not enhance cognitive performance in healthy, intact animals [8].

Glucose is thought to improve some aspects of cognitive performance, however, the results are inconsistent. Some reports showed that caffeine and glucose, alone and together, reduced reaction time and together improved sustained attention and verbal memory [9].

Giles et al [10], studied the relative effects of caffeine, taurine and their combination; and reported that caffeine had improvements on cognitive performance, while taurine intake opposed caffeine effects on mood, including reducing feelings of vigor and increasing caffeine withdrawal symptoms. However, the specific brain regions associated with main effects and interactions of ingredients on cognitive performance were not studied.

The primary objective of this study was to evaluate the main effects and interactions of caffeine, taurine and glucose on mood, working memory and to identify the specific brain regions involved. We hypothesized that if working memory could be evaluated using the N-Back task and activate anterior cingulate, parietal, temporal and frontal regions; then a multivariate repeated measurements of functional magnetic resonance imaging (fMRI) could identify brain regions associated with main effects and interactions of the three main ED ingredients.

I. Torres et al. (eds.), *VII Latin American Congress on Biomedical Engineering CLAIB 2016, Bucaramanga, Santander, Colombia, October 26th -28th, 2016*,
IFMBE Proceedings 60,
DOI: 10.1007/978-981-10-4086-3_158

II. MATERIALS AND METHODS

A. Subjects

Twelve healthy volunteers (6 male, 6 female; mean age 21.38±3.4 years, mean Body Mass Index 22.56±1.9) participated in this double-blind, mixed factor, repeated measures fMRI study. All participants were undergraduate students, non-regular caffeine consumers, non nicotine users, in good health, and did not use prescription medication. Written informed consent was obtained, and all experimental procedures were approved by the Ethics Committee of the Research Center (CI3M).

B. Experimental design

A double-blind, mixed factor, repeated measures design was used with caffeine and taurine treatments as within subjects factors and glucose treatment as a between subjects factors. All participants completed the four caffeine and taurine treatments 1) 0 mg caffeine, 0 mg taurine; 2) 0 mg caffeine, 2000 mg taurine; 3) 200 mg caffeine, 0 mg taurine and 4) 200 mg caffeine, 2000 mg taurine. These treatments were crossed with glucose, such that half of participants were administered glucose, 11 g, and half were administered placebo, 3 g of Stevia. The values chosen for glucose, caffeine and taurine approximate those found in the commonly consumed energy drink Monster® in Mexico city (160 mg caffeine, 2000 mg taurine, 52 g sugar) [11]. Caffeine and taurine were administered in identical capsule form with water. Glucose or Stevia was administered as a 350-ml sparkling drink.

The Profile of Mood States (POMS) questionnaire was used to evaluate subjective mood and arousal states [12]. Participants were asked to rate a series of 65 mood related adjectives on a five-point scale, using the response set "how are you feeling right now?" The adjectives factor into six mood subscales (tension, depression, anger, vigor, fatigue, and confusion).

During the study sessions, participants first completed baseline measures of the POMS questionnaire. They consumed their assigned treatment capsule and drink. They then took a second POMS questionnaire after 45 minutes break in order to have an evaluation while treatment constituents remained at near-maximal plasma levels [13]. Seventy minutes after treatment, functional images were acquired while participants undertake the N-Back task, which main emphasis is on monitoring and constant updating of working memory and evaluate the reaction time performance. After 80 minutes of baseline a third-final POMS evaluation was completed.

C. N-back task

A 3-step letter version of the N-back task was used [14]. Letters used as test stimuli were presented one at a time. Each letter appeared for 1000 ms and the inter-stimulus interval was 500 ms allowing a maximal reaction time of 1500 ms. In the 0-back phase, subjects were asked to press a push-button when letter "X" appeared. In the 1-back phase participants were asked to press a push-button when the letter presented was identical to 1 letter preceding it. For the 2-back phase subjects had to press the push-button whenever the letter presented was identical to the letter presented 2 positions back (see Fig. 1).

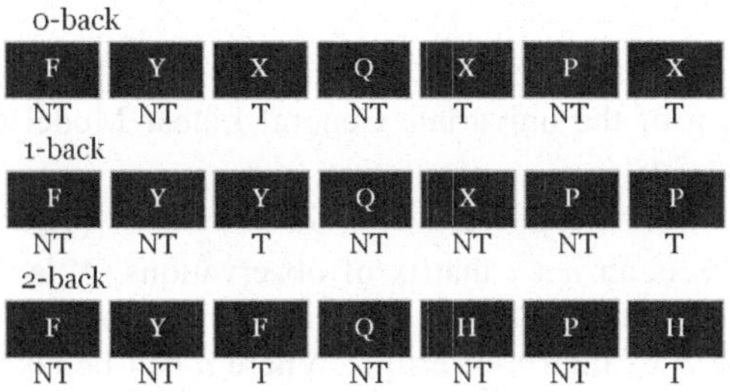

Fig. 1 Schematic diagram of the N-back paradigm where NT= non target and T= target.

Each N-back phases were presented individually and repeated three times. In each block 11 letters were presented, containing 3 targets for 0-back, 1-back and 2-back. There was an 8 s instruction screen before each task block. The paradigm was programmed in Matlab (2007a, The MathWorks Inc) with the Cogent 2000 toolbox developed by the Cogent 2000 team at the FIL and the ICN and Cogent Graphics developed by John Romaya at the LON at the Wellcome Department of Imaging Neuroscience(http://www.vislab.ucl.ac.uk/cogent.php). Reaction time data were acquired with a 2-Button Fiber Optic Response Pad (Current Designs®).

D. fMRI acquisition

Functional images were acquired on a 3T MRI scanner (Philips Medical Systems, Achieva®) using a SENSE head coil of 8 elements. A gradient echo EPI sequence (TE=35 ms, TR=2000 ms, flip angle=90°, matrix=128) and a T1 weighted 3D (FOV=250x250x181, slices=301) were used.

The task was electronically synchronized with the MRI scanner. Task stimuli were projected onto a special fMRI screen (BOLDscreen, Cambridge Research Systems) with subjects viewing the screen via a mirror system located in the head coil. None of the subjects presented trouble with visualization within the scanner. 480 scans were acquired per subject for each session.

E. fMRI preprocessing

All images were pre-processed with SPM8 software (Functional Imaging Laboratory at University College London, http://www.fil.ion.ucl.ac.uk/spm/software/spm8/). Functional volumes per subject were realigned to the mean functional volume of each session (EST&WRITE), then data were normalized (EST&WRITE) and smoothed (Gaussian kernel, [8 8 8]). Three contrasts of first level analysis per subject were computed (0-back, 1-back, 2-back) and considered as a repeated measurement condition.

F. Analysis of multivariate repeated measurements of fMRI data

As was described by McFarquhar et al [15], the multivariate form of the univariate General Linear Model could be expressed as:

$$\boldsymbol{Y} = \boldsymbol{XB} + \boldsymbol{E} \quad (1)$$

where **Y** is an $n \times t$ matrix of observations, $\boldsymbol{X}$ is the $n \times k$ design matrix, $\boldsymbol{B}$ is the $k \times t$ matrix of model parameters, and **E** is the $n \times t$ matrix of errors. Where n can be taken as the number of subjects, t as the number of *repeated measurements,* and k as the number of independent variables or predictors. It is assumed that $\boldsymbol{Y}_i \sim N(\boldsymbol{X}_i\mathbf{B}, \boldsymbol{\Sigma})$ so that each ith row of $\boldsymbol{Y}$ is considered drawn from a multivariate normal distribution with a mean vector given by $\boldsymbol{X}_i\mathbf{B}$, and an unstructured covariance matrix $\boldsymbol{\Sigma}$.

Hypothesis testing in the multivariate GLM is based on the contrast

$$\boldsymbol{ABC}' = \mathbf{0} \quad (2)$$

The hypotheses testing approach is conceptualized as combining hypotheses about groups using $\boldsymbol{A}$, and hypotheses about repeated measures using $\boldsymbol{C}$.

The test statistics from the multivariate GLM was carried out using the MRM toolbox [15] considering the Wilks'lambda statistics, and a voxel-level permutation test to provide a FEW correction, thresholding the images at $p_{FWE} < 0.05$ with 5000 permutations.

III. RESULTS AND DISCUSSIONS

A. Profile of mood states (POMS) questionnaire

Table 1 show the main significant interactions from POMS questionnaire analysis. Results from the POMS shown that Glucose is the main responsible for lower feelings of Tension, Depression, Anger and Fatigue and higher feeling of Vigor; and combination of Glucose with Caffeine trends to increase these differences.

Table 1 Results of the ANOVA test with Caffeine, Taurine and Time as within-participants variables and Glucose as a between-participant variable.

Subscale/Interaction	DM	SEM	CI	*p*
Tension				
(St+Ta) - (Gl+Caf+Ta)	4.21	1.3	(0.21, 8.22)	0.031
(St+Ta) - (Gl+Ta)	4.33	1.3	(0.32, 8.33)	0.024
Depression				
(St+Caf) - (Gl+Caf)	2.079	0.629	(0.14, 4.01)	0.026
(St+Caf) - (Gl+Caf+Ta)	2.357	0.629	(0.41, 4.29)	0.006
(St+Caf) - (Gl+Pl)	2.524	0.629	(0.58, 4.46)	0.003
(St+Caf) - (Gl+Ta)	2.468	0.629	(0.52, 4.40)	0.003
Anger				
(St+Bas) - (Gl+45m)	2.143	0.694	(0.13, 4.14)	0.029
(St+Bas) - (Gl+80m)	2.351	0.694	(0.34, 4.35)	0.012
Vigor				
(St+Caf+Ta) - (Gl+Caf)	-8.11	2.07	(-14.48, -1.74)	0.003
(St+Ta) - (Gl+Caf)	-6.44	2.07	(-12.81, -0.08)	0.045
(St+Caf+Ta) - (Gl+Caf+Ta)	-8.33	2.07	(-14.70, -1.97)	0.002
(St+Ta) - (Gl+Caf+Ta)	-6.67	2.07	(-13.03, -0.30)	0.033
(St+Caf+Ta) - (Gl+Pl)	-9.17	2.07	(-15.53, -2.80)	0.001
(St+Ta) - (Gl+Pl)	-7.5	2.07	(-13.87, -1.13)	0.009
(St+Caf+Ta) - (Gl+Ta)	-7.89	2.07	(-14.26, -1.52)	0.005
(St+Ta) - (Gl+Ta)	-6.22	2.07	(-12.59, 0.14)	0.06
Fatigue				
(St+Pl) - (Gl+Caf)	6.12	1.57	(1.27, 10.97)	0.004
(St+Pl) - (Gl+Pl)	6.01	1.57	(1.16, 10.85)	0.005
(St+Pl) - (Gl+Ta)	6.23	1.57	(1.38, 11.08)	0.003

DM= Difference of Means, SEM = Standard Error of Means, CI= 95% Confidence Interval, St = Stevia, Ta= Taurine, Gl = Glucose, Ca = Caffeine, Pl = Placebo, Bas = baseline time, 45m = 45 minutes after baseline time, and 80m = 80 minutes after baseline time.

Our findings are inconsistent with report of Giles et al [10], who states that feelings of tension were higher following caffeine intake when consumed with glucose but not without glucose; and that taurine intake lowered feelings of vigor in absence of caffeine intake.

B. N-Back reaction time

These preliminary results have shown significant statistical differences only for task difficulty, with $F(2,22)= 16.3$, $p<0.0001$. Pairwise comparison tests revealed that 0-back was significantly different from both 1-back ($p<0.01$) and 2-back ($p<0.0001$). However, 1-back and 2-back were not significantly different from each other. Although not significant, Glucose has found to be somewhat better than Stevia in terms of reaction time. Furthermore, the ingestion of only Taurine seemed to negate this difference. The best combination that achieved the lowest reaction time seemed to be either Glucose alone, or in a combination with the Caffeine and Taurine. According to task type, Placebo was the best option for the lowest difficulty (0-back), the full treatment treatment seemed to work best for 1-back and just Taurine seemed to be most effective in 2-back.

C. fMRI multivariate data results

Main activations of brain regions having a working memory task were located at frontal gyrus (superior, middle and inferior), temporal gyrus, cingulate gyrus and precentral gyrus as has been reported in a meta-analysis studies [16]. In this preliminary results we can observe (see Table 2) that statistical effects and interactions of brain regions are explained mainly by the differences of Task difficulty (0-Back vs 1-Back vs 2-Back) and none significant for Treatment or Drink.

Table 2. Results of brain regions with significant N-Back main effect and interactions Group x N Back and Group x Treatment x N-Back.

Region		Cluster size	F-value	x	y	z
N-back						
R	Superior Temporal Gyrus	6	330.73	46	-14	-6
R	Inferior Frontal Gyrus	1	110.25	34	26	-4
L	Insula	11	254.84	-32	22	2
L	Sub-Gyral	1	103.75	-32	-50	40
R	Cingulate Gyrus	1	118.46	2	-10	46
R	Middle Frontal Gyrus	1	115.23	28	12	48
R	Superior Frontal Gyrus	9	196.42	34	14	54
R	Sub-Gyral	1	183.80	22	-42	54
L	Medial Frontal Gyrus	1	110.04	-6	-12	54
R	Postcentral Gyrus	3	146.03	24	-36	64
Group x N-back						
L	Insula	1	110.476	-42	-2	8
R	Precentral Gyrus	1	125.006	38	-14	60
Grouop x Treatment x N-back						
L	Middle Frontal Gyrus	1	234-68	-20	32	40

IV. CONCLUSIONS

In this study we evaluated the main effects of caffeine, taurine and glucose on mood, working memory and the brain regions involved. Analysis of POMS shows that Glucose had the most consistent effects on mood, decreasing feelings of Tension, Depression, Anger and Fatigue and increasing feeling of Vigor. However, reaction time analysis revealed significant differences only for task difficulty without interactions due to the Treatment or Glucose intake. Functional images identified statistical changes of activation in regions that has been described by a meta-analysis study, however the repeated measurement analysis found that significant main effects and interactions with Treatment and Glucose were explained only for the N-Back task.

One concern of these preliminary results was number of subjects analyzed, because previous studies that found mood and cognitive effects for treatment and glucose intake typically had larger samples. However, these findings are not enough to provide overwhelming evidence for the beneficial effects of Glucose on mood and discard the effects of Taurine on cognitive performance. A stronger design experiment is needed to further elucidate a solid conclusion.

ACKNOWLEDGMENT

To the biomedical and technical staff of the CI3M, Mexico City, Mexico.

REFERENCES

1. "New Market Research Report: Energy Drinks in Mexico," ed: SyndiGate Media Inc., 2016
2. K. A. Sepkowitz, "Energy drinks and caffeine-related adverse effects," JAMA, vol. 309, pp. 243-4, Jan 16 2013.
3. S. Attila and B. Çakir, "Energy-drink consumption in college students and associated factors," Nutrition, vol. 27, pp. 316-322, 2011
4. J. E. James and P. J. Rogers, "Effects of caffeine on performance and mood: withdrawal reversal is the most plausible explanation," Psychopharmacology, vol. 182, pp. 1-8, 2005.
5. F. Koppelstaetter, T. D. Poeppel, C. M. Siedentopf, A. Ischebeck, C. Kolbitsch, F. M. Mottaghy, et al., "Caffeine and cognition in functional magnetic resonance imaging," J Alzheimers Dis, vol. 20 Suppl 1, pp. S71-84, 2010
6. K. Ito, M. Arko, T. Kawaguchi, M. Kuwahara, and H. Tsubone, "The effect of subacute supplementation of taurine on spatial learning and memory," Experimental Animals, vol. 58, pp. 175-180, 2009
7. F. Junyent, L. De Lemos, J. Utrera, S. Paco, F. Aguado, A. Camins, et al., "Content and traffic of taurine in hippocampal reactive astrocytes," Hippocampus, vol. 21, pp. 185-197, 2011
8. B. P. Vohra and X. Hui, "Improvement of impaired memory in mice by taurine," Neural Plast, vol. 7, pp. 245-59, 2000
9. M. B. Gilsenan, E. A. de Bruin, and L. Dye, "The influence of carbohydrate on cognitive performance: a critical evaluation from the perspective of glycaemic load," Br J Nutr, vol. 101, pp. 941-9, Apr 2009
10. G. E. Giles, C. R. Mahoney, T. T. Brunye, A. L. Gardony, H. A. Taylor, and R. B. Kanarek, "Differential cognitive effects of energy drink ingredients: caffeine, taurine, and glucose," Pharmacol Biochem Behav, vol. 102, pp. 569-77, Oct 2012
11. P. F. del Consumidor, "Bebidas saborizadas, hidratantes y para deportistas," Revista del consumidor, vol. 281, 2000
12. D. M. McNair, M. Lorr, and L. F. Droppleman, Profile of mood states: Educational and Industrial Testing Service San Diego, CA, 1981
13. J. L. Donovan and C. L. DeVane, "A primer on caffeine pharmacology and its drug interactions in clinical psychopharmacology," Psychopharmacol Bull, vol. 35, pp. 30-48, Summer 2001
14. L. E. Nystrom, T. S. Braver, F. W. Sabb, M. R. Delgado, D. C. Noll, and J. D. Cohen, "Working memory for letters, shapes, and locations: fMRI evidence against stimulus-based regional organization in human prefrontal cortex," Neuroimage, vol. 11, pp. 424-46, May 2000.
15. M. McFarquhar, S. McKie, R. Emsley, J. Suckling, R. Elliott, and S. Williams, "Multivariate and repeated measures (MRM): A new toolbox for dependent and multimodal group-level neuroimaging data," NeuroImage, vol. 132, pp. 373-389, 2016
16. D. E. Nee, J. W. Brown, M. K. Askren, M. G. Berman, E. Demiralp, A. Krawitz, et al., "A meta-analysis of executive components of working memory," Cereb Cortex, vol. 23, pp. 264-82, Feb 2013

Design of a Validation Protocol for Medical Technology According to Current Standards

J. Acevedo[1], N. Saldarriaga[1], J.S. Orozco[1], and J.H. García[1]

[1]Bioinstrumentation and Clinical Engineering Research Group - GIBIC, Bioengineering Department, Engineering Faculty, University of Antioquia UdeA; Calle 70 No. 52-21, Medellín, Colombia

Abstract—**Currently, every company of designs and manufactures medical devices should have a solid quality assurance system focused on product and service quality with minimal manufacturing errors, meeting the specifications proposed by the company and the norms of the country, looking to benefit patients and health authorities.**

This article presents a methodology for designing validation protocols for medical technology; starting with a review of national and international standards and processes, continuing with the protocol design for tests according to a particular device and finally analyzing the results and making necessary adjustments to improve the quality of patients life through safe technologies, looking to improve services in different hospital environments, making sure that a device works properly and introduce to the market in accordance with the stipulated standards and specification.

Keywords— **Standards, Biomedical equipment, quality designs, Health and safety, Hospitals, Organizations.**

I. Introduction

Currently in many developing countries, the evaluation of medical technology is minimal and regulatory controls are only to prevent the importation or use of irregular medical devices. Governments need to develop policies to regulate medical devices and all elements related to these, ensuring access to high quality products, establishing safety guides proper use and disposal of these[1].

In Colombia, the development of standardized protocols for validation is necessary by different laboratories that specialize in medical devices. This triggers the addition of activities under a measurement system that allows it to be applied to different types of devices. All medical devices used in patient contact will carry some degree of risk and may cause problems in specific circumstances. Some of the problems of these devices are not being able to detect them until these are in the market. Currently the orientation of a device security is to estimate the potential that it can become harmful, which could lead to security problems [2].

According to the International Standard Organization (ISO) a standard is a document that provides the requirements, specifications, guidelines or characteristics that can be used consistently to ensure that the materials, products, processes and services are fit for their purpose [3].

The development of specific standards for each medical device in particular is not easy, given the great diversity of these and the continuous changes introduced by technological advances [4]. For this reason it is necessary to have more validation protocols for medical equipment based on current standards.

A protocol for the quality of new medical devices helps to improve the effectiveness and efficiency in the development of them, in order to protect the operability of institutions providing health services. Likewise, there is a control in each of the procedures for the release of a medical device into the market [5].

To carry out the validation protocol, the following requirements are taken into account: current national and international standards for medical devices, current norms, process validation, experimental design, and statistical analysis, among others.

II. Methodology

In order to develop the validation protocol for biomedical technology, a searching of regulatory standards for the validation of medical equipment as the basis for the protocol was performed. Then, a relevant test for the validation of different biomedical devices like the modules of a vital sign monitors was designed and evaluated. These tests were modified to improve the protocol.

A. Recollecting standards

Databases, empiric information and experts in regulations and normativity were necessary to accomplish a collection of the required standards that performed to validate the equipment, and the corresponding specifications of quality in usability and ergonomics were selected, both hardware and software.

B. Design of the checklist accordance with the norms.

A list of questions was made based on the selected standards to evaluate the performance of each biomedical device. To

I. Torres et al. (eds.), *VII Latin American Congress on Biomedical Engineering CLAIB 2016, Bucaramanga, Santander, Colombia, October 26th -28th, 2016*,
IFMBE Proceedings 60,
DOI: 10.1007/978-981-10-4086-3_159

evaluate the efficiency tests of Table 1, the following should be considered:

1. An expert in equipment operation performed each test and the time spent is recorded in order to evaluate the efficiency with respect to a non-expert.

2. The following ranges are taken into account based on the times recorded for both the expert and non-expert:

Time expert: Te, Time non-expert: Ti

- If Ti ≤2Te, an X should be put in box 4, table 1.
- If 2Te<Ti<2.5Te, an X should be put in box 3.
- If 2.5Te<Ti<3Te, an X should be put in box 2.
- If 3Te<Ti<3.5Te, an X should be put in box 1.
- If Ti>3.5Te, an X should be put in box 0.

3. Make a list of tasks to perform on the device and determine a time limit to perform them. The expert should perform the tasks to take as a reference when evaluating the non-expert. During this time, the number of errors committed by both the expert and the non-expert should be recorded.

Number of tasks performed by the expert: Ne

Number of tasks performed by the non-expert: Ni

- If Ni >0.8Ne, an X should be put in box 4, table 1.
- If 0.5Ne<Ni<0.8Ne, an X should be put in box 3.
- If Ni=0.5Ne, an X should be put in box 2.
- If 0.1Ne<Ni<0.5Ne, an X should be put in box 1.
- If Ni=0, an X should be put in box 0.

Number of errors made by the expert: Ee

Number of errors made by the non-expert: Ei

- If Ei >0.8Ee, an X should be put in box 0, table 1.
- If 0.5Ee<Ei<0.8Ee, an X should be put in box 1.
- If Ei=0.5Ee, an X should be put in box 2.
- If 0.1Ee<Ei<0.5Ee, an X should be put in box 3.
- If Ei=0, an X should be put in box 4.

4. The device correctly measures was determinate by the calibration process.

Based on the preceding paragraphs the qualification of the device is given and finally according to order of importance previously established the final state is determined. After the data were taken, the evaluation and statistics analysis of results were made to implement improvements. Then test to selected technologies were applied, in this case was a prototype of vital sign monitor, to determine the status of this device. Finally the validation report was designed and filled with the objective of delivering the evaluation reports to the maker of the device.

III. Results

A. Recollecting standards

According to the searching criteria the following standards were selected: ISO 1138-1, ISO 18472, ISO 14161 and ISO 15882, the electrical safety among which are IEC 60601, IEC 60950-1, IEC 01/01/1000, usability, software and interference of medical equipment, especially those relating to this case were oximeters, blood pressure and EKG. Also, some were selected based on international standards, such as ANSI / AAMI EC 12-1991, ANSI / AAMI EC53-RC01, ANSI / AAMI EC 11R-1991.

B. Design of the checklist accordance with the standards.

To assess the efficiency, depending on the results of paragraphs 2 and 3 referred in the methodology, in time and number of tasks, a score is assigned to this item, which will only have the following values: 0: No efficient, 4: Efficient (results are similar to those of the expert).

Table 1. Checklist model based on the collection of standards. The table shows some of the implemented tests. For reasons of space only a fraction is shown. Own source.

	TESTS	Score 0	1	2	3	4	N/A
Usability	Efficiency: resources spent (time and effort)						
	Relative efficiency with a quick guide compared to an expert user.						
	Effectiveness: ease of learning, no system errors, easiness of system to be remembered						
	Relevant functions used						
	Performance: determine device performance						
	The device correctly measures values in the range described in the technical manual						

C. Tests performed on the oximetry module of the poet plus 8100 device (criticare).

For tests 1 and 2, 6 random samples were taken for 30 percentages of saturation within the important range in order to eliminate randomization errors and be able to confirm the status of operation. This was done with a constant heart rate 75b pm so that these variations would not affect the saturation measurements, each data in 15 s intervals for stabilization. (See Fig 1)

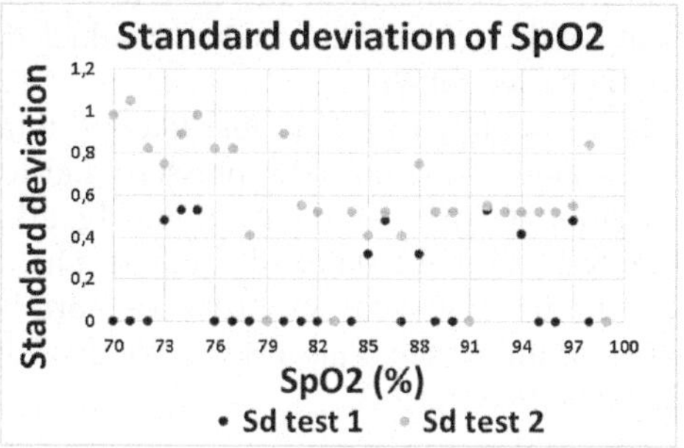

Fig. 1. Comparison between standard deviation (Sd) of test1 and test2. Own source

D. Tests performed on an automatic blood pressure monitor.

❖ For the first test, 6 samples of 26 different blood pressure values were recorded in descending order, all within the range allowed by both the simulator and the equipment being tested, in order to corroborate the functionality of the device (See fig. 2a). This test was performed with the following constants:
Pulse volume: 1.00 ml, Wave: Adult, Wave: -4%, heart rate: 60 bpm. Nomenclature results: systolic / diastolic pressure.

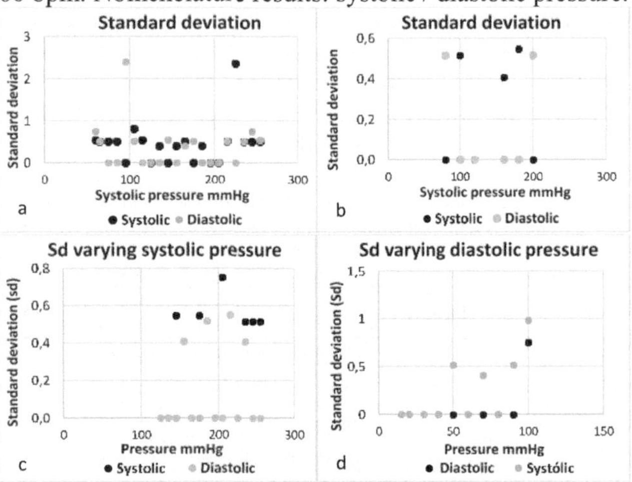

Fig. 2. Tests, a: First, b: Second, c: Third, d: Fourth. Own source.

❖ For the second test, 6 sample for 6 different blood pressure values were recorded in descending order with a difference between the value of systolic and diastolic pressure of 40mmHg, all within the range allowed by both the standard equipment as well as the equipment to be analyzed, in order to corroborate the functionality of the device. This test was conducted under the same constants (See fig. 2b).

❖ For the third test, six samples for 14 different blood pressure values were recorded in descending order by varying the value of the systolic pressure and leaving the value of the diastolic pressure fixed on 80 mmHg, all within the range allowed by both the standard equipment as well as the equipment to be analyzed, in order to corroborate the functionality of the device. This test was conducted under the same constants (See fig. 2c).

❖ For the fourth test was done just like the third test but varying the value of the diastolic pressure and leaving the value of systolic pressure fixed on 120 mmHg, all within the range allowed by both the standard equipment as well as the equipment to be analyzed, in order to corroborate the functionality of the device. This test was conducted under the same constants (See fig. 2d).

E. Alarm Test

Seven subjects were taken for the alarm test. Initially, they underwent a pressure test with the monitor with an inflation limit of 270 mmHg to determine the range in which the pressure of each fell under. From these results, the alarms were programmed, specifically the upper ranks which were the ones who allowed us to carry out the test. The pressures of the subjects were once again measured and were taken two times every 2 minutes to make sure the alarms were working. According to the registered results, it is clear that the equipment operates correctly, since the subjects where the programmed superior range falls below the registered range for the tests subjects, the alarm sounded correctly. And for the specific case of subject 6, wherein the table records that the alarm did not sound, it was expected since the programmed range was higher than that recorded by the subject (See table 2).

Table 2. Results obtained from the alarm test

Alarm: superior range	Alarm: inferior range	Test 1	Sound	No sound	Test 2	Sound	No sound
100-110	90-50	101-70	X		114-79	X	
135-110	90-50	136-91	X		138-90	X	
106-110	90-50	109-72	X		111-70	X	
106-110	90-50	136-99	X		133-87	X	
106-110	90-50	135-77	X		131-81	X	
120-110	90-50	112-74		X	102-65		X
120-110	90-50	145-99	X		149-105	X	

F. Results: devices in the pre-market stage

After designing the validation protocol, it was run on a vital signs monitor prototype that is in pre-market stage in order to check the restrictions and functionality that the manufacturer specifies, in addition to evaluating the protocol on a device that is not yet commercial. For the vital signs monitor, the test performed on the EKG module consisted of taking 6 samples for 7 different values generated by the simulator device. In the figure 3a the error rate presented is shown. Errors obtained did not exceed 4.5%.

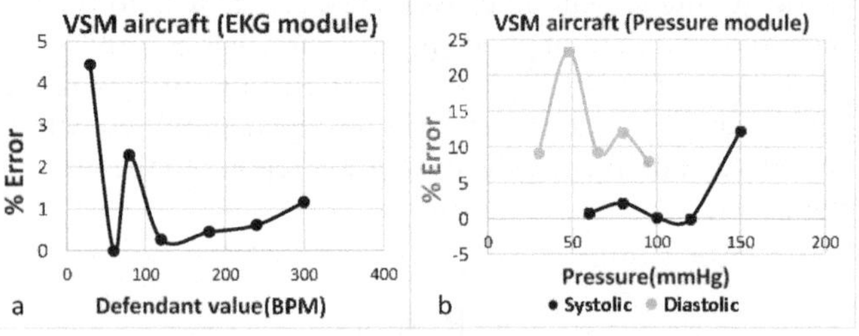

Fig. 3.VSM aircraft, a: EKG module, b: Pressure module

For the VSM, the test performed on the pressure module consisted in taking 6 samples for 5 different pressure values generated by the simulator equipment, in figure 3b the error

percentage that occurred between the generated value and the measured shown versus each of the pressure values.
The error calculation and processing of the graph were performed separately for systolic and diastolic pressure, in order to know if there was a considerable difference between errors. The highest error rate was on the diastolic pressure test (approximately between 10% a 25%) showing the highest values at low pressures. Unlike the diastolic, the systolic pressure shows errors between 0% and 10% and also higher values of this are obtained at higher pressures.

G. Validation report

At a point where an opinion is given on whether or not to validate a medical technology device that is on the market or is in the pre-market stage, it's necessary to document the entire process of validation in a report in order to sustain the testing and the final results, specifically for the manufacturers because they become very important when prototypes need improvements, with information that shows that the device works properly.

The validation report includes information about the device and information about the applicant who may be the manufacturer; where the validation will take place, who performs it, standard equipment specifications such as their last calibration and environmental conditions under which they will develop the tests. In addition, for each item of the tests there is a box detailing clearly whether or not the equipment or technology fulfilled the expectations; to validate the latter a photographic and graphic record is left with the analysis of the data supporting the work done in the test and therefore the results, because if the device is not validated you need to provide the manufacturer or applicant documentation.

IV. Conclusions

In this paper it was proposed a protocol for the quality of new medical devices and according to the results the objective was gotten successfully.
According to figures 1, 2a, 2b, 2c, and 2d, we can conclude that the results of the second test are in concordance with the results obtained in the first test, which satisfactorily confirms that the experimental designs used for retrieving the data and analysis it are reliable when implementing the protocol. According to the results obtained for both tests, it is clear that the data's are not so distance and they adjust to a media value. It was possible to design and to test validation protocols for medical devices in various technologies with successful results. The improvements implemented and the comparison of the results showed that this project has validity, because they are of vital importance at the time of minimizing risks for anyone who it is in constant contact with medical technology, for caregivers as well as patients.

The validation of medical devices that are in the pre - market stage is essential for professionals and for the public in general because of the growing importance of these in clinical practice in a near future. Considering that there are not existing protocols available to the public for the validation of devices soon to enter the market, creating a protocol provides a simplification on the method and is an effort to facilitate the validation of new technologies or the existing ones.

Conflict of interests

The authors declare that there is no conflict of interest regarding the publication of this paper.

Acknowledgment

This work was supported partially by Sistema General de Regalías, Republic of Colombia through the project Fortalecimiento de la plataforma tecnológica para la formación especializada en el área de la salud y el desarrollo de tecnología biomédica", code RutaN-139C. Authors thanks to Bioengineer Juan Camilo Guerrero for his collaboration and his knowledge.

References

[1] World Health Organization. Medical device regulations: Global overview and guiding principles, pp.5 Y 6. WHO Library Cataloguing-in-Publication Data. 2003

[2] CRUZ, Antonio Miguel. A view of clinical engineering from scientific publications. Biomédica, vol. 30, no 2, p. 188-198, 2010.

3] D. Rubio, S. Ponce, A. Abril, and E. Fascioli. Acreditación de Laboratorios de Ensayos de Equipamiento Electromédico. IV Latin American Congress on Biomedical Engineering 2007, Bioengineering Solutions for Latin America Health. Springer Berlin Heidelberg, 2007. p. 873-876.

4] A. V Kaplan, D. S. Baim, J. J. Smith, D. a Feigal, M. Simons, D. Jefferys, T. J. Fogarty, R. E. Kuntz, and M. B. Leon, "Medical device development: from prototype to regulatory approval.," Circulation, vol. 109, no. 25, pp. 3068–72, Junio 2004.

5] L. Marín, C. López, and A. Herrera, "Incidencia de eventos adversos asociados a dispositivos médicos en una Institución de salud en Colombia," Revista Ingeniería Biomédica, vol. 4, no 8, p. 71-84, 2010.

Author address

Author: J. H. García.
Institute: University of Antioquia
Street: Calle67 No. 53 – 108
City: Medellín
Country: Colombia
Email:hernando.garcia@udea.edu.co

Analysis of a Nurse Call System Implementation using a Wireless Sensors Network

R.A. Zapatán[1], E.E Armijos[1] L. Serpa-Andrade[1], Eduardo Pinos[1]

[1] GI-IATa, Investigation Center, Development and Innovation in Engineering Universidad Politécnica Salesiana, Cuenca-Ecuador

Abstract— **A nurse call system is used to improve the patient assistance and to make efficient use of the hospital staff. The aim of this project is to analyze the use of a Wireless Sensor Network (WSN) that works as an alarm center, emergency calls and as abnormal environmental parameters. The system allows knowing the temperature, relative humidity (RH) and the level of CO2, in different hospital areas, which will ease the preventive management of environmental emergencies. Moreover, the system enables to know the response time and the nurses' attention, providing statistics that will help to control the efficiency of the hospital staff and the quality level of the care provided. During the functionality testing made to the developed nurse call system, the results are encouraging, the reached data transmission reliability is 93%, from the farthest room to the nursing center. The system successfully provides relevant data to the environmental parameters of the hospital, the attention time and the staff response are successfully controlled. In the testing phase, it was also observe that the system is highly scalable, due to the easy integration of new devices to the network, thanks to the use of the ZigBee protocol and to the mesh topology. The implementation of this system is simple because it is wireless and it does not cause damage to the physical structure of the health center.**

Keywords— **Nurse call, Mesh, WSN, ZigBee.**

I. Introduction

Nurse call systems are used in hospitals to facilitate the communication between the inpatients and the nurses on duty [1]. The use of an integral nurse call system is essential to optimize the use of nursing staff, and to improve the patients' care [1, 2].

Optimizing the efficiency of the nursing staff allows reducing the health care costs, but this finality is of secondary importance in relation to the main objective, to provide a high level of medical attention. If the nursing staff is reduced, without simplifying functions and responsibilities, the level of patient care would decrease [2]. For this reason, it is necessary the implementation of a system that facilitate the work of nurses, through visual and auditory signals. Besides, knowing exactly who is the patient that requires nursing care.

Given the above, the purpose of this work is to design, develop and implement a system that maximize the efficiency of nurses, and that the design system should be scalable, reliable and easy to implement in the health center, without neglecting the principal objective of providing a high level of medical care. It is also part of this work to take advantage of the existing network to measure temperature, relative humidity and CO2 level within the health center in order to help the preventive management of environmental emergencies.

The remainder of this paper is organized as follows: Section 2, provides a brief description of the related work. Section 3, presents the proposed methodology. Section 4, shows the obtained results with our proposal. Finally, in Section 5, the conclusions of this work are presented.

II. Related work, a brief overview

There are several proposals to improve the nurse call systems, for example, wiring and wireless systems [3-8]. In [4] a system that uses the RS-485 protocol is observed, it operates with the command of a button and shows visual and auditory alarms, however the wireless technology is an option that has many advantages compared to wired systems, this is why it is proposed as a future work to implement the same system through a wireless network.

In [5] a system, able to communicate through wireless protocol, is found. This system was developed jointly with patients and nursing staff, who have participated at each point of its progress. The system counts with identification cards that allow localizing the patient quickly and easily. The assistance button of the patient and the sensors for the compilation of environmental parameters, the implemented methodology, increases the degree of acceptance by users of the system.

III. Proposed approach

With the aim of improving the service for the patient in the health centers, a nurse call system is proposed, which use a Wireless Sensors Network (WSN) under a mesh to-

I. Torres et al. (eds.), *VII Latin American Congress on Biomedical Engineering CLAIB 2016, Bucaramanga, Santander, Colombia, October 26th -28th, 2016*,
IFMBE Proceedings 60,
DOI: 10.1007/978-981-10-4086-3_160

pology, the devices employed for the communication are the xbee s2 that manage the ZigBee communication protocol. In the Fig.1 the used topology is shown.

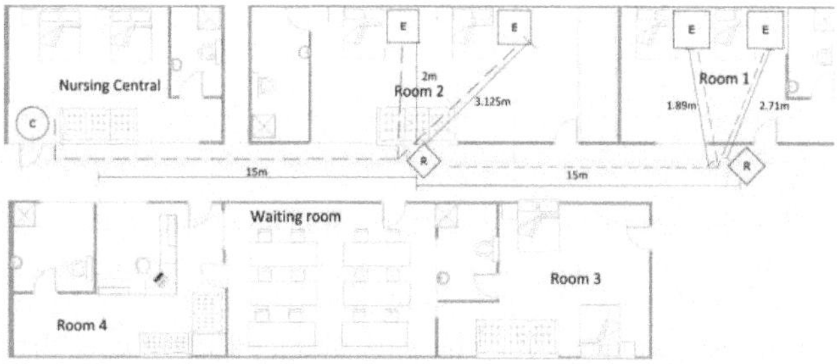

Fig.1. Network used for testing the system performance

The reach of the XBee S2 devices in indoor environments is 30 meters [9], since a mesh configuration is used, the number of modules can be easily expand [10], since the communication will be perform through jumps through the devices with a higher Received Signal Strength Indication (RSSI).

The RSSI level, allows to measure the power of signals received by a device, in the application allows this establish the best connection path between the modules that make up the system.

The performance of the nurse call system is described in Fig.2

The nurse call system consists of:

A. Patient Module

This module is placed near the patient. The module consists of a button that when activated sends a message to the central indicating the need for attention, additionally a light-indicator that turns on after the delivery of the message to the central and a battery that ensures an approximate autonomy of about 4 days.

B. Hallway Module

This device will be responsible for performing the extension of the network's reach, it will be located out of the room and will sense the environmental variables such as temperature, CO2 and humidity, which will be displayed on a touch screen and transmitted to the Central. Furthermore, when a call from a patient is received the system will provide visual and sound signals.

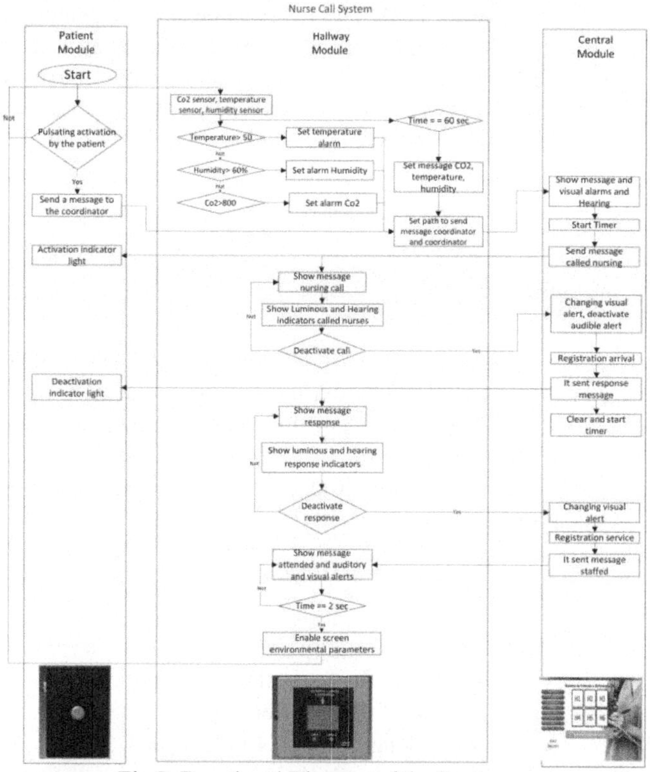

Fig.2. Functional Diagram of the System

C. Central Module

The central module maintains communication with the patient and hallway modules, it is in charge of the register of attention times from nursing, shows visual and audible indicators, different for each patient, and allows the visualization of the temperature, humidity and Co2 parameters. It also incorporates an alarm if the normal values are exceed. The central module has an application developed in JAVA, where the system statistics (response time and attention) are shown. Moreover, it allows the register of patients to the health center.

IV. Preliminary Results

In order to ensure the reliability of the network it has proceeded to measure the RSSI of the signal, from the central to the patient rooms.

The RSSI indicates the intensity of the received signal, it is a parameter to measure the signal intensity from the receiver, it is expressed in dBm and can be used to find the distance between devices [11], the advantage of this method of localization through RSSI is that it does not require additional hardware [12]. The RSSI values are presented in the Table 1, these values are taken from the nursing Central to the rooms of patients. The propagation model used is the one that depends in the partitions of a site, which is used in the indoor environments [13] and is given by the equation 1

$$L_P = L_O + 20\log d + \sum\nolimits_{tipo} m_{tipo} W_{tipo} \qquad (1)$$

where:

- m_{tipo}: Number of partitions of a particular type.
- w_{tipo}: Loss associated to a particular type.
- L_O : Loss of a signal path at a distance of 1 meter.

The signal loss at a distance of 1 meters calculated by the equation 2.

$$L_O = -10\log G_T G_R \left(\frac{\lambda}{4\,\pi}\right)^2 \qquad (2)$$

where:

- G_T: Gain of the transmitter antenna
- G_R: Gain of the receiver antenna
- λ: Wavelength for a 2.4 GHz signal.

The gains of the transmitting antennas of the XBee modules are calculated with the value of 1 or 0 dBi. The power of the received signal is given by equation 3.

$$P_{rx} = P_{tx} - L_P \qquad (3)$$

Table 1 Obtained measured values

Node	Remote RSSI	% Transmission success	Distance (meters)
Room 1	-69	99	15
Bed 1	-75	96	17
Bed 2	-77	93	19
Room 2	-66	99	31.2
Bed 1	-71	96	34
Bed 2	-73	98	37

The Table 2 shows a comparison between the RSSI calculated and measured values. The highest percentage of error is 5.79%, which means that the performance of xbee modules in the developed nurse call system is satisfactory.

Table 2 Comparison between measured and calculated obtained values.

Node	Calculated RSSI	Real RSSI	% error
Room 1	-65	-69	5.79
Bed 1	-72	-75	4
Bed 2	-74	-77	3.89
Room 2	-65	-66	1.51
Bed 1	-70	-71	1.4
Bed 2	-72	-73	1.36

The minimum success rate of transmission data is 93%, which ensures network reliability, it is verified that the calculated RSSI level is similar to the real one; the maximum variation is 5%. The data obtained certificate that the proposed nurse call system works satisfactorily.

V. Conclusions

The remote RSSI levels prove that using a WSN under the ZigBee protocol, as a nurse call system is highly satisfactory, thanks to a minimum percentage in the data transmission of 93%. Moreover, the use of a ZigBee mesh network, has great advantages over other technologies and topologies, where the ones that stand out are: the low current consumption by the final devices, and the establishment of new routes if a node moves from one place to another, or is disconnected from the network.

The developed nurse call system helps to ensure timely, efficient and quality care and attention by health centers. This is possible because of the statistics provided by the application where it can be observed in what time the nurse call are answered and attended.

Taking advantage of the deployed network to monitor environmental parameters in rooms increases the quality of the medical care, and optimizes the use of network resources.

References

1. Miller, E. T., Deets, C., & Miller, R. V. (1997). Nurse call systems: impact on nursing performance. Journal of nursing care quality, 11(3), 36-43.

2. Gallant, D., Harnden, J. C., Myers, J. E., & Ulrich, D. J. (1998). U.S. Patent No. 5,838,223. Washington, DC: U.S. Patent and Trademark Office.

3. Wu, Q., & Kai, S. (2011, October). Design of medical ward call system based on power line carrier technology. In *System Science, Engineering Design and Manufacturing Informatization (ICSEM), 2011 International Conference on* (Vol. 2, pp. 101-104). IEEE.

4. Sharma, C., & Gautam, D. K. (2015, October). Design development and implementation of wired Nurse calling system. In Green Computing and Internet of Things (ICGCIoT), 2015 International Conference on (pp. 1258-1262). IEEE

5. Ongenae, F., Duysburgh, P., Verstraete, M., Sulmon, N., Bleumers, L., Jacobs, A., y ... De Turck, F. (2012, mayo). Diseño guiado por el usuario de una aplicación sensible al contexto: un sistema de llamada de enfermera ambiente inteligente. En *2012 6ª Conferencia Internacional sobre Tecnologías de la informática generalizada de Salud (PervasiveHealth) y Talleres* (pp. 205-210).IEEE.

6. Aswin, S., Gopalakrishnan, N., Jeyender, S., Prasanna, R. G., & Kumar, S. P. (2011, December). Design development and implementation of wireless nurse call station. In *2011 Annual IEEE India Conference* (pp. 1-6). IEEE.

7. Şahin, Ö., & Şahin, H. (2010, December). Nurse call system design integrated with the code-blue application. In *Electrical, Electronics and Computer Engineering (ELECO), 2010 National Conference on* (pp. 531-535). IEEE.

8. Wyffels, J., De Brabanter, J., Crombez, P., Verhoeve, P., Nauwelaers, B., & De Strycker, L. (2014). Distributed, signal strength-based indoor localization algorithm for use in healthcare environments. *IEEE journal of biomedical and health informatics*, *18*(6), 1887-1893.

9. Kioumars, A. H., & Tang, L. (2011, December). ATmega and XBee-based wireless sensing. In Automation, Robotics and Applications (ICARA), 2011 5th International Conference on (pp. 351-356). IEEE.

10. Dhaka, H., Jain, A., & Verma, K. (2010, February). Impact of Coordinator Mobility on the throughput in a Zigbee Mesh Networks. In *Advance Computing Conference (IACC), 2010 IEEE 2nd International* (pp. 279-284). IEEE.

11. Mukhopadhyay, B., Sarangi, S., & Kar, S. (2015, February). Performance evaluation of localization techniques in wireless sensor networks using RSSI and LQI. In *Communications (NCC), 2015 Twenty First National Conference on*(pp. 1-6). IEEE.

12. Vimala, T., & Rajaram, U. (2014, February). Analysis of RSSI in diverse field. In Information Communication and Embedded Systems (ICICES), 2014 International Conference on (pp. 1-3). IEEE. ISO 690

13. Saavedra Arancibia, C. A., & Ordonez Riofrio, D. M. (2015). Diseño e Implementación de una Red de Sensores Inalámbricos para el Monitoreo de Temperatura Corporal.

Multichannel Planar Microelectrode Platform for Recording Extracellular Field Potentials

J.C. Franco[1,2], M.A. Portela[1,2] and H. Andrade-Caicedo[1]

[1] Centro de Bioingeniería, Universidad Pontificia Bolivariana, Medellín, Colombia
[2] These authors contributed equally to complete this work

Abstract— **This paper presents the implementation of a Microelectrode Array Platform for registering until 60 channels of cellular bio-potentials in real-time. This electrophysiological system was set by using home-made and commercial elements acquired from Multichannel Systems (Germany). Some of them are, the planar microelectrode arrays, the housing system, and the amplifier. Additionally, we used a data-acquisition card (NI USB-6225, from National Instruments) for the digitalization of the signals. We developed the monitoring and acquisition software through a high-level programming environment (LabView, National Instruments). The system has additional synchronization features, which allow it to interact with external devices. For instance, an electrical stimulator, an optical video camera, etc. The result is a completely modular electrophysiological system, which allows monitoring the bio potentials from living cells cultured onto the planar microelectrodes. The data registered by the acquisition card can be visualized and stored on the computer by a friendly graphic user interface.**

Keywords— **Microelectrode Array, User Interface, LabView, Electrophysiological System, Celluar Biopotentials.**

I Introduction

The complexity of the biological processes involved in the genesis and sustenance of human pathologies requires the development of novel bioengineering methods of diagnosis, monitoring and therapy. Therefore, it is important to perform multi-factorial investigations, by combining systems that provide information about the biological phenomena [1, 2]. Cardiovascular system is a particular challenge due to the complexity of the simultaneous electrical, chemical and mechanical interactions. There is a particular interest in the electrophysiological behavior at different scales: the cell, the tissue and the heart as a whole organ.

Currently, there are different commercial systems available to carry out electrophysiological investigations. The most widespread systems among the scientific community are the Patch Clamp and Microelectrode Array Systems (MEA) [3]. Patch Clamp systems allow to visualize electrophysiological phenomena at a sub-cellular scale. They provide data about single or whole-cell ionic currents, as well as other important electric parameters for characterizing cardiac cells [4, 5]. However, they show high complexity and cost, require user's skills, and a substantial training time.

MEA systems instead, show a more general panorama of the electrophysiological behavior of cardiac tissue. They allow to obtain extracellular potentials of multiple locations uniformly distributed. Depending on the Microelectrode Array, it is possible to have different configurations [6]. For instance, the conventional 60 electrodes setup, 4096 electrodes in newer devices, etc. [7]. MEA systems are more attainable for a wide number of users. They are low cost, and do not depend, too much, on the researcher skills. However, the signal acquisition module is only compatible with a proprietary software. It limits the inclusion of new features, and brings low flexibility; which causes disadvantages with simultaneous experiments that use combined systems.

This paper presents the implementation of an electrophysiological system based on planar Micorelectrode Arrays. The system can register cellular biopotential in real-time. We used commercial devices combined with home-made technology. Its main components are: a planar micro-electrode array (60 channels), a housing [8], an acquisition card [9], and an acquisition, recording and visualization software. The system guarantees flexibility and modularity, enabling the interaction and synchronization with external systems. For instance, it can be linked to an optical microscope or an electrical stimulation system. Besides, under this architecture, it can be used independently of the kind of data-acquisition module. Moreover, since it is an open-source software, it is possible to add new modules that allow to perform signal processing, characteristic's extraction, parameter measurements, etc.

II Methods

A Hardware

The Hardware consists of five main elements: a planar Microelectrode Array (60MEA200/30iR-Ti), where cells are seeded. A housing (MEA1060-Inv), which is the interface be-

I. Torres et al. (eds.), *VII Latin American Congress on Biomedical Engineering CLAIB 2016, Bucaramanga, Santander, Colombia, October 26th -28th, 2016*,
IFMBE Proceedings 60,
DOI: 10.1007/978-981-10-4086-3_161

tween the cell culture and the electronic components. There is a filter/amplifier (FA64), with a gain of 1 000 V/V and bandwidth range between 0.1 Hz – 10 000 Hz. A signal divider (SD-MEA), where each of the 60 channels is accessible independently, by using BNC connectors. Finally, a low-noise power source (PS40W)[10].

It was necessary to develop an electronic circuit board to overcome the incompatibility between the MEA systems connectors and the data-acquisition card. The cable used to connect the acquisition-card (NISH68-68-EPM), and the electronic card is a 68 pin shielded connector. It ends in two female connector series 0.050, type D.

The complete system is connected through a power source via three banana-banana cables to the signal divider. The amplifier is connected to a male-male MCS-68 pin connector. The amplifier is also connected to the MEA housing (MEA 1060-Inv) with the same kind of connector. The acquisition-card is connected to the electronic card (e-card) through a NI SH68-68-EPM cable, and the e-card is connected to the signal divider using a MCS-68 pin connector.

To register the cell's bio-potentials, we used a data acquisition card NI USB-6225. It has 16-bit resolution, sample rate of 250 kS/s *National Instruments* [9].

Figure 1 shows an image of the complete connected system.

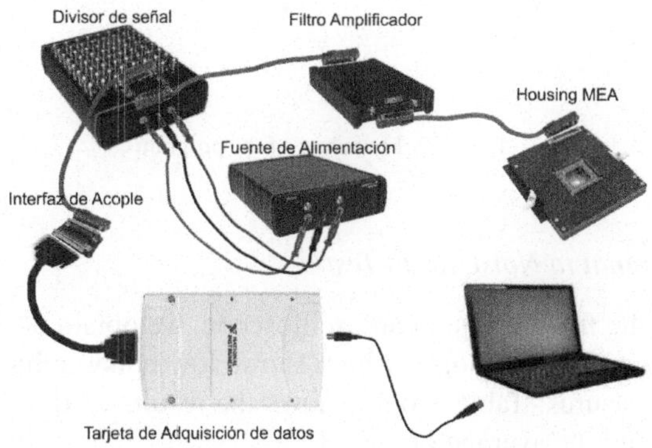

Figure 1: MEA system fully connected

The MEA signal generator (60MEA-SG) acts as a substitute for expensive biologic samples. That means that you can test your amplifier and data-acquisition settings without using valuable supplies. The simulator generates sinusoidal waves or prerecorded real signals in a digitized form. It is possible to simulate 12 different types of signals, ranging from Hippocampal Neurons Spikes to Cardiomyocyte Ventricle Field Potentials.

B Software

We programmed the software for visualizing and registering cell's bio potentials in LabView. It allows simultaneous visualization of 60 channels in real-time. We used a designing user interface guide during the development process of the software *Guidelines for Designing User Interface Software* [11]. As a result, the application is divided into three main components: input data, output data and program flow.

Input data: The following guidelines are considered for designing the user interface:

- input data should be entered only once;
- the application provides feedback during data entry and at the end of it;
- the answer speed must be swift during this process;
- the software uses only one method for data input;
- areas for data input are particularly specified.

The application has an automatic data validation method. It accepts correct entries, while no mistakes are detected. Then, it shows an error message box at the end of the entry, instead of interrupting the process. Besides, default values are predefined to speed up the data entry.

The main entry sections are: project folder's name, sample rate, number of samples, channel configuration, file's path, channels selected for visualization, name of the configuration file to save or to load, and vertical axis scale of graphs. Furthermore, there are flow control entries, which are buttons to navigate between application windows, and for accepting particular actions.

Output data: Due to the system characteristics, the information is presented mainly through waveform graphs. Therefore, the visual appearance of the graphs is very relevant. We based the design of the software on the following guidelines:

- the software allows to compare data, directly through the graphs;
- it is possible to monitor the data changes directly on the graph;
- there should be consistency in the information displayed;
- only display necessary information;
- axis scales should be linear;
- have the option for zooming into the graphics;
- display the scale changes;
- have conventions and consistency in the scales;
- to label the axes;
- to have marks in scales separated at standard intervals;
- the grids need to be distinguishable from the waveform.

The data output is displayed through graphs and text labels.

Program flow: The design goals for this application are, consistency in actions, minimizing the need for control actions, and diminish the memory load to the user.

C Performance Test and Signal to Noise Ratio (SNR)

We conducted a general performance test, in order to establish the influence of the different noise sources on the electronic system. This test intended to evaluate the SNR parameter when the whole MEA system is connected and ready to get measures. It includes the electronic card of the interface, the connection cables, the data acquisition card, and the personal computer. We used the signal generator (60MEA-SG, Multichannel Systems) to simulate different analog waveforms. As a first step, we set up the simulator to generate a prerecorded sinusoidal waveform of 1.25 Hz, then we recorded different channels by using a digital oscilloscope. We obtained access to individual channels from the signal divider [10]; each channel was recorded in a USB stick to be analyzed afterwards *off-line* by using the software Matlab. We chose 23 random channels to carry out an evaluation of Signal to Noise Ratio (SNR). Subsequently, we set a second signal on the 60MEA-SG. We recorded six files, each of them with 10, 20, 30, 40, 50 and 60 channels respectively, we calculated the signal to ratio (SNR) to all of them.

D Usability, performance and interaction with external system

Usability: To evaluate the system usability, we design a survey with specific questions about size and distribution of controls and graphics, execution velocity of actions, application stability, etc. Besides, we asked questions about the perception about comprehensibility of the software, easy of use, the relevance of the information, manageability, etc. Each question was scored from 1 to 5, where 1 is the lowest score and 5 is the highest [12].

Performance: The system performance was assessed in the most critical scenarios, that occurs when visualizing and recording 60 channels simultaneously. The parameters evaluated were: the use of CPU, the access velocity to the hard drive, and the amount of RAM loaded [12].

Interaction with External Systems: We developed an electronic card with four analog inputs and four digital outputs. The card allows synchronizing the system actions to external signals triggers. Additionally, to visualize events coming from external devices through the analog inputs.

III RESULTS

A Software

The first window named *"creating project"* allows creating a new work file, or to load one created previously. The second window named *"set up"* allows the user to set parameters such as the sample rate, the number of channels to acquire, and the time frame to show. The third window, named *"Reading"* shows the signals being captured in real time. This representation is a matrix of 60 plots where each channel has a correspondence with the spatial distribution of the microelectrodes in the array. In Figure 2 is shown the full program flow diagram when is explained in detail more application features.

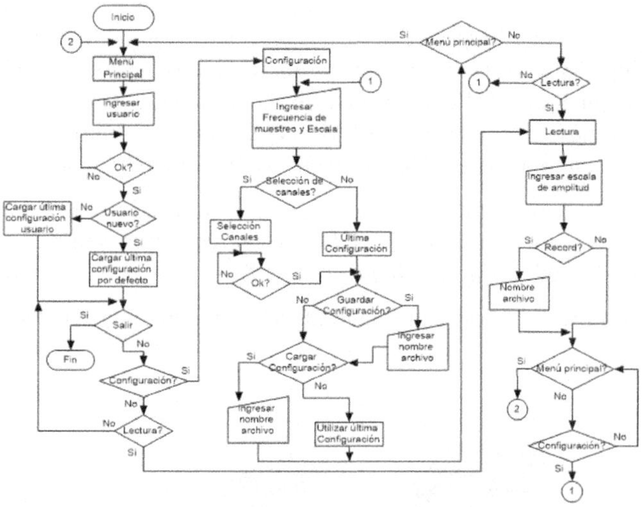

Figure 2: Flowchart of the program

B Signal to Noise Ratio Test

In the first test,using an oscilloscope, we obtain 34.792 of average signal to noise, which shows low noisy influence in the measures. Table 1 only shows the results of three channels, and the average of the 23.

Table 1: SNR of 23 randomized MEA channels

Channels	SNR (dB)
1	38.849
2	32.598
3	32.930
Average	34.792

In the second test we obtain 36.945 of fully average of the seven files created.

Table 2: Average SNR calculated from 6 files - second test

Trial	SNR (dB)
1	36.629
2	36.991
3	36.575
4	36.337
5	38.195
6	35.990
Average	36.945

These results demonstrate that additional home-made hardware elements used to built the system does not introduce relevant noise signals. The average SNR parameter in both tests is very similar.

IV Discussion and Conclusions

Bioengineering experiments require multifactorial designs to obtain the greatest possible amount of data from a single phenomenon. The main advantage of developing home-made systems is the possibility of integrating technologies. We propose an electrophysiological platform based on commercial elements, but with a philosophy of open source either from the hardware and software components. We implemented an Electrophysiological Platform for measuring Extracellular Field Potentials of excitable cells, using specialized commercial elements, components present in the laboratory and other designed and manufactured locally. The platform is flexible, due to its hardware design that allows it to interact with other systems and devices easily. The software is sensitive to improve its capabilities, by developing new modules of signal processing and data analysis.

Acquisition modules from *National Instruments* turned out very helpful prototyping tools. They allow the use of generic software libraries provided by the company, so the process of programming the application components was straightforward and efficient. However, the sampling rate of the acquisition card is a key quality factor, since it defines the maximum number of channels that can be read simultaneously. In this project, this factor limits the maximum sampling rate to 4 000 samples per second to capture 60 channels simultaneously.

Computer performance was not affected significantly when the application demanded a greater amount of resources. The CPU had an increase of 6%, memory was not affected and the access time to the hard disk had an increase of 52 *s*, which is due to the time it takes for the software to create the file with the 60 data channels. It means that the software can be installed and used in a generic personal computer with no special requirements. We used the following specifications for running the tests, Intel Core i7 2.6 GHz, 6 GB RAM, and a hard drive of 1 TB.

The software application obtained a high rating from the survey. Users think that the software is intuitive (score 3.8/5.0), it is easy to learn how to use it (4.2/5.0), it is well designed (4.8/5.0), among the principal issues evaluated. It means that the user experience regarding the usability was high. However, it is a dynamic process and the platform can be improved with additions of software that allows the users a friendly approach with the technology.

References

1. Hales Chadwick M, Rolston John D, Potter Steve M. How to culture, record and stimulate neuronal networks on micro-electrode arrays (MEAs) *J Vis Exp.* 2010;39.
2. Wilders Ronald. Dynamic clamp: a powerful tool in cardiac electrophysiology. *The Journal of physiology.* 2006;576:349–59.
3. Natarajan A, Molnar P, Sieverdes K, Jamshidi A, Hickman JJ. Microelectrode array recordings of cardiac action potentials as a high throughput method to evaluate pesticide toxicity *Toxicology in Vitro.* 2006;20:375–381.
4. Sakmann Bert, Neher Erwin. Patch clamp techniques for studying ionic channels in excitable membranes *Annual review of physiology.* 1984;46:455–472.
5. Lehmann-Horn Frank, Fauler Michael, Holzherr Boris, Jurkat-Rott Karin. Principles of Electrophysiological In Vitro Measurements in *Cardiac Electrophysiology Methods and Models* (Sigg Daniel C., Iaizzo Paul A., Xiao Yong-Fu, He Bin. , eds.)ch. 6, :119–134Springer 2010.
6. Fejtl Michael, Stett Alfred, Nisch Wilfried, Boven Karl-Heinz, Möller Andreas. On Micro-Electrode Array Revival: Its Development, Sophistication of Recording, and Stimulation in *Advances in Network Electrophysiology* (Taketani Makoto, Baudry Michel. , eds.)ch. 2, :24–37Springer US 2006.
7. Reppel Michael, Pillekamp Frank, Lu Zhong Ju, et al. Microelectrode arrays: a new tool to measure embryonic heart activity *Journal of electrocardiology.* 2004;37:104–109.
8. Extracellular Recording with Microelectrode Arrays for all Applications at http://www.multichannelsystems.com 2015.
9. Manual LabVIEW User. National Instruments *Austin, TX.* 1998.
10. Multi Channel Systems at http://www.multichannelsystems.com 2015.
11. Smith Sidney L, Mosier Jane N. *Guidelines for designing user interface software.* Mitre Corporation Bedford, MA 1986.
12. Sommerville Ian, Galipienso María Isabel Alfonso. *Ingeniería del software.* Pearson Educación 2005.

Author:Henry Hermel Andrade Caicedo
Institute:Universidad Pontificia Bolivariana
Street:Circular 1 No. 73-76, Bloque 22C
City: Medelln
Country: Colombia
Email: henry.andrade@upb.edu.co

Modeling and simulation of ciprofloxacin pharmacokinetics: electric circuits approach

J.D. Otálvaro[1], A.F. Zuluaga[2] and A.M. Hernández[1]

[1] Bioinstrumentation and Clinical Engineering Research Group - GIBIC, Bioengineering Department, Engineering Faculty, Universidad de Antioquia UdeA; Calle 70 No. 52-21, Medellín, Colombia
[2] GRIPE (Grupo Investigador de Problemas en Enfermedades Infecciosas), Department of Pharmacology and Toxicology, Medical School, Universidad de Antioquia UdeA; Calle 70 No. 52-21, Medellín, Colombia

Abstract— **The development of models that fit to behavior of the drug pharmacokinetics and quantitatively describe the concentration-time relationship plus its integration with the development of tools that allow to implement them in clinical practice to model and simulate the better clinical outcomes are recommended to promote the rational use of drugs. It relies on the development of models that describe the body as a set of compartments and transitions. These models are then represented by a set of ordinary differential equations.**

This article discusses the development of a pair of circuital models that describes the pharmacokinetics of ciprofloxacin, a broad spectrum antibiotic and their fitting to experimental data previously published. As a result, there is no relevant difference between the RLC model and the RC model, the fitting algorithm used found that L should be zero reducing the RLC model to a RC model, showing that a first order model is sufficient for modeling pharmacokinetics of ciprofloxacin.

Keywords—**Drug modeling, Pharmacokinetics, Antibiotics**

I. INTRODUCTION

According to the World Health Organization (WHO), the irrational use of medicines causes between the 8% and the 10% of clinical admissions and near of 100.000 deaths per year worldwide [1]. The non-rational use of antibiotics entails other problems, such as the bacterial resistance or the probability to select the subpopulation of resistant bacteria during a treatment, leading to regrowth despite the use of high drug concentration originally effective.

According to the WHO, the antimicrobial resistance is an increasing serious threat to global public health that demands attention. Without urgent, coordinated action, the world is heading towards a post-antibiotic era, in which common infections and minor injuries in the community and hospitals, which have been treatable for decades, can once again kill [2].

Ciprofloxacin is an antibiotic used to treat or prevent certain infections and in the treatment and prevention of anthrax in people who may have been exposed to anthrax germs in the air, a misuse of this antibiotic may increase the risk of bacterial resistance at the same time that may promote the development of tendinitis and other kinds of disorders [10].

The pharmacometrics is the science of quantitative pharmacology [6], relying on the pharmacokinetic and pharmacodynamic studies of the drug, provides mathematical tools to model and simulate clinical outcomes, that help to support the generation and optimization of the dosing schedules to contain bacterial resistance [9].

As a part of pharmacometrics, the pharmacokinetics (PK) is focused to determine the concentration-time relationship of a drug through the development of the model that describe the process of absorption, distribution and elimination of the drug [5]. These PK models are currently based on compartmental systems with diffusion and removal constants as shown in Figure 1 [5], varying the number of compartments or the location where happens the infusion and the elimination.

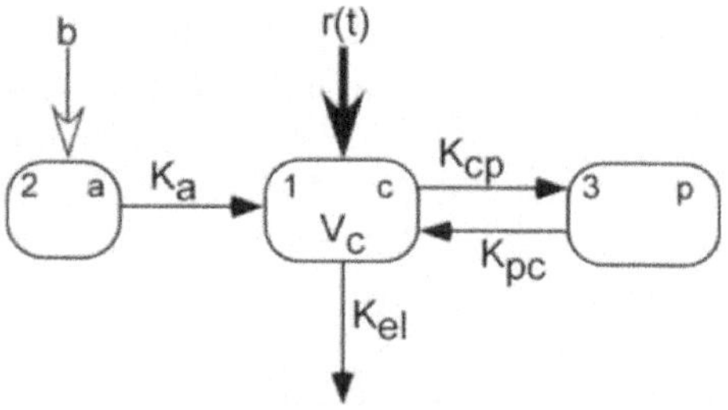

Fig. 1 Example of a pharmacokinetic model with 3 compartments and 4 diffusion variables.

The pharmacokinetic data of ciprofloxacin have been published elsewhere [8], the raw data used in this paper were adjusted to a first order resistor-capacitor (RC) circuit

I. Torres et al. (eds.), *VII Latin American Congress on Biomedical Engineering CLAIB 2016, Bucaramanga, Santander, Colombia, October 26th -28th, 2016*,
IFMBE Proceedings 60,
DOI: 10.1007/978-981-10-4086-3_162

model and a second order resistor-inductor-capacitor (RLC) circuit model using its representation in the state space. The parameters of the resulting circuital models were subsequently adjusted following the methodology described in Figure 2 [3], using the mean squared error as a cost function and a regression of the data as a representation of the physiological system. Finally, the results of both models were compared and the findings were presented.

The aim of this paper is to describe the pharmacokinetic behavior of ciprofloxacin using an electric circuits approach that can be used to simulate in different platforms, mobile devices included. These models and their respective tests were developed in MATLAB / Simulink MathWorks R2013a X64

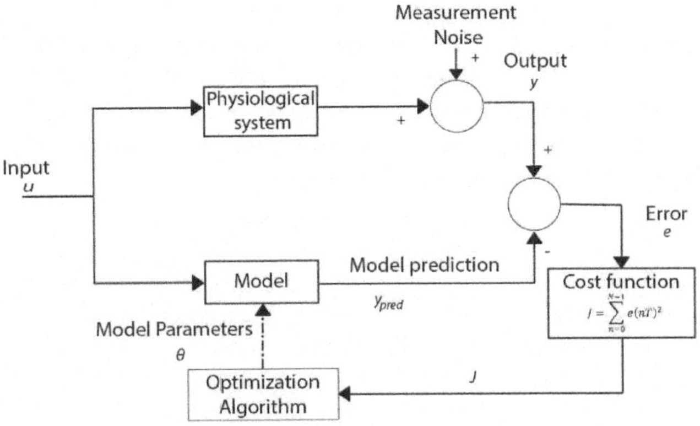

Fig. 2 Model fitting approach, modified from [3]

II. METHODOLOGY AND DEVELOPMENT

Pharmacokinetic data were published elsewhere by Rodriguez et al. [8]. The Table 1 shows the compiled data of 12 mice (A to L), they had a weight of 25 ± 2 g and received subcutaneously a single 16 mg/kg dose of ciprofloxacin [8]. Since the maximum number of blood samples which can be performed in mice is 3 [7], the 12 mice were divided into 3 groups (N=4 mice), the concentrations were measured in the first group at the 5 min, 1 h and 3 h, in the second group at the 15 min, 1.5 h and 4 h, finally the third group at the 30 min, 2 h and 5 h, the measured concentrations have units of mg/L.

Table 1 Measured levels in mg/L obtained from [8]

Mouse	0.083 h	0.25 h	0.5 h	1 h	1.5 h	2 h	3 h	4 h	5 h
A	0.53	-	-	1.5	-	-	0.24	-	-
B	0.63	-	-	1.32	-	-	0.15	-	-
C	0.88	-	-	1.37	-	-	0.13	-	-
D	0.68	-	-	1.47	-	-	0.17	-	-
E	-	2.98	-	-	0.6	-	-	0.09	-
F	-	0.95	-	-	0.87	-	-	0.09	-
G	-	1.11	-	-	0.98	-	-	0.09	-
H	-	1.1	-	-	0.75	-	-	0.09	-
I	-	-	1.54	-	-	0.75	-	-	0.09
J	-	-	1.91	-	-	0.28	-	-	0.09
K	-	-	1.74	-	-	0.45	-	-	0.09
L	-	-	1.99	-	-	0.42	-	-	0.09

As show in Figure 2, the block representing the physiological system must be continuous in order to be compared with the model output in the cost function, it is necessary to find a mathematical expression that describes the behavior of the discrete data of Table 1.

The first step consisted in represent all data looking for some kind of pattern, taking into account that all the data represents a single behavior, a scatter plot was made, this scatter plot is shown in the Figure 3.

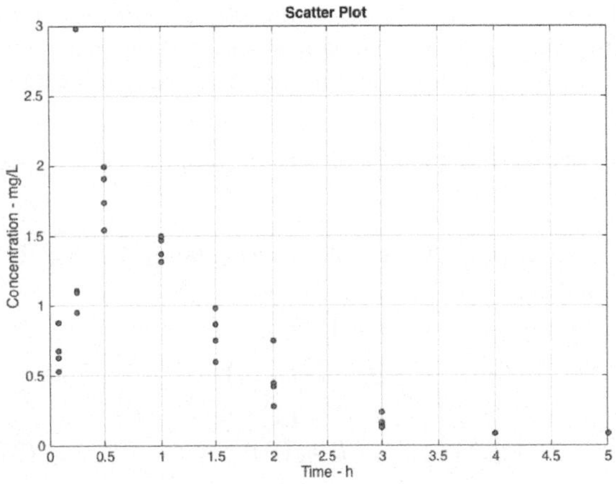

Fig. 3 Scatter plot of the pharmacokinetic data

When all data are put together the pattern becomes visible, then a regression may represent the data, following the exponential equation shown at the Eq. 1.

$$A.\,e^{B.x} + C.\,e^{D.x} \tag{1}$$

III. RESULTS

Figure 4 shows the non-linear concentration-time relationship described by the Eq. 1 that allows us to analyze the behavior of the data through a continuous expression with the electric circuits approach. The result of that regression was the Eq. 2.

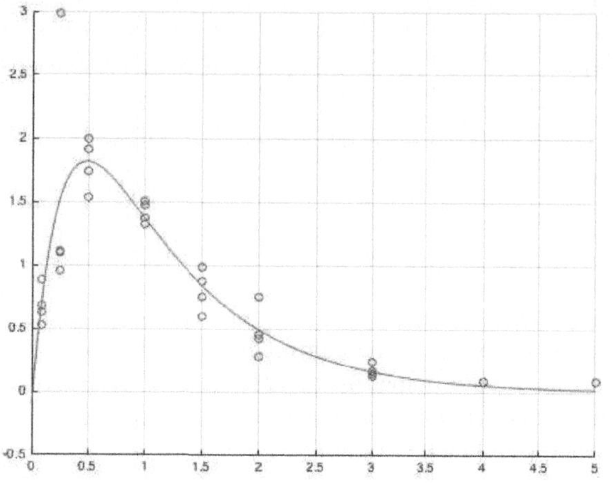

Fig. 4 Regression of the mice data

Implementing the Nelder-Mead simplex algorithm (direct search approach) using squared error as the cost function, the Eq. 2 as representation of the physiological system and two mathematical representations in the state space as models: (i) a first-order RC circuit and (ii) a second-order RLC circuit.

$$-4.67 * exp(-3.52 * x) + 4.57 * exp(-1.11 * x) \quad (2)$$

The calculated space state representation for the RC circuit was:

$$A = \left(-1/_{R.C}\right)$$
$$B = \left(1/_{R.C}\right)$$
$$C = (1)$$
$$D = (0) \quad (3)$$

And the RLC one was:

$$A = \begin{pmatrix} 0 & 1 \\ -C/_L & -R/_L \end{pmatrix}$$
$$B = \begin{pmatrix} 0 \\ C/_L \end{pmatrix}$$
$$C = (1 \quad 0)$$
$$D = (0) \quad (4)$$

In the case of the first order representation, the R and C parameters can not be determined individually, its representation on the space state allows only to determine the RC parameter as a whole, in the second case as shown in the Eq. 4 the R, C and L parameters can be determined individually, allowing to compare the results on both of them.

Both models used the same physiological system equation, the same cost function and the same optimization algorithm in the case of the RC model, the value of the combined parameter RC was found and for the case of the RLC model the R, L and C models were found.

When the optimization ended the following results were obtained: The RC approximation got an RC value of 0.7558 and a J value of 3.8901 (see Fig. 2), otherwise the RLC model got an R value of=2.336, a C value of 0.3236 and a L value of 0 with a J value of 3.8901.

The results show that both approaches are equally accurate, the L value of the RLC approximation is equal to zero reducing the second-order approximation to a first-order, finally the R*C value in both electric circuits equal to 0.7558. In consequence, the simplest model was chosen to be excited with the same input using the data from mice and the result is shown at the Figure 5.

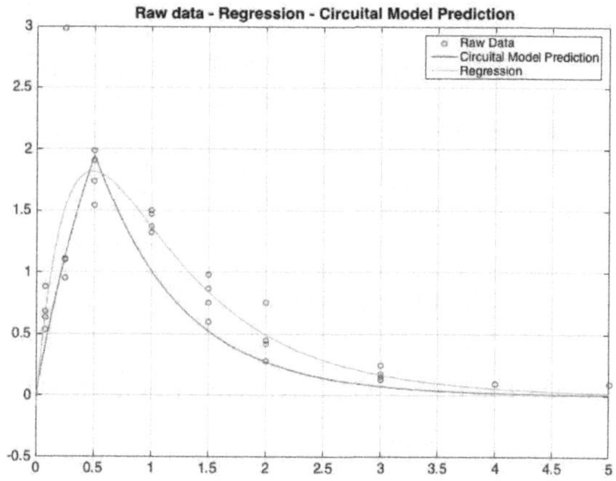

Fig. 5 Comparison between the Raw data, the regression and the prediction obtained with the circuit model.

Trying to obtain a different result of the RLC approximation all the seeds of R, L and C were changed multiple times, although the result always was different, the parameter L still was 0 and the product of R and C kept constant.

IV. DISCUSSION

Although the regression result from the electric circuits approach doesn't perfectly fit to data, it is mathematically valid and potentially suitable to describe real data, including some experimental error, because our model fitting approach (Figure 2) assumes the insertion of an error in the output of the physiological system.

One possible explanation of why the model does not perfectly predict the regression is that the input signal used to excite the electrical system assumes that the input signal is a perfectly square pulse which experimentally does not have to be true in pharmacokinetic studies, a first step to address this problem is taken into account in the model as the body metabolizes the drug before it reaches the circulatory system.

An important result is that this methodology has shown that the pharmacokinetics of ciprofloxacin has a first-order behavior, when was attempted to model it as a second-order system, the third parameter became 0, reducing it to first-order one again.

For future studies it's proposed to try different circuit configurations and restricting the solution to real roots of equations. Concerning the order reduction, facing the identification problem with an input signal similar to that one that excited the system will provide suitable results. Additionally, the evaluation of other drugs that allow comparisons will help in the challenge of the reduction of compartmental models.

ACKNOWLEDGMENT

This work has been supported in part by "Convocatoria 727 – Programa de Becas de Doctorado Nacionales 2015 de Colciencias", the University of Antioquia, the project "Fortalecimiento de plataforma tecnológica para la formación especializada en el área de la salud y el desarrollo de tecnología biomédica", code Ruta N 139C of the *Fondo Nacional de Regalías*, Republic of Colombia, and *Sistema General de Regalías de Colombia Gobernación de Antioquia* (BPIN: 2013000100168)–Ruta N and the project "Desarrollo en soluciones en ciencia, tecnología e innovacción para el asesoramiento especializado en farmacología y toxicología en Medellín, Antioquia, Occidente", code Ruta N 138C-2014 of the *Fondo Nacional de Regalías*, Republic of Colombia, and *Sistema General de Regalías de Colombia Gobernación de Antioquia* (BPIN: 2013000100183)–Ruta N. Doctor Carlos A. Rodriguez for its contribution sending us the raw data of ciprofloxacin, published elsewhere [8].

CONFLICT OF INTEREST

The authors declare that there is no conflict of interest regarding the publication of this paper.

REFERENCES

1. Johnson JA, Bootman JL, Drug related Morbidity and Mortality. Arch Intern Med,1995. Ernst FR, Grizzle AJ, Drug-related morbidity and mortality: updating the cost-of-illness model. J AM Pharm Assoc 2001;41:192-9
2. WHO. (2015). Resistencia a los antimicrobianos. Retrieved from http://www.who.int/mediacentre/factsheets/fs194/es
3. Hernández AM (2014), Sistema de control respiratorio ante estímulos y patologías: Análisis, Modelado y Simulación (Spanish Edition). Publicia.
4. D'Argenio David Z., Schumitzky Alan, W. X. (2009). ADAPT 5 User's Guide: Pharmacokinetic/Pharmacodynamic Systems Analysis Software, 316.
5. Katzung, B. G., Masters, S. B., & Trevor, A. J. (2012). Basic & clinical pharmacology. New York: McGraw-Hill Medical
6. Ette E.I., &Williams P.J. John Wiley & Sons, Inc., Hoboken, New Jersey; 2007. Pharmacometrics: The Science of Quantitative Pharmacology
7. National Centre for the Replacement Refinement & Reduction of Animals in Research. (2016). Decision tree. Retrieved from https://www.nc3rs.org.uk/mouse-decision-tree-blood-sampling
8. Rodriguez CA, et al. Antimicrob Agents Chemother. 2015 Jan;59(1):53-8
9. Trivedi A. et al. Expert Rev Clin Pharmacol. 2013 Mar; 6(2):
10. National Institute of Health. (2016). a688016 @ www.nlm.nih.gov. Retrieved from https://www.nlm.nih.gov/medlineplus/druginfo/meds/a688016.html

Corresponding author: A.M. Hernández.
Institute: University of Antioquia
Street: Calle 67 No. 53 - 108
City: Medellín
Country: Colombia
Email: alher.hernandez@udea.edu.co

Gait Kinematics of Load Carriage in Healthy College Students

Fátima G. Tapia-Rodríguez[1], Daniela M. Castro[1], Favia P. Aviles H.[1], J. Daniel Moreno-González[1], L. Fernando Andrade-Heredia[1], and Arturo Vega-González[1]

[1] División de Ciencias e Ingenierías DIQEB, Universidad de Guanajuato, León, Guanajuato, México

Abstract— **Daily use of backpacks have raised some concerns and it has been an issue of research on children and teenagers. This paper shows results of the gait kinematics of load carriage in healthy college students. The analysis was done with videography on the sagittal plane. Three loads were tested (4.1 kg, 6.8 kg and 10.5 kg). It was observed that the angles of ankle and knee do not change, whereas the initial hip angle and the maximum flexion angle showed an increase. Thus all of the observed differences were at the hip level.**

Keywords— **load carriage, gait, backpack.**

I. Introduction

The impact of a load on the human gait is a function of the relation between load mass and body mass of each person [1]. College students usually wore a backpack or a similar device to transport all the materials needed for their academic activities, such as notebooks, books, and laptop. In some cases the backpack weight can be substantially heavy and it can cause discomfort to the users [2].

One approach to analyze how the load carriage affects us is through gait analysis. Human motion analysis and particularly gait analysis has become a useful tool in research and clinical settings [3]. Although there has been some research reports on the use of backpacks. Most of them are focused on children [4], teenagers [2,5], industrial or military use [6]. It has been establish that there is an association between the use of backpacks and back pain and muscle fatigue [3,7]. This has raised concerns on health and it has influenced the design of backpacks [8,9].

However, despite the biomechanical-designed backpacks, the use of a backpack or similar in the college environment has not changed. Thus there is a variety of backpacks, their use is inappropriate and the load is not constant trough the term. These factors contribute to complaint such as back pain and muscle fatigue. The aim of this pilot study was to establish a protocol to evaluate the gait under different load carriage in college students.

II. Materials and Methods

A. Participants

Five healthy subjects were included in the study: 2 males with heights 173.8 ± 0.4 cm and weights 83.7 ± 15.6 kg, and 3 females with heights 163.3 ± 3.5 cm and weights 58.1 ± 10.9 kg. Participants (aged between 20-21 years) signed an informed consent prior to start the measurements. Measurement protocol is in line with the Helsinki declaration.

B. Measurements

Subjects were evaluated through kinematic analysis using videography in the sagittal plane. Figure 1 and 2 show the setting and the processing blocks for this study. Briefly, subject wore a set of markers on: base rib cage, greater trochanter, lateral epicondyle of thigh, head of fibula, lateral malleolus, heel and 5th metatarsal, as recommended by [10,11]. A commercial video camera (Sony-HDR-CX250, 60 fps) was used for recording. The camera was placed on the sagittal plane of the subject at 90cm over the floor and 4.5m from the subject's track.

Participants were asked to perform three trials for each load condition (without load, 4.1 kg, 6.8 kg and 10 kg), using a commercial bilateral backpack. Heavier load represent on average 15.3 % of body mass (table 1). For all trials, subjects were barefoot, wore jeans and the arm was flexed towards the opposite shoulder.

C. Kinematic Analysis

Trajectories for every one of the markers were obtained using the software Tracker. Joint angles were calculated for hip, knee and ankle as show in figure 3. Then, the angles were filtered with an 8-order Butterworth low-pass filter with a cutoff frequency of 10 Hz.

Table 1 Relation between participant's weights and load carriage

	Men			Women		
Load (kg)	4.1	6.8	10.5	4.1	6.8	10.5
% of body Mass	4.9	8.1	12.5	7.1	11.7	18.1

I. Torres et al. (eds.), *VII Latin American Congress on Biomedical Engineering CLAIB 2016, Bucaramanga, Santander, Colombia, October 26th -28th, 2016*,
IFMBE Proceedings 60,
DOI: 10.1007/978-981-10-4086-3_163

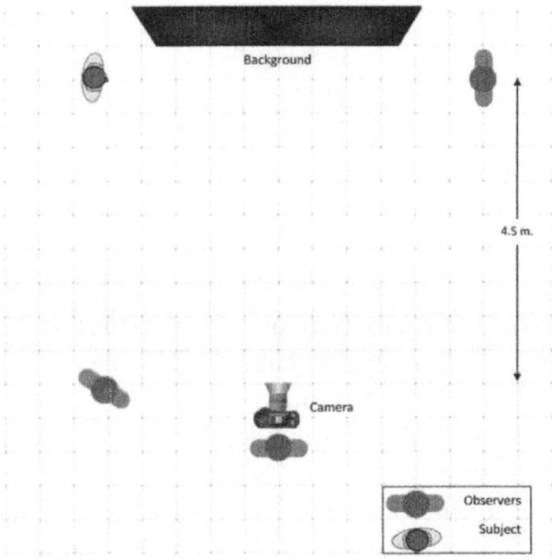

Fig. 1 Setting configuration. The camera was placed on the sagittal plane of the subject at 90cm over the floor and 4.5m from the subject's track.

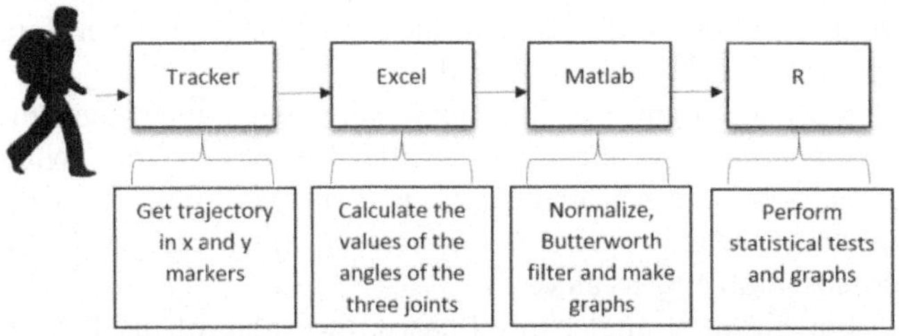

Fig. 2 Kinematic Analysis. The steps involved on the kinematic analyses were as follow; video recording, computation of trajectories and joint angles.

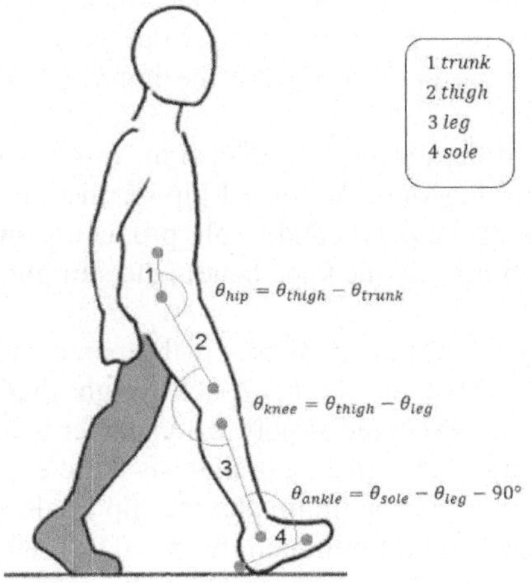

Fig. 3 Joint Angles. Calculation of joint angles considering the markers settings.

D. Statistical Analysis

Data of normalized joint angles were compared using R software and applying a T test for paired samples with α=0.05.

III. Results

Figure 4 shows the joint angles for hip, knee and ankle during the four load-carriage conditions. Figure 5 shows the comparison of the percentages (gait cycle). Figure 6 shows the statistical comparison for the hip angles (initial angle and maximum flexion), which were the only joint angles that showed a statistical significant difference.

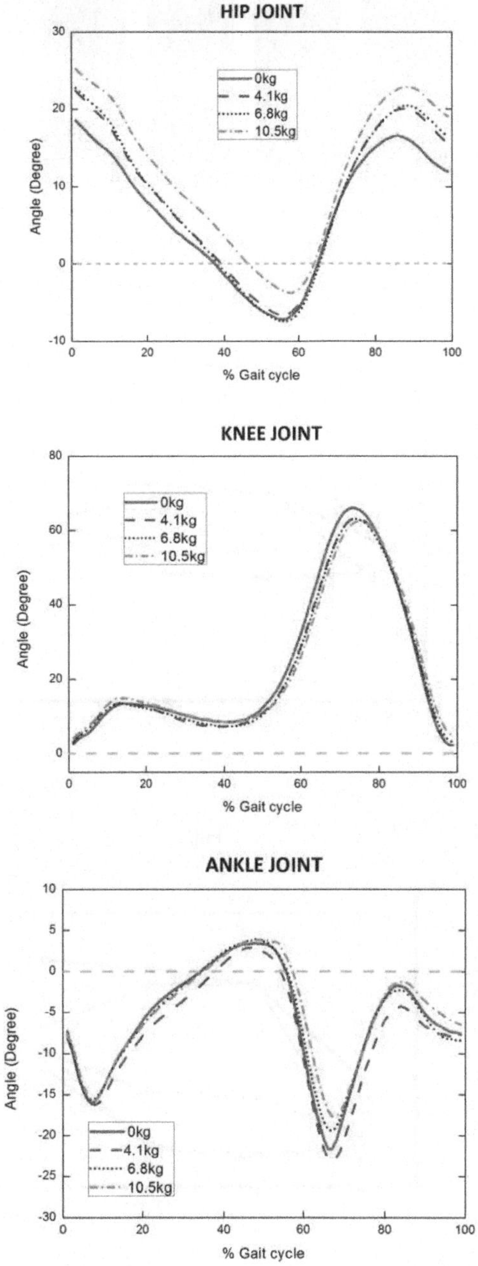

Fig. 4 Joint Angles. Average of hip, knee and ankle angles during the four load-carriage conditions.

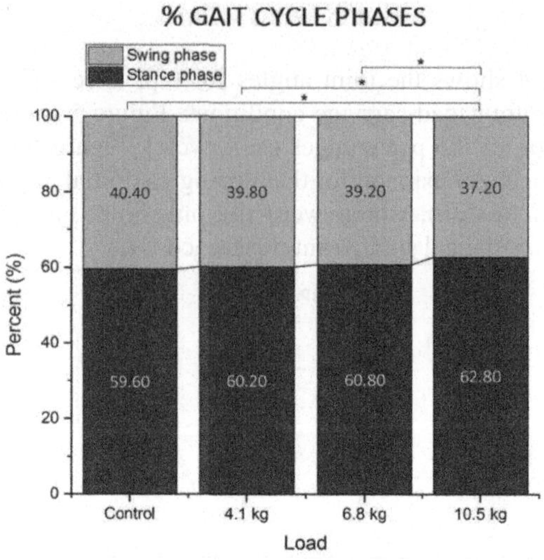

Fig. 5 Percentage of gait cycle during the four load-carriage conditions.

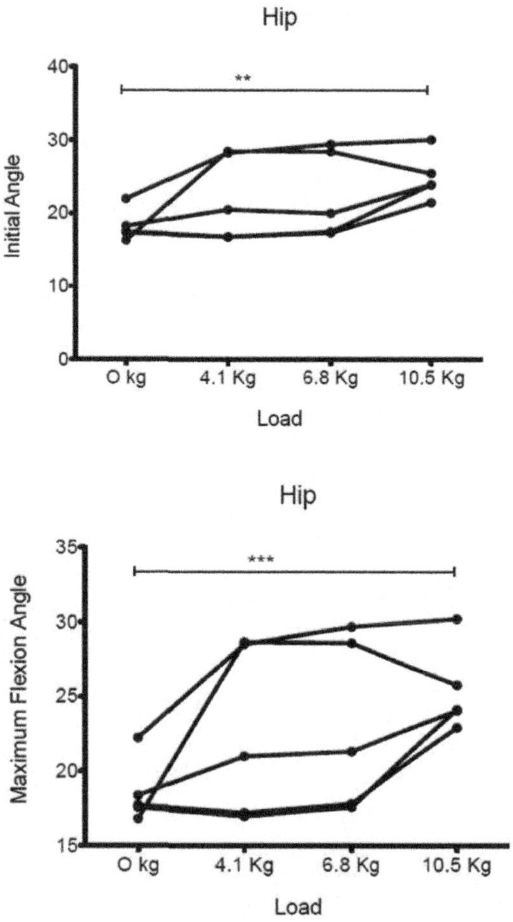

Fig. 6 Statistical comparison for hip angles during the four load-carriage conditions.

IV. DISCUSSION

For this study we decided to take in account these three loads: 4.1 kg, 6.8 kg and 10.5 kg. It is because they represent the most common cases bag's weight that college students carry during the university term. The first load represented a common day with a laptop and a book or notebook inside bag. The second one could be the weight of school bag in a project delivery day with two more books (that we had to give back to library) than the first case. Finally the third load represented a day at final season when school bags have more books, notebooks, folders, final deliveries and laptop, just for giving an example.

In general terms, along the gait cycle, knee flexion angles were decreased meanwhile the extension angles were increased.

In knee joint was observed a decrease in the flexion angle and an offset to the right side of the gait cycle, it means that the percentage of the stance phase was increased, therefore the swing phase was decreased.

For ankle joint, the mean gait cycle began under the neutral position, i.e. in plantar-flexion. Besides the pre-swing phase was increased at gait cycle percentage. For the test with the larger load, in general, an increase was presented at the dorsiflexion angle value.

For hip joint angle, in all cases, the value of the flexion was increased at the initial contact in order to keep the gravity center. In the hip joint angle, it is shown that the flexion angle was increased for all load-carriage conditions.

During the first half of the stance phase, an increase was observed in the knee flexion, it is due to act as protective measure by the body to absorb the impacts and reduce the risk of injury.

A leak in the proper hip extension causes a reduction of the opposite leg step. A limited hip extension modifies the alignments of the pelvis and thigh, producing anterior tilt of the pelvis, trunk and the knee bending to straighten the pelvis and trunk.

Figure 5 shows an increase in the percentage of stance phase with increasing the backpack weight that means than this phase was executed slowly with heavier load.

According to the statistical analysis there exists a significant difference in maximum flexion hip angle ($p < 0.001$), like it happens in initial hip angle ($p < 0.01$), when load carriage is heavier. It is showed in figure 6.

Unfortunately mostly college population is not informed or is not interested in how bad using of load affect their health. Despite every day there exist improvements in design and comfort in backpacks, the selling of backpacks is still more influenced by fashion than by ergonomics.

V. CONCLUSIONS

The observed changes in ankle joint and hip joint kinematics, as well as increase in percentage of stance phase of gait cycle in tests with heavier loads are consistent with results in [12,13]. In these studies, participants had similar age and they walked with loads which represents around 15% of their body mass.

The hip joint angle it is mostly affected from the effects of the 10.5 kg load. The angle is gradually changing to keep stable the gravity center and reduce the energy consumption. When the heel touches the floor, the increase of the flexion angle enlarges the damping between the heel and the sole. In general, we can see that in cases with very heavy loads, the hip dispenses of the extension. It sounds logical to think that becomes natural to walk curved performing a greater effort in the abdominal muscles especially the abdominal rectus to keep the gravity center.

This pilot study gives good results which are consistent with previous researches, but now studying a lost population: college students. This study could be improve using others devices as a force platform; obtaining physiological parameters like energy cost; or measuring time-space variables as cadence.

ACKNOWLEDGMENT

DAIP-UGTO 1,146/2016 and PAPIME PE104715.

CONFLICT OF INTEREST

The authors declare that they have no conflict of interest.

REFERENCES

1. N. A. S. Taylor, G. E. Peoples, and S. R. Petersen, (2016), "Load carriage, human performance, and employment standards," Appl Physiol Nutr Metab, vol. 41, pp. S131–47.
2. Sheir-Neiss, G.I., Kruse, R.W., Rahman, T., Jacobson, L.P., Pelli, J.a., 2003. The association of backpack use and back pain in adolescents. Spine 28 (9), 922–930.
3. Lu, T.-W. and Chang, C.-F. (2012). Biomechanics of human movement and its clinical applications. Kaohsiung J Med Sci, 28(2 Suppl):S13–25.
4. Hong, Y., Li, J.-X., Fong, D.T.-P., 2008. Effect of prolonged walking with backpack loads on trunk muscle activity and fatigue in children. J. Electromyogr. Kinesiol. 18 (6), 990–996.
5. Kirk, J., & Schneider, D. (1992). Physiological and perceptual responses to load-carriage in female subjects using internal and external frame backpacks. Ergonomics, 35, 445-55.
6. Knapik. J., Reynolds, K., & Harman, E. (2004). Soldier load carriage: historical, physiological, biomechanical, and medical aspects. Military and Medicine, 169, 45-46.
7. Brackley, H.M., Stevenson, J.M., Selinger, J.C., 2009. Effect of backpack load placement on posture and spinal curvature in prepubescent children. Work, 32 (3), 351–360
8. Pau, M., Mandaresu, S., Leban, B., Nussbaum, M.A., 2015. Short-term effects of backpack carriage on plantar pressure and gait in schoolchildren. J. Electromyogr. Kinesiol. 25 (2), 406–412.
9. Legg, S. J. and Cruz, C. O. (2004). Effect of single and double strap backpacks on lung function. Ergonomics, 47(3):318–23.
10. X. Yang, G. Zhao, D. Liu, W. Zhou, and H. Zhao, (2015), "Biomechanics analysis of human walking with load carriage," Technol Health Care, vol. 23 Suppl 2, pp. S567–75.
11. David A. Winter, (2009) Biomechanics and Motor Control of Human Movement, Fourth Edition, John Wiley & Sons, Inc. New Jersey USA.
12. M. F. Vieira, G. C. Lehnen, M. Noll, F. B. Rodrigues, I. S. de Avelar, and P. H. L. da Costa, (2016), "Use of a backpack alters gait initiation of high school students," J Electromyogr Kinesiol, vol. 28, pp. 82–9.
13. Kevin D. Dames and Jeremy D. Smith, (2015) "Effects of load carriage and footwear on spatiotemporal parameters, kinematics, and metabolic cost of walking", Gait & Posture, vol. 42(2), pp. 122-126.,

Corresponding author:

Author: Arturo Vega-González
Institute: Universidad de Guanajuato, División de Ciencias e Ingenierías, DIQEB
Street: Loma del Bosque 103
City: León, Guanajuato, 37150
Country: Mexico
Email: a.vega@ugto.mx

An approach to emotion recognition in single-channel EEG signals using StationaryWavelet Transform

A. Gómez[1], L. Quintero[1], N. López[2], J. Castro[3], L. Villa[4] and G. Mejía[5]

[1] Mathematical Modeling Research Group, GRIMMAT, Universidad EAFIT, Medellín, Colombia
[2] Medical Technology Laboratory, GATEME, Universidad Nacional de San Juan, San Juan, Argentina
[3] Psychology, Education and Culture Research Group, Institución Universitaria Politécnico Grancolombiano, Bogotá, Colombia
[4] System Engineering Research Group, ARKADIUS, Universidad de Medellín, Medellín, Colombia
[5] Functional Analysis and Aplications Research Group, Universidad EAFIT, Medellín, Colombia

Abstract— **In this work, we perform an approach to emotion recognition from Electroencephalography (EEG) single channel signals extracted in four (4) mother-child dyads experiment in developmental psychology. Single channel EEG signals are decomposed by several types of wavelets and each subsignal are processed using several window sizes by performing a statistical analysis. Finally, three types of classifiers were used, obtaining accuracy rate between 50% to 87% for the emotional states such as happiness, sadness and neutrality.**

Keywords— **EEG, Emotion, Wavelet, Features, KNN, QDA, RFC**

I INTRODUCTION

Define exactly the emotions is a difficult matter [1, 2], but for convenience, we consider the emotions as states, and these emotional states are neurological answer to external stimulus. The study of these responses to external stimuli are extremely important, because it allows to establish a psychological profile of the person, which it is useful for the diagnosis of possible psychopathology of the individual, allowing the creation of optimal treatment strategies. [3, 4] However, the study of emotional states has had a growing development in improving the interaction between human and computer, developing artificial intelligence systems that can interact and react depending on the emotional state of the human user.[5, 6] Over the last years different approaches have been reported to determining emotional states, implementing face recognition techniques, discourse analysis, and interpretation of body language. [7, 8] A different approach that has been studied in recent years for detection of emotional states is by analyzing the electroencephalographic signals, these signals being the interaction between neurons when an action or a thought happen, in this case when an emotional state is given. Detect emotions using EEG signals is very important, because these signals are directly related to brain activity, allowing a mapping from emotional states to the different areas of neuronal activity. This can allow psychologists validate or determine emotional states, or allowing an artificial intelligence system takes decisions based on a human user's emotional state; without requiring the human user performs a phonetic action or gestural expression, which is quite useful in infants.

Since signals were taken from a group of dyads, it is convenient to define what is a dyad. Kenny define the dyad as "The fundamental unit of interpersonal interaction and interpersonal relations" [9]. This is basically a social group of 2 persons who interact with each other, in this case the group corresponds to a mother and her child. The importance of this dyad is in the high impact of the behavior of the mother in the psychosocial and emotional development of her child. [10] It is important to focus on studying this interactions of the dyad mother-child, based on, that this will build a complete mindset of infant development through the time. In this case the mother-child dyad was analyzed for the evaluation of the interaction mother-child into a developmental psychological study. The study evaluated the capabilty of the emotional induction over a child from the mother without verbal communication. Also, detecting emotions evoked in the mother and its consequence on the evocation of emotions in her child[11, 12]. Therefore, this research seeks to validate the previous results by a classifier system for estimating between three emotional states, the state of happiness, the state of sadness and the state of neutrality, developing an analysis method for EEG single channel signals.The raw EEG signal generally has a range of nearly μV and with frequency components higher than 300 Hz[13], in several studies report a range for the detection of emotions in EEG signals in different frequency bands. This bands are: Delta band (0.5 to 4 Hz), Theta band (4 to 8 Hz), Alpha band (8 to 13 Hz), Beta band (13 to 30 Hz), and Gamma band (30 to 70 Hz) [13, 14]. In the last years has been studied the detection of emotional states in EEG signals, Pentrantonakis used the Delta, Theta, Alpha and Beta band with a HOC analysis for six emotions: happiness, surprise, anger, fear, disgust, and sadness with a minimal rate of classification for Fear with 75% and a maximum rate for Happiness with 99.373% [15]; Lee used Theta, Alpha, Beta, and Gamma band for demonstrated different pat-

I. Torres et al. (eds.), *VII Latin American Congress on Biomedical Engineering CLAIB 2016, Bucaramanga, Santander, Colombia, October 26th -28th, 2016*,
IFMBE Proceedings 60,
DOI: 10.1007/978-981-10-4086-3_164

terns for differents emotional state, with different combination of this bands, performing a complete analysis of correlation, coherence, and phase synchronization index of each band in different emotional states, [16]; Varun executes a more conservative analysis as to the frequency range, taking all the frequency bands previously exposed in a range of 0.5 Hz to 100 Hz, implementing a classification between happiness, neutral, sadness, and fear based on multiwavelet transform, with a mean accuracy of 80%[17]; Murugappan proposes the implementation of a Surface Laplacian filter to remove noises and artifacts and used an extraction method based on multi-resolution analysis of Wavelet Transform, for determinated between disgust, happiness, surprise, fear, and neutral with different classifier as LDA and KNN, comparing the number of channels of EEG from 62 to 8, with a mean accuracy of 85% [18]; and we previously proposed the classification of two emotions, sadness and happiness performing a statistical analysis over features in the time and frequency domains with a mean accuracy of 99%.[19] This paper relates a study carried out on a small cohort of mother-child "Dyads" in order to determine between three emotional states: sadness, happiness and neutrality. The work revolves around using some standard statistical measures applied to the subsignals obtain from Stationary Wavelet Transform of a single channel from an EEG signal obtained from the subjects adult and child alike, after conducting an experiment of evocation of emotions. Different window sizes were used to perform the statistical analyses and these were then used features in three different classifiers: KNN, RFC and QDA to classify incomes into happiness, sadness and neutrality. In this work we propose an approach to a methodology for emotion recognition in EEG signals from one channel information. Also, presents an easily and understandable solution with preprocessing of the signal, decomposition of the signal, and statistical features extration and therefore three kinds of classifiers. No specific EEG frequency bands were taken into account and for simplicity statistical measurements were performed to extract relevant information from features. Signals are not only sampled from adults but also from children and emotional states were confirmed by human experts. This work is organized as follows: Section 2 presents the methodology, Section 3 presents the results obtained and finally the conclusions.

II METHODOLOGY

The proposed methodology, consists of five main stages: experimental protocol, pre-processing, feature extraction, classification and finally analysis of the results. As it is shown in Figure 3.

A Experimental Protocol

The experimental protocol includes the design of the experiment, the validation by expert triangulation and execution of the experiment with taking electroencephalographic data. To perform the experiment the following protocol was used: First, each mother was asked to make a recording in a room, the happiest moment of her life for the stimulus of happiness, and the saddest moment of her life to the stimulus sadness, then every dyad was placed face to face; the mother had headphones and listen in each case (happy or sad) the story that her previously recorded, evoking the feeling of happiness or sadness, as the case, creating an evocation of emotions on her child who was staring at her. The state of neutrality is recorded before each session evocation of emotions. It was counted with a team of psychologists who verified the validity of this protocol that meets the ethical principles of Helsinksi [11, 12], and the single channel, it was located between the F3 and F4 position and it was taken at a sampling rate of 1000 Hz.

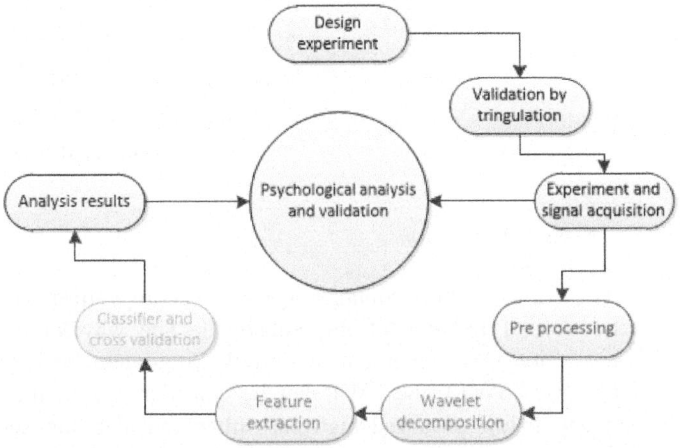

Fig. 3: Proposed Methodology

B Pre-processing

The pre-processing includes filtering and windowing signals. The signal filtering process was implemented with three Butterworth filters of 10th order: a high pass filter at 0.5Hz, a low pass filter at 70Hz, and a notch filter at 60Hz. Since the EEG signal is considered stationary in short intervals [13], a windowing process is implemented with a Hanning window. An overlap of 50% was done, and three window sizes: 64 ms, 128 ms and 254 ms; each windows sizes were essayed, in order to decomposed the signal as detalied below.

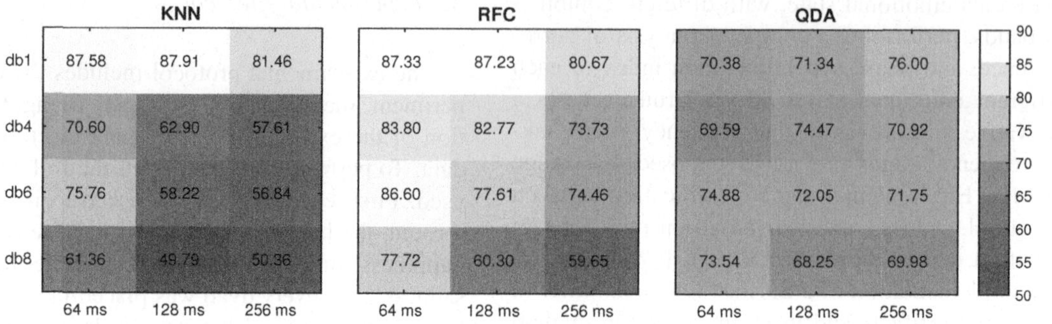

Fig. 1: Classification scores using different types of wavelet and window sizes for each classifier.

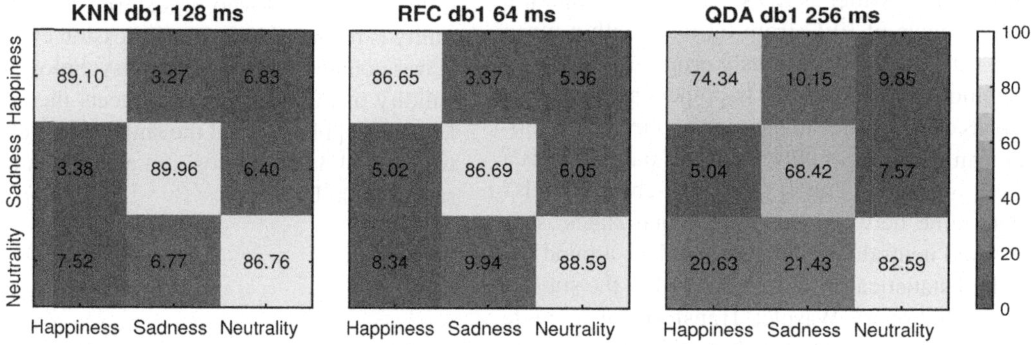

Fig. 2: Confusion Matrices for the best classifier of each type.

C Feature Extraction

Feature extraction contains: wavelet decomposition and extraction of features of this subsignal. The wavelet decomposition was implemented using the Stationary Wavelet Transform (SWT), the SWT can be considered as a filter bank, with low pass and high pass filter, that decomposed the input signal into two subsignals: approximation and detail; for the following levels of decomposition, the approximation coefficients of the previous level are the input of the filter bank for the next level [20]. Each filter bank is related to a mother wavelet, in this case the mother wavelet is the Daubechies wavelet, particularly we worked with the wavelet db1, db4, db6 and db8 [21]. Each segment taken from the signal on the windowing process, is decomposed decomposed at 5 levels with each type of wavelet; for analyzing this signal a matrix was constructed for each type of wavelet and for each window size, with 5 coefficients of detail and the 5th coefficient approximation of the decomposed signal. [20, 22] Each subsignal obtained by the SWT was analyzed using three descriptive characteristics. For this work we use the mean, the variance [15] and the kurtosis[23] of each subsignal in order to properly describe the original signal relating to an emotional state.

D Classification

Classification includes the training of different classifiers, with a cross-validation. Eigth classification processes were employed for this work, but the three best are presented: 15-Nearest Neighbor (KNN), Random Forest (RFC) and Quadratic Discriminant Analysis (QDA) [24, 25]. Every classification process was implemented with cross-validation using the holdout method and random subsampling for a total of thirty experiments.

E Analysis of Results

The analysis of the results contain the psychological analysis and validation of emotional states [11, 12]; in addition the evaluation of emotion recognition using supervised learning techniques that is presented in the following section.

III Results and Discussion

The emotion detection of happiness, sadness and neutrality was performed with 3 different sizes of windows and 4 different wavelet decomposition types. From each sub-signal we extracted 3 features and built a matrix resulting from the extraction of descriptive characteristics. We proceed to train a KNN, RFC and QDA classifiers with 50% of the dataset with a Cross Validation using the Holdout method and random subsampling for thirty experiments for each group of sub-signals and the 3 different sizes of window, which gives us a total of 36 classifiers and after the Cross Validation gives us a total of 1080 classifiers. Once each classifier was trained, we calculated the average percentage of correct classification for the 36 classifiers(Figure 1). Notice that in general as windows size increases, the average percentage of accuracy decreases. On the other hand, as wavelet vanishing moments increase (db1 - db8), the average percentage of accuracy decrease for the KNN and RFC Classifiers, indicating that is nontrivial selecting the wavelet for the decomposition. Notice that the QDA classifier has the most stable behavior regarding to the windows size and the type of wavelet, and all of the three classifiers obtain their best scores when the signal decomposition was carried out using the db1. But the higher scores correspond to the RFC and KNN classifiers without highly significant changes between window sizes. To give an further look of the behavior of these classifiers, the confusion matrix of the best classifier of each type is calculated, and it is shown in the figure 2. Notice, the good performance of RFC and KNN classifiers, over the three emotional states, whereas that QDA classifier is notoriously difficut to separate the emotional states of happiness and sadness, so indicating that it may not be the best classifier for this case.

IV Conclusion

The recognition between the happiness, sadness and netraulity emotional states, is possible through an analysis of a single channel of EEG. The study of the signal using Stationary Wavelet Transform, evaluating the effect of the signal decomposition with different types of wavelets and statistical analysis of each subsignal, allows the classification of the emotional states with a high rate with different kinds of classifiers. Moreover, the effect of window size on the results was also evaluated. Given the high classification rates, it can be concluded that the size of the window is important but stable. Allowing the selection of larger window sizes to reduce computational costs, and trying to reduce the small effect of misclassification of the signal, assesing the segments before and after the emotional states obtained by the classifier. This analysis can be used to study the mental mappings during the mother-child dyads interactions. As future work, we will study the detection of a greater number of emotions, and different number of channels.

References

1. Cabanac Michel, Cabanac Michel. What is emotion? *Behavioural processes.* 2002;60:69–83.
2. Scherer K. R.. What are emotions? And how can they be measured? 2005.
3. Affrunti Nicholas W., Woodruff-Borden Janet. The effect of maternal psychopathology on parentchild agreement of child anxiety symptoms: A hierarchical linear modeling approach *Journal of Anxiety Disorders.* 2015;32:56 - 65.
4. Kleiman Evan M., Ammerman Brooke, Look Amy E., Berman Mitchell E., McCloskey Michael S.. The role of emotion reactivity and gender in the relationship between psychopathology and self-injurious behavior *Personality and Individual Differences.* 2014;69:150 - 155.
5. Martnez-Miranda Juan, Aldea Arantza. Emotions in human and artificial intelligence *Computers in Human Behavior.* 2005;21:323 - 341.
6. Kreps Gary L., Neuhauser Linda. Artificial intelligence and immediacy: Designing health communication to personally engage consumers and providers *Patient Education and Counseling.* 2013;92:205 - 210.
7. Milton A., Selvi S. Tamil. Class-specific multiple classifiers scheme to recognize emotions from speech signals 2014;28:727 - 742.
8. Feng Xiaoyan, Watada Junzo. Building a Recognition System of Speech Emotion and Emotional States *2013 Second International Conference on Robot, Vision and Signal Processing.* 2013;0:253-258.
9. Kenny D.A., Kashy D.A., Cook W.L.. *Dyadic Data Analysis.* Methodology in the social sciencesGuilford Press 2006.
10. Killeen Lauren A.. Understanding parenting as a process: Frontal EEG alpha asymmetry as a measure of "online" maternal responsiveness to infant cues
11. Castro Martinez Jaime Alberto. Neurodinámica y autoorganización en la interacción socioemocional madre-hijo: aproximación de los sistemas dinámicos a los principios del desarrollo emocional infantil ;2:17.
12. Castro Martinez Jaime Alberto. Sistemas dinámicos en la interacción emocional madre-hijo: primera fase ;9:129-138.
13. Sanei Saeid, Chambers Jonathon. a.. *EEG Signal Processing*;1 . 2007.
14. Teplan Michal. Fundamentals of EEG measurement *Measurement Science Review.* 2002;2:1–11.
15. Petrantonakis Panagiotis C, Hadjileontiadis Leontios J. Emotion recognition from EEG using higher order crossings. *IEEE transactions on information technology in biomedicine : a publication of the IEEE Engineering in Medicine and Biology Society.* 2010;14:186–197.
16. Lee You Yun, Hsieh Shulan. Classifying different emotional states by means of eegbased functional connectivity patterns *PLoS ONE.* 2014;9.
17. Varun Bajaj Ram Bilas Pachori. Classification of human emotions based on multiwavelet transform of EEG signals 2012.
18. Murugappan Murugappan. Classification of human emotion from EEG using discrete wavelet transform 2010.
19. Gomez Alejandro, Quintero Lucia, Lopez Natalia, Castro Jaime. An approach to emotion recognition in single-channel EEG signals: a mother child interaction. *XX Congreso Argentino de Bioingeniera, SABI 2015.* 2015.
20. Nason G. P., Silverman B. W.. The Stationary Wavelet Transform and some Statistical Applications :281–300Springer-Verlag 1995.
21. Campo D, Quintero O L, Bastidas M. Multiresolution analysis (discrete wavelet transform) through Daubechies family for emotion recognition in speech. *Journal of Physics: Conference Series.* 2016;705:12034.
22. Bustamante P A, Celani N M Lopez, Perez M E, Montoya O L Quintero. Recognition and regionalization of emotions in the arousal-valence plane *Conf Proc IEEE Eng Med Biol Soc.* 2015:6042–6045.
23. Kevin P. Balanda H. L. MacGillivray. Kurtosis: A Critical Review *The American Statistician.* 1988;42:111-119.
24. Dougherty Geoff. *Pattern Recognition and Classification*;53 . 2013.
25. James Gareth, Witten Daniela, Hastie Trevor, Tibshirani Robert. *An Introduction to Statistical Learning*;103 . 2013.

Processing of thermal images oriented to the automatic analysis of hand thermoregulation

N. Zapata-Osorio[1], S. Orrego-Serna[1], L. Ramirez-Arbelaez[2], A. Castro-Ospina[3] and H. Fandiño-Toro[3]

[1] Instituto Tecnológico Metropolitano, Medellín, Colombia
[2] Instituto Tecnológico Metropolitano, Grupo de Investigación e Innovación Biomédica, Medellín, Colombia
[3] Instituto Tecnológico Metropolitano, Grupo de Automática, Electrónica y Ciencias Computacionales, Medellín, Colombia

Abstract— **In this work we propose a methodology for segmenting and labeling regions of interest (ROI) corresponding to the palm and the five from infrared thermal images of the hand, acquired from a immersion test. The segmentation relies on measuring distances between the pixels in the contour of the hand and its center of mass, and involves the consideration of very few free parameters. The results are promising and suggest that the proposed methodology could serve as a component of an automatic scheme for the analysis of the hand thermoregulation.**

Keywords— **Fingertip, Hand thermoregulation, Interdigital landmarks, Thermal image.**

I INTRODUCTION

Thermoregulation refers to the set of physiological processes that aim to keep constant the body temperature values, both at the core and as at skin surface. For the scientific community, analysis of thermoregulation is relevant because some pathologies can affect it and as a result, the affected subjects can not recover the temperature values exhibited before experiencing an external thermal stimulus. Some published studies have shown evidence about the relationship between alterations in thermoregulation and the presence of pathologies such as Raynaud's phenomenon [1], cold intolerance [2] and carpal tunnel syndrome [3]. It is possible to observe the thermoregulation by thermal imaging, a completely non-invasive technique that does not involve application of ionizing emissions to the surface of hands. One way in which researchers evaluate thermoregulation with thermal images is performing immersion tests on subjects under analysis hands in water at controlled temperature. Concerning to the analysis of thermoregulation with thermal images, there are some published studies showing curves with profiles of hand temperature recovery in immersion tests.

It is known that the response to a cold stimulus is not uniform in the whole hand, because of differences in the structures involved in the microcirculation and vasoconstriction of the palm and fingers. As a result, the analysis of this profiles usually involve the manual segmentation of specific ROI, and there are comparatively few studies that have explored the automatic segmentation of the ROI corresponding to the palm hand and fingers [4, 5]. At both works it is sequentially applied a morphological opening to the thresholded image of the hand with different versions of structuring elements, aiming to segment the five fingers from the palm region and wrist. In [5] distal phalanges are segmented by using morphological dilation with a circular structuring element. It is worth to state that the main disadvantage of methods based on mathematical morphology is the need to tune the size of structuring element, which highly relies on the amount of pixels of the foreground/background of image under analysis. Under the hypothesis that the automatic analysis of thermoregulation can benefit from the automatic segmentation of relevant ROI in the whole hand, a methodology of thermal imaging processing is proposed, which could be used to analyze temperature time plots of the segmented ROI.

II EXPERIMENTAL SETUP

The protocol for the acquisition of the images used in this work is described in detail in [6]. The complete database comprises of 32 sequences of images acquired at a frame rate of one frame per second, during approximately 840 seconds that takes the cold stimulation test of the hand (of a volunteer subject) with water at 15 degrees Celsius. In order to avoid occlusion or interference due to movement, from each image sequence we extract a single image consisting of the arithmetic average of the first five frames of the sequence.

A Finding the fingertips in the thresholded thermal images

To segment each finger two representative points are required for defining its base. But to find these representative points for the index, the middle and third finger, it is necessary to previously find their corresponding fingertips. An additional processing step is required to segment the thumb and the little finger. The details of such processing is described as

I. Torres et al. (eds.), *VII Latin American Congress on Biomedical Engineering CLAIB 2016, Bucaramanga, Santander, Colombia, October 26th -28th, 2016,*
IFMBE Proceedings 60,
DOI: 10.1007/978-981-10-4086-3_165

follows. First we obtain the Cartesian coordinates of the pixels belonging to the contour of the thresholded image. Then we generate a vector whose elements are the distances between these coordinates and the Cartesian coordinates of the center of mass of the hand region, in a process inspired by the work presented in [7]. Henceforth, we will refer to this vector of distances as $1DTS$. In order to obtain the center of mass of the region corresponding to the hand, we calculate the inverse of the distance function for the foreground pixels in the thresholded image, and then we segment the region enclosing pixels with the ten percent of highest values within this inverse distance function. We use the coordinates of the center of mass of this latter region, as the coordinates for the center of mass of the hand region. The calculation of $1DTS$ allows the association of the linear location of each of its elements with the spatial location of a pixel belonging to the contour of the hand in the thresholded image. From $1DTS$ we extract their two extreme points: its starting point and its end point. We will refer to the extreme point closest to the origin (of the image) as **EPCTO**, and to the extreme point farthest to the origin as **EPFTO**. We calculate the proximity of these extreme points to the origin, via Euclidean distance. Depending on the orientation of the hand in the thermal image, the first element of $1DTS$ can be **EPCTO** or **EPFTO**. In any case, the elements in $1DTS$ are sorted from left to right, so that their distances are increasing relative to the first element.

Our first hypothesis is that convex regions in $1DTS$ correspond to regions of the hand contour close to the fingertips. Therefore it is useful to extract peaks in $1DTS$ in order to localize the fingertips. There is a considerable number of methodologies available for automatic extraction of peaks in one-dimensional signals; such methodologies often involve the adjustment of free parameters [8]. We propose to iteratively filter out $1DTS$ with a moving average filter, until we obtain five peaks related to the five fingertips. The span of the moving average is three for all iterations. As we iteratively filter $1DTS$, we consider that an element b in this signal is a peak, if its y coordinate is greater than the y coordinate for elements a and c immediately before and after that b, respectively. In other words, given three contiguous elements: (x_a, y_a), (x_b, y_b) and $(x_c, x_c) \in 1DTS$, b is a peak if $y_a \leq y_c < y_b$, for $x_a < x_b < x_c$, three consecutive elements in $1DTS$. Once we obtain five peaks in $1DTS$ after the iterative filtering, each peak is associated with its corresponding fingertip in the thresholded image. For this association we use a distance matrix D where each element d_{ij} represent the Euclidean distance between the coordinates of two peaks p_i and p_j present in $1DTS$, as presented in Eq. (1). In this equation n is the number of peaks in $1DTS$, in this case five.

$$D = \begin{bmatrix} d_{11} & d_{12} & \dots & d_{1n} \\ d_{21} & d_{22} & \dots & d_{2n} \\ \vdots & \vdots & \ddots & \vdots \\ d_{n1} & d_{n2} & \dots & d_{nn} \end{bmatrix} \quad (1)$$

In this equation $d_{ij} = d(p_i, p_j) = \sqrt{(p_i - p_j)^2}$, $\forall\, i, j \leq n$. Considering Eq. (1), we formulate a second hypothesis: the cumulative sum of the Euclidean distances between the coordinates of each fingertip and the coordinates of the points of the other fingers, is greater for the case of the thumb fingertip. After identifying the thumb fingertip with this criterion, the identification of the other fingertips is straightforward, by sorting them from smallest to largest distance from the thumb fingertip. We will refer to the fingertips of the thumb, index, middle, third and little, as **FT1**, **FT2**, **FT3**, **FT4** and **FT5**, respectively.

B Finding approximations to the bases of the fingers

In order to find approximations to the bases of the fingers, we found the lowest point between each pair of consecutive peaks in $1DTS$, which allow us to obtain the coordinates of four inter-digital landmarks: **IDLM1**, **IDLM2**, **IDLM3** y **IDLM4**. After drawing the line segments: $\overrightarrow{\textbf{IDLM1 IDLM2}}$, $\overrightarrow{\textbf{IDLM2 IDLM3}}$ and $\overrightarrow{\textbf{IDLM3 IDLM4}}$, we delimit the regions of the image comprising the fingers: index, middle and third, respectively. Given that this approach does not allow thumb and little finger segmentation, we propose it as follows:

To segment the thumb, we consider one of the two extreme points **EPCTO** or **EPFTO** (It depends on the hand orientation in the image) and the fingertip **FT1**. Straight line segments are drawn, starting from points that lie in the closed interval that includes the selected extreme point and **FT1**, and ending in **IDLM1**. Moreover, a line segment is drawn between the points **IDLM1** and **IDLM2**, and its slope is calculated. Finally, from the points in the closed interval used to draw the straight line segments, we extract the coordinates of a point **MLM-T**, for which the line segment $\overrightarrow{\textbf{MLM-T IDLM1}}$ have a slope as close as possible to the slope of the line segment $\overrightarrow{\textbf{IDLM1 IDLM2}}$.

To segment the little finger, **FT5**, and one of the extreme points **EPCTO** or **EPFTO** are used. Straight line segments are drawn, starting from points that lie in the closed interval which includes **FT5** and the selected extreme point, and ending in **IDLM4**. A straight line segment is drawn between the points **IDLM3** and **IDLM4**, and its slope is calculated. Finally, from the points in the closed interval used to draw the straight line segments, we extract the coordinates of a point

MLM-L, for which the line segment $\overrightarrow{\textbf{MLM-L IDLM4}}$ have a slope as close as possible to the slope of the line segment $\overrightarrow{\textbf{IDLM3 IDLM3}}$.

C ROI labeling

Image segmentation is performed when six ROIs are obtained for the following regions of the hand: a region covering the palm and the wrist, and other five related to the fingers, i.e. thumb, index, middle, third and little. These regions are named: **ROI1**, **ROI2**, **ROI3**, **ROI4**, **ROI5** and **ROI6**, respectively. A separation of the ROIs is done, by iteratively performing a morphological erosion with a circular structuring element, with a radius that is increased from an initial value of 1 until the number of connected components in the image is equal to six. The ROI with the highest number of pixels is named **ROI1** and the remaining five ROIs are ordered by the reasoning introduced in Section II-*B*. ROIs mass centers are used as reference points to compute distances.

III Results

An example of a thermogram processed in this work is shown in Fig. 1.

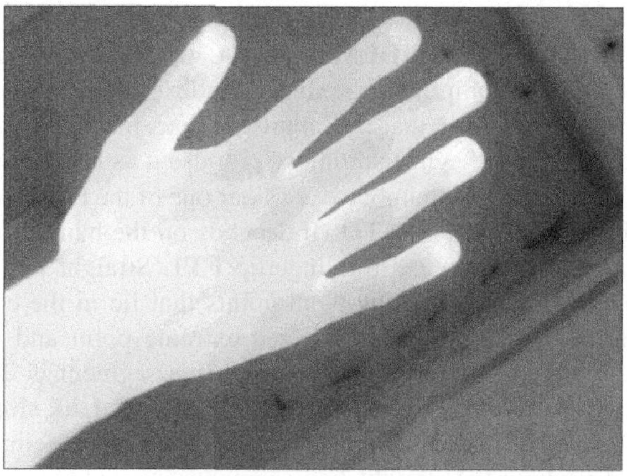

Fig. 1 Original infrared image.

In Fig. 2 is highlighted the hand contour of the image shown in Fig. 1. Inside the hand is shown the result of the inverse of the distance function. It can be seen how the pixels intensities values increases as the points are distant from the contour. The delimited region in blue has an illustrative purpose; it is used to emphasize the pixels within the 10% of the highest values obtained by the inverse of the distance function. Finally, the red dot inside the blue contour depicts its mass center, hence the hand mass center.

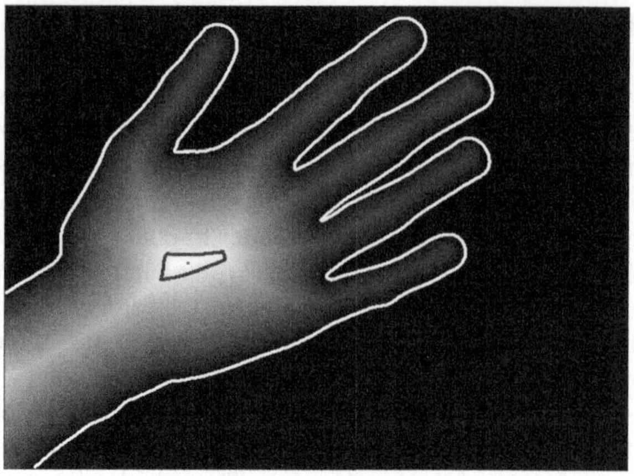

Fig. 2 Hand contour and inverse of distance function.

Distance vector between hand contour and hand mass center (Fig. 2) is shown in Fig. 3. (a) before and (b) after an iterative average filtering until obtain five peaks corresponding to each fingertip. It is also shown the four valleys located between the peaks. From Fig. 3 (a) is clear that without the iterative filtering of $1DTS$ can be found a lot of peaks from such vector, which makes it difficult to locate each fingertip. The correspondence of fingertips and reference points found in Fig. 3(b) of the thermal image is shown in Fig. 4. In addition to these points it is also depicted **EPCTO** and **EPFTO** points. It can be seen how the reference points allow to obtain suitable approximations of the base of each finger. Finally, segmented ROIs are shown in Fig. 5, after a final separation through a morphological erosion. Such erosion has an additional purpose which is to ensure that each ROI is enclosed by its corresponding region in the original image, avoiding the consideration of background pixels.

IV Conclusions and Future works

A methodology for automatic segmentation and labeling of ROI from infrared thermal images of the hand is proposed, achieving suitable results for future work. Proposed methodology is simple yet effective; it allows hand segmentation with minimal *a-priori* information of the images to process and only employs two free parameters, whose tuning are done by a high interpretability criterion. As future work a comparison against related segmentation techniques must be performed to quantify achieved results. Moreover, a temporal analysis of the thermoregulation of each segmented ROI should be done. Also from such analysis it can be generated relevant features for discriminate among pathologies related to neurological and vascular impairments.

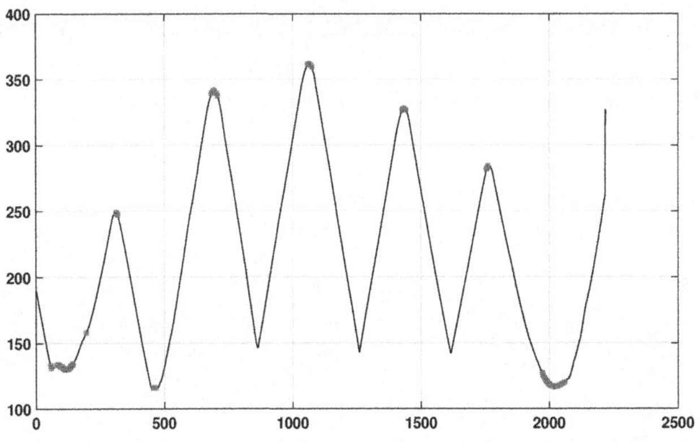

(a)

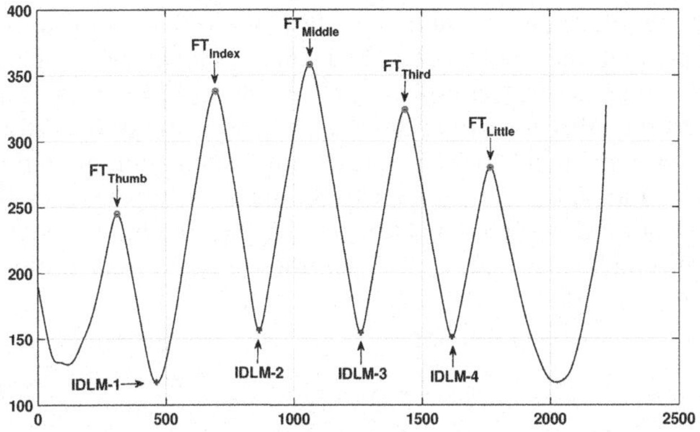

(b)

Fig. 3 (a) Original distances vector; (b) The five peaks remaining in the distances vector after its processing.

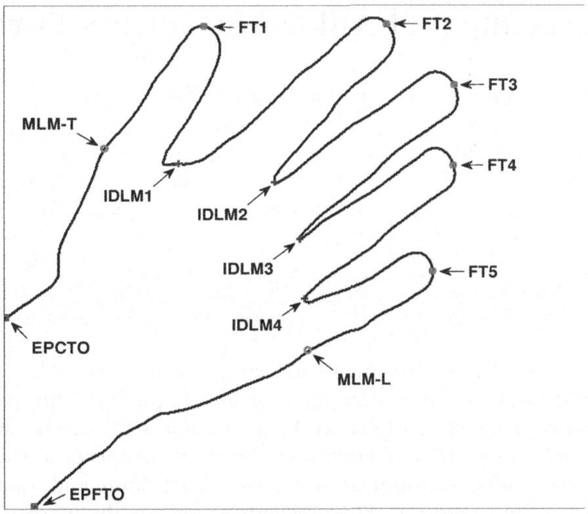

Fig. 4 Landmarks obtained from the distances vector.

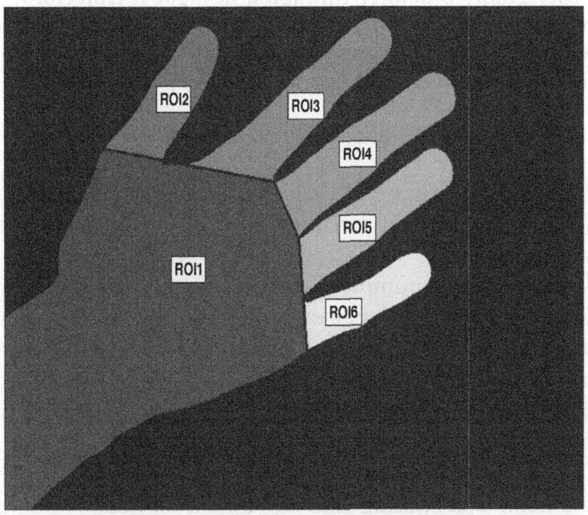

Fig. 5 The final six segmented ROI.

Conflict of Interest

The authors declare that they have no conflict of interest

References

1. Schlager O, Gschwandtner M, Herberg K et al. Correlation of infrared thermography and skin perfusion in Raynaud patients and in healthy controls *Microvasc Res.* 2010;80:54–57.
2. Smits E, Nijhuis T, Huygen F et al. Rewarming patterns in hand fracture patients with and without cold intolerance *J Hand Surg Am.* 2011;36:670–676.
3. Ming Z, Zaproudina N, Siivola J et al. Sympathetic pathology evidenced by hand thermal anomalies in carpal tunnel syndrome *Pathophysiology.* 2005;12:137–141.
4. Blank M, Kargel C. Infrared Imaging to Measure Temperature Changes of the Extremities caused by Cigarette Smoke and Nicotine Gums in *2006 IEEE Instrumentation and Measurement Technology Conference Proceedings*:794–799 2006.
5. Zhang H, He Y, Wang X et al. Dynamic infrared imaging for analysis of fingertip temperature after cold water stimulation and neurothermal modeling study *Comput Biol Med.* 2010;40:650–656.
6. Ramírez L, Jiménez K, Correa A et al. Protocolo de adquisición de imágenes diagnósticas por termografía infrarroja *Medicina & Laboratorio.* 2015;21:161–178.
7. EtehadTavakol M, Ng E, Lucas C et al. Nonlinear analysis using Lyapunov exponents in breast thermograms to identify abnormal lesions *Infrared Phys Techn.* 2012;55:345–352.
8. Scholkmann F, Boss J, Wolf M. An efficient algorithm for automatic peak detection in noisy periodic and quasi-periodic signals *Algorithms.* 2012;5:588–603.

Author: Hermes Alexander Fandiño Toro
Institute: Instituto Tecnológico Metropolitano
Street: Calle 31 # 54-22
City: Medellín
Country: Colombia
Email: hermesfandino@itm.edu.co

Assessment of Ankle Movements Through a Game-Based Sphere: Proof of Concept

Alan Meana[1], David Negrete-Rojas[1], Rafael A. Nava-Gomez[1], Jose A. Ruiz-Diaz[1], Arturo Vega-Gonzalez[2], Juan M. Gomez-Gonzalez[1]

[1] Facultad de Ingeniería, DCR, Universidad Nacional Autónoma de México, México DF,
[2] División de Ciencias e Ingenierías DIQEB, Universidad de Guanajuato, León, Guanajuato, México

Abstract— **Ankle movements are complex and play a key role in displacement and manipulation of equipment. This paper presents a novel game-based device to assess the ankle movements. The system is composed by a hemisphere, a smartphone, a software application for the smartphone and another one for the computer. The hemisphere acts as the main interface between the foot and the software application. Through the hemisphere, which acts as a pointer or jositick for the foot, the user guides a pointer through a hexagonal trajectory. This allows assessing the ankle movements. The developed system was tested with 11 subjects. Preliminary result shows the potential of the system.**

Keywords— **ankle, kinematics, rehabilitation, game-based.**

I. Introduction

The ankle is a complex joint. It is used for different activities such as displacement, manipulation of equipment and sports among others [1]. Due to its extensive use, it is common that it suffer from a lession or injury either on the ankle itself or indirectly such as patients who suffer from a stroke.

The aim of the proposed system is to assess simple movements considering fine movement on healthy subjects. Nonetheless the system could be used to complement a rehabilitation intervention, where the gross and fine movements could be practiced and monitored [2,3].

The ankle plays a key role in gait and therefore it is required to have a precise control [1]. It is well known that the nervous terminals are less for the ankle, compared to those dedicated to the fingers or wrist. Therefore it is important that the developed system translate accurately the position of the joint to coordinates x and y in a computer display [1,4].

Other factor that should be considered is the dominance among lower limbs. Thus, people usually perform a variaety of activities with one or other limb in a daily basis [1,4,5,6]. This “unconscious training” helps to develop abilities that can make a difference when performing Activities of Daily Living. Therefore for some activities it is not possible to change the role assigned to each upper or lower limb.

The developed system was designed to assess all movements of the ankle (Fig 1). Table 1 shows the range of movements for a non-phatological subjects [1]. It has been observed that for a right dominat subject is easier to perform such movements (flexion/extension, supination/pronation, aduction/abduction), in clockwise direction whereas for the left-dominant subject is easier to perform the movements in anti-clockwise direction. Therefore a proposed protocol must include both feet and both directions (clockwise and anti-clockwise) [1,4,7,8]

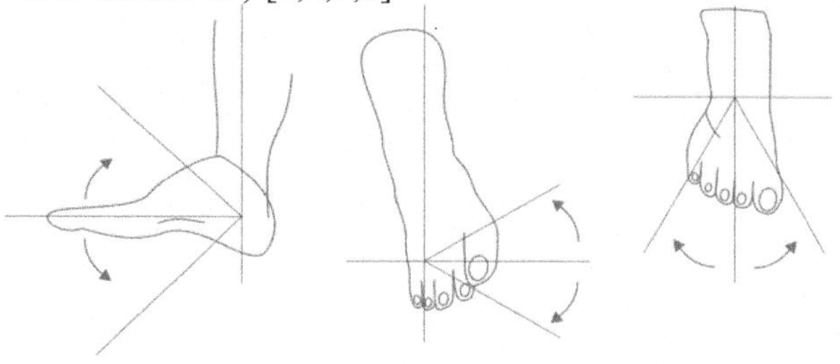

a b c

Fig. 1 Movements of the ankle joint. a) Flexion/extension, b) Supination/pronation and c) abduction/aduction. The system assesses all of these movements.

Table 1 Angles of the joint ankle. Range of movements for non-phatolofical subject [1]

Movement	Angle		
Flexion	20.3°	to	29.8°
Extension	37.6°	to	45.8°
Supination	14.5°	to	22.0°
Pronation	10.0°	to	17.0°
Abduction	15.4°	to	25.9°
Aduction	22.0°	to	36.0°

I. Torres et al. (eds.), *VII Latin American Congress on Biomedical Engineering CLAIB 2016, Bucaramanga, Santander, Colombia, October 26th -28th, 2016*,
IFMBE Proceedings 60,
DOI: 10.1007/978-981-10-4086-3_166

II. Metodos

A. Display

The kinematic trial requires that the subject must be seated in a comfortable position. Figure 2 shows the suggested postion for the subject. The foot to be tested should be inside the hemisphere and the attention of the subject should be on a display monitor or wall. The display area will show the path and the pointer that the user has to guide (Fig 4, 5).

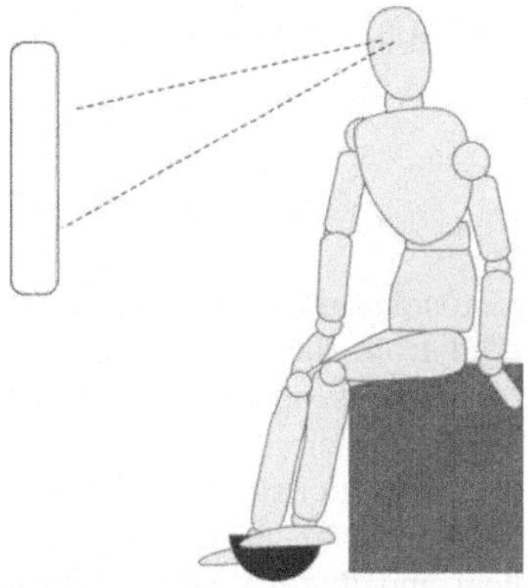

Fig. 2 Position of the subject at trial time. The position should be comfortable and focus on the display area.

The subject hast to interact with the hemisphere to move a pointer through a hexagonal trajectory. The trajectory can be followed either in a clockwise or anti-clockwise direction (fig 4). The subject will have to perform turns of 120°. The precission of the movements should assure that the pointer do not go outside the track. If this occurs the subject should manipulate the sphere to return the pointer to the track and continue with the trajectory.

The view that the user has will be only a section of the hexagon (Fig 5) so that the sphere is always in the center and is seen as a walking line. The application was created in the graphical development engine Unity [9], using as lenguage development c-sharp . The path that the user sees has a width of 10 graphic units while the pointer that will go through the trajectory has a diameter of 6 graphic units, each side that make up the hexagon has 102 units.

The mobility of the sphere is controlled via a mobile device (smartphone), Through the gyroscope and integrated accelerometers, so that the axes used are (x , y) (Fig. 3).

The application stores all the coordinates of the position of the sphere, and the corresponding time at approximately 60 frames per second, in a file whose name is updated automatically. This allows to perform a off-line processing and to recreate the movements and trajectory.

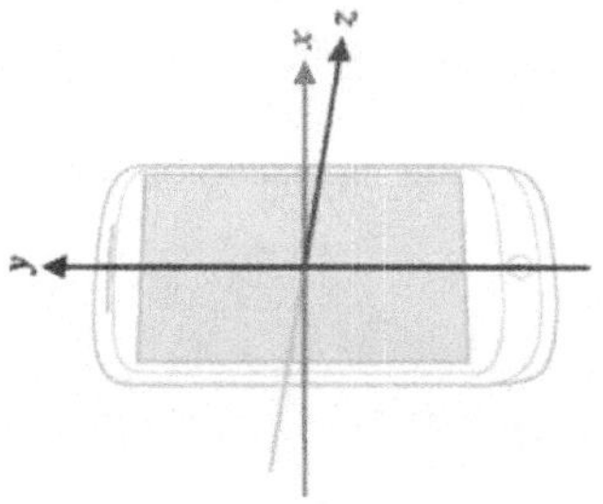

Fig. 3. The axes of the device are (x, y), These will be matched with the movement of the sphere.

The application has a section with which the user can calibrate the position and motion allowing to get used to the preset sensitivity before starting the trial. This allows always at the beginning of each trial that sphere start in a controlled point by the user. This indicates the start of the trial.

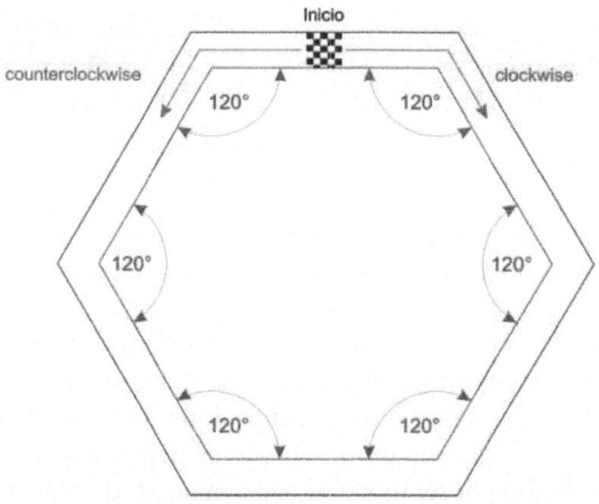

Fig. 4 Trial Trajectory. The user has to guide a pointer trough the trajectory using a hemisphere with the foot.

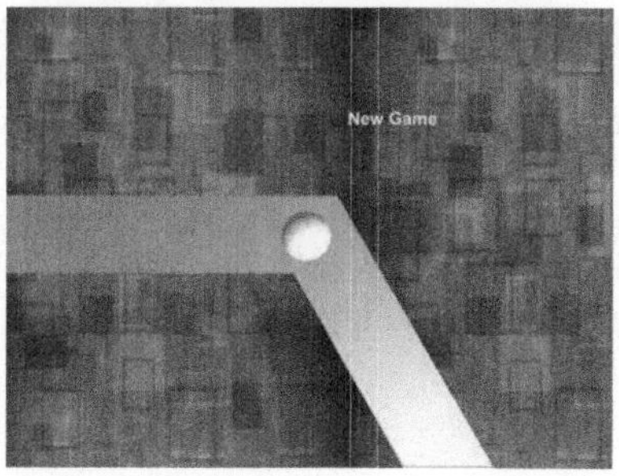

Fig. 5 Application View. The user has to guide a ball pointer, but only a section of the trajectory is visible (users view).

B. Control

To record the movements of the foot, a custom made joystick named podal hemisphere (Fig. 6), in which the foot (left or right) is placed while an electronic device, that is responsible for sensing, is placed on top (fig 3).

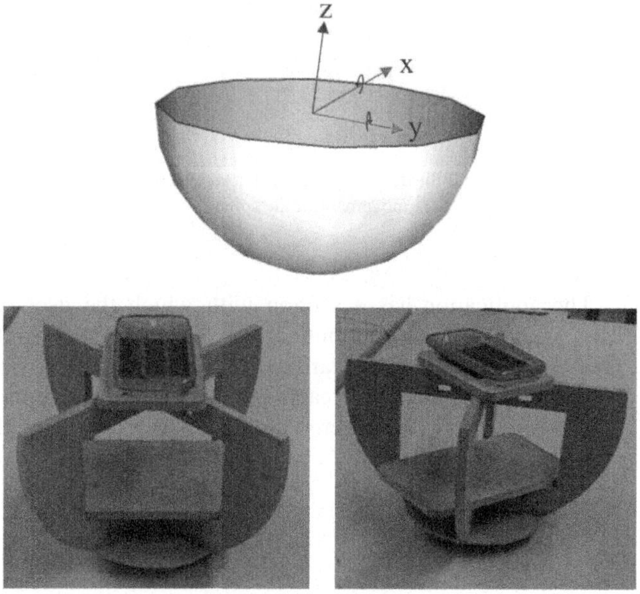

Fig. 6 Podal hemisphere

The hemisphere was manufactured in chipboard with a base of a polymer avoid slipping on the floor.

As the position is registered with goroscopios of a smart phone, the operation is designed to function as a joystick, it is not necessary to have a continuous movement to achieve progress in any direction. However depending on the angle of the hemisphere, this will cause a small displacement (Fig. 7) in the virtual sphere in the same direction at a constant speed, if the angle increases the speed will do it proportionally. Thus the rest position occurs when the angle of both axes is equal to 0 °, the mimina graphics speed is 0.5 units per second and maximum of 3 units. As for the range of angles for maximum speed is 7 ° for all movements of the ankle.

The device is coupled to the podal hemispherical with a sheath (Fig. 6) which keeps it completely stable, and at the same time connects to the computer .

Fig. 7 Podal hemispherical in use.

III. Trials

The trial protocol was provided to healthy volunteers, so it was discarded from the study people with hip or ankle injuries. All participants signed an informed consent.

Eleven subjects participated on the test trials. A cuestionary was applied to all participants. It include information on physical and intellectual activities, such as sports, playing a musical instrument, table games, video games, languages, etc. Frecuency and level of these activities also was checked with the participants.

Each participant was asked to perform ten trails for each foot and for each direction (clockwise and anticlockwise).

IV. Results

Figures 6 and 7 show the hemisphere developed, figures 8 and 9 show same graphical and numerical data from the system developed

At all times the opinion of volunteers was considered, so it was insisted that in case of discomfort or difficult to achieve control to the system, the test could be aborted, which did not happen over the application of 440 tests.

Figure 8 shows the first and last of the trajectories made by the volunteers with each foot. It is observed that it is possible to achieve a good performance with few variations. The total time to performa a trajectory did not show any difference between the first and last trajectory. Nonetheless the number of times the pointer go outside the path and the time remainded outside were decressing as the user gained experience with the hemisphere.

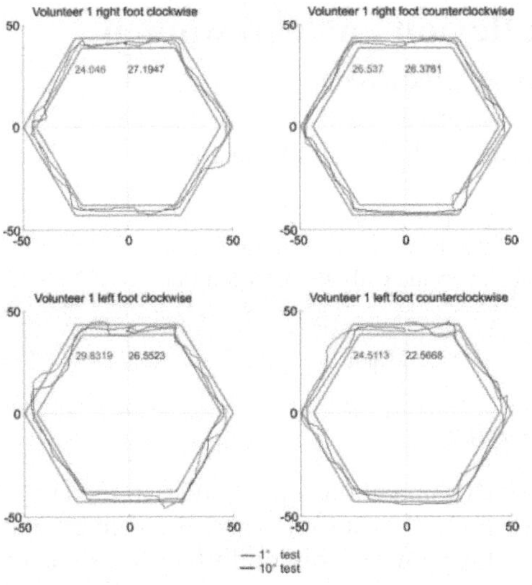

Fig. 8 Result of the first (red line) and last trials (blue line), for left and rigth foot and both directions.

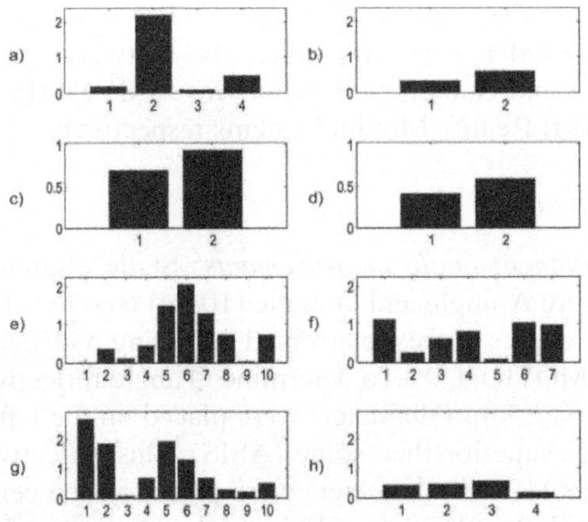

Fig. 9 Number of times the pointer was outside the trajectory vs the time it remained outside. a) and b) right foot clockwise, c) and d) right foot anti-clockwise., e) and f) left foot clockwise, g) and h) left foot anti-clockwise.

Conclusions

The results shown are consistent. Only one volunteers attempted to perform the trial at maximum speed. The average for this participant was 14 s per trial. This participant also showed the maximum numer of errors.

The system store valuable information and it was possible to match the fine movemnts of the ankle to the movements of the poiter in a trajectory. All stored information need to be exploited to match the requirements of different applictions such as rehabilitation intervention, training, communication with feet, etc.

Acknowledgment

DAIP-UGTO 1,146/2016 and PAPIME PE104715.

Conflict of Interest

The authors declare that they have no conflict of interest.

References

1. Lamm, B. M., Stasko, P. A., Gesheff, M. G., and Bhave, A. (2016). Normal foot and ankle radiographic angles, measurements, and reference points. Journal of Foot and Ankle Surgery, .
2. Lubetzky-Vilnai, A., McCoy, S. W., Price, R., and Ciol, M. A. (2015). Young adults largely depend on vision for postural control when standing on a bosu ball but not on foam. J Strength Cond Res, 29(10):2907–18.
3. Williams, G., Banky, M., and Olver, J. (2015). Ankle plantarflexor spasticity does not restrict the recovery of ankle plantarflexor strength or ankle power generation for push-off during walking following traumatic brain injury. J Head Trauma Rehabiltation, vol. 31, No. 4, E52-E58.
4. Natsakis, T., Burg, J., Dereymaeker, G., Jonkers, I., and Vander Sloten, J. (2016). Foot-ankle simulators: A tool to advance biomechanical understanding of a complex anatomical structure. Proc Inst Mech Eng H, 230(5):440–9.
5. Deschamps, K., Dingenen, B., Pans, F., Van Bavel, I., Matricali, G. A., and Staes, F. (2016). Effect of taping on foot kinematics in persons with chronic ankle instability. J Sci Med Sport, 19(7):541–6.
6. Cho, K. H., Jeon, Y., and Lee, H. (2016). Range of motion of the ankle according to pushing force, gender and knee position. Ann Rehabil Med, 40(2):271–8.
7. Gribble, P. A., Bleakley, C. M., Caulfield, B. M., Docherty, C. L., Fourchet, F., Fong, D. T.-P., Hertel, J., Hiller, C. E., Kaminski, T. W., McKeon, P. O., Refshauge, K. M., Verhagen, E. A., Vicenzino, B. T., Wikstrom, E. A., and Delahunt, E. (2016). Evidence review for the 2016 international ankle consortium consensus statement on the prevalence, impact and long-term consequences of lateral ankle sprains. Br J Sports Med.
8. Kim, T., Kim, E., and Cho, H. (2016). Effects of a 6-week neuromuscular rehabilitation program on ankle evertor strength and postural stability in elite women field hockey players with chronic ankle instability. J Sport Rehabil.
9. Unity Technologies at http://www.unity3d.com

Corresponding author:

Author: Juan Manuel Gómez-González
Institute: Facultad de ingeniería, UNAM,
Street: Circuito escolar S/N, Ciudad Universitaria
City: Mexico DF
Country: Mexico
Email: juan@dctrl.fi-b.unam.mx

Analysis of the alignment angles and flexion angle in women with patellofemoral pain syndrome

N.F. Diaz-Martinez[1], J.D. Pulgarin-Giraldo[2], L.E. Vinasco-Isaza[3] and W. Agredo[2]

[1] Faculty of Engineering, Universidad Libre Cali, Cali, Colombia
[2]G-Bio Research Group, Universidad Autonoma de Occidente, Cali, Colombia
[3]Civil and Industrial Department, Pontificia Universidad Javeriana Cali, Cali, Colombia

Abstract— **Patellofemoral pain syndrome (PFPS) is one of the most com-mon disorders in the knee that occurs with a higher incidence in women than men. Q-angle and A-angle as well as alignment and flexion angle of bending were considered in this study as kinematic variables. The hypothesis is that patients with changes in alignment variables are more likely to have pain in the patellofemoral joint. To verify this, we worked with patients with PFPS which was compared with control patients. No significant difference in angle A and the flexion angle of the knee were found, but a significant difference was found in Q-angle.**

Keywords— **Gait analysis, A-angle, Q-angle, knee flexion angle, one-way ANOVA, patellofemoral pain.**

I. Introduction

The gait analysis is a biomechanical study used as a fundamental tool in rehabilitation engineering [1], which facilitates diagnosis, treatment and prevention plans of pathologies associated with human movement. Besides, it gets a quantitative evaluation that allows comparing pathological patterns with normal patterns [2].

Among the pathologies of the lower limb, there are the generated by deformation of the bone structure. These are manifested by poor alignment of the lower extremities [3,4]. One of the diseases that have the greatest impact on the knee is the anterior knee pain known as patellofemoral pain syndrome (PFPS) [5,6].

This is why the gait analysis provides biomechanical variables to quantify, mainly alignment variables and its effects on the kinematic variables, temporo-spatial and kinetic gait patterns [7,8].

In this article, they are determined and analyzed static alignment variables such as *A* angle and *Q* angle, and knee flexion angle as a kinematic variable, using one-way ANOVA [9]. Besides, these variables show the modifications generated by PFPS in patients. Cases (PFPS) and control (no pain) women in the city of Santiago de Cali were considered.

II. Experimental setup

A. Participants

The study population were women in the city of Cali between 18 and 50 years, of which 12 women had PFPS (6 women with pain and discomfort in his right knee and 6 women with pain and discomfort in the left knee), and control group consisted of 14 women with no history of congenital or traumatic deformity of the lower limbs. All participants explicitly volunteered to participate in the study, approved by the Ethics Committee of the Universidad del Valle. The data were collected with 4 video cameras JVC and two force platforms Kistler 9286 AA type. The software for data processing and camera calibration was SIMI MOTION 3D ™ and SIMI Reality Motion Systems respectively.

B. Methods

Alignment angle measurements: Static alignment variables were A-angle and Q-angle [10-14] (see Fig. 1a and 1b). These angles were determined by using videography and SIMI MOTION ™. To determine Q angle reflective circular markers (25 mm diameter) were placed on the left and right anterior superior iliac spine (ASIS), plus reflective circular markers (12 mm diameter) were placed in the center of the patella (left and right) and tibial tubercle (left and right). To determine A angle reflective circular markers (12 mm diameter) were placed on the center and lower pole the patella (left and right) and on tibial tubercle.

Kinematic measurements: To properly quantify the kinematic variables are needed records of static and dynamic videography women who participated in the project. Temporary and fixed markers are listed in Table 1.

For measurements of the kinematic variables two registers, a static and a dynamic record registration were recorded.

I. Torres et al. (eds.), *VII Latin American Congress on Biomedical Engineering CLAIB 2016, Bucaramanga, Santander, Colombia, October 26th -28th, 2016*,
IFMBE Proceedings 60,
DOI: 10.1007/978-981-10-4086-3_167

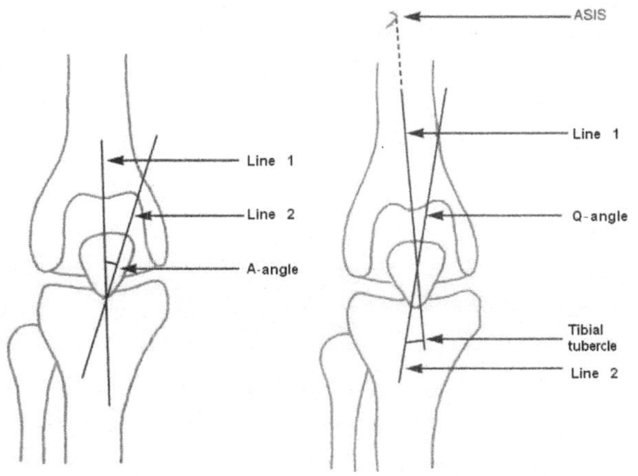

a) A-angle **b) Q-angle**

Fig. 1 Static alignment angles. a) A-angle. Line 1 connects the center of the patella to the lower pole of the patella. Line 2 connects tibial tubercle to the inferior pole of the patella. b) 2 Q-angle. Line 1 connects anterior superior iliac spine to the center of the patella. Line 2 connects tibial tubercle to the center of the patella (Adapted from [10]).

Table 1 Anatomical markers location (Adapted from [15])

Temporary markers
Distal phalanx of the second finger right
Distal phalanx of the second left finger
Right medial malleolus
Left medial malleolus
Right medial condyle
Left medial condyle
Right major trochanter
Left major trochanter
Fixed markers
Second right metatarsal
Second left metatarsal
Right heel
Left heel
Right lateral malleolus
Left lateral malleolus
Right middle third tibial
Left middle third tibial
Right lateral condyle
Left lateral condyle
Right anterior superior iliac spine
Left anterior superior iliac spine
L4 (called S2)

Women (patients) did not use shoes or socks. Also in the dynamic registration, patients conducted locomotion activities walking naturally. Static images are displayed records in Figure 2, while images of dynamic records are displayed in Figure 3.

Fig 2 Static record. Biomechanics laboratory. Universidad Autonoma de Occidente. Cali – Colombia.

Fig. 3 Dynamic record. Biomechanics laboratory. Universidad Autonoma de Occidente. Cali – Colombia.

Starting from the positions of the external markers placed on the skin, the information is processed in SIMI MOTION ™. The linear kinematic and angular patterns for each lower limb segment obtained are: displacements, velocities, and accelerations both linear and angular. In this way, the behavior of the angular displacement of the knee in flexion and extension along the gait cycle is obtained in the sagittal plane.

Kinematic data normalization: To facilitate comparison of results between the group with PFPS symptomatic (painful) and asymptomatic control group (no pain), the gait cycle was normalized to 100% (by turning the time percentage of gait cycle) [7].

The locomotion activities were measured in women walking naturally- This to avoid variations in the flexion angle of the knees and in the reaction force of both feet. These changes can be generated by an increase in walking speed of women (patients) [2].

III. STATISTICAL ANALYSIS AND RESULTS

Means comparison was performed for all variables: alignment angles (Q-angle and A-angle) and kinematics (knee flexion angle) variables. Descriptive statistics of all variables, graphical representation, and validation of assumptions hypothesis with one-way ANOVA was performed, as is necessary when several variables are studied in a group. The quantified data were obtained with a confidence level of 95%, as required by hypothesis testing in clinical cases [16].

The null hypothesis was that there is no variation between the patient population control and patient cases, considering normal distribution and equal variance. As an alternative hypothesis, we stated that there is variability between the characteristics of both populations. If the P-value is less than 0.05, it means that there are significant differences between patients in the control and case evaluated with a confidence level of 95% feature.

A. Alignment variables results

The ANOVA did not reveal the presence of significant differences in static angular alignment variables of the right limbs. But significant differences in the static angular alignment variables of the left limbs were observed (Table 2). The mean Q-angle and A-angle on the right limbs were slightly higher for control patients with respect to case patients. But the Q-angle on the left limbs revealed a significant difference. Its angular average was higher for control patients regarding the case patients (13.2° vs. 24.2°). Also, as in the right limbs, the average A-angle in left limbs was higher for control patients regarding the case patients.

Table 2 Mean values (±standard deviation) of static angular measurements

Static angular measurements	Control patients	Case patients	P-value
Right limbs A-angle	16.6 (11.9)	11.6 (12.1)	0.41
Right limbs Q-angle	10.7 (7.2)	8.7 (5.4)	0.54
Left limbs A-angle	13.7 (13.2)	4.8 (5.7)	0.13
Left limbs Q-angle	24.2 (8.9)	13.2 (8.4)	0.020

B. Kinematic variables results

ANOVA did not reveal the presence of significant differences between the control patients regarding case patients with pain in the right knee or pain in the left knee (with PFPS) for the angular displacement of the knee joint (P value> 0.05).

However, data from the flexion angle of the right knee are lower in case patients (who have PFPS) with respect to control patients in the support period of the gait cycle and can be viewed in Figure 4.

Data from flexion angle of the left knee are higher in case patients (who have PFPS) with respect to the control patients in the support period of the gait cycle as can be visualized in Figure 5.

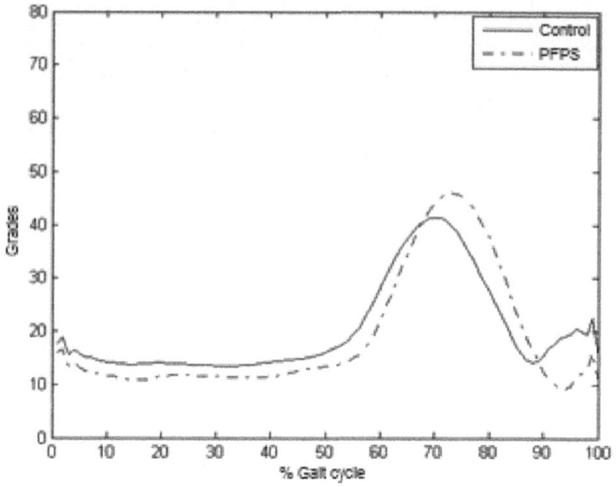

Fig. 4 Mean flexion angle from right knee during gait cycle.

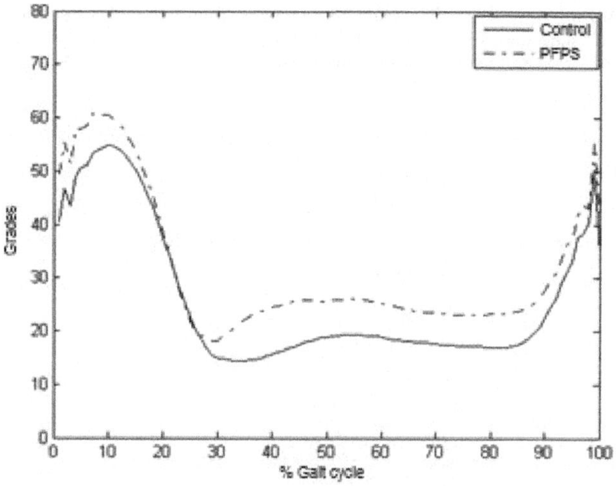

Fig. 5 Mean flexion angle from left knee during gait cycle.

The mean flexion angle of the right knee was slightly lower for the case patients than control patients during important moments of gait cycle support period (GC). These moments were 10% G.C. and 20% G.C. [3] and can be viewed in Table 3.

Table 3 Mean values (±standard deviation) of the behavior of flexion angle on the right knee

Knee flexion angle	Control patients (n=14)	Case patients (n=6)	P-value
10 % G.C. (°)	14.1 (5.99)	11.5 (4.38)	0.35
20 % G.C. (°)	14.0 (4.81)	11.7 (3.40)	0.29

The mean flexion angle in the left knee was higher for case patients with respect to the control patients during the support period 50% G.C. and 60% G.C. This can be visualized in Table 4.

Table 4 Mean values (±standard deviation) of the behavior of flexion angle on the left knee

Knee flexion angle	Control patients (n=14)	Case patients (n=6)	P-value
50 % G.C. (°)	19.1 (9.68)	25.8 (11.7)	0.20
60 % G.C. (°)	18.7 (10.71)	24.8 (12.28)	0.27

IV. DISCUSSION AND CONCLUSIONS

A. Alignment angles variables

Q-angle. Right lower limbs. In this study, all case patients with pain in the right knee had a Q-angle below average (Q-angle ≤20°), indicating no significant change in this variable due to PFPS. Similarly, the Q-angle does not show significant differences between case and control patients. Although no statistically significant differences are presented, it was observed that the average Q-angle was slightly more pronounced in control patients.

Q-angle. Left lower limbs. Q-angle presented significant differences between case patients with pain in the left knee and control patients. The average Q-angle in the control patients was higher than those who frequently been reported in the literature, while the average Q-angle in case patients was also lower than that reported [14,17]. In this study, 78.6% of control patients had a high Q-angle without presenting PFPS therefore showed a trend of higher Q-angles in the study population.

A-angle. Right and left lower limbs. All case patients with knee pain, either left or right, had an A-angle below average (angle A ≤35 °), indicating no significant changes exists in this variable alignment due to PFPS.

Although no statistically significant differences occur, the mean A-angles was slightly more pronounced in control patients than in case patients.

B. Kinematic variables

Behavior patterns for flexion angle on the right knee between case patients and control patients revealed no statistically significant differences in both phases of initial contact and loading response as the rest of the gait cycle, indicating in our study that PFPS does not alter this variable.

Similarly flexion angle in the left knee revealed no statistically significant differences. However, it was observed that case patients with pain in left knee revealed a greater knee flexion than control patients in the left lower limbs, although there is not statistically significance.

ACKNOWLEDGMENT

This work was supported by the project *Estudio biomecánico de mujeres colombianas para el análisis del síndrome de dolor patelofemoral* code 09INTER-108 funded by Dirección de Investigaciones y Desarrollo Tecnológico from Universidad Autonoma de Occidente.

CONFLICT OF INTEREST

The authors declare that they have no conflict of interest.

REFERENCES

1. Winter D.A (2009) Biomechanics and motor control of human movement. 4th Edition. John Wiley & Sons, New Jersey
2. Kirtley C (2006) Clinical gait analysis theory and practice. Elsevier Churchill Livingstone, Edinburgh.
3. Perry J (1992) Gait Analysis Normal and Pathological function.. SLACK Incorporated, Thorofare, NJ
4. Levangie P, Norkin, C (2005) Joint Structure and Function: A comprehensive analysis. Fourth edition. F. A. Davis Company, Philadelphia.
5. Nuttal C, Winters B (2015). Understanding Anterior Knee Pain: Patellofemoral Pain Syndrome. The Journal for Nurse Practitioners – JNP 11, 1032 – 1035.
6. Rothermich M, Glaviano N, Li J, Hart J (2015) Patellofemoral Pain Epidemiology, Pathophysiology, and Treatment Options. Clin Sports Med 34, 313–327.
7. Nadeau S, Gravel D, Hébert L, Bertrand Arsenault A, Lepage Y (1997). Gait study with patellofemoral pain syndrome. Gait Posture 5, 21-27.
8. Levinger P, Gilleard W (2005) The heel strike transient during walking in subjects with patellofemoral pain syndrome. Physical Therapy in Sport 6, 83-88.
9. Vinasco-Isaza L.E (2012) Estadística Descriptiva con Minitab versión 15,0. Editorial Pontificia Universidad Javeriana, Cali, Colombia.
10. Selfe J, Gillard D, Marshall S (1996) Patello-femoral pain: is the A-angle a clinically useful measurement? Physiotherapy 82, 324-330.
11. Chaitow L, Walker J (2011) Clinical Application of Neuromuscular Techniques, Volume 2: The Lower Body. 2nd Edition. Churchill Livingstone, Edinburgh.
12. Emami M, Ghahramani M, Abdinejad F, Namazi H (2007) Q-angle: an invaluable parameter for evaluation of anterior knee pain. Arch Iranian Med 10, 24-26
13. Freedman B, Brindle T, Sheehan F (2014) Re-evaluating the functional implications of the Q-angle and its relationship to in-vivo patellofemoral kinematics. Clinical Biomechanics 29, 1139–1145.
14. Green S (2005) Patellofemoral syndrome. Journal of bodywork and movement therapies 9, 16-26.
15. Simi Motion™ (2015) 3D Inverse Dynamics manual at http://www.simi.com/fileadmin/user_upload/Dokumente/Downloads/Simi_Clinical_and_Sports_Catalogue.pdf
16. Dawson B, Trapp R, Arias G, De la Garza V (2005) Bioestadística médica. Cuarta edición. Editorial El Manual Moderno, México, D. F.
17. Behnke R (2012) Kinetic Anatomy, 3rd Edition. Human Kinetics, Champaign.

Managing Heterogeneous Medical Data: Learning from Experiences in Telemedicine

J.C. Vanegas-Serna[1], J.J. Perez[1] and H. Andrade-Caicedo[1]

[1] Universidad Pontificia Bolivariana/Centro de Bioingeniería, Facultad de Ingeniería Eléctrica y Electrónica, Medellín, Colombia

Abstract— **The Internet of Things is one of the most promising technical revolutions of our age. The idea of integrating sensors to common objects and linking to the Internet offers a wide range of new possibilities. However, establishing standard communication protocols, storing heterogeneous data and guaranteeing security to the transmitted information need to be studied. As a support, web services have been released to establish the way how web applications interchange information by using XML, JSON or other plain formats. In this paper, we describe a general Back-End solution to store heterogeneous data collected from different sources. We used the stack named "MEAN" to program web services, and JSON as the data model to interchange information. On this study, we show the advantage of using JSON and a NoSQL database to integrate different solutions developed in projects that involve the Internet of Things philosophy.**

Keywords— **Health Informatics, Heterogeneous Medical Data, Telemedicine, Internet of Medical Things.**

I Introduction

The Internet has changed our daily life and has permitted interactions worldwide in ways that we do not even know. The concept of smart things represents a new manner to think about ourselves and our environment. The Internet of Things (IoT) comes out as a formalization of the way how the "Things" are connected to the conventional Web 2.0 [1]. Its importance relies on the impact that it generates on the experience of users in their daily life and the activities that can be influenced by it, for instance: smart cities, smart environments, security, healthcare, etc. [1, 2].

Additionally, the rapid spread of smartphones has opened new opportunities where continuous and remote connectivity is required. Health services based on this technology have been benefited through the implementation of strategies that improve processes that involve education, patient care, and preventive actions. All these strategies are expected to guarantee access to specialized clinical services to patients living in rural areas [3]. Medicine is not only concerned about using smartphones for healthcare (mHealth), but also regards the use of biomedical devices connected to the Internet (wearables, Body Area Networks, and clinical devices). Such specialization can be considered as a branch of IoT: *The Internet of Medical Things* (IoMT). It seeks for integrating medical devices with healthcare systems. It is also a synonymous for telemedicine, a well-known field of study where the medicine and the Information and Communications Technology meets to manage remote clinical data [4].

One of the technical challenges of implementing real telemedicine applications is the management of heterogeneity of the medical data [5]: each telemedicine-based service has its own data model, for both transmission (local or remote) and storage. There are many protocols used in data transmission in the context of the IoT: DDS, MQTT, CoAP, REST, etc. [6, 7]. Selecting one of these protocols depends on the understanding of the requirements of each data frame and the architecture of each solution [6].

The most common solution among clinical institutions for the storage of clinical and medical data are based on relational databases. Nevertheless, many fields on tables remain empty, since they are not needed in all cases (e.g. not all the medical observations need the electrocardiogram). Besides, relational databases have shown to be inefficient and have poor performance under these situations [8]. In this paper, we propose an architecture that involves devices, gateways, and servers using JSON notation as the data model. We report the results of using this methodology over three projects of telemedicine in our research group.

II Related work

The emergence of low-cost platforms like Raspberry, BeagleBone, and Arduino favors the development of projects under the IoT concept. Recent works have focused on developing innovative tools that improve the way how healthcare is exerted. For instance, the smartphones apps are invaluable tools for healthcare professionals. However, in some cases they still remain reluctant to adopt them for clinical use [9].

Population aging is becoming a major problem for governments. It is reflected in the increasing number of people suffering from chronic diseases, disability, and dependency

I. Torres et al. (eds.), *VII Latin American Congress on Biomedical Engineering CLAIB 2016, Bucaramanga, Santander, Colombia, October 26th -28th, 2016*,
IFMBE Proceedings 60,
DOI: 10.1007/978-981-10-4086-3_168

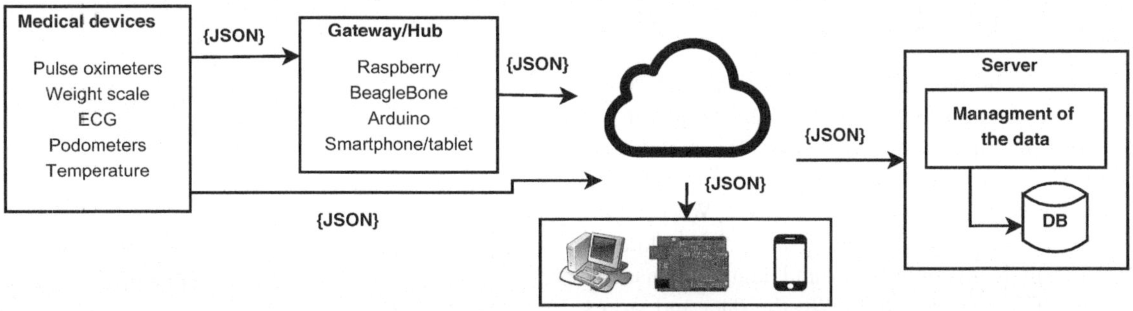

Fig. 1: General architecture of the solution

[10]. Literature evidences different studies where the development of IoMT applications is reported, and web platforms designed to mitigate this problem [11, 12, 10, 13, 14]. Other efforts address the treatment of addictive conditions [15], ambulance monitoring [16], chronic diseases [11, 12, 17, 14], and support to healthcare professionals [18]. Many of these works make use of the smartphones as a gateway/hub to collect the medical/clinical data [11, 15, 12, 10, 17, 14] and direct it to the cloud using web services based on XML [12, 10] and JSON [11, 12].

For the storage of the data, due to the dynamic and heterogeneity of clinical data, an attractive solution are the NoSQL databases [8]. These databases are focused on dealing with high data volumes [19], with a good grade of scalability/flexibility [8] and solving the traditional performances issues of relational databases with heterogeneous data [20].

III System architecture

For telemedicine projects in our research group, the hypothesis was: there is a scalable solution that is able to offer significative performance no matter what platform (Raspberry, BeagleBone, Arduino, smartphone, etc.) or communication technology (WiFi, Bluetooth, ZigBee, USB, UART, etc.) is used. The data model was kept simple as possible for its handling at the server side. We selected JSON as the data model for all levels of communication due to any type of data can be stored following a pre-established model. One of the advantages of this strategy is that developers do not need to be concerned about the communication protocol, which provides them time to concentrate on the main objective of their investigation. All the data is received by a REST service (HTTP) or REST-like service over WS and will be stored in *MongoDB*. A diagram of the architecture is presented in Fig. 1.

A Generic telemedicine platform

The objective of this project is to acquire medical-clinical data from patients through different communication technologies. We used a BeagleBone Black board as a Hub. It was connected by different communication technologies, WiFi, Bluetooth Low Energy (BLE), XBee S2C and UART to read the data acquired from patients. Then, data is sent to the servers using JSON and HTTP to a specific REST service for this project.

B Wireless Heart Arrhythmia Monitoring - WHAM

WHAM is a system based on the Arduino DUE platform, it uses a GSM/GPRS module and Bluetooth V 2.0 module. Its purpose is to monitor ECG signal, in order to detect arrhythmias. When an event happens, it stores the signal before and after the arrhythmia period on an SD card and sends the data over WS to a server. The data collected can be examined by a cardiologist using a web-based application to determinate the current condition of the patient. The device is able to connect to smartphones by Bluetooth through an Android app. It allows the professional to check the correctness of the electrode placement. All the communication process (device to smartphone and device to server) uses JSON as the data model.

C Patients monitoring in ambulances

A real time application to monitor a patient in an ambulance was developed. The hub on the ambulance acquires data from the different medical devices, creates a WS connection to the server and redirects the data in another WS to a web page where a clinician could help a paramedic to stabilize the patient. The platform manages data from ECG (3-lead), SpO_2 and mechanical ventilation. An Android app is available to see some variables and set alarms for them.

IV RESULTS

We developed a solution at the server side (Back-End) that receives and stores data in JSON notation, independently of the platform or communication technology. We are able to use the same system for different projects, or create a copy and deploy a unique Back-End if it is needed. We used an end-to-end MEAN stack (*MongoDB*, *Express.js*, *Angular.js* and *Node.js*) to handle all connections and data transmission. There exists a generic service that takes the data and stores it in the database. Each project can create a specific REST service for additional processing. In Table 1 we present the technologies used on each project.

A Generic telemedicine platform

This project has a BeagleBone Black with an electronic system designed by our research group. It has different wired and wireless communication technologies. Data is sent to a hub in a JSON notation, and it is automatically transmitted to the server for storage using Ethernet. The data stored in the database can be retrieved using REST services. We created this project to use the hub as a telemedicine platform. However, it can be integrated with any telemetry project.

B Wireless Heart Arrhythmia Monitoring - WHAM

WHAM Technology includes three elements: *WHAM device* to acquire the ECG data, *WHAM app* to show ECG signal locally, and *WHAM Web* to display events from ECG trace.

WHAM device uses an Arduino DUE, Bluetooth V 2.0 and GSM/GPRS. All data sent from both Bluetooth and GSM/GPRS modules is in JSON format. The Bluetooth communication is for local use through the WHAM app. The device analyzes a *D-II* recorded ECG signal and distinguishes eight types of arrhythmias. When a cardiac event is detected, the device creates a connection with the server and sends the ECG from a time before and after the event, allowing the clinicians to monitor the trace.

The WHAM app was developed using *PhoneGap* for Android OS. This app allows the clinicians check if the electrodes are placed properly and configure the parameters to the WHAM device.

The web page was developed using the *MEAN* stack. The web page allows the clinicians to see the events reported by the devices and configure the parameters if necessary.

C Patients monitoring in ambulances

The device on the ambulance acquires the information given by the medical equipment, which is then sent through

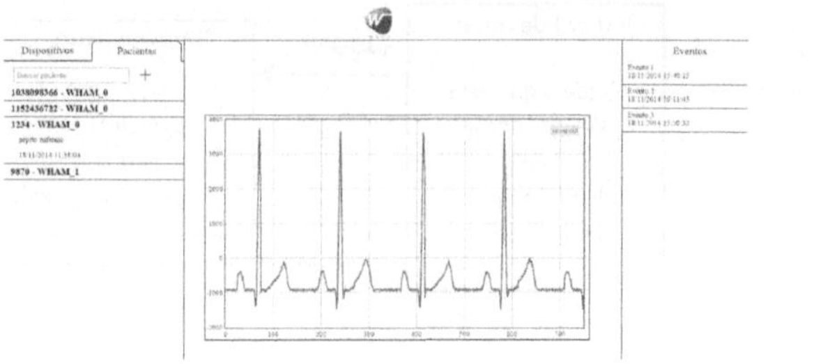

Fig. 2: WHAM Web

a 4G LTE connection using WS. It can acquire four kinds of data: ECG, SpO_2, Non-Invasive Blood Pressure (NIBP) and mechanical ventilation.

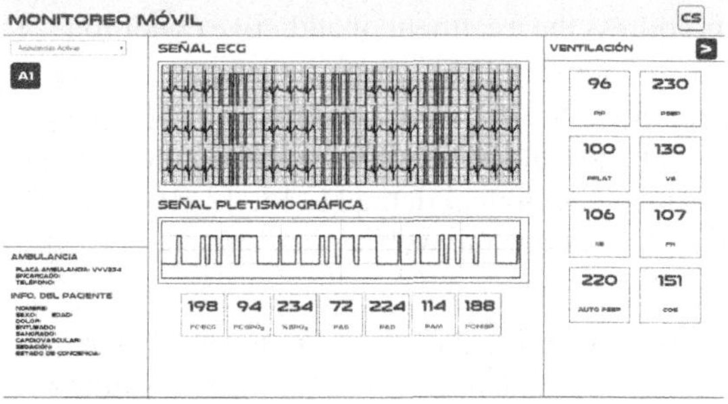

Fig. 3: Web app for monitoring patients in ambulances

A web page was developed using the *MEAN* stack, where the clinicians at the hospital can see in real time the vital signs of the patient (Fig. 3). It also can search for old registers taken by the device on each ambulance.

The mobile app was developed with *PhoneGap*. The clinicians can use it to see some parameters of the patient on the ambulance, call the paramedic if it is necessary to assign him a specific task or set automatic alarms. If some of the parameters are out of boundaries, the clinician will receive a notification on the smartphone where the mobile app is installed.

V CONCLUSION

JSON is a data model that facilitates the handling of the data at the server side. This feature allows it to manage heterogeneous data coming from the telemetry processes. JSON and *MongoDB* allow saving the information directly into the database, both of them use the same object model (JSON), thereby a map between the data and the object to save is not necessary. However, the implementation of JSON in embedded systems can be difficult since object-oriented programming is not available. Platforms using Linux embedded O.S. is recommended. They do not have this limitation considering

Table 1: Technologies used on each project

Project	Communication technology	Board	Server communication	Web application	Mobile application
Generic telemedicine platform	Bluetooth Low Energy, WiFi, XBee, Serial, Ethernet	BeagleBone Black	HTTP	-	-
WHAM	Bluetooth V 2.0 GSM/GPRS	Arduino DUE	HTTP/WebSocket	MEAN stack	PhoneGap
Patients monitoring in ambulances	Serial 4G LTE	-	HTTP/WebSocket	MEAN stack	PhoneGap

that Linux provides the necessary tools for the development of complex IoT and IoMT applications.

We developed a Back-End solution that can be integrated with a system based on the JSON data model. We programmed the web server and the communication with the database through the MEAN stack. One of the main advantages of using this strategy is that it can be implemented as a general instance of the Back-End or copied and deployed for a specific project. In both cases, specific REST services should be programmed. However, this kind of solutions allows the user to integrate and prototype new IoT projects faster since it is not necessary to create a Back-End system from the beginning. JSON and *MongoDB* give the opportunity to create a scalable and flexible solution due to its "schemaless" nature.

CONFLICT OF INTEREST

The authors declare that they have no conflict of interest.

REFERENCES

1. Atzori Luigi, Iera Antonio, Morabito Giacomo. The Internet of Things: A survey *Computer Networks.* 2010;54:2787–2805.
2. Botta Alessio, De Donato Walter, Persico Valerio, Pescapé Antonio. Integration of Cloud computing and Internet of Things: A survey *Future Generation Computer Systems.* 2016;56:684–700.
3. Ozdalga Errol, Ozdalga Ark, Ahuja Neera. The Smartphone in Medicine: A Review of Current and Potential Use Among Physicians and Students *Journal of Medical Internet Research.* 2012;14:e128.
4. Fong Bernard, Fong a. C. M., Li C. K.. *Telemedicine Technologies: Information Technologies in Medicine and Telehealth.* Chichester: John Wiley & Sons, Ltd 2010.
5. Wasan Siri Krishan, Bhatnagar Vasudha, Kaur Harleen. The impact of data mining techniques on medical diagnostics *Data Science Journal.* 2006;5:119–126.
6. Foster Andrew (PrismTech). Messaging Technologies for the Industrial Internet and the Internet of Things Tech. Rep. MarchPrismTech Corp.Woburn 2014.
7. Talaminos-Barroso Alejandro, Estudillo-Valderrama Miguel A., Roa Laura M., Reina-Tosina Javier, Ortega-Ruiz Francisco. A Machine-to-Machine protocol benchmark for eHealth applications - Use case: Respiratory rehabilitation *Computer Methods and Programs in Biomedicine.* 2016;129:1–11.
8. Lee Ken Ka Yin, Tang Wai Choi, Choi Kup Sze. Alternatives to relational database: Comparison of NoSQL and XML approaches for clinical data storage *Computer Methods and Programs in Biomedicine.* 2013;110:99–109.
9. Ventola C Lee. Mobile devices and apps for health care professionals: uses and benefits. *P & T : a peer-reviewed journal for formulary management.* 2014;39:356–64.
10. García-Sánchez Pablo, González Jesús, Mora Antonio M., Prieto Alberto. Deploying intelligent e-health services in a mobile gateway *Expert Systems with Applications.* 2013;40:1231–1239.
11. Serhani Mohamed Adel, Menshawy Mohamed El, Benharref Abdelghani. SME2EM: Smart mobile end-to-end monitoring architecture for life-long diseases *Computers in Biology and Medicine.* 2016;68:137–154.
12. Skałkowski Kornel, Zieliński Krzysztof. Applying formalized rules for treatment procedures to data delivered by personal medical devices *Journal of Biomedical Informatics.* 2013;46:530–540.
13. Gill Asif Qumer, Phennel Nathan, Lane Dean, Phung Vinh Loc. IoT-enabled emergency information supply chain architecture for elderly people: The Australian context *Information Systems.* 2016;58:75–86.
14. Pires Pedro, Mendes Luís, Mendes Jorge, Rodrigues Rúben, Pereira António. Integrated e-Healthcare System for Elderly Support *Cognitive Computation.* 2016;8:368–384.
15. Capon Hannah, Hall Wayne, Fry Craig, Carter Adrian. Realising the technological promise of smartphones in addiction research and treatment: An ethical review *International Journal of Drug Policy.* 2016.
16. Almadani Basem, Bin-Yahya Manaf, Shakshuki Elhadi M.. E-AMBULANCE: Real-time integration platform for heterogeneous medical telemetry system *Procedia Computer Science.* 2015;63:400–407.
17. Gómez Jorge, Oviedo Byron, Zhuma Emilio. Patient Monitoring System Based on Internet of Things *Procedia Computer Science.* 2016;83:90–97.
18. Ogescu C., Plaisanu C., Bistriceanu D.. Web based platform for management of heterogeneous medical data in *2008 IEEE International Conference on Automation, Quality and Testing, Robotics*;3:257–260IEEE 2008.
19. Arefin Mohammad Shamsul, Hossain Khondoker Nazibul, Morimoto Yasuhiko. Efficiency of NoSQL Databases under a Moderate Application Load :213–227 2015.
20. Dharmasiri H. M L, Goonetillake M. D J S. A federated approach on heterogeneous NoSQL data stores in *2013 International Conference on Advances in ICT for Emerging Regions (ICTer)*no. December:234–239IEEE 2013.

Author: Henry H. Andrade
Institute: Universidad Pontificia Bolivariana
Street: Circular 1 No. 73-86, Bloque 22C
City: Medellin
Country: Colombia
Email: henry.andrade@upb.edu.co

Fibrosis evaluation of animal liver tissue by thermal conduction

N. Alemán-García[1,2], A. Pérez-García[3], J. Sánchez-Melecio[3], F. Silva-Aguilera[3], E. Gutiérrez-Herrera[4], M.R. Ortiz-Posadas[1], J. Hernández-Ruiz[3], D. Kershenobich [5] and C. Sánchez-Pérez[2].

[1] Universidad Autónoma Metropolitana Iztapalapa. Electrical Engineering Department, Mexico City, Mexico
[2] Universidad Nacional Autónoma de México, UNAM, CCADET. Mexico City, Mexico
[3] Unidad de Medicina Experimental, HIPAM, UNAM, Hospital General de México "Dr. Eduardo Liceaga"
[4]Unidad de Investigación y Desarrollo Tecnológico, CCADET-UNAM-Hospital General de México "Dr. Eduardo Liceaga"
[5]Dirección General, Instituto Nacional de Ciencias Médicas y Nutrición "Salvador Zubirán"

Abstract— **Hepatic fibrosis is a condition that alters normal structure and composition of the liver. Long term, it leads to complications such as cirrhosis; an advanced stage of chronic liver diseases, which in turn can drive to hepatocarcinoma (liver cancer). Hence, knowing the fibrosis degree is an important clinical data that can help in the prognosis and treatment of liver diseases. This work is focused in evaluating the thermal performance of 16 samples of rat liver tissue graded in different fibrosis degrees, as an alternate method to know how these properties change as the disease progresses. A heat sensor based on photo-thermal deflection was used. The thermal response was compared to the fibrosis degree in the METAVIR score, as evaluated by a group of experimented observers.**

Keywords— **Liver fibrosis, Thermal behavior, METAVIR score.**

I. Introduction

Hepatic fibrosis is a progressive change in the density and composition of the extracellular matrix in the liver parenchyma, which if persists leads to cirrhosis. Environmental and genetic factors have been described to influence its progress, as well as different methods to estimate it [1]. Cirrhosis is a condition where the structure and function of the liver are severely compromised [2]. Hepatic cirrhosis is a disorder with high mortality and morbidity rates. In Mexico, the reported cases of liver cirrhosis increased 7.8% from 2005 to 2006; morbidity is also reported to have increased 6%, hence becoming the third cause of death in the country. Cirrhosis is the second cause of death in the age group of 15-64 years, and the sixth in persons over 65 years [3]. There for the importance of developing auxiliary tools for evaluation and timely and precise progress of liver fibrosis in patient diagnosis.

Characterization of hepatic tissue by its thermal properties is a useful method to study fibrosis, since it alters normal liver structure and function [4]. Thermal properties of biological tissues evaluated by physical parameters can be associated to structural and compositional alterations caused by degenerative processes. Thus, assessing the relevance of the physical properties that define an organ's condition is very helpful to enhance our knowledge on some pathologies and to develop new resources to analyze them. Regarding hepatic fibrosis, recognition of the advance degree is very important to properly establish the prognosis and the therapeutic strategy. Due to the aforementioned, the focus of this work was to evaluate the thermal response of liver tissue from rats with different degrees of fibrosis, in order to better understand how hepatic tissue becomes progressively modified by this condition, so that it can contribute to a more precise diagnostic of the phases of liver diseases.

II. Methodology

A. Animal model of fibrosis induction

20 male Wistar rats, 6 weeks of age, 190-220 g. were used. 5 groups of 4 rats each were formed randomly. Fibrosis was induced with a 33% carbon tetrachloride (CCl_4) solution in olive oil at an initial dose of 250µl, given intraperitoneally twice a week [5]. After the rats gained 20% of their initial weight, the dose was adjusted weekly to 1 µl/g. The groups were sacrificed after different periods of induction as described in Table 1.

B. Sampling

After sacrifice, the livers were collected whole and put in 3.8% formaline. Then, a 1 cm wide, 3 cm long and 0.1 to 0.2 cm thick cut was taken from the right medial lobe. This cut was used to perform the heat conduction measurements. Slices of the obtained tissue were histochemically processed with Masson Trichrome stain. The slides were photographed using an optical microscope at 4x. 10 photographs were made of different zones of each slide, meaning that 40 images were made per rat, and that 160 images were analyzed. The analysis was not performed in the control group, since no CCl_4 treatment was given, and therefore, no histopathological findings were expected.

I. Torres et al. (eds.), *VII Latin American Congress on Biomedical Engineering CLAIB 2016, Bucaramanga, Santander, Colombia, October 26th -28th, 2016*,
IFMBE Proceedings 60,
DOI: 10.1007/978-981-10-4086-3_169

Table 1 Groups of the fibrosis induction model

Group	n	Time of treatment
0 (control)	4	No treatment
T1	4	4 weeks
T2	4	8 weeks
T3	4	12 weeks
T4	4	18 weeks

C. Assessment of fibrosis using the METAVIR score

Each image underwent 3 fibrosis assessments with the METAVIR score [6] which defines five degrees: F0, no fibrosis, F1, portal fibrosis (mild); F2, portal fibrosis with few septa (moderate); F3, fibrosis with numerous septa forming incomplete nodules, (severe), F4, cirrhosis with complete regeneration nodules. This score yields an ordinal qualitative variable, so the mean and the standard deviation (σ) of the assessments was calculated, progressively ranking the values F1=1 to F4=4 in order to have an approach on how scattered the diagnostic could be. The assessments were blinded and independently made by three experienced observers ($\kappa > 0.8$) from the Laboratory of Liver, Pancreas and Motility (HIPAM), of the Experimental Medicine Unit at the General Hospital of Mexico. The final evaluation for each image was obtained by the mode of the three assessments. The conclusive evaluation reported for each rat was made by estimating the mode of the 10 assessed images of each rat's lobe.

D. Assessment of heat conduction

To evaluate the heat conduction of the liver samples, a photo-thermal deflection based heat flow sensor of the Optical Sensors Laboratory of the CCADET-UNAM was used [7]. Photo-thermal effect is achieved when a light beam is deflected from its trajectory to a certain angle by a temperature gradient. An indirect evaluation of the heat flow is obtained from the decrease of the measured power (P_{out}) resulting from the angular deflection of the laser beam. It is important to say that this sensor has been previously used for experimental calibration tests with different biological tissues. [7].

Once the heat conduction responses of the samples were obtained, their changes were analyzed for the different fibrosis stages, comparing them by the time of fibrosis induction and the reported degree of the METAVIR score.

III. Results

A. Assessment of fibrosis using the METAVIR score

The results of the fibrosis assessment by three different observers show that from of 160 images, in 70% the three observers concurred in their grading, whereas in 28.7%, one of the experts dissented from the other two, and only in 1.3% of the cases, none of the observers agreed, and a fourth opinion was needed to grade the image.

The grouping and weight at sacrifice of the sampled individuals, as well as the fibrosis assessment by the METAVIR score are shown in Table 2. Noticeably, rat B of the T1 group had the highest standard deviation (σ) among all the samples, yet it was graded F3 with the METAVIR score, despite being inducted for cirrhosis for only 4 weeks, so that it was expected to be F1. The average standard deviation $\bar{\sigma}$ for the groups shows that T4 had a more homogeneous development of fibrosis than the rest. This was obtained because two of the individuals of the group had a σ equal to zero; this is to say that all their images were graded as cirrhosis. Hence, the progression of fibrosis in these individuals was more attached to the expected prognosis at the beginning of the induction process.

Table 2. Grouping, weight at sacrifice and METAVIR assessment of the rats in the heat flow experimental animal model

			METAVIR score			
Group	Individual	Weight [g]	Mode	Mean	σ	$\bar{\sigma}$
	A	311	F1	1.1	0.32	
T1	B	321	F3	2.5	0.71	0.55
	C	335	F1	1.3	0.48	
	D	345	F1	1.3	0.68	
	A	311	F2	1.7	0.48	
T2	B	287	F2	1.7	0.48	0.49
	C	315	F1	1.2	0.42	
	D	340	F2	2.1	0.57	
	A	300	F2	2	0.47	
T3	B	352	F4	3.9	0.32	0.44
	C	378	F3	3.3	0.32	
	D	342	F4	3.7	0.48	
	A	372	F4	4.0	0	
T4	B	382	F4	4.0	0	0.20
	C	386	F3	2.9	0.32	
	D	384	F4	3.7	0.48	

On the other hand, it was observed that in the induction time 4, T4, individuals with the least σ were those with lower weight in this group. However, this is not overwhelming, due to the fact that in other groups we found no relation between weight and the METAVIR evaluation.

B. Assessment of heat conduction

The measurements of thermal conduction were grouped according to the obtained METAVIR score (Table 2). Four sets of graphics were made, according to the fibrosis grade observed. Then, the curves of the previously described groups were averaged. As observed in Fig.1, the curves do not define a predominant state of fibrosis based on its heat conducting features; thus, this parameter is not determinant to distinguish the phases of the disorder. Likewise, the difference of values between extreme phases was calculated; F1 for mild fibrosis and F4 for cirrhosis, resulting in 0.5%, which is non-significant to discriminate between phases of the disease.

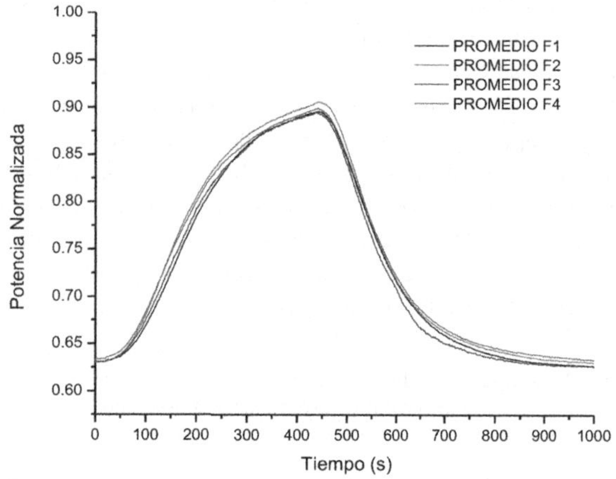

Fig. 1 Averange of the sensor response to heat conduction for all the samples at each METAVIR grading.

On the other hand, the individuals with minor standard deviations in the METAVIR assessment were selected using the previously described measures of dispersion (Table 1). Thus, the selected samples were T1A for the F1 phase, T3A for F2, T3C for F3 and T4B for F4 (Figure 2). Same as in the previous case, an analysis of the maximums revealed a difference of 3% between the curves of phase 1 and 4. Moreover, when grouping the individuals with minor dispersion, it can be seen that the highest curve corresponds to the measurement of an F4 cut; followed by one in F3, then in F2 and finally in F1, meaning that the more the fibrosis increases, the more normalized power increments, as a consequence of a greater heat flow. It is very important to point out that the method for normalizing the measured power (P_{out}) takes into account the power monitoring of the laser source (P_{in}) in order to discard possible variations in the excitation signal. Due to this, the normalized power in the shown curves increases with the heat flow. In the equation (1), the term P_{out} in the numerator appears as a remainder of P_{in}. If the out power decrements proportionally to the heat flow, the numerator increases. This explains the response of the normalized curves, because the greater difference between P_{in} and P_{out}, the more decrement of power and the value of normalized power. This being, the hepatic structure and composition resulting of fibrosis progression appears an increment in the heat flow. In this regard, it is noticeable that the appearance order of the curves follows the order of the METAVIR grading.

$$P_{nor} = \frac{P_{in}-P_{out}}{P_{in}+P_{out}} \quad (1)$$

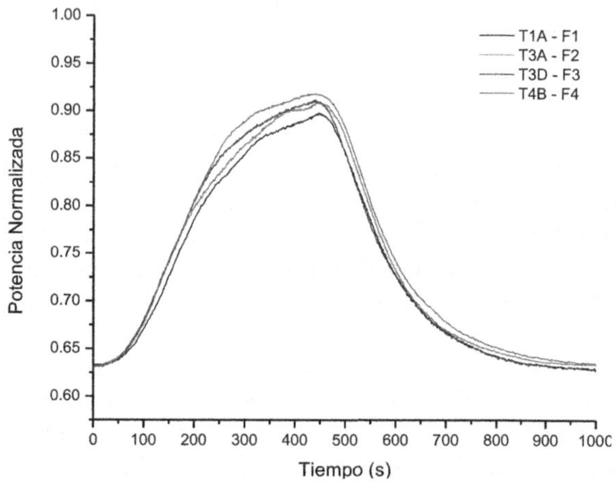

Fig. 2 Sensor response to heat conduction of the samples with lowest *METAVIR* dispersion.

IV. CONCLUSIONS

The fibrosis assessment method supported by three experienced observers showed congruence in 70% of the cases, and even when the procedure reduced the subjectivity of the grading, these results point out the need of developing new quantitative tools to properly determine the stage of the disorder.

In regard to the heat flow measurements, they were not conclusive to classify the biopsies. However, comparing the results of the groups in distinct stages, differences can be appraised for the cases with a lower dispersion of fibrosis. Grouping and comparing the curves in this manner, a directly proportional relation was found between fibrosis and thermal conduction; meaning that the more fibrosis in the hepatic tissue, the higher thermal conductivity of the sample. However, it is desireable to increase the sensitivity of the heat flow sensor in order to continue evaluating this method as a tool to analyze hepatic samples.

Continuing to study the samples with other methods is important. In this matter, calculation of the thermal conductivity is an important tool.

Acknowledgment

Authors Sánchez-Pérez C. and Gutiérrez-Herrera E. acknowledge financial support of research grants PAPIIT IT-100515 and PAPIIT TA-100215 from Dirección General de Asuntos del Personal Académico from Universidad Nacional Autónoma de México.

Conflict of Interest

The authors declare that they have no conflict of interest.

References

1. Domíguez M., Colmenero J. & Bataller R. (2009). Tratamiento de la fibrosis hepática. Gastroenterología y Hepatología 32 (9), 10-16.
2. Anthony P., Ishak K., Nayak N. et al (1978). The morphology of cirrhosis. Recommendations on definition, nomenclature, and classification by a working group sponsored by the World Health Organization. Journal of Clinical Pathology, 31, 395-414.
3. Salud: México 2006. Informe para la rendición de cuentas. Secretaría de Salud, 46-48. http://www.salud.gob.mx/unidades/evaluacion/saludmex2006/SM06.pdf
4. Valvano J., Cochran J. & Diller K. (1985). Thermal conductivity and diffusivity of biomaterials measured with self-heated thermistors. International Journal of Thermophysics, 6 (3), 301-311.
5. García L., González M. & Moreno R. (2012). Cirrosis hepática. Medicine - Programa de Formación Médica Continuada Acreditado, 11: 625-633. Julio, 2015, Elsevier, Base de datos.
6. Ausina V. & Moreno S. (2005). Tratado SEIMC de enfermedades infecciosas y microbiología clínica. Buenos Aires: Médica Panamericana.
7. Gutiérrez A., Sánchez C. & Alemán N. (2013). Optical sensor for heat conduction measurement in biological tissue. Journal of Physic: Conference Series, 450.

Corresponding author:

Author: Celia A. Sánchez Pérez
Institute: Universidad Nacional Autónoma de México
Street: Av. Universidad No. 3000, C. P. 04510
City: Mexico city
Country: Mexico
Email: celia.sanchez@ccadet.unam.mx

Clinical Evaluation of Inductive Spectrometer to detect Breast Cancer

C.A. González-Díaz [1*], M.C. Uscanga-Carmona[2], L.M. Lozano-Trenado[1], J.L. Ortíz[2], J.A. González[2], and C.I. Guerrero-Robles[1]

[1] Universidad del Ejército y Fuerza Aérea/Centro Militar de Ciencias de la Salud, Esc.Mil.Gds.Snd., México City, México
[2] Hospital Militar de Especialidades de la Mujer y Neonatología/Centro de Detección Oportuna de Cáncer de Mama, México City, México

Abstract—**Breast Cancer (BC) is the most common malignancy in women worldwide. Impedance measurements based on Magnetic Induction Spectroscopy (MIS) has been proposed for detection of malignant patho-physiological conditions associated to tissue electrical conductivity changes. In particular; we propose the use of Inductive Phase Shift (ΔΘ) measurements based on MIS to detect BC.** Objective. **Evaluate clinically the use of ΔΘ measurements trough an Inductive Spectrometer device to detect BC.** Methods. **Female volunteer patients (600) randomly enrolled as they assisted to its programmed screening were separated in two experimental groups (Control; n=589 and BC; n=11), and were measured by an experimental inductive spectrometer designed to estimate volumetric ΔΘ in human breast tissue at 136 frequencies spaced logarithmically in a whole bandwidth (0.001 to 100 MHz). All patients were examined with conventional mastography by experienced radiologists. ΔΘ data were homogenized in every patient with respect its contra-lateral breast measurement in such a way that the effects of age, corporal mass index even hormonal cycle phase factors were neutralized. The histopathology result obtained from ultrasound-guided core biopsy or operation excisions were used as reference standard. A parametric t-student test was used for comparing mean values of ΔΘ in Control *vs* BC groups.** Results. **Increments in volumetric inductive phase shift measurements at specific frequencies were evident in BC with respect to control conditions (p<0.05).** Conclusion. **Inductive phase shift increments at frequencies in the Beta dielectric dispersion region could be associated to changes in the cell structure and has the potential to detect pathological conditions in breast tissue associated to cancer. Further complementary studies are warranted to confirm the observations.**

Keywords— **Breast, Cancer, Magnetic, Induction.**

I. Introduction

Breast Cancer (BC) is the main cause of cancer death and the most frequently diagnosed cancer among women in 140 of 184 countries worldwide. While the incidence rates are higher in more developed regions the mortality is relatively higher in less developed countries due to a lack of early detection and access to treatment facilities [1]. The hyper-vascularization of malignancies promotes changes in the electrical impedance (EI) of the tissue. Given this phenomena it has been suggested the use of bioimpedance measurements for detection of cancer in different organs and tissues. Bioelectrical measurements through magnetic induction at a single frequency with noninvasive coils have been proposed as a valuable alternative to monitor, non-invasively, the health of organs and tissues [2], [3], [4] and [5]. Our group has proposed the use of tissue bioimpedance measurements by multiple frequency magnetic fields as a valuable alternative to detect non-invasively BC [6], [7].

Human breast tissue tumor shows an increase in electrical conductivity from 4 to 8 mS/cm compared to normal tissue [8], [9]. We have developed an inductive spectrometer device of low intensity and non-ionizing magnetic fields, ergonomically adapted to the anatomy of the breast. The system uses the Magnetic Induction Spectroscopy (MIS) to measure the electrical properties of tissue through magnetic fields at multiple frequencies. Basically two coupled coils of different radius in an inductor-sensor arrangement were coaxially centered and spaced in such a manner that a typical breast tissue volume is placed along the center line as shown figure 1.

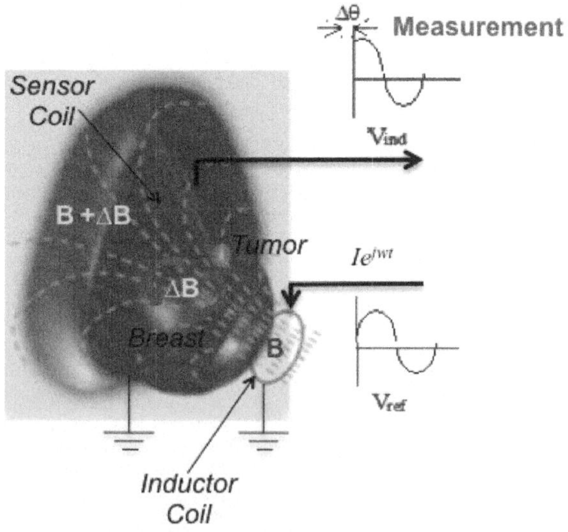

Fig. 1. Biophysical concept for non-invasive detection of malignant neoplasms in breast tissue by MIS measurements.

I. Torres et al. (eds.), *VII Latin American Congress on Biomedical Engineering CLAIB 2016, Bucaramanga, Santander, Colombia, October 26th -28th, 2016*,
IFMBE Proceedings 60,
DOI: 10.1007/978-981-10-4086-3_170

The aim of this study was to evaluate clinically the use of MIS for detection of breast cancer tissue conditions associated to its volumetric electrical conductivity changes through inductive phase shift measurements. In particular to determine the sensitivity of an Inductive Spectrometer designed and manufactured for BC detection.

II. MATERIAL AND METHODS

A. Biophysical Concept

The Inductive Spectrometer device consists of two coils of different radii in an inductor-sensor arrangement. The coils are coaxially centered. The breast volume is placed between the coils. An alternating current Ie^{jwt} is injected to the inductor coil to generate a primary magnetic field **B** and to induce Eddy currents in the tissue as well in the sensor coil. Induced currents promote a perturbation of the primary magnetic field (**ΔB**) as a function of the bulk conductive electrical properties of the breast and eventual tumour volumes. Therefore, an "inductive phase shift" (Δθ) is generated, which is detected as a composite magnetic field **B+ΔB** in the sensor coil (figure 1). The phase shift of the total magnetic field is obtained through the voltage phase difference between inductor and sensor coils [6]. The Inductive spectrometer used here was designed to acquire data from measurements through the entire volume of the breast, thus the study hypothesis is that electrical conductivities changes given by the presence of tumour tissue promotes changes in the spectra of Δθ estimated at multiple magnetic fields frequencies, and not volumetric resolution is expected.

B. Inductive Spectrometer device

An experimental inductive spectrometer to use MIS measurements in a typical breast tissue volume was designed and constructed. The system consists of five modules: digital synthesizer, transceiver, phase detector, data acquisition and data processing. A general description of the different modules has been previously documented in [7] and [8]. as follows: the digital synthesizer was implemented by a signal generator AD9958 (Analog Device Inc. Norwood, MA, USA), which supplies a signal Icos(ωt) of approximately 10 mA in the range of 0.001–100 MHz at preprogrammed steps controlled by PC. The transceiver consists of an arrangement of inductor and sensor coils coaxially centered as the concept shown in figure 1. Both coils were built from five turns of magnet wire AWG32 rolled on a coil holder ergonomically adapted to the anatomy of the breast. The inductor-sensor coils geometrical disposition correspond to diameters $\phi 1=5.5$ and $\phi 2=13.5$ cm respectively, both coils spaced a distance d=5cm as shown in figure 1. The coil inductances, as calculated on the basis of Faraday's law, were 14.9 and 89.9 μH for inductor and sensor respectively. The estimated mutual inductance coefficient was M=0.01μH approximately. A resonance frequency of the digital synthesizer-inductor coil system was determined experimentally around 90 KHz; this frequency is excluded from the analysis. To avoid inductive pickup the leads of the coils are twisted. A phase detector as the voltage phase difference between the inductor and sensor coils estimates the inductive phase shift. The phase detector was designed on the basis of the AD8302 (Analog Devices Inc. Norwood, MA, USA). The AD8302 is a fully integrated RF IC for measuring differences in phase between two signals with a resolution of 10 mV/degree. The data acquisition (A/D) module uses a 10-Bit Analog-to-Digital module of the microcontroller 18F4550 (Microchip Technology Inc. Chandler, Arizona, USA). The experimental inductive spectrometer prototype manufactured is shown in figure 2.

Experimental Design

Female volunteer patients (600) randomly enrolled as they assisted to its programmed mastography screening were separated in two experimental groups (control, n=589 and BC, n=11). All patients were measured by the experimental inductive spectrometer designed to estimate volumetric ΔΘ in human breast tissue at 136 frequencies spaced logarithmically in a whole bandwidth (0.001 to 100 MHz). All patients were examined with conventional mastography by experienced radiologists. As original conception, ΔΘ data were homogenized in every patient with respect its contralateral breast measurement in such a way that the effects of age, corporal mass index even hormonal cycle phase factors were neutralized or minimized, , even though the implicit statistical risk of not detect small specific variations. The histopathology result obtained from ultrasound-guided core biopsy or operation excisions were used as reference standard. A parametric t-student test was used for comparing mean values of ΔΘ in Control *vs* BC groups.

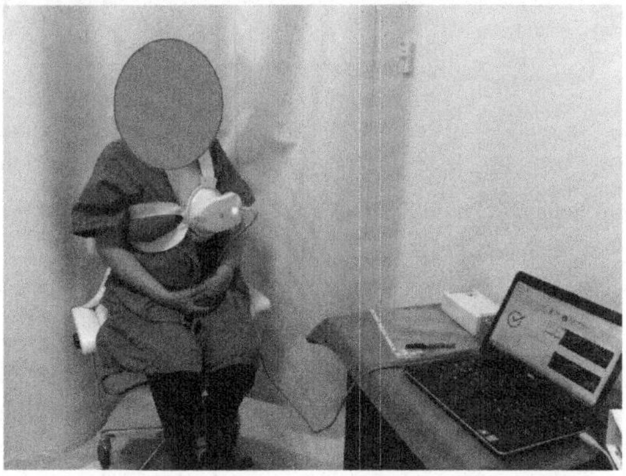

Fig. 2 Inductive Spectrometer System in a female volunteer.

The study was conducted according to the principles expressed in the Declaration of Helsinki and was approved by the "Research" and "Bioethics" Committees of the Institution. The volunteer patients were gently informed and signed an informed consent.

III. RESULTS

Figure 3 shows the Inductive Phase Shift Spectra for Control and BC groups. The obvious observation suggests that increments of inductive phase shift as a function of frequency are more evident above 10 MHz; nevertheless specific frequencies below 2 KHz show significant differences between groups. A t-student analysis in the whole bandwidth shows differences at specific magnetic fields frequencies and the most significant points have been showed in figure 3, in addition; a close-up of those mean values as well as its dispersion bars are shown in figure 4.

Table 1 *t-student* test for independent samples.

KHz	Mean Control	Mean BC	t-value	fd	P
14065	3.574499	8.968275	-3.63569	598	0.000301
21544	0.198148	0.623834	-2.21956	598	0.026826

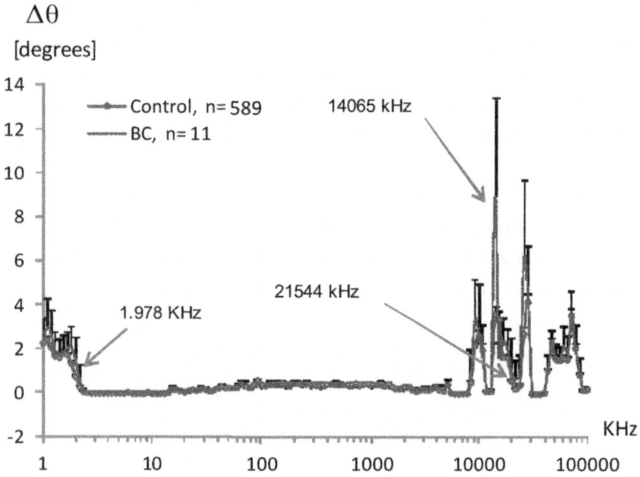

Fig. 3 Inductive Phase Shift Spectra for both Control and BC groups.

IV. DISCUSION

The observations suggest that Inductive phase shift increments at frequencies in the Beta dielectric dispersion region could be associated to changes in the cell structure and has the potential to detect pathological conditions in breast tissue associated to cancer, such assumption is on the basis of tissue properties have strong variation according to morphological patterns, structural lines of muscle fibers, tendons, nerve fibers, and blood vessels [10]. An analysis of the radiology reports (not shown in this paper in order to keep simplicity) indicates that breast tissue has multi-factorial conditions such as the ratio fat/glandular tissue density, calcifications/micro-calcifications, as well as the size and morphological shape of benign and malignant nodules, which might be associated to changes in the volumetric electrical properties of the breast tissue; it might explains in part the highest dispersion observed in the inductive phase shift spectrum of BC condition. Thus is expected that an eventual algorithm for automatic pattern recognition of inductive phase shift spectra associated to BC demand specific training under such multi-factorial conditions.

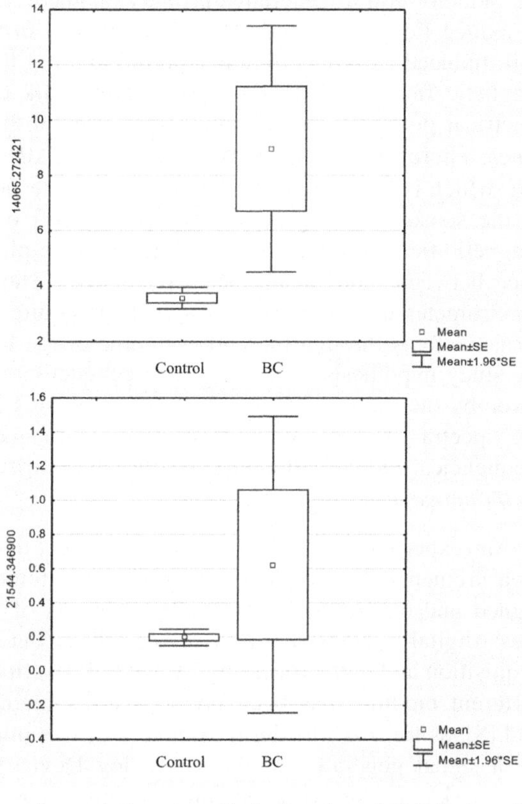

Fig. 4 Mean values and dispersion of inductive phase shift (Δθ) at specific magnetic fields frequencies at the most significant differences.

V. CONCLUSIONS

Inductive phase shift increments at frequencies in the Beta dielectric dispersion region have the potential to detect pathological conditions in breast tissue associated to cancer. Further complementary studies are warranted to confirm the observations.

Acknowledgment

This research was supported by "Centro de Investigación y Desarrollo del Ejército y Fuerza Aérea" of the Mexican Army, trough the "Programa de Investigación y Desarrollo en Colaboración con Instituciones de Educación Superior A022".

Conflict of Interest

The study was developed in the "Centro de Diagnóstico Oportuno de Cáncer de Mama" from the "Hospital Militar de Especialidades de la Mujer y Neonatología", México. And was approved for research in humans by the "Ethic in Research Committee" of the Institution and the "Hospital Militar Regional de San Luis Potosí-SLP", México, as well as the "Comisión Federal de Protección contra Riesgos Sanitarios", COFEPRIS - México.

References

1. Cancer statistics, 2015, Rebecca L. Siegel, MPH; Kimberly D. Miller, MPH, Ahmedin Jemal, DVM, PhD, CA CANCER J CLIN 2015;65:5-29.
2. Scharfetter H, Ninaus W, Puswald B, Petrova G I Kovachev D and Hutten H 1999 Inductively coupled wideband transceiver for bioimpedance spectroscopy (IBIS) Ann N Y Acad Sci 873:322-34
3. Griffiths H, Steward W R and Gough W 1999 Magnetic induction tomography - a measuring system for biological materials Ann N Y Acad Sci 873:335-45
4. Griffiths H 2001 Magnetic induction tomography Meas Sci Tech 12:1126-31
5. Watson S, Williams R J, Gough W A and Griffiths H 2008 A magnetic induction tomography system for samples with conductivities less than 10s m−1 Meas Sci Tech 19:045501-11
6. González C A, Lozano L M, M C Uscanga, Silva J G and Polo S M. "Theoretical and Experimental Estimations of Volumetric Inductive Phase Shift in Breast Cancer Tissue" Journal of Physics: Conference Series 434 (2013) 012004.
7. González C A, Silva J G, Lozano L M and Polo S M. "Simulation of Multifrecuency Induced Currents in Biophysical Models and Agar Phantoms of Breast Cancer". Journal of Electromagnetic Analysis and Applications. August 2012. Vol. 4 No.8. pp 317-325.
8. Burdette E C 1982 Electromagnetic and acoustic properties of tissues in pyisical Aspects of Hyperthermia G H Nussbaum ed (American Association of Phycisist in Medicine) 8: 105-50
9. Duck F A 1990 Physical properties of tissue Academic Press Limited (London NW1) 197-204
10. *Bioimpedance and Bioelectricity Basics*, 3rd Edition. Sverre Grimnes, Orjan G. Martinsen. Academic Press, 2014.

• corresponding author:

Author: César A González
Institute: Universidad del Ejército y Fuerza Aérea-Esc.Mil.Gds.Snd.
Street: Cda. Palomas s/n Col Lomas de sotelo
City: Ciudad de México
Country: México
Email: gonzalezantoni@hotmail.com

Kinematic Soccer Kick Analysis Using a Motion Capture System

Juan Pablo Ángel-López[1], Belarmino Segura-Giraldo[1], Luz Dary Rodríguez-Sotelo[2] and Karol Bibiana García-Solano[3]

[1] Universidad Autónoma de Manizales (UAM®), Grupo de Investigación en Automática, Colombia
[2] Universidad Autónoma de Manizales (UAM®), Grupo de Investigación Neuroaprendizaje, Colombia
[3] Universidad Autónoma de Manizales (UAM®), Grupo de Investigación Cuerpo Movimiento, Colombia

Abstract— **This paper shows a kinematic analysis of the penalty and free kick in soccer in order to write an algorithm to identify the phases of movement when executing said kicks. At the beginning of the paper, there is a short review of similar and previous investigations, which serve as a foundation for this work. Then, the motion capture (MoCap) system used to develop this research and the phases detection methodology are described, and finally, the results from the MoCap process for kicking and their mathematical descriptions are shown. Those results aim to a trend on the execution of the kick in terms of movement phases, sorted as: approaching, counter-movement, swinging of the attacking leg, foot-ball impact, and post-impact. An algorithm to detect the exact moments those phases start, using motion capture info, was written.**

Keywords— **Kinematics, biomechanics, motion capture, soccer, free kick.**

I. Introduction

One of the main interests for the scientific community is to identify movement patterns associated to sports. The need to understand said movement implies that the researchers must quantify the motor actions related to those movements. This quantification requires the use of measure systems based on photo-grams (2D videos) [1], systems limited by the subjectivity of the evaluator and the perspective of the video.

There can be found several movement analysis using video for identifying kinematic features, by means of generic algorithms that detect probabilistic events [2] and identify mobile points on each captured frame from the video at different times with different motion magnitudes [1]. Nowadays, motion capture systems are used for detecting specific dots placed over the body, in real time and also in 3D ((x, y, z) spatial coordinates) [3], which can be used as a reference for identifying kinematic features.

Most of the motion analysis systems with standard video cameras are susceptible to external agents like light variations, particular characteristics of the analysed event, occlusions, gasps of information, image quality, among others [4]. For this reason, it is necessary to get access to different methodologies that allow the researchers to identify features like trajectory, velocity and acceleration, using algorithms for filtering and data subtraction and correlation [5] [6].

On the other hand, there are also some researches that compare different analysis methods, using different types of cameras, resolutions and recording file extensions [5] [7], taking into count software for 3D graphics creation from 2D frames and movement simulation [5]. However, the main obstacles are found during the 3D reconstruction for being dependant of the used cameras quantity [7]. At the same time, there are also studies about movement analysis from a series of several still images, being the blind spots between the images the obstacles for achieving data fidelity [8].

Soccer could be considered as the most important show of the human kind because of its worldwide physical, technical, economic and social implications. It is the most popular game in the world, there are more than 50 million registered players and many others as amateur players [9] [10]. Its success depends on a great variety of strategies and basis that allows the team to achieve the goal target, and some of these factors depend on the biomechanics. Nowadays, research in this field has only been approached in advanced countries and professional clubs.

In this sense, this research aims to develop an algorithmic approach for analysing the biomechanic gesture of the soccer kick, based on motion capture information, specially with data from the kicking foot. By detecting the phases of the kick described by [9], it would be able to set bases for determining if the soccer player has executed a technically correct kick.

II. Equipment and Methods

The main goal of this research is to develop an strategy for detecting presented phases during a soccer kick execution by tracking several markers placed over the body of a soccer player. In this study, twelve male college soccer players were analysed and each one had to simulate a Free Kick inside a controlled environment. Each player was asked to execute, on their own criteria, the best kick possible in terms of technique,

I. Torres et al. (eds.), *VII Latin American Congress on Biomedical Engineering CLAIB 2016, Bucaramanga, Santander, Colombia, October 26th -28th, 2016*,
IFMBE Proceedings 60,
DOI: 10.1007/978-981-10-4086-3_171

for achieving an imaginary goal. The tests took place inside a Biomechanics Laboratory, equipped with a twelve infra-red cameras Mocap system.

A. MoCap System

This research took place thanks to a Motion Capture System composed by infrared Flex 3 cameras and the software Motive, both manufactured by OptiTrack^{TM}. In this study, twelve cameras were used, for which it was necessary to design a specific layout configuration that would capture each marker placed on the hand with the least possible amount of occlusions. Fig.1 shows the setup used, where it is possible to appreciate a rectangular frame and the capture area where the cameras are aiming to. This capture area is 2 by 3 meters approx..

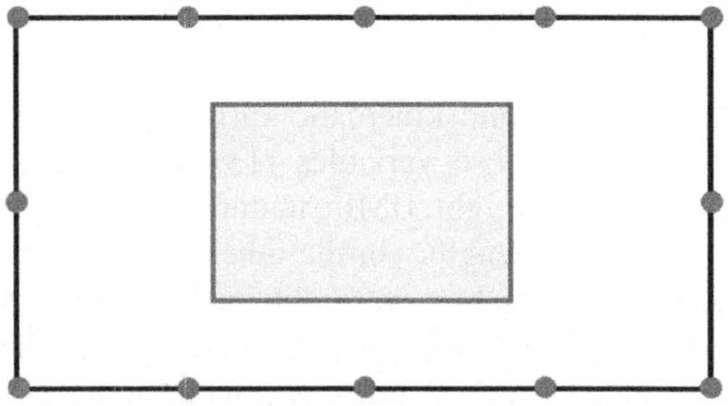

Fig. 1: The external filled circles represent the cameras and the inner rectangular area represents the capture volume.

The cameras used for this study have a capture rate of 100 *FPS*[1], a video resolution of 640x480 pp and a maximum range of detection of $11m$ [3]. On the other hand, the software which the cameras where connected to, gathers the information sent by the cameras to create a 3D data set. Motive© builds a tridimensional matrix that contains the *xyz* coordinates from each one of the markers placed inside the capture volume [3].

B. Free Kick Phases

According to [9], the soccer kick biomechanical gesture presents the following phases, also shown in Fig.2:

1. **Approaching:** In this phase, the soccer player starts the acquisition of speed towards the ball, and ends at the moment the supporting foot touches the ground (Fig2(a)).
2. **Counter-Movement:** Starts when the supporting foot touches the ground and is executed during an one-foot supporting, and goes until the kicking foot reaches its maximum altitude behind the body (Fig.2(b)).

[1]Frames per second

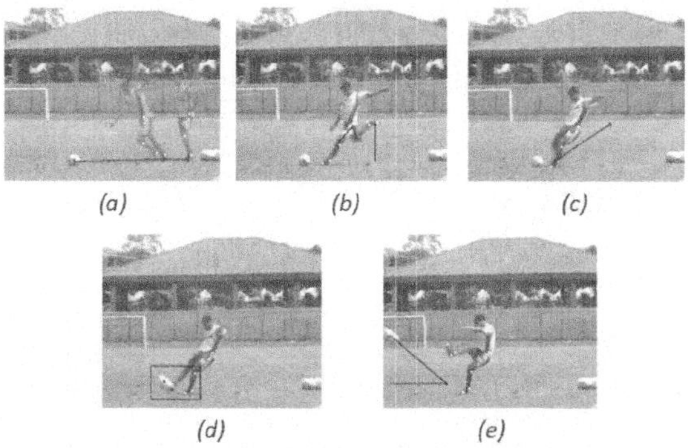

Fig. 2: Soccer kick biomechanical phases. (a) Approaching. (b) Counter-Movement. (c) Balancing. (d) Foot-Ball Impact. (e) Post-Impact. Adapted from [9]

3. **Balancing:** All the movements executed when the kicking leg goes from behind the body towards the ball and ends at the exact moment the kicking foot makes contact with the ball (Fig.2(c)).
4. **Foot-Ball Impact:** The moment when the foot instep makes contact with the ball surface and it starts its trajectory (Fig.2(d)).
5. **Post-Impact:** Are all the body movements executed after the ball impact until the body recovers its equilibrium (Fig2(e)).

C. Data Analysis

For each player, the analysed data was a 3D matrix (Fig.3) containing spacial information of a total of 34 markers placed over the player's body. Those matrices were composed by three different sub-matrices, where each one contained the coordinates on the x, y and z coordinate axes, respectively. Each sub-matrix had a row number equal to the total recorded frames, so, for example, if the recording lasted five seconds, the number of rows would be 500. The column number was equal to the number of markers placed on the body, that in this case was 34 markers.

Given that the MoCap system shows some noise, related to the infra-red rays of the cameras, it was necessary to smooth all the captured data inside the Matlab® environment using a *Savitsky-Golay* filter [12].

D. Sample

During this study, 35 soccer players from different college soccer teams from the city of Manizales (Colombia) voluntarily participated on the motion capture sessions, performing three takes each for a simulation of a free kick inside a

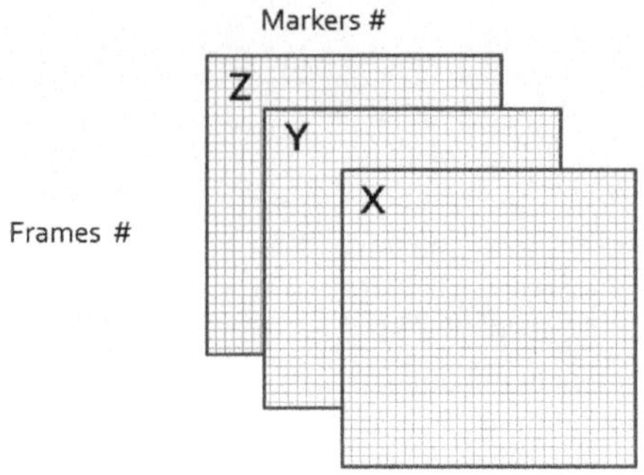

Fig. 3: Schematic representation of the data given by the motion capture system.

controlled environment. Between those 35 players, 12 were selected randomly in order to identify their kick's gesture. All players were from 17 to 21 years old and their predominant foot was the right.

III. RESULTS

After recording all the executed tests by the twelve selected players, it was noted that the way the markers were sorted by the motion capture system in each take was random, so it was necessary to write a visualization algorithm under the Matlab® environment in order to identify the assigned indexes for each marker to build connectivity matrices for every body segment on every player, aiming to achieve a good visualization of the data. An example of this representation is shown in Fig.4.

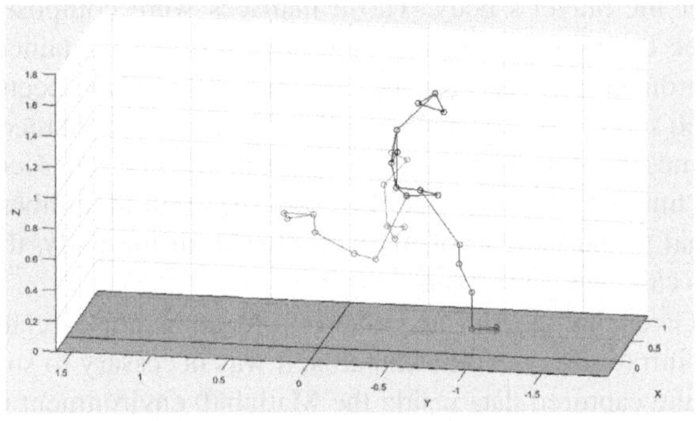

Fig. 4: Schematic representation of the data given by the motion capture system.

When all the markers' indexes where identified, all the recordings taken from the players where analysed under the same algorithm, also written under the Matlab® environment, for detecting the kick phases by reading the Z axis trajectory of the marker placed on the kicking foot. With all the analysed soccer players, it was able to detect said phases, and an example of said detection is shown in Fig.5.

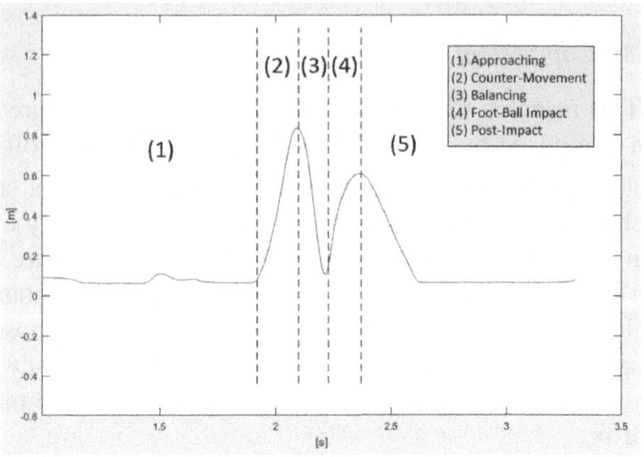

Fig. 5: Automatic phases detection using a Matlab® algorithm.

Anthropometric and kinematic variables (154) were analyzed. Some of those variables (13) are anthropometric, which are: height, weight, BMI[2], femur and leg length, thighs perimeter and feet length. On the other hand, the remaining 141 (kinematic) variables were obtained using the Matlab® environment, separated by the detected phases mentioned above: traveled distance, speed, acceleration and angles between body segments, as defined by [9]. All this information was stored on a single data base to which a descriptive statistic analysis was applied in order to identify the most relevant variables.

After performing a bivariate correlation between the kinematic variables, using a non parametric test of SRCC[3] [13], with a significance level lower than 0.05 and degrees of correlation: high (> 0.8), medium (between 0.5 and 0.79), or low (< 0.05), the most relevant variables per phase, comapred to [9], were identified as showed in Table 1.

IV. CONCLUSIONS

During MoCap sessions, it is imperative to avoid occlusions between cameras and markers in order to decrease the presence of noise and gaps in the recorded data. When gaps occur, the data analysis turns into a seriously difficult task when the motion capture doesn't count with the proper tools for editing data, resulting most of the times in recording several, sometimes dozens of takes for a single player, until a perfect take is achieved.

[2]Body Mass Index

[3]Spearman's rank correlation coefficient

Table 1: Most relevant variables obtained from the SRCC.

Phase	Variables
Approaching	Number of steps
	Body approach mean speed
	Last step length
	Last step's body speed
	Angle between the trunk and the horizontal
Counter-Movement	Angle between the kicking foot and the horizontal
Balancing	Angle between each thigh and the trunk
	Angle between each thigh and leg (knee)
	Angle between between both thighs
Foot-Ball Impact:	Speed of each thigh
	Speed of each leg
	Speed of each foot
Post-Impact	Angle between the trunk and the horizontal
	Angle between the kicking thigh and the trunk
	Angle between supporting side's thigh and leg (knee)

By detecting maximums, minimums and turning points in the Z axis trajectory curve, a functional algorithm for automatically detecting the frames where each phase started, was achieved. This approach allows future researches to make further analysis of the biomechanic gesture of the soccer kick in order to identify patterns and to create mathematical models for kicking actions in soccer.

Even when all the players were asked to execute a kick under the same circumstances, inside the same controlled area, all of them showed different features like the number of steps taken before kicking the ball, the amplitude of the kick and the duration of the movements. This behaviour implies that further kinematic analysis in soccer should take into count these variables in order to obtain more accurate mathematical models about this certain action.

According to the SRCC, the most relevant phases for describing a kicking pattern are Counter-Movement, Balancing and Foot-Ball Impact. Further analysis should be done in order to determine a proper kicking pattern.

V. Acknowledgement

The authors thank the support given by Universidad Autónoma de Manizales and their Research Committee, providing the facilities of the Biomechanics Laboratory, under the researh project "Caracterización del Gesto Deportivo del Pateo en el Lanzamiento de Tiro Libre y Desde el Punto Penal en Futbolistas Juveniles".

References

1. Martínez A, Espí M, Jover E, Prats G, Pons I, Martínez B, "El Video Digital como recurso didáctico para el estudio cinemática del movimiento", Actes VII Jornades de la Curie, pp. 53–65, 2003.
2. Ekin A, Tekalp AM, editors, "Generic event detection in sports video using cinematic features", Computer Vision and Pattern Recognition Workshop, 2003.
3. Optitrack Motion Capture Systems, Motive Software tutorials and Flex 3 Cameras specifications. `http://www.naturalpoint.com/optitrack/`.
4. D'Orazio T, Leo M, "A review of vision-based systems for soccer video analysis", Pattern recognition, 2010.
5. Shin T, Kasuya N, Kitahara I, Kameda Y, Ohta Y, editors, "A comparison between two 3d free-viewpoint generation methods: Player-billboard and 3d reconstruction", 3DTV-Conference: The True Vision-Capture, Transmission and Display of 3D Video (3DTV-CON), 2010.
6. Poppe R, Van Der Zee S, Heylen DK, Taylor PJ, "AMAB: Automated measurement and analysis of body motion", Behavior research methods, 2013.
7. Chavan A, Sasi S, editors, "Vision-based real-time speed tracking", Intelligent Signal Processing, IEEE International Workshop, 2005.
8. Duan L-Y, Jin JS, Tian Q, Xu C-S, "Nonparametric motion characterization for robust classification of camera motion patterns", IEEE Transactions on Multimedia, 2006.
9. Acero, J.A. and Albarracín, J., "Biomecánica bidimensional (2D) del tiro libre preferencial en fútbol: un modelo integral experimental". [Online] 2005. Avalaible: `http://publicacion05.unipamplona.edu.co/unipamplona02/hermesoft/portal/home_1/rec/arc_12428.pdf`.
10. Suárez, R., "Biomecánica deportiva y control del entrenamiento", Universidad de Antioquia, 2009.
11. Cano, O., Rodríguez, J., Chavarría, F., Isaza, P., "Análisis biomecánico del gesto del pateo. Especialidad en entrenamiento deportivo", Universidad de Antioquia. 2008.
12. Matlab Documentation Center. Savitzky-Golay filtering "sgolayfilt". `http://www.mathworks.com/help/signal/ref/sgolayfilt.html`. 2013.
13. Lehman, A., O'Rourke, N., Hatcher, L., Stepanski, E.J., "JMP® for Basic Univariate and Multivariate Statistics Methods for Researchers and Social Scientists", Second Edition, Cary, NC: SAS Press. p. 123. 2013.

Measurement of the strength of upper limbs in cervical injury users to design a propulsion wheelchair mechanism

Norma Araceli Coral Hernández[1], Cuitláhuac Osornio Correa[1], María del Carmen Mercado Chaparro[1], Ana María Vásquez Gallego[1], Jorge Letechipia Moreno[1] y Andrés Torres Velásquez[2]

[1] Universidad Iberoamericana/Posgrado en Ciencias de la Ingeniería, Ciudad de México, México.
[2] Universidad EIA-Universidad CES/Ingeniería Biomédica, Medellín, Colombia.

Abstract— **It is known that the propulsion technique of manual wheelchair it is mechanically inefficient, due to the relationship given by the propulsion force and the distance obtain, and the number of repetitions put the user at risk in developing cumulative trauma compromising joint and tissue. At least 70% of user will suffer secondary injuries because of constant and prolonged used of their manual wheelchair.**

The users with cervical injury, are exposed in higher physical risk of not getting prescribe technological assistive; due to the fact to the consequences of this type of injuries they present weakness muscles required for the propulsion such as: triceps brachii, middle deltoid, serratus anterior, extensor carpi radial and ulnar extensor carpi, furthermore they have no grip because they have damage at dermatome C6, C7, C8-T1, taking into account that in regular users 50% of the necessary force is lost at the moment they grip the rim to move forward, which represent for this type of user a heavy physical risk that can cause cumulative trauma disorders.

We are looking to develop assistive technology for a better propulsion technique that suits their remaining capabilities, empower them and make them self-sufficient as possible. In order to do that, a mechanism that allow us to measure the force in a difference range of motions of the upper limbs and force hands grip has been created, taking into consideration for the final design we are measuring anthropometric dimensions of the Mexican users and their personal records to better satisfy their needs. With all this information we will be able to have a better understanding of the problematic by applying the quantitative and qualitative data in a methodology of design.

Keywords— **Assistive Technology, Cervical lesion, Propulsion wheelchair.**

I. INTRODUCCIÓN

A spinal cord injury refers to any type of damage to the spinal cord as a result of trauma, illnesses, or tissue degradations [1]. An injury close to the cervical segments implicates a total loss (complete spinal cord injury) or partial (incomplete spinal cord injury) of the motor and/or sensitive function of the four extremities of the body; this level is actually known as tetraplegia or quadriplegia [2].

According to the World Health Organization, an estimate of 250 000 to 500 000 people suffer muscular injuries, which stem from several causes: traffic accidents, collapses, or violent acts [1]. In Mexico, the annual incidence of muscular injuries is approximately 2 500 people, from whom 50 % correspond to quadriplegia and 55 % of these correspond to incomplete spinal cord injuries [3].

People with spinal cord injury require technological assistance that allows them to augment, keep, or enhance their functional capacities, such as promoting or improving their physical health [4]; therefore, 90 % of the people who experience spinal cord injury use wheelchairs for movement [5]. Nevertheless, the relation between the user and the technological assistance is so narrow that it is necessary to have access to appropriate technology as a pivotal factor for the development of daily life activities [4].

The WHO's World Report on Disability, establishes that people with disabilities are more vulnerable to preventive secondary sicknesses [6]; such vulnerability is compromised to the lack of knowledge of the capability of movement in the functional extremities of people with disabilities; namely, an inadequate prescription of the technological assistant [4]. As for the wheelchairs, Goutam indicates that donated equipment oftentimes do not satisfy the necessities of the users, generating pain, fatigue, and uncomfortably or incompatibility use with the users' surroundings [7].

Several research studies that focus on the satisfaction of the use of technological assistance, oriented to wheelchairs, revel that users are not feeling comfortable with the mechanical characteristics of their wheelchairs owing to the power needed and the speed obtained [8]. Moreover, studies have found a direct relation between the use of wheelchairs and body pain [9]; and a relation between maneuver and pain, fatigue, and comfort [10]. According to the statistic data provided by the National Institution of Security and Health (NIOSH), 32 % of the injuries and sicknesses that occur nowadays are as a result of repetitive movements and great efforts, which 7 % affect shoulders and 55 % wrists [11]. Some studies report that the levels spinal cord injury affects remarkably the shoulder muscles, which are indis-

I. Torres et al. (eds.), *VII Latin American Congress on Biomedical Engineering CLAIB 2016, Bucaramanga, Santander, Colombia, October 26th -28th, 2016*,
IFMBE Proceedings 60,
DOI: 10.1007/978-981-10-4086-3_172

pensable for the wheelchair propulsion [12]. Rankin acknowledges that 70 % of wheelchair users suffer injuries due to overuse [13]. The assessment of functional clinic of hands and arms in users quadriplegics is extremely imperative, because just as much movement and control of the superior members is the level of dependence that the user can achieve when doing their regular, life activities; thus, it has been found that 75% of people with quadriplegia prefer to recover the functionality of their superior members to any other part of their bodies [2].

II. METHODS

A. Equipment

Hardware: To gather or collect data, BIOPAC's data-acquisition system of four channels MP36/35 was used. When putting forth two BIOPAC's dynamometers SS25LA, a system calibration was done, using an INSTRON 4482 machine (Norwood, Massachusetts).

In Fig 1, there is an assembly of the system that comprises two steady, vertical rigid bars with an oscillating angle from 90 ° to 110 ° at a step of 5 °.

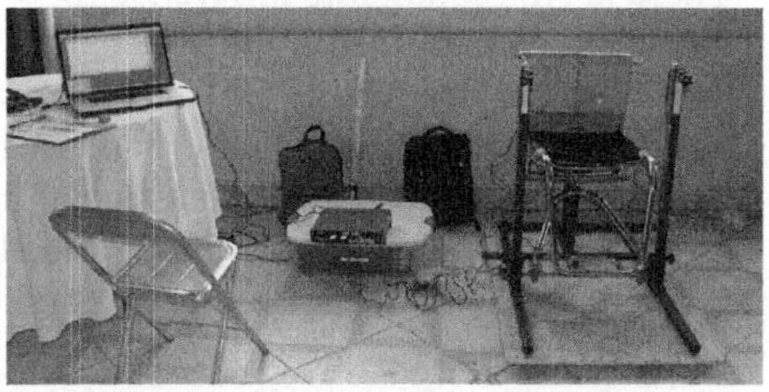

Fig 1 Mounting the force measuring system.

The dynamometers are located in the bar upper section, whose height varies with accordance to the user. The employed structure for the study, on which the Activa® wheelchair (Vida independiente model) is entrenched, is adaptable to the dimensions of the chair.

Software: The system calibration was done with software Bluehill® (Norwood, Massachusetts). The acquisition of data was done with Biopack M35®. The analysis of the data was performed in Microsoft Excel® 2013 (Redmond, Washington) and Minitab ® 17.

B. Methodology

Selecting participants: To conduct the testing, three groups were established: non-users of wheelchairs, paraplegic users and quadriplegic users. The criteria for selection was as follows:

- Physically, healthy and active people.
- Every wheelchair user was able to propel its own wheelchair without assistance.
- Average age of the participants was 32 years.
- Average length, in time, of the injury: 18 years.
- People with pain symptoms in every part of their bodies were excluded from the research, and if the participant had performed any intensely physical activity 12 hours previous to the study, he or she was also disregarded.

The group of non wheelchair users was composed of 10 men and 9 women, the group of paraplegics was composed of 14 men and 9 women; lastly, the quadriplegics was of 8 men and 4 women.

The collected data allowed the researchers to compare the capabilities of each group and helped to establish the design criteria, which was oriented to contemplate an integral rehabilitation based upon the user.

Informed consent signature: To conduct the study, each participant signed an informed consent.

Development of the testing: To lead the testing, handles were 90 ° positioned, in accordance to the floor, with an initial angle of the elbow to 90 °, researchers proceeded to take anthropometrics measures of the user and adapted a system to the user, registration the force in witch the propulsion was exerted on the handles. The procedure of collecting force was registered every 5 ° to a maximum of 110 °, in accordance to the floor, *Fig 2* shows the assembling of the system with the user, finally, researchers gauged the grabbing force through the dynamometer SS25LA de BIOPAC®.

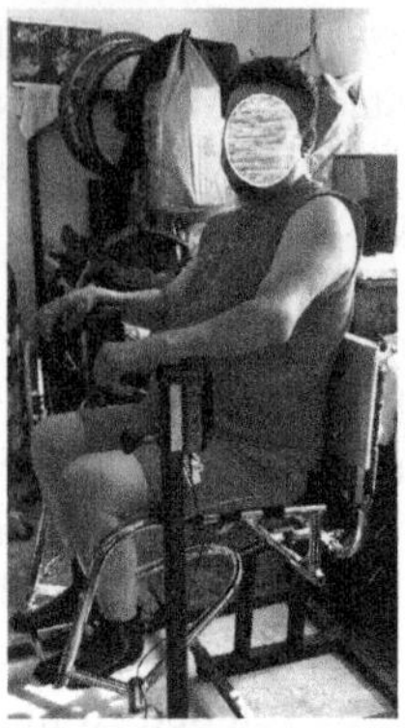

Fig 2 Assembling of the system with the user

Gathering data: With interval breaks of two minutes, three repetitions were performed in each group in every position (angle) of the system.

The data was tabulated on Microsoft Excel® 2013 and thereafter, to conduct the comparison among groups, a statistic ANOVA software test was used on software Minitab ® 17.

III. Resultados

- The quadriplegic men users (Fig 3) obtained 39.54 % less force in regards to that of paraplegic men, and 51.88 % less force compared to that of non-user men.
- The quadriplegic women users (Fig 4) obtained 48.32 % less force in regards to that of paraplegic women, and 23.40 % less force compared to that of non-user women.

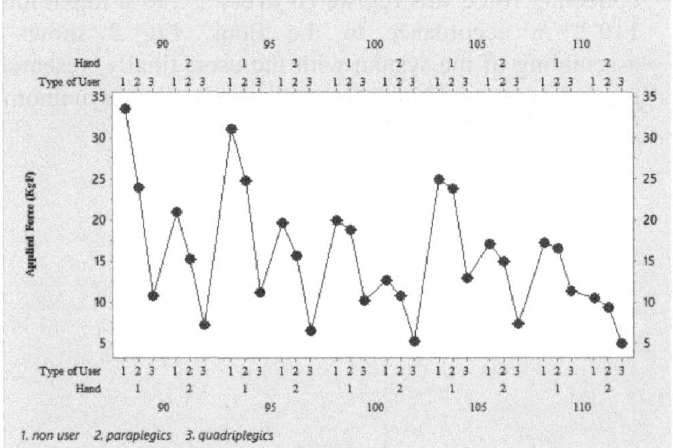

Fig 3 Strength differences between the three groups of men (90 °, 95 °, 100°, 105 ° y 110 °)

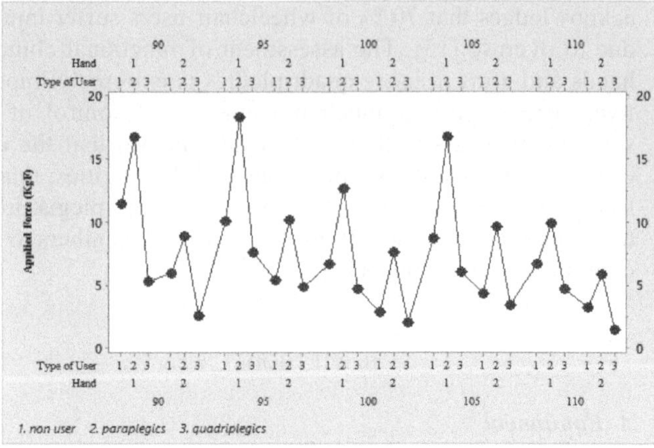

Fig 4 Strength differences between the three groups of women (90 °, 95 °, 100 °, 105 ° y 110 °)

- In the quadriplegic men group, it was found that the arithmetic median of force was 17.22 Kgf; with an angle of 100 °, a greater force was obtained on both hands.
- In the quadriplegic women group, it was found that the arithmetic median of force was 10.77 Kgf in an angle of 95 ° with the right hand.
- There is No remarkable difference ($p>0.05$) in both men and women (right and left hand). Such consistency was due to the lack of estimation of the propelled force applied to the handles in each angle. The graph above in the analysis of these data is a platykurtic graph. The arithmetic medians of the applied force on the different studied angles are virtually identical.
- There is a significant difference ($p<0.05$) between each of the medians for every angle in the paraplegic user group and the non user wheelchair one, in both men and women (right and left hand)

IV. Conclusions

The unremarkable differences in the quadriplegic men and women group allow researchers to establish that the crucial factor, as a principle to consider, is not an initial angle of inclination of the handles; rather, it is the force that requires each user to employ for moving the propulsion mechanism of the wheelchair. In an attempt to design and implement an assisting technology that can be adaptable to the user and his or her environment, instead of compelling the user to adapt to a bestowed technology, it is fundamental to characterize the intended population, comprehend the problematic, and analyze the limits and strengthen the capabilities of the user. The obtained forces will be represented through mathematical models that allow a simulation of the system, adapting them to the users' necessities.

Acknowledgment

Every author of the investigation thanks to each of the participants of the investigation; likewise, they thank to Fundaction Vida Independiente, to Fundaction Grupo Altia, the Iberoamericana University and Consejo Nacional de Ciencia y Technología.

Conflict of Interest

The authors declare that does not exist an interest conflict.

References

1. Organización Mundial de la Salud. (2013, Noviembre). Centro de prensa. Retrieved Febrero 2, 2015, from Lesiones Medulares: http://www.who.int/mediacentre/factsheets/fs384/es/
2. Duruöz, Mehemet Tuncay, Hand Function: A Practical Guide to Assesment, Ed. Springer, 2014.
3. Estrada Mondaca, S., Carreón Rodríguez, A., Parra, M. d., Ibarra Ponce de León, C., Velasquillo Martínez, C., Vacanti, C., et al. (2007, Noviembre-Diciembre). Lesión de médula espinal y medicina regenerativa. Salud Pública México, 49(6), 437-444.
4. Gutierrez Brezmes, J. L. (2011). Accesibilidad: Personas con discapacidad y Diseño Arquitectónico. Ciudad de México DF: Universidad Iberoamericana.
5. Choromanski, W., Dobrzynski, G., & Fiok, K. (2010). Optimization of Lever-Driven Wheelchairs. In 6th World Congress of Biomechanics (WCB 2010). August 1-6, 2010 Singapore (pp. 671-674). Springer Berlin Heidelberg.
6. Organización Mundial de la Salud. (2011, Junio 9). Discapacidades. Retrieved Mayo 28, 2015, from Informe mundial sobre la discapacidad: http://www.who.int/topics/disabilities/es/
7. Goutam, M., & Amalendu, S. (2005). Wheelchair charity: A useless benevolence in community-based rehabilitation. Taylor & Francis. Disability and Rehabilitation, 27(10), 591-596.
8. The Quebec User Evaluation of Satisfaction with Assistive Technology (QUEST 2.0): An Over view and Recent Progress. 2002, Canadá.
9. Leung, V., Colantonio, A., & Santaguida, P. (2005). Wheelchair use, pain and satisfaction with life in a national simple of older adults. International Journal on the fundamental aspects of technology to serve the ageing society, 3(3), 156-164.
10. Mukherjee , G., & Samanta , A. (2005, Mayo 20). Wheelchair charity: A useless benevolence in community-based rehabilitation. Disability Rehabilitation, 27(10), 591-596.
11. Qadri, M. J., & Peterson, D. R. (2011). A Method for modeling long-duration joint motion. Bioengineering Conference (NEBEC), 2011 IEEE 37th Annual Northeast (pp. 1-2). Nueva York: IEEE Xplore Digital Library.
12. Mulroy, S. J., Farrokhi, S., Newsam, C. J., & Perry, J. (2004). Effects of spinal cord injury level on the activity of shoulder muscles during wheelchair propulsion: an electromyographic study. Archives of physical medicine and rehabilitation, 85(6), 925-934.
13. Rankin, J. W., Kwarciak, A. M., Richter, W. M., & Neptune, R. R. (2010). The influence of altering push force effectiveness on upper extremity demand during wheelchair propulsion. Journal of Biomechanics, 43(14), 2771-2779.

Author: Norma Araceli Coral Hernández.
Institute: Universidad Iberoamericana.
Street: Prolongación Paseo de la Reforma 880, Lomas de Santa Fe, C.P. 01219.
City: Ciudad de México.
Country: México.
Email: ruacoral@gmail.com

Knee Joint Angle Monitoring System Based on Inertial Measurement Units for Human Gait Analysis

J.J. Castañeda, A.F. Ruiz-Olaya, C.N. Lara-Herrera and F.Z. Roldán

Faculty of Electronics and Biomedical Engineering, Antonio Nariño University, Colombia

Abstract— **Gait Analysis involves analyzing human walking, particularly its temporal parameters. Gait parameters (cadence, step length, trajectory of center of mass, etc) require measurements of lower-limb joint angles. The most commonly used techniques for gait analysis are optical systems. However, these systems have several drawbacks which include: markers need to be attached to the human body; it is required a considerable amount of time for setting up the experimental session; High cost; Bulk and space requirements; finally, they are restricted to a limited place or laboratory. Inertial measurement units (IMU) can be used in body-worn biomechanical monitoring devices for movement data collection in daily life. This paper presents a system based on 2 IMUs for knee joint angle monitoring to be used for long-time periods in out-of-lab daily life. For validation, the system was compared to a vision-based motion capture system, in a set of experiments involving walking on a treadmill under three different velocities (slow-speed, normal-speed, high-speed). Obtained results showed high correlation (>0.94) between measurements from the developed device and the vision-based motion capture system, according to the obtained concordance correlation coefficients.**

Keywords— **Biomechanical Measurements, Gait Analysis, Inertial Measurement Units, Knee Joint Angle.**

I. Introduction

Recently, measurements using inertial sensors (accelerometers and gyroscopes) have been applied in many fields including navigation systems, camera stabilization, games, and are being embedded in portable electronic devices such as smartphones and tablets [1]. Taking into account that inertial sensors are low-power, small, and accurate, they have been used for wearable applications such as rehabilitation and therapy for patients, human-machine interaction, falling detection for medical alerts in the elderly or disability, and sport training support [2]. One of the fields of application is biomechanical analysis, where inertial measurements can supplement other means of measurement or assessment for clinical diagnosis like optical motion analysis and assessment of other symptoms by visual observation [3]. Optical motion capture is not suitable for ambulatory or real scenarios or low-cost systems. Inertial sensors can be fixed to body segments to acquire data related to the motion. They can be used in body-worn biomechanical monitoring devices for movement data collection in daily life.

In a normal human walking activity, the movement of the upper and lower limbs can be seen as a quasi-periodic activity [4]. The term "gait" is used to describe the way of walking and consists of consecutive gait cycles, defined as "the duration from one event, usually foot contact, to the next occurrence of the same event on the same limb." [4]. Instrumented gait analysis is becoming of special interest for gait event detection and biomechanical monitoring.

Gait measurements techniques can be divided into kinematics and kinetics. In kinematic gait analysis, the kinematic parameters such as velocity, acceleration and joint angles are measured. The most commonly used techniques are video digitizing systems and marker systems (video based, optoelectronic) [5]. However, these systems have several drawbacks which include: markers need to be attached to the human body; it is required a considerable amount of time for setting up the experimental session; High cost; Bulk and space requirements; finally, they are restricted to a limited place or laboratory.

The inertial and magnetic sensors based devices called inertial measurement units (IMUs) can be used in most scenarios without any limitations (i.e. illumination, space, etc.). IMUs base on the integration of accelerometers, gyroscopes and magnetometers. They calculate the orientation of the body segments where they are attached by using multi-sensor information through specific sensor fusion algorithms mainly based on Kalman filtering. This paper presents a system based on IMUs for knee joint angle monitoring to be used for long-time periods in out-of-lab daily life.

Paper is organized as follows: next section describes experimental methods including hardware and software, and protocol. Section 3 presents obtained results in the validation stage. Finally, last section presents conclusions.

II. Materials and methods

A. IMU-based Motion Capture System

The motion capture system based on inertial sensors comprises two inertial measurement units (IMU). It was used IMU MPU9150 from Invensense, which is composed

I. Torres et al. (eds.), *VII Latin American Congress on Biomedical Engineering CLAIB 2016, Bucaramanga, Santander, Colombia, October 26th -28th, 2016*,
IFMBE Proceedings 60,
DOI: 10.1007/978-981-10-4086-3_173

of a tri-axis accelerometer, a tri-axis gyroscope and tri-axis magnetometer, which provide information of linear acceleration, angular velocity and angular position relative to the Earth's magnetic field, respectively. Data are sent through the I2C digital protocol. In addition, there is a controller based on Max32 chipKIT card from Digilent, which contains a microcontroller Microchip PIC32MX795F512L, a 32-bit controller that manages an operating speed of 80MHz. This card is programmed using an Arduino-based IDE environment.

The accelerations, angular velocities, and magnetic signals are collected and processed in the controller by an Euler-based fusion algorithm that calculate an absolute orientation estimation (roll, pitch and yaw) relative to a reference system formed by the gravitational acceleration and the Earth's magnetic field, which is shown in Fig.1.

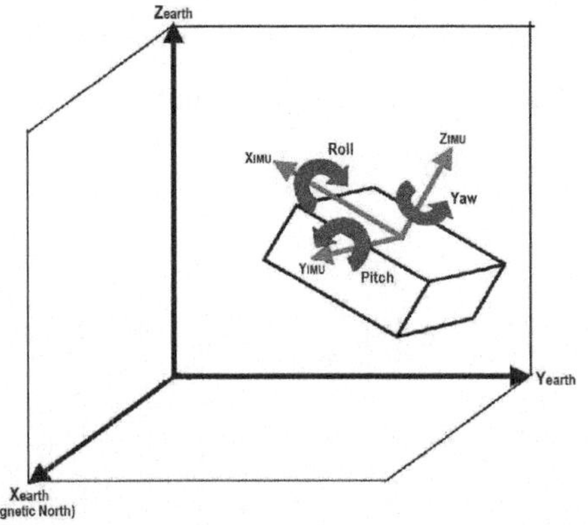

Fig. 1 Absolute orientation estimation (roll, pitch and yaw) of IMU relative to a reference system (earth's gravity and magnetic field)

The Euler angles are three angles to describe the orientation of a rigid body, and are commonly used to describe robotic arm movements. Euler's rotation theorem shows that any orientation can be described as three consecutive rotations. For instance, Fig. 2 presents a sequence of rotations that changes the orientation of a rigid body from the coordinate system OXY to Oxyz. The Euler angles (θ, φ, ψ) are also called "pitch", "roll", and "yaw" angles.

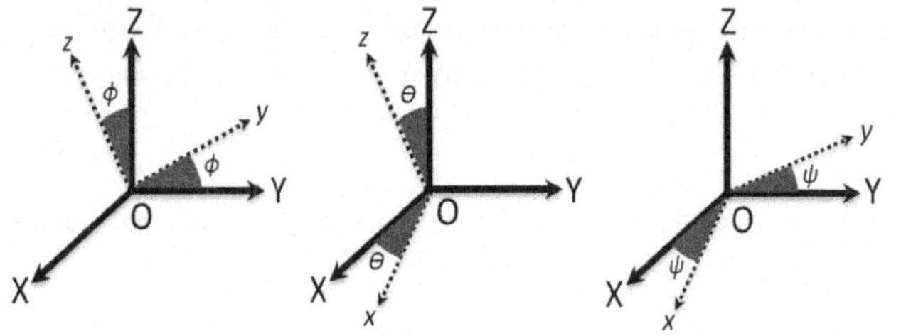

Fig. 2 Euler angles (θ, φ, ψ) definition.

It is possible to configure an acquisition frequency until 50 Hz, which is sufficient for normal human gait analysis. It was developed a software application that is responsible for reading the status of sensors and observing real-time data angles, angular velocity and linear velocity in an interface that puts the data in a graph and also export them to a text file for further off-line analysis. This interface was designed in Processing, which is a programming language based on Java that allows you to design graphical applications for Windows and Linux operating systems.

B. Validation

For validation, the IMU-based system was compared to a vision-based motion capture system. These systems are widely used to obtain joint kinematic data for human gait. The vision-based motion capture system used for validation consists of high speed cameras, software to capture and analyze the collected information and additionally, they can be passive, when retroreflective markers are used; active which use LED markers; or markerless when software is required to track the subject's movement. Although passive optical motion capture is the most accurate, flexible and common type, useful information can be lost due to marker occlusions [6] [7]. Optical motion capture systems should be calibrated before collecting a set of measurement, since the precision and repeatability of the data depends on this.

The vision-based motion capture system for acquiring joint kinematics data on the sagittal plane combines 2 cameras Basler SCA640-70 gc GIGE, passive reflective markers and commercially-available motion analysis software. The software MaxTraq 2D/3D from Innovision Systems Inc was used to capture and digitize the data and calibrate the system. Six reflective markers were attached on specific anatomic positions of the subject's right lower limb as shown in Fig. 3. These markers defined a system with 3 rigid segments, thigh, leg and foot, according to Helen Hayes protocol [8].

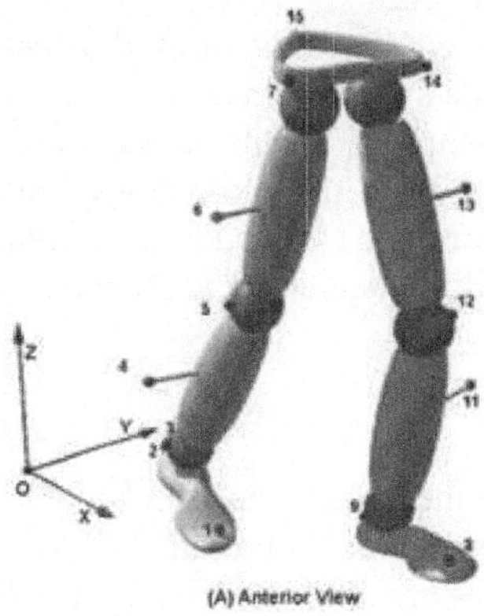

Fig. 3 Six reflective markers attached on specific anatomic positions of the subject's right lower limb according to Helen Hayes protocol [8].

C. Protocol

Reflective markers and IMUs were attached on the subject's right lower limb (see Fig. 4). One healthy subject participated in the experimental protocol. Two IMUs were fixed to right lower limb segments: IMU1 was attached to the thigh; IMU2 was attached to the shank. Test were conducted on a treadmill for different velocities: 1, 2 and 3 miles/hour which were called in this paper slow, normal and fast, respectively. Both systems were initialized simultaneously. Both systems were recorded during walking movements, with a sampling frequency of 30 Hz.

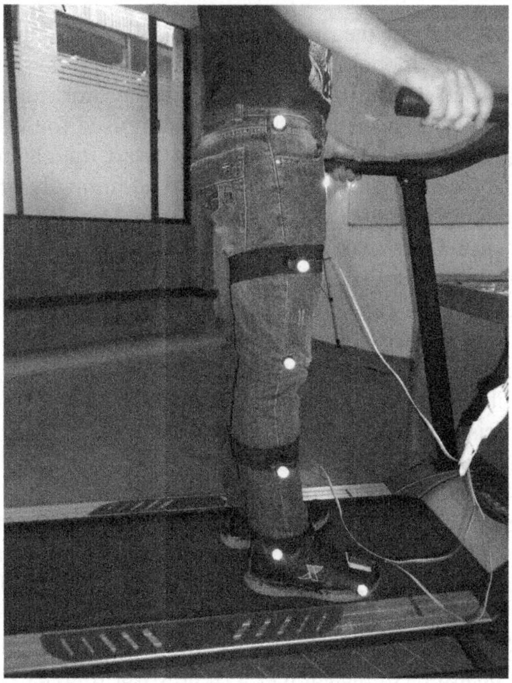

Fig. 4 Reflective markers and IMUs attached on the subject's right lower limb. IMU1: thigh; IMU2: shank.

It was aligned the axis from the vison-based motion capture reference system to axes to be measured from IMU-based motion capture reference system. With the information obtained from this axis, it was possible to find the relative angle for knee (sagittal plane) from calculated absolute angles for thigh and shank segments.

III. RESULTS

The concordance correlation coefficient ρC [9] was used to compare measurements provided by the IMU-based motion capture system and the vision-based motion capture system. The concordance correlation coefficient evaluates the agreement between two readings from the same sample by measuring the variation from the 45° line through the origin (the concordance line). It evaluates the degree to which these pairs fall on the 45° line. This method allows knowing whether a new method or device agrees well enough with other, as a suitable surrogate. A value of 1 denotes perfect concordance; a value of zero denotes its complete absence. McBride suggests a descriptive scale for values of the concordance correlation coefficient (see Table 1) [10].

Table 1. Descriptive scale for values of the concordance correlation coefficient.

Value of ρc	Strength of agreement
< 0.90	Poor
0.90 - 0.95	Moderate
0.95 - 0.99	Substantial
>0.99	Almost perfect

Fig. 5 presents data of knee joint angle of a high speed walking test.

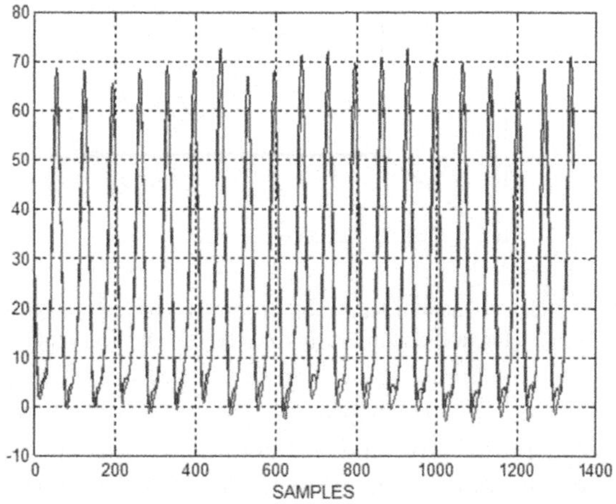

Fig. 5 Knee joint angle of a high speed walking test on treadmill, where red graph is from IMU and blue graph is from cameras.

The concordance correlation coefficient ρC found for slow-speed, normal-speed and high-speed walking are showed in Table 1. Results obtained from the developed system are consistent with those from the vision-based system.

Table 2 Concordance correlation coefficient ρC for slow-speed, normal-speed and high-speed walking tests

Movement	**ρc:** concordance correlation coefficient	Cb: bias factor correction	r: Pearson correlation coefficient

Slow-speed walking	0.9858	0.9912	0.9946
Normal-speed walking	0.9475	0.9612	0.9857
High-speed walking	0.9697	0.9936	0.9759

The concordance correlation coefficient ρc contains a measurement of precision ρ and accuracy Cb:

$$\rho c = \rho \ . \ Cb \qquad (1)$$

where ρ is the Pearson correlation coefficient, which measures how far each observation deviates from the best-fit line, and is a measure of precision, and Cb is a bias correction factor that measures how far the best-fit line deviates from the 45° line through the origin, and is a measure of accuracy.

IV. CONCLUSIONS

A low-cost IMU-based system for human knee joint angle monitoring is presented. The proposed device is suitable for gait analysis in ambulatory applications, taking into account that it is wearable, easy to use and shows a promising performance in terms of joint range of motion measurements in a dynamic working mode for normal velocities of human walking.

The system was validated with a set of experiments to monitor the knee flexion/extension angles for sagittal plane during walking on a treadmill for three velocities (low-speed, normal-speed, high-speed). Obtained results showed high correlation (>0.947) between measurements from the developed device and a vision-based motion capture systems, according to the concordance correlation coefficients obtained.

ACKNOWLEDGMENT

This research has been supported by Universidad Antonio Nariño under project 20141096.

CONFLICT OF INTEREST

The authors declare that they have no conflict of interest.

REFERENCES

1. Tamura T (2014) Chapter: Wearable Inertial Sensors and Their Applications. In the book Wearable Sensors: Fundamentals, Implementation and Applications, edited by Edward Sazonov, Michael R Neuman, Elsevier
2. Bonato P (2005) Advances in wearable technology and applications in physical medicine and rehabilitation. Journal of NeuroEngineering and Rehabilitation 2(2):1-4
3. Tik-Pui Fong D, Chan Y (2010) The Use of Wearable Inertial Motion Sensors in Human Lower Limb Biomechanics Studies: A Systematic Review. Sensors 10(12): 11556–11565
4. Troje NF (2002) Decomposing biological motion: A framework for analysis and synthesis of human gait patterns. Journal of Vision 2(5):371-387
5. Mündermann L, Corazza S, Andriacchi TP (2006) The evolution of methods for the capture of human movement leading to markerless motion capture for biomechanical applications. Journal of NeuroEngineering and Rehabilitation 3(6):1-11
6. Kirk AG, O'Brien JF, Forsyth DA (2005) Skeletal parameter estimation from optical motion capture data. Proc. vol. 2, IEEE Computer Society Conference on Computer Vision and Pattern Recognition, 2005, pp. 782-788
7. Kurihara K, Hoshino SI, Yamane K, Nakamura Y (2002) Optical motion capture system with pan-tilt camera tracking and realtime data processing. Proc. IEEE Int. Conference on Robotics and Automation, Washington, USA, 2002, pp. 1241-1248
8. Pribani T, Peharec S, Medved V (2009) A comparison between 2d plate calibration and wand calibration for 3d kinematic systems. Kinesiology 41(2):147-155
9. Lin LI (1989) A concordance correlation coefficient to evaluate reproducibility. Biometrics 45: 255-268
10. McBride GB (2005) A proposal for strength-of-agreement criteria for Lin's Concordance Correlation Coefficient. NIWA Client Report: HAM2005-062

Author: Andres Felipe Ruiz Olaya
Institute: Universidad Antonio Nariño
Street: Cra. 3 Este # 47A-15
City: Bogotá
Country: Colombia
Email: andresru@uan.edu.co

Case Study: Audiometry and ECR profile time convergence.

A.K. Quintana; Ma.P. Granados & J.M. Cornejo

Universidad Autónoma Metropolitana – Izt. Electrical Engineering Department. Audiology Laboratory. México

Abstract— **Through the follow up of a pediatric implanted patient in a period of time of eighteen months after CI initial activation, comparison of patient hearing threshold obtained by means standard audiometry and an objective method was made. Eighteen months after implantation Electric Cochlear Response (ECR) test and audiometry profile showed time convergence. This make think in the possibility to obtain an implanted patient audiometry estimation through ECR test. Because of its objective nature ECR test offers some advantages over standard behavioral audiometry. Not being patient health condition dependent, make possible to obtain an estimation of the patient hearing threshold since the very first time of the cochlear implant turning on. This feature allows to reduce the necessary time extension to reach a suitable electrical stimulation that provide a safe, useful and comfortable audition to implanted patient.**

Keywords— **Electric Cochlear Response, hearing threshold, pediatric patient.**

I. Introduction

Once initial activation of the Cochlear Implant (CI) to pediatric patient has been perform, audiometric evaluation is necessary to assure an audible and comfortable stimulation. Pediatric implanted patient audiometric evaluation may be a time consuming and difficult task. At that time patient has not experience enough in the use of the CI, proper training for a behavioral audiometry and has not yet alternative communication resources to provide proper feedback, among other limitations due to health condition.

Standard audiometry frequency range goes from 250 to 8 kHz for a sound intensity level range that goes from 0 to 120 dB_{HL} [1,2]. As a behavioral test audiometry needs the conscious patient collaboration, so its result is reliable for patient elder than 4 y.o. [3].

In the other hand standard audiometry frequency set do not match to the frequency band intracochlear electrode assignment made in the sound processor. This may result in an auditory threshold underestimation up to 3dB, if the sound stimulus frequency fall dawn outside the band-pass filter wide band for a given intracochlear electrode.

Audiologist adjust T/M current level of the electrode array looking for to provide comfortable, suitable and safe audition to patient, however standard audiometry result is the only clinical means to know if the implanted patient reach the audiogram normal audition zone [4].

To get a reliable audiogram for the implanted pediatric patient, it is necessary to await patient has gain experience and abilities enough to perform the test, besides become familiar with the CI use.

As an alternative way to overcome audiometry limitations to obtain pediatric implanted patient hearing threshold, we´ve been using an objective technique denominated Electric Cochlear Response (ECR) [5], in order to get feedback to assist the current level T/M adjustment of the electrode array.

In a previous work a number of 16 implanted children, average age of 4.4 y.o. was observed that ECR threshold value profile follows standard audiometry profile within a ± 5 dB band [6].

This work shows the follow up of the behavioral audiometry, ECR threshold values and the successive adjustment of the T/M current level of the electrode array, in a period of time of eighteen months after initial CI turning on for an individual patient.

At the end of this period of time an audiometry and ECR threshold value profile convergence was observed. This result make think the possibility to adjust T/M current level of the electrode array base on the ECR threshold value since the very first time CI turning on or when patient health condition do not allow to obtain a reliable standard audiometry.

II. Metodology

Population: Female, 1.8 y.o., patient of the Rehabilitation National Institute (RNI) in México City. Bilateral sensorineural hearing loss, analogical hearing aid used for six months previous to the implant surgery. Complete insertion of the electrode array. CI turning on one month posterior to implantation surgery.

A set of ECR test, behavioral audiometry and auditory performance index, Infant Toddler - Meaningful Auditory Integration Scale, (IT-MAIS) [7], provided by parents, were made four times (T_n; n=1,..., 4) within an observation period of eighteen months after implantation.

Test: Patient located inside an anechoic and sound proof room, 2.5x2.5x3.0 m, at the RNI. Sound field calibration

I. Torres et al. (eds.), *VII Latin American Congress on Biomedical Engineering CLAIB 2016, Bucaramanga, Santander, Colombia, October 26th -28th, 2016,*
IFMBE Proceedings 60,
DOI: 10.1007/978-981-10-4086-3_174

by substitution method according ISO 389-7 standard by using sound meter B&K 2235, filter B&K 1625 y un sound calibrator B&K 4230.

For ECR test an *ECR Monitor ECR* was used [5, 6]. Sound Intensity Level varies from 10 to 90 dB_{HL}. Sound stimulus frequency matched central frequencies in the electrode array. Stimulus presented to patient by means of loudspeaker located one meter away from patient's CI, 0° azimuth. Asleep and lying down patient. Two EEG recording channels, A1(–), A2(–), Cz(+) and FPz(GND), electrode impedance < 5kΩ. EEG filtering 300 Hz LPF 12dB/Octave and zero phase, 20 kHz sampling rate. ECR obtaining by 100 EEG epochs averaging time locked with sound stimulus presentation in a 50 ms time window.

For sound field Audiometry, an *Interacoustics* AC-30 Clinical Audiometer was used. Patient auditory threshold was established by calculating Pure Tone Average (PTA) for audiometry and ECR threshold value, PTA-A and PTA-RCE respectively in a frequency range from 500 to 2000 Hz. According PTA [dB_{HL}] value hearing level was classified as normal (0-20dB_{HL}), and hearing loose as mild (20-40dB_{HL}), moderate (40-60dB_{HL}), severe (60-80dB_{HL}) and profound (+ 80dB_{HL}) [2].

To determinate a possible relationship between ECR hearing threshold change and audiometry due to T/M adjustment and the IT-MAIS score, Pearson product correlation was used.

III. Results

Refers to female, 1.8 y.o., Advanced Bionics cochlear implant user, model Hires 90K/HiFocus and sound processor Harmony.

Table 1 shows patient's chronological and auditory ages, and T and M levels average current units (CU) in the electrode array, corresponding to T_n along a period of 18 months. Elapsed time between T_n and T_{n+1} was four months on average.

Table 1 Patient's chronological and auditory ages at T_n. T and M average level values in the electrode array.

ECR record		T_1	T_2	T_3	T_4
Chronological age	[years]	1.8	2.1	2.6	3.0
Auditory age	[years]	0.3	0.6	1.1	1.4
Average M level	[CU]	224±29	230±32	250±32	261±33
Average T level	[CU]	22±3	23±3	25±4	26±3

Figure 1 shows a decreasing ECR threshold value profile, meaning an improvement of the auditory threshold for all intracochlear electrodes as T and M level increases, see bottom row in Table 1. Additionally audiometric frequency data points show lack of coincidence to intracochlear electrode central frequency. From T_1 to T_2 audiometry does not exhibit any significant change in the hearing threshold profile, notice that figure 1 depicts just a few frequency data points at T_1 (500, 1000, 2000 y 4000 Hz) and three data points at T_2 (500, 2000 y 4000 Hz), all this due to non reliable audiometry. In contrast from T_3 to T_4 audiometry shows an auditory threshold improvement and more frequency data points. In the other hand in figure 1 is shown that ECR profile includes frequency data points that audiometry does not take into account.

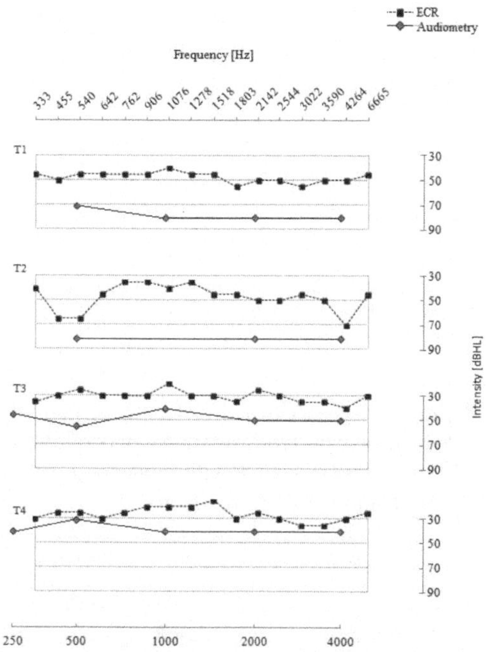

Fig. 1. ECR (■) and audiometry (♦) profiles for T_1 to T_4 observation time after CI initial programming, T_0 , expressed in years. $T_1= T_0 + 0.3$ y., $T_2= T_0+ 0.6$ y., $T_3= T_0+1.1$ y., and $T_4= T_0+1.4$ y. Upper horizontal axis shows ECR test frequencies, bottom horizontal axis shows audiometry test frequencies and right vertical axis intensity sound level [dB_{HL}].

In time PTA-ECR and PTA-A show a reduction value indicating a better patient hearing threshold as can be seen from T_1 to T_4 in Table 2. Also its difference in a given T_n observation is decreasing, however hearing threshold value obtained by ECR was always better than audiometric.

Table 2 PTA-ECR and PTA-A values for T_n.

		T_1	T_2	T_3	T_4
PTA-ECR	[dB_{HL}]	46±4	41±6	29±5	24±4
PTA-A	[dB_{HL}]	78±4	80±0	47±4	38±4

Figure 2 shows PTA intensity threshold level change exhibited by ECR and audiometry and their relationship to T and M current level adjustment. Another result of the gradual PTAs value decreasing from T_1 to T_4 is that PTA-ECR profile shows better correlation to T level profile (r^2 = 1.0) and M level profile (r^2 = 0.99) than PTA-A profile to T level profile (r^2 = 0.93) and M level profile (r^2 = 0.96).

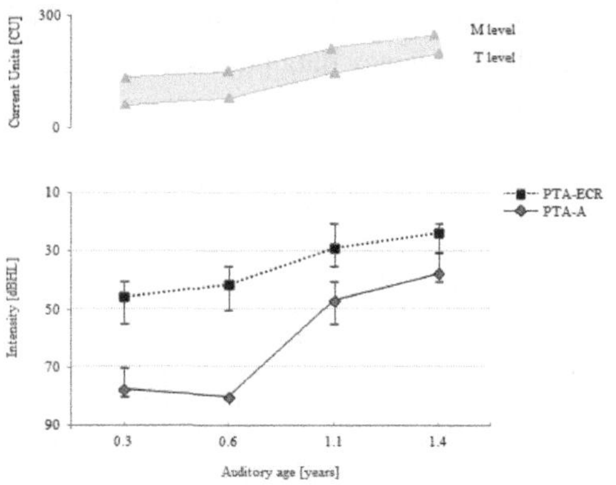

Fig. 2. Top: T and M current averaging level increasing graph from T_1 (after 0.3 years of CI use) to T_4 (after 1.4 years of CI use). Bottom: ECR and audiometry average thresholds differences from T_1 to T_4.

According to IT-MAIS score, figure 3, patient auditory performance improves from T_1 a T_4. Correlation between IT-MAIS score and PTA-ECR value was better than PTA-A value, r2 = 0.81 and r_2 = 0.59 respectively.

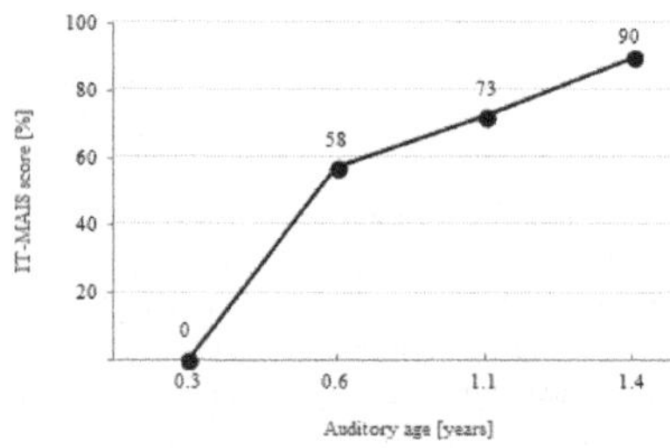

Fig. 3. IT-MAIS score improves from T_1 (after 0.3 years of CI use) to T_4 (after 1.4 years of CI use).

IV. DISCUSION

To estimate the benefit provided by a CI to patient is necessary to know patient hearing threshold as a result of CI operation parameters settings including T and M current level. Probably this benefit will be grater if the patient hearing threshold could be known since CI initial activation.

Pediatric patient audiometric evaluation is restricted by patient cooperation. This is because of age, general health condition and standard audiometry intrinsic methodological limitations when applied to implanted patient, along with a not knowledgeable CI use. This can be seen in figure 1 where audiometry for T_1 and T_2 only includes four and three frequency data points respectively. Considering residual hearing system sensibility to current stimulation change along cochlea, audiometry in T_1 and T_2 does not make sense because it is unlikely that a given sound intensity lead to the same hearing threshold for same T current level in two different intracochlear electrodes.

Patient audiometry gets better from T_1 to T_4 however high and low frequency information might be affected by error because frequency sound stimulus could be out of the intracochlear electrode or CI wide band, which is the case for 250 Hz frequency, figure 1. In contrast ECR provided information concerning hearing threshold for every single intracochlear electrode since T_1.

Similarity is observed between PTA-ECR and T/M profile, from T_1 to T_3 (figure 2). This suggest that T/M adjustment is related to moderate to mild hearing loss, as established by ECR instead of severe to moderate hearing loss as audiometry indicates.

In T_4 ECR and audiometry point out to mild hearing loss, however a PTA-A= 38 dB is closer to a moderate hearing loss meantime PTA-ECR= 24 dB is closer to a normal hearing.

Comparison between hearing threshold and patient auditory performance at time T_1 (after 3.6 months of CI use) shows a better relationship between severe hearing loss as reported by audiometry (PTA-A = 78 dB) and the IT-MAIS score = 0%.

At T_2 (after 7.2 months of CI use) audiometry classification indicate a severe hearing loss, however with an IT-MAIS score = 58%, is better related to a moderate hearing loss as indicated by PTA-ECR = 41 dB.

In T_3 and T_4, when difference between PTA-ECR and PTA-A decrease to 46% and 36 % compared to T_2, an IT-MAIS of 78% and 90% is better related to a mild hearing loss, as indicated by PTA-ECR = 29 dB and 27 dB, respectively, than PTA-A.

Correlation coefficients between hearing threshold changes PTA-ECR and the T/M levels adjustment, and IT-MAIS score; $r^2_{\mathrm{RCE-nivel\,T}} = 1.0$, $r^2_{\mathrm{RCE-nivel\,M}} = 0.99$ and $r^2_{\mathrm{RCE-ITMAIS}} = 0.81$, confirmed the relationship between ECR and true patient hearing threshold.

V. CONCLUSIONS

As objective test ECR is not affected by the patient health condition and it is possible to use it to make an objective audiometry estimation, allowing an individual T/M adjustment of the electrode array, improving the benefit provided by the cochlear implant to the patient in a short period of time.

Conflict of Interest

"The authors declare that they have no conflict of interest".

References

1. Dobie R A and Van Hemel S 2004 Hearing Loss: Determining Eligibility for Social Security Benefits., ed R A Dobie and S Van Hemel (Washington (DC): National Academies Press)
2. Katz J 2009 Hand book of clinical audiology
3. Gifford R H (2013) Adult Implant Recipients. Cochlear Implant Patient Assessment: Evaluation of Candidacy, Performance, and Outcomes, 85. (San Diego, CA: Plural Publishing)
4. Baudhuin J, Cadieux J, Firszt J B, Reeder R M and Maxson J L 2012 Optimization of programming parameters in children with the advanced bionics cochlear implant J Am Acad Audiol 23 302-12
5. Cornejo Cruz J M and Granados Trejo M d P 2012 Systems and methods for detecting and using an electrical cochlear response ("ECR") in analyzing operation of a cochlear stimulation system US Patent Number,165,687
6. Quintana A, Beltran N, Granados M, Chamlati E, Mena M and Cornejo J 2015 Objective Approach to Audiometry in the Pediatric Implanted Patient. In: VI Latin American Congress on Biomedical Engineering CLAIB 2014, Paraná, Argentina 29, 30 & 31 October 2014: Springer) pp 707-10
7. Monteiro E., Feitosa C., Catanhede P. (2008) Comparison between the ITMAIS and MUSS questionnaires with video-recording for evaluation of children who may receive a cochlear implantation, Rev Bras Otorrinolaringol, 74(1):91-8.

Author: Agar Karina Quintana López
Institute: Universidad Nacional Autónoma Metropolitana - Iztapalapa
Street: Prolongación Canal de Miramontes 3855, Col, Exda. San Juan de Dios
City: Mexico City
Country: Mexico
Email:agar_k@hotmail.com

CONACYT Scholarship no. 225855

Computer models for ions under electric and magnetic fields: random walks and relocation of calcium in dendrites depends on timing and population type

J.F. Gomez-Molina[1*], M. Corredor[2], A.A. Restrepo-Velasquez[3] and U.M. Ricoy[4]

[1] International Group of Neuroscience, IGN (Colombia, USA, Europe, India), address: Cra 64c #48-94 (603) Medellin, Colombia
[2] Genetics and Biochemistry of Microorganisms group (GEBIOMIC), University of Antioquia, Biology Institute, FCEN, address: calle 67 No. 53 – 108, block 7, office: 318, Medellín, Colombia
[3] Informática y Sistemas, EAFIT University, address: Carrera 49 N° 7 Sur-50, Medellín, Colombia
[4] Department of Biology, Chemistry and Environmental Science, Northern New Mexico College, Española, address: 921 N. Paseo de Oñate Española, NM 87532 NM, USA

Abstract— **In this computational study we analyze segregated population of ions in biological tissues and how electric or magnetic fields can relocate them. The appropriate definition of a segregated population of ions is justified for its physiological relevance, algorithmic simplicity and biophysical realism. Although this study can be valid for several ions and cell compartments, we focus on calcium ions in parallel dendrites of neurons. Computer simulations are presented as calcium flux visualizations showing the final position of each ion in different conditions in neurons: in the absence of an electric field and in the presence of it at different timings in relation to the initial release event of calcium into the dendrites. The simulation suggests that it is possible to relocate (probabilistically) aggregations of calcium ions in the spaces of the dendrites, allowing neuromodulation of synaptic connections. In conclusion, the maximal response to endogenous electric fields and the efficient way to design "friendly" devices for electrical field stimulation of neurons for relocating calcium ions close to their targets (e.g. vesicle sensors, proteins in membranes, or cytosol) depends on the geometry of dendrites, the duration and timing of the field (respect to ongoing activity), and the selection of the appropriate subpopulation we want to relocate.**

Keywords— **Electric Fields, Calcium dynamics, Brownian motion, computational neuroscience, dendritic electrophysiology.**

I. Introduction

Ions move in a *random walk* (sequences of steps in which they change direction in a probabilistic form) *toward certain targets* (e.g. proteins). Brownian motion is the random motion of particles. The mathematical model of Brownian motion has numerous applications in ions transport in neurons. However, the movement of ions in neurons is a probabilistic phenomenon affected by fields. Field effects might not be always modeled as terms (slight bias) added to the random motion of the ion only in certain regions. It was initially assumed, for instance, that Calcium dynamics was passive in the dendrites. Now it is known that relocation of calcium ions produce effects that depend on dendritic geometry and is guided by active properties [1-4]. A stable resting membrane potential (down state) and a metastable, or transiently stable depolarized state (up state) have been described in neurons [5] and these states interact with the dynamics of calcium. The last has been incorporated in deterministic, 1D-models of dendrites [6-8].

Endogenous electric fields from groups of neurons can also affect other neurons directly [9]. Long-range electric signaling and mesoscopic wave-phenomena play an important role [10-13]. The effects of fields in the whole trajectory of ions are clearer for exogenous fields (e.g. applied by *transcranial electric and/or magnetic stimulation*, TEMS, [14]). How to stimulate the excitable tissue in a non-invasive and "friendly" way? If endogenous electric fields guide neural activity [15], how and when can exogenous fields be applied in order to enhance (or attenuate) the natural/healthy (or pathological) effects of the first ones?

In this work we visualize, with a computer simulation, how the spatial location of calcium ions, after entering one compartment, can be modified by electric fields. A probabilistic, 2D-approach for individual ions is proposed. Unlike other stochastic models, we simplify the dynamics by segregating the population of calcium ions in 3 subpopulations. We analyze the subpopulation with the simplest behavior. It is proposed that, for the other subpopulations, the action of reactants and buffers can be described by making zero (or different than zero) the probabilities of location of the ions, as they bind to a buffer (or are released from them). We focus on dendrites, but the relative simplicity of the model suggests that some conclusions can be valid to synaptic boutons and extracellular spaces.

II. Materials and methods

In order to simplify the model, we define two fields: 1. *a general field* representing the random collision of particles, the action of restricted spaces, membranes, in homogeneities, buffers or even "macromolecular networks" [13, 16],

I. Torres et al. (eds.), *VII Latin American Congress on Biomedical Engineering CLAIB 2016, Bucaramanga, Santander, Colombia, October 26th -28th, 2016*,
IFMBE Proceedings 60,
DOI: 10.1007/978-981-10-4086-3_175

and 2. *An electric or magnetic field* [9, 11], which can be endogenous or exogenous. The general field involves, then, any non-electric (or magnetic) agent that affects the probabilistic variables of the ion: direction, length or duration of each step in the random walk of the ion [17, 18]. This scenario follows the modeling philosophy of a "toy model" [12, 18].

A. Model and Material

One important strategy is to segregate the whole population of calcium ions in a dendritic compartment into three (sub)populations of ions:

P1-ions. They bind to buffers in few steps.

P2-ions. They stay free for the whole simulation.

P3-ions. They are released from buffers.

Inspired by conventional models of Brownian motion, we simulate the behavior of P1, P2 and P3 as sequences of random positions of either ion coordinates, $(x, y) = \mathbf{r}$, or probability of location of ions, $p(\mathbf{r}, t)$.

We focus here on the position of P2-ions. P1/P3-ions can phenomenologically "disappear or appear": $p(\mathbf{r}, t)$ depends on the probability of presence of reactants. P2-ions do not present this behavior (they do not interact with reactants). Diffusible agents [6] can then be modeled by $p(\mathbf{r}, t)$ with (anti)particle-like properties (i.e. particles able to create and/or annihilate free-calcium charges; not shown here).

Finally -after capturing individual single ion behavior- the final ion positions we claim to simulate correspond to the average of multiple ion subpopulations of synchronized dendrites with similar cell-membrane orientation. The expected increase with the field in the apparent speed of $p(\mathbf{r}, t)$, $dp(\mathbf{r}, t)/dt$, depends on average population levels (columnar states) and it was analyzed elsewhere [16, 19].

B. Computational resources

Simulation was elaborated mainly in Matlab (MathWorks). An equivalent version was developed also almost entirely with a combination of Python and hoc, in order to use the powerful tools of NEURON to define region-objects and reaction-diffusion modules (e.g. by writing *rxd.Region*, *rxd.Species*, etc [6,8]). Modeling approaches [1-8] were reinterpreted as explained in this section. Numerical simulations and methods were compared.

III. Results

To preserve all the spatial information, outcomes are presented as representative visualizations (instead of statistical graphs) showing the final position of each calcium ion in

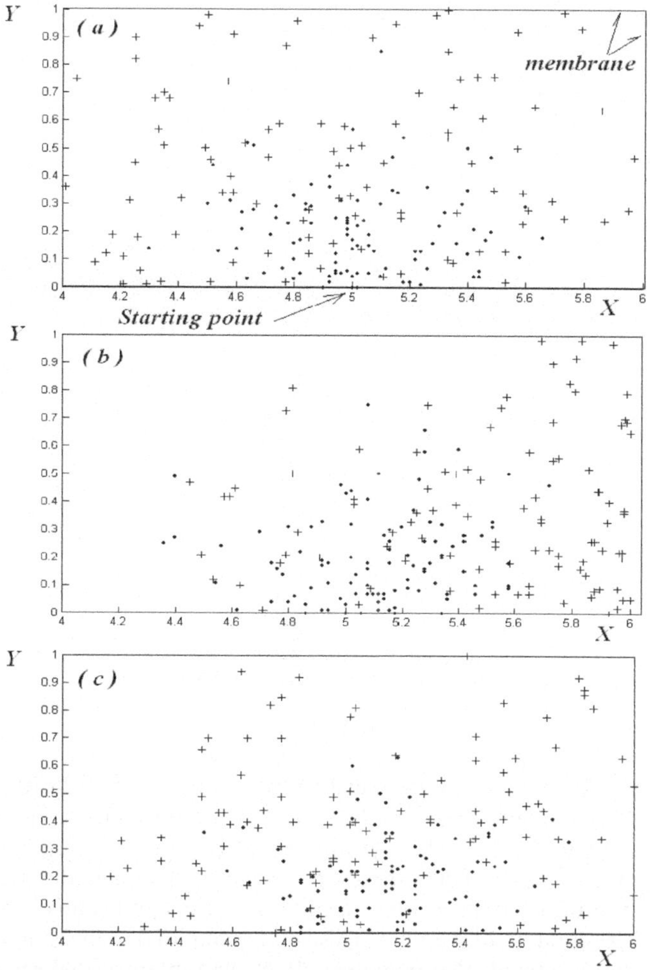

Fig. 1. For the three graphs, the final position of calcium ions in dendrites (P2-population) is represented at time $t=T/3$ by dots (.) and at $t=T$ by cruses (+). T is the number of steps of the simulation. (a) Without electrical field. (b) With an electrical field applied in the interval between 0 and T. (c) With an electrical field only between 0 and T/3. The targets (e.g. proteins) are located at probabilistic positions in the right wall.

different conditions: in absence of an electric field and in presence of it at different time intervals (figure 1).

In Fig. 1, the final position of P2-ions at time $t=T/3$ (.) and at $t=T$ (+) is shown. The electric field is positive, toward the right in all cases, except for (a) where it is zero for the whole time interval. The electrical field is applied between 0-T (for the whole interval) in (b) and only between the subinterval 0-T/3 in (c).

In Fig. 2, the available space is reduced by half. (a) shows the final position of calcium ions without electrical field. (b) shows the field applied from 0 to T. (c) the field is applied only the first T/5 msec. In (d), the field is applied for the same duration but at the end (from T/5 to T).

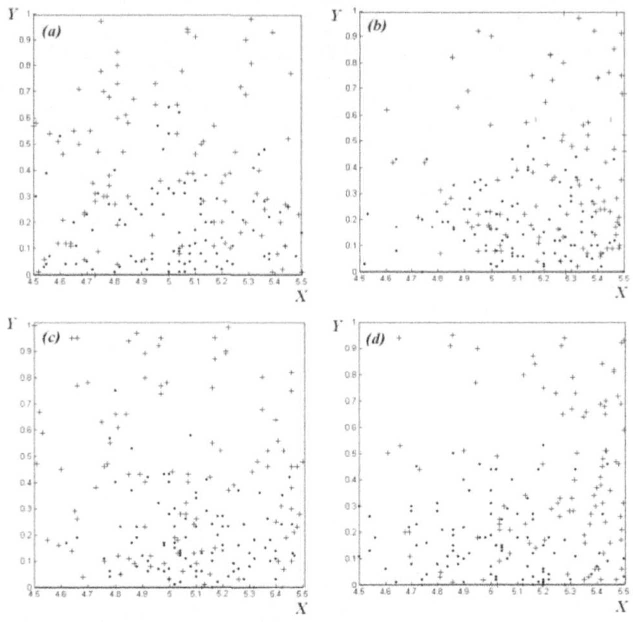

Fig. 2. Same as in Fig.1 but reducing the space by half. (a) Final position of calcium ions without electrical field. (b) Field applied from 0 to T. (c) Field applied only the first T/5 msec (early subinterval). (d) Field applied only from T/5 to T (late subinterval).

After running the simulation several times, the following robust relationships can be intuitively proposed from visual inspection: 1. The *spatial heterogeneity in Calcium-distribution:* (i) decreases with geometry (area) and the time for diffusion but (ii) not seem to increase significantly with the duration of the field. This is important to design stimulation protocols with maximal effects and minimal exposure. 2. The longer the dimension of the compartment in the dimension of the field, the more capacity the system shows to respond to a field. 3. The strength (efficiency) and sign (additive or subtractive) of the effects of diffusion and the electric field depends on the subinterval in which the field is applied. For ions with *atypical random walks:* Late subintervals (closest to the time T at which calcium is approaching the target) are more efficient than late subintervals (with respect to t=0, when calcium is released from a channel or from a starting point; compare Fig. 2(c) vs. 2 (d)).

IV. DISCUSSION

When the number of molecules is low (e.g. P2), stochastic simulation may be more appropriate than deterministic simulation [6-8]. The longer the simulation time, the shorter P2, and therefore the stronger the need for stochastic simulation. NEURON and GENESIS [4] uses calcium concentration averages and it is relatively easy to segregate the population of calcium using these powerful packages. In this way, their opposite effects and drifts do not cancel out prematurely between them. Weak electric fields can guide subpopulations of ions *with atypical trajectories* but not gross concentrations (averages) of them.

NEURON has a GUI tool (ChannelBuilder) that makes it easy to specify voltage-gated calcium channels. To solve differential equations for channels and diffusion, numerical integration methods (e.g. Crank-Nicholson) are prone to numerical oscillations if *dt* is too long but implicit Euler provides robust stability. Random-walk algorithms do not have these issues. The slight bias term of the random walk can be interpreted as the action of endogenous and/or TEMS-fields synchronized with the expected time of calcium release in multiple parallel dendrites of a principal cell (e.g. Purkinje cell or Pyramidal cell). An arriving wave of activity in axonal inputs can generate such synchronized calcium release. Friendly brain electrical stimulation can then mimic (amplify) endogenous field instead of impose them and reduce (attenuate) unwanted fields [12, 16].

An important concern in brain electrical stimulation is to produce maximum positive effects with minimal adverse effects [14]. We have shown, using a relatively simple toy model, that by changing the timing of the field (subinterval position) a change in the distribution of calcium is evident. Using EEG we can estimate the time at which ion channels open ("EEG-triggered TEMS" [10, 12, 16]).

For future clinical applications of TEMS, hybrid algorithms (stochastic and deterministic, [17]) should use both the biological realism and efficient numerical integration methods of NEURON or GENESIS (e.g. by adjusting integration order, time step and discretization parameters of the cable equation as necessary to satisfy a local error criterion, [6-8]) and the advantages of stochastic models. The last show $p(\mathbf{r}, t')$ for atypical individual ions. They are the target ions to relocate, using TEMS at a subinterval $a < t' < b$.

Although the sophistication of other stochastic methods [17, 18] is recognized, the intuitive understanding of mechanisms might be sacrificed. Instead, toy models allow, for its simplicity, formulation of robust relationships –at the expenses of biological realism.

Additional considerations can further be explored: 1. The strength of the field (and perhaps their frequency [10, 16, 20]) determines the population of calcium ions that is going to respond more strongly. 2. As the angle between the *electric field* and the *buffer-target vector* increases the probability that P3-ions find the target decreases (for angles less than 180 degrees). 3. Fast charge transients of Ca-ions can exist in Ca-binding and Ca-Current-phenomena [12]. Charge and spin unbalances of hydrogen ions can be important in magnetic resonance and TEMS. If quantum-like models need to be used for these cases, $p(\mathbf{r}, t)$ and the notion of (re)location probability can be more computationally powerful than the more conventional and deterministic concept of calcium concentration.

V. Conclusions

1. Although highly simplified, the simulation suggests that it is possible to relocate aggregations of calcium ions in the spaces of the dendrites, allowing neuromodulation of synaptic connections. Also, the efficient way to apply TEMS depends strongly on the timing respect to outgoing physiological activity (i.e. the moment of calcium release). With the use of repetitive TEMS [14] any small difference in calcium is expected to be clinically important.

2. Single calcium ions produce important physiological effects [4, 6-8, 17-20]. The number of active *calcium ions is also generally small.* Therefore, probabilistic algorithms for single-particles present unique properties respect to collective approaches. Intriguingly, fuzzy and exotic probabilities [10] might describe charge relocation and charge transients [12] as a new computational property of cells analogous to state superposition in quantum computing.

3. Weak Electric fields can change the *final location probabilities of P2-ions* and *atypical* calcium ions, allowing neuromodulation of synaptic connections. An appropriate interpretation of simple probabilistic models ("toy models") can unravel how weak fields might amplify significantly–if applied at the right conditions- the unusual behavior of few calcium ions.

Acknowledgment

In this independent IGN-initiative ("science for friendly interaction") we are thankful to international advisors of IGN, UdeA, EAFIT (Colombia), UTSA and NNMC(USA).

Conflict of Interest

The authors declare that they have no conflict of interest.

References

1. Anwar H, Roome CJ, Chen W *et al.* (2015) Dendritic diameters affect the spatial variability of intracellular calcium dynamics in computer models. Front. Cell. Neurosci., 23 July

2. Llinas R (2014) Intrinsic electrical properties of mammalian neurons and CNS function: a historical perspective Front. Cell. Neurosci., 04 November | http://dx.doi.org/10.3389/fncel..00320

3. Santamaria F, Wils S, De Schutter E *et al* (2011) The diffusional properties of dendrites depend on the density of dendritic spines. Eur. J. Neurosci. 1460-9568.

4. Bower JM (2015) The 40-year history of modeling active dendrites in cerebellar Purkinje cells: emergence of the first single cell "community model" Front Comput Neurosci. Oct 20;9:129. doi: 10.3389/fncom.2015.00129.

5. Wilson CJ (2008) Up and down states. Scholarpedia 3:1410.

6. McDougal RA, Hines ML, Lytton WW (2013) Reaction-diffusion in the NEURON simulator. Front Neuroinform. Nov 15;7:28. DOI: 10.3389/fninf.2013.00028.

7. Neymotin SA, McDougal RA, Sherif MA *et al.* (2015) Neuronal calcium wave propagation varies with changes in endoplasmic reticulum parameters: a computer model. Neural Comput. Apr;27(4):898-924. DOI: 10.1162/NECO_a_00712. Epub Mar 3.

8. McDougal RA, Morse TM, Hines ML *et al.* (2015) ModelView for ModelDB: Online Presentation of Model Structure. Neuroinformatics. Oct;13(4):459-70. DOI: 10.1007/s12021-015-9269-2.

9. Anastassiou CA1, Perin R, Markram H et al. (2011) Ephaptic coupling of cortical neurons. Nat Neurosci. Feb;14(2):217-23. DOI: 10.1038/nn.2727. Epub Jan 16.

10. Gomez JF (2003) Ionic channels and long-range electrical signals: a probabilistic interaction. Med Hypotheses. Vol 60, 4, 463-467

11. Weiss SA, Faber DS (2010) Field effects in the CNS play functional roles. Front Neural Circuits. May 18;4:15. doi: 10.3389/fncir..00015. eCollection

12. Gomez-Molina JF, Restrepo-Velazques A, and Botero-Posada LF (2015) Field Generated by Waves, Sequential Activations and Apparent Motion: Effects and Typical Patterns Revista Ingenieria Biomedica 9:17 http://repository.eia.edu.co/revistas/index.php/BME/article/view/732

13. Scholkmann F (2015) Two emerging topics regarding long-range physical signaling in neurosystems: Membrane nanotubes and electromagnetic fields. J Integr Neurosci. Jun;14(2):135-53. doi: 10.1142/ S0219 635215300115. Epub May 11.

14. Fregni F, Pascual-Leone A (2007) Technology Insight: noninvasive brain stimulation in neurology—perspectives on the therapeutic potential of rTMS and tDC Nat Rev Neurol 3, 383-393 DOI:10.1038/ncpneuro0530

15. Fröhlich F, McCormick DA (2010) Endogenous electric fields may guide neocortical network activity. Neuron. Jul 15;67(1):129-43. DOI: 10.1016/j.neuron 06.005

16. Gomez-Molina JF, Ricoy UM, J. Velez-M.D., D. Sepulveda-Falla (2014) Reestablishing Ca2+ amplitude and speed during deep sleep in Alzheimer´s disease with EEG-triggered TES/TMS: Neuromodulation of abnormal columnar states. Abstract Society for Neuroscience Meeting. DOI: 10.13140/RG.2.1.3568.8403 Program 93.11/BB36

17. Wieder N, Rainer HA, Wegner F (2011) Exact and Approximate Stochastic Simulation of Intracellular Calcium Dynamics J Biomed Biotechnol. 2011; 572492. PMCID: PMC3216318

18. Blackwell KT (2013) Approaches and tools for modeling signaling pathways and calcium dynamics in neurons. J Neurosci Methods. 220(2):131-140

19.Gomez-Molina JF, Corredor M, Restrepo A, Ricoy U (2015) Computer Simulation for ions under electric and magnetic fields: from random walks in aqueous solutions to stochastic manifolds for calcium location probabilities in microbes and neurons. III Colombian Congress on Computational Biology and Bioinformatics Organized by University of Antioquia, EAFIT Univ, CIB, ITM, UPB, CES. DOI: 10.13140/RG.2.1.4097.4805 http://www.researchgate.net

20. Ricoy UM1, Frerking ME (2014) Distinct roles for Cav2.1-2.3 in activity-dependent synaptic dynamics. J Neurophysiol. Jun 15;111(12):2404-13. DOI: 10.1152/jn.00335.2013. Epub Feb 12.

Address of the corresponding author*:

Author: J.F. Gomez-Molina., Institute, member: International Group of Neuroscience, IGN (Colombia, USA, Europe, India)
Street: Cra. 64c #48-94 (603)
City: Medellin
Country: Colombia, South America.
Email: gomezmoljfmedcol@gmail.com

Structural and functional genomics analysis of methyltransferase genes and networks associate to understand antibiotic resistance inside the pangenome of *Pseudomonas aeruginosa*

Diana Carolina Castaño[1], Jeanneth Mosquera-Redón[1], Mauricio Corredor[1*]

[1] University of Antioquia, Biology Institute, FCEN, Genetics and Biochemistry of Microorganisms group (GEBIOMIC), address: calle 67 No. 53 – 108, block 7, office: 318, Medellín, Colombia

Abstract—**Multi-drug resistant *Pseudomonas aeruginosa* has multiple issues, all associated with the diverse genetic and epigenetics traits present in this pathogen, ranging from multidrug resistant genes to the molecular machinery for the biosynthesis of biofilms developed by *quorum sensing*. In this work was established bioinformatics strategy to study resistance inside the pangenome and compared with Colombian resistant strains. Using the information hosted in PATRIC database of *P. aeruginosa* from 200 genomes, was very important to understand the genomics structure and the relationship between antibiotic resistance associated to *rsmH*, *rsmG* and *rsmI* methyltransferases genes and proteins. This bioinformatics analysis and the sequencing of *rsmG*, *rsmH* and *rsmI* in *P. aeruginosa* from Colombia, was useful, establishing a correlation inside the pangenome from this bacteria. There is an association between genomic structure and function to draw together antibiotic resistance with those genes in the pangenome.**

Keywords— **Bioinformatics, pangenome, methyltransferase, antibiotic resistance, multi-drug resistant, structural and functional genomics.**

I. Introduction

Major issues in the development of a successful *Pseudomonas aeruginosa* vaccine arise from the probable genotypic variation at the strain level, making *P. aeruginosa* a presumably antigenically variable organism. Results supporting this assumption have been reported, yielding genetic information from the *P. aeruginosa* genome. For example, genetic variability explored in multiple *P. aeruginosa* isolates from different regions of the world indicated that *pcrV*, a member of the type III secretion system, exhibits limited genetic variation in terms of non-synonymous substitutions [1].

Previously, it was constructed a pangenome of *P.aeruginosa* [2]. The *P. aeruginosa* pangenome was estimated to contain more than 16,000 non-redundant genes, and approximately 15 % of these constituted the core genome. Functional analyses of the accessory genome indicated a wide presence of genetic elements directly associated with pathogenicity. Now we want to emphasize in antibiotic resistance.

In recent years have become important scrutinize on the molecular function of the ribosome and its importance in the processes of antibiotic resistance and post-translational modifications that occur in the rRNAs and tRNAs were involved in their function [3]. Of particular interest is the study of the methylations present in the bacterial 16S rRNA and 23S, because mechanisms are established in several human pathogens such as *P. aeruginosa* to acquire resistance to antibiotics. One interesting group of proteins acting as RNA-modifying enzymes is composed of AdoMet- (or S-adenosyl-L-methionine) dependent RNA methyltransferases (MTases). Globally, enzymes that methylate RNA comprise two major classes of MTases according to their structure core: i) Rossmann-Fold MTases (RFM) including almost all the N and C methylases and modify nucleobases; ii) SPOUT MTases consisting of 2'-*O*-methylases which act essentially in tRNAs with very few exceptions [4,5].

This bio-computational study, was supported with new molecular data in order to deepen in antibiotic resistance in the *P. aeruginosa* pangenome using the *rsmG*, *rsmH* and *rsmI* genes from Colombian strains resistant to aminoglycosides.

II. Material and methods

A. *Downloaded genomes*

Genome information from *P. aeruginosa* strains was downloaded via the ftp server from the PATRIC database [6]. A set of 200 available genomes (*ffn* files) was retrieved from the PATRIC database, April 2016 release. Estimation of the *Pseudomonas aeruginosa* pangenome size was assessed in a similar manner to that previously reported as genome- and gene-oriented methods using iterative and combinatorial approaches [7, 8, 9]. Briefly, a BLAST-based iterative method was used to extract the full set of non-redundant genes representing the *P. aeruginosa* pangenome. A single iteration consisted in a random selection of a strain as pangenome primer, then the remaining set of strain was randomly incorporated to the pangenome. The above process was calculated over

I. Torres et al. (eds.), *VII Latin American Congress on Biomedical Engineering CLAIB 2016, Bucaramanga, Santander, Colombia, October 26th -28th, 2016*,
IFMBE Proceedings 60,
DOI: 10.1007/978-981-10-4086-3_176

200 iterations with random permutation of the strain order in every iterative step. At the other hand, the *rsmG*, *rsmH* and *rsmI* genes were sequenced in *P. aeruginosa* Colombian strains. These strains are listed in table 1, where only 6 over 200 strains are shown there.

Table 1. Some strains inside this study. For the remained strains see [1]

Strain	Host	Country	Resistance G	A	Ce	Ci	Co	I	L	P	T
P. aeruginosa 6077	Human*	USA	R	S	S	S	S	S	S	S	R
P. aeruginosa BL17	Human*	USA	R	R	S	S	S	S	In	S	S
P. aeru. WHPSA001	Human*	USA	S	S	R	R	In	S	R	R	S
P. aeru. WHPSA008	Human*	USA	R	R	S	S	S	S	In	S	S
P. aeruginosa E2	Tomato*	USA	S	S	S	S	S	S	S	S	S
P. aeruginosa PAO1	Human*	USA	U	U	U	U	U	U	U	U	U
P. aeruginosa S1	Human	Colombia	R	R	U	U	U	U	U	U	U
P. aeruginosa S2	Human	Colombia	R	R	U	U	U	U	U	U	U
P. aeruginosa S3	Human	Colombia	R	R	U	U	U	U	U	U	U
P. aeruginosa S4	Human	Colombia	R	R	U	U	U	U	U	U	U
P. aeruginosa S5	Human	Colombia	R	R	U	U	U	U	U	U	U
P. aeruginosa S6	Human	Colombia	R	R	U	U	U	U	U	U	U

* Genome available, G: gentamicin, A: amikacin, Ce: ceftazidime, Ci: ciprofloxacin, Co: colistin, I : imipenem, L: levofloxacin, P: piperacillin, T : tobramycin, R: resistant, S: susceptible, In: Intermediate and U: Unknown.

B. *Alignments of Genome*

Some interesting genomes were visualized with UGENE 1.16 and Artemis 16.0.0 tools [10, 11] and the structures were aligned with ACT and Mauve suites [12, 13]. This analysis was used not only for search the orthologous and paralogous genes but also to find non-redundant genes.

C. *Methyltransferase genes and proteins*

Functional annotation of genes was performed using the KEGG Automatic Annotation Server for KEGG Orthology [14]. KEGG functional modules and ontologies were explored in the KEGG BRITE database [15]. On the other hand, The DNA fragments of *rsmG*, *rsmH* and *rsmI* genes obtained by PCR were sequenced both strands by automatic Sanger method on an ABI 3730 XL sequencer. The sequences were aligned with respective genes from *P. aeruginosa* PAO1 using the package Jalview 2.8.1 [16]. The alignments were carried out with MUSCLE [17], inside the program.

D. *Functional genomics analysis*

RsmG, RsmH and RsmI proteins were modeled with portal Phyre2 web-based services (http://www.sbg.bio.ic.ac.uk/phyre2/) [18] and their structure were analyzed with Chmera suite 1,1 [19]. In addition, the network of those proteins were downloaded from STRING database (http://string-db.org/cgi/input.pl) [20] and reanalyzed by Cytoscape 3.2.0 program [21].

III. Results

A. *Alignments of genomes*

The alignment of genomes exhibits strong variability in structure as shown Fig. 1. The host is irrelevant between animals and plants and it is complicated to see the relationship with antibiotic resistance or susceptibility. The arrangements of some specific regions are constant but specific regions in *P. aeruginosa* are conserved, however specific region in *P. aeruginosa* 6077, the first in Fig. 1. (2.000.000-2.500.000 bp) is dispersed in BL17, WHPSA001, WHPSA008, E2 and PAO1strains.

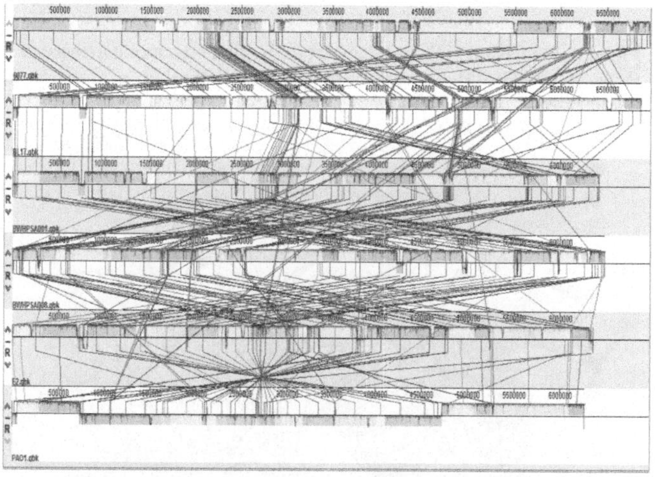

Fig. 1 Alignments of 6 *P. aeruginosa* genomes compared by Mauve, from 200 genomes of pangenome analyzed. Each genome corresponds to the table 1 in the same order: 6077, BL17, WHPSA001, WHPSA008, E2 and PAO1strains.

B. *Mutations of methyltransferase genes and proteins*

The whole mutations occurred in the analyzed *P. aeruginosa* S1, S2, S3, S4, S5 and S6 isolates; a total of 35 mutations of which 60% were synonymous mutations and 40% synonymous mutations were mostly in non-conserved enzyme sites were identified. It is noteworthy that such changes in the variants of RsmG and RsmH proteins occurred in non-conserved regions, what makes us think that these mutations do not have impact on the function of these enzymes (Fig. 2: A and B). The alignment of RsmG structure with *Thermus thermophilus* putative protein shows the main architecture conserved despite the amino acids variability (Fig. 3). In contrast, four changes (S20T, D37H, E162Q and S191N) present in *P. aeruginosa* S1, S2, S3, S4, S5 and S6 clinical isolates for RsmI protein were found in highly conserved sites (data not show).

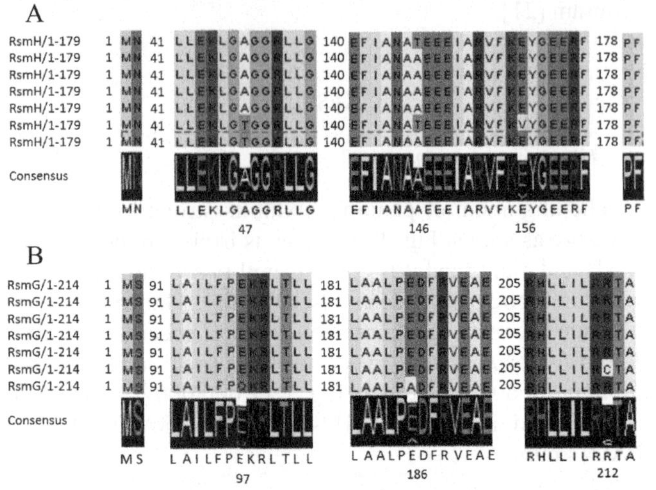

Fig. 2 A RsmH and B RsmG sequences. The first sequence corresponds to *P. aeruginosa* PAO1, second *P. aeruginosa* S1, third *P. aeruginosa* S2, fourth *P. aeruginosa* S3, fifth *P. aeruginosa* S4, sixth P. aeruginosa S5, seventh *P. aeruginosa* S6.

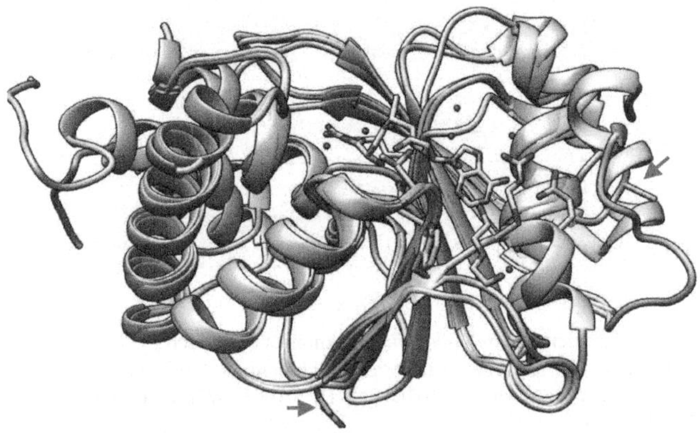

Fig. 3 Structure alignment of RsmG proteins in blue 3G88 *T. thermophiles*, in brown RSMG from *P. aeruginosa* PAO1 and violet *P. aeruginosa* S6 obtained by Phyre2 suite. Amino acid mutations are marked with red arrows: E97Q and E186A.

C. Methyltransferase networks analyzed

The ontologies of network from RsmH and RsmG networks of *P. aeruginosa* PAO1 are well studied, but RsmI is still unclarified for the species. The main proteins that interact with RsmG protein are DnaI, CycB, DnaN, GibB, MnmA, MnmG. RecF, RpmH, RpsL, Soj, and SpoOJ. The proteins that interact or co-express with RsmH are shown in the Fig. 4 and is interesting the association with RsmI.

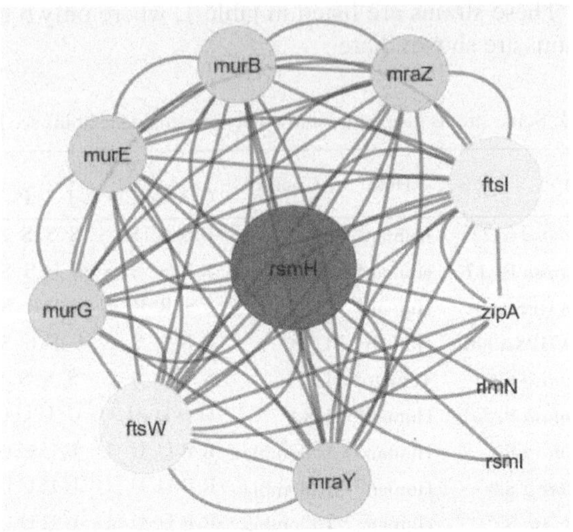

Fig. 4 Network of RsmH protein, obtained from STRING and analyzed by Cytoscape. In red main protein, in green main proteins associated, in yellow secondary associations. The big ball have good evidence.

IV. Discussion

The *P. aeruginosa* chromosome contains 6175 genes on average, with a distribution ranging from 5382 to 7170 genes per genome, indicating a variation of 13–16 % in terms of gene content among all strains analyzed. By using the genome-centered approximation to define the *P. aeruginosa* pangenome, more than of 17,000 non-redundant genes were retrieved from those 200 genomes analyzed. We found a similar distribution to that reported by Lapierre and Gogarten when they estimated the pangenome for more than 500 different bacterial genomes [22].

The RsmG, RsmH and RsmI proteins methylate 16S rRNA generating antibiotic resistance in bacteria [2]. The synonym and non-synonym mutations of *rsmG*, *rsmH* and *rsmI* genes in sequenced genes and pangenome seem do not affect the proteins structure in RsmG, the only gene completely sequenced, but *rsmH* and *rsmI* were sequenced partially. Looking inside the 200 genomes there are variation in gene that seem the same case for the Colombian strains, but we found more mutations that published on alignments published in NCBI for 100 strains from rsmG (https://www.ncbi.nlm.nih.gov/sutils/alnview.cgi?job_Key=JSID_01_160037_130.14.24.130_9034__muscle_fasta&key=&filter_string=Min Score:100).

Methyltransferases protein interaction supports the robust machinery of multi-drug resistance in *P. aeruginosa*. Our work demonstrate that mutation in methyltransferase genes are not alone mechanism to develop antibiotic resistance, but are important.

V. Conclusions

This report described a closed *P. aeruginosa* pangenome at very large scale, including 200 bacterial genomes from this human pathogen and performing a pangenome-scale molecular antibiotic resistance analysis, corroborating a relationship with *rsmG*, *rsmH* and *rsmI* genes. We found more no-synonym mutations in Colombian resistant strains that reported today in databases for RsmG. Theoretical structure of RsmG from Colombian mutant has important variations, which probably contribute to antibiotic resistance.

Acknowledgment

We want to thanks to GEBIOMIC group from the University of Antioquia.

Conflict of Interest

The authors declare that they have no conflict of interest.

References

1. Lynch, S. V., Flanagan, J. L., Sawa, T., Fang, A., Baek, M. S., Rubio-Mills, A, & Misset, B. (2010). Polymorphisms in the Pseudomonas aeruginosa type III secretion protein, PcrV–implications for anti-PcrV immunotherapy. *Microbial pathogenesis*, *48*(6), 197-204.
2. Mosquera-Rendón, J., Cárdenas-Brito, S., Pineda, J. D., Corredor, M., & Benítez-Páez, A. (2014). Evolutionary and sequence-based relationships in bacterial AdoMet-dependent non-coding RNA methyltransferases. *BMC research notes*, *7*(1), 1.
3. Mosquera-Rendón, J., Rada-Bravo, A. M., Cárdenas-Brito, S., Corredor, M., Restrepo-Pineda, E., & Benítez-Páez, A. (2016). Pangenome-wide and molecular evolution analyses of the Pseudomonas aeruginosa species. *BMC genomics*, *17*(1), 1.
4. Anantharaman, V., Koonin, E. V., & Aravind, L. (2002). SPOUT: a class of methyltransferases that includes spoU and trmD RNA methylase superfamilies, and novel superfamilies of predicted prokaryotic RNA methylases. *Journal of molecular microbiology and biotechnology*, *4*(1), 71-76.
5. Anantharaman, V., Koonin, E. V., & Aravind, L. (2002). Comparative genomics and evolution of proteins involved in RNA metabolism. *Nucleic acids research*,*30*(7), 1427-1464.
6. Wattam, A. R., Abraham, D., Dalay, O., Disz, T. L., Driscoll, T., Gabbard, J. L., & Machi, D. (2013). PATRIC, the bacterial bioinformatics database and analysis resource. *Nucleic acids research*, gkt1099.
7. Tettelin, H., Masignani, V., Cieslewicz, M. J., Donati, C., Medini, D., Ward, N. L.,& DeBoy, R. T. (2005). Genome analysis of multiple pathogenic isolates of Streptococcus agalactiae: implications for the microbial "pan-genome".*Proceedings of the National Academy of Sciences of the United States of America*, *102*(39), 13950-13955.
8. Tettelin, H., Riley, D., Cattuto, C., & Medini, D. (2008). Comparative genomics: the bacterial pan-genome. *Current opinion in microbiology*, *11*(5), 472-477.
9. Lapierre, P., & Gogarten, J. P. (2009). Estimating the size of the bacterial pan-genome. *Trends in genetics*, *25*(3), 107-110.
10. Okonechnikov, K., Golosova, O., & Fursov, M. (2012). Unipro UGENE: a unified bioinformatics toolkit. *Bioinformatics*, *28*(8), 1166-1167.
11. Rutherford, K., Parkhill, J., Crook, J., Horsnell, T., Rice, P., Rajandream, M. A., & Barrell, B. (2000). Artemis: sequence visualization and annotation. *Bioinformatics*, *16*(10), 944-945.
12. Carver, T. J., Rutherford, K. M., Berriman, M., Rajandream, M. A., Barrell, B. G., & Parkhill, J. (2005). ACT: the Artemis comparison tool. *Bioinformatics*, *21*(16), 3422-3423.
13. Darling, A. C., Mau, B., Blattner, F. R., & Perna, N. T. (2004). Mauve: multiple alignment of conserved genomic sequence with rearrangements. *Genome research*, *14*(7), 1394-1403.
14. Moriya, Y., Itoh, M., Okuda, S., Yoshizawa, A. C., & Kanehisa, M. (2007). KAAS: an automatic genome annotation and pathway reconstruction server.*Nucleic acids research*, *35*(suppl 2), W182-W185.
15. Aoki-Kinoshita, K. F., & Kanehisa, M. (2007). Gene annotation and pathway mapping in KEGG. *Comparative Genomics*, 71-91.
16. Waterhouse, A. M., Procter, J. B., Martin, D. M., Clamp, M., & Barton, G. J. (2009). Jalview Version 2—a multiple sequence alignment editor and analysis workbench. *Bioinformatics*, *25*(9), 1189-1191.
17. Edgar, R. C. (2004). MUSCLE: multiple sequence alignment with high accuracy and high throughput. *Nucleic acids research*, *32*(5), 1792-1797.
18. Kelley, L. A., Mezulis, S., Yates, C. M., Wass, M. N., & Sternberg, M. J. (2015). The Phyre2 web portal for protein modeling, prediction and analysis. *Nature protocols*, *10*(6), 845-858.
19. Pettersen, E. F., Goddard, T. D., Huang, C. C., Couch, G. S., Greenblatt, D. M., Meng, E. C., & Ferrin, T. E. (2004). UCSF Chimera—a visualization system for exploratory research and analysis. *Journal of computational chemistry*, *25*(13), 1605-1612.
20. Szklarczyk, D., Franceschini, A., Kuhn, M., Simonovic, M., Roth, A., Minguez, P., & Jensen, L. J. (2011). The STRING database in 2011: functional interaction networks of proteins, globally integrated and scored. *Nucleic acids research*, *39*(suppl 1), D561-D568.
21. Shannon, P., Markiel, A., Ozier, O., Baliga, N. S., Wang, J. T., Ramage, D., & Ideker, T. (2003). Cytoscape: a software environment for integrated models of biomolecular interaction networks. *Genome research*, *13*(11), 2498-2504.
22. Lapierre, P., & Gogarten, J. P. (2009). Estimating the size of the bacterial pan-genome. *Trends in genetics*, *25*(3), 107-110.

Address of the corresponding author*:

Author: Mauricio Corredor
Institute: Biology Institute, Universidad de Antioquia
Street: Calle 67 No. 53 – 108, block 7, office 318
City: Medellín
Country: Colombia
Email: mauricio.corredor@udea.edu.co

Descriptive study of the courses offered in degree programs in Biomedical Engineering in Mexico

A. Pliego-Carrillo[1], E. Del Hierro-Gutierrez[1],

[1]Facultad de Medicina, Universidad Autónoma del Estado de México, Estado de México

Abstract— **This work provides an overview on undergraduate programs on Biomedical Engineering (BME) in 37 Mexican Higher Education institutions recognized by the Mexican Society of Biomedical Engineering (SOMIB). In addition to school affiliation, denomination and duration of each program, we classified 2,392 courses, according to the categories established by the Accreditation Council for Teaching Engineering (CACEI, Mexico). A second classification was made according to the topics of interest of the 38th Annual International Conference (2016) IEEE-EMBS. We are confident that this paper will serve as a guide to show the inclination of each of the undergraduate programs and point out its contents on topical issues.**

Keywords— **Education, Formation, Academic Programs, Biomedical Engineering.**

I. Introduction

In Mexico, the undergraduate program to obtain a Bachelor of Science in Biomedical Engineering (BME) degree was first offered in 1973 by the Universidad Iberoamericana (UIA), a private educational institution, and a year later by the Metropolitan Autonomous University (UAM-I), a public one [1]. On 2016, the Mexican Society of Biomedical Engineering (SOMIB)[2], founded in 1978, recognizes on its official website 37 higher education institutions (IES), both public and private, distributed on eleven out of thirty two states in Mexico [2]. Currently, the Accreditation Council for Teaching Engineering (CACEI) has certified only seven BME undergraduate programs. This represents 18.9% of the programs included in this study [3].

In the table 1 shows the relationship of IES that offer the BME by state.

II. Methodology

It classified 2,392 courses of 37 Mexican universities with BME undergraduate programs, in four categories: Basic Science, Engineering Science and Applied Engineering, Social Science and Humanities, and Others, as suggested by the CACEI. We also mention the correspondence between the offered courses and the topics of interest of the 38th IEEE-EMBS Conference (2016). Total duration of the program and the name of the professional degree were also considered.

Table 1. Relationship of IES that offer the BME by state

State	Nr. of IES	State	Nr. of IES
Baja California	2	Michoacán	1
Sonora	2	Estado de México	1
Chihuahua	3	Ciudad de México	7
Sinaloa	1	Guanajuato	3
Nuevo León	2	Querétaro	1
Tamaulipas	1	Hidalgo	1
Durango	1	Puebla	1
San Luis Potosí	1	Yucatán	2
Aguascalientes	2	Tabasco	1
Jalisco	3	Chiapas	1
		TOTAL	37

III. Results

It found out that, from 2,392 courses, 24.7% are of Basic Science, 51% belong to Engineering Science and Applied Engineering, 17.4%% are of Social Science and Humanities and 6.9% belong to Others. On the other hand, 9.7% of the courses are related to the top topics of the 38th EMBS Annual Conference, being "Therapeutic & Diagnostic Systems, Devices and Technologies, Clinical Engineering" (22%) and "Bio-Robotics, Surgical Planning and Biomechanics" (13%) the most frequent among all programs. Besides, we found out that 56.7% of the programs belong to a Faculty of Engineering or related engineering areas, 21.6% belong to a Faculty of Medicine and Biomedical Sciences or a Health Division, and 21.6% are part of a Technological Institute. The name of the degree also differs between IES, representing Bachelor in Science of Biomedical Engineering 83.8%, Electromedical Engineering 8.1%, Bioengineering 5.4%, and Bachelor of Science in Engineering of Biomedical Systems 2.7%. The duration of the program is of 9 semesters for 48.6%, 8 semesters for 32.4%, 9 four-month periods for 8.1%, 10 semesters for 5.4%, 10 four-month periods and 12 four-month periods for 2.7%, respectively.

The original version of this chapter has been revised: Incorrect chapter title has been corrected. The erratum to the chapter is available at https://doi.org/10.1007/978-981-10-4086-3_193

I. Torres et al. (eds.), *VII Latin American Congress on Biomedical Engineering CLAIB 2016, Bucaramanga, Santander, Colombia, October 26th -28th, 2016,*
IFMBE Proceedings 60,
DOI: 10.1007/978-981-10-4086-3_177

The Figure 1, shows the BME Assigment in the University, the denomination degree and the duration time.

Fig.1

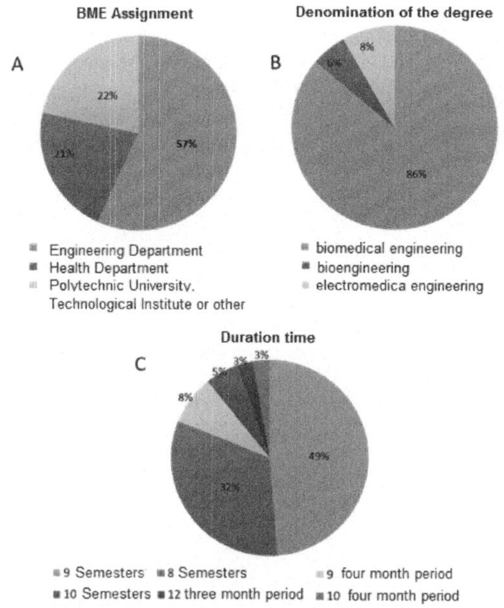

IV. Conclusions

According to the categories defined by the CACEI 30.77% (800 hours) program should be devoted to the study of Basic Sciences, 50% (1300 hours) Engineering Sciences and Applied Engineering, 11.53% (300 hours) Social Sciences and Humanities and 7.7% (200 hours) to Other courses. After comparing with the results, we found out that all undergraduate programs should reduce the percentage of subjects dedicated to the study of the Social Sciences and Humanities (17.4%) and increase courses on Basic Science (24.7%) on approximately 6%. For Engineering Science and Applied Engineering the percentage is 51%. Our study shows that although the percentage of courses which include avant-garde topics according to EMBS-IEEE is only 10%, all higher education institutions include in their offer at least one of the topics of interest. We noted that none of the courses is related to the main topic of this Annual Conference "Empowering Individual Health Care Decisions through Technology". We propose, for a future study, to compare the total number of credits per program, per subject, and compare graduate profiles. We also consider relevant to determine whether the absence of courses on top topics directly impacts the performance of graduates, their income, or their job opportunities. We strongly suggest this paper to serve as a guide in order to know the direction of the undergraduate program in Mexico towards specific areas. For instance, before applying to a university, candidates should look up for the program that better suits their orientation and interests.

Acknowledgment

We thank the SOMIB and the Universities that kindly provided information not available on their website.

Conflict of Interest

It has no conflict of interest in this study, regardless of whether the researchers are teaching in one of the universities which curricula were analyzed, it is not intended to say if there is any better than another, only used to study the state art in Biomedical Engineering in the country. The reason for conducting this article is for academic purposes with the resources of the authors.

References

1. M. R. Ortiz Posadas, M. J. Gaitán González (2001) La ingeniería biomédica y el Sector Salud, 2nd ed. Universidad Autónoma Metropolitana, México.
2. SOMIB at http://www.somib.org.mx/
3. CACEI at http://www.cacei.org/
4. CACEI framework at http://cacei.org.mx/index.php/acreditacion/formatos-y-manuales/manual-de-marco-de-referencia-tsu
5. CACEI programs accredited at http://cacei.org.mx/index.php/acreditacion/programas-acreditados-vigentes
6. EMBS at http://www.embs.org/
7. EMBS 2016 at http://embc.embs.org/2016/
8. INSTITUTO TECNOLOGICO DE TIJUANA at http://tectijuana.edu.mx
9. UABC at http://www.uabc.mx
10. ULSA-NOROESTE at http://www.ulsa-noroeste.edu.mx
11. ITH at http://www.ith.mx
12. UACH at http://www.uach.mx
13. UACJ at http://www.uacj.mx/Paginas/Default.aspx
14. ULSA-CHIHUAHUA at http://ulsachihuahua.edu.mx
15. ITESM at http://www.itesm.mx/wps/portal?WCM_GLOBAL_CONTEXT
16. UDEM at http://www.udem.edu.mx/Esp/Paginas/default.aspx
17. ULSA-VICTORIA at http://www.ulsavictoria.edu.mx/
18. UASLP at http://www.uaslp.mx/Spanish/Paginas/default.aspx
19. ULSA-LAGUNA at http://www.ulsalaguna.edu.mx
20. UPSIN at http://www.upsin.edu.mx
21. ITESM-AGUASCALIENTES at www.itesm.mx/wps/wcm/connect/Campus/AGS/Aguascalientes
22. UAA at http://www.uaa.mx
23. UAQ at http://www.uaq.mx
24. AUG at http://www.uag.mx
25. ITESM-GUADALAJARA at http://www.itesm.mx/wps/wcm/connect/Campus/GDA/Guadalajara
26. UDG at http://www.udg.mx
27. ULSA-BAJIO at http://bajio.delasalle.edu.mx/
28. UGTO at http://www.ugto.mx/campusleon/
29. UIA-LEON at http://www.leon.uia.mx

30. ITS-PUREPECHA at http://www.its-purhepecha.edu.mx
31. UPP at http://www.upp.edu.mx/front/
32. ULSA at http://lasalle.mx
33. UIA at http://www.uia.mx
34. UAM at http://www.uam.mx
35. UNAM at http://unam.mx
36. IPN at http://www.ipn.mx/Paginas/aviso.aspx
37. ANAHUAC at http://www.anahuac.mx/Pages/NvoHome.aspx
38. ITESM-CCM at http://www.itesm.mx/wps/wcm/connect/Campus/CCM/Ciudad+de+Mexico/
39. UAEMEX at http://www.uaemex.mx
40. UP-CHIAPAS at http://www.upchiapas.edu.mx
41. UNIVERSIDAD OLMECA at http://www.olmeca.edu.mx
42. UNIVERSIDAD MODELO at http://www.unimodelo.edu.mx
43. INSTITUTO TECNOLOGICO DE MERIDA at http://www.itmerida.mx
44. UNIVERSIDAD DE LAS AMERICAS at http://www.udlap.mx/inicio.aspx
45. (2-44 consulted until March 2016)

Author: E.G. Del Hierro-Gutiérrez
Institute: Universidad Autónoma del Estado de México
Street: Paseo Tollocan y Jesús Carranza
City: Toluca
Country: México
Email: ed.hierro@hotmail.com

Motor Imagery BCI System with Visual Feedback: Design and Preliminary Evaluation

L.C. Carrere and C.B. Tabernig

Laboratorio de Ingeniería en Rehabilitación e Investigaciones Neuromusculares y Sensoriales. Facultad de Ingeniería. Universidad Nacional de Entre Ríos, Oro Verde, Argentina

Abstract— **Nowadays, strokes are a growing cause of mortality and many people remain with motor sequelae and troubles in the daily activities. To treat these sequelae, alternative rehabilitation techniques are needed. This article describes the design, development and preliminary evaluation of a system based on Brain Computer Interfaces (BCI) by Motor Imagery, with visual feedback for lower limb rehabilitation of people post stroke. The system consists of three modules: Sensing and Conditioning; Control Signal Generator; and Visual Feedback. The first module acquires, filters and segments 5 channels of EEG. The second module performs spatial filtering using a Laplacian, estimates the signal power spectral density, extracts and selects EEG features which are then used by the classifier to detect event related desynchronization. The command signal generated by the BCI is inputted into the third module, which simulates the movement of foot dorsiflexion of an avatar displayed on a screen. For the implementation, the BCI2000, V-REP platforms and MATLAB software were used. Performance evaluation of the system was done in a healthy volunteer by estimating the sensitivity and specificity, and through interviews with specialists. Average values for sensitivity and specificity were 0,67 and 0,70 respectively, and professional opinions were very good. These results are encouraging for deepening the performance evaluation system and taking steps for clinical implementation.**

Keywords—**Neuro-rehabilitation, BCI System, Motor Imagery, Visual Feedback, Stroke**

I. Introduction

Some rehabilitation techniques for people with stroke are based on Hebbian learning principles, which propose an explanation for the adaptation of neurons in the brain during the learning process. They establish that an increase in synaptic efficacy arises from the presynaptic cell's repeated and persistent stimulation of the postsynaptic cell. This kind of neural stimulation could be achieved by the modulation of Sensory Motor Rhythms (SMR) and sensory feedback. SMRs are oscillations in the electric or magnetic fields recorded over sensorimotor cortices which can be detected on the scalp electroencephalography (EEG). These are the rhythms μ and β of the EEG, between 8 and 12 Hz and between 18 and 30 Hz, respectively [1].

On other hand, the advancement of technology offers Systems based on Brain Computer Interfaces (BCI) by Motor Imagery (MI) for therapeutic use. The main BCI elements are shown in Figure 1: user, the BCI and the external device or actuator.

One of the most used neurophysiological phenomena in BCI-MI Systems is the reduction in rhythmic activity of the SMR related to the real or imaginary movement. This SMR´s decrease is known as Event-Related Desynchronization (ERD) [2].

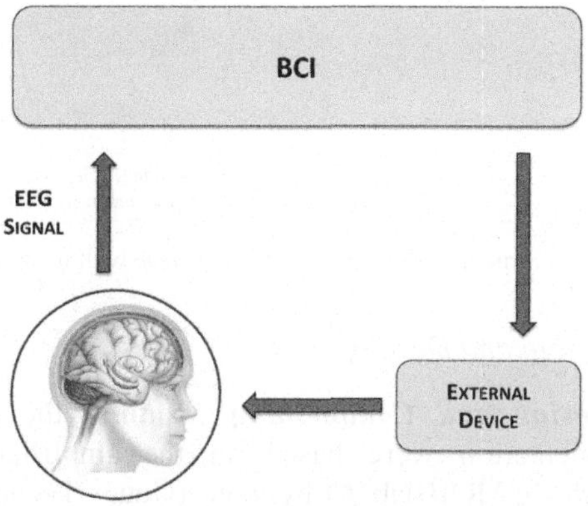

Fig. 1. A system based on BCI.

The application of such facts in therapies to restore motor function in patients post stroke has been used to improve neuroplasticity, mainly in upper limb [3]. But, since these patients has foot drop as a chronic and serious sequelae, it is necessary also to explore therapeutic applications of BCI-MI systems in lower limb and to develop appropriate technology for such therapeutic strategy. On the other hand it is necessary to research about the actuator device, such that the sensory feedback complies with the Hebbian principles in order to facilitate the neuroplasticity.

This article describes the design, development and preliminary evaluation of a system for clinical studies in neuro-rehabilitation based on BCI by MI with Visual Feedback (*BCI-MI+VF System)*. It detects the ERD in the EEG signal during the intention to move the affected foot

I. Torres et al. (eds.), *VII Latin American Congress on Biomedical Engineering CLAIB 2016, Bucaramanga, Santander, Colombia, October 26th -28th, 2016*,
IFMBE Proceedings 60,
DOI: 10.1007/978-981-10-4086-3_178

and gives, through simulation, visual feedback of the movement.

II. MATERIALS AND METHODS

A. General Design Requirements

It was proposed that the system consists of a synchronous BCI-MI triggered by an auditive cue and detects the ERD related to the imagination of a foot movement. The motor imagery task will have 2 seconds of a maximum duration, according to reports about the temporal ERD behavior for motor imagery guided by a cue [1]. To detect the ERD, the system will compare the power of the EEG recorded during the motor imagery task with a threshold indicated by the user: *ERD_threshold*; which should be lower than the signal power in the resting state.

The structure of the *BCI-MI+VF System* was defined in 3 modules, as shown in Figure 2: *Sensing and Conditioning*; *Control Signal Generator*; and the *Visual Feedback.*

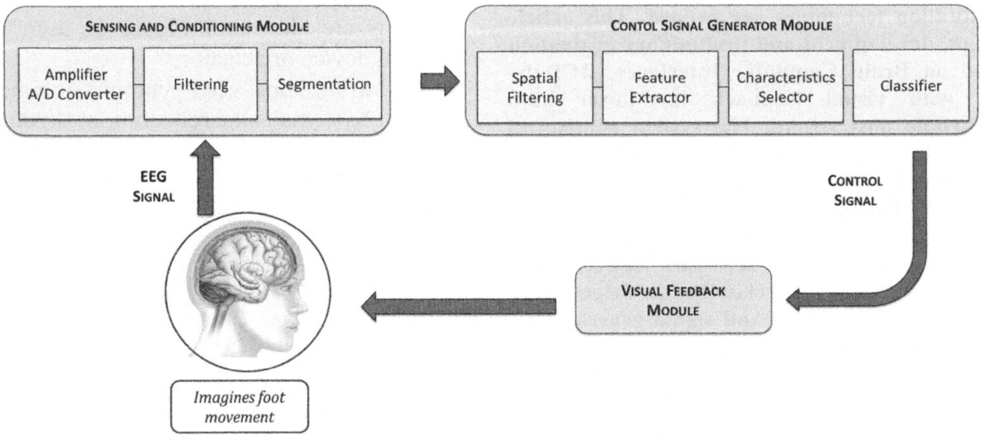

Fig. 2. Structure of the *BCI-MI+VF System*, with its modules: *Sensing and Conditioning*; *Control Signal Generator*; and *Visual Feedback.*

B. Design and Development

***Sensing and Conditioning* Module**: the design and implementation were based on the amplifier and A/D converter g.MOBIlab+® by g.tec (Guger Technology) and on the BCI2000 platform [5], with the following blocks:

Amplifier and A/D converter: consists of the amplifier and A/D converter g.MOBIlab+® by g.tec (Guger Technology); pasives Ag/AgCl electrodes positioned based on extended 10/20 system, and a cap designed ad hoc. The block acquires 5channels of EEG, which are digitized with a resolution of 16 bits and sampling frequency of 256 Hz. In order to implement a Laplacian spatial filter and taking into account the area of the motor cortex where the feet are represented [1], the *C3, C4, Pz, Fz* and *Cz* positions were selected. The ground and reference electrodes were placed on the right and left mastoids, respectively

Filtering: It consists of a low pass filter and frequency power line filter. It besides multiplies the signal by a gain factor.

Segmentation: this block segments the acquired signal on each channel, with the aim of reducing the computational processing time in the *Control Signal Generator* Module. Each segment contains 8 signal samples. Thus, there are 5 segments (one for each channel) of 31.25 msec of duration; which enter the next module.

***Control Signal Generator* Module:** in order to identify the characteristics associated with the ERD related to the imagination of foot movement, in this module the following blocks were developed:

Spatial Filtering: It is a Laplacian spatial filter to improve the signal to noise ratio and to enhance the feet cortical position, *Cz*. This filter applies a linear transformation to the input signals, as shown in equation 1. Thus the average of the signals on the cortical EEG positions *F3, F4, C3, C4, Pz* is subtracted from *Cz*, generating a unique output signal of this block: *Cz'*.

$$Cz' = Cz - (C3 + C4 + Pz + Fz)/4 \quad (1)$$

The implementation of this filter was performed through the BCI2000 platform.

Feature Extractor: It performs the power spectral density (PSD) estimation, also through BCI2000 platform. The PSD is estimated in 500 msec signal epochs via an AR model because these models exhibit good frequency resolution in short segments of the EEG signal. An AR model of order 16 was selected [1]. The signal epochs for estimating the PSD were constructed with segments of 8 samples and updated every 31.25 msec. The model coefficients were determined by the Burg method [1]. The output of this block is the amplitude of the PSD, which is defined as the square root of the signal power. The PSD was estimated in the range of 0-30 Hz, discretly in bins of 3 Hz. Thus, the output of this block consists of a feature vector that has the 11 PSD's amplitude values (Amp_{1x11}) for each signal epoch.

In order to classify the features to generate a visual feedback command signal, the following blocks were developed.

Features Selector: It selects, in the Amp_{1x11} vector, the PSD's amplitude value for the frequency of maximum ERD, f_{maxERD}. The determination of f_{maxERD} was performed in a calibration session following the methodology reported in the literature [6]. During the 2 sec, which takes the task assigned by the auditve cue, these PSD values are incorporated into a new vector called $Amp\text{-}f_{maxERD}$; so this new vector has 64 elements. Each element of the vector $Amp\text{-}f_{maxERD}$ is calculated as a linear combination of the vector (Amp_{1x11}) and a weight vector (P_{11x1}) which consists of a single non-zero element in the position corresponding to the f_{maxERD}, as shown in the following equation:|

$$y = \sum_{i=1}^{11} Amp_{1xi} P_{ix1} \tag{2}$$

where y is the element of the $Amp\text{-}f_{maxERD}$ vector.

Classifier: this block processes and classifies data in the $Amp\text{-}f_{maxERD}$ vector to generate the control signal. For this, it was developed an application in MATLAB 7.10 (R2010a) which performs a unidirectional communication channel with BCI2000 through a UDP data transmission protocol, and transmits the control signal to the *Visual Feedback* Module.

The application reads the $Amp\text{-}f_{maxERD}$ vector and calculates the average of the vector's elements. Then, it compares this average with *ERD_ threshold*; if the average is greater than *ERD_ threshold*; then the user is in the state of "rest" and the *Visual Feedback* Module is not activated. In the opposite case, it interprets that the user has achieved the ERD related to the imagination of the movement of a foot and activates feedback through the *Visual Feedback* Module. The value of *ERD_ threshold* will be configured by the operator as the mean of the $Amp\text{-}f_{maxERD}$ vector in a previous system calibration. The developed application also informs the mean $Amp\text{-}f_{maxERD}$ vector and if the *Visual Feedback* Module was activated or not.

***Visual Feedback* Module**: its development is based on the V-REP PRO EDU 3.0 software as simulation platform. This module generates an avatar that is displayed on a screen and is controlled by the output of the previous module. It is a human model which consists of mobile elements. To control the movement of the avatar's joints, all of its elements were selected as static ones except those representing feet. The right and left ankle's joints were defined as mobile, while the others were specified as immobile. In order to simulate the movement of each foot dorsiflexion, an angle of 45° was assigned as the maximum ankle joint angle, measured from horizontal. The module is controlled by the external application through its API (Application Programming Interface), also developed in MATLAB 7.10 (R2010a).

C. Design Evaluation

Sensitivity (*Se*) and specificity (*Sp*) were estimated because of these metrics are adequate to evaluate the performance in each of two classes or tasks [4]. Two sessions of three series of use of the *BCI-MI+VF System* were performed with a healthy volunteer of 22 years old.

During the test, two neuro-rehabilitation professionals were present: a physician and a therapist. They were asked about the viability of using the *BCI-MI+VF System* in a clinical environment and about the *Visual Feedback*.

III. RESULTS

In Figure 4, the developed *BCI-MI+VF System* is presented. It can be seen the visual interface running in a PC screen and the user located at a distance of about 1m. Moreover, the operator controlling the system operation is observed.

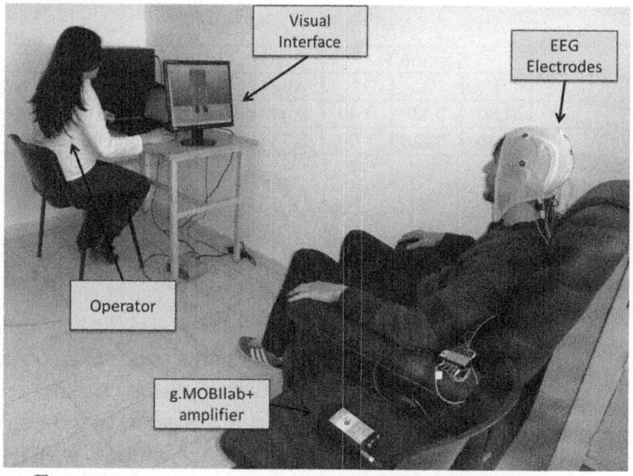

Fig. 4. Photography of the developed *BCI-MI+VF System*

Figure 5 presents the visual interface, in which the avatar in a sitting position is shown. It is observed that the interface's image is a simulation of a person sitting in front of a mirror.

In Table 1 results of *Se* and *Sp* are shown. It can be observed that the values of the metrics present variation between series in the same session.

Neuro-rehabilitation professionals said in the interview that *BCI-MI+VF System* can be used in clinical practice as a tool in rehabilitation therapy. Also they commented that the placement and cleaning of EEG electrodes could be a possible disadvantage for daily use. Regarding the developed visual interface, they confirmed that it simulated properly the volunteer's imagined action. But they emphasized the importance of providing a simulation target related to an everyday life movement of the lower limb.

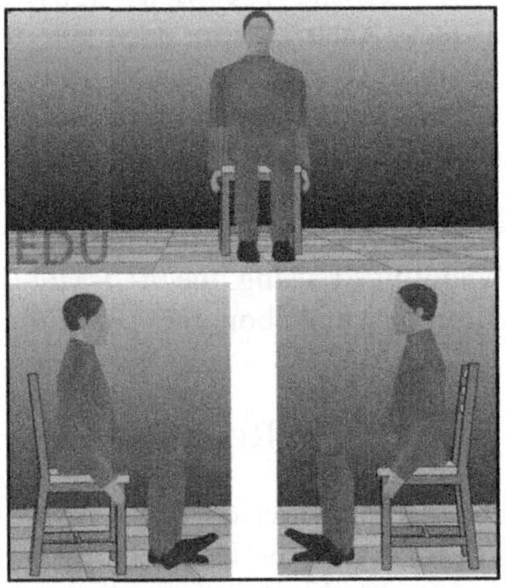

Fig. 5. Visual interface

Table 1. *Sp* and *Se* of performance testing of the *BCI-MI+VF System* with a healthy volunteer.

	Session I				Session II			
Serie	1	2	3	**Avg**	1	2	3	**Avg**
Se	0,80	0,80	0,40	**0.67**	0,60	0,80	0,60	**0,67**
Sp	0,40	0,80	1,00	**0.73**	0,40	0,80	0,80	**0,67**

IV. Discussion y conclusions

The designed and developed *BCI-MI+VF System* fulfilled successfully the requirements set. The obtained average values of *Se* and *Sp* were similar to the results reported by Leeb et al. [7]. Furthermore it is expected to achieve improvement on these metrics values with user's training.

In our and in the interviewed professionals opinions, the placement of the electrodes for EEG recording during use of the system is a drawback. Then, it is necessary to develop new technologies that reduce the time electrode placement and at the same time ensure the system stability.

It was evident that the time involved in communication between the *Classifier* block (developed in MATLAB) and BCI2000 platform is strongly dependent on the hardware. According to reports from different authors and Hebbian theory, a very fast sensorial feedback promotes neuronal plasticity and facilities recovery of motor function [4]. In this context, the short communication time of the developed *BCI-MI+VF System* could positively influence the neuro-rehabilitation therapy.

The results of this preliminary study allow us to think it is possible to continue evaluating the *BCI-MI+VF System* performance in more healthy volunteers and people with neurological sequelae; and, in a closely future, to investigate the effects of using the *BCI-MI+VF System* as a therapeutic tool to promote neuronal plasticity in people post stroke.

Conflict of interests

The authors declare no conflict of interest.

References

1. G. Pfurtscheller y D. Mcfarland. BCIs that use sensorimotor rhythms. En: Brain-computer interfaces: principles and practice, J. R. Wolpaw and E. Wolpaw, Eds. Oxford University Press; 2012; pp. 227–240.
2. G. Pfurtscheller y F. H. Lopes Da Silva. Event-related EEG/MEG synchronization and desynchronization: Basic principles. Clin. Neurophysiol.1999;110(11): 1842–1857.
3. C. Brunner, N. Birbaumer, B. Blankertz, C. Guger, A. Kübler, D. Mattia, J. D. R. Millán, F. Miralles, A. Nijholt, E. Opisso, N. Ramsey, P. Salomon, y G. R. Müller-Putz. BNCI Horizon 2020: towards a roadmap for the BCI community. Brain-Computer Interfaces.2015; 2 (1): 1-10.
4. F. Lotte, J. Faller, C. Guger, Y. Renard, G. Pfurtscheller, L. Anatole, y R. Leeb. Combining BCI with Virtual Reality: Towards New Applications and Improved BCI. En: B. Z. Allison, editor, Towards Practical Brain-Computer Interfaces, Berlin: Springer-Verlag Berlin Heidelberg; 2013; pp. 197–220.
5. Gerwin Schalk y J. Mellinger. A Practical Guide to Brain–Computer Interfacing with BCI2000. 2010.
6. L. C. Carrere y C. B. Tabernig. Detection of Foot Motor Imagery Using the Coefficient of Determination for Neurorehabilitation Based on BCI Technology. En: IFMBE Proceedings- VI Latin American Congress on Biomedical Engineering CLAIB 2014; 2014; pp. 944–947.
7. Leeb, R., Lee, F., Keinrath, C., Scherer, R., Bischof, H., y Pfurtscheller, G.. Brain–computer communication: motivation, aim, and impact of exploring a virtual apartment. IEEE Transactions on Neural Systems and Rehabilitation Engineering, 15(4), 473-482. 2007

Author: Carrere Lucía Carolina
Institute: Facultad de Ingeniería.Universidad Nacional de Entre Ríos
Street: Ruta 11, km 10
City: Oro Verde
Country: Argentina
Email: carrerecarolina@bioingenieria.edu.ar

Linear and non-linear methods for analysis Center Pressure and its application in Diabetic Peripheral Neuropathy: A systematic review

D.Toloza[1] y M. Zequera[2]

[1]Pontificia Universidad Javeriana/Department of Electronic, PhD student, Bogotá, Colombia
[2] Pontificia Universidad Javeriana/ Department of Electronic, Associate Professor, Bogotá, Colombia

Abstract— **Body stability problems can limit mobility, daily activities and is one of the risk factors for falls in elderly people, increasing in Diabetic Peripheral Neuropathy (DPN) population. Methods of stability analysis focus on Center of Pressure (CoP), using different parameters focused on a linear analysis, but currently several studies demonstrate that CoP can also present a nonlinear behavior that gives rise to other information to complement the stability evaluation. Currently, very few studies have done this type of analysis in DPN, for this reason, this systematic review has been conducted aiming to identify the most reliable and efficient parameters by linear and non-linear methods for CoP characterization in the antero-posterior direction (AP) under static conditions. In order to achieve this objective, a total of 16 articles were selected. Included articles examined CoP in diabetic patients with or without DPN. These studies suggest that linear and non-linear parameters to CoP analysis should be: Power Spectral Density, Range, Root Mean Square, Average Velocity, Detrended Fluctuation Analysis, Entropy, Hurst Exponent and Wavelet Transform. All parameters show reliability and statistical significance between groups. This approach may facilitate earlier detection of fall risk.**

Keywords — **Diabetes, Diabetic Peripheral Neuropathy, Center of pressure, linear and non-linear analysis, Body stability**

I. Introduction

The International Diabetes Federation Diabetes Atlas Seventh Edition 2015 showed that Colombia has 3.0487 million adult people with diabetes, about 19.000 die every year and 1.22 million are undiagnosed. These people have an increased risk of developing a number of serious health problems as Diabetic Peripheral Neuropathy (DPN). DPN is the major complications of this type of diabetes and affects up to 50% of diabetics, is asymptomatic early-stage and it is the most common cause of foot ulceration and amputation [1]. This disease decreased sensory and motor function in lower limbs reflected by a marked decrease in proprioception, loss of ankle reflexes, reducing vibration sense and plantar pressure on the plantar area of the foot, reduced muscle strength and pain in advanced stages characterized by numbness, tingling, and burning. The loss of proprioception and vibration sensation at the toes is also associated with CoP displacement increased and higher velocity in AP direction in a quiet standing position. These findings indicate that CoP in the AP direction evaluation provides more information on postural control. CoP is the quantitative parameter most commonly used to evaluate stability in static standing position, which reflects the orientations of the segments and body movements. Several investigations, such research by Palmieri et al. (2002), Lafond et al. (2004) and Raymakers et al. (2005) have concluded that large CoP displacements may be associated with decreased body control, an indicator of instability and loss of balance [2]. The data obtained from CoP during static standing position have been used to calculate different parameters to analyze body stability behavior. Usually, researchers have been using parameters in the time and frequency domain (Traditional Parameters TP), but recently, studies are using nonlinear or stochastic analysis (Non-Linear Parameters NLP) [3]. TP is a statistical summary of data, while that NLP identify the underlying structural and dynamic properties of body stability. This systematic review aims to analyze the TP and NLP used to study of CoP behavior and identify which of them provide useful information to characterize CoP-AP data in DPN.

II. Methods

A systematic review of research to assess CoP-AP direction under static conditions. Types of participants: healthy elderly people and/or any disease. Studies were obtained from multiple electronic databases: EMBASE, IEEE, MEDLINE, PubMed, Science Direct, Scopus and Web of Science. The subject headings in databases were used in conjunction with key word variants to build gold-standard search strategies, which were then run on March 7, 2015. Variant terms used in the search included "elderly", "center of pressure", "diabetes", "body stability" and "DPN". Inclusion criteria included studies that assessed CoP-AP on a rigid surface with eyes open and CoP parameters (TP and/or NLP). Exclusion criteria were: gait analysis, questionnaires, case studies, dissertations and gestational diabetes. In total, 42 studies were published in 1995-2016. We made a data extraction form or table to organize the information extracted from each reviewed study: authors (publication year), population studied (number of participants), TP and NLP, trial period (time),

I. Torres et al. (eds.), *VII Latin American Congress on Biomedical Engineering CLAIB 2016, Bucaramanga, Santander, Colombia, October 26th -28th, 2016*,
IFMBE Proceedings 60,
DOI: 10.1007/978-981-10-4086-3_179

sample frequency (Hz) and statistical significance test. Review of studies: Titles and abstracts of all references identified were screened and non-relevant studies excluded. Screening of titles and abstracts was performed by two authors with at least two assessing every title and abstract. Potentially relevant articles were then retrieved for a final assessment based on full text assessment by at least two authors. Disagreement in all cases was resolved by consensus and articles that did not fulfil inclusion criteria were excluded.

III. Results

Table 1 shows the most relevant parameters used for the CoP analysis in diabetic people with or without DPN and control people, which values were reported in the selected articles and correspond to measures of CoP in the AP time series. In column two was registered: number of participants, statistical test (level of significance) and scientific journal. The following columns on table were: parameters (linear –as time and frequency- and nonlinear), test period and sample frequency. Next, we discuss the most important issues to keep in mind when recording and analyzing CoP signal into AP direction.

A. CoP registration

For CoP registration, environmental conditions must be appropriate for the tests, such as, adequate lighting and without sounds and/or noise that may cause distractions during measurements. It is also important to standardize the parameters for signal acquisition, such as sampling frequency and acquisition time. Regarding the sample frequency, Winter (1995) in his book found that CoP signal frequency is below 10 Hz, therefore, an acquisition frequency of 20 Hz (Nyquist theorem) is sufficient for perfect fidelity. Based on data reported in Table 1, it can be seen that sample frequency is between 20 to 500 Hz, with predominant frequency of 100 Hz. Acquisition time, Lafond et al. (2004) and Corriveau et al. (2000) suggested 1 to 2 minutes[4]. However, Rugelj et al. (2007) and Pinsault et al. (2008) suggested that 30 seconds is sufficient for elderly people or individuals with a particular disease affecting body stability. In Table 1, researchers used several measurement times from 20 to 120 seconds, although prevailed 30 seconds to CoP signal acquisition.

B. Linear and nonlinear parameters

TP are commonly used in body stability studies and it reports statistical values in time and frequency domain, but emerging research suggests that this kind of analysis does not fully characterize CoP properties. It is because, the body system has a strong nonlinearities due to elastic and damping muscles properties involved in maintenance of upright posture and it has nonlinear feedback control created by nervous system, due to the time delay caused by reception and transmission of nerve impulses from the foot to the brain and vice versa. This time delay was estimated by Peterka (2002) from 100 to 200 msec in normal people, increasing with some diseases such as DPN [5]. According to the above, CoP analysis as a stochastic process will be useful to assess body stability in normal and pathological stages.

More recent evidence NLP analysis, supports the idea that, with advancing age and with any disease, temporal structure complexity is broken reflecting irregular movements in CoP time series, suggesting that changes in fluctuations regularity would indicate impaired sensory feedback on postural control. For this purpose, researchers have implemented some algorithms for dynamic stability analysis: Entropy (En) (approximate, sample and multiscale) to quantify the regularity over time series data; Hurst exponent (H) to detect the existence of long-range correlations reflecting the influence of past events in the future evolution of signal; Wavelet transform (WT) to identify frequencies of CoP signal on different time scales; Stabilogram diffusion analysis to represents the time series as a Brownian motion; Detrended fluctuation analysis (DFA) to detects of long-term correlations and Lyapunov Exponent used to determine periodic or chaotic behavior. Therefore, these tools have begun to be used to describe and understand CoP signal, but actually, very few investigations have been conducted in subjects with DPN, as recorded in Table 1. As research by Fioretti et al. (2010) used Lyapunov Exponent, but they did not find any significant differences between diabetics with and without DPN and Morrison et al. (2012) used approximate Entropy finding significant differences between controls and diabetics [6].

IV. Discussion

Body Stability studies in DPN found that these are significantly less stable with respect to healthy control subjects and even diabetics without DPN. Similarly, authors say it is due to disturbance in lower limb proprioceptive feedback suggesting that visual information can not completely compensate proprioception impaired [2], [7]. This sensorimotor integration deficits increased ankle stiffness, which forced to adopt different postural strategies, either ankle or hip, according to disease stage. Time parameters in all studies reported in Table 1, indicated that CoP-AP in Diabetes and/or DPN have increased: Range, excursion, RMS, velocity and amplitude, compared to controls, but for this systematic review, we will focus on those that presented statistically significant differences between groups. Therefore, the most important parameters were, Range: 2.6±0.3 cm, RMS: 0.26±0.21 mm, and Average Amplitude: 4.4 mm. Fioretti et al. (2010) indicated that AP power (dB/Hz) in asymptomatic DPN is approximately 20 and symptomatic DPN around 13.

Table 1. Parameters used in the assessment of CoP-AP in DPN

Authors	Population/Statistical test /Scientific journal	Parameters	T (s)	S-F (Hz)
Bergin et al. (1995) [8]	32 HC and 25 D-DPN Mann-Whitney (p≤0.01) Journal of Neurology, Neurosurgery, and Psychiatry	AvgAmp (linear): [-]	60	-
Boucher et al. (1995) [9]	27 Diabetics and 12 HC ANOVA (p<0.001) Diabetes Care	Range* (linear): [13,15.5 mm]	30	500
Katoulis et al. (1997) [10]	20 N-D Control 20 D without DPN 20 D-DPN without ulcer 20 D-DPN with ulcer ANOVA (p<0.0001) Diabetic Medicine	AvgAmp * (linear): 5.3 mm IC 95% (4.6, 6.2)	30	-
Dickstein et al. (2001) [11]	8 D-DPN and 8 HC MANOVA (p<0.0001) Gait and Posture	RMS (linear): [0.6 ± 01 cm]	40	240
Yamamoto et al. (2001) [7]	123 D without DPN 32 D-DPN and 55 HC Tukey–Kramer Test (p=0.05) Diabetes Research and Clinical Practice	AvgAmp (linear): [−0.62±1.45 cm]	60	-
Horak et al. (2002) [12]	12 HC 13 D-DPN ANOVA (p< 0.003) Somatosensory & Motor Research	Range * (linear): [2.6±0.3 cm]	40	-
Dickstein et al. (2003) [13]	10 HC and 8 D-DPN ANOVA (p < 0.004) J Neurol Neurosurg Psychiatry	AvgAmp (linear): [-] VelP (linear): [-]	-	240
Lafond et al. (2004) [14]	11 D-DPN and 20 HC ANOVA (p < 0.05) Diabetes Care	RMS* (linear): [0.27±0.10 mm]	120	20
Kanade et al. (2008) [15]	23 D-DPN 23 D-DPN and ulcer 16 D-DPN and P-Amp 22 D-DPN and T-Amp ANOVA (p < 0.01) Clinical Biomechanics	Exc (linear): [0.67±0.09 m]	30	-
Schilling et al. (2009) [16]	39 HC 10 D without ND 22 N-D ND 28 D-DPN ANOVA (p=0.0038) IEEE	Max value: AvgAmp * (linear): [18.3 mm] RMS* (linear): [23.7 mm] VelP* (linear): [627 mm/s] PotP* (linear): [562 Hz]	20	100
Fioretti et al. (2010) [17]	12 D without DPN 25 D-DPN Linear discriminant analysis (p < 0.0001) Gait and Posture	AvgAmp* (linear)[4.4 mm] VelP* (linear): [-] Pot* (linear): [20 Hz] F95 (linear): [-] Lya (nonlinear): [-]	60	100
Morrison et al. (2012) [18]	14 Controls WoF 7 Controls WF 9 Diabetics WoF 7 Diabetics WF Repeated measures test (p<0.05) Gait and Posture	Range * (linear): [25.46 mm] ApE* (nonlinear) [High value]	30	-
Salsabili et al. (2013) [19]	19 D-DPN ANOVA (p<0.05) Journal of Diabetes & Metabolic Disorders	CFr (linear): [0.08 ± 0.04 Hz] Pot (linear): [8.73 ± 4.37 Hz] F99 (linear): [1.65 ± 0.32] PSD (linear): [4.33 ± 2.14 Hz]	30	50
Silva et al. (2015) [20]	20 Diabetics ANOVA (p<0.001) Journal of Bodywork & Movement Therapies	AvgAmp (linear): [2.17±0.72 cm)]	30	100
Dixit et al. (2015) [21]	61 D-DPN RANOVA (p<0.05) Indian Journal of Medical Research	AvgAcc (linear): [8.1±2.5 mm/s²] AvgAmp (linear): [0.39±0.09 mm]	30	-
Dixit et al. (2016) [6]	46 HC and 36 D-DPN RANOVA (p<0.05) American Journal of Physical Medicine & Rehabilitation	VelP (linear): [2.31±1.16 mm/s] AvgAmp (linear): [2.86±1.25 mm]	30	-

*: statistical significance with other groups; AvgAcc: Average acceleration; AvgAmp: Average amplitude; ApE: Approximate entropy; cm: Centimeters; CMT1: Charcot-Marie-Tooth 1A; CMT2: Charcot-Marie-Tooth 2; D: Diabetics; D-DPN: Diabetics with DPN; Exc: Excursion; F95, F99: Frequency for which 95%, 99% of the total power of CoP; HC: Healthy controls; CFr: Centroidal frequency; S-F: Sample Frequency; Lya: Lyapunov exponent; m: Meters; mm: Millimeters; N-D: Nondiabetic; N-L: Nonlinear; P-Amp: Partial amputation; PotP: Average amplitude of power; Pot: Total power; PSD: Power spectral density; T-Amp: Trans tibialis amputation; WF: with falls; WoF: without falls; RMS: Root mean square; T: Time; VelP: Average velocity.

Concerning to CoP-AP analysis by using nonlinear methods, Morrison et al. (2012) presented significant differences between controls and diabetics by using approximate Entropy, suggesting that CoP signal in Diabetes is less predictable with a complex structure for its high value in entropy. However, regardless of the available studies in Table 1, other research, such as those by Boucher et al. (1995) and Uccioli et al. (1995, 1997) finding in DPN greater velocity and sway area. Horak et al (2002), indicated that DPN subjects exhibit 40-50% higher sway than control subjects during static standing position on a rigid surface with eyes open and closed [12]. LaFond et al (2004) concluded DPN subjects presented a higher sway area, average velocity, and CoP range in both AP and ML directions, compared with healthy controls [14].

V. Conclutions

CoP-AP recording and analysis in DPN is relevant to understand the body stability behavior and that information should contribute to clinical area to identify patients at risk for falls. Therefore, we suggest some specifications for signal recording and TP-NLP parameters to analyze the CoP-AP behavior in DPN. For data acquisition, we suggested 100Hz sampling frequency because according to the previous study by Schilling et al. (2009), Fioretti et al. (2010) y Silva et al.

(2015), found that frequency optimal to CoP analysis and they reported significant difference between groups. For measured time with eyes open, we proposed 30 seconds, used by Rugelj et al. (2007) and Pinsault et al. (2008). They reported that this time is adequate to CoP data acquisition in elderly and on the other hand, a long time, the person may be fatigued and consequently alterations in sway body. For TP parameters, we recommended measurements that presented statistically significant differences in studies: Range, Average velocity and the frequency band that contains up to 80% of the power spectrum (Table 2) and for NLP parameters, we analyzed studies to other populations (young and older adults) because only a few studies have examined CoP signal in DPN, as mentioned earlier. The selected NLP parameters were: WT, H, En and DFA, as shown in Table 2. Wang et al. (2012), Kirchner et al. (2013) and Yeh et al. (2014) suggested that TW, H, En and DFA are useful tools to study changes in AP CoP fluctuations; because these nonlinear techniques analyzes CoP time series as a stationary or nonstationary signal and thus identify new information that contribute to characterization. Also, the analysis showed that is necessary to quantify other important related variables feature in DPN and that impact on stability, such as: diabetes duration, age and changes in foot sensation on plantar pressure (monofilament test), as shown by Dickstein et a. (2003), Lafond et al. (2004), Schilling et al. (2009) and Morrison et al. (2012). Therefore, further research should include additional information about.

Table 2. Suggested parameters in the assessment of CoP-AP in DPN

Time	Frequency	Nonlinear
Range Average Velocity RMS	F80	Wavelet transform Hurst exponent Approximate entropy Multiscale entropy DFA

DFA: Detrended fluctuation analysis; F80: Power spectral density containing up to 80% of the power spectrum. RMS: root mean square value.

Conflict of Interest

The authors declare that they have no conflict of interest.

References

1. IDF, International Diabetes Federation (2015) Atlas – 7th Edition, at http://www.idf.org
2. Lafond D, Corriveau H, Prince F (2004) Postural Control Mechanisms During Quiet Standing in Patients With Diabetic Sensory Neuropathy. Diabetes Care 27:173–178
3. Loughlin PJ, Redfern MS, Furman JM (2003) Nonstationarities of Postural Sway. IEEE Eng Med Biol Mag 22(2):69-75 DOI:10.1109/MEMB.2003.1195699
4. Lafond D, Corriveau H, Prince F (2004) Postural control mechanisms during quiet standing in patients with diabetic sensory neuropathy. Diabetes Care 27(1):173-178 DOI:10.2337/diacare.27.1.173
5. Peterka RJ (2002) Sensorimotor integration in human postural control. J Neurophysiol 88(3):1097-1118
6. Dixit S, Maiya A, Shastry BA, Guddattu V (2016) Analysis of Postural Control During Quiet Standing in a Population with Diabetic Peripheral Neuropathy Undergoing Moderate Intensity Aerobic Exercise Training. Am J Phys Med Rehabil 1 DOI:10.1097/PHM.0000000000000426
7. Yamamoto R, Kinoshita T, Momoki T, et al (2001) Postural sway and diabetic peripheral neuropathy. Diabetes Res Clin Pract 52:213-221
8. Bergin PS, Bronstein AM, Murray NMF, Sancovic S, Zeppenfeld K (1995) Body Sway and Vibration Perception Thresholds in Normal Aging and in Patients With Polyneuropathy. J Neurol Neurosurg Psychiatry 58:335-340 DOI:10.1136/jnnp.58.3.335
9. Boucher P, Teasdale N, Courtemanche R, Bard C, Fleury M (1995) Postural Stability in Diabetic Polyneuropathy 18(5):638-645
10. Katoulis EC, Ebdon-Parry M, Hollis S, et al (1997) Postural instability in diabetic neuropathic patients at risk of foot ulceration. Diabet Med 14(4):296-300
11. Dickstein R, Shupert CL, Horak FB (2001) Fingertip touch improves postural stability in patients with peripheral neuropathy. Gait Posture 14(3):238-247 DOI10.1016/S0966-6362(01)00161-8
12. Horak FB, Dickstein R, Peterka RJ (2002) Diabetic neuropathy and surface sway-referencing disrupt somatosensory information for postural stability in stance. Somatosens Mot Res 19(4):316-326 DOI:10.1080/0899022021000037782
13. Dickstein R, Peterka RJ, Horak FB (2003) Effects of light fingertip touch on postural responses in subjects with diabetic neuropathy. J Neurol Neurosurg Psychiatry 74(5):620-626
14. Kanade R V., Van Deursen RWM, Harding KG, Price PE (2008) Investigation of standing balance in patients with diabetic neuropathy at different stages of foot complications. Clin Biomech 23(9):1183-1191 DOI:10.1016/j.clinbiomech.2008.06.004
15. Schilling RJ, Bollt EM, et al (2009) A quiet standing index for testing the postural sway of healthy and diabetic adults across a range of ages. IEEE Trans Biomed Eng 56(2):292-302 DOI:10.1109/TBME.2008.2003270
16. Fioretti S, Scocco M, Ladislao L, Ghetti G, Rabini RA (2010) Identification of peripheral neuropathy in type-2 diabetic subjects by static posturography and linear discriminant analysis. Gait Posture 32(3):317-320 DOI:10.1016/j.gaitpost.2010.05.017
17. Morrison S, Colberg SR, Parson HK, Vinik AI (2012) Relation between risk of falling and postural sway complexity in diabetes. Gait Posture 35(4):662-668 DOI:10.1016/j.gaitpost.2011.12.021
18. Salsabili H, Bahrpeyma F, Esteki A, Karimzadeh M, Ghomashchi H (2013) Spectral characteristics of postural sway in diabetic neuropathy patients participating in balance training. J Diabetes Metab Disord 12(1):29 DOI:10.1186/2251-6581-12-29
19. Silva P, Figueredo P, De Oliveira E, et al (2015) Long-term benefits of somatosensory training to improve balance of elderly with diabetes mellitus. J Body Mov Ther 19(3):453-457 DOI:10.1016/j.jbmt.2014.11.002
20. Dixit S, Maiya A, Shasthry BA, D SK, Guddattu V (2015) Postural sway in diabetic peripheral neuropathy among Indian elderly 713-720 DOI:10.4103/0971-5916.174562

[1] Daissy Carola Toloza Cano. Pontificia Universidad Javeriana. Carrera 7 No. 40 - 62, Bogotá, Colombia. Email: dtoloza@javeriana.edu.co
[2] Martha Lucía Zequera Díaz. Pontificia Universidad Javeriana. Carrera 7 No. 40 - 62, Bogotá, Colombia. Email: mzequera@javeriana.edu.co

Optimal design of a mechanism for children foot guiding

P.A. Niño Suarez[1], F. Calderon-Romero[1], M.B. Calva-Yañez[1], E.A. Portilla-Flores[2] and O.A. Aviles-Sánchez[3]

[1]Instituto Politécnico Nacional, ESIME Unidad Azcapotzalco, Av. de las Granjas No. 682, Col. Santa Catarina, C.P. 02250, Delegación Azcapotzalco, Ciudad de México, México
fercalderon02@gmail.com, b_calva@hotmail.com,pninos@ipn.mx
[2]Instituto Politécnico Nacional, CIDETEC, Av. Juan de Dios Bátiz s/n esq. Miguel O. Mendizábal, Col. Nueva Industrial Vallejo, C.P. 07700, Ciudad de México, México
aportilla@ipn.mx
[3]Universidad Militar Nueva Granada,Carrera. 11 #101-80, Bogotá, Colombia
oscar.aviles@unimilitar.edu.co

Abstract— **In this paper the result of a binational research, Mexico-Colombia, developed for the last four years is presented. This project aims to develop a robotic system to gait training for children 2 to 12 years with cerebral palsy. The concept design system and the methodology used during the development of the design process is presented. And finally, the optimal design of the ankle-foot stage is showed. The design of this stage was resolved as a dynamic optimization problem using a differential evolution algorithm with a constraint handling mechanism.**

Keywords— **ankle trajectory, gait rehabilitation, optimal design.**

I. Introduction

Cerebral palsy (CP) is a postural and movement disorder, it is a frequent cause of pediatric physical disabilities. According to the severity of involvement, infants have functional limitations in the musculoskeletal system, such as learning to walk by themselves. The use of robotic devices in active progress for rehabilitation of disorders in the central nervous system has shown significant improvement over conventional functional therapies, [1] and [2]. However, there are still critical factors to improve because none of the equipment used was created to train children with CP, and them were not designed taking into account the specific characteristics of their rehabilitation. Therefore, the development of systems specific to the needs of the CP gait training resulting in a technical and scientific challenge and opens a window of opportunity so that they are not a few rehabilitation centers and health institutions, those who can afford the technology allowing for optimal quality rehabilitation for infants.

Up to now, the exoskeleton designs proposed are generally oriented to adult patients. A review of the gait rehabilitation devices designed for patients with central nervous system disorders is presented in [3] and [4]. Only few gait rehabilitation devices have been developed for children with cerebral palsy, [5]. In addition, there are few research project that used the numerical optimization problem for design those devices. In [6] an optimum kinematic design of a planar cable-driven parallel robot with wrench-closure gait trajectory is presented.

In this article the design of the concept of the rehabilitation children system is presented in the section II. In the section III, the optimal developed design is shown. The manufacturing design is presented in the section IV. In the section V, the conclusions of the work performed are presented.

II. Desing methodology

This article is product of the research binational project trying to contribute with the children with CP rehabilitation. Its aim is developing a robotic system for rehabilitation, focused on children with CP between 2- 12 years old allowing the development of the necessary neural connections that allows them, in the future, perform the gait process by themselves. Since the project is too ambitious, it has been decided to divide it into three stages. The first stage is the conceptual design of the system. According with the results of the Quality Function Deployment, QFD methodology, it was achieved to consider the needs given by the user and medical experts into the technical requirements that should be covered by the system at the end of the design process, [7]. The design concept obtained is shown in Figure 1.

The second stage consists on implementing a concurrent design methodology, beginning with biomechanics of human gait to develop the different stages of the system structure. It began with ankle-foot stage. The first analysis of the lower limb is performed to obtain a mathematical model, determining its kinematics and the path followed by the ankle during the gait cycle. The goal is to find a mechanism being able of tracking natural trajectory of the ankle in the sagittal plane efficiently, with a maximum error of 1%. In this case a four-

I. Torres et al. (eds.), *VII Latin American Congress on Biomedical Engineering CLAIB 2016, Bucaramanga, Santander, Colombia, October 26th -28th, 2016*,
IFMBE Proceedings 60,
DOI: 10.1007/978-981-10-4086-3_180

bar mechanism was chosen. The methodology includes the formulation of a numerical optimization problem to find the dimensional synthesis of four-bar mechanism using a metaheuristic algorithm and the best solution obtained was implemented in CAD, whose performance analysis by simulation, allowed evaluate its functionality as rehabilitation device and determine the feasibility of building the system. From the results, it was concluded that the mechanism designed allows to reproduce the natural path of the ankle in the sagittal plane efficiently. Due to its reconfigurable dimensions, the system can be used for children of 2 to 12 years.

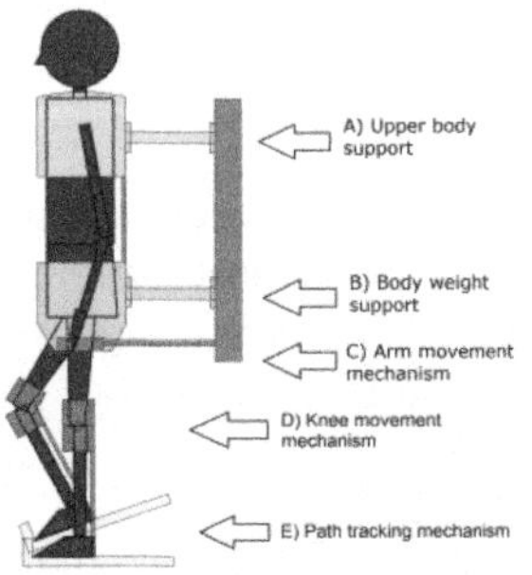

Fig. 1 System Design Concept

The third stage consists in the development of detailed design for the manufacture of the prototype. For manufacturing the mechanism a finite element analysis (FEA) tools are used for geometry evaluation and material selection.

III. OPTIMAL DESIGN

For determining the motor mechanism for foot guiding, is necessary to establish the trajectory of the lower limb from a reference point. For getting this trajectory, the kinematic model of the leg was obtained.

A. Kinematic model leg

Taking the coxofemoral joint located in the base of the hip as that reference and considering the leg like a planar mechanism with 2 degrees of freedom and 2 links from the hip to the ankle, analyzed on sagittal plane, as shown in Figure 2. It is possible describe the forward kinematics of a leg from (1) and (2),

$$X = l_1 \cos(q_1) + l_2 \cos(q_1 + q_2) \tag{1}$$

$$Y = l_1 \sin(q_1) + l_1 \sin(q_1 + q_2) \tag{2}$$

where l_1 and l_2 are the distance from hip to knee and knee to ankle respectively. Meanwhile q_1 and q_2 are the angle of rotation of X around the Z axis of each joint, [8].

B. Ankle trajectory

From forward kinematics and angles q_1 and q_2 defined during the gait cycle by hip bone and knee, ankle trajectory is obtained, [8]. In the Figure 3, ankle trajectory of a child of seven years is shown. It is noted that, this is a trajectories type drop.

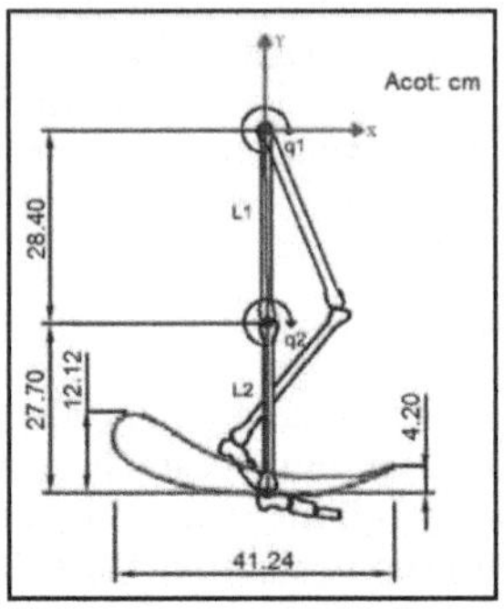

Fig. 2. Kinematic model leg of a child of seven years

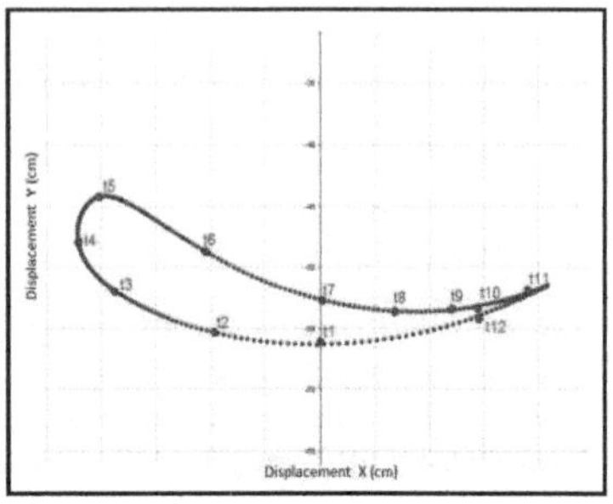

Fig. 3. Ankle trajectory of a child of seven years

C. Selection mechanism

Therefore, a four-bar mechanism as the basis for the design of the ankle-foot stage was selected, Figure 4. Because its topology allows the generation of type drop trajectories.

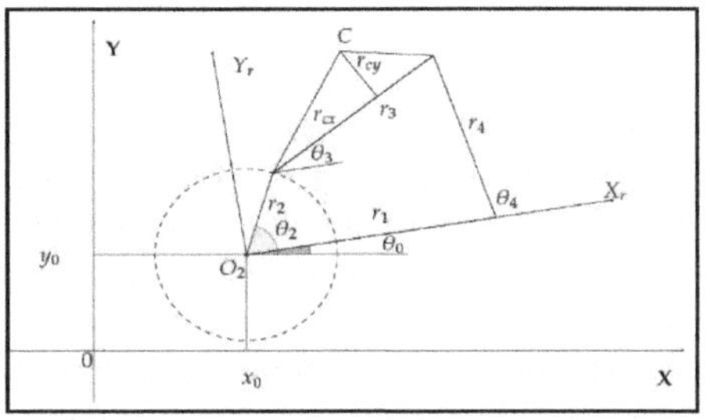

Fig. 4. Four-bar mechanism

Where r_1 is the ground link, r_2 is the input link, r_3 is the coupler y r_4 is the output link, θ_i is the angular displacement for links $i=1,2,3,4$. The mechanism has an actuator coupled to the link input shaft. The point C located in the coupler has to follow an N point sequence, which are the track for the mechanism, [9].

D. Static Optimization Problem

The dimensions of the mechanism selected are obtained by formulating the problem as a static optimization problem. At the description of four bars mechanism in Figure 4, in the global coordinate system OXY, the "i" precision point is written by (3),

$$C_d^i = [C_{xd}^i, C_{yd}^i] \quad (3)$$

And the N set precision points are defined as it is shown in (4),

$$\Omega = \left\{ C_d^i \middle| i \in N \right\} \quad (4)$$

In the first design case, the given problem is to determinate the value of the length of links (r_1, r_2, r_3, r_4, r_{cx}, r_{cy}) of the mechanism, the angular offset θ_0, the distance R_0 and the angular movements that make possible to reach the N defined precision points. Thus, given a set of link values and R_0, θ_0, it can be written each point of the coupler as follow in (5),

$$C^i = [C_x(\theta_2^i), C_y(\theta_2^i)]^T \quad (5)$$

The position of the coupler in the OX_rY_r system reference it is given as follow in (6),

$$\begin{aligned} C_{xr} &= r_2 \cos\theta_2 + r_{cx} \cos\theta_3 - r_{cy} \sin\theta_3 \\ C_{yr} &= r_2 \sin\theta_2 + r_{cx} \sin\theta_3 - r_{cy} \cos\theta_3 \end{aligned} \quad (6)$$

Thus, from the *OXY* global coordinate system, it is shown that the coupler is placed by (7),

$$\begin{aligned} C_x &= \cos\theta_0 C_{xr} - \sin\theta_0 C_{yr} + X_0 \\ C_y &= \sin\theta_0 C_{xr} - \cos\theta_0 C_{yr} + X_0 \end{aligned} \quad (7)$$

Thereby, the optimal design of the four bars mechanism was made proposing an optimization numeric problem which was solved by a differential evolution algorithm (DE), [8].

This solution is only valid for a particular trajectory. A new optimization problem should be performed, to obtain a four-bar mechanism that can follow at least three different trajectories, (for 2 years, 7 years and 12 years). In this way the designed system can be reconfigurable. For this purpose, dimensional parameters that can be held constant in the three dimensional synthesis are chosen. The remainder parameters were calculated using the optimization numeric problem.

Analyzing four-bar mechanism, r_1 is the ground link, it has no movement relative with other components of the mechanism. It is a link that is not constructed physically. It represents the distance between the points where the input link and output link can rotate. Another parameter that does not change, is the rotation of the mechanism with respect to the global coordinate system, Θ_0. Therefore, r_1 and Θ_0 were chosen as the constant parameters.

The problem of optimal design is rewritten using a dimensional parameters r_1 and Θ_0 that were obtained as a result of the best solution of the optimization problem for the course of an infant 12 years.

Be the numerical optimization problem mono-objective:

$$Min \quad f(x) = \sum_{i=1}^{N} \left[\left(C_{xd}^i - C_x^i\right)^2 + \left(C_{yd}^i - C_y^i\right)^2 \right] \quad x \in \mathbb{R}^{15} \quad (8)$$

subject to,

$$\begin{aligned} g_1(x) &= x_1 + x_2 - x_3 - x_4 \le 0 \\ g_2(x) &= x_2 - x_3 \le 0 \\ g_3(x) &= x_3 - x_4 \le 0 \\ g_4(x) &= x_4 - x_1 \le 0 \\ g_5(x) &= x_{10} - x_{11} \le 0 \\ g_6(x) &= x_{11} - x_{12} \le 0 \\ g_7(x) &= x_{12} - x_{13} \le 0 \\ g_8(x) &= x_{13} - x_{14} \le 0 \\ g_9(x) &= x_{14} - x_{15} \le 0 \\ x_i &= 47.880 \qquad i = 1 \\ 0 \le x_i &\le 60 \qquad i = 2,3,4 \\ -60 \le x_i &\le 60 \qquad i = 5,6,8,9 \\ x_i &= 2.244 \qquad i = 5,6,8,9 \\ 0 \le x_i &\le 2\pi \qquad i = 10,11,12,13,14,15 \end{aligned}$$

And a N set precision points are defined for each of the trajectories, where Ωa is obtained from the trajectory of infants under 2 years, Ωb for infants under 7 years and Ωc for 12 years. The dimensions for the paths were taken from [10].

Again, the optimization numeric problem was solved using a differential evolution algorithm. In the present work, a set of 30 independent runs was carried out. A population of 100 individuals, a random crossover rate (0.8 , 1), a random mutation rate (0.3 , 0.9) and 2000 generations was used. The DE algorithm was coded in Matlab ® R2008a and was run in a Laptop computer with 6 GB RAM, Intel ® Core i5 processor @ 2.5 GHz, and Microsoft Windows ® 7 OS. From simulations implemented for the optimal dimensional synthesis, the best for its implementation in CAD were selected, figure 5. And its dimensions are presented in Table 1.

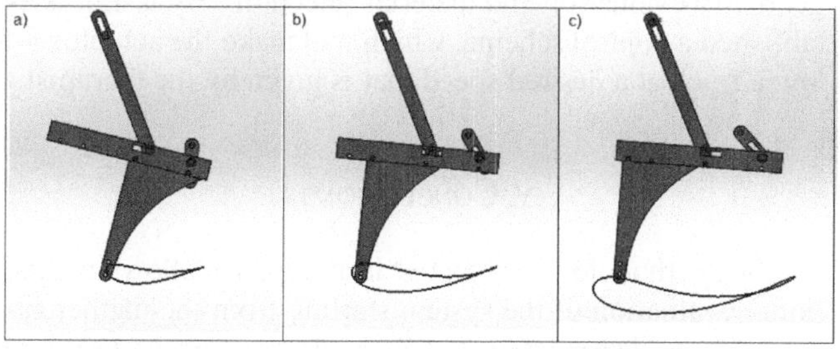

(a) (b) (c)

Fig. 5 Four-bar mechanisms CAD. (a). 2 years, (b). 7 years (c). 12 years

Table 1. Dimensions of the length of links

Parameters	Two years	Seven years	Twelve years
r_1	0.47880 m.	0.4788 m.	0.4788 m.
r_2	0.05.663 m.	0.08605 m.	0.1112 m.
r_3	0.14756 m.	0.16876 m.	0.1939 m.
r_4	0.42334 m.	0.46660 m.	0.4788 m.
r_{cx}	0.18826 m.	0.37068 m.	0.4497 m.
r_{cy}	0.43640 m.	0.40227 m.	0.4497 m.
Θ_0	2.243 rad.	2.243 rad.	2.243 rad.
X_0	0.56996 m.	0.61708 m.	0.6558 m.
Y_0	0.57461 m.	0.63850 m.	0.6946 m.

IV. Desing for Manufacturing

With the dimensions resulted by the best solution of the optimization problem, a CAD model to evaluate the dimensions and operation for a correct manufacturing was generated. In this way, a virtual prototype was turned out as shown in figure 6a; where it is seen the four bar mechanism with the coupler and a tracker in a manufacturing configuration which helps to save material during machine process. In figures 6b, it is shown the patient use concept, where each lower extremity uses an individual mechanism; and both mechanisms are synchronized to reproduce the gait cycle path.

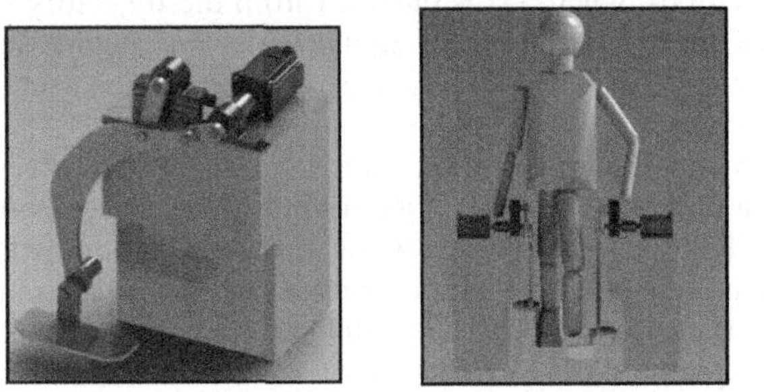

Fig. 6 Patient use concept rehabilitation children system

A physical prototype is needed to start the manufacturing process; a behavior of the system load response is needed too, and hence, finite element analysis (FEA) tools are used for geometry evaluation and material selection. By last, it is established a control scheme, which will make the actuator follow a track at a desired speed that is given by the therapist.

V. Conclusions

The methodology presented in this paper allows a functional evaluation of the system starting from the mathematical model and simulation its behavior. Concurrent design results in the development of intelligent and flexible mechatronic system. Design features, the optimal synthesis mechanism allows reconfiguration, providing great flexibility to the system. The design CAD mechanism allows, without the need to build, verify that operation is the desired and determine if the results obtained dimensional synthesis makes possible the manufacture of the mechanism.

Acknowledgment

The first and second author acknowledges support from CONACYT through a scholarship to pursue graduate studies. Authors acknowledge support from AMEXID and APC through binational research project program "Cooperacion Sur-Sur".

Conflict of Interest

The authors declare that they have no conflict of interest.

References

1. Borggraefe I., Schaefer J. S., Klaiber M., *et al.*, (2010). Robotic-assisted treadmill therapy improves walking and standing performance in children and adolescents with cerebral palsy. European Journal of Pediatric Neurology, Vol. 14, pp. 496-502.
2. Meyer-Heim A., Ammann-Reiffer C., Schmartz A., Schaefer J., et al. (2009). Improvement of walking abilities after robotic-assisted locomotion training in children with cerebral palsy. Archives of disease in childhood, Vol. 94, pp. 615-620.
3. Kubo K., Miyoshi T., Kanai A. and Terashimal K. (2011). Gait Rehabilitation Device in Central Nervous System Disease: A Review. Journal of Robotics. Vol. 2011, pp. 2-14.
4. Hesse S., Waldner A., Tomelleri C. (2010). Research Inno-vative gait robot for the repetitive practice of floor walking and stair climbing up and down in stroke patients. Journal Neuro Engineering and Rehabilitation, Vol 7, No. 30, pp. 1-10.
5. Canela M., Del Alma A. J., Pons J. L. (2013). Design of a pediatric exoskeleton for the rehabilitation of the physical disabilities caused by C.P. Converging Clinical & Engineering Research on NR. BIOSYSROB 1, pp. 255–258.
6. Ghasem Abbasneiad, Jungwon Yoon, Hose Lee. (2016). Optimum kinematic design of a planar cable-driven parallel robot with wrench-closure gait trajectory. Mechanism and Machine Theory, Elsevier, Vol. 99, pp. 155–162.
7. Figueroa Amador M. T. (2016). Diseño de un sistema robótico para rehabilitación de marcha enfocado a niños con parálisis cerebral. Master's Thesis in Ingeniería de Manufactura, ESIME Azcapotzalco, IPN, Ciudad de México, México.
8. Calva-Yáñez M. B., Calderón-Romero F., Portilla-Flores E. A., Niño-Suarez P. A., Vega E. (2015). Diseño de mecanismo para seguimiento de trayectoria de tobillo en rehabilitación de marcha. In Proceeding MECAMEX, Queretaro, México.
9. Calva Y. M. B., Niño S. P. A., Villarreal C. M. G., et al. (2013). Differential Evolution for the Control Gain's Optimal Tuning of a Four-bar Mechanism. POLIBITS, Vol. 47, pp. 67-73.
10. Ávila R., Prado L., and González E. (2001). Dimensiones antropométricas de población latinoamericana, Universidad de Guadalajara: Centro de Investigaciones en Ergonomía, pp. 24-276.

Thematic profile of the e-health field, based on mainstream scientific production

P. Romero[1], Y. Piedra[1] and M. Nyssen[2]

[1] University of Havana, Havana, Cuba
[2] Vrije Universiteit Brussel, Brussels, Belgium

Abstract— **The study and analysis of knowledge areas using metrics and visualization tools represent an important scientific, technical and social achievement. Applying appropriate tools of the Information Sciences to knowledge spaces, as little explored as the e-health field constitutes a step forward with regard to the level of recognition of these fields. The application of indicators to the products of the (formal and informal) scientific communication allows detection of research fronts and new trends, among other things. Applying this analysis to the field of e-health , monitoring and evaluation of its scientific production (mainly literature and scientific events) will help obtain a better insight into the development of the related field. In this article, the scientific production of e-health in the "Web of Science" in the period 2010-2015 is analyzed to provide the thematic field profile.**

Keywords— **e-health, thematic profile, web of science, scientific production, domain analysis, metric studies**

I. Introduction

After the Internet breakthrough in the 90's' and when users became aware of all the services, offered by the network of networks, so-called "e-terms" started to emerge, as a way to refer to user friendly applications in public life.

The development of e-health aims at improving the health system, via information and communication technology, taking a prominent place in the e-developments, and an interesting field of study. For this purpose, metrical studies concerning disciplines provide valuable information about the fields based on the analysis of the produced publications and texts, regarding authors, countries, journals, institutions. This way we can get an idea of the characteristics and the development parameters characterizing the e-health field.

Traditionally this type of study is developed from the compilation, processing and analysis of the information contained in national and international databases to which metric techniques are applied in order to build quantitative and qualitative indicators of scientific development and thus generate new Scientific information.

Applying this analysis to the field of e-health, by monitoring and evaluating scientific production will help to get a broader view on this field's development.

In recent years the situation of computer medicine has been poorly explored from a metric point of view. While studies stand out as (Moser, 2004 [1]; Demiris and Tao, 2005 [2]; Whan, 2006 [3]; Liang, 2010 [4]; Fateh and Wootton, 2012 [5]; Armfield 2014 [6]); there is a need for a study analyzing how the terms have changed with the evolution of time and the publication rate of articles in the medical field it is high, so it is possible to conduct such a study.

But it is essential to note that 80% of relevant publications are posted outside medical journals. Another important aspect is that the first publications on this topic focused mainly on technical issues, and now this focus has shifted to the application and medical evaluation.

In the early years, these studies identified that the term "imaging" was often used to describe applications of telemedicine involving radiological images. But tele-radiology has become essential in medical treatment and this may be an explanation for the absence of this term. Metrical studies provide new information regarding publications in the field of tele-medicine and tele-health. As a consequence of the expansion of this field in recent years, further studies are essential.

These studies allow to visualize new information regarding the nature of publications and to understand this field that has revolutionized medical treatment. Production in health issues has moved from technical problems related to applications and clinical evaluations; and home begins to have more coverage as a place for quality health treatments, of course, through the Internet.

Thus, this research aims to examine the field in question from the scientific production available in international databases and with tools applied to traditional metrics.

II. Materials and Methods

The research methodology used here is mixed, as qualitative analyzes will be conducted from quantitative results. This is a metric and descriptive study.

The classic documentary analysis, which involves the selection, collection, analysis and processing of the consulted documentary sources that support the development of the study, to collect the necessary data from each of the chapters is used. information display software that will allow the re-

I. Torres et al. (eds.), *VII Latin American Congress on Biomedical Engineering CLAIB 2016, Bucaramanga, Santander, Colombia, October 26th -28th, 2016*,
IFMBE Proceedings 60,
DOI: 10.1007/978-981-10-4086-3_181

alization of statistical procedures for the analysis will be used. These include the Ucinet Bibexcel and software for frequency counts and generating matrices that allow the analysis of the references. VosViewer and Netdraw for reticular representation of the results.

The data source used was the Web of Science of Thomson Reuters, integrated into ISI Web of Knowledge. To search the database the terms "e-health", "ehealth", "eHealth", "electronic health", "health informatics" and "telemedicine" were used.

Downloading the database was exported directly to an EndNote library. Each of the fields used were duly standardized to export the data necessary to create arrays through Bibexcel software, these matrices allow the representation and visualization of data through Ucinet, Netdraw and VosViewer, creating figures that facilitate the presentation of the study results.

Analysis results are then presented with their graphical representation. We worked on a total of 6538 records from the Web of Science, pertaining to the period from 1979 to 2015. These items were normalized in order to adjust the period of time relevant to this investigation. Of the total, 3111 documents were recovered in the time range from 1979 to 2009, these were separated from the sample. Later in the normalization process 626 records were discarded because they were present more than once. After this process, this study was conducted on the remaining sample of 2901 records.

Once normalized the fields, proceeded to the creation of the lists of topics, which were exported to text files that, later on, were processed with the program Microsoft Excel to generate the charts and graphic corresponding. It was also used the program Bibexcel (Olle Persson, University of Umeå, Sweden) to carry out the general counts of frequency and for the matrices generation of collaboration networks at different levels, main topics and abstracts co-occurrence. Later on they were processed to achieve their reticular representation by means of the use of the program VoSViewer 1.6.4

As theoretical methods were used the analytic-synthetic, deductive - inductive, the mensuration, documentary investigation and statistical methods relatives to metric information studies. Visualization techniques based on social networks for the representation of the analysis. The co-occurrence of topics and words of the summary, were the techniques mainly used for the realization of the present contribution.

III. Results

A. Frequency of occurrence of keywords

Keyword analysis lead to a total of 8604 keywords. Of this total, a fraction of 68% appear only once, 27 keywords have a frequency of occurrence greater than 100 representing 0.31% of the sample. Only 6 (0.06%) appear 200 times or more.

The keywords were split-up between words assigned by the author, and terms assigned by the Wos data base. In this section, we first analyze the frequency of occurrence of keywords provided by the author and then those coming from the database. All keywords can be found in the annexes.

Table 1 Keyword occurrences

Keyword	**Occurrence**
Electronic health records	373
Telemedicine	252
Internet	210
health	189
eHealth	182
Informatics	137

As shown above, the principal term used to describe the field of work is "Electronic Health Records", a methodology to store the relevant health records, pertinent to patients from hospitals or other health care entities, where health workers (physicians and others) will deposit and get access to all the relevant information they may need for the treatment of patients, possibly accessible via the network. Electronic management and updating of medical records has become widespread and gives rapid access to pertinent information such as lab tests and patient parameters to assist in diagnosing and appropriate treatment of the patients.

Electronic health records potentially allow entitled health care professionals, with medical links to the patient, to obtain up to date information without the need for searching paper based physical records. The second most used term by the authors is "telemedicine". This term is very important within the e-health field, because it refers to the first popular widespread e-health application, to the point that it was confused with the whole field. Telemedicine allows to perform some medical acts from a distance via the network, for example to render health care to remote rural areas, lacking personnel with medical expertise.

Internet is the third most common word in this field is the means through which it is possible to provide the e-health-services, as the e- prefix indicates.

The terms health, e-health and informatics consecutively follow in the list. Clearly the relationship between these three terms, e-health, the field studied and computer as the tool through which they can provide health services through internet. Following were the most occurring keywords from the Wos database:

Table 2 Author assigned keyword occurrences

Keyword	Occurrence
Care	477
Systems	254
Technology	222
Information	206
Medical Informatics	192

The most frequent was "care” with a frequency of 477 occurrences, here, the term refers to medical care, understandable as the subject under study is in the Health domain.

Then there is "systems" (254) referring to the new e-health developments in support of the health care field, these new systems are based on health technologies (222). "Information" is next with 206 appearances. Finally, "medical informatics" with a frequency of 192, this term is often used to refer to the e-health field, as there is no clear consensus on the name of this scientific field but clearly linked to the 6 frequently used keywords of the first table.

B. Co-occurrence of keywords

In order to understand the figure below, it is necessary to explain that the first two co-occurrence maps correspond to the keywords given by the authors and their linkages to the words from the database.

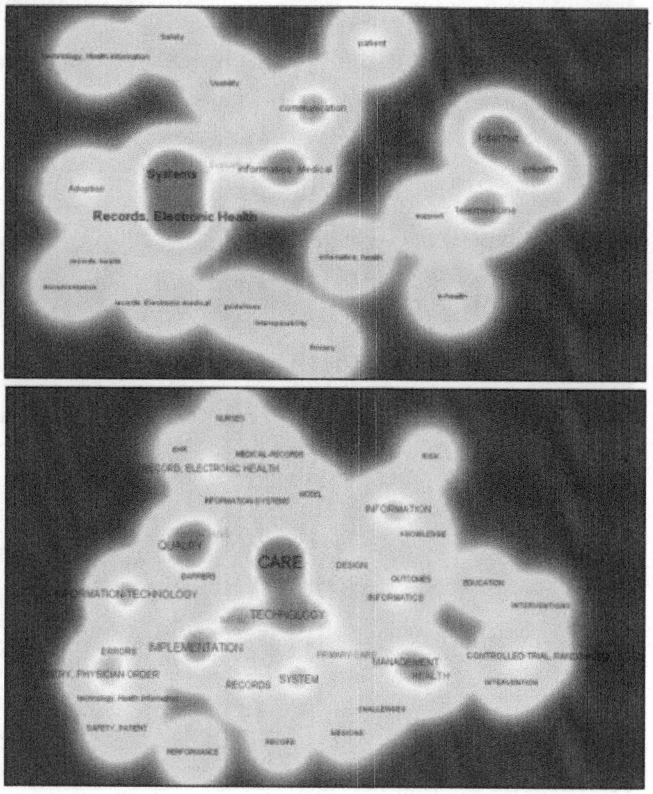

Fig. 1 Keyword density maps

The figures above represent the density of keywords, based on the co-occurrence reached by the elements, the density is represented by red, yellow, green and blue. The intensity of the color corresponds to the density level of each term, the higher the density, the closer to red, however, if the co-occurrence is less, the color will be closer to blue. The proximity to red determines the terms that are addressed in the records analyzed in this study, close to the blue color, the less frequent or perhaps emerging terms. Returning to the density maps, as you can see in both cases the words which are closer to the red and matching keywords have increased frequency of appearance. In the first figure you can also observe other terms closely related with the e-health field that are close to the red zone, such as: "medical informatics" (24), "communication" (136) "patient" (81) and "health informatics" (86).

In the second case, we rediscover the theme "Electronic health record" in a strong position as well as terms assigned by the database, this can be due to the great interest in the subjects, related to the processing and storage of medical records. Other terms are "information technology" (27), "health" (58), "quality" (182).

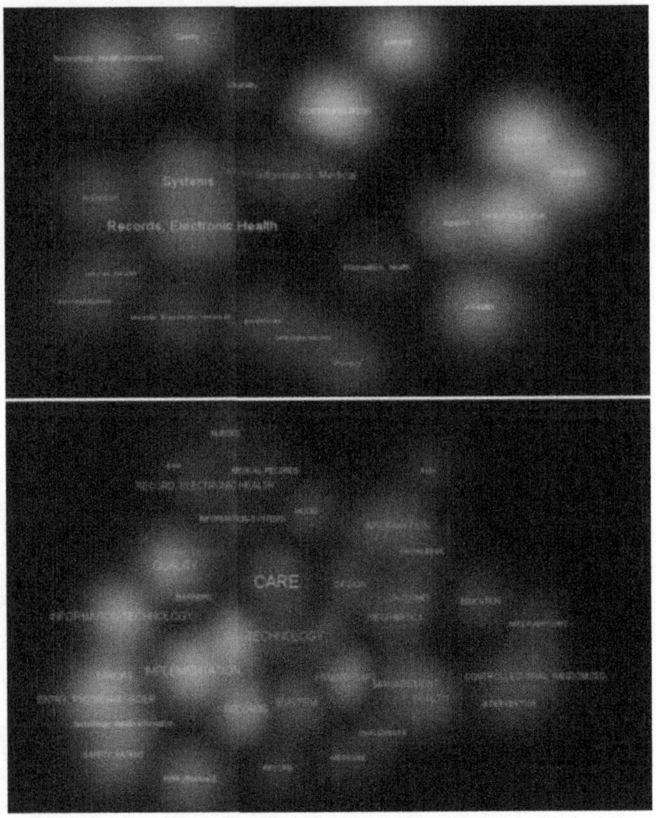

Fig. 2 Keyword co-occurrence maps

In the second series of 2 figures, the terms are grouped depending on the number of times these terms were found joined together. On the upper one, groups can clearly be identified:

Red: all having to do with "records" and "health systems".
Green: "e-health" and "Internet".
Blue: "Technologies".
Yellow: "Information and communication".

On the bottom figure, somewhat differently:
Red: "information systems" and "health management".
Green: "Patient care" and "health technologies".
Blue: "electronic records", "technology and health care".

IV. CONCLUSIONS

The Term e-health can still be considered as new and sometimes unknown to many developing countries, however in North America, Europe and other developed countries it is becoming very popular.

In a world where the daily lives of people is impacted by Internet, impact of this network on the health system is not uncommon.

'This phenomenon has opened doors to an easier and more efficient health care. ICT's allow reaching far away places and offer many different applications, to share medical knowledge and experiences improving medical practice for all.

Advancing technologies increase their impact on the health system. Not many studies were found that conduct metric type analyses in relation to the e-health field, but those that were performed are rich in information and offer insight on the subject from a different point of view.

ACKNOWLEDGMENT

The authors are endebted to the "University of Havana" and the "Vrije Universiteit Brussel" for enabling this joint research.

CONFLICT OF INTEREST

The authors declare that they have no conflict of interest.

REFERENCES

1. Moser, P.L., Hauffe, H., Lorenz, I.H. et al, Publication output in telemedicine during the period January 1964 to July 2003. J. Telemed. Telecare. 2004;10:72–77

2. Demiris, G., Tao, D. An analysis of the specialized literature in the field of telemedicine. J. Telemed. Telecare. 2005;11:316–319

3. Whan, P., Brown, N.A., Wootton, R. A bibliographic snapshot of the telemedicine citation literature. J. Telemed. Telecare. 2006;12:95–102

4. Liang, H.-N. (2010). Overview of the Health Informatics Research Field: A Bibliometric Approach. IFIP AICT, 335, 37-48

5. Fatehi, F Wootton, R. (2012). Telemedicine, telehealth or e-health? A bibliometric analysis of the trends in the use of these terms. Journal of Telemedicine and Telecare, 18(8), 460-464

6. Armfield, N. (2014). Telemedicine – A bibliometric and content analysis of 17,932 publication records. International Journal of Medical Informatics., 83, 715-725

corresponding author:

Author: Marc Nyssen
Institute: Vrije Universiteit Brussel
Street: Laarbeeklaan 103
City: B-1090 Brussels
Country: Belgium
Email: mnyssen@vub.ac.be

Thermoregulation of the hand: assessment with infrared thermography

L.C. Ospina-Restrepo[1], L.M. Herrera-Velasquez[1], C.J. Barrera-Causil[1], H.A. Fandiño-Toro[2] and L.M. Ramirez-Arbelaez[1]

[1] Instituto Tecnológico Metropolitano, Grupo de Investigación e Innovación Biomédica, Medellín, Colombia
[2] Instituto Tecnológico Metropolitano, Grupo de Automática, Electrónica y Ciencias Computacionales, Medellín, Colombia

Abstract— **Clinical infrared thermography is based on measuring the skin surface temperature. In this study, thermographic images were used to analyze the thermoregulation process after a cold stimulus, applied to regions of interests distributed along the hand in healthy subjects. Two parameters were assessed by statistical methods: acclimation percentage and** 60% **acclimation time. The averaging values for all subjects in acclimation percentage parameter, ranges between** 94.06% – 99.02% **and** 60% **acclimation time parameter ranges between** 2.08 – 3.58 **minutes. Statistically significant differences between women and men were not found in the parameters evaluated, nevertheless, statistically significant differences were found between regions of interest. The infrared thermography is useful to assess the thermoregulation processes of the hand.**

Keywords— **Thermoregulation, infrared thermography, microcirculation.**

I INTRODUCTION

The internal temperature of the body is one of the most rigorously controlled physiological parameters of the organism. Its regulation is a thermodynamic process in which heat is added or removed from the body, allowing variations of up to 0.4°C around 37°C to sustain normal metabolic functions [1]. In 1934 Hardy associated physiology with infrared emission from the human body, proposing that human skin can be considered as a radiator, establishing the temperature as a parameter of great importance in clinical diagnosis, whilst opening the way for applications with infrared thermography (IRT) [2]. The skin blood flow and microcirculation affect skin temperature [3]. Whilst changes in cutaneous flow modify skin temperature at a gross level, local fluctuations are more challenging to accurately diagnose due to the complexity of the microvasculature distribution patterns [4]. Indeed, with the complexity of the underlying tissue of several anatomical areas, it is plausible that changes in the underlying tissue composition could modify skin temperature within a region [5].

Clinical thermography is based on measurement and quantification of the temperature on the skin surface, which depends on the blood flow at it outer layers. This flow is subject to complex regulation processes by the nervous system and local factors. Therefore, the nervous and vascular reactions to different stimuli can be visualized and studied as changes in heat distribution patterns by using thermography [2]. Infrared thermography has many applications in the medical field: breast cancer [6], detection of pain [7], Raynaud's phenomenon [8], peripheral neuropathies [9], carpal tunnel syndrome [10], among others; nevertheless, these researches are mainly in exploratory stages. Raynaud's phenomenon is one of the most researched disorders by IRT, where in previous studies positive results were found to visualize thermoregulation patterns in patients and healthy volunteers [11]. The same principle applies to assess peripheral neuropathies because in these disorders the peripheral nerves are affected, which alters vasodilation and vasoconstriction functions and changes the temperature distribution pattern. In this paper the thermoregulation process of the hand is analyzed by infrared thermography in healthy volunteers, by evaluating parameters such as temperature recovery and reacclimation times, by subjecting the hand to a cold stimulus and comparing patterns of temperature distribution between men and women.

II MATERIALS AND METHODS

A Subjects

Sixty healthy volunteers were recruited; twenty-eight women and thirty-two men (age: 20 – 40, mean: 26.14 years), who met the following criteria for inclusion and exclusion. Inclusion criterion: Adult between 20 and 40 years old. Exclusion criterion: previous diagnosis of peripheral neuropathy or Raynaud's phenomenon, nor previous lesions in upper extremity. Ethical approval was obtained from ethics committee of Instituto Tecnológico Metropolitano and informed consent was used.

B *Image acquisition and processing*

The images are acquired by a FLIR thermographic camera (model A655sc), which has a spatial resolution of 640 × 480

I. Torres et al. (eds.), *VII Latin American Congress on Biomedical Engineering CLAIB 2016, Bucaramanga, Santander, Colombia, October 26th -28th, 2016*,
IFMBE Proceedings 60,
DOI: 10.1007/978-981-10-4086-3_182

pixels and accuracy of 2°C or 2% of the reading. The software FLIR ResearchIR was used for image processing. The image acquisition frequency is 1Hz, and the ResearchIR software shows the acquired images sequentially as a film. A previously developed diagnostic protocol [12] was used; this protocol establishes a preparation period for subject acclimation, before image acquisition; then, image acquisition begins with 2 minutes of measurements; a $15 \pm 0.5°C$ cold stimulus is applied for 30 seconds, and image acquisition continues during 10 minutes after stimulus to record the acclimation process. An acrylic water tank of $32 \times 22 \times 6$cm was used to position and offer support to the hand during measurement. It also provides interdigital rods to prevent movement of the fingers during the acquisition. Thirteen regions of interest (ROI) were manually defined in the hand (Fig. 1), each ROI includes a number of pixels, and each pixel of the image has associated a temperature value.

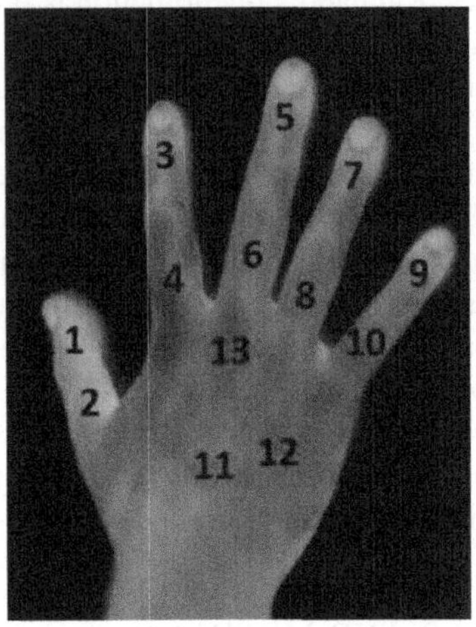

1. DP of thumb.
2. PP of thumb.
3. DP of index.
4. PP of index.
5. DP of middle.
6. PP of middle.
7. DP of ring.
8. PP of ring.
9. DP of little.
10. PP of little.
11. Dorsum of the hand.
12. Basilic vein.
13. Third Metacarpo-phalangeal joint.

Fig. 1 ROI selected in the hand. Distal phalanx (DF) and proximal phalanx (PP).

For each ROI, the mean temperature was calculated and a *temperature vs time* curve was built; in this curve, points correspond to the mean temperature of the ROI in each of the images of the acquisition sequence, i.e. one point per second during the stages of measurement: before, during and after of the cold stimulus.

C Parameters and statistical analysis

The *temperature vs time* curves were built using averaging temperature of the ROI; two parameters were obtained from the curves for all the ROI and all subjects: Parameter No. 1: Acclimation percentage. Percentage of temperature that ROI has rewarmed after 10 minutes of the cold stimulus. Parameter No. 2: 60% acclimation time. Time spent to rewarm up to 60% of initial temperature for the ROI. It is important highlight that acclimation percentage parameter only correlates the first and the last period of measurements in the thermoregulation process, while 60% acclimation time parameter provides information about how the thermoregulation process occurs. The parameters acclimation percentage and 60% acclimation time for thirteen ROI in sixty subjects were analyzed using the R software [13]. A non-parametric test was applied to find differences between parameters measured in the ROI. This procedure was made because the normality test (Shapiro-Wilk test) was rejected. A Kruskal-Wallis test and pairwise comparisons between ROI (using the Wilcoxon Rank Sum Tests with corrections) were used for multiple testing. An implementation of the False Discovery Rate (FDR) [14] was used to check and detect differences between ROI.

III Results

One of the assumptions of an anova model is the normality of the residuals. This assumption was rejected using a Shapiro-Wilk test with p-value of 2.2^{-16}. So, a non-parametric test was implemented to check the differences between ROI. Kruskal-Wallis test allowed us to detect a significant difference among the population means of the ROI with p-value of 2.2^{-16}. Then, to find which ROI are different, the Wilcoxon Rank Sum Tests with corrections for multiple testing, implementing the FDR method, was used to compare ROI with $\alpha = 0.05$. This test was applied to acclimation percentage parameter for thirteen ROI and sixty subjects, and 60% acclimation time parameter for thirteen ROI and sixty subjects. The comparison was made with all ROI each other, in all voluntaries. Statistically significant differences between women and men were not found in acclimation percentage parameter (Table 1). For this parameter, the comparisons between ROI showed a different behavior of the ROI 13 related to the other regions (statistically significant), except to ROI 4 and 6 which are closer. Averaging acclimation percentage for all subjects in ROI 13 was 91,69%, and averaging acclimation percentages for all ROI were in the interval 94,06% and 99,02% (Fig. 2).

Statistically significant differences between women and men were not found in 60% acclimation time parameter (Table 1), except to ROI 10, 11 and 12. For this parameter, the comparisons between ROI showed a different behavior between all of them (statistically significant). Among distal phalanxes (corresponding to ROI 1, 3, 5, 7 and 9) discrepancies were not found; these findings also were observed among proximal phalanxes (corresponding to ROI 2, 4, 6, 8 and 10),

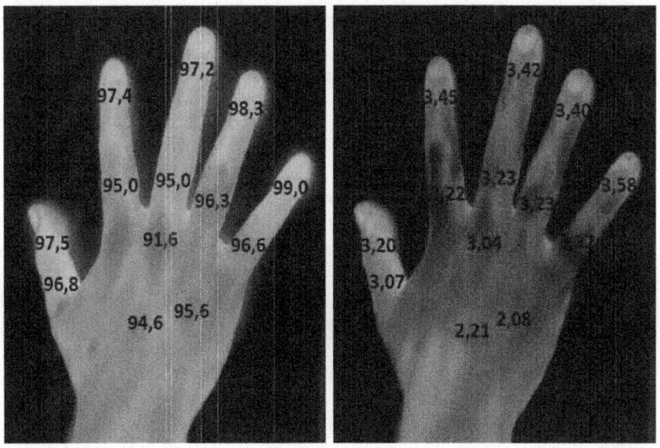

(a) (b)

Fig. 2 (a) Acclimation percentage (%); (b) 60% acclimation time (min).

which showed variations in the thermoregulation process in the fingers. In the regions located in dorsum of the hand (ROI 11 and 12) were observed the shortest 60% acclimation times, 2,22 and 2,08 minutes respectively; in the distal phalanxes regions were observed the longest 60% acclimation times. Averaging 60% acclimation times for all ROI were in the interval 2,08 y 3,58 minutes (Figure 2.B). Comparing the parameters, acclimation percentage and acclimation time, it was observed that ROI 11 and 12 were faster in 60% acclimation time but 10 minutes after cold stimuli their acclimation percentage is similar, even lower than other ROI. Whilst ROI located in distal phalanxes were slower in 60% acclimation time but 10 minutes after cold stimuli their acclimation percentage is higher than other ROI. Comparing regions located on the dorsum of the hand, ROI 11 (center of the dorsum) and ROI 12 (Basilic vein), were not found statistically significant differences neither in the 60% acclimation time nor in the acclimation percentage.

IV Discussion

In this study, the thermoregulation process of hand after a cold exposure showed that was not statistically significant differences between women and men. As reported by Charkoudian y Stachenfeld, physiological thermoregulation response to exercise did not show differences between women and men, although sex hormones effects theoretically can suggest the opposite response [15] [16]. Comparative cold responses studies of women and men were not found. In this study, discrepancies were found in ROI, they could be explained by the underlying tissue structure below each ROI. Acclimation percentage parameter in ROI (1, 3, 5, 7 and 9), corresponding to distal phalanxes, showed similar behavior; likewise, acclimation percentage parameter in ROI (2, 4, 6, 8 and 10), corresponding to proximal phalanxes, showed similar behavior; and acclimation percent-age parameter in ROI on the dorsum of the hand also exhibited similar behavior; these similar values were statistically significant. Nevertheless, ROI 13, corresponding to third Metacarpo-phalangeal joint (knuckle), presented a statistically significant difference in its acclimation percent-age compared to ROI 1 to 12; this discrepancy is presumably due to a variation in the underlying tissue composition, the bone is immediately below the skin surface in the ROI 13 hile muscle tissue is presented below the skin in the ROI 1 to 12.

Table 1 Parameter values: acclimation percentage, and 60% acclimation time. General value (G), Women value (W) and Men value (M).

	Acclimation percentage (%)			60% Acclimation time (min)		
	G	M	W	G	M	W
R1	97.5	98.3	96.6	3.2	3.21	3.19
R2	96.8	97.1	96.3	3.07	3.07	3.07
R3	97.4	97.6	97.2	3.45	3.49	3.42
R4	95.0	94.8	95.2	3.22	3.25	3.19
R5	97.2	97.4	97.0	3.42	3.48	3.35
R6	95.0	94.5	95.6	3.23	3.26	3.2
R7	98.3	98.5	98.1	3.41	3.44	3.37
R8	96.3	96.0	96.7	3.23	3.26	3.2
R9	99.0	99.3	98.6	3.58	3.56	3.61
R10	96.6	96.9	96.3	3.32	3.34	3.31
R11	94.6	94.3	94.9	2.22	1.79	2.7
R12	5.63	95.5	95.7	2.08	1.49	2.77
R13	91.6	91.0	92.3	3.04	3.08	3.0

In the 60% acclimation time parameter discrepancies among all ROI each other were presented, it may be related to variation in the vascular system along the fingers and the dorsum of the hand, even when the underlying tissue composition of the ROI could be similar. Braverman found that skin blood microcirculation is composed by two plexuses, the first one is 1-1,5 mm below the skin surface and the second one is at the dermal-subcutaneous junction; arterioles and venules connect the plexuses, and are ex-tended along skin in 1 mm^3 volumes which could be mainly composed by arteriolar, venular or devoid of all microvascular elements [4]. Ludwing et al analyzed methods to extract a representative temperature value of a body area [5]; skin temperature is affected by anatomical factor such as irregularities on the surface and different underlying tissue [17], these factors can change the spread of the skin temperature of a determined body area, thus it is frequent to observe a non-Gaussian thermal distribution inside a ROI. The maximum temperature method (T_{max}) extracts the highest values of temperature and can be useful for analyzing infrared images during patients in movement and when fast response is required [5].

Thermography is promoted by the American College of Clinical Thermology (ACCT) (18). The information provided by this institution mainly concerns thermography applied to the diagnosis of breast cancer and do not concern studies similar to the study presented in this work. The journal Infrared Physics & Technology published in April this year an article, about a thermographic study, using a cold stimulus to see the effects of age on the body thermoregulation processes; the difference with this study is that they do not use a protocol previously established which prevents the reproducibility study [18]. In future studies, comparing *temperature vs time* curves built using (T_{max}) values and *temperature vs time* curves built using mean values; in healthy volunteers and patients with pathologies affecting microcirculation and thermoregulation of hands, as peripheral neuropathies and Raynaud's phenomenon.

Limitations of the present study: in the present study we assessed the thermoregulation process after cold response but we did not assess the thermoregulation process after heat response. Gagnon et al found that are differences in perspiration between the sexes, but it is evident at very high exercise intensities joined to very high requirements for heat loss (high ambient temperatures). Finally, there were no variations in skin blood flow responses between women and men [19].

V CONCLUSION

Measurement of skin temperature of the hand using a thermal camera has been shown to be feasible and reproducible. The parameters, acclimation percentage and 60% acclimation time showed that there are not significant differences between women and men in the thermoregulation of hand. Dynamic IRT studies correlate with the vascular and microvascular system physiology of the skin and underlying tissue, it suggests new medical applications for diagnostic imaging by IRT where skin, vascular system or peripheral nervous system (which control vasodilation and vasoconstriction processes) are affected. Disorders such as peripheral neuropathy are likely to be diagnosed by IRT, e.g. diabetic neuropathy, medications induced neuropathy and others. The database built in this study will be the reference images for IRT images of patients in future studies.

CONFLICT OF INTEREST

The authors declare that they have no conflict of interest

REFERENCES

1. Biazzotto C, Brudniewski M, Schmidt A et al. Perioperative hypothermia *Rev. Bras. Anestesiol.* 2006;56:89–106.
2. Domínguez E, Priego F, Jiménez A et al. Capítulo 4. Termografía Médica y Neuro-termografía
3. Maley M, Eglin C, House J et al. The effect of ethnicity on the vascular responses to cold exposure of the extremities *Eur. J. Appl. Physiol.* 2014;114:2369–2379.
4. Braverman I. The cutaneous microcirculation in *J. Invest. Dermatol. Symp. Proc*;5:3–9 2000.
5. Ludwig N, Formenti D, Gargano M et al. Skin temperature evaluation by infrared thermography: Comparison of image analysis methods *Infrared Phys. Techn.* 2014;62:1–6.
6. Pérez M, Conci A, Aguilar A et al. Detección temprana del cáncer de mama mediante la termografía en Ecuador *Maskana.* 2016:111–123.
7. Robaina F, Miranda G. Advances in the diagnosis and treatment of sympathetic pain *Rev. Soc. Esp. Dolor.* 2000;7:390–402.
8. Correa M, Mariz H, Andrade L et al. Oral N-acetylcysteine in the treatment of Raynaud's phenomenon secondary to systemic sclerosis: A randomized, double-blind, placebo-controlled clinical trial *Rev. Bras. Reumatol.* 2014;54:452–458.
9. Pardo J, García T, Costa E. Neuropatías, radiculopatías y plexopatías *Medicine.* 2015;11:4698–4710.
10. Roldan K, Piedrahita M, Benitez H. Design and implementation of a protocol for acquisition and processing of infrared images obtained from hands in *Symposium of Signals, Images and Artificial Vision: STSIVA*:1–7 2013.
11. Schlager O, Gschwandtner M, Herberg K et al. Correlation of infrared thermography and skin perfusion in Raynaud patients and in healthy controls *Microvasc. Res.* 2010;80:54–57.
12. Ramírez L, Jiménez K, Correa A et al. Protocolo de adquisición de imágenes diagnósticas por termografía infrarroja *Medicina & Laboratorio.* 2015;21:161–178.
13. Team RCore. R: A language and environment for statistical computing 2013.
14. Benjamini Y, Hochberg Y. Controlling the false discovery rate: a practical and powerful approach to multiple testing *J. R. Stat. Soc. Ser. B. Met.* 1995:289–300.
15. Charkoudian N, Stachenfeld N. Sex hormone effects on autonomic mechanisms of thermoregulation in humans *Auton. Neurosci. Basic Clin.* 2016;196:75–80.
16. Charkoudian N, Stachenfeld N. Reproductive hormone influences on thermoregulation in women *Compr. Physiol.* 2014.
17. Deng A, Liu D, Gu C. Active skin perfusion and thermoregulatory response in the hand following nerve injury and repair in human upper extremities *Brain Res.* 2016;1630:38–49.
18. Lahiri B, Bagavathiappan S, Nishanthi K. Infrared thermography based studies on the effect of age on localized cold stress induced thermoregulation in human *Infrared Phys. Techn.* 2016;76:592–602.
19. Gagnon D, Crandall C, Kenny G. Sex differences in postsynaptic sweating and cutaneous vasodilation *J. Appl. Psychol.* 2013;114:394–401.

Author: Laura Cristina Ospina Restrepo
Institute: Instituto Tecnológico Metropolitano
Street: Calle 573 # 76A - 354
City: Medellín
Country: Colombia
Email: lauraospina137431@correo.itm.edu.co

Exhaled Breath analysis of smokers and nonsmokers using sensors based ultrapure Organically Modified Gold Nanoparticles

O.E. Gualdrón[1*], T.G. Welearegay[3], A.L. Jaimes[1], J.M. Cáceres[1], C.M. Durán[1], R. Ionescu[3], M. Maestre[2], G. Pugliese[3]

[1]Universidad de Pamplona, Facultad de Ingenierías, EEST, Pamplona, Colombia
[2] Universidad de Pamplona, Facultad de Económicas, Pamplona, Colombia
[3]MiNoS, Universitat Rovira i Virgili,DEEA, Tarragona, Spain

Abstract— **In this study we present a new technological approach for the fabrication of ultrapure organically modified gold nanoparticles (AuNPs) for chemical sensing applied to exhaled breath analysis. To achieve a high purity level of the sensing films, we combined Advanced Gas Deposition (AGD) technique to deposit ultrapure monodispersed AuNPs, and dip coating process for functionalization of the AuNPs with thiolated organic ligands. Morphology and surface analysis revealed the deposition of ultrapure isolated AuNPs after the first processing step, and a network of nanoparticle–ligand nanoassemblies after the second processing step. Gas sensing measurements were performed with exhaled breath samples collected from a group of smokers, a group of non-smokers, and ambient air. Sensors responses towards these samples demonstrated characteristic responses for each study group. PCA analysis further revealed samples classification in three distinct characteristic clusters, which indicates the suitability of the molecularly modified AuNPs presented in this communication for breath analysis applications.**

Keywords: Monolayer capped AuNPs; advanced Deposition; Breath analysis.

I. INTRODUCTION

The search for non-invasive diagnostic methods and diseases monitoring on the basis of volatile organic compounds (VOCs) detected in exhaled breath has led to the development of sensors arrays sensitive to low VOCs concentrations. In this regard, molecularly modified gold nanoparticles have proven excellent potential for detecting very low VOCs concentrations in biofluid samples [1]. The conventional fabrication method of these nanomaterials is based on wet-chemistry processes, which can leave traces of impurities and residual compounds in the sensing film [2]. In order to counteract this negative effect, we propose in this study an innovative methodology that employs the Advanced Gas Deposition technique to fabricate ultrapure, size-controlled, dispersed AuNPs in the first processing step, followed by AuNPs functionalization with thiolated organic ligands in the second processing step. A chemical sensors array based on monolayer-capped AuNPs fabricated employing this innovative technique was tested for exhaled breath analysis.

II. EXPERIMENTAL

a. AuNPs fabrication and functionalization

AuNPs were deposited on sensing devices fabricated on one side polished p-type <100> Si substrates (13 mm x 8 mm) with two parallel Au electrodes, 15 μm gapped, patterned on top by rf-sputtering. Nanoparticles growth and deposition was achieved employing the AGD equipment (ULVAC, Japan) configured in a two-chambers set-up for ultrapure AuNPs generation and deposition [3] – see **Fig.** 1. A piece of pure gold metal (99.99% purity) was placed on a carbon crucible inside the lower evaporation chamber, and a copper induction coil surrounding the crucible provided the heating necessary to melt the gold metal piece. The sensing substrates were mounted onto a movable stage in the upper deposition chamber. The two chambers of AGD were connected via a narrow transfer pipe (3 mm inner diameter), positioned in the vapor zone where melted gold atoms are evaporated.

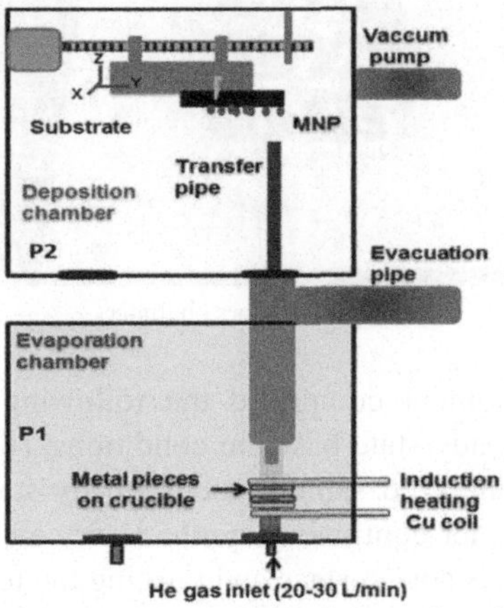

Fig. 1. Advanced Gas Deposition (AGD) technique adapted for AuNPs fabrication and deposition.

I. Torres et al. (eds.), *VII Latin American Congress on Biomedical Engineering CLAIB 2016, Bucaramanga, Santander, Colombia, October 26th -28th, 2016*,
IFMBE Proceedings 60,
DOI: 10.1007/978-981-10-4086-3_183

High purity He gas with a laminar flow rate of 20 L/min introduced underneath the heated gold piece, with upward direction, transported the Au atoms through the transfer pipe from the evaporation chamber to the deposition chamber. Because of the pressure difference between the two AGD chambers (0.88 mbar and 90 mbar in the upper and lower chamber, respectively), the Au atoms were propelled at high speed and got deposited onto the sensing substrates when impinging into them. Induction heating power, pressure in AGD chambers, speed of the movable stage and number of deposition cycles were optimized to obtain ultrapure monodispersed and crystalline AuNPs.

Next, the as-deposited AuNPs films were coated with various thiolated molecular ligands (4-methoxy-α-toluenethiol, methyl-3-mercaptopropionate, and 1-decanethiol) by dip coating the substrates in a solution containing 100–200 μL ligand dissolved in 20 ml ethanol for one hour, followed by drying at 50 °C for one hour for evaporating the solvent.

III. RESULTS AND DISCUSSION

a. *Exhaled breath samples analysis*

The responses of the sensors were analyzed upon exposure to exhaled breath samples of a group of 5 smoker and 5 nonsmoker volunteers, and to ambient air samples. For performing the sensing tests, the sensors were placed inside a Teflon test chamber (26 cm^3 inner volume), Fig 2.

Fig 2. Sensors chamber.

Each measurement comprised the following cycles: (i) 5 min under steady state baseline conditions; (ii) 5 min exposure to the analyzed sample under steady state conditions; (iii) 5 min under continuous synthetic dry air flow used for cleaning the sensors surface and purging the test chamber.

The samples were acquired employing the BioVOC^{TM} sampling system and immediately introduced to the sensors chamber, Fig 3. 5V constant voltage was alternately applied for intervals of 10 sec to each sensor during operation, and the current through the sensors was recorded for further analysis.

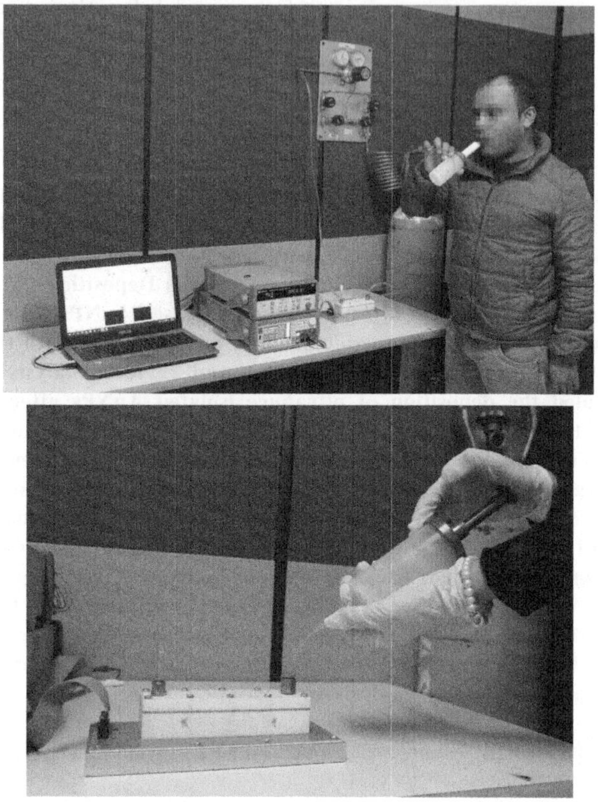

Fig 3. BioVOC, sensor chamber and Acquisition system

The sensors responded to the exhaled breath samples of both smoker and non-smoker volunteers, displaying different characteristic responses towards each study group and a good recovery after samples evacuation, as it is remarked from the investigation of **Fig**. 4. In this kind of sensing films, the AuNPs act as electrical conductors, while the molecular component of the nanoassembly provides the sensing surface that interacts with the VOCs from the sample analyzed. The smaller increase in sensor's response towards the smokers breath samples compared with the non-smokers breath samples could be related to the lower concentration of VOCs adsorbed.

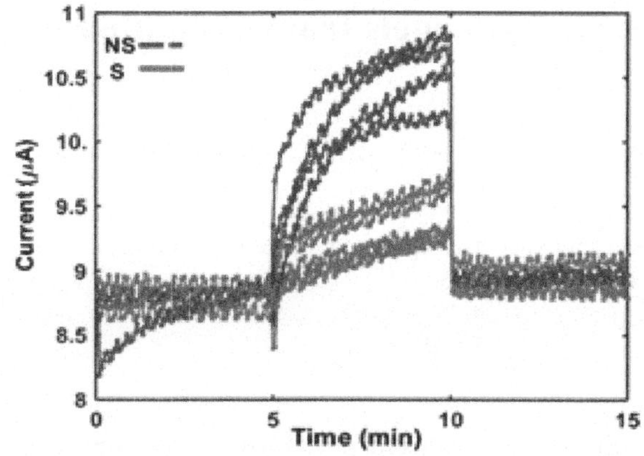

Fig 4. Responses of the 4-methoxy-α-toluenethiol capped AuNPs sensor to the exhaled breath of the 5 smokers and 5 non-smokers volunteers

Sensors 3, 4, 8 and 9 placed in the chamber are best responded to the presence of the breath of smokers and non-smokers. Table 1 shows the relationship of each of these sensors with its organic components ligands.

Table 1. Relationship of each of these sensors with its organic components ligands

Sensor	AuNP + Ligand
Sensor 3	D: 4-methoxy -α-toluenethiol
Sensor 4	M: Methyl-3-mercaptopropionate
Sensor 8	B: 1-decanethiol
Sensor 9	B: 1-decanethiol

Fig. 5. Shows the results of the Principal Component Analysis (PCA) performed with the overall responses of four sensors (1 sensor with 4-methoxy-α-toluenethiol ligand – $R_{baseline}$≈190kΩ; 1 sensor with methyl-3-mercaptopropionate ligand – $R_{baseline}$≈100kΩ; and 2 sensors with 1-decanethiol lugand – $R_{aseline}$≈11MΩ and ≈450kΩ), which demonstrates excellent discrimination between smokers, non-smokers, and ambient air.

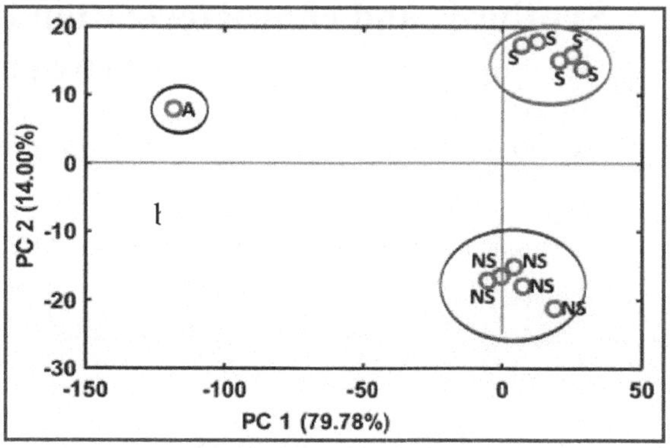

Fig 5.PCA plot showing excellent discrimination of exhaled breath samples of smokers, non-smokers, and ambient air.

IV. CONCLUSION

A new technological approach that combines the advanced gas deposition technique for the fabrication of ultrapure monodispersed AuNPs and single step functionalization of the AuNPs with thiolated organic ligands, was demonstrated for gas sensing applications in the present study. The morphological and surface analysis of the sensing films demonstrated a high purity level. Sensors responses towards ambient air and exhaled breath samples of smokers and non-smokers showed characteristic responses for each study group. PCA analysis revealed the excellent discrimination potential between different samples achieved by the sensors array, which suggests the suitability of the developed sensors for breath analysis applications.

ACKNOWLEDGEMENTS

This work has been supported by EC's H2020 Program H2020-MSCA-RISE-2014, Grant Agreement No. 645758 ("TROPSENSE"), and by COLCIENCIAS (Colombia), Project Number 112165741516.

REFERENCES

[1] U. Tisch, H. Haick, "Nanomaterials for cross-reactive sensor arrays", MRS Bulletin. Vol 35, pp .797-803, 2010.

[2] M. Brust, J. Fink, D. Bethell, D. Schiffrin, C. Kiely, "Synthesis and reactions of functionalized gold nanoparticles", J. Chem. Soc. Chem. Commun, Vol. 16, pp. 1655-1656, 1995.

[3] C.G. Granqvist, R.A. Buhrman., "Ultrafine metal particles", J. Appl. Phys. Vol 47, pp. 2200–2219, 1976.

Author: Oscar Eduardo Gualdron Guerrero
Institute: University of Pamplona.
Street: Km 1 way Bucaramanga
City: Pamplona-Colombia
Email: oscar.gualdron@unipamplona.edu.co

Synthesis and characterization of nanofibroin hydrogels from Colombian silkworm *Bombyx Mori* L.

A. Zuluaga-Velez[1], R Buitrago-Sierra[2], J.F. Santa[2], F.A. Tabares-Villa[1], E. Aguilar[1], J.C. Sepúlveda-Arias[1]

[1] Grupo Infección e Inmunidad. Facultad de Ciencias de la Salud. Universidad Tecnológica de Pereira, Pereira, Colombia.
[2] Grupo de Materiales Avanzados y Energía-MATyER. Instituto Tecnológico Metropolitano, Medellín, Colombia.

***Abstract*— Fibroin is a natural insoluble protein that are used as a scaffolding material in regenerative medicine. In this work, fibroin hydrogels obtained from the silkworm *Bombyx Mori* L. were synthesized by two different methods (vortex and sonication). Samples were characterized by Scanning Electron Microscopy, Infrared and Thermogravimetric analysis. Cytotoxicity was evaluated by the MTT method on adherent and non-adherent cells. The results indicate that nanofibers can be obtained by both vortex and sonication methods. Vortex samples have higher thermal stability. The MTT test revealed that hydrogels have low cytotoxicity in the cells lines evaluated.**

***Keywords*— biomaterials, silk fibroin, regenerative medicine.**

I. Introduction

Biomaterials science includes fundamental aspects of physical, mechanical, chemical and biological (compatibility) properties of synthetic and natural biomaterials per se [1]. Scaffolds play a key role in regenerative medicine, as they provide a suitable environment to promote proliferation, differentiation and cell regeneration [2]. The structural characteristics of the material influence the quality of the graft [3], so the scaffold must exhibit appropriate mechanical and physicochemical properties [4].

Hydrogels are attractive scaffolding materials due to their highly swollen network structure, their ability to encapsulate cells and bioactive molecules, and their efficient mass transfer [5]. Hydrogels are interconnected networks of hydrophilic blocks that become insoluble due to the presence of crosslinks and the absorbtion of large amounts of water [6]. Different polymers, including natural, synthetic and natural/synthetic hybrid polymers, have been used to obtain hydrogels via chemical or physical crosslinking [7]. In general, hydrogels based on polymers from natural sources are advantageous in regenerative medicine applications due to their intrinsic biological recognition characteristics of biological recognition, including receptor-binding ligands presentation and susceptibility to cell-triggered proteolytic remodeling and degradation. However, the use of natural component-based hydrogels has shown some drawbacks, such as complexities associated with purification, immunogenicity and pathogen transmission [6].

Silk fibroin is a fibrillar protein from the silk produced by some arthropods such as silkworms [8], consisting of a heavy and a light chain linked by disulfide bonds [9]. Fibroin hydrogels have been widely used in regenerative medicine [10, 11], however it is still unclear how manufacturing methods affect the biomaterial structural characteristics. Therefore, the aim of this study was to assess the influence of the methods to synthesize fibroin hydrogels on their structural properties and cytotoxicity toward adherent and non-adherent cell lines.

II. Materials and methods

A. Materials

Bombyx mori L. silkworm cocoons were collected at the experimental farm "El Pílamo" (Universidad Tecnológica de Pereira, Colombia). Reagents were purchased at Sigma-Aldrich. All solutions were prepared with ultrapure water (UPW) generated by a Millipore Direct- Q 5 purification system.

A.1. Silk fibroin solution preparation

Silk fibroin solutions were prepared using a process described by Rockwood et al [12] with some modifications. 5 g of silk cocoons were cut into small pieces and degummed by boiling in an aqueous solution of 0.02 M Na_2CO_3 for 60 min, in order to remove the glue-like sericin protein. After three ambient UPW rinses, the silk fibroin was dried in an oven at 45 °C for 8 hours. The dry fibroin was solubilized at 20% (w / w) in a 9.3 M aqueous LiBr solution at 60°C for 8 hours. This solution was dialyzed with ultrapure water for 48 hours using 6000-8000 MWCO regenerated cellulose dialysis tubing (Fisherbrand). Silk concentrations were determined by compar-

I. Torres et al. (eds.), *VII Latin American Congress on Biomedical Engineering CLAIB 2016, Bucaramanga, Santander, Colombia, October 26th -28th, 2016*,
IFMBE Proceedings 60,
DOI: 10.1007/978-981-10-4086-3_184

ing the mass of the solution to the mass of dry silk after storage at 70°C for 12 hours. Silk concentration after dialysis was about 6.5 % (w/w).

A.2. Sonication-Induced Gelation of Silk Fibroin

A 5 mL aliquot of 4% (w/v) fibroin solution was transferred to a 15 mL tube and sonicated with a Microson XL 2000. The sonication time was 30 seconds at 20 W. The solution was incubated at 37°C after sonication and the sol-gel transition time was 12 hours [13].

A.3. Vortex-Induced Gelation of Silk Fibroin

A 1 mL of 4% (w/v) fibroin solution was mixed for 7 minutes at 3500 rpm using a vortex mixer (Fisher Scientific, Hampton, NH) to induce fibroin self-assembly and hydrogelation. Then, the solution was incubated at 37°C with a sol-gel transition time of 12 hours [14].

B. Characterization

Fibroin morphology was characterized in a JEOL 7100F FEG-SEM at the Microscopy Laboratory of Instituto Tecnológico Metropolitano (ITM), Medellín, Colombia. The obtained gel was fractured and coated with gold in order to obtain a conducting material enabling the morphological study via SEM. The plasma coating was done in a dual head Quorum Q300TD sputtering system available at the ITM. All images where obtain using the SE-detector. FTIR analysis was done using a Shimadzu IRTracer-100 spectrometer. Spectra were acquired in ATR mode with 16 scans at a resolution of 4 cm^{-1} over the 4000–600 cm^{-1} spectral region. Thermal properties were studied by simultaneous TGA-DSC analysis using a TA Instruments Q600. Samples (10–15 mg) were loaded into an alumina pan. The samples were heated in a nitrogen atmosphere from 25 to 1000 °C at a heating rate of 10 °C min^{-1}.

C. Cytotoxicity evaluation of silk fibroin hydrogel

C.1. Cell culture

HEK-293 and THP-1 cells were obtained from ATCC (American Type Culture Collection). THP-1 cells were maintained in RPMI-1640 medium and HEK-293 cells were cultured in DMEM medium. Both media were supplemented with 10% fetal bovine serum (Gibco) and 100 U/mL penicillin-streptomycin. Cells were cultured at 37 °C in a 5% CO_2 humidified atmosphere.

C.2. Cytotoxicity assay

Cytotoxicity assessment was carried out using vortex and sonicated hydrogels as scaffolds. Fibroin solutions were sterilized by autoclaving for 15 minutes at 121°C before polymerization. The purpose of the MTT assay was to evaluate cytotoxicity [15]. HEK-293 cells were seeded at 3 x 10^4 cells/mL, and THP-1 cells were seeded at 6 x 10^4 cells/ mL in 24-well plates containing the hydrogels. Cells cultured without hydrogels were used as control. After treatment for 24, 48 and 72 h, 500 μL of MTT (5 mg/mL) were added into a culture media without phenol red, afterwards cells were cultured for another 4 h. Thereafter, the medium with MTT were aspirated, and 500 μL of DMSO were added into each well. Finally, the samples were read at 570 and 630 nm on a microplate reader (Thermo Fisher Scientific). In the THP-1 assay it was necessary to centrifuge the cell suspension at 1500 RPM for 5 min in each step. Statistical analysis was performed using a two-way ANOVA followed by Tukey's post hoc test performed using GraphPad Prism 6 software (GraphPad Software, San Diego, CA, USA). A p value less than 0.05 was considered statistically significant.

III. Results and discussion

Figure 1a shows the silkworm cocoons from the Colombian silkworm *Bombyx Mori* L. These cocoons have a typical elongated egg shape form with a length of 33 mm and a width of 16 mm. The wall thickness is near to 1 mm. All the cocoons used were homogenous and had a typical white colour. Figure 1b shows the photograph by the two different methods: sonication (right) and vortex agitation (left).

a

b

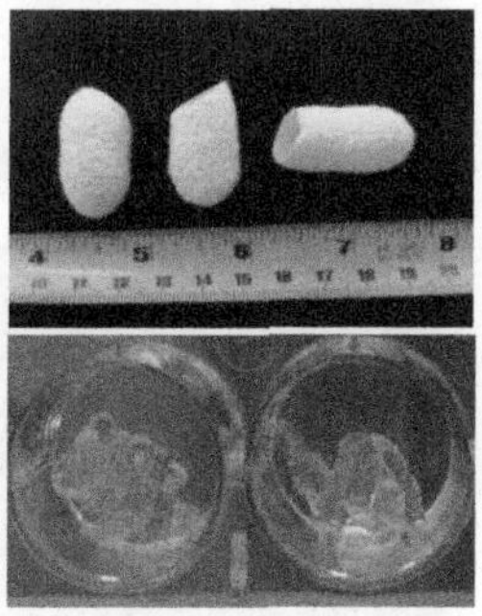

Fig 1. (a) Photograph of Colombian silkworm *Bombyx Mori* L. cocoons. (b) Photograph of silk fibroin hydrogels.

Figure 2 presents the SEM images of the fibroin obtained by the two different methods: sonication and vortex agitation. In both cases layer-shaped hydrogels were obtained. The gels show a flat homogeneous surface formed by the agglomeration of the fibroin nanofibers. This structure can be observed in the borders of the layers in both cases. Similar results were obtained by Zhong *et al* [16] in SF/Copolymer hydrogels synthesized for drug release purposes.

The morphology of fibers is important for the regeneration purpose. When the fibers are nanostructured, they are suitable for cell culturing [17] which is one of the aims of this investigation in order to use hydrogels as biomaterials. The fibers can be described as nanofibers since their diameter is under 100 nm.

Fibers have an average diameter of 24 ± 3 nm when they are processed by the sonication method and 40 ± 6 nm by the vortex method. In all cases, fibers obtained by the sonication method were thicker. The fibers are obtained by hundreds of molecular chains gathered after the drying process as it is described in the literature [18].

a

b

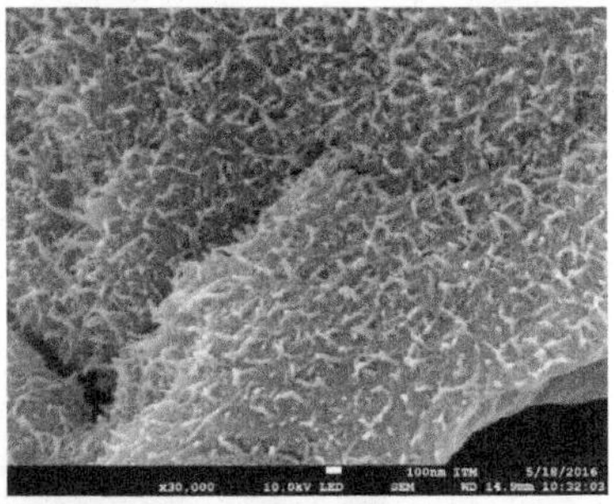

c

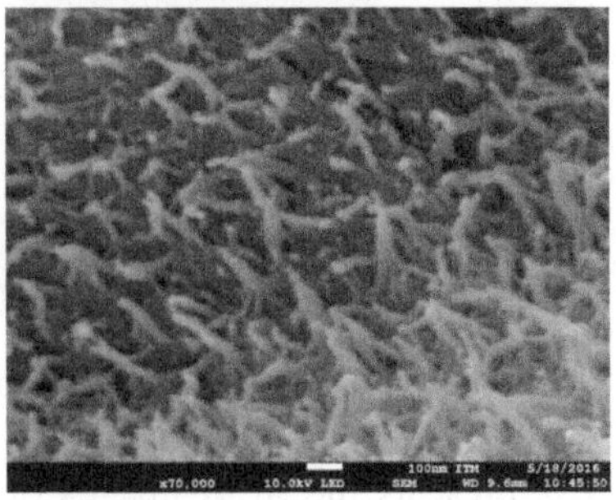

d

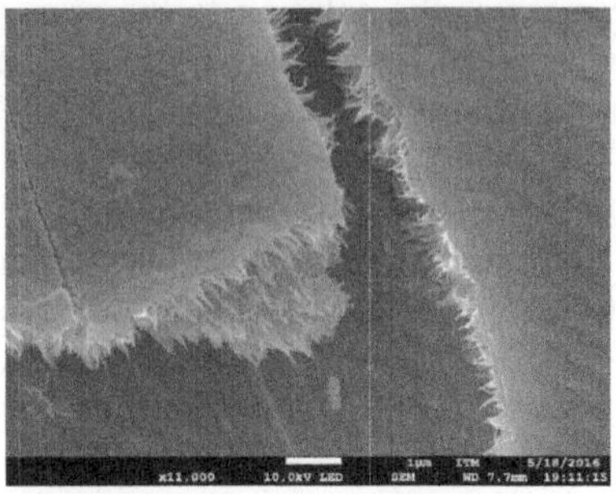

e

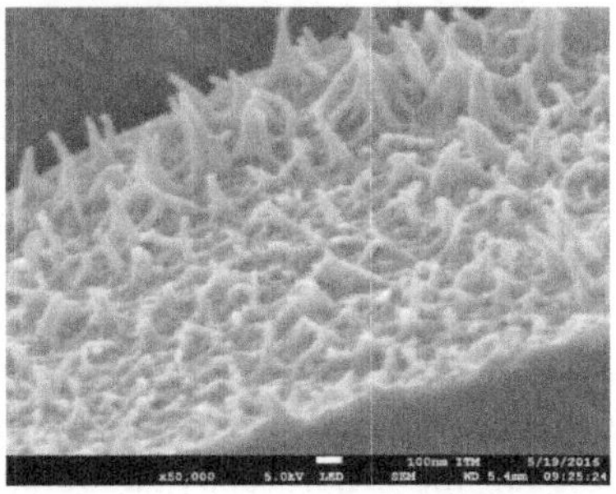

f

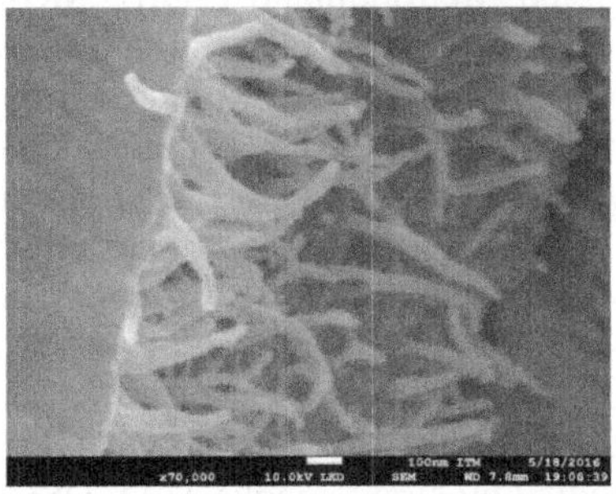

Fig 2. SEM images of fibroin obtained by (a) 10,000X, (b) 30,000X, (c) 70,000X sonication method. (d) 11,000X, (e) 50,000X, (f) 70,000X vortex method. All images were obtained with 10.0 kV LED.

Fibroin samples were studied by FTIR. Figure 3, shows the vibrations at 1800-800 cm^{-1} in both vortex and sonicated samples. Typical fibroin spectra were obtained in the two cases: vibrations related to fibroin molecules are observed. Amide I (C=O stretching) and II (C-N stretching) absorption bands at 1683 and 1540 cm^{-1}, respectively, were observed. Typical absorption of β-sheet structures of silk fibroin was observed at 1246 cm^{-1} [19, 20].

In the samples prepared by sonication, a small shift of the spectrum to lower wave numbers was observed. The shift could be related to the presence of a β-sheet structure [20, 21].

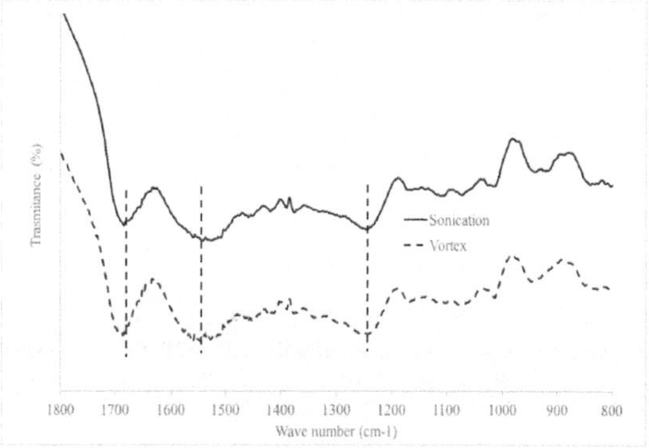

Fig 3. FTIR analysis of Fibroin hydrogels.

The results of TGA analysis of fibroin gels are shown in Figure 4. Vortex sample is represented by dashed lines and sonicated sample by a solid line. For both samples, three thermal events are clearly observed.

The first weight lost could be related to moisture and water molecules weakly ligated to the gel system. The weight loss occurs at 40ºC and in both cases represents 20% of weight losses of Vortex and 35% of sonicated sample. The second weight loss was observed at 110ºC in the vortex and 90ºC in the sonicated sample. This event can be explained by the loss of humidity from the gel. The weight losses of vortex samples were higher since the content of humidity is higher.

Finally, at 290ºC material decomposition was observed. This is normally attributed to the rupture of side-chain groups and amino acid residues breakdown as well as the peptide bonds cleavage. This behaviour implies that the fibroin has an extended crystallinity and it also has a clear molecular orientation of the proteins. The fibroin with amorphous structures has a typical breakdown at 285ºC [20, 22]. On the other hand, the humidity losses from the vortex sample occurs at lower temperature when compared to sonicated sample. However, the total amount of water eliminated at the end of the test is the same in both samples. This behaviour could be related to the density of the samples.

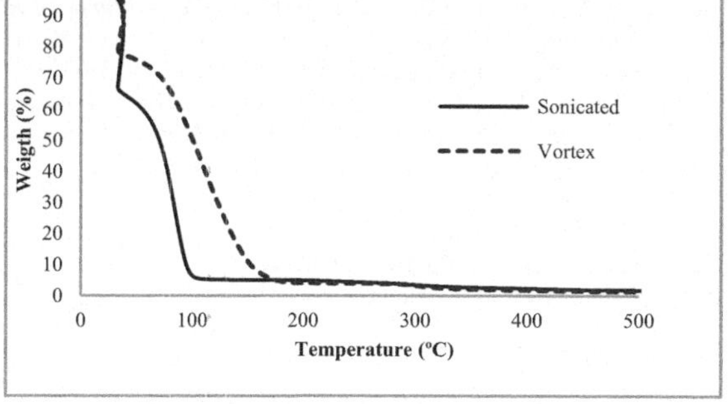

Fig 4. TGA analysis of Fibroin hydrogels.

When designing biomaterials it is necessary to characterize their cytotoxicity and biocompatibility. *In vitro* cell culture studies are very useful at this respect. Regarding cytotoxicity assessment of any biomaterial, *in vitro* tests should be conducted prior to *in vivo* tests [23]. Materials dispaying low cytotoxicity may be useful for medical applications such as regenerative medicine, where cell-scaffold unions are required. The results of the MTT assay of fibroin hydrogels are shown in Figures 5 and 6. Both types of hydrogels exhibited no differences in terms of cytotoxicity in the cell lines tested. It is clear that both, sonication and vortexing methods could be used to generate fibroin materials with low toxicity, although the hydrogels showed a decrease in cell proliferation relative to control. For HEK-293 cells this could be explained by a limited cell adhesion, as previously reported in the literature [24, 25]. In THP-1 cells (non-adherent cells), cell proliferation was higher at 48 h in control wells, this behavior depends on cell mobility and the geometry of the well-plate [26]. This perhaps because the cells were into the scaffold, leading to a decrease in cellular connectivity and therefore low cell growth.

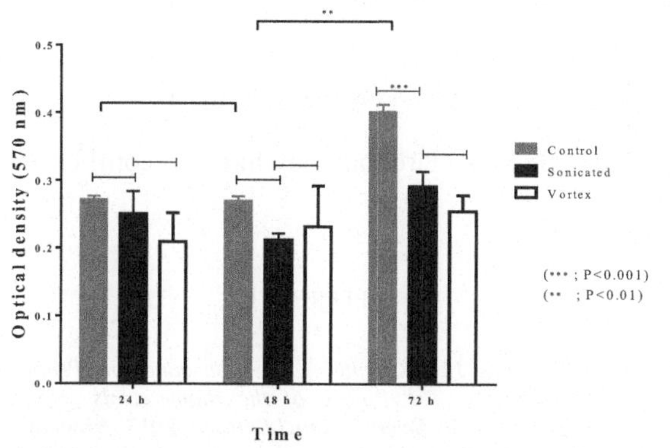

Fig 5. Influence of vortex (white) and sonicated (black) hydrogels on HEK-293 cell density after an incubation period of 24, 48 and 72 h as determined by the MTT assay.

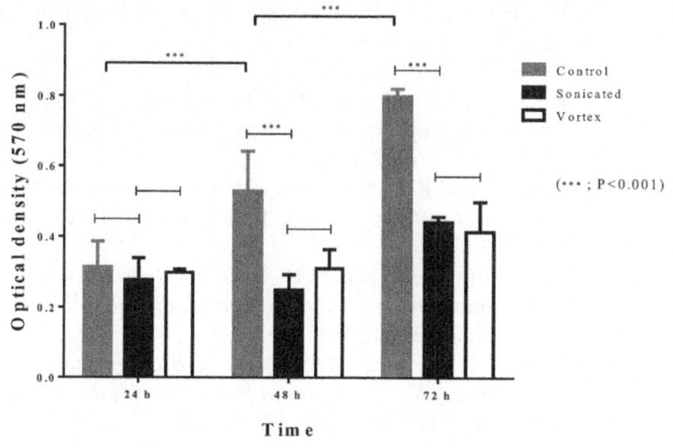

Fig 6. Influence of vortex (white) and sonicated (black) hydrogels on THP-1 cell density after an incubation period of 24, 48 and 72 h as determined by the MTT assay.

IV. Conclusions

The vortex and sonicated method allowed the generation of fibroin hydrogels from silkworm *Bombyx Mori* L. The nanofibers have a diameter of 24 ± 3 nm and 40 ± 6 nm when they are obtained by sonication and vortex method, respectively. These differences could be explained by the physical crosslinking level, however, marked differences at the structural level were not observed.

Fibroin hydrogels can be used as scaffolds for tissue regeneration and can be used to release drugs or molecules in media where there are suspension cells or adherent cells, without affecting cell viability.

Acknowledgment

The authors would like to acknowledge to Electron Microscopy Laboratory of Instituto Tecnológico Metropolitano for the SEM images. This work was funded by, Sistema General de Regalias de Colombia and Universidad Tecnológica de Pereira.

Conflict of Interest

The authors declare that they have no conflict of interest.

References

1. Lemons, J.E., *Chapter I.1.1 - Introduction: Properties of Materials: The Palette of the Biomaterials Engineer*, in *Biomaterials Science (Third Edition)*. 2013, Academic Press. p. 5.
2. Silva, R., B. Fabry, and A.R. Boccaccini, *Fibrous protein-based hydrogels for cell encapsulation.* Biomaterials, 2014. **35**(25): p. 6727-38.
3. Hendriks, J.A.A., et al., *The effect of scaffold-cell entrapment capacity and physico-chemical properties on cartilage regeneration.* Biomaterials, 2013. **34**(17): p. 4259-4265.
4. Keane, T.J. and S.F. Badylak, *Biomaterials for tissue engineering applications.* Seminars in Pediatric Surgery, 2014. **23**(3): p. 112-118.
5. Zhu, J. and R.E. Marchant, *Design properties of hydrogel tissue-engineering scaffolds.* Expert Rev Med Devices, 2011. **8**(5): p. 607-26.
6. El-Sherbiny, I.M. and M.H. Yacoub, *Hydrogel scaffolds for tissue engineering: Progress and challenges.* Glob Cardiol Sci Pract, 2013. **2013**(3): p. 316-42.
7. Peppas, N.A., et al., *Hydrogels in Biology and Medicine: From Molecular Principles to Bionanotechnology.* Advanced Materials, 2006. **18**(11): p. 1345-1360.
8. Yonemura, N., et al., *Conservation of silk genes in Trichoptera and Lepidoptera.* J Mol Evol, 2009. **68**(6): p. 641-53.
9. Kundu, B., et al., *Silk fibroin biomaterials for tissue regenerations.* Adv Drug Deliv Rev, 2013. **65**(4): p. 457-70.
10. Melke, J., et al., *Silk fibroin as biomaterial for bone tissue engineering.* Acta Biomater, 2016. **31**: p. 1-16.
11. Murphy, A.R. and D.L. Kaplan, *Biomedical applications of chemically-modified silk fibroin.* J Mater Chem, 2009. **19**(36): p. 6443-6450.
12. Rockwood, D.N., et al., *Materials fabrication from Bombyx mori silk fibroin.* Nature Protocols, 2011. **6**(10): p. 1612-1631.
13. Wang, X., et al., *Sonication-induced gelation of silk fibroin for cell encapsulation.* Biomaterials, 2008. **29**(8): p. 1054-64.
14. Yucel, T., P. Cebe, and D.L. Kaplan, *Vortex-induced injectable silk fibroin hydrogels.* Biophys J, 2009. **97**(7): p. 2044-50.
15. Mandal, B.B., S. Kapoor, and S.C. Kundu, *Silk fibroin/polyacrylamide semi-interpenetrating network hydrogels for controlled drug release.* Biomaterials, 2009. **30**(14): p. 2826-36.
16. Zhong, T., et al., *Silk fibroin/copolymer composite hydrogels for the controlled and sustained release of hydrophobic/hydrophilic drugs.* Int J Pharm, 2015. **494**(1): p. 264-70.
17. Chan, B.P. and K.W. Leong, *Scaffolding in tissue engineering: general approaches and tissue-specific considerations.* Eur Spine J, 2008. **17 Suppl 4**: p. 467-79.
18. Lu, Q., et al., *Silk self-assembly mechanisms and control from thermodynamics to kinetics.* Biomacromolecules, 2012. **13**(3): p. 826-32.
19. Jin, Y., et al., *Bio-inspired mineralization of hydroxyapatite in 3D silk fibroin hydrogel for bone tissue engineering.* Colloids Surf B Biointerfaces, 2015. **134**: p. 339-45.
20. Motta, A., et al., *Fibroin hydrogels for biomedical applications: preparation, characterization and in vitro cell culture studies.* Journal of Biomaterials Science, Polymer Edition, 2004. **15**(7): p. 851-864.
21. Guziewicz, N., et al., *Lyophilized silk fibroin hydrogels for the sustained local delivery of therapeutic monoclonal antibodies.* Biomaterials, 2011. **32**(10): p. 2642-50.
22. Weska, R.F., et al., *Effect of freezing methods on the properties of lyophilized porous silk fibroin membranes.* Materials Research, 2009. **12**(2): p. 233-237.
23. Cheddadi, M., et al., *Cytotoxicity and biocompatibility evaluation of a poly(magnesium acrylate) hydrogel synthesized for drug delivery.* Int J Pharm, 2011. **413**(1-2): p. 126-33.
24. Bray, L.J., et al., *Incorporation of Exogenous RGD Peptide and Inter-Species Blending as Strategies for Enhancing Human Corneal Limbal Epithelial Cell Growth on Bombyx mori Silk Fibroin Membranes.* J Funct Biomater, 2013. **4**(2): p. 74-88.
25. Gil, E.S., et al., *Helicoidal multi-lamellar features of RGD-functionalized silk biomaterials for corneal tissue engineering.* Biomaterials, 2010. **31**(34): p. 8953-63.
26. Nienow, A.W., *Reactor engineering in large scale animal cell culture.* Cytotechnology, 2006. **50**(1-3): p. 9-33.

Author: J. C. Sepúlveda-Arias
Institute: Universidad Tecnológica de Pereira
Street: Cra 27 No. 10-02 Los Álamos
City: Pereira
Country: Colombia
Email: jcsepulv@utp.edu.co

Human-Computer Interaction for Image Guided Surgery Systems using Physiological Signals: Application to Deep Brain Stimulation Surgery

D.A. Jimenez[1], J.B. Padilla[1], R. Arango[1] and H.F. Garcia[2]

[1] Universidad del Quindío/Research group GIDET, Armenia-Quíndio, Colombia
[2] Universidad tecnológica de Pereira/Research group in Automatics, Pereira-Risaralda, Colombia

Abstract— **Deep brain stimulation (DBS) is a surgical procedure used to treat Parkinson disease, the success of this surgery relies in the correct location of the neurostimulator device. A 3D Brain atlas allows to interactively visualizing the brain structures over a 3D medical image volume giving a view over the real area of surgery where the neurostimulator device has to be allocated. The movement of the 3D image volume is commonly made by a computer mouse or touch screens, which are not the best for a sterile environment as a surgery room and have some limitations as distance and move sensibility of the computer cursor. Hence in this paper we proposed a Human-Computer Interaction (HCI) to navigate over the 3D brain atlas structure and improve the limitations that the common computer peripherals have. For the development of the HCI was used the bluetooth sensor MYO of ThalamicLabs TM. The HCI developed were tested in the planning of DBS surgery under the NEURONAV software [1] giving a suitable navigation and control over 3D atlas.**

Keywords— **Deep brain stimulation, human computer interaction, image guided surgery (IGS), machine learning, physiological signals.**

I INTRODUCTION

Deep brain stimulation (DBS) in Parkinson disease is a surgical procedure used to treat a variety of disabling neurological symptoms such as rigidity, tremor, walking problems and stiffness. Here, the neurosurgeon has to implant a neurostimulator device into the thalamic area with high accuracy. The procedure of DBS surgery involve the implantation of a neurostimulator into the brain.

Due to the complexity of some surgeries an image guided surgery (IGS) is very useful some applications [3]. In the particular case of DBS surgery an IGS software as NEURONAV allows the user to generate a 3D image volume of the brain structures of the atlas, place the neurostimulator in one of the three possible zones (VIM, GPi and STN) [2] and also the sofrware shows to the user the volume of tissue activated (VTA) giving some parameters such active pulse, kind of neurostimulator and voltage [1].

In the Neuronav environment the 3D image volume movement is done by the computer cursor which is commonly a mouse, touch screen or kinect camera but they have an issue that it is mainly the wired connection (Mouse), the limited distance of use and the line of view with the user (Kinect). So this kind of devices implies that if they are used in an operating room a change of surgery gloves and sterilization is needed before and after the use of them (mouse or touch screen) and this involves a loss of time. Hence a Human-Computer interaction (HCI) is proposed allowing the user to: move 3D brain atlas over X-Y-Z axis, zoom in, zoom out and to move X-Y-Z axis. The HCI features are done by using orientation signals and superficial electromyography signals (sEMG) provided by MYO of ThalamicLabs TM.For the sEMG signals there are many applications in literature such as human-computer interfaces (HCI), Wheelchair handling, emotion recognition, classification of hand movements [4] [5] [6] [7] [8]. The success of this applications relies on the features extraction, but as told in [4] "the nature of EMG signal is complex and highly nonlinear that makes it difficult to have an explicit relation between the measured signals and a motion command. To distinguish and identify the functionality of different sEMG signals from their extracted features, pattern recognition techniques are very important" is not easy to extract features from sEMG signals without a pre-processing of them so there are different methods to extract features from sEMG signals in time and frequency domain[9].

II MATERIALS AND METHODS

A Database

In order to take the pertinent actions into the IGS NEURONAV software were defined three (3) main gestures and some transition moves showed in figure 1 and figure 2 respectively.

From figure 1 (a) denotes the gesture to rotate the 3D image volume in the space X-Y-Z (Right Click), (b) is the gesture to zooming (in or out) (Left Click) and (c) the gesture to translation the axis X-Y-Z (Scroll Click).

I. Torres et al. (eds.), *VII Latin American Congress on Biomedical Engineering CLAIB 2016, Bucaramanga, Santander, Colombia, October 26th -28th, 2016*,
IFMBE Proceedings 60,
DOI: 10.1007/978-981-10-4086-3_185

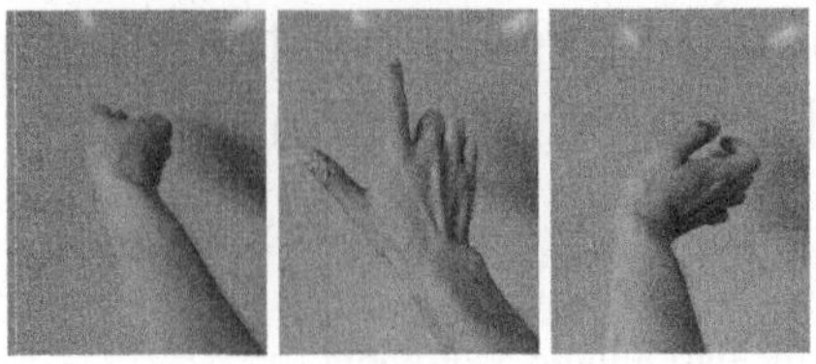

(a) (b) (c)

Fig. 1: Defined gestures to control the IGS software NEURONAV.

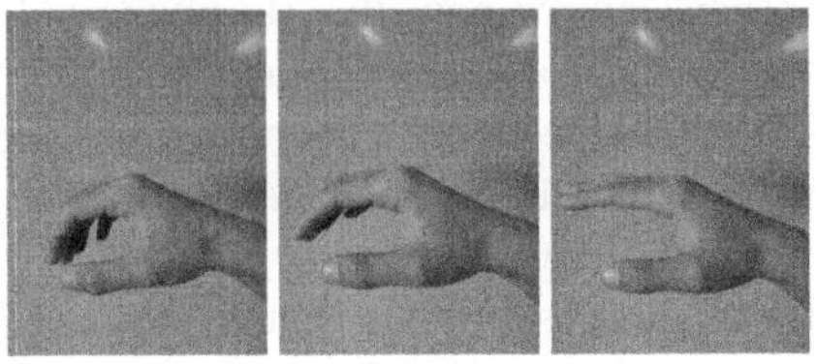

(a) (b) (c)

Fig. 2: Transition moves from one main gestures to another.

Once defined the main gestures (see figure 1) and transition moves (see figure 2) it proceeds to generate the database. The generation of the database counted with the participation of ten (10) people, the methodology used to acquire the data is described in the steps showed below:

1. Putting the MYO armband of ThalamicLabs ^{TM} on the dominant arm (right or left).
2. Enabling the movement of computer cursor by using MYO armband of ThalamicLabs ^{TM}.
3. For rotation gesture (R. click), the participant was asked to make and hold the gesture (see figure 1 (a)) while he/she was moving the computer cursor in any direction or form (circles, diagonal, and so on possible moves) around the whole computer screen.
4. For zoom gesture (L. Click), it proceeds in the same way of step three (3) but the participant was asked only to move the computer cursor from upper part to the bottom part of the screen and vice versa.
5. For traslation gesture (S. Click) and transition moves, it proceeds the same way as step (3).

The time spent to acquire the data for each gesture was approximately of one (1) minute, the signals acquired were the superficial electromyography (sEMG) sampled at 200Hz and the computer used was a Hewlett Packard (HP) with intel core i5 processor of second generation of 3.2GHz and memory RAM of 8GB.

B MYO Armband

The hardware used to build the HCI was the MYO armband of ThalamicLabs ^{TM}. The reason to choose the MYO device was that it uses the Bluetooth protocol and has four (4) sensors such accelerometer, gyroscope, orientation signals (Pitch, Yaw and Roll) and Eight(8) superficial electromyography (sEMG) sensors.[1]

C HCI Development

C.1 Cursor movement

To get the movement of the computer cursor were used the orientation signals (Pitch, Roll and Yaw), those signals represent the movement made by the arm where the MYO is worn as show in the figure 3.

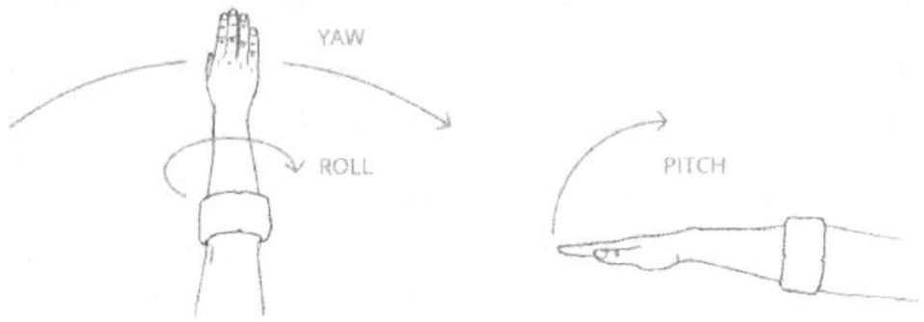

Fig. 3: Representation of orientation signals in arm movements.

Of the three orientation signal Pitch, Roll and Yaw were only used the Pitch signal as the displacement of X axis and Yaw as the displacement of Y axis.

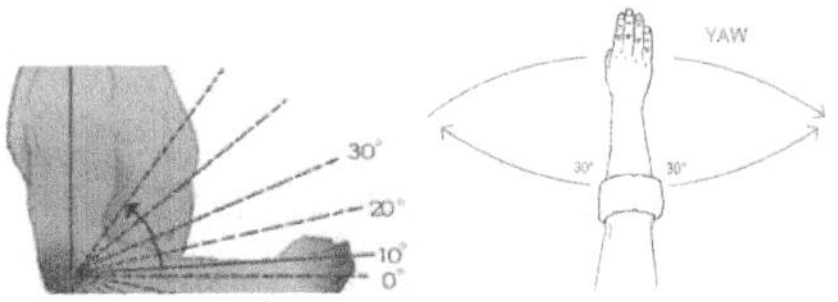

(a) (b)

Fig. 4: Initial point of Pitch and Yaw respectively.

The figure 4 denotes for (a) the moves take in account for the user to get the Y coordinate and (b) the moves to get the X coordinate. So the way to get the screen coordinates is by using a linear transformation from measures taken by orientation signal into screen coordinates.

C.2 Features extraction of gestures

In order to recognize the gestures made by the neurosurgeon (user) a pre-processing of the electromyographic signals

[1] https://www.myo.com/techspecs

is needed. This pre-processing is done by a normalization of sEMG signals and a second order discrete wavelet decomposition (DWT) for a time window of 0.15 seconds. This detail coefficients of the second order DWT are used to extract features such as root mean square (RMS), standard deviation (STD) and wave length (WL) of the gestures [9].

C.3 Gesture classifier

The features extracted from the sEMG signals from the database are the input of a supported vectorial machine (SVM), which was trained with two hundred (200) data per gesture giving a total of eight hundred (800) training data. The SVM classifier uses the one vs all multi-class methodology to identify four (4) classes that are:

- Rotate (Right click) (See figure 1 (a)).
- Zoom (Left click) (See figure 1 (b)).
- Translation (Scroll click) (See figure 1 (c)).
- Transition moves (Scroll click) (See figure 2).

Once the SVM classifier is trained it proceeds to make a cross-validation this results are showed in section IV.

C.4 Avoiding fake gestures and lock functions

To avoid fake gestures a thresholding is implemented by using the integration of sEMG signals. This threshold ensure that the movement of the cursor (arm movement) does not provoke an unwanted action. Also to support the thresholding explained above both an automatic and manual lock [2] function and a unique manual unlock function are implemented.

- **Automatic lock:** The automatic lock is activate when the Pitch signal is considerably lower than the initial point (see figure 4 (a)) indicating that the arm is in a relax position.
- **Manual lock:** The manual lock is activate when the user make the default gesture of Wave In of MYO.
- **Unlock:** If the user wants to unlock the application it must do the default gesture Double Tap of MYO twice, the first one to enable the mouse movement and the second one to enable the gestures detection.

C.5 HCI functionality

If the HCI is not locked it reads the orientation signals (pitch and yaw) and transform them into screen coordinates. Also reads the Pitch level to lock the application if the arm is down. If the HCI is not locked and the move made by the user overcomes the threshold, features (RMS, STD and WL) are extracted and classified by a SVM to make the respective action, which are Rotation, Zoom or Translation. The MYO classifier only take action when the wave in (lock gesture) or the double tap gesture is made. When the HCI is locked waits for the unlock gesture.

[2]unable the detection of gestures and cursor movement

III NEURONAV

The Neuronav is an open-source application developed by the research group in Automatics of the Universidad Tecnológica de Pereira (UTP). This application allows interactive neurosurgical planning, volume rendering, rigid and non-rigid registration, 3D image viewer, flexible layouts and slice viewers, atlas registration, and on-line tracking of microelectrode devices during DBS.
Neuronav main application is to show the volume of tissue activated (VTA) [1].

IV EXPERIMENTAL RESULTS AND DISCUSSION

A Features extraction results

Table 1: Mean value of RMS feature extracted

Gesture	RMS							
	RMS1	RMS2	RMS3	RMS4	RMS5	RMS6	RMS7	RMS8
R. Click	0.1603	0.1624	0.1610	0.1535	0.0838	0.0404	0.0526	0.0978
Zoom	0.3054	0.2260	0.5622	0.3160	0.0967	0.0842	0.0824	0.0662
S. Click	0.0796	0.3502	0.7173	0.4888	0.1621	0.0758	0.0512	0.0453

Due to a large data is only presented the RMS feature results showed in table 1. It denotes that feature extracted is very usefull to establish a differentiable factor between all gestures (transition and main gestures), in other words this features allows the use of a "simple" machine learning and therefore improves the time spent in the classification step. Because this application runs in parallel with the IGS software NEURONAV, this implies a soft HCI in terms of process made (computational cost) but a robust result in the application is need, and that is what we reflect on the results shown in table 2.

B Classifier results

The concussion matrix and accuracy of the train made to the SVM is shown in table 2.

Considering the low complexity of the HCI, the percentage of accuracy is acceptable to use the HCI in a medical appli-

Table 2: Classifier cross validation

Gesture	R.Click	Zoom	Scroll click	Transition moves
R.Click	195	2	0	3
Zoom	2	195	3	0
Scroll Click	2	5	193	0
Transition moves	3	0	0	197
Accuracy	97.5%	97.5%	96.5%	98.5%
Total Accuracy			**97.5%**	

cation as it is an IGS software (Neuronav) for the plannig of parkinson surgery,

C Results in Operating room

The HCI was tested in the **planning** of Deep Brain Stimulation surgery (DBS) under the IGS software NEURONAV. The tasks evaluated was first the accuracy of classifier, second the complexity of the usage of the HCI and third the robustness of HCI avoiding fake gestures. After evaluated this task named before the HCI was very usefull in the planing of DBS allowing the neurosurgeon to perform a better control over the 3D image volume, making the pertinent actions in the IGS software NEURONAV and finally bringing a wearable computer cursor making the manage of NEURONAV more suitable for doctors. Of course the HCI is not perfect sometimes it confuses some main gestures but it reflects the result of table 2 so it still a good HCI for a medical application.

C.1 Neurosurgeon feedback

The Neurosurgeon who tested it give to us this feedback: "Using this wereable device MYO in NEURONAV makes the control of software more suitable to us (Doctors) and it is not wire attached so we can move and still have control of it."

V CONCLUSION

In this paper a Human-Computer Interaction was presented, the goal achieved with this HCI was to give a better control over 3D image volume of the NEURONAV software to the neurosurgeon, making it more suitable in term of interactivity and avoiding issues that other peripherals have.

The problems presented during the process was the classification, because there has to be a balance between the kind and amount of features extracted.

As future work is expected to improve the classification accuracy by using another techniques of machine learning, the feature extraction could be improved by using probabilistic methods like the characteristic function of each gesture or a distribution function to reach at least 98% of accuracy to implement the HCI in surgery.

ACKNOWLEDGEMENTS

This research is developed under the project financed by COLCIENCIAS with code 111065740687. H.F. García is funded by Colciencias under the program: *Formación de alto nivel para la ciencia, la tecnología y la innovación - Convocatoria 617 de 2013*. David A. Jimenez is funded by Colciencias under the program: *Convocatoria 706 del 2015 - Jóvenes investigadores e Innovadores*. Thanks for the support to the program of master in electrical engineering of the Universidad Tecnológica de Pereira.

REFERENCES

1. Padilla José Bestier, Arango Ramiro, García Hernán F., et al. NEURONAV: A Tool for Image-Guided Surgery - Application to Parkinson's Disease in *Advances in Visual Computing 11th International Symposium, ISVC 2015, Las Vegas, NV, USA, December 14-16, 2015, Proceedings, Part I*:349-358 2015.
2. Chou Kelvin L., Grube Susan, Patil Parag G.. *Deep Brain Stimulation: A New Life for People with Parkinson's, Dystonia, and Essential Tremor*. New York: Demos Medical Publishing 2011.
3. Manea S. M., Kamblib R. A., Kazic F. S., Singhc N. M.. From clinical imaging and computational models to personalised medicine and image guided interventions *Medical Image Analysis*. 2016;33:50-50.
4. Manea S. M., Kamblib R. A., Kazic F. S., Singhc N. M.. Hand Motion Recognition From Single Channel Surface EMG Using Wavelet and Artificial Neural Network *Procedia Computer Science*. 2015;49:58-65.
5. Chen Yumiao, Yang Zhongliang, Wang Jiangping. Eyebrow emotional expression recognition using surfaceEMG signals *Neurocomputing*. 2015;168:871-879.
6. Ahsan Md. R., Ibrahimy Muhammad I., Khalifa Othman O.. Hand motion detection from EMG signals by using ANN based classifier for Human Computer Interaction
7. Moon Inhyuk, Lee Myungjoon, Chu Junuk, Mun Museong. Wearable EMG-based HCI for Electric-Powered Wheelchair Users with Motor Disabilities in *Proceedings of the 2005 IEEE International Conference on Robotics and Automation*:2649-2654 2005.
8. Moon Inhyuk, Kim Kyunghoon, Ryu Jeicheong, Mun Museong. Face Direction-based Human-Computer Interface using Image Observation and EMG Signal for The Disabled in *Proceedings of the 2003 IEEE International Conference on Robotics and Automation*:1515-1520 2003.
9. Phinyomark Angkoon, Limsakul Chusak, Phukpattaranont Pornchai. A Novel Feature Extraction for Robust EMG Pattern Recognition *JOURNAL OF COMPUTING,*. 2009;1:71-80.

Author: David Alejandro Jimenez Sierra
Institute: Universidad del Quindío
Street: Carrera 15 Calle 12 Norte
City: Armenia, Quindío
Country: Colombia
Email: dajimenezs_1@uqvirtual.edu.co

Programs in Biomedical Engineering Education: How to Improve it

G. Avendaño[1] and A. Rienzo[1]

[1] Escuela de Ingeniería Civil Biomédica, Universidad de Valparaíso, Valparaíso, Chile

Abstract— **This work collects information on educational models applied in the teaching of biomedical engineering, in several countries. There is no universal uniformity in teaching subjects and contents of the courses. The result internationally is great variability in the curriculum, the analysis is based in areas covers and didactic modalities. We used information available on the Web sites of careers and we use the AWP model as valuation method. Some optimization proposals are made, based on the results of professional performance of graduates and the demands of the professional field.**

Keywords—**Biomedical Engineering, programs, models.**

I. Introduction and teaching models

In several countries, programs in BME (Bio Medical Engineering) are based on the requirements of the professional field in which graduates perform, so some models have a high content of basic science because the main demand comes from biomedical or bioengineering research institutions, on the other hand there are careers heavily involved in device design because the ties with the medical equipment industry that is a priority; while in most Latin American countries, the professional field of higher demand is related to the management of biomedical technologies in the hospital world, that is known as clinical engineering. Based in the study of the different programs we can establish that, there are the following training models:

A. Scientific Model based in Biological Sciences: This first model is introspective, with teachers focused on the high scientific research. The professional field is limited

B. Management Model based on Clinical Engineering: This second model has focused on generic problems of hospital world.

C. Model focused in Technology for service: It is derived from careers of electronic, industrial and mechanical engineering, maintenance is the core in formation.

D. Mixed model Technical scientific: It is the most suitable, allows the formation of competent professionals, to solve hospital problems and clinical engineering.

E. Hospital Model: Is inclined to meet the demands of hospital design, is focused on hospital engineering and its relationship to architecture and civil engineering.

II. How to analize it

To analyze how teaching is done in various careers in Latin American, were studied the curricula of two by country, only those in which there is a form of undergraduate [1] [2] [3] [4] [5] [6] [7] [8]. The study of different training strategies in Biomedical Engineering It is required to take into consideration the knowledge that constitute the "core" of the specialty, which are strongly linked to actual medical devices that relate to patients in a variety of ways, (diagnostic, therapy, prosthetic, telemedicine) as shown in the debate held in Madrid 2002 [9] .

We use the curriculum distribution, AWP (analysis of work place) in which the magnitude of specific thematic content distribution is displayed. As shown in Figure 1. On continuous line, the classical curriculum distribution of a career based on scientific model is shown, while the dashed line represents a career with a mixed approach, which is more universal because clinical engineering and technical work with medical devices becomes highly important.

III. The main problems of training in BME

A. Theory and Practice

The problematic relationship defined as theory and practice is ancient cross in the formation of any engineering, however in the case of BME, is particularly important because of the nature of professional work to be developed by university graduates [10].

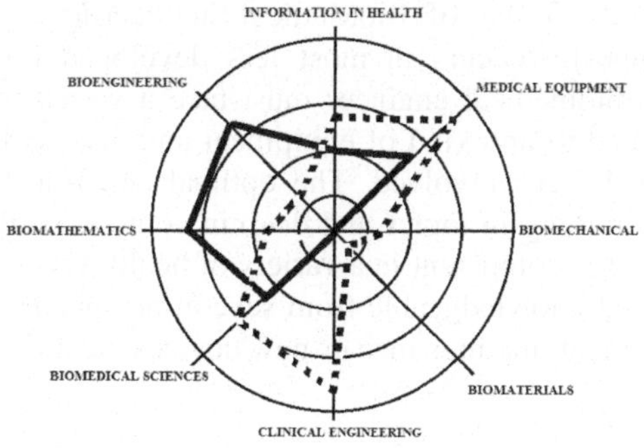

Fig. 1 AWP Diagram applied to BME Curriculum

I. Torres et al. (eds.), *VII Latin American Congress on Biomedical Engineering CLAIB 2016, Bucaramanga, Santander, Colombia, October 26th -28th, 2016*,
IFMBE Proceedings 60,
DOI: 10.1007/978-981-10-4086-3_186

This contradiction affect their working career from the beginning, in such a way that tends to think (wrongly in most cases), that many traditional theoretical aspects are useless or unnecessary, and should be removed from the curriculum, several graduates in addition, complain that should have been trained or at least warned they would find in real working life, some issues that nobody mentioned in class [8] [11].

Coverage of some disciplines in different careers can be explained by external demands detected [12] and by the availability of teachers in these aspects.

It is appreciated that Clinical Engineering, is not a strong activity in the universities analyzed, despite being an activity of relative lower cost of implementation and a large external necessity. In this type of careers, studies high-lighted that the secondary priority are issues like biomaterials, tissue engineering, allegedly by lower demand in the labor field and the need for few specialists.

B. General or specific knowledge

Most professionals in any discipline, have the known dilemma about their cognitive capacity related with their professional performance, this means, being a professional with deep knowledge in few matters ignoring others, or learn to have an adaptive general culture in order to solve general requirements of practice, without being a great specialist into a specific topic.

The classic picture of professional competence is in the case of BME, very indicative of the relationship between university education and obtained skills for better action in practice; we see that as an engineer has more knowledge on general subjects while reduce their level of specialization, aspect that needs in areas of work such as corresponding to those of highly developed countries, where a professional can cover a specific activity in a context of many devices of the same type, thus achieving a deep knowledge in all aspects of these devices and increasingly reducing its information in other issues of BioMedical Engineering.

The opposite occurs in most less developed countries where the biomedical engineer must face a variety of problems related to any kind of equipment and also with a big variety of devices involved. The optimal situation is with extensive training in topics and deep in content, so that the professional is competent in a variety of health-related technologies and knowledgeable from several perspectives such as management, maintenance or new devices creation.

IV. How to get the best program

In some countries the demand for labor and professional performance of graduates determines an inclination to work in the management of biomedical technology, due to the great need for specialists to the hospital world, while in others countries increased labor demand in creating new technological developments and research in bioengineering, so, the training in clinical engineering deteriorates and even disappears as professional option, generating a lamentable lack of specialists in a context of increasing technological complexity of hospital systems.

In Figure 2 we shows the distribution of content more common in BME careers in Latin American countries, where professional requirements are highly variable, but mostly focused in Medical devices, Information in Health and Clinical engineering. At universities in most developed countries the study delves into topics more related to research [13] such as biological and biochemistry engineering, molecular and cellular biology, biochemistry, biotechnology, pharmaceutical sciences or other areas related with bioengineering.

The contents of the Biomedical engineering courses and BME technology labs include theoretical and practical aspects about medical equipment, informatics in health, experiments in design and creation of new types of devices.

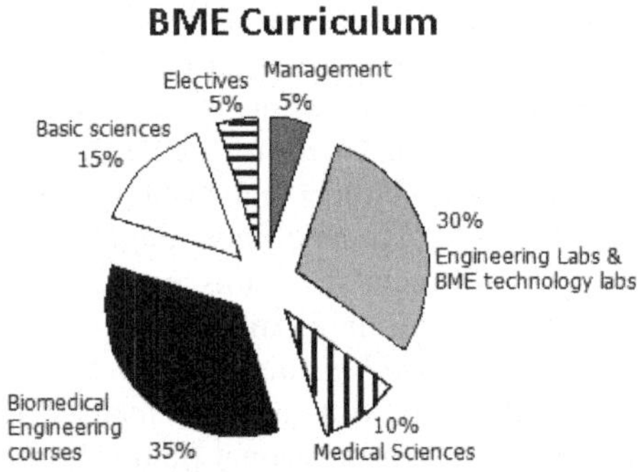

Fig. 2 Distribution of matters in the curriculum

The requirements that the real world makes to the University in order to forming the best engineers, are complex and not always possible to implement, even worse, there is not always channels to detect which are the real needs and which is best program to satisfy a better professional profile [8] .

It is known that PAHO (Pan American Health organization) [14] has determined that 50% of expensive medical equipment in Latin America, is not able to be used, this for a variety of reasons, ranging from the donation of equipment without information for maintenance, and the inadequate use of medical devices by lack of knowledge of users. This indicates that is necessary to have adequate professionnals with the required tools, to make a much-needed sustained action of technology management, tending to optimization and savings in these resources, which will have an impact beneficially on public health in general.

Moreover, the Latin American presence in the global market for medical equipment is marginal, very few countries have an industry in this area and it is clear that it would be necessary to count with a growing development in this sense, at least to achieve import substitution on some devices of high-cost.

The inclusion of professionals in hospitals, the need for development of medical equipment, support for biomedical, pharmacological and scientific research in general, are lines of work of the graduates in Biomedical engineering, and their effect is highly productive and effective.

Is necessary a feedback information from the professional world (medical centers, companies, ministries, laboratories, etc.) to the University, which increase efficiency in engineers formation, in order to obtain a greater relevance of the matter studied and an accurate definition of the field of responsibility of the professional performance. Figure 3 shows the distribution of the professional field in last ten years of biomedical engineers graduates from two Chilean Universities [15].

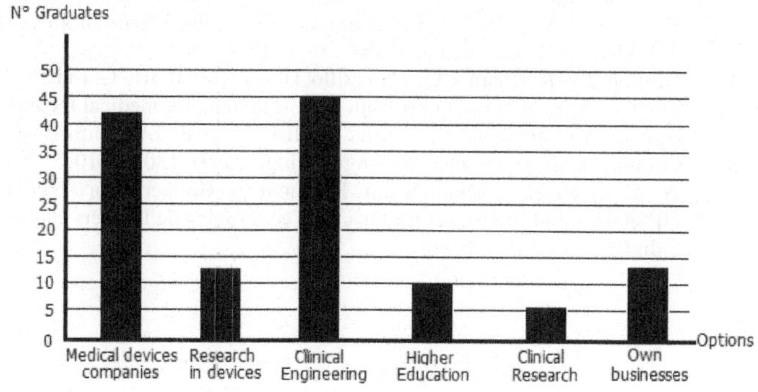

Fig. 3 Professional field of BME graduates in Chile

V. Proposal for Attempting to Resolve the Challenges

To meet the objective of designing a strategy to improve training in biomedical engineering careers, the following proposals were raised. The first proposals we make are aimed at overcoming the contradiction theory practice, which occurs frequently in the university, this problem is of great importance in biomedical engineering.

a) The preferential hiring of more experienced teachers in the professional activity (equipment industries, trading companies and health institutions), which in the exclusive teaching in universities.

b) The implementation of various laboratories, if possible for each and every one of the technical subjects of the curriculum. Not expecting to have optimal technological facilities, acting with the view that something is always better than nothing.

c) Multivariate equipment development; obtaining medical equipment donated for educational purposes, together with the creation of self-constructed equipment and the acquisition of essential medical equipment, prioritized according to the specific professional focus [16].

d) The effective implementation of the links between universities and health institutions or enterprises through visits of students to institutions and vice versa, professionals in various fields coming to classrooms for discussions, symposia, conferences and seminars themed around real world of engineering Biomedical.

e) The orientation of the work of qualification to the study, research or implementation of technology related to specific problems of hospital or technological world of Biomedical Engineering and Clinical Engineering.

f) The use in basic science subjects, such as mathematics, statistics, physics, biophysics, chemistry, examples of theoretical problems especially made in relation to biomedicine, bioengineering, process simulation, software application, biosignals, biomaterials and other specific disciplines, rather than conventional abstract examples of problems applied to mechanical, electrical or chemical.

g) The completion of assignments, projects and exams most lectures, with applications close to real situations taken from the hospital problematic, engineering requirements or management of health institutions, as opposed to the classic problems taken from texts.

h) The realization of lectures about classical issues of Clinical Engineering in hospital campuses (clinical fields), in order to permeate students in both, attitudinal and cognitive aspects of specific situations of health facilities. So, the students do homework and projects solving specific problems of the institution that receives them.

i) The conversion of the introductory courses and general training character, (example, "Introduction to Engineering"), in discussion topics, experimenting and searching for the students themselves, about aspects of hospital life.

j) The intensive use of computer systems and educational software simulation of organ functions or physiological processes for teaching purposes (LabView, PASCO, SIMULINK, BIOPAC, etc.).

k) The transverse incorporation of Hospital Safety thematic, throughout the whole career with concrete and real references on the different risk involved in all types of health facilities, always bearing in mind that international experience shows that the risk real that run in hospitals is determined by inexperience, the action with urgency stress and lack of protocols in the procedures [17]. The increasing use of technology in all its forms, does increase the risk related to how to install, use and maintain the biomedical equipment.

l) Early incorporation of students to research and development of the Faculty, as well as participation in working groups for technical competitions and advisory companies or health institutions.

m) The creation of teaching materials aimed at the definition given to the programs; this results in the creation of collections of texts of theoretical material, laboratory manuals, series of PowerPoint presentations, maintenance manuals, installation and management protocols, formats practices, formats and research tasks, etc. Material that have a double educational value, on the one hand, content organized systematically and continuously generated and other the learning achieved in the organization of written forms with documentation style.

VI Conclusions

Biomedical engineering is complex profession and unlike other engineering have in their curriculum specific materials that link to other knowledge disciplines such as biology, health management, including relationship with patients and not only with objects, so that training should consider these differences and incorporate them in teaching in a creative and appropriate form related with the reality of each country.

The idea It is form skilled professionals who know the real problems and the tools necessary to properly solve the problems of all existing technological systems in the world of human and animal health; it is necessary to get a correct curriculum structure and based in reality and not in purely theoretical abstraction; is of great importance to have a positive impact on job performance of graduates.

Conflict of Interest

The authors declare that they have no conflict of interest.

References

[1] Carrera de Ingeniería Civil Biomédica, Universidad de Valparaíso; Chile: http://www.biomedica.uv.cl

[2] Carrera de Ingeniería Biomédica, Universidad de Concepción; Chile: http://biomedica.udec.cl/?Page_id=53

[3] Carrera de Ingeniería Biomédica, Universidad Manuela Beltrán; Colombia: http://www.umb.edu.co/programas/pregrados/ingenieria-biomedica.html

[4] Carrera de Ingeniería Biomédica, Universidad Politécnica del Perú; Perú: http://utp.edu.pe/sites/default/files/ingenieria_biomedica.pdf

[5] Carrera de Ingeniería Biomédica, Universidad del Valle; Bolivia: http://www.univalle.edu/index.php/facultades/informatica/biomedica

[6] Carrera de Ingeniería Biomédica, Universidad Don Bosco; El Salvador: http://www.udb.ed.sv/udb/archivo/pensum/ing_biomedica.pdf

[7] Carrera de Ingeniería Biomédica; Universidad Nacional de Entre Ríos, Argentina: http://www.bioingenieria.edu.ar

[8] A. Rios, A. Cruz et al.: La ingeniería biomédica en Colombia: una perspectiva desde la formación del pregrado; Revista Ingeniería Biomédica, volumen 4, número 7, enero-junio 2010, págs. 23-34 ; Medellín, Colombia.

[9] E. J. Gomez A., "La formación del Bioingeniero en Nuestro Entorno: La realidad Europea" en *Aula Datex Ohmeda Junio,* pp.14-17. 2002

[10] Designing a Career in Biomedical Engineering, IEEE Engineering in Medicine & Biology, PP 1-12 2015

[11] E. Berjano, Transferencia de Resultados de Investigación desde el Ámbito Académico en Ingeniería Biomédica: Deseos, Realidades y Desatinos, Revista Mexicana de Ingeniería Biomédica Vol. 35, No. 1, 7-12 Abril 2014

[12] Lin KP, Kao T, Wang JJ, Chen MJ, Su FC. Education of biomedical engineering in Taiwan, Conf Proc IEEE Eng Med Biol Soc. 2014;2014:5153-5. doi: 10.1109/EMBC.2014.6944785 2014.

[13] Area de Bioingeniería; Fundación para la Ciencia y la Tecnología, MIT: http://www.mitportugal.org/bioengineering-systems/

[14] Dispositivos Médicos: La gestión de la Discordia: Los problemas de los dispositivos Médicos, pp 45-55, Informe OMS, ISBN 978 92 4 356404 3 , Switzerland, 2012 .

[15] G. Avendaño: Análisis Crítico sobre la Formación en Ingeniería Biomédica; V Congreso Latinoamericano de Ingeniería Biomédica (CLAIB 2011); La Habano, Cuba.

[16] Brinton TJ, Kurihara CQ, Camarillo DB, Pietzsch JB, Gorodsky J, Zenios SA, et al.: Outcomes from a postgraduate biomedical technology innovation training program: the first 12 years of Stanford Biodesign, Annals of Biomedical Engineering,; 41(9): 1803-1810. 2013

[17] A. Hernandez, "Diagnóstico Regional y Cooperación Técnica OPS/OMS en Infraestructura Física y Tecnológica de los Servicios de Salud. OPS. 2011.

Video Tracking Analysis System for Forelimb Akinesia test in the Rat Parkinson model

M.P Bonaccorso Marinelli[1], M.J Ledesma[2], F.E Nieto Grimalt[2] and R.J Cabrera[1]

[1] Instituto de Investigaciones Biomedicas (INBIOMED- IMBECU-CONICET), Universidad de Mendoza, Mendoza, Argentina
[2] Area de Bioingenieria, Facultad de Ingenieria, Universidad de Mendoza, Mendoza, Argentina

Abstract— **This study analyzes forelimb akinesia with stepping test in adult male rats with nigrostriatal cell loss and striatal dopamine depletions, similar to levels reported in the clinical setting of Parkinson Disease (PD), and determines if subjects who received a pharmacological induction with apomorphine exhibit more robust parkinsonian effects. With manual testing only one parameter (adjusting steps) is evaluated; the software presented increases the number of parameters as it estimates adjusting steps, stepping time, step length and initiation time. Manual and automated adjusting steps scores are correlated to show software validity. Adult male rats received infusions of 6-hydroxydopamine (6-OHDA) in left corpus striatum. Two months after lesion they were tested for forelimb akinesia using stepping test, before and after an apomorphine induction. Video tracking analysis system developed consists in two stages: 1) where animals performance is recorded and their trajectories are acquired and processed using open source software; and 2) wherein post processing methods are conducted to measure stepping test parameters. Based on these measurements, the program has the ability to make statistical analysis and present data in different formats, including matrixes and graphs. Results show high correlations between manual and automated adjusting steps scores. What is more, this computerized system decreases inter/intra operator errors due to direct observation and handling, expands the amount of parameters assessed, early and acutely detects those parameters and increases data acquisition speed and quality.**

Keywords— **Hemiparkinsonism, forelimb akinesia, object identification, tracking system, video processing.**

I. Introduction

Parkinson Disease (PD) was first described by James Parkinson in 1817. The brains of patients suffering from PD show a profound deficit in dopamine levels, because of the loss of neurons in the substantia nigra. Accordingly, authors consider the disease primary as a disorder of movement [1]. It is a relatively common and serious impairment of health in developing countries all around the world, affecting 1% of the population over 55 years, with a mean onset of 60 years [2]. This neurological disorder is degenerative and progressive, and it is expected to become a serious public health problem [3].

Several reports inform that unilateral 6-OHDA lesioned rats are a suitable animal model for behavioral and biochemical evaluation of PD [4][5]. Here, we use forelimb stepping test in this model to evaluate motor deficits before and after a pharmacological induction with apomorphine. This test is a useful tool to monitor lesion changes in forelimb akinesia adjusting steps, stepping time, step length and initiation time; which may be analogous to limb akinesia and gait problems seen in patients with PD [6][7].

Actually, many behavioral studies used in our institute are analyzed directly by the experimenters. It is our primary goal and necessity to abolish inter and intra operator observations and to establish automatized technology in the practices. Nowadays, there are many types of software available that track objects and extract their positions and orientations. Most of them apply computing processing to images obtained by an analogical camera, as well as they required special equipment to function [8][9].

In this report we present a video tracking software system to early detect and analyze forelimb akinesia modifications; which also improves data amount, precision, speed, flexibility and costs. Adjusting steps is the only stepping parameter that can be manually measured so, to determine software validity we correlate adjusting steps manual scores with the automated ones. The other three parameters which are impossible to quantify with manual recording or direct observation, are automatically detected and measured by this system. What is more, we implement two very useful post processing tools, statistics to analyze data by parametric or non parametric tests and graphic interface to export outputs.

II. Materials and Methods

A. Animals

We used male Sprague Dawley rats form our breeding colony. They were 60-80 days old at the beginning of the

I. Torres et al. (eds.), *VII Latin American Congress on Biomedical Engineering CLAIB 2016, Bucaramanga, Santander, Colombia, October 26th -28th, 2016*,
IFMBE Proceedings 60,
DOI: 10.1007/978-981-10-4086-3_187

study, and their weight was 280-340 g. Experimental subjects were housed under controlled conditions of temperature (22 ± 3 °C) and lighting (12 hour cycle beginning at 7:00 a.m.), with food and water made available *ad libitum*. Animals were kept and handled according to the Guide for the Care and Use of Laboratory Animals of the National Research Council (National Academies, U.S.A, 8th edition, 2011). All efforts were made to minimize animal suffering.

B. Surgical procedures

In order to achieve unilateral lesions of the nigrostriatal system, 10 rats received 6-OHDA injections into the left striatum. Animals were anesthetized with an intraperitoneal injection ketamine (100 mg/kg) and xylazine (20 mg/kg) and placed into a stereotaxic frame (David Kopf, USA). 6-OHDA was dissolved at a concentration of 2 μg/μl saline in 0.1% ascorbic acid. The lesion was performed by injecting the neurotoxic or vehicle with a Hamilton syringe at the following coordinates: AP: +1.2 mm; ML: +2.5 mm; DV: -5.0 mm; TB at 0 mm. The injection was conducted at a rate of 0.5 μl/min and the needle was left in place for another 5 min before it was slowly drawn back. Groups were conformed as: HP (lesion with 6-OHDA, n=5); and Sham (lesion with vehicle, n=5).

C. Forelimb akinesia test

The procedure was similar to the one described by Olsson et al [6] with some modifications. The test monitoring stepping test parameters was performed on a smooth-surfaced wooden table with a length of 0.9 m. All testing was performed during daytime.

- Testing sequence.

  The rat was held by the experimenter with one hand fixing the handlimb and slightly raising the hind part above the surface. The other hand fixed the forelimb not to be monitored. Then it was moved slowly sideways (5 s for 0.9 m) by the experimenter; the testing sequence was right paw forehand and backhand adjusting steps (RP-F, RP-B), followed by left paw forehand and backhand directions (LP-F, LP-B)(Fig. 1).

- Stepping test parameters.

  Initiation time [s] was measured until the rat initiated each forelimb movement. Stepping time [s] was measured from movement initiation until the rat reached the other extreme of the table. The number of adjusting steps was counted for both paws in the backhand and forehand directions of movement. Step length [cm] is the magnitude of stepping movement. The test was repeated three times each day.

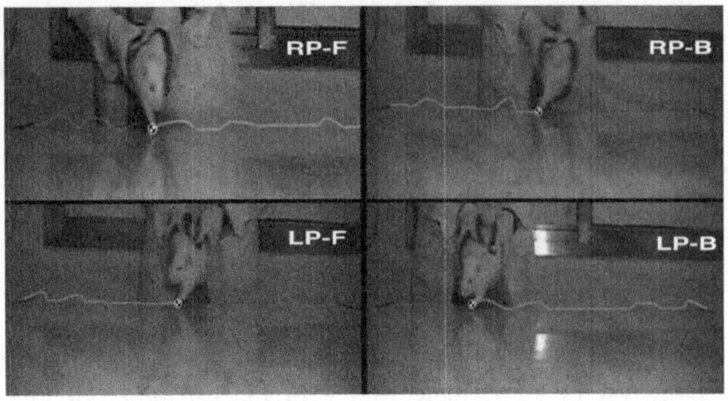

Fig. 1: Testing sequence

- Handling and training.

  For three days prior to the test, rats were handled by the experimenter to become familiar with the experimenters grip and trained for the testing sequence described above. Scores were not recorded.

- Drug induced forelimb akinesia.

  Forelimb akinesia test was used to examine and measure locomotive functionality before and after stimulation with apomorphine (2 mg/kg s.c.). This pharmacological induction is useful to enhance deficiencies in laterality as bradykinesia, time between footsteps, footsteps length and stiffness.

- Activity analysis.

  Behavioral records were all performed by an observer blinded to the experimental condition of the group. Each experimental group began with at least 7 animals. Five animals were finally assigned to each group. Procedures were performed according to the following outline: a) On day one, one hour before treatment all animals were taken to the behavioral testing room to get used to the environment, after that the sequence of testing was carried on. At the end of it, subjects were returned to their home cages. b) On the second day all animals were handled as in day one but, previous to the test, they received an apomorphine injection.

  On both days, activity was assessed manually (direct observation for adjusting steps) and video monitored (with a bioinformatic system for all the parameters).

- Statistical analysis

  Results are expressed as means ± SD. For statistical evaluation, average data form manual and automated recordings were subjected to a correlation analysis. Statistical significance level was set at $p < 0.05$.

D. Bioinformatic systems

- Video tracking procedure stages.

  Room setting is crucial before starting performing behavioral tests. Light and temperature must be set according to standard conditions to avoid animals stress. Camera settings also must be determined at the beginning of the experiment.

  Data acquisition process is where animals performance is video recorded and their trajectories are tracked and processed. For this stage we use Kinovea 0.8.15 [10], a biomechanical analysis video software. This tool adds value for behavioral analysis and helps to determine failure or abnormal movements related to coordination, technique or strategy.

  Data processing software methods are conducted to measure stepping test parameters; and data layout consists in statistical analysis and results presentation. For both, we use computational intelligence strategies developed in software language Python 2.7.9 [11] which allows us to optimize the accuracy, speed and flexibility in manipulating data detection, calculation and analysis. Such implementation increases the amount of behavioral information comparisons between groups and among animals within each group.

- Software development.

  Fig. 2 shows a full diagram of the software development. Briefly, Kinoveas output for each animal is a data sheet (Excel file) which is imported into the software. After that, the algorithm written starts calculating all the parameters listed above per animal and then by groups. Statistical analysis starts testing normal distribution, depending on the result it applies tests for equal arithmetic means and variances or for medians. It works at a significance level of $p < 0.05$, for every *p-value* there is a Sidak correction available which exactly controls family wise error rate (FWER).

  To begin the graphic user interface (GUI), the program needs the folder location where animals tracked data is saved. In experiment settings, the program requests the user to specify the number and amount of animals within each group. With these parameters this block begins to work; data layout can be graphs (barcharts), matrixes of raw data and/or Excel sheets.

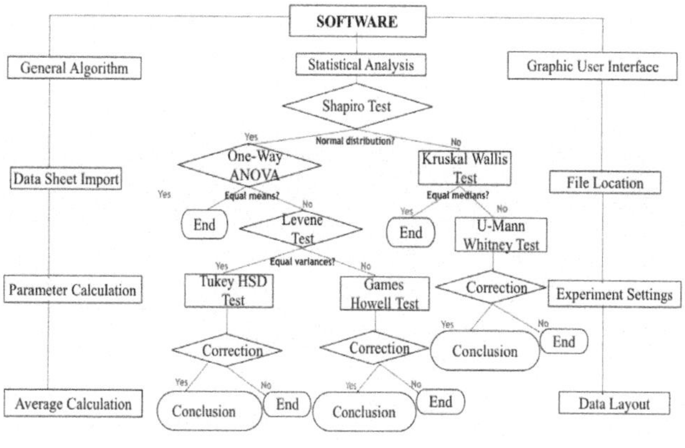

Fig. 2: Software development diagram

III. Results

Adjusting steps is the only parameter that can be correlated between manual and automated results. On day one, 6-OHDA degenerative action on the dopaminergic system is shown as a misbalance in adjusting steps for lesioned (HP) animals compared with controls (Sham). Linear regression coefficient (r) between manual and automated adjusting steps was high for both groups; r=0.9952 for HP and r=0.9966 for Sham, $p<0.01$ was statistically significant.

Apomorphine is a non selective postsynaptic dopamine receptor agonist that enhances adjusting steps difference between both groups. The purpose of administrating this drug on the second day, was to intensify the misbalance evaluated on the first day. Manual and automated measurements under this enhanced situation also correlate; for HP r=0.9445 with $p < 0.05$ and for Sham r=0.9975 , with $p < 0.01$.

Correlation analysis between manual and automated data acquisition methods shows that: 1) Forelimb akinesia adjusting steps correspond correctly with and without apomorphine induction; 2) Bioinformatic method increments the amount, quality and speed of parameters assessed and there is better error detection performance and lower subjective workload in the automatic mode.

Fig. 3 presents GUI output for adjusting steps, stepping time and initiation time parameters in HP and Sham on both testing days.

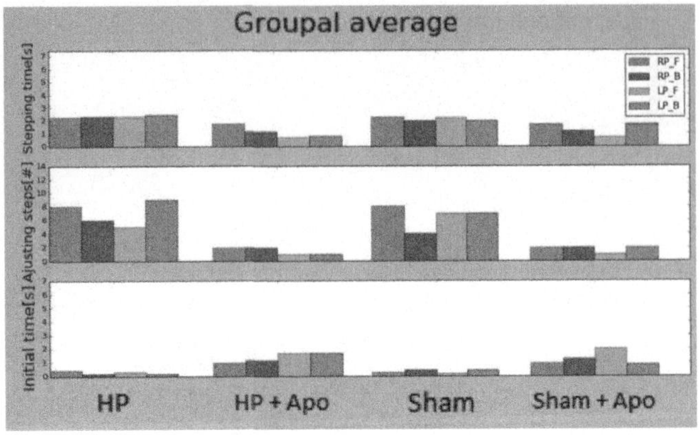

Fig. 3: GUI output

IV. Conclusions

Forelimb akinesia test is an advantageous implement to diagnose motor neurodegenerative disorders. It enriches early diagnosis and improves lesion and post treatment evolution monitoring trough different stages of the disease. Results indicate that 6-OHDA induced marked and long lasting deficits in forelimb stepping in the side contralateral to the lesion, while changes on the ipsilateral side were more subtle. What is more, apomorphine induction enhanced those deficits.

Behavioral evaluation is a powerful tool for animal models *in vivo*. To name some advantages, it allows to establish inclusion/exclusion criteria, demonstrates the efficiency of treatments, identifies sedative, stimulant or toxic influences of different substances, among others. However, experimenter effects are responsible for a significant proportion of the laboratory effects in behavioral research. As this test in particular requires the interaction between the experimenter and the animal, there is an introduction of an inevitable subjective element that makes inter operator comparisons difficult to establish. The operator requires patience, experience and skills, and although it is potentially stressful performance for the animal, to standardize the test conditions as far as possible in room settings, handling, training and the test itself is the way for acquiring consistent results.

It is a fact that bioinformatic benefits animal behavior assessment. Our development provides free software tools to detect and measure behavioral parameters related to motor deficits (impossible to quantify with manual recording or direct observation as stepping time, step length and initiation time) with precision and over a longer period of time. What is more, we implemented two very useful post processing tools, statistics to analyze data by parametric or non parametric tests and graphic interface to export outputs. Scores from automated behavior correlate highly with scores of independent observers, but cost-benefit analysis favors the automated system. Currently, we are working on two more projects to track animals and analyze their behavior by extracting their positions and orientations in water maze and open field maze. As this advances will make our results more accurate, the developments are low cost and time saving.

Conflict of Interest

The authors declare that they have no conflict of interest.

Acknowledgements

This work was partly supported by a grant from the National Research Council of Argentina (CONICET), and a grant from Universidad of Mendoza, to RC.

References

1. Dauer W Przedborski S. Parkinson's disease: mechanisms and models. *Neuron.* 2003:889-909.
2. Hayes M Kimber T O'Sullivan J. Current concepts in the management of Parkinson disease. *Med. J. Aust.* 2010:9-144.
3. Menken M Janca A. Parkinson's disease and public health: educational and management implications. in *WHO Working Group on Parkinson's Disease. Meeting*(Geneva, Switzerland) :1997. www.who.int/iris/handle/10665/63572.
4. Ahmad A Ahmad M Saleem S Yousuf S M Hoda. Neuroprotection by crocetin in a hemiparkinsonian rat model. *Pharmacol., Biochem. and Behav.*. 2005:13-805.
5. Saravanan K Mohanakumar K. Acute intranigral infusion of rotenone in rats causes progressive biochemical lesions in the striatum similar to Parkinson's disease. *Brain Res.*. 2005:55-147.
6. Olsson M Bentlage C Bjrklund A. Forelimb akinesia in the rat parkinson model: differential effects of dopamine agonists and nigral transplants as assessed by a new stepping test. *The Journal of Neuroscience.* 1995:3863-3875.
7. Larramendy C Saborido M Ferrario J Murer M Gershanik O. Cabergoline and pramipexole fail to modify already established dyskinesias in an animal model of parkinsonism. *Behav. Brain Res.*. 2008:44-51.
8. Mukhina TV Lermontova NN Zefirov NS. Versatile computerized system for tracking and analysis of water maze tests. *Beh. Research Methods, Inst. and Computers.* 2011:371-380.
9. Noldus LPJJ Tegelenbosch RAJ. Etho Vision: A versatile video trackins system for automation of behavioral experiments. *Beh. Research Methods, Instruments and Computers.* 2011:398-414.
10. Kinovea at http://www.kinovea.org/.
11. Python at https://www.python.org/.

Author: Maria Paula Bonaccorso Marinelli
Institute: Instituto de Biologia Expermiental de Cuyo
Street: Av. Ruiz Leal S/N - Parque General San Martin
City: Ciudad de Mendoza
Country: Argentina
Email: paulabonaccorso@gmail.com

Rat spinal cord transection injury progression: an MRI study

M. Flores-Leal[1], A. Morales-Guadarrama[2], H. Salgado-Ceballos[3], E. Sacristán-Rock[1]

[1]Departamento de IngenieríaEléctrica, UniversidadAutónomaMetropolitantaUnidadIztapalapa, Ciudad de México.
[2] Centro Nacional deInvestigaciónen Imagenología e InstrumentaciónMédica (Ci[3]m), Ciudad de México
[3]Unidad de InvestigaciónMédicaenEnfermedadesNeurológicas, Centro Médico NacionalSiglo XXI, Ciudad de México.

Abstract— **Transection Spinal Cord Injury (SCI) model in rats has been used to evaluate the efficacy of numerous visionary techniques aimed to preserve and regenerate damaged neural tissue and improve functional recovery. In recent years quantitative Magnetic Resonance Imaging (qMRI) has shown the potential to visualize and extend SCI treatment efficacy assessment in a non-invasive manner. In this study, we used conventional and qMRI to follow rat transection SCI progression through a two-month period study, on a weekly basis, in order to identify crucial injury progress stages, vital for in vivo treatment assessment. This study validates the use of a 7T MRI scanner to visualize and quantify SCI damage on animal model, necessary for any forthcoming treatment approach.**

Keywords— **Spinal Cord Injury, Quantitative MRI, transection, 7T, in vivo**

I. Introduction

Due to the complexity of secondary injury mechanism in spinal cord and the ineffective spontaneous regeneration capabilities of neural cells, repairing central nervous system after SCI remains being the biggest challenge for any therapeutic strategy that aims to reestablish proper neurologic function.

Recent investigation has focused in develop new strategies to increase functional recovery, including the use of stem cells for replacing damaged tissue and promoting neuroregenerative repair [1], biological and synthetic implants that endorse neuroprotective response [2] and magnetic stimulation therapies that combined with exercise have shown promising results [3].

In this regard, different laboratory animal models of SCI have been created to explore the secondary injury cascade effects of traumatic SCI [4]. These models play also a very important role in treatment validation as researchers are able to test and obtain quantitative behavioral, electrophysiological, imaging or biochemical outcomes.

Although histology or immunohistochemistry techniques are commonly used to evaluate SCI treatments effects, animal sacrifice at different stages of therapy are necessary to obtain significant histological evidence.

MRI offers an in vivo approach to follow traumatic injury evolution in a noninvasive manner, providing a unique opportunity to examine a single subject through the entire study and adding quantitative information that can be correlated with treatment efficacy.

In this study, we used MRI to assess complete transection SCI development through a standard eight week period study in pursuit of key anatomical and quantitative changes that can help us to better understand the rat spinal cord transection overall damage and ultimately identify critical time points of injury development, indispensable for any forthcoming treatment approach.

II. Material and methods

An adult female Long-Evans rat (250g) was used in the present study. The animal was housed under standard conditions in the National Center for Imaging and Medical Instrumentation Research (Ci^{3}m). Animal care and study protocol was approved by local ethics committee.

Surgical procedure: The animal underwent a spinal cord transection at 9th thoracic vertebra (T9). The surgery was performed under general anesthesia induced by intramuscular injection of ketamine (77 mg/kg) and xylazine (12.5 mg/kg). After dissecting the paravertebral muscles, two spinous processes were removed, T8 and T9, and a 2-level laminectomy was carried out while keeping the meninges intact. A 5mm longitudinal incision was made to expose the spinal cord and carry on a complete transection. A microsurgery hook was used to corroborate that no nerve pathways remained connected. The meninges were closed using a 10-0 suture. Finally, overlying muscle and skin were sutured.

Following surgery, animal received anti-inflammatory treatment (5 mL/2 L acetaminophen into drinking water) and an antibiotic dose (200 μL benzathine penicillin) to avoid infection. Daily assistance in bladder emptying was procured until spontaneous micturition reflex was recovered.

Magnetic Resonance Imaging: MRI studies were carried out in a 7T Varian scanner located at Ci^{3}m. All images were

I. Torres et al. (eds.), *VII Latin American Congress on Biomedical Engineering CLAIB 2016, Bucaramanga, Santander, Colombia, October 26th -28th, 2016*,
IFMBE Proceedings 60,
DOI: 10.1007/978-981-10-4086-3_188

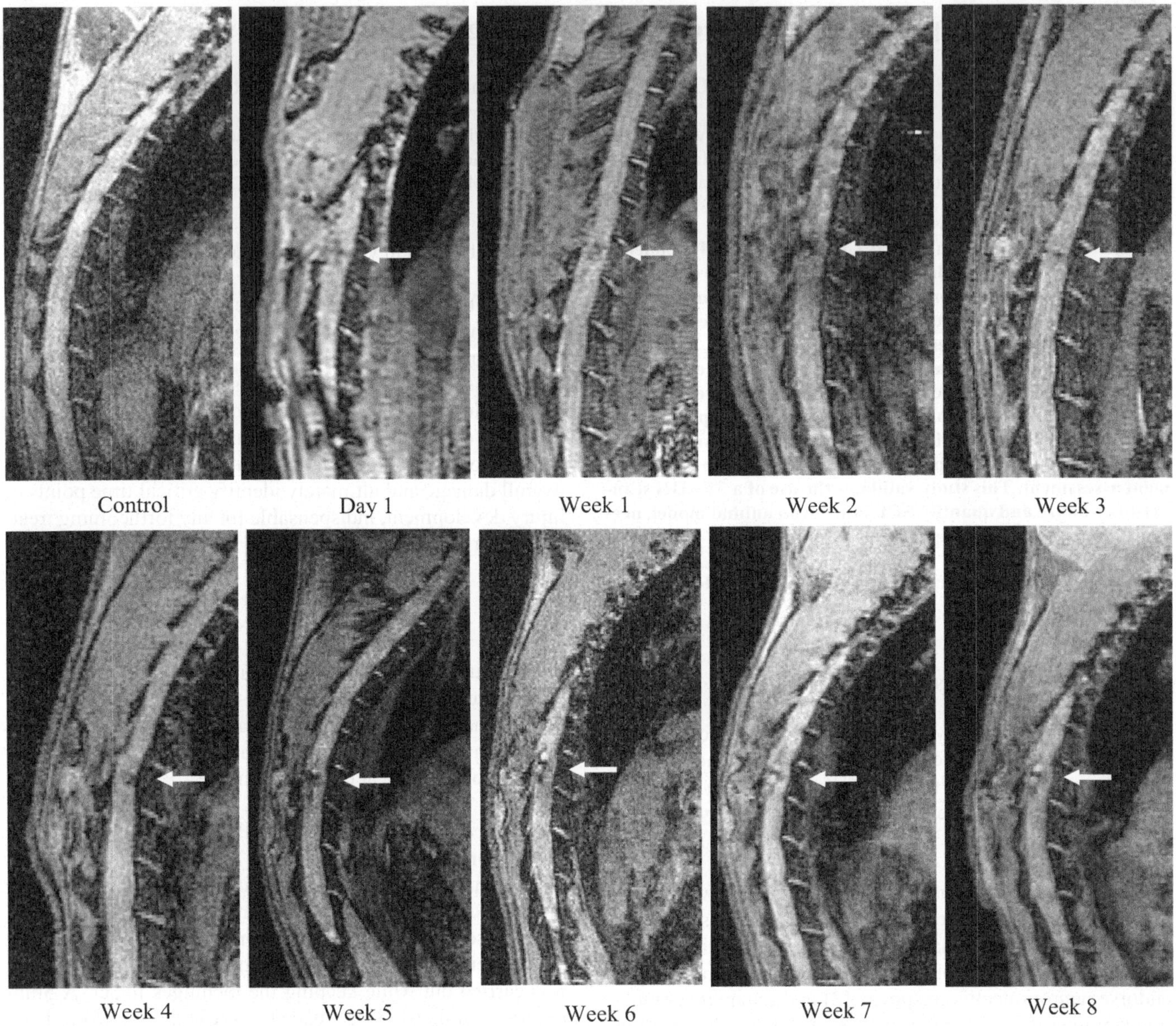

Figure 1. - T_1-weighted MR images in sagittal plane view show the location and progress of the injury through the eight week period. Control image before surgery was included to improve trauma appreciation. Injury site is signaled with an arrow.

acquired under general anesthesia beginning a day after surgery and later on a weekly basis for a total of 8 acquisitions during the evaluation period. Each image trial consisted on 3-D Gradient Echo sequence (TR = 100 ms; TE= 10 ms, FOV = 50 X 60 mm, angle = 20°, matrix size = 256x256) followed by Fast Spin Echo sequence (TR = 2500; TE = 20, FOV = 50 X 60 mm, slices = 20, matrix size = 192x192) used for anatomical evaluation. Finally a Diffusion Weighted Echo Planar Imaging (TR = 2500 ms; TE = 43 ms; FOV = 40 x 40 mm, Matrix = 64 x 64; b value = 1000 s/mm^2) sequence was acquired for quantitative assessment for a total of 45 minutes image acquisition length.

Image processing: T_1 and T_2 weighted images slices were obtained at injury epicenter and displayed in chronological order (Figure 1 and 2).

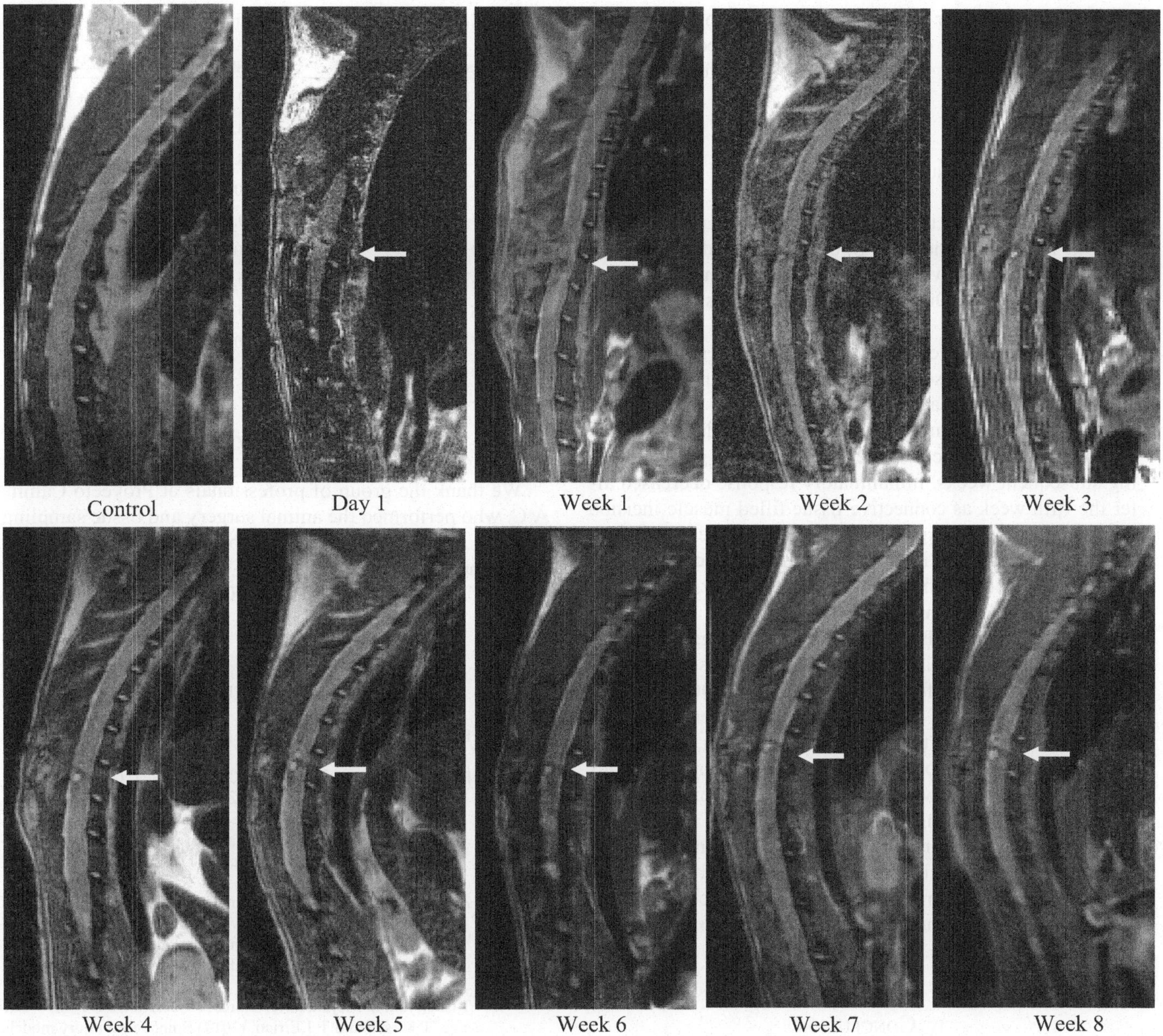

Figure 2. – T_2-weighted MRI acquisitions in sagittal plane view showing injury location and development through the eight week period. Injury site is signaled with an arrow.

DWI were processed using DSI Studio (http://dsi-studio.labsolver.org). Spinal cord tissue was segmented using a threshold mask. Apparent Diffusion Coefficient (ADC) and Fractional Anisotropy (FA) maps were generated from each image round. An average 50 mm^3 Region of Interest (ROI) surrounding the SCI epicenter was created in order to calculate statistics. Averaged FA and ADC were obtained from each ROI and listed in table 1.

Functional recovery assessment: The animal was evaluated a day after surgery to corroborate the absence of movement and each subsequent week before each image acquisition using the Basso, Beattie and Bresnahan (BBB) motor scale [5], which has been developed to measure motor recovery in spinal cord injured rats, ranging from 0 to 21 score depending on the level of motor activity complexity in the hind limbs.

Sacrifice: A day after the last image acquisition the animal was sacrificed and a 2 cm spinal cord segment containing the lesion and surrounding tissue was obtained for future ex-vivo work.

III. Results and Discussion

The animal went through the eight-week period study without major complication. Spontaneous bladder function was recovered four weeks after surgery and no defecation disorder was found. Outcome BBB scale scores are included in Table 1.

Conventional T_1-weighted images (Figure 1) showed a notorious signal decrease at the transection injury site. Main signal loss area increased substantially during the first two weeks as severe inflammatory response prevailed. Cyst formation began after second week and continued growing during subsequent weeks. Inflammatory response decreased after the fifth week as connective tissue filled muscle incision area. Similarly, T2-weighted images (Figure 2) exhibited noticeable signal loss at injury site. Starting on second week, cyst can be noticed with a clear center boundary.

Quantitative MRI outcome measures (Table 1) showed a decrease tendency in FA values after SCI, suggesting progressive neuronal damage. Similarly, ADC presented an increase value tendency following SCI, higher ADC values are correlated with loss of tissue integrity and increase in water content, probably due to cyst formation among other tissue damaging process involved in the injury development. This results are consistent with previous imaging which evaluate similar SCI model using qMRI [6].

Although more animals are needed for significant data generalization, we believe this single-subject run provides new and valuable information about rat spinal cord transection injury course of action and potential treatment assessment in the near future.

IV. Conclusion

This preliminary study can lay the groundwork for the use of Quantitative and conventional MRI techniques in SCI development and treatment assessment. Although weekly image acquisition delivers detailed information of injury development, we think that three image acquisitions (D1, W2 and W5) can provide sufficient data for a satisfactory SCI treatment assessment. Further work includes histological evaluation in order to correlate results and proper data integration.

Table 1 Functional recovery and quantitative MRI outcome measures

Time-lapse	BBB	FA ± SD	ADC ± SD
CTRL	21	0.625 ± 0.201	1.113 + 0.391
D1	0	0.617 ± 0.235	1.175 ± 0.408
W1	1	0.729 ± 0.166	1.38 ± 0.507
W2	1	0.619 ± 0.15	1.665 ± 0.573
W3	2	0.523 ± 0.257	1.114 ± 0.504
W4	2	0.602 ± 0.17	1.325 ± 0.392
W5	2	0.567 ± 0.237	1.228 ± 0.52
W6	3	0.508 ± 0.245	1.298 ± 0.602
W7	3	0.449 ± 0.28	1.281 ± 0.802
W8	3	0.457 ± 0.305	1.384 ± 0.942

Acknowledgment

We thank the group of professionals at Proyecto Camina A.C. who performed the animal surgery and tissue sampling. We also thank Ci^{3}m team for helping with animal care and image acquisition preparations. The first author thanks CONACYT for the scholarship (406006) granted during his studies.

References

1. B. E. Trapp, (2001) Spinal Cord Injury: Hope Through Research, J. Consum. Health Internet, vol. 14, no. 2, pp. 167–174.
2. R. Olayo, C. Ríos, H. Salgado-Ceballos, G. J. Cruz, J. Morales, M. G. Olayo, M. Alcaraz-Zubeldia, A. L. Alvarez, R. Mondragon, A. Morales, and A. Diaz-Ruiz, (2008) Tissue spinal cord response in rats after implants of polypyrrole and polyethylene glycol obtained by plasma, *J. Mater. Sci. Mater. Med.*, vol. 19, no. 2, pp. 817–826.
3. Z. Ahmed, M. Wagdy, M. Benjamin, S. Mohamed, H. Mohamed, S. Ahmed, B. Kanjilal, and A. Wieraszko, (2011) Therapeutic effects of acrobatic exercise and magnetic field exposure on functional recovery after spinal cord injury in mice, *Bioelectromagnetics*, vol. 32, pp. 49–57.
4. T Cheriyan, D J Ryan, J H Weinreb, J Cheriyan, J C Paul, V Lafage, T Kirsch and T J Errico, (2014) Spinal cord injury models: a review, Spinal Cord, vol 52, pp. 588-595.
5. Basso DM , Beattie MS , Bresnahan JC, (1996) Graded histological and locomotor outcomes after spinal cord contusion using the NYU weight-drop device versus transection, Exp Neurol, vol. 139, pp. 244 – 56 .
6. Mondragon-Lozano F, Diaz-Ruiz A, Ríos C, Olayo Gonzalez R, Favila R, Salgado-Ceballos H, Roldan-Valadez E, (2013) Feasibility of In Vivo Quantitative MRI With Diffusion Weighted Imaging, T2-Weighted Relaxometry, and Diffusion Tensor Imaging in a Clinical 3T MR Scanner for the Acute Traumatic Spinal Cord Injury of Rats, Spine, vol 38, pp. 1242-1249.

Mobile Application: Assistance in Mathematics Basic Operations in Children with Learning Disabilities

F.D.L. Abreu[1], F.P. Silva[1], P.B. Neto[1], M.A.S. Bissaco[1] e S.C.M. Rodrigues[1]

[1] University of Mogi das Cruzes/Biomedical Engineering Mogi das Cruzes, Brasil.

Abstract— **This research aims to develop a mobile application(app) in the Android system, to aid in mathematics learning for in autistic children. We used the MIT APP Inventor Tools Beta, Google Chrome and Inkspace, including images and sounds with free. In addition, ISO/IEC9126 standard was used. The application has five screens: loading, avatar selection, the application scenario (questions and answers) and authorship information. The scenario screen is composed by three questions with four answers for each arithmetic operation. The application assessment was made by three experts: 1 in software development, 1 in mobile interface design and 1 math teacher focused on special education, which analyzed through a satisfaction questionnaire. At the end the application was considered by experts as playful, colorful and pleasant sounds, there by can influence the autistic child in uses it, so practicing simulated exercises. In the next step the application will be tested on autistic children.**

Keywords— **Mobile application, mathematics learning, special education.**

I. INTRODUCTION

Mathematics is an area quite prized in academia and much needed for the autonomy of the subject in the society in which he lives. However knows how difficult that few students present regarding discipline, this reality who wins bigger proportions when it involves students with disabilities. For people with disabilities, for example, the autistic the special educator plays a very important role in mediating the development of math skills [1].

Research in the school of medicine, Stanford University, United States, noted that autistic children with an IQ within the overall average perform better in solving mathematical problems than children without the disorder. To the researchers, this may be explained by the unique way in which the brains of people with autism is organized, that is, in people with the disorder there is a major activity in brain areas that favor this ability [2].

The Declaration of Salamanca, proposes that every child has the fundamental right to education and must have the opportunity to achieve and maintain an acceptable level of learning, where the same have characteristics, interests, abilities and learning needs own [3].

The use of technological resources in teaching enhancement tools has shown positive results in learning, especially the virtual environments for on-site support [4][5].

There are several math apps for kids in the Google Play store, such as Monkey Mathematician; Children mathematics; Mathematicians tatters, among others, but none aimed at children with disabilities, for example, autism.

This paper describes the development of a mathematical application containing basic operations of mathematics, such as add, multiply, divide and subtract; to be used in autistic children, aged 7 and 8 years old. This application can be used as assistance or encouragement in the teaching of mathematics for autistic children with severe level or in early treatment and also for any child with learning difficulties. It is important to report that the application is not intended to replace the teacher, special paper and, yes, help you, because it can also be used in another environment outside school. This application developed could be used as an aid or encouragement in the teaching of mathematics for autistic children, in a severe level or in the initial phase of treatment and, also, for any child with learning difficulties.

II. METHODS

After the definition of the educational content to be approached, defined a detailed roadmap on the structure of the application and construction planning respectively in programming and design suitable for child.

A. Application development

The first step of the development process was the definition of the document of the application architecture. The same is used for locating faults/bugs, thus avoiding the rework of the programmer. To ensure the quality of the

I. Torres et al. (eds.), *VII Latin American Congress on Biomedical Engineering CLAIB 2016, Bucaramanga, Santander, Colombia, October 26th -28th, 2016*,
IFMBE Proceedings 60,
DOI: 10.1007/978-981-10-4086-3_189

application followed-if the ISO/IEC9126 [6]. Figure 1 illustrates one of the artifacts used in the software architecture document.

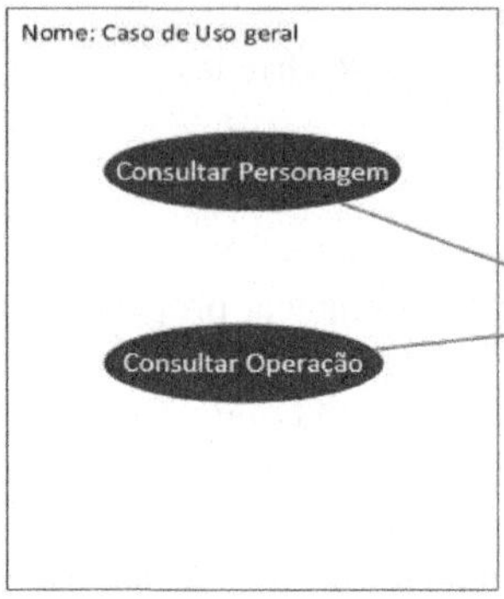

Figure 1: General application use case.

The tools used for programming were: the MIT APP Inventor Beta [7], which is an online platform for open source and free for the development of applications in the Android system, the same can be executed in the web browser Google Chrome. To emulate the application was required to use the software: MIT Companion or AI2 MIT APP Inventor Tools. The images and icons have been searched in IconArchive image bank, free license for personal or academic use, and for editing the same using InkScape.

Settled three blocks of questions with different difficulty levels (easy, medium and difficult). In each arithmetic operation there are three questions (P1, P2, P3), being that in each operation contains a randomization of questions and answers (R1, R2, R3, R4)

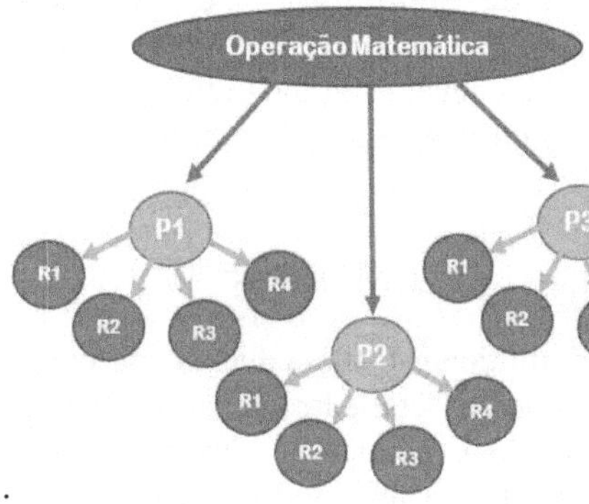

Figure 2 – application Flowchart.

B. *Development of interfaces*

- Splash Screen: Used to load the application modules. While this occurs is shown the child application logo.
- Choosing the Character screen: After the loading of the application process the child shall be directed to the choice screen character, where, it is characterized by two characters (figures), named for Pipa and Pepe.
- User ID screen: When choosing your child's character will be forwarded to the identification screen, which consists of a text field for entering your name.
- Main Menu Screen: This screen is represented by five buttons, four arithmetic operations and one for application authoring information. Arithmetic operations were implemented between the screens: main Menu and application Scenario. On the main menu screen has been set the choice of arithmetic options (addition, subtraction, Division and multiplication) and the user action, the same will be directed to the application scenario.
- Scenario questions screen: There will be three random math questions and each question four alternatives, answers. In the first 10 seconds the child has the possibility to respond without alternatives, after that time the alternatives will appear and the child will have the time to find it necessary to answer the math question. The screen will be modified at the time the child answer the question. When responding to correct the alternative will be green in color and the image button to continue will be displayed. Otherwise, the child will remain in the same step until no matter the setting, however the incorrect alternatives will be red in color and the same will become blocked. Upon arriving at the last arrangement of issues according to the mathematical operation you have chosen will be displayed the button-Choose option, which has direct action the user menu screen to do again the choice of mathematical operation you want
- Authoring Information screen: This screen has information for people and the financier of the application.

C. *Evaluation of the Application*

The assessment of the application was made in two distinct ways:

1. Satisfaction questionnaire based on Likert scale applied in three experts (one in software development, a mobile interface design and a professor in special education), which responded to questions about the application.

2. Test plan document – Used to evaluate the logical paths with your expected result, i.e. the four mathematical operations (add, subtract, multiply and divide) three questions for each and every question

there are four possible answers, so was generated 48 test cases.

III. RESULTS

Figures 3, 4 and 5 illustrate the operation of the application, execution mode in a cell phone with the Android system 5.0- Lollipop.

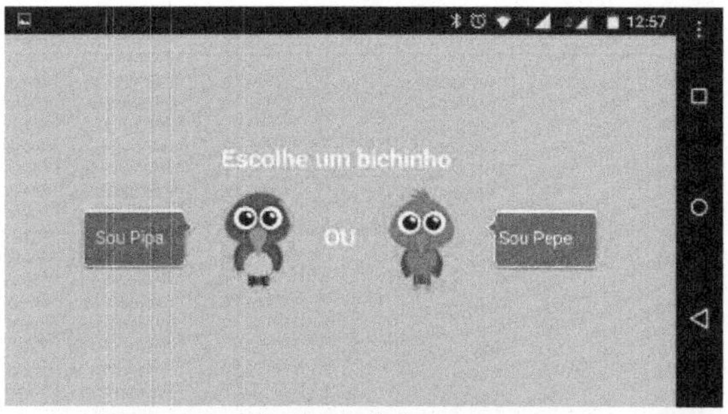

Figure 3: Choice of character screen.

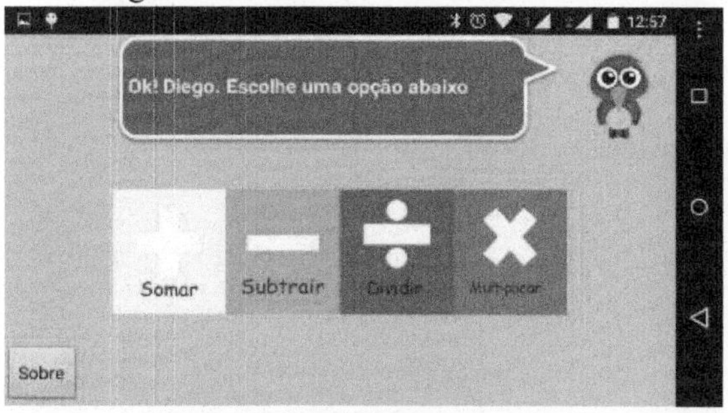

Figure 4: Main Menu screen.

Figure 5: Scenario questions screen.

As the evaluations:

• Experts: in the assessment of the experts the application obtained the value sets = 4.0 equal in all the requirements (functionality, performance, usability, interface and configuration) which represents agree partially Linkert scale. Suggestions will be incorporated in the next version.

•Software: All executions were carried out through the application's emulator, which simulates a phone or tablet in any screen size or Android operating system version. On the emulator execute test plans and complete the application efficiency in its portability to other versions of the Android system, including on different screens of smartphones and tablets, thus allowing a better reproduction without loss of quality or pixels.

IV. CONCLUSIONS

Through tests conducted at MIT App Inventor Tools [7] (emulator), it was possible to complete the application efficiency in its portability to other versions of Android, including on different screens, allowing the application to be responsive to your layout, this it is playing on any screen without loss of quality or pixels. The same did not show flaws in execution, and is easy to install and adapt to other devices. The next step is to use the application for disabled children or not and adults with difficulty in learning math.

V. REFERÊNCIAS

1. Bertazzo, Joíse de Brum; RAMBURGER, Janice Saratt et al (2011). Autismo e matemática: práticas que fazem a diferença e viabilizam a inclusão social at http://www.educasul.com.br/2011/anais/formacao/Joíse de Brum Bertazzo.pdf.
2. Supekar, K., Swigart, A. G., Tenison, C., Jolles, D. D., Rosenberg-Lee, M., Fuchs, L., Menon, V. (2013) Neural predictors of individual differences in response to math tutoring in primary-grade school children. Proceedings of the national academy of sciences of the United States of America 2013; 110 (20): 8230-8235 https://med.stanford.edu/profiles/vinod-menon?tab=publications
3. DECLARAÇÃO DE SALAMANCA: Sobre Princípios, Políticas e Práticas na Área das Necessidades Educativas Especiais, 1994, Salamanca-Espanha. http://unesdoc.unesco.org/images/0013/001393/139394por.pdf
4. Parrish K.L et al (1996).Using interactive multimedia to teach the cognitive e psychomotor skills of cardiopulmonary resuscitation (CPR) and advanced cardiac life support (ACLS).
5. Castro MVd, Bissaco MAS, Panccioni BM, Rodrigues SCM, Domingues AM (2014) Effect of a Virtual Environment on the Development of Mathematical Skills in Children with Dyscalculia. PLoS ONE 9(7): e103354. doi:10.1371/journal.pone.0103354.
6. ABNT at http://www.google.com.br/url?sa=t&rct=j&q=&escr=s&source=web&cd=1&ved=0CFgQFjAA&url=http%3A%2F%2Fprof.swai.com.br%2Fdownload.php%3Ffile%3D1012%26idProf%3D4&ei=wQe1T4zzKYyO8wTew_TqDw&usg=AFQjCNGgEWfb2IGBVhlcOG7tMomPpOGS8Q
7. MIT App Inventor BETA. at http://appinventor.mit.edu/explore

Corresponding author:
Author: Francisco Douglas Lima Abreu
Institute: University of Mogi das Cruzes
Street: A. Dr. Cândido Xavier de Almeida Souza, 200 City: Mogi das Cruzes - SP
Country: Brazil
Email: franciscodougllas@outlook.com

Dispersive Raman Analysis of Sacha Inchi Ozonated Oil

H.C. Carvalho[1-2], D.A.B. Palacios[1-2], C.J. Lima[2-3], M.S. Melo[2], L. Silveira Jr[2-3] and R.A. Zângaro[2-3]

[1] UNICASTELO – Universidade Camilo Castelo Branco, R. Carolina Fonseca, 235, Itaquera, CEP 08230-030, São Paulo, SP, Brasil.
[2] CITÉ – Associação Cidade da Ciência Tecnologia e Educação, Estrada Dr. Altino Bondesan, 500, Eugênio de Melo, CEP 12247-016, São José dos Campos, SP, Brasil.
[3] UAM – Universidade Anhembi Morumbi, Rua Casa do Ator, 294, 7º. andar, Unidade 5, Campus Vila Olímpia, Itaim Bibi, CEP 04546-001, São Paulo, SP, Brasil.

Abstract— **The sacha inchi oil has major advantages when compared to other vegetable oils used in the pharmaceutical, food and cosmetics due to its high concentration of organic compounds. Raman spectroscopy is an analytical technique that allows real-time analysis and nondestructive way the physical-chemical composition of vegetable oils. In this work the Raman spectrum of sacha inchi oil ozonated was characterized by correlating the respective functional groups by PCA. Knowledge of the spectral changes of sacha inchi ozonated oil is important to study and obtain new therapeutic processes.**

Keywords— **ozone, sacha inchi oil, Raman spectroscopy.**

I. Introduction

The sacha inchi (*Plukenetia volubilis* L.) oleaginous plant originated in the Amazon forest is rich in protein and vegetable oil [1, 2]. The oil is extracted from the pressing of their minds up, it is mainly characterized by a high content of polyunsaturated fatty acids (omega 3, 6 and 9) and the high concentration of various organic compounds when compared to conventional oil seed [2]. Recently vegetable oils also gained prominence due to its application as a cosmetic and therapeutic use [1-5].

Ozonation of vegetable oils is a promising alternative aimed at treating various diseases, particularly in dermatology [6, 7]. In addition to the intrinsic properties of the oil, ozonation may confer various antimicrobial properties, making oil-ozone therapy quite interesting.

For a better assessment of the effects of ozonation on sacha inchi, it is important to use a non-destructive technique, showing the results in real time, without requiring sample preparation. Such features are possible using Raman spectroscopy technique, which allows to evaluate the spectral components of the constituents of vegetable oils [8-11].

This study aims to evaluate the composition of ozonated sacha inchi oil, aiming future applications in ozone therapy.

II. Materials and Methods

The sacha inchi oil (*P. volubilis* L.), obtained by pressing of dried seeds, was donated by farmers in Antioquia region (Colombia).

The ozonation was performed using 10 ml of weeding sacha inchi oil, placed in a beaker using a stainless steel diffuser attached to a silicone tube with an internal diameter of 4.5 mm. The hose was connected to the output of an ozone generator (OzonLife, Medical Systems, Brazil) set to generate 66 mg/l of ozone, oxygen flow $\frac{1}{8}$ l/min. The experiment was carried out with the aid of a spectrometer Ra-man dispersive (Raman Dimension P-1, Lambda Solutions, USA), operating in the near infrared, with 830 nm excitation wavelength, optical power of 230 mW, resolution 2 spectrograph cm^{-1}, and spectral band range of 600-1800 cm^{-1}. The exposure time for obtaining the spectra was 1 second and 10 accumulations, totalizing 10 sec. These parameters were defined based on previous work by our group considering the optical scattering characteristics of the oil. The spectral analysis was obtained from the average of three signals collected from each of the ozonated oil samples.

Of each sample were extracted three aliquots of 50 μl, they were placed in three different wells of a sample holder of aluminum. The spectrum of sample from each well was collected in five conditions: 0, 5, 10, 20 and 40 minutes of ozonation, with the distal Raman probe positioned 1 cm from the sample surface. To ensure the reproducibility of

I. Torres et al. (eds.), *VII Latin American Congress on Biomedical Engineering CLAIB 2016, Bucaramanga, Santander, Colombia, October 26th -28th, 2016*,
IFMBE Proceedings 60,
DOI: 10.1007/978-981-10-4086-3_190

the pickup signal, the probe central axis was aligned with the center of the sample.

The signal processing of the Raman spectra was performed using Matlab routine developed that enables normalizing the signal according to the area, to identify and cancel the intrinsic fluorescence, filter the high frequency noise and remove cosmic rays.

In order to show the possible spectral changes induced by the ozonation process, data were analyzed using multivariate statistics based on Principal Components Analysis-PCA.

III. Results

During the ozonation period is observed saturation of oil evidenced by some spectral changes as shown in the figures below.

Figure 1 presents the Raman spectrum corresponding to the vibrational bands characteristics of sacha inchi ozonized oil in function of time.

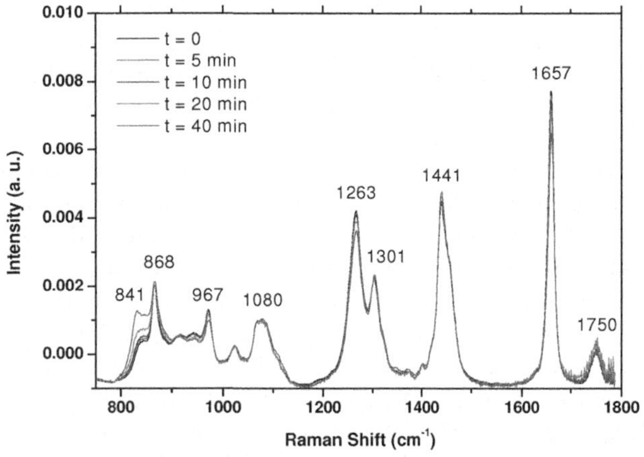

Fig. 1 Raman Spectrum of sacha inchi ozonized oil at 0, 5, 10, 20 and 40 minutes

The Raman scattering of vegetable oils is widely described in the literature. Based on other papers [8-12] are shown in Table 1 the main Raman shifts observed in other vegetable oils. These displacements were also observed in the sacha inchi oil ozonated in this work, for which the functional groups and the corresponding vibrational mode are presented.

Table 1 Assignment of major Raman peaks in sacha inchi ozonated oil.

Raman Shift (cm^{-1})	Functional Group	Vibrational Mode
1750	C═O	C═O stretching
1657	*cis* C═C	C═C stretching
1441	$-CH_2$	C—H bending (scissor)
1301	$-CH_2$	C—H bending (twist)
1263	*cis* HC═CH	═C—H bending (scissor)
1080	$-(CH_2)_n-$	C—C stretching
967	*trans* HC═CH	C═C bending
868	$-(CH_2)_n-$	C—C stretching
841	—O—O—	O—O stretching

Figure 2 presents the spectral vectors obtained from the PCA, and only the Score 2 can be seen relevant spectral information, spectrum showing variation over time. In Scores 1 and 3, this condition is not repeated.

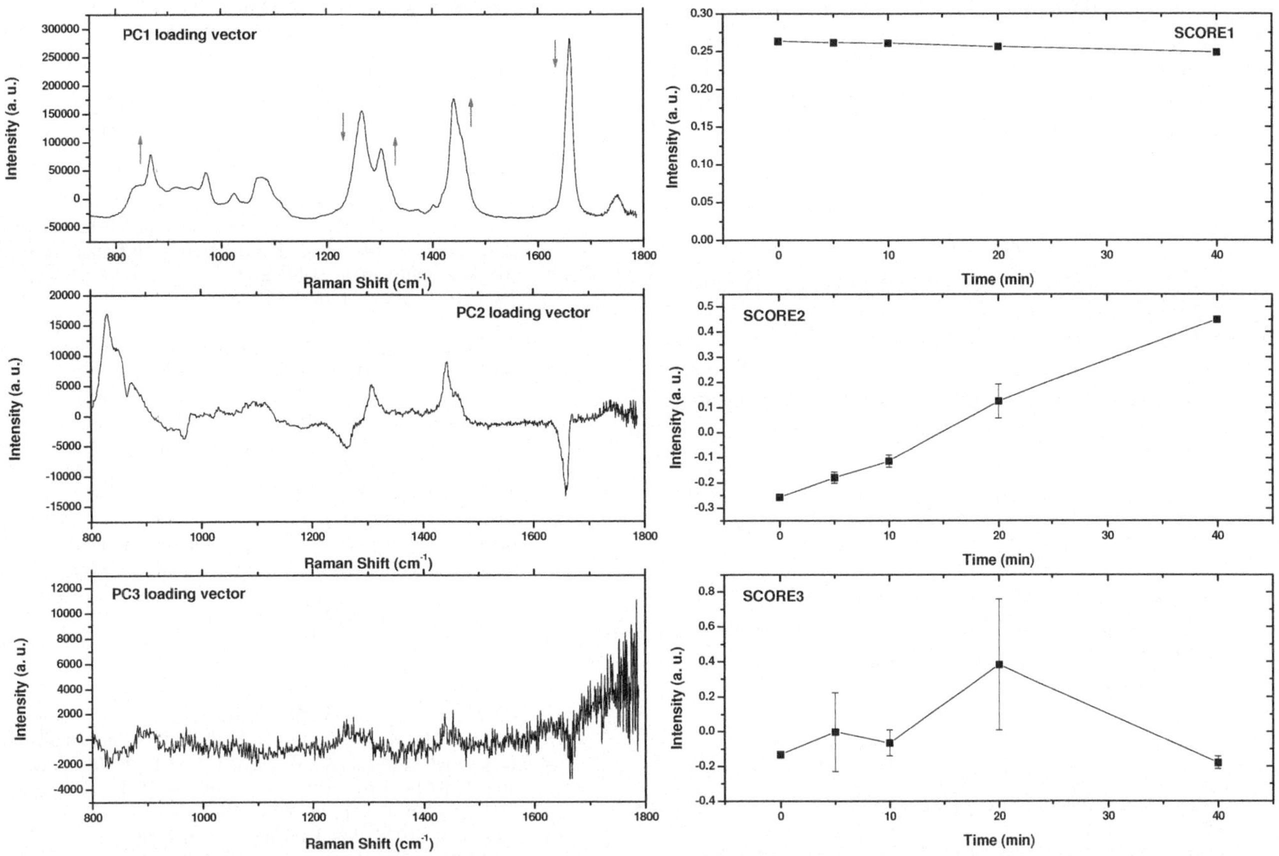

Fig. 2 Plotting vector of the main components PC1, PC2 and PC3 calculated using the spectra of sacha inchi ozonated oil.

IV. DISCUSSION

Ozone reacts chemically with double bond, especially carbon found in unsaturated fatty acids present in triglycerides of vegetable oils [8, 12], this mechanism known as Criegee Mechanism [14]. The reactions with ozone produce different products, which are responsible for providing the ozonated oil features that make it interesting for healthcare applications.

The Raman spectrum of sacha inchi oil ozonated shows high intensity peaks characteristic of the molecular structure of the oil that is rich in polyunsaturated acyl groups. The non-polarity of double bonds also favors the high intensity peaks of the Raman bands.

The PCA revealed that mainly peaks at 967, 1263 and 1657 cm^{-1} decreased in intensity while the peaks at 868, 1301, 1441 cm^{-1} increased indicating that the ozonization produces an increase of the unsaturated components in the sacha inchi oil.

The decrease of the intensity in the Raman peaks is related to oxidative attack to carbon double bond which decrease in intensity which occurs as oxidative attack [15]. This behavior is consistent with that reported in other scientific papers which expose vegetable oils at different oxidative agents, such as linseed oil in the cobalt catalysts action [16], in sesame oil by ozone [8] and coconut oil [17] and other oils [18] by heat.

The increase of intensity in the peaks evidenced by PC2 occurs due to simple bonds connecting with skeletal chain and combined with conjugated ramifications. The functional groups of these radicals are expressed more strongly due to oxidative damage to the double bonds produced by the ozone.

In the region between 800-1000 cm^{-1} are single bonds stretches in sterols, which is a region of little spectral relevance without large variations between different groups of vegetable oils [10, 11], but in this study there was a significant shift band around 841 cm^{-1}, with a consequent increase in band intensity as is the ozonation process. These simple oxygen bonds are susceptible to ozonation and related to displacement of Raman bands in the low wavenumber region generated by the formation of peroxides, such as OH and O-O [18, 19].

The Raman scattering can also allow the improvement of the characterization process ozonated oil by quantifying physical-chemical parameters such as peroxide contents, acidity and iodine, which are also capable of evaluating the properties of the vegetable oil.

V. Conclusions

The Raman spectroscopy technique was able to analyze in real-time and non-destructive way the spectral differences generated by ozonation in sacha inchi oil that may contribute to therapeutic processes in healthcare.

Acknowledgment

To CAPES (Coordenação de Aperfeiçoamento de Pessoal de Nível Superior) for the financial support to H. C. Carvalho and FAPESP (Fundação de Amparo a Pesquisa do Estado de São Paulo).

Conflict of Interest

The authors declare that they have no conflict of interest.

References

1. Chirinos R, Pedreschi R, Domínguez G et al. (2015) Comparison of the physico-chemical and phytochemical characteristics of the oil of two *Plukenetia* species. Food Chem 173:1203-06 DOI 10.1016/j.foodchem.2014.10.120
2. Chirinos R, Necochea O, Pedreschi R et al. (2016) Sacha inchi (*Plukenetia volubilis* L.) shell: an alternative source of phenolic compounds and antioxidants. Int J Food Sci Tech 51(4):986-93 DOI 10.1111/ijfs.13049
3. Cirlini M, Caligiani A, Palla G et al. (2012) Stability studies of ozonized sunflower oil and enriched cosmetics with a dedicated peroxide value determination. Ozone-Sci Eng 34(4):293-99 DOI 10.1080/01919512.2012.692992
4. Hernandez F, Hernandez D, Zamora Z et al. (2009) *Giardia duodenalis*: effects of an ozonized sunflower oil product (Oleozon) on *in vitro* trophozoites. Exp Parasitol 121(3):208-12 DOI 10.1016/j.exppara.2008.10.009
5. Kim HS, Noh SU, Han YW et al. (2009) Therapeutic effects of topical application of ozone on acute cutaneous wound healing. J Korean Med Sci 24(3):368-74 DOI 10.3346/jkms.2009.24.3.368
6. Zanardi I, Burgassi S, Paccagnini E et al. (2013) What is the best strategy for enhancing the effects of topically applied ozonated oils in cutaneous infections? Biomed Research Int 27:702949 DOI 10.1155/2013/702949
7. Sega A, Zanardi I, Chiasserini L et al. (2013) Properties of sesame oil by detailed ^{1}H and ^{13}C NMR assignments before and after ozonation and their correlation with iodine value, peroxide value, and viscosity measurements. Chem Phys Lipids 163(2):148-56 DOI 10.1016/j.chemphyslip.2009.10.010
8. Barthus RC, Poppi RJ (2001) Determination of the total unsaturation in vegetable oils by Fourier transform Raman spectroscopy and multivariate calibration. Vib Spectrosc 26(1):99-105 DOI 10.1016/S0924-2031(01)00107-2
9. Palacios DAB, Silveira Jr L, Zângaro RA (2011) Estudio analítico de aceite de girasol ozonizado por espectroscopía Raman dispersiva, IFMBE Proc. vol. 33, V Latin American Congress on Biom. Eng., Habana, Cuba, 2011, pp. 987-990.
10. Guzmán E, Baeten V, Pierna JAF (2011) Application of low-resolution Raman spectroscopy for the analysis of oxidized olive oil. Food Control 22(12):2036-40 DOI 10.1016/j.foodcont.2011.05.025
11. Silveira FL, Silveira Jr L, Villaverde AB et al. (2010) Use of dispersive Raman spectroscopy in the determination of unsaturated fat in commercial edible oil- and fat-containing industrialized foods. Instrum Sci Technol 38(1):107-23 DOI 10.1080/10739140903430339
12. Yang H, Irudayaraj J (2001) Comparison of near-infrared, fourier transform-infrared, and fourier transform-raman methods for determining olive pomace oil adulteration in extra virgin olive oil. JAOCS 78:889-95 DOI 10.1007/s11746-001-0360-6
13. Guillén MD, Ruiz A, Cabo N et al. (2003) Characterization of sacha inchi (*Plukenetia volubilis* L.) oil by FTIR spectroscopy and ^{1}H NMR. Comparison with linseed oil. JAOCS 80(8):755-62 DOI 10.1007/s11746-003-0768-z
14. Criegee R (1975) Mechanism of ozonolysis. Angew Chem Int Ed 14:745-52 DOI 10.1002/anie.197507451
15. Oyman ZO, Ming W, van der Linde R (2005) Oxidation of drying oils containing non-conjugated and conjugated double bonds catalyzed by a cobalt catalyst. Prog Org Coat 54:198-204 DOI 10.1016/j.porgcoat.2005.06.004
16. Rohman A, Che Man YB, Ismail A et al. (2011) Monitoring the oxidative stability of virgin coconut oil during oven test using chemical indexes and FTIR spectroscopy. Int Food Res J 18:303-10.
17. Muik B, Lendl B, Molina-Díaz A et al. (2005) Direct monitoring of lipid oxidation in edible oils by Fourier transform Raman spectroscopy. Chem Phys Lipids 134:173–82 DOI 10.1016/j.chemphyslip.2005.01.003
18. Lee AKY, Chan CK (2007) Single particle Raman spectroscopy for investigating atmospheric heterogeneous reactions of organic aerosols. Atmos Environ 41(22):4611-21 DOI 10.1016/j.atmosenv.2007.03.040
19. Jia J, Zhang P, Chen L (2016) The effect of morphology of α-MnO_2 on catalytic decomposition of gaseous ozone. Cat Sci Tec 6(15):5841-47 DOI 0.1039/C6CY00301J

Author: Renato Amaro Zângaro
Institute: Biomedical Engineering Institute (UNICASTELO)
Street: Parque Tecnológico de São José dos Campos - Estrada Dr. Altino Bondesan, 500, Eugênio de Melo, CEP 12247-016
City: São José dos Campos
Country: Brazil
Email: razangaro@gmail.com

Design of Chassis and Adjustable Elements to Support Posture for Pediatric Wheelchair

Torres-Pérez Y.[1], Caballero-Reyes C.[1]

[1] Faculty of Mechanical Engineering, Saint Thomas University (sectional Tunja), Boyacá, Colombia

Abstract— **Wheelchairs are devices designed to help with the transportation of people in motor disability condition; are elements used by patients of different ages, weights, heights and diseases. However, most have the same problem as the chassis and postural support elements are fixed which creates several problems: poor posture, overruns to suit the measurements of each person, deformation in the column and in the case of growing children, wheelchair chance not comply with the new measures. In this Project, works in the design of a chassis and the postural support elements of a wheelchair for children in motor disability condition in their lower extremities. The design was developed considering the state of the art of wheelchairs, anthropometric measures of Colombian population of children 5 to 9 years, determining functional requirements (needs) of users, using design methodologies of product, generating design proposals, performing mechanical calculations, mathematical analysis of loads and stresses supported by the chassis and accessories, by the finite element method and through a dynamic simulation of the operation of the device using Autodesk Inventor software. It got to the design of an adjustable chassis width, height and depth, and their postural support elements: headrests, armrests, trunk support, leg supports, footrests, graduating to position and provide body support the user and he reached the conclusion is that design is not only functional for the construction of more ergonomic and functional wheelchairs but also serves as medical or orthopedic custom tool prescription wheelchair.**

Keywords—**Disability, Inclusion, Mechanical Design, Support Elements, Wheelchair.**

I. Introduction

According to statistics from the National Administrative Department of Statistics (DANE), for March 2010 in the Tunja city, there were 545 children who do not study because of their disability. The main causes of disability are accidents, general illness and hereditary genetic alteration and the highest incidence is in the primary education level, where the mostly are students of head municipal. [1][2]. A critical problem is permanently present in cities or small towns in Colombia and in many other countries, is that the health service companies deliver standard technical assistance to people in disability condition or the exact physical condition of the patient, or anthropometric measures, or mobility conditions in your home or environment and much less to the functional requirements of the patient, so sometimes it is not used by the patient or the patient's family has to spend money to make adjustments and make the device work under certain restrictions.

Although international level, children have a wide range of wheelchair with various setting. Electrical pediatric wheelchairs, standard manual pediatric wheelchairs) [3], [4], [5], [6]. In Tunja; wheelchairs are mostly fixed chassis and some folding or removable; fixed chassis are more robust to improve stability and better resist external overload. Chassis are not as strong, conform well to irregular land by joints that allow it's folding and have more manageable sections for transport and storage [7], [8], [9]. Although there are several models of wheelchairs on trade (Fig 1), almost all handled only three sizes for all people and are few manufacturers that consider the voice of the customer in terms of their needs and functional requirements. [10].

Fig. 1. Pediatric wheelchair models

In this document, information on the design of a chassis and its elements adjustable postural support is provided (arm rests, foot rests, head support and chest support) of a wheelchair for children on condition of motor disability. Also include the calculating and modelling of 3D parts by Autodesk Inventor software 2015®.

II. Materials y Methods

A. Anthropometric Groundwork

Ergonomic principles and some anthropometric measurements of potential users, were key parameters for the proper design of the chassis and postural support elements of the wheelchair proposal, as are those governing the sizes, shapes and which they are factors that greatly affect the device.

To determine the appropriate dimensions of the chassis and postural support elements, information from a study on

I. Torres et al. (eds.), *VII Latin American Congress on Biomedical Engineering CLAIB 2016, Bucaramanga, Santander, Colombia, October 26th -28th, 2016*,
IFMBE Proceedings 60,
DOI: 10.1007/978-981-10-4086-3_191

anthropometric dimensions of the Latin American population was taken, where different anthropometric measures (weights, heights, body lengths, perimeters, lengths and heights bipedal and seated positions, etc.) of Colombian children at differences ages are present. The data reported for children between 5 years (minimum measurements) and 9 years (maximum measurements) in bipedal and seated position was taken, such as: (weight, height, lengths, heights to different parts of the body, scope, wide body, perimeters, etc., see table 1) [11].

The maximum and minimum dimensions (height, length, depths backrest height, seat depth, leg length, seat width, etc.) that should have each of the parts of the chassis and postural support elements (Table 1), as the ranks of gradualness thereof, so that the device conforms to the dimensions of each child (a) growing between 5 and 9 years, they were established according to the information reported in the study of anthropometric measurements of Colombian children. [11]

Table 1. Dimensional requirements for the design of the chassis and postural support elements.

Measures relating to the chair (cm), weight (kg)	Maximum 9 years	Minimum 5 years
Seat width	0,38	0,26
Seat depth	0,48	0,33
Seat height	--	0,4
Armrest height	0,23	0,11
Armrest length	0,24	--
Back height	0,48	0,35
Footrest length	0,24	0,38
Along the foot platform	0,23	--
Width of the foot platform	0,9	--
Total width	0,6	--
Total length	0,9	--
Weight load		0,4
Headrest height	1,17	0,98

B. Funcional requirements

The functional requirements were determined through interviews with children and adult users of wheelchairs, and some caregivers of such patients in the city of Tunja, to which they were asked things like, if you like your chair would fit the dimensions of your body, if one can manipulate or need an assistant, that improve using the chair, which need accessories, etc.) All were informed of the purpose of this survey and signed informed consent (children, adults and caregivers). Some needs that should satisfy the proposed design according to these usurious are: it is folding; with reclining backrest; with change of height position; easy to transport; safe; comfortable; low birth weight; useful for the inclined topography of Tunja; width, height, length and depth adjustable; resistant strains, with minimal maintenance; durable, etc.

The chassis is a key element in any wheelchair, because it is responsible for supporting all the time to the user and other components, so it is vital that is strong, reliable and comfortable, were analysed by a morphological picture by studying Fortitude, Opportunities, Weaknesses, and Threats, based on the needs of potential users, and taking into account parameters such as comfort and support seating, mobility. Seat adjustability, durability, low cost, manufacturability in Tunja, a house of quality (QFD method) was built three chassis designs and postural support elements were generated) and by the method of weighted objectives with which it came to identifying the best design proposal.

C. Mechanical Calculations Chassis and Postural Elements Support

According to theoretical study with the lengths and weights of body segments found for anthropometric of children 5 to 9 years of age [11], the mass and inertial properties of body parts of interest were determined; They were calculated their centres of mass where is located the mass loads for static calculations of the chair. The preliminary selection of material for performing calculations and simulation of stress and strain of the chassis and postural support elements, was performed by consulting manufacturers catalogues, material resistance books, and by rating weighted objectives, where the material selection criteria were: mechanical strength and corrosion resistance, weldability, weight, among others. 6061 T6 aluminium being the material selected.

To make the mechanical design in detail, was taken as reference the ISO 7176, where it states that wheelchairs bearing static loads, since the movements of the chair are considered slow and insignificant accelerations [12].

The calculations for the chassis and the seat base structure were performed considering two different load conditions. The first, under conditions of use and normal load and the second under maximum load conditions in concentrated points of the structure. Furthermore, it is assumed that the weight of the person is distributed over the chassis and other components of support.

Then a summary exemplifies the calculation methodology of stress, strain, moments, modules section materials, etc., which is applied to each of the chassis components, the support structure and postural support elements to ensure proper material selection manufacturing and parts support the loads to which shall be subject to probability of failure.

Calculation of profile for the seat support structure.

The load to support the chassis, associated with the weight of a 9-year-old is 40 kg (392.4N) (see Table 1). This value was used throughout the study to calculate all chassis components. For calculating the maximum moment of the seat support, it was considered the most critical situation that may occur, which is when the child sits on the edge of the seat. In Fig.2, the modelling of loads in Autodesk Inventor software version 2015 is observed.

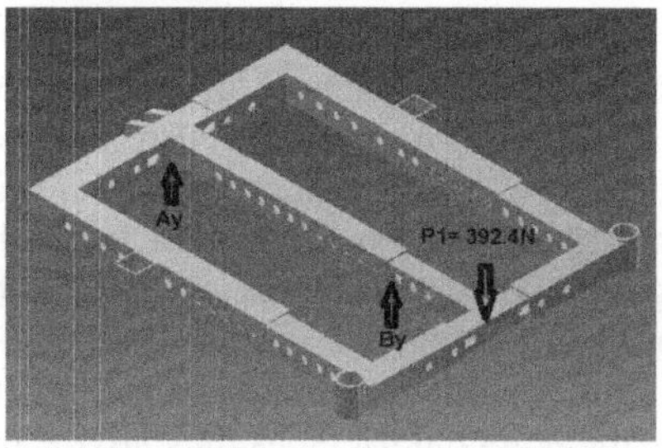

Fig. 2. Seat structure under critical load (part 2 of the Fig. 4)

Where P1 = load exerted by the total weight of the child (or girl) at the vertex of the seat and Ay and By = reactions based on the chassis structure.

Then free diagrams each device body elements were made and calculated the maximum bending moment by software MDsolids 3.5. The maximum moment in the most critical load condition in the seat support is 58.86Nm. In this condition the maximum load on the chassis base is 570.76 N, corresponding to the reaction in By. A safety factor of 3 was estimated, because it is designing a device for human use. [13].

The calculation of tensile stress (maximum) allowable by the structure, it was calculated according to equation 1.

$$\sigma'_{maximum} = \frac{S_y}{n} \quad (1)$$

$$\sigma'_{maximum} = \frac{240MPa}{3} = 80MPa$$

Then, the section modulus of the square profile according to equation 2 was calculated.

$$s = \frac{M}{\sigma'_{maximum}} = \frac{58.86Nm}{80MPa} \quad (2)$$

$$s = 0.735x10^{-6}m^3 = 0.735cm^3$$

This methodology was used to calculate each of the chassis components, the support structure and elements of postural support.

D. Analysis of Stress and Strain by Means of Finite Element.

Subsequently, according to section modulus calculated under critical loads; the technical specifications of the materials and according to the availability in the market, was selected: the best material that met the mechanical requirements, the best profile, suitable wall thickness, etc. Then with these specifications was calculated again for each of the pieces, Von Mises stress (normal and critical) using Autodesk Inventor software 2015®, as well as normal and critical security factor of the structure.

Then, this process is presented for selection and calculation of the effort profile for manufacturing of base of chassis's structure (part 1 of Fig 4) according to section modulus calculated, circular aluminium alloy 6061 T6 of 30mm outside diameter and wall thickness 33mm, with module section profile 1.565cm^3 of chassis.

According to equation 2, the tensile strength allowable was calculated.

$$\sigma'_{máximo} = \frac{58{,}6Nm}{1{,}565x10^{-6}\, m^3}$$

$$\sigma'_{máximo} = 37{,}4MPa$$

The safety factor is that this structure provided was calculated, according to equation 1.

$$n = \frac{240MPa}{37{,}4MPa} = 6.4$$

Through finite element analysis, could be determined the maximum tension Von Mises existing in the structure (Fig. 3), the magnitude and location of the maximum tensile stress is observed.

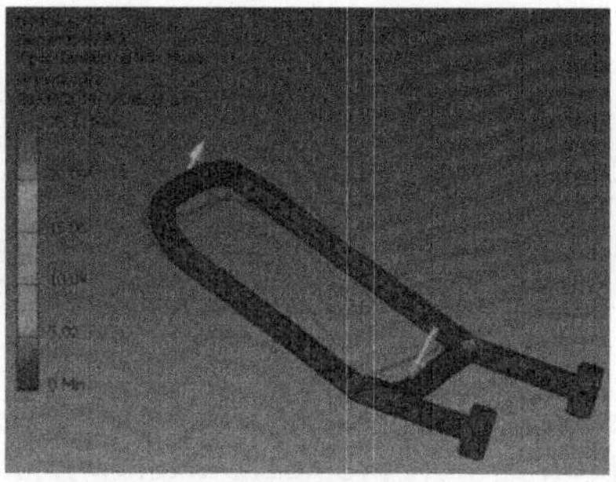

Fig. 3. Von Mises stress to the chassis base.

According to equation 1, the safety factor was calculated again:

$$n = \frac{240MPa}{25.11MPa} = 9.5$$

The maximum displacement suffering each part of chassis structures was calculated by the same software and same methodology.

III. Results

Drawings, circulation of stress, strain, time, and materials section modules, etc., and selecting appropriate for each of the chassis components materials, the support structure and postural support elements were performed for ensure proper performance of the chair (see fig 4).

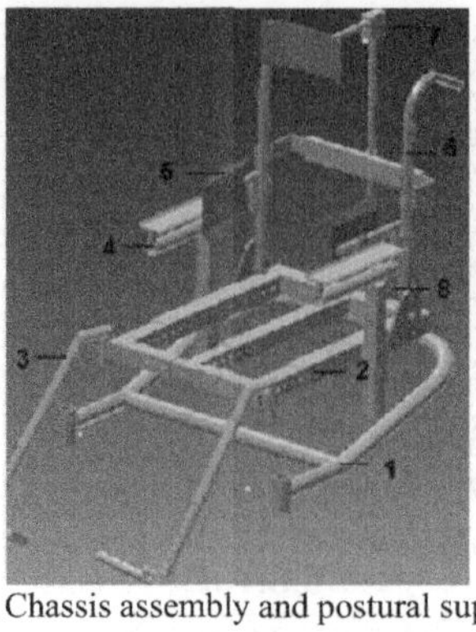

Part Number	Assembly Parts
1	Based chassis structure
2	Seat support structure
3	Footrest structure
4	Armrest structure
5	Thoracic support structure
6	Back support structure
7	Support structure cephalic
8	Position of the seat pivot axis

Fig. 4. Chassis assembly and postural support elements of the chair.

Design, dynamic simulation, construction and assembly of the pediatric chassis and the support elements was achieved, the significant advantage is that most of its parts are adjustable and removable by means of knobs in width, height, length and depth. This will help a child of 5 years, that growth has not changed every year or two wheelchair, but the chair can graduate and respond to anthropometric patient conditions of 5 to 9 years and older to be possible. (Fig. 4).

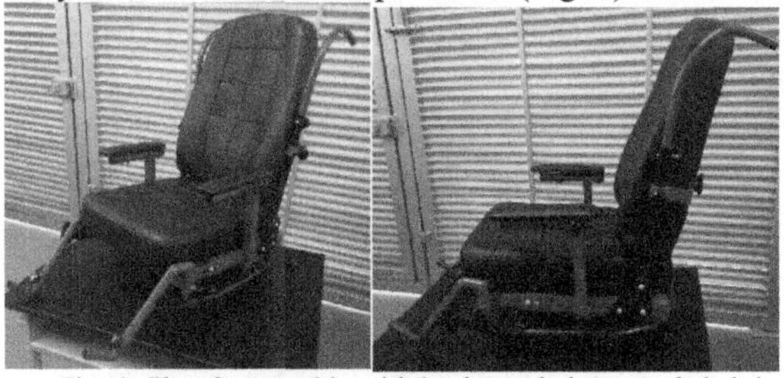

Fig. 4. Chassis assembly with leather upholstery and chair base

For the chair is adjustable in the seat support using three different profiles were proposed, to allow relative movement between profiles and thus ensure and facilitate graduation (height, width and depth) of the structure according to the anthropometric measurements patient.

Ideal design for improved posture, breathing, feeding and overall comfort design was achieved. They are compact and portable yet strong, durable and reliable. Offer versatility, security, functionality, manoeuvrability to the growing child.

IV. Conclutions

The design propose allows multiple adjustments as needed by the user, for example, it allows to graduate the backrest angle vs the seat; the width, height and depth of the seat base; they can be scaled up and along the armrests, as the foot rest and head support and can change the position of the base of the back.

If the chassis is adapted for an electric wheelchair, it should also take into account the location and weight of all constituent elements such as engines, wheels, batteries, suspension, electronic control, etc., as these added other charges, generating efforts and different deformations. The entire chassis is properly calculated to be no structural failure easily and that patient safety is not affected.

This chassis model with elements of postural support and a strategically placed rules, may also be used by doctors and orthopedists surgeons to prescribe custom wheelchairs, making a fitting controlled of all measures of the elements of the chair to suit the morphology required for each person, thus improving their posture, ergonomics and comfort.

For a product to use population is functional and commercial, it should be based on the functional requirements expressed by the potential users, as they are the primary source that has clearly needs to be met by the device.

Conflict of Interest

The authors declare that no conflict of interest.

References

1. Sev, S. D. Dgee, Direction General of Special Education, Online http://eespecial.sev.gob.mx. Accessed 2014 February 10

2. Dirección de Censos y Demografía, DANE, Online http://www.dane.gov.co, Accessed 2014 February 7.

3. Richard Simpson, Edmund LoPresti, Steve Hayashi, Illah Nourbakhsh, David Miller,(2004) The Smart Wheelchair Components System, Journal of Rehabilitation Research & Development, Vol 41, No. 3B, pp 429-442

4. Krey CH, Calhoun CL. Utilizing research in wheelchair and seating selection and configuration for children with injury/dysfunction of the spinal cord. J Spinal Cord Med 2004;27 Supply 1:S29-37.

5. Technical MFG Group.(1968). Adjustable Wheelchair Device. U.S.Patent 3, 379, 450.

6. Zipfel E, Cooper RA, Pearlman J, Cooper R, McCartney M. New design and development of a manual wheelchair for India. Journal of disability and rehabilitation 2007; 29(11):949-962.

7 McLaurin Colin. Future Developments. Current Directions in Wheelchair Re-search. In JRRD Clinical Supplement no2: Choosing a Wheelchair System pp 88-99 URL http://www.rehab.research.va.gov/mono/wheelchair/mclaurin.pdf Accessed 2016 July 17

8. Dale, P.(1994). Power Stand-up and Reclining Wheelchair. U.S. Patent 5, 366, 036.

9. Lin Zhang, Xi Feihong, 2012, An Optimization Design for the Stair-Climbing Wheelchair, thesis of Master of Science in Mechanical Engineering, Blekinge Institute of Technology, Karlskrona, Sweden. Online http://www.diva-portal.org/smash/get/diva2:829385/FULLTEXT01.pdf, (fecha de consulta 15 mayo 2016)

10. Pieter Desmet, Eva Dijkhuis, 2003, A wheelchair can be Fun: A Case of emotion-drive design, Conference: Proceedings of the 2003 International Conference on Designing Pleasurable Products and Interfaces, Pittsburgh, PA, USA, pp.23-26, DOI: 10.1145/782896.782903

11. Anthropometric measures of Colombian children in Anthropometric Dimensions of Population Latinoamerican. Online: http://biblio3.url.edu.gt/Libros/DA2/5/5.1.2.pdf, Accessed 2015 march 10

12. Falcón Víctor. 2004. Concurrent Design Wheelchairs for Disabled. Online-http://saber.ucv.ve/jspui/bitstream/123456789/645/1/Dise%C3%B1o%20Concurrente%20de%20Sillas%20de%20Ruedas.pdf. Accessed 2015 April 25.

Prosthetic Alignment and Biomechanical Parameters in Transtibial Amputees due Landmines

L.A. Luengas[1], G. Sanchez[2], K. Novoa[1]

[1] Universidad Distrital Francisco Jose de Caldas/Ingeniería en Control, Docente, Bogotá, Colombia
[2] Universidad de San Buenaventura/Facultad de Ingeniería, Docente, Bogotá, Colombia

Abstract— **The lack of objective information concerned to prostheses alignment is a common phenomenon in the area. When an inadequate angular location of the transtibial prosthetic socket appears, the biomechanical parameters in the joint and the location of the center of pressure (COP) in the amputees are affected; it also affects the amputee's comfort in the standing position. The aim of this study was to investigate the effect of transtibial prosthesis malalignment on biomechanical parameters. The effects of transtibial prosthesis alignment on biomechanical parameters were measured in terms of lower limb angles and the location of the center of pressure (COP) when the subject is in the static standing position. Amputees who were regularly attending in the Hospital Militar Central, Bogotá, Colombia, were invited to participate in this study. Seven subjects with transtibial amputation were recruited from the community. Seven insertion alignment conditions were measured in each of the 6 participants of the study, whereas participant seven (7) only six conditions were located, for a total number of 144 tests. The socket angle, the joint angles and the COP were recorded in computer files. The statistical analysis of the measurements showed variation intra and inter subjects. All angle perturbations revealed statistically significant differences on joint angles and the COP compared to the nominal alignment. The sagittal angle perturbations of the socket led to statistically significant changes in the antero-posterior direction of COP and the joint angles. Therefore, malalignment affected the lower limb angles and the location of the center of pressure (COP) in amputees with transtibial prostheses.**

Keywords— **Amputees, Alignment, Standing up.**

I. Introduction

Limb loss is one of the most physically and psychologically devastating events that can happen to a person. Not only does lower limb amputation cause major disfigurement, it renders people less mobile and at risk for loss of independence. The high incidence of amputation both global and national levels has made the rehabilitation and treatment of people with amputations of lower limbs is one of the main concerns of the current health system. There are nearly 10 million people living with limb loss in worldwide [1]. In Colombia the number of amputees is 46,200 people, being the most common transtibial amputation [2], and one of the causes is the action of landmines [3], [4].

In affected subjects the goal of rehabilitation is to improve bipedal locomotion function (gait), amputee needs stability and balance in static standing position. Amputee rehabilitation must provide adequate training to walk comfortably and safely without physical or mental exertion. Visually, the aim of the gait training is to have prosthetic gait patterns similar to non-amputees as much as possible, with little or no demonstrable asymmetry.

For safe and independent walking, it is necessary that a person develop the movement abilities to standing up and walk with a prosthesis. This is possible with an adequate rehabilitation process and a successful alignment of the prosthesis. Static and dynamic prosthetic alignment has a significant influence on general performance of a patient. Amputee must be able to stand comfortably [5], [6]. The process and measured orientation of the prosthetic socket relative to the foot is called Alignment. It is important for proper function of a transtibial prosthesis because to optimize the dynamic balance and biomechanical function exhibited by the user. Alignment plays an important role in the function of the entire prosthesis and general performance of a patient and is one of the critical elements for optimal limb function. Manufacturers' guidelines for static prosthetic alignment are not individualised for each amputee. As a result, prosthetic alignment must be optimised for each individual during iterative dynamic alignment [6]. In the clinical setting, it is challenging to align a prosthesis because prosthetists rely on observational gait analysis and patients' feedback [7]. Optimisation of prosthetic alignment is a time-consuming and subjective process requiring many years of experience. It is inevitable that this subjective method results in a wide variation in acceptable definitive alignments [8]. Current prosthetic alignment methods require many years of experience from the prosthetist and clear feedback and assessment of the functionality of the prosthesis by the amputee. The effect of alignment change on the kinematic

I. Torres et al. (eds.), *VII Latin American Congress on Biomedical Engineering CLAIB 2016, Bucaramanga, Santander, Colombia, October 26th -28th, 2016*,
IFMBE Proceedings 60,
DOI: 10.1007/978-981-10-4086-3_192

symmetry in amputees wearing transtibial prostheses revealed that they could use compensatory mechanisms that would work to maintain symmetry of kinematic parameters. It seems attractive to measure limbs angles and ground reaction forces as a valid measure of limb loading and the alignment [9].

The researchers reported in the literature have predominately focused on determining the effect of prosthetic alignment on measured gait parameters; few researchers have analyzed the effect of alignment in the static position standing.

The aim of this study was to investigate the effect of transtibial prosthesis alignment on lower limb angles and the location of the center of pressure (COP) when the subject is in the static standing position. These biomechanical parameters were expected to be influenced by the alignment of the socket relative to the foot. Through improved understanding of prosthetic alignment, one may improve the efficiency of the prosthetist in optimizing alignment and minimize mistakes in alignment of prostheses.

In Section 2, the materials and experimental methodology used in this study are described. Sections 3 and 4 contain a discussion of the results and conclusions from the study, respectively.

II. Materials and Methodology

2.1. Subjects inclusion criteria

Inclusion criteria of this study were using prosthesis for at least 1 year, using a modular endoskeletal prosthesis, unilateral amputee, bilateral normal range of motion, ability to walk without additional assistive devices and ability to provide informed consent. All the subjects had energy storage and return prostheses (Flex-foot), total surface bearing and pin suspension. Prosthesis alignment was establishing clinically.

Seven volunteers, table 1, males, aged 33,29 ± 3,77 years; weight 71,43 ± 4,58 kg; height 1,71 ± 0,054 m, were recruited for this pilot study. In addition, subjects were excluded if: they using any assistive device, having experienced osteomuscular symptoms, injuries, peripheral neuropathy, vasculopathy, or other systemic pathological conditions which could affect measurements.

2.2. Apparatus

The study used seven electro-goniometers, in which high accuracy and repeatability were demonstrated, to measure the joint angles (Biometrics) [10]. The goniometers were placed on the socket, (prosthetic) ankle, (prosthetic) knee and hip of both limbs. We calibrated the goniometers by placing the subject in an erect position with hips and knees in extension and the feet in a plantigrade position. Subjects wore their own shoes. We recorded the COP with soles of Novel Pedar [11].

Table 1: Patient characteristics

SUBJECT	AMPUTEE SIDE	WEIGHT (kg)	HEIGHT (m)	AGE (years)
S1	Right	67	1,70	30
S2	Left	75	1,74	29
S3	Left	74	1,70	31
S4	Left	65	1,60	35
S5	Left	75	1,76	33
S6	Left	68	1,73	40
S7	Right	76	1,75	35

2.3. Procedure

Alignment of each subject's transtibial prosthesis was dynamically tuned by a prosthetist to the satisfaction of himself and the subject. From this nominally aligned condition, controlled angle perturbations of 2, 4, 6° (flexion, extension) were induced. Including the nominally aligned condition, there were put 7 alignment conditions measured for each subject. Joint angles and COP were recorded. Three test sessions were conducted with each subject.

The system was mounted in the Hospital Militar Central, Bogotá, Colombia, with constant access to the complete experimental setup: a computer and the guides (Fig. 1). At the beginning of every session, a calibration procedure was performed.

One technique was used for feet and posture positioning during the test sessions. The method consisted of a 2D frame placed on the floor to control foot and body position, following the anatomical position.

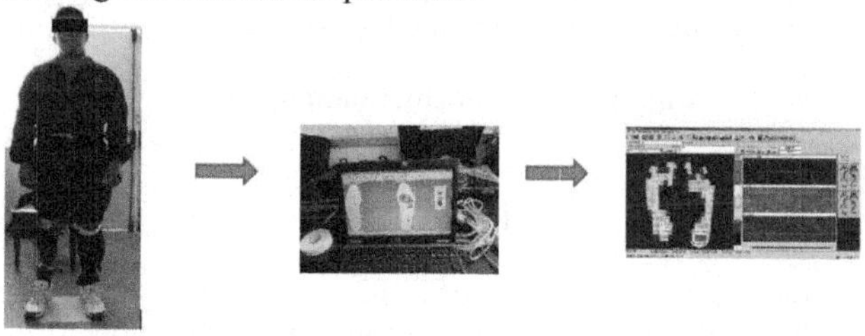

Fig. 1: Procedure to assess the joint angles and COP.

The data files were post-processed with custom software. Mean and standard deviation of measurements per subject were calculated.

2.5. Statistical analysis

Levene's test was performed to confirm the homogeneity of variance for all the collected variables. They were subsequently compared between nominally aligned and malaligned conditions. Statistical significance was set to an alpha = 0.05.

For each limb condition, the mean (X') and standard deviation (SD) value of the outcome variables was calculated. They were subsequently compared between nominally aligned and misaligned conditions using Spearman test. Spearman's rank correlation coefficient uses the ranks of the data to compute a correlation coefficient. The effect of sagittal plane angle malalignment was analyzed in the sagittal plane.

III. RESULTS

Lower limb angles and the location of the center of pressure (COP) in response to socket alignment perturbations produced systematic and some statistically significant changes.

Seven subjects with unilateral transtibial amputation were recruited at Hospital Militar Central (Table 1). All subjects had amputations due to trauma. Amputation technique varied and was not controlled. All the subjects used a clinically prescribed prosthesis. The Ethics Committee of Hospital Militar Central, Bogotá, Colombia, approved the study and informed consent was obtained from each patient.

Table 2 summarizes the range, minimum, maximum, mean and standard deviations of each parameter under sagittal plane alignment perturbation conditions.

Table 2: Effect of sagittal perturbations on joint angles, body weight and COP. Socket is position angular. CNA refers to hip no-amputee side. RNA, knee no-amputee side. TNA, ankle no-amputee side. CA, hip amputee side. RA, knee amputee side. TA, ankle amputee side. PNA, body weight on no-amputee side. PA, body weight on amputee side. XNA and YNA, COP no-amputee side. XA and YA, COP amputee side.

VARIABLE	RANGE	MIN	MAX	$\bar{X}$	(S)
SOCKET (°)	12	-6	6	.00	4.000
CNA (°)	4.7	-2.0	2.7	-.118	1.49
RNA (°)	5.1	-1.0	4.1	2.00	1.43
TNA (°)	7.4	-.9	6.5	3.16	2.02
CA (°)	4.2	-2.8	1.4	-.64	1.26
RA (°)	10.1	-1.8	8.3	2.16	3.88
TA (°)	2.8	-1.3	1.5	.05	.80
PNA (%)	14.45	40.41	54.85	48.6	2.9
PA (%)	14.44	45.14	59.58	51.3	2.9
X NA (mm)	8.75	51.86	60.61	55.7	1.9
Y NA (mm)	65.85	66.43	132.28	95.9	19.09
X A (mm)	7.86	26.01	33.88	28.9	1.56
Y A (mm)	46.86	113.29	160.15	138.7	14.14

The homogeneity of data subjects was calculated. The Levene's test for equality of variances was used to assess homogeneity. Large intersubject differences in variables were found. Ipsilateral hip joint angles, Fig. 2, showed high variability inter-subjects. Levene's test showed that variances in joint ranges and COP were no equal. The level of significance was set on $p \leq 0.01$.

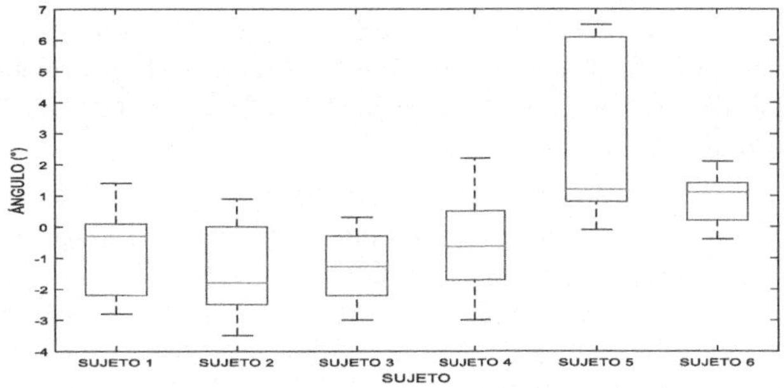

Fig. 1: Boxplot hip ipsilateral side. Large intersubject differences in variables was founded.

In amputees, the body weight was similar on both limb 50% Isakov et al. [12] and Viton et al. [13], but variability weight distribution between the limbs occurs while the subject is standing (SD 7,28%). The body weight was minimum affected by the alignment perturbations in sagittal plane (SD ± 9%). Spearman's Correlation analysis was utilized to investigate the relationship between socket alignment perturbations (angles) and the body weight. In ipsilateral, the Spearman's correlation analysis resulted in a weak positive correlation (r_s=.025, p<.01) and a weak (non-significant) positive correlation in contralateral (r_s=.081, p<.01).

In alignment, COP position was anterior on ipsilateral side versus contralateral. The effect of sagittal alignment perturbations on COP position was sensitive in angle perturbations. Ipsilateral side: the angular misalignment induced a posterior and medial COP shift. Contralateral side: the angular misalignment not induced a medial COP shift, but induced an anterior shift. The Spearman's correlation analysis between socket position and COP axis-X resulted in a weak correlation (r_s=.03, p<.01) whereas a significant positive correlation is shown in axis-Y (r_s=.7, p>.01).

Joint angles limb showed variations. Knee variation in the prosthetic limb was lower than in the non- affected. Hip variation in the non- affected limb was lower than in the prosthetic [14]. The effect of sagittal alignment perturbations on joint angles was sensitive. Except for ankle in the contralateral side, which had a weak correlation with the socket position (r_s=.07, p<.01), significant correlations could be found. Contralateral side: hip r_s=.2, p<.01, knee r_s=.74, p<.01. Ipsilateral side: hip r_s=.46, p<.01, knee r_s=.92, p<.01, ankle rs=.77, p<.01.

IV. Conclusions

Alignment is a complex process and not the measurement of the resulting alignment was the focus of this research. The importance of prosthetic alignment to the success of prosthetic fitting is well known. Prosthetic alignment is performed by prosthetists using visual gait observation and amputees' feedback. Thus, the procedure of prosthetic alignment is very time-consuming and subjective. Objective, reproducible and faster techniques are desirable [7], [9].

This study demonstrated that the alignment perturbations in sagittal plane would have a systematic influence, though not all conditions revealed significant differences from the nominally aligned condition, on the body weight, joint angles and COP in subjects with transtibial prostheses. It was therefore suggested that joint angles and COP could be a biomechanical effect of prosthetic alignment.

Individuals are likely to accommodate for sagittal plane alignment changes to a certain extent, by applying or enduring greater forces on their residual limb using active control of the knee joint. In the sagittal plane, it was clear that the extension of the socket would lead to changes in the limb angles and COP position.

Amputees loaded 45–50% of their body weight on their prosthetic limb. Hence, loading of the no-amputee limb seems to increase slightly when alignment is perturbed.

Socket angular position affected COP. Alignment of the socket changes COP in the anteroposterior (flexion–extension) direction.

The findings reported here are specific to transtibial prostheses. While this constitutes the majority of lower-limb prostheses, it does not account for the significant differences one would expect at either higher or lower levels of amputation.

The participants in this study had all used modular endoskeletal prosthesis. Future work should explore other type of prosthesis. Another limitation is the gender of the sample, because our sample included only male.

Referencias

[1] M. Leblanc, "Give Hope - Give a Hand - The LN-4 Prosthetic Hand," 2008. [Online]. Available: http://www.stanford.edu/class/engr110/2011/LeBlanc-03a.pdf. [Accessed: 01-Oct-2013].

[2] DANE and CENSO 2005, "Discapacidad Personas con limitaciones permanentes," 2006. [Online]. Available: http://www.dane.gov.co/files/censo2005/discapacidad.pdf. [Accessed: 03-Oct-2013].

[3] Dirección Contra Minas, "Víctimas de Minas Antipersonal," 2014. [Online]. Available: http://www.accioncontraminas.gov.co/estadisticas/Paginas/victimas-minas-antipersonal.aspx. [Accessed: 03-Oct-2014].

[4] E. Camargo, L. A. Luengas, and M. Balaguera, "RESPUESTA A CARGA DE UNA PRÓTESIS TRANSTIBIAL CON ELEMENTOS INFINITOS DURANTE EL APOYO Y BALANCEO," *Visión Electrónica*, vol. 6, pp. 82–92, 2012.

[5] M. A. González, O. Cohí, and F. Salinas, *Amputación de extremidad inferior y discapacidad: Prótesis y rehabilitación*. Barcelona: Masson, 2005.

[6] M. S. Pinzur, W. Cox, J. Kaiser, T. Morris, A. Patwardhan, and L. Vrbos, "The effect of prosthetic alignment on relative limb loading in persons with trans-tibial amputation: a preliminary report.," *J. Rehabil. Res. Dev.*, vol. 32, no. 4, pp. 373–378, 1995.

[7] D. Boone, T. Kobayashi, T. G. Chou, A. K. Arabian, K. L. Coleman, M. S. Orendurff, and M. Zhang, "Influence of malalignment on socket reaction moments during gait in amputees with transtibial prostheses.," *Gait Posture*, vol. 37, no. 4, pp. 620–626, Nov. 2012.

[8] S. Blumentritt, T. Schmalz, R. Jarasch, and M. Schneider, "Effects of sagittal plane prosthetic alignment on standing trans-tibial amputee knee loads.," *Prosthet. Orthot. Int.*, vol. 23, no. 3, pp. 231–238, 1999.

[9] S. Blumentritt, T. Schmalz, R. Jarasch, and M. Schneider, "Effects of sagittal plane prosthetic alignment on standing trans-tibial amputee knee loads.," *Prosthet. Orthot. Int.*, vol. 23, no. 3, pp. 231–8, Dec. 1999.

[10] Biometrics Ltd, "Biometrics Ltd." [Online]. Available: http://www.biometricsltd.com/. [Accessed: 01-Oct-2013].

[11] Novel.de, "The pedar® system." [Online]. Available: http://www.novel.de/novelcontent/pedar. [Accessed: 11-May-2014].

[12] E. Isakov, J. Mizrahi, Z. Susak, I. Ona, and N. Hakim, "Influence of prosthesis alignment on the standing balance of below-knee amputees," *Clin. Biomech.*, vol. 9, no. 4, pp. 258–262, 1994.

[13] J. Viton, L. Mouchnino, M. L. Mille, M. Cincera, a. Delarque, a. Pedotti, a. Bardot, and J. Massion, "Equilibrium and movement control strategies in trans-tibial amputees," *Prosthet. Orthot. Int.*, vol. 24, no. 2, pp. 108–116, Jan. 2000.

[14] L. A. Luengas, E. Camargo, and G. Sánchez, "Estudio de los rangos articulares en la bipedestación estática en personas normales vs . Amputados transtibiales," *Tecnura*, vol. 17, pp. 60–68, 2013.

Erratum to: Descriptive study of the courses offered in degree programs in Biomedical Engineering in Mexico

A. Pliego-Carrillo, E. Del Hierro-Gutierrez

Erratum to:
Chapter "Descriptive study of the courses offered in degree programs in Biomedical Engineering in Mexico" in: I. Torres et al. (eds.), *VII Latin American Congress on Biomedical Engineering CLAIB 2016, Bucaramanga, Santander, Colombia, October 26th -28th, 2016*, IFMBE Proceedings 60, https://doi.org/10.1007/978-981-10-4086-3_177

The original version of the book was inadvertently published with incorrect title for Chapter 177 in XML and book PDF (TOC), which has to be now corrected. The erratum book has been updated with the change.

The updated online version of this chapter can be found at https://doi.org/10.1007/978-981-10-4086-3_177

I. Torres et al. (eds.), *VII Latin American Congress on Biomedical Engineering CLAIB 2016, Bucaramanga, Santander, Colombia, October 26th -28th, 2016*,
IFMBE Proceedings 60,
DOI: 10.1007/978-981-10-4086-3_193

Author Index

The manufacturer's authorised representative in the EU is Springer Nature Customer Service Centre GmbH, Europaplatz 3, 69115 Heidelberg, Germany. If you have any concerns regarding our products, please contact ProductSafety@springernature.com

Printed and bound by CPI Group (UK) Ltd, Croydon, CR0 4YY
02/10/2026
02226684-0003